Soil Biology Guide

Soil Biology Guide

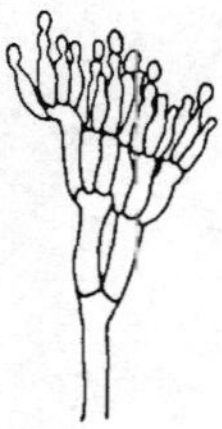

Edited by

DANIEL L. DINDAL

Distinguished Teaching Professor of Soil Ecology
State University of New York College of Environmental Science and Forestry
Syracuse, New York

WILEY

A Wiley-Interscience Publication

JOHN WILEY & SONS

Library of Congress Cataloging in Publication Data:

Soil biology guide.
"A Wiley-Interscience publication."
Bibliography: p.
Includes index.
1. Soil Biology. 1. Dindal, Daniel L.
QH84.8.S632 1990 574.909'48 88-28016
ISBN 978-0-471-04551-9

Printed and bound by CPI Group (UK) Ltd, Croydon, CR0 4YY
C9780471045519_230226

The manufacturer's authorized representative according to the EU General Product Safety Regulation is Wiley-VCH GmbH, Boschstr. 12, 69469 Weinheim, Germany, e-mail: Product_Safety@wiley.com.

To future students of soil ecology,
young and old;
may this aid their education in basic science,
and hasten their solving of the applied
problems of our day.

Contributors

B. D. AINSCOUGH, (Former Associate Curator), Entomology Division, Provincial Museum, Victoria, British Columbia, Canada

PATRICIA A. ARCHIBALD, Department of Biology, Slippery Rock University, Slippery Rock, Pennsylvania

JAMES G. BALDWIN, Department of Nematology, University of California, Riverside, California

IAN R. BALL, Department of Biology, Memorial University of Newfoundland, St. Johns, Newfoundland, Canada

STUART S. BAMFORTH, Department of Biology, Tulane University, New Orleans, Louisiana

ROSS T. BELL, Department of Zoology, University of Vermont, Burlington, Vermont

J. B. BURCH, Museum of Zoology, Department of Biology, The University of Michigan, Ann Arbor, Michigan

DONALD S. CHANDLER, Department of Entomology, University of New Hampshire, Durham, New Hampshire

KENNETH A. CHRISTIANSEN, Department of Biology, Grinnell College, Grinnell, Iowa

T. P. COPELAND, Department of Biological Sciences, East Tennessee State University, Johnson City, Tennessee

CLIFFORD S. CRAWFORD, Department of Biology, University of New Mexico, Albuquerque, New Mexico

MADHAB C. DASH, School of Life Sciences, Sambalpur University, Orissa, India

DANIEL L. DINDAL, Department of Environmental and Forest Biology, State University of New York College of Environmental Science and Forestry, Syracuse, New York

C. D. DONDALE, Biosystematics Research Institute, Agriculture Canada, Ottawa, Ontario, Canada

The late **HENRY S. DYBAS,** Division of Insects, Field Museum of Natural History, Chicago, Illinois

LINDA DYBAS, Department of Biology, Knox College, Galesburg, Illinois

ARLAN L. EDGAR, Department of Biology and Chemistry, Alma College, Alma, Michigan

CLIVE A. EDWARDS, Department of Entomology, The Ohio State University, Columbus, Ohio

DOUGLAS E. EVELEIGH, Department of Biochemistry and Microbiology, Rutgers—The State University of New Jersey, New Brunswick, New Jersey

WILLIAM M. FENDER, Soil Biology Associates, McMinnville, Oregon

LYNN M. FERGUSON, Department of Natural Sciences, Longwood College, Farmville, Virginia

DIANA W. FRECKMAN, Dry Lands Research Institute and Department of Nematology, University of California, Riverside, California

ALFONSO NERI GARCÍA ALDRETE, Instituto de Biología-Departmento de Zoología, Universidad Nacional Autónoma de México, Mexico, D. F. Mexico

T. R. G. GRAY, Department of Biology, University of Essex, Colchester, England

ROBERT P. HIGGINS, Department of Invertebrate Zoology, National Museum of Natural History, Smithsonian Institution, Washington, D.C.

RICHARD L. HOFFMAN, Virginia Museum of Natural History, Martinsville, Virginia

J. C. HUNTER-CEVERA, Cetus Corporation, Emeryville, California

G. W. KRANTZ, Department of Entomology, Oregon State University, Corvallis

GENTARO IMADATE, Biology Department, Konodai College, Tokoya Medical and Dental University, Ichikawa, Japan

SAMUEL W. JAMES, Department of Biology, Maharishi International University, Fairfield, Iowa

W. BRYCE KENDRICK, Department of Biology, University of Waterloo, Waterloo, Ontario, Canada

JOHN KETHLEY, Division of Insects, Field Museum of Natural History, Chicago, Illinois

G. W. KRANTZ, Department of Entomology, Oregon State University, Corvallis, Oregon

J. DANIEL LOUSIER, Research Branch, Ministry of Forests and Lands, Province of British Columbia, Victoria, British Columbia, Canada

J. F. McALPINE, Biosystematics Research Institute, Agriculture Canada, Ottawa, Ontario, Canada

DOROTHY McKEY-FENDER, Soil Biology Associates, McMinnville, Oregon

WILLIAM B. MUCHMORE, (Professor Emeritus), Department of Biology, University of Rochester, Rochester, New York

PETER MUNDEL, Department of Entomology, Smithsonian Institution-MSC, Washington, D.C.

DIANE R. NELSON, Department of Biological Sciences, East Tennessee State University, Johnson City, Tennessee

ALFRED F. NEWTON, JR., Division of Insects, Field Museum of Natural History, Chicago, Illinois

ROY A. NORTON, Department of Environmental and Forest Biology, State University of New York College of Environmental Science and Forestry, Syracuse, New York

W. L. NUTTING, Department of Entomology, The University of Arizona, Tucson, Arizona

DENNIS PARKINSON, Biology Department, University of Calgary, Calgary, Alberta, Canada

TIMOTHY A. PEARCE, Museum of Zoology, Department of Biology, The University of Michigan, Ann Arbor, Michigan

STEWART B. PECK, Department of Biology, Carleton University, Ottawa, Ontario, Canada

JAMES R. PHILIPS, Science Department, Babson College, Babson Park (Wellesley), Massachusetts

ROY T. SAWYER, Biopharm (UK) Limited, Swansea, Wales; Biopharm (USA) Limited, Charleston, South Carolina

ULF SCHELLER, Lundsberg, Sweden

DONALD P. SCHWERT, Geology Department, North Dakota State University, Fargo, North Dakota

RONALD SLUYS, Institute of Taxonomic Zoology, University of Amsterdam, The Netherlands

HARUO TASHIRO, Entomology Department, New York State Agricultural Experiment Station, Cornell University, Geneva, New York

H. J. TESKEY, Biosystematics Research Institute, Agriculture Canada, Ottawa, Ontario, Canada

GEORGE C. WHEELER, (Professor Emeritus), Department of Biology, University of North Dakota, Grand Forks, North Dakota

JEANETTE WHEELER, (Adjunct Research Associate), Desert Research Institute, University of Nevada, Reno, Nevada

Preface

Upon discovery, soil organisms fascinate many people. Thus, this book was organized to support a broad audience of users. It is my hope to encourage the keen interests of the beginner as well as to assist and provide scientific breadth and depth for the professional. For the first time, most of the major representative microbial and invertebrate inhabitants of North American soils are drawn together, in one book, as they are found in nature. All the specialists have presented what they believe to be the most valuable mass of knowledge available to date on each taxon. This book, therefore, represents the "state of the art" of the biology, taxonomy, and ecology for each soil biotic group. In general, it is assumed that the user of the *Soil Biology Guide* has been exposed to basic biology and may have used an identification key. Even though that may not be the case, I am convinced that anyone with a burning quest in soil science will be able to master any part, given the time. Once mastered, each chapter offers the foundation from which further study may progress. New literature which has been published subsequently to that given here should be sought as future studies continue.

Chapters are diverse in size and content, reflecting (1) the number of researchers who have worked on a given taxon, and (2) how much the body of knowledge has been developed. Taxa that have received little attention to date are fertile areas for future taxonomic studies. Each taxon also has its value in the total ecological picture, so each deserves to be studied from that viewpoint. The introductory chapter concentrates on an ecological approach to soil studies. An extensive bibliography is provided of ecological publications having both scientific and historical significance. An ecological approach can facilitate compilation and multivariate treatment of data obtained from use of any or all chapters. In contrast to the general coverage of the Introduction, the remaining 42 chapters emphasize specific information on each of the taxonomic groupings. All contributing authors of the chapters or their representatives would welcome communications regarding any aspects of their interest areas.

Identification keys are similarly diverse in their type and organization. Individual keys vary with the authors. Versatility of pattern and form of the keys may provide some alternative ideas of future taxonomic key presentation. Also, the variable degree of hierarchy is a function of the complexity to which each taxon has evolved compared to others. Some keys are based on functional morphological delineations; others lead the user to classical identification at family, genus, or species levels. Descriptions of new genera and new species combinations are incorporated in other keys.

A special feature of this book is the use of illustrated keys whenever possible. Once again, diversity is obvious in the presence or absence of illustrations as well as the simplicity or level of detail with which illustrations are presented. These decisions were the authors' perogative based on the developmental stage of the given taxonomic unit.

This guide is applicable for use with mesic, xeric, and hydric terrestrial sites. For those working in wet histosols, submerged soils, and aquatic or marine substrates, I recommend using this book in conjunction with it's companion volume, Edmonson's *Ward and Whipple's Fresh-Water Biology.**

The style and format of the references in each chapter also differ with each taxon discipline. In some cases reference sections represent only the publications cited in the text, whereas in others a more complete bibliography of additional readings of merit is included. As noted before, it must be realized that a constant commitment to finding and reading contemporary literature is a necessity in order to remain fluent in any taxonomic group.

Numerous people were responsible and supportive during the production of this monumental treatise. First, I offer my thanks to all of the 54 contributing authors. Each one labored over the chapters, constantly updating them while waiting during the eleven years of preparation as all parts of this document were put into place. Those who were forever optimistic and patient deserve extra gratitude.

With a project of this magnitude and longevity, various traumatic complications of daily life often crept into our way. Most tragic was the loss of two of our colleagues, Dr. Henry S. Dybas and Dr. Pedro Wygodzinsky. Dr. Dybas had his chapter almost completed at the time of his death, and he had instructed his daughter, Linda Dybas, Department of Biology, Knox College, Galesburg, Illinois, to see that his contribution was submitted. I am most grateful to her and the colleagues at the Field Museum of Natural History for having aided in this request. Dr. Wygodzinsky had contem-

* *Fresh-Water Biology*, by Henry Baldwin Ward and George Chandler Whipple, 2nd edition, edited by W. T. Edmondson (John Wiley, New York, 1959).

plated some revised keys for a chapter on Thysanura but was never able to complete them to his satisfaction before his passing. Dr. Lynn M. Ferguson graciously agreed to prepare the chapter. For this he deserves our recognition.

My appreciation is extended to Mary M. Conway, Isabel Stein, and Rosalyn Farkas of John Wiley and Sons, Inc., for their help and patience in this endeavor. I am appreciative of the able assistance provided to me by Professors Robert L. Burgess and Neil H. Ringler. I also thank Ruth T. Piatoff for her very helpful assistance in typing manuscripts and all the correspondence during this project. Many graduate and undergraduate students deserve credit for the stimulating conversations on soil biota which enlightened me during all phases of book preparation. I also acknowledge with appreciation my graduate students, who had to patiently share my time while I was deeply involved in this enterprise.

Moreover, I could never have accomplished my part of this major project without the considerate support and constant help of my wife, Anna Jean. She provided continual encouragement, understanding, and long months of patience while I completed this task. In addition to her invaluable moral support, she readily assisted in the typing of manuscripts, editing, proofreading, and indexing. My wife shared unselfishly in the much-needed discussion and advice in her specialty of information studies as related to this book. I am thankful that we can work together so harmoniously. Finally, I wish to recognize the rest of my family, Christopher, Ann Marie, Mark, Daphne, Leanna, and Mom and Paul, for their caring presence, advice, and understanding.

DANIEL L. DINDAL

Syracuse, New York
September 1989

Contents

Errata for

SOIL BIOLOGY GUIDE

Edited by

Daniel L. Dindal

page 96, add reference:

194. **Archibald, P. A. 1988.** *Chlorococcum pamirum* and *C. salinum*, two new species of the Chlorophyceae from Central Asia. *Br. Phycol J.* 23:121–128.

page 567, couplet 3b, line 5, should read:

with six, seven, or eight setae, . . .

page 642, couplet 2b (Pachylaelapidae), should read:

tibiae III and IV each with 8 setae $\left(2\,\frac{1}{1}\,\frac{2}{1}\,1\right)$ . . .

page 654, couplet 4a (Polyaspididae), line 1, should read:

Idiosoma subtriangular; . . .

page 987, couplet 102 or 103, should read:

Venter of manubrium of adult specimens with 12 or more setae . . .

page 987, couplet 103 or 102, should read:

Venter of manubrium of adult specimen with 10 or fewer ventral setae . . .

page 990, couplet 126 or 127, add:

(Fig. 32.71)

Corrections to Index

page 1322: Methods:
collection: Pselaphidae, 1177

page 1324: Microhabitats:
feces: Diplopoda, 840, 842

page 1326: Microhabitats:
soil litter: leaf: Opiliones, 532

Micropeplinae, 1138, *1147*, 1159, 1162, *1163*

Microplana:
terrestris (=*Geodesmus terrestris* and *Orthodemus terrestris*), **143–144,** *145–146*, 149–151

page 1328: Mycelium, 22, 33–35, 61, *66*, 952, 1038; *see also* Fungi: Hyphae

page 1329: Nitrogen cycle, *see also* Excretion, Excretory systems

page 1332: Periodicity:
nocturnal: Opiliones, 530

page 1335: *Promecognathus*, 1054, *1072*

page 1336: *Pselaphus, 1181: fustifer,* 1176

Pselaptus, 1184: belfragei, **1184**

page 1338: *Rhinoscepsis, 1184: bistriatus,* **1185**

page 1340: Setae. *See also* Rutella (blade-like setae); Trichobotheria (sensory setae)

page 1341: Soils:
aggregates: Sphecidae, desert, 468–469
subsurface: Actinomycetes, 35, 41

Soil Biology Guide

1

Introduction

DANIEL L. DINDAL

Soils of the world usually abound with active individuals and propagules representing a multitude of microbial and invertebrate taxa. Few humans are aware of the existence of such cryptic life forms. These soil biota are organized both structurally and functionally into specific microcommunities. Such organized microcommunities are modified by human activities with the degree of modification being dependent on the nature, intensity, and longevity of human disturbance. In disturbed soils, as well as undisturbed soils, the types and numbers of soil organisms and their specific functions reflect the pedogenic status, vegetative successional patterns, and environmental perturbations of given sites.

Functional Roles of Soil Biota

The invaluable role played by soil microorganisms—bacteria, actinomycetes, fungi, algae and protozoa—includes the dynamic decomposition of organic matter and production of humus, cycling of nutrients and energy and elemental fixation, soil metabolism, and the production of complex compounds that cause soil aggregation. These active organisms with their high surface area-to-volume ratio are living tissue pools of chemical elements, compounds, and energy. Many soil microbes are functionally related in a symbiotic way to plants and animals. Microbes of the soil have relationships with higher plants as nitrogen-fixers (Smith and Rice 1986), mycorrhizae (Smith 1980), and other rhizosphere associates (Rovira 1965, Gray and Williams 1971). Microorganisms are associated with soil invertebrates as gut microbes (Martin 1987), mycetangial and cecal symbionts (Buchner 1965), and as cultures in fungal gardens (Weber 1972). Microbial–vertebrate symbioses are common among rumen and gut microorganisms (Hungate 1966, 1978), fecal and carcass decomposers (Stevenson and Dindal 1987b), and nest–burrow symbionts (Phillips and Dindal 1977). In a variety of ways, soil microorganisms are the very bases of the heterotrophic microcommunities that dominate soil ecosystems.

Soil invertebrates are also extremely important agents in the generation and maintenance of the biological, chemical, and physical character of the soil ecosystem. These invertebrates function as regulators of microbial processes through comminution, translocation, defecation, and inoculation of microbial propagules. Soil invertebrates are significant living tissue pools of energy, nitrogen, and other nutrients. By their burrowing activity they make low-resistance channels through which air and water pass within the soil profile. Jenny (1980) refers to these cryptic invertebrates as mechanical blenders of soil components. Mechanical blending is caused by burrowing and complex feeding activities of soil invertebrates. For instance, the nutritional ecology of soil arthropods involves symbiotic interactions with microorganisms, predator–prey transfers, phoresy–food relationships, and coprophagy which directly or indirectly cause physical–chemical modifications of soil (Werner and Dindal 1987). Also, the role of soil invertebrates in modifying humification rates and in distribution of organic materials into the mineral soil matrix via fecal pellet deposition affects soil aggregate size and frequency (Dindal 1985). Finally, the increases in the presence and diversity of water-stable soil aggregates cause (a) the enhancement of soil aeration and nutrient and water-holding capacities, and (b) the reduction of bulk density and susceptibility of soils to erosive forces of water and wind.

Ecologic and/or Taxonomic Approach?

Since microorganisms and invertebrates function naturally in unison within the soil, both should be investigated together using an ecological approach. For example, analyses of the components of food webs in heterotrophic microhabitats can illuminate the total structural and functional involvement of microorganisms and invertebrates (Pimm 1982). Investigations of this type require that one evaluates the entire community. These complete works can provide qualitative and quantitative assessments of both natural soil systems, such as forest and desert soils and litter, and human-modified systems, such as compost piles and landspread waste recycling sites (Dindal 1978). However, consideration of the soil invertebrates and microbes as a community has too often been neglected in past research. Therefore, when pursuing research objectives dealing with entire soil ecosystems or when modeling dynamic edaphic conditions, the influences and quantification of all soil biota must be evaluated.

Furthermore, some soil ecological research has been hampered by claims that the taxonomy of soil organisms is incomplete, limited, or subject to disagreement among taxonomists. Even though the systematics of many soil taxa are ever changing, ecological studies must persist to answer the questions about soils of the world. The answers will lead to understanding of patterns in natural soil ecosystems, will aid in interpretation of impacts of human activities, and will stimulate the production of new soil management techniques. The taxonomic classification for any group, no matter what stage of development, is a most important tool, and it is imperative that it be used in ecological studies. Any formal classification scheme can be used in ecological works as long as it is dated and documented. In some studies, soil life forms may be classified and quantified based on functional trophic levels, such as producers, herbivores, predators, parasites, and decomposers. This book is a source of such information and documentation for either taxonomic or functional categorization.

For those solely interested in taxonomy, this volume provides contemporary thought and the level of taxonomic elegance at the time of publication. Future taxonomists are encouraged to start with this foundation and further develop the classification of any taxon. Work of this type will be appreciated by those in systematic disciplines as well as those in ecological pursuits.

Book Objectives

Considering the comments above, this book was written and compiled with the following specific objectives in mind:

1. To produce one book devoted to the taxonomy, biology, and ecology of representative North American soil biota as a terrestrial companion volume to *Ward and Whipple's Fresh-Water Biology* (Edmondson 1959).
2. To provide an illustrated means of identification of most soil forms for use by anyone with an interest or a problem in soil biology.
3. To demonstrate the diversity of soil organisms that comprise microcommunities associated with characteristic soils and ecosystems.
4. To elucidate some of the adaptations of microorganisms and invertebrates to their terrestrial existence.
5. To prepare a source book as a basis on which reference collections can be made.
6. To provide a teaching–research resource concentrating on taxonomy, biology, and ecology of soil biota.
7. To stimulate the imagination with diverse ideas toward furthering education and research in soil ecosystems.

EARLY CONTRIBUTIONS IN SOIL ECOLOGY

Soil Biology in Europe and Asia

In a detailed global history of soil zoology, Kevan (1985) traces back to ancient times where images of burying beetles, other scarabs, and various soil fauna were found on artifacts of the Magdalenian cultures (25,000–30,000 years ago) and Sumarian and Egyptian cultures (2000–4000 years ago). Little scientific interest was shown until 1881, when Darwin stressed the value of earthworms in vegetable mold formation, followed in 1887 by Drummond's statement that termites were the tropical ecological equivalent to earthworms in temperate zones. In 1889, Müller coined the terms "mull" and "mor" to describe patterns of humus development; in these definitions Müller emphasized the role of soil biota in humus formation. Some of the other scientists who published important details on soil microhabitats and soil microcommunity ecology were Bornebusch (1930), Ghilarov (1947), Kubiëna (1953), Haarløv (1960), Kühnelt (1961), Burges and Raw (1967), Wallwork (1970, 1976), Edwards and Lofty (1972), Dickinson and Pugh (1974), Edwards and Veeresh (1978), Swift et al. (1979), Satchell (1983), and Lee (1985). Of great importance since 1955, eight International Colloquia of Soil Zoology have been held in Europe and Asia.

Soil Ecological Activities in North America

People with Original Ideas

Early research and writings in soil ecology within North America were accomplished by isolated individuals. Jacot (1935, 1936, 1940), working for the U.S. Forest Service in North Carolina and Connecticut, carried out extensive research on oribatid mites. He was one of the first in this country to publish the ecological role of these decomposers. Also in the 1930s and 1940s, Pearse (1946), working at Duke University, investigated soil fauna and arthropod symbionts in forest litter. Another Forest Service researcher in North Carolina, L. J. Metz, was instrumental in compiling the taxonomic findings of Jacot (Metz et al. 1977). He also carried out further ecological studies concentrating on the effects of forestry practices on litter decomposition and on microcommunity structure of soil arthropods (Metz and Farrier 1973, Metz and Dindal 1975). Contemporaries of Jacot and Pearse, Eaton and Chandler (1942), conducted ecological research on soil organisms at the Cornell University Arnot Forest. At about the same time, microorganism ecology was being related to soil conservation by Waksman and Martin (1939) as a part of a long-term microbial research program at Rutgers University. With his book *Forest Soils,* Wilde (1958) of the University of Wisconsin presented a thorough classification of organic matter and humus based on a soil ecological approach. Englemann (1961) of the University of Michigan was among the first to evaluate the bioenergetics of soil invertebrates. In Iowa at Grinnell College, Christiansen (1964) developed a strong program in taxonomy, biology, and ecology of Collembola. In addition to preparing a much needed treatise on soil fauna, Kevan (1968) of McGill University has developed a program in soil animals (Kevan and Hill 1979).

The individual work of soil microbiologists Alexander (1971) and Stotzky (1972) has provided valuable knowledge with which biologists can pursue total soil ecosystems. All facets of soil microbiology and ecology have been approached with thoroughness by Alexander at Cornell University. At New York University, Stotzky has provided a wealth of information on cation exchange capacity and clay mineral chemistry as related to soil microbial ecology. Both of these scientists have developed their efforts into major research programs at their respective institutions. Primary work on oribatid mites and their symbioses with tapeworms was conducted at the State University of New York College at Buffalo by Sengbusch (1977). Most of the contributing authors of this book were among the first to do research related to North America in their specialized taxonomic groups, and they have thus cited many important works in their chapters.

Much-Needed Group Efforts

Building on the initiative of the individual North American pioneers in soil ecology, educational and research programs have been further promoted and maintained by groups working in various institutions. A variety of conferences and symposia was held in North America, which has aided in the development of soil ecology as a discipline. Early in the 1950s, an intense research effort in soil ecology was initiated at Oak Ridge National Laboratory by Auerbach (1958) and many colleagues, including Crossley and Witkamp (1964) and Reichle (1970). Since 1963, the Ohio State University Acarology Program has been presenting courses in general acarine taxonomy and agricultural, soil, medical, and veterinary acarology. This successful program was originally founded by G. W. Wharton, whose early works included the symbiosis of litter-related mites at Duke University (Pearse et al. 1936). Michigan State University was the site of a symposium on chemical perturbations, "Pesticides in the Soil" (Guyer 1970). Butcher, Snider, and Snider (1971) and colleagues have maintained a soil ecology program ever since. From 1970 to 1985 at the Natural Resources Ecology Laboratory at Colorado State University, research emphasis by Coleman (1976) was on interactions between soil microbes, invertebrates, plant roots, and nutrient cycling. Also, in the early 1970s, D. A. Crossley, Jr., joined the Institute of Ecology at the University of Georgia, where he promoted much research in various areas of soil ecology, especially nutrient cycling, soil ecosystem modeling, and agroecosystem studies (Crossley et al. 1984); in

1985, D. C. Coleman joined the soil ecologists at the University of Georgia. A workshop on the role of earthworms in stabilizing wastes was organized by Applehof (1981) and cosponsored by Western Michigan University and the Kalamazoo Nature Center. A two-volume proceedings and bibliography was a helpful product of that meeting. Freckman (1982), from the University of California–Riverside, edited a useful proceedings from her conference on "Nematodes in Soil Ecosystems." The soil microbiology division of the Soil Science Society of America sponsored several symposia on interactions of soil microbes with plants (Todd and Giddens 1984) and interactions of microorganisms with organic matter and soil minerals (Huang and Schnitzer 1986). The international symposium on "Faunal Influences on Soil Structure" was organized by Spence (1985) and his colleagues at the University of Alberta, Edmonton, Canada.

From 1971 through 1978, I organized six "Soil Microcommunities Conferences" (Dindal 1973a) and a "Biology of Oribatid Mites" symposium (Dindal 1977), which were sponsored by the State University of New York College of Environmental Science and Forestry, Syracuse. The main objective of these conferences was to bring together the isolated researchers of soil taxa so that they could meet each other, discuss their specialities, and develop common ecological concepts related to soil. These meetings were followed by the college hosting the 7th International Soil Zoology Colloquium (Dindal 1980). Starting in 1980 through 1983, I taught the National Science Foundation/American Association for Advancement of Science Chautauqua course "Ecology of Terrestrial Microcommunities" at 10 college campuses throughout the United States. My colleagues, Mitchell and Nakas (1986), compiled and edited a fine book on interactions of soil microbes and invertebrates in natural and agroecosystems, also as a function of the ongoing soil ecology program. In 1987, Edwards et al. (1988) and colleagues organized a soil ecology symposium sponsored by the Ohio State University and Ohio Agricultural Research and Development Center. At that meeting the U.S. Soil Ecology Society was formed. Many Ohio State faculty members from different departments are supporting this concentration in basic and applied soil ecology.

From this brief overview it is apparent that many productive centers of soil ecology exist in North America. With the scientists from these centers, the contributing authors of this book, and all the soil ecologists of the future using our common knowledge, most of the basic and applied problems related to the biology of soil should be within our grasp.

MICROSCOPIC TO MACROSCOPIC SCOPE

Biotic Coverage

Organism Size

Soil life forms presented in this book range in size from ultramicroscopic to the very obvious large invertebrates, such as earthworms, beetles, and terrestrial slugs and snails. In general, as the chapters progress from the front of the book onward, the size of the biota represented increases. Identification of smaller forms will therefore require the use of a light microscope with oil-immersion objectives and phase optics. The smaller forms include all of the microbes, turbellarians, nematodes, tardigrades, pauropods, symphylids, proturans, thysanurans, diplurans, collembolans, pseudoscorpions, mites, and psocopterans. The light microscope will be helpful with the remaining taxa, and a good dissecting microscope will be most important with the larger forms.

Of the larger creatures, small mammals such as mice, voles, shrews, moles, gophers, and squirrels have a major influence on soil ecosystems. Because of their direct impact, small mammals should be included in soil ecological studies. However, since many taxonomic keys and field guides are readily available to mammals of North American regions (Burt 1952, Blair et al. 1968), they have not been included in this volume.

Collection and Preservation of Soil Life

Specific methods of collecting, culturing, fixing, and preserving soil biota are presented in each chapter. For ecological studies, additional references, such as Klute (1986), Allen (1974), and Phillipson (1970), are very helpful since they deal with all aspects of the microbial, zoological, chemical, and physical components of the soil ecosystem.

Numerous chemical compounds that have been used for many years as preservatives and fixatives are now classed as controlled substances, suspected carcinogens, and/or suspected allergens or environmental pollutants. Since some of these compounds are ingredients in effective, classical procedures, they are presented in this book. Before using any of these substances, contact the local department of health, environmental departments, extension agencies, and controlled substances bureaus to obtain the most recent classification, handling, and licensing procedures, if necessary. Be extremely cautious of potential human health hazards in the laboratory, classroom, and in the field. If you find a given compound banned or decide that it is too hazardous, adopt an existing alternative procedure or develop and test a new method that may be more safe.

Formalin, an aqueous solution of formaldehyde, is an example of a widely used, age-old preservative, fixative, and extracting fluid. Recently, it has become the subject of concern as a causal agent of allergies in some people (Bureau of Toxic Substances Management 1981) and it is a suspected human carcinogen (Marshall 1987). In addition to its preservative characteristics, extremely weak formalin solutions are used on soil as an earthworm expulsion agent. Some workers have been concerned that this use of formalin causes soil pollution. Behrens and Hannes (1984) have shown that formaldehyde is degraded rapidly by soil bacteria, which use it as a carbon and energy source. Leonova et al. (1985) further demonstrated that the ubiquitous genus of soil bacteria, *Pseudomonas,* readily uses formaldehyde as a substrate.

These are the kinds of facts and hypotheses that must be weighed by each of us when confronted with using certain chemical compounds. In this specific case with formalin, hand digging, hand sorting, and sifting for earthworms are existing techniques that are fine alternatives.

Geographic Coverage

This book was organized to emphasize organisms of North America. However, researchers worldwide should be able to use the ecological and biological information and the taxonomic keys, especially for cosmopolitan taxa and for those that have been introduced outside their endemic areas.

Soil biota of the deserts of the southwestern United States are featured in the chapter on Scorpionida, Solpugida, and associated taxa. Other desert fauna are included in a number of chapters, such as those on Isoptera and Formicidae. Desert biotic information is applicable to a degree to desert regions of Mexico. In contrast, in the chapter on Opiliones, the key does not usually apply to areas of Mexico.

Some very unique soil taxa are found in caves. Chapters on Thysanura, Diplura, Collembola, Pseudoscorpionida, and Isopoda mention some of these terrestrial life forms. This is of special ecological interest since the cave ecosystem is considered by Margalef (1968) to be one of the most stable.

Microcommunity Structure and Classification

I compiled the definitions and literature on microcommunities for the purpose of refining the community concept (Dindal 1973b). The concept of a microcommunity, whether artificial or not, provides a manageable unit that is easy to study. Reed (1958) first defined microcommunities as the individual subunits of organisms comprising the conventional macrocommunity. More specifically, a microcommunity must include characteristic life forms that reproduce and/or spend a major part of their life cycle within a given microhabitat. Usually, a microcommunity is a small, ephemeral grouping of organisms (Elton 1966), which also must exhibit complex food webs, many intricate inter- and intraspecific associations, and highly specialized strategies for organism transport from one microhabitat to another, similar one.

In a temporal, areal, and spatial sense, each macrocommunity is comprised, structurally and functionally, of many characteristic microcommunities. Microsuccession occurs within each microcommunity; Allee et al. (1949) referred to this as *intracommunity succession.* When considered collectively through time, all patterns of intracommunity succession ultimately lead to macrocommunity change or succession. Each microcommunity is characteristic of a microhabitat with its own microclimate, thus making a microecosystem. Finally, microcosms are experimental microecosystems (Stevenson and Dindal 1987a, Cruzan 1988).

The classification of specific microcommunities is based on a judgment of the relative autotrophy–heterotrophy, reflecting community metabolism after Odum (1971). A number of terrestrial conditions fit the definition and characteristics of a microcommunity as given. With these factors in mind, microhabitat sites that support specific terrestrial microcommunities are classified as follows:

Heterotrophic (Unstable)

- Vertebrate carrion and regurgitated pellets (Reed 1958, Philips and Dindal 1979)
- Cadavers (Smith 1986)
- Invertebrate carcasses and exuvia
- Cercopid spittle (Whittaker 1970)
- Ecto- and endoparasitic sites
- Feces (dung, manure, frass) (Hutchinson 1950, Stevenson and Dindal 1987b, Mohr 1943)
- Vertebrate nests and burrows (Philips and Dindal 1977)
- Invertebrate nests and burrows, vermispheres (Hamilton and Dindal 1983)
- Phytotelmata (Maguire 1971), tree holes (Fashing 1979)
- Cortex (bark)
- Decaying wood; standing dead timber, rotten logs
- Sawdust piles
- Fungal fruiting bodies
- Soil litter; leafy, woody
- Composts
- Soil pedons (Buol et al. 1980); humus, mineral
- Rocks, stones, gravel
- Decaying fruit
- Tidal debris, beach wrack, strandlines, drift (Norton and Dindal 1976)
- Human shelters (Ordish 1981, Nero 1988)

Heterotrophic–Autotrophic Intermediates

- Galls
- Some flowers
- Seeds (Winston 1956, McMillian and Wiseman 1972)
- Stored products; grain, meal, dry fruit
- Rhizospheres (Rovira 1965), roots
- Spermatosphere, spermospheres (Slykhuis 1947, Verona 1963)
- Laimospheres (Magyarosy and Hancock 1972)
- Herbaceous plant stems (Hudson 1965)
- Ligules (Birkby and Preece 1987)
- Epiphytes
- Subcortex (underbark)
- Phytotelmata

Rosette angiosperms

Nodal pools, Teasel *Dipsacus laciniatur* L.

Pitcher plant reservoirs, Sarraceniaceae

Bromeliad tanks (Laessle 1961, Benzing 1980)

Autotrophic Unstable

Aeolian environments? (Parker 1970)

Moss, liverworts

Lichens (Andre 1979)

Living fruit (fresh)

Phyllospheres (Preece and Dickinson 1971, Ruinen 1961); living foliage, dead foliage

Various methods of community quantification permit an evaluation and comparison of microcommunity structure. Quantification methods include taxon diversity, taxon richness, taxon equitability, ordination, dendrogram analysis, similarity coefficient analysis, and interspecific association analysis. These types of microcommunity structural assays have been completed on agricultural soils (Clemen and Pedigo 1970), on DDT-treated old field herbaceous sites (Dindal et al. 1975a) and on sewage effluent disposal sites (Dindal et al. 1975b). Even with these available data, many more microcommunities need to be assessed in this manner. The microcommunity approach, linked with quantitative comparisons, provides much information on insular ecology and allows an easily controlled analysis of basic ecological trends and of the degree of human disturbances.

Soil Defined

In this volume, soil is defined in the broadest sense of the term. The basic classification of soil profiles of the United States is described very well by the Soil Survey Staff (1975) of the U.S. Department of Agriculture. Each of the soil taxonomic subgroups within the soil families represents a unique physical–chemical condition, which makes a unique microhabitat for soil organisms.

Soil of the Earth's Mantle

Although some type of biotic activity is associated with all the soil horizons, the organic horizons are usually the places of highest populations and greatest species diversity of soil forms. Unfortunately, organic horizons are fragile and may quickly be destroyed by soil manipulation. When present, organic layers include (a) the litter (L or O) horizon, an accumulation of recognizable plant materials and animal residues; (b) the fermentation (F or O_1) layer, a mass of partially decomposed mixed leaf and animal litter that is conspicuously permeated with hyphae of fungi and actinomycetes; and (c) the humus (H or O_2) horizon, the dark amorphous products of decomposition with no identity as to source.

The L, F, and H horizons all collect over time on the undisturbed surface of mineral soils. Ultimately, the organic matter becomes incorporated within the soil profile of (a) the A mineral horizons, a leached, eluvial, or extraction zone; (b) the B mineral horizons, the enriched, illuvial, or concentration zone; and (c) the C mineral horizons, the physically and chemically weathered parent material.

Aerial Arboreal Soil

High above ground level within the arboreal strata (or tree canopies), organic soil layers form in the crotches of tree branches, in tree holes, in nests of vertebrates and invertebrates, and among epiphytes. Litter accumulates in these arboreal sites, and fermentation and other decomposition processes occur just as they do on the soil surface. Furthermore, soil biota that colonize from the ground are predominantly responsible for the soil formation in arboreal microhabitats. For instance, epiphytic sites of organic soil genesis involve the specific microcommunities in microhabitats formed by the Bromeliaceae, various bryophytes, and numerous forms of lichens. True mosses and lichens are represented from the arboreal zone down the trunks of trees and stems of plants, thus connecting the aerial strata directly with the rock and soil surfaces below.

Litter—Innocuous or Offensive?

Leaves, seeds, fruits, and other plant parts make up a large mass of the components of the litter layer. They have a relatively high C/N ratio, and as a part of the soil organic horizon they are easy and innocuous to work with. Animal residues in the organic horizons are very important to the soil ecosystem mainly because of their relatively low C/N ratio and available nutrient content. Vertebrate and invertebrate nests and the more noxious carrion and fecal matter are certainly significant components of organic layers of the soil which are frequently, and perhaps conveniently, overlooked.

In summary, many soil researchers think of only mineral soil pedons as falling under the definition of soil. However, since the organic O, O_1, and O_2 layers are presented as distinct horizons under the most recent soil classification (Soil Survey Staff 1975), all the functional components of each organic horizon are components of soil. Also, the arboreal formation of soil must be realized and accepted, just as is the character of the organic and mineral horizons.

References

Alexander, M. 1971. *Microbial Ecology.* Wiley, New York. 511 pp. **Allee, W. C., A. E. Emerson, O. Park, T. Park and K. P. Schmidt. 1949.** *Principles of Animal Ecology.* W. B. Saunders, Philadelphia. 837 pp. **Allen, S. E. (ed.). 1974.** *Chemical Analysis of Ecological*

Materials. Wiley, New York. 565 pp. **Andre, H. 1979.** Notes on the ecology of corticolous epiphyte dwellers. 1. The mite fauna of fruticose lichens. Pp. 551–557 in J. G. Rodriguez (ed.). *Recent Advances in Acarology*. Academic Press, New York. **Applehof, M. (ed.). 1981.** *Workshop on the Role of Earthworms in the Stabilization of Organic Residues,* Vols. I, II. Beach Leaf Press, Kalamazoo, Mich. **Auerbach, S. I. 1958.** Soil ecosystem and radioactive waste disposal to the ground. *Ecology* 39(3):522–529. **Behrens, U. and J. Hannes. 1984.** Zum Abbau von Formaldehyde durch adaptierte Bacterien. *Acta Hydrochim. Hydrobiol.* 12(1):39–45. **Benzing, D. H. 1980.** *The Biology of the Bromelaids.* Mad River Press, Eureka, Calif. 305 pp. **Birkby, K. M. and T. F. Preece. 1987.** The ligules of green leaves of cocksfoot grass, *Dactylis glomerata* L. are a microhabitat for bacteria and fungi. *J. Appl. Bacteriol.* 63(6):505–512. **Blair, W. F., A. P. Blair, P. Brodkorb, F. R. Cagle and G. A. Moore. 1968.** *Vertebrates of the United States.* McGraw-Hill, New York. 616 pp. **Bornebusch, C. H. 1930.** The fauna of forest soils. *Forstl. Forsogsvaes. Dan.* 11:1–224. **Buchner, P. 1965.** *Endosymbiosis of Animals with Plant Microorganisms.* Wiley-Interscience, New York. 909 pp. **Buol, S. W., F. D. Hole and R. J. McCracken. 1980.** *Soil Genesis and Classification,* 2nd ed. Iowa State University Press, Ames, Iowa. 406 pp. **Bureau of Toxic Substances Management. 1981.** *Chemical Fact Sheet: Formaldehyde.* New York State Dept. of Health. 1 p. **Burges, N. A. and F. Raw (eds.). 1967.** *Soil Biology.* Academic Press, New York. 532 pp. **Burt, W. H. 1952.** *A Field Guide to the Mammals.* Houghton Mifflin, Boston. 200 pp. **Butcher, J. W., R. M. Snider and R. J. Snider. 1971.** Bioecology of edaphic Collembola and Acarina. *Annu. Rev. Entomol.* 16:249–288. **Christiansen, K. 1964.** Bionomics of Collembola. *Annu. Rev. Entomol.* 9:147–178. **Clemen, R. B. and L. P. Pedigo. 1970.** Collembola populations from selected arable fields. *Proc. North Cent. Branch Entomol. Soc. Am.* 25(2):115–119. **Coleman, D. C. 1976.** A review of root production processes and their influence on soil biota in terrestrial ecosystems. Pp. 417–434 in J. M. Anderson and A. Macfadyen (eds.). *The Role of Terrestrial Aquatic Organisms in Decomposition Processes.* Blackwell Scientific, Oxford. **Crossley, D. A., Jr. and M. Witkamp. 1964.** Forest soil mites and mineral cycling. *Acarologia (Paris) Fasc.h.s.* 137–145. **Crossley, D. A., Jr., G. J. House, R. M. Snider, R. J. Snider and B. R. Stinner. 1984.** The positive interactions in agroecosystems. Pp. 73–81 in R. R. Lawrence, B. R. Stinner and G. J. House (eds.). *Agricultural Ecosystems: Unifying Concepts.* Wiley, New York. **Cruzan, J. 1988.** Teaching ecology with microcosms. *Am. Biol. Teach.* 50(4):226–228. **Darwin, C. 1881.** *The Formation of Vegetable Mould through the Action of Worms, with Observations on Their Habits.* Murray, London. 326 pp. **Dickinson, C. H. and G. J. F. Pugh (eds.). 1974.** *Biology of Plant Litter Decomposition,* 2 vols. Academic Press, New York. 775 pp. **Dindal, D. L. (ed.). 1973a.** *Soil Microcommunities I.* USAEC, CONF-711076. NTIS, Springfield, Va. 285 pp. **1973b.** Microcommunities defined. Pp. 2–6 in D. L. Dindal (ed.). *Soil Microcommunities I.* USAEC, CONF-711076. NTIS, Springfield, VA. **(ed.). 1977.** *Biology of Oribatid Mites.* SUNY College of Environmental Science and Forestry, Syracuse, N.Y. 122 pp. **1978.** Soil organisms and stabilizing wastes. *Compost Sci./Land Util.—J. Waste Recycling* 19(4):8–11. **(ed.). 1980.** *Soil Biology as Related to Land Use Practices.* Proc. 7th International Colloquium of Soil Zoology, Syracuse, 1979. USEPA-560/13-80-038. U.S. Environmental Protection Agency, Washington, D.C. 880 pp. **1985.** Soil animals and soil fabric production: facts and perceptions. *Quaest. Entomol.* 12(4):587–594. **Dindal, D. L., D. W. Folts and R. A. Norton. 1975a.** Effects of DDT on community structure of soil microarthropods in an old field. Pp. 505–513 in Jan Vanek (ed.). *Progress in Soil Zoology.* Dr. W. Junk, The Hague, The Netherlands. **1975b.** Effects of sewage effluent disposal on community structure of soil invertebrates. Pp. 419–427 in Jan Vanek (ed.). *Progress in Soil Zoology.* Dr. W. Junk, The Hague, The Netherlands. **Drummond, H. 1887.** On the termite or the tropical analogue of earth-worm. *Proc. R. Soc. Edinb.* 13(1884–85):137–146. **Eaton, T. H. and R. F. Chandler. 1942.** The fauna of forest-humus layers in New York. *Cornell Univ. Agric. Exp. Stn. Mem.* 247:1–26. **Edmondson, W. T. (ed.). 1959.** *Ward and Whipple's Fresh-Water Biology,* 2nd ed. Wiley, New York. 1248 pp. **Edwards, C. A. and J. R. Lofty. 1972.** *Biology of Earthworms.* Chapman & Hall, London. 283 pp. **Edwards, C. A. and G. K. Veeresh (eds.). 1978.** *Soil Biology and Ecology in India.* UAS Tech. Ser. 22. University of Agricultural Science, Hebbal, Bangalore, India. 352 pp. **Edwards, C. A., B. R. Stinner, D. A. Stinner and S. Rabatin (eds.). 1988.** *Biological Interactions in Soil.* Elsevier, Amsterdam. 380 pp. **Elton, C. S. 1966.** *The Pattern of Animal Communities.* Wiley, New York. 432 pp. **Engleman, M. D. 1961.** The role of soil arthropods in the energetics of an old field community. *Ecol. Monogr.* 31:221–238. **Fashing, N. J. 1979.** *Naiadacarus arboricola,* a

"K-selected" acarid mite. Pp. 525–529 in J. G. Rodriguez (ed.). *Recent Advances in Acarology*, Vol. I. Academic Press, New York. **Freckman, D. (ed.). 1982.** *Nematodes in Soil Ecosystems.* University of Texas Press, Austin, Tex. 206 pp. **Ghilarov, M. S. 1947.** Distribution of humus, root-systems and soil invertebrates within the soil of the walnut forests of the Ferhana mountain range. *Dokl. Akad. Nauk SSSR (N.S.)* 55:49–52. **Gray, T. R. G. and S. T. Williams. 1971.** *Soil Microorganisms.* Hafner Press, New York. 240 pp. **Guyer, G. E. (ed.). 1970.** *Pesticides in the Soil: Ecology, Degradation and Movement.* Proceedings of the International Symposium, Michigan State University, East Lansing, Mich. 144 pp. **Haarløv, N. 1960.** Microarthropods from Danish soils: ecology, phenology. *Oikos Suppl.* 3:1–176. **Hamilton, W. E. and D. L. Dindal. 1983.** The vermisphere concept: earthworm activity and sewage sludge. *BioCycle—J. Waste Recycling* 24(5):54–55. **Huang, P. M. and M. Schnitzer (eds.). 1986.** *Interactions of Soil Minerals with Natural Organics and Microbes.* SSA Spec. Publ. 17. Soil Science Society of America, Madison, Wis. 606 pp. **Hudson, H. J. 1965.** The ecology of fungi on plant remains above the soil. *New Phytol.* 67:837–874. **Hungate, R. E. 1966.** *The Rumen and Its Microbes.* Academic Press, New York. 533 pp. **1978.** Gut microbiology. Pp. 258–264 in M. W. Loutit and J. A. R. Miles (eds.). *Microbial Ecology.* Springer-Verlag, Berlin. **Hutchinson, G. E. 1950.** Survey of existing knowledge of biogeochemistry. 3. The biogeochemistry of vertebrate excretion. *Bull. Am. Mus. Nat. Hist.* 96:1–554. **Jacot, A. P. 1935.** Wild life and the forest carpet. *Sci. Mon. N.Y.* 40:425. **1936.** Soil Structure and soil biology. *Ecology* 17:359–379. **1940.** The fauna of the soil. *Q. Rev. Biol.* 15:28–58. **Jenny, H. 1980.** *The Soil Resource, Origin and Behavior,* Springer-Verlag, New York. 377 pp. **Kevan, D. K. McE. 1968.** *Soil Animals.* Witherby, London. 244 pp. **1985.** Soil zoology, then and now—mostly then. *Quaest. Entomol.* 12(4):371.7–472. **Kevan, D. K. McE. and S. B. Hill. 1979.** Fauna. Pp. 153–161 in R. W. Fairbridge and C. W. Finkl (eds.). *The Encyclopedia of Soil Science,* Part 1. (Encyclopedia of Earth Science, 12) **Klute, A. (ed.). 1986.** *Methods of Soil Analysis,* 2nd ed., Part 1. Agron. Ser. Monogr. 9. [A. L. Page, R. H. Miller and D. R. Keeney (eds.). Part 2.] Soil Science Society of America and Agronomy Society of America, Madison, Wis. 2400 pp. **Kubiëna, W. L. 1953.** *The Soils of Europe.* Mirby, London. 317 pp. **Kühnelt, W. 1961.** *Soil Biology with Special Reference to the Animal Kingdom* (trans. N. Walker). Faber & Faber, London. 397 pp. **Laessle, A. M. 1961.** A microlimnological study of Jamaican bromeliads. *Ecology* 42(3):499–517. **Lee, K. E. 1985.** *Earthworms: Their Ecology and Relationships with Soils and Land Use.* Academic Press, New York. 411 pp. **Leonova, V. E., A. V. Abramov and V. F. Karpukhin. 1985.** Kinetics of formaldehyde biodegradation. *Microbiology* 54(1):124–128. **Maguire, B., Jr. 1971.** Phytotelmata: biota and community structure determination in plant-held waters. *Annu. Rev. Ecol. Syst.* 2:439–464. **Magyarosy, A. and J. G. Hancock. 1972.** Microbial populations of the laimosphere of squash (*Cucurbita maxima*). *Plant Soil* 37(1):187–190. **Margalef, R. 1968.** *Perspectives in Ecological Theory.* University of Chicago Press, Chicago. 111 pp. **Marshall, E. 1987.** EPA indicts formaldehyde, 7 years later. *Science* 236(4800):381. **Martin, M. M. 1987.** *Invertebrate–Microbial Interactions.* Cornell University Press. Ithaca, N.Y. 176 pp. **McMillian, W. W. and B. R. Wiseman. 1972.** Insect species present on sorghum heads of various stages of maturity. *J. Ga. Entomol. Soc.* 7(3):179–182. **Metz, L. J. and D. L. Dindal. 1975.** Collembola populations and prescribed burning. *Environ. Entomol.* 4(4):583–587. **Metz, L. J. and M. H. Farrier. 1973.** Prescribed burning and populations of soil mesofauna. *Environ. Entomol.* 2(3):433–440. **Metz, L. J., R. M. Reeves, and R. A. Norton. 1977.** The works of A. P. Jacot. Pp. 1–31 in D. L. Dindal (ed.). *Biology of Oribatid Mites.* SUNY College of Environmental Science and Forestry, Syracuse, N.Y. **Mitchell, M. J. and J. P. Nakas (eds.). 1986.** *Microfloral and Faunal Interactions in Natural and Agro-ecosystems.* Martinus Nijhoff/Dr. W. Junk, Dordrecht, The Netherlands. 505 pp. **Mohr, C. O. 1943.** Cattle droppings as ecological units. *Ecol. Monogr.* 13(3):275–298. **Müller, P. E. 1889.** Recherches sur les formes naturelles de l'humus. *Ann. Sci. Agron. Fr. Etrang.* (*Nancy*) 6:85–423. **Nero, A. V. 1988.** Controlling indoor air pollution. *Sci. Am.* 258(5):42–48. **Norton, R. A. and D. L. Dindal. 1976.** Structure of the microarthropod community in Lake Ontario beach debris. *Environ Entomol.* 5(4):773–779. **Odum, E. P. 1971.** *Fundamentals of Ecology.* 3rd ed. W. B. Saunders, Philadelphia. 574 pp. **Ordish, G. 1981.** *The Living American House.* William Morrow, New York. 320 pp. **Parker, B. C. 1970.** Life in the sky. *Nat. Hist.* 79(8):54–59. **Pearse, A. S. 1946.** Observations on the microfauna of Duke Forest. *Ecol. Monogr.* **Pearse, A. S., M. T. Patterson, J. S. Rankin and G. W. Wharton. 1936.** The ecology of *Passalus carnutus* Fabricius. *Ecol. Monogr.* 6:456–490. **Philips, J. R. and D. L.**

Dindal. 1977. Raptor nests as a habitat for invertebrates: a review. *Raptor Res.* 11(4):86–96. **1979.** Decomposition of raptor pellets. *Raptor Res.* 13(4):102–111. **Phillipson, J. (ed.). 1970.** *Methods of Study in Soil Ecology.* UNESCO, Paris. 303 pp. **Pimm, S. L. 1982.** *Food Webs.* Chapman & Hall, London. 219 pp. **Preece, T. F. and C. H. Dickinson (eds.). 1971.** *Ecology of Leaf Surface Microorganisms.* Academic Press, New York. 640 pp. **Reed, H. B. 1958.** A study of dog carcass communities in Tennessee, with special reference to the insects. *Am. Midl. Nat.* 59(1):213–245. **Reichle, D. (ed.). 1970.** *Analysis of Temperate Forest Ecosystems.* Springer-Verlag, Berlin. 304 pp. **Rovira, A. D. 1965.** Interactions between plant roots and soil microorganisms. *Annu. Rev. Microbiol.* 19:241–266. **Ruinen, J. 1961.** The phyllosphere. I. An ecologically neglected milieu. *Plant Soil* 15(2):81–109. **Satchell, J. E. (ed.). 1983.** *Earthworm Ecology from Darwin to Vermiculture.* Chapman & Hall, London. 495 pp. **Sengbusch, H. G. 1977.** Review of oribatid mite–anoplocepholan tapeworm relationships. Pp. 86–102 in D. L. Dindal (ed.). *Biology of Oribatid Mites.* SUNY College of Environmental Science and Forestry, Syracuse, N.Y. **Slykhuis, J. T. 1947.** Studies on *Fusarium culmorum* blight of crested wheat and brome grass seedlings. *Can. J. Res. Sect. C* 25:155–180. **Smith, S. S. E. 1980.** Mycorrhizas of autotrophic higher plants. *Biol. Rev. Camb. Philos. Soc.* 55:475–510. **Smith, K. G. V. 1986.** A Manual of Forensic Entomology. Cornell University Press, Ithaca, N.Y. 224 pp. **Smith, M. S. and C. W. Rice. 1986.** The role of microorganisms in the soil nitrogen cycle. Pp. 245–284 in M. J. Mitchell and J. P. Nakas (eds.). *Microfloral and Faunal Interactions in Natural and Agro-Ecosystems.* Martinus Nijhoff/Dr. W. Junk, Dordrecht, The Netherlands. **Soil Survey Staff. 1975.** *Soil Taxonomy.* Agric. Handb. 436. U.S. Dept. of Agriculture, Washington, D.C. 754 pp. **Spence, J. R. (ed.). 1985.** Faunal influences on soil structure. *Quaest. Entomol.* 21(4):371.1–700. **Stevenson, B. G. and D. L. Dindal. 1987a.** Insect effects on decomposition of cow dung in microcosms. *Pedobiologia* 30(2):81–92. **1987b.** Functional ecology of coprophagous insects: a review. *Pedobiologia* 30(3):285–297. **Stotzky, G. 1972.** Activity, ecology and population dynamics of microorganisms in soil. *CRC Crit. Rev. Microbiol.* 2:59–137. **Swift, M. J., O. W. Heal and J. M. Anderson. 1979.** *Decomposition in Terrestrial Ecosystems.* University of California Press, Berkeley, Calif. 372 pp. **Todd, R. L. and J. E. Giddens (eds.). 1984.** *Microbial–Plant Interactions.* ASA Spec. Publ. 47. Soil Science Society of America, Madison, Wis. 68 pp. **Verona, O. 1963.** Interaction entre la graine en germination et les microorganismes telluriques. *Ann. Inst. Pasteur (Paris)* 105:75–84. **Waksman, S. A. and J. P. Martin. 1939.** The role of microorganisms in the conservation of soil. *Science* 9:304–305. **Wallwork, J. A. 1970.** *Ecology of Soil Animals.* McGraw-Hill, New York. 283 pp. **1976.** *The Distribution and Diversity of Soil Fauna.* Academic Press, New York. 355 pp. **Weber, N. A. 1972.** The fungus-culturing behavior of ants. *Am. Zool.* 12:577–587. **Werner, M. R. and D. L. Dindal. 1987.** Nutritional ecology of soil arthropods. Pp. 815–836 in F. Slansky, Jr., and J. G. Rodriguez (eds.). *Nutritional Ecology of Insects, Mites and Spiders.* Wiley, New York. **Whittaker, J. B. 1970.** Cercopid spittle as a microhabitat. *Oikos* 21(1):59–64. **Wilde, S. A. 1958.** *Forest Soils: Their Properties and Relation to Silviculture.* Ronald Press, New York. 537 pp. **Winston, P. W. 1956.** The acorn microsere, with special reference to arthropods. *Ecology* 37(1):120–132.

2

Soil Bacteria

T. R. G. GRAY

It is probably true to say at the time of writing (1980) that almost any known bacterium and a good many more can be found in soil. If it is necessary to identify an isolate, the only safe recourse is to start by using one of the standard works of identification. Perhaps the most useful of these are *Bergey's Manual of Determinative Bacteriology* (Buchanan and Gibbons 1974, Krieg and Holt 1984, Sneath, Mair, Sharpe and Holt (1986) and Cowan and Steel (1965), although the latter is concerned primarily with bacteria of medical importance.

However, whereas almost any bacterium might exist in soil for short periods, a more restricted range of organisms live and grow and attain relatively high population levels (i.e., about 10^4 or 10^5 per gram of soil). It is, therefore, only a more restricted range of organisms that are likely to appear frequently on isolation plates or in selective liquid media.

The organisms that are isolated often do not conform exactly with published descriptions of organisms. This phenomenon is due to the enormous diversity of slightly differing microenvironments that exist in soil, permitting the survival and growth of a much larger number of variants than would be likely in a homogeneous environment. For this reason, it is unwise in preliminary investigations or in teaching exercises to attempt identification at the specific level.

To understand many of the microbial transformations that occur in soil, it is necessary to know the composition of its microbial community, the functions of members of the predominant taxa, and the environmental conditions that modify the nature of the populations and their activities. As often as not, precise ecological work rests on stable classification and reliable and workable identification of the predominant bacteria (Gray 1969).

Until recently, the classification of bacteria indigenous to soil was based on a small number of subjectively weighted, morphological, nutritional, and physiological characters (Goodfellow 1969, Goodfellow et al. 1976). Such artificial classifications contributed to the formation of heterogeneous groups and were unable to accommodate new discoveries. It is now generally accepted that the most useful classifications are those based on different types of taxonomic criteria. Thus the consistency found between different types of taxonomic information is a measure of the confidence that can be placed in the classification (Sneath 1976, Jones and Sackin 1980).

In the past 20 years the introduction of new, more exacting biochemical, chemical, genetical, and numerical methods has led to changing views on how bacteria ought to be classified and identified (see Goodfellow and Board 1980). Established genera have been circumscribed more carefully, new species reduced to synonyms of established species, and misclassified strains and tests of presumptive diagnostic value highlighted. Modern taxonomic methods have generally made their greatest impact on the classification of groups where reliance on form and function proved most inadequate. Thus major revisions have taken place in the classification and identification of coryneform bacteria (see Bousfield and Callely 1978) and actinomycetes (Minnikin and Goodfellow, 1980), organisms that form an integral part of many balanced microbial communities in soil (Goodfellow and Cross 1974, Gyllenberg and Eklund 1974). The impact of some of these developments is reflected in the diagnostic key presented below. In the future, however, identification methods will reflect advances in disciplines such as chemotaxonomy (Minnikin and Goodfellow 1980) and developments in computer-assisted methods of identification (Sneath 1978).

The keys that follow attempt to place an organism in one or another of the common genera that have been encountered on general isolation plates or in specialized selective media. Since the key is not comprehensive, it must be borne in mind that the organisms being examined may not be dealt with in the key, in which case it must be labeled as unidentifiable or referred to a more specialized key. If an organism does not conform exactly with the descriptions given, other checks must be made.

It is not always easy to decide when a soil becomes a mud, so a few organisms that are characteristics of muds and which might be encountered have been included. Most animal and plant pathogens and marine forms have been excluded, as well as most forms found in body cavities, the intestine and rumen, and in fermenting animal and plant products. If contamination from one of these sources is likely, the key should not be used.

The genera that have been included are all to be found in the eighth edition of *Bergey's Manual of Determinative Bacteriology* (Buchanan and Gibbons 1974) and the classification used there has also been adopted in most cases. However, the steps involved in the identification of these genera are not always the same. Since nonsoil forms are omitted, it has been possible to use tests that would not distinguish between genera and groups if organisms from other sources were present. Further genera maybe found in Krieg and Holt (1984) and subsequent volumes of *Bergey's Manual of Systematic Bacteriology*, published while this article was in press.

KEY TO THE IDENTIFICATION OF THE MAIN GROUPS OF SOIL BACTERIA

1a Phototrophic, usually only in mud or stagnant water . . . . **Group A**
1b Chemotrophic **2**
2a Chemoorganotrophic **4**
2b Chemolithotrophic **3**
3a Derive energy from oxidation of nitrogen, sulfur, or iron compounds, do not produce methane from carbon dioxide . . . . **Group H**
3b Not as above, do produce methane from carbon dioxide . . . . . . **Group I**
4a Cells glide . . . . **Group B**
4b Cells do not glide **5**
5a Have appendages other than flagella or pili or reproduce by budding . . . **Group C**
5b Not as above. **6**
6a Gram negative. **7**
6b Gram positive **9**
7a Strict aerobes, straight rods or cocci, usually oxidase positive . . . **Group D**
7b Strict aerobes, very short rods or cocci, oxidase negative . . . . . **Group G**
Anaerobic **8**
8a Facultative anaerobes . . . . **Group E**
8b Strict anaerobes . . . . **Group F**
9a Cocci . . . . **Group J**
9b Rods, at least in 12- to 24-hour cultures **10**
10a Endospores produced . . . . **Group K**
10b No endospores produced **11**
11a Straight rods. . . . . **Group L**
11b Irregular rods, often becoming coccoid in older cultures . . . . . **Group M**
11c Filamentous in 24-hour-old cultures . . . . **Chapter 3**

DESCRIPTIONS AND KEYS OF MAIN GROUPS OF BACTERIA

Group A: The Phototrophic Bacteria

These photosynthetic bacteria all contain bacteriochlorophylls and carotenoids. They require light, are anaerobic, and use external electron donors other than water (cf. **Cyanobacteria** and higher plants). They are, therefore, found only in habitats where these conditions coincide and are unlikely to be found in many soils They are more characteristic of muds and sediments with surface water. They are unlikely to be isolated except on selective media.

1a Cells contain bacteriochlorophyll *a* and *b* (generally purple, pink, or orange-brown in color) **2**

1b Cells contain bacteriochlorophyll *c* and *d* (green or brown in color) . . . . . **Chlorobiaceae**

2a Cells able to grow with sulfide as sole electron donor for photosynthesis . . **Chromatiaceae**

2b Usually unable to grow with sulfide as sole electron donor for photosynthesis; photoassimilate simple organic substances; most strains require vitamins for growth . . . . **Rhodospirillaceae**

Unfortunately, there are exceptions to this simplified key and confirmation will require more elaborate investigations and the use of special keys.

Group B: The Gliding Bacteria

These bacteria are generally motile but do not possess flagella. They move only when in contact with a solid surface or an air–water interface. They have gram-negative, rod-shaped cells, often with tapered ends. In many genera, the cells are able to aggregate and form fruit bodies in which the rods are transformed into cysts or spores. These bacteria are able to attack a wide range of polymers and may also lyse other microorganisms, under strictly anaerobic conditions. To isolate them, agar media containing dead bacterial cells or cellulose or other suitable polymers are used. Fruit body formation occurs mainly on dilute, nutrient-poor media (e.g., water agar). Many of the forms that form fruit bodies grow on dung rather than in the soil itself.

1a May form fruit bodies **2**
1b Incapable of forming fruit bodies, often produce carotenoids **4**
2a Vegetative cells with blunt ends. . . . . **Polyangiaceae**
2b Vegetative cells with tapered ends, microcysts produced **3**
3a Microcysts rod-shaped . . . . **Archangiaceae or Cystobacteriaceae**
3b Microcysts spherical or oval . . . . **Myxococcaceae (*Myxococcus*)**
4a Produce microcysts. . . . . ***Sporocytophaga***
4b Do not produce microcysts **5**
5a Attack cellulose or chitin . . . . ***Cytophaga***
5b Do not attack cellulose or chitin. . . . . ***Flexibacter***

Group C: Budding and/or Appendaged Bacteria

This group includes bacteria that have appendages other than flagella or pili (i.e., stalks or filaments) and/or reproduce by budding. In some, the appendages are formed from the cell wall (prosthecae), while in others the appendage resembles a capsular secretion. They are Gram-negative.

1a Cells prosthecate **2**

1b Cells not prosthecate, rod-shaped, spirally twisted cells often in star-shaped rosettes, reproduces by budding . . . . ***Seliberia***

2a Appendages derived from the cell wall. **3**

2b Appendages excreted, cells kidney shaped on long stalks, often covered with iron or manganese deposits . . . . ***Gallionella***

3a Prosthecae have reproductive function, cells often oval with filamentous outgrowths and flagellate swarm cells **4**

3b Prosthecae have no reproductive function **5**

4a Ferric or manganese deposits on the cells . . . . *Pedomicrobium*
4b No such deposits on the cells . . . . *Hyphomicrobium*
5a Prostheca extends from one pole and attaches cell to substrate. . . . . *Caulobacter*
5b Prostheca arises excentrically from one pole or laterally, does not attach cell to substrate . . . . *Asticcacaulis*

Group D: Gram-Negative, Aerobic Rods and Cocci, usually Oxidase Positive

This group of bacteria, which has a respiratory metabolism, is very difficult to differentiate into genera. There are no truly satisfactory characters that will distinguish *Pseudomonas* from *Xanthomonas,* or *Alcaligenes* from *Agrobacterium.* Characters such as pigmentation and motility are notoriously unreliable, as they are easily lost in culture or are capable of misinterpretation. They may also be confused with *Acinetobacter* in group G.

1a Fix atmospheric nitrogen 2
1b Do not fix atmospheric nitrogen 7
2a Use only one-carbon compounds (e.g., methane, methanol) as sole carbon and energy sources . . . . *Methylomonas, Methylococcus*
2b Utilize a variety of organic compounds as carbon and energy source. . . . 3
3a Fix nitrogen in symbiotic relationship with legumes. . . . . *Rhizobium*
3b Fix nitrogen in free-living condition, aerobically 4
4a Large ovoid cells, often with extracellular slime 4
4b Small rods 5
5a Produce cysts . . . . *Azotobacter*
5b Do not produce cysts . . . . *Azotomonas*
6a Catalase positive, bipolar lipoid bodies. . . . . *Beijerinckia*
6b Catalase negative, many lipoid bodies . . . . *Derxia*
7a Straight or curved rods, generally polar flagellate 8
7b Rods, generally peritrichously flagellate 9
8a Usually yellow carotenoids present, usually plant pathogenic. . . . . *Xanthomonas*
8b Diffusible fluorescent pigments sometimes present, often nonpigmented, sometimes have non-diffusible yellow, orange, brown or pink pigments . . . *Pseudomonas*
9a Rods, generally with 1–4 peritrichous flagella, mostly cause stem hypertrophies on plants; produce 3-ketolactose . . . . *Agrobacterium*
9b Rods, 1–4 peritrichous flagella, saprophytic . . . . *Alcaligenes*

Group E: Gram-Negative, Facultatively Anaerobic Rods

Unlike group D, these bacteria are facultatively anaerobic. Acid is produced from the fermentation of glucose and many of the strains also produce a mixture of gases. There are two main groups, one usually peritrichously flagellate and oxidase negative (Enterobacteriaceae) and the other polar flagellate oxidase positive and sensitive to 2,4-diamino-6,7-diisopropyl pteridine (0/129) (Vibrionaceae), but there are other important genera, such as *Chromobacterium* and *Flavobacterium,* which are not clearly related to either group. *Flavobacterium* is a genus that is particularly difficult to classify, especially as it produces little if any acid from glucose, and fermentation is not conspicuous. They are easily confused with gliding bacteria.

1a Produce violet-colored colonies on agar media . . . . . . ***Chromobacterium***
1b Not as above. **2**
2a Produce insoluble yellow, orange, red, or brown pigments, metabolism respiratory or weakly fermentative . . . . ***Flavobacterium***
2b Strong fermentative metabolism, usually nonpigmented but may have yellow or pink pigments **3**
3a Peritrichously flagellate, oxidase negative **4**
3b Polar flagellate, oxidase positive, and sensitive to 2,4-diamino-6,7-diisopropyl pteridine . . . . **Vibrionaceae**
(usually in water, not soil)
4a Grow poorly at 37°C, rarely produces urease. . . . . ***Erwinia***
4b Grow well at 37°C, may or may not produce urease. **5**
5a Deaminate phenylalanine . . . . ***Proteus***
5b Do not deaminate phenylalanine **6**
6a Produce prodigiosin, a pink pigment. . . . . ***Serratia***
6b Do not produce prodigiosin **7**
7a Vosges–Proskauer positive **8**
7b Vosges–Proskauer negative **11**
8a Nonmotile rods, do not decarboxylate ornithine . . . . ***Klebsiella***
8b Motile rods, decarboxylate ornithine **9**
9a Produce acid from sorbitol. . . . . ***Enterobacter***
9b Do not produce acid from sorbitol. **10**
10a Produce DNase. . . . . ***Serratia***
10b Do not produce DNase . . . . ***Hafnia***
11a Use citrate as sole carbon source. **12**
11b Do not use citrate as sole carbon source **13**
12a Decarboxylate lysine . . . . ***Salmonella***
12b Do not decarboxylate lysine . . . . ***Citrobacter***
13a Produce gas from glucose at 37°C, decarboxylate lysine . . . . ***Escherichia***
13b Do not produce gas from glucose at 37°C, do not decarboxylate lysine . . . ***Shigella***

Group F: Gram-Negative, Anaerobic Rods

Most of the organisms in this group are isolated from natural cavities of humans or animals or the alimentary canal or the rumen, and some species are pathogenic. One genus, *Desulphovibrio,* may be encountered in soil. The cells are rod-shaped, polar flagellate, and obtain energy by anaerobic respiration. Sulfates are the usual terminal electron acceptors and these are reduced to sulfides. When grown in media containing sulfates and iron salts, the cultures turn black. They are only likely to be isolated on selective media.

Group G: Gram-Negative Cocci and Coccobacilli, Oxidase Negative

Most genera in this group are parasitic, and only *Acinetobacter* spp. are saprophytic. Acinetobacters are rod-shaped in the exponential phase of growth but become spherical in the stationary phase (cf. *Arthrobacter* spp., which are Gram-positive or Gram-variable). They divide in one plane, grow on simple defined media, are resistant to penicillin, are oxidase negative, and fail to reduce nitrates. The genus includes some strains which used to be termed *Achromobacter,* although others of this genus have been transferred to *Alcaligenes* (see group D).

Group H: Gram-Negative Chemolithotrophic Bacteria

These bacteria are only likely to be encountered on selective media containing no organic matter and having suitable reduced inorganic substances which can be oxidized to obtain both energy and reducing power for the incorporation of carbon dioxide. In recent years, the number of recognized genera of these organisms has increased, but many of these are marine forms and are excluded here. They participate in important processes such as nitrification in agricultural soils and sulfur oxidation, which may have important applications in the extraction of minerals from ores. Organisms not grown in pure culture are excluded from the keys.

1a Oxidize ammonia or nitrite 2
1b Oxidize sulfur or sulfur-containing compounds, motile rods . . ***Thiobacillus***
2a Oxidize nitrite to nitrate, short rods . . . . ***Nitrobacter***
2b Oxidize ammonia to nitrite 3
3a Straight rods . . . . ***Nitrosomonas***
3b Spiral cells . . . . ***Nitrosospira***
3c Spherical cells . . . . ***Nitrosococcus***
3d Lobed, pleomorphic cells . . . . ***Nitrosolobus***

Group I: Methane-Producing Bacteria

These strictly anaerobic organisms are morphologically diverse but are all able to form methane. This is done by the reduction of carbon dioxide with reducing power obtained from the oxidation of hydrogen or formate, or by the fermentation of acetate or methanol. They frequently exist in association with other bacteria, which led to the belief that other organic compounds were utilized by these organisms. They are, therefore, unlikely to be isolated except on selective media, and in any case are found mainly in anaerobic muds.

1a Rod-shaped cells or cocci in chains . . . . ***Methanobacterium***
1b Cocci not in chains 2
2a Cocci occur singly, in pairs, or in irregular clumps . . . . . . ***Methanococcus***
2b Large cocci in irregular packets . . . . ***Methanosarcina***

Group J: Gram-Positive Cocci

Characteristically, the majority of the soil representatives of this group, placed in the family Micrococcaceae, divide in either two or three planes, so that they form either irregular clumps or regular packets. They can be respiratory or fermentative but never produce gas from glucose. The soil forms are aerobic, with the exception of *Sarcina* (family Peptococcaceae), which is a strict anaerobe. Some other forms may be found in soil (e.g., strains of *Streptococcus faecalis*). Other streptococci are excluded from the key.

1a Cells form only pairs or chains of cocci, cells elongated in direction of the chain . . . . ***Streptococcus***
1b Cells form clusters or packets 2
2a Strictly anaerobic . . . . ***Sarcina***
2b Aerobic or facultatively anaerobic 3
3a Acid from glucose under aerobic conditions only, sensitive to erythromycin (0.4 μg/ml) . . . . ***Micrococcus***
3b Acid from glucose under both aerobic and anaerobic conditions (the latter sometimes weakly), resistant to erythromycin (0.4 μg/ml) . . . ***Staphylococcus***

Group K: Endospore-Forming Bacteria

All these organisms produce endospores which may be distinguished from the vegetative cells that produce them, as they are highly refractile, difficult to stain, and have considerable resistance to heat. They are mostly Gram-positive rod-shaped organisms which are either nonmotile or peritrichously flagellate. They are among the commonest of organisms on general isolation plates.

1a Rod-shaped 2
1b Spherical . . . . ***Sporosarcina***
1c Mycelial (see also Chapters 3 and 4) . . . . ***Thermoactinomyces***
2a Aerobic, catalase positive . . . . ***Bacillus***
2b Anaerobic, usually catalase negative 3
3a Reduce sulfate to sulfide . . . . ***Desulfotomaculum***
3b Do not reduce sulfate to sulfide . . . . ***Clostridium***

Group L: Gram-Positive, Asporogenous, Rod-Shaped Bacteria

These bacteria are not generally encountered in soil samples, but some members of the genus *Lactobacillus,* characteristic of fermenting animal and plant products (e.g., milk, silage, or cavities of warm-blooded animals), are occasionally reported from soil. They are fermentative and have complex nutritional requirements, although growth generally occurs in the presence of air. They produce lactate from carbohydrate metabolism and often grow at pH 5. They are catalase negative.

Group M: Coryneform Bacteria

The term "coryneform bacterium" delineates a broad morphological group, sometimes imperfectly, and does not imply relatedness within it (Keddie 1978). The organisms generally included in this group have respiratory metabolism and are Gram-positive or Gram-variable and non-acid fast. Their morphology often changes during the growth cycle from rod to short rod or coccus. The rods are often irregular in shape and may be clubbed, bent, or curved, sometimes giving rise to angular formations in the shape of a "V". Some cells may branch but a mycelium is not formed. Some anaerobic strains have been included in the group from time to time.

It is now generally recognized that there is no sharp morphological division between corynebacteria and actinomycetes such as *Mycobacterium* and *Rhodococcus,* where mycelium formation is rarely advanced. The genera in this group are being redefined, often on the basis of chemotaxonomic data. Thus it has been suggested (Keddie 1978, Minnikin et al. 1978) that *Corynebacterium* be restricted to organisms having *meso*-diaminopimelic acid, arabinose, and galactose in the cell wall, guanine plus cytosine ratios in the range 48–59%, dihydrogenated menaquinones (eight or nine isoprene units), relatively low molecular weight mycolic acids (22–23 carbon atoms), and which are facultative anaerobes. Similarly, *Arthrobacter* should probably be restricted to strains that possess peptidoglycans based on L-lysine, although this property is not restricted to the genus. All *Cellulomonas* strains would contain ornithine and lack galactose in the cell wall, whereas the type species of *Brevibacterium* would contain *meso*-diaminopimelic acid and ribose. Unfortunately, proposals such as these leave some strains unallocated to genera. Furthermore, this information cannot be used satisfactorily in keys at the present time, as there are no simple tests for these properties that are available in most laboratories. For most purposes, it is probably desirable to leave identification at the group level.

1a Cells change from rods to cocci during the cell cycle, do not attack cellulose Probably ***Arthrobacter***
1b Cells shorten during the life cycle, not many cocci produced 2
2a Attack cellulose . . . . ***Cellulomonas***
2b Do not attack cellulose 3
3a Facultatively anaerobic . . . . ***Corynebacterium***
3b Obligate aerobes . . . . ***Brevibacterium* or *Corynebacterium***

Isolation Media

The media listed are those that have proved useful for isolating either a wide range of organisms, or else are necessary for the selective isolation of special groups. If solid media are required, agar at the maker's recommended concentration should be used (usually 1–2%).

Soil Extract Medium (James 1958). Autoclave 1 kg soil with 1 liter of water for 20 minutes at 1 kg/cm^2. Strain off the solid residues and restore the volume to 1 liter. Add a small amount of calcium sulfate if the medium is cloudy, allow to stand, and then filter. This extract may be used directly for isolation or diluted or have nutrients added (e.g., K_2HPO_4, 0.025%; glucose, 0.1%; yeast extract, 0.5%; K_2HPO_4, 0.02%). Incorporation of actidione (50 μg/ml) and nystatin (50 $\mu g\ ml^{-1}$) will suppress most fungal growth.

Peptone Yeast Extract Medium (Goodfellow et al. 1968). Peptone, 5.0 g; yeast extract (spray dried), 1.0 g; ferric phosphate, 0.01 g; water 1 liter; pH 7.2.

Sulfur-Photosynthetic Bacteria (Whittenbury 1971). Dissolve $MgSO_4 \cdot 7H_2O$ (1.0 g), $CaSO_4 \cdot 2H_2O$ (0.5 g), $CaCO_3$ (0.05 g), $FeSO_4 \cdot 7H_2O$ (10.0 mg), and either $(NH_4)_2SO_4$ or ammonium acetate in 700 ml of distilled water. Autoclave and cool. Dissolve KH_2PO_4 (1.0 g) in 100 ml of distilled water. Add to foregoing cooled medium. Prepare 10 ml of a trace element solution ($FeSO_4 \cdot 7H_2O$, 200 mg; $ZnSO_4 \cdot 7H_2O$, 10 mg; $MnCl_2 \cdot 4H_2O$, 3 mg; H_3BO_3, 30 mg; $CoCl_2 \cdot 6H_2O$, 20 mg; $CaCl_2 \cdot 2H_2O$, 1 mg; $NiCl_2 \cdot 6H_2O$, 2 mg; $Na_2MoO_4 \cdot 2H_2O$, 500 mg; water, 1 liter). Autoclave and cool and add to the foregoing medium. Prepare 10 ml of growth factor solution (biotin, 0.1 mg; calcium pantothenate, 0.1 mg; *p*-aminobenzoic acid, 1.0 mg; nicotinic acid, 2.0 mg; thiamine hydrochloride, 2.0 mg; water, 1 liter). Filter sterilize and add to cooled medium. Add 0.5–0.05 g of Na_2S (former for green and purple forms, latter for gas-vacuole-forming types) after partial neutralization with sterile HCl. Magnetic stirring and pH checks between additions to the basal medium are required. Adjust the pH to 6.8 for *Chlorobium,* 7.2 for *Chromatium,* and 7.0 for gas-vacuole-forming types. Dispense into sterile bottles leaving 2–3 ml of gas space. Put in a cold room for 2–3 days before use. Incubation conditions: high light, 540–2150 lux

(50–200 footcandles), 25–30°C (*Chlorobium*); high light and infrared filter (*Chromatium*); low light, 54–215 lux (5–20 footcandles), 10–20°C (gas-vacuole-forming types).

Non-Sulfur-Photosynthetic Bacteria (Whittenbury 1971). Dissolve NH_4Cl (0.5 g), $MgSO_4 \cdot 7H_2O$ (0.4 g), NaCl (0.4 g), $CaCl_2 \cdot 2H_2O$ (0.05 g), electron donor (e.g., succinate) (2.0–0.2 g), yeast extract (if required) (1.0 g). Dispense into bottles after melting the agar, autoclave, cool, and add heat-labile materials if necessary. Dissolve KH_2PO_4 (1.0 g) in 10 ml of trace element mixture (see sulfur photosynthetic bacteria), filter sterilize, and add to autoclaved medium. Adjust the pH before autoclaving. The following is one of several possible selective versions of this medium for *Rhodomicrobium vanielii.* Make the basal medium with no yeast extract or growth factors and add 0.03% $Na_2S \cdot 9H_2O$. Replace chlorides with sulfates. Add 0.2% acetate, pH 6.8. For *Rhodopseudomonas palustris,* use the basal medium with yeast extract and 0.1–2% (w/v) ethanol, methanol, glutarate, or benzoate. Adjust the pH to 6.8.

Thiobacillus Species—Aerobic Species Add sulfur (10 g) or $Na_2S_2O_3 \cdot 5H_2O$ (5 g), $(NH_4)_2SO_4$ (2–4 g), KH_2PO_4 (2–4 g), $CaCl_2$ (0.25 g), $MgSO_4 \cdot 7H_2O$ (0.5 g), and $FeSO_4$ (10 mg) to 1 liter of water. If sulfur is used, sterilize by steaming for 30 minutes on three successive days. Dispense in conical flasks to give shallow aerobic cultures. Adjust to pH 5.0 for *T. thiooxidans* and pH 7.5 for *T. thioparus.*

For *T. ferrooxidans,* dissolve $(NH_4)_2SO_4$ (0.15 g), KCl (0.05 g), $MgSO_4 \cdot 7H_2O$ (0.5 g), KH_2PO_4 (0.05 g), and $Ca(NO_3)_2$ (0.01 g) in 1 liter of water. Sterilize and add 10 ml of separately sterilized 10% $FeSO_4 \cdot 7H_2O$ solution. Adjust to pH 3.5. This species may be grown in sulfur or thiosulfate media but requires a pH of 4.5–5.0 to do so.

Thiobacillus (Anaerobic Species) (e.g., *T. denitrificans*) (Baalsrud and Baalsrud 1954). Add NH_4Cl (0.5 g), $MgCl_2 \cdot 6H_2O$ (0.5 g), $Na_2S_2O_3 \cdot 5H_2O$ (5 g), and KNO_3 (2 g) to water and sterilize. Add separately sterilized solutions of KH_2PO_4 (2 g), $NaHCO_3$ (1 g), and $FeSO_4 \cdot 7H_2O$ (10 mg). Make up to a total volume of 1 liter. A trace element mixture is sometimes useful and may be added to the media above. If added, do not duplicate the iron requirements.

Nitrosomonas (Soriano and Walker 1968). Add $(NH_4)_2SO_4$, 0.5 g: KH_2PO_4, 0.2 g; $CaCl_2 \cdot 2H_2O$, 0.04 g; $MgSO_4 \cdot 7H_2O$, 0.04 g; Fe (as ferric citrate), 0.5 mg and phenol red, 0.5 mg to distilled water to give 1 liter. After sterilization, adjust pH to 7.5–8.0 with sterile aqueous 5% sodium carbonate solution. For solid media use 1–1.5% Noble (Difco) agar.

Nitrobacter (Smith and Hoare 1968). Add $NaNO_2$ (for *Nitrobacter*), 1.38 g, Na_2HPO_4 (2.55 g), KH_2PO_4, 0.27 g, $MgSO_4 \cdot 7H_2O$ (20 mg), iron solution (10 ml), and trace salts solution (1 ml) to distilled water to make 1 liter. The iron solution contains $FeSO_4 \cdot 7H_2O$ (1.0 g), Na-EDTA (1.0 g), and distilled water (1 liter). The trace salts solution contains H_3BO_3 (0.02 g),

$CaSO_4 \cdot 5H_2O$ (0.1 g), $MnSO_4 \cdot 2H_2O$ (0.02 g), $(NH_4)_6Mo_7O_{24} \cdot 4H_2O$ (0.02 g), $ZnSO_4 \cdot 7H_2$) (0.15 g) $CoCl_2$ (0.01 g), Na-EDTA (1.0 g), and distilled water 1 liter). Growth may be detected by the production of acid and phenol red (0.5 mg/liter) is a useful indicator. Growth may be inhibited by volatile materials from rubber.

Methanogenic Bacteria (Zeikus et al. 1975). $(NH_4)_2SO_4$, 0.45 g; NaCl, 0.9 g; $MgSO_4 \cdot 7H_2$), 0.18 g; $CaCl_2 \cdot 2H_2O$, 0.1 g; NH_4Cl, 0.5 g; KH_2PO_4, 1.5 g; K_2HPO_4, 2.19 g; trace mineral solution (Wolin et al. 1963); trace vitamin solution, 5 ml (Wolin et al. 1963); 5% solution of $FeSO_4$, 0.3 ml; 0.2% solution of resazurin, 1 ml; and Na_2S, 1.0 g, added after filter sterilization of the medium. Incubate under an atmosphere of 80% H_2: 20% CO_2 (v/v). For acetate-fermenting methanogens, add 10% (w/v) sodium acetate to the medium above and incubate under an N_2/CO_2 atmosphere.

Rhizobium (from Nodules) Mannitol, 10.0 g; $K_2HPO_4 \cdot 3H_2O$, 0.65 g; $MgSO_4 \cdot 7H_2O$, 0.2 g; NaCl, 0.1 g; yeast extract (spray dried), 1.0 g; and distilled water, 1 liter.

Free-Living, Aerobic, Nitrogen-Fixing Bacteria K_2HPO_4, 0.5 g; $MgSO_4 \cdot 7H_2O$, 0.2 g; NaCl, 0.2 g; $FeCl_3$, trace; $MnSO_4$, trace; and distilled water, 1 liter. To every 50 ml of sterile medium, add 1 ml of 10% mannitol solution (filter sterilized).

Methylotrophic Bacteria (Foster and Davis 1966). $NaNO_3$, 2.0 g; $MgSO_4 \cdot 7H_2O$, 0.2 g; KCl, 0.04 g; $CaCl_2$, 0.015 g; Na_2HPO_4, 0.21 g; NaH_2PO_4, 0.09 g; $FeSO_4 \cdot 7H_2O$, 1 mg; $CuSO_4 \cdot 5H_2O$, 5 μg; H_3BO_3, 10 μg; $MnSO_4 \cdot 5H_2O$, 10 μg; $ZnSO_4 \cdot 7H_2O$, 70 μg; MoO_3, 10 μg; and distilled water, 1 liter; Incubate in a desiccator under a 50% air–50% methane mixture at 1 atm pressure.

Desulphovibrio (Pankhurst 1967). K_2HPO_4, 0.5 g; NH_4Cl, 1.0 g; $CaSO_4$, 1.0 g; $MgSO_4 \cdot 7H_2O$, 2.0 g; 70% (w/v) sodium lactate solution, 5.0 g; Oxoid ionagar No. 2, 10.0 g; and distilled water, 930 ml. Adjust to pH 8.1. Autoclave for 20 minutes at 115°C and cool rapidly to prevent resolution of oxygen. Add 50 ml of supernatant from 1% (w/v) aqueous solution of $FeSO_4(NH_4)_2SO_4 \cdot 6H_2O$ steam sterilized for 1 hour on three successive days; 10 ml of 10% (w/v) membrane-filtered yeast extract; and 10 ml of aqueous solution of sodium thioglycollate (filter sterilized) to give 0.01% (w/v) in the final medium. Disperse sediment with medium into culture vessels. Completely fill the vessels.

Selected Tests

Stains

Gram Reaction (Hucker and Conn 1923). Prepare a suspension of bacteria in demineralized water which is faintly opalescent. Place a drop on a slide and spread this out. Allow to dry. If it is not dry within 30 seconds, too much material has been used. Gently and quickly, heat fix the film.

Stain with ammonium oxalate crystal violet for 1 minute (crystal violet, 2 g; ammonium oxalate, 0.8 g; ethanol, 20 ml; distilled water, 80 ml). Wash and drain off the excess water. Stain with Lugol's iodine solution for 1 minute (iodine, 1 g; potassium iodide, 2 g; distilled water, 300 ml). Wash, drain, and blot dry. Treat with 95% ethanol for 30 seconds. Wash and drain and then stain with safranin for 30 seconds (safranin, 10 ml of a 2.5% solution in 95% alcohol, distilled water, 100 ml). Wash, drain, and blot dry. Gram-positive bacteria appear purple and Gram-negative bacteria appear red when examined with a 100× oil-immersion lens. Spores remain unstained.

Spore Stain (Conklin 1934). Prepare a heat-fixed film as for the Gram stain. Stain with 5% aqueous malachite green for 5 minutes, heating the slide over a low flame until it steams. Wash the slide, drain, and counterstain with 5% aqueous mercurochrome for 30 seconds. Wash and blot dry and observe under a 100× oil-immersion lens. Spores stain green, vegetative cells red.

Motility and the Flagella Stain (modified from Rhodes 1958). Motility is best observed in a hanging-drop preparation. Roll out a piece of Plasticine to equal the circumference of a coverslip. Place the Plasticine on a slide in the form of a ring or square. Place a small drop of bacterial suspension centrally on a coverslip. Press the Plasticine against the coverslip and invert the slide quickly, leaving a discrete drop hanging from the coverslip. Observe the edge of the film with the iris diaphragm stopped down at 40× magnification. Motile bacteria are obvious, although they must be distinguished from cells showing Brownian movement.

To stain flagella, prepare a bacterial suspension as for the Gram stain but leave the suspension at room temperature for 10 minutes and then place a drop at the end of a very clean slide. Allow the drop to run down the slide onto a piece of blotting paper and leave the slide to air dry. Do not heat fix it. Flood the dried film with a filtered mordant (10 ml of one-fourth saturated aqueous picric acid, 5 g of tannic acid, 7.5 g of ferrous sulfate, and 90 ml of distilled water). Wash after 5 minutes with distilled water. Drain and treat the slide with hot sensitized silver nitrate solution for 5 minutes (2% $AgNO_3$ in distilled water; add 0.88 ammonia drop by drop until the precipitate redissolves; add more silver nitrate solution drop by drop under the solution becomes faintly turbid; boil). Wash and blot dry. Observe under a 100× oil-immersion lens. Flagella appear as gray or brown threads attached to the cells.

Type of Metabolism (Hugh and Leifson 1953)

Inoculate tubes of the following medium by stabbing (peptone, 0.2 g; NaCl, 0.5 g; K_2HPO_4, 0.03 g; bromothymol blue, 0.3 g; glucose, 1.0 g;

water, 100 ml; agar, 0.1 g; pH 7.1). The bromothymol blue is best made as a 1% solution and 0.3 ml of this added to the medium. The glucose is sterilized by filtration and added aseptically to the autoclaved medium. After inoculation, cover the surface of one agar culture with 1–2 cm of sterile molten Vaseline. Observe the color change after 24 hours, 48 hours, and sometimes longer. A yellow color indicates acid production. If this occurs only in the open tube, the bacterium has a respiratory metabolism. If it occurs in both tubes, the metabolism is fermentative or fermentative and respiratory. If bacteria are aerogenic (i.e., produce gas), the Vaseline will be pushed up the tube or large bubbles will form in the medium. Any soluble carbohydrate can be used instead of glucose if further acid production tests are needed (e.g., sorbitol).

Use of Sole Carbon Sources (e.g., Citrate)

Use of citrate can be detected in the following medium, designed by Koser (1924): KH_2PO_4, 1.0 g; $MgSO_4 \cdot 7H_2O$, 0.2 g; $NaNH_4PO_4 \cdot 4H_2O$, 1.5 g; sodium citrate, 2.0 g; bromothymol blue (as for type of metabolism test); distilled water, 1 liter; pH 6.7–6.9. A positive reaction can be detected by the development of turbidity and by the alkaline reaction that develops, turning the indicator blue. The use of other organic acids can be detected in a similar way, although lower concentrations of the acids may be necessary if they are toxic (e.g., 0.1%).

Production of Enzymes

Catalase (Skerman 1967). Pour 1 ml of H_2O_2 (10 vol %) over the surface of a 1-day-old agar slope culture. Release of oxygen bubbles indicates the presence of catalase.

Cellulase (Eggins and Pugh 1961). Prepare an agar medium containing $NaNO_3$, 0.5 g; K_2HPO_4, 1.0 g; $MgSO_4 \cdot 7H_2O$, 0.5 g; $FeSO_4$, 0.01 g; ball-milled cellulose, 12 g; and distilled water 1 liter. Pour the medium as a thin layer over solidified tap water agar in a petri dish. Streak the organism across the plate and incubate for at least 1 week. Clearing of the medium indicates cellulase production.

Chitinase (Baxby and Gray 1968). Prepare an agar medium containing K_2HPO_4, 1.0 g; $MgSO_4 \cdot 7H_2O$, 1.0 g; and 1% ball-milled suspension of chitin in distilled water, 1 liter. Pour the medium as a thin layer over solidified tap water agar in a petri dish. Streak the organism across the plate and incubate for at least 1 week. Clearing of the medium indicates chitinase production. A small amount of yeast extract may also be added to the medium to promote growth.

Deaminase (Phenylalanine) (Ewing 1962). Grow bacteria on the following medium: yeast extract, 3.0 g; DL-phenylalanine, 2.0 g; Na_2HPO_4, 1.0 g; NaCl, 5.0 g; distilled water 1 liter; and agar, 12.0 g. Incubate at 37°C for 24 hours and then flood the slope culture with 4–5 drops of ferric chloride solution (10% w/v). A green color indicates the presence of phenyl pyruvate.

Decarboxylases (Ornithine and Lysine) (Moeller 1955). Add 1% L-ornithine hydrochloride or 1% of L-lysine dihydrochloride (if DL acids are used, use 2%) to the following basal medium: peptone, 5 g; beef extract, 5 g; bromocresol purple (1.6%), 0.625 ml; cresol red (0.2%), 2.5 ml; glucose, 0.5 g; pyridoxal, 5 mg; and distilled water, 1 liter; adjust to pH 6.0. Autoclave 4-ml amounts for 10 minutes. Inoculate, together with a control. Add a 4- to 5-mm layer of sterile paraffin oil to all tubes. Examine after 4 days. A positive reaction is shown by a color change from yellow to violet or reddish violet.

DNase (Jacobs et al. 1964). Prepare an 8% solution of the sodium salt of thymus gland DNA in sterile distilled water. Add 10 ml of this to a cooled sterile nutrient agar base to give a final concentration of 0.2% DNA and pour. A suitable base is the Difco heart infusion broth solidified with agar. Note that the DNA has not been sterilized. Inoculate by streaking and incubate at 37°C for 18 hours and then flood the plates with 1.5 *N* HCl. DNA is precipitated, so DNase is detected if a zone of clearing is evident.

Nitrogenase. Inoculate a small quantity of a culture into a nitrogen-free medium: for example, K_2HPO_4, 0.5 g; $MgSO_4 \cdot 7H_2O$, 0.2 g; NaCl, 0.2 g; $MnSO_4$, trace; $FeCl_3$, trace; and distilled water 1 liter. To every 50 ml of medium add 1 ml of 10% mannitol or glucose solution (filter sterilized).

If growth occurs, transfer a small amount to fresh medium and repeat this two or three times. If growth develops quickly, it is possible that nitrogen-fixing organisms are present, but many bacteria are able to scavenge minute traces of combined nitrogen, and more elaborate tests may be required (e.g., acetylene reduction).

Oxidase (Kovacs 1950). Put 2–3 drops of fresh 1% aqueous tetramethyl-*p*-phenylenediamine dihydrochloride on a piece of filter paper. Smear a colony onto the paper. If it turns purple in 5–10 seconds, oxidase is present.

Urease (Christensen 1946). Prepare the following medium: peptone, 1.0 g; NaCl, 5.0 g; glucose, 1.0 g; K_2HPO_4, 2.0 g; phenol red, 0.012 g (6 ml of a 1 : 500 solution); urea, 20.0 g; and distilled water 1 liter. Adjust to pH 6.8 and filter sterilize. Dissolve 15 g of agar in 900 ml of water and autoclave. Cool to 50°C and add 100 ml of the medium above. Mix and distribute in tubes to form agar slopes. Inoculate the entire surface of the slant. Incubate at 37°C and observe at regular intervals for up to 4 days. Urease positive cultures produce an alkaline reaction and a red color in the medium.

Production of Characteristic Compounds

Production of Acetyl Methyl Carbinol from Glucose, the Vosges–Proskauer Test (Eddy 1961). Prepare a medium containing 0.5% glucose, 0.5% peptone, and 0.5% K_2HPO_4 in distilled water. Inoculate and incubate for 3 days at 37°C. Then add 5 ml of 40% KOH and a trace of creatine to 2 ml of the culture, shake, and leave for up to 1 hour. The culture is positive if a cherry red color develops.

Production of 3-ketolactose (Bernaerts and De Ley 1963). Grow a culture for 1–2 days at 25°C on an agar slope (yeast extract, 1 g; glucose, 2 g; $CaCO_3$, 2g; agar, 1.5 g; distilled water, 100 ml). Transfer a loopful of the organism to a solid medium in a petri dish (1 g of lactose; 0.1 g of yeast extract; 2 g of agar; 100 ml of water) as a small heap about 0.5 cm in diameter and incubate at 28°C for 1–2 days. Flood the plate with Benedict's reagent and leave at room temperature for up to 1 hour. A yellow ring of Cu_2O develops around the cell mass if 3-ketolactose is present.

Production of Nitrite or Nitrogen Gas from Nitrate (Society of American Bacteriologists 1957). Prepare a nutrient broth with 1% KNO_3. Dispense into Universal vials, add a Durham tube, and autoclave. Inoculate the broth and after a few days test the culture for a production of nitrite or nitrogen gas. Nitrogen gas will accumulate in the Durham tube. Nitrite is detected by the addition of a few drops of sulfanilic acid reagent (sulfanilic acid, 8 g; 5 *N* acetic acid, 1 liter), followed by a few drops of dimethyl-α-napthylamine reagent (dimethyl-α-napthylamine, 6 ml; 5 *N* acetic acid, 1 liter) to a portion of the culture. A red color, sometimes transient, indicates the presence of nitrite.

If no nitrite is detected, it is essential to check for residual nitrate, for sometimes nitrite *is* produced but further reduced and therefore not detected. To do this, add a small quantity of zinc dust to another portion of the culture to chemically reduce nitrate to nitrite. Then test for nitrite as before. If the original test and this test are both negative, it follows that all the nitrite had been reduced by the bacterium to another product. Thus the bacterium does possess nitrate reductase.

Sensitivity Tests

Sensitivity to Erythromycin. Incorporate filter-sterilized erythromycin in a cooled agar medium that will support growth at a concentration of 0.4 μg/ml. Also inoculate the medium and pour at once into a petri dish. Absence of growth indicates sensitivity.

Sensitivity to Pteridine Derivative (0/129). Soak sterile filter paper strips in a filter sterilized, saturated, aqueous solution of 0/129 2,4-diamino-6,7-diisopropyl pteridine and place them on inocula streaked on yeast extract agar. Sensitivity is detected by failure to grow over the strips.

ACKNOWLEDGMENT

I wish to thank Dr. M. Goodfellow for his helpful comments on the construction of these keys.

References

Baalsrud, K. and K. Baalsrud. 1954. Studies on *Thiobacillus denitrificans*. *Arch. Mikrobiol.* 20:34–62. **Baxby, P. and T. R. G. Gray. 1968.** Chitin decomposition in soil. I. Media for isolation of chitinoclastic micro-organisms from soil. *Trans. Br. Mycol. Soc.* 51:287–292. **Bernaerts, M. J. and J. De Ley. 1963.** A biochemical test for crown gall bacteria (*Agrobacterium tumefaciens* and *Agrobacterium radiobacter*). *Nature* (*Lond.*) 197:406. **Bousfield, I. J. and A. G. Callely. 1978.** *Coryneform Bacteria.* Spec. Publ. Soc. Gen. Microbiol. 1. Academic Press, London. **Buchanan, R. E. and N. E. Gibbons. 1974.** *Bergey's Manual of Determinative Bacteriology,* 8th ed. Williams & Wilkins, Baltimore. **Christensen, W. B. 1946.** Urea decomposition as a means of differentiating *Proteus* and para-colon cultures from each other and from *Salmonella* and *Shigella* types. *J. Bacteriol.* 52:461–466. **Conklin, M. E. 1934.** Mercurochrome as a bacteriological stain. *J. Bacteriol.* 27:30. **Cowan, S. T. and K. J. Steel. 1965.** *Manual for the Identification of Medical Bacteria.* Cambridge University Press, Cambridge. **Eddy, B. P. 1961.** The Vosges–Proskauer reaction and its significance: a review. *J. Appl. Bacteriol.* 24:27–41. **Eggins, H. O. W. and G. J. F. Pugh. 1961.** Isolation of cellulose decomposing fungi from soil. *Nature* (*Lond.*) 193:94–95. **Ewing, W. H. 1962.** *Enterobacteriaceae. Biochemical methods for group differentiation.* U.S. Dept. of Health, Education, and Welfare, Public Health Serv. Publ. 734 (revised). **Foster, J. W. and R. H. Davis. 1966.** A methane dependent coccus, with notes on classification and nomenclature of obligate, methane-utilizing bacteria. *J. Bacteriol.* 91:1924–1931. **Goodfellow, M. 1969.** Numerical taxonomy of some heterotrophic bacteria isolated from a pine forest soil. Pp. 83–105 in J. G. Sheals (ed.). *The Soil Ecosystem.* Systematics Association, London. **Goodfellow, M. and R. G. Board. 1980.** *Microbiological Classification and Identification.* Academic Press, London. **Goodfellow, M. and T. Cross, 1974.** Actinomycetes. Pp. 269–302 in C. H. Dickinson and G. J. F. Pugh (eds.). *Biology of Plant Litter Decomposition,* Vol. 2. Academic Press, London. **Goodfellow, M., I. R. Mill, and T. R. G. Gray. 1968.** Bacteria in a pine forest soil. Pp. 500–515 in T. R. G. Gray and D. Parkinson (eds.). *The Ecology of Soil Bacteria.* University of Liverpool Press, Liverpool, England. **Goodfellow, M., B. Austin, and D. Dawson. 1976.** Classification and identification of phylloplane bacteria using numerical taxonomy. Pp. 275–292 in C. H. Dickinson and T. F. Preece (eds.). *Microbiology of Aerial Plant Surfaces.* Academic Press, London. **Gray, T. R. G. 1969.** The identification of soil bacteria. Pp. 73–82 in J. G. Sheals (ed.). *The Soil Ecosystem.* Systematics Association, London. **Gyllenberg, H. G. and E. Eklund. 1974.** Bacteria. Pp. 245–268 in C. H. Dickinson and G. J. F. Pugh (eds.). *Biology of Plant Litter Decomposition,* Vol. 2. Academic Press, London. **Hucker, G. J. and H. J. Conn. 1923.** Methods of Gram staining. *N.Y. State Agric. Exp. Stn. Tech. Bull.* 93. **Hugh, R. and E. Leifson. 1953.** The taxonomic significance of fermentative versus oxidative metabolism of carbohydrates by various gram-negative bacteria. *J. Bacteriol.* 66:24–26. **Jacobs, S. I., A. T. Willis and G. M. Goodburn. 1964.** Pigment production and enzymatic activity of staphylococci: the differentiation of pathogens from commensals. *J. Pathol. Bacteriol.* 87:151–156. **James, N. 1958.** Soil extract in soil microbiology. *Can. J. Microbiol.* 4:363–370. **Jones, D. and M. J. Sackin. 1980.** Numerical methods in the classification and identification of bacteria with especial references to the Enterobacteriaceae. Pp. 73–106 in M. Goodfellow and R. G. Board (eds.). *Microbiological Classification and Identification.* Academic Press, London. **Keddie, R. M. 1978.** What do we mean by coryneform bacteria? Pp. 1–12 in I. J. Bousefield and A. G. Callely (eds.). *Coryneform Bacteria.* Academic Press, London. **Koser, S. A. 1924.** Correlation of citrate utilization by members of the colon-aerogenes groups with other differential characteristics and with habitat. *J. Bacteriol.* 9:59–77. **Kovacs, N. 1950.** Identification of *Pseudomonas pyocyanea* by the oxidase reaction. *Nature* (*Lond.*) 178:703. **Krieg, N. R. and J. G. Holt (eds.). 1984.** *Bergey's Manual of Systematic Bacteriology,* Vol. 1. 1st ed. Williams & Wilkins, Baltimore. **Minnikin, D. E. and M.**

Goodfellow. 1980. Lipid composition in the classification and identification of acid-fast bacteria. Pp. 189–256 in M. Goodfellow and R. G. Board (eds.). *Microbiological Classification and Identification.* Academic Press, London. **Minnikin, D. E., M. Goodfellow and M. D. Collins. 1978.** Lipid composition in the classification and identification of coryneform and related taxa. Pp. 85–160 in I. J. Bousfield and A. G. Callely (eds.). *Coryneform Bacteria.* Academic Press, London. **Moeller, V. 1955.** Simplified test for amino acid decarboxylases and the arginine dehydrolase system. *Acta Pathol. Microbiol. Scand.* 36:158–172. **Pankhurst, 1967.** A simple culture tube for anaerobic bacteria. *Lab. Pract.* 16:58–59. **Rhodes, M. E. 1958.** The cytology of *Pseudomonas* spp. as revealed by a silver-plating staining method. *J. Gen. Microbiol.* 18:639–648. **Skerman, V. B. D. 1967.** *A Guide to the Identification of the Genera of Bacteria,* 2nd ed. Williams & Wilkins, Baltimore. **Smith, A. J. and D. S. Hoare. 1968.** Acetate assimilation by *Nitrobacter agrilis* in relation to its obligate autotrophy. *J. Bacteriol.* 95:844–855. **Sneath, P. H. A. 1978.** An evaluation of numerical taxonomic techniques in the taxonomy of *Nocardia* and allied taxa. Pp. 74–101 in M. Goodfellow, G. H. Brownell, and J. A. Serrano (eds.). *The Biology of the Nocardiae.* Academic Press, London. **Sneath, P. H. A., N. S. Mair, M .E. Sharpe and J. G. Holt. 1986.** *Bergey's Manual of Systematic Bacteriology,* Vol. 2. 1st ed. Williams & Wilkins, Baltimore. **Society of American Bacteriologists. 1957.** *Manual of Microbiological Methods.* McGraw-Hill, New York. **Soriano, S. and N. Walker. 1968.** Isolation of ammonia-oxidizing autotrophic bacteria. *J. Appl. Bacteriol.* 31:493–497. **Whittenbury, R. 1971.** Enrichment and isolation of photosynthetic bacteria. Pp. 241–249 in D. A. Shapton and R. G. Board (eds.). *Isolation of Anaerobes.* Academic Press, New York. **Wolin, E. A., M. J. Wolin and R. S. Wolfe. 1963.** Formation of methane by bacterial extracts. *J. Biol. Chem.* 238:2882–2886. **Zeikus, J. G., P. J. Weimer, D. R. Nelson, and L. Daniels. 1975.** Bacterial methanogenesis: acetate is a methane precursor in pure culture. *Arch. Microbiol.* 104:129–134.

3

Actinomycetes

J. C. HUNTER-CEVERA

DOUGLAS E. EVELEIGH

Actinomycetes represent a diverse group of bacteria that share the common property of a "filamentous" and often branched growth habit. Their morphology includes single or branched cells, short rudimentary hyphae, and extensively branched, narrow, mycelium (0.05–2.0 μm in diameter). They derive their name from their radial and fungal-like colony form: sun's ray—*actino*, fungi—*mycos* (1). Asexual reproduction can occur by fragmentation of hyphae to yield bacillary and coccoid elements and/or the formation of a wide range of spore types, including true endospores in the genus *Thermomonospora*. The range of spore types is the basis of one of the most practical morphological classifications.

The actinomycetes are of major ecological importance. In general, they are aerobic, saprophytic mesophiles, although a small percentage are pathogenic for humans, animals, and plants (2). Thermophilic strains are also well known (3). This major group of soil microbes are remarkable for their broad degradative capabilities. Thus, although often slow growing, they are extremely important in the mineralization of a wide range of complex soil organic matter (e.g., cellulose) (4, 5). The actinomycetes are particularly well known for their ability to produce antibiotics (e.g., streptomycin, tetracycline) and have been widely used industrially for this purpose (6, 7). They also make numerous other secondary metabolites—for example, geosmin, the material responsible for the "earthy" odor of soil (8).

As procaryotes, they have peptidoglycan in the cell wall, 70S ribosomes, monofibrous flagella, a single chromosome, and plasmids. They lack nuclear membranes and organelles and their membranes lack sterols. Despite these broad generalities, the actinomycetes are a diverse group, as illustrated by the wide guanine-plus-cytosine ratio among them (62–78.5%) (9).

In overview, this concatenation of genera are successful members of diverse terrestrial and aquatic habitats. Their success can be attributed in part to their broad nutritional capabilities, their ramifying mycelial growth habit, and their ability to form spores for both survival and dispersal. More detailed aspects regarding the wide range of actinomycetes can be found in several excellent review articles (2, 6, 7, 10–14).

Ecology

Actinomycetes are numerous and widespread in all soils, river muds, and lake bottoms (15–18). They apparently occur less frequently than bacteria but in greater numbers than fungi, with estimates ranging between 1.0×10^5 and 1.0×10^8 per gram of dry soil in temperate zones, based on plate counts. This concept should be qualified in that the plate count method emphasizes the sporing ability of an organism rather than the cellular mass. Other estimates of microbial numbers, based on biomass, indicate mycelial fungi as the dominant microbial group (18). In a similar vein, actinomycete mycelial filaments are readily observed on root fragments, leaves and twigs, and soil particles (19, 20) and thus represent a considerable portion of the soil microbiotal biomass. Even so, soil actinomycetes are believed to exist for long periods, as dormant spores until optimal conditions are provided for growth (18, 19, 21).

The size of the actinomycete population in soils is dependent on moisture content and aeration, organic material, pH, rhizosphere and vegetation, soil depth, and soil type (7, 10, 15). The most abundant actinomycetes isolated from soils using standard plate dilution techniques are *Streptomyces* spp., followed by *Nocardia* spp. and *Micromonospora* spp. (11). However, the application of selective isolation techniques, such as use of antibiotics (22) or heat treatment (22a), results in recovery of a greater species diversity.

Actinomycete spores are resistant to desiccation (10, 15, 22b), and they can germinate in a low relative humidity range (91.5–99%) (23). On these bases the survival and growth of actinomycetes are favored by lower soil moisture content in comparison to bacteria (10). In contrast, poor aeration of soils results in a general decrease or even complete absence of streptomycetes (24). This is in part due to anaerobiosis (25), although microaerophilic actinomycetes have been isolated from soil in numbers that were 10–100 times greater than the total plateable soil microorganisms (26). Although it is thought that most actinomycetes will not survive in waterlogged conditions, *Micromonospora* (25, 27, 28) and certain other genera (*Actinoplanes, Nocardia, Streptomyces,* and *Streptosporangium*) are reproducibly obtained from lake muds (29, 30).

Many actinomycetes are intolerant of a pH below 5.0. Thus in acid environments, actinomycetes comprise a small proportion of the microbiota. Exceptions occur; *Streptomyces griseus* can compose 20% of the microbiota in pine litter (31). Occasionally, in alkaline locations actinomycetes are recorded to be the dominant component of the microbial population (e.g., the Bikini Atoll; pH 9) (32).

The number of actinomycetes is higher in soils that are rich in organic matter or humus (10, 15). Amendments to soils with organic material such as farmyard manure and crop residues, increase the total number of actinomycetes isolated but only following initial colonization by bacteria and fungi (33). Perhaps in a similar vein, the number of actinomycetes is usually higher in the rhizosphere than in soil, but they differ qualitatively depending on the vegetation present (34, 35).

In relation to all these factors, it is interesting to note that actinomycetes occur predominantly in subsurface soil layers. It is concluded that such layers provide the optimal balance of the range of factors affecting growth (36).

Classification and Key

Classically, actinomycetes have been grouped according to their morphologic and physiologic characteristics when cultured on agar media. Characteristics used in classification include morphology of the colony, spore orientation, the color and form of the aerial spores and of the substrate mycelium, and the type of diffusable pigments (37, 38). A basic classification scheme is presented in Fig. 3.1. Cultures are grown on standardized agar media and then examined microscopically, preferably *in situ* using a long-working-distance objective (45×). The major microscopic morphological characteristics include mycelial form, spore chain, spore type, and the presence of sporangia.

Mycelium

The majority of actinomycetes form varying amounts of mycelium (sparse to abundant), which develops deep in the substrate agar (*Oerskovia*), forms aerially (*Sporichthya*), or has both substrate and aerial components (*Streptomyces*). The majority of actinomycetes form narrow hyphae (less than 2 μm), but extremely "fine" hyphae are exhibited by *Micromonospora*. Mycelial forms range from rudimentary (*Rothia*) to extensively branched hyphae (*Streptomyces*), those with sclerotia (*Chainia*), those with fragmenting cells (*Nocardia*), or even a filamentous form that divides both longitudinally and transversely (*Geodermatophilus*) (Fig. 3.1).

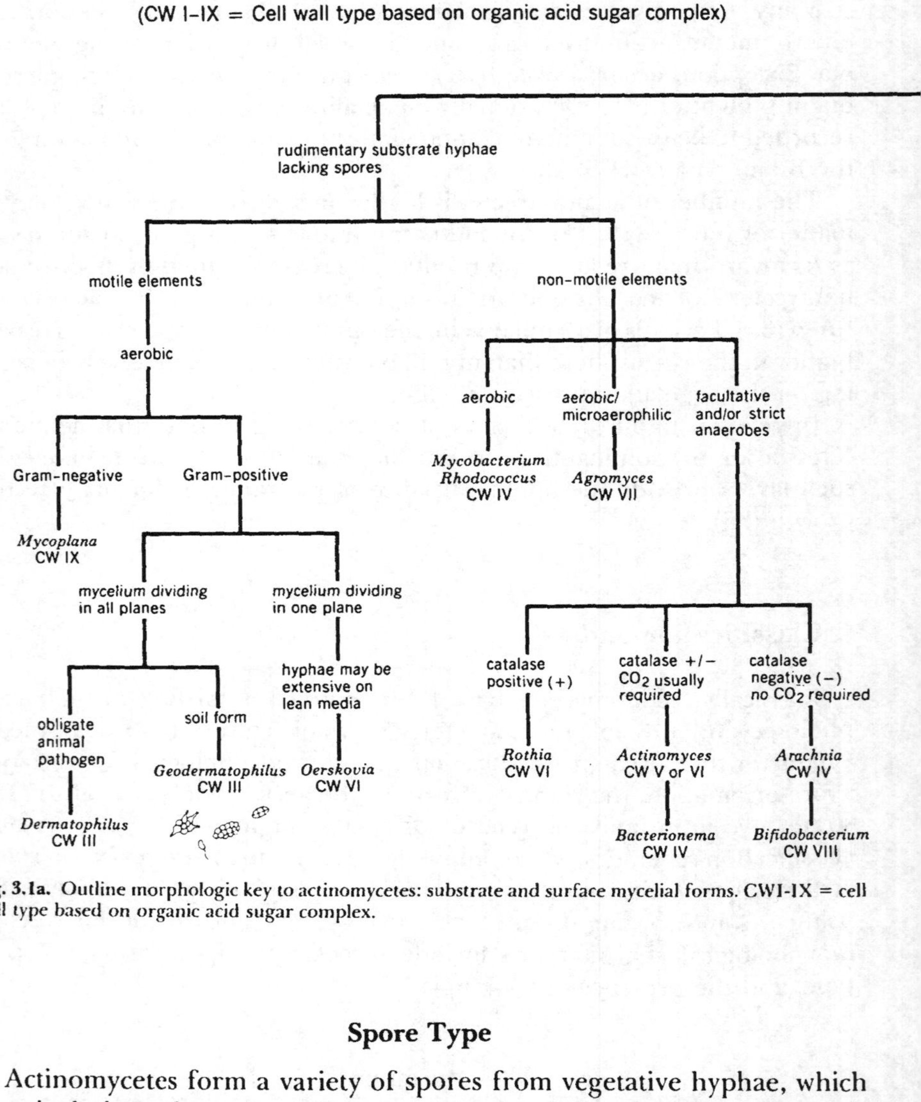

Fig. 3.1a. Outline morphologic key to actinomycetes: substrate and surface mycelial forms. CWI-IX = cell wall type based on organic acid sugar complex.

Spore Type

Actinomycetes form a variety of spores from vegetative hyphae, which may include arthrospores (*Streptomyces*), aleuriospores (*Micromonospora*), zoospores (*Spirillospora*), and endospores (*Thermoactinomyces*) (Fig. 3.1). Spores may be developed only on aerial mycelium (*Planomonospora*), or conversely on substrate mycelium only (*Micromonospora*). *Micropolyspora* as well as some other genera form spores on both aerial and substrate mycelia. In some genera (*Arachnia*), "elements" rather than spores develop from a fragmenting substrate mycelium. The spore morphology includes rod-shaped, cylindrical, elliptical, and spherical forms. The spore surface (smooth, warty, spiny, or hairy) is a further useful taxonomic criterion but requires the use of an electron microscope for its determination. Spores are borne on hyphae in characteristic numbers: single spores (*Micromonospora*),

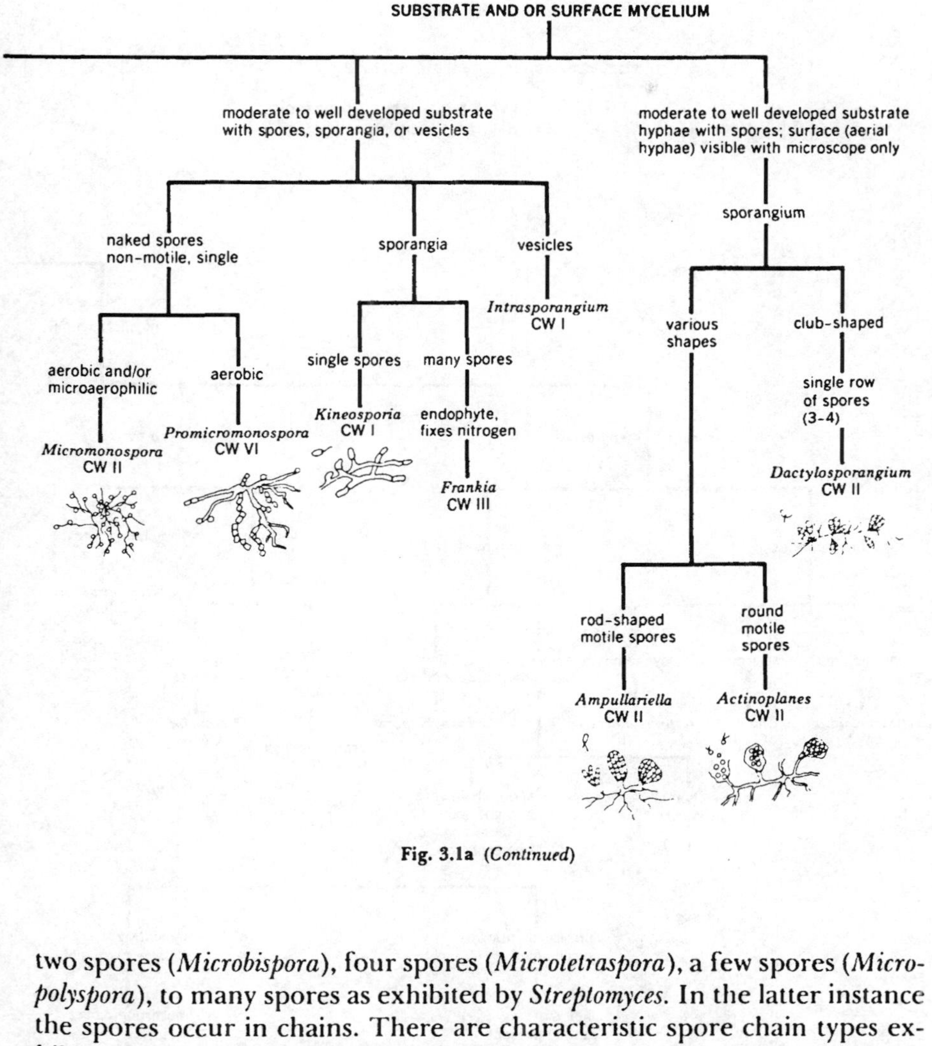

Fig. 3.1a (*Continued*)

two spores (*Microbispora*), four spores (*Microtetraspora*), a few spores (*Micropolyspora*), to many spores as exhibited by *Streptomyces.* In the latter instance the spores occur in chains. There are characteristic spore chain types exhibited by streptomycetes; the spores may be arranged in (a) straight (rectus), flexing, wavy (flexibilis), or open *loops, hooks, or extended spirals* of "wide diameter" (retinaculum–apertum); or (b) simple *spiral* (spira) *chains* that are long, extended, open, short, or compact (Fig. 3.1). Spore chains may also appear as whorls attached on a long axis (*Streptoverticillium*). Single spores are either dispersed on the mycelium (*Thermoactinomyces*) or can be either single or clustered together in *Micromonospora.*

Spore Sporangia and Vesicles

Some actinomycetes have spores contained within a sporangium or vesicle. The shapes of the sporangia can be used as an aid in classification:

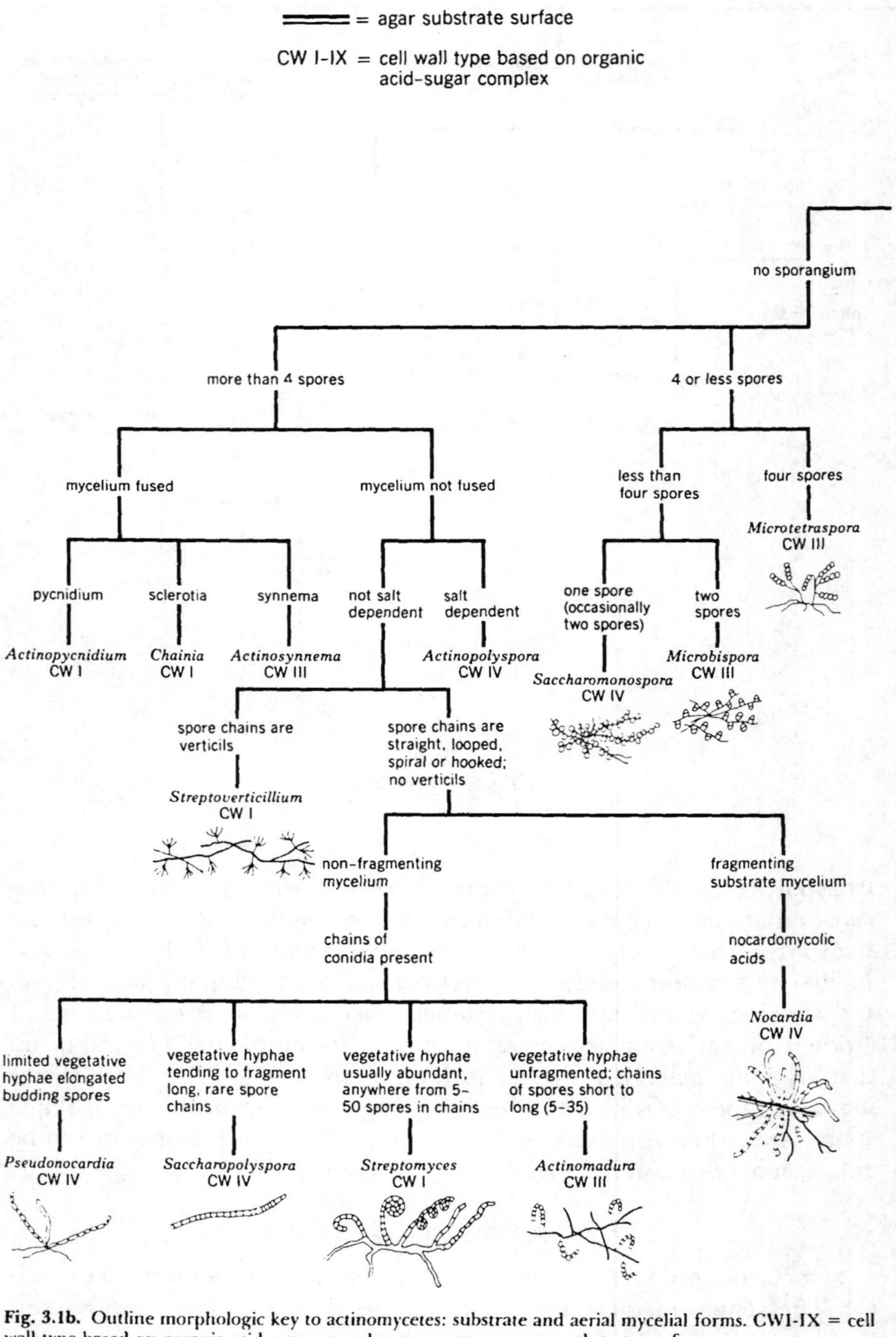

Fig. 3.1b. Outline morphologic key to actinomycetes: substrate and aerial mycelial forms. CWI-IX = cell wall type based on organic acid sugar complex. ═══ = agar substrate surface.

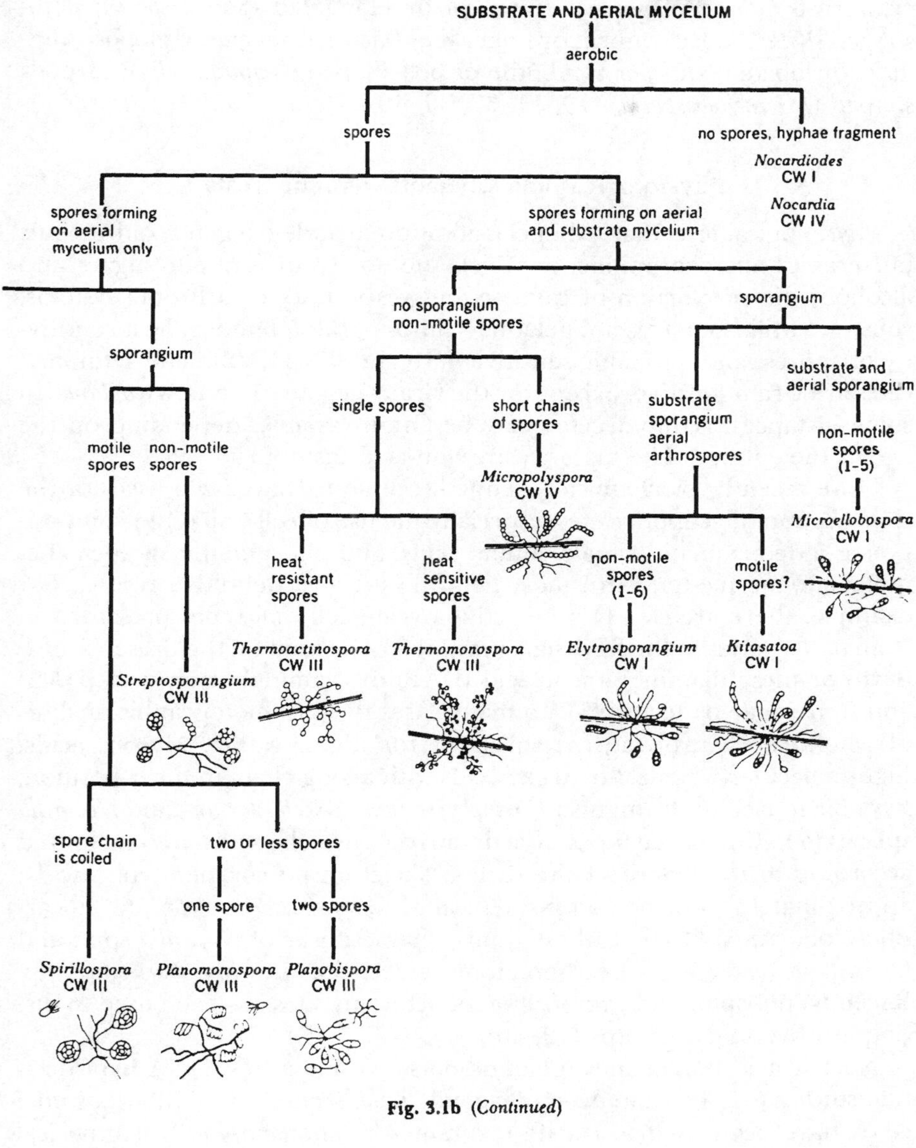

Fig. 3.1b (*Continued*)

cylindrical (*Planomonospora*); spherical or elongated (*Spirillospora*); club-shaped (*Kitasatoa*); claviform or fingerlike (*Dactylosporangium*); globose, digitate, or lobate (*Actinoplanes*); bottle or bell-shaped (*Ampullariella*); or pod-shaped (*Microellobosporia*) (10, 11, 37, 39, 40).

Physiological and Chemotaxonomic Tests

Physiological tests that aid in classification include oxygen requirements (anaerobic, microaerophilic, aerobic), the ability to ferment sugars and alcohols, the production of catalase and respiratory cytochrome systems, substrate utilization (casein, gelatin, cellulose), salt tolerance, heat sensitivity of spores, and antibiotic sensitivity (10, 12, 39, 41, 42). The actinomycetes are Gram-positive, except for the Gram-negative genus *Mycoplana*. In some instances, actinomycetes may be Gram-variable, depending on the age of the culture. The majority are non-acid fast (39).

More recently, biochemically refined techniques have been used for the classification of actinomycetes, especially the use of cell wall (CW) composition as a determinant (sugars, amino acids, and phospholipids). Such chemotaxonomic methods are most helpful and give definitive results. For example, there are clearly nine actinomycete cell wall types presented in Roman numerals in the flowchart key (Fig. 3.1), based on the presence of L (levo) or meso-diaminopimelic acid (DAP) or diaminobutyric acid (DAB) and four sugar patterns (43). In many cases the chromotographic analysis of whole cell hydrolysates is sufficient for identification. Mycolic acids, high-molecular-weight β-hydroxy acids with a long chain in the α-position, have been used to distinguish *Corynebacterium, Mycobacterium,* and *Nocardia* species (44, 45). Three types of actinomycete mycolic acids are recognized according to the length of the carbon skeleton: (a) corynemycolic acids; approximately 30 carbon atoms, typical of *Corynebacterium* spp.; (b) nocardomycolic acids; 50–60 carbon atoms, characteristic of *Nocardia* spp.; and (c) mycolic acids; 80–90 carbon atoms, associated with *Mycobacterium* spp. Recently, phospholipid composition of actinomycetes has been used to distinguish five major groups (12, 46).

Nucleic acid homologies [guanine-plus-cytosine ratios, DNA hybridization studies (47, 48)] and even nucleotide base sequences of 16S ribosomal RNA have been used to clarify taxonomic relationships (49). Numerical taxonomic approaches have similarly been used most effectively (50).

Unfortunately, these chemotaxonomic classifications require specialized equipment that is not available in many laboratories. Therefore, a practical scheme is presented here based on morphologic and culture characteristics (Fig. 3.1).

The classification scheme shown in Fig. 3.1 represents a simple identification of actinomycetes by morphology alone. Key morphological characters are listed or described that aid taxonomists in distinguishing between genera. The more common soil forms are illustrated by simple drawings.

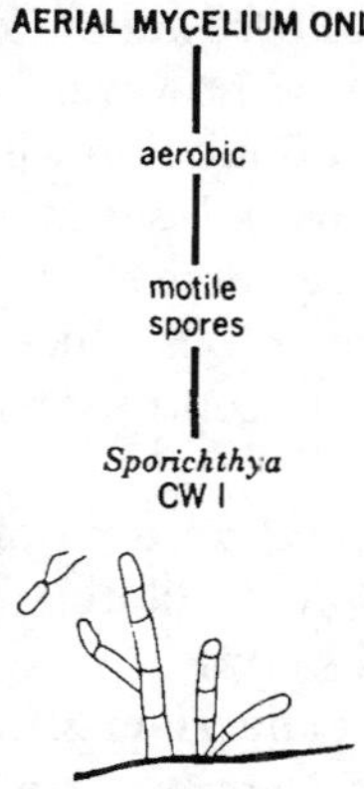

Fig. 3.1c. Outline morphologic key to actinomycetes: aerial mycelial form. CWI = cell wall type based on organic and sugar complex.

This type of identification is referred to as "tentative identification" by taxonomists. For a more precise identification of a particular genus, chemotaxonomy (e.g., cell wall analysis) is generally required. Species identification, which is sometimes very difficult, would entail physiological tests. Note that cell wall types are listed for those students wishing to do a more exact identification (see Refs. 12 and 43 to 46).

Isolation

A range of routine cultural procedures can be used for the isolation of soil actinomycetes. However, to isolate a representative population from a particular soil, especially those species occurring in unique microenvironmental niches, considerable attention should be paid to sampling procedures and cultural methods. Aseptic technique should be used during sampling. Samples should be representative of a site, for example, of a particular soil and its horizons, a leaf litter, or a rhizosphere surface. For general soil sampling, in order to recover the greatest number and variety of species, the top surface layer (1–2 cm) is used as one site sample and second samples are taken from the subsurface layer (3–10 cm deep). Leaf litter and dried plant vegetation may also yield high numbers and a variety of actinomycetes (16, 51). Samples are readily transported in sterile Whirlpak bags to the laboratory and should be processed directly on return or stored briefly at 1–5°C. Seasonal and temporal aspects of collecting should be considered, as a true autochthonous population may occur transiently. As an example, the actinomycete population can change dramatically within a week, following heavy rainfall (H. B. Woodruff, personal communication).

A wide range of routine culture media can be used for growth and isolation of actinomycetes (21). If however, a highly alkaline soil is being studied, a more representative population will be isolated by use of alkaline culture media. Other environmental parameters that must be considered include high or low temperature, ionic strength, or even substrate concentration. For example, certain microbes which are most efficient at utilizing substrates available at very low concentrations may not be isolated using the "routine" rich growth media.

The most frequently used method for isolation of soil actinomycetes is the dilution plate method. Soil is diluted ($\times\ 10^2$–10^7) in sterile saline (0.85% NaCl) and then 0.1-ml samples are spread plated onto "Actinomycete" media (21). This method characteristically yields mainly *Streptomyces* spp., with smaller numbers of *Nocardia* and *Micromonospora*. It is emphasized that this method is indicative of the spore numbers in the soil rather than absolute actinomycete biomass. A greater variety of species can be isolated using modified plating procedures. "Stamping" is one such rapid and facile method (52). The soil sample is first air dried by spreading as a thin layer in a sterile petri dish and left at room temperature for 2–7 days. This considerably reduces the bacterial population. The soil is then gently powdered in a sterile mortar. A small circular sponge (Dispo-culture plug, 16 mm) is pressed onto the soil powder, removed, and excess soil shaken off. A stack of 10 plates is now inoculated. The agar surface of each plate is successively "stamped" 13 times (10 stamps on the outside perimeter and three in the middle). A dilution effect is achieved and clear isolated colonies result. Other treatments include the prior heating of soils (70°C, 10 minutes) in order to obtain the endospore producing *Thermomonospora* (3).

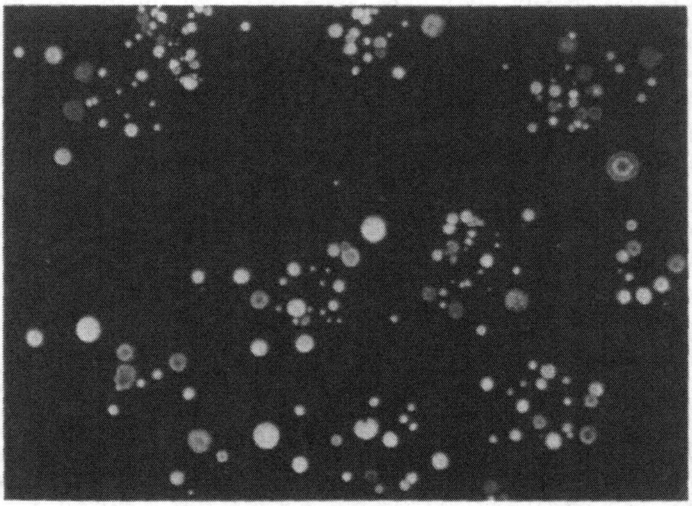

Fig. 3.2. Actinomycete colonies on agar plate.

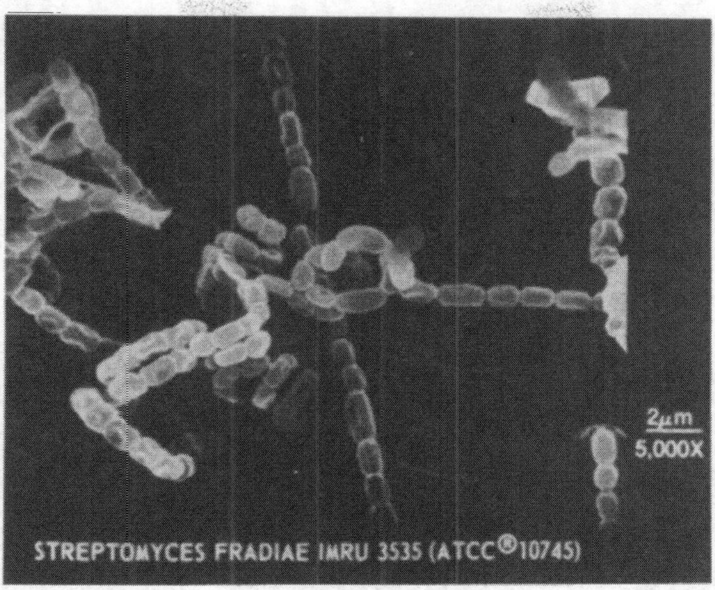

Fig. 3.3. Scanning electron micrograph of streptomycete spore chains. (Courtesy of A. Dietz, SIM Publication 6, *Actinomycete Taxonomy*, 1980.)

Baiting aids can be used in enrichment procedures prior to isolation; for example, the use of wet pollen grains facilitates isolation of actinoplanes from water samples (20). Amendments to soil often increase or enhance the number and/or type of actinomycetes isolated (53); for example, prior addition of chitin to soil enhances the *Micromonospora* population (29).

Media

The majority of agar media employed to isolate aerobic actinomycetes are nutritionally poor, rather than rich in carbon and nitrogen substrates. For example, water (54) and chitin agars (55) have each been employed successfully to isolate actinomycetes. Water agar is simply agar prepared with tap water and without additional nutrients. Chitin agar is a colloidal chitin (1–5%) in a salts medium. Two widely used agars that yield a representative number and variety of actinomycete colonies are arginine–glycerol salts (AGS) (54) and starch casein (SC) media (56) (Table 3.1). Many other agar media have been developed for the isolation of actinomycetes (21).

To retard the development of fungi on the plates, antifungal antibiotics are incorporated into the media: cycloheximide (Actidione) or a polyene (nystatin). The antibiotics are filter sterilized separately and added to the sterilized culture media to a final concentration of 75 μg/ml. In addition, antibiotics are often added to agars for the selective isolation of specific genera such as *Nocardia* (57) or thermophiles (3).

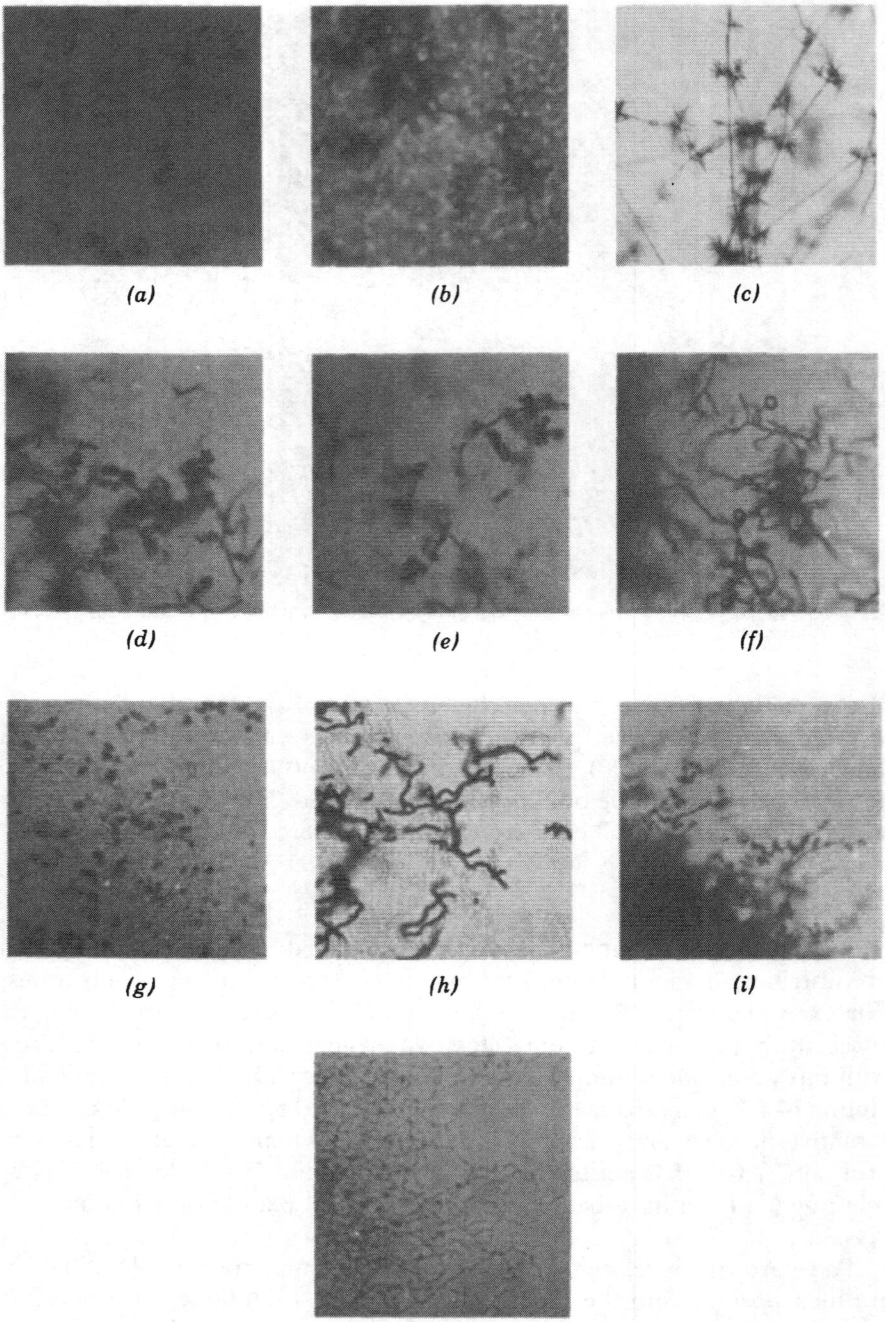

Fig. 3.4. Different actinomycete genera growing on casein–starch agar as viewed through a 45× long-working-distance objective (Lietz): (*a*) *Actinoplanes* sp.; (*b*) *Microtetraspora* sp.; (*c*) *Streptoverticillium* sp.; (*d*) *Streptomyces* sp.; (*e*) *Planomonospora* sp.; (*f*) *Streptosporangium* sp.; (*g*) *Dactylosporangium* sp.; (*h*) *Nocardia* sp.; (*i*) *Microbispora* sp.; (*j*) *Micromonospora* sp. Magnification 56×.

Table 3.1. Culture Media

Arginine–Glycerol Salts Medium (54)		Starch–Casein Medium (56)	
Ingredient	Amount	Ingredient	Amount
Arginine monohydrochloride	1.0 g	Soluble starch	10.0 g
Glycerol	12.5 ml	Casein, vitamin-free	0.3 g
K_2HPO_4	1.0 g	KNO_3	2.0 g
NaCl	1.0 g	NaCl	2.0 g
$MgSO_4 \cdot 7H_2O$	0.5 g	K_2HPO_4	2.0 g
$Fe(SO_4)_3 \cdot 6H_2O$	0.010 g	$MgSO_4 \cdot 7H_2O$	0.05 g
$CuSO_4 \cdot 7H_2O$	0.001 g	$CaCO_3$	0.02 g
$ZnSO_4 \cdot 7H_2O$	0.001 g	$FeSO_4 \cdot 7H_2O$	0.01 g
$MnSO_4 \cdot 7H_2O$	0.001 g	Agar	20.0 g
Agar	15.0 g	Distilled water	1 liter
Distilled water	987.5 ml		

ACKNOWLEDGMENT

We appreciate the work of Dr. M. A. Lechevalier, who reviewed the detailed content of this chapter.

References

1. **Harz, C. O. 1977–1978.** *Dtsch. Z. Tiermed.* 5:125–140.
2. **Sykes, G. and F. A. Skinner** (eds.). **1973.** *Actinomycetes: Characteristics and Practical Importance,* 2nd Symposium of the Society for Applied Bacteriology, London. Academic Press, New York. 339 pp.
3. **Cross, T. 1968.** *J. Appl. Bacteriol.* 31:36–53.
4. **Ishaque, M. and D. Kluepfel. 1980.** *Can. J. Microbiol.* 26:183–189.
5. **Crawford, D. L. and R. L. Crawford. 1980.** *Enzyme Microb. Technol.* 2:11–22.
6. **Porter, J. N. 1971.** *Adv. Appl. Microbiol.* 14:73–92.
7. **Waksman, S. A. 1967.** *The Actinomycetes: A Summary of Current Knowledge.* Ronald Press, New York. 280 pp.
8. **Turner, W. D. 1973.** Pp. 209–218 in G. Sykes and F. A. Skinner (eds.). *Actinomycetales: Characteristics and Practical Importance.* Academic Press, New York.
9. **DeLey, J. 1970.** Pp. 317–327 in H. Prauser (ed.). *The Actinomycetales.* Gustav Fischer, Jena, East Germany.
10. **Küster, E. 1972.** Pp. 111–127 in A. Burges and F. Raw (eds.). *Soil Biology.* Academic Press, New York.
11. **Lechevalier, H. A. and M. P. Lechevalier. 1967.** *Annu. Rev. Microbiol.* 21:71–100.
12. **Lechevalier, H. A. and L. Pine. 1979.** Pp. 360–381 in A. L. Laskin adn H. A. Lechevalier (eds.). *Bacteria,* Vol. I. CRC Press, Cleveland, Ohio.
13. **H. Prauser (ed.). 1970.** *The Actinomycetales.* Gustav Fischer, Jena, East Germany.
14. **Williams, S. T. 1966.** *Int. Biodeterior. Bull.* 2:125–133.
15. **Alexander, M. 1977.** Actinomycetes. Pp. 36–50 in *Introduction to Soil Microbiology,* 2nd Ed. Wiley, New York.
16. **Goodfellow, M. and T. Cross. 1974.** Actinomycetes. Pp. 269–302 in C. H. Dickinson and G. J. F. Pugh (eds.). *Biology of Plant Litter Decomposition.* Academic Press, New York.

17. **Gray, T. R. G. and S. T. Williams. 1971.** *Soil Microorganisms.* Oliver & Boyd, Edinburgh. p. 235.
18. **Gray, T. R. G. and S. T. Williams. 1971.** Pp. 255–286 in D. E. Hughes and S. H. Rose (eds.). *Microbes and Biological Productivity.* Cambridge University Press, Cambridge.
19. **Mayfield, C. I., S. T. Williams, D. M. Ruddick and N. L. Hatfield. 1972.** *Soil Biol. Biochem.* 4:79–91.
20. **Willoughby, L. G. 1971.** *Freshwater Biol.* 1:23–27.
21. **Williams, S. T. and T. Cross. 1971.** *Meth. Microbiol.* 4:295–334.
22. **Williams, S. T. and F. L. Davies. 1965.** *J. Gen. Microbiol.* 38:251–261.
22a. **Nonomura, H. and Y. Ohara. 1969.** *J. Ferment. Technol.* 47:463–469.
22b. **Meiklejohn, J. 1957.** *J. Soil Sci.* 8:240–247.
23. **Jagnow, G. 1957.** *Arch. Mikrobiol.* 26:175–191.
24. **Waksman, S. A. and E. R. Purvis. 1932.** *Soil Sci.* 34:95–113.
25. **Williams, S. T., M. Shameemullah, E. R. Watson and C. I. Mayfield. 1972.** *Soil Biol. Biochem.* 4:215–225.
26. **Gledhill, W. E. and L. E. Casida. 1969.** *Appl. Microbiol.* 18:114–121.
27. **Chen, A. W. C. and D. M. Griffin. 1966.** *Trans. Br. Mycol. Soc.* 49:419–426.
28. **Kouyeas, V. 1964.** *Plant Soil* 20:351–363.
29. **Johnston, O. W. 1972.** *Actinomycetes in Aquatic Habitats.* Ph.D. thesis, University of Bradford, England. 216 pp.
30. **Willoughby, L. G. 1969.** *Hydrobiologia* 34:465–485.
31. **Ruddick, S. M. and S. T. Williams. 1972.** *Soil Biol. Biochem.* 4:93–103.
32. **Johnston, D. B. 1947.** *Soil Sci.* 64:453–458.
33. **Waksman, S. A., W. W. Embreit and T. C. Cordon. 1939.** *Soil Sci.* 47:37–61.
34. **Abraham, T. A. and L. J. Herr. 1964.** *Can. J. Microbiol.* 10:281–285.
35. **Webley, D. M., D. J. Eastwood and C. H. Gimingham. 1952.** *J. Ecol.* 40:168–178.
36. **Davies, F. L. and S. T. Williams. 1970.** *Soil Biol. Biochem.* 2:227–238.
37. **Küster, E. 1959.** *Int. Bull. Bacteriol. Nomencl. Taxon.* 9:98–104.
38. **Tresner, H. D. and E. J. Backus. 1963.** *Appl. Microbiol.* 11:335–338.
39. **Cross, T. and M. Goodfellow. 1973.** Pp. 11–112 in G. Sykes and F. A. Skinner (eds.). *Actinomycetales: Characteristics and Practical Importance.* Academic Press, New York.
40. **Waksman, S. A. 1961.** *The Actinomycetes.* Vol. II. *Classification, Identification, and Description of Genera and Species.* Williams & Wilkins, Baltimore. 363 pp.
41. **Lechevalier, M. P., C. Bievre and H. A. Lechevalier. 1977.** *Biochem. Syst. Ecol.* 5:249–260.
42. **Tresner, H. D., J. A. Hayes and E. J. Backus. 1968.** *Appl. Microbiol.* 16:1134–1136.
43. **Lechevalier, M. P. and H. A. Lechevalier. 1980.** Pp. 225–291 in A. Dietz and P. W. Thayer (eds.). *Actinomycete Taxonomy.* SIM Publ. 6.
44. **Lechevalier, M. P., A. C. Horan and H. A. Lechevalier. 1971.** *J. Bacteriol.* 105:313–318.
45. **Mordarska, H., M. Mordaski and M. Goodfellow. 1972.** *J. Gen. Microbiol.* 71:77–86.
46. **Lechevalier, M. P., C. Bievre and H. A. Lechevalier. 1977.** *Biochem. Syst. Ecol.* 5:249–260.
47. **Trejo, W. 1980.** Pp. 292–326 in A. Dietz and D. W. Thayer (eds.). *Actinomycete Taxonomy.* SIM Publ. 6.
48. **Bradley, S. G. 1975.** *Adv. Appl. Microbiol.* 19:59–70.
49. **G. E. Fox, E. Stackebrandt, R. B. Hespell, J. Gibson, J. Maniloff, T. A. Dyer, R. S. Wolfe, W. E. Balch, R. S. Tanner, L. J. Magrum, L. B. Zablen, R. Blakemore, R. Gupta, L. Bonen, B. J. Lewis, D. A. Stahl, K. R. Luehrsen, K. N. Chen and C. R. Woese. 1980.** *Science* 209:457–463.
50. **Goodfellow, M. 1971.** *J. Gen. Microbiol.* 69:33–80.
51. **Hunter, J. C. 1978.** *Actinomycetes of a Salt Marsh.* Ph.D. thesis, Rutgers University, New Brunswick, N.J.

52. **Tresner, H. D. and J. A. Hayes. 1970.** *Appl. Microbiol.* 19:186–187.
53. **Porter, J. N. and J. J. Wilheim. 1961.** *Dev. Ind. Microbiol.* 2:293–260.
54. **El-Nakeeb, M. A. and H. A. Lechevalier. 1963.** *Appl. Microbiol.* 11:75–77.
55. **Lingappa, Y. and J. L. Lockwood. 1962.** *Phytopathology* 52:317–323.
56. **Küster, E. and S. T. Williams. 1964.** *Nature* (*Lond.*) 202:928–929.
57. **Orchard, V. A. and M. Goodfellow. 1974.** *J. Gen. Microbiol.* 89:160–162.

Bibliography

Additional references of value to researchers and students are listed below.

Isolation References

Cross, T. 1982. *Dev. Ind. Microbiol.* 23:1–18. **Hunter, J. C., D. E. Eveleigh and G. Casella. 1981.** In K. P. Schaal and G. Pulverer (eds.). *Actinomycetes.* Zentralbl. Bakteriol. Parasitenkd. Infektionskr. Hyg. Suppl. 11. Gustav-Fischer, Stuttgart, West Germany. **Hunter, J. C., M. E. Fonda, L. Sotos, R. Toso and A. Belt. 1984.** *Dev. Ind. Microbiol.* 25:247–266. **Hunter-Cevera, J. C., M. E. Fonda and A. Belt. 1986.** Pp. 3–23 in A. L. Demain and N. A. Solomon (eds.). *Manual of Industrial Microbiology and Biotechnology.* ASM, Washington, D.C. **Williams, S. T. 1986.** Pp. 693–700 in G. Szabó, S. Biró and M. Goodfellow (eds.). *Biological, Biochemical and Biomedical Aspects of Actinomycetes.* Akadémiai Kiadó, Budapest, Hungary. **Williams, S. T. and E. M. H. Wellington. 1982.** Pp. 9–26 in J. D. Bulock, L. J. Nisbet and D. J. Winstanley (eds.). *Bioactive Microbial Products: Search and Discovery.* Academic Press, London. **Williams, S. T., S. Lanning and E. M. H. Wellington. 1984.** Pp. 481–528 in M. Goodfellow, M. Mordarski and S. Williams (eds.). *Biology of the Actinomycetes.* Academic Press, London.

Identification References

Goodfellow, M. 1986. Pp. 487–496 in G. Szabó, S. Biró and M. Goodfellow (eds.). *Biological, Biochemical and Biomedical Aspects of Actinomycetes.* Akadémiai Kiadó, Budapest, Hungary. **Stackebrandt, E. 1986.** Pp. 498–506. in G. Zsabó, S. Biró and M. Goodfellow (eds.). *Biological, Biochemical and Biomedical Aspects of Actinomycetes.* Akadémiai Kiadó, Budapest, Hungary. **Stackebrandt, E., B. Wunner-Füssl, V. Fowler and K. H. Schleifer. 1981.** *Int. J. Syst. Bacteriol.* 31:420–431.

Soil Fungi

W. BRYCE KENDRICK

DENNIS PARKINSON

The functional roles of the fungi in soil are determined by their physiological and morphological characters. Being heterotrophic organisms the fungi must be saprobes, parasites, or symbionts, and as such they play, directly or indirectly, vital roles in affecting primary production. Primary production is directly affected by the actions of parasitic and symbiotic (mycorrhizal) fungi, and is indirectly affected by the actions of soil and litter saprobes as decomposers of organic matter and hence as important agents in nutrient cycling.

In this chapter only the saprobic, filamentous soil fungi will be considered. For information on pathogenic and mycorrhizal fungi, Garrett (1970), Baker and Cook (1974), Marks and Kozlowski (1973), and Saunders et al. (1975) should be consulted and used as initial entrées to those aspects of soil mycology.

A large number of taxa have been isolated from different soil types, these taxa being from all the major groups of the fungi. The distribution of fungi in soil is affected by various factors, such as the distribution of suitable organic substrates, by available moisture, and by temperature. Within an individual soil, fungal distribution frequently varies with soil depth (i.e., in different soil horizons). The heterogeneous nature of the soil will therefore have great effects on the distribution of specific fungi.

The cosmopolitan nature of many species of soil fungi, postulated by Waksman and others in the 1920s, was emphasized by Domsch and Gams (1972). However, considerable differences have been shown between the fungal species most frequently isolated from different biomes. In grassland mineral soils, *Fusarium* spp., *Paecilomyces* spp., *Cylindrocarpon* spp., and *Rhizoctonia* sp. might be most frequently isolated, whereas from the mineral soil of coniferous forests, *Trichoderma* spp., *Penicillium* spp., and *Mortierella* spp. would be most commonly isolated.

Since the detailed work of Kendrick and Burges (1962), considerable attention has focused on fungi associated with plant leaf material at different stages of decomposition. The ideas developing from these studies (i.e., ideas on various successional patterns) have been reviewed by Hayes (1979) and Swift (1976). In many of these studies one must be struck by the absence of species of basidiomycetes from the lists of fungi isolated. This is particularly unfortunate in studies of forest litter layers, where even casual observation of the forest floor in general reveals the presence of a diversity of fruit bodies or microscopic examination of litter samples reveals the presence of clamp-bearing (Basidiomycete) hyphae.

It must be emphasized that when detailed studies on spatial and temporal variations of fungi in soil or decomposing organic matter are being attempted, critical consideration must be given to the proper design of the sampling program, to proper statistical analysis of the data obtained, and to appropriate "banking" of data. In such detailed studies each investigator will be faced with numerous taxonomic problems, and we hope these will be attacked critically.

Over the last 20 years there has been increasing interest in determining fungal biomass and standing crop in soils. This interest developed from the need to assess the roles of the soil fungi in energy and nutrient flows through litter and soil systems. While assessments of soil fungal biomass and standing crop are possible, effective methods for assessing fungal production are still awaited.

Fungi exist in soil in a variety of morphological forms, that is, as spores of various types or other dormant propagules (e.g., sclerotia), or as hyphae in various physiological states. The hyphae, with their large surface area/volume ratio, are the important structures for organic matter decomposition. They are morphologically adapted for ramifying over and penetrating into suitable organic substrates, and for the absorption of soluble nutrients. Hence it is the aim of many ecological studies to assess (qualitatively and quantitatively) the fungi present in litter or soil in an active hyphal state.

Various types of hyphal structures occur in soil. These range from individual branched hyphal systems to various types of hyphal strands (i.e., twined, branched hyphae) within which various degrees of differentiation can be observed, culminating in the true rhizomorph. Even where hyphal systems are produced with no strand formation, different types of hyphal structures can be seen (Harley 1971). These structures range from ephemeral hyphae growing rapidly on simple carbon substrates to relatively long-lived (even perennial) hyphae growing on such complex substrates as cellulose or lignin. The latter are generally septate hyphae among which hyphal fusions are frequent—a phenomenon that can both produce hyphal networks and introduce the possibility of heterokaryosis.

Apart from the different types of hyphae present in soil, different species of fungi exhibit different growth forms in litter and soil systems. Burges (1960) recognized five growth patterns exhibited by soil fungi: dense colonization of the substrate with no hyphal growth into surrounding soil, hyphal growth into surrounding soil (with chlamydospore production) following substrate exploitation, growth through soil (with utilization of organic material in the soil solution) with no obvious colonization of organic fragments, mycelial strands or rhizomorphs, and the well-defined mycelial zone in the soil typical of "fairy ring" fungi.

Harley (1971) in considering fungi in ecosystems pointed out that the peculiar difficulty in studying soil fungi stems from the fact that field identification is very rarely possible, and that in studying the ecology of soil fungi both direct observation and isolation methods must be used. In the following sections we discuss some of these methods and address the problems of identifying the large variety of taxa that may be isolated from soil and organic debris.

Methods for Qualitative and Quantitative Studies of Soil Fungi

Choice of Nutrient Media for Isolating Soil Fungi

A number of detailed reviews are available on the range of methods available for isolating soil fungi (Parkinson et al. 1971, Johnson and Curl 1972, Parkinson 1986). All methods involve the isolation onto and growth of the fungal isolates on a nutrient medium. Therefore, it is important to choose a nutrient medium appropriate for the study being undertaken. At one time a number of efforts were made to develop nonselective media (i.e., media that would allow isolation of all fungi present in a soil sample under study), and media such as water agar (i.e., devoid of added nutrients) were used for the primary isolation of soil fungi. Naturally, fungi growing on such a medium had to be transferred to a nutrient medium for further study (identification).

The idea of a single all-purpose medium has now been abandoned as impractical, and recently it has become common to use a medium that will allow the isolation of the maximum number of fungal taxa from the soil being studied (i.e., it is accepted that not all species of fungi present in the soil samples will be isolated). Given sufficient resources (time and personnel) the use of a wide range of selective media for studies on a single soil would provide valuable data, but this will be a rare occurrence.

It is dangerous, because of differences in soil types and the range in possible investigations, to recommend any particular isolating medium. If a long-term, detailed investigation on the ecology of fungi in a particular soil is being planned, a properly designed, comparative study on a range of nutrient media should be carried out to determine the most appropriate medium for that ecological investigation.

As can be imagined, a considerable number of different media have been developed for the isolation of fungi from soil. Among the most frequently used have been Czapek–Dox agar (with or without small additions of yeast extract), malt extract agar (with or without additions of peptone, dextrose, and yeast extract), and dextrose–peptone agar (with or without additions of such substances as rose bengal, oxgall, or sodium propionate). Although soil extract media have been used for studies on soil fungi, this has not been so frequent as for studies on soil bacteria. Detailed recipes for a range of media are given in Parkinson et al. (1971); however, details on three commonly used media are given below:

Czapek–Dox Medium:

Sucrose, 30.0 g; $NaNO_3$, 3.0 g; K_2HPO_4, 1.0 g; KCl, 0.5 g; $MgSO_4 \cdot 7H_2O$, 0.5 g; $FeSO_4$, trace; agar, 15.0 g; distilled water, 1000 ml; and 0.5–1.0 g of yeast extract, if desired. Dissolve inorganic constituents separately, add $FeSO_4$ last. Add sucrose just before final sterilization.

Malt Extract Agar:

Variants of this medium range from malt extract, 20.0 g; agar, 15.0 g; distilled water, 1000 ml, to malt extract, 20.0 g; dextrose, 20.0 g; peptone, 1.0 g; agar, 15.0 g; and distilled water, 1000 ml. The amount of malt extract can be decreased if desired and small amounts of yeast extract can be added (0.5–1.0 g/liter).

Dextrose–Peptone Agar (Martin 1950):

Dextrose, 10.0 g; peptone, 5.0 g; KH_2PO_4, 1.0 g; $MgSO_4 \cdot 7H_2O$, 0.5 g; rose bengal (1%), 3.3 ml; streptomycin, 30.0 mg; agar, 20.0 g; and distilled water, 1000 ml. All materials except rose bengal and streptomycin are dissolved in water and heated, then rose bengal is added. The medium is autoclaved and, following cooling, streptomycin is added to the still liquid agar medium.

Dextrose–Peptone Agar (Papavizas and Davey 1959):

Dextrose, 5.0 g; peptone, 1.0 g; yeast extract, 2.0 g; NH_4HO_3, 1.0 g; K_2HPO_4, 1.0 g; $MgSO_4 \cdot 7H_2O$, 0.5 g; $FeCl_3 \cdot 6H_2O$, trace; oxgall, 5.0 g; sodium propionate, 1.0 g; agar, 20.0 g; aureomycin, 30.0 mg; streptomycin, 30.0 mg; and distilled water, 1000 ml. Once again, antibiotics are added following autoclaving and cooling of the medium.

If it is decided to use an isolation medium that incorporates soil extract, the usual method to prepare this extract is to autoclave 1 kg of soil with 1 liter of water (20 minutes at 15 or 20 lb pressure). The autoclaved material is filtered and the filtrate made up to 1 liter. If the filtrate is cloudy, a little $CaSO_4$ is added and, after standing, filtration is repeated.

When dealing with soils containing large numbers of bacteria, it may be necessary to reduce competition from these organisms by adding antibacterial substances to the isolation medium (thereby increasing the likelihood of isolating slow-growing fungi).

While acidification of the isolation medium (to about pH 5.0) has been used frequently for this purpose, it must be remembered that this may add yet another selective factor for the isolation of the fungi in the soil. Substances such as crystal violet, oxgall, rose bengal, sodium deoxycholate, and sodium propionate have been used to restrict bacterial development on nutrient media.

The use of antibacterial antibiotics is now common (some examples have been given above). The addition of aureomycin (30 mg/liter medium) with or without streptomycin (30 mg/liter medium) is widely used. It must be remembered to add antibiotics after autoclaving and cooling the otherwise complete nutrient medium.

For studies on a particular species or allied species of fungi (i.e., autecological studies) selective media are used. Many such media have been, and continue to be, developed. Examples of such selective media are given in Johnson and Curl (1972).

One group of the soil fungi that has been relatively infrequently isolated but which has great importance in organic matter decomposition and ectomycorrhizal associations is the basidiomycetes. Many of these species seem to be fastidious in their nutritional requirements and hence are not isolated on conventional nutrient media; however, it is possible that for many species, their slow growth on commonly used media makes them susceptible to competition from faster-growing soil fungi during attempts at their isolation. Hale and Savory (1976) reviewed the use of substances that inhibit groups of fungi other than basidiomycetes in conjunction with isolation media. Benomyl (5 ppm active ingredient) has been found useful in suppressing growth of hyphomycetes during attempts to isolate basidiomycetes from ectomycorrhizal roots.

Methods for Isolating Soil Fungi

A range of methods have been described for the isolation of fungi from soil (Parkinson 1986). Only a restricted number of methods are presented here.

Soil Dilution Plates. In the early period of studies on soil fungi the soil dilution plate method was used almost exclusively. Warcup (1957) demonstrated that this method selectively isolates fungi present in soil samples as spores, that is, it is inappropriate for use in studies where knowledge of fungi present in soil as hyphae is required. Nevertheless, it can be valuable in studies where information is desired on the types and numbers of fungal spores in soil samples. Details on this method can be found in Parkinson (1986).

Soil Plates. A simple variant of the soil dilution plate method was described by Warcup (1950). In this method (the soil plate method), very small weights of soil (0.005–0.015 g) are dispersed in known volumes of an appropriate nutrient medium. A very small subsample of soil is taken from the soil sample using a sterile needle with a flattened tip and placed in the base of a sterile petri dish. This small amount of soil is then thoroughly broken up and spread over the base of the dish (addition of a drop of sterile water frequently aids this process). Cooled (45°C) molten nutrient agar (8–10 ml) is then added to the dish, which is gently rotated to disperse the soil fragments in the nutrient medium. Because of its simplicity, replicate soil plates from a considerable number of soil samples can be prepared in a relatively short time. Soil plates are incubated (usually, 22–25°C) and fungi developing from the soil fragments are isolated into pure culture for subsequent identification.

While the soil plate method suffers from the same defect as soil dilution plating (i.e., it selectively isolates fungi present in soil as spores), because of its simplicity and speed of execution, it is frequently used in preliminary studies on soil mycotas prior to applying a more critical isolation method (see below).

Direct Hyphal Isolation. Warcup (1955) described a method of direct hyphal isolation from soil which would appear ideal for studies on fungi present as hyphae in soil samples. Here small soil samples or soil crumbs are saturated with sterile water and are then broken up with a fine jet of water. Heavier soil particles sediment and fine particles are decanted off. Repeated application of this procedure allows separation of the heavier particles, which are then spread in a film of sterile water and are examined under a dissecting microscope. Any fungal hyphae observed are picked out using sterile needles or fine forceps. To remove debris or spores attached to the hyphae, each fragment of hyphae is drawn through semisolid agar prior to plating onto a nutrient agar plate. It is important to mark on the base of the petri dish the position of each piece of hyphal inoculum. Plates prepared in this way are incubated (22–25°C) and the plated hyphal fragments are observed (under the microscope) at least once per day to make sure that any fungal growth which occurs does originate from the plated hyphae and not from spores still attached to the hyphae.

Despite its obvious value in restricted studies (particularly of specific taxa), this method is not practical for large-scale studies on soil fungi (i.e., studies involving isolations of fungi from replicate isolations from large numbers of soil samples on a regular basis) because of the time and skill required in its execution. If the method is used, care must be taken to avoid bias in selecting hyphae in the samples under study. There is a tendency for workers to select hyphae of large diameter, hyphae that do not fragment easily, and darkly pigmented hyphae.

Washing Techniques. The development of various washing techniques for isolation of fungi from soil, roots, decaying organic matter of various types has provided a simpler, less-skill-dependent approach to the isolation of fungi present as hyphae in these substrates. The essence of these methods is the serial vigorous washing of the material (soil, roots) under study with sterile water (2 minutes per wash), the washing water being carefully decanted (or otherwise removed) after each washing, the aim of this procedure being to remove as many fungal spores as possible from the soil sample. Following the washing procedure the washed particles are placed on sterile filter paper to remove excess water and are then individually placed onto nutrient agar medium (two to four particles per plate). The plates are incubated and fungi developing from the plated particles are subcultured for subsequent identification.

The optimal number of washings required to achieve efficient washing of the substrate under study must be determined by preliminary investigation (see Harley and Waid 1955). The efficiency of soil washing is related to the physical nature of the soil being studied (e.g., sandy soils can be washed more efficiently than soils with high clay content).

Washing methods range from simple washing–decanting methods (Watson 1960) to the use of automated washing machines (Hering 1966, Bissett and Widden 1972), where numerous samples can be washed simultaneously and where separation of soil samples into different particle sizes can be achieved.

Damp Chambers. In investigations on the mycofloras of such substrates as leaves, twigs, and dung, the observation of the development of fungi and the succession of fungi can be followed by placing the substrates in moist sterile chambers and by regular microscopic (dissecting and compound) observation of the fungi sporulating on the surface of the substrate. This method complements the conventional isolation studies described earlier when dealing with various types of plant and animal debris.

In this method sterile petri dishes or sterile deeper glass dishes with lids are used, damp conditions are maintained by enclosing either moist sterile filter paper on the base of the dishes or surrounding the walls of the dishes with sterile water-saturated asbestos string. The substrates under study are placed in the dishes and following incubation are observed regularly for fungal development. Care should be taken to avoid getting the substrates too wet.

Isolation of fungi growing directly on the substrate can be attempted by picking off small portions of hyphae or small numbers of spores and plating these onto nutrient agar (see later). Following the primary isolation of fungi from soil or organic debris by any of the methods described above, individual isolates must be subcultured for identification. A wide range of media are available for the maintenance of pure cultures of fungi (e.g., potato extract, potato–dextrose, potato–carrot, and malt extract). When specific identification of individual taxa is being undertaken, it is essential to grow the fungi on the specific media described in the taxonomic monographs being used for identification purposes (see later).

Methods for Quantifying Soil Fungi

In view of the need in many ecological investigations to achieve determination of fungal standing crop and/or biomass, various methods have been developed for such determinations. In earlier studies on soil fungi the soil dilution plate method was used to achieve so-called quantitative estimates of soil fungi, but it is now known that colony counts on dilution plates only allow an estimate of the number of fungal spores in soil samples. Assessment of absolute amounts of fungal hyphae per unit weight or volume of substrate is now attempted by one of three types of methods: direct observation methods, chemical methods, or physiological methods.

A detailed description and discussion of these various methods has been given by Parkinson (1986). In view of the fact that direct observation methods have been most used and are most straightforward in application, only this type of approach will be considered here.

Soil–Agar Film Method. Jones and Mollison (1948) described a soil–agar film method for quantitatively studying microorganisms in mineral soil samples. This method has been modified as follows for mineral soil or organic matter samples: a known weight of soil (or decaying litter), the weight having been predetermined by preliminary testing, is blended (Waring blender or similar apparatus) at 10,000 rpm for 3 minutes in 100 ml of water. Swift (1973) emphasized the need for testing both maceration time and degree of dilution when using this method in detailed studies. The suspension prepared is then decanted into a beaker, care being taken to ensure that all material is decanted off, and an equal volume of double strength water agar is added, thoroughly mixed, and the soil–water–agar suspension is kept at 50–60°C to prevent gelling.

The suspension is thoroughly agitated and a small amount of the suspension is pipetted from a known (constant) depth in the suspension onto the platform of a hemocytometer slide. A weighted (5 g) coverslip is immediately placed over the platform, and the enclosed suspension is allowed to solidify. Then the slide is immersed in distilled water, the coverslip removed, agar in the moats of the slide is cut away, and the agar film remaining is floated onto an ordinary microscope slide. The agar film on the slide is dried at room temperature and can then be stained if desired (1 h phenolic aniline blue, then three to four washings with 98% alcohol) and mounted (Euparal or other suitable mountant). Thus a soil–agar film of known thickness (0.1 mm, the conventional depth of the well of a hemocytometer slide) is mounted on a microscope slide. The prepared film is observed under the microscope at known magnification (400–1000×) and hyphal lengths in each of a number of random fields are assessed using the intersect method (Olson 1950). Replicate observations of each of a number of replicate soil films are essential (Frankland et al. 1978). One of the problems of this method is the difficulty of distinguishing between live and dead hyphae in the agar films. The use of fluorescent dyes with agar films is impractical, but observation of films stained with phenolic aniline blue has allowed estimations of live (stained) and dead (unstained) hyphae. Observation of films under phase contrast microscopy allows distinction between hyphae with cell contents and empty hyphae (Frankland 1975, Visser and Parkinson 1975).

Knowing the length of hyphae per microscope field, the area of the field, the depth of the film, and the dilution factor in preparing the soil–water–agar suspension, the length of hyphae per unit weight of soil can be calculated. From these data hyphal standing crop (dry weight) can be calculated given data on the average radius of observed hyphae, an average specific gravity of hyphae, and an average value for hyphal moisture content (Parkinson et al. 1971, Frankland et al. 1978).

Membrane Filter Method. In an attempt to reduce the perceived tediousness of the soil–agar film method and to allow the use of fluorescent staining, Hansen et al. (1974) developed a membrane filtration method. In this method 0.5–1.0 g of soil is suspended in 500 ml of water, homogenized (30–90 seconds, depending on soil type) in a Waring blender, and known volumes of the suspension are membrane filtered (0.45 μm Oxoid standard grade). Each filter is placed on a slide, allowed to dry, and cleared (liquid paraffin or decahydronaphthalene for Oxoid filters). If fluorescence microscopy using incident light is available, clearing is unnecessary. The filters are observed and hyphal lengths recorded (as described for the soil–agar film method).

Paul and Johnson (1977) described a variation of this method where filters are stained with water-soluble aniline blue (1% w/v in 0.2 *M* K_2HPO_4, pH 8.0), the filters being observed using bright-field illumination. Soderstrom (1977) has described a vital method (fluorescein diacetate) in an attempt to distinguish clearly between live and dead hyphae. Assessment of these methods is given in Parkinson (1986).

Nylon Mesh Method. A simple method for studying fungal growth in soil was described by Waid and Woodman (1957) and modified by Nagel-de-Boois and Jansen (1971) for detailed studies of fungi in deciduous forest soils. In this method pieces of nylon gauze (125 meshes per centimeter) are buried in soil and replicate gauzes are sampled at appropriate time intervals, returned to the laboratory, stained, and examined under the microscope. By examining 200 meshes per gauze and recording the presence or absence of fungi in each gauze, an expression of fungal growth can be obtained. Actual lengths of mycelium on each gauze can also be measured directly. Of course, it is difficult to relate data obtained by this method to absolute weights or volumes of soil, and it must be remembered that the placing of pieces of gauze into soil may create physical conditions which stimulate fungal growth.

Pure Culturing of Fungal Isolates and Their Preservation

Detailed accounts of methods for obtaining pure cultures of fungal isolates, of media on which fungi may be grown, and of methods for preservation of fungal cultures are given in Booth (1971), Hawksworth (1974), Stevens (1974), and Gams et al. (1975). Therefore, the following comments must be viewed as a brief summary of commonly used methods.

Obtaining Pure Cultures of Fungi Isolated from Soil

In all the isolation methods described above, where particles of soil or organic debris (washed or unwashed) are plated onto nutrient agar, it is necessary to obtain pure cultures of all isolates in order to achieve specific identification of these isolates. This is made difficult because frequently more than one species of fungi grow from each plated particle and overgrow each other, and frequently fast-growing species rapidly cover isolation plates, thereby masking slower growing forms. Therefore, when isolation of fungi into pure culture is to be attempted, the primary isolation plates should be observed under a dissecting microscope to allow clearer definition of the colonies of isolated fungi—with experience, identification of some taxa sometimes can be made at this stage, but normally, subculturing must be achieved to allow detailed study of pure isolates.

Subculturing is effected by taking small numbers of spores or small amounts of hyphae (preferably hyphal tips) and either streaking or plating the fungal inoculum onto a suitable nutrient agar medium in a petri dish. Subcultured fungi are observed regularly to ensure that pure cultures of the desired fungi have been obtained, and to observe all stages of the life history of each isolate (detailed regular microscopic study being required). If pure cultures have not been obtained during the initial subculturing process, it is essential that further, careful subculturing be performed to obtain pure cultures of the fungi to be identified.

Frequently, fungal isolates do not sporulate on the nutrient agar medium used for subculturing. In this case the sterile isolates must be grown on a range of media and under a range of incubation conditions. Gams et al. (1975) give a thorough description of media and methods for growth and identification of fungi isolated from soil and plant debris.

Slide Preparations of Fungi

Identification of fungi obtained in pure culture requires first, careful observation under the dissecting microscope of the growth form of the fungus on the agar medium and of any spore-bearing structures, and second, the preparation of wet mounts of each isolate. The usual method is to place a small drop of mounting liquid (e.g., lactic acid, lactophenol, Shear's mounting liquid) with or without an added stain (e.g., cotton blue, trypan blue) on a slide. Using either fine forceps or a mounted needle, a *small* amount of the fungus material is placed in the drop of mounting liquid and is then carefully teased out in the drop using two needles and preferably using a dissecting microscope to observe the degree of separation. A coverslip is then carefully placed over the preparation, avoiding the trapping of air bubbles. A variant of this is to place a coverslip on a clean piece of paper, quickly invert the slide so that the drop containing the fungus is on the lower surface, and then lower the slide until the drop touches the coverslip. Surface tension draws the coverslip to the slide and the mountant will spread evenly, usually without trapping air bubbles. The preparation can then be observed under different magnifications of the compound microscope.

If the fungus under study has very delicate spore-bearing structures which are disrupted by the mounting procedure described above, it may be necessary to prepare either slide or coverslip cultures of the fungus under study. Here a small block of agar is placed on a sterile microscope slide, each side of the block is inoculated with the fungus, and the block is then covered with a sterile coverslip. The culture is then kept in a sterile moist chamber until the fungus has grown substantially both over the coverslip and on the slide. Then wet mounts can be made of the material on both the slide and the coverslip.

It is important when transferring fungi for purification and subculturing purposes to use proper aseptic technique, that is, the use of flame-sterile equipment, sterile glassware, and sterile nutrient media. The bench on which transferences are being made should be kept very clean.

Maintenance of Pure Cultures

Where mass isolation of fungi from soil is being attempted it will be obvious that pure cultures must be kept for varying periods prior to identification. Once identification of isolates has been achieved (and confirmed) it is necessary to keep a reference collection of the named isolates. Therefore, appropriate preservation of pure cultures is required and can be achieved in various ways. The growth of the fungi on agar slants at low temperature (2–5°C) is a common method, but periodic transfers onto fresh slants are necessary. If long-term storage of fungi is required, such methods as freeze-drying, growth in sterile soil, slant cultures under sterile paraffin oil, or under glycerol and frozen (liquid N_2) can be used.

Identification

Let us assume that you have isolated some microorganisms from the soil. You think they are fungi, and you want to identify them. Our key is intended to set you on the right path but does not presume to guide you all the way to your destination—the genus and species of your unknown. The scope of this chapter is simply too restricted to permit such a detailed treatment. We point you instead to some recently published, specialized literature, which should give you an excellent chance of arriving at a correct identification.

In constructing our key, we have tried to keep specialized terminology to a minimum and to assume minimal familiarity with the fungi on the part of our readers. But there is a practical limit to this, and those who are totally unacquainted with the group should refer to a textbook such as Mueller and Loeffler (1976), Webster (1980) or Kendrick (1985) before attempting to use the key, so that they will better understand the conceptual framework involved.

Microorganisms isolated from soil commonly include representatives of several major groups: bacteria, true fungi of the phylum Zygomycota and phylum Dikaryomycota (mostly Ascomycetes), and finally, a heterogeneous assemblage called yeasts.

Although we know that many protoctistan fungi of the phyla Chytridiomycota and Oomycota, and many true fungi of the subphylum Basidiomycotina, are also common and very active in many soils, they are not isolated by the techniques and media usually employed to survey the soil mycota. It is apparent that they are outcompeted by members of other groups, and simply do not grow on these media. If you want to study the occurrence of these three groups in soil, you will have to use specialized techniques and selective media. These are discussed by Stevens (1974) and Booth (1971).

The primary features by which the major groups of fungi are delineated have to do with the rather different structures (grouped under the general heading of teleomorphs) that they produce during sexual reproduction. This emphasis can create major problems for the would-be identifier, because it is difficult to persuade many fungi to exhibit their sexual characteristics in culture. In fact, many Ascomycetes apparently no longer reproduce sexually, and exist entirely as anamorphs (placed in the anamorph-classes Hyphomycetes and Coelomycetes) which produce only mitospores (conidia).

Fortunately, even when sexual structures are lacking, other characters enable us to narrow our search. We begin our key with vegetative features, since these are always present. And since mitospores are much more frequently encountered than meiospores, we tend to use these at the second level of our key. Only when these are absent or taxonomically inadequate do we fall back on the more rarely observed teleomorph.

Dichotomous keys are unforgiving—if you make one error, you go off on the wrong track. Such keys also have a habit of asking for information that may not be present in the specimen you are looking at. For example, most keys to flowering plants assume the presence of flowers. How one is supposed to identify the plants during the perhaps 11 months of the year when they are not in flower is rarely made clear. Similar things happen in mycological keys. In the fungi, if reproductive structures are absent, we are often left with no other avenue to explore. We cannot count chromosomes in most cases, and the taxonomic potential of vegetative morphology or biochemical tests has not been fully exploited in most groups.

So the plain truth is that unless sporulation occurs spontaneously, or can be induced, it may prove impossible to identify a fungus to genus or species: although, as our key shows, it is often possible to assign the sterile mycelium to one of the major taxonomic divisions of fungi.

When a soil fungus grows in culture, it produces what we call a colony. This may arise from a spore or from a hypha, but the end result is the same: a mass of hyphae growing more or less radially outward from their point of origin, and usually branching frequently in order to explore the substrate fully. As the number and extent of the hyphae increase, they become visible to the naked eye as a mycelium, forming a circular, often more or less disklike aggregation. Some fungi grow very slowly and form extremely restricted colonies. Others may cover a petri dish in a day or so, and a few weedy fungi will even grow out of a dish and probe the surroundings with their hyphae.

Zygomycetes are among the fastest-growing fungi, and members of this phylum can usually be recognized by the unaided eye of the experienced observer. Similarly, the ascomycetous anamorph genera *Penicillium* and *Aspergillus* are easily pinpointed by the practiced eye; here it is not the growth rate or the appearance of the vegetative hyphae that are diagnostic, but rather the formation of spores (conidia in this case) in some shade of green on characteristic conidiophores.

Determination to genus is usually not so easy, and identification of even *Penicillium* and *Aspergillus* to species calls for special media and large monographic treatments of each genus (Raper and Thom 1949; Samson et al. 1976, 1981, Raper and Fennell 1965; Samson 1979).

Nevertheless, as you examine more and more isolates, some will quickly become familiar by their unique appearance (often as difficult to describe as it is easy to recognize), which may involve such features as color, texture, growth rate, zonation, exudates, and secretion of pigment into the medium. Before we present our keys, it might be a good idea for us to consider the likelihood of your meeting any particular group. Table 4.1 is

Table 4.1. Fungal Taxa Isolated from Soil[a]

Taxa	Genera	Species
Oomycetes	3 (1.8%)	13 (2.8%)
Zygomycetes	12 (7%)	46 (10%)
Ascomycetes	56 (32.8%)	111 (24.1%)
Ascomycetous anamorphs	92 (53.8%)	282 (61.2%)
Basidiomycetes	3 (1.8%)	4 (0.9%)
Basidiomycetous anamorphs	2 (1.2%)	2 (0.4%)
Filamentous yeasts	3 (1.8%)	3 (0.7%)
Total	171 genera	461 species

[a] Note that this table does not take into consideration the numerous sterile isolates that are normally encountered.

derived from the list of taxa covered in a very useful (although expensive) book, *Compendium of Soil Fungi* (Domsch et al. 1980). We have no figures for actual numbers of individual colonies derived from each major group, but Table 4.1 shows the relative frequencies of occurrence of genera and species belonging to different groups.

Almost 94% of the genera you are likely to isolate from soil belong to two groups, the Zygomycetes and the Ascomycetes, and the Ascomycetes with their anamorphs make up almost 87% of the total. This last figure is, in one way, spuriously high, because some of the ascomycete genera listed are the teleomorphs of some of the anamorphic genera listed. Nevertheless, teleomorph and anamorph always look different, often have different nutritional and other ecological requirements, and thus develop at different times on different substrates. This is not to say that anamorph and teleomorph never occur together: both develop simultaneously in, for example, some of the Eurotiales. But in terms of the diversity of forms encountered and of the literature required for identification, the numbers and groups quoted are a reasonable reflection of reality.

It seems, then, that when you examine any particular soil fungus, it is very likely to fall into one of three categories: ascomycetous anamorphs, ascomycetous teleomorphs, or Zygomycetes, in that order. If we now shift the discussion to the species level, it appears that almost two-thirds will be ascomycetous anamorphs—fungi producing conidia on hyphae or specialized conidiophores, and mostly belonging to the anamorph class Hyphomycetes—while another one-fourth would be ascomycetous teleomorphs producing spores inside specialized meiosporangia called asci, usually in some kind of fruit body. The last important group, containing one-tenth of the total species isolated, would be Zygomycetes, almost all of them in the anamorphic phase, and therefore producing spores usually inside some kind of delicate mitosporangia on long stalks. Some zygomycetous anamorphs produce single-spored sporangia that can easily be mistaken for conidia, and this possibility is noted in the key. Only homothallic Zygomycetes can be reasonably expected to produce zygosporangia.

Faced with the statistical evidence presented above, it is only logical that we should make our key as efficient as possible, diagnosing the commonest groups first, and others in order of frequency of occurrence. The opening couplets ensure that you are dealing with a filamentous fungus, and after that we go immediately to the main sequence outlined above.

Finally, before you begin to use the key, it is vital that you check the illustrations. Familiarize yourself with the appearance of an ascus, a zygosporangium, a characteristic penicillate conidiophore of *Penicillium*, and other commonly encountered structures. If you miss out this step and go straight to the key, we will only point you back to the pictures anyway. You should have a very clear image in your mind of the various diagnostic characters, because the slide preparations you make will contain a lot of distracting elements—loose spores, broken and disoriented hyphae—a whole spectrum of discordant and misleading material. The illustrations will help you to see through this microscopic smoke screen and pick out the "signal" from the "noise."

KEY TO MAJOR GROUPS OF FUNGI ISOLATED FROM SOIL

1a Colony containing filamentous elements (hyphae) wider than 2 μm . . 2

1b Colony lacking filamentous elements, or these extremely narrow (less than 1.5 μm). 9

2a (1) Hyphae with cross-walls (septa) at more or less regular intervals (Dikaryomycota) (Figs. 4.10, 4.11). 3

2b (1) Hyphae usually lacking cross-walls (Zygomycota) 7

3a (2) Spores or other differentiated propagules (bulbils, sclerotia; Figs. 4.1–4.9) present 4

3b (2) Spores and other propagules absent 8

4a (3) Spores or other propagules produced singly or in indeterminate numbers; on undifferentiated hyphae (Figs. 4.10, 4.11) or on individual, differentiated, simple or branched conidiophores (Figs. 4.12, 4.13) or on conidiophores aggregated into characteristic fructifications (conidiomata; Figs. 4.14–4.16)—ascomycetous and basidiomycetous anamorphs . . . 5

4b (3) Spores produced inside saclike cells (asci; Figs. 4.17–4.20), usually 8 per ascus (1, 2, 4, or multiples of 8 ascosphores occur in some taxa)—ascomycetous teleomorphs and some yeasts 6

4c (3) Spores borne singly at the tips of (usually) four outgrowths arising from a cell called a basidium (Fig. 4.21), basidia often developing only on characteristic macroscopic fructifications (basidiomata)—basidiomycetous teleomorphs 8

4d (3) Anamorph and teleomorph both present (identify through appropriate pathway in the key; check connection in Kendrick and DiCosmo 1979, Kendrick and Watling 1979 or Carmichael et al. 1980)

5a (4) Conidia not formed inside a conidioma (Figs. 4.10–4.15) . . . . **HYPHOMYCETES**

See Carmichael et al. (1980), Ellis (1971, 1976), Barron (1968), Domsch and Gams (1972), Domsch et al. (1980), von Arx (1981).

COELOMYCETES pro parte

See Sutton (1980), Morgan-Jones et al. (1972 or 1982) (Fig. 4.29).

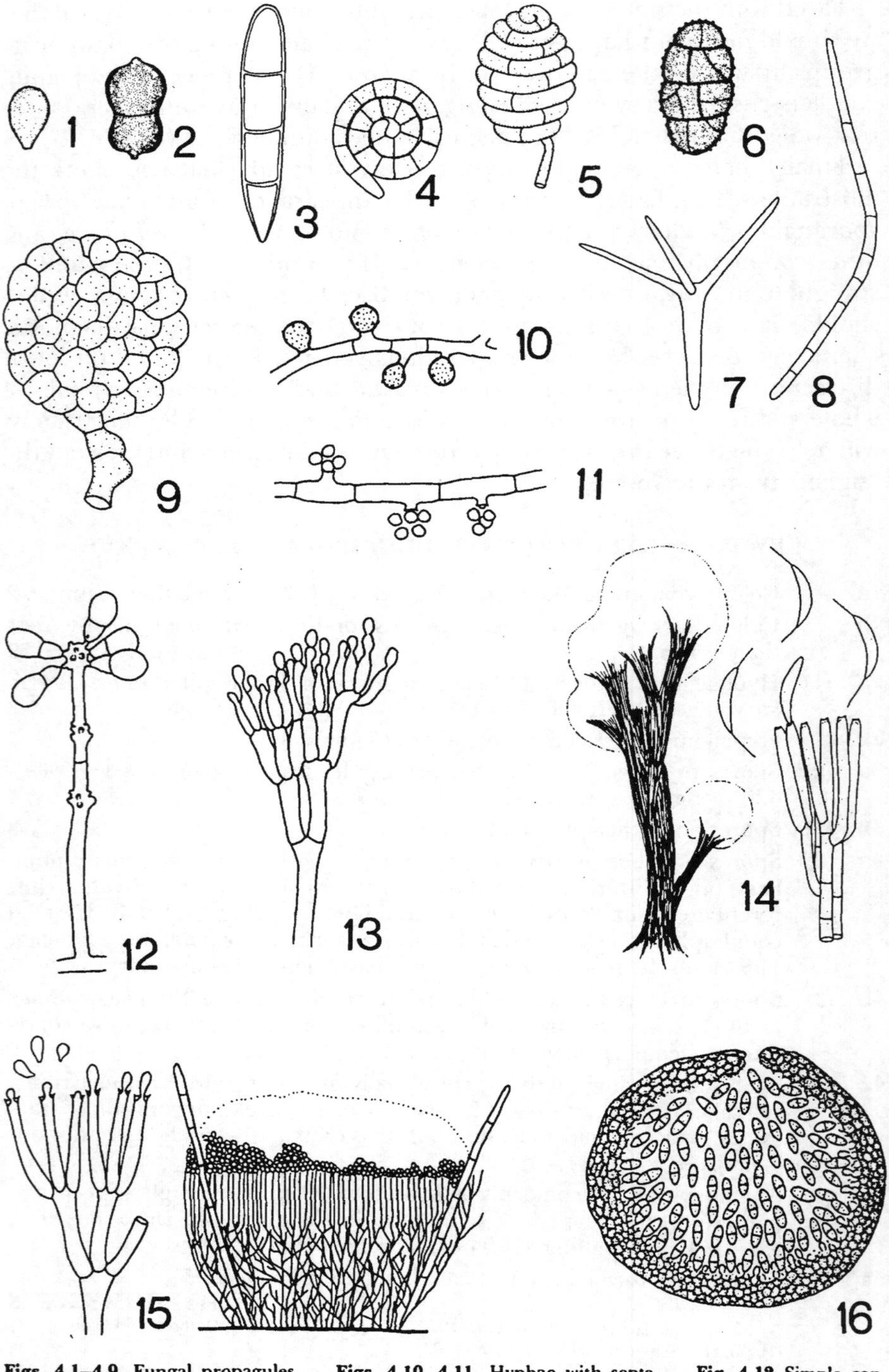

Figs. 4.1–4.9. Fungal propagules. **Figs. 4.10, 4.11.** Hyphae with septa. **Fig. 4.12** Simple conidiophore. **Fig. 4.13** Branched conidiophore. **Figs. 4.14–4.16.** Conidiomata.

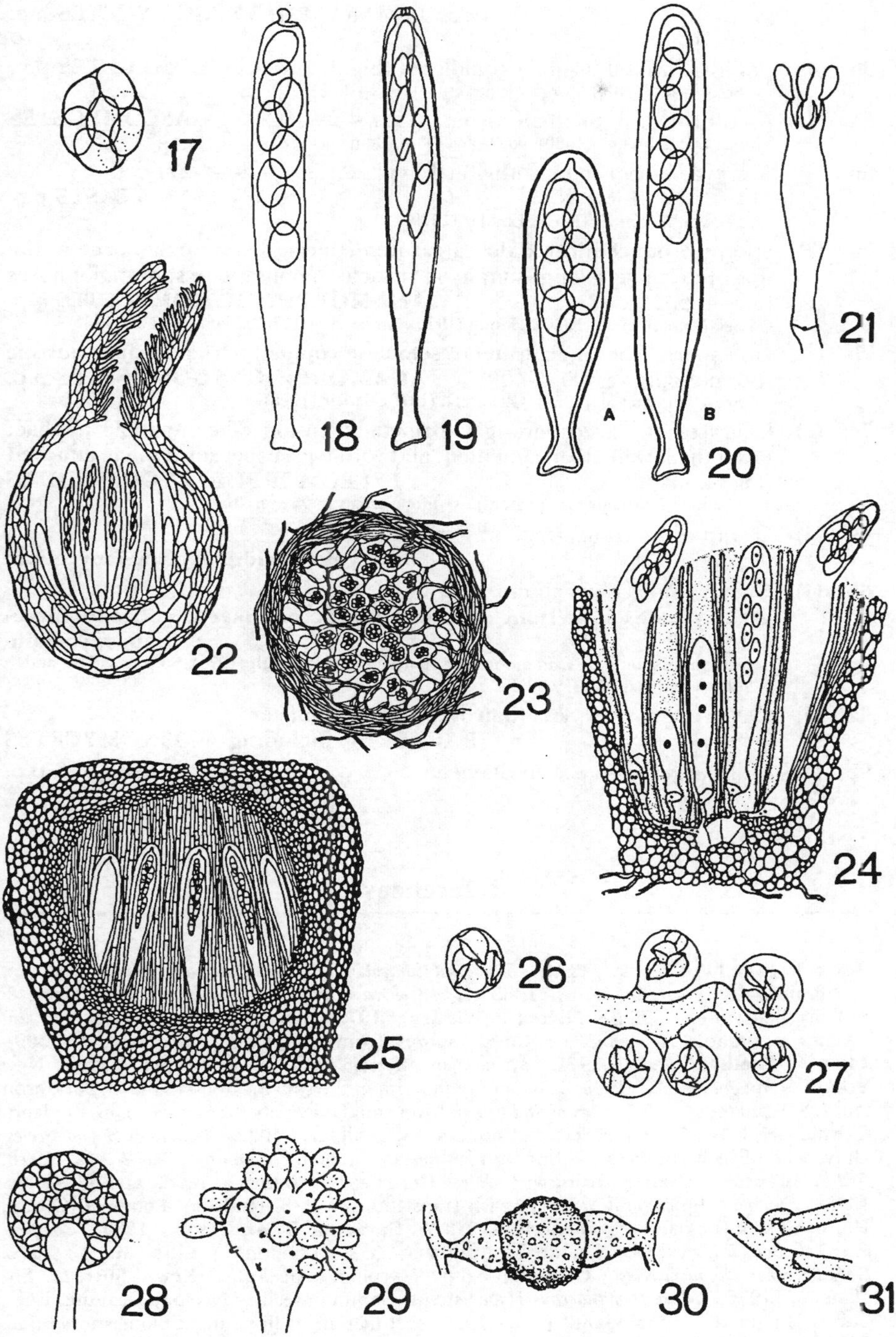

Figs. 4.17–4.20. Asci. **Fig. 4.21.** Basidium. **Figs. 4.22–4.25.** Asci on or in an ascoma. **Fig. 4.26.** Free asci. **Fig. 4.27.** Hyphal-borne asci. **Fig. 4.28.** Spores within spherical sporangia. **Fig. 4.29.** Single-spored sporangia on characteristic sporangiophore. **Fig. 4.30.** Zygosporangium. **Fig. 4.31.** Clamp connections.

or ANAMORPHIC ZYGOMYCETES p.p.
See **7b**.

5b (4) Conidia formed inside a conidioma (Fig. 4.16) . **COELOMYCETES** p.p.
See Sutton (1980), Morgan-Jones et al. (1972–1982)

6a (4) Asci formed on or in an ascoma (Figs. 4.22–4.25) . . **ASCOMYCETES**
See Domsch et al. (1980), von Arx (1981), Dennis (1978).

6b (4) Asci free or borne on individual hyphae (Figs. 4.26–4.27) . . . .
YEASTS p.p.
See von Arx et al. (1977), Lodder (1979).

7a (2) Spores produced inside delicate, often spherical sporangia borne at the tips of often very long, simple or branched, nonseptate sporangiophores (Fig. 4.28) . . . . **ANAMORPHIC ZYGOMYCETES** p.p.
See Domsch et al. (1980), Hanlin (1973), Zycha et al. (1969), Samson et al. (1981).

7b (2) One-spored sporangia (often resembling conidia) borne on characteristic sporangiophores (Fig. 4.29) . . **ANAMORPHIC ZYGOMYCETES** p.p.
See Domsch et al. (1980), O'Donnell (1979), Fuller (1978).

7c (2) Characteristic zygosporangia formed by fusion of compatible hyphae; often thick-walled, ornamented, and with two suspensor hyphae attached (Fig. 4.30) . . . . **TELEOMORPHIC ZYGOMYCETES**
Anamorph usually also present; use literature as in **7a** and **7b**.

8a (4) Clamp connections (Fig. 4.31) present at septa . . . .
dikaryotic basidiomycetous mycelium

8b (4) Clamp connections absent . . . .
ascomycetous mycelium or monokaryotic or dikaryotic basidiomycetous mycelium
Attempts should be made to induce sporulation; see Booth (1971), Stevens (1974) and Kendrick and Nag Raj (1979).

9a (1) Cells very narrow, less than 1.25 μm in diameter . . . .
BACTERIA, including ACTINOMYCETES

9b (1) Cells larger, 1–5 μm in diameter . . . . **YEASTS** p.p.
See **6b**.

References

Baker, K. F. and R. J. Cook. 1974. *Biological Control of Soil-borne Pathogens.* W. H. Freeman, San Francisco. **Barron, G. L. 1968.** *The Genera of Hyphomycetes from Soil.* Williams & Wilkins, Baltimore. **Bissett, J. and P. Widden. 1972.** An automatic, multichamber soil washing apparatus for removing fungal spores from soil. *Can. J. Microbiol.* 18:1399–1409. **Booth, C. (ed.). 1971.** *Methods in Microbiology,* Vol. 4. Academic Press, New York. **Burges, N. A. 1960.** Dynamic equilibria in the soil. Pp. 185–191 in D. Parkinson and J. S. Waid (eds.). *The Ecology of Soil Fungi.* Liverpool University Press, Liverpool, England. **Carmichael, J. W., B. Kendrick, I. L. Conners and L. Sigler. 1980.** *Genera of Hyphomycetes.* University of Alberta Press, Edmonton, Alberta, Canada. **Dennis, R. W. G. 1978.** *British Ascomycetes.* Cramer, Brunswick, West Germany. **Domsch, K. H. and W. Gams. 1972.** *Fungi in Agricultural Soils.* (English translation by P. S. Hudson). Longman Group, Harlow, Essex, England. **Domsch, K. H., W. Gams and T. Anderson. 1980.** *Compendium of Soil Fungi,* Vols. 1 and 2. Academic Press, London. **Ellis, M. B. 1971.** *Dematiaceous Hyphomycetes.* Commonwealth Mycology Institute, Kew, Surrey, England. **1976.** *More Dematiaceous Hyphomycetes.* Commonwealth Mycology Institute, Kew, Surrey, England. **Frankland, J. C. 1975.** Estimation of live fungal biomass. *Soil Biol. Biochem.* 7:339–340. **Frankland, J. C., D. K. Lindley and M. J. Swift. 1978.** A comparison of two methods for the estimation of mycelial biomass in leaf litter. *Soil Biol. Biochem.*

10:323–333. **Fuller, M. S. (ed.). 1978.** *Lower fungi in the Laboratory.* Dept. of Botany, University of Georgia, Athens, Ga. **Gams, W., H. A. van der Aa, A. J. van der Plaats-Niterink, R. A. Samson and J. A. Stalpers. 1975.** *CBS Course of Mycology.* Centraalbureau voor Schimmelcultures, Baarn, The Netherlands. **Garrett, S. D. 1970.** *Pathogenic Root-Infecting Fungi.* Cambridge University Press, London. **Hale, M. D. C. and J. G. Savory. 1976.** Selective agar media for the isolation of basidiomycetes from wood—a review. *Int. Biodeterior. Bull.* 12:112–115. **Hanlin, R. T. 1973.** *Keys to the Families, Genera and Species of the Mucorales.* Cramer, Brunswick, West Germany. **Hansen, J. F., T. F. Thingstad and J. Goksoyr. 1974.** Evaluation of hyphal length and fungal biomass in soil by a membrane filtration technique. *Oikos* 25:102–107. **Harley, J. L. 1971.** Fungi in ecosystems. *J. Ecol.* 59:653–668. **Harley, J. L. and J. S. Waid. 1955.** A method for studying active mycelia on living roots and other surfaces in soil. *Trans. Br. Mycol. Soc.* 38:104–118. **Hawksworth, D. L. 1974.** *Mycologists Handbook.* Commonwealth Mycology Institute, Kew, Surrey, England. **Hayes, A. J. 1979.** The microbiology of plant litter decomposition. *Sci. Prog. (Oxf.)* 66:25–42. **Hering, T. F. 1966.** An automatic soil washing apparatus for fungal isolation. *Plant Soil* 25:195–200. **Johnson, L. F. and E. A. Curl. 1972.** *Methods for Research on the Ecology of Soil-borne Pathogens.* Burgess, Minneapolis, Minn. **Jones, P. C. T. and J. E. Mollison. 1948.** A technique for the quantitative estimation of soil micro-organisms. *J. Gen. Microbiol.* 2:54–69. **Kendrick, B. 1985.** *The Fifth Kingdom.* Mycologue Publications, Waterloo, Ontario, Canada. **Kendrick, W. B. and N. A. Burges. 1962.** Biological aspects of the decay of *Pinus sylvestris* litter. *Nova Hedwigia* 4:313–342. **Kendrick, B. and F. DiCosmo. 1979.** Teleomorph–anamorph connections in ascomycetes. In B. Kendrick (ed.). *The Whole Fungus,* Vol. 1. National Museum of Canada, Ottawa. **Kendrick, B. and T. R. Nag Raj. 1979.** Morphological terms in fungi imperfecti. In B. Kendrick (ed.). *The Whole Fungus,* Vol. 1. National Museum of Canada, Ottawa. **Kendrick, B. and R. Watling. 1979.** Mitospores in Basidiomycetes. In B. Kendrick (ed.) *The Whole Fungus.* Vol. 2. National Museum of Canada, Ottawa. **Lodder, J. 1979.** *The Yeasts, a Taxonomic Study.* North-Holland, Amsterdam. **Marks, G. C. and T. T. Kozlowski (eds.). 1973.** *Ectomycorrhizae.* Academic Press, New York. **Martin, J. P. 1950.** Use of acid, rose bengal and streptomycin in the plate method for estimating soil fungi. *Soil Sci.* 69:215–232. **Morgan-Jones, G., T. R. Nag Raj, F. DiCosmo and B. Kendrick. 1972–1982.** *Icones generum coelomycetum.* Fasc. I–XIII. Univ. Waterloo Biol. Ser. Ontario, Canada. **Mueller, E. and W. Loeffler. 1976.** *Mycology* (English translation by B. Kendrick and F. Baerlocher). Georg Thieme, Stuttgart, West Germany. **Nagel-de-Boois, H. M. and E. Jansen. 1971.** The growth of fungal mycelium in forest soil layers. *Rev. Ecol. Biol. Sol* 8:509–520. **O'Donnell, K. L. 1979.** *Zygomycetes in Culture.* Dept. of Botany, University of Georgia, Athens, Ga. **Olson, F. C. W. 1950.** Quantitative estimates of filamentous algae. *Trans. Am. Microsc. Soc.* 59:272–279. **Papavizas, G. C. and C. B. Davey. 1959.** Evaluation of various media and antimicrobial agents for isolation of soil fungi. *Soil Sci.* 88:112–117. **Parkinson, D. 1986.** Filamentous fungi. In J. L. Page (ed.). *Methods of Soil Analysis,* Part 2, 2nd ed. Agronomy Handb. 9. American Society of Agronomy, Madison, Wis. **Parkinson, D., T. R. G. Gray and S. T. Williams. 1971.** *Methods for Studying the Ecology of Soil Microorganisms.* IBP Handb. 19. Blackwell Scientific, Oxford. **Paul, E. A. and R. L. Johnson. 1977.** Microscopic counting and ATP measurement in determining microbial growth in soils. *Appl. Environ. Microbiol.* 34:263–269. **Raper, K. B. and D. I. Fennell. 1965.** *The Genus Aspergillus.* Williams & Wilkins, Baltimore. **Raper, K. B. and C. Thom. 1949.** *A Manual of the Penicillia.* Williams & Wilkins, Baltimore. **Samson, R. A. 1979.** A compilation of the aspergilli described since 1965. *Stud. Mycol. (Baarn)* 14:18–40. **Samson, R. A., A. C. Stolk and R. Hadlok. 1976.** Revision of the subsection fasciculata of *Penicillium* and some allied species. *Stud. Mycol. (Baarn)* 11:1–47. **Samson, R. A., E. S. Hoekstra and C. A. N. van Oorschot. 1981.** *Introduction to Food-borne Fungi.* Centraalbureau voor Schimmelcultures, Baarn, The Netherlands. **Saunders, F. E., B. Mosse and P. B. Tinker (eds.). 1975.** *Endomycorrhizas.* Academic Press, New York. **Soderstrom, B. E. 1977.** Vital staining of fungi in pure cultures and in soil with fluorescein diacetate. *Soil Biol. Biochem.* 9:59–63. **Stevens, R. B. (ed.). 1974.** *Mycology Guidebook.* University of Washington Press, Seattle, Wash. **Sutton, B. C. 1980.** *The Coelomycetes.* Commonwealth Mycology Institute, Kew, Surrey, England. **Swift, M. J. 1973.** The estimation of mycelial biomass by determination of hexosamine content of wood tissue decayed by fungi. *Soil Biol. Biochem.* 5:321–332. **1976.** Species diversity and structure of microbial communities in terrestrial habitats. Pp. 185–222 in J. M. Anderson and A. Macfadyen (eds.). *The Role of Terrestrial and Aquatic Organisms in Decomposition Processes.* Blackwell Scientific, Oxford. **Visser, S. and D. Parkinson. 1975.** Litter respiration and fungal growth under low temperature conditions. Pp. 88–97 in G. Kilbertus, O. Reisinger, A. Mourey and J. A. Cancela da Fonseca (eds.). *Biodegradation et Humification.* Pierron, Sarreguemines, France. **von Arx, J. A. 1981.** *The Genera of Fungi Sporulating in Pure Culture.* Cramer, Brunswick, West Germany. **von Arx, J. A., L. Rodriques de Miranda, M. T. Smith and D. Yarrow. 1977.** The general of yeasts and the yeast-like fungi. *Stud. Mycol* (*Baarn*) 14:1–42. **Waid, J. S. and M. J. Woodman. 1957.** A method for estimating hyphal activity in soil. *Pedologie* 7 (no. spec.):155–158. **Warcup, J. H. 1950.** The soil plate method for isolation of fungi from soil. *Nature* (*Lond.*) 116:117. **1955.** Isolation of fungi from hyphae present in soil. *Nature* (*Lond.*) 175:953–954. **1957.** Studies on the occurrence and activity of fungi in a wheat field soil. *Trans. Br. Mycol. Soc.* 40:237–262. **Watson, R. D. 1960.** Soil washing improves the value of the soil dilution and plate count method of estimating populations of soil fungi. *Phytopathol.* 50:792–794. **Webster, J. 1980.** *Introduction to Fungi,* 2nd ed. Cambridge University Press, Cambridge. **Zycha, H., R. Siepmann and G. Linnemann. 1969.** *Mucorales.* Cramer, Brunswick, West Germany.

5

Soil Algae

PATRICIA A. ARCHIBALD

Development of the Study of Soil Algae

Lyngbye (1) credits Paracelus, an ancient Greek, as the first researcher to describe an algal isolate from soil. Metting (2) summarizes a series of isolated reports of edaphic genera recorded during the nineteenth and early twentieth centuries. Soil phycologists, however, date the beginning of the serious study of edaphic forms to the publication of Petersen's classical report in 1935 (3). Another study from the same date is that of James (4). Other important early publications in the field are those of Fritsch and John (5), John (6), and Lund (7) from Great Britain in the 1940s. Broady (8) probably is the most active soil phycologist working on British soils at this time. The major works of Hollerbach and Shtina (9, 10), Novichkova-Ivanova (11), and Rodin, dating from the 1950s, contribute to our knowledge of the algae of Russian soils; Forest (12) and Cameron (13) list translations of the Russian bibliographies. Reisigl's (14–16), Pitschmann's (17), Trenkwalder's (18), Vinatzer's (19), and Schwarz's (20) studies of the 1970s characterize the alpine soil flora of Austria and the surrounding areas. Akiyama's (21, 22) and Nakano's (23, 24) research detail the genera of algae common to Japanese soils.

American edaphic phycological literature of the past 30 years reports the extensive work of Bold and his students (25–27). The Phycological Studies Series published at the University of Texas–Austin by Bold and his students describe techniques in detail and contain important taxonomic revisions for genera of Cyanophyta (28, 29) and Chlorophyta (30–39).

Deason's initial study (30) outlines the basic procedures followed in all subsequent studies. Brown's (34) and Groover's (37) research illustrate the important differences between the Chlorococcales and Chlorosarcinales, those orders of Chlorophyta that contain so many genera common to soil. Archibald's (38) monograph of *Chlorococcum* Meneghini cites specific criteria for delimiting species in the unicellular greens. Deason and Archibald (40–42, 194) continue to study the edaphic chlorococcalian and chlorosarcinalian genera. George's 1976 summary (43) of algal keys cites their revision of genera in these orders.

Distribution

Edaphic algal distributional studies most frequently list the taxa of a particular biome, relate taxa to specific types of soil, or compare the populations of undisturbed and disturbed areas. A few studies are general rather than specific (3–6, 8, 44, 45). Forest's (46) analysis of different soil algal communities summarizes generalized patterns of distribution. Metting (2) lists references to algal distribution by specific geographic areas. Alpine and coniferous forest soils reveal a wealth of algae (15, 16, 18–20, 47–49). The alpine forest flora of Europe exhibits a larger number of Xanthophyceae (15, 16, 18–20) than the North American flora, which shows a higher percentage of Chlorophyceae (47, 48). Studies of deciduous forests, which produce fewer Xanthophyceae, are fewer in number (50–53). Although supporting a rich and varied algal flora, bogs rarely have been studied (54–57). Unicellular greens dominate the flora of both bogs and the immediately surrounding areas (57, 58). Polderman and Polderman-Hall (59) identify distinctly different communities distributed throughout salt marshes.

Semiarid biomes are well studied compared to many other areas. French works (60–62) of the middle 1930s reports surveys of the Sahara. Fletcher and Martin in 1948 (63) were among the first to investigate the algal composition of surface crusts formed after desert rains. The literature contains several studies of both the southwestern desert region of the United States (31, 33, 64–72), the Israeli desert (73–76), and the Soviet deserts of central Asia (77–81). A bibliography of Russian studies of this area are listed by Novichkova-Ivanova (11) and Metting (2). Other projects are known to have been undertaken at scattered, isolated sites (82–84). Papers referring to surface crusts (63, 80, 84) list primarily filamentous cyanophycean genera. Zoosporic chlorophycean taxa occur in palmolloid stages (80). Cameron (85) indicates a decrease in abundance from cold desert, hot desert, and arid high mountain areas, to barren antarctic sites. Techniques employed varied so greatly that it is difficult to draw correlations between various arid studies and between arid and nonarid investigations. In several incidents identification was made from mixed algal cultures after only short periods of incubation (see the section "Techniques of Study"). Even different taxonomic schemes were used in identification of the Cyanophyta. Certainly, more careful attention to taxonomic detail is evidenced in some reports (31, 33, 71, 72, 76, 82–84) than in others.

Literature on the tropics and the polar regions is sparse and appears in a variety of journals. Durrell (86) and MacEntee (87) describe a variety of tropical soils, while other reports relate to small geographic sections (88–91). Akiyama's (92), Novichkova-Ivanova's (93) and Cameron's (94) papers represent the work in the northern polar regions. Studies concerned with the edaphic algae of Antarctica appear in the literature from 1966 (95) through 1970 (96–101). Cameron summarizes the ecological aspects of much of this early work (102–104). From the mid-1970s, the literature contains references to chasmoendolithic, epilithic, and terrestrial algae in Antarctica. It is Broady's work that includes studies of the truly edaphic taxa (105–108). His detailed work on Signy Island in the South Orkney Islands (105) is the definitive terrestrial study for the south polar region.

The early works of Petersen (3), Fritsch and John (5), and John (6) relate distribution of genera to soil types. Again workers, particularly those in the USSR (11, 109–114), are realizing the importance of this. Studies in the United States enumerate the genera occurring in saline prairie soils (115) and silt loam (116). Polderman and Polderman-Hall discuss salt marsh diversity in Scotland (117). Phycologists need, and nearly all lack, training in the basic knowledge of soil structure and development. Even the latest papers characterize soil types in the broadest of terms. The discipline needs to advance beyond the stage of stating that alkaline soils support profuse growth of Cyanophyceae and that acidic soils are rich in Chlorophyceae. Ground cover, texture, moisture content, ion availability, pH, and parent material of the soil are important factors to consider in a discussion of the distribution and occurrence of algae in the world's soils. Cameron's early studies of arid regions considers some of the interactions of physical factors to algal distribution (118–122). Metting (2) summarizes the sparse literature in this area for other habitats.

Quantitative studies of soil algae are the exception rather than the rule. This may be related to the lack of a completely satisfactory method of measurement. Three basic techniques of counting are reported in the literature: direct microscopic count, culture dilution and plate count, and spectrophotometric measurement of chlorophyll extract. Early quantitative Russian studies are discussed in Lund (123). Russian workers have revised their count methods (124, 125). Various workers suggest suspension of algal cells in either agar (126) or resin (127) for counting. Major papers relative to observation of chlorophyll fluorescence are few in number (128–130). Examination of the literature indicates that environmentalists prefer estimation of chlorophyll extract, while taxonomists select the dilution culture plate count. Two quantitative articles in the English literature, in which the dilution culture plate count technique was used, are those by

King and Ward (131) and Broady (105). Broady reports the method highly effective for use with the unicellular green and yellow-green algae; however, some workers believe that cell count is usually lower with this method. Akiyama (97) prefers to vary the count method according to the type of soil being studied. Metting (2) reports that the average count of algal cells per gram of soil is 10^3–10^4 cells. His Table VII reports the count range from a diverse collection of soil types.

Quantitative studies may compare productivity among soil types (3–5, 54, 110, 125) or between undisturbed and disturbed soils (111, 131). Noskova (110) and Pomelova (111) report similar results for studies comparing soil types (110) and disturbed and undisturbed soils (111). Production is higher in spring and autumn and where there is a high concentration of nitrogen, phosphorus, and potassium. Broady sampled 23 sites from six distinct areas with a pH range of 4.1–7.4. King and Ward sampled 51 sites in a 7188-ha (17,760-acre) area with a pH range from 5.6 to 8.5. The Texas study (131) indicates that pH and soil disturbances were the major factors affecting productivity, while phosphorus is also a limiting factor. Undisturbed soils exhibit greater diversity but produce smaller standing crops. Little growth is evidenced below 15 cm. In Broady's English study, productivity was patchy, and at least in one site, was higher in undisturbed soil. Both Broady and Kulikova (54) report compacted soils to support larger standing crops than loose soils. The U.S. soils show a lower productivity than the English soils. Depth of collection and range of pH are major variables between the two studies.

Methods reported in the literature to date are so diverse that it is questionable if valid comparisons of results can be made for formulating statements concerning ecological relationships. Studies reporting both qualitative and quantitative distribution of algal taxa in soil need to follow a set of basic procedures if data are to be extrapolated between projects with any accuracy. The author would suggest that standard methodology should include (a) broad random sampling, at specific depths, throughout the area being described; (b) cultivation in various media exhibiting a range of pH or salinity and nutrient level to eliminate selection of species; (c) cultivation over a temperature range of 10–35°C; (d) examination of original inoculum over a period of time (1–4 weeks) so that succession of species will not prevent isolation of some genera; and (e) isolation and identification of unialgal cultures under defined conditions of growth.

Environmental Relationships

The cosmopolitan occurrence of edaphic genera indicates that there must be few physical or chemical factors that are of prime operational importance in limiting their ability to survive dormancy over extended periods of time and under extreme conditions. Lund's 1967 article (123) is a major paper discussing physical and chemical environmental effects on growth, biomass production, of edaphic algae. A summary of the major findings to that date are given. Metting's 1981 summary of the literature (2) still indicates a paucity of work related directly to physical and chemical requirements in nature. The 1981 reference (2) lists a series of studies that attempt to correlate abiotic factors to rate of growth, diversity of species, distribution of specific taxa in soils, and regulation of developmental stages in life cycles.

With few exceptions (116, 132–134), research related to the chemical effects of pH, ion balance, and inorganic and organic nutrient needs is primarily either of field studies in environments in which it is difficult to evaluate all the indirect influences operating (2) or of culture studies developed as taxonomic tools. The tests, though, are not definitive enough for taxonomic purposes except as supplementary attributes (38). *Chlorella* Beijernick, because of the availability of strains, has been studied in this manner more than any other genus (134–138). Organic nitrogen and carbon requirements related to autotrophic and heterotrophic growth have been reported for other selected genera of the Chlorophyta (30–39, 42) and Cyanophyta (28, 29). The author excludes here the number of applied agricultural articles on the Cyanophyta relative to rice and cotton production. Metting (2) cites some of this research plus additional agricultural studies. Three major recent editorial works concerned with physiological and ecological attributes of Cyanophyta are available (139–141). Genera and species in both the Chlorophyta and Cyanophyta exhibit a vast range of nutritional requirements.

Perhaps the physiological tests that identify enzyme and secondary carotenoid production in edaphic forms may reveal their role in the environment. It is known that amylase, gelatinase (29, 37, 38, 42), catalase (35), nitratase (29, 37, 38), and hydrogenase (2, 142–148) are produced by some soil algae. Techniques and references for these and other tests can be found in Hellebust and Craigie (149), Kessler (150), and Ben-Amotz et al. (151). Soil taxa exhibit both mineral and sulfate (152) uptake, and release of organic materials (2, 153) into the soil. Laboratory studies show that metabolism may be inhibited by metabolite analogs, chelators, thiol reagents (154), and antibiotics (32, 38, 42, 155). Edaphic species of a genus are not as likely to be inhibited by antibiotics produced from soil organisms as are aquatic species of the same genus (155).

With the increased use of pesticides, one must consider their effect on the edaphic algae. Metting (2) summarizes much of the literature through 1980. Torres's work (156) considers soil flora; however, some of the studies examine aquatic species. Few workers culture the algae in soil when pesticides are applied; Atkins' and Tchan's (157) and Audus's (158) papers are exceptions. Anderson's (159) and Cullimore's (160) reports examine the effect of pesticides on the nontarget organisms in the environment. Syngerism is not addressed in the studies summarized by Metting (2), and some of the pesticides are no longer used. References by Gaunt and Evans (161), Kirkwood and Fletcher (162), Arvik et al. (163), Anderson and Drew (164), Cullimore and McCann (165), Komaromy and Mihaly (166), and Metting and Rayburn (167) apply to pesticides still being used. Kiss (168) reports that the inhibitory effect of some herbicides lasts for only 3 months. Texture (169) and cultivation of the soil affect the degree of inhibition experienced, and the degree of inhibition varies from one taxon to another (170). Few authors report on both the uptake of pesticides by algae and the degradation of pesticides by algae. McCann and Cullimore (171) and Lal and Saxena (172) present information on the cytological and biochemical effects of pesticides. An important approach in the study of the effect of pesticides on algae is the one that Cain and Cain (173) take in their study comparing the sensitivity of vegetative cells and zygospores. Pesticide toxicity to the higher plants varies considerably with temperature, moisture content of air and soil, and soil texture. It is reasonable to assume that the same may be true for the algae.

Techniques of Study

Basic and specialized techniques are described in detail by Bold (174), Baker (29), Archibald (38), and Nichols (175). Only a brief description of basic procedure is given here. Aseptic random collections are made at the site of study. Both surface and subsurface samples, to a few centimeters, should be taken. Soil can be transported and stored in a variety of containers for extended periods of time (176).

Samples require culturing in a variety of media to eliminate the problem of species exclusion by medium selection, to test for specific growth requirements, and to reveal structural details used in taxonomy. A modification of inorganic Bristol's solution such as Bold's Basal Medium (38), a frequently employed control growth medium suitable for the screening of most edaphic algae, can be prepared and adjusted to a pH range from acid–neutral–alkaline for initial screening. Isolates from saline soil may require a medium showing a salinity range for the first screening tests. Medium can be used in either liquid or 1.5% agarized form.

Portions of 5 g should be inoculated into 125 ml of medium and cultured under controlled conditions of light and temperature. Temperature at 18–22°C and cool white fluorescent light at 2500–3000 lux (4.37 W/m^2) on a diurnal light–dark cycle of 12–12 or 16–8 is sufficient. Constant illumination can alter the morphology of the plastid, thickness of wall development, and development of the various stages in a taxon's life cycle (177–180).

When sufficient growth has appeared in the culture, a portion of the algal suspension is removed by a sterile pipette and placed into a sterile centrifuge tube. A uniform suspension is obtained after a minute of sonication. The suspension of cells is sprayed or streaked onto agarized plates of medium and incubated for approximately 2 weeks. On the appearance of individual colonies, each different type of colony needs to be isolated (by using sterile pipettes of a fine bore) into small tubes of liquid medium and incubated to obtain unialgal cultures. A slurry of soil and water may be introduced onto agar plates and colonies isolated directly from the initial inoculation. Soil particles, however, may present problems with this technique.

Accurate identification of a taxon, even to the level of genus, should be undertaken by the study of unialgal cultures, time consuming though it is. Periodical examinations of the initial inoculum must be made to reveal the rate of growth and succession of types of genera. Some filamentous forms may be identified tentatively, with a certain degree of accuracy, by observation of the inoculum. These initial identifications then should be verified by observation of an unialgal culture to check for the presence or absence of sheaths and for the presence and arrangement of specialized cells (e.g., akinetes, heterocysts, holdfasts, oogonium, and sporangium).

Starr's 1955 work (181) outlines the criteria for delimiting those microscopic, unicellular, zoosporic, chlorophycean genera of the Chlorococcales most frequently isolated from soil. They are the shape of plastid, type of zoospore, and presence or absence of a pyrenoid. Later workers (34, 37, 42) applied the same criteria for identification of genera of the chlorosarcinales. Identification of such unicellular genera in a mixed culture simply is impossible. The characteristic plastid types of parietal, reticulate, stellate, and axial are distinguished easily in young vegetative cells, but starch deposits in older cells obscure plastid shape. Each genus has a specific type of zoospore that must be identified. In mixed cultures autospores can be mistaken for *Chlorella* Beijernick, and zoospores for *Chlamydomonas* Ehrenberg or *Euglena* Ehrenberg. Species are delimited on the criteria of vegetative cell size in log and stationary phases of growth, thickness of cell wall, and size and morphology of zoospore (38, 42).

Classification

A review of the major taxonomic soil studies starting with Petersen's 1935 work (3) and continuing through 1985 was conducted to determine those genera that are common to soils. The system of classification followed in the key is that recognized by Bourrelly (182–184) except for a few genera of the Chlorophyta. Genera are assigned to the Chlorococcales, Chlorosarcinales, and Tetresporales according to Bold's and Wynne's recent revision (185). I do not accept Drouet's and Daily's revisions of the Cyanophyta (186) and Ettl's and Komarek's revision of the Chlorophyta (187). Both Deason (188) and Melkonian (189) elaborate on the inaccuracy in Ettl's and Komarek's scheme. The most recent detailed, phylogenic and taxonomic work on the unicellular green algae common to soil is that by Mattox and Stewart (190); however, the scheme requires examination of cells at the transmission electron microscope level.

Metting (2) lists 172 edaphic genera exclusive of the Bacillariophyceae (diatoms). His list includes taxa that have been reported only once, and some that are as frequent in aquatic as in edaphic environments. A taxon may be listed both as named in Bourrelly (182–184) and as named by Drouet and Daily (186) and the "German schools." What Metting refers to as the German schools is literature written in German primarily by Austrian–Czech workers, some of whom closely follow Psacher's pre–World War II papers to the exclusion of subsequent literature. Metting's list excludes some common soil genera listed in my key. The following families, genera, and authors of each genus are represented in my key:

CYANOPHYTA: CHROOCOCCALES (Chroococcaceae) *Chroococcus* Nägeli; *Dactylococcopsis* Hansgirg; *Synechococcus* Nägeli; *Synechocystis* Sauvageau. PLEUROCAPSALES (Chroococcidiaceae) *Myxosarcina* Printz (Hyellaceae) *Pleurocapsa* Thuret and Hauck. STIGONEMATALES (Stigonemaceae) *Fischerella* Bornet and Flahault; *Hapalosiphon* Nägeli. NOSTOCALES (Microchaetaceae) *Microchaete* Thuret (Nostocaceae) *Anabaena* St. Vincent; *Anabaenopsis* Woloszynsha; *Cyclindrospermum* Kützing; *Nodularia* Mertens; *Nostoc* Vaucher (Oscillatoriaceae) *Borzia* Cohn; *Lyngbya* Agardh; *Microcoleus* Desmozieres; *Oscillatoria* Vaucher; *Porphyrosiphon* Kützing (Rivulariaceae) *Calothrix* Agardh; *Gloeotrichia* J. Agardh; *Rivularia* Roth (Scytonemaceae) *Plectonema* Thuret; *Scytonema* C. A. Agardh; *Tolypothrix* Kützing. **RHODOPHYTA:** PORYPHYRIDIALES (Phorphyridiaceae) *Porphyridium* Nägali. **CHROMOPHYTA-XANTHOPHYCEES:** MISCHOCOCCALES (Pleurochloridaceae) *Botrydiopsis* Borzi; *Chloridella* Pascher; *Chlorocloster* Pascher; *Ellipsoidion* Pascher; *Monodus* Chodat; *Nephrodiella* Pascher; *Pleurochloris* Pascher; *Vischeria* Pascher (Chlorobotrydaceae) *Chlorellidium* Vischer and Pascher; *Chlorobotrys* Bohlin; *Gloeobotrys* Pascher (Chlorotheciaceae) *Bumilleriopsis* Printz. TRIBONEMATALES (Tribonemataceae) *Bumilleria* Borzi; *Heterothrix* Pascher; *Heterotrichella* Reisigl; *Tribonema* Derbis and Solier (Heterodendraceae) *Heterococcus* Chodat. VAUCHERIALES (Botrydiaceae) *Botrydium* Wallroth. **EUGLENOPHYTA:** EUGLENALES (Euglenaceae) *Euglena* Ehrenberg. **CHLOROPHYTA:** VOLVOCALES (Volvocaceae) *Gonium* Muller; *Volvox* Linné (Chlaymdomonadaceae) *Carteria* Diesing; *Chlamydomonas* Ehrenberg; *Sphaerellopsis* Korchikott. TETRASPORALES (Tetrasporaceae) *Tetraspora* Link (Palmellaceae) *Gloeocystis* Nägeli; *Hormotila* Borzi; *Hormotilopsis* Trainor and Bold; *Palmella* Lyngbye (Hypnomondaceae) *Sphaerocystis* Chodat. CHLOROCOCCALES (Chlorococcaceae) *Bracteacoccus* Tereg; *Characium* A. Braun; *Chlorococcum* Menighini; *Dictyochloris* Vischer; *Dictyococcus* Gerneck; *Kentrosphaera* Borzi;

Myrmecia Printz; *Nautococcus* Korchikoff; *Neochloris* Starr; *Neospongiococcum* Deason; *Planktosphaeria* G. M. Smith; *Pulchrasphaera* Deason; *Radiosphaera* Snow; *Rhopalocystis* Schussnig; *Spongiochloris* Starr; *Trebouxia* de Puymaly emend Archibald (Coccomyxaceae) *Coccomyxa* Schmidle (Dictyosphaeriaceae) *Dictyosphaerium* Nägeli (Oocystaceae) *Ankistrodesmus* Corda; *Chlorella* Beijernick; *Pseudochlorella* Lund emend Reisigl; *Dactylococcus* Nägeli; *Muriella* B. Petersen; emend Vischer; *Planktosphaerella* Reisigl; *Pseudochlorococcum* Archibald; *Oocystis* Nägeli; *Scotiella* Fritsch; *Selenastrum* Reinsch; *Trochiscia* Kützing (Radiococcaceae) *Coenochloris* Korchikoff; *Coenocystis* Korchikoff; *Phacomyxa* Skuja (Scenedesmaceae) *Coelastrum* Nägeli; *Crucigenia* Morren; *Scenedesmus* Meyer. CHLOROSARCINALES (Chlorosarcinaceae) *Axilosphaera* Cox and Deason; *Borodinella* Miller; *Borodinellopsis* Dykstra; *Botryokoryne* Reisigl; *Chlorosarcina* Gerneck emend Vischer; *Chlorosarcinopsis* Herndon; *Chlorosphaeropsis* Vischer emend Herndon; *Fasciculochloris* McLean and Trainor; *Fernandinella* Chodat; *Friedmannia* Chantanachat and Bold; *Heterotetracystis* Cox and Deason; *Planophila* Gerneck; *Pseudotetracystis* Arneson; *Pseudotrebouxia* Archibald; *Spongiococcum* Deason; *Tetracystis* Brown and Bold. ULOTRICHALES (Microsporaceae) *Microspora* Thuret (Ulotrichaceae) *Geminella* Turpin; *Klebsormidium* Silva, Mattox and Blackwell; *Koliella* Hindák; *Stichococcus* Nägeli; *Ulothrix* Kützing. ULVALES (Ulvaceae) *Prasiola* Agardh. CHAETOPHORALES (Chaetophoraceae) *Desmococcus* Brand; *Leptosira* Borzi; *Microthamnion* Nägeli; *Pleurastrum* Chodat; *Protoderma* Kützing; *Pseudopleurococcus* Snow emend Vischer. OEDOGONIALES (Oedogoniaceae) *Oedogonium* Link. SIPHONALES (Protosiphonaceae) *Protosiphon* Klebs. TRENTEPOHLIALES (Trentepohliaceae) *Trentepohlia* Martius. ZYGNEMATALES (Desmidiaceae) *Closterium* Nitzsch; *Cosmarium* Chorda (Mesotaeniaceae) *Cylindrocystis* Meneghini; *Mesotaenium* Nägeli.

In the following key, diatoms are excluded since fixation of the cell is required for accurate identification, and it is difficult to maintain cultures for prolonged study. If an organism appears in the literature under more than one name, only the genus cited as valid by Bourrelly is listed. Nine additional edaphic genera have been described since prepration of the latest editions of Bourrelly (19, 87, 191, 192). George's review (43) lists 121 references for monographs and keys to species. Andreyeva's monograph on *Chlorella* Beijernick (193) should be added to the references in George. Drawings are composites illustrating the morphological traits noted in the key.

ACKNOWLEDGMENTS

Dr. Harold C. Bold and Dr. Temd R. Deason thoroughly reviewed this manuscript, and their suggested modifications and valuable time involvement are greatly appreciated.

KEY TO THE MORPHOLOGICAL TRAITS OF THE COMMON GENERA OF EDAPHIC ALGAE

1a Pigment not in a plastid but diffuse in the cell **(CYANOPHYTA)** 2
1b Pigment in a plastid or plastids of specific pattern 32
2a Unicellular 3
2b Colonial or filamentous. 5
3a Cells solitary or in pairs (Fig. 5.6) . . . . *Synechocyctis*
3b Cells solitary or in colonies of few cells. 4
4a Cells or colony lacking gelatinous sheath (large, oval bright blue cells) (Fig. 5.7) . . . . *Synechococcus*
4b Cells or colony exhibiting gelatinous sheath (one layer of cells in common mucilage) (Fig. 5.8) . . . . *Myxosarcina*
5a Colonial 6
5b Filamentous 11
6a Cells fusiform and in sheath that is not always definite (Fig. 5.9) . . . . . *Dactylococcopsis*
6b Cells not fusiform and in globular arrangement 7
7a Colony of compressed filaments appearing as encrustment with rhizoidal extensions (Fig. 5.10) . . . . *Pleurocapsa*
7b Colony not of compressed filaments 8
8a Mucilage surrounding a few cells. 9
8b Mucilage surrounding numerous cells 10
9a Mucilage not arranged in concentric rings (Fig. 5.11). . . . . *Chroococcus*
9b Mucilage arranged in concentric rings (Fig. 5.12) . . . . *Gleocapsa*
10a Cells regularly spaced, spherical, often in pairs (Fig. 5.13) . . *Aphanocapsa*
10b Cells irregularly spaced, elongated cylinders (Fig. 5.14). . . . *Aphanothece*
11a Heterocyst present 12
11b Heterocyst not present 19
12a Sheath missing or if present obscure 13
12b Sheath appearing pronounced 15
13a Cross walls missing, coiled in regular fashion (Fig. 5.15). . . . . *Spirulina*
13b Cross walls appearing distinct, not coiled 14
14a Filament short, usually not more than 4 cells, not exceeding 20 cells (Fig. 5.16) . . . . *Borzia*
14b Filament long, slightly tapered toward anterior end . . . . . . *Oscillatoria*
15a Sheath enclosing one trichome 16
15b Sheath enclosing more than one trichome 18
16a Sheath firm, trichome rounded at apex (Fig. 5.17) . . . . *Lyngbya*
16b Sheath soft and sticky, adjacent plants adhering 17
17a Apical cell distinct from other cells (Fig. 5.18) . . . . *Phormidium*
17b Apical cell similar to other cells (Fig. 5.19) . . . . *Symploca*
18a Many intertwined trichomes crowded within sheath (Fig. 5.20) . . . . . . *Microcoleus*
18b Few trichomes within sheath . . . . *Schizothrix*
19a Trichome not branching 20

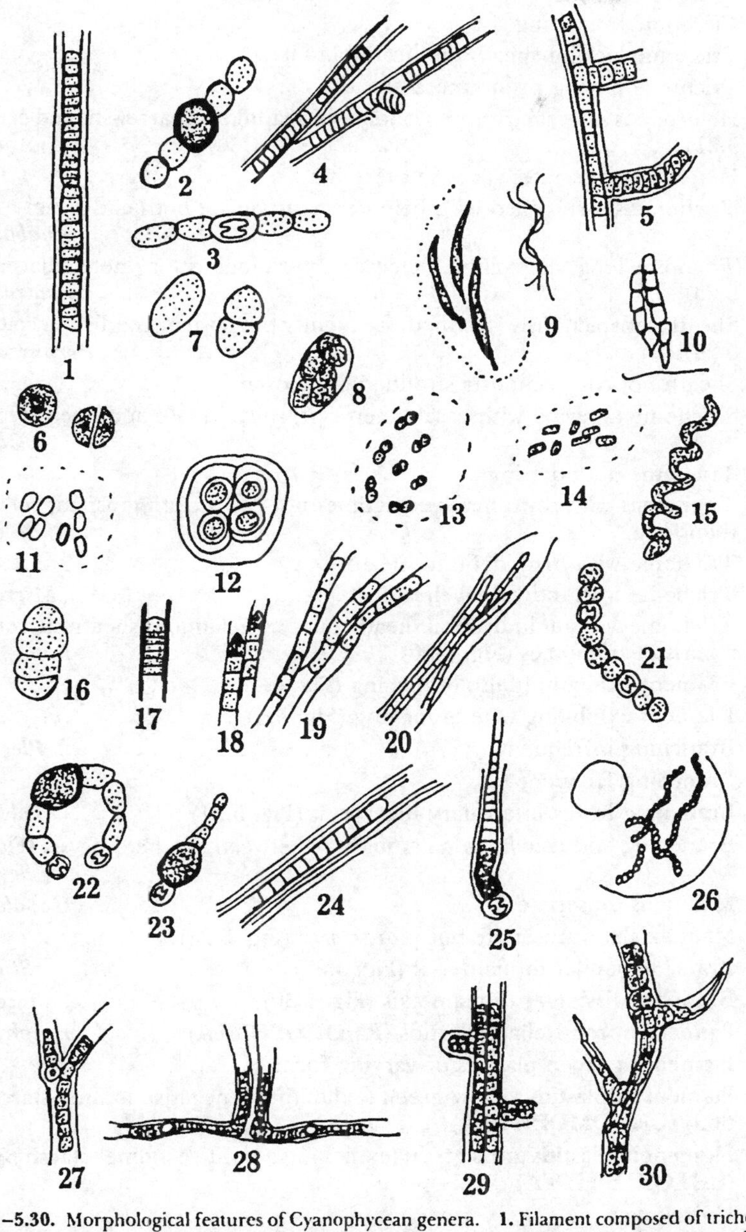

Figs. 5.1–5.30. Morphological features of Cyanophycean genera. **1.** Filament composed of trichome and sheath. **2.** Trichome exhibiting thick walled akinete. **3.** Trichome exhibiting clear heterocyst with polar nodules. **4.** False branching with hormogonium emerging from ruptured sheath. **5.** True branching with sheath intact. **6.** *Synechocyctis* sp. (600×). **7.** *Synechococcus* sp. (150×). **8.** *Myxosarcina* sp. (300×). **9.** *Dactylococcopsis* sp. (300×). **10.** *Pleurocapsa* sp. (540×). **11.** *Chrococcus* sp. (540×). **12.** *Gleocapsa* sp. (540×). **13.** *Aphanocapsa* sp. (540×). **14.** *Aphanothece* sp. (540×). **15.** *Spirulina* sp. (540×). **16.** *Borzia* sp. (150×). **17.** *Lyngbya* sp. (210×). **18.** *Phormidium* sp. (210×). **19.** *Symploca* sp. (210×). **20.** *Microcoleus* sp. (60×). **21.** *Anabaena* sp. (150×). **22.** *Anabaenopsis* sp. (150×). **23.** *Cylindrosperum* sp. (150×). **24.** *Porphyrosiphon* sp. (210×). **25.** *Calothrix* sp. (150×). **26.** *Nostoc* sp. (45×). **27.** *Tolypothrix* sp. (150×). **28.** *Scytonema* sp. (90×). **29.** *Stigonema* sp. (90×). **30.** *Fischerella* sp. (240×).

19b Trichome branching **27**
20a Trichome lacking sheath or sheath obscure **21**
20b Trichome having pronounced sheath **23**
21a Heterocysts intercalary and trichome of beadlike or barrel-shaped cells (Fig. 5.21) . . . . ***Anabaena***
21b Heterocysts not intercalary but terminal **22**
22a Trichome short, coiled with heterocysts usually at both ends (Fig. 5.22) . . ***Anabaenopsis***
22b Trichome long, not coiling, heterocysts terminal with akinete adjacent (Fig. 5.23) . . . . ***Cylindrosperum***
23a Sheath conspicuously stratified, frequently purplish or reddish in color (Fig. 5.24) . . . . ***Porphyrosiphon***
23b Sheath not conspicuously stratified or colored **24**
24a Trichome tapering with basal heterocyst, sheath loose and open (Fig. 5.25) ***Calothrix***
24b Trichome not tapering **25**
25a Trichomes with thin sheath sometimes not easily distinguished, cells wider than long . . . . ***Nodularia***
25b Trichome with firm, definite sheath. **26**
26a Trichome with individual sheath . . . . ***Microchaete***
26b Trichome without individual sheath, but one gelatinous sheath surrounding a mass of trichomes (Fig. 5.26) . . . . ***Nostoc***
27a Filament exhibiting false branching (Fig. 5.4) **28**
27b Filament exhibiting true branching (Fig. 5.5) **30**
28a Branching infrequent . . . . ***Plectonema***
28b Branching frequent **29**
29a Branching below intercalary heterocyst (Fig. 5.27) . . . . ***Tolypothrix***
29b Branching and usually in pairs midway between two heterocysts (Fig. 5.28) ***Scytonema***
30a Main axis uniseriate . . . . ***Hapalosiphon***
30b Main axis not uniseriate but pluriseriate (Fig. 5.28) **31**
31a Branches similar to main axis (Fig. 5.29) . . . . ***Stigonema***
31b Branches dissimilar to main axis (Fig. 5.30) . . . . ***Fischerella***
32a Pigment in red stellate plastids **(RHODOPHYTA)** . . . . . ***Porphyridium***
32b Pigment in green plastids of varying form **33**
33a Pigment in plastids yellow-green (exhibiting a negative iodine–starch reaction) **(CHROMOPHYTA)** **34**
33b Pigment in plastids grass green (exhibiting a positive iodine–starch reaction) **(CHLOROPHYTA)** **52**
34a Unicellular **35**
34b Colonial or filamentous **44**
35a Large balloon unicell (to 2 mm) with rhizoidlike extension (Fig. 5.31) . . . ***Botrydium***
35b Cells less than 100 μm in diameter **36**
36a Cell form irregular (Fig. 5.32) . . . . ***Vischeria***
36b Cell form not irregular **37**
37a Cells spherical. **38**
37b Cells not spherical. **40**

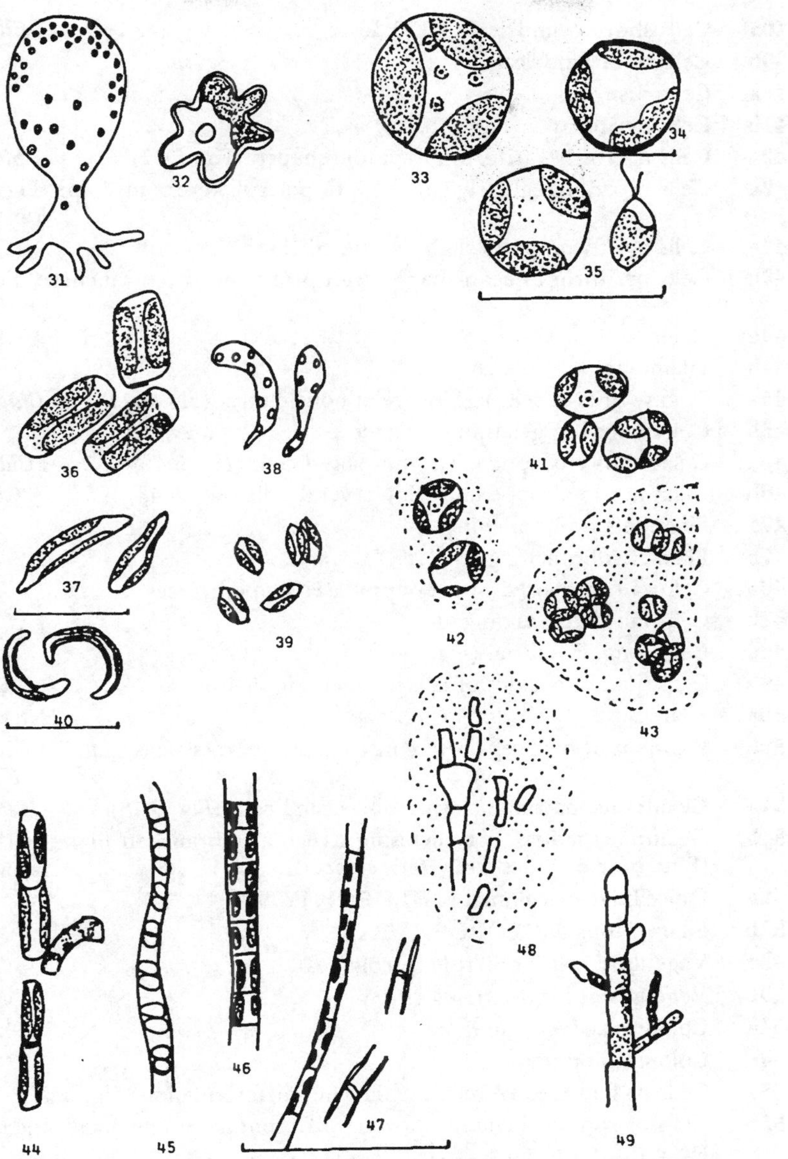

Figs. 5.31–5.49. Morphological features of Chrysophycean genera. **31.** *Botrydium* sp. (18×). **32.** *Vischeria* sp. (720×). (After Bourrelly.) **33.** *Botridiopsis* sp. (1440×). **34.** *Pleurochloris* sp. (720×). **35.** *Chloridella* sp. (720×). **36.** *Ellipsoidion* sp. (600×). **37.** *Chlorocloster* sp. (300×). **38.** *Bumilleriopsis* sp. (300×). **39.** *Monodus* sp. (300×). **40.** *Nephrodiella* sp. (300×). **41.** *Chlorellidium* sp. (900×). **42.** *Chlorobotrys* sp. (900×). **43.** *Gloeobotrys* sp. (900×). **44.** *Heterotrichella* sp. (300×). **45.** *Heterothrix* sp. (180×). **46.** *Bumilleria* sp. (180×). **47.** *Tribonema* sp. (180×). **48.** *Heterococcus* sp. (300×). **49.** *Monocilia* sp. (300×).

38a Cells large with many discoid plastids and nuclei (Fig. 5.33) . . . . ***Botrydiopsis***

38b Cells small with few discoid plastids. 39

39a Producing zoospores (Fig. 5.35) . . . . ***Pleurochloris***

39b Lacking zoospores (Fig. 5.34). . . . . ***Chloridella***

40a Cells short, cylindrical (Fig. 5.36) . . . . ***Ellipsoidion***
40b Cells not cylindrical **41**
41a Cells fusiform **42**
41b Cells pyriform. **43**
42a Cells narrowly fusiform or spindle-shaped (Fig. 5.37) . . . . ***Chlorocloster***
42b Cells broad, irregularly curved with poles different in shape (Fig. 5.38) . . ***Bumilleriopsis***
43a Cells pyriform, poles slightly pointed (Fig. 5.39) . . . . ***Monodus***
43b Cells pyriform to bean-shaped with pronounced central curve (Fig. 5.40) . ***Nephrodiella***
44a Colonial **45**
44b Filamentous **47**
45a Colony of 2–4 cells lacking gelatinous matrix (Fig. 5.41). . . ***Chlorellidium***
45b Colony having gelatinous matrix **46**
46a Gelatinous envelope enclosing paired cells (Fig. 5.42). . . . . ***Chlorobotrys***
46b Gelatinous envelope enclosing several cells (Fig. 5.43) . . . . . ***Gloeobotrys***
47a Filaments not branching **48**
47b Filaments branching **51**
48a Cell wall composed of overlapping H-shaped pieces **49**
48b Cell wall not so arranged **50**
49a Cells long, cylindrial (Fig. 5.47). . . . . ***Tribonema***
49b Cells quadrate, sides of cell parallel (Fig. 5.46) . . . . ***Bumilleria***
50a Filament 2–3 cells long (Fig. 5.44) . . . . ***Heterotrichella***
50b Filament of oval cells with constrictions at cross-walls (Fig. 5.45). . . . . . ***Heterothrix***
51a Gelatinous matrix enclosing dissociated cells (Fig. 5.48) . . . ***Heterococcus***
51b Gelatinous matrix not enclosing irregularly branched filament (Fig. 5.49) (Bourrelly includes with ***Heterococcus***). . . . . ***Monocilia***
52a Unicellular or colonial **(CHLOROPHYTA)** **53**
52b Filamentous **(CHLOROPHYTA)** **121**
53a Vegetative cells occurring in colonies **54**
53b Vegetative cells occurring singly **69**
54a Colonies actively motile **55**
54b Colonies nonmotile **56**
55a Cells of flattened colony not showing differentiation (Fig. 5.51) . . ***Gonium***
55b Cells of spherical colony showing differentiation into small vegetative and large reproductive cells (Fig. 5.52) . . . . ***Volvox***
56a Cells in a prominent, readily observable amorphous matrix **57**
56b Cells in a colony of distinct form **68**
57a Cells in gelatinous matrix exhibiting flagella or pseudoflagella **58**
57b Cells in gelatinous matrix lacking flagella **59**
58a Cells exhibiting pseudoflagella (stiff flagella-like processes) (Fig. 5.56) . . . ***Tetraspora***
58b Cells exhibiting flagella (Fig. 5.57). . . . . ***Gloeococcus***
59a Cells at apices of stalklike gelatinous tubes (Fig. 5.55). **60**
59b Cells not so arranged **61**
60a Zoospores biflagellated . . . . ***Hormotila***

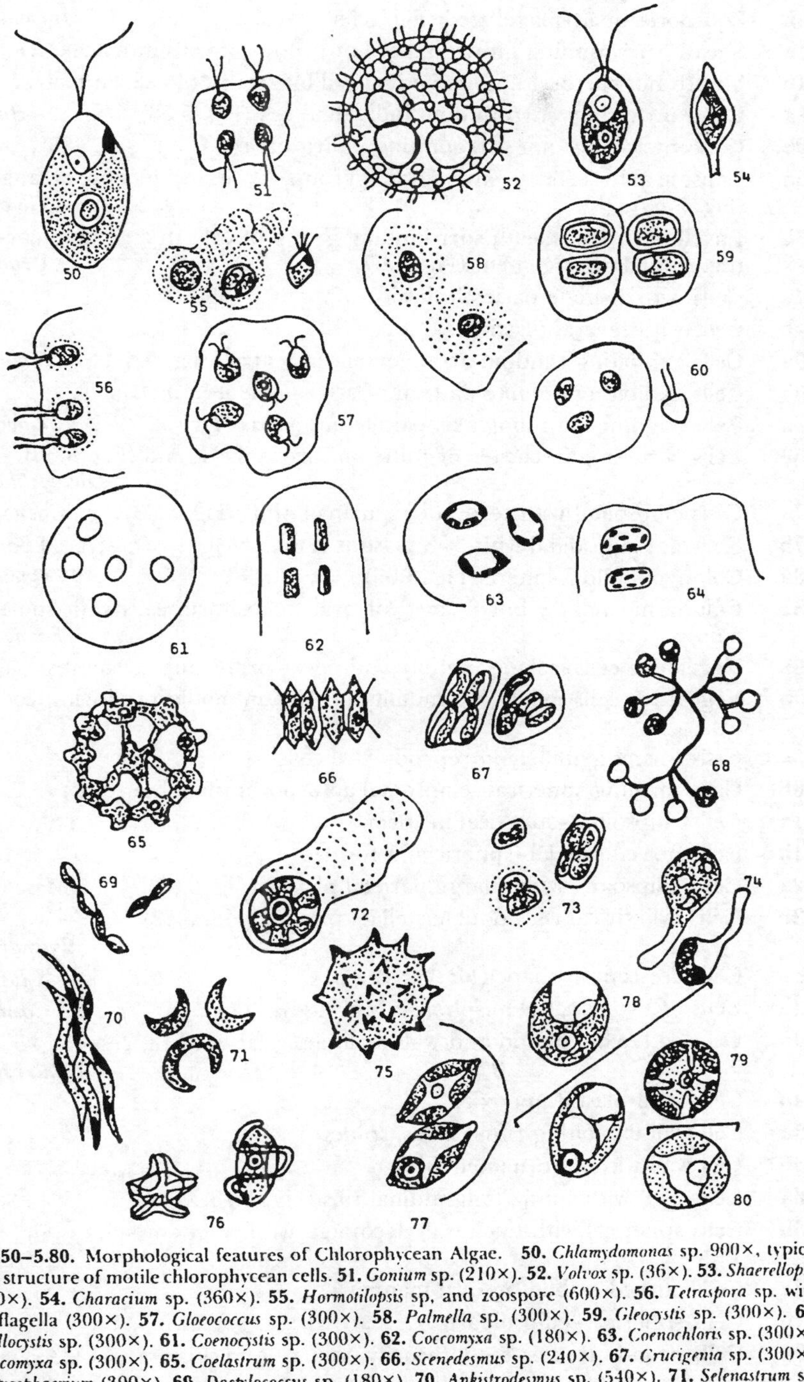

Figs. 5.50–5.80. Morphological features of Chlorophycean Algae. **50.** *Chlamydomonas* sp. 900×, typical cellular structure of motile chlorophycean cells. **51.** *Gonium* sp. (210×). **52.** *Volvox* sp. (36×). **53.** *Shaerellopsis* sp. (900×). **54.** *Characium* sp. (360×). **55.** *Hormotilopsis* sp, and zoospore (600×). **56.** *Tetraspora* sp. with pseudoflagella (300×). **57.** *Gloeococcus* sp. (300×). **58.** *Palmella* sp. (300×). **59.** *Gleocystis* sp. (300×). **60.** *Sphaerellocystis* sp. (300×). **61.** *Coenocystis* sp. (300×). **62.** *Coccomyxa* sp. (180×). **63.** *Coenochloris* sp. (300×). **64.** *Phacomyxa* sp. (300×). **65.** *Coelastrum* sp. (300×). **66.** *Scenedesmus* sp. (240×). **67.** *Crucigenia* sp. (300×). **68.** *Dictyosphaerium* (300×). **69.** *Dactylococcus* sp. (180×). **70.** *Ankistrodesmus* sp. (540×). **71.** *Selenastrum* sp. (540×). **72.** *Kentosphaera* sp. (600×). **73.** *Pseudochlorella* (300×). **74.** *Protosiphon* sp. (21×). **75.** *Trochiscia* sp. (600×). **76.** *Scotiella* sp. (600×). **77.** *Oocystis* sp. (release from autosporangium); (600×). **78.** *Chlorella* sp. (1200×). **79.** *Pseudochlorococcum* sp. (1200×)—young cell showing large vacuoles. **80.** *Muriella* sp. (1200×).

60b Zoospores quadriflagellated (Fig. 5.55) . . . . ***Hormotilopsis***
61a Sheath surrounding individual cell or groups of cells in matrix. **62**
61b Sheath not surrounding individual cell or groups of cells in matrix . . **64**
62a Concentric rings surrounding individual cell (Fig. 5.58) . . . . . ***Palmella***
62b Concentric rings not surrounding individual cell **63**
63a Capsule a thick sheath surrounding group of cells in a globose arrangement (Fig. 5.59). . . . . ***Gloeocystis***
63b Envelope a thin sheath surrounding group of cells in a rectangular–oval–trapezoidal arrangement (Fig. 5.67) . . . . ***Crucigenia***
64a Cells with a single parietal plastid. **65**
64b Cells with several plastids **67**
65a Cells exhibiting random arrangement in matrix (Fig. 5.61) . . ***Coenocystis***
65b Cells exhibiting definite pattern of arrangement in matrix **66**
66a Cells aligning with long axis parallel in matrix (Fig. 5.62) . . . ***Coccomyxa***
66b Cells occurring in cluster of fours on threadlike strand (Fig. 5.68). . . . . ***Dictyosphaerium***
67a Cells ellipsoidal with several irregular plastids (Fig. 5.63). . . ***Coenochloris***
67b Cells not ellipsoidal with 3–8 plastids (Fig. 5.64) . . . . ***Phacomyxa***
68a Colony a hollow sphere (Fig. 5.65) . . . . ***Coelastrum***
68b Colony not hollow but formed of oval to fusiform cells with spines (Fig. 5.66) . . . . ***Scenedesmus***
69a Vegetative cells lacking motility and never producing zoospores **70**
69b Vegetative cells exhibiting motility, or if non-motile producing zoospores **85**
70a Cells forming amoeboid reproductive cells **71**
70b Cells forming spherical–ellipsoidal autospores (daughter cells) **74**
71a Cells ellipsoidal–spherical in form **72**
71b Cells not ellipsoidal–spherical in form **73**
72a Cells ellipsoidal with a large parietal plastid (Fig. 5.81). . . . ***Mesotaenium***
72b Cells cylindrical with double stellate plastids (Fig. 5.82) . . . ***Cylindrocystis***
73a Cells crescent in shape (Fig. 5.83). . . . . ***Closterium***
73b Cells of two equal hemispheres (Fig. 5.84) . . . . ***Cosmarium***
74a Cells with axial plastid and wall forming a flattened process (Fig. 5.72) . . ***Kentrosphaera***
74b Cells with plastid not axial **75**
75a Cell wall exhibiting ribs, ridges, spines **76**
75b Cell wall without ornamentation **77**
76a Cells oval with spiral, longitudinal ribs (Fig. 5.76). . . . . ***Scotiella***
76b Cells spherical with thick wall decorated with ridges or spines (Fig. 5.75) . ***Trochiscia***
77a Cells curved or bent in oval manner **78**
77b Cells spherical–ellipsoidal. **79**
78a Cells pronounced crescent, usually occurring in clusters of 4–32 (Fig. 5.71) ***Selenastrum***
78b Cells slightly curved to straight and loosely entangled (Fig. 5.70) . . . . . ***Ankistrodesmus***

79a Cells ellipsoidal and frequently in loose chains (Fig. 5.69) . . . . *Dactylococcus*
79b Cells spherical. 80
80a Cells may be surrounded by indistinct gelatinous matrix or may occur in packets (Fig. 5.73) . . . . *Pseudochlorella*
80b Individual cells not in close contact 81
81a Cells with single parietal plastid. 82
81b Cells with more than one plastid and ellipsoidal 83
82a Plastid filling half or more of a spherical cell (Fig. 5.78) . . . . . . *Chlorella*
82b Plastid in young cell a thin rim next to wall with most of cell filled with large vacuoles, pronounced pyrenoid (Fig. 5.79) . . . . . . *Pseudochlorococcum*
83a Cells ellipsoidal with several plastids and polar nodes producing a lemon-shaped cell (Fig. 5.77) . . . . *Oocystis*
83b Cells spherical with several plastids 84
84a Plastids lacking pyrenoid (Fig. 5.80) . . . . *Muriella*
84b Plastids exhibiting pyrenoid . . . . *Planktosphaerella*
85a Vegetative cells flagellated 86
85b Vegetative cells nonmotile, producing zoospores 89
86a Cells ellipsoidal–spherical and having a firm wall, so maintaining shape (Fig. 5.50) 87
86b Cells ellipsoidal–spherical and lacking a firm wall, so changing shape **(EUGLENOPHYTA)** . . . . *Euglena*
87a Cells quadriflagellated . . . . *Carteria*
87b Cells biflagellated 88
88a Cell wall lacking an envelope (Fig. 5.50). . . . . *Chlamydomonas*
88b Cell wall having an envelope (Fig. 5.53). . . . . *Sphaerellopsis*
89a Cells large, coenocytic, bulbous with reticulate sheetlike plastid (Fig. 5.74). *Protosiphon*
89b Cells small, noncoenocytic, spherical–ellipsoidal 90
90a Cells ellipsoidal, attaching to substrate by a stalk (Fig. 5.54) . . *Characium*
90b Cells not attaching to a substrate 91
91a Cells exhibiting eleutheroschisis (nonvegetative cell division) and zoosporogenesis (Fig. 5.92). 92
91b Cells exhibiting desmoschisis (vegetative cell division) in addition to zoosporogenesis (Fig. 5.88, 5.89). 108
92a Cells pyriform–ovate in shape 93
92b Cells ellipsoidal–spherical in shape 96
93a Cells with a single parietal plastid. 94
93b Cells with irregularly lobed plastid(s) and at times showing pronounced thickening of the cell wall. 95
94a Cells more pyriform than ovate and lacking a gelatinous sheath (Fig. 5.85) *Rhopalocystis*
94b Cells more ovate than pyriform and exhibiting a thick gelatinous sheath (Fig. 5.60) . . . . *Sphaerellocystis*
95a Plastid irregularly lobed without a pyrenoid . . . . *Myremecia*
95b Plastid massive, axial with a central pyrenoid, gelatinous cap forming with age or in liquid culture (Fig. 5.86) . . . . *Nautococcus*

96a Cells with parietal plastid (Fig. 5.78) **97**
96b Cells with other than pareital plastid **99**
97a Plastid with one or more pyrenoids **98**
97b Plastid without a pyrenoid . . . . ***Pulchrasphaera***
98a Zoospores remaining ellipsoidal for a period after quiescence . . . . ***Chlorococcum***
98b Zoospores becoming spherical immediately upon quiescence. . . . . ***Neochloris***
99a Cells with reticulate or spongy plastid (Fig. 5.87) **100**
99b Cells with other than reticulate or spongy plastid **102**
100a Plastid with one or more pyrenoids **101**
100b Plastid without a pyrenoid . . . . ***Dictyochloris***
101a Zoospores remaining ellipsoidal for a period after quiescence . . . . ***Neospongiococcum***
101b Zoospores becoming spherical immediately upon quiescence. . . . . ***Spongiochloris***
102a Cells with numerous plastids appearing as discs (Fig. 5.91) or papillate projections. **103**
102b Cells with stellate or axial plastid (Figs. 5.86, 5.90) **105**
103a Plastids disc shaped **104**
103b Plastids papillate in shape . . . . ***Dictyococcus***
104a Plastids with one or more pyrenoids . . . . ***Planktosphaera***
104b Plastids without a pyrenoid . . . . ***Bracteacoccus***
105a Plastid stellate (Fig. 5.86) zoospores remaining ellipsoidal for a period after quiescence . . . . ***Radiosphaera***
105b Plastid axial (Fig. 5.90) zoospores becoming spherical immediately upon quiescence . . . . ***Trebouxia***
106a Cells pyriform in shape with a parietal plastid and grouped in tetrads. . . **107**
106b Cells ellipsoidal–spherical in shape **108**
107a Biflagellated zoospores (Fig. 5.93) . . . . ***Fernandinella***
107b Quadriflagellated zoospores . . . . ***Botryokoryne***
108a Cells with reticulate or spongy plastid (Fig. 5.87) **109**
108b Cells with other than reticulate or spongy plastid **110**
109a Plastid with one or more pyrenoids. . . . . ***Spongiococcum***
109b Plastid without one or more pyrenoids . . . . ***Chlorosphaeropsis***
110a Cells with stellate or axial plastid **111**
110b Cells with other than stellate or axial plastid. **114**
111a Cells with stellate plastid (Fig. 5.86) **112**
111b Cells with axial plastid (Fig. 5.90) **113**
112a Zoospores remaining ellipsoidal for a period after quiescence . . . . ***Bordinellopsis***
112b Zoospores becoming spherical immediately upon quiescence. . . . . ***Borodinella***
113a Zoospores remaining ellipsoidal for a period upon quiescence. . . . . . . ***Axillosphaera***
113b Zoospores becoming spherical immediately upon quiescence. . . . . ***Pseudotrebouxia***

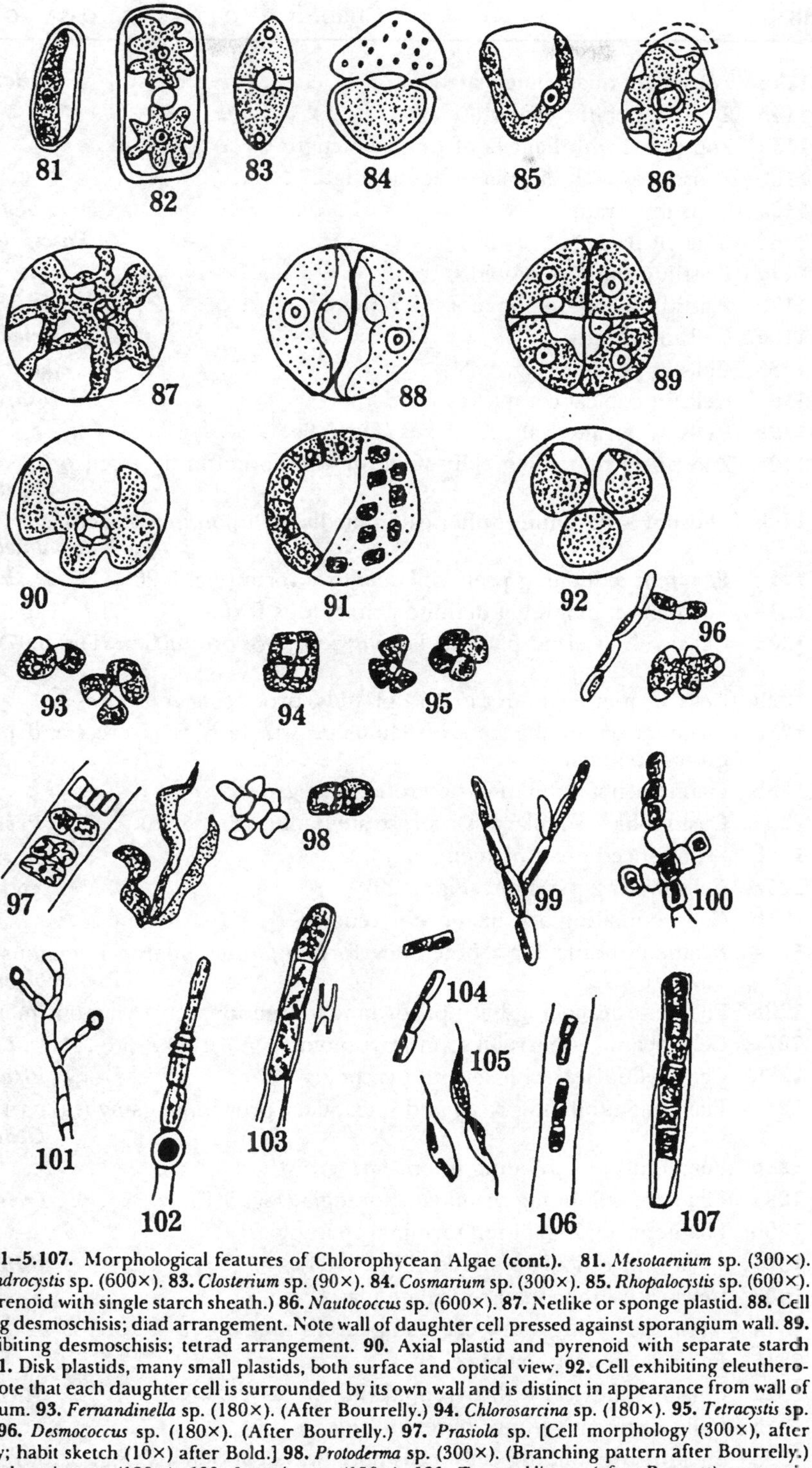

Figs. 5.81–5.107. Morphological features of Chlorophycean Algae **(cont.)**. **81.** *Mesotaenium* sp. (300×). **82.** *Cyclindrocystis* sp. (600×). **83.** *Closterium* sp. (90×). **84.** *Cosmarium* sp. (300×). **85.** *Rhopalocystis* sp. (600×). (Note pyrenoid with single starch sheath.) **86.** *Nautococcus* sp. (600×). **87.** Netlike or sponge plastid. **88.** Cell exhibiting desmoschisis; diad arrangement. Note wall of daughter cell pressed against sporangium wall. **89.** Cell exhibiting desmoschisis; tetrad arrangement. **90.** Axial plastid and pyrenoid with separate starch grains. **91.** Disk plastids, many small plastids, both surface and optical view. **92.** Cell exhibiting eleutheroschisis; note that each daughter cell is surrounded by its own wall and is distinct in appearance from wall of sporangium. **93.** *Fernandinella* sp. (180×). (After Bourrelly.) **94.** *Chlorosarcina* sp. (180×). **95.** *Tetracystis* sp. (180×). **96.** *Desmococcus* sp. (180×). (After Bourrelly.) **97.** *Prasiola* sp. [Cell morphology (300×), after Bourrelly; habit sketch (10×) after Bold.] **98.** *Protoderma* sp. (300×). (Branching pattern after Bourrelly.) **99.** *Microthamnion* sp. (120×). **100.** *Leptosira* sp. (150×). **101.** *Trentepohlia* sp. (after Prescott) sporangia arrangement. **102.** *Oedogonium* sp. (150×). Note dark oogonium, rings of intercalary cell division.) **103.** *Microspora* sp. (180×). (Note H-shaped pieces of cell wall.) **104.** *Stichococcus* sp. (300×). **105.** *Koliella* sp. (33×). (After Bourrelly.) **106.** *Geminella* sp. (420×). **107.** *Ulothrix* sp. (120×).

114a Zoospores quadriflagellated. ***Planophila***
114b Zoospores biflagellated. **115**
115a Zoospores with flagella of unequal length **116**
115b Zoospores with flagella of equal length **117**
116a Cells in tetrads . . . . ***Heterotetracystis***
116b Cells in packets . . . . ***Fasciculochloris***
117a Plastid lacking pyrenoid **118**
117b Plastid having one or more pyrenoids **119**
118a Cells in tetrads . . . . ***Friedmannia***
118b Cells in packets (Fig. 5.94). . . . . ***Chlorosarcina***
119a Cells in cubical complexes . . . . ***Chlorosarcinopsis***
119b Cells in tetrahedral complexes (Fig. 5.95) **120**
120a Zoospores remaining ellipsoidal for a period after quiescence . . . . ***Tetracystis***
120b Zoospores becoming spherical immediately upon quiescence. . . . . ***Pseudotetracystis***
121a Filament exhibiting pseudofilamentous form (Fig. 5.96) . . . ***Desmococcus***
121b Filament exhibiting a definite filamentous form **122**
122a Cells with asteroid plastids, lacking zoospore production (Fig. 5.97) . . . . ***Prasiola***
122b Cells with other than asteroid plastids, producing zoospores **123**
123a Filament exhibiting heterotrichous growth (a distinct erect and prostrate growth pattern) **124**
123b Filament not exhibiting heterotrichous growth **128**
124a Cushionlike, circular disk, prostrate thallus (Fig. 5.98). . . . . ***Protoderma***
124b Pronounced erect filament. **125**
125a Cells lacking pyrenoid (Fig. 5.99) . . . . ***Microthamnion***
125b Cells exhibiting one or more pyrenoids **126**
126a Filament branching, sometimes forming pseudoparenchymatous matrix ***Pseudopleurococcus***
126b Filament branching but not forming a pseudoparenchymatous matrix **127**
127a Cells beadlike–barrellike with pyrenoid (Fig. 5.100) . . . . ***Leptosira***
127b Cells cylindrical with several pyrenoids . . . . ***Pleurastrum***
128a Filament exhibiting basal and intercalary growth, oogamy (Fig. 5.102). . . ***Oedogonium***
128b Filament not exhibiting intercalary growth **129**
129a Filament exhibiting terminal sporangia (Fig. 5.101) . . . . . ***Trentepohlia***
129b Filament not exhibiting terminal sporangia **130**
130a Cells with reticulate plastid (Fig. 5.103) . . . . ***Microspora***
130b Cells with other than reticulate plastid. **131**
131a Filament having a few cells, single cells may occur. **132**
131b Filament having many cells **133**
132a Cells cyclindrical with pyrenoid (Fig. 5.104). . . . . ***Stichococcus***
132b Cells fusiform without pyrenoid (Fig. 5.105). . . . . ***Koliella***
133a Sheath surrounding cells, cells not always contiguous (Fig. 5.106) . . . . . ***Geminella***

133b Sheath not surrounding cells, cells always contiguous **134**

134a Differentiation of growth in filament exhibiting formation of unicellular holdfast (Fig. 5.107) . . . . *Ulothrix*

134b No differentiation of growth in filament, presence of H-shaped cell wall sections in culture . . . . *Klebsormidium*

References

1. **Lyngbye, H. C. 1819.** *Tentamen Hydrophytologiae Danicae.* Hafinae.
2. **Metting, B. 1981.** The systematics and ecology of soil algae. *Bot. Rev.* 47:195–321.
3. **Petersen, J. B. 1935.** Studies on the biology and taxonomy of soil algae. *Dan. Bot. Ark.* 8:1–183.
4. **James, E. J. 1935.** An investigation of the algal growth in some naturally occurring soils. *Beih. Bot. Zentralbl.* 53(A):519–553.
5. **Fritsch, F. E. and R. P. John. 1942.** An ecological and taxonomic study of the algae of British soils. II. Consideration of the species. *Ann. Bot.* (*N.S.*) 6:371–395.
6. **John R. P. 1942.** An ecological and taxonomic study of the algae of British soils. I. The distribution of surface-growing algae. *Ann. Bot.* (*N.S.*) 6:323–349.
7. **Lund, J. W. G. 1947.** Observations on soil algae. II. Notes on groups other than diatoms. *New Phytol.* 46:35–60.
8. **Broady, P. A. 1979.** Qualitative and quantitative observations on green and yellow-green algae in some English soils. *Br. Phycol. J.* 14:151–160.
9. **Hollerbach, M. M. and E. A. Shtina. 1969.** *Soil Algae.* Nauka, Leningrad (in Russian).
10. **Shtina, E. A. 1974.** The principle directions of experimental investigation in soil algology with emphasis on the U.S.S.R. *Geoderma* 20:151–156.
11. **Novichkova-Ivanova, L. N. 1968.** Principal trends and problems of soil algology in the U.S.S.R. In D. F. Jackson (ed.), *Algae, Man, and the Environment.* Syracuse University Press, Syracuse, N.Y. 554 pp.
12. **Forest, H. S. 1965.** The soil algal community. II. Soviet soil studies. *J. Phycol.* 4:164–171.
13. **Cameron, R. E. 1974.** A review of the Russian book *Soil Algae* by M. M. Hollerbach and E. A. Shtina, including a translation of the bibliography. *Phycologia* 13:109–120.
14. **Reisigl, H. 1963.** Uber die Verteilung der Bodenalgen in der Gipfelstufe der Otztaler Alpen. *Ber. Naturwiss.-Med. Ver. Innsb.* 53:163–172 (Festscrift H. Gams).
15. **Reisigl, H. 1964.** Zur Systematik and Ökologie alpiner Bodenalgen, *Ost. Bot. Z.* 111:402–498.
16. **Reisigl, H. 1969.** Bodenalgenstudien II. *Ost. Bot. Z.* 116:492–506.
17. **Pitschmann, H. 1963.** Vorarbeiten zu einer Monographie der Gattung *Heterococcus. Nova Hedwigia* V:487–531.
18. **Trenkwalder, H. 1975.** Neue Bodenalgen aus Föhrenwäldern im Raum von Brixen (Südtirol, Italien). *Ber. Naturwiss.-Med. Ver. Innsb.* 62:7–19.
19. **Vinatzer, G. 1975.** Neue Bodenalgen aus den Dolomiten. *Plant Syst. Evol.* 123:213–235.
20. **Schwarz, K. 1979.** *Neue Bodenalgen aus Dalmatien.* Plant Syst. *Evol.* 131:193–209.
21. **Akiyama, M. 1965.** Some soil algae from Japan. *Bull. Shimane Univ.* (*Nat. Sci.*) 15:96–117.
22. **Akiyama, M. 1970.** Some aerial and soil algae from the Ryukyu Islands. *Mem. Fac. Educ. Shimane Univ.* 3:24–45.
23. **Nakano, T. 1971.** Some aerial and soil algae from the Ishizuchi Mountains. *Hikobia* 6:139–152.
24. **Nakano, T. 1977.** *Hormotilopsis gelatinosa* Trainor et Bold, a new member of the Japanese soil algae. *Hikobia* 8:117–123.

25. **Herndon, W. R. 1958.** Some new species of chlorococcalean algae. *Am. J. Bot.* 45:308–323.

26. **Herndon, W. R. 1958.** Studies on chlorosphaeracean algae from soil. *Am. J. Bot.* 45:298–308.

27. **Arce, G. and H. C. Bold. 1958.** Some Chlorophyceae from Cuban soils. *Am. J. Bot.* 45:492–503.

28. **Kantz, T. and H. C. Bold. 1969.** *Phycological Studies IX.* Morphological and taxonomic investigations of *Nostoc* and *Anabaena* in culture. Univ. Tex. Publ. 6924. Austin, Tex. 67 pp.

29. **Baker, A. F. and H. C. Bold. 1970.** *Phycological Studies X.* Taxonomic studies in the Oscillatoriaceae. Univ. Tex. Publ. 7004. Austin, Tex. 105 pp.

30. **Deason, T. R. and H. C. Bold. 1960.** *Phycological Studies I.* Exploratory studies of Texas soil algae. Univ. Tex. Publ. 6022. Austin, Tex. 72 pp.

31. **Chantanachat, S. and H. C. Bold. 1962.** *Phycological Studies II.* Some algae from arid soils. Univ. Tex. Publ. 6218. Austin, Tex. 75 pp.

32. **Mattox, K. R. and H. C. Bold. 1962.** *Phycological Studies III.* The taxonomy of certain Ulotrichalean algae. Univ. Tex. Publ. 6222. Austin, Tex. 67 pp.

33. **Bishoff, H. W. and H. C. Bold. 1963.** *Phycological Studies IV.* Some algae from Enchanted Rock and related algal species. Univ. Tex. Publ. 6318. Austin, Tex. 95 pp.

34. **Brown, R. M., Jr. and H. C. Bold. 1964.** *Phycological Studies V.* Comparative studies of the algal genera *Tetracystis* and *Chlorococcum,* Univ. Tex. Publ. 6417. Austin, Tex. 213 pp.

35. **Smith, R. L. and H. C. Bold. 1966.** *Phycological Studies VI.* Investigations of the algal genera *Eremosphaera* and *Oocystis.* Univ. Tex. Publ. 6612. Austin, Tex. 121 pp.

36. **Cox, E. R. and H. C. Bold. 1966.** Taxonomic investigations of *Stigeoclonium. Univ. Tex. Publ.* 6612. Austin, Tex. 167 pp.

37. **Groover, D. R. and H. C. Bold. 1969.** *Phycological Studies VIII.* The taxonomy and comparative physiology of the Chlorosarcinales and certain edaphic algae. Univ. Tex. Publ. 6907. Austin, Tex. 165 pp.

38. **Archibald, P. A. and H. C. Bold. 1970.** *Phycological Studies XI.* The genus *Chlorococcum* Meneghini. Univ. Tex. Publ. 7015. Austin, Tex. 155 pp.

39. **Lee, K. W. and H. C. Bold. 1974.** *Phycological Studies XII. Characium* and some *Characium*-like algae. Univ. Tex. Publ. 7403. Austin, Tex. 127 pp.

40. **Deason, T. R. 1967.** *Pulchrasphaera,* a new chlorococcalean genus. *J. Phycol.* 3:19–21.

41. **Cox, E. R. and T. R. Deason. 1968.** *Axilosphaera* and *Heterotetracystis,* a new chlorosphaeracean genus from Tennessee soil. *J. Phycol.* 4:240–249.

42. **Archibald, P. A. 1970.** *Pseudochlorococcum,* a new chlorococcalean genus. *J. Phycol.* 6:127–132.

43. **George, E. A. 1976.** A guide to algal keys. *Br. Phycol. J.* 11:49–55.

44. **Lowe, C. W. and A. V. Moyse. 1948.** An investigation of some Manitoba soils for the presence of soil algae. *Trans. R. Soc. Can. Sect. 5* 28:119–152.

45. **Lanciani, G. D. and J. M. Kingsbury. 1965.** Soil algae of 40 ponds under construction at Ithaca, N.Y. *Rhodora* 67:242–254.

46. **Forest, H. S. 1962.** Analysis of the soil algal community. *Trans. Am. Microsc. Soc.* 81:189–198.

47. **MacEntee, F. J., G. Schreckenberg and H. C. Bold. 1972.** Some observations on the distribution of edaphic algae. *Soil Sci.* 114:171–179.

48. **MacEntee, F. J. and H. C. Bold. 1974.** Some observations of edaphic algae in Pike County, Pa. *Soil Sci.* 117:66–69.

49. **Chanlygina, O. Ya. 1976.** Soil algae of pine and spruce forests of the Moscow oblast. *Bot. Zh. (Leningr.)* 61:1077–1088.

50. **Komaromy, Z. P. 1969.** Algological investigations of Hungarian forest soils. I. The life cycle of three soil algae. *Ann. Hist.-Nat. Mus. Natl. Hung.* 61:147–150.

51. **Hindak, F. 1970.** *The Microphytoedaphon of the Oak–Hornbeam Forest at B&B Res.* Project B&B, Prog. Rep. I, pp. 59–63.

52. **Kabirov, R. R. and R. G. Minibaev. 1975.** Pecularities of seasonal dynamics of algae populations in gray forest soil of the Ural foothills in the Bashlin Autonomous S.S.R. *Biol. Nauki (Mosc.)* 11:114–117.

53. **Novichkova-Ivanova, L. N. 1978.** Soil algae of the mixed coniferous–broadleaved forest subzone of the Amur-Zeya Watershed, Amurskaya Taiga. *Izv. Akad. Nauk SSSR* 32.

54. **Kulikova, R. M. 1965.** Development of algae in peat-bog soils of varying degrees of cultivation. *Sov. Soil Sci.* 10:1023–1207.

55. **Kusel-Fitzmann, E. and W. Url. 1965.** Das Schwinggrasenmoor am Goggausee und seine Algengesellschaften, *Sitzungs beneti (Abt. 1) Ot. A.K. der Urss.* 174:315.

56. **Compere, P. 1966.** Observations on the algae in the peat bogs of the Hautes-Fagner of Belgium, Brussels. *Bull. Jard. Bot. Etat. Brux.* 36:5–51.

57. **Archibald, P. A. 1969.** *A Study of the Algal Flora of Bogs with Special Reference to the Genus* Chlorococcum. Unpubl. Ph.D. thesis, University of Texas, Austin, Tex. 204 pp.

58. **Busygina, E. A. 1976.** Some algal communities on ameliorated cutover peatlands. *Pochvovedenie* 12:103–109.

59. **Polderman, P. J. G. and R. A. Polderman-Hall. 1080.** Algal communities in Scottish salt marshes. *Br. Phycol. J.* 15:59–71.

60. **Killian, C. and D. Fehér. 1935.** Recherches sur les phénomènes microbiologiques des sols Sahariens. *Ann. Inst. Pasteur (Paris)* 55:573–622.

61. **Killian, C. and D. Fehér. 1939.** Recherches sur la microbiologie des sols désertiques. Première partie. Recherches sur les phénomènes microbiologiques des sols Sahariens. Pp. 1–44 in P. Lechevalier (ed.). *Encyclopedia of Biology*, Vol. 21. Paris.

62. **Killian, D. and D. Fehér. 1939.** Recherches sur la microbiologie des sols désertiques. Deuzième partie. Résultats de la mission Saharienne Killian-Fehér. Pp. 45–127 in P. Lechevalier (ed.). *Encyclopedia of Biology*, Vol. 21. Paris.

63. **Fletcher, J. E. and W. P. Martin. 1948.** Some effects of algae and molds in the rain-crust of desert soils. *Ecology* 29:95–100.

64. **Durrell, L. W. 1959.** Algae in Colorado soils. *Am. Midl. Nat.* 61:322–328.

65. **Cameron, R. E. 1960.** Communities of soil algae occurring in the Sonoran desert in Arizona. *J. Ariz. Acad. Sci.* 1:85–88.

66. **Cameron, R. E. 1963.** Algae of southern Arizona. I. Introduction—blue green algae. *Rev. Algol (N.S.)* 6:282–318.

67. **Cameron, R. E. 1964.** Terrestrial algae of southern Arizona. *Trans. Am. Microsc. Soc.* 133:212–218.

68. **Cameron, R. E. 1964.** Algae of southern Arizona. II. Algal flora (exclusive of blue-green algae). *Rev. Algol (N.S.)* 7:151–177.

69. **Cameron, R. E., F. A. Morelli and G. B. Blank. 1968.** *Soil Studies: Desert Microflora in an Area of Soil at White Mt. Range, CA.* J.P.L. Space Program Summary 4. Jet Propulsion Laboratories, California Institute of Technology, Pasadena, Calif., pp. 212–214.

70. **Cameron, R. E. 1969.** *Abundance of Microflora in Soils of Desert Regions.* J.P.L. Tech. Rep. Jet Propulsion Laboratories, California Institute of Technology, Pasadena, Calif., pp. 32–1378.

71. **Shields, L. M. and L. W. Durrell. 1967.** Distribution of terrestrial algae within the Nevada test site. *Am. J. Bot.* 49:547–554.

72. **Dykstra, R. F., F. J. MacEntee and H. C. Bold. 1975.** Some edaphic algae of the Texas coast. *Tex. J. Sci.* 26:171–177.

73. **Friedmann, E. I. 1961.** *Chroococcidiopsis kashau* sp. nov. and the genus *Chroococcidiopsis*. *Ost. Bot. Z.*, 108:354–367.

74. **Friedmann, E. I. 1964.** *Xerophytic Algae in the Negev Desert.* 10th International Botany Congress, pp. 290–291.

75. **Friedmann, E. I. and R. Ocampo-Paus. 1966.** *Bracteacoccus minor* (Chodat) Petrova var. *desertorum* n. var., a remarkable alga from the Negev. *Nova Hedwigia* 10:481–494.

76. **Friedmann, E. I., Y. Lipkin and R. Ocampo-Paus. 1967.** Desert algae of the Negev (Israel). *Phycologia* 7:185–200.

77. **Sdobnikova, N. V. 1959.** Algae which inhabit the sands of middle Asia. *Bot. Mater. Otd. Sporovykh Rast. Bot. Inst. Akad. Nauk SSSR* 12.

78. **Melnikova, V. V. 1962.** Algal flora on serozems soils of Tadzhikistan. *Tr. Bot. Inst. Akad. Nauk Tadzh. SSR* 18.

79. **Shtina, E. A. 1969.** The distribution and ecological significance of algae in arable soils. *Pedobiologia* 9:226–242.

80. **Novichkova-Ivanova, L. N. 1972.** Soil algae of middle Asia deserts. In L. E. Rodin (ed.). *Eco-Physiological Foundation of Ecosystem Productivity in Arid Zone.* International Symposium, U.S.S.R., Nauka, Leningrad. 232 pp.

81. **Gayel, A. G. and E. A. Shtina. 1974.** Algae on the sands of arid regions and their role in soil formation. *Sov. Soil Sci.* 6:311–319.

82. **Akiyama, M. 1961.** Aerial and terrestrial algae in San-In region of Honshu, Japan. *Bull. Shimane Univ. (Nat. Sci.)* 10:75–89.

83. **Akiyama, M. 1973.** Ecological studies on the terrestrial algae found in the coastal sand dune in San-In region, Japan. *Mem. Fac. Educ. Shimane Univ.* 7:33–42.

84. **Cameron, R. E. and G. B. Blank. 1966.** *Desert Algae: Soil Crusts and Diaphanous Substrata as Algal Habitats.* NASA Tech. Rep. 32-971. Jet Propulsion Laboratories, California Institute of Technology, Pasadena, Calif., pp. 1–41.

85. **Cameron, R. E. 1972.** A comparison of soil microbial ecosystem in hot, cold, and polar desert regions. In L. E. Rodin (ed.). *Eco-Physiological Foundation of Ecosystem Productivity in Arid Zone,* International Symposium, U.S.S.R. Nauka, Leningrad. 232 pp.

86. **Durrell, L. W. 1964.** Algae in tropical soils. *Trans. Am. Microsc. Soc.* 83:79–85.

87. **MacEntee, F. J., H. C. Bold and P. A. Archibald. 1977.** Notes on some edaphic algae of the South Pacific and Malaysian areas, with special reference to *Pseudotetraedron polymorphum* gen. et spec.-nov. *Soil Sci.* 124:161–166.

88. **Flint, E. A. 1969.** Biology of the main soils. III. Algae on the surface of some New Zealand soils. *N.Z. Dep. Sci. Ind. Res. Bull.* 26:183–190.

89. **Akiyama, M. 1971.** On some Brazilian species of Trentepohliaceae. *Mem. Fac. Sci. Shimane Univ.* 5:81–95.

90. **Archibald, P. A. 1972.** A preliminary survey of the edaphic algae of Costa Rica and San Andreas Isle. *Soil Sci.* 113:63–64.

91. **Archibald, P. A. 1975.** Notes on the edaphic algae of the Galapagos. *Soil Sci.* 120:400–402.

92. **Akiyama, M. 1970.** Some soil algae from the Arctic: Alaska, Canada and Greenland. *Mem. Fac. Educ. Shimane Univ.* 4:53–75.

93. **Novichkova-Ivanova, L. N. 1972.** Soil and aerial algae of polar deserts and arctic tundra. In F. E. Wielgolaski and T. Rosswall (eds.). *I.B.P. Tundra Biome,* Proceedings of the 4th International Meeting on the Biological Productivity of Tundra, Tundra Steering Committee, Stockholm. 265 pp.

94. **Cameron, R. E. 1970.** Soil microbial ecology of Valley of 10,000 Smokes, Alaska. *J. Ariz. Acad. Sci.* 6:11–40.

95. **Cameron, R. E. 1966.** *Soil Studies-Desert Microflora. XIII. Identification of Some Algae from Antarctica.* J.P.L. Space Program Summary 37-40, IV. Jet Propulsion Laboratories, California Institute of Technology. Pasadena, Calif., pp. 123–133.

96. **Benoit, R. E. and R. E. Cameron. 1967.** *Microbial Ecology of Some Dry Valley Soils of Antarctica,* Proc. 3. Space Program Jet Propulsion Laboratories, California Institute of Technology, Pasadena, Calif.

97. **Akiyama, M. 1967.** On some Antarctic terrestrial and subterranean algae. *Mem. Fac. Educ. Shimane Univ.* 1:36–56.

98. **Kol, E. 1968.** Algae from the Antarctica. *Ann. Hist.-Nat. Mus. Natl. Hung. Perspect. Bot.* 60:71–77.

99. **Akiyama, M. 1968.** A list of terrestrial and subterranean algae from the Ongul Islands, Antarctica. *Antarct. Rec. (Nat. Sci. Mus. Tokoyo)* 32:71–77.

100. **Akiyama, M. 1969.** Cold desert problems and characteristics relevant to other arid lands.

In W. McGinnies and B. Goldman (eds.). *Arid Lands in Perspective.* University of Arizona Press, Tucson, Ariz.

101. **Cameron, R. E. 1970.** Soil microbial and ecological investigations in the antarctic interior. *Antarct. J. U.S.* 5:87–88.

102. **Cameron, R. E. and J. King. 1970.** Soil microbial ecology of Wheeler Valley, Antarctica. *Soil Sci.* 109:110–120.

103. **Cameron, R. E. 1971.** Antarctic soil microbiological and ecological investigations. In L. Quam (ed.). *Research in Antarctica.* American Association for the Advancement of Science, Washington, D.C.

104. **Cameron, R. E. 1972.** Ecology of blue-green algae in Antarctic soils. In T. V. Desikachary (ed.). *The Taxonomy and Biology of Blue-Green Algae,* Symposium at Madras, India.

105. **Broady, P. A. 1979.** *The Terrestrial Algae of Signy Island, South Orkney Islands.* Br. Antarctic Survey Sci. Rep. 98. National Research Council, Cambridge. 117 pp.

106. **Broady, P. A. 1982.** New records of chlorophycean micro-algae cultured from Antarctic terrestrial habitats. *Nova Hedwigia* 36:445–484.

107. **Broady, P. A. 1984.** Taxonomic and ecological investigations of algae on steam-warmed soil on Mt. Erebus, Ross Island, Antarctica. *Phycologia* 23:257–271.

108. **Broady, P. A. 1986.** Ecology and taxonomy of the terrestrial algae of the Vestfold Hills. In J. Pickard (ed.). *Antarctic Oasis. Terrestrial Environments and History of the Vestfold Hills.* Academic Press, Sydney, Australia. 367 pp.

109. **Prasad, B. N. and P. N. Srivastava. 1968.** Systematic and ecological studies on algae of alkaline (usar) soils. *Phykos* 7:102–111.

110. **Noskova, T. S. 1968.** *Bot. Zh. (Leningr.)* 53:983–987.

111. **Pomelova, G. I. 1970.** Number of soil algae in different fields of a rotation on slightly podzolic/moderately loamy soil. *Pochvovedenie* 8:70–74.

112. **Rüs, 1972.** The number and floristic composition of algae in cultivated soddy calcareous soil. *Izv. Akad. Nauk Est. SSR Ser. Biol.* 21:366–373 (Engl. Sum.).

113. **Busygina, E. A. 1976.** Soil algae communities on ameliorated cutover peatlands. *Pochvovedenie* 12:103–109. (Engl. Sum.).

114. **Nordin, R. N. and D. W. Blinn. 1972.** Analysis of a saline tallgrass prairie ecosystem. IV. Preliminary investigations of soil algae. Proc. N.D. Acad. Sci. 25:8–17.

115. **Balezina, L. S. 1975.** Effect of mineral and organic fertilizers on the development of algae in a sod-podzolic soil. *Microbiology* 44:306–309.

116. **Metting, B. and W. Rayburn. 1979.** Algal communities and soil microenvironment in an eastern Washington silt loam. *Soil Sci.* 127:74–78.

117. **Polderman, P. J. G. and R. A. Polderman-Hall. 1980.** Algal communities in Scottish saltmarshes. *Br. Phycol. J.* 15:59–71.

118. **Cameron, R. E. and G. B. Blank. 1966.** *Soil studies: desert microflora. XI. Desert Soil Algae Survival at Extremely Low Temperatures.* J.P.L. Space Programs Summary 37-37, IV. Jet Propulsion Laboratories, California Institute of Technology, Pasadena, Calif. pp. 174–181.

119. **Cameron, R. E., C. N. David and J. King. 1968.** Soil toxicity in antarctic dry valleys. *Antarct. J. U.S.* 3:164–166.

120. **Cameron, R. E. and H. P. Conrow. 1969.** Antarctic dry valley soil microbial incubation and gas composition. *Antarct. J. U.S.* 4:28–33.

121. **Cameron, R. E. and H. P. Conrow. 1969.** Soil moisture, relative humidity, and microbial abundance on dry valleys of southern Victoria Land. *Antarct. J. U.S.* 4:23–28.

122. **Cameron, R. E. 1972.** A comparison of soil microbial ecology in hot, cold, and popular desert regions. In L. E. Rodin (ed.). *Ecophysiological Foundation of Ecosystem Productivity in Arid Zones.* Nauka, Leningrad.

123. **Lund, J. W. G. 1967.** Soil algae. In A. Burges and F. Raw (eds.). *Soil Biology.* Academic Press, London. 532 pp.

124. **Kabirov, R. R. 1978.** Method for studying the number and biomass of soil algae. *Pochvovedenie* 5:125–129.

125. **Nokrasova, K. A. and E. A. Busygina. 1977.** Some corrections to the method of quantitative account of soil algae. *Bot. Zh. (Leningr.)* 62:214–222.

126. **Jones, C. P. T. and J. E. Mollison. 1948.** A technique for the quantitative estimation of soil microorganisms. *J. Gen. Microbiol.* 2:54–69.

127. **Jones, D. and E. Griffiths. 1964.** The use of thin soil sections for the study of soil microorganisms. *Plant Soil* 20:232–240.

128. **Tchan, Y. T. 1952.** Study of soil algae. I. Fluorescence microscopy for the study of soil algae. *Proc. Linn. Soc. Lond.* 77:265–269.

129. **Sharabi, N. E. D. and D. Pramer. 1973.** A spectrophotofluorometric method for studying algae in soil. *Bull. Ecol. Res. Comm. (Stockholm)* 17:77–84.

130. **Drew, E. A. and J. R. Anderson. 1977.** Studies on the survival of algae added to chemically treated soils. I. Methodology. *Soil Biol. Biochem.* 9:207–215.

131. **King, J. M. and C. H. Ward. 1977.** Distribution of edaphic algae as related to land usage. *Phycologia* 16:23–30.

132. **Hunt, M. E., G. L. Floyd and B. B. Stout. 1979.** Soil algae in field and forest environments. *Ecology* 60:362–375.

133. **Metting, B. 1979.** *A Comparative Study of Algal Communities on Cultivated and Uncultivated Portions of a Schumacher Silt Loam.* Unpubl. Ph.D. dissertation, Washington State University, Pullman, Wash.

134. **Darley, W. M., C. L. Montague, F. G. Plumley, W. W. Sage and A. T. Psalidas. 1981.** Factors limiting edaphic algal biomass and productivity in a Georgia saltmarsh. *J. Phycol.* 17:122–128.

135. **Shihira, I. and R. W. Krauss. 1965.** *Chlorella: Physiology and Taxonomy of 41 Isolates.* University of Maryland Press, College Park, Md. 97 pp.

136. **Kessler, E. 1976.** Comparative physiology, biochemistry and the taxonomy of *Chlorella* (Chlorophyceae). *Plant Syst. Evol.* 125:125–138.

137. **Kessler, E. 1965.** Physiologische und biochemische Beiträge zur Taxonomie der Gattung *Chlorella.* I. Säueresistenz als taxonomisches Merkmal. *Arch. Mikrobiol.* 52:291–296.

138. **Kessler, E. 1967.** Physiologische und biochemische Beiträge zur Taxonomie der Gattung *Chlorella.* III. Merkmale von 8 autotrophen Orten. *Arch. Mikrobiol.* 55:346–357.

139. **Carr, N. G. and B. A. Whitton. 1973.** *The Biology of Blue-Green Algae,* Botany Monographs, Vol. 9. University of California Press, Berkeley, Calif. 676 pp.

140. **Stewart, W. D. P. (ed.). 1974.** *Algal Physiology and Biochemistry,* Botany Monographs, Vol. 10, University of California Press, Berkeley, Calif. 989 pp.

141. **Granhall, U. (ed.). 1976.** *Environmental Role of Nitrogen-Fixing Blue-Green Algae and Asymbiotic Bacteria.* Ecol. Bull. 26. Swedish Natural Science Research Council, Uppsala, Sweden. 391 pp.

142. **Kessler, E., I. Ludwig and H. Wiechmann. 1963.** Bildung von Sekindär-Carotinoiden bei Stickstoffmangel und Hydrogenase Aktivität als taxonomishe Merkmale in der Gattung *Chlorella.* Pp. 7–20 in *Studies on Microalgae and Photosynthetic Bacteria.* Ed. Japanese Society Plant Physiology, Tokoyo.

143. **Kessler, E., W. Langer, I. Lusewig and H. Wiechman. 1963.** Bildung von Sekundär-Carotenoiden bei Stickstoffmangel und Hydrogenase Aktivität als taxonomische Merkmale in der Gattung *Chlorella.* In *Studies on Microalgae and Photosynthetic Bacteria.* Ed. Japanese Society Plant Physiology, Tokoyo.

144. **Kessler, E. and F. C. Czygen. 1967.** Physiologische und biochemische Beiträge zur Taxonomie der Gattungen *Ankistrodesmus* und *Scenedesmus.* I. Hydrogenase, Sekundär-Carotenoide und Gelatine-Verflüssigung. *Arch. Microbiol.* 55:320–326.

145. **Czygan, F. C. 1968.** Sekundär-Carotinoide in Grünalgen I. Chemie, Vorkomen und Faktoren, welche die Bildung dieser Polëne beinflussen. *Arch. Microbiol.* 61:81–102.

146. **Czygan, F. C. 1970.** Untersuchungen über die Bedeutung der Biosynthese von Sekundär-Carotenoiden als Artmerkmal vei Grünalgen. *Arch. Microbiol.* 74:77–81.

147. **Weber, A. and F. C. Czygan. 1972.** Chlorophylle und Carotinoide der Chaetophonineae (Chlorophyceae-Ulotrichales) Siphonoxanthin in *Microthamnion* Kützingianum. *Arch. Microbiol.* 84:243–253.

148. **Kessler, E. 1978.** Physiological and biochemical contribution to the taxonomy of the genus *Prototheca*. II. Starch hydrolysis and base complements of DNA. *Arch. Microbiol.* 116:105–107.

149. **Hellebust, J. A. and J. S. Craigie (ed.). 1978.** *Handbook of Phycological Methods: Physiological and Biochemical Methods.* Cambridge University Press, New York. 512 pp.

150. **Kessler, E. 1982.** Chemotaxonomy in Chlorococcales. In F. E. Round and D. J. Chapman (eds.). *Progress in Phycological Research,* Vol. 1. Elsevier, New York.

151. **Ben-Amotz, A., T. G. Tornabene and W. H. Thomas. 1985.** Chemical profile of selected species of microalgae with emphasis on lipids. *J. Phycol.* 21:72–81.

152. **Twang, M., R. C. Hodson and J. A. Schiff. 1978.** Sulfate uptake. In J. A. Hellebust and J. S. Craigie, (eds.) *Handbook of Phycological Methods: Physiological and Biochemical Methods.* Cambridge University Press, New York. 512 pp.

153. **Nalewajko, C. 1978.** Release of organic substances. In J. A. Hellebust and J. S. Craigie (eds.). *Handbook of Phycological Methods: Physiological and Biochemical Methods.* Cambridge University Press, New York. 512 pp.

154. **Badour, S. S. 1978.** Inhibitors used in studies of algal metabolism. In J. A. Hellebust and J. S. Craigie (eds.). *Handbook of Phycological Methods: Physiological and Biochemical Methods.* Cambridge University Press, New York. 512 pp.

155. **Archibald, P. A. and V. J. Smith. 1987.** Notes on variation in physiological attributes between aquatic and edaphic species of the chlorophycean algal genus *Neochloris. Trans Am. Microsc. Soc.* 106:179–182.

156. **Torres, A. and L. O'Flaherty. 1976.** Influence of pesticides on *Chlorella, Chlorococcum, Stigeoclonium,* (Chlorophyceae) *Tribonema, Vaucheria* (Xanthophyceae) and *Oscillatoria* (Cyanophyceae). *Phycologia* 15:25,36.

157. **Atkins, C. A. and Y. T. Tchan. 1967.** Study of soil algae. VI. Bioassay of atrazine and the prediction of its toxicity in soils using algal growth method. *Plant Soil* 27:432–442.

158. **Audus, L. J. 1970.** *The Action of Herbicides on the Microflora of the soils,* Proceedings of the 10th British Weed Control Conference, pp. 1036–1051.

159. **Anderson, J. R. 1978.** Pesticide effects on non-target soil microorganisms. In I. R. Hill and S. J. L. Wright (eds.). *Pesticide Microbiology.* Academic Press, London.

160. **Cullimore, D. 1971.** Interaction between herbicides and soil microorganisms. *Residue Rev.* 35:65–80.

161. **Gaunt, J. and W. Evans. 1961.** Metabolism of 4-chloro-2-methylsphenoxyacetic acid by a soil microorganism. *Biochem. J.* 79:25.

162. **Kirkwood, R. and W. F. Fletcher. 1970.** Factors influencing the herbicidal efficiency of MCPA and MCPB in three species of microalgae. *Weed Res.* 10:3–10.

163. **Arvik, J., D. Willson and L. Darlington. 1970.** Response of soil algae to picloram-2,4-D mixtures. *Weed Soil* 19:276–278.

164. **Anderson, J. R. and E. A. Drew. 1976.** Effects of pure paraquat dichloride, "Granoxone W" and formulation additives on soil microbiological activities. I. Estimation of soil biomass in laboratory treated soil. *Zentralbl. Bakteriol. Abt. 2* 131:125–135.

165. **Cullimore, D. and A. McCann. 1977.** Influence of four herbicides on the algal flora of a prairie soil. *Plant Soil* 46–499–510.

166. **Komaromy, A. P. and K. Mihaly. 1978.** Diqual-dibromide sensitivity of some blue-green and green soil algae. *Stud. Bot. Hung.* 12:31–40.

167. **Metting, B. and W. Rayburn. 1979.** The effects of the pre-emergence herbicide di-allate and the post-emergence herbicide MCPA on the growth of some soil algae. *Phycologia* 18:269–272.

168. **Kiss, A. 1967.** *Agrokem. Talajtan* 16:111–124.

169. **Panters, H. 1970.** *Meded. Rijksfac. Landbouwwet. Gent* 35:847–854.

170. **Mikhailova, P. and Yu. V. Kruglov. 1973.** *Pochvovedenie* 8:81–85.

171. **McCann, A. E. and D. R. Cullimore. 1979.** Influence of pesticides on the soil algal flora. *Residue Rev.* 72:1–31.

172. **Lal, R. and D. M. Saxena. 1980.** Cytological and biochemical effects of pesticides on microorganisms. *Residue Rev.* 73:49–86.

173. **Cain, J. R. and R. K. Cain. 1983.** The effects of selected herbicides on zygospore germination and growth of *Chlamydomonas moewusii* (Chlorophyceae, Volvocales). *J. Phycol.* 19:301–305.

174. **Bold, H. C. 1970.** Some aspects of the taxonomy of soil algae. *Ann. N.Y. Acad. Sci.* 175:601–616.

175. **Nichols, H. W. 1973.** 1: Growth media—freshwater. In J. R. Stein (ed.). *Handbook of Phycological Methods: Culture Methods and Growth Measurements.* Cambridge University Press, Cambridge. 448 pp.

176. **Starr, R. C. 1973.** Special methods—dry soil samples. In J. R. Stein (ed.). *Handbook of Phycological Methods: Culture Methods and Growth Measurements.* Cambridge University Press, Cambridge. 448 pp.

177. **Andreva, V. M. and T. V. Dedova. 1965.** On the variability of the taxonomic characters of the unicellular Chlorophyta under the conditions of culture. *Bot. Zh.* (*Leningr.*) 50:954–961.

178. **Brown, R. M., Jr. and R. Wilson. 1968.** Electron microscopy of the lichen *Physcia aerpolia* (Ehrh.) Nyl. *J. Phycol.* 4:230–240.

179. **Laudi, G., P. Bonnatti and L. D. Trovatelli. 1970.** Differenze ultrastructurali di alcume species di *Trebouxia* poste in condizione di illüminazine differenti. *G. Bot. Ital.* 103:79–107.

180. **McCarthy, J. and N. J. Lang. 1971.** Comparative ultrastructure of culture species of *Trebouxia, J. Phycol.* 7:155–165.

181. **Starr, R. C. 1955.** A Comparative Study of Chlorococcum *Meneghini and Other Spherical, Zoospore-Producing Genera of the Chlorococcales.* Indiana Univ. Publ. Sci. Ser. 20. Indiana University Bloomington, Ind.

182. **Bourrelly, P. 1972.** *Les Algues d'eau douce. Initiation á la systématique.* 1. *Les algues vertes,* rev. ed. Boubée, Paris. 569 pp.

183. **Bourrelly, P. 1968.** *Les Algues d'eau douce. Initiation á la systématique. II. Les algues jaunes et brunes, Chrysophycees, Xanthophycees et Diatomées,* Boubée Paris. 438 pp.

184. **Bourrelly, P. 1970.** *Les Algues d'eau douce. Initiation á la systématique. III. Les algues bleues et rouges, les Eulgéniens, Peridiniens et Cryptomonadines.* Boubée, Paris. 512 pp.

185. **Bold, H. C. and M. J. Wynne. 1985.** *Introduction to the Algae: Structure and Reproduction,* 2nd ed. Prentice-Hall, Englewood Cliffs, N.J. 720 pp.

186. **Drouet, F. and W. A. Daily. 1981.** Revision of the Stigonemataceae with a summary of the classification of the blue-green algae. *Beih. Nova Hedwigia* 66 221 pp.

187. **Ettl, H. and J. Komarek. 1982.** Was versteht man unter dem Perguff, coccale Gruñalgen? *Arch. Hydrobiol. Suppl.* 60:345–374.

188. **Deason, T. R. 1984.** A discussion of the classes Chlamydophyceae and Chlorophyceae and their subordinate taxa. *Plant Syst. Evol.* 146:75–86.

189. **Melkonian, M. 1982.** Systematics and evolution of the algae. In H. Ellenberg, K. Esser, K. Kubitzki, E. Schnepf, and H. Ziegler (eds.). *Progress in Botany,* Vol. 44. Springer, New York.

190. **Mattox, K. R. and K. D. Stewart. 1984.** Classification of the green algae: a concept based on comparative cytology. Pp. 29–72 in D. John and D. Irvene (eds.). *Systematics of the Green Algae,* Systematics Assoc. Spec. Vol. 27. Academic Press, Orlando, Fla.

191. **Bold, H. C. and F. J. MacEntee. 1974.** Phycological Notes III. Two new saccate unicellular Chlorophyceae. *J. Phycol.* 10:189–193.

192. **Broady, P. A. 1977.** A new genus and two new species of terrestrial chlorophycean algae from Signy Island, South Orkney Islands, Antarctica. *Br. Phycol. J.* 12:7–15.

193. **Andreyeva, V. M. 1975.** *The Genus* Chlorella: *Morphology, Taxonomy, Principles of Classification.* Nauka, Leningrad 88 pp. (in Russian).

6

Soil Protozoa

J. DANIEL LOUSIER

STUART S. BAMFORTH

Common Groups of Soil Protozoa

Protozoa constitute a level of animal organization or a way of life rather than a "natural" group of organisms. The individual protozoan cell is an entire organism that must perform all the functions requisite for the survival and perpetuation of the species. Thus it maintains a higher degree of differentiation than metazoan cells. In terrestrial ecosystems, free-living protozoa are most commonly and characteristically associated with the surface layers of the soil, particularly with decomposing plant material.

The free-living protozoa of litters and soils belong to two phyla, Sarcomastigophora and Ciliophora (104). The former includes a spectrum of flagellated organisms, some amoeboflagellates and amoeboid types, the latter including unicellular naked amoebae, other naked amoebae and plasmodial forms that reproduce by sporangia (slime molds), radiate Heliozoa, and shelled amoebae (testacea). The Ciliophora are a better defined group that possess cilia at some time during their life cycle and usually throughout their life histories. Pragmatically and ecologically, soil protozoa may be arranged into four groups: flagellates, naked amoebae, testacea, and ciliates (Fig. 6.1). The first three groups are polyphyletic (104). Because of differing ecological requirements, each group varies in distribution within soil layers and among sites. Community structure and pattern are closely related to ecological conditions of the habitat, so that some protozoa can be considered indicator species in their respective habitats.

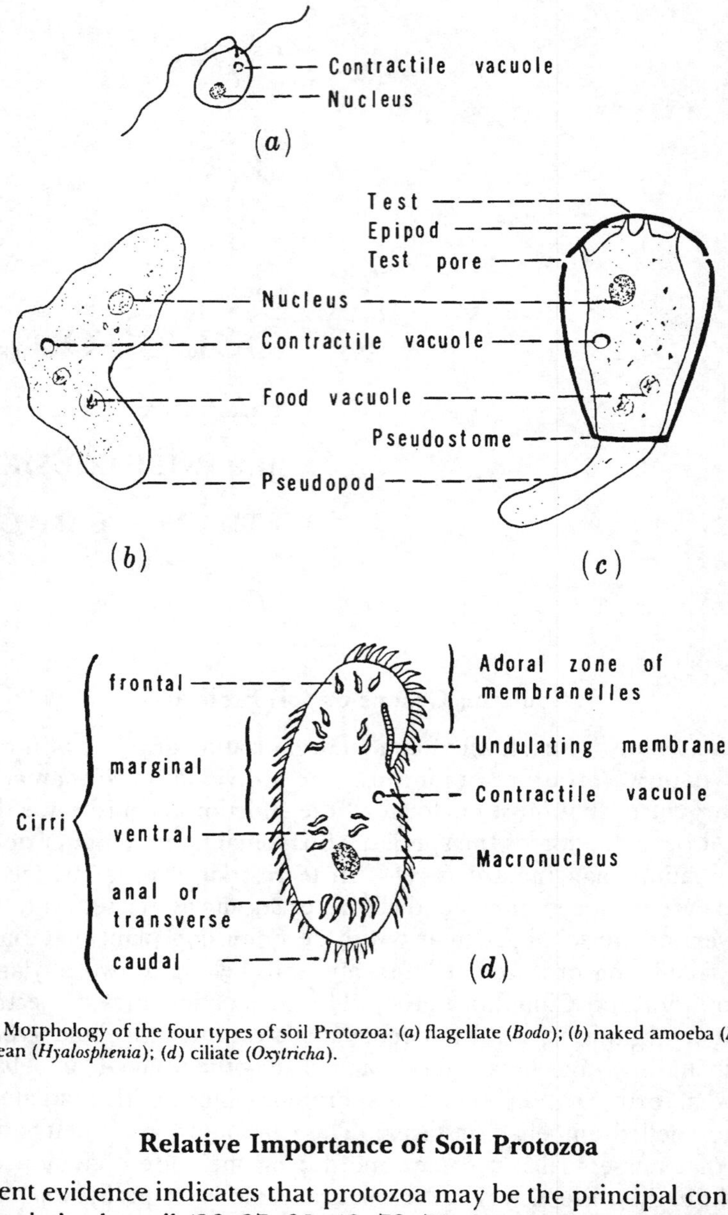

Fig. 6.1. Morphology of the four types of soil Protozoa: (*a*) flagellate (*Bodo*); (*b*) naked amoeba (*Naegleria*); (*c*) testacean (*Hyalosphenia*); (*d*) ciliate (*Oxytricha*).

Relative Importance of Soil Protozoa

Recent evidence indicates that protozoa may be the principal consumers of bacteria in the soil (26, 37, 38, 49, 73, 79, 85, 159–161, 163). Through this consumption, soil protozoa help regulate and modify the size and character of the bacterial community, thus accelerating the turnover of the soil biomass and, consequently, the turnover of soil organic matter.

The comparative roles of the various groups of soil protozoa do not appear to be characterized by differences in kind, but rather in degree. The flagellates, naked amoebae, and small ciliates appear to be the most widespread forms and to have the largest biomass and highest turnover rates, thus exerting greater impact on the organic cycle of soil than do the large ciliates and testacea. Many testacea may be facultative primary decomposers; for example, when soil conditions have deteriorated and edible bacteria have become less available, some species of testacea were observed to ingest primarily detrital particles (96).

There have been few studies in North America that compare the impact of protozoa on the organic cycle of the soil to that of other fauna. For a Canadian deciduous woodland soil, Lousier and Parkinson (116) reported a mean annual dry biomass of 0.07 g dry weight per square meter for testacea, which was considerably less than that for bacteria (23 g dry weight per square meter) and fungi (40 g dry weight per square meter), less by a factor of 10 than the estimate for Enchytraeidae (52) and greater than that for Carabidae (0.02 g dry weight per square meter) (30). The testacean secondary production, 21 g dry weight per square meter per year, was also 10–13 times more than the estimated production of the Enchytraeidae (52). If the testacea are the *least* abundant and productive of the four major groups of soil protozoa (163), it is then evident that the potential for productivity and organic cycling by soil protozoa as a whole is tremendous.

Soil Protozoa in North America

Although protozoa are a major component of the terrestrial organic cycle, they are a poorly known and understood group of soil fauna in North America. While reports of freshwater protozoa in North America had begun in the mid-1800s (4), the first paper to mention soil protozoa was that of Leidy in 1879 (102). The early part of this century was characterized by a more active interest in the species composition and distribution of protozoa in various soils (2, 44–46, 68, 69, 143, 144, 170, 171). While qualitative studies continue to appear (6, 7, 21, 107, 108, 120, 139), the emphasis now has shifted to include basic biology (14, 26, 121, 122), community structure (8, 13, 19, 107, 109), estimates of abundance (5, 8, 10, 13, 105, 106, 109–113, 116) and biomass (67, 105, 106, 109–113, 116), population dynamics (67, 105, 106, 109–113, 116), and effects on the dynamics of energy and nutrients (3, 26, 40, 67, 105, 106, 109–113, 116).

Morphological Adaptations of Soil Protozoa

The soil protozoa lack the large size and structurally elaborate morphologies that are commonly found in protozoa of freshwater habitats (35, 70–72). Among common soil species, the flagellates are tiny, only 5–20 μm in length; amoebae are small, usually less than 50 μm; testacea usually range only from 45 to 60 μm; and ciliates are generally less than 100 μm long, small for ciliates. Besides smaller size, the chief morphological features of the naked protozoa in soils are a relatively simple form and a high degree of flexibility. Even for the larger species of soil ciliates, this flexibility enables the cell to utilize very thin water films that encompass soil particles and allows them to occupy the smaller pores where moisture usually persists (72, 163). Many of the ciliates are flattened and all of the common species, with the exception of *Vorticella* spp. and Suctoria, are free-swimming throughout their active life histories.

The testacea show considerable morphological variation within species in relation to variations of their microhabitats, but except for ubiquitous species, such as *Euglypha rotunda, E. laevis, Trinema lineare,* and *Difflugiella oviformis,* the tests of the commoner soil testacea are subspherical with the ventral surface being slightly or completely flattened, particularly in the area of the pseudostome (shell opening). The pseudopodia are extended out through the pseudostome (Fig. 6.1*c*), and the commoner species, which show a tendency toward bilateral symmetry of the test, are better able to utilize the thin water films. The morphological character of the ventral surface and of the pseudostomal area is a reflection of the degree of adaptation a species has made to life in the soil (19, 22) and is an important taxonomic and evolutionary criterion (22).

The small species of protozoa present in soils are usually the more abundant, even in congeneric species. The larger congeneric species are restricted, either in distribution and/or in number, or do not occur in soils. When the larger species of amoebae, testacea, and ciliates are present in soil habitats, they appear to be confined to areas with better environmental conditions (i.e., higher moisture content), such as litter layers, or they show morphological adaptations similar to those described above.

Physiology

Nutrition

Most soil protozoa are heterotrophic; that is, they must rely on preformed organic matter for their nutrition, either by saprobic feeding (obtaining their nourishment from soluble organic and inorganic substances) or by phagotrophy (direct ingestion of microbial cells or other particulate material). Many flagellates can feed osmotrophically (131, 166), and many species of flagellates, amoebae, and ciliates have been cultured axenically on soluble nutrients (159). The significance of saprobic feeding by soil protozoa is unknown, however, because it is likely that available soluble nutrients would be readily metabolized by the soil microflora which produce extracellular enzymes (1, 159). The dominant mode of protozoan nutrition in the soil is therefore considered here to be phagotrophic (159–

161, 167), with bacteria, fungi, yeasts, algae, other protozoa (36), and detrital particles comprising most of the material ingested (42, 138). It has been shown that even in axenic cultures, particulate material must be present to stimulate formation of food vacuoles by some protozoa (32, 159). Others, such as *Glaucoma,* may not require the particulate "starter" materials.

Axenic culture of soil protozoans has revealed a diversity of complex and exacting nutritional requirements (163), but the diversity of soil microorganisms is such that ingestion of bacteria, fungi, yeasts, and algae, for example, can ordinarily meet these requirements. While there is some evidence that some free-living soil protozoa readily ingest humic substances (96, 145, 148) and can digest cellulose (165), it is generally accepted that most soil protozoa are not capable of efficiently metabolizing the decomposing plant residues (159–161). The studies on feeding cited in Stout and Heal (163) have indicated that most soil protozoa can feed on a wide variety of bacteria but can also select one species of bacteria from mixed populations.

Excretion

There is limited evidence identifying excretion products or processes of soil protozoa. Protozoa in general appear to be ammoniotelic and do not excrete complex nitrogenous or phosphatic compounds (159). The carbohydrate metabolism of protozoan cells is comparable to that in other metazoan cells and may result in the excretion of organic acids (159).

Respiration

Stout and Heal (163) concluded that respiratory activity of soil protozoa is a function of oxygen tension, food supply, and the physiological health of the particular population. When these factors are favorable in cultures, soil protozoa maintain a high rate of respiration (98–100, 117, 118, 122, 173). The respiratory activity of ciliates is a little greater than that for flagellates, amoebae, or testacea (135, 140, 172).

In natural soil environments, however, conditions fluctuate and are generally less favorable than in aquatic environments and protozoan cells face harsher conditions, less food, lower oxygen and higher carbon dioxide concentrations, and more often will be either starving, resting, or encysted. Field respiration rates over longer periods of time should therefore be substantially lower than laboratory rates.

Encystment

When environmental conditions become unfavorable (e.g., decreased availability of edible bacteria, desiccation, low oxygen tensions), the active protozoan cell can enter a nonmotile inactive stage, or cyst. Encystment is a distinctive and common feature of soil protozoa and permits survival. Among species of soil protozoa there is great variation in the degree of cellular differentiation of the encysted organism, thickness of the cyst membrane, the readiness with which cyst formation and reactivation occur, the tolerance of the cyst to adverse environmental conditions, and the viability of the cell after long periods of encystment (1, 163).

Colpoda steini, the most widespread of the soil ciliates and perhaps of all soil protozoa, forms two types of resistant cysts: resting and unstable (157). Resting cysts generally form after a period of growth and encystment is stimulated by exhaustion of the food supply and crowding. Unstable cysts form in response to adverse environmental conditions that apparently inhibit growth and cellular division. Cysts of small amoebae are typically simple and spherical, with the cyst membrane varying in thickness and structure (136, 152); some are with and some without plugged exit pores.

Soil testacea appear to employ several types of cyst formation (169), three of which may be related to environmental conditions. The simplest "cyst" is the formation of a wrinkled membrane or pellicle, which covers the cytoplasm and produces a "pre-cyst" in only a few minutes (19).

In the second case, a true cyst is formed inside the test, with the cyst membrane varying in thickness and shape. A third form involves the plugging of the pseudostome (the plug is called an epiphragm and consists of organic debris cemented together with a cellular secretion). The latter two types of cysts are seen commonly in soil testacea (107, 108) and such encysted organisms can constitute a prominent portion of testacean populations (96, 109–113, 116). No detailed information is available on cyst structure and formation by flagellates found in soils.

The number of viable cysts formed will diminish with time after encystment (74). Cysts of the common, well-adapted soil protozoa tend to survive extremes in environmental factors (e.g., drought), but cysts of the more aquatic protozoa are not as viable (1). Once living conditions become more amenable, the cell is able to excyst (i.e., leave the cyst and return to the active form). Generally, excystment of soil protozoa is closely related to edibility of bacteria (66, 95, 156), presence of microbial products (53, 154–156), or improving physical conditions [especially the availability of water (163)].

Lousier (105, 106) postulated an inverse relationship between numbers of cysts and numbers of active forms (i.e., many cysts and few active testacea during poor soil conditions and the converse during improved conditions). He found, however, that for most species of testacea in a deciduous woodland soil, the number of cysts tended to fluctuate directly with the number of active forms (110–113, 116). For some of the other species, cyst numbers did increase during decreases in active numbers, even to the point of exceeding significantly the number of active individuals. Lousier (110–113, 116) also showed that a population of soil testacea, despite the complications of having a test and more complex mechanisms of encystment, can excyst and encyst quickly.

Reproduction

Under favorable culture conditions, soil protozoans have been observed to reproduce rapidly—within hours for many species. However, under natural conditions, reproduction is slower and much less frequent. Short periods (hours to days) of favorable conditions, characterized by relatively high rates of reproduction, are interspersed for most soil protozoa with longer periods (weeks to months) of much less favorable conditions when the protozoa reproduce infrequently or not at all (117).

Reproduction usually occurs asexually by binary fision. In most groups, division is parallel to the anteroposterior axis of the body, but in ciliates the division is at right angles (i.e., transverse). Reproduction in testacea is complicated by the presence of the test and the mode of division is dependent on the test's rigidity (16, 86–89, 119, 123–130).

Sexual reproduction of flagellates, amoebae, and testacea is by syngamy, where two cells fuse. Meiosis may occur before or after fusion, depending on the species. Sexual reproduction of ciliates occurs by conjugation, where the micronucleus undergoes meiosis in two individuals; the two cells join at the region of the cytostome and exchange haploid "gametic" nuclei. The two cells separate and undergo binary fision to produce individuals that are genetically different from the pre-conjugant parents.

Some soil testacea and ciliates are known to form cysts that may be associated with reproductive processes (19, 34 157). Such cysts either are locations of ordinary binary fision (e.g., *Colpoda cucullus*) or are involved with sexual stages in the organisms' life cycles.

Stout and Heal (163) maintained that flagellates, small amoebae, and ciliates can divide rapidly, perhaps once or twice a day under favorable conditions, but that testacea divide more slowly. Support for this was furnished by Lousier, who showed that mean generation times for the testacean community in a deciduous soil in Canada were 6–10 days (105, 106) and 6–8 days (110–113, 116). Generation times for some individual populations (e.g., *Euglypha laevis, E. rotunda, E. scutigera, E. cuspidata, Trinema enchelys, Difflugiella oviformis, Phryganella acropodia, Hyalosphenia minuta, H. subflava, Heleopera petricola*) were found, however, to be as short as 0.9–1.0 day at various times during the year. Schönborn (146, 147, 149) also found generation times of 6–10 days for testacea in some East German soils.

Relationships to Humification and Mineralization

The reviews of Stout and Heal (163), Bamforth (9, 12), Stout (159–161), and Foissner (72) provide detailed assessments of the roles of soil protozoa, but show that despite the ubiquity and abundance of protozoa, little is known of their specific functions in the soil. Protozoan activities in the dynamic biological equilibrium of terrestrial systems have often been extrapolated from experimental evidence.

Nutrients taken from the soil by plants are returned in the litter as dead leaves, roots, stems, reproductive structures, and the feces and carcasses of animals. This detritus contains nitrogen, phosphorus, carbon, and a number of other chemical elements generally in a form unavailable to growing plants. Decomposition of this litter releases the chemical constituents in forms that plants can use, hence materials are recycled. Where decomposition ceases or slows and where debris tends to accumulate, soil fertility becomes depleted. Thus an efficient decomposition cycle is essential to maintain soil fertility and structure. Protozoa are thought to affect the cycling of soil organic matter indirectly through their continual predation on the bacteria, in particular, and other microflora. This grazing modifies bacterial populations, accelerates turnover of soil biomass and consequently of organic matter, and expedites the recycling of elements essential for plant nutrition (26, 37–39, 73, 159–161). Other ways in which protozoa affect the organic cycle include symbiosis (e.g., termite flagellates and oligochaete ciliates), parasitism on and in soil metazoans, and serving as food for other organisms in the food web (15, 50). Some free-living protozoa, particularly testacea, are thought to affect the soil organic cycle directly by ingesting plant residues or humic substances (96, 145, 148).

In estimating the importance of protozoa to the soil organic cycle, population turnover must be considered. In general, population studies have measured the standing crop of soil protozoa and have not attempted an estimation of reproductive rate, production, predation losses, or mortality (114, 141). Lousier and Parkinson (116) estimated that the mean annual standing crop of testacea in a Canadian deciduous woodland soil was 0.72 g/m^2 (wet weight), that total annual production was 206 g/m^2 (wet weight), and that the total weight of bacteria they consumed annually was 1377 g/m^2. The mass of bacteria consumed was about 60 times the annual mean standing crop of bacteria in the soil, and 25 times the estimated consumption by amoebae in a Colorado grassland soil (67). The annual secondary production of the testacea was 290 times the standing crop, more than double the estimate (100 : 1) for naked amoebae from agricultural soils (48, 82, 163), slightly less than that (300 : 1) for flagellates from the same soils, and 70–100 times the estimate derived for the above-mentioned Colorado soils (67). The estimated carbon loss through respiration (40 g/m^2 per year) amounted to 60% of the total annual carbon input through litterfall (116). This estimate seems disproportionately high and appears to indicate a significant recycling of organic matter within the microbial–microfaunal biomass.

The substantive production and rapid turnover of various protozoan groups suggest that despite having a lower biomass than other soil fauna, the protozoa exert a greater impact on the organic cycle (159–161, 163).

Ecology

Litter–Soil Environmental Conditions

Litters and soils contain irregular surfaces and interstices that hold small volumes of water in surface films and in pore spaces (43). Each film constitutes a microcosm (microhabitat) and is connected to other microcosms when moisture is abundant and is joined or separated by animal movements (e.g., burrowing) and microbial action (by flocculation and by disintegration of humus particles).

The abiotic factor that appears to affect litter–soil protozoa most directly is available moisture (19, 26, 67, 96, 105, 106, 167). Loss of water due to evaporation and evapotranspiration increases the relative concentration of salts, producing osmotic stress. Heavy or extended precipitation can cause dilution stress and may introduce anaerobic conditions by filling up air spaces between particles (waterlogging).

Temperatures can rise to 50°C in litters and the upper layers of soil; hence the most successful protozoa tolerate moisture and temperature fluctuations by developing resistant cyst membranes and by this ability to change quickly from active to latent metabolism. Lousier (110–113) found that all the testacean populations occurring in a deciduous woodland soil reproduced and remained active throughout the winter period, even when temperatures were below 0°C for several months and the soil and litter layers were frozen. Some species maintained higher densities than at other times in the year and showed seasonal peaks in abundance and biomass, and/or had higher than normal rates of production and turnover during this frozen period. Rogerson and Berger (142) recorded high total protozoan populations and rapid excystment rates on numerous brief occasions throughout the winter months and suggested that significant protozoan activity occurred at these times.

Terrestrial ecosystems are dominated by higher plants that produce leachates, and dying and dead organic matter. These organic products are supplemented by fecal material, dead animals, and microbial protoplasm, thus adding a greater amount of nutrients than in most aquatic systems. Decomposition of these detrital materials, together with root respiration in soils, can produce high concentrations of carbon dioxide, adding further stress to that which can be caused by the water and temperature conditions described above.

Three soil-determining factors—climate, vegetation, and parental material—are critical in developing soil habitats and thus influence the types and abundance of protozoa present. The first two influence the development of deserts, grasslands, peat lands, and forests; and the third distinguishes acid from basic soils, mull from mor humus types, and fens from bogs.

Composition and Distribution

Over 300 species of protozoa have been reported from litters and soils, but due to the unpredictable and changeable conditions in these habitats, most species must be considered transient rather than resident. The presence of high numbers of less edaphic species in a litter or soil, however, indicates that the substrate has a high moisture content and provides a more stable environment. The thin moisture films present a "benthic" surface of large area relative to its volume of water, thus favoring interface locomotion typical of rhizopods (naked amoebae and testacea).

The protozoan fauna of litters and soils is generally cosmopolitan (163), although species may occur with different frequencies under differing soil conditions and may often be restricted in distribution within a particular soil (150).

Testacea are generally better represented in forest litters and zonal peats (favoring a slow organic turnover, particularly characteristic of raw humus) than they are in grasslands or lowland peats. Naked amoebae, flagellates, and ciliates frequent habitats characterized by a high base status (e.g., grasslands) and a mellow humus with a high rate of mineralization (e.g., mulls).

Naked amoebae, including "limax" forms (e.g., *Naegleria, Vahlkampfia, Hartmannella*), *Acanthamoeba, Thecamoeba,* and *Mayorella*), are ubiquitous and can comprise 50–90% of the protozoa in soils and litters. They are flexible, enabling them to penetrate the tiny spaces unavailable to other soil protozoa. Several large amoebae (e.g., *Vampyrella, Biomyxa,* and *Gephyramoeba*) have pseudopodia that can penetrate small spaces. The slime molds (Acrasea and Eumycetazoa) are more typical of wet litters and masses of vegetation. Several genera of amoebae, such as *Naegleria* and *Tetramitus,* can transform into flagellates, becoming more motile when moisture is abundant; that is, the dilution of the soil water content may stimulate the amoebae to flagellate transformation.

Other naked protozoa are swimmers, and the best adapted to the litter/soil habitat are the ubiquitous zooflagellates, *Oicomonas termo, Pleuromonas jaculans,* and species of *Bodo, Cercobodo* and *Mastigamoeba,* common also to organically rich waters. They are most abundant in forest litters (13, 166), which are comparable to hay infusions, in which these flagellates are the first protozoa to appear.

The larger size of the soil ciliates limits the size of the pores they can enter (51). The genus *Colpoda* has efficient encystment mechanisms, as do the two preceding groups, and can comprise 50–95% of the litter–soil ciliates. Several of the following genera and species can also be expected to occur in a litter or soil collection: *Chaenia, Enchelys, Keronopsis, Pleurotricha, Oxytricha minor, Leptopharynx sphagnetorum, Cyrtolophosis elongata, C. mucicola, Vorticella microstoma, Cyclidium glaucoma, Chilodonella cucullus,* and *C. uncinata.* The last five, together with *Colpoda cucullus* and *C. steini,* possess tolerances to higher temperatures and usually to higher carbon dioxide levels than do most common limnetic species (17), thus emphasizing the rigorous demands on edaphic forms of life.

Other species are characteristic of more aquatic habitats; hence the diversity of ciliates reflects the degree of moisture availability and stability in the litter–soil habitat. Increased moisture may encourage the appearance of one or more predaceous forms, such as *Litonotus, Dileptus, Breslaua,* and *Sphaerophrya,* and these forms may be accompanied by aquatic nonciliates such as colorless euglenoid-flagellates and Heliozoa.

Testacea are soil interface organisms that can withdraw into a test to supplement encystment ability. As with the ciliates, testacea occur as a wide range of edaphic or aquatic forms. Bonnet (19, 22) has developed an elaborate classification of test forms and has described certain species as indicators of various types of soils. Four basic morphotypes emerge from his classification scheme: (a) flattened "*Arcella*"-like species are typical of aquatic vegetation and moist soils and include *Arcella, Microchlamys,* and *Diplochlamys;* (b) "acrostome" species have high-vaulted tests (such as *Euglypha, Assulina, Tracheleuglypha, Nebela, Heleopera,* and *Hyalosphenia*) and reside in aquatic vegetation, mosses, and forest litters; (c) wedge-shaped "plagiostome" species, such as *Centropyxis, Plagiopyxis, Bullinularia, Corythion,* and the ubiquitous *Trinema,* are prominent in litters and drier soils where the water film is thinner and more transitory; (d) globose (spherical) "axial" species, such as *Phryganella, Cyclopyxis, Trigonopyxis,* and *Distomatopyxis,* are truly edaphic forms that can withstand considerable dryness. These morphotypes contain both aquatic and edaphic species, but the relative proportions of the morphotypes in a testacean community characterize the relative amount of moisture in the litter or soil.

Relative Numbers and Biomass

Flagellates and amoebae are, under most soil conditions, the more numerous protozoa, ranging from 10^2 per gram in desert soils to 10^6 per gram in humid forest litters and soils (11, 163). Up to 50% of the amoebae and flagellates may be encysted at any given time (27, 154). Testacea have been recorded in the range of 10^2 per gram in desert and agricultural soils to 10^5 per gram in forest soil humus (116, 158, 163). The encysted proportion of a testacean population may exceed 50% at a given time (109–113). The same range of abundance has been found for ciliates (11), but the inactive portion of the ciliate population may exceed 70% (27). All protozoan groups appear to decrease in abundance with soil profile depth (97, 107, 116, 163, 168).

The smaller protozoan species are usually the more common (48, 96, 109–113). Many previous estimates of field biomass must be used cautiously, because most protozoa have been measured in cultures and are usually larger there than in the natural soil environment. Very few estimates of field biomass are available for all groups of protozoa, but Stout and Heal (163) suggested a relatively small biomass (<5 g/m^2) in woodland or cold grassland soils, where organic matter is accumulating and larger species of rhizopods are more frequent. Lousier and Parkinson (116) reported a biomass of 0.7 g/m^2 for 2.6×10^8 testaceans per square meter in a Canadian deciduous woodland soil.

Methodology

Methods for the study of soil protozoa have been well reviewed and evaluated by Heal (83, 84) and Stout et al. (162). Basically, there are two useful techniques available: direct observation and culturing.

Culturing has been the most versatile and widely used method; it is essential for qualitative and detailed taxonomic studies and for the initiation of pure culture studies. Cultures are selective and may be chosen and adapted to reveal either a wide range of species or a particular group. The choice of culture medium is dependent on the objectives of the study, but liquid over an agar (plain or enriched) surface in a petri dish and liquid soil extract appear to be the most suitable media (81). In such cultures, small flagellates and *Colpoda* normally appear first, followed in a few days by small naked amoebae, and then several days later by other ciliates and testacea. The dilution-culture technique (47, 151, 153, 158, 162) has been used extensively for estimating protozoan abundance.

Techniques for direct observation have proven to be successful mostly for soil testacea because their tests can withstand the rigors involved in sampling and extraction. The soil sample may be examined microscopically in an undisturbed condition (93), in prepared sections (28, 78, 81), in a soil agar suspension (81), or concentrated on a Millipore filter (41, 115). These methods, particularly the Coûteaux Millipore filter technique, allow counting of testacea, and with the use of stains, allow a distinction between live cells, empty tests, and encysted forms, important information when studying population dynamics and activity (115).

All these methods are subject to error, and all must be adapted and tested in relation to the populations being studied, and soil being sampled, and the intent of the study (115, 162). It is also essential to recognize distinctions between the biological and statistical significance of the data produced by a particular method.

Startling and realistic observations of protozoa have been made by electron microscopy: transmission (TEM) for the internal structure of protozoan cells, and scanning (SEM) for the external morphology of the cells and tests. Continued improvement in methods of electron microscopy will produce tremendous advances in the knowledge of every aspect of soil protozoology.

Classification

Characteristics of Soil Protozoa

Protozoan taxonomy is continually changing as new information alters phylogenetic concepts. The traditional division of the "phylum Protozoa" into four classes based on locomotion is no longer tenable, and the new system of Levine et al. (104) recognized that many morphologically similar groups are polyphyletic. The new system lists seven phyla, two of which, the Sarcomastigophora and Ciliophora, contain the free-living soil protozoa.

The Sarcomastigophora include protozoa that move by flagella or by protoplasmic flow or by both. On the basis of the predominant type of locomotion, the phylum is subdivided into the Mastigophora (or flagellates) and the Sarcodina (or amoeboid protozoa).

The Mastigophora possess one or several body-length (or longer) whiplike appendages—flagella—arising from the anterior portion of the cell. Most soil flagellates have one or two flagella; if two, the second usually trails posteriorly (Fig. 6.1*a*). Each flagellum has a cablelike ultrastructure of microtubules arising from a basal body just below the cell membrane. The Mastigophora contain two groups, the Phytomastigophorea, typically chlorophyll bearing but also including colorless forms that are clearly related to certain pigment-bearing forms; and the Zoomastigophorea, a polyphyletic, colorless group defined by lack of similarity to the first group.

Few chlorophyll-bearing protozoa occur in soils. (Most chlorophyll-bearing protists found in soils belong to the ubiquitous *Chlorococcum* and other Chlorophyceae, diatoms, and blue-green algae.) Colorless representatives of the Phytomastigophorea in soils include the ubiquitous *Oicomonas termo* and several euglenoids—*Scytomonas, Anisonema,* and *Peranema*—in moist soils. Most soil flagellates belong to the Zoomastigophorea and are typically characterized by two flagella, one of which is trailing (e.g., as in *Bodo, Pleuromonas*), or by a single flagaellum and pseudopodia (e.g., *Mastigamoeba* and *Mastigella*).

Members of the subphylum Sarcodina move by means of temporary extensions of protoplasm, called pseudopodia, or the whole body flows in one direction. Because the naked sarcodines have no rigid external surface, their body shapes change continually as the cytoplasm streams or as pseudopodia are extended and withdrawn.

There are two superclasses of Sarcodina: Rhizopoda, which move by nonaxopodal pseudopodia or by protoplasmic flow without the production of pseudopodia; and Actinopoda, which possess spherical bodies and axopodia, the latter being fine pseudopodia supported by internal cores of microtubules (Fig. 6.2*a*). The latter group is represented occasionally in moist soils by certain Heliozoa (Fig. 6.8*e,f*).

Six of the eight classes of rhizopods are found in soils. Two classes contain test-bearing forms as well as "naked amoebae." Formerly, testate forms were considered a distinct taxonomic group (see, e.g., Ref. 90), but studies of pseudopodial structure and function have shown the testacea to be a polyphyletic assemblage. However, testacea constitute an ecologically distinct group (evident from the preceding ecological discussions), hence they are separated in this key.

Among the naked amoebae, protoplasmic streaming manifests itself in a variety of ways (Fig. 6.2*b–g*). The entire body may flow in one direction as a cylinder ("limax" locomotion) or expand in a fanlike manner. Discrete pseudopodia may be lobose or fingerlike, conical, filose, and elongate, and the latter may anastomose to form a reticulum. Some amoeboid organisms consist of a large mass that may be multicellular or multinucleate—a plasmodium (Fig. 6.2*g*). Most amoeboid organisms can be recognized by pseudopod type and manner of locomotion, but many species can display more than one type of locomotion.

Slime molds are naked rhizopods characterized by an amoeboid feeding stage which later differentiates into a reproductive stage where the organism forms a fruiting body (usually on a stalk) containing spores. This mode of reproduction resembles that of fungi, from which they are distinguished by their amoeboid trophic stage and the lack of filamentous connections to the sporangium. The large amoeboid stage of many myxogastric slime molds is a plasmodium often visible to the naked eye, but in other true slime molds the amoeboid stage is small and difficult to observe. The amoeboid stage of the cellular slime molds consists of tiny uninucleate amoebae

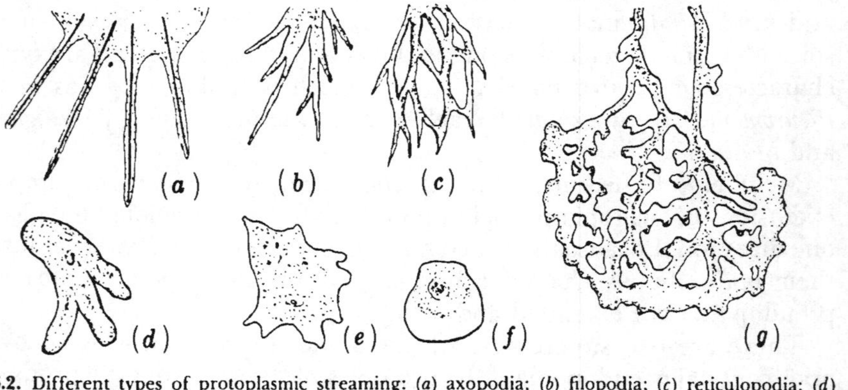

Fig. 6.2. Different types of protoplasmic streaming: (*a*) axopodia; (*b*) filopodia; (*c*) reticulopodia; (*d*) amoeba with lobopodia; (*e*) amoebae with conical pseudopodia; (*f*) amoeba with flattened sheetlike body; (*g*) plasmodium.

which later coalesce to form a multicellular "slug" that moves monopodially and then differentiates into a stalk and sporangium. Slime molds are best identified by their fruiting bodies.

The testacea secrete a test which, along with the type of pseudopodia, provides the basis for identification (Fig. 6.1*c*). There are two basic types of tests: those entirely composed and secreted by the cell itself, and those constructed of extraneous sedimental materials in a secreted binder. The first type may be more or less flexible or rigid. The rigid tests may have scales or platelets (generally siliceous in composition) of only one type or of several types (pseudostomal, body, and spine scales). The second type of test involves the specific selection by the cell of materials (grains of sand or quartz of similar or differing sizes, diatom shells, remnants of other tests) from its surroundings. Spines, horns, and other projections found on some species are generally variable in form, number, and location, but usually occur within a range that is characteristic of the species. The form and border of the pseudostome are important taxonomic characteristics and are usually constant for each species. Generally, within a soil or site, test size and shape for a particular species are fairly uniform, but it has been known for many years that tests are highly susceptible to habitat-induced morphological variation. This has often resulted in the description in the literature of varieties, subspecies, and in some cases, species that cannot readily be distinguished and identified by other workers.

The Ciliophora, or ciliates, are protozoa that move by movement of many slender hairlike processes, cilia, arising from basal bodies (kinetosomes) just below the cell membrane. The ultrastructure of cilia is similar to that of flagella, but cilia are smaller and more numerous. In the Suctoria, the cell is ciliated only in the larval stage. The adult is often spheroid and extends tentacles out to capture prey. The suctorians found in soils resemble the Heliozoa, but may be distinguished by the tiny knobs on the distal ends of their tentacles.

Cilia may be distributed in rows, with some rows shortened and fused, to form membranelles, or undulating membranes, or into tufts called *cirri* (Fig. 6.3*a–d*). The latter function as walking appendages in hypotrichous ciliates, enabling the organism to dart back and forth along surfaces.

Ciliates ingest food at only one place in the cell surface, the cytostome (cell mouth), which may occur on the cell surface; in the bottom of a small cavity, the vestibulum; or at the base of a large buccal cavity, lined by membranelles (Fig. 6.3*f–g*).

ACKNOWLEDGMENTS

The authors are grateful for the following contributions: the testacean drawings by Ms. Susan Kerr; the typing of the manuscript by Ms. Helen Mequet, Ms. Elizabeth Steele, Ms. Jean Stringer, and Mrs. Louise Gronmyr; the support of MacMillan Bloedel Limited and the British Columbia Ministry of Forests and Lands; and the review by Dr. Eugene Bovee.

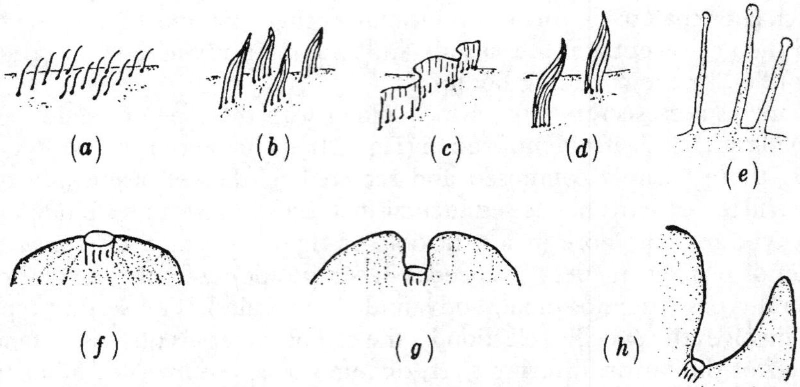

Fig. 6.3. Different kinds of ciliary structures (*a–d*), tentacles (*e*), and position of the cytostome (*f–h*) in ciliates: (*a*) cilia; (*b*) membranelles; (*c*) undulating membrane; (*d*) cirri; (*e*) tentacles of a suctorian; (*f*) cytostome on cell surface; (*g*) cytostome in vestibulum; (*h*) cytostome in buccal cavity.

KEY TO COMMON NORTH AMERICAN SOIL PROTOZOA

The following key includes genera reported as common in various litters and soils in North America. For genera with commonly occurring species, the species are usually illustrated, though not described in the key. The following taxonomic references were used: flagellates (29, 96, 101, 103, 137, 143); naked amoebae (18, 91, 101, 134, 136); testacea (20, 22–25, 31, 33–35, 54–65, 75–77, 164); ciliates (70, 71, 92, 94, 101, 132, 133, 143).

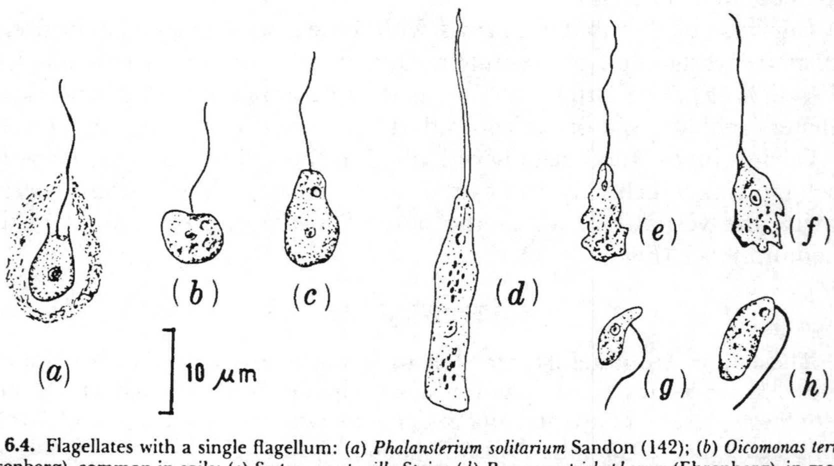

Fig. 6.4. Flagellates with a single flagellum: (*a*) *Phalansterium solitarium* Sandon (142); (*b*) *Oicomonas termo* (Ehrenberg), common in soils; (*c*) *Scytomonas pusilla* Stein; (*d*) *Peranema trichophorum* (Ehrenberg), in moist soils; (*e*) *Mastigamoeba reptans* Stokes (29); (*f*) *Mastigella simplex* Kent (29); (*g*) *Sainouron mikroteron* Sandon (142); (*h*) *Allantion tachyploon* Sandon (142).

KEY TO MAJOR GROUPS OF SOIL PROTOZOA

1a		Motile unicellular colorless protists	2
1b		Sporangia not attached to the end of filaments or a filamentous network; fruiting bodies of slime molds	**4a**
2a	(1)	Movement by flagella or cilia	3
2b		Movement by protoplasmic flow	4
3a	(2)	Naked organisms	**Flagellates**
3b		Movement by cilia	**Ciliates**
4a	(2)	Naked organisms	**Naked amoebae**
4b		Test (shell) present	**Testacea**

KEY TO GENERA AND COMMON SPECIES OF FLAGELLATES

1a		One flagellum	2
1b		More than one flagellum	9
2a	(1)	Flagellum directed anteriorly	3
2b		Flagellum directed posteriorly	8
3a	(2)	Cell ovoid, 10–15 μm, usually in a gelatinous sheath (Fig. 6.4*a*)	***Phalansterium solitarium*** Sandon
3b		Cell not surrounded by sheath	4
4a	(3)	Ovoid, 5–20 μm; often swims in large slow circles (Fig. 6.4*b*)	***Oicomonas termo*** (Ehrenberg)
4b		Body more elongate; creeping movement	5
5a	(4)	No pseudopodia formed; flagellum is body length	6
5b		Pseudopodia formed; flagellum two to four times body length	7
6a	(5)	Ovate, 10–20 μm; organism moves straight ahead, then angles in another direction (***Copromonas*** of Dobell) (Fig. 6.4*c*)	***Scytomonas pusilla*** Stein
6b		Elongate, rounded to truncate, body 20–70 μm; only the anterior end of the thick flagellum vibrates; slow, gliding movement (Fig. 6.4*d*)	***Peranema trichophorum*** (Ehrenberg)
7a	(5)	Flagellum connects through fibril to nucleus (May be difficult to see); several species of variable forms, usually 15–25 μm (Fig. 6.4*e*)	***Mastigamoeba*** Schulze
7b		Flagellum not connected to nucleus; several species, 12–40 μm (Fig. 6.4*f*)	***Mastigella*** Frenzel
8a	(2)	Small, about 5 μm, crescent shape; locomotion sinuous but not spiral (Fig. 6.4*g*)	***Sainouron mikroteron*** Sandon
8b		Ovoid, 8–14 μm; with thick flagellum; has rigid gliding motion (Fig. 6.4*h*)	***Allantion tachyploon*** Sandon
9a	(1)	Two flagella	10
9b		Four flagella; pyriform cell with pointed anterior end; 10–30 μm (Fig. 6.5*a*); flagellate phase of	***Tetramitus*** Perty
10a	(9)	Both flagella anteriorly directed	11
10b		One flagellum trailing	13
11a	(10)	Flagella of equal length; flagellate phase of limax amoebae	12
11b		Second flagellum very short, barely visible; oval body, 5–20 μm, often confused with ***Oicomonas*** (Fig. 6.5*b*)	***Monas*** (Ehrenberg) Stein

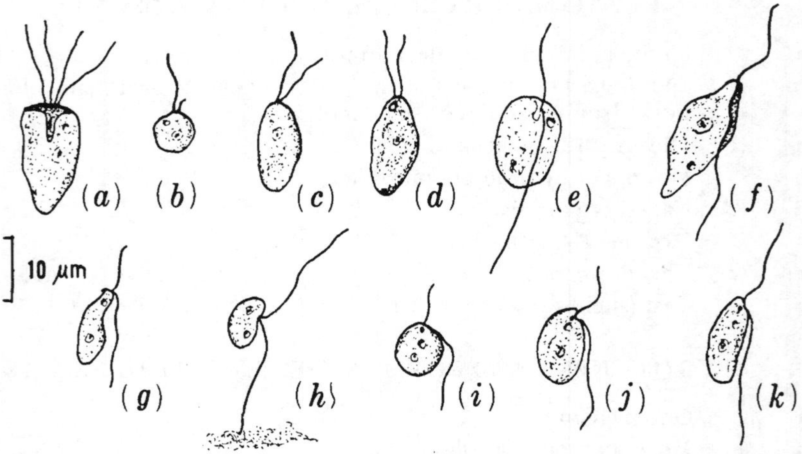

Fig. 6.5. Flagellates with two or four flagella and flagellate stages of limax amoebae: (*a*) *Tetramitus rostratus* (Perty), flagellate phase (136); (*b*) *Monas minima* H. Meyer; (*c*) *Naegleria gruberi* (Schardinger), flagellate phase; (*d*) *Adelphamoeba galeacystis* Napolitano, Wall and Ganz, flagellate phase (136); (*e*) *Anisonema ovale* Klebs; (*f*) *Cercobodo crassicauda* (Alexieff) (29); (*g*) *Spiromonas angusta* (Dujardin) (29); (*h*) *Pleuromonas jaculans* Perty, common in soils; (*i*) *Bodo globosus* Klebs, common in soils; (*j*) *B. edax* Klebs; (*k*) *B. mutabilis* Klebs.

12a	**(11)**	No cytostome (mouthlike) opening near flagella; body 10–20 μm (Fig. 6.5*c*); flagellate phase of . . . . ***Naegleria*** Alexieff
12b		Small anterior cytostomelike opening near flagella; body 9–19 μm (Fig. 6.5*d*); flagellate phase of . . . . ***Adelphamoeba*** Napolitano, Wall and Ganz
13a	**(10)**	Cell oval, flat, rigid, 10–15 μm; trailing flagellum passes across body in a thin groove (Fig. 6.5*e*) . . . . ***Anisonema ovale*** Klebs
13b		Trailing flagellum attached to side of body or free **14**
14a	**(13)**	Posterior flagellum usually attached to side of body; body 12–16 μm; genus includes ***Cercomonas*** of older authors; several species (Fig. 6.5*f*) ***Cercobodo*** Krassilotchick
14b		Flagellum not attached to body **15**
15a	**(14)**	Body narrows anteriorly; cell 10 μm; swims actively with anterior flagellum; common (Fig. 6.5*g*) . . . . ***Spiromonas angusta*** (Dujardin) Alexieff
15b		Body spherical, oval, or bean-shaped. **16**
16a	**(15)**	Flagella emerge laterally; organism vibrates violently, usually attached to substrate by trailing flagellum; bean-shaped cell, 6–10 μm (Fig. 6.5*h*) . . . . ***Pleuromonas jaculans*** Perty
16b		Flagella emerge at anterior end; forward locomotion; includes ***Heteromita;*** many species, 3 common in soil . . . ***Bodo*** Ehrenberg **17**
17a	**(16)**	Cell spherical to broadly pyriform, 7–13 μm; swims with steady rocking motion (Fig. 6.5*i*). . . . . ***Bodo globosus*** Klebs
17b		Cell more oval **18**

18a (17) Cell oval, 11–15 μm, with anterior pointed end; trailing flagellum twice body length (Fig. 6.5*j*) . . . . ***Bodo edax*** Klebs

18b Elongate, 8–14 μm; jerky swimming motion (Fig. 6.5*k*). . . . . ***Bodo mutabilis*** Klebs

KEY TO COMMON GENERA OF NAKED AMOEBAE AND HIGHER TAXA OF SLIME MOLDS

1a Amoeboid, changing body, or spherical with fine radiating pseudopods 2

1b Spore mass (sporangium) not attached to a filamentous mycelium . . 27

2a (1) Locomotion by protoplasmic flow with or without production of discrete pseudopodia . . . . **Rhizopoda, 3**

2b Spherical body with fine radiating pseudopods . . . . **Actinopoda, Heliozoa, 34**

3a (2) Protoplasmic flow in one direction 4

3b Plasmodium usually conspicuous, with back and forth (shuttle) streaming . . . . **Plasmodial stage of Myxogastria, 26**

4a (3) Amoebae behave independently of each other and remain in amoeboid stage. 5

4b Amoebae eventually change to form sporangium, usually by aggregating together . . . . **Amoeboid stage of slime molds, 25**

5a (4) Very fine pseudopodia (filopodia) 24

5b Lobose or conical pseudopodia 6

6a (5) Body not branched 7

6b Body a reticulum of coarse branches; advancing tips are lobose 22

7a (6) Body a flowing cylinder which may be monopodal or produce several pseudopods 8

7b Body flat, discoid or globose, often with small pseudopodia 15

8a (7) Polypodal, or monopodal with posterior bulb 9

8b Small, monopodal . . . . **"Limax" amoebae, 10**

9a (8) Monopodal, with posterior villous-like uroid (bulb); 80–150 μm (Fig. 6.6*a*) . . . . ***Trichamoeba*** Fromentel

9b One pseudopod dominant during locomotion, with several shorter pseudopodia developing from a stem of flow; 15–200 μm (Fig. 6.6*b*) ***Amoeba*** Ehrenberg

10a (8) Locomotion by steady flow 11

10b Locomotion by succession of eruptive waves; amoeboid stages appear similar to all four genera, although depicted separately; genus identification of the three genera possessing flagellate stage depends on inducing stage by dilution technique (see Ref. 136) 12

11a (10) Hyaline cap (clear area) at advancing anterior end; body length 12–35 μm (Fig. 6.6*c*) . . . . ***Hartmannella*** Alexieff

11b Hyaline cap usually lacking; posterior villus bulb (uroid) present; length 30–70 μm (Fig. 6.6*d*) . . . . ***Saccamoeba*** Frenzel

12a (10) No flagellate stage; length 10–40 μm (Fig. 6.6*e*) . . . . ***Vahlkampfia*** Chatton and Lalung-Bonnaire

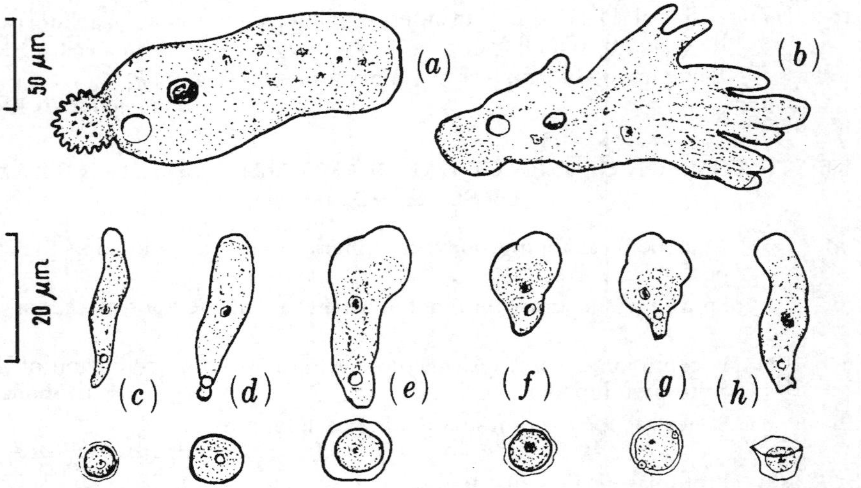

Fig. 6.6. Lobose amoebae and limax amoebae and their cysts: (*a*) *Trichamoeba villosa* (Wallich); (*b*) *Amoeba proteus* (Leidy); (*c*) *Hartmannella vermiformis* Page; (*d*) *Saccamoeba stagnicola* Page; (*e*) *Vahlkampfia russeli* (Singh); (*f*) *Tetramitus rostratus* (Perty) (136); (*g*) *Naegleria gruberi* (Schardinger); (*h*) *Adelphamoeba galeacystis* Napolitano, Wall and Ganz (136).

12b Flagellate stage can be induced by diluting culture medium 13
13a (12) Two anterior flagella; body ovoid 14
13b Four flagella; pyriform body with pointed posterior end; 10–30 µm (Figs. 6.5*a*, 6.6*f*) . . . . ***Tetramitus*** Perty
14a (13) Lozenge-shaped; no cytostome (mouthlike) opening near flagella; 10–20 µm (Figs. 6.5*c*, 6.6*g*) . . . . ***Naegleria*** Alexieff

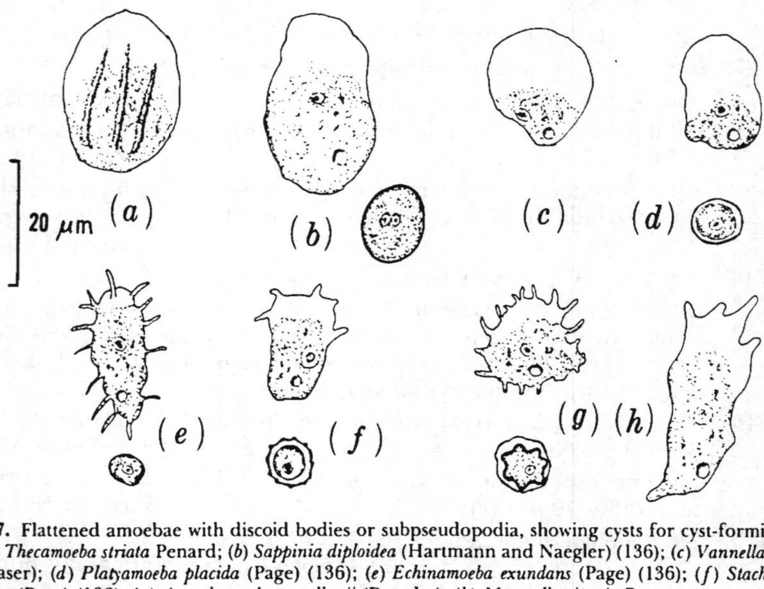

Fig. 6.7. Flattened amoebae with discoid bodies or subpseudopodia, showing cysts for cyst-forming genera: (*a*) *Thecamoeba striata* Penard; (*b*) *Sappinia diploidea* (Hartmann and Naegler) (136); (*c*) *Vannella platypodia* (Glaser); (*d*) *Platyamoeba placida* (Page) (136); (*e*) *Echinamoeba exundans* (Page) (136); (*f*) *Stachyamoeba lipophora* (Page) (136); (*g*) *Acanthamoeba castellanii* (Douglas); (*h*) *Mayorella riparia* Page.

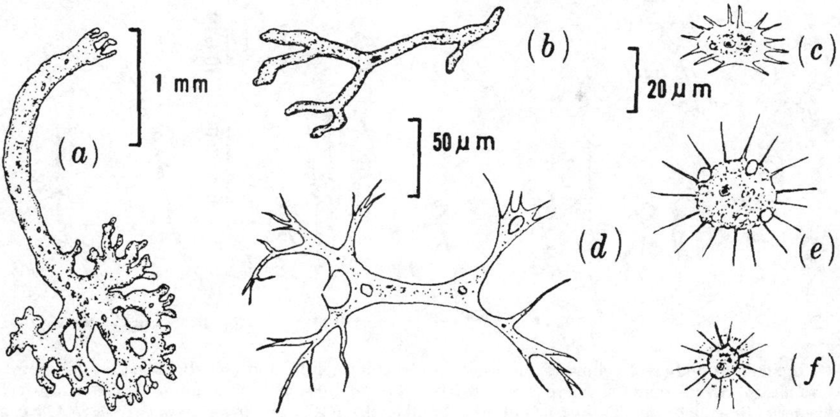

Fig. 6.8. Filose and branched amoebae and Heliozoa: (*a*) *Leptomyxa reticulata* Goodey (136); (*b*) *Gephyramoeba delicatula* Goodey; (*c*) *Nuclearia simplex* Cienkowski; (*d*) *Biomyxa vagans* Leidy; (*e*) *Actinophrys sol* Ehrenberg; (*f*) *Heterophrys glabrescens* Penard.

14b Oval to pyriform; small cytostomelike opening near flagella; 9–19 μm (Figs. 6.5*d*, 6.6*h*) . . . . . ***Adelphamoeba*** Napolitano, Wall and Ganz

15a (7) Flattened; no pseudopodia produced during locomotion **16**

15b Small pseudopodia produced during locomotion. **19**

16a (15) Ovoid oblong body; pellicle with stiff appearance; clear anterior (hyaline) area extends backward along sides; circular granular mass not thick **17**

16b Body broader; wide thin hyaline region anterior to thicker granular mass; floating forms have pointed pseudopodia **19**

17a (16) Longitudinal ridges in pellicle; 30–200 μm (Fig. 6.7*a*) . . . . ***Thecamoeba*** Fromentel

17b No (or faint) ridges; 40–85 μm (Fig. 6.7*b*) . . . . ***Sappinia*** Dangeard

18a (16) Fanlike body, usually tapering posteriorly; floating forms with long arms; 10–80 μm (Fig. 6.7*c*) . . . . ***Vannella*** Bovee

18b Oval or tonguelike shape; length equals or exceeds width; floating forms with short arms; 15–35 μm (Fig. 6.7*d*) . . . ***Platyamoeba*** Page

19a (16) Short spinelike pseudopods from distinct hyaline zone **20**

19b Conical or tapering pseudopods **21**

20a (19) Triangular body that may be elongated; no limaxlike locomotion; 7–21 μm (Fig. 6.7*e*) . . . . ***Echinamoeba*** Page

20b Thickly flattened irregular body; may sometimes display eruptive limax-type locomotion; 15–33 μm (Fig. 6.7*f*) . . . ***Stachyamoeba*** Page

21a (19) Short pseudopods taper to fine tip; 10–30 μm (Fig. 6.7*g*). . . . . . . ***Acanthamoeba*** Volkonsky

21b Pseudopodia conical or rodlike; 12–200 μm (Fig. 6.7*h*) . . . . ***Mayorella*** Schaeffer

22a (6) Protoplasm flows in one direction **23**

22b Protoplasmic streaming reverses in alternations **30**

23a (22) Radially branching with many anastomoses; up to 3 mm (Fig. 6.8*a*) ***Leptomyxa*** Goodey

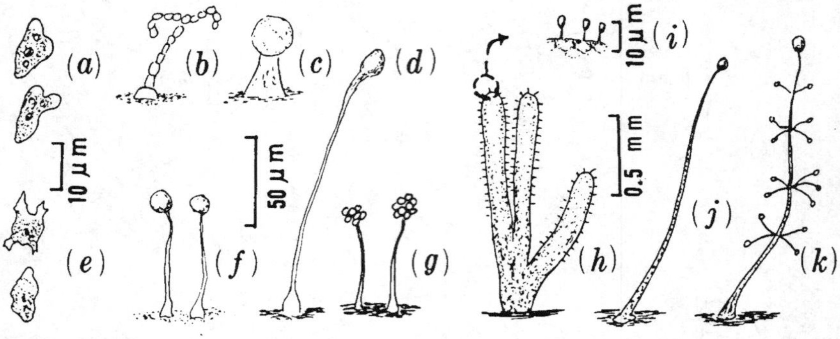

Fig. 6.9. Fruiting bodies of cellular slime molds and Echinostelida, and amoeboid stages of *Acrasis* and *Protostelium:* (*a*) *Acrasis rosea* Olive and Stoianovitch, amoebae; (*b*) same, sporangium; (*c*) *Guttinalopsis* (18); (*d*) *Pocheina* (Loeblich and Tappan) Guttalina Cienkowsky) (18); (*e*) *Protostelium mycophaga* Olive and Stoianovitch, amoebae; (*f*) same, sporangia; (*g*) *Echinostelium minutum* DeBary (134); (*h*) *Ceratiomyxa fructiculosa* (Mueller) (134); (*i*) same, enlargement of surface showing several sporangia (134); (*j*) *Dictyostelium mucoroides* Brefeld (18); (*k*) *Polysphondylium violaceum* Brefeld (18).

23b Body divided into slender armlike lobes; 60–250 µm (Fig. 6.8*b*) . . . ***Gephyramoeba*** Goodey

24a (5) Body oval, but spindle-shaped or triangular in locomotion; very fine hyaline pseudopodia arise from all parts of body; 10–30 µm (Fig. 6.8*c*) ***Nuclearia*** Cienkowski

24b Variable body continually extending filopodia in any direction; filopodia often branch and anastomose; up to 100 µm (Fig. 6.8*d*) . . . . ***Biomyxa*** Leidy

25a (4) Limax-type amoebae, 10 µm, with eruptive locomotion (Fig. 6.9*a–d*) **Acrasea**

25b Filose amoebae **26**

26a (25) Aggregating amoebae form plasmodium with back-and-forth (shuttle) streaming prior to sporangium formation . . . . . . **Myxogastria, 30**

26b Plasmodium without shuttle streaming; elongate ("slug") moving in one direction . . . . **Cellular slime molds, 28**

27a (1) One cell, several cells in ovoid group, or several chains of spores at end of stalk . . . . **.28**

27b Mass of spores in a mesh (capitillum) surrounded by membrane (epitheca); sporangium may be at end of stalk or lie on substrate as an oval or irregular mass . . . . **Myxogastria, 30**

28a (27) A few spores in spheroid group or chains of spores radiating from cellular stalk (Fig. 6.9*a–d*) . . . . **Acrasea**

28b One to many spores in spherical mass at end of noncellular stalk. . . **Cellular slime molds, 29**

29a (28) Single or 1–4 spores at end of short stalks (Fig. 6.9*e,f,h,i*) . . . . . . **Protostelida**

29b Spherical to ovoid cluster of spores at end of stalk, 0.5 mm or longer (Fig. 6.9*j,k*) . . . . **Dictyostelida**

30a (27) Spores light color to white **31**

30b Spores dark purple to black **33**

31a (30) Sporangium minute, nonreticulate (Fig. 6.9*g*) . . . . . **Echinostelida**

31b Larger sporangia 32
32a (31) True capitallum lacking (Fig. 6.10*a,b*). . . . . **Liceida**
32b Threadlike capitillum present (Fig. 6.10*c–e*) . . . . **Trichiida**
33a (31) Sporangium without lime (Fig. 6.10*f,g*) . . . . **Stemonitida**
33b Sporangium with lime (Fig. 6.10*h,i*) . . . . **Physarida**
34a (2) Without mucilaginous envelope; protoplasm highly vacuolated, especially periphery (ectoplasm); 30–50 μm (Fig. 6.8*e*) . . . . ***Actinophrys*** Ehrenberg
34b Mucilaginous envelope around body, with tiny spicules radiating at edge of envelope between pseudopodia; 10–15 μm (Fig. 6.8*f*) . . . . ***Heterophyrys*** Archer

KEY TO COMMON GENERA OF SOIL TESTACEA

1a Pseudopods lobose, fingerlike, rarely anastomosing . . . . Class **Lobosea** 2
1b Pseudopods filoform, pointed, branched and sometimes anastomosing Class **Filosea** 16
2a (1) Pseudopods fingerlike, not anastomosing. . . Superorder **Eulobosa** 3
2b Pseudopods often pointed, anastomosing . . . . Superorder **Reticulobosa** 15
3a (2) Test membranous 4
3b Test with mineral or organic particles or plates, sometimes very minute. 7
4a (3) Test semirigid, flexible near the pseudostome. . . . . Family **Microcoryciidae** 5
4b Test rigid, distinctly and densely punctuated, often yellow or brown (Fig. 6.11*a–d*) . . . . Family **Arcellidae, *Arcella*** Ehrenberg
5a (4) Pseudostome with membranous frame (Fig. 6.11*e*) . . . . ***Microcorycia*** Cockerell

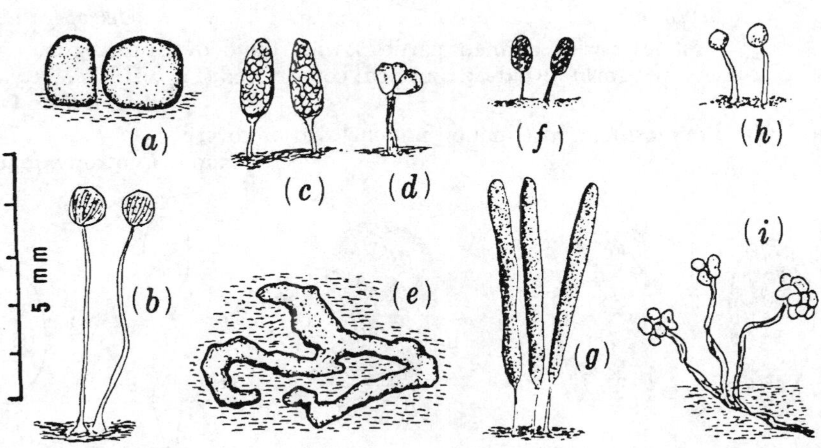

Fig. 6.10. Fruiting bodies of myxogastric slime molds (134). (*a*) *Lycogala epidendron* (Linneaus); (*b*) *Dictydium cancellatum* (Batsch) MacBride; (*c*) *Arcyria cinerea* (Bull.) Persoon; (*d*) *Metatrichia vesparium* (Batsch) Nennenga-Bremekamp; (*e*) *Hemitrichia serpula* (Scop.) Rostafinski; (*f*) *Diachea leucopodia* (Bull.) Rostafinski; (*g*) *Stemonitis* sp.; (*h*) *Didymium iridis* (Ditmar); (*i*) *Physarum polycephalum* Schweikoff.

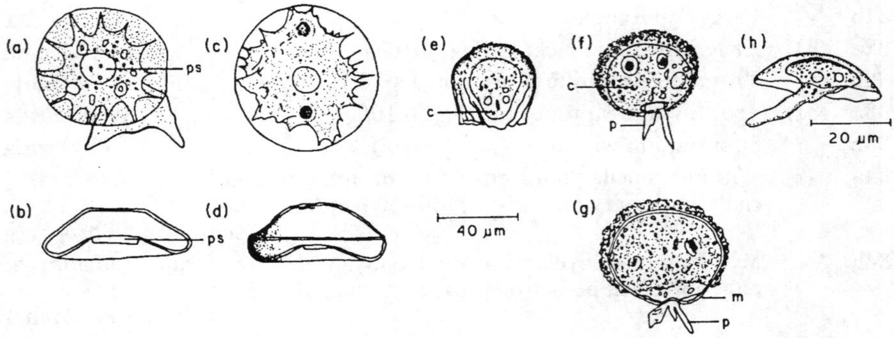

Fig. 6.11. Testacea: family Arcellidae. (*a*) *Arcella catinus* Penard, ventral view (61); (*b*) same, lateral view (61); (*c*) *Arcella arenaria* Greeff, ventral view (61); (*d*) same, lateral view (61); (*e*) *Microcorycia flava* (Greeff) Cockerell, lateral view (75); (*f*) *Diplochlamys fragilis* Penard, lateral view (75); (*g*) *Diplochlamys timida* Penard, lateral view (80); (*h*) *Microchlamys patella* Claparede and Lachman (75). *c*, Cell; *m*, test membrane; *ps*, pseudostome.

5b Pseudostome without membranous frame **6**

6a (5) Test membrane double; hemispherical to cup-shaped; loosely coated with small particles (Fig. 6.11*f*,*g*) . . . . ***Diplochlamys*** Greeff

6b Test membrane simple; flattened in side view (Fig. 6.11*h*) . . . . . . ***Microchlamys*** Cockerell

7a (3) Test with platelets or scales, rounded or angular, sometimes with sediment particles as well . . . . Family **Nebelidae 8**

7b Test with sediment particles, without platelets secreted by the cytoplasm **10**

8a (7) Scales or platelets indistinct, membrane appears structureless (Fig. 6.12*d–f*) . . . . ***Hyalosphenia*** Stein

8b Scales or platelets very apparent **9**

9a (8) Test with a small amount of sediment material at fundus of test (Fig. 6.12*a–c*) . . . . ***Heleopera*** Leidy

9b Test never with sediment particles; with round, oval, or irregular platelets; pyriform, ovoid or rounded; compressed (Fig. 6.13) . . . . . . ***Nebela*** Leidy

10a (7) Pseudostome invagination not enclosed or covered . . . . Family **Centropyxidae 11**

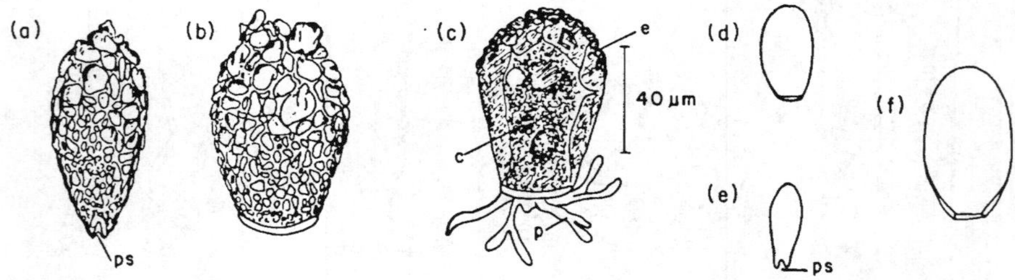

Fig. 6.12. Testacea: family Nebelidae. (*a*) *Heleopera petricola* Leidy, profile showing shape of pseudostome (64, 65); (*b*) same, lateral view (64, 65); (*c*) *Heleopera rosea* Penard, lateral view (75); (*d*) *Hyalosphenia minuta* Cash, lateral view (77); (*e*) same, profile showing shape of pseudostome (77); (*f*) *Hyalosphenia subflava* Cash, lateral view (77). *e*, Epipod; *ps*, pseudostome.

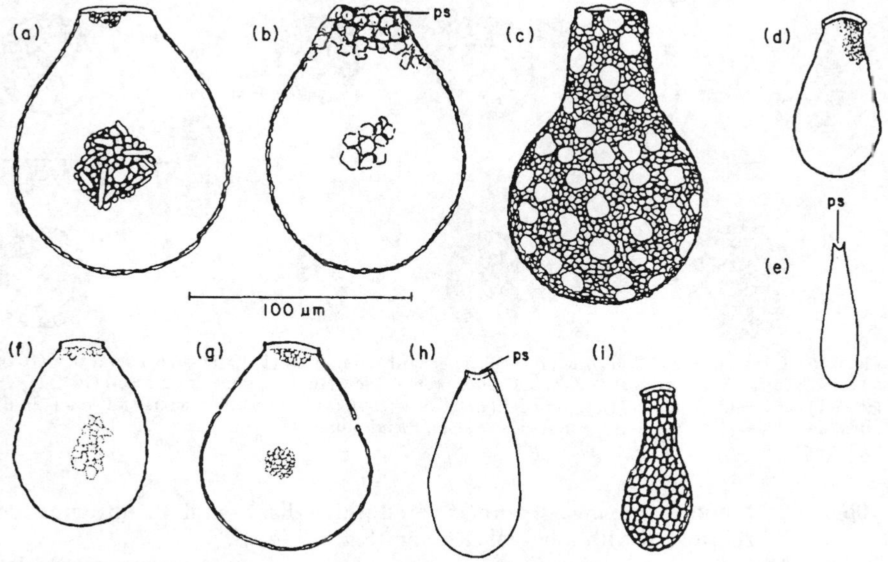

Fig. 6.13. Testacea: genus *Nebela* (63). (*a*) *N. bohemica* Taranek, lateral view; (*b*) *N. dentistoma* Penard, lateral view; (*c*) *N. lageniformis* Penard, lateral view; (*d*) *N. militaris* Penard, lateral view; (*e*) same, profile; (*f*) *N. parvula* Cash, lateral view; (*g*) *N. tincta* Leidy (Awarinzew), lateral view; (*h*) same, profile; (*i*) *N. tabulata* Brown, lateral view. *ps*, pseudostome.

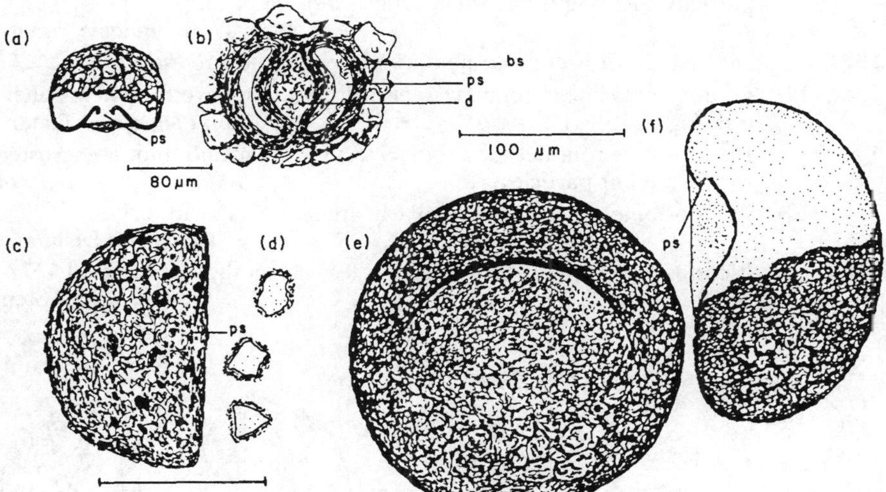

Fig. 6.14. Testacea; family Distomatopyxidae (*a,b*) and Centropyxidae (*c–f*). (*a*) *Distomatopyxis couillardi* Bonnet, lateral view (20); (*b*) same, oral view (20); (*c*) *Trigonopyxis arcula* (Leidy) Penard, lateral view (25); (*d*) same, various pseudostomal shapes (25); (*e*) *Bullinularia indica* Penard, ventral view (25); (*f*) same, lateral view (25). *bs*, Buccal skeleton; *d*, diaphragm; *ps*, pseudostome.

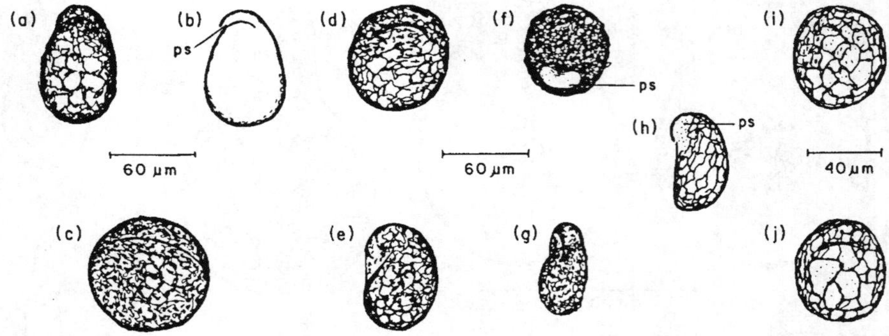

Fig. 6.15. Testacea: genus *Plagiopyxis*. (*a*) *P. callida* Penard, lateral view (164); (*b*) same, lateral view (164); (*c*) same, ventral view (164); (*d*) *P. declivis* Thomas, ventral view (164); (*e*) same, lateral view (164); (*f*) *P. labiata* Penard, ventral view (164); (*g*) same, ¾ lateral view (164); (*h*) *P. minuta* Bonnet, lateral view (19); (*i*) same, dorsal view (19); (*j*) same, ventral view (19). *ps*, Pseudostome.

10b Pseudostome invagination covered with a diaphragm; the invagination reinforced with a buccal skeleton (Fig. 6.14*a,b*) . . . . Family **Distomatopyxidae** ***Distomatopyxis*** Bonnet

11a (10) Pseudostome triangular or irregularly trilobed or quadrangular; test hemispherical (Fig. 6.14*c,d*) . . . . ***Trigonopyxis*** Penard

11b Pseudostome not triangular **12**

12a (11) Inferior lip of the pseudostome extends to superior lip, sometimes partially covering it; test hemispherical or elliptical (Fig. 6.14*e,f*) . . . ***Bullinularia*** Penard

12b Inferior lip of pseudostome does not extend to superior lip **13**

13a (12) Pseudostome linear, lunate; superior lip without pores; test hemispherical (Fig. 6.15) . . . . ***Plagiopyxis*** Penard

13b Pseudostome rounded or angular; test mostly membranous, encrusted with sediment particles **14**

14a (13) Pseudostome central; test regularly arched (Figs. 6.16, 6.17*a,b*) . . . . ***Cyclopyxis*** Deflandre

14b Pseudostome usually not central; test generally compressed (Figs. 6.17*c–f*, 6.18, 6.19*a–f*) . . . . ***Centropyxis*** Stein

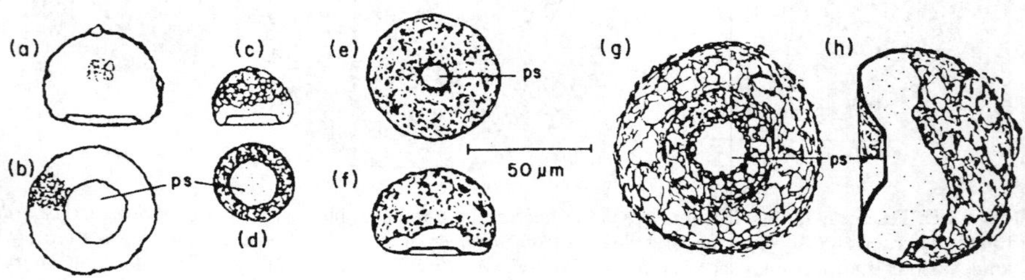

Fig. 6.16. Testacea: genus *Cyclopyxis*. (*a*) *C. eurystoma* Deflandre, lateral view (64); (*b*) same, ventral view (64); (*c*) *Cyclopyxis eurystoma* var. *parvula* Bonnet and Thomas, lateral view (25); (*d*) same, ventral view (25); (*e*) *C. ambigua* Bonnet and Thomas, ventral view (25); (*f*) same, lateral view (25); (*g*) *C. kahli* Deflandre, ventral view (64); (*h*) same, lateral view (64). *ps*, Pseudostome.

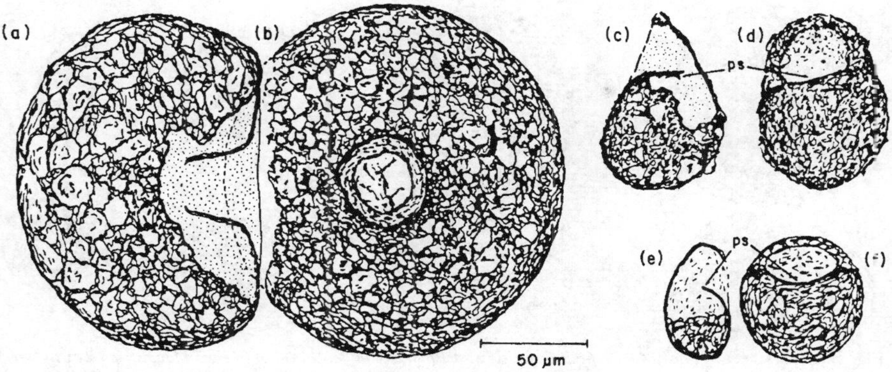

Fig. 6.17. Testacea: genera *Cyclopyxis* (cont'd) and *Centropyxis* (25). (*a*) *Cyclopyxis puteus* Thomas, lateral view; (*b*) same, ventral view; (*c*) *Centropyxis aerophila* Deflandre, lateral view; (*d*) same, ventral view; (*e*) *Centropyxis aerophila* var. *sphagnicola* Deflandre, lateral view; (*f*) same, ventral view. *ps*, Pseudostome.

15a (2) Test membranous; densely covered with sand grains (Fig. 6.19*g,h*) . . Family **Phryganellidae** ***Phryganella*** Penard

15b Test membranous, ovoid, transparent; few or no sand or sediment particles (Fig. 6.20*a–c*). . . . . Family **Cryptodifflugiidae** ***Difflugiella*** Cash

16a (1) Test with circular or elliptical scales . . . . . Family **Euglyphidae 17**

16b Test without distinct plates, often with foreign elements (Fig. 6.20*d*) Family **Gromiidae** ***Pseudodifflugia*** Schlumberger

17a (16) Pseudostome terminal 18

17b Pseudostome subterminal 20

18a (17) Pseudostome circular or oval. 19

18b Pseudostome elliptical, elongate, or linear; border of pseudostome very thin and finely dentate; elongate–elliptical plates; usually brown; test without a neck (Fig. 6.20*e,f*) . . . . ***Assulina*** Ehrenberg

19a (18) Pseudostome bordered by scales; scales round or oval; margin of pseudostome with prominent denticles; spines often develop (Figs. 6.21–6.23). . . . . ***Euglypha*** Dujardin

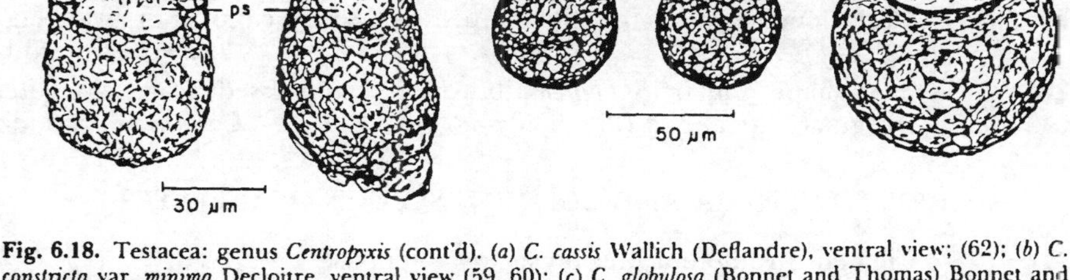

Fig. 6.18. Testacea: genus *Centropyxis* (cont'd). (*a*) *C. cassis* Wallich (Deflandre), ventral view; (62); (*b*) *C. constricta* var. *minima* Decloitre, ventral view (59, 60); (*c*) *C. globulosa* (Bonnet and Thomas) Bonnet and Thomas, ¾ lateral view (25); (*d*) same, lateral view (25); (*e*) *C. orbicularis* Deflandre, ventral view (25). *ps*, Pseudostome.

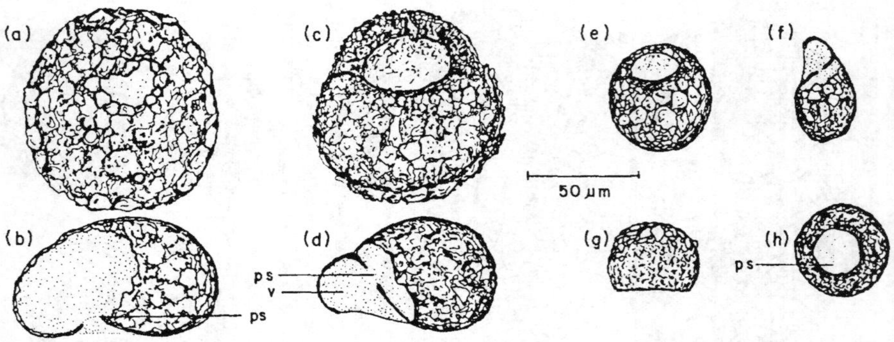

Fig. 6.19. Testacea: genera *Centropyxis* (cont'd) and *Phryganella* (25). (*a*) *Centropyxis plagiostoma* Bonnet and Thomas, ventral view; (*b*) same, lateral view; (*c*) *Centropyxis sylvatica* (Deflandre) Thomas, ventral view; (*d*) same, lateral view; (*e*) *Centropyxis sylvatica* var. *minor* Bonnet and Thomas, ventral view; (*f*) same, lateral view; (*g*) *Phryganella acropodia* (Hertwig and Lesser) Hopkinson, lateral view; (*h*) same, ventral view. *ps*, Pseudostome; *v*, vestibulum (for pseudostome).

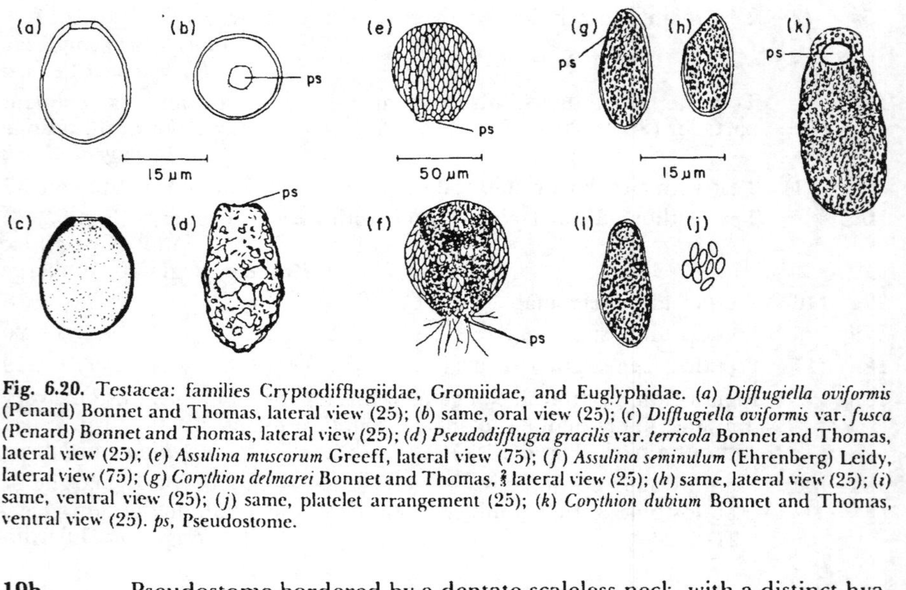

Fig. 6.20. Testacea: families Cryptodifflugiidae, Gromiidae, and Euglyphidae. (*a*) *Difflugiella oviformis* (Penard) Bonnet and Thomas, lateral view (25); (*b*) same, oral view (25); (*c*) *Difflugiella oviformis* var. *fusca* (Penard) Bonnet and Thomas, lateral view (25); (*d*) *Pseudodifflugia gracilis* var. *terricola* Bonnet and Thomas, lateral view (25); (*e*) *Assulina muscorum* Greeff, lateral view (75); (*f*) *Assulina seminulum* (Ehrenberg) Leidy, lateral view (75); (*g*) *Corythion delmarei* Bonnet and Thomas, lateral view (25); (*h*) same, lateral view (25); (*i*) same, ventral view (25); (*j*) same, platelet arrangement (25); (*k*) *Corythion dubium* Bonnet and Thomas, ventral view (25). *ps*, Pseudostome.

19b Pseudostome bordered by a dentate scaleless neck, with a distinct hyaline collar (denticulate or laciniate) (Fig. 6.24*a,b*) . . . . ***Tracheleuglypha*** Deflandre

20a **(17)** Test elongate–oval, usually compressed; scales circular or oval (Fig. 6.24*c–f*) . . . . ***Trinema*** Dujardin

20b Test shape similar to ***Trinema*** but more compressed; scales elongate, irregularly arranged (Fig. 6.20*g–k*). . . . . ***Corythion*** Taranek

KEY TO GENERA AND COMMON SPECIES OF CILIATES

1a Ciliated throughout life; no suctorial tentacles **2**

1b Suctorial tentacles radiating from spherical body, 30–50 µm, in adult (Fig. 6.25*a*) . . . . ***Sphaerophrya magna*** Maupas

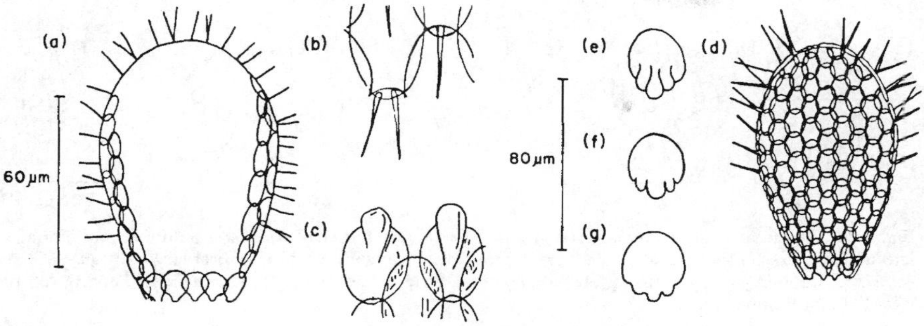

Fig. 6.21. Testacea: genus *Euglypha* (54, 55). (*a*) *E. ciliata* (Ehrenberg) Leidy, lateral view (*E. ciliata* forma *glabra*, the spineless form of this species, has a similar test shape and similar pseudostomal denticles); (*b*) same, test platelets and attachment of spines; (*c*) same, pseudostome platelets; (*d*) *E. compressa* Carter, lateral view (*E. compressa* forma *glabra*, the spineless form of this species, has a similar test shape and similar pseudostomal denticles); (*e*) same, pseudostome platelet types; (*f*) same, pseudostome platelet types; (*g*) same, pseudostome platelet types.

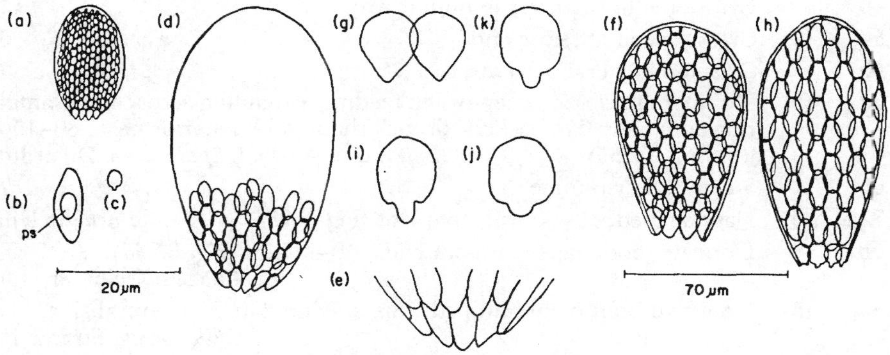

Fig. 6.22. Testacea: genus *Euglypha* (cont'd) (54, 55). (*a*) *E. cuspidata* Bonnet, lateral view; (*b*) same, oral view; (*c*) same, pseudostome platelets; (*d*) *E. denticulata* Brown, lateral view; (*e*) same, pseudostome platelets; (*f*) *E. laevis* Perty, lateral view; (*g*) same, pseudostome platelets; (*h*) *E. rotunda* Wailes, lateral view; (*i*) same, pseudostome platelet types; (*j*) same, pseudostome platelet types; and (*k*) same, pseudostome platelet types. *ps*, Pseudostome.

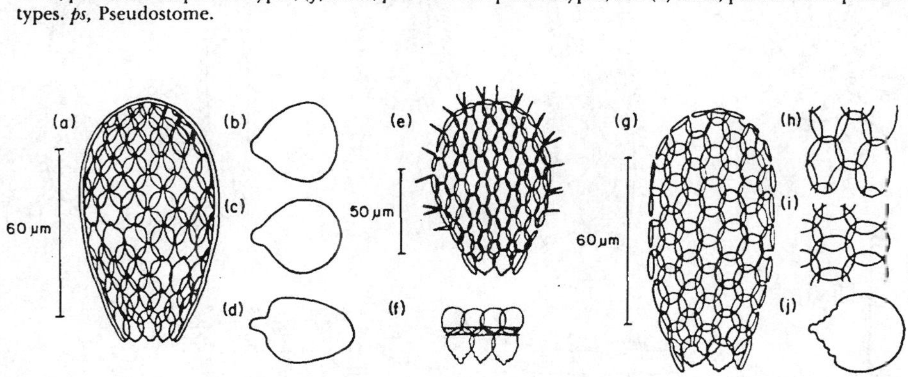

Fig. 6.23. Testacea: genus *Euglypha* (cont'd) (54, 55). (*a*) *E. scutigera* Penard, lateral view; (*b*) same, test platelet forms; (*c*) same, test platelet forms; (*d*) same, test platelet forms; (*e*) *E. strigosa* (Ehrenberg) Leidy, lateral view; *E. strigosa* forma *glabra;* (*f*) same, pseudostome platelets; (*g*) *E. tuberculata* Dujardin, lateral view; (*h*) same, test platelet forms; (*i*) same, test platelet forms; (*j*) same, pseudostome platelet form.

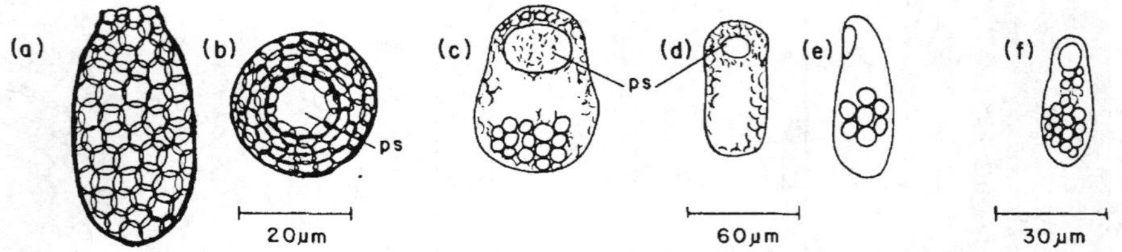

Fig. 6.24. Testacea: genera *Tracheleuglypha* and *Trinema.* (*a*) *Tracheleuglypha acolla* Bonnet and Thomas, lateral view (25); (*b*) same, oral view (25); (*c*) *Trinema complanatum* Penard, ventral view (90); (*d*) *Trinema enchelys* (Ehrenberg) Leidy, ventral view (90); (*e*) same, ¾ lateral view (90); (*f*) *Trinema lineare* Penard, ventral view (90). *ps,* Pseudostome.

2a	**(1)**	Without adoral zone of membranelles	3
2b		With adoral zone of membranelles	21
3a	**(2)**	Cytostome on body surface or in small vestibulum	4
3b		Cytostome in membrane-lined buccal cavity	17
4a	**(3)**	Cytostome on body surface	5
4b		Cytostome in small vestibulum (cavity)	11
5a	**(4)**	Cytostome at anterior end	6
5b		Cytostome lateral or ventral	8
6a	**(5)**	Mouth a slit, closed except when feeding, extending across entire anterior end; body flat, flasklike or sack shape, with anterior neck; 80–150 µm (Fig. 6.25*b*)	***Spathidium*** Dujardin
6b		Small oval cytostome	7
7a	**(6)**	Flask-shaped, 30–60 µm; frequent (Fig. 6.25*c*)	***Enchelys*** Hill
7b		Elongate; body tapers at both ends; 50–80 µm (Fig. 6.25*d*)	***Chaenia*** Quennerstedt
8a	**(5)**	Mouth ventral; body humped; cilia only on flattened ventral side	***Chilodonella*** Strand **10**
8b		Mouth lateral; body elongate with narrow flexible anterior end	9

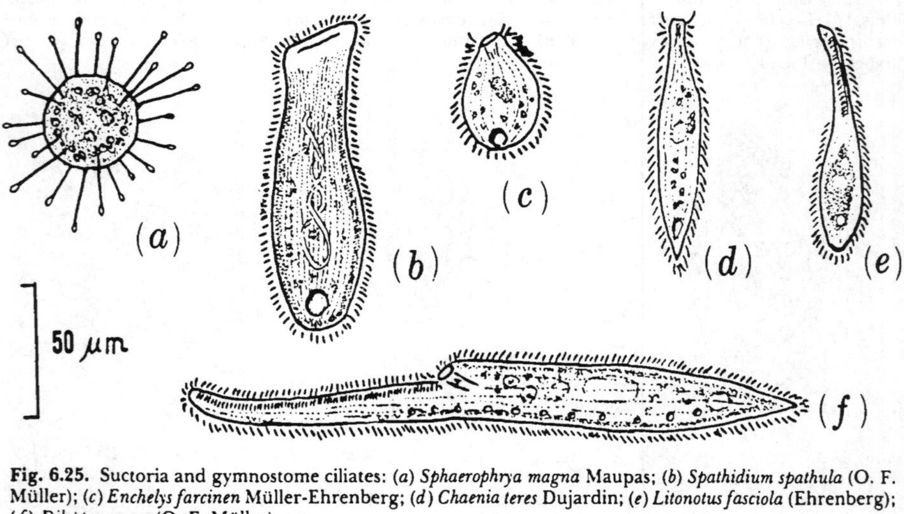

Fig. 6.25. Suctoria and gymnostome ciliates: (*a*) *Sphaerophrya magna* Maupas; (*b*) *Spathidium spathula* (O. F. Müller); (*c*) *Enchelys farcinen* Müller-Ehrenberg; (*d*) *Chaenia teres* Dujardin; (*e*) *Litonotus fasciola* (Ehrenberg); (*f*) *Dileptus anser* (O. F. Müller).

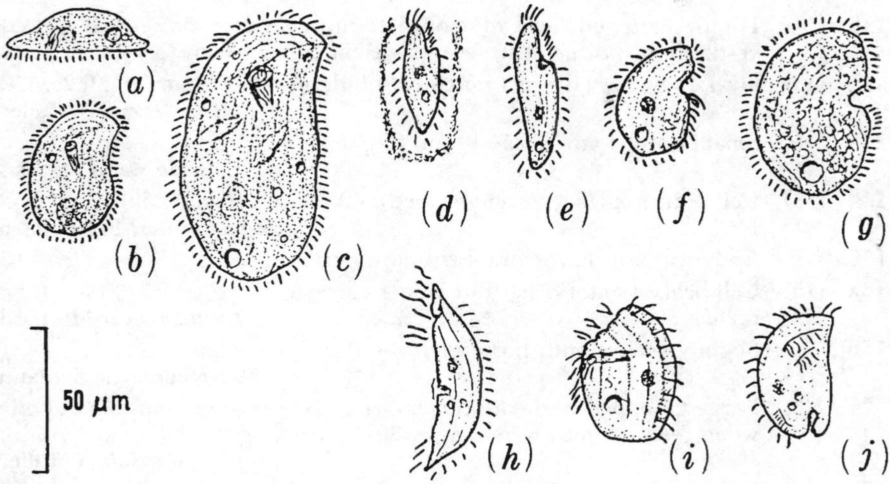

Fig. 6.26. Vestibulate ciliates: (*a*) *Chilodonella uncinata* (Ehrenberg), lateral view; (*b*) same, ventral view; (*c*) *Chilodonella uncinatus* (Müller); (*d*) *Cyrtolophosis mucicola* Stokes; (*e*) *Cyrtolophosis elongata* Kahl; (*f*) *Colpoda steini* Maupas; (*g*) *Colpoda cucullus* Müller; (*h*) *Drepanomonas dentata* Fresenius; (*i*) *Leptopharynx sphagnetorum* (Mermod); (*j*) *Microthorax pusillus* Engelmann.

9a (8) Slitlike mouth, evident only when feeding; 60–100 μm (Fig. 6.25*e*). . ***Litonotus*** Wrzesniowski

9b Visible mouth, halfway down side of body at base of flexible proboscis; 150–200 μm (Fig. 6.25*f*) . . . . ***Dileptus*** Dujardin

10a (8) Small, 35–50 μm (Fig. 6.26*a,b*) . . ***Chilodonella uncinata*** (Ehrenberg)

10b Larger, more elongate (Fig. 6.26*c*) . . . . ***C. cucullus*** (Müller)

11a (4) Flattened body; vestibulum lateral or posterior **13**

11b Ovoid, elongate; vestibulum at anterior end. . . . . ***Cyrtolophosis*** Stokes **12**

12a (11) Small, 25–30 μm; usually inhabits and darts back and forth in a gelatinous envelope (Fig. 6.26*d*) . . . . ***Cyrtolophosis mucicola*** Stokes

12b Longer, 40–50 μm; narrow body; no envelope (Fig. 6.26*e*) . . . . . . ***Cyrtolophosis elongata*** Kahl

13a (11) Flat oval, kidneylike body, indented on one side by easily seen vestibulum . . . . ***Colpoda*** Müller **14**

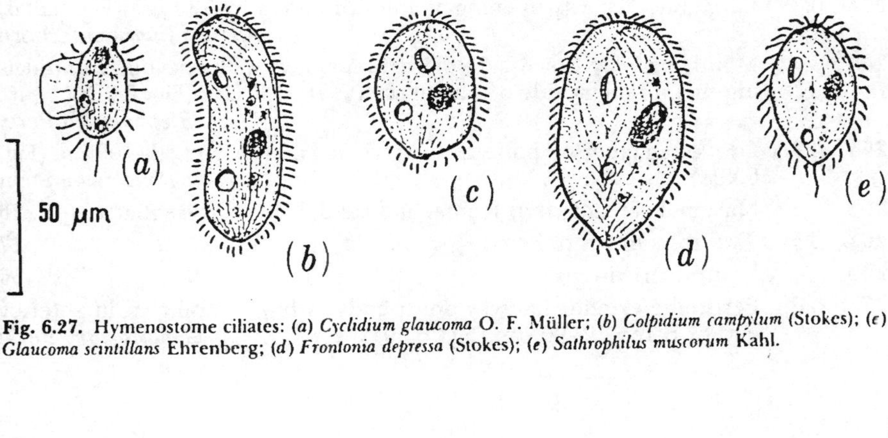

Fig. 6.27. Hymenostome ciliates: (*a*) *Cyclidium glaucoma* O. F. Müller; (*b*) *Colpidium campylum* (Stokes); (*c*) *Glaucoma scintillans* Ehrenberg; (*d*) *Frontonia depressa* (Stokes); (*e*) *Sathrophilus muscorum* Kahl.

13b Highly flattened, oval or elongate; the lateral or posteriorly placed vestibulum is difficult to see; sparse ciliation on body **15**

14a **(13)** Oval body, very distinct notch (vestibulum); 40–80 μm (Fig. 6.26*g*). . ***Colpoda cucullus*** Müller

14b Smaller, not so oval; 25–60 μm (Fig. 6.26*f*). . . . . ***Colpoda steini*** Maupas

15a **(13)** Cell semicircular (crescent-shaped), 40–70 μm (Fig. 6.26*h*) . . . . . . ***Drepanomonas*** Fresenius **6**

15b Body without sharp contours; more oval. **16**

16a **(15)** Cell beaked anteriorly; four ridges curve to one side, 25–50 μm (Fig. 6.26*i*) . . . . ***Leptopharynx*** Mermod

16b Slightly oval; mouth posterior, 30–40 μm (Fig. 6.26*j*). . . . . ***Microthorax*** Englemann

17a **(3)** Large C-shaped undulating membrane extends out from side of body when feeding; moves by jumps; 20–25 μm (Fig. 6.27*a*). . . . . ***Cyclidium*** Müller

17b Inconspicuous undulating membrane within buccal cavity **18**

18a **(16)** Mouth axis oblique to body axis **19**

18b Mouth axis parallel to body axis **20**

19a **(18)** Elongate, reniform (beanlike), small triangular mouth near side; 50–70 μm (Fig. 6.27*b*) . . . . ***Colpidium*** Stein

19b Oval, with oval mouth toward midline of body; 45–60 μm (Fig. 6.27*c*) ***Glaucoma*** Ehrenberg

20a **(18)** Large, about 100 μm (Fig. 6.27*d*). . . . . ***Frontonia*** Ehrenberg

20b Small, 35–40 μm (Fig. 6.27*e*). . . . . ***Sathrophilus*** Corliss

21a **(2)** Stalked body; circular adoral zone of membranelles at anterior end of inverted bell-shaped body; body otherwise unciliated (Figs. 6.28*a,b*) ***Vorticella*** Linnaeus

21b No stalk **22**

22a **(21)** Cilia sparse on diamond-shaped body, 25–50 μm; saltatorial bouncing movement (Fig. 6.28*c*) . . . . ***Halteria grandinella*** Müller

22b Elongated body, uniformly covered with cilia or with groups of cirri **23**

23a **(22)** Cirri on flattened ventral surface; rows of cilia also present in some genera; often show darting locomotion . . . . **Hypotrichs 25**

23b Body covered by cilia. . . . . **Heterotrichs 24**

24a **(23)** Large buccal cavity opening at anterior end; 100–130 μm (Fig. 6.28*d*) ***Bursaridium*** Lauterborn

24b Mouth halfway down side of slender body; well-developed undulating membrane at side of buccal cavity; 70–130 μm (Fig. 6.28*e*). . . . ***Blepharisma*** Perty

25a **(23)** Oval often tapering posteriorly; cirri long and sparse; 80–100 μm (Fig. 6.29*a*) . . . . ***Psilotricha*** Stein

25b Marginal rows of cirri; frontal and caudal cirri differentiated . . . **26**

26a **(25)** Two rows of ventral cirri. **27**

26b Ventral cirri in groups **30**

27a **(26)** Peristome extends halfway down body; a few ventral cirri in anterior end of body; 70–90 μm (Fig. 6.29*b*). . . . . . . ***Gonostomum*** Sterki

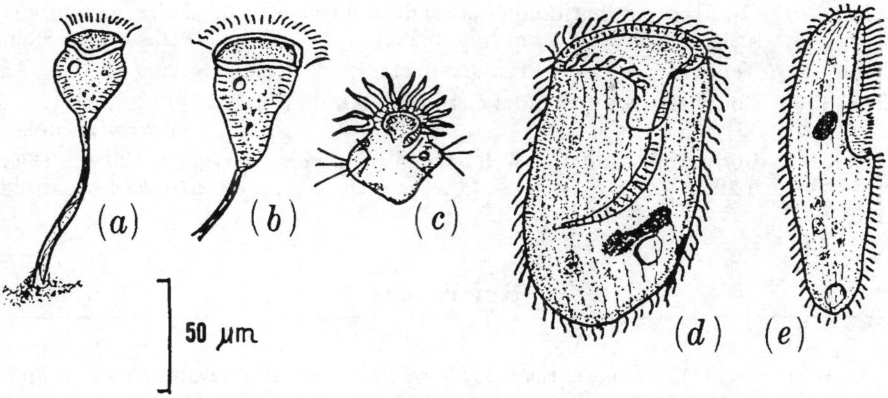

Fig. 6.28. Peritrich, oligotrich, and heterotrich ciliates: (*a*) *Vorticella microstoma* Ehrenberg; (*b*) *Vorticella nebulifera* O. F. Müller; (*c*) *Halteria grandinella* (Müller); (*d*) *Bursaridium difficile* Kahl; (*e*) *Blepharisma steini* Kahl.

27b Peristome in anterior one-third or one-fourth of body. **28**

28a (27) No anal (transverse) cirri; elongate body, tapering into a tail-like portion; 150–200 μm (Fig. 6.29*c*). . . . . ***Uroleptus*** Ehrenberg

28b Anal (transverse) cirri present; body does not taper posteriorly **29**

29a (28) Frontal cirri prominent; 120–160 μm (Fig. 6.29*d*) . . . . ***Holosticha*** Wrzeniowski

29b Frontal cirri not prominent; anal (transverse) cirri close to posterior end of body; large, 160–250 μm (Fig. 6.29*e*) . . . ***Keronopsis*** Penard

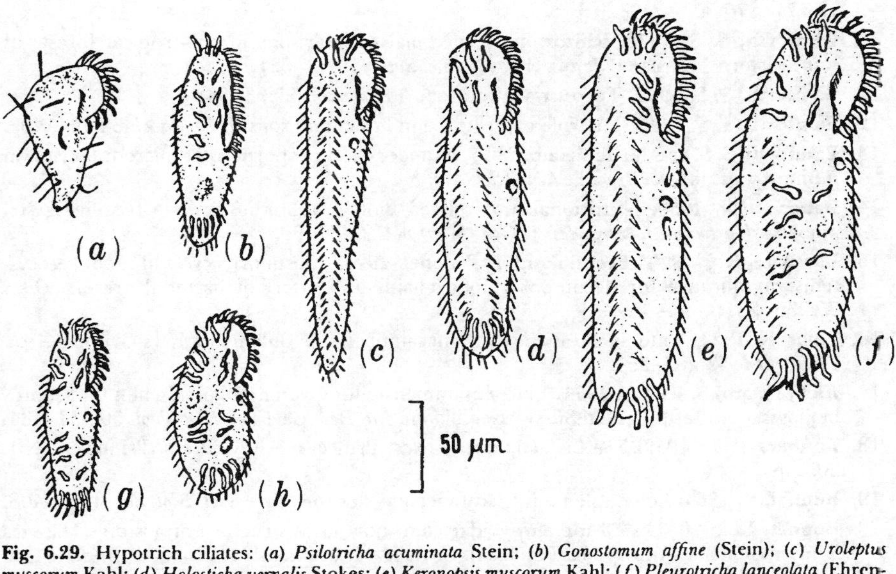

Fig. 6.29. Hypotrich ciliates: (*a*) *Psilotricha acuminata* Stein; (*b*) *Gonostomum affine* (Stein); (*c*) *Uroleptus muscorum* Kahl; (*d*) *Holosticha vernalis* Stokes; (*e*) *Keronopsis muscorum* Kahl; (*f*) *Pleurotricha lanceolata* (Ehrenberg); (*g*) *Tachysoma pellionella* (Müller-Stein); (*h*) *Oxytricha minor* Kahl.

30a (26) Ovoid, somewhat pointed at end; one row of ventral cirri with others scattered; 120–180 μm (Fig. 6.29*f*) . . . . ***Pleurotricha*** Stein

30b Ventral cirri scattered in groups **31**

31a (30) Body narrow; no caudal cirri; 80–100 μm (Fig. 6.29*g*) . . . . ***Tachysoma*** Stokes

31b Body oval; caudal as well as anal (transverse) cirri; 80–120 μm (Fig. 6.29*h*) . . . . ***Oxytricha*** Ehrenberg

References

1. **Alexander, M. 1977.** *Introduction to Soil Microbiology,* 2nd ed. Wiley, New York. 300 pp.
2. **Allison, R. V. 1924.** A note on the protozoan fauna of the soils of the United States. *Soil Sci.* 18:339–352.
3. **Anderson, R. V., E. T. Elliott, J. F. McClellan, D. C. Coleman, C. V. Cole and H. W. Hunt. 1978.** Trophic interactions in soils as they affect energy and nutrient dynamics. III. Biotic interactions of bacteria, amoebae, and nematodes. *Microb. Ecol.* 4:361–371.
4. **Bailey, J. W. 1879.** *Smithsonian Contributions, Washington, 1850.* Through J. Leidy, U.S. Geol. Surv. Terr. Monogr. 12, p. 299.
5. **Bamforth, S. S. 1967.** A microbial comparison of two forest soils of southeastern Louisiana. *Proc. L. Acad. Sci.* 30:7–16.
6. **Bamforth, S. S. 1968.** Forest soil protozoa of the Florida parishes of Louisiana. *Proc. L. Acad. Sci.* 31:5–15.
7. **Bamforth, S. S. 1969.** Protozoa and algae of the Mississippi deltaic soils. *Proc. L. Acad. Sci.* 32:68–77.
8. **Bamforth, S. S. 1971.** The numbers and proportions of testacea and ciliates in litters and soils. *J. Protozool.* 18:24–28.
9. **Bamforth, S. S. 1973.** Population dynamics of soil and vegetation protozoa. *Am. Zool.* 13:171–176.
10. **Bamforth, S. S. 1976.** Rhizosphere microbial comparisons in sub-tropical forests of southeastern Louisiana. *Trans. Am. Microsc. Soc.* 95:613–621.
11. **Bamforth, S. S. 1980.** Terrestrial Protozoa. *J. Protozool.* 27:33–36.
12. **Bamforth, S. S. 1985.** The role of Protozoa in litters and soils. *J. Protozool.* 32:404–409.
13. **Bamforth, S. S. and N. de Pelaez. 1977.** Numbers and proportions of microorganisms in humid forest litters. *Proc. L. Acad. Sci.* 40:33.
14. **Band, R. N. 1959.** Nutritional and related biological studies on the free-living soil amoeba, *Hartmannella rhysodes. J. Gen. Microbiol.* 21:80–95.
15. **Barsdate, R. J., R. T. Prentki and T. Fenchel. 1974.** Phosphorus cycle of model ecosystems: significance for decomposer food chains and effect of bacterial grazers. *Oikos* 25:239–251.
16. **Belar, K. 1921.** Untersuchungen über Thecamöben der Chlamydophrys-Gruppe. *Arch. Protistenkd.* 53:287–354.
17. **Bick, H. and S. Kunze. 1971.** Eine Zusammenstellung von autökologischen und saprobiollogischen Befunden an Süsswasserciliaten. *Int. Rev. Gesamten Hydrobiol.* 56:337–384.
18. **Bonner, T. L. 1959.** *The Cellular Slime Molds.* Princeton Univ. Press, Princeton, N.J. 205 pp.
19. **Bonnet, L. 1964.** Le peuplement thécamoebiens des sols. *Rev. Ecol. Biol. Sol* 1:123–408.
20. **Bonnet, L. 1970.** Les Distomatopyxidae fam. nov. et la structure diplostome chez les Thécamoebiens (Rhizopoda Testacea). *C. R. Hebd. Seances Acad. Sci. Ser. (Paris) D Sci. Nat.* 271:1189–1191.

21. **Bonnet, L. 1974.** Quelques aspects du peuplement thécamoebien des sols de la province de Québec (Canada). *Can. J. Zool.* 52:29–41.

22. **Bonnet, L. 1975.** Types morphologiques, ecologie et évolution de la théque chez les Thécamoebiens. *Protistologica* 11:363–378.

23. **Bonnet, L. and R. Thomas. 1955.** Étude sur les Thécamoebiens du sol (I). *Bull. Soc. Hist. Nat. Toulouse* 90:411–428.

24. **Bonnet, L. and R. Thomas. 1960.** Étude sur les Thécamoebiens du sol (II). *Bull. Soc. Hist. Nat. Toulouse* 95:339–349.

25. **Bonnet, L. and R. Thomas. 1960.** *Thécamoebiens du sol, Faune terrestre et d'eau douce des Pyrénées-Orientales.* Fascicule 5, Vie et Milieu (Suppl.). Hermann, Paris, pp. 1–103.

26. **Bryant, R. J., L. E. Woods, D. C. Coleman, B. C. Fairbanks, J. F. McClellan and C. V. Cole. 1982.** Interactions of bacterial and amoebal populations in soil micro-cosms with fluctuating moisture content. *Appl. Environ. Microbiol.* 43:747–752.

27. **Bunt, J. S. and Y. T. Tchan. 1955.** Estimation of protozoan populations in soils by direct microscopy. *Proc. Linn. Soc. N.S.W.* 80:148–153.

28. **Burges, A. and D. P. Nicholas. 1961.** Use of soil sections in studying amount of fungal hyphae in soil. *Soil Sci.* 92:25–29.

29. **Calaway, W. T. and J. Lackey. 1962.** *Waste Treatment Protozoa: Flagellata.* Eng. Ser. 3. University of Florida, Gainesville, Fla. 140pp.

30. **Carter, A. 1975.** *Aspects of the Population Biology of* Agonum retractum *Leconte (Coleoptera: Carabidae) and Its Role and That of Some Other Soil Arthropods in the Cycling of Chemical Elements in an Aspen Woodland Ecosystem.* Ph.D. thesis, University of Calgary, Calgary, Alberta, Canada, 111 pp.

31. **Cash, J., G. H. Wailes and J. Hopkinson. 1905–1921.** *The British Fresh-Water Rhizopoda and Heliozoa.* 5 parts. Ray Society, London.

32. **Chapman-Andresen, C. 1973.** Endocytic processes. Pp. 319–348 in K. W. Jeon (ed.). *The Biology of Amoeba.* Academic Press, New York.

33. **Chardez, D. 1962.** Sur *Centropyxis sylvatica* (Deflandre) Thomas. *Biol. Jaarb.* 30:316–320.

34. **Chardez, D. 1965.** Sur un mode particulier et peu connu de reproduction chez les Thécamoebiens aquatiques. *Bull. Inst. Agron. Stn. Rech. Gembloux* 33:26–34.

35. **Chardez, D. 1967.** *Histoire naturelle de Protozoaires thécamoebiens.* Les Naturalistes Belges, Brussels, Belgium, pp. 1–100.

36. **Chardez, D. 1985.** Protozoaires prédateurs de Thécamoebiens. *Protistologica* 21:187–194.

37. **Clarholm, M. 1981.** Protozoan grazing of bacteria in soil—impact and importance. *Microb. Ecol.* 7:343–350.

38. **Clarholm, M. 1985.** Interactions of bacteria, Protozoa, and plants leading to mineralization of soil nitrogen. *Soil Biol. Biochem.* 17:181–187.

39. **Cole, C. V., E. T. Elliott, H. W. Hunt and D. C. Coleman. 1978.** Trophic interactions in soils as they affect energy and nutrient dynamics. V. Phosphorus transformations. *Microb. Ecol.* 4:381–387.

40. **Coleman, D. C., R. V. Anderson, C. V. Cole, E. T. Elliott, L. Woods and M. K. Campion. 1978.** Trophic interactions in soils as they affect energy and nutrient dynamics. IV. Flows of metabolic and biomass carbon. *Microb. Ecol.* 4:373–380.

41. **Coûteaux, M.-M. 1967.** Une technique d'observation des Thécamoebiens du sol pour l'estimation de leur densité absolue. *Rev. Ecol. Biol. Sol* 4:593–596.

42. **Coûteaux, M.-M. and J. Dévaux. 1983.** Effet d'un enrichissement en champignons sur la dynamique d'un peuplement thécamoebien d'un humus. *Rev. Ecol. Biol. Sol* 20:519–545.

43. **Coûteaux, M.-M. 1985.** The effects of compressing forest litter (H layer) on the populations of testate amoebae. *Rev. Ecol. Biol. Sol* 22:447–461.

44. **Cunningham, A. 1914.** Studies on soil Protozoa: some of the activities of protozoa (I). *Zentralbl. Bakteriol. Abt. 2* 42:8–27.

45. **Cunningham, A. 1915.** Studies on soil Protozoa. *J. Agric. Sci.* 7:47–74.

46. **Cunningham, A. and F. Lohnis. 1914.** Studies on soil Protozoa: some of the activities of Protozoa (I). *Zentralbl. Bakteriol. Abt. 2* 39:596–610.

47. **Cutler, D. W. 1920.** A method for estimating the number of active Protozoa in the soil. *J. Agric. Sci. Camb.* 10:135–143.

48. **Cutler, D. W., L. M. Crump and H. Sandon. 1921.** A quantitative investigation of the bacterial and protozoan population of the soil. *Philos. Trans. R. Soc., Lond. B Biol. Sci.* 211:317–350.

49. **Danso, S. K. A. and M. Alexander. 1975.** Regulation of predation by prey density: the protozoan–*Rhizobium* relationship. *Appl. Microbiol.* 29:515–521.

50. **Darbyshire, J. F. 1972.** Nitrogen fixation by *Azotobacter chroococcum* in the presence of *Colpoda steini.* I. The influence of temperature. *Soil Biol. Biochem.* 4:359–369.

51. **Darbyshire, J. F. 1976.** Effect of water suctions on the growth in soil of the ciliate *Colpoda steini,* and the bacterium *Azotobacter chroococcum. J. Soil Sci.* 27:369–376.

52. **Dash, M. C. and J. B. Cragg. 1972.** Ecology of Enchytraeidae (Oligochaeta) in Canadian Rocky Mountain soils. *Pedobiologia* 12:323–335.

53. **Datta, T. and J. Kaur. 1978.** Effect of fungal extracts on excystment of soil amoebae. *Protistologica* 14:121–123.

54. **Decloitre, L. 1962.** Le genre *Euglypha* Dujardin. *Arch. Protistenkd.* 106:51–100.

55. **Decloitre, L. 1976.** Le genre *Euglypha.* Compléments à jour au 31 décembre 1974 de la monographie du genre parue en 1962. *Arch. Protistenkd.* 118:18–33.

56. **Decloitre, L. 1976.** Le genre *Arcella* Ehrenberg. Compléments à jour au 31 décembre 1974 de la monographie du genre parue en 1928. *Arch. Protistenkd.* 118:291–309.

57. **Decloitre, L. 1977.** Le genre *Cyclopyxis.* Compléments à jour au 31 décembre 1974 de la monographie du genre parue en 1929. *Arch. Protistenkd.* 119:31–53.

58. **Decloitre, L. 1977.** Le genre Nebela. Compléments à jour au 31 décembre 1974 de la monographie du genre parue en 1936. *Arch. Protistenkd.* 119:325–352.

59. **Decloitre, L. 1978.** Le genre *Centropyxis* I. Compléments à jour au 31 décembre 1974 de la monographie parue en 1929. *Arch. Protistenkd.* 120:63–85.

60. **Decloitre, L. 1979.** Le genre *Centropyxis* II. *Arch. Protistenkd.* 121:162–192.

61. **Deflandre, G. 1928.** Le genre *Arcella* Ehrenberg. *Arch. Protistenkd.* 64:152–287.

62. **Deflandre, G. 1929.** Le genre *Centropyxis* Stein. *Arch. Protistenkd.* 67:323–375.

63. **Deflandre, G. 1936.** Étude monographique sur le genre *Nebela* Leidy (Rhizopoda–Testacea). *Ann. Protistol.* 5:201–286.

64. **Deflandre, G. 1953.** Thécamoebiens. Pp. 92–148 in P. P. Grasse (ed.). *Traité de Zoologie,* Tome 1, Fascicule, 2. Masson, Paris.

65. **Deflandre, G. 1959.** Rhizopoda and Actinopoda. Pp. 232–264 in W. T. Edmondson (ed.). *Freshwater Biology,* 2nd ed., Wiley, New York.

66. **Dudziak, B. 1955.** Kielkowanie i wzorst ameb w zaleznosci od pokarmu bakteryjnego. *Acta Microbiol. Pol.* 4:115–125.

67. **Elliott, E. T. and D. C. Coleman. 1977.** Soil protozoan dynamics in a shortgrass prairie. *Soil Biol. Biochem.* 9:113–118.

68. **Feher, D. and L. Varga. 1929.** Untersuchungen über die Protozoen. *Zentralbl. Bakteriol. Parasitenkd. Abt. II* 77:524–542.

69. **Fellers, C. R. and F. E. Allison. 1920.** The protozoan fauna of the soils of New Jersey. *Soil Sci.* 9:1–25.

70. **Foissner, W. 1982.** Ökbologie und Taxonomie der Hypotrichida (Protozoa : Ciliophora) einiger österreicher Böden. *Arch. Protistenkd.* 126:19–143.

71. **Foissner, W. 1985.** Klassification und Phylogenie de Colpodea (Protozoa : Ciliophora). *Arch. Protistenkd.* 129:239–290.

72. **Foissner, W. 1986.** Soil protozoa: fundamental problems, ecological significance, adaptation, indicators of environmental quality, guide to the literature. *Prog. Protist.* 2:69–212.

73. **Frey, J. S., J. F. McClellan, E. R. Ingham and D. C. Coleman. 1985.** Filter-out-grazers (FOG): a filtration experiment for separating protozoan grazers in soil. *Biol. Fertil. Soils* 1:73–79.

74. **Giese, A. C. (ed.). 1973.** *Blepharisma: The Biology of a Light-Sensitive Protozoan.* Stanford University Press, Stanford, Calif. 366 pp.

75. **Grospietsch, T. 1958.** *Wechseltierchen (Rhizopoden).* Kosmos, Stuttgart, West Germany, pp. 1–80.

76. **Grospietsch, T. 1964.** Die Gattungen *Cryptodifflugia* and *Difflugiella* (Rhizopoda, Testacea). *Zool. Anz.* 172:243–257.

77. **Grospietsch, T. 1965.** Monographische Studie der Gattung *Hyalosphenia* Stein. *Hydrobiologia* 26: 211–241.

78. **Haarlov, M. and T. Weis-Fogh. 1953.** A microscopical technique for studying the undisturbed texture of soils. *Oikos* 4:44–57.

79. **Habte, M. and M. Alexander. 1975.** Protozoa as agents responsible for the decline of *Xanthomonas campestris* in soil. *Appl. Microbiol.* 29:159–164.

80. **Harnish, O. 1958.** Rhizopoda. Pp. 1–75 in P. Brohmer, P. Ehrmann, and G. Ulmer (eds.). *Die Tierwelt Mitteleuropas.* Leipzig, East Germany.

81. **Heal, O. W. 1964.** The use of cultures for studying Testacea (Protozoa: Rhizopoda) in soil. *Pedobiologia* 4:1–7.

82. **Heal, O. W. 1967.** Quantitative studies on soil amoebae. Pp. 120–125 in O. Graff and J. E. Satchell (eds.). *Progress in Soil Biology.* North-Holland, Amsterdam.

83. **Heal, O. W. 1970.** Methods of study of soil Protozoa. Pp. 119–126 in J. Phillipson (ed.). *Methods of Study in Soil Ecology.* UNESCO, Paris.

84. **Heal, O. W. 1971.** Protozoa. Pp. 51–64 in J. Phillipson (ed.). *Methods of Study in Quantitative Soil Ecology: Population, Production and Energy Flow.* IBP Handbook 18. Blackwell Scientific, Oxford.

85. **Heal, O. W. and M. J. Felton. 1970.** Soil amoebae: their food and their reaction to microflora exudates. Pp. 145–162 in A. Watson (ed.). *Animal Populations in Relation to Their Food Resources.* Br. Ecol. Symp. 10. Blackwell Scientific, Oxford.

86. **Hedley, R. H. and C. G. Ogden. 1973.** Biology and fine structure of *Euglypha rotunda* (Testacea : Protozoa). *Bull. Br. Mus. (Nat. Hist.) Zool.* 25:119–137.

87. **Hedley, R. H. and C. G. Ogden. 1974.** Observations on *Trinema lineare* Penard (Testacea : Protozoa). *Bull. Br. Mus. (Nat. Hist.) Zool.* 26:185–199.

88. **Hedley, R. H. and C. G. Ogden. 1974.** Adhesion plaques associated with the production of a daughter cell in *Euglypha* (Testacea : Protozoa). *Cell Tissue Res.* 153:261–268.

89. **Hedley, R. H., C. G. Ogden and J. I. Krafft. 1974.** Observations on clonal cultures of *Euglypha acanthophora* and *Euglypha strigosa* (Testacea : Protozoa). *Bull. Br. Mus. (Nat. Hist.) Zool.* 27:103–111.

90. **Hoogenraad, H. R. and A. A. de Groot. 1940.** Zoetwaterrhizopoden enheliozoen. Pp. 1–303 in H. Boschma (ed.). *Fauna van Nedeland.* Sijthoff's Uiteversm, Alphen aan der Rijn, The Netherlands.

91. **Jahn, T. L., E. C. Bovee and F. F. Jahn. 1979.** *How to Know the Protozoa,* 2nd ed. Wm. C. Brown, Dubuque, Iowa. 279 pp.

92. **Kahl, A. 1932–1935.** *Urtiere oder Protozoa.* I. *Wimpertiere oder Ciliata (Infusoria),* Parts 18, 21, 25, 30. Gustav Fischer, Jena, East Germany.

93. **Kubiena, W. 1938.** *Micropedology.* Collegiate Press, Iowa State College, Ames, Iowa, pp. 1–243.

94. **Kudo, R. R. 1966.** *Protozoology,* 5th ed. Charles C Thomas, Springfield, Ill. 1174 pp.

95. **Kunicki-Goldfinger, W., W. Drozanski, D. Blaszczak, J. Mazur and J. Skibinski. 1957.** Bakterie jako pokarm pelzakow. *Acta Microbiol. Pol.* 6:331–334.

96. **Laminger, H. 1978.** The effect of soil moisture fluctuations on the testacean species *Trinema enchelys* (Ehrenberg) Leidy in a high mountain brown-earths-podsol and its feeding behaviour. *Arch. Protistenkd.* 120:446–454.

97. **Laminger, H., G. Kainz and G. Horvath. 1982.** Populationsdynamik terrestrischer Protozoen (Testacea, Rhizopoda) in zentralalpinen Lagen Tirols. IV. Distribution der Testaceenzönosen in Hochgebirgsboden. *Zool. Anz. Jena* 208:339–351.

98. **Laybourn, J. 1975.** Respiratory energy losses in *Stentor coeruleus* Ehrenberg (Ciliphora). *Oecologia (Berl.)* 21:273–278.

99. **Laybourn, J. 1976.** Energy budgets for *Stentor coerulus* Ehrenberg (Ciliophora). *Oecologia (Berl.)* 22:431–437.

100. **Laybourn, J. E. M. and J. M. Stewart. 1975.** Studies on consumption and growth in the ciliate *Colpodium campylum* Stokes. *J. Anim. Ecol.* 44;165–174.

101. **Lee, J. J., S. H. Hutner and E. D. Bovee. 1985.** *An Illustrated Guide to the Protozoa.* Society of Protozoologists, Lawrence, Kans.

102. **Leidy, J. 1879.** *Fresh-Water Rhizopods of North America.* U.S. Geol. Surv. Terr. Monogr. 12, pp. 1–323.

103. **Lemmermann, E. 1914.** Protomastiginae. In A. Pascher (ed.). *Susswasserflora Deutschlands, Österreichs, und der Schweiz,* Heft 1. Gustav Fischer, Jena, East Germany.

104. **Levine, N. D., J. O. Corliss, F. E. G. Cox, D. Deroux, J. Grain, B. M. Honigberg, G. F. Leedale, R. Loeblich III, J. Lom, D. Lynn, E. G. Merinfeld, F. C. Page, G. Poljansky, V. Sprague, J. Vavra, F. G. Wallace and J. Wieser. 1980.** A new revised classification of the protozoa. *J. Protozool.* 27:36–58.

105. **Lousier, J. D. 1974.** Effects of experimental soil moisture fluctuations on turnover rates of Testacea. *Soil Biol. Biochem.* 6:19–26.

106. **Lousier, J. D. 1974.** Response of soil Testacea to soil moisture fluctuations. *Soil Biol. Biochem.* 6:235–239. (Errata: **1975.** *Soil Biol. Biochem.* 7:185.)

107. **Lousier, J. D. 1975.** Relationship between distribution of Testacea (Protozoa, Rhizopoda) and the soil habitat. *Nat. Can. (Que.)* 102:57–72.

108. **Lousier, J. D. 1976.** Testate amoebae (Rhizopoda, Testacea) in some Canadian Rocky Mountain soils. *Arch. Protistenkd.* 118:191–201.

109. **Lousier, J. D. 1982.** Colonization of decomposing deciduous leaf litter by Testacea (Protozoa, Rhizopoda): species succession, abundance and biomass. *Oecologia (Berl.)* 52:381–388.

110. **Lousier, J. D. 1984.** Population dynamics and production studies of species of Euglyphidae (Testacea, Rhizopoda, Protozoa) in an aspen woodland soil. *Pedobiologia* 26:309–330.

111. **Lousier, J. D. 1984.** Population dynamics and production studies of *Phryganella acropodia* and *Diffugiella oviformis* (Testacea, Rhizopoda, Protozoa) in an aspen woodland soil. *Pedobiologia* 26:331–347.

112. **Lousier, J. D. 1984.** Population dynamics and production studies of species of Nebelidae (Testacea, Rhizopoda, Protozoa) in an aspen woodland soil. *Acta Protozool.* 23:145–159.

113. **Lousier, J. D. 1985.** Population dynamics and production studies of species of Centropyxidae (Testacea, Rhizopoda, Protozoa) in an aspen woodland soil. *Arch. Protistenkd.* 130:165–178.

114. **Lousier, J. D. and D. Parkinson. 1981.** Disappearance of empty tests in soil and litter Testacea (Protozoa, Rhizopoda). *Arch. Protistenkd.* 124:312–336.

115. **Lousier, J. D. and D. Parkinson. 1981.** Evaluation of a membrane filter technique to count soil and litter Testacea. *Soil Biol. Biochem.* 13:209–213.

116. **Lousier, J. D. and D. Parkinson. 1984.** Annual population dynamics and production ecology of Testacea (Protozoa, Rhizopoda) in an aspen woodland soil. *Soil Biol. Biochem.* 16:103–114.

117. **Lüftenegger, G., W. Foissner and H. Adam. 1985.** *r*- and *K*-selection in soil ciliates: a field and experimental approach. *Oecologia (Berl.)* 66:574–579.

118. **Madoni, P. 1986.** Abundance, productivity and energy flow of a population of *Coleps hirtus* NITZSCH (Protozoa, Ciliophora) during the first stage of colonization in an experimental ricefield. *Arch. Protistenkd.* 132:43–51.

119. **Moraczewski, J. 1971.** Structure et formation de la coque d'*Arcella. Acta Protozool.* 8:423–443.

120. **Mote, R. F. 1954.** A study of soil Protozoa on an Iowa virgin prairie. *Proc. Iowa Acad. Sci.* 61:570–592.

121. **Neff, R. J. 1957.** Purification, axenic cultivation, and description of a soil amoeba, *Acanthamoeba* sp. *J. Protozool.* 4:176–181.

122. **Neff, R. J., R. H. Neff and R. E. Taylor. 1958.** The nutrition and metabolism of a soil amoeba, *Acanthamoeba* sp. *Physiol. Zool.* 31:73–91.

123. **Netzel, H. 1971.** Die Schalenbildung bei der Thekamöben-Gattung *Arcella* (Rhizopoda, Testacea). *Cytobiologie* 3:89–92.

124. **Netzel, H. 1972.** Die Bildung der Gehäusewand bei der Thekamöbe *Centropyxis discoiaes* Rhizopoda, Testacea). *Z. Zellforsch.* 135:45–54.

125. **Netzel, H. 1972.** Die Schalenbildung bei *Difflugia oviformis* (Rhizopoda, Testacea). *Z. Zellforsch.* 135:55–61.

126. **Netzel, H. 1972.** Morphogenese des Gehäuses von *Euglypha rotunda* (Rhizopoda, Testacea). *Z. Zellforsch.* 135:63–69.

127. **Netzel, H. 1976.** Die Abschiedung der Gehäusewand bei *Centropyxis discoides* (Rhizopoda, Testacea). *Arch. Protistenkd.* 118:53–91.

128. **Netzel, H. 1977.** Die Bildung des Gehäuses bei *Difflugia oviformis* (Rhizopoda, Testacea). *Arch. Protistenkd.* 119:1–30.

129. **Netzel, H. 1977.** Die Struktur des Plättchen-Gehäuses von *Euglypha rotunda* (Rhizopoda, Testacea). *Arch. Protistenkd.* 119:191–216.

130. **Netzel, H. 1977.** Die Herstellung des Plättchen-Gehäuses bei *Euglypha rotunda* (Rhizopoda, Testacea). *Arch. Protistenkd.* 119:301–324.

131. **Nikoljuk, B. 1956.** *Soil Protozoa and Their Role in Cultivated Soils of Uzbekistan.* Izdatel stvo Akad. Nauk. Uzbekskoi SSR. Tashkent, USSR. 144 pp. (in Russian).

132. **Nikolyuk, B. and J. Geltzer. 1972.** *Pochvennye Proteyshie CCCP* (*Soil Protozoa of the USSR*). Tashkent, USSR. 310 pp. (in Russian).

133. **Noland, L. E. 1959.** Ciliophora. Pp. 265–297 in W. T. Edmondson (ed.). *Fresh-Water Biology,* 2nd ed. Wiley, New York.

134. **Olive, L. A. 1970.** *The Mycetozoans.* Academic Press, New York, 292 pp.

135. **Ormsbee, R. A. 1942.** The normal growth and respiration of *Tetrahymena geleii. Biol. Bull. Mar. Biol. Lab. Woods Hole* 82:423–437.

136. **Page, F. C. 1976.** *An Illustrated Key to Fresh-Water and Soil Amoebae.* Freshwater Biol. Assoc. Publ. 34. Titus Wilson, Kendal, England.

137. **Pascher, A. 1913.** Chrysomonadinae, Cryptomonadinae, Eugleninae, Chloromonadinae. In *Susswasserflora Deutschlands, Jena. Osterreichs und der Schweiz,* Heft 2. Gustav Fischer, Jena, East Germany.

138. **Petz, W., W. Foissner, E. Wirnsberger, W. D. Krautgartner and H. Adam. 1986.** Mycophagy, a new feeding strategy in autochthonous soil ciliates. *Naturwissenschaften* 73:560.

139. **Puytorac, P. de, J. P. Mignot, J. Grain, C. A. Groliere, L. Bonnet and P. Couillard. 1972.** Premier relevé de certains groupes de Protozoaires libres sur le territoire de la Station de biologie de l'Université de Montréal (Saint Hippolyte, Comté de Terrebonne, Québec). *Nat. Can.* (*Que.*) 99:417–440.

140. **Reich, K. 1948.** Studies on the respiration of an amoeba, *Mayorella palestinensis. Physiol. Zool.* 21:390–412.

141. **Rogerson, A. 1981.** The ecological energetics of *Amoeba proteus* (Protozoa). *Hydrobiologia* 85:117–128.

142. **Rogerson, A. and J. Berger. 1981.** The effects of cold temperature and crude oil on the abundance and activity of Protozoa in a garden soil. *Can. J. Zool.* 59:1554–1560.

143. **Sandon, H. 1927.** *The Composition and the Distribution of the Protozoan Fauna of the Soil.* Oliver & Boyd, Edinburgh, pp. 150–219.

144. **Sandon, H. 1928.** A study of the Protozoa of some American soils. *Soil Sci.* 25:107–120.

145. **Schönborn, W. 1965.** Untersuchungen über die Ernährung bodenbewohnender Testaceen. *Pedobiologia* 5:205–210.

146. **Schönborn, W. 1975.** Ermittlung der Jahresproduktion von Boden-Protozoen. I. Euglyphidae (Rhizopoda, Testacea). *Pedobiologia* 15:415–424.

147. **Schönborn, W. 1977.** Production studies on Protozoa. *Oecologia* (*Berl.*) 27:171–184.

148. **Schönborn, W. 1978.** Untersuchungen zur Produktion der Boden-Testaceen. *Pedobiologia* 18:373–377.

149. **Schönborn, W. 1983.** Beziehungen zwischen Produktion, Mortalität und Abundanz terrestrischer Testaceen Gemeinschaften. *Pedobiologia* 25:403–412.

150. **Schönborn, W. 1983.** Modifikabilität und Evolutionstrends bei Protozoen. *Biol. Rundsch.* 21:225–235.

151. **Singh, B. N. 1946.** A method of estimating the numbers of soil Protozoa, especially amoebae, based on their differential feeding on bacteria. *Ann. Appl. Biol.* 33:112–119.

152. **Singh, B. N. 1952.** Nuclear division in nine species of small free-living amoebae and its bearing on the classification of the order Amoebida. *Philos. Trans. R. Soc. Lond. B Biol. Sci.* 236:405–461.

153. **Singh, B. N. 1955.** Culturing soil Protozoa and estimating their numbers in soil. Pp. 403–411 in D. K. McE. Kevan (ed.). *Soil Zoology*. Butterworth, London.

154. **Singh, B. N. and L. M. Crump. 1953.** The effect of partial sterilization by steam and formalin on the numbers of amoebae in field soil. *J. Gen. Microbiol.* 8:421–426.

155. **Singh, B. N., T. Datta and G. P. Dutta. 1971.** Factors inducing excystation in free living amoebae. *Indian J. Exp. Biol.* 9:350.

156. **Singh, B. N., S. Mathew and N. Anand. 1958.** The role of *Aerobacter* sp., *Escherichia coli,* and certain amino acids in the excystment of *Schizopyrenus russelli. J. Gen. Microbiol.* 19:104–111.

157. **Stout, J. D. 1955.** Environmental factors affecting the life history of three soil species of *Colpoda* (Ciliata). *Trans. R. Soc. N.Z.* 82:1165–1188.

158. **Stout, J. D. 1962.** An estimation of microfaunal populations in soils and forest litter. *J. Soil Sci.* 13:314–320.

159. **Stout, J. D. 1973.** The relationship between protozoan populations and biological activity in soils. *Am. Zool.* 13:193–201.

160. **Stout, J. D. 1980.** Role of Protozoa in nutrient cycling and energy flow. Pp. 40–44 in T. K. Crosby and R. P. Pottinger (eds.). *Second Australasian Conference on Grassland Invertebrate Ecology*. Government Printer, Wellington, New Zealand.

161. **Stout, J. D. 1980.** The role of Protozoa in nutrient cycling and energy flow. *Adv. Microb. Ecol.* 4:1–50.

162. **Stout, J. D., S. S. Bamforth and J. D. Lousier. 1982.** Protozoa. Pp. 1103–1120 in R. H. Miller, (ed.). *Methods of soil analysis.* Am. Soc. Agron. Monogr. 9.

163. **Stout, J. D. and O. W. Heal. 1967.** Protozoa. Pp. 149–195 in A. Burges and F. Raw (eds.). *Soil Biology*. Academic Press, London.

164. **Thomas, R. 1958.** Le genre *Plagiopyxis* Penard. *Hydrobiologia* 10:198–214.

165. **Tracey, M. V. 1955.** Cellulase and chitinase in soil amoebae. *Nature* (*Lond.*) 175:815.

166. **Varga, L. 1957.** Untersuchungen über die Mikrofauna der Waldstreu einiger Waldtypen im Bukkgebugt (Ungarn). *Acta Zool. Hung.* 4:443–478.

167. **Vargas, R. and T. Hattori. 1986.** Protozoan predation of bacterial cells in soil aggregates. *FEMS Microbiol. Ecol.* 38:233–242.

168. **Vergeiner, J., H. Laminger, C. Jaitner-Knaden and H. Bayer. 1982.** Populationsdynamik terrestrischer Protozoen (Testacea, Rhizopoda) in zentralalpinen Lagen Tirols. II. Untersuchungen gestörter und relativ ungestörter Böden im Raume Patsch (Tirol, Österreich). *Arch. Protistenkd.* 126:173–201.

169. **Volz, P. 1929.** Studien zur Biologie der bodenbewohnenden Thekamöben. *Arch. Protistenkd.* 68:349–406.

170. **Waksman, S. A. 1916.** Studies on soil Protozoa. *Soil Sci.* 1:135–152.

171. **Wilson, C. W. 1916.** On the life-history of a soil amoeba. *Univ. Calif. Publ. Zool.* 16:241–292.

172. **Zeuthen, E. 1943.** A Cartesian diver micro-respirometer, with a gas volume of 0.1 ml. *C.R. Trav. Lab. Carlsberg Ser. Chim.* 24:479–518.

173. **Zeuthen, E. 1953.** Oxygen uptake as related to body size in organisms. *Q. Rev. Biol.* 28:1–12.

7

Turbellaria: Tricladida: Terricola

IAN R. BALL

RONALD SLUYS

The large acoelomate phylum Platyhelminthes comprises the well-known parasitic flatworms such as flukes (Trematoda and Cercomeromorphae) and tapeworms (Cestoda), and a highly diverse group of predominantly free-living forms commonly known as the Turbellaria, although the latter class name does not feature in modern classifications (Ehlers 1985). A good general account of the phylum was given by Hyman (1951). A pictorial key to the families and genera of the Turbellaria has been published by Cannon (1986).

Most turbellarians are aquatic in habit but a few have adapted themselves to life in wet conditions close to permanent waters and one group, the planarians or triclads, contains many species which have become fully adapted to life on land or in the soil. Since it is the terrestrial planarians that are the most frequently encountered of the soil Turbellaria, and since they are the best known, this chapter deals principally with them. We know almost nothing of the soil-dwelling lower turbellarians in North America, but general accounts have been provided by Reisinger (1924) for the Neorhabdocoela, and Steinbock (1928) and Porter (1936) for the Prorhynchidae, and for these forms the keys to the freshwater Turbellaria provided by Ruebush (1941), Jones (1959) and Pennak (1978) are sufficient, at least to the generic level.

Published work on the terrestrial planarians includes mainly descriptive or distributional studies; very little is known of their biology and thus there is here a wealth of opportunity for students of soil organisms. Most of the species occur in tropical areas, but there are some indigenous to the north temperate zone. Moreover, many of the larger tropical species have attained wider distributions through human activities. It is not unusual to find them in greenhouses and other horticultural establishments, and some of them have been able to establish themselves outdoors both in North America and Europe.

Most of our knowledge of the North American terrestrial planarians is due to the work of Libbie H. Hyman and it must be admitted that many of her descriptions are based on a few (sometimes one) specimens and derived from serial sections of such poor quality as to prohibit an accurate account. It is therefore not surprising that there are errors and omissions in her descriptions, particularly of the native rhynchodemids. Furthermore, Hyman undertook nomenclatural revisions of the Rhynchodemidae that led to much confusion. This is not the place to discuss such problems and full accounts are easily available elsewhere (Hyman 1954, Minelli 1977). In this chapter the classificatory and nomenclatural decisions of Pantin (1953), Marcus (1953), and Minelli (1977) are in large part accepted.

Table 7.1 lists and classifies all the known North American land planarians, giving both the current nomenclature and the most commonly encountered synonyms. In most of the existing literature the two species of *Kontikia* are assigned to the genus *Geoplana*. But Sluys (1983) showed that *G. mexicana* should be transferred to the genus *Kontikia,* while the diagnostic characteristics of this genus also apply to *G. vaga* (Ogren, in litt.). The species of *Bipalium, Dolichoplana,* and *Kontikia* are, no doubt, exotics, whereas most of the Rhynchodemidae are indigenous. Two species of the latter family are also widespread in Europe. *Microplana terrestris* is the common European land planarian, and *Rhynchodemus sylvaticus* is now considered to be conspecific with the European *R. bilineatus* (see Minelli 1977). Furthermore, it is likely that *R. sylvaticus* and *R. americanus* are also conspecific because the variation found in the former in both Canadian and European populations encompasses the condition found in the latter. They are here maintained separately so as to encourage further critical comparison.

Large areas of North America remain unexplored as far as the lower invertebrates are concerned, and since terrestrial planarians in particular are mostly cryptic and difficult to find, there can be little doubt that the species list given here is far from complete.

General Morphology

Terrestrial planarians form a diverse group varying widely in size (<10 mm to about 600 mm), appearance, and structure. This diversity is reflected in the difficulty of achieving a stable higher classification of the group (de Beauchamp 1961, Froehlich 1967). They may be readily distinguished from other sorts of worms by their smooth unsegmented bodies, which lack spines, appendages, or bristles. The gut is blind, lacking an

Table 7.1. Known North American Species of Terrestrial Planarians, Showing Current Nomenclature and (in Parentheses) the Most Commonly Encountered Synonyms

Family Bipaliidae von Graff, 1896

Genus Bipalium Stimpson, 1857
1. *Bipalium kewense* Moseley, 1878
   (*Placocephalus kewense*)
2. *Bipalium adventitium* Hyman, 1943
3. *Bipalium pennsylvanicum* Ogren, 1987

Family Geoplanidae Stimpson, 1857

Genus Kontikia Froehlich, 1955
4. *Kontikia mexicana* (Hyman, 1939)
   (*Geoplana mexicana*)
5. *Kontikia vaga* (Hyman, 1943)
   (*Geoplana vaga*)

Family Rhynchodemidae von Graff, 1896

Subfamily Rhynchodeminae Correa, 1947
(= Dolichoplaninae Hyman, 1943)

Genus Rhynchodemus Leidy, 1851
6. *Rhynchodemus sylvaticus* (Leidy, 1851)
   (*Rhynchodemus bilineatus*)
7. *Rhynchodemus americanus* Hyman, 1943

Genus Dolichoplana Moseley, 1877
8. *Dolichoplana striata* Moseley, 1877
   (*Dolichoplana feildeni*)

Subfamily Microplaninae Pantin, 1953
(= Geodesminae Hyman, 1943)

Genus Microplana Vejdovsky, 1890
9. *Microplana terrestris* (Muller, 1773)
   (*Rhynchodemus terrestris, Geodesmus terrestris, Orthodemus terrestris*)
10. *Microplana atrocyanea* (Walton, 1912)
   (*Rhynchodemus atrocyaneus, Geodesmus atrocyaneus*)
11. *Microplana rufocephalata* Hyman, 1954

Genus Diporodemus Hyman, 1938
12. *Diporodemus indigenus* Hyman, 1943

anus, and forms three diverticulated branches. The first branch extends forward from the posteriorly located pharynx, and the second and third branches are directed posteriorly alongside and behind the pharynx. The body is usually covered with mucus, and at least some part of the body surface, especially the creeping sole, is ciliated.

The native terrestrial planarians are usually small (10–70 mm), more or less rounded, dull gray, black, or brown animals, with not more than two eyes. The introduced exotics, however, may be much larger (up to 600 mm), are flatter, and more brightly colored with stripes of various types. Most possess many eyes. Some of the species may be characterized by sensory tracts or grooves on the sides of the head; often, however, these are visible only in sections.

Since the precise identification of species of land planarians depends on analysis of the reproductive system it is necessary to give a brief account of this (Fig. 7.2). Almost all planarians are hermaphroditic, even though cross-copulation is the rule, but the male and female systems are not necessarily at the same stage of development in a given individual. Isolated descriptions of protandrous or protogynous individuals may well be one of the causes of the present taxonomic confusion in the group.

The female gonads are the paired ovaries. These are ovoid bodies lying ventrally, above the nerve cords, just behind the brain. From each an oviduct runs ventrally caudad to the copulatory apparatus, receiving on the way the openings of numerous vitelline follicles which are scattered throughout the body. These ducts thus function as ovovitelline ducts. The male gonads are the testes and these take the form of numerous spherical or ovoid follicles which are distributed from the level of the ovaries to the pharynx, to the copulatory apparatus, or to the tail, depending on the species. They may lie ventrally or dorsally to the intestine, or the follicles may be so large as to fill the entire dorsoventral space.

The copulatory apparatus is situated between the pharynx and the posterior end of the animal. It consists of a genital atrium, with ventral genital pore, into which both the male and female ducts enter. The penis occupies the anterior part, the male atrium. There is a muscular hemispherical bulb, embedded in the anterior wall, and the penis papilla, with ejaculatory duct, projects posteriad into the atrium. In some species the penis papilla is reduced, or absent (Fig. 7.3*A,B*). Within the penis bulb there may be an enlarged cavity, the bulbar cavity or seminal vesicle, which leads into the ejaculatory duct. From the testes the vasa deferentia run alongside the pharynx, becoming enlarged and distended with sperm to form spermiducal vesicles or false seminal vesicles. Behind the pharynx they curve upward and open into the seminal vesicle within the penis bulb.

The female genital duct extends from the posterior wall of the genital atrium, receiving numerous shell glands, and ultimately the openings of the two oviducts either separately, or after their union to form a short common oviduct. This part of the female genital canal is also referred to as the vagina or the glandular duct. Opening into the terminal part of the vagina, or connected with it by a duct or canal, there may be a sacciform organ, the bursa copulatrix, which serves to receive the sperm during copulation. In some species the bursa may have a separate opening to the ventral surface. Also, it is not uncommon for the bursa, or for the female genital canal if a bursa is not present, to connect with the intestine via a genitointestinal duct (Fig. 7.2), and in certain species there may be two or even three such communications.

The functions of the various parts of the reproductive apparatus are discussed below. Other organ systems are less important in the identification and classification of the Terricola, and adequate accounts of these will be found in Hyman (1951) or de Beauchamp (1961). A general introduction to all aspects of planarian structure, biology, and taxonomy is given by Ball and Reynoldson (1981). The classic monograph on Terricola is that of von Graff (1899).

Collection and Preservation

Terrestrial planarians are usually to be found underneath logs, stones, or boards lying on loose soil in damp shady places. The native rhynchodemids are particularly difficult to see and are easily overlooked; to the practiced eye the glisten of their mucus sheath may reveal them. Nevertheless, there is much truth in Hubricht's dictum (in Hyman 1954) that one finds them only when searching for other animals.

Planarians are fairly fragile animals, they fragment easily, and consequently they are unlikely to be recovered by normal soil extraction techniques. However, Ogren (1955) has described the use of the gentle heat from a 100-W bulb to induce *Rhynchodemus* species to leave samples of topsoil. The animals must be handled individually and carefully with clean moist implements, and protected from extremes of temperature and from desiccation. The larger exotic forms are more robust and more amenable to laboratory culture.

The best killing and fixing agents are those containing mercuric chloride. For the smaller forms Steinmann's fluid (1 part distilled water, 1 part concentrated nitric acid, 1 part 5% sodium chloride in saturated mercuric chloride) is satisfactory. Susa's fixative may also be effective, as is also a simple saturated solution of mercuric chloride used at 60°C. Larger tropical forms may be killed directly with boiling water. In all cases the specimens are best treated individually as they are crawling normally. After killing they should be transferred to 80% ethanol to which a little glycerine has been added, and when they have hardened they may be used for the preparation of serial sections.

Identification

Introduction to the Key

The recognition of the North American land planarians can be accomplished to the generic level on the basis of external appearance. The living worms may be observed when they are crawling in a petri dish lined with moistened filter paper. Features to note are the normal gliding length, coloration of both the dorsal and ventral surfaces, the shape of the head, and the number and position of the eyes. Intensity of color may be affected by preservatives. With this information most of the forms can be provisionally keyed out, at least to the generic level, and often to the species.

Accurate specific identification, however, is a hazardous venture for the nonspecialist and normally requires the preparation of serial sections. The problem is further compounded by the lack of adequate reference material in museums and by the fact that our knowledge of the North American forms is so scanty. It follows that caution must be exercised in attempting to fit unusual specimens into a key that is valid only for the fauna as at present known. The account below must be considered highly provisional, for there is still much basic work to be done on this group.

Key to the North American Species of Terrestrial Planarians

Where a couplet is not reached immediately from the one preceding it, the number of the couplet from which the direction came is given in parentheses; characters requiring serial sections are given in brackets.

1a Eyes two, anteriorly situated (Fig. 7.1*A–D,H,I*) **2**

1b Eyes many, scattered over the anterior margin and along the sides of the body (Fig. 7.1*E–G*) **8**

2a Dorsal surface light brown, with six longitudinal stripes (Fig. 7.1*H*) . . ***Dolichoplana striata***

This distinctive species has a flattened ribbonlike body of size up to 200 × 35 mm. The anterior end is rounded with two small eyes. The ventral creeping sole is white; the dorsal surface is a light olive- or yellow-brown and bears three pairs of dark longitudinal stripes which begin just behind the eyes. The medial stripes are usually narrow, whereas the lateral and marginal stripes are broader and usually darker. The species is rarely found in the sexual state and reproduces by fragmentation. A native of Indo-Malaya, it is one of the commonest tropical species to have become widely distributed by humans. In temperate regions it is normally associated with greenhouses, but in Florida populations have become established outdoors and have proved to be a nuisance in commercial earthworm-rearing beds as a result of its predatory activity. Distribution: Missouri, greenhouses (Hyman 1943); Florida (Hyman 1954).

2b Dorsal surface uniform gray, black, or blue-black, sometimes with two longitudinal stripes **3**

3a Eyes terminal, or nearly so (Fig. 7.1*A*) [penis papilla present in mature specimens] **4**

3b Eyes subterminal, head with proboscislike region (Fig. 7.1*B*) [penis papilla absent in mature specimens]. **7**

4a Dorsal surface black, anterior tip reddish brown [vagina and genitointestinal canal very long] (Fig. 7.2*c*) . . . . ***Microplana rufocephalata***

35 mm long preserved and of elongate cylindroid shape. Color black, with a lighter creeping sole, and with a distinctive reddish anterior tip bearing two small eyes. The testes of the holotype are small, ventrally located, and apparently prepharyngeal. The penis comprises a massive bulb, with spacious seminal vesicle, and a large conical papilla. The female genital canal is a long duct extending backward from the atrium and distally it receives the openings of the two oviducts; it then curves forward once more as a long genitointestinal canal, opening into the gut wall without forming a bursa. This species is known only from the type material, which is in poor condition. Distribution: Kentucky (Hyman 1954); North Carolina (Darlington 1979, in litt.).

4b Dorsal surface totally black, blue-black, or gray [vagina and genitointestinal canal of normal proportions] **5**

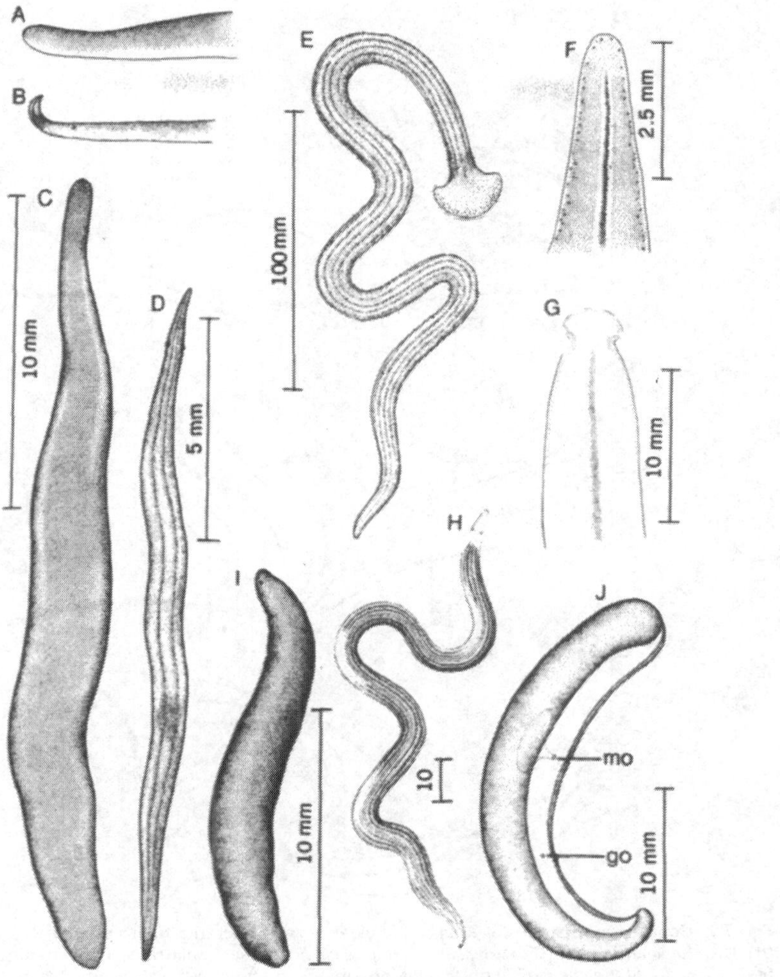

Fig. 7.1. North American terrestrial planarians: (*A*) *Microplana terrestris*, lateral view of head; (*B*) *Rhynchodemus sylvaticus*, lateral view of head; (*C*) *Microplana terrestris;* (*D*) *R. sylvaticus;* (*E*) *Bipalium kewense;* (*F*) *Kontikia mexicana*, head; (*G*) *Bipalium adventitium*, head; (H) *Dolichoplana striata;* (*I*) *Microplana atrocyanea*, preserved specimen; (J) *Diporodemus indigenus*, preserved specimen. [(*F*) After Hyman (1939); (*G,J*) after Hyman (1943).]

5a Up to 70 mm long, dorsal surface slaty gray, two sensory cephalic tracts ventrally situated on the head [bursa present, often with a separate ventral pore to the exterior] (Figs. 7.1*J*, 7.3*C*) . . . . **_Diporodemus indigenus_**

Slender worm reaching a length of 70 mm. Color dark slaty gray with a lighter, or white, creeping sole. Head rounded, with two deep-set eyes which may be discernible only in sections. Characteristic of this species are two cephalic sensory tracts on the ventral side of the head, one on each side of the creeping sole. Testes numerous, ventral, extending no farther than the pharynx. Penis large, with a strong muscular bulb and stout conical papilla. The female genital canal runs posteriad, receiving the oviducts separately, and then continues as a narrow sinuous tube which functions as a genitointestinal canal. It empties into a large stromatic muscular bursa which itself communicates directly with the

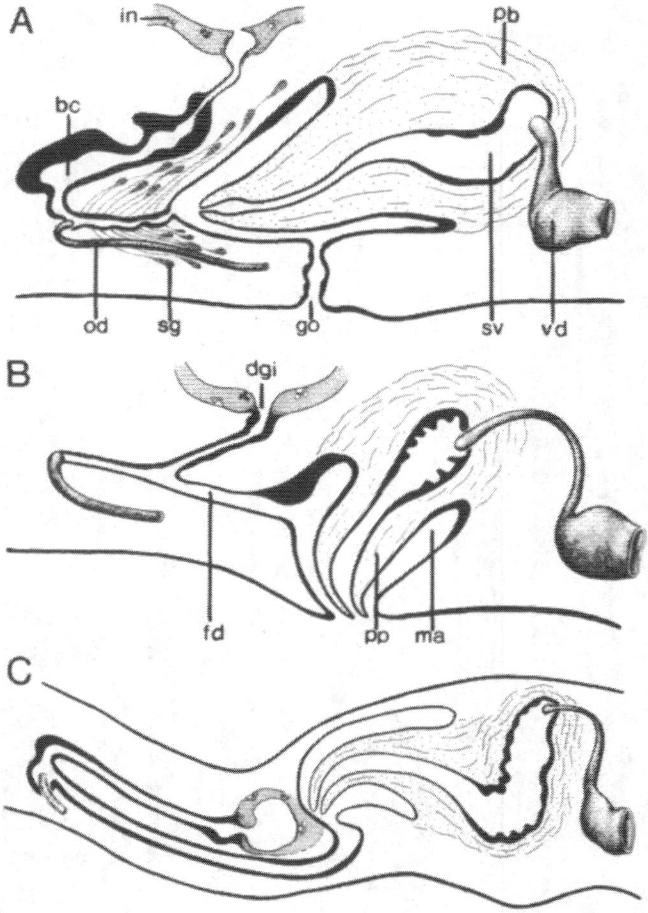

Fig. 7.2. Copulatory apparatus of Rhynchodemidae, viewed from the right side: (*A*) *Microplana terrestris;* (*B*) *Microplana atrocyanea;* (*C*) *Microplana rufocephalata. bc,* bursa copulatrix; *dgi,* genito-intestinal duct; *fa,* female atrium; *fd,* female genital duct; *go,* gonopore; *in,* intestine; *ma,* male atrium; *od,* oviduct; *pb,* penis bulb; *pp,* penis papilla; *sg,* shell glands; *sv,* seminal vesicle; *vd,* vas deferens. [(B) after Hyman (1954).]

intestine and may, in mature specimens, also open to the exterior by a separate ventral pore, behind and lateral to the common genital pore. Distribution: Maryland, Virginia, West Virginia, Washington D.C., Michigan; North Carolina, Kentucky, Tennessee? (Hyman 1943, 1954); Louisiana (Conella and Stern 1969); New York (specimen in American Museum of Natural History); Quebec (Ogren and Ball 1989). A suspected second species of the genus is reported, but not described, from North Carolina (Darlington and Place 1966).

5b Up to 30 mm long, dorsal surface usually blue-black or dark gray, sensory cephalic tracts absent [bursa present or absent, never with a separate pore to the ventral surface] 6

6a [Bursa copulatrix absent; one genitointestinal connection between vagina and intestine] (Fig. 7.1*I*, 7.2*B*). . . . . . . ***Microplana atrocyanea***

Cylindroid, up to 25 mm long. Color black, with a blue or purplish sheen, white or light gray below. Head rounded or truncate with two eyes near the tip. Testes numerous, ventral, prepharyngeal. Penis normal; female genital canal extends posteriorly and throws off a branch that curves forward and opens into the intestine without forming a well-developed bursa. At the distal end of the vagina the two oviducts are received separately. Distribution: Ohio (Walton 1904); Illinois, Missouri, Kentucky, Tennessee, Arkansas (Hyman 1943); Virginia, North Carolina (Hyman 1954); New York (specimens in the American Museum of Natural History), Pennsylvania (Ogren 1981, 1982). According to Darlington (1979, in litt.), this is the most common species in North Carolina.

6b [Bursa copulatrix present usually with two, or three, communications between it and the intestine on each side] (Figs. 7.1*A*,*C*, 7.2*A*) . . . . . ***Microplana terrestris***

Cylindroid, up to 25 mm long. Dorsal surface a uniform shade of gray or purplish black, ventral surface paler. Anterior end truncate, with two tiny eyes at the tip. Testes numerous, ventral, distributed to the pharyngeal region. Penis well developed. Female genital canal receives the two oviducts separately and also leads into a bursa copulatrix which may have one, two, or three genitointestinal connections. One of the most widespread species in Europe, it is possible that ***M. terrestris*** is an introduction into North

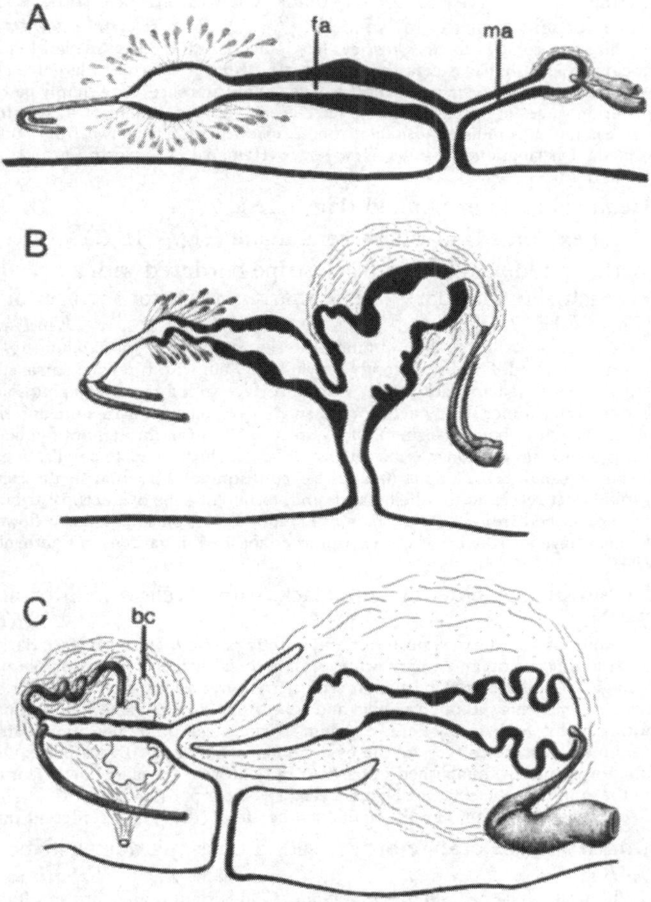

Fig. 7.3. Copulatory apparatus of Rhynchodemidae, viewed from the right side: (*A*) *Rhynchodemus americanus;* (*B*) *Rhynchodemus sylvaticus;* (*C*) *Diporodemus indigenus.* [(C) after Hyman (1943.] For identification of parts, see Fig. 7.2.

America. The relationships between this species and ***M. atrocyanea*** need elucidation. Distribution: Wisconsin (Hyman 1939); New York (Ogren 1984a); Ontario, Quebec (Ogren and Ball 1989). Ogren (1988) has provided sound evidence that ***M. atrocyanea*** and ***M. Terrestris*** are conspecific. The latter has priority.

7a (3) Dorsal surface gray or brown with two darker longitudinal stripes; male atrium larger than the female atrium (Figs. 7.1*B,D*, 7.3*B*) . . . . ***Rhynchodemus sylvaticus***

Up to 15 mm long. Most of the body rounded but the anterior region flattened, and usually lifted from the substratum, thus having a proboscis-like appearance. Two large eyes situated a little back from the anterior tip. The color is variable, usually some shade of gray or brown, but occasionally dirty white, purplish brown, or almost black. There are usually two dark dorsal longitudinal stripes united with a dark saddle in the pharyngeal region; but these stripes on rare occasions are absent. The ventral surface is pale. The testes are large, few in number, and extend postpharyngeally. The copulatory complex lacks a penis papilla and the female genital canal lacks a bursa and a genitointestinal duct. This species is doubtfully distinct from ***R. americanus*** (q.v.) Distribution: Pennsylvania (Ogren 1981); Rhode Island (Leidy 1858); Ohio (Walton 1912); Missouri (Hyman 1943); Illinois, Wisconsin, New York (Ogren 1955); Louisiana (Conella and Stern 1969); Ontario (Ogren and Ball 1989).

7b Dorsal surface gray or brown-black, without stripes, male atrium smaller than female atrium (Fig. 7.3*A*) . . . . ***Rhynchodemus americanus***

Size and color as for ***R. sylvaticus*** (q.v.), but lacks the dorsal stripes. Reproductive apparatus also similar except that the female atrium is very much larger than the male, whereas in ***R. sylvaticus*** the reverse is true. The differences may simply be reflections of differing stages of sexual maturity (see Ogren 1956); specimens of ***R. sylvaticus*** without stripes, and thus indistinguishable from ***R. americanus***, are known from both Europe and Canada. Distribution: Missouri, New Jersey (Hyman 1943); North Carolina (Darlington 1979, in litt.).

8a (1) Head rounded or pointed (Fig. 7.1*F*). **9**

8b Head expanded, spatulate, semilunate (Fig. 7.1*E,G*) **10**

9a With a middorsal brown-black stripe bordered with a broad light stripe on each side and dark lateral margins; ventral surface faintly banded (Figs. 7.1*F*, 7.4*A*) . . . . ***Kontikia mexicana***

Flat, elongate, up to 20 mm long. Anterior end narrow with a blunt tip, posterior end abruptly pointed. Color apparently brown-black, but with two light dorsal stripes bordering a narrow median black stripe. Ventral surface with a raised creeping sole and a faint banded appearance. Many eyes form a single irregular row across the anterior tip and along the entire body margin. Testes numerous, ventral, throughout the body length. Penis comprises an elongate muscular bulb and a short pluglike papilla. Seminal vesicle an S-shaped canal. Female apparatus simple, consisting of a backwardly directed female genital duct receiving the shell glands and, terminally, the oviducts. Distribution: originally described from material taken at Laredo, Texas, on shipments of flowers from Mexico (Hyman 1939), the species is now established in gardens in California (Hyman 1943).

9b Dorsal surface blue, or blue-black, with a yellow middorsal stripe (Fig. 7.4*C*). . . . . ***Kontikia vaga***

Anterior end narrow, posterior end bluntly pointed. Dorsal surface dark blue with a whitish anterior tip and a narrow yellow middorsal stripe. Ventral surface pale. Eyes form a single irregular row along the anterior margin but do not cross the anterior tip. Testes not known. Penis without a papilla and comprising an elongated hollow muscular organ with a highly folded wall. Female atrium spacious, extending posteriorly, and leading to a glandular duct which receives the oviducts and shell glands. Distribution: Original habitat unknown but now established in gardens in California (Hyman 1954); Florida (specimen in the American Museum of Natural History); Georgia? (Olewine 1972). The unnamed ***Geoplana*** species from Florida mentioned by Mead (1973) may represent this species.

10a (8) Dorsal surface orange-brown with a dark middorsal stripe (Figs. 7.1*G*, 7.4*B*) . . . . ***Bipalium adventitium***

Elongate, flattened, up to 75 mm long. Color bright orange-brown with a single narrow, dark middorsal stripe which does not extend onto the head. Anterior characteristically lunate, or fan shaped, bearing numerous eyes around the edge and extending along the anterior body margins. Ventrally, the eyes may continue over the whole length of the body. Testes not known. Penis comprises an elongate muscular bulb and a long

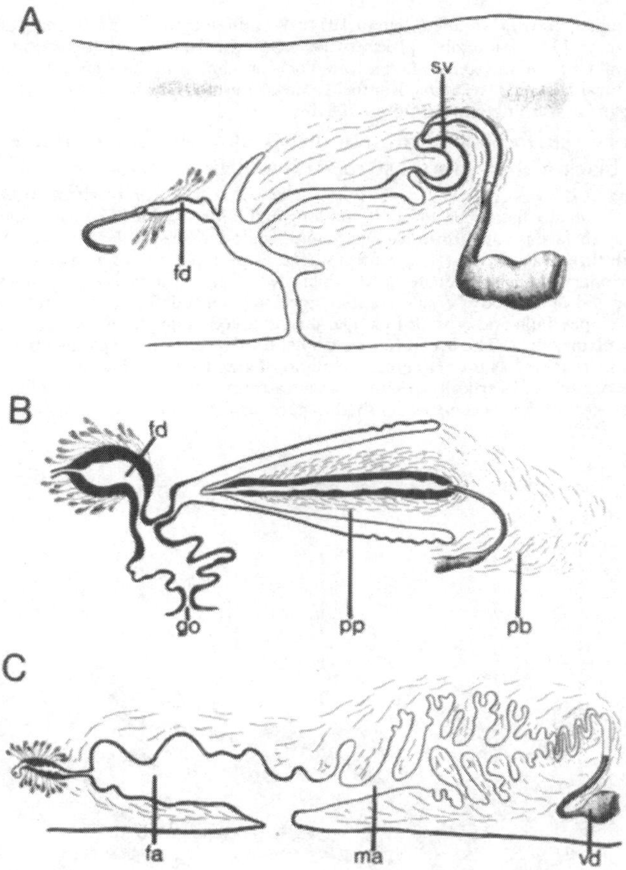

Fig. 7.4. Copulatory apparatus of Bipaliidae and Geoplanidae, viewed from the right side: (*A*) *Kontikia mexicana;* (*B*) *Bipalium adventitium;* (*C*) *Kontikia vaga.* [(*B,C*) After Hyman (1943).] For identification of parts, see Fig. 7.2.

conical papilla containing a spacious ejaculatory duct. Female genital duct broad, bursa-like, receiving numerous glands as well as the oviducts. The original habitat of this species is unknown but it has now become established in a number of places in North America. Distribution: California (Hyman 1943); New York (Hyman 1954, Klots 1960, Dindal 1970); Massachusetts (Klots 1960); Pennsylvania (Ogren 1981, 1984b, 1985).

10b Dorsal surface yellowish with five purple-black longitudinal stripes which converge to give a dark patch at the neck (Fig. 7.1*E*). . . . . ***Bipalium kewense***

Elongate, up to 350 mm long. Basic color yellow-ochre on which there are five black or violet dorsal longitudinal stripes. These stripes run together just behind the head to give a characteristic dark patch. The ventral surface is paler. There are numerous eyes scattered over the anterior end. This species is rarely found in the sexual state and reproduces by fragmentation. A native of Indo-Malaya, ***B. kewense*** has become widely distributed by humans. In North America it is usually to be found in greenhouses, but in the southern United States, populations have become established outdoors, even as far north as North Carolina (Ogren 1984b). Distribution: Louisiana, Alabama, Mississippi (Hyman 1943, 1954; Dundee and Dundee 1963; Conella and Stern 1969); Florida (Hyman 1943); Texas (Tramel and Campbell 1971); California (Hyman 1943, 1954); Tennessee (Chandler 1974, 1976); Georgia (Hyman 1943, 1954; Morton and Kleinginna 1971); South

Carolina, North Carolina (Hyman 1954); Washington D.C., New Jersey, Ohio, Illinois (Hyman 1943); Manitoba (Jackson 1940); Ontario (Judd 1957; also specimens in the Royal Ontario Museum); Maine, New York (specimens in the American Museum of Natural History); Arkansas, Kentucky, Massachusetts, Oklahoma, Pennsylvania, Utah, Virginia, Washington (see Ogren 1984b).

10c Dorsal surface pale brown or dirty yellow with some orange, with purple to black median dorsal stripe and a diffuse lateral stripe on each side of the body (Fig. 7.5) . . . . ***Bipalium pennsylvanicum***

Elongate, flattened, up to 80 mm long and 3.5 mm wide. The basic color of the dorsal body surface is pale brown to dirty yellow, while the head is dark brown. Dorsal surface with three dark stripes. The ventral surface is pale. Numerous eyes occur along the dorsal periphery of the semilunate head, while rows of eyes extend along the sides to the posterior end of the body. Testes prepharyngeal and ventral. The penis consists of a well-developed bulb and a conical papilla, the latter containing a seminal vesicle and a spacious ejaculatory duct. The broad female genital duct receives the separate openings of the oviducts as well as the numerous openings of glands. The female genital duct opens into the narrow and vertically oriented female atrium. The species is not indigenous to North America but has become established in gardens. Distribution: Pennsylvania (Ogren 1987).

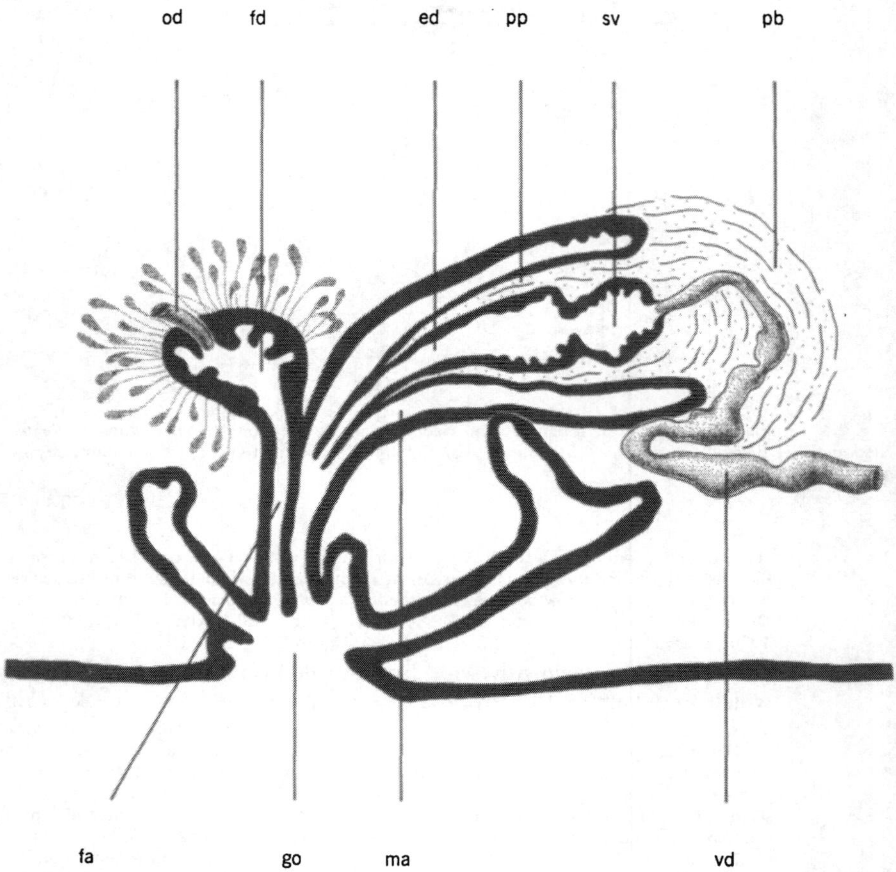

Fig. 7.5. Copulatory apparatus of *Bipalium pennsylvanicum;* sagittal reconstruction. [After Ogren (1987).] For identification of parts, see Fig. 7.2.

General Biology

Ecology and Distribution

Since terrestrial planarians lack special mechanisms for water conservation, they are more affected than most other organisms by the moisture of their environment. Too little and they desiccate; too much and they deliquesce (Froehlich, 1955; Barnwell, 1969). Because of these limitations they tend to remain hidden during the day in humid, but not wet, refuges, emerging at night when the relative humidity of the air is high. Their favorite resting places are under stones, logs, boards, bricks, refuse, leaf litter, and the like, lying in damp shady places. Occasionally, they may be found deeper in the soil, and Ogren (1955) is of the opinion that *Rhynchodemus sylvaticus* must be regarded as a true soil dweller since it lives in the top 50 mm of the soil and deposits its cocoons there. *Bipalium adventitium* may be found in rich mesic woodland, in the burrows of the earthworms upon which it preys (Klots 1960), and Ogren (1985) has recorded specimens of this species in the upper 20 cm of the soil. Only rarely are land planarians found crawling in daytime, and this usually after heavy rains which probably have rendered their refuges waterlogged.

The introduced exotic species of *Bipalium, Kontikia,* and *Dolichoplana* show a scattered distribution, as might be expected. They seem always to be associated with humans and probably are transported around in the soil surrounding the roots of potted plants. Nevertheless, *Bipalium kewense, B. adventitium,* and *B. pennsylvanicum* have established themselves outdoors in areas subject to winter freezing (Hyman 1954, Klots 1960). *B. kewense* survives temperatures near 0°F (−18°C), probably by going underground (Barnwell 1969). Further notes on the biology of this species have been given by Chandler (1976).

Native rhynchodemids are so rarely found that little can be said of their ecology and distribution. Published records all lie east of a line from Wisconsin to Louisiana, and the west of North America is unexplored for these animals. Species such as *Rhynchodemus sylvaticus, Microplana atrocyanea,* and *M. terrestris* range as far north as the Canadian border, and thus must be able to withstand severe cold. The latter species is puzzling in that it is so widespread in Europe, yet there are only three records from North America. *Diporodemus indigenus* is especially widespread, although never numerous, and seems to be an Appalachian species. It, too, ranges as far as Ontario.

It is noteworthy that although land planarians are typically free-living soil invertebrates, they have been known to invade human body openings and to assume a pseudoparasitic mode of life. The three known cases, including one from Atlanta, Georgia, all involved species of the genus *Bipalium* (Walton and Yokogawa 1972).

Feeding and Digestion

Despite their apparent fragility, planarians are active and successful predators on other invertebrates. They can tackle living snails, slugs, earthworms, isopods, insect larvae, springtails, and even other land planarians. The larger introduced species, such as *Dolichoplana striata* and *Bipalium adventitium,* have become pests in parts of the United States because of their predatory activities in commercial earthworm beds (Hyman 1954, Dindal 1970).

In attacking large prey, such as a slug or earthworm, the planarians are vulnerable during the early stages. But once they have mounted the prey, and taken a firm hold with the pharynx, the prey is literally eaten alive (Froehlich 1955, Dindal 1970). Small slugs and arthropods may be swallowed whole. Species of *Rhynchodemus* are known to use their anterior proboscislike region (Fig. 7.1*A*) to catch their prey. Since it is covered with secretions when it touches such as a springtail, the prey adheres to the tip, is lifted off the ground, and is transferred to the mucus current of the creeping sole, and then the planarian creeps over it until it is within reach of the pharynx (Froehlich 1955).

Once large prey have been trapped, they are pierced by the extensible and protrusible pharynx of the planarian. The secretion of subepidermal glands may assist in disintegrating the prey. It seems that in some species external digestion occurs and the pharynx is used to suck up the more or less liquefied tissues of the prey. In *Microplana terrestris,* however, digestion appears to be intracellular, following rapid phagocytosis by the gut epithelium, but some intraluminar digestion of proteins may occur (Jennings 1959, 1974).

Since planarians do not possess an anus, the elimination of solid waste is via the mouth assisted by the action of the pharynx and by convulsions of the entire body.

Reproduction and Life Cycle

Reproduction may be sexual and asexual, and in the latter case cross-copulation is the rule, even though these animals are hermaphrodite. In general, it is the introduced tropical forms that reproduce asexually. They do this by fragmentation, yielding pieces just a few centimeters in length, which then regenerate to form new worms. No doubt the successful colonizing habit of these species is due to this rapid mode of reproduction. In Louisiana, sexually reproducing populations of *Bipalium kewense* have become established (Conella and Stern 1969).

In sexually reproducing species sperm is introduced during copulation into the bursa copulatrix (or female genital canal). From here it is transferred to the oviducts, where it moves anteriorly to be stored just behind the ovaries. Thus as eggs are released they are fertilized and then collect the products of the vitelline follicles as they travel to the atrium, where the cocoon is formed.

Few data on egg capsules and life cycles are available. The cocoons are usually attached to objects in the humid refuges of the flatworms. Those of *Bipalium kewense* are 5–7 mm in diameter, bright red when laid, changing to black. They contain few (about three) young (Conella and Stern 1969). The cocoons of *Rhynchodemus sylvaticus* show a similar color change but are much smaller when laid, being about 1.2–1.5 mm in diameter (Ogren 1957). They hatch in about 30 days, yielding four or five immature replicas of their parents.

It is probable that most of the rhynchodemids are either protogynous or protandrous, and thus one cannot expect to find all parts of the reproductive system of one individual at the same stage of development. This is evident from the work of Ogren (1957, 1986) on *Rhynchodemus sylvaticus* and from our own work on the same species and *Microplana terrestris* in both Canada and Europe.

Movement and Locomotion

For the most part terrestrial planarians are surface dwellers and their principal mode of locomotion is gliding on a layer of mucus that is secreted from special glands. Consequently, they leave behind them a slime trail, and sometimes mucus threads are used to bridge gaps in the substratum or to enable the worm to lower itself from one level to another. During locomotion the head may be raised and waved from side to side; this is particularly obvious in species of *Bipalium* and *Rhynchodemus*. The main locomotory forces are provided by both ciliary and muscular action.

According to Pantin (1950) and Jones (1978) *Microplana terrestris* glides at about 20–30 mm/min, and the entire length of the creeping sole is in contact with the substratum. In this species cilia apparently provide the propulsive forces during normal gliding. But under certain circumstances stationary waves, involving muscular action, may be used, especially under conditions of low humidity (Jones 1978).

Stationary waves are more usual in *Rhynchodemus sylvaticus*, which can glide at speeds up to 50 mm/min. In this species locomotion is achieved by three or four ventral protruberances which remain stationary relative to the substratum. Both the muscles and cilia are involved because the stationary waves are caused by the superimposition of peristaltic waves and ciliary locomotion. Thus, during gliding, *R. sylvaticus* leaves mucus "footprints" rather than a continuous slime trail. This conserves mucus, and therefore water, enabling *R. sylvaticus* to live in drier habitats than *Microplana terrestris*.

While it is true that most planarians inhabit the surface, leaf litter, or loose soil, some species have been able to take advantage of their muscularity and extensibility and may live somewhat deeper. In Britain, for example, one species of *Microplana* can burrow into clay to a depth of 45 cm (Percival 1925).

ACKNOWLEDGMENTS

I. R. Ball is grateful to Dr. E. Kirsteuer (American Museum of Natural History) for granting access to the collections of the late Dr. Libbie Hyman and for lending him many of her slides for detailed study, and he thanks also Mr. Harold Feinberg for facilitating his work in New York. Fieldwork and laboratory studies in Ontario and Quebec were carried out under the auspices of the Royal Ontario Museum, Toronto. Professor J. T. Darlington (Southwestern University at Memphis) generously supplied much unpublished information and Professor T. B. Reynoldson (University College of North Wales, Bangor) read and criticized a draft of the manuscript. All of the illustrations were prepared by Maria Tran Thi Vinh-Hao.

References

Ball, I. R. and T. B. Reynoldson. 1981. *British Planarians.* Cambridge University Press, Cambridge. 141 pp. **Barnwell, G. M. 1969.** The adaptability and distribution of the land planarian *Bipalium kewense. J. Biol. Psychol.* 11(2):30–32. **Cannon, L. R. G. 1986.** *Turbellaria of the World—A Guide to Families and Genera.* Queensland Museum, South Brisbane, Australia. 136 pp. **Chandler, C. M. 1974.** The land planarian *Bipalium kewense* Moseley, 1878, in middle Tennessee. *J. Tenn. Acad. Sci.* 49(4):128–129. **1976.** Field Observations on a population of the land planarian *Bipalium kewense* (Turbellaria, Tricladida) in middle Tennessee. *J. Tenn. Acad. Sci.* 51(2):73–75. **Connella, J. V. and D. H. Stern. 1969.** Land planarians: sexuality and occurrence. *Trans. Am. Microsc. Soc.* 88(2):309–311. **Darlington, J. T. and J. G. Place. 1966.** A new terrestrial triclad from the Highlands, North Carolina Region. *ASB Bull.* 13(2):33 (abstract). **de Beauchamp, P. 1961.** Classe des Turbellaires. Pp. 35–212 in P. P. Grasse (ed.). *Traite de Zoologie,* Vol. IV. Masson et Cie, Paris. **Dindal, D. L. 1970.** Feeding behavior of a terrestrial Turbellarian *Bipalium adventitium. Am. Midl. Nat.* 83(2):635–637. **Dundee, D. S. and H. A. Dundee. 1963.** Observations on the land planarian *Bipalium kewense* Moseley in the Gulf coast. *Syst. Zool.* 12:36–37. **Ehlers, U. 1985.** *Das phylogenetische System der Plathelminthes.* Gustav Fischer, Stuttgart, West Germany. 317 pp. **Froehlich, C. G. 1955.** On the biology of land planarians. *Univ. São Paulo Fac. Filos. Cienc. Let. Bol. Zool.* 20:263–271. **1967.** A contribution to the zoogeography of neotropical land planarians. *Acta Zool. Lilloana* 23:153–162. **Hyman, L. H. 1939.** New species of flatworms from North, Central, and South America. *Proc. U.S. Nat. Mus.* 86(3055):419–439. **1943.** Endemic and exotic land planarians in the United States with a discussion of necessary changes of names in the Rhynchodemidae. *Am. Mus. Novit.* 1241:1–21. **1951.** The acoelomate Bilateria. In *The Invertebrates.* Vol. II. *Platyhelminthes and Rhynchocoela.* McGraw-Hill, New York. 550 pp. **1954.** Some land planarians of the United States and Europe, with remarks on nomenclature. *Am. Mus. Novit.* 1667:1–21. **Jackson, V. W. 1940.** *Placocephalus kewensis* in Manitoba. *Can. Field-Nat.* 54:75. **Jennings, J. B. 1959.** Observations on the nutrition of the land planarian *Orthodemus terrestris* (O. F. Muller). *Biol. Bull.* (*Woods Hole*) 177:119–124. **1974.** Digestive physiology of the Turbellaria. Pp. 173–197 in N. W. Riser and M. P. Morse (eds.). *Biology of the Turbellaria.* McGraw-Hill, New York. 530 pp. **Jones, E. R. 1959.** Catenulida, Macrostomida, Neorhabdocoela, Alloeocoela. Pp. 334–365 in W. T. Edmonson (ed.). *Ward and Whipple's Freshwater Biology,* 2nd ed. Wiley, New York. 1248 pp. **Jones, H. D. 1978.** Observations on the locomotion of two British terrestrial planarians (Platyhelminthes, Tricladida). *J. Zool.* (*Lond.*) 186:407–416. **Judd, W. W. 1957.** A terrestrial turbellarian flatworm collected at London, Ontario. *Can. Field-Nat.* 71:154. **Klots, A. B. 1960.** A terrestrial flatworm well established outdoors in the northeastern United States. *Syst. Zool.*

9:33–34. **Leidy, J. 1858.** Remarks on *Rhynchodemus terrestris. Proc. Acad. Nat. Sci. Phila.* 10:171–172. **Marcus, E. 1953.** *Turbellaria Tricladida. Exploration du Parc National de l'Upemba. Mission G. F. Witte.* Institut des Parcs Nationaux du Congo Belge, Afl. 21. 61 pp. **Mead, A. R. 1973.** New outbreaks in the Florida giant African snail infestation. *Bull. Am. Malacol. Union Inc.* 38:19 (abstract). **Minelli, A. 1977.** A taxonomic review of the terrestrial planarians of Europe. *Boll. Zool.* 44:399–419. **Morton, R. and P. Kleinginna. 1971.** Running speed and mortality of *Bipalium kewense* as a function of different levels of illumination. *J. Biol. Psychol.* 13(1):25–26. **Ogren, R. E. 1955.** Ecological observations on the occurrence of *Rhynchodemus,* a terrestrial turbellarian. *Trans. Am. Microsc. Soc.* 74:54–60. **1956.** Physiological observations on movement and behavior of the land planarian *Rhynchodemus sylvaticus* (Leidy). *Proc. Pa. Acad. Sci.* 30:218–225. **1957.** Developmental observations, the egg capsule and sexual maturity of the land planarian *R. sylvaticus. Trans. Am. Microsc. Soc.* 76(1):45–52. **1981.** Land planarians in Pennsylvania. *Proc. Pa. Acad. Sci.* 55:52–56. **1982.** The identification and reproductive apparatus of the land planarian, *Microplana atrocyaneus* (Walton) from Pennsylvania (Platyhelminthes; Turbellaria). *Proc. Pa. Acad. Sci.* 56:39–43. **1984a.** The land planarian *Microplana terrestris* (Platyhelminthes: Turbellaria) from the United States. *Trans. Am. Microsc. Soc.* 103(2):130–136. **1984b.** Exotic land planarians of the genus *Bipalium* (Platyhelminthes: Turbellaria) from Pennsylvania and the Academy of Natural Sciences, Philadelphia. *Proc. Pa. Acad. Sci.* 58:193–201. **1985.** The human factor in the spread of an exotic land planarian in Pennsylvania. *Proc. Pa. Acad. Sci.* 59:117–118. **1986.** Development of the reproductive apparatus of the land planarian *Rhynchodemus sylvaticus* (Turbellaria: Tricladida) and its significance for classification in the genus. *Hydrobiologia* 132:175–180. **1987.** Description of a new three-lined land planarian of the genus *Bipalium* (Turbellaria: Tricladida) from Pennsylvania, U.S.A. *Trans. Am. Microsc. Soc.* 106:21–30. **1988.** Developmental and morphological evidences that the land planarians *Microplana terrestris* and *Microplana atrocyanea* are conspecific. *Progress in Zoology* 36:395–398 in P. Ax, U. Ehlers and B. Sopott-Ehlers (eds.). *Free-Living and Symbiotic Platyhelminthes.* Gustav Fischer Verlag, Stuttgart and New York. 529 Pp. **Ogren, R. E. and I. R. Ball. 1989.** Land Planarians from Ontario and Quebec. *Trans. Am. Microsc. Soc.* (in press). **Olewine, D. A. 1972.** Further observations in Georgia on the land planarians *Bipalium kewense* and *Geoplana vaga* (Turbellaria, Tricladida, Terricola). *ASB Bull.* 19(2):88 (abstract). **Pantin, C. F. A. 1950.** Locomotion in British terrestrial nemertines and planarians: with a discussion on the identity of *Rhynchodemus bilineatus* (Mecznikow) in Britain, and on the name *Fasciola terrestris* O. F. Muller. *Proc. Linn. Soc. Lond.* 162:23–37. **1953.** On the name of the grand fluke *Fasciola terrestris* O. F. Muller, on *Othelosoma sysmondsi* Gray, and on the genus *Amblyplana* von Graff. *J. Linn. Soc. Lond. Zool.* 42:209–218. **Pennak, R. W. 1978.** *Fresh-Water Invertebrates of the United States,* 2nd ed. Wiley, New York. 803 pp. **Percival, E. 1925.** *Rhynchodemus brittanicus,* n. sp. A new British terrestrial triclad, with a note on the excretion of calcium carbonate. *Q. J. Microsc. Soc.* 69:344–355. **Porter, R. J. 1936.** *The Family Prorhynchidae in America.* M.A. thesis, University of Virginia, Charlottesville, Va. 25 pp. + 2 plates. **Reisinger, E. 1924.** Die terricolen Rhabdocoelen Steiermarks. *Zool. Anz.* 59:33–48, 72–86, 128–143. **Ruebush, T. K. 1941.** A key to the American freshwater turbellarian genera, exclusive of the Tricladida. *Trans. Am. Microsc. Soc.* 60:29–40. **Sluys, R. 1983.** A new species of land planarian from Madeira (Platyhelminthes, Turbellaria, Tricladida). *J. Zool.* (*Lond.*) 201:433–443. **Steinbock, O. 1928.** Monographie der Prorhynchidae (Turbellaria). *Z. Morphol. Oekol. Tiere* 8:538–662. **Tramell, P. R. and J. W. Campbell. 1971.** Carbamyl phosphate synthesis in invertebrates. *Comp. Biochem. Physiol. B Comp. Biochem.* 40:395–406. **von Graff, L. 1899.** *Monographie der Turbellarien.* Vol. II. *Tricladida Terricola.* Wilhelm Engelman, Leipzig, East Germany. 574 pp. (plus Atlas of 57 plates). **Walton, L. B. 1904.** A land planarian in Ohio. *Proc. Ohio Acad. Sci.* 4:178. **1912.** The land planarians of North America, with a note on a new species. *Science* 35:940. **Walton, B. C. and M. Yokogawa. 1972.** Terrestrial turbellarians (Tricladida, Bipaliidae) as pseudoparasites of man. *J. Parasitol.* 58(3):444–446.

8

Nematoda

DIANA W. FRECKMAN

JAMES G. BALDWIN

Nematodes (= Nemata *sensu* Maggenti 1981) are a phylum of aquatic, bilaterally symmetric, unsegmented, triploblastic, elongated roundworms. They range in length from about 0.3 mm to 8 m. Nematodes are ubiguitous, occurring in such diverse habitats as soil, fresh water, oceans, and as parasites of vertebrates, invertebrates, and plants. They cannot be classified on the basis of habitat alone. Some nematodes found in the soil are actually free-living (nonparasitic) stages of vertebrate or invertebrate parasites and resemble commonly found terrestrial nematodes. For example, first-, second-, and third-stage juveniles of vertebrate parasites in Strongylida feed on bacteria in the soil and are morphologically similar to some Rhabditida. Differences between stages of nematodes parasitic on invertebrates found in soil and free-living terrestrial forms are also difficult to discern. Freshwater free-living species occur in moist or marshy soils and along the sides and bottoms of ponds or streams. Nematodes that are terrestrial are also aquatic, because they require a small film of water around a soil particle in which to survive, move, and reproduce. In this chapter, we limit our discussion to nematodes commonly found in soil.

The study of terrestrial nematodes has increased greatly in recent years, partly as a result of the acceptance of nematodes as economic pests and instigators of disease on major crops, and the impetus of the International Biological Program. Since Nielsen's (1967) chapter on the Nematoda, the number of taxonomic descriptions, life histories, and biological details on nematodes has doubled. However, the science of nematology is still in an embryological state, as compared to, for example, entomology because (a) there are many undescribed species due to a young taxonomy and the relatively few nematologists throughout the world; (b) most nematodes are transparent and microscopic, and require special techniques in handling in order to discern minute taxonomic characters; and (c) three-fourths of the taxonomic work involves economically important parasitic nematodes of vertebrates, invertebrates, and plants, so that there is a paucity of biological description of the free-living terrestrial nematodes.

Fundamental Morphology

The capacity of nematodes to exploit virtually every habitat available to life is reflected in diverse morphological adaptations. Yet many fundamental characteristics are shared throughout the phylum. Nematodes are elongate and in cross section they are round (Figs. 8.1, 8.2). Most species are *fusiform,* tapering at the ends, but others are *vermiform,* with little taper. Certain terrestrial nematodes, for example, females of the plant parasites, Heteroderoidea, become swollen as they mature, approaching a *globose* (spherical) or *pyriform* shape (Fig. 8.9*E*).

Since the nematode body is predominantly bilaterally symmetric, they have a *dorsal* and *ventral* side and move on one of two *lateral* sides (Fig. 8.2*C*). The orientation of a given specimen can be readily identified since most body openings are ventral. The anterior end of nematodes, as viewed *en face,* typically shows a high degree of radial symmetry, with the pattern repeated in multiples of three (Fig. 8.2*B*).

Nematodes vary in size but most terrestrial forms are less than 4 mm long. They are not segmented and body organs are said to occur in a nonpartitioned "fluid-filled" cavity. Nematodes consist of a body wall, digestive, nervous, excretory, and reproductive system, but lack a typical circulatory, respiratory, and perhaps, endocrine system (Fig. 8.1).

The nematode body wall includes an outermost layer, the cuticle, the hypodermis which produces the cuticle, and an inner layer of somatic muscles (Fig. 8.2*C*). The cuticle is a flexible, multilayered exoskeleton, which is supported by internal hydrostatic pressure. It extends inward at all body openings, including the stoma and esophagus lumen, the anus, vulva, excretory pore, and sensory openings (Fig. 8.1). Superficial markings of the cuticle are important for identification of nematodes (Fig. 8.2). Most species have transverse indentations. Fine patterns are typically termed *striations,* whereas coarse indentations might be more specifically called *annulations.* In some groups, such as certain Criconematoidea (Tylenchida), transverse markings are very pronounced and termed *apparent segmentation.* In the latter case, the bottom edge of each "segment" projects over the adjacent segment. Some nematodes have longitudinal markings. Certain Dorylaimidae have *ridges* that extend the length of the body and are present on all sides. When such longitudinal markings occur simulta-

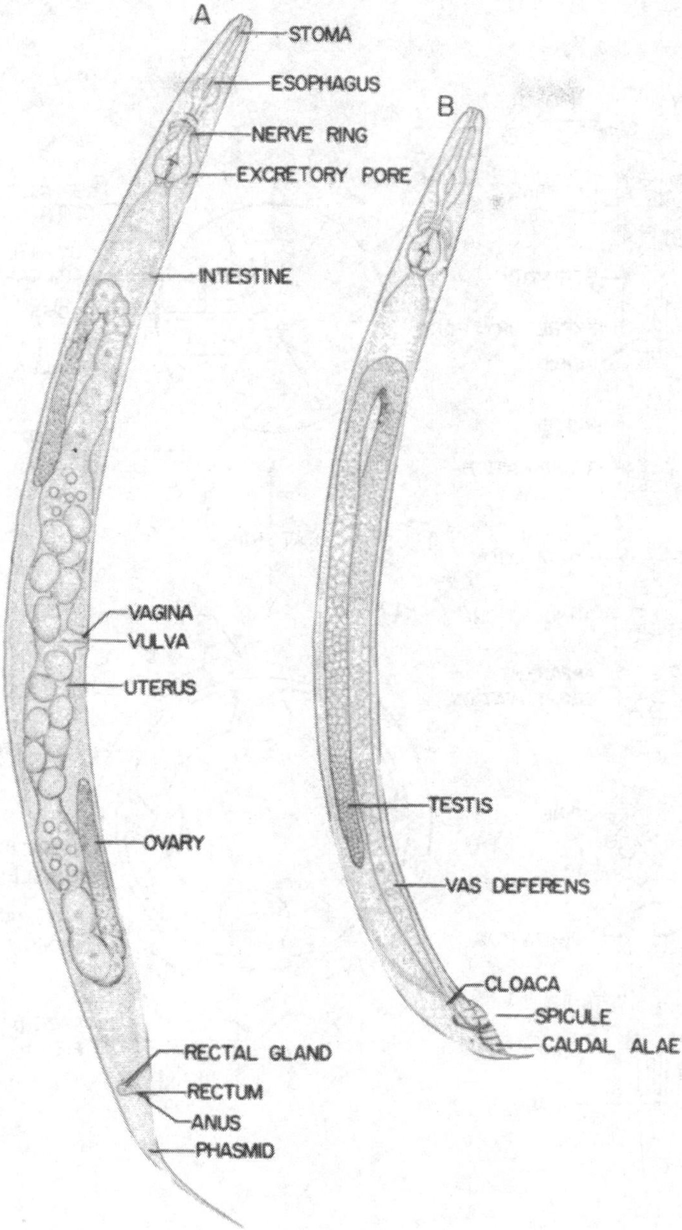

Fig. 8.1. *Rhabditis* sp.: (*A*) female; (*B*) male.

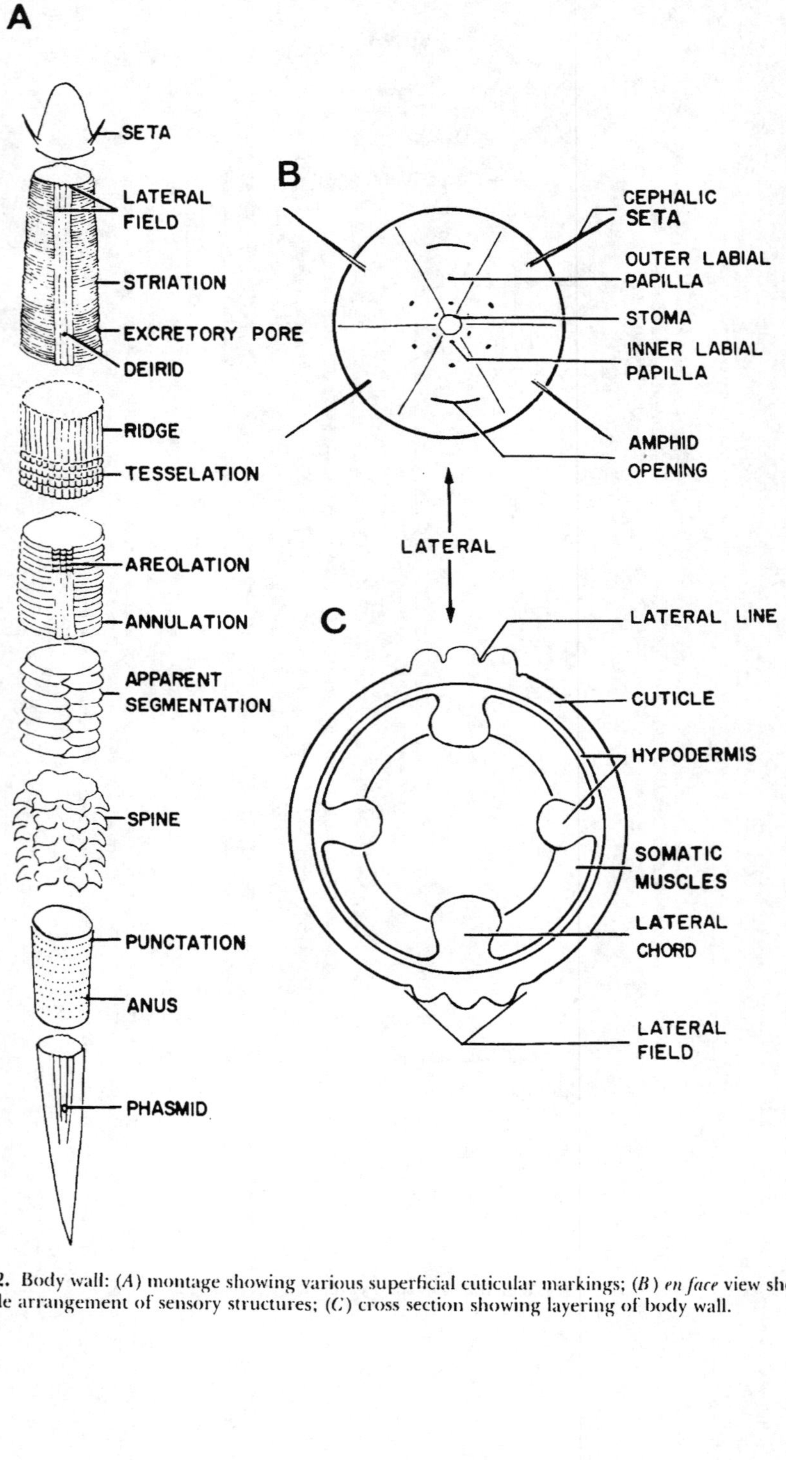

Fig. 8.2. Body wall: (*A*) montage showing various superficial cuticular markings; (*B*) *en face* view showing example arrangement of sensory structures; (*C*) cross section showing layering of body wall.

neously with transverse markings as in certain Tylenchorhynchidae (Tylenchida), the pattern is *tesselate*. Many nematodes have *alae* which are localized longitudinal projections. The most common are those that extend the length of each lateral side forming a *lateral field* (Fig. 8.2*A*,*C*). The lateral field consists of a variable number of incisures or lateral lines, and the number of lines is often taxonomically important. When the field is crossed by transverse lines, it is said to be *areolated*. In some nematodes certain alae are limited to the anterior or posterior region. Caudal alae (bursae) are a pair of lateral flaps of cuticle in the tail region, and are common among males (Fig. 8.8*A*). Cervical alae, flaps of cuticle positioned near the anterior end, are rare among terrestrial nematodes; however, anterior dorsal and ventral projections on *Wilsonema* spp. are often interpreted as alae (Fig. 8.15*F*). The cuticle of some nematode species is marked by numerous dots or *punctations*, which are sometimes arranged in rows or other specific patterns (Fig. 8.2*A*). Elongate extensions or projections or cuticle occur on many nematode species and are taxonomically important. *Spines*, which are present on *Criconema* spp., are nonsensory and project from the posterior edge of each apparent segment. Sensory projections include *papillae* and *setae*. Although papillae are minute elevations and setae are longer, the demarcation may not be clear cut. Sensory projections may occur anywhere on the body but are typically concentrated in the head and sometimes the tail region. The basic pattern on the head, as viewed *en face*, is two inner concentric circles of six projections each and an outer circle of four projections (Fig. 8.2*B*). Sometimes the two outer circles join, forming a circle of ten projections. In some forms, not all sensory projections are expressed; in others, some projections may occur as papillae, whereas others appear as setae.

Deirids are paired papillae-like structures occurring laterally near the level of the nerve ring (see below) (Fig. 8.2*A*).

Amphids occur on the lateral sides of the head region and apparently are chemoreceptive. Amphids are usually difficult to see in the light microscope, particularly in most terrestrial forms. In Rhabditida and Tylenchida the external openings are inconspicuous pores or slits, which are usually found so far anterior that they are best seen *en face* (Figs. 8.2*B*, 8.3*A*). In Dorylaimida, Mononchida, and Enoplida, the amphids usually occur sufficiently posterior in the head region to observe them in lateral mount. Amphid openings in these orders are slitlike or oval, but extending posterior to the opening is a broad cup-shaped cavity or pouch; thus the amphid is termed *cyathiform* (Figs. 8.3*B*, 8.14*D*). Some of the most elaborate amphids occur in orders with few terrestrial species such as Monhysterida, Chromadorida, Araeolaimida, and Desmodorida, in which the pouch is circular or spiral (Figs. 8.3*C*,*D*, 8.16). Inconspicuous pore or slitlike openings occur in certain species of all the orders.

Probolae are cuticular modifications of lips, projecting forward in threes or sixes, and occur among many Cephaloboidea (Rhabditida) and other

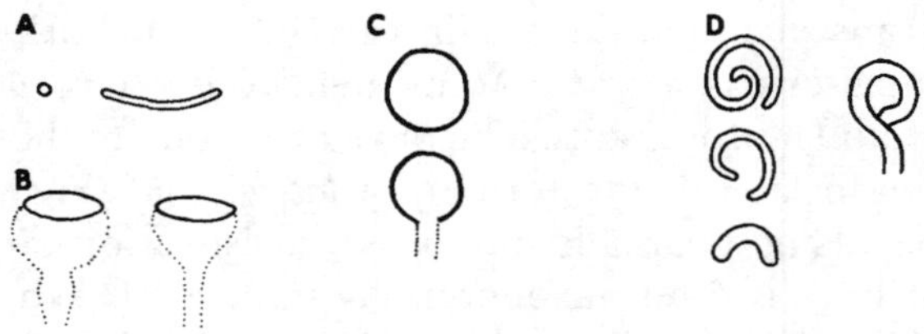

Fig. 8.3. Amphid openings: (*A*) minute pore or slit; (*B*) cyathiform variations; (*C*) circular variations; (*D*) spiral variations. [(*B*) Redrawn from Hirschmann (1960); (*D*) partially redrawn from Hyman (1951).]

superfamilies (Fig. 8.10*I,J*). Often, as in *Acrobeles* spp., there is an inner circle of three (labial) and an outer circle of six (cephalic). Probolae may be shaped as simple triangles or they may be elaborate with featherlike projections. The labial projections of Chambersielloidea (Rhabditida) are probolae, although the finely cut anterior ends are specifically described as *cirri* (Fig. 8.10*J*). Similarly, the pointed lips of *Teratocephalus* and *Euteratocephalus* (discussed below with Araeolaimida) (Fig. 8.15*G*) are similar to the probolae of certain Cephaloboidea (Fig. 8.10*I,J*). Some nematodes have broad cuticle elevations, or *warts,* which on Bunonematoidea usually are limited to one lateral side (Fig. 8.10*D*). Although these structures apparently do not have a sensory function in Rhabditida, similar structures of other species may be innervated. In some cases wartlike structures occur in the precloacal position (anterior to the cloacal opening), on the ventral side, and by their location and association with glands, are defined as *supplementary organs.* Other supplementary organs may be expressed on the surface as papillae or suckers. Glands of the supplements are associated with the hypodermis (Fig. 8.14*F*). However, hypodermal glands with openings through the cuticle are not restricted to supplementary organs, but occur in many Adenophorea, usually as longitudinal rows on the lateral side. Some hypodermal glands are restricted to the caudal region. Certain caudal glands may terminate in a single modified cuticular duct to the exterior, forming a *spinneret* (Fig. 8.15*H*). The spinneret and associated caudal glands are common among aquatic species and apparently function in secreting an adhesive for anchorage.

Phasmids occur as a pair on the lateral sides and usually are limited to the posterior region; they characterize Secernentea (Figs. 8.2, 8.4). Their surface structure typically occurs in the center of the lateral field and is usually pore or papillae-like. In some forms, a fine channel extends inward from the surface, and thus the phasmid appears tubular (Fig. 8.4*B*). Sometimes the phasmid includes a large superficial cavity filled with secretion and may be termed a *scutellum* (Fig. 8.4*C*). The position of the phasmids relative to the anus is an important taxonomic character.

The hypodermis underlies and secretes the cuticle (Fig. 8.2*C*). Except for its association with glands and sensory structures with cuticular manifestations, it is insignificant with respect to identification. The hypodermis consists of a thin layer which is interrupted by two lateral and two or more additional chords. The hypodermal chords extend the length of the nematode and are associated with major nerve bundles. In nematodes with lateral alae, the lateral chords directly underlie these cuticle structures (Fig. 8.2*C*).

Somatic muscles are the innermost layer of the body wall, but the layer is discontinuous where hypodermal chords occur (Fig. 8.2*C*). The muscles are longitudinally spindle-shaped with contractile elements adjacent to the hypodermis. A noncontractile region extends toward the interior of the body. Early nematologists attempted to develop a system of classification based on numbers of muscle cells, but differences in size, numbers, and types of cells may be more a function of nematode size than fundamental relationships.

The morphological diversity of the nematode digestive system reflects adaptations to almost every habitat and food source. The fundamental structure includes a *stoma* (= buccal cavity or mouth), *esophagus* (= pharnyx), *intestine,* and *rectum* (Figs. 8.1, 8.6*A*), but the stoma and esophagus are the most variable and taxonomically important. Variations in the stoma include overall shape and modifications of the cuticular lining into diverse thickened structures (Fig. 8.5). The esophagus also varies in shape and specialization of regions with muscles, valves, and glands (Fig. 8.6).

In some genera demarcation between stoma and esophagus is indistinct. The posterior portion of the "stoma" may be surrounded by esophagus and might even be of esophageal origin; in other cases the stoma may be greatly reduced. For example, in *Tripyla* spp., the long tubular stoma tends to collapse, and except for a small tooth, the lining is thin and surrounded by tissue of the esophagus (Fig. 8.13*B*). Conversely, the stoma of Rhabditida is a broad relatively rigid tube, which is lined by thickened cuticular structures called rhabdions (Figs. 8.5*A*, 8.10). Separate rhabdions are particularly distinct in *Cephalobus* (Fig. 8.5*A*). The anteriormost structures that line the *cheilostom* (anterior lip cavity) are *cheilorhabdions. Prostom, mesostom,* and *metastom* make up the protostom and are lined by corresponding rhabdions (respectively, prorhabdion, mesorhabdion, metarhabdion). The *telorhabdions* line the posterior end of the stoma, which is the *telostom* (Fig. 8.5*A*) (Goodey 1963a, Andrássy 1976). In some genera, various rhabdions are fused (Fig. 8.10*E*). Rhabdion terms apply only to Rhabditida. However,

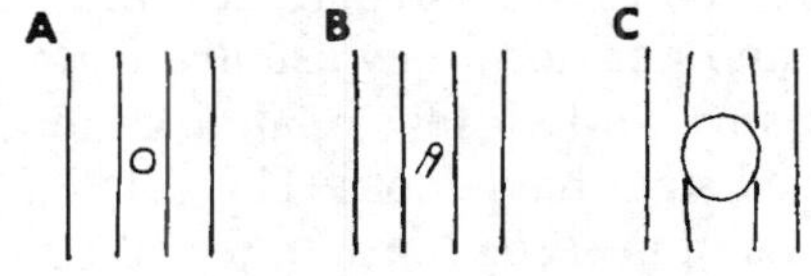

Fig. 8.4. Phasmids in relation to lateral field; (*A*) pore-like; (*B*) pore with conspicuous channel (tubular); (*C*) scutellum.

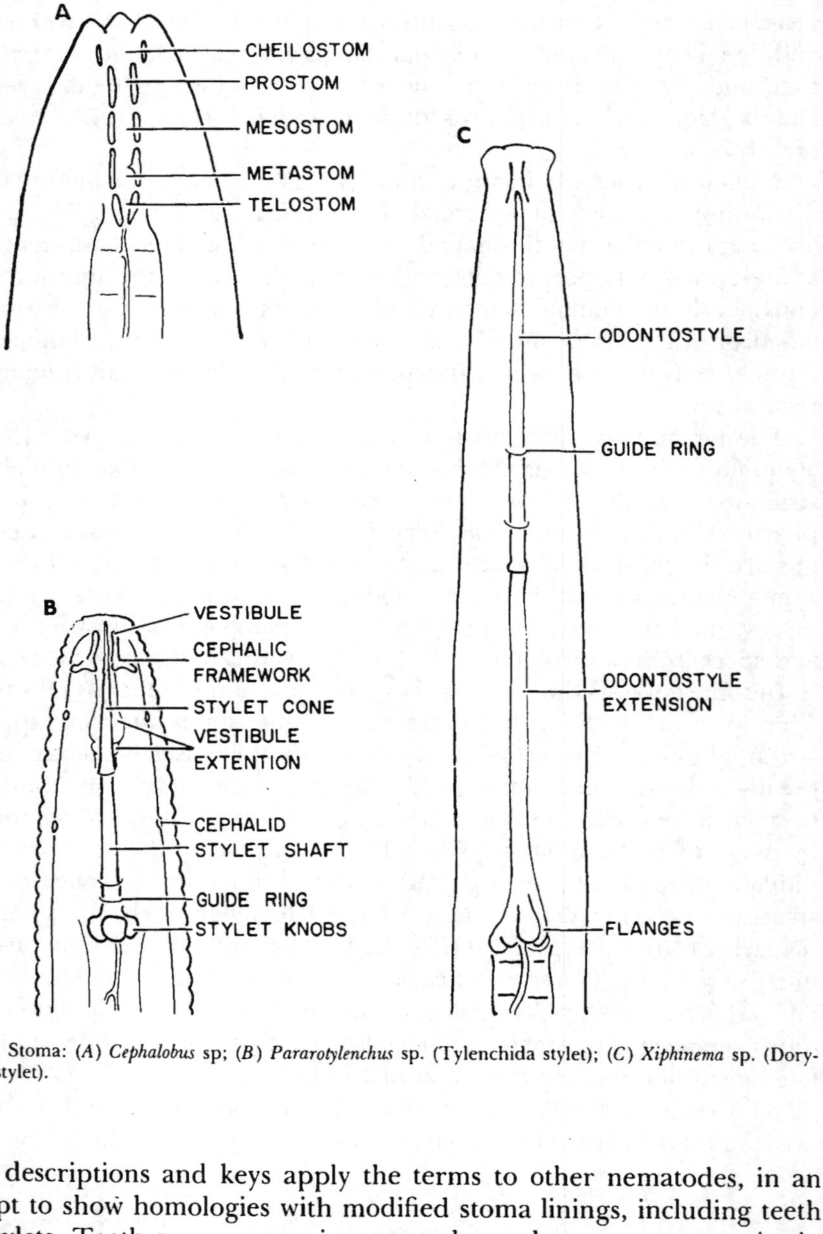

Fig. 8.5. Stoma: (*A*) *Cephalobus* sp; (*B*) *Pararotylenchus* sp. (Tylenchida stylet); (*C*) *Xiphinema* sp. (Dorylaimida stylet).

many descriptions and keys apply the terms to other nematodes, in an attempt to show homologies with modified stoma linings, including teeth and stylets. Teeth are common in nematodes and vary among species in size, shape, number, and position; they usually characterize predators. For example, *Mononchus* spp. and closely related genera have a single large *onchium* or broad-based stationary tooth (Fig. 8.12*B*,*C*). In addition, many mononchids have rows of *denticles* (minute teeth) or *rasps* lining the stoma (Fig. 8.12*B*,*C*). In some nematodes, such as *Dorylaimus,* a long narrow protrusible *axial spear* dominates the narrow stoma (Fig. 8.5*C*). Perhaps the most complex armature is the protrusible *stylet* of Tylenchida, which is interpreted as derived from the fused lining of the stoma (Fig. 8.5*B*). The stoma cavity is thus reduced to a fine lumen which extends the length of the stylet. Anteriorly, the stylet consists of a *cone,* which is fused posteriorly with a *shaft,* and three knobs provide points of protractor muscle attachment.

The stylet moves in a cuticularized channel. Anteriorly, the channel consists of a cylinder, the *vestibule.* The vestibule is the central portion of the cuticularized *cephalic framework,* which underlies the lip region (Fig. 8.5*B*). Posteriorly, the vestibule merges with a *vestibule extension,* which surrounds the cone area of the retracted stylet. Folds in membranes of

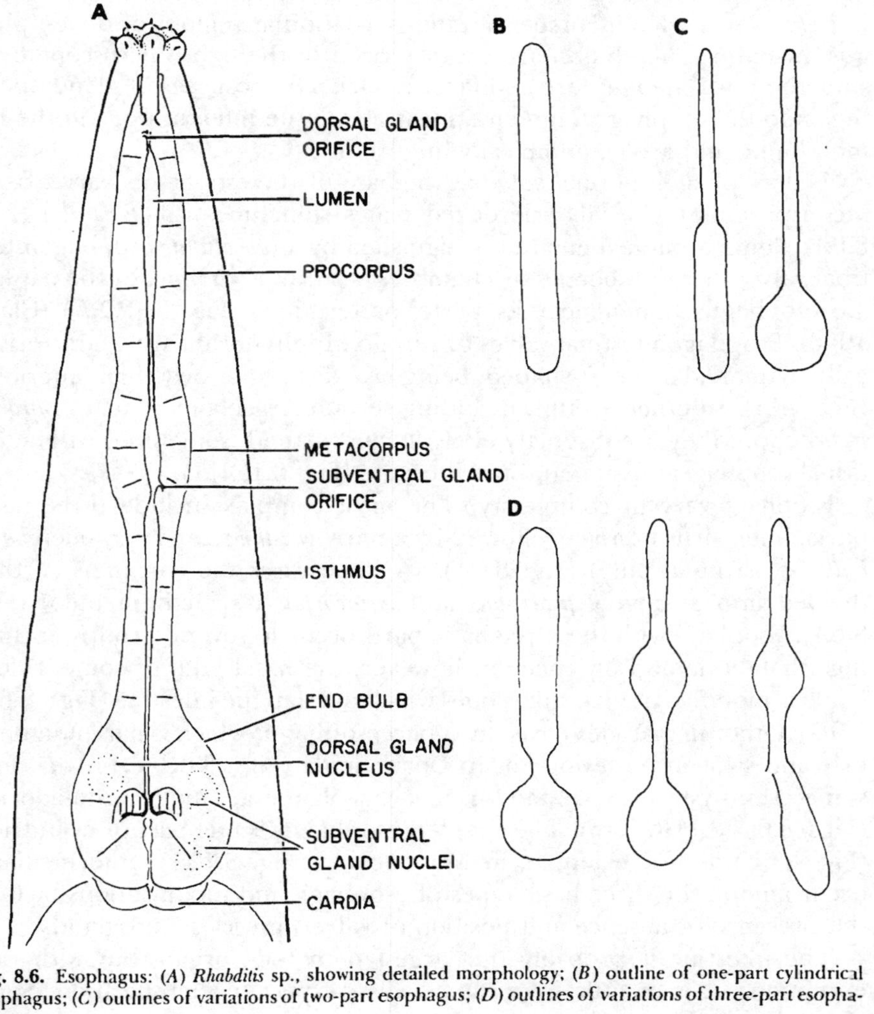

Fig. 8.6. Esophagus: (*A*) *Rhabditis* sp., showing detailed morphology; (*B*) outline of one-part cylindrical esophagus; (*C*) outlines of variations of two-part esophagus; (*D*) outlines of variations of three-part esophagus.

tissue that surround the stylet may give the impression of rings, and are termed "*guide rings*" (Fig. 8.5*B,C*). The hollow stylet of many plant-parasitic Dorylaimida superficially resembles that of Tylenchida (Fig. 8.5*B,C*). The anterior portion, the *odontostyle,* is similar to the cone and the *odontostyle extension* and *flanges* may resemble the shaft and knobs, respectively. The odontostyle of the plant-parasitic dorylaim, *Paratrichodorus* is ventrally curved and lacks flanges (Fig. 8.14*D*).

The esophagus is variably shaped with a cuticle-lined lumen which extends from the base of the stoma, posteriorly, to an esophagointestinal valve (Fig. 8.6*A*). Muscular regions of the esophagus usually appear as fine transverse striations in the microscope. They are generally associated with a broadened triradiate and more heavily cuticularized lumen. Sometimes, the esophagus broadens to form a muscular *bulb,* but bulbs may also occur in regions with glandular specializations. Glandular regions of the esophagus, including such bulbs, are characterized in the light microscope by a smooth nonstriated texture and large nuclei. The positions of gland openings into the esophageal lumen appear as minute interruptions in the lumen lining and are taxonomically important.

The esophagointestinal valve at the base of the esophagus varies from inconspicuous to heavily sclerotized plates sometimes called *cardia* (Fig. 8.6*A*). Some Mononchida are distinguished by a *tuberculate* esophagointestinal valve; that is, tubercles or chambers occur at each apex of the triradiate esophagus lumen near its posterior terminus (Fig. 8.12*D,E*) (Clark 1960). Esophagointestinal valves of certain Monhysterida are characteristically pyramidal or heart-shaped, being broader posteriorly than anteriorly (Fig. 8.11). In other groups, including certain Araeolaimida (e.g., *Chronogaster* spp.), they are distinctly elongate and extend, sometimes with additional esophageal tissue, into the intestine (Fig. 8.15*A,B*).

Esophagi vary in complexity. The most complex include three basic parts: anteriorly a *corpus,* followed by a narrow *isthmus* and a swollen *basal bulb* (= terminal bulb) (Fig. 8.6*A*). In some cases the corpus is further divided into a narrow *procorpus* and *metacorpus* (= medium bulb) (Fig. 8.6*A*). Esophagi with the three basic parts occur in several groups, including Rhabditida and Tylenchida; however, the basal bulb of some Tylenchida is modified as glandular lobes which overlap the intestine (Figs. 8.6*D*, 8.9*D*). Other nematodes have a two-part esophagus which is narrow anteriorly and swollen posteriorly as in Dorylaimida (Figs. 8.6*C*, 8.14*A–D*) and some Monhysterida, Desmodorida, Araeolaimida, and Chromadorida (Figs. 8.6*C*, 8.11*B*, 8.16). The simplest esophagus is one part or cylindrical (Fig. 8.6*B*) as, for example, in Mononchida (Fig. 8.12*A*). Intermediates occur among the three basic types of esophagi, and modifications include the presence or absence and position of valves, muscles, and glands.

The intestine is probably the largest nematode organ, but with rare exceptions, it is of little importance with respect to identification. Some older descriptions refer to the number of cells in the intestine, but this appears to be a function of nematode size and probably has little relationship to phylogeny. The intestine typically terminates posteriorly in juveniles and females as a *rectum* and *anus*, and in males as a *cloaca* (Figs. 8.2, 8.8*A*). The position of body openings may be taxonomically important; for example, the character "tail length" refers to the distance from the anus (or cloacal opening) to the posterior terminus, and the shape of the tail is also important to identification. The cloaca is associated with important structures which are best considered with the reproductive system (Fig. 8.8*A*).

The most visible part of the nematode nervous system, excluding the cuticular portion of sense organs, is the nerve ring (Fig. 8.1). The *nerve ring*, a commissure, is a concentration of nerves that encircle a portion of the esophagus. Bundles of longitudinal nerves extend anteriorly and posteriorly from the ring and innervate the entire body.

Peripheral nerves may occur within the thin layer of hypodermis between the cuticle and somatic muscles. These nerves often appear in lateral mounts as small lens-shaped structures, although they encircle or partially encircle the specimen. *Cephalids* refer to such structures in the head region and may appear as an anterior and posterior "pair." A *hemizonion* or *hemizonid* may occur just anterior or posterior to the excretory pore, respectively, and *caudalids* are confined to the tail region.

In addition to sensory organs considered in relation to cuticular structure, some aquatic nematodes may have subsurface pigmented sensory structures, *ocelli.* Ocelli occur as a lateral pair in the esophagus region and are thought to be light sensitive.

The most conspicuous part of the excretory system is a cuticle-lined duct which opens ventrally to the exterior, generally in the region of the esophagus (Figs. 8.1, 8.2*A*). In Adenophorea the system may be obscure or absent, but where a duct occurs it originates proximally from a simple gland. In Secernentea the system includes one or more glands as well as tubes or canals. The canals may extend throughout the body length, often appearing to be closely associated with the intestine. The structure of the excretory system is taxonomically important in distinguishing between Adenophorea and Secernentea, and variations in the position of the excretory pore can also be useful for identification.

The reproductive system is generally important to identification of nematodes. In both sexes, the reproductive tract is a tube, which is divided into a series of specialized regions. The distal end, which includes the gonad, is straight (outstretched) or reflexed (Figs. 8.7, 8.8). A female may be *monodelphic,* with a single reproductive tract (Fig. 8.7*B,C*) or *didelphic,* with two similar tracts joined at the vagina (Fig. 8.7*D,E*). The system is *amphidelphic* if the two tracts extend in opposite directions (Fig. 8.7*D*), *prodelphic* if anteriorly directed (Fig. 8.7*C,D*), or *opistodelphic* (rare) if posteriorly directed. Males of Secernentea are typically *monorchic,* with one gonad (testis) (Fig. 8.8*A,B*), and most Adenophorea are *diorchic,* with two testes which join proximally at a common duct (Fig. 8.8*C*). The male reproductive tract

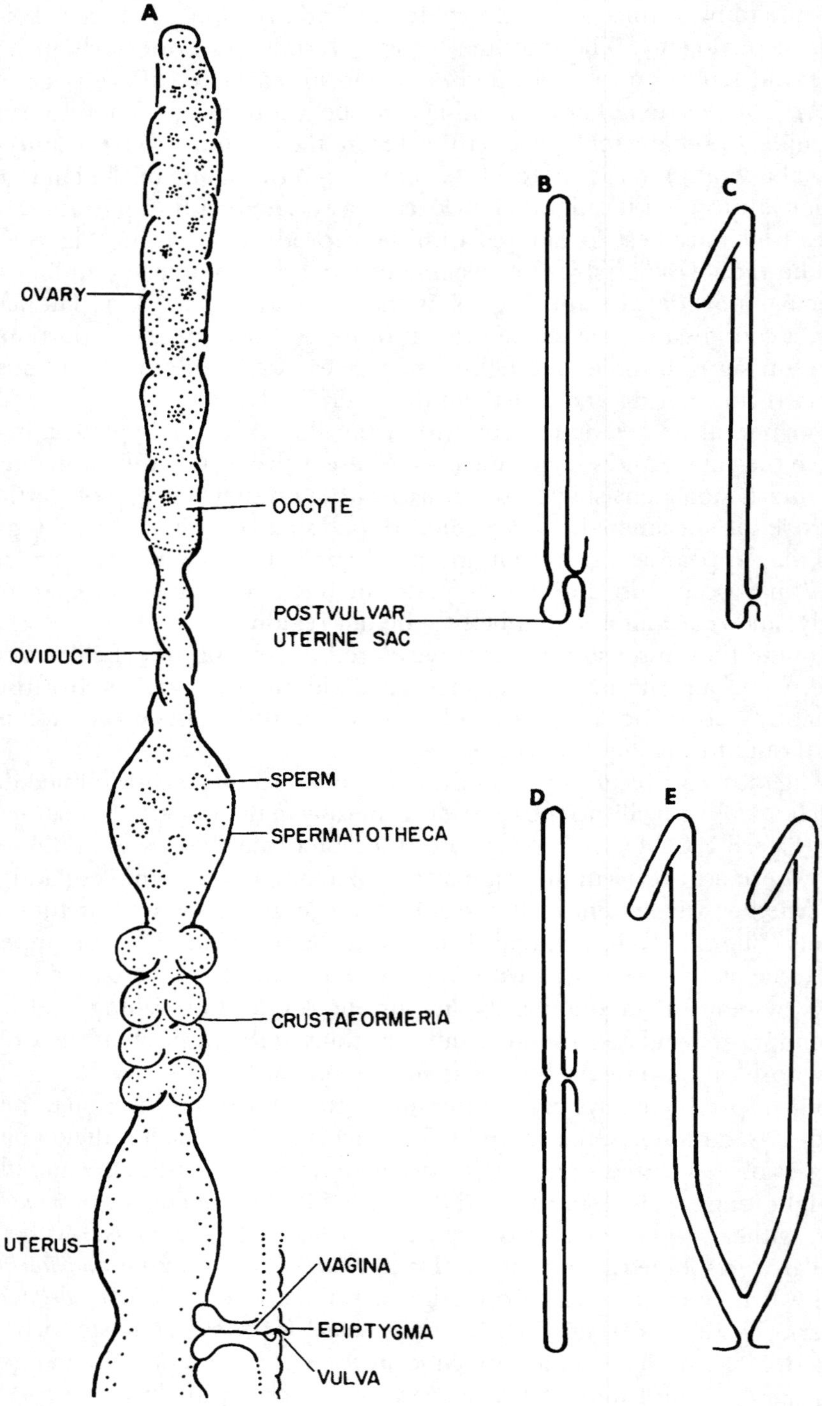

Fig. 8.7. Female reproductive tract: (*A*) detailed morphology; (*B*) diagrammatic outline of monodelphic, prodelphic tract; (*C*) diagrammatic outline of monodelphic, prodelphic, reflexed tract; (*D*) diagrammatic outline of didelphic, amphidelphic tracts; (*E*) diagrammatic outline of didelphic, prodelphic, reflexed tract.

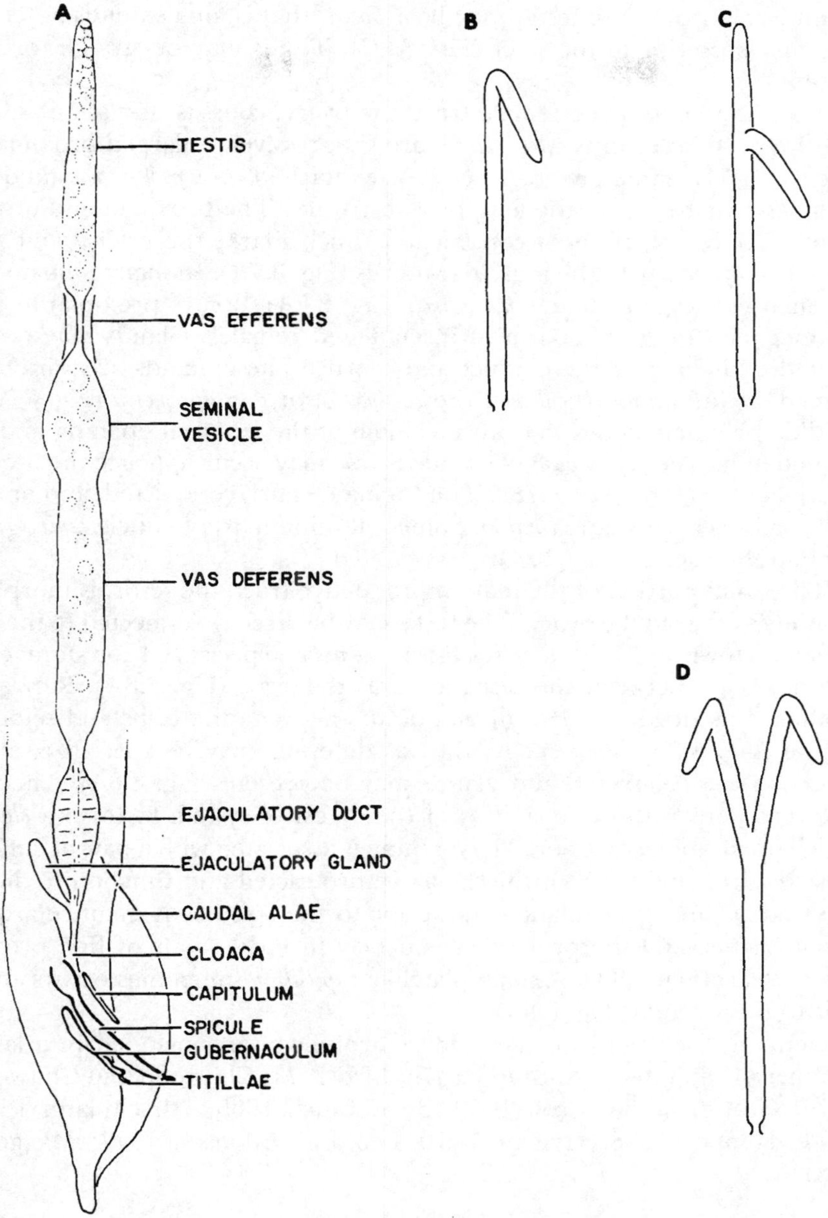

Fig. 8.8. Male reproductive tract: (*A*) detailed morphology; (*B*) diagrammatic outline of monorchic reflexed tract; (*C*) diagrammatic outline of diorchic outstretched tracts; (*D*) diagrammatic outline of diorchic anteriorly directed, reflexed tracts.

extends anteriorly, but testes may be *opposed;* that is, one extends anterior and one posterior to the duct (Fig. 8.8*C*). Testes may be straight or reflexed.

The distal end of the female tract, the ovary, consists of a sac of small tightly packed cells or *oocytes,* which are progressively enlarged proximally (Fig. 8.7*A*). In some genera, oocytes are attached to a *rachis,* a strand of cytoplasm that extends the length of the ovary. The proximal end of the ovary may be set off by a constriction which marks the *oviduct* and the oviduct connects with the muscular uterus (Fig. 8.7*A*). In many nematodes an enlarged region, the *spermatotheca* (Fig. 8.7*A*), occurs proximal to the oviduct for storage of sperm in inseminated females. Glands may occur near the junction of the oviduct and uterus. These glands are variously named, including *quadricolumella, tricolumella,* or *curstaformeria* (Fig. 8.7*A*). In didelphic nematodes the two uteri join at the external opening and in monodelphic species, a *postvulvar–uterine sac* may occur opposite the uterus and join the vagina (Fig. 8.7*B*). The vagina is short, constricted, and opens externally as a slit-shaped *vulva.* Sometimes fine flaps of cuticle, *epiptygma,* overlap the vagina (Fig. 8.7*A*).

The distal portion of the male reproductive tract, the testis, is morphologically similar to the ovary. The testis may be directly connected to the *vas deferens.* However, in some cases a *seminal vesicle* specialized for storage of sperm occurs between the testis and vas deferens (Fig. 8.8*A*). Rarely, a distinct constriction, or *vas efferens,* occurs between the testis and seminal vesicle (Fig. 8.8*A*). Proximally, the vas deferens may be specialized as a muscular *ejaculatory duct* and glands may be present (Fig. 8.8*A*). The vas deferens joins with the terminus of the digestive system to form a *cloaca* which opens to the exterior. The opening is associated with a pair of cuticularized *spicules* (Fig. 8.8*A*) which may be protracted and function to dilate the vagina during copulation. Posterior to the spicules is a cuticularized *gubernaculum,* and the protruded end may include a pair of fine projections, *titillae* (Fig. 8.8*A*). A small platelike *capitulum* sometimes occurs anterior to the spicules (Fig. 8.8*A*).

General discussions of nematode morphology are included in a large number of references (Andrássy 1976; Bird 1971; Chitwood and Chitwood 1977; Crofton 1966; Croll 1976; de Coninck 1965; Hirschmann 1960, 1971; Hooper 1978; Hyman 1951; Lee and Atkinson 1977; Maggenti 1981).

Digestive Characteristics—Feeding

Nematodes feed on a wide variety of foods, but most species have only one kind of nutrition at a particular stage of development. The anatomy and physiology of their digestive system varies with species and can be related to different types of feeding and food. Generally, nematodes ingest food through the stoma (buccal) opening of an alimentary canal that terminates in the anus. The nematode body cavity has a high turgor pressure which keeps the intestine lumen collapsed except when it is full of food. Food is pumped into the intestine against the high turgor pressure by the esophagus, which is a powerful sucking and pumping organ.

Microbivorous nematodes draw bacterial suspensions into the alimentary canal by the sucking action of the esophagus. *Acrobeles* has probolae (Fig. 8.10*I*) which might function as sievelike structures to restrict the size of the ingested particles. When feeding, the esophagointestinal valve closes and the lumen of the esophagus fills with food which is mixed with secretions from three or more digestive glands. As the anterior end of the esophagus closes, the esophagointestinal valve opens, and the food is forced into the intestine by action of the esophageal muscles. In some species the valve of the terminal bulb appears to crush particles, aiding in digestion. In one aquatic species, the movement of food from the stoma to the intestines takes place in about 0.3 second with the whole intestine filling in about 5 minutes (Jennings and Colam 1970).

Penetration of cells by phytophagous or fungivorous nematodes is a fascinating phenomena. The worm searches the surface of the cells, occasionally probing with the stylet. The nematodes seem to prefer a more rigid cell surface, and once a suitable surface has been found, the stoma and stylet are placed at right angles to the cell, while the rest of the body remains stationary although it is sometimes braced against soil particles or the surface of the plant or mycelium (Doncaster and Seymour 1973). The cell is penetrated by powerful thrusts of the stylet, and due to the pumping action of the esophagus, the contents are ingested through the hollow stylet. Some members of Tylenchida can digest the contents of a cell externally by injecting secretions from the esophageal glands into the cell (McElroy and Van Gundy 1968). The injection may last minutes and is followed by a period of ingestion, which may continue for minutes to hours. Many nematodes bring about destructive cellular changes. For example, *Aphelenchus avenae* removes the cell contents of fungal hyphae in 15 seconds, while the feeding *Pratylenchus* affects the permeability of tissues at a distance from the feeding site (Lee and Atkinson 1977). Heteroderoidea induce morphological changes in host cells (i.e., the formation of giant cells or syncytia) which provides a sustained localized supply of nutrients for the sedentary females.

Predaceous nematodes often possess a stylet, teeth, denticles, or various combinations. They feed on protozoa, rotifers, tardigrades, and small oligochaetes, as well as other nematodes. Many predators with teeth can ingest a small animal whole, ripping it as it passes through the stoma. Nematodes can often be seen whole or partially digested in the intestine of mononchids, together with parts of stylets and spicules or chaetae from oligochaetes. Predaceous nematodes with a stylet feed much like the phytophagous forms, piercing their prey and sucking out the contents.

Digestive Characteristics—Energy Metabolism

Barrett (1976) and Atkinson (1980) reviewed the metabolism and respiratory physiology of nematodes, respectively. They hesitated on drawing general conclusions for nematodes since much of the work was on such a diverse range of species, which were primarily animal parasites in different developmental stages. The data for catabolic pathways in free-living and plant parasitic nematodes are limited, and it is assumed that their metabolism is similar to that of other aerobes. It appears, however, that lipids, rather than carbohydrates, are the major energy reserve. Detection of enzymes have indicated that nematodes probably have the normal sequence of the thyrocalcitonin and glycolytic enzyme cycles. Under anaerobic conditions, activity is greatly reduced, glycogen is catabolized, and ethanol and lactic acid appear to be the main end product of glycolysis (Cooper and Van Gundy 1971). Little is known about the electron transport system in nematodes, although several cytochrome enzymes have been isolated.

Excretory Characteristics

Variations in structure and physiology of the excretory system suggest that it does not have a universal function among groups of nematodes. In free-living and phytophagous nematodes, osmoregulation or removal of nitrogenous waste does not appear to be the only function of the excretory system. The structure suggests secretion function, but many of these data are derived from animal parasites. One example in phytophagous nematodes is *Tylenchulus,* where swollen sedentary females secrete an external gelatinous matrix through the excretory pore; this exudate may have a protective function for eggs. Other metabolic products are secreted from the rectum and in some Dorylaimida the prerectum. *Caenorhabditis briggsae,* a free-living nematode, which is usually studied on artificial media, secretes amino acids, glycerol, sugars, and other metabolites into surrounding medium (Rothstein 1965). Myers and Krusberg (1965) noted similar results in five species of Tylenchida. Ammonia is also a major excretory product in both Rhabditida and Tylenchida, although this could be excreted through the cuticle (Wright and Newell 1976).

Osmoregulation in nematodes has not been directly examined. Evidence is based on changes in body length (Stephenson 1942) because direct methods [i.e., using tritiated water for H_2O content (Marks et al. 1968, Castro and Thomason 1973) and measurement of wet weight (Myers 1966)] require large numbers of nematodes. The ability of nematodes to maintain a constant volume under osmotic stress is particularly important because of their hydrostatic skeleton. Freshwater and terrestrial nematodes conform to their expected behavior, both in their relative permeabilities to water and salts and in their ability to regulate hypotonic solutions. Water and salts move through the cuticle to restore equilibrium. A *Rhabditis* sp., a microbivore, shrank and swelled during gradual hypertonic and hypotonic regulation, but burst with rapid changes in osmotic pressure (Stephenson 1942). There is a difference in tolerance among species, with some phytophages more tolerant of osmotic change than *Rhabditis spp.* (Nicholas 1975). Many nematodes can survive periods of osmotic stress or dehydration in a dormant state (see the section "Ecology—Behavior").

Reproduction Characteristics

Life cycles, including embryogenesis and postembryogenesis, are similar among terrestrial nematodes, but gametogenesis and therefore the mode of reproduction, are variable. Hyaline eggs are usually laid in the one-celled stage. However, in some forms, especially Rhabditida, embryogenesis may be completed within the female and second-stage juveniles emerge from the vulva, but this is often associated with "old" females or mutants who have lost the ability to lay eggs. According to Crofton (1966), the lineage of cells in nematode embryology is the best documented in the animal kingdom. Cleavage is total since there is no yolk, and determinate since even after the first division most cells are destined to give rise to specific parts of the adult. Since nematodes are triploblastic, the body systems develop from ectoderm, endoderm, and mesoderm.

Postembryogenesis includes four juvenile stages plus the adult. The stages are separated by molts of the entire cuticle, including linings of body openings and all cuticularized structures, such as stoma parts and spicules. Typically, in Secernentea the first juvenile stage and molt occurs within the egg; the juvenile hatches in the second stage, whereas in Adenophorea molts occur outside the egg. Since the nematode exoskeleton is not rigid, growth occurs between molts, including during the adult stage. Hatching and molting are usually triggered by environmental conditions which favor completion of the life cycle. For example, eggs of Heteroderidae are stimulated to hatch in the presence of exudates from plant hosts, and a limited food supply favors suspended development of certain nematodes after the second stage as resistant, nonfeeding "dauer" juveniles.

Reproduction—Specific Modes

Most terrestrial nematodes are bisexual, having male and female individuals, and reproduce by cross-fertilization or amphimixis (Triantaphyllou 1971, Triantaphyllou and Hirschmann 1980). Amphimixis is generally inferred as common, when the numbers of males and females in a given population are approximately equal. However, many amphimictic species also have the capacity to reproduce by parthenogenesis, or in hermaphrodism, by automixis or self-fertilization. In parthenogenic species, males may be rare and even sexually nonfunctional. Parthenogenic species may reproduce by pseudogamy; that is, sperm stimulates the development of oocytes, but there is no fusion of male and female pronuclei (Triantaphyllou 1971). Hermaphrodism and pseudogamy are common in some populations of *Rhabditis sensu lato* spp. Hermaphrodism is syngonic and protandric; a single gonad (ovotestis) first produces sperm, then eggs, and the individual has the appearance of a female.

Modes of reproduction have been most thoroughly examined in Heteroderoidea (Tylenchida), including *Meloidogyne* spp. (root-knot nematodes) (Triantaphyllou 1979, Triantaphyllou and Hirschmann 1980). The most common species—*M. javanica, M. incognita, M. arenaria,* and a cytological race of *M. hapla*—reproduce by mitotic parthenogenesis; meiosis never occurs during oogenesis. However, certain other species, including additional races of *M. hapla,* reproduce by meiotic parthenogenesis. In these species meiosis occurs and $2N$ is reestablished by fusion of the oocyte pronuclei with chromatin of a polar body or by doubling of chromosomes as the oocyte develops. Meiotic parthenogenic species are often facultative amphimictic, but males also occur where there is no amphimixis.

Sex expression in *Meloidogyne* spp. and genera of some other orders is affected by environment during postembryogenesis. Generally, environmental stress results in production of males, but environmental changes during critical times in development can result in sex reversal, including males with two gonads (normally, *Meloidogyne* females have two gonads and males have one) or male intersexes. Intersexes have also been reported for other nematode species.

Ecology—Importance

Nematodes are the most abundant metazoa in soils and interact closely with other soil organisms. They are usually divided into four or five trophic groups, based on schemes of Banage (1963), Wasilewska (1971), and Yeates (1979). The trophic groups commonly used are:

1. Phytophages
    a. Endoparasites (feed within roots or rarely, plant shoots)
        1. Migratory
        2. Sedentary
    b. Ectoparasites (feed on surface of roots)
        1. Migratory
        2. Sedentary
2. Microbivores (feed on bacteria and microflora)
3. Fungivores (feed on mycelium of fungi)
4. Omnivores (feed on several sources, i.e., fungi, bacteria, algae, protozoans, rotifers)

5. Predators (feed on other nematodes, enchytraeids, tardigrades, and protozoa)

Often the morphology of the nematode stoma indicates its affinity to a particular trophic group. Nematodes with protrusible hollow stylets are typically phytophagous (Tylenchida and some Dorylaimida), but some are fungivores, predators, or parasites of invertebrates. Fungivores of Tylenchida and Aphlenchida usually have small stylets and reduced esophageal musculature. Most nematodes with an open, unarmed stoma are considered microbivores, while those with a large onchium or axial tooth are probably predaceous. Most Dorylaimida are omnivores and have an onchiostyle. Often a given trophic group can be inferred if a higher taxonomic category such as superfamily is known.

Nematodes feed on living protoplasm and thus their activity affects many processes, such as (a) primary production as phytophages feeding on plant roots and aboveground plant parts, (e.g., seed head); (b) decomposition as fungivores and microbivores feeding on the primary decomposers; and (c) other soil fauna.

As primary consumers, the phytophagous nematodes (herbivores) are directly dependent on the biological stability of the root. Yeates (1979) reported a positive correlation between average nematode abundance and primary production in terrestrial ecosystems. Examples of damage to crops due to feeding by obligate parasitic nematodes, as well as their interactions with other nematodes or microflora, are abundant, and a major impetus has been established for the study of nematodes in agriculture, which includes the determination of economic thresholds in nematology. In terms of soil community structure, phytophagous nematodes may represent 98% of the population in an agriculture field but only 20–30% in uncultivated soils (Yeates 1979). When root-fungal feeders were compared to other organisms among various sites in Sweden and Poland, their importance was high when abundance was considered, but low in terms of biomass and only slightly higher for respiration (Sohlenius 1980). This was due to the smaller average size of the nematodes; smaller animals have a higher respiratory rate.

Nematodes are involved in decomposition by consuming primary decomposers, bacteria, fungi, and other microflora and microfauna. In many communities, microbivores and fungivores are the most abundant consumer of primary decomposers and thus have a major effect on nutrient cycling (Yeates 1979). Fungivores are known to feed on pathogenic and beneficial fungi (mycorrhizae) and microbivores feed on pathogenic and beneficial bacteria. For example, *Acrobeloides* feeding on *Rhizobium* nodules of soybeans was found to affect N_2 fixation (Westcott and Barker 1976). Microcosm studies on the relationship between microbivores and microbial populations may aid in delineating the role of nematodes in nutrient cycling (Coleman et al. 1977, Rosswall et al. 1977). Nematodes can also be disseminators of bacteria, fungi, actinomycetes, and mycoplasma by transmitting propagules internally and externally. Some species transmit plant pathogenic viruses (Norton 1978).

Nematodes are a major reserve of nutrients which upon death are returned to the cycle and used by other soil invertebrates as a source of nutrition. Nielsen (1967) has suggested that nematode movement expends little energy and therefore most assimilated energy is converted to biomass. Although nematodes are abundant in soil ecosystems, their importance in total soil respiration has been estimated at only 0.6–0.9% (Sohlenius 1980). Sohlenius estimated, however, that if only heterotrophic soil respiration was considered, soil animals would be responsible for about 5–10%. Nematodes would then be responsible for about 10–15% for the total soil animal respiration. Consumption, another measure of importance, gives nematodes a more significant position in the soil ecosystem. Smolik (1974) estimated that root-feeding nematodes consumed as much plant material as cattle in a Colorado grassland ecosystem. However, Sohlenius (1980) has cautioned that at this point, we have too few results in either C or N_2 cycling or energetics to determine the general role of nematodes in ecosystems.

Ecology—Abiotic Factors

The physical structure of soil and the abiotic factors that interact within soil are of great importance to nematode ecology. The size of particles directly affects movement of the nematodes, since they can only move in the water film within pores between particles. Only a few larger nematodes (Dorylaimida) can actually displace soil particles. If particles are small, such as in a compact clay, nematode movement may be limited by the small channels. Optimum movement occurs when the diameter of pores is narrower than the nematodes' diameter (Wallace 1971).

Changing the physical environment of soil by adding organic amendments has been shown to affect the nematode community. Mankau and Minteer (1962) showed that of eight organic materials added to soil infested with the citrus nematode, *Tylenchulus semipenetrans,* only cattle manure decreased the population. Johnson (1962) found that populations of root-knot nematode, *Meloidogyne,* were adversely affected in the presence of organic amendments. Nonpredaceous nematodes increased in an experiment of Mankau (1962b), whereas predaceous *Dorylaimus* spp. did not.

Next to soil structure, moisture is most important in affecting nematode biology. The amount of water in soil influences the general nematode biology, including movement, reproduction, and egg hatch; it is also responsible for active and passive dispersal. Nematodes are carried passively great distances in irrigation canals, streams, and rivers and can be spread within a smaller distance by rainfall. Maximum mobility of nematodes is obtained in a shallow water film around soil particles when the pores are almost empty. Too little soil moisture can trigger nematodes to enter a dormant state of anhydrobiosis, although phytophagous nematodes, (i.e., *Meloidogyne*) can remain active up to the wilting point of the plant. Conversely, when there is too much soil water and pores are filled, nematode movement is very slow. Since nematodes live in a water film, they are affected by gases dissolved in the water and not the atmosphere. The diffusion of gases is slower in water than air. In wet soils where the amount of air is reduced, anaerobic conditions increase, decreasing nematode activity. Bacteria, plants, and other organisms use available oxygen quickly, decreasing the amount available for nematodes. In the absence of oxygen, some nematodes can survive in a dormant state of anoxybiosis (Wallace 1971).

Nematodes are affected by a wide variety of chemicals in soil. Soil water acts as a medium for transport of host exudates and pheromones, both of which trigger specific behavior responses. Second-stage juveniles of the phytophagous nematode *Heterodera schachtii* are attracted to roots over 4.5 cm away, and hatching is stimulated when the eggs are placed 3 cm from cress roots (Wallace 1958). Prot and Van Gundy (1981) have shown that *Meloidgyne javanica* migrates vertically up to 45 cm in 5 days when placed in the presence of tomato roots. Gradients of O_2 and CO_2 are known to occur in water, and it is certain that gradients also exist for root exudates. Pheromones occur in a range of soil nematodes (i.e., *Panagrolaimus, Pelodera*) and *Heterodera* males are attracted to a concentration gradient of water-soluble chemicals produced by sedentary females (Green 1966, 1971).

Other chemicals affect nematodes. Wallace (1971) notes that there is little evidence that pH is an important factor to soil nematodes. However, it is difficult to separate the effects of pH from other influences within soil, and thus far nematodes appear to tolerate wide pH ranges (Dropkin 1980). Soil is dynamic and fluctuations in moisture are accompanied by changes in pH, suction, and osmotic pressure. In most agricultural soils osmotic pressures rarely exceed 2 atm and phytophages can tolerate up to 10 atm. That nematodes vary in their tolerance to salts is suggested by species that occur only in habitats of relatively high osmotic pressure and suction. Wyss (1970) showed that some members of Dorylaimida were more susceptible to increasing desiccation of soil and high osmotic pressures than some species of Tylenchida. Decreases in hatch and mobility with increasing osmotic pressures have been noted in several studies (Dropkin et al. 1958, Viglierchio et al. 1969). Viglierchio et al. (1969) have suggested three means of injury related to osmotic pressure: (a) irreversible dehydration of tissue, (b) toxicity of the chemicals in the solution, and (c) excessive uptake of distilled water, after regulation in hypotonic solutions, resulting in body rupture.

Since nematodes are poikilothermic, temperature affects their activity. Response to temperature extremes by nematodes may be by metabolic compensation, dormancy, or behavior (Lee and Atkinson 1976). In many nematodes the rates of change in oxygen consumption with alterations in temperature is usually at a constant rate, although some nematodes (i.e., *Ditylenchus dipsaci*) vary in oxygen uptake within their environmental range (Bhatt and Rhode 1970). Variations in optimum temperature affects all life processes of the nematode, including hatching, development, reproduction, and length of life cycle.

Ecology—Biotic Factors

Nematodes interact with many organisms in the soil and some, such as obligate parasites, depend on plant roots for nutrition and reproduction. Parasitism in nematodes is often specialized and a given species often has a specific host range. *Meloidogyne hapla* generally will not penetrate the roots of corn, yet corn is a good host for other species of *Meloidogyne*. The soybean, *Glycine max* var. Centennial, is resistant to *M. incognita* race 1, but susceptible to *M. javanica*. *Resistance* implies a degree of interaction between host and parasite, in which the parasite's capacity to exploit the host is reduced, relative to *susceptibility*. In *immunity* there is no interaction between potential host and parasite. Some hosts sustain large numbers of parasites without a pathogenic effect and are thus *tolerant,* whereas others are intolerant and are affected by only a few parasites. The genetic variability that exists in economically important plants of the same species has been used by plant breeders to develop nematode-resistant or nematode-tolerant varieties.

The effect of nematodes on disease may be enhanced by interactions with other organisms. For example, *Longidorus* spp. and other plant parasitic Dorylaimida transmit specific plant viruses. A certain nematode species can only transmit a certain virus, and the nematode and virus are apparently unaffected by the transfer. Interactions of nematodes with fungi or bacteria are not as highly specific, and the effect on the plant host is often synergistic; that is, the combined pathogenic effect on the host is greater than the additive effect of the two organisms acting independently. For example, when NC 95 tobacco (resistant to *M. incognita*) was infected by *M. javanica* and the black shank fungus, there was a 66% black shank infection versus 27% when the fungus was added alone. Examples of nematode, nematode–fungus, and nematode–bacterial interactions as related to plant disease are listed in Norton (1978) and Webster (1972).

Nematodes have parasites and predators; thus interest has been stimulated in these agents as a means of biocontrol for plant parasites. Although little is known about viral diseases of nematodes, studies with bacteria, protozoa, fungi, and predators are hopeful. The nematode-trapping fungi (Mankau 1962a), bacteria that rupture nematodes by producing toxins, amoebas that ingest nematodes, and the fatal sporozoan disease of plant parasitic nematodes are all currently being investigated as alternatives to chemical control. Predation of nematodes by mites (Muraoka and Ishibashi

1976), tardigrades, oligochaetes, and turbellarians have not been investigated as thoroughly (Norton 1978).

Ecology—Behavior

The behavior of nematodes is limited by their simple nervous system. Nematodes have been studied as to feeding, aggregation, host and sexual attraction, as well as avoidance reactions. Techniques used for these studies include neurophysiology, system analysis, tracking with microcinematography and CCTV recording, and microsurgery. Only about 0.01% of all nematodes have been studied with respect to behavior, mainly, the infective stages of plant and animal parasites. Most of these studies are analyses of locomotion and number or types of body waves. Evidence that nematodes respond to gravity or moisture gradient (since they live in water and cannot be hydrotactic) is not available. However, soil nematodes are sensitive and orient themselves toward or away from many sources of stimuli. Attraction and response to roots, opposite sexes, and chemicals have been discussed.

Some nematodes, as well as rotifers, tardigrades, and some protozoa are capable of withstanding adverse environmental conditions by entering into a dormant state (cryptobiosis) at any time in their life cycle. Conditions that induce this (and their respective responses) are low O_2 (anoxybiosis), dehydration (anhydrobiosis), high salt concentrations (osmobiosis), and freezing (cryobiosis). The main characteristics of these cryptobiotic states are a nondetectable metabolism and/or coiling of the body. Nematodes from desert soils have been found to be anhydrobiotic during hot dry months, thus affecting the total oxygen consumption per year (Freckman 1978). Nematodes exist in these states for years and yet can be revived to normal activity. Anhydrobiotic nematodes can withstand freezing in liquid N_2 and are more resistant to chemicals, such as nematicides (Freckman et al. 1980). References on this subject include Crowe and Clegg (1978), Evans and Womersley (1980), and Demeure and Freckman (1981).

Sampling and Extraction

There are many problems in estimating nematode population density and community structure in soils (Freckman et al. 1986). One of the most critical, and the basis on which reliable nematode quantitative evaluations are made, is an adequate and representative sampling program (Bird 1974, Goodell and Ferris 1980, Proctor and Marks 1975, Southey 1970, Wasilewska and Paplinska 1976). For nematode community analysis, samples should be taken at least monthly for a year, with a minimum of three times per year. Hopefully, the year will be a typical year (Petersen and Luxton,

1982). No one method of collecting nematode samples for assay (Goodell and Ferris 1981) or extracting nematodes can be used in all situations. There should be an adequate number of samples which give repeatable assays and are representative of the plot; also, these samples must be mixed and extracted by appropriate techniques. The objective of sampling and extraction is to collect and extract a sample that represents the nematode population in the field at any given time. Nematode distribution varies horizontally and vertically and is influenced by plant roots, litter, and soil types. For example, a nematode species may be more abundant in light textured soil than in clay.

The Society of Nematologists (Barker 1978) has listed recommendations for (a) the number, diameter, and depth of cores needed to provide an adequate sample; (b) representative pattern of sampling to provide reliable data on population densities; (c) the sampling frequency to reflect population densities at critical stages; and (d) proper handling and storage of samples. The recommendations are results of many investigations and are applicable to nematode ecology studies. Of course, adjustments in sampling procedures must be made for various soil ecosystems. Recommendations are as follows: a 2.0-cm sampling tube or cone-shaped tube should be used to a minimum depth of 20 cm. Each core (2.0 cm × 2.0 cm) provides about 62 cm^3 of soil. The minimum number of cores to be collected in a systematic manner for one soil sample are (a) small plots (1–5 m^2), 10 cores; (b) medium plots (5–100 m^2), 20 or more cores; and (c) large plots (greater than 100 m^2), 30 or more cores. Recommendations for a tree are five cores taken close to the plant and five cores at the drip line of the tree. If there are differences in soil type of moisture characteristics within the plot, the plot may need to be subdivided to reduce sampling error. Samples should be immediately placed in plastic bags and in an insulated cooler in the shade. Exposure to temperatures of 30–40°C, even for a short time, can kill certain species, which will interfere with subsequent recovery and identification. If the nematodes are not extracted in 2 days the soil should be stored at 10–15°C to keep the nematode morphology intact for identification. Samples from arid, nonirrigated areas can be stored safely in an air-conditioned building.

The next step in analysis of the soil nematode population is adequate mixing of the bulked samples if the entire sample (>500 cm^3) cannot be processed. The samples should be screened first and then mixed thoroughly by either running them through a soil sample splitter 2–4 times, or mixing and quartering the sample and then removing a 100-cm^3 aliquot. The mixing apparatus should be cleaned before mixing each sample. Mixing eliminates variation within a sample caused by nematode aggregation.

Extraction methods vary in efficiency and ability to extract various species equally. Without knowledge of the deficiencies and efficiencies of the method employed, both density and trophic structure may be misrepresented. Several extraction methods should be tested before proceeding with a nematode study. For example, ecological studies have largely ignored the extraction of endoparasites of roots, since this often requires a separate method and a longer incubation period.

Soil extraction methods are usually based on (a) the motility of the nematode or (b) flotation of the nematode from the soil. Motile nematode extraction techniques include Baermann funnels or trays, semiautomatic elutriator, Seinhorst elutriator, Oostenbrink elutriator, and the mist chamber. For flotation of the nematodes there are centrifugal flotation, sugar flotation sieving, semiautomatic elutriator, and Cobb's sieving and decanting (Barker 1978, Goodey 1963b). Recently, a highly efficient and new method for collecting nematodes from litter (dried leaves, pine needles) and peat moss was developed by W. A. Coolen. For information, write Dr. W. A. Coolen, Rijksstation voor Nematologie en Entomologie, Burg. van Gansberghelaan 96, B-9220 Merelbeke, Belgium. A partial description of a similar technique is given in Coolen (1979). We discuss only the Baermann funnel (BF) technique, for three reasons: It is the most commonly used by soil ecologists; the BF has a low efficiency and problems of which scientists must be aware; and with proper use, the BF can be inexpensive without sacrificing efficiency.

The BF was first devised for collection of juvenile stages of vertebrate parasites such as hookworm, and is now used widely for plant and soil nematodes. The equipment consists of 10 cm diameter glass funnels with rubber tubing, pinch clamps, 20-mesh plastic or stainless steel screen (copper is nematicidal), and facial tissues or cheesecloth. Soil is mixed and a 50-cm^3 aliquant is placed on the tissue, which is supported by the screen. Water is placed in the funnel, and the screen containing the tissue and sample is placed over the funnel until water just covers the soil. The samples are incubated at 21–24°C, and water is added as needed. The nematodes become active, swim out of the soil, and sink to the bottom of the funnel. The nematodes are collected at 24-hour intervals for 3 days. This method and modifications was widely used for roots and soil during the International Biological Program and extraction efficiencies varied from 35 to 90% (Sohlenius 1980). Reasons for variable efficiencies are (a) nematode motility varies with species, the more active being microbivores and the least active ectoparasites; (b) nematodes must be in an active state; (c) temperature controls nematode motility; (d) nematodes confined too long at the bottom of the funnel may die due to lack of oxygen and increased microbial activity; (e) eggs of some species can hatch during 72 hours, giving a skewed distribution of juveniles; (f) excessive water over soil or too much soil makes a poor recovery; and (g) use of deionized water decreases nematode extraction. Careful attention to the setting up of this apparatus and a knowledge of associated problems will assure increased extraction efficiency.

To determine the quantitative importance of soil organisms, it is necessary to have an estimation of their abundance and biomass, which is also useful for energy and material flow studies. The more common procedures for quantitative estimates of biomass and energy flow for nematodes are outlined below. For additional information consult the excellent reviews of Sohlenius (1980) and Yeates (1979).

Procedures for determining individual nematode weight and biomass are based on a formula of Andrássy (1956),

$$G = \frac{w^2 \times l}{1.6 \times 10^6}$$

where G is the weight in μg, w the nematode width, and l the length in μm. This formula approximates fresh-weight values. A conversion factor of 0.25 dry weight is used (Yeates 1979). Procedures for measuring oxygen consumption vary. However, the formula of Klekowski et al. (1972), $R = 1.40G^{0.72}$, where R is μl O_2 consumed/individual per hour, is the most commonly used. Recently, however, Atkinson (1980) reviewed the relationship between oxygen consumption and body weight of nematodes using data from single animal techniques. He concluded that the best approximation for estimating the requirements for the maintenance of nematode communities after assessing body size and abundance was $y/x = 1.6x^{-0.216}$, where y is mm^3 O_2 per individual per hour and x is the wet weight in milligrams. Respiration rates are corrected for soil temperature (Winberg 1971) and a calorific equivalent of 4.8 cal/ml O_2 is used (Yeates 1973). Croll and de Soyza (1980) determined calorie values for a wide range of nematodes and found a mean of 5.095 kcal/g dry weight. Production is calculated according to McNeill and Lawton's equation (1970) for short-lived poikilotherms—$\log P = 0.8262 \log R - 0.0948$. For a discussion of the problems involved in determining nematode energetics by these procedures, see Sohlenius (1980).

Preparation of Specimens

Some nematodes can be identified to superspecific taxonomic categories just by using a dissecting microscope and transmitted light. Specific identification requires carefully prepared slides of adult specimens and examination with an oil-immersion lens on a compound microscope. The specimens must be inactivated and preserved for the time needed to observe them; in some cases a few hours is adequate, whereas for other purposes, specimens should be preserved indefinitely. Nematodes are usually inactivated by killing with heat, with fixative, or by various methods combining both; heat and fixative; direct mounting in fixative at room temperature results in distortion. A mass collection containing large numbers of specimens may be killed and subjected to fixative in a single step. Nematodes are transferred to a few milliliters of water at room temperature and an equal volume of boiling 10% formalin is then added to the water. A more precise method is to transfer them in a vial and place the vial in a water bath at 57°C for 15 minutes. In our laboratory, a small number of specimens (less than 30) are handpicked into a drop of water and killed on a slide. Killing is by carefully applying minimum heat from an alcohol lamp until specimens "relax" or stop motion in an open or loosely coiled position. Temporary mounts are fixed by immediately transferring relaxed specimens to a second slide and mounting them in aqueous 2% formalin (2 ml of commercial 38% formaldehyde solution plus 98 ml of water). For permanent mounts, the nematodes are fixed in 5% formalin for at least 48 hours. Other fixatives recommended for nematodes include various formulas for FAA (formalin, ethanol, acetic acid), TAF (triethanolamine, ethanol, acetic acid) and lactophenol (Thorne 1961, Goodey 1963a). Specimens for temporary slides are mounted in a hydrated state, but permanent preparations must be dehydrated and infiltrated with glycerin. Infiltration can be facilitated by a minute prick in the intestine region of a partially fixed specimen, using a sharp instrument, such as an oculist's scalpel. In our laboratory, fixed specimens are transferred to a small open dish of 2.5% aqueous glycerin until the water is evaporated at room temperature (at least 1 week). Other dehydration and infiltration techniques use glycerin solutions in ethanol (Thorne 1961).

Specimens must be mounted, with the coverslip supported by finely drawn glass rods, glass wool (Christmas "angel hair") or for temporary mounts, with a ring of Zut® (Thorne 1961) or Gurr Glyceel®. Inadequately supported coverslips result in pressed distorted specimens. On the other hand, if supports are too thick, nematodes "float" and cannot be focused or kept in the field of view at high magnifications. Coverslips are generally sealed in place with a ring of Zut® or Gurr Glyceel®. Zut® may be difficult to obtain, but Glyceel is available from Atomergic Chemetals Corp. (100 Fairchild Avenue, Plainville, NY 11803). Many laboratories permanently mount specimens on "Cobb slides." These 75 × 25 mm aluminum frames have an 18-mm-diameter central hole over which a 25-mm square No. 1 coverslip is attached. Specimens are mounted on the square coverslips in glycerin and covered and sealed under an 18-mm round coverslip. This type of slide permits viewing of specimens on both sides. Cobb slide frames may be available from some laboratories with large collections (Table 8.1). A device for making these frames can be built (Mason and Bosher 1963) or purchased from Arthur H. Thomas Co.

Identification of certain nematode species may require specialized dissecting and mounting methods, including *en face* views, perineal patterns, or hand cut sections. Techniques for these procedures are discussed in Thorne (1961) and Goodey (1963a). Dissecting specimens is facilitated by tools, including an oculist's scalpel; similarly, needles such as a finely honed dental canal file or bamboo split are helpful in manipulating and transferring individual specimens.

Table 8.1. Locations of Some Nematode Slide Collections[a]

Australia	
Collection of the CSIRO, Division Horticultural Research, Merbein, Victoria 3505	
Belgium	
Nematode Collection, Instituut voor Dierkunde, Ledeganckstraat 35, B-9000 Gent	
Canada	
Canadian National Collection of Nematodes, Biosystematics Research, Institute, Central Experimental Farm, Ottawa, Ontario K1A OC6	
Denmark	
Zoological Museum, University of Copenhagen, Denmark	
France	
Collection du Museum National d'Historie Naturelle, 61 rue de Buffon, 75231 Paris	
Japan	
Lab of Nematology, Forestry and Forest Products Research Institute, P.O. Box 16, Tsukuba Norin Kenkyu Danchi-nae, Ibaraki, 305	
Netherlands	
Lab voor Nematologie, Landbouwhogeschool, Binnenhaven 10, P.O. Box 8123, 6700 ES Wageningen	
New Zealand	
Entomology Division, Mount Albert Research Centre, 120 Mt. Albert Rd., Auckland	
United Kingdom	
Rothamsted Experimental Station, Department of Nematology, Harpenden, Herts AL5 2JQ	
Commonwealth Institute of Helminthology, St. Albans, Herts AL4 0XU	
United States	
ARS, USDA Nematology Section, Plant Protection Institute, Beltsville, Maryland 20705	
Museum of Natural History, Department of Invertebrate Zoology, Washington, D.C. 20560	
Purdue University Nematode Collection, Department of Entomology, West Lafayette, Indiana 47907	
University of California, Division of Nematology, Davis, California 95616	
University of California, Department of Nematology, Riverside, California 92521	
USSR	
The Zoological Institute, Academy of Sciences, Leningrad	

[a] See also Nicholas 1974 and Lichtenfels and Pritchard 1982.

Measurement and Identification

The type of classification that is efficient and satisfactory for a given ecology project must be determined before beginning the difficult task of specific taxonomic identification of large numbers of specimens. Many ecological studies list terrestrial nematodes as numbers of individuals or numbers of genera by trophic groups. This can often be inferred from morphology of the stoma. For some ecological studies, it is satisfactory to identify terrestrial nematodes according to higher taxonomic categories, such as order or superfamily; often as much biological information can be inferred from a higher taxon as from a little-studied species. For example, the biology of a specimen identified as the order Mononchida is readily inferred as a predaceous nematode.

Some ecological studies require specific taxonomic identification of terrestrial nematodes. The person carrying out such a task should be familiar with taxonomically important characteristics as well as measurements, ratios, and percentages which are commonly encountered in keys (Table 8.2).

Identification to genus usually requires adult specimens, sometimes of both sexes, and specimens must be carefully preserved and mounted. Since nematology is a relatively young science, surveys of terrestrial habitats will frequently include new species. These should be submitted to an expert (see cited authorities), or specimens can be sent to one of the institutions with large collections of preserved nematodes (Table 8.1). In addition, it is important that voucher specimens from ecological studies be retained and deposited in recognized repositories for future reference.

Terrestrial nematodes predominate in the large orders Tylenchida, Rhabditida, Mononchida, and Dorylaimida. Complete keys to genera of

Table 8.2. Symbols for Measurements, Ratios, and Percentages Commonly Used in Nematode Descriptions

a	*L*/greatest body width
b	*L*/distance from anterior end to esophageal valve
b'	*L*/distance from anterior end to base of esophagus
c	*L*/tail length
c'	Tail length/body width at anus
DGO	Distance from base of stoma or stylet to dorsal gland orifice
h	Length of hyaline terminus of tail
L	Total body length
m	Length of stylet cone/length of stylet shaft plus knobs (also length of stylet cone as percent of total stylet length)
MB	Distance from anterior end to medium bulb as percent of total length of esophagus
o	Distance from base of stoma (or stylet knobs) to dorsal esophageal gland orifice/spear length (may be expressed as percent)
R	Total number of body annules
R an	Number of annules on tail (posterior end to anus)
R ex	Number of annules from anterior end to excretory pore
R V	Number of annules from posterior end to vulva
R Van	Number of annules from vulva to anus
s	Stylet length/body width at base of stylet
T	Distance from anterior end of testis to cloaca as percent of *L*
V	Distance from anterior end to vulva as percent of *L*

these orders is beyond the scope of this chapter. Therefore, we have limited our keys for these groups to identification of higher categories such as superfamilies, and references are given for aids to more specific identification. On the other hand, aquatic nematodes predominate in the orders Monhysterida, Araeolaimida, Enoplida, Desmodorida, and Chromadorida and we have included keys to a few terrestrial genera. Samples from marshy areas, muck, or tidal flats may include forms from the above and other orders which are essentially aquatic and several keys and other guides are available for identification of these nematodes, including Allen and Noffsinger (1978), Ferris et al. (1973), Filipjev (1968, 1970), Gerlach and Riemann (1973, 1974), Hope (1971), Mulvey (1969), Pennak (1978), Tarjan (1980), Tarjan et al. (1977), Wieser (1959), and Wieser and Hopper (1967). General keys for, or including, phytophagous and free-living terrestrial nematodes are Golden (1971), Goodey (1963b), Hopper and Cairns (1959), Mai and Lyon (1975), Southey (1978a), and Thorne (1961). General systems of classification of nematodes are included in many of the references above, as well as other sources: Andrássy (1976), Chitwood and Chitwood (1977), Hooper (1978), Hooper and Southey (1978), Maggenti (1981, 1982), and Poinar (1979). The systems presented in Grassé (1965) is summarized as a table in Nicholas (1975). Genera of free-living nematodes commonly found in soil samples are discussed by Southey (1978b).

The key characters employed herein were revised and updated to February 1984. Readers are urged to examine current literature to supplement their identifications and studies of soil nematodes.

KEY TO CLASSES AND ORDERS OF TERRESTRIAL NEMATODES

Two Classes

1a Phasmids present; esophagus two- or three-part; setae extremely rare **(Secernentea)** **2**

1b Phasmids absent; esophagus one-, two-, or rarely three-part; setae common **(Adenophorea)** **3**

Seven Orders

2a Stoma modified to stylet; esophagus three-part (Fig. 8.9) . . . . **Tylenchida**

2b Stoma without stylet; esophagus two- or three-part (Fig. 8.10) . . **Rhabditida**

3a Esophagus one-part cylindrical; some tending toward two-part, with large pyramidal esophagointestinal valve **4**

3b Esophagus two-part (rarely three-part), without large pyramidal esophagointestinal valve **6**

4a Amphids typically circular; cephalic setae usually present, stoma lining weakly sclerotized without large teeth (Fig. 8.11) . . . . **Monhysterida**

4b Amphid openings indistinct slits or cyathiform; presence of cephalic setae and teeth, variable. **5**

5a Stoma lining cuticularized with broad dorsal tooth (onchium), esophageal lu-

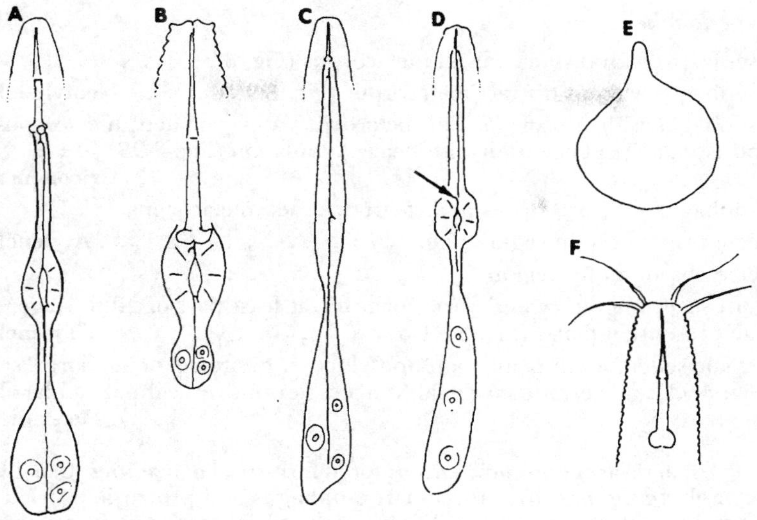

Fig. 8.9. Diagnostic characteristics of superfamilies of Tylenchida: (*A*) esophagus of Tylenchoidea (*Tylenchorhynchus* sp.); (*B*) esophagus of Criconematoidea (*Criconemoides* sp.); (*C*) esophagus of Neotylenchoidea (*Nothotylenchus* sp.); (*D*) esophagus of Aphelenchoidea (*Aphelechoides* sp.). Arrow indicates position of dorsal gland orifice; (*E*) outline showing form of female of Heteroderoidea (*Meloidogyne* sp.); (*F*) head region of Atylenchoidea (*Atylenchus* sp.).

men interrupted, posterior to nerve ring, by openings of gland ducts (often difficult to observe); body without setae (Fig. 8.12) . . . . **Mononchida**

5b Stoma variable; esophageal glands open into stoma; and lumen interrupted by openings to two gland ducts posterior to nerve ring (often difficult to observe); body usually with setae (Fig. 8.13) . . . . **Enoplida**

6a Esophagus narrow anteriorly, broad and usually elongate and cylindrical posteriorly; amphids typically cyathiform, (Fig. 8.14). . . . . **Dorylaimida**

6b Esophagus narrow anteriorly, bulboid posteriorly; amphid variable but rarely cyathiform **7**

7a Cuticle in head region thickened and hyaline, forming "helmet"; distinct annulation posterior to helmet (Fig. 8.16*A*). . . . . **Desmodorida**

7b Cuticle in head region not forming helmet; cuticular markings variable . . . 8

8a Cuticle without punctations; amphid variable, often spiral or round (Fig. 8.15) **Araeolaimida**

8b Cuticle ornamented, generally with punctations; amphid variable, often spiral or single loop (doughnut-shaped), (Fig. 8.16*B*) . . . . **Chromadorida**

TYLENCHIDA: KEY TO SUPERFAMILIES

Two Suborders

1a Outlet of dorsal esophageal gland orifice in procorpus, usually near base of stylet (Tylenchina; Fig. 8.9*A–C*) **2**

1b Outlet of dorsal gland in large–medium bulb (Aphelenchina, with one superfamily; Fig. 8.9*D*) . . . . **Aphelenchoidea**

Five Superfamilies

2a Esophagus with distinct valved metacorpus (Fig. 8.9*A,B*) 3
2b Esophagus without distinct metacorpus (Fig. 8.9*C*). . . . . **Neotylenchoidea**
3a Esophagus with procorpus and metacorpus amalgamated; metacorpus large and elongate; cuticle often with heavy annulation (Fig. 8.9*B*) . . . . **Criconematoidea**
3b Esophagus with procorpus and clearly defined metacorpus. 4
4a Setae present on lip region (Fig. 8.9*F*) . . . . **Atylenchoidea**
4b Setae absent on lip region 5
5a Both sexes migratory and vermiform for at least part of adult stage; caudal alae present on males (Fig. 8.8*A*) . . . . **Tylenchoidea**
5b Female sedentary in plant roots and globose, pyriform, or similarly "swollen" (Fig. 8.9*A*); no vermiform females; males vermiform without caudal alae . . **Heteroderoidea**

Tylenchida are commonly encountered terrestrial nematodes. Usually, they are easily recognized by a three-part esophagus and protrusible stylet. Since the order includes economically important plant parasites, its taxonomy is better developed than that of most orders. However, Golden (1971) estimates that less than 10% of the total species thought to exist have been described. The five superfamilies included here are distinguished primarily by the morphology of the esophagus (Figs. 8.6*D*, 8.9*A–D*). Some classifications include species of Heteroderoidea and Atylenchoidea within Tylenchoidea (Maggenti 1981, 1982). In addition, Raski and Maggenti (1983) reject the validity of Neotylenchoidea as a separate subfamily and consider the lack of a metacorpus valve to be a character that occurs repeatedly throughout several related groups of Tylenchida. Atylenchoidea, with only a few rare species, is distinctive among Secernentea because it has setae (Fig. 8.9*F*). Not all Tylenchida are plant parasites. Insect associates, predators, and fungus feeders are included in genera of Aphelenchoidea, Neotylenchoidea, and Tylenchoidea. Some recent classifications include aphelenchids in Aphelenchida, separating it from Tylenchida (Siddiqi 1980; Maggenti 1981, 1982).

Keys to genera of Tylenchoidea include Golden (1971) and Mai and Lyon (1975). Classification and/or keys to some groups are included in Skarbilovich (1959), Thorne (1961), Goodey (1963b), Ritter and Theodorides, (1965), Siddiqi (1971, 1980), Andrássy (1976), and Hooper (1978). Keys to many specific groups of Tylenchida are given by several authors in Southey (1978b).

RHABDITIDA: KEY TO SUPERFAMILIES

1a Esophagus with muscular basal bulb with a broad platelike valve (Fig. 8.10*A,B*); teeth rare or if present, small (Fig. 8.10*E–G*) 2
1b Esophagus with glandular basal bulb without valve (Fig. 8.10*C*); teeth, including one or more large onchia, typically present (Fig. 8.10*H*) . . . . **Diplogasteroidea**
2a Wartlike structures usually present on one side (Fig. 8.10*D*) . . . . **Bunonemoidea**
2b Wartlike structures absent 3
3a Stoma narrow appearing collapsed for most of length; rhabdions largely separated; probolae often present, without cirri (Fig. 8.10*F,I*) . . . . **Cephaloboidea**

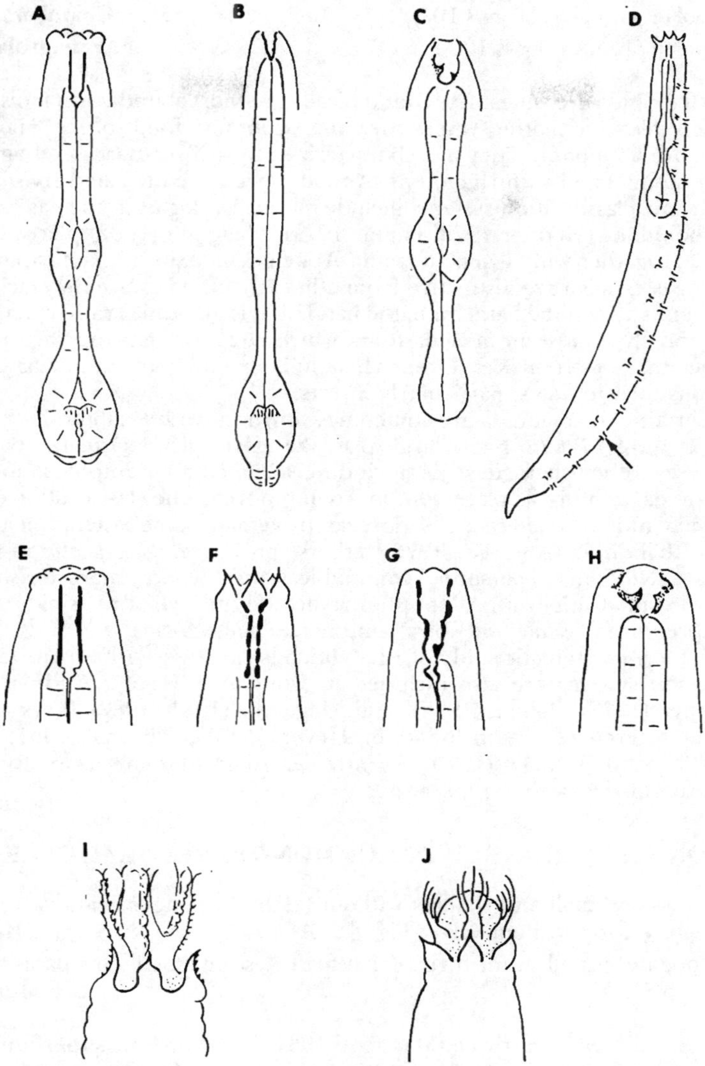

Fig. 8.10. Diagnostic characteristics of superfamilies of Rhabditida: (*A*) esophagus of Rhabditoidea (*Rhabditis* sp.); (*B*) esophagus of Panagrolaimoidea (*Panagrolaimus* sp.); (*C*) esophagus of Diplogasteroidea (*Diplogaster* sp.). (*D*) outline of Bunonemoidea (*Bunonema* sp.). Arrow indicates position of a wart; (*E*) stoma of Rhabditoidea (*Rhabditis* sp.); (*F*) stoma of Cephaloboidea (*Eucephalobus* sp.); (*G*) stoma of Panagrolaimoidea (*Panagrolaimus* sp.); (*H*) stoma of Diplogasteroidea (*Diplogaster* sp.); (*I*) probolae (*Acrobeles* sp.); (*J*) probolae showing cirri of Chambersielloidea (*Chambersiella* sp.).

3b Stoma mostly broad tubular or funnel shaped; rhabdions partially fused (Fig. 8.11*E*,*G*); probolae, if present, with fine cirri (Fig. 8.10*J*) 4

4a Stoma cylindrical tube; walls not heavily cuticularized (Fig. 8.10*E*); rhabditoid esophagus (Fig. 8.10*A*) . . . . **Rhabditoidea**

4b Stoma broad anteriorly, but tapering posteriorly; walls heavily cuticularized (Fig. 8.10*G*); panagrolaimoid esophagus (Fig. 8.10*B*) 5

5a Probolae with cirri (Fig. 8.10*J*) . . . . **Chambersielloidea**
5b Probolae absent (Fig. 8.10*G*) . . . . **Panagrolaimoidea**

Rhabditida are widespread but typically are most abundant in moist organic soil on decaying matter, where their most common food source, microorganisms, are abundant. They are characterized by a three-part esophagus and a large stoma lined with rhabdions. Although we have indicated six superfamilies, some classification systems include all but Diplogasteroidea as families of Rhabditiodea (Hooper 1978, Poinar 1979). Maggenti (1982) places Diplogasterida, together with Tylenchida and Aphelenchida, in subclass Diplogasteria. Diplogasteroidea are distinctive from other Rhabditida since they lack a broad platelike valve in the basal bulb, the basal bulb is glandular rather than muscular, and they have an armed stoma which suggests that they are predators rather than microbivores. Other rhabditids may be parasites or associates of various invertebrates, particularly insects.

Certain Araeolaimida are sometimes confused with Rhabditida (Fig. 8.15). For example, *Plectus* has a large platelike valve (grinder) in the basal bulb; however, other characteristics, including setae, circular amphids, and lack of phasmids, indicate its placement in Adenophorea. The classification of *Teratocephalus* and *Euteratocephalus* is more controversial. Some systems include these with Rhabditida (Andrássy 1976); others consider *Eutatocephalus* in Chromadorida (Noffsinger, personnel communications). Several characteristics which indicate an affinity with Adenophorea include the lack of phasmids, as well as the presence of setae and spiral amphids in some forms.

A recent systematic study of Rhabditida is Andrássy (1983). However, classification systems are also included in Dougherty (1955), Andrássy (1976), Hooper (1978), Poinar (1979), and Maggenti (1981, 1982). Keys and additional references are included in Heyns (1971), Thorne (1961), Goodey (1963b), and Bernard (1979) recently described new species of Bunonematoidea and reviewed the literature.

MONHYSTERIDA: KEY TO COMMON TERRESTRIAL GENERA

Stoma anteriorly small and shallow without teeth; esophagus cylindrical, but may broaden posteriorly without valve (Fig. 8.11*A*) . . . . ***Monhystera***
Stoma elongate, usually with three small teeth; esophagus forms basal bulb with valve (Fig. 8.11*B*) . . . . ***Monhystrella***

Monhysterida includes three (Maggenti 1981, 1982) or four superfamilies (Andrássy 1976) which are entirely aquatic except for a few terrestrial genera in Monhysteroidea. Monhysterida is characterized by a cylindrical esophagus (bulboid rare), and four or more setae in the head region. The stoma is usually small without teeth, although denticles may be present. Amphids are typically circular. The genera *Monhystera* and *Monhystrella* are probably both microbivores and are distinguished by the shallow and elongate stoma, respectively. Classification of Monhysterida include Poinar (1979) and Maggenti (1981). Andrássy (1981) has recently published a revision of the order.

MONONCHIDA: KEY TO SUPERFAMILIES

Stoma wide; most of stoma walls not surrounded by esophageal tissue (Fig. 8.12*A*,*B*). . . . . **Mononchoidea**

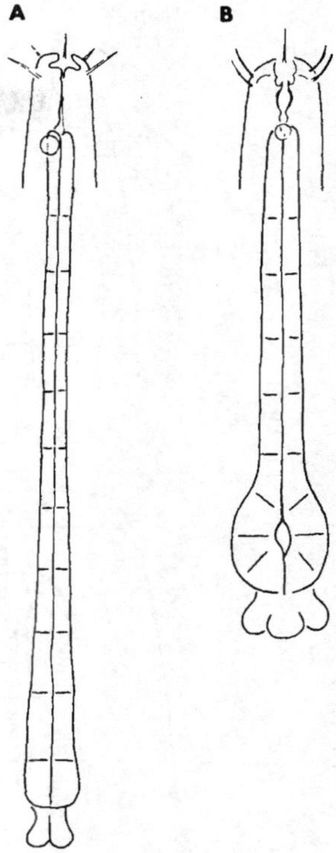

Fig. 8.11. Stoma and esophagi of Monhysterida: (*A*) *Monhystera* sp.; (*B*) *Monhystrella* sp. [(*B*) Redrawn from Goodey (1963b).]

Stoma narrow; most of stoma walls surrounded by esophageal tissue (Fig. 8.12*C*) **Bathyodontoidea**

Mononchida are widely distributed in uncultivated soils, although a few species are aquatic. They are distinguished by a one-part or cylindrical esophagus with five gland cells (Fig. 8.12*A*) and a stoma cuticularized with at least one large tooth. The esophageal glands open into the lumen posterior to the nerve ring. The two superfamilies are identified by the length-to-width proportions of the stoma; other important characters include the morphology of various stoma parts, and the esophagointestinal valve. The valve may be tuberculate with the lumen ending in three swollen hollow tubercles, or it may be nontuberculate (Fig. 8.12*D,E*) Monochida are predators.

Some taxonomic systems include the mononchids as a taxon within the Dorylaimida (Andrássy 1976, Goodey 1963b, Poinar 1979), but others elevate them to a separate order (Hooper 1978; Maggenti 1981, 1982). In addition to the above, keys to genera of Mononchida include Mulvey (1961a,b 1962, 1963), Jensen and Mulvey (1968), and Mulvey and Jensen (1967). Bathyodontina is discussed by Coomans and Loof (1970). More recently, Jairajpuri and Khan (1982) reviewed the order.

ENOPLIDA: KEY TO COMMON TERRESTRIAL GENERA

1a Stoma walls heavily cuticularized; barrel or tubular-shaped (Fig. 8.13*A,D*) . **2**

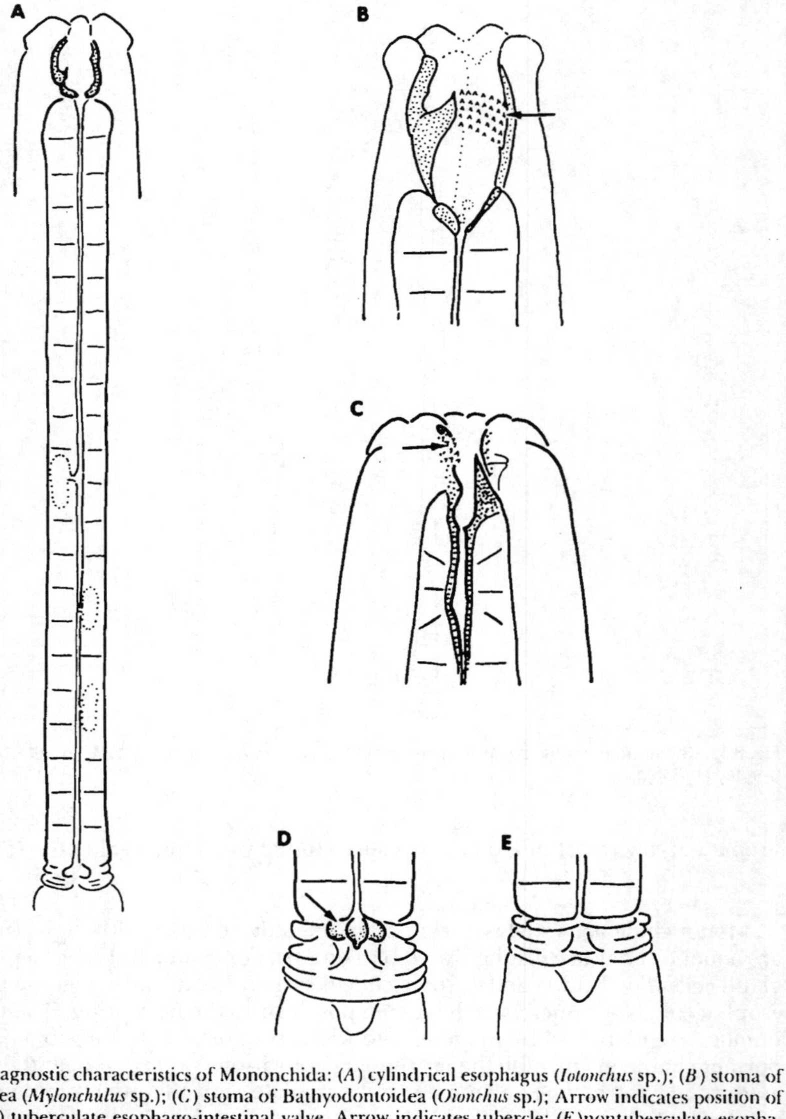

Fig. 8.12. Diagnostic characteristics of Mononchida: (*A*) cylindrical esophagus (*Iotonchus* sp.); (*B*) stoma of Mononchoidea (*Mylonchulus* sp.); (*C*) stoma of Bathyodontoidea (*Oionchus* sp.); Arrow indicates position of denticles; (*D*) tuberculate esophago-intestinal valve. Arrow indicates tubercle; (*E*)nontuberculate esophagointestinal valve. [(*D,E*) Redrawn from Clark (1960).]

1b Stoma walls weakly cuticularized; collapsed or funnel-shaped (Fig. 8.13*A,D*). **3**

2a Stoma barrel-shaped; denticles at posterior end of stoma, sometimes with small dorsal tooth; six long cephalic setae, and outer row of four slightly shorter setae; amphid opening a transverse slit, posterior to stoma (Fig. 8.13*A*) . . . . ***Prismatolaimus***

2b Stoma long cylindrical tube, with three teeth anteriorly; four cephalic setae; amphid cyathiform, opening near level of setae (Fig. 8.13*D*) . . . . . ***Ironus***

3a Stoma collapsed with small dorsal tooth, three lips each with circlet of very short blunt setae; amphid opening inconspicuous (Fig. 8.13*B*) . . . . ***Tripyla***

3b Stoma funnel-shaped without teeth, six lips with 10 or more blunt setae; amphid opening, cyathiform (Fig. 8.13*C*). . . . . ***Tobrilus***

Enoplida is almost exclusively marine and freshwater. In addition to relatively few genera with terrestrial species, some essentially aquatic forms may be found in very wet or marshy soils. Enoplida species have a one-part cylindrical esophagus (Fig. 8.6*B;* bulboid rare) which resembles that of Mononchida, but differs in that three esophageal gland openings are in or near the stoma and two open just anterior to the nerve ring. In Mononchida all the gland openings are posterior to the nerve ring (Maggenti 1982). The stoma is highly variable with or without teeth. Setae are usually present. Andrássy (1976) includes

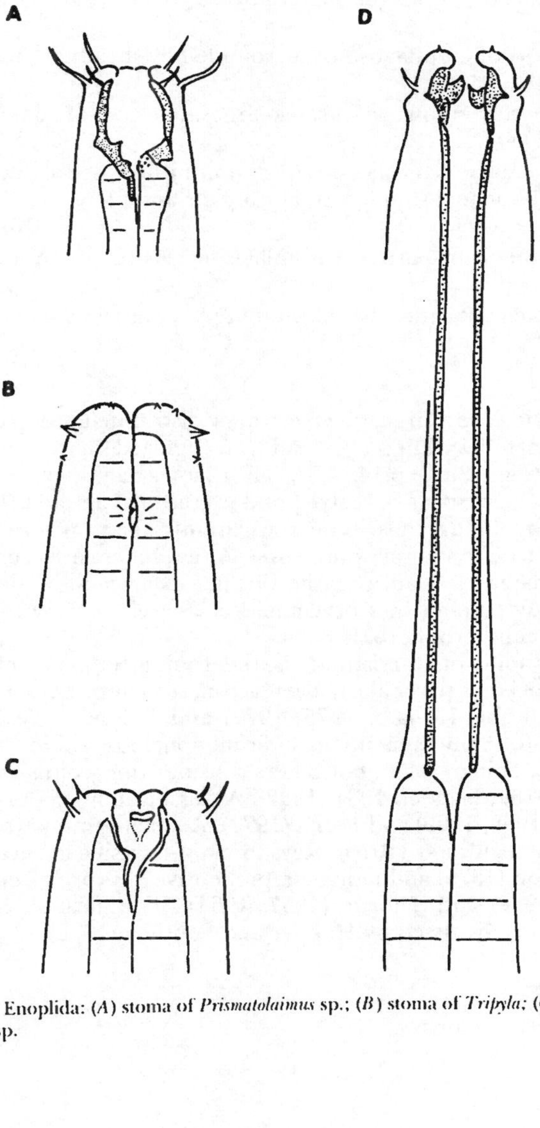

Fig. 8.13. Stomas of Enoplida: (*A*) stoma of *Prismatolaimus* sp.; (*B*) stoma of *Tripyla;* (*C*) stoma of *Tobrilus;* (*D*) stoma of *Ironus* sp.

eight superfamilies in Enoplida; Maggenti (1981) includes six. Genera found in soil are likely to be in Triphyloidea (*Tripyla, Tobrilus, Prismatolaimus*) or closely related Ironoidea (*Ironus*). Little is known about the biology, but they are probably predaceous. Goodey (1963b) suggests that *Tobrilus* (syn *Trilobus*) may also feed on diatoms or algae. Classifications of Enoplida include Andrássy (1976) and Maggenti (1981, 1982).

DORYLAIMIDA: KEY TO SUPERFAMILIES†

1a Enlarged part of esophagus includes posterior one-third or more (Fig. 8.14*A*,*B*) **2**

1b Enlarged part of esophagus includes less than posterior one-third (Fig. 8.14*C*) or may form pyriform bulb (Fig. 8.15*D*) **3**

2a Enlarged part of esophagus surrounded by a sheath of muscles (Fig. 8.14*A*). **Belondiroidea**

2b Enlarged part of esophagus not surrounded by sheath of muscles (Fig. 8.14*B*) **4**

3a Amphid apertures obscure slits, two testes, one pair of adanal supplementary organs present . . . . **Leptonchoidea**

3b Amphid aperatures conspicuous ovals (cyathiform; see arrow Fig. 8.14*D*), one testis, pair of adanal supplementary organs absent . . . . **Diphtherophoroidea**

4a Walls of stoma cuticularized with plates or ribs surrounding axial spear (Fig. 8.14*E*) . . . . **Actinolaimoidea**

4b Walls of stoma not cuticularized without plates or ribs surrounding axial spear or stylet. . . . . **Dorylaimoidea**

Dorylaimida are a large, diverse, widely distributed group of terrestrial and aquatic nematodes. They are readily recognized by a two-part cylindrical esophagus (Figs. 8.6*C*, 8.14*A–D*). They lack setae, and most forms have a conspicuous axial spear or a stylet and cyathiform amphid (Fig. 8.3*B*). Many dorylaims are predaceous; others apparently feed on lower plants such as algae and mosses; many are omnivores. A few genera are economically important plant parasites. Plant parasites include *Longidorus, Xiphinema,* and other closely related genera in Dorylaimoidea, as well as *Trichodorus* and related genera in Diphtherophoroidea.

The taxonomy of Dorylaimida is undergoing frequent revisions, and considerable work has particularly been accomplished in Leptonchoidea (including Goseco et al. 1974a,b, 1975, 1977) and Belondiroidea (Jairajpuri and Ahmad 1980). Some systems include mononchids within Dorylaimida (Andrássy 1976, Poinar 1979), but others consider the groups as separate orders (Hooper 1978; Maggenti 1981, 1982). A classification of Dorylaimida is given by Siddiqi (1968, 1969) and Ferris (1971). A system with keys to higher taxa is presented by Andrássy (1976). Keys to plant-parasitic genera are included in Mai and Lyon (1975) and more comprehensive keys are given by Thorne and Swanger (1936) and Thorne (1957, 1964). Identification of plant-parasitic Dorylaimida is discussed by Hooper and Southey (1978).

† Adapted from Thorne (1957).

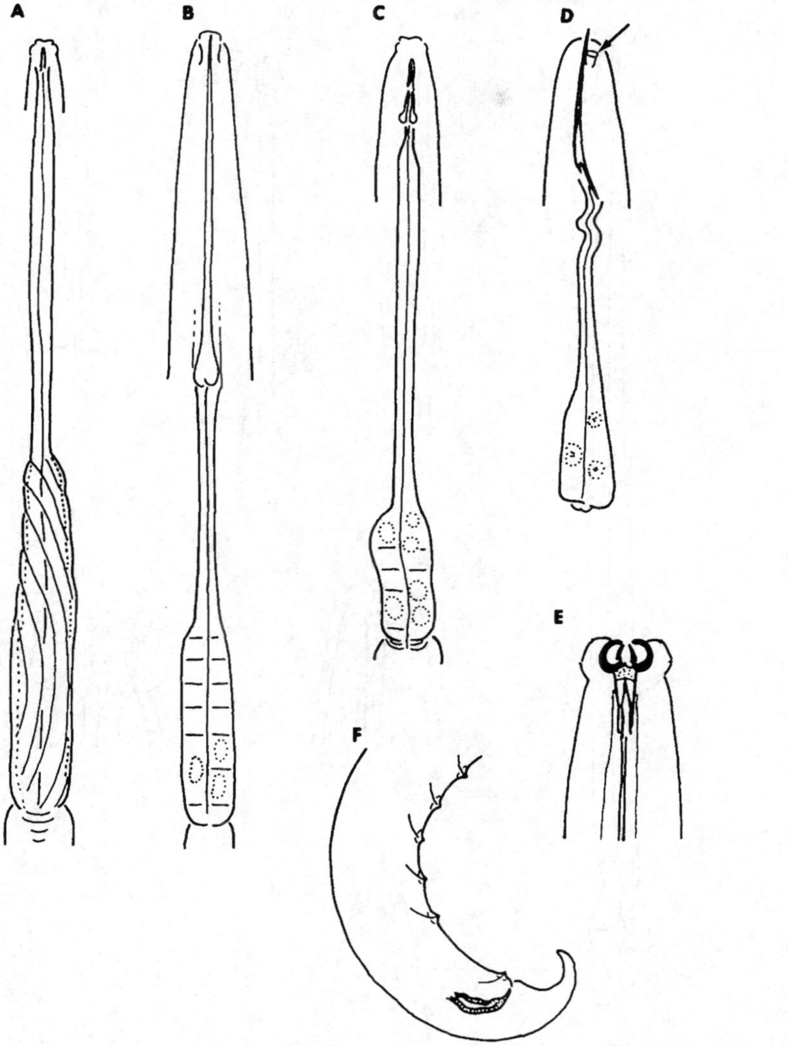

Fig. 8.14. Diagnostic characteristics of superfamilies of Dorylaimida: (*A*) esophagus of Belondiroidea (*Dorylaimellus* sp.); (*B*) esophagus of Dorylaimoidea (*Xiphinema* sp.); (*C*) esophagus of Leptonchoidea (*Tylencholaimellus* sp.); (*D*) esophagus of Diphtherophoroidea (*Paratrichodorus* sp.) (arrow indicates position of cyathiform amphid opening); (*E*) stoma of Actinolaimoides (*Carcharolaimus* sp.); (*F*) male tail region showing supplements (*Dorylaimus* sp.).

ARAEOLAIMIDA: KEY TO COMMON TERRESTRIAL GENERA

1a Terminal bulb of esophagus with broad "butterfly" valve (with conspicuous striated plates) (Fig. 8.15*B*) 2

1b Terminal bulb of esophagus without valve, or if present, smooth and narrow (without conspicuous striated plates) (Fig. 8.15*A,D*) 7

2a Head region with six pointed lips curved inwardly (Fig. 8.15*G*). 3

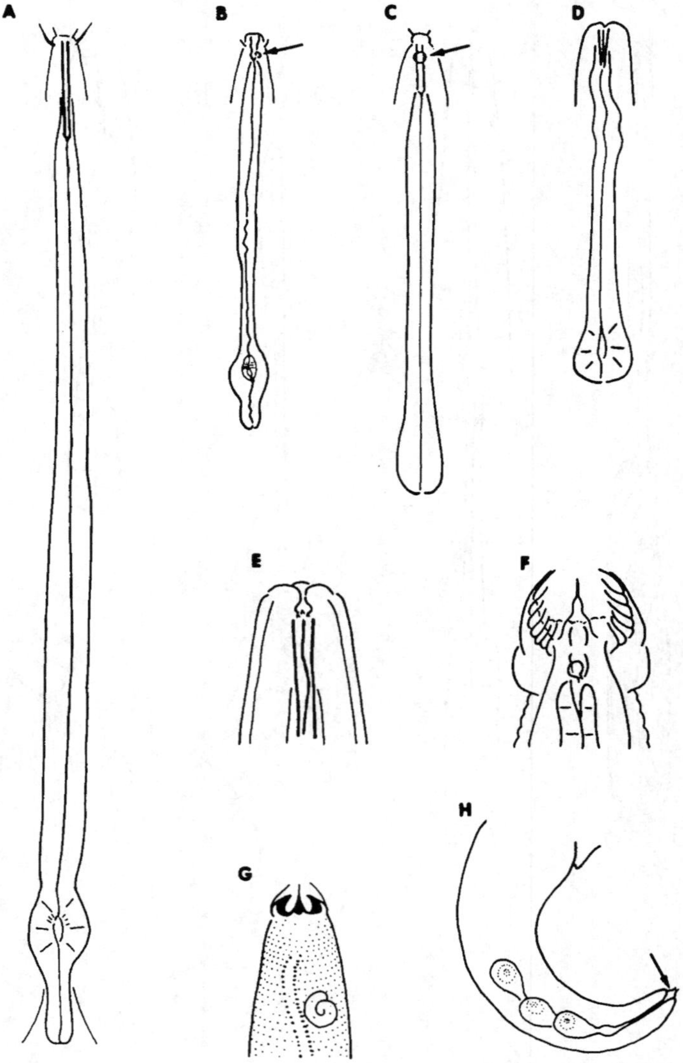

Fig. 8.15. Diagnostic characteristics of Araeolaimida: (*A*) esophagus of *Chronogaster* sp.; (*B*) esophagus of *Plectus* sp. (arrow indicates position of amphid opening); (*C*) esophagus of *Cylindrolaimus* sp. (arrow indicates position of amphid opening); (*D*) esophagus of *Rhabdolaimus* sp.; (*E*) stoma of *Rhabdolaimus* sp.; (*F*) head region of *Wilsonema* sp.; (*G*) head region of *Euteratocephalus* sp.; (*H*) tail region of *Plectus* sp. (arrow indicates spinneret which is a common duct to three caudal glands). [(*D*,*E*) Redrawn from Goodey (1963) and Ferris et al. (1973).]

2b Head region without pointed lips **4**

3a Body with distinct annules; amphid opening porelike, cephalic setae absent; one ovary . . . . ***Teratocephalus***

3b Body with fine horizontal rows of punctations; amphid spiral; cephalic setae present; two ovaries (Fig. 8.15*G*) . . . . ***Euteratocephalus***

4a Head with four to six winglike structures on sides (Fig. 8.15*F*) **5**

4b Head without winglike structures **6**
5a Head with four winglike structures . . . . *Tylocephalus*
5b Head with six winglike structures (Fig. 8.15*F*) . . . . *Wilsonema*
6a Body with very fine striations; amphid circular or elliptical (Fig. 8.15*B*) . . *Plectus*
6b Body with distinct annules; amphid transverse slitlike or obscure . . . . . . *Anaplectus*
7a Female with one ovary; esophagus barely enlarged posteriorly (Fig. 8.15*C*). *Cylindrolaimus*
7b Female with two ovaries; esophagus variable (Fig. 8.7*E*) **8**
8a Gonads prodelphic (Fig. 8.7*E*); clearly defined terminal bulb of esophagus with long "neck" (including esophagointestinal valve) (Fig. 8.15*A*) . . . . . *Chronogaster*
8b Gonads amphidelphic (Fig. 8.7*D*); anterior part of esophagus gradually enlarges posteriorly and merges with terminal bulb **9**
9a Terminal bulb of esophagus with valve in lumen (Fig. 8.15*D*); three longitudinal thickenings in stoma; amphids inconspicuous (Fig. 8.15*E*) . . . . . . *Rhabdolaimus*
9b Terminal bulb of esophagus reduced without valve; stoma reduced and inconspicuous **10**
10a Large circular amphid located near lip region (Fig. 8.3*C*) . . *Aphanolaimus*
10b Spiral or crescent-shaped amphid posterior to lip region (Fig. 8.3*D*) . . . . *Eastiania*

Araeolaimida is a large order, but only a few genera occur in soil. Araeolaimida are distinguished by the esophagus, which typically broadens posteriorly to form a basal bulb which may or may not include a valve. The stoma is usually narrow and tubular, and teeth or denticles are rare. Typically, 4 or 10 setae occur on the lip regions, and amphids vary and may be spiral, circular, or inconspicuous transverse slits. Araeolaimida includes a number of superfamilies, but systems of classification are highly variable, and there is some disagreement as to what genera should be included in the order. The most common soil forms occur in Plectoidea, including the genera *Plectus, Anaplectus,* and *Wilsonema. Euteratocephalus* and *Teratocephalus* are less common. Probably most of these genera are microbivores. Recent classifications of Araeolaimida include Andrássy (1976) and Maggenti (1981, 1982).

DESMODORIDA

Desmodorida includes five superfamilies (Maggenti 1982) and like Chromadorida is almost exclusively aquatic. One exception is *Prodesmodora* (Fig. 8.16*A*), which occurs in moist soil; feeding habits are unknown. *Prodesmodora* has strong annulations which contrast sharply with the smooth hyaline helmet. The helmet, which has four cephalic setae, is composed of thick layers of cuticle and terminates near the base of the stoma. The stoma is elongate and includes a distinct dorsal tooth and very small subventral teeth. The amphid is a single loop slightly posterior to the stoma, and the esophagus is two-part with an end bulb.

CHROMADORIDA

Chromadorida includes four superfamilies (Maggenti 1981, 1982) and is primarily aquatic. Only a few genera, including *Achromadora* (Fig. 8.16*B*), occur in

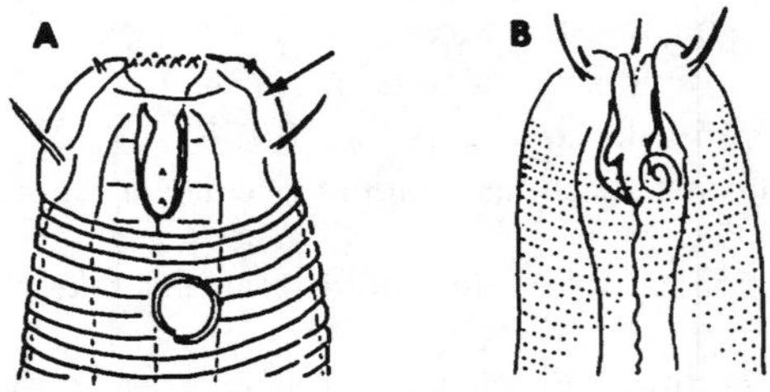

Fig. 8.16. Terrestrial genera of Desmodorida and Chromadorida: (*A*) head region of *Prodesmodora* showing thickened cuticle of helmet (arrow); (*B*) head region of *Achromadora*.

moist soil; feeding habits are unknown. *Achromadora* has striations composed of punctations and a spiral amphid, near or just posterior to the funnel-shaped stoma. The stoma has a conspicuous dorsal tooth, and the esophagus is two-part with an end bulb. Ten setae occur on the head region.

References

Allen, M. W. and E. M. Noffsinger. 1978. *A Revision of the Superfamily Draconematoidea Filipjev, 1918 (Nematoda: Draconematina).* Univ. Calif. Publ. Zool. 109. University of California Press, Berkeley, Calif. 133 pp. **Andrássy, I. 1956.** Die Rauminhalts—und Gewichtsbestimmung der Fadenwürmer (Nematoden). *Acta Zool. Acad. Sci. Hung.* 2:1–15. **1976.** *Evolution as a Basis for the Systematization of Nematodes.* Pitman, London. 288 pp. **1981.** Revision of the order Monhysterida (Nematoda) inhabiting soil and inland waters. *Opusc. Zool. (Budap.)* 17 and 18:13–47. **1983.** *A Taxonomic Review of the Suborder Rhabditina (Nematoda: Secernentea.)* Eotvos Lorand University, Budapest, Hungary. 241 pp. **Atkinson, H. J. 1980.** Respiration in nematodes. Pp. 101–142 in B. M. Zuckerman (ed.). *Nematodes as Biological Models,* Vol II. Academic Press, London. **Banage, W. B. 1963.** The ecological importance of free-living nematodes with special reference to those of moorland soil. *J. Anim. Ecol.* 32:133–140. **Barker, K. R.,** Chairman, SON/ASTM E 35.16 Task Force on Nematode Population Response Committee. **1978.** Determining nematode population responses to control agents. Pp. 114–125 in E. I. Zehr (ed.). *Methods for Evaluating Plant Fungicides/ Nematicides and Bactericides.* American Phytopathological Society, St. Paul, Minn. **Barrett, J. 1976.** Energy metabolism in nematodes. Pp. 11–70 in N. A. Croll (ed.). *The Organization of Nematodes.* Academic Press, London. **Bernard, E. C. 1979.** New species of Bunonematoidea (Rhabditida) from Georgia and Tennessee. *J. Nematol.* 11:344–358. **Bhatt, B. D. and R. A. Rohde. 1970.** The influence of environmental factors on the respiration of plant-parasitic nematodes. *J. Nematol.* 2:277–285. **Bird, A. F. 1971.** *The Structure of Nematodes.* Academic Press, New York. 318 pp. **Bird, G. W. 1974.** *Nematode Detection.* MSU Ag. Facts 9. Cooperative Extension Service, East Lansing, Mich. 1 p. **Castro, C. E. and I. J. Thomason. 1973.** Permeation dynamics and osmoregulation in *Aphelenchus avenae. Nematologica* 19:100–108. **Chitwood, B. G. and M. B. Chitwood. 1977.** *Introduction to Nematology.* University Park Press, Baltimore. 344 pp. **Clark, W. C. 1960.** *The oesophago-intestinal junction in the Mononchidae (Enoplida, Nematoda).* Nematologica 5:178–183. **Coleman, D. C., C. V. Cole, R. V. Anderson, M. Blaha, M. K. Campion, M. Clarholm, E. T. Elliott, H. W. Hunt, B. Schaefer and J. Sinclair. 1977.** An analysis of rhizosphere-saprophage interactions in terrestrial ecosystems. Pp. 299–309 in U. Lohm and T. Persson (eds.). *Soil Organisms as Component of Ecosystems.* Ecol. Bull. (Stockholm) 25. **Coolen, W. A. 1979.** Methods for the extraction of *Meloidogyne* spp. and other nematodes from roots and soil. Pp. 317–339 in F. Lamberti and C. E. Taylor (eds.). *Root-Knot Nematodes (Meloidogyne Species): Systematics Biology and Control.* Academic Press, London. **Coomans, A. and P. A. A. Loof. 1970.** Morphology and taxonomy of Bathyodontina

(Dorylaimida). *Nematologica* 16:180–196. **Cooper, A. F. and S. D. Van Gundy. 1971.** Ethanol production and utilization by *Aphelenchus avenae* and *Caenorhabditis* sp. *J. Nematol.* 3:205–214. **Crofton, H. D. 1966.** *Nematodes.* Hutchinson University Library, London. 160 pp. **Croll, N. A. (ed.) 1976.** *The Organization of Nematodes.* Academic Press, New York, 440 pp. **Croll, N. A. and K. de Soyza. 1980.** Comparative calorie values of nematodes. *J. Nematol.* 12:132–135. **Crowe, J. and J. Clegg. 1978.** *Dry Biological Systems.* Academic Press, New York. 357 pp. **de Coninck, L. 1965.** Classe généralités des nematodes. Pp. 3–217 in P. P. Grasse (ed.). *Traité de Zoologie Anatomie, Systématique, Biologie,* Vol. IV, No. 2. Masson, Paris. **Demeure, Y. and D. W. Freckman. 1981.** Recent advances in anhydrobiosis. Pp. 205–226 in B. M. Zuckerman and R. A. Rohde (eds.). *Plant Parasitic Nematodes,* Vol. III. Academic Press, New York. **Doncaster, C. C. and M. K. Seymour. 1973.** Exploration and selection of penetration site by Tylenchida. *Nematologica* 19:137–145. **Dougherty, E. C. 1955.** The genera and species of the subfamily Rhabditinae Micoletzky, 1922 (Nematoda): a nomenclatorial analysis—including an addendum on the composition of the family Rhabditidae, Örley, 1880. *J. Helminthol.* 29:105–132. **Dropkin, V. H. 1980.** *Introduction to Plant Nematology.* Wiley, New York. 293 pp. **Dropkin, V. H., G. C. Martin and R. W. Johnson. 1958.** Effect of osmotic concentration on hatching of some plant parasitic nematodes. *Nematologica* 3:115–126. **Evans, A. A. F. and C. Womersley. 1980.** Longevity and survival in nematodes. Pp. 193–212 in B. M. Zuckerman (ed.). *Nematodes as Biological Models,* Vol. II. Academic Press, New York. **Ferris, V. R. 1971.** Taxonomy of Dorylaimida. Pp. 163–189 in B. M. Zuckerman, W. F. Mai, and R. A. Rohde (eds.). *Plant Parasitic Nematodes,* Vol. I. Academic Press, New York. **Ferris, V. R., J. M. Ferris and J. P. Tjepkema. 1973.** Genera of freshwater nematodes (Nematoda) of eastern North America. Water Poll. Control. Ser. Biota of Freshwater Ecosystems. Indent. Man. No. 10. 38 pp. **Filipjev, I. N. 1968.** *Free-Living Marine Nematodes of the Sevastopal Area,* Issue I (translated by M. Raveh) Wiener Bindery, Jerusalem. (Available from the U.S. Dept. of Commerce, Springfield, Va.) 225 pp. **1970.** *Free-Living Marine Nematodes of the Sevastopol Area, Issue II (translated by M. Raveh). Wiener Bindery, Jerusalem.* (Available from the U.S. Dept. of Commerce, Springfield, Va.) 203 pp. **Freckman, D. W. 1978.** Ecology of anhydrobiotic nematodes. Pp. 345–347 in J. Crowe and J. Clegg (eds.). *Dried Biological Systems.* Academic Press, New York. 357 pp. **Freckman, D. W., Y. Demeure, D. Munnecke and S. D. Van Gundy. 1980.** Resistance of anhydrobiotic *Aphelenchus avenae* to methyl bromide fumigation. *J. Nematol.* 12:19–22. **Freckman, D. W. K. Cromack, Jr. and J. A. Wallwork. 1986.** Recent advances in quantitative soil Biology. Pp. 399–442 in M. J. Mitchell (ed.). *Microfloral and Faunal Interactions in Natural and Agroecosystems.* Martinus Nijhoff, Hingham, Mass. 505 pp. **Gerlach, S. A. and F. Riemann. 1973, 1974.** *The Bremerhaven Checklist of Aquatic Nematodes; A Catalogue of Nematoda Adenophorea excluding the Dorylaimida,* Vols. I and II. Ditzen, Breman, West Germany. **Golden, A. M. 1971.** Classification of the genera and higher categories of the order Tylenchida (Nematoda). Pp. 191–232 in B. M. Zuckerman, W. F. Mai and R. A. Rohde (eds.). *Plant Parasitic Nematodes,* Vol. I. Academic Press, New York. **Goodell, P. B. and H. Ferris. 1980.** Plant-parasitic nematode distributions in an alfalfa field. *J. Nematol.* 12:136–140. **1981.** Sample optimization of five plant-parasitic nematodes in an alfalfa field. *J. Nematol.* 13:304–313. **Goodey, J. B. 1963a.** *Laboratory Methods for Work with Plant and Soil Nematodes.* Tech. Bull. 2. Her Majesty's Stationery Office, London. 72 pp. **Goodey, T. 1963b.** Rev ed. by J. B. Goodey) *Soil and Freshwater Nematodes.* Wiley, New York. 544 pp. **Goseco, C. G., V. R. Ferris and J. M. Ferris. 1974a.** Revisions in Leptonchoidea (Nematoda: Dorylaimida). *Leptonchus, Proleptonchus, Funaria,* and *Meylis* n. gen. in Leptonchidae, Leptonchinae. *Purdue Univ. Agric. Exp. Stn. Res. Bull.* 911. 32 pp. **1974b.** Revisions in Leptonchoidea (Nematoda: Dorylaimida). *Tyleptus* in Leptonchidae, Tyleptinae; *Basirotyleptus* in Leptonchidae, Belonenchinae; and *Loacharionema* n. gen. in Leptonchidae, Xiphinemmellinae. *Purdue Univ. Agric. Exp. Stn. Res. Bull.* 913. 25 pp. **1975.** Revisions in Leptonchoidea (Nematoda: Dorylaimida). *Tylencholaimellus, Doryllium, Gerthus* n. gen, and *Agmodorus* in Tylencholaimedellidae; and *Aulolaimoides adenolaimus* in Aulolaimoididae. *Purdue Univ. Agric. Exp. Stn. Res. Bull.* 928. 40 pp. **1977.** Revisions in Leptonchoidea (Nematoda: Dorylaimida). *Dorylaimoides* in Dorylaimoididae, Dorylaimoidirae; *Calolaimus* and *Timmus* n. gen. in Dorylaimoididae, *Calolaiminae;* and *Miranema* in Miranematidae. Purdue Univ. Res. Bull. 941:1–146. **Grassé, P. P. 1965.** *Traité de Zoologie, Anatomie, Systematique, Biologie,* Vol. IV, Nos. 2, 3. Masson, Paris. **Green, C. D. 1966.** Orientation of male *Heterodera rostochiensis* Woll. and *H. schactii* Schm to their females. *Ann. Appl. Biol.* 58:327–339. **1971.** Mating and host finding behaviour of plant nematodes. Pp. 247–266 in B. M. Zuckerman, W. F. Mai, and R. A. Rohde (eds.). *Plant Parasitic Nematodes,* Vol. II. Academic Press, New York. **Heyns, J. 1971.** *A Guide to the Plant and Soil Nematodes of South Africa.* A. A. Balkema, Cape Town, South Africa. 233 pp. **Hirschmann, H. 1960.** Part III. Morphology and anatomy. Pp. 125–167 in J. N. Sasser and W. R. Jenkins (eds.). *Nematology Fundamentals and Recent Advances with Emphasis on Plant Parasitic and Soil Forms.* University of North Carolina Press, Chapel Hill, N.C. 480 pp. **1971.** Comparative morphology and anatomy. Pp. 11–63 in B. M. Zuckerman, W. F. Mai and R. A. Rohde (eds.). *Plant Parasitic Nematodes,* Vol. I. Academic Press, New York. **Hooper, D. J. 1978.** Structure and classification of nematodes. Pp. 3–45 in J. F. Southey (ed.). *Plant Nematology.* Ministry of Agriculture, Fisheries and Food, Her Majesty's Stationery Office, London. 440 pp. **Hooper, D. J. and J. F. Southey. 1978.** Plant-parasitic Dorylaimida: morphology and identification. Pp. 207–231 in J. F. Southey (ed.). *Plant Nematology.* Ministry of Agriculture, Fisheries and Food, Her Majesty's Stationery Office, London. 440 pp. **Hope, D. 1971.** Class Nematoda (Nemata). Nematodes or roundworms. Pp. 200–209 in K. L. Gosner (ed.). *Guide to Identification of Marine and Estuarine Invertebrates.* Wiley-Interscience, New York. **Hopper, B. E. and E. J. Cairns. 1959.** *Taxonomic Keys to Plant, Soil, and Aquatic Nematodes.* Alabama Polytechnic Institute, Auburn, Ala. 176 pp. **Hyman, L. H. 1951.** The pseudocoelomate bilateria. In *The Invertebrates.* Vol. III. *Acanthocephala, Aschelminthes, and Entoprocta.* McGraw-Hill, New York. 572 pp. **Jairajpuri, M. S. and M. Ahmad. 1980.** Revised classification of the superfamily Belondiroidea Thorne, 1964 with notes on the systematics of *Dorylaimellus* Cobb 1913 (Nematoda: Dorylaimida) *Indian J. Nematol.* 10:9–22. **Jairajpuri, M. S. and W. U. Khan. 1982.** *Predatory nematodes (Mononchida).* Associated Publishing, New Delhi, India. 131 pp. **Jennings, J. B. and J. B. Colam. 1970.** Gut structure, digestive physiology, and food storage in *Pontonema vulgaris* (Nematoda: Enoplida). *J. Zool.* 161:211–221. **Jensen, H. J. and R. H. Mulvey. 1968.** *Predaceous Nematodes (Mononchidae) of Oregon.* Oregon State University Press, Corvallis, Oreg. 58 pp. **Johnson, L. F. 1962.** Effect of addition of organic amendments to soil on root-knot of tomatoes. II. Relation of soil temperature, moisture and pH. *Phytopathology* 52:410–413. **Klekowski, R. Z., L. Wasilewska and E. Paplinska. 1972.** Oxygen consumption by soil-inhabiting nematodes. *Nematologica* 18:391–403. **Lee, D. L. and H. J. Atkinson. 1976.** *Physiology of Nematodes.* Columbia University Press, New York. 215 pp. **Lichtenfels, J. R. and M. H. Pritchard (eds.). 1982.** *A Guide to the Parasite Collections of the World.* The American Society of Parasitologists. 79 pp. **Maggenti, A. 1981.** *General Nematology.* Springer-Verlag, New York. 372 pp. **1982.** Nemata. Pp. 879–929 in S. P. Parker (ed.). *Synopsis and Classification of Living Organisms.* McGraw-Hill, New York. **Mai, W. F. and H. H. Lyon. 1975.** *Pictorial Key to Genera of Plant-Parasitic Nematodes.* Cornell University Press, Ithaca, N.Y. 223 pp. **Mankau, R. 1962a.** Soil fungistasis and nematophagous fungi. *Phytopathology* 52:611–615. **1962b.** The effect of some organic additives upon a soil nematode population and associated natural enemies. *Nematologica* 7:65–73. **Mankau, R. and R. J. Minteer. 1962.** Reduction of soil populations of the citrus nematode by the addition of organic materials. *Plant Dis. Rep.* 46:375–378. **Marks, C. F., I. J. Thomason and C. E. Castro. 1968.** Dynamics of the permeation of nematodes by water, nematicides, and other substances. *Exp. Parasitol.* 22:321–337. **Mason, E. B. B. and J. E. Bosher. 1963.** Combination die for making aluminum micro slides. *Proc. Helminthol. Soc. Wash.* 30:19–20. **McElroy, F. D. and S. D. Van Gundy. 1968.** Observations on the feeding processes of *Hemicycliophora arenaria. Phytopathology* 58:1558–1565. **McNeill, S. and J. H. Lawton. 1970.** Annual production and respiration in animal populations. *Nature (Lond.)* 225:472–474. **Mulvey, R. H. 1961a.** The Mononchidae: a family of predaceous nematodes. I. Genus *Mylonchulus* (Enoplida: Mononchidae). *Can. J. Zool.* 39:665–696. **1961b.** The Mononchidae: a family of predaceous nematodes. II. Genus *Anatonchus* (Enoplida: Mononchidae). *Can. J. Zool.* 39:807–826. **1962.** The Mononchidae: a family of predaceous nematodes. III. Genus *Miconchus* (Enoplida: Mononchidae). *Can. J. Zool.* 40:65–81. **1963.** The Mononchidae: a family of predaceous nematodes. V. Genera *Sporonchulus, Granonchulus,* and *Prionchuloides.* N. Gen-(Enoplida: Mononchidae). *Can. J. Zool.* 41:763–774. **1969.** Soil inhabiting nematodes of the orders Areolaimida, Chromadorida, Enoplida and Monhysterida from the Canadian high Arctics. *Can. J. Zool.* 47:365–382. **Mulvey, R. H. and H. J. Jensen. 1967.** The Mononchidae of Nigeria. *Can. J. Zool.* 45:667–727. **Muraoka, M. and N. Ishibashi. 1976.** Nematode-feeding mites and their feeding behaviour. *Appl. Entomol. Zool.* 11:1–7. **Myers, R. F. 1966.** Osmoregulation in *Panagrellus redivivus* and *Aphelenchus avenae.*

Nematologica 12:579–586. **Myers, R. F. and L. R. Krusberg. 1965.** Organic substances discharged by plant-parasitic nematodes. *Phytopathology* 55:429–437. **Nicholas, W. R. 1974.** *The Biology of Free-Living Nematodes.* Clarendon Press, Oxford. 219 pp. **Nielsen, C. O. 1967.** Nematoda. Pp. 199–211 in A. Burges and F. Raw (eds.). *Soil Biology. Academic Press, London.* **Norton, D. C. 1978.** *Ecology of Plant-Parasitic Nematodes.* Wiley, New York. 268 pp. **Pennak, R. W. 1978.** *Fresh-Water Invertebrates of the United States,* 2nd ed. Wiley, New York. 803 pp. **Petersen, H. and M. Luxton. 1982.** A comparative analysis of soil fauna populations and their role in decomposition processes. *Oikos* 287–386. **Poinar, G. O. 1979.** *Nematodes for Biological Control of Insects* C.R.C. Press, Boca Raton, Fla. 277 pp. **Proctor, J. R. and C. F. Marks. 1975.** The determination of normalizing transformations for nematode count data from soil samples and of efficient sampling schemes. *Nematologica* 20:395–406. **Prot, J. C. and S. D. Van Gundy. 1981.** Effect of soil texture and the clay component on migration of *Meloidogyne incognita* second-stage juveniles. *J. Nematol.* 13:213–216. **Raski, D. J. and A. R. Maggenti. 1983.** Pp. 131–142 in A. R. Stone, H. M. Platt and L. F. Khalil (eds.). *Concepts in Nematode Systematics.* Academic Press, London. 388 pp. **Ritter, M. M. and J. Theodorides. 1965.** Systématique des nematodes. Ordre des tylenchides. Pp. 803–868 in P. P. Grassé. (ed.). *Traité de Zoologie Anatomie, Systématique, Biologie,* Vol. IV, No. 2. Masson, Paris. **Rosswall, T., U. Lohm and B. Sohlenius. 1977.** Développement d'un micrososme pour l'étude de la minéralisation et de l'absorption radiculaire de l'azote dans l'humus d'une forêt de conifères (*Pinus sylvestris L.*). *Lejeunia* 84:1–24. **Rothstein, M. 1965.** Nematode biochemistry. V. Intermediary metabolism and amino acid interconversions in *Caenorhabditis briggsae. Comp. Biochem. Physiol.* 14:541–552. **Siddiqi, M. R. 1968.** Five new species of Belondiroidea (Nematoda) from Sinsagar' India, with a revised classification of the superfamily. *Proc. Helminthol. Soc. Wash.* 35:248–258. **1969.** *Crateronema* n. gen. (Crateronematidae n. fam.), *Poronemella* n. gen. (Lordellonematinae n. sub.-fam.) and *Chrysonemoides* n. gen. (Chrysonematidae n. fam.), with a revised classification of Dorylaimoidea (Nematoda). *Nematologica* 15:81–100. **1971.** Structure of the oesophagus in the classification of the superfamily Tylenchoidea (Nematoda). *Indian J. Nematol.* 1:25–43. **1980.** The origin and phylogeny of the nematode orders Tylenchida Thorne, 1949 and Aphelenchida N. Ord. *Helminthol. Abstr. B. Plant Nematol.* 49:143–170. **Skarbilovich, T. S. 1959.** On the structure of systematics of the nematodes order Tylenchida Thorne, 1949. *Acta Parasitol. Pol.* (*Warsaw*) VII:117–132. **Smolik, J. D. 1974.** *Nematode Studies on the Cottonwood Site.* USIBP Grassland Biome Tech. Rep. 251. Colorado State University, Ft. Collins, Colo. **Sohlenius, B. 1980.** Abundance, biomass, and contribution to energy flow by soil nematodes in terrestrial ecosystems. *Oikos* 34:186–194. **Southey, J. F. 1970.** Principles of sampling for nematodes. Pp. 1–4 in J. F. Southey (ed.). *Laboratory Methods for Work with Plant and Soil Nematodes.* Ministry of Agriculture, Fisheries and Food, Her Majesty's Stationery Office, London. **1978a.** *Plant Nematology.* Ministry of Agriculture, Fisheries and Food, Her Majesty's Stationery Office, London. 440 pp. **1978b.** Some free-living nematodes found in plant material and soil samples. Pp. 244–266 in J. F. Southey (ed.). *Plant Nematology.* Ministry of Agriculture, Fisheries and Food, Her Majesty's Stationery Office, London. **Stephenson, W. 1942.** The effect of variations in osmotic pressure upon a free-living soil nematode. *Parasitology* 34:253–265. **Tarjan, A. C. 1980.** *An Illustrated Guide to the Marine Nematodes.* IFAS, Gainesville, FL. 135 pp. **Tarjan, A. C., R. P. Esser and S. L. Chang. 1977.** An illustrated key to nematodes found in fresh water. *J. WPCF* 2318–2319. **Thorne, G. 1957.** A monograph of the nematodes of the superfamily Dorylaimoidea. Reprinted from *Capita Zoologica,* Vol. VIII, No. 5. Martinus Nijhoff, The Hague, The Netherlands. 263 pp. **1961.** *Principles of Nematology.* McGraw-Hill, New York. 553 pp. **1964.** Nematodes of Puerto Rico: Belondiroidea new superfamily, Leptonchidae, Thorne, 1935, and Belonenchidae new family (Nemata, Adenophorea, Dorylaimida). *Univ. Puerto Rico Agric. Exp. Stn. Tech. Pap.* 39. 51 pp. **Thorne, G. and H. H. Swanger. 1936.** A monograph of the nematode genera *Dorylaimus* Dujardin, *Aporcelaimus* n.g., *Dorylaimoides* n.g., and *Pungentus* n.g. *Capita Zoologica,* Vol. VI, No. 5. Martinus Nijhoff, The Hague, The Netherlands. 223 pp. **Triantaphyllou, A. C. 1971.** Genetics and cytology. Pp. 1–34 in B. M. Zuckerman, W. F. Mai and R. A. Rohde (eds.). *Plant Parasitic Nematodes,* Vol. I. Academic Press, New York. 345 pp. **1979.** Cytogenetics of root-knot nematodes. Pp. 85–109 in F. Lamberti and C. E. Taylor (eds.). *Root-Knot Nematodes* (Meloidogyne Species) *Systematics, Biology and Control.* Academic Press, New York. 477 pp. **Triantaphyllou, A. C. and H. Hirschmann. 1980.** Cytogenetics and morphology in relation to evolution and speciation of plant-parasitic nematodes. *Annu. Rev.*

Phytopathol. 18:333–359. **Viglierchio, D. R., N. A. Croll and J. H. Gortz. 1969.** The physiological response of nematodes to osmotic stress and an osmotic treatment for separating nematodes. *Nematologica* 15:15–21. **Wallace, H. R. 1958.** Observations on the emergence from cysts and the orientation of larvae of three species of the genus *Heterodera* in the presence of host plant roots. *Nematologica* 3:236–243. **1971.** Abiotic influences in the soil environment, Pp. 257–280 in B. M. Zuckerman, W. F. Mai and R. A. Rohde (eds.). *Plant Parasitic Nematodes,* Vol. II. Academic Press, New York. 347 pp. **Wasilewska, L. 1971.** Nematodes of the dunes in the Kampinos forest. II. Community structure based on number of individuals, state of biomass and respiratory metabolism. *Ekol. Pol.* 19:651–688. **Wasilewska, L. and E. Paplinska. 1976.** Method of soil sampling and estimations of numbers of biomass and the community structure of soil nematodes. *Ekol. Pol.* 24:593–606. **Webster, J. M. 1972.** *Economic Nematology.* Academic Press, London. 563 pp. **Westcott, S. W. and K. R. Barker. 1976.** Interaction of *Acrobeloides buetschlii* and *Rhizobium leguminosarum* on Wando pea. *Phytopathology* 66:468–472. **Wieser, W. 1959.** *Free-Living Nematodes and Other Small Invertebrates of Puget Sound Beaches.* University of Washington Press, Seattle, Wash. 179 pp. **Wieser, W. and B. Hopper. 1967.** Marine nematodes of the east coast of North America. I. Florida. *Bull. Mus. Comp. Zool.* 135:239–344. **Winberg, G. G. 1971.** *Methods for the Estimation of Production of Aquatic Animals.* Academic Press, London. 175 pp. **Wright, D. J. and D. R. Newell 1976.** Nitrogen excretion, osmotic, and ionic regulation in nematodes. Pp. 163–210 in N. A. Croll (ed.). *The Organization of Nematodes.* Academic Press, London. 440 pp. **Wyss, U. 1970.** Zur Toleranz wandernder Wurzelnematoden gegenüber zunehmender Austrocknung des Bodens und hohen osmotischen Drucken. *Nematologica* 16:63–73. **Yeates, G. W. 1973.** Nematoda of a Danish beech forest. II. Production estimates. *Oikos* 24:179–185. **1979.** Soil nematodes in terrestrial ecosystems. *J. Nematol.* 11:213–229.

9

Terrestrial Gastropoda

J. B. BURCH

TIMOTHY A. PEARCE

A major event in molluscan evolution occurred when gastropods were able to invade land and had the opportunity to radiate into the many available terrestrial habitats. As terrestrial inhabitants, snails achieved a great measure of success, being able to occupy nearly every type of land environment. Subsequently, the terrestrial gastropod faunas of the world have become rich and diversified, and those of North America are no exception. Nearly a thousand species occur in North America north of Mexico.

Native land snails and slugs are found almost everywhere, but in general they prefer habitats offering shelter, adequate moisture, and an abundant food supply, and for snails generally, an available source of lime. Forested river valleys usually supply such habitats, and those with outcrops of limestone generally show the most abundant and varied mollusk faunas. Introduced species often tend to be somewhat more urban. For example, in its native Europe the slug *Milax gagates* (Draparnaud) lives in woodlands and on damp rocks on open hillsides and is not a slug of cultivated areas. But where it has been introduced into North America, it is mostly found around human dwellings and in greenhouses and gardens.

The various gastropod species differ in their physical requirements. Some species may require calcareous or damp soils, while others tolerate a wide range of moisture or soil types. Factors important in a particular snail's distribution may not always be easy to identify. Long term studies on morph frequencies (Cain 1983) suggest that factors influencing gastropod distribution may be in some cases imperceptible to researchers.

Gastropods may have differing tolerances to environmental conditions at various life history stages. Most gastropod eggs have little desiccation resistance, so they must be deposited in sites that will remain moist. Environmental conditions also can influence life cycles. For example, snails reach maturity more slowly in drier circumstances.

Gastropods require moisture. Water is needed for basic physiological processes; it is also expended for locomotion and reproduction, and may be lost through evaporation from external tissues (Riddle 1983). Gastropods make mucous trails as they crawl, and since mucus is composed largely of water, this crawling in itself expends much water. Yet, terrestrial mollusks are able to survive considerable water loss. The shell-free weight of a partially dehydrated snail may be 50% or less of the fully hydrated weight. Because slugs do not have an external shell that helps reduce evaporative water loss, a source of moisture is especially important for them. Slugs survive unfavorable conditions by crawling into small crevices or cracks, or by burrowing into the humus and soil.

Most shelled snails are able to reduce water loss and survive dry periods by estivating. During estivation, prosobranch snails close the shell aperture with an operculum and pulmonate snails seal the aperture by an epiphragm (a calcareous or non-calcareous mucous sheet). Both types of apertural coverings may be quite efficient. For example, snails in deserts may remain sealed in their shells for several years, only to emerge to feed and reproduce when moisture becomes available. Small snail species are usually absent from arid areas, perhaps because they have a higher surface area to volume ratio and cannot tolerate even minimal water loss during estivation.

Snails require a source of calcium for shell production. For this reason, many snail species are most abundant on limestone, soils derived from limestone, or soils otherwise high in calcium carbonate content (Burch 1954). Limestone is also especially favorable for snail distribution, because it weathers into deep cracks, providing especially convenient refuges.

Vegetation can influence terrestrial gastropod distribution and abundance by its effect on soil chemistry and quality of leaf litter. Snails requiring calcium can live in areas of calcium poor soil if local trees concentrate calcium in their leaves. Tree species produce leaf litter having different moisture-holding or draining characteristics. Fallen leaves that curl rather than lie flat provide spaces for gastropods in the leaf litter. Most species of gastropods prefer deciduous vegatation, although some prefer conifer-dominated habitats (Kralka 1986).

Slope of the land is correlated with drainage. Some snails may be confined to moist depressions, while other species may be found on well drained, steeper slopes. Slope aspect, or orientation to the sun, may be especially important to snails at the climatic edge of their range (Cain 1983).

Microclimate influences the microenvironment such as soil moisture, humidity of the air, and temperature. While on a broad scale snails may occur in a certain geographical region, on a small scale they may occur only in certain microhabitats. For example, native snails tend to be most abundant in soil humus and leaf litter, under and in rotting logs, and under and around the base of stones. Introduced species tend to be more urban and may be found under old boards, bricks and stones, and beneath damp refuse and litter in towns and cities. Another aspect of climate, insolation, may influence shell color. Shells in areas of high insolation tend to have lighter colors which reflect away some of the heat, causing a reduction in evaporative water loss (Riddle 1983). Visual predators may also affect color or color patterns by bringing about selection for cryptically colored shells.

Most terrestrial gastropods are generalist herbivores. Many consume fungus. Most are poor at digesting their food. Many plants contain toxins or digestibility reducing compounds such as tannins, so some gastropods prefer to eat senescent plant material that has lost many of those secondary compounds. A few gastropod species are carnivorous, eating soil invertebrates, including other gastropods.

Terrestrial gastropods are preyed upon by a variety of organisms besides carnivorous gastropods. Ducks consume slugs, and other birds (such as trushes), rodents and other small mammals eat snails. Terrestrial gastropods are also preyed upon by lampyrid bettle larvae and other insects.

Refuges may be the most important factor limiting terrestrial gastropod abundance. This is in contrast to so many vertebrate and arthropod groups, whose populations seem to be regulated by food or predation. Although food and predation may be implicated in regulating gastropod populations in some instances, abundance and distribution of refuges has a more general effect on gastropods. Refuges provide shelter from cold, heat and desiccation, and protection from predators. Slugs have been known to "fight over" refuges (Rollo and Wellington 1979).

Mucus appears to be a mechanism of population density regulation. Snails can detect mucous trails of their own and other species. Experiments have shown that snails grow more slowly and attain smaller adult sizes in the presence of conspecific mucous trails (Cain 1983).

Activity patterns of terrestrial gastropods are influenced primarily by light intensity, relative humidity, and temperature. Gastropods are primarily nocturnal. A rise in relative humidity and a drop in temperature usually causes snails and slugs to become more active.

The panmictic unit of a gastropod population may be small relative to those of other animal groups because snails tend to be less vagile. For example, desert conditions provide such little opportunity for migration that panmictic units are very small, and speciation may occur relatively easily.

Long distance dispersal of terrestrial gastropods is undoubtedly passive (adventitious) (H.B. Baker 1958). Gastropods can actively migrate by crawling relatively slowly, over relatively short distances, when weather conditions are favorable. Passively, snails can be carried (as phoretics) more quickly over longer distances on terrestrial mammals, and over barriers such as water on birds and on insects. It has been speculated that snails may be carried through the air by wind or rafted across water on floating objects (Vagvolgyi 1975). Through human agency, especially over the last century, snails have been dispersed to many parts of the world accidently on plants, on heavy machinery and sometimes intentionally. The hermaphroditic reproductive systems of pulmonate snails may enhance their chances of establishing new populations. Indeed, some terrestrial gastropods, especially slugs, may routinely fertilize themselves.

The land snail faunas of eastern and western North America, and of the southern tips of Florida and Texas, are fundamentally different. The more northern Nearctic fauna is by far the largest. Its species are divided nearly equally into eastern and western groups by the Continental Divide. The southernmost fauna is part of a much larger Middle American fauna, which also inhabits the Caribbean area and Middle America. The physical barriers separating these regions are not impassable, and their boundaries are not distinct lines but in general rather broad transitional zones.

The Eastern and Western divisions are each subdivided into several molluscan provinces (Fig. 9.1), less distinctly and sharply defined than the divisions, but nevertheless each having its own more or less distinctive land snail fauna. The Eastern Division extends from the eastern limit of the Rocky Mountains to the Atlantic coast. It consists of the Northern, Interior, Cumberland, Texan, and Austroriparian provinces. The Western Division consists of the Rocky Mountain, Southwestern, Washingtonian, Oregonian, and Californian provinces. For a discussion of the mollusks characterizing various of these zoogeographic areas, the reader is referred to the publications of Binney (1869, 1878, 1885), and to Henderson (1931), Pilsbry (1948), and Bequaert and Miller (1973).

The land snail fauna of North America north of Mexico is a rich and diverse one, containing 140 genera, belonging to 37 families. These snails belong to four quite dissimilar major groups. The most primitive group is the prosobranch snails, which, when living, can be distinguished readily by their possession of an operculum on the posterior dorsum of the foot. This operculum covers and seals the aperture when the snail is withdrawn into its shell. These operculated snails are related to the aquatic gill-breathing prosobranchs, which also have an operculum. Compared to the non-operculated snails, the land prosobranchs are relatively few.

The other three groups of land snails are pulmonates and are related to the prosobranchs only in that they are also gastropods. Otherwise, they are distinctly different. Two of these groups, the Acteophila and the Gymnophila, are small, each containing only one family and a few species in North America.

The tiny acteophiles, the Carychiidae, are more closely related to the freshwater pulmonate snails, the Lymnophila, and like them, have only one pair of tentacles, with eyes at their bases (rather than having two pairs of

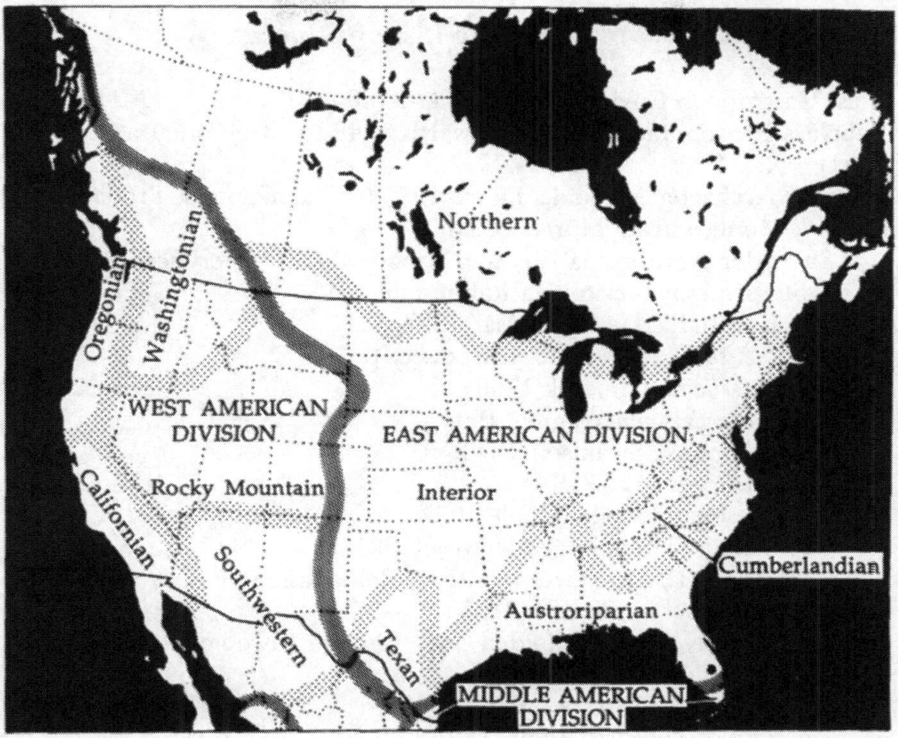

Fig. 9.1. Molluscan divisions and provinces in central North America. The approximate divisional boundaries are delineated by fine parallel diagonal lines and the provincial boundaries by rows of parallel dots. [After Binney (1873, 1878, 1885), Henderson (1931), Pilsbry (1948).]

tentacles, with eyes at the tips of the larger upper pair, as in the other pulmonate land snails).

The gymnophile gastropods, the veronicellid slugs, have their own set of peculiar characteristics, which make them a problematic group, difficult to relate easily to the other gastropods. They have been placed at various times and by different authors with the pulmonates, the opisthobranchs, or in their own order (by themselves or combined with the Rathouisiidae and Onchidiidae), variously called the Ditremata, Soleolifera, Teletremata, Digonophora, or Systellommatophora.

By far the largest group of land snails belong to the order Geophila. In North America north of Mexico, the geophile snails are grouped into 31 families and 128 genera.

Because of the magnitude of the North American land snail fauna and the many taxa involved in its classification, we are including a systematic list of genera for convenience to those using the accompanying identification keys. The taxonomic names and arrangement follows that of Baker (1956a,b, 1957, 1960, 1962b, 1963, 1964) Solem (1976, 1983) and Wurtz (1955).

Systematic List of Genera

Class Gastropoda Cuvier 1797 (Duméril 1806)†
Subclass Prosobranchia Milne Edwards 1848 (= Streptoneura Spengel 1881)
Order Archaeogastropoda Thiele 1927 (= Taenioglossa Troschel 1848; Monotocardia Mörch 1865)
Suborder Neritopsina Cox and Knight 1960 (= Neritacea auct.)
Superfamily Neritoidea Rafinesque 1815
HELICINIDAE Férussac 1822
Alcadia Gray 1840 [introduced]
Helicina Lamarck 1799
Hendersonia Wagner 1905
Lucidella Swainson 1840
Oligyra Say 1818
Order Mesogastropoda Thiele 1927
Superfamily Pomatioidea Newton 1891
POMATIIDAE Newton 1891 (= Pomatiasidae Pilsbry 1919)
Chondropoma Pfeiffer 1847
Superfamily Truncatelloidea Gray 1840 (= Rissooidea H. and A. Adams 1854)
TRUNCATELLIDAE Gray 1840
Truncatella Risso 1820
POMATIOPSIDAE Stimpson 1865
Pomatiopsis Tryon 1862

Subclass Pulmonata Cuvier 1817 (Fleming 1882) (= Euthyneura Spengel 1881, in part)
Order Acteophila Férussac (= Gehydrophila Férussac 1821; Basommatophora Keferstein 1864, in part)
Superfamily Auriculoidea Férussac 1821
CARYCHIIDAE Jefferys 1829
Carychium Müller 1774
Order Gymnophila Baker 1955 (= Systellommatophora Pilsbry 1948)
Suborder Soleolifera Simroth 1890
Superfamily Veronicelloidea Gray 1840
VERONICELLIDAE Gray 1840 (= Vaginulidae Gill 1871)
Angustipes Colosi 1922 [introduced]
Diplosolenodes Thomé 1975 [introduced]
Leidyula H. B. Baker 1925 [introduced, in part]
Order Geophila Férussac 1812 (= Stylommatophora Schmidt 1856)
Suborder Orthurethra Pilsbry 1900

† Genera in each family or subfamily are listed in alphabetical order. Subgenera are placed in parentheses below their genera. Introduced genera (i.e., snails not native to North America but brought by human commerce) are noted.

Superfamily Cionelloidea Clessin 1879
CIONELLIDAE Clessin 1879 (= Cochlicopidae Pilsbry 1900)
Cionella Jefferys 1829 (= *Cochlicopa* Risso 1826)
Superfamily Pupilloidea Turton 1831
VALLONIIDAE Morse 1864
Planogyra Morse 1864
Vallonia Risso 1826
Zoogenetes Morse 1864
STROBILOPSIDAE Pilsbry 1918
Strobilops Pilsbry 1893
PUPILLIDAE Turton 1831
Vertigininae Fitzinger 1833
Bothriopupa Pilsbry 1898
Columella Westerlund 1898
Pupisoma Stoliczka 1873
Sterkia Pilsbry 1898
Vertigo Müller 1774
Pupillinae Turton 1831
Pupilla Leach (in Fleming) 1828
Pupoides Pfeiffer 1854
Chondrininae Steenberg 1925
Chaenaxis Pilsbry and Ferriss 1906
Gastrocopta Wollaston 1878
Suborder Mesurethra H. B. Baker 1955
Superfamily Cerioidea Pilsbry 1901
CERIIDAE Fleming 1818 (= Cerionidae Pilsbry 1901)
Cerion Röding 1798
Suborder Heterurethra Pilsbry 1900
Superfamily Succineoidea Beck 1837
SUCCINEIDAE Beck 1837
Catinella Pease 1870
Oxyloma Westerlund 1885
Succinea Draparnaud 1801
Suborder Sigmurethra Pilsbry 1900
Infraorder Holopodopes H. B. Baker 1962
Superfamily Achatinoidea Swainson 1840
ACHATINIDAE Swainson 1840
Achatininae Swainson 1840
Achatina Lamarck 1799 [introduced; ? not established]
Ferussaciinae Bourguignat 1883
Cecilioides Férussac 1814
Subulininae Crosse and Fischer 1877
Beckianum H. B. Baker 1961 [? introduced]
Lamellaxis Strebel and Pfeffer 1882 [introduced]
Opeas Albers 1850 [introduced]

Rumina Risso 1826 [introduced]
Subulina Beck 1837 [introduced]
SPIRAXIDAE Baker 1955
Euglandina Fischer and Crosse 1870
Pseudosubulina Strebel and Pfeffer 1882
Superfamily Streptaxoidea Gray 1860
STREPTAXIDAE Gray 1860
Gulella Pfeiffer 1856 [introduced]
Superfamily Rhytidoidea Pilsbry 1893
HAPLOTREMATIDAE H. B. Baker 1925
Haplotrema Ancey 1881
Superfamily Orthalicoidea Albers-Martens 1860
UROCOPTIDAE Pilsbry and Vanatta 1898
Cochlodinella Pilsbry and Vanatta 1898
Holospira Martens 1860
Metastoma Strebel 1880
Microceramus Pilsbry and Vanatta 1898
ORTHALICIDAE Albers-Martens 1860 (= Bulimulidae Tryon 1896)
Bulimulus Leach 1815
Drymaeus Albers 1850
Liguus Montfort 1810
Naesiotus Albers 1850
Orthalicus Beck 1837
Rabdotus Albers 1850
Infraorder Aulacopoda Pilsbry 1896
Superfamily Arionoidea Gray (in Turton) 1840
PUNCTIDAE Morse 1864
Paralaoma Iredale 1913
Punctum Morse 1864
CHAROPIDAE Hutton 1884
Radiodiscus Pilsbry and Ferriss 1906
HELICODISCIDAE Pilsbry 1927
Helicodiscus Morse 1864
Polygyriscus Pilsbry 1948
Speleodiscoides Smith 1957
DISCIDAE Thiele 1931
Anguispira Morse 1864
Discus Fitzinger 1833
PHILOMYCIDAE Gray 1847
Megapallifera Hubricht 1956
Pallifera Morse 1864
Philomycus Rafinesque 1820
ARIONIDAE Gray (in Turton) 1840
Anadenulus Cockerell 1890

Ariolimax Mörch 1860
Arion Férussac 1819 [introduced]
Binneya Cooper 1863
Hemphillia Bland and Binney 1872
Hesperarion Simroth 1891
Magnipelta Pilsbry 1953
Prophysaon Bland and Binney 1873
Zacoleus Pilsbry 1903
Superfamily Limacoidea Rafinesque 1815
EUCONULIDAE Strebel and Pfeffer 1879 (= Helixarionidae Bourguignat 1883)
Dryachloa Thompson and Lee 1980
Euconulus Reinhardt 1883
Guppya Mörch 1867
VITRINIDAE Fitzinger 1833 (= Zonitidae Mörch 1864)
Vitrininae Fitzinger 1833
Vitrina Draparnaud 1801
Zonitinae Mörch 1864
Gastrodonta Albers 1850
Glyphyalinia Martens 1892
Hawaiia Gude 1911
Mesomphix Rafinesque 1819
Nesovitrea Cooke 1921
Oxychilus Fitzinger 1833 [introduced]
Paravitrea Pilsbry 1898
Pilsbryna Baker 1926
Pristiloma Ancey 1887
Striatura Morse 1864
Vitrea Fitzinger 1833 [introduced]
Vitrinizonites Binney 1879
Zonitoides Lehmann 1862
(*Ventricallus* Pilsbry 1946)
(*Ventridens* Binney 1863)
(*Zonitoides* s.s. Lehmann 1862)
LIMACIDAE Rafinesque 1815
Deroceras Rafinesque 1820 [introduced, in part]
Lehmannia Heynemann 1862 [introduced]
Limax Linnaeus 1758 [introduced]
MILACIDAE Ellis 1926
Milax Gray 1855 [introduced]
Superfamily Testacelloidea Gray 1840
TESTACELLIDAE Gray 1840
Testacella Draparnaud 1801 [introduced]
Infraorder Holopoda Pilsbry 1896
Superfamily Mesodontoidea Tryon 1866

THYSANOPHORIDAE Pilsbry 1926
Megomphicinae H. B. Baker 1930 (= Ammonitellidae Pilsbry 1930)
Ammonitella Cooper 1818
Glyptostoma Bland and Binney 1873
Megomphix H. B. Baker 1930
Polygyrella Binney 1863
Polygyroidea Pilsbry 1930
Oreohelicinae Pilsbry 1939
Oreohelix Pilsbry 1904
Radiocentrum Pilsbry 1905
Thysanophorinae Pilsbry 1926
Hojeda Baker 1926
Microphysula Cockerell and Pilsbry 1926
Thysanophora Strebel and Pfeffer 1880
MESODONTIDAE Tryon 1866 (= Polygyridae Pilsbry 1895)
Mesodontinae Tryon 1866
Mesodon Rafinesque 1821
(*Appalachina* Pilsbry 1939)
(*Inflectarius* Pilsbry 1939)
(*Mesodon* s.s. Rafinesque 1821)
(*Patera* Albers 1850)
Polygyra Say 1818
(*Daedalochila* Beck 1837)
(*Polygyra* s.s. Say 1818)
Praticolella Martens 1892)
Stenotrema Rafinesque 1819)
(*Euchemotrema* Archer 1939)
(*Stenotrema* s.s. Rafinesque 1819)
Triodopsinae Pilsbry 1940
Allogona Pilsbry 1939
(*Allogona* s.s. Pilsbry 1939)
(*Dysmedoma* Pilsbry 1939)
Cryptomastix Pilsbry 1939
Trilobopsis Pilsbry 1939
Triodopsis Rafinesque 1819
(*Neohelix* Ihering 1892)
(*Triodopsis* s.s. Rafinesque 1819)
(*Xolotrema* Rafinesque 1831)
Vespericola Pilsbry 1939
Ashmunellinae Webb 1954
Ashmunella Pilsbry and Cockerell 1899
Superfamily Oleacinoidea Adams 1855
SAGDIDAE Pilsbry 1895
Lacteoluna Pilsbry 1926

OLEACINIDAE Adams 1855
Varicella Pfeiffer 1855
Superfamily Camaenoidea Pilsbry 1895
CAMAENIDAE Pilsbry 1895
Zachrysia Pilsbry 1894 [introduced]
Superfamily Helicoidea Rafinesque 1815
XANTHONYCHIDAE Strebel and Pfeffer 1879
Humboldtianinae Pilsbry 1939
Humboldtiana Ihering 1892
Sonorellinae Pilsbry 1939
Sonorelix Berry 1943
Sonorella Pilsbry 1900
Helminthoglyptinae Pilsbry 1939
Eremarionta Pilsbry 1913
Eremariontoides Miller 1981
Helminthoglypta Ancey 1887
Micrarionta Ancey 1880
Mohavelix Berry 1943
Monadenia Pilsbry 1895
Plesarionta Pilsbry 1939
Xerarionta Pilsbry 1913
Cepolinae Hoffman 1928
Cepolis Montfort 1810
Eulotinae Wenz 1923 (= Bradybaeninae Pilsbry 1939)
Bradybaena Beck 1837 [introduced, in part (?)]
HYGROMIIDAE Tryon, 1866 (= Helicellidae Wenz 1923)
Candidula Kobelt 1871 [introduced]
Cochlicella Férussac 1820 [introduced]
Helicella Férussac 1821 [introduced]
Monacha Fitzinger 1833 [introduced]
Trichia Hartmann 1840 [introduced]
Trochoidea Brown 1827 [introduced]
HELICIDAE Rafinesque 1815
Cepaea Held 1837 [introduced, in part (?)]
Eobania Hesse 1913 [introduced]
Helix Linnaeus 1758 [introduced]
Otala Schumacher 1817 [introduced]
Theba Risso 1826 [introduced]

Identification

All land snails and most slugs possess a calcareous shell. In the snails, it is a hard, rigid external structure that covers the soft parts of the animal, providing protection (Fig. 9.2). It is twisted in a continually increasing

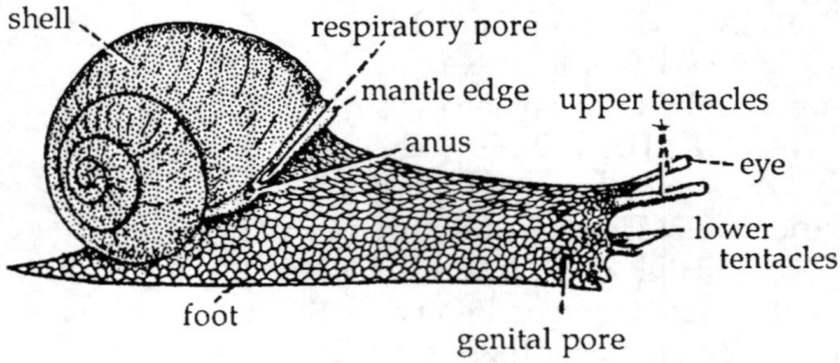

Fig. 9.2. Active crawling snail, showing the position and orientation of its protective shell and external features of its body.

spiral, and in nearly all North American snails it coils to the right or clockwise. The characteristics of this shell are different for each species, but within each species the appearance of the shell is constant, except for slight individual variations, differences due to age of the snail, and sometimes minor diversity exhibited between different local populations. Because of the constant appearance of the shell for each species, its characteristics are very important in species recognition, and usually for generic and familial placement as well. For that reason it is important to know the terminology for the various parts of the snail's shell (Fig. 9.3). Especially useful are the adult size (Fig. 9.4) and general form of the shell. Among the many species, the shell may take various shapes (Figs. 9.5, 9.6).

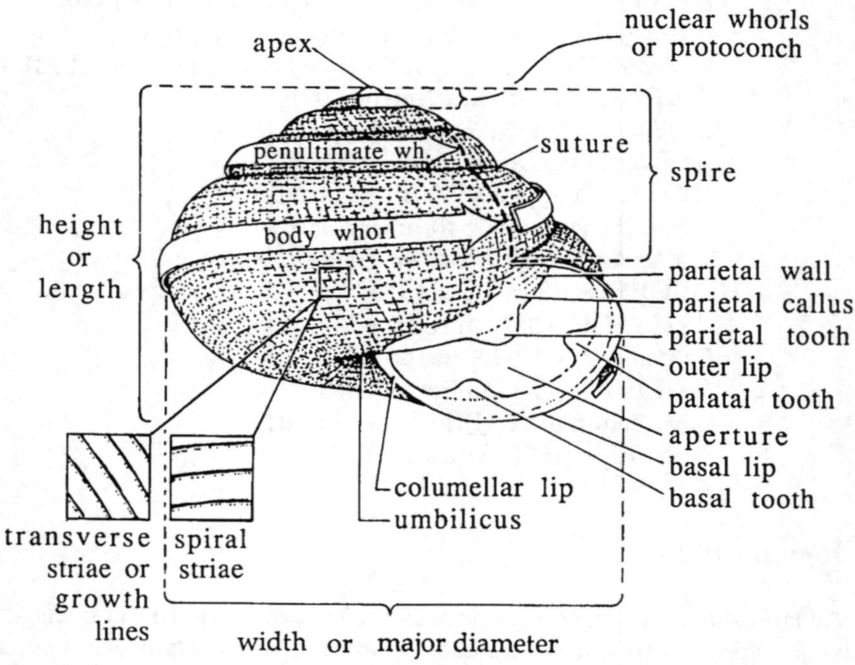

Fig. 9.3. Shell terminology.

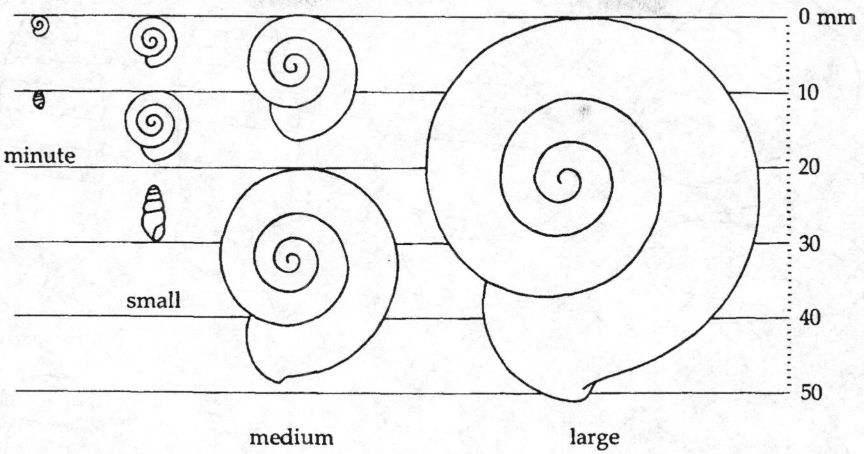

Fig. 9.4. Range of sizes and shell-size nomenclature of North American land snails. Less than 3 mm in longest dimension = *minute;* 3–10 mm = *small;* 11–30 mm = *medium;* more than 30 mm = *large.*

The shells of the different species may vary from very elongate to globose, depressed, and discoidal. Elongate shells may be oval, cylindrical, or conical, and those with conical shapes vary from narrowly conic to widely conic. The shell may be longer than wide, or wider than long [the columella determining the length/width axis]. Its coils may be rounded, angular, shouldered, or flattened (Fig. 9.7), and have shallow or impressed sutures (Fig. 9.8). The shell may have few or many whorls (Fig. 9.9), may lack an opening (umbilicus) at its base, or may have either a narrow or a wide umbilical opening (Fig. 9.10). The columella or central axis of the shell may be either twisted or straight and may or may not end abruptly (Fig. 9.11). The outer lip of the shell may be either straight ("sharp") or variously curved and is sometimes turned back or reflected (Fig. 9.12). The surface of the shell may be marked in various ways (i.e., differently colored or

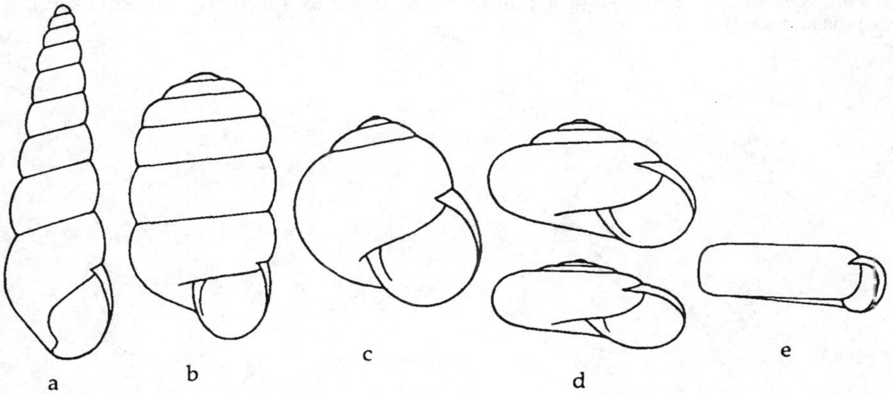

Fig. 9.5. Shell shapes: (*a*) elongate conic; (*b*) elongate cylindric; (*c*) globose; (*d*) depressed; (*e*) discoidal.

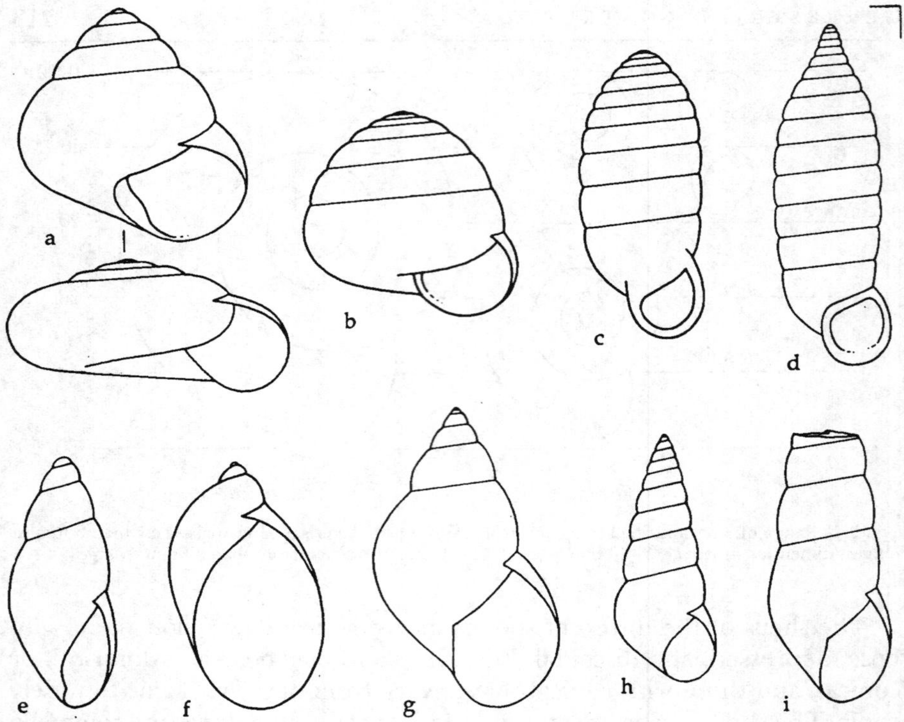

Fig. 9.6. Shell shapes: (*a*) heliciform; (*b*) dome or bee-hive-shaped; (*c*) pupilliform or pupa-shaped; (*d*) urocoptiform; (*e*) fusiform of spindle-shaped; (*f*) succiniform; (*g*) bulimoid; (*h*) conical; (*i*) decollated.

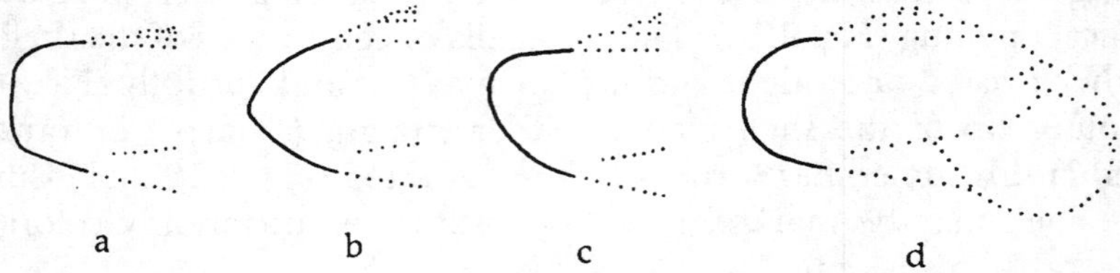

Fig. 9.7. Shell terminology: shell with (*a*) flattened whorls; (*b*) angular whorls; (*c*) shouldered whorls; (*d*) well-rounded whorls.

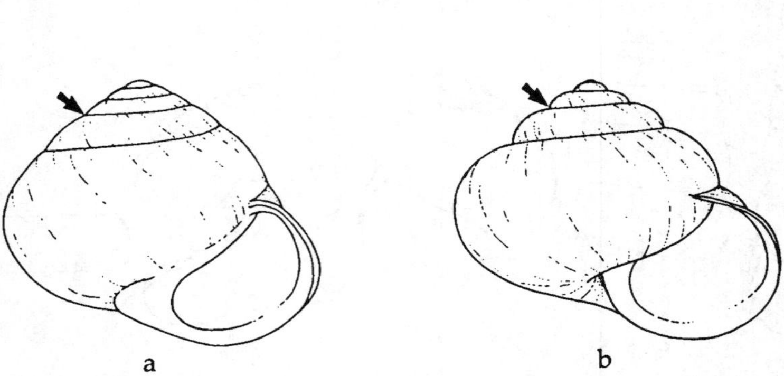

Fig. 9.8. Shell terminology: sutures (*a*) shallow; (*b*) indented or impressed.

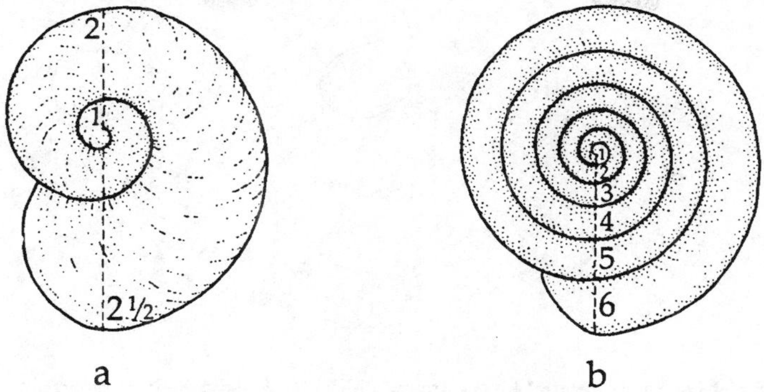

Fig. 9.9. Shell with (*a*) few, rapidly increasing whorls; (*b*) many, slowly increasing whorls. The dashed lines on the whorls illustrate the method of counting whorls; the shell in (*a*) has 2½ whorls, the shell in (*b*) has 6 whorls.

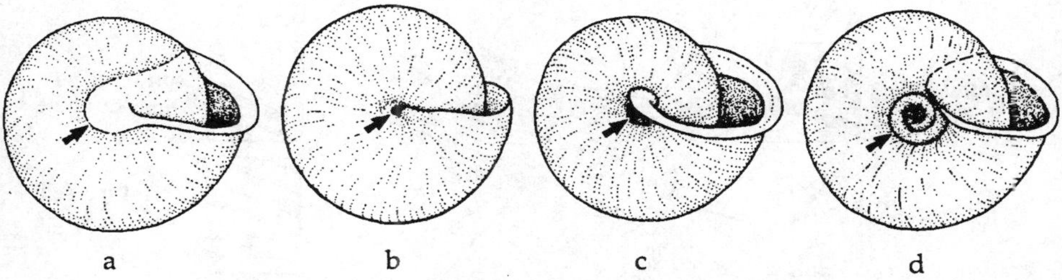

Fig. 9.10. Shell (*a*) imperforate; (*b*) perforate; (*c*) rimate; (*d*) umbilicate.

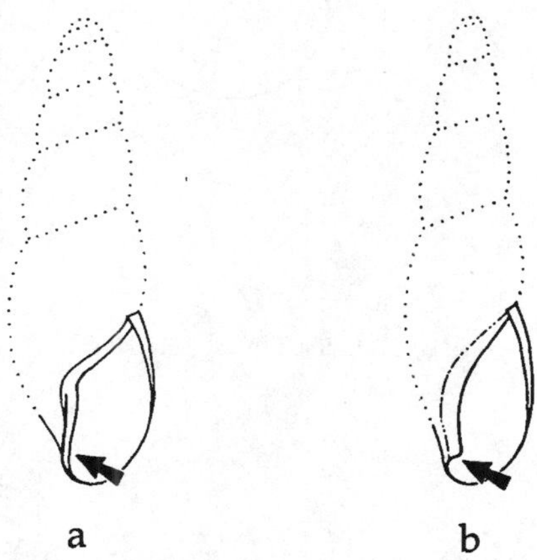

Fig. 9.11. Shell with (*a*) a straight columella; (*b*) a truncate columella.

sculptured) (Fig. 9.13), or simply may be unicolored and smooth. The outline of the shell aperture ("mouth") may take many forms due to the shape and relation of the whorls to each other. The aperture may (Fig. 9.14) or may not (Fig. 9.15) be closed by a cover (operculum), which itself has important recognition characters (Fig. 9.16).

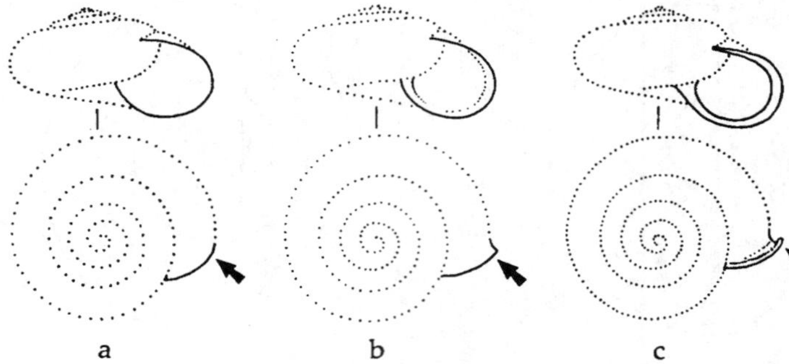

Fig. 9.12. Shell apertural lip (*a*) straight ("sharp"); (*b*) expanded; (*c*) reflected.

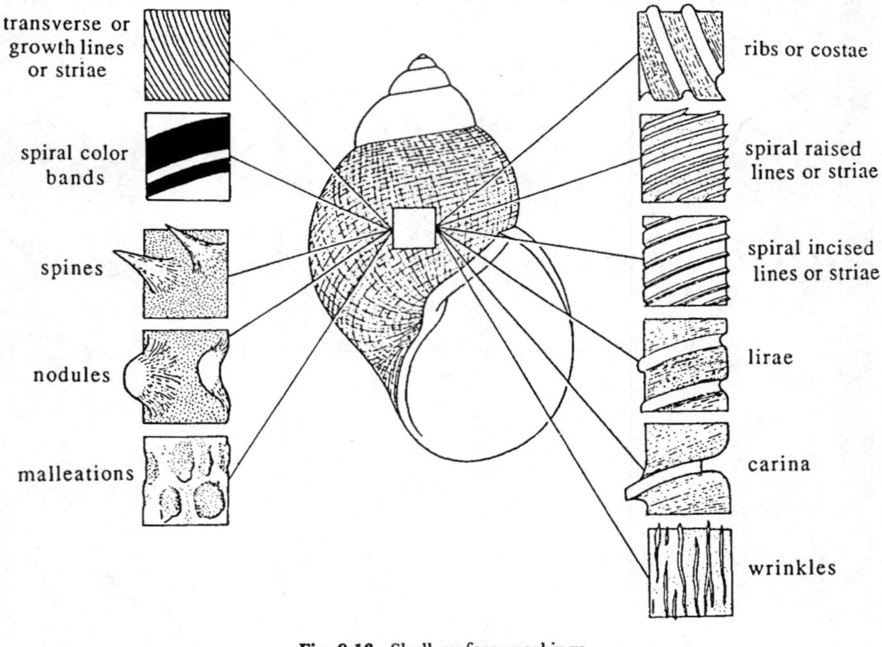

Fig. 9.13. Shell surface markings.

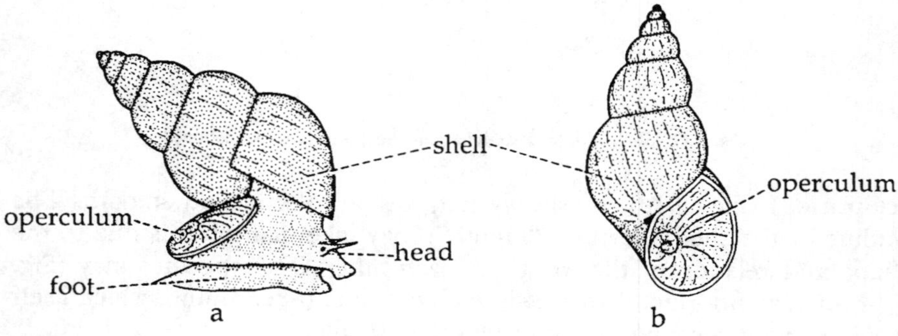

Fig. 9.14. An operculated snail, i.e., one that carries an operculum attached to its dorsal posterior foot. Position of the operculum when the snail (*a*) is active; (*b*) has withdrawn into its shell.

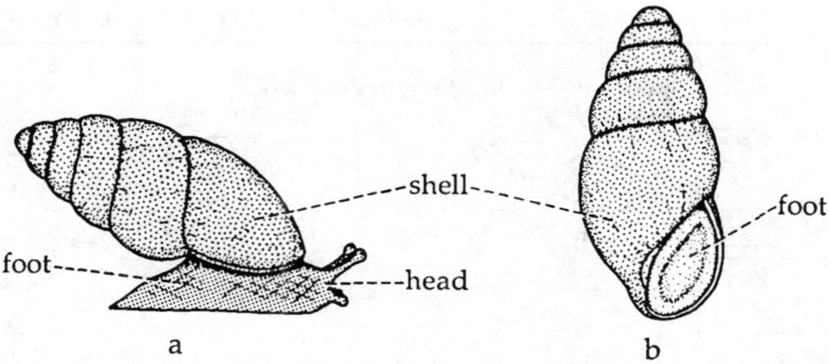

Fig. 9.15. A non-operculated (i.e., a pulmonate) snail. It does not have a protective operculum to seal the shell aperture when the animal has withdrawn into its shell. (*a*) Active snail; (*b*) Inactive snail withdrawn into its shell, with only the surface of its foot showing.

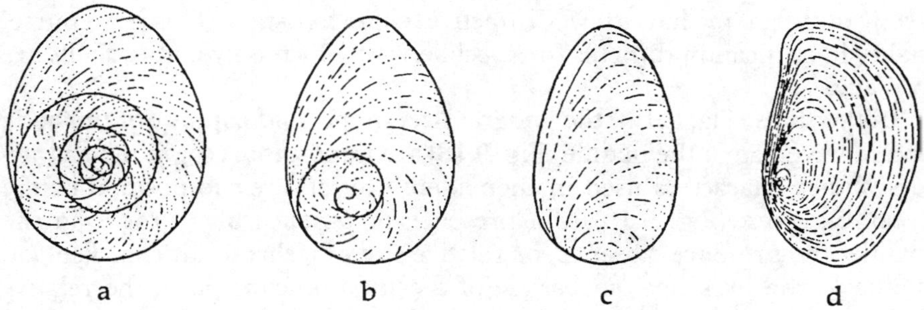

Fig. 9.16. Types of opercula: (*a*) multispiral; (*b*) paucispiral; (*c*) subspiral; (*d*) concentric.

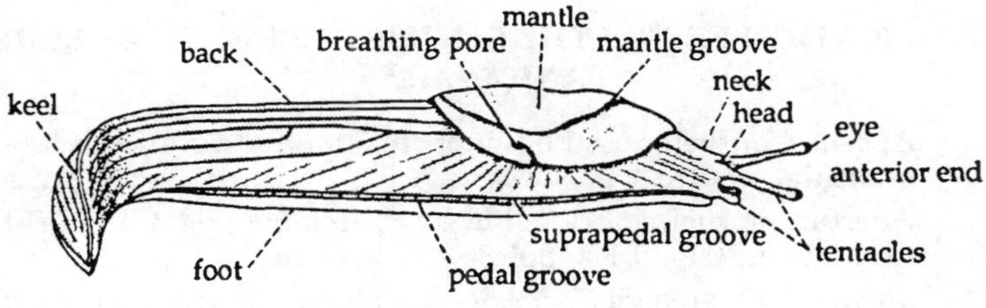

Fig. 9.17. Slug body terminology.

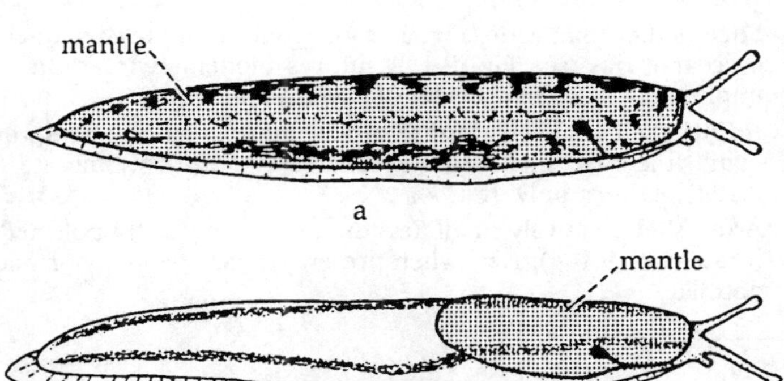

Fig. 9.18. Mantle covering (*a*) the entire back of the slug; (*b*) only an anterior portion.

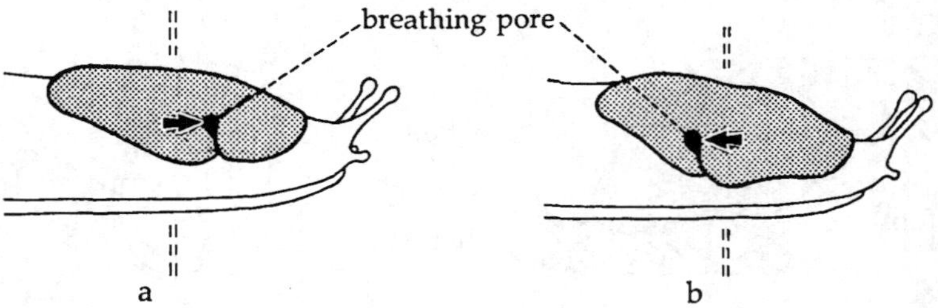

Fig. 9.19. Breathing pore (*a*) in the anterior half of the right side of the mantle; (*b*) in the posterior half.

In some snail groups, aspects of soft anatomy are essential for identification, because the various taxa in these groups have shells which are relatively uniform or have few distinctive characteristics. This, of course, makes identification difficult for specimens of which only empty shells are available.

In the slugs (Fig. 9.17), the general size, shape, and contour of the body and relative size of the mantle (Fig. 9.18) are important recognition characters. Other characters used in their identification are the position of the breathing pore (Fig. 9.19); the presence or absence of a groove in the mantle; the presence, absence, or relative size of a sharp ridge, or keel, on the back; the presence or absence of a caudal mucous pore; the relative development of the suprapedal groove; the color and pigmentation of the body; and the color of the mucus.

IDENTIFICATION KEY TO THE FAMILIES OF NORTH AMERICAN LAND SNAILS

1a Aperture of shell closed by an operculum (Fig. 9.14); Subclass **Prosobranchia** 2

1b Aperture of shell not closed by an operculum (Fig. 9.15), or shell entirely absent (Fig. 9.17); Subclass **Pulmonata** 5

2a (1) Shell wider than high (Fig. 9.5*c–e*), without ribs; sutures not impressed (Fig. 9.8*a*); Order **Archaeogastropoda,** Superfamily **Neritoidea;** eastern only . . . . **HELICINIDAE**

2b Shell higher than wide (Fig. 9.5*a,b*), usually sculptured with either fine or coarse ribs (see Fig. 9.13); sutures moderately to rather strongly impressed (Fig. 9.8*b*); Order **Mesogastropoda** 3

3a (2) Adult† shell relatively large (8 mm or more in length), sculptured with many fine ribs; shell often decollate (Fig. 9.6*i*); umbilicate (see Fig. 9.10*d*); eastern only . . . . **POMATIIDAE**

3b Adult shell relatively small (less than 8 mm in length) polished, imperforate (Fig. 9.10*a*); ribs, when present, thick; shell may or may not be decollate 4

† The keys in this chapter are based mostly on characters of adult shells. A key that would also include immature shell characters, or mainly anatomical characters, would be almost impossible to construct or too cumbersome to use.

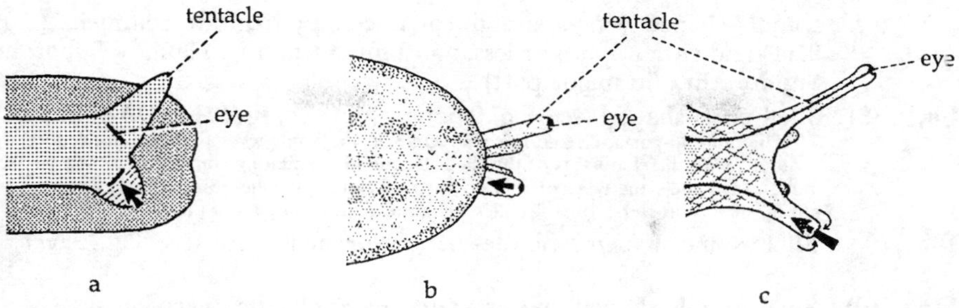

Fig. 9.20. Snail with (*a*) *one* pair of contractile tentacles, eyes at the *bases:* (*b*) *two* pairs of *contractile* tentacles, eyes at the *tips* of the upper, larger pair; (*c*) two pairs of *invaginable* tentacles, eyes at the tips of the upper, larger pair.

4a (3) Shell decollate (see Fig. 9.6*i*); i.e., the first several whorls broken or worn off; operculum subspiral (Fig. 9.16*c*). . . **TRUNCATELLIDAE**

4b Shell not decollate, the first several whorls clearly evident; operculum paucispiral (Fig. 9.16*b*). . . . . **POMATIOPSIDAE**

5a (1) Head of animal with one pair of tentacles (Fig. 9.20*a*), with eyes at their bases; shell very small to minute, less than 3 mm in length, elongate (but not pupa-shaped), translucent white; shell aperture with a reflected or expanded lip; entire columella with a lamella; Order **Actoephila** (= **Basommatophora,** in part) . . . . **CARYCHIIDAE**

5b Head with two pairs of tentacles (Fig. 9.20*b,c*), with eyes at the tips of the upper pair; shell, when present, of various shapes, but if very small or minute, elongate and with a reflected lip, the shell is not white and the internal columella lacks a lamella **6**

6a (5) Tentacles contractile (Fig. 9.20*c*); animal without a shell; Order **Gymnophila** (= **Systellommatophora**), Suborder **Soleolifera** (in part); eastern only . . . . **VERONICELLIDAE**

6b Tentacles retractile (Fig. 9.20*b*); animal with or without a shell; Order **Geophila** (= **Stylommatophora**). **7**

7a (6) Shell elongate, succiniform (Fig. 9.6*f*), very thin, with large aperture, its length more than half the shell length; Suborder **Heterurethra** **SUCCINEIDAE**

7b Shell, when present, of various shapes, but not succiniform; in shells higher than wide, aperture is less than half the shell height **8**

8a (7) Shell pupa-shaped (Fig. 9.6*c*), 20 mm or more in length; Suborder **Mesurethra;** southeastern only . . . . **CERIIDAE**

8b Shell, when present, may or may not be pupa-shaped, but if so, it is small, 6 mm or less in length **9**

9a (8) Shell rather small, less than 8 mm in length; pupa-shaped, or when heliciform usually having ribs or when dome shaped (Fig. 9.6*b*) usually having a reflected lip (Fig. 9.12*c*), or when spindle-shaped (Fig. 9.6*e*), being 5 mm or longer but less than 8 mm and with a twisted, but not abruptly truncate, columella (Fig. 9.11*a*); Suborder **Orthurethra** (in major part). **10**

9b Shell minute to large; not pupa-shaped; generally without ribbing when heliciform, without a reflected lip when dome-shaped, and when spindle-shaped with a straight or decidedly truncate columella (Fig. 9.11) and either 5 mm or less, or 10 mm or more, in length; Suborder **Sigmurethra** (in major part) **14**

10a (9) Shell pupa-shaped (see Fig. 9.6*c*). . . . . . . **PUPILLIDAE** (in part)

The more important diagnostic characters at the family level are often in the soft anatomy, with the characters of the shells frequently exhibiting convergence. Accordingly, nearly the same types of shells occur in different families. Since this key uses mostly shell characters, some families will appear in the key more than once

10b Shell dome-shaped, spindle-shaped, or heliciform (see Fig. 9.6*a,b,e*) **11**

11a (10) Shell spindle-shaped, imperforate (see Fig. 9.10*a*), very glossy . . . . **CIONELLIDAE**

11b Shell dome-shaped or heliciform, perforate to umbilicate (see Fig. 9.10*b–d*), dull to moderately glossy. **12**

12a (11) Shell with one or more lamellae in the aperture, usually dome-shaped and with rather stout ribs; usually dark brown or reddish brown; eastern only . . . . **STROBILOPSIDAE** (in part)

12b Shell without lamellae in the aperture, usually depressed heliciform, but without ribs or with rather thin ribs when dome-shaped or globose-heliciform; usually olive-brown, light tan, or white. **13**

13a (12) Shell depressed, with or without ribs, or if globose then with ribs; olive-green, olive-brown, light tan, or white . . . **VALLONIIDAE** (in part)

13b Shell globose, without ribs; dark or light reddish brown . . . . **PUPILLIDAE** (in part) [***Pupisoma***]

14a (9) Pedal grooves of the foot conspicuous and well above the angle of the lateral and ventral foot margins (Fig. 9.21*a*); external shell, when present, usually distinctly wider than high, sometimes colored with irregular color markings, but rarely banded; aperture lip not reflected; Infraorder **Aulacopoda** **15**

14b Pedal grooves of the foot inconspicuous and in or close to the angle of the lateral and ventral foot margins (Fig. 9.21*b*); external shell always present, either wider than high or higher than wide, and sometimes marked with color bands; aperture lip may or may not be reflected;

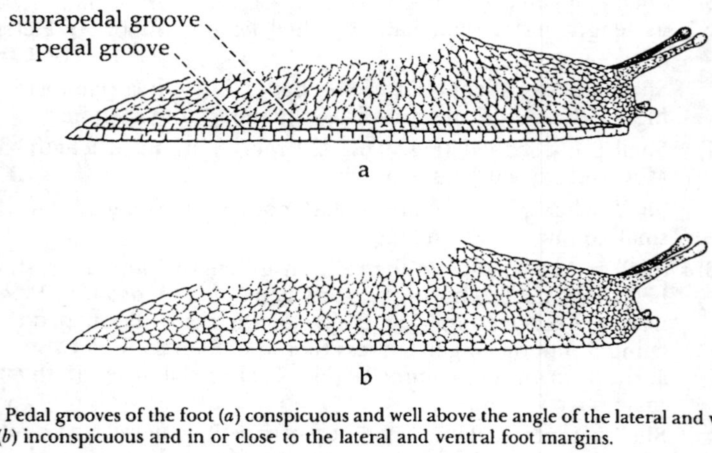

Fig. 9.21. Pedal grooves of the foot (*a*) conspicuous and well above the angle of the lateral and ventral foot margins; (*b*) inconspicuous and in or close to the lateral and ventral foot margins.

Infraorders **Holopodopes** and **Holopoda** [these two suborders are distinguished by internal characters (Baker 1955, 1962a)] 29

The foot in **Megomphix** (Holopoda, Mesodontoidea, Thysanophoridae, Megomphicinae) is aulacopod in appearance and has a caudal foss and tripartite sole. However, its other anatomical characters are similar to those of **Glyptosoma,** another genus of its subfamily and more clearly a holopod.

15a **(14)** Animals slug-like, external shell lacking (Fig. 9.17), or, if present, then shell reduced and much too small to accommodate the animal. . . 16

15b Animals snail-like with shell well developed, coiled, large enough to completely or nearly accommodate the contracted animal 24

16a **(15)** Slugs with a small external shell, either at the extreme posterior end of the body or on an elevated mantle hump 17

16b Shell entirely internal or lacking 18

17a **(16)** With a small ear-shaped external shell at the posterior end of the body (Fig. 9.148); largely subterranean; carnivorous; introduced from Europe . . . . **TESTACELLIDAE**

17b Small external shell present on a centrally located elevated hump **ARIONIDAE** (in part) [***Hemphillia, Binneya***]

18a **(16)** Mantle covering nearly the entire dorsum of the animal, more than two-thirds the length of the body (Fig. 9.18*a*) 19

18b Mantle covering only a central or anterior portion of the animal, less than one-half the length of the body (Fig. 9.18*b*) 20

19a **(18)** Mantle shortly free or not free anteriorly; breathing hole located in anterior fourth of mantle; eastern North America from Canada to Florida, west to Iowa and northeastern Texas. . . **PHILOMYCIDAE**

19b Mantle free anteriorly about one-fourth its length, and shortly free posteriorly; breathing hole located slightly anterior to middle of right side of mantle; Idaho and Montana. . . . . **ARIONIDAE** (in part) [***Magnipelta***]

20a **(18)** Breathing hole located in the anterior half of the right side of the mantle (Fig. 9.19*a*); body not keeled . . . . **ARIONIDAE** (in part) [***Arion, Prophysaon***]

20b Breathing hole located in the posterior half (Fig. 9.19*b*) or rarely at the middle of the right side of the mantle; body may or may not be keeled 21

21a **(20)** Animal with a conspicuous caudal mucous pore at the posterior end of the body at the meeting of the pedal grooves; sole not distinctly divided; foot fringe (the area between the latero-ventral foot margin and the pedal groove) moderately to quite wide . . . . **ARIONIDAE** (in part) [***Ariolimax, Hesperarion***]

21b Caudal mucous pore lacking; sole longitudinally divided into three sections; foot fringe narrow to moderately wide 22

22a **(21)** Posterior part of body not keeled or only weakly keeled; mantle lacking concentric wrinkles; jaw ribbed; small slugs (14–28 mm long); native slugs in southern California [***Anadenulus***] or Idaho and western Montana [***Zacoleus***] . . . . **ARIONIDAE** (in part)

22b Posterior part of body with a strong dorsal keel which may or may not reach the posterior margin of the mantle; mantle with or without concentric wrinkles; jaw smooth; small to large mostly introduced European slugs (15–100 mm long) 23

23a **(22)** Back keeled posteriorly all the way to the posterior mantle margin; mantle with a U-shaped or circular groove (Fig. 9.17); mantle not concentrically wrinkled . . . . **MILACIDAE**

23b Back keeled only near the posterior end; mantle without a U-shaped or circular groove; mantle concentrically wrinkled . . . . **LIMACIDAE**

24a **(15)** Shell usually opaque, generally dull and with prominant growth lines or low ribs, umbilicate to widely umbilicate, brown, reddish brown, tan, or almost white, sometimes with darker reddish color markings (patches or bands) **25**

24b Shell usually translucent, glossy, smooth, or without prominant growth lines or ribs, umbilicate, perforate, or occasionally imperforate, white or light tan to dark olive-brown, often containing some tint of green, but without reddish patches or bands **28**

Megomphix has a foot that appears aulacopod, but based on other characters the snail is classified with Holopoda. It has an umbilicus one fourth or more the diameter of the shell; the shell is greenish and more than 8 mm in diameter.

25a **(24)** Shell lirate, smooth, or with faint growth lines, but not ribbed . . . . **HELICODISCIDAE**

25b Shell ribbed or with well-developed (generally coarse) growth lines **26**

26a **(25)** Shell minute (less than 2 mm in diameter with about 4 whorls), surface usually sculptured with major and minor riblets . . . . **PUNCTIDAE**

26b Shell small to medium, or when minute (less than 2 mm in diameter and with $3\frac{1}{2}$–$4\frac{1}{2}$ whorls), the shell surface is smooth and without sculpturing of major and minor riblets **27**

27a **(26)** Shells without color markings; nuclear (embryonic) whorls with distinct spiral striae; jaw composed of about 20 loosely connected square plates; western in distribution . . . . **CHAROPIDAE**

27b Shells with or without color bands or blotches; nuclear (embryonic) whorls smooth or nearly so; jaw of one elongate, arcuate piece; widely distributed . . . . **DISCIDAE**

28a **(24)** Shell minute or small, less than 4 mm in diameter, depressed or with a conspicuously raised spire (often dome-shaped); penis with a digitiform appendix (Fig. 9.22*a*). . . . . **EUCONULIDAE**

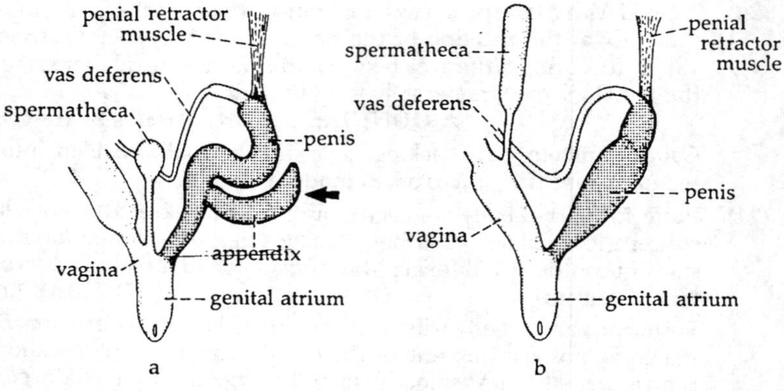

Fig. 9.22. Lower genitalia of (*a*) an euconulid snail [e.g., *Euconulus fulvus* (Müller)] and (*b*) a vitrinid snail [e.g., *Nesovitrea electrina* (Gould)]. The arrow points to the penial appendix. The penis and penial appendix are shaded. (Modified from H. B. Baker, 1928, *Proc. Acad. Nat. Sci. Phila.* 80, pl. 1, fig. 7; pl. 3, fig. 2.)

28b Shell minute, small, medium, or large (rarely), and may or may not have a conspicuously raised spire (species with raised spires have shells larger than 4 mm in diameter); penis without a digitiform appendix (Fig. 9.22*b*) . . . . **VITRINIDAE**

29a (14) Shell higher than wide **30**

29b Shell wider than high **48**

30a (29) Aperture with several lamellae or teeth; lip reflected **31**

30b Aperture without lamellae or teeth; lip sharp or reflected **32**

31a (30) Shell with short radial ribs at the sutures; shell at least three times higher than wide; Gulf coast states . . . . **STREPTAXIDAE**

31b Shell not ribbed [except for ***Vertigo californica,*** which occurs in California]; shell height less than three times its width . . . . **PUPILLIDAE (Orthurethra)** (in part)

32a (30) Shell urocoptiform (Fig. 9.6*d*) or cylindrical (Fig. 9.5*b*) **33**

32b Shell conical, fusiform, or bulimoid (see Fig. 9.6*e,g,h*); lip may or may not be reflected **34**

33a (32) Shell urocoptiform, more than 5 mm high, with more than 7½ whorls; lip of shell expanded, often reflected (see Fig. 9.12*b,c*); restricted in North America (north of Mexico) to the southern United States . . . **UROCOPTIDAE**

33b Shell cylindrical, less than 4 mm high, with fewer than 7 whorls; lip of shell sharp, not expanded or reflected . . . . **PUPILLIDAE (Orthurethra)** (in part) [***Columella*** (in part)]

34a (32) Aperture lip reflected (Fig. 9.12*c*) . . . . **PUPILLIDAE (Orthurethra)** (in part)

34b Aperture lip not reflected **35**

35a (34) Shell nearly globose, height less than 1.4 times the width **36**

35b Shell elongate, height more than 1.4 times the width **42**

36a (35) Shell large, 14 mm or more in width, usually banded **37**

36b Shell small, 4 mm or less in width, not banded **38**

37a (36) Shell less than 25 mm in diameter . . . . **XANTHONYCHIDAE** (in part) [***Cepolis*** (in part)]

37b Shell greater than 25 mm in diameter . . . . **HELICIDAE** (in part) [***Helix*** (in part)]

38a (36) Shell very small, 2 mm or less in diameter; surface pitted, granulose or striate, but not ribbed (see Fig. 9.13) . . . . **PUPILLIDAE (Orthurethra)** (in part) [***Pupisoma,*** in part]

38b Shell somewhat larger, more than 2 mm in diameter; surface not pitted, granulose, or striate, but either smooth or with fine, low ribs **39**

39a (38) Shell sculptured with thin periostracal ribs more oblique than the growth lines . . . . **THYSANOPHORIDAE** (in part) [***Thysanophora,*** in part]

39b Shell smooth, without ribs **40**

40a (39) Shell surface glossy and transparent; Mountain states, Washington, and Vancouver Island in British Columbia . . . . **THYSANOPHORIDAE** (in part) [***Microphysula*** (in part)]

40b Surface of shell dull, translucent whitish or pale brown; southern Florida **41**

41a **(40)** Whorls rounded . . . . . **THYSANOPHORIDAE** (in part) [***Hojeda***]
41b Whorls strongly shouldered (Fig. 9.7*c*) . . . . . **SAGDIDAE** (in part)
42a **(35)** Shell bulimoid (Fig. 9.6*g*) **43**
42b Shell conical (Fig. 9.6*h*) or fusiform (Fig. 9.6*e*). **45**
43a **(42)** Shell relatively small, 13 mm or less in length. . . . . **HYGROMIIDAE (in part)**
43b Shell relatively large, 15 mm or more in length **44**
44a **(43)** Shell less than 71 mm long, with or without spiral bands but shells greater than 60 mm long always with spiral bands . . . . **ORTHALICIDAE**
44b Shell 60 to 208 mm long, without spiral bands; Subfamily **Achatininae** **ACHATINIDAE** (in part)
45a **(42)** Shell marked with either color bands or alternating transverse opaque and translucent bands . . . . **HYGROMIIDAE** (in part) [***Cochlicella***]
45b Shell without color bands or alternating opaque and translucent bands **46**
46a **(45)** Shell either fusiform and medium in size (more than 20 mm long), or if conical then with radial ribs, and if conical and in Florida then with ribs and flattened whorls **47**
46b Shell either fusiform and small (less than 8 mm long), or if conical then either without ribs or if in Florida then with ribs and whorls rounded; Subfamilies **Ferussaciinae, Subulininae**. . . . . **ACHATINIDAE** (in part)
47a **(46)** Shell fusiform, medium to large (20 mm or more in length), without ribs . . . . **SPIRAXIDAE**
47b Shell conical with ribs and flattened whorls; in Florida . . . . **OLEACINIDAE**
48a **(29)** Outer aperture lip not reflected (Fig. 9.12*a*). **49**
48b Outer aperture lip reflected (Fig. 9.12*c*) **72**
49a **(48)** Shell large, more than 13 mm in diameter. **50**
49b Shell relatively small, 13 mm or less in diameter. **60**
50a **(49)** Shell distinctly and often widely umbilicate (Fig. 9.10*d*), umbilicus greater than one-fifteenth of the shell diameter; western and Rocky Mountain states [except ***Haplotrema concavum,*** which occurs in the east and ***Helicella obvia*** which has been introduced from Europe] . . . **51**
50b Shell perforate or imperforate (Fig. 9.10*a,b*), umbilicus less than one-fifteenth of the shell diameter **54**
51a **(50)** Shell surface roughly sculptured, sometimes radially or spirally striate; shell typically with two spiral bands, but bands vary from many to none; embryonic whorls and sometimes adult whorls carinate; embryonic whorls with radial striae, not granulose; found in Rocky Mountain states, Great Basin, and Santa Catalina Island in California. . . . . . **THYSANOPHORIDAE** (in part) [***Oreohelix*** (in part), ***Radiocentrum*** (in part)]
51b Shell surface smooth or of weak growth wrinkles, sometimes sculptured with spiral lines but never spiral striae; shell with or without spiral bands, but usually not two-banded [may be two-banded in ***Monadenia***]; usually not carinate as adult [may be carinate in ***Monadenia***]; embryonic whorls smooth or granulose, without radial striae **52**

52a **(51)** Umbilicus one-fifth or less of shell diameter; shell usually with one spiral band . . . . **XANTHONYCHIDAE** (in part) [***Monadenia*** (in part), ***Eremarionta*** (in part), ***Eremariontoides*** (in part), ***Sonorelix, Sonorella*** (in part)]

52b Umbilicus one-fourth or more of shell diameter; shell without spiral bands [***Helicella*** in **HYGROMIIDAE** has a large umbilicus and has spiral lines]. **53**

53a **(52)** Shell glossy; upper lip of aperture convex upward; shell surface smooth or with weak growth wrinkles, and shells have minute spiral lines on the parietal wall of the aperture (in ***Megomphix*** present in only young shells); shell dark or light colored, sometimes greenish; jaw strong, weakly and closely rib-striate; reproductive system with an oviducal cecum or accessory sac near junction of the free oviduct and the vagina (Fig. 9.23*a*) . . . . **THYSANOPHORIDAE** (in part) [***Glyptostoma, Megomphix*** (in part)]

53b Shell usually not glossy; upper lip of aperture straightened or dipping downward; shell surface of growth wrinkles or striations; shell usually greenish; jaw reduced to a narrow crescentic plate with a median point on the cutting edge; reproductive system without an oviducal cecum nor accessory sac near the junction of the free oviduct and the vagina (Fig. 9.23*b*). . . . . **HAPLOTREMATIDAE** (in part)

54a **(50)** Found east of the Continental Divide **55**

54b Found west of the Continental Divide **59**

55a **(54)** Shell large, 25–50 mm in diameter; shell horn colored or yellow and either spirally striate with pale brown spiral bands, or sculptured with fine growth wrinkles, lacking spiral striae, and with brown spiral bands interrupted by yellow flecks or streaks . . . . **HELICIDAE** (in part) [***Helix*** (in part)]

55b Shell less than 34 mm in diameter, with or without spiral bands, but not with the foregoing combination of characters **56**

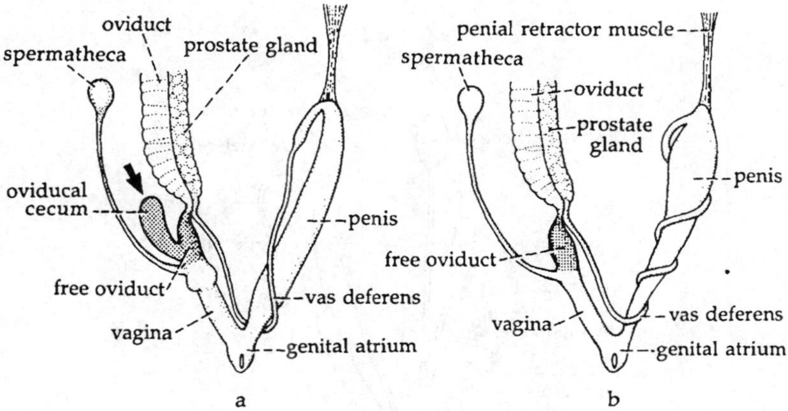

Fig. 9.23. Lower genitalia of (*a*) a thysanophorid snail (e.g., *Megomphix hemphilli* (W. G. Binney) (modified from H. B. Baker, 1930, *Nautilus* 43, pl. 5, fig. 3) and (*b*) a haplotrematid snail (e.g., *Haplotrema durant continentis* H. B. Baker) (modified from H. B. Baker, 1930, *Proc. Acad. Nat. Sci. Phila.*, 82, pl. 33, fig. 6). The arrow points to the oviducal cecum. The free oviduct and the oviducal cecum are shaded.

56a **(55)** Shell sculpture of growth ridges crossed by fine spiral striae; shell with $5\frac{1}{2}$–6 whorls. . . . . **HELICIDAE** (in part) [***Theba*** (in part)]

56b Shell sculpture of growth ridges or radial wrinkles, without spiral striae [a few interrupted spiral threads may be present in ***Sonorella micromphala***]; shell with 3–6 whorls **57**

57a **(56)** Found in Texas, New Mexico, Arizona, and Mexico; shell with 4–$4\frac{3}{4}$ whorls. . . . . **XANTHONYCHIDAE** (in part) [***Humboldtiana, Sonorella micromphala***]

57b Found in Florida; shell with 3–6 whorls **58**

58a **(57)** Shell 14–16 mm in diameter, white, usually with spiral color bands; shell with 3–6 whorls; reproductive system with dart apparatus (Fig. 9.24) . . . . **XANTHONYCHIDAE** (in part) [***Cepolis*** (in part)]

58b Shell 25–29 mm in diameter, yellow to horn color, lacking color bands; shell with about $4\frac{1}{2}$ whorls; reproductive system lacking dart apparatus **CAMAENIDAE**

59a **(54)** Shell typically with more than one band; introduced snails; shell either sculptured with fine growth wrinkles, lacking spiral striae, colored pale brown to yellow and having brown spiral bands interrupted by yellow flecks or streaks [***Helix aspersa***]; or shell sculpture of growth ridges crossed by fine spiral striae, white to ginger in color and usually with darker spiral bands [***Theba pisana***]; mucous glands of reproductive system either tubular and branched, or club-shaped and unbranched **HELICIDAE** (in part)

59b Shell usually with one band at or above the periphery; native snails, either found along the coast of southern California, Baja California, and the southern California islands [shell typically with a dark often interrupted band at or above the periphery, ***Plesarionta*** and ***Xerarionta***], or found in Arizona [shell buff-colored with one thin brown spiral band above the shell periphery, ***Sonorella micromphala***]; mucous glands of reproductive system either expanded, flattened and not branched, or absent . . . . **XANTHONYCHIDAE** (in part)

60a **(49)** Shell less than 6 mm in diameter **61**

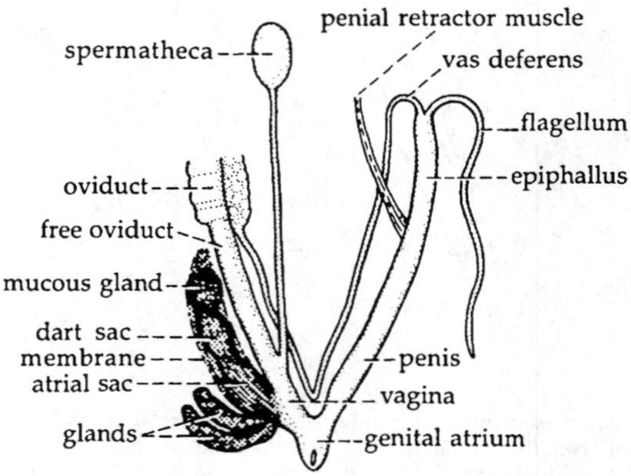

Fig. 9.24. Terminal genitalia of a xanthonychid snail [e.g., *Cepolis varians* (Menke)]. Modified from Pilsbry (1939). The dart apparatus and accessory structures are shaded.

60b Shell greater than 6 mm in diameter. **66**

61a **(60)** Shell less than 2 mm in diameter . . . . **PUPILLIDAE (Orthurethra)** (in part) [***Pupisoma***]

61b Shell greater than 2 mm in diameter. **62**

62a **(61)** Umbilicus wide (1/2.7–1/3.2 of the shell diameter); shell diameter 5 mm or greater; found in southern and central coastal California . . . **HAPLOTREMATIDAE** (in part)

62b Umbilicus narrower (one-third to one-tenth of the shell diameter); found in various parts of North America (north of Mexico), but not in California **63**

63a **(62)** Shell sculptured with thin periostracal ribs more oblique than the growth lines . . . . **THYSANOPHORIDAE** (in part) [***Thysanophora*** (in part)]

63b Shell smooth, without ribs **64**

64a **(63)** Shell surface glossy and transparent; Rocky Mountain states, Washington, and on Vancouver Island in British Columbia . . . . **THYSANOPHORIDAE** (in part) [***Microphysula*** (in part)]

64b Surface of shell dull, translucent whitish or pale brown; southern Florida **65**

65a **(64)** Whorls rounded . . . . . **THYSANOPHORIDAE** (in part) [***Hojeda***]

65b Whorls strongly shouldered . . . . **SAGDIDAE** (in part)

66a **(60)** Shell perforate (see Fig. 9.10*b*), umbilicus less than one-fifteenth of the shell diameter . . . . **HELICIDAE** (in part) [***Theba*** (in part)]

66b Shell umbilicate (see Fig. 9.10*c,d*), umbilicus greater than one ninth of the shell diameter **67**

67a **(66)** Umbilicus greater than 1/3.5 of shell diameter; shell without bands **68**

67b Umbilicus less than 1/3.5 of shell diameter; shell with or without spiral bands. **70**

68a **(67)** Shell either with spire concave and aperture lacking teeth or barriers (Fig. 9.25*a*), or spire flat or convex and aperture with teeth or barriers (Fig. 9.25*b*); western . . . . **THYSANOPHORIDAE** (in part) [***Ammonitella, Polygyrella, Polygyroidea***]

68b Shell with spire convex, aperture without teeth or barriers **69**

69a **(68)** Shell usually not glossy; upper lip of aperture straightened or dipping downward; shell surface of growth wrinkles or striations; shell usually greenish; jaw reduced to a narrow crescentic plate with a median point on the cutting edge; reproductive system without an oviducal cecum nor accessory sac near the junction of the free oviduct and the vagina (Fig. 9.23*b*) . . . . **HAPLOTREMATIDAE** (in part)

69b Shell glossy; upper lip of aperture convex upward; shell surface smooth or with weak growth wrinkles, and young shells have minute spiral lines on the parietal wall of the aperture; shell light colored, sometimes greenish; jaw strong, weakly and closely rib-striate; reproductive system with an oviducal cecum or accessory sac near junction of the free oviduct and the vagina (Fig. 9.23*a*). . . . . **THYSANOPHORIDAE** (in part) [***Megomphix*** (in part)]

70a **(67)** Eastern United States, not montane. . . . . **HYGROMIIDAE** (in part) [***Trochoidea, Candidula***]

70b Western United States and Rocky Mountain states. **71**

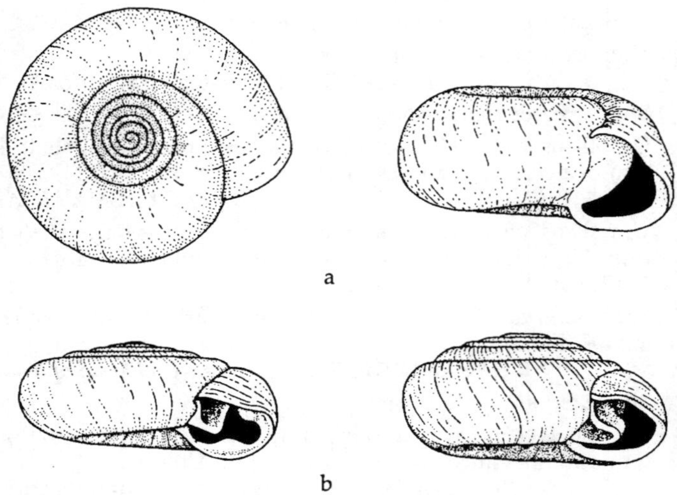

Fig. 9.25. Shell(s) with (*a*) concave spire and lacking apertural barriers; (*b*) convex spire and apertural barriers.

71a **(70)** Shell surface smooth or of weak growth wrinkles; shell with one spiral band located above the periphery; shell not carinate . . . .
XANTHONYCHIDAE (in part)
[***Eremarionta*** (in part), ***Eremariontoides*** (in part), ***Sonorella micra*** and ***Mohavelix***]

71b Shell surface striate or roughly sculptured; shell with or without spiral bands, shell may or may not be carinate; Subfamily **Oreohelicinae** . . . . **THYSANOPHORIDAE** (in part)
[***Oreohelix*** (in part), ***Radiocentrum*** (in part)]

72a **(48)** Shell less than 3 mm in diameter **73**

72b Shell greater than 4 mm in diameter. **74**

73a **(72)** Shell without lamellae in the aperture; growth lines not prominent; whitish to light tan-colored . . . .
VALLONIIDAE (***Orthurethra***) (in part) [***Vallonia***]

73b Shell with one or more lamellae in the aperture; growth lines prominent; chestnut brown . . . .
STROBILOPSIDAE (***Orthurethra***) (in part)

74a **(72)** Found west of Continental Divide **75**

74b Found east of Continental Divide **81**

75a **(74)** Shell diameter greater than 25 mm; aperture without teeth or barriers **76**

75b Shell diameter less than 25 mm; aperture with or without teeth or barriers **79**

76a **(75)** Umbilicus narrow (less than one-fifteenth of the shell diameter) to closed **77**

76b Umbilicus wider than one-fifteenth of the shell diameter **78**

77a **(76)** Shell usually with one brown spiral band, located above the shell periphery, and usually with a narrower, paler band adjacent to each side of the brown band; native species . . . . **XANTHONYCHIDAE** (in part) [***Helminthoglypta*** (in part)]

77b Shell usually with more than one brown spiral band (up to five); introduced species . . . . **HELICIDAE** (in part) [***Helix*** (in part), ***Cepaea*** (in part), ***Eobania*** (in part), ***Otala*** (in part)]

78a (76) Shell without bands; whorls rounded . . . . **MESODONTIDAE** (in part) [***Allogona*** (***Dysmedoma***) (in part)]

78b Shell usually with bands, whorls rounded to carinate . . . . **XANTHONYCHIDAE** (in part) [***Monadenia*** (in part), ***Helminthoglypta*** (in part)]

79a (75) Shell without spiral bands; aperture often with teeth or barriers; reproductive system without mucous glands (Fig. 9.26*a*) . . . . **MESODONTIDAE** (in part) [***Cryptomastix, Vespericola, Allogona*** (***Dysmedoma***) (in part), ***Trilobopsis, Ashmunella***]

79b Shell usually with spiral bands; aperture without teeth or barriers; reproductive system with mucous glands (Fig. 9.26*b*) present [except mucous glands lacking in ***Micrarionta opuntia***] **80**

80a (79) Shell usually tan to brown with one darker spiral band located above the shell periphery, often with a thinner, paler band adjacent to either side of the darker band; or shell dark brown with a yellowish band above the periphery and other spiral color bands present or absent **XANTHYNYCHIDAE** (in part) [***Monadenia*** (in part), ***Helminthoglypta*** (in part), ***Micrarionta, Eremarionta*** (in part)]

80b Shell usually of a white or pale base color with up to five darker spiral bands . . . . **HELICIDAE** (in part) [***Cepaea*** (in part) and ***Theba*** (in part)]

81a (74) Umbilicus greater than one-twelfth of diameter **82**

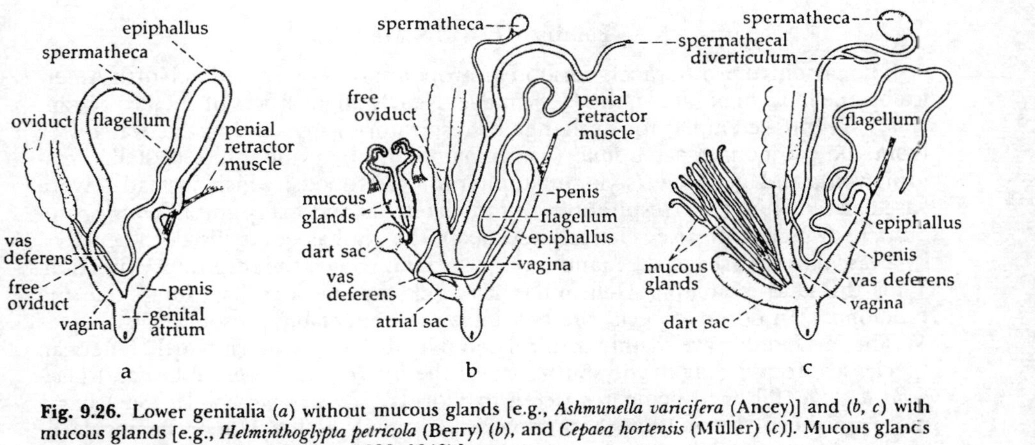

Fig. 9.26. Lower genitalia (*a*) without mucous glands [e.g., *Ashmunella varicifera* (Ancey)] and (*b, c*) with mucous glands [e.g., *Helminthoglypta petricola* (Berry) (*b*), and *Cepaea hortensis* (Müller) (*c*)]. Mucous glands are shaded. [Modified from Pilsbry (1939, 1940).]

81b Umbilicus narrow (less than one-twelfth of the diameter) to closed . . . **83**

82a **(81)** Aperture often with one or more teeth or barriers; reproductive system without dart sac or mucous glands (Fig. 9.26*a*) . . . . **MESODONTIDAE** (in part)
[***Polygyra, Stenotrema*** (***Euchemotrema***) (in part), ***Mesodon*** (in part), ***Triodopsis* s.s., *Allogona, Ashmunella***]

82b Aperture without teeth or barriers; reproductive system with mucous glands and usually with a dart sac. . . . . **HYGROMIIDAE** (in part)
[***Monacha, Trichia***]

83a **(81)** Shell aperture often with one or more teeth or barriers; reproductive system without dart sac or mucous glands (e.g., see Fig. 9.6*a*) . . . . **MESODONTIDAE** (in part)
[***Stenotrema, Stenotrema*** (***Euchemotrema***) (in part), ***Praticolella, Mesodon*** (in part), ***Triodopsis*** (***Xolotrema*** + ***Neohelix***)]

83b Shell aperture without teeth or barriers; introduced and tropical species; reproductive system with a dart sac and mucous glands (e.g., see Fig. 9.26*b,c*) **84**

84a **(83)** Shell diameter 17 mm or greater . . . . **HELICIDAE** (in part)
[***Helix*** (in part), ***Cepaea*** (in part), ***Otala*** (in part), ***Theba*** (in part)]

84b Shell diameter less than 17 mm **85**

85a **(84)** Shell higher, ratio of height to width greater than 0.85; lip only very slightly reflected; southern Florida . . . . **XANTHONYCHIDAE** (in part) [***Cepolis*** (in part)]

85b Shell more depressed, ratio of height to width less than 0.85; lip slightly to moderately reflected. **86**

86a **(85)** Shell with a single spiral band located above the periphery; introduced species in Gulf coast states . . . . **XANTHONYCHIDAE** (in part) [***Bradybaena***]

86b Shell usually with more than one spiral band; species introduced throughout the United States, usually farther north than Gulf coast states. . . . . **HELICIDAE** (in part)
[***Cepaea*** (in part), ***Theba*** (in part)]

Family HELICINIDAE

The primitive prosobranch neritoid gastropods are represented in North America by one land snail family, the Helicinidae. Nearly all members of the Prosobranchia have gills (ctenidia) for exchange of respiratory gases, and in this way differ from the pulmonate acteophile (basommatophoran), gymnophile (systellommatophoran), and geophile (stylommatophoran) gastropods, which instead have a vascularized "lung" for respiration. But in terrestrial and many amphibious prosobranchs (which comprise relatively few taxa), the gill has gradually been replaced by a vascularized area of the mantle roof, which takes over the respiratory function. Thus the land-inhabiting Helicinidae lack gills, having a pallial "lung" like the Pulmonata. In other respects, the helicinids are undeniably prosobranchs.

The Helicinidae are mainly tropical and five of the six eastern North American species are found only in the south, two of them [*Helicina clappi* Pilsbry and *Lucidella tantilla* (Pilsbry)] being restricted to Florida. *Hendersonia occulta* (Say) has a more northern distribution, occurring as far north as the Upper Peninsula of Michigan. *Alcadia striata* (Lamarck) has been introduced from Puerto Rico into southern Florida.

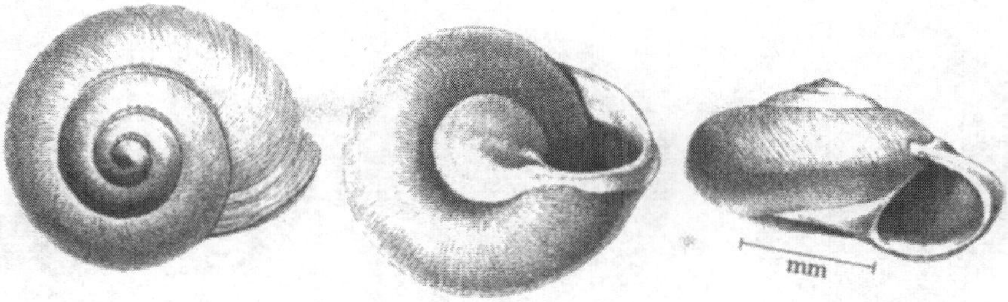

Fig. 9.27. *Lucidella tantilla* (Pilsbry).

Helicinid snails are usually wider than high and all are imperforate, the umbilical region being occupied by a callous pad. As is typical for nearly all prosobranch snails, the shell aperture is closed by an operculum when the animal is withdrawn into the shell. The opercula are either paucispiral or concentric, depending on the genus, and are layered with a rather thin calcareous deposit.

Key to Genera

1a Shell small, 3 mm or less in diameter, strongly depressed; operculum concentric (Fig. 9.16*d*); southern Florida (Fig. 9.27) . . . . . . ***Lucidella***

1b Shell larger, 5 mm or more in diameter, not depressed, or only moderately depressed; operculum paucispiral or concentric **2**

2a (1) Lower edge of aperture adjacent to columella with a distinct notch (Fig. 9.28). . . . . ***Alcadia***

2b Lower edge of aperture without a distinct notch. **3**

3a (2) Shell sculptured dorsally, and usually ventrally, with coarse growth lines or fine transverse riblets, and often with fine, close spiral striae between the growth lines or riblets; operculum paucispiral (Fig. 9.16*b*); Virginia west to eastern Oklahoma, north to Wisconsin and Michigan (Fig. 9.29) ***Hendersonia***

3b Shell smooth, with very weak growth lines, and only occasionally with a few widely spaced striae; operculum concentric **4**

4a (3) Shell 7–12 mm in diameter; no tooth at base of columella (Fig. 9.30) ***Helicina***

4b Shell 5–8.5 mm in diameter; a blunt projection present at the base of the columella (Fig. 9.31) . . . . ***Oligyra***

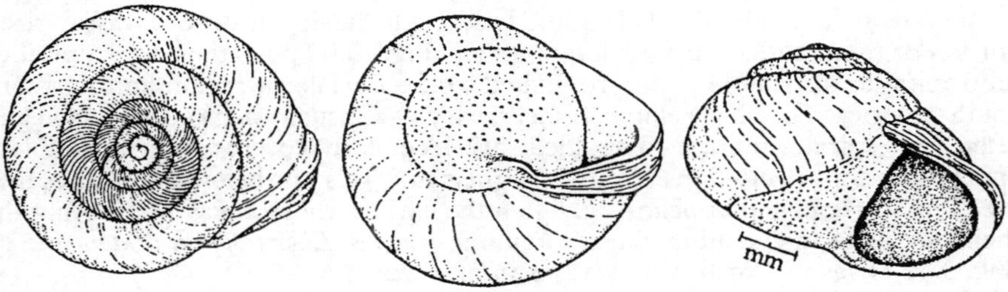

Fig. 9.28. *Alcadia striata* (Lamarck).

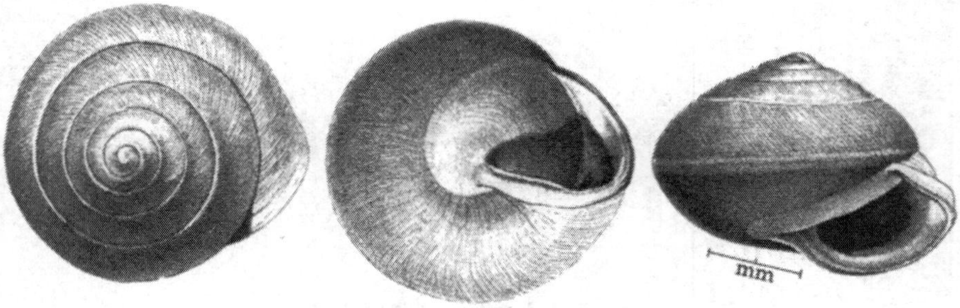

Fig. 9.29. *Hendersonia occulta* (Say).

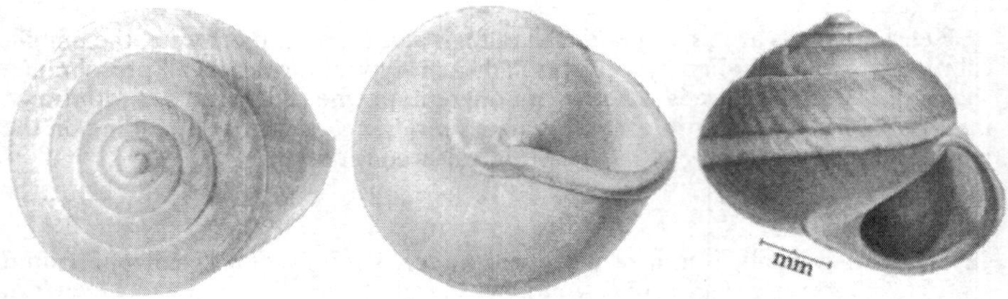

Fig. 9.30. *Helicina clappi* Pilsbry.

Fig. 9.31. *Oligyra orbiculata* (Say).

Family POMATIIDAE

The Pomatiidae Gray 1852 (= Pomatiasidae Pilsbry 1919, Chondropomatidae Henderson and Bartsch 1920, and Choanopomidae Thiele 1929) are terrestrial prosobranchs, and like the helicinids, lack a gill, having instead a vascularized portion of the mantle wall functioning as a "lung." The pomatiids are tropical or subtropical in distribution and are limited to the Caribbean region and northern South America. They have elongate shells, which are usually marked with axial and often with spiral sculpture. The shells are frequently covered with fine riblets. Small calcareous projections at rather regular intervals just below the sutures give the shells a crenulated appearance. Often the earlier whorls break off, leaving the shells decollate in the adult stage. Only one species, *Chondropoma dentatum* (Say) (Fig. 9.32), lives in North America (north of Mexico).

The shell of *Chondropoma dentatum* has about four whorls remaining in decollated adults and measures from 10 to 12 mm in length. It is dull red to light tan and

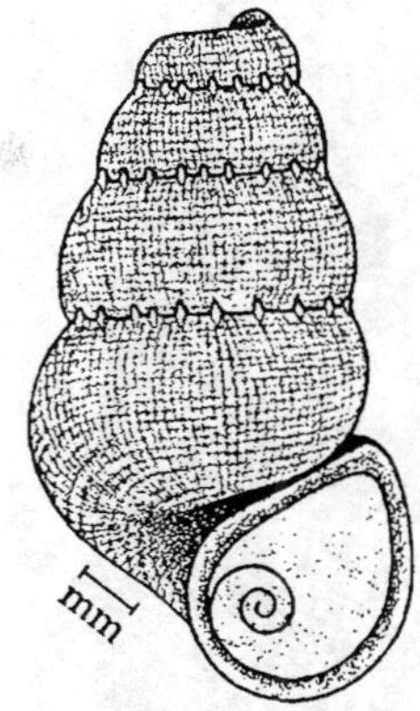

Fig. 9.32. *Chondropoma dentatum* (Say).

sometimes has several narrow interrupted bands or spots. The shell is umbilicate, the outer surface has both transverse and spiral sculpture, and the inner edge of the aperture is smooth and rounded. The operculum is spiral and cartilaginous.

Family TRUNCATELLIDAE

The Truncatellidae comprise species with small (3–9 mm in length), cylindrical shells, which are imperforate, usually decollate, and often heavily ribbed. The operculum is subspiral (Fig. 9.16*c*) and corneous, often covered with a calcareous

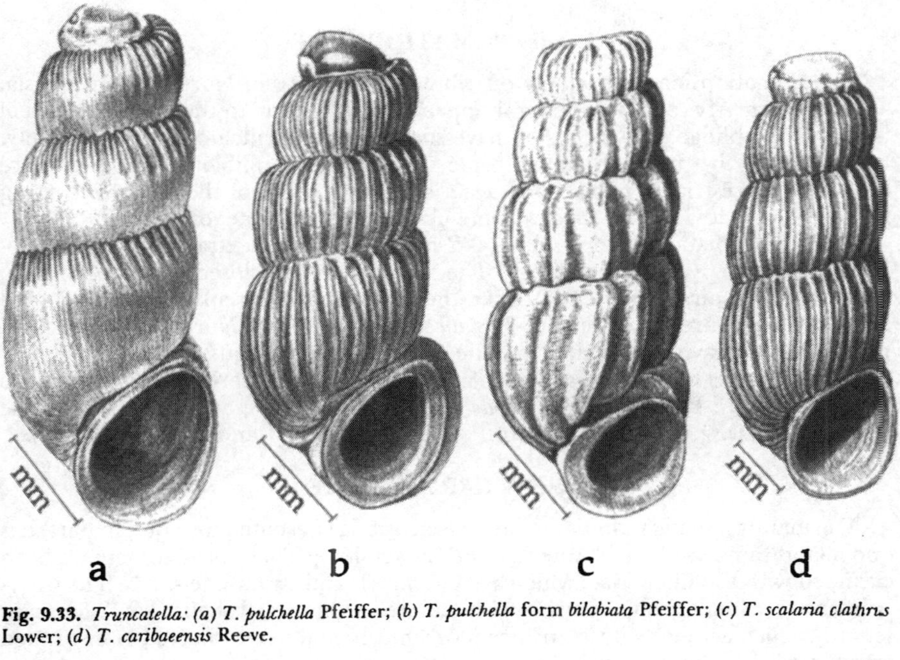

Fig. 9.33. *Truncatella:* (*a*) *T. pulchella* Pfeiffer; (*b*) *T. pulchella* form *bilabiata* Pfeiffer; (*c*) *T. scalaria clathrus* Lower; (*d*) *T. caribaeensis* Reeve.

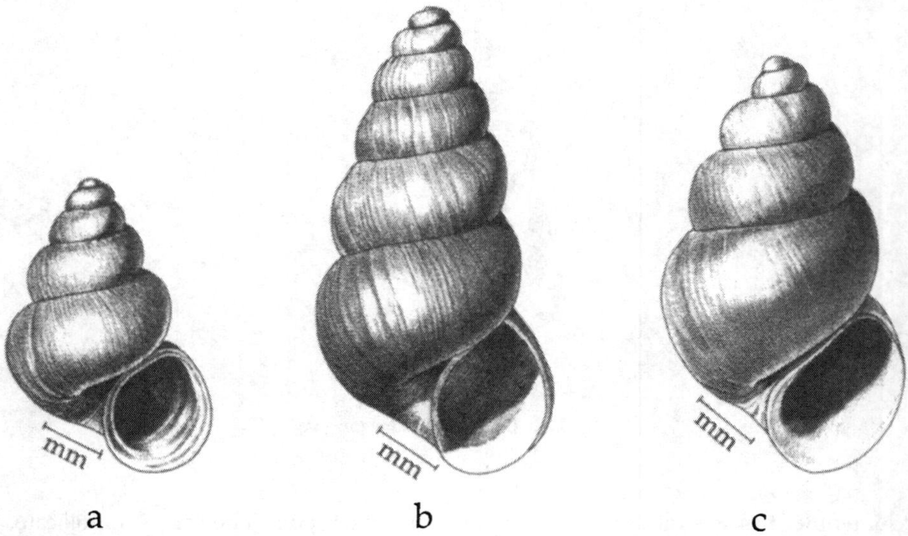

Fig. 9.34. *Pomatiopsis: (a) P. cincinnatiensis* (Lea); *(b) P. lapidaria* (Say); *(c) P. hinkleyi* Pilsbry.

layer. The family and its main genus, *Truncatella,* get their names from the characteristically truncated shell spire, which results from the upper spire being broken off at about the fourth or fifth whorl as the snails reach maturity. The truncatellids are widely distributed in tropical and subtropical areas, usually living on beaches above high tide level, often on rocks or under stones or debris. Only one genus occurs in the United States, *Truncatella* (Fig. 9.33), which has several species in the east and two, *T. californica* Pfeiffer and *T. stimpsoni* Stearns, in the west.

Family POMATIOPSIDAE

The Pomatiopsidae are widely distributed, occurring in North America, Asia, and southern Africa. Their general appearance is similar to the freshwater snail family Hydrobiidae, in which they have sometimes been included as a subfamily. Many species live in fresh water, some species are amphibious, and a few are essentially terrestrial in habit. *Pomatiopsis* is the only genus of the family occurring in North America, and its species come under the latter category.

Shells of pomatiopsids are small (4–7 mm), elongate, dextral, drab and unicolored, and have five to seven whorls. The shell aperture is closed by an operculum, which is paucispiral (Fig. 9.16*b*). Like the rest of the operculate snails included here, the sexes are separate. But *Pomatiopsis,* unlike other North American land prosobranchs, have gills in their mantle cavity for respiration.

Three species of *Pomatiopsis* occur in eastern North America [*P. cincinnatiensis* (Lea), *P. hinkleyi* Pilsbry, and *P. lapidaria* (Say)] (Fig. 9.34), and three species are found in the west (*P. binneyi* Tryon, *P. californica* Pilsbry, and *P. chacei* Pilsbry).

Family CARYCHIIDAE

The mainly aquatic pulmonate order Acteophila (Basommatophora, in part) has one inland family in North America, the Carychiidae. These have sometimes been classified with the Ellobiidae, which are salt marsh and strandline snails. The Carychiidae are a small group of snails with minute to very small shells (1.2–2.5 mm in length), which are elongate, translucent or whitish, and have reflected or expanded apertural lips. The columella has a lamella, which is visible through translucent

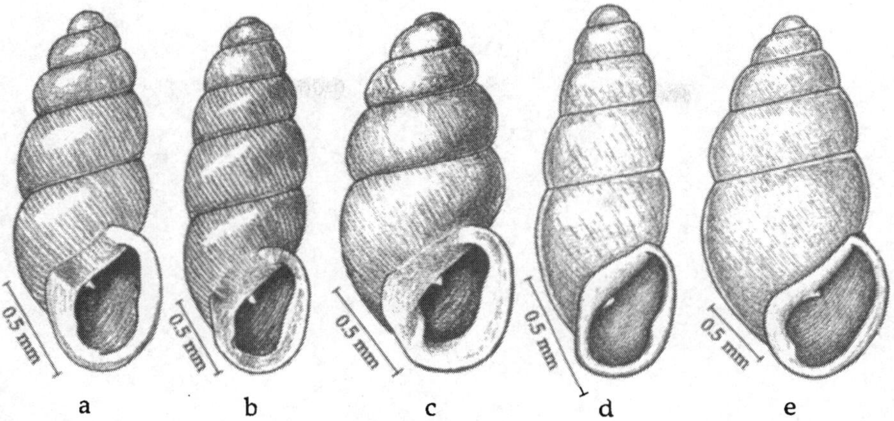

Fig. 9.35. *Carychium;* (*a*) *C. exile* H. C. Lea; (*b*) *C. exile canadense* Clapp; (*c*) *C. floridanum* Clapp; (*d*) *C. nannodes* Clapp; (*e*) *C. stygium* Call.

juvenile and adult shells. The end of this lamella appears as a small tooth on the parietal wall of the aperture. Like the lymnophile snails, the eyes of carychiid snails are sessile at the bases of the single pair of contractile tentacles. Only one genus, *Carychium*, is found in the United States (Fig. 9.35). It has several or more species, but their taxonomy is still somewhat confused. Several nominal species seem to be clearly distinct, yet others are difficult to distinguish, despite seemingly contrary statements in the literature. The various species usually frequent moist to wet places, often rotting logs or decaying leaves.

Family VERONICELLIDAE (Vaginulidae)

The Veronicellidae Gray 1840 (= Vaginulidae Gill 1871) are a tropical family of primitive slugs that have their eyes on contractile (not inversible) stalks or tentacles. The mantle (notum) covers the entire back of the animal and contains neither an external nor an internal shell. The body is characterized by a flattened, elongate, or elongate–oval form, hiding the sidewalls of the body. Differing from other terrestrial slugs, the anus and renal opening in veronicellids are at the posterior end of the animal (Fig. 9.36). This is because veronicellids are detorted gastropods; that is, the mantle cavity has migrated to the posterior end during development, carrying

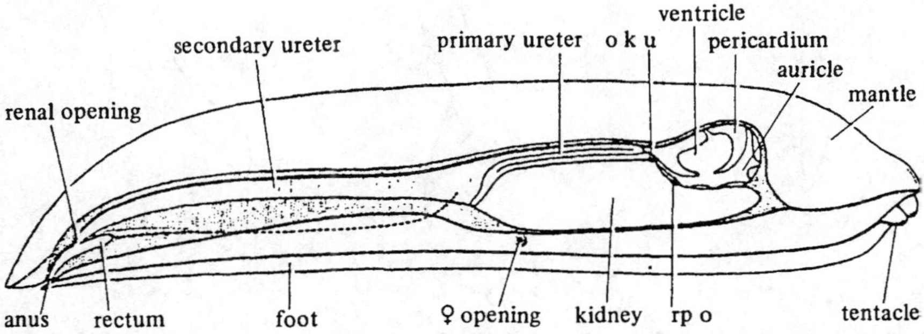

Fig. 9.36. Longitudinal partial section of a veronicellid slug. *o k u*, opening from the kidney to the ureter; *rp o*, renopericardinal opening. (From P. Pelseneer, 1901, *Mem. Acad. Roy. Sci., Lett., Beaux-Arts Belgique*, 54, pl. 6, fig. 51.)

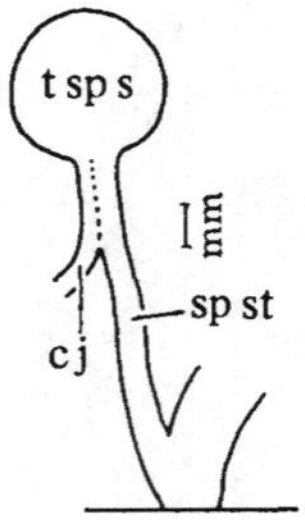

37

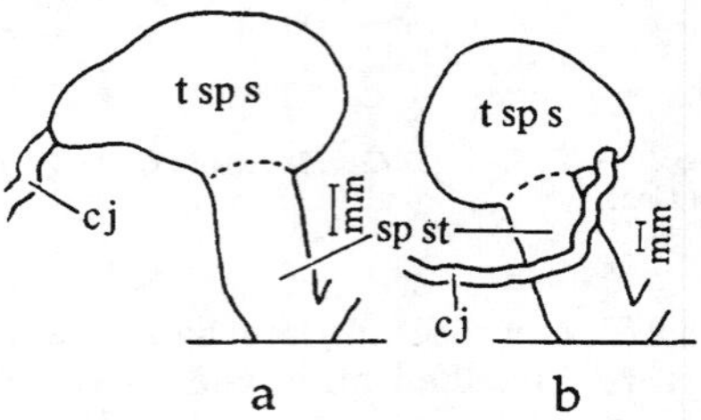

38

Fig. 9.37. Spermatheca and ducts of *Leidyula*. **Fig. 9.38.** Spermatheca and ducts of (*a*) *Diplosolenodes* and (*b*) *Angustipes*. *cj*, canalis junctor; *sp st*, spermathecal stalk; *t sp s*, terminal spermathecal sac. (Adapted from H. B. Baker, 1925, pls. 3–6.)

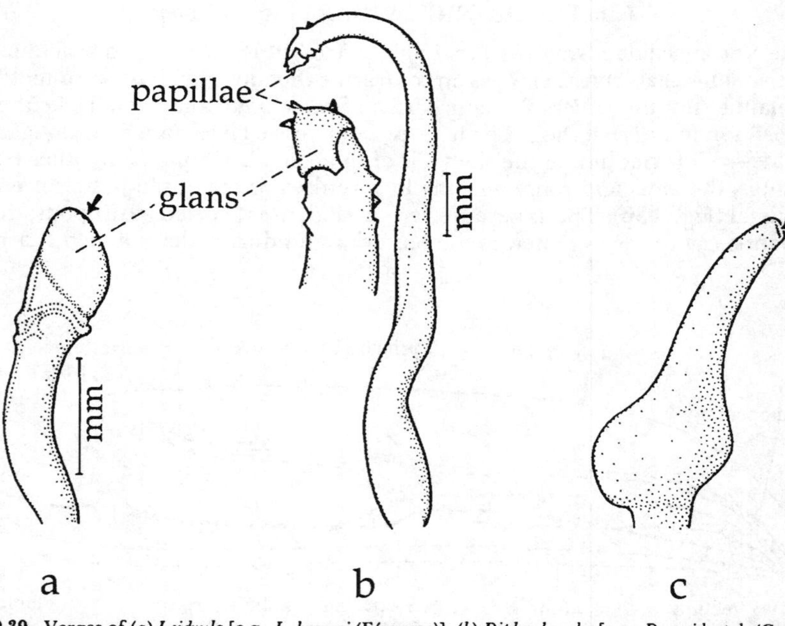

Fig. 9.39. Verges of (*a*) *Leidyula* [e.g., *L. kraussi* (Férussac)]; (*b*) *Diplosolenodes* [e.g., *D. occidentale* (Guilding)] (inset: apex, much enlarged) (after H. B. Baker, 1925, pls. 5, 6); (*c*) *Angustipes ameghini* (Gambetta) [after L. Gambetta, 1923, *Boll. Mus. Zool. Anat. Comp. Univ. Torino* N.S., 38(11), pl. 8, fig. 6]. Arrows indicate position of seminal pore.

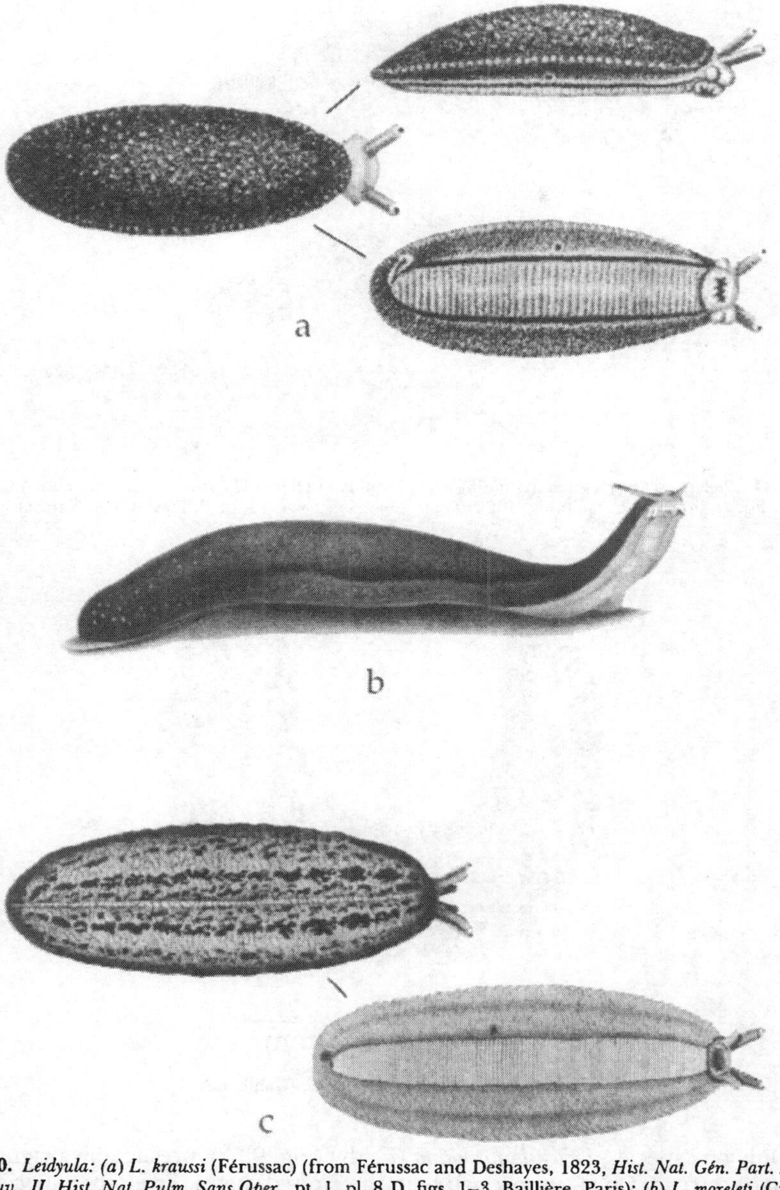

Fig. 9.40. *Leidyula:* (*a*) *L. kraussi* (Férussac) (from Férussac and Deshayes, 1823, *Hist. Nat. Gén. Part. Moll. Terr. Fluv., II, Hist. Nat. Pulm. Sans Oper.*, pt. 1, pl. 8 D, figs. 1–3, Baillière, Paris); (*b*) *L. moreleti* (Crosse and Fischer) (from P. Fischer and H. Crosse, 1878, *Etud. Moll. Terr. Fluv. Mex. Am. Centr. . . . , Rech. Zool. . . .* , pt. 7, vol. 1, pl. 24, fig. 14); (*c*) *L. floridana* (Leidy) (from W. G. Binney, 1857, *Terr. Air-Breath. Moll. U.S., Adj. Terr. N. Am.*, vol. 3, pl. 67, Little, Brown, Boston). Natural size.

with it the rectal and renal openings. In adults, the mantle cavity is completely obliterated; veronicellids lack a respiratory cavity.

The Philomycidae and the arionid *Magnipelta* also bear a mantle covering the entire dorsum, or nearly so, but they differ from veronicellid slugs in having invaginable tentacles (see Fig. 9.20) and a breathing hole in the anterior half of the right side of the mantle.

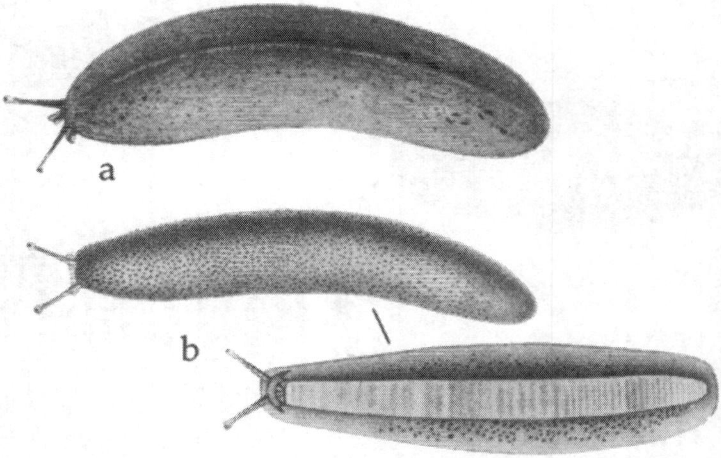

Fig. 9.41. *Diplosolenodes occidentale* (Guilding). [(*a*) From P. Fischer, 1871, *Nouv. Arch. Mus. Hist. Nat. Paris* 7, pl. 11, fig. 2); (*b*) from L. Guilding, 1823, *Trans. Linn. Soc. Lond.* 14, pl. 9, Figs. 9, 10). Natural size.

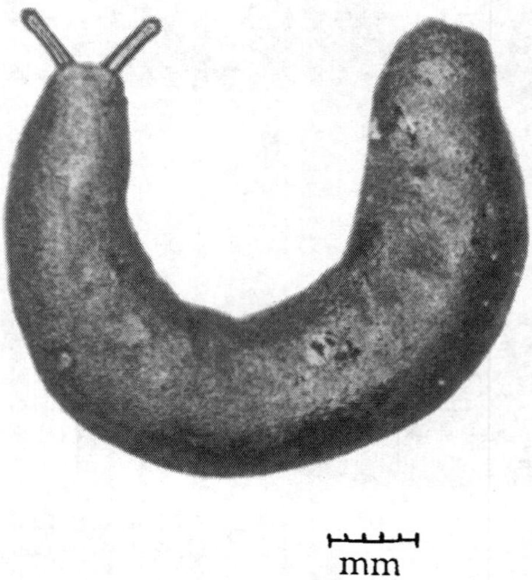

Fig. 9.42. *Angustipes ameghini* (Gambetta).

Only one species, *Leidyula floridana* (Leidy) (Fig. 9.37), presumably originally an immigrant from Cuba, is now native to the United States. Five other species have been reported as recent introductions: *L. kraussi* (Férussac), *L. moreleti* (Crosse and Fischer), *L. sloanei* (Cuvier), *Angustipes ameghini* (Gambetta) and *Diplosolenodes occidentale* (Guilding).

***Key to Genera*†**

1a Oval anus protected ventrally by an anterior flap so that the superficial opening is a transverse crescentic slit which usually extends beyond the

† Adapted from Baker (1925).

right edge of the retracted foot; dorsum with or without two longitudinal dark bands; spermathecal stalk elongate and cylindrical, joined by the canalis junctor, which enters near the base of the terminal sac (Fig. 9.37); female opening posterior to the middle; hindgut and genitalia enter the body wall in close juxtaposition; verge simply cylindrical or with subspiral ridges, with the orifice at its tip and surrounded by a granulate, chalky-white glans (Fig. 9.39*a*); verge retractor short and stout; dart-papilla small with paired retractors (Fig. 9.40) . . . . *Leidyula*

1b Roundish anus almost median, usually protected ventrally by a dextral flap, so that the superficial crescentic slit is more or less longitudinal and usually hidden under end of even retracted foot; spermathecal stalk short and usually swollen; canalis junctor enters spermathecal sac rather than joining the spermathecal stalk (Fig. 9.38); verge without granulate, chalky-white glans (Fig. 9.39*b,c*) 2

2a (1) Verge long and cylindrical (Fig. 9.39*b*), quite symmetrical, except at apex; canalis junctor enters apex of spermathecal sac (Fig. 9.38*a*); tip of verge (Fig. 9.39*b*) with papillae and with prominent, reflected emarginate cap (glans) which hides the pseudacrocaul opening; hyponotum with black spots (Fig. 9.41). . . . . *Diplosolenodes*

2b Verge short and stout, usually symmetrical (Fig. 9.39*c*); canalis junctor enters near base of the spermathecal sac (Fig. 9.38*b*); dorsum black and hyponotum light gray in adults (Fig. 9.42) . . . . *Angustipes*

Family CIONELLIDAE (Cochlicopidae)

Snails of this family have small shells (5–7.5 mm in length), which are very smooth and glossy and are transparent or translucent brown or yellow-brown in color. The shells are elongate, spindle-shaped, with unreflected but internally thickened lips, and shells of adults have 5½–6 whorls. Only one genus, *Cionella* (= *Cochlicopa*) (Fig. 9.43), occurs in North America. It has been reported from all the continental United States except Florida and Texas.

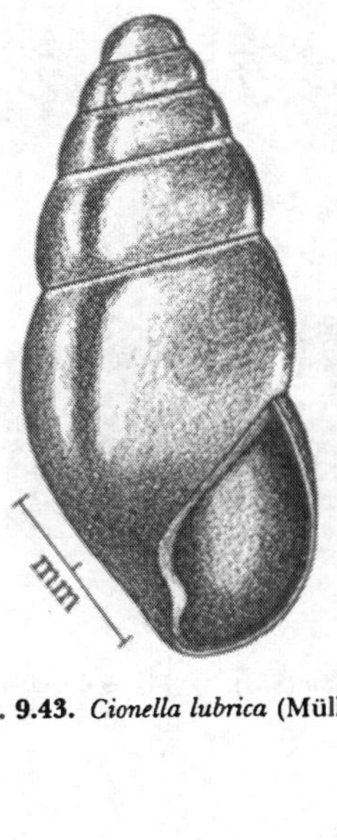

Fig. 9.43. *Cionella lubrica* (Müller).

Family VALLONIIDAE

Members of this family are minute to very small snails (shells 1.7–3 mm in diameter) with depressed (most species) or globose [*Zoogenetes harpa* (Say)] shells, which are often sculptured with fine transverse costae. Of the three genera, *Zoogenetes* contains the single species *Z. harpa,* and *Planogyra* contains the two species *P. asteriscus* and *P. clappi. Vallonia* has two species restricted to the east [*V. costata* (Müller) and *V. parvula* Sterki] and six species occurring in both regions [*V. albula* Sterki, *V. cyclophorella* Sterki, *V. excentrica* Sterki, *V. gracilicostata* Reinhardt, *V. perspectiva* Sterki and *V. pulchella* (Müller)]. *Vallonia* and *Planogyra* are oviparous; *Zoogenetes* is ovoviviparous.

Key to Genera

1a Shell strongly depressed, with or without ribs **2**

1b Shell globose, last several whorls with thin periostracal ribs (Fig. 9.44) ***Zoogenetes***

2a **(1)** Shell thin, less calcified, brownish, sculptured with thin periostracal ribs; aperture lip thin, not reflected (Fig. 9.45). . . . . ***Planogyra***

2b Shell moderately calcified, rather sturdy, with or without ribs; aperture lip reflected and may or may not be thickened (Fig. 9.46) . . . ***Vallonia***

Family STROBILOPSIDAE

In the United States, the snails of this family have dome-shaped, ribbed shells (except for *Strobilops hubbardi* Brown, which has a depressed shell). The shell aperture contains one or more lamellae or folds (Fig. 9.47). These snails are usually found on decaying or dead leaves or wood in moderately humid forests. The single genus in the United States, *Strobilops,* contains five species [*S. aenea* Pilsbry, *S. affinis*

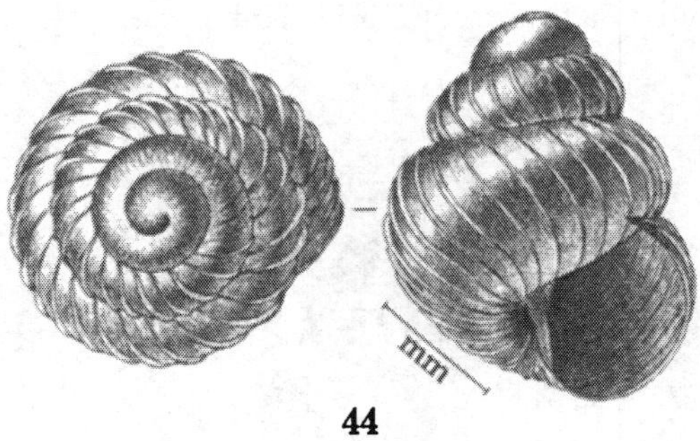

44

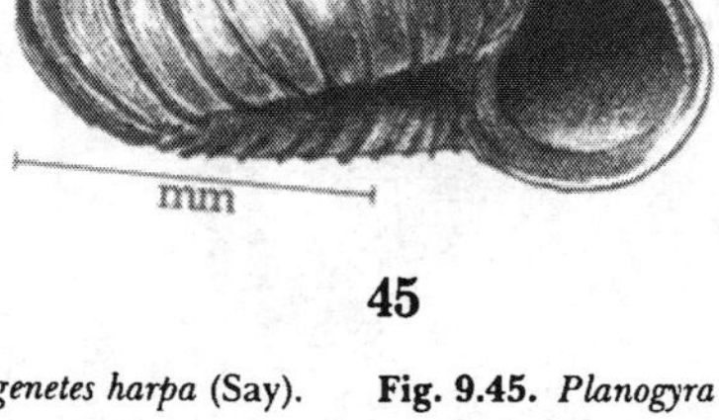

45

Fig. 9.44. *Zoogenetes harpa* (Say). **Fig. 9.45.** *Planogyra asteriscus* (Morse).

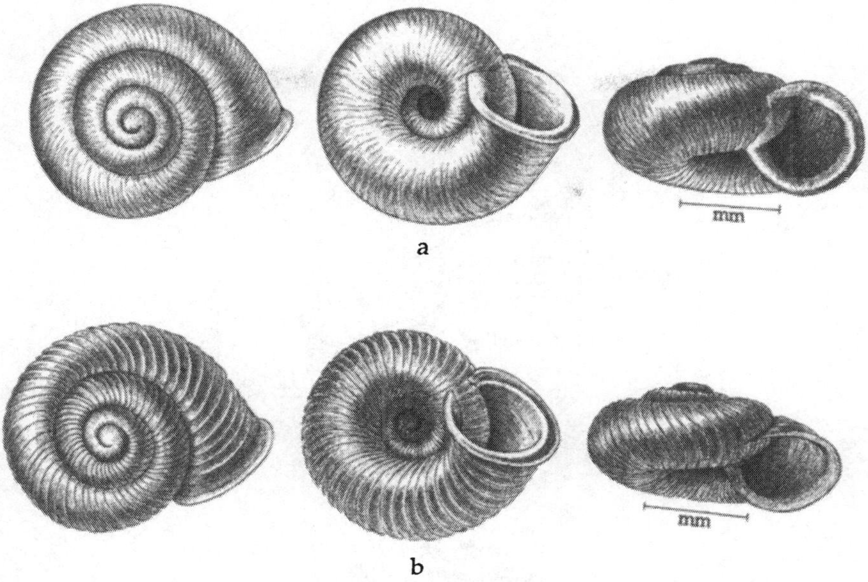

Fig. 9.46. *Vallonia:* (*a*) *V. excentrica* Sterki; (*b*) *V. costata* (Müller).

Pilsbry, *S. hubbardi* Brown (Fig. 9.48*a*), *S. labyrinthica* (Say) (Fig. 9.48*b*) and *S. texasiana* Pilsbry and Ferriss]. All of these species are found in the eastern part of the country.

Family PUPILLIDAE

This is a large family of small-to-minute snails, which are generally pupa-shaped and frequently possess apertural lamellae and folds (Fig. 9.49). These apertural "teeth" are important in taxonomy and in species recognition. Also useful in identifying species is the general shape of the adult shells (Fig. 9.50). The family is nearly worldwide in its distribution. The subfamilies have sometimes been used as families. Three subfamilies occur in the United States: Pupillinae (*Pupilla* and *Pupoides*), Chondrininae (*Gastrocopta* and *Chaenaxis*), and Vertigininae (*Vertigo, Columella, Bothriopupa, Sterkia* and *Pupisoma*). Nine genera and many species and subspecies occur in the United States. *Gastrocopta* and *Vertigo* are both genera that contain numerous species; the other seven genera have few species.

Key to Genera

1a		Shell aperture without "teeth" (lamellae, folds, or plicae)	**2**
1b		Aperture with one or more "teeth" (see Fig. 9.49)	**5**
2a	**(1)**	Aperture lip thickened and noticeably reflected (Fig. 9.51)	***Pupoides*** (in part)
2b		Aperture lip not thickened, and although it may be expanded, it is not reflected	**3**
3a	**(2)**	Adult shell dome-shaped (Fig. 9.52)	***Pupisoma***
3b		Adult shell cylindrical	**4**

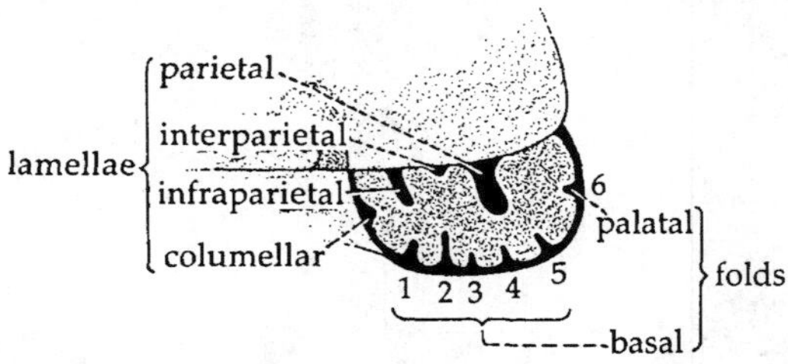

47

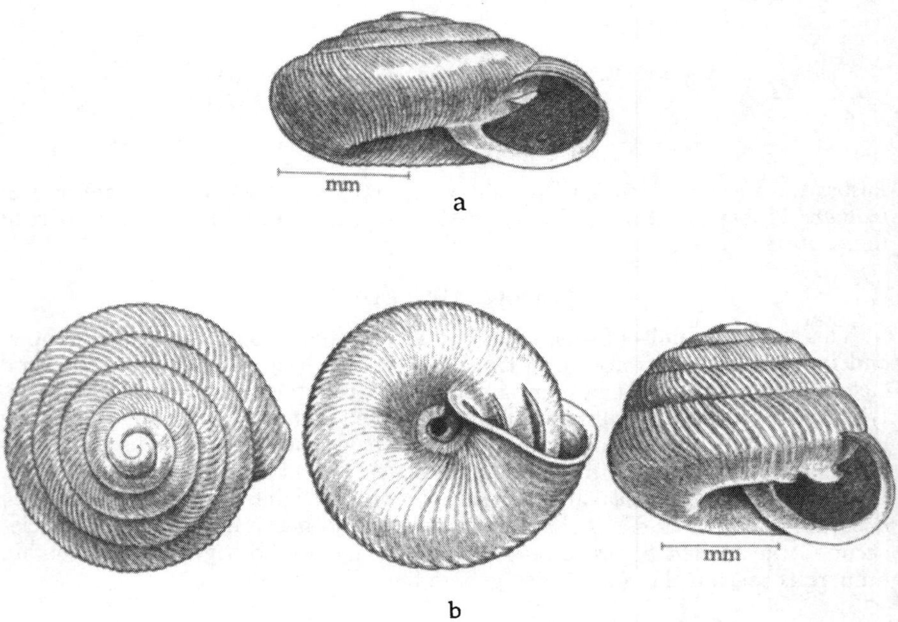

48

Fig. 9.47. Terminology of strobilopsid teeth. Only the parietal lamellae are usually seen in frontal apertural view of mature shells. [After Pilsbry (1948).] **Fig. 9.48.** (*a*) *Strobilops hubbardi* Brown; (*b*) *S. labyrinthica* (Say).

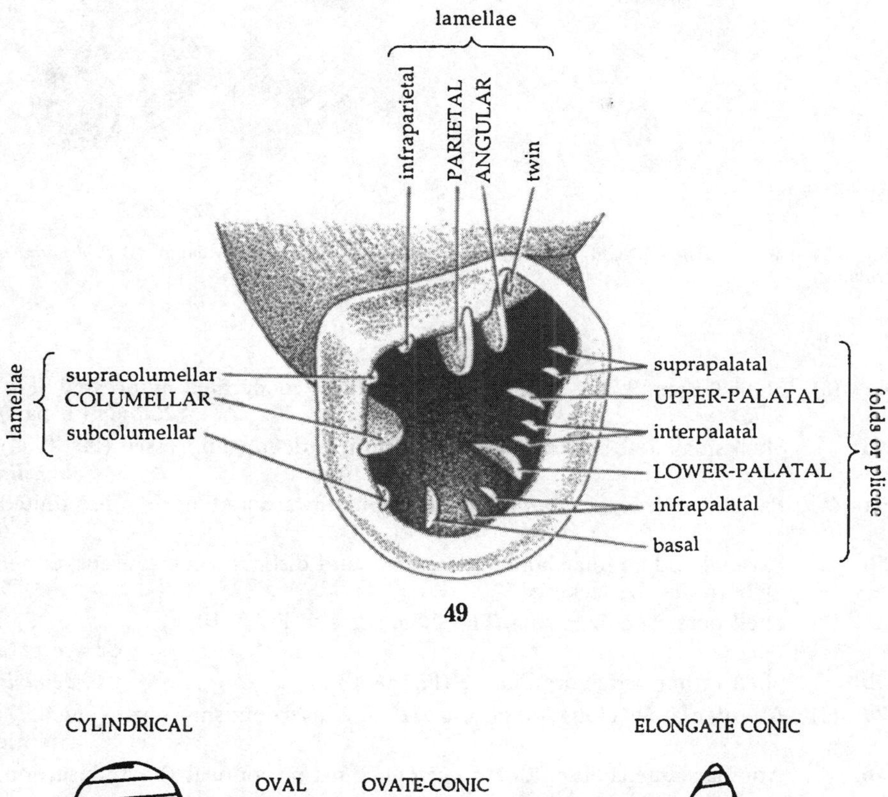

49

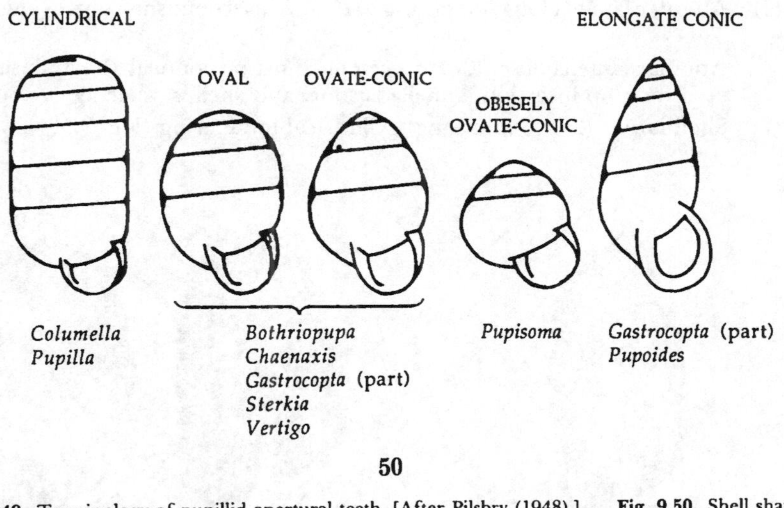

50

Fig. 9.49. Terminology of pupillid apertural teeth. [After Pilsbry (1948).] **Fig. 9.50.** Shell shapes in the Pupillidae.

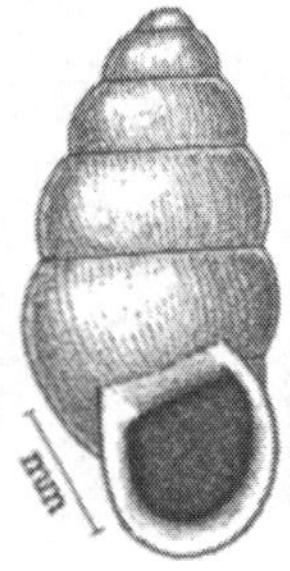

51

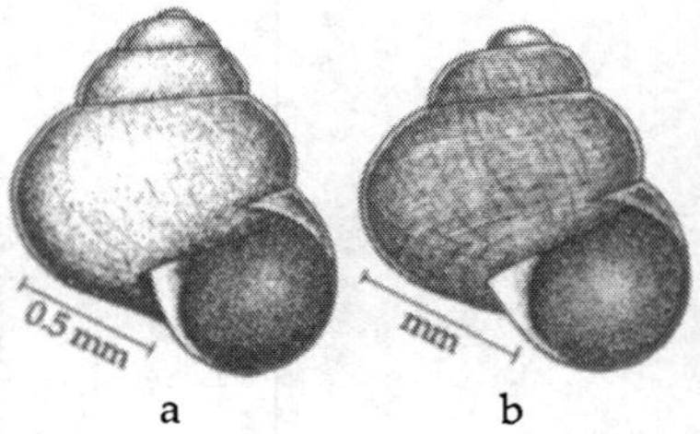

52

Fig. 9.51. *Pupoides modicus* (Gould).

Fig. 9.52. *Pupisoma:* (*a*) *P. macneilli* (Clapp); (*b*) *P. dioscoricola* (Adams).

4a (3) Shell larger, 3.2–4 mm in length; sutures moderately impressed (Fig. 9.53) . . . . ***Pupilla*** (in part)

4b Shell smaller, 1.5–3 mm in length; sutures deeply impressed (Fig. 9.54) ***Columella***

5a (1) Parietal and angular lamellae converging inward and more or less united **6**

5b Parietal and angular lamellae separate and distinct when present; either or both may be lacking **7**

6a (5) Shell perforate or rimate (Fig. 9.55; see also Fig. 9.10*b,c*) . . . . ***Gastrocopta***

6b Shell rather widely umbilicate (Fig. 9.56) . . . . ***Chaenaxis***

7a (5) Angular lamella long, emerging to the lip insertion; shell ovate (Fig. 9.57) ***Sterkia***

7b Angular lamella absent, or if present, short or not near the lip insertion, or if near lip insertion, lamella is short and shell is tapering. **8**

8a (7) Shell larger (2.8–6 mm long) cylindrical to tapering (Fig. 9.50) **9**

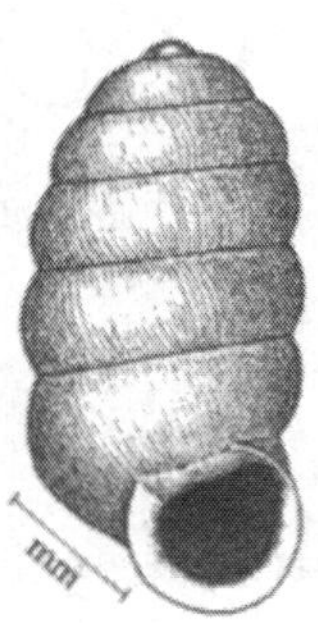

53

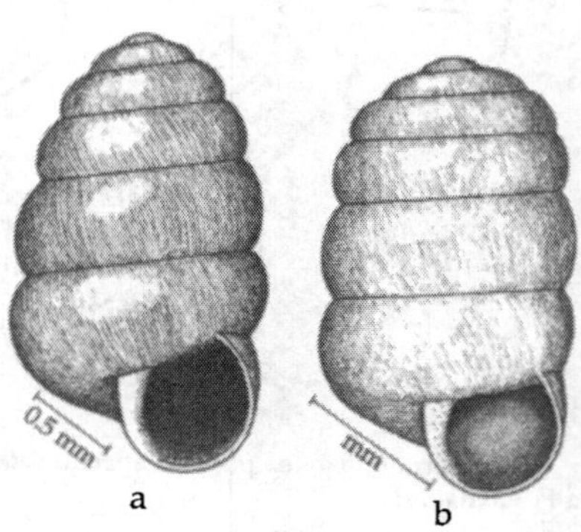

54

Fig. 9.53. *Pupilla muscorum* (Linnaeus).

Fig. 9.54. *Columella:* (*a*) *C. edentula* (Linnaeus); (*b*) *C. alticola* (Ingersoll).

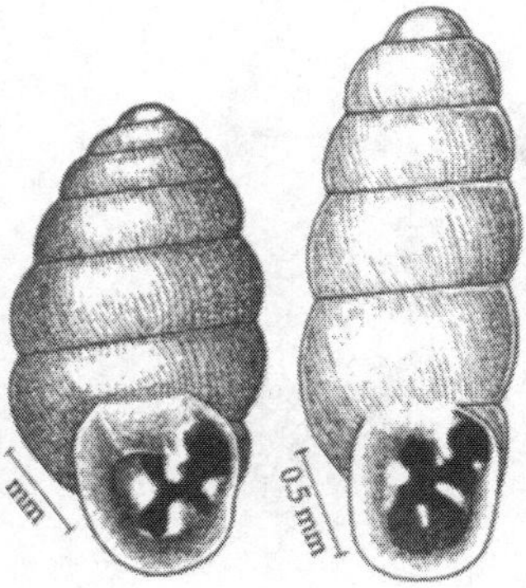

a b
55

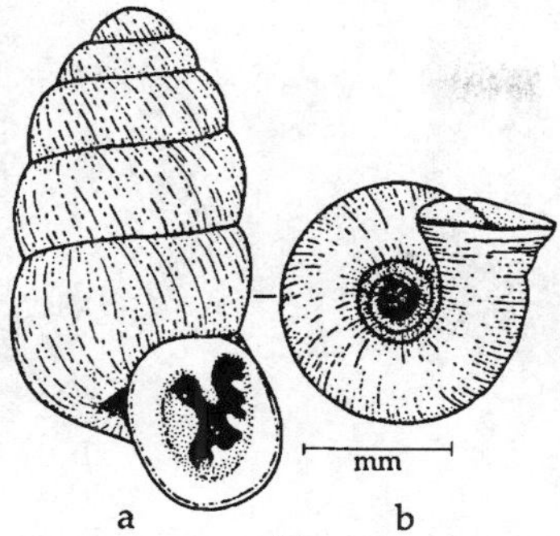

a b
56

57

Fig. 9.55. *Gastrocopta:* (*a*) *G. armifera* (Say); (*b*) *G. pellucida hordeacella* (Pilsbry). **Fig. 9.56.** *Chaenaxis tuba* (Pilsbry and Ferriss): (*a*) apertural view; (*b*) umbilical view. **Fig. 9.57.** *Sterkia eyriese rhoadsi* (Pilsbry).

8b		Shell smaller (1.5–3 mm long), oval to ovate–conic (Fig. 9.50) **10**
9a	(8)	Shell cylindrical; without angular lamellae (Fig. 9.58) ***Pupilla*** (in part)
9b		Shell tapering; an angular lamella in the lip insertion is the only tooth (Fig. 9.59). . . . . ***Pupoides*** (in part)
10a	(8)	Shell surface pitted or granulose (Fig. 9.60*a*). . . . . ***Bothriopupa***
10b		Shell surface smooth or striate (Fig. 9.60*b,c*). . . . . ***Vertigo***

Family CERIIDAE (Cerionidae)

This is a tropical and subtropical family restricted to the western hemisphere and represented by only one species in the United States, *Cerion incanum* (Binney) of the Florida Keys. The shell of *C. incanum* (Fig. 9.61) is medium to large, 20–40 mm in length, solid, pupilliform, and its color is white, sometimes with a faint bluish tint, or pale gray, sometimes marked with brown stripes. The shell aperture has a reflected lip and usually has a parietal and a columellar lamella. *Cerion incanum* is confined to the sea coast above the high-tide level.

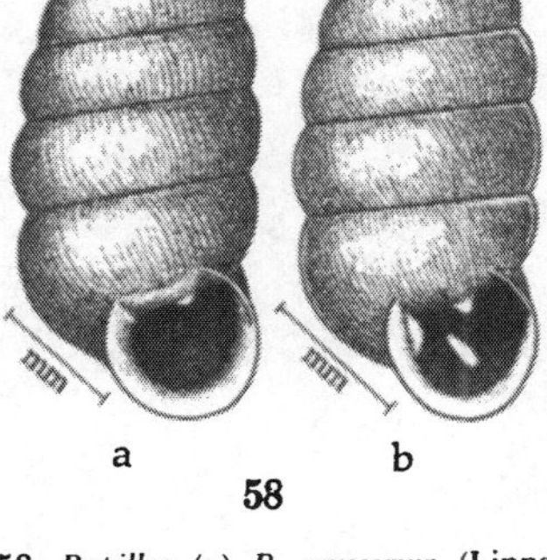

a b
58

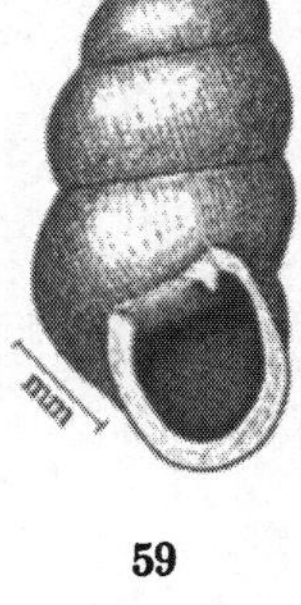

59

Fig. 9.58. *Pupilla:* (*a*) *P. muscorum* (Linnaeus); (*b*) *P. blandi* Morse. **Fig. 9.59.** *Pupoides albilabris* (Adams).

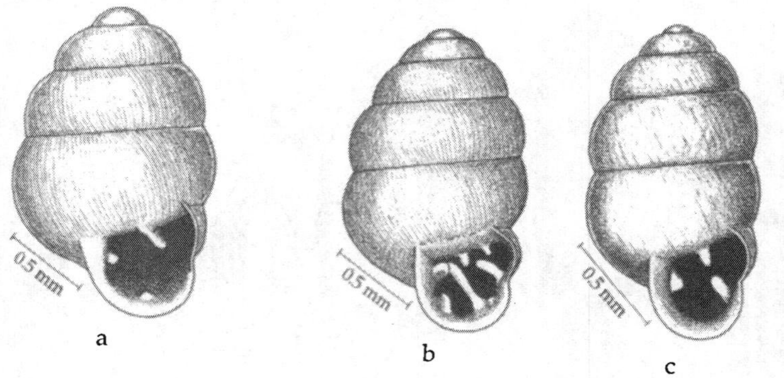

Fig. 9.60. (*a*) *Bothriopupa variolosa* (Gould); (*b*) *Vertigo milium* (Gould); (*c*) *V. parvula* Sterki.

Family SUCCINEIDAE

The shells of snails belonging to this primitive family of land pulmonates are very thin, with unusually large apertures, and are often amber-colored (giving them the common name "amber snails"). Succineids are most commonly found close to bodies of water, along streambanks, at the edges of ponds or lakes, and in or near marshes. About 30 species have been recorded in the eastern United States and another 10 from the west. Identification is difficult and future much-needed revisions will rest mainly on anatomical characters, such as genitalia (Fig. 9.62). Three genera are recognized in the United States: *Succinea*, *Oxyloma*, and *Catinella*.

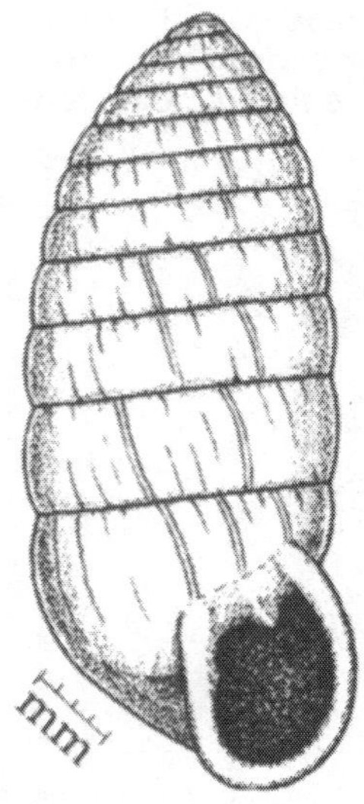

Fig. 9.61. *Cerion incanum* (Binney).

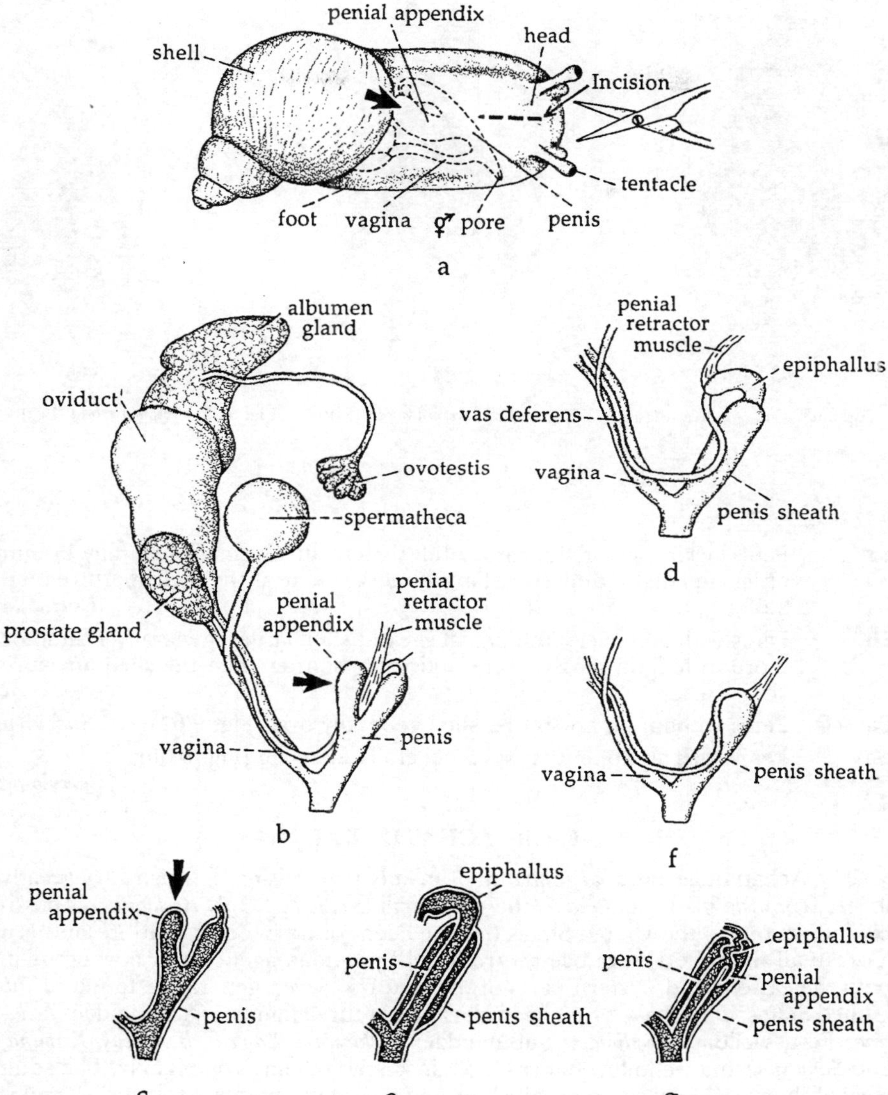

Fig. 9.62. Generic characters of terminal male genitalia in succineid snails. (*a–c*) *Catinella avara* (Say): (*a*) dorsal view of animal showing natural position of terminal genitalia and place for incision to open the dorsal head-foot to observe the lower reproductive system; (*b*) complete reproductive system; (*c*) longitudinal section through penis and penial appendix; arrows point to penial appendix. (*d,e*) *Succinea ovalis* (Say): (*d*) lower genitalia; (*e*) longitudinal section through penis, penial sheath, and epiphallus. (*f,g*) *Oxyloma retusa* (Lea): (*f*) lower genitalia; (*g*) longitudinal section through penis, penial appendix, penial sheath, and epiphallus.

63

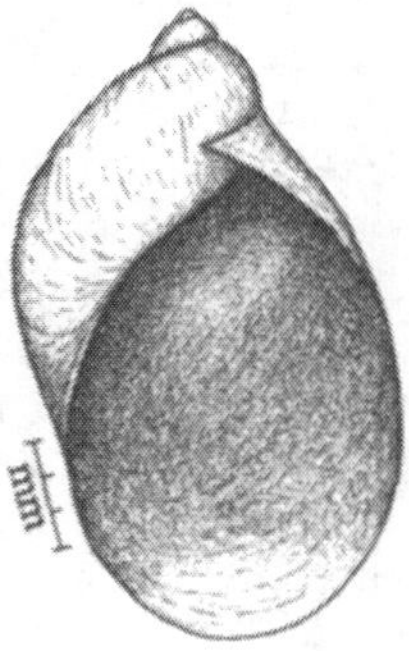

64

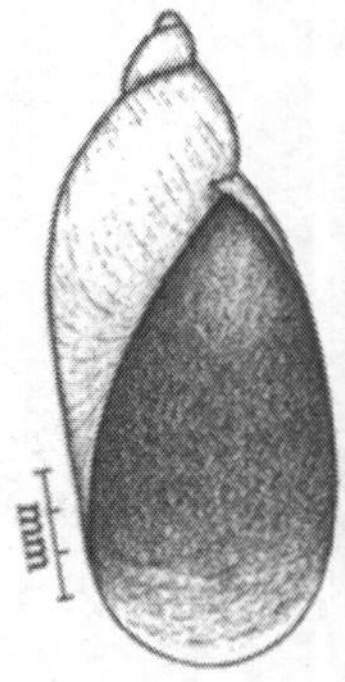

65

Fig. 9.63. *Catinella avara* (Say). **Fig. 9.64.** *Succinea ovalis* (Say). **Fig. 9.65.** *Oxyloma retusa* (Lea).

Key to Genera

1a Penis lacking a penial sheath; adult shell relatively small, generally 11 mm or less in length, dull, spire long, almost as long as the shell aperture (Fig. 9.63) . . . . **Catinella**

1b Penis with an enveloping penial sheath; shell larger, generally 12 mm or more in length, glossy, spire noticeably shorter than the shell aperture **2**

2a (1) Penis without an appendix; shell generally oval (Fig. 9.64). . . ***Succinea***

2b Penis with an appendix; shell generally elongate (Fig. 9.65) . . . . ***Oxyloma***

Family ACHATINIDAE

The Achatinidae are a large group of mainly tropical snails, which vary greatly in size from the giant *Achatina* to the very small *Cecilioides.* Only *Beckianum* is possibly indigenous to the United States (having been found in beach drift in southern Texas); all the others have been introduced by human agency and now occur in both the eastern and western parts of the country. Seven genera are found in the United States. They have often been classified in three families (Achatinidae: *Achatina;* Ferussaciidae: *Cecilioides;* Subulinidae: *Beckianum, Lamellaxis, Opeas, Rumina,* and *Subulina*), but we follow Baker (1963) in keeping them as one family. The adult shell of the various species is usually long and thin, without internal lamellae, and is either perforate or imperforate. The columella may be either straight or truncate.

Key to Genera

1a Shell very large, 50–200 mm in height (Fig. 9.66) . . . . ***Achatina***

1b Shell small to large, less than 45 mm high **2**

2a (1) Shell medium to large (25–45 mm in length); spire partly broken off in adults (Fig. 9.67) . . . . ***Rumina***

2b Shell medium to small (less than 20 mm in length); spire not normally broken off **3**

3a (2) Base of columella truncate (Fig. 9.11*b*); shell imperforate **4**

3b Base of columella straight, not truncate (Fig. 9.11*a*); shell perforate **5**

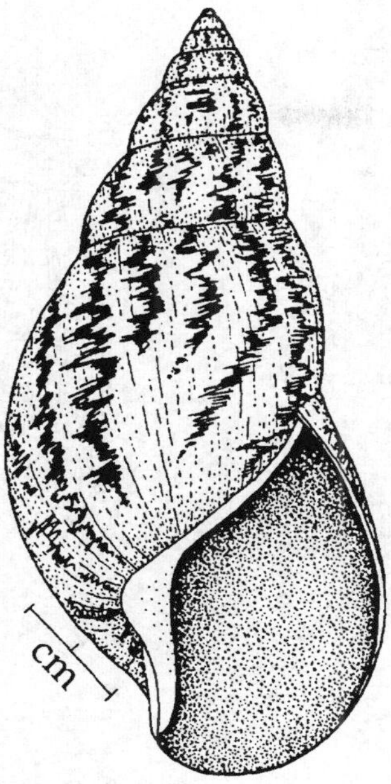

66

Fig. 9.66. *Achatina fulica* Bowdich.

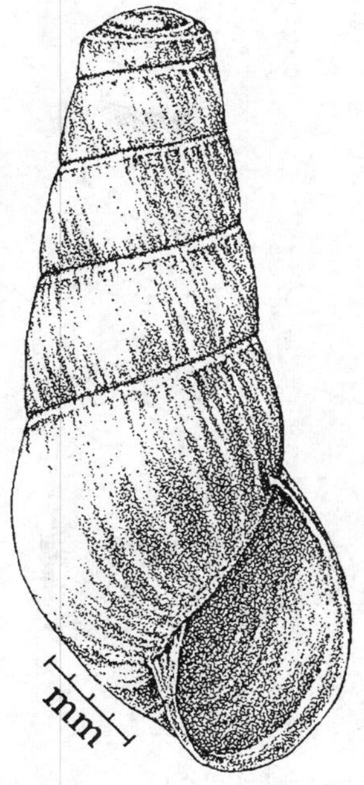

67

Fig. 9.67. *Rumina decollata* (Linnaeus).

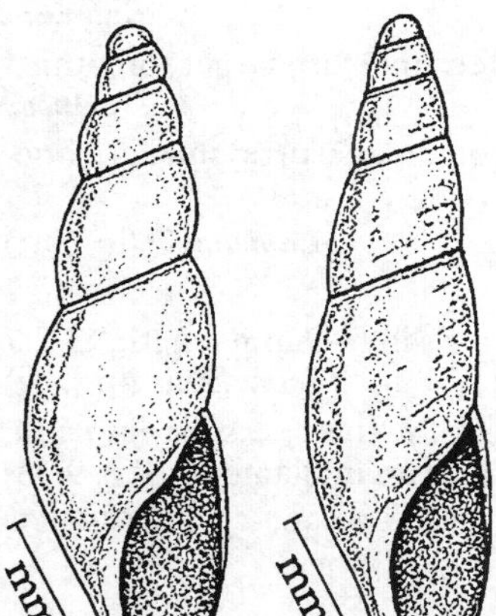

a b

68

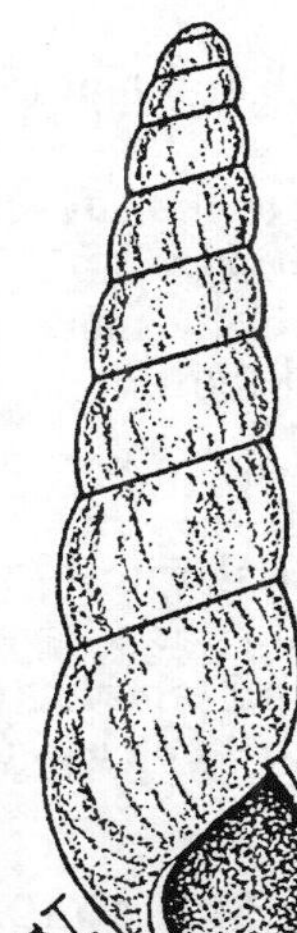

69

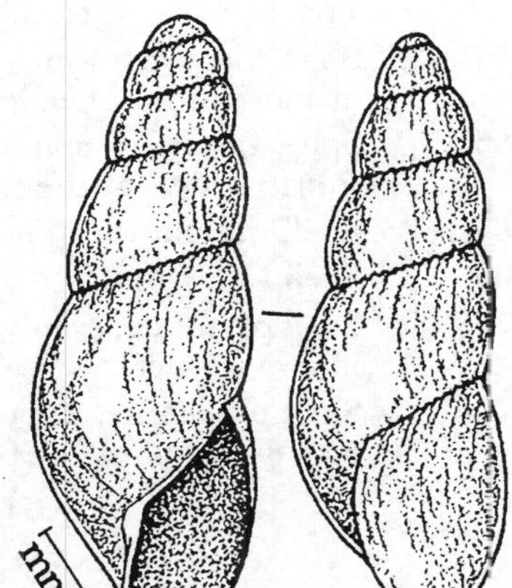

a b

70

Fig. 9.68. *Cecilioides:* (*a*) *C. aperta* (Swainson); (*b*) *C. acicula* (Müller). **Fig. 9.69.** *Subulina octona* (Bruguière). **Fig. 9.70.** *Opeas pyrgula* Schmacker and Boettger: (*a*) apertural view; (*b*) right-side view.

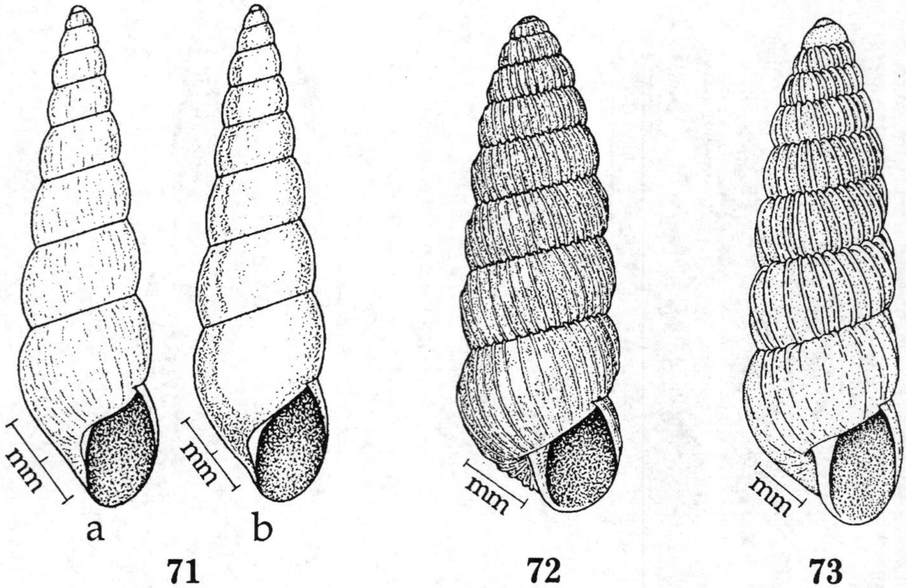

Fig. 9.71. *Lamellaxis:* (*a*) *L. gracilis* (Hutton); (*b*) *L. mauritianus* (Pfeiffer). **Fig. 9.72.** *Lamellaxis micra* (Orbigny). **Fig. 9.73.** *Beckianum beckianum* (Pfeiffer).

4a (3) Shell small, adults (with 5–6 whorls) less than 6 mm in length; very glossy; shell surface smooth or only very weakly striate (Fig. 9.68) . . . . ***Cecilioides***

4b Shell larger, adults more than 6 mm in length; moderately glossy; shell surface striate (Fig. 9.69) . . . . ***Subulina***

5a (3) Spire somewhat wider and terminally blunter; aperture larger, one-third or more of the shell length (Fig. 9.70) . . . . ***Opeas***

5b Spire narrower and terminally more pointed; aperture smaller, one-fourth or less of the shell length 6

6a (5) Shell without ribs (Fig. 9.71) . . . . ***Lamellaxis*** (in part)

6b Shell with ribs 7

7a (6) Shell perforate, evenly conical, with 7–8 whorls in 6–9 mm length, 2–2.7 mm in diameter (Fig. 9.72) . . . . ***Lamellaxis*** (in part)

7b Shell umbilicate, upper third tapering conically, the rest subcylindrical, with 8–9½ whorls in a 6.2- to 9-mm length, 3.5 mm in diameter (Fig. 9.73) ***Beckianum***

Family SPIRAXIDAE

This is a family of tropical and subtropical America and the Mediterranean region. Its species have mostly large shells, which are elongate, imperforate, and have rather flattened but rounded whorls, truncate collumellae, and fusiform or narrowly ovate–lunate apertures. Two genera occur in the United States. *Euglandina* has three species, ranging in distribution along the Gulf coast states, Georgia, South Carolina, and North Carolina. They are all active, aggressive predators of other land snails. One species, *E. rosea* (Férussac), has been widely distributed in the

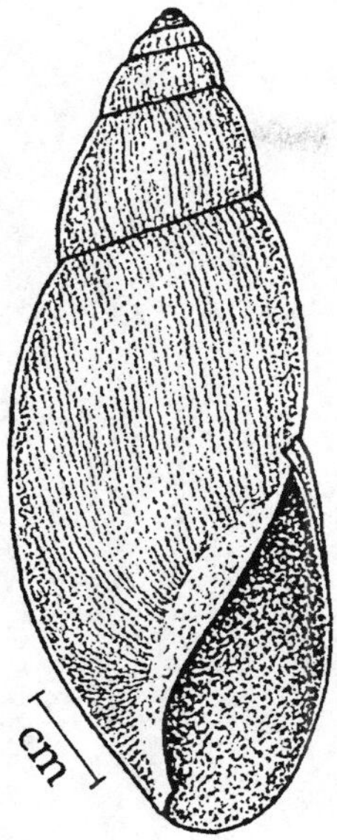

74

Fig. 9.74. *Euglandina rosea* (Férussac).

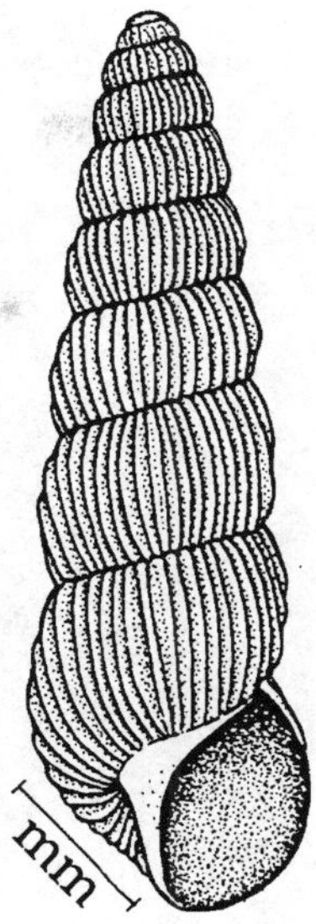

75

Fig. 9.75. *Pseudosubulina cheatumi* Pilsbry.

Pacific area, and to Thailand, by the Hawaiian Department of Agriculture in persistent and consistently futile attempts to control the introduced snail pest, *Achatina fulica* Bowdich. *Pseudosubulina cheatumi* Pilsbry, found in southern Texas, is the only species of that genus occurring in the United States.

Key to Genera

1a Shell fusiform, large, 20–80 mm high (Fig. 9.74) . . . . . . ***Euglandina***

1b Shell narrowly conical or turret-shaped, smaller, 5–7 mm high (Fig. 9.75) ***Pseudosubulina***

Family STREPTAXIDAE

This family is not native to North America, but is represented in our fauna by the introduced *Gulella bicolor* (Hutton), a species originally from South Africa. Shells of this family are heliciform, cylindrical, or sometimes pupa-shaped, the aperture lip is expanded, and the columella is always thickened and usually toothed.

Gulella bicolor (Fig. 9.76) has been introduced into the Gulf coast states and South Carolina. It has a small shell, about 7 mm in length, with 6–7 whorls, and is white, polished, and smooth (except at the sutures, where the shell is ribbed). The aperture contains four teeth: a parietal lamella, a subconically shaped lamella on the right margin, a small basal tooth, and a deep-seated columellar lamella.

Family HAPLOTREMATIDAE

This is mainly a family of North America north of Mexico, but a few members inhabit Middle America, the West Indies, and northern South America. They can

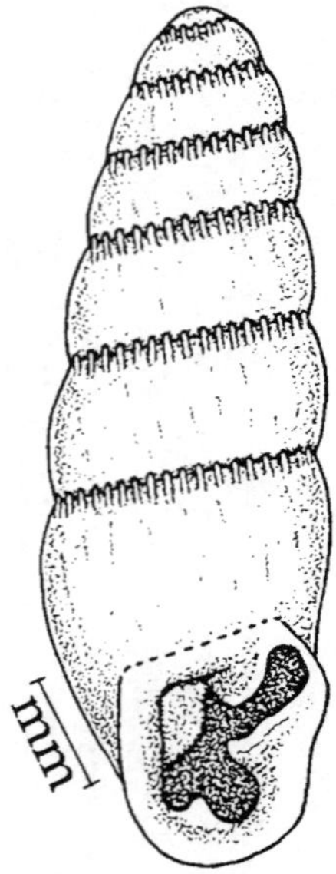

Fig. 9.76. *Gulella bicolor* (Hutton).

be carnivorous, feeding on other land snails, but also are known to eat plants. Some species are oviparous and others are ovoviviparous. The family contains but one genus, *Haplotrema,* which is best represented in the western United States. In the east, there is only one species, *H. concavum* (Say), but it has a wide distribution. About 10 species occur in the west.

The shells of *Haplotrema* are very depressed (planispiral or nearly so), generally light-colored (often light greenish yellow), and are umbilicate or widely umbilicate (Fig. 9.77). The shells are generally sturdy and opaque, in contrast to the thinner and often translucent shells of the Vitrinidae.

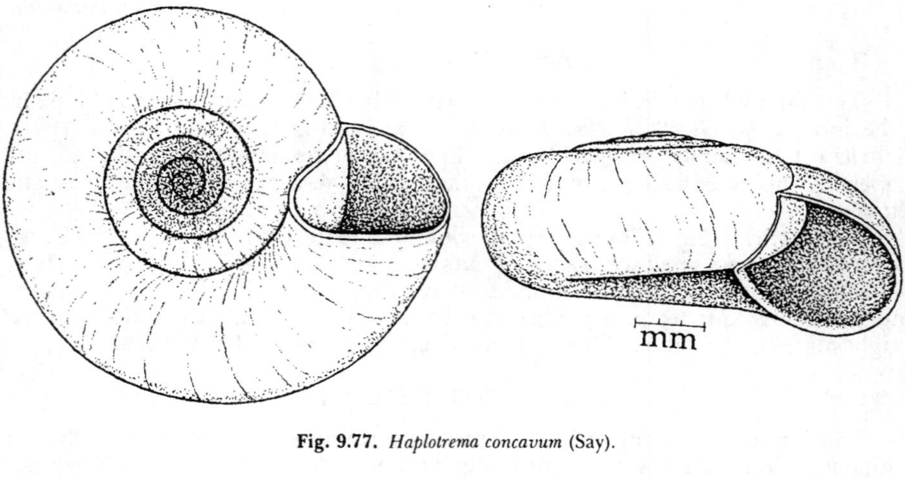

Fig. 9.77. *Haplotrema concavum* (Say).

Family UROCOPTIDAE

This is a large, mainly tropical family with relatively few species in the United States. The family is native to the western hemisphere, occurring in northern South America, Middle America, the Antilles, and, in the United States, in southern Florida, Texas, and New Mexico. The shell is very elongate, with a characteristically tapering spire which changes its taper near the apex. The apertural lip is expanded, often strongly reflected. Four genera occur in the United States: *Holospira, Metastoma, Microceramus,* and *Cochlodinella,* the latter with a single species, *C. poeyana* (Orbigny), in the United States.

Key to Genera

1a Shell imperforate to rimate **2**

1b Shell narrowly umbilicate **3**

2a (1) Shell slender, the early whorls broken off in adult shells (Fig. 9.78). . . . ***Cochlodinella***

2b Shell wider, the early whorls retained in adult shells (Fig. 9.79) . . . . . ***Microceramus***

3a (1) Last whorl distorted, sometimes making the shell appear whorled to the left; surface relatively smooth (Fig. 9.80). . . . . ***Metastoma***

3b Last whorl not distorted, shell obviously dextral; surface ribbed (Fig. 9.81) . . . . ***Holospira***

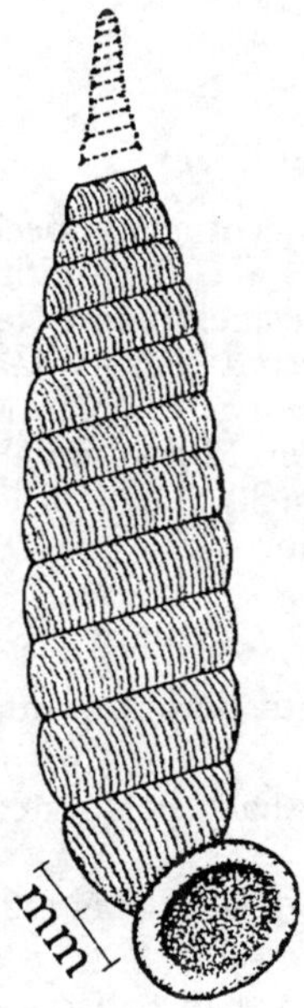

78

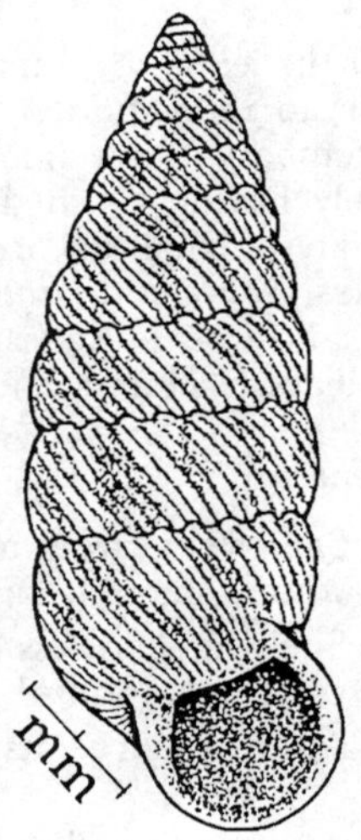

79

Fig. 9.78. *Cochlodinella poeyana* (Orbigny). **Fig. 9.79.** *Microceramus pontificus* (Gould).

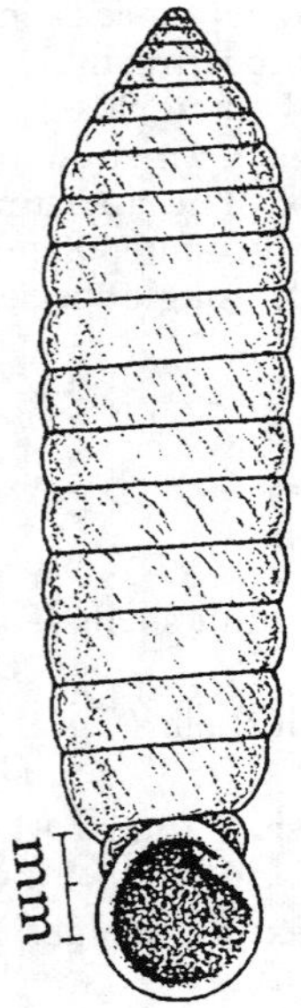

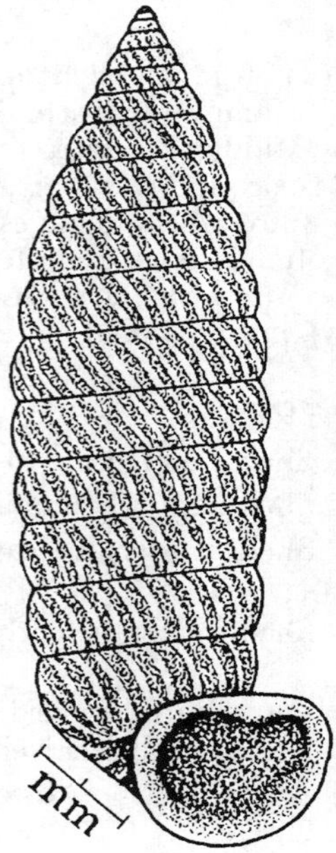

Fig. 9.80. *Metastoma roemeri* (Pfeiffer). **Fig. 9.81.** *Holospira goldfussi* (Menke).

Family ORTHALICIDAE (Bulimulidae)

The Orthalicidae are a mainly South American family of many species, a few of which spread into Middle America, the West Indies, and the southern United States. Members of the family occur in Australia, Melanesia, and New Zealand, presumably having reached these regions via Antarctica (Pilsbry 1946). In the United States, orthalicids occur in Florida, Texas, Arizona, and New Mexico, and one species, *Rabdotus dealbatus* (Say), is distributed from Illinois and Kentucky to Alabama, and west to Kansas and New Mexico. The shells of orthalicids are often large, brilliantly colored, and have capacious body whorls.

***Key to Genera*†**

1a Shell very large, more than 40 mm in length; imperforate; embryonic whorls smooth; Orthalicinae **2**

1b Shell smaller, less than 40 mm in length; usually perforate; embryonic whorls sculptured; Bulimulinae **3**

2a **(1)** Aperture large, its length nearly half the shell length (Fig. 9.82) . . . . ***Orthalicus***

2b Aperture smaller, its length considerably less than half the shell length (Fig. 9.83) . . . . ***Liguus***

3a **(1)** Transverse and spiral sculpture of embryonic whorls equally strong,

† From Breure (1979) in part.

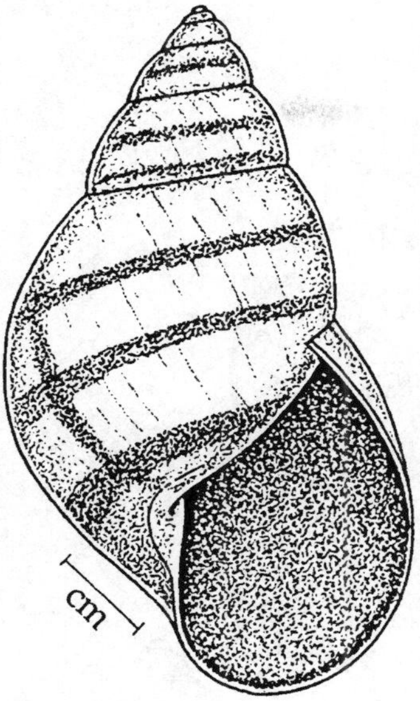

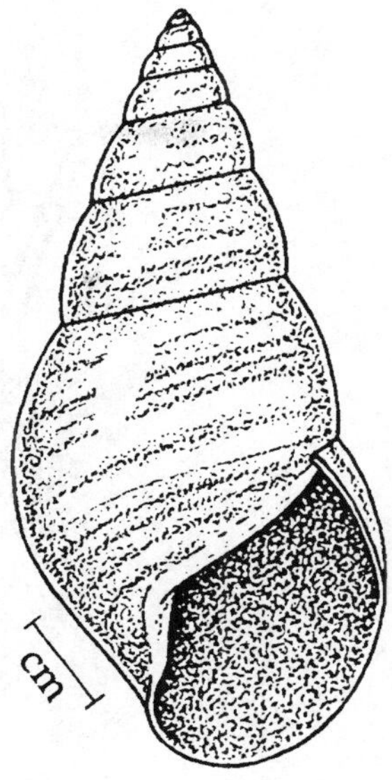

Fig. 9.82. *Orthalicus floridensis* Pilsbry. **Fig. 9.83.** *Liguus fasciatus* (Müller).

transverse sculpture consisting of straight ribs; shell uniform in color or with spiral color bands (Fig. 9.84) . . . . *Drymaeus*

3b Protoconch with only transverse sculpture, or with fine spiral sculpture between the more prominent transverse riblets **4**

4a **(3)** Transverse sculpture of embryonic whorls consisting of riblets (Fig. 9.85) **5**

4b Transverse sculpture of embryonic whorls consisting of rather weakly developed wrinkles; introduced species (Fig. 9.86) . . . . . . ***Bulimulus***

5a **(4)** Embryonic whorls with fine spiral threads in the interstices between transverse riblets . . . . ***Naesiotus***

5b Embryonic whorls lacking spiral threads between transverse riblets (Fig. 9.85) . . . . ***Rabdotus***

Family PUNCTIDAE

The endodontoid snails have recently undergone a significant revision (Solem 1976, 1983) and previously used subfamilies have been raised one level in the taxonomic hierarchy to the status of full families. Thus the Endodontidae (in the more restricted sense) are inhabitants of the Pacific islands and do not occur in North America. The six genera of the United States are placed in four families: *Paralaoma* and *Punctum* in the Punctidae, *Helicodiscus* in the Helicodiscidae, *Discus* and *Anguispira* in the Discidae, and *Radiodiscus* in the Charopidae (subfamily Rotadiscinae).

The Punctidae get their name from the minute size of their species. Adult shells of Punctidae measure less than 2 mm in diameter, rarely exceeding 2 mm, and are sculptured with delicate riblets. They are umbilicate and vary in shape from subdis-

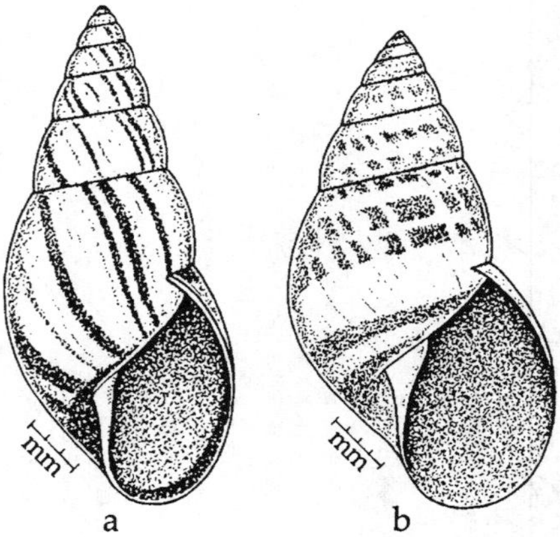

84

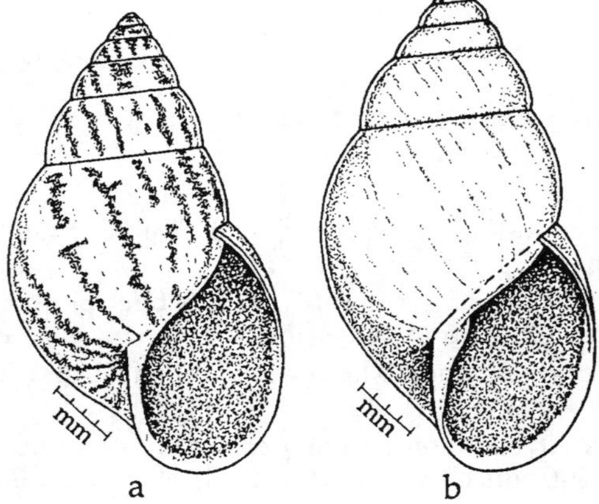

85

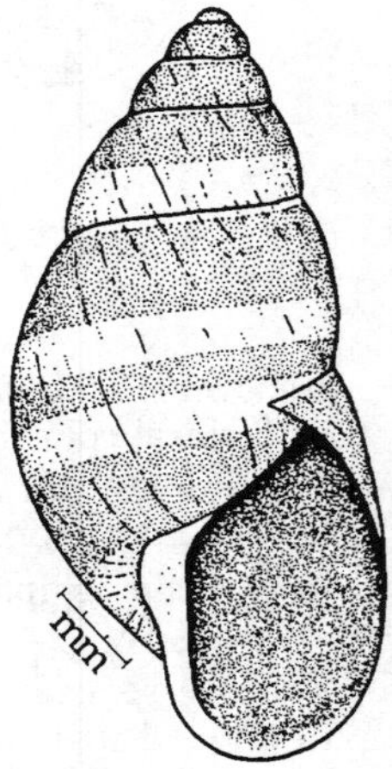

86

Fig. 9.84. *Drymaeus:* (*a*) *D. multilineatus* (Say); (*b*) *D. dormani* (Binney). **Fig. 9.85.** *Rabdotus:* (*a*) *R. dealbatus* (Say); (*b*) *R. alternatus* (Say). **Fig. 9.86.** *Bulimulus guadalupensis* (Bruguière).

coidal to low helicoid, depending on the species. As in *Radiodiscus* of the endodontoid family Charopidae, the jaw is composed of numerous separate plates, which are connected loosely by a thin membrane. In North America north of Mexico, three species of *Paraloama* and five species of *Punctum* have been recognized.

Key to Genera

1a Shell with major riblets widely spaced and prominent, usually with five or more minor riblets in each interval; embryonic whorls with fine spiral lirae (Fig. 9.87*b*) . . . . ***Paralaoma***

1b Shell with major riblets low and closely spaced, usually with less than 5 minor growth threads in each interval; embryonic whorls smooth (Fig. 9.87*a*) . . . . ***Punctum***

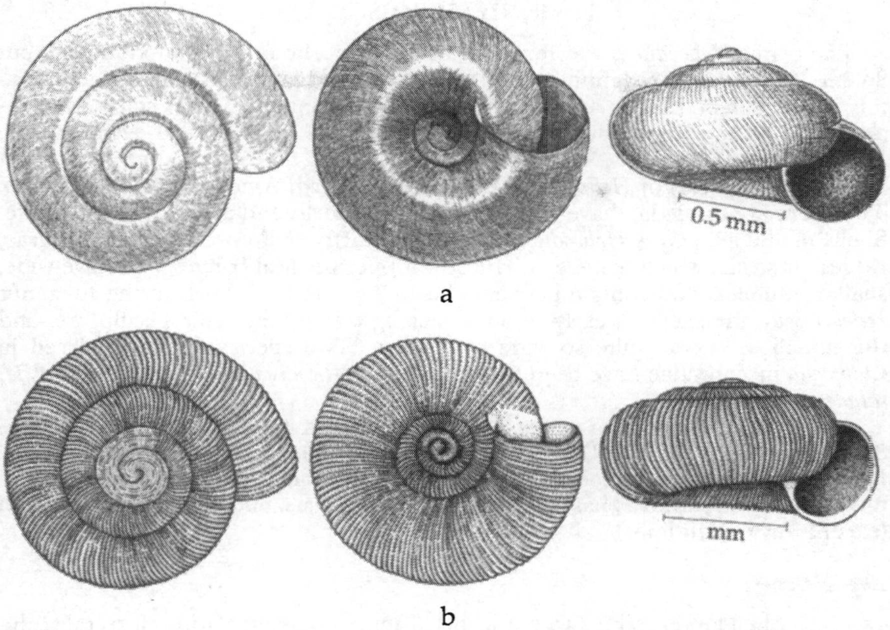

Fig. 9.87. (*a*) *Punctum minutissimum* (Lea); (*b*) *Paralaoma hannai* Roth. (From Roth, 1985, *Malacol. Rev.* 18:52, fig. 1.)

Family CHAROPIDAE

The Charopidae, as most recently defined (Solem 1983), contain a large number of taxa on islands of the Pacific Ocean, in South and Central America, and in southern Africa. Only one genus, *Radiodiscus,* is found in North America north of Mexico. Here it has two species, *R. millecostatus* Pilsbry and Ferriss of Arizona, New Mexico, and Mexico (Fig. 9.88) and *R. abietum* (H. B. Baker) of Oregon and Idaho. These are small shells, just over 2 mm (*R. millecostatus*) to nearly 7 mm (*R. abietum*) in diameter. The shells are subdiscoidal, umbilicate, sculptured with numerous close-set riblets, and have prominently striate nuclear whorls. Previously, *Radiodiscus* was included in the Endodontidae, a family now (in its restricted sense) limited in distribution to islands of the Pacific Ocean (Solem 1976).

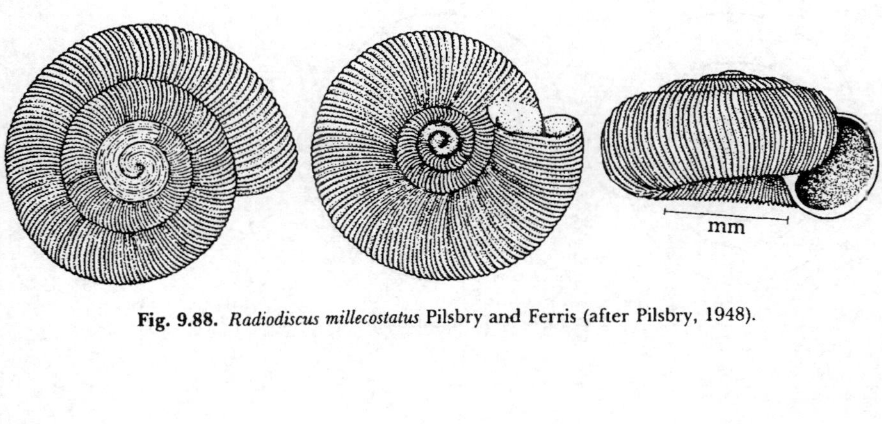

Fig. 9.88. *Radiodiscus millecostatus* Pilsbry and Ferris (after Pilsbry, 1948).

Family HELICODISCIDAE

The genus *Helicodiscus* was formerly included in the family Endodontidae, but Solem's (1976, 1983) redefinitions of the endodontoid taxa restricted the endodontids to the Pacific islands and raised the Helicodiscinae to family status, and included in it the rare, isolated, *Polygyriscus virginianus* (Burch), and extralimitally, *Stenopylis*.

About 18 species of *Helicodiscus* are found in North America north of Mexico. The shells are discoidal, have slowly increasing whorls, and are widely umbilicate. Shells of the subgenus *Helicodiscus s.s.* are distinctly sculptured with spiral lirae, ridges, or striae, which may be enhanced with periostracal fringes, and have wide, shallow umbilici, and contain barriers ("teeth") in the last whorl. In the subgenus *Hebetodiscus*, the shell is nearly smooth, lacking conspicuous spiral sculpture, and the umbilicus is not quite so wide or shallow. Two species formerly placed in *Clappiella* in Zonitidae have been transferred to *Helicodiscus: H. aldrichiana* and *H. saludensis*.

A third North American genus, *Speleodiscoides*, with its single species, *S. spirellum* Smith, has been considered to be an endodontoid somewhat related to *Helicodiscus*. It occurs in a few isolated localities in central California. The shell is small, 7 mm in diameter with 5½ whorls, has a wide, shallow umbilicus, and is glossy and smooth (except for growth lines).

Key to Genera

1a Shell larger, adults with 5 whorls 7 mm in diameter, without spiral sculpture, with barriers in the body whorl (may be lacking in empty, worn shells); central California (Fig. 9.89) . . . . ***Speleodiscoides***

1b Shells small, adults with 4 or more whorls 5 mm or less in diameter, with or without spiral sculpture, body whorl with or without internal barriers 2

2a (1) Aperture and last part of body whorl not detached and without prominent deflection; with or without well-developed spiral sculpture; widely distributed (Fig. 9.90). . . . . ***Helicodiscus***

2b Aperture and last part of body whorl prominently deflected and detached; without well-developed spiral striae; found at only one locality in Montgomery County, Virginia (Fig. 9.91). . . . . ***Polygyriscus***

Family DISCIDAE

This is a family with small to medium shells. The shells of some species have distinctive color markings, and some species are sculptured with rather coarse ribs.

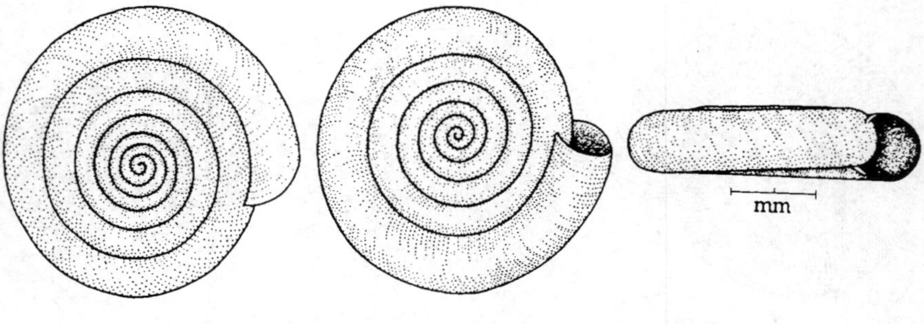

Fig. 9.89. *Speleodiscoides spirellum* Smith.

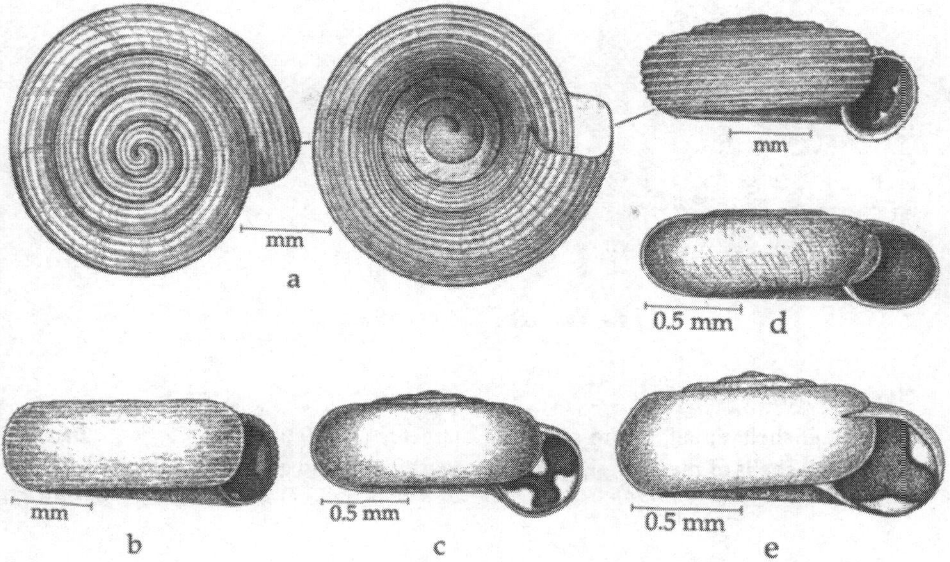

Fig. 9.90. *Helicodiscus:* (*a*) *H. parallelus* (Say); (*b*) *H. saludensis* (Morrison); (*c*) *H. tridens* (Morrison); (*d*) *H. nummus* (Vanatta); (*e*) *H. roundyi* Morrison.

Species without ribbed shells have coarse or well-developed transverse striae. All shells are umbilicate, and most species of *Discus* are widely umbilicate. None of the shells have reflected lips or apertural barriers.

The genus *Discus* is widely distributed and has about 10 naturally occurring species in North America north of Mexico. Most of these species are found only in the east, two are western in distribution. One species, *D. shimeki* (Pilsbry), is largely western, but occurs as far east as South Dakota, and one species, *D. cronkhitei* (Newcomb), occurs from the Atlantic Ocean to the Pacific. Another species, the European and North African *Discus rotundatus* (Müller), which has brown or reddish-brown color blotches, has been introduced into the eastern states and into California.

The genus *Anguispira* contains species of substantial size and their shells are marked with reddish-brown color bands or patches on a lighter horn, yellow, or tan background. The genus is mostly eastern in distribution, but several species occur in the west.

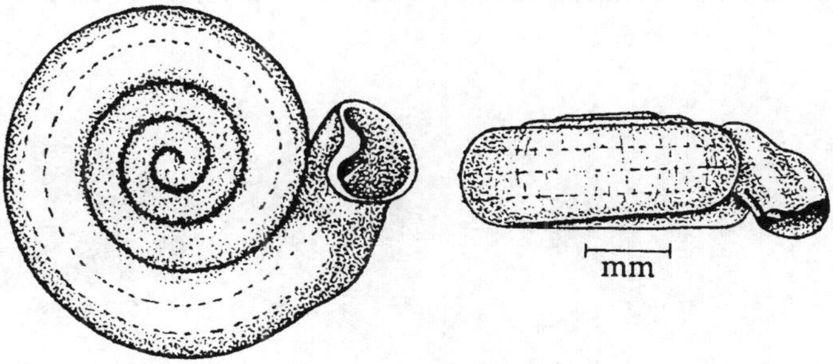

Fig. 9.91. *Polygyriscus virginianus* (Burch).

Fig. 9.92. *Discus cronkhitei* (Newcomb).

Key to Genera

1a Adult shells small, 8 mm or less in diameter (Fig. 9.92) . . . . ***Discus***

1b Adult shells of medium size, adults at least 9 mm and generally 15 mm or more in diameter (Fig. 9.93) . . . . ***Anguispira***

Family PHILOMYCIDAE

The slugs of this family are related to the Arionidae. In the United States, the philomycids are found only in the eastern part, but the family's extralimital distribution includes part of Canada, humid temperate and tropical Middle and South America, and parts of Eastern Asia. One species has been introduced into Hawaii. The philomycids differ from other geophile (stylommatophoran) slugs of the United States by the long mantle, which covers their entire back. The philomycids should not be confused with the veronicellid slugs (which have contractile tentacles and no breathing pore in the right side of the mantle) or the arionid slug *Magnipelta* (which has the breathing hole near the middle of the right side of the mantle; the philomycids have the breathing hole in the anterior fourth of the right side of the mantle). A large shell sac is located inside the mantle. Two genera were formerly recognized in North America, *Philomycus* and *Pallifera*. A third genus, *Megapallifera*, has been added recently. Species of *Megapallifera* are larger than those of *Pallifera s.s.* and have chevron-shaped pigment markings on their backs. *Megapallifera* is difficult to distinguish from *Philomycus* on external characters.

Key to Genera

1a Animal in general relatively small, usually less than 30 mm in length; mantle not covering the head region and either lacking spots or with

Fig. 9.93. *Anguispira alternata* (Say).

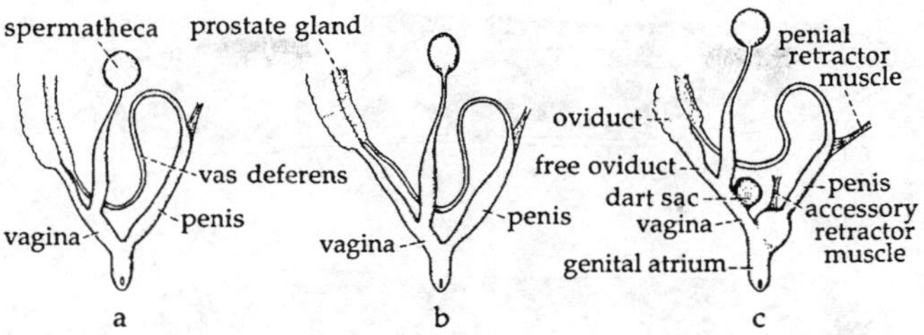

Fig. 9.94. Lower genitalia of (*a*) *Pallifera dorsalis* (W. G. Binney), (*b*) *Megapallifera wetherbyi* (W. G. Binney), and (*c*) *Philomycus carolinianus* (Bosc). The dart sac is shaded. [(*a,c*) Modified from Pilsbry (1948); (*b*) modified from G. R. Webb, 1950, *Trans. Am. Microsc. Soc.* 69, pl. 1, fig. 9a.]

poorly developed ones and a middorsal line; slime clear (Figs. 9.94*a*, 9.95) . . . . ***Pallifera***

1b Animal relatively large, more than 50 mm in length; mantle covering entire body and covered with spots, which may coalesce to form bands; slime milky. **2**

2a (1) Jaw ribbed; dart sac absent in reproductive system; a pilaster is present from the lower vagina to the upper atrium (Figs. 9.94*b*, 9.96*a*) . . . . . ***Megapallifera***

2b Jaw striate; dart sac present in reproductive system; vagina without pilaster (Fig. 9.94*c*, 9.96*b*) . . . . ***Philomycus***

Family ARIONIDAE

The Arionidae are slugs (*Anadenulus, Ariolimax, Arion, Hesperarion, Magnipelta, Prophysaon,* and *Zacoleus*) or semislugs (*Binneya* and *Hemphillia*). In the western hemisphere, the native species are restricted to western North America. Extralimitally, the arionids are native to Europe, Asia, North Africa, and South Africa. Species of the introduced European genus *Arion* are common in parts of eastern North America, and several species of *Arion* also have been introduced into California, Oregon, Washington and British Columbia.

The Arionidae are aulacopod gastropods with an oval mantle on the anterior part of the back. The mantle has a short slit on the right side which ends posteriorly in a breathing pore (pneumostome). The mantle extends forward as a free lobe under which the head can be retracted during rest or when the animal is disturbed. The shell is completely embedded in the mantle in most species, but in *Hemphillia* it is partly exposed, and in *Binneya* it is wholly external. In species in which the shell is embedded, the shell is usually simply a nonspiral oval plate, or sometimes it is reduced to only a few granules. The sides of the foot are well marked with pedal grooves and with "suprapedal" grooves as well.

The Arionidae superficially resemble slugs of the family Limacidae, but they are more closely related to the shelled family Punctidae and to the philomycid slugs,

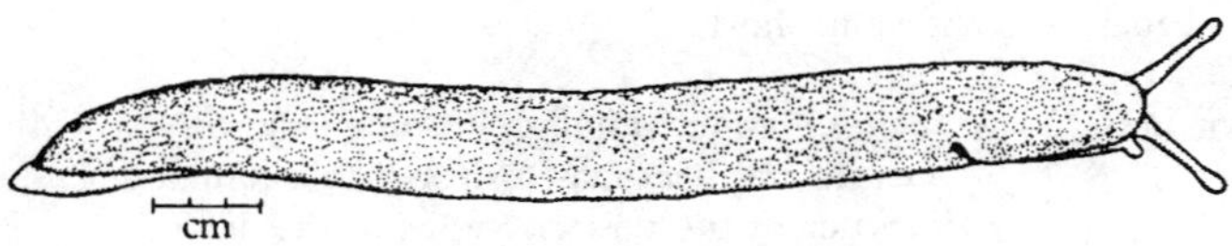

Fig. 9.95. *Pallifera hemphilli* (W. G. Binney).

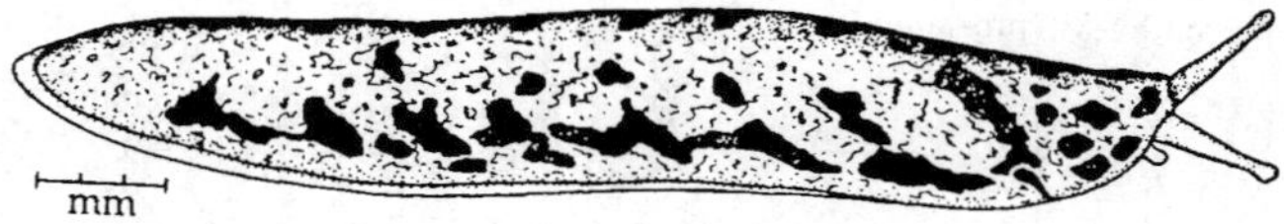

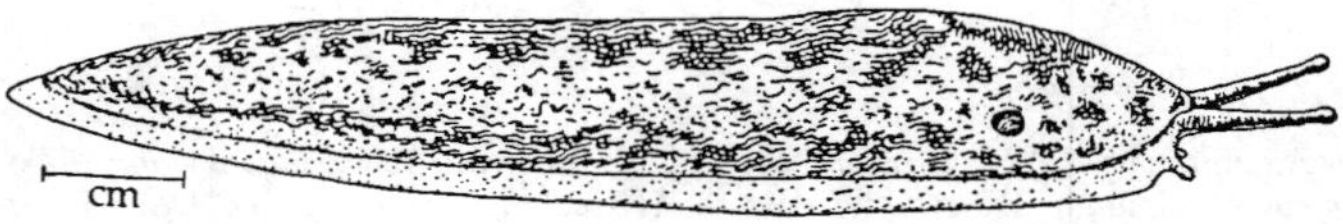

Fig. 9.96. (*a*) *Megapallifera ragsdalei* (Webb): (*b*) *Philomycus carolinianus* (Bosc). [From F. C. Baker (1939).]

although the arionids probably evolved independently of the latter, but from the same shelled ancestors. The arionids differ from the limacids in many details, such as the ridged jaw, the wide, short basal plates of the marginal teeth, the posterior separation of the free retractor muscles, the broadness of the band below the pedal groove, and the arhythmic or faint waves of progression of the foot while crawling (Table 9.1).

Arionid slugs flourish only in regions of high or moderate humidity, and in the eastern United States, they are found almost entirely in the northern states.

Table 9.1. Comparison of the Slug Families Arionidae and Limacidae

Character	Arionidae	Limacidae
Breathing pore in mantle	Anterior (*Arion, Magnipelta Prophysaon*) or posterior	Posterior
Caudal pit	Present (*Arion, Ariolimax, Hemphillia, Hesperarion*) or absent (*Anadenulus, Binneya, Magnipelta, Prophysoan, Zacoleus*)	Absent
Viscera	In foot (most genera), or in dorsal hump (*Binneya, Hemphillia*)	In foot
Sole of foot	Uniform (*Arion, Ariolimax, Hesperarion, Prophysaon, Hemphillia, Magnipelta)* or tripartite [*Anadenulus, Arion* (weakly tripartite), *Binneya, Zacoleus*]	Tripartite
Foot fringe	Wide	Narrow
Vertical lines on foot fringe	Often present	Usually absent
Muscular movements of the foot sole while crawling	Arhythmic, or with faint waves in the central area of the sole	With distinct waves
Jaw	Ridged	Smooth
Basal plates of marginal radular teeth	Wide and short	Narrow
Free retractor muscles	Remain separate, although in some taxa some retractor muscles may converge to a common insertion at the posterior edge of the diaphragm	All united into a single band posteriorly

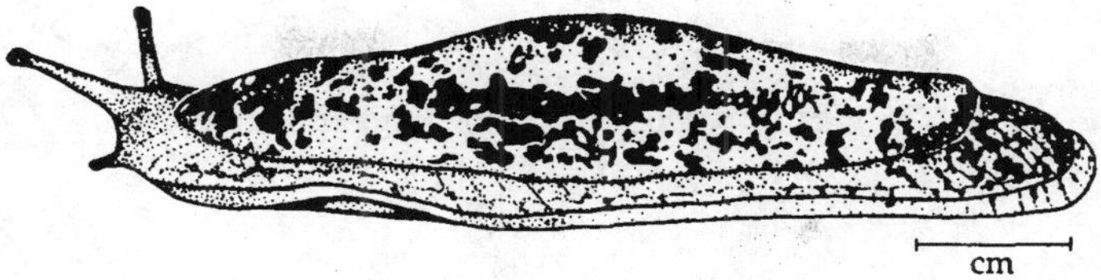

97

Fig. 9.97. *Magnipelta mycophaga* Pilsbry. (After H. A. Pilsbry, 1954, *Not. Nat.* 262, fig. 1.)

Key to Genera

1a Shell completely embedded in and covered by the mantle; viscera contained entirely in the foot, not in a dorsally elevated hump **2**

1b Shell completely or partially exposed; viscera contained in a dorsally elevated hump, behind which the foot is solid (i.e., does not contain a cavity for the viscera). **8**

2a (1) Mantle covering nearly the entire dorsum, more than two-thirds the length of the animal; breathing hole located near the middle of the right side of the mantle; found in Idaho and Montana (Fig. 9.97) . . . . . . ***Magnipelta***

2b Mantle covering only a central or anterior portion of the animal, less than one-half the length of the animal **3**

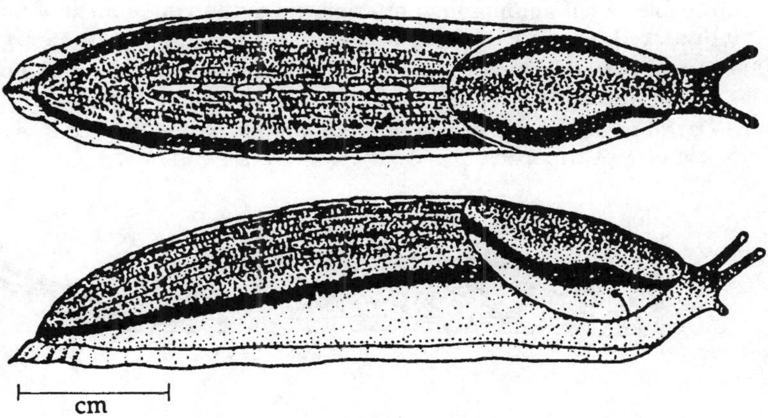

98

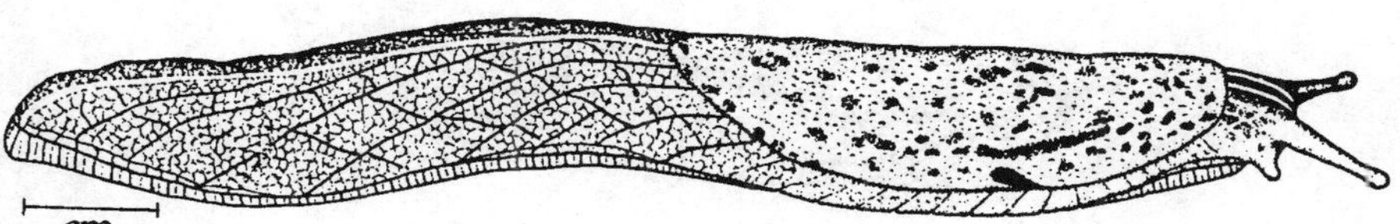

99

Fig. 9.98. *Arion fasciatus* (Nilsson). [From Chichester and Getz (1973, fig. 4a).] **Fig. 9.99.** *Prophysaon foliolatum* (Gould).

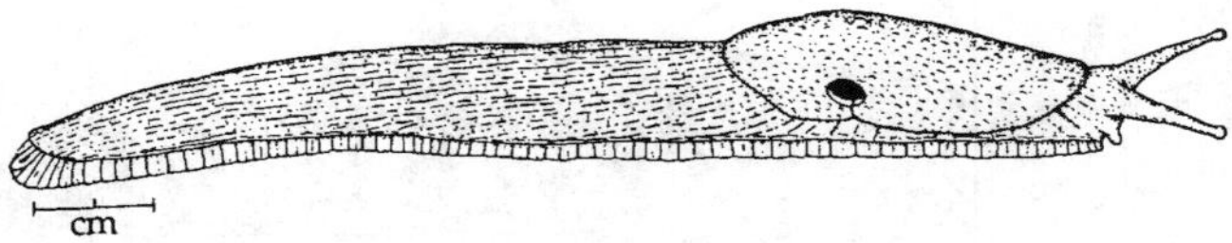

100

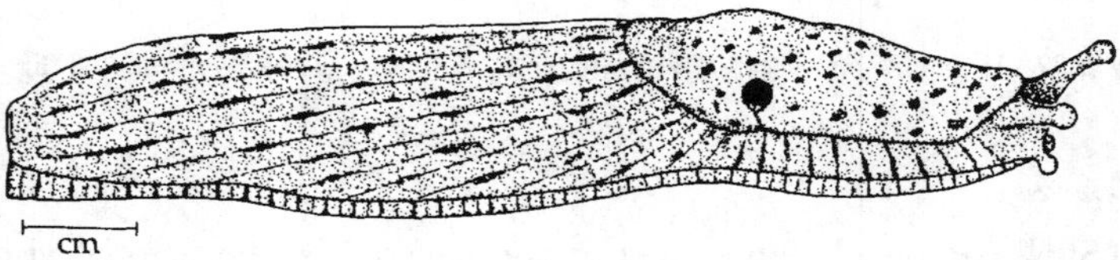

101

Fig. 9.100. *Ariolimax columbianus* (Gould). **Fig. 9.101.** *Hesperarion niger* (J. G. Cooper).

3a (2) Breathing pore in the anterior half of the right side of the mantle . . **4**

3b Breathing hole in the posterior half (or rarely in the middle) of the right side of the mantle **5**

4a (3) Dorsal posterior end of foot with a median caudal mucous gland pit; introduced throughout the United States (Fig. 9.98) . . . . ***Arion***

4b Dorsal posterior end of foot without a caudal mucous gland pit; animal capable of self-amputating posterior one-sixth to one-fourth of tail when disturbed; western United States (Fig. 9.99). . . . . ***Prophysaon***

5a (3) Dorsal posterior end of foot with a median caudal mucous gland pit; sole of foot undivided **6**

5b Dorsal posterior end of foot without a median caudal mucous gland pit; sole of foot tripartite. **7**

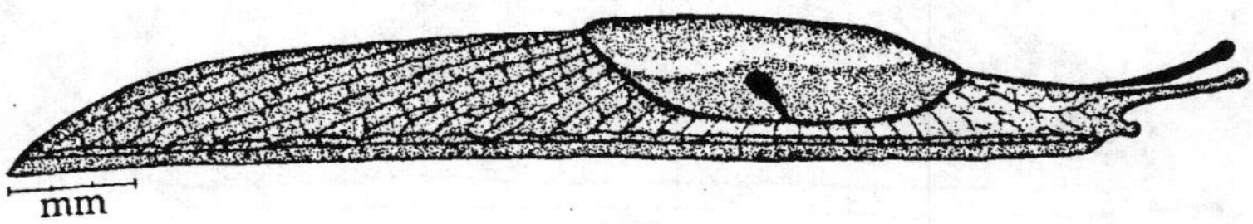

102

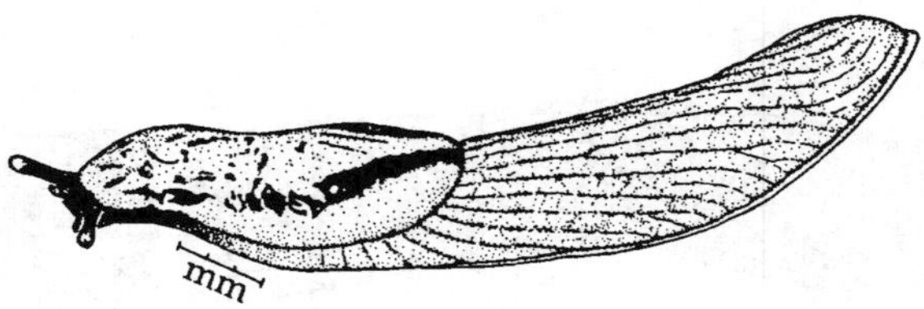

103

Fig. 9.102. *Anadenulus cockerelli* (Hemphill). [After W. G. Binney, 1892, *Bull. Mus. Comp. Zool. Harvard* 22(4), pl. 1, fig. 1); and H. A. Pilsbry and E. G. Vanatta, 1898, *Proc. Acad. Nat. Sci. Phila.* 50, pl. 9, figs. 13, 14.] **Fig. 9.103.** *Zacoleus lyrata russelli* (Webb). [After G. R. Webb, 1980, *Gastropodia* 2(1), pl. 2, fig. 1.]

104

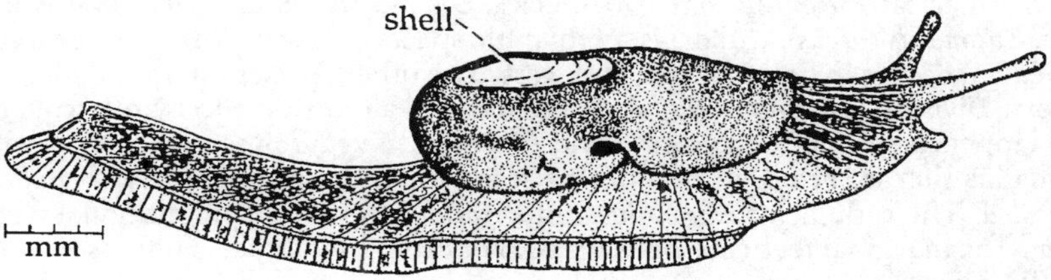

105

Fig. 9.104. *Binneya.* (*a*) shell of *B. notabilis* (Cooper), dorsal (on left) and ventral (on right) views; (*b*) preserved, somewhat contracted animal of *B. guadalupensis* Pilsbry with shell removed. [From Pilsbry (1948).] **Fig. 9.105.** *Hemphillia dromedarius* Branson.

6a (5) Animal very large, adults over 10 cm long when extended; yellow to ochre brown to olive green, with or without black blotches; west coast, southern California to Alaska (Fig. 9.100). . . . . ***Ariolimax***

6b Animal medium or small, adults less than 6 cm when extended; tan to gray to brown-black, with or without black spots; middle California (Fig. 9.101). . . . . ***Hesperarion***

7a (5) Breathing pore is submedian, near the middle of the right side of the mantle; southern California (Fig. 9.102). . . . . ***Anadenulus***

7b Breathing pore in the posterior third of the mantle; Idaho and western Montana (Fig. 9.103) . . . . ***Zacoleus***

8a (1) Shell coiled, wholly external, with 1½–2 rapidly increasing whorls, sole tripartite; southern California islands and Baja California (Fig. 9.104) ***Binneya***

8b Shell not coiled, flat, partly covered by the mantle; sole of foot undivided; animal may jump by thrashing its tail from side to side when disturbed; Pacific northwest and British Columbia (Fig. 9.105) . . . . . ***Hemphillia***

Family EUCONULIDAE (Helixarionidae, Helicarionidae)

The Euconulidae found in North America are small species, all belonging to the subfamily Euconulinae. This subfamily has a wide distribution in the world. Its members have small to minute shells, with depressed, convex, or dome-shaped spires, and are perforate or imperforate. They differ anatomically from the Vitrinidae by the possession of a digitiform appendix on the penis. Few species occur in

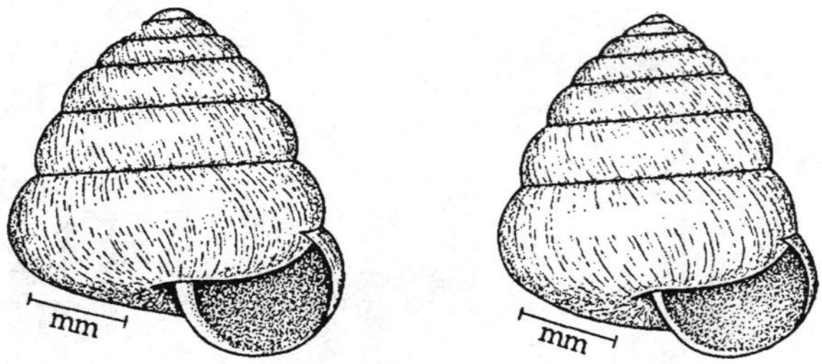

Fig. 9.106. *Euconulus:* (*a*) *E. fulvus* (Müller); (*b*) *E. chersinus* (Say).

North America: *Euconulus* has two species, *Guppya* three, and *Dryachloa* is monotypic. *Euconulus fulvus* (Müller) is a Holarctic species, and in North America (north of Mexico) occurs in both the east and west. The other species are found only in the Eastern Division. In *Euconulus,* shell lappets are absent on the mantle collar, but such lappets occur on both *Guppya* and *Dryachloa. Guppya* possesses a penis with an epiphallus that is cecoid at the end; such an appendix is absent in *Euconulus* and *Dryachloa.* The radulae of *Euconulus* and *Dryachloa* have bicuspid marginal teeth; in *Guppya* the marginal teeth are polycuspid. For other details, see Thompson and Lee (1980).

Key to Genera

1a Shell spire high, dome-shaped; embryonic whorl sculptured with spiral striations, crossed by axial riblets (Fig. 9.106) . . . . ***Euconulus***

1b Shell spire depressed; embryonic whorls sculptured with spiral pits **2**

2a (1) Neanic whorls† with axial striations; umbilical plug present (Figs. 9.107, 9.108*a*) . . . . ***Dryachloa***

2b Neanic whorls without axial striations; umbilical plug absent (Fig. 9.108*b*) ***Guppya***

Family VITRINIDAE (Zonitidae)

This family of medium to small snails is almost worldwide in its distribution, and in North America it is one of the dominant land snail families. The shells of the Vitrinidae usually have a depressed spire, are perforate or umbilicate, and are generally smooth and glossy. The apertural lip is thin and not reflected. The animal has the margin of the foot defined by pedal and suprapedal grooves, like other aulacopod snails. Several of the North American genera are rather large ones, each containing a number of species: *Glyphyalinia, Mesomphix, Paravitrea, Pristiloma,* and *Zonitoides. Ventridens* (including *Ventricallus*) has often been used as a separate genus; we follow Baker (1962b) in including it in *Zonitoides.* Three genera are probably monotypic: *Gastrodonta, Hawaiia,* and *Vitrinizonites. Gastrodonta, Mesomphix, Paravitrea, Pilsbryna,* and *Vitrinizonites* are restricted to the eastern region. *Glyphyalinia* is also found only in the east, except for *G. indentata paucilirata,* which extends into Arizona, New Mexico, and Utah. *Hawaiia, Nesovitrea, Striatura, Vitrina,* and

† In land snails, neanic whorls are those formed by a young snail from birth or hatching up to completion of spiral growth of the shell.

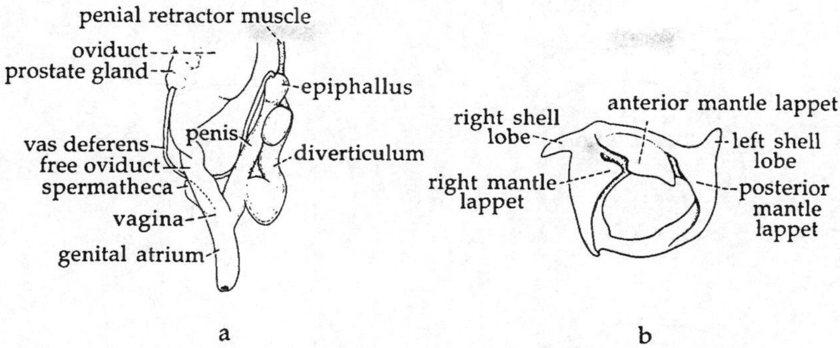

Fig. 9.107. (*a*) Lower genitalia of *Dryachola dauca* Thompson and Lee; (*b*) mantle collar of *D. dauca*. (From Thompson and Lee, 1980).

Zonitoides are found in both eastern and western North America. *Pristiloma* occurs only in the west. Members of the genus *Oxychilus,* usually found in gardens and in and around greenhouses and cellars, have been introduced from Europe into both the eastern and western United States. *Vitrea contracta* (Westerlund) has been introduced into the western United States.

Key to Genera

1a Shell with few (3 or less) rapidly enlarging whorls, usually imperforate, very thin; aperture unusually large, larger than the rest of the shell when seen in side view (Fig. 9.109*a*) **2**

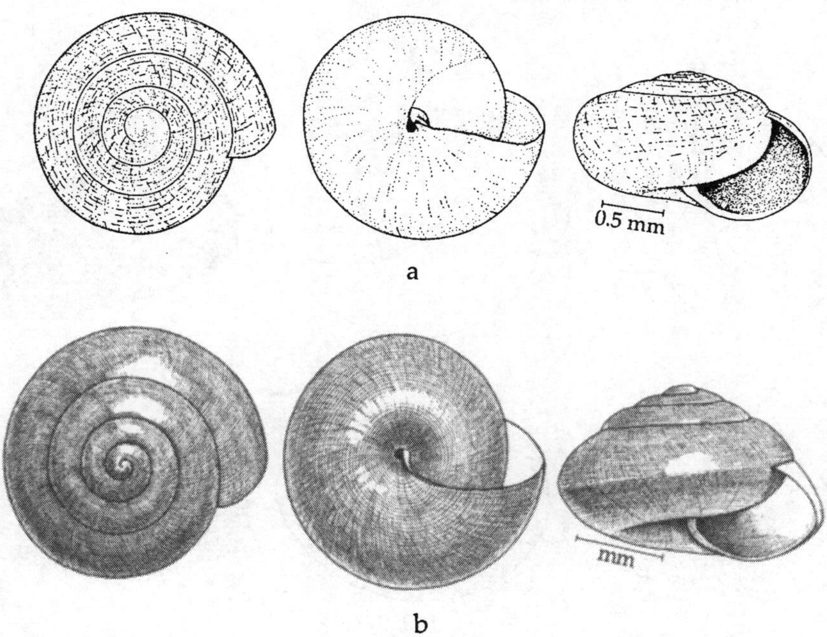

Fig. 9.108. (*a*) *Dryachloa dauca* Thompson & Lee; (*b*) *Guppya gundlachi* (Pfeiffer).

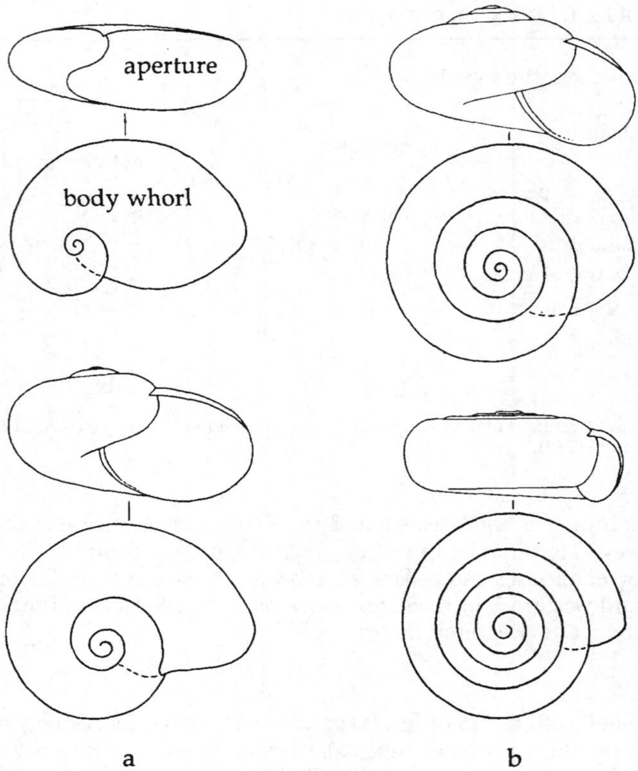

Fig. 9.109. Shells with (*a*) very wide body whorls and apertures; (*b*) narrower body whorls and apertures. Dashed lines mark end of the body whorl.

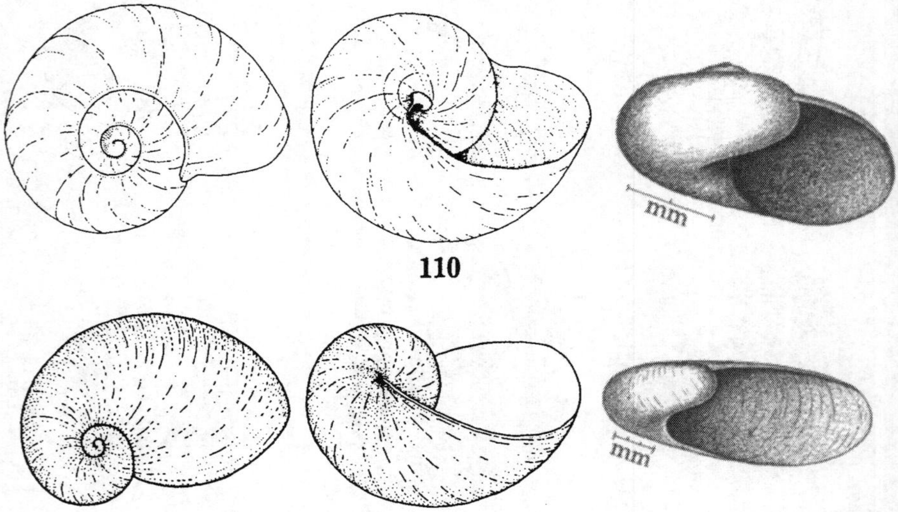

Fig. 9.110. *Vitrina limpida* Gould. **Fig. 9.111.** *Vitrinizonites latissimus* (Lewis).

1b Shell with 3 or more whorls which increase more slowly in size; shell thicker; aperture smaller than rest of shell when seen in side view (Fig. 9.109*b*) **3**

2a (1) Shell small, 6 mm or less in diameter, colorless or with a pale greenish or yellowish tint (Fig. 9.110). . . . . *Vitrina*

2b Shell medium, 16 mm or more in diameter; dark olive-green or dark tannish brown (Fig. 9.111) . . . . *Vitrinizonites*

3a (1) Shell sculptured with ribs or riblets **4**

3b Shell without ribs or riblets **5**

4a (3) Adult shell small, more than 6 mm in diameter, upper surface sculptured with ribs, base smooth, spiral striae absent (Fig. 9.112) . . . . . *Gastrodonta*

4b Adult shell small to minute, less than 3.5 mm in diameter, upper and lower surfaces sculptured with distinct or indistinct riblets and fine spiral striae (Fig. 9.113) . . . . *Striatura* (in part)

5a (3) Whorls increase regularly in size, the last very much wider (Fig. 9.114*a*); without teeth in the shell aperture or body whorl **6**

5b Whorls narrow, tightly coiled, increasing very slowly in size (Fig. 9.114*b*); with or without teeth in the shell aperture and body whorl **19**

6a (5) Adult shell 16 mm or more in diameter. **7**

6b Adult shell smaller, 15 mm or less in diameter **8**

7a (6) Shell opaque, brown or dark olive in color, smooth or striate (Fig. 9.115) . . . . *Mesomphix*

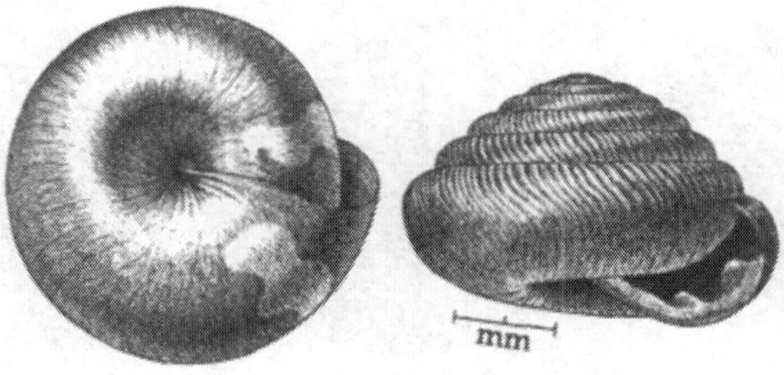

112

113

Fig. 9.112. *Gastrodonta interna* (Say). **Fig. 9.113.** *Striatura exigua* (Stimpson).

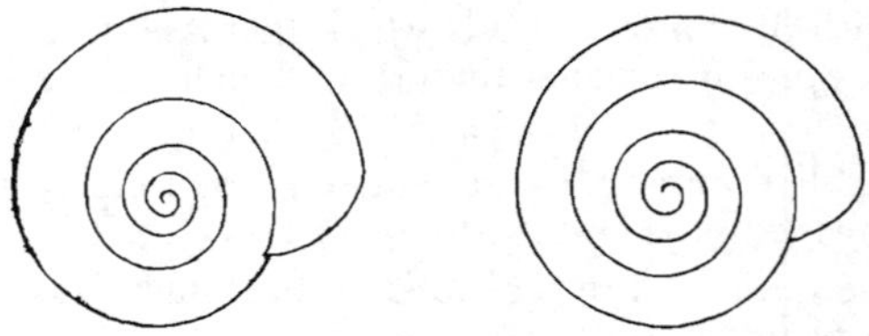

a b

Fig. 9.114. (*a*) Whorls increasing regularly in size, the last much wider; (*b*) whorls narrow, tightly coiled, increasing very slowly in size.

7b Shell translucent, tan or yellowish in color, smooth (Fig. 9.116). . . . ***Oxychilus*** (in part)

8a **(6)** Adult shell 6 mm or more in diameter, without radiating indented lines (Figs. 9.117, 9.118*a*) . . . . ***Oxychilus*** (in part)

8b Adult shell usually less than 6 mm in diameter, or if larger, with radiating indented lines (Fig. 9.118*b*) **9**

9a **(8)** Shell without radiating indented lines in addition to its ordinary growth lines or wrinkles [these genera are difficult to separate by shell characters] **10**

9b Shell with radiating indented lines in addition to its ordinary growth lines or wrinkles **11**

10a **(9)** Sole of foot tripartite; penis lacking a papilla; jaw smooth; throughout United States (Fig. 9.119). . . . . ***Nesovitrea*** (in part)

Nesovitrea and ***Glyphyalinia*** were formerly considered North American subgenera of the European ***Retinella,*** but more recently have been elevated to full genera.

115

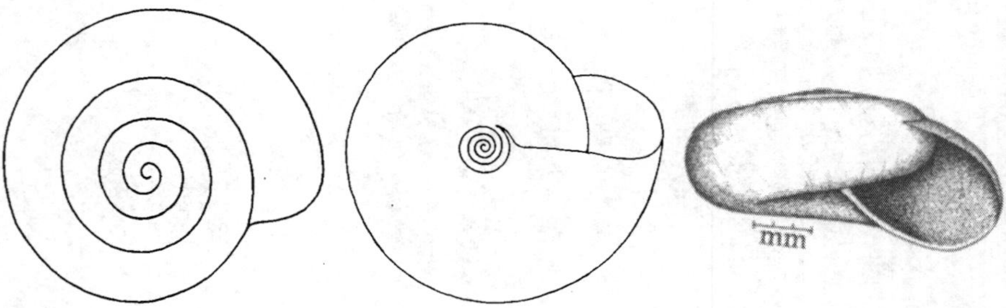

116

Fig. 9.115. *Mesomphix cupreus politus* (Pilsbry). **Fig. 116.** *Oxychilus draparnaudi* (Beck).

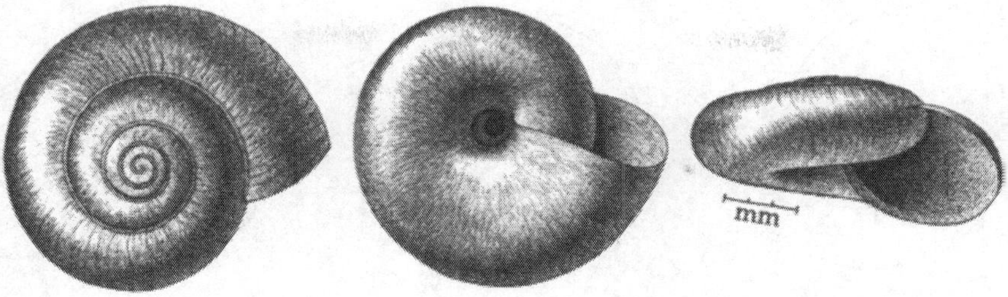

117

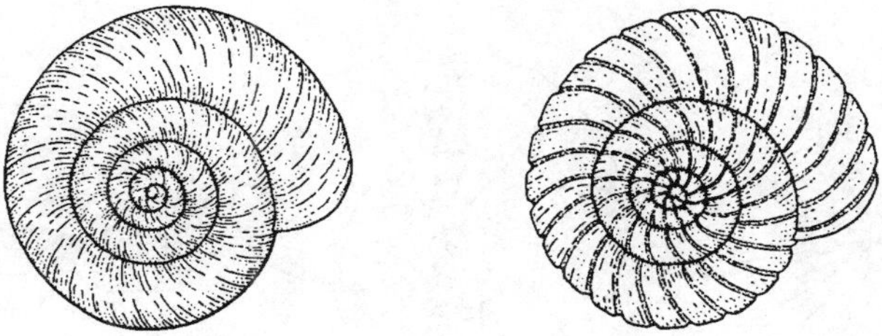

a b

118

Fig. 9.117. *Oxychilus cellaruis* (Müller).

Fig. 9.118. Shell (*a*) without radiating indented lines; (*b*) with radiating indented lines.

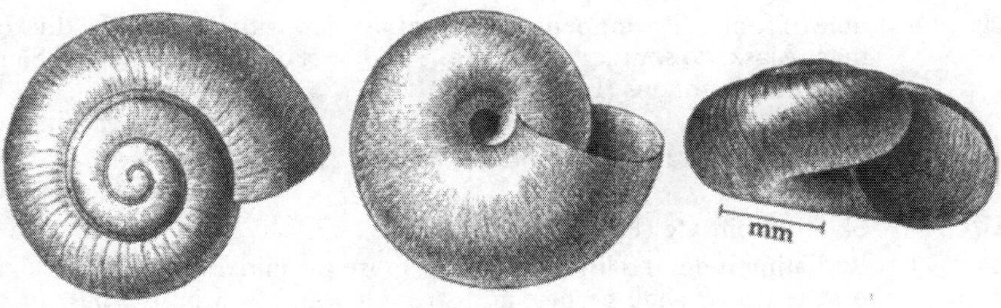

119

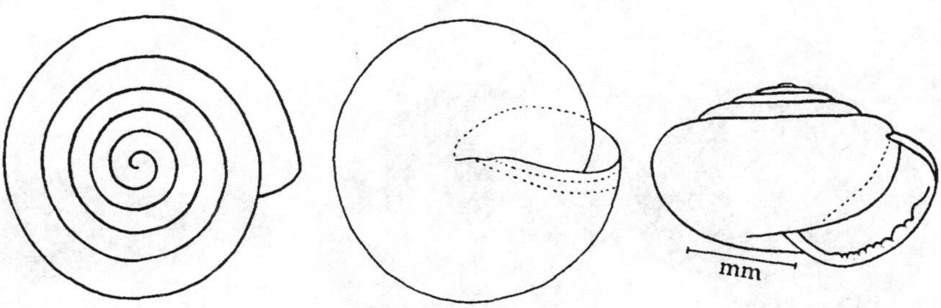

120

Fig. 9.119. *Nesovitrea binneyana* (Morse).

Fig. 9.120. *Pristiloma lansingi* (Bland). [After Pilsbry (1946).]

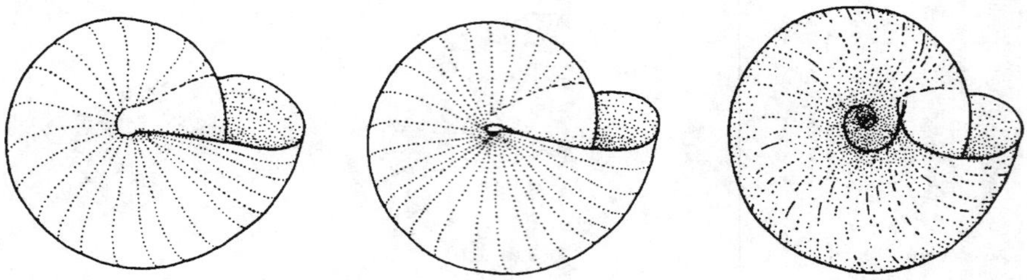

a b c

121

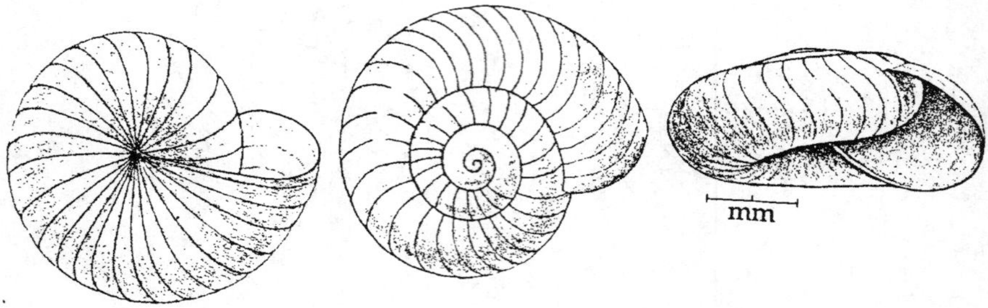

Fig. 9.121. Shell (*a*) imperforate; (*b*) perforate; (*c*) umbilicate.

Fig. 9.122. *Glyphyalinia indentata* (Say).

10b Sole of foot uniform; penis with a papilla; jaw with flat, rib-like thickenings; Alaska to southern California, and west of the Cascade and Sierra Nevada Mountains (Fig. 9.120) . . . . **_Pristiloma_** (in part)

11a **(9)** Shell imperforate or perforate (Figs. 9.121*a,b;* 9.122). . . . . **_Glyphyalina_** (in part)

11b Shell umbilicate (Fig. 9.121*c*). **12**

12a **(11)** Radiating indented lines noticeably more prominent than growth lines on the upper shell surface and strongly impressed on the base of the shell **13**

12b Radiating indented lines only slightly more prominent than growth

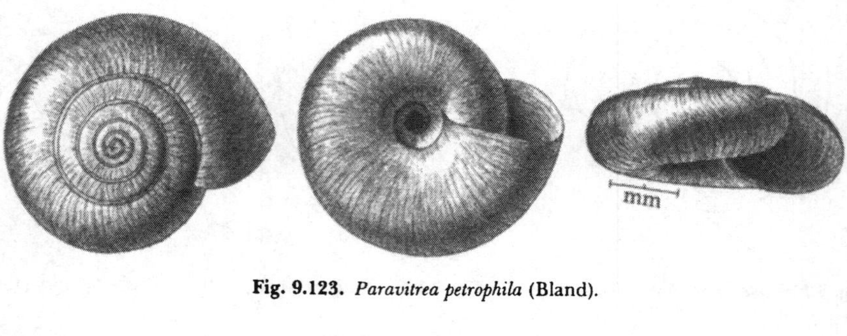

Fig. 9.123. *Paravitrea petrophila* (Bland).

124 125

Fig. 9.124. *Pilsbryna castanea* H. B. Baker. **Fig. 9.125.** *Glyphyalinia rhoadsi* (Pilsbry).

		lines on the upper shell surface and weakly impressed on the base of the shell	**15**
13a	**(12)**	Shell larger than 5.5 mm in diameter, whitish (Fig. 9.123) . . . . . . ***Paravitrea*** (in part)	
13b		Shell smaller than 5.5 mm in diameter, yellowish or brownish . . .	**14**
14a	**(13)**	Shell with about 4–5 whorls less than 4 mm in diameter (Figs. 9.124, 9.129, 9.135) . . . . ***Pilsbryna*** (in part)	
14b		Shell with about 4–5 whorls 4.5 mm or more in diameter (Fig. 9.125) ***Glyphyalinia*** (in part)	
15a	**(12)**	Radiating indented lines closely and regularly spaced	**16**
15b		Radiating indented lines widely and irregularly spaced	**17**
16a	**(15)**	Umbilicus more than 20% of the shell diameter (Fig. 9.126) . . . . . ***Glyphyalinia*** (in part)	
16b		Umbilicus less than 20% of the shell diameter (Fig. 9.127) . . . . . . ***Nesovitrea*** (in part)	
17a	**(15)**	Umbilicus narrow, less than one-eighth of the shell diameter; shell color greenish horn (Fig. 9.128) . . . . ***Glyphyalinia*** (in part)	
17b		Umbilicus wide, one-sixth to one-third of the shell diameter; shell color yellowish, reddish, or brownish	**18**
18a	**(17)**	Shell 3.6–3.7 mm in diameter (Figs. 9.124, 9.129, 9.135) . . . . ***Pilsbryna*** (in part)	
18b		Shell more than 4 mm in diameter (Fig. 9.130) . . . . ***Glyphyalinia*** (in part)	

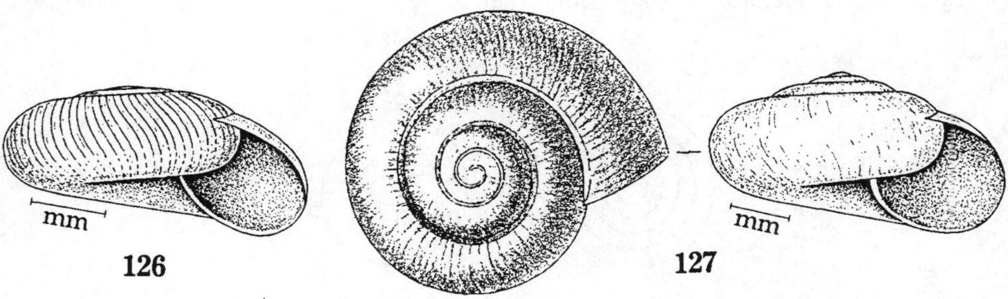

126 127

Fig. 9.126. *Glyphyalinia raderi* (Dall). **Fig. 9.127.** *Nesovitrea electrina* (Gould).

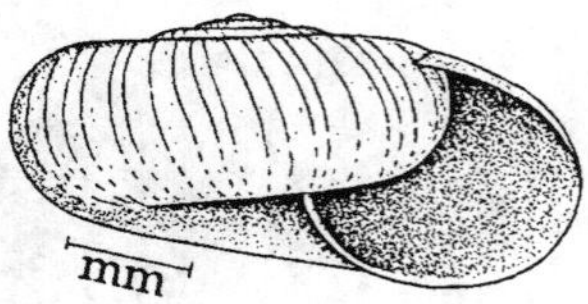

128

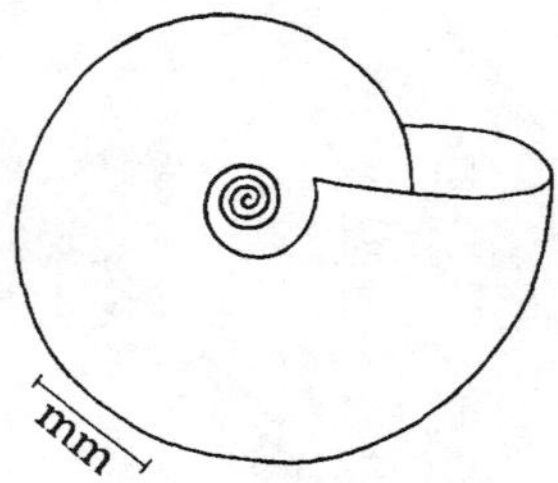

129

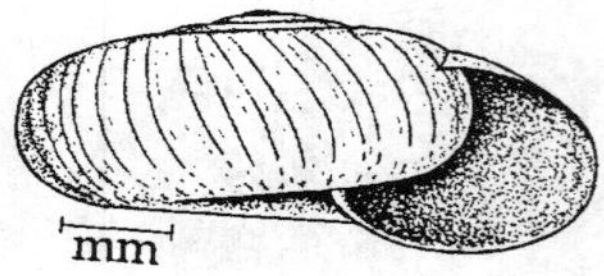

130

Fig. 9.128. *Glyphyalinia clingmani* (Dall). (After B. Walker and H. A. Pilsbry, 1902, *Proc. Acad. Nat. Sci. Phila.* 54, pl. 23, figs. 7–9.) **Fig. 9.129.** *Pilsbryna castanea* H. B. Baker. **Fig. 9.130.** *Glyphyalinia virginica* (Morrison).

19a **(5)** Shell small, less than 4 mm in diameter **20**

19b Shell larger, more than 4 mm in diameter **26**

20a **(19)** Shell very tightly coiled (Fig. 9.131*a*), usually with 6 or more whorls, which are flattened or shortened, last whorl with or without teeth or lamellae (eastern); or with 5 or more whorls, which may be either shouldered or rounded, and without teeth or lamellae [***Pristiloma lansingi*** (Bland), a western species, has a lamella just inside the basal and outer aperture lip (Fig. 9.120)]. **21**

20b Shell less tightly coiled (Fig. 9.131*b*), usually with less than 6 whorls, which are rounded; last whorl usually without teeth **22**

21a **(20)** East of the Continental Divide (Fig. 9.132) . . . . ***Paravitrea*** (in part)

21b West of the Continental Divide (Fig. 9.133). . . . ***Pristiloma*** (in part)

22a **(20)** Shell white, opaque (Fig. 9.134) . . . . ***Hawaiia***

22b Shell corneous, translucent. **23**

23a **(22)** Whorls more rapidly expanding, the last one rather prominent (Figs. 9.124, 9.129, 9.135) . . . . ***Pilsbryna*** (in part)

23b Whorls more gradually expanding. **24**

24a **(23)** Umbilicus deep and steep-sided, penultimate whorl not prominent within the umbilicus; $4\frac{1}{2}$–6 whorls **25**

24b Umbilicus relatively shallow, prominently showing the penultimate whorl; $3\frac{1}{2}$–4 whorls (Fig. 9.136) . . . . ***Striatura*** (in part)

25a **(24)** Apertural lamellae present (Fig. 9.137) . . . . . . ***Pilsbryna*** (in part)

25b Apertural lamellae absent (Fig. 9.138) . . . . ***Vitrea***

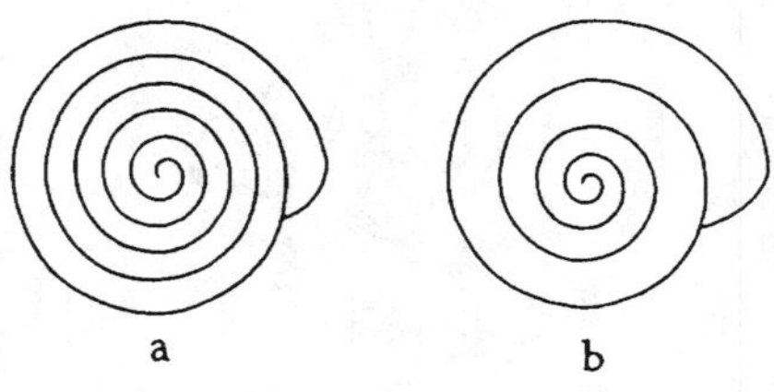

Fig. 9.131. Shell (*a*) very tightly coiled; (*b*) less tightly coiled.

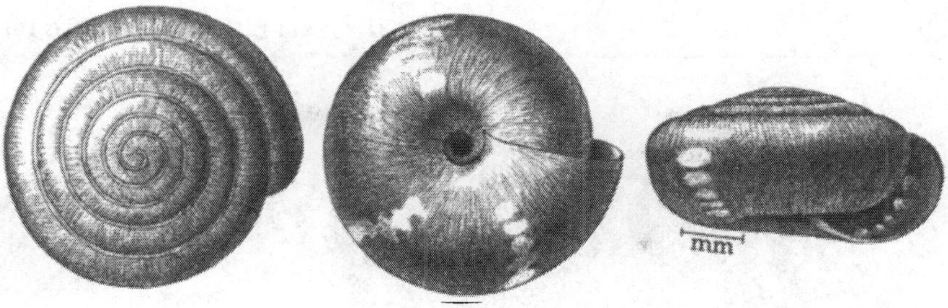

132

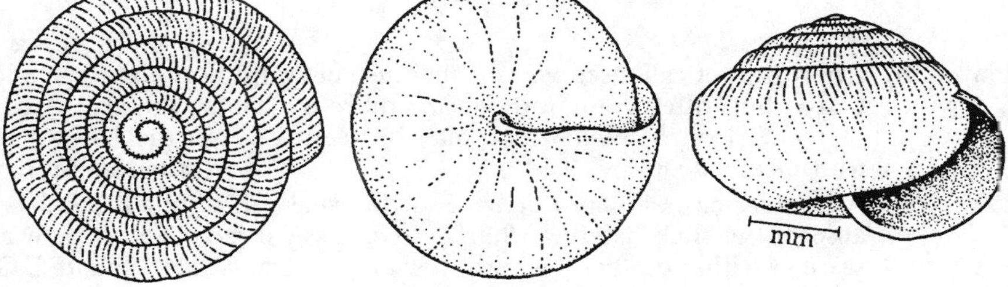

133

Fig. 9.132. *Paravitrea multidentata* (A. Binney). **Fig. 9.133.** *Pristiloma stearnsi* (Bland). [After Pilsbry (1946).]

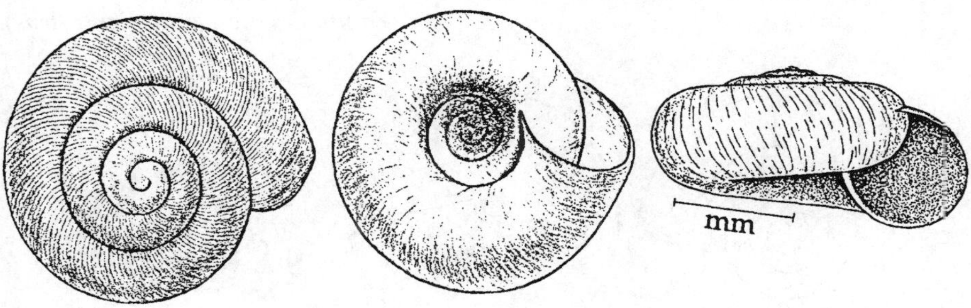

134

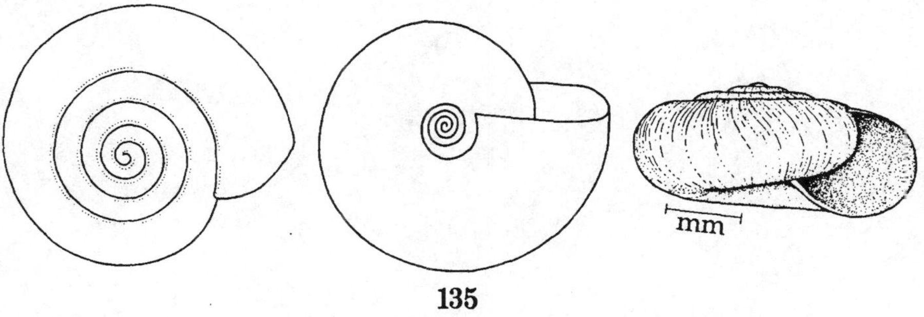

135

Fig. 9.134. *Hawaiia minuscula* (A. Binney). **Fig. 9.135.** *Pilsbryna castanea* H. B. Baker.

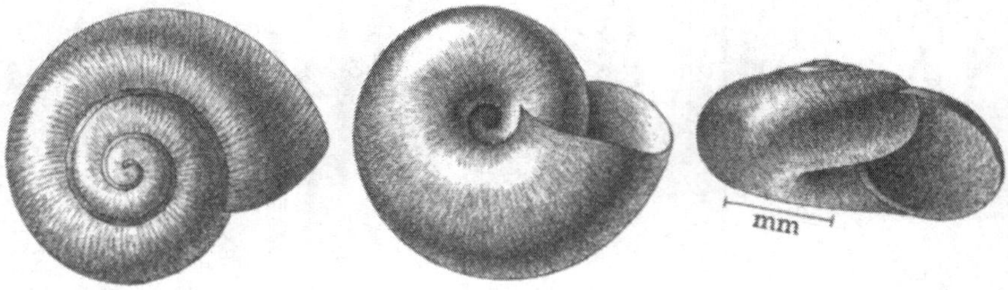

Fig. 9.136. *Striatura ferrea* Morse.

26a (19) Aperture not calloused within; shell usually with more than 5 whorls; body whorl often containing one or more transverse lamellae or rows of teeth (Fig. 9.139*b*), which may be seen through the base of the translucent shell (Fig. 9.140) . . . . ***Paravitrea*** (in part)

26b Aperture may or may not have an internal callus, but if a callus is absent, the shell has few whorls (4½ or less); if lamellae or teeth are present within the body whorl, they are not transversely situated. Genus ***Zonitoides*** **27**

27a (26) Aperture without a callous pad, lamellae, or teeth; shell with 5 whorls or less (Fig. 9.141). . . . . ***Zonitoides*** (Subgenus ***Zonitoides s.s.***)

27b Aperture with a callous pad (Fig. 9.139*a*), lamellae, or teeth; shell with 5 or more whorls. **28**

28a (27) Aperture in adults with lamellae or teeth (Fig. 9.142). . . . . ***Zonitoides*** (Subgenus ***Ventridens***)

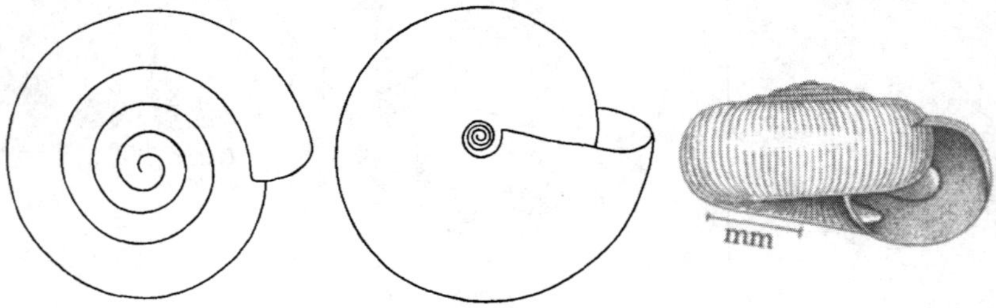

137

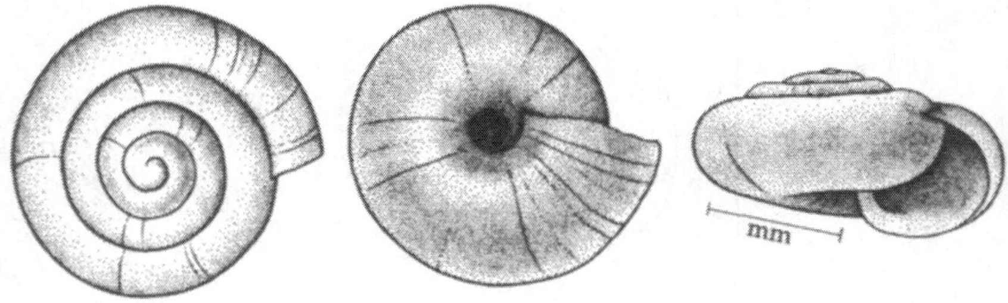

138

Fig. 9.137. *Pilsbryna aurea* H. B. Baker. **Fig. 9.138.** *Vitrea contracta* (Westerlund).

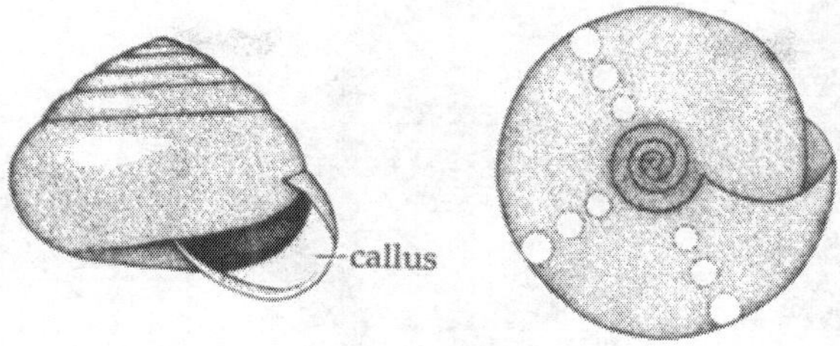

a b

139

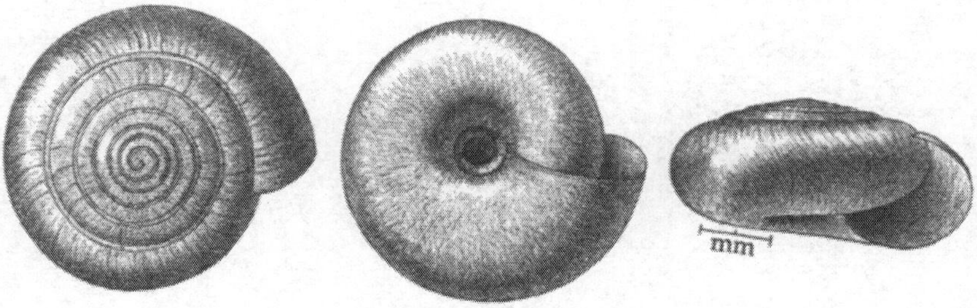

140

Fig. 9.139. (*a*) Apertural view of shell showing callus; (*b*) basal view of shell with "teeth" visible through the last whorl. **Fig. 9.140.** *Paravitrea capsella* (Gould).

28b Aperture in adults with a callous pad, but without teeth or lamellae (Fig. 9.143) . . . . *Zonitoides* (Subgenus *Ventricallus*)

Family LIMACIDAE

The Limacidae are a family of slugs native to Europe and adjacent parts of Asia and Africa; some species of one of its genera, *Deroceras,* also occur naturally in northeastern Asia and North America. All other members of this family in North America (north of Mexico) have been introduced, mainly from Europe. *Deroceras reticulatum* (Müller), native to Europe, has been introduced throughout the world and is a destructive pest in gardens and vegetable fields. Some workers have put *Deroceras* in the separate family Agriolimacidae. Besides *Deroceras,* species of *Limax* and *Lehmannia* occur in North America, and like *Deroceras,* they occur in both the Eastern and Western divisions.

***Key to Genera*†**

1a Tail truncated in side view; posterior mantle margin rounded; center of concentric mantle ridges on the right side of the mantle (Fig. 9.144) ***Deroceras***

1b Tail pointed in side view; posterior mantle margin obtusely angular; center of concentric mantle ridges medially located on the mantle **2**

† Adapted, in part, from Chichester and Getz (1973).

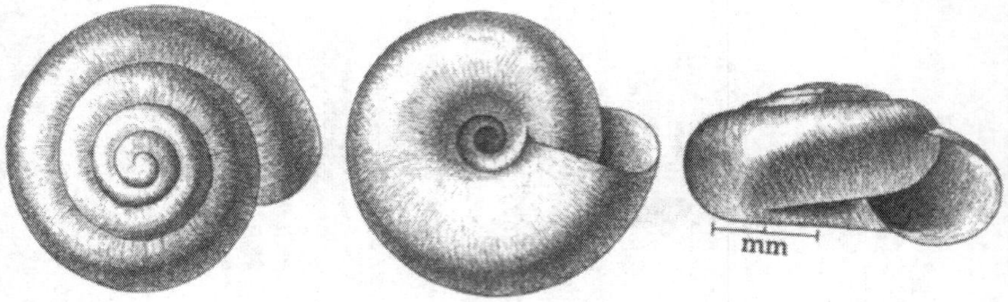

141

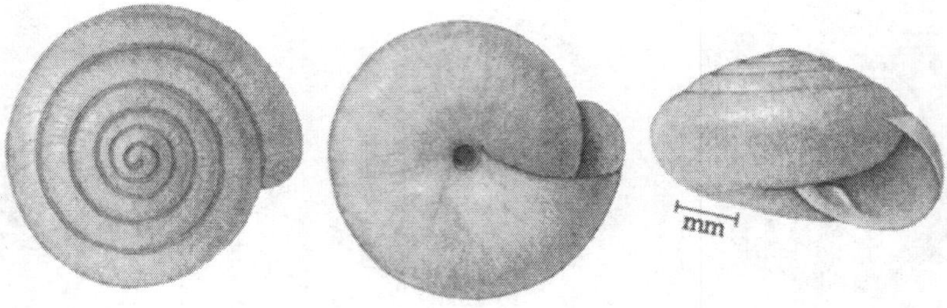

142

143

Fig. 9.141. *Zonitoides* (*Zonitoides*) *arboreus* (Say). **Fig. 9.142.** *Zonitoides* (*Ventridens*) *suppressus* (Say). **Fig. 9.143.** *Zonitoides* (*Ventricallus*) *demissus* (A. Binney).

2a **(1)** Mantle with one or more pairs of dark longitudinal bands; penial appendix present (Fig. 9.145) . . . . ***Lehmannia***

2b Mantle plain, or with yellow spots, or black spots or patches; penial appendix absent (Fig. 9.146) . . . . ***Limax***

Family MILACIDAE

The Milacidae are mainly a family of southern Europe, but several species have been widely dispersed throughout the world by human agency. The respiratory pore is on the right side, in the posterior half of the mantle. These slugs lack a caudal mucous gland. Externally, the milacids can be distinguished from the Limacidae by having a dorsal median keel running from the posterior tip of the foot anteriorly to the posterior edge of the mantle, and by the groove running around the dorsal centrum of the mantle. The introduced *Milax* is the only genus occurring

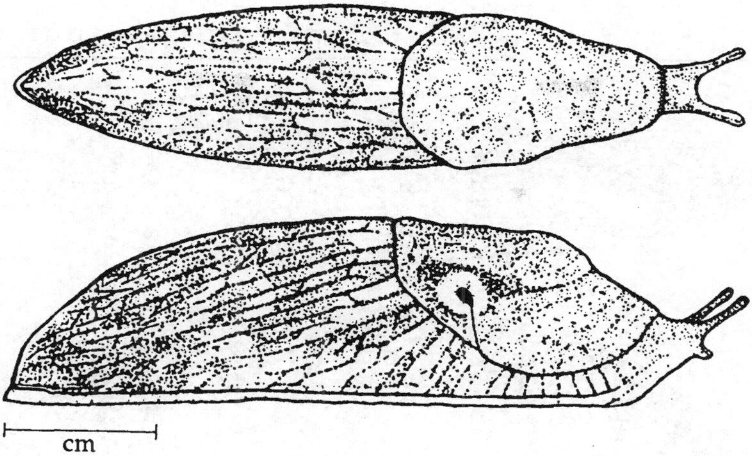

Fig. 9.144. *Deroceras reticulatum* (Müller). [From Chichester and Getz (1973, fig. 7c).]

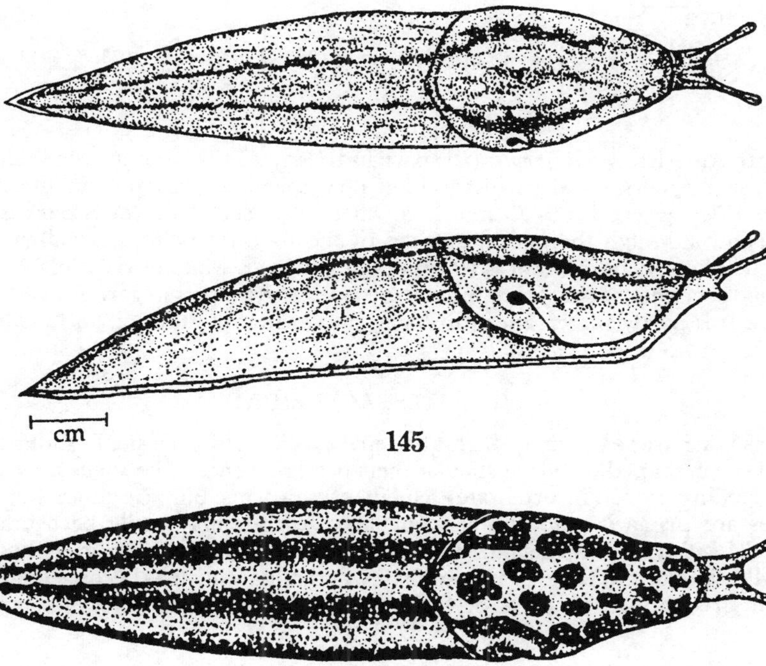

145

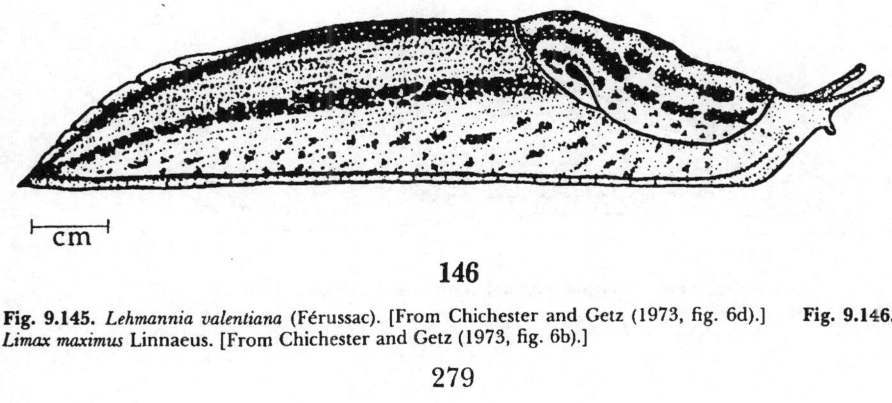

146

Fig. 9.145. *Lehmannia valentiana* (Férussac). [From Chichester and Getz (1973, fig. 6d).] **Fig. 9.146.** *Limax maximus* Linnaeus. [From Chichester and Getz (1973, fig. 6b).]

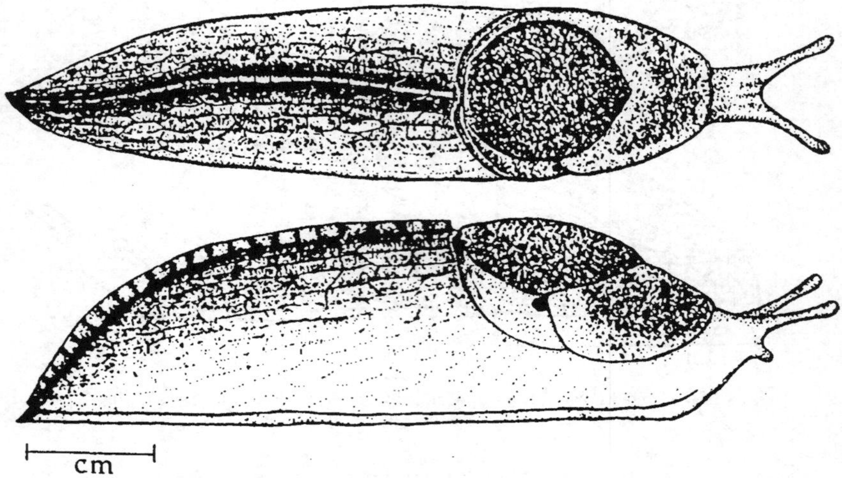

Fig. 9.147. *Milax gagates* (Draparnaud). [From Chichester and Getz (1973, fig. 8c).]

in North America. *Milax gagates* (Draparnaud) (Fig. 9.147) was introduced long ago into North America and its distribution now covers a great part of the United States. *Milax gagates* is 50–70 mm long when extended. Its color is dark gray or black, and although the lateral grooves of the foot are paler, the animal is not banded. The sole of the foot is whitish. The mucus is white or colorless.

This "greenhouse slug" is reputed to be a destructive pest in gardens and greenhouses. It is partly subterranean in habit, burrowing into the soil and feeding on roots.

Family TESTACELLIDAE

This is a European family of sluglike snails which possess a small, rudimentary, cap-like, ear-shaped, external shell at their posterior ends. The snails are carnivorous, feeding on soil invertebrates, chiefly earthworms, but also other slugs. The species are primarily subterranean in habit, so they may easily be overlooked. *Testacella haliotidea* Draparnaud (Fig. 9.148) has been reported in greenhouses in both the eastern and western United States, and apparently has become established outside greenhouses in parts of Oregon. This snail has an anteriorly tapering,

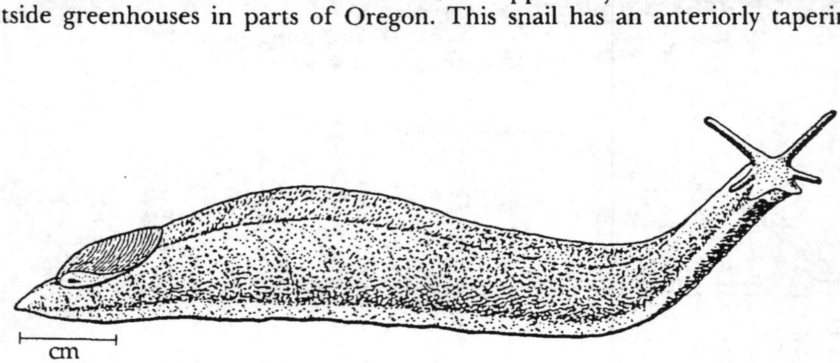

Fig. 9.148. *Testacella haliotidea* Draparnaud. [From F. C. Baker (1939).]

grayish-brown or pale yellow body and attains a length of up to 120 mm. Its body wall is scored by longitudinal, branching lateral grooves.

Family THYSANOPHORIDAE

The Thysanophoridae Pilsbry 1926, as treated here, contain three subfamilies, the Megomphicinae H. B. Baker 1930 (= Ammonitellidae Pilsbry 1930), Oreohelicinae Pilsbry 1939, and Thysanophorinae *s.s.* The Megomphicinae are medium to large discoidal snails with umbilicate shells in which the lip is neither expanded nor reflected. In the reproductive system, the dart apparatus is lacking, and the penis is without a verge. The subfamily occurs only in the western United States and northern Baja California. *Megomphix* has three species, *Glyptostoma* two, and *Ammonitella, Polygyrella,* and *Polygyroidea* are monotypic genera.

The Oreohelicinae are medium to large helicoidean snails which lack a dart apparatus in the female reproductive system and lack a penial verge in the male system. They are restricted to North America and are the dominant land snail group in the Rocky Mountain states. They do not occur in the eastern United States. The shells are umbilicate and have sharp apertural lips that are neither reflected nor expanded. The subfamily contains two genera, *Oreohelix* and *Radiocentrum.* Shells of the discid snail *Anguispira kochi* are like those of *Oreohelix. Oreohelix* can be distinguished from *Anguispira* by its radially or spirally striate embryonic whorls. In *Anguispira,* the embryonic whorls are smooth. Also, the suprapedal groove which is so noticeable in *Anguispira,* is absent in *Oreohelix.*

Three genera of Thysanophorinae occur in the United States, *Hojeda, Microphysula* and *Thysanophora.* Their small umbilicate shells vary from globose to depressed and are unicolored, either white or brown. The outer lip of the shell aperture is thin and unexpanded, while the columellar lip is generally reflected. *Thysanophora* is easily distinguished by its periostracal threads, which are peculiarly more oblique than the shell growth lines. The shells of *Hojeda* and *Microphysula* look like those of some of the Vitrinidae, especially *Pilsbryna,* but the animal of these thysanophorids is characteristically holopod (see Fig. 9.21*b*), not aulacopod, and the shell lacks internal barriers and has a wider umbilicus than most vitrinids, especially *Pilsbryna. Hojeda* is found in the Florida Keys. *Microphysula* is widespread in the Rocky Mountain region, distributed from British Columbia to southern Arizona and New Mexico. *Thysanophora* is found from Arizona to Venezuela, and also occurs in Florida and the West Indies.

Key to Genera

1a Shell small, less than 6 mm in diameter. Subfamily **Thysanophorinae 2**

1b Shell small to large, more than 7 mm and up to nearly 40 mm in diameter. Western in distribution **4**

2a (1) Shell brownish, having thin periostracal threads more oblique than

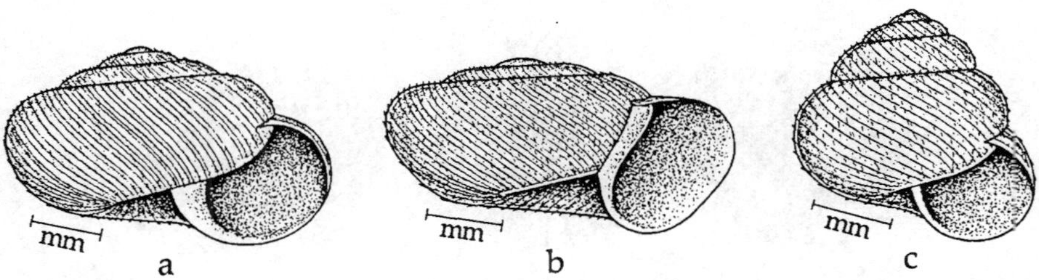

Fig. 9.149. *Thysanophora; (a) T. horni* (Gabb); (*b*) *T. incrustata* (Poey); (*c*) *T. plagioptycha* (Shuttleworth).

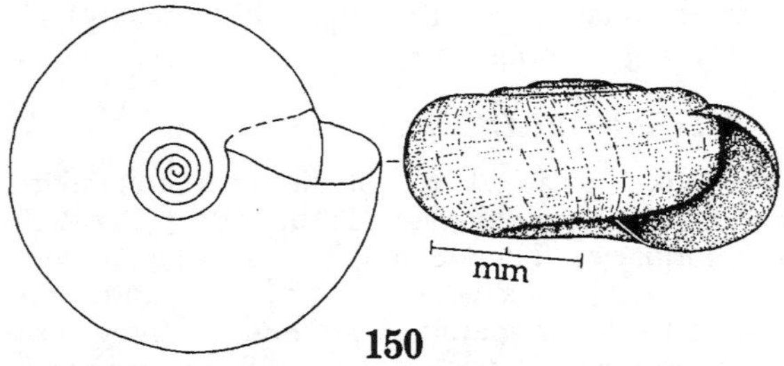

Fig. 9.150. *Microphysula ingersolli* (Bland).

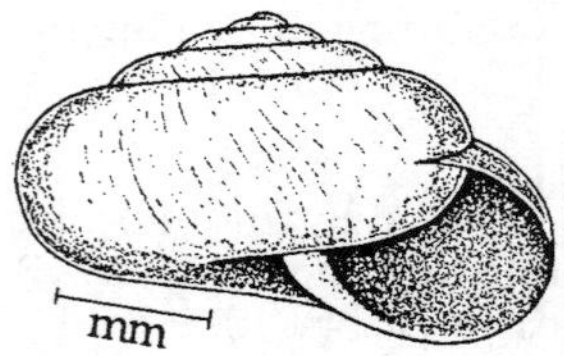

Fig. 9.151. *Hojeda inaguensis* (Weinland).

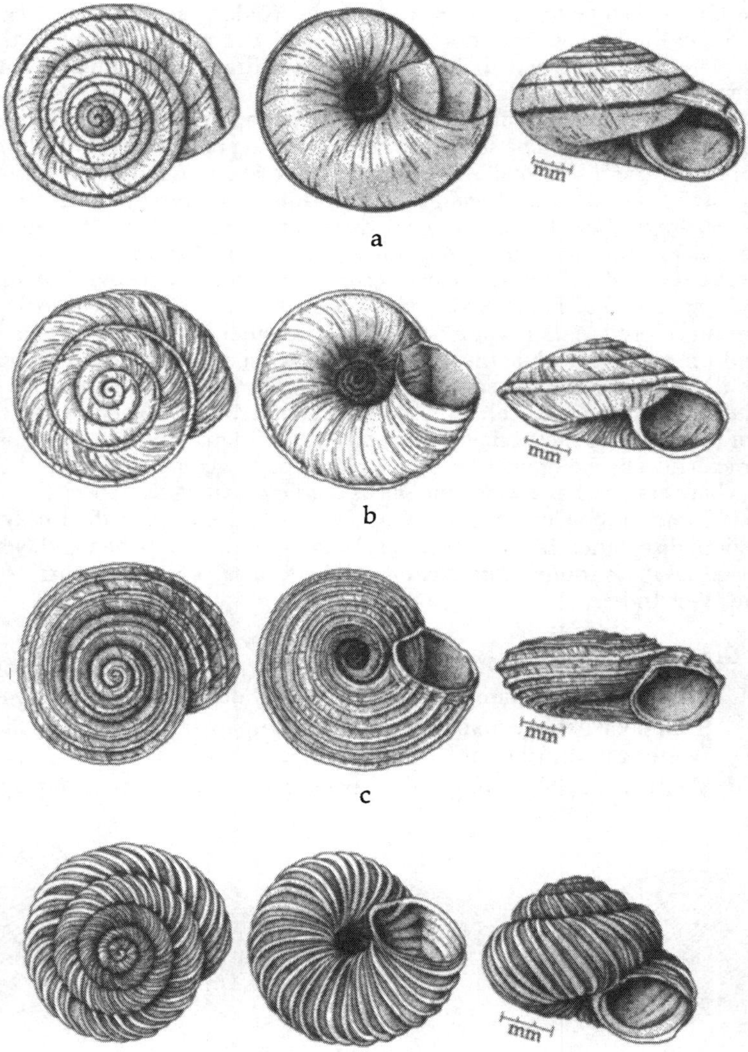

Fig. 9.152. *Oreohelix:* (*a*) *O. strigosa* (Gould); (*b*) *O. metcalfi acutidiscus* Pilsbry and Ferriss; (*c*) *O. haydeni* (Gabb); (*d*) *O. idahoensis* (Newcomb).

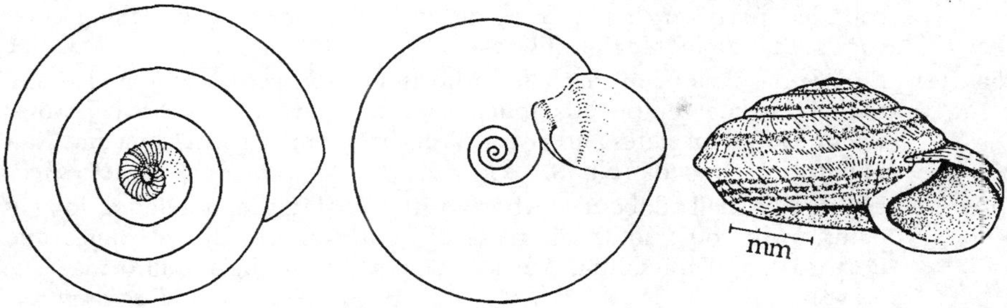

Fig. 9.153. *Radiocentrum chiricahuana* (Pilsbry).

growth lines; occurring in southern Florida, Texas, New Mexico, and Arizona (Fig. 9.149). . . . . ***Thysanophora***

2b Shell without periostracal threads **3**

3a **(2)** Surface of shell glossy and clear; occurring in mountain states, Washington, and on Vancouver Island (Fig. 9.150) . . . . ***Microphysula***

3b Surface of shell dull, translucent whitish or pale brown; occurring in the Florida Keys (Fig. 9.151). . . . . ***Hojeda***

4a **(1)** Shell helicoid to depressed, surface roughly sculptured, often radially or spirally striate; periphery of young shells either angular or carinate; with (generally) or without color bands; Subfamily **Oreohelicinae** **5**

4b Shell very depressed to discoidal, rather smooth, often glossy, with

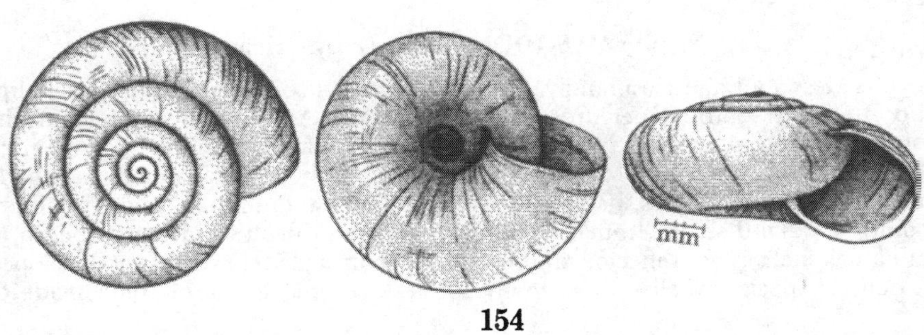

154

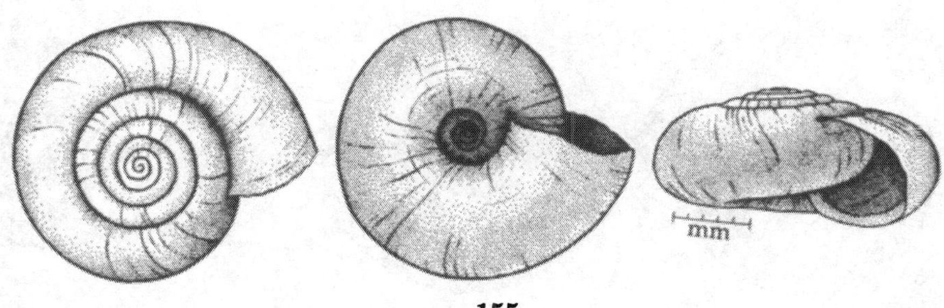

155

Fig. 9.154. *Glyptostoma newberryanum* (W. G. Binney). **Fig. 9.155.** *Megomphix hemphilli* (W. G. Binney).

rounded periphery in both young and adult shells, and lacking color bands; Subfamily **Megomphicinae** 6

5a (4) Embryonic shell of more than 2 whorls at birth, variously wrinkled and radially striate; epiphallus much shorter than the penis; viviparous; Rocky Mountain states and Great Basin from northern Arizona and New Mexico into Canada (Fig. 9.152). . . . . ***Oreohelix***

5b Embryonic shell of about $1\frac{1}{2}$ whorls, with radial ribs; epiphallus as long as penis; oviparous; southern parts of southwestern United States and northeastern Chihuahua, Mexico, and southern Baja California (Fig. 9.153) . . . . ***Radiocentrum***

6a (4) Shell dark or pale; apertural lip thin; parietal wall with minute spiral lines (in ***Megomphix*** these are visible only in immature shells). 7

6b Shell pale; apertural lip thickened within; parietal wall without spiral lines 8

7a (6) Shell dark; adult shells 25 mm or more in diameter; southern California (Fig. 9.154) . . . . ***Glyptostoma***

7b Shell pale, adult shells 20 mm or less in diameter; Oregon and Washington (Fig. 9.155) . . . . ***Megomphix***

8a (6) Shell spire concave (inverted); shell aperture narrow and lacking barriers ("teeth"); California (Fig. 9.156) . . . . ***Ammonitella***

8b Shell spire flat or convex; shell aperture moderately wide and having a parietal tooth 9

9a (8) Apertural lip without teeth, but with one or two radial series of three teeth each within the last whorl and visible through the shell; upper surface of shell with costae (Fig. 9.157) . . . . ***Polygyrella***

9b Apertural lip with two teeth; teeth lacking within the last whorl; upper surface of shell striate (Fig. 9.158) . . . . ***Polygyroidea***

Family MESODONTIDAE (Polygyridae)

The Mesodontidae are a family of small to large helicoid snails with reflected lips and often with toothed apertures. Endemic to North America, they are one of the continent's dominant land snail families. Although mainly a family of temperate climates, the range of the Mesodontidae extends from Alaska to eastern Canada and south into the tropics. Its species are most abundant in humid regions and are mainly woodland snails. Immature individuals are difficult to identify; the lip of immature shells is not reflected, and usually apertural teeth have not yet developed in dentate species. Shells of the many species are unicolored, of some shade of

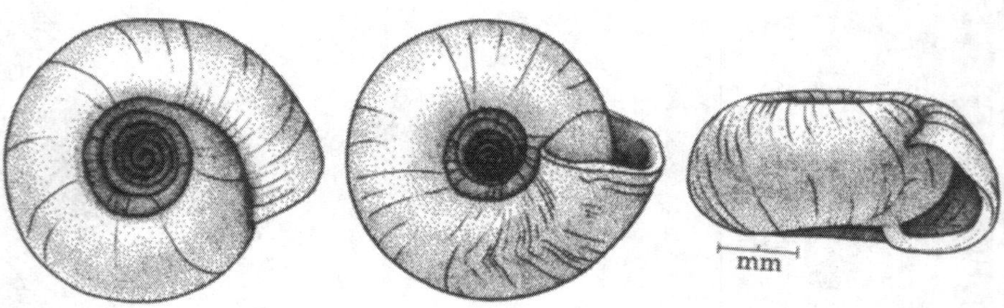

Fig. 9.156. *Ammonitella yatesi* J. G. Cooper.

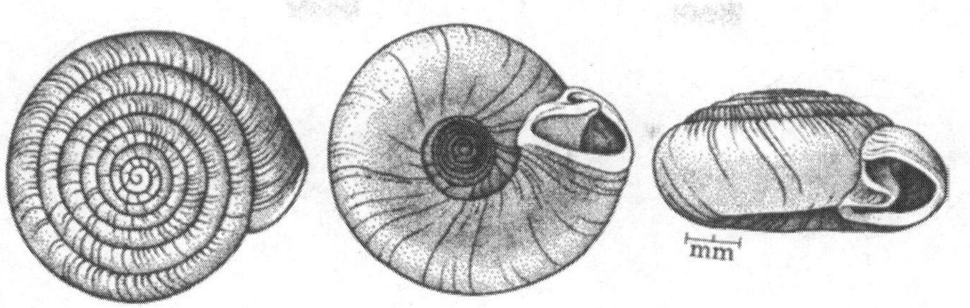

157

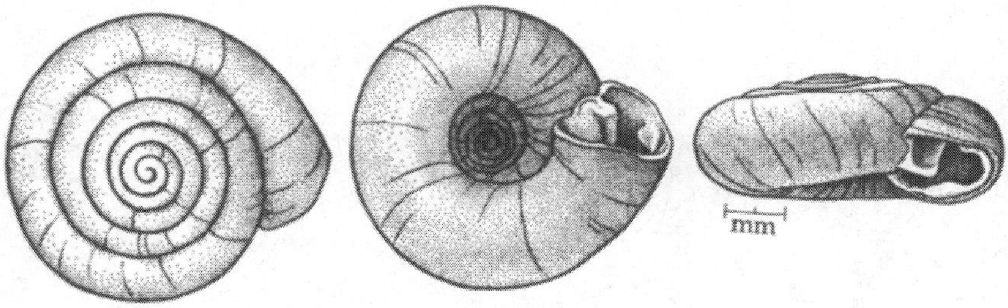

158

Fig. 9.157. *Polygyrella polygyrella* (Bland and J. G. Cooper). **Fig. 9.158.** *Polygyroidea harfordiana* (J. G. Cooper).

yellow, tan, or brown, but several of the species of *Praticolella* have spiral color bands, as do *Allogona profunda* (Say), *Polygyra texasiana* (Moricand), *Stenotrema fraternum fasciatum* Pilsbry, *Triodopsis multilineata* (Say), and rarely *Mesodon elevatus* (Say) (Pilsbry 1940). There are 10 genera of mesodontids in the United States. *Allogona s.s., Mesodon, Polygyra, Praticolella,* and *Stenotrema* are restricted to the east, and *Allogona* (*Dysmedoma*), *Ashmunella, Cryptomastix, Trilobopsis* and *Vespericola* are restricted to the west.

Members of the genus *Stenotrema* are clearly distinguishable by their characteristic shells, and some of the species of *Polygyra* are quite distinct as a group. But much convergence or parallelism is exhibited by the shells of the other mesodontid genera, and some of the *Polygyra* species as well, which makes a key to the family's genera based on shell characters rather difficult to construct. The two keys below are adapted from Pilsbry (1940) and Burch (1962; for the eastern mesodontids). The first key is based on shell characters; the second is based mainly on anatomical characters.

The Mesodontidae are divided into three subfamilies according to characteristics of the terminal male genitalia. The Mesodontinae contain the genera *Mesodon, Polygyra, Praticolella,* and *Stenotrema.* The Triodopsinae includes *Allogona, Cryptomastix, Trilobopsis, Triodopsis,* and *Vespericola.* The Ashmunellinae is a monogeneric family, containing only *Ashmunella.*

Key to Shells of the Genera

1a		Distribution east of the Continental Divide (see Fig. 9.1) **2**
1b		Distribution west of the Continental Divide **13**
2a	(1)	Found in the arid mountains of New Mexico, Arizona, and Mexico (Fig. 9.175). *Ashmunella* (in part)

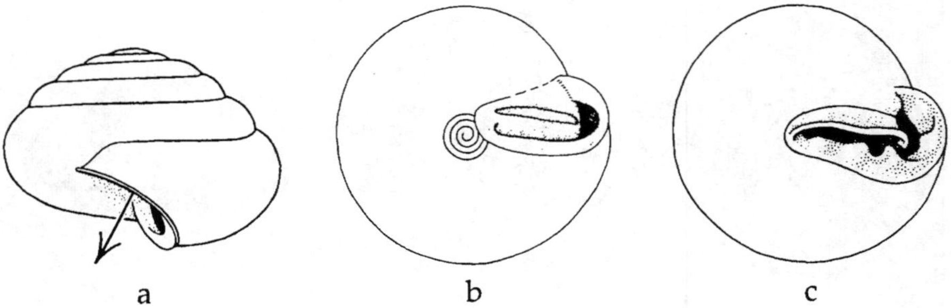

159

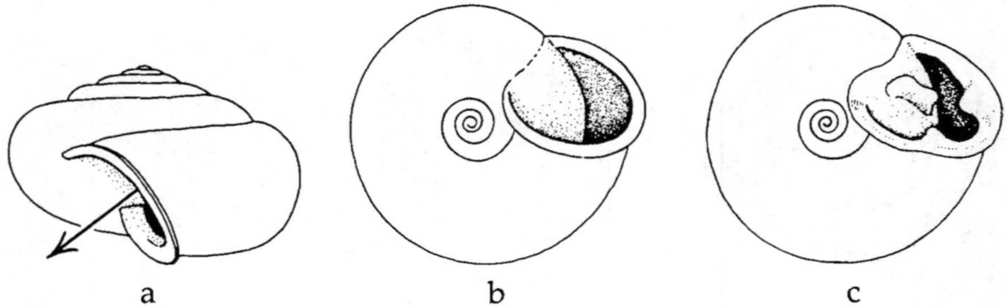

160

Fig. 9.159. Shell aperture and parietal teeth in *Stenotrema:* (*a*) aperture directed basally; (*b*) aperture more broadly open, parietal tooth moderately long; (*c*) aperture more slit-like, parietal tooth very long. **Fig. 9.160.** Shell apertures in other mesodontid snails: (*a*) Aperture directed laterally; (*b*) aperture ovate; (*c*) aperture ovate, with "teeth."

2b Found in other areas of eastern North America. 3

3a (2) Aperture basal (Fig. 9.159*a*), rather narrow (Fig. 9.159*b*) to very narrow (Fig. 9.159*c*); parietal tooth unusually long, bladelike, basal lip often with a notch. Genus ***Stenotrema*** 4

3b Aperture not directed basally, generally roundly or ovately lunate (Fig. 9.160); parietal tooth, when present, not bladelike; basal lip without a notch 5

4a (3) Aperture long and slit-like; parietal tooth very long (Fig. 9.161) . . . ***Stenotrema*** (Subgenus ***Stenotrema***)

4b Aperture more broadly open; parietal tooth shorter (Fig.9.162) . . . ***Stenotrema*** (Subgenus ***Euchemotrema***)

5a (3) Aperture relatively small in relation to overall size of shell (Fig. 9.163*a–c*), or aperture nearly closed by teeth or with the parietal callus often biramose and having a raised proximal edge. Genus ***Polygyra*** . . . 6

5b Aperture usually relatively large, parietal callus not biramose and without a raised proximal edge (Fig. 9.163*d,e*). 7

6a (5) Shell with relatively low whorls; umbilicus shallow and very wide; apertural lip without teeth (Fig. 9.164) . . . ***Polygyra*** (Subgenus ***Polygyra***)

6b Shell with relatively high whorls; umbilicus, if present, narrower and deeper; aperture lip with or without teeth (Fig. 9.165) . . . . ***Polygyra*** (Subgenus ***Daedalochila***)

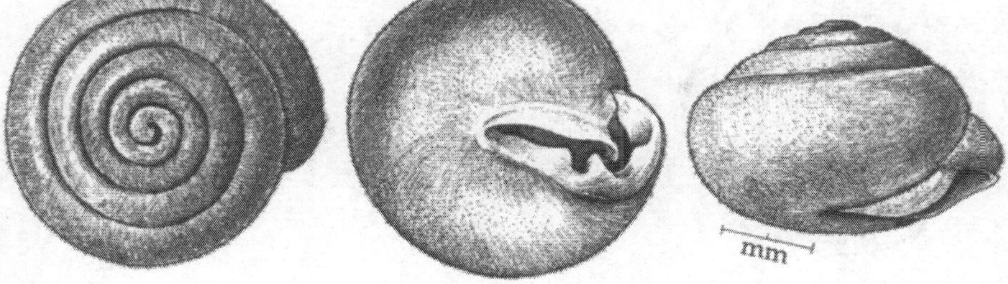

161

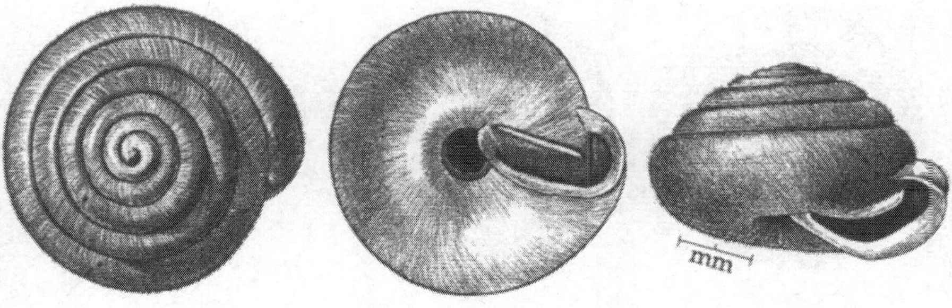

162

Fig. 9.161. *Stenotrema (Stenotrema) deceptum* (Clapp).

Fig. 9.162. *Stenotrema (Euchemotrema) leai* (A. Binney).

7a	(6)	Aperture with three teeth	**8**
7b		Aperture with two, one, or no teeth	**10**
8a	(7)	Shell umbilicate (Fig. 9.166) . . ***Triodopsis*** (Subgenus ***Triodopsis s.s.***)	
8b		Shell imperforate or nearly so	**9**
9a	(8)	Shell periphery rounded to slightly shouldered; palatal tooth at the distal curvature of the outer lip (Fig. 9.167). ***Mesodon*** (Subgenus ***Inflectarius***)	
9b		Shell periphery subangular to carinate; palatal tooth on the upper curvature of the outer lip (Fig. 9.168) ***Triodopsis*** (Subgenus ***Xolotrema***)	

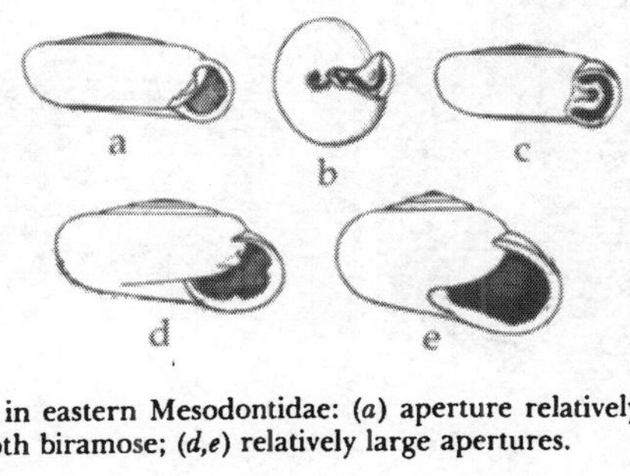

Fig. 9.163. Shell apertures in eastern Mesodontidae: (*a*) aperture relatively small; (*b*) edge of parietal callus raised; (*c*) parietal tooth biramose; (*d,e*) relatively large apertures.

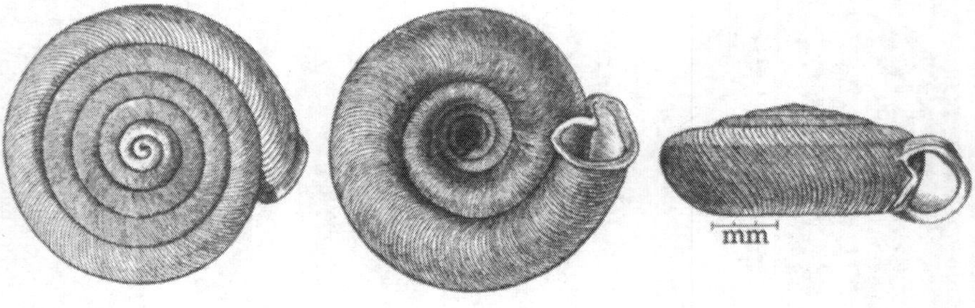

164

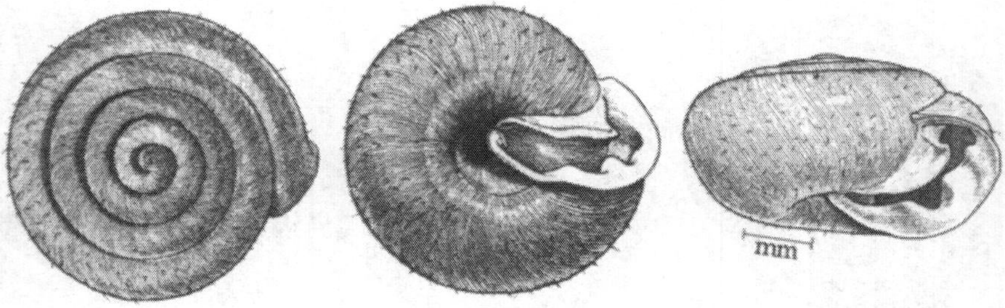

165

Fig. 9.164. *Polygyra* (*Polygyra*) *septemvolva* Say. **Fig. 9.165.** *Polygyra* (*Daedalochila,* section *Lobosculum*) *pustula* (Férussac).

10a (7) Shell depressed, imperforate, with a parietal tooth; the basal lip is widened by a marginal lamina (Fig. 9.169) ***Mesodon*** (Subgenus ***Patera***)

10b Shell depressed to subglobose, imperforate or umbilicate, with or without a parietal tooth; the basal lip is usually without a marginal lamella, but if one occurs, the shell is subglobose, not very depressed . . . **11**

11a (10) Shell depressed, with a rather large umbilicus; aperture rounded, toothless, or with a small parietal tooth (one species, ***M. sayanus*** (Pilsbry) has a basal tooth, in addition to a parietal tooth) (Figs. 9.170, 9.171) . . . . . . ***Mesodon*** (Subgenus ***Appalachina***), ***Allogona*** (in part)

11b Shell depressed-globose to globose-conic, capacious, imperforate or with a very narrow umbilicus; toothless or with a small parietal and sometimes a blunt prominence on the basal lip near the columella **12**

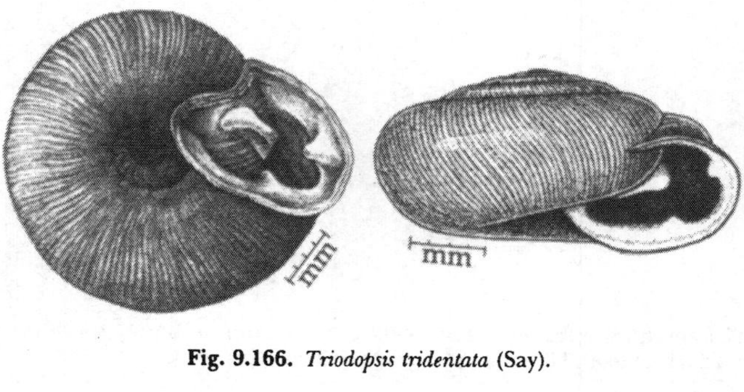

Fig. 9.166. *Triodopsis tridentata* (Say).

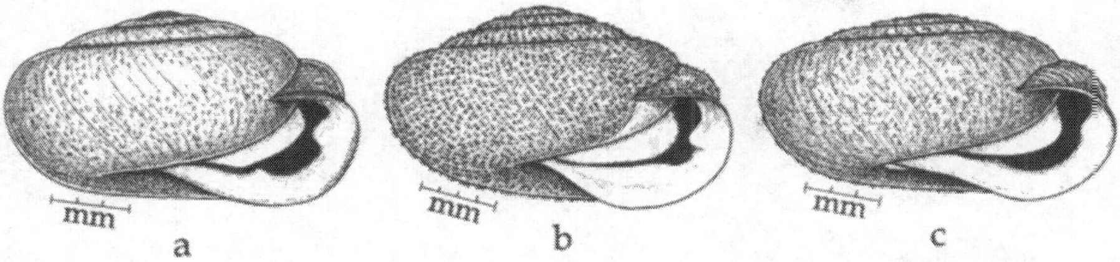

167

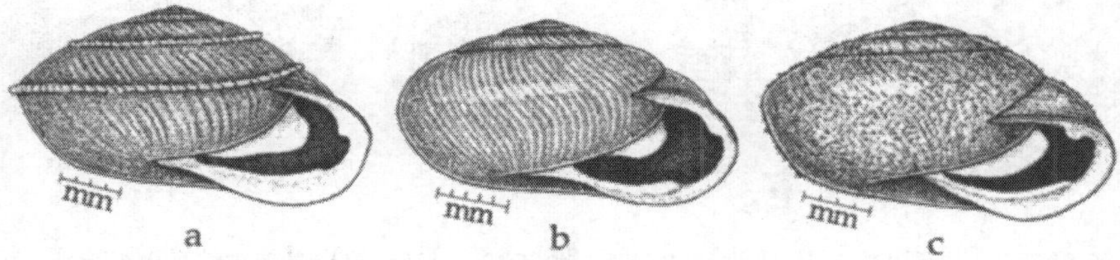

168

Fig. 9.167. *Mesodon,* subgenus *Inflectarius:* (*a*) *M.* (*I.*) *inflectus* (Say); (*b*) *M.* (*I.*) *smithi* (Clapp); (*c*) *M.* (*I.*) *magazinensis* (Pilsbry and Ferriss). **Fig. 9.168.** *Triodopsis,* subgenus *Xolotrema:* (*a*) *T.* (*X.*) *obstricta* (Say); (*b*) *T.* (*X.*) *fosteri* (F. C. Baker); (*c*) *T.* (*X.*) *denotata* (Férussac).

12a **(11)** Shell large, usually more than 14 mm in diameter (Figs. 9.172, 9.173) ***Mesodon*** (in part), ***Triodopsis*** (Subgenus ***Neohelix***)

12b Shell smaller, usually less than 14 mm in diameter (Fig. 9.174) . . . . ***Praticolella***

13a **(1)** Distribution New Mexico, Arizona, and Mexico (Fig. 9.175) . . . . . ***Ashmunella*** (in part)

13b Distribution Alaska to California, Idaho, and Montana **14**

14a **(13)** Last whorl not constricted behind lip; shell with strong radial striations or wrinkles; shell width 10 to 32 mm; without teeth in the aperture and hairs on the shell; shell openly umbilicate (Fig. 9.176) . . . . ***Allogona*** (Subgenus ***Dysmedoma***)

14b Last whorl distinctly constricted behind lip; shell not distinctly striate; shell less than 24 mm in width; aperture with up to three teeth; shell with or without hairs, umbilicate to imperforate **15**

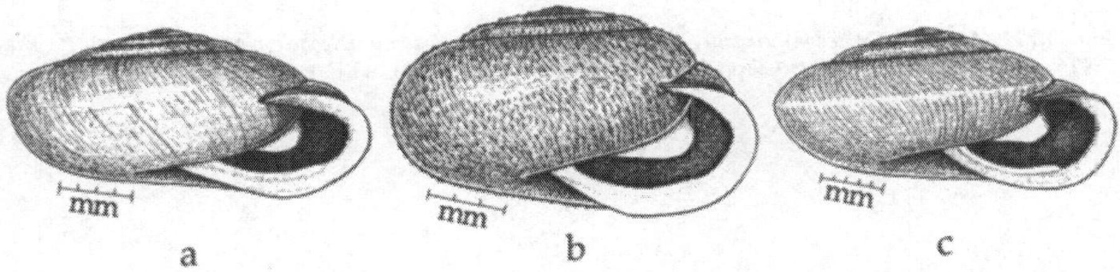

Fig. 9.169. *Mesodon,* subgenus *Patera:* (*a*) *M.* (*P.*) *appressus* (Say); (*b*) *M.* (*P.*) *jonesianus* (Archer); (*c*) *M.* (*P.*) *sargentianus* (Johnson and Pilsbry).

170 171

Fig. 9.170. *Mesodon (Appalachina) sayanus* (Pilsbry). **Fig. 9.171.** *Allogona profunda* (Say).

172 173

a b c

174

Fig. 9.172. *Mesodon (Mesodon) elevatus* (Say). **Fig. 9.173.** *Triodopsis (Neohelix) albolabris* (Say). **Fig. 9.174.** *Praticolella:* (*a*) *P. campi* Clapp & Ferriss; (*b*) *P. lawae* (Lewis); (*c*) *P. griseola* (Pfeiffer).

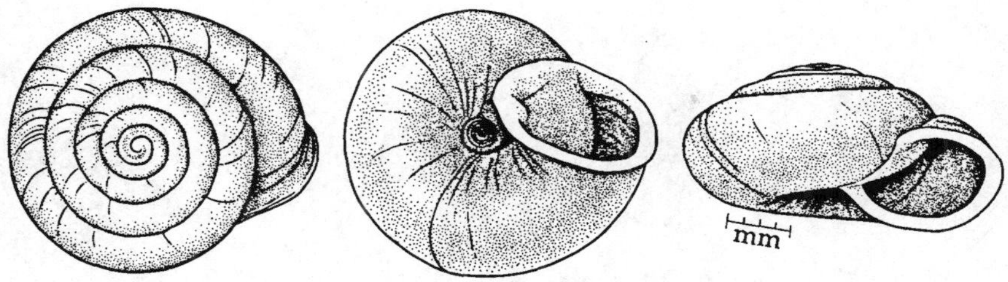

175

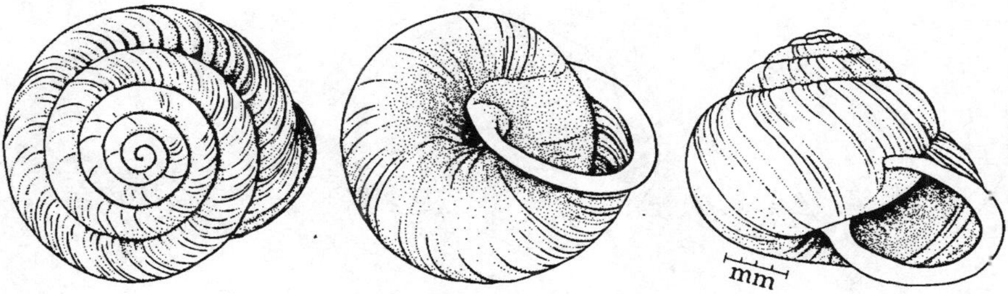

176

Fig. 9.175. *Ashmunella chiricahuana* (Dall). **Fig. 9.176.** *Allogona (Dysmedoma) ptychophora* (Brown).

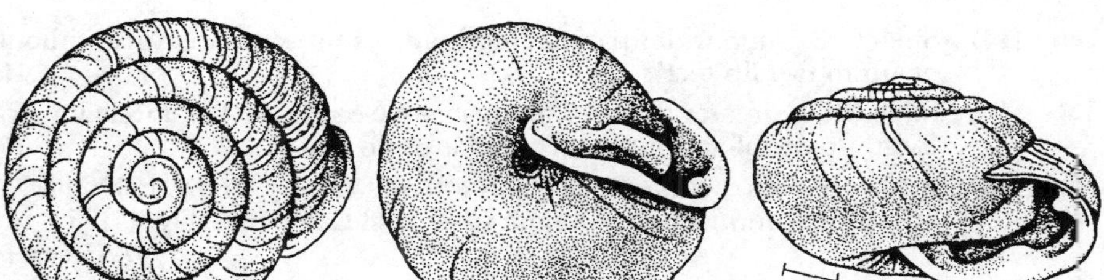

177

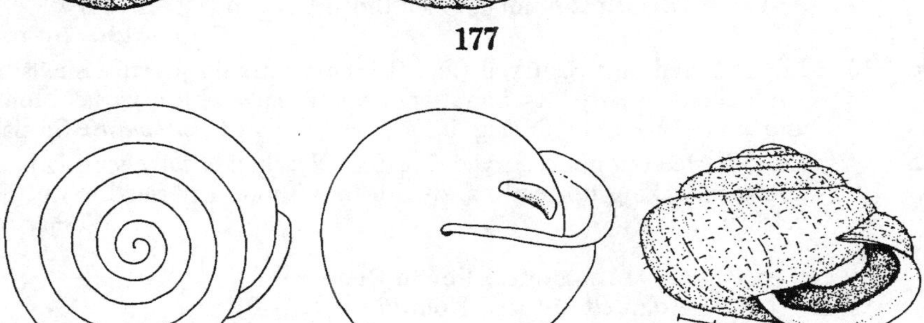

178

Fig. 9.177. *Trilobopsis loricata* (Gould). **Fig. 9.178.** *Cryptomastix germana* (Gould).

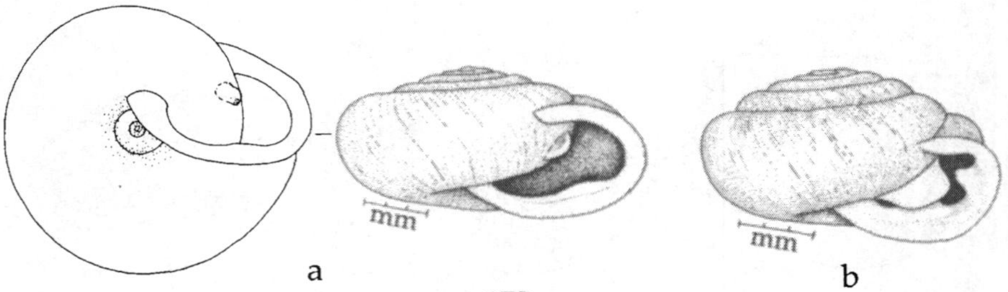

179

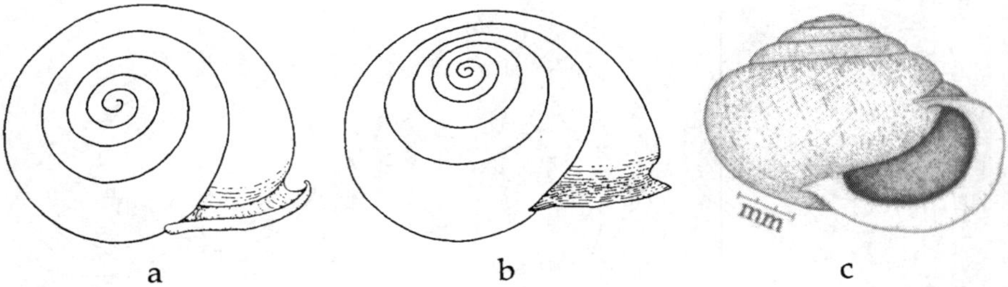

180

Fig. 9.179. *Cryptomastix:* (*a*) *C. mullani* (Bland and J. G. Cooper); (*b*) *C. sanburni* (W. G. Binney). **Fig. 9.180.** (*a*) Aperture lip reflected and recurved: (*b*) lip reflected, but not recurved; (*c*) *Vespericola columbiana* (I. Lea).

15a **(14)** Shell 6–8.5 mm wide; parietal tooth always present; aperture without or up to two lip teeth **16**

15b Shell width greater than 8.5 mm; rimate (see Fig. 9.10*c*), rarely imperforate; parietal and lip teeth present or absent **17**

16a **(15)** Aperture with a curved parietal tooth and two lip teeth; shell 6.0–8.2 mm wide; southwestern Oregon to central California (Fig. 9.177) . . ***Trilobopsis***

16b Aperture with a long parietal tooth but lip teeth lacking; shell 6.5–8.5 mm wide; British Columbia to southern Oregon (Fig. 9.178) . . . . . ***Cryptomastix*** (in part)

17a **(15)** Lip reflected and recurved (Fig. 9.180*a*); hairs on juvenile shells but adult shells usually lack hairs; aperture usually with a parietal tooth, and up to two lip teeth (Fig. 9.179) . . . . . . ***Cryptomastix*** (in part)

17b Lip reflected but not recurved (Fig. 9.180*b*); shell usually bearing hairs; aperture lacking lip teeth, and with or without a parietal tooth (Fig. 9.180*c*) . . . . ***Vespericola***

Anatomical Key to Genera

[Adapted, in part, from Pilsbry (1940)]

1a A short flagellum is present; spermatheca of medium length, not divided into bulb or duct; penial sheath imperfect (Fig. 9.175); **Ashmunellinae** ***Ashmunella***

1b Flagellum absent or vestigial; spermatheca divided into a distinct bulb and duct (Figs. 9.181–184). **2**

2a (1) Penis a rather simple tube, without a sheath (Fig. 9.181, 9.182) **Mesodontinae** **3**

2b Lower penis covered with a complete or partial sheath (Figs. 9.183, 9.184) **7**

3a (2) Male reproductive system with a well-developed epiphallus (Figs. 9.181*c*, 9.182*b*) **4**

3b Male reproductive system without an epiphallus. **5**

4a (3) Spermathecal duct slender (Fig. 9.181*c*); shell aperture narrow, basal; parietal tooth usually long, blade-like (Fig. 9.159*b, c*); eastern in distribution (Figs. 9.161, 9.162) . . . . ***Stenotrema***

4b Spermathecal duct wide (Fig. 9.182*b*); shell aperture ovate-lunate, not directed basally; parietal tooth not blade-like; western in distribution (California and Oregon). (Fig. 9.177) . . . . ***Trilobosis***

5a (3) Penis with an appendix; shell subglobose, without apertural teeth or with a parietal tooth (Figs. 9.174, 9.182*a*). . . . . ***Praticolella***

5b Penis without an appendix, or if one is present (in ***Polygyra,*** Section ***Lobosculum***), the shell is depressed and the aperture has three or four teeth, including teeth on the lip **6**

6a (5) Vagina very elongated, longer than the penis (Fig. 9.181*b*); shell with a small, D-shaped aperture and a biramose parietal tooth, or with the edge of the parietal callus raised (Fig. 9.164) . . . . ***Polygyra***

6b Vagina shorter than the penis (Fig. 9.181*a*); aperture larger; shell lacks a biramose parietal tooth, and is without a raised edge on the parietal callus (Figs. 9.163*d, e,* 9.167, 9.169, 9.170, 9.172). . . . . ***Mesodon***

7a (2) Penis with a large stimulator; penial sheath imperfect, largely adnate (Figs. 9.171, 9.176, 9.183*a*) . . . . ***Allogona***

7b Penis with a stimulator; penial sheath well developed (Fig. 9.183*b*, 9.184) . **8**

8a (7) Verge present at apex of the penial cavity (Fig. 9.183*b*) . . . ***Vespericola***

8b Verge absent (Fig. 9.184) **9**

9a (8) The concealed vestige of a flagellum is present at the junction of the epiphallus and vas deferens (Figs. 9.178, 9.179, 9.184*a*) . . . ***Cryptomastix***

9b No trace of a flagellum (Figs. 9.166, 9.168, 9.184*b*). . . . . . ***Triodopsis***

Family SAGDIDAE

The Sagdidae are a family of mostly small helicoid snails of the tropical and subtropical Middle American region. The one species found in the United States, *Lacteoluna selenina* (Gould) (Fig. 9.185), occurs in southern Florida. Its shell is small (5.0–5.5 mm in diameter), has 4½–5 whorls, is depressed, strongly shouldered, umbilicate, bluish white, and dull (not glossy). The outer lip is thin and not expanded, while the columellar lip is generally reflected.

Family OLEACINIDAE

The Oleacinidae have a large number of species in the West Indies, one of which reaches the United States, the west Cuban *Varicella gracillima* (Pfeiffer), of southern Florida. The shell of the American subspecies, *V. gracillima floridana* Pilsbry (Fig. 9.186), has more ribs and more transverse striae between the ribs than its Cuban relatives. *Varicella gracillima floridana* is a small snail, its adult shell measuring 6–7.5 mm in length with 8–8½ whorls. The shell is elongate, thin, imperforate, and

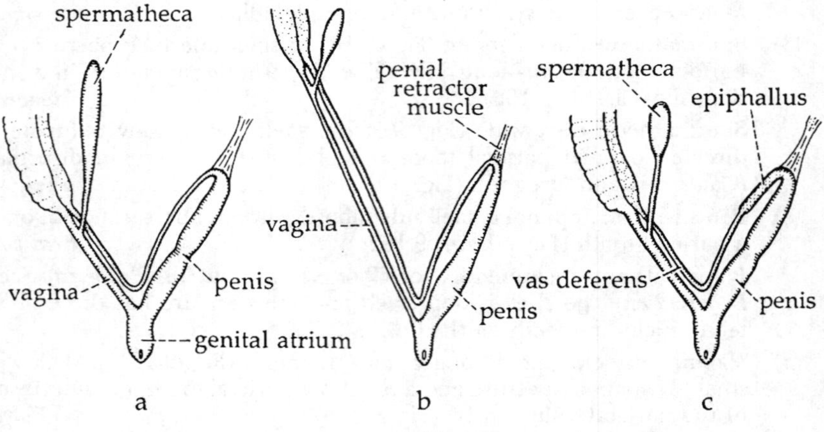

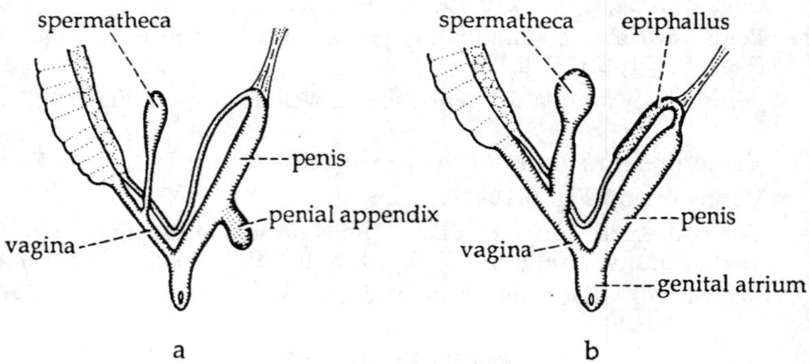

Fig. 9.181. Lower genitalia of Mesodontinae: (*a*) *Mesodon;* (*b*) *Polygyra;* (*c*) *Stenotrema.* The male system in all three genera is rather simple, lacking a penial sheath, appendix (except in *Polygyra,* section *Lobosculum*), stimulator, flagellum and verge. *Stenotrema* has an epiphallus (shaded). [Figs. 181–184 are modified from Pilsbry (1940).] **Fig. 9.182.** Lower genitalia of Mesodontinae: (*a*) *Praticolella* (the penial appendix is shaded); (*b*) *Trilobopsis* (the epiphallus is shaded). Both genera lack a penial sheath, stimulator, flagellum and verge.

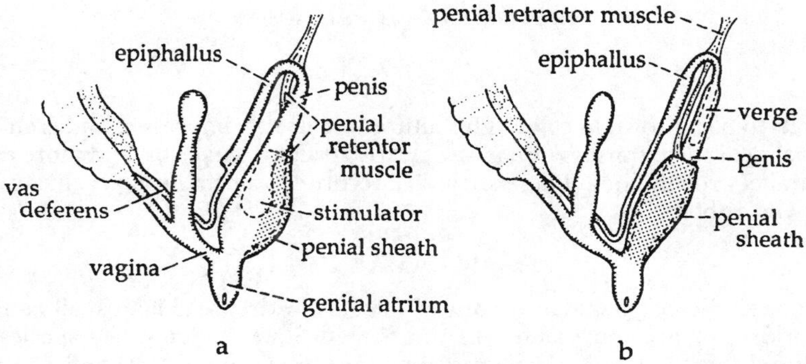

183

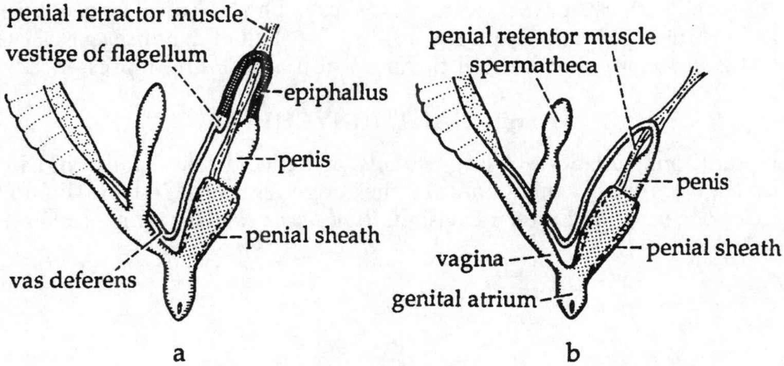

184

Fig. 9.183. Lower genitalia of Triodopsinae. The penial sheath is shaded. (*a*) *Allogona* (the stimulator is outlined in the bulbous lower penis); (*b*) *Vespericola* (the verge is outlined in the upper penis). **Fig. 9.184.** Lower genitalia of Triodopsinae. The penial sheath is lightly shaded. (*a*) *Cryptomastix* (the epiphallus is darkly shaded); (*b*) *Triodopsis.*

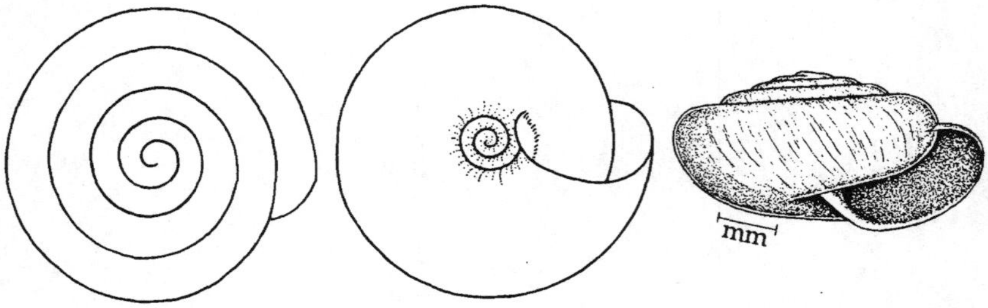

Fig. 9.185. *Lacteoluna selenina* (Gould).

corneous to pale brown in color. The sutures are deeply impressed and crenulated. The shell sculpture consists of narrow, nearly straight, axial ribs, and more numerous parallel growth lines. These snails are recorded as being solitary, living under stones and rubbish.

Family CAMAENIDAE

Although the Camaenidae are one of the most widespread land snail families in the world, they are represented in the United States by only two species, both introduced from Cuba: *Zachrysia provisoria* (Pfeiffer) (Fig. 9.187*b*) and *Z. auricoma* (Férussac). The camaenids are medium-sized to large helicid snails which lack a dart apparatus in the reproductive system. *Zachrysia provisoria* has a shell 25–29 mm in diameter with about 4½ whorls, is yellow to horn colored, somewhat glossy, solid and sculptured with rather even transverse striae. The apertural lip is not reflected, but is rather thick, particularly on the inside. The shell of *Z. auricoma* is very similar to that of *Z. provisoria,* but it is a little larger and a bit more depressed.

Family XANTHONYCHIDAE

The Xanthonychidae are a large family of helicoid land snails, and in North America they are mostly inhabitants of the zoogeographic Western Division. Only two species occur in the Eastern Division, *Bradybaena similaris* (Férussac) and *Cepolis varians* (Menke).

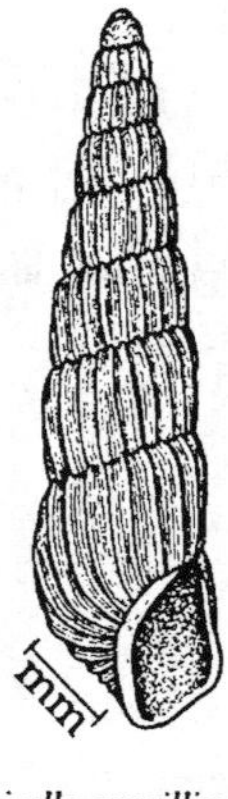

Fig. 9.186. *Varicella gracillima floridana* Pilsbry.

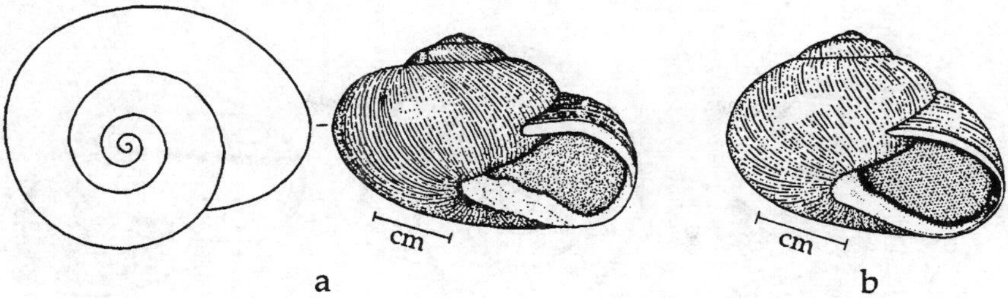

Fig. 9.187. *Zachrysia:* (*a*) *Z. auricoma* (Férussac); (*b*) *Z. provisoria* (Pfeiffer).

It is convenient to divide the family into smaller subfamilies, some of which have been considered families by various authors. The North American species have been grouped into five such subfamilies: Cepolinae, Eulotinae, Helminthoglyptinae, Humboldtianinae, and Sonorellinae. The Cepolinae are found in the southern United States (one species), the Greater Antilles, and the Bahamas. *Cepolis varians* Menke (Fig. 9.188*a*), the only species of that genus in the United States, occurs in the Florida Keys. It has a globosely conic, white shell, usually marked with brown color bands (one band at the periphery, and one band above and one below it). Its shell is 14–16 mm in diameter, with 4–6 whorls. The apertural lip is not reflected, but is usually thickened along the inner margin.

The Eulotinae are a subfamily of medium-sized to small, depressed shells, native to eastern Asia. One species, *Bradybaena similaris* (Férussac) (Fig. 9.188*b*), has been introduced into the southeastern United States by commerce. Its shell is 12–16 mm in diameter, with about 5½ whorls. Its light brown shell often bears a single spiral chestnut band. The shell is sculptured with fine, irregular growth lines and fine spiral striae. The lip is reflected, with the columellar portion partly covering the umbilicus.

The large subfamily Helminthoglyptinae is native to North, Middle, and South America. In the western United States, the species are grouped into eight genera: *Eremarionta, Eremariontoides, Helminthoglypta, Micrarionta, Mohavelix, Monadenia, Plesarionta,* and *Xerarionta.* They have large to medium-sized shells that vary in shape from globose or conic to very depressed, and may be umbilicate, perforate, or imperforate. The shells are usually banded. The apertural lip is generally expanded or reflected and lacks teeth, although there may be lamellae within the last whorl. Several of the genera of the subfamily are difficult to identify on shell characters alone, so their anatomy must be examined (Fig. 9.189*b*). In the reproductive system, there is an epiphallus and flagellum, and the penis usually has a verge. In the female reproductive system, there are one or more dart sacs and one or two mucous glands (the dart sac and one mucous gland is lacking in *Micrarionta opuntia* and the dart sac and both mucous glands are lacking in *Mohavelix* and *Eremariontoides*).

The subfamily Sonorellinae contains four genera found in New Mexico, Arizona, and eastern California: *Eremariontoides, Mohavelix, Sonorelix,* and *Sonorella.* Their shells are depressed and umbilicate and have expanded apertural lips. The reproductive system lacks dart sacs and mucous glands.

The subfamily Humboldtianinae has but one genus in the United States, *Humboldtiana,* which occurs in the Mexican plateau and nearby Texas. The shells are medium-sized to large (20–34 mm in diameter) with few (3–4⅔) rather rapidly enlarging whorls, three spiral color bands, granulose microsculpture, and unexpanded apertural lips.

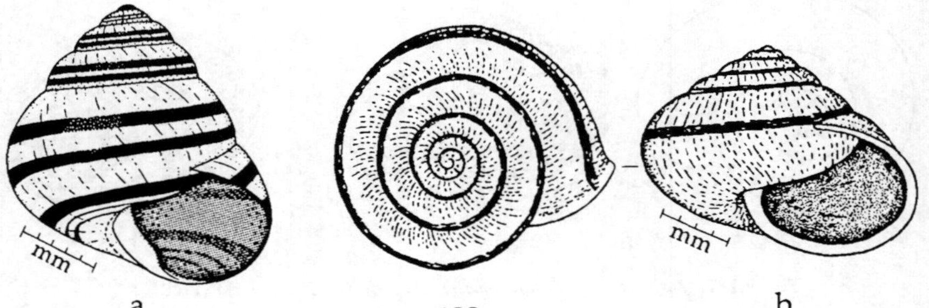

a 188 b

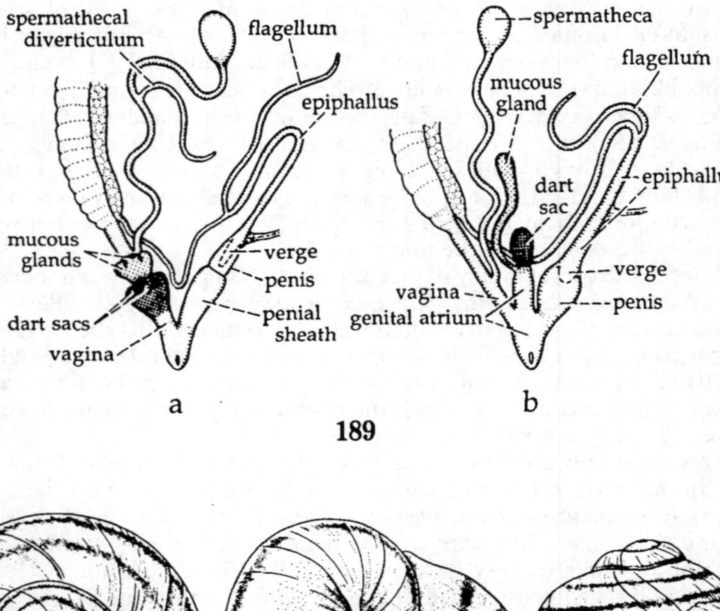

a b
189

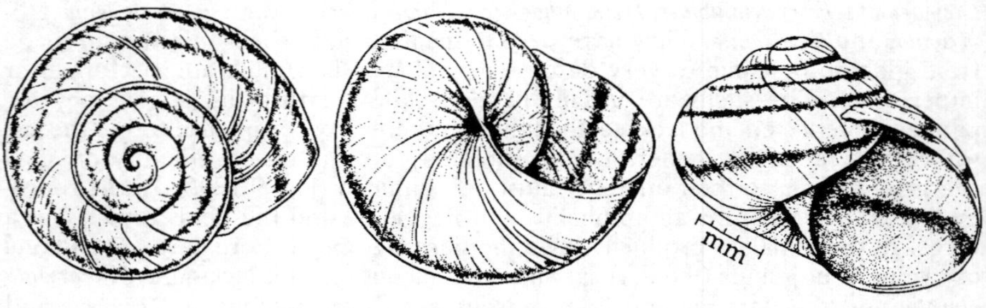

190

Fig. 9.188. (*a*) *Cepolis varians* (Menke); (*b*) *Bradybaena similaris* (Férussac). **Fig. 9.189.** Lower genitalia of (*a*) *Humboldtiana* and (*b*) *Monadenia*. Dart sacs are darkly shaded and mucous glands lightly shaded. **Fig. 9.190.** *Humboldtiana chisosensis* Pilsbry.

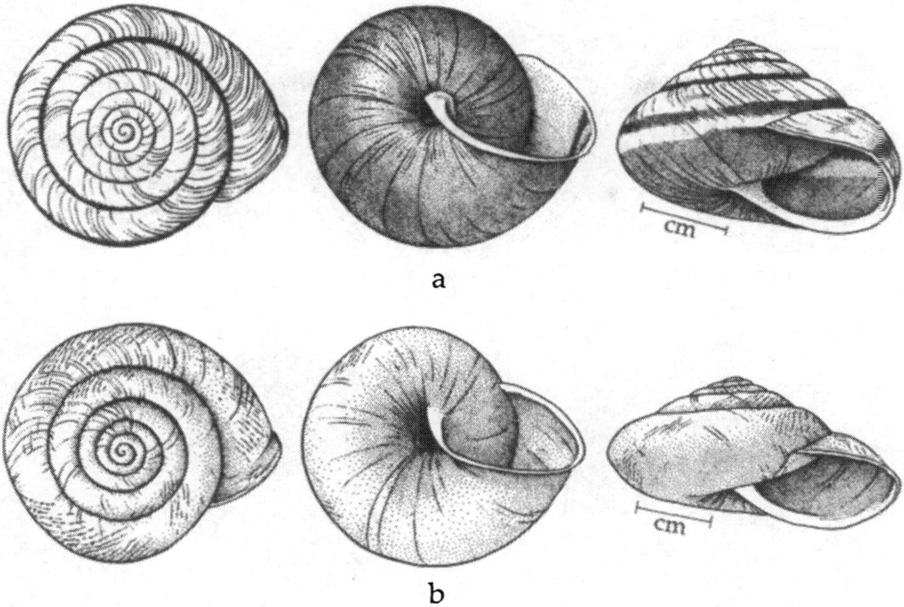

Fig. 9.191. *Monadenia:* (*a*) *M. fidelis* Gray; (*b*) *M. infumata* (Gould).

***Key to Xanthonychid Genera*†**

1a Eastern (southeastern states) in distribution **2**

1b Western (Rocky Mountain and Pacific states) in distribution **3**

2a **(1)** Shell depressed helicoid (height/width ratio less than 0.85), light brown, with reflected lip and fine spiral striae; Texas to Florida (Fig. 9.188*b*); Subfamily **Eulotinae** . . . . ***Bradybaena***

2b Shell globosely conic (height/width ratio greater than 0.85), white, without spiral striae, lip with little or no reflection (Fig. 9.188*a*); Subfamily **Cepolinae** . . . . ***Cepolis***

3a **(1)** Shell large (23–34 mm in diameter), subglobose, adults with 3½–4 whorls and three spiral color bands; lip thin, not expanded; reproductive system with four dart sacs and four mucous glands (Figs. 9.189*a*, 9.190); Subfamily **Humboldtianinae** . . . . ***Humboldtiana***

3b Shell size and shape variable, adults have more than four whorls; one spiral color band is usually present above the periphery, but other bands may be present or absent; lip thin or expanded; reproductive system with one dart sac or none, and two, one, or no mucous glands. Subfamilies **Helminthoglyptinae** and **Sonorellinae** **4**

4a **(3)** Embryonic whorl sculpture irregularly granulose, not papillose; embryonic whorls usually carinate and adult shell may be carinate; spiral color band not far above the shell periphery; shell large; reproductive

† Adapted, in part, from Pilsbry (1939).

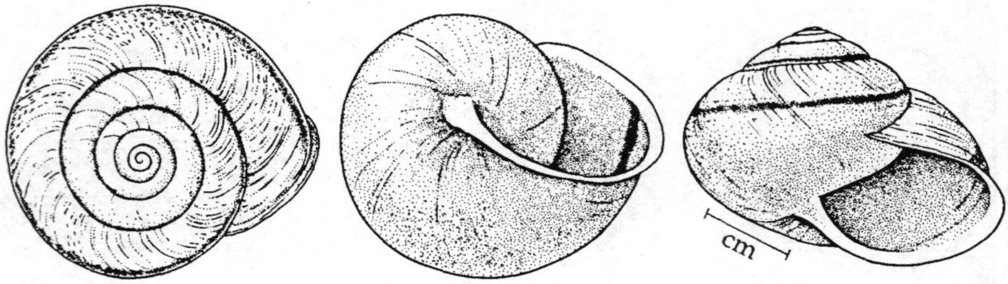

192

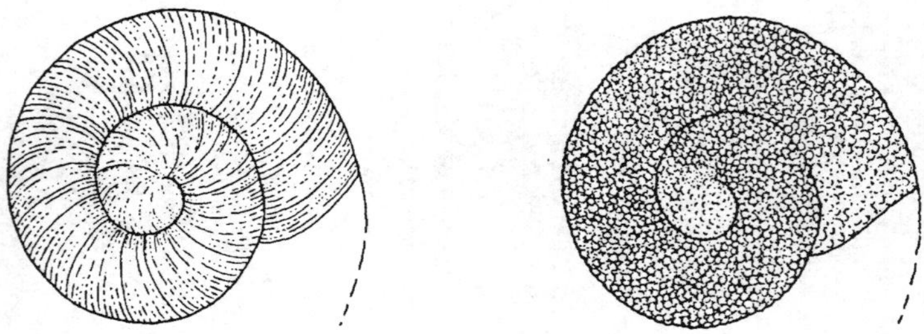

a **193** b

Fig. 9.192. *Helminthoglypta tudiculata* (A. Binney). **Fig. 9.193.** Embryonic shell sculpture: (*a*) radial (transverse) striae; (*b*) papillose.

system with one club shaped mucous gland (Fig. 9.189*b*); Pacific states (Fig. 9.191) . . . . ***Monadenia***

4b Embryonic whorl sculpture of radial lines or of papillae; embryonic whorls and adult shell not carinate; spiral color band well above the shell periphery; shell size variable; mucous glands two or none (rarely one), not club shaped **5**

5a **(4)** Shell globose to depressed, umbilicate to imperforate, usually larger with more than 5 whorls; reproductive system with a spermathecal diverticulum, flagellum long (Fig. 194*b*) **6**

5b Shell depressed, umbilicate, usually smaller with less than 5 whorls; reproductive system with or without a spermathecal diverticulum; flagellum variable, usually short **8**

6a **(5)** Shell depressed to globose; umbilicus open or closed; shell with spiral band but without additional color streaks; embronic whorls papillose (Fig. 9.193*b*); shell surface sculpture variable, including malleations, elongate papillae, and smooth, without incised spiral lines; dart sac inserts on an extension of the atrium; mucous gland ducts unite at base and insert on dart sac; California to southwest Oregon (Fig. 9.192) ***Helminthoglypta***

6b Shell globose; umbilicus closed; shell with color streaks in addition to the spiral band; embryonic whorls with radial sculpture (Fig. 9.193*a*); shell surface sculpture smooth or of fine growth lines with or without incised spiral lines; dart sac inserts on vagina; mucous gland ducts remain separate; mucous gland ducts insert in crotch between dart sac and vagina; southern California coast and islands **7**

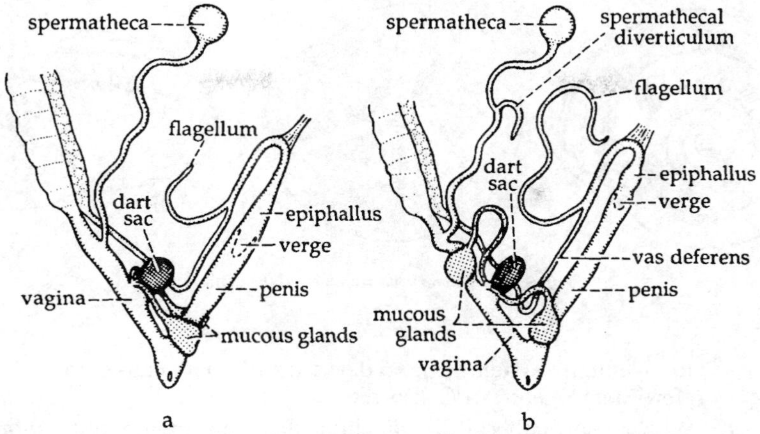

194

Fig. 9.194. Lower genitalia of (*a*) *Eremarionta indioensis* (Yates); (*b*) *Plesarionta stearnsiana* (Gabb). Dart sacs are darkly shaded and mucous glands lightly shaded. [Modified from Pilsbry (1939).]

7a (6) Cavity of the penis separated by a verge from that of the epiphallus (Figs. 9.194*b*, 9.197) . . . . ***Plesarionta***

7b Cavity of the penis continuous with that of the epiphallus (Fig. 9.198) ***Xerarionta***

8a (5) Lip strongly reflected, lip at upper part of aperture usually reflected; spiral band not far above shell periphery; embryonic whorls radially sculptured (Fig. 9.193*a*); umbilicus open or closed; flagellum moderate to long; one mucous gland descends, the other ascends; southern California islands (Fig. 9.196) . . . . ***Micrarionta***

8b Lip moderately to scarcely reflected, lip at upper part of aperture not reflected; spiral band well above shell periphery; embrynonic whorls papillose; shell umbilicate; in reproductive system, flagellum is short, mucous glands, if present, both descend; desert snails of Arizona, New Mexico, eastern California. **9**

9a (8) Reproductive system with a dart sac and two descending mucous glands (Fig. 9.194*a*); eastern California and western Arizona (Fig. 9.195) . . . . ***Eremarionta***

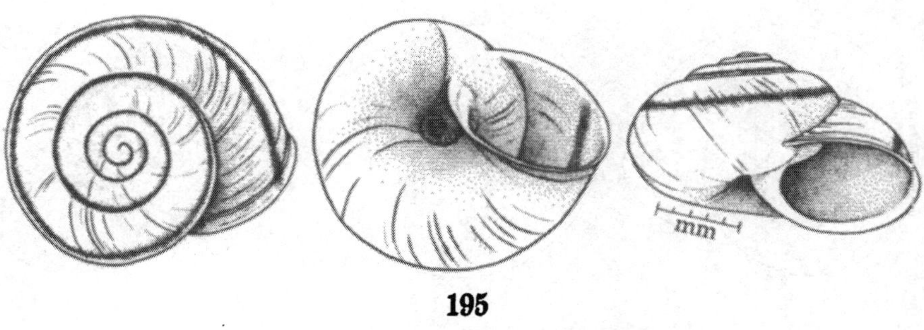

195

Fig. 9.195. *Eremarionta indioensis* (Yates).

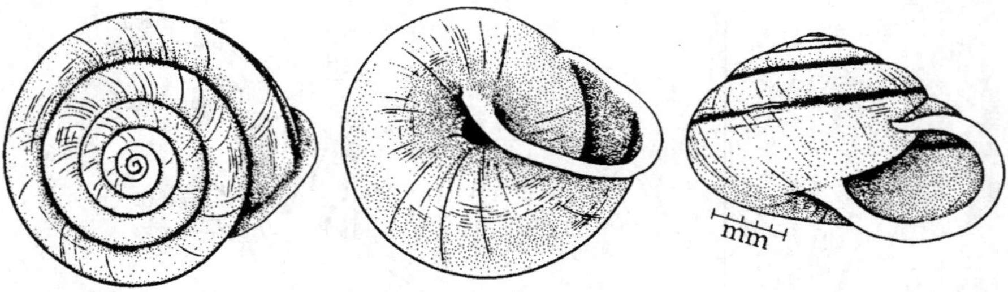

Fig. 9.196. *Micrarionta rufocincta* (Newcomb).

9b Reproductive system with no dart sac and no mucous glands; Arizona, New Mexico, eastern California **10**

10a (9) With a spermathecal diverticulum; flagellum moderate or short (Fig. 9.199*a*) **11**

10b No spermathecal diverticulum; flagellum short (Fig. 199*b*). **12**

11a (10) Flagellum of moderate length (Fig. 9.199*a*); Arizona, New Mexico, eastern California (Fig. 9.200). . . . . ***Sonorelix***

11b Flagellum short; eastern California (Fig. 9.201) . . . . ***Eremariontoides***

12a (10) Shell greater than 12 mm wide, umbilicus less than one-fifth the width of the shell; shell surface smooth or of weak growth lines, occasionally with spiral granules or hairs; Arizona and New Mexico (Fig. 9.202) ***Sonorella***

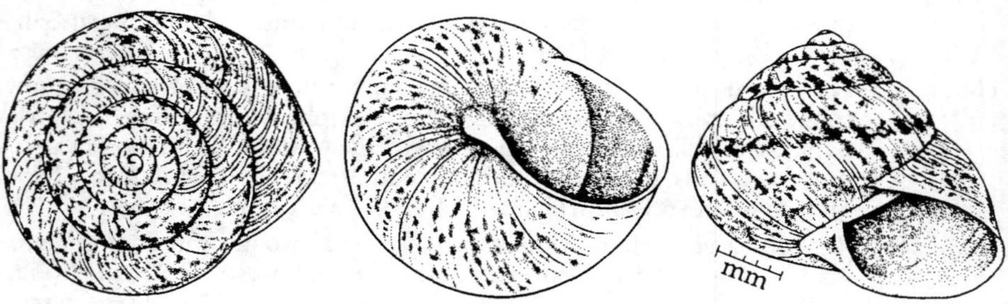

197

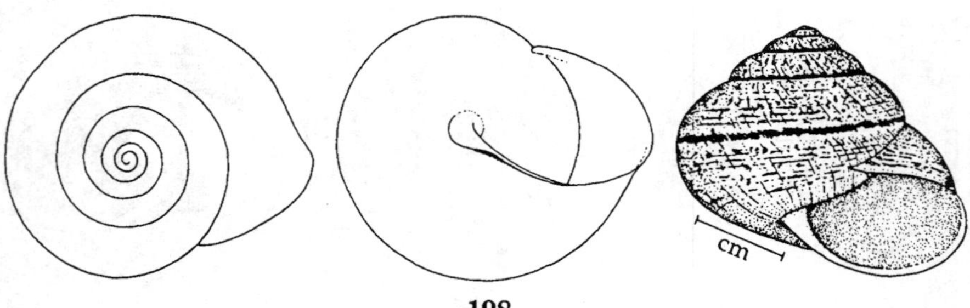

198

Fig. 9.197. *Plesarionta strearnsiana* (Gabb). **Fig. 9.198.** *Xerarionta kelletti* (Forbes).

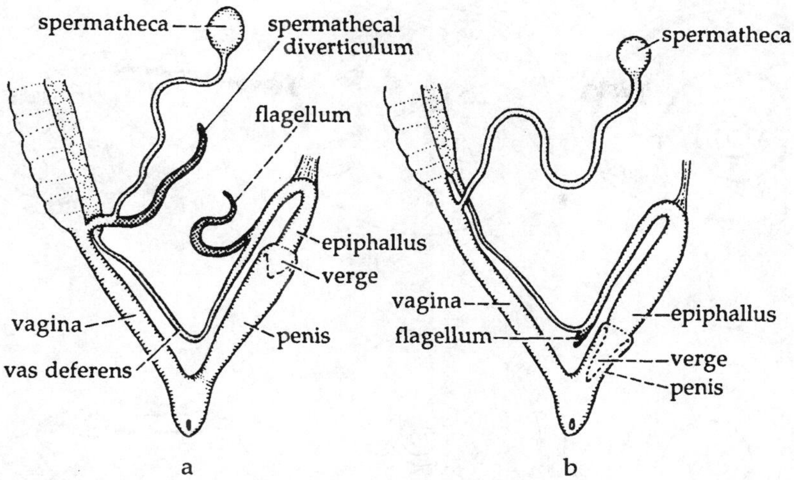

Fig. 9.199. Lower genitalia of (*a*) *Sonorelix borregoensis* (Berry) (after S. S. Berry, 1943, *Trans. San Diego Soc. Nat. Hist.* 10(1), fig. 3); (*b*) *Sonorella hachitana* (Dall) [after Pilsbry (1939)]. The spermathecal diverticulum and the flagella are shaded.

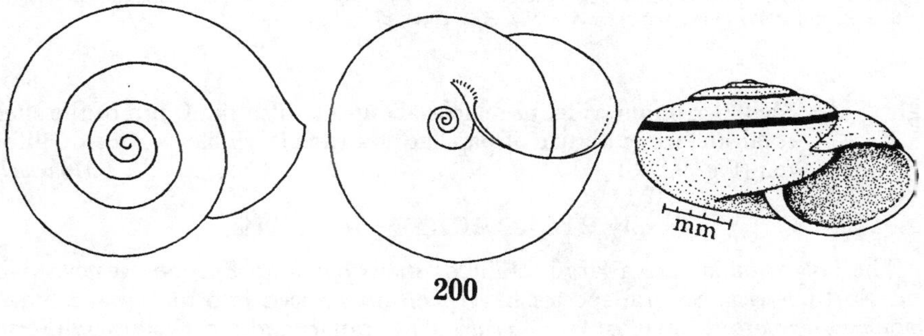

200

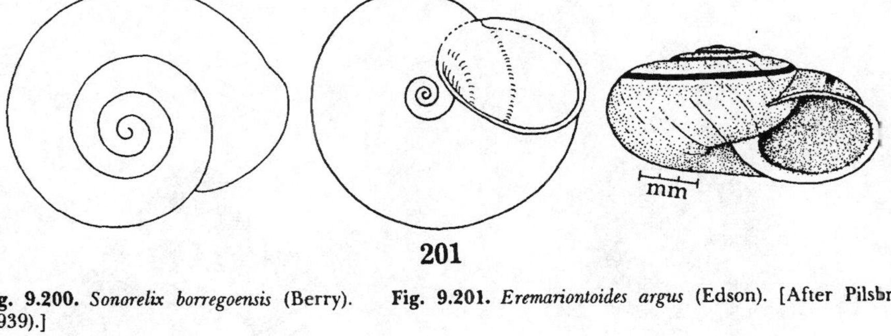

201

Fig. 9.200. *Sonorelix borregoensis* (Berry). **Fig. 9.201.** *Eremariontoides argus* (Edson). [After Pilsbry (1939).]

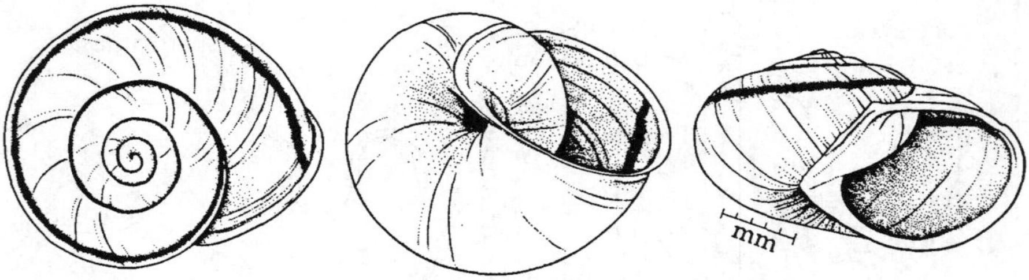

202

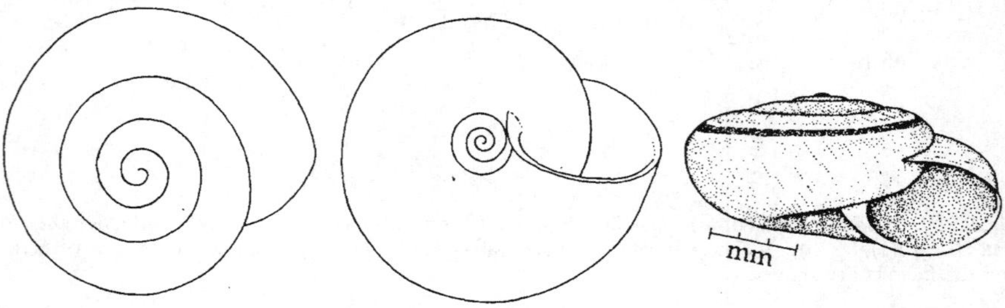

203

Fig. 9.202. *Sonorella baboquivariensis* Pilsbry & Ferriss. **Fig. 9.203.** *Mohavelix micrometalleus* (Berry). (After S. S. Berry, 1943, *Trans. San Diego Soc. Nat. Hist.* 10(1), pl. 1, figs. 1–3.)

12b Shell 9–10 mm wide, umbilicus about one-fifth the width of the shell; shell surface sculpture of papillae and radial wrinkles; eastern California (Fig. 9.203) . . . . **_Mohavelix_**

Family HYGROMIIDAE (Helicellidae)

The Hygromiidae are a large family of snails native to Europe, western Asia, and North Africa. Several species have been introduced into the United States: *Cochlicella barbara* (Linnaeus) [= *C. ventricosa* (Draparnaud); = *C. ventrosa* (Férus-

204

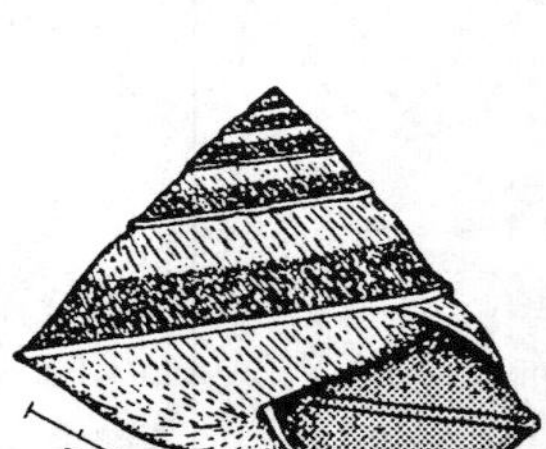

205

Fig. 9.204. *Cochlicella barbara* (Linnaeus). **Fig. 9.205.** *Trochoidea elegans* (Gmelin).

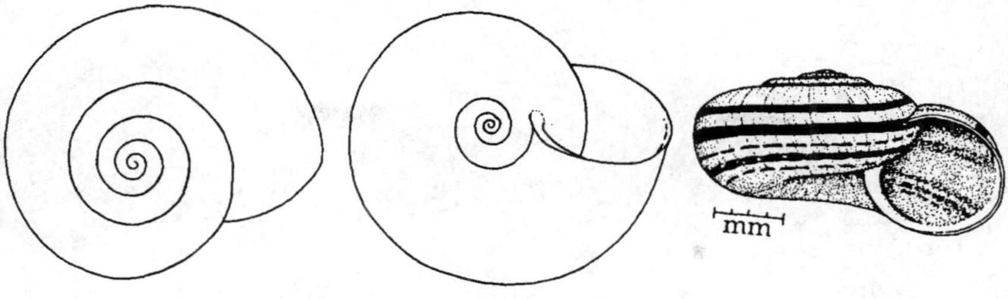

206

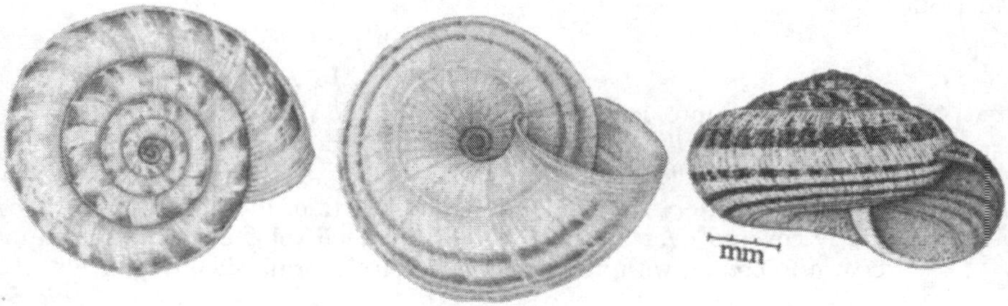

207

Fig. 9.206. *Helicella obvia* (Menke). **Fig. 9.207.** *Candidula intersecta* (Poiret).

sac)], *Helicella obvia* (Menke), *Trochoidea elegans* (Gmelin) (= *Helicella elegans*), *Candidula intersecta* (Poiret) [= *Helicella caperata* (Montagu)], *Monacha cantiana* (Montagu), *Trichia hispida* (Linnaeus) (= *Hygromia hispida*), and *T. striolata* (Pfeiffer) (= *Hygromia striolata*). They are all either umbilicate or perforate and of medium to small size. Their shapes vary from long and rather narrow in *C. barbara* to broad and flat in *T. striolata*. Anatomically, the hygromiids are characterized by having one or two dart sacs, or they may be absent, and often with accompanying accessory sacs. When a dart is present it is simple or has two blades and lacks a fluted base. There are usually more than two mucous glands, which are tubular and may be simple or branching, situated well above the dart sac. The spermathecal duct is short or medium in length, never branched (Pilsbry 1939).

Key to Genera

1a		Shell higher than wide (Fig. 9.204)	***Cochlicella***
1b		Shell wider than high	**2**
2a	**(1)**	Shell with spiral color bands or keeled periphery	**3**
2b		Shell without spiral color bands or keeled periphery	**5**
3a	**(2)**	Shell conical, periphery of whorls sharply keeled (Fig. 9.205)	***Trochoidea***
3b		Spire convex, not conical, periphery rounded	**4**
4a	**(3)**	Shell background color white; last whorl broad, about 2.5 times the width of the penultimate whorl (Fig. 9.206)	***Helicella***
4b		Shell background color reddish to yellowish, ribbing distinct, last whorl less than twice the width of the penultimate whorl (Fig. 9.207)	***Candidula***

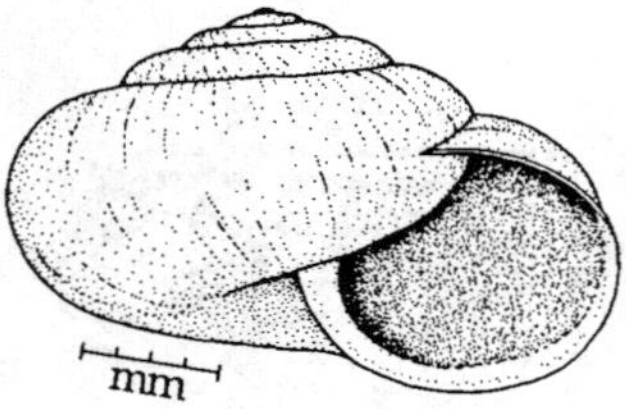

208

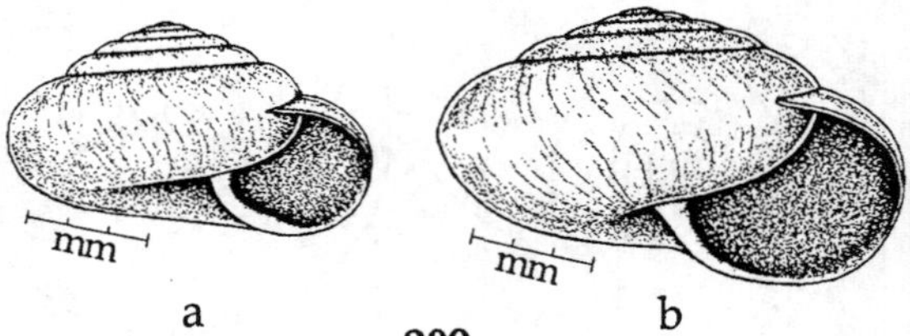

209

Fig. 9.208. *Monacha cantiana* (Montagu). **Fig. 9.209.** *Trichia: (a) T. hispida* (Linnaeus); *(b) T. striolata* (C. Pfeiffer).

5a **(2)** Umbilicus minute, less than one-tenth of the shell diameter, and partly covered over by the reflected lip; shell color creamy white; without hairs on the adult shell (Fig. 9.208) . . . . ***Monacha***

5b Umbilicus moderate, one-eighth to one-fourth of the shell diameter, not partly covered over by the reflected lip; shell color dark red-brown or brown to cream; with or without hairs on the adult shell (Fig. 9.209) . . ***Trichia***

Family HELICIDAE

The Helicidae are medium-sized to large snails of European or west Asian origin and include the edible snails (escargots) of Europe [*Helix pomatia* Linnaeus, *H. aspersa* Müller, *Otala lactea* (Müller), and *Eobania vermiculata* (Müller)]. Helicid shells are commonly banded, generally wider than high, and are somewhat loosely coiled, so that the central axial column is hollow or umbilicate. However, in the adult shell the umbilicus is often closed over by an extension of the columellar lip.

Anatomically, the family is characterized by having one dart sac which contains a two- or four-bladed dart. Between the vagina and dart sac are two tubular mucous glands, which may vary from simple to many- branched. The penis contains a papilla or a short verge. An epiphallus is present and usually there is a flagellum (Pilsbry 1939).

Helix aspersa, regarded as a garden pest, has become well established in California and in other states in the east and west, and is spreading. The same holds true

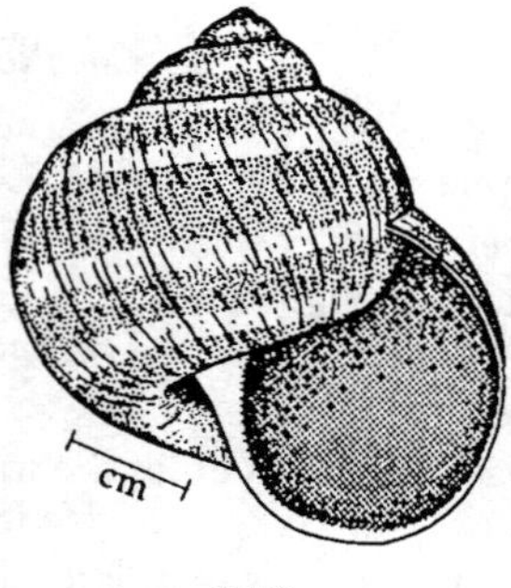

210

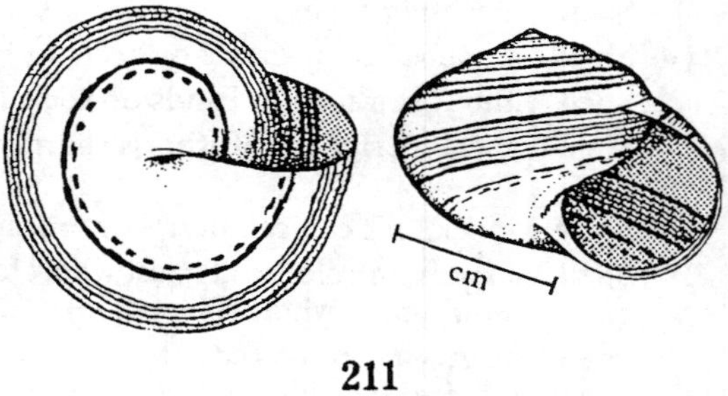

211

Fig. 9.210. *Helix pomatia* Linnaeus. **Fig. 9.211.** *Theba pisana* (Müller).

212

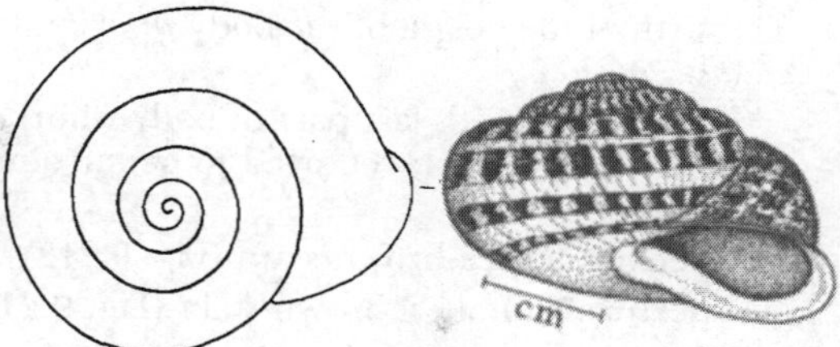

213

Fig. 9.212. *Otala lactea* Müller. **Fig. 9.213.** *Eobania vermiculata* (Müller).

for *H. pomatia* in Michigan. Other members of the family apparently have spread relatively little from points of introduction and generally are found close to sites of human activity. *Theba pisana* (Müller), a serious pest to the California citrus crop, once eradicated, has returned.

Key to Genera

1a		Aperture lip not reflected or rarely slightly expanded (Fig. 9.210) . . .	***Helix*** (in part)
1b		Aperture lip reflected or noticeably expanded (see Fig. 9.12*b,c*)	**2**
2a	**(1)**	Shell with microscopic spiral sculpture.	**3**
2b		Shell without microscopic spiral sculpture	**5**
3a	**(2)**	Shell narrowly umbilicate; last part of body whorl descends less than one-	

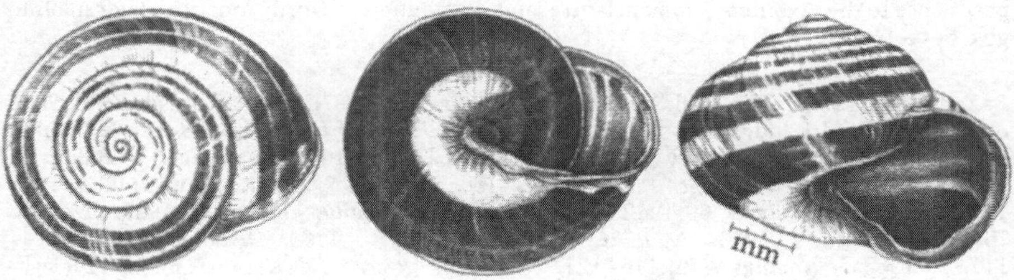

214

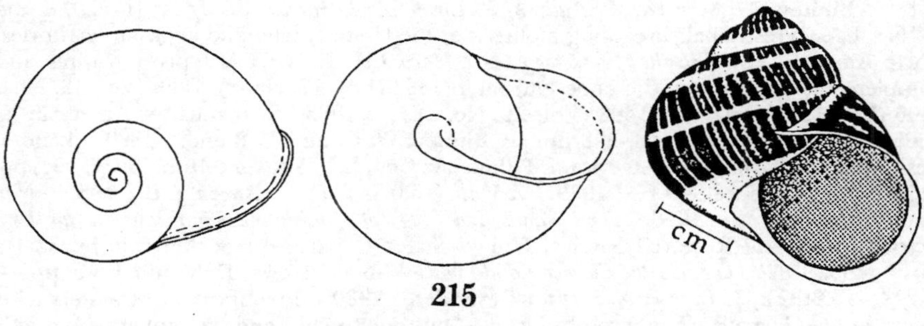

215

Fig. 9.214. *Cepaea nemoralis* (Linnaeus). **Fig. 9.215.** *Helix aspersa* Müller.

		fourth of the height of the body whorl; introduced in southern California (Fig. 9.211)	***Theba***
3b		Umbilicus closed; last part of body whorl descends about half the height of the body whorl; reported to be introduced in California and Arizona	**4**
4a	**(3)**	Aperture with a brown stain (Fig. 9.212)	***Otala***
4b		Aperture without a brown stain (Fig. 9.213)	***Eobania***
5a	**(2)**	Shell medium in size, less than 25 mm in diameter; color bands continuous but not always present (Fig. 9.214)	***Cepaea***
5b		Shell large, generally more than 30 mm in diameter; color bands always present and interrupted by yellow flecks or streaks (Fig. 9.215)	***Helix*** (in part)

Figure Credits

Figs. 27, 29–31, 33, 34, 44, 48, 63, 87a (left two figures), 90 (left two figures), 92 (left two figures), 93 (left two figures), 108, 122, 134 (left two figures), 143 (left two figures), 164 and 165 were drawn by John L. Tottenham; Figs. 138, 152, 154–158, 175–177, 190–192, 195–197 and 202 by Margaret Van Bolt; Figs. 35c, 45, 161, 162, and 170–173 by Arthur Cushman; Figs. 112, 113, 115, 117, 119, 123, 125 (left figure), 127 (left figure), 132, 136, and 140–142 by Daniel S. Cutler; and Figs. 67, 166 (left figure), 204, 207, 212 and 214 by Bonnie Hall. All other figures, except where indicated in the legends, are by the first author.

References

A few of the more comprehensive publications on the North American (north of Mexico) fauna, or regional parts of the fauna, are listed below. (There are too many references of pertinence to the taxonomy, nomenclature, and distribution of North American land snails to give but a few.)

Baker, F. C. 1902. The Mollusca of the Chicago area. Part II. *Chic. Acad. Sci. Nat. Hist. Surv. Bull.* 3(2):137–410, index, pls. 28–36. **1939.** *Fieldbook of Illinois Land Snails.* Manual 2. State of Illinois Natural History Survey Division. pp. i–xi, 1–166, frontispiece. **Baker, H. B. 1925.** *Proc. Acad. Nat. Sci. Phila.* 77:157–184; pls. 3–6. **1955.** *Nautilus* 68(4):109–112. **1956a.** *Nautilus* 69(4):128–139. **1956b.** *Nautilus* 70(1):34. **1957.** *Nautilus* 70(4):141–142. **1958.** *Nautilus* 71(4):141–148. **1960.** *Nautilus* 73(3):114–117. **1962a.** *Nautilus* 75(3):116–121. **1962b.** *Proc. Acad. Nat. Sci. Phila.* 114(1):1–21. **1963.** *Proc. Acad. Nat. Sci. Phila.* 115(8):191–259. **1964.** *Proc. Acad. Nat. Sci. Phila.* 116(4):149–193. **Bequaert, J. C. and W. B. Miller. 1973.** *The Mollusks of the Arid Southwest, with an Arizona Checklist.* University of Arizona Press, Tucson, Ariz., pp. i–xvi, 1–271. **Binney, W. G. 1873.** *Bull. Mus. Comp. Zool. Harvard Col.* 3:191–220, 1 map. **1878.** The terrestrial air-breathing mollusks of the United States and adjacent territories of North America, Vol. 5. *Bull. Mus. Comp. Zool. Harv. Col.* 4:i–v, 1–449, pls. 1–74, pls. 1–16. [Supplements appeared in the same journal in 1883 (Vol. 11, No. 8), 1886 (Vol. 13, No. 2), 1890 (Vol. 19, No. 4), and 1892 (Vol. 22, No. 4).] **1885.** A manual of American land snails. *U.S. Nat. Mus. Bull.* 28. 528 pp. **Binney, W. G. and T. Bland. 1869.** Land and fresh water-shells of North America. I. Pulmonata Geophila. *Smithson. Misc. Collect.* 194, pp.i–xii, 1–316. **Breure, A.S.H. 1979.** *Z. Verh.* (168):1–215. **Burch, J. B. 1955.** *Nautilus* 69(2):62–69. **1960.** *Some Snails and Slugs of Quarantine Significance to the United States.* Agricultural Research Service, United States Department of Agriculture, 81(1):1–73. **1962.** *How to Know the Eastern Land Snails.* Wm. C. Brown, Dubuque, Iowa, pp. i–vi, 1–214. **Burch, J. B. and A. S. Van Devender. 1980.** Identification of eastern North American land snails. The Prosobranchia, Opisthobranchia and Pulmonata (Actophila). *Walkerana* 1(2):33–80. **Cain, A. J. 1983.** In *The Mollusca,* vol. 6, Ecology, pp. 597–647. Academic Press. **Call, R. E. 1900.** *A Descriptive Illustrated Catalogue of the Mollusca of*

Indiana. 24th Annual Report of the State Geologist for 1899, Department of Geology and Natural Resources of Indiana, pp. 335–535, pls. 1–78. **Cheatum, E. P. and R. W. Fullington. 1971.** The aquatic and land Mollusca of Texas. 1. The Recent and Pleistocene members of the gastropod family Polygyridae in Texas. *Dallas Mus. Nat. Hist. Bull.* 1, pp. i–vi, 1–74. **1973.** The aquatic and land Mollusca of Texas. 2. The Recent and Pleistocene members of the Pupillidae and Urocoptidae (Gastropoda) in Texas. *Dallas Mus. Nat. Hist. Bull.* 1, pp. i–vi, 1–67 pp. **Chichester, L. F. and L. L. Getz. 1973.** The terrestrial slugs of northeastern North America. *Sterkiana* (51):11–42. **Dall, W. H. 1910.** *Land and Fresh Water Mollusks of Alaska and Adjoining Regions.* In C. Hart Merriman (ed.). *Alaska,* Harriman Alaska Series, Vol. 13. Smithsonian Institution Publ. 2000, pp. i–xii, 1–171, pls. 1, 2. **DeKay, J. E. 1843.** Part V. Mollusca. In *Zoology of New York, or the New-York fauna; comprising detailed descriptions of all the animals hitherto observed within the state of New York; with brief notices of those occasionally found near its borders: and accompanied by appropriate illustrations.* Carroll and Cook, Albany, N.Y. 271 pp. **Fullington, R. W. and W. L. Pratt, Jr. 1974.** The aquatic and land Mollusca of Texas. Part 3. The Helicinidae, Carychiidae, Achatinidae, Bradybaenidae, Bulimulidae, Cionellidae, Haplotrematidae, Helicidae, Oreohelicidae, Spiraxidae, Streptaxidae, Strobilopsidae, Thysanophoridae, Valloniidae (Gastropoda) in Texas. *Dallas Mus. Nat. Hist. Bull.* 1, pp. i–vi, 1–48. **Goodrich, C. 1932.** *The Mollusca of Michigan.* Michigan Handbook Series 5. University Museums, University of Michigan, Ann Arbor, Mich., pp. 1–120, pls. 1–7. **Goodrich, C. and H. van der Schalie. 1944.** A revision of the Mollusca of Indiana. *Am. Midl. Nat.* 32(2):257–326. **Hanna, G. 1966.** Introduced mollusks of western North America. *Occas. Pap. Calif. Acad. Sci.* (48):1–108, pls. 1–4. **Henderson, J. 1924.** Mollusca of Colorado, Utah, Montana, Idaho and Wyoming. *Univ. Colo. Stud.* 13(2):65–223. **1929.** Non-marine Mollusca of Oregon and Washington. *Univ. Colo. Stud.* 17(2):47–190. **1931.** Molluscan provinces in the western United States. *Univ. Colo. Stud.* 18(4):177–186. **1936a.** Mollusca of Colorado, Utah, Montana, Idaho and Wyoming—Supplement. *Univ. Colo. Stud.* 23(2):81–145, pls. 1, 2. **1936b.** Non-marine Mollusca of Oregon and Washington—Supplement. *Univ. Colo. Stud.* 23(4):251–280, pls. 1, 2. **Hubricht, L. 1985.** The distributions of the native land mollusks of the eastern United States. *Fieldiana Zool.* (N.S.) (24):i–viii, 1–191. **Keep, J. and J. L. Baily, Jr. 1935.** *West Coast Shells. A description in familiar terms of the principal marine, fresh-water, and land mollusks of the United States. British Columbia, and Alaska, found west of the Sierra.* Stanford University Press, Stanford, Calif., pp. i–xi, 1–350. **LaRocque, A. 1970.** Terrestrial Gastropoda. Chapter 7 in *Pleistocene Mollusca of Ohio.* Division of Geological Survey, Department of Natural Resources, State of Ohio. Bulletin 62, Part 4, pp. xxvii–xxxiv, 555–800, pls. 15–18. **Leonard, A. B. 1959.** Handbook of gastropods in Kansas. *Univ. Kans. Mus. Nat. Hist. Misc. Publ.* 20, pp. 1–224, map. **MacMillan, G. K. 1949.** The land snails of West Virginia. *Ann. Carnegie Mus.* 31(7):89–207, pls. 1–15, index pp. i–iv. **Morse, E. S. 1864.** Observations on the terrestrial Pulmonifera of Maine, including a catalogue of all the species of terrestrial and fluviate Mollusca known to the state. *J. Portland Soc. Nat. Hist.* 1(1):1–63. **Oughton, J. 1948.** A zoogeographical study of the land snails of Ontario. *Univ. Toronto Stud. Biol. Ser.* (37):i–xi, 1–126, frontispiece, table. **Pilsbry, H. A. 1939–1948.** Land Mollusca of North America (north of Mexico). *Monogr. Acad. Nat. Sci. Phila.* 3. Vol. 1, No. 1, pp. i–xvii, 1–573, index pp. i–ix, 1939; Vol. 1, No. 2, pp. i–viii, 575–997, index pp. i–ix, 1940; Vol. 2, No. 1, pp. i–viii, 1–520, 1946; Vol. 2, No. 2, pp. i–xlvii, 521–1113. **Riddle, W. A. 1983.** In *The Mollusca,* vol. 6, Ecology, pp. 431–461. Academic Press. **Robertson, I. C. S. and C. L. Blakeslee. 1948.** The Mollusca of the Niagara Frontier Region. *Bull. Buffalo Soc. Nat. Sci.* 19(3):i–xi, 1–191, frontispiece, map. **Rollo C. D. and W. G. Wellington, 1979.** Can. J. Zool. 57:846–855. **Smith, A. G. 1957.** Snails from California caves. *Proc. Calif. Acad. Sci. Ser. 4* 29((2):21–46. **Solem, A. 1976.** *Endodontoid Land Snails from Pacific Islands,* Part 1. Field Museum of Natural History, Chicago. **1983.** *Endodontoid Land Snails from Pacific Islands,* Part 2. Field Museum of Natural History, Chicago. **Sterki, V. 1907.** A preliminary catalogue of the land and freshwater molluscs of Ohio. *Proc. Ohio State Acad. Sci.* 4(8):367–402 (Special Paper 12). **Taft, C. 1961.** The shell-bearing land snails of Ohio. *Bull. Ohio Biol. Surv.* (N.S.) 1(3):i–xii, 1–108. **Thompson, F. G. and H. G. Lee. 1980.** *Malacol. Rev.* 13:37–44. **Vagvolgyi, J. 1975.** *Syst. Zool.* 24:465–488. **Walker, B. 1906.** *An Illustrated Catalogue of the Mollusca of Michigan.* Part I. *Terrestrial Pulmonata (Land Snails).* Michigan State Board of Geological Survey, Report for 1905. Wynkoop Hallenbeck Crawford Co., Lansing, Mich., pp. 427–531, frontispiece, table. **1928.** The terrestrial shell-bearing Mollusca of Alabama. *Misc. Publ. Mus. Zool. Univ. Mich.* (18):1–180. **Wurtz, C. B. 1955.** *Proc. Acad. Nat. Sci. Phila.* 107:99–143.

10

Oligochaeta: Enchytraeidae

MADHAB C. DASH

The Enchytraeidae is a family of Oligochaetes that occur in terrestrial, littoral, and aquatic habitats. They are usually pale-colored worms. The body size is generally 10–20 mm long and anatomically, they form a relatively simple and uniform group. Some 600 species of Enchytraeidae are now known from different parts of the world. They have been recorded from every continent, but because of their abundance in the arctic and north temperate zones and sensitiveness to drought, they have been supposed to be of arctic origin. They, however, reach greatest abundance in acid soils with high organic matter content. In tundra, moorlands, and temperate coniferous forest soils, the enchytraeids form a large percentage of microfaunal biomass. They form some 35.6–92.2% ($\bar{x}$ = 73.1%) of the total microfaunal biomass in some nine sites of Barrow, Alaska (Maclean 1974). Peachey (1963), Dash and Cragg (1972b), and Kitazawa (1977) have recorded densities up to 290 × 10^3 per square meter, 22 × 10^3 per square meter, and 180 × 10^3 per square meter in Pennine moorland, England; aspen woodland, Alberta, Canada; and in Japanese subalpine coniferous forest soils, respectively. Their role in the breakdown of organic matter in sewage beds has been demonstrated by Reynoldson (1939a, 1947a,b, 1948). Zachariae (1963, 1964) examined the form and distribution of invertebrate feces in many soils and concludes that the so-called collembolan soils (Kubiena 1953) have been formed by Enchytraeidae. He claims that enchytraeids take part in soil formation.

The relative importance of an enchytraeid population in a community can be calculated as the percentage of the total annual rate of energy input into the habitat released by the Enchytraeidae population. An assessment of energy utilization of the Enchytraeidae population is impossible from respiratory data alone. Information on population structure, distribution of mortality between age classes, growth rates of individuals, and calorific values of cocoons and worms must be available. O'Connor (1967) has calculated that enchytraeids release 11% of the total annual energy input of a Douglas fir plantation in North Wales. The total annual energy input into an aspen forest floor of the Kananaskis area of Alberta, Canada, was estimated to be 2360 kcal/m^2 and the respiratory activity of enchytraeids released about 2.2% of the total annual energy input. The total energy utilized by an Enchytraeidae population would amount to about 3–4% of the total energy input if energy expenses for body building and reproduction are included (Dash and Cragg 1972b). Satchell (1971) has calculated the

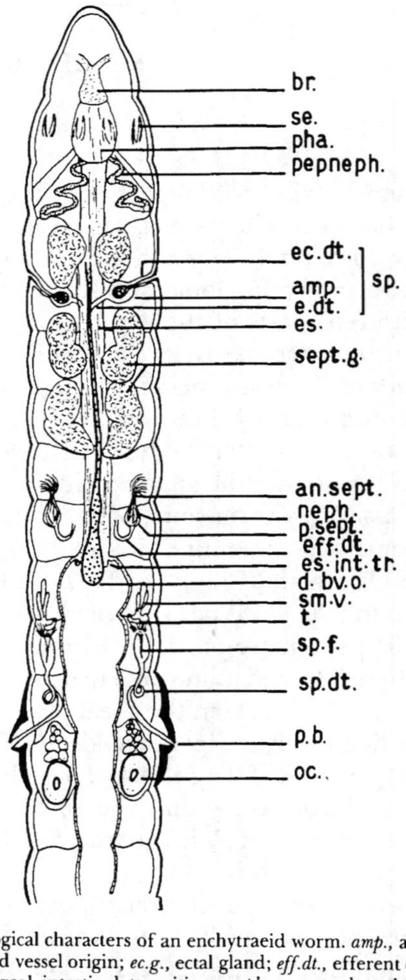

Fig. 10.1. Typical morphological characters of an enchytraeid worm. *amp.*, ampulla; *an. sept.*, ante-septal; *br.*, brain; *d.bv.o.*, dorsal blood vessel origin; *ec.g.*, ectal gland; *eff.dt.*, efferent duct; *e.op.*, ental opening; *es.*, esophagus; *es.int.tr.*, esophageal intestinal transition; *m.pha.*, muscular pharynx; *neph.*, nephridia; *oc.*, oocyte; *pha.*, pharynx; *p.b.*, penial bulb; *pepneph.*, peptonephridia; *p.sept.*, postseptal; *se.*, setae; *sept.g.*, septal gland; *sm.v.*, seminal vesicle; *sp.*, spermatheca; *sp.dt.*, sperm duct; *sp.f.*, sperm funnel; *t.*, testes.

respiration of soil organisms in a broad-leaved forest of England and found that the values of fungi, bacteria plus actinomycetes, and soil fauna (enchytraeids, nematodes, and lumbricids) are 1727, 77, and 361 kcal/m^2 per year, respectively. Among the soil fauna, the value for enchytraeids was 167 kcal/m^2 per year. These studies indicate the importance of Enchytraeidae in the decomposer system.

External Morphology

Morphological characters described here can be observed in live material as presented in the typical body form of an enchytraeid worm (Fig. 10.1).

Size

The body size of worms may be less than 10 mm to about 50 mm long. A live worm when fully contracted may be reduced to half of its fully extended size. The average length somewhere between maximum extension and minimum contraction of a worm may be recorded while it moves in a film of water. The length of preserved worms depends on the fixation technique: cold Bouin's fixative produces coiled worms, hot Bouin's fixative produces straight but slightly contracted worms.

Segments

The newly hatched worms possess about 15–20 segments and the number increases steadily with age. Worms become mature when the reproductive system becomes fully differentiated. Segment counts for taxonomic purposes refer to fully matured worms with developed reproductive systems containing mature ova and spermatozoa in the receptacle. The prostomium plus peristomium is counted as the first segment, which does not bear any setae.

Segment counts show some intraspecific variation and usually the range is given when describing species. The segments are indicated by Indo-Arabic (Roman) numerals; thus IV–V means segment numbers 4 and 5. The range of segment counts is often significantly different in populations from different habitats. It is not known whether this is due to environmental or genetic factors.

Color

Most species are either pale white or gray. Color is often determined by the gut contents. Chloragogen cells present in the intestine also give a certain tinge specially marked in some greenish *Lumbricillus* sp. The presence of large lymphocytes rich in colorless, refractile globules causes an intensively white color. This is observed in some *Marionina* sp. and *Cernosvitoviella* sp. Distinctly yellow lymphocytes give a distinct yellow color in some species (*Mesenchytraeus* sp.). A yellow or brown pigment may be present in the tissues instead of lymphocytes. This gives the worms a yellow or brown tinge.

Cutaneous Glands

These glands may be rectangular, oval, or with irregular outline and with the long axis on the transverse position on the segments. These glands are present in most of the Enchytraeidae, but their conspicuousness varies from species to species. They are usually arranged in transverse rows between the setal bundles in the middle of the segments. Yellow, red, orange, and brown cutaneous glands may occur in some *Marionina* sp. and *Fridericia* sp.

Setae

Setae are present in all Enchytraeidae except the genus *Achaeta*. Some species of *Achaeta* may possess oval or spherical setal follicles, which are believed to occupy the positions of former setae. Four bundles of setae per segment of the body are usually present. Two bundles are lateral (dorsolateral) and two bundles are ventrolateral and are usually called dorsal and ventral bundles. Any reduction of the typical four bundles of setae should be mentioned in the description of a species.

The shape of the setae shows variation. They are, however, usually simple. The shape may be straight with or without ental hook, curved with or without ental hook, or sigmoid with or without nodules. The genus *Propappus* (so far not discovered in North America) has sigmoid setae with forked distal ends. Intermediary types occur frequently. Straight setae are commonly found near the center of the setal bundles in the genera *Henlea* and *Enchytraeus*. Straight setae with ental hooks are typical of the genera *Enchytraeus, Hemienchytraeus, Enchytronia, Fridericia,* and *Hemifridericia*. Curved setae occur in the genera *Enchytraeus* and *Henlea*. In all Enchytraeidae, the setae are slightly sigmoid, and therefore it is important to realize that to be classified as sigmoid, the setae must possess a distinct double curvature. Sigmoid setae are typical of the genera *Propappus, Buchholzia, Mesenchytraeus, Lumbricillus, Cernosvitoviella,* and *Cognettia*. In some *Mesenchytraeus* sp. and *Cernosvitoviella* sp., swelling is well defined and called a nodulus.

Setal arrangements may vary from genus to genus. The simplest arrangement occurs in *Enchytraeus, Hemienchytraeus,* and in bisetose species of *Fridericia*. In *Fridericia* sp. except the bisetose species, the setae are disposed pairwise, the central pair being shortest. A similar pattern is often found in

Henlea sp. In *Buchholzia, Bryodrilus,* and in some species of *Lumbricillus* the setae may be arranged fanwise.

The number of setae varies between bundles of the same segment, between segments of one individual, between individuals of the same species, and between species. Dorsal bundles have usually fewer setae than ventral bundles. The number of setae in the four bundles is written in a sequence: dorsal anteclitellar–ventral anteclitelar: dorsal postclitellar–ventral postclitellar (i.e., 2,3,4,5–3,4,5 : 3,4,5,6–4,5,6).

Head Pore

This links the coelom and the exterior. It lies either on the anterior end of the prostomium or at the transition between pro- and peristomium. It may be absent in *Propappus, Stercutus,* and *Aspidodrilus.*

Dorsal Pore

Dorsal pores are present in *Fridericia* sp. They form an open, dorsal communication between coelom and exterior from VII onward. They do not provide distinguishing characters at species level.

Clitellum

The clitellum extends over a varying number of segments which contain male and female pores. It usually occurs in XII–XIII. The cutaneous glands are modified and usually occur in greater abundance in clitellar segments than in other segments.

Digestive Characteristics

Gut

The gut consists of mouth cavity, pharynx, esophagus, and intestine. Salivary glands known as peptonephridia of unknown function may arise from the pharynx and extend backward freely floating in the coelom in *Enchytraeus* sp. and *Fridericia* sp. Three pairs of primary septal glands of unknown function are present at IV–VII. The glands may merge dorsally and may possess ventral lobes. The esophagus merges gradually with the intestine in many enchytraeids (*Mesenchytraeus, Fridericia, Enchytraeus, Lumbricillus, Cognettia, Cernosvitoviella, Hemienchytraeus, Stercutus*). In others the esophageal–intestinal transition is sudden and the intestine is usually wider. Esophageal appendages may occur at IV or VI of some species. Intestinal diverticula often occur in the transitional region. These diverticula are usually sac or pouch shaped and directed forward. The peptonephridia, esophageal–intestinal traction, and intestinal diverticula are important generic characteristics.

Feeding

A knowledge of the food requirements of Enchytraeidae is essential to assess their role in the decomposer system. Jegen (1920) [quoted in O'Connor (1967)] observed that enchytraeids ingested plant remains and silica. Clark (1949) [quoted in O'Connor, (1967)] reported that in Australian forest soil, enchytraeids ingest finely divided plant remains together with large quantities of fungal mycelium. Reynoldson (1939a) reported that enchytraeids of sewage beds fed on algae, fungi, and bacteria. Dougherty and Solberg (1960, 1961) have succeeded in rearing *Enchytraeus fragmentosus* (Bell) in monogenic cultures with *E. coli* growing on agar slants and also in axenic culture on nutrient agar supplemented with heated lamb liver extract. Observing a large number of enchytraeids in and around the roots of nematode-infested strawberry plants, Jegen (1920) [quoted in O'Connor (1967)] suggested that enchytraeids can kill root-feeding nematodes, and in pot experiments he found that nematode infestations of strawberry plants could be checked by introducing enchytraeids during the early stage of infection.

O'Connor (1967) studied squash preparations of three species of enchytraeids from a Douglas fir site in North Wales. His result indicated that *Achaeta eiseni* and *Cognettia cognetti* consumed twice the amount of fungus that would have been expected if they had fed on their surrounding substrates in a random manner. Furthermore, he observed selective feeding by enchytraeids and found that *Achaeta* avoided plant material, and *Cognettia* fed selectively on fungi in the litter layer.

Dash and Cragg (1972a) examined the food preferences of four species of North American Enchytraeidae using three techniques: fungal baits, squash preparations, and culturing of gut contents in an aspen woodland site in Alberta, Canada. Four fungal species common in the aspen forest soil were tested as baits for the soil fauna. About 3000 animals, comprising Collembola, mites, insect larvae, nematodes, and enchytraeids, were collected from 500 baits.

Of the total Enchytraeidae, 53% occurred on *Cladosporium* baits, 42% on *Ceuthospora,* 4% on *Pencillium,* and 1% on *Paecilomyces.* The analysis of gut contents showed that the worms ingested decomposed plant material with a high content of fungal mycelium. Isolation experiments produced one *Pencillium* sp. and one *Cladosporium* sp. from the preclitellar portion of the gut of the worms, and isolations from the postclitellar region gave rise to fungal colonies. Some species of enchytraeids showed a tendency to select certain fungal material, and ingested nematodes were recorded in the gut of some enchytraeids. Recently, using similar techniques, Dash et al. (1980) observed that three species of tropical Enchytraeidae grazed over *Rhizopus nigricans, Trichoderma viride,* and *Syncephalastraum recemosum* but could not digest *Penicillium steckii* and *Aspergillus niger.* Standen and Latter (1977) and Springett and Latter (1977) working on *Cognettia sphagnetorum* found that the worm prefers *Calluna* and *Eriophorum* litters as food. Gut content observations showed that worms ingested decomposing litter containing fungal mycelium, but there was no evidence for the actual selection of fungi. Dozsa-Farkas (1973a) found that in Hungarian forests *Stercutus niveus* feeds largely on decaying litter leaves.

Nielsen (1962) examined four species of Enchytraeidae (*Cognettia sphagnetorum, Enchytraeus albidus, Mesenchytraeus glandulosus,* and *Fridericia hegemon*) for carbohydrases in their gut. Some of them had the requisite enzymes to digest disaccharides and amylose. But they did not possess enzymes which would enable them to digest cellulose, pectin, xylan, and chitin of higher plants. Dash et al. (1980) examined three species of tropical enchytraeids (*Hemifridericai khallikotosus, Enchytraeus berhampurosus,* and *Fridericia kalinga*) for amylase, invertase, protease, urease, and cellulase. They found these enzymes present, thus indicating that they can digest disaccharides, proteins, urea, and cellulose.

Relationship to Humification

Laboratory studies on the role of enchytraeids in humification are lacking. Our knowledge of their role in humification is based chiefly on field studies. Zachariae (1963) has reported that Enchytraeidae consumed the droppings of litter-feeding Collembola along with loose particles of leaf litter in coniferous and deciduous forests. He also reported that enchytraeids produced crumblike droppings of finely divided remains in which there were no recognizable cellulose residues. He has examined the form and distribution of invertebrate feces in many soils and concludes that so-called collembolan soils (Kubiena 1953) were formed by Enchytraeidae.

Burges (1967) reported that after litter fall extensive microbial development occurs. There is a rapid loss of both cellulose and lignin. At this stage the soil fauna completely destroys the mesophyll tissue of the leaves. The amount of black fecal material increases, and initially this appeared to be due to the activity of mites and Collembola, but as the pellets accumulate the Enchytraeidae become more noticeable and their excreta in the form of amorphous black material becomes more abundant.

It seems that the role of Enchytraeidae is (a) to ingest decomposing plant material that has a high microbial content and digest part of it, and the partial breakdown of such materials may be important in making food material available for other decomposer organisms, (b) to produce fecal material that is added to the soil to provide sites for microbial activity, and (c) to encourage a high level of activity by the microflora, which they do in their grazing activity by preventing aging of the microflora.

Information on the rate and amount of fecal matter production by Enchytraeidae in different ecosystems is lacking and studies on their gut enzyme components are inadequate to make useful generalizations.

Excretory Characteristics

The organs of excretion are nephridia. Usually, one pair of nephridia occurs per segment starting from VII. Each nephridium has three parts: (a) the anteseptale part, consisting of funnel and sometimes one or more coils of the nephridial canal (the shape of the anteseptale portion is variable, i.e., in some species, it is long and tubular and in others large and oval or small and squarish); (b) the postseptale part, consisting mainly of nephridial canal and interstitial tissue; and (c) the efferent duct, whose location may vary from anteroventral to terminal. The structure of nephridia provide some taxonomic clues for identification of genera. Information on the composition and rate of nephridial excretion is not available. The nephridia probably produce mucilage which is richly nitrogenous.

Reproductive Characteristics

Reproductive Organs

The male reproductive organs consist of paired seminal vesicles in the X segment and may look conspicuous in some species due to the brownish or reddish color of the later developmental stages of the spermatozoa. The vesicle may also be lobed as in *Lumbricillus* or may extend two to three segments as in *Fridericia.* A pair of sperm funnels is present in the XI segment and usually two to three times longer than wide. The vas deferens (sperm ducts) open through the penial bulb in the clitellar segments. The enlargement of the sperm duct adjoining the penial bulb is called the atrium. Atrial glands may be present along with the sperm duct. Accessory glands may be found in *Mesenchytraeus.* A pair of spermathecae are present in the IV segment and may communicate with the esophagus at IV–V or extend freely, usually up to segment V. They store the sperm and each spermathecum opens to the outside by an ectal duct. Parthenogenetic forms usually do not possess spermathecae.

The female reproductive organs consist of a pair of ovaries. Usually, 1–2 mature eggs occur at a time in the clitellar segments. In some species of *Enchytraeus,* 5–10 mature eggs may be present at a time. In *Mesenchytraeus* an egg sac may be present.

Normal Method of Reproduction

The normal method of reproduction in the Enchytraeidae is by mutual transfer of spermatozoa between hermaphrodites. The penes of one worm are inserted into the spermathecae of the other. Ova are passed into a mucilaginous cocoon secreted by the clitellum. The spermatozoa are passed into the cocoon and fertilize the ova while the cocoon passes forward over the spermatbecae. The eggs develop into worms of about 1 mm length, which hatch out by rupture of the cocoon wall. Dozsa-Farkas (1973a) found that newly hatched worms of *Stercutus niveus* contain 20 segments. Reynoldson (1939b, 1943), Ivleva (1953), Williams et al. (1969), Kirk (1971), and Dozsa-Farkas (1973a) have done pioneering work on the breeding biology of Enchytraeidae. Table 10.1 gives information on the reproductive biology of four species of Enchytraeidae.

Typical Life Cycle

Christensen (1956) used glass observation chambers to study cocoon production in some species of Enchytraeidae. Each cocoon may contain 1 or more eggs, depending on the species and the individual. In *Enchytraeus buchholzia,* some 5–10 eggs, and in *Stercutus niveus,* some 1–15 eggs may be present in a cocoon. The genera *Enchytraeus, Fridericia,* and *Henlea* actively cover their cocoons with a layer of sand grains and organic debris. The type of cocoon covering may be of ecological significance in relation to drought resistance. Some believe that Enchytraeidae are of arctic origin and a greater proportion of individuals in Enchytraeidae populations become sexually mature in winter in temperate climate. Reynoldson's (1939b, 1943) study of the life cycle of *Lumbricillus lineatus* suggests that there is no obligatory winter breeding. Kirk (1971) found peak cocoon production of *Lumbricillus rivalis* in late winter/early spring.

Cocoon production continues throughout the life of the worm but is more rapid in young mature than in old worms. Temperature has an important influence on the reproduction of Enchytraeidae (Reynoldson 1943). Reproduction and multiplication in worms of temperate region is possible over the range 7–20°C for *Lumbricillus lineatus* and 5–25°C for *Enchytraeus albidus* (Reynoldson 1939b, 1943; Ivleva 1953). The incubation and maturation periods decline with increasing temperature up to a lethal limit between 20 and 25°C. However, in tropical worms normal cocoon production can occur at higher temperatures, 15–25°C (Dash and Nanda, unpublished data). The maximum longevity of worms may be 1 year and the total life cycle from cocoon hatching to maturity may be 65–120 days. Field census on different age classes over 2 years indicate that some species of North American terrestrial Enchytraeidae mature in 60–90 days (Dash 1970a).

Dash and Cragg (1972b) observed two periods of high densities in spring (May) and autumn (October) with two low-density periods in summer

Table 10.1. Summary of Data on Reproductive Biology of Four Species of Enchytraeidae

	L. lineatus (Reynoldson 1939b, 1943)	*L. rivalis* (Kirk 1971)	*Enchytraeus albidus* (Reynoldson 1943)	*Enchytraeus albidus* (Ivleva 1953)	*Stercutus niveus* (Dozsa-Farkas 1973a)
Cocoon size (mm) (length × breadth)	1.18 × 0.84 (average)		0.59 × 0.55 (average)	0.5–1.8 (length)	0.71–1.35 × 0.49–0.64
Eggs/cocoon	6–11	17.1–47.8 (average)	2–6	1–35 (average 10)	1–15
Incubation period (days)	15.7 ± 0.9 (at 10°C)		23.8 ± 1.7 (at 10°C)	12 (at 18°C)	
Hatching success (%)	74	19–92	97		
Size at hatching (mm)	1.0 (13 segs.)			1.5–3.0	20 (segs.)
Total life cycle (days)	117 (at 10°C)		68.3 (at 10°C)	261 (mean max. longevity)	>365
Temperature range (°C) for reproduction	5–20	10 ± 1	1.5–25.5	18	13–18

(July) and winter (January) in an aspen woodland site of Alberta, Canada. The high-density periods were largely dominated by juveniles, indicating that two generations per year occurred for most of the species of Enchytraeidae. The winter population was dominated by both nonmature and mature worms. This does not completely rule out winter breeding, although the main breeding period was early spring and autumn.

Asexual Reproduction and Parthenogenesis

Some species of the Enchytraeidae are parthenogenetic and some reproduce by asexual method (Christensen 1959). *Enchytraeus fragmentosus* breaks into 3 to 14 fragments, each of about five segments, and the fragments regenerate into complete worms in about 10 days (Bell 1959). *Cognettia sphagnetorum, C. glandulosa,* and *Buchholzia appendiculata* also reproduce by fragmentation. *C. glandulosa* and *B. appendiculata* can produce eggs which usually develop parthenogenetically. Out of 88 cytotypes studied by Christensen (1961), 41 polyploids were recorded. It seems that Enchytraeidae are evolving rapidly by polyploidy.

Unique Ecological Characteristics

Distribution

The Enchytraeidae reach their greatest abundance in temperate climates, particularly in mor-type soils. They are mainly terrestrial but frequently occur in littoral and aquatic habitats. Thambi and Dash (1973) and other studies by Dash et al. (1974, 1978) show that they occur in tropical grasslands and forest habitats in India. The smallest number of records of the occurrence of Enchytraeidae in the southern hemisphere is probably due to lack of extensive studies on them. A detailed study on their geographical distribution in the tropics has yet to be made. Recent studies on the distribution of Enchytraeidae in European soils have been made by O'Connor (1967), Nurminen (1967), and Abrahamsen (1968, 1972). *Fridericia* and *Achaeta* are mainly terrestrial species; *Lumbricillus, Mesenchytraeus,* and *Cognettia* tend to occur in wet places; *Lumbricillus* mainly in marine littoral habitats; *Mesenchytreaus* and *Cognettia* occur in limnic or wet places such as fens and bogs; *Marionina, Henlea,* and *Enchytraeus* occur in a wide variety of habitats.

The Harriman Alaska Expedition did not cross the Arctic Circle. This expedition, however, brought together a large amount of material from areas very near it (Eisen 1904). Southern parts of North America have been surveyed extensively for the Enchytraidae fauna. Eisen (1904), Welch (1914a,b, 1916, 1917a–c, 1919a,b), Altman (1931, 1936), and Bell (1936, 1942, 1945, 1947, 1949, 1958, 1959, 1962) reported new species from the western part of North America. Recent studies on terrestrial Enchytraeidae from the western part of North America based on the modern taxonomy of Nielsen and Christensen (1959, 1961, 1963) are by Tynen (1969) and Dash (1972). Although there has been no comprehensive revision of the Enchytraeidae fauna of the western Rocky Mountains, the Pacific coast, and Alaska, it seems that the fauna of this region is to a large extent different from that of North America east of the Rocky Mountains. About 80 species of Enchytraidae are known from this region, and of these species only *Henlea nasuta* (Eisen 1878) is known from Europe, and a few others are known from Siberia (Nurminen 1973c). The genera *Mesenchytraeus, Henlea, Marionina,* and *Fridericia* are well represented in this region particularly in Alaska (S. F. Maclean, personal communication). The Canadian deciduous and coniferous forests are inhabited by communities with dominant species of *Fridericia, Buchholzia, Enchytraeus,* and *Henlea.*

Aggregation

Nielsen (1954), O'Connor (1957), Peachey (1963), Nurminen (1967), Abrahamsen (1972), Dash and Cragg (1972b), Thambi and Dash (1973), Standen and Latter (1977), and Nakamura (1979), all working on Enchytraeidae in different habitats under different environmental conditions, have observed aggregated distributions of Enchytraeidae. The major factors to cause aggregation are (a) the destruction of animals in unfavorable microhabitat by desiccation and other adverse conditions; (b) the movement of worms toward a favorable area such as a suitable food source, favorable site for copulation and cocoon deposition, and areas of optimum humidity and temperature; (c) the slow dispersal of the worms from an egg batch or a favored area, increased susceptibility of young worms to adverse environmental conditions, and subsequent slow dispersal from a favored area; and (d) gregariousness.

Vertical Distribution

Peachey (1963) found in Moorland soils over 75% of the worms in the top 6 cm of the 12 cm deep sample, taken in January and May. Under drying conditions he took 30 cm deep samples and extracted 20% of the worms below 12 cm depth. Nurminen (1967) working on coniferous forest soil in Finland observed that the majority of the worms were confined to the uppermost 9 cm of the soil profile. He does not mention the effect of drought. Nielsen (1955) observed that there was a tendency of the worms to migrate deeper during drought, but the movement was confined to the top 10 cm of the soil profile. In the two Canadian fen and forest sites more than 70% of Enchytraeidae were concentrated in the top 3 cm of the soil profile (Dash and Cragg 1972b). In the warmer months of the year, some 82% of the worms in the fen site and 74% in the aspen site were concentrated in the top 3 cm of the soil profile, whereas in winter in both sites some 65% were found in that zone. More than 90% of all species were concentrated in the top 6 cm of the profile. These studies indicate that the optimum conditions for life for enchytraeids occur in the top layers of the soil profile.

Field experiments on the vertical movement of enchytraeids in relation to rising soil temperature and laboratory experiments on vertical movement in relation to moisture showed that enchytraeids moved downward against rising soil temperature and decreasing humidity, but movement was confined to the top 6 cm of the profile (Springett et al. 1970, Dash and Cragg 1972b).

Population Changes

Population studies of Enchytraeidae have been made and a variety of habitats in Denmark (Nielsen 1955), in moorland soils in England (Peachey 1962), in coniferous forest soils in Finland (Nurminen 1967), in aspen forest soil and fen soil in Alberta, Canada (Dash and Cragg 1972b), in Scandinavian soils (Abrahamsen 1972), in tropical soil from India (Thambi and Dash 1973), and in Japanese soil (Kitazawa 1977, Nakamura 1979). These habitats cover a wide variety of environmental conditions. Table 10.2 summarizes the density and biomass and metabolic activity of Enchytraeidae in different habitats. The highest population of enchytraeids have been recorded by O'Connor (1957) and Peachey (1963). The number of enchytraeids in the North American aspen study site at Kananaskis, Alberta, Canada, which experiences a continental-type climate with a hot summer and long cold winter is only one-tenth of the number recorded by O'Connor (1957) from coniferous forest soil in north Wales, but is approximately the same as recorded by Nurminen (1967) from Finnish coniferous forest soil, which also experiences a continental climate. In all probability these low-density populations may be characteristic of northern continental areas. In the Canadian fen site there was one peak density in late summer or early autumn and one low-density period in winter (Dash and Cragg 1972b). This population trend is of the same type as described by Peachey (1963) from permanently moist habitat at Moor House, England. Temperature and soil moisture have been found to be important environmental factors regulating enchytraeid population changes (O'Connor 1967, Dash and Cragg 1972b, Abrahamsen 1972). Predation or parasitism are not considered as important in regulating the seasonal trend of enchytraeid population (Nielsen 1955, Dash 1973).

The summer and winter populations in the Canadian study were dominated by immature and mature worms and the spring and autumn populations were dominated by juveniles and immature worms. Most of the species had two generations per year. However, in the permanently moist habitat (fen site), most species showed only one generation and the summer

Table 10.2. Density, Biomass, and Metabolic Activity of Enchytraeidae in Different Habitats

	Nielsen (1955), Denmark, Permanent Pasture Station			O'Connor (1957), North Wales	Peachey (1962, 1963), England Pennine Moorland			Abrahamsen (1972), Norway,	Dash and Cragg (1972b), Alberta, Canada		Thambi and Dash (1973), India,
	1	4	18	Coniferous Woodland	Bare Peat	*Nardus* Grass-land	*Juncus* Moor	Norwegian Coniferous Forests	Aspen Woodland	Fen	Tropical Pasture and Grassland
Altitude (m)	3.2	3.1		500	633		564	430	1380		5
Annual precipitation (mm)						1780		866–1297	600–700		1250
Annual mean air temperature (°C)				5.5–15.5		5.6		4.1–5.5	2.5		30
Soil moisture		Humid		Moist	Water-logged	Moist	Water-logged	Humid-moist	Dry		Moist
Dominant vegetation				*Pseudotsuga taxifolia*	No veg.	*Nardus-stricta*	*Juncus squarosus*	*Pices abies*, *Pinus silvestris*, and others	*Populus tremuloides*	*Betula* sp. moss cover	*Cynodon dactylon* and *Hygrorhiza* sp.
Density (mean or range) ($\times 10^3$ m^2)	44	30	74	134.3	12–50	37–200	130–290	20–55	2–22	0.5–10	0.5–7
Mean biomass (g m^{-2})	2.97	3.03	10.50	10.79	10	35	53		3	0.5	0.14
Secondary production (g m^{-2})									8	1.1	0.25
Metabolic activity ml O_2m^{-2} h^{-1}	0.84	1.10	3.70	3.61					1.05	0.175	
liters O_2m^{-2} yr^{-1}	7	10	32	31.6					9.2	1.53	

population was dominated by juveniles (peak density reached in August–September).

Biomass and Metabolism

The biomass value for the Canadian sites (Dash and Cragg 1972b) are lower than these recorded elsewhere (Table 10.2). The average monthly standing crop (wet weight) of worms was 3 g/m^2 in the aspen site and 0.5 g/m^2 in the fen site. From calculations on the observed population turnover, production must exceed three times the average biomass.

Using the length/volume regression (Abrahamsen 1973), Maclean (1974) estimated the biomass of enchytraeids in nine sites near Barrow, Alaska, from late June to late August in 1972. The biomass values range from 291 to 2851 mg/m^2 ($\bar{x}$ = 1400 mg/m^2) and enchytraeids made up between 35.6 and 92.2% ($\bar{x}$ = 73.1%) of the microfaunal biomass. Oxygen consumption by Enchytraeidae has been measured by Nielsen (1961) using a Cartesian diver respirometer and by O'Connor (1963) using both the diver and Warburg respirometer. Laboratory measurements of oxygen uptake underestimate the actual respiratory rates of worms in soil (O'Connor 1967). Although less than ideal, these measurements help us to get an overall picture of community metabolism.

O'Connor (1967) has given general information on levels of oxygen consumption at 20°C by five genera and 21 species of Enchytraeidae. The genera of Enchytraeidae can be placed in ascending order with regard to increasing rate of oxygen uptake: *Fridericia, Mesenchytraeus, Henlea, Enchytraeus, Lumbricillus.* The ecophysiological basis of these intergeneric differences in oxygen consumption is not fully understood. O'Connor (1967) pointed out that the genera studied so far form a series of decreasing adaptation to the terrestrial environment, and the high respiratory rate is associated with greater mobility in water and in observation slides. The measurement of respiratory quotients of Enchytraeidae may help to explain the differential respiratory activity with regard to the utilization of different substrates.

O'Connor (1967) has calculated that the respiratory activity of enchytraeids (average monthly biomass 10.8 g wet weight per square meter) in a Douglas fir plantation in North Wales accounted for 11% of the total annual energy input to the forest floor (1350 kcal/m^2). This estimate takes no account of the energy consumed in growth and reproduction. Thus the total metabolic activity is likely to exceed the estimate by at least 20% if the values given by Englemann (1961) for oribatid mite populations can be taken as a guide.

The annual litter input to a North American aspen forest site at Kananaskia, Alberta, was high, consisting of 318 g dry weight per square meter aspen leaves and at least 103 g dry weight per square meter as understory vegetation (Cragg et al. 1977). Applying the Peterson et al.

(1970) calorific determination of 4.720 kcal/g dry weight for aspen leaves to the total litter input (approx. 420 g/m^2), the total annual energy input would amount to about 2000 kcal/m^2. Assuming that the enchytraeids have the same degree of metabolic efficiency (mean temperature, $<5°C$; mean individual weight, <0.5 mg) as those quoted by O'Connor (1967), the average monthly biomass of 3 g/m^2 of enchytraeids will account for about 2% (about 3%, including growth and reproduction) of the total energy input to the aspen litter layer.

Collection and Extraction Technique

A steel corer is generally used to collect soil samples for a population study of Enchytraeidae. Tufnol rings help to maintain the soil samples in a relatively undisturbed condition. Abrahamsen (1969) found that stratified random sampling is highly preferable to simple random sampling. He also found that the coefficient of variation increased rapidly when the mean number of specimens per soil core fell noticeably below 40 and the precision did not markedly improve even if the mean was noticeably above 40. Thus the choice of the size of sample units depends on the objectives of the study.

Extraction

Two techniques, both dependent on the activity of worms, have been described for the large-scale extraction of enchytraeids from litter and soil. Both methods depend on the movement of worms in response to physical stimuli but differ in their mode of operation. In Nielsen's (1952–1953) extractor, the soil sample is heated in a glazed earthenware vessel immersed in a water bath, and the worms are driven upward into a cooled layer of sand or fine gravel, placed on top of the soil sample. The earthenware cylinder is 14.5 cm high and 7.5 cm in diameter (internal dimension) and has a hole (7 mm in diameter) in its base. The bottom is filled to a depth of 3 cm with gravel, covered with wire gauze (1 mm mesh) fastened to the wall of the vessel with plaster of paris. The extractor was designed to take soil cores of 6 cm deep and 7.3 cm in diameter. The soil core is placed with the upper surface upward in the extraction vessel so that it rests on the ring of plaster of paris. The soil is covered with a layer of fine gravel or sand (2.5 cm deep) of particle size 2–4 mm. A cooling coil (4 mm in internal diameter) is connected to the cold water supply put at a depth of 0.5 cm in the upper gravel layer and is supported by two wires. The extraction vessel is immersed in a water bath so that the water comes to the upper surface of the gravel, with the upper 6 cm of the vessel protruding above the ring of the water bath. The water bath is heated at a rate to provide, after 2 hours, a water temperature of 60°C. At the same time, water evaporates from the lower gravel surface and condenses in the soil, thus establishing both a temperature and a moisture gradient, high at the base and low at the top of the sample. When the upper 1 cm of the soil core reaches 25°C after 1 hour, cold water is circulated through the cooling coil, resulting in an abrupt decrease in temperature in this layer. Heating is continued until the water reaches 85°C and the upper 1 cm of the sample about 35°C. The temperature level is maintained for about 1 hour when the upper gravel layer is removed and the worms are recovered from repeated washing and decantation. The complete process takes 3 hours.

The alternative method is the O'Connor (1955) wet funnel method. Soil core samples ≤6.3 cm in diameter and ≤3 cm deep can be processed in this apparatus. Each soil sample or subsample is placed surface downward on a sieve plate in the funnel. The worms are collected periodically in petri dishes. The heating of the funnel (60-W electric bulbs on a rheostat) is increased gradually in steps, and when the sample surface has reached a temperature of 45°C after 3 hours, extraction is completed.

Both methods have been used widely for population studies of Enchytraeidae. Peachey (1962) compared the two methods using samples from four sites ranging from peat to mineral soil. His results indicated that the wet funnel was more efficient than the Nielsen extractor for extracting enchytraeids from wet peat soil, but there was little to choose between the two methods when the mineral soil was used. Nielsen's method can readily be modified for larger-diameter soil cores since the temperature gradient is not affected by soil core diameter, and this is not the case in the wet funnel method. The wet funnel method is simple and easy to operate, heat control is easier, and the extraction is a single process. It is to be preferred for soils with a high organic matter content.

Observation Technique

Worms after extraction are left in a petri dish in a cool place for about 6–12 hours for clearance of gut contents. Sexually mature worms are transferred, usually one at a time, to a drop of water on a slide and covered by a cover glass. The most suitable pressure for examination is given by using different sizes of cover glasses and varying the amount of water. Blotting paper may be used to soak up excess water. The examination should start at about 100× magnification, and once some experience is gained in identifying external characteristics, high-power 400× magnification can be used to study in detail the external and internal structures. A good standard research microscope with good lighting facilities should be used. Most of the structures are fairly transparent, and therefore to achieve good constrast, the diaphragm and condenser must be used effectively.

Cytotaxonomic Technique

Nielsen and Christensen (1959) have found that mitosis are easily obtained from worms in which spermatogenesis has just begun. The worm is placed in a drop of orcein on a slide and examined under a binocular microscope in transmitted light. The worm is cut near segment IX and just behind the clitellum. Other portions are removed and the cut piece is torn into small pieces with a pair of needles. The coverslip is then placed on the preparation, which is heated continuously over a flame (50°C) for about 10 minutes. The surplus orcein is removed by filter paper. The slide is now ready for examination for chromosomes.

The first meiotic metaphase in eggs can also be studied using an orcein preparation. When a mature worm is examined in transmitted light, it is possible to know whether the egg nucleus shows a prophase or a metaphase. In the former case, the nucleus shows up as a light circle among the dark yolk granules, while the metaphase spindle is seen as an oval light area. The worm is cut across the clitellum immediately in front of the egg(s). The contraction of the worm will often force out the egg(s). If not, the eggs can easily be forced out by pressing just behind them with the needle. When the eggs are isolated, the remains of the animal are removed. The eggs are kept moist and are punctured with a needle. The yolk flows through the ruptured membrane and when the spindle is about to pass through the rupture, a drop of orcein is added and a cover glass applied. The drop of orcein must be added at the right moment; otherwise, the staining will either be weak or the chromosomes scattered over a wide area. Cytotaxonomical techniques are described by Christensen (1960, 1961).

Identification

The published descriptions of Enchytraeidae vary considerably with regard to taxonomical criteria. The first major critical revision of this family was made by Cernosvitov (1937). He produced a world list of 303 species with 69 doubtful species. The species described by him dated from 1850 to 1937. It is now generally agreed that most of the taxonomic difficulties encountered arise from intraspecific variation, which led some of the earlier investigators (1850–1920) to give species status to varieties of the same species, and in some cases the same species was placed in different genera. The standard current monograph is that of Nielsen and Christensen (1959, 1961, 1963), who have provided a critical revision of the family with descriptions of different genera and a list of about 407 species, including 75 doubtful species.

Taxonomic literature on North American Enchytraeidae is fairly sparse and mostly recent. Authors have therefore been able to scan previous descriptions before publishing new species; thus synonyms are few. However, because of certain generic revisions by Cernosvitov (1937) and Nielsen and Christensen (1959), many species have been transferred to genera different from those in which they were placed by their original authors. For example, many species placed in the genus *Marionina* are now placed in other genera by Nielsen and Christensen (1959). A revision of the North American species of this family would be desirable, particularly verifying descriptions based on limited material. Many North American type specimens have been lost, and vast areas of North America have never been examined for Enchytraeidae. This would be a major undertaking and would involve years of detailed survey and study.

Tynen (1975) gives a checklist of 143 species of North American Enchytraeidae placed in 15 genera, including the very doubtful genus *Chirodilus*. The genera *Grania* and *Hemigrania* are not terrestrial and they are now considered to be synonyms (Erseus and Lasserre 1976). A key of North American genera of Enchytraeidae along with the number of species of each genus and area of North American records is given in Table 10.3.

A fixed procedure for the description of species of Enchytraeidae has not yet been established, and therefore gross morphological features which are sufficiently constant to acquire taxonomic importance were discussed in earlier sections. Considerable importance is given to the form and number of setae, structure of the gut and peptonephridia, and structure of the

Table 10.3. North American Species List and Distribution of Enchytraeidae

Genera and Species	Area of North American Records
***Achaeta* Vedjovsky, 1877**	
camerani (Cognetti, 1889)	E
eiseni (Vedjovsky, 1877)	E
Silvatica Nurminen, 1973a	E
***Bryodrilus* Ude, 1892**	
arctica (Bell, 1962)	W
ehlersi Ude, 1892	A
novaescotiac (Bell, 1962)	E
parvus Nurminen, 1970	A
p. kananaskis Dash, 1970b	E
***Buchholzia* Michaelsen, 1887**	
appendiculata (Buchholz, 1862)	E
***Cernosvitoviella* Nielsen and Christensen, 1959**	
atrata (Bretscher, 1903)	E
celere Nurminen, 1973a	E
christenseni Dash, 1970b	E
immota (Knollner, 1935)	E
pusilla Nurminen, 1973b	A

Table 10.3. (*Continued*)

Genera and Species	Area of North American Records
***Cognettia* Nielsen and Christensen, 1959**	
glandulosa (Michaelsen, 1888)	E
***Enchytraeus* Henle, 1837**	
alaskae Eisen, 1904	W
albidus Henle, 1837	E, C
buchholzi Vedjovsky, 1879	E
capitatus Bulow, 1957	E
cryptosetosus Tynen, 1969	W
fragmentosus Bell, 1959	C
gillettensis Welch, 1914a	E
kincaidi Eisen, 1904	W
minutus Nielsen and Christensen, 1961	A
***Fridericia* Michaelsen, 1889**	
agilis Smith, 1895	E
alba Moore, 1895b	E
anglica Friend, 1912a	E
agaricola Moore, 1895b	E
berkelyensis Bell, 1936	W
bulboides Nielsen and Christensen, 1959	E
bulbosa Rosa, 1887	E, M
californica Eisen, 1904	W
canadensis Dash, 1972	W
caprensis Bell, 1947, augm. Nielsen and Christensen, 1959	E
firma Smith and Welch, 1913	E
fuchsi Eisen, 1904	W
galba (Hoffmeister, 1845)	E
harrimani Eisen, 1904	W
Johnsoni Eisen, 1904	W
kalffi Nurminen, 1973a	E
longa Moore, 1895b	E
losangelensis Bell, 1936	W
macgregori Eisen, 1904	W
nielseni Dash, 1972	W
oconeensis Welch, 1914a	E
papofiana Eisen, 1904	W
paroniana Issel, 1904	E
parva Moore, 1895b	E
ratzeli (Eisen, 1872)	E
sacculata Bell, 1936	W
santaebarbarae Eisen, 1904	W
sima Welch, 1914a	E
sonorae Eisen, 1904	W
sylvestris (Leidy, 1882)	E
syracussa Dash and Mitchell, 1981	E
tenera Smith and Welch, 1913	E
whatcomae Altman, 1936	W

Table 10.3. (*Continued*)

Genera and Species	Area of North American Records
***Grania* Southern, 1913**	M
americana Kennedy, 1966	E
atlantica Coates and Erseus, 1985	E
incerta Coates and Erseus, 1980	W
levis Coates and Erseus, 1985	E
maricola Southern, 1913	W
paucispina Eisen, 1904)	W
postclitellochaeta (Knollner, 1935). This species was removed to *Hemigrania* from *Marionina* by Lasserre (1971). *Hemigrania* is now found by Erseus and Lasserre (1976) to be a synonym for *Grania.*	*E*
reducta Coates and Erseus, 1985	E
***Henlea* Michaelsen, 1889**	
californica Eisen, 1904	W
c. monticola Eisen, 1904	W
eiseni Bell, 1942	W
erhorni Eisen, 1904	W
glabra Altman, 1936	W
helenae Cernosvitov, 1937	W
heleotropha Stephenson, 1922, augm. Nielsen and Christensen, 1959	A
moderatoidea Altman, 1936	W
nasuta Eisen, 1878	A
ochracea (Eisen, 1878)	A, E
(= *arctica* Welch, 1912a)	
(= *moderata* Welch, 1914a)	
(= *tenella* Eisen, 1878)	
(= *tubulifera* Welch, 1914b)	
parva Friend, 1912b	E
perpusilla Friend, 1911, Cernosvitov, 1837	A
scharffi Southern, 1910	E
udei (Eisen, 1904), Holmquist, 1968	
(= *Bryohenlea udei*)	A, W
urbanensis Welch, 1914a	E
ventriculosa Udekem, 1854	A
welchi Bell, 1942	E
***Lumbricillus* Orsted, 1844**	
arenarius (Michaelsen, 1889), augm. Knollner, 1935	A
annulatus Eisen, 1904	W
belli Tynen, 1969	W
charae (Tynen, 1970)	A
codensis Lasserre, 1971; (Coates and Erseus, 1985: *Randidrilus*)	E
curtus Coates, 1981	W
franciscanus Eisen, 1904	W
f. borealis Eisen, 1904	W
f. unalaskae Eisen, 1904	W
georgiensis Tynen, 1969	W
kalatdlitus Nurminen, 1970	A
lineatus (Müller, 1774 (= *agilis* Moore, 1905)	A, E

Table 10.3. (*Continued*)

Genera and Species	Area of North American Records
Merriami Eisen, 1904	W
m. elongatus Eisen, 1904	W
mirabilis Tynen, 1969	W
pagenstecheri (Ratzel, 1869), augm. Coates and Ellis, 1981	W
qualicumensis Tynen, 1969	W
ritteri Eisen, 1904	W
rupertensis Coates, 1981	W
rutilus Welch, 1914a	E
santaeclarae Eisen, 1904	W
tsimpseanis Coates, 1981	W
tuba Stephenson, 1911, augm. Coates and Ellis, 1981	W
vancouverensis Tynen, 1969	W
***Marionina** Michaelsen, 1889**	
achaeta Lasserre, 1964	E
alaskae Eisen, 1904	W
americana Eisen, 1904	W
appendiculata Nielsen and Christensen, 1959, augm. Coates and Ellis, 1981	W
argentea (Michaelsen, 1889)	E
canadensis Dash, 1970b	E
charlottensis Coates, 1980	W
citrina (Eisen, 1904)	W
craggi Dash, 1970b	E
diazi Coates and Erseus, 1985	E
dirksi Bell, 1942	E
forbesae Smith and Welch, 1913	E
klaskisharum Coates, 1983	E
modesta Eisen, 1904)	W
neroutsensis Coates, 1980	W
nevisensis (Right and Kanner, 1979)	W
paucispina (Eisen, 1904)	W
preclitellochaeta Nielsen and Christensen, 1963	E
pugetensis (Altman, 1931)	W
saxicola (Eisen, 1904)	W
sjaelandica Nielsen and Christensen, 1959, augm. Coates and Ellis, 1981	W
southerni (Cernosvitov, 1937)	E, W
spicula (Leuckart, 1847)	E
subterranea (Knollner, 1935)	E
trevori Coates, 1980 (Coates, 1983: *Stephensoniella*)	W
vancouverensis Coates, 1980	W
welchi Lasserre, 1971	E
Mesenchytraeus Eisen, 1878**	
altus, Welch, 1917b	E
americanus Bell, 1942	E
arcticus Bell, 1942	W
argentctus Nurminen, 1973b	A, E
armatus (Levinsen, 1884)	

Table 10.3. (*Continued*)

Genera and Species	Area of North American Records
a. kananaskis Dash, 1970b	E
atriophorus Altman, 1936	W
beringenis Eisen, 1904	W
chromophorus Altman, 1936	W
diplobulbosus Bell, 1949	W
eastwoodi (Eisen, 1904	W
fontinalis Eisen, 1904	W
franciscanus Eisen, 1904	W
fuscus Eisen, 1904	W
f. inermis Eisen, 1904	W
gelidus Welch, 1916	W
grandis Eisen, 1904	W
harrimani Eisen, 1904	W
hydrius Welch, 1919b	W
johanseni Welch, 1919a	A
kincaidi Eisen, 1904	W
macnabi Bell, 1942	W
maculatus Eisen, 1904	W
magnus Altman, 1936	W
minimus Altman, 1936	W
monothecatus Bell, 1945	W
nanus Eisen, 1904	W
obscurus Eisen, 1904	W
orcae Eisen, 1904	W
pedatus, Eisen, 1904	W
penicillus Eisen, 1904	W
setchelli Eisen, 1904	W
solifugus (Emery, 1898)	W
s. rainierensis Welch, 1916	W
unalaskae Eisen, 1904	W
***Randidrilus* Coates and Erseus, 1985**	M
codensis (Lasserre, 1971)	E
quadrithecatus Coates and Erseus, 1985	E
†*Stephensoniella* Cernosvitov, 1934	
marina (Moore, 1902)	E
trevori (Coates, 1980) (= *Marionina trevori*)	W

* Some species (e.g., *Enchytraeus saxicola*) have been transferred to *Marionina* from the genus *Enchytraeus* because they lacked peptonephridia (Nielsen and Christensen, 1959) and those from the invalidated genus *Michaelsena* are also placed here. *Marionina americana* is retained, although Nielsen and Christensen (1959) refer it to *Lumbricillus* (Tynen, 1975). Coates and Ellis (1981) have provided very useful informations on a number of species of marine enchytraeids of the genera *Marionina, Lumbricillus, Enchytraeus,* and *Granic.*

** There are five new species of *Mesenchytraeus* from taiga and tundra habitats of the northeastern USSR. The enchytraeid fauna of the northeastern USSR is similar to that of subarctic and arctic Alaska (Emery 1898, 1900 and Moore 1899).

† Coates (1983a) has reviewed the genus *Stephensoniella* and has transferred *Lumbricillus sterreri,* Lasserre and Erseus, 1976 to *Stephensoniella.* It is named as *S. sterreri* (Erseus and Lasserre, 1976) and found in Bermuda.

Key: A: North of Arctic Circle. C: known from cultures. E: east of Rockies. W: west of Rockies. M: marine.

reproductive organs, particularly sperm funnels, vas deferens, and spermathecae for taxonomic identification. Dash (1983) has reviewed the biology of enchytraeids and has given a world list of 23 genera consisting of 533 species. At present some 28 genera (including some doubtful) consisting of more than 600 species are known from different parts of the world. Two genera, *Pelmatodrilus* Moore, 1943 and *Sinenchytraeus* Liang and Hus, 1979, reported from Jamaica and Tibet, respectively, are not described in this review, as they are not reported from North America.

ACKNOWLEDGMENT

Professor J. B. Cragg of University of Calgary, Canada, read an earlier version of the manuscript and I am thankful to him for his helpful criticisms.

KEY TO THE GENERA OF THE ENCHYTRAEIDAE

This key was prepared considering the following characters as important at the generic level. Generic characters: (a) presence and absence of peptonephridia, (b) nature of the transitional zone of esophagus and intestine, (c) setal structure, (d) origin of dorsal vessel, and (e) structure of the reproductive organs.

I. Peptonephridia (salivary gland) present

(A) Unpaired peptonephridia, spermathecae free, dorsal vessel arising usually in XII–XIII, dorsal pores absent, straight setae usually two in number . . . . ***Hemienchytraeus*** Cernosvitov, 1935 (not yet reported from North America)

(B) Paired peptonephridia, spermathecae usually attached or communicating to the esophagus (free in ***Fridericia sonorae*** and in two non-North American species; Dash, Nanda and Thambi (1978) and Dozsa-Farkas, 1973b). Dorsal vessel arising in XIV–XXV, dorsal pores present from VII. Two types of lymphocytes, large nucleated, small hyaline anucleated. Setae straight, without nodulus, and the innermost setae remain pairwise in the bundle and shorter than the outer ones; the setal numbers may vary from 2 to 16 . . . . ***Fridericia*** Michaelsen, 1889 [= ***Distichopus*** Leidy, 1882 (see Brinkhurst and Jamieson 1971)]

(C) Paired peptonephridia. Spermathecae attached or communicating to esophagus. Dorsal vessel arising in XII–XVII. Dorsal pores absent. Only nucleated lymphocytes. Setae straight and without nodulus. . . . . . . ***Enchytraeus*** Henle, 1837

(D) Paired unbranched peptonephridia present. Dorsal vessel arising in VII. Dorsal pores absent. Sudden expansion of intestine at VII/VIII, esophageal appendages absent but intestinal diverticula present in VII. Spermathecae communicating with esophagus. Setae straight and of equal sizes in the bundle and without nodulus . . . . ***Enchylea*** Nielsen and Christensen, 1963 (not yet reported in North America)

II. Peptonephridia absent:

(A) Setae sigmoid, simple with nodulus. Gradual transition between esophagus and intestine.

(1) Egg sac present. Postseptal part of nephridium bilobed, efferent duct arising between the lobes. Dorsal vessel arising in the intraclitellar or postclitellar segments. The ampullae of spermathecae usually possess diverticulum . . . . ***Mesenchytraeus*** Eisen, 1878

(2) Small worms, egg sac absent. Postseptal part of nephridium entire, efferent duct terminal, spermathecae simple. . . . .
Cernosvitoviella Nielsen and Christensen, 1959

(B) Setae sigmoid and without nodulus. Gradual transition between esophagus and intestine.

(1) Dorsal vessel arising in the anteclitellar segments.

(a) Esophageal and intestinal diverticula absent. Spermathecae without opening to esophagus. Dorsal vessel arising usually in IX. Only nucleate lymphocytes present . . . .
Stercutus Michaelsen, 1888
(not yet reported in North America)

(b) Esophageal and intestinal diverticula absent or may be present. Spermathecae very long and extend to many posterior segments without opening to the esophagus. Dorsal vessel arising in X . . . . *Tupidrilus* Righi, 1974
(reported only from South America, not yet reported in North America)

(2) Dorsal vessel arising in the intraclitellar or postclitellar segments.

(a) Spermathecae free. Seminal vesicle compact, genital organs often displaced forward. Esophageal diverticula absent. Efferent duct of nephridia long, narrow, arising anteroventrally. Usually white slender worms . . . .
Cognettia Nielsen and Christensen, 1959

(b) Spermathecae attached to esophagus. Seminal vesicles usually lobed. Efferent duct of nephridia short, stout, arising mid- or posteroventrally. No esophageal diverticula. Usually reddish or greenish worms . . . . *Lumbricillus* Orsted, 1844

(c) Spermathecae attached to esophagus (occasionally free), seminal vesicle present or absent. Efferent duct of nephridia terminal. No esophageal diverticula . . . .
Marionina Michaelsen, 1889

(C) Setae simple, straight or bent, and without nodulus. Usually gradual transition between esophagus and intestine. Dorsal vessel arising in intraclitellar or postclitellar segments.

(1) Two types of lymphocytes present, large nucleated and small hyaline anucleated lymphocytes present. No esophageal appendages.
Hemifridericia Nielsen and Christensen, 1959
(not yet reported in North America)

(2) Esophageal appendages and intestinal diverticula absent. Anteseptal of nephridia consists of funnel only. Paired voluminous seminal vesicles. Sperm duct usually confined to XII. Thick muscular preclitellar septa are present . . . . . *Stephensoniella* Cernosvitov, 1934

(3) Body wall generally transversely striated with thick cuticula. Proximal end of setae broad or curved, setae totally absent in the most anterior segments and from XII, dorsal setae beginning posterior to the first segment bearing ventral setae or totally absent; supplementary setae may be present. Esophageal appendages and intestinal diverticula absent. Sperm duct large, narrow, and convoluted extending backward as far as XIX. Spermathecae communicating to esophagus . . . .
Grania Southern, 1913 (see Erseus and Lasserre 1976)

(4) Single seta, ventrolateral usually in II and lateral (dorsal) setae absent. Dorsal lobes of pharyngeal glands usually join together; ventral lobes may extend up to VII. No secondary glands, but separate

anterior and posterior glandular masses are present. Dorsal vessel postclitellar in origin and lymphocytes are ovoid and nucleate. Seminal vesicle simple, dorsally unlobed, and may extend from IX anteriorly and posteriorly. One or two pairs of spermathecae in V or in V and VI with glands at ectal orifice and the ampullae are connected to esophagus. Sperm lodged in ampullar cavity. Colorless marine worms . . . . ***Randidrilus*** Coates and Erseus, 1985

(5) Straight and anteriorly directed setae occur only in II and III. Lateral (dorsal) setae absent. Penial bulb compact and spermathecal ducts are elongated. Usually worms occur in freshwater habitats. ***Barbidrilus*** Loden and Locy, 1981

(6) Setae slightly curved and without nodulus. Dorsal vessel arising in XII. Usually, two pairs of intestinal diverticula in VIII. Seminal vesicle absent, sperm funnel cylindrical, and penial bulb compact. Spermathecae simple, join together before communicating with esophagus . . . . ***Bryohenlea*** Cernosvitov, 1934

Note: Nielsen and Christensen (1959) included one species (***B. udei*** Eisen, 1904) under this genus. Holmquist (1968) and Tynen (1975) have included this species in the genus ***Henlea***. Thus ***Bryohenlea*** is a very doubtful genus.

(D) Usually sudden transition between esophagus and intestine.

(1) Setae absent, usually ovoid or pear-shaped follicles or glandular sacs occupy the position of some or all of the missing setal bundles ***Achaeta*** Vejdovsky, 1877

(2) Setae sigmoid and forked at distal end. . . . .
Propappus Michaelsen, 1905
(not yet reported from North America)

(3) Setae sigmoid, simple, and without nodulus.

(a) Dorsal vessel arising in the anteclitellar segments.

(i) Sudden transition between esophagus and intestine at VII/VIII, esophageal appendages in IV, one or two large intestinal diverticula at VII/VIII. Dorsal vessel arising from the anterior end of the intestinal diverticula. Nucleated lymphocytes and small hyaline anucleated corpuscles present. Spermathecae attached to the esophagus . . . .
Buchholzia Michaelsen, 1887

(b) Dorsal vessel originating in the intra- or postclitellar segments.

(i) Four esophageal diverticula in VI. Spermathecae merging into a common narrow canal which communicates with the esophagus dorsally. Lymphocytes usually discoid . . . . .
Bryodrilus Ude, 1892

(c) Dorsal vessel originating intra- or anteclitellar segments. Two setae per bundle. One pair of intestinal diverticula at esophageo-intestinal transition. Paired esophageal appendages in VI (peptonephridia?). Seminal vesicle present. Spermathecae simple and free. . . . . ***Guaranidrilus*** Cernosvitov, 1937
(a doubtful genus, not yet reported from North America)

(E) Setae straight or bent, simple without nodulus.

(1) Dorsal vessel originating in the anteclitellar segments.

(a) Dorsal vessel arising in VIII or IX with heartlike pulsating expansions in two to three segments near the origin. Zero, two, or four intestinal diverticula in VII. Spermathecae attached to the esophagus. . . . . ***Henlea*** Michaelsen, 1889

(b) Dorsal vessel arising in the anteclitellar region, usually in VII or IX with pulsating expansion, blood usually red or colorless, intestinal diverticula present or absent. Spermathecae simple with or without gland at the ectal orifice and attached to the esophagus. . . . . ***Punahenlea*** Nurminen, 1980

(c) Dorsal vessel arising in VIII. One pair of intestinal diverticula in VII. Esophageal appendages absent (?). Spermathecae simple, not communicating with the esophagus. . . . . ***Aspidodrilus*** Baylis, 1914 (not yet reported from North America and a doubtful genus)

(2) Dorsal vessel originating in intra- or postclitellar segments.

(a) Two setae per bundle, dorsal bundles absent in VIII–IX. Dorsal vessel usually arising in XIII. No pulsating heartlike expansions. Two lateral intestinal diverticula in VI . . . . ***Enchytronia*** Nielsen and Christensen, 1959 (not yet reported from North America)

References

Abrahamsen, G. 1968. *Records of Enchytraeidae (Oligochaeta) in Norway.* Publ. 89, Bind XXV. Norwegian Forest Research Institute, Vollebekk, Norway. **1969.** Sampling design in studies of population densities in Enchytraeidae (Oligochaeta). *Oikos* 20:54–66. **1972.** Ecological study of Enchytraeidae (Oligochaeta) in Norwegian coniferous forest soils. *Pedobiologia* 12:26–32. **1973.** Studies on body-volume, body-surface area, density and live weight of Enchytraeidae (Oligochaeta). *Pedobiologia* 13:6–15. **Altman, L. C. 1931.** *Enchytraeus pugetensis,* a new marine enchytraeid from Puget Sound. *Trans. Am. Microsc. Soc.* 50:154–163. **1936.** Oligochaeta of Washington. *Univ. Wash. Publ. Biol.* 4:1–137. **Baylis, H. A. 1914.** Preliminary account of *Aspidodrilus,* remarkable epizoic Oligochaeta. *Ann. Mag. Nat. Hist. Ser.* 8:14, 145. **Bell, A. W. 1936.** Three new species of *Fridericia* (Enchytraeidae) from California. *Univ. Calif. Publ. Zool.* 41(12):145–164. **1942.** Some new enchytraeid worms (Oligochaeta) from North America. *Trans. Am. Microsc. Soc.* 61:404–429. **1945.** A new species of *Mesenchytraeus* (Oligochaeta) from Oregon. *Trans. Am. Microsc. Soc.* 64:29–33. **1947.** Some new enchytraeids (Oligochaeta) from the Old World. *Trans. Am. Microsc. Soc.* 66:190–202. **1949.** *Mesenchytraeus diplobulbosus,* a new enchytraeid worm (Oligochaeta) from southern California with notes on the so-called "cement gland." *Trans. Am. Microsc. Soc.* 68:240–244. **1958.** The anatomy of the oligochaeta, *Enchytraeus albidus,* with a key to the species of the genus *Enchytraeus. Am. Mus. Novit.* (1902):1–13. **1959.** *Enchytraeus fragmentosus,* a new species of naturally fragmenting oligochaeta worm. *Science (Wash. D. C.)* 129:1278. **1962.** Enchytraeids (Oligochaeta) from various parts of the world. *Trans. Am. Microsc. Soc.* 81(2):158–178. **Bretscher, K. 1903.** Zur Biologie und Faunistch der wasserbewohnenden Oligochaeten der Schweiz. *Biol. Zentralbl.* 23. **Buchholz, R. 1862.** Beiträge zur Anatomie der Gattung *Enchytraeus.* Konigsberg. *Schriftenr. Phys. Oekon. Ges.* 3:93–130. **Bulow, T. Von. 1957.** Systematisch-autokologische Studien an Eulitoralen Oligochaeten der Kimibrischen Halbinsel. *Kiel. Meeresforsch.* 13:69–116. **Burges, A. 1967.** The decomposition of organic matter in the soil. Pp. 479–492 in A. Burges and F. Rew (eds.). *Soil Biology.* Academic Press, New York. **Cernosvitov, L. 1934.** Zur Kenntnis der Enchytraeiden. I. Über *Enchytraeus marinus* Moore 1902; Über die Beziehung zwischen den Gattungen *Enchytraeus* und *Pachydrilus. Zool. Anz.* 105:233–247. **1937.** System der Enchytraeiden. *Bull. Ass. Russ. Rech. Sci. Prague* 5:263–295. **Christensen, B. 1956.** Studies on Enchytraeidae. 6. Technique for culturing Enchytraeidae, with notes on cocoon types. *Oikos* 7:302–307. **1959.** Asexual reproduction in the Enchytraeidae (Olig.). *Nature (Lond.)* 184:1159–

1160. **1960.** A comparative cytological investigation of the reproductive cycle of an amphimictic diploid and a parthenogenic triploid from *Lumbricillus lineatus* (O.F.M.) (Oligochaeta, Enchytraeidae). *Chromosoma (Berl.)* 11:365–379. **1961.** Studies on cytotaxonomy and reproduction in the Enchytraeidae with notes on parthenogenesis and polyploidy in the animal kingdom. *Hereditas* 396–450. **Coates, K. 1981.** New species of Lumbricillus (Oligochaeta, Enchytraeidae) from littoral habitats of British Columbia. *Can. J. Zool.* 59:1302–1311. **1983.** A contribution to the taxonomy of the Enchytraeidae (Oligochaeta). A review of *Stephensoniella,* with new species records. *Proc. Biol. Soc. Wash.* 96:411–419. **Coates, K. and D. V. Ellis. 1981.** Taxonomy and distribution of marine Enchytraeidae (Oligochaeta) in British Columbia. *Can. J. Zool.* 59:2129–2150. **Coates, K. and C. Erseus. 1985.** Marine enchytraeids (Oligochaeta) of the coastal northwest Atlantic (northern and mid U.S.A.). *Zool. Scr.* 14:103–116. **Cognetti, L. 1899.** Descrizione dell '*Anachaeta camerani,*' nuova specie della famiglia degli Enchitredi. *Boll. Mus. Zool. Univ. Torino* 14(354). **Cragg, J. B., A. Carter, C. Leischner, E. B. Peterson and G. N. Sykes. 1977.** Litter fall and chemical cycling in an aspen (*Populus tremuloides*) woodland ecosystem in the Canadian Rockies. *Pedobiologia* 17:428–443. **Dash, M. C. 1970a.** *Ecology of Enchytraeidae (Oligochaeta) in Rocky Mountain* forest soils. Unpublished Ph.D. thesis, University of Calgary, Calgary, Alberta, Canada. **1970b.** A taxonomic study of Enchytraeidae (Oligochaeta) from Rocky Mountain forest soil of the Kananaskis region of Alberta, Canada. *Can. J. Zool.* 48(6):1429–1435. **1972.** New Enchytraeidae (Oligochaeta) from the Penny Forest soil of British Columbia, Canada. *Ann. Zool. Fenn.* 9:15–16. **1973.** Note on nematodes occurring in Enchytraeidae (Olig.). *Megadrilogica* 1(8):1–2. **1983.** *The Biology of Enchytraeidae.* International Book Distributors, Dehradun, India. 171 pp. **Dash, M. C. and J. B. Cragg. 1972a.** Selection of microfungi by Enchytraeidae (Oligochaeta) and other members of the soil fauna. *Pedobiologia* 12:282–286. **1972b.** Ecology of Enchytraeidae (Oligochaeta) in Canadian Rocky Mountain soils. *Pedobiologia* 12:323–335. **Dash, M. C. and M. J. Mitchell. 1981.** *Fridericia syracussa* (Oligochaeta: Enchytraeidae), a new species found in sewage sludge. *Rev. Ecol. Biol. Sol.* 18:259–261. **Dash, M. C., U. C. Patra and A. V. Thambi. 1974.** Comparison of primary production of plant material and secondary production of Oligochaetes in a tropical grassland of southern Orissa, India. *Trop. Ecol.* 15(1,2):16–21. **Dash, M. C., B. Nanda and A. V. Thambi. 1978.** A new species of Enchytraeidae (Oligochaeta) from forest and grassland soils of Orissa, India. *Indian Biol.* 10:35–37. **Dash, M. C., B. Nanda and N. Behera. 1980.** Fungal feeding by Enchytraeidae (Oligochaeta) in a tropical woodland from Orissa, India. *Oikos* 34:202–206. **Dougherty, E. C. and B. Solberg. 1960.** Monoxenic culture of *Enchytraeus fragmentosus. Nature (Lond.)* 186:1067–1068. **1961.** Axenic culture of an enchytraeid annelid. *Nature (Lond.)* 192:184–185. **Dozsa-Farkas, K. 1973a.** Ananeosis, a new phenomenon in the life history of the Enchytraeids (Oligochaeta). *Opusc. Zool. (Budap.)* 12(1,2):43–55. **1973b.** A new *Fridericia* species (Oligochaeta: Enchytraeidae). *Acta. Zool. Acad. Sci. Hung.* 20:27–32. **Eisen, G. 1878.** Redogorelse for Oligochaeter, samlade under de Svenska expeditionerna till arkiske trakter. *Ofvers. K. Vet. Akad. Forh. (Stockholm)* 3:63–79. **1904.** Enchytraeidae. In *Harriman Alaska Expedition. Alaska XII.* Doubleday, Page, New York. **Emery, C. 1898.** *Sur un oligochaeta noir des glaciers de l'Alaska.* Bull. Soc. Zool. Suisse. Assemblée Générale de Berne, Geneva, Switzerland. **1900.** On *Mesenchytraeus solifugus.* In *The Ascent of Mt. St. Elias (Alaska) by H. R. H. Prince Luigi Amedeo di Savoia.* Appendix D, pp. 224–241. **Englemann, M. D. 1961.** The role of soil arthropods in the energetics of an old field community. *Ecol. Monogr.* 31:221–238. **Erseus, C. and P. Lasserre. 1976.** Taxonomic status and geographic variation of the marine Enchytraeid genus *Grania* Southern (Oligochaeta). *Zool. Scr.* 5(3,4):121–132. **Friend, H. 1912a.** British Enchytraeids. III. The genus *Fridericia. J. R. Microsc. Soc.* 1912:9–27. **1912b.** British Enchytraeids. IV. The genus *Henlea. J. R. Microsc. Soc.* 1912:577–598. **Henle, F. 1837.** Über *Enchytraeus,* eine neue Anneliden-Gattung. *Arch. Anat. Physiol. Med.* 1837:74. **Hoffmeister, W. 1845.** *Die bis jetzt bekannten Arkenaus der Familie der Regenwurmer.* Braunschweig, West Germany. **Holmquist, C. 1968.** On Eisen's species *Bryodrilus udei* and the genus *Bryohenlea* Cernosvitov (Oligochaeta: Enchytraeidae). *Ark. Zool. Ser. 2* 22(4):149–153. **Issel, R. 1904.** Due nuove *Fridericia. Atti Soc. Ligust. Sci. Nat. Geogr. Genova* 15:31–39. **Ivleva, I. V. 1953.** Rosti razmnozhenie gorshechnogo cherv'a (*Enchytraeus albidus* Henle). *Zool. Zh.* 32:394–404. **Kennedy, C. R. 1966.** A taxonomic revision of the genus *Grania* (Oligochaeta; Enchytraeidae). *J. Zool.* (1966) 148:399–407. **Kirk, R. G. 1971.** Reproduction of *Lumbricillus rivalis* (Levinsen) in laboratory cultures and in decaying seaweed. *Ann. Appl. Biol.* 67:255–264. **Kitazawa, R.**

1977. Numbers and biomass of Enchytraeidae. Pp. 110–114 in Y. Kitazawa (ed.). *Ecosystem Analysis of the Subalpine Coniferous Forest of the Shigayama IBP Area, Central Japan.* JIBP Synthesis 15. 199 pp. 110–114. **Knollner, F. H. 1935.** Ökologische und systematische Untersuchungen über litorale und marine Oligochaeten der Kieler Bucht. *Zool. Jahrb. Abt. Syst. Oekol. Geogr. Tiere* 66:425–563. **Kubiena, W. L. 1953.** *The Soils of Europe.* Thomas Murty, London. 318 pp. **Lasserre, P. 1964.** Notes sur quelques Oligochetes enchytraeides, présents sur les plages du bassin d'Arcachon. *P.-V. Soc. Linn. Bordeaux* 101:1–5. **1971.** The marine Enchytraeidae (Annelida, Oligochaeta) of the eastern coast of North America with notes on their geographical distribution and habitat. *Biol. Bull. (Woods Hole)* 140:440–460. **Leidy, J. 1882.** On *Enchytraeus, Distichopus* and their parasites. *Proc. Acad. Nat. Sci. Phila.* 145–148. **Leuckart, F. 1847.** Beiträge zur Kenntnis der wirbellosen Tiere. **Levinsen, G. M. R. 1884.** Systematisk-geografisk Oversigt over de nordiske Annulata, Gephyrea, Chaetognathi og Balanoglossi. *Vidensk. Medd. Dan. Naturhist. Foren.* 1883:92–350. **Loden, M. S. and S. M. Locy. 1980.** *Barbidrilus paucisetus,* new genus, new species (Oligochaeta : Enchytraeidae), from eastern North America. *Proc. Biol. Soc. Wash.* 93:1173–1176. **Maclean, S. F., Jr. 1974.** Primary production, decomposition, and the activity of soil invertebrates in tundra ecosystems: a hypothesis. Pp. 197–206 in A. J. Holding et al. (eds.). *Soil Organisms and Decomposition in Tundra.* Tundra Biome Steering Committee, Stockholm. **Michaelsen, W. 1888.** Beiträge zur Kenntnis der deutschen Enchytraeide Fauna. *Stud. Arch. Mikr. Anat.* 31:483–498. **1889.** Synopsis der Enchytraeiden. *Abh. Naturwiss. Ver. Hamb.* 11. **1905a.** Die Oligochäten Deutsch-Ostafrikas. *Z. Wiss. Zool.* 82:288–367. **1905b.** Die Oligochaeten der deutschen Südpolar-Expedition 1901–1903. Nebst Erörterung der Hypothese über einen früheren Grösses die Südspelzen der Kontinente verbindenden antarktischen Kontinent. *Deutsch. Sudpol Exped.* 9:1–58. **Moore, J. P. 1895a.** The characters of the Enchytraeid genus *Distichopus. Am. Nat.* 29:754–56. **1895b.** Notes on American Enchytraeidae. 1. New species of *Fridericia* from the vicinity of Philadelphia. *Proc. Acad. Nat. Sci. Phila.* 341–345. **1899.** A snow-inhabiting enchytraeid [*Mesenchytraeus solifugus* (Emery)] collected by Mr. Henry G. Bryant on the Malaspina Glacier, Alaska. *Proc. Acad. Nat. Sci. Phila.* 125–144. **1902.** Some Bermuda Oligochaeta with a description of new species. *Proc. Acad. Nat. Sci. Phila.* 54:80–84. **1905.** Some marine Oligochaeta of New England. *Proc. Acad. Nat. Sci. Phila.* 57: 373–399. **1943.** *Pelmatodrilus planariformis,* a new oligochaete (Enchytraeidae) modified for epizootic life on Jamaican earthworms. *Not. Nat. (Phila.)* 128:1–7. **Müller, O. F. 1774.** *Vermium terrestrium et fluviatilum.* Heineck and Faber, Havniae and Lipsiae. **Nakamura, Y. 1979.** Micro-distribution of Enchytraeidae in zoysia grassland. *Bull. Natl. Grassl. Res. Inst.* 14:21–27. **Nielsen, C. O. 1952–1953.** Studies on Enchytraeidae. 1. A technique for extracting Enchytraeidae from soil samples. *Oikos* 4:187–196. **1954.** Studies on Enchytraeidae. 3. The microdistribution of Enchytraeidae, *Oikos* 5:167–178. **1955.** Studies on Enchytraeidae. 5. Factors causing seasonal fluctuations in numbers. *Oikos* 6:153–169. **1961.** Respiratory metabolism of enchytraeid worms and free living nematodes. *Oikos* 12:17–35. **1962.** Carbohydrates in soil and litter invertebrates. *Oikos* 13:200–215. **Nielsen, C. O. and B. Christensen. 1959.** Studies on Enchytraeidae VII. The Enchytraeidae. Critical revision and taxonomy of European species. *Nat. Jutl.* 8,9:1–160. **1961.** Studies on Enchytraeidae VII. Critical revision and taxonomy of European species. Supplement 1. *Nat. Jutl.* 10:1–23. **1963.** Studies on Enchytraeidae VII. Critical revision and taxonomy of European species. Suppl. 2. *Nat. Jutl.* 10:1–19. **Nurminen, M. 1967.** Ecology of Enchytraeidae (Oligochaeta) in Finish coniferous forest soils. *Ann. Zool. Fenn.* 4:147–157. **1970.** Records of Enchytraeidae (Oligochaeta) from the west coast of Greenland. *Ann. Zool. Fenn.* 7:199–209. **1973a.** Enchytraeidae (Oligochaeta) from the vicinity of Montreal, Canada. *Ann. Zool. Fenn.* 10:399–402. **1973b.** Enchytraeidae (Oligochaeta) from the Arctic archipelago of Canada. *Ann. Zool. Fenn.* 10:403–411. **1973c.** Distribution of northern enchytraeids (Oligochaeta). *Ann. Zool. Fenn.* 10:483–486. **O'Connor, F. B. 1955.** Extraction of Enchytraeid worms from a coniferous forest soil. *Nature (Lond.)* 175:815,816. **1957.** An ecological study of the enchytraeid worm from a coniferous forest soil. *Oikos* 8:161–169. **1963.** Oxygen consumption and population metabolism of Enchytraeidae. Pp. 32–48 in J. Doeksen, and J. Van der Drift (eds.). *Soil Organisms.* North-Holland, Amsterdam. **1967.** The Enchytraeide. Pp. 213–256 in A. Burges and F. Raw (eds.). *Soil Biology.* Academic Press, New York. **Orsted, A. 1844.** De regionibus marinis. Kobenhavn. **Peachey, J. E. 1962.** A comparison of two techniques for extracting Enchytraeidae from moorland soil. Pp. 286–293 in P. W. Murphy (ed.). *Progress in Soil Zoology.*

Butterworth, London. **1963.** Studies on Enchytraeidae (Oligochaeta) of moorland soils. *Pedobiologia* 2:81–95. **Peterson, E. B., Y. H. Chan and J. B. Cragg. 1970.** Above ground standing crop, leaf area, and caloric value in an aspen clone near Calgary, Alberta. *Can. J. Bot.* 48:1459–1469. **Reynoldson, T. B. 1939a.** Enchytraeid worms and the bacteria bed method of sewage treatment. *Ann. Appl. Biol.* 26:139–164. **1939b.** On the life history and ecology of *Lumbricillus lineatus* Mull. (Oligochaeta). *Ann. Appl. Biol.* 26:782–799. **1943.** A comparative account of the life cycles of *Lumbricillus lineatus* Mull. and *Enchytraeus albidus* Henle in relation to temperature. *Ann. Appl. Biol.* 30:60–66. **1947a.** An ecological study of the enchytraeid worm population of sewage bacteria beds. Field investigations. *J. Anim. Ecol.* 16:26–37. **1947b.** An ecologican study of the enchytraeid worm population of sewage bacteria beds. Laboratory experiments. *Ann. Appl. Biol.* 34:331–345. **1948.** An ecological study of the enchytraeids worm population of sewage bacteria beds. Synthesis of field and laboratory data. *J. Anim. Ecol.* 17:27–38. **Right, G. and E. Kanner. 1979.** Marine Oligochaeta (Tubificidae and Enchytraeidae) from the Caribbean Sea. *Stud. Fauna Curacao Other Caribb. Isl.* 58:44–68. **Rosa, D. 1887.** 11 *Neoenchytraeus bulbosus* n.sp. *Boll. Mus. Zool. Univ. Torino* 2(29). **Satchell, J. E. 1971.** Feasibility study of an energy budget for Meathop wood. Pp. 619–630 in Duvigneaus (ed.). *Productivity of Forest Ecosystem.* Ecology and Conservation 4. Proc. Brussels Symposium, 1969. **Smith, F. 1895.** Notes on species of North American Oligochaeta. *Ill. State Lab. Nat. Hist. Bull.* 4:285–297. **Smith, F. and P. S. Welch. 1913.** Some new Illinois Enchytraeidae. *Ill. State Lab. Nat. Hist. Bull.* XII:615–636. **Southern, R. 1910.** A new species of enchytraeid worm from the White Mountains. *Proc. Acad. Nat. Sci. Phila.* 62:18–20. **1913.** Clare Island Survey. 48. Oligochaeta. *Proc. R. Ir. Acad. Sect. B Biol. Geol. Chem. Sci.* 31:sect.3. **Springett, J. A. and P. M. Latter. 1977.** Studies on the micro-fauna of blanket bog with particular reference to Enchytraeidae. I. Field and laboratory tests of microorganisms as food. *J. Anim. Ecol.* 46:959–974. **Springett, J. A., J. E. Brittain and B. P. Springett. 1970.** Vertical movement of Enchytraeidae (Oligochaeta) in moorland soils. *Oikos* 21:16–21. **Standen, V. and P. M. Latter. 1977.** Distribution of a population of *Cognettia sphagnetorum* (Enchytraeidae) in relation to microhabitats in a blanket bog. *J. Anim. Ecol.* 46:213–229. **Stephenson, J. 1922.** The Oligochaeta of the Oxford University Spitsburgen Expedition. *Proc. Zool. Soc.* 1922:1109–1138. **Thambi, A. V. and M. C. Dash. 1973.** Seasonal variation in numbers and biomass of Enchytraeidae (Oligochaeta) populations in tropical grassland soils from India. *Trop. Ecol.* 14(2):228–237. **Tynen, M. J. 1969.** New Enchytraeidae (Oligochaeta) from the east coast of Vancouver Island. *Can. J. Zool.* 47:387–393. **1970.** *Marionina charae,* a new species of enchytraeid (Oligochaeta) from the Canadian Arctic. *Can. J. Zool.* 48(6):1359–1361. **1975.** A checklist and bibliography of the North American Enchytraeidae (Annelida : Oligochaeta). *Syllogeus* 9:1–14. **Ude, H. 1892.** Ein neues Enchytraeiden-Genus. *Zool. Anz.* 401. **Udekem, D. 1854.** Description d'une nouvelle espèce d'*Enchytraeus. Bull. Acad. R. Sci. Belg.* 21:583–864. **Vedjovsky, F. 1877.** Zur Anatomie und Systematik der Enchytraeiden. *Sitzungsber. Bohem. Ges. Wiss. Prague* 1877. **1879.** Beiträge zur vergleichenden Morphologie der Anneliden. 1. *Monogr. Enchytraeiden. Prague.* **Welch, P. S. 1914a.** Studies on the Enchytraeidae of North America. *Ill. State Lab. Nat. Hist. Bull.* 10(3):123–211. **1914b.** A new Henlea (Enchytraeidae) from northern Michigan. *Trans. Am. Microsc. Soc.* 33:155–164. **1916.** Snowfield and glacier Oligochaeta from Mt. Rainier, Washington. *Trans. Am. Microsc. Soc.* 35:85–124. **1917a.** The Enchytraeidae of the Woods Hole region, Mass. *Trans. Am. Microsc. Soc.* 36:119–138. **1917b.** Enchytraeidae (Oligochaeta) from the Rocky Mountain region. *Trans. Am. Microsc. Soc.* 36:67–81. **1917c.** *Alaskan Glacier Worms (Oligochaeta).* Bionomical Leafl. 2. McGill University, Montreal, Canada. **1919a.** The Enchytraeidae. *Rep. Can. Arctic Exped. 1913–18* 9(A)11:9–19. **1919b.** Further studies on North American mesenchytraeids (Oligochaeta). *Trans. Am. Microsc. Soc.* 38:175–188. **Williams, N. V., J. F. Dell, G. Solbe and R. W. Edwards. 1969.** Aspects of the distribution, life history and metabolism of the enchytraeid worms *Lumbricillus rivalis* (Levinsen) and *Enchytraeus coronatus* (N. & C.) in a percolating filter. *J. Appl. Ecol.* 6:171–183. **Zachariae, G. 1963.** Was leisten Collembolen für den Waldhumus? Pp. 109–123 in J. Van der Drift and J. Doeksen (eds.). *Soil Organisms.* Amsterdam. **1964.** Welche Bedentung haben *Enchytraus* in Waldboden? Pp. 57–68 in A. Jongerius (ed.). *Soil Micromorphology* Elsevier, Amsterdam.

11

Oligochaeta: Lumbricidae

DONALD P. SCHWERT

A majority of North Americans might depict an "earthworm" as a member of the family Lumbricidae and, particularly, as the species *Lumbricus terrestris*. Remarkably, *L. terrestris* is not one of the more common species of Lumbricidae in North America, and, furthermore, about 10 other families of earthworms inhabit this continent. Most of the dominant earthworm species of North America, however, especially those in regions outside of the southern United States, are members of Lumbricidae; included in this family are those species of greatest agricultural and economic importance.

The Lumbricidae include over 200 species worldwide, of which 34 are known from the United States and Canada. Several of these species are peregrine and, with humans as the agent, have spread throughout the world. The anatomical features of Lumbricidae, such as *L. terrestris*, are familiar to almost any student with a basic background in general zoology; except for taxonomically important characters, they are not reviewed here. For more specific discussions of oligochaete and lumbricid morphology, one should refer to Stephenson (1930), Avel (1959), Støp-Bowitz (1969), Bouché (1972), Gates (1972a), Edwards and Lofty (1972), or Reynolds (1977).

Reproduction

Most lumbricids are hermaphroditic and biparental, the sperm being exchanged between partners during copulation into the spermathecae of the partner. After copulation, the clitellum secretes a chitinoid cocoon, which loosens upon full development and passes anteriorly over the earthworm; ova are deposited into it as it slips past the female pores, then the stored sperm of the other parent is forced into it from the spermathecae. As the cocoon slips completely off the earthworm, it closes to form a tough capsule in which the embryo(s) develop.

Stalklike spermatophores, containing mature sperm, are commonly found attached in pairs to preclitellate segments of some lumbricid earthworms. This probable mechanism of sperm transfer has been poorly investigated. Because the distances between the attached spermatophore pairs are approximately equal to the distances between the male pore openings, the spermatophores may be deposited from the male pore atria onto the surface of a partner of the same species. Possibly, the spermatophores, as well as the ova, are then included within the cocoon as it passes over the anterior segments, a mechanism similar to one proposed by Bouché (1975a) for the Albanian lumbricid *Spermophorodrilus albanianus* Bouché, 1975.

Spermatophores are particularly common on some species of *Bimastos,* a genus whose members have long been considered to be parthenogenetic because of the absence of spermathecae. Moore (1895) observed the formation of spermatophores within the atrial invaginations of the male pores of *Bimastos palustris.* Investigation is still needed to determine if these spermatophores also contain mature sperm; if so, at least some species of *Bimastos* could be biparental.

Ecology

Lumbricidae occur in a variety of habitats, ranging from grasslands of the marine coast to alpine meadows of montane regions, from open pasture to dense forest, from acid bogs to alkaline upland soils. Because all lumbricids share a requirement for adequate moisture in their environment, they rarely occur in dry or desert regions. Some, such as *Eiseniella tetraedra* and *Eisenoides lonnbergi,* are even subaquatic, preferring partially or wholly submerged habitats.

The ecology of Lumbricidae in North America is poorly known. Fortunately, most members of the family, excluding those in the genera *Eisenoides* and *Bimastos,* occur in Europe and other continents, and we can refer to the often-detailed research from those regions. Useful summaries of lumbricid ecology have been presented by Bouché (1972), Edwards and Lofty (1972), Satchell (1983) and Lee (1985).

Ecological Categories

Gates (1961) classified lumbricids of Maine into three ecological groups, based principally on dietary preferences:

1. *Geophagous:* those that pass much soil through their intestines.
2. *Limicolous (or limiphagous):* those that inhabit mud or saturated soils.

3. *Litter feeders:* those that feed on various types of organic material, such as leaf litter, compost, or manure.

A second system of ecological classification, proposed by Bouché (1972), is based not only on habitat, but on morphological features common within the groups:

1. *Épigé:* small, red-colored earthworms found, not in the mineral zones of soil, but in decaying plant debris such as compost, litter, and wood; morphologically, they are poorly adapted for burrowing.
2. *Anécique:* brownish, medium-to-large size earthworms inhabiting vertical burrows into the mineral zone, but feeding at or near the surface, especially at night; morphologically, adapted for burrowing and rapid movement therein.
3. *Endogé:* unpigmented, small-to-medium-sized earthworms which continuously inhabit the organic-mineral zones of the soil, feeding especially on dead roots; morphologically, they are adapted for burrowing, but their burrows are more or less horizontal.

Nutrient and Organic Recycling

The burrowing activity of lumbricids serves not only for movement within a protected environment, but for the gathering and ingestion of nutrient material. They derive their nutrients from organic matter ingested either from the soil or, for several species, from plant material on the soil surface. Solid wastes are eliminated through a terminal anus in the form of castings.

Edwards and Lofty (1972) have summarized research, indicating that much of the mineral component ingested by members of the Lumbricidae is returned in these castings to the soil in forms more readily available to plants. Through lumbricid activity, the amounts of exchangeable calcium, magnesium, and potassium and of available potassium and molybdenum are increased, enhancing soil fertility. In addition, castings have been shown to be higher than soil in microorganisms and in organic matter of forms more available to plants. The production of castings varies with species and season, with up to 100 MT/ha being produced each year by healthy populations.

Although the principal excretory organs are the paired nephridia of each segment, nearly half of the nitrogenous waste is excreted from the body as proteinaceous mucus. The nitrogenous waste from the nephridia is mainly urea and ammonia, both simple forms of nitrogen readily available for use by plants. Large amounts of available nitrogen may, likewise, be added to the soil upon decay of earthworm corpses.

With the exception of several species, most lumbricids inhabiting North American forest and woodlands are consistently associated with mull humus, and undoubtedly, they are major agents contributing to the breakdown of organic matter and its subsequent mixture with the mineral soil. Humification processes are enhanced not only by the size reduction and movement of organic matter within the soil, but also by microbial activity within the intestines of the earthworms and by aeration of the soil through earthworm movement. As Bouché (1972) notes, épigé lumbricids are most notably active in these decomposition processes.

Distribution and Dispersal

The North American Fauna

Members of the Lumbricidae inhabit every state or province of the United States and Canada. North American species of the genera *Allolobophora, Aporrectodea, Eisenia, Dendrobaena, Eiseniella, Lumbricus,* and *Octolasion* are not only widely distributed across this continent, but are widespread in Europe and several other landmasses (Gates 1972a). The endemic North American genera, *Bimastos* and *Eisenoides,* are most common in the eastern United States but are not restricted to this region.

Although the Lumbricidae of northeastern North America have been well surveyed [see Reynolds (1977) for a review], the fauna of the southeast, Great Plains, Rocky Mountain, west coast, and far northern regions is still poorly known, and the fauna of Mexico has never been surveyed. Intensive collecting in the states of Kansas, Oklahoma, Missouri, Arkansas, and states bordering the Gulf of Mexico should yield new species, particularly within the genus *Bimastos.*

With the exception of *Bimastos* and *Eisenoides,* all North American Lumbricidae are considered to be "exotic" by Gates (1976a), the source continent for these lumbricids presumably having been Europe. According to this hypothesis, transport of these species was probably accidental, such as lumbricids trapped in ship ballast or among the roots of imported plants. Based on distribution data, Gates (1970, 1976a,b) further asserts that earthworms were completely absent from northern North America from the time of glaciation until the time of their presumed, accidental introduction by European settlers.

Although Gates's hypotheses may be justified for some North American species of Lumbricidae, he has failed to account for the often extensive distribution of *Eisenoides, Bimastos,* and species of the endemic acanthodrilid earthworm genus *Diplocardia* in several northeastern states; according to his hypotheses, except for small colonies obviously the result of northward human transport, these species should presently be restricted to the southern states.

Fossil evidence (Schwert 1979) further indicates that not only were lumbricids present in northern North America over 10,000 years ago (therefore, well before the arrival of European settlers), but that the fauna included at least one "exotic" "European" species, *Dendrobaena rubida.* As is the case with many holarctic taxa within other phyla, neither a mechanism nor a date of immigration can yet be confidently proposed for this species.

Dispersal

Although lumbricids seemingly are sedentary animals, most lumbricids in northern North America are remarkably active throughout the spring and autumn, and in the southern regions, even in winter. The dense and widespread distribution of many Lumbricidae across North America have benefited not only from humans and from burrowing activity, but from their own extraordinary abilities at natural dispersal. These have been discussed by Schwert (1980) and are summarized as follows:

Stream Drift. The movement of cocoons and individuals was measured in a small stream for 5 months by Schwert and Dance (1979). Of the over 300 lumbricid cocoons recovered from drift subsamples, over 92% were found to be viable. The small size and tough spheroidal walls of these cocoons are ideally suited for long and rigorous transport downstream, where they could eventually settle and hatch. Individual earthworms, however, are more subject to predation and physical damage during drift and are, therefore, of minor significance to downstream dispersal.

Mass Emergence. It is common for certain members of the Lumbricidae to emerge from soil in high numbers. This phenomenon occurs during rain or heavy dew in cooler seasons of the year. Such surfacing has been variously proposed to be the result of illness, drowning, or a chemotactic response to gaseous conditions in the burrows. However, because this emergence occurs only during a combination of proper soil temperature and sufficient surface moisture, it could represent an evolved behavioral response, triggering the earthworms to enter the loose litter zone, where free and rapid surface dispersal can occur in relative safety. Such a response would be favorable to those species involved, as it no longer limits the active outward dispersal of geophagous species to burrowing activity. Also, it decreases population pressure during periods of peak reproduction, it rapidly expands the range of the species involved, and it enhances the possibility of genetic exchange among scattered populations of a species.

Accidental Transport by Other Animals. The possibility of cocoons or individuals being transported by other creatures is difficult to investigate and probably is of marginal significance to distribution. Birds, dropping live individuals captured as food or transporting cocoons trapped in mud on their feet, could, for example, result in sporadic range expansions of some lumbricid species.

Sampling Techniques

There are several qualitative and quantitative methods of sampling populations of Lumbricidae, using hand sorting, chemical extraction, or electrical extraction. None of these techniques is perfect for sampling every habitat or for extracting every species. The effectiveness of the various procedures has been critically tested or discussed by Satchell (1955, 1969), Bouché (1969, 1972, 1975b), Nordström and Rundgren (1972), and Reynolds (1977).

Hand Sorting

This is generally the most efficient sampling technique, although small immatures can easily be overlooked, and damaged specimens must be expected. Unless care is taken to dig deeply, burrowing species, particularly *Lumbricus terrestris*, may escape downward. Hand sorting is usually laborious and inevitably disturbs the habitat. An effective mechanical system for removing earthworms and cocoons from excavated soil was designed by Bouché and Beugnot (1971), but the machine is, unfortunately, not commercially available and is difficult to construct.

Chemical Extraction

Although a widely utilized method for sampling lumbricids, its effectiveness varies with species, habitat, soil conditions, and even time of day. Compared with hand sorting, sampling time and habitat disturbance are substantially reduced, and deep burrowing species can often be forced to the surface. The technique, however, is ineffective for sampling subaquatic species, and aestivating individuals will not surface.

An extraction method using formalin was described by Raw (1959) and is probably the best of the chemical techniques. Add 25 ml of 37% formalin to 5 liters of water and pour slowly over an area of about 0.5 m^2. Collect all the earthworms that surface in 10 minutes, and for quantitative measurements, repeat the application. Adjustments in the time and strength of these applications may be necessary, dependent on sampling conditions and the type of experiment. If the worms are to be kept alive after extraction, they must be washed in fresh water immediately after surfacing.

Electrical Extraction

This method is seldom applied today in experimental work, although it is still used by some bait dealers. Earthworms are expelled from the soil by electric current, usually applied from a generator. Like chemical extraction, its effectiveness varies with species and habitat conditions. The method is generally too costly, cumbersome, and time consuming for most research work.

Preservation Techniques

Careless preservation of specimens can be the greatest problem encountered during earthworm identification. Preparation of the earthworms in a straight, robust condition is easily accomplished with minimal time and equipment: two shallow pans, one pair of blunt-tipped forceps, bottles of 70% ethanol solution and 10% formalin solution, vials (with caps or stoppers that will not discolor the preservative), and some cloth or paper toweling.

Kill the earthworms in a pan of cool 70% ethanol. As soon as all movement ceases, place the specimens into a second pan containing cloth or paper toweling soaked with 10% formalin. Quickly straighten the earthworms in the pan without stretching them. For larger specimens, lightly sprinkle their upper surfaces with formalin. After several minutes, the earthworms will harden and will be in a contracted, but straight condition.

Place the specimens in a vial of 10% formalin solution, without overcrowding. After 3–5 days, replace the solution with 5% formalin; formalin should also be replaced whenever the solution becomes discolored. For permanent collections, the earthworms should be gently cleaned of all debris. Label each vial with the locality, collection date, name of collector, and important habitat information. Avoid contact of the preserved specimens with sunlight, as loss of coloration may occur; even the green color of *Allolobophora chlorotica* may be retained indefinitely if the specimens are exposed to a minimum of all types of light.

Many taxonomists, particularly those in Europe, store their specimens in 70% ethanol after a brief, initial hardening in formalin. Specimens prepared in this manner are usually softer, easier to dissect, and will not adversely affect the hands of the examiner. However, unless consistent care is given to an alcohol collection, tissue maceration of the specimens can occur within a short period, often rendering the earthworms useless for further study. The adverse effects of formalin-preserved earthworms on the examiner can be minimized if the specimens are allowed to soak, preferably up to an hour, in cool, fresh water before examination.

Classification and Identification

Classification

The earliest descriptions of lumbricid earthworms were based chiefly on external features, such as color and the form of the body, position of the clitellum, male pores, and tubercula pubertatis. Later classification schemes (Rosa 1893, Michaelsen 1900, Pop 1941, Omodeo 1956, Perel 1976, and others) are based principally on internal organization, such as on the number and arrangement of seminal vesicles, on the numbers, shape, and positions of spermathecae, and on musculature; considerable controversy still exists as to which one of these proposed schemes, if any, is the better.

Because many of the Lumbricidae presently domiciled in North America also inhabit other landmasses, the nomenclature applied in this section is intended to provide both a means of identifying lumbricids and a way of relating one's research to studies of the same peregrine species elsewhere; it is not intended as a defense or an assessment of any taxonomic research. In the key, the names applied are footnoted for purposes of important synonymy.

Two distinctly North American genera, *Eisenoides*, and *Bimastos*, are recognized as defined by Gates (1969). *Eisenoides*, with two species, was reviewed by Gates (1972b). *Bimastos*, which is in great need of review and probably of revision, undoubtedly includes more than the nine species listed in the key.

Important Morphological Features

Although the various proposed classification schemes for the Lumbricidae are based chiefly on internal organization, all of those species inhabiting the United States and Canada can be identified on external characters alone, provided that the specimens are well-preserved adults with well-developed clitella. Because many other families (notably the Acanthodrilidae, Megascolecidae, and Glossoscolecidae) are represented in some regions of North America, one must first ascertain whether all of the specimens being examined share the following basic external characteristics before using the key: clitellum beginning on or behind segment 19; male pores present in segment 15 (segment 13 for the common morph of *Eiseniella tetraedra*); female pores on segment 14; eight setae per segment, in regular longitudinal ranks.

The key uses a minimum of external features for identifying a species, and these are defined as follows:

Clitellum: a regional swelling of the epidermis associated with cocoon production and often obscuring the intersegmental furrows; it usually differs from the rest of the body in color (Figs. 11.1, 11.3). The convention 26,27,28–33,34 indicates that the clitellum generally occurs on segments 28–33, but it may also overlap onto segments 26, 27, and/or 34.

Female pores: the external openings of the oviducts, usually present in the region of the A and B setae on segment 14 of the Lumbricidae and often difficult to see (Fig. 11.1).

Genital tumescences: glandular swellings, usually of paler color than the rest of the body, through which follicles of the genital setae open (Figs. 11.1, 11.4).

Intersegmental furrows: the border between two consecutive segments. In regions where the furrow boundaries are confused by the clitellum or by secondary annulation, the segment positions can usually be distinguished by the presence of the four setal pairs.

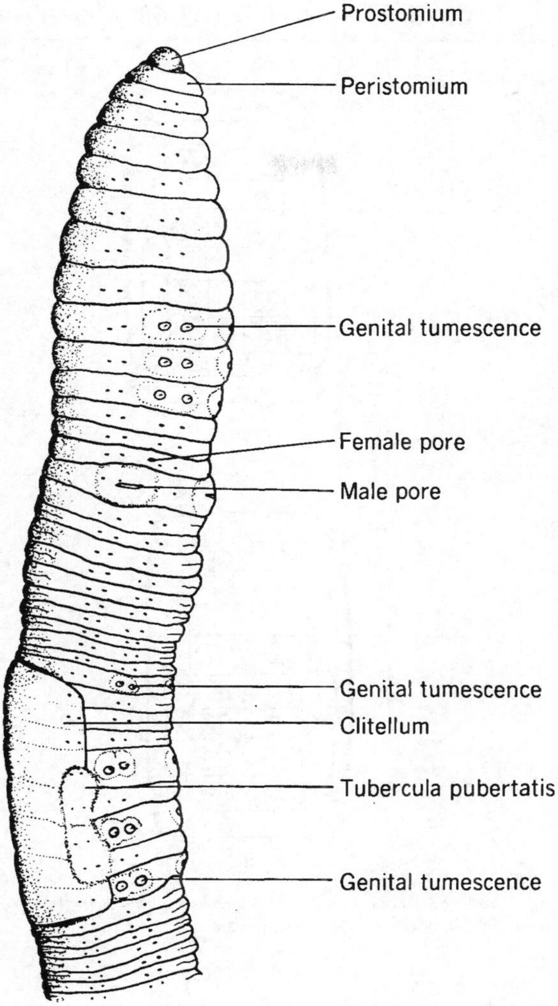

1

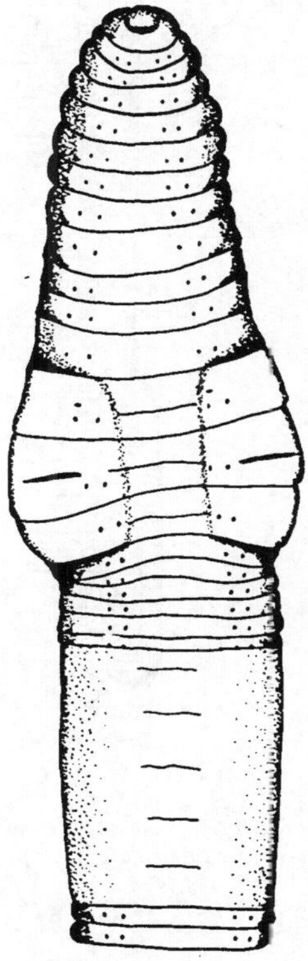

3

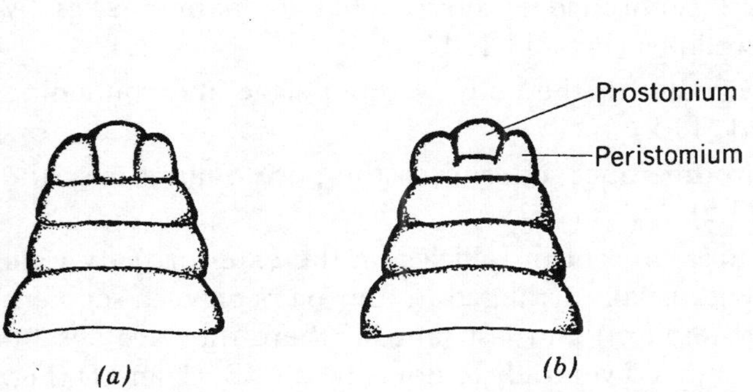

2

Fig. 11.1. Longitudinal view of *Aporrectodea tuberculata*. **Fig. 11.2.** Dorsal view of segments 1–4: (*a*) peristomium completely divided by prostomium; (*b*) peristomium incompletely divided by prostomium. **Fig. 11.3.** *Bimastos palustris:* ventral view of segments 1–29.

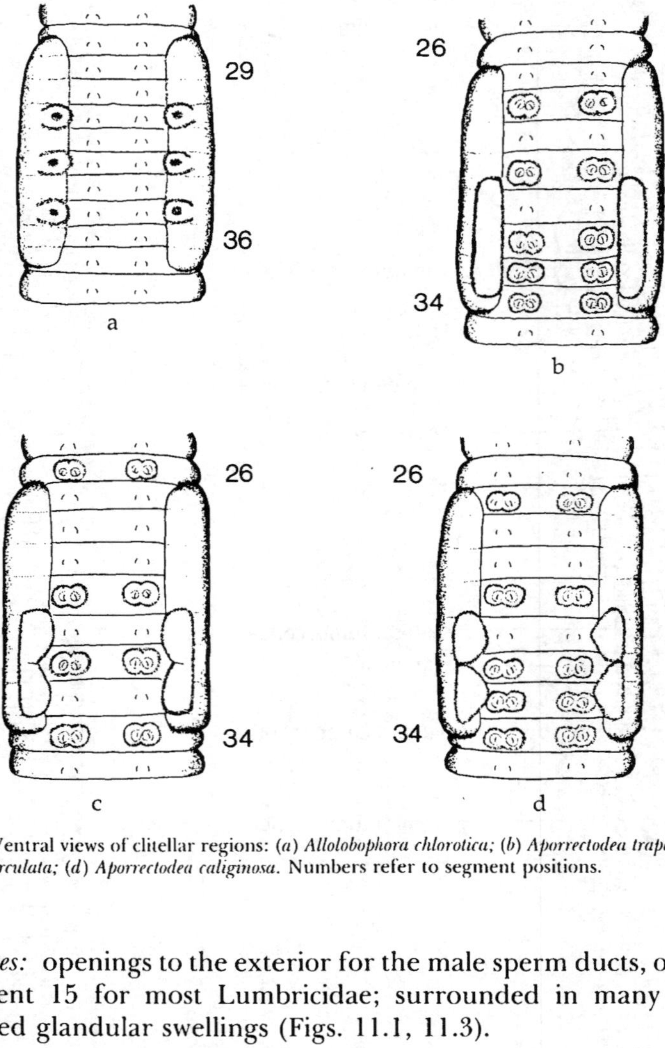

Fig. 11.4. Ventral views of clitellar regions: (*a*) *Allolobophora chlorotica;* (*b*) *Aporrectodea trapezoides;* (*c*) *Aporrectodea tuberculata;* (*d*) *Aporrectodea caliginosa.* Numbers refer to segment positions.

Male pores: openings to the exterior for the male sperm ducts, occurring in segment 15 for most Lumbricidae; surrounded in many species by marked glandular swellings (Figs. 11.1, 11.3).

Peristomium: the first segment of the body, without setae and containing the mouth (Figs. 11.1, 11.2).

Prostomium: a lobate protuberance anterior to the peristomium over the mouth (Figs. 11.1, 11.2).

Setae: bristlelike structures present in follicles on the exterior body wall. Eight setae are present, usually arranged in four pairs on each segment except the first (peristomium) and last (anus), where they are absent. Individual setae are referred ventrally to dorsally as A, B, C, and D (Fig. 11.5), and the relative distances between them can be of importance to species identification.

Tubercula pubertatis: glandular swellings on each side of the clitellum, usually just above the B setae (Figs. 11.1, 11.4); morphologically, they aid

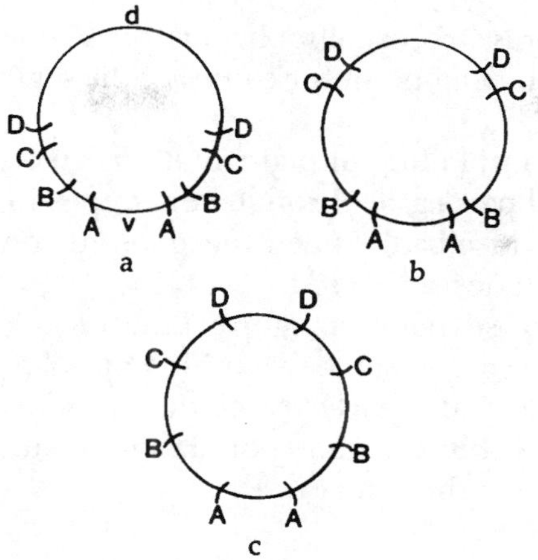

Fig. 11.5. Setal positions of Lumbricidae: (*a*) closely paired (*d*, dorsum; *v*, venter); (*b*) widely paired; (*c*) equidistant.

the forcing of sperm into the spermathecae of the partner during copulation (Bouché 1972).

Additional information or detailed illustrations of lumbricid morphology can be found in Stephenson (1930), Bouché (1972), and Reynolds (1977). A catalog has been published by Reynolds and Cook (1976) of names, descriptions, and type specimens of Oligochaeta. Gates (1972a,c, 1975, 1978a,b) discusses his acceptance of certain nomenclature as well as provides detailed accounts on the anatomy of some important groups of Lumbricidae. Standards for modern taxonomic description of earthworms were proposed by Ljungström (1970) and are widely accepted today; for reasons of simplicity and clarity, the conventional use of roman numerals to designate segment numbers is omitted in the key.

Methods of Identification

Lumbricidae are not difficult to identify, provided that the specimens are well preserved. All of the features used in the key should be readily visible under a low-power binocular microscope. Live Lumbricidae can also be identified with this key, although some frustration is inevitable due to movement of the earthworms during examination. Generally, once one becomes proficient in identifying preserved material, the identification of live specimens on sight becomes easier.

For each specimen being examined, one should first note its pigment, the nature of its prostomium, the positions of its clitellum, any tubercula pubertatis, any genital tumescences, and the relative position of its setae behind the clitellum. On all species except the common morph of *Eiseniella tetraedra,* the male pores are usually obvious on segment 15, and these provide a convenient starting point for counting the segmental positions of other features.

Despite all precautions, fading of pigmentation can occur through time. Most specimens should pass easily through key couplets 11, 14, or 21, but it may be necessary to retrace badly faded specimens through the key until a suitable anatomical diagnosis is found.

The identification of sexually immature Lumbricidae is difficult, even for the specialist. Some species can be identified on color, setal patterns, or on adolescent specimens, the early traces of secondary sexual features. Recognition of adults within the same populations could be an aid in providing possible names for the immatures.

ACKNOWLEDGMENTS

Dr. Gabriel Comita and Dr. Laurent LeSage deserve recognition for their thorough review of this manuscript.

KEY TO THE SEXUALLY MATURE SPECIES OF LUMBRICIDAE OF NORTH AMERICA

1a Prostomium completely divides the peristomium (Fig. 11.2*a*) **2**

1b Prostomium incompletely divides the peristomium (Fig. 11.2*b*) . . . **6**

2a **(1)** Clitellum begins before segment 30 . . . . **3**

2b Clitellum begins behind segment 30. . . . . **5**

3a **(2)** Tubercula pubertatis present **4**

3b Tubercula pubertatis absent; clitellum on segments 23, 24–32, 33; pigment brown to blackish . . . . ***Eisenia eiseni*** (Levinsen, 1884)
Syn: ***Lumbricus eiseni*** Levinsen, 1884

4a **(3)** Clitellum on segments 26, 27–32; tubercula pubertatis on segments 28–31; pigment violet to reddish-brown . . . .
Lumbricus rubellus Hoffmeister, 1843

4b Clitellum on segments 28–33; tubercula pubertatis on segments 29–32; marked pale genital tumescence in region of setae A and B usually present on segment 10; pigment red to reddish brown . . . .
Lumbricus castaneus (Savigny, 1826)

5a **(2)** Clitellum on segments 32–37; tubercula pubertatis on segments 33–36; length usually greatly exceeding 100 mm; pigment violet to reddish brown . . . . ***Lumbricus terrestris*** Linnaeus, 1758

5b Clitellum on segments 34–39, 40; tubercula pubertatis on segments 35–37 (occasionally extending onto 38); length seldom exceeding 100 mm; pigment reddish brown . . . ***Lumbricus festivus*** (Savigny, 1826)

6a **(1)** Male pores on segment 15 . . . . **7**

6b Male pores on segment 13; clitellum on segments 22, 23–26, 27; tubercula pubertatis on segments 23–25 (occasionally onto 26); pigment variable, yellow to brown . . . .
Eiseniella tetraedra (Savigny, 1826) (in part)

7a **(6)** Tubercula pubertatis absent. **8**

7b Tubercula pubertatis present **13**

8a (7) Clitellum saddle-shaped, not completely encircling body **9**

8b Clitellum annular, completely encircling body on segments 22, 23–27, 28; ventral region surrounding male pores often markedly swollen (Fig. 11.3) . . . . ***Bimastos palustris*** Moore, 1893

9a (8) Clitellum beginning before segment 26 . . . . **10**

9b Clitellum beginning on or behind segment 26 . . . . **11**

10 (9) **a.** Clitellum on segments 20–29, 30, 31, 32; pigment red anterodorsally . . . . ***Bimastos gieseleri*** (Ude, 1895)

b. Clitellum on segments 22–29, 30; pigment reddish anterodorsally ***Bimastos tumidus*** (Eisen, 1874)

c. Clitellum on segments 23, 24–32, 33; pigment nearly absent . . . ***Bimastos longicinctus*** Smith and Gittins, 1915

d. Clitellum on segments 24, 25–30; pigment red dorsally . . . . ***Bimastos parvus*** (Eisen, 1874)

e. Clitellum on segments 24, 25–31; pigment red anterodorsally . . . ***Bimastos beddardi*** (Michaelsen, 1894)

f. Clitellum on segments 25–31, 32; pigment reddish dorsally . . . . ***Bimastos heimburgeri*** Smith, 1928

g. Clitellum on segments 25–35; pigment absent (known only from holotype collected in Kansas). . . . . . ***Bimastos welchi*** (Smith, 1917)

11a (9) Pigment (formalin preservation) present, at least anterodorsally, red, deep violet, or brown **12**

11b Pigment (formalin preservation) absent; clitellum on segments 27–33; four pairs of pits present at level of A and B setae between segments 28/29, 29/30, 30/31, and 31/32 . . . . ***Aporrectodea muldali*** (Omodeo, 1956)

12a (11) Clitellum on segments 26, 27–31, 32; length less than 90 mm (rare morph) . . . . ***Dendrobaena rubida*** (Savigny, 1826) (in part)

Syn: ***Dendrodrilus rubidus*** (Savigny, 1826)

12b Clitellum on segments 27–36, 37; length generally greater than 90 mm ***Bimastos zeteki*** Smith and Gittins, 1915

13a (7) Tubercula pubertatis elliptical, suckerlike disks on segments 31, 33, and 35 (Fig. 11.4*a*); clitellum on segments 28, 29–37; color of live specimens usually green to greenish yellow . . . . ***Allolobophora chlorotica*** (Savigny, 1826)

13b Tubercula pubertatis not elliptical and suckerlike; color of live specimens not green **14**

14a (13) Red to reddish-brown pigment (formalin preservation) present, at least dorsally **15**

14b Pigment (formalin preservation) either absent or, if present, yellow, slate, deep violet, pure brown, but not pinkish or reddish **21**

15a (14) Tubercula pubertatis beginning on or before segment 28 . . . . . **16**

15b Tubercula pubertatis beginning behind segment 28 . . . . **17**

16a (15) Tubercula pubertatis on segments 27–29; clitellum on segments 24, 25–30, 31 . . . . ***Eisenoides carolinensis*** (Michaelsen, 1910) (in part)

16b Tubercula pubertatis on segments 28–30; clitellum on segments 24, 25, 26–32; pigment often dorsally in transverse bands alternating with colorless intersegmental furrows . . . . ***Eisenia fetida*** (Savigny, 1826)

Dugès' (1837) emended spelling of ***foetida*** for this species is unjustified.

17a (15) Clitellum beginning on or before segment 29 . . . . **18**

17b Clitellum on segments 31–36; tubercula pubertatis on segments 33–34 . . . . ***Dendrobaena mammalis*** (Savigny, 1826)

18a **(17)** Pigment dorsally with little or no colorless intersegmental banding **19**

18b Pigment dorsally in obvious transverse bands alternating with colorless intersegmental furrows; clitellum on segments 26–33 (rarely limited to 27–33); tubercula pubertatis on segments 30–31 (sometimes extending onto 29 or 32) . . . . ***Eisenia zebra*** Michaelsen, 1902

19a **(18)** Setae equidistant, at least behind clitellum (Fig. 11.5*c*), distances between setae A and A, A and B, C and D, and D and D approximately equal to one another; clitellum on segments 27, 28–33, 34; tubercula pubertatis on segments 31–33. ***Dendrobaena octaedra*** (Savigny, 1826)

19b Setae obviously paired (Fig. 11.5*a,b*) **20**

20a **(19)** Tubercula pubertatis on segments 29–30; clitellum on segments 26, 27–31, 32 . . . . ***Dendrobaena rubida*** (Savigny, 1826) (in part)
Syn: ***Dendrodrilus rubidus*** (Savigny, 1826)

20b Tubercula pubertatis on segments 30–31; clitellum on segments 26, 27, 28–32, 33, 34; marked colorless area often present around C and D setae of segment 11 (or 10–11) . . . . ***Eisenia hortensis*** (Michaelsen, 1890)

21a **(14)** Pigment (formalin preservation) present, yellow, brown, deep violet, or slate **22**

21b Pigment (formalin preservation) absent **25**

22a **(21)** Clitellum clearly begins on or in front of segment 24 . . . . **23**

22b Clitellum begins on or behind segment 25 (rarely infringing forward onto segment 24). **24**

23a **(22)** Clitellum on segments 23, 24–30; tubercula pubertatis on segments 26–28; length generally greater than 60 mm; pigment deep violet to slate . . . . ***Eisenoides lonnbergi*** (Michaelsen, 1894)

23b Clitellum on segments 22, 23–26, 27; tubercula pubertatis on segments 23–25 (occasionally onto 26); length less than 60 mm; pigment variable, yellow to brown (rare morph) . . . . ***Eiseniella tetraedra*** (Savigny, 1826) (in part)

24a **(22)** Tubercula pubertatis on segments 31–33, bandlike in shape; in region of clitellum, genital tumescences on segments 30, 32–33 and often on 28 or 34; clitellum on segments 27, 28–33, 34 (Fig. 11.4*b*) . . . . . ***Aporrectodea trapezoides*** (Dugès, 1828) (in part)

24b Tubercula pubertatis on segments 32–34; in region of clitellum, genital tumescences on segments 31, 33, and 34; clitellum on segments 27, 28–34, 35 . . . . ***Aporrectodea longa*** (Ude, 1885)

25a **(21)** Tubercula pubertatis beginning on or before segment 29 . . . . . **26**

25b Tubercula pubertatis beginning behind segment 29 . . . . **27**

26a **(25)** Tubercula pubertatis on segments 27–29; clitellum on segments 24, 25–30, 31 . . . . ***Eisenoides carolinensis*** (Michaelsen, 1910) (in part)

26b Tubercula pubertatis on segments 29–31; clitellum on segments 24, 25, 26–32. ***Eisenia rosea*** (Savigny, 1826)

27a **(25)** Tubercula pubertatis beginning on or before segment 31 . . . . . **28**

27b Tubercula pubertatis beginning behind segment 31 . . . . **32**

28a **(27)** Setae widely paired (Fig. 11.5*b*) **29**

28b Setae closely paired (Fig. 11.5*a*) **30**

29a **(28)** Clitellum on segments 29–34; tubercula pubertatis on segments 30–33 . . . . ***Octolasion cyaneum*** (Savigny, 1826)

29b Clitellum on segments 30–35; tubercula pubertatis on segments 31–34 . . . . ***Octolasion tyrtaeum*** (Savigny, 1826)

30a **(28)** Genital tumescences present in region of A and B setae on segment 33, among other segments in region of clitellum. **31**

30b Genital tumescences absent from region of A and B setae on segment 33, but present on segments 30, 32, 34, and often 26; clitellum on segments 27–34; tubercula pubertatis on segments 31–33 (occasionally extending onto 30 or 34) (Figs. 11.1, 11.4*c*). . . . .
Aporrectodea tuberculata (Eisen, 1874)

31a **(30)** Tubercula pubertatis on segments 31–33, bilobate or crescentic in shape; clitellum on segments 27, 28, 29–34, 35; genital tumescences in region of clitellum on segments 30, 32–34, and often 27 (Fig. 11.4*d*); length generally less than 85 mm . . . .
Aporrectodea caliginosa (Savigny, 1826)
Syns: ***Allolobophora caliginosa*** (Savigny, 1826); ***Aporrectodea turgida*** (Eisen, 1873)

31b Tubercula pubertatis on segments 31–33, bandlike in shape; clitellum on segments 27, 28–33, 34; genital tumescences in region of clitellum on segments 30, 32–33, and often 28 or 34 (Fig. 11.4*b*); length generally greater than 85 mm. . . . .
Aporrectodea trapezoides (Dugés, 1828) (in part)

32a **(27)** Tubercula pubertatis on segments 33–34; clitellum on segments 28, 29, 30–35, 36. . . . . ***Aporrectodea limicola*** (Michaelsen, 1890)

32b Tubercula pubertatis on segments 35–42; clitellum on segments 33, 34–43; marked genital tumescence present in region of C and D setae on segment 9 . . . . ***Aporrectodea icterica*** (Savigny, 1826)

References

Avel, M. 1959. Classe des Annélides Oligochètes (Oligochaeta, Huxley, 1875). Pp. 224–470 in P. P. Grassé (ed.). *Traité de Zoologie,* Vol. 5. Masson, Paris. **Bouché, M. B. 1969.** Comparaison critique de méthodes d'évaluation des populations de Lombricidés. *Pedobiologia* 9:26–34. **1972.** *Lombriciens de France, écologie et systématique.* Institut National de Recherche Agronomique, Paris. 671 pp. **1975a.** La reproduction de *Spermophorodrilus albanianus* nov. gen., nov. sp. (Lumbricidae): explique-t-elle la fonction des spermatophores? *Zool. Jahrb. Abt. Syst. Oekol. Geogr. Tiere* 102(1):1–11. **1975b.** Fonctions des lombriciens. IV. Corrections et utilisations des distorsions causées par les méthodes de capture. Pp. 571–582 in J. Vaněk (ed.). *Progress in Soil Zoology.* Dr. W. Junk, The Hague, The Netherlands. 512 pp. **Bouché, M. B. and M. Beugnot. 1971.** Contribution à l'approche méthodologique de l'étude des biocénoses. II. L'extraction des macroéléments du sol par lavagetamisage. *Ann. Zool. Ecol. Anim.* 4:537–544. **Dugès, A. 1837.** Nouvelles observations sur la zoologie des annélides abranches sétigères. Pt. 1. *Ann. Sci. Nat.* Ser. 2, 8:17–24. **Edwards, C. A. and J. R. Lofty. 1972.** *Biology of Earthworms.* Chapman & Hall, London. 283 pp. **Gates, G. E. 1961.** Ecology of some earthworms with special reference to seasonal activity. *Am. Midl. Nat.* 66:61–86. **1969.** On two American genera of the earthworm family Lumbricidae. *J. Nat. Hist.* 9:305–307. **1970.** Miscellanea megadrilogica VII. *Megadrilogica* 1(2):1–14. **1972a.** Burmese earthworms—an introduction to the systematics and biology of megadrile oligochaetes with special reference to southeast Asia. *Trans. Am. Philos. Soc.* 62(7):1–326. **1972b.** Contributions to North American earthworms (Annelida) no. 4. On American earthworm genera. I. *Eisenoides* (Lumbricidae). *Bull. Tall Timbers Res. Stn.* (13):1–17. **1972c.** Contributions to North American earthworms (Annelida : Oligochaeta) no. 3. Toward a revision of the earthworm family Lumbricidae IV. The *trapezoides* species group. *Bull. Tall Timbers Res. Stn.* (12):1–146. **1975.** Contributions to a revision of the earthworm family Lumbricidae XII. *Enterion mammale* Savigny, 1826 and its position in the family. *Megadrilogica* 2(1):1–5. **1976a.** More on earthworm distribution in North America. *Proc.*

Biol. Soc. Wash. 89(40):467–476. **1976b.** More on oligochaete distribution in North America. *Megadrilogica* 2(11):1–6. **1978a.** The earthworm genus *Lumbricus* in North America. *Megadrilogica* 3(6):81–116. **1978b.** Contributions to a revision of the earthworm family Lumbricidae XXII. The genus *Eisenia* in North America. *Megadrilogica* 3(8):131–147. **Lee, K. E. 1985.** *Earthworms: Their Ecology and Relationships with Soils and Land Use.* Academic Press, New York. 411 pp. **Ljungström, P.-O. 1970.** Introduction to the study of earthworm taxonomy. *Pedobiologia* 10:265–285. **Michaelsen, W. 1900.** Oligochaeta. *Das Terreich* 10:xxix + 575 pp. **Moore, H. F. 1895.** On the structure of *Bimastos palustris*, a new oligochaete. *J. Morphol.* 10:473–496. **Nordström, S. and S. Rundgren. 1972.** Methods of sampling lumbricids. *Oikos* 23:344–352. **Omodeo, P. 1956.** Contributo alla revisione dei Lumbricidae. *Arch. Zool. Ital.* 41:129–212. **Perel, T. S. 1976.** A critical analysis of the Lumbricidae genera system (with key to the USSR fauna genera). *Rev. Ecol. Biol. Sol* 13(4):635–644. **Pop, V. 1941.** Zur Phylogenie und Systematik der Lumbriciden. *Zool. Jahrb.* 74:487–522. **Raw, F. 1959.** Estimating earthworm populations by use of formalin. *Nature (Lond.)* 184:1661–1662. **Reynolds, J. W. 1977.** *The Earthworms (Lumbricidae and Sparganophilidae) of Ontario.* Life Sci. Misc. Publ. Royal Ontario Museum, Toronto, Canada. 141 pp. **Reynolds, J. W. and D. G. Cook. 1976.** *Nomenclatura Oligochaetologica, A Catalogue of Names, Descriptions, and Type Specimens of the Oligochaeta.* University of New Brunswick, Fredericton, Canada. 217 pp. **Rosa, D. 1893.** Revisione dei Lumbricidi. *Mem. Accad. Torino Ser. 2* 43:399–476. **Satchell, J. E. 1955.** An electrical method of sampling earthworm populations. Pp. 356–364 in P. Murphey (ed.). *Soil Zoology.* Proc. Nottingham School of Agricultural Science. **1969.** Methods of sampling earthworm populations. *Pedobiologia* 9:20–25. **(ed.). 1983.** *Earthworm Ecology from Darwin to Vermiculture.* Chapman & Hall, London. 495 pp. **Schwert, D. P. 1979.** Description and significance of a fossil earthworm (Oligochaeta : Lumbricidae) cocoon from postglacial sediments in southern Ontario. *Can. J. Zool.* 57(7):1402–1405. **1980.** Active and passive dispersal of lumbricid earthworms. Pp. 182–189 in D. L. Dindal (ed.). *Soil Biology as Related to Land Use Practices.* EPA-560/13-80-038. U.S. Environmental Protection Agency, Washington, D.C. **Schwert, D. P. and K. W. Dance. 1979.** Earthworm cocoons as a drift component in a southern Ontario stream. *Can. Field-Nat.* 93:180–183. **Stephenson, J. 1930.** *The Oligochaeta.* Clarendon Press, Oxford. 978 pp. **Støp-Bowitz, C. 1969.** A contribution to our knowledge of the systematics and zoogeography of Norwegian earthworms (Annelida : Oligochaeta : Lumbricidae). *Nytt Mag. Zool. (Oslo)* 17:169–280.

12

Oligochaeta: Megascolecidae

and Other Earthworms from Western North America

WILLIAM M. FENDER

DOROTHY McKEY-FENDER

The terrestrial earthworm fauna of the western United States is composed primarily of two groups, the native species and the introduced Lumbricidae (Fender 1984). The native earthworms are of the family Megascolecidae in the broad sense (Acanthodrilidae and Octochaetidae *sensu,* Gates 1959), and include several genera. The bulk of these species were traditionally placed in *Plutellus* Perrier but were removed to *Argilophilus* Eisen, 1893 (Jamieson 1972, Gates 1977b). Two other genera, *Arctiostrotus* (McKey-Fender and Fender 1982) and *Kincaidodrilus* (McKey-Fender 1982), have recently been described to contain several species that did not fit well into *Argilophilus.* Many new species are being described in a revision of the argilophiline earthworms which is presently being compiled by us. We consider all of these species to belong to a single tribe, herein called the Argilophilini. Brief descriptions of several new genera and a more detailed taxonomic discussion are given later in this chapter. The second group, the adventive Lumbricidae, now predominate in most agricultural and urban areas, whereas the native Megascolecidae predominate where native vegetation is still dominant.

In addition to the two groups above, there are a few species of exotic genera of the Megascolecidae (*Pheretima* and its segregates, *Microscolex* and *Dichogaster*) and Ocnerodrilidae (*Eukerria* and *Ocnerodrilus*). These taxa are not especially abundant anywhere in the western United States, but are more commonly encountered in southern areas.

Range

The Argilophilini occur in a narrow strip of Pacific coastal North America. In British Columbia they appear to be limited to the coastal islands, a single species (*Arctiostrotus perrieri*) having been described from the Queen Charlotte Islands and that species and two others having been collected from Vancouver Island. In Washington, they are present along the entire coast, but in the Puget lowlands and Cascade Range only south of the Puget Sound. Two species are found in the Columbia Gorge in Washington. *Driloleirus americanus* is found in the area around Pullman, Washington, and Moscow, Idaho, and in the hills west of Ellensburg, Washington, but is not known from the arid land between. Several species of *Arctiostrotus* are now known from central Idaho. In Oregon, the argilophiline worms are present in great diversity west of the Cascade Range but eastward have been found only in the Dalles area (two species) and in the Ochoco Mountains (two species?) and Blue Mountains (one species). The apparent paucity of species in that area could be a collecting artifact, but most of central and eastern Oregon is too dry for earthworms, and pumice soils inimical to earthworms blanket most of the east slope of the Cascade Range. Southward, argilophiline worms have been found in northern and central California in the mountains and along the coast (although not on the dry floor of the Sacramento Valley) as far south as Santa Cruz and Tulare counties. The species diversity drops rapidly southward, only two species, *Argilophilus sierrae* and *A. papillifer,* having been collected south of San Francisco Bay.

No argilophiline worms are known south of Tulare County in the United States, but a single specimen, described as *Argilophilus hyalinus* Eisen, 1900, was collected from Guatemala. This specimen, in poor shape to begin with, was lost in the San Francisco earthquake and fire of 1906, but of its general affinities there can be little doubt. No native earthworm species are known from western North America outside Washington, Idaho, Oregon, California, and the islands of coastal British Columbia. The nearest relatives of the Argilophilini are to be found in southeast Asia (Gates 1972), and although there are no congeners in the two areas, some of the Asian *"Plutellus"* species could possibly fit within the Argilophilini.

Anatomy and Function

Peregrine Lumbricidae can usually be identified to species simply by external features, but in the Megascolecidae, specimens must be dissected to identify the genus, and identification to species requires an intimate knowledge of earthworm anatomy.

The argilophiline worms are cylindrical in cross section except in the area of the clitellum; the posterior end is never flattened as in the Lumbricidae, but is usually expanded to form a "caudal bulb." The ambulatory setae are in eight regular rows and are generally not closely paired. The worms are, with few exceptions, unpigmented and the coelomic fluid is a milky white due to the presence of abundant coelomocytes. They often have a marbled appearance posteriorly where the milky coelomic fluid and the dark chloragogen alternatively show through the thin body wall.

The clitellum is closely associated with both the female and male pores, covering at least segments 13–18, and may extend onto segment 12 anteriorly and/or to segment 22 posteriorly (*Driloleirus* spp.), whereas in the Lumbricidae the clitellum is always posterior to 18. The clitellum may be a complete ring, or may be divided ventrally. Round, oval, or transversely elongate suckerlike genital markings are present on the anterior ventral surface of the body (not just in the clitellum), usually intersegmentally. These aid in attachment during copulation, since no slime tube is formed by these worms. Tubercula pubertatis and genital tumescences such as are found in the Lumbricidae are absent in the Megascolecidae. The male pores are a single pair on segment 18 (15 in the Lumbricidae) and are in glandular porophores in line with *A* and *B* setal rows. They are combined with the pores of cylindrical "prostates" and are associated with modified "penial" setae, which are generally longer and more slender than the ambulatory setae and lack the nodulus of the latter. These setae apparently act as an intromittent organ, entering the spermathecal duct and guiding the passage of sperm; similar penial setae and prostates are present in *Diplocardia* but in the segments adjacent to that bearing the male pores. Female pores are on 14. Spermathecal pores are in line with setal row *B* and are usually in or near intersegments 7/8 and 8/9, although some species possess only one pair in 7/8 or 8/9 and others have as many as five pairs, the most posterior being in 8/9. In some genera these pores are nearly invisible. The corresponding spermathecae are in the segment posterior to the intersegment and typically have a thick duct containing or bearing a lateral diverticulum in which the spermatozoa are stored, and a terminal ampulla. In one group of *Argilophilus* spp. the diverticulum has disappeared and spermatozoa are stored in the ampulla. Gonads and their funnels are as in the Lumbricidae; testes in x and xi, ovaries in xiii.† Ovaries are composed of numerous egg strings rather than a single egg string as in the Lumbricidae. Seminal vesicles are postseptal from 10/11 and 11/12 only (i.e., none are anteseptal). Ovisacs are postseptal on 13/14, and small ovisac-like rudiments are postseptal on 12/13. "Testis sacs" such as in *Lumbricus terrestris* are not found in the western natives, although a similar arrangement has arisen in *Arctiostrotus perrieri,* where the testes of xi have disappeared and those of x have become enclosed by the adherence of septa 9/10 and 10/11. While the oviducts pass from the ovifunnels directly to the wall in xiv, the

† We use the convention of referring to internal segmental locations by Roman numerals, external ones by Arabic numerals, and septa by two Arabic numerals separated by a slash.

sperm ducts travel backward just under the peritoneum to segment xviii, where they connect to the ental ends of the prostate ducts. The prostates are tubular and coiled or folded into a mass and are composed of two parts: the glandular ental portion, and a much shorter duct which is covered with shining circular muscle. The sperm ducts connect to the prostate at or very near the juncture of the duct and gland.

Gastrointestinal System

The alimentary canal is divided into a series of functional parts. This series is similar to that of *Diplocardia* and differs strongly from that in the Lumbricidae. Just inside the mouth is the pharynx, which contains a dorsal glandular pad, or tongue, which can be extruded from the mouth and aids in feeding, probably also providing the first digestive enzymes to be mixed with the food. The "saliva" of most Argilophilini has a definite flowery or spicy odor, characteristic of the species, and is perhaps involved in defense since it may be copiously produced in response to a mechanical stimulus. Behind the pharynx, of segment iv or iv–v, is a nonmuscular crop which serves for storage of food until it is processed by the muscular gizzard, which is in v, v and vi, vi, or vi and vii. Even though it may occupy more than one segment, the large muscular gizzard is a single structure, not double as in *Diplocardia,* and is derived from segment v. The gizzard of the Lumbricidae is in segment xvii and is not a homologous structure. In segments (ix)x–xiii(xvii) is a highly vascularized region of the esophagus which produces spherules of solid $CaCO_3$. This region of the esophagus may be a simple tube (as in most species) or may have portions organized into distinct calciferous "glands" communicating with the lumen of the esophagus only by ducts (*Kincaidodrilus*), but it always has abundant blood-filled flaps, or lamellae, on its inner surface and it is from these flaps that the $CaCO_3$ is produced (compare to the calciferous gland of lumbricids and *Diplocardia;* Smith 1924). The floor of the calciferous esophageal region may be flat or with a single or double ridge (the "ventral esophageal typhlosole"). Between the calciferous region and the expanded intestine is a valvular constriction composed of one to three segments.

The intestine of the Argilophilini has four parts: pretyphlosolar (peristaltic) region, anterior typhlosolar (secretory) region, posterior typhlosolar (absorptive) region, and rectum. The peristaltic region extends from the intestinal origin (xvi–xix) to the typhlosole origin and may comprise 2–10 segments. As the name implies, this portion of the intestine is a mixing region with muscular (though thin) walls that in life are in constant motion. This region is in the same area and is homologous with the gizzard and crop of the Lumbricidae. The typhlosole begins in the region of xix–xxix, usually abruptly. Unlike the lumbricid typhlosole, which is an infolding of the entire intestine dorsum, the megascolecid typhlosole involves only the inner intestine wall and contains a vascular sinus. It may be variously bifurcate, forming a T- or Y-shaped or more complex structure (viewed in cross section), but it has none of the transverse grooves or flaps characteristic of many lumbricid typhlosoles. All argilophiline species have a typhlosole, but in a few species it is reduced to a vein in the intestine dorsum. In these cases it may be difficult or impossible to find the typhlosole in poorly preserved specimens (see note on preparation of specimens). The anterior typhlosolar region is nearly cylindrical, has relatively thick walls with a dense coating of brown or black chloragogen, and is secretory in function, as can be observed in anesthetized worms dissected in Ringer's solution. In *Drilochaera* the anterior intestine is covered by a pearly white layer of unknown function. Several (apparently unrelated) taxa possess ceca in this region of the intestine. The ceca appear to be secretory. Most species have a longitudinal groove midventral in the anterior typhlosolar intestine. This groove is flanked by ridges bearing long, dense cilia, which can be observed in living specimens to beat backward, while similar shorter cilia on the adjacent intestine wall beat toward the midventrum. The posterior typhlosolar region has a thinner wall and is more slender, but is tortuously arranged. It is this area that gives the "marbled" appearance (see above). The rectum is the final portion of the alimentary canal and may comprise the last 10–20 segments or so [i.e., the portion in the caudal bulb (see above)]. It lacks any trace of the typhlosole and may be empty or contain the rather dry pelletlike castings of this group. These firm castings are quite characteristic and probably contribute to the well-granulated character of the soil in some areas the worms inhabit.

Excretory System

Most Argilophilini have a single pair of nephridia in each segment starting with segment ii; there is always a set of nephridia which can be seen to derive from segment ii. However, only in *Arctiostrotus* are the nephridial funnels functional outside the rectal region. The remaining genera have small nonfunctional stalked mass of cells, except in the rectal area, where the funnels are large and functional. The nephridia are distinctly tubular in shape (i.e., without a thick glandular investment covering the tubules). In *Arctiostrotus,* the subterminal portion of the duct is expanded into a bladder. In all Argilophilini, the ducts enter the body wall at or near the site of the nephropore; in the Lumbricidae the nephridial ducts enter the body wall in *B* and pass subepidermally to the site of the pore. *Driloleirus* spp. are meronephric (i.e., having numerous nephridia in each segment). Most of the nephridia are blind entally, but the medianmost nephridia have the characteristic preseptal funnels of normal holonephridia, although these are nonfunctional mass of cells outside the rectal area.

There is a definite, though poorly developed tendency toward buccal enteronephry in the Argilophilini. This is the condition where the anterior nephridia empty into the buccal cavity rather than to the body surface.

Only in *Driloleirus americanus* has this reached the stage of an actual bladder emptying directly into the buccal cavity from the anterior nephridia, but in *Arctiostrotus* the anteriormost nephropores are associated with the ends of the prostomial grooves, allowing fluid to flow to the opening of the mouth along those grooves. We hypothesize that this has the effect of encouraging microbial growth within the intestine, by adding nitrogen to the otherwise nitrogen-poor food supply of these worms.

Circulatory System

The vascular system, as in the Lumbricidae, has a pulsing dorsal vessel which carries the blood forward, and a nonpulsing ventral vessel which distributes blood to the various tissues. These are connected by a series of commissures in the anterior. Anterior to segment x these commissures are noncontractile and attach to the dorsal vessel, the ventral vessel, and the parietes. In segments x–xii(xiii) are "lateroesophageal hearts." These commissures are pulsing and have a large connection to the esophagus dorsum, a small connection to the dorsal vessel, and feed directly into the ventral vessel; lumbricid hearts connect only to the ventral and dorsal vessels. The ventral vessel in these segments is unbranched. These "hearts" have a series of valves that impart a beaded appearance, but the dorsal part is generally much longer than the others.

Nervous System

There are no subneural nor lateral neural vessels in the Argilophilini. Instead, two pairs of parietal trunks are present at about the level of the nephridia, and these appear to perform the same function of connecting the parietal circulation between the posterior segments.

Reproduction

The Pacific coast states are dominated by a maritime climate with mild, rainy winters and hot, dry summers and the native earthworms have developed a reproductive cycle that mirrors this fact. Nearly all species have a peak of cocoon production in the late winter and early spring months, and in most species this is followed by a summer-long estivation. Especially in *Argilophilus* spp. the sexual organs regress, and in *A. garloughi* they disappear entirely. Only those taxa that live in the moist coastal forests and in the High Cascades will maintain activity throughout the summer. Various Lumbricidae, notably *Aporrectodea trapezoides,* will stay active as long as the ground is sufficiently moist, so it is quite possible, by collecting late in the season, to collect nothing but Lumbricidae in areas that a month earlier would have been dominated by the native species. Activity resumes after the ground becomes thoroughly moist in the fall, but the clitellum and other sexual structures will generally not regain full size until midwinter.

Parthenogenesis

To judge by morphological and ecological evidence, parthenogenesis is not nearly as common among the Argilophilini as has been suggested (Gates 1977b), but it does occur. *Arctiostrotus perrieri, Chetcodrilus umbellulariae,* and *Argilophilus sierrae* are all commonly, perhaps usually parthenogenetic. All three are widespread "weedy" species and can be found in fairly disturbed situations. Features that we use as indicators of parthenogenesis include consistent lack of mature sperm on funnels and in spermathecae of mature individuals, frequent aberrations of morphology and/or number of male sexual structures, extreme parasitism of the seminal vesicles and male segments, and lack of the genital papillae which function in copulation.

Habitats

Much of the ecological literature on earthworms, especially from the temperate zones, treats Lumbricidae only, and the other families, when they are recognized as existing at all, are considered to be on the way to being exterminated by the more agressive Lumbricidae. There is some justification for this picture in the western United States, at least on the surface. If one looks for earthworms in lawns, gardens, fields, roadside ditches, irrigated areas, pastureland or grazed woodlots, coarse-textured soils, grassy areas, or areas dominated by European weeds, one will almost certainly find nothing but Lumbricidae. *Argilophilus garloughi, Toutellus toutellus, A. sierrae,* and a few other natives have been found to persist in disturbed areas such as grazed hillsides with oaks and in wheat fields, but this is usually (not always) in reduced numbers and in specialized microhabitats. Although *T. toutellus* and *A. sierrae* have been found to feed on the underside of aged cattle dung, this is an extremely unusual practice for this group. Further, the native taxa rarely travel over the surface as the Lumbricidae so commonly do. In consequence of the above, very few people, including biologists, ever see the native species.

Optimal Soil and Organic Matter Characteristics

In the extensive areas dominated by native forest, the Argilophilini are the predominant worms, and Lumbricidae are very rare, existing only in a few specialized habitats. Abundance of the natives can range from quite sparse (0.5 m^{-2}) in the Cascade Range to levels more nearly like those of agricultural soils (e.g., 30+ m^{-2} at Castle Crags State Park, California).

Driloleirus macelfreshi, D. wellsi, and *D. cascadensis,* all of which can reach 60 cm or more in length, can be found every 1–2 m along trails in favorable sites. Obviously, these worms can and do exert a dominating influence on soil structure and chemistry in much of the west. There are areas (e.g., under *Ceanothus* sp. on lower Table Rock, Jackson County, Oregon) where the entire A_1 horizon can be seen to be composed of the pelletlike castings of the local earthworms. On Vancouver Island, Graeme Spiers et al. (personal communication) found 40–50 large specimens per square meter of an undescribed *Arctiostrotus* sp. and observed that the acid, very deep (0.5–1.5 m) organic layer in virgin forests of this region was composed entirely of castings of this species.

The preferred habitats of the various species are quite diverse but fall into several groups. Members of the genus *Argilophilus* are largely soil feeders, living in the A_1 horizon of fine-textured soils and feeding indiscriminately on a mixture of soil and organic fragments. Most *Argilophilus* spp. are found in areas of mixed forest, including both hardwood and coniferous trees. This is a large genus, with many subgroupings, some species of which are litter feeders, but the rule will hold for most.

Most *Driloleirus* spp. make permanent burrows in fine-textured, compact, deep soils. They feed exclusively on organic fragments at or near the surface, including mosses, decaying conifer needles, bits of wood, grass seeds and stems, and an occasional dead insect. (A single undescribed species lives in and feeds on red clay soil.) Their burrows can be quite deep, having been traced down almost 5 m in Palouse soils near Pullman, Washington (Smith 1897). Several species will cast on the surface. Concentrations of these worms often occur where the water table is reachable but have not been found in waterlogged soil. Specimens of *D. macelfreshi* and *D. wellsi* kept in glass containers in the laboratory spent most of their time at the surface of an artificial water table (W. M. Fender, personal observation).

Arctiostrotus spp. live in concentrations of fairly raw organic matter, (i.e., upper litter layers of coniferous forest, soft rotting wood of old logs, in moss layers, and in well-aerated muck of cold seeps and small streams). Several species are specialists, while *Arctiostrotus perrieri,* which ranges from Coos Bay, Oregon, to the Queen Charlotte Islands and inland to the Cascade Range, takes advantage of all the foregoing habitats. These taxa are extremely intolerant of heat and will decompose quickly in a room that is comfortably warm to humans.

Species of *Toutellus* also live in organic matter, but they tend to be found in the lower organic layers (O_2) and into the upper mineral soil layers. One (undescribed) species is a true limicole and most can be found in or near aquatic situations, but more are not limited to such situations. *T. toutellus* is often found on dry slopes covered sparsely by *Quercus garryana,* the Oregon white oak, where they are found in the litter or organic-rich soils under the trees.

Chetcodrilus will inhabit both litter and soil, and at least one species lives in well-rotted wet wood and in the mud of streams. These taxa are limited to within a few miles of the coast between Reedsport, Oregon, and Trinidad, California, and are probably dependent on the combination of high rainfall and moderate temperatures characteristic of that region. *Chetcodrilus umbellulariae* is the most widespread species, being extremely opportunistic, and can be found in a wide variety of habitats [e.g., soil under redwoods, deep litter under *Lithocarpus* (tan oak), extremely acid litter on Blacklock soil (spodosol) under shore pine, or in rotting wood].

Most species of *Nephrallaxis* inhabit litter only, but unlike other litter inhabitants are found in the interior valleys of southwestern Oregon, although *N. davisi* and an undescribed species extend to the south Oregon coast and one, *N. blacki,* gets up to the central Willamette valley. One species, undescribed, from Lower Table Rock, near Medford, Oregon, was abundant in soil under *Ceanothus,* and the A_1 horizon consisted almost entirely of its castings. The soil between the shrubs, on the other hand, was nearly structureless. Another species was found within Oregon caves, but normally inhabits the litter of mixed conifer–hardwood forests.

Kincaidodrilus spp. are restricted to within a few kilometers of the coast in Washington and northwestern Oregon. Their ecology is poorly known but they seem to be primarily soil feeders and can stand a considerable amount of disturbance. An undescribed species has been found under grasses and herbs along a major highway and in a plowed field. Many of the known localities for *K. kincaidii* had been disturbed by logging or road building.

A common feature of native earthworms is for them to be found in higher numbers along trails and in other areas of compacted soil. Many of our best collections have been from old logging skid roads and within forest camps. Some species seem to have a reticulated distribution, determined largely by the game trails which are quite abundant in our forests.

Within the forested regions, there are certain kinds of sites which are much more likely than others to be inhabited by lumbricids to the exclusion of natives: streambanks and floodplains, especially if the associated soils are coarse-textured silts or sands; areas near roadsides, especially if dominated by hardwoods such as red alder (*Alnus rubra*) and with silty soils; campsites; open areas with blackberries or grasses, meadows; former pastureland regrown to forest (especially if with silty soil); and areas near fishing streams. Factors such as pH, calcium availability, and history of introduction also being important, not all of these areas will actually support lumbricids, but it is more likely that they will.

Serpentine Soils

In southwestern Oregon and northern California are large areas dominated by serpentinitic soils. These areas have a peculiar flora dominated by shrubs and are much more open than surrounding areas. The soils are rich in clays and magnesium and often have high levels of toxic heavy metals (White 1971, Whittaker 1954). Native megascolecid earthworms (all undescribed species) are abundant in these soils but the Lumbricidae are absent. Interestingly, those local species are much more resistant than is normal to our technique of anesthetization with $MgSO_4$.

Moisture

The lower limit of precipitation tolerated by argilophiline species appears to be about 15 in. (38 cm) annual precipitation. This is about the edge of moist forests in our area, although the range of *Driloleirus americanus* extends into treeless areas. While some argilophiline species are semiaquatic, none seems to take advantage of moist situations in otherwise arid environments, such as do some Lumbricidae. The native taxa usually live in fine-textured soils, generally clays, clay loams, or silt loams, and those few species found in areas of sandy soil actually live in litter layers overlying or interbedded within the sand. Most species live in moderately acid soils (pH 6.5–4.5), but some are found in high abundance in organic soils of less than pH 3.0. Our only collections of native taxa from basic soil (8.0) were from serpentinitic clays. Soils inhabited by Argilophilini average lower in pH, finer in texture, and higher in woody (or otherwise hard to digest) organic matter than soils inhabited by Lumbricidae.

Preservation

Although it is possible to identify specimens that are poorly preserved, it is a difficult task. Also, we have found that the most valuable specimens and those easiest to study are those which have been relaxed prior to fixing. The following method has been developed over a period of nearly 40 years by James A. Macnab and the authors and if carefully followed can yield excellent specimens with good color and excellent internal detail. It is a modification of a method used initially with marine invertebrates (Hedgpeth, personal communication).

This method involves four separate steps: anesthetization, killing, fixing and preservation.

1. After the worms have been rinsed (in cold water), place them in a $MgSO_4$ solution which contains 3–4 parts water to 1 part saturated $MgSO_4$ solution. It is generally best to start with a more dilute solution and add the saturated solution gradually. After a period of 30 minutes to 3 hours (usually 1 hour) the worms will be relaxed and unresponsive to mechanical stimulus, at which point they can be killed and fixed (but see note on serpentine soils above). If they become wrinkled, they must immediately be taken to the next stage before they decompose.

2. Dip the worms briefly (2–5 seconds) in Bouin's picro-formol, then lay them straight, extended but not stretched, on a sheet of paper for about 1 minute. This time is very approximate, but must be long enough for the tissues to be killed; otherwise, the longitudinal muscles will contract irreversibly in the next step. Bouin's picro-formol is the only solution we have found which, while killing them, will still reliably leave the worms extended. Various methods that work well for Lumbricidae are unreliable with the Argilophilini. Immersion in hot water, for example, will cause some argilophiline worms to fall apart immediately. Some of the very large *Driloleirus* spp. can be killed in hot water, but the water must be hot enough to coagulate proteins. Contracted megascolecids are much more difficult to work with than lumbricids because they contract much more strongly. The internal organs end up lying on top of each other and segmental positions are obscured.

3. Put the worms in a V-shaped container, such as a dissecting tray elevated on one edge, and add enough Lavdowsky's solution (a mixture of formalin, alcohol, and acetic acid) to cover the specimens. At present, we are using a modified formula with half the acetic acid of the formula given by Guyer (1936). The worms will be stiff and ready to put in tubes in about half an hour but should be kept in the Lavdowsky's solution for at least several hours. Put the specimens in stoppered glass culture or test tubes and lay the tubes flat; the specimens can be folded to fit them in the tubes.

4. Replace the Lavdowsky's solution with 4% formaldehyde solution (9 parts water to 1 part commercial formalin) and afterward keep the tubes upright. If cork stoppers are used, care must be taken to ensure that the fluid, specimens, and labels do not touch the cork, or they will be blackened beyond use. The fluid should be replaced within a week and in any event should be replaced if it becomes cloudy, if the tubes are opened frequently, or after several years of storage. Care should also be taken that there is more volume of preservative than specimens. (It is astonishing how often this simple principle is ignored.)

Do not place unfixed earthworms directly into 70–80% alcohol and expect to get usable specimens. They become extremely flaccid and are of little or no scientific value. The internal organs dissolve to almost nothing recognizable, external features disappear, and all pigments are usually washed away. Formalin is essential for the fixation of delicate earthworm tissues. If there is no time to preserve them properly, worms can be placed directly in Lavdowsky's solution or Bouin's picro-formol and be usable (although badly contracted), but even then the fixing fluid should be replaced with 4% formaldehyde after several hours or overnight.

An efficient method for use during collecting trips is to lay the worms which have been dipped in Bouin's on a 3 by 5 in. file card and then slip the whole card, worms and all, into a small self-sealing polyethylene bag. Enough Lavdowsky's solution to cover the specimens when the bag is held by the top is then poured directly into the bags, the air forced out, and the bag sealed. After several hours the fixative is strained out and replaced with 4% formaldehyde as above, the bags resealed, rolled up, and placed in glass-top canning jars, which are then filled with 4% formaldehyde. Many collections can thus be put into a small space and transported efficiently, not to mention the savings on tubes and chemicals. Collection data can be written directly on the cards, preferably in pencil or some resistant ink, and thus not get separated from the specimen.

Fixative and Preservation Solutions

Lavdowsky's Mixture (FAA) (Modified)

Water	40 parts
95% ethyl alcohol	50 parts
Formalin	10 parts
Acetic acid (glacial)	1 part

Bouin's Picro-formol

Picric acid, sat. aq. sol.	75 parts
Formalin	25 parts
Acetic acid (glacial)	5 parts

Note: Picric acid should never be allowed to become perfectly dry, as it then becomes a dangerous explosive. For this reason it is always sold with a sufficient moisture content. We recommend keeping the powder under water and using the resulting saturated solution. Rinse the cap of the container with water before attempting to open it so that no dry powder is left around the lip.

KEY TO THE MEGADRILE FAMILIES OF THE UNITED STATES*

1a Hearts 11 pairs, in xi–xxi; ovaries in xxiii; testes 10 pairs xii–xxi (large mud-dwelling worms of southern United States) . . . . **Lutodrilidae (*Lutodrilus multivesiculatus*)***

1b Hearts fewer, only anterior to xiv; ovaries in xiii; testes one to two pairs in x–xi **2**

2a An intestinal gizzard present in or behind segment xvii; calciferous gland intramural, consisting of longitudinal tubules within the esophagus wall, opening anteriorly; typhlosole present, involving all layers of gut wall . . . . . . **Lumbricidae**

2b Gizzard not intestinal, if present; calciferous glands various, but not as above; typhlosole present or absent, if present then involving only inner gut wall and enclosing a vascular sinus **3**

3a With one or two (not more) pairs of "prostates" in or adjacent to the male pore

* Asterisks indicate species not yet reported from the western United States; included in this key for increased usefulness.

segments, these prostates having a glandular, rough-textured ental portion and a basal duct portion with shining circular muscle 4

3b Prostates present or absent, but never as described above 5

4a Extramural calciferous glands present in ix and x; last hearts in segment xi. . . . . **Ocnerodrilidae**

4b Calciferous glands various, but never extramural in ix and x only; last hearts in segment xii or xiii . . . . **Megascolecidae (s. lat.)***†

5a Prostates a pair in xvii, extending backward under the intestine, shining, covered with longitudinal (apex to base) muscle; female apparatus complex, ovaries in sacs connected to "spermathecae," which open into the oviduct; esophageal gizzard present; typhlosole absent . . . **Eudrilidae (*Eudrilus eugeniae*)***

5b Prostates, if present, a long series of small glands associated with the ventral setae; ovaries, oviducts, and spermathecae all separate, not joined 6

6a Calciferous glands present, extramural in vii–ix; typhlosole present; gizzard present, esophageal . . . . **Glossoscolecidae (*Pontoscolex corethrurus*)***

6b Calciferous glands absent; typhlosole absent; gizzard present or absent . . 7

7a Gizzard absent. . . . . **Sparganophilidae**

7b Gizzard present, esophageal. . . . . **Komarekionidae (*Komarekiona eatoni*)***

Megascolecidae s. lat.

The family-level classification of the megascolecid earthworms is in chaos. The western endemic taxa, as an example, cannot be satisfactorily fitted into the schemes of either Gates (1959), who recognizes three distinct families in this group, or Jamieson (1972), who places these same taxa into two subfamilies. Sims's (1980) treatment does not provide a place for the western endemics. For these reasons, we choose for the present to consider only an inclusive family Megascolecidae, which has well-defined limits, without subdivisions, rather than using subdivisions that are unnatural. Whereas according to the classifications of Gates (1959) and Sims (1980), the western natives would be placed in two separate families, Acanthodrilidae and Octochaetidae, they are a homogeneous group that is referable to a single tribe by our concepts. For this and other reasons we place them only in the inclusive family Megascolecidae. [For an overview of the dispute over family-level classifications in the megascolecoid earthworms, see Jamieson (1971), Gates (1959), and Sims (1980).]

KEY TO GENERA OF WESTERN MEGASCOLECIDAE‡

Due to time constraints and the presence of so many undescribed species in the western United States, no attempt will be made to key to the species level.

1a Setae perichaetine (very many per segment); prostate gland branched, racemose 2

1b Setae lumbricine (eight per segment, in regular rows); prostate glands unbranched 3

2a Holonephric (one pair of nephridia per segment) . . . . ***Perionyx excavatus****

2b Meronephric, nephridia a great many tiny twists on the peritoneum, giving it a

† See discussion under the family.

‡ Includes three genera not yet reported from the West.

furry appearance, especially anteriorly; gizzard in viii, but often seeming to include ix and x as well . . . . ***Pheretima* (s. lat.)***

Sims and Easton (1972) subdivided this unwieldly genus. Those interested in pursuing this problem should consult that reference as well as Gates (1972 and 1982).

3a Prostates a single pair, opening in the same segment as the male pores; esophageal gizzard single or lacking in our species **4**

3b Prostates one or two pair, opening in segments adjacent to that containing the male pore; gizzard double **6**

4a Male pores on 17, opening separately from the prostates in the same depression; gizzard weak, in v; nephridia vesiculate, a single pair per segment, all pores of a side in single longitudinal rank . . . . ***Microscolex***

4b Male pores on 18, combined with prostate pore; gizzard various; nephridia various **5**

5a Gizzard weak or absent; nephridia absent anterior to segment xiii, glandular (saline, highly organic situations; tropical). . . . . ***Pontodrilus****

5b Gizzard strong, in v, vi, v and vi, or vi and vii; nephridia present from ii, holonephric or meronephric, nephridia tubular . . . .
Argilophilini (see separate key)

6a Usually holonephric (but sometimes with two or three pairs of nephridia in individual segments, occasionally throughout), nephridia tubular, without extramural calciferous glands (refer to Gates 1977a) . . . . . . ***Diplocardia****

6b Meronephric, nephridia discoidal; calciferous glands extramural and stalked in xv–xvii, opening by a common duct into esophagus of xvi . . . .
Dichogaster

Argilophilini

Type Genus: Argilophilus Eisen

Description: Circular in cross section. Setae in eight regular rows. Clitellum including at least segments xiv–xviii. Nephropores arrayed in two or more poorly to strictly defined levels, arrangement varying from totally random to very regular or, in meronephric species, in several regular longitudinal rows.

Septa all present from 4/5, strengthened in anterior esophageal region. Gizzard single, strong, in v, v and vi, vi, or vi and vii but derived from segment v. Typhlosole always present. Intestine origin behind xv, before xx. Supraintestinal glands absent. Calciferous glands, if present, sessile, never stalked. Nephridia tubular, unbranched (always?) although sometimes fused, holonephric or meronephric throughout, with a single pair of stomate nephridia adjacent to the nerve cord in all segments behind xv. Funnels functional in rectal region, nonfunctional in all but one genus (*Arctiostrotus*) anterior to rectal region. Exonephric or with buccal enteronephry (one species). Seminal vesicles postseptal in xi and xii or xi only (proandric species). Prostates tubular, one pair in xviii. Vasa deferentia join prostate at juncture of duct and gland. Dorsal vessel single. Ventral vessel unbranched in "heart" segments [x–xii(xiii)]. No subneural or lateral neural vessels, their place taken by lateral parietal connections or trunks. Lateroesophageal "hearts" in x–xii(xiii).

The species listed under this heading are a tightly knit group but include several recognizable genera, three of which (*Argilophilus, Kincaidodrilus, Arctiostrotus*) have been described previously. Except for two *Argilophilus* species (*A. marmoratus, A. papillifer*), all the species in this group were originally described as *Plutellus* or *Megascolides,* which at the time were considered to be widespread genera with many species in Australia, New Zealand, Asia, and North America. *Argilophilus* Eisen,

1893 was synonymized with *Plutellus* by Michaelsen (1900). According to modern concepts, *Plutellus* and *Megascolides* are now each restricted to a few Australian species. Accordingly, Jamieson (1972) resurrected *Argilophilus* for the American *Plutellus* but did not formally transfer any of these species. Gates (1977b) formally transferred all of the described western *Plutellus* to *Argilophilus* except for *Plutellus* (= *Kincaidodrilus*) *kincaidii,* which has extramural calciferous glands and therefore did not fit his concept of *Argilophilus* (nor ours!), and *P. panulirus,* which was apparently overlooked. (In addition to the name changes, Gates gave brief "aids to the study of the genus" in the form of comments on the species, most of which, however, are either erroneous or misleading, due to his lack of experience in the group.) Two more plutelline genera have recently been described, *Kincaidodritus* McKey-Fender, 1982 and *Arctiostrotus* McKey-Fender, 1982 (McKey-Fender and Fender 1982), but several remaining species do not belong in *Argilophilus* and new genera for these are described in this chapter.

1a Meronephric (with many pairs of nephridia in each segment) . . . . **Driloleirus, gen. nov.**

1b Holonephric (with only a single pair of nephridia in each segment) **2**

2a Intestine with dorsolateral caeca adnate to the anterior septa starting in region of xxiii–xxv . . . . **Toutellus, gen. nov.**

2b Intestine without such ceca **3**

3a Prostomium tanylobic with nephropores of segment ii at or near ends of prostomial grooves; last hearts in xiii, nephrostomes functional through length of body. . . . . **Arctiostrotus**

3b Nephropores of segment 2 not at ends of prostomial grooves; if prostomium tanylobic, then last hearts in xii, nephrostomes reduced to small stalked clumps of cells in all but rectal region **4**

4a Commissure of v on posterior gizzard, septum 5/6 fairly well developed; gizzard entirely in segment v **5**

4b Commissure of vi on posterior gizzard, of v on anterior gizzard margin, septum 5/6 weak, closely investing gizzard or gizzard at least partly in segment vi or posterior **7**

5a Calciferous region with extramural calciferous glands in all or some of xiv–xvii, at least dorsolateral sacculations (pouches), at most large paired glands with narrow openings; longitudinal lamellae are continuous into these sacculations. . . . . **Kincaidodrilus**

5b Calciferous region with or without enlargements but not segmental paired glands. **6**

6a Intestine with mV cecum near typhlosolar origin. . . . . **Macnabodrilus gen. nov.**

6b Intestine without ceca of any kind; covering of anterior intestine a white, pearly layer, removable in thin sheets; dorsointestinal vessels large, pulsating in life. . . . . **Drilochaera**

7a Spermathecal pores obvious, one to four pairs; prostate duct well developed, sometimes huge. **8**

7b Spermathecal pores small, although often associated with obvious glandular papillae; prostate duct small; calciferous lamellae all longitudinal or a mixture of longitudinal and transverse; spermathecae always two pairs . . . . **Argilophilus**

8a Prostate ducts very long, strong, extending most of height of worm; calciferous lamellae all transverse; spermathecae two to four pairs; axis of typhlosole evenly corrugated vertically, like a washboard; typhlosole often T-shaped in cross section . . . . *Chetcodrilus*, **gen. nov.**

8b Prostate ducts only somewhat less strong; calciferous lamellae longitudinal; typhlosole simple; spermathecae one to two pairs. . . ***Nephrallaxis*** **gen. nov.**

Driloleirus gen. nov.

Type Species: Megascolides americanus Smith, 1897
Other Included Species: *Megascolides wellsi* Altman 1936, *Megascolides macelfreshi* Smith 1937, *Megascolides cascadensis* Smith 1937, *Megascolides eiseni* Smith 1937, *Megascolides michaelseni* Smith 1937, and three or more undescribed species
Diagnosis: Anterior and posterior ends bulbous, anterior septa nested. Unpigmented. Prostomium in juveniles epilobous with parallel grooves nearly to one-half, with cross groove near the anterior of segment, wide anteriorly. In older worms this portion appearing prolobous or short epilobous closed behind or tanylobous. Clitellum saddle $\frac{1}{2}$xii–$\frac{1}{2}$xix(xxiii) varying specifically. Dorsal pores not conspicuous, present in all postclitellar segments, posterior one to five pores enlarged. Setae ventral, widely paired. Meronephric throughout. Nephridia behind gizzard separate, exoic, in regular longitudinal rows, a single transverse series per segment; nearly constant in number per segment behind prostates. Median nephridia possessing (functionless) funnels except in some ante-prostatic segments; enlarged in caudal bulb. Buccal enteronephry in one species. Genital markings intersegmental. Spermathecae paired in viii and ix or in ix. Pores conspicuous; duct short, thick-walled with one or more intramural to sessile diverticula; duct expanded ectally (vaginal?). Prostate gland long but tightly coiled, rarely entering adjacent segments, duct short. Penial setae stout, sculptured, reserve setae often fascicled. Penisetal and male pores not united, on paired, round glandular porophores. Gizzard in v, commissures of v on posterior third of gizzard. Calciferous region dilated to sacculated; some lamellae transverse midsegmentally (varying specifically). Intestinal origin xvii to xix, usually xviii, preintestinal constriction two segments. Typhlosole always present, a simple lamellar ridge or with free edge once or more bifurcate. Ventral ciliated intestinal groove absent (at most sometimes rudimentary?) Last hearts xiii. Vascular systems with posterior longitudinal intersegmental connections of both afferent and efferent systems present but always on lateral parietes, never midventral. Posterior lateroparietal vessels having branches both of subesophageal vessels and directly to esophagus. Lymph glands inconspicuous, not enveloping dorsal vessel (a characteristic of *Megascolides* s. str.).
Range: West of Cascade Range from Cowlitz and Skamania counties in Washington, south to Lane County, Oregon. East of Cascade Range in vicinity of Pullman and Clarkston, Washington; Moscow, Idaho; and Ellensberg, Washington.

Arctiostrotus McKey-Fender, 1982

Type Species: Plutellus perrieri Benham, 1892
Other Included Species: *Plutellus altmani* Gates, 1942, and seven undescribed species
Diagnosis: Pigmented or unpigmented, pigment in muscle layers. Prostomium tanylobic. Clitellum xiii–xviii, annular. Dorsal pores lacking except a few in caudal bulb. Nephropores alternating at two levels, first pair at or near ends of prostomial grooves. Holonephric exonephridia, vesiculate (vesicles may be inconspicuous), terminal ducts of one or more successive anterior segments confluent bilaterally in some species. Genital markings segmental, intersegmental, or a combination (rarely absent). All genital pores small, superficial. Spermatheca three to five pairs, last pair in ix. Prostates with slender ducts, confined to xviii; prostate pores and penisetal pores united. Penial setae weak, small, but present. Moderate gizzard in v, v and vi,

or vi. Commissures of v on anterior of gizzard, of vi on posterior. Esophagus simple, calciferous lamellae low, predominantly longitudinal, or partly transverse, short. Intestinal origin xvii, without ceca or ventral ciliated groove. Typhlosole present, a simple lamella or nearly obsolete, reduced to a dorsal longitudinal vein. Last hearts xiii, dorsal part of hearts of x–xiii long. Posterior lateroparietal trunks join bifurcation of continuous subesophageal vessels in xiii. Anterior lateroparietal trunks originate from small vessels below ventral valves of commissures of vii, viii, and especially ix, slender. Both lateroparietals extend to posterior end of body. Vascularization of intestinal wall quadrangular with dorsointestinals visible as slender, paired vessels perpendicular in lateral view. Musculature of body wall thicker ventrally, ventral setae large, those of spermathecal segments (vi–ix) very large. Dorsal interval narrow to wide, $DD < \frac{1}{3}C$ to $\frac{1}{2}C$ (specific variation).
Range: Islands of western British Columbia, south along the coast and in the Puget Trough of Washington; Oregon west of the Cascade Range summits south to Coos Bay, exclusive of the Willamette Valley floor; and in the mountains of central Idaho (Clearwater and Latah counties).

Macnabodrilus gen. nov.

Type Species: Plutellus macnabi McKey-Fender, 1957
Other Included Species: *Plutellus hopsonae* McKey-Fender, 1970
Diagnosis: Unpigmented. Prostomium epilobic. Clitellum saddle xiii–xviii. Dorsal pores large, beginning anterior to clitellum. Holonephric, exoic, heteropore, avesiculate. Genital markings intersegmental. Quadrithecal in viii and ix. Seminal chambers anteromedially intramural in duct. Penial setae stout (may be fascicled), sculptured with rows of spines. Male and penisetal pores close or united. Gizzard in v, commissures of v on posterior one-third of gizzard. Calciferous region broad, constricted septally, some lamellae transverse. No extramural calciferous glands. Typhlosole lamellar or T-shaped (specific variation), double ventral esophageal typhlosole. Intestine beginning xviii. Intestinal groove distinct, wall uniformly thick throughout, a midventral cecum opposite beginning of typhlosole, involving all layers of intestinal wall. Last hearts xii. Anterior lateroparietal trunks from ventral part of hearts of ix. Posterior lateroparietal trunks join paired adnate subesophageal vessels in xiii. Both lateroparietal trunks extend to end of body. Vascularization of intestinal wall a finely quadrangular sinus; dorsointestinal veins perpendicular in lateral view. Dorsal interval very wide, $DD > \frac{1}{2}C$.
Range: Washington south of Puget Sound in Puget Trough and Wind River area; in Oregon just south of the Columbia River in Clatskanie area, Columbia County.

Kincaidodrilus McKey-Fender, 1982

Type Species: Plutellus kincaidii Altman, 1936
Diagnosis: Unpigmented. Prostomium small, epilobic closed behind or nearly prolobous. Clitellum saddle, xiii–xviii. Dorsal pores large, beginning anterior to clitellum. Nephropores very inconspicuous, alternating at two distinct levels. Holonephric exonephridia, avesiculate; postclitellar tubules not parietal, on a mesentery. Genital markings intersegmental. Diagonal glandular bars form male field. Bithecal or quadrithecal in ix or viii and ix (specific variation). Pores small slits with tumid lips; ample ampulla and stout duct with multilocular sessile diverticulum anteriorly or anterolaterally. Prostates confined to xviii, very long, bulging septa, duct moderate. Penial setae very long and fine, capillary; sculptured with fine transverse streaks. Male and penisetal pores united on conical porophores. A strong rounded gizzard in v, commissures of v on posterior one-third of gizzard. Esophagus complex, in region of (xiii)xiv–xvii with dorsal paired pouches with

strong longitudinal lamellae, ranging from simple outpocketings to basally constricted sessile calciferous glands with narrow openings into esophageal lumen. Intestine beginning xviii or xix, without ceca and with simple shallow intestinal groove. Typhlosole present, very large, T-shaped in cross section. Last hearts xii. Anterior extraesophageal trunks form slender, paired subesophageals in vii or viii to xi and xiii (not found in xii). Posterior lateroparietal trunk joins esophagus directly in xiii. Anterior (ventral) lateroparietals from ventral part of commissures of ix, conspicuous anterior to clitellum, but posteriorly traceable only as slender connectives in BC between circumferential vessels on parietes. Most vascular elements of intestine wall vertical. Musculature of body wall not divided. Setae regular, paired, none especially enlarged. Dorsal interval wide, DD > $\frac{1}{3}$C.
Range: Washington, northwest Olympic Penninsula (Clallam, Jefferson, Gray's Harbor, and Mason counties) south along the coast in Willapa Bay area and Pacific County. Oregon in the Columbia River lowland forest (Clatsop County) near Astoria and along coast as far south as Cape Meares (central Tillamook County).

Toutellus gen. nov.

Type Species: Plutellus toutellus Altman, 1936
Other Included Species: *Plutellus oregonensis* Smith, 1937, *Plutellus oregonensis swiftae* Smith, 1937, *Plutellus hyoiedes* Macnab and McKey-Fender 1959, *Plutellus adecus* Macnab and McKey-Fender, 1959, and (at least?) four undescribed species
Diagnosis: Bulbous anteriorly, slender posteriorly, phallic in contraction. Usually, unpigmented or immatures with red epidermal alcohol-soluble chromophores (specific variation). Epilobic open behind. Clitellum annular xiii–xviii ($\frac{1}{2}$xix dorsally). Dorsal pores present behind clitellum (uncommonly beginning 14/15). Holonephric exonephridia, avesiculate; pores alternating irregularly, very inconspicuous. Genital markings intersegmental. Quadrithecal, ducts stout, seminal chambers intramural in anterior wall of duct, bulging duct walls. Prostates confined to xviii, moderately short to long, ducts moderately stout. Penial setae strong, curved, ornamented with rows of teeth or scales. Male and penisetal pores separate on low, raised porophores. Strong gizzard in v, v and vi, or vi. Commissures of v on anterior of gizzard, of vi on posterior or farther forward. Calciferous lamellae longitudinal, low to high, continuous in (xiv)xv–xvi. Intestine beginning xviii (rarely xvii?), with ceca and ventral intestinal groove. Ceca paired dorsolaterally and/or laterally on intestine, anterior several pairs adnate to and supported by anterior septa of their segments; saccate to digitate, a series in region of xxii to lx. Typhlosole lamellar to branched T-shape in cross section. Last hearts xii or xiii (specific or individual variation). Anterior (ventral) lateroparietal trunks originate below ventral valve of commissures of ix. Posterior (dorsal) lateroparietals enter esophagus in xiii. Subesophageals not continuous. Both lateroparietal trunks extend to rear of body. Epidermal capillaries longitudinal. Musculature of body wall continuous, thickness nearly uniform dorsally and ventrally. Setae regular. None especially enlarged. Widely paired to separate. Dorsal interval wide DD > $\frac{1}{3}$C.
Range: British Columbia, Vancouver Island; Washington west of the Cascade Range and in the central Puget Trough, south in Oregon west of the Cascade Summits, in the Cascade Range and central valleys to the Douglas–Jackson county line, and along the coast to southernmost Curry County. Not yet known from the Washington coast or California. Ranging in elevation from sea level to at least 1280 m (4200 ft).

Chetcodrilus gen. nov.

Type Species: Plutellus umbellulariae Macnab and McKey-Fender, 1958
Other Included Species: *Plutellus fenderi* Macnab and McKey-Fender, 1958, *Plutellus exutus* Macnab and McKey-Fender, 1958 and two(?) undescribed species

Diagnosis: Unpigmented. Epilobic. Clitellum xiii–xviii, annular over several anterior segments. Dorsal pores present behind clitellum. Nephropores modified heteropore. Holonephric exonephridia, avesiculate. Genital markings intersegmental. Quadrithecal; (quadrithecal to octothecal in the type species); spermathecal duct ectally shining, strengthened with circular muscles. Pores small, distinct round pits, in adults closed with mucous plugs; seminal chambers intramural, bulging duct, or sessile on duct. Prostates large, drawn back into xix, gland plicate, ducts strongly muscular. Penial setae spined. Male and penisetal follicles united or nearly so, even in species with strong penial setae. Gizzard in v and vi or vi, commissures of v on anterior of gizzard. Calciferous lamellae simple, rippled, transverse, sharply defined posteriorly, esophagus not enlarged, ventral esophageal typhlosole double, extending posteriorly into first intestinal segment beyond other lamellae of preintestinal constriction. Intestinal origin xvii. No intestinal ceca. Distinct ventral intestinal groove opposite typhlosole. Typhlosole lamellar (with square or grooved face) to broadly T-shaped. Axis with many strong vertical veins. Last hearts xii or xiii. Posterior lateroparietals join bifurcations of continuous paired subesophageal trunks in xiii. Vascularization of intestine a rectangular sinus grid, neither anterior nor posterior dorsointestinals especially prominent. Musculature of ventrum slightly stronger. Ventral pairs of ambulatory setae slightly enlarged. Dorsal interval DD = $\frac{1}{3}$ to $\frac{1}{2}$C (specific variation).
Range: Limited to coastal and near coastal areas from Coos County, Oregon, south to Patrick's Point near Trinidad, Humboldt County, California.

Nephrallaxis gen. nov.

Type Species: Plutellus davisi Macnab and McKey-Fender, 1953
Other Included Species: *Plutellus blacki* Macnab and McKey-Fender, 1953, and at least seven undescribed species
Diagnosis: Mostly small rather stout worms. Unpigmented or with red epidermal chromophores (one species). Prostomium long-epilobic to tanylobic. Clitellum annular, xiii–xviii. Dorsal pores present beginning 12/13 (usually evident only behind clitellum). Heteropore. Holonephric exonephridia, avesiculate. Postclitellar tubules compactly arranged, low on parietes (always?). Genital markings segmental or intersegmental or a combination, or absent. Bithecal in viii or ix or (one species) quadrithecal in viii and ix, with expanded pores in 7/8 or 8/9 or 7/8 and 8/9. Seminal chambers intramural, sessile on duct, or sperm contained in duct lumen. Pores of mature individuals closed by mucous plugs. Prostates with strong to very strong ducts entrant into xix. Lateral musculature vicinity of prostates very strong. Penial setae strong, curved; ornamented with rows or groups of sharp spines. Prostate and penisetal pores separate. Gizzard vi or vi and vii. Commissures of v on anterior of gizzard, of vi on posterior of gizzard or farther forward. Esophagus tubular, calciferous lamellae low, discontinuous, mainly longitudinal, a few scattered transverse folds between main lamellae. Intestinal origin xvii; lacking ceca, ventral groove simple. Typhlosole lamellar with rounded free edge containing a sinus. Preintestinal constriction short (parts of two segments). Last hearts xii. Hearts long above valves. Posterior lateroparietal trunks join bifurcation of continuous paired subesophageals in xiii. Anterior lateroparietals from below ventral valve of commissures of ix. Both continuous to end of body. Vascularization of intestine wall rectangular, anterior dorsointestinals visible. Musculature of body wall thicker ventrally. Ventral setae slightly enlarged. Interval DD < $\frac{1}{3}$C (always?). Setae regular, separate.
Range: Eola Hills (southern Yamhill and Polk counties), Salem (Marion County) south through Douglas and Jackson counties, Oregon, into northern California, Smith River (Del Norte County) north along the Oregon coast and Coos Bay in Curry and Coos counties and in western Douglas and Lane counties.

Drilocheira gen. nov.

Type Species: Plutellus chenowithensis McKey-Fender, 1970
Other Included Species: three undescribed species
Diagnosis: Slender worms with thin body wall. Unpigmented. Broadly epilobic. Clitellum pale, saddle (xii)xiii–($\frac{1}{2}$xviii)xviii. Dorsal pores visible behind clitellum, may be present in clitellum, but obscured; last one or two enlarged. Holonephric, exoic, avesiculate. Nephropores irregular, inconspicuous. Genital markings intersegmental, visible internally through the body wall. Quadrithecal in viii and ix; pores small. Seminal chambers intramural or forming extramural lateral, dorsolateral or paired median and lateral, sessile or short-stalked diverticula. Prostate gland length moderate, slightly flattened, ducts slender. Penial setae moderately stout to very slender, with toothed ornament, curved at least proximally (rarely absent?). One or three pores on each male porophore (correlated with degree of reduction of penial setae). Gizzard v, commissures of v on or behind middle of gizzard. Esophagous nearly cylindrical or very slightly broadened, not segmentally enlarged. Calciferous lamellae longitudinally oriented short ridges, fine longitudinal ridges or a combination, in (xiii)xiv–xvi (or in homologous segments preceding the preintestinal constriction and valve), moderate to high with tendency toward fusion medially. Flat or double ventral esophageal typhlosole extending into first intestinal segment. Intestine begins xvi, xvii, or xviii (specific variation). Pretyphlosolar region of intestine short, involving only one or two segments. Typhlosole lamellar to inverted T-shape in cross section (specific variation). Intestinal ceca absent. Ventral intestinal groove wide, shallow, flanked by low ciliated ridges. Last hearts xii. Dorsal vessel behind hearts broad, in anteriormost (two or three) intestinal segments very broad; dorso-intestinals in these segments also widened near dorsal vessel; naked dorsally and with heavy chloragogen laterally, forming a prominently pulsing organ in life, a crudely stellate pattern in the preserved state. Each posterior lateroparietal joining a bifurcation of the continuous paired subesophageals in xiii. Typhlosolar region of intestine, including dorsal vessel and dorsointestinals, covered by a distinct, glistening (white to yellow) pearly layer closely investing and masking structures beneath. This is distinct from the chloragogen and covers it. Typical chloragogen mostly on esophagus and aggregated along dorsal vessel. Vascularization of intestinal wall longitudinal. Musculature not divided into separate bands, narrowly interrupted at setae. Dorsal setal interval moderate to wide, DD = $\frac{2}{5}$ to $\frac{2}{3}$C (specific variation).
Range: The type species is from the Dalles area (east of the Cascade Range). Other representatives are from southwestern Washington south along the coast and in the Coast Range to Lane County, and in the Willamette Valley.

Argilophilus Eisen, 1893

Type Species: Argilophilus marmoratus Eisen, 1893
Other Included Species: *Argilophilus papillifer* Eisen, 1893, *Argilophilus collinus* Eisen, 1900, *Plutellus sierrae* Michaelsen, 1921, *Plutellus garloughi* Smith, 1937, *Plutellus panulirus* Macnab and McKey-Fender, 1959, *Plutellus hammondi* McKey-Fender, 1970 and at least 30(?) undescribed species
Diagnosis: Mostly slender worms. Unpigmented. Epilobic. Clitellum saddle or indefinitely or distinctly annular (xii)xiii–xvii. Dorsal pores present, sometimes very small. Nephridia holoic, exoic, avesiculate. Pores irregularly alternating with tendency in some species to form series in the same line. Genital markings intersegmental. Quadrithecal; spermathecal diverticula distinct, multilocular on lateral, anterior, or median face of duct, intramural or absent (then sperm received in ampulla). Pores small, though distinguishable. Prostate ducts weak in relation to glandular portion. Penial setae various. Male and penisetal pores separate or

united. Gizzard in v, v and vi, or vi. Commissures of v anterior on gizzard, of vi on posterior. Calciferous region not expanded, lamellae essentially longitudinal. Intestinal origin xvi–xviii (specific variation). Typhlosole distinct, lamellar to T-shaped (specific variation). Dorsolateral intestinal ceca if present shallow, not involving septa. Several species have a midventral cecum involving all layers of intestine wall, near origin of typhlosole. Midventral intestinal groove distinct, often broad, in one of the groups often with thin floor. Last hearts xii or xiii. Parietal trunks various; posterior lateroparietals join a bifurcation of continuous subesophageals, enter esophagus directly, or send branches both entering esophagus and joining subesophageals (specific and group variation). Vascularization of gut wall longitudinal, may be diagonal (group variation?). Musculature various. Setal intervals various. This is an extremely variable genus containing some groups that may be of generic or at least of subgeneric status.

Range: In Oregon throughout area west of Cascade Range, except extreme northwest coast and mid to high elevation in the Cascade Range. They have been found in two areas of the Ochoco and Blue mountains of eastern Oregon. In California they are known from as far south as Tulare County.

Described Species

Driloleirus americanus (Smith, 1897) comb. nov.
Driloleirus wellsi (Altman, 1936) comb. nov.
Driloleirus macelfreshi (Smith, 1937) comb. nov.
Driloleirus cascadensis (Smith, 1937) comb. nov.
Driloleirus eiseni (Smith, 1937) comb. nov.
Driloleirus michaelseni (Smith, 1937) comb. nov.
Arctiostrotus perrieri (Benham, 1892) McKey-Fender and Fender, 1982
Arctiostrotus altmani (Gates, 1942) McKey-Fender and Fender, 1982
Macnabodrilus macnabi (McKey-Fender, 1957) comb. nov.
Macnabodrilus hopsonae (McKey-Fender, 1970) comb. nov.
Kincaidodrilus kincaidii (Altman, 1936) McKey-Fender, 1982
Toutellus toutellus (Altman, 1936) comb. nov.
Toutellus oregonensis (Smith, 1937) comb. nov.
Toutellus hyoiedes (Macnab and McKey-Fender, 1959) comb. nov.
Toutellus adecus (Macnab and McKey-Fender, 1959) comb. nov.
Chetcodrilus umbellulariae (Macnab and McKey-Fender, 1958) comb. nov.
Chetcodrilus fenderi (Macnab and McKey-Fender, 1958) comb. nov.
Chetcodrilus exutus (Macnab and McKey-Fender, 1958) comb. nov.
Nephrallaxis davisi (Macnab and McKey-Fender, 1953) comb. nov.
Nephrallaxis blacki (Macnab and McKey-Fender, 1953) comb. nov.
Drilochaera chenowithensis (McKey-Fender, 1970) comb. nov.
Argilophilus marmoratus Eisen, 1893
Argilophilus papillifer Eisen, 1893
Argilophilus collinus Eisen, 1900
Argilophilus sierrae (Michaelsen, 1921) Gates, 1977
Argilophilus garloughi (Smith, 1937) Gates, 1977
Argilophilus panulirus (Macnab and McKey-Fender, 1959) comb. nov.
Argilophilus hammondi (McKey-Fender, 1970) Gates, 1977

For detailed discussion of the species see the references by Altman (1936), Eisen (1894, 1896), Gates (1942, 1962), Macnab and McKey-Fender (1947–1960), McKey-Fender (1957–1982), McKey-Fender and Fender (1982), and Smith (1897, 1937).

References

Altman, L. C. 1936. Oligochaeta of Washington. *Univ. Wash. Publ. Biol.* 4(1):1–137. **Benham, W. B. 1892.** Description of three new species of earthworms. *Proc. Zool. Soc. Lond.* 1892(x):136–141. **Eisen, G. 1894.** On the California Eudrilidae. *Mem. Calif. Acad. Sci.* 2(3):21–62. **1895–1896.** Pacific coast Oligochaeta. I, II. *Mem. Calif. Acad. Sci.* 2(4):63–122. **1900.** Researches on American Oligochaeta, with special reference to those of the Pacific coast and adjacent islands. *Proc. Calif. Acad. Sci. Ser. 3* 2(2):82–277. **Fender, W. M. 1984.** Earthworms of the western United States. I. Lumbricidae. *Megadrilogica* 4(5):93–129. **Gates, G. E. 1942.** Notes on a California earthworm, *Plutellus papillifer* (Eisen, 1893). *Proc. Calif. Acad. Sci.* 23(30):443–452. **1959.** On a taxonomic puzzle and the classification of the earthworms. *Bull. Mus. Comp. Zool.* 121(6):229–261. **1962.** On some earthworms of Eisen's collection. *Proc. Calif. Acad. Sci.* 31(8):185–225. **1972.** Burmese earthworms. *Trans. Am. Philos. Soc.* 62(7):1–326. **1977a.** More on the earthworm genus *Diplocardia. Megadrilogica* 3(1):1–47. **1977b.** On the correct generic name for some west coast native earthworms, with aids for a study of the genus. *Megadrilogica* 3(2):54–60. **1982.** Farewell to North American Megadriles. *Megadrilogica* 4(1,2):12–80. **Guyer, M. F. 1936.** *Animal Micrology.* University of Chicago Press, Chicago. 331 pp. **Jamieson, B. G. M. 1971.** A review of the Megascolecoid earthworm genera (Oligochaeta) of Australia, I. Reclassification and checklist of the Megascolecoid genera of the world. *Proc. R. Soc. Queensl.* 82(6):75–86. **1972.** The Australian earthworm genus *Spenceriella* and the description of two new genera (Megascolecidae : Oligochaeta). *Mem. Nat. Mus. Vic.* 33:73–88. **Macnab, J. A. and D. McKey-Fender. 1946.** Aids to the teaching and study of earthworm anatomy. *Turtox News* 24(6):3 pp. **1947.** An introduction to Oregon earthworms with additions to the Washington list. *Northwest Sci.* 21(2):69–75. **1948.** North American *Plutellus* and *Megascolides* with synonymical notes (Annelida, Oligochaeta). *Am. Midl. Nat.* 39(1):160–164. **1952.** A northwestern endemic earthworm, *Plutellus perrieri* Benham (Oligochaeta : Megascolecidae). *Wasmann J. Biol.* 10(1):67–73. **1953.** Description of two new species of *Plutellus* (Oliochaeta : Megascolecidae). *Wasmann J. Biol.* 10(3):343–354 (dated 1952, actually published 1953). **1958.** Descriptions of new Pacific coast earthworms of the genus *Plutellus* (Oligochaeta : Megascolecidae). *Northwest Sci.* 32(4):103–121. **1959a.** A new species of *Plutellus* from western Oregon (Oligochaeta : Megascolecidae). *Northwest Sci.* 33(2):69–75. **1959b.** Two new species of *Plutellus* from Coos County, southwestern Oregon (Oligochaeta : Megascolecidae). *Northwest Sci.* 33(4):157–159. **1959c.** Corrected setal measurements of *Diplocardia* and *Plutellus* (Oligochaeta : Megascolecidae). *Wasmann J. Biol.* 17(1):159–160. **1960.** Subspecific variation in *Plutellus davisi* (Oligochaeta : Megascolecidae). *Northwest Sci.* 34(4):127–133. **McKey-Fender, D. 1957.** A new species of *Plutellus* from southwestern Washington (Oligochaeta : Megascolecidae). *Wasmann J. Biol.* 15(1):49–59. **1970a.** Concerning native earthworms from southwestern Washington and northwestern Oregon (Oligochaeta, Acanthodrilidae). *Northwest Sci.* 44(4):225–234. **1970b.** Description of endemic earthworms from eastern Oregon (Oligochaeta, Acanthodrilidae). *Northwest Sci.* 44(4):235–243. **1982.** Dr. Kincaid's other worm. *Megadrilogica* 4(3):86–88. **McKey-Fender, D. and W. M. Fender. 1982.** *Arctiostrotus* (gen. nov.). I. The identity of *Plutellus perrieri* Benham, 1892 and its distribution in relation to glacial refugia. *Megadrilogica* 4(3):81–86. **Michaelsen, W. 1900.** Oligochaeta. In *Das Tierreich,* Lief 10. von Friedländer, Berlin. 575 pp. **1921.** Neue und wenig bekannte Oligochäten aus skandinavischen Sammlungen. *Ark. Zool. (Stockholm)* 13(19):1–25, 1 plate. **Sims, R. W. 1980.** A classification and the distribution of earthworms, suborder Lumbricina (Haplotaxida : Oligochaeta). *Bull. Br. Mus. (Nat. Hist.) Zool.* 39(2):103–124. **Sims, R. W. and E. G. Easton. 1972.** A numerical revision of the earthworm genus *Pheretima* auct. (Megascolicidae : Oligochaeta) with the recognition of new genera and an appendix on the earthworms collected by the Royal Society North Borneo Expedition. *Biol. J. Linn. Soc.* 4:169–268. **Smith, F. 1897.** Upon an undescribed species of *Megascolides* from the United States. *Am. Nat.* 31:202–204. **1924.** The calciferous glands of Lumbricidae and *Diplocardia. Ill. Biol. Monogr.* 9(1):1–78. **1937.** New North American species of earthworms of the family Megascolecidae. *Proc. U.S. Natl. Mus.* 84(3009):157–181. **White, C. D. 1971.** *Vegetation–Soil Chemistry Correlations in Serpentine Ecosystems.* Unpublished Ph.D. thesis, University of Oregon, Eugene, Oreg. 274 pp. **Whittaker, R. H. 1954.** The ecology of serpentine soils. IV. The vegetational response to serpentine soils. *Ecology* 35:275–288.

13 Oligochaeta: Megascolecidae and Other Earthworms from Southern and Midwestern North America

SAMUEL W. JAMES

Native earthworms of the United States and Canada consist of members of the following families: Megascolecidae, Komarekionidae, Lumbricidae, Lutodrilidae, and Sparganophilidae. There is only one native terrestrial genus of Lumbricidae, *Bimastos.* All other lumbricid genera were probably introduced.

The Komarekionidae and Lutodrilidae are monospecific and relatively recently described [Gates (1974) and McMahan (1976), respectively]. Both are inhabitants of wet soils, and both are morphologically quite distinct from other earthworms of North America. In the United States, *Komarekiona eatoni* is the only nonred earthworm with a tanylobic prostomium (see Fig. 13.1), no dorsal pores, and a saddle-shaped clitellum from segment xix to xxv or xxvi (Reynolds 1977c). Presently it is known to occur in the southeastern states, Illinois, and Indiana.

Lutodrilus multivesiculatus, presently known from southeast Louisiana, is unusual in color, dimensions, and internal organization. It is green to blue green, changing to brown in the posterior portions, and mature specimens are 20–40 cm long when contracted (McMahan 1976). The striking feature of its internal anatomy is its having 12 pairs of seminal vesicles and as many or more spermathecae. As such, it should present few difficulties to the nonspecialist seeking an identification. For further details, see McMahan (1976).

The Sparganophilidae are another mud-dwelling (or limicolous) group, and may be considered outside the bounds of this volume. However, they have been found in the same soil with terrestrial oligochaetes, so they should be mentioned. The best sources of information on the habits and distributions of *Sparganophilus* are Reynolds (1977a,b). Distinguishing characteristics of the family are a clitellum beginning in segments xv–xix, absence of gizzard and calciferous gland, zygolobic prostomium (Fig. 13.1), and their typically long and thin bodies. For instance, *S. eiseni* is 150–200 mm long and 2.5 mm in diameter (Reynolds 1977a).

Diplocardia

This megascolecid genus is largely restricted to regions south of the limit of Wisconsinan glaciation. It is the opinion of Gates (1966) that in addition to glacial removal of soil, climatic factors eliminated earthworms from the northern portions of the United States and all of Canada. However, Murchie (1958) has found members of this group in southern Michigan on sites covered with Wisconsinan drift, and they are also known from other states in the Great Lakes region. This means they must have made some progress since the furthest advance of Wisconsinan ice 18,000 years ago (Flint 1957). Areas covered by Kansan and Illinoian glaciations, but not the Wisconsinan, have been ice-free for over 100,000 years, and do have native earthworms.

In the past, considerable information on the genus *Diplocardia* was published under Acanthodrilidae as presented by Reynolds and Cook (1976). Therefore, literature on this family may be searched for details on the genus.

Distribution

The genus *Diplocardia* consists of some 40 or more species, of which several are at this writing undescribed. With this in mind, and the presence of large areas of the central United States which have not been collected thoroughly, it is likely that the systematics of the group will be in flux for some time. The group is best known in the southern states, including Texas and Oklahoma. The distribution of the genus is from Nebraska east to Delaware, the northern limit roughly at the latitude of Detroit, Michigan, and from Nebraska south to Texas. There is also a record of *Diplocardia*

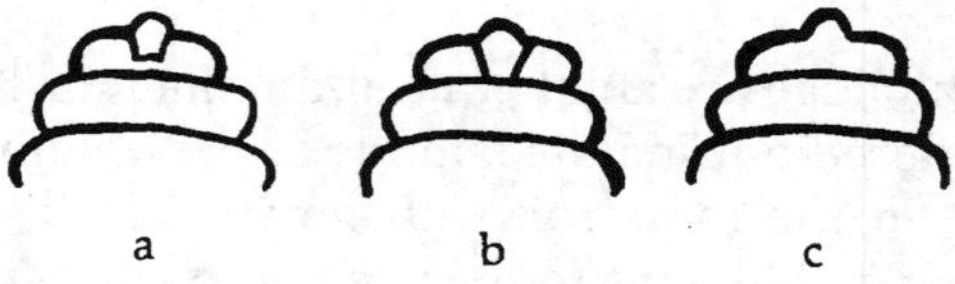

Fig. 13.1. Prostomium types: (*a*) epilobic; (*b*) tanylobic; (*c*) zyglobic.

verrucosa from New Mexico (Gates 1967), but this observation may have been the result of human transportation outside the natural range of the species.

Ecology

Relatively little is known of the ecology of these worms. The morphological correlates of niche developed in Bouché (1977) and Perel (1977) are not as well developed in *Diplocardia* as in the Lumbricidae, but the general trends are similar. Lee's (1985) ecological categorizations were developed on New Zealand Megascolecidae, and may also be applied in a general way to obtain an educated guess of the ecology of a species.

With the exception of *D. eiseni,* which is found in wet or submerged soils, all known species are terrestrial. The *D. riparia* is found in moist river banks, where it is very abundant. A small bait industry collecting the "sodworm" or "riverworm" (both refer to *D. riparia*) has developed in Kansas and Missouri.

Diplocardia fusca and *D. biprostatica* are two large (40 cm long) species occurring in Texas and Oklahoma, respectively. Preliminary indications are that they feed on surface plant litter from a home burrow system, as does the nightcrawler, *Lumbricus terrestris.*

The species with calciferous lamellae, a structure that secretes calcium carbonate into the esophagus, are restricted to the southern states. The structure seems to have arisen only once in the evolutionary history of the genus (S. W. James, unpublished). The function of calciferous glands in earthworms is not clearly understood (see Lee 1985), but probably is related to aspects of dietary quality. There may be some ecological factor peculiar to the southern range, or the distribution of this characteristic may reflect a recent evolutionary origin.

Temperature tolerances of *Diplocardia* are higher than those of the introduced lumbricid *Aporrectodea turgida* (James 1983). Anecdotal reports and the ability of *D. prosenteris* and other species to survive temperatures in excess of 30°C indicate that temperature tolerances are generally higher than in Lumbricidae.

James (1988) demonstrated a positive effect of prairie burning on *D. smithii* and *D. verrucosa,* due to increased root production of burned prairie. These species feed primarily on soil organic matter (James and Cunningham 1989).

Seasonal patterns of reproduction can be estimated from the presence of fully clitellate adults recorded in collection data. An example is the winter–spring reproduction of *D. fusca* inferred from information in Gates (1977). In Kansas tallgrass prairie, most species reproduce in the spring, as seen in the appearance of adults, and then small individuals, in the populations

(S. W. James, unpublished). Some are capable of reproducing whenever conditions permit; if rainfall is sufficient, *D. smithii* may reproduce continuously. On the other hand, *D. rugosa* reproduces in the autumn only.

Potential Areas of Research

An area of ecological concern is the interaction between introduced Lumbricidae and Megascolecidae and native species. It has been hypothesized that these introduced species, which are quite aggressive colonizers, are displacing the native forms (Stebbings 1962). It appears that this is not entirely without foundation in fact, but that the reasons may be incorrectly perceived. The colonizing species tend to be those which tolerate human-made habitats or flourish as a result of human disturbance. In the case of North American Megascolecidae, habitat destruction, rather than competition, is probably the critical factor. Our observations of *Diplocardia* spp. abundant in lawns inhabited by Lumbricidae indicate that peaceful coexistence is possible. An experimental approach would be useful.

Modes of reproduction are another area where research is needed. Earthworms are known to reproduce sexually and by parthenogenesis (Muldal 1952, Jaenike and Selander 1979). Polyploidy and parthenogenesis are interrelated, and polyploidy has been a factor in the evolution of *Diplocardia* (Murchie 1967a).

Some of the intraspecific variation in *D. communis* and *D. singularis* may be related to the formation of polyploid and/or parthenogenetic races. Both are suspected to have parthenogenetic morphs (Gates 1977). On theoretical grounds, earthworms are expected to have high rates of endemism and local variation (Easton 1979), so the study of intraspecific variation could be rather complicated. Work in this area should be careful to consider ploidy, reproductive mode, and geographic variation simultaneously, in order to reach definitive conclusions.

Identification

Taxonomy of *Diplocardia* species is not possible from external characteristics. Even those which are readily distinguished while alive can be very hard to tell apart once they are preserved.

The key in Gates (1977) has been rearranged slightly here. Further information on the worms can be found in Gates (1977), including descriptions of most of the species. Terminological questions raised in further reading may be answered in Gates (1972) and Reynolds (1977a).

As discovery of new, undescribed worm species is very likely, care should be taken to record the locations and circumstances of collection, and to conduct detailed and careful dissections.

Collecting is best done by digging, since not all species can be relied upon to respond to vibrational, electrical, or chemical means of extraction. The author has had no success with the formalin method of extraction, and recommends that it not be used for *Diplocardia.* Many species are small as adults and care must be taken to examine closely for the presence of a clitellum. Otherwise, species will be missed.

Worms should be killed in ethanol or very hot water, and preserved in 4% formalin. After they have hardened, pin a specimen to a firm substrate in a flat dish suitable for use under a low-power dissecting microscope. I use petri dish bottoms with a layer of silicone rubber caulking compound (clear or white) spread as smoothly as possible. The rubber is firm and shows little sign of wear after hundreds of pinnings. Examine dorsal and ventral sides for features used in the key, and then pin it dorsal side up for dissection. A dorsal cut, or in the case of larger species, a more lateral cut, should be made from the third segment to the fortieth. With the body wall spread out and pinned down, the worm will be ready for identification. Conducting further probing and examination under water will be easier. Loss of segment count is a problem at first, and can be avoided by noting the locations of septa with respect to an external marker (i.e., posterior edge of clitellum) during the spreading of the body wall. Segment numbering can be recovered by searching for the ovaries under the esophagus in segment xiii. They resemble a cluster of tiny white grapes and are always present in that segment.

KEY TO SPECIES OF *DIPLOCARDIA*†

1a Spermathecae one pair **2**
1b Spermathecae more than one pair. **3**
2a Spermathecae, spermathecal pores in vii. . . . . ***D. biprostatica***
2b Spermathecae, spermathecal pores in ix . . . . ***D. bitheca***
3a Spermathecae in viii and ix **4**
3b Spermathecae in vii and viii, or vii–ix **32**
4a Intestinal origin in xiv or xv. **5**
4b Intestinal origin behind xvi **6**
5a Intestinal origin in xv, male grooves centered on xxi no typhlosole ***D. keyesi***
A typhlosole is a longitudinal infolding of the dorsal intestinal wall, having the effect of increasing the surface area of the intestine.
5b Intestinal origin in xiv, male field grooves centered on segment xviii . . . . ***D. koebeli***
6a Calciferous lamellae (Fig. 13.2) absent. **7**
6b Calciferous lamellae present in any of segments xiv–xvi **17**
7a Intestinal origin in xvii **8**
7b Intestinal origin behind xvii **14**
8a Last hearts in xii. **9**

† Modified from Gates (1977).

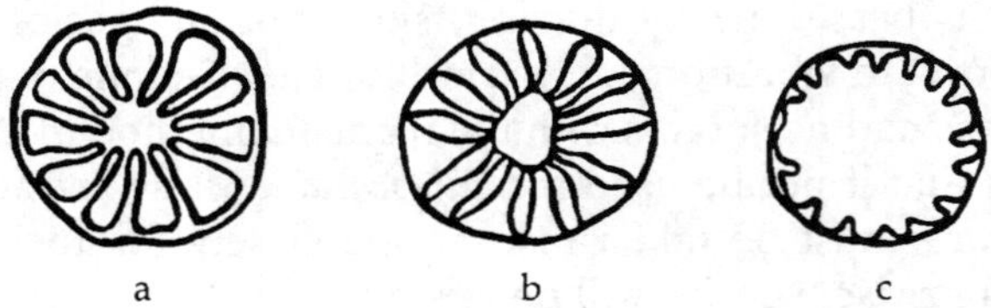

Fig. 13.2. Esophageal cross sections showing various stages of development of calciferous glands or lamellae: (*a*) calciferous lamellae free mesially; (*b*) calciferous lamellae united mesially; (*c*) no calciferous lamellae, a condition that grades from a smooth esophagus to one having deeper folds.

8b Last hearts in xiii 13
9a Spermathecal ducts (Fig. 13.3) cryptate 10
9b Spermathecal ducts not cryptate 11

A crypt is an outpocketing of the spermathecal duct lumen (being *cryptate*), which may be entirely within a nearly cylindrical duct, or *intramural.* If the crypt causes a bulge in the duct wall, it is not intramural. The crypt is not to be confused with the diverticulum.

10a Crypt mostly intramural . . . . ***D. conoyeri***
10b Crypt markedly protuberant from duct . . . . ***D. longiseta***
11a Clitellum annular (uninterrupted at midventral line) . . . . ***D. invecta***
11b Clitellum not annular 12
12a Prostomium epilobous, clitellum saddle-shaped . . . . ***D. smithii***
12b Prostomium tanylobous, clitellum interrupted at midventral line, ventral setae of viii and ix often missing . . . . ***D. rugosa***
13a Genital markings or genital tumescences all anterior to xv . . . . ***D. fuscula***
13b Some or all genital markings posterior to xv . . . . ***D. prosenteris***
14a Intestinal origin in xviii 15
14b Intestinal origin in xix . . . . ***D. texensis***
15a Last hearts in xii 16
15b Last hearts in xiii . . . . ***D. riparia***
16a Male field grooves centered on xix . . . . ***D. gracilis***
16b Male field grooves centered on xx . . . . ***D. verrucosa***

The male field is on the ventral side behind or under the posterior portion of the clitellum, and consists of two faint longitudinal parallel grooves extending over three segments.

17a Calciferous lamellae free mesially 18
17b Calciferous lamellae united mesially 27

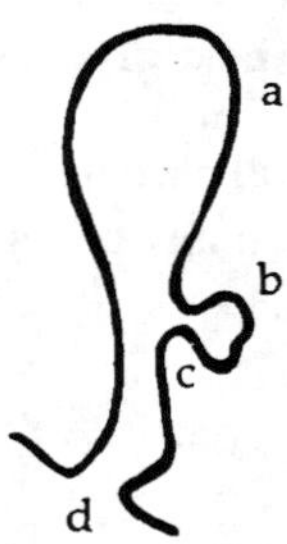

Fig. 13.3. Spermathecal parts: A, ampulla; B, diverticulum; C, duct; D, pore.

18a Calciferous lamellae in xiv–xv 19
18b Calciferous lamellae in xv–xvi . . . . *D. bivesiculata*
19a Intestinal origin in xvii 20
19b Intestinal origin behind xvii 26
20a Dorsal blood vessel single throughout 21
20b Dorsal blood vessel doubled in part 23
21a Spermathecal duct without a diverticulum. . . . . *D. udei*
21b Spermathecal duct with a diverticulum. 22
22a Seminal vesicles in ix and xii . . . . *D. farmvillensis*
22b Seminal vesicles in xi and xii . . . . *D. alabamana*
23a Dorsal vessel doubled only posteriorly . . . . *D. gatesi*
23b Dorsal vessel doubled anteriorly 24
24a Last hearts in xii 25
24b Last hearts in xiii. . . . . *D. eiseni*
25a Prostomium tanylobous . . . . *D. michaelseni*
25b Prostomium epilobous . . . . *D. macdowelli*
26a Last hearts in xii. . . . . *D. varivesicula*
26b Last hearts in xiii . . . . *D. alba*
27a Dorsal blood vessel single throughout 28
27b Dorsal blood vessel doubled in part 29
28a Last hearts in xii . . . . *D. minima*
28b Last hearts in xiii . . . . *D. vaili*
29a Last hearts in xii 30
29b Last hearts in xiii. 31
30a Prostomium tanylobous . . . . *D. pettibonae*
30b Prostomium epilobous . . . . *D. floridana*
31a Prostomium tanylobous. . . . . *D. komareki*
31b Prostomium epilobous . . . . *D. mississippiensis*
32a Setae lacking in segments i–ix or more . . . . *D. glabra*
32b Setae present in all segments except i 33
33a Spermathecae in vii and viii . . . . *D. fusca*
33b Spermathecae in vii–ix 34
34a Dorsal blood vessel single throughout 35
34b Dorsal vessel doubled at least in part. 42
35a Last hearts in xii 36
35b Last hearts in xiii. 41
36a Prostomium epilobous 37
36b Prostomium tanylobous 38
37a Setae widely paired, no copulatory setae in spermathecal segments, may be pigmented . . . . *D. singularis*
37b Setae closely paired, copulatory setae present, unpigmented . *D. caroliniana*
38a Spermathecal duct with a crypt . . . . *D. ornata*
38b Spermathecal duct without a crypt 39
39a Clitellum annular. 40
39b Clitellum interrupted (saddle-shaped) . . . . *D. sylvicola*

40a Spermathecal diverticulum attached without a stalk next to the ampulla, segments >200 . . . . ***D. meansi***
40b Spermathecal diverticulum attached by a stalk at the middle of the duct, segments about 100 . . . . ***D. hulberti***
41a Female pores in xiii and xiv, and ovaries in xii and xiii . . . . . ***D. sandersi***
41b Female pores in xiv and ovaries in xiii only . . . . ***D. longa***
42a Dorsal blood vessel doubled throughout . . . . ***D. communis***
42b Dorsal blood vessel doubled in part . . . . ***D. nova***

References

Bouché, M. B. 1977. Strategies Lombriciennes. In U. Lohm and T. Persson (eds.). *Ecol. Bull. (Stockholm)* 25:122–133. **Easton, E. G. 1979.** A revision of the "acaecate" earthworms of the *Pheretima* group (Megascolecidae: Oligochaeta): *Metapheretima, Planapheretima, Pleinogaster,* and *Polypheretima. Bull. Br. Mus. (Nat. Hist.) Zool.* 35:1–126. **Flint, R. F. 1957.** *Glacial and Pleistocene Geology.* Wiley, New York. 553 pp. **Gates, G. E. 1966.** Requiem for megadrile utopias. A contribution to our understanding of the earthworm fauna of North America. *Proc. Biol. Soc. Wash.* 79:239–254. **1967.** On the earthworm fauna of the Great American desert and adjacent areas. *Great Basin Nat.* 27(3):142–176. **1972.** Burmese earthworms. *Trans. Am. Philos. Soc.* 62(7):1–326. **1974.** On a new species of earthworm in a southern portion of the United States. *Bull. Tall Timbers Res. Stn.* 15:1–13. **1977.** More on the earthworm genus *Diplocardia. Megadrilogica* 3:1–48. **Jaenike, J. and R. K. Selander. 1979.** Evolution and ecology of parthenogenesis in earthworms. *Am. Zool.* 19:729–737. **James, S. W. 1983.** Factors affecting the distribution and abundance of three earthworm species in a tall grass prairie. Unpublished Ph.D. thesis, University of Michigan. Ann Arbor, Mich. **1988.** The postfire environment and earthworm populations in tallgrass prairie. *Ecology* 69:476–483. **James, S. W. and M. Cunningham. 1989.** Feeding ecology of some earthworms in tallgrass prairie. *Am. Midl. Nat.* (in press). **Lee, K. E. 1985.** *Earthworms: Their Ecology and Relationships with Soils and Land Use.* Academic Press, New York. 411 pp. **McMahan, M. L. 1976.** Preliminary notes on a new megadrile species, genus, and family from the southeastern United States. *Megadrilogica* 2:6–8. **Muldal, S. 1952.** The chromosomes of earthworms. I. The evolution of polyploidy. *Heredity* 6:55–76. **Murchie, W. R. 1958.** A new megascolecid earthworm from Michigan with notes on its biology. *Ohio J. Sci.* 58(5):270–272. **1967.** Chromosome numbers of some diplocardian earthworms (Megascolecidae- Oligochaeta). *Am. Midl. Nat.* 78:534–537. **Perel, T. S. 1977.** Differences in lumbricid organization connected with ecological properties. In U. Lohm and T. Persson (eds.). Soil organisms as components of ecosystems. *Ecol. Bull. (Stockholm)* 25:56–63. **Reynolds, J. W. 1977a.** *The Earthworms (Lumbricidae and Sparganophilidae) of Ontario.* Life Sci. Misc. Publ. Royal Ontario Museum, Toronto, Canada. 144 pp. **1977b.** The earthworms of Tennessee (Oligochaeta). II. Sparganophilidae, with the description of a new species. *Megadrilogica* 3:61–64. **1977c.** The earthworms of Tennessee (Oligochaeta). III. Komarekionidae, with notes on distribution and biology. *Megadrilogica* 3:65–69. **Reynolds, J. W. and D. G. Cook. 1976.** *Nomenclatura Oligochaetologica.* University of New Brunswick, Fredericton, Canada. 217 pp. **Stebbings, J. H. 1962.** Endemic–exotic earthworm competition in the American midwest. *Nature (Lond.)* 196:905–906.

14

Terrestrial Hirudinea

ROY T. SAWYER

General Characteristics

The Hirudinea or leeches are segmented worms closely allied to the Oligochaeta. They differ from the latter group in having suckers and in having external rings (annuli) which do not correspond to the internal segments. Most leeches are parasitic, but many forms are predaceous on earthworms, snails, or even other leeches. Although most leeches live in fresh water or in the sea (1,2,3), there are terrestrial species that live in many parts of the world. This habit is confined to the Arhynchobdellida, a large group with varied amphibious tendencies. Most arhynchobdellids, for example, have the habit of lying at the water's edge with the anterior part of the body exposed to the air, a behavior that probably facilitates oxygen uptake. Also, during the reproductive season most species in this group leave the water to deposit their cocoons in soil above the water's edge. A few species [e.g., *Haemopis marmorata* (1,2,3)] sometimes feed a few centimeters above the waterline on earthworms or snails; whereas others may inadvertently, but commonly, leave the water with the host while feeding. Finally, every year in some localities arhynchobdellids survive prolonged periods in which the pond or stream completely dries up (4). These annual dry periods are most predominant in the tropics, which has alternating dry and wet seasons, and in the floodplains of larger rivers.

Morphological Adaptations and Ecology

These amphibious tendencies that so characterize the Arhynchobdellida could be considered as ecological overtures leading to truly terrestrial adaptation. The successful transition from water to land has, in fact, evolved in two different directions in the Arhynchobdellida, in the surface dwelling bloodsucking forms on the one hand, and in the burrowing predaceous forms on the other. Both groups are most prevalent in areas of perpetually high humidity, especially in tropical mountains, but each has adapted to its particular life-style with its own morphological specializations.

The bloodsucking land leeches (Haemadipsidae) of southeast Asia have for centuries been the source of many horrible stories involving attacks on unsuspecting people. These highly adapted terrestrial species occur from India, China, and Japan to Australasia, including Madagascar and other islands of the Indian and Pacific oceans (5). The systematically interesting, relict form, *Mesobdella,* lives in the southern tip of South America (6). Two additional bloodsucking genera, *Xerobdella* (7) from southern Europe and *Diestecostoma* (8,9) from Mexico and Guatemala, are less well known primarily because they feed on salamanders rather than mammals.

The haemadipsids are terrestrial annelids *par excellence.* They live on the surface of the ground or in some cases even on bushes, and are notoriously capable of quickly homing in on a host. Unfortunately, their many morphological, physiological, and behavioral adaptations to terrestrial life have never been investigated in detail. The specializations of their segmental nephridia are especially remarkable in that each nephridium has an enlarged bladder which serves as a spacious reservoir for precious fluids. The first and last pairs open conveniently onto the oral and caudal suckers to keep them moist. The sensory and locomotory systems of hemadipsids are also highly specialized.

The predaceous species are less well known because they are, for the most part, burrowers and feed on earthworms, which they somewhat resemble. Morphological specializations associated with the burrowing habit include large body size, firmness of the body, prolonged head region, and reduction of the caudal sucker. The burrowing land leeches are especially common and diverse in South America, where they are represented by *Blanchardiella, Cylicobdella* (10,11), *Americobdella* (12), and *Lumbricobdella* (13). The Chilean species *Americobdella valdiviana* is one of the largest leeches in the world (12). The Trinidad species *Lumbricobdella schaefferi* is remarkable in completely lacking a caudal sucker (13). Predaceous land leeches are also well represented in southeast Asia [e.g., *Mimobdella, Orobdella,* and *Gastrostomobdella* (14–17)]. A related American species, *Gastrostomobdella quinqueannulata,* lives in Punaluu, Hawaii (9). An effort should be made to rediscover this land leech, which is known only from the original account.

The existence of two large land leeches in the southern and central United States usually comes as a surprise, even to soil biologists. In fact, the two predaceous species involved are among the largest terrestrial invertebrates in America. This being the case, it is especially surprising that the second species was not described until 1976. Both species are currently placed in the genus *Haemopis,* to which they are obviously allied, but eventually a revision may be in order. It should be emphasized that neither is a bloodsucking species.

KEY TO SPECIES

1a Gonopores on midventral line separated by 5 annuli . . . . **Haemopis oenops** (Grube, 1871), new combination

1b Gonopores on midventral line separated by 6½–7 annuli . . . . **Haemopis septagon** Sawyer and Shelley, 1976

Haemopis oenops (Grube, 1871), new combination

Not *Hirudo lateralis* Say, 1824.

Aulastoma oenops Grube, 1871: 100–101. Type locality: "Mississippi." Holotype: no. 1396, Zoologisches Museum für Naturkunde der Humboldt-Universität, Berlin.

Semiscolex terrestris Forbes, 1890a,b.

Haemopis lateralis: Moore, 1898.

Percymoorensis lateralis: Richardson, 1969.

Haemopis terrestris: Sawyer, 1972.

Systematic Note: To add further to the somewhat tortuous nomenclatural history of this species, I feel compelled to accept the priority of Grube's (18) specific name *oenops.* Upon examination of a collection of leeches from the Berlin Museum, I ran across a specimen of *Aulastoma oenops* from Mississippi which was indistinguishable externally from *Haemopis terrestris.* A reading of the original description confirmed this view: "obtained from Mr. Sallé of Paris from the state of Mississippi . . . of considerable size . . . the genital openings lie somewhat further than usual for they are situated on complete annuli 27 and 32" (quoting from translations of passages).

Description: Hirudinidae; uniform dark ground color with a conspicuous middorsal black stripe and yellowish lateral bands; body very firm; on the ventral side the male gonopore located 27 annuli from the oral cavity; annuli VIIa3 and VIIIa1 completely subdivided ventrally; female gonopore not raised; size large, 8–15 cm.

Biology: *Haemopis oenops* lives in soil, usually under rocks and logs, well away from the water. This life-style is reflected by the large, firm body and increased annulation of the head region. Little information is known on the biology of this species apart from the fact it feeds on earthworms (2,19). To this I can add from personal observation that *H. oenops* in water is capable of graceful swimming, and is typically hirudinid in depositing spongy cocoons in soil. In the laboratory they spend most of their time in formed tunnels, but occasionally enter the water.

This species is distributed along the Gulf states from northern Florida to Louisiana (20) and up the Mississippi–Ohio River valleys as far as the Great Lakes region (1). *H. oenops* belongs basically to the southern fauna and in this respect differs from all other *Haemopsis,* with the exception of the recently discovered *H. septagon* discussed below.

Haemopis septagon Sawyer and Shelley, 1976

Haemopis septagon Sawyer and Shelley, 1976; 85–88, figs. 6E,7. Type locality: Durham County, North Carolina. Holotype: no. 51539, U.S. National Museum.

Description: Hirudinidae; uniform dark olive-green with faint longitudinal dark stripe along dorsal midline, with numerous small irregularly scattered black flecks; body not especially firm; on the ventral side male gonopore located in furrow

between annuli 24/25 from the oral cavity; annuli VIIa3 and VIIIa1 not subdivided ventrally; female gonopore raised, nipplelike; size medium, about 7 cm.
Biology: Little is known about the habits of this species other than it feeds on earthworms (23). It lives under logs in the swamps and moist floodplains in coastal North and South Carolina and southeastern Virginia (20–23). In general, *H. septagon* is less adapted to terrestrial life than its congenitor *H. oenops.* Its body is more flaccid than that of the latter species and, accordingly, is not especially adapted to burrowing. The peculiar arrangement of the gonopores is unique and reinforces the impression that this species is relict.

References

1. **Sawyer, R. T. 1986.** *Leech Biology and Behaviour.* Vols. 1, 2, 3. Oxford Univ. Press, England.
2. **Sawyer, R. T. 1972.** *North American Freshwater Leeches, Exclusive of the Piscicolidae, with a Key to All Species.* University of Illinois Press, Urbana, Ill. 154 pp.
3. **Klemm, D. J. 1985.** Freshwater leeches (Annelida: Hirudinea). Pp. 70–194 in D. J. Klemm (ed.). *A Guide to the Freshwater Annelida (Polychaeta, Naidid and Tubificid Oligochaeta, and Hirudinea) of North America.* Kendall/Hunt Publishing Co., Dubuque, Iowa.
4. **Sawyer, R. T. 1970.** Observations on the life history and behavior of *Erpobdella punctata* (Annelida : Hirudinea). *Am. Midl. Nat.* 83(1):65–80.
5. **Richardson, L. R. 1975.** A contribution to the general zoology of the land leeches (Hirudinoidea : Haemadipsoidea superfam. nov.). *Acta Zool. Acad. Sci. Hung.* 21:119–152.
6. **Richardson, L. R. 1971.** The relationship of the terrestrial jawed sanguivorous g. *Mesobdella* to the neotropical hirudiniform leeches (Hirudinoidea). *Proc. Linn. Soc. N.S.W.* 95(3):215–220.
7. **Moosbrugger, G. and E. Reisinger. 1971.** Zur Kenntnis des europäischen Landblutegels *Xerobdella lecomtei* (Frauenfeld). *Z. Wiss. Zool.* 183(1/2):1–50.
8. **Moore, J. P. 1945.** Two new leeches (Hirudinea) in the collection of the United States National Museum. *J. Wash. Acad. Sci.* 35:261–265.
9. **Moore, J. P. 1946.** Leeches (Hirudinea) from the Hawaiian Islands, and two new species from the Pacific region in the Bishop Museum collection. *Occas. Pap. Bernice P. Bishop Mus.* 18(11):171–191.
10. **Ringuelet, R. A. 1968.** Llave o clave para el reconocimiento de las sanguijuelas conocidas de la República Argentina (Hirudinea) y apuntamientos sobre la Hirudofauna neotropical y transicional Mexicana. *Physis (B. Aires)* 27(75):367–390.
11. **Ringuelet, R. A. 1974.** Los hirudineos terrestres del género *Blanchardiella* Weber del páramo nor-andino de Colombia. *Physis (B. Aires)* 33(86):63–69.
12. **Moore, J. P. 1924.** The anatomy and systematic position of the Chilean terrestrial leech, *Cardea valdiviana* (Philippi). *Proc. Acad. Nat. Sci. Phila.* 76:29–48.
13. **Kennel, J. 1886.** Über einige Landblutegel des tropischen America (*Cylicobdella* Grube und *Lumbricobdella* n.g.). *Zool. Jahrb. (Syst.)* 2:37–64.
14. **Moore, J. P. 1929.** Leeches from Borneo with descriptions of new species. *Proc. Acad. Nat. Sci. Phila.* 81:267–295.
15. **Moore, J. P. 1935.** Leeches from Borneo and the Malay peninsula. *Bull. Raffles Mus.* 10:67–79.
16. **Gilyarov, M. S., E. I. Lukin and T. S. Perel. 1969.** The first terrestrial leech in fauna of the USSR, *Orobdella whitmani* Oka (Hirudinea, Herpobdellidae), a tertiary relict of forests of the southern maritime region. *Dokl. Acad. Nauk SSSR* 188(1):235–237.
17. **Richardson, L. R. 1971.** Gastrostomobdellidae f. nov. and a new genus for the gastroporous *Orobdella octonaria* Oka 1895 of Japan (Hirudinoidea : Arhynchobdellae). *Bull. Natl.*

Sci. Mus. (*Tokyo*) 14(4):585–602. **1975.** A new species of terricolous leeches in Japan (Gastrostomobdellidae, *Orobdella*). *Bull. Natl. Sci. Mus.* (*Tokyo*) *Ser. A.* (*Zool.*) 1(1):39–56.

18. **Grube, A. E. 1871.** Beschreibungen einiger Egelarten. *Arch. Naturgesch.* 37:87–121.

19. **Forbes, S. A. 1890.** An American terrestrial leech. *Bull. Ill. Lab.* 3:118–122. **1890.** An American terrestrial leech. *Am. Nat.* 24:646–649.

20. **Sawyer, R. T. and R. M. Shelley. 1976.** New records and species of leeches (Annelida: Hirudinea) from North and South Carolina. *J. Nat. Hist.* 10:65–97.

21. **Shelley, R. M. 1975.** Distributional patterns of leeches in North and South Carolina (Annelida: Hirudinea). *Bull. Assoc. Southeast. Biol.* 22(2):79.

22. **Shelley, R. M. 1977.** North Carolina's terrestrial leech. *Wildl. N.C.* 41(9):24.

23. **Shelley, R. M., A. L. Braswell and D. L. Stephan. 1979.** Notes on the natural history of the terrestrial leech, *Haemopis septagon* Sawyer and Shelley (Gnathobdella: Hirudinidae). *Brimleyana* 1:129–133.

15

Tardigrada

DIANE R. NELSON

ROBERT P. HIGGINS

General Characteristics

Tardigrades (Fig. 15.1) are micrometazoans found in a diversity of marine, freshwater, and terrestrial habitats. They range in size from 50 μm, the smallest juvenile, to 1200 μm, the largest adult. Mature individuals that have attained full growth average 250–500 μm; few species exceed 500 μm. Considerable size variation is exhibited by many species.

The first recorded observation of a tardigrade was in 1773, when J. A. E. Goeze (1), a naturalist and parish priest, described a freshwater tardigrade which he called a "Kleiner Wasser Bär" (little water bear). In 1774, Bonaventura Corti (2), an Italian abbot and professor of physics, referred to tardigrades found in gutter sand as "Brucolini" (little caterpillars), and he noted their ability to revive after desiccation. J. C. Eichorn (3) published his observation of a "Wasser-Bär" in freshwater algae in 1775. In 1776, an Italian abbot and professor of natural history, Lazzaro Spallanzani (4), named the animal "Il Tardigrado" (slow stepper) because its slow crawling movement resembled that of a small tortoise. Spallanzani also investigated the phenomenon of anhydrobiosis, which he referred to as resurrection after death. Later naturalists provided more detailed observations, and the first genera (*Macrobiotus, Echiniscus,* and *Milnesium*) were described in the nineteenth century.

Since their discovery, the Tardigrada have been placed in various systematic categories, with affinities to the Infusoria, Gastrotricha, Rotifera,

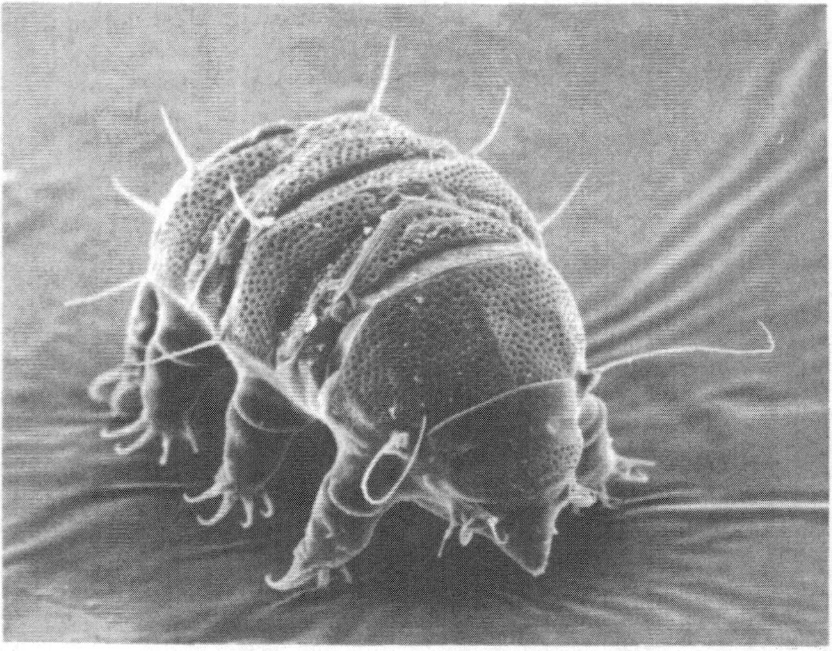

Fig. 15.1. Heterotardigrada, *Echiniscus spiniger,* from New Zealand. (SEM photograph by D. R. Nelson.)

Nematoda, Annelida, Onychophora, Linguatulida, and Arthropoda (Acarina, Crustacea, Insecta). Early workers recognized similarities of tardigrades to both the onychophoran–arthropod complex and the aschelminth complex. Marcus (5) considered the Tardigrada as a class of the phylum Arthropoda between the classes Onychophora and Eutracheata. Today, although their phylogenetic position is still debatable, they are regarded as a phylum with affinities more closely allied with either the aschelminth complex or the Arthropoda.

The ability of tardigrades to revive after extended periods of dehydration was noted by early naturalists. This latent state, anhydrobiosis, also occurs in some nematodes and rotifers and enables these organisms to withstand periodic drying of their habitat. The active state requires a thin film of water surrounding the body. As this water evaporates, the organism undergoes anhydrobiosis. Normal appearance and activity resume only when appropriate moisture conditions return.

In addition to prolonging life, anhydrobiosis may contribute to the distribution of tardigrades throughout the world. Dissemination of active tardigrades, eggs, cysts, and anhydrobiotic specimens by wind, water, and other animals may explain the broad geographic distributions of many cosmopolitan species. As yet, few well-defined zoogeographic distributions have been established.

Tardigrades occupy a variety of microhabitats. Marine and freshwater species are components of benthic communities inhabiting the interstices of sediments or living on the surface of algae, mosses, and other aquatic plants. These tardigrades feed primarily on algae, protozoans, organic detritus, and the cell contents of submerged plants. The terrestrial tardigrades, which also must be enveloped by a film of water to remain active, are found in soil, mosses, lichens, liverworts, and other vegetation. Many species feed on algae, organic detritus, bacteria, and other plant cells. Some large species of tardigrades are predaceous, feeding on protozoans, nematodes, and even other tardigrades.

In some tardigrade species, the sexes are separate but there is little external sexual dimorphism. In many species, males are rare or unknown and reproduction occurs by parthenogenesis. One instance of hermaphroditism in tardigrades has been reported (6). Eggs may be laid singly, in groups attached to a substrate, or in the molted cuticle. Growth is accompanied by a series of ecdyses (molts).

Two classes, formerly orders, have been established: Heterotardigrada (Figs. 15.1, 15.2), the marine and armored terrestrial tardigrades, and Eutardigrada (Fig. 15.3), the unarmored freshwater and terrestrial tardigrades. A dubious third class, established by Rahm (7) in 1937, was based on a single species, *Thermozodium esakii,* from a thermal spring in Japan. The species, as described, is difficult to assess; no type specimens exist and the habitat has been destroyed by an earthquake. Approximately 600 species of tardigrades have been described, and the potential for finding undescribed species remains high.

The taxonomy of tardigrades is based primarily on the morphological characters of the cuticle, claws, and buccal apparatus. In general, marine tardigrades are divided into many genera with few species in each genus. Samples collected in marine habitats generally exhibit high species diversity but low numbers of individuals. Some genera of both Heterotardigrada and Eutardigrada consist of many species. Samples of these tardigrades more commonly contain large numbers of individuals of one or more dominant species, and usually species diversity is lower than in marine habitats.

External Morphology

Tardigrades are bilaterally symmetrical animals with four pairs of legs terminated by claws of various numbers and shapes. The body, flat on the ventral side and convex on the dorsal side, is divided into five, somewhat indistinct, body segments: a cephalic segment, three trunk segments bearing the first three pairs of legs, and one terminal segment with the last pair of legs. In some cases the body appears divided into nine parts by cuticular folds.

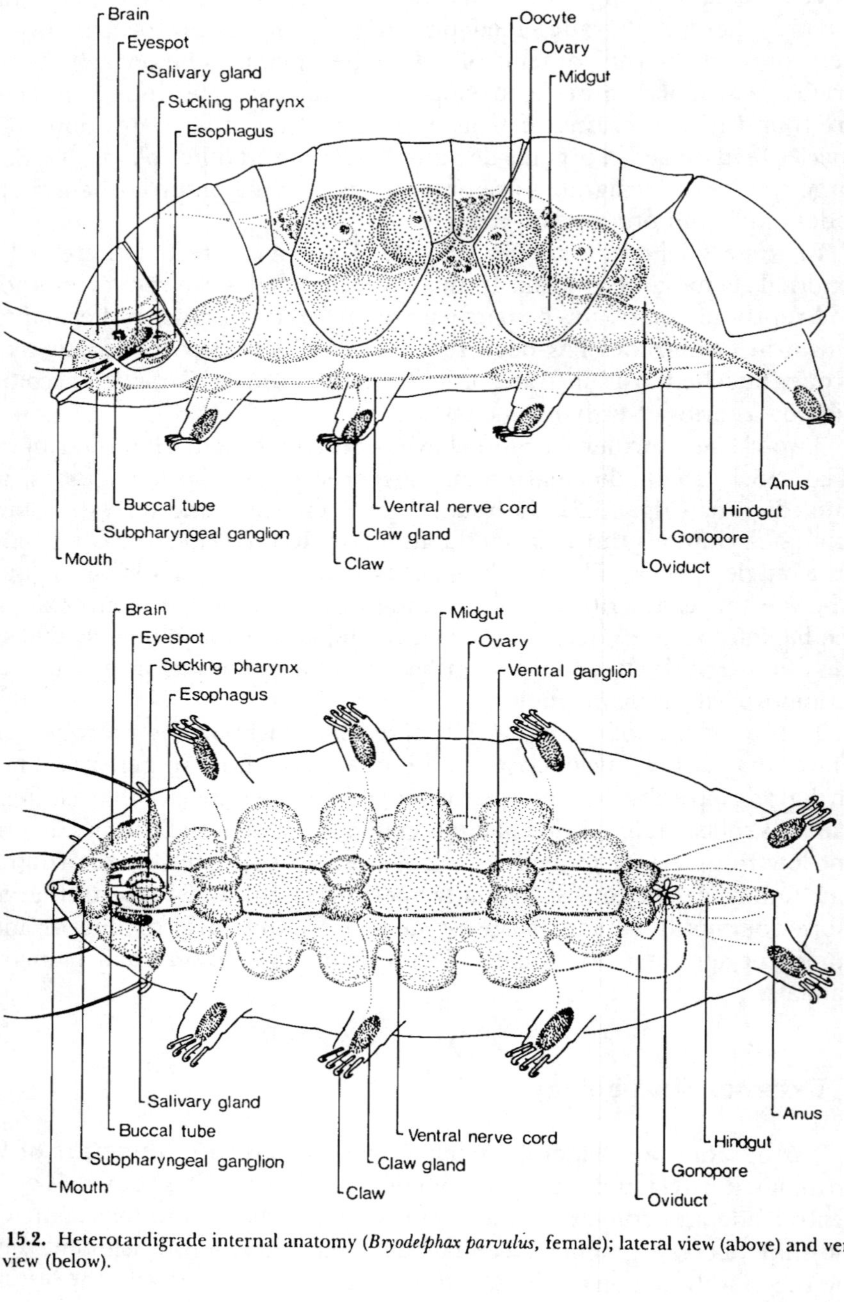

Fig. 15.2. Heterotardigrade internal anatomy (*Bryodelphax parvulus,* female); lateral view (above) and ventral view (below).

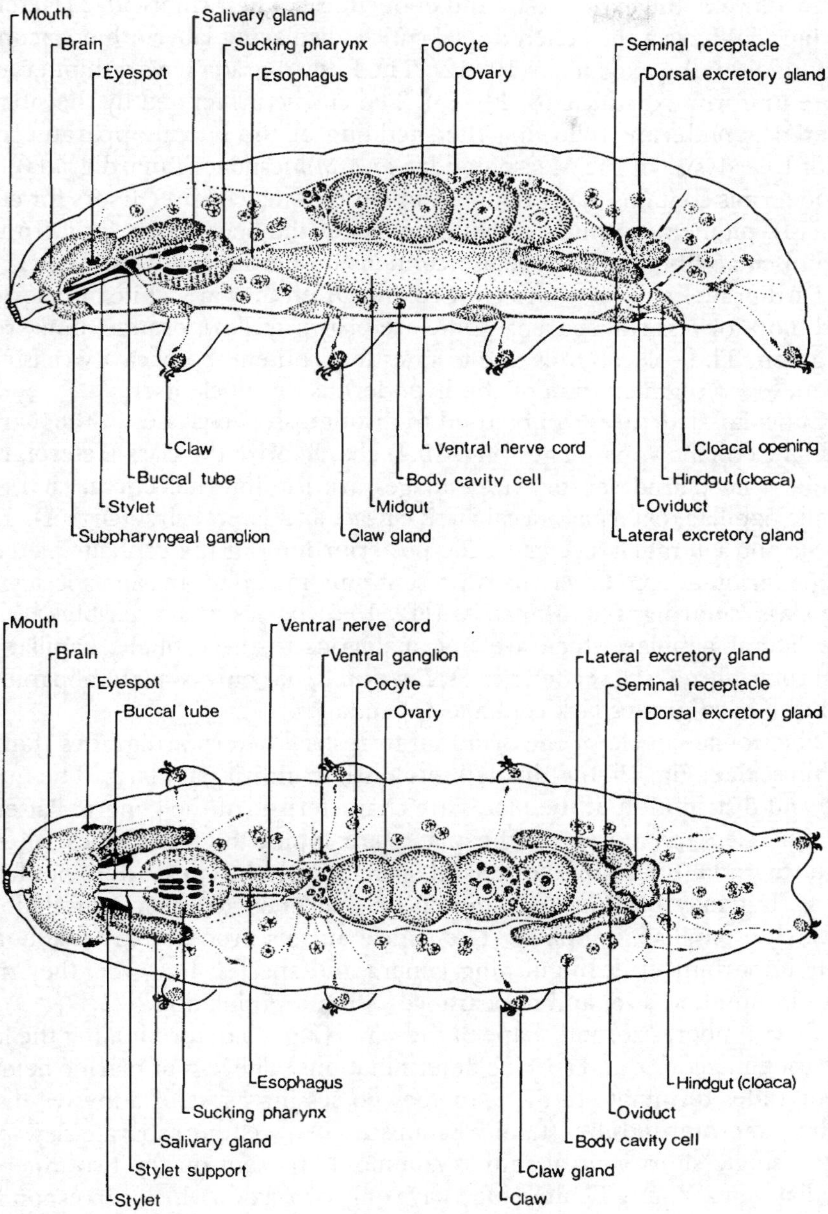

Fig. 15.3. Eutardigrade internal anatomy (*Macrobiotus hufelandii*, female); lateral view (above) and dorsal view (below).

The cuticle of tardigrades has been reported both to lack chitin (8) and contain chitin (9). It is a complex multilayered structure, usually with three distinct layers, the exo-, meso- and endocuticle. The Echiniscidae (Heterotardigrada) have a thickened dorsal cuticle displaying labyrinthine lacunae and canals in the exocuticle (10–12). The Eutardigrada have a thinner and more uniform exocuticle (8, 11–13). The cuticle is secreted by the single-layered hypodermis following the shedding of the buccal apparatus and prior to ecdysis. In the Macrobiotidae and Milnesiidae (Eutardigrada) the hypodermis contains many dense granules that may be precursors for cuticle (14), pigment granules or components of the body cavity fluid. In the Echiniscidae the hypodermis also contains numerous lipid droplets (11).

Tardigrades exhibit a variety of colors. Often they are white or colorless and more or less transparent; some are brown, yellow, orange, pink, red, or green. The coloration is due to intestinal content, body cavity cells and granules, or pigmentation of the hypodermis or cuticle itself.

Cuticular structures can be used to distinguish two classes of the Tardigrada. Cephalic appendages (Fig. 15.4) characterize the class Heterotardigrada. The paired sensory appendages are the internal buccal cirri, cephalic papillae, external buccal cirri, clavae, and lateral cirri (cirri A). The clavae and lateral cirri A mark the posterior limit of the cephalic plate. A single terminal median cirrus is present only in certain marine species. In the class Eutardigrada, *Milnesium* (Fig. 15.5) possesses six peribuccal and two lateral papillae which are not analogous to the cephalic papillae of heterotardigrades; a related genus, *Limmenius*, has only two lateral papillae. Other eutardigrades lack cephalic appendages.

The dorsal cuticle of the armored terrestrial heterotardigrades (family Echiniscidae, Fig. 15.4) is divided into distinct individual plates. The number and distribution of the plates are characteristic of the genera. Patterns of tubercles and pores on the plates are typical of the species. In the Eutardigrada (unarmored or naked tardigrades), the cuticle does not form plates but may be thin and smooth or thickened to form granulations, tubercles, papillae, or spines. The cuticle and its processes are taxonomically important in distinguishing genera and species; however, they may vary in number, size, and appearance within a population.

The number, size, and shape of the claws (Fig. 15.6) terminating the legs are used in generic and specific determinations. The legs of marine heterotardigrades terminate in digits or toes possessing claws of adhesive discs. Other heterotardigrades (family Echiniscidae) have four separate claws per leg; a single short ventral spur is commonly present on the two inner or medial claws. Young Echiniscidae have only two claws, which correspond to the medial claws in adults. The Eutardigrada usually possess two double claws per leg. The claws of the first three pairs of legs are anteriorly directed and generally arranged as external and internal claws. Compared with these claws, the claws of the fourth legs are rotated nearly 180° and are thus posteriorly directed. Therefore, the external claw of the third leg

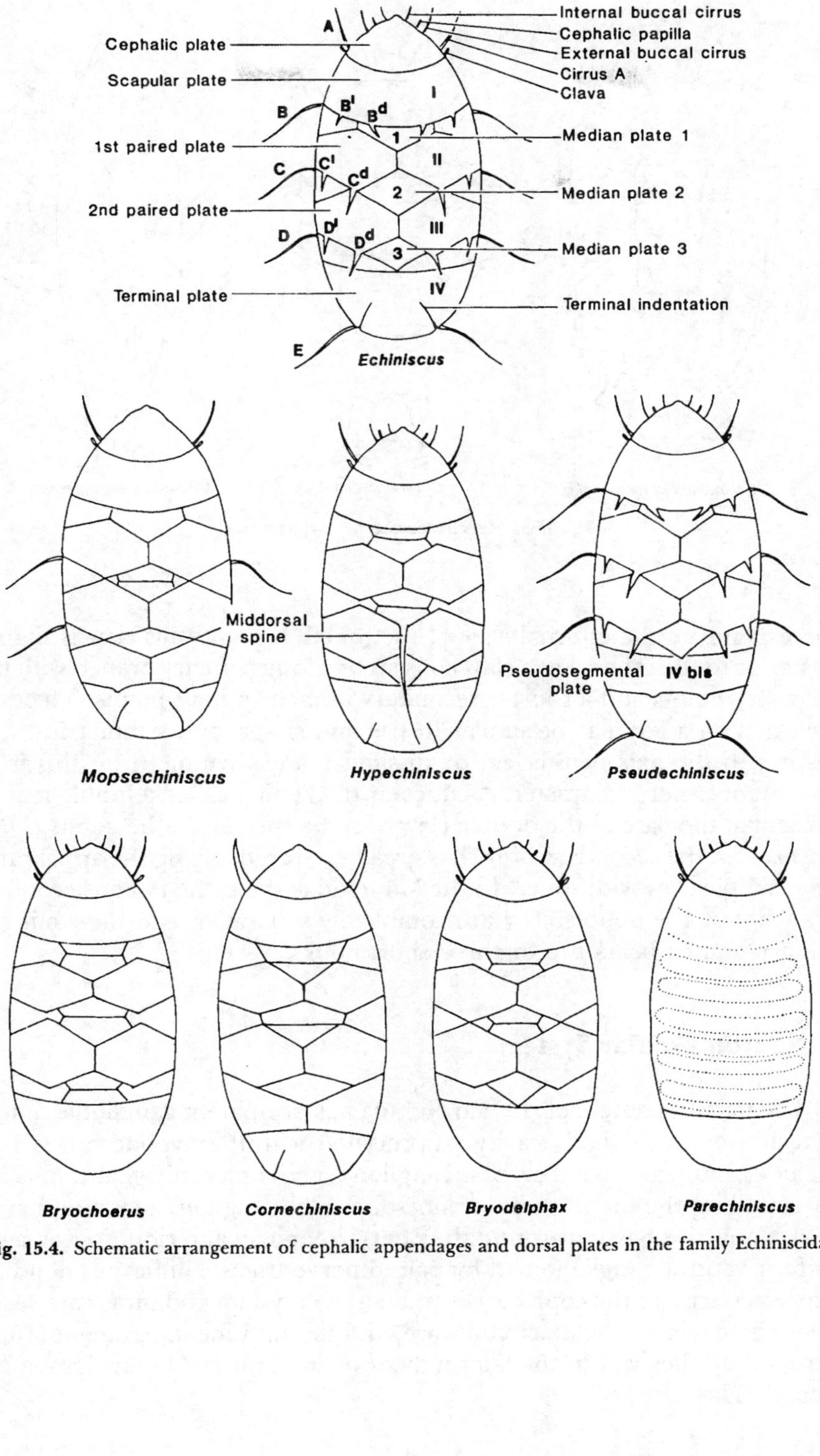

Fig. 15.4. Schematic arrangement of cephalic appendages and dorsal plates in the family Echiniscidae.

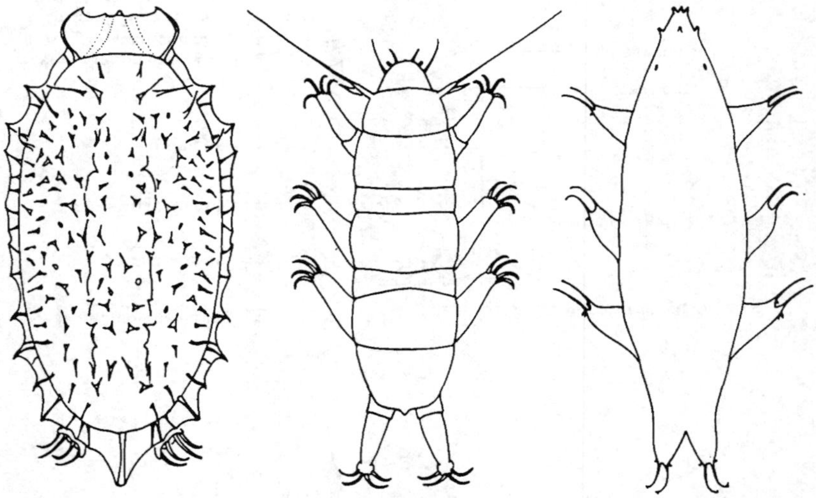

Fig. 15.5. Miscellaneous genera, dorsal view.

corresponds to the internal claw of the fourth leg, and this claw is slightly posterior to the other. Each claw consists of a long primary branch with two accessory points and a shorter secondary branch without points. The double claws on a leg may be similar in size and shape, and symmetrical with respect to the axis of the leg, or dissimilar and asymmetrical; this is an important generic character. A cuticular thickening called a lunule may be present at the base of the double claw of some species. In the genus *Haplomacrobiotus,* the secondary branch is greatly reduced, giving the appearance of a "simple claw with spur." In the Milnesiidae the primary and secondary branches of the double claw are completely separated, and the stout secondary branch bears two or three short hooks.

Neuromuscular System

The nervous system of the Tardigrada has prominent ganglionic centers that lie free in the body cavity unassociated with the hypodermis (15). It includes a dorsal brain (cerebral ganglion), a circumesophageal connective surrounding the mouth tube, a subesophageal ganglion, a circumpharyngeal connective passing around the pharynx, and an anterioposterior series of four ventral ganglia joined by paired nerve tracts. Numerous dendritic processes arise in the cephalic central nervous system and innervate receptive regions on the head, circum-oral, and mouth tube integument (16). A cerebral eye lies within the lateral lobes of the brain in Eutardigrada and Echiniscidae.

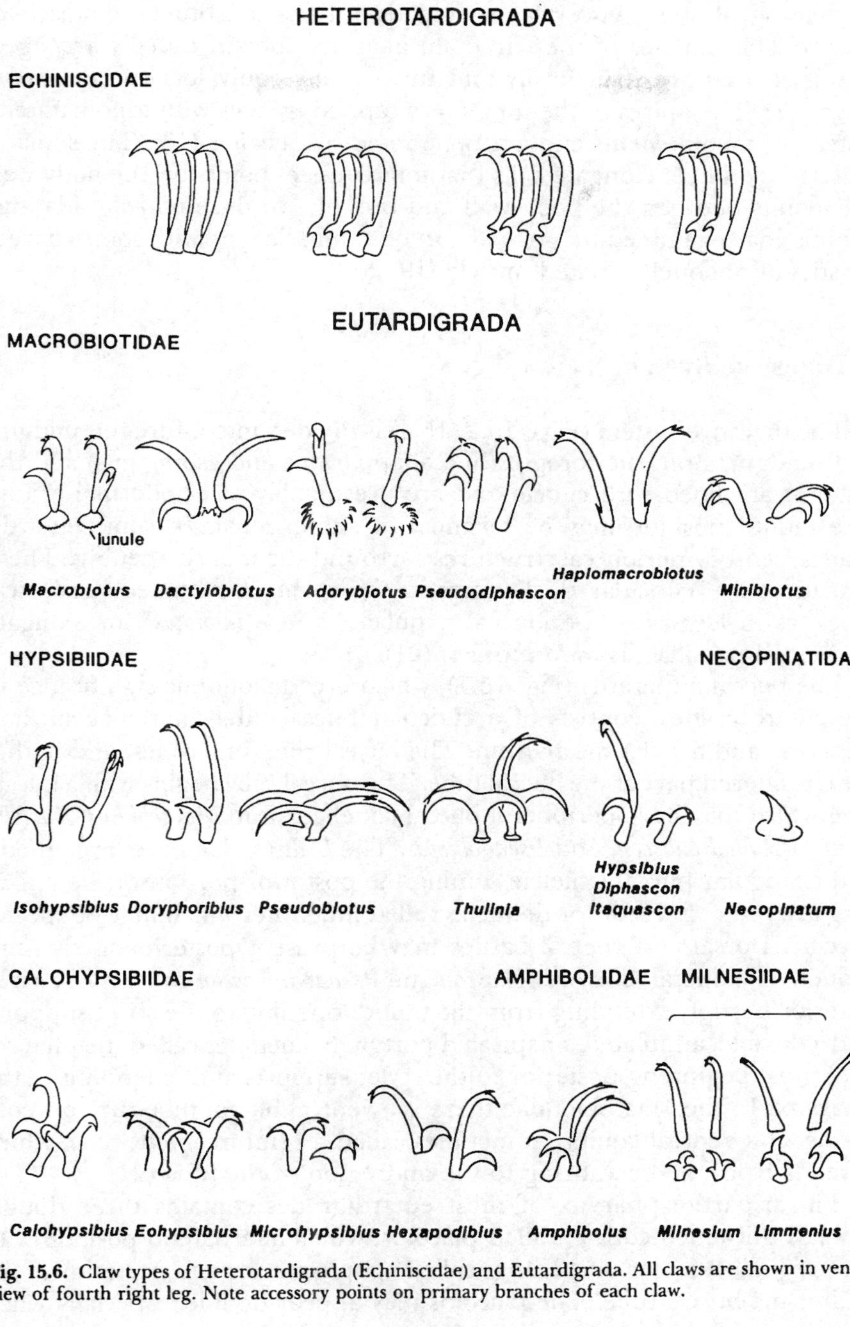

Fig. 15.6. Claw types of Heterotardigrada (Echiniscidae) and Eutardigrada. All claws are shown in ventral view of fourth right leg. Note accessory points on primary branches of each claw.

Pharyngeal, stylet, visceral, and somatic muscles constitute the muscular system. The muscles of the triradiate pharynx contain radially arranged myofibrils that are structurally and functionally equivalent to one sarcomere (17). The apices of the lumen are capped by cells with tonofilaments instead of myofilaments and are penetrated by neurites (18). The somatic muscles are single elongate cells that interconnect points on the body wall and points between the body wall and the gut. In the Eutardigrada and Echiniscidae examined they are of a smooth muscle type with some characteristics of obliquely striated muscle (19, 20).

Digestive System

The digestive system (Figs. 15.2, 15.3) is divided into a foregut, midgut, and hindgut. Both the foregut (buccal apparatus and esophagus) and the hindgut are lined with cuticle and are presumably of ectodermal origin. The mouth position may be terminal or subterminal. In some eutardigrades, sensory peribuccal structures surround the mouth opening. These structures are cuticular thickenings in the form of flattened peribuccal lobes (as in *Hypsibius*), peribuccal papulae (as in *Minibiotus*), or elongate peribuccal papillae (as in *Milnesium*) (21).

The buccal apparatus (Fig. 15.7), which is of taxonomic significance in the eutardigrades, consists of a cuticular buccal tube, muscular sucking pharynx, and a stylet mechanism. The buccal ring, mucrones, and baffles are considered part of the buccal tube. The buccal tube begins with a buccal ring which may have peripheral buccal lamellae, as in *Macrobiotus, Dactylobiotus, Pseudodiphascon,* and *Pseudobiotus*. The Milnesiidae have large modified triangular buccal lamellae. Lining the posterior portion of the buccal ring are rows of "teeth" or denticles called mucrones which may be species specific. Dorsal and ventral baffles may be present posterior to the mucrones. The buccal tube of *Diphascon* and *Pseudodiphascon* consists of a rigid anterior portion extending from the mouth opening to the stylet supports and a flexible, annulated, or spiraled portion (sometimes called the pharyngeal tube) beginning posterior to the stylet supports and extending to the pharynx. In the Macrobiotidae there is a ventral buccal tube support consisting of a ventral lamina (sometimes called a reinforcement rod), which extends from the buccal ring to the midregion of the tube (21).

The tripartite pharynx of most eutardigrades contains three double rows of thickened cuticle, called placoids, which alternate in position with three equal cuticular thickenings called apophyses, located posterior to the end of the buccal tube. The placoids may appear double, especially when the pharynx is flattened by coverslip pressure. The large anterior placoids (macroplacoids) occur in two or three transverse rows, numbered posteriorly from the buccal tube. Posterior to the last macroplacoids, a row of small placoids (microplacoids) may be present. A median septulum may

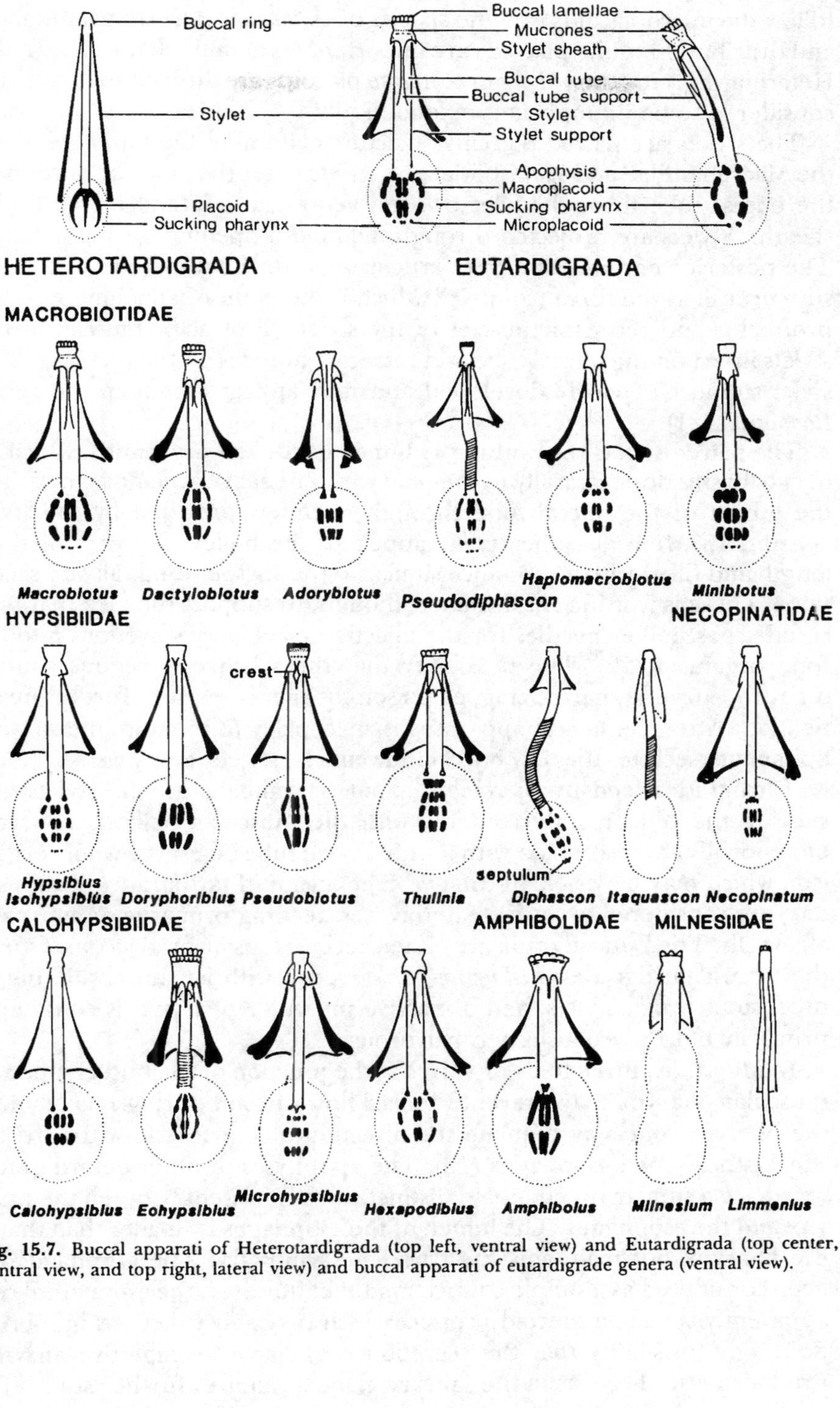

Fig. 15.7. Buccal apparati of Heterotardigrada (top left, ventral view) and Eutardigrada (top center, ventral view, and top right, lateral view) and buccal apparati of eutardigrade genera (ventral view).

follow the microplacoids in some species of *Diphascon*. The shape, number, and distribution of the placoids are important taxonomic characters. In the Heterotardigrada, the very thin elongate placoids are difficult to detect and considered to be unimportant taxonomically.

The stylets are paired piercing structures lateral to the buccal tube. In the Macrobiotidae and Hypsibiidae the stylets enter the anterior portion of the buccal tube through stylet orifices, ventral to a lateral flange (stylet sheath). Stylets are projected through the mouth opening during feeding. The posterior end of each stylet articulates with the distal end of a stylet support called the furca (condyle), which is the point of attachment of the protractor and retractor muscles of the stylets. Protractor muscles of the stylets insert on the buccal tube and retractor muscles on the pharynx. The stylet support is poorly developed and may appear absent in the genus *Itaquascon* (21).

The paired stylets rest within the lumen of the salivary glands, which lie in two masses dorsolaterally to the pharynx. The salivary glands empty into the lumen of the buccal tube through openings provided by the stylet sheaths. In *Milnesium,* the stylets appear to be hollow for part of their length and filled with a substance similar to the surrounding salivary secretions. There is insufficient evidence, though, to support the idea that they could act as hollow needles for the injection of salivary secretions into the food organism (22). The possibility that the salivary secretions contain digestive enzymes, narcotizing, or poisoning agents requires further investigation. After the buccal apparatus is shed prior to molting, the salivary epithelium secretes the new buccal tube cuticle, stylets, and stylet supports.

Tardigrades feed by piercing a plant or animal with the stylets and sucking the cell or body contents with the muscular bulbous pharynx. Occasionally, Eutardigrada with a wide buccal tube ingest the whole organism, which may be visible in the gut. The armored tardigrades (Echiniscidae) are considered exclusively herbivores, feeding primarily on moss and algal cells. The Eutardigrada are either herbivorous or predaceous. One of the largest tardigrades, *Milnesium tardigradum,* with its short wide buccal tube, stout broad stylets, and a massive muscular pharynx, is considered primarily but not exclusively carnivorous.

In *Milnesium,* three folds or valves at the junction of the buccal tube and muscular pharynx may prevent the backflow of food during relaxation of the pharynx and may regulate the direction of nutrient flow toward the short cuticle-lined esophagus (18). The apophyses of other eutardigrades may have a similar function. No distinct valve is present between the pharynx and the esophagus. The lumen of the esophagus is smaller than that of the pharynx but maintains the triradiate symmetry. The esophagus has been considered as a simple connecting tube; however, the esophageal cells apparently secrete a mucoid product which probably functions in lubrication. The possibility that the secretion could activate digestive enzymes produced elsewhere, as in the salivary glands, requires further study. The cuticular lining of the esophagus extends a short distance into the intestinal lumen.

Food is digested in the midgut (intestine), which is endodermal in origin and not lined with cuticle. The epithelial cells are organized into a convoluted single layer that has an apical microvillous border. There is no connective tissue or muscle layer surrounding the midgut, although in the Eutardigrada, eight small muscles that extend from the esophagus to the junction of the excretory glands circumscribe the midgut (51). The midgut of the Heterotardigrada exhibits five or six lateral diverticula (Fig. 15.2), which are not evident in the Eutardigrada. The intestinal content of herbivorous tardigrades is usually yellow or green just after feeding and becomes browner and turns dark prior to defecation. Transparent intestinal contents are characteristic of either predatory or starved tardigrades. Food reserves are stored in lipid droplets or lipoprotein granules in body cavity cells. These cells float freely in the pseudocoelom (thought to be a persistent blastocoel) and may have functions in addition to food storage. Cell numbers vary dramatically and are positively correlated with nutritional status. Autophagic and heterotrophic vacuoles are frequently present (23).

The midgut empties into the hindgut, which is cuticle-lined and presumably ectodermal in origin. In the Eutardigrada the hindgut is a true cloaca since it junctions with the genital ducts. The hindgut of *Milnesium tardigradum* consists of two morphologically distinct sections, the anterior hindgut and the cloaca (24). The anterior hindgut receives the contents of the midgut and excretory glands (Malpighian tubules). It is specialized for osmoregulation and may be important in regulating the transition from active to cryptobiotic (anhydrobiotic) states. The cloaca junctions with the anterior hindgut and the reproductive tract. The mechanism for the passage and expulsion of eggs, feces, and excretory products of the excretory glands is not clear, but two prominent muscles insert on the dorsal body wall of the cloaca and may facilitate expulsion. The cloacal opening is a transverse slit that opens ventrally between the fourth pair of legs. In the Heterotardigrada, the oviduct leads to a rosette-shaped preanal gonopore, and the separate anus is a longitudinal slit between the fourth pair of legs. Young Echiniscidae (those with only two claws on each leg) have a mouth but no gonopore or anus.

Excretory System

Ramazzotti (25) listed four possible methods of excretion in tardigrades: (a) through the salivary glands when the buccal apparatus is expelled prior to ecdysis; (b) through the shedding of the cuticle, which accumulates excretory granules; (c) through the wall of the midgut into the body fluid; and (d) through organs called excretory glands or Malpighian tubules, if present, into the hindgut.

The Heterotardigrada lack excretory glands (Fig. 15.2), but the midgut has five or six lateral diverticula, which presumably provide considerable surface area for excretion. The midgut of the Eutardigrada (Fig. 15.3) is not lobed and may play only a minor role in excretion, which occurs primarily through excretory glands (Malpighian tubules). Marcus (15) described two lateral and one dorsal excretory glands (Fig. 15.3) which empty into the junction of the midgut and hindgut. The lateral glands are either the "long type" or the "short type." The dorsal gland is always short. The shape of the glands (lobed or smooth) varies with the species and the nutritional state of the individual.

In addition to the three glands described by Marcus, a fourth ventral excretory gland has been found in *Milnesium tardigradum* (24). In this species the glands empty into the gut via the pylorus, a valvelike structure that is "open" in the active state and "closed" in the anoxybiotic state.

The basal surface of the gland cells is increased by interdigitations of modified foot processes resting on the basement membrane. The lumen of the tubules is small until it enlarges into a Malpighian ductule adjacent to the pylorus. Large round granules within the tubule cytoplasm and spherites having concentric laminations within the ductule lumen suggest both secretory and excretory activity (26).

Reproduction and Developmental Life History

Sexual reproduction, parthenogenesis, and hermaphroditism are exhibited in the Tardigrada. In many tardigrade species, the sexes are separate but there is very little external sexual dimorphism. Generally, males are smaller than females, and in some species the claws of the first pair of legs are modified in the male. Females comprise the majority of populations throughout the year, with males reported to reach peak abundance during the winter or early spring. In bisexual populations of some species of *Macrobiotus,* the sex ratio is 1 : 1 and the diploid chromosome number ranges from 10 to 16, depending on the species and the population (27).

A single gonad (Figs. 15.2, 15.3), dorsal to the intestine and attached to the body wall by one or two ligaments, is present in both sexes. Morphological variations in the gonad are associated with species, sex, age, and reproductive activity. In the Eutardigrada, the male has two sperm ducts which open into the posterior hindgut (cloaca). The female has a single oviduct on either the right or the left side. In some females of *Macrobiotus* and *Hypsibius,* a small seminal receptacle (Fig. 15.3), which also opens into the cloaca, is present during the autumn and winter. In the Heterotardigrada, the genital ducts terminate in a ventral preanal gonopore, often with rosette-shaped cuticular folds (Fig. 15.2).

Fertilization has been observed for very few species of tardigrades. In species that deposit eggs in the cuticle, males release sperm into the cloacal opening of the female just before the female molts. Fertilization occurs as the eggs are deposited in the old cuticle as it separates from the new one. In other species, fertilization is independent of molting. Sperm are deposited in the cloaca and may be stored in the seminal receptacle or may fertilize the eggs in the ovary. Eggs, however fertilized, may be laid singly, in groups attached to the substrate, or in the molted cuticle. Eggs deposited freely and singly or in small groups are usually ornamented with pores, reticulations, or processes and are of taxonomic importance especially in *Macrobiotus* (Fig. 15.8). Smooth eggs, varying in number from 1 to 60, have a thin shell and are generally deposited in the molted cuticle but may also be laid freely. The method of egg deposition is unknown for many species and some genera.

Parthenogenesis also occurs in the Tardigrada, especially in the Echiniscidae, in which males are rare or unknown. The eutardigrades undergo meiotic parthenogenesis in diploid biotypes and ameiotic parthenogenesis in triploid or tetraploid biotypes (27). Some species of *Macrobiotus* and *Hypsibius* undergo sexual reproduction when males are present and parthenogenesis when males are absent. Males are sometimes frequent in the spring but always absent in the other seasons. Parthenogenesis has been confirmed for *Milnesium tardigradum* and *Hypsibius dujardini* in artificial cultures maintained in the laboratory; however, the reproductive mode of the Echiniscidae remains an unsolved problem due to difficulties in rearing the organisms.

Until recently, all tardigrades except some of the Echiniscidae were considered gonochoristic (dioecious) with both bisexual and unisexual populations. Bertolani (6) reported hermaphroditic populations of *Isohypsibius* in which the gonad contained both male and female gametes in various stages of development. Further investigation of hermaphroditism in natural populations of tardigrades is needed.

The time span between egg deposition and hatching varies considerably with the species and environmental conditions. Development of the embryo requires from 5 to 40 days; eggs with thin shells in the old cuticle may develop more rapidly than eggs with thicker shells with projections. Eggs generally can be found at any time of the year, but are more numerous in the winter in soil-inhabiting species.

When development is complete, the tardigrade emerges from the egg, which is ruptured by an increase in hydrostatic pressure or by the action of the stylets. Immature tardigrades are smaller and may differ slightly from the adult in reproductive structures, buccal apparatus, and number or size of claws and spines. Young Echiniscidae have only two claws per leg instead of four, cuticular structures are incompletely developed, and they have a mouth but no gonopore or anus. An increase in body size through growth of the individual cells rather than through the production of new cells is the primary change that occurs in the successive molts from juvenile to adult.

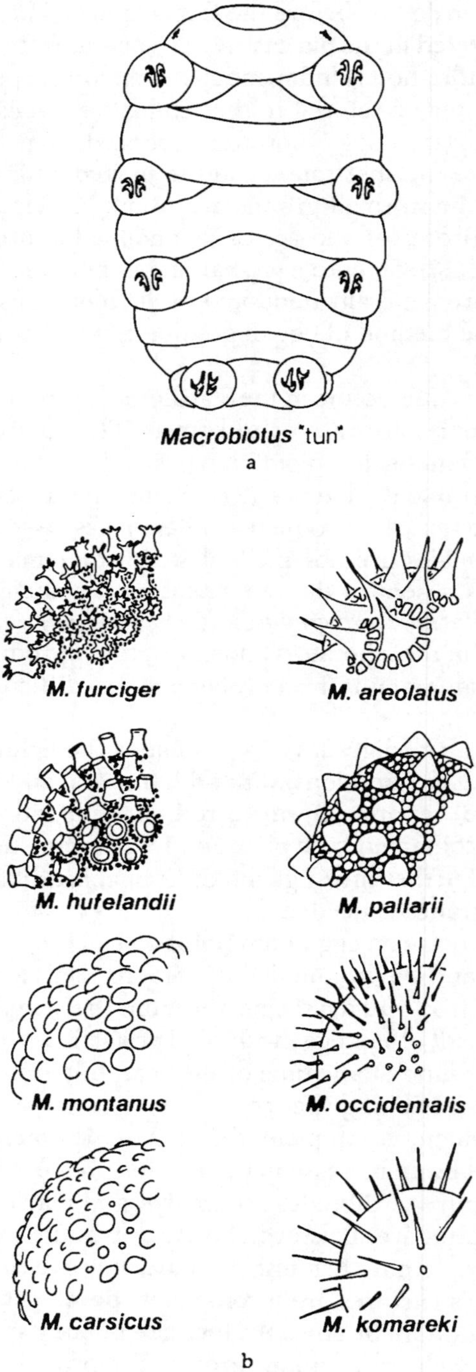

Fig. 15.8. (a) *Macrobiotus* "tun" or anhydrobiotic state and (b) various ornamentations of eggs.

Prior to molting, the cuticular lining of the foregut is ejected and a new buccal apparatus begins to be formed by the salivary glands. The simplex stage is characterized by the absence of the buccal apparatus and therefore animals in this stage are unable to feed. The reformation of the buccal apparatus is followed by molting of the body cuticle, claws, and lining of the hindgut. New claws are formed by claw glands (Figs. 15.2, 15.3) in the legs prior to the shedding of the exoskeleton. Normally, growth occurs with each molt until the maximum size is attained; however, a decrease in size after molting has been noted when food was lacking.

The number of molts in the life history has been determined for only a few species of tardigrades (25). Frequency distributions of body length and length of the buccal apparatus indicate 4–12 molts, but considerable variation and overlapping of the stages occur. After the second or third molt, the individual is sexually mature and egg production continues throughout life. The active life span of the tardigrade is generally from a few months to 1 or 2 years, but the total life span may be extended greatly by anhydrobiosis and/or encystment.

Ecology

Regardless of the habitat (marine, freshwater, or terrestrial), all tardigrades are essentially aquatic. The distinction between the aquatic and the terrestrial fauna is a tenuous one since activity (metabolism) depends on the presence of a film of water surrounding the body. As the water in terrestrial or temporary aquatic habitats evaporates, tardigrades undergo anhydrobiosis, a phenomenon that enables them to survive periods of desiccation. The ability of obligate marine and freshwater species to become anhydrobiotic is unknown. Generally, the environments in which tardigrades live are divided into (a) marine and brackish water, (b) freshwater, and (c) terrestrial habitats, although there may be considerable overlap between these environments. Marine and brackish water species will not be discussed further in this section.

Certain tardigrade species are hydrophilous and found only in permanent freshwater habitats. Although these are occasionally found in plankton samples, they are benthic invertebrates either living in the interstitial spaces of sediments or on aquatic plants in ponds, lakes, rivers, and streams. Although usually littoral, some specimens have been collected at depths of 150 m on lake bottoms (25). The aquatic habitat may also yield hygrophilous tardigrades, which typically inhabit mosses which remain moist, and eurytopic species, which live in a range of very dry to very wet habitats.

Habitats of terrestrial tardigrades must provide (a) sufficient aeration since tardigrades are hypersensitive to the lack of oxygen, (b) alternate periods of wet and dry conditions, and (c) sufficient food. Soil-inhabiting tardigrades are found in the upper porous strata where the oxygen concentration is greatest and the interstices allow movement. Tardigrades are not likely to inhabit impermeable clay substrates, permanently wet ground or moss, dense cushion mosses, or mosses with thick cellulose cell walls.

Many mosses, lichens, liverworts, rosette angiosperms, and other plants at times contain large tardigrade populations. These habitats may also occasionally yield aquatic and semiaquatic species. Population densities of tardigrades in mosses vary widely. Mihelcic (28) reported for Austrian mosses a minimum of 100–500 to a maximum of 500–2000 specimens per 10 cm^2 of the surface. Ramazzotti (25) considered Mihelcic's minimum number to be a more common population density; however, much lower densities may frequently be encountered, including sometimes, none at all. Marcus (15) reported a maximum of 22,000 tardigrades in 1 g (dry weight) of moss. Lichens and other plants are usually less densely populated than mosses, although lichens from northern regions may have very high numbers of tardigrades. Species diversity within a habitat generally ranges from two to six, but 10 or more species may be found in a single moss or lichen sample.

In 1916, Heinis (29) was the first to report tardigrades and their eggs in soil, but Marcus (15) believed soil was not their normal habitat and explained their presence by passive transportation. Soil and leaf litter as a normal habitat for tardigrades was later confirmed by various workers (30–41).

The more common species of Eutardigrada reported from soil and leaf litter include *Macrobiotus harmsworthi, M. hufelandii, M. richtersi, Hypsibius convergens, H. dujardini, H. oberhaeuseri, H. pallidus, Isohypsibius prosostomus, I. franzi, I. tuberculatus, Diphascon bullatum, D. scoticum,* and *Milnesium tardigradum.* Within the Heterotardigrada, only a single family, Echiniscidae (Fig. 15.4), is a common component of the soil fauna; *Echiniscus granulatus, E. trisetosus,* and *Pseudechiniscus suillus* are frequently found. Only one other genus of heterotardigrades is nonmarine: *Oreella* (family Oreellidae) has been found in mosses. *Echinursellus* (family Halechiniscidae) was reported in moist soil on the beach of a freshwater lake (Fig. 15.5); however, Kristensen (personal communication) stated that "the only known specimen is a cyst of a cyclomorphic stage of a eutardigrade closely related to *Pseudobiotus.*" Typical moss-inhabiting tardigrades are also frequently found in the soil. Two species have been collected only from the soil: *Macrobiotus terricola* and *Isohypsibius tuberculoides.* Few tardigrade specialists have investigated the soil and leaf litter habitat; therefore, many more species are likely to be found in this habitat than have been reported.

Tardigrades exhibit a patchy distribution in the soil. Densely populated zones are irregularly spaced in the upper strata of the soil, especially in the first few centimeters, but gradually decrease and disappear at a depth of 5–10 cm. Aeration (ventilation), oxygen concentration, light, temperature, chemical composition, pH, humidity, desiccation cycle, particle size, food, egg production, and other factors affect both distribution and density of tardigrade populations. The degree of compaction of the soil, which varies considerably, is probably one of the primary factors affecting distribution. Compaction decreases both the oxygen content and the interstitial space. Soil tardigrades feed on algae, fungi, bacteria, protozoa, rotifers, nematodes, organic debris, and other tardigrades. In turn, they are preyed on by some ameboid protozoans, nematodes, fungi, other tardigrades, and perhaps other organisms such as mites, spiders, and insect larvae.

Population densities of soil tardigrades show a high degree of variation and patchy distributions. Franz (31) reported 300 per 10 cm^2, Mihelcic (28), 50–200 per 10 cm^2, and Ludi (32) 20–5000 per kilogram of soil. Ramazzotti (25) found an average corresponding to 400 tardigrades, 15 cysts, and 1200 eggs per 10 cm^2 in field soil in Pallanza, Italy; cysts were often very numerous.

Leaf litter in deciduous and coniferous forests also provides a suitable habitat for tardigrades. The distinction between litter and soil fauna is often obscure since the zones are continuous. Populations are more dense in the upper layers; tardigrades rarely penetrate strata deeper than 3–5 cm. Apparently, the presence of tardigrades is dependent on a well-established, permanent, or long-lasting litter in addition to other factors, such as proper humidity, oxygen, and food.

Minimum and optimum conditions for soil and leaf litter tardigrades are unknown. Two apparently identical habitats may exhibit considerable variation in population density and species diversity. Further investigation is essential to determine the precise ecological requirements for a given species. Factors other than those listed above that may affect population density include competition, predation, and parasitism (especially by protozoa and fungi).

Ecological classifications of tardigrade species have been proposed, but sufficient data on environmental conditions are not available. One classification proposed by Mihelcic (39) divides tardigrade species into four groups based on moisture requirements: (a) xerophilous species, in dry habitats; (b) eurytopic species, in a wide range of moisture conditions; (c) hygrophilous species, in moist habitats; and (d) hydrophilous species, in wet habitats, including standing and running water. Mihelcic (39) also proposed a more complex classification which is not practical in light of our limited knowledge of the ecological requirements of tardigrade species. He classified species on the basis of plasticity (stenoplastic, euryplastic, eu-euryplastic), humidity (stenokous, eurokous, eu-eurokous), sunlight (mesostenoheliophilous, polystenoheliophilous, mesoeuryheliophilous), oxygen requirements (stenaerobic, euryaerobic), feeding, geophilia, and other factors. Each of these main categories has further subdivisions.

Although they are cosmopolitan in distribution as a phylum, little is known about the biogeography of the Tardigrada. They are found from the Arctic to the Antarctic and from the slopes of the Himalayas to the depths of the oceans. For some species the ecological requirements appear to be quite broad, resulting in a worldwide distribution. Other species, perhaps with more narrow tolerances, appear to be highly restricted in their distribution. Dissemination of active tardigrades, eggs, cysts, and anhydrobiotic specimens by wind, water, and other animals may result in some broad geographic distributions. However, more data are essential before generalizations can be made about the biogeography of the group.

Crowe (42) discussed five types of latency found among the tardigrades: (a) encystment, (b) anoxybiosis, (c) cryobiosis, (d) osmobiosis, and (5) anhydrobiosis. These phenomena play an integral role in the ecological interactions or characteristics of the organisms. The latent states are characterized by reduction (or cessation) in metabolism, temporary cessation of growth and reproduction, reduction in senescence, and enhanced resistance to environmental extremes such as cold, heat, drought, chemicals, and ionizing radiation. Extreme latency has been called "anabiosis" or "abiosis" (without life) and most recently "cryptobiosis" (hidden life).

Encystment, most common in aquatic tardigrades that live in permanent freshwater pools, is induced by unknown ecological factors. A gradual worsening of environmental conditions is apparently one inducing factor. Before encystment, a tardigrade takes in large amounts of food which is stored in body cavity cells for use during the encystment period. The buccal apparatus is ejected, the animal enters the simplex stage, and the gut is emptied by defecation. The organism molts but remains in the old cuticle in a contracted immobile state. The old outer cuticle is sloughed off after the new inner cuticle becomes dark and hardened. Reasons for the emergence of the tardigrade from the cyst state have not been established. Encystment is important for survival since the cyst can last for over a year in nature without depleting food reserves.

Anoxybiosis is considered a type of cryptobiosis, although the metabolic aspects are unknown. This phenomenon, also called the "asphyxial state" or "asphyxia," is induced by low oxygen tensions. When the environmental water surrounding the organism is lacking in oxygen, the tardigrade becomes turgid due to the loss of osmotic control and the consequential absorption of water. The organism is extended, immobile, and appears lifeless. Because of the high degree of transparency in this state, anatomical structures can be observed more easily. Tardigrades in the asphyxial state for less than 4 or 5 days can be revived by the addition of oxygen to the water.

Cryobiosis is a type of cryptobiosis induced by low temperatures, which allows tardigrades to survive freezing and thawing. Slow cooling rates and the formation of tuns (Fig. 15.8) enhance survival after freezing. A tun is the result of contraction of the head, legs, and body which produces a rounded barrel shape. Normal activity occurs during temperatures that range from near freezing to 25–30°C. Some species inhabit warm springs where the temperature is a constant 40°C.

Osmobiosis is a cryptobiotic tun state which is induced by exceptionally high osmotic pressure when the animal is subjected to abnormal salt concentrations. This enables freshwater species to survive short periods of exposure to seawater. One marine tardigrade, *Echiniscoides sigismundi*, becomes turgid when placed in fresh water but resumes activity when returned to seawater within 3 days.

Anhydrobiosis is a cryptobiotic state induced by removal of water from semiaquatic and terrestrial tardigrades by evaporation. Marine species and true aquatic freshwater tardigrades may lack the ability to survive dehydration. Anhydrobiotic animals remain viable for as long as 120 years (43) and become active when free water in the environment becomes available. Survival depends on dehydration with slow evaporation rates at high humidities and the formation of tuns. Tun formation apparently slows the rate of evaporative water loss and increases preparation time for protection against gross mechanical damage and damage at the molecular level.

Cryptobiosis is an adaptation that enhances survival by increasing the longevity of the organism and by enabling the organism to survive harsh or hazardous environmental conditions. Further investigations are obviously needed in this area.

Techniques for Collection, Extraction, and Microscopy

Qualitative collections of tardigrades from aquatic habitats with sandy bottoms can be made by stirring the sand in a container of water and decanting immediately after the sand settles (44). The decanted water is passed through a sieve (No. 325) with pores of about 44 μm, and the specimens are rinsed from the screen to a sample jar. Tardigrades in large volumes of water can be preserved by the addition of formalin (5%) or gluteraldehyde.

Collections of tardigrades from vegetation can be made by removing the plant (moss, lichen, liverwort) from the substrate (tree, rock, soil) and placing the sample in a small paper bag for dry storage until processing (45). Then the sample is placed in a container of water for several hours to reverse anhydrobiosis. The sample is agitated and squeezed to remove specimens, eggs, and cysts. After the excess water is decanted, the remaining sediment containing the live tardigrades can be examined with a dissecting microscope at a magnification of 30× or greater. Boiling water or boiling alcohol (85% or greater) added to specimens in small amounts of water is adequate for tardigrade fixation. The sample may also be washed through sieves to remove particles smaller than 50 μm or larger than 1200 μm.

Core samples of soil and leaf litter can also be placed in paper bags for dry storage and processed according to the traditional procedure for vegetation. The sample is thoroughly washed in water for several hours and examined in small quantities of thin aqueous suspensions since eggs and cysts are easily overlooked. The technique is extremely laborious and time consuming.

Modification of techniques to extract soil nematodes by centrifugation in water have been used to remove tardigrades by many workers (46–49). A watery suspension of the sample is prepared and filtered. The portion containing particles of the appropriate size range is centrifuged with water or a sucrose solution. The supernatant is filtered and the residue washed into a petri dish for examination with a dissecting microscope.

After fixation, individual specimens are mounted on microscope slides and observed with phase microscopy. Hoyer's mounting medium (50 cc of distilled water, 30 g of crude gum arabic, 150 g of chloral hydrate, 20 g of glycerine, 1 g of potassium iodide, and 2 g of iodine; dissolved in sequence) produces cleared, distended specimens, but the coverslip must be sealed to ensure permanence. The amount of chloral hydrate may be reduced to decrease the extent of clearing of specimens. Various types of commercially available media are useful for temporary mounts; however, excessive stain may obscure important structures, and the specimens may clear completely after a period of time. More permanent mounts with balsam or synthetic resins are not as useful for phase microscopy.

For scanning electron microscopy, fixed tardigrades are dehydrated to absolute alcohol, transferred to an intermediary, and dried by the critical point method (50). Individuals are mounted on SEM stubs and coated with 200–300 nm of gold or gold–palladium.

Identification

Identification of species in the phylum Tardigrada is based primarily on the morphology of the cuticle, claws, buccal apparatus, and eggs, which have been discussed in previous sections. Ramazzotti (25) published the definitive monograph, in Italian, which provides keys and illustrated diagnoses of the species, thorough sections on morphology and biology, and an extensive bibliography. A supplement (51) was produced in 1974, and a revision of the monograph was published in 1983 by Ramazzotti and Maucci (52). Earlier monographs by Marcus (5, 15, 53) in German and Cuénot (54) in French formed the basis of tardigrade systematics for many years. Useful references in English include those by Pennak (55), Riggin (56), Morgan and King (57), and Nelson (58).

The system of classification presented in Ramazzotti's monograph (25) reflected previous changes in tardigrade systematics and was generally accepted, although the need for a thorough revision was recognized. Pilato (59) revised the systematics of the Eutardigrada and based genera and families on a single character, claw structure. Subsequently, Schuster et al. (21) presented another revision of Eutardigrada systematics based on a multicharacter analysis of the complex buccal apparatus, and they proposed two orders (Parachela and Apochela), three families (Macrobiotidae, Hypsibiidae, and Milnesiidae), and three new genera (*Dactylobiotus, Minibiotus,* and *Pseudobiotus*). Ramazzotti and Maucci (52) incorporated parts of both schemes, retaining the orders Parachela and Apochela and designating six families and 20 genera based primarily on claw structure rather than buccal apparatus. Figures 15.6 and 15.7 illustrate the claw types and buccal apparati of 21 genera of eutardigrades, according to Ramazzotti and Maucci's classification system, with the inclusion of *Minibiotus,* which we feel is a valid genus. The figures illustrate the range of variation in morphological structures that are considered of taxonomic importance in defining genera and also the need for further analysis.

The following key includes genera previously found and likely to be found in soil and leaf litter habitats. Positive identification with the key may be difficult because of inadequate generic descriptions and the need for further revision of tardigrade systematics.

KEY TO GENERA

1a Lateral cirri A present (Fig. 15.2) **2**

1b Lateral cirri A absent (Fig. 15.3) **9**

2a **(1)** Cuticle without dorsal plates; body indistinctly subdivided into eight segments (Fig. 15.5) . . . . ***Oreella***

2b Cuticle with complete or partial dorsal plates (Fig. 15.4) **3**
[See Kristensen (60) for revision of genera in Fig. 15.4]

3a **(2)** Cuticle with thin transverse dorsal plates, better defined caudodorsally; no plates with longitudinal divisions forming paired plates; terminal plate clearly defined; legs without dorsal leg spines (Fig. 15.4) . . . . ***Parechiniscus***

3b Cuticle with well-defined dorsal plates; longitudinal divisions form paired plates **4**

4a **(3)** Second paired plate (III) or third median plate (3) directly followed by terminal plate (Fig. 15.4) **5**

4b Second paired plate (III) or third median plate (3) followed by another double or single transverse plate (pseudosegmental plate, IV), then by a terminal plate (Fig. 15.4) **8**

5a **(4)** Median plates 1, 2, and 3 transversely divided into anterior and posterior section so that a total of six median plates are visible; indentations absent on terminal plate (Fig. 15.4). . . . . ***Bryochoerus***

5b Median plate 3 never transversely divided, median plates 1 and 2 may or may not be divided so that maximum of five median plates apparent (Fig. 15.4) **6**

6a **(5)** Median plates 1 and 2 transversely divided, indentations absent on terminal plate (Fig. 15.4) . . . . ***Bryodelphax***

6b Total of three or five median plates visible, indentations present on terminal plate (Fig. 15.4) **7**

7a **(6)** Median plates 1 and 2 transversely divided, so five median plates visible (Fig. 15.4) . . . . ***Hypechiniscus***

7b Median plates 1, 2, and 3 not transversely divided, so three median plates visible (Fig. 15.4) . . . . ***Echiniscus***

8a **(4)** Internal and external buccal cirri absent (Fig. 15.4) ***Mopsechiniscus***

8b Internal and external buccal cirri present (Fig. 15.4) ***Pseudechiniscus***

9a **(1)** Head with cephalic papillae (Fig. 15.5). **10**

9b Head without cephalic papillae (Fig. 15.3) **11**

10a **(9)** Head with six peribuccal papillae and two lateral papillae (Fig. 15.5) ***Milnesium***

10b Head without peribuccal papillae; two lateral papillae present . . . . ***Limmenius***

11a **(9)** Single unbranched claws without secondary branches **12**

11b Double claws, with primary and secondary branches. **13**

12a **(11)** One unbranched conical claw on leg I, no claws on legs II, III, IV (Fig. 15.6). . . . . ***Necopinatum***

12b Two unbranched claws on legs I–III, small basal spurlike claws on legs IV (Fig. 15.6). ***Haplomacrobiotus***

13a **(11)** Two double claws on each leg similar in size and shape and symmetrical with respect to the median plane of the leg; claw branch sequence is 2112 (secondary, primary, primary, secondary) **14**

13b Two double claws on each leg usually dissimilar in size and shape and asymmetrical with respect to the median plane of the leg; claw branch sequence is 2121 (secondary, primary, secondary, primary) **18**

14a **(13)** Buccal tube of distinctly greater length between stylet supports and placoids than between stylet supports and buccal opening; posterior part of tube of spiral composition (Fig. 15.7) . . . . ***Pseudodiphascon***

14b Buccal tube between stylet supports and placoids not longer than distance between stylet supports and buccal opening; tube without spiral composition (Fig. 15.7). **15**

15a **(14)** Stylet supports attach near middle of buccal tube; with 10 peribuccal papulae (not lamellae) (Fig. 15.7) . . . . ***Minibiotus***

15b Stylet supports attach in posterior one-half of buccal tube **16**

16a **(15)** Branches of claws at right angles so that tips are remote; two claws structurally connected at base; claws without lunules; cuticle without pores (Fig. 15.6). . . . . ***Dactylobiotus***

16b Both branches of claw follow similar axis so that tips are close together; two claws not structurally connected at base; claws with lunules; cuticle with pores **17**

17a **(16)** Ventral lamina (reinforcement bar) on buccal tube well developed; no anterior crests on buccal tube; 10 peribuccal lamellae (Fig. 15.7) . . . ***Macrobiotus***

17b Ventral lamina narrow or absent; unequal dorsal and ventral crests on buccal tube; lamellae unknown (Fig. 15.7) . . . . ***Adorybiotus***

18a **(13)** Buccal tube with median longitudinal anteroventral lamina (ventral tube support; Fig. 15.7) **19**

18b Buccal tube without median longitudinal anteroventral lamina (ventral tube support; Fig. 15.7) **20**

19a **(18)** Legs IV reduced or absent, with no claws or one or two minute double claws (Fig. 15.6). . . . . ***Hexapodibius***

19b Legs IV normally developed (Fig. 15.6) . . . . ***Doryphoribius***

20a	**(18)**	Claw of *Calohypsibius* type, primary and secondary branches rigidly connected, claws equal in size	**21**
20b		Claw with flexible connection between primary and secondary branches	**23**
21a	**(20)**	Posterior part of buccal tube of spiral composition, peribuccal lamellae present (Fig. 15.7)	***Eohypsibius***
21b		Buccal tube rigid, without spiral composition	**22**
22a	**(21)**	Cuticle sculptured, two macroplacoids (Fig. 15.7). . . .	***Calohypsibius***
22b		Cuticle smooth; three macroplacoids (Fig. 15.7). . . .	***Microhypsibius***
23a	**(20)**	Buccal tube rigid, without spiral composition	**24**
23b		Buccal tube with posterior part of spiral composition	**28**
24a	**(23)**	Each double claw with three distinct parts, 14 peribuccal lamellae, ornamented eggs deposited freely (Fig. 15.6, 15.7)	***Amphibolus***
24b		Each double claw with two parts	**25**
25a	**(24)**	Claws of dissimilar size and shape; primary branch with thin, flexible articulation; secondary branch continuous with base; no lunules; no buccal lamellae; cuticle smooth (Figs. 15.6, 15.7)	***Hypsibius***
25b		Claws of similar size and shape, with secondary branch forming right angle with base.	**26**
26a	**(25)**	Dorsal and ventral crests on buccal tube absent; lamellae and/or lunules rarely present; cuticle smooth or with granulations or gibbosities (Fig. 15.7)	***Isohypsibius***
26b		Dorsal and ventrals crests on buccal tube present	**27**
27a	**(26)**	With 12 peribuccal lamellae (Fig. 15.7)	***Thulinia***
27b		With 30 peribuccal lamellae (Fig. 15.7)	***Pseudobiotus***
28a	**(23)**	Pharynx with normal placoids; buccal tube with obvious style supports (Fig. 15.7) . . . . [See Pilato (61) for revision of genus]	***Diphascon***
28b		Pharynx with placoids and stylet supports reduced or absent (Fig. 15.7).	***Itaquascon***

References

1. **Goeze, J. A. E. 1773.** Uber den kleinen Wasserbar. In K. Bonnet (ed.). *Abhandl. A. D. Insektologie* (aus d. Franzos. übers. usw). 2. *Beobachtg.*, pp. 67, 145–210.
2. **Corti, B. 1774.** *Osservazióni Microscòpiche sulla Tremella e sulla Circolazióne del Flúido in una Piànta Acquajuola* (*Chara*) (208 pp., 3 tab.), 8. Lucca, pp. 97, 100.
3. **Eichorn, J. C. 1775.** Beyträge zur Natur-Geschichte der kleinsten Wasser-Thierchen (8 Kupfer-Tafeln, p. 74, tab. 7, fig. E). Der Wasser-Bar. Danzig.
4. **Spallanzani, L. 1776.** *Opúscoli di Física Animàle, e Vegetabile.* Vol. 2. *Il Tardigrado etc.*, Opusc. 4, sez. spec., p. 222ff. Modena.
5. **Marcus, E. 1936.** Tardigrada. In F. Schulze (ed.). *Das Tierreich.* Walter de Gruyter, West Berlin. 340 pp.
6. **Bertolani, R. 1979.** Hermaphroditism in tardigrades. *Int. J. Invertebr. Reprod.* 1:67–71.
7. **Rahm, G. 1937.** A new order of tardigrades from the hot springs of Japan (Furu-yu Section, Unzen). *Annot. Zool. Jpn.* 16:345–352.
8. **Crowe, J. H., I. M. Newell and W. W. Thomson. 1971.** Fine structure and chemical composition of the cuticle of the tardigrade. *Macrobiotus areolatus* Murray. *J. Microsc. (Paris)* 11:107–120.

9. **Bussers, J. C. and C. Jeaniaux. 1973.** Chitinous cuticle and systematic position of Tardigrada. *Biochem. Syst.* 1:77–78.
10. **Crowe, J. H., I. M. Newell and W. W. Thomson. 1970.** *Echiniscus viridis* (Tardigrada): fine structure of the cuticle. *Trans. Am. Microsc. Soc.* 89:316–325.
11. **Greven, H. 1972.** Vergleichende Untersuchungen am Integument von Hetero- und Eutardigraden. *Z. Zellforsch.* 135:517–538.
12. **Greven, H. 1975.** New results and considerations regarding the fine structure of the cuticle in tardigrades. *Mem. Ist. Ital. Idrobiol. (Suppl.)* 32:113–131.
13. **Baccetti, B. and F. Rosati. 1971.** Electron microscopy on tardigrades. III. The integument. *J. Ultrastruct. Res.* 34:214–243.
14. **Crowe, J. H., I. M. Newell and W. W. Thomson. 1971.** Cuticle formation in the tardigrade, *Macrobiotus areolatus* Murray. *J. Microsc.* (*Paris*) 11:121–132.
15. **Marcus, E. 1929.** Tardigrada. In H. G. Bronn (ed.). *Klassen und Ordnungen des Tierreichs,* Band 5, Abt. IV, Buch 3. Leipzig, East Germany. 608 pp.
16. **Walz, B. 1978.** Electron microscopic investigation of cephalic sense organs of the tardigrade *Macrobiotus hufelandi* C.A.S. Schultze. *Zoomorphologie* 89:1–19.
17. **Walz, B. 1973.** Zur Feinstruktur der Muskelzellen des Pharynx-Bulbus von Tardigraden. *Z. Zellforsch.* 140:389–399.
18. **Dewel, R. A. and W. H. Clark. 1973.** Studies on the tardigrades. II. Fine structure of the pharynx of *Milnesium tardigradum* Doyere. *Tissue & Cell* 5:147–159.
19. **Walz, B. 1974.** The fine structure of somatic muscles of Tardigrada. *Cell Tissue Res.* 149:81–89.
20. **Walz, B. 1975.** Ultrastructure of muscle cells in *Macrobiotus hufelandi. Mem. Ist. Ital. Idrobiol.* (*Suppl.*) 32:425–443.
21. **Schuster, R. O., D. R. Nelson, A. A. Grigarick and D. Christenberry. 1980.** Systematics of the Eutardigrada. *Trans. Am. Microsc. Soc.* 99(3):284–303.
22. **Dewel, R. A. and W. H. Clark. 1973.** Studies on the tardigrades. I. Fine structure of the anterior foregut of *Milnesium tardigradum* Doyere. *Tissue & Cell* 5:133–146.
23. **Weglarska, B. 1975.** Studies on the morphology of *Macrobiotus richersi* Murray 1911. *Mem. Ist. Ital. Idrobiol.* (*Suppl.*) 32:445–464.
24. **Dewel, R. A. and W. C. Dewel. 1979.** Studies on the tardigrades. IV. Fine structure of the hindgut of *Milnesium tardigradum* Doyere. *J. Morphol.* 161:79–110.
25. **Ramazzotti, G. 1972.** Il Phylum Tardigrada, Secónda Edizióne, *Mem. Ist. Ital. Idrobiol.* 28:1–732.
26. **Greven, H. 1979.** Notes on the structure of vasa Malpighii in the eutardigrade *Isophysibius augusti* (Murray, 1907). *Zesz. Nauk. Uniw. Jagiellonsk. Pr. Zool.* 25:87–95.
27. **Bertolani, R. 1975.** Cytology and systematics in Tardigrada. *Mem. Ist. Ital. Idrobiol.* (*Suppl.*) 32:17–35.
28. **Mihelcic, F. 1953.** Contribución al conocimiento de los tardigrados con especial consideración de los tardigrados de Osttirol, II. *An. Edafol. Fisiol. Veg.* 12:431–479.
29. **Heinis, F. 1916.** Ueber die Mikrofauna am Bolchen. *Tatbericht Naturforsch. Ges. Baselland* 1911–1916, pp. 81–98.
30. **Delamare-Deboutteville, C. 1954.** Eaux souterraines littorales de la côte catalane française. *Vie Milieu* 5:408–451.
31. **Franz, H. 1952.** État de nos connaissances sur la microfaune de sol. *Colloq. Int. Cent. Natl. Rech. Sci. Ecol.* 33:241–252.
32. **Lüdi, W. 1948.** Die Pflanzengesellschaften der Schingeplatte bei Interlaken und ihre Beziehungen zur Umbelt. Pp. 1–400 in Hans Huber (ed.). *Veröffentl. Geobot. Inst. Rübel.,* Zurich.
33. **Ramazzotti, G. 1959.** Tardigradi in terréni prativi. *Atti Soc. Ital. Sci. Nat.* 98:199–210.
34. **Mihelcic, F. 1949.** Neuvos biotopos de tardigrados: contribución al conocimiento de la ecología de los tardigrados. *An. Edafol. Fisiol. Veg.* 8:511–526.
35. **Mihelcic, F. 1952.** Contribución al estudio de la ecología de los tardigrados que hábitan suelos de humus, I. *An. Edafol. Fisiol. Veg.* 11:407–446.

36. **Mihelcic, F. 1952.** Contribución al estudio de la ecología de los tardigrados que hábitan suelos de humus, II. *An. Edafol. Fisiol. Veg.* 11:651–680.
37. **Mihelcic, F. 1953.** Contribución al conocimiento de los tardigrados de Osttirol, I. *An. Edafol. Fisiol. Veg.* 12:243–274.
38. **Mihelcic, F. 1954.** Zur Ökologie der Tardigraden. *Zool. Anz.* 153:250–257.
39. **Mihelcic, F. 1957.** Contribución a la ecología de los tardigrados de suelos humedos. *An. Edafol. Fisiol. Veg.* 16:651–671.
40. **Mihelcic, F. 1962.** Können Tardigraden im Boden leben? *Pedobiologia* 2:96–101.
41. **Mihelcic, F. 1967.** Der Boden als Wohnraum für Tardigraden. *An. Edafol. Agrobiol.* 26:145–157.
42. **Crowe, J. H. 1975.** The physiology of cryptobiosis in tardigrades. *Mem. Ist. Ital. Idrobiol.* (*Suppl.*) 32:37–59.
43. **Crowe, J. H. and A. F. Cooper, Jr. 1971.** Cryptobiosis, *Sci. Am.* 225(6):30–36.
44. **Schuster, R. O., E. C. Toftner and A. A. Grigarick. 1978.** Tardigrada of Pope Beach, Lake Tahoe, California. *Wasmann J. Biol.* 35:115–136.
45. **Nelson, D. R. and D. S. Horning, Jr. 1979.** Tardigrada of the Kowhai Bush, New Zealand. *Zesz. Nauk. Uniw. Jagiellonsk. Pr. Zool.* 25:125–142.
46. **Fleegar, J. W. and W. D. Hummon. 1975.** Distribution and abundance of soil Tardigrada in cultivated and uncultivated plots of an old field pasture, *Mem. Ist. Ital. Idrobiol.* (*Suppl.*) 32:93–112.
47. **Martin, N. A. and G. W. Yeates. 1975.** Effect of four insecticides on the pasture ecosystem. 3. Nematodes, rotifers, and tardigrades. *N.Z. J. Agric. Res.* 18:307–312.
48. **Hallas, T. E. and G. W. Yeates. 1972.** Tardigrada of the soil and litter of a Danish beech forest. *Pedobiologia* 12:287–304.
49. **Hallas, T. E. 1975.** A mechanical method for the extraction of Tardigrada. *Mem. Ist. Ital. Idrobiol.* (*Suppl.*) 32:153–158.
50. **Horning, D. S., Jr., R. O. Schuster and A. A. Grigarick. 1978.** Tardigrada of New Zealand. *N.Z. J. Zool.* 5:185–280.
51. **Ramazzotti, G. 1974.** Supplemento A, Il Phylum Tardigrada, Secónda Edizióne, 1972. *Mem. Ist. Ital. Idrobiol.* 31:69–179.
52. **Ramazzotti, G. and W. Maucci. 1983.** Il Phylum Tardigrada, Tèrza Edizióne. *Mem. Ist. Ital. Idrobiol.* 41:1–1012.
53. **Marcus, E. 1928.** Spinnentiere oder Arachnoidea. IV. Bartierchen (Tardigrada). Pp. 1–230 in F. Dahl (ed.). *Die Tierwelt Deutschlands,* Part 12. Gustav Fischer, Jena.
54. **Cuénot, L. 1932.** Tardigrades. Pp. 1–96 in P. Lechevalier (ed.). *Faune de France 24.*
55. **Pennak, R. 1979.** Tardigrada. Pp. 239–253 in *Fresh-Water Invertebrates of the United States,* 2nd ed. Wiley, New York.
56. **Riggin, G. T. 1962.** Tardigrada of southwest Virginia: with the addition of a description of a new marine species from Florida. *Va. Agr. Expt. Sta. Tech. Bull.* 152:1–145.
57. **Morgan, C. I. and P. E. King. 1976.** *British Tardigrada. Tardigrada Keys and Notes for the Identification of the Species.* Synopses of the British Fauna 9. Academic Press, London, pp. 1–132.
58. **Nelson, D. R. 1982.** Developmental Biology of the Tardigrada. Pp. 363–398 in F. Harrison and R. Cowden (eds.). *Developmental Biology of Freshwater Invertebrates.* Alan R. Liss, New York.
59. **Pilato, G. 1969.** Schema per una nuova sistemazione delle famiglie e dei generi degli Eutardigrada. *Boll. Sedute Accad. Gioenia Sci. Nat. Catania* 10:181–193.
60. **Kristensen, R. M. 1987.** Generic revision of the Echiniscidae (Heterotardigrada), with a discussion of the origin of the family. Pp. 261–335 in R. Bertolani (ed.). *Biology of Tardigrades.* Selected Symposia and Monographs U.Z.I., 1, Mucchi, Modena.
61. **Pilato, G. 1987.** Revision of the genus *Diphascon* Plate, 1889, with remarks on the family Itaquasconinae (Eutardigrada, Hypsibiidae). Pp. 337–357 in R. Bertolani (ed.). *Biology of Tardigrades.* Selected Symposia and Monographs U.Z.I., 1, Mucchi, Modena.

16

Scorpiones, Solifugae, and Associated Desert Taxa

CLIFFORD S. CRAWFORD

Unlike other chapters in this book, the present one deals with soil-associated invertebrates inhabiting an entire geographic region, the deserts and peripheral arid lands of North America. Although many "desert species" are also found in other biomes, the fauna treated here seem sufficiently distinct, both taxonomically and ecologically, to warrant separate consideration.

The biology of the animals in question is in general poorly understood. Except for pioneering studies made earlier in this century, research on desert soil invertebrates has only recently been given much emphasis. Because of this uneven progression of knowledge it is difficult to treat each group with much consistency. Nevertheless, whenever possible I have tried to address the functional roles of these interesting animals by combining information from a variety of sources, not all of which concern North American species. In doing so I may have written more of a review than a synthesis; however, the reader should be able to derive certain ecological patterns from the following passages in spite of the rather broad approach taken.

Excluded from the chapter are desert nematodes and social insects (termites and ants), which are discussed in separate parts of the book. Ectoparasites of burrowing vertebrates are also ignored since their story would take as much space as the one to be presented. Protozoans are omitted because of our monumental ignorance of their biology in deserts.

North American Deserts

Distribution and General Characteristics

Four subdivisions of North American deserts (Fig. 16.1) are generally recognized (1–3). Of these the Great Basin desert is the most northern, being an example of a "continental interior desert" (4). Although its boundaries are debatable, particularly to the north and east, its coverage of nearly all of Nevada, the western half of Utah, part of southeastern Oregon, a strip of southern Idaho, and small portions of eastern California is usually accepted.

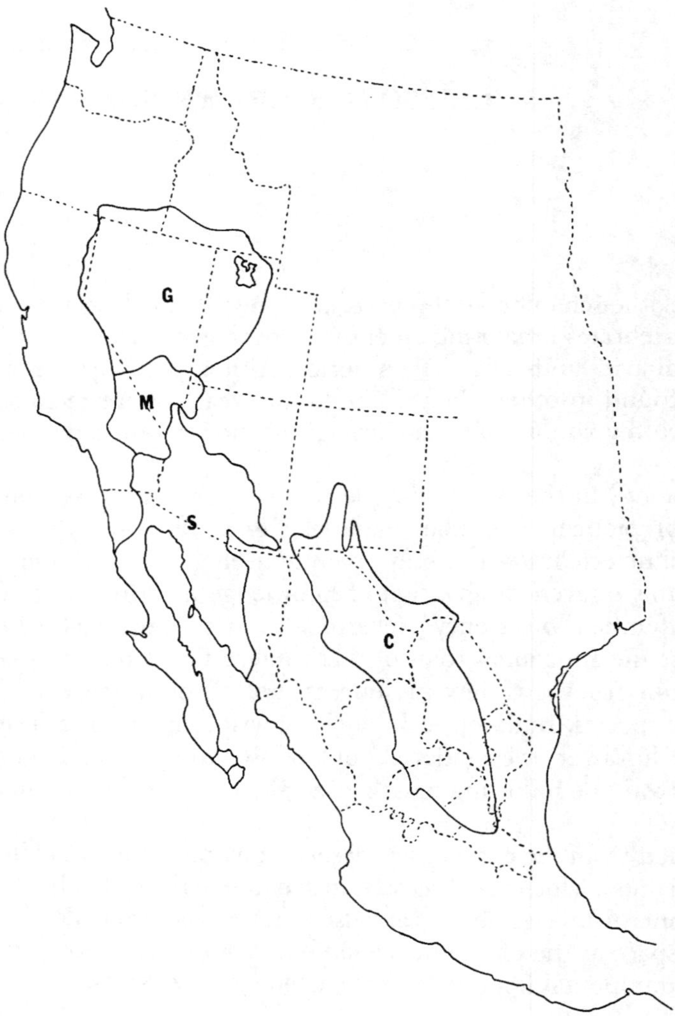

Fig. 16.1. Subdivisions of North American deserts as used in this chapter. C, Chihuahuan; G, Great Basin; M, Mojave; S, Sonoran.

The Great Basin can be extremely cold in winter when frigid air masses arrive from the Canadian interior. Mountain barriers to the west and southeast prevent the ingress of moderating maritime air (4). Otherwise, light snowfall accumulates, especially on the western slopes of higher ranges. In summer, rain-shadow effects and thermal instability caused by the highly uneven topography produce erratic and sometimes severe thunderstorms.

About half of the Great Basin is included in the Basin and Range Province, where mountain chains generally separate valleys in a north–south pattern. Major alluvial basins occur in the north and to the south (Death Valley). Aside from alluvial deposits and Lithosols associated with steep terrain, most Great Basin soils are well-structured Aridosols (5). Sierozems predominate to the north. Gray Desert soils (with more B-horizon clay) and Red Desert soils, together with saltier Aridosols, comprise soils of valleys and low-lying basins to the south, as well as soils of the more southern deserts.

Immediately to the south the smaller, upland Mojave desert occupies about equal portions of southern California and Nevada, in addition to the extreme northwest corner of Arizona. Precipitation in the Mojave occurs with moderate intensity and takes place almost entirely in the winter, when moisture-laden air moves in from the Pacific. The Mojave and the following two deserts have much Basin and Range topography, but their winters are warmer than those of the Great Basin.

The Sonoran desert abuts the Mojave's southern and southeastern borders. It extends in a rough arc across Arizona to the east and south, and into Mexico in the form of two projections surrounding most of the Gulf of California. One covers nearly all of Baja California, the other much of the western half of Sonora. Seven regional subdivisions were erected by Shreve and Wiggins (6) on the basis of climate, topography, and vegetation; clearly the physiography of the Sonoran desert is highly variable. Since much of it borders marine waters its winter precipitation is not unexpected. Prevailing summer winds, however, bring moist air off the Gulf of Mexico and result in brief but torrential rains.

To the east of the Sonoran desert lies the Chihuahuan desert. It includes the Rio Grande drainage of southern New Mexico and western Texas, but its main location is in Mexico. There it stretches south through the eastern half of Chihuahua, the western half of Coahuila, and parts of Durango, Zacatecas, Nuevo Leon, and San Luis Potosi (6). Relatively high valleys and block-faulted ranges are common in the north and south. Summer rainfall usually dominates winter precipitation.

Major Biotic Associations

All four North American deserts exhibit tremendous biotic heterogeneity, but within each there are somewhat characteristic assemblages of plants and animals. A useful recent synthesis of these by MacMahon (3) serves as a basis for the following comments.

In the Great Basin big sagebrush, *Artemisia tridentata,* is very common in the north, while associations of shadscale, *Atriplex confertifolia,* are especially evident on drier and more saline soils to the south. Even though these shrubs often dominate their communities, perhaps a dozen others also are present, as are a variety of grasses and annuals. Many of the annuals are exotics.

The creosote bush, *Larrea tridentata,* is an "indicator" species on a large scale because its distribution covers the Mojave, Sonoran, and Chihuahuan deserts. Stands of creosote and white bursage, *Ambrosia dumosa,* dominate large expanses of the Mojave and Sonoran deserts; in the Chihuahuan desert tarbush, *Fluorensia cernia,* replaces bursage as a codominant. In terms of diversity of prominent plant species in all four deserts, the cacti, subtrees, and shrubs of the Sonoran desert are the most striking. Diversity of summer annuals in the three warm deserts increases to the southeast and is maximal in the Sonoran desert, while that of winter annuals shows the opposite pattern (7).

The major Mojave subtree is the Joshua tree, *Yucca brevifolia.* That of the Chihuahuan desert is mesquite, *Prosopis glandulosa,* which in shrub form as well is one of many woody perennials in the Sonoran desert. There, a diversity of cacti (tallest in the south) comprises the succulent vegetation. Yuccas and agaves share this distinction with cacti in the Chihuahuan desert, but few of each exist in the Mojave. Grasses are moderately abundant in parts of all four deserts.

Vertebrate faunas in the Great Basin are less diverse (except for mammals) than they are in the three warm deserts, which together display remarkable similarities in vertebrate guild structure. Invertebrates in each desert are difficult to compare because of high species numbers. Extensive (8) and intensive (9, 10) studies of invertebrate communities are still too rare to produce robust generalizations. However, some individual species are potential "indicators" of biotic associations and of whole deserts. These are discussed in ensuing sections.

Soil Invertebrate Fauna (Exclusive of Protozoa, Nematodes, Social Insects, and Ectoparasites of Burrowing Vertebrates)

Definition

A precise definition of "soil invertebrates" in deserts is difficult to make because desert soils at times shelter practically all invertebrate species present in one stage or another. In this chapter I employ the following criteria in determining whether a taxon should be included: (a) its use of the soil as a habitat, and (b) its capacity to occupy another habitat. Even then some arbitrary decisions are inevitable, especially among insects occupying desert soil only as immatures.

All invertebrates treated below are either mollusks or arthropods. Oligochaete annelids are found in North America desert regions (11), but seem to be introduced and are very restricted in range. Enchytraeid annelids have recently been collected in southern New Mexico, but as yet nothing is known about them (W. G. Whitford, personal communication). Since a certain basic familiarity with invertebrates by the reader is assumed, and since major soil-inhabiting groups are dealt with in detail in other chapters and other publications (11a–c), no attempt is made here to describe the characteristics of the following major taxa: Gastropoda, Isopoda, Arachnida, Diplopoda, Chilopoda, and Insecta. Each of these is discussed below at a level commensurate with our present knowledge of its ecology in deserts. Keys are given only to the major orders of arachnids and to the families of scorpions, since most interested persons are less familiar with these arthropods than with insects, the other great class about which some desert biology is known. Keys in the other four classes are not justified in view of our ignorance of their biology in deserts.

Gastropoda

The considerable diversity of pulmonate snails in the southwestern deserts of North America has been documented by Bequaert and Miller (12). Many species occur in the Southwestern Molluscan Province proposed by these authors, an area that includes much of the Mojave, Sonoran, and Chihuahuan deserts. Despite its limited number of genera and species, the province's snail fauna presents a complexity of origins and distributions. It also contains an unusually high proportion of "precinctive" taxa that evolved in and are restricted to the province. Otherwise, fewer taxa presumably arrived from points south, and even lesser numbers came from other regions.

Species richness in the province is actually high in comparison with that of adjacent regions, and is probably due to diversified climate and topography (12). Distinctive local genera include *Sonorella, Ashmunella, Oreohelix, Holospira,* and *Chaenaxis.* To the east, in the Chihuahuan desert of west Texas and nearby areas, an encroachment by central Texas genera such as *Rabdotus, Polygyra,* and *Helicina* has occurred. Metcalf (13) relates elevation and habitat in the northern Chihuahuan desert to the distribution of both living and fossil mollusks. There, in mountains of medium elevation, accumulations of talus limestone provide effective refugia. Snail faunas of lower mountains consist mainly of small xeric-adapted species. Jaeger (2) also refers to small species of land snails living beneath rocks in arid mountain ranges, especially in the Death Valley area.

The mollusk fauna of Baja California south of the province's boundary

(28°N) is also rich and contains a good number of precinctive genera, including *Berendtia, Spartocentrum,* and subgenera of *Rabdotus* (12). Living *Greggelix indigena* (14) and *Rabdotus laevapex* (15) have been collected from rock slides in that region. Apparently, land snails are unstudied or at least not common in the Great Basin Desert; none are listed in the extensive survey of biotic communities undertaken by Fautin (8).

What little is known of the phenology and adaptations of North American desert snails suggests adaptive patterns not unlike those of species in the Middle East (16, 17) and Australia (18, 19). Activity seems confined to rainy seasons, during which feeding is nocturnal or restricted to cool daytime hours (2, 20–22). In the southwest no carnivorous species are known (12); rather they are herbivores (2) or feed on some combination of detritus, algae, and fungi (20, 22). Gilbertson's (20) brief report on the biology of *Sonorella odorata* (actually, a xeric-forest representative of this widespread genus) is apparently the only one of its kind in North American deserts and reveals that a number of fungus species are present in the gut of that animal. A similar symbiotic association is known for other desert detrivores (23) and implies rate control of decomposition.

As with desert snails in other parts of the world, reproduction probably also occurs in rainy seasons, provided that temperatures are not too low. In *S. odorata* mating in the field takes place in summer, and oviposition occurs just below the soil surface during heavy rains several months later. Incubation may take at least 3 weeks. Walton (24) inferred that clutch size of North American snails should be relatively low in marginal environments.

Dormancy occupies most of the lives of desert snails. Increased longevity (probably up to 15–20 years in *Sonorella* spp.) and slow maturation appear to correlate with extended dormancy (24). In summer, sometimes even when activity is expected, dormant *S. odorata* are affixed by calcerous mucus to a variety of objects. In winter dormancy this species remains free of attachment but with the aperture facing up, and is found in crevices, under rocks, in debris, and in soil. Several epiphragms may then be formed, and nearly 60% of the body weight may be lost in 8 months of dormancy (20).

Metabolic adaptations to prolonged periods of drought and nonfeeding are well developed in desert snails (25). Riddle (22) showed in *Rabdotus schiedeanus* from the Chihuahuan desert that increased activity during moist periods was reflected in elevated levels of oxygen consumption in humid air. Metabolic rates over a temperature range consistent with field conditions were generally much lower than in a common garden snail during dormancy as well as during activity, indicating a well-adapted system of energy conservation while exposed to dry desert conditions (26). As in *R. schiedeanus,* oxygen consumption in *R. dealbatus* drops markedly during dormancy, and dormant snails become active at elevated humidities (21). Energy depletion rather than water loss seems to be the most pressing physiological problem during dormancy in the arid-region snail from Texas, and is apparently countered by enhanced catabolism of tissue protein.

Isopoda

Terrestrial isopods are present but not common in North American deserts. Introduced species occurring beneath rocks near riparian habitats in xeric regions will not be considered here. The only well-studied desert species in North America is *Venezillo arizonicus* (Armadillidae), which was investigated in the Santa Rita Mountains of southern Arizona by Warburg (27) and reported (as *Armadillo arizonicus*) from the southern Great Basin desert by Allred and Mulaik (28). The relative rareness of this species seems to typify the isopod fauna of arid North America. Desert species are much more common and their biology as a whole much better known in North Africa (29, 30), the Middle East (31–33), and Australia (34).

Venezillo arizonicus is active on the soil surface between May and November in southern Nevada, where its preferred habitat is dominated by *Lycium* shrubs but where "suitable" sheltering rocks are lacking (28). In contrast, Warburg (35) found the species beneath stones and logs in the Santa Ritas. There the isopods move on the open surface at night, and also during the day when relative humidities and temperatures exceed those of potential shelters. Diurnal locomotion may be an adaptation for changing hiding places or for exploiting the higher surface-humidity conditions.

As with other desert isopods, *V. arizonicus* uses a combination of behavior and relatively great resistance to water loss (35) to survive the physical stresses of its environment. Associated adaptations include conglobation (rolling up), pseudotracheal respiration, negative photoreaction during high temperatures, and positive hygroreaction except at high temperatures (36).

Practically nothing is known of the seasonal biology and life history of *V. arizonicus* except that immatures numerically exceeded adults collected in Nevada can traps between July and September (28). This may mean that birth occurs in early summer following a gestation period beginning during the wet winter season, not unlike the situation in a species of *Hemilepistus* from Afghanistan (37).

Arachnida

Four of the arachnid orders, the Scorpiones, Solifugae, Araneae, and Acari, contain many soil-associated desert species in North America. The Pseudoscorpionida, Opiliones, and Uropygi are also represented, but not as well. With the exception of numerous acarines, desert arachnids are essentially carnivores. Predaceous arachnids are generally abundant in deserts, but their relative impact is difficult to assess.

A discussion of the orders is preceded by the following key, which should help readers unfamiliar with arachnids.

KEY TO THE ORDERS OF DESERT ARACHNIDS

1a Body divided into a leg-bearing prosoma and a posterior opisthoma; species highly variable in size 2

1b Body appears to be undivided, except for an anterior gnathosoma containing mouthparts; most species small to minute . . . . **Acari**

2a (1) Pedipalps enlarged, chelate. **3**

2b Pedipalps not enlarged **5**

3a (2) Last five segments of opisthoma form tail with terminal sting . . . . . **Scorpiones**

3b Opisthoma not so constructed **4**

4a (3b) Opisthoma bearing terminal whiplike flagellum; front legs long and antenniform; adults among the larger arachnids . . . . **Uropygi**

4b Opisthoma rounded posteriorly; all species very small . . . . **Pseudoscorpionida**

5a (2b) Chelicerae greatly enlarged . . . . **Solifugae**

5b Chelicerae not greatly enlarged. **6**

6a (5b) Prosoma and opisthosoma broadly joined; opisthoma segmented . . . . **Opiliones**

6b Prosoma and opisthosoma joined by narrow pedicel; opisthoma unsegmented . . . . **Araneae**

Scorpiones. Scorpions are widely distributed throughout the tropics and subtropics; some occur at higher latitudes as well. Many are desert species. The following key separates the four families found in North American deserts. See also the scorpion key by Francke and Soleglad (37a).

1a Two pedal spurs in the articular membrane between protarsus and tarsus (Fig. 16.2a) **2**

1b One pedal spur in the articular membrane between protarsus and tarsus (Fig. 16.2*b*) . . . . **Diplocentridae**

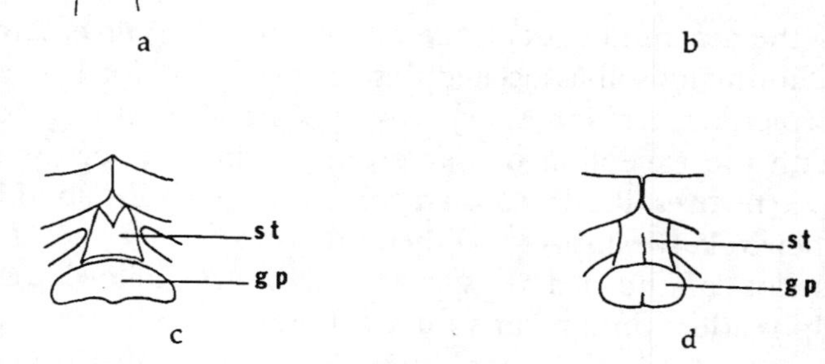

Fig. 16.2. (*a,b*) Pedal spurs (*ps*) in articular membrane between protarsus (*p*) and tarsus (*t*): (*a*) drawn from *Paruroctonus utahensis* (Vaejovidae); (*b*) drawn from *Diplocentrus peloncillensis* (Diplocentridae). (*c,d*) sternum (*st*) anterior to genital plate (*gp*) and between coxae: (*c*) drawn from *Centruroides sculpturatus* (Buthidae); (*d*) drawn from *Paruroctonus utahensis* (Vaejovidae).

2a	**(1a)**	Three to five pairs of lateral eyes	**3**
2b		No more than two pairs of lateral eyes	**Chactidae**
3a	**(2a)**	Sternum roughly triangular (Fig. 16.2*c*); tubercle may be present at base of stinger	**Buthidae**
3b		Sternum with nearly parallel sides (Fig. 16.2*d*)	**Vaejovidae**

A combination of behavioral, physiological, and metabolic traits adapts desert scorpions to their unpredictable environments (37b). This principle also applies to other desert arthropods, even though specific responses are emphasized differently in different taxa. Adaptations to the physical environment are keyed to the maintenance of optimal water and heat balance, the dynamics of which are intimately related because both are functions of radiant energy.

Water balance in desert animals is achieved by regulating the ratio of water uptake and loss. In scorpions, water uptake during the warmer months seems mainly due to fluids provided by tissues of their prey (38). Some North American species are also known to imbibe water directly or from moist substrates (39, 40). Unpublished studies show that water absorption can also occur at very high relative humidities (E. C. Toolson, personal communication). Metabolic water produced as a consequence of respiration completes the positive side of the water-balance ledger; however, since scorpions are comparatively inactive arthropods the contribution from this source is probably minimal.

Water loss in these and other desert arthropods occurs largely across the body surface. In their ability to resist cuticular transpiration desert scorpions have few equals, the large desert vaejovid, *Hadrurus arizonensis,* being especially well adapted in this respect (41). This capacity appears related to habitat in scorpions and other desert invertebrates (40, 42), with species in very arid habitats having the most resistance. It is also associated with season, since transcuticular water loss is greater in the desert buthid, *Centruroides sculpturatus,* during winter than in summer (43). Much of the resistance appears to be a function of high levels of long-chain branched hydrocarbons and/or long-chain fatty acids in the epicuticle (44, 45).

Water is also lost across the respiratory surface in arthropods and other animals. In scorpions these surfaces consist of "book lungs" opening to the exterior via small abdominal spiracles. Some evidence of the spiracular control of gas exchange exists (40, 46), but in general low ventilation rates (reflecting low metabolism) make respiratory transpiration an unimportant avenue for water loss except at high temperatures (40, 41). During starvation, which is probably common in desert scorpions, metabolic rates are depressed (47) and loss of moisture across respiratory surfaces should be inconsequential.

Another form of regulation associated with metabolism in scorpions and other arachids is the production of guanine as an excretory product. This highly insoluble nitrogenous waste is similar to the uric acid of insects in that it can be excreted in essentially dry form following extensive water reabsorption by portions of the gut and excretory system. Studies with a single species suggest that when extensive desiccation does occur, desert scorpions simply tolerate the resulting osmotic imbalance (48).

Escape from high surface temperatures and desiccating air is effected in most desert scorpions by nocturnal and shelter-seeking patterns of behavior. Another explanation for nocturnal activity may be avoidance of daytime predators (49). It should be noted, however, that nocturnal predators do exist; these include shrews and owls (50) as well as bats (O. F. Francke, personal communication). Whatever the selective agents for nocturnality may have been, most scorpions exhibit a pronounced activity increase with the onset of darkness (51). Often this is associated with an endogenous circadian rhythm (49), but in *Diplocentrus peloncillensis* (Fig. 16.3), which lives beneath rocks in southwestern New Mexico, a direct, noncircadian response occasions the same result (52). Recent work with a Saharan desert species reveals the exquisite sensitivity of scorpion eyes to extremely low light intensities (53), which perhaps explains why the surface activities of several North American species are reduced during periods of full moon (51, 54).

Fig. 16.3. *Diplocentrus peloncillensis* at burrow entrance following removal of sheltering rock. Peloncillo Mountains, New Mexico. (Photo by author.)

Burrows provide a common form of shelter for desert scorpions. This is well illustrated in the case of *H. arizonensis,* which moves in such a way within its burrow that it regularly avoids exposure to hot daytime temperatures (55). Nonburrowing scorpions resort instead to hiding beneath rocks, bark and other plant parts, and surface debris (38, 56, 57). They also make use of burrows dug by other animals.

Survival during the winter is also a function of appropriate shelter. Crawford and Riddle (58) estimated that when *D. peloncillensis* inhabiting north-facing slopes remained close to the mouths of their burrow openings beneath rocks, they risked 50% mortality due to freezing an average of one day a year. The chances of death due to cold diminished with depth and changed with site locations. While this species is moderately tolerant of repeated freezing, its supercooling point remains seasonally unchanged to any significant degree (59). In contrast, Riddle and Pugach (60) showed that the burrowing desert–grassland vaejovid, *Paruroctonus utahensis,* is killed by a single freezing but presumably avoids this by lowering its supercooling point as winter approaches.

The population biology of desert scorpions has only recently come under scrutiny. In North America that of a single species, *Paruroctonus mesaensis* (Fig. 16.4), which is very abundant in the Sonoran and Mojave deserts, has been thoroughly investigated in the Mojave, although many admirable studies have been made on selected aspects of scorpion life history elsewhere. The following comments are based in part on the work of Polis and Farley (62, 63) with *P. mesaensis.*

Male reproductive activity in some North American desert scorpions first involves an obvious increase in surface locomotion (61, 62) during the warmer months. Termed "migratory" or "vagrant" activity by various authors, this presumably facilitates finding females. An elaborate courtship dance ensues, in which the female is pulled over a spermatophore previously deposited on the substrate by the male. Insemination results, and is followed by a gestation period of roughly half a year to a year. In populations of desert vaejovids at least, birth may often be a synchronous event in any given year (61). Polis and Farley (63) consider such synchrony a probable consequence of springtime "pulses" of prey available to gravid females. A number of scorpion species are known to bear young more than once in a lifetime (O. F. Francke, personal communication). *Centruroides sculpturatus* can definitely have two clutches in a year (61), and some marked *P. mesaensis* females were gravid for two to three consecutive years (63).

Newborn scorpions ascend the mother's back and remain there for a week to a month until they molt to the second instar. At this time they become independent from the female and are able to capture their own food. Although overt brooding past the first instar seems not to take place, it is common to find early instars in the vicinity of their mother and their birth site.

Scorpions average about six molts in a lifetime (63). Potential longevity

Fig. 16.4. *Paruroctonus mesaensis* preying on adult myrmeleontid (ant lion). Mojave desert. (Photo courtesy of G. A. Polis.)

seems to range from about 1 year to over 6 years (62), being shorter in buthids than in diplocentrids and scorpionids (O. F. Francke, personal communication). Actual longevity, however, depends on mortality induced by abiotic and biotic factors. The effects of winter cold have already been mentioned, and *D. peloncillensis* apparently also drowns in its burrows if spring rains are unusually heavy (personal observations). The lethal influence of excessive heat and dryness on scorpion populations appears not to have been quantified. Carnivory by large invertebrates such as centipedes, Solifugae, spiders, and other species of scorpions has been recorded, and a wide variety of vertebrate predators are known (64). Although knowledge of cannibalism among scorpions is not new, the effects of this form of behavior have only recently been estimated. Polis (65) finds that in *P. mesaensis* it accounts for the largest amount of prey biomass; the capacity for density-dependent population regulation in this desert species, and perhaps in others as well, therefore appears to be well developed (66).

In addition to feeding on conspecifics, desert scorpions utilize a broad assortment of prey (64, 54). In *P. mesaensis* prey selection is somewhat age specific, and over 100 species have been recorded as prey (67). Brownell (68) recently demonstrated that *P. mesaensis* detects prey as distant as 50 cm on sandy substrates by using sensory structures on its legs to perceive low-velocity compressional and surface waves. The work of Hadley and Williams and of Polis suggests that relatively few individuals in desert populations forage at any one time, and that prey capture and prey consumption take on different patterns in different species. In a very broad sense scorpions may be classified either as mobile hunters or as "sit-and-wait" predators, or may be assigned to intermediate categories. Utilization of a common habitat by more than one species is well known (52, 69, 70) and implies mechanisms of resource utilization that allow for coexistence.

Pseudoscorpionida and Uropygi. Neither of these orders is common in deserts, but in North America several relatively arid-adapted species are known. Two pseudoscorpion species were collected in the northern

Fig. 16.5. *Mastigoproctus giganteus*. Peloncillo Mountains, New Mexico. (Photo by author.)

reaches of the Chihuahuan desert by Hoff (71). *Parachernes nubilis* (Chernetidae) was found in dry, sand-mixed litter of shinnery oak, *Quercus havardii,* and mesquite, *Prosopis juliflora. Levichelifer fulvopalus* (Cheliferidae) occurred only in shinnery oak litter. Although pseudoscorpions are small arachnids, both of these species appear fairly well adapted against desiccation because they are comparatively large and possess heavily sclerotized cuticles.

Only one uropygid or whipscorpion seems to be desert-adapted, and it inhabits the Chihuahuan desert. *Mastigoproctus giganteus* (Thelyphonidae) (Fig. 16.5) actually ranges in distribution from Florida to Arizona, consistent with its preference for moist, warm microhabitats regardless of macroclimate. Ahearn (72) collected it from beneath large rocks and logs on the eastern slopes of the Chiracahua Mountains in Arizona. New Mexico locations known to me include dry oak woodland in the Peloncillo Mountains, rocky areas and stone walls near Elephant Butte Reservoir (northern Chihuahuan desert), and the central Rio Grande valley.

Summer rains coincide with the surface appearance of these large arachnids in the southwest. Unlike desert scorpions, they are only moderately resistant to cuticular transpiration, but compensate by imbibing water readily from moist substrates (72, 73). Although nocturnal, *M. giganteus* shows no preference between light and darkness. On the other hand, when desiccated, it is sensitive to moist air and burrows against objects or sides of surface depressions if the soil is damp. Detection of nonhorizontal surfaces is accomplished by probing with the antenniform front legs (73), while most of the digging and soil carrying is done with the massive pedipalps (74).

Reproduction in *M. giganteus* commences with a nocturnal courtship dance differing from that of scorpions (and pseudoscorpions that dance) in that both partners face the same direction during spermatophore construction (75). Another difference is that male *M. giganteus* use palpal fingers to push the two sperm carriers of the spermatophore into females' gonopores. The morphology of these carriers is seen as an adaptation to dry environments in that it should prevent sperm masses from desiccating during the relatively slow process of insemination (76). Birth may occur about the time of summer rains in the southwest, judging from personal observations of probable first instars on the surface following heavy rains. Four instars occur in this species; molting takes place in early summer in an underground chamber, and reproductive maturity begins after the last molt (76). Adults can be kept alive for several years in the laboratory, but longevity in the field is not known.

A variety of prey, including small vertebrates, is captured by *M. giganteus,* which uses its front legs to detect movement and its pedipalps to grasp prey after a usually cautious stalking period (64). As a defense against potential predators it accurately directs a deterrent spray of 84% acetic acid, 5% caprylic acid, and water from glands at the base of its flagellum

(77). Feeding intensity presumably diminishes with the relatively dry winter months in arid regions. What appear to be dormant whipscorpions are then occasionally found in soil hibernacula (personal observations).

Solifugae. These are highly mobile arachnids (Fig. 16.6) that inhabit mostly warm, arid environments (78). In North America members of the families Ammotrechidae and Eremobatidae occur in the four major deserts, being more abundant in some habitats than in others. For example, Allred and Muma (79) found that high population densities in the northern Great Basin were associated with plant communities dominated by sagebrush, *Artemisia tridentata,* and rabbitbrush, *Chrysothamnus* spp. Since some of the same Solifugae were associated with other dominant shrub species in southern Nevada, the replacement role of the plant cover was felt to have greater ecological significance than its taxonomic composition (80).

Of 29 species collected in desert sites of Nevada, Idaho, and Utah, none were found in all three places (81). In each geographic area these arachnids were most abundant in plant communities containing relatively low populations of spiders and scorpions. In New Mexico, Muma (82) found that populations of Solifugae were two to four times larger in arid grassland than in the pinyon–juniper zone. He surmised that flooding depresses while heavy grazing promotes some species populations in that general

Fig. 16.6. Solifugae (Eremobatidae) exiting from burrow. Big Bend National Park, Texas. (Photo by author.)

region. However, Cloudsley-Thompson (78) implies that Solifugae are adapted to flash-flood conditions because of their ability to survive submersion.

The physiological ecology of North American Solifugae has not been investigated; nevertheless, certain inferences can be made from Cloudsley-Thompson's (78) recent review of the adaptional biology of this order. Thus if adaptations of North American species are at all similar to those of North African species, the former should also be resistant to high temperatures and low relative humidities. Yet despite their presumed abilities to cope with heat and dryness, nearly all North American species are nocturnal (83). Nocturnalism may therefore have evolved as an antipredator strategy. During the day individuals remain in burrows excavated with chelicerae, pedipalps, and legs. This behavior has been described by Muma (84), who noted that eremobatids may dig 40 or more burrows in a lifetime (85).

As the reproductive season approaches for a given species-group in North America, the first individuals to appear are mature males. Females mature at about the same time, but are not abundant until some time later. Muma (82) considers that near-simultaneous maturation within a species-group, regardless of the group's geographical distribution, prevents interbreeding among sympatric populations.

Each sex can mate more than once (86, 87). Unlike scorpions, pseudoscorpions, and whipscorpions, the Solifugae do not deposit a spermatophore on the ground. Instead, following various versions of a courtship ritual, the spermatophore is inserted directly by the male's chelicerae into the female's aperture. Eggs are laid at the terminus of a burrow during the summer (80), and in North America, the adult female then leaves after first plugging the burrow entrance, an act that probably stabilizes the burrow microclimate (85).

Eggs hatch in a matter of weeks, the newborns being essentially immobile. In another week or so the offspring molt into an active second instar that feeds and burrows. Although little is known about the lives of developing individuals, Muma (87) recorded nine instars in one species. The entire life cycle appears to occupy not more than a year in this order; in that respect the Solifugae differ greatly from the less metabolically active scorpions and whipscorpions.

In keeping with their high rate of energy use, specimens in the field seldom seem to slow down. Their exploratory movements are extremely rapid as they search for their prey mainly on the ground. With their powerful chelicerae they are quite able to subdue and ingest a captured food item.

As with most other carnivorous desert arachnids, the impact of these animals on their prey populations has not been adequately measured. The immature stages feed extensively on termites, particularly in North America (88). Since termites may be the most populous of the southwestern desert detritivores (23), a case can be made for a possible regulatory influence of the Solifugae on these important social insects. Adults have a much broader arthropod diet than that of immatures (88).

Araneae. Many species of desert spiders in North America are not directly associated with the soil. For example, Chew (89) suggested that only in the winter do many of the shrub-inhabiting species in southeastern Arizona probably move into soil litter, as well as into burrows and soil crevices [see also Fowler and Whitford (89a)]. Hence for these species soil may be a valuable temporary refuge, but not necessarily a feeding site. Similarly, although the retreats of the funnel-web spider *Agelenopsis aperta* (Agelenidae) sometimes extend into protected spaces in the substrate and vegetation of New Mexico's south-central Malpais Lava Beds (90), actual use of the soil habitat appears facultative.

The pattern of using soil and litter as a refuge is probably widespread among North American desert spiders. While published evidence of this seems to be meager, quite a few species crawl upon the surface, burrow into it, and sometimes build webs against soil clods (D. W. Lowrie, personal communication). Among these spiders are members of the families Gnaphosidae, Clubionidae, Hahniidae, Micryphantidae, Amaurobiidae, Dictynidae, Oonopidae, and Pholcidae, as well as some rare Agelenidae other than *Agelenopsis*.

The most consistent association with soil is generally seen among the tarantulas (Fig. 16.7) and wolf spiders (Fig. 16.8). Exceptions do occur, of course, as in *Syspira* spp. (Clubionidae), a southwestern desert group that looks and behaves very much like the wolf spiders (91). The following paragraphs, however, emphasize only the life histories and adaptations of common desert tarantulas (Theraphosidae) and wolf spiders (Lycosidae), arachnids that spend most of their lives in the soil or on its surface. The trap-door spiders (Ctenizidae) belong to the same primitive suborder (Orthognatha) as the tarantulas and also occur in North American deserts. Nonetheless, aside from their unique nest construction, the little information available on the ecology and physiology of these species suggests that their life-history patterns are similar to those of tarantulas.

Tarantulas have fascinated naturalists and biologists for years, and highly readable accounts of these large arachnids are found in Comstock (92), Gertsch (91), and Baerg (93). Recently, Minch (94) wrote a fairly detailed report on the biology of *Aphonopelma chalcodes* in the Sonoran desert. Much of the following discussion is based on Minch's work.

Overwintering ceases and activity begins with the removal of soil plugs from burrow entrances. This happens in May or in the summer, depending on the individual. Those with early activity retire again briefly in June, replugging the burrow in the process. Surface activity lasts until October. At various intervals during this time span egg sacs are produced, and adult females are known to hold them with the chelicerae and pedipalps at the

Fig. 16.7. Tarantula (Theraphosidae). Big Bend National Park, Texas. (Photo by author.)

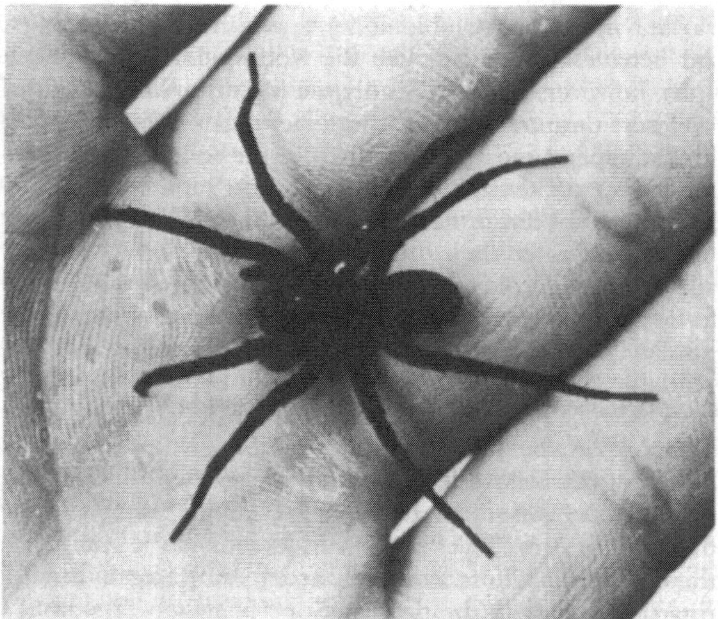

Fig. 16.8. *Geolycosa rafaelana.* Central New Mexico. (Photo courtesy of M. C. Matlack.)

burrow entrance. Female tarantulas can live roughly a decade after reaching sexual maturity, and prior development involves an equivalent duration (91). Since only a small proportion of *A. chaleodes* females bears egg sacs each year, they may be iteroparous. Having more than one generation of offspring in a lifetime occurs in other tarantulas (93).

Fecundity is high in tarantulas; *A. chalcodes* can have more than 550 offspring per egg sac (94). Egg eclosion takes place in the protected sac, juveniles remaining there for over a month. After hatching the juveniles are very gregarious for awhile, but eventually leave the noctural burrows and disperse to nearby shelters. Presumably, mortality is heavy in the early months and years of their development.

Mating, molting, and the consumption of prey also take place in the warmer months. Following the last molt males leave their burrows and wander in search of females. Such exposure, which occurs between July and late autumn, surely has a pronounced additional effect on the mortality pattern of desert tarantulas. Mating in *A. chalcodes* occurs at night (95). Molting in this species takes place in June, following a period of fasting. Adult females kept in the laboratory for over a year under constant conditions exhibited an annual molting cycle that peaked between March and May (95); this suggests a circannual rhythm, a phenomenon seldom demonstrates in the invertebrates. Early instars probably molt more frequently, as they do in other tarantula species.

Southwestern desert tarantulas seem to be highly territorial. According to Gertsch (91), they become attached to their own burrows or to one abandoned by rodents, and remain there indefinitely. Hunting is accomplished within a small space (perhaps 1 m in diameter) around the burrow entrance. Insects are captured and eaten with great regularity, and small vertebrates are potential prey. Contrary to earlier opinion, *A. chalcodes* can hunt in the daytime (96), although aktograph studies showed that several Sonoran desert species are strictly nocturnal (97). Daytime temperatures on the desert floor far exceed the upper lethal limit (43°C) of an *Aphonopelma* sp. studied by Seymour and Vinegar (98), so returning to the burrow or some other shaded retreat before midday becomes a necessity. These authors found that temperatures at burrow depths of about 20 cm on hot days were well below the lethal limit.

A thorough cataloging of prey items remains to be done for desert tarantulas; it is known, however, that a large percentage of tenebrionid beetles and spirostreptid millipedes are rejected because of their defensive secretions (99). Beetles and grasshoppers are among the preferred prey of North American species (91).

Desert tarantulas are well adapted to their habitat partly because of their ability to endure overwintering without feeding. Prolonged starvation is withstood in these arachnids (91) and in other spiders (100). The metabolism-depressing effects of starvation in tarantulas can be inferred from the results of scorpion and spider studies previously mentioned.

Resistance to water loss in tarantulas is accomplished by a combination of size (i.e., low surface-to-volume ratio) and effective citucular waterproofing (97, 98, 101). Resting metabolism is relatively low (98, 102), which means that little water is lost via ventilation at moderate temperatures. Water uptake occurs, of course, when prey are ingested, but also comes about due to direct drinking when water is available. Perhaps like other arachnids (see Ref. 103; see also citations of whipscorpion and scorpion behavior) tarantulas can extract water from moist substrates. Stewart and Martin (101) present indirect evidence that drinking may maintain blood volume in *Dugesiella hentzi* following bleeding or egg laying.

A variety of vertebrate predators, including rodents, birds, reptiles, and amphibians (91), eat tarantulas. Pompilid wasps ("tarantula hawks" in the genus *Pepsis*) specialize in provisioning their developing young with tarantulas and trap-door spiders. A partial defense against such attacks is a rapid retreat to the burrow. Short dashes to escape predation and to capture prey are therefore part of the adaptive repertoire of these desert spiders, while steady running is not. The finding by Seymour and Vinegar (98) that *Aphonopelma* fatigues easily is consistent with these observations.

Wolf spiders are smaller than tarantulas and belong to the suborder Labidognatha, which contains the "true" spiders. Members of the genera *Lycosa* and *Geolycosa* are especially well represented in North American deserts. *Geolycosa rafaelana* (Fig. 16.8) is a common burrower in soils of the desert southwest (91); however, very little is known of the biology of this genus in that region. On the other hand, Shook (104) recently elucidated the life history of *Lycosa carolinensis,* which in western deserts may well be a distinct species or at least a subspecies in comparison to the eastern form (D. W. Lowrie, personal communication). The burrows of *Geolycosa* and *Lycosa* are like those of tarantulas, and are also lined with silk at the top. A characteristic turret of webbed debris usually surrounds the burrow entrance and probably has many functions, including acting as an "early warning system" against potential enemies (104).

In the lower Sonoran desert *L. carolinensis* is active between March and October. Its egg sacs appear most abundantly in late July and late August, suggesting two broods a year as occurs in some other lycosids (104). The species is less fecund than are the desert tarantulas, having about 150 eggs per sac. This may be associated with a greater degree of maternal care exercised by wolf spiders (D. W. Lowrie, personal communication). Eggs hatch after about 2 weeks, whereupon the hatchlings ascend the mother's body and are carried about by her for another week or so. They do not feed at this stage (91). After molting they descend and disperse.

Molting normally takes place within the burrow. Development to reproductive maturity takes about 2 years. Adult males are seen in peak numbers a month before females and apparently live not more than 1 year after that, while females live an additional year. Immatures of *L. carolinensis* are most obvious in March and October, months corresponding to the times of main burrow construction by *G. rafaelana* in the central Rio Grande valley of New Mexico.

Although wolf spiders are often diurnal hunters, *L. carolinensis* is completely nocturnal in the lower Sonoran desert. Like *A. chalcodes,* it waits for prey in a predetermined home range, which averages just over 1 m^2 (104). Burrows are randomly dispersed with regard to each other and have adjacent home ranges that do not overlap, a situation consistent with the well-known cannibalistic tendencies of spiders. For some reason the burrows of *L. caroliensis* are uniformly distributed with respect to shrubs (104).

Adaptations to abiotic stresses differ somewhat between this species and the desert tarantulas. One difference is metabolic, the smaller lycosid having a somewhat higher weight-specific metabolic rate at equivalent temperatures (98, 102, 105). A poorly developed tracheal system is present in *Lycosa* spp. and is felt by Anderson (102) to be of significance mainly in the prevention of water loss during daytime activity. Perhaps this attribute is useful in habitats where *L. carolinensis* is not nocturnal.

The upper lethal limit of this wolf spider was first inferred from burrow temperatures to be about 45°C, just exceeding the thermal conditions of burrows near Phoenix (105). These authors later found that acclimation to high and cycling temperatures raises the upper thermal limit, an adaptation not seen in *A. chalcodes* (98). Furthermore, respiratory acclimation to several different temperatures increases occurred in only a few days, an adjustment also not observed in the tarantulas. Finally, the lateral displacement of the seasonal rate–temperature curve of respiration was shallower in January than in June, indicating a switch to an energy-conserving mechanism with the advent of high summer temperatures. Both species have relatively low respiratory Q_{10} values within the range of temperatures expected to occur in microhabitats occupied during their active seasons.

Acari. Mites (Fig. 16.9) make up the large majority of the desert soil microarthropod fauna, which also consists of Collembola and other small or immature insects. We know little about the taxonomic array and ecological roles of any of these animals in arid lands. Fortunately, the few comprehensive North American studies deal both with broad assemblages as well as with major species components. This approach is reflected to some extent in the following discussion, in which Collembola are mentioned relative to the spatial and temporal patterns of soil mites. Following a somewhat general ecological review, we shall look more closely at a unique group of acarines, the giant velvet mites.

Wallwork's (106, 107) nine-month examination of soil microarthropods in the southern Mojave (Joshua Tree National Monument) was the first of its kind on the continent. He, and others later, examined population attributes of the following groups: Cryptostigmata (Oribatida or Oribatei), Prostigmata, Astigmata, Mesostigmata, and Collembola. In comparing the relative abundances of these and associated pterygote insects in a consist-

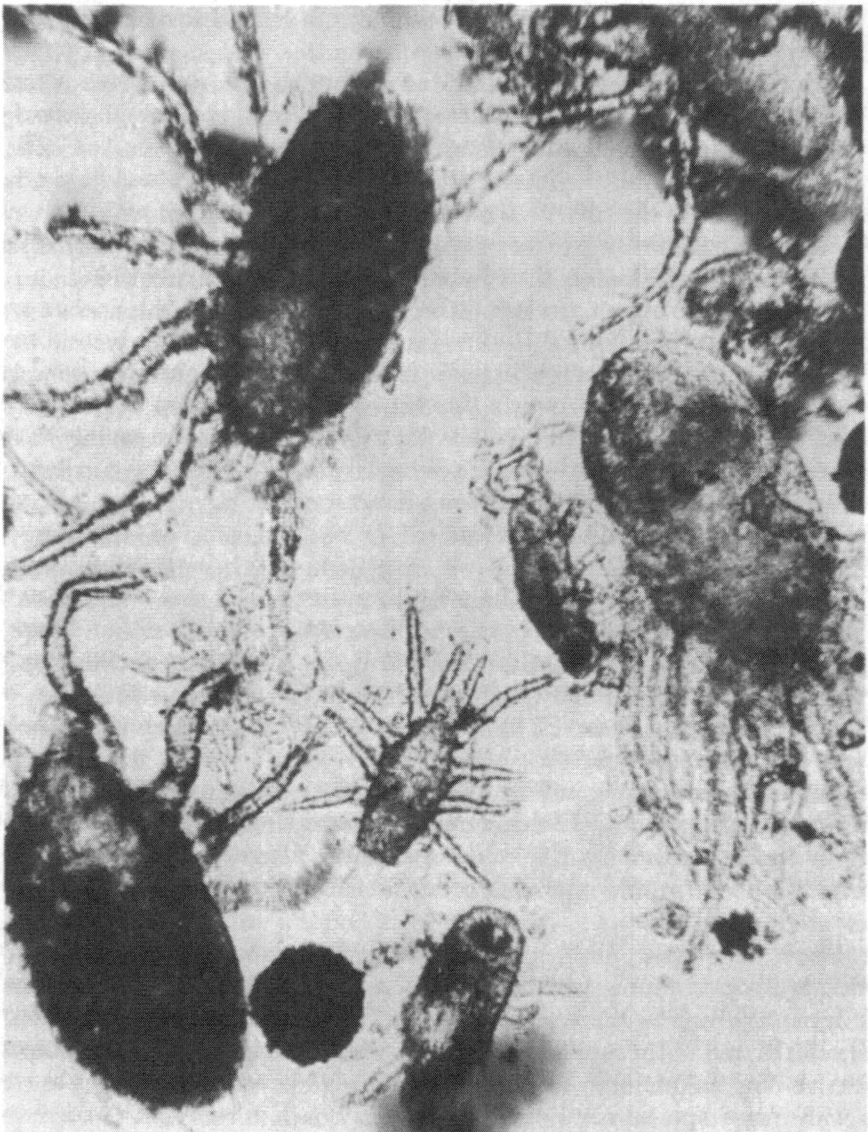

Fig. 16.9. Assorted soil mites from shrub litter. Jornada Experimental Ranch, New Mexico State University, south-central New Mexico. (Photo courtesy of P. F. Santos.)

ently studied dry site (juniper litter) with single collections from rotting wood (agave, *Nolina parryi*) and an oasis (in debris beneath *Equisetum*), Wallwork found that among the mites the major litter species were oribatids. This contrasts with an observation by Whitford, cited in Santos et al. (108) that oribatids made up only 12% of the microarthropods under *Juni-*

perus monospermae in central New Mexico. Wallwork (106) also noted that several mite genera from juniper litter were widely dispersed in arid and semiarid parts of North America, yet he found no evidence of faunal interchanges among relatively close study sites. He concluded that the distribution of these organisms in deserts is influenced by ecological rather than geographical factors.

A comprehensive study of soil microarthropods beneath desert shrubs in the northern Mojave desert's Rock Valley was reported by Edney et al. (109, 110) and Franco et al. (111), who found that shrub size and species did not influence faunal densities. However, decreases in density did occur with depth and distance from shrub base. Prostigmatids, not oribatids, were the most abundant mites in this location (110), although at times the latter accounted for more than half of the total arthropods present (111). These findings are generally consistent with a tendency for oribatid/prostigmatid ratios to increase as soils become drier and more mineralized (108, 112, 113).

Increased rainfall and moderate temperatures during late autumn and early winter months in the Mojave appear to increase microarthropod populations (111). Samples taken at 10 cm depth intervals showed that except for June and September (hot, dry months), densities were highest near the surface (110, 111). Densities also correlated positively with concentrations of soil carbon. While vertical distribution was associated with temperature, artificial watering confirmed that moisture was an important factor controlling density (110).

Seasonal changes in vertical distribution are probably caused by a number of events. The argument that optimal temperature levels are achieved by vertical migration makes sense intuitively, as does an explanation invoking seasonally varying birth/death ratios at different depths (110). Wallwork (107) inferred from the presence of gravid female mites in mineral soil and from generally greater abundance of immature stages in the litter above, that juveniles disperse upward, in a species-specific manner, to the warmer litter where development rates are enhanced. Migration as an explanation for vertical distribution was further supported by observations that litter mites were comparatively mobile species, and that higher mineral-soil densities coincided with midday sampling (107).

In the Chihuahuan desert Santos et al. (108) obtained prostigmatid: oribatid ratios that corresponded with most previous desert studies, although in one part of a topographical transect oribatids were most abundant. Weight of surface litter proved to be an excellent predictor of density.

The roles of soil microarthropods in desert ecosystems are surely geared to climatic and seasonal changes, and also to the population fluctuations and trophic activities of different species. Oribatids, astigmatids, and Collembola are usually considered primary consumers (mainly saprovores and fungivores), and mesostigmatids are most likely carnivores (113a). However, while Wallwork (107) considered prostigmatids to be carnivorous, Edney et al. (109, 110) classified them in the approximate nonpredator/predator ratio 5 : 1. With the addition of Collembola and oribatids [as calculated from Edney et al. (109, Table 7)], and ignoring mesostigmatids which were not treated as a discrete group, the nonpredator/predator ratio at Rock Valley becomes 8 : 1. This corresponds roughly to Wallwork's (107) overall ratio of 4.6 : 1, but not to his analysis for litter, where approximately equal values occurred, nor for that of mineral soil, where the ratio was 19 : 1.

Annual expenditure of respiratory energy among Rock Valley microarthropods apparently does not reflect trophic-group ratios. Edney et al. (110) estimated metabolism in cal/m^2 as follows: predatory mites, 151.64; saprophagous mites, 24.80; Collembola, 67.74. Two prostigmatid families dominated the collective metabolism at their respective trophic levels. The predatory Caligonellidae contributed 46%, and the saprophagous Nanorchestidae—a very widespread group in arid regions—accounted for 51% (compared to 20% by oribatids). These percentages were derived from Table 7 in Edney et al. (109), from which it was also determined that calligonellids and nanorchestids, respectively, comprised only 8.3% of the total mite density. Although aware of potential errors in their estimates, the authors tentatively assessed respiration by soil microarthropods (and tenebrionid beetle larvae) at about 0.2% of the annual energy input into the Rock Valley ecosystem. If this figure is reasonably accurate and representative, the ecological significance of soil microarthropods in deserts may well have more to do with their control of decomposition rates than with their actual consumption of food materials.

Giant red velvet mites (Prostigmata : Trombidiidae) are striking arachnids that appear on desert surfaces after rains (Fig. 16.10). Their brilliant crimson or crimson and white coloration may be aposematic because they are rejected by invertebrate predators (14), although caged scrub jays will eat them (personal observations).

Several species of these broadly distributed acarines were described from the Southwest by Newell and Tevis (115), who reported densities approximating 4000 per square meter in about a hectare of the Sonoran desert near Tucson. An intensive study by Tevis and Newell (116) of *Dinothrombium pandorae* in the Mojave desert forms the basis for the following treatment of velvet mite life history.

Whereas summer storms trigger surfacing of species in the Sonoran (115) and Chihuahuan (personal observation) deserts, adult *D. pandorae* in the Mojave respond only to winter and spring rains of at least 0.6 mm. Most individuals emerge once in a season and remain on the surface, in bright sunlight, for 2.5 hours at the most. Emergence coincides with swarms of termites, which serve as prey for adults.

Courtship, but not insemination has been observed on the surface. Burrows are dug only in fine sandy soil. Vertical in structure, each burrow

Fig. 16.10. Giant velvet mite (probably *Dinothrombium* sp.) on soil surface following rain. Big Bend National Park, Texas. (Photo by author.)

houses a single mite. Digging is done with the pedipalps. Mites exhibit differential vertical movement in burrows according to temperature. The vast majority remain at the bottom (mean depth 17 cm, range 11–20 cm) during summer, when deep burrow temperatures reach 42–46°C. Mites move closer to the surface during winter, reducing the size of the overhead sand plug as they ascend.

Several thousand eggs per female begin to appear in burrows toward the end of April, following an estimated gestation of 11 weeks. Eggs incubate for just over a month, then hatch into larvae that remain in the burrow for a short time before crawling out to search for grasshoppers, which they parasitize. Eventually larvae probably drop off their hosts and enter the soil for further development. Postadult molts may occur; adult females stay alive after oviposition and may emerge the following year. Iteroparity may therefore apply to these spectacular mites.

Diplopoda

Millipedes can be abundant in certain North American desert habitats, but like many other soil-associated invertebrates, they are infrequently seen. At least three orders are represented. A single species of Cambalidae

(Leioderidae) was collected at the Nevada Test Site by Chamberlin (117). In the Spirobolida the family Atopetholidae is apparently endemic to the Sonoran region of Mexico and to the southwestern United States (118). Species in four subfamilies are reported from regions that should be generally arid, but microhabitats have seldom been disclosed. Exceptions in this regard are three atopetholid species from the Nevada Test Site; they were winter-active and found in various desert shrub communities (117, 119). Two of them were reported as being numerous. Fairly large numbers of moderately large, gray atopetholids can also be seen on the surface after summer rains in the northern Chihuahuan desert (west Texas); similar millipedes occur on the surface after summer rains in central New Mexico. They seem most abundant under shrubs in heavily vegetated ravines, and I have also found them infrequently in the soil.

The largest and most common desert millipedes in North America are in the order Spirostreptida. They belong to the genus *Orthoporus* (Spirostreptidae), with ranges south to Argentina (120). Causey (121) refers to four *Orthoporus* species in North American deserts; however, the ranges of two of these are confined to comparatively mesic regions. Another spirostreptid, *O. flavior,* appears restricted to parts of the eastern Chihuahuan desert. By far the most widespread and abundant species is *O. ornatus* (Fig. 16.11), which occurs throughout the Chihuahuan and Sonoran deserts. It is the only desert millipede to have been studied from an ecological standpoint. A résumé of its biology, which is summarized elsewhere in greater detail (122), is now presented.

Most of the life of *O. ornatus* is spent underground. For perhaps two-thirds of each year the millipedes are tightly coiled in soil, often beneath

Fig. 16.11. *Orthoporus ornatus* feeding on detritus. Central New Mexico. (Photo by author.)

rocks or in harvester ant (*Novomessor cockerelli*) nests, depending on population location. Frequently they are highly aggregated, a tendency that carries over to the behavior of surface-active animals when not feeding.

Metabolism is reduced in dormant specimens, especially just prior to annual molt (123). Molting takes place in June and July in New Mexico, many individuals molting within a week of each other. A large proportion of the population then tunnels to the surface following the first heavy summer rains, which usually arrive a few weeks later and wet the soil sufficiently to make this upward movement possible. In Big Bend National Park (Chihuahuan desert), where mature individuals are much larger and populations are polymorphic in coloration, emergence coincides with heavy rains in May. Second-instar larvae encased in fecal egg pellets (about 100–500 pellets to an underground egg chamber) also emerge from these structures as summer rains begin, but these and other early instars seldom surface.

Like most other millipedes *O. ornatus* feeds on detritus. In doing so it also ingests soil and fungi, which along with bacteria contribute to its bulky gut contents. With its relatively high rate of assimilation (124) it makes efficient use of ingested food, and by the end of the feeding season has increased in dry weight by about one and a half times (125). A relatively dense population at Big Bend may consume about 0.2% of the annual net primary production of its ecosystem, a value equivalent to that estimated for soil microarthropods and tenebrionid larvae in the Mojave desert (110).

During its active months *O. ornatus* is exposed periodically to the physical stresses of the desert surface, yet it avoids temperatures above 35°C by differential movements in vegetation, on the surface, and in mammal burrows (126). By such behavior potential desiccation is also circumvented. Maintenance of optimal water balance at this time is enhanced by exceptional resistance (for a millipede) to evapotranspiration and by rectal and oral imbibition (127). Adult females in particular cope with desiccation in the summer by fairly strong osmoregulation (48).

Over an entire year water volumes in the gut and body proper undergo measurable variation (128). Gut water in summer greatly exceeds that of winter. As spring approaches body water increases markedly, presumably because of transcuticular uptake against a vapor-pressure gradient. Hemolymph osmolality increases by half again in the winter, and the proportions of major hemolymph ions, as well as of total hemolymph animo acid and protein, shift throughout the year (129).

Reproductive and demographic patterns are poorly known in *O. ornatus*. Mating is seldom seen but occurs in summer. Ova that were immature at time of surface emergence are ready for oviposition about a month later. Iteroparity is probable because gravid females span a rather large size range. Neither longevity nor instar number have been estimated precisely, but size increases at molt suggest that individuals may live for about a decade. Population mortality rates may be highest within the egg pellet, judging from the large numbers of pellets that contain egg or larval remains but have no emergence holes. Although spirostreptid millipedes secrete defensive benzoquinones from their repugnatorial glands (130), they are nevertheless preyed upon by a variety of vertebrates, a phengodid beetle, and probably by certain ants as well (unpublished data).

Larvae probably hatch several months after oviposition. Following a molt some time before the next summer the second instar feeds on its fecal covering, most likely gaining a gut flora in the process. Fecal pellets can reduce moisture gradients between larvae and the surrounding soil, but osmoregulation seems to be the most effective way for a larva to protect itself against fluctuations in soil moisture (131).

Chilopoda

Not usually considered desert animals, centipedes are indeed found in xeric regions. Distributional and taxonomic treatments of North American taxa, some of which are desert inhabiting, are available (132–135). Unfortunately, these publications seldom refer to actual habitats. Much of the following brief discussion is therefore based on what is known of centipedes elsewhere, on a few actual studies of desert centipede biology, and on personal observations and anectodal information.

Of the four chilopod orders the Lithobiomorpha contains the smallest individuals. Occurring beneath rocks and under bark of fallen logs in arid southwestern woodlands, few lithobiomorphs are found in true deserts, although several species occur in shrub communities at the Nevada Test Site (119). Whether members of the rapidly moving Scutigeromorpha should be termed desert centipedes is debatable. Species in this order are known from dry parts of west Texas (133), but habitat details were not reported. We also know little about the slender, wormlike Geophilomorpha, essentially a deep-soil group, some of which live beneath desert shrubs in North Africa (136).

Some species in the Scolopendromorpha qualify as desert inhabitants. Members of the family Cryptopidae are found under rocks in dry mountains of the southwest (personal observations), while larger-bodied species in the Scolopendridae occur in true desert soils. In the latter family, *Scolopendra heros, S. polymorpha* (Fig. 16.12), and perhaps *S. viridis* are among the most representative in North America (135). The first reaches 15 cm in length and is obviously well adapted to ambient dryness. I have seen it walking on the desert surface during hot afternoons in New Mexico. While the biology of *S. heros* has not been studied, the ecology of *S. polymorpha* is now slightly understood, and is summarized below.

This wide-ranging species is smaller in size than *S. heros,* and according to aktograph studies demonstrates circadian control over its strictly nocturnal activity (137). Compared with west African scolopendrids (138) it is relatively resistant to water loss. Its distribution in regions with cold winters

Fig. 16.12. *Scolopendra polymorpha* after removal of sheltering rock. Peloncillo Mountains, New Mexico. (Photo by author.)

suggests that it should become cold-hardy; however, it appears unable to adapt physiologically to seasonal cold, although single supercoolings are survived (139). Its mortality due to winter cold should be similar to that of the scorpion *Diplocentrus peloncillensis* (see above) when *S. polymorpha* overwinters at comparable soil depths (58).

Like some other centipedes from temperate regions (138) this species probably matures in 2 years and may live several years longer. The number of broods it has in a season or in its lifetime is unknown. Personal observations and studies of other scolopendromorphs (140, 141) suggest relatively low fecundity, as well as brooding by the female parent. The survival curve of *S. polymorpha* should therefore be rather gradual compared to that of desert millipedes. In New Mexico its predators include burrowing owls (137) and scorpions, the latter encountered because of its daytime habit of crawling beneath rocks and other shelters (personal observations). Its own prey items are probably quite varied, but have not been carefully documented in the field.

Insecta

Soils of North American deserts contain members of the apterygote, orthopteroid, hemipteroid, and holometabolous orders. These insects contribute greatly to the taxonomic, morphological, and ecological diversity of desert soil fauna.

Apterygota. Two orders dominate the primitively wingless insects in desert soils. These are the Collembola and the Thysanura. Collembola were mentioned earlier in connection with mites because together these constitute the great body of desert soil microarthropods. In general, the springtails are less numerous than mites in deserts. Moderate numbers were collected in the Great Basin by Fautin (8) and at the Nevada Test Site by Allred and Beck (142). Wallwork (107) obtained an average of 110 per square meter (three species) in juniper litter at Joshua Tree National Monument between October and June. Edney et al. (110) found that winter densities can exceed 800 per square meter at Rock Valley. Scott (143) made a thorough distributional study in New Mexico and found relatively few species associated with plants in arid soils. In a transect collection, Collembola were extracted only from the unvegetated center of a large wash and from beneath shrubs at its edge at the Jornada Site in southern New Mexico (108).

Broadly classified as detritivores, springtails are also known as algavores and even more so as fungivores (144). One might therefore expect their population densities to increase with relatively moist conditions; organic debris should then be more decomposable, and algal and mycelial growth more rapid than when soil is dry. Wet winter months at Rock Valley in the Mojave are in fact associated with greatest annual densities of Collembola there (109). Preliminary evidence suggests that this pattern also applies in southern and central New Mexico (P. F. Santos, personal communication; also personal observations). Abundance in these arid regions may therefore be a function of relatively low temperature and moist soil.

Population metabolism, in contrast, is maximal at Rock Valley in August, when abundance is well below that of winter. A dramatic biomass increase between June and August helps to resolve this seeming paradox, as do high summer temperatures that promote increased metabolic rates. The Rock Valley studies may indicate that reproduction intensifies with the onset of winter rains, and that consumption rates depend strongly on temperature. Seasonal activity patterns of desert Collembola may also reflect the effects of carnivorous mites, which according to Christiansen (144), are their main predators. More precise explanations of these trends are needed if the roles of desert springtails are to be well understood.

Despite their small size, some Collembola are comparatively resistant to water loss (145, 146). It has recently been reported that, like nematodes, some Collembola can undergo anhydrobiosis (147). The somewhat larger, but often more surface-active Thysanura are also well adapted to xeric environments, and in southern Africa's Namib Desert *Ctenolepisma terebrans* more than compensates for water loss in dry air by absorbing water vapor from relative humidities above 47.5% (148). Comparable studies in North America are not available.

Desert thysanurans are nevertheless widespread on this continent. Beck and Allred (149) report them from the Great Basin desert and Wygodzinsky (150) discusses species of Lepismatidae that occur in the southwest. Two of the southwestern species are considered native, while others are probably introduced. Members of the Machilidae also occur in the region's soils and litter (personal observations). *Leucolepisma arenaria* is an example of a North American desert lepismatid that is cryptic by day but becomes surface-active at sunset. This behavior is not uncommon among machilids and lepismatids in general; however, it is restricted to nonmyrmecophilous and nontermitophilous species (P. Wygodzinsky, personal communication). In central Asia, desert lepismatids occupy a variety of feeding niches; their densities peak in the fall and they may live 4–5 years (151). Their ubiquity in North American deserts, where they are encountered in a variety of microsites, including harvester ant nests (personal observations), warrants comprehensive ecological study.

The Orthopteroid Orders. Members of orders discussed in this section typically spend most of their lives in soil (often sand) or beneath rocks. Grasshoppers and locusts (Orthoptera : Acrididae, Tetrigidae, Tanaoceridae, Eumasticidae, and Tettigoniidae) are therefore omitted; they oviposit in soil and sometimes use burrows and crevices for shelter, but they feed and develop on or above the soil surface. Also excluded from serious consideration are the earwigs (Dermaptera), webspinners (Embioptera), and booklice (Psocoptera). While in theory these qualify for discussion, there is really little to say about them. Helfner (152) lists a single species of *Spongovostox* earwig from the desert southwest. Beck and Allred (149) report webspinners from the Nevada Test Site and I have collected them from beneath rocks in southwestern New Mexico. Borror et al. (153) state that a species of *Chelicera,* a large webspinner genus in Mexico (154), occurs in the region. Pscocoptera are not uncommon in soil and litter under shrubs in central New Mexico (personal observations), and were collected at the Nevada Test Site (149). They may well be a significant food-web component.

Soil-associated desert orthopteroids qualifying for more intensive treatment include the sand and camel crickets (Gryllacrididae : Stenopelmatinae and Rhapidophorinae, respectively), the true crickets (Gryllidae), and the cockroaches (Blattoidea). Remarkably little is known about the ecology of these insects in desert ecosystems.

The sand cricket *Stenopelmatus cahuilaensis* inhabits dunes in California's Coachella Valley (Mojave desert). Tinkham (155) suggests that it takes $2\frac{1}{2}$ years for this species to reach reproductive maturity. Other species of these

Fig. 16.13. *Stenopelmatus* sp. digging in heavy soil. Central New Mexico. (Photo by author.)

robust and rapidly burrowing omnivores are common underground and also beneath rocks elsewhere in the southwest (Fig. 16.13).

Somewhat more seems to be known about camel-cricket biology. Species called "sand treaders" have a "sand basket" of long spines on the hind tibia. This structure facilitates their swift digging as they move through the substrate. By day these insects live in sand dunes; at night they emerge and feed (152). Ovipositon occurs in spring (155). Species of *Ammobaenetes* are common in gypsum sands and other dune systems in central New Mexico (Fig. 16.14). *Macrobaenetes velgum* lives in dunes in the Coachella Valley and *Daihinoides larvale* is a whitish species without a sand basket that inhabits White Sands, New Mexico, and feeds on fallen *Ephedra* cones (152). Several other species may be found under rocks and wood, and in mammal burrows in arid North America. The genus *Ceuthophilus* is widespread in the southwest, *C. variegatus* being a large species often seen at night on the desert floor at Big Bend National Park (156).

Encountered in many of the same microhabitats as camel crickets are gryllids comprising field crickets (Gryllinae), ground crickets (Nemobiinae), and short-winged crickets (Mogoplistinae). Tinkham (156) lists species of these gryllid subfamilies in the Trans Pecos region and suggests that ant-loving crickets (Myrmecophilinae) should be found there too. Helfner (152) states that *Myrmecophila manni* occurs in ant nests throughout much of the arid West.

Cockroaches are far more abundant in deserts than was formerly realized, particularly in North Africa (136). Not often observed because of their nocturnal surface activity, species of Blattidae and Polyphagidae are well known in deserts of North America. Endemic desert blattids, however, including *Parcoblatta desertae* from the Chihuahuan desert (156), have received comparatively little attention. The arid-mountain polyphagid *Arenivaga grata* is found under rocks (152), as is *P. desertae*. *Eremoblatta subdiaphana* is a widespread and relatively hairy desert polyphagid (152, 156).

Among North American desert cockroaches, the genus *Arenivaga* (Fig. 16.15) has been the most seriously studied. A number of species are found in or adjacent to the Chihuahuan, Sonoran, and Mojave deserts (156–158). In the Mojave, members of the genus spend most of their lives burrowing in sand dunes, where at least one species feeds on decaying leaves and clumps of adventitious *Prosopis* roots with mycorrhizae and fungal hyphae attached (159). Leaves eaten by *A. investigata* include those of *Larrea, Dalea,* and *Atriplex* (160), and the peculiar subterranean "swimming" of this cockroach may allow it to locate them when they accumulate near the sand surface.

Seasonally and diurnally defined vertical locomotion is typical of *Arenivaga* spp. in the Mojave. Such behavior enables *A. investigata* to remain continually exposed to subsurface humidities greater than 82% relative

Fig. 16.14. *Ammobaenetes* sp. digging in sand dune. Sevilleta National Wildlife Refuge, central New Mexico. (Photo by author.)

Fig. 16.15. *Arenivaga* sp. about to dig in sandy loam soil. Central New Mexico. (Photo by author.)

humidity (160), under which conditions both nymphs and adult females can absorb water vapor (161). Absorption is probably accomplished by protrusible bladders on either side of the mouth (162). Orientation to substrate moisture is by orthokinesis and is mediated by antennal sense organs (163), while spatial positioning in sand is mediated by equilibrium receptors on the cerci (164).

Edney (165) found that *Arenivaga investigata* osmoregulates effectively when desiccated. Dead specimens lose water at lower rates than the cosmopolitan species *Periplaneta americana,* especially at elevated temperatures (166). When the population biology of *Arenivaga* spp. becomes documented, this genus will serve as a model for the research efforts of many desert ecologists.

The Hemipteroid Orders. Desert soils provide habitats for relatively few hemipteroid insects, including the order Hemiptera. The burrower bugs (Cydnidae), however, are exceptions. Most of these small, stout insects feed on plant roots (167), and *Microporus obliquus* is said to inhabit roots in sandy areas, including dunes. It has been collected from the Great Basin, Sonoran, and Chihuahuan deserts, while *Melanaethus noctivagus* is known from the Mojave as well (168). Species in at least five genera are recorded by Froeschner (168) from North American deserts, but aside from the assumption that all are burrowers, nothing seems to be known of their biology.

In the order Homoptera, species of Cicadidae are long-term inhabitants of desert soils. A rich cicada fauna is found in western North America and is described in a long series of papers by W. T. Davis in the *Journal of the New York Entomological Society* (1915 to 1941). On hot days in southwestern deserts the shrill singing of adult males in trees and shrubs is not unusual, depending on the area and time of day. Unfortunately, what little is known of the biology of these interesting insects comes from studies of the adults. Examples are those of Heath and Wilkin (169) with *Diceroprocta apache* and of Heath et al. (170). with *Cacama valvata.* Immature stages in deserts probably feed on roots of perennial plants and live at least 4 years, these traits being characteristic of the Cicadidae in general (153).

Neuroptera. Several neuropteran families are associated with soils in arid parts of North America. Included here are the mantidflies (Mantispidae), which as hypermetamorphosing larvae parasitize spider egg sacs and probably solitary-bee larvae as well. Mantidflies are known from the Nevada Test Site (149), central New Mexico (personal observations), and undoubtedly other arid parts of the continent. Also included are the owlflies (Ascalaphidae), whose larvae resemble ant lions although they remain on the soil surface while hunting (153). However, the most appropriate family in the context of this chapter is easily the Myrmeleontidae (see Fig. 16.4), whose ant lion larvae actually live in the soil. A key to the larvae of North American species is provided by Stange (171).

Several predatory patterns are used by myrmeleonid larvae. Best known is that of digging conical pits, which are characteristic of the genus *Myrmeleon,* a widely distributed group in much of North America and elsewhere. Pits are dug in sandy soil. In southeastern Arizona *M. immaculatus* accomplishes this by flicking sand as it circles along the bottom of a "digging groove" (172). In South Africa *M. obscurus* possesses a circadian pit-building activity rhythm that peaks at dusk, when surface temperatures drop sufficiently to permit larvae buried below the pit base to come to the surface, reconstruct eroded pits, and feed (173). Following pit construction larvae lie concealed at the base and wait for prey to fall over the edge.

The South African study probably applies broadly within the Myrmeleontidae, as do two others from a dry region in Costa Rica. There, Wilson (174) determined that both ant lion size and maximum prey size were directly related to pit diameter. This observation tends to reinforce the finding of Youthed and Moran (175) that in general the biggest pits are constructed by the largest ant lions. McClure (176) demonstrated that *Myrmeleon* sp. in Costa Rica optimizes prey capture by constructing pits at maximum distances from neighboring pits when densities are high (five larvae per 100 m^2); this results in a uniform distribution of pits.

The largest group of North American myrmeleontids consists of small-to-medium-sized ant lions that are most abundant in the arid southwest (171). They belong to the tribe Brachynemurini which includes five genera comprising 69 species. The following comments are based largely on Stange's (171) account of *Brachynemurus* in the Mojave and Sonoran deserts.

Eggs are laid at dusk just beneath the sand surface. Larvae do not make pits, instead they lie with spread mandibles (actually, united mandibles and maxillae) directly below the surface where a weak depression is usually evident around the hidden mouthparts. Sometimes larvae emerge, run rapidly to a new location, and quickly dig into the sand. As Wilson (174) noted, presumably with *Myrmeleon, Brachynemurus* larvae are also indiscriminate feeders as regards prey variety. They are very sensitive to vibrations made by potential prey, often surfacing quickly and running toward prey when vibrations are detected. Capture is facilitated by a paralytic secretion transmitted through ducts in the sickle-shaped mouthparts.

Larval *Brachynemurus* respond to low temperature and excessive moisture by burrowing 7–8 cm and then becoming inactive, sometimes for many weeks. Activity quickly resumes when less severe conditions return. Pupation occurs in a silken cocoon beneath the sand, and adults emerge in about 3 weeks. Most species appear to be univoltine, but *B. sackeni,* which is the first of the year to appear at light traps near Palm Springs, may have two annual generations.

Coleoptera. As adults, desert beetles usually feed on or above the soil surface, although they often seek shelter beneath surface debris, under rocks, or in burrows. In contrast, the immature stages of some desert Coleoptera develop in soil; in this section we treat such taxa by dividing them roughly into three trophic categories: herbivores, carnivores, and detritivores.

Most subterranean herbivores are root feeders, except for those that consume algae and fungi. While the literature on this subject is scanty for desert Coleoptera, it can be surmised from more general accounts that the root-feeding habit applies to a number of the click beetles (Elateridae) that feed so extensively on roots of grasses and grains, certain leaf beetles (Chrysomelidae: Gallerucinae) that specialize on roots of curcubits, and the weevils (Curculionidae), many of which burrow. Some weevils are apparently endemic to certain western sand-dune systems (177). The few instances of documented root feeding by North American desert Coleoptera include those of larvae in the small family Monomidae, which bore into root stalks of *Agave lecheguilla* (178), and larvae in the genera *Mecas* and *Crossidius* (Cerambycidae), which consume roots of the desert shrubs *Artemisia* and *Chrysothamnus* (179). Many other subterranean beetle larvae undoubtedly have similar feeding niches in North American deserts.

Carnivorous beetles are better known, but in most instances their bionomics are poorly documented in desert soils. A good example of our ignorance involves the enormous ground-beetle family (Carabidae), species of which are probably abundant in all of the world's nonpolar deserts. Most carabids are predatory as larvae and adults, the former being found frequently in burrows or beneath debris (153). In parts of the western United States, carabid larvae can destroy at least 3% of the grasshopper egg pods in a given area (180). However, their collective role in desert ecosystems remains obscure.

Tiger beetles (Cicindelidae) are carnivores in the same suborder (Adephaga) as carabids and also inhabit the major North American deserts. The genus *Cicindela* (Fig. 16.16) predominates in the southwest (181). Some tiger-beetle larvae require moist soil for their vertical burrows, within which they wait just below the entrance for their prey. *Cicindela* spp. of this type are found near wet playas (182). Another ecological group lives in sand dunes and includes *C. limbata albissima* as well as morphologically similar species. The former species burrows into the lee sides of very large dunes in southeastern Utah, according to Rumpp (182), who provides some life-history information. Overwintering adults of this species emerge from hibernation and probably mate in spring. Since a new generation of adults emerges in the fall, larval life must occur only in the warmer months when prey are obtainable.

Among carnivores in the larger suborder Polyphaga, the blister beetles (Meloidae) (Fig. 16.17) stand out as an important component of the coleop-

Fig. 16.16. Tiger beetle (probably *Cicindela* sp.) on sandy bank of Rio Grande. Central New Mexico. (Photo by author.)

Fig. 16.17. Meloid beetle (probably *Cysteodemus wislizeni*). Central New Mexico. (Photo by author.)

terous fauna in xeric regions of the world (183). Unlike carabids and cicindelids, only the larvae of meloids are predaceous, adults being usually specialized herbivores. The account of the hypermetamorphic life history of meloids in the southwest by Werner et al. (180) serves as a partial basis for the following description of their desert biology.

Active triungulin (first-instar) larvae are the host-seeking stage. Probably all triungulins in the tribe Epicautini locate grasshopper egg pods on which they feed by first wandering over the soil surface. Other groups feed on eggs and stored food of wild bees that nest in the ground. This is accomplished in the Lyttini also by wandering, but in other groups the triungulins attach themselves to flower-visiting bees and are transported aerially to the nests. Subsequent molting to a grublike stage occurs after a certain amount of feeding at the host site has taken place. Energy is now spent for growth rather than for dissemination, and further molts also result in immobile instars except for the sixth, which has a hardened cuticle and becomes dormant. In dry soil the "pseudopupa" may well persist for several years. Eventually it molts again to a grub form, and later it pupates. Adults are short-lived (1–2 months), like much of their plant food.

Larvae of the wedge-shaped beetles (Rhipiphoridae) undergo a hypermetamorphosis similar to that of the Meloidae. Larvae are parasitic on various wasps and bees (153) and should be expected in the burrows of these insects in North American deserts.

Rarer than the Meloidae and less widely distributed are members of the glow-worm family (Phengodidae). In deserts and other parts of western North America larvae of *Zarhippus* spp. (Fig. 16.18) live in soil and prey on millipedes, which they kill and eat in a behaviorally unique manner. I have observed them hunting *Orthoporus* millipedes in west Texas and southern New Mexico (Chihuahuan desert), but never in central New Mexico. Tiemann's (184) description of *Zarhippus* life history mentions that first instar larvae are active and may not feed. Their longevity is not known, but larvae in the absence of food become quiescent for indefinite periods. The last larval instar in females is much larger and more strikingly pigmented than in males. While males undergo pupation and become winged adults, females become colorless larvaform pupae that molt into large and perhaps nonfeeding larvaform adults.

Larvae of several well-known families of soil-associated Coleoptera range in feeding behavior from broadly defined detritivory to more restrictive feeding. The latter includes necrophagy, coprophagy, fungivory, and algavory. Conceivably the rove beetles (Staphylinidae) should be included

Fig. 16.18. Last instar larva or adult female of *Zarhippus* sp. Big Bend National Park, Texas. (Photo by author.)

in this group, although they are usually considered litter dwellers, rather than true soil forms. The same applies to the Lathridiidae, a family of fungivores known from litter in desert areas east of the Sierra Nevada (185). More appropriate families include those of the scarab (Scarabaeidae) and darkling beetles (Tenebrionidae). These and two smaller groups will now be discussed in the context of their known biology in North American deserts.

Among the Scarabaeidae several genera in the subfamily Aphodiinae are psammophilous, or sand-loving (186). These include a few probable detritivorous *Aphodius* species from Colorado and Utah, and from Chihuahua (187). Other members of the genus from the Great Basin and more southern locations inhabit rodent burrows (188, 189). Still other scarabs are myrmecophiles in nests of desert harvester ants (C. N. Slobodchikoff, personal communication). Quite a few scarabaeids are called "dung beetles" because they roll and then bury balls of vertebrate fecal material, on which they oviposit. Resulting larvae are simultaneously provisioned and sheltered. Gordon and Cartwright (190) suggest that prairie dog dung may be the excrement preferred by *Boreocanthon praticola* (Scarabaeinae); cow dung is definitely used by other members of this subfamily in desert regions. Preformed dung pellets serve as food for species in some genera, but large-bodied species of *Canthon* construct their own dung balls instead. Members of this genus are also attracted to fungi and dead animals (190).

Adults of many other scarabaeids have been collected from western sand dunes [see Hardy and Andrews (191) for a partial list of pertinent literature]. Habits of their larvae remain unknown, but detritivory is suspected. Some species seem to be endemic to certain dune systems (191).

A ruteline scarab, *Paracotalpa puncticollis,* is suspected of feeding on juniper litter since its adults emerge from such a substrate throughout the southwest (192). In the related stag-beetle family (Lucanidae) several species apparently develop and pupate in wood of juniper (in southeastern Utah), paloverde, or mesquite (in southern California) buried by sand dunes (193).

Carrion beetles (Silphidae) should also be mentioned as desert detritivores in the broad sense because species of *Nicrophorus* are occasionally seen in southwestern arid lands. Adults excavate beneath dead vertebrates, allowing each carcass to sink into the ground. Eggs are then laid on the dead animals, which are eventually eaten by developing larvae and adults (153).

Tenebrionid beetles (Fig. 16.19) are among the best "indicator species" of deserts (194). Large numbers populate most deserts, including those of North America, where a number of species are either subsurface dwellers (many Tentyrinae) or live in vegetation as adults (C. N. Slobodchikoff, personal communication). Yet while a rapidly growing literature now reveals much about the taxonomy, distribution, physiology, behavior, and ecology of these insects, it applies mainly to adult beetles, which are essen-

Fig. 16.19. *Eleodes caudifera* entering a burrow. Central New Mexico. (Photo by author.)

tially surface-active. Furthermore, few complete life-history studies of desert tenebrionids have been published, although certain trends can be pieced together from some of the more ecologically oriented papers. One broad tendency is that of habit preference by individual species (195–197). Another is the seasonal succession of these flightless beetles on the desert surface over the course of a year (198–202). This pattern is not one of simple replacement; rather, it is complicated by varying lengths and numbers of annual activity periods. As measured by trap collections, collective abundance tends to peak in spring and fall, although a few species are caught all year long.

Longevity of both immatures and adults varies considerably, depending on the species and perhaps on microclimate and food. One or two years of larval life have been recorded for some species (203, 204), while adults can live between a month (204) and at least 3 years (205). Doyen (203) suggests that a long adult life span is typical of the family, and C. N. Slobodchikoff (personal communication) asserts that species' life spans may be short (2–4 weeks), intermediate (2–6 months), or very long (1–5 years). Many members of the family may be *K*-selected (196).

Reproduction in tenebrionids seems geared to endogenous patterns as well as to climate and topography. In general, their fecundity may be low compared with desert spiders and millipedes (see above); judging from estimates based on dissections or laboratory observations (202, 203), it may average about 50 eggs per female. However, Wakeland (206) gives a range of 240–367 for numbers of eggs found in *Eleodes hispilabris*, while *Philolithus* (= *Pelecyphorus*) *densicollis* is assumed to lay a single large egg each day, according to K. W. Brown (204). Eggs of several species with low mean clutch sizes (2–5 eggs per clutch) are longer than those of other species with larger clutches (C. N. Slobodchikoff, personal communication).

Periods of adult activity encompassing mating and oviposition sometimes shift at Rock Valley in the Mojave desert because of climate (202). There, reproductive success in several species is correlated with soil moisture produced by ephemeral summer rains. The influence of topography—and perhaps of biotic factors as well—is evidenced by the tendency of certain species to oviposit around the fringes of harvester-ant nests in parts of New Mexico and Texas (personal observations).

Some desert species may estivate (199), and winter dormancy is characteristic of many populations. In New Mexico I have found larvae overwintering in harvester-ant nests, rodent burrows (nesting material of kangaroo rats often contains larvae), and in soil beneath shrubs and rocks. Overwintering by adults was recorded for *Eleodes* in the Mojave desert (205), for *Philolithus* and *Stenomorpha* in the shrub steppe region of southeastern Washington (195), and for *Coelocnemis* from arid mountains in California (203).

Precipitation has a variety of effects on the activity and abundance of adults, and must therefore indirectly influence demography in entire populations. Rainfall is significantly correlated with abundance of some species (adults) in the Mojave during a given year (202), while its effect appears postponed for 2 years in *Philolithus* in southeastern Washington (204). That the effect of precipitation is not via primary production is evident from Thomas's (202) study, which suggests that it acts by increasing soil and detrital moisture.

Both larval and adult tenebrionids seem to have generalized nutritional requirements (203). Many authors consider the larvae to be detritivores, and several workers have clearly demonstrated that a variety of organic material eaten in the laboratory results in successful development. An interesting relationship between use of defensive secretions by adults and their feeding on harvester-ant trash heaps was shown recently by Slobodchikoff (207). He found that three *Eleodes* species with defensive secretions were much more abundant on *Pogonomyrmex barbatus* mounds than were two sympatric *Stenomorpha* spp. in northern Arizona. Apparently, the ability to repel attacking ants effectively allows the secreting beetles to utilize this rich source of detritus.

If the "detritivore" designation is valid, very dense populations can be expected to have ecologically significant effects on their desert environments. While densities of adults alone are difficult to estimate with reliability (208) and can change dramatically over a decade (209), there is little question that populations can indeed be large. For example, *Cryptoglossa* (adults) can achieve a biomass greater than the sum of all vertebrates in its community at Rock Valley (202). Even this value (275 g/ha) is several orders of magnitude less than those reported for other genera in desert steppe vegetation in Washington (204, 210).

Lepidoptera. Plant roots in deserts present a considerable biomass that can be utilized by herbivores. Reference has already been made to the relatively few known desert arthropods associated directly or indirectly with these structures. In the present section we see that species of Lepidoptera can be added to the list of root feeders, although some of the comments below are admittedly speculative. A great deal remains to be learned about subterranean herbivores in North American and other deserts.

Among the moths of this continent larvae of at least four families contribute to the root-feeding category. Genera of Pyralidae, especially in the subfamily Crambinae where grass crown and root feeding is common (210, 212), include desert grassland species of *Thaumatopsis* that become periodically abundant in parts of New Mexico (personal observations). In the Phyctinae numerous species attack cactus, and at least one, the widely distributed *Olycella subumbrellum,* sometimes extends its feeding tunnel into the underground bulb of *Opuntia junctolineela* (213). Mares et al. (214) report *Acrolophus* (Acrolophidae) from roots of *Prosopis.*

A number of cutworm genera in the Noctuidae are known for their destruction of plant roots. Extrapolation from distribution records and from some published information (215; R. W. Holland, personal communication), suggests that probable desert root-feeding genera include species of *Agrotis, Chorizagrotis, Copablepheron, Crymodes, Luperina,* and *Peridroma.* Huge swarms of *Agrotis ypsilon* moths occasionally appear in late spring and early summer in New Mexico. Scorpions capture what are presumed to be their late-instar larvae during the early spring (W. A. Riddle, personal communication).

Larvae of the clearwing moths (Sesiidae) have a predilection for boring, and those of some desert-distributed species do so in roots. Duckworth and Eichlin (216) provide the following information: *Sphaeralcea* (Malvaceae) roots are hosts for *Zenodoxus* spp. in Arizona; *Oenothera* (Onagraceae) roots are eaten by *Euhagena nebraskae* in the Mojave; *Curcubita* (Curcubitaceae) roots are bored into by *Melitta* spp., which are probably found in all four major deserts; and *Pentstemon* (Scrophulariaceae) roots are hosts for *Penstemonia* spp., which, like *Melitta,* may take 2 years to develop.

Butterflies in the family of giant skippers (Megathymidae) also contribute some species to the guild of desert root borers. Roever (217) indicates that larvae of *Megathymus* spp. burrow into the root systems of *Yucca* in parts of southwestern North America. At least the last three instars are spent in roots; most species are univoltine.

Life-history patterns of root-feeding Lepidoptera in deserts probably vary considerably. Pyralid moths are well known for having from one to at least four generations per year, and we have seen that it may take several years for a single generation of sesiid larvae to develop. Average fecundities are also quite variable. Mature eggs of crambine grass moths probably average below 300 per female (218), while values for many noctuid species are two to three times as high (219). Crambines live at most for a few weeks as adults; in contrast the noctuid *Agrostis ypsilon* estivates during the summer in New Mexico and presumably reproduces in the fall. More life histories will have to be documented before generalizations can be made for desert species.

Diptera. Many kinds of flies are found in deserts, and undoubtedly many more than the comments below imply develop in desert soils. At present, however, substantial information about most soil-associated desert Diptera is very limited. This is especially true of small-bodied groups, such as the Chloropidae and other acalypterate muscoid flies that breed in friable soils. The recent discovery of a subterranean gall midge (Cecidomyiidae) on *Artemisia ludoviciana* in the Great Basin (220), supports the need to examine the trophic contribution of roots to dipteran consumers in deserts.

Ecologically, the two most extensively studied families of concern here are the bee flies (Bombyliidae (Fig. 16.20) and the robber flies (Asilidae) (Fig. 16.21). Both are large and diverse taxa in warm deserts (221, 222). In summarizing the larval habits of beeflies, Hull (222) noted that all are parasites and that their hosts involve six insect orders. One feeding unit composed of 19 genera attacks "soil-hidden" larvae of scarabaeid and tenebrionid beetles and of noctuid moths, as well as egg pods of grasshoppers. Another group of genera parasitizes insects that are "soil miners," such as ant lion larvae and mining bees. All of these hosts are of course common in deserts. The "miner" hosts are parasitized after the female bee fly flips an egg into the victim's burrow. It may be that after consuming their prey, immature bombyliids can postpone final development until conditions occur that are conducive for eventual breeding. This is inferred from Hull's (222) observation that the nectar-feeding adults often appear in large numbers about a week following desert rains. Allred et al. (223) collected most bee fly adults at the Nevada Test Site in May, probably around the end of the rainy season. These authors found that adult abundance varied considerably in different plant communities; presumably this also reflected both larval and prey abundances.

Larvae of many robber-fly genera also develop in the soil, but their role as predators of insects located there is disputed (222). Evidence seems more conclusive that vegetative matter is their main food, with insect consumption being secondary or sporadic. The predaceous adults are particularly common in vegetation along dry streambeds where relatively lush

Fig. 16.20. Bee fly (Bombyliidae). Central New Mexico. (Photo by author.)

Fig. 16.21. Robber fly (Asilidae) with grasshopper prey. Central New Mexico. (Photo by author.)

growth ensures the presence of prey and perhaps provides nutrition for the subterranean larvae as well.

Snipe flies (Rhagionidae) also have soil-inhibiting larvae. The rhagionid genus *Vermileo* occurs in "the Western mountains" (178), and since it is well adapted to a xeric existence (224) might also be expected from regional deserts, although no species were reported from the Nevada Test Site insect survey (149). Larvae of *Vermileo* and other genera in the same subfamily are convergent with *Myrmeleon* ant lion larvae in morphology and predaceous behavior; they too construct conical pit traps in sandy soil (224).

The larval role, and even the presence of other soil-associated dipteran families in North American deserts, is conjectural. Groups meriting further consideration in this regard include the Anthomyiidae, Tipulidae, Bibionidae, Dolichopodidae, and Empididae, partly because these are generally common soil dwellers (225), and also because all (except for the empidids) have been collected at the Nevada Test Site (149). Some anthomyiid larvae are root feeders, tipulid and bibionid larvae are detritivores, and dolichopodid and empidid larvae are probably carnivores (153). If larvae of the last four families are true desert inhabitants, they may be restricted to habitats where the soil remains moist, since they are typically found in mesic locations.

Hymenoptera. In the brief treatment here accorded to the nonsocial Hymenoptera it is impossible to do justice to the extraordinary diversity of species that spend parts of their lives underground. Rather than attempting to do so, I instead emphasize conspicuous assemblages of soil-nesting, provisioning wasps and bees that inhabit deserts of North America. These groups belong to the superfamilies Scolioidea, Vespoidea, Sphecoidea, and Apoidea. Because their developing stages are so frequently consumed by other desert arthropods, occasional mention of the latter will both reinforce and add to discussions presented earlier in this chapter.

Two scolioid families, the Tiphiidae and Mutillidae, are relatively prominent in the region concerned. In the southwest the genus *Brachycistis* is representative of tiphiid wasps. Most of its species inhabit the Sonoran desert, although others occur in all major deserts on the continent (226). Adult females in the subfamily (Brachycistidinae) are wingless; strongly fossorial, those of *Brachycistis* probably spend most of their lives underground (227). Little is known of the lives of immatures in the subfamily. However, since larvae of other typhiids parasitize scarab beetle larvae, as do larvae of the Scoliidae (153), the possibility of a similar habit in *Brachycistis* remains open. In general, female scoliids burrow until they find a scarabaeid grub, which they sting and paralyze. The host is then buried deeper still, together with an egg, in a cell the wasp constructs. The larvae then develop as external parasites.

Mutillids are widespread in deserts. Wingless females of "velvet ants" (*Dasymutilla* spp.) are often wooly or velvety as well as brightly colored; they are commonly seen during early morning and late afternoon in southwestern deserts as they run over the sandy areas where their host bees and wasps are nesting (E. G. Linsley, personal communication). Similarly, the nocturnal southwestern genus *Photopsis* typically selects hosts from larvae of gregarious nesting bees and other solitary aculeate (with a sting) Hymenoptera (228). In *Photopsis,* optimal conditions for selection of larva-containing host cells include hard-packed soil at depths of about 10 cm. Oviposition takes place through a small hole made in the cell surrounding the larva, which is then entirely eaten after the mutillid egg hatches. Provided that this happens in a region with cold winters, the parasite larva next enters diapause. In such cases one annual generation is typical, but up to three per year are possible in regions where summers are long and hot.

In the Vespoidea the family Eumenidae comprises many southwestern desert species. Of these there is only one large genus of ground-nesting forms, *Pterocheilus,* the species of which provision their nests with caterpillars (E. G. Linsley, personal communication).

In the vespoid family of spider wasps (Pompilidae) (Fig. 16.22), members of the genus *Pepsis* are quite large and usually have bright red wings. Their behavior while attacking tarantulas has been well documented (91).

Fig. 16.22. Spider wasp (*Pepsis* sp.) emerging from freshly dug burrow. Jornada Experimental Ranch, New Mexico State University, south-central New Mexico. (Photo by author.)

After paralyzing the prey, a female frequently constructs a subterranean cell into which she drags the spider. She then oviposits on the arachnid and the resulting larva is thereby provisioned for its period of development.

The only sphecoid family is the Sphecidae. It comprises many species of mostly ground-nesting solitary wasps. Among its several subfamilies some genera are termed "mud-daubers" because they use mud to construct nests in crevices or on surfaces. In the subfamilies Philanthinae and Crabroninae the many species called "digger wasps" provision their respective underground nests mainly with beetles or flies. The thread-waisted wasps (Sphecinae) also have ground-nesting members; species of one tribe (Sphecini) capture orthopterans, while those of another (Ammophilini) provision with caterpillars (153). Many digger and thread-waisted wasps occur in North American deserts.

The remaining sphecid subfamily consists of the sand wasps (Nyssoninae). Prey assortment in this large and diverse group is variable on the whole, but somewhat more specific at the tribal level. Nearly all nyssonines nest in sandy soil. Details of their burrowing and provisioning behavior are given by Evans (229), who also lists and discusses their parasites and predators. Nesting aggregations or colonies involving more than one genus are common. In the tribe Bembicini some colonies are composed of hundreds of individuals; at times their nest entrances are separated by only a few centimeters.

Most families of bees (Apoidea) are composed of solitary species. According to Linsley (230), their greatest abundance is in warm and arid regions, including the southwestern deserts of North America. In the Avra Valley of Arizona a total of 188 species of solitary bees (and no social species) were recorded (231, Appendix E). The percentage contribution made by each family was Andrenidae (30.9%), Megachilidae (22.3%), Anthophoridae (20.7%), Halictidae (15.4%), Colletidae (5.8%), Melittidae (0.5%), and Oxaeidae (0.5%). Since 50 of the 58 andrenid species were in the subfamily Panurginae, and since most desert panurgines belong to the genus *Perdita* (153), which is the second largest among solitary bees in America north of Mexico (230), the potential significance of that genus in the southwest must be considerable.

With the exception of carpenter bees and most megachilids, the majority of solitary bees nest in the ground. Their nests are sometimes close together with entrances numbering in the thousands (230), although far fewer may also occur at a given site (232, 233) and some construct isolated burrows (E. C. Linsley, personal communication). In the Halictidae many bees use a common passageway to the outside (153); otherwise, entrances are usually connected to cells beneath by separate burrows. Nesting sites are variously located, depending on the substrate and nutrient requirements of each taxon. Anthophorine digger bees tend to nest close to a water source (230); *Emphoropsis pallida,* for example, utilizes dry watercourses in the Sonoran and Mojave deserts (234). Soil-nesting megachilids require the proximity of certain plant parts, mud, or pebbles for their nest-cell linings, and cells of *Hereranthidium larreae* are impregnated with plant resin (232). Most other groups line their cells with waxlike or varnishlike waterproofing material (230).

Located at the ends of burrow shafts that vary in length, depth, and slope, cells in the solitary Apoidea (except for parasitic species) are provisioned with pollen and nectar. Oviposition usually follows provisioning, then cells are sealed with mud. Larval development takes from 8 days to several weeks (230). In many species this is followed by a period of diapause lasting most of the year. Pupation generally occurs after overwintering, but in the large genera *Andrena* (Andrenidae) and *Osmia* (Megachilidae) it takes place in fall and the adult stage overwinters. Drouth conditions tend to limit or prevent adult emergence in arid regions, in which case subterranean diapause may last 2 or more years.

Linsley (230) gives a summary of categories and kinds of organisms detrimentally associated with larvae of solitary bees. A review of his summary provides an appropriate ending to this chapter because it reminds us not only of arthropods previously dealt with, but also of the need to consider the complexity of biotic interactions when studying the biology of soil-associated desert invertebrates.

One category is restricted to parasites that as first-instar larvae prey on the bee egg, subsequent larval instars feeding on the stored pollen and nectar. Some parasitic bees and meloid beetles are examples. A second unit concerns predators and parasites of the developing and grown larva. Some of these (e.g., larvae of rhipiphorid beetles, stylopid strepsipterans, and a variety of parasitoid wasps) are endoparasites; others include larvae of clerid and meloid beetles, mantispid neuropterans, bombyliid and asilid flies, and mutillid wasps. A third group is that of "food depredators," which consume the provisioned food and either directly harm the bee larva or starve it to death. Involved are larvae of parasitic megachilid bees, ptinid beetles, and phorid flies, in addition to certain mites and fungi. Finally, there are scavengers of old cells and burrow refuse that accidentally affect the developing bee. Larvae of dermestid beetles, anthomyiid flies, and certain Lepidoptera (*Plodia*: Gelechiidae; *Tineola*: Tineidae) fall into this final category.

ACKNOWLEDGMENTS

I am most grateful to the following persons for directing me to or providing me with useful literature and information: E. U. Balsbaugh, Jr., E. B. Edney, R. T. Gordon, A. R. Hardy, R. W. Holland, M. C. Matlack, W. A. Riddle, P. F. Santos, D. P. Schwert, D. B. Thomas, and C. A. Triplehorn. Valuable references also came from reviewers of portions of the chapter. Many thanks for their critical reviews are due to O. F. Franke, E. G. Linsley, D. W. Lowrie, W. B. Miller, G. A. Polis, C. N. Slobodchikoff, E. C. Toolson, and P. Wygodzinsky. The editorial contribution of D.

L. Dindal is greatly appreciated, as is the technical help with illustrations provided by J. M. Evans and G. Minion. Photographs were kindly contributed by M. C. Matlack, P. F. Santos, and G. A. Polis.

References

1. **Shreve, F. 1942.** *Bot. Rev.* 8:195.
2. **Jaeger, E. G. 1957.** *The North American Deserts.* Stanford University Press, Stanford, Calif. 308 pp.
3. **MacMahon, J. A. 1979.** North American deserts. In D. W. Goodall and R. A. Perry (eds.). *Arid-Land Ecosystems,* Vol. I. Int. Biol. Programme 16. Cambridge University Press, Cambridge. 881 pp.
4. **Logan, R. F. 1968.** Causes, climates, and distribution of deserts. In W. G. Brown, Jr. (ed.). *Desert Biology,* Vol. I. Academic Press, New York.
5. **Fuller, W. H. 1974.** Desert soils. In W. G. Brown, Jr. (ed.). *Desert Biology,* Vol. II. Academic Press, New York. 601 pp.
6. **Shreve, F. and I. L. Wiggins. 1964.** *Vegetation and Flora of the Sonoran Desert,* Vol. I. Stanford University Press, Stanford, Calif. 840 pp.
7. **Mulroy, T. W. and P. W. Rundel, 1977.** *BioScience* 27:109.
8. **Fautin, R. W. 1946.** *Ecol. Monogr.* 16:251.
9. **Chew, R. M. 1961.** *J. N.Y. Entomol. Soc.* LXIX:5.
10. **Gertsch, W. J. and S. E. Riechert. 1976.** *Am. Mus. Novit.* 2604:1.
11. **Gates, G. E. 1967.** *Great Basin Nat.* 27:142.

11a. **Crawford, C. S. 1981.** *Biology of Desert Invertebrates.* Springer-Verlag, Berlin. 314 pp.

11b. **Crawford, C. S. 1986.** The role of invertebrates in desert ecosystems. Pp. 73–91 in W. G. Whitford (ed.). *Patterns and Processes in Desert Ecosystems.* University of New Mexico Press, Albuquerque, N. Mex.

11c. **Schoenly, K. and W. Reid. 1983.** Community structure of carrion arthropods in the Chihuahuan desert. *J. Arid Environ.* 6:253–263.

12. **Bequaert, J. C. and W. B. Miller. 1973.** *The Mollusks of the Arid Southwest.* University of Arizona Press, Tucson, Ariz. 271 pp.
13. **Metcalf, A. L. 1977.** Some quaternary molluscan faunas from the northern Chihuahuan desert and their paleoecological implications. In R. H. Wauer and D. H. Riskind (eds.). *Transactions of the Symposium on the Biological Resources of the Chihuahuan Desert Region.* U.S. Dept. of Interior, Washington, D.C.
14. **Miller, W. B. 1972.** *Nautilus* 85:128.
15. **Christensen, C. C. and W. B. Miller. 1975.** *Nautilus* 89:43.
16. **Yom-Tov, Y. 1972.** *J. Anim. Ecol.* 41:17.
17. **Shachak, M., E. A. Chapman and Y. Orr. 1976.** *Isr. J. Med. Sci.* 12:887.
18. **McMichal, D. F. and T. Iiredale. 1959.** The land and freshwater mollusks of Australia. In A. Kest, R. L. Crocker, and C. S. Christian (eds.). *Biogeography and Ecology in Australia.* Dr. W. Junk, The Hague, The Netherlands. 640 pp.
19. **Warburg, M. R. 1965.** *Proc. Malacol. Soc. Lond.* 36:297.
20. **Gilbertson, L. H. 1969.** *Nautilus* 83:29.
21. **Horne, F. R. 1973.** *Biol. Bull. (Woods Hole)* 144:321.
22. **Riddle, W. A. 1975.** *Comp. Biochem. Physiol. A Comp. Physiol.* 51:579.
23. **Crawford, C. S. 1979.** *J. Arid Environ.* 2:31.
24. **Walton, M. L. 1963.** *Nautilus* 76:127.
25. **Schmidt-Nielsen, K., C. R. Taylor and A. Shkolnik. 1972.** *Symp. Zool. Soc. Lond.* 31:1.
26. **Riddle, W. A. 1977.** *Comp. Biochem. Physiol. A Comp. Physiol.* 56:369.

27. **Warburg, M. R. 1965.** *Am. Midl. Nat.* 73:363.
28. **Allred, D. M. and S. Mulaik. 1965.** *Great Basin Nat.* 25:43.
29. **Cloudsley-Thompson, J. L. and M. J. Chadwick. 1964.** *Life in Deserts.* Dufour Editions, Philadelphia. 218 pp.
30. **Linsenmair, K. E. and C. Linsenmair. 1971.** *Z. Tierpsychol.* 29:134.
31. **Shachak, M., E. A. Chapman and Y. Steinberger. 1976.** *Oecologia (Berl.)* 24:57.
32. **Warburg, M. R., D. Rankevich and K. Chasanmus. 1978.** *J. Arid Environ.* 1:157.
33. **Kheirallah, A. M. 1979.** *J. Arid Environ.* 2:51.
34. **Warburg, M. R. 1965.** *Crustaceana (Leiden)* 9:302.
35. **Warburg, M. R. 1965.** *Physiol. Zool.* 38:99.
36. **Warburg, M. R. 1968.** *Am. Zool.* 8:545.
37. **Schneider, P. 1975.** *Zool. Anz. Jena* 195:155.
37a. **Francke, O. F. and M. E. Soleglad. 1981.** The family Iuridae Thorell (Arachnida, Scorpiones). *J. Arachnol.* 9:233–258. (Contains a new key to the families of scorpions.)
37b. **Williams, S. C. 1987.** Scorpion bionomics. *Annu. Rev. Entomol.* 32:275–295.
38. **Hadley, N. F. 1974.** *J. Arachnol.* 2:11.
39. **Hadley, N. F. 1971.** *Southwest. Nat.* 15:495.
40. **Crawford, C. S. and R. C. Wooten, Jr. 1973.** *Physiol. Zool.* 46:218.
41. **Hadley, N. F. 1970.** *J. Exp. Biol.* 53:547.
42. **Edney, E. B. 1977.** Water balance. In *Land Arthropods.* Springer-Verlag, Berlin. 282 pp.
43. **Toolson, E. C. and N. F. Hadley. 1979.** *J. Comp. Physiol.* 129:319.
44. **Hadley, N. F. and L. L. Jackson. 1977.** *Insect Biochem.* 7:85.
45. **Toolson, E. C. and N. F. Hadley. 1977.** *Physiol. Zool.* 50:323.
46. **Hadley, N. F. and R. D. Hill. 1969.** *Comp. Biochem. Physiol.* 29:217.
47. **Riddle, W. A. 1978.** *J. Arid Environ.* 1:243.
48. **Riddle, W. A., C. S. Crawford and A. M. Zeitone. 1976.** *J. Comp. Physiol.* 112:295.
49. **Cloudsley-Thompson, J. L. 1978.** *Bull. Br. Arachnol. Soc.* 4:184.
50. **Williams, J. C. 1966.** *Proc. Calif. Acad. Sci.* 34:419.
51. **Polis, G. A. 1980.** *J. Anim. Ecol.* 49:1.
52. **Crawford, C. S. and R. C. Krehoff. 1975.** *J. Arachnol.* 2:195.
53. **Fleissner, G. 1977.** *J. Comp. Physiol.* 118:109.
54. **Hadley, N. F. and S. C. Williams. 1968.** *Ecology* 49:726.
55. **Hadley, N. F. 1970.** *Ecology* 51:434.
56. **Stahnke, H. L. 1966.** *Bull. South. Calif. Acad. Sci.* 65:65.
57. **Frohlich, D. R. and W. H. Clark. 1976.** *J. Idaho Acad. Sci.* 12:87.
58. **Crawford, C. S. and W. A. Riddle. 1974.** *Oikos* 25:86.
59. **Crawford, C. S. and W. A. Riddle. 1975.** *Physiol. Zool.* 48:84.
60. **Riddle, W. A. and S. Pugach. 1976.** *Cryobiology* 13:248.
61. **Williams, S. C. 1969.** *Proc. Calif. Acad. Sci.* 37:1.
62. **Polis, G. A. and R. D. Farley. 1979.** *J. Arachnol.* 7:33.
63. **Polis, G. A. and R. D. Farley. 1979.** *J. Zool. (Lond.)* 187:517.
64. **Cloudsley-Thompson, J. L. 1968.** *Spiders, Scorpions, Centipedes and Mites.* Pergamon Press, Oxford. 278 pp.
65. **Polis, G. A. 1980.** *Behav. Ecol. Sociobiol.* 7:25.
66. **Polis, G. A. and R. D. Farley. 1980.** *Ecology* 61:620.
67. **Polis, G. A. 1979.** *J. Zool. (Lond.)* 188:333.
68. **Brownell, P. H. 1977.** *Science* 197:479.
69. **Williams, S. C. 1970.** *Pan-Pac. Entomol.* 46:254.
70. **Allred, D. M. and W. J. Gertsch. 1976.** *J. Arachnol.* 3:87.
71. **Hoff, C. C. 1959.** *The Ecology and Distribution of the Pseudoscorpions of North-Central New Mexico.* University of New Mexico Press, Albuquerque, N. Mex. 68 pp.

72. **Ahearn, G. A. 1970.** *Comp. Biochem. Physiol. A Comp. Physiol.* 35:339.
73. **Crawford, C. S. and J. L. Cloudsley-Thompson. 1971.** *Entomol. Exp. Appl.* 14:99.
74. **Pergande, T. 1886.** *Proc. Entomol. Soc. Wash.* 1:42.
75. **Weygoldt, P. 1970.** *Behaviour* 36:1.
76. **Weygoldt, P. 1971.** *J. Zool. (Lond.)* 164:137.
77. **Eisner, T., J. Meinwald, A. Monroy and R. Ghent. 1961.** *J. Insect Physiol.* 6:272.
78. **Cloudsley-Thompson, J. L. 1977.** *Bull. Br. Arachnol. Soc.* 4:61.
79. **Allred, D. M. and M. H. Muma. 1971.** *Great Basin Nat.* 31:164.
80. **Muma, M. H. 1963.** *Brigham Young Univ. Sci. Bull. Biol. Ser.* 3:1.
81. **Allred, D. M. 1975.** *Great Basin Nat.* 35:405.
82. **Muma, M. H. 1974.** *Fla. Entomol.* 57:385.
83. **Muma, M. H. 1966.** *Fla. Entomol.* 49:23.
84. **Muma, M. H. 1966.** *Psyche (Camb.)* 73:251.
85. **Muma, M. H. 1967.** *Fla. Entomol.* 50:115.
86. **Muma, M. H. 1966.** *Anim. Behav.* 14:346.
87. **Muma, M. H. 1966.** *Fla. Entomol.* 49:233.
88. **Muma, M. H. 1966.** *Fla. Entomol.* 49:199.
89. **Chew, R. M. 1961.** *J. N.Y. Entomol. Soc.* LXIX:5.
89a. **Fowler, H. G. and W. G. Whitford. 1985.** Structure and organization of a winter community of cavity-inhabiting, web-building spiders (Pholocidae and Theriidae) in a Chihuahuan desert habitat. *J. Arid Environ.* 8:57–65.
90. **Riechert, S. E., W. G. Reeder and T. A. Allen. 1973.** *J. Anim. Ecol.* 42:19.
91. **Gertsch, W. J. 1979.** *American Spiders,* 2nd ed. Van Nostrand Reinhold, New York. 274 pp.
92. **Comstock, J. H. 1971.** *The Spider Book,* third printing, revised by W. J. Gertsch. Cornell University Press, Ithaca, N.Y., p. 729.
93. **Baerg, W. J. 1958.** *The Tarantula.* University of Kansas Press, Lawrence, Kans. 88 pp.
94. **Minch, E. W. 1979.** *Novit. Arthropodae* 1:1.
95. **Minch, E. W. 1977.** *J. Arachnol.* 5:133.
96. **Minch, E. W. 1978.** *Bull. Br. Arachnol. Soc.* 4:231.
97. **Cloudsley-Thompson, J. L. 1967.** *Entomol. Mon. Mag.* 103:217.
98. **Seymour, R. S. and A. Vinegar. 1973.** *Comp. Biochem. Physiol. A Comp. Physiol.* 44:83.
99. **Minch, E. W. 1978.** *Bull. Br. Arachnol. Soc.* 4:275.
100. **Anderson, J. F. 1974.** *Ecology* 55:576.
101. **Stewart, D. M. and A. W. Martin. 1970.** *Z. Vgl. Physiol.* 70:223.
102. **Anderson, J. F. 1970.** *Comp. Biochem. Physiol.* 33:51.
103. **Parry, D. A. 1954.** *J. Exp. Biol.* 31:218.
104. **Shook, R. S. 1978.** *J. Arachnol.* 6:53.
105. **Moeur, J. E. and C. H. Eriksen. 1972.** *Physiol. Zool.* 45:290.
106. **Wallwork, J. A. 1972.** *J. Zool. (Lond.)* 168:91.
107. **Wallwork, J. A. 1972.** *J. Anim. Ecol.* 41:291.
108. **Santos, P. F., E. DePree and W. G. Whitford. 1978.** *J. Arid Environ.* 1:41.
109. **Edney, E. B., J. F. McBrayer, P. J. Franco and A. W. Phillips. 1975.** *US/IBP Desert Biome Research Memorandum 75-29,* p. 39.
110. **Edney, E. B., P. J. Franco and J. F. McBrayer. 1976.** *US/IBP Desert Biome Research Memorandum 76-24.* 11 pp.
111. **Franco, P. J., E. B. Edney and J. F. McBrayer. 1979.** *J. Arid Environ.* 2:137.
112. **Deboutteville, C. D. and E. Rapoport. 1968.** *Biologie de L'Amerique Australe,* Vol. 9. Centre National de la Recherche Scientifique, Paris. 472 pp.
113. **Wood, T. G. 1971.** *Pedobiologia* 11:446.

113a. **Wallwork, J. A., B. W. Kamil and W. G. Whitford. 1985.** Distribution and diversity patterns of soil mites and other microarthropods in a Chihuahuan desert site. *J. Arid Environ.* 9:215–231.

114. **Cloudsley-Thompson, J. L. 1962.** *Entomol. Exp. Appl.* 5:69.

115. **Newell, I. M. and L. Tevis, Jr. 1960.** *Ann. Entomol. Soc. Am.* 53:293.

116. **Tevis, L., Jr. and I. M. Newell. 1962.** *Ecology* 43:497.

117. **Chamberlin, R. V. 1962.** *Proc. Biol. Soc. Wash.* 75:53.

118. **Hoffman, R. L. and B. S. Orcutt. 1960.** *Proc. U.S. Natl. Mus.* 111:95.

119. **Allfred, D. M. 1971.** *Great Basin Nat.* 31:161.

120. **Crawford, C. S. 1979.** Desert millipedes: a rationale for their distribution. In M. Camatini (ed.). *Proceedings of the 4th International Congress of Myriapodology.* Academic Press, London.

121. **Causey, N. B. 1975.** *Occas. Pap. Mus. Tex. Tech Univ.* 35:1.

122. **Crawford, C. S. 1979.** A year in the life of *Orthoporus ornatus*, a desert millipede. Pp. 393–396 in R. M. Linn (ed.). *Proceedings of the First Conference on Scientific Research in the National Parks,* Vol. 1. National Park Service Trans. Proc. Ser. 5.

123. **Wooten, R. C., Jr. and C. S. Crawford. 1974.** *Oecologia (Berl.)* 17:179.

124. **Wooten, R. C., Jr. and C. S. Crawford. 1975.** *Oecologia (Berl.)* 20:231.

125. **Crawford, C. S. 1976.** *Oecologia (Berl.)* 24:265.

126. **Wooten, R. C., Jr., C. S. Crawford and W. A. Riddle. 1975.** *Zool. J. Linn. Soc.* 57:59.

127. **Crawford, C. S. 1972.** *Comp. Biochem. Physiol. A Comp. Physiol.* 42:521.

128. **Crawford, C. S. 1978.** *Ecology* 59:996.

129. **Pugach, S. and C. S. Crawford. 1978.** *Can. J. Zool.* 56:1460.

130. **Eisner, T., J. J. Hurst, W. T. Keeton and Y. Meinwald. 1965.** *Ann. Entomol. Soc. Am.* 58:247.

131. **Crawford, C. S. and M. C. Matlack. 1979.** *Pedobiologia* 19:48.

132. **Chamberlin, R. V. and S. Mulaik. 1940.** *Entomol. News* 107.

133. **Chamberlin, R. V. 1943.** *Proc. Biol. Soc. Wash.* 56:97.

134. **Chamberlin, R. V. 1943.** *Bull. Univ. Utah* 33:1.

135. **Crabill, R. E. 1960.** *Proc. U.S. Natl. Mus.* 111:1.

136. **Ghabbour, S. I., W. Mikhail and M. A. Rizk. 1977.** *Rev. Ecol. Biol. Sol* 14:429.

137. **Cloudsley-Thompson, J. L. and C. S. Crawford. 1970.** *Entomol. Exp. Appl.* 13:187.

138. **Lewis, J. G. E. 1972.** *J. Zool. (Lond.)* 167:399.

139. **Crawford, C. S., W. A. Riddle and S. Pugach. 1975.** *Physiol. Zool.* 48:290.

140. **Lewis, J. G. E. 1969.** *Bull. Mus. Natl. Hist. Nat. (Paris)* 41:85.

141. **Brunhuber, B. S. 1970.** *Zool. J. Linn. Soc.* 49:225.

142. **Allred, D. M. and D. E. Beck. 1964.** *Brigham Young Univ. Sci. Bull. Biol. Ser.* 5:16.

143. **Scott, H. G. 1957.** *The Collembola of New Mexico with Especial Reference to the North-Central Part of the State.* Ph.D. dissertation, University of New Mexico, Albuquerque, N. Mex., p. 400.

144. **Christiansen, K. 1964.** *Annu. Rev. Entomol.* 9:147.

145. **Vannier, G. 1973.** *C. R. Acad. Sci. (Paris) Ser. D Sci. Nat.* 277:85.

146. **Vannier, G. 1978.** *Bull. Soc. Ecophysiol.* 3:13.

147. **Poinsot-Balaguer, N. 1976.** *Pedobiologia* 16:1.

148. **Edney, E. B. 1971.** *Physiol. Zool.* 44:61.

149. **Beck, D. E. and D. M. Allred. 1968.** *Great Basin Nat.* 28:132.

150. **Wygodzinsky, P. 1972.** *Am. Mus. Novit.* No. 2481, p. 26.

151. **Kaplin, V. G. 1978.** *Entomol. Rev.* 56:72.

152. **Helfner, J. R. 1963.** *How to Know the Grasshoppers, Cockroaches and Their Allies.* Wm. C. Brown, Dubuque, Iowa. 353 pp.

153. **Borrer, D. J., D. M. DeLong and C. A. Triplehorn. 1976.** *An Introduction to the Study of Insects,* 4th ed. Holt, Rinehart and Winston, New York. 852 pp.
154. **Ross, E. S. 1970.** *Annu. Rev. Entomol.* 15:157.
155. **Tinkham, E. R. 1968.** *Great Basin Nat.* 28:124.
156. **Tinkham, E. R. 1948.** *Am. Midl. Nat.* 40:521.
157. **Hebard, M. 1920.** *Trans. Am. Entomol. Soc.* 46:197.
158. **Friauf, J. F. and E. B. Edney. 1969.** *Proc. Entomol. Soc. Wash.* 71:1.
159. **Hawke, S. D. and R. D. Farley. 1973.** *Oecologia* (*Berl.*) 11:263.
160. **Edney, E. B., S. Haynes and D. Gibo. 1974.** *Ecology* 55:420.
161. **Edney, E. B. 1966.** *Comp. Biochem. Physiol.* 19:387.
162. **O'Donnell, M. J. 1977.** *Proc. Natl. Acad. Sci.* 74:1757.
163. **Edney, E. B., P. Franco and R. Wood. 1978.** *Physiol. Zool.* 51:241.
164. **Hartman, H. B., W. W. Walthall, L. P. Bennett and R. A. Stewart. 1979.** *Science* 205:503.
165. **Edney, E. B. 1968.** *Comp. Biochem. Physiol.* 25:149.
166. **Edney, E. B. 1974.** *Physiol. Zool.* 47:1.
167. **Slater, J. A. and R. M. Baranowski. 1978.** *How to Know the True Bugs* (*Hemiptera–Heteroptera*). Wm. C. Brown, Dubuque, Iowa. 256 pp.
168. **Froeschner, R. C. 1960.** *Proc. U.S. Natl. Mus.* 111:337.
169. **Heath, J. E. and P. J. Wilkin. 1970.** *Physiol. Zool.* 43:145.
170. **Heath, J. E., P. J. Wilkin and M. S. Heath. 1972.** *Physiol. Zool.* 45:238.
171. **Stange, L. A. 1970.** *Univ. Calif. Publ. Entomol.* 55:192.
172. **Tuculescu, R., H. Topoff and S. Wolfe. 1975.** *Ann. Entomol. Soc. Am.* 68:719.
173. **Youthed, G. J. and V. C. Moran. 1969.** *J. Insect Physiol.* 15:1103.
174. **Wilson, D. S. 1974.** *Biotropica* 6:187.
175. **Youthed, G. J. and V. C. Moran. 1969.** *J. Insect Physiol.* 15:867.
176. **McClure, M. S. 1976.** *Biotropica* 8:179.
177. **Hardy, A. R. and F. G. Andrews. 1976.** *A Final Report to the Office of Endangered Species on Contract 14-16-0008-966.* Calif. Dept. of Food and Agriculture, Sacramento, Calif., 42 pp.
178. **Peterson, A. 1953.** *Larvae of Insects—Coleoptera, Diptera, Neuroptera, Siphonaptera, Mecoptera, Trichoptera,* Part II. Ohio State University, Columbus, Ohio. 416 pp.
179. **Linsley, E. G. 1957.** *J. Kans. Entomol. Soc.* 30:81.
180. **Werner, F. G., W. R. Enns and F. H. Parker. 1966.** *The Meloidae of Arizona.* Univ. Calif. Agric. Exp. Stn. Tech. Bull. 175, Tucson, Ariz. 95 pp.
181. **Cazier, M. A. 1948.** *Am. Mus. Novit.* 1382:28.
182. **Rumpp, N. L. 1961.** *Bull. South. Calif. Acad. Sci.* 60:165.
183. **Cohen, A. C. and J. D. Pinto. 1977.** *Ann. Entomol. Soc. Am.* 70:741.
184. **Tiemann, D. L. 1967.** *Proc. Calif. Acad. Sci.* 35:235.
185. **Andrews, F. G. 1976.** *Pan-Pac. Entomol.* 52:339.
186. **Gordon, R. D. 1977.** *Proc. Entomol. Soc. Wash.* 79:275.
187. **Gordon, R. D. 1977.** *Proc. Biol. Soc. Wash.* 90:232.
188. **Gordon, R. D. 1973.** *Proc. Entomol. Soc. Wash.* 75:435.
189. **Gordon, R. D. 1976.** *Proc. Entomol. Soc. Wash.* 78:458.
190. **Gordon, R. D. and O. L. Cartwright. 1974.** *Entomol. News* 85:181.
191. **Hardy, A. R. and F. G. Andrews. 1978.** *Pan-Pac. Entomol.* 54:1.
192. **Cheary, B. S., A. R. Hardy and J. A. Robertson. 1972.** *Coleopt. Bull.* 26:97.
193. **Hardy, A. R. and F. G. Andrews. 1974.** *Pan-Pac. Entomol.* 50:124.
194. **Schmoller, R. R. 1970.** *Biologist* 52:77.
195. **Rickard, W. H. 1970.** *Northwest Sci.* 44:107.
196. **Doyen, J. T. and W. T. Tschinkel. 1974.** *Ann. Entomol. Soc. Am.* 67:617.

197. **Slobodchikoff, C. N. 1978.** *Am. Midl. Nat.* 99:45.
198. **Tanner, V. M. and W. A. Packham. 1965.** *Brigham Young Univ. Sci. Bull. Biol. Ser.* 6:44.
199. **Ahearn, G. A. 1971.** *Am. Midl. Nat.* 86:385.
200. **Krehoff, R. C. 1975.** Ph.D. dissertation, University of New Mexico, Albuquerque, N. Mex. p. 64.
201. **Smith, W. E. and W. G. Whitford. 1976.** *Southwest. Entomol.* 1:161.
202. **Thomas, D. B., Jr. 1979.** *Environ. Entomol.* 8:568.
203. **Doyen, J. T. 1973.** *Univ. Calif. Publ. Entomol.* 73:p. 110.
204. **Hinds, W. T. and W. H. Rickard. 1973.** *J. Anim. Ecol.* 42:341.
205. **Kramm, R. A. and K. R. Kramm. 1972.** *Southwest. Nat.* 16:341.
206. **Wakeland, C. 1926.** *Univ. Idaho Agric. Exp. Stn. Res. Bull.* 6:52.
207. **Slobodchikoff, C. N. 1979.** *Environ. Entomol.* 8:770.
208. **Thomas, D. B., Jr. and E. L. Sleeper. 1977.** *Ann. Entomol. Soc. Am.* 70:242.
209. **Rogers, L. E. and W. H. Rickard. 1975.** *Ann. Entomol. Soc. Am.* 68:1069.
210. **Rickard, W. H. and L. E. Haverfield. 1965.** *Ecology* 46:873.
211. **Felt, E. P. 1894.** *Cornell Univ. Agric. Exp. Stn. Bull.* 64:47.
212. **Crawford, C. S. and R. F. Harwod. 1964.** *Wash. Agric. Exp. Stn. Tech. Bull.* 44:p. 25.
213. **Mann, J. 1969.** *Cactus-Feeding Insects and Mites.* U.S. Natl. Mus. Bull. 256. Smithsonian Institution, Washington, D.C., 158 pp.
214. **Mares, M. A., F. A. Enders, J. M. Kingslover, J. L. Neff and B. B. Simpson. 1977.** *Prosopis* as a niche component. P. 250 in B. B. Simpson (ed.). *Mesquite—Its Biology in Two Desert Scrub Ecosystems.* US/IBP Synthesis Series. Dowden, Hutchinson & Ross, Stroudsburg, Pa.
215. **Crumb, S. E. 1958.** *The Larvae of the Phalaenidae.* Tech. Bull. 1135. U.S. Dept. Agriculture, Washington, D.C., p. 356.
216. **Duckworth, W. D. and T. D. Eichlin. 1978.** *The Clearwing Moths of California (Lepidoptera: Sesiidae).* Occas. Pap. Entomol. 27. Calif. Dept. Food and Agriculture, Sacramento, Calif., p. 80.
217. **Roever, K. 1975.** Family Megathymidae (the giant skippers). P. 633 in W. H. Howe (ed.). *The Butterflies of North America.* Doubleday, Garden City, N.Y.
218. **Crawford, C. S. 1971.** *Ann. Entomol. Soc. Am.* 64:52.
219. **Walkden, H. H. 1950.** *Circular 849.* U.S. Dept. Agriculture, Washington, D.C., p. 52.
220. **Ranasinghe, M. A. S. K. 1977.** *Great Basin Nat.* 37:429.
221. **Hull, F. M. 1962.** *Robber Flies of the World—The Genera of the Family Asilidae,* Part I. U.S. Natl. Mus. Bull., 224. Smithsonian Institution, Washington, D.C. 430 pp.
222. **Hull, F. M. 1973.** *Bee Flies of the World—The Genera of the Family Bombyliidae.* U.S. Natl. Mus. Bull. 286. Smithsonian Institution, Washington, D.C. 687 pp.
223. **Allred, D. M., D. E. Johnson and D. E. Beck. 1965.** *Great Basin Nat.* 25:5.
224. **Oldroyd, H. 1964.** *The Natural History of Flies.* W. W. Norton, New York. 324 pp.
225. **Frost, S. W. 1959.** *Insect Life and Insect Natural History,* 2nd rev. ed. Dover, New York. 526 pp.
226. **Wasbauer, M. S. 1966.** *Univ. Calif. Publ. Entomol.* 43:96.
227. **Wabauer, M. S. 1973.** *Great Basin Nat.* 33:109.
228. **Ferguson, W. E. 1962.** *Univ. Calif. Publ. Entomol.* 27:82.
229. **Evans, H. E. 1966.** *The Comparative Ethnology and Evolution of the Sand Wasps.* Harvard University Press, Cambridge, Mass. 526 pp.
230. **Linsley, E. G. 1958.** *Hilgardia* 27:543.
231. **Orians, G. H. and O. T. Solbrig (eds.). 1977.** *Convergent Evolution in Warm Deserts.* US/IBP Synth. Ser. I. Dowden, Hutchinson & Ross, Stroudsburg, Pa. 333 pp.
232. **MacSwain, J. W. 1946.** *Pan-Pac. Entomol.* 22:159.
233. **Hurd, P. D., Jr. and J. A. Powell. 1958.** *Pan-Pacific Entomologist* 34:147.
234. **Bohart, G. E., P. F. Torchio, Y. Maeta and R. W. Rust. 1972.** *J. Kans. Entomol. Soc.* 45:381.

17

Litter Araneae (Araneida)

C. D. DONDALE

The thin connected spaces within the litter, or L layer, of the soil are occupied by a living community of scavengers, mold feeders, and predators. The Araneae, or true spiders, belong to the predatory component of this fauna.

With few exceptions, spiders are solitary, predatory, and terrestrial. They reproduce by sexual pairing and the deposition of eggs, and development is comparable with that of the orthopteroid insects. Family groupings of spiders generally comprise either species that build and use silken webs of various types in obtaining prey or species that hunt their prey on the surface of soil or plants.

Most analytical studies of litter spiders involve hand sorting, sifting, or heat extracting of litter samples (1–4), or they involve the simpler but often debated pitfall trapping of motile animals on the soil surface (5–16). Experimental studies designed to elucidate the role of spiders in litter are those by Duffey (17) and by Kajak and Jakubczyk (18). Norton (19) reviewed most of the ecological literature through 1971.

This chapter presents keys for the identification of 16 families and 95 genera of litter-inhabiting spiders in North America. The main criterion for inclusion is persistent occurrence in litter in significant numbers, as reported by investigators. Workers who require more inclusive keys to families may consult Kaston (20), Roth (20a), or Borror et al. (21); those requiring more inclusive keys to genera may consult Kaston (20), Roth (20a), Chamberlin and Gertsch (22) for Dictynidae, Leech (23) for Amaurobiidae, Roth and Brame (24) for Agelenidae, or Levi and Randolph (25) for Theridiidae. Generic revisions, when available, are cited in the keys to genera given here. Basic anatomical terms are explained by Kaston (20).

ACKNOWLEDGMENT

My appreciation is extended to Dr. E. E. Lindquist and Dr. H. J. Teskey for their thorough review of the manuscript.

KEY TO FAMILIES OF LITTER SPIDERS

1a Spider with six or fewer eyes **2**

1b Spider with eight eyes **4**

2a **(1)** Abdomen with cribellum (*cri*, Fig. 17.1); leg tarsi with three claws (Fig. 17.2) . . . . **Dictynidae** (in part)

2b Abdomen without cribellum; leg tarsi with two claws (Fig. 17.10) **3**

3a **(2)** Sternum (*st*) extending around bases of leg coxae (Fig. 17.7) (Genus ***Dysdera***) . . . . **Dysderidae**

3b Sternum not extending around bases of leg coxae (Genus ***Oonops***) . . **Oonopidae**

4a **(1)** Leg tarsi with three claws, with middle claw small, hooklike, separated from pair dorsal to it (Figs. 17.4–17.6) **9**

4b Leg tarsi with two claws (*cl*), often with scopula (*scop*) and tufts (*clt*) of setae (Fig. 17.3) **5**

5a **(4)** Legs prograde (Fig. 17.8) **7**

5b Legs laterigrade (Figs. 17.11, 17.19). **6**

6a **(5)** Leg I (and usually II) much longer and thicker than III and IV; tarsi without claw tufts (Fig. 17.10) . . . . **Thomisidae**

6b Leg I (and usually II) of approximately same length and thickness as III and IV (Figs. 17.11, 17.19); tarsi with claw tufts (*clt*, Fig. 17.3) **Philodromidae**

7a **(5)** Eyes in one curved row along front and sides of carapace (Fig. 17.9); anterior median eyes (*ame*) two or more times as large as anterior lateral eyes (*ale*), and posterior median eyes (*pme*) minute (Fig. 17.9). **Salticidae**

7b Eyes in two transverse rows (Figs. 17.15, 17.16); anterior median eyes (*ame*) approximately equal in size to anterior lateral eyes (*ale*), and posterior median eyes (*pme*) approximately equal to or larger than posterior lateral eyes (*ple*) (Figs. 17.15, 17.16). **8**

8a **(7)** Palp-coxal lobes (*pcxl*) with oblique or transverse depression on ventral surface (Fig. 17.17); anterior spinnerets (*aspin*) cylindrical, separated at base (Fig. 17.18) . . . . **Gnaphosidae**

8b Palp-coxal lobes (*pcxl*) without depression on ventral surface (Fig. 17.14); anterior spinnerets (*aspin*) conical, not separated at base (Fig. 17.20) . . . . **Clubionidae**

9a **(4)** Abdomen with cribellum (*cri*, Figs. 17.1, 17.22); basitarsus IV usually with calamistrum (*cal*, Figs. 17.2, 17.21) **10**

9b Abdomen without cribellum; basitarsus IV without calamistrum **11**

10a **(9)** Eyes all appearing light in color; cribellum (*cri*) often with dark longitudinal band at midline (Fig. 17.22); calamistrum (*cal*) usually less than one-half length of basitarsus IV (Fig. 17.21). . . . . . **Amaurobiidae**

10b Eyes in part or whole dark; cribellum (*cri*) without dark longitudinal

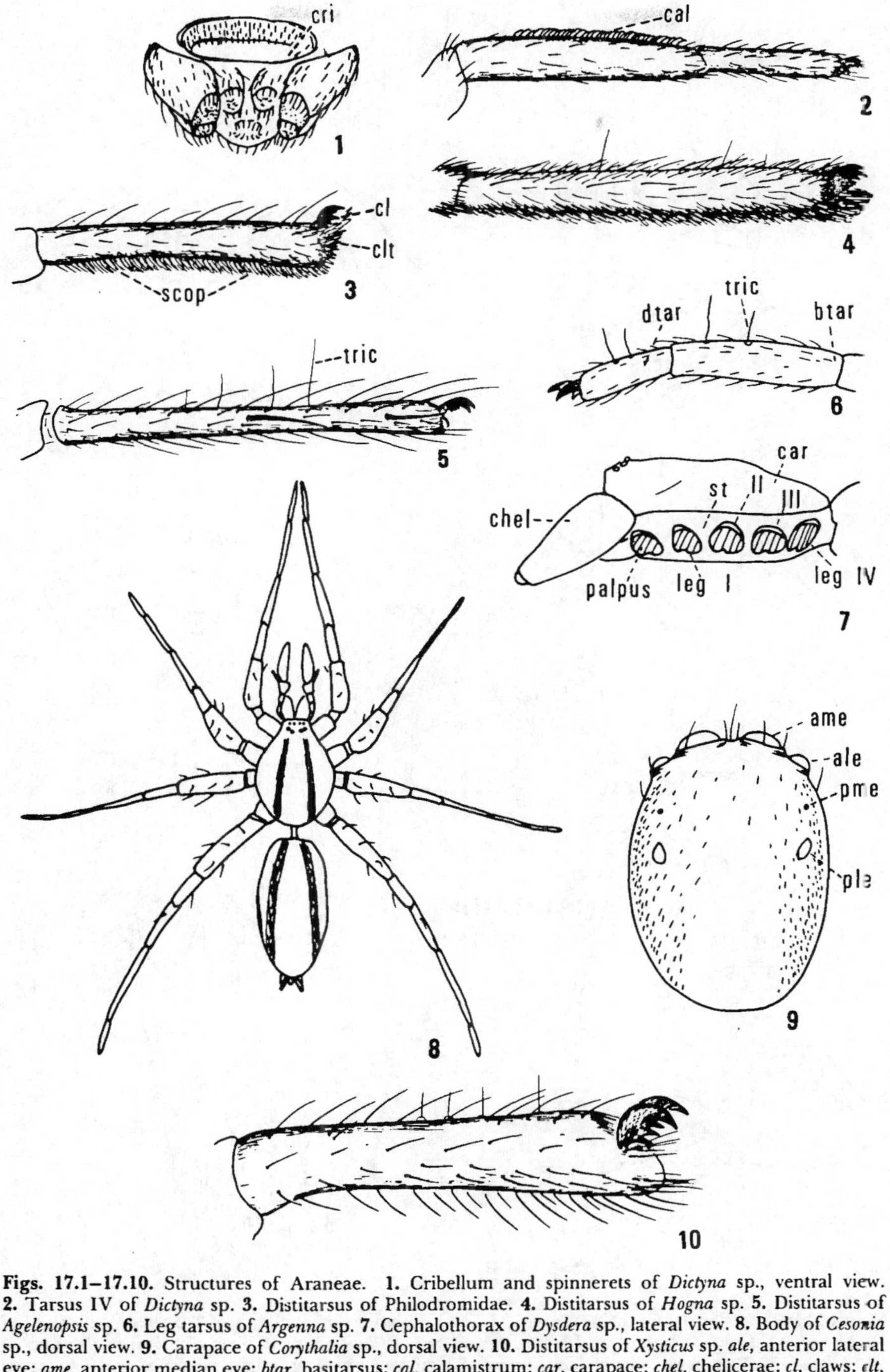

Figs. 17.1–17.10. Structures of Araneae. **1.** Cribellum and spinnerets of *Dictyna* sp., ventral view. **2.** Tarsus IV of *Dictyna* sp. **3.** Distitarsus of Philodromidae. **4.** Distitarsus of *Hogna* sp. **5.** Distitarsus of *Agelenopsis* sp. **6.** Leg tarsus of *Argenna* sp. **7.** Cephalothorax of *Dysdera* sp., lateral view. **8.** Body of *Cesonia* sp., dorsal view. **9.** Carapace of *Corythalia* sp., dorsal view. **10.** Distitarsus of *Xysticus* sp. *ale,* anterior lateral eye; *ame,* anterior median eye; *btar,* basitarsus; *cal,* calamistrum; *car,* carapace; *chel,* chelicerae; *cl,* claws; *clt,* claw tufts; *cri,* cribellum; *dtar,* distitarsus; *ple,* posterior lateral eye; *pme,* posterior median eye; *scop,* scopula; *st,* sternum; *tric,* trichobothrium.

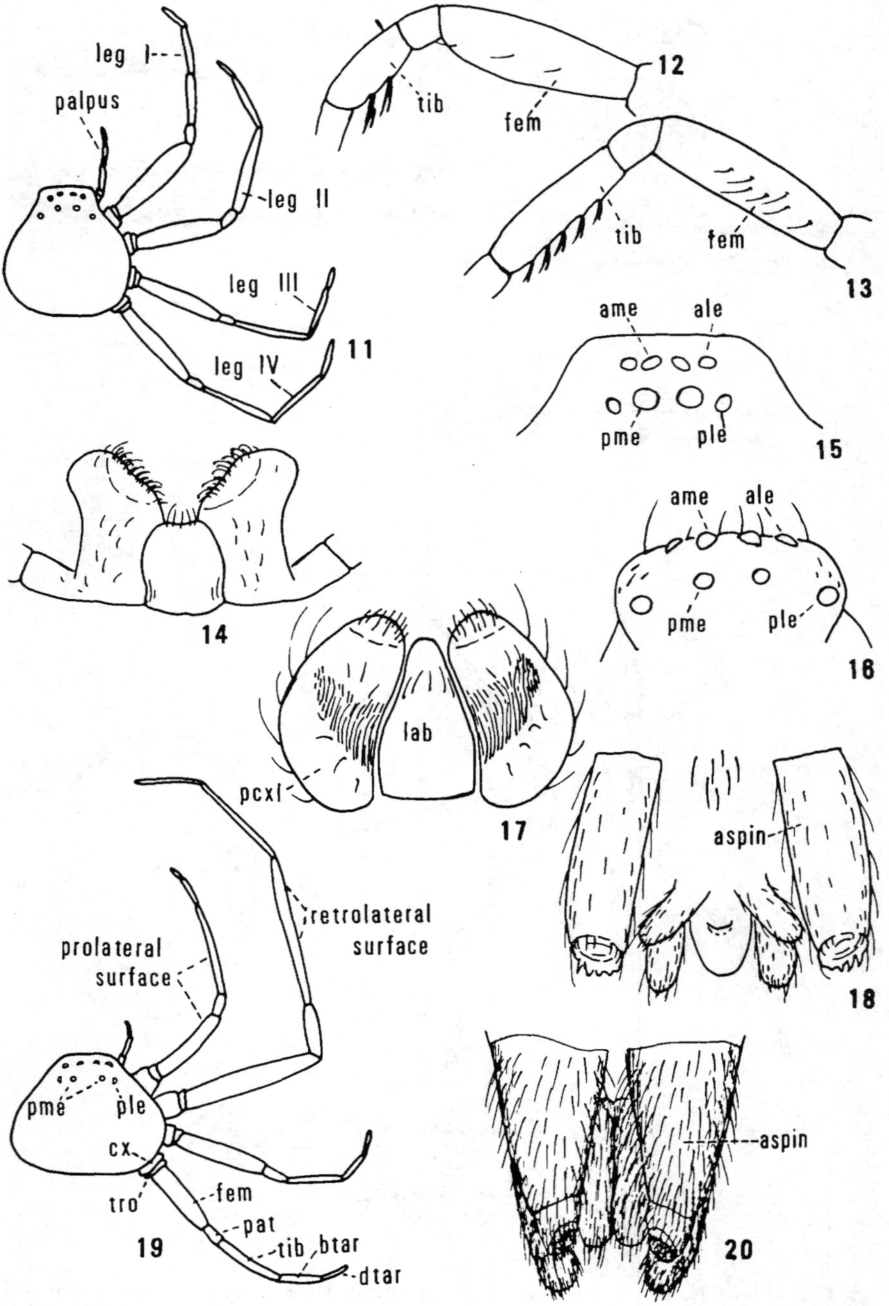

Figs. 17.11–17.20. Structures of Araneae. **11.** Cephalothorax and abdomen of *Thanatus* sp. **12.** Leg I of *Ozyptila* sp. **13.** Leg I of *Xysticus* sp. **14.** Palp-coxal lobes of *Clubiona* sp. **15.** Eyes of *Zelotes* sp. **16.** Eyes of *Trachelas* sp. **17.** Palp-coxal lobes of *Gnaphosa* sp. **18.** Spinnerets of *Gnaphosa* sp., ventral view. **19.** Cephalothorax and legs of *Ebo* sp. **20.** Spinnerets of *Clubiona* sp., ventral view. *ale,* anterior lateral eye; *ame,* anterior median eye; *aspin,* anterior spinneret; *btar,* basitarsus; *cx,* coxa; *dtar,* distitarsus; *fem,* femur; *lab,* labium; *pat,* patella; *pcxl,* palp-coxal lobe; *ple,* posterior lateral eye; *pme,* posterior median eye; *tib,* tibia; *tro,* trochanter.

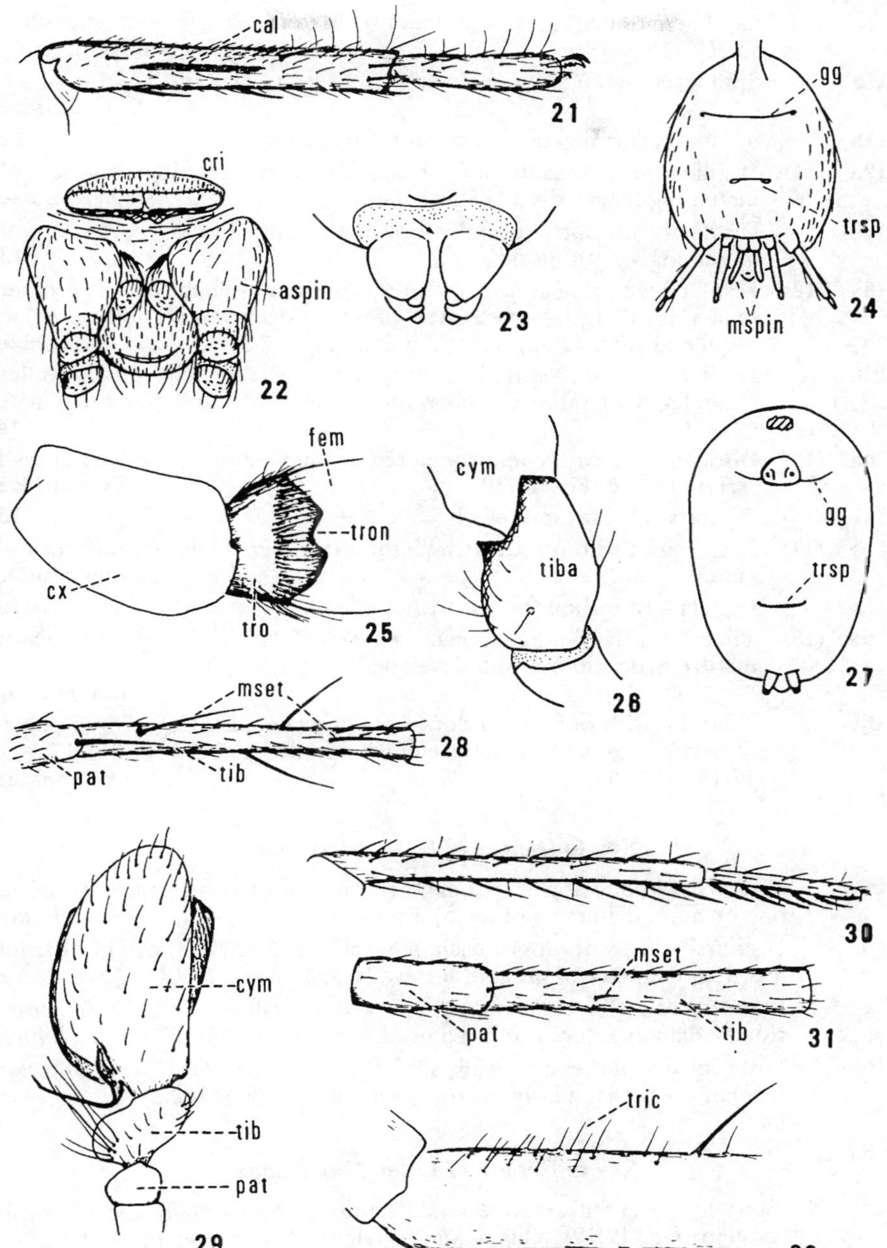

Figs. 17.21–17.32. Structures of Araneae. **21.** Tarsus IV of *Callobius* sp. **22.** Cribellum and spinnerets of *Callobius* sp. **23.** Spinnerets of *Theridion* sp. **24.** Abdomen of *Antistea* sp., ventral view. **25.** Leg trochanter of *Hogna* sp., ventral view. **26.** Palpal tibia of male *Pocadicnemis* sp. **27.** Abdomen of *Tennesseellum* sp., ventral view. **28.** Tibia IV of Linyphiidae, dorsal view. **29.** Palpus of male *Neriene* sp., dorsal view. **30.** Tarsus IV of *Theridion* sp. **31.** Tibia IV of *Erigone* sp., dorsal view. **32.** Leg femur of Tetragnathidae. *aspin,* anterior spinneret; *cal,* calamistrum; *cri,* cribellum; *cx,* coxa; *cym,* cymbium; *fem,* femur; *gg,* genital groove; *mset,* macroseta; *mspin,* median spinneret; *pat,* patella; *tib,* tibia; *tiba,* retrolateral apophysis of palpal tibia; *tric,* trichobothrium; *tro,* trochanter; *tron,* notch on leg trochanter; *trsp,* tracheal spiracle.

band at midline (Fig. 17.1); calamistrum (*cal*) nearly as long as basitarsus IV (Fig. 17.2) . . . . **Dictynidae** (in part)

11a (9) Spinnerets arranged in essentially transverse row (Fig. 17.24) . . . . **Hahniidae**

11b Spinnerets arranged in tight cluster (e.g., Fig. 17.27) **12**

12a (11) Distitarsi with dorsal trichobothria (*tric*) aligned in a single row and of increasing length distad (Fig. 17.5) . . . . **Agelenidae**

12b Distitarsi with dorsal trichobothria not arranged in single row, not of increasing length distad **13**

13a (12) Distitarsi with dorsal trichobothria projecting well above tips of other setae (Fig. 17.4); leg trochanters (*tro*) with deep ventral notch (*tron*) at distal end (Fig. 17.25); legs scopulate (Fig. 17.4) . . . . . . **Lycosidae**

13b Distitarsi with dorsal trichobothria not projecting above tips of other setae; leg trochanters usually without ventral notch; legs not scopulate **14**

14a (13) Distitarsus IV with comb composed of midventral row of long curved serrated setae (Fig. 17.30) . . . . **Theridiidae**

14b Distitarsus IV without comb **15**

15a (14) Leg femora with dorsal trichobothria (*tric*, Fig. 17.32) (genus ***Pachygnatha***) . . . . **Tetragnathidae**

15b Leg femora without dorsal trichobothria **16**

16a (15) Tibia IV with two dorsal macrosetae (*mset*, Fig. 17.28); palpal tibia of mature male without well-developed apophysis (Fig. 17.29) . . . . . . **Linyphiidae**

16b Tibia IV with one or no dorsal macrosetae (*mset*, Fig. 17.31); palpal tibia of male with well-developed apophysis (Figs. 17.26, 17.126, 17.137–17.139) . . . . **Erigonidae**

Key to Genera of Litter Dictynidae

1a Distitarsi without dorsal trichobothria; height of front usually equal to one or more diameters of an anterior eye (Fig. 17.33) (22)*. . . ***Dictyna***

1b Distitarsi (*dtar*) with dorsal trichobothria (*tric*, Fig. 17.6); height of front less than diameter of an anterior eye (Fig. 17.34) **2**

2a (1) Posterior row of eyes strongly procurved (dorsal view, Fig. 17.37); anterior median eyes (*ame*) reduced or absent (Fig. 17.34) (22) . . . . ***Lathys***

2b Posterior row of eyes essentially straight (dorsal view, Fig. 17.38); anterior median eyes (*ame*) nearly as large as other eyes (Fig. 17.35) (22) ***Argenna***

Key to Genera of Litter Thomisidae

1a Body and legs with clavate setae; femur I (*fem*) usually conspicuously swollen (Fig. 17.12); tibia 1 (*tib*) usually with only two pairs of ventral macrosetae (Fig. 17.12) (26). . . . . ***Ozyptila***

1b Body and legs usually without clavate setae; femur I (*fem*) not conspicuously swollen (Fig. 17.13); tibia I (*tib*) with more than two pairs of ventral macrosetae (Fig. 17.13) (27–30). . . . . ***Xysticus***

* Note: Lightface numerals in parens are citations to reference list.

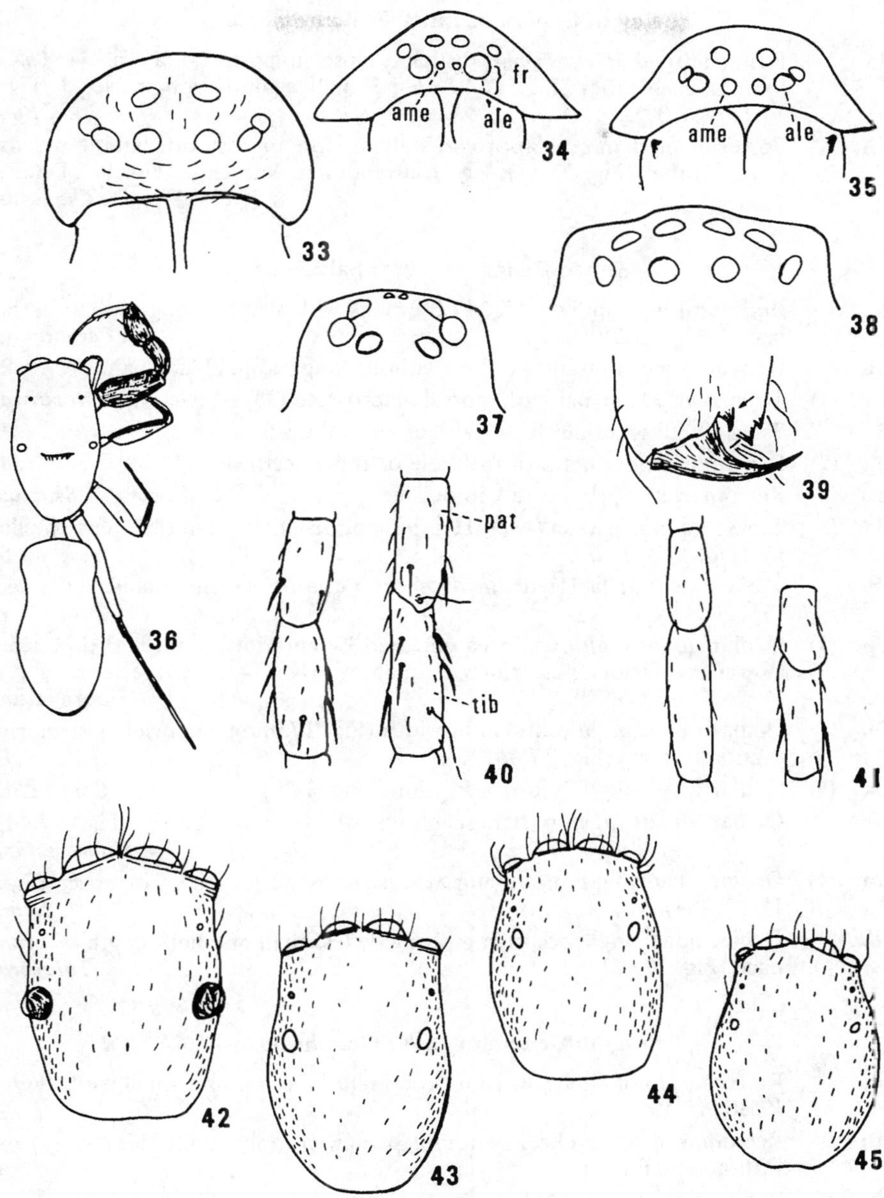

Figs. 17.33–17.45. Structures of Dictynidae and Salticidae. **33.** Eyes of *Dictyna* sp., anterior view. **34.** Eyes of *Lathys* sp., anterior view. **35.** Eyes of *Argenna* sp., anterior view. **36.** Body of *Peckhamia* sp., dorsal view. **37.** Eyes of *Lathys* sp., dorsal view. **38.** Eyes of *Argenna* sp., dorsal view. **39.** Chelicera of *Habronattus* sp., retrolateral view. **40.** Legs III (right) and IV (left) of *Habronattus* sp., dorsal view. **41.** Legs III (right) and IV (left) of *Neon* sp., dorsal view. **42.** Carapace of *Neon* sp., dorsal view. **43.** Carapace of *Talavera* sp., dorsal view. **44.** Carapace of *Habrocestum* sp., dorsal view. **45.** Carapace of *Habronattus* sp., dorsal view. *ale,* anterior lateral eye; *ame,* anterior median eye; *fr,* front of carapace; *pat,* patella; *tib,* tibia.

Key to Genera of Litter Philodromidae

1a Posterior median eyes (*pme*) distinctly closer to posterior lateral eyes (*ple*) than to each other (Fig. 17.19); leg II half again as long as leg I (Fig. 17.19) (30,31) . . . . ***Ebo***

1b Posterior median eyes approximately as close to posterior lateral eyes as to each other (Fig. 17.11); leg II little longer than leg I (Fig. 17.11) (32) ***Thanatus***

Key to Genera of Litter Salticidae

1a Body and legs antlike (Fig. 17.36); legs with slender longitudinal black bands (Fig. 17.36) . . . . ***Peckhamia***

1b Body and legs not antlike; legs without longitudinal black bands . . . **2**

2a (1) Tibia I with four pairs of ventral macrosetae (33) . . . . ***Marpissa***

2b Tibia I with less than four pairs of ventral macrosetae **3**

3a (2) Retromargin of chelicera with one or more teeth (Fig. 17.39) **4**

3b Retromargin of chelicera without teeth . . . . ***Sitticus***

4a (3) Tibia (*tib*) plus patella (*pat*) III as long as or longer than tibia plus patella IV (Fig. 17.40) **5**

4b Tibia plus patella III distinctly shorter than tibia plus patella IV (Fig. 17.41) **7**

5a (4) Ocular quadrangle (i.e., area enclosed by anterior lateral and posterior lateral eyes) wider posteriorly than anteriorly (Fig. 17.45) (33a) . . . . . ***Habronattus***

5b Ocular quadrangle uniform in width (Fig. 17.9) or narrower posteriorly than anteriorly (Fig. 17.44) **6**

6a (5) Ocular quadrangle uniform in width (Fig. 17.9) . . . . ***Corythalia***

6b Ocular quadrangle narrower posteriorly than anteriorly (Fig. 17.44) ***Habrocestum***

7a (4) Ocular quadrangle occupying at least one-half length of carapace (Fig. 17.42) (34) . . . . ***Neon***

7b Ocular quadrangle occupying distinctly less than one-half length of carapace (Fig. 17.43) . . . . ***Talavera***

Key to Genera of Litter Gnaphosidae

1a Retromargin of chelicera with prominent lobe, without small teeth (Figs. 17.46, 17.49) **2**

1b Retromargin of chelicera without prominent lobe, with (Fig. 17.47) or without small teeth **3**

2a (1) Lobe on retromargin of chelicera serrated (Fig. 17.46) (35) . . . . ***Gnaphosa***

2b Lobe on retromargin of chelicera smooth (Fig. 17.49) (36) . . . . ***Callilepis***

3a (1) Body antlike, with slender legs (Fig. 17.55); abdomen (and usually carapace) covered with iridescent scales; femur I usually with one dorsal macroseta (36c) . . . . ***Micaria***

3b Body not antlike, with stout legs; abdomen and carapace without iridescent scales; femur I with two or three dorsal macrosetae **4**

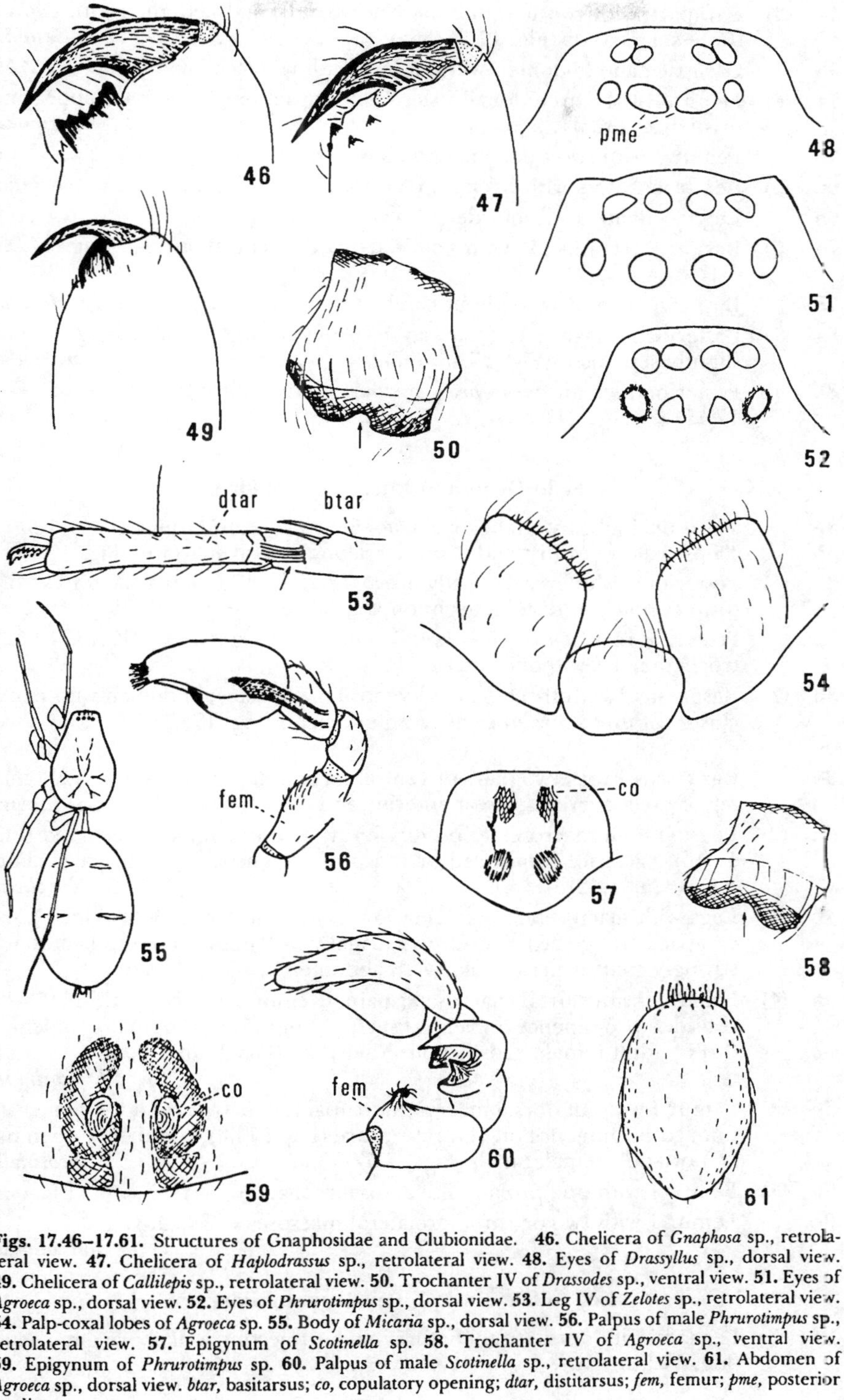

Figs. 17.46–17.61. Structures of Gnaphosidae and Clubionidae. **46.** Chelicera of *Gnaphosa* sp., retrolateral view. **47.** Chelicera of *Haplodrassus* sp., retrolateral view. **48.** Eyes of *Drassyllus* sp., dorsal view. **49.** Chelicera of *Callilepis* sp., retrolateral view. **50.** Trochanter IV of *Drassodes* sp., ventral view. **51.** Eyes of *Agroeca* sp., dorsal view. **52.** Eyes of *Phrurotimpus* sp., dorsal view. **53.** Leg IV of *Zelotes* sp., retrolateral view. **54.** Palp-coxal lobes of *Agroeca* sp. **55.** Body of *Micaria* sp., dorsal view. **56.** Palpus of male *Phrurotimpus* sp., retrolateral view. **57.** Epigynum of *Scotinella* sp. **58.** Trochanter IV of *Agroeca* sp., ventral view. **59.** Epigynum of *Phrurotimpus* sp. **60.** Palpus of male *Scotinella* sp., retrolateral view. **61.** Abdomen of *Agroeca* sp., dorsal view. *btar,* basitarsus; *co,* copulatory opening; *dtar,* distitarsus; *fem,* femur; *pme,* posterior median eye.

4a **(3)** Carapace with two longitudinal black bands, and abdomen with two or three such bands (Fig. 17.8) (36a) . . . . ***Cesonia***

4b Carapace and abdomen without longitudinal black bands **5**

5a **(4)** Femur I with three dorsal macrosetae (distal one sometimes appearing bristlelike) (36b) . . . . ***Sergiolus***

5b Femur I with two dorsal macrosetae **6**

6a **(5)** Leg trochanters with deep notch (Fig. 17.50) (37). . . . . . . ***Drassodes***

6b Leg trochanters without deep notch **7**

7a **(6)** Basitarsi III and IV with comb at tip on ventral surface (Fig. 17.53) **8**

7b Basitarsi III and IV without comb (38) . . . . ***Haplodrassus***

8a **(7)** Posterior median eyes (*pme*) large, ovoid or angular, touching or only slightly separated (Fig. 17.48) (38a) . . . . ***Drassyllus***

8b Posterior median eyes (*pme*) smaller, round, distinctly separated (Fig. 17.15) (38b) . . . . ***Zelotes***

Key to Genera of Litter Clubionidae

1a Palp-coxal lobes straight or convex along lateral margins (Fig. 17.54) **2**

1b Palp-coxal lobes distinctly concave along lateral margins (Fig. 17.14) **6**

2a **(1)** Posterior row of eyes distinctly procurved (Fig. 17.51); trochanter IV with distinct notch in distal margin on ventral surface (Fig. 17.58) **3**

2b Posterior row of eyes recurved or nearly straight (Figs. 17.16, 17.52); trochanter IV without notch **4**

3a **(2)** Basitarsus I with three pairs of ventral macrosetae; abdomen with dense cluster of erect curved setae at anterior end (Fig. 17.61) (39,39a) . . . . ***Agroeca***

3b Basitarsus I with two pairs of ventral macrosetae; abdomen without cluster of erect curved setae at anterior end (39a,40) . . . . . . ***Castianeira***

4a **(2)** Legs without macrosetae; posterior row of eyes distinctly recurved (Fig. 17.16); carapace bright red or red brown, contrasting with paler abdomen (39a,41,42) . . . . ***Trachelas***

4b Legs with macrosetae; posterior row of eyes nearly straight (Fig. 17.52); carapace orange brown or brown, usually with network of black lines, not strongly contrasting in color with abdomen **5**

5a **(4)** Femur I with dorsal macroseta; palpal femur (*fem*) of adult male with rounded prominence on ventral surface (Fig. 17.56); copulatory opening (*co*) of adult female conspicuous, cavitylike (Fig. 17.59) (39a) . . . . . . ***Phrurotimpus***

5b Femur I without dorsal macroseta; palpal femur (*fem*) of adult male with hooked prominence on ventral surface (Fig. 17.60); copulatory openings (*co*) of adult female small (Fig. 17.57) (39a) . . . . ***Scotinella***

6a **(1)** Femur I with one prolateral macroseta (39a,43) . . . . ***Clubiona***

6b Femur I with two or three prolateral macrosetae (39a,43). . . . . ***Clubionoides***

Key to Genera of Litter Amaurobiidae

1a Leg tarsi with dorsal trichobothria long, projecting well above other setae (Fig. 17.21) **2**

1b Leg tarsi with dorsal trichobothria short, not projecting above other setae (23) . . . . *Titanoeca*

2a (1) Dorsal process on palpal tibia of adult male ridgelike (Fig. 17.62) or subdivided; epigynum of adult female with angular median plate lying within excavation at posterior margin (Fig. 17.64) (23) . . . *Amaurobius*

2b Dorsal process on palpal tibia of adult male tonguelike (Fig. 17.63) or minute and pointed (Fig. 17.66); epigynum of adult female with two broad lateral lobes that converge posteriad (Figs. 17.65, 17.67) 3

3a (2) Dorsal process on palpal tibia of adult male tonguelike (Fig. 17.63); epigynum of adult female with two lobes that enfold small triangular median plate (Fig. 17.67) (23) . . . . *Callobius*

3b Dorsal process on palpal tibia of adult male minute, pointed (Fig. 17.66); epigynum of adult female with two lobes not enfolding small median plate (Fig. 17.65) (23) . . . . *Callioplus*

Key to Genera of Litter Hahniidae

1a Tracheal spiracle (*trsp*) closer to base of spinnerets than to genital groove (*gg*) (Fig. 17.77); anterior median eyes (*ame*) distinctly smaller than anterior lateral eyes (*ale*) (Fig. 17.68) (44) . . . . *Hahnia*

1b Tracheal spiracle (*trsp*) closer to genital groove (*gg*) than to base of spinnerets (Fig. 17.74) or equidistant (Fig. 17.24); anterior median eyes (*ame*) slightly larger than or equal to anterior lateral eyes (*ale*) (Figs. 17.69, 17.70) 2

2a (1) Tracheal spiracle (*trsp*) nearly two-thirds the distance from base of spinnerets to genital groove (*gg*) (Fig. 17.74); median spinnerets (*mspin*) separated at bases (Fig. 17.74) (44) . . . . *Neoantistea*

2b Tracheal spiracle (*trsp*) equidistant between base of spinnerets and genital groove (*gg*) (Fig. 17.24); median spinnerets (*mspin*) touching at bases (Fig. 17.24) (44) . . . . *Antistea*

Key to Genera of Litter Agelenidae

1a Spider with both rows of eyes strongly procurved (Fig. 17.71); posterior spinnerets (*pspin*) with distal segment longer than basal segment (Fig. 17.76) (45) . . . . *Agelenopsis*

1b Spider with neither row of eyes strongly procurved (Fig. 17.72). Posterior spinnerets (*pspin*) with distal segment shorter than basal segment (Fig. 17.73) 2

2a (1) Tibia I with two unpaired ventral macrosetae in addition to two pairs (Fig. 17.75) . . . . *Cryphoeca*

2b Tibia I with one or no unpaired ventral macroseta in addition to two (or three) pairs 3

3a (2) Femur I with two dorsal macrosetae (46) . . . . *Circurina*

3b Femur I with one dorsal macroseta (47) . . . . *Cybaeus*

Key to Genera of Litter Lycosidae

1a Retromargins of both chelicerae with two teeth (Fig. 17.79) 2

1b Retromargins of one or of both chelicerae with three teeth (Fig. 17.78) 4

2a (1) Femur I with one prolateral macroseta 3

2b Femur I with two prolateral macrosetae (47a) . . . . *Trochosa* (in part)

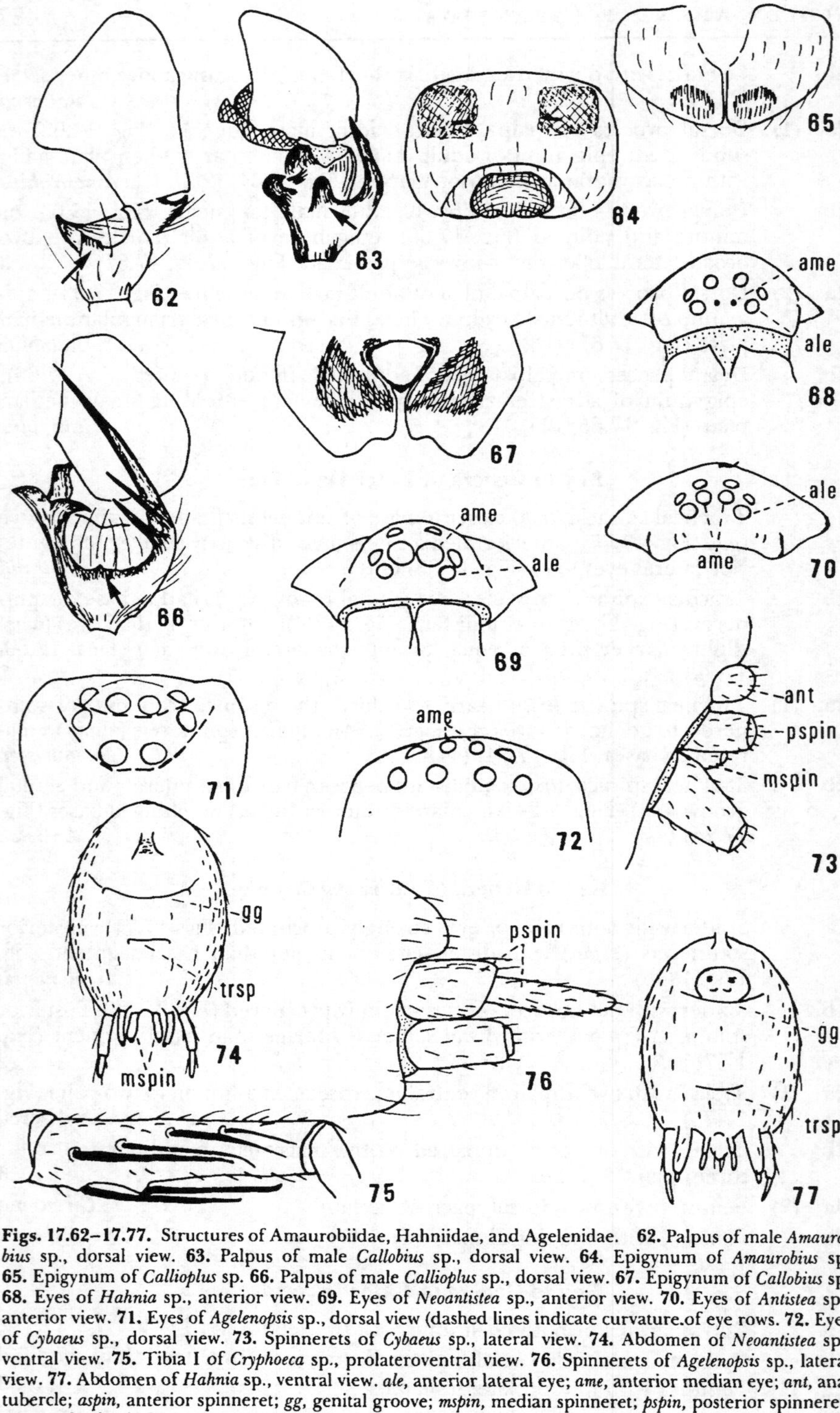

Figs. 17.62–17.77. Structures of Amaurobiidae, Hahniidae, and Agelenidae. **62.** Palpus of male *Amaurobius* sp., dorsal view. **63.** Palpus of male *Callobius* sp., dorsal view. **64.** Epigynum of *Amaurobius* sp. **65.** Epigynum of *Callioplus* sp. **66.** Palpus of male *Callioplus* sp., dorsal view. **67.** Epigynum of *Callobius* sp. **68.** Eyes of *Hahnia* sp., anterior view. **69.** Eyes of *Neoantistea* sp., anterior view. **70.** Eyes of *Antistea* sp., anterior view. **71.** Eyes of *Agelenopsis* sp., dorsal view (dashed lines indicate curvature.of eye rows. **72.** Eyes of *Cybaeus* sp., dorsal view. **73.** Spinnerets of *Cybaeus* sp., lateral view. **74.** Abdomen of *Neoantistea* sp., ventral view. **75.** Tibia I of *Cryphoeca* sp., prolateroventral view. **76.** Spinnerets of *Agelenopsis* sp., lateral view. **77.** Abdomen of *Hahnia* sp., ventral view. *ale,* anterior lateral eye; *ame,* anterior median eye; *ant,* anal tubercle; *aspin,* anterior spinneret; *gg,* genital groove; *mspin,* median spinneret; *pspin,* posterior spinneret; *trsp,* tracheal spircle.

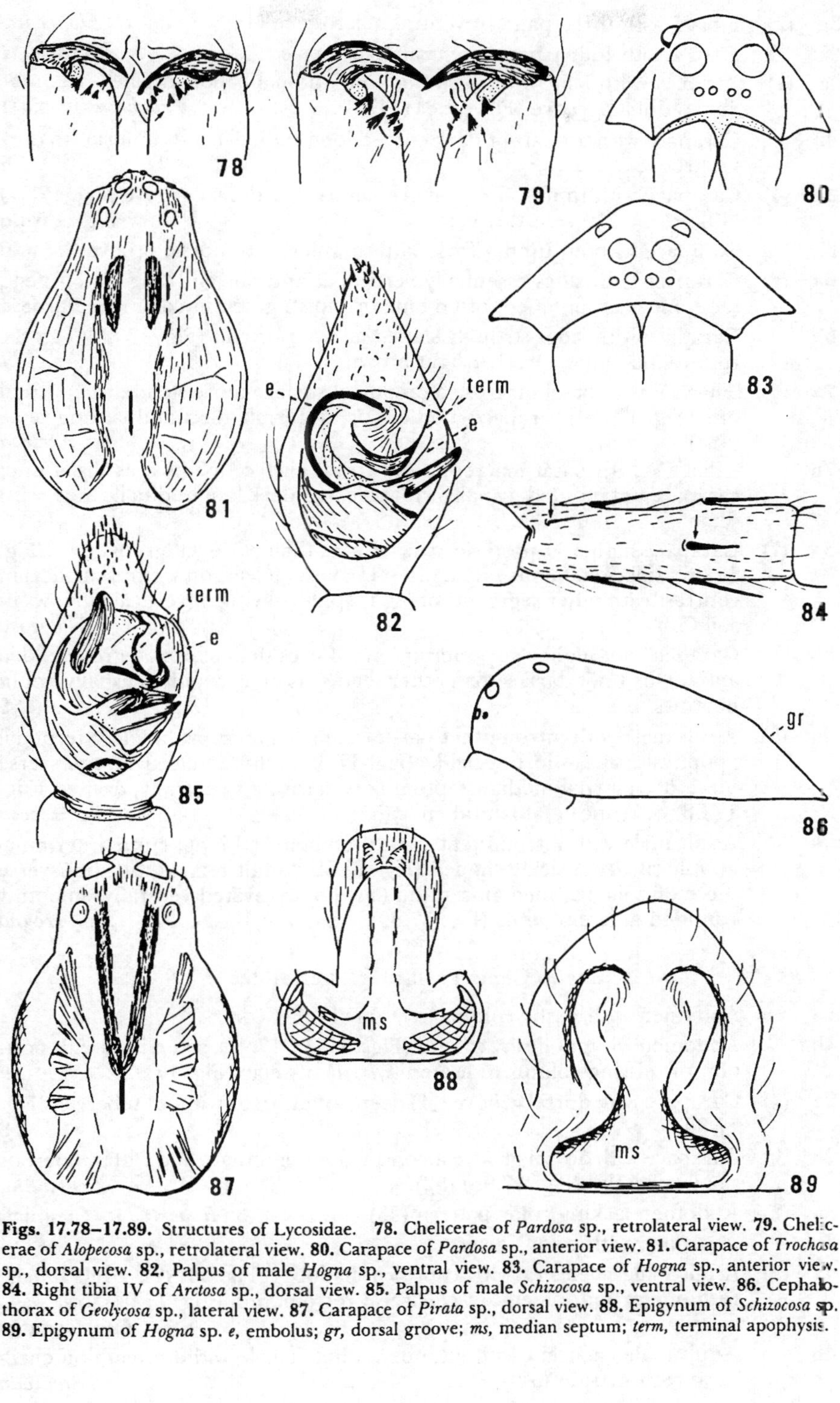

Figs. 17.78–17.89. Structures of Lycosidae. **78.** Chelicerae of *Pardosa* sp., retrolateral view. **79.** Chelicerae of *Alopecosa* sp., retrolateral view. **80.** Carapace of *Pardosa* sp., anterior view. **81.** Carapace of *Trochosa* sp., dorsal view. **82.** Palpus of male *Hogna* sp., ventral view. **83.** Carapace of *Hogna* sp., anterior view. **84.** Right tibia IV of *Arctosa* sp., dorsal view. **85.** Palpus of male *Schizocosa* sp., ventral view. **86.** Cephalothorax of *Geolycosa* sp., lateral view. **87.** Carapace of *Pirata* sp., dorsal view. **88.** Epigynum of *Schizocosa* sp. **89.** Epigynum of *Hogna* sp. *e,* embolus; *gr,* dorsal groove; *ms,* median septum; *term,* terminal apophysis.

3a (2) Tibia I with three pairs of ventral macrosetae (48) . . . . . . ***Alopecosa***
3b Tibia I with four pairs of ventral macrosetae . . . . ***Trabeops***
4a (1) Carapace with pair of dark curved longitudinal bands in pale area anterior to dorsal groove (Fig. 17.81) (47a) . . . . ***Trochosa*** (in part)
4b Carapace without pair of dark curved longitudinal bands anterior to dorsal groove **5**
5a (4) Carapace with tuning fork pattern anterior to dorsal groove (Fig. 17.87) (49) . . . . ***Pirata***
5b Carapace without tuning fork pattern anterior to dorsal groove **6**
6a (4) Carapace with sides essentially vertical at anterior end (Fig. 17.80); egg sac somewhat flattened, often blue or bluish green in color . . ***Pardosa***
6b Carapace with sides strongly sloped at anterior end (Fig. 17.83); egg sac round and white, gray, or light brown **7**
7a (6) Tibia IV with basal macroseta on dorsal surface more slender than distal one (Fig. 17.84); carapace mottled, or uniformly dark and lacking setae (49a) . . . . ***Arctosa***
7b Tibia IV with basal macroseta on dorsal surface as thick as distal one; carapace not mottled, usually with pale and dark longitudinal bands, with setae **8**
8a (7) Carapace higher at posterior row of eyes than at dorsal groove (*gr*) (Fig. 17.86); tibia and tarsus I (and often II) dark (at least on ventral surface) in contrast with other segments of leg I; spiders living in vertical burrows in soil (50) . . . . ***Geolycosa***
8b Carapace not higher at posterior row of eyes than at dorsal groove; tibia and tarsus I not darker than other segments of leg; spiders usually not in burrows **9**
9a (8) Adult male with intromittent part of embolus (*e*) short, angled; terminal apophysis (*term*) short, scalelike (Fig. 17.85); adult female with transverse piece of epigynal median septum (*ms*) excavated ventrally, excised (Fig. 17.88) or truncate at lateral ends (51) . . . . ***Schizocosa***
9b Adult male with intromittent part of embolus (*e*) long, curved; terminal apophysis (*term*) sickle-shaped (Fig. 17.82); adult female with transverse piece of epigynal median septum (*ms*) not excavated ventrally, smoothly rounded at lateral ends (Fig. 17.89) . . . . ***Hogna***

Key to Genera of Litter Theridiidae

1a Abdomen with fleshy colulus (*col*, Fig. 17.90) **2**
1b Abdomen without fleshy colulus (Figs. 17.23, 17.92), sometimes with position of missing colulus indicated by pair of setae (Fig. 17.92) **5**
2a (1) Carapace with dorsal groove (*gr*) deep, and with or without tubercles (Fig. 17.96) **3**
2b Carapace with dorsal groove inconspicuous or absent, and with numerous small tubercles (Fig. 17.95) (52). . . . . ***Crustulina***
3a (2) Abdomen lacking color pattern (53) . . . . ***Robertus*** (= ***Ctenium***)
3b Abdomen with color pattern. **4**
4a (3) Adult male chelicera with conspicuous lobes (Fig. 17.97); adult female with one or more teeth on cheliceral retromargin (Fig. 17.94) (54) . . . ***Enoplognatha***
4b Adult male chelicera without lobes; adult female without teeth on cheliceral retromargin (52) . . . . ***Steatoda***

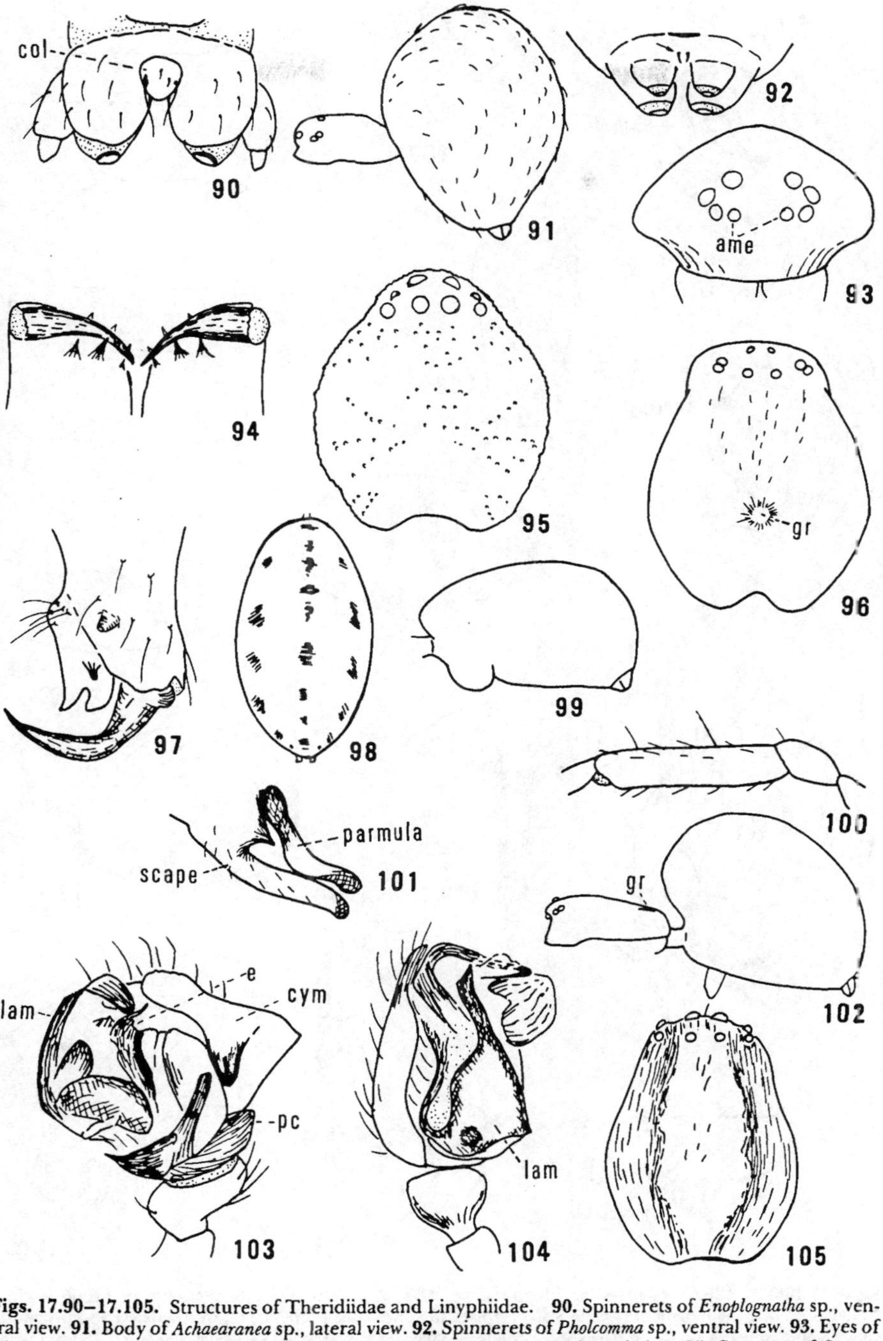

Figs. 17.90–17.105. Structures of Theridiidae and Linyphiidae. **90.** Spinnerets of *Enoplognatha* sp., ventral view. **91.** Body of *Achaearanea* sp., lateral view. **92.** Spinnerets of *Pholcomma* sp., ventral view. **93.** Eyes of *Pholcomma* sp., anterior view. **94.** Chelicerae of *Enoplognatha* sp., retrolateral view. **95.** Carapace of *Crustulina* sp., dorsal view. **96.** Carapace of *Enoplognatha* sp., dorsal view. **97.** Chelicera of *Enoplognatha* sp., prolateral view. **98.** Abdomen of *Stemonyphantes* sp., dorsal view. **99.** Abdomen of *Neriene* sp., lateral view. **100.** Leg I of *Centromerus* sp. **101.** Epigynum of *Bathyphantes* sp., lateral view. **102.** Body of *Allomengea* sp., lateral view. **103.** Left palpus of male *Meioneta* sp., retrolateral view. **104.** Left palpus of male *Bathyphantes* sp., mesal view. **105.** Carapace of *Tapinopa* sp., dorsal view. *ame,* anterior median eye; *aspin,* anterior spinneret; *col,* colulus; *cym, cymbium; e,* embolus; *gr,* dorsal groove; *lam,* lamina; *pc,* paracymbium.

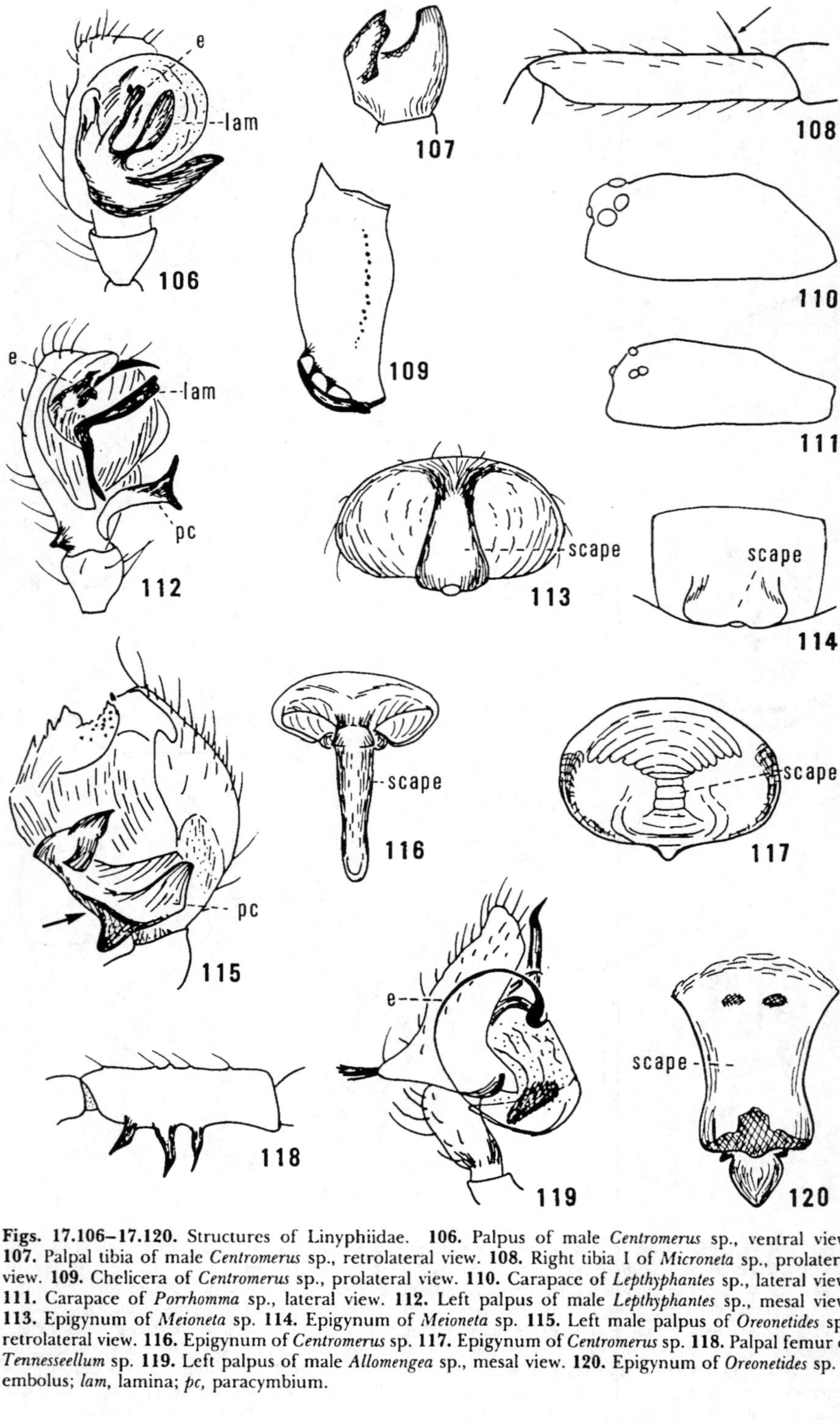

Figs. 17.106–17.120. Structures of Linyphiidae. **106.** Palpus of male *Centromerus* sp., ventral view. **107.** Palpal tibia of male *Centromerus* sp., retrolateral view. **108.** Right tibia I of *Microneta* sp., prolateral view. **109.** Chelicera of *Centromerus* sp., prolateral view. **110.** Carapace of *Lepthyphantes* sp., lateral view. **111.** Carapace of *Porrhomma* sp., lateral view. **112.** Left palpus of male *Lepthyphantes* sp., mesal view. **113.** Epigynum of *Meioneta* sp. **114.** Epigynum of *Meioneta* sp. **115.** Left male palpus of *Oreonetides* sp., retrolateral view. **116.** Epigynum of *Centromerus* sp. **117.** Epigynum of *Centromerus* sp. **118.** Palpal femur of *Tennesseellum* sp. **119.** Left palpus of male *Allomengea* sp., mesal view. **120.** Epigynum of *Oreonetides* sp. *e*, embolus; *lam*, lamina; *pc*, paracymbium.

5a **(1)** Position of missing colulus indicated by pair of setae (Fig. 17.92) . . . **6**
5b Position of missing colulus not indicated by setae (Fig. 17.23) **7**
6a **(5)** Anterior median eyes (*ame*) distinctly smaller than other eyes (Fig. 17.93); adult less than 1.8 mm in length (55) . . . . ***Pholcomma***
6b Anterior median eyes equal to other eyes or larger; adult more than 1.8 mm in length (56) . . . . ***Dipoena***
7a **(5)** Abdomen higher than long (Fig. 17.91) (57) . . . . ***Achaearanea***
7b Abdomen not higher than long (54) . . . . ***Theridion***

Key to Genera of Litter Linyphiidae

1a Femur I with a dorsal and/or a prolateral macroseta **2**
1b Femur I without macrosetae **8**
2a **(1)** Dorsum of abdomen with longitudinal rows of small black spots (Fig. 17.98) (58) . . . . ***Stemonyphantes***
2b Dorsum of abdomen without longitudinal rows of small black spots, with or without other pattern **3**
3a **(2)** Tibia I with one or more ventral macrosetae; abdomen usually about as thick at posterior end as at anterior end (lateral view, Fig. 17.99) (59) ***Neriene***
3b Tibia I without ventral macrosetae (Fig. 17.100); abdomen distinctly thicker at anterior end than at posterior end (lateral view Fig. 17.102) **4**
4a **(3)** Carapace with broad white median band contrasting with dark lateral areas (Fig. 17.105) . . . . ***Tapinopa***
4b Carapace essentially uniform in color **5**
5a **(4)** Male palpus with lamella (*lam*) large, shieldlike (Fig. 17.104); female epigynum with distinct parmula (Fig. 17.101) (60) . . . ***Bathyphantes***
5b Male palpus with lamella (*lam*) smaller, more slender (Figs. 17.103, 17.106, 17.112); female epigynum without parmula. **6**
6a **(5)** Tibia I with one or more retrolateral macrosetae; male chelicera lacking longitudinal row of small tubercles **7**
6b Tibia I without retrolateral macrosetae; male chelicera with longitudinal row of small tubercles (Fig. 17.109) (61). . . . . ***Centromerus*** (in part)
7a **(6)** Basitarsi with one or more macrosetae; eyes located on prominence (Fig. 17.110) (62) . . . . ***Lepthyphantes***
7b Basitarsi without macrosetae; eyes not located on prominence (Fig. 17.111) . . . . ***Porrhomma***
8a **(1)** Tracheal spiracle (*trsp*) located at one-third the distance from spinnerets to genital groove (*gg*) (Fig. 17.27); male palpal tibia with row of ventral spurs (Fig. 17.118) . . . . ***Tennesseellum***
8b Tracheal spiracle located immediately anterior to spinnerets; male palpal tibia without ventral spurs **9**
9a **(8)** Basitarsus IV with trichobothrium **10**
9b Basitarsus IV without trichobothrium **13**
10a **(9)** Abdomen with dorsal pattern; male embolus (*e*) long, threadlike (Fig. 17.119). **11**
10b Abdomen without dorsal pattern; male embolus (*e*) short, not threadlike (Figs. 17.103, 17.106, 17.112) **12**

11a **(10)** Carapace thicker at level of posterior row of eyes than at dorsal groove (*gr*) (lateral view, Fig. 17.102); chelicerae with four or five teeth on promargin; tibia I with prolateral, retrolateral, and ventral macrosetae (63) . . . . ***Allomengea***

11b Carapace approximately as thick at level of posterior row of eyes as at dorsal groove (lateral view); chelicerae with three teeth on promargin; tibia I without prolateral, retrolateral, or ventral macrosetae (64) . . . ***Wubana***

12a **(10)** Tibia I with basal dorsal macroseta located at 0.14 length of segment (Fig. 17.108) . . . . ***Microneta***

12b Tibia I with basal dorsal macroseta located at 0.25 length of segment ***Macrargus***

13a **(9)** Male palpus with paracymbium folded along margin (Fig. 17.115); female epigynum with long broad scape (Fig. 17.120). . . . . ***Oreonetides***

13b Male palpus with paracymbium not folded at margin (Fig. 17.107); female epigynum with scape platelike (Fig. 17.114), long and slender (Fig. 17.116), curled (Fig. 17.113), or short and wrinkled (Fig. 17.117) **14**

14a **(13)** Male palpus with lamella (*lam*) long, bent prolaterad or dorsad near tip (Fig. 17.103); male chelicera without longitudinal row of small tubercles; female epigynum with scape platelike (Fig. 17.114) or curled (Fig. 17.113) . . . . ***Meioneta***

14b Male palpus with lamella (*lam*) short, somewhat sickle-shaped, not bent near tip (Fig. 17.106); male chelicera with longitudinal row of small tubercles (Fig. 17.109); female epigynum with scape long and slender (Fig. 17.116) or short and wrinkled (Fig. 17.117) (61). . . . . ***Centromerus*** (in part)

Key to Genera of Litter Erigonidae

1a Adult male **2**

1b Adult female **24**

2a **(1)** Carapace with pit on each side of eye area (Figs. 17.121, 17.122, 17.125) **3**

2b Carapace without pits **8**

3a **(2)** Embolus (*e*) long, threadlike (Fig. 17.125). **6**

3b Embolus (*e*) shorter, not threadlike (Fig. 17.129) **4**

4a **(3)** Carapace (and sternum) coarsely rugose (Fig. 17.123) (65) ***Ceratinops***

4b Carapace smooth **5**

5a **(4)** Palpal tibia with apophysis drawn into two points (Fig. 17.126) (65) ***Eridantes***

5b Palpal tibia with apophysis drawn into one point (Fig. 17.124) or none (65) . . . . ***Tapinocyba***

6a **(3)** Posterior median eyes (*pme*) situated at base of carapace hump (Fig. 17.121) (65,65a). . . . . ***Scotinotylus***

6b Posterior median eyes (*pme*) situated on carapace hump (Fig. 17.125) **7**

7a **(6)** Retrolateral apophysis on palpal tibia long, slender, tapered to point (Fig. 17.144) (65) . . . . ***Minyriolus***

7b Retrolateral apophysis (*tiba*) on palpal tibia short, broad, little tapered (Fig. 17.26), sometimes forked (66) . . . . ***Pocadicnemis***

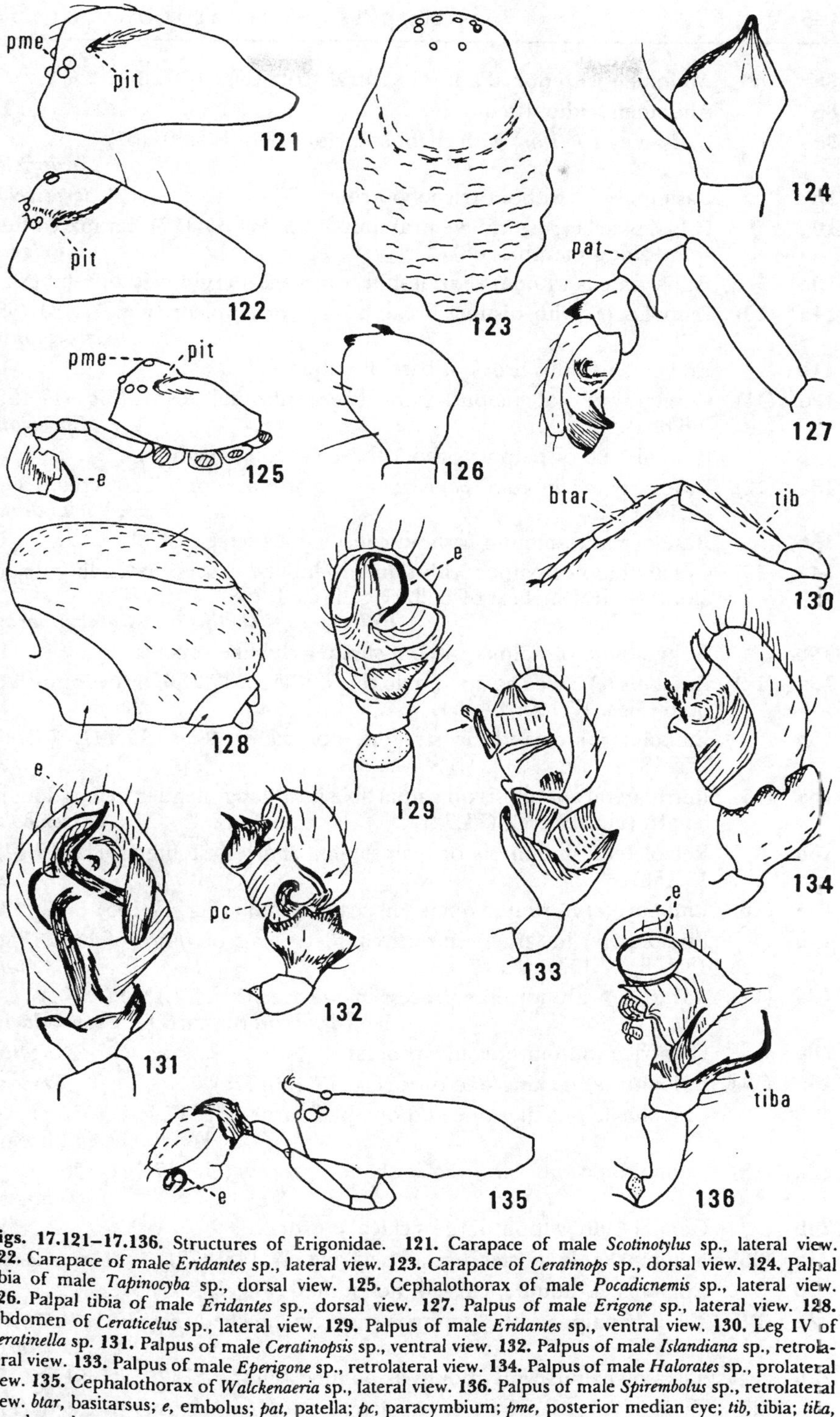

Figs. 17.121–17.136. Structures of Erigonidae. **121.** Carapace of male *Scotinotylus* sp., lateral view. **122.** Carapace of male *Eridantes* sp., lateral view. **123.** Carapace of *Ceratinops* sp., dorsal view. **124.** Palpal tibia of male *Tapinocyba* sp., dorsal view. **125.** Cephalothorax of male *Pocadicnemis* sp., lateral view. **126.** Palpal tibia of male *Eridantes* sp., dorsal view. **127.** Palpus of male *Erigone* sp., lateral view. **128.** Abdomen of *Ceraticelus* sp., lateral view. **129.** Palpus of male *Eridantes* sp., ventral view. **130.** Leg IV of *Ceratinella* sp. **131.** Palpus of male *Ceratinopsis* sp., ventral view. **132.** Palpus of male *Islandiana* sp., retrolateral view. **133.** Palpus of male *Eperigone* sp., retrolateral view. **134.** Palpus of male *Halorates* sp., prolateral view. **135.** Cephalothorax of *Walckenaeria* sp., lateral view. **136.** Palpus of male *Spirembolus* sp., retrolateral view. *btar*, basitarsus; *e*, embolus; *pat*, patella; *pc*, paracymbium; *pme*, posterior median eye; *tib*, tibia; *tiba*, retrolateral apophysis of palpal tibia.

8a (2) Abdomen with dorsal and/or ventral scuta (Fig. 17.128) **9**
8b Abdomen without scuta **10**
9a (8) Basitarsus IV (*btar*) with trichobothrium (Fig. 17.130) (67) . . . . . . ***Ceratinella***
9b Basitarsus IV without trichobothrium (67) . . . . ***Ceraticelus***
10a (8) Palpal patella (*pat*) with ventral spur at tip (Fig. 17.127); carapace often spiny along margins (68) . . . . ***Erigone***
10b Palpal patella without ventral spur; carapace margins without spines **11**
11a (10) Embolus (*e*) with distinct break at base of tailpiece (Fig. 17.131 (69) ***Ceratinopsis***
11b Embolus without break at base of tailpiece. **12**
12a (11) Genital bulb of palpus with three subequal teeth (Fig. 17.133) (68,68a). . . . . ***Eperigone***
12b Genital bulb of palpus without three subequal teeth **13**
13a (12) Base of paracymbium (*pc*) bearing two or more conspicuous setae (Fig. 17.132) (70) . . . . ***Islandiana***
13b Base of paracymbium without conspicuous setae. **14**
14a (13) Genital bulb of palpus with slender clublike process extending distad along ventral surface of bulb (Fig. 17.134) (68) . . . . ***Halorates*** (= ***Collinsia*** = ***Catabrithorax***)
14b Genital bulb of palpus without slender clublike process **15**
15a (14) Embolus (*e*) coiled at tip of bulb (Figs. 17.135, 17.136) or over much of bulb (Figs. 17.150, 17.151) **16**
15b Embolus (*e*) curved or straight, not coiled (Figs. 17.141, 17.147, 17.148). **20**
16a (15) Retrolateral apophysis on palpal tibia (*tiba*) long, slender throughout its length (Fig. 17.136) (71,72). . . . . ***Spirembolus***
16b Retrolateral apophysis on palpal tibia shorter, stouter (Figs. 17.142, 17.150). **17**
17a (16) Embolus (*e*) in simple open coil at tip of bulb (Fig. 17.135) **18**
17b Embolus (*e*) in large coil encircling one-half or more of bulb (Figs. 17.150, 17.151). **19**
18a (17) Carapace with hornlike process in eye area (Fig. 17.135) (73,74) . . . ***Walckenaeria*** (in part) (= ***Cornicularia***)
18b Carapace without hornlike process (75) . . . . ***Sisicottus***
19a (17) Embolus (*e*) making one turn (Fig. 17.150) (75) . . . . ***Scironis***
19b Embolus (*e*) making one and one-half turns (Fig. 17.151) (73,74). . . ***Walckenaeria*** (in part)
20a (15) Genital bulb with large sickle-shaped process (Fig. 17.141) (76). . . . ***Tunagyna***
20b Genital bulb without large sicklelike process **21**
21a (20) Embolus (*e*) long, slender, arched (Fig. 17.148) (77). . . ***Grammonota***
21b Embolus (*e*) minute, curved, not arched (Fig. 17.147) **22**
22a (21) Tibia of palpus with apophysis truncate, notched (Fig. 17.138) (75,76) ***Sciastes***
22b Tibia of palpus with apophysis pointed (Figs. 17.137, 17.139) . . . **23**
23a (22) Tibia of palpus with short retrolateral apophysis bearing two teeth (Fig. 17.137) (75,76). . . . . ***Diplocentria*** (= ***Scotoussa***)

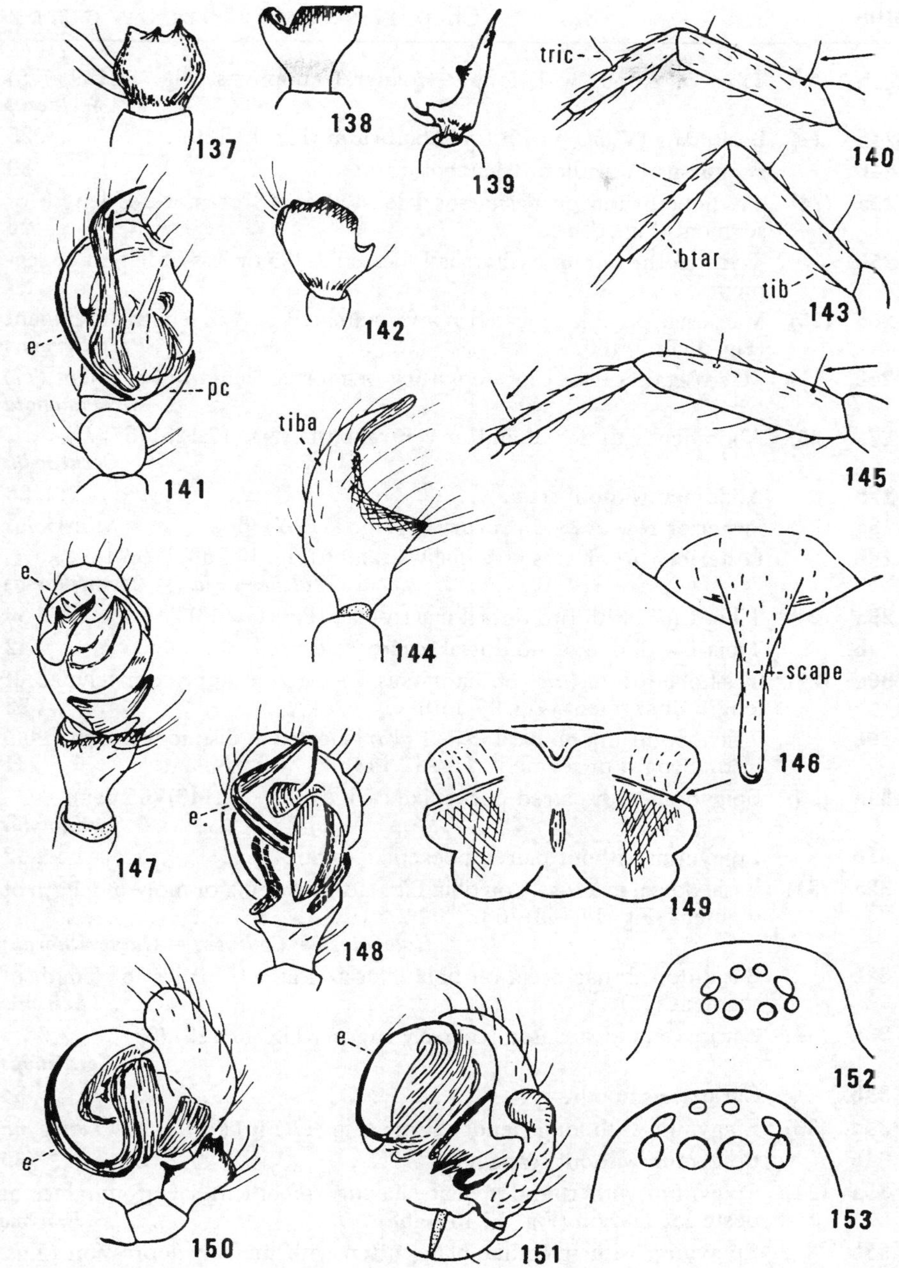

Figs. 17.137–17.153. Structures of Erigonidae. **137.** Palpal tibia of male *Diplocentria* sp., dorsal view. **138.** Palpal tibia of male *Sciastes* sp., dorsal view. **139.** Palpal tibia of male *Scylaceus* sp., dorsal view. **140.** Leg I of *Eridantes* sp. **141.** Palpus of male *Tunagyna* sp., ventral view. **142.** Palpal tibia of male *Sisicottus* sp., dorsal view. **143.** Leg I of *Halorates* sp. **144.** Palpal tibia of male *Minyriolus* sp., dorsal view. **145.** Leg I of *Pocadicnemis* sp. **146.** Epigynum of *Tunagyna* sp. **147.** Palpus of male *Sciastes* sp., ventral view. **148.** Palpus of male *Grammonota* sp., ventral view. **149.** Epigynum of *Scotinotylus* sp. **150.** Palpus of male *Scironis* sp., retrolateral view. **151.** Palpus of male *Walckenaeria* sp., retrolateral view. **152.** Eyes of *Minyriolus* sp., dorsal view. **153.** Eyes of *Walckenaeria* sp., dorsal view. *e,* embolus; *pc,* paracymbium; *tiba,* retrolateral apophysis of palpal tibia; *tric,* trichobothrium.

23b Tibia of palpus with long retrolateral apophysis (Fig. 17.139) (75) ***Scylaceus***

24a (1) Basitarsus IV (*btar*) with trichobothrium (Fig. 17.130) **25**

24b Basitarsus IV without trichobothrium **29**

25a (24) Trichobothrium on basitarsus I located at 0.8 or more of length of segment (Fig. 17.145) **26**

25b Trichobothrium on basitarsus I located at 0.5 or less of length of segment **27**

26a (25) Macroseta on tibia I located at approximately 0.14 of length of segment (Fig. 17.145) (66) . . . . ***Pocadicnemis***

26b Macroseta on tibia I located at 0.2 or more of length of segment (77) ***Grammonota***

27a (25) Abdomen with dorsal and/or ventral scuta (Fig. 17.128) (67) . . . . . ***Ceratinella***

27b Abdomen without scuta **28**

28a (27) Posterior row of eyes procurved (Fig. 17.152) (65). . . . . ***Minyriolus***

28b Posterior row of eyes essentially straight (Fig. 17.153) (73,74). . . . . ***Walckenaeria*** (= ***Cornicularia***)

29a (24) Tibia I (*tib*) with two dorsal macrosetae (Fig. 17.143) **30**

29b Tibia I with one or no dorsal macrosetae **42**

30a (29) Trichobothrium (*tric*) on basitarsus I located at approximately 0.5 of length of segment (Fig. 17.140) **33**

30b Trichobothrium on basitarsus I (*btar*) located at distinctly less than 0.5 of the length of segment (Fig. 17.143) **31**

31a (30) Epigynum with paired pocketlike cavities (Fig. 17.149) (65,65a) . . . ***Scotinotylus***

31b Epigynum without paired pocketlike cavities. **32**

32a (31) Basal dorsal macroseta on tibia I (*tib*) located at 0.2 or more of length of segment (Fig. 17.143) (68). . . . . ***Halorates*** (= ***Collinsia*** = ***Catabrithorax***)

32b Basal dorsal macroseta on tibia I located at 0.15 or less of length of segment (75) . . . . ***Scironis***

33a (30) Carapace (and sternum) coarsely rugose (Fig. 17.123) (65) . . . . . . ***Ceratinops***

33b Carapace smooth. **34**

34a (33) Epigynum with long slender scape (Fig. 17.146) (76). . . . ***Tunagyna***

34b Epigynum without scape **35**

35a (34) Epigynum with rounded plate having smooth mesal prominence at posterior margin (Fig. 17.154) (68) . . . . ***Erigone***

35b Epigynum with irregular plate, often with median depression (Figs. 17.155, 17.157, 17.158) **36**

36a (35) Posterior row of eyes distinctly procurved (Fig. 17.160) (65) . . . . . ***Eridantes***

36b Posterior row of eyes essentially straight (Fig. 17.156) **37**

37a (36) Dorsal macroseta on tibia IV equal in length to thickness of tibia (71,72) . . . . ***Spirembolus***

37b Dorsal macroseta on tibia IV distinctly greater in length than thickness of tibia **38**

38a	**(37)**	Basal dorsal macroseta on tibia I located at 0.15 of length of segment (75,76). . . . . ***Diplocentria*** (= ***Scotoussa***)
38b		Basal dorsal macroseta on tibia I located at 0.2 or more of length of segment **39**
39a	**(38)**	Eyes of posterior row small, separated by more than their diameter (Fig. 17.159) (75,76) . . . . ***Sciastes***
39b		Eyes of posterior row larger, separated by their diameter or less (Fig. 17.161). **40**
40a	**(39)**	Epigynum with plate projecting conelike over genital groove (Fig. 17.155) (70) . . . . ***Islandiana***
40b		Epigynum with plate projecting in two lobes over genital groove (*gg*) (Fig. 17.158), or not projecting (Fig. 17.157). **41**
41a	**(40)**	Epigynum with plate projecting in two lobes over genital groove (*gg*) (Fig. 17.158) (68,68a). . . . . ***Eperigone***
41b		Epigynum with plate not projecting over genital groove (*gg*) (Fig. 17.157) (75) . . . . ***Sisicottus***

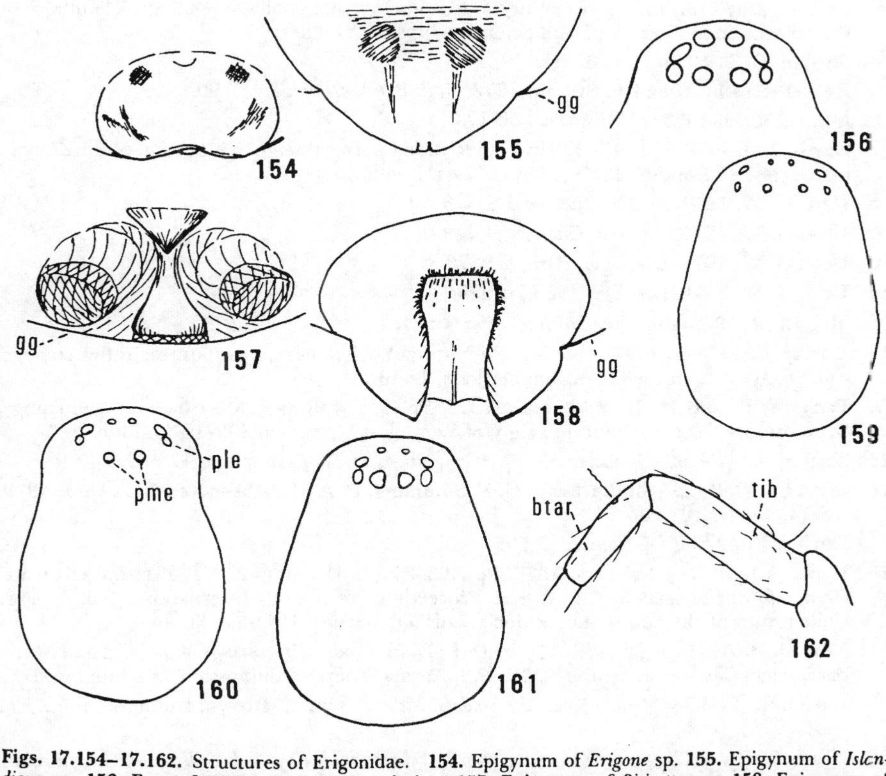

Figs. 17.154–17.162. Structures of Erigonidae. **154.** Epigynum of *Erigone* sp. **155.** Epigynum of *Islandiana* sp. **156.** Eyes of *Diplocentria* sp., dorsal view. **157.** Epigynum of *Sisicottus* sp. **158.** Epigynum of *Eperigone* sp. **159.** Carapace of *Sciastes* sp., dorsal view. **160.** Carapace of *Eridantes* sp., dorsal view. **161.** Carapace of *Islandiana* sp., dorsal view. **162.** Leg I of *Tapinocyba* sp. *btar,* basitarsus; *gg,* genital groove; *ple,* posterior lateral eye; *pme,* posterior median eye; *tib,* tibia.

42a **(29)** Trichobothrium on basitarsus I (*btar*) located at 0.4–0.6 of length of segment (as in *Ceratinella* sp., Fig. 17.130) **43**
42b Trichobothrium on basitarsus (*btar*) located at 0.31–0.35 of length of segment (Fig. 17.162) (65) . . . . ***Tapinocyba***
43a **(42)** Abdomen with dorsal and/or ventral scuta (Fig. 17.128) (67) . . . . . ***Ceraticelus***
43b Abdomen without scuta **44**
44a **(43)** Carapace (and sternum) coarsely rugose (Fig. 17.123) (65) ***Ceratinops***
44b Carapace smooth. **45**
45a **(44)** Leg tibiae each with one dorsal macroseta (75) . . . . ***Scylaceus***
45b Leg tibiae without dorsal macrosetae (69) . . . . ***Ceratinopsis***

References

1. **Moulder, B. C. and D. E. Reichle. 1972.** *Ecol. Monogr.* 42:473.
2. **Christophe, T. and P. Blandin. 1977.** *Bull. Br. Arachnol. Soc.* 4:132.
3. **Koponen, S. 1977.** Pp. 515–517 in U. Lohm and T. Persson (eds.). *Soil Organisms as Components of Ecosystems.* Proceedings of the 6th International Soil Zoology Colloquium of the International Society of Soil Science, Uppsala, Sweden.
4. **Strojan, C. L. 1978.** *Oikos* 31:41.
5. **Russell-Smith, A. and P. Swann. 1972.** *Bull. Br. Arachnol. Soc.* 2:99.
6. **Muma, M. H. 1973.** *Fla. Entomol.* 56:173.
7. **Curtis, D. J. 1978.** Pp. 149–159 in P. Merritt (ed.). *Arachnology.* Symposium of the Zoological Society of London 42. Academic Press, London.
8. **Uetz, G. W. 1974.** *Environ. Entomol.* 3:313.
9. **Uetz, G. W. 1975.** *Environ. Entomol.* 4:719.
10. **Uetz, G. W. 1976.** *Oecologia (Berl.)* 22:373.
11. **Uetz, G. W. 1979.** *Oecologia (Berl.)* 40:29.
12. **Merritt, P. 1976.** *Bull. Br. Arachnol. Soc.* 3:214.
13. **Duffey, E. 1978.** Pp. 109–123 in P. Merritt (ed.). *Arachnology.* Symposium of the Zoological Society of London 42. Academic Press, London.
14. **Peck, W. B. and W. H. Whitcomb. 1978.** Pp. 131–138 in P. Merritt (ed.). *Arachnology.* Symposium of the Zoological Society of London 42. Academic Press, London.
15. **Thaler, K., J. Ausserlechner and F. Mungenast. 1977.** *Pedobiologia* 17:389.
16. **Uetz, G. W., K. L. Van der Laan, G. F. Summers, P. A. K. Gibson and L. L. Getz. 1979.** *Am. Midl. Nat.* 101:286.
17. **Duffey, E. 1975.** *Biol. Conserv.* 7:255.
18. **Kajak, A. and H. Jakubczyk. 1977.** Pp. 493–496 in U. Lohm and T. Persson (eds.). *Soil Organisms as Components of Ecosystems.* Proceedings of the 6th International Soil Zoology Colloquium of the International Society of Soil Science, Uppsala, Sweden.
19. **Norton, R. A. 1973.** Pp. 138–156 in D. L. Dindal (ed.). *Proceedings of the First Soil Microcommunities Conference, Syracuse, N.Y.* U.S. Atomic Energy Commission, Washington, D.C.
20. **Kaston, B. J. 1978.** *How to Know the Spiders,* 3rd ed. Wm. C. Brown. Dubuque, Iowa. 272 pp.

20a. **Roth, V. D. 1985.** *Spider Genera of North America with Keys to Families and Genera and a Guide to Literature.* auth. publ. Portal, Ariz.

21. **Borror, D. J., D. M. DeLong and C. A. Triplehorn. 1976.** *Introduction to the Study of Insects,* 4th ed. Holt, Rinehart and Winston, New York. 852 pp.
22. **Chamberlin, R. V. and W. J. Gertsch. 1958.** *Bull. Am. Mus. Nat. Hist.* 116(1).

23. **Leech, R. E. 1972.** *Mem. Entomol. Soc. Can.* 84.
24. **Roth, V. D. and P. L. Brame. 1972.** *Am. Mus. Novit.* 2505.
25. **Levi, H. W. and D. E. Randolph. 1975.** *J. Arachnol.* 3:31.
26. **Dondale, C. D. and J. H. Redner. 1975.** *J. Arachnol.* 2:129.
27. **Gertsch, W. J. 1939.** *Bull. Am. Mus. Nat. Hist.* 76:277.
28. **Gertsch, W. J. 1953.** *Bull. Am. Mus. Nat. Hist.* 102(4).
29. **Schick, R. X. 1965.** *Bull. Am. Mus. Nat. Hist.* 129(1).
30. **Dondale, C. D. and J. H. Redner. 1978.** *The Insects and Arachnids of Canada.* Part 5. *The Crab Spiders of Canada and Alaska (Araneae: Philodromidae and Thomisidae).* Canada Department of Agriculture, Ottawa, Canada. 255 pp.
31. **Sauer, R. J. and N. I. Platnick. 1972.** *Can. Entomol.* 104:35.
32. **Dondale, C. D., A. L. Turnbull and J. H. Redner. 1964.** *Can. Entomol.* 96:636.
33. **Barnes, R. D. 1958.** *Am. Mus. Novit.* 1867.
33a. **Griswold, C. E. 1987.** *Univ. Cal. Publns. Ent.* 107.
34. **Gertsch, W. J. and W. Ivie. 1955.** *Am. Mus. Novit.* 1743.
35. **Platnick, N. I. and M. U. Shadab. 1975.** *Bull. Am. Mus. Nat. Hist.* 155(1).
36. **Platnick, N. I. 1975.** *Am. Mus. Novit.* 2573.
36a. **Platnick, N. I. and M. U. Shadab. 1980.** *Bull. Am. Mus. Nat. Hist.* 165:1.
36b. **Platnick, N. I. and M. U. Shadab. 1981.** *Am. Museum Novit.* 2717.
36c. **Platnick, N. I. and M. U. Shadab. 1988.** *Am. Mus. Novit.* 2916.
37. **Platnick, N. I. and M. U. Shadab. 1976.** *Am. Mus. Novit.* 2593.
38. **Platnick, N. I. and M. U. Shadab. 1975.** *Am. Mus. Novit.* 2583.
38a. **Platnick, N. I. and M. U. Shadab. 1982.** *Bull. Am. Mus. Nat. Hist.* 173:1.
38b. **Platnick, N. I. and M. U. Shadab. 1983.** *Bull. Am. Mus. Nat. Hist.* 174:1.
39. **Kaston, B. J. 1938.** *Am. Midl. Nat.* 20:562.
39a. **Dondale, C. D. and J. H. Redner. 1982.** *The Insects and Arachnids of Canada.* Part 9. *The Sac Spiders of Canada and Alaska (Araneae: Clubionidae and Anyphaenidae).* Canada Department of Agriculture, Ottawa, Canada. 194 pp.
40. **Reiskind, J. 1969.** *Bull. Mus. Comp. Zool. Harv. Univ.* 138:163.
41. **Platnick, N. I. and M. U. Shadab. 1974.** *Am. Mus. Novit.* 2553.
42. **Platnick, N. I. and M. U. Shadab. 1974.** *Am. Mus. Novit.* 2560.
43. **Edwards, R. J. 1958.** *Bull. Mus. Comp. Zool. Harv. Univ.* 118:365.
44. **Opell, B. D. and J. A. Beatty. 1976.** *Bull. Mus. Comp. Zool. Harv. Univ.* 147:393.
45. **Chamberlin, R. V. and W. Ivie. 1941.** *Ann. Entomol. Soc. Am.* 34:585.
46. **Chamberlin, R. V. and W. Ivie. 1940.** *Bull. Univ. Utah Biol. Ser.* 5(9).
47. **Chamberlin, R. V. and W. Ivie. 1932.** *Bull. Univ. Utah Biol. Ser.* 2(1).
47a. **Brady, A. R. 1979.** *Psyche (Camb.)* 86:167.
48. **Dondale, C. D. and J. H. Redner. 1979.** *Can. Entomol.* 111:1033.
49. **Wallace, H. K. and H. Exline. 1978.** *J. Arachnol.* 5:1.
49a. **Dondale, C. D. and J. H. Redner. 1983.** *J. Arachnol.* 11:1.
50. **Wallace, H. K. 1942.** *Am. Midl. Nat.* 27:1.
51. **Dondale, C. D. and J. H. Redner. 1978.** *Can. Entomol.* 110:143.
52. **Levi, H. W. 1957.** *Bull. Mus. Comp. Zool. Harv. Univ.* 117:367.
53. **Kaston, B. J. 1946.** *Am. Mus. Novit.* 1306.
54. **Levi, H. W. 1957.** *Bull. Am. Mus. Nat. Hist.* 112(1).
55. **Levi, H. W. 1957.** *Trans. Am. Microsc. Soc.* 76:105.
56. **Levi, H. W. 1953.** *Am. Mus. Novit.* 1647.
57. **Levi, H. W. 1955.** *Am. Mus. Novit.* 1718.
58. **van Helsdingen, P. J. 1968.** *Zool. Meded. Rijksmus. Nat. Hist. Leiden* 43:117.
59. **van Helsdingen, P. J. 1969.** *Zool. Verh. Rijksmus. Nat. Hist. Leiden* 105.

60. **Ivie, W. 1969.** *Am. Mus. Novit.* 2364.
61. **van Helsdingen, P. J. 1973.** *Zool. Verh. Rijksmus. Nat. Hist. Leiden* 124.
62. **Zorsch, H. M. 1937.** *Am. Midl. Nat.* 18:856.
63. **van Helsdingen, P. J. 1974.** *Zool. Meded. Rijksmus. Nat. Hist. Leiden* 46:295.
64. **Chamberlin, R. V. and W. Ivie. 1936.** *Ann. Entomol. Soc. Am.* 29:85.
65. **Crosby, C. R. and S. C. Bishop. 1933.** *Ann. Entomol. Soc. Am.* 26:105.
65a. **Millidge, A. F. 1981.** *J. Arachnol.* 9:167.
66. **Millidge, A. F. 1975.** *Bull. Br. Arachnol. Soc.* 3:145.
67. **Crosby, C. R. and S. C. Bishop. 1925.** *N.Y. State Mus. Bull.* 264.
68. **Crosby, C. R. and S. C. Bishop. 1928.** *N.Y. State Mus. Bull.* 278.
68a. **Millidge, A. F. 1987.** *Am. Mus. Novit.* 2885.
69. **Bishop, S. C. and C. R. Crosby. 1930.** *J. N.Y. Entomol. Soc.* 38:15.
70. **Ivie, W. 1965.** *Am. Mus. Novit.* 2221.
71. **Chamberlin, R. V. and W. Ivie. 1945.** *Trans. Conn. Acad. Arts Sci.* 36:215.
72. **Millidge, A. F. 1980.** *J. Arachnol.* 8:109.
73. **Crosby, C. R. and S. C. Bishop. 1931.** *J. N.Y. Entomol. Soc.* 39:359.
74. **Millidge, A. F. 1983.** *J. Arachnol.* 11:105.
75. **Bishop, S. C. and C. R. Crosby. 1938.** *J. N.Y. Entomol. Soc.* 46.
76. **Millidge, A. F. 1984.** *J. Arachnol.* 12:121.
77. **Bishop, S. C. and C. R. Crosby. 1932.** *J. N.Y. Entomol. Soc.* 40:393.

18

Pseudoscorpionida

WILLIAM B. MUCHMORE

The pseudoscorpions, or false scorpions, are small arachnids, giving the general impression of scorpions, but without the long, narrow tail and sting. They are more or less dorsoventrally compressed and their appendages extend out to the front and sides. They are drab in color, ranging from light tan to reddish brown to almost black.

External Morphology

The body of a pseudoscorpion is, as in all arachnids, divided into an anterior *prosoma*, or *cephalothorax*, and a posterior *opisthosoma*, or *abdomen*, which are broadly connected to one another (Fig. 18.1).

The dorsal surface of the cephalothorax is covered by a single, large *carapace*, which shows little or no evidence of segmentation (Fig. 18.3). The carapace is variously shaped, from square to triangular as viewed from above. There may be one or two transverse furrows, one near the middle and the other toward the posterior margin; in nymphs, the area behind the posterior furrow may be clearly separate from the rest of the carapace. There may be two pairs of *eyes* laterally toward the front, or there may be only one pair or none. As with most other parts of the body, the carapace may be smooth or roughened with granules, and is provided with a number of *setae*. The actual number, size, shape, and placement of the setae are rather varied and are often of diagnostic value.

Ventrally, the cephalothorax is covered by the expanded proximal segments, or *coxae*, of the palps and legs (Fig. 18.2). Occasionally, a small, apparent vestige of a sternum, the *intercoxal tubercle*, may be present in

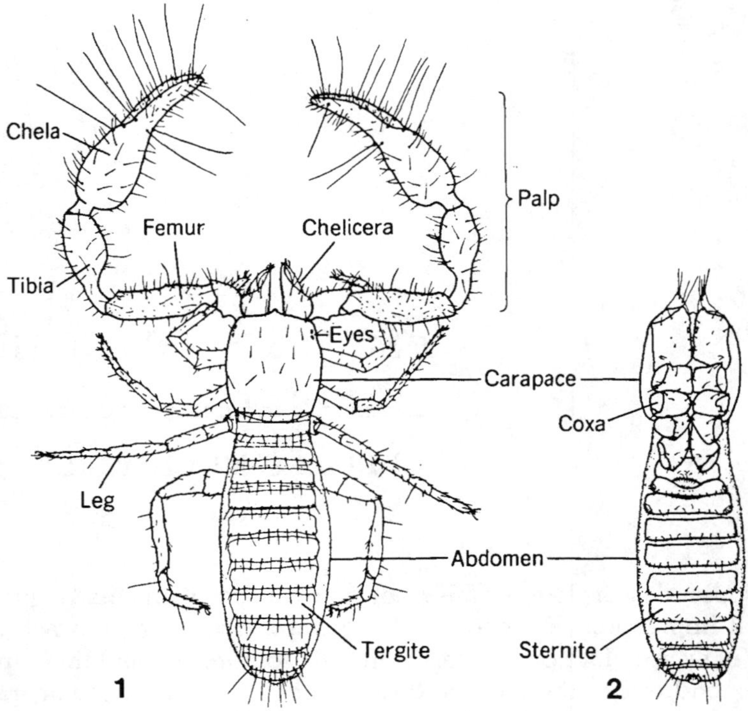

Figs. 18.1, 18.2. *Microcreagris* sp., dorsal and ventral.

some Chthonioidea. In the small family Sternophoridae a *pseudosternum* is produced by the thinning of the medial edges of the pedal coxae.

The mouthparts surround the mouth at the anterior end of the cephalothorax. Internally are several sclerotized structures which are difficult to see and not well understood. The first pair of appendages, the *chelicerae,* and the coxae of the second pair of appendages, the *palps,* serve as accessory mouthparts.

The *chelicerae* are two-segmented, chelate structures (Fig. 18.4) attached close together under the anterior margin of the carapace and are used mainly to grasp and macerate food organisms. Each chelicera bears several setae, some of which are closely grouped together on the medial side as the *flagellum;* there are frequently five, long, conspicuous setae on the lateral side: the exterior seta (*es*) near the articulation of the movable finger, the basal seta (*b*) and subbasal seta (*sb*) on the surface of the hand, and the interior seta (*is*) and laminal seta (*ls*) on the base of the fixed finger; and on the movable finger, usually in the distal half, is the smaller, galeal seta (*gs*). There is a comblike or membranous *serrula* along the side of each finger.

And near the end of the movable finger is a *spinneret* or *galea,* which carries the openings of *silk glands* located in the cephalothorax.

The succeeding five pairs of appendages all have six or seven segments, the proximal ones of which, the *coxae,* are enlarged and usually meet in the midline on the ventral side. The first of these are the long, conspicuous *palps,* or *pedipalps,* the coxae of which serve as accessory mouthparts. Beyond the coxa, each palp has five segments; the trochanter, the femur, the tibia, and the chela with fixed and movable fingers. The *chela* (Fig. 18.5) is used mainly for food gathering and defense. It is the possession of these chelate palps that provides the strongest resemblance of pseudoscorpions to scorpions. Usually, the palpal chela has numerous *teeth* along the inner margins of the fingers and the sharp tip (*venedens*) of one or both fingers has an opening for the venom gland; the venom is used to subdue prey. The palpal chela also bears long, thin, sensory setae called *trichobothria*—

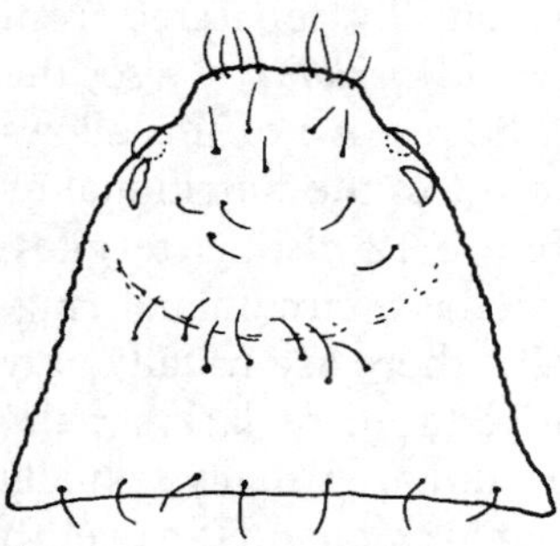

3

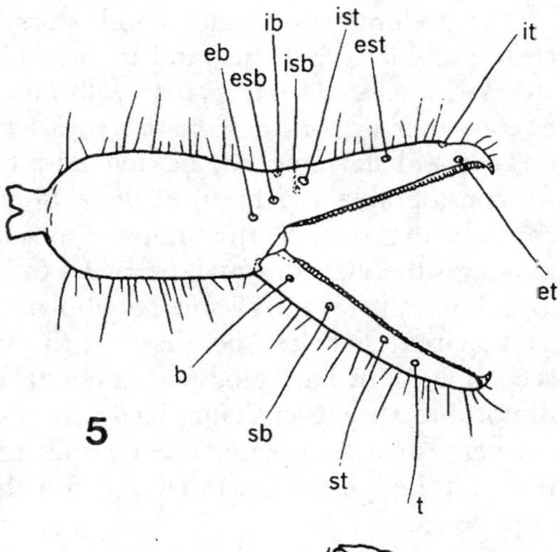

5

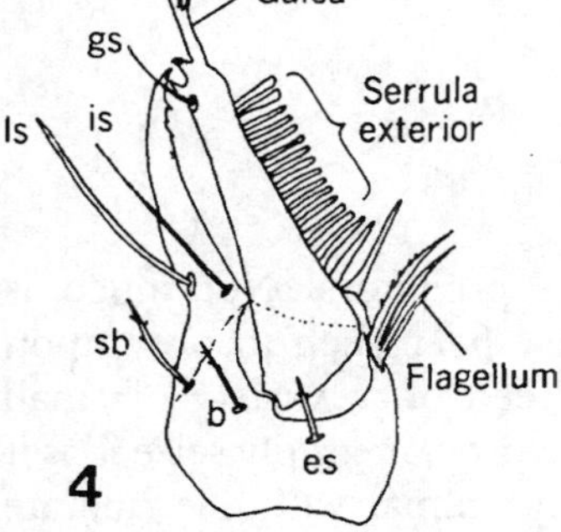

4

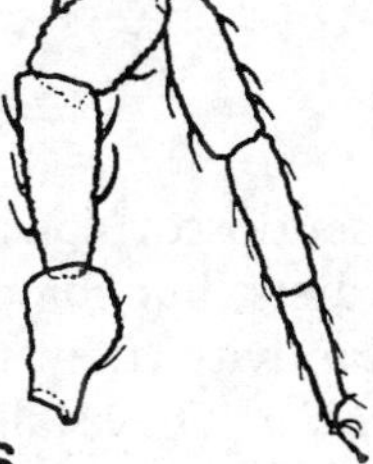

6

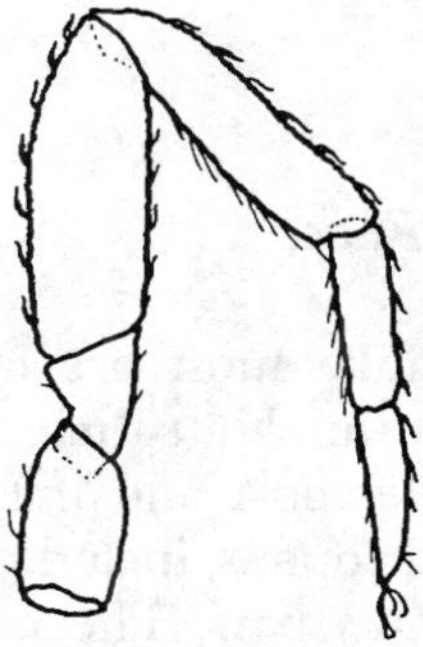

7

Figs. 18.3–18.7. 3. *Larca granulata.,* carapace. **4.** *Lamprochernes* sp., chelicera. **5.** *Microcreagris* sp., palpal chela, lateral view. **6.** *Larca granulata,* leg I. **7.** *Larca granulata,* leg IV.

usually, on the lateral surface of the movable finger there is a series of four trichobothria named (from base to tip) the basal (*b*), subbasal (*sb*), subterminal (*st*) and terminal (*t*); and usually, on the fixed finger there are both a lateral or exterior series of four and a medial or interior series of four, the exterior basal (*eb*), exterior subbasal (*esb*), exterior subterminal (*est*) and exterior terminal (*et*), and the interior basal (*ib*), interior subbasal (*isb*), interior subterminal (*ist*), and interior terminal (*it*), respectively. Rarely, the chela may bear specialized bumps or processes, the functions of which are unknown.

The remaining four pairs of appendages are the walking *legs,* each consisting of a coxa, trochanter, basifemur, telofemur, tibia, and one or two tarsal segments. Usually, the anterior two pairs of legs are quite different from the posterior two pairs in details of structure, posture, and movement (Figs. 18.6, 18.7). The two femoral segments may be movable on one another or not, and there may be one or two tarsal segments. At the end of each tarsus is a small pretarsus with two *claws* and an *arolium,* which allow the animal to hold on to surfaces.

The abdomen is divided, both dorsally and ventrally, into 11 or 12 clearly defined segments, and is smoothly rounded posteriorly. It is provided with sclerotized *tergites* dorsally and *sternites* ventrally, separated from one another by soft, flexible *intersegmental* and *pleural membranes;* also, the tergites and sternites may be divided into lateral halves. All of this allows for considerable expansion of the abdomen to adjust to the nutritional or reproductive state of the animal. Dorsally, there are 12 distinct tergites, although the eleventh may be fused to the sternite as a circumanal ring, and the twelfth is a very small anal plate. Ventrally, there are usually only 11 apparent sternites, because the first is much reduced or lacking; the second and third are modified as genital opercula and are usually sexually dimorphic; the eleventh may be a part of a circumanal ring; and the twelfth is a very small anal plate. The *spiracles,* or tracheal openings, are found at the lateral edges of the third and fourth sternites. The abdomen has no appendages.

Food

Like most arachnids, pseudoscorpions are predators. Not much is known about their natural diets, but some have been seen to prey upon collembola and mites. In captivity, they will feed on a variety of small arthropods, including *Drosophila.* They stalk the prey or simply seize it as it moves past. The palpal chelae grasp the prey animal with the dentate fingers; when venom apparatus is present, poison is injected to immobilize the prey quickly. Then the food organism is pressed against the chelicerae, which pierce and manipulate it so that the body juices can be sucked up by the pharynx. Presumably, enzymes are secreted to dissolve some body tissues.

Reproduction and Life History

The life cycle is straightforward—adults mate to produce young which go through three nymphal stages to attain adulthood. Mating is often preceded by courtship of the female by the male; various dancing steps and posturing and grasping with the palps are common, and one instance of production of a silken pathway has been reported (see Weygoldt 1969). The male then produces a complex stalked spermatophore, which is attached to the substratum, and the female subsequently takes up the sperm from that structure. The fertilized eggs are passed from the body but are retained in a secreted pouch attached to the genital opercula. The eggs are fairly large and number from 5 to 20 or more. They develop slowly, nourished by their own yolk and by genital secretions of the female (see Weygoldt 1969). The young are retained in the brood pouch until they reach a protonymphal stage, quite recognizable as little pseudoscorpions. Three ecdyses produce deutonymphs, tritonymphs, and adults. In a few species, notably of *Microbisium,* parthenogenesis appears to be the rule, as males are exceedingly rare while females and young are common.

In many pseudoscorpions, females protect themselves during brooding by spinning a silken chamber with the spinnerets on the chelicerae. Similarly, nymphs often produce silken chambers to protect themselves during ecdysis.

Ecology

Pseudoscorpions are essentially cryptic animals. Their small size and bodily construction facilitate movement through the small spaces in soil and litter and under logs and rocks. Their sensory apparatus is adjusted to this environment. The eyes, four at most, consist of simple ocelli and are relatively small; often, they are reduced in both size and number, and many pseudoscorpions are blind. The major sense appears to be tactile, mediated by long tactile setae, or trichobothria, located on the palpal chela and posterior abdominal segments.

From the soil and ground litter, pseudoscorpions have easily moved into trees, living under bark; into caves, where they often move about on mud or rock surfaces, increasing in size and slenderness and losing eyes and color; and onto the seashore, where they are found under rocks and debris between tidemarks. Also, they have frequently found their way into the nests of larger animals, such as ants, bees, termites, birds, and mammals, as commensals or inquilines. And some pseudoscorpions engage in phoresy, where they attach themselves to a larger animal and are carried from place to place; this has been most frequently documented with respect to flies and beetles, but certainly also occurs with other insects and birds and mammals.

Collection and Study

Pseudoscorpions should be sought wherever there is organic material to serve as food for their prey, mainly small herbivores and detritivores. They may be found by direct search or by sifting litter or debris and picked up with a wet toothpick or brush or sucked up with an aspirator. They may be more easily separated from litter or debris by the use of a Berlese or Tullgren funnel. Phoretic individuals will be found on their hosts or retrieved from traps used to collect the host animals.

They should be collected and preserved in 70–80% alcohol; it is permissible to add a little glycerol to prevent complete desiccation in case the alcohol should be lost.

Most of the characters used in the key may be seen on whole specimens observed under the higher powers of a dissecting microscope. However, the detailed study necessary for discrimination of many specific characters can be done only on dissected, slide-mounted specimens viewed with a compound microscope. Specimens can be mounted temporarily in glycerol, lactic acid, or Hoyer's medium; for permanent mounts, they are best dehydrated, cleared, and mounted in Canada balsam (see Hoff 1949, 1959).

ACKNOWLEDGMENT

Appreciation is extended to Dr. Ellen M. Benedict for the very thorough review of the scientific content of this chapter.

KEY TO GENERA

This key is designed to identify an adult specimen to genus, and covers all genera known to be represented in North America north of Mexico. It will usually not work with nymphs, which are smaller, of different proportions, and with fewer setae than adults. Further, it may not work with females in groups where the main diagnostic characters are sexually dimorphic (as in some Chernetidae, Withiidae, and Cheliferidae).

1a Legs III and IV with more segments (six beyond coxa) than legs I and II (five beyond coxa) (Figs. 18.9, 18.8); Superfamily **Chthonioidea** **2**

1b All legs with same number of segments (either five or six beyond coxa) **14**

2a **(1)** Carapace with more than 50 setae; spiracles oblique and with distinct sclerites (Fig. 18.10); coxal spines present on legs I and II; Family **Tridenchthoniidae** . . . . *Verrucaditha* Chamberlin

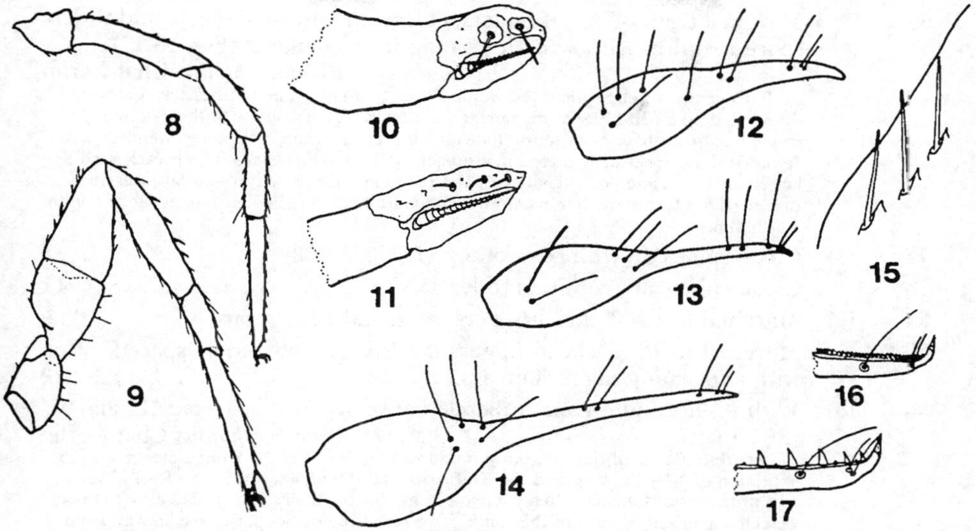

Figs. 18.8–18.17. 8. *Apochthonius* sp., leg I. **9.** *Apochthonius* sp., leg IV. **10.** *Verrucaditha spinosa,* spiracular plate. **11.** *Chthonius tetrachelatus,* spiracular plate. **12.** *Lechytia hoffi,* Palpal chela, dorsal view. **13.** *Chthonius tetrachelatus,* palpal chela, dorsal view. **14.** *Pseudotyrannochthonius* sp., palpal chela, dorsal view. **15.** *Apochthonius* sp., coxal spines on leg I. **16.** *Apochthonius* sp., end of finger of palpal chela. **17.** *Kleptochthonius* sp., end of finger of palpal chela.

Only one representative of the genus and family, *V.* ***spinosa*** (Banks), is known from the United States, although species of ***Tridenchthonius*** are found in Mexico and the West Indies. This species is fairly common in the southeastern quarter of the country; in litter and under bark of dead stumps. Body length 1–1.5 mm.

2b Carapace with fewer than 30 setae; spiracles transverse and without distinct sclerites (Fig. 18.11); coxal spines variably present on legs I, II, III, or lacking; Family **Chthoniidae** 3

3a (2) Without coxal spines; four trichobothria on dorsum of chelal hand (Fig. 18.12) . . . . ***Lechytia*** Balzan

This genus is represented in North America by three species, two from Florida and the Gulf coast, and one widespread through the Rocky Mountain region; mainly in dry litter. Body length about 1 mm.

3b With coxal spines; two trichobothria on dorsum of chelal hand or base of fixed finger 4

4a (3) Transverse pair of trichobothria on dorsum of chelal hand near or proximal to middle (Fig. 18.13) 5

4b Transverse pair of trichobothria on dorsum of chela near base of fixed finger (Fig. 18.14) . . . . ***Pseudotyrannochthonius*** Beier

Three species of this genus are known from the western states of California, Oregon, Washington, Idaho, and Utah; found in leaf litter and soil. Body length 1.5–2.5 mm.

5a (4) Coxal spines only on coxa of leg I (Fig. 18.15) 6

5b Coxal spines not confined to leg I. 7

6a (5) Marginal teeth of chelal fingers small and contiguous (Fig. 18.16) ***Apochthonius*** Chamberlin

This genus is widespread throughout the country, except in very dry regions; usually in moist deciduous litter. It has also successfully invaded caves in Virginia, West Virginia, Georgia, Alabama, Indiana, Missouri, Arkansas, Oregon, and California. The cave-dwelling forms are better known taxonomically than the epigean ones. Body length 1–2 mm.

6b Marginal teeth of chelal fingers, at least in part, tall, sharp, and widely spaced, and usually with alternating microdenticles (Fig. 18.17) . . . ***Kleptochthonius*** Chamberlin

This genus is well represented within its range, most of the southeastern quarter of the United States and the Pacific northwest; usually in moist litter. In the east, many troglobitic forms have been found in West Virginia, Virginia, Tennessee, Kentucky, and Indiana; these are usually placed in a subgenus, ***Chamberlinochthonius,*** characterized by large size, attenuation of appendages, and reduction of eyes. As in ***Apochthonius,*** the cave-dwelling forms are better known taxonomically than the epigean ones. Body length 1.5–3 mm.

7a (5) Coxal spines only on coxa of leg II (Fig. 18.18). **8**

7b Coxal spines not confined to leg II. **11**

8a (7) Marginal teeth of chelal fingers small and contiguous. **9**

8b Marginal teeth of chelal fingers tall, sharp, and widely spaced, often with alternating microdenticles. **10**

9a (8) With bisetose intercoxal tubercle between bases of coxae III and IV (Fig. 18.19) . . . . ***Mundochthonius*** Chamberlin

Representatives of this genus are found through most of the United States, with the exception of New England and the Gulf coast states; it also occurs in the Rocky Mountain provinces of Canada; often common in moist, cool litter. The species have not been well characterized in most of the range. Two rare, troglobitic species are known, from Virginia and Illinois. Body length 0.75–1.5 mm.

9b Without an intercoxal tubercle . . . . . ***Malcolmochthonius*** Benedict

Three species are presently known, from California and Oregon; they seem to prefer relatively dry leaf litter. Body length 0.85–1.25 mm.

10a (8) Inner surface of chelal hand with a single, conspicuous, spinelike seta (Fig. 18.20) . . . . ***Tyrannochthonius*** Chamberlin

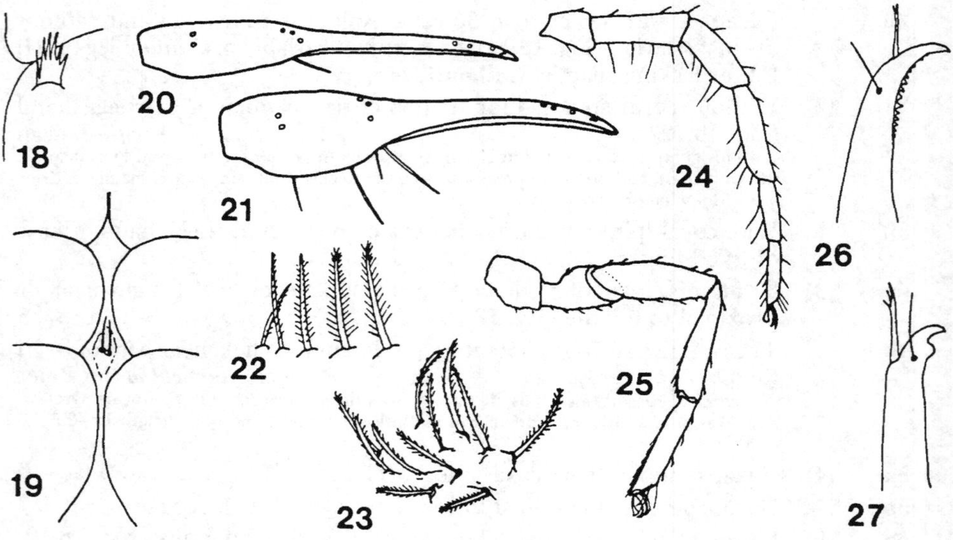

Figs. 18.18–18.27. 18. *Mundochthonius* sp., coxal spine on leg I. **19.** *Mundochthonius* sp., intercoxal tubercle. **20.** *Tyrannochthonius* sp., palpal coxa, dorsal view. **21.** *Paraliochthonius weygoldti*, palpal coxa, dorsal view. **22.** *Neochthonius stanfordianus*, coxal spines on leg II. **23.** *Kewochthonius* sp., coxal spines on leg II. **24.** *Chitrella* sp., leg I. **25.** *Pselaphochernes* sp., leg I. **26.** *Microcreagris* sp., movable finger of chelicera. **27.** *Pachyolpium* sp., movable finger of chelicera.

This is a widespread tropical and subtropical genus. Only two species have been described from the United States, ***T. floridensis*** Malcolm and Muchmore from Florida and Georgia, and ***T. troglodytes*** Muchmore from Texas, but others are known as far north as Kentucky and Arkansas; found mainly in moist litter. Several (as yet undescribed) cave-adapted forms occur in Alabama and Tennessee. Body length 1–3 mm.

10b Inner surface of chelal hand with more than one spinelike seta (Fig. 18.21) . . . . ***Paraliochthonius*** Beier

This tropical and subtropical genus is represented in the United States by a single species, ***P. weygoldti*** Muchmore, from Florida; it is a marine littoral form, to be found in wrack and beneath debris at the edge of the water. Body length about 1.2 mm.

11a **(7)** Coxal spines on coxae of legs I and II . . . . ***Aphrastochthonius*** Chamberlin

In the United States this genus is represented only by cavernicolous species: two in Alabama, one in New Mexico and two in California. It is also known from Middle America and Cuba, where tiny litter inhabiting forms exist. Body length about 1.5 mm.

11b Coxal spines on coxae of legs II and III **12**

12a **(11)** Marginal teeth of both chelal fingers small and contiguous **13**

12b Marginal teeth of at least the fixed chelal finger acute and distinctly spaced . . . . ***Chthonius*** C. Koch

One species, ***C. tetrachelatus*** (Preyssler), has been reported from most of the eastern states. The five other species known from the United States are of spotty distribution, two on the east coast and three on the west coast. They all prefer moist litter and debris. Body length 1.5–2.5 mm.

13a **(12)** Coxal spines of leg II in a transverse row across middle of coxa (Fig. 18.22) . . . . ***Neochthonius*** Chamberlin

This genus is known only from California where three species, including one troglobite, have been described. They are litter dwelling forms of small size, about 1.5 mm in length.

13b Coxal spines of leg II irregularly placed in a rounded patch (Fig. 18.23). ***Kewochthonius*** Chamberlin

Representatives of this genus are known from California, Colorado, Utah, and the southeastern United States; they probably occur also in the Gulf coast states, where collecting has been sparse. They are small forms, usually less than 1.5 mm long, and live in moist litter and soil.

14a **(1)** Tarsus of each leg divided into two parts; each leg with six segments beyond coxa (Fig. 18.24). **15**

14b Tarsus of each leg undivided; each leg with five segments beyond coxa (Fig. 18.25). **41**

15a **(14)** Chelicera with inner margin of movable finger distinctly dentate (Fig. 18.26); Superfamily **Neobisioidea** **16**

15b Chelicera with inner margin of movable finger smooth (Fig. 18.27); Superfamily **Garypoidea** **27**

16a **(15)** Only fixed finger of palpal chela with prominent venedens and venom duct (Fig. 18.28), movable finger without (Fig. 18.29); chela with 12 trichobothria or fewer **17**

16b Both fingers of chela with prominent venedens and venom duct; chela with many more than 12 trichobothria; Family **Ideoroncidae**. . . . . ***Albiorix*** Chamberlin

This genus is widespread in Mexico and extends into southwestern United States as far north as Utah. Four species have been recorded from the United States, but they are poorly defined. They are found under stones and debris, often in very dry areas. Body length up to 3 mm.

17a **(16)** Pleural membranes of abdomen granulate (Fig. 18.30); Family **Neobisiidae** **18**

17b Pleural membranes of abdomen longitudinally striate (Fig. 18.31); Family **Syarinidae** **23**

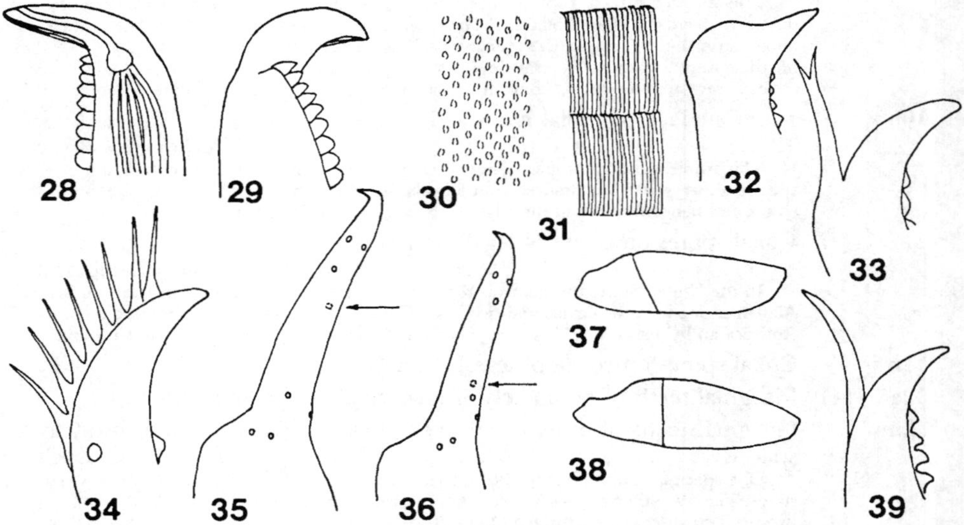

Figs. 18.28–18.39. 28. *Microcreagris rufula,* tip of fixed finger of chela. **29.** *Microcreagris rufula,* tip of movable finger of chela. **30.** *Microcreagris* sp., pleural membrane. **31.** *Syarinus* sp., pleural membrane. **32.** *Novobisium* sp., tip of movable finger of chelicera with spinneret. **33.** *Microcreagris* sp., tip of movable finger of chelicera with spinneret. **34.** *Halobisium* sp., tip of movable finger of chelicera with spinneret. **35.** *Novobisium* sp., fixed chelal finger showing positions of trichobothria (arrow indicates *ist*). **36.** *Parobisium utahensis,* fixed chelal finger showing positions of trichobothria (arrow indicates *ist*). **37.** *Syarinus* sp., entire femur of leg IV. **38.** *Chitrella cavicola,* entire femur of leg IV. **39.** *Ideoblothrus floridensis,* tip of movable finger of chelicera with spinneret.

18a (17) Spinneret on movable finger of chelicera a long, usually branched or multipartite galea **19**

18b Spinneret in the form of a low rounded knob or absent (Fig. 18.32). (One known species of ***Microcreagris*** will key out to this point, but may be recognized by the strongly pedicellate palpal tibia.) **20**

19a (18) Spinneret consisting of a single, usually branched galea (Fig. 18.33) ***Microcreagris*** Balzan

Forms assigned to this genus are widespread through the southern half of the United States and extend as far north as British Columbia on the west coast; some 35 species have been described. Their natural habitat is cool, moist soil and litter, and they are often found on the undersides of large stones and rocks. Many have entered caves and become highly modified troglobites in Virginia, Tennessee, Alabama, Texas, Nevada, California, and Oregon. They range in size from 2 to 5.5 mm in body length, and the appendages of troglobites are often very long and slim. [It has been shown recently that the genus ***Microcreagris, s. str.,*** is restricted to northern Asia. Curcic (1978–1984) has placed some of the North American forms in new genera, but the taxonomic situation is still not stable.]

19b Spinneret in the form of 8–10 simple galeae in a row (Fig. 18.34) ***Halobisium*** Chamberlin

Only a single species, ***H. occidentale*** Chamberlin, occurs in the United States, as a littoral form on the Pacific coast from Alaska to California. It is found among rocks and in litter on rocky coasts and in salt marshes. Body length 3–4 mm.

20a (18) With 4 eyes **21**

20b With 2 eyes . . . . ***Roncus*** L. Koch

Only the European species ***R. lubricus*** L. Koch has been reported in the United States, from the vicinity of Rochester, New York, where it occurs in greenhouses and outdoors in moist litter. Body length 2–2.5 mm.

21a (20) Palpal chela with normal complement of four trichobothria on movable finger and eight on fixed finger **22**

21b Palpal chela with only three trichobothria on movable finger and seven on fixed finger . . . . ***Microbisium*** Chamberlin

This genus is widespread, probably represented in every state and southern Canada. Two species seem to be represented: ***M. brunneum*** (Hagen) in boggy situations in the north and ***M. parvulum*** (Banks) in woodland litter throughout the region. These animals are apparently parthenogenetic, males being very rare. Body length 1.5–2 mm.

22a (21) Trichobothrium *ist* near middle of fixed chelal finger, much nearer to *est* than to *isb* (Fig. 18.35) . . . . ***Novobisium*** Muchmore

The three species of this genus are confined to the southern Appalachian Mountain region, from West Virginia to Alabama. Species distinctions are not clear, but ***N. carolinense*** (Banks) seems to be the most common. They live in moist litter. Body length 2.5–4 mm.

22b Trichobothrium *ist* near base of fixed chelal finger, much nearer to *isb* than to *est* (Fig. 18.36) . . . . ***Parobisium*** Chamberlin

Several species have been described from Oregon, California, Colorado, Utah, and Idaho; one of these, ***P. charlotteae*** Chamberlin, is cave adapted. Most can be found in forest litter. Body length 2.5–3 mm for most forms, 6 mm for ***P. charlotteae.***

23a (17) Suture between basifemur and telofemur of leg IV strongly oblique to long axis of leg (Fig. 18.37); spinneret on chelicera in form of a branched process. **24**

23b Suture between basifemur and telofemur of leg IV weakly oblique or perpendicular to long axis of leg (Fig. 18.38); spinneret a slender simple process or absent. **25**

24a (23) Marginal teeth of movable chelal finger contiguous; movable finger with four trichobothria . . . . ***Syarinus*** Chamberlin

This genus is represented in the colder parts of the United States and the warmer parts of Canada, being known from Washington, Oregon, California, New Mexico, Utah, Colorado, Wyoming, Montana, Minnesota, Wisconsin, Michigan, Pennsylvania, New York, and New Hampshire, and British Columbia, Saskatchewan, Ontario, and Newfoundland. They are found under rocks and in damp litter, and in soil to a depth of several centimeters. Body length 2.5–4 mm.

24b Marginal teeth of movable chelal finger distinctly spaced; movable finger with only three trichobothria. . . . . ***Hyarinus*** Chamberlin

Only a single specimen is known, taken from the stomach of a toad at Santa Barbara, California. It is named ***H. hesperus*** Chamberlin. Body length 2.7 mm.

25a (23) No spinneret evident on chelicera **26**

25b Spinneret a slender, simple process (Fig. 18.39). . . . . ***Ideoblothrus*** Balzan

This genus is widespread through Middle America and the West Indies. However, only one species is known from the United States, ***I. floridensis*** (Muchmore) from the Florida Keys. It is found in litter. Body length about 1.75 mm.

26a (25) Trichobothrium *ib* near center of dorsum of chelal hand; *ist* near middle of fixed finger, clearly distant from *est* . . . . ***Aglaochitra*** Chamberlin

One species is known in this genus: ***A. rex*** Chamberlin from Monterey County, California. It was found under large rocks, perhaps in association with pack rats. Body length about 4 mm.

26b Trichobothrium *ib* on dorsum of chelal hand clearly distal to center; *ist* at distal third of fixed finger, at same level as *est* . . . ***Chitrella*** Beier

Epigean forms are known from California, Colorado, New Mexico, Pennsylvania, Kentucky, West Virginia, Virginia, and District of Columbia, while cave-adapted forms have been found in Texas, West Virginia, Virginia, and Tennessee. The epigean habitat is moist litter and under large rocks. Body length 2–4 mm.

27a (15) Venedens and venom duct present in both fingers of palpal chela **28**

27b Venedens and venom duct only in fixed finger of chela; Family **Menthidae** . . . . ***Menthus*** Chamberlin

Only one species of this desert-adapted genus has been found in the United States: ***M. californicus*** Chamberlin in southern California. They are usually encountered under rocks. Body length about 2.5 mm.

28a (27) Coxal area widened posteriorly (Fig. 18.40); Family **Garypidae** **29**

28b Coxal area parallel-sided (Fig. 18.41) Family **Olpiidae**. **31**

29a (28) Arolia of pedal tarsi longer than claws. **30**

29b Arolia of pedal tarsi shorter than claws . . . . ***Garypus*** L. Koch

Only two species of this circumtropical genus are known from the United States; ***G. californicus*** Banks is a common littoral form on the Pacific coast as far north as Marin County, California, while ***G. floridensis*** Banks has been reported only once, from St. Lucie County, Florida. The species are all adapted for shore life, occurring under stones and in wrack at high-tide line. Body length 4–6 mm.

30a (29) Movable finger of palpal chela with four trichobothria . . . . ***Archeolarca*** Hoff and Clawson

Five species are known in this genus, which is apparently restricted to the Rocky Mountain region of the United States. An epigean form, ***A. rotunda*** Hoff and Clawson, is found associated with pack rats in New Mexico, Utah, and Oregon; and four cavernicolous forms are known, from California, Arizona, and Texas. Body length 2.5–3 mm.

30b Movable chelal finger with two or three trichobothria. . . . . ***Larca*** Chamberlin

This genus includes a species, ***L. notha*** Hoff, with three trichobothria on the movable chelal finger, known from Colorado, Oregon, and Saskatchewan. The other three species have only two trichobothria on the movable finger; ***L. granulata*** (Banks) from many northeastern and north-central states, and epigean and cave-dwelling species from California. The epigean forms are found in litter and frass, sometimes quite dry, and often in association with the nests of small mammals. Body length about 2 mm.

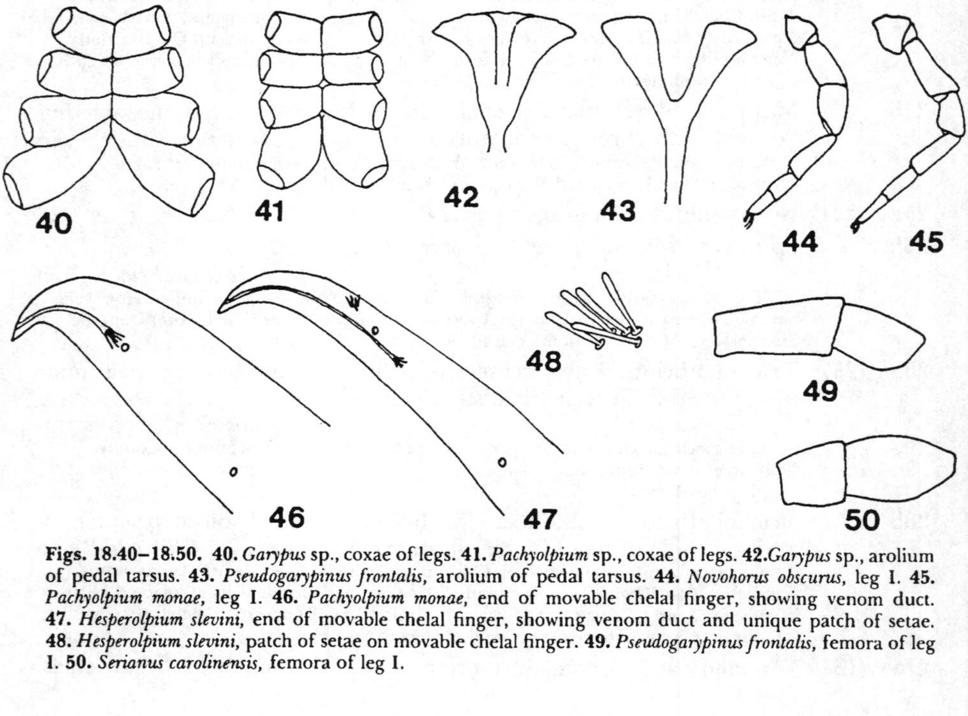

Figs. 18.40–18.50. 40. *Garypus* sp., coxae of legs. **41.** *Pachyolpium* sp., coxae of legs. **42.***Garypus* sp., arolium of pedal tarsus. **43.** *Pseudogarypinus frontalis,* arolium of pedal tarsus. **44.** *Novohorus obscurus,* leg I. **45.** *Pachyolpium monae,* leg I. **46.** *Pachyolpium monae,* end of movable chelal finger, showing venom duct. **47.** *Hesperolpium slevini,* end of movable chelal finger, showing venom duct and unique patch of setae. **48.** *Hesperolpium slevini,* patch of setae on movable chelal finger. **49.** *Pseudogarypinus frontalis,* femora of leg I. **50.** *Serianus carolinensis,* femora of leg I.

31a (28) Arolia of pedal tarsi entire (Fig. 18.42); tergites and sternites entire; Subfamily **Olpiinae** **32**

31b Arolia of pedal tarsi bifurcate (Fig. 18.43); at least some anterior tergites divided; Subfamily **Garypininae** **37**

32a (31) Leg I with basifemur and telofemur about equal in length (Fig. 18.44) **33**

32b Leg I with basifemur distinctly longer than telofemur (Fig. 18.45) **34**

33a (32) All trichobothria except *et* confined to proximal halves of chelal fingers ***Oreolpium*** Benedict and Malcolm

Only a single species is known to represent this genus: ***O. nymphum*** Benedict and Malcolm, from western Oregon. It is found mainly in the rotted bark of old standing conifers. Body length about 2 mm.

33b Trichobothria *et, it,* and *t* in distal halves of chelal fingers . . . . . . ***Novohorus*** Hoff

One species of this Caribbean genus is found in the United States; ***N. obscurus*** (Banks) from peninsular Florida. It is usually found in litter of sand pine. Body length 2–3 mm.

34a (32) Nodus ramosus of venom duct in movable chelal finger is proximal to trichobothrium *t* (Fig. 18.47) **35**

34b Nodus ramosus of venom duct in movable finger is distal to trichobothrium *t* (Fig. 18.46) . . . . ***Pachyolpium*** Beier

One species, ***P. monae*** (Hoff), of this neotropical genus has been reported from the Florida Keys. It is often found in rather dry litter and under rocks. Body length 2–2.5 mm.

35a (34) Movable chelal finger with a small group of short, club-shaped setae on the exterior side, distal to trichobothrium *t* (Figs. 18.47, 18.48). . . . ***Hesperolpium*** Chamberlin

Two species of this genus have been described: ***H. slevini*** Chamberlin from southern California, Arizona, and Baja California, and ***H. andrewsi*** Muchmore from central California. They are found under stones in dry situations. Body length 3–4.5 mm.

35b Movable chelal finger without such setae. **36**

36a (35) Vestitural setae long and conspicuous . . . . ***Aphelolpium*** Hoff

One species of this Caribbean genus is found on the Florida Keys: ***A. cayanum*** Muchmore. It lives in fairly dry litter and under rocks. Body length 2–2.5 mm.

36b Vestitural setae short and inconspicuous . . . . . ***Planctolpium*** Hoff

Two species of this Caribbean genus are known from the United States: ***P. peninsulae*** Muchmore from peninsular Florida and ***P. suteri*** Muchmore from the Gulf coast of Florida and Alabama. They live in various litter and under objects on the ground. Body length 1.5–2 mm.

37a (31) Leg I with articulation between basifemur and telofemur partially mobile, and basifemur distinctly longer than telofemur (Fig. 18.49) **38**

37b Leg I with femoral articulation virtually immobile, and telofemur longer than basifemur (Fig. 18.50). **39**

38a (37) Trichobothrium *isb* on fixed chelal finger much closer to *ist* than to *ib* ***Neoamblyolpium*** Hoff

Only two species of this genus are known: ***N. alienum*** Hoff from New Mexico and Colorado, and ***N. giulianii*** Muchmore from central California. They live in forest litter. Body length 2–2.5 mm.

38b Trichobothrium *isb* closer to *ib* than to *ist* . . . ***Pseudogarypinus*** Beier

Three species are known from the western United States. ***P. frontalis*** (Banks) ranges widely in New Mexico, Colorado, Utah, Idaho, Washington, Oregon, and California. The other two species are apparently restricted to Colorado and California, respectively.

They are all found in rather dry situations, in litter, under rocks, and in mammals' or birds' nests. Body length 2.5–4.5 mm.

39a (37) With two trichobothria on movable chelal finger. **40**

39b With four trichobothria on movable finger . . . ***Serianus*** Chamberlin

This tropical and subtropical genus is represented by several species in the southern United States, ranging as far north as West Virginia in the east and Utah in the Great Basin. They are poorly understood taxonomically. They inhabit litter, often in sandy places. Body length 2–2.5 mm.

40a (39) Hand of palpal chela very heavy; fixed chelal finger with seven trichobothria . . . . ***Aldabrinus*** Chamberlin

Only one species of this uncommon genus has been encountered in the United States, ***A. floridanus*** Muchmore on Key Largo, Florida. Body length about 3 mm.

40b Hand of chela normal; fixed finger with only five trichobothria . . . ***Solinellus*** Muchmore

The only known species of this genus, ***S. simberloffi*** Muchmore, is from the Florida Keys, where it is found under the bark of red mangrove trees. Body length 1.5–2 mm.

41a (14) With four prominent eyes and anterolateral processes on carapace (Fig. 18.51); Family **Pseudogarypidae** . . . . ***Pseudogarypus*** Ellingsen

This genus is widely distributed in the cooler parts of the United States and in southern Canada; reported from Wyoming, Idaho, Utah, Colorado, Arizona, California, Oregon, Michigan, New York, New Hampshire, and Quebec. Six species are known, three from caves in California and Arizona. The epigean forms are usually found in dry situations, among rocks, or in frass in tree holes. Body length 2.5–3.5 mm.

41b With two eyes or none; no processes on carapace **42**

42a (41) Femora of all legs similar (Figs. 18.52, 18.53) **43**

42b Femora of legs I and II quite different from those of legs III and IV (Figs. 18.54, 18.55); Superfamily **Cheliferoidea** **47**

43a (42) Femoral articulations well developed (Fig. 18.56) **44**

43b Femoral articulations obscure (Figs. 18.52, 18.53); Family **Cheiridiidae** **46**

44a (43) Medial ends of pedal coxae modified as a distinct, thin pseudosternum (Fig. 18.57); Family **Sternophoridae** **45**

44b No such pseudosternum present; fourth coxae expanded posteriorly (Fig. 18.58); Family **Pseudochiridiidae** . . . . ***Pseudochiridium*** With

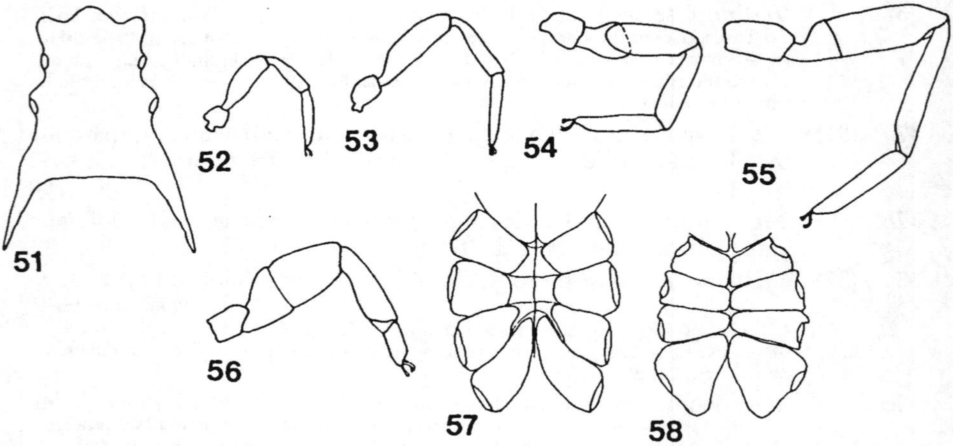

Figs. 18.51–18.58. 51. *Pseudogarypus banksi*, carapace. **52.** *Cheiridium* sp., leg I. **53.** *Cheiridium* sp., leg IV. **54.** *Dinocheirus* sp., leg I. **55.** *Dinocheirus* sp., leg IV. **56.** *Garyops depressa*, leg IV. **57.** *Garyops depressa*, coxae of legs, showing pseudosternum. **58.** *Pseudochiridium* sp., coxae of legs.

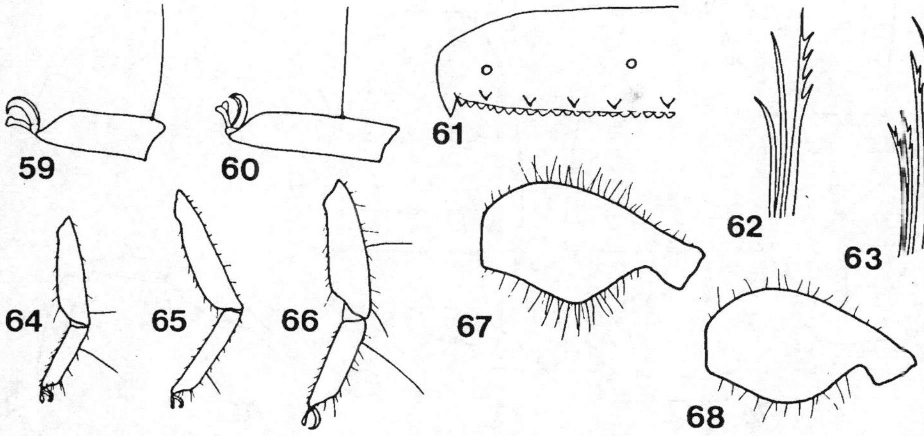

Figs. 18.59–18.68. 59. *Paratemnus elongatus,* tarsus of leg IV. **60.** *Diplotemnus rothi,* tarsus of leg IV. **61.** *Dinocheirus* sp., end of fixed chelal finger, showing marginal teeth and accessory teeth. **62.** *Parachernes* sp., flagellum of chelicera. **63.** *Dinocheirus* sp., flagellum of chelicera. **64.** *Lamprochernes* sp., tibia and tarsus of leg IV. **65.** *Dinocheirus* sp., tibia and tarsus of leg IV. **66.** *Lustrochernes* sp., tibia and tarsus of leg IV. **67.** *Cordylochernes* sp., palpal tibia. **68.** *Lustrochernes* sp., palpal tibia.

Only one representative of this genus and family is known from the United States, namely, ***P. insulae*** Hoff from the Florida Keys. It is found in rather dry litter and in pack rat nests. Body length about 1 mm.

45a **(44)** Movable chelal finger with three trichobothria . . . . ***Garyops*** Banks

Two species are known from the United States, both from Florida: ***G. depressus*** Banks and ***G. pumilus*** Hoff. They are usually found beneath the loose bark of living trees. Body length 2.5–3.5 mm.

45b Movable chelal finger with two trichobothria . . . . ***Idiogaryops*** Hoff

Only one species of this genus is known. ***I. paludis*** Chamberlin apparently ranges along the Gulf coast from Florida to Texas, up the Atlantic coast to New Jersey, and up the Mississippi valley to Illinois. It seems to be a lowland form, found mostly under the bark of living trees. Body length 2–2.5 mm. [A different concept of ***Garyops*** and ***Idiogaryops*** is presented by Harvey 1985].

46a **(43)** Movable chelal finger with two trichobothria . . . ***Cheiridium*** Menge

Three species are known in the United States, one of them, ***C. museorum*** (Leach), an anthropophilic form having been introduced into Massachusetts from Europe. The others have been found rarely in Illinois and in California, in tree hole litter and in pack rat nests, respectively. Body length about 1 mm.

46b Movable chelal finger with one trichobothrium . . . . ***Apocheiridium*** Chamberlin

The genus is widespread, being represented in the west by six species from California, Oregon, and Utah, and in the east by one named species ranging from Michigan to Massachusetts and south to Florida. They live mainly under the bark of living trees and have been taken from the bodies of noctuid moths. Body length 1–1.5 mm.

47a **(42)** Venedens and venom duct present only in fixed finger of palpal chela.. . . . . **48**

47b Venedens and venom duct well developed in both chelal fingers or only in movable finger **49**

48a **(47)** Tactile seta on tarsus of leg IV basal in position (Fig. 18.59); surface of carapace smooth; Family **Atemnidae** . . . . ***Paratemnus*** Beier

This circumtropical genus is represented in the United States by one species, ***P. elongatus*** (Banks), known from Florida and Alabama. It is found, often in numbers, under the bark of both living and dead trees. Body length 3–4.5 mm.

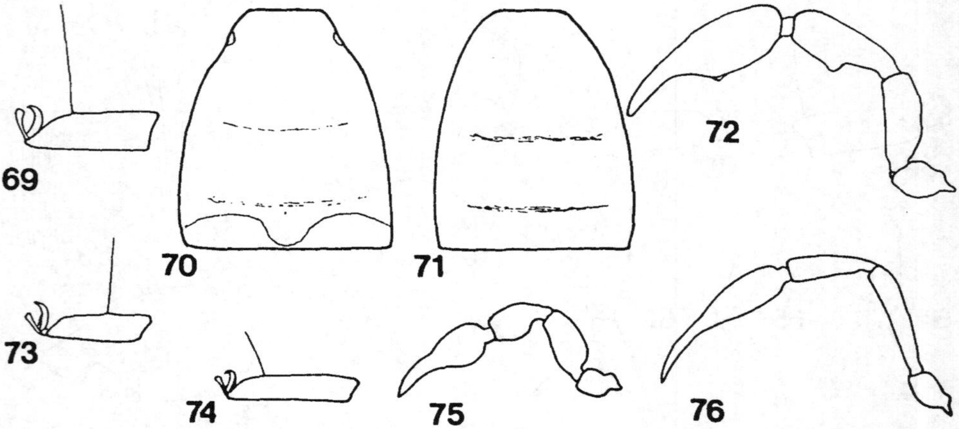

Figs. 18.69–18.76. 69. *Parachernes* sp., tarsus of leg IV. **70.** *Parachernes* sp., carapace. **71.** *Dinocheirus* sp., carapace. **72.** *Bituberochernes mumae*, ♂, palp, dorsal view. **73.** *Pselaphochernes scorpioides*, tarsus of leg IV. **74.** *Tychochernes* sp., tarsus of leg IV. **75.** *Tejachernes* sp., palp, dorsal view. **76.** *Zaona biseriatum*, palp, dorsal view.

48b Tactile seta on tarsus of leg IV near middle of segment (Fig. 18.60); surface of carapace distinctly granulate; Family **Miratemnidae**. . . . ***Diplotemnus*** Chamberlin

A single species of this wide-ranging genus is known from the United States, ***D. rothi*** Muchmore, from southeastern Arizona. It apparently lives in dry litter. Body length 4.5–6.5 mm.

49a (47) Venedens and venom duct well developed in movable chelal finger, poorly developed in or absent from fixed finger; both chelal fingers with accessory teeth (Fig. 18.61); Family **Chernetidae** **50**

49b Venedens and venom duct well developed in both chelal fingers; no accessory teeth on either finger **73**

50a (49) Flagellum on chelicera composed of three setae (Fig. 18.62). . . . **51**

50b Flagellum of four setae (Fig. 18.63) **65**

51a (50) Tibia of leg IV with one or two prominent tactile setae on outer margin (Fig. 18.64). **52**

51b Tibia of leg IV without conspicuous tactile setae (Fig. 18.65) . . . **55**

52a (51) Two tactile setae on tibia of leg IV, one at middle and one at distal end (Fig. 18.66). **53**

52b One tactile seta on tibia of leg IV very near distal end of outer margin (Fig. 18.64) . . . . ***Lamprochernes*** Tömösvary

This genus is probably widespread through the United States and southern Canada; but its precise distribution and taxonomy here is unknown, mainly because it has until recently been confused with ***Pycnochernes*** Beier (which is a synonym) and ***Americhernes*** Muchmore (which is distinct). They apparently live mostly in damp organic debris and are often found clinging to the legs of insects in a phoretic relationship. Body length 2–3 mm.

53a (52) Trichobothrium *it* on fixed chelal finger at least as close to fingertip as the distance between *ist* and *isb*. **54**

53b Trichobothrium *it* clearly farther from the fingertip than the distance between *ist* and *isb*. . . . . ***Americhernes*** Muchmore

This newly recognized genus is represented in the United States by at least five species. ***A. oblongus*** (Say) is widespread through the eastern and central states. The others are known only from restricted areas of Florida, Colorado, and Utah. They are usually found under bark of dead tree stumps and logs, but occasionally under bark of living trees. Body length 2–3.5 mm.

54a **(53)** Palpal tibia of male with a conspicuous dorsomedial protuberance (Fig. 18.67) . . . . ***Cordylochernes*** Beier

This and the following genus are not yet sharply defined. It is possible that representatives of ***Cordylochernes,*** which is widespread through Middle and South America, may occur in the United States near the Mexican border. They are often found under the elytra of large cerambycid beetles. Body length 3–4.5 mm.

54b Palpal tibia of male without such a protuberance (Fig. 18.68). . . . . ***Lustrochernes*** Beier

Two species of this widespread Neotropical genus are recognized in the United States: ***L. pennsylvanicus*** (Ellingsen) in the southeast and ***L. grossus*** (Banks) in the west. They have been found beneath the bark of trees and logs and under the elytra of large cerambycid beetles. Body length 2.5–3.5 mm.

55a **(51)** Tarsus of leg IV with a conspicuous tactile seta (Fig. 18.69) **56**

55b Tarsus of leg IV without a conspicuous tactile seta **61**

56a **(55)** All trichobothria on fixed chelal finger except *et* confined to proximal half of finger. **57**

56b Trichobothria of fixed chelal finger not so confined **59**

57a **(56)** Carapace with a thickened "keel" near middle of posterior margin (Fig. 18.70) . . . . ***Parachernes*** Chamberlin

This genus is predominantly Neotropical but is represented in the United States by at least nine species. They are most common in the southern states, but have been taken as far north as Delaware on the east coast, Michigan in the Mississippi valley, and Colorado in the Great Plains. They are usually found beneath the bark of standing trees, but also in logs and litter. Body length 2–4 mm.

57b Carapace without a conspicuous "keel" (Fig. 18.71) **58**

58a **(57)** Leg I of male with many long, conspicuous setae on outer margins of tibia and tarsus; palpal chela of female only moderately heavy . . . . ***Pachychernes*** Beier

One uncommon species of this Neotropical genus occurs in Florida, ***P. shelfordi*** Hoff, which is also known from two localities in Mexico. It has been found in moist litter. Body length 2.5–3.5 mm.

58b Male unknown; palpal chela of female conspicuously short and heavy ***Dinochernes*** Beier

In the United States there is one rare species of this genus, which is otherwise known only from Baja California, Mexico. ***D. wallacei*** Muchmore has been found in Alachua and Dade counties, Florida, on a tree trunk and in litter. Body length about 5 mm.

59a **(56)** Palpal tibia and chelal hand of male each with a medial protuberance (Fig. 18.72) . . . . ***Bituberochernes*** Muchmore

One species of this Caribbean genus is known from the United States. ***B. mumae*** Muchmore has been found in Dade County, Florida, under bark. Body length 3–3.5 mm.

59b Palpal tibia and chela of male without protuberances **60**

60a **(59)** Tactile seta on tarsus of leg IV median in position (Fig. 18.73) . . . . ***Pselaphochernes*** Beier

This genus is widespread in the United States from the Mississippi valley eastward; it is also known to occur in Utah and Washington. The species are not yet well defined. They are commonly found in moist litter and compost. Body length 1.5–2.5 mm.

60b Tactile seta on tarsus of leg IV distal in position (Fig. 18.74) . . . . . ***Tychochernes*** Hoff

Only one species is known in this genus: ***T. inflatus*** Hoff, from New Mexico and Arizona. It is found in pack rat nests. Body length 2.5–3.5 mm.

61a **(55)** Palp relatively stout, *l/w* ratio of femur less than 4.0 (Fig. 18.75) **62**

61b Palp slender, femur with *l/w* ratio greater than 4.0 (Fig. 18.76). . . . *Zaona* Chamberlin

Only one species has been described in this genus: *Z.* ***biseriatum*** (Banks). It is known from Florida and Texas, where it occurs in the axils of palm trees. Body length about 2.5 mm.

62a (61) Hand of chelicera with five setae. **63**

62b Hand of chelicera with only four setae . . . . ***Tejachernes*** Hoff

This genus is widely represented in caves in the southwestern United States and Middle America. However, only one species has been described, ***T. stercoreus*** (Turk), from caves in Texas. It is found on guano below the roosts of bats. Body length 2.5–3 mm.

63a (62) Fingers of palpal chela with a few accessory teeth. . . . . ***Allochernes*** Beier

This palearctic genus has been detected only once in the United States. A single specimen referable to ***A. peregrinus*** Lohmander was found in Cheshire County, Massachusetts, in beech litter. Body length 2 mm.

63b Fingers of chela without accessory teeth **64**

64a (63) Movable chelal finger with trichobothrium *st* nearer to *sb* than to *t*; vestigial venom duct present in fixed finger. . . . . ***Wyochernes*** Hoff

This genus is known only by ***W. hutsoni*** Hoff, from Albany County, Wyoming, at an elevation of 3240 m (10,800 ft). It was found on the undersides of rocks. Body length 2–3 mm.

64b Movable chelal finger with trichobothrium *st* nearer to *t* than to *sb*; no venom duct in fixed finger . . . . ***Chrysochernes*** Hoff

This genus is known only by ***C. elatus*** Hoff, from Torrance County, New Mexico, at an elevation of 2040 m (6800 ft). It was found in dry juniper litter. Body length about 2.5 mm.

65a (50) Tarsus of leg IV with a conspicuous tactile seta (Fig. 18.77) **66**

65b Tarsus of leg IV without a conspicuous tactile seta (Fig. 18.78) . . **71**

66a (65) Vestitural setae very numerous (about 200 on carapace); most setae long, acuminate, and with a single, small, lateral spinule (Fig. 18.79) ***Mucrochernes*** Muchmore

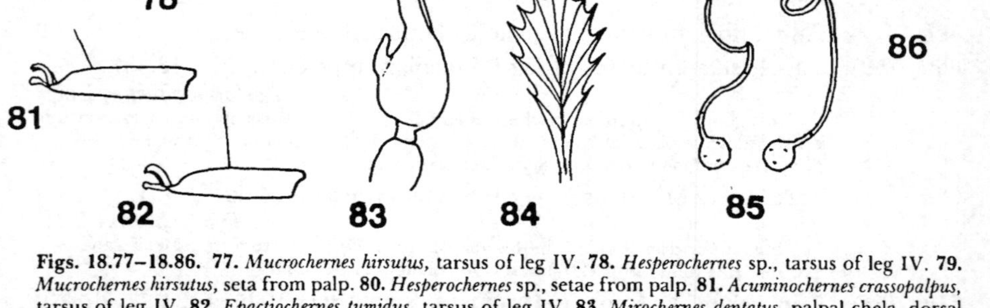

Figs. 18.77–18.86. 77. *Mucrochernes hirsutus,* tarsus of leg IV. **78.** *Hesperochernes* sp., tarsus of leg IV. **79.** *Mucrochernes hirsutus,* seta from palp. **80.** *Hesperochernes* sp., setae from palp. **81.** *Acuminochernes crassopalpus,* tarsus of leg IV. **82.** *Epactiochernes tumidus,* tarsus of leg IV. **83.** *Mirochernes dentatus,* palpal chela, dorsal view. **84.** *Illinichernes distinctus,* seta from palp. **85.** *Hesperochernes* sp., ♀, spermathecae. **86.** *Chernes* sp., ♀, spermathecae.

This genus is known only by ***M. hirsutus*** (Banks), from Orange and Los Angeles counties, California. They are found at the shore in cracks between rocks. Body length 2.5–3.5 mm.

66b Vestitural setae normal in number, each rather short and terminally dentate or clavate (Fig. 18.80) **67**

67a **(66)** Movable chelal finger with trichobothrium *st* nearer to *t* than to *sb*. **68**

67b Movable chelal finger with trichobothrium *st* nearer to *sb* than to *t* . . ***Dendrochernes*** Beier

This Holarctic genus is fairly widespread in North America, known from New Mexico, Colorado, Arizona, California, Oregon, Montana, and Saskatchewan in the west and from Michigan, New York, New Jersey, Maryland, Tennessee, North Carolina, and Florida in the east. However, it is only rarely encountered; under bark or phoretic on insects. Body length 3–3.5 mm.

68a **(67)** Tactile seta on tarsus of leg IV distinctly distal to middle of segment (Fig. 18.81). **69**

68b Tactile seta on tarsus of leg IV median in position (Fig. 18.82) . . . . ***Epactiochernes*** Muchmore

This genus includes mainly littoral forms from the east coast of the United States and the West Indies. ***E. tristis*** (Banks) from New York, Connecticut, and Massachusetts and ***E. tumidus*** (Banks) from Florida and North Carolina are found under debris on the shore. Body length 1.5–2.5 mm.

69a **(68)** Hand of chelicera with seta *b* acuminate and *sb* terminally denticulate . **70**

69b Hand of chelicera with both *b* and *sb* simple and acuminate . . . . . ***Acuminochernes*** Hoff

This Nearctic genus (= ***Phoberocheirus*** Chamberlin) includes two species, ***A. crassopalpus*** Hoff and ***A. tacitus*** Hoff. They have been found in Texas, Colorado, Utah, Oregon, Kansas, Michigan, Arkansas, Tennessee, Illinois, New York, Virginia, and Florida. They are usually encountered in tree holes or decayed logs, often in association with birds or small mammals. Body length 2–3 mm.

70a **(69)** Male with a prominent, anvil-shaped process on medial side of chelal hand (Fig. 18.83). . . . . ***Mirochernes*** Beier

Only one species is known, ***M. dentatus*** (Banks), from Florida, North Carolina, Virginia, Connecticut, Indiana, Illinois, Michigan, Arkansas, and Oklahoma. It is found in decayed trees, sometimes in association with small mammals. Body length 2–3 mm.

70b Male without a process on the chelal hand (females are very much like those of ***Mirochernes***) . . . . ***Dinocheirus*** Chamberlin

This is the largest and most varied genus of pseudoscorpions in North America, with 16 species known here. It probably is represented in every state of the United States and the southern parts of Canada. They are found in a wide variety of debris; many species seem strongly associated with animal nests and shelters. Common are ***D. sicarius*** Chamberlin in the west and ***D. pallidus*** (Banks) in the east. Body length 2–4.5 mm.

71a **(65)** Setae of palps and tergites large and bilaterally pinnate and leaflike (Fig. 18.84); with a few, long, clavate setae near center of outer margin of fixed chelal finger . . . . ***Illinichernes*** Hoff

Two species of this indigenous genus are known: ***I. distinctus*** Hoff from Illinois, Indiana, Michigan, Maryland, and Alabama and ***I. stephensi*** Benedict and Malcolm from Oregon. They occur in tree holes and decayed logs. Body length 1.7–2 mm.

71b Setae of palps and tergites not bilaterally pinnate and leaflike; without unusual setae on palpal chela **72**

72a **(71)** Hand of chelicera usually with setae *b* and *sb* both denticulate; spermathecae of female terminally expanded (Fig. 18.85). . . . . ***Hesperochernes*** Chamberlin

Next to ***Dinocheirus,*** this is the most widespread and varied genus in North America, with about 12 described species. It has been recorded from most of the states and provinces. Species are found in a variety of organic debris, often in association with animals; and two are widespread on bat guano in caves from Indiana and Ohio to Alabama and Georgia. Body length 2.5–3.5 mm.

72b Hand of chelicera with seta *b* acuminate and *sb* denticulate; spermathecae of female not terminally expanded (Fig. 18.86). . ***Chernes*** Menge

This genus is probably well represented across southern Canada and northern United States. However, its taxonomy and distribution are not known in detail because it has long been confused with ***Hesperochernes***. They apparently live in moist forest litter. Body length 2–3 mm.

73a (49) Flagellum of chelicera composed of four setae; males without ramshorn organs (Fig. 18.87); Family **Withiidae** **74**

73b Flagellum of chelicera composed of three setae; males with ramshorn organs (Fig. 18.88); Family **Cheliferidae**. **75**

74a (73) Sternites 4–9 of male with median patches of small, sensory setae (Fig. 18.89) . . . . ***Withius*** Kew

This genus has been reported from around the world in tropical and subtropical areas. One species, ***W. piger*** (Simon), is anthropophilic and has been recognized in Florida, South Carolina, Arizona, and California. Other species have been described from southwestern United States and Mexico, but their status is not clear. They are found in dry debris and in grain stores. Body length 2–3 mm.

74b Sternites of male without patches of sensory setae . ***Juxtachelifer*** Hoff

Only one species is known, ***J. fructuosus*** Hoff, found in pack rat nests in New Mexico and California. Body length 3–4 mm.

75a (73) Male with genital sacs in coxae of legs IV (Fig. 18.90); female with median cribriform plates smaller in diameter than diameter of anterior tracheal trunk (Fig. 18.91) **76**

75b Male without genital sacs in coxae of legs IV; female with median cribriform plates as large in diameter as diameter of anterior tracheal trunk (Fig. 18.92) . . . . ***Haplochelifer*** Chamberlin

One species is known, ***H. philipi*** (Chamberlin); found in the western United States as far north as Oregon and Idaho, where it lives in leaf litter. Body length 1.75–3.5 mm.

76a (75) Coxal sacs of male with well-defined atria (Fig. 18.93) **77**

76b Coxal sacs of male without atria (Fig. 18.90) **82**

77a (76) Atrium of coxal sac with conspicuous cribrate surface; females with setae on the pleural membranes. . . . . ***Levichelifer*** Hoff

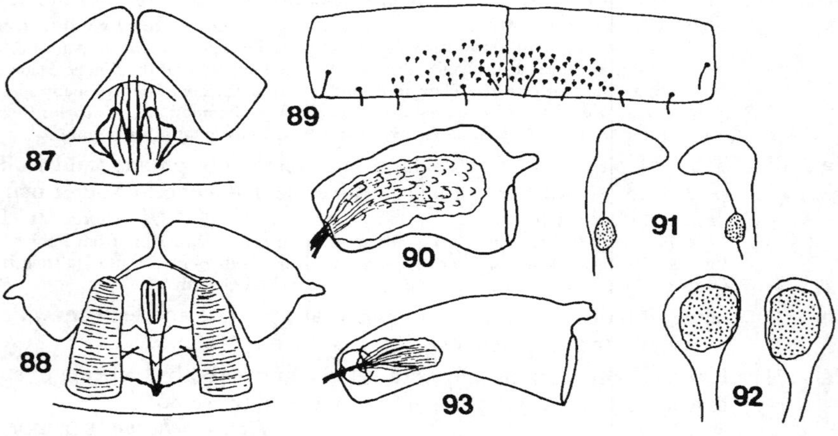

Figs. 18.87–18.93. 87. *Withius piger*, ♂, coxae of legs IV and internal genitalia. **88.** *Chelifer cancroides*, ♂, coxae of legs IV and internal genitalia, with large ramshorn organs. **89.** *Withius piger*, ♂, sternite 7, showing sensory setae. **90.** *Chelifer cancroides*, ♂, coxa of leg IV with coxal sac. **91.** *Chelifer cancroides*, ♀, spermathecae and median cribriform plates. **92.** *Haplochelifer philipi*, ♀, spermathecae and median cribriform plates. **93.** *Tyrannochelifer* sp., coxa of leg IV with coxal sac.

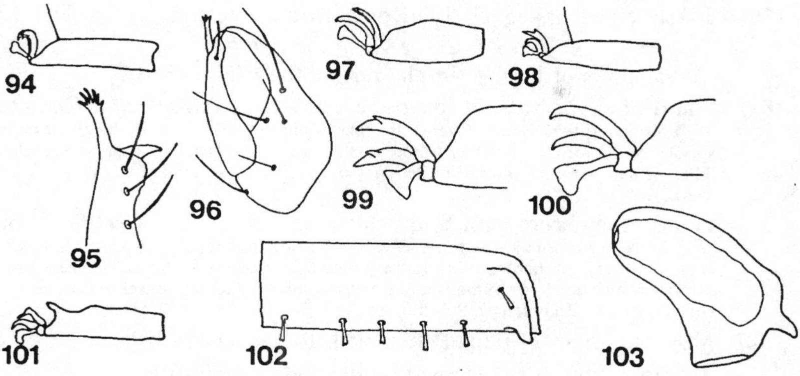

Figs. 18.94–18.103. 94. *Tyrannochelifer* sp., tarsus of leg IV. **95.** *Ellingsenius* sp., end of movable finger of chelicera, showing 3 galeal setae. **96.** *Dactylochelifer silvestris,* chelicera. **97.** *Dactylochelifer copiosus,* ♂, tarsus of leg I. **98.** *Florichelifer aureus,* ♂, tarsus of leg I. **99.** *Parachelifer* sp., end of tarsus of leg IV, showing claws. **100.** *Hysterochelifer* sp., end of tarsus of leg IV, showing claws. **101.** *Hysterochelifer* sp., ♂, tarsus of leg I, with apical spine. **102.** *Hysterochelifer* sp., ♂, right half of tergite 4, showing lateral spur. **103.** *Hysterochelifer* sp., ♂, coxa of leg IV with coxal sac and lateral spur.

This genus (= ***Ocalachelifer*** Chamberlin) includes two species: ***L. fulvopalpus*** Hoff from Texas and New Mexico and ***L. cribratus*** (Chamberlin) from Florida and Alabama. They are found in rather dry litter. Body length 2–2.5 mm.

77b Atrium of coxal sac with smooth surface; females without setae on the pleural membranes. **78**

78a (77) Tarsus of leg IV without a conspicuous tactile seta **79**

78b Tarsus of leg IV with a tactile seta, distal in position (Fig. 18.94) . . . ***Tyrannochelifer*** Chamberlin

One species of this Neotropical genus occurs in south Florida, ***T. floridanus*** (Banks). It is found beneath the bark of sand pine or mangrove trees. Body length 2.5–3.5 mm.

79a (78) Movable finger of chelicera with three galeal setae (Fig. 18.95) . . . . ***Ellingsenius*** Chamberlin

An African species ***E. sculpturatus*** (Lewis) has been reported once from California, where it may have been introduced with honeybees, in the hives of which it lives. Body length 3–3.5 mm.

79b Movable finger of chelicera with one galeal seta **80**

80a (79) Hand of chelicera with five setae (Fig. 18.96) **81**

80b Hand of chelicera with four setae. . . . . ***Pugnochelifer*** Hoff

The only species of this genus, ***P. amoenus*** Hoff, is known from Florida, Louisiana, and North Carolina. It lives under the bark of various trees. Body length 1.8–2 mm.

81a (80) Tarsus of leg I of male thickened in middle, and claws enlarged and strongly modified (Fig. 18.97); coxa of leg IV without a lateral spur . ***Dactylochelifer*** Beier

This Holarctic genus is represented in the United States and Canada by two species: ***D. copiosus*** Hoff in the central and eastern states as far north as Nova Scotia, and ***D. silvestris*** Hoff in the Rocky Mountains from New Mexico to British Columbia. They are found in litter, often rather dry. Body length 2.5–3.5 mm.

81b Tarsus of leg I of male normal and claws little modified (Fig. 18.98); coxa of leg IV of male with a distinct lateral spur (see Fig. 18.93) . . ***Florichelifer*** Hoff

This monotypic genus is known only by ***F. aureus*** Hoff from Florida. It is found under bark of slash pines. Body length 2–3 mm.

82a (76) Tarsal claws of leg IV bifid or with an accessory tooth (Fig. 18.99) . . . **83**

82b Tarsal claws of leg IV simple, not toothed (Fig. 18.100) **84**

83a (82) Hand of chelicera with five setae . . . . . . ***Parachelifer*** Chamberlin

This widespread Nearctic genus is represented in North America by about eight species. Some form is probably found in every state but New York and New England. They usually are encountered under bark of living or dead trees. Body length 2.5–4.5 mm.

83b Hand of chelicera with four setae . . . . ***Chelifer*** Geoffroy

This genus is apparently represented by a nearly cosmopolitan species, ***C. cancroides*** (Linnaeus). It is probably found in every state and province in North America, usually in close association with humans and domestic animals (i.e., in houses, barns, and chicken coops). Body length 2.5–3.5 mm.

84a (82) Movable finger of palpal chela with four trichobothria **85**

84b Movable finger of chela with only two trichobothria . . . . ***Xenochelifer*** Chamberlin

This monotypic genus is known by ***X. davidi*** Chamberlin, from southern California. The few specimens were found on cottonwood trees. Body length 3–4 mm.

85a (84) Tarsus of leg I of male with a well-developed apical spine (Fig. 18.101) . . . **86**

85b Tarsus of leg I of male lacking an apical spine (see Fig. 18.98) . . **88**

86a (85) Male with lateral spurs of anterior abdominal tergites fairly well developed (Fig. 18.102) **87**

86b Male without lateral spurs on abdominal tergites . . . . ***Aspurochelifer*** Benedict and Malcolm

This newly described genus, based on ***A. littlefieldi*** Benedict and Malcolm, has been recognized in California, Nevada, Oregon, and Washington. It is found in forest litter. Body length 2–3 mm.

87a (86) Male with a lateral spur on coxa of leg IV (Fig. 18.103) . . . . ***Hysterochelifer*** Chamberlin

This Holarctic genus is known in the United States by three species, the detailed relations and distribution of which are not yet understood. Representatives have been found in California, Oregon, Washington, Wyoming, Colorado, Arizona, New Mexico, and Texas. They are found in a variety of habitats, mostly dry. Body length 2–3 mm.

87b Male without a lateral spur on coxa of leg IV . . . ***Phorochelifer*** Hoff

This monotypic genus has only the species ***P. mundus*** Hoff, known from two specimens from New Mexico at altitudes of 2100–2700 m (7000–9000 ft). Body length 2.5–3.5 mm.

88a (85) Male with a lateral spur on coxa of leg IV (see Fig. 18.103) . . . . . ***Idiochelifer*** Chamberlin

This monotypic genus is represented by ***I. nigripalpus*** (Ewing), found in the midcentral United States, usually under the bark of trees, both living and dead. Body length 2.5–3 mm.

88b Male without a lateral spur on coxa of leg IV . . . ***Paisochelifer*** Hoff

This indigenous genus includes two species: ***P. callus*** Hoff from eastern and central United States and ***P. utahensis*** Hoff from Utah. They are found in moist litter and in birds' nests. Body length 1.5–3 mm.

References

References to the early work on pseudoscorpions of North America may be found in Beier (1932a,b), Chamberlin (1931), and Hoff (1958). Below are listed most of the significant papers on pseudoscorpions published since 1957.

Aitchison, C. W. 1979. Low temperature activity of pseudoscorpions and phalangids in southern Manitoba. *J. Arachnol.* 7:85–86. **Beier, M. 1932a.** *Pseudoscorpionidea I.* Subord. Chthoniinea et Neobisiinea. *Tierreich* 57:1–258. **1932b.** *Pseudoscorpionidea II.* Subord. C. Cheliferinea. *Tierreich* 58:1–294. **Benedict, E. M. 1978a.** False scorpions of the genus *Apocheiridium* Chamberlin from western North America (Pseudoscorpionida, Cheiridiidae). *J. Arachnol.* 5:231–241. **1978b.** A new pseudoscorpion genus *Malcolmochthonius,* n.g., with three new species from the western United States. *Trans. Am. Microsc. Soc.* 97:250–255. **1979.** A new species of *Apochthonius* Chamberlin from Oregon (Pseudoscorpionida, Chthoniidae). *J. Arachnol.* 7:79–83. **Benedict, E. M. and D. R. Malcolm. 1970.** Some pseudotyrannochthoniine false scorpions from western North America. (Chelonethida, Chthoniidae). *J. N.Y. Entomol. Soc.* 78:38–51. **1973.** A new cavernicolous species of *Apochthonius* (Chelonethida : Chthoniidae) from the western United States with special reference to troglobitic tendencies in the genus. *Trans. Am. Microsc. Soc.* 92:620–628. **1974.** A new cavernicolous species of *Mundochthonius* from the eastern United States (Pseudoscorpionida, Chthoniidae). *J. Arachnol.* 2:1–4. **1977.** Some garypoid false scorpions from western North America (Pseudoscorpionida : Garypidae and Olpiidae). *J. Arachnol.* 5:113–132. **1978.** The family Pseudogarypidae (Pseudoscorpionida) in North America with comments on the genus *Neopseudogarypus* Morris from Tasmania. *J. Arachnol.* 6:81–104. **1979.** Pseudoscorpions of the family Cheliferidae from Oregon (Pseudoscorpionida, Cheliferoidea). *J. Arachnol.* 7:187–198. **1982.** Pseudoscorpions of the family Chernetidae newly identified from Oregon (Pseudoscorpionida, Cheliferoidea). *J. Arachnol.* 10:97–109. **Brach, V. 1978.** Social behavior in the pseudoscorpion *Paratemnus elongatus* (Banks) (Pseudoscorpionida : Atemnidae). *Insectes Soc.* 25:1–11. **1979.** Species diversity and distributional relationships of pseudoscorpions from slash pines (*Pinus elliottii* Eng.) in Florida (Arachnida : Pseudoscorpionida). *Bull. South. Calif. Acad. Sci.* 78:32–39. **Chamberlin, J. C. 1931.** The arachnid order Chelonethida. *Stanford Univ. Publ. Biol. Sci.* 7:1–284. **1962.** New and little-known false scorpions, principally from caves, belonging to the families Chthoniidae and Neobisiidae (Arachnida, Chelonethida). *Bull. Am. Mus. Nat. Hist.* 123:299–352. **Chamberlin, J. C. and D. R. Malcolm. 1960.** The occurrence of false scorpions in caves with special reference to cavernicolous adaptation and to cave species in the North American fauna. *Am. Midl. Nat.* 64:105–115. **Curcic, B. P. M. 1978.** *Tuberocreagris,* a new genus of pseudoscorpions from the United States (Arachnida, Pseudoscorpiones, Neobisiidae). *Fragm. Balc. Mus. Macedonici Sci. Nat.* 10:111–121. **1981.** A revision of some North American pseudoscorpions (Neobisiidae, Pseudoscorpiones). *Bull. Mus. Hist. Nat. Belgr.* 36B:101–107. **1982a.** *Trisetobisium* (Pseudoscorpiones, Neobisiidae), a new genus of pseudoscorpions based on *Microcreagris fallax* Chamberlin. *Bull. Mus. Hist. Nat. Belgr.* 37B:57–61. **1982b.** *Americocreagris,* a new genus of pseudoscorpions from the United States. *Bull. Acad. Serbe Sci. Arts Cl. Sci. Math. Nat. Sci. Nat.* 22:47–50. **1984.** A revision of some North American species of *Microcreagris* Balzan, 1892 (Arachnida : Pseudoscorpiones : Neobisiidae). *Bull. Br. Arachnol. Soc.* 6:149–166. **Dunkle, S. W. 1984.** First record of pseudoscorpions phoretic on dragonflies. *Not. Odonatol.* 2:48. **Harvey, M. S. 1985.** The systematics of the family Sternophoridae (Pseudoscorpionida). *J. Arachnol.* 13:121–209. **Hoff, C. C. 1949.** The pseudoscorpions of Illinois. *Illinois Nat. Hist. Surv. Bull.* 24:409–498. **1957.** *Tejachernes* (Arachnida-Chelonethida, Chernetidae, Chernetinae), a new genus of pseudoscorpions based on *Dinocheirus stercoreus. Southwest. Nat.* 2:83–88. **1958.** List of the pseudoscorpions of North America north of Mexico. *Am. Mus. Novit.* 1875:1–50. **1959.** The ecology and distribution of the pseudoscorpions of north-central New Mexico. *Univ. N.M. Publ. Biol.* No. 8:1–68. **1961.** Pseudoscorpions from Colorado. *Bull. Am. Mus. Nat. Hist.* 122:409–464. **1963a.** Pseudoscorpions from the Black Hills of South Dakota. *Am. Mus. Novit.* 2134:1–10. **1963b.** Sternophorid pseudoscorpions, chiefly from Florida. *Am. Mus. Novit.* 2150:1–14. **1964a.** A new pseudochiridiid pseudoscorpion from Florida. *Trans. Am. Microsc. Soc.* 83:89–92. **1964b.** Atemnid and cheliferid pseudoscorpions, chiefly from Florida. *Am. Mus. Novit.* 2198:1–43. **Hoff, C. C. and D. T. Jennings. 1974.** Pseudoscorpions phoretic on a spider. *Entomol. News* 85:21,22. **Lee, V. F. 1979.** The maritime pseudoscorpions of Baja California, Mexico (Arachnida : Pseudoscorpionida). *Occas. Pap. Calif. Acad. Sci.* 131:1–38. **Lloyd, J. E. and W. B. Muchmore. 1974.** Pseudoscorpions phoretic on fireflies. *Fla. Entomol.* 57:381. **Lloyd, J. E., S. Correale and W. B. Muchmore. 1975.** Pseudoscorpions phoretic on fireflies II. *Fla. Entomol.* 58:241,242. **Malcolm, D. R. and W. B. Muchmore. 1985.** An unusual species of *Tyrannochthonius* from Florida (Pseudoscorpionida, Chthoniidae). *J. Arachnol.* 13:403–405. **Manley, G. V. 1969.** A pictorial key and annotated list of Michigan pseudoscorpions (Arachnida : Pseudoscorpionida). *Mich. Entomol.* 2:2–13. **Muchmore, W. B. 1962.** A new cavernicolous pseudoscorpion belonging to the genus *Microcreagris. Postilla* 70:1–6. **1963a.** Redescription of some cavernicolous pseudoscorpions (Arachnida, Chelonethida) in the collection of the Museum of Comparative Zoology. *Breviora* 188:1–16. **1963b.** Two European arachnids new to the United States. *Entomol. News* 74:208–210. **1965.** North American cave pseudoscorpions of the genus *Kleptochthonius,* subgenus *Chamberlinochthonius* (Chelonethida, Chthoniidae). *Am. Mus. Novit.* 2234:1–27. **1966a.** Two new species of *Kleptochthonius* (Arachnida, Chelonethida) from a cave in Tennessee. *J. Tenn. Acad. Sci.* 41:68,69. **1966b.** A cavernicolous pseudoscorpion of the genus *Microcreagris* from southern Tennessee. *Entomol. News* 77:97–100. **1966c.** The use of chloropicrin with a Berlese funnel. *Turtox News* 44:278–279. **1967a.** New cave pseudoscorpions of the genus *Apochthonius* (Arachnida, Chelonethida). *Ohio J. Sci.* 67:89–95. **1967b.** Pseudotyrannochthoniine pseudoscorpions from the western United States. *Trans. Am. Microsc. Soc.* 86:132–139. **1967c.** Two new species of the pseudoscorpion genus *Paraliochthonius. Entomol. News* 78:155–162. **1967d.** *Novobisium* (Arachnida, Chelonethida, Neobisiidae, Neobisiinae), a new genus of pseudoscorpions based on *Obisium carolinensis* Banks. *Entomol. News* 78:211–215. **1968a.** A new species of the pseudoscorpion genus *Aphrastochthonius* (Arachnida, Chelonethida), from a cave in Alabama. *Bull. Natl. Speleol. Soc.* 30:17,18. **1968b.** Two new species of chthoniid pseudoscorpions from the western United States (Arachnida : Chelonethida : Chthoniidae). *Pan-Pac. Entomol.* 44:51–57. **1968c.** Redescription of the type species of the pseudoscorpion genus *Kewochthonius* Chamberlin. *Entomol. News* 79:71–76. **1968d.** A cavernicolous species of the pseudoscorpion genus *Mundochthonius* (Arachnida, Chelonethida, Chthoniidae). *Trans. Am. Microsc. Soc.* 87:110–112. **1968e.** A new species of the pseudoscorpion genus *Syarinus* (Arachnida, Chelonethida, Syarinidae) from the northeastern United States. *J. N.Y. Entomol. Soc.* 76:112–116. **1968f.** A new species of the pseudoscorpion genus *Serianus* (Arachnida, Chelonethida, Olpiidae) from North Carolina. *Entomol. News* 79:145–150. **1968g.** A new species of the pseudoscorpion genus *Parobisium* from Utah (Arachnida, Chelonethida, Neobisiidae). *Am. Midl. Nat.* 79:531–534. **1969a.** The pseudoscorpion genus *Neochthonius* Chamberlin (Arachnida, Chelonethida, Chthoniidae) with description of a cavernicolous species. *Am. Midl. Nat.* 81:387–394. **1969.** A population of a European pseudoscorpion established in New York. *Entomol. News* 80:66. **1969c.** New species and records of cavernicolous pseudoscorpions of the genus *Microcreagris* (Arachnida, Chelonethida, Neobisiidae, Ideobisiinae). *Am. Mus. Novit.* 2392:1–21. **1970.** New cavernicolous *Kleptochthonius* spp. from Virginia (Arachnida, Pseudoscorpionida, Chthoniidae). *Entomol. News* 81:210–212. **1971a.** On phoresy in pseudoscorpions. *Bull. Brit. Arachnol. Soc.* 2:38. **1971b.** Phoresy by North and Central American pseudoscorpions. *Proc. Rochester Acad. Sci.* 12:79–97. **1971c.** The identity of *Olpium obscurum* (Pseudoscorpionida, Olpiidae). *Fla. Entomol.* 54:241–243. **1971d.** A new *Lamprochernes* from Utah. *Entomol. News* 82:327–329. **1972a.** European pseudoscorpions from New England. *J. N.Y. Entomol. Soc.* 80:109,110. **1972b.** The pseudoscorpion genus *Paraliochthonius* (Arachnida, Pseudoscorpionida, Chthoniidae). *Entomol. News* 83:248–256. **1972c.** The unique, cave-restricted genus *Aphrastochthonius* (Pseudoscorpionida, Chthoniidae). *Proc. Biol. Soc. Wash.* 85:433–444. **1973a.** A new genus of pseudoscorpions based upon *Atemnus hirsutus* (Pseudoscorpionida, Chernetidae). *Pan-Pac. Entomol.* 49:43–48. **1973b.** Ecology of pseudoscorpions—a review. *Proc. 1st Soil Microcommunities Conf.,* D. L. Dindal (ed.), pp. 121–127. **1973c.** The genus *Chitrella* in America (Pseudoscorpionida, Syarinidae). *J. N.Y. Entomol. Soc.* 81:183–192. **1974a.** Pseudoscorpions from Florida. 1. The genus *Aldabrinus* (Pseudoscorpionida : Olpiidae). *Fla. Entomol.* 57:1–7. **1974b.** Pseudoscorpions from Florida. 2. A new genus and species *Bituberochernes mumae.* (Chernetidae). *Fla. Entomol.* 57:77–80. **1974c.** Pseudoscorpions from Florida. 3. *Epactiochernes,* a new genus based upon *Chelanops tumidus* Banks (Chernetidae). *Fla. Entomol.* 57:397–407. **1974d.** Clarification of the genera *Hesperochernes* and *Dinocheirus* (Pseudoscorpionida, Chernetidae). *J. Arachnol.* 2:25–36. **1974e.** New cavernicolous species of *Kleptochthonius* from Virginia and West Virginia (Pseudoscorpionida, Chthonoiidae). *Entomol. News* 85:81–84. **1975a.** The genus *Lechytia* in the United States (Pseudoscorpionida, Chthoniidae). *Southwest. Nat.* 20:13–27. **1975b.** Use of the spermathecae in the taxonomy of chernetid pseudoscorpions. *Proc. 6th Int. Arachn. Congr.,* pp. 17–20. **1975c.** Two miratemnid pseudoscorpions from the western hemisphere (Pseudoscorpionida, Miratemnidae). *Southwest. Nat.* 20:231–239. **1975d.** Pseudoscorpions from Florida. 4. The genus *Dinochernes* (Chernetidae). *Fla. Entomol.* 58:275–279. **1976a.** New species of *Apochthonius,* mainly from caves in central and eastern United States (Pseudoscorpionida, Chthoniidae). *Proc. Biol. Soc. Wash.* 89:67–80. **1976b.** *Aphrastochthonius pachysetus,* a new cavernicolous species from New Mexico (Pseudoscorpionida, Chthoniidae). *Proc. Biol. Soc. Wash.* 89:361–364. **1976c.** Pseudoscorpions from Florida and the Caribbean area. 5. *Americhernes,* a new genus based upon *Chelifer oblongus* Say (Chernetidae). *Fla. Entomol.* 59:151–163. **1976d.** New cavernicolous species of *Kleptochthonius* and recognition of a new species group within the genus (Pseudoscorpionida, Chthoniidae). *Entomol. News* 87:211–217. **1979.** Pseudoscorpions from Florida and the Caribbean area. 7. Floridian diplosphyronids. *Fla. Entomol.* 62:193–213. **1980a.** A new species of *Apochthonius* with paedomorphic tendencies (Pseudoscorpionida, Chthoniidae). *J. Arachnol.* 8:87–90. **1980b.** A new cavernicolous *Apochthonius* from California (Pseudoscorpionida, Chthoniidae). *J. Arachnol.* 8:93–95. **1980c.** Three new olpiid pseudoscorpions from California (Pseudoscorpionida, Olpiidae). *Pan-Pac. Entomol.* 56:191–169. **1981a.** Cavernicolous species of *Larca, Archeolarca,* and *Pseudogarypus* with notes on the genera (Pseudoscorpionida, Garypidae and Pseudogarypidae). *J. Arachnol.* 9:47–60. **1981b.** Redescription of *Chthonius virginicus* Chamberlin (Pseudoscorpionida, Chthoniidae). *J. Arachnol.* 9:110–112. **1981c.** New pseudoscorpion synonymies (Pseudoscorpionida, Chernetidae and Cheliferidae). *J. Arachnol.* 9:335,336. **1981d.** The identity of *Olpium minutum* Banks (Pseudoscorpionida, Olpiidae). *J. Arachnol.* 9:337,338. **1981e.** Cavernicolous pseudoscorpions in North and Middle America. *Proc. 8th Int. Congr. Speleol.,* Bowling Green, Ky., pp. 381–384. **1982a.** Pseudoscorpionida. Pp. 96–102 in *Synopsis and Classification of Living Organisms,* Vol. 2. McGraw-Hill, New York. **1982b.** The genera *Ideobisium* and *Ideoblothrus,* with remarks on the family Syarinidae (Pseudoscorpionida). *J. Arachnol.* 10:193–221. **1984a.** Further data on *Mucrochernes hirsutus* (Banks) (Pseudoscorpionida, Chernetidae). *Pan-Pac. Entomol.* 60:20–22. **1984b.** New cavernicolous pseudoscorpions from California (Pseudoscorpionida, Chthoniidae and Garypidae). *J. Arachnol.* 12:171–175. **1984c.** Additional pseudoscorpions, mostly from caves, in Mexico and Texas (Arachnida : Pseudoscorpionida). *Texas Mem. Mus., Speleol. Monogr.* 1:17–30. **1986.** Redefinition of the genus *Olpiolum* and description of a new genus *Banksolpium* (Pseudoscorpionida, Olpiidae). *J. Arachnol.* 14:83–102. **Muchmore, W. B. and C. H. Alteri. 1969.** *Parachernes* (Arachnida, Chelonethida, Chernetidae) from the coast of North Carolina. *Entomol. News* 80:131–137. **1974.** The genus *Parachernes* (Pseudoscorpionida, Chernetidae) in the United States, with descriptions of new species. *Trans. Am. Entomol. Soc. (Phila.)* 99:477–506. **Muchmore, W. B. and E. M. Benedict. 1976.** Redescription of *Apochthonius moestus* (Banks), type of the genus *Apochthonius* Chamberlin (Pseudoscorpionida, Chthoniidae). *J. N.Y. Entomol. Soc.* 84:67–74. **Nelson, S., Jr. 1975.** A systematic study of Michigan Pseudoscorpionida (Arachnida). *Am. Midl. Nat.* 93:257–301. **1982.** The external morphology and life history of the pseudoscorpion *Microbisium confusum* Hoff. *J. Arachnol.* 10:261–274. **1984.** The genus *Microbisium* in North and Central America (Pseudoscorpionida, Neobisiidae). *J. Arachnol.* 12:341–350. **Nelson, S., Jr. and G. V. Manley. 1972.** *Dinocheirus horricus* n. sp., a new species of pseudoscorpions from Michigan. *Trans. Am. Microsc. Soc.* 91:217–221. **Schulte, G. 1976.** Litoralzonierung von Pseudoscorpionen an der nord-amerikanischen Pazifikküste (Arachnida : Pseudoscorpiones : Neobisiidae, Garypidae). *Entomol. Ger.* 3:119–124. **Sweeney, M. P., R. W. Taylor and C. C. Counts III. 1977.** Non-cavernicolous pseudoscorpions from West Virginia. *Entomol. News* 88:55–66. **Sweeney, S. J., S. L. Stryker, M. P. Sweeney, R. W. Taylor and C. L. Counts III. 1977.** New record for a non-cavernicolous pseudoscorpion in West Virginia. *Entomol. News* 88:98. **Weygoldt, P. 1966a.** Mating behavior and spermatophore morphology in the pseudoscorpion *Dinocheirus tumidus* Banks (Cheliferinea, Chernetidae). *Biol. Bull. (Woods Hole)* 130:462–467. **1966b.** Spermatophore web formation in a pseudoscorpion. *Science* 153:1647–1649. **1969.** *The biology of Pseudoscorpions.* Harvard University Press, Cambridge, Mass. 145 pp.

19

Opiliones (Phalangida)

ARLAN L. EDGAR

Opilio means "a shepherd" in Latin. This may liken the elevated body position afforded by long legs of opilionids to the fact that in some countries shepherds walk about on stilts, the better to count their flocks. The names "shepherd spiders" and "harvestman" in England suggest that they frequently are associated with grazed vegetation and the harvest season. In France they are known as "reapers" because they give the appearance of reaping, as with a sickle or scythe, when they walk. The names "daddy longlegs" and "phalangids" in North America probably derive from the comparison of the numerous tarsal articles of their conspicuously long legs with the phalanges of tetrapod digits.

External Morphology

Opilionids may be distinguished from other arachnid orders by the broadly fused cephalothorax (prosoma) and abdomen (opisthosoma); the body appears oval in outline. Usually, there is evidence of some of the 9 or 10 segments of the opisthosoma. The prosoma probably consists of six somites and carries six pairs of appendages. A pair of simple eyes perch atop a tubercle in the dorsal midline near the anterior prosomal margin in all except a few primitive forms. Near the bases of the second pair of walking legs are the openings of repugnatorial glands, which secrete an odoriferous liquid demonstrated to be offensive to predators (Blum and Edgar 1971, Eisner et al. 1971, Juberthie 1976). It is likely to be useful in intraspecies recognition (Holmberg 1986). The body length in North

American species is from 3 to 10 mm, although a few soil and litter species are never more than 2 mm and resemble mites.

The 12 appendages consist of a pair of chelicerae, a pair of pedipalpi, and four pairs of walking legs. The three-segmented chelicerae have a movable opposable finger which is used for tearing and crushing food and for defense and feeding. Posterior to the chelicerae are six-segmented pedipalpi used for manipulating food and for sensory reception. These segments, named in order from the body, are coxa, trochanter, femur, patella, tibia, and tarsus. The coxae carry gnathobases, extensions toward the mouth, which help convey food to the mouth. Some are sharp and help crush food and others are lobed and soft.

Each walking leg is composed of seven segments, the same segments as make up pedipalpi plus the insertion of a metatarsus distal to the tibia. Gnathobases occur on the first walking leg and on the second in some genera. All walking legs end in a single or double claw. Cuticular sensillae of three configurations are abundant on the legs of harvestmen; lyriform organs, single-slit organs, and circular campaniform structures are placed strategically to provide information on cuticular deformation as a result of the pull of gravity and attack or any other interference with normal movement (Edgar 1963).

Ecology

The leg length relating to the body size reflects where opilionids live. The familiar small-bodied, long-legged forms occupy surfaces where there is access to relatively large spaces, such as forest floor and tree surfaces. Species with shorter legs may be found in loose litter and among dense ground vegetation; smaller, often flat-bodied forms with short legs live in smaller interstices of litter, under rocks, and in soil crevices. The small-bodied, mitelike forms may be found with other small arthropods, such as Collembola, mites, and microdiptera. They are easily overlooked.

Feeding in nature is done when and where the animal finds food. More food is consumed at night than in daylight because there is more movement by both predator and prey at that time (Phillipson 1960b, Edgar and Yuan 1968). Living food is taken cautiously and usually only by larger-bodied, stronger species such as *Leiobunum calcar* and *L. vittatum* (Edgar 1971). Examination of fecal pellets reveals that food consists almost exclusively of animal matter and that plant material comprises only a very small portion of the diet (Phillipson 1960a). Mayflies, caddis flies, dipterans, and small lepidopterans are taken readily. Harder-bodied forms are eaten if the body wall is broken or can be pierced. Food is assimilated more efficiently by rapidly growing instars than by adults (Phillipson 1960b), with the possible exception of females at the time of oocyte production (Phillipson 1962).

Life History Patterns

Palpatorid opilionids in North America, those commonly seen in forests, thickets, gardens, and around buildings, can be categorized into three different life-cycle patterns. All three are more pronounced in northern latitudes because of the distinct seasons produced by cold winters. Most long-legged species [e.g., *Leiobunum aldrichi* (= *L. longipes*)] reach adulthood in summer, mate and oviposit in the fall, then die. The eggs produced hatch next spring and then grow to maturity to repeat the process. A second approach, demonstrated by *Leiobunum nigripes* and a few others, shows them reaching maturity in the spring, mating, and ovipositing in summer. The eggs hatch and the young molt several times before growth is slowed by cold weather. Wintering over as late instars sets the stage for early maturation the next spring. The third life-history type combines the other two. The best example is *Phalangium opilio,* the species that frequents areas disturbed by humans. It overwinters both as eggs and immatures; therefore, animals in two stages of maturity may be found at any one time in a population.

The life-history characteristics of cyphophthalmids and laniatorids do not reflect seasonal changes as conspicuously as palpatorids. They usually are small-bodied, short-legged forms whose physical environment is shielded against both short- and long-term fluctuations. Many of them live longer than one year.

Mating behavior observed ranges from simple insemination characteristic of most long-legged forms to rather complicated pre- and postcopulatory interactions seen in a few shorter-legged species. Martens (1969) describes a courtship behavior pattern in *Ischyropsalis* in which the male presents the female with a secretion coming from the base of its chelicera. Only after the mouth organs of the female contact the extruded secretion does she assume a posture favorable for copulation. Among long-legged forms the mating behavior of *Leiobunum aldrichi* is exceptional. In the presence of a receptive female, single males copulate readily and repeatedly. If several males are present, one must dominate the others via a peculiar, vigorous leg-pulling contest. Using chelicerae, one male grasps a leg of another near its base and violently jerks up and down or rapidly moves the body in a circle. These tussles go on until a winner emerges; he then claims the female. After insemination the male hovers over or near the female in a manner which suggests that of shepherding the female to oviposit then and there. Eggs always appear to be inserted into some appropriate small space supplied with moisture. Protective structures and parental guarding of the eggs such as is provided by spiders and many insects is known but unusual. Upon hatching, the young remain confined to small spaces both for protection from predators and for maintenance of a requisite high humidity. Subsequent peregrination from the hatching site depends on the species. In some, adaptations are very specialized and tolerance limits narrow. All phases of living may take place within the physical influence of a single rock or decaying log. Several species of the phalangodid, *Sitalcina,* in California are found associated exclusively with serpentine rock (Briggs 1968). In dry weather, the cyphophthalmid, *Siro,* may descend into the soil as far as 1 m, utilizing cracks and passageways of worms (Juberthie 1964).

Parthenogenesis is known to occur in opilionids (Tsurusaki 1986). Phillipson (1959) found populations of *Megabunus diadema* in England to be almost all females. Virgin females, isolated in the laboratory, produced eggs that hatched and developed normally.

Molting can occur successfully almost anywhere provided that there is adequate relative humidity and no disturbance. An opilionid ready to molt will most frequently hang upside down in a secluded space such as a curled leaf, spider web, or twig. The entire molting process, from the initial split at the anterior lateral margins separating dorsum from venter to the withdrawal of the longest pair of legs, takes about 20 minutes for the last of the seven molts in a long-legged species. An additional 40 minutes of inactivity is necessary for exoskeleton hardening so that muscle contractions can be supported. During this period the animal is vulnerable to attack.

Characteristic Habitats

The majority of opilionids are located in damp places. All are terrestrial and breathe air via spiracles and a tracheal system. Principal habitat types for adults are three: (a) soil and cave inhabitants, (b) litter dwellers, and (c) those found on surfaces. All three groups utilize the soil milieu, broadly defined, for the early stages of the life cycle. A small assemblage of true cave opilionids occurs in North America (Goodnight and Goodnight 1967, Briggs 1974a,b). All cavernicolous species are presumably derived from modern surface species. Intermediate or transition forms may be found near the entrance to caves where true cavernicoles reside. The latter have eyes reduced or absent, longer and more robust legs, pigmentation lost or reduced, and an exoskeleton with reduced sclerotization (Goodnight and Goodnight 1960).

Virtually all species of opilionids encountered by the casual observer belong to the family Phalangiidae. The life-history stage most frequently seen is the adult. This is because it is the largest, its markings are less cryptic, and mating activities occasionally result in its being found in unusual locations. There are both seasonal and diurnal vertical migrations characteristic of many long-legged species. Seasonally, the majority of such species hatch in the spring. The young remain in the smaller litter interstices until molts occur that produce legs of increasing length; these require greater space for movement. Juveniles soon spend more time moving about on litter and eventually ascend low herbs and shrubs, fallen limbs, and trunks of trees. Upon the onset of adulthood and mating there is a marked increase in distances traveled. Some species ascend high into trees. Some exhibit diurnal vertical migration, while others appear unaffected by changing light conditions. All species tested for movement in relationship to light and time of day were found to be most active at night and in the dark (Edgar and Yuan 1968).

Long-legged epigenean opilionids in temperature and relative humidity gradients in the laboratory show behavior that matches up closely with where they exist in nature. For example, *Phalangium opilio* is an anthrophile of worldwide distribution whose usual habitat in lawns, gardens, and along roadsides makes it relatively exposed to broad temperature and humidity fluctuations. This species has a mean habitat temperature preference of 27.6°C and survives an average of 97.8 hours in dry air. By comparison, *Leiobunum politum,* seldom found far from deep shade and damp soil, has a temperature preference of 24.3°C and survival time in dry air of 28.5 hours (Edgar 1971). The physiology involved in the rate of loss of body fluids and other environmental tolerance characteristics is of basic importance to that species in determining its success or failure in adapting to a wide variety of habitats.

Collection

The long-legged and medium-to-large-bodied opilionids are most easily collected by hand, using the thumb and first two fingers. This occasionally causes damage to the specimen; if animals in perfect condition are needed, usually they can be persuaded to step onto a platform of cardboard and be transferred to a live jar, or can be "herded" into a large-mouthed container. Sweeping of vegetation with an insect net yields specimens readily. Other methods for collecting, especially small forms, are (a) pit traps, (b) Tullgren funnel sorting of litter, and (c) sifting of organic debris onto a white cloth or enamel pan.

Several genera of opilionids may be maintained quite easily in the laboratory by providing adequate space, food, water, and isolation from predators. A shell vial with a mouth diameter of approximately 10 mm stoppered with a piece of plastic sponge, laying horizontally, provides a reservoir of available water. A variety of foods can be used. In my experience, the mix of insects captured in a sweep net is best. Dry them thoroughly, then pulverize in a mortar and pestle. Serve a "pinch" with forceps every few days. In the absence of fresh insects, obtain very lean beef. Dry and pulverize it thoroughly. Others have used dried bacon and cornmeal (Klee and Butcher 1968) and a mixture of dried egg, wholemeal flour, and yeast (Todd 1949).

Both field and permanent collections may be preserved in 70–80% alcohol. Some curators add 1–2% glycerol to prevent hardening should the alcohol evaporate, but this causes some natural colors to be lost. When alcohol becomes discolored it should be replaced. Relatively permanent storage of specimens can be done using either shell vials stoppered with cotton and stored in larger-mouthed screw-cap bottles of approximate capacity ½–1 liter or in homeopathic (shouldered) glass vials of 4–6 dram capacity closed with neoprene stoppers. Labels should be good-quality bond paper and written with soft pencil, carbon ink, or typed using a fresh ribbon.

For taxonomic study a good stereoscopic binocular microscope with magnifications from 10 to 100 diameters should be available. An ocular grid, calibrated to the objectives of that microscope, is useful in both measurements of structures and making drawings. Occasionally, magnifications afforded by the compound microscope are necessary to see the detail of genitalia, tarsal claws, setae, denticulation, and other very small structures.

Preface to the Key

Characteristics that distinguish one species from all others generally are better displayed on adult reproductive males than on females or immatures. In those species exhibiting sexual dimorphism (many do not), males' colors usually are more vivid and their body shape less variable than those of females. The female abdomen may show varying degrees of distention due to the egg production phase she was in when preserved. Males frequently show greater and more exaggerated development in a variety of structures. Examples are spines, tubercles, length of appendages, shape of appendage segments, body ornamentation, and genitalia. The male penis probably is the single most distinctive structure for species identification in groups where penial morphology has been studied. The morphology of the female ovipositor occasionally is helpful in species determination. Both penis and ovipositor may be exposed by tearing along one side of the genital operculum and bending it out of the way.

Considerable utilization of color, pattern, pigment intensity, and shading is made in the following key. Frequently, confusion and indecision reign when determinations involving these characteristics are called for. Part of the problem can be understood and remedied if the condition of the specimen under scrutiny can be ascertained. Sometimes the same individual can hardly be recognized as belonging to the same species if its appearance just before molting is compared with its appearance just after molting. Prior to molting the body appears bloated with the enclosed molting fluid, colors are dark and dull, and patterns are indistinct. The opposite is the case just after the molt case has been shed; colors are bright and patterns are sharp. In fact, pigmentation may not be developed fully. This results in some specimens appearing pale and albinolike. The exoskeleton is weak, pliable, and easily torn, irrespective of whether the species is hard or soft between molts. Specimens preserved a week before or after molting will show conditions intermediate between these extremes.

Postreproductive adults decrease slightly in body size from dehydration and tissue loss. They exhibit greater melanization. Patterns and color bandings that were bright and in contrast at the time of mating become dark, dull, and obscure. Appendages and other parts break more readily from handling and preservation when specimens are very old.

Juveniles and subadults without penes or ovipositors are very difficult to identify. Occasionally, they can be determined to the genus level. It is therefore advantageous to work with adults whenever possible. The presence of adults in the field can be learned by putting gentle pressure on the specimen's abdomen with thumb and forefinger. This usually causes the penis or ovipositor to be exserted if the animal is adult.

Taxonomic studies on opilionids in North America have been going on for over a century. Early studies naturally were concerned with description of new genera and species. H. C. Wood (1871), C. M. Weed (1897–1917), and N. Banks (1892–1911) described many of the larger and accessible forms. C. R. Crosby and S. C. Bishop published studies on eastern United States species over a period of nearly a half century (1904–1949). N. W. Davis reported on the large and complex genus *Leiobunum* in 1934. Beginning in the 1940s, Clarence and Marie Goodnight published numerous papers dealing predominantly with the fauna of the United States and Mexico.

Until the work of T. S. Briggs in the late 1960s and 1970s, only limited attention was given to the relationships of North American forms to those on other major landmasses. He made many revisions in suborder Laniatores at several taxonomic levels (Briggs 1968, 1969, 1971a,b; Briggs and Hom 1966, 1967; Briggs and Ubick 1981). Revisionary work primarily at the generic level in suborder Palpatores began to appear in the 1970s and continues actively to the present time. J. Gruber published on *Hesperonemastoma* (1970); W. A. Shear on *Sabacon* (1975a), family Caddidae (1975b) and family Ceratolasmatidae (1986), suborder Cyphophthalami, 1980; Shear and Gruber on subfamily Orthlasmatinae, 1983; C. R. McGhee on part of *Leiobunum* (1977), and J. C. Cokendolpher on *Eumesosoma* (1980a), *Trachyrhinus* (1981a), *Liopilio* (1981b), *Dalquestia* (1984a, 1986), *Leptobunus* (1985a), *Leiobunum* (1982, 1984b, 1985b), and *Homolophus* (1987). At least four species have been introduced from Europe (Bragg and Holmberg 1975, Lesage 1972, Muchmore 1963).

As in all groups not thoroughly researched, the opilionids still need revisionary studies in several genera and at other taxonomic levels. *Taracus, Eurybunus,* and *Leiobunum* are prime candidates. *Leiobunum,* the largest genus in North America with 26 species, perhaps is polyphyletic, an aggregate of several genera.

It should be recognized that any key to species may be misleading in that it suggests completeness. It does not accommodate undescribed species or provide evidence that a species is or is not in its appropriate genus. This key, therefore, may be viewed as a taxonomic halfway house, perhaps bridging the large gap that presently exists between the miscellany of described species and ultimately an understanding of how the North American fauna articulates with that of the rest of the world. Literature concerning almost all original species descriptions and other papers are given at the end to facilitate further study.

ACKNOWLEDGMENTS

Substantive revisions to the key and other helpful suggestions were made by J. C. Cokendolpher and W. A. Shear. Those contributions are deeply appreciated and hereby acknowledged.

KEY TO THE RECENT GENERA AND SPECIES OF ORDER OPILIONES (= PHALANGIDA) IN NORTH AMERICA, EXCLUSIVE OF MEXICO

Section 1: Key to Suborders *Cyphophthalmi, Laniatores,* and *Palpatores;* Key to Genera and Species of *Cyphophthalmi* and *Laniatores*[1]

1a Small, mitelike animals with scent gland pores elevated on small projections at anterolateral margin of cephalothorax (Fig. 19.1); leg tarsi usually one segmented; gonopore lacks an operculum; Suborder **Cyphophthalmi,** Family **Sironidae** (females cannot easily be identified) **3**

1b Generally larger, not mitelike in appearance; scent glands not on elevated projections; leg tarsi with more than one article; gonopore covered by an operculum **2**

2a Terminal claws on walking legs III and IV either double or if single, claw usually is divided into three parts (Fig. 19.2); heavily spined palpi, tipped with a strong claw; leg IV usually longest; genital operculum usually separates fourth coxae only (Fig. 19.3); Suborder **Laniatores** **7**

2b Only a single undivided claw on tarsi of legs III and IV; slender palpi with the terminal claw weak or occasionally lacking; leg II longest; genital operculum usually separates all four leg coxae (Fig. 19.4) . . . . Suborder **Palpatores** (*Section 2*)

3a Adenostyle of leg IV distally brush-tipped and located near the articulation with metatarsus on dorsal side (Fig. 19.5); Florida, Georgia, South Carolina *__Metasiro americana__* (Davis, 1933)

3b Adenostyle of male leg IV not distally brush-tipped and not located near articulation with metatarsus (Fig. 19.6) *__Siro__*[2] Latreille, 1804. . . . . **4**

4a Tarsus IV of males clearly divided distal to adenostyle; Washington and Idaho. . . . . *__Siro kamiakensis__* (Newell, 1943)

4b Tarsus IV of males not divided **5**

5a Male tarsus IV modified with ventral spur and excavated region (Fig. 19.6B); San Francisco Bay region . . . . *__Siro sonoma__* Shear, 1980

---

[1] See Figs. 19.11–19.12 for certain general structures.
[2] Key adapted from Shear (1980).

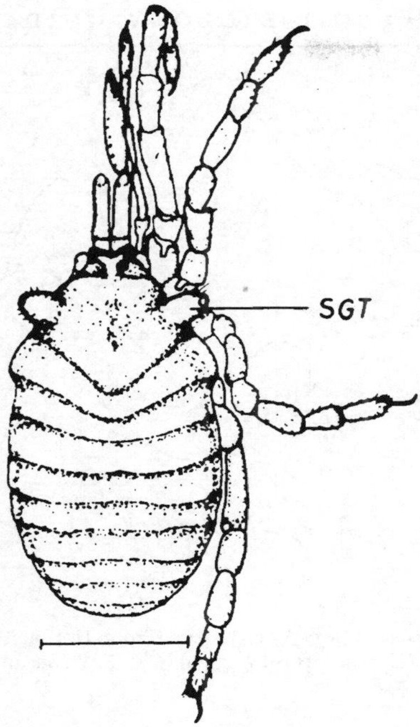

1

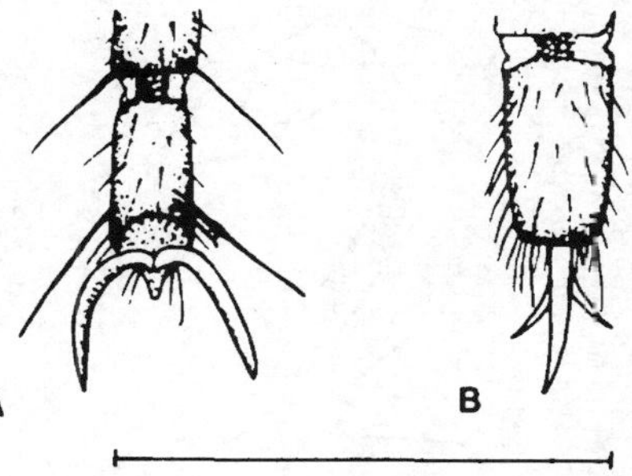

2

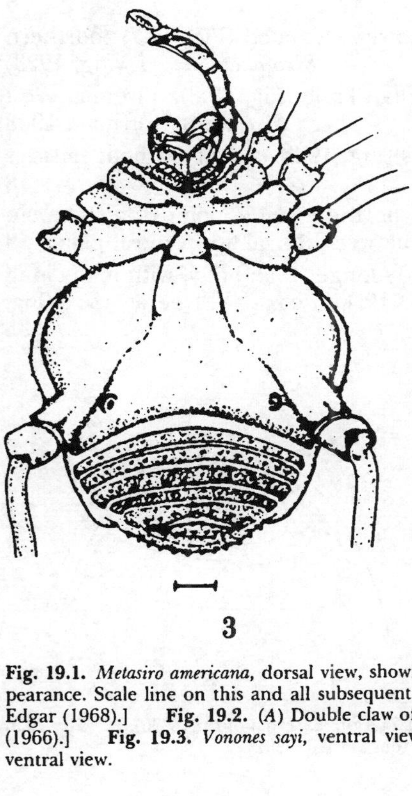

3

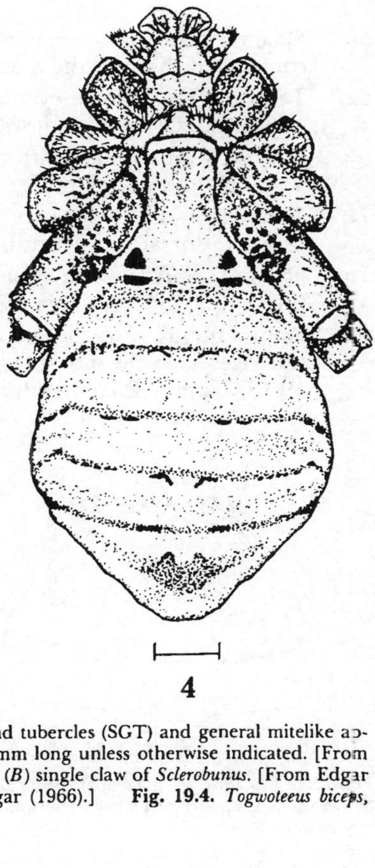

4

Fig. 19.1. *Metasiro americana,* dorsal view, showing scent gland tubercles (SGT) and general mitelike appearance. Scale line on this and all subsequent figures is 1 mm long unless otherwise indicated. [From Edgar (1968).] **Fig. 19.2.** (*A*) Double claw of *Phalangodes;* (*B*) single claw of *Sclerobunus.* [From Edgar (1966).] **Fig. 19.3.** *Vonones sayi,* ventral view. [From Edgar (1966).] **Fig. 19.4.** *Togwoteeus biceps,* ventral view.

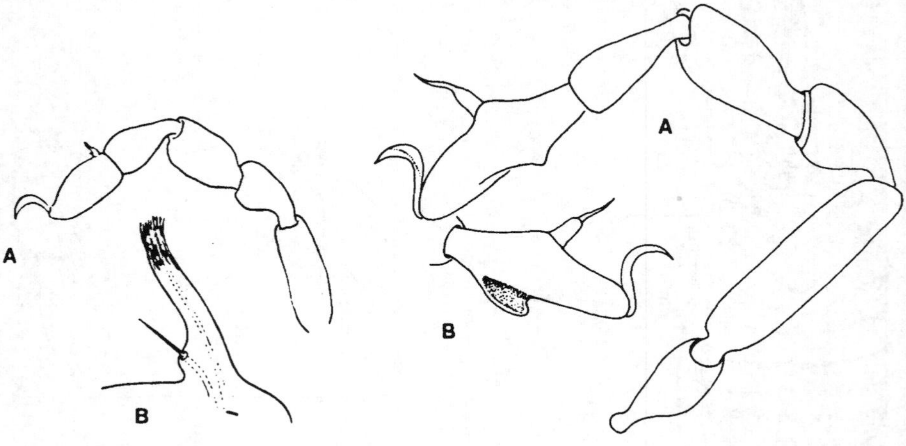

Fig. 19.5. *Metasiro americana:* (*A*) leg IV of male; (*B*) tarsus of leg IV enlarged. [From Hoffman (1963).] **Fig. 19.6.** *Siro sonoma:* (*A*) male leg IV, ectal view; (*B*) male tarsus IV, mesal view. [From Shear (1980).]

5b Male tarsus IV not strongly modified, more or less cylindrical or somewhat constricted distal to the adenostyle **6**

6a Tergite VIII of male evenly and shallowly excavated (Fig. 19.7); northern California, Oregon, and Washington . . . . . ***Siro acaroides*** (Ewing, 1923)

6b Tergite VIII of male with a small median knob (Fig. 19.8); Virginia, West Virginia, and Maryland . . . . ***Siro exilis*** Hoffman, 1963

7a Tarsi III and IV with two simple claws (Fig. 19.2*A*); penis without muscles between dorsal and ventral plates **8**

7b Tarsi III and IV with a complex claw having a single point of attachment (Figs. 19.9, 19.10); penis with muscles between dorsal and ventral plates . **9**

8a Palpi large, robust, rounded, and usually longer than body, with many long tubercles bearing movable spines (Fig. 19.11); coxa IV free at apex (Fig. 19.12); Family **Phalangodidae** **35**

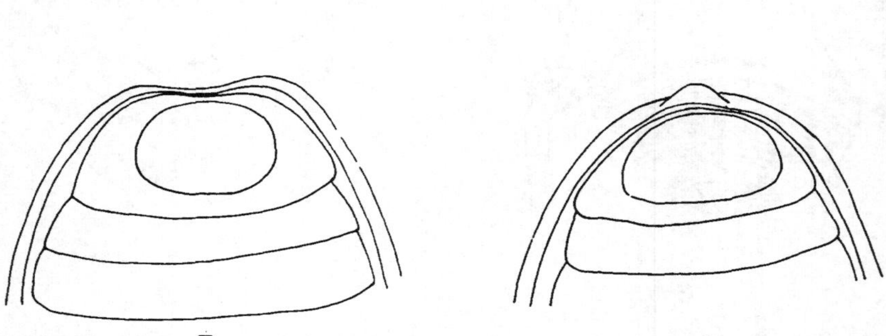

Fig. 19.7. *Siro acaroides,* male anal region showing terminal tergite 8. [From Shear (1980).] **Fig. 19.8.** *Siro exilis,* male anal region showing terminal tergite 8. [From Shear (1980).]

9

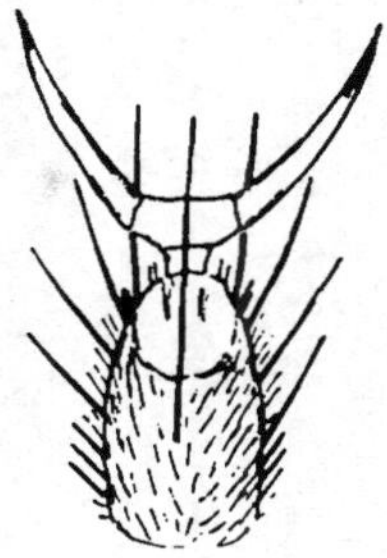

10

Fig. 19.9. Claw of leg III of *Erebomaster flavescens,* dorsal view. [From Briggs (1969).] **Fig. 19.10.** Claw of leg IV of *Cryptomaster leviathan,* dorsal view. [From Briggs (1969).]

8b Palpi shorter than body, dorsoventrally flattened; coxa IV wholly united to venter; spiracles distinct (Fig. 19.3); Family **Cosmetidae,** ***Vonones*** Simon, 1879 . . . . **33**

9a Tarsi III and IV with a claw bearing a median prong, usually with paired branches (Fig. 19.13) (use 100–500× magnification) **10**

9b Tarsi III and IV with a bifurcate claw having no median prong (Fig. 19.2A); Family **Cladonychidae** **23**

10a Two or three pairs of lateral abdominal sclerites present; free tergite nine present (Fig. 19.14). **11**

10b Lateral abdominal sclerites absent; free tergite nine absent; Family **Triaenonychidae** **12**

11a Tarsi III and IV with claw terminating in a broad median plate having scalelike branches (Fig. 19.15); Family **Travuniidae** . . . .
Speleonychia sengeri Briggs, 1974

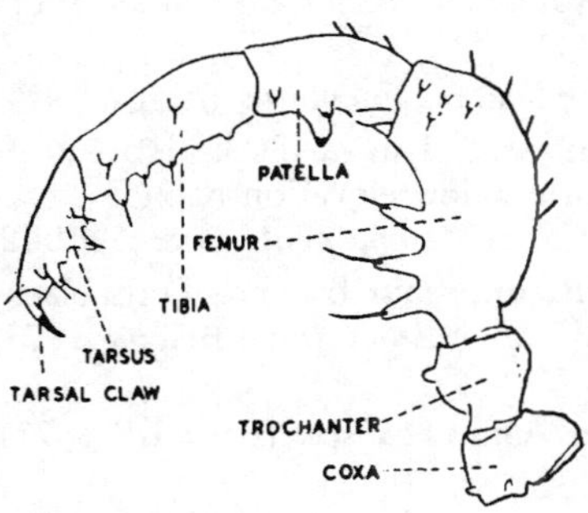

11

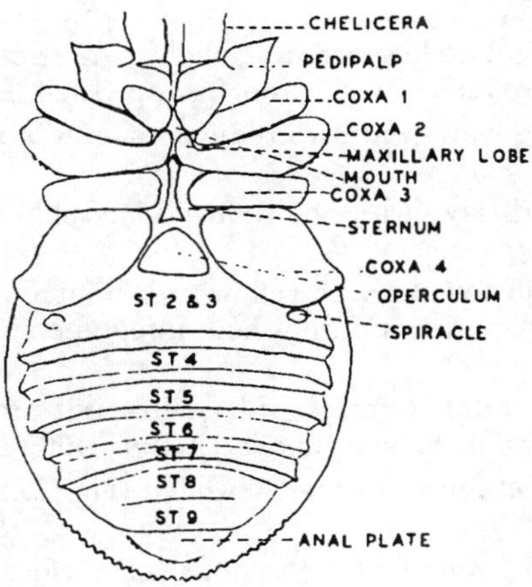

12

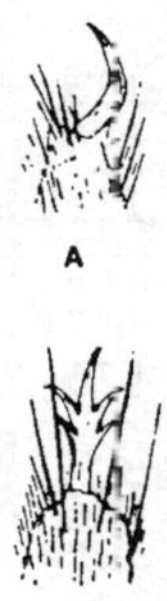

13

Fig. 19.11. Palpus of *Sitalcina madera,* lateral view. [From Briggs (1968).] **Fig. 19.12.** Body of *Sitalcina cockerelli,* ventral view. [From Briggs (1968).] **Fig. 19.13.** Claw of leg IV of *Isolachus spinosus:* (*A*) lateral view; (*B*) dorsal view. [From Briggs (1971a).]

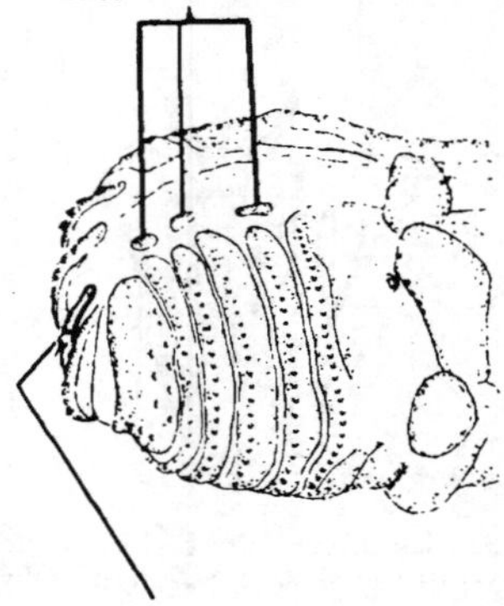

Fig. 19.14. Lateral abdominal sclerites in *Pentanychus*. [From Briggs (1971a).]

14

11b Tarsi III and IV with a continuous median prong bearing four uniformly tapering branches (Fig. 19.13); Family **Pentanychidae** **28**

12a Hind claws with two pairs of branches on median prong (Fig. 19.16) . . **19**

12b Hind claws with one pair or fewer branches on median prong. **13**

13a Leg femora, patellae, and tibiae with anterior and posterior rows of spined tubercles, twice width of segment on leg I (Fig. 19.17) . . . .
Fumontana deprehendor Shear, 1977

13b Leg tubercles shorter than above. **14**

14a Hind claws without ventral tooth, branches small or absent (Fig. 19.18); ***Cyptobunus*** Banks, 1905 . . . . **16**

14b Hind claws with ventral tooth and two branches (Fig. 19.19). **15**

15a Eye tubercle conical, on anterior margin of scute (Fig. 19.20); ***Zuma*** G & G, 1942[3]. **17**

15b Eye tubercle subconical, slightly recessed from anterior margin (Fig. 19.21); ***Sclerobunus*** Crosby and Bishop, 1924 (key to males) **18**

16a Hind claws with two reduced lateral branches. . . . .
Cyptobunus cavicolens Banks, 1905

16b Hind claws with one reduced lateral branch. . . . .
Cyptobunus ungulatus Briggs, 1971

17a Adult females with yellow-brown integument underlain by black pigment; if black pigment diminished, integument colorless; low elevation habitat . . .
Zuma acuta G & G, 1942

17b Live adult females with yellow integument underlain by very faint black pigment; habitats above 1219 m (4000 ft) . . . . . ***Zuma tioga*** Briggs, 1971

18a Palpal femur of males swollen (Fig. 19.21) . . . .
Sclerobunus robustus (Packard, 1877)

18b Palpal femur of males not swollen (Fig. 19.22) . . . .
Sclerobunus nondimorphicus Briggs, 1971

---

[3] G & G refers to Goodnight and Goodnight.

15

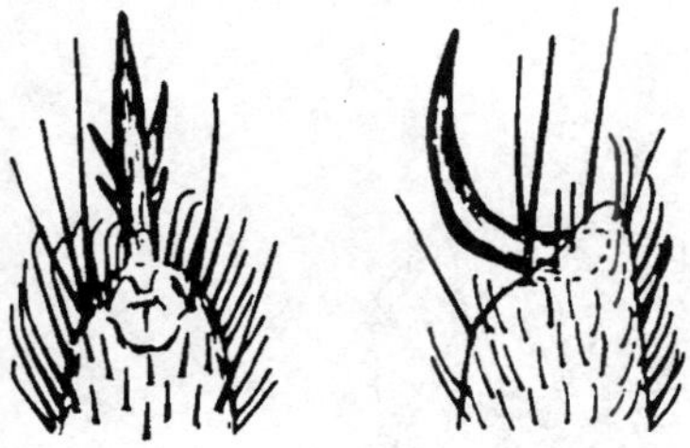

A B

16

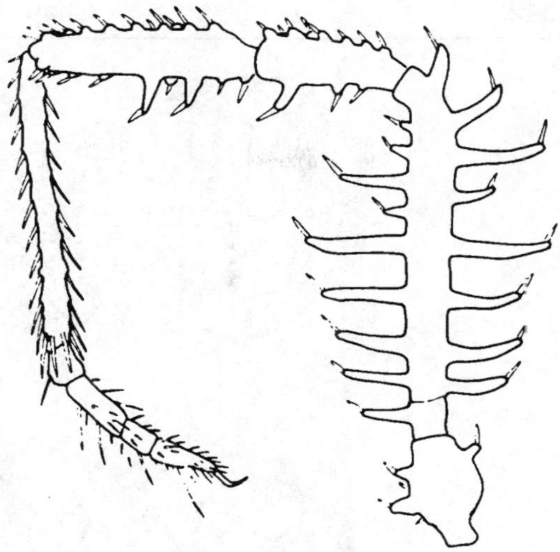

17

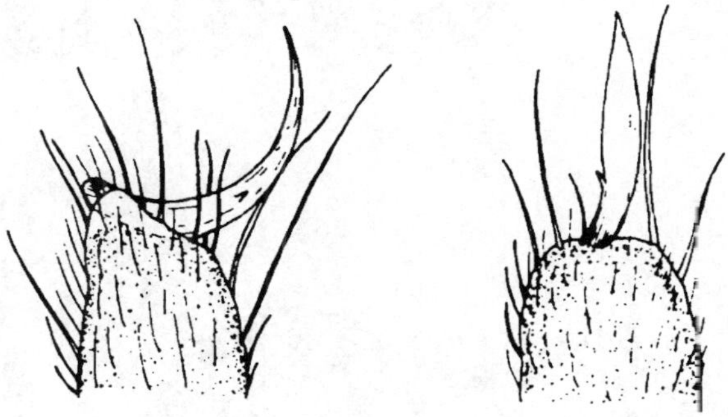

A B

18

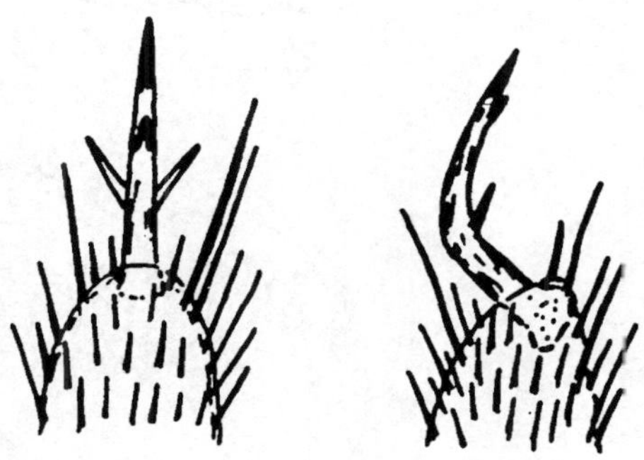

A B

19

Fig. 19.15. Claw of leg IV of *Spelenonychia sengeri*, lateral view. [From Briggs (1974a).] **Fig. 19.16.** Claw of leg IV of *Paranonychus brunneus: (A)* dorsal view; *(B)* lateral view. [From Briggs (1971b).] **Fig. 19.17.** Leg I of male *Fumontana deprehendor*. [From Shear (1977).] **Fig. 19.18.** Claw of leg IV of *Cytobunus ungulatus madhousensis:* (*A*) lateral view; (*B*) dorsal view. [From Briggs (1971b).] **Fig. 19.19.** Claw of leg IV of *Zuma acuta:* (*A*) ventral view; (*B*) lateral view. [From Briggs (1971b).]

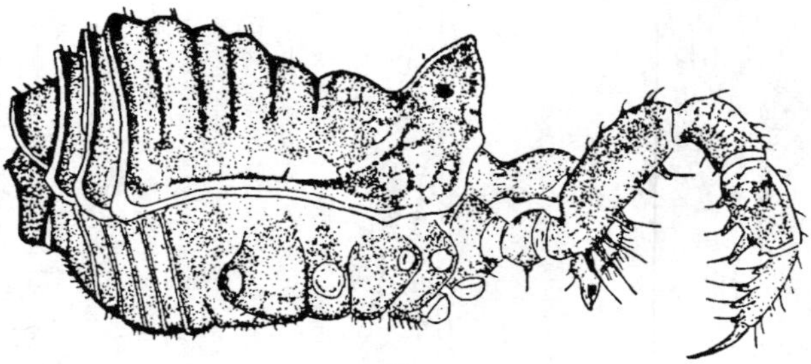

20

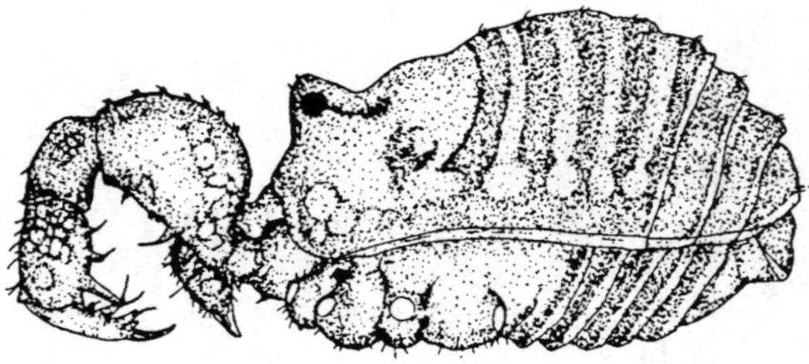

21

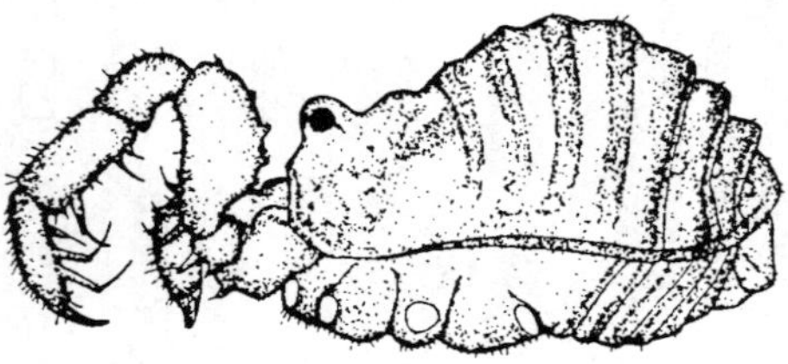

22

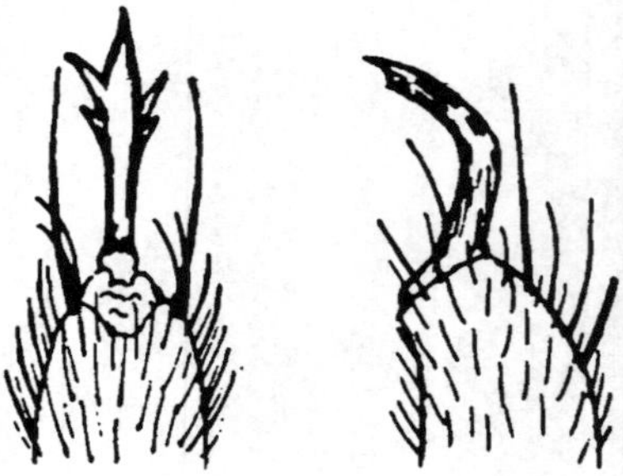

A B

23

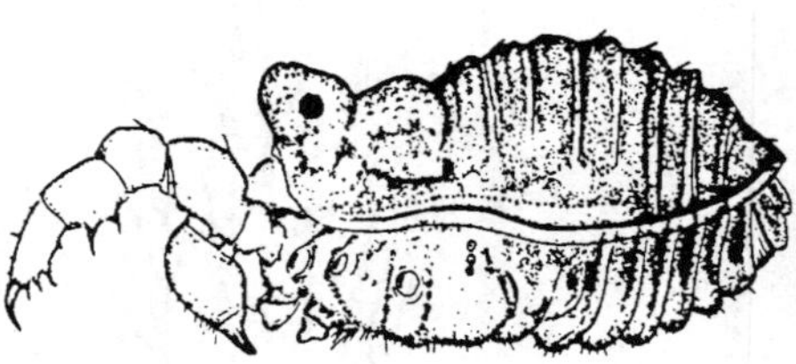

24

Fig. 19.20. *Zuma acuta,* lateral view. [From Briggs (1971b).] **Fig. 19.21.** *Sclerobunus robustus robustus,* lateral view. [From Briggs (1971b).] **Fig. 19.22.** *Sclerobunus nondimorphicus,* lateral view. [From Briggs (1971b).] **Fig. 19.23.** Claw of leg IV of *Metanonychus nigricans oregonus:* (*A*) dorsal view; (*B*) lateral view. [From Briggs (1971b).] **Fig. 19.24.** *Paranonychus brunneus,* lateral view. [From Briggs (1971b).]

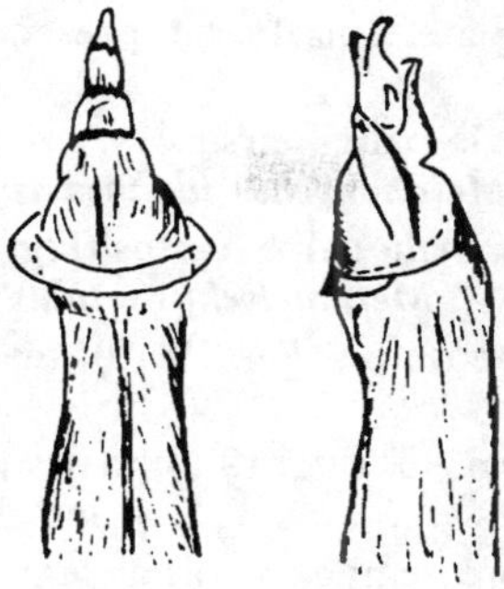
A B
25

A B
26

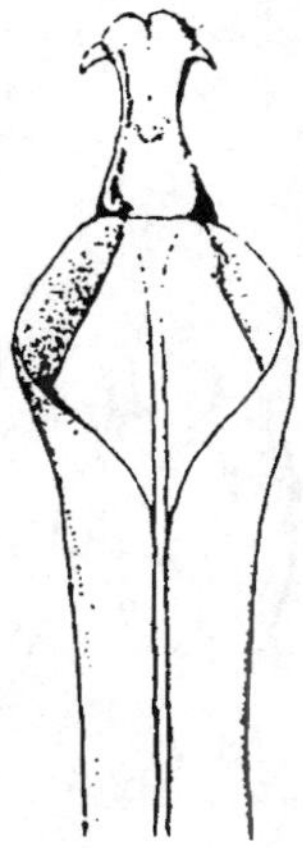
27

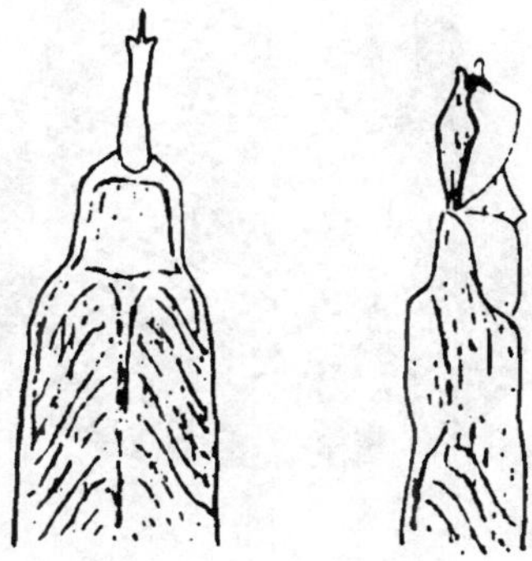
A B
28

Fig. 19.25. Penis of *Metanonychus nigricans nigricans:* (*A*) ventral view; (*B*) lateral view. [From Briggs (1971b).] **Fig. 19.26.** Penis of *Metanonychus setulus mazamus:* (*A*) ventral view; (*B*) lateral view. [From Briggs (1971b).] **Fig. 19.27.** Penis of *Speleomaster lexi,* ventral view. [From Briggs (1974a).] **Fig. 19.28.** Penis of *Cryptomaster leviathan:* (*A*) ventral view; (*B*) lateral view. [From Briggs (1969).]

19a Hind claws with ventral tooth on median prong (Fig. 19.23); ***Metanonychus*** Briggs, 1971 (key to males). **21**

19b Hind claws without ventral tooth on median prong (Fig. 19.16); ***Paranonychus*** Briggs, 1971 (key to males) **20**

20a Dorsal process of penis tubular, with acute apex; eye tubercle extends over anterior margin of cephalothorax (Fig. 19.24). . . . . ***Paranonychus brunneus*** (Banks, 1893)

20b Dorsal process of penis flattened, with beveled apex; eye tubercle does not extend over anterior margin of cephalothorax . . . . ***Paranonychus concolor*** Briggs, 1971

21a Dorsal process of penis a sclerotized spine, apical setae absent (Fig. 19.25) ***Metanonychus nigricans*** Briggs, 1971

21b Dorsal process of penis a sclerotized plate, apical setae present (Fig. 19.26) **22**

22a Ventral swelling or spur on base of male palpal tarsus . . . . ***Metanonychus idahoensis*** Briggs, 1971

22b No ventral swelling or spur on base of male palpal tarsus (Fig. 26) . . . . . ***Metanonychus setulus*** Briggs, 1971

23a Prongs on hind claws meet in 180° opposition (Fig. 19.10); male with distal swelling on tibia II **24**

23b Prongs on hind claws meet at angle near 90° (Fig. 19.2); males without distal swelling on tibia II **26**

24a Distitarsus of leg II with four or more segments; distal segment of penis broad in ventral view and with dorsoventral flattening and basal stem seg-

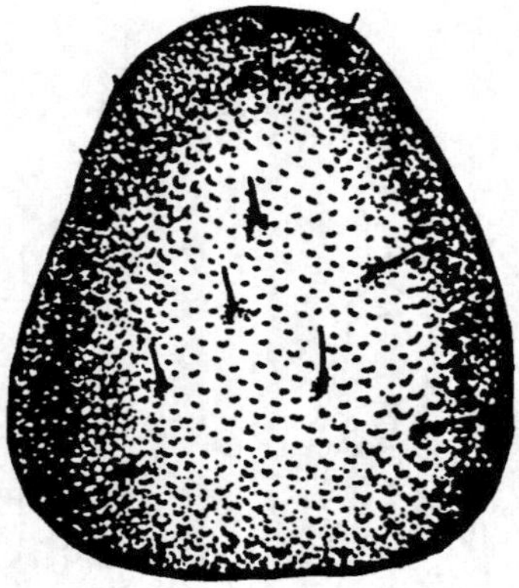

29

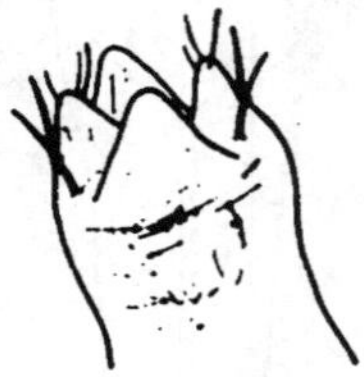

30

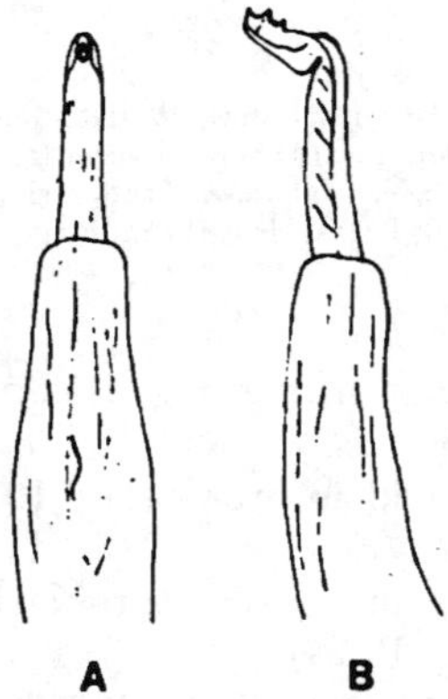

A B

31

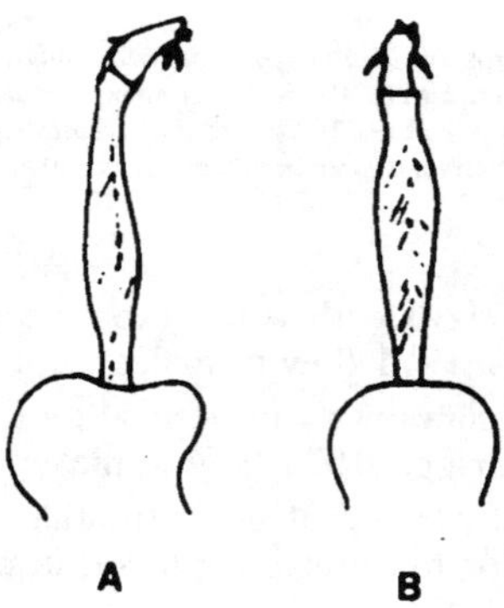

A B

32

Fig. 19.29. Operculum of *Speleomaster pecki,* female. [From Briggs (1974a).] **Fig. 19.30.** Ovipositor of *Erebomaster acanthina,* ventral view. [From Briggs (1969).] **Fig. 19.31.** Penis of *Erebomaster acanthina:* (*A*) ventral view; (*B*) lateral view. [From Briggs (1969).] **Fig. 19.32.** Penis of *Thermomaster brunnea:* (*A*) lateral view; (*B*) ventral view. [From Briggs (1969).]

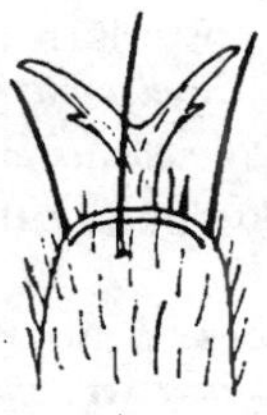

33

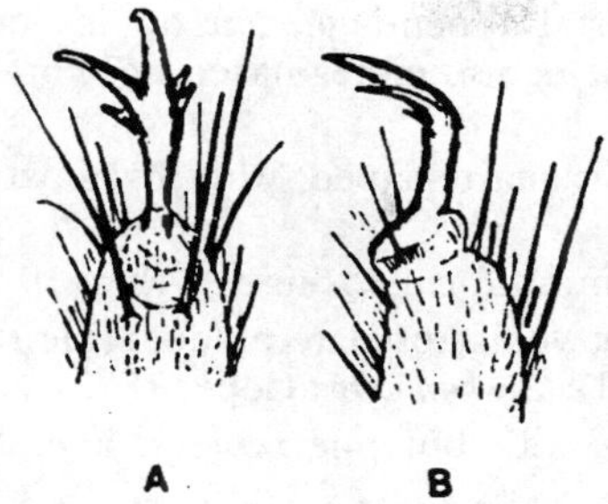

34

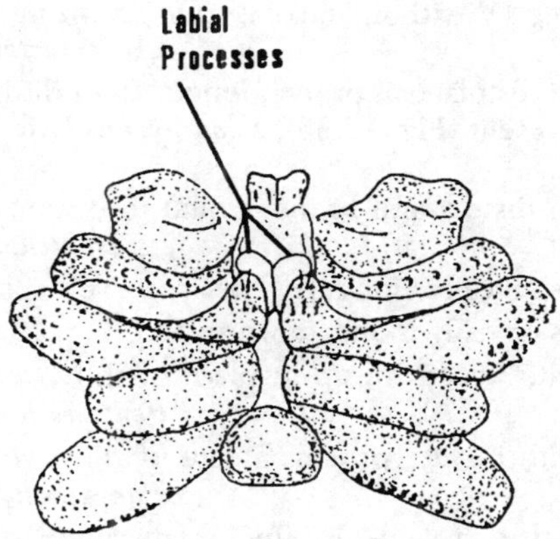

35

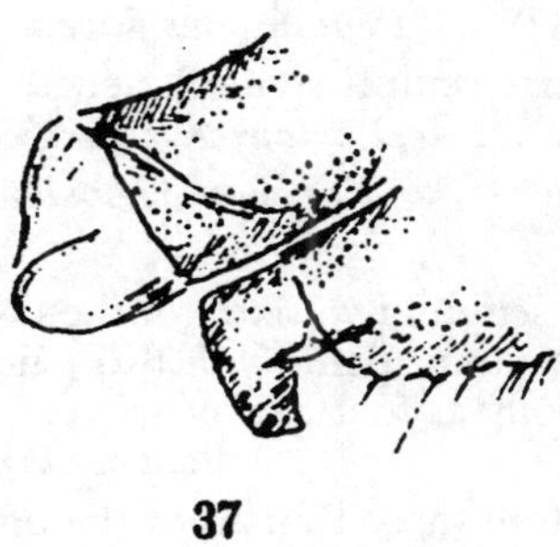

37

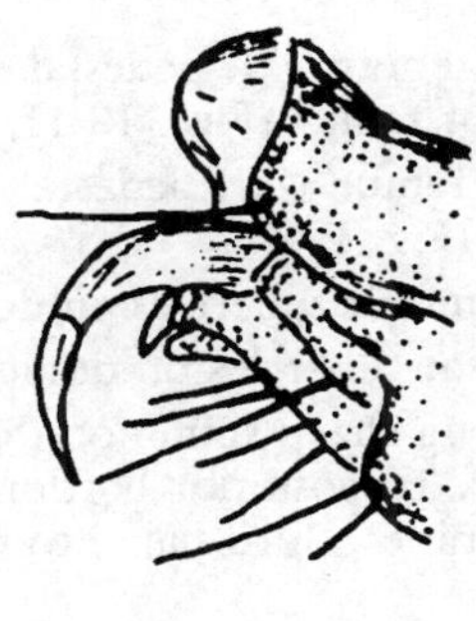

36

Fig. 19.33. Claw of leg IV of *Erebomaster acanthina,* dorsal view. [From Briggs (1969).] **Fig. 19.34.** Claw of leg IV of *Pentanychus flavescens:* (*A*) dorsal view; (*B*) lateral view. [From Briggs (1971a).] **Fig. 19.35.** Sternal view of typical *Pentanychus.* [From Briggs (1971a).] **Fig. 19.36.** Labial process of *Pentanychus hamatus* showing recurvid spine. [From Briggs (1971a).] **Fig. 19.37.** Labial process of *Pentanychus flavescens.* [From Briggs (1971a).]

ment of penis flared and cup-shaped at apex (Fig. 19.27); ***Speleomaster*** Briggs, 1974 . . . . **25**

24b Distitarsus of leg II with three segments; penis with cultriform or knife-shaped distal appendage, narrow in ventral view (Fig. 19.28); basal stem segment of penis neither flared nor cup-shaped . . . . ***Cryptomaster leviathan*** Briggs, 1969

25a Operculum heart-shaped, with invagination posteriorly. . . . . ***Speleomaster lexi*** Briggs, 1974

25b Operculum uniformly rounded (Fig. 19.29) ***Speleomaster pecki*** Briggs, 1974

26a Ovipositor with setose lateral lobes (Fig. 19.30), penis cultriform or knifelike (Fig. 19.31); ***Erebomaster*** Cope, 1872. . . . . **27**

26b Ovipositor with but one seta on lateral lobes; penis not cultriform (Fig. 19.32). ***Theromaster brunnea*** (Banks, 1902)

27a Tarsal claw of leg IV with medial spur on each of the two prongs (Fig. 19.33) ***Erebomaster acanthina*** (Crosby and Bishop, 1924)

27b Tarsal claw of leg IV without medial spur on prongs . . . . ***Erebomaster flavescens*** Cope, 1872

28a Hind claws with distolateral branch longer than distomesial (Fig. 19.34); labial processes present (Fig. 19.35); ***Pentanychus*** Briggs, 1971 (key to males)[4] **29**

28b Hind claws with distolateral branch equal to distomesial (Fig. 19.13); labial processes absent. . . . . ***Isolachus spinosus*** Briggs, 1971

29a Labial processes apically elongated into recurved spines (Fig. 19.36). . . **30**

29b Labial processes apically blunt, downturned (Fig. 19.37) **31**

30a Palpal femur with broad hump in place of third ventral spine (Fig. 19.38) ***Pentanychus hamatus*** Briggs, 1971

30b Palpal femur with clavate process at site of third ventral spine (Fig. 19.39) ***Pentanychus clavatus*** Briggs, 1971

31a Palpal femur with first ventral spine on laterally flattened pedestal . . . **32**

31b Palpal femur with first ventral spine on rounded pedestal (Fig. 19.40) . . . ***Pentanychus bilobatus*** Briggs, 1971

32a Palpal femur with pedestal of first ventral spine flattened perpendicular to plane of femur (Figs. 19.41, 19.42). . . ***Pentanychus flavescens*** Briggs, 1971

32b Palpal femur with pedestal of first ventral spine flattened 45° to plane of femur (Fig. 19.43). . . . . ***Pentanychus pacificus*** Briggs, 1971

33a No spines or tubercles on dorsum . . . . ***Vonones bimaculata*** (Banks, 1893)

33b Spines or tubercles on dorsum **34**

34a Two long sharp spines on dorsal scutum five-sixths of the distance from eye tubercle to posterior border of scutum; penis with two pairs of well-developed spines, third pair $\frac{1}{2}$ to $\frac{2}{3}$ as long as first pair of spines . . . . ***Vonones ornata*** (Say, 1821)

34b Two low tubercles on dorsal scutum three-fourths of the distance from eye tubercle to posterior border of scutum; penis with three pairs of equally well-developed spines (Fig. 19.3). . . . . ***Vonones sayi*** (Simon, 1879)

35a Eyes widely separated, not on tubercle, median spine between (Fig. 19.44) . ***Stygnomma spinifera*** (Packard, 1888)

---

[4] Key adapted from Briggs (1971a).

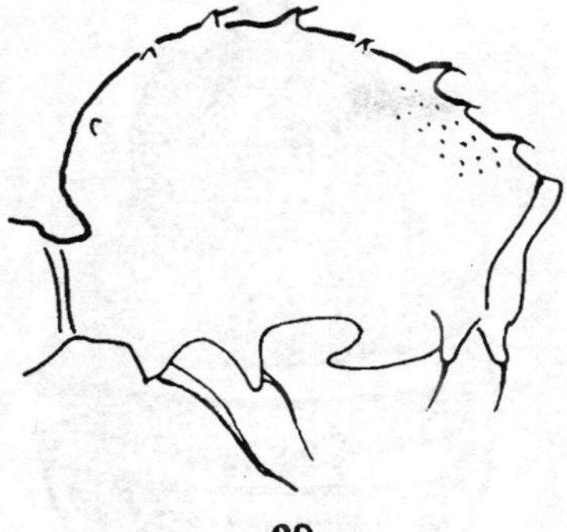

38

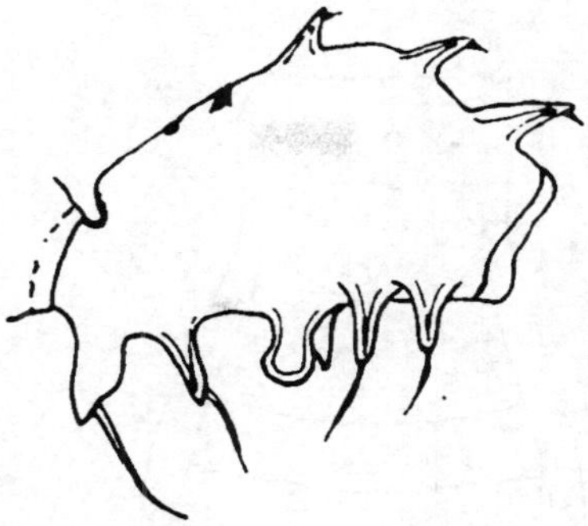

39

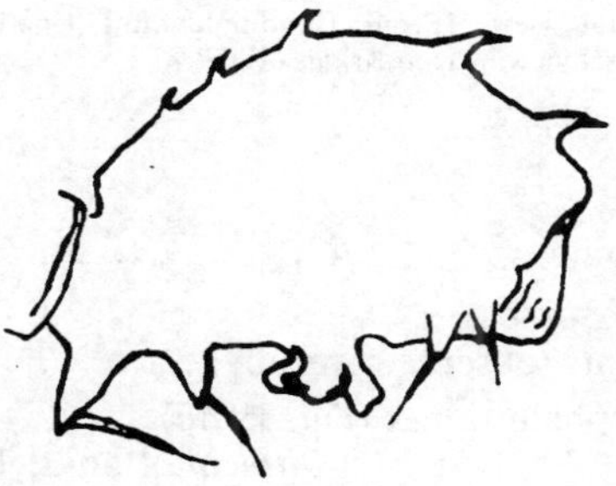

40

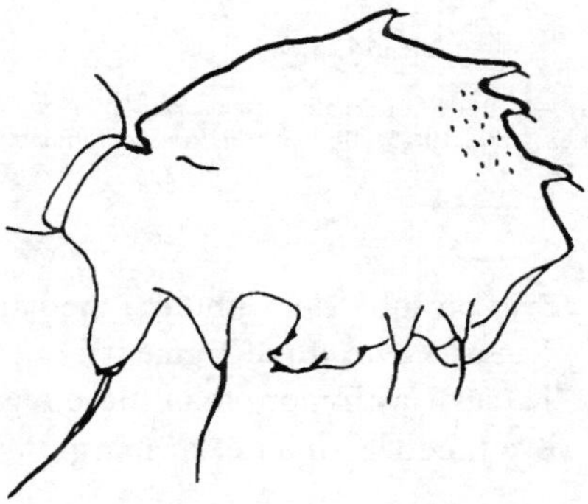

41

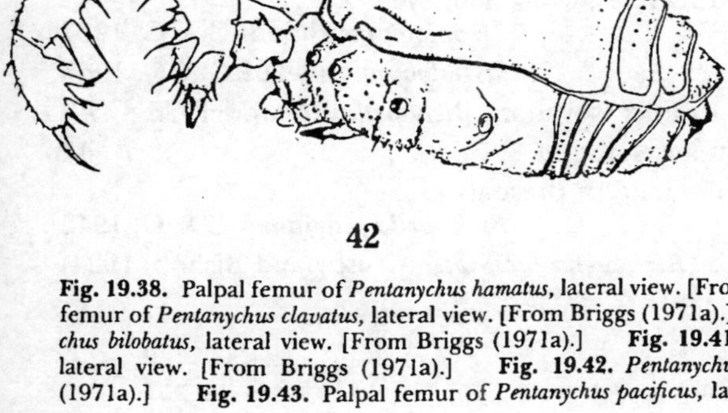

42

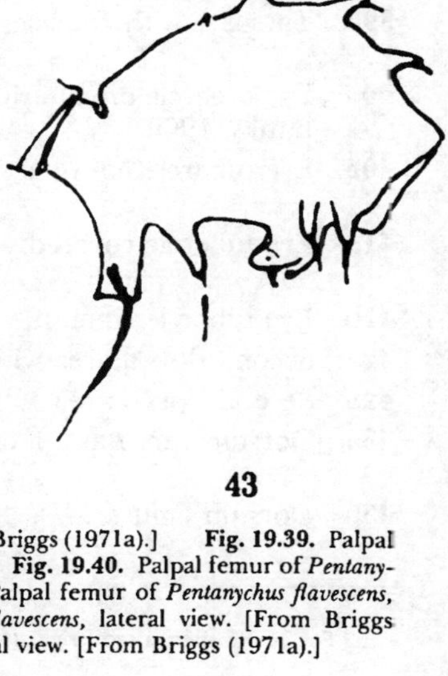

43

Fig. 19.38. Palpal femur of *Pentanychus hamatus,* lateral view. [From Briggs (1971a).] **Fig. 19.39.** Palpal femur of *Pentanychus clavatus,* lateral view. [From Briggs (1971a).] **Fig. 19.40.** Palpal femur of *Pentanychus bilobatus,* lateral view. [From Briggs (1971a).] **Fig. 19.41.** Palpal femur of *Pentanychus flavescens,* lateral view. [From Briggs (1971a).] **Fig. 19.42.** *Pentanychus flavescens,* lateral view. [From Briggs (1971a).] **Fig. 19.43.** Palpal femur of *Pentanychus pacificus,* lateral view. [From Briggs (1971a).]

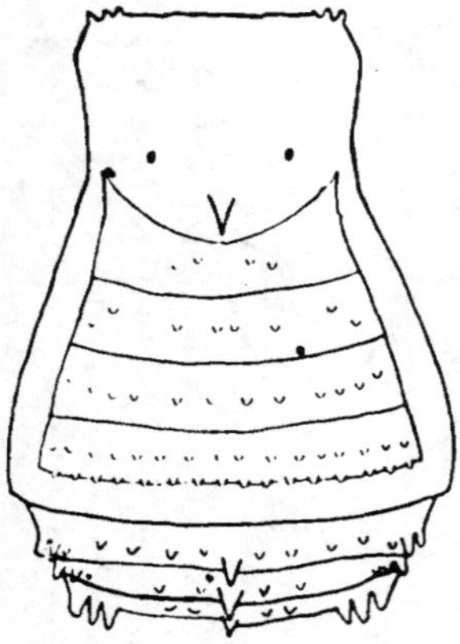

44

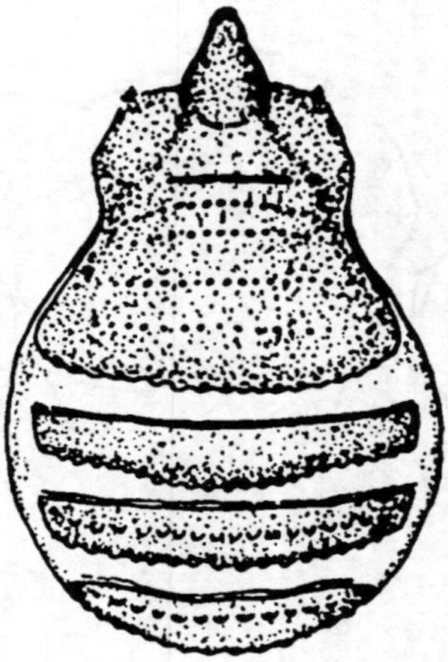

45

Fig. 19.44. Body of male *Stygnomma spinifera,* dorsal view. [From Goodnight and Goodnight (1951).] **Fig. 19.45.** Body of *Sitalcina californicus,* dorsal view. [From Briggs (1968).]

35b Eyes on tubercle, without a median spine **36**
36a Tarsus I with three segments. **37**
36b Tarsus I with more than three segments (observe carefully) **39**
37a Eye tubercle on anterior margin of cephalothorax (Fig. 19.45) . . . . *Sitalcina* Banks, 1911[5]
37b Eye tubercle not on anterior margin of cephalothorax **38**
38a Tarsus III with five segments. . . ***Phalangodes arkansasensis*** (G & G, 1942)
38b Tarsus III with four segments. . . . ***Phalangodes formosus*** (G & G, 1942)
39a Tarsus I with five or more segments **40**
39b Tarsus I with four segments (may be a false articulation on metatarsus I) **42**
40a Eye tubercle on anterior margin of cephalothorax, Texas caves; ***Hoplobunus*** Banks, 1900 . . . . **41**
40b Eye tubercle not on anterior margin of cephalothorax . . . . ***Phalangodes appalachius*** (G & G, 1942)
41a Eye tubercle covered with tubercles, with small eyes . . . . ***Hoplobunus russelli*** G & G, 1967
41b Eye tubercle smooth, no eyes . . . . ***Hoplobunus madlae*** G & G, 1967
42a Second dorsal area with a distinct elevation; ***Bishopella*** Roewer, 1926 . . **43**
42b Second dorsal area without an elevation **44**
43a Dorsum very dark, a definite pattern present . . . . ***Bishopella marianna*** G & G, 1942
43b Dorsum light . . . . ***Bishopella laciniosa*** (Crosby and Bishop, 1924)

[5] For key to species see Briggs (1968).

44a Eye tubercle on anterior margin of the cephalothorax, genital operculum of male with ventrally pointing projection; ***Texella*** G & G, 1942 . . . . . . **45**

44b Eye tubercle not on anterior margin of the cephalothorax; genital operculum of male without ventrally point projection **46**

45a Tarsus II with six or seven segments; trochanter IV with distal projection bearing several tubercles on posterior margin in male; eyes absent . . . . . ***Texella mulaiki*** G & G, 1942

45b Tarsus II with five segments; trochanter IV with large, distal, bluntly rounded projection on distal posterior margin in male; eyes present . . . . ***Texella reddelli*** G & G, 1942

46a Palpal femur with fewer than six dorsal spines or none (Fig. 19.46) . . . **47**

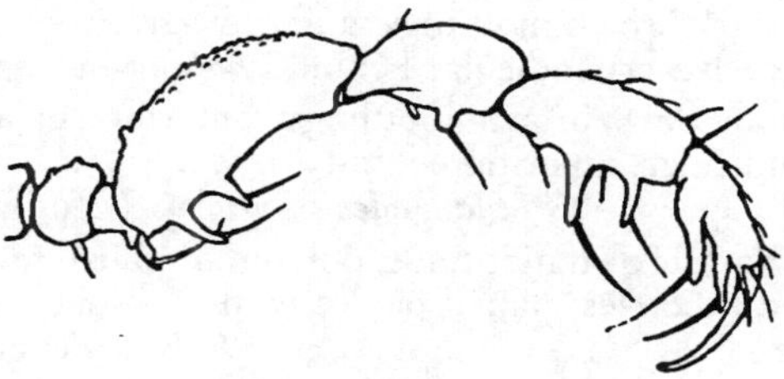

46

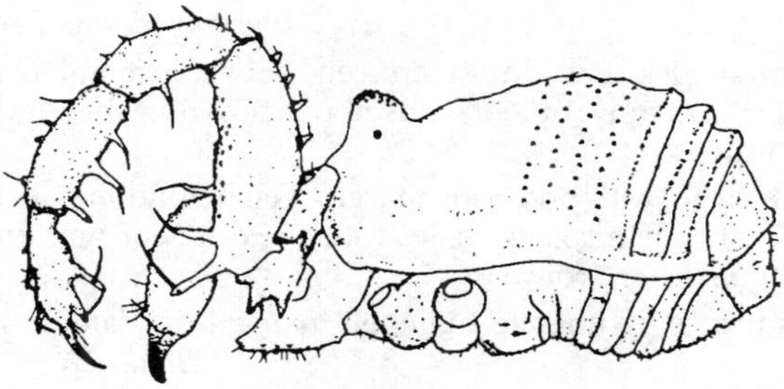

47

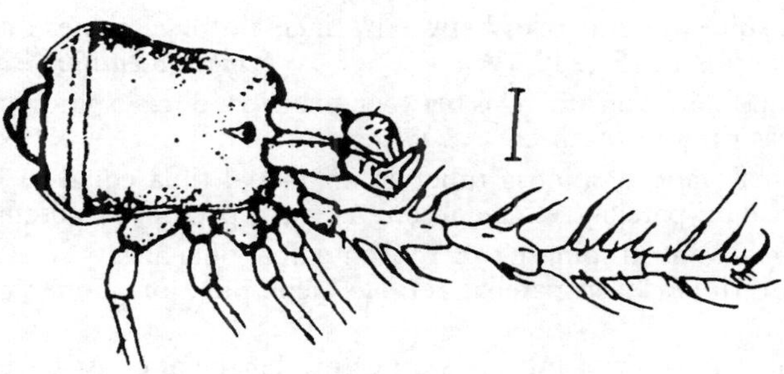

48

Fig. 19.46. Right palpus of *Phalangodes spinturnix*, lateral view. [Crosby and Bishop (1924).] **Fig. 19.47.** *Banksula galilei*, lateral view. [From Briggs (1974b).] **Fig. 19.48.** *Phalangodes armata*, dorsal view. [From Edgar (1966).]

46b Palpal femur with six to eight spines dorsally (Fig. 19.47); ***Banksula*** Roewer, 1949[6] **53**

47a Without eyes, tarsus II with six or more segments; ***Phalangodes*** Tellkampf, 1844 . . . . **48**

47b With eyes, tarsus II with six segments **49**

48a Palpal femur stout, nearly twice as wide in middle as at ends . . . . ***Phalangodes distincta*** (G & G, 1942)

48b Palpal femur slender, only slightly thicker in middle than at ends (Fig. 19.48) ***Phalangodes armata*** Tellkampf, 1844

49a Dorsum heavily tuberculate . . . . . ***Phalangodes tuberculata*** G & G, 1942

49b Dorsum slightly tuberculate or smooth **50**

50a Dorsum of palpal femur smooth or granulated but not spined or coarsely tuberculated (Fig. 19.46) **51**

50b Dorsum of palpal femur spined or coarsely tuberculated **52**

51a Dorsum with dark brown mottlings; anterior surface of eye tubercle slanted posteriorly, nearly perpendicular behind ***Phalangodes montana*** G & G, 1942

51b Dorsum without dark brown mottlings; anterior surface of eye tubercle nearly perpendicular, tapering behind . . . . ***Phalangodes spinturnix*** (Crosby and Bishop, 1924)

52a Dorsum with dark brown mottlings; dorsum of palpal femur with more than three rounded tubercles; palpal patella with two spine-bearing tubercles located distomesially . . . . ***Phalangodes roeweri*** G & G, 1942

52b Dorsum without dark brown mottlings; dorsum of palpal femur with three pointed tubercles; palpal patella with one spine-bearing tubercle located distomesially . . . . ***Phalangodes archeri*** (G & G, 1942)

53a Operculum small, held almost entirely between mesal margins of fourth coxae; males with apex of aedeagus not enclosed in sheath (Fig. 19.49) (***californica*** group) **54**

53b Operculum large, with posterior margin well behind mesal extreme of posterior margin of fourth coxae; males with apex of aedeagus enclosed in sheath (Fig. 19.50) (***melones*** group) **59**

54a Retinae entire; eye tubercle obliquely truncate in lateral view (Fig. 19.47) ***Banksula galilei*** Briggs, 1974

54b Retinae absent or incomplete; if retinae present, eye tubercle a rounded cone (Fig. 19.51) **55**

55a Small spines or tubercles between principal dorsal spines on palpal femur; corneas absent (Fig. 19.51) . . . . ***Banksula californica*** (Banks, 1900)

55b No small spines or tubercles between principal dorsal spines on palpal femur; corneas present or absent **56**

56a Proximal ventral spinose tubercle on palpal tibia equal to ventral spinose tubercle on patella; retinae absent; corneas present or absent **57**

56b Proximal ventral spinose tubercle on palpal tibia always smaller than ventral spinose tubercle on patella; retinae incomplete or absent; corneas present **58**

57a Submarginal row of tubercles on eighth tergite and last sternite (Fig. 19.52) ***Banksula rudolphi*** Briggs and Ubick, 1981

---

[6] Key adapted from Briggs and Ubick (1981).

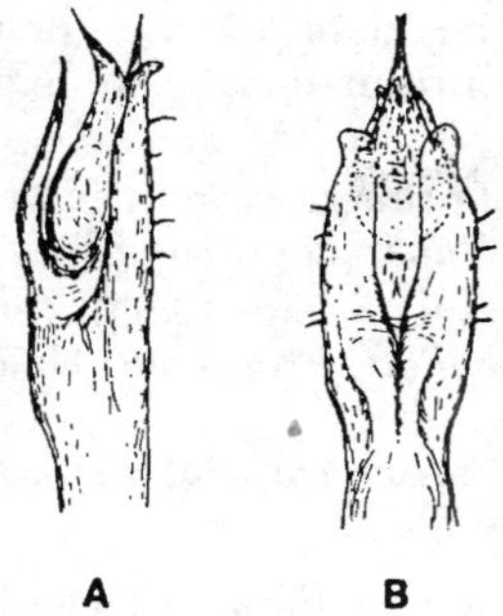

49

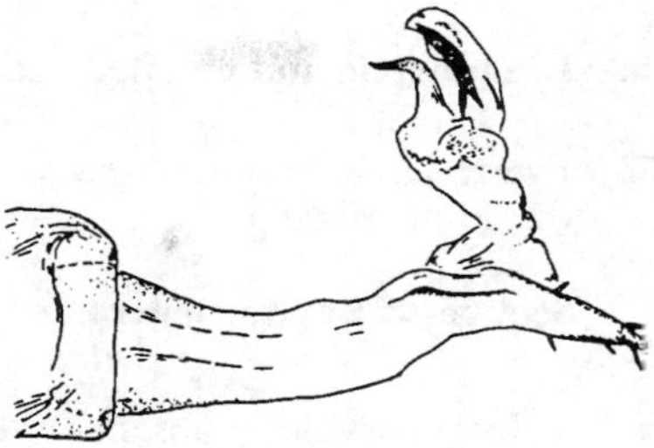

50

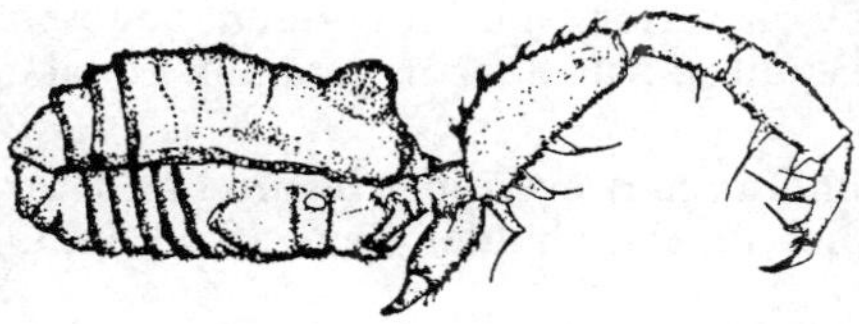

51

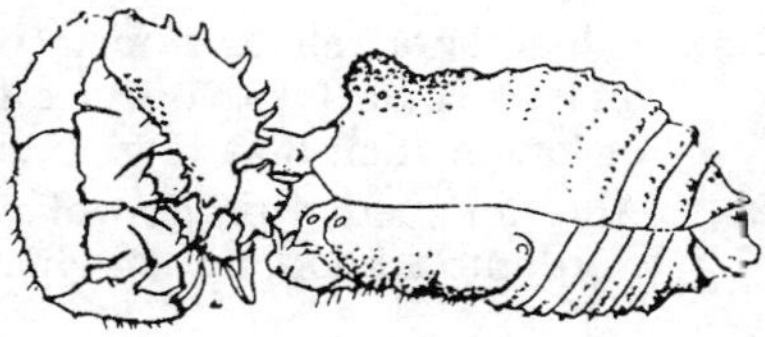

52

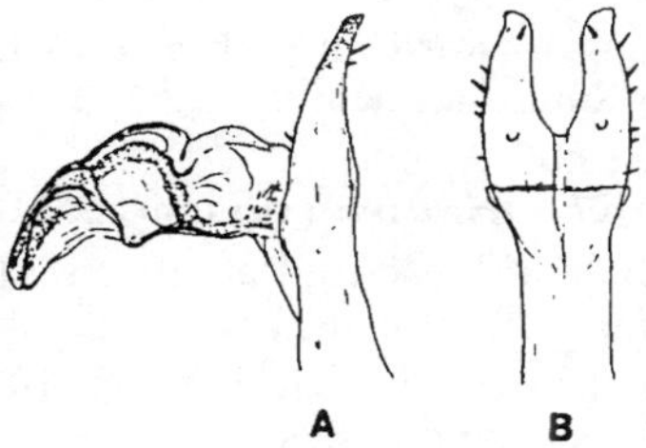

53

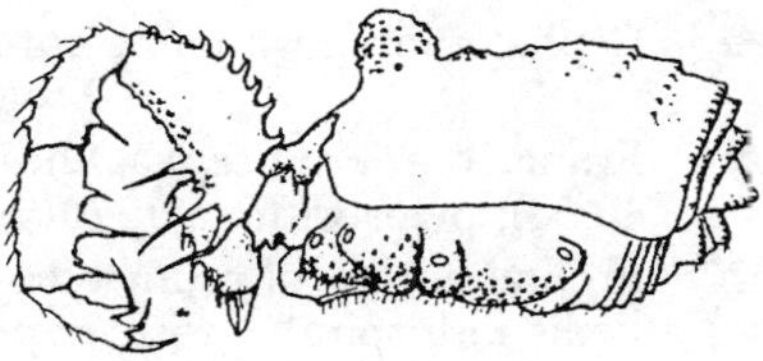

54

Fig. 19.49. Penis of *Banksula californica,* (*A*) lateral view, (*B*) ventral view. [From Briggs (1974b).] **Fig. 19.50.** Penis of *Banksula melones,* lateral view. [From Briggs (1974b).] **Fig. 19.51.** Body, chelicera, and palpus of *Banksula californica.* [From Briggs (1974b).] **Fig. 19.52.** Body, chelicera, and palpus of *Banksula rudolphi.* [From Briggs and Ubick (1981).] **Fig. 19.53.** Penis of *Banksula grahami:* (*A*) lateral view; (*B*) ventral view. [From Briggs (1974b).] **Fig. 19.54.** Body, chelicera, and palpus of *Banksula elliotti.* [From Briggs and Ubick (1981).]

57b No submarginal rows of tubercles on tergites or sternites . . . .
Banksula martinorum Briggs and Ubick, 1981

58a Eye tubercle a rounded cone; small spur ectal to proximal ventral spine on palpal femur; incomplete retinae usually present . . . .
Banksula tuolumne Briggs, 1974

58b Eye tubercle obliquely truncate in lateral view; no spur ectal to proximal ventral spine on palpal femur; retinae absent . . . .
Banksula grubbsi Briggs and Ubick, 1981

59a Leg I longer than 4 mm; retinal diameter greater than 0.03 mm; aedeagus and enclosing sheath fold anterior to recurved prongs of dorsal plate (Fig. 19.50) . . . . ***Banksula melones*** Briggs, 1974

59b Leg I shorter than 4 mm; retinal diameter less than 0.03 mm or missing; aedeagus and enclosing sheath held within bifurcate dorsal plate (Fig. 19.53) **60**

60a Eye tubercle and elevated area behind without tubercles; retinae present. . *Banksula grahami* Briggs, 1974

60b Eye tubercle and elevated area behind tuberculate (Fig. 19.54); retinae reduced or missing . . . . *Banksula elliotti* Briggs and Ubick, 1981

Section 2: Key to Genera and Species of Suborder Palpatores, except *Leiobunum* (see Section 3)

1a Palpal claws either absent or very weak; no accessory spiracles on legs; coxa II with endites small, almost fixed or absent; palpal tarsus shorter than tibia (Fig. 19.55) **2**

1b Palpal claws well developed (Fig. 19.56); accessory spiracles present or absent on legs; coxa II with distinct, flexible endites; palpal tarsus generally equal to or longer than tibia (Fig. 19.57) **34**

2a Anterior medial portion of cephalothorax and eye tubercle produced forward into a hood, under which the chelicerae are hidden (Figs. 19.58, 19.59) **3**

2b Anterior medial portion of cephalothorax and eye tubercle either produced forward in the form of a cone or not produced forward **9**

3a Cephalothorax with stout spines projecting forward on each side of hood (Figs. 19.55, 19.58); **Nemastomatidae : Ortholasmatinae**[7] **4**

3b Cephalothorax without stout spines (Fig. 19.59). . . . . **Trogulidae:** ***Trogulus tricarinatus*** (Linneaus, 1758)

4a Frontal border of cephalothorax on each side with two conical processes (Fig. 19.58); penis with an apical hook; ***Ortholasma*** Banks, 1894 . . . . **5**

4b Frontal border of cephalothorax with one process on each side (Fig. 19.60); penis with spirally twisted apex; ***Dendrolasma*** Banks, 1894 . . . . **8**

5a Femora of legs banded, body with median light stripe; Mendocino County, California, north to coastal British Columbia ***Ortholasma pictipes*** Banks, 1911

5b Femora of legs not banded **6**

6a Femora of legs without false articulations, appearing bumpy in outline; Coast Ranges north and south of San Francisco Bay. . . . . ***Ortholasma rugosum*** Banks, 1894

6b At least femora II with false articulations, all femora appearing smooth in outline **7**

7a Microsculpture of leg femora appearing short-hairy, false articulations usually lacking on femora I and III; southern California . . . . ***Ortholasma setulipes*** Shear and Gruber, 1983

7b Microsculpture of leg femora granulate, all femora with false articulations; San Francisco Bay region south to Los Angeles, foothills of Sierra Nevada . ***Ortholasma levipes*** Shear and Gruber, 1983

8a Scute with dense network of small cells (Fig. 19.60); legs short, banded; male palp with distal patellar tooth (Fig. 19.61); northwestern California. . . . . ***Dendrolasma dentipalpe*** Shear and Gruber, 1983

[7] Key adapted from Shear and Gruber (1983).

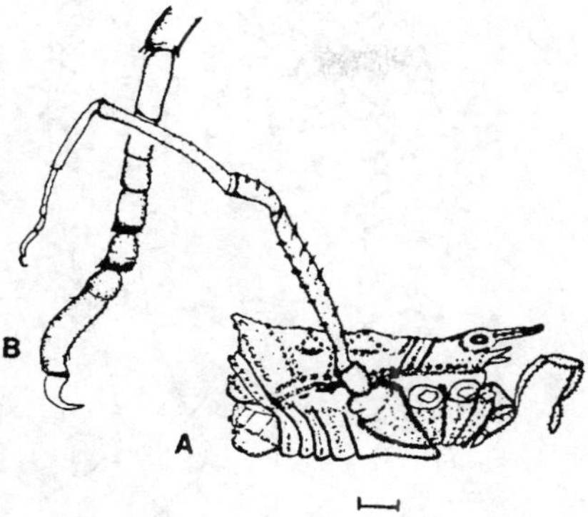

55

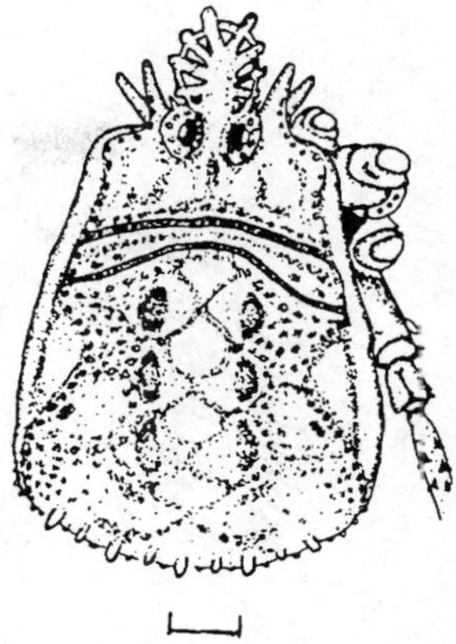

58

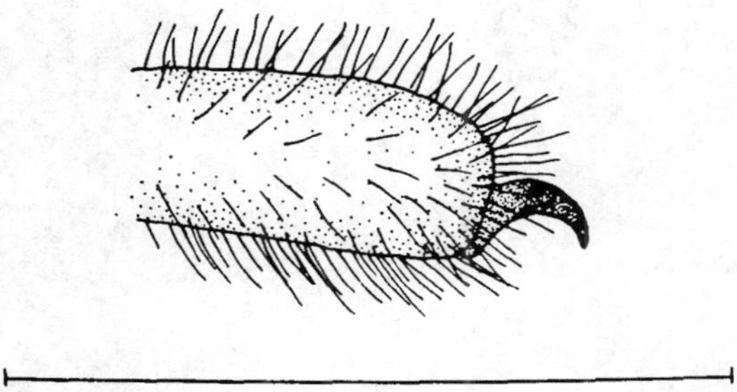

56

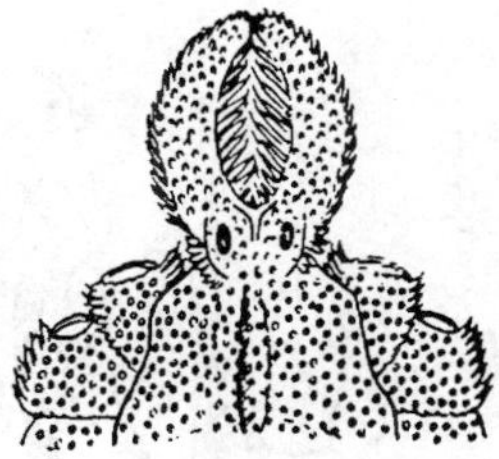

59

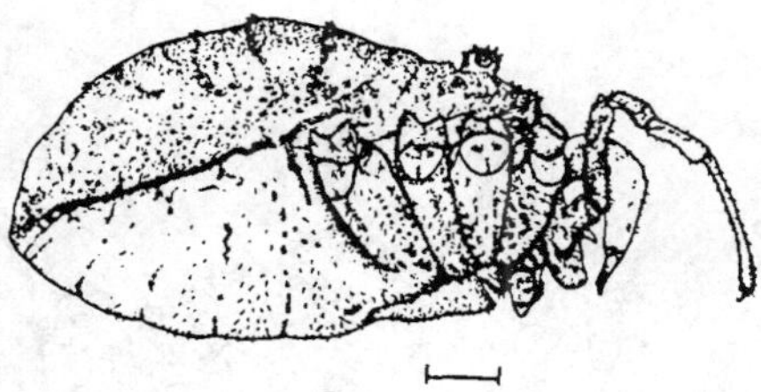

57

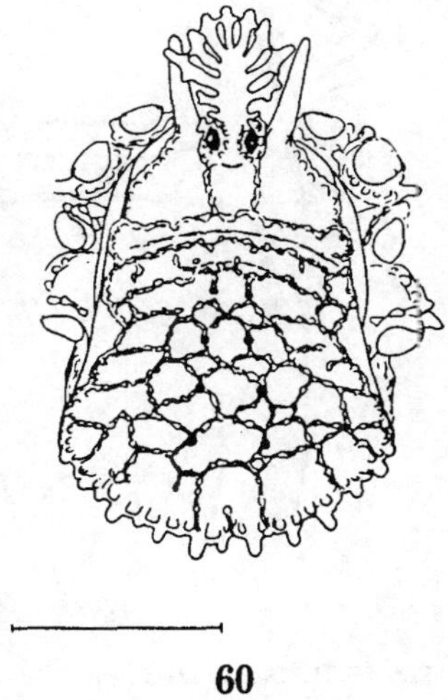

60

Fig. 19.55. *Ortholasma pictipes: (A)* body, palpus, and leg IV, lateral view; *(B)* tarsus of leg IV. **Fig. 19.56.** Palpal claw of *Dalquestia formosa,* lateral view. **Fig. 19.57.** *Phalangiulm opilio* female, lateral view. **Fig. 19.58.** *Ortholasma pictipes,* dorsal view. **Fig. 19.59.** Anterior portion of body of *Trogulus tricarinatus,* dorsal view. [From Roewer (1923).] **Fig. 19.60.** *Dendrolasma mirabile,* dorsal view of male. Scale line = 2.0 mm. [From Shear & Gruber (1983).]

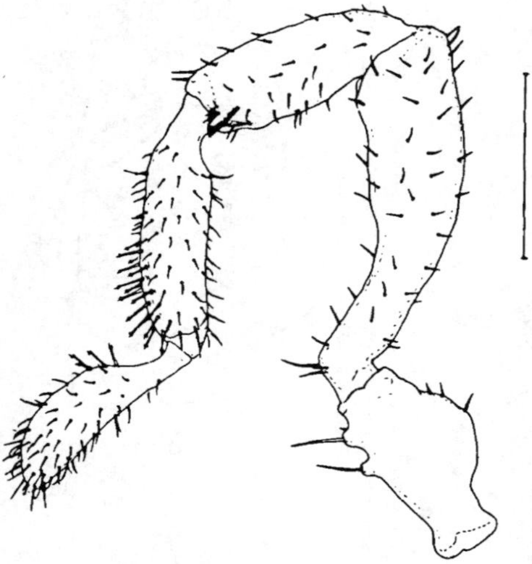

61

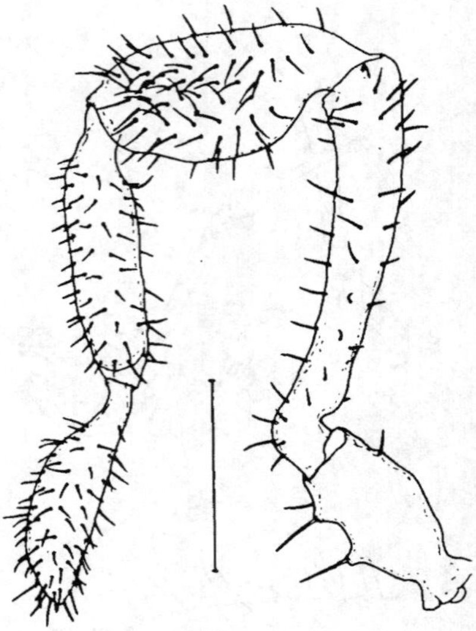

62

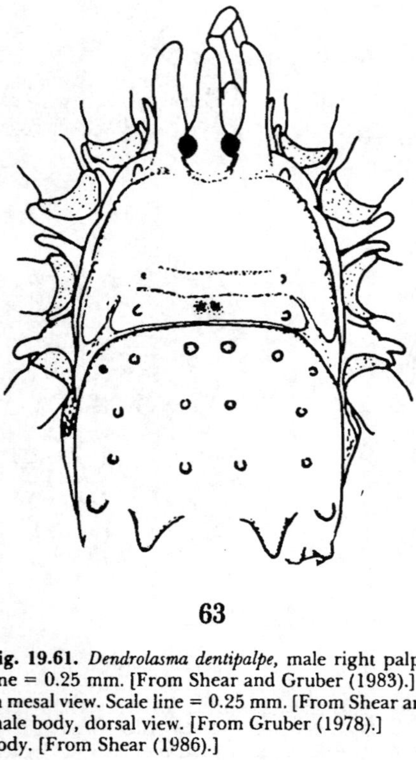

63

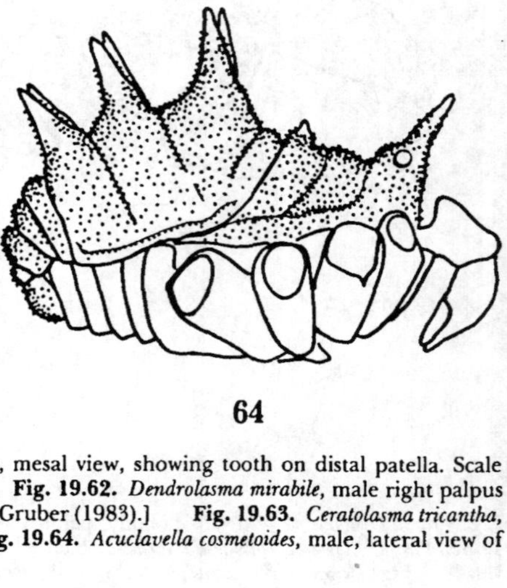

64

Fig. 19.61. *Dendrolasma dentipalpe*, male right palpus, mesal view, showing tooth on distal patella. Scale line = 0.25 mm. [From Shear and Gruber (1983).] **Fig. 19.62.** *Dendrolasma mirabile*, male right palpus in mesal view. Scale line = 0.25 mm. [From Shear and Gruber (1983).] **Fig. 19.63.** *Ceratolasma tricantha*, male body, dorsal view. [From Gruber (1978).] **Fig. 19.64.** *Acuclavella cosmetoides*, male, lateral view of body. [From Shear (1986).]

8b Scute with larger cells; legs long, concolorous; male palp with enlarged patella (Fig. 19.62) but no tooth; along the coast from southern Oregon to southern Alaska . . . . ***Dendrolasma mirabile*** Banks, 1894

9a Eye tubercle in the form of a forward-pointing cone; . . **Ceratolasmatidae** **10**

9b Eye tubercle low, not in form of cone **15**

10a Cone-shaped projection present on both sides of the cephalothorax, which are over the chelicarae (Fig. 19.63) . . . ***Ceratolasma tricantha*** G & G, 1942

10b Anterior border of cephalothorax without cone-shaped processes (Fig. 19.64) ***Acuclavella*** Shear, 1986[8] **11**

11a Dorsum of abdomen with prominent acute spines (Fig. 19.64) **12**

11b Without such spines, but paired, low, rounded tubercles on all five dorsal areas . . . . females of ***Acuclavella shoshone*** Shear, 1986

12a A single pair of abdominal spines, on area 2 ***Acuclavella merickeli*** Shear, 1986

12b More than one pair of abdominal spines **13**

13a Abdominal spines on area 1 . . . . **14**

13b No spines on area 1, paired spines on areas 2, 3, and 4 (Fig. 19.64) . . . . ***Acuclavella cosmetoides*** Shear, 1986

14a Four pairs of abdominal spines (on areas 1–4) . . . . ***Acuclavella shoshone*** Shear, 1986

14b Two pairs of abdominal spines (on areas 1 and 2) . . . . females of ***Acuclavella quattuor*** Shear, 1986

15a Metapeltidium (small shield on posterior cephalothorax) with one or two centrally placed sensory cones or spines **16**

15b Metapeltidium smooth, or at least without sensory spines **28**

16a A single sensory cone or spine present on metapeltidium (Fig. 19.65); chelicerae enlarged and elongated, often greatly so, the sum of the lengths of the two segments equal to or exceeding the length of the body (Figs. 19.65, 19.66); basal segment of chelicerae with spine-bearing tubercles . . . . . . **Sabaconidae: *Taracus*** Simon, 1879 (7 species)[9]

16b A pair of sensory cones present on metapeltidium (Fig. 19.67); chelicerae not enlarged, or if enlarged then without spine bearing tubercles (Fig. 19.68); **Sabaconidae: *Sabacon*** Simon, 1879[10] **17**

17a Palpi stout, heavily set with stiff bristles (Fig. 19.69); cuticle mostly leathery, but well sclerotized in males of some species **18**

17b Palpi more slender, not heavily set with stiff bristles, but with more scattered, often glandular, hairs (Fig. 19.70), region of Mt. St. Elias, Alaska . . . . . . ***Sabacon (= Tomicomerus) bryanti*** (Banks, 1898)

18a Males[11] **19**

18b Females[12] **24**

19a Femora of legs with false articulations; northern coastal Oregon (females unknown) . . . . ***Sabacon astoriensis*** Shear, 1975

---

[8] Key adapted from Shear (1986).

[9] For descriptions of species, see Banks (1894b), Goodnight and Goodnight (1942b, 1943, 1945), Roewer (1930), and Simon (1870).

[10] Key adapted from Shear (1975a).

[11] Males of *S. briggsi* Shear, 1975a are unknown.

[12] Females of *S. astoriensis* Shear, 1975a and *S. sheari* Cokendolpher, 1984 are unknown.

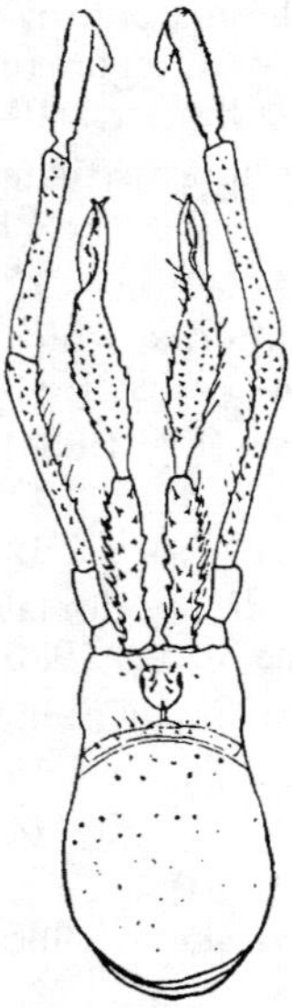

65

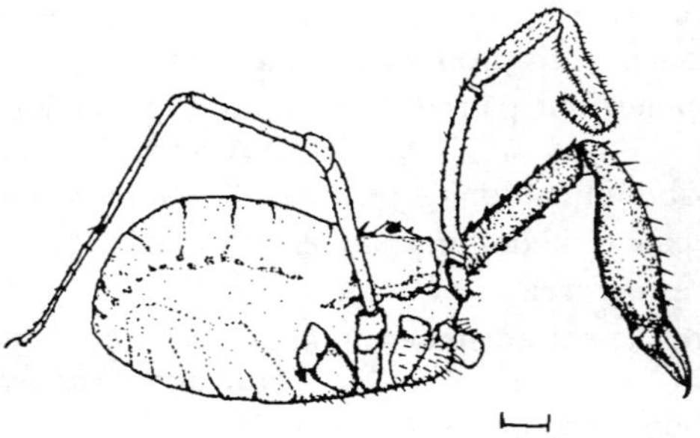

66

67

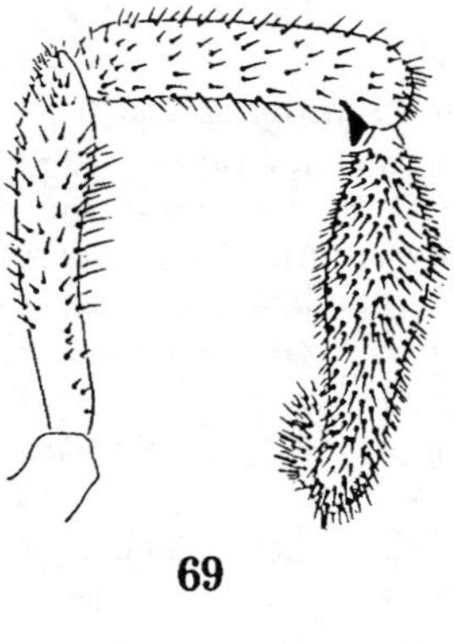

69

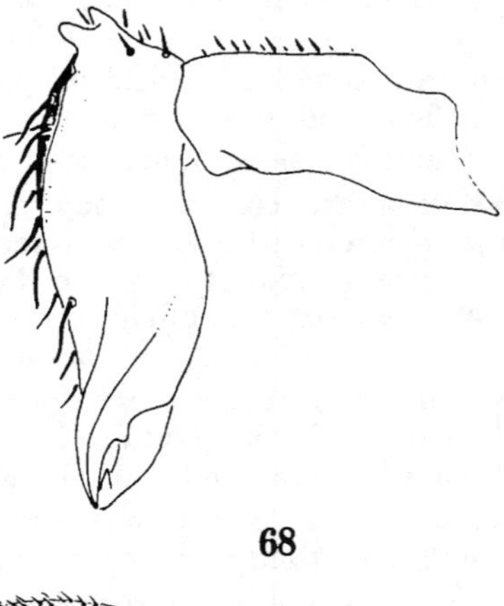

68

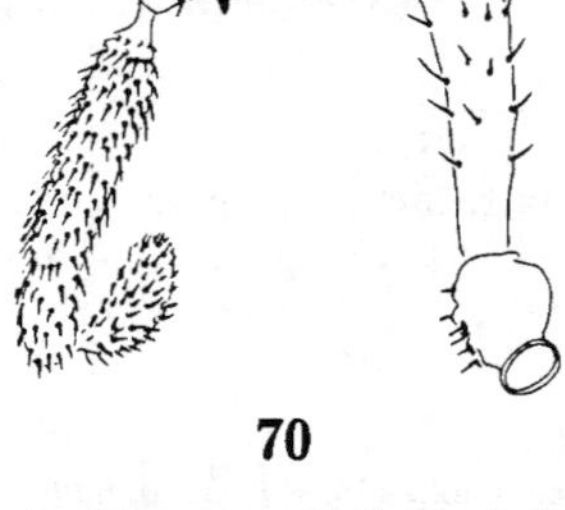

70

Fig. 19.65. Chelicera, palpus and body of *Taracus malkani*, dorsal view. [From Goodnight and Goodnight (1945).] **Fig. 19.66.** *Taracus packardi*, lateral view, showing body, palpus, chelicera, and leg III. **Fig. 19.67.** Body of male *Sabacon mitchelli*, dorsal view. [From Shear (1975a).] **Fig. 19.68.** Right chelicera of *Sabacon* (= *Tomicomerus*) *bryanti*, mesal view. [From Shear (1975a).] **Fig. 19.69.** Left palpus of *Sabacon cavicolens*, male, lateral view. [From Shear (1975a).] **Fig. 19.70.** Right palpus of *Sabacon* (= *Tomicomerus*) *bryanti*, lateral view. [From Shear (1975a).]

19b Femora of legs without false articulations **20**

20a Penis slender, straight, with weak setation (Fig. 19.71); palpal patella with two large denticles and a single small denticle on mesal margin, Oregon . . . . ***Sabacon sheari*** Cokendolpher, 1984

20b Penis and palpal patella with other than the foregoing characters **21**

21a Basal segment of chelicera without a knoblike gland (Fig. 19.72); high mountains of North Carolina. . . . . ***Sabacon mitchelli*** Crosby and Bishop, 1924

21b Basal segment of chelicera with a knoblike gland (Figs. 19.73, 19.74) . . **22**

22a Tip of penis with fingerlike projections (Fig. 19.75); British Columbia south to northern California. . . . . ***Sabacon occidentalis*** (Banks, 1894)

22b Penis otherwise. **23**

23a Cheliceral gland often large, narrowed at base (Fig. 19.73); if small, then shaft of penis with at least two pair of stiff spines (Fig. 19.76); California and Oregon . . . . ***Sabacon siskiyou*** Shear, 1975

23b Cheliceral gland always smaller, rounded, not narrowed at base; northeastern and midwestern United States (Fig. 19.72) . . . . ***Sabacon cavicolens*** (Packard, 1884)

24a Genital operculum pointed (Fig. 19.77); west-central California. . . . . . . ***Sabacon briggsi*** Shear, 1975

24b Genital operculum evenly rounded. **25**

25a Only first two abdominal segments sclerotized; high peaks in North Carolina (Fig. 19.67) . . . . ***Sabacon mitchelli*** Crosby and Bishop, 1924

25b All abdominal segments with sclerotized patches **26**

26a Ovipositor short, sparsely setose; eastern United States (Fig. 19.78). . . . . ***Sabacon cavicolens*** (Packard, 1884)

26b Ovipositor longer, densely setose; British Columbia south to northern California (Fig. 19.79) **27**

27a Legs unbanded; femora II short (about 2.5 mm); generally with large metapeltidial sensory spines . . . . ***Sabacon siskiyou*** Shear, 1975

27b Legs banded; femora II longer (about 4.5 mm); metapeltidial sensory spines small . . . . ***Sabacon occidentalis*** (Banks, 1894)

28a Length of body about 1 mm or less (Fig. 19.80); legs covered with curly, decumbent setae; northeastern and midwestern United States; **Ceratolasmatidae** . . . . ***Crosbycus dasycnemus*** (Crosby, 1911)

28b Length of body greater than 1 mm; legs without decumbent curly setae. **29**

29a Leg femora with pseudosegments, penis with pointed tubercles in two lateral rows (Fig. 19.81); **Nemastomatidae : Nemastomatinae**. . . . . ***Nemastoma bimaculatum*** (Fabricius, 1775)

29b Leg femora without pseudosegments, penis with tubercles encircling the tip (Fig. 19.82); **Ceratolasmatidae :** ***Hesperonemastoma*** Gruber, 1970) **30**

30a Tibia of palpus more than twice as long as tarsus **31**

30b Tibia of palpus less than twice as long as tarsus **33**

31a Two diverging rows of four curved spines on dorsum near union of cephalothorax and abdomen; first segment of chelicera not constricted near base **32**

31b Spines not arranged as above, first segment of chelicera strongly constricted near base (Fig. 19.83) . . . . ***Hesperonemastoma kepharti*** (Crosby and Bishop, 1924)

32a Palpal femur swollen in distal third (Fig. 19.84). . . . . ***Hesperonemastoma packardi*** (Roewer, 1914)

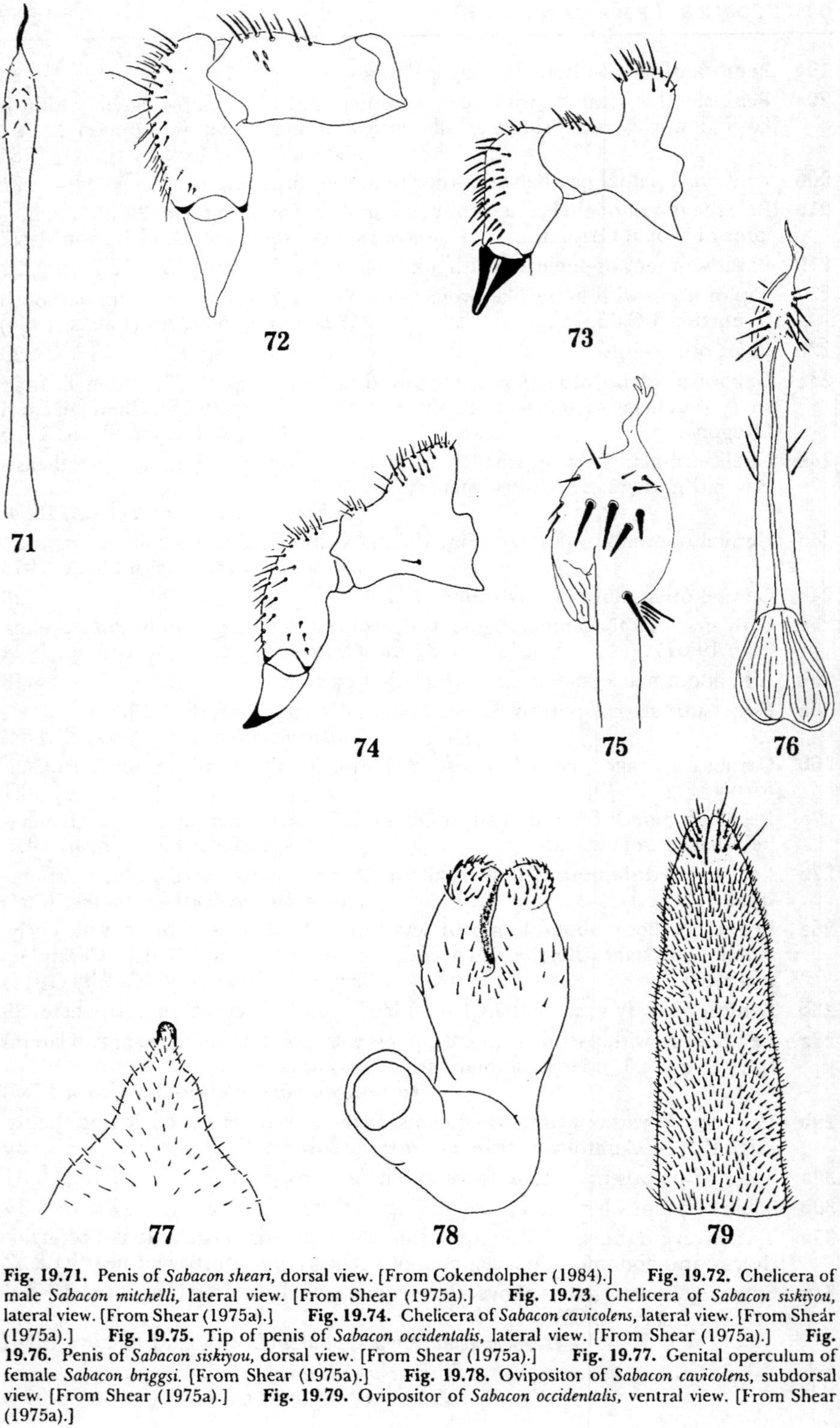

Fig. 19.71. Penis of *Sabacon sheari,* dorsal view. [From Cokendolpher (1984).] **Fig. 19.72.** Chelicera of male *Sabacon mitchelli,* lateral view. [From Shear (1975a).] **Fig. 19.73.** Chelicera of *Sabacon siskiyou,* lateral view. [From Shear (1975a).] **Fig. 19.74.** Chelicera of *Sabacon cavicolens,* lateral view. [From Shear (1975a).] **Fig. 19.75.** Tip of penis of *Sabacon occidentalis,* lateral view. [From Shear (1975a).] **Fig. 19.76.** Penis of *Sabacon siskiyou,* dorsal view. [From Shear (1975a).] **Fig. 19.77.** Genital operculum of female *Sabacon briggsi.* [From Shear (1975a).] **Fig. 19.78.** Ovipositor of *Sabacon cavicolens,* subdorsal view. [From Shear (1975a).] **Fig. 19.79.** Ovipositor of *Sabacon occidentalis,* ventral view. [From Shear (1975a).]

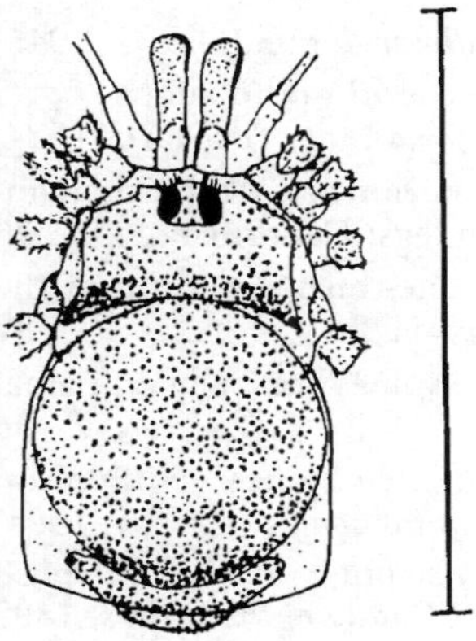

80

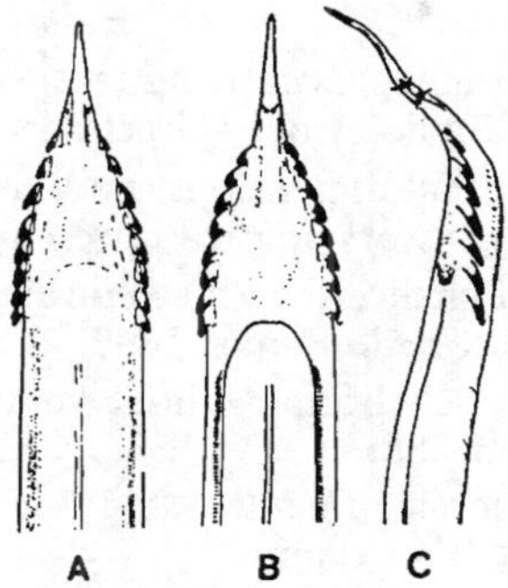

81

82

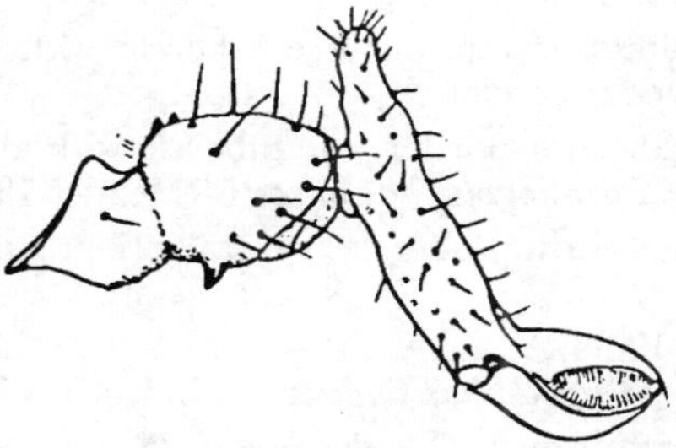

83

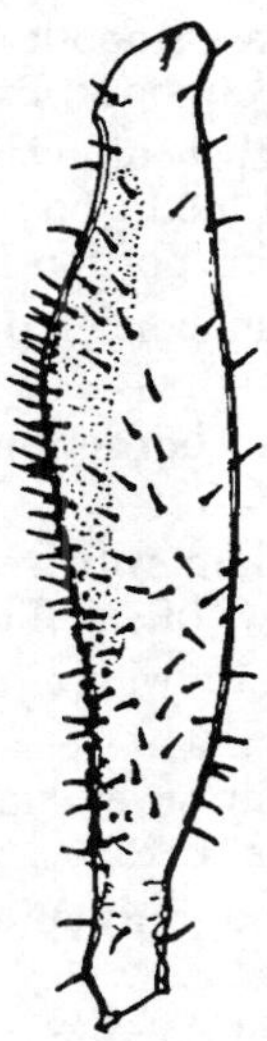

84

85

Fig. 19.80. Body of *Crosbycus dasycnemus*, dorsal view. **Fig. 19.81.** *Nemastoma bimaculatum*, glans of penis: (*A*) ventral view; (*B*) dorsal view; (*C*) lateral view. [From Gruber and Martens (1968).] **Fig. 19.82.** Apical portion of *Hesperonemastoma packardi* penis, lateral view. [From Gruber (1970).] **Fig. 19.83.** Right chelicera of male *Hesperonemastoma kepharti*, lateral view. [From Crosby and Bishop (1924).] **Fig. 19.84.** Right palpal femur of *Hesperonemastoma packardi*, mesal view. [From Gruber (1970).] **Fig. 19.85.** Right palpal femur of *Hesperonemastoma modesta*, mesal view. [Modified from Gruber (1970).]

32b Palpal femur nearly parallel sided and not swollen in distal third (Fig. 19.85) ***Hesperonemastoma modesta*** (Banks, 1894)

33a Row of conspicuous light, shiny spots on lateral and posterior margins of dorsal scute; dorsal scute, including eye tubercle and cephalothorax, covered with small round granulations. . . . . ***Hesperonemastoma pallidimaculosum*** (G & G, 1945)

33b Without row of light spots and covering of small round granulations. . . . ***Hesperonemastoma inops*** (Packard, 1884)

34a Eye tubercle enormous, occupying most of the anterior half of the cephalothorax (Fig. 19.86); accessory spiracles lacking on legs; **Caddidae** **35**

34b Eye tubercle of usual dimensions, no more than one-third the width of the cephalothorax; accessory spiracles present on legs **37**

35a Basal segment of chelicera with a ventral spur; no spines ventrally on palpal tibia (Fig. 19.87); ***Caddo*** Banks, 1892. . . . . **36**

35b Basal segment of chelicera without ventral spur; two spines ventrally on palpal tibia (Fig. 19.88) . . . . ***Acropsopilio boopis*** (Crosby, 1904)

36a Width of eye tubercle of *adults* 1.3–1.5 mm; body strongly marked purplish brown and silver (Fig. 19.86) . . . . ***Caddo agilis*** Banks, 1892

36b Width of eye tubercle of *adults* 0.6–0.8 mm; body more or less evenly colored medium brown . . . . ***Caddo pepperella*** Shear, 1975

37a Penis in lateral view strongly bent at glans–shaft junction (Figs. 19.89, 19.90); palpal claw always smooth (Fig. 19.56); body generally soft, genital operculum flexible throughout; **Phalangiidae**. **38**

37b Penis not strongly bent in lateral view (Figs. 19.91, 19.92); palpal claw with or without ventral teeth; body more sclerotized; genital operculum generally solid, the anterior portion forming a lip (Fig. 19.4) **52**

38a Anterior median border of cephalothorax with one to three conspicuous spines (Fig. 19.93) **39**

38b Anterior median border smooth, granulated, or with one to several small spines **43**

39a Genital operculum with rounded anterior margin; eye tubercle with two longitudinal rows of small denticles or tubercles **40**

39b Genital operculum with a notch on anterior border; eye tubercle with a few hairs only . . . . ***Paroliogolophus agrestis*** (Meade, 1855)

40a Palpal femur with several large spine-bearing tubercles ventrally (Fig. 19.93) ***Odiellus*** Roewer, 1923 . . . . **41**

40b Palpal femur with only spines (Fig. 19.94). . . . . ***Oligolophus tridens*** (C. L. Koch, 1836)

41a Central figure of dorsum mottled with brown. ***Odiellus pictus*** (Wood, 1868)

41b Central figure of dorsum not conspicuously mottled but uniform dark brown or black, except for a few small white spots **42**

42a Lateral borders of central figure nearly parallel on abdomen, diverging anteriolaterally on cephalothorax. . . . . ***Odiellus argentus*** Edgar, 1971

42b Lateral borders of central figure narrowest at abdominal segments 2 and 5, giving an hourglass or figure-eight appearance . . . . ***Odiellus nubivagus*** Crosby and Bishop, 1924

43a Eye tubercle with spines **49**

43b Eye tubercle smooth **44**

44a Palpal femur and patella with distal ends expanded on inner margin and male palpal tarsus toothed ventrally; ***Liopilio*** Schenkel, 1951 . . . . . . **48**

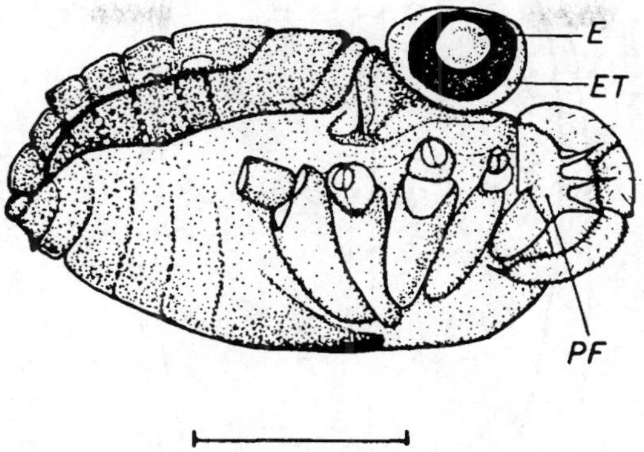

86

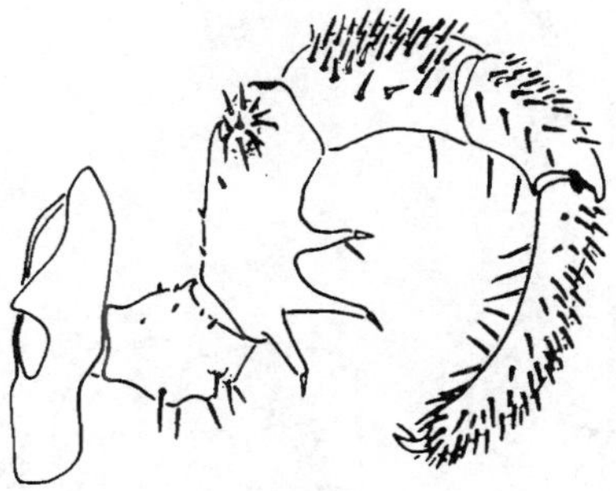

87

88

Fig. 19.86. *Caddo agilis,* lateral view. *E,* Eye; *ET,* eye tubercle; *PF,* palpal femur. [From Edgar (1966).] **Fig. 19.87.** Left palpus of female *Caddo agilis,* mesal view. [From Shear (1975b).] **Fig. 19.88.** Left palpus of female *Acropsopilio boopis,* mesal view. [From Shear (1975b).]

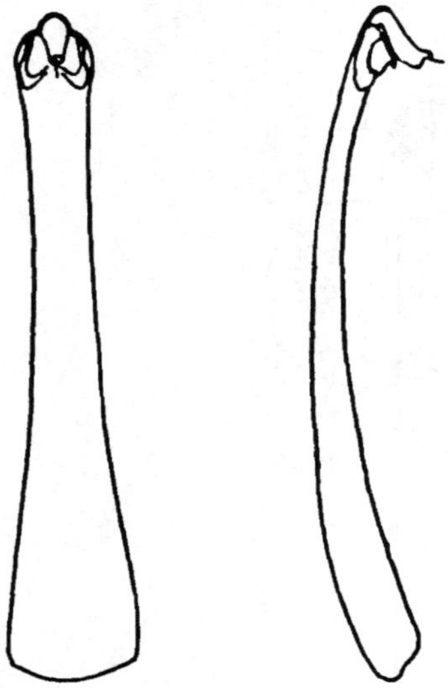

A B

89

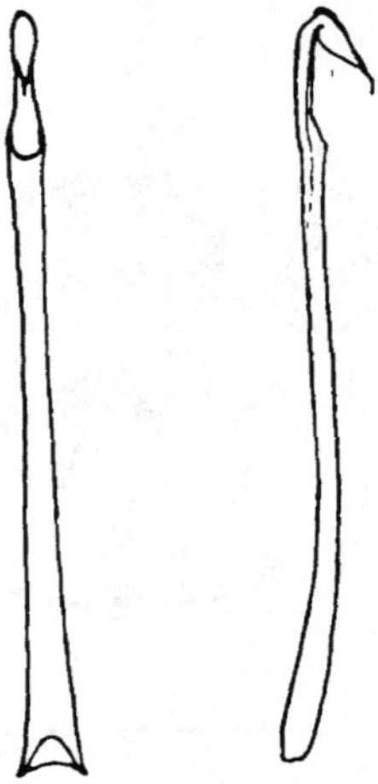

A B.

90

91

92

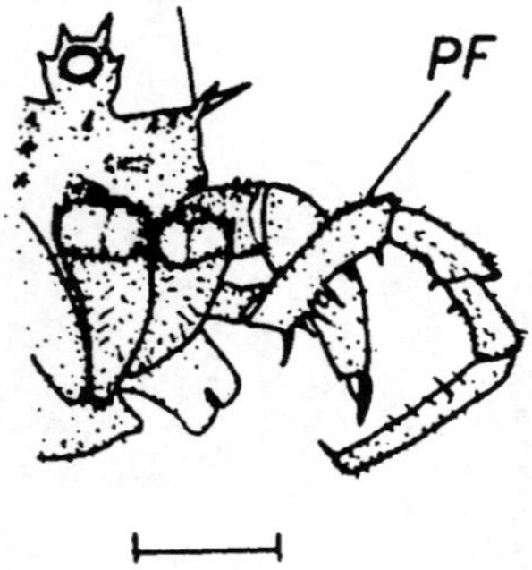

93

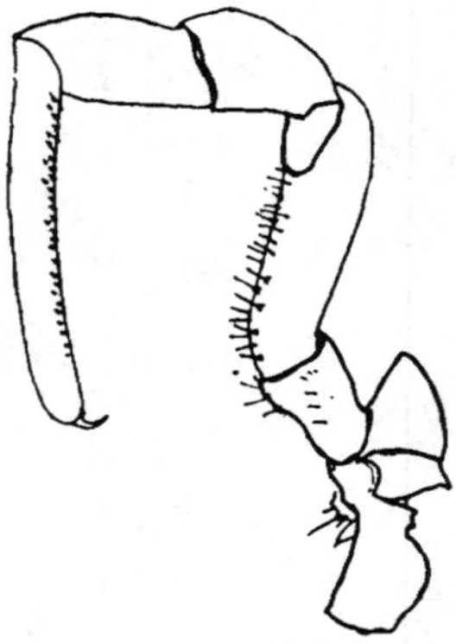

94

Fig. 19.89. Penis of *Opilio parientinus:* (*A*) dorsal view; (*B*) lateral view. [From Bishop (1949).] **Fig. 19.90.** Penis of *Odiellus pictus:* (*A*) dorsal view; (*B*) lateral view. [From Bishop (1949).] **Fig. 19.91.** Penis of *Hadrobunus maculosus,* dorsal view. [From Bishop (1949).] **Fig. 19.92.** Penis of *Eumesosoma roeweri,* dorsal view. [From Cokendolpher (1980a).] **Fig. 19.93.** Anterior portion of *Odiellus pictus,* lateral view. Trident of spines (ST) palpal femur (PF). [From Edgar (1966).] **Fig. 19.94.** Palpus of *Oligolophus tridens,* mesal view. [From Meinertz (1962).]

44b Palpal femur and patella not extended; male palpal tarsus untoothed ventrally; ***Leptobunus*** Banks, 1893[13] **45**

45a Male chelicera with distinct apophysis on movable finger (Fig. 19.95); stylus of penis equal to or shorter than primary setae of glans (Fig. 19.96); primary loop of seminal receptacle short, less than three times as long as wide (Fig. 19.97) (***parvulus*** group) **46**

45b Chelicera without apophysis on movable finger (Fig. 19.98); stylus of penis much longer than primary setae of glans (Fig. 19.99); primary loop of seminal receptacle long, more than four times as long as wide (Fig. 19.100) (***californicus*** group) **47**

46a Body color golden, legs speckled; primary setae of glans separated more than the length of the stylus (Fig. 19.96) ***Leptobunus aureus,*** Cokendolpher, 1985

46b Body color brown, gray, or bronze, legs banded; primary setae of glans closely spaced . . . . ***Leptobunus parvulus*** (Banks, 1894)

47a Male palpal tibiae and tarsi noticeably expanded on mesal margins (Fig. 19.101B); dorsum of penis glans slightly depressed (Fig. 19.99); basal loops of seminal receptacles consisting of three coils or loops (Fig. 19.100) . . . .
Leptobunus californicus Banks, 1893

47b Male palpal tibiae and tarsi only slightly expanded (Fig. 19.101A); dorsum of penis glans straight; basal loops of seminal receptacles consisting of two coils or loops . . . . ***Leptobunus borealis*** Banks, 1899

48a Dorsal abdominal pattern distinct (Fig. 19.102B); male palpal patella and tibia armed ventrally with small denticles . . . .
Liopilio yukon Cokendolpher, 1981

48b Dorsal abdominal pattern obscured or absent (Fig. 19.102A); male palpal patella and tibia unarmed . . . . ***Liopilio glaber*** Schenkel, 1951

49a First (basal) segment of chelicera ventrally with one sharp, forward-pointed tooth (observe carefully) (Fig. 19.103) . . . ***Mitopus morio*** (Fabricius, 1779)

49b First segment without a tooth (Fig. 19.104). **50**

50a Supracheliceral laminae with two anteriorly directed spines visible below anterior margin of cephalothorax; distal segment of male chelicera prolonged dorsally as a spur (Figs. 19.57, 19.105, 19.106) . . . .
Phalangium opilio Linnaeus, 1761

50b Superacheliceral laminae smooth without spines; male chelicerae normal **51**

51a Eye tubercle strongly canaliculate and about one-third the width of the cephalothorax (Fig. 19.107) . . . .
Rilaena (= Platybunus) triangularis (Herbst, 1799)

51b Eye tubercle not canaliculate and about one-fifth the width of the cephalothorax (Fig. 19.108) . . . . ***Opilio parietinus*** (DeGeer, 1778)

52a Leg femora with one or two pseudoarticulary nodules (Fig. 19.109); male coxae III and IV enlarged (Fig. 19.110); **Gagrellidae: Gagrellinae; *Trachyrhinus*** Weed, 1892[14] **53**

52b Leg femora never with nodules; male coxae normal **60**

53a Palpal patella with an apophysis (Fig. 19.111) **58**

53b Palpal patella without apophyses (Fig. 19.112) **54**

---

[13] Key adapted from Cokendolpher (1985).
[14] Key adapted from Cokendolpher (1981a).

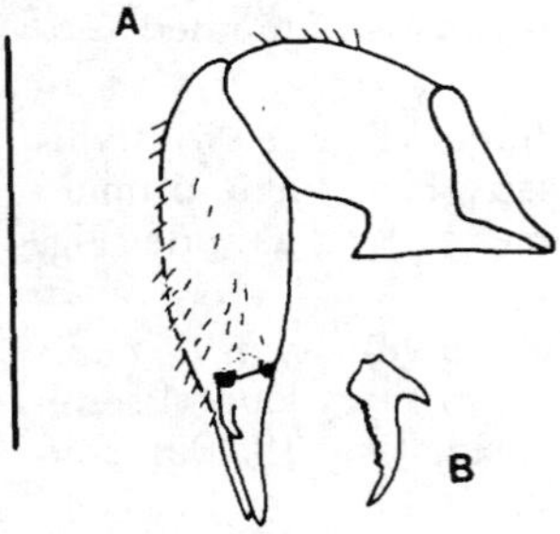

95

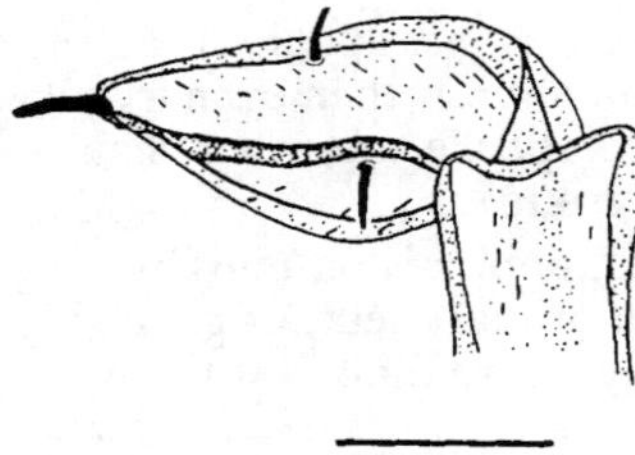

96

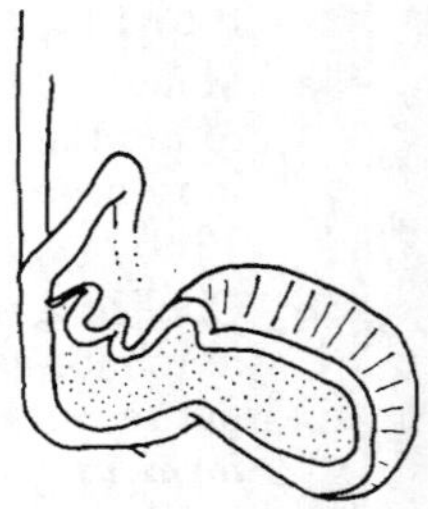

97

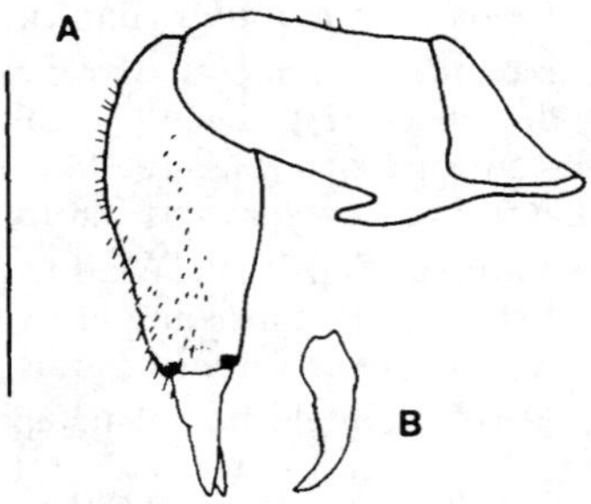

98

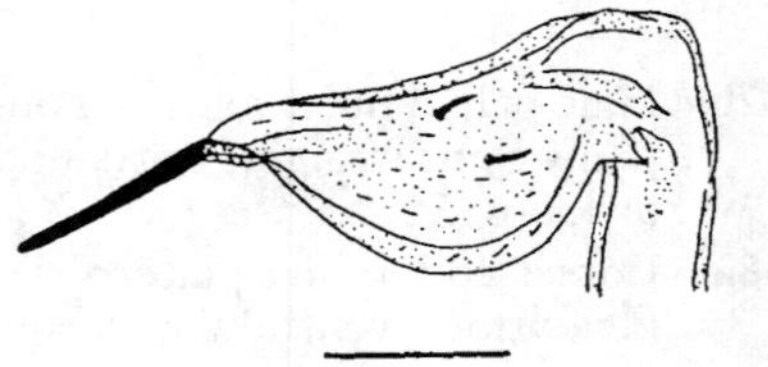

99

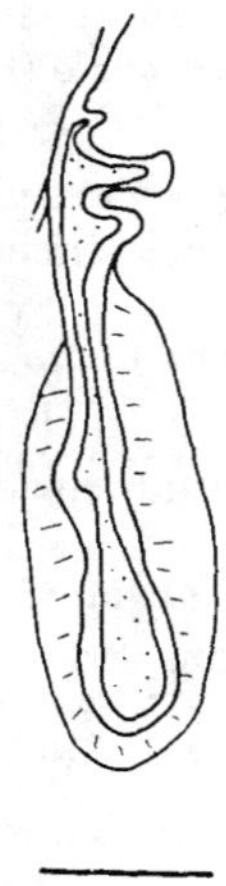

100

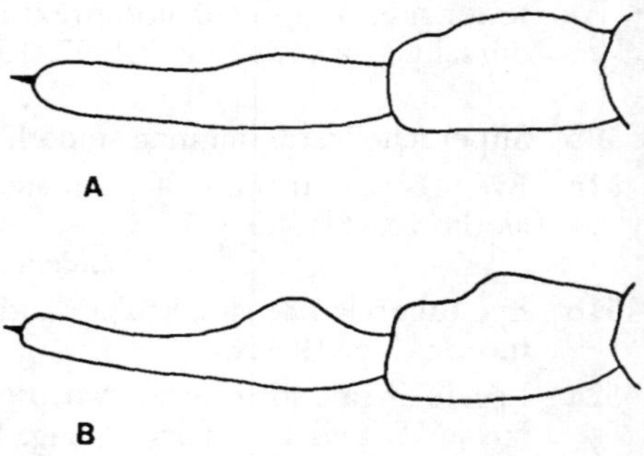

101

Fig. 19.95. *Leptobunus parvulus:* (*A*) male chelicera in lateral view; (*B*) moveable finger without apophysis. [From Cokendolpher (1985a).] **Fig. 19.96.** Distal end of *Leptobunus aureus* penis, lateral view. Scale line = 0.12 mm. [From Cokendolpher (1985a).] **Fig. 19.97.** Seminal receptacle of *Leptobunus parvulus.* [From Cokendolpher (1985a).] **Fig. 19.98.** *Leptobunus californicus:* (*A*) male chelicera; (*B*) moveable finger without apophysis. [From Cokendolpher (1985a).] **Fig. 19.99.** Distal end of *Leptobunus californicus* penis, lateral view. Scale line = 0.12 mm. [From Cokendolpher (1985a).] **Fig. 19.100.** Seminal receptacle of *Leptobunus californicus.* Scale line = 0.05 mm. [From Cokendolpher. (1985).] **Fig. 19.101.** Dorsal views of palpal tibia and tarsus of male: (*A*) *Leptobunus borealis;* (*B*) *L. californicus.* [From Cokendolpher (1985a).]

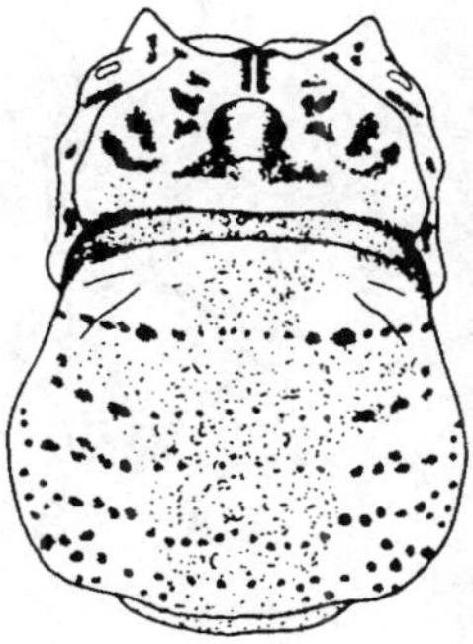

102 A

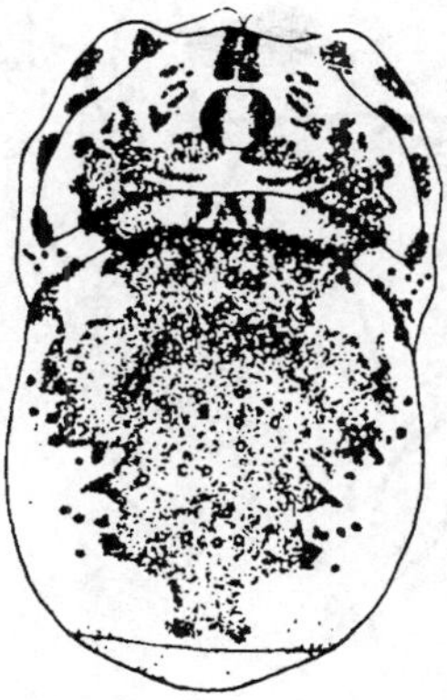

102 B

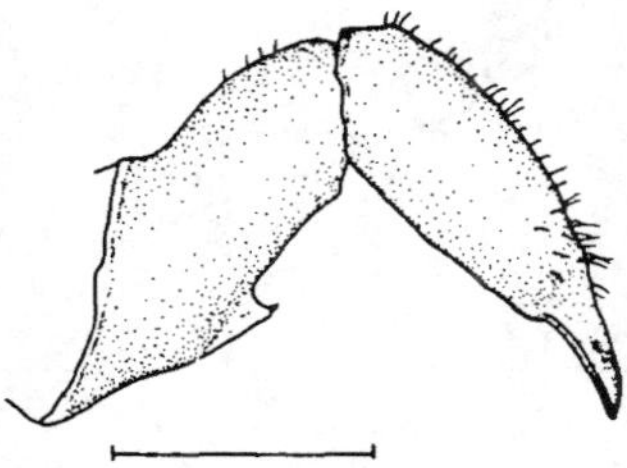

103

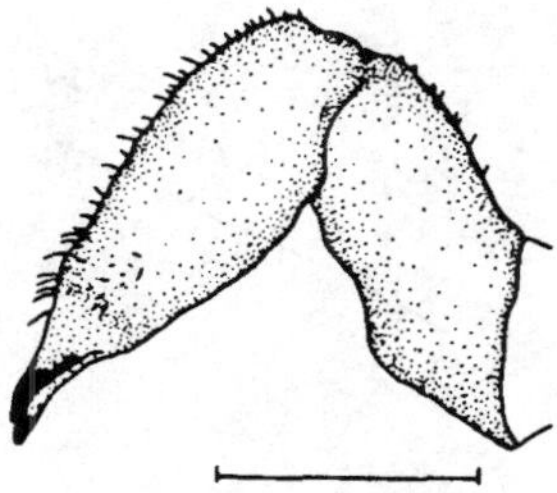

104

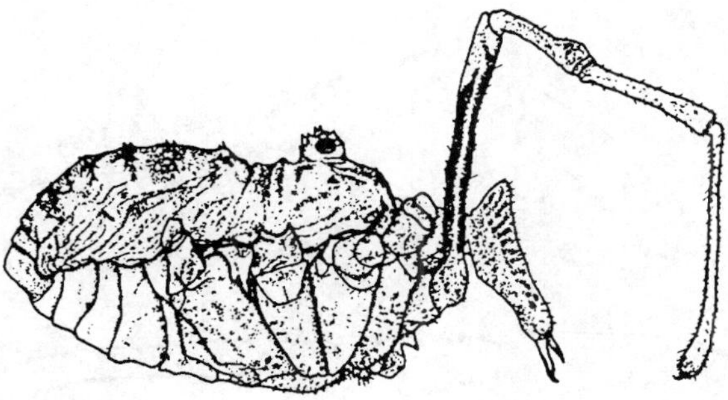

105

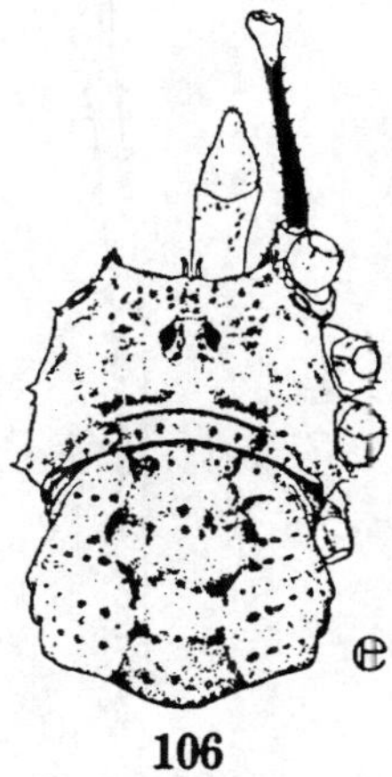

106

Fig. 19.102. Dorsal view of male body: (*A*) *Liopilio yukon;* (*B*) *Liopilio glaber.* [From Cokendolpher (1981b).] **Fig. 19.103.** Chelicera of *Mitopus morio,* lateral view. **Fig. 19.104.** Chelicera of *Opilio parientinus,* lateral view. **Fig. 19.105.** *Phalangium opilio* male with short-spurred chelicera and long palpus. **Fig. 19.106.** *Phalangium opilio* male, dorsal view. [From Bishop (1949).]

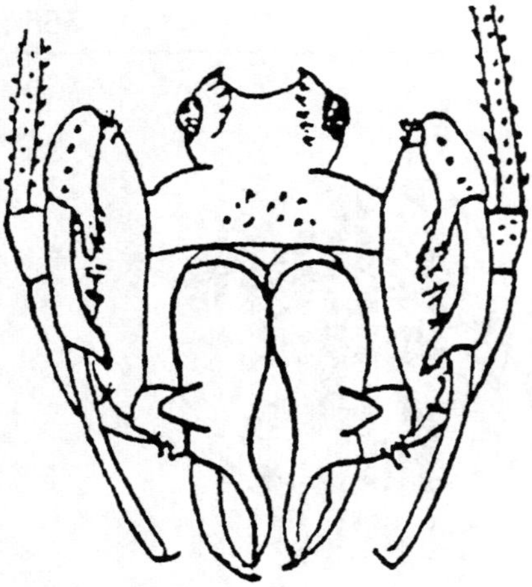

107

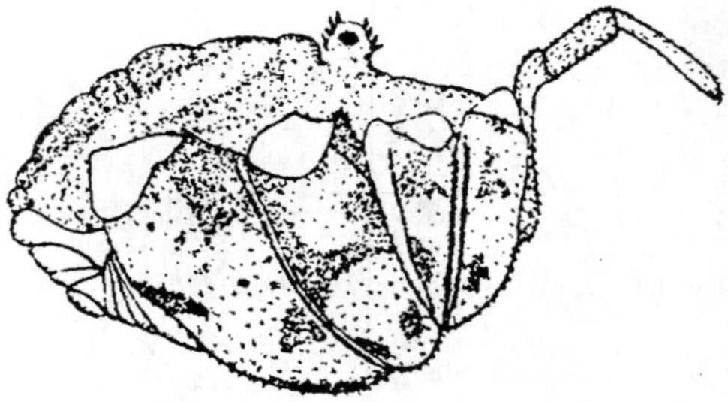

110

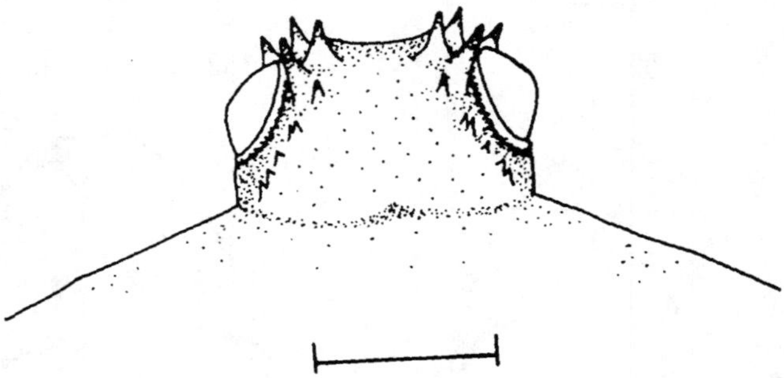

108

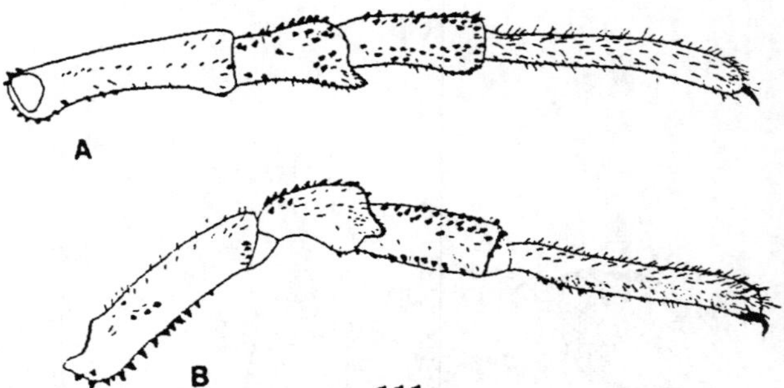

111

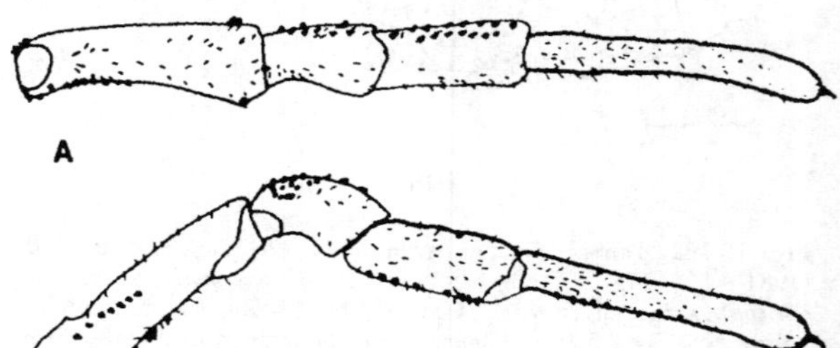

109

112

54a Males 55
54b Females 56
55a Femur II less than 6.5 mm in length . . . . *Trachyrhinus horneri* Cokendolpher, 1981
55b Femur II greater than 7.0 mm in length . . . . *Trachyrhinus rectipalpus* Cokendolpher, 1981
56a Femur II with two pseudoarticulary nodules (Fig. 19.109), less than 6.5 mm in length. . . . . *Trachyrhinus horneri* Cokendolpher, 1981
56b Femur II with one pseudoarticulary nodule, greater than 8.0 mm in length . . . 57
57a Femur II less than 10.5 mm in length. *Trachyrhinus rectipalpus* Cokendolpher, 1981
57b Femur II greater than 11.5 mm in length. *Trachyrhinus mesillensis* Cokendolpher, 1981
58a Patellar apophysis greater than 0.23 mm in length . . . . *Trachyrhinus dicropalpus* Cokendolpher, 1981
58b Patellar apophysis less than 0.18 mm in length. 59
59a Palpal tarsi with only ventral row of denticles, bare in females . . . . *Trachyrhinus marmoratus* Banks, 1894
59b Palpal tarsi with many small obtuse tubercles ventrally, reduced in females, as well as ventral row of denticles in males. *Trachyrhinus favosus* (Wood, 1868)
60a Female palpal patella with an apophysis; male palpal femur and tibia enlarged; palpal tarsus generally equal to or shorter than tibia (Fig. 19.113); **Sclerosomatidae:** ***Protolophus*** Banks, 1893[15] 74
60b Palpus normal, without apophyses in adults, tarsus longer than tibia; **Gagrellidae: Leiobuninae** 61
61a Preocular area with an elevated area or hump that is often covered with many short spines; femora and tibiae of male legs I and III generally enlarged in diameter 62
61b Preocular area not elevated, no more rugose than remainder of cephalothorax; male leg femora and tibiae normal. 64
62a Supracheliceral laminae with spines visible below anterior margin of carapace; ocular tubercle about as high as long . . . . *Togwoteeus*[16] **(=** ***Homolophus*****)** ***biceps*** (Thorell, 1877)
62b Supracheliceral laminae without spines; ocular tubercle only about one-half as high as long 63

[15] Modified from Goodnight and Goodnight (1942a).
[16] See Cokendolpher (1987).

Fig. 19.107. *Rilaena* (= *Platybunus*) *triangularis*, anterior aspect. [From Roewer (1923).] **Fig. 19.108.** Eye tubercle of *Opilio parientinus*. [From Edgar (1966).] **Fig. 19.109.** Femur II of female *Trachyrhinus horneri*. [From Cokendolpher (1981a).] **Fig. 19.110.** *Trachyrhinus marmoratus* males, lateral view. [From Cokendolpher (1981a).] **Fig. 19.111.** Male palpus of *Trachyrhinus favosus*: (*A*) dorsal view; (*B*) lateral view. [From Cokendolpher (1981a).] **Fig. 19.112.** Male palpus of *Trachyrhinus rectipalpus*: (*A*) dorsal view; (*B*) retrolateral view. [From Cokendolpher (1981a).]

63a Ocular tubercle smooth . . . . ***Eurybunus*** Banks, 1893[17]

63b Ocular tubercle with spines or tubercles **81**

64a Abdominal tergites 1–6, sometimes tergites 7 and 8 fused together in a hard scutum, body hard and granulate; femur I shorter than width of body; ***Eumesosoma*** Cokendolpher 1980[18]. **65**

64b At most, tergites 1–5 of abdomen fused in a hard scutum, but usually all freely distinguishable, body usually soft or subcoriaceous; femur I often longer than body width **71**

65a Denticles in lateral rows on coxae blunt and rounded **66**

65b Denticles in lateral rows on coxae reduced to either tripointed or single spines **68**

66a Males. **67**

66b Females . . . . ***Eumesosoma arnetti*** Cokendolpher, 1980
Eumesosoma roeweri (G & G, 1943)[19]

67a Alate portion of penis forked anteriorly (Fig. 19.114). . . . .
Eumesosoma arnetti Cokendolpher, 1980

67b Alate portion of penis convergent anteriorly (Fig. 19.115) . . . .
Eumesosoma roeweri (G & G, 1943)

68a Venter of abdomen with only lateral obtuse tubercles, genital operculum and coxae with tripointed spines . . . ***Eumesosoma ephippiatum*** (Roewer, 1923)

68b Venter with many obtuse tubercles, genital operculum and coxae with obtuse tubercles or simple spines **69**

69a Genital operculum and coxae with few or no obtuse tubercles; if present, not as large as tubercles on venter of abdomen **70**

69b Genital operculum and coxae with many obtuse tubercles (Fig. 19.115). . .
Eumesosoma ocalense Cokendolpher, 1980

70a Alate portion of penis forked anteriorly (Fig. 19.116), without median sheath . . . . ***Eumesosoma nigrum*** Say, 1821

70b Alate portion of penis convergent anteriorly (Fig. 19.117), median sheath present. . . . . ***Eumesosoma sayi*** Cokendolpher, 1980

71a Femur I longer than length of body or in some females, a little shorter than length of body, but longer than width of body **72**

71b Femur I shorter than length of body and, in females, shorter than body width (Fig. 19.118) ***Hadrobunus*** banks, 1900 . . . . **73**

72a Palpal claws smooth, or at most with two to three small teeth at base, coxae never with lateral rows of denticles . . ***Leuronychus pacificus*** (Banks, 1894)

72b Palpal claws denticulate (Fig. 19.119), some or all coxae with at least anterior row of denticles . . . . ***Leiobunum*** C. L. Koch, 1839 (**Section 3**)

73a Dorsum finely spinulate, legs of female poorly marked with brown at tips of segments, northeastern North America (Figs. 19.118, 19.91) . . . .
Hadrobunus maculosus (Wood, 1868)

73b Dorsum smooth or nearly so with small round pale spots on dorsum, southeastern United States. . . . . ***Hadrobunus grandis*** (Say, 1821)

---

[17] For species descriptions, see Banks (1893), Goodnight and Goodnight (1943), Schenkel (1951), and Cokendolpher (1980b).

[18] Key adapted from Cokendolpher (1980a).

[19] See Cokendolpher (1980a) to distinguish females of these two species.

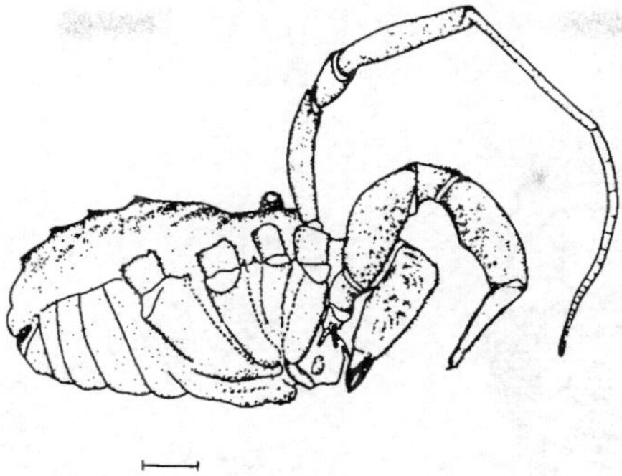

113

114

115

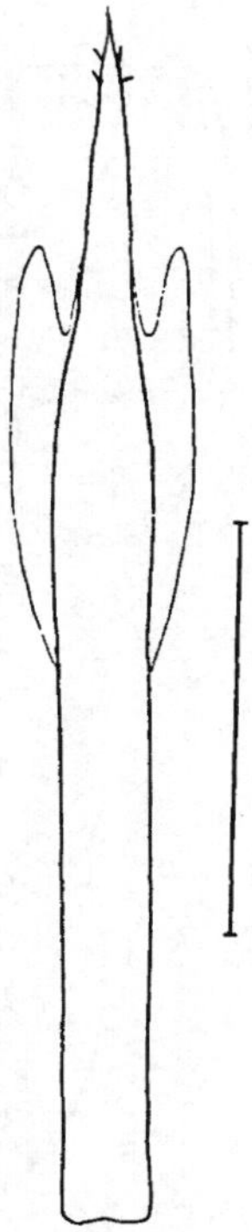

116

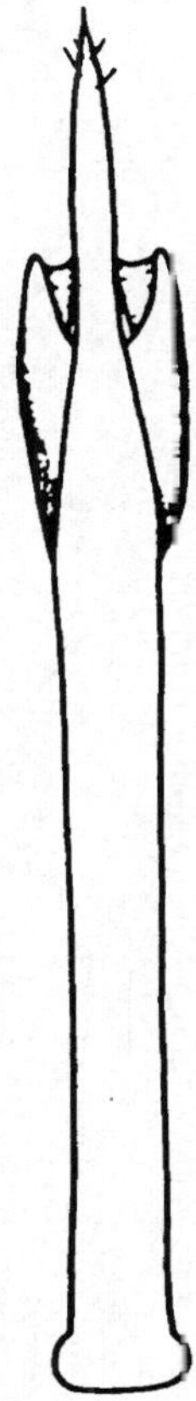

117

Fig. 19.113. *Protolophus singularis* male, lateral view. **Fig. 19.114.** Penis of *Eumesosoma arnetti*, dorsal view. [From Cokendolpher (1980a).] **Fig. 19.115.** Penis of *Eumesosoma ocalense*, dorsal view. [From Cokendolpher (1980a).] **Fig. 19.116.** Penis of *Eumesosoma nigrum*, dorsal view. Scale line = 2 mm. [From Cokendolpher (1980a).] **Fig. 19.117.** Penis of *Eumesosoma sayi*, dorsal view. [From Cokendolpher (1980a).]

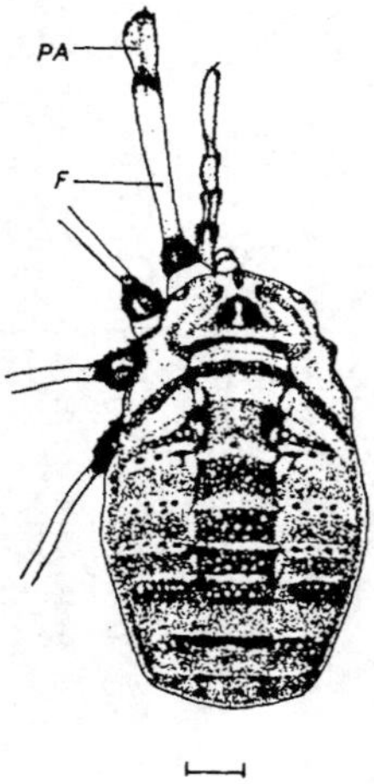

118

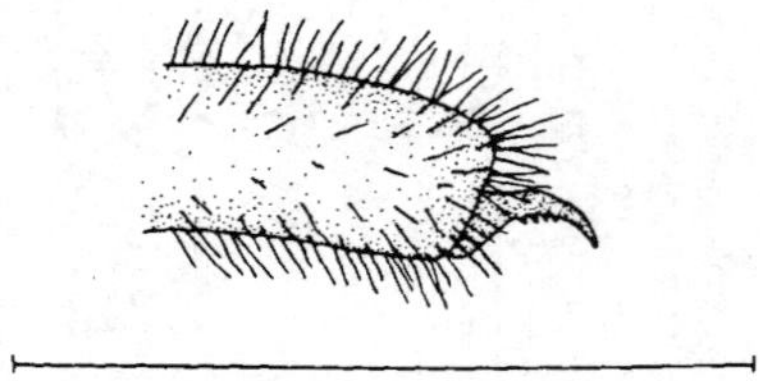

119

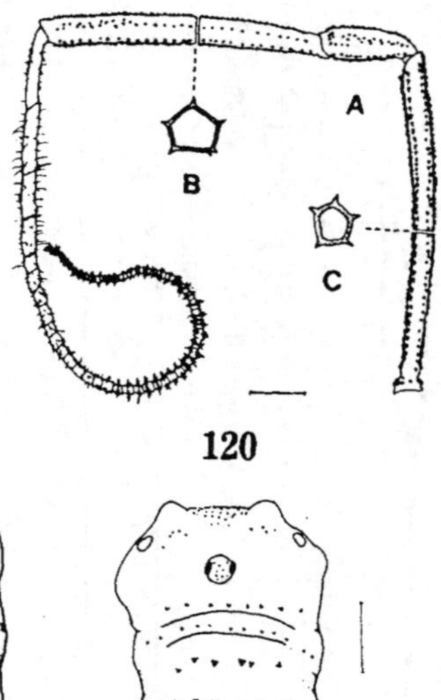

120

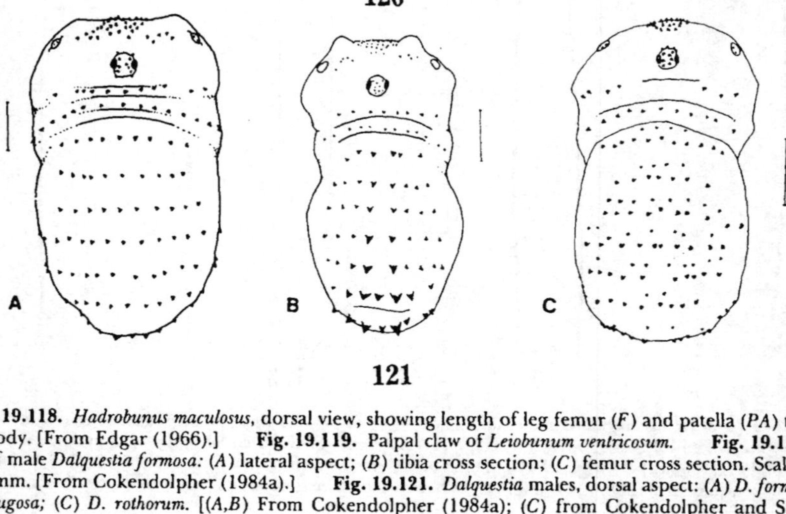

121

Fig. 19.118. *Hadrobunus maculosus*, dorsal view, showing length of leg femur (*F*) and patella (*PA*) to width of body. [From Edgar (1966).] **Fig. 19.119.** Palpal claw of *Leiobunum ventricosum*. **Fig. 19.120.** Leg II of male *Dalquestia formosa:* (*A*) lateral aspect; (*B*) tibia cross section; (*C*) femur cross section. Scale line = 0.5 mm. [From Cokendolpher (1984a).] **Fig. 19.121.** *Dalquestia* males, dorsal aspect: (*A*) *D. formosa;* (*B*) *D. rugosa;* (*C*) *D. rothorum.* [(*A,B*) From Cokendolpher (1984a); (*C*) from Cokendolpher and Stockwell (1986).]

74a Median abdominal tubercles bearing spines (Fig. 19.113) **75**

74b Median abdominal tubercles without spines **77**

75a Femur of male palpus very heavy, 1.1 mm in width (Fig. 19.113). . . . . . .
Protolophus singularis Banks, 1893

75b Femur of male palpus normal, 0.7 mm or less in width **76**

76a Legs with a definite annulate appearance, color of animal brownish . . . .
Protolophus dixiensis Chamberlin, 1925

76b Legs mottled, but not annulate, color of animal grayish. . . . .
Protolophus differens G & G, 1942

77a Median abdominal tubercles scarcely elevated above the surface of the dorsum **78**

77b Median abdominal tubercles distinctly elevated above the surface of the dorsum **80**

78a Dorsum black, trochanters orange on top . . . .
Protolophus longipes Schenkel, 1951

78b Dorsum brown, trochanters brown **79**

79a Dorsum reddish brown, trochanters dark brown . . . .
Protolophus niger G & G, 1942

79b Dorsum yellowish brown, with darker brown median strips, trochanters light brown . . . . ***Protolophus rossi*** G & G, 1943

80a Femur, patella and tibia of legs dusky brown . . . .
Protolophus tuberculatus Banks, 1893

80b Femur, patella and tibia of legs brown or yellow . . . .
Protolophus cockerelli G & G, 1942

81a Abdominal tubercles present on dorsum; tibia II angular in cross section (Fig. 19.120) ***Dalquestia***[20] Cokendolpher, 1984 . . . . **83**

81b Abdomen is without conspicuous dorsal tubercles; tibia II round in cross section; ***Globipes*** Banks, 1893 . . . . **82**

82a Tibia II with three pseudosegments . . . . ***Globipes spinulatus*** Banks, 1893

82b Tibia II with four pseudosegments . . . . ***Globipes simplex*** Schenkel, 1951

83a Abdominal tubercles arranged in seven single transverse rows (Figs. 19.121A,B) **84**

83b Abdominal tubercles arranged in seven double transverse rows or 14 weakly defined rows (Fig. 19.121C); Arizona . . . .
Dalquestia rothorum Cokendolpher and Stockwell, 1986

84a Tibiae of all legs with distal white bands; abdominal tubercles approximately uniform in size (Fig. 19.121A); Texas . . ***Dalquestia formosa*** (Banks, 1910)

84b Tibiae of all legs uniformly brown or brown mottled with white; abdominal tubercles noticeably larger toward the posterior midline (Fig. 19.121B); southwestern California . . . . ***Dalquestia rugosa*** (Schenkel, 1951)

Section 3: Key to Species of Genus *Leiobunum* C. L. Koch, 1839

1a Males (male of ***L. ephippiatum*** is unknown) **2**

1b Females (female of ***L. crassipalpe*** is unknown) **28**

[20] Key adapted from Cokendolpher (1986).

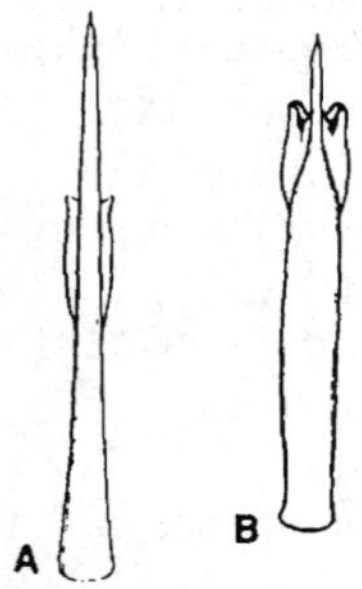

122

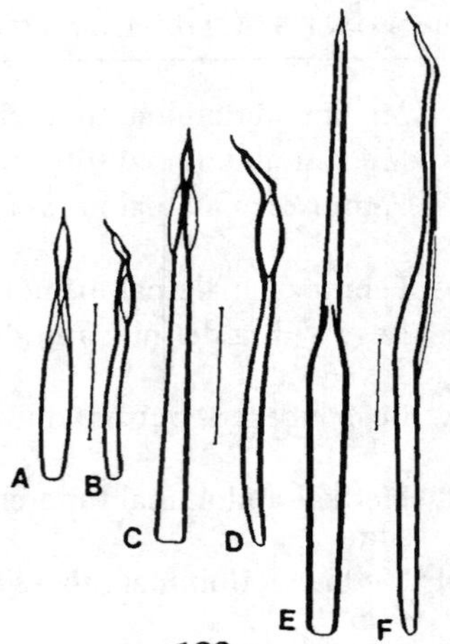

123

124

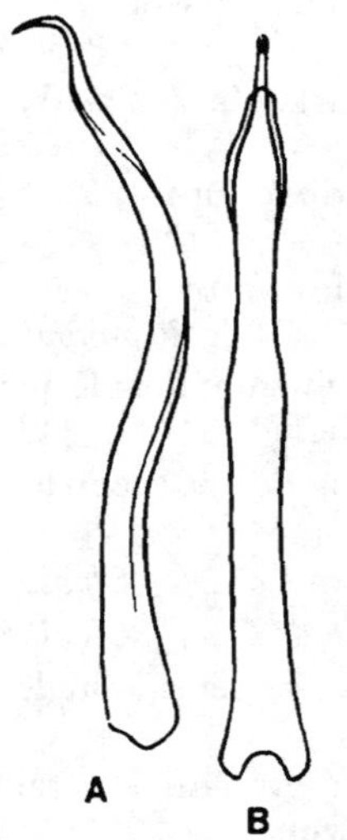

125

126

127

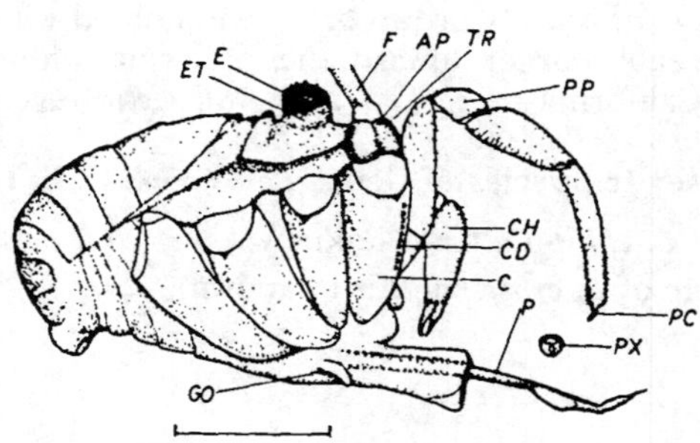

128

2a Penial shaft alate with a thin-walled sacculus or bulbus which may or may not open to the exterior at its distal end (Fig. 19.122–19.124). **3**

2b Penial shaft alate or not but with no lateral thin-walled accessory structure (Figs. 19.125, 19.126). **20**

3a Sacculus very prominent, projecting noticeably beyond the lateral edge of the penial shaft in dorsal view (Fig. 19.122) **8**

3b Sacculus usually inconspicuous, projecting only slightly or not at all beyond lateral edge of the penial shaft in dorsal view (Figs. 19.123, 19.124, 19.127) **4**

4a White banding on false articulations of metatarsus and on proximal portion of tarsi. **5**

4b Tarsi and metatarsi not banded as above **6**

5a From western North America, penis as in Fig. 19.124. . . . . *L. paessleri* Roewer, 1910

5b From eastern North America, penis as in Fig. 19.127 . . . . *L. elegans* Weed, 1889 (= *L. bicolor* (Wood, 1868))[21]

6a Penial bulb double, located laterally or ventrolaterally along the distal end of the shaft; may or may not protrude beyond the lateral margin of the shaft (Fig. 123A–D); penis length less than 5.0 mm. **7**

6b Penial bulb single, located on ventral side of shaft, may lie close to the shaft wall and be rather inconspicuous; penial shaft very long and slender and laterally flattened along approximately the distal one-half (Fig. 19.123E,F); penis length 6.0–7.0 mm . . . . *L. holtae* McGhee, 1977

7a Penial bulb rather large, conspicuous and extending slightly beyond the edge of the shaft when viewed from the dorsal or ventral side (Figs. 19.123C,D, 19.128); penis length 2.5–4.0 mm; genital operculum length 1.2–1.8 mm *L. politum* Weed, 1889

7b Penial bulb not as conspicuous, relatively narrow and almost flush with the lateral edge of the shaft when viewed from the dorsal or ventral side (Fig. 19.123A,B); penis length 1.5–2.5 mm; genital operculum length 0.9–1.2 mm. . . . . *L. bracchiolum* McGhee, 1977

8a Dorsum brown or black with conspicuous white to yellow spots near union of carapace with abdomen; trochanters brown, penis as in Fig. 19.122B . . . . *L. bimaculatum* Banks, 1893

8b Dorsum without such markings; trochanters brown or black **9**

9a Conspicuous banding on some or all leg femora or tibiae **10**

[21] See Cokendolpher (1984b).

Fig. 19.122. Penis of (*A*) *Leiobunum aurugineum,* (*B*) *Leiobunum bimaculatum,* (*C*) *Leiobunum cretatum,* (*D*) *Leiobunum depressum,* dorsal views. [From Davis (1934).] **Fig. 19.123.** Penis of (*A*) *Leiobunum bracchiolum,* dorsal view; (*B*) *Leiobunum bracchiolum,* lateral view; (*C*) *Leiobunum politum,* dorsal view; (*D*) *Leiobunum politum,* lateral view; (*E*) *Leiobunum holtae,* dorsal view; (*F*) *Leiobunum holtae,* lateral view. [From McGhee (1977).] **Fig. 19.124.** Penis of *Leiobunum paessleri,* dorsal view. [From Davis (1934).] **Fig. 19.125.** Penis of *Leiobunum calcar:* (*A*) lateral view; (*B*) dorsal view. [From Davis (1934).] **Fig. 19.126.** Penis of *Leiobunum uxorium,* dorsal view. [From Davis (1934).] **Fig. 19.127.** Penis of *Leiobunum elegans* (= *L. bicolor*), dorsal view. [From Davis (1934).] **Fig. 19.128.** Lateral view of *Leiobunum politum,* with penis exserted. *AP,* autotomy plane; *C,* coxa; *CD,* coxal denticle; *CH,* chelicera; *E,* eye; *ET,* eye tubercle; *F,* femur; *GO,* genital operculum; *P,* penis; *PC,* palpa claw; *PP,* patella of palpus; *PX,* penis, cross section showing sclerotized shaft of nonalate type; *TR,* trochanter [From Edgar (1966).]

9b Legs relatively uniform in coloration **12**

10a Apex of tibia II only banded with white. . . . .
L. aldrichi (Weed, 1893) (= *L. longipes* Weed, 1890)[22]

10b Leg femora banded with white distally **11**

11a Dark bands distally on tibiae preceded by white bands; southwestern United States . . . . *L. townsendi* Weed, 1893

11b White bands on tibiae extend to distal margin; Alabama, Illinois, and Missouri . . . . *L. gordoni* G & G, 1945

12a Body of adult less than 4 mm long (Fig. 19.129) . . . .
L. cretatum Crosby & Bishop, 1924 (= *L. lineatum* Edgar, 1962)

12b Body of adult 4 mm long or longer **13**

13a Femur, patella, and tibia of palpus black; tarsus yellow; west coast . . . . .
L. exilipes (Wood, 1868)

13b Palpus brown or yellow. **14**

14a Trochanters of legs black contrasting with coxae and much darker than dorsum. **15**

14b Trochanters of legs brown or yellow, not contrasting with coxae and or no darker than dorsum **16**

15a Legs light brown, dorsum brown contrasting with yellowish-white venter; penis length 2.9 mm (Fig. 19.130) . . . . *L. verrucosum* (Wood, 1868)

15b Legs uniform dark brown to black, dorsum orange-yellow, only slightly darker than venter; penis length 3.7–4.1 mm . . . *L. nigripes* Weed, 1887

16a Dorsal surface of abdomen strongly arched as seen from side, covered thickly with sturdy retrorse tubercles . . . *L. aurugineum* Crosby and Bishop, 1924

16b Dorsal surface of abdomen nearly flat, very few or no tubercles present **17**

17a Dorsal surface of penis arched to form a medial longitudinal keel (Fig. 19.122D); southwestern United States. . . . . . . *L. depressum* Davis, 1934

17b Dorsal surface of shaft of penis without medial longitudinal keel **18**

18a Dorsum brown or golden contrasting with yellowish-white venter. . . . . .
L. peninsulare Davis, 1934

18b Dorsum concolorous with or slightly darker than venter **19**

19a Abdomen long and pointed behind (Fig. 19.131A) . . . .
L. ventricosum (Wood, 1868)

19b Abdomen truncate or bluntly rounded behind . . . . *L. flavum* Davis, 1934

20a Femur of palpus with a distinct ventrolateral spur or raised area with large black tubercles (Figs. 19.132, 19.133) . . . . *L. calcar* (Wood, 1868)

20b Femur of palpus without spur or area with black tubercles **21**

21a Femur, patella and tibia of palpus thickened (Fig. 19.134) . . . .
L. crassipalpe Banks, 1900

21b Not as above **22**

22a Femur of palpus very long, curved, extending above surface of carapace a distance equal to three times height of eye tubercle (Figs. 19.135, 19.136E) **23**

---

[22] See Cokendolpher (1984b).

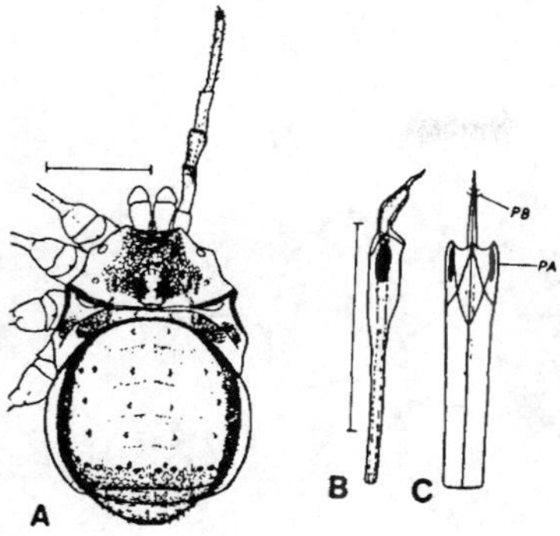

129

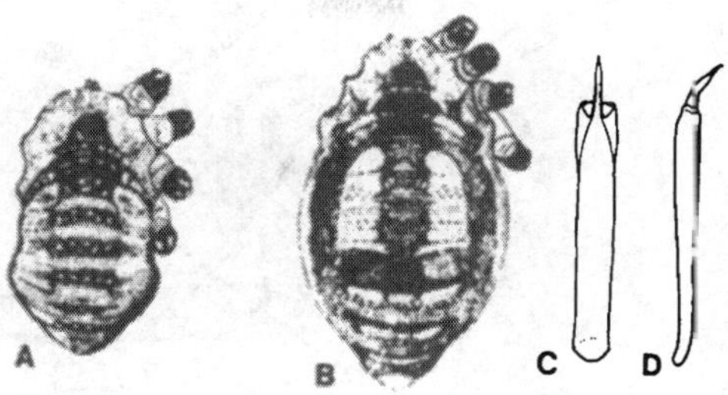

130

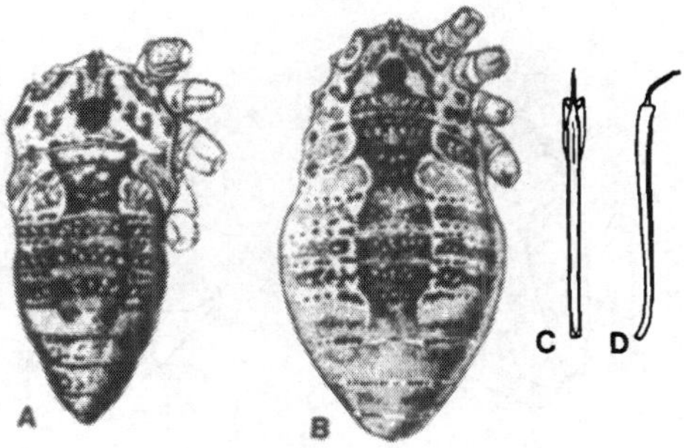

131

Fig. 19.129. *Leiobunum cretatum* (= *lineatum*): (*A*) dorsal view of body; (*B*) lateral view; (*C*) dorsal view of penis. *PA*, alate portion of penis; *PB*, penial bristle. [From Edgar (1962).] **Fig. 19.130.** *Leiobunum verrucosum:* (*A*) body of male, dorsal view; (*B*) body of female, dorsal view; (*C*) penis, dorsal view; (*D*) penis, lateral view. [From Bishop (1949).] **Fig. 19.131.** *Leiobunum ventricosum:* (*A*) body of male, dorsal view; (*B*) body of female, dorsal view; (*C*) penis, dorsal view; (*D*) penis, lateral view. [From Bishop (1949).]

22b Femur of palpus extending above surface of carapace a distance equal to less than three times the height of eye tubercle (Fig. 19.137). **24**

23a Abdomen long and pointed behind; central figure distinct black stripe . . . ***L. vittatum*** (Say, 1821)

23b Abdomen bluntly rounded behind, dorsum yellowish, central marking obsolete (Fig. 19.136A) . . . . ***L. speciosum*** Banks, 1900

24a Legs with white annulations; Oklahoma. . . . . . . ***L. relictum*** Davis, 1934

24b Legs with no white annulations, often with brown or black annulations **25**

25a Trochanters black contrasting strongly with coxae **26**

25b Trochanters yellow to golden concolorous or nearly concolorous with coxae ***L. flavum*** Davis, 1934

26a Femur banded with black or dark brown distally . . . . ***L. formosum*** (Wood, 1868)

26b Femur dark brown or black but not banded **27**

27a Distal end of femur and all of patella and tibia of palpus black (Fig. 19.138A–C). . . . . ***L. nigropalpi*** (Wood, 1868)

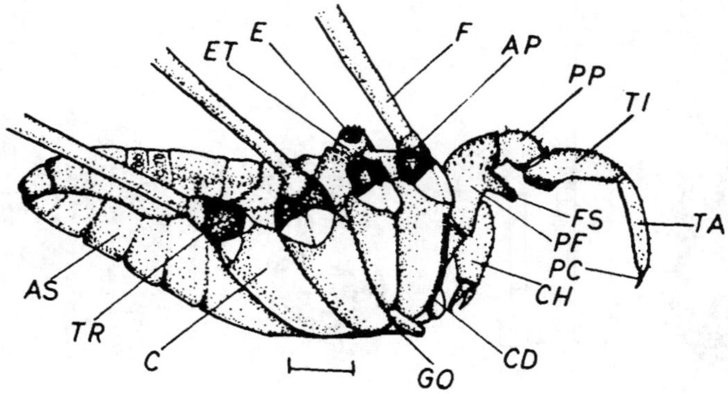

132

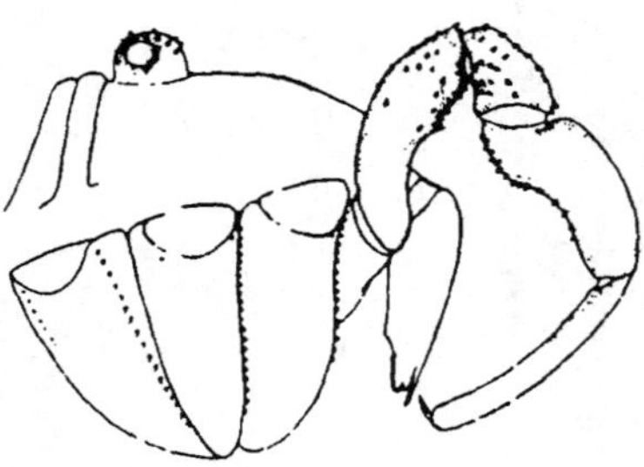

133

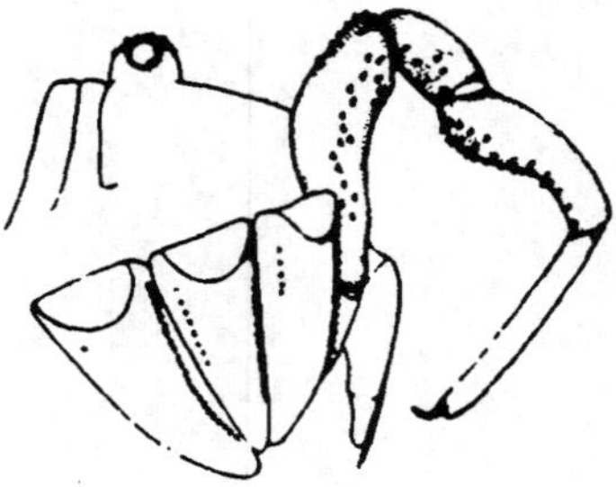

134

Fig. 19.132. Lateral view of *Leiobunum calcar,* male. *AS,* abdominal segment; *AP,* autotomy plane; *C,* coxa; *CD,* coxal denticle; *CH,* chelicera; *E,* eye; *ET,* eye tubercle; *F,* femur; *FS,* spur of femur; *GO,* genital operculum; *PC,* palpal claw; *PF,* femur of palpus; *PP,* patella of palpus; *TA,* tarsus of palpus, *TI,* tibia of palpus; *TR,* trochanter. [From Edgar (1966).] **Fig. 19.133.** Cephalothorax and palpus of *Leiobunum calcar* male showing variation in form of palpal femur, lateral view. [From Davis (1934).] **Fig. 19.134.** Cephalothorax and palpus of *Leiobunum crassipalpe,* lateral view. [From Davis (1934).]

27b Palpus golden yellow (Fig. 19.137) . . . . **L. uxorium** Crosby and Bishop, 1924

28a Legs with white annulations (sometimes with only a faint band distally on tibia II, especially in preserved and very mature specimens) **29**

28b Legs without white annulations on femora and tibiae, often with brown or black annulations **39**

29a All femora and tibiae banded distally with white **30**

29b Banding on legs not as above **32**

30a Leg tibiae with dark distal bands preceded by white bands. **31**

30b Leg tibiae with only distal white band; Alabama, Illinois, and Missouri . . . **L. gordoni** G & G, 1945

31a Ocular tubercle smooth; southwestern United States . . . . **L. townsendi** Weed, 1893

31b Ocular tubercle with well-developed rows of tubercles; Oklahoma. . . . . . **L. relictum** Davis, 1934

32a Trochanters of legs black. **33**

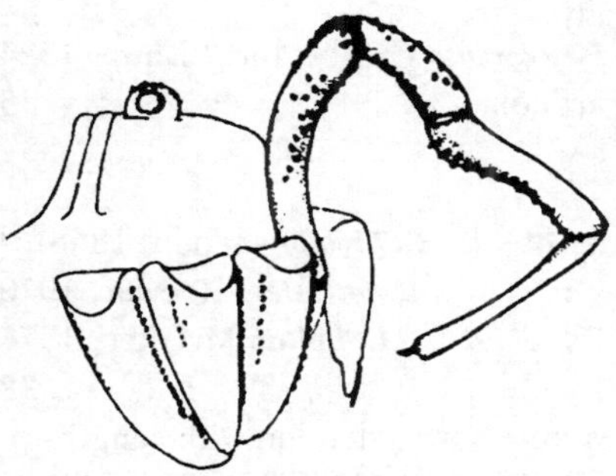

135

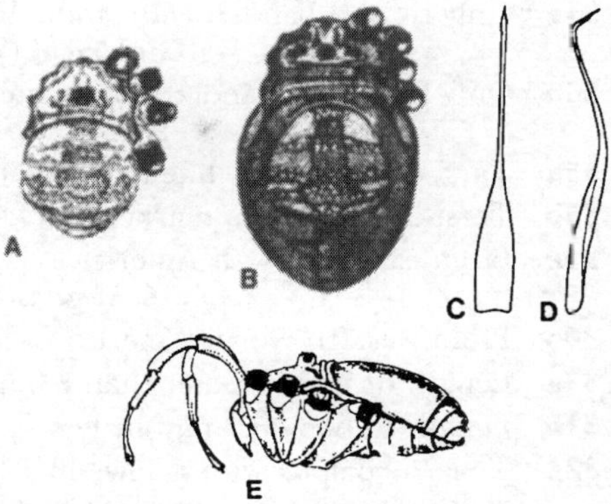

136

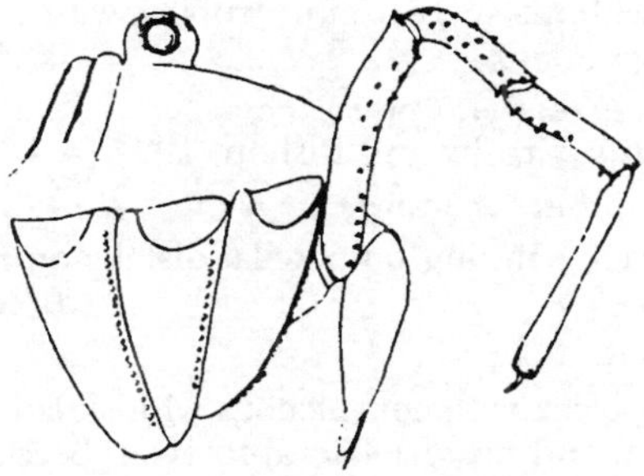

137

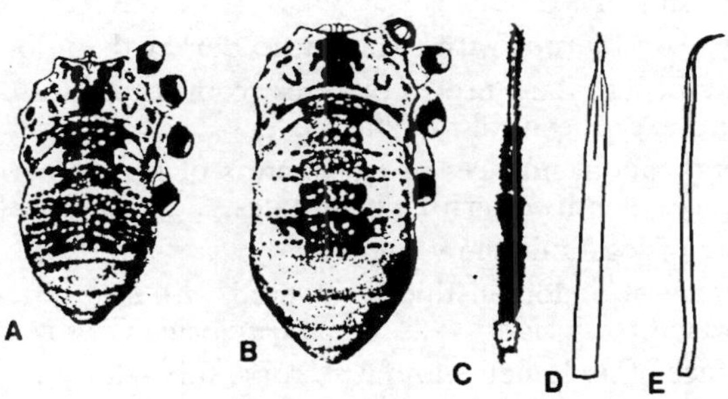

138

Fig. 19.135. Cephalothorax and palpus of *Leiobunum vittatum,* lateral view. [From Davis (1934).] **Fig. 19.136.** *Leiobunum speciosum:* (*A*) body of male, dorsal view; (*B*) body of female, dorsal view; (*C*) penis, dorsal view; (*D*) penis, lateral view; (*E*) body of male, lateral view. [From Bishop (1949).] **Fig. 19.137.** Cephalothorax and palpus of *Leiobunum uxorium,* lateral view. [From Davis (1934).] **Fig. 19.138.** *Leiobunum nigropalpi:* (*A*) body of male, dorsal view; (*B*) body of female, dorsal view; (*C*) palpus, dorsal view; (*D*) penis, dorsal view; (*E*) penis, lateral view. [From Bishop (1949).]

32b Trochanters of legs yellow **35**
33a Palpus marked with black; west coast . . . . ***L. exilipes*** (Wood, 1868)
33b Palpus not marked with black **34**
34a Only tibia II banded with white distally . . . .
L. aldrichi (Weed, 1893) (= ***L. longipes*** Weed, 1890)[22]
34b Only femur IV banded with white distally. . . . .
L. uxorium Crosby and Bishop, 1924
35a Tarsi of legs with white annulations at articulations **36**
35b Tarsi of legs not so marked **37**
36a From eastern North America . . . .
L. elegans Weed, 1889 [= ***L. bicolor*** (Wood, 1868)][21]
36b From western North America . . . . ***L. paessleri*** Roewer, 1910
37a Length of body greater than 8.0 mm. . . . . ***L. holtae*** McGhee, 1977
37b Length of body 8.0 mm or less. **38**
38a Tibia of palpus with a few small spines anteriorly, slender with length-to-width ratio of approximately 3.5 : 1 (maximum length vs. diameter at middle of segment). . . . . ***L. bracchiolum*** McGhee, 1977
38b Tibia of palpus without spines, more robust with length-to-width ratio of approximately 3.2 : 1. . . . . ***L. politum*** Weed, 1889
39a Length of adult less than 4.5 mm . . . .
L. cretatum Crosby and Bishop, 1924 (= ***L. lineatum*** Edgar, 1962)
39b Length of adult 4.5 mm or more. **40**
40a Dark brown or black banding on patella, distally on femur and tibia . . . .
L. formosum (Wood, 1868)
40b Legs not so banded. **41**
41a Dorsum brown or black with conspicuous whitish lateral markings especially pronounced lateral and posteriolateral to eye tubercle . . . .
L. bimaculatum Banks, 1893
41b Dorsum not so marked. **42**
42a Trochanters concolorous with coxae or no darker than dorsum **43**
42b Trochanters of legs dark brown or black, contrasting with coxae or darker than dorsum except central marking. **48**
43a Cephalic and caudal surfaces of trochanters of legs somewhat darker than lateral surfaces; southwestern United States. . . . ***L. depressum*** Davis, 1934
43b Trochanters of legs uniformly colored **44**
44a Dorsal surface of abdomen thickly covered with sturdy retrorse tubercles, strongly arched from side . . . . . ***L. aurugineum*** Crosby and Bishop, 1924
44b Dorsal surface of abdomen without retrorse tubercles **45**
45a Venter contrasting sharply with entire dorsum or central figure of dorsum; found in eastern United States, Florida . . . . . ***L. peninsulare*** Davis, 1934
45b Venter only slightly lighter than dorsum. **46**
46a Body length 8.0 mm or greater **47**
46b Body length less than 8.0 mm, Oregon. . . . ***L. ephippiatum*** Roewer, 1910
47a Abdomen long and pointed behind (Fig. 19.131B) . . . .
L. ventricosum (Wood, 1868)
47b Abdomen rounded behind . . . . ***L. flavum*** Davis, 1934
48a Venter nearly concolorous with dorsum, femur of palpus extending above

surface of carapace a distance equal to twice the height of the eye tubercle (Fig. 19.136B). . . . . **L. speciosum** Banks, 1900

48b Venter distinctly lighter than dorsum in color, femur of palpus extending above surface of carapace a distance less than twice height of eye tubercle . **49**

49a Palpal femur with a ventrolateral swelling armed with several short black spines; tibia with a group of similar spines localized on the ventrolateral aspect . . . . **L. calcar** (Wood, 1868)

49b Palpal femur without the above swelling and spination **50**

50a Coxa, expanded proximal portion of femur, eye tubercle and distal palpal femur and tibia all of a similar dark brown to black coloration; blunt, anteriomedial projection on palpal patella (Fig. 19.138B) . . . . **L. nigropalpi** (Wood, 1868)

50b Coloration not of the foregoing combination. **51**

51a Dorsal central figure distinct, predominantly black, flanked by whitish to tan margins . . . . **L. vittatum** (Say, 1821)

51b Dorsal central figure distinct or indistinct but not predominantly black . **52**

52a Venter and coxae concolorous, femur I less than length of body . . . . . . **L. nigripes** Weed, 1887

52b Venter lighter than dorsum except in old, darkened specimens; femur I at least as long as body . . . . **L. verrucosum** (Wood, 1868)

References

Banks, N. 1892. A new genus of Phalangiidae. *Proc. Entomol. Soc. Wash.* 2(2):249–251. **1893a.** The Phalangida Mecostethi of the United States. *Trans. Am. Entomol. Soc. (Phila.)* 20:149–152. **1893.** The Phalanginae of the United States. *Can. Entomol.* 25(8):205–211. **1894a.** The Nemastomatidae and Trogulidae of the United States, I and II. *Psyche (Camb.)* 7:11–12, 51–52. **1894b.** Notes on Phalangidae. *J. N.Y. Entomol. Soc.* 2(4):145–146. **1894c.** Washington Phalangida, with description of a new southern *Liobunum. Can. Entomol.* 26:160–164. **1895.** Two California phalangids. *J. N.Y. Entomol. Soc.* 3(2):66. **1898.** Arachnida from the Malaspina Glacier, Alaska. *Entomol. News* 9(1):16. **1899.** Arachnida. Pp. 347–350 in W. H. Ashmead (ed.). *Reports upon the Insects, Spiders, Mites and Myriopods Collected by Dr. L. Stejneger and Mr. G. E. H. Barrett-Hamilton on the Commander Islands.* Report on Fur-Seal Investigations 1896–1897, Part IV, App. C. Government Printing Office, Washington, D.C. **1900a.** New genera and species of American phalangida. *J. N.Y. Entomol. Soc.* 8:199–201. **1900b.** Some Arachnida from Alabama. *Proc. Acad. Nat. Sci. Phila.* 52:529–543. **1901.** Synopsis of North-American invertebrates. XVI. The Phalangida. *Am. Nat.* 35:669–679. **1902.** A new phalangid from the Black Mountains, N.C. *J. N.Y. Entomol. Soc.* 10:142. **1905.** A new genus and species of Phalangida. *Entomol. News* 16:251–253. **1910.** A handsome species of Phalangidae. *Entomol. News* 21:156. **1911.** *The Phalangida of California. Pomona Coll. J. Entomol. Claremont* 3(1):412–421. **Bell, R. T. 1975.** A European harvestman in North America (Phalangida, Phalangiidae). *Entomol. News* 85:(5,6):154. **Bishop, S. C. 1949.** The Phalangida (Opiliones) of New York. *Proc. Rochester Acad. Sci.* 9(3):159–235. **Blum, M. S. and A. L. Edgar. 1971.** 4-Methyl-3-heptanone: identification and role in opilionid exocrine secretions. *Insect Biochem.* 1(2):181–188. **Bragg, P. D. and R. G. Holmberg. 1975.** *Platybunus triangularis* and *Paroligolophus agrestis:* two phalangids introduced to North America (Arachnida, Opiliones). *J. Arachnol.* 2(2):127. **Briggs, T. S. 1968.** Phalangids of the laniatorid genus *Sitalcina* (Phalangodidae : Opiliones). *Proc. Calif. Acad. Sci. Ser. 4* 36(1):1–32. **1969.** A new Holarctic family of laniatorid phalangids (Opiliones). *Pan-Pac. Entomol.*

45(1):35–50. **1971a.** Relict harvestmen from the Pacific northwest. *Pan-Pac. Entomol.* 47(3):165–178. **1971b.** The harvestmen of family Triaenonychidae in North America (Opiliones). *Occas. Pap. Calif. Acad. Sci.* 90:1–43. **1974a.** Troglobitic harvestmen recently discovered in North American lava tubes (Travuniidae, Erebomastridae, Trianonychidae, Opiliones). *J. Arachnol.* 1(3):205–214. **1974b.** Phalangodidae from caves in the Sierra Nevada (California) with a redescription of the type genus (Opiliones: Phalangodidae). *Occas. Pap. Calif. Acad. Sci.* (108):1–15. **Briggs, T. S. and K. Hom. 1966.** Five new species of Phalangodidae from California (Opiliones). *Pan-Pac. Entomol.* 42(4):262–269. **1967.** New Phalangodidae from the Sierra Nevada mountains (Opiliones). *Pan-Pac. Entomol.* 43(1):48–52. **Briggs, T. S. and D. Ubick. 1981.** Studies on cave harvestmen of the Sierra Nevada with descriptions of new species of *Banksula. Proc. Calif. Acad. Sci.* 42(11):315–322. **Cokendolpher, J. C. 1980a.** Replacement name for *Mesosoma* Weed, 1892, with a revision of the genus (Opiliones, Phalangiidae, Leiobuninae). *Occas. Pap. Tex. Tech. Univ.* 66:1–19. **1980b.** Comments on Opiliones described from western North America by Schenkel. *Entomol. News* 91(4):133–135. **1981a.** Revision of the genus *Trachyrhinus* Weed (Opiliones, Phalangioidea). *J. Arachnol.* 9(1):1–18. **1981b.** The harvestmen genus *Liopilio* Schenkel (Opiliones: Phalangiidae). *J. Arachnol.* 9(3):309–316. **1982.** Comments on some *Leiobunum* species in the U.S.A. (Opiliones: Palpatores, Leiobunidae). *J. Arachnol.* 10(1):89–90. **1984a.** A new genus of North American harvestmen (Arachnida: Opiliones: Palpatores). Pp. 27–43 in N. V. Horner (ed.). *Festschrift for Walter W. Dalquest.* Midwestern State University, Wichita Falls, Tex. 163 pp. **1984b.** Homonyms of American and European *Leiobunum* (Opiliones, Palpatores, Leiobuninae). *J. Arachnol.* 12(1):118,119. **1985.** Revision of the harvestman genus *Leptobunus* and dismantlement of the Leptobunidae (Arachnida: Opiliones: Palpatores). *J. N.Y. Entomol. Soc.* 92(4):371–402. **1987.** On the identity of the genus *Homolophus:* a senior synonym of *Euphalangium* (Opiliones: Phalangiidae). *Acta Arachnol.* 35. (In press). **Cokendolpher, J. C. and W. F. Rapp. 1985.** *Leiobunum lineatum:* a synonym of *Leiobunum cretatum* (Opiliones, Gagrellidae). *J. Arachnol.* 13(3):347–354. **Cokendolpher, J. C. and S. A. Stockwell. 1986.** A new species of harvestman from Arizona (Arachnida: Opiliones: Palpatores: *Dalquestia*), *Southwest. Nat.* 31(1):49–53. **Crosby, C. R. and S. C. Bishop. 1924.** Notes on the Opiliones of the southeastern United States with descriptions of new species. *J. Elisha Mitchell Sci. Soc.* 40:8–26. **Davis, N. W. 1934.** A revision of the genus *Leiobunum* (Opiliones) of the United States. *Am. Midl. Nat.* 15(6):662–705. **Edgar, A. L. 1963.** Proprioception in the legs of Phalangida. *Biol. Bull. (Woods Hole)* 124(3):262–267. **1971.** Studies on the biology and ecology of Michigan Phalangida (Opiliones). *Misc. Publ. Mus. Zool. Univ. Mich.* 144:1–64. **Edgar, A. L. and H. A. Yuan. 1968.** Daily locomotry activity in *Phalangium opilio* and seven species of *Leiobunum* (Arthropoda: Phalangida). *Bios* 39(4):167–176. **Eisner, T., A. F. Kluge, J. E. Carrel and J. Meinwald. 1971.** Defense of phalangid: liquid repellent administered by leg dabbing. *Science* 173:650–652. **Goodnight, C. J. and M. L. Goodnight. 1942a.** The genus *Protolophus* (Phalangida). *Am. Mus. Novit.* (1157):1–7. **1942b.** New American phalangida. *Am. Mus. Novit.* (1164):1–4. **1943.** New and little known phalangids from the United States. *Am. Midl. Nat.* 29(3):643–656. **1945.** Phalangida from the United States. *J. N.Y. Entomol. Soc.* 53:239–245. **1960.** Speciation among cave opilionids of the United States. *Am. Midl. Nat.* 64(1):34–38. **1967.** Opilionids from Texas caves (Opiliones, Phalangodidae). *Am. Mus. Novit.* (2301):1–8. **Gruber, J. 1970.** Die "Nemastoma" Arten nordamerikas (Ischyropsalididae, Opiliones, Arachnida). *Ann. Naturhist. Mus. Wien.* 74:129–144. **Holmberg, R. G. 1986.** *The scent glands of Opiliones: A review of their function.* Pp. 131–133 in Proc. 9th Internat. Cong. Arachnol., Panama, 1983, Smithsonian Instit. Press. **Juberthie, C. 1964.** Recherches sur la biologie des Opilions. *Ann. Speleol.* 19(1):1–245. **1976.** Chemical defence in soil opiliones. *Rev. Ecol. Biol. Sol* 13(1):155–160. **Klee, G. E. and J. W. Butcher. 1968.** Laboratory rearing of *Phalangium opilio* (Arachnida: Opiliones). *Mich. Entomol.* 1(8):275–278. **Lesage, L. 1972.** Premiere mention de *Nemastoma bimaculatum* (Fabricius) pour l'amerique du nord (Opiliones: Nemastomatidae). *Naturaliste Canadian* 104:485. **Martens, J. 1969.** Die Sekretdarbietung wahrend des Paarungsverhaltens von *Ischyropsalis* C. L. Koch (Opiliones). *Z. Tierpsychol.* 26:513–523. **McGhee, C. R. 1977.** The *politum* group (bulbate species) of *Leiobunum* (Arachnida: Phalangida: Phalangiidae) of North America. *J. Arachnol.* 3(3):151–163. **Muchmore, W. B. 1963.** Two European arachnids new to the United States. *Entomol. News* 74(8):208–210. **Phillipson, J. 1959.** The seasonal occurrence, life histories and fecundity of harvest-spiders (Phalangida, Arachnida) in the neighborhood of Durham

City. *Entomol. Mon. Mag. Ser. 4* 95:134–138. **1960a.** A contribution to the feeding biology of *Mitopus morio* (F) (Phalangida). *J. Anim. Ecol.* 29:35–43. **1960b.** The food consumption of different instars of *Mitopus morio* (F.) (Phalangida) under natural conditions. *J. Anim. Ecol.* 29:299–307. **1962.** Respirometry and the study of energy turnover in natural systems with particular reference to harvestspiders (Phalangiida). *Oikos* 13(2):311–322. **Roewer, C. F. 1926.** *Abh. Naturwiss. Ver. Bremen* XXVI:289. **1930.** Zwei nordamerikanische höhlenopilioniden. *Boll. Labor. Zool.* 23:10–12. **Schenkel, E. 1951.** Spinnentiere aus dem westlichen Nordamerika, gesammelt von Dr. Hans Schenkel-Rudin. *Verh. Naturforsch. Ges. Basel* 62:28–62. **Shear, W. A. 1975a.** The opilionid genera *Sabacon* and *Tomicomerus* in America (Opiliones, Troguloidea, Ischyropsalidae). *J. Arachnol.* 3(1):5–29. **1975b.** The opilionid family Caddidae in North America, with notes on species from other regions (Opiliones, Palpatores, Caddoidea). *J. Arachnol.* 2(1):65–88. **1980.** A review of the Cyphophthalmi of the United States and Mexico, with a proposed reclassification of the suborder (Arachnida, Opiliones). *Am. Mus. Novit.* (2705):1–34. **1986.** A cladistic analysis of the opilionid superfamily Ischyropsalidoidea, with description of the new family Ceratolasmatidae, the new genus *Acuclavella,* and four new species. *Am. Mus. Novit.* (2844):1–29. **Shear, W. A. and J. Gruber. 1983.** The opilionid subfamily Ortholasmatinae (Opiliones, Troguloidea, Nemastomatidae). *Am. Mus. Novit.* (2757):1–65. **Simon, E. 1879.** Descriptions d'Opiliones nouveaux. *Ann. Soc. Entomol. Belg.* 22:70–75. **Todd, V. 1949.** The habits and ecology of the British harvestmen (Arachnida, Opiliones), with special reference to those of the Oxford district. *J. Anim. Ecol.* 18(2):209–229. **Tsurusaki, N. 1986.** Parthenogenesis and geographic variations on sex ratio in two species of *Leiobunum* (Arachnida, Opiliones). *Zool. Sci. (Tokyo)* 3:517–532. **Weed, C. M. 1889.** A description catalogue of the Phalangiinae of Illinois. *Ill. State Lab. Nat. Hist. Bull.* 3:79–97. **1893.** The cinnamon harvest-spider and its variations. *Am. Nat.* 27(318):534–541. **1897.** *Life Histories of American Insects.* Macmillan Co., Ltd. 272 pp. **Wood, N. C. 1868.** On the Phalangeae of the United States of America. *Commun. Essex Inst.* 6(1):10–40.

20

Acarina: Mesostigmata (Gamasida)

G. W. KRANTZ

B. D. AINSCOUGH

The Mesostigmata, or Gamasida, is a large and cosmopolitan group of acarines that has successfully adapted to a remarkably wide range of habitats. Many are free-living predators in soil, in litter, or on plants where they prey on other small invertebrates. Others are internal or external parasites of vertebrates or invertebrates, while still others are fungivores or detritivores. A number of mesostigmatid mites have established phoretic associations with other animals, some of which may involve a high degree of specificity.

The edaphic Mesostigmata of North America comprise a rich and varied fauna, including representatives of at least 120 genera in approximately 30 families. Literature on this group is quite comprehensive for a number of families, but a treatise on Nearctic soil Mesostigmata such as those compiled for the Palaearctic fauna by Stammer et al. (1963), Karg (1971a), and Ghilyarov et al. (1977), is not presently available. Habitat and distributional records are sparse and frequently reflect highly selective or arbitrary collecting patterns rather than real distribution.

External Morphology

Edaphic mesostigmatids may range in length from 200 to over 2000 μm and usually possess a number of distinctive characteristics which readily identify them as members of the suborder. The gnathosoma (Figs. 20.1–20.3) is a prognathous tube with a sclerotized base (*basis capituli*) composed

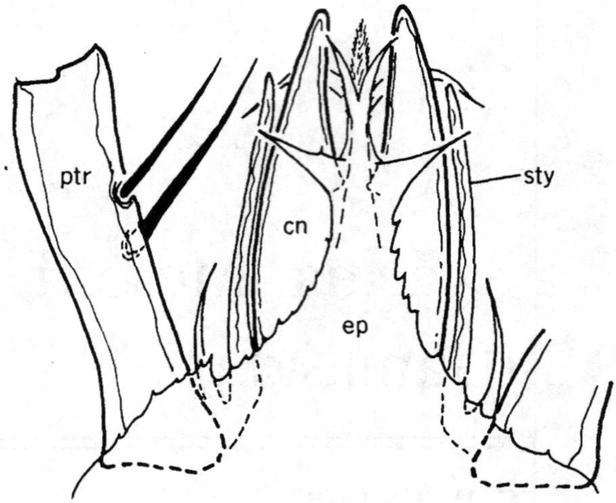

1

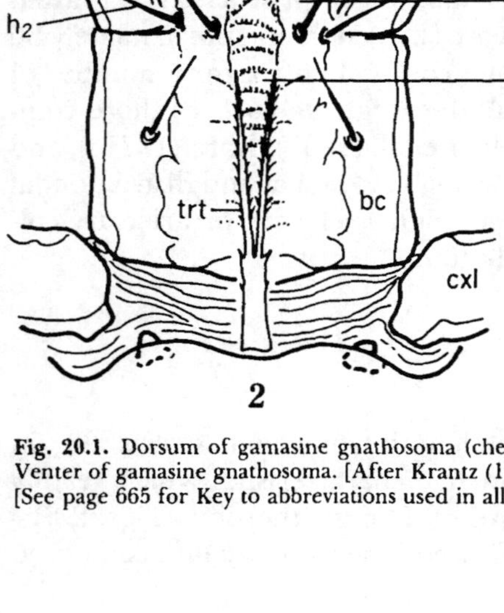

2

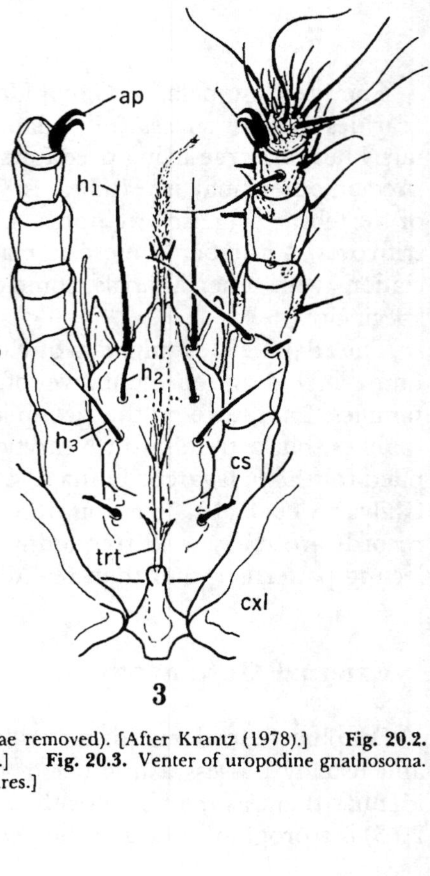

3

Fig. 20.1. Dorsum of gamasine gnathosoma (chelicerae removed). [After Krantz (1978).] **Fig. 20.2.** Venter of gamasine gnathosoma. [After Krantz (1978).] **Fig. 20.3.** Venter of uropodine gnathosoma. [See page 665 for Key to abbreviations used in all figures.]

largely of fused primitive palpcoxal elements. The anterior dorsal extremity of the gnathosomatic tube may be produced into a simple or ornamented *epistome* (tectum capituli), while the remnants of the palpcoxal endites form a substantial portion of the gnathosomatic floor. The ventral endites are separated medially by the *deutosternal groove,* considered by some authors to represent the remnants of the palpal sternites. The groove is ornamented with a series of transverse rows of denticles and terminates anteriorly in paired membranous processes (*internal malae*). The palpcoxal endites extend anteriorly to form the *hypostome,* which bears a pair of *corniculi* on its anterolateral angles. The corniculi usually are strongly sclerotized and often assume a hornlike appearance. In addition to the corniculi, the hypostome has three pairs of setae (two pairs in larvae), whose insertions may form a triangle (Fig. 20.2) or a straight line (Fig. 20.3) on each hypostomatic lobe. A pair of *capitular setae* is inserted behind the hypostome in all but the larval stase. A pair of *salivary styli* is often visible lateral to the corniculi, sometimes lying in grooves that extend the entire length of each corniculus. The styli represent the terminal elements of a salivary gland system that originates in the idiosoma.

The *palpi* (Fig. 20.3) are simple four- or five-segmented structures articulated from the anterolateral angles of the basis capituli. A two- or three-tined claw (*apotele*) arises from the paraxial surface of the palptarsal base. The tarsus itself terminates in a profusion of sensory setae. The *chelicerae* (Fig. 20.4) are of the chelate type with digits that are usually well developed and dentate. A single seta may be found dorsally at the base of the fixed digit and a *pilus dentilis* on the ventral face. *Arthrodial processes* may be

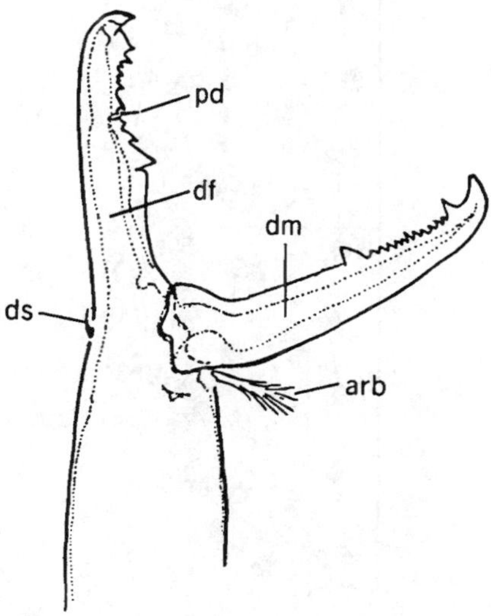

Fig. 20.4. Female chelicera.

present at the basal angle of the movable digit. The male often carries a free or partially fused sperm transfer process (*spermatodactyl*) on the movable digit (Fig. 20.170). Other gnathosomatic features are discussed by Hughes (1959) and Evans and Till (1965, 1979).

The idiosoma of edaphic mesostigmatids bears a series of sclerites (shields or plates) which generally illustrate progressive fusion during ontogeny. The three major dorsal shield components are the anterior *podonotal,* the median *mesonotal,* and the posterior *pygidial* (Fig. 20.12). The mesonotal shield may be fragmented (Fig. 20.106) or incorporated into the podonotal shield (Fig. 20.181). Amalgamation of mesonotal and pygidial elements creates an *opisthonotal shield* which is generally subequal to the anterior podonotal shield (Fig. 20.113). Fusion of podonotal, mesonotal, and pygidial elements results in the formation of a *holonotal shield* which may be laterally incised (Fig. 20.77) or entire (Fig. 20.83). A partial or entire *marginal shield* and/or independent marginal scutella are widespread among uropodine mesostigmatids (Figs. 20.177, 20.182, 20.185). Most gamasines lack marginal sclerites, having instead posterolateral zones of striated unsclerotized integument in which submarginal (UR) setae often may be found. The dorsal chaetotaxy (Fig. 20.5) includes *dorsal* (j, J), *median*

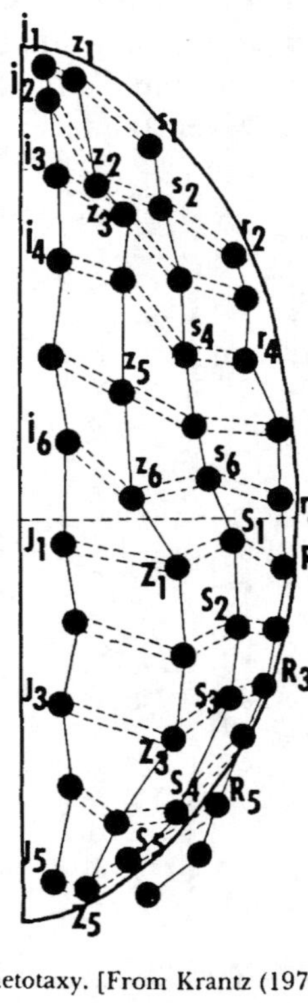

Fig. 20.5. Generalized dorsal chaetotaxy. [From Krantz (1978), after Lindquist and Evans (1965).]

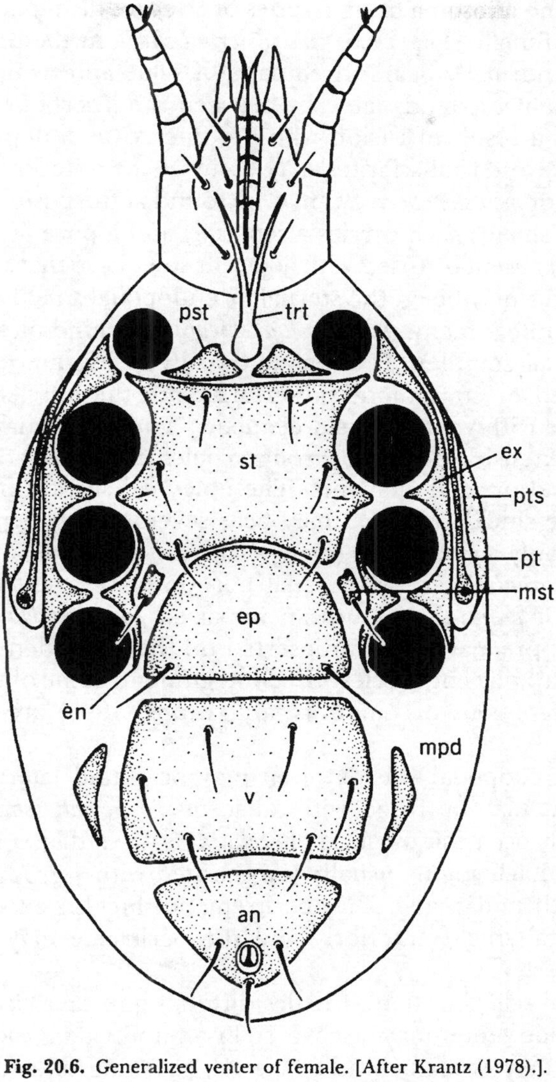

Fig. 20.6. Generalized venter of female. [After Krantz (1978).].

(z, Z), *lateral* (s, S), and *marginal* (r, R) setae arranged in ranks and files characteristic for both stase and taxon. Ontogenetic acquistion of setae occurs from the larval to the deutonymphal stase, with the adult chaetotaxy generally remaining unchanged from that of the deutonymph. The complete complement of dorsal setae (44 pairs) may or may not occur in the adult; in some cases accessory setae may be present (hypertrichy). The dorsum often is ornamented with a series of pores that mark the positions of subcuticular glands. Distinctive patterns of muscle insertion scars (sigillae) may also be seen (Athias-Henriot 1975).

Ventrally, the idiosoma bears a series of sclerites distinctive to both stase and sex. The female (Fig. 20.6) has a *sternal shield* in the propodosomatic region which normally bears three pairs of setae and two pairs of pores. The shield usually extends laterally between the insertions of coxae I–II and II–III as a result of fusion with the intercoxal endopodal elements. Independent presternal sclerites may occur at the anterior margin of the shield. A pair of *metasternal shields* may be found at the posterolateral angles of the sternal shield; each carries a seta (*st*4) and a pore. The metasternal shields may be absent or fused with adjacent shields, so that *st*4 and/or their associated pores may be on the sternal or endopodal shields, or free in the unsclerotized integument. An *epigynial shield* lies behind or within the sternal–metasternal complex. It opens outwardly at its anterior margin and may be hinged or continuous posteriorly with the unsclerotized ventral integument or with ventral shield elements. The shield may be expanded into a more extensive complex through coalescence with adjacent ventral, metapodal, and/or anal elements. The epigynial shield (or the epigynial portion of the shield complex) may bear several setae (Sejina), but more often carries only a single pair (many Gamasina) or none at all (most Uropodina). The *anal shield,* or the anal portion of the ventrianal complex, bears a pair of *paranal setae* and an unpaired *postanal seta.* A number of opisthogastric preanal setae are inserted on the expanded ventrianal elements or in the adjacent unsclerotized integument. Some of the opisthogastric setal designations of Lindquist and Evans (1965) are shown in Fig. 20.102.

Sclerotized exopodal shields often may be found lateral to the coxal insertions and may be fused with adjacent *peritrematic shields.* A pair of *stigmata* usually occurs at the level of coxae III–IV and lateral to the exopodal elements. Each stigma usually is associated with a grooved *peritreme* of variable length and shape. The peritrematic shields generally are fused with the dorsal shield anteriorly and the peritreme may extend to the idiosomatic vertex.

The ventral sclerotization of males differs significantly from that of females. The male venter may be covered by an undivided (*holoventral*) shield or a division may occur behind coxae IV, resulting in a separate anal or ventrianal shield. The chaetotaxy of the male venter generally parallels that of the female. The male genital opening may be located at the anterior margin of the sternal region (most Gamasina; Fig. 20.7) or in the metapodosomatic region.

Both males and females usually have a *tritosternum* inserted between, or anterior to, the insertions of coxae I. The tritosternal base carries one or more variously ornamented laciniae which extend anteriorly to lie along the deutosternal groove of the gnathosoma.

The legs of edaphic Mesostigmata are seven-segmented and possess a distinctive chaetotaxy that is useful not only in identification but also as an indicator of systematic relationships within the suborder (Evans 1963a,

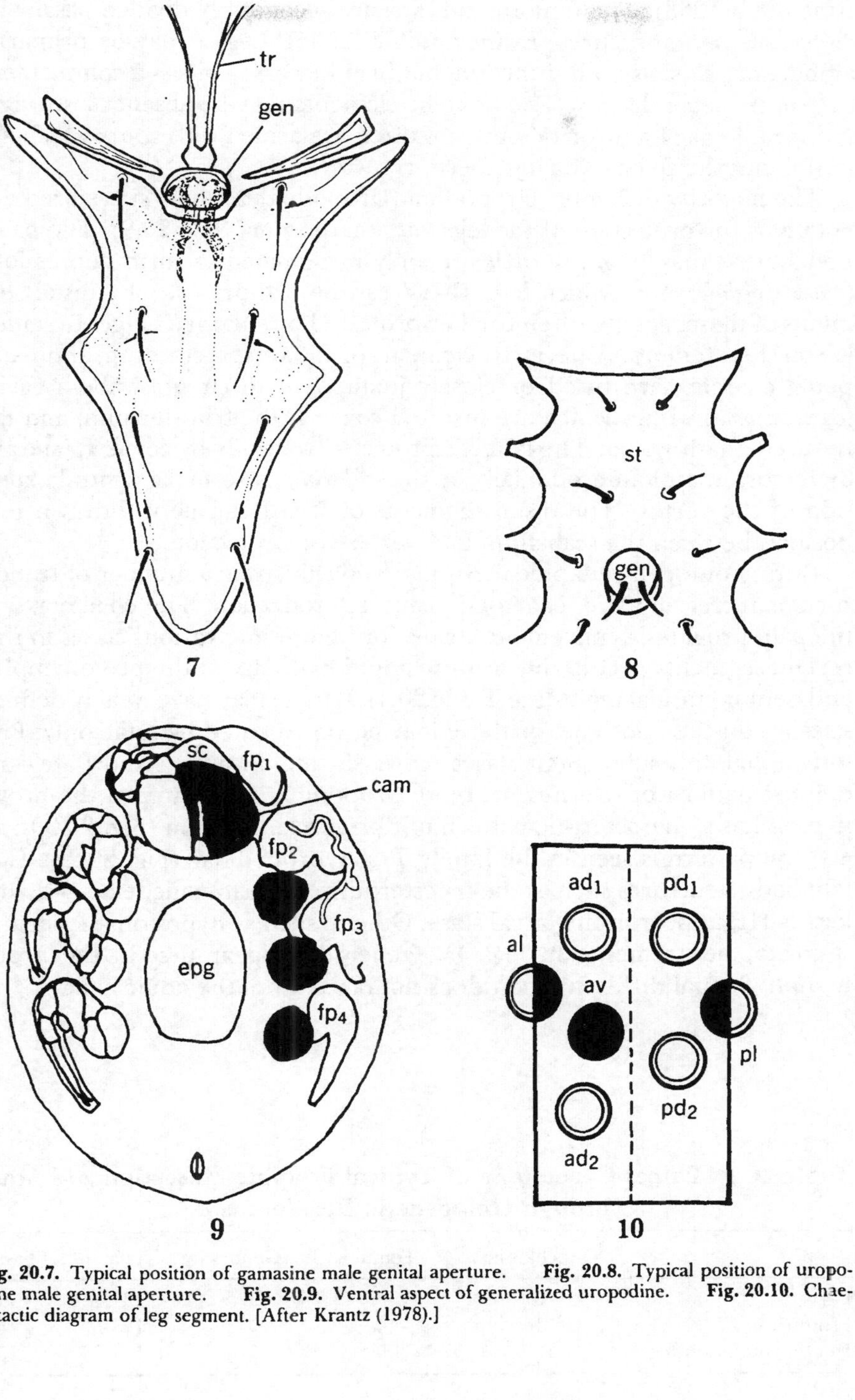

Fig. 20.7. Typical position of gamasine male genital aperture. **Fig. 20.8.** Typical position of uropodine male genital aperture. **Fig. 20.9.** Ventral aspect of generalized uropodine. **Fig. 20.10.** Chaetotactic diagram of leg segment. [After Krantz (1978).]

Ainscough 1963). The femora and tarsi are secondarily divided basally by circumsegmental fissures (Evans and Till 1965). Legs I may be primarily ambulatory or sensory in function, but in either case possess a complement of sensory setae distally. The apotelic elements may be absent or strongly reduced. Legs II and/or IV of males often are armed with spurs which aid in holding the female during sperm transfer.

The majority of higher Uropodina (Uropodoidea) show specialized adaptations for protection of the legs and gnathosoma (Fig. 20.9). The parapodal areas may be more or less deeply invaginated to form depressions (*foveae pedales*) into which legs II–IV can be withdrawn. The distal segments of these legs are often further protected by enlarged ridges or lamellae on their femora. Anteriorly, elements of the vertex, sternum, and parapodal complex are fused or closely juxtaposed to circumscribe a cavity (*camerostome*) within which are inserted coxae I, the tritosternum, and the reduced gnathosoma. The latter is protected ventrally by coxae I, laterally by femora I, and anterodorsally by the *scabellum,* a scalelike ventral extension of the vertex. The distal segments of legs I can be withdrawn to a position between the scabellum and vertex for protection.

During ontogeny edaphic mesostigmatids illustrate a number of trends, such as increasing size, chaetotaxy, and sclerotization. The chaetotaxy of the palp provides a convenient means for identifying various stases in that certain segments predictably acquire additional setae at the protonymphal and deutonymphal molts (see Table 20.1). Larvae may have weakly defined sclerites on the idiosoma, or these may be represented by setae only. Protonymphal and subsequent stases, however, almost always show a clearly defined pattern of sclerites. In most Uropodina, for example, the protonymph has a characteristic pattern of dorsal sclerotization (Fig. 20.30); an atypical pattern is seen in the family Trachyuropodidae (Fig. 20.29). Certain body structures such as the tritosternum, hypostomatic setae 1–2, and legs I–III, appear in the larval stase. Others, such as hypostomatic setae 3, stigmata, peritremes, and legs IV, generally appear first in the protonymph. Sexual differentiation does not occur until the adult molt.

Table 20.1. Palpal Chaetotaxy of Typical Edaphic Mesostigmatid Mites Through Ontogenetic Development

	Trochanter	Femur	Genu	Tibia	Tarsus
Larva	0	4	5	12	11
Protonymph	1	4	5	12	15
Deutonymph, adult	2	5	6	14	15

Internal Morphology

The digestive system of edaphic Mesostigmata is typical of a generalized predatory arthropod (Michael 1892, Hughes 1959, Ainscough 1960, Coons 1978). The mouth opens into a triradiate pharynx which leads to an esophagus. The esophagus passes through the dorsal and ventral elements of the central nervous system and joins the ventriculus via a cardiac valve. The ventriculus has a series of projecting gastric ceca (as many as four pairs) which provide additional digestive surface. The ventriculus opens posteriorly into a hindgut consisting of an anterior rectal tube ("intestine") and a posterior rectal bulb, which discharges via the anus. The Malpighian tubules arise from the rectal bulb and extend anteriorly into the body cavity. The tubules collect and concentrate nitrogenous wastes for elimination via the hindgut. In phoretic deutonymphs of higher Uropodina a median unpaired dorsal gland releases secretions into the hindgut which are discharged to form the anal pedicel (Faasch 1967).

The male reproductive system consists of fused or paired testes and vasa deferentia, one or more accessory glands, and an ejaculatory duct (Krantz 1978). An extrusible ejaculatory organ has been observed in some families (Lee 1974, Krantz and Wernz 1979) but an aedeagus per se is absent. Whereas sperm transfer in the Uropodina appears to be accomplished by direct genital contact, transfer in many Gamasina is indirect. Here, the sperm is transferred by the male cheliceral spermatodactyl to female extragenital pores (Athias-Henriot 1968, Lee 1970). The female reproductive system generally comprises an unpaired ovary, paired oviducts, a short uterus, and a vagina.

Other internal systems seen in edaphic Mesostigmata include a typical parasitiform central nervous system (Ainscough 1960) and a glandular system of paired acinous glands in the dorsal region of the propodosoma. These empty into ducts that serve the salivary stylets. A pair of globose glands of uncertain function opens into gastric ceca III of certain Uropodina (Woodring and Galbraith 1976).

Reproduction, Biology, and Ecology

Reproduction in edaphic Mesostigmata may be either sexual or facultatively parthenogenetic, with the latter being either arrhenotokous or thelytokous (Oliver 1971). The apparent absence of males in certain species of Uropodina suggests that obligate parthenogenesis may occur in the cohort (B. D. Ainscough, personal observation).

Like other Mesostigmata, edaphic forms generally pass through five stases: egg, larva, protonymph, deutonymph, and adult. Ovoviviparity and larviparity are known to occur in the gamasine family Macrochelidae (Filipponi and Francaviglia 1963), but oviparity is typical in edaphic representatives of the suborder. One or more juvenile stases are commonly nonfeeding (Edlers and Solomon 1979). Edaphic forms tend to be multivoltine, with development from egg to adult requiring from $\frac{1}{2}$ to 3 weeks. Developmental time, however, may vary greatly within a given species at different temperatures (Karg 1971a).

Many mesostigmatid inhabitants of litter, soil, and humus are predaceous on nematodes, enchytraeids, or on a variety of microarthropods or their eggs. A few groups are considered fungivorous (e.g., some Ameroseiidae and Uropodidae), bacteriophagous (certain Uropodina), or facultatively polleniphagous or phytophagous (some Phytoseiidae), while the food preferences of some others remain essentially unknown (e.g., the Halolaelapidae and Zerconidae). Karg (1961) has noted that there is a correlation between cheliceral structure and the type of food taken: mites with short dentate cheliceral digits feed primarily on nematodes, whereas those with long, slender digits and retrorse teeth feed on other mites or on Collembola.

The deutonymphal or adult stases of many edaphic mesostigmatid families have established phoretic relationships with a variety of arthropods (Krantz 1978). Whereas phoresy in the edaphic gamasines entails no obvious adaptations for attachment to the carrier, phoretic uropodine deutonymphs often attach to their arthropod associates by means of an anal pedicel, a liquid strand drawn from the anus that hardens on contact with the air. A similar strategy may be used also by certain sejines (Evans and Till 1979). Two deutonymphal morphotypes may occur in certain uropodines, only one of which is phoretic (Ainscough 1963, Faasch 1967, Athias 1975).

Data on the distribution of edaphic Mesostigmata indicate that a substantial proportion of the fauna may occur in the mineral soil beneath the litter–humus layers. Price (1973) found that only 16.5% of the edaphic mesostigmatid mites in a California pine forest were encountered in litter and humus, and that the highest percentage of individuals was found at depths of 4–6 in. In some situations, however, the mesostigmatid fauna may occur primarily at the humus–soil interface, rarely penetrating farther than 2–3 cm into the underlying soil (Evans et al. 1961). Smaller species (e.g., members of the Rhodacaridae and Digamasellidae) often are able to move interstitially into the deeper soil layers. Coineau (1974) observed that the size of edaphic mesostigmatids diminishes with increase in the depth at which they occur. Thus species of *Veigaia* (Veigaiidae) are common in the litter–humus layers, the smaller *Dendrolaelaps* (Digamasellidae) inhabit the humus–soil interface, and the diminutive *Rhodacarellus* (Rhodacaridae) live in the soil itself. Edaphic forms often move vertically in response to seasonal fluctuations in rainfall (Price 1975), but show little preference for particular soil types. On the other hand, they may follow a predictable species succession in decomposing forest litter (Metz and Farrier 1969).

The density of mesostigmatid populations in edaphic habitats varies considerably with locality and season; at times, their numbers may be quite substantial. McBrayer et al. (1977) showed that the density of edaphic Mesostigmata in an east Tennessee hardwood forest varied during a 12-month period from a February low of 7077 mites per square meter (biomass = 1906.6 mg/m^2) to a July high of 16,775 mites per square meter (biomass = 4532.4 mg/m^2). The average monthly mesostigmatid biomass over the year was 3073 mg/m^2.

Collection and Study Technique

Like most edaphic arthropods, mesostigmatid mites are best collected by extraction of habitat samples. Samples of litter, soil, or humus are collected, labeled, and transported to the laboratory for separation. A partial removal of extraneous material may be accomplished by preliminary mechanical sieving. A common qualitative extraction method utilizes a modified Tullgren apparatus which concentrates the mites by means of progressive desiccation of the sample and forces them to fall into a collecting container (Krantz 1978). Mites may be collected alive by placing moistened pieces of absorptive paper in the collecting container, or they may be preserved directly by partially filling the container with a suitable preservative. Flotation offers another useful collecting method. Here, extraneous soil and organic debris are precipitated by use of a liquid of appropriate specific gravity and the mites are decanted or collected manually from the surface film (Edwards and Fletcher 1971, Gridelet-de-Saint-Georges 1975). Quantitative sampling methods devised by Macfadyen (1953) and by Merchant and Crossley (1970) provide high gradient extraction of mites from relatively small core samples. Other extraction methods suitable for edaphic Mesostigmata are described by Edwards and Fletcher (1971).

Microscopic study of external features of edaphic mesostigmatids often requires the use of a clearing agent. Immersion of mites in lactophenol or aqueous solutions of lactic acid for 1–7 days usually provides satisfactory clearing of internal organs. Nesbitt's fluid is recommended for clearing old alcohol-stored specimens that do not yield to clearing in lactophenol or lactic acid. Dissection of specimens prior to study often reveals structural details that may be obscure in whole specimens. Techniques for dissection are discussed by Krantz (1978).

Observation of prepared specimens may be accomplished by means of permanent slide mounts using Hoyer's medium or similar water-soluble mountants. Alternatively, temporary preparations may be made using lactive acid in a cavity slide. Mites placed in a drop of lactic acid near the sidewall of the cavity may be manipulated into any position by moving the coverslip which partially covers the cavity, or by inserting a needle probe under the coverslip (Grandjean 1949, Coineau 1974).

Key Characters

Identification of edaphic mesostigmatid families and genera depends to a great extent on qualitative characters. Features such as comparative setal lengths or shield size are, with some exceptions, of value only at the specific level. Major character states used in the keys to edaphic North American Mesostigmata are described below.

Gnathosoma

The epistome may be smooth marginally or may be distinctively produced in some families and genera (Figs. 20.47, 20.136, 20.146). Development of deutosternal denticles and the position of the hypostomatic setae on the venter of the gnathosoma (Figs. 20.48, 20.66, 20.195) also provide useful indicators. The variation in number of tines on the palpal apotele is an important character at generic and suprageneric levels throughout the suborder. Cheliceral dentition is a useful generic character in some families (e.g., the Parholaspididae and Uropodidae), as is development of the corniculi and labrum. The form of the male spermatodactyl (Fig. 20.170) often is distinctive at the familial or generic level.

Idiosoma

The development of dorsal and ventral sclerites and setae comprise a major source of variation among the genera, families, and higher taxa of the edaphic Mesostigmata. The dorsal shield may be entire or variously divided (Figs. 20.12, 20.14, 20.40); dorsal setation may be complete, deficient, or hypertrichous, and setal form may vary at the generic level from simple to highly modified (Fig. 20.121). Ventral sclerites of females may depart in many ways from the generalized pattern shown in Fig. 20.6, and often are distinctively amalgamated or reduced. The female sternal shield typically bears three pairs of setae and two pairs of pores, but sternals 3 are off the shield in some genera (Fig. 20.123) and the metasternal setae and/or pores may be amalgamated into the sternal shield in others (Fig. 20.147). Males have a medioventral or anteroventral genital opening located in either a sternal, sternoventral, or holoventral shield complex. A separate ventrianal or anal shield occurs in the males of a number of supraspecific taxa. Peritremes in both sexes extend well beyond coxae I in many genera, but may be reduced or absent in others. The stigmata generally are lateroventral but assume a laterodorsal position in the Epicriidae (Fig. 20.19).

The structure of the tritosternal base and laciniae are useful for identifying a number of taxa (Fig. 20.6).

Legs

The development of the apotele on legs I is highly variable throughout the suborder but in some cases it provides a useful character at the generic level. Chaetotactic patterns on particular leg segments provide valuable criteria for determination of cohorts, families, and genera (Evans 1963a, 1972; Ainscough 1963, Evans and Till 1979), and are used extensively in the keys to follow. Evans' system of setal nomenclature permits characterization of a given segment by means of the following formula:

$$\text{anterolaterals}\,\frac{\text{anterodorsals}}{\text{anteroventrals}}\,\frac{\text{posterodorsals}}{\text{posteroventrals}}\,\text{posterolaterals}$$

Thus a leg segment with one anterolateral and one posterolateral seta, two anterodorsals and two posterodorsals, one anteroventral, and no posteroventrals, would be characterized as follows:

$$al\,\frac{ad}{av}\,\frac{pd}{pv}\,pl = 1\,\frac{2}{1}\,\frac{2}{0}\,1$$

These relationships are shown diagramatically in Fig. 20.10.

Identification

The keys to families and genera presented below include taxa which are commonly edaphic in character, and which are known by the authors to occur in North America. It is important to note, however, that additional genera undoubtedly remain to be discovered in the Nearctic edaphon.

A variety of basically nonedaphic Mesostigmata may occasionally be encountered in soil or litter habitats in North America. These include groups that are generally associated with mammals or insects (e.g., the laelapid genera *Haemogamasus* and *Laelaspis*) or that are usually found on plants (e.g., the phytoseiid genera *Phytoseius* and *Typhlodromus*) or in dung (e.g., the macrochelid genus *Glyptholaspis*). Due to limitations of space, only the more commonly encountered of these genera are included in the keys and illustrations that follow. Taxa that are only rarely found in North American edaphic substrates are identified in the keys with an asterisk.

ACKNOWLEDGMENTS

We are grateful to Dr. E. E. Lindquist, Biosystematics Research Centre, Ottawa, for his comments regarding the family Ascidae and for his kindness in providing records and representative specimens of a number of mesostigmatid genera. We also wish to thank Dr. C. Blaszak, Institute of Biology, A. Mickiewicz University, Poznan, for his key to Nearctic genera of Zerconidae. We have borrowed freely from Dr. Blaszak in writing the zerconid key included in this chapter. Our thanks also to Dr. J. B. Kethley, Field Museum of Natural History, Chicago, for his comments regarding the distribution of Eviphididae in North America.

It should be noted that many of the generic keys were derived from important works by a number of our colleagues, including Micherdzinski (1969), Karg (1978a), and Evans and Till (1979) for the Parasitidae, Lindquist (1975) for the Digamasellidae, Lindquist and Evans (1965) for the Ascidae, Lee (1970) for the Ologamasidae and Rhodacaridae, Karg (1971a) for the Ameroseiidae and Halolaelapidae, Karg (1976) and Evans and Till (1979) for the Eviphididae, and Karg (1971a, 1979a) and Evans and Till (1966) for the Laelapidae. We are pleased to acknowledge their contributions to this chapter.

KEY TO FAMILIES

1a Females with one primary genital cover (epigynial shield), with or without a pair of associated setae; movable cheliceral digit without medially or terminally inserted ventral excrescenses, but often with brushlike or coronetlike outgrowths in the arthrodial angle at the base of the digit; males often with a sperm transfer apparatus originating on the movable digit; tarsi II–IV typically with 18 setae $\left(3\ \frac{3}{2}\frac{1}{1}\frac{3}{2}\ 3\right)$; Supercohort **Monogynaspides**. **2**

1b Females with three primary genital covers (two latigynials and one mesogynial element), variously coalesced or reduced; movable cheliceral digit with medial or terminal dendritic, brushlike, or filamentous excrescences, occasionally with additional excrescence at base of digit; males without sperm transfer apparatus on movable digit; tarsi II–III of deutonymph and adult with 19 setae $\left(4\ \frac{3}{1}\frac{1}{1}\frac{3}{2}\ 3\right)$, and tarsus IV with 21 setae $\left(4\ \frac{3}{3}\frac{1}{1}\frac{3}{3}\ 3\right)$; Supercohort **Trigynaspides;** Cohort **ANTENNOPHORINA**. **23**

2a Epigynial shield truncate, drop-shaped, or rounded posteriorly, often extending beyond the podosomatic region as a fused genitiventral or genitiventrianal entity; shield usually with a pair of setae in the podosomatic region, or setae are inserted in the integument laterad of the shield; usually with one or two dorsal shields, without separate marginal shields or platelets; femur IV of adult and deutonymph typically with six setae; hypostomatic setae 2–3 inserted at approximately the same level; male genital aperture at anterior edge of sternal shield or within it; sperm transfer structures present on chelicerae of those forms with anterior genital opening; males often with leg armature; Cohort **GAMASINA**. **6**

2b Epigynial shield variously shaped, generally confined to the podosomatic region, rarely displaced posteriorly and occasionally fused with ventral elements; with or without epigynial setae; with one to several dorsal shields, marginal shields or platelets generally present; hypostomatic setae 3 at more or less the same level as hypostomatic setae 2, or well behind them; femur IV of adult and deutonymph usually with seven setae (if only six setae are present, hypostomatic setae 3 are always behind hypostomatic setae 2); male genital aperture within sternal region, never at anterior shield margin; cheliceral sperm transfer structures absent; males may have specialized spines or apophyses on legs II. **3**

3a Sternal shield fragmented (Fig. 20.16) or simply divided behind sternals 2; epigynial shield typically with one, three, or several pairs of setae in the podosomatic region; femur IV of adults and deutonymphs with seven setae; hypostomatic setae 2 and 3 inserted at more or less the same level; Cohort **SEJINA,** Superfamily **SEJOIDEA** 4

3b Sternal shield entire, often fused with adjacent podal and peritrematic shields; epigynial shield truncate posteriorly, occasionally displaced behind podosomatic region (***Metagynella*** Berlese, Fig. 20.187), without setae (except **THINOZERCONOIDEA** and ***Polyaspis*** (***Dyscritaspis***)); femur IV of adults with six or seven setae, hypostomatic setae 3 always well behind hypostomatic setae 2, describing a nearly straight line with hypostomatic setae 1 (Fig. 20.3); Cohort **UROPODINA** 25

4a With one pair of epigynial setae; male genital aperture posterior to coxae II (Figs. 20.11, 20.12); Family **MICROGYNIIDAE** . . ***Microgynium*** Trägårdh

4b With more than one pair of epigynial setae; male genital aperture between coxae II 5

5a Epigynial shield subquadrangular, extended anteriorly so as to overlap posterior sternal shield elements, ornamented with at least 20 setae; with one to four major dorsal shields (Figs. 20.13, 20.14); Family **ICHTHYOSTOMATOGASTERIDAE** . . . . ***Uropodella*** Berlese

5b Epigynial shield more or less reniform, not overlapping posterior sternal shield elements, with three (occasionally two) pairs of laterally inserted setae; with two to seven dorsal shields (Figs. 20.15, 20.16); Family **SEJIDAE** . . . . ***Sejus*** Koch

6a Sternal and metasternal shields of female fused, with four pairs of setae, sternals 1 may be inserted in a weakly defined anteromarginal extension; epigynial shield usually rounded anteriorly, separated posteriorly from ventrianal shield; Superfamily **RHODACAROIDEA** 13

The metasternal setae (sternals 4) may be separated from the sternal shield in certain genera of the Family **OLOGAMASIDAE** (some ***Periseius*** Womersley, for example). These forms are distinguished by their distinctive tubuli annulati, combined with the presence of four ventral setae on tibia I.

6b Sternal shield of female usually with three or fewer pairs of setae; metasternal shields free or absent, occasionally contiguous with sternal shield or incorporated into adjacent endopodal elements; epigynial shield rounded or truncate anteriorly, fused posteriorly to ventral or ventrianal elements (Figs. 20.154, 20.165) or freely articulated 7

The sternal shield of ***Pachylaelaps*** Berlese extends posteriorly to incorporate both the metasternal and endopodal elements, and carries sternal setae 4 (the metasternals). The fused metasternal–endopodal elements of the pachylaelapid genera ***Pachyseius*** and ***Zygoseius*** are only tenuously fused to the sternal shield at their anterior extremities (Figs. 20.148, 20.150).

7a With the leg chaetotactic combination of five dorsal setae on tibia I, two posterolateral setae on genu II, and seven or eight setae on tibia IV (if only four dorsal setae on tibia I, then the segment has only two ventral setae and the palptarsal apotele is three-tined); Superfamily **EVIPHIDOIDEA** . . 20

7b Without the above combination of characters, usually with four or six dorsal setae on tibia I and one posterolateral seta on genu II; tibia IV may have only six or seven setae $\left(1\ \frac{1\,2}{1\,0}\ 1 \text{ or } 1\ \frac{1\,2}{1\,1}\ 1\right)$ or, more commonly, 9–11 setae 8

8a Sternal shield well developed and entire, with three pairs of setae and two or three pairs of pores; peritremes present and extending at least to level of coxae II; dorsal shield entire or divided, occasionally reduced posteriorly;

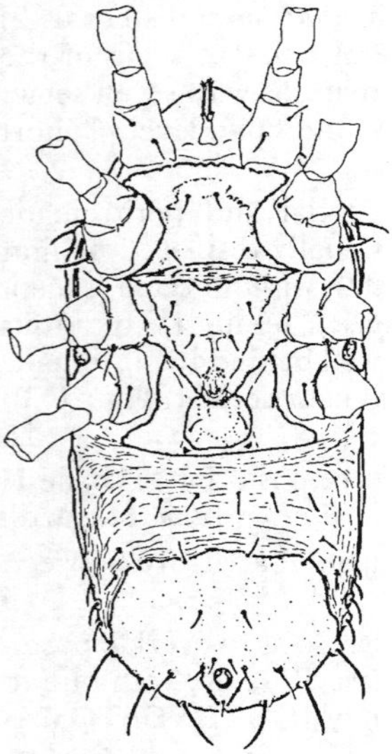
11

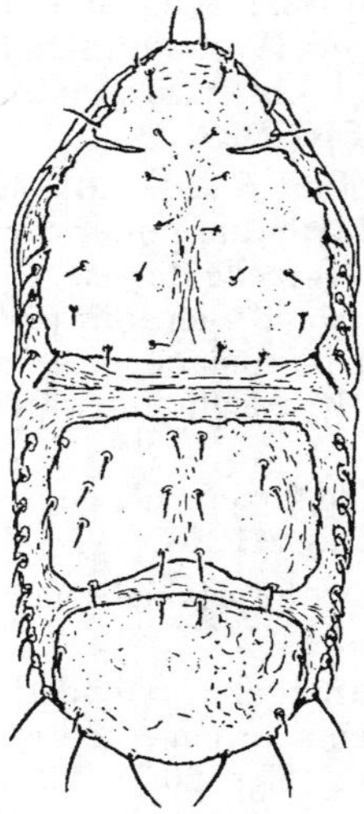
12

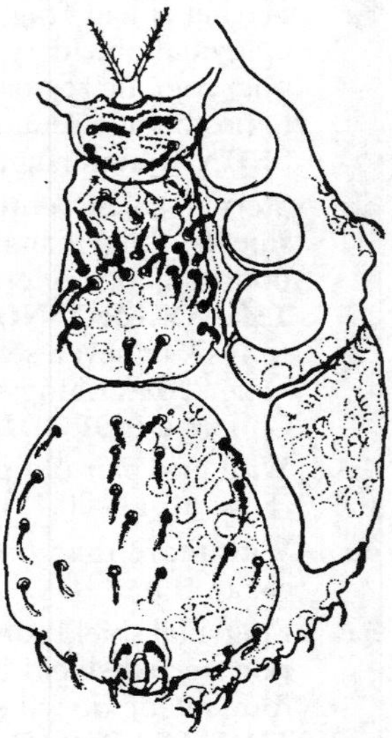
13

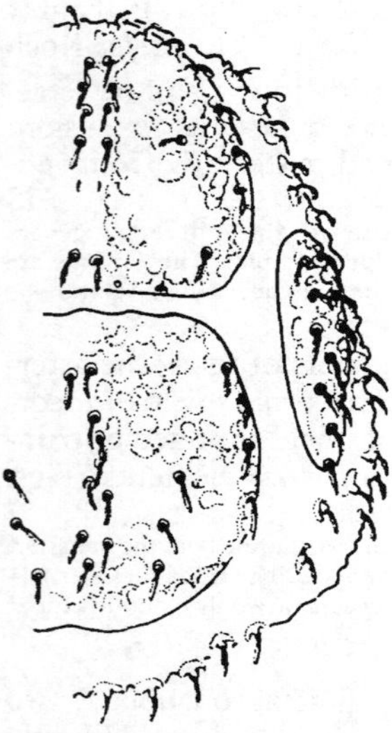
14

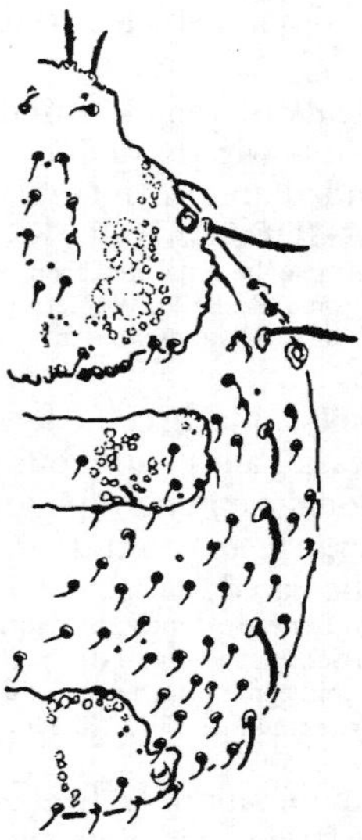
15

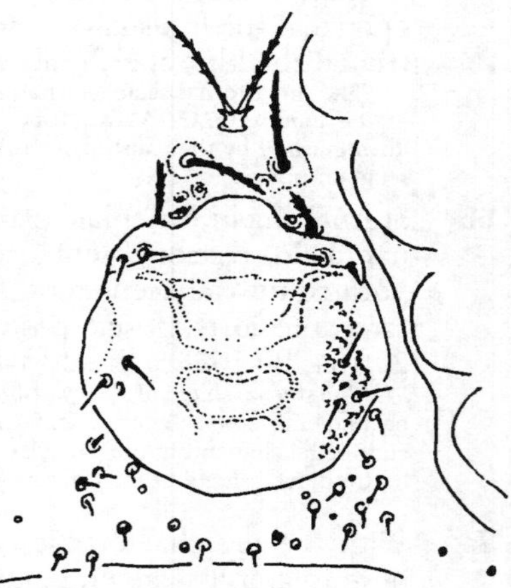
16

Fig. 20.11. *Microgynium incisum* Krantz (Oregon), female venter. [From Krantz (1961).] **Fig. 20.12.** *Microgynium incisum* Krantz (Oregon), female dorsum. [From Krantz (1961).] **Fig. 20.13.** *Uropodella laciniata* Berlese (Ohio), female venter. **Fig. 20.14.** *Uropodella laciniata* Berlese (Ohio), female dorsum. **Fig. 20.15.** *Sejus* sp. (Oregon), female dorsum. **Fig. 20.16.** *Sejus* sp. (Oregon), intercoxal region.

palptarsal claw always three-tined; with one *pv* seta on genu III $\left(2\,\frac{2}{1}\,\frac{2}{1}\,1 \text{ or } 2\,\frac{2}{1}\,\frac{2}{1}\,2\right)$ and two on tibia I $\left(2\,\frac{3}{2}\,\frac{3}{2}\,2\right)$; tibia IV with three *pd* setae $\left(2\,\frac{2}{1}\,\frac{3}{1}\,2\right)$; Superfamily **PARASITOIDEA** **11**

8b Without the foregoing chaetotactic combination. Palptarsal claw two- or three-tined. **9**

9a Epigynial shield truncate or weakly convex posteriorly and widely separated from anal shield, sometimes abutting a ventrianal shield (if epigynial shield is rounded posteriorly, then anal shield usually is not triangular in shape) **10**

9b Epigynial shield rounded posteriorly, widely or narrowly separated from triangular anal shield, occasionally expanded so as to include the anal shield or to abut it, in the latter case assuming a flattened or invaginated aspect posteriorly . . . Superfamily **DERMANYSSOIDEA,** Family **LAELAPIDAE**

10a Deutonymphs and adults with less than 20 pairs of dorsal shield setae (if more than 20 pairs are present, then the peritremes are absent or tibia IV has two *ad* setae $\left(2\,\frac{2}{1}\,\frac{2}{1}\,1\right)$ Superfamily **PHYTOSEIOIDEA** **17**

10b Deutonymphs and adults with more than 20 pairs of dorsal shield setae; peritremes present, tibia IV with only one *ad* seta; Superfamily **ASCOIDEA** **15**

11a Epigynial shield triangular and flanked by large discrete metasternal shields; male spermatodactyl coalesced distally with movable digit to form a median atrium (Fig. 20.33); males with highly developed spurs and apophyses on legs II, femoral spur often as long or longer than femur; tritosternum of male often reduced, occasionally absent . . . . Family **PARASITIDAE**

11b Epigynial and metasternal shields not as above; male spermatodactyl free, not coalesced distally with movable digit; apophyses may be present on legs II, but rarely developed as described above; tritosternum of male normally produced **12**

12a Distal extensions of internal malae of hypostome broadly produced, often moustache-like; palpal claw with a hyaline, membranous excrescence arising at the apotelic base; epigynial and ventral shields usually separated by incomplete suture; with claws on tarsus I; male genital opening at anterior margin of sternal shield . . . . Family **VEIGAIIDAE**

12b Internal malae terminate in unmodified, lightly fringed elements; palpal claw without adjacent hyaline element; epigynial shield rounded posteriorly, without a discrete adjacent or attached ventral shield as above; tarsus I terminates in a series of short hairs, claws present or absent; male genital aperture well within sternal region (Fig. 20.17); Family **ARCTACARIDAE** ***Arctacarus*** Evans

The position of the male genital aperture in the **ARCTACARIDAE** is similar to that in the Families **ZERCONIDAE (Ascoidea)** and **EPICRIIDAE (Phytoseioidea).** The coincident occurrence of this unusual character in the three families has led to their clustering in some treatments.

13a Dorsal shield divided; adults usually with distinctive sclerotic nodules (scleronoduli, Fig. 20.62) between anterior dorsal shield setae j_4 and j_5 (absent in ***Digamasellus***); anterior portion of sternal shield weakly defined, usually carrying sternal setae 1; spermatodactyli of male often recurved basally **14**

13b Dorsal shield divided or entire, without scleronoduli; anterior portion of sternal shield undifferentiated from more posterior portions; spermatodactyli of male not recurved basally. . . . . Family **OLOGAMASIDAE**

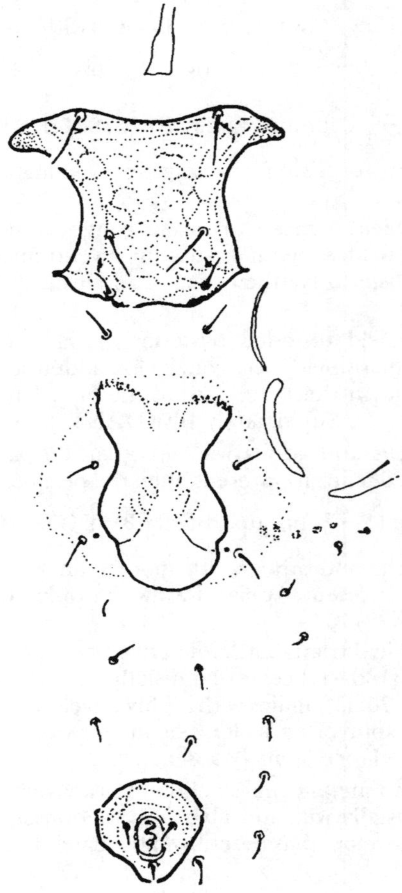

17

Fig. 20.17. *Arctacarus* sp. (Oregon), female venter.

14a Palptarsal claw two-tined; usually with seven setae on genu IV $\left(1\,\frac{2}{1}\,\frac{2}{0}\,1\right)$ and seven on tibia IV $\left(1\,\frac{1}{1}\,\frac{2}{1}\,1\right)$; male with sternal setae 5 ("epigynial" setae) on separate platelets . . . . Family **DIGAMASELLIDAE**

14b Palptarsal claw three-tined; with 10 setae on genu and tibia IV $\left(2\,\frac{2}{1}\,\frac{3}{1}\,1\right.$ and $\left.2\,\frac{1}{1}\,\frac{3}{1}\,2\right)$; male with sternal setae 5 on sternitigential shield . . . .
Family **RHODACARIDAE**

15a Dorsal shield divided medially, with a transverse row of four posteromarginal fossae usually clearly defined on opisthonotal shield, medial fossae may be fused; peritremes reduced, never extending anteriorly beyond coxae III;

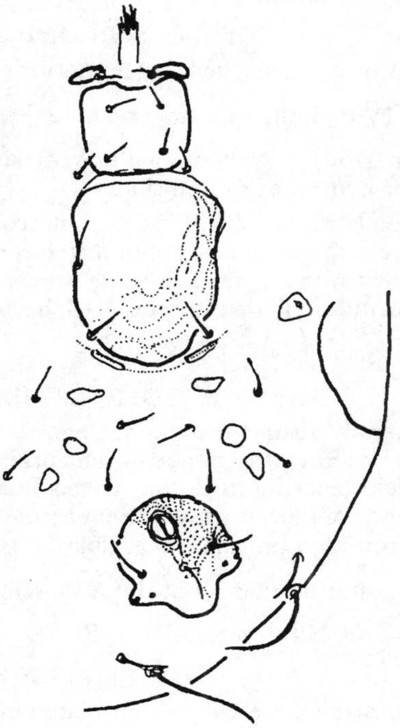

18

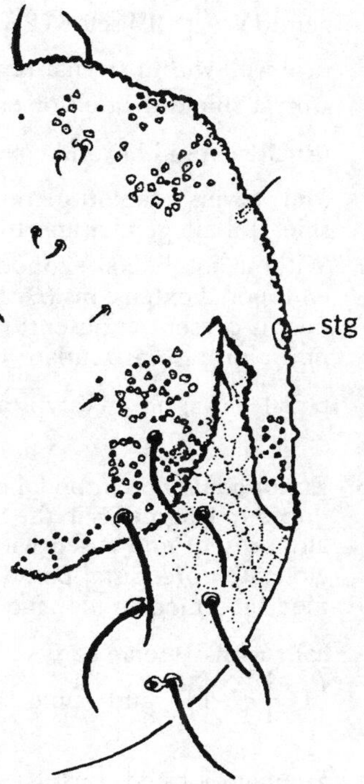

19

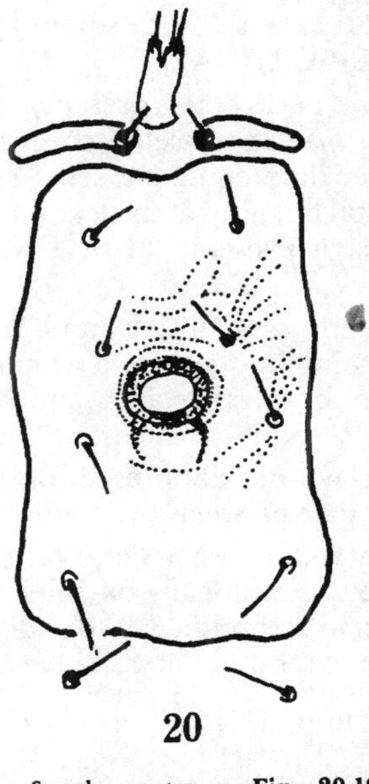

20

Fig. 20.18. *Epicrius* sp. (Oregon), female venter. **Fig. 20.19.** *Epicrius* sp. (Oregon), female dorsum. **Fig. 20.20.** *Epicrius* sp. (Oregon), male intercoxal region.

genu IV with 10 setae $\left(2\,\frac{2}{1}\,\frac{3}{1}\,1\right)$; with a ventrianal shield; male genital aperture well within sternal region . . . . Family **ZERCONIDAE**

15b Dorsal shield divided or entire, without posteromarginal fossae; peritremes usually extend beyond coxae III; genu IV typically with nine setae $\left(2\,\frac{2}{1}\,\frac{3}{0}\,1\right)$ (only seven setae on genu IV of certain groups); with an anal or ventrianal shield; male genital aperture at anterior margin of sternal shield **16**

16a Without intercoxal exopodal development, sternal shield lacking intercoxal endopodal extensions (***Halodarcia incideta*** Karg is an exception), free endopodals present or absent; tarsi II–IV often with a pair of elongate, membranous, distally acuminate dorsodistal paradactyli that extend well beyond tarsal claws; tibia IV typically with eight setae $\left(2\,\frac{1}{1}\,\frac{2}{1}\,1\right)$ . . . .
Family **HALOLAELAPIDAE**

16b Endopodal and exopodal elements generally distinct (exopodal elements reduced to fragments in the Subfamily **Arctoseiinae**), endopodal elements usually present and fused with sternal shield (exceptions in the **Arctoseiinae**); elongate dorsodistal paradactyli present or absent, if present then accompanied by a median elongate or broadly rounded pulvillar lobe; tibia IV typically with 10 setae $\left(2\,\frac{1}{1}\,\frac{3}{1}\,2\right)$ [only seven setae on tibia IV of the Arctoseiinae $\left(1\,\frac{1}{1}\,\frac{2}{1}\,1\right)$, and nine on tibia IV of ***Protogamasellus*** $\left(2\,\frac{1}{1}\,\frac{3}{1}\,1\right)$]
Family **ASCIDAE**

17a Stigmata located dorsolaterally in a distinctive tuberculate–reticulate dorsal shield which may cover the dorsum, or which may be abbreviated or weakened posteriorly and variously incised laterally (Figs. 20.18–20.20); peritremes absent; sternal setae 1 on presternal shield or paired jugularia; setae Zv_1 often inserted on epigynial shield; male genital opening within sternal shield, spermatodactyli absent; tarsus I with some dorsal setae minutely clubbed, claws absent; Family **EPICRIIDAE** . . . . ***Epicrius*** Canestrini and Fanzago

17b Stigmata ventrolateral, peritremes present; dorsal shield variously ornamented, generally entire and completely covering the dorsum; sternal setae 1 on sternal shield or free in integument; setae Zv_1 never inserted on epigynial shield; male with genital opening at or near anterior margin of sternal shield element, spermatodactyli present; tarsus I without clubbed setae, with or without claws. **18**

18a Legs I greatly elongated; genual, tibial and tarsal segments attenuate, subequal, tarsus I terminating in one or two whiplike setae; with a pair of large, distinctive dorsal pores between setae J_4 and Z_4 (Figs. 20.21, 20.22); Family **PODOCINIDAE** . . . . ***Podocinum*** Berlese

18b Legs I not as above, claws usually present on tarsus I; dorsal pores present but without a distinct pair between the J and Z series **19**

19a Sternal shield typically with two pairs of setae (st_1–st_2), st_3 free in integument or on adjacent platelets, occasionally on shield proper; corniculi may be divided distally or otherwise modified in the distal portion; posterior row of deutosternal denticles extending laterally beyond insertions of capitular setae; chaetotaxy of femur II is $\left(2\,\frac{4}{3}\,1\right)$, that of tibia IV is $\left(2\,\frac{2}{1}\,\frac{2}{1}\,1\right)$
Family **AMEROSEIIDAE**

19b Sternal shield typically with three pairs of setae, st_3 occasionally on sepa-

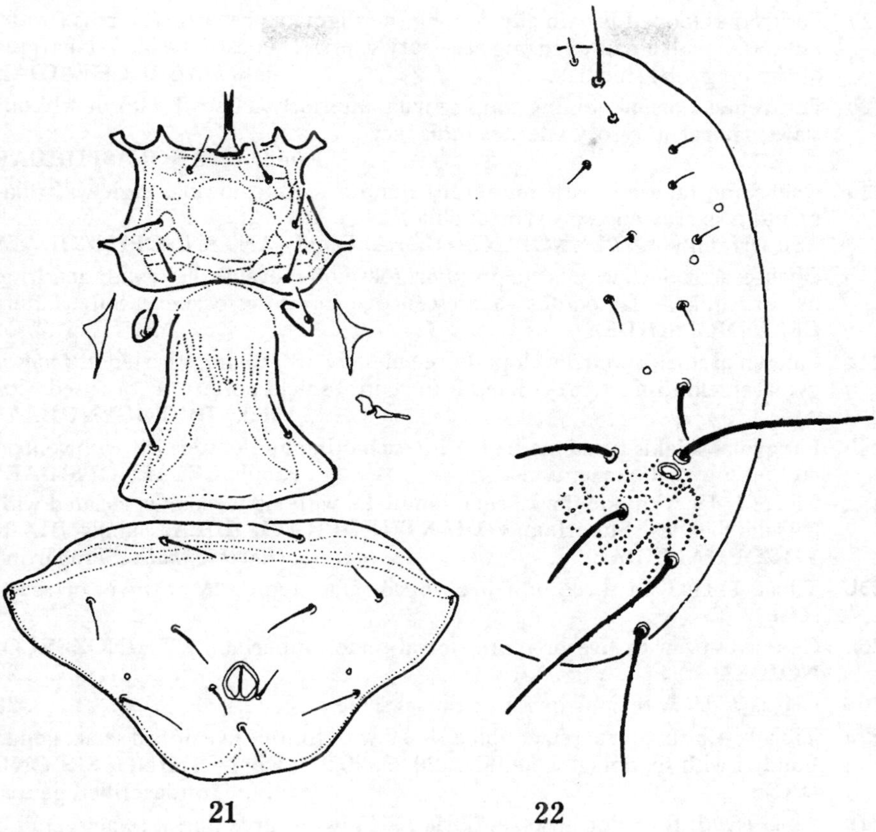

Fig. 20.21. *Podocinum pacificum* Berlese (Oregon), female venter. **Fig. 20.22.** *Podocinum pacificum* Berlese (Oregon), female dorsum.

rate platelets; corniculi without distal modifications, posterior row of deutosternal denticles more or less confined to deutosternal groove; chaetotaxies of femur II and tibia IV not as above, tibia IV with only one *al* seta Family **PHYTOSEIIDAE**

20a Tibia and genu I each with one *al* seta . . . . . . .Family **EVIPHIDIDAE**

20b Tibia and genu I each with two *al* setae. **21**

21a Chelicerae with one or two plumose or filamentous arthrodial processes, or with an arthrodial brush and an adjacent setulose coronet; sternal shield of female with three pairs of setae, typically with a ventrianal shield (if anal shield is present, it is widely separated from genitiventral elements) . . . **22**

21b Arthrodial process, when apparent, rarely forming a distinct brush; if brush is present, then leg II of female distinctly thickened and carrying one to two greatly enlarged terminal tarsal setae; sternal shield complex with four pairs of setae, sternal shield insensibly or tenuously fused to metasternal–endopodal elements; with an anal shield closely bordered by a gentitiventral shield, or with a fused genitiventrianal shield; peritrematic, exopodal and metapodal elements fused, often extending posteriorly beyond coxae IV. . . . . Family **PACHYLAELAPIDAE**

22a Peritremes looped proximally, joining the stigmata posteriorly; tarsus I without claws; with a pair of strong accessory sclerites beneath the lateral margins of the epigynial shield . . . . Family **MACROCHELIDAE**

22b Peritremes normal, joining the stigmata anteriorly; tarsus I with or without claws; lateral accessory sclerites indistinct . . . . Family **PARHOLASPIDIDAE**

23a Chelicerae tapered, with numerous minute teeth; movable digit with filamentous excrescences; Myrmecophiles . . . . Superfamily **ANTENNOPHOROIDEA,** Family **ANTENNOPHORIDAE***

23b Chelicerae robust, with large proximal tooth on movable digit; with dendritic or brushlike filamentous excrescences; insect associates; Superfamily **CELAENOPSOIDEA** **24**

24a Latigynial shields well developed, free mesally over their entire length; mesogynial shield distinct, articulated with ventral shield posteriorly or fused with it . . . . Family **DIPLOGYNIIDAE***

24b Latigynial shields fused medially with each other and posteriorly with ventral and mesogynial elements . . . . Family **CELAENOPSIDAE***

25a Tibiae I–IV with one dorsal seta, femur IV with eight setae; associated with passalid beetles; Superfamily **DIARTHROPHALLOIDEA,** Family **DIARTHROPHALLIDAE** . . . . *Diarthrophallus* Trägårdh*

25b Tibiae I–IV with three or more dorsal setae, femur IV with six or seven setae **26**

26a Genua I–IV with five or more dorsal setae; Superfamily **THINOZERCONOIDEA** **27**

26b Genua I–IV with four or fewer dorsal setae **28**

27a Tibia I with six dorsal setae; tibiae II–IV with four or five dorsal setae; genua I and II with six dorsal setae (Figs. 20.23, 20.24); Family **THINOZERCONIDAE** . . . . **(undescribed genus)**

27b Tibia I with four dorsal setae, tibiae II–IV with three dorsal setae; genua I and II with five dorsal setae (Figs. 20.25, 20.26); Family **PROTODINYCHIDAE** . . . . *Protodinychus* Evans

28a Genua II and III with two *al* and two *pl* setae; tibia I usually with four dorsal setae; foveae pedales lacking; Superfamily **POLYASPIDOIDEA** **29**

28b Genua II and III with one *al,* and one or no *pl,* setae; tibia I with three dorsal setae; foveae pedales usually present; Superfamily **UROPODOIDEA** **30**

29a Genu I with two *al* and two *pl* setae; femur IV with two ventral setae. . . . Family **POLYASPIDIDAE**

29b Genu I usually with two *al* and one *pl* setae; femur IV usually with two ventral setae (if femur IV with one seta, then genu I with two *pl* setae). . . . . . . Family **DITHINOZERCONIDAE**

30a Chelicerae with internal apodeme at base of movable digit (Fig. 20.27); foveae pedales present; tritosternal base narrow, longer than broad; apex of leg II directed posteriorly when withdrawn; internal malae usually fimbriate or plumose (Fig. 20.28); protonymph with podonotal and mesonotal shields insensibly fused (Fig. 20.29); myrmecophiles . . . . Family **TRACHYUROPODIDAE***

30b Chelicerae with or without internal apodeme at base of movable digit; foveae pedales usually present; tritosternal base variable in shape; apex of leg II directed anteriorly or posteriorly; internal malae variable in form; protonymph with separate podonotal and mesonotal shields (Fig. 20.30) . . . . . . Family **UROPODIDAE**

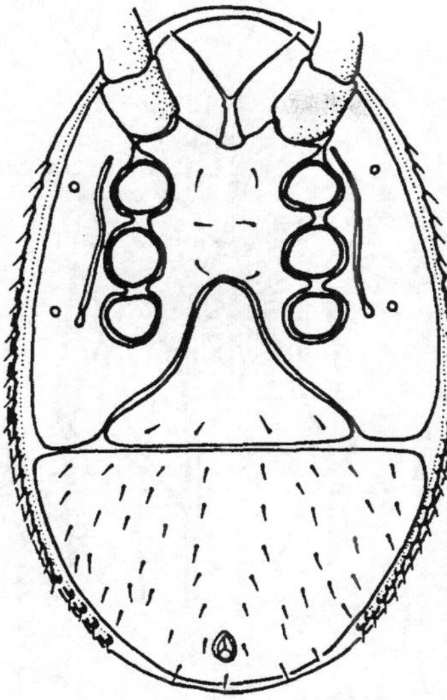

23

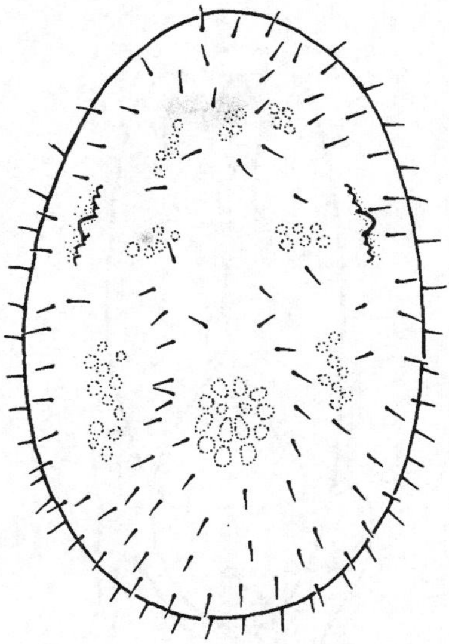

24

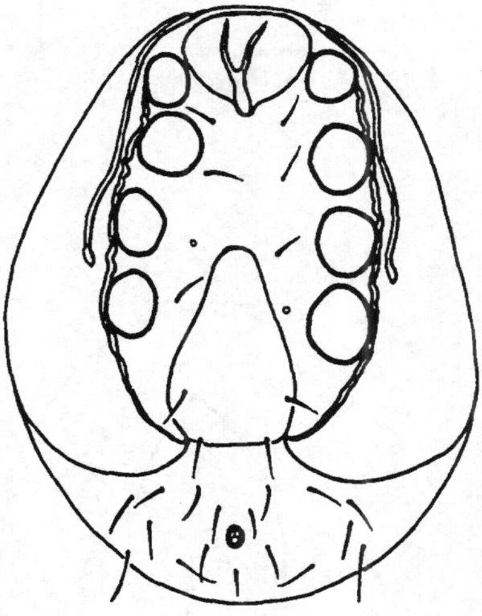

25

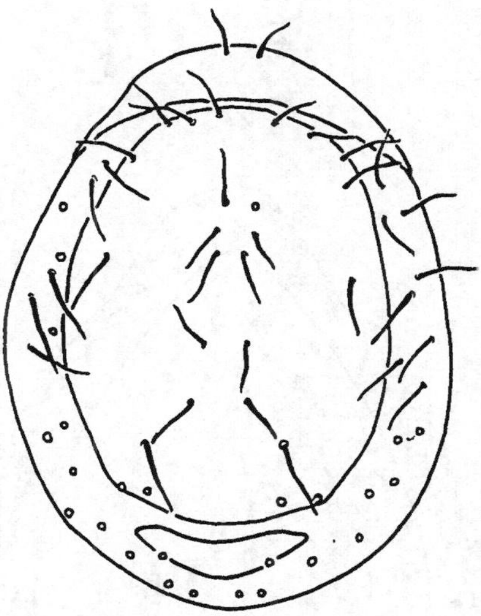

26

Fig. 20.23. Thinozerconid (undescribed genus) (Oregon), female venter. **Fig. 20.24.** Thinozerconid (undescribed genus) (Oregon), female dorsum. **Fig. 20.25.** *Protodinychus* sp. (Ontario), female venter. **Fig. 20.26.** *Protodinychus* sp. (Ontario), female dorsum.

KEYS TO GENERA

Parasitidae Oudemans, 1901

1a Dorsal shield of female and male entire, without lateral incisions; some males with a median notal line; male with tritosternal base strongly reduced and obscured by lamina of genital orifice (Fig. 20.32); dorsal setae j_5, j_6, and z_5 similar in form and length; Subfamily **Pergamasinae** **2**

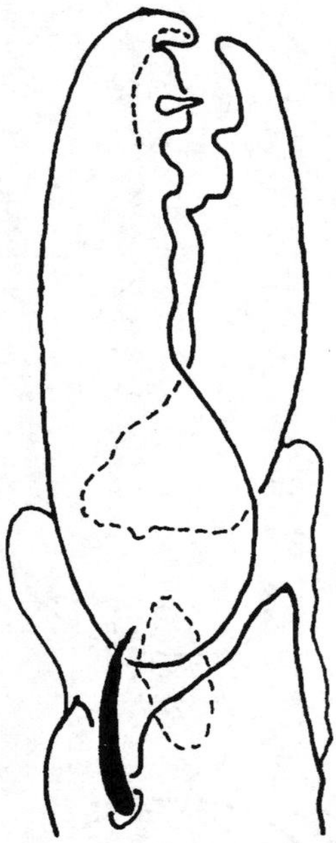

27

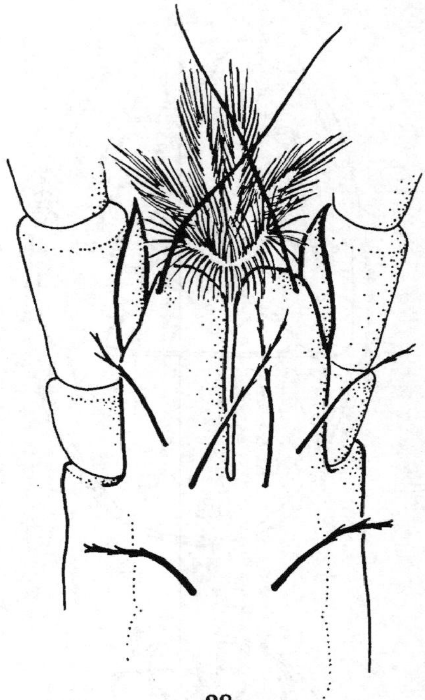

28

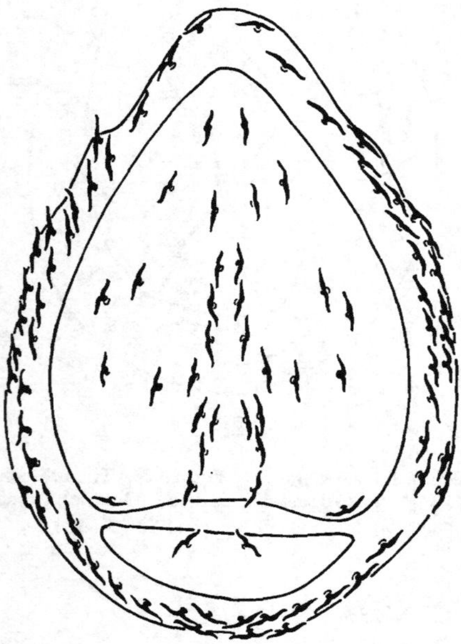

29

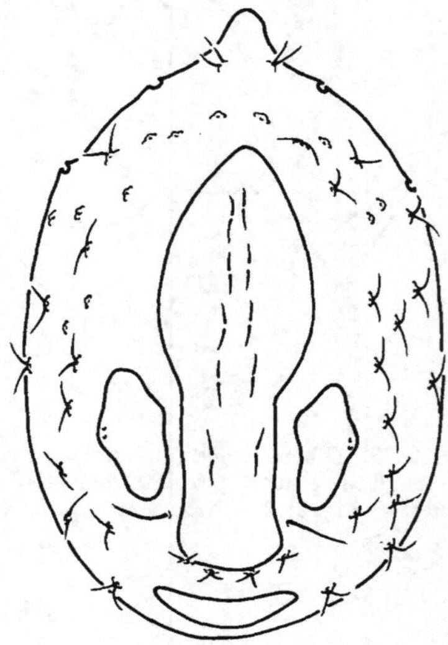

30

Fig. 20.27. Chelicera of trachyuropodid. **Fig. 20.28.** Gnathosoma of trachyuropodid. **Fig. 20.29.** Dorsum of trachyuropodid protonymph. **Fig. 20.30.** Dorsum of uropodid protonymph.

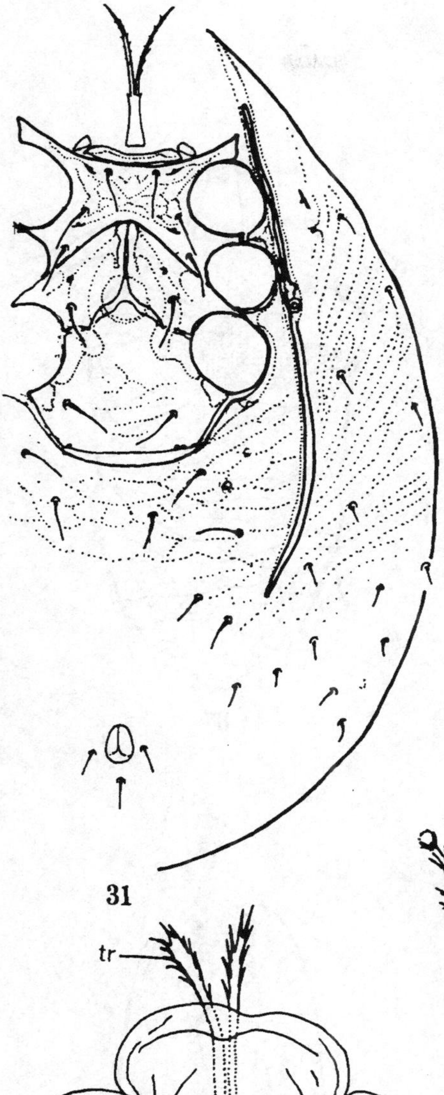

31

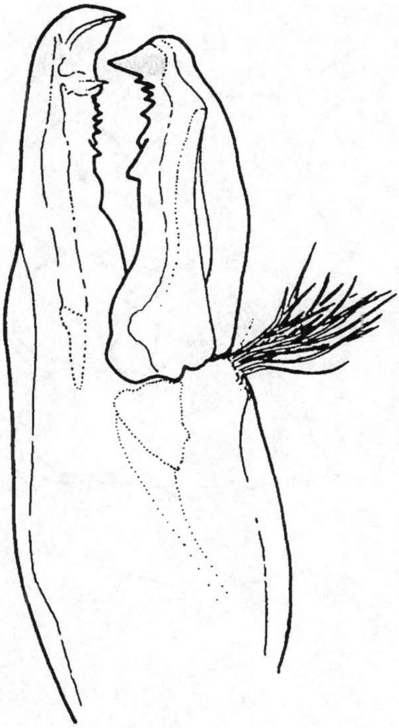

33

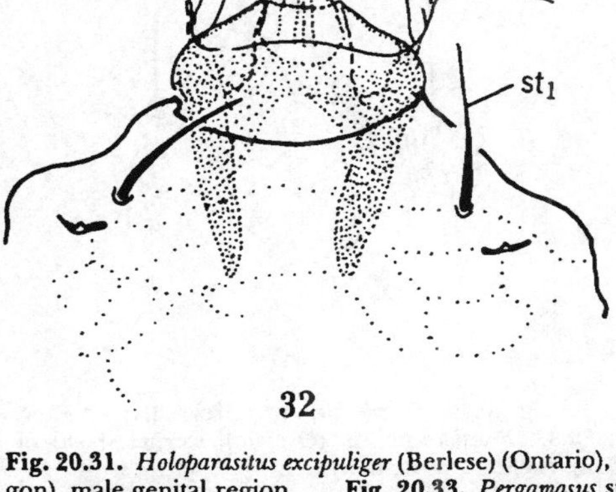

32

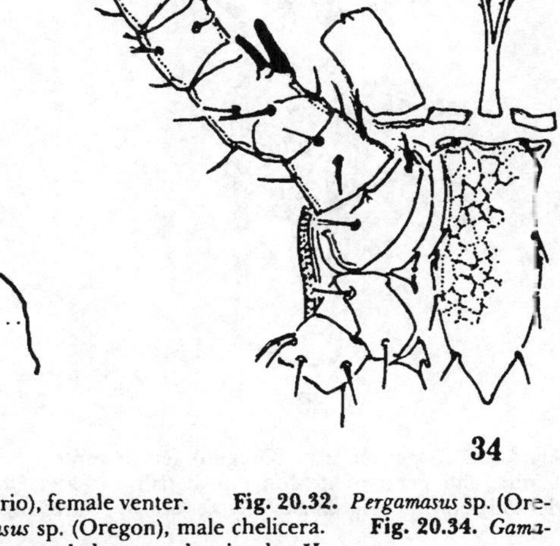

34

Fig. 20.31. *Holoparasitus excipuliger* (Berlese) (Ontario), female venter. **Fig. 20.32.** *Pergamasus* sp. (Oregon), male genital region. **Fig. 20.33.** *Pergamasus* sp. (Oregon), male chelicera. **Fig. 20.34.** *Gamasodes bispinosus* (Halbert) (New York), part of deutonymphal venter showing leg II.

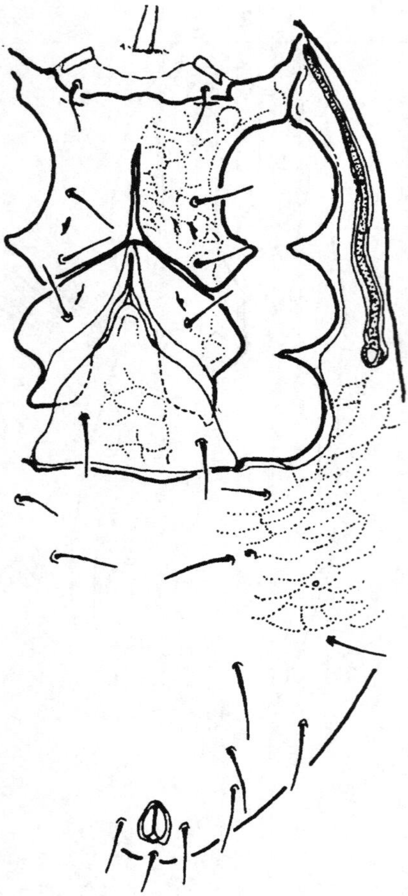

35

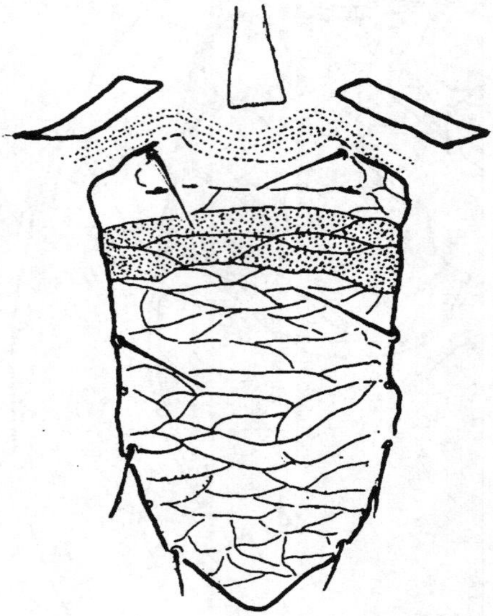

37

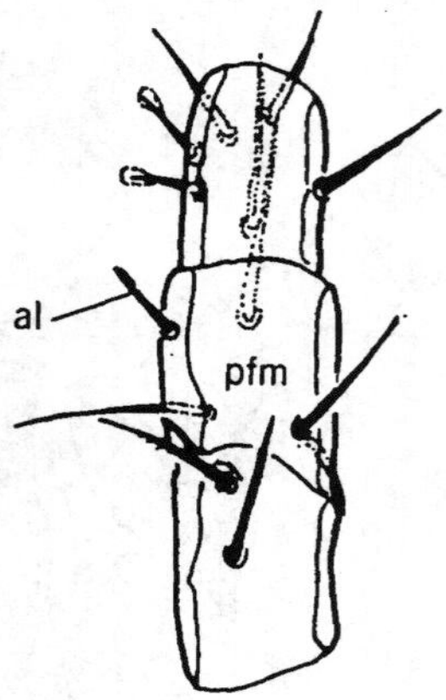

36

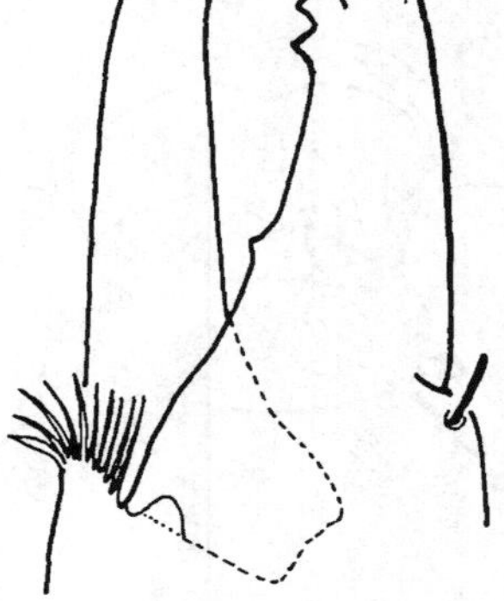

38

Fig. 20.35. *Poecilochirus* sp. (Oregon), female venter. **Fig. 20.36.** *Poecilochirus* sp. (Oregon), trochanter, femur, and genu of deutonymphal palp. **Fig. 20.37.** *Poecilochirus* sp. (Oregon), sternal shield of deutonymph. **Fig. 20.38.** *Poecilochirus* sp. (Oregon), chelicera of deutonymph.

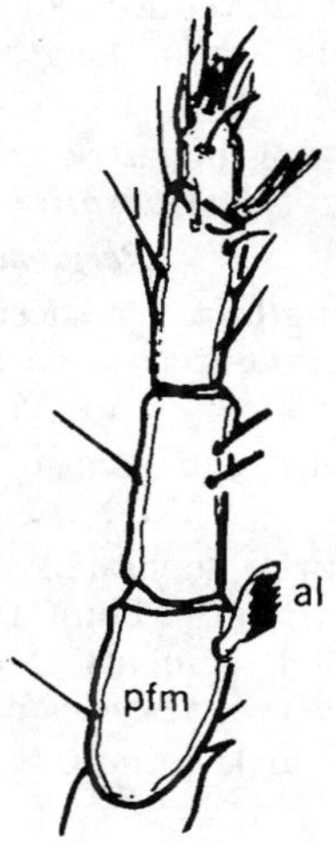

39

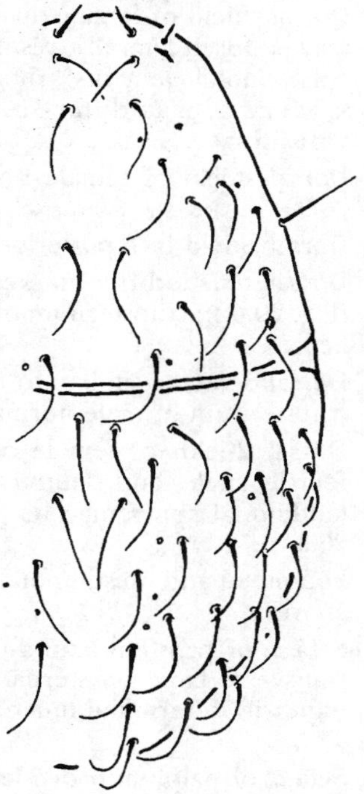

40

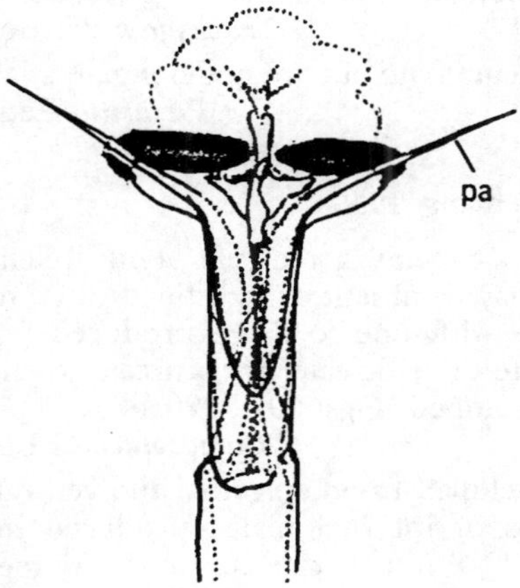

41

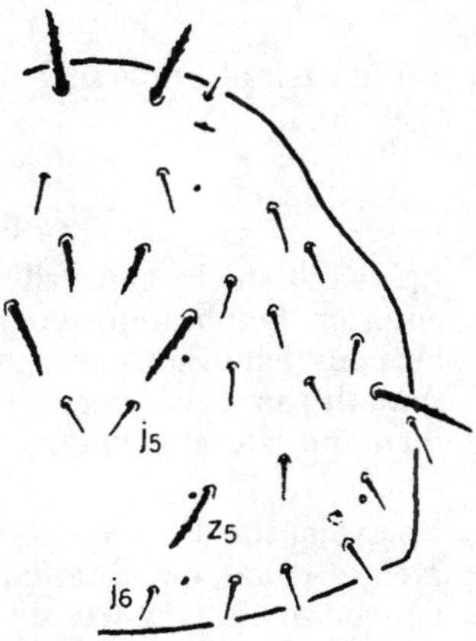

42

Fig. 20.39. *Vulgarogamasus* sp. (Oregon), female palp. **Fig. 20.40.** *Vulgarogamasus* sp. (Oregon), female dorsum. **Fig. 20.41.** *Trachygamasus* sp. (Quebec), dorsal aspect of pretarsus IV of deutonymph. **Fig. 20.42.** *Parasitus coleoptratorum* (L.), female podonotum. [After Micherdzindki (1969).]

1b Dorsal shield of female divided, rarely fused medially; dorsal shields of male may be fused laterally, resulting in a median separation of the podonotal and opisthonotal elements; tritosternum of male normal, reduced, or absent; dorsal setae z_5 may differ from setae j_5–j_6 in form and/or length; Subfamily **Parasitinae** **3**

2a Dorsal shield of female fused posteriorly with opisthogastric complex (Fig. 20.31) . . . . ***Holoparasitus*** Oudemans

2b Dorsal shield free posteriorly (Figs. 20.32, 20.33) . . . . ***Pergamasus*** Berlese

3a Dorsal setae z_5 differ markedly in form and/or length from adjacent setae j_5–j_6 (Fig. 20.42); tritosternum of male may be absent, reduced, or variously modified **4**

3b Dorsal setae z_5 similar to setae j_5 and j_6 in form and length (Fig. 20.40); tritosternum of male normally produced **5**

4a Dorsal shields of female fused medially or entirely separated; femur II of female, male, and deutonymphs with one or two large blunt ventral spurs (additional spurs may be present on more distal segments of leg II) (Fig. 20.34) . . . . ***Gamasodes*** Oudemans

4b Podonotal and opisthonotal shields separate; femur II without spurs as noted above **6**

5a Seta *al* of palpfemur undivided distally, spatulate; deutonymph with a dark transverse band on sternal shield between sternals 1 and 2, fixed cheliceral digit with a terminal mucro (Figs. 20.35–20.38) . . . .
Poecilochirus G. & R. Canestrini

5b Seta *al* of palpfemur divided distally into a series of narrow processes; deutonymphal sternal shield and chelicerae not as above (Figs. 20.39, 20.40) . . .
Vulgarogamasus Tichomirov

6a Paradactyli of pretarsi II–IV greatly elongated, acuminate (Fig. 20.41) . . .
Trachygamasus Berlese

6b Paradactyli of pretarsi II–IV may be acuminate but are not elongate as above (Fig. 20.42) . . . . ***Parasitus*** Latreille

Veigaiidae Oudemans, 1939

1a Epigynial shield generally represented by only a reduced ventral element, anterior membranous epigynial flap may be absent or indistinct; distal membranous extensions of internal malae with one to three produced points, smooth throughout; epistome bifurcate or trifurcate; if trifurcate, then median and lateral elements similarly developed (Figs. 20.43, 20.44) . . . . . .
Gamasolaelaps Berlese

1b Epigynial shield comprised of well-developed fused epigynial and ventral elements; distal membranous extensions of internal malae produced into a fringed moustache-like structure (Fig. 20.48); trifurcate epistome with median element longer than lateral elements, often divided distally (Figs. 20.45–20.48). . . . . ***Veigaia*** Oudemans

Digamasellidae Evans, 1957

1a Podonotal shield lacking setae z_3, marginal setae R_1 absent in nymphs and adults; without scleronoduli between setae z_5 in adult; anal aperture enlarged both in adults and immatures; femur II with 10 setae, ad_3 absent $\left(2\,\frac{2}{2}\,\frac{2}{1}\,1\right)$ (Fig. 20.49). . . . . ***Digamasellus*** Berlese

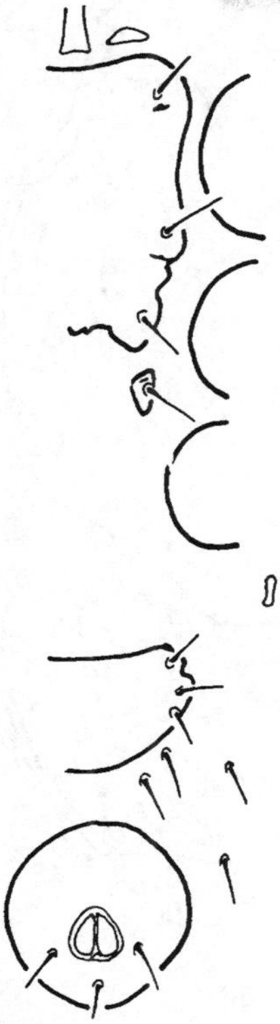

43

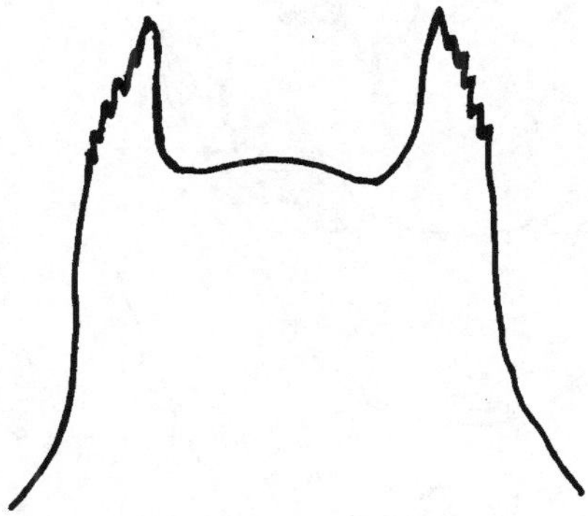

44

Fig. 20.43. *Gamasolaelaps whartoni* (Farrier) (North Carolina), female venter. [After Farrier (1957).]
Fig. 20.44. *Gamasolaelaps whartoni* (Farrier) (North Carolina), female epistome. [After Farrier (1957).]

1b Setae z_3 present on podonotal shield, with setae R_1 present in nymphs and adults, with two pairs of conspicuous scleronoduli between setae z_5 of adult; anal aperture of nymphs and adults not enlarged; femur II with 11 setae, ad_3 present $\left(2\,\frac{3}{2}\,\frac{2}{1}\,1\right)$ (Figs. 20.50, 20.51) . . . . ***Dendrolaelaps*** Halbert

Ologamasidae Ryke, 1962

1a Dorsal and ventrianal shields broadly or narrowly fused posteriorly in both sexes (Fig. 20.53) **2**

1b Dorsal and ventrianal shields not fused in female; if fused in male, then ventrianal is also fused to peritrematic shields **4**

2a Peritrematic shields extend posteriorly well beyond coxae IV, not fused to exopodals; male with reduced, edentate chelicerae (Fig. 20.52). . . . .
Stylochirus G. & R. Canestrini

2b Peritrematic shields either abbreviated or extending posterior to coxae IV, generally fused to exopodals IV; male chelicerae dentate, not reduced. . . **3**

3a Tibia I with three ventral setae, sternal–metasternal complex not fused with

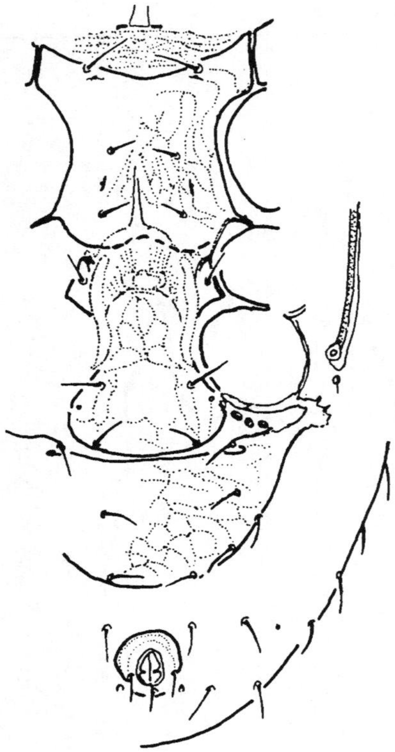

45

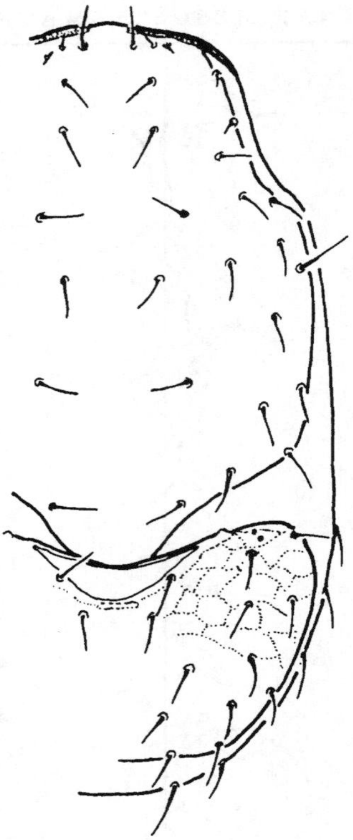

46

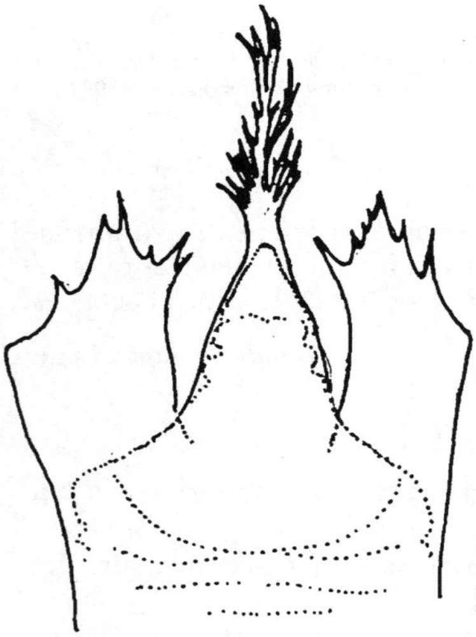

47

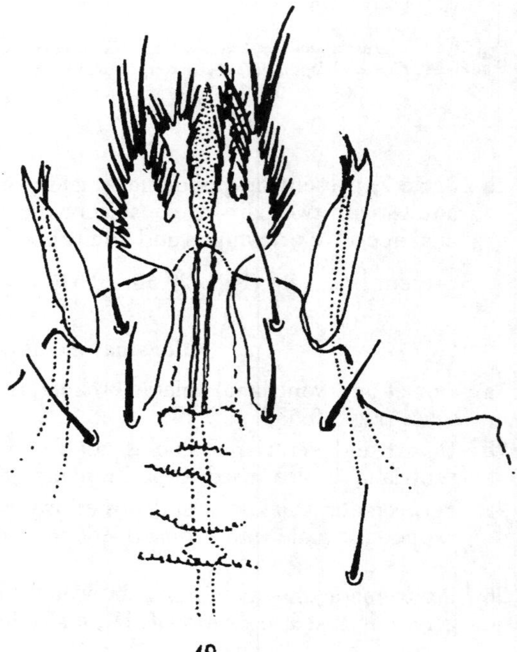

48

Fig. 20.45. *Veigaia serrata* Willmann (Oregon), female venter. **Fig. 20.46.** *Veigaia serrata* Willmann (Oregon), female dorsum. **Fig. 20.47.** *Veigaia* sp. (Oregon), female epistome. **Fig. 20.48.** *Veigaia* sp. (Oregon), ventral aspect of gnathosoma.

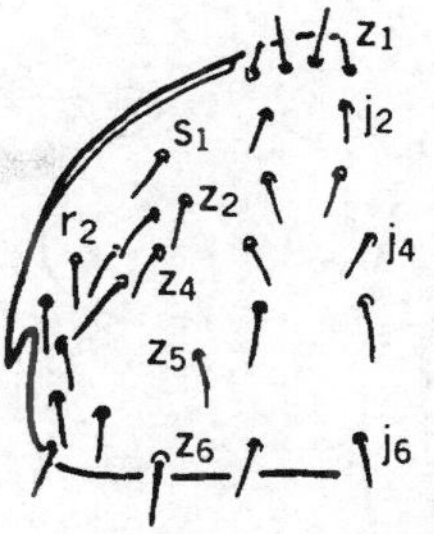

49

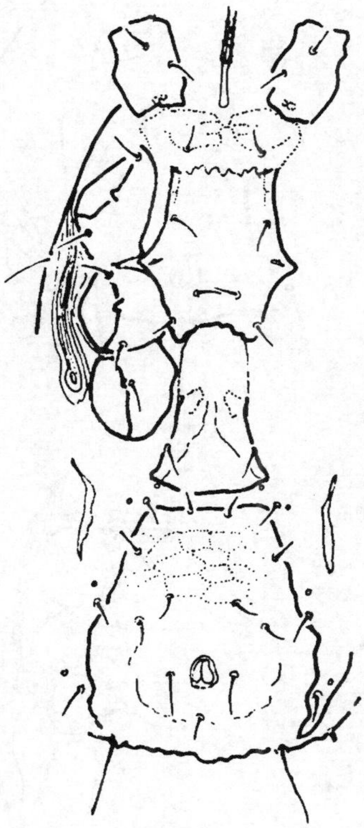

50

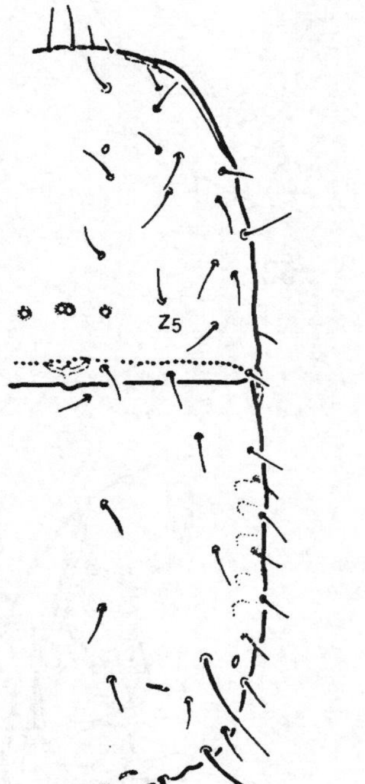

51

Fig. 20.49. *Digamasellus perpusillus* (Berlese), female podonotum. [After Hirschmann (1960).] **Fig. 20.50.** *Dendrolaelaps* sp. (Oregon), female venter. **Fig. 20.51.** *Dendrolaelaps* sp. (Oregon), female dorsum (z5).

endopodal shields IV; sternal setae 5 of male on ventrianal shield (Fig. 20.53) ***Hydrogamasus*** Berlese

3b Tibia I with four ventral setae, sternal–metasternal complex fused with exopodal shields IV; sternal setae 5 of male on fused sternitigenital shield (Figs. 20.54, 20.55) . . . . ***Gamasiphis*** Berlese

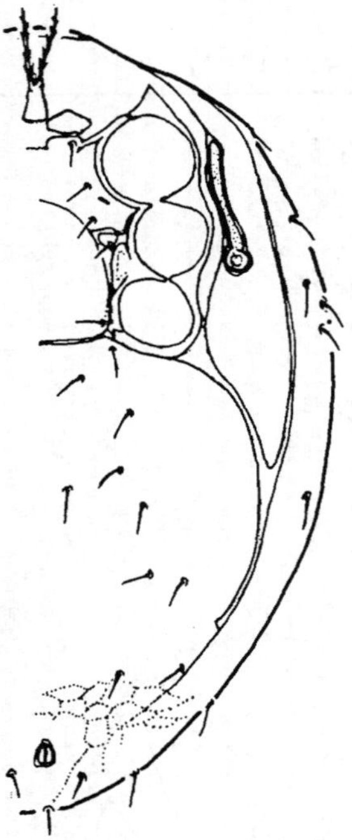

52

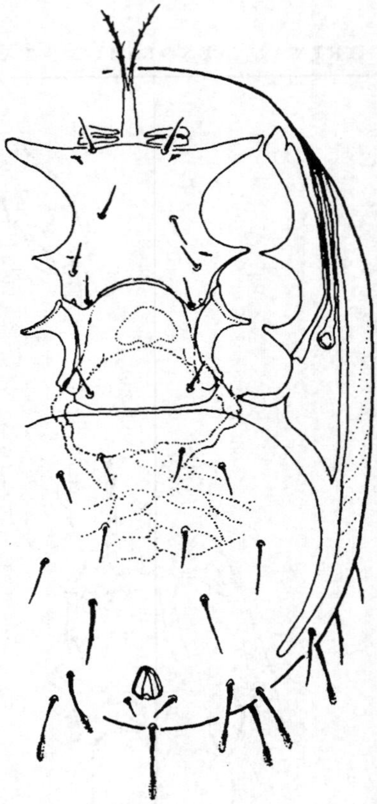

53

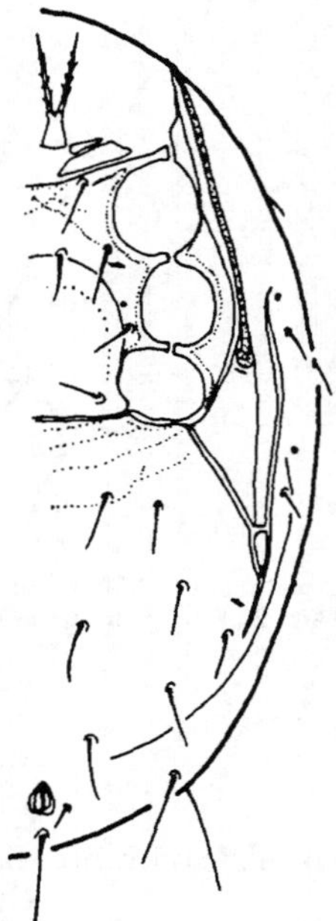

54

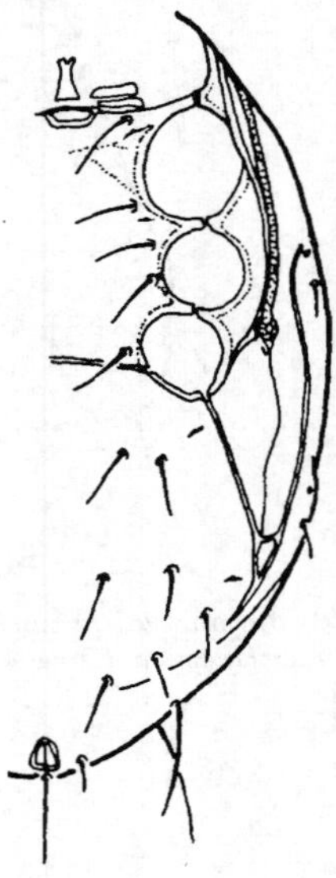

55

Fig. 20.52. *Stylochirus rarior* (Berlese) (Missouri), female venter. [After Lee (1970).] **Fig. 20.53.** *Hydrogamasus* sp. (Oregon), female venter. **Fig. 20.54.** *Gamasiphis pulchellus* (Berlese), female venter. [After Lee (1970).] **Fig. 20.55.** *Gamasiphis pulchellus* (Berlese), male venter. [After Lee (1970).]

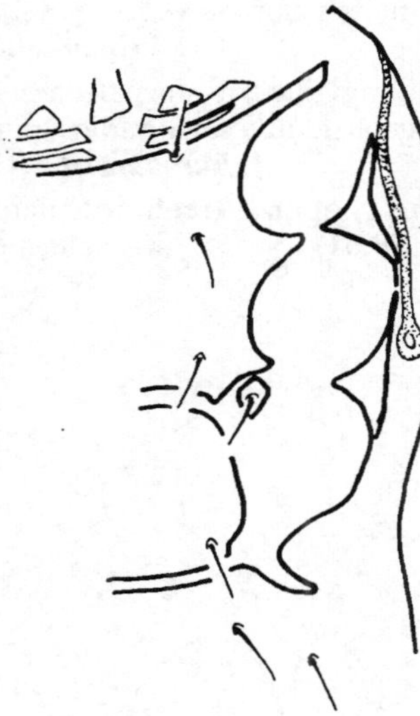

56

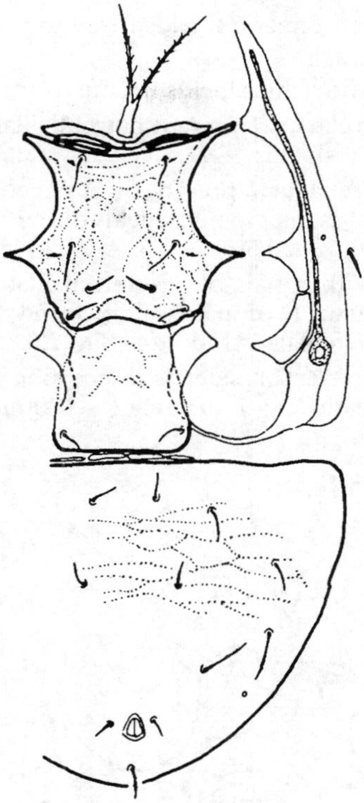

57

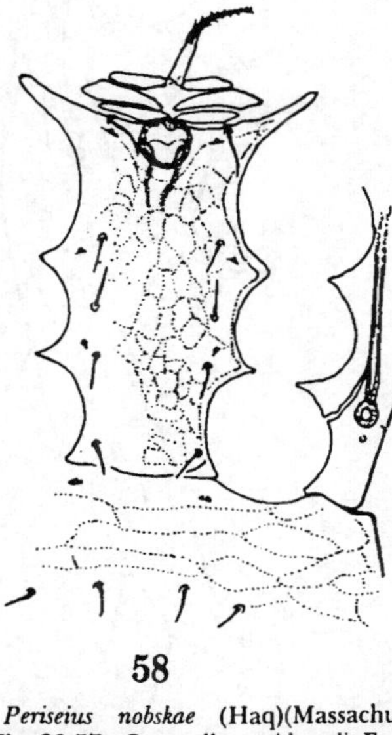

58

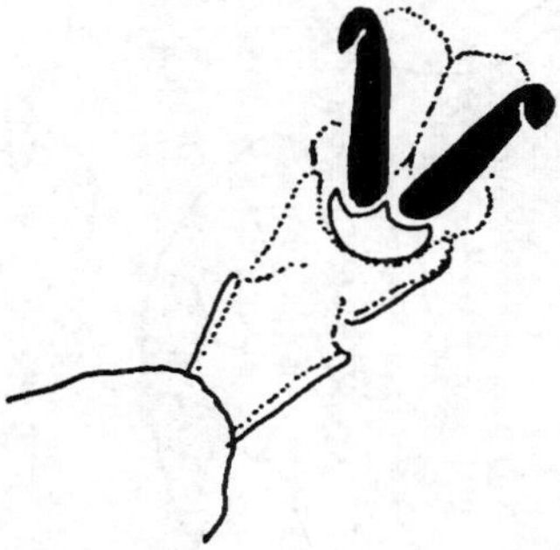

59

Fig. 20.56. *Periseius nobskae* (Haq)(Massachusetts), podogastric region of female. [After Haq (1965).] **Fig. 20.57.** *Gamasellus occidentalis* Emberson (Oregon), female venter. **Fig. 20.58.** *Gamasellus* sp. (Oregon), male venter. **Fig. 20.59.** *Gamasellus* sp. (Oregon), ventral aspect of pretarsus I.

4a Peritrematic shields fused to ventrianal shield (Fig. 20.56) . . . . ***Periseius*** Womersley

4b Peritremic shields not fused to ventrianal shield **5**

5a Pretarsus I present but pullvillar lobes reduced; female with discrete metapodal shields **6**

5b Pretarsus I present or absent; if present, then with well-developed pullvillar lobes; metapodal shields of female contained in a broad ventrianal shield (Figs. 20.57–20.59) . . . . ***Gamasellus*** Berlese

6a With a pair of presternal platelets flanking the tritosternal base; tibia and tarsus II of male rigidily fused; extremely large mites, with idiosomatic length of female >1500 μm (Fig. 20.60) . . . . ***Euryparasitus*** Oudemans

6b Presternal platelets absent; tibia and tarsus II of male freely articulated; idiosomatic length of female <1000 μm (Fig. 20.61) . . . . . ***Cyrtolaelaps*** Berlese

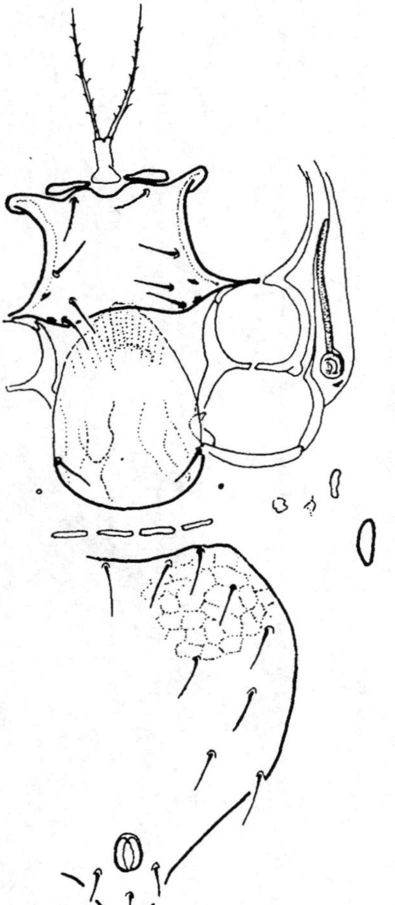

60

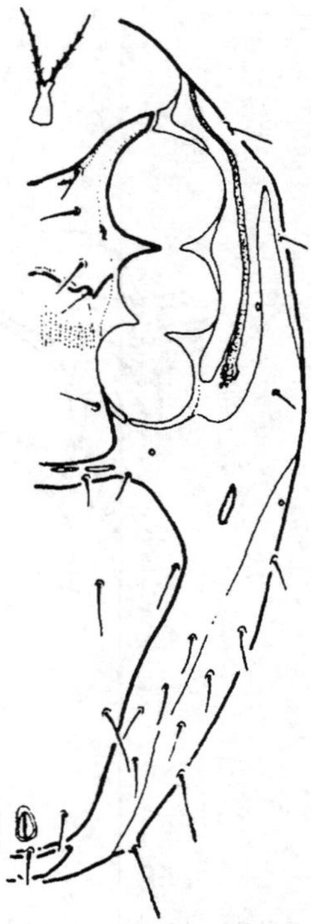

61

Fig. 20.60. *Euryparasitus* sp. (Oregon), female venter.

Fig. 20.61. *Cyrtolaelaps rectus* (Berlese), female venter. [After Lee (1970).]

Rhodacaridae Oudemans, 1902

1a Median pair of podonotal scleronoduli fused (Fig. 20.65) (***Rhodacarus minimus*** Karg is an exception), flanked by a more lateral pair; hypostomatic setae 3 displaced posteriorly, inserted between hypostomatic setae 2 and the capitular setae (Fig. 20.66) **2**

1b With four separate podonotal scleronoduli; hypostomatic setae 2 and 3 at approximately the same level (Fig. 20.62) . . . . . . ***Rhodacarellus*** Willmann

2a Pretarsus I present, consisting of paired sessile claws; with two pairs of presternal shields (Fig. 20.63) . . . . ***Rhodacaropsis*** Willmann

2b Pretarsus I normally absent (if present, consisting of paired sessile claws); without typical presternal shields (Fig. 20.64–20.66) . . . . ***Rhodacarus*** Oudemans

Ascidae Oudemans, 1905

1a Median pulvillar lobe of pretarsi II–IV slender and acute (Fig. 20.72); femora I and II with 11 and 10 setae, respectively, lacking proximoventral seta (v_3); paranal setae inserted on a level with or posterior to hind margin of anal aperture, usually longer than postanal seta (Fig. 20.71); hypostomatic setae 1 and paraxial palptrochanteral setae extremely long, whiplike, tapering only near tip; Subfamily **Platyseiinae**. **2**

1b Median pulvillar lobe of pretarsi II–IV broadly rounded (Fig. 20.79), femora I and II with 12 and 11 setae, respectively, v_3 present; paranal setae inserted anterior to hind margin of anal aperture, usually shorter than postanal seta (Fig. 20.80); hypostomatic setae 1 and paraxial palptrochanteral setae not whiplike and generally not elongate, tapering along entire length. **3**

2a Opisthonotum with two to four pairs of setae in *J* series; sternal setae 1 distinctly shorter than sternals 3; tibia II with nine setae, pd_1 absent $\left(2\,\frac{2}{1}\,\frac{1}{1}\,2\right)$ (Figs. 20.67, 20.68). . . . . ***Platyseius*** Berlese

2b Opisthonotum with five pairs of *J* setae; sternal setae 1 and 3 subequal in length; tibia II with 10 setae, pd_1 present $\left(2\,\frac{2}{1}\,\frac{2}{1}\,2\right)$ (Figs. 20.69–20.72) ***Cheiroseius*** Berlese

3a Opisthonotum usually with four pairs of lateral (*S*) setae, S_2 generally absent (Fig. 20.77); genu and tibia I without seta pd_3 $\left(2\,\frac{3}{2}\,\frac{2}{1}\,2\right)$, genu and tibia II each with only one *pl* seta; genua and tibiae III and IV with only seven setae, al_2 absent $\left(1\,\frac{2}{1}\,\frac{2}{0}\,1 \text{ and } 1\,\frac{1}{1}\,\frac{2}{1}\,1\right)$; rows of deutosternal denticles moderately wide, multidenticulate (Fig. 20.78); Subfamily **Arctoseiinae** **4**

3b Opisthonotum usually with five pairs of *S* setae, S_2 present (Fig. 20.81); genu and tibia I with pd_3 present, genu and tibia II with two *pl* setae; genua and tibiae III and IV with 8–10 setae, al_2 present; rows of deutosternal denticles variously developed, often narrow and each with few (2–10) denticles (Fig. 20.99); Subfamily **Ascinae**. **7**

4a Female with one to six pairs of opisthogastric setae on a ventrianal shield; genu III with eight setae, pv_1 present $\left(1\,\frac{2}{1}\,\frac{2}{1}\,1\right)$; tarsi II–IV each with elongate and curved dorsoproximal setae (ad_2 and pd_2). **5**

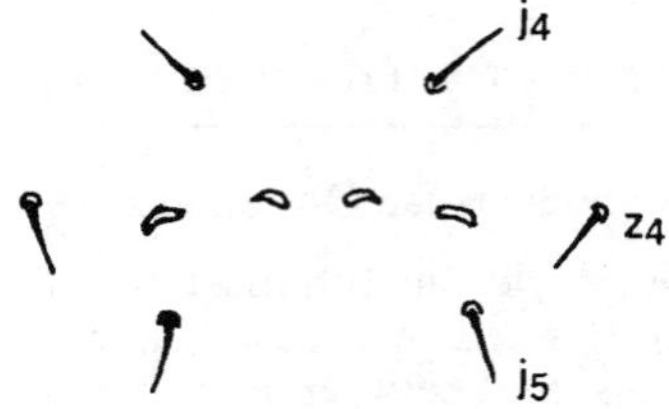

62

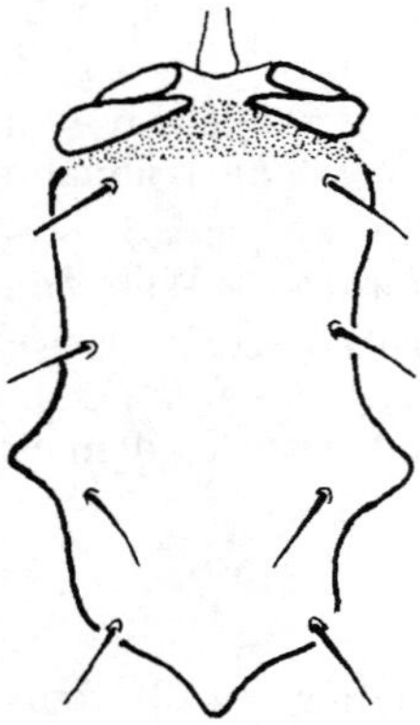

63

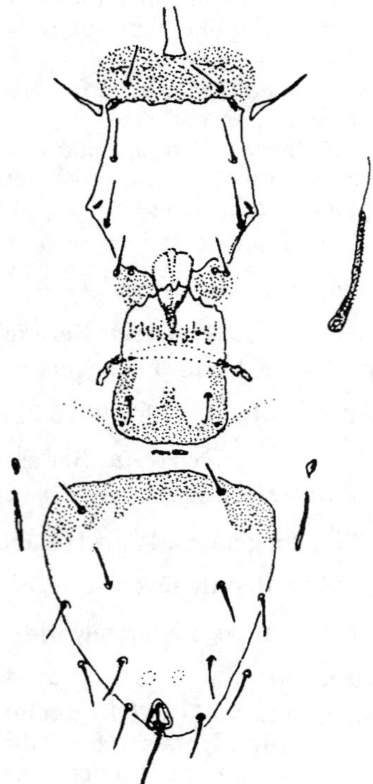

64

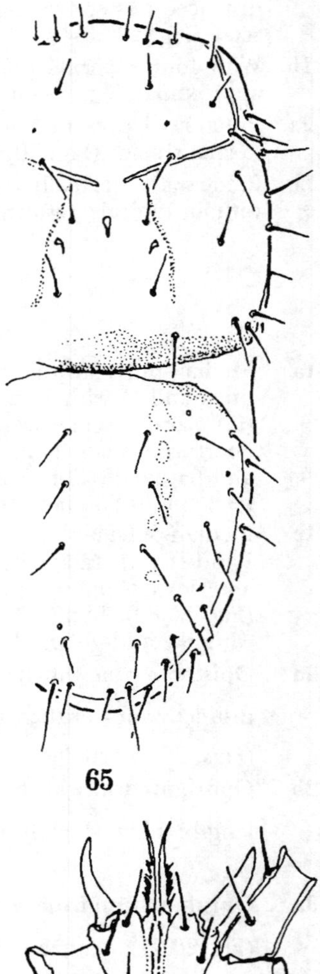

65

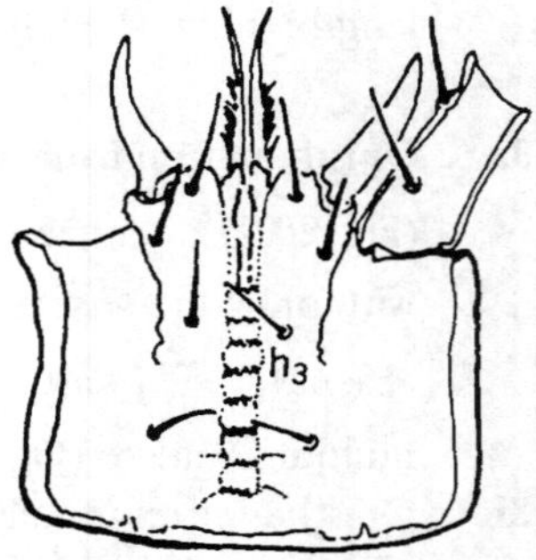

66

Fig. 20.62. *Rhodacarellus,* portion of podonotum showing position of scleronoduli. **Fig. 20.63.** *Rhodacaropsis inexpectatus* Willmann, female sternal shield. [After Ghilyarov et al. (1977).] **Fig. 20.64.** *Rhodacarus roseus* Oudemans (Oregon), female venter. **Fig. 20.65.** *Rhodacarus roseus* Oudemans (Oregon), female dorsum. **Fig. 20.66.** *Rhodacarus roseus* Oudemans (Oregon), ventral aspect of female gnathosoma.

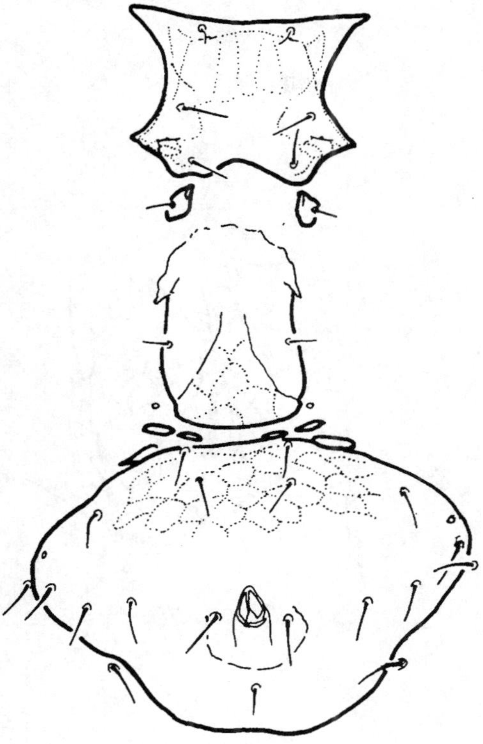

67

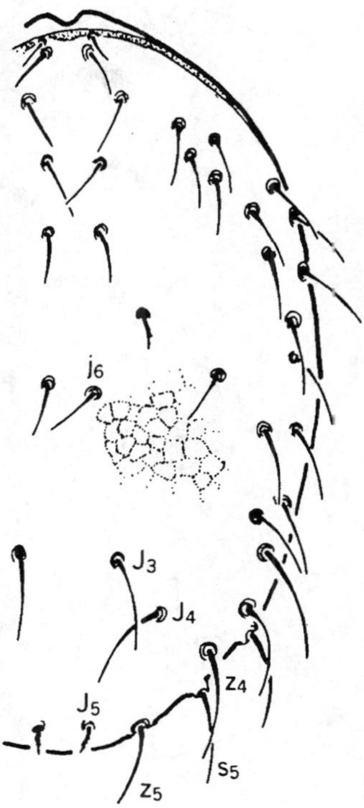

68

Fig. 20.67. *Platyseius* sp. (Idaho), female venter. **Fig. 20.68.** *Platyseius* sp. (Idaho), female dorsum.

4b Female with an anal shield carrying only the typical adanal and postanal setae; genu III with seven setae, pv_1 absent $\left(1\ \frac{2}{1}\ \frac{2}{0}\ 1\right)$; ad_2 and pd_2 of tarsi II–IV neither elongate nor curved **6**

5a Tarsus I without pretarsus, all dorsal setae simple; epistome triramous, with a greatly elongate and apically bifid median process (Fig. 20.73). . . . . *Xenoseius* Lindquist and Evans

5b Tarsus I with pretarsus; some dorsal shield setae (always s_4 and Z_5) paddle-shaped; epistome triramous but with median process usually similar to lateral ones (Figs. 20.74, 20.75). . . . . ***Zerconopsis*** Hull

6a Tarsi II–IV with al_1, or al_1 and pl_1, very slender and elongate, palptarsus with macroseta; vertex of dorsal shield strongly arched downward, concealing setae j_1; anterior extremities of elongate peritremes sharply recurved posteroventrally (Fig. 20.76). . . . . ***Iphidozercon*** Berlese

6b Setae al_1 and al_2 of tarsi II–IV neither slender nor elongate; palptarsus without macroseta; vertex of dorsal shield not strongly arched; peritremes short or long, their anterior extremities not recurved (Figs. 20.77–20.80) ***Arctoseius*** Thor

7a Dorsal shield of deutonymph and adult entire, lateral incisions of deutonymphal dorsal shield retained or obliterated in adult. **8**

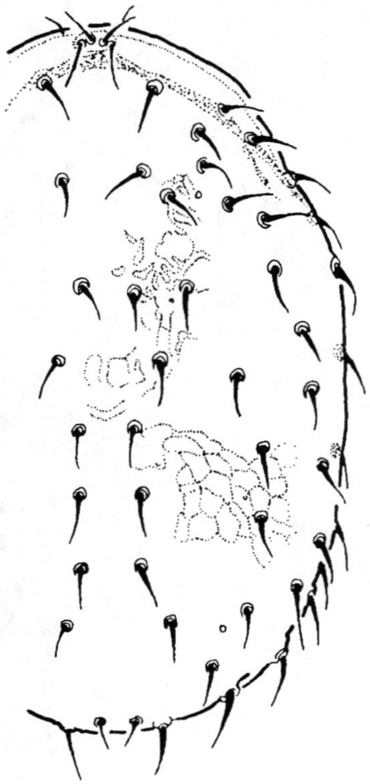

69

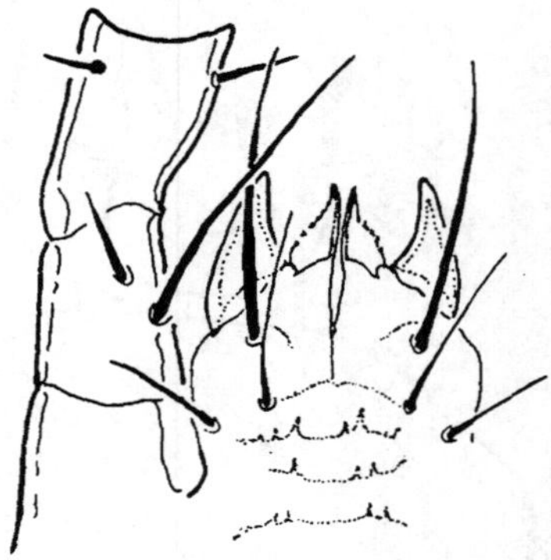

70

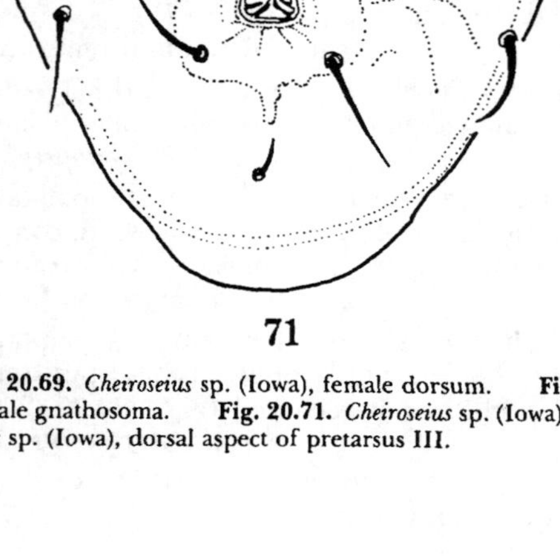

71

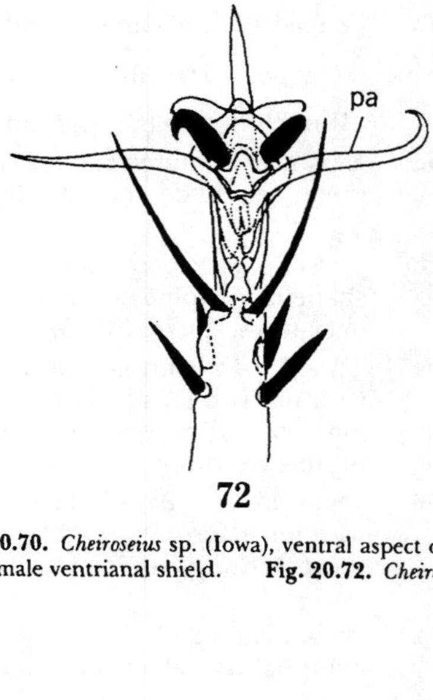

72

Fig. 20.69. *Cheiroseius* sp. (Iowa), female dorsum. **Fig. 20.70.** *Cheiroseius* sp. (Iowa), ventral aspect of female gnathosoma. **Fig. 20.71.** *Cheiroseius* sp. (Iowa), female ventrianal shield. **Fig. 20.72.** *Cheiroseius* sp. (Iowa), dorsal aspect of pretarsus III.

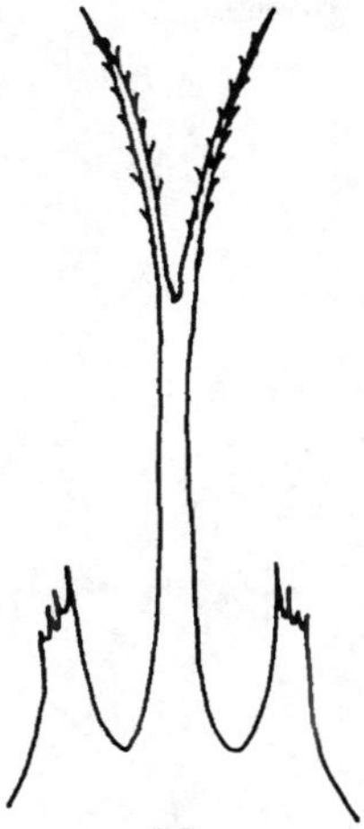

73

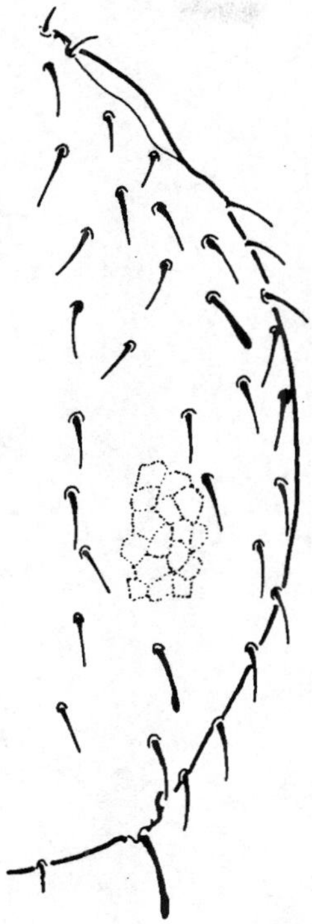

74

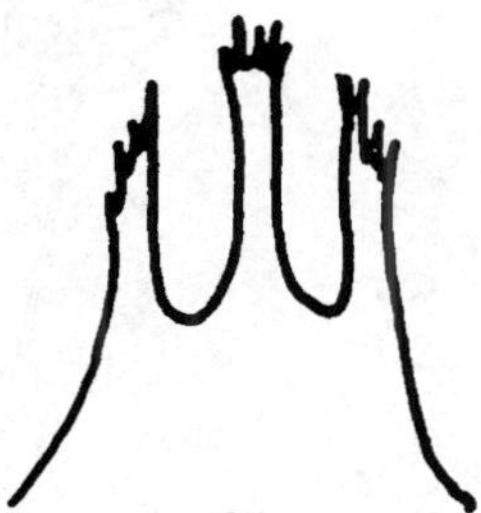

75

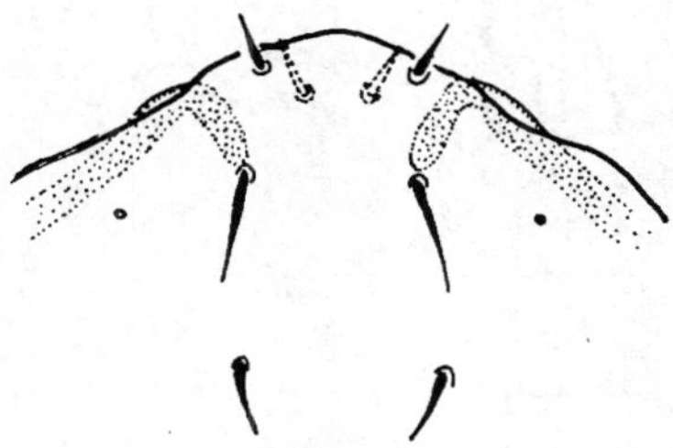

76

Fig. 20.73. *Xenoseius clayi* Evans and Hyatt, epistome. [After Evans and Hyatt (1960).] **Fig. 20.74.** *Zerconopsis remiger* (Kramer), female dorsum. [After Westerboer (1963).] **Fig. 20.75.** *Zerconopsis remiger* (Kramer), epistome. **Fig. 20.76.** *Iphidozercon californicus* Chant (California), anterior region of female podonotum. [After Lindquist and Evans (1965).]

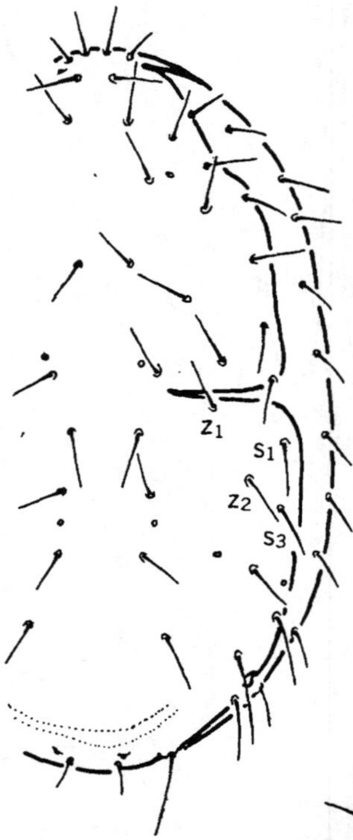

77

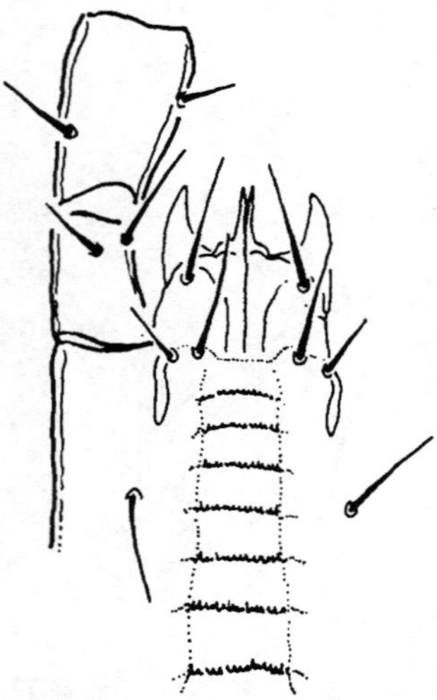

78

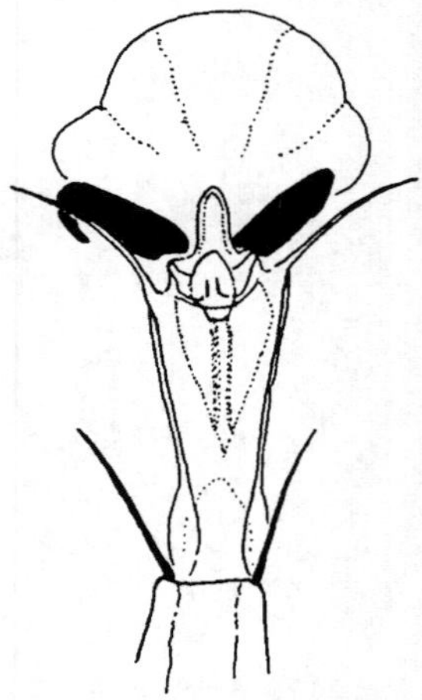

79

Fig. 20.77. *Arctoseius* sp. (Oregon), female dorsum. **Fig. 20.78.** *Arctoseius* sp. (Oregon), ventral aspect of female gnathosoma. **Fig. 20.79.** *Arctoseius* sp. (Oregon), dorsal aspect of pretarsus III.

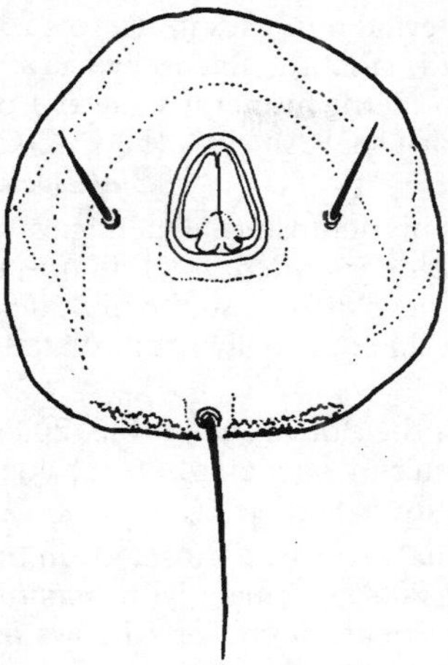

80

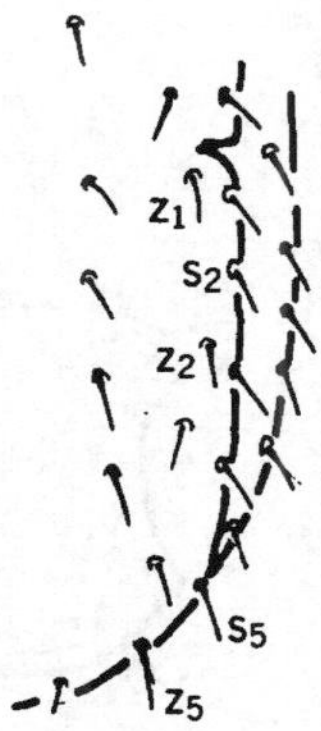

81

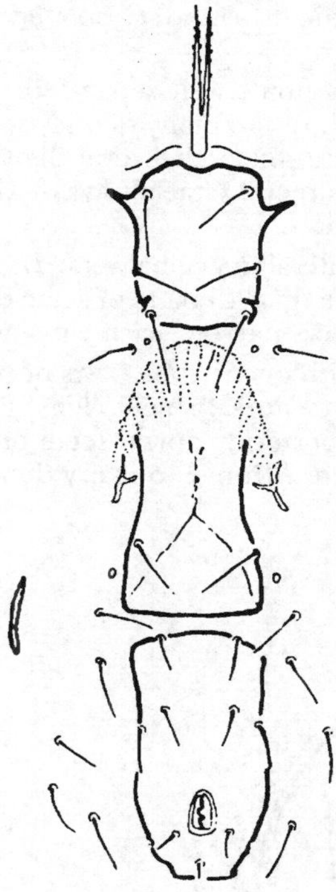

82

Fig. 20.80. *Arctoseius* sp. (Oregon), female anal shield. **Fig. 20.81.** *Leioseius minusculus* (Berlese) (Connecticut), dorsal aspect of female opisthosoma. **Fig. 20.82.** *Blattisocius tarsalis* (Berlese) (Oregon), female venter.

7b Dorsal shield of deutonymph and adult complete divided in two (Fig. 20.101). **14**

8a Lateral dorsal shield incisions retained in adult (Fig. 20.81). ***Leioseius*** Berlese

8b Lateral dorsal shield incisions obliterated in adult. **9**

9a Epigynial shield of female usually truncate posteriorly (Fig. 20.82), with a ventrianal shield carrying two to seven pairs of opisthogastric setae; peritrematic shield of adults broadly joined posteriorly to exopodal shield which curves behind coxa IV. **10**

9b Epigynial shield of female rounded posteriorly (Fig. 20.102), with an anal shield which rarely carries one pair of opisthogastric setae; peritrematic shield of adults free or narrowly joined posteriorly to exopodal shield adjacent to coxa IV. **17**

10a Corniculi slender, approximal; deutosternal denticles in narrow rows, denticles few in number; epistome rounded, smooth; fixed cheliceral digit well developed or reduced, with few or no teeth; humeral setae (r_3) usually off dorsal shield, without isolated metasternal shields (Figs. 20.82–20.84) ***Blattisocius*** Keegan

10b Corniculi stout, well separated; rows of deutosternal denticles narrow or wide, with few to many denticles; epistome rounded or triramous, smooth or denticulate; fixed cheliceral digit well developed, usually multidentate (Fig. 20.96); humeral setae always on dorsal shield, usually with isolated metasternal shields. **11**

11a Adults with all marginal setae (*r*, *R*) on the dorsal shield (Fig. 20.86), female with more than 15 pairs of setae on posterior portion of dorsal shield; deutosternal rows narrow, each with two to four denticles. **12**

11b Adults with one to nine pairs of marginal setae (*r*, *R*) inserted in integument adjacent to dorsal shield (Figs. 20.89, 20.92), female with maximum of 15 pairs of posterior dorsal setae on the shield, deutosternal rows moderately wide, each with five to many denticles. **13**

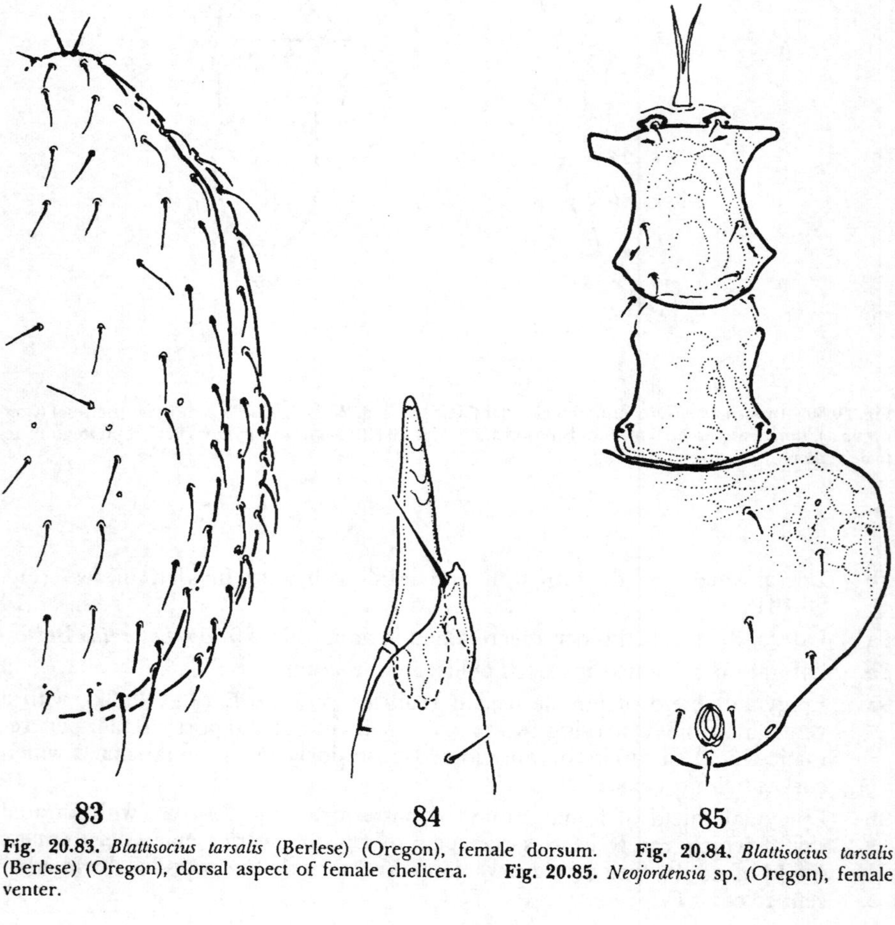

83 84 85

Fig. 20.83. *Blattisocius tarsalis* (Berlese) (Oregon), female dorsum. **Fig. 20.84.** *Blattisocius tarsalis* (Berlese) (Oregon), dorsal aspect of female chelicera. **Fig. 20.85.** *Neojordensia* sp. (Oregon), female venter.

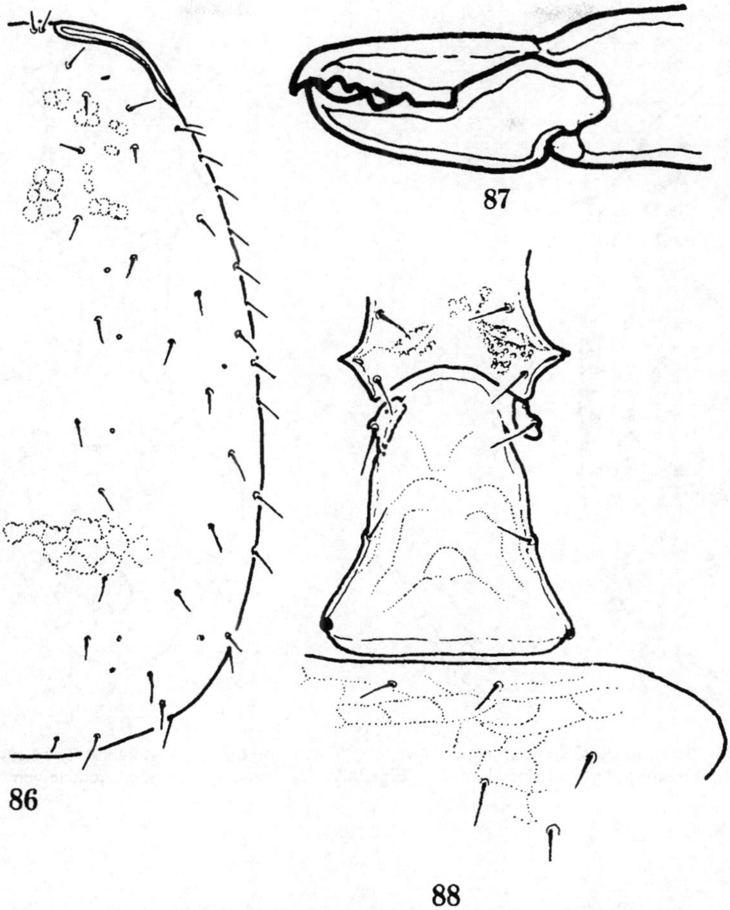

Fig. 20.86. *Neojordensia* sp. (Oregon), female dorsum. **Fig. 20.87.** *Neojordensia* sp. (Oregon), female chelicera. **Fig. 20.88.** *Zercoseius spathuliger* (Leonardi) (Oregon), female venter.

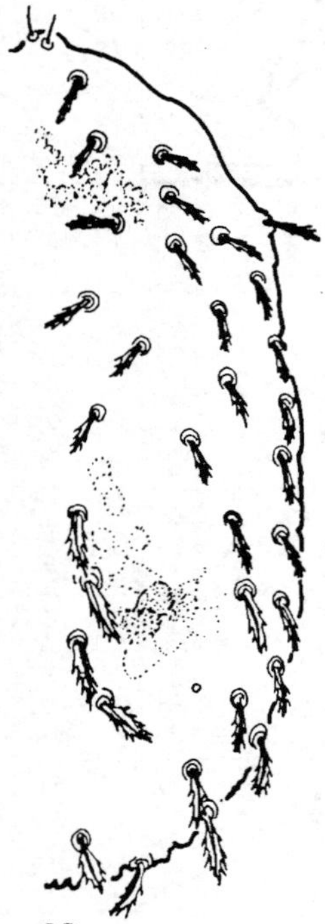

89

90

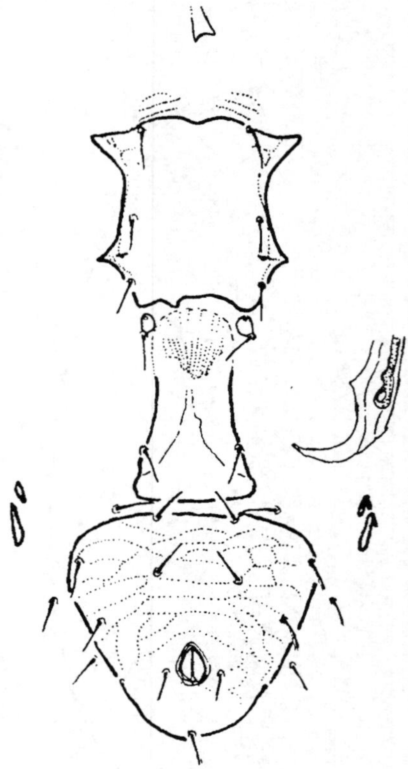

91

Fig. 20.89. *Zercoseius spathuliger* (Leonardi) (Oregon), female dorsum. **Fig. 20.90.** *Zercoseius spathuliger* (Leonardi) (Oregon), detail of dorsal seta. **Fig. 20.91.** *Lasioseius* sp. (Oregon), female venter.

12a Female with third pair of sternal pores on sternal shield, metasternal setae without associated shields (Figs. 20.85–20.87). . . . . . . ***Neojordensia*** Evans

12b Female with third pair of sternal pores and metasternal setae on metasternal shields. . . . . ***Orthadenella*** Athias-Henriot

13a Adults with only one pair of marginal setae (R_1) inserted in integument adjacent to dorsal shield; setae z_1, s_1–s_2, r_2 not present on anterior portion of dorsal shield; epigynial shield strongly widened posteriorly, bearing paragenital pores; male with holoventral shield; genu III with 10 setae, pl_2 present $\left(2\ \frac{2}{1}\ \frac{2}{1}\ 2\right)$ (Figs. 20.88–20.90). . . . . ***Zercoseius*** Berlese

13b Adults with three to nine pairs of marginal setae (r, R) inserted in integument; anterior portion of dorsal shield usually holotrichous or, if lacking several setae, j_1 absent; epigynial shield not strongly widened posteriorly, paragenital pores in adjacent integument; males with ventral shield complexes; genu III usually with nine setae, pl_2 absent (Figs. 20.91, 20.92). . . . ***Lasioseius*** Berlese

14a Pretarsus I typically absent, movable cheliceral digit with a basal pointed ventral projection; female with an inversely triangular anal or small ventrianal shield bearing up to two pairs of opisthogastric setae; epigynial shield strongly rounded posteriorly (Fig. 20.93–20.96). . . . . ***Antennoseius*** Berlese

14b Pretarsus I present, movable cheliceral digit without ventral projection; female with a nontriangular ventrianal shield bearing two to six pairs of opisthogastric setae (Fig. 20.97); epigynial shield truncate posteriorly. **15**

15a Dorsal setae Z_4 and S_5 usually approximal and borne on a pair of prominent posterolateral tubercles; setae z_1 absent, j_2 well behind level of j_1; peritrematic shield of adult broad, truncate posteriorly; genu I with 12 setae, av_2 absent $\left(2\,\frac{3}{1}\,\frac{3}{1}\,2\right)$ (Figs. 20.97–20.99). . . . . ***Asca*** Heyden

15b Dorsal setae Z_4 and S_5 normally spaced, not borne on posterolateral tubercles; setae z_1 present, j_2 tending to be displaced anteriorly, often in approximately transverse row with setae z_1 and $j1$ (Fig. 20.101); peritrematic shield of adult narrow or of moderate width, tapered posteriorly; genu I with 13 setae, av_2 present **16**

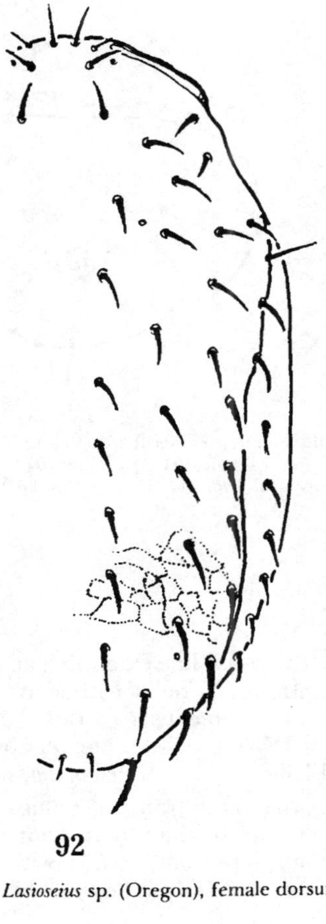

92

Fig. 20.92. *Lasioseius* sp. (Oregon), female dorsum.

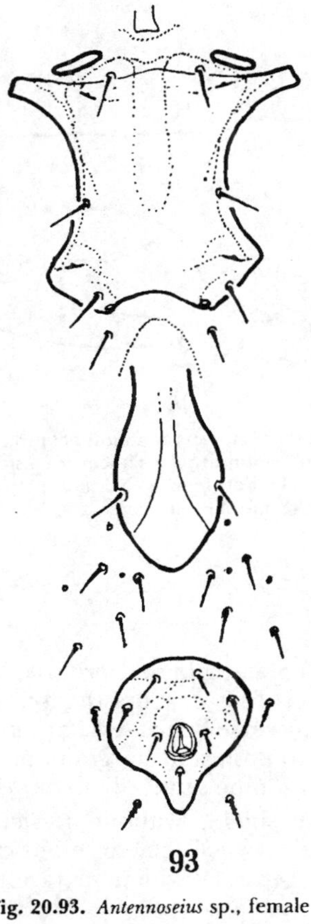

93

Fig. 20.93. *Antennoseius* sp., female venter

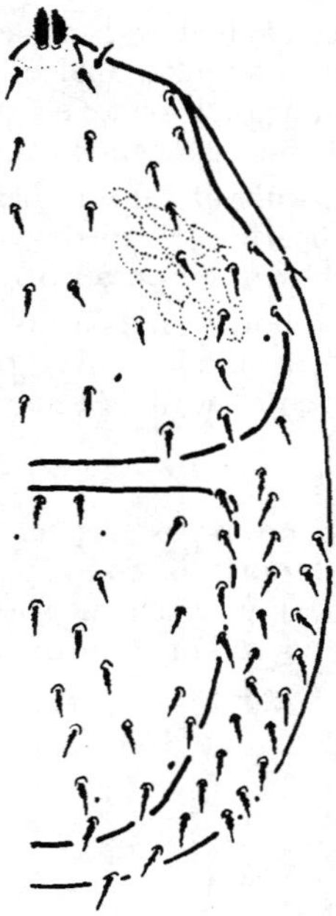

94

95

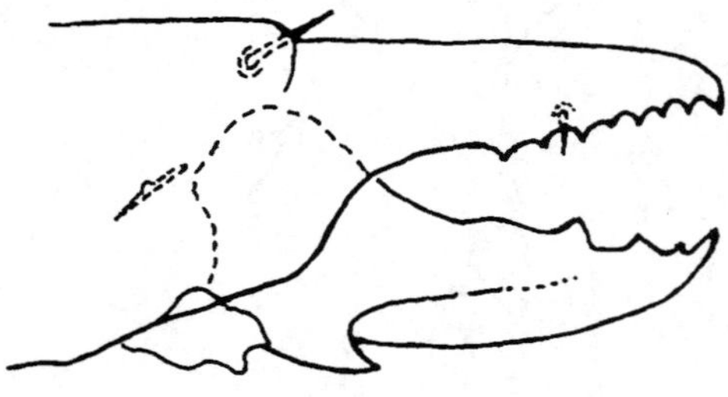

96

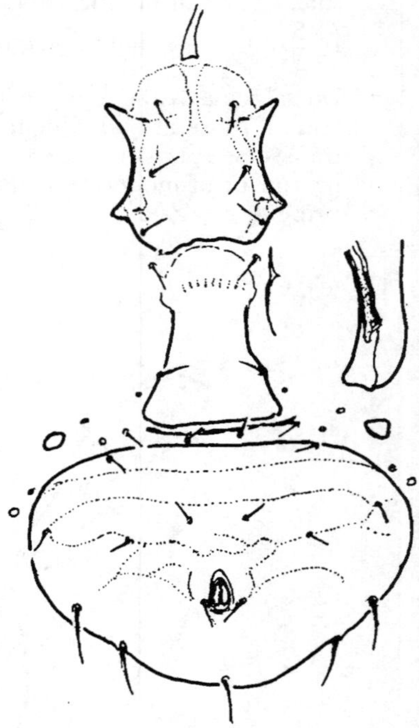

97

Fig. 20.94. *Antennoseius* sp., *borussicus* Sellnick, female dorsum. [After Bernhard (1963).] **Fig. 20.95.** *Antennoseius* sp., *borussicus* Sellnick, female epistome. [After Weis-Fogh (1948).] **Fig. 20.96.** *Antennoseius bacatus* Athias-Henriot, female chelicera. [After Lindquist and Evans (1965).] **Fig. 20.97.** *Asca nova* (Willmann) (Oregon), female venter.

16a Podonotal shield desclerotized behind transverse line extending across shield at level of setae z_6, opisthonotal shield with similar desclerotized area anterior to transverse line at level of setae J_1; paranal setae inserted closer to anterior than to posterior margin of anus; genu IV with eight setae, pl_1 absent; tibia IV with nine setae, pl_2 absent (Fig. 20.100). . . . . . . ***Protogamasellus*** Karg

16b Dorsal shields without desclerotized areas and transverse lines as noted above; paranal setae inserted closer to posterior than to anterior margin of anus; genus IV with nine (rarely six) setae, pl_1 present; tibia IV with 10 (rarely eight) setae, pl_2 present (Fig. 20.101). . . . . . ***Gamasellodes*** Athias-Henriot

17a Adults with 7–13 pairs of marginal setae (*r*, *R*) inserted in integument adjacent to dorsal shield (rarely fewer in cases where some marginals are lacking); female with maximum of 15 pairs of setae on posterior portion of dorsal shield (Figs. 20.102, 20.103). . . . . ***Melichares*** Hering

17b Adults with up to three pairs of the most posterior marginal setae inserted in integument adjacent to dorsal shield, others on shield margin; female with 18–22 pairs of setae on posterior portion of dorsal shield (Fig. 20.104) ***Proctolaelaps*** Berlese

Halolaelapidae Karg, 1965

1a Dorsum with a holonotal shield (Fig. 20.105). . . . . ***Halodarcia*** Karg

1b Dorsum with two or more shields. 2

2a Dorsum of female with a podonotal shield anteriorly, and four mesonotal shields and a pygidial shield posteriorly; setae Z_4 and S_5 borne on a pair of posterolateral protuberances (Fig. 20.106). . . . . ***Saprosecans*** Karg

2b Dorsum with a divided dorsal shield; hysterosomatic shield entire, commonly with an anterior longitudinal separation between anterior *J* setae. 3

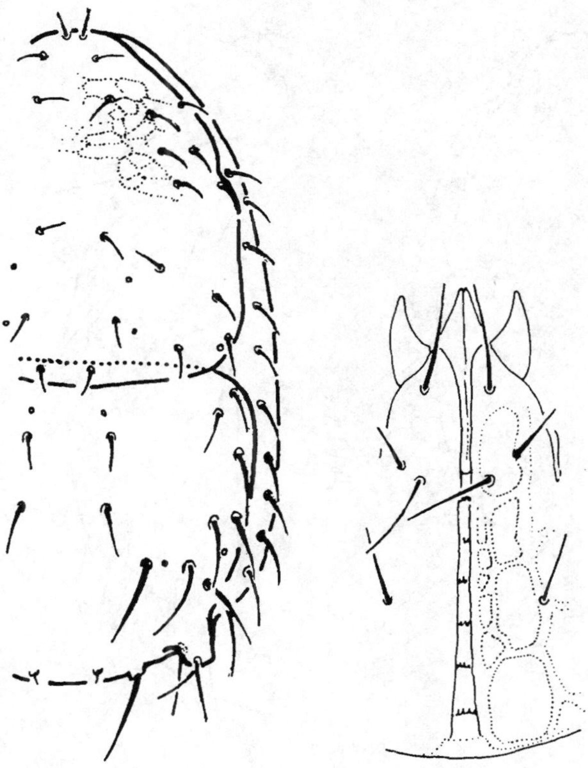

98 99

Fig. 20.98. *Asca nova* (Willmann) (Oregon), female dorsum. **Fig. 20.99.** *Asca* sp. (Oregon), ventral aspect of female gnathosoma.

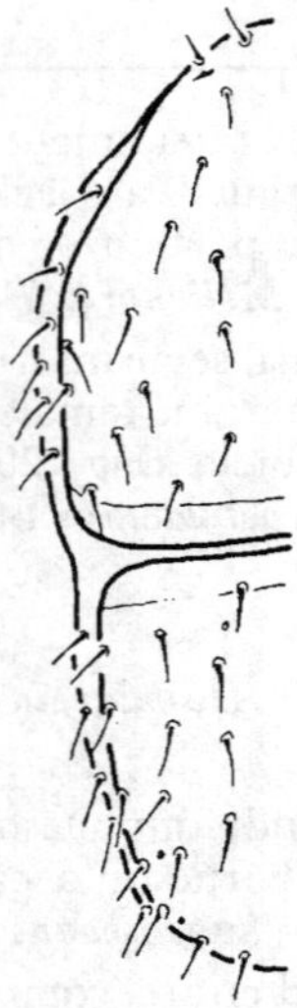

100

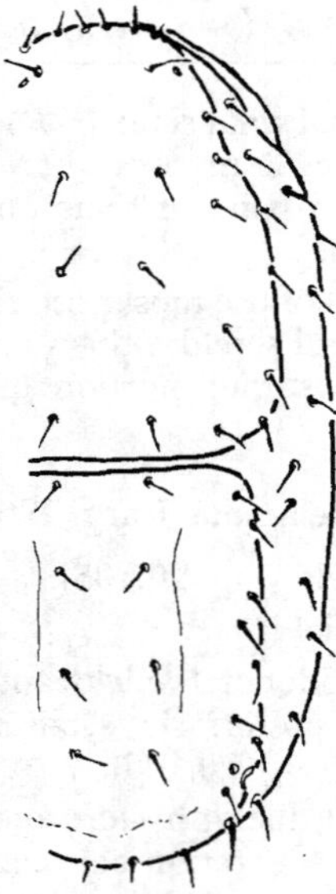

101

102

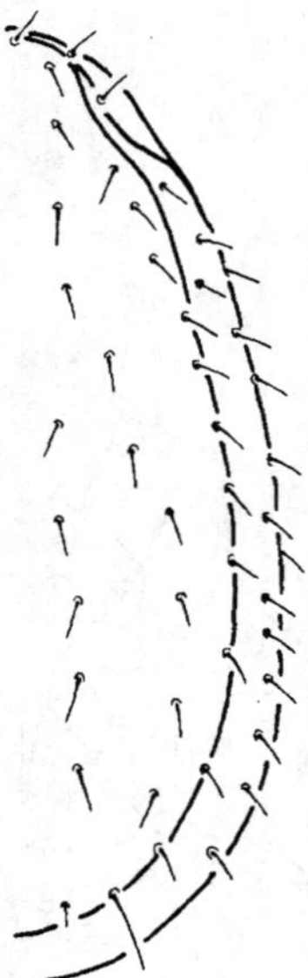

103

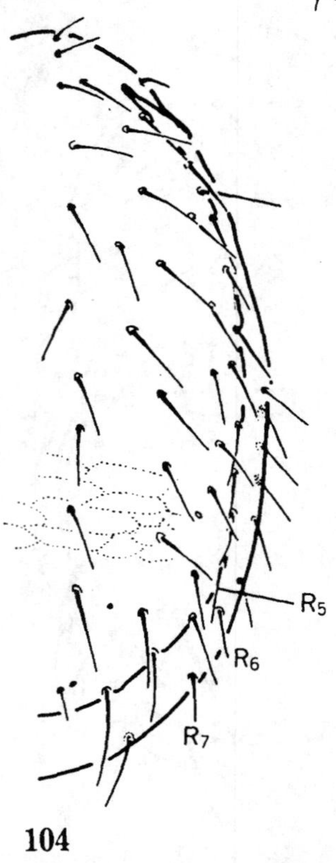

104

Fig. 20.100. *Protogamasellus*, female dorsum (composite). **Fig. 20.101.** *Gamasellodes* sp. (Oregon), female dorsum. **Fig. 20.102.** *Melichares agilis* Hering, female venter. [After Evans (1958).] **Fig. 20.103.** *Melichares agilis* Hering, female dorsum. [After Evans (1958).] **Fig. 20.104.** *Proctolaelaps hypudaei* (Oudemans) (Oregon), female dorsum.

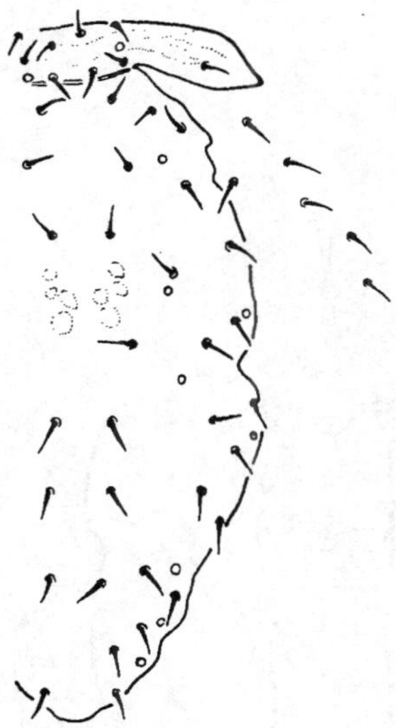
105

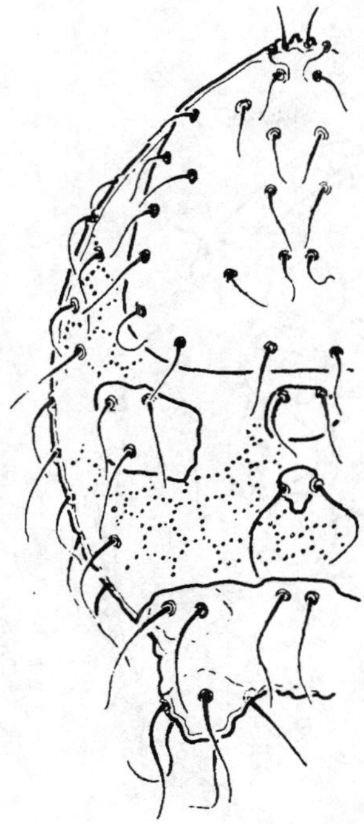
106

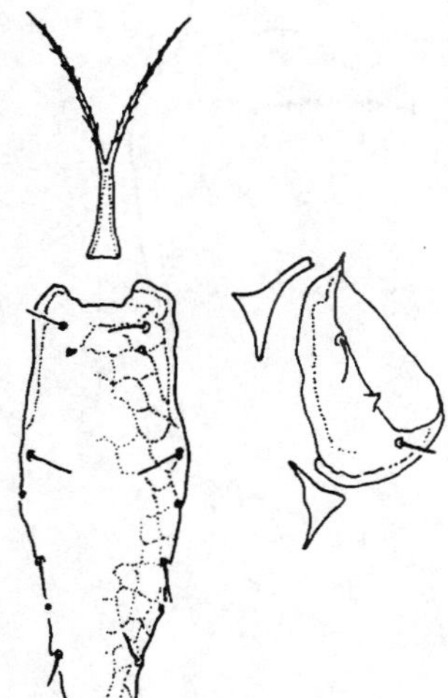
107

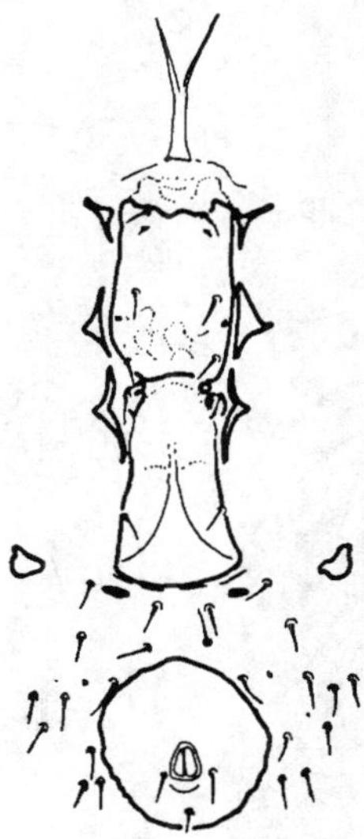
108

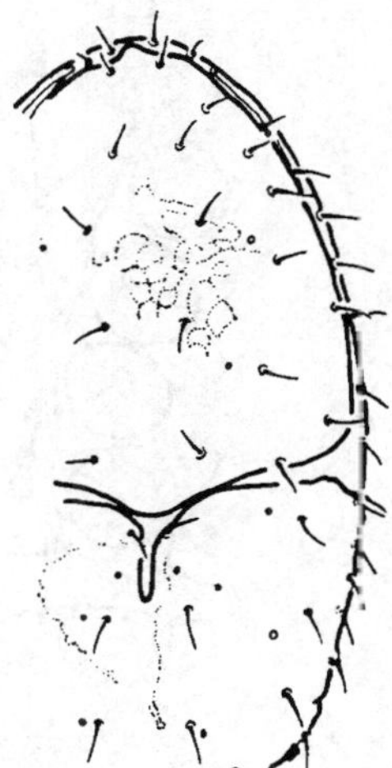
109

Fig. 20.105. *Halodarcia* sp. (Alberta), female dorsum. **Fig. 20.106.** *Saprosecans* sp. (Oregon), female dorsum. **Fig. 20.107.** *Saprolaelaps* sp. (Ontario), coxa II and intercoxal region of deutonymph. **Fig. 20.108.** *Halolaelaps* sp. (Oregon), female venter. **Fig. 20.109.** *Halolaelaps* sp. (Oregon), female dorsum.

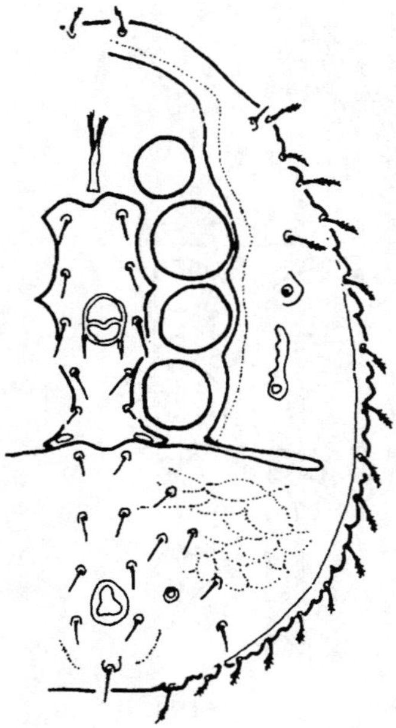

110

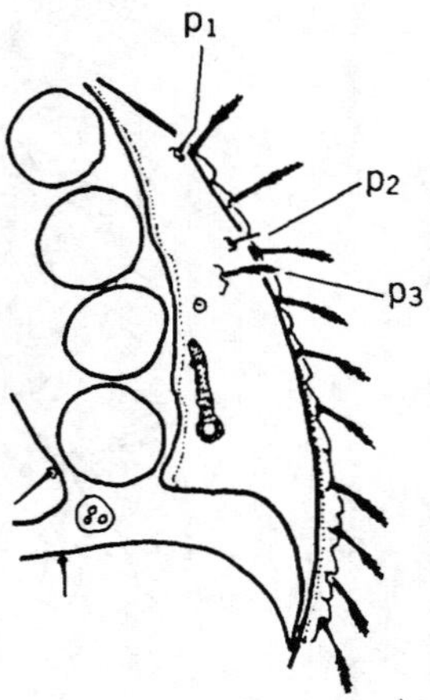

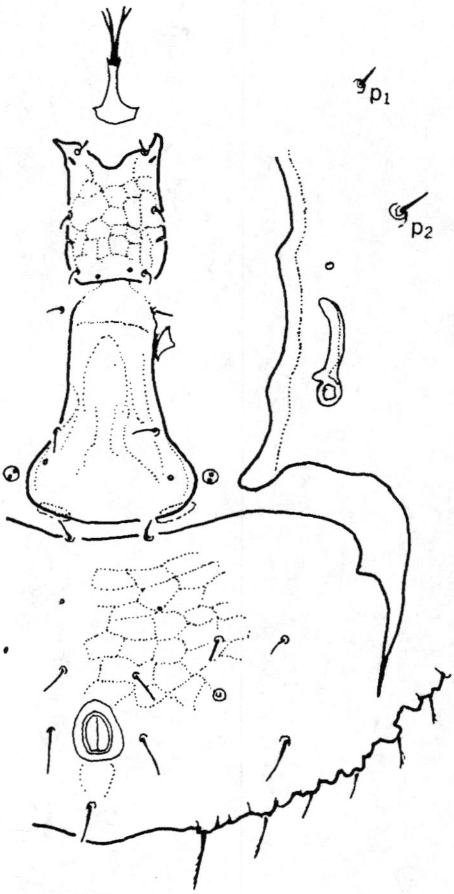

112

111

Fig. 20.110. *Amerozercon suspiciosus* Halašková, male venter. [After Halašková (1969).] **Fig. 20.111.** *Parazercon* sp., female parapodal region. [After Blaszak (1975).] **Fig. 20.112.** *Microzercon* sp. (Oregon), female venter.

3a Genu III with eight setae; coxae II each with a distinctive anterior spine; male with holoventral shield (Fig. 20.107). . . . . ***Saprolaelaps*** Leitner

3b Genu III with nine setae; coxae II without anterior spines; male with separate ventrianal shield (Figs. 20.108, 20.109). . . . . ***Halolaelaps*** Berlese and Trouessart

Zerconidae Canestrini, 1891 (Females Only)[1]

1a Peritrematic shields fused to ventrianal shield (Fig. 20.112). 2

1b Peritrematic shields not fused to ventrianal shield. 3

2a Peritrematic shield with three setae (p_1–p_3), the most posterior of which (p_3) is pilose and distinctly longer than more anterior simple p_1 and p_2 (Fig. 20.111) ***Parazercon*** Trägårdh

2b Peritrematic shield with only two setae (p_1 and p_2), both setae simple and short (Figs. 20.112, 20.113). ***Microzercon*** Blaszak

3a Peritrematic shield with three setae (p_1–p_3). 4

3b Peritrematic shield with only two setae. 5

4a Adgenital pore plates present, each with one or two pores; with one pair of setae (Jv_1) on anterior margin of ventrianal shield (Fig. 20.114). . . . . . . ***Skeironozercon*** Halašková 1977

4b Adgenital pore plates absent; with two pairs of setae (Jv_1 and $Zv1$) on anterior margin of ventrianal shield (as in Fig. 20.119) (Fig. 20.115). ***Aspar*** Halašková

5a Adgenital pore plates present, each with two to four pores (Fig. 20.119). . . 6

5b Adgenital pore plates absent (Fig. 20.116). . . . . ***Mixozercon*** Halašková

6a Ventrianal shield with deep lateral incisions which extend posteriorly to the level of marginal setae R_6 (Fig. 20.117). . . . . ***Macrozercon*** Blaszak

6b Ventrianal shield entire, not as above. 7

7a Peritrematic shields abbreviated posteriorly, extending only to a level just behind coxae IV (Fig. 20.119). 8

7b Peritrematic shields extend posteriorly behind coxae IV to a level beyond marginal setae R_3 (Fig. 20.118). . . . . ***Caurozercon*** Halašková

8a Opisthonotal margin with seven pairs of setae; peritrematic setae p_1 and p_2 of different lengths (Fig. 20.119, 20.120). . . . . ***Zercon*** C. L. Koch

8b Opisthonotal margin with eight pairs of setae; p_1 and p_2 equal in length. . . . ***Bledas*** Halašková

Ameroseiidae Evans, 1963

1a Some of the dorsal setae greatly elongated, often approaching the length of the strongly tuberculate dorsal shield; genu III with eight setae, *pv* and pl_2 absent $\left(2\,\frac{2}{1}\,\frac{2}{0}\,1\right)$ (Figs. 20.121, 20.122). . . . . ***Epicriopsis*** Berlese

1b Dorsal setae variously developed but not elongate as above, often broadened and with marginal serrations; dorsal shield often sculptured with distinctive areolate patterns, not tuberculate; genu III with 10 setae $\left(2\,\frac{2}{1}\,\frac{2}{1}\,2\right)$ (Figs. 20.123, 20.124). . . . . ***Ameroseius*** Berlese

[1] The genus ***Amerozercon*** Halašková 1969, a monotypic taxon described only from the male (Fig. 20.110), cannot be accommodated in this key.

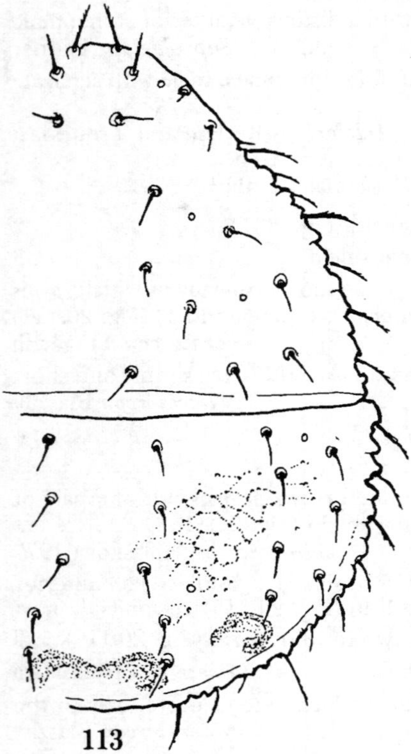

113

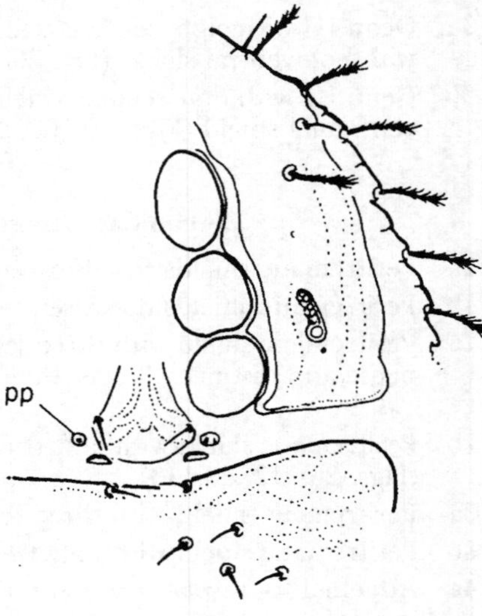

114

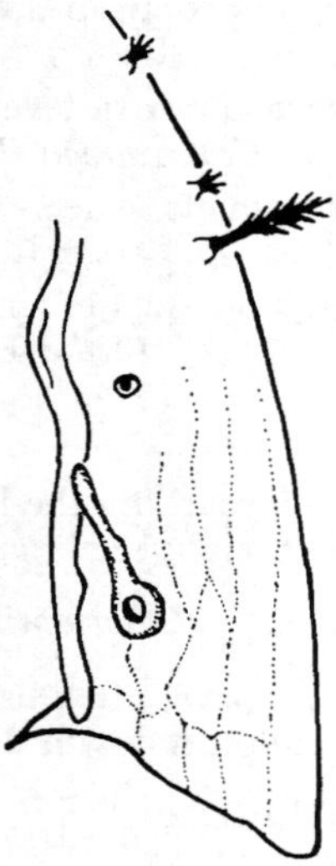

115

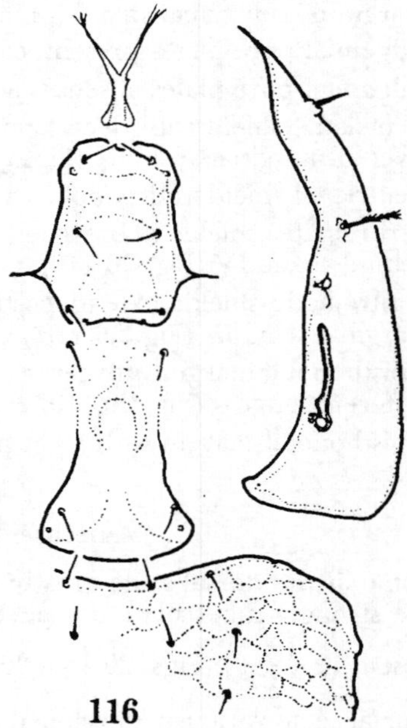

116

Fig. 20.113. *Microzercon* sp. (Oregon), female dorsum. **Fig. 20.114.** *Skeironozercon embersoni* Halašková (British Columbia) portion of female venter. **Fig. 20.115.** *Aspar anisotrichus* Halašková (British Columbia), female parapodal region. **Fig. 20.116.** *Mixozercon stellifer* Halašková (British Columbia), female venter.

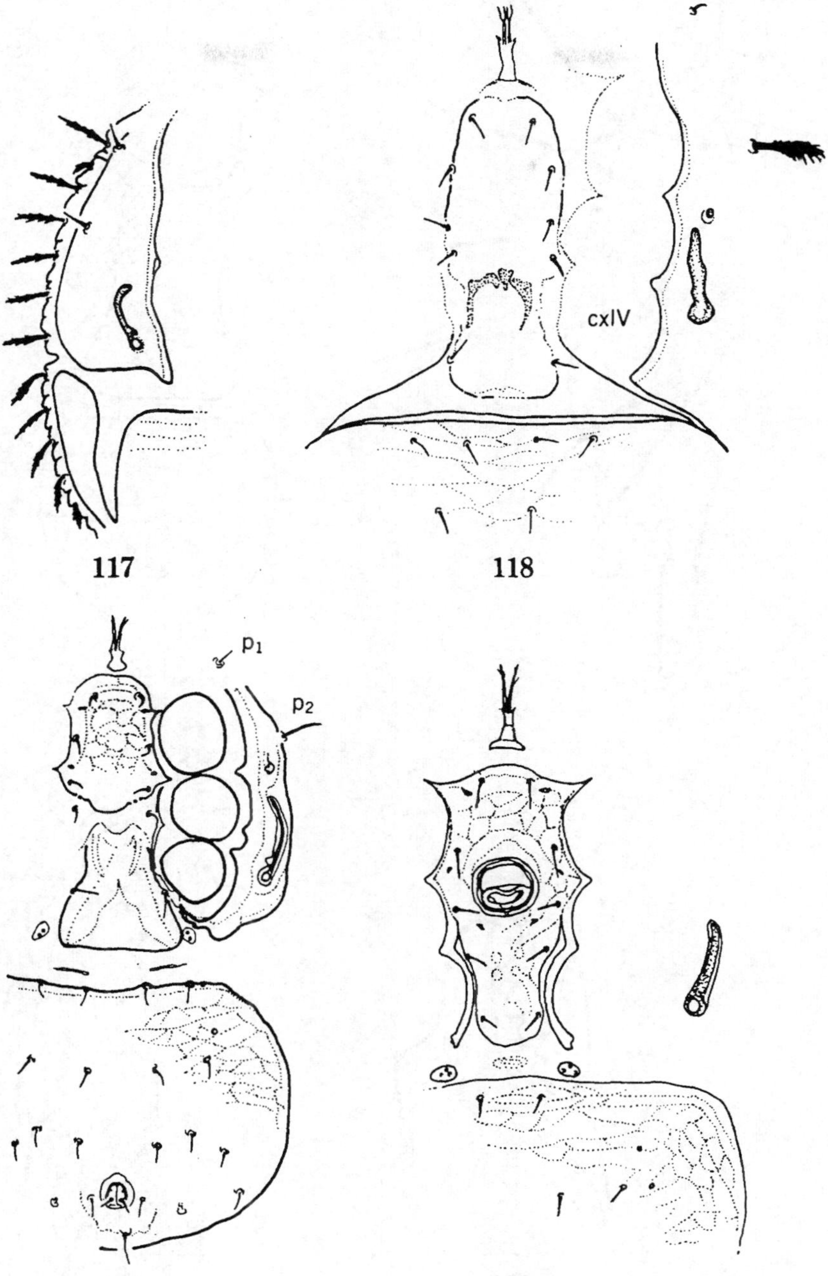

Fig. 20.117. *Macrozercon praecipuus* (Sellnick) female parapodal region. [After Sellnick (1958).] **Fig. 20.118.** *Caurozercon duplex* Halašková (Quebec) female venter. **Fig. 20.119.** *Zercon* sp. (Oregon), female venter. **Fig. 20.120.** *Zercon* sp. (Oregon), male venter.

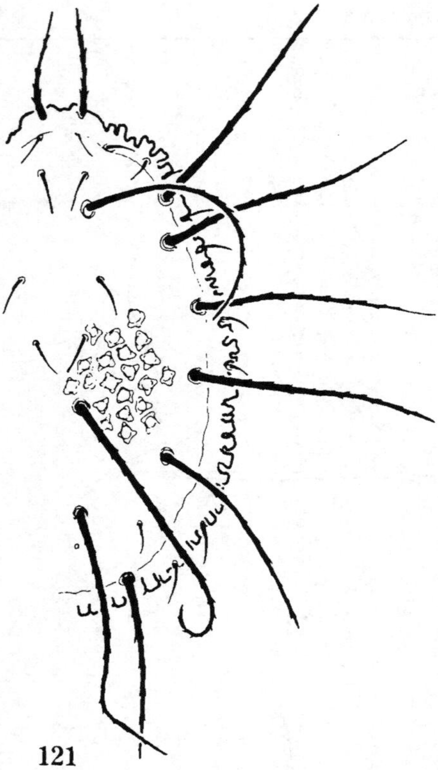

121

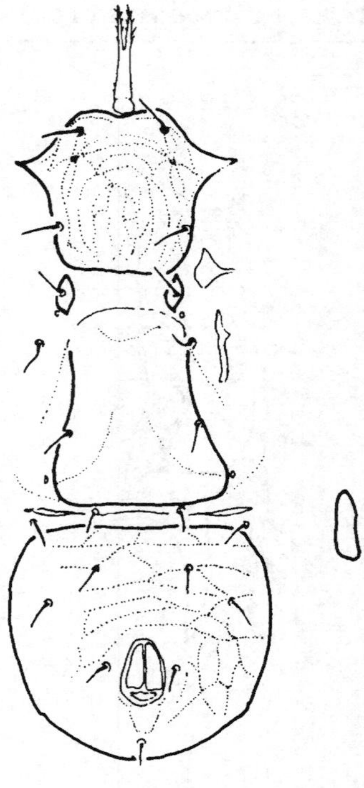

123

122

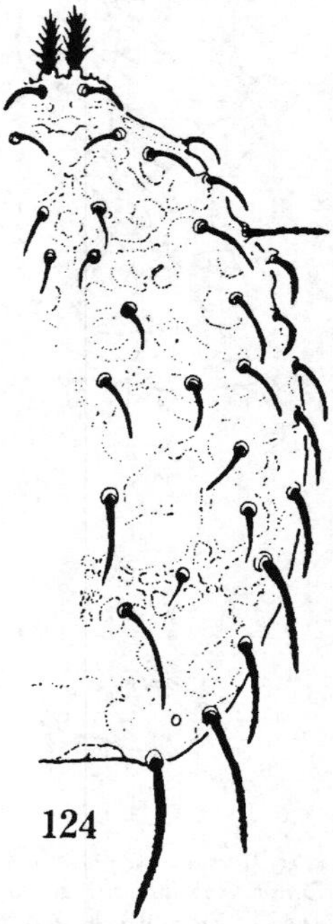

124

Fig. 20.121. *Epicriopsis horridus* Kramer (Oregon), female dorsum. **Fig. 20.122.** *Epicriopsis horridus* Kramer (Oregon), female chelicera. **Fig. 20.123.** *Ameroseius* sp. (Oregon), female venter. **Fig. 20.124.** *Ameroseius* sp. (Oregon), female dorsum.

125

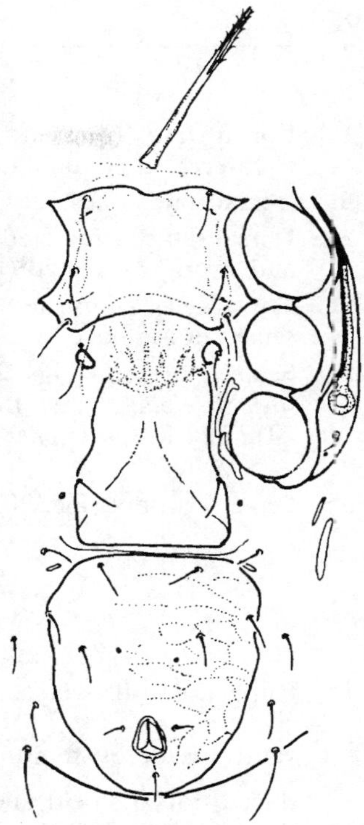

126

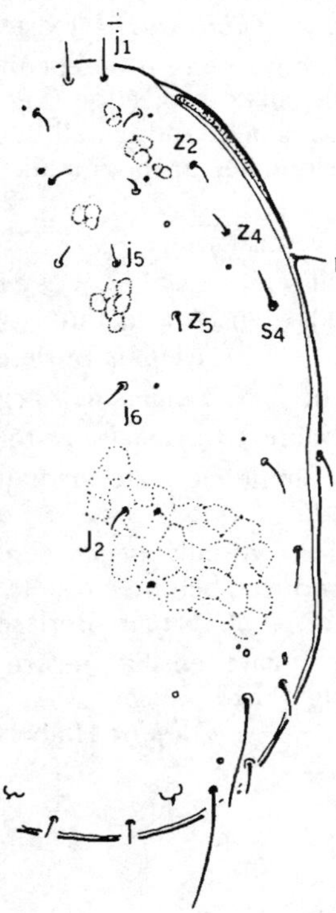

127

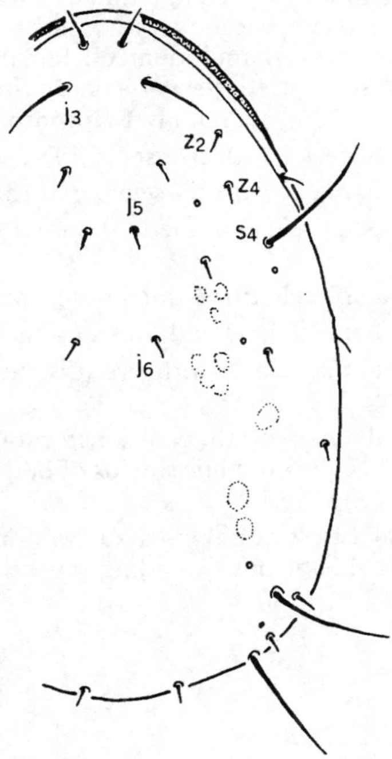

128

Fig. 20.125. *Phytoseius macrophilis* (Banks) (Maryland), female dorsum. **Fig. 20.126.** *Amblyseius* sp. (Oregon), female venter. **Fig. 20.127.** *Amblyseius* sp. (Oregon), female dorsum. **Fig. 20.128.** *Proprioseiopsis rotundus* (Muma), female dorsum. [After Muma and Denmark (1970).]

Phytoseiidae Berlese, 1916[1]

1a Dorsal state z_3 present (Fig. 20.125); plant-inhabiting forms occasionally encountered in edaphic habitats. **2**

lb Dorsal state z_3 absent; Commonly edaphic. **3**

2a Dorsal setae r_3 inserted on dorsal shield, with several dorsal setae thickened and strongly serrated (Fig. 20.125). . . . . ***Phytoseius*** Ribaga*

2b Dorsal setae r_3 inserted in lateral integument adjacent to dorsal shield; dorsal setae not as above. ***Typhlodromus*** Scheuten*

3a Setae J_2 present (Figs. 20.126, 20.127). . . . .***Amblyseius*** Berlese (***Amblyseiella*** Muma, ***Chelaseius*** Muma and Denmark, ***Fundiseius*** Muma and Denmark, ***Iphiseiodes*** De Leon, ***Neoseiulus*** Hughes, ***Typhlodromalus*** Muma, ***Typhlodromips*** De Leon)

3b Setae J_2 absent (Fig. 20.128) . . . . ***Proprioseiopsis*** Muma, (***Amblyseiulus*** Muma, ***Neoledius*** Muma and Denmark, ***Proprioseiulus*** Muma and Denmark)

Eviphididae Berlese, 1913

1a Palpgenu with five setae; femur II with only one *al* seta $\left(1\ \frac{5}{3}\ 1\right)$, with two ventral setae on femur III $\left(1\ \frac{3}{2}\ 1\right)$; epistome multidentate; female with reduced dorsal shield and partially or completely unsclerotized sternal shield (Figs. 20.129–20.131); Subfamily **Thinoseiinae**. . . . . . ***Thinoseius*** Halbert*

lb Palgenu with six setae; femur II with two *al* setae, with only one ventral seta on femur III; epistome typically with a long, median lanceolate projection (Fig. 20.136), rarely multidentate; female with a holodorsal shield and a well defined sclerotized sternal shield; chelicerae may be elongate, uropodine-like (Fig. 20.135); Subfamily **Eviphidinae**. **2**

2a Trochanter I with five setae, genu III with only one ventral seta **3**

2b Trochanter I with six setae, genu III with two ventral setae. **4**

3a Palpal apotele two-tined, idiosoma strongly arched (Figs. 20.132–20.136). . . ***Eviphis*** Berlese

3b Palpal apotele three-tined, idiosoma not strongly arched. . . . ***Evimirus*** Karg

4a Genu and tibia II with one *al* seta, epistome multidentate. ***Crassicheles*** Karg*

4b Genu and tibia II with two *al* setae, epistome with a single elongate median process. **5**

5a Palpal tarsus with two conspicuously elongate sensilla distally (as in Fig. 20.134), peritrematic shields of both sexes strongly produced behind coxae IV (as in Fig. 20.132). . . . . ***Copriphis*** Berlese

5b Distal aspect of palpal tarsus without conspicuously elongate sensilla; peritrematic shields not extending posterior to coxae IV (Fig. 20.137). . . . . ***Alliphis*** Halbert

[1] This is a simplified treatment based on characters utilized by Evans and Till (1979) in a key that follows the generic concepts of Karg (1971a). Taxa listed as synonyms of ***Amblyseius*** and ***Proprioseiopsis*** are known to contain species that occur in North American edaphic habitats.

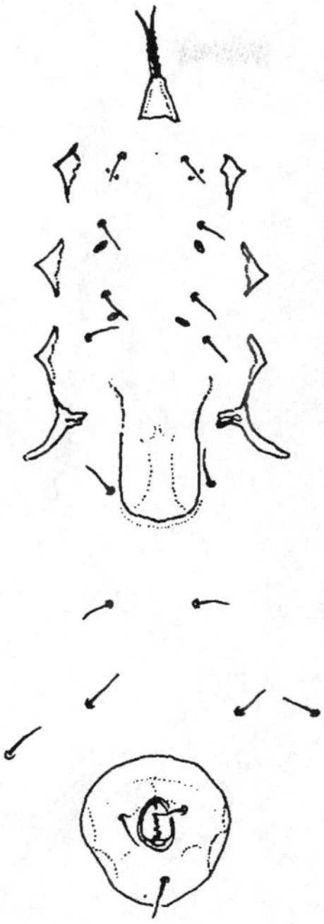

129

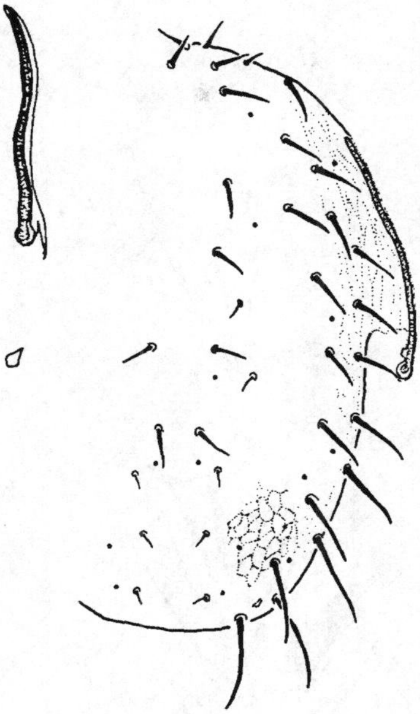

130

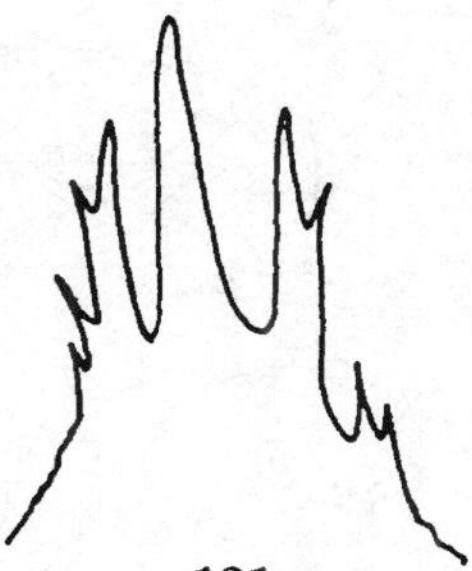

131

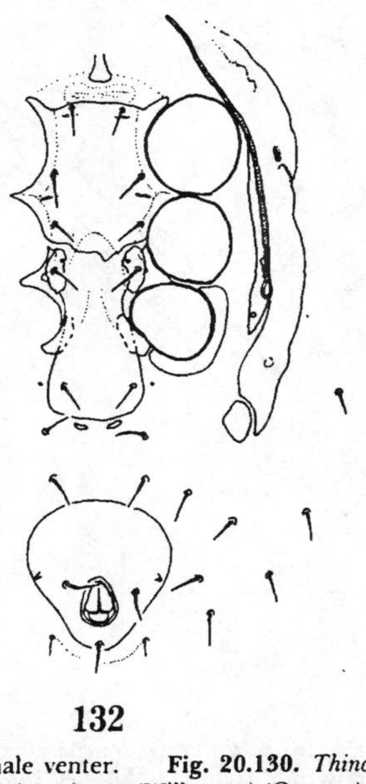

132

Fig. 20.129. *Thinoseius spinosus* (Willmann) (Oregon), female venter. **Fig. 20.130.** *Thinoseius spinosus* (Willmann) (Oregon), male dorsum. **Fig. 20.131.** *Thinoseius spinosus* (Willmann) (Oregon), female epistome. **Fig. 20.132.** *Eviphis* sp. (Ontario), female venter.

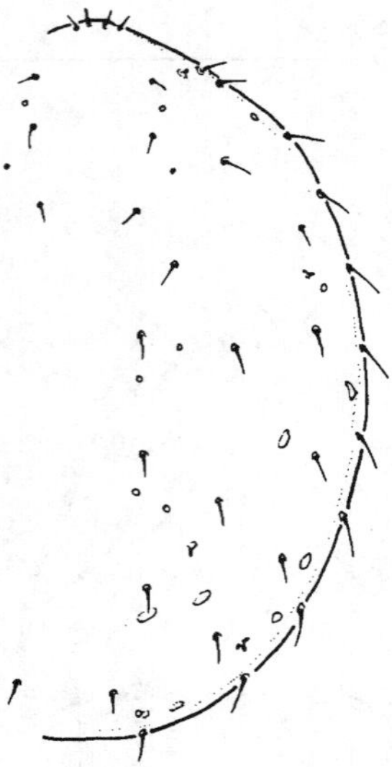

133

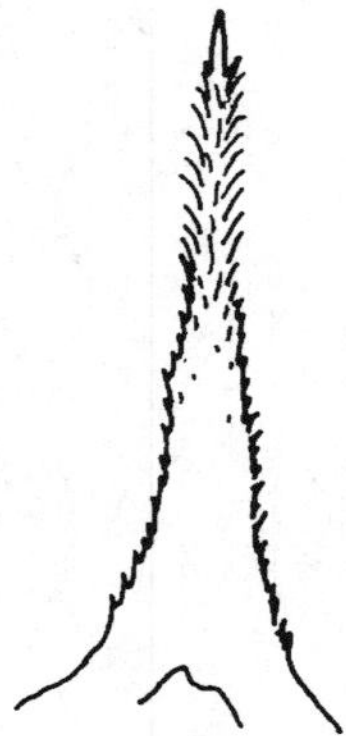

136

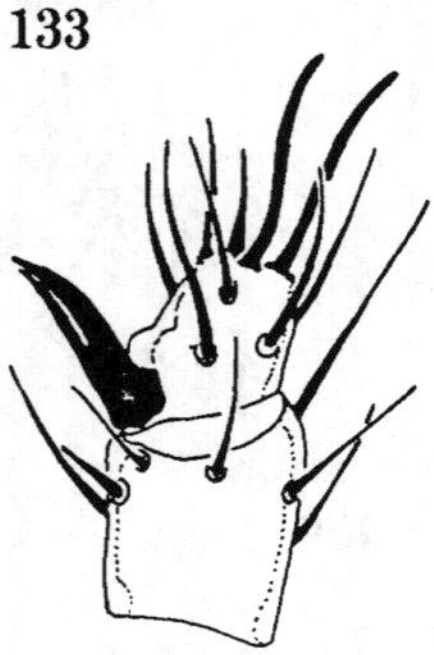

134

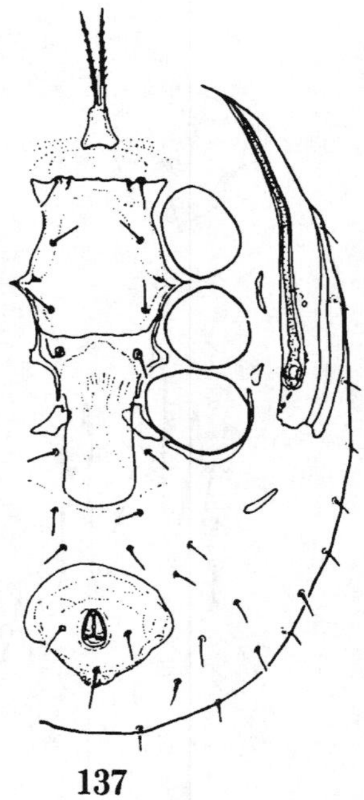

137

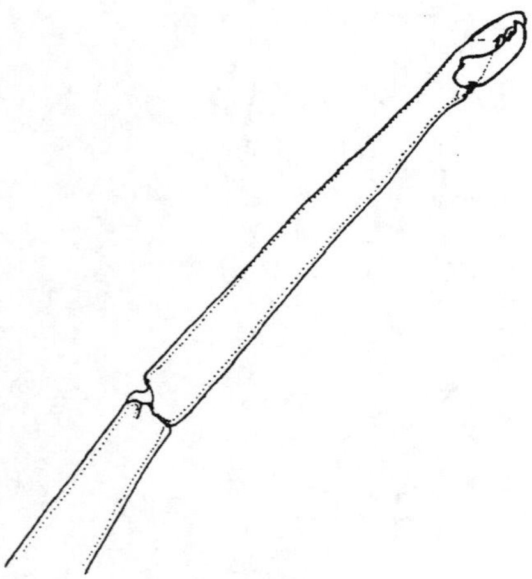

135

Fig. 20.133. *Eviphis* sp. (Ontario), female dorsum. **Fig. 20.134.** *Eviphis* sp. (Ontario), ventral aspect of female palptibia and tarsus. **Fig. 20.135.** *Eviphis* sp. (Ontario), female chelicera. **Fig. 20.136.** *Eviphis* sp. (Ontario), female epistome. **Fig. 20.137.** *Alliphis* sp. (Oregon), female venter.

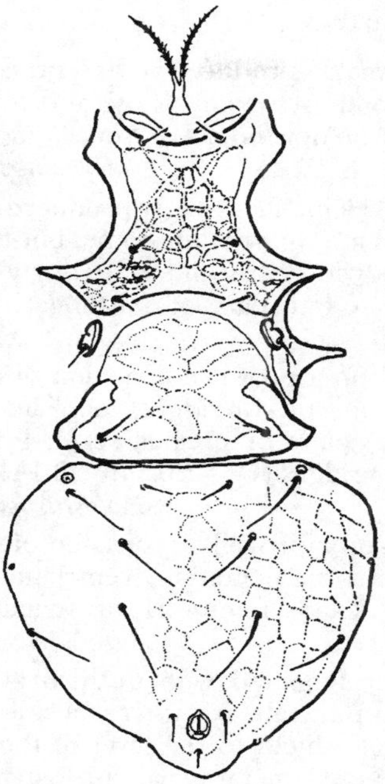

138

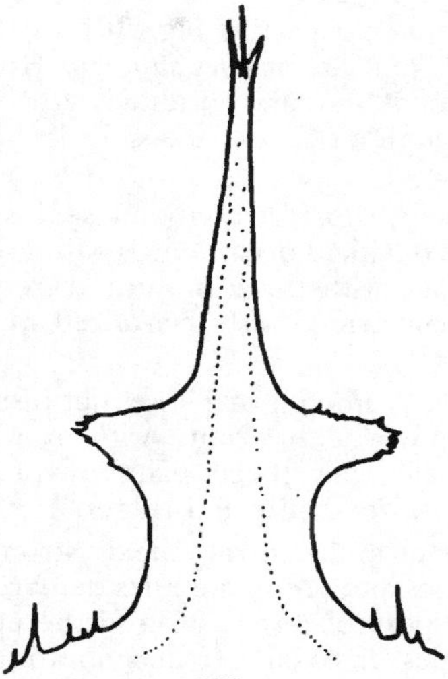

139

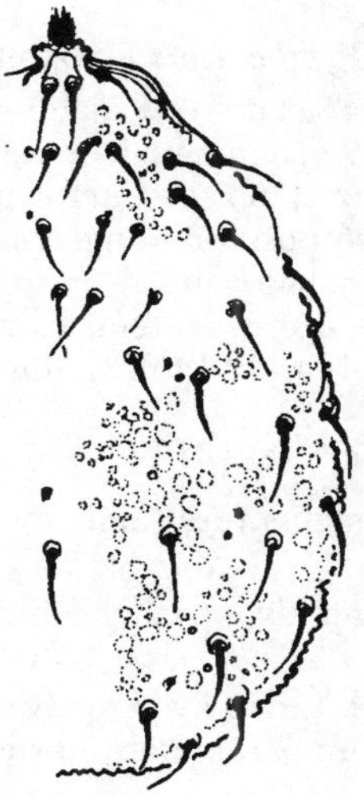

140

141

Fig. 20.138. *Geholaspis longispinosa* (Kramer) (Oregon), female venter. **Fig. 20.139.** *Geholaspis longispinosa* (Kramer) (Oregon), female epistome. **Fig. 20.140.** *Holostaspella ainscoughi* Krantz (Oregon), female dorsum. **Fig. 20.141.** *Holostaspella ainscoughi* Krantz (Oregon), femur and genu II of female.

Macrochelidae Vitzthum, 1930

1a Epistome unipartite, lateral elements weakly produced or absent; central element elongate, uniformly narrow, smooth or weakly pilose and divided distally; ventrianal shield with five pairs of opisthogastric setae (Figs. 20.138, 20.139). . . . . ***Geholaspis*** Berlese

1b Epistome tripartite (Fig. 20.146), lateral elements strongly produced and often flaglike (***Macrocheles superbus*** Hull has a unipartite epistome, but the central element narrows gradually to an undivided tip and is strongly fimbriate throughout); ventrianal shield with two to four pairs of opisthogastric preanal setae. **2**

2a Setae j_1 short, strongly pilose, inserted on a prowlike projetion of the dorsal shield; shield ornamented with areolations, pits or ridges; ventrianal shield of female with three or four (rarely two) pairs of preanal setae; femur II of female armed with sclerotized spurs or ridges (Figs. 20.140, 20.141). . . . . ***Holostaspella*** Berlese

2b Setae j_1 smooth or pilose, not inserted on a prowlike projection of the dorsal shield; shield various, with or without ornamentation; ventrianal shield of female with three pairs of preanal setae; femur II of female without sclerotized spurs and ridges. **3**

3a Dorsal and ventral shields strongly ornamented with distinctive polygonal ridges and crenulate lines demarcating punctate areas; sternal shield extending poserolaterally around the epigynial shield to the level of the posterior angles of coxae III and abutting rounded metasternal shields; males with holoventral shield; species of great size, with females averaging 12–1300 μm in length; coprophiles (Figs. 20.142, 20.143). . . . . ***Glyptholaspis*** Filipponi and Pegazzano

3b Dorsal shield commonly punctate or punctate–reticulate, occasionally with ridges but not as above, ventral shields ornamented in various patterns of reticulations, punctations, and areolations; posterolateral angles of sternal shield usually not extending to level of posterior angles of coxae III and abutting metasternal shields; males with holoventral or separate ventrianal shield; species of small or intermediate size, with females rarely attaining a length of more than 1000 μm (Figs. 20.144–20.146). . . ***Macrocheles*** Latreille

Pachylaelapidae Berlese, 1913

1a Female with metasternal-endopodal elements fused only tenuously to sternal shield proper (Fig. 20.148). **2**

1b Female with an anal shield, or a fused genitiventrianal complex (Fig. 20.147) ***Pachylaelaps*** Berlese

2a Leg II of female distinctly thickened, with two to five greatly enlarged terminal tarsal setae; peritrematic shields free from parapodals; tibiae III and IV each with seven setae $\left(1\ \frac{1}{1}\,\frac{2}{1}\ 1\right)$ (Figs. 20.148, 20.149). . . . . . ***Pachyseius*** Berlese

2b Leg II of female similar to other legs, terminal tarsal setae thickened but not greatly enlarged; peritrematic and parapodal shields fused; tibiae III and IV each with 10 setae $\left(2\ \frac{2}{1}\,\frac{2}{1}\ 2\right)$ (Fig. 20.150). . . . . ***Zygoseius*** Berlese

Parholaspididae Evans, 1956

1a Peritrematic–exopodal shield complex fused posteriorly with ventrianal shield which covers the entire opisthogaster; usually with a pair of distinctive inguinal solenostomes behind coxae IV (Figs. 20.151, 20.152). . . ***Neparholaspis*** Evans

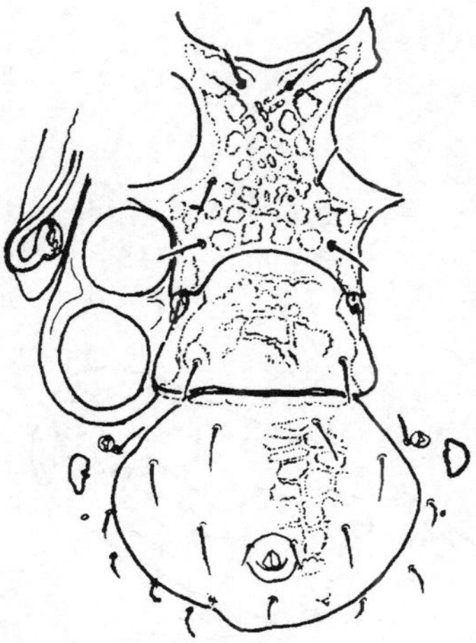

142

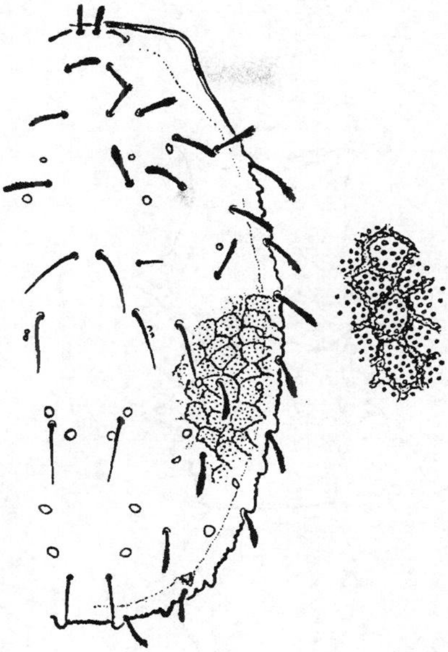

143

Fig. 20.142. *Glyptholaspis americana* (Berlese) (Oregon), female venter.

Fig. 20.143. *Glyptholaspis americana* (Berlese) (Oregon), female dorsum with detail of ornamentation.

1b Peritrematic–expodal shield complex separate from ventrianal shield, ventrianal shield variously developed but usually not covering the entire opisthogaster, sometimes fused anteriorly to epigynial shield; inguinal solenostomes behind coxae IV indistinct.. 2

2a Metasternal setae and pores on sternal shield; ventrianal shield with more than four pairs of preanal setae (Fig. 20.153). . . . . **Calholaspis** Berlese

2b Metasternal setae and pores either on free metasternal shields or on fused metasternal–endopodal element, not on sternal shield; ventrianal shield with three or four pairs of preanal setae. 3

3a Epistome trifurcate (Fig. 20.156); ventrianal shield with three pairs of preanal setae. 4

3b Epistome produced medially, lateral processes absent (Fig. 20.160); ventrianal shield with three or four pairs of preanal setae. 5

4a Presternal sclerites fragmented, consisting of eight or more elements, metasternal shields free; dentition of movable cheliceral digit consisting only of a median and a subterminal tooth; trochanter IV without spur (Figs. 20.154–20.156). . . . . **Holaspina** Berlese

4b With two unfragmented presternal sclerites, metasternal shields fused with endopodals; movable cheliceral digit multidentate, with 12–14 small teeth between the large median and subterminal teeth; trochanter IV with a ventral spur (Figs. 20.157, 20.158). . . . . **Snaveolaspis** Johnston

5a Cheliceral digits short, not as long as the corniculi, dentition of movable digit consisting only of a median and a subterminal tooth; epistome produced medially but without an elongate median process; claws present on pretarsus I (Figs. 20.159, 20.160). . . . . **Lattinella** Krantz

5b Cheliceral digits as long or longer than corniculi, movable digit multidentate,

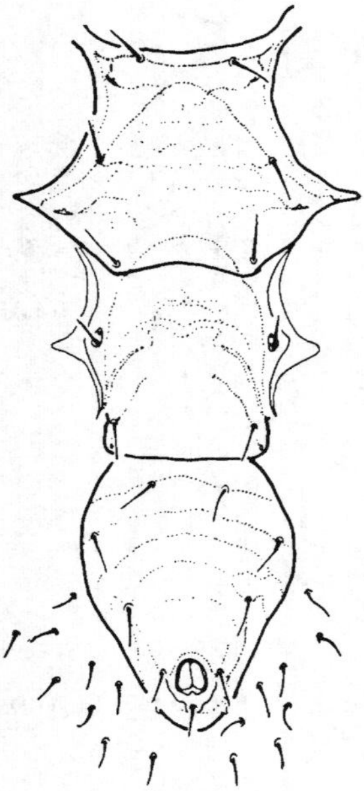

144

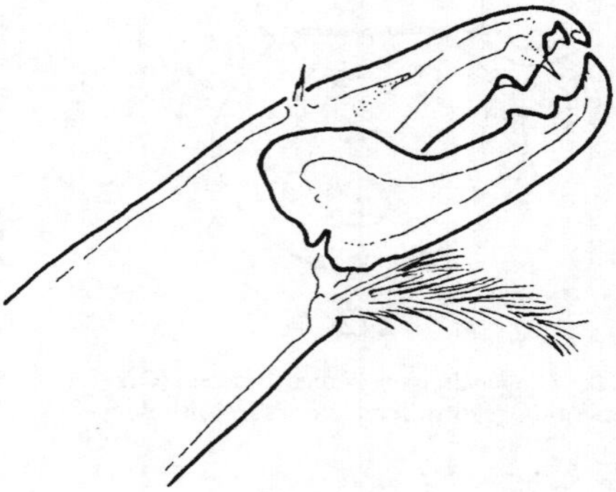

145

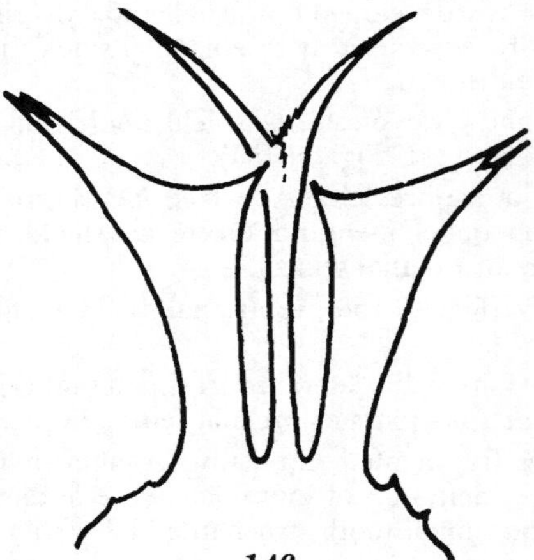

146

Fig. 20.144. *Macrocheles merdarius* (Berlese) (Quebec), female venter. **Fig. 20.145.** *Macrocheles penicilliger* Berlese (Oregon), female chelicera. **Fig. 20.146.** *Macrocheles perglaber* Filipponi and Pegazzano, female epistome. [After Filipponi and Pegazzano (1962).]

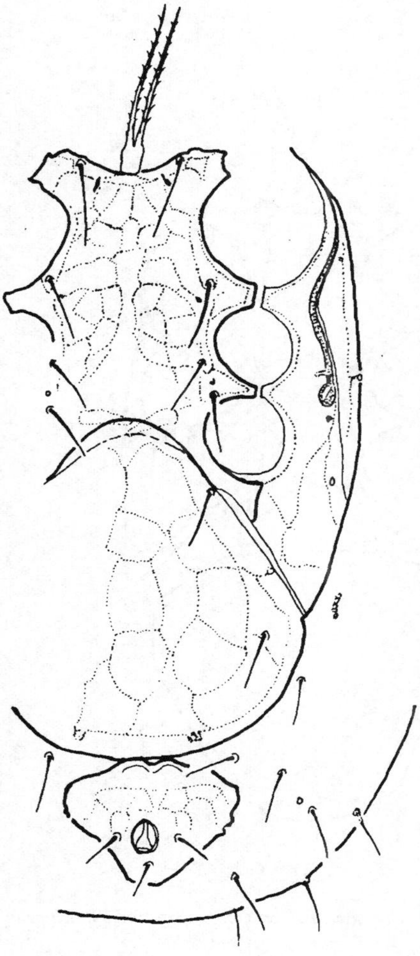

147

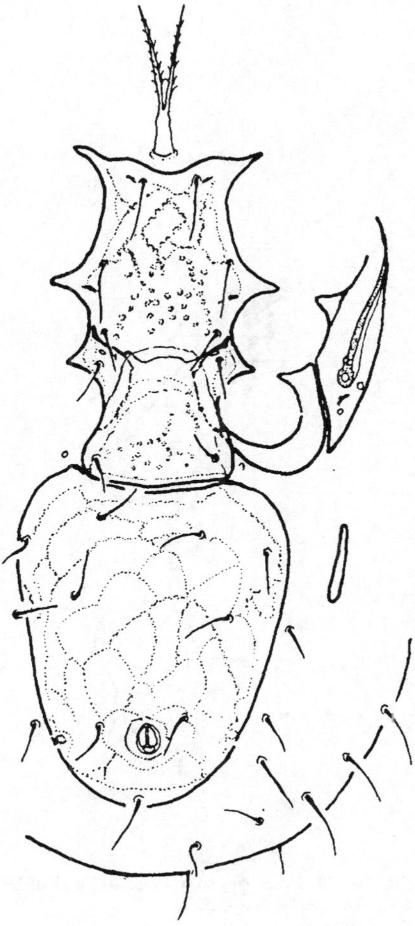

148

Fig. 20.147. *Pachylaelaps* sp. (Oregon), female venter.

Fig. 20.148. *Pachyseius* sp. (Oregon), female venter.

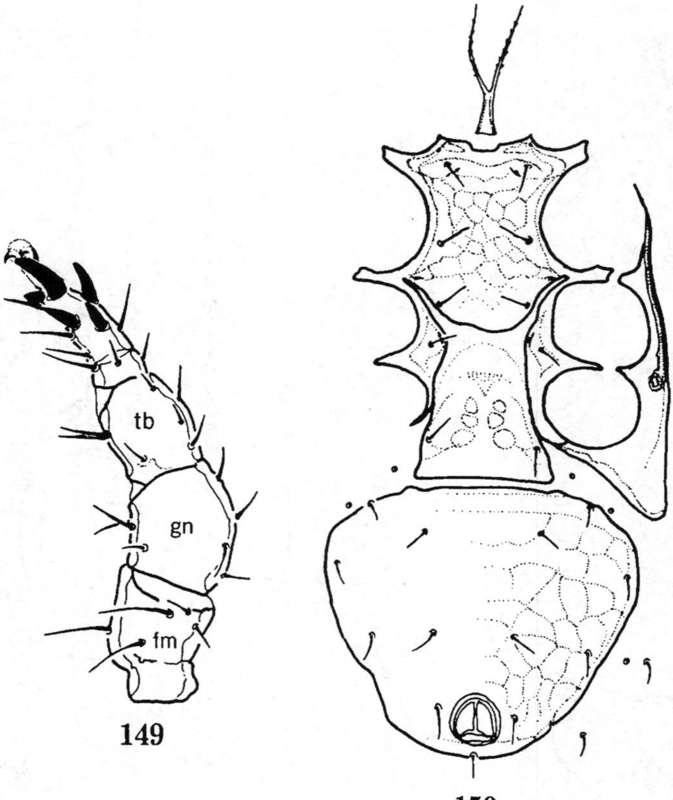

149

150

Fig. 20.149. *Pachyseius* sp. (Oregon), leg II of female (less coxa and trochanter). **Fig. 20.150.** *Zygoseius* sp. (Florida), female venter.

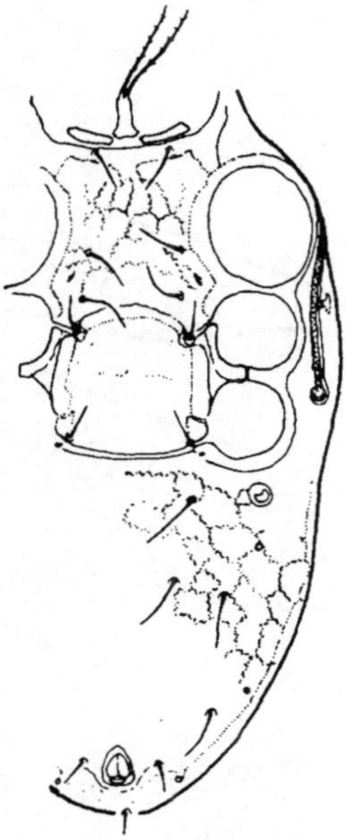

151

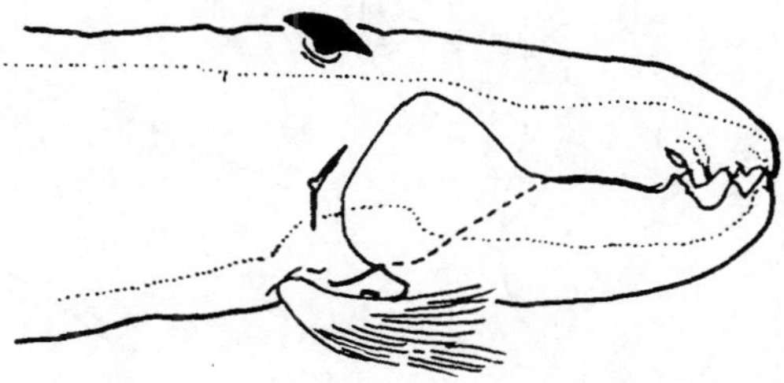

152

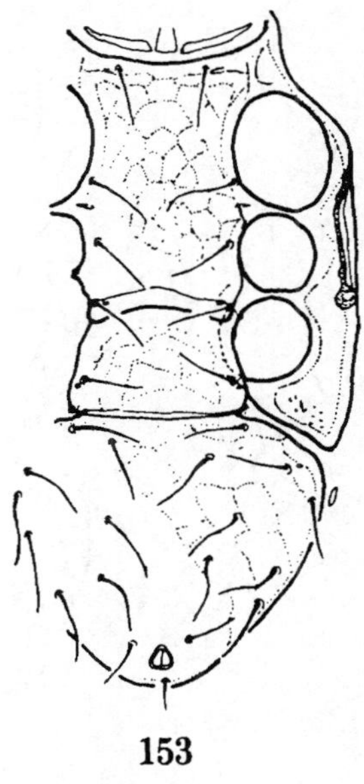

153

Fig. 20.151. *Neparholaspis evansi* Krantz (Oregon), female venter. **Fig. 20.152.** *Neparholaspis evansi* Krantz (Oregon), female chelicera. **Fig. 20.153.** *Calholaspis superba* Berlese (Quebec), female venter.

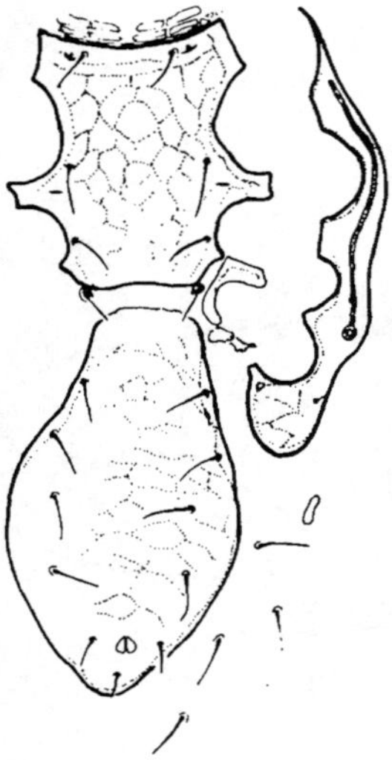

154

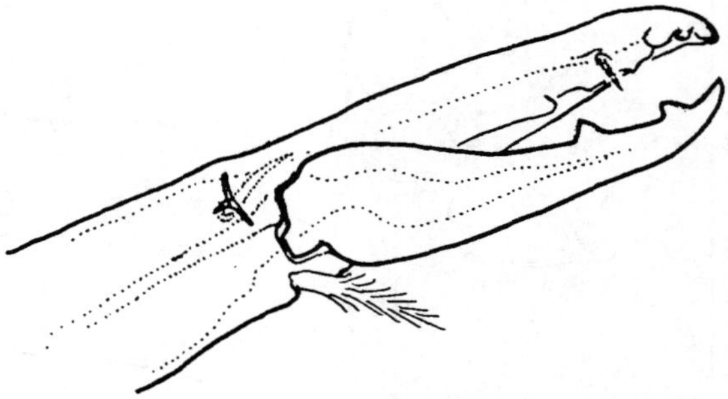

155

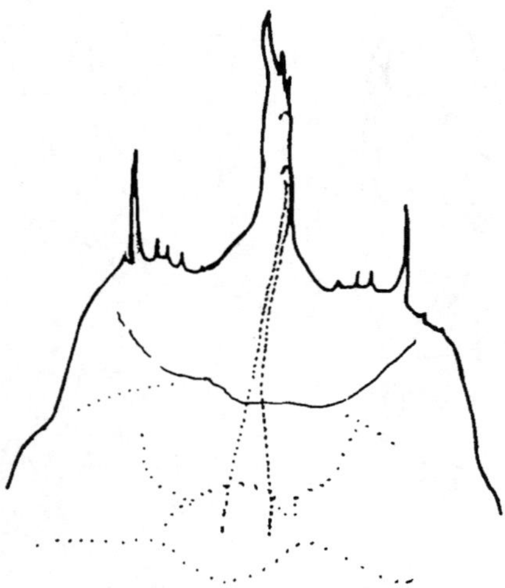

156

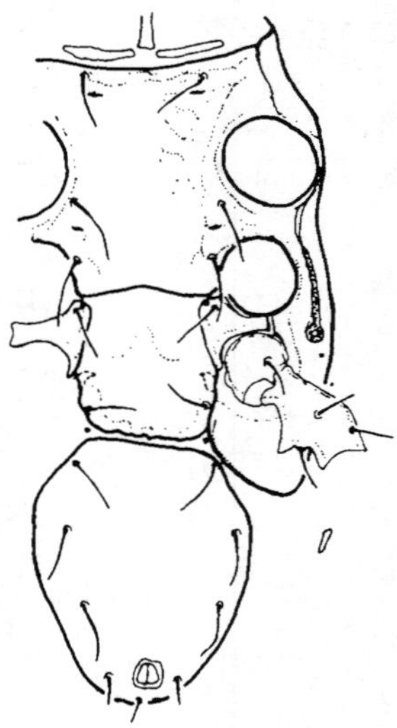

157

Fig. 20.154. *Holaspina coalescens* (Krantz) (Louisiana), female venter. **Fig. 20.155.** *Holaspina hurlbutti* (Krantz) (Maryland), female chelicera. **Fig. 20.156.** *Holaspina hurlbutti* (Krantz) (Maryland), female epistome. **Fig. 20.157.** *Snaveolaspis lobata* (Krantz) (Oregon), female venter.

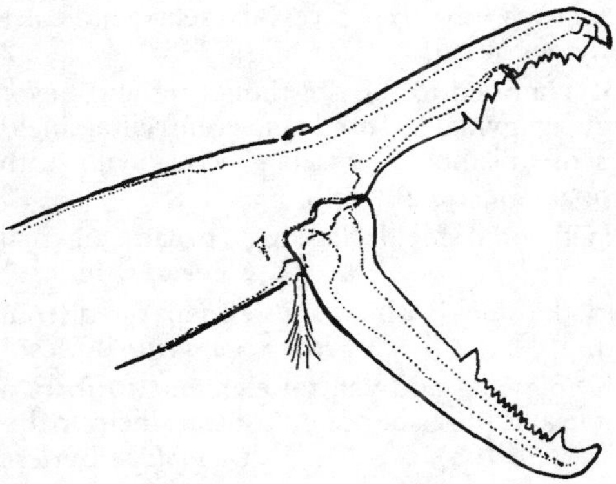

158

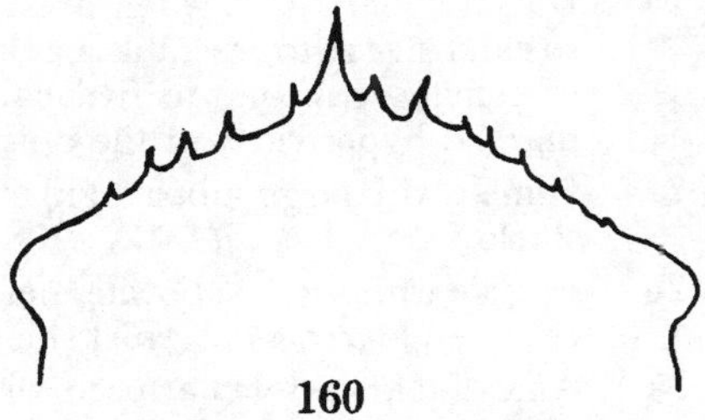

160

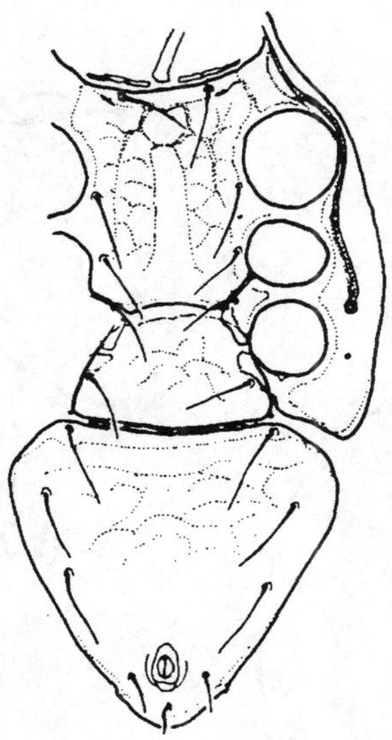

159

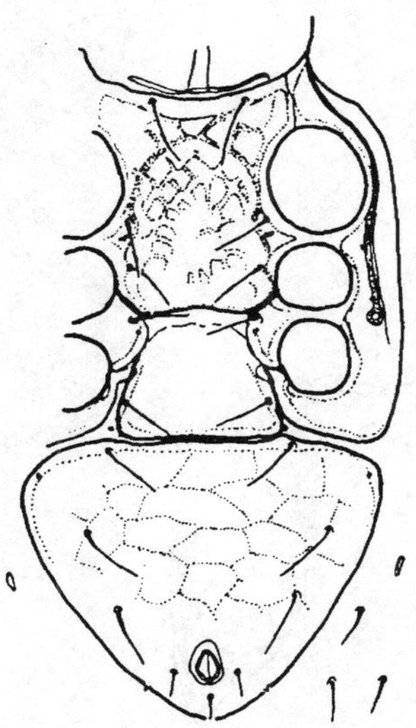

161

Fig. 20.158. *Snaveolaspis parvilobata* (Krantz) (Oregon), female chelicera. **Fig. 20.159.** *Lattinella capizzii* Krantz (Oregon), female venter. **Fig. 20.160.** *Lattinella capizzii* Krantz (Oregon), female epistome. **Fig. 20.161.** *Parholaspella spatulata* Krantz (Oregon), female venter.

with 9–13 small teeth interposed between a larger median and a subterminal tooth; epistome with an elongate median process; claws absent on partarsus I (Fig. 20.161). . . . . ***Parholaspella*** Krantz

Laelapidae Berlese, 1892

1a Dorsal shield of adult and deutonymph strongly hypertrichous, resembling a pelage; epigynial shield with numerous unpaired accessory setae; nidicoles and facultative parasites, primarily of rodents. **2**

1b Dorsal hypertrichy, when present, confined to the opisthonotum and never so extensive as to resemble a pelage; epigynial or fused genitiventrianal shield of female with one to five pairs of opisthogastric setae, occasionally with marked hypertrichy of the opisthogastric integument. **3**

2a Female with large subtriangular metapodal shields flanking a broad epigynial shield. . . . . ***Eulaelaps*** Berlese*

2b Female with small subovate metapodal shields which are well separated from relatively narrow epigynial shield. . . . . ***Haemogamasus*** Berlese*

3a Anal shield of female fused with epigynial and ventral elements to form a hologastric shield, metasternal setae and pores borne on sternal shield; palptarsal claw three-tined (Figs. 20.162, 20.163). . . . . ***Ololaelaps*** Berlese

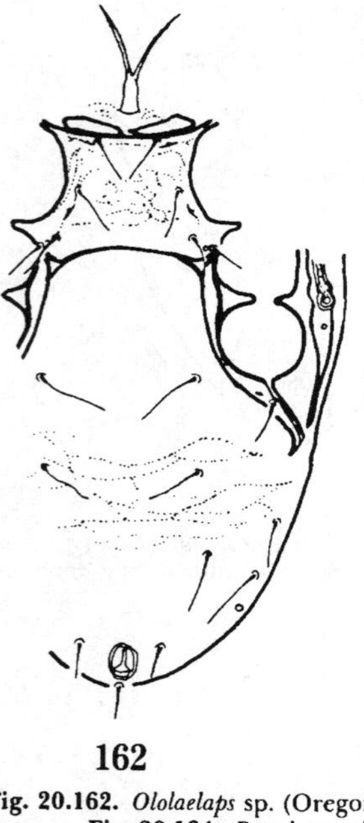
162

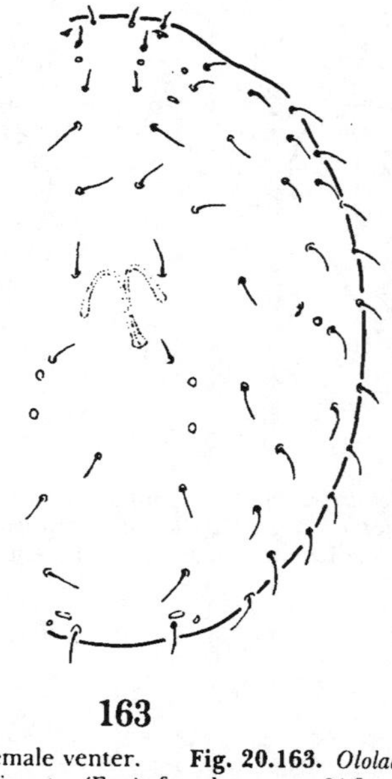
163

164

Fig. 20.162. *Ololaelaps* sp. (Oregon), female venter. **Fig. 20.163.** *Ololaelaps* sp. (Oregon), female dorsum. **Fig. 20.164.** *Pseudoparasitus stigmatus* (Fox), female venter. [After Hunter (1966).]

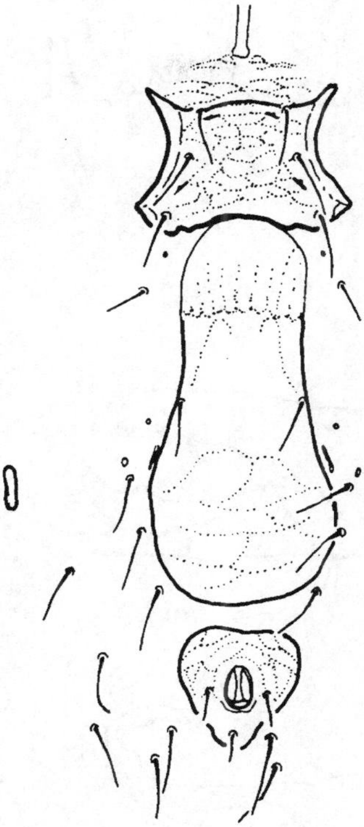

165

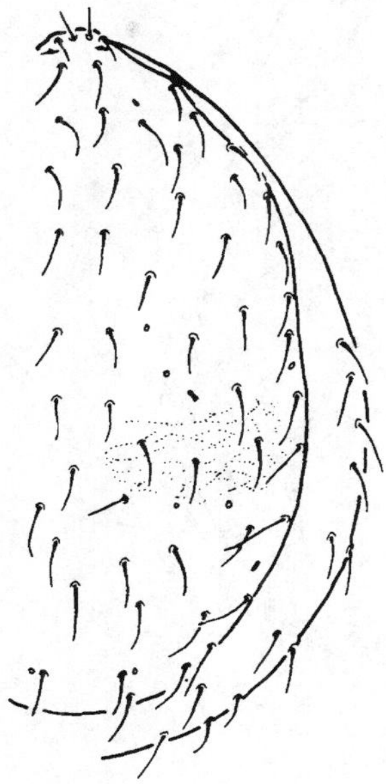

166

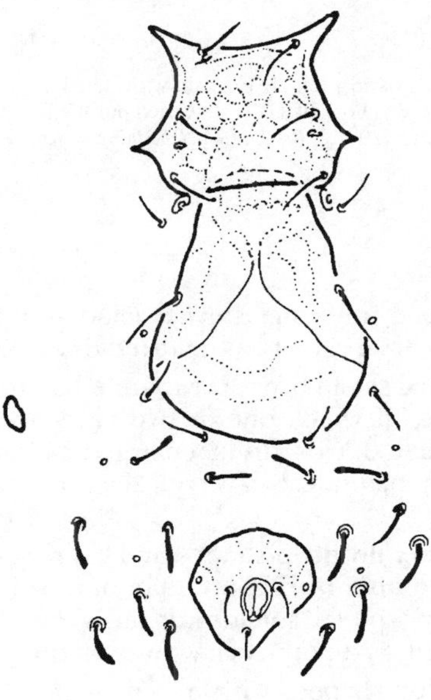

167

Fig. 20.165. *Androlaelaps* sp. (Oregon), female venter. **Fig. 20.166.** *Androlaelaps* sp. (Oregon), female dorsum. **Fig. 20.167.** *Cosmolaelaps* sp. (Maryland), female venter.

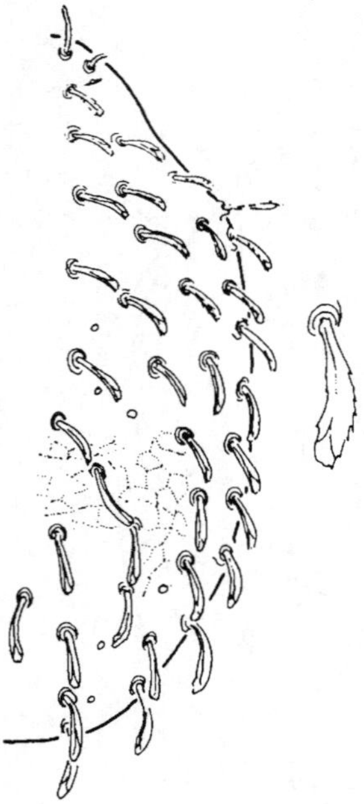

168

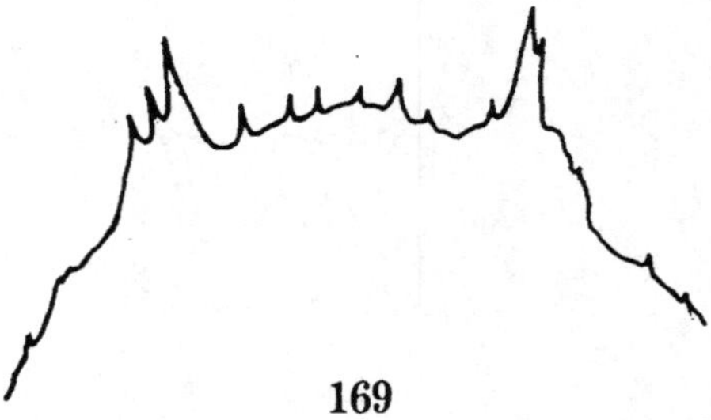

169

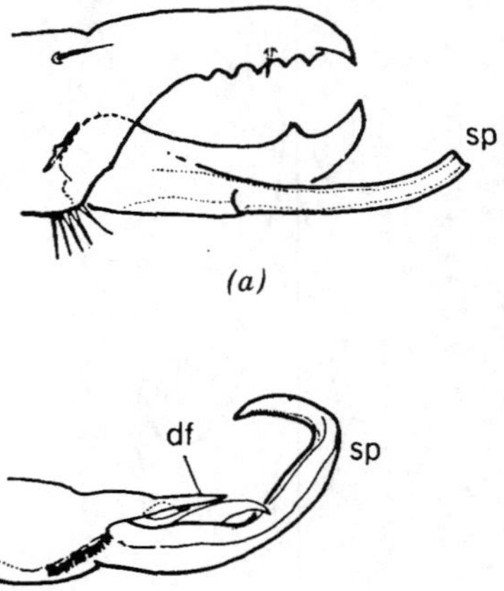

(a)

(b)

170

Fig. 20.168. *Cosmolaelaps* sp. (Maryland), female dorsum with detail of seta. **Fig. 20.169.** *Cosmolaelaps* sp. (Maryland), female epistome. **Fig. 20.170.** Male laelapid cheliceral types: **(a)** *Cosmolaelaps claviger* Berlese, male chelicera [after Karg (1971)]; **(b)** *Androlaelaps fahrenholzi* (Berlese), male chelicera [after Till (1963)].

3b Anal shield of female free; metasternal setae and pores usually not on the sternal shield; palptarsal claw two- or three-tined. **4**

4a Female with epigynial shield broadly expanded behind coxae IV, with four or more pairs of setae, of which one or two pairs are well removed from the shield margin; podal shields often enlarged behind coxae IV, free from peritrematic shields; palptarsal claw typically three-tined, proximal tine small (Fig. 20.164). . . . . ***Pseudoparasitus*** Oudemans

4b Epigynial shield generally drop-shaped and bearing one to four pairs of setae (if shield is enlarged and bears more than one pair of setae, all of them are inserted marginally); podal shields variously developed, often fused with peritrematic elements; palptarsal claw two- or three-tined. **5**

5a Epigynial shield drop-shaped, with one pair of setae; genu IV with two *pl* setae $\left(2\,\frac{2}{1}\,\frac{3}{0}\,2\right)$; pilus dentilis (as in Fig. 20.4, *pd*) long, sometimes inflated; chelicerae of male strongly modified; fixed digit reduced, edentate (Figs. 20.165, 20.166, 20.170*b*). . . . . ***Androlaelaps*** Berlese

5b Epigynial shield variously developed, often with more than one pair of setae; genu IV typically with only one *pl* seta $\left(1(2)\ \frac{2}{1}\,\frac{3}{0}\ 1\right)$; pilus dentilis short, setiform; chelicerae of male normally produced, chelate–dentate (Fig. 20.170*a*). . . . .(***Hypoaspis* s.l.**) **6**

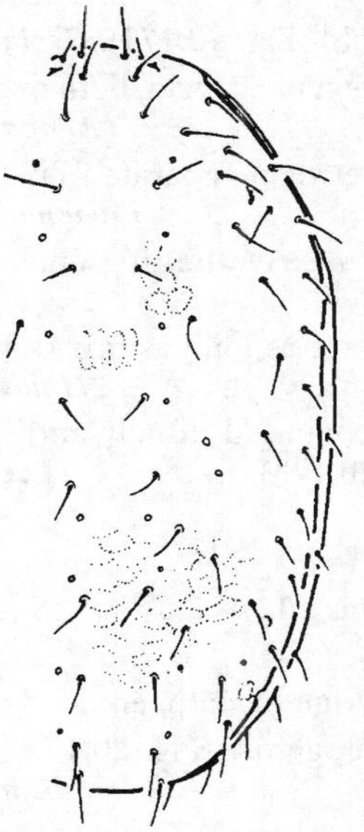

171

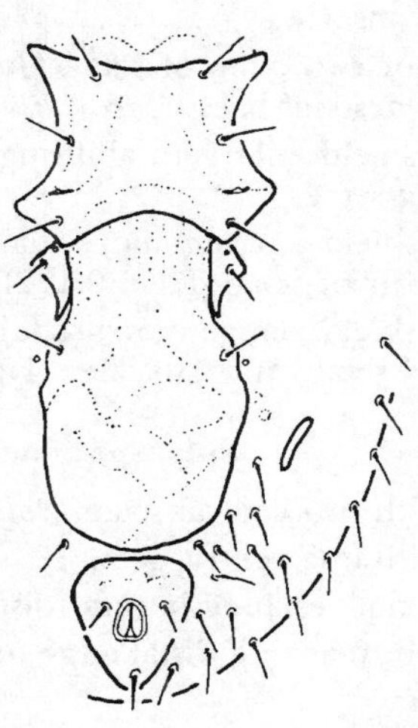

172

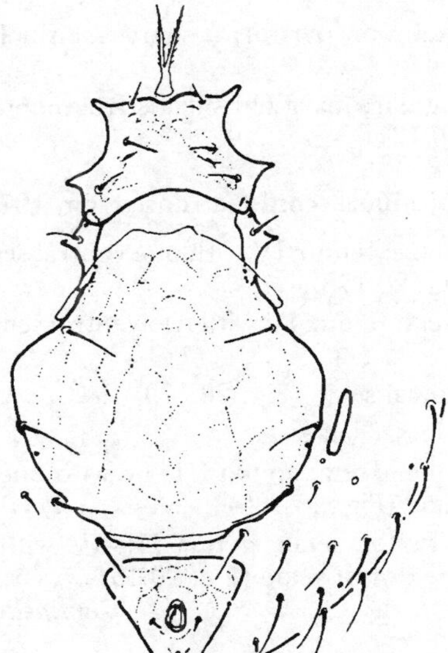

173

Fig. 20.171. *Geolaelaps aculeifer* (Canestrini) (Oregon), female dorsum. **Fig. 20.172.** *Holostaspis acuta* (Michael), female venter. [After Karg (1971).] **Fig. 20.173.** *Laelaspis* sp. (Oregon), female venter.

6a Anterior margin of epistome serrated (Fig. 20.169) or denticulate; rows of deutosternal denticles mostly with more than six teeth. **7**

6b Anterior margin of epistome smooth; rows of deutosternal denticles usually with only two to six teeth. **9**

7a Dorsal shield setae spatulate, cuneiform or knifelike; some with basal dilations (Figs. 20.167–20.170[*a*]. . . . . ***Cosmolaelaps*** Berlese

7b Dorsal shield setae simple or pilose. **8**

8a With one pair of setae on epigynial shield (Fig. 20.171). ***Geolaelaps*** Trägårdh

8b With at least two pairs of setae on the epigynial shield, both or all of which are marginally inserted. . . . . ***Alloparasitus*** Berlese

9a With one or two pairs of setae (*Jv*) between separated epigynial and anal shields; in nests of bees. . . . . ***Pneumolaelaps*** Berlese*

9b Epigynial shield enlarged, abutting or nearly abutting anal shield; without intervening setae. **10**

10a Epigynial shield typically more than twice as long as broad, but always with only one pair of setae (Fig. 20.172). . . . . ***Holostaspis*** Kolenati

10b Epigynial shield elongate or broadly expanded distally and with more than one pair of setae; ant associates (Fig. 20.173). . . . . . . ***Laelaspis*** Berlese*

Polyaspididae Berlese, 1917

1a Genu IV with two ventral setae; ***Polyaspis* s.l.**. **2**

1b Genu IV with one ventral seta. **4**

2a Corniculi long, reaching beyond distal edge of palpgenua. **3**

2b Corniculi not reaching distal edge of palpgenua (Fig. 20.174). . . . . ***Dipolyaspis*** Berlese

3a Tibia I with one *pl* seta (Fig. 20.175). . . . . ***Polyaspis*** Berlese

3b Tibia I with two *pl* setae. . . . . ***Dyscritaspis*** Berlese

4a Isiosoma subtriangular or pyriform; marginal shields entire laterally (Fig. 20.176). . . . . ***Trachytes*** Michael

4b Idiosoma subrectangular; marginal shields fragmented into numerous platelets laterally (Fig. 20.177). . . . . ***Polyaspinus*** Berlese

Dithinozerconidae Ainscough, 1979

1a Genu I with two *pl* setae, femur IV with one ventral seta, and tibia I with two *al* and two *pl* setae (Fig. 20.178). . . . . ***Apionoseius*** Berlese

1b Genu I with one *pl* seta; femur IV with two ventral setae, and tibia I with one *al* and one *pl* seta. **2**

2a Tibia I with four dorsal setae (Fig. 20.179). . . . . . ***Dithinozercon*** Berlese

2b Tibia I with three dorsal setae. **3**

3a Tibiae II–IV with one *pl* seta, genua III–IV with one ventral seta, and femur III with 4 dorsal setae (Fig. 20.180). . . . . ***Iphidinychus*** Berlese

3b Tibiae II–IV with two *pl* setae, genua III–IV with two ventral setae, and femur III with three dorsal setae (Fig. 20.181). . . . . ***Caminella*** Krantz and Ainscough

Uropodidae Kramer, 1881

1a Foveae pedales lacking. **2**

1b Foveae pedales more or less well developed. **6**

174

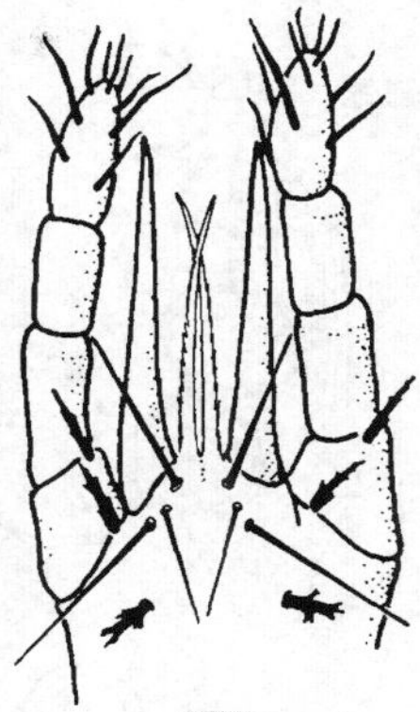

175

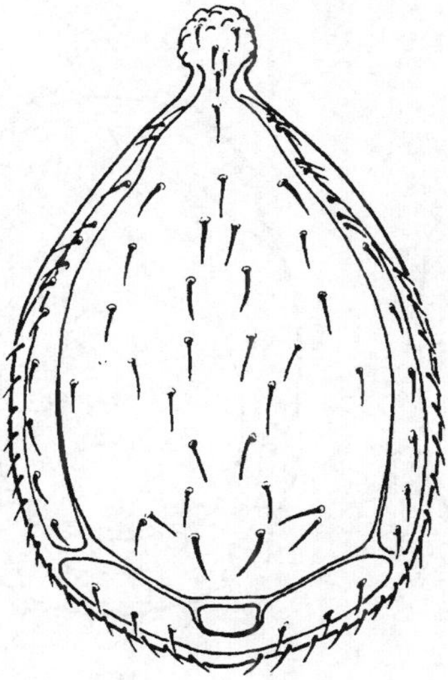

176

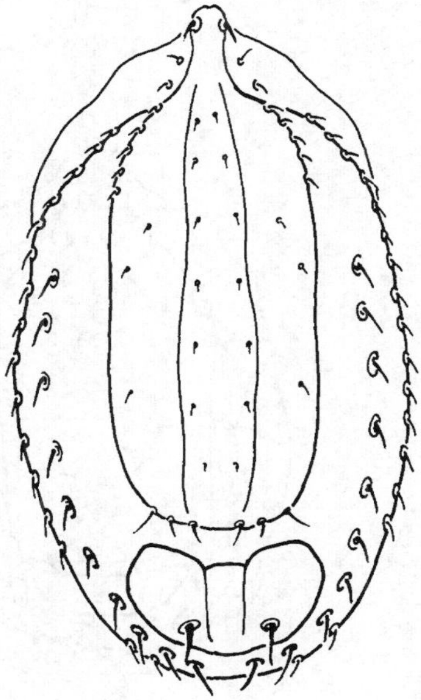

177

Fig. 20.174. *Polyaspis (Dipolyaspis)* sp. (Oregon), female dorsum. **Fig. 20.175.** *Polyaspis (Polyaspis)* sp., ventral aspect of female gnathosoma. **Fig. 20.176.** *Trachytes* sp. (Oregon), female dorsum. **Fig. 20.177.** *Polyaspinus* sp. (Washington), female dorsum.

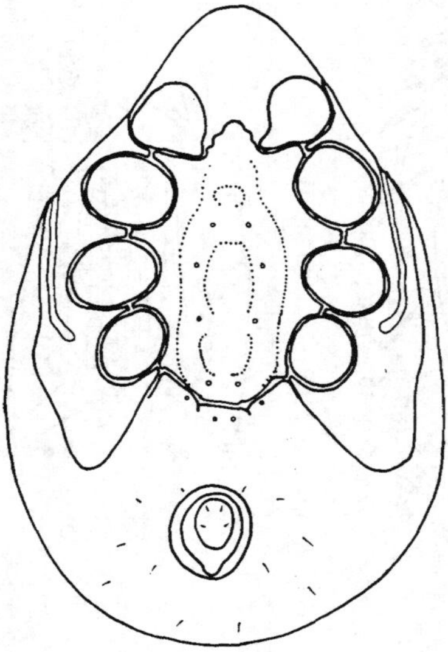

178

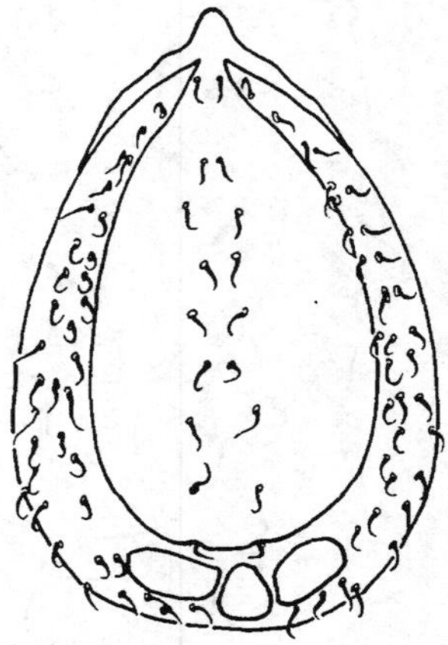

179

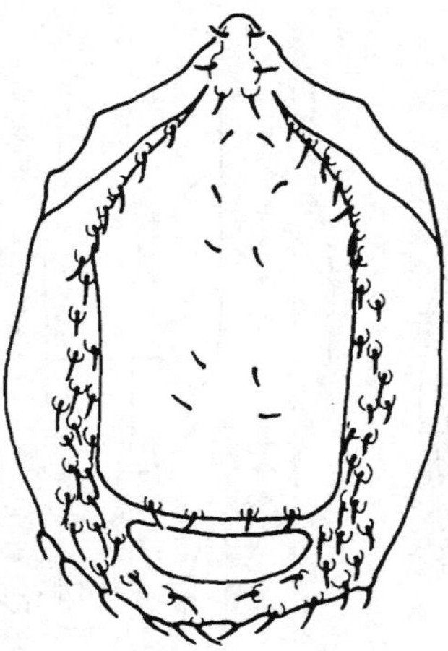

180

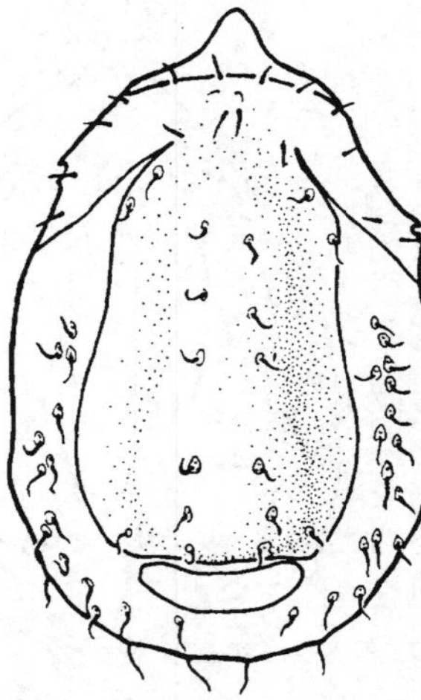

181

Fig. 20.178. *Apionoseius lagenaiformis* (Berlese) (Kentucky), deutonymphal venter. **Fig. 20.179.** *Dithinozercon halberti* (Berlese) (Ontario), female dorsum. **Fig. 20.180.** *Iphidinychus* sp. (North Carolina), female dorsum. **Fig. 20.181.** *Caminella peraphora* Krantz and Ainscough (Oregon), female dorsum.

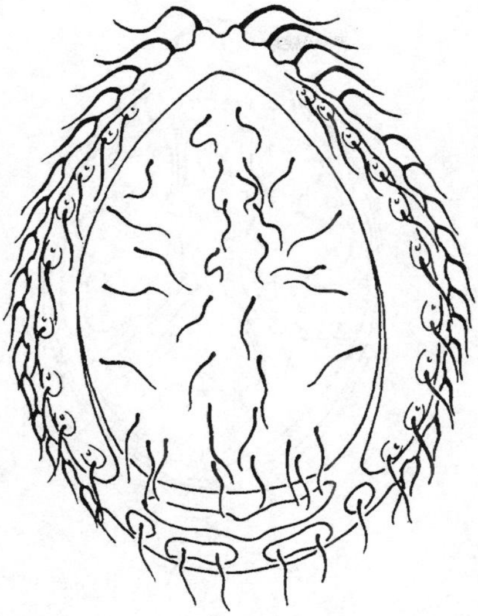

182

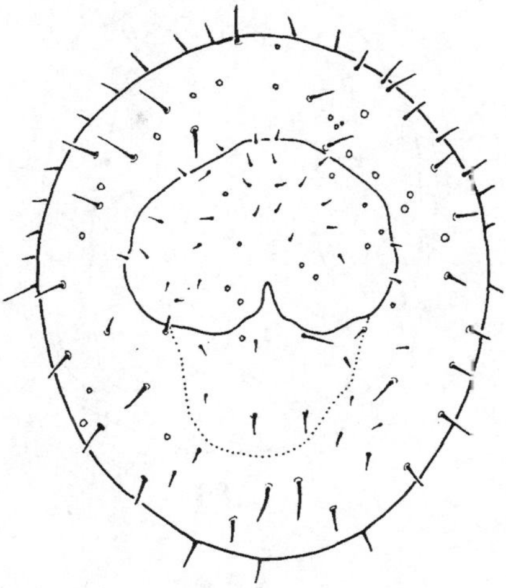

184

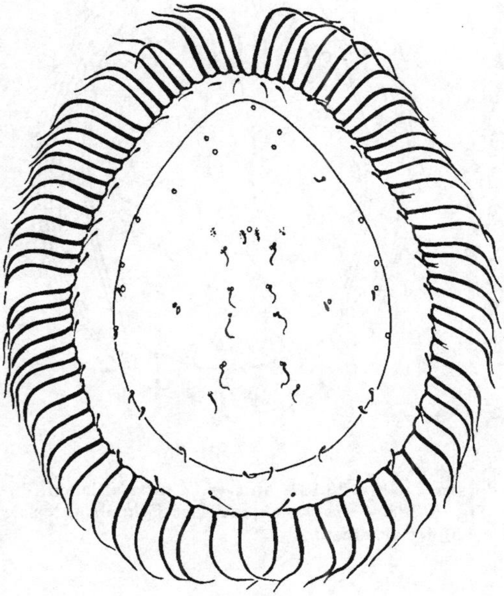

183

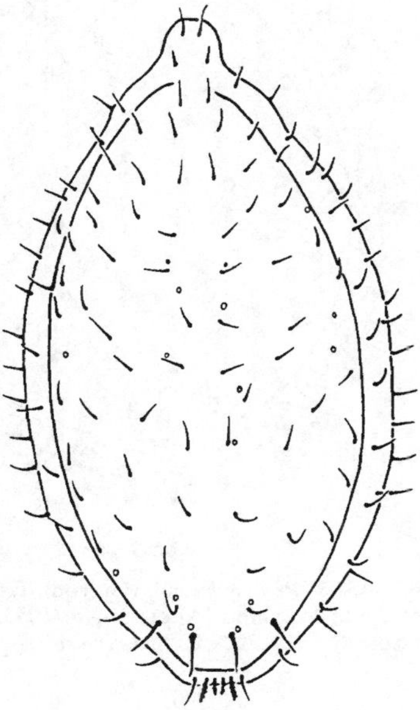

185

Fig. 20.182. *Discourella modesta* (Leonardi) Missouri), female dorsum. **Fig. 20.183.** *Clausiadinychus* sp. (Florida), female dorsum. **Fig. 20.184.** *Uroseius* sp., female dorsum. **Fig. 20.185.** *Dinychus* sp. (Oregon), female dorsum.

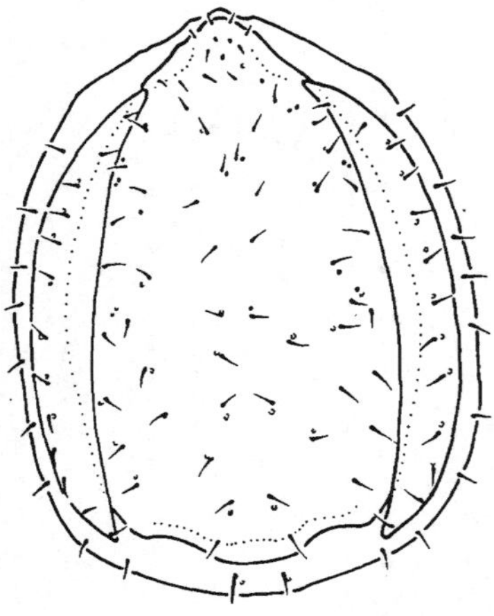

186

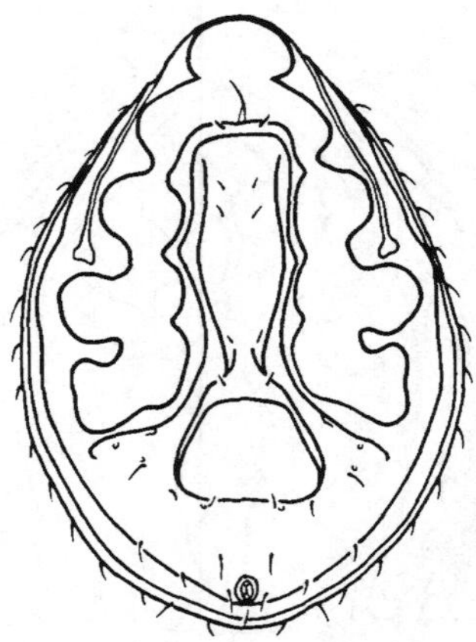

187

188

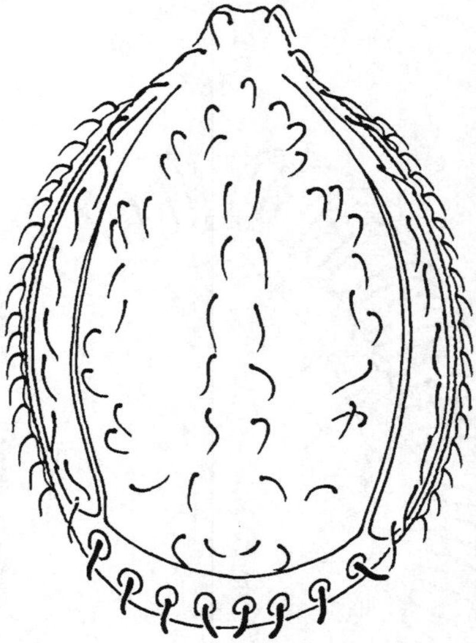

189

Fig. 20.186. *Prodinychus* sp. (Oregon), female dorsum. **Fig. 20.187.** *Metagynella parvula* Camin (Illinois), female venter. [After Camin (1953).] **Fig. 20.188.** *Uropolyaspis* sp. (British Columbia), female dorsum. **Fig. 20.189.** *Phaulodinychus* sp. (Oregon), female dorsum.

2a Tritosternal base narrow, longer than broad. **3**

2b Tritosternal base broad, at least as broad as long (Fig. 20.182). . . . . ***Discourella*** Berlese

3a Marginal setae greatly elongated (Fig. 20.183). . . . ***Clausiadinychus*** Sellnick

3b Marginal setae of normal length. **4**

4a Legs I without claws; dorsal sclerotization greatly reduced posteriorly (Fig. 20.184). . . . . ***Uroseius*** Berlese

4b Legs I with claws; dorsal sclerotization well developed posteriorly. **5**

5a Posterior region of dorsal shield and postmarginal platelets each with a transverse row of four stout, pilose or serrated setae (Fig. 20.185). . . . . **_Dinychus_** Kramer

5b Posterior dorsal region without such setae (Fig. 20.186). **_Prodinychus_** Berlese

6a Apex of leg II directed anteriorly when withdrawn. **7**

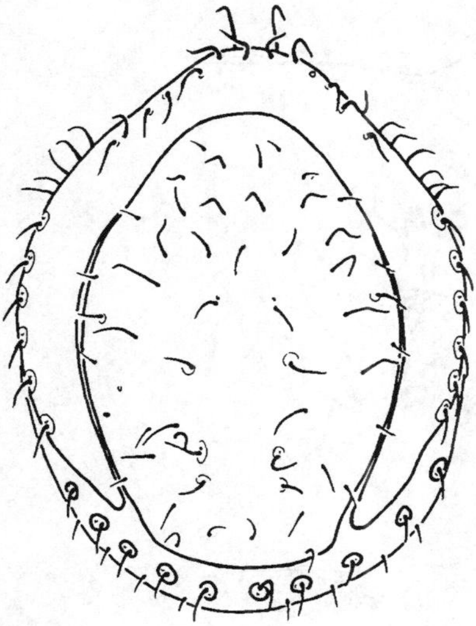

190

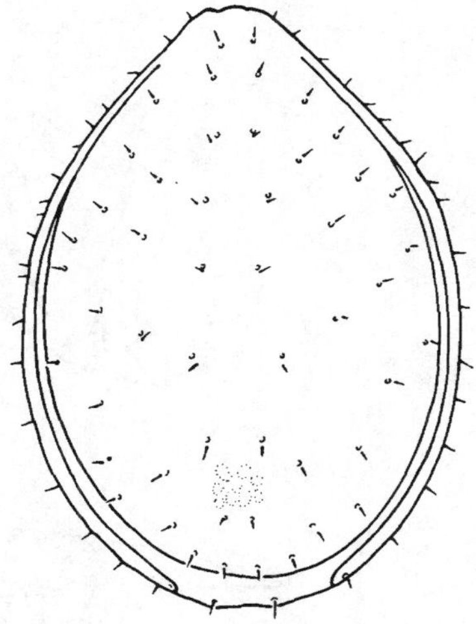

191

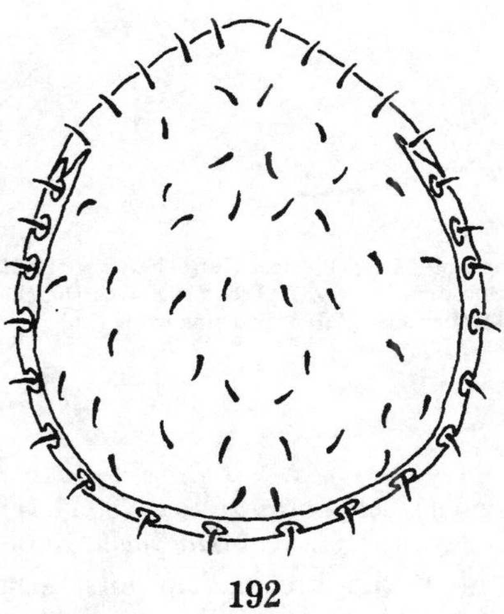

192

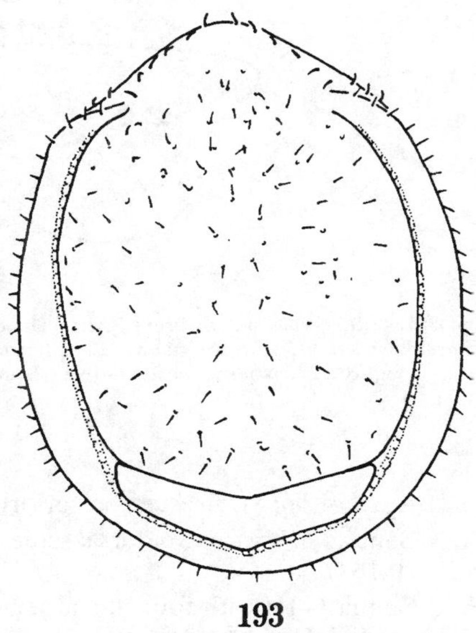

193

Fig. 20.190. *Neodiscopoma* sp. (Oregon), female dorsum. **Fig. 20.191.** *Olodiscus* sp. (Washington), female dorsum. **Fig. 20.192.** *Uropoda* sp. (Oregon), female dorsum. **Fig. 20.193.** *Urodiaspis* sp. (British Columbia), female dosum.

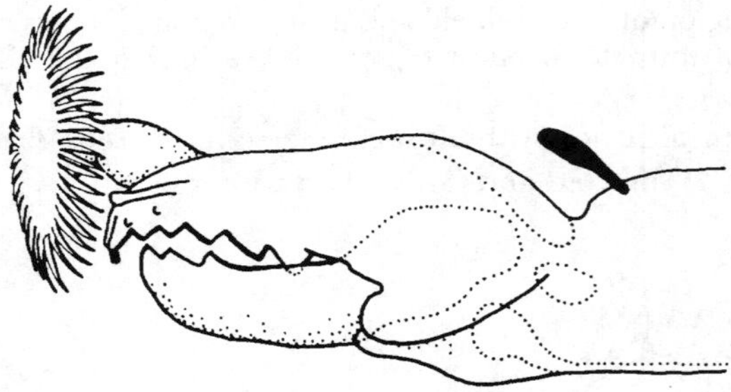

194

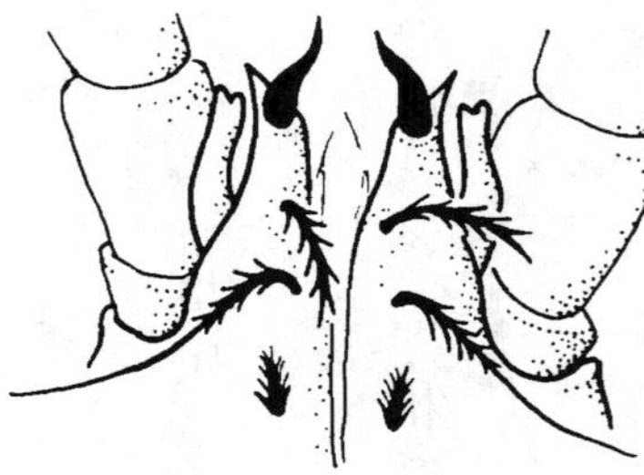

195

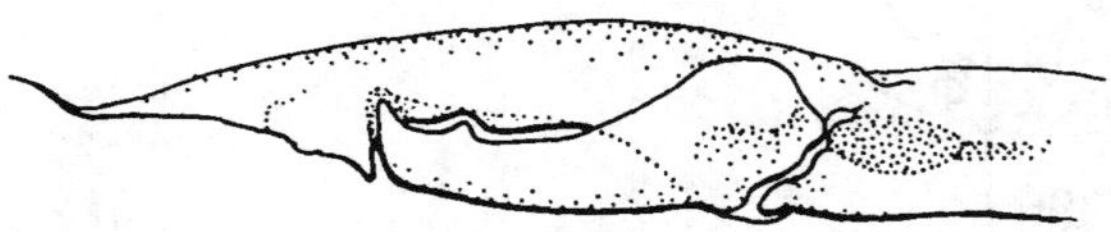

196

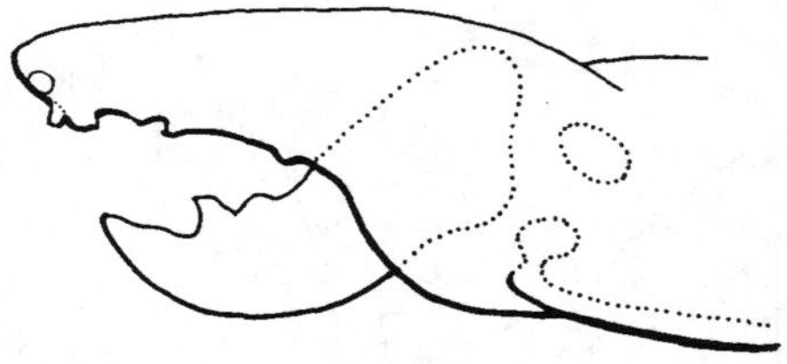

197

Fig. 20.194. *Uroactinia* sp., chelicera. [After Hirschmann and Zirngiebl-Nicol (1961).] **Fig. 20.195.** *Trichouropoda* sp. (Oregon), ventral aspect of female gnathosoma. **Fig. 20.196.** *Uroobovella* sp., chelicera. **Fig. 20.197.** *Nenteria canadensis* (New Brunswick), chelicera. [After Hirschmann (1978).]

6b Apex of leg II directed posteriorly. **12**

7a Genua I–IV with two dorsal setae; epigynial shield posterior to coxae IV (Fig. 20.187). . . . . ***Metagynella*** Berlese

7b Genua I–III with four dorsal setae, genu IV with three or four dorsal setae; epigynial shield between coxae II and IV. **8**

8a Legs I with well-developed claws. **9**

8b Legs I without claws. **10**

9a Marginal shield fragmented into platelets throughout its length; body setae clavate (Fig. 20.188). . . . . ***Uropolyaspis*** Berlese
9b Marginal shield fragmented into platelets only posteriorly; body setae not clavate (Fig. 20.189). . . . . ***Phaulodinychus*** Berlese
10a Dorsal surface more or less strongly sculptured; marginal shield fragmented posteriorly (Fig. 20.190). . . . . ***Neodiscopoma*** Vitzthum
10b Dorsal surface smooth; marginal shield variable. **11**
11a Idiosoma less than 500 μm in length; marginal shield reduced posteriorly (Fig. 20.191). . . . . ***Olodiscus*** Berlese
11b Idiosoma more than 500 μm in length; marginal shield variable (Fig. 20.192). ***Uropoda*** Latreille
12a Dorsal shield divided into anterior and posterior elements (Fig. 20.193). . . ***Urodiaspis*** Berlese
12b Dorsal shield entire. **13**
13a Chelicerae each with actiniform appendage on apex of fixed digit (Fig. 20.194). . . . . ***Uroactinia*** Sellnick
13b Chelicerae without actiniform appendages. **14**
14a Corniculi curved mesad, with two to four teeth or branches distally (Fig. 20.195). . . . . ***Trichouropoda*** Berlese
14b Corniculi smooth and hornlike, unbranched distally. **15**
15a Movable digit of chelicera unidentate (Fig. 20.196). . . . ***Uroobovella*** Berlese
15b Movable digit of chelicera multidentate (Fig. 20.197). . ***Nenteria*** Oudemans

References

Since the submission of this chapter, a number of papers have appeared which should be consulted for further information on the systematics, biology and distribution of North American edaphic Mesostigmata. The more significant of these recent contributions are included below.

Ainscough, B. D. 1960. *The Internal Morphology of Caminella peraphora Krantz and Ainscough, with descriptions of the Immature Stages.* Unpublished M.Sc. thesis, Oregon State University, Corvallis, Oreg. 53 pp. [DITHINOZERCONIDAE] **1963.** *Some Aspects of the Morphology of the Uropodina (Acarina: Mesostigmata).* Unpublished Ph.D. thesis, Birkbeck College, University of London, London. 244 pp. **1979.** Recent advances in the classification of the Uropodina. Pp. 451–454 in J. G. Rodriguez (ed.). *Recent Advances in Acarology,* Vol. II. Academic Press, New York. **1981.** Uropodine studies. I. Suprageneric classification in the cohort Uropodina Kramer, 1882 (Acari: Mesostimata). *Inter. J. Acarol.* 7:47–56. **Athias, F. 1975.** Observations morphologiques sur *Polyaspis patavinus* Berlese 1881 (Acariens: Uropodides). I. Morphologie de l'idiosoma au cours du développement postembryonnaire. *Acarologia (Paris)* 17(3):410–435. [POLYASPIDIDAE] **Athias-Henriot, C. 1968.** L'appareil d'insemination laelapoïde (Acariens anactinotriches: Laelapoidea, ♀♀). Premières observations. Possibilité d'emploi à des fins taxonomiques. *Bull. Sci. Bourgogne* 25:229–274. **1969.** Notes sur la morphologie externe des gamasides (Acariens anactinotriches). *Acarologia (Paris)* 11(4):609–629. **1975.** Nouvelles notes sur les Ambyseiinae. II. Le relevé organotaxique de la face dorsale adulte (Gamasides protoadeniques, Phytoseiidae). *Acarologia (Paris)* 17(1):20–29. **1982.** *Schizosthetus* n.g. (type *Eugamasus lyriformis* McGr. & Farr., 1969) avec deux espèces nouvelles (Parasitiformes, Parasitidae). *Acarologia (Paris)* 23(3):207–214. **Bernhard, F. 1963.** Die Familie Ascaidae (Oudemans 1905) Bernhard nov. comb. In H.-J. Stammer (ed.). *Beiträge zur Systematik and Ökologie mitteleuropäischer Acarina* 2. Geest und Portig, Leipzig, East Germany. *Mesostigmata* 1(3):33–177. [ASCIDAE] **Blas-**

zak, C. **1975.** A revision of the family Zerconidae (Acari, Mesostigmata) (Systematic studies on family Zerconidae I). *Acarologia (Paris)* 17(4):553–569. **1982.** New species of the genera *Echinozercon* Blaszak and *Skeironozercon* Halašková (Acari, Mesostigmata: Zerconidae) from United States of America. *Bul. Acad. Polonaise Sci.* 30(1–12):61–68. **1984.** Three new genera of Zerconid mites (Acari, Mesostigmata, Zerconidae) from the United States of America with a key to the American genera. *Bull. Entomol. Pologne* 53:587–601. **Bregetova, N. G. and E. V. Koroleva 1960.** The macrochelid mites (Gamasoidea, Macrochelidae) in the USSR. *Parazitol. Sb.* 19:32–154. **Camin, J. H. 1953.** Metagynellidae, a new family of uropodine mite, with the description of *Metagynella parvula,* a new species from tree holes. *Bull. Chic. Acad. Sci.* 9(18):391–409. **Camin, J. H. and F. Gorirossi 1955.** A revision of the suborder Mesostigmata (Acarina) based on new interpretations of comparative morphological data. *Chic. Acad. Sci. Spec. Publ.* 11:70 pp. **Chant, D. A. 1959.** Phytoseiid mites (Acarina : Phytoseiidae). I. Bionomics of seven species in southeastern England. II. A toxonomic review of the family Phytoseiidae, with descriptions of 38 new species. *Can. Entomol.* 91(Suppl. 12), 164 pp. **1965.** Generic concepts in the family Phytoseiidae (Acarina : Mesostigmata). *Can. Entomol.* 97:351–374. **Coineau, Y. 1974.** Introduction a l'Étude des Microarthropodes du Sol et de ses Annexes. *Documents pour l'enseignement pratique de l'Écologie.* DoIN, Paris. 118 pp. **Coons, L. B. 1978.** Fine structure of the digestive system of *Macrocheles muscaedomesticae* (Scopoli) (Acarina : Mesostigmata). *Int. J. Insect Morphol. Embryol.* 7(2):137–153. **Edlers, A. and L. Solomon. 1979.** Morphology of preadult stages of *Laelaps agilis* Koch, 1836 with notes on life history and feeding habits (Acari/Mesostigmata : Laelapidae). *Entomol. Scand.* 10(2):81–90. **Edwards, C. A. and K. E. Fletcher. 1971.** A comparison of extraction methods for terrestrial arthropods. Pp. 150–185 in J. Phillipson (ed.). *Methods of Study in Quantitative Soil Ecology; Population, Production and Energy Flow.* IBP Handb. 18. Blackwell Scientific, Oxford. **Evans, G. O. 1955.** A revision of the family Epicriidae (Acarina : Mesostigmata). *Bull. Br. Mus. (Nat. Hist.) Zool.* 3(4):169–200. **1957.** An introduction to the British Mesostigmata (Acarina) with keys to families and genera. *J. Linn. Soc. Lond.* 43:203–259. **1958.** A revision of the British Aceosejinae (Acarina : Mesostigmata). *Proc. Zool. Soc. Lond.* 131(2):177–229. [ASCIDAE]. **1963a.** Observations on the chaetotaxy of the legs in the free-living Gamasina. *Bull. Br. Mus. (Nat. Hist.) Zool.* 10(5):275–303. **1963b.** Some observations on the chaetotaxy of the pedipalps in the Mesostigmata (Acari). *Ann. Mag. Nat. Hist.* 6(13):513–527. **1969.** A new mite of the genus *Thinoseius* Halbert (Gamasina : Eviphididae) from the Chatham Islands. *Acarologia (Paris)* 11(3):505–514. **1972.** Leg chaetotaxy and the classification of the Uropodina (Acari : Mesostigmata). *J. Zool. (Lond.)* 167:193–206. **Evans, G. O., J. G. Sheals, and D. Macfarlane. 1961.** *The Terrestrial Acari of the British Isles. Vol. I, Introduction and Biology.* Adlard & Son, Bartholemew Press, Dorking, U.K.: 219 pp. + v. **Evans, G. O. and W. M. Till. 1965.** Studies on the British Dermanyssidae (Acari : Mesostigmata). I. External morphology. *Bull. Br. Mus. (Nat. Hist.) Zool.* 13(8):247–294. [LAELAPIDAE] **1966.** Studies on the British Dermanyssidae (Acari : Mesostigmata). II. Classification. *Bull. Br. Mus. (Nat. Hist.) Zool.* 14(5):107–370. [LAELAPIDAE] **1979.** Mesostigamatic mites of Britain and Ireland (Chelicerata : Acari-Parasitiformes). An introduction to their external morphology and classification. *Trans. Zool. Soc. Lond.* 35:139–270. **Faasch, H. 1967.** Beitrag zur Biologie der einheimischen Uropodiden *Uroobovella marginata* (C. L. Koch 1839) und *Uropoda orbicularis* (O. F. Müller 1776) und experimentelle Analyse ihres Phoresieverhaltens. *Zool. Jahrb. (Syst.)* 94:521–608. [UROPODIDAE] Farrier, M. H. 1957. A revision of the Veigaiidae (Acarina). *N. C. Agric. Exp. Stn. Tech. Bull.* 124. 103 pp. **Filipponi, A. and G. Francaviglia. 1963.** Oviparitá e larviparitá in *Macrocheles peniculatus* Berl. (Acari : Mesostigmata) regolate da fattori ecologici. *Riv. Parassitol.* 24(2):81–104. [MACROCHELIDAE] **Ghilyarov, M. S. and N. G. Bregetova (eds.). 1977.** *A Key to the Soil-Inhabiting Mites Mesostigmata.* Akademii Nauk SSSR, Zoological Institute, Leningrad. 719 pp. (in Russian). **Grandjean, F. 1949.** Observation et conservation des très petits arthropodes. *Bull. Mus. Natl. Hist. Nat. Paris Ser. 2* 21:363–370. **Gridelet-de-Saint-Georges, D. 1975.** Techniques d'extraction applicables a l'étude écologique des acariens des poussières de maison. Comparaison qualitative et quantitative de divers types de poussières. *Acarologia (Paris)* 17(4):693–708. **Halašková, V. 1977.** *A Revision of the Genera of the Family Zerconidae (Acari : Gamasides) and Descriptions of New Taxa from Several Areas of Nearctic Region.* Studie CSAV 7. Československý Akadémie Vied, Bratislava, Czechoslovakia. 74 pp. **Hirschmann, W. 1960.** Gangsystematik der Parasitiformes. 3. Die Gattüng *Dendrolaelaps* Halbert 1915. *Acarologie (Nuernb.)—Schriftenr. Vergl. Milbenkd.* 3:1–27, Taf. 1–28. [DIGAMASELLI-

DAE] **1978.** Gangsystematik der Parasitiformes. 276. Ventralbestimmungstabelle von 17 *Nenteria*-Arten. Teilgang, Stadien von 14 neuen *Nenteria*-Arten der Breviunguiculata-Moseri-Gruppe aus Kanada, Mexiko, Argentinien und das Männchen von *Nenteria eulaelaptis* (Vitzthum 1930) aus Ungarn (Trichouropodini, Uropodinae). *Acarologie (Nuernb.)—Schriftenr. Vergl. Milbenkd.* 24:48–57, Taf. 7–8. **Hirschmann, W. and W. Krauss 1965.** Gangsystematik der Parasitiformes. Teil 8. Bestimmungstafeln von 55 *Pachylaelaps*-Arten. *Acarologie (Nuernb.)—Schriftenr. Vergl. Milbenkd.* 7:1–5, Taf. 1–28. [PACHYLAELAPIDAE] **Hirschmann, W. and J. Wisniewski. 1982.** Weltweite Revision der Gattung *Dendrolaelaps* Halbert 1915 und *Longoseuis* Chant 1961 (Parasitiformes). *Acarologie (Nuernb.)—Shriftenr. Vergl. Milbenkd. Bd. I* 29:1–190; *Bd. II* 29:1–46, Taf. 1–94 + XIII. **Hirschmann, W. and I. Zirngiebl-Nicol. 1961.** Gangsystematik der Parasitiformes. 4. Die Gattung *Trichouropoda* Berlese 1916 nov. comb., die Cheliceren und das System der Uropodiden. *Acarologie (Nuernb.)—Schriftenr. Vergl. Milbenkd.* 4:1–41, Taf. 1–14. **Hughes, T. E. 1959.** *Mites, or the Acari.* University of London, Athlone Press, London. 255 pp. **Hurlbutt, H. W. 1984.** A study of North American *Veigaia* (Acarina : Mesostigmata) with comparisons of habitats of unisexual and bisexual forms. *Acarologia (Paris)* 25(3):207–222. **Hyatt, K. H. 1956.** British mites of the genus *Pachyseius* Berlese, 1910 (Gamasina, Neoparasitidae). *Ann. Mag. Nat. Hist.* 9(12):1–6. [PACHYLAELAPIDAE] **Johnston, D. E. 1968.** I. *The Families of Parasitiformes and Opilioacariformes.* Publ. 172. Acarology Laboratory, Ohio State University Columbus, Ohio. 110 pp. **1982.** Mesostigmata. Pp. 112–117 in S. B. Parker (ed.). *Synopsis and Classification of Living Organisms,* Vol. 2. McGraw-Hill, New York. **Karg, W. 1961.** Ökologische Untersuchungen von edaphischen Gamasiden (Acarina, Parasitiformes). *Pedobiologia* 1(1,2):53–74, 77–98. **1965a.** Larvalsystematische und phylogenetische Untersuchung sowie Revision de Systems der Gamasina Leach, 1915 (Acarina, Parasitiformes). *Mitt. Zool. Mus. Berl.* 41(2):193–340. **1965b.** Neue Erkenntnisse zum System der Gamasina (Acarina, Parasitiformes) durch larvalsystematische Untersuchungen. *Zesz. Probl. Postepow Nauk Roln.* 65:89–114. **1965c.** Die Anwendung Systematisch-Phylogenetischer Arbeits methoden bei einer Bearbeitung der Gamasina (Acarina, Parasitiformes). *Zesz. Probl. Postepow Nauk Roln.* 65:115–138. **1965d.** Entwickslungsgeschichtliche Betrachtung zur Ökologie der Gamasina (Acarina. Parasitiformes). *Zesz. Probl. Postepow Nauk Roln.* 65:139–155. **1971a.** Die freilebenden Gamasina (Gamasides), Raubmilben. *Tierwelt Deutschl.* 59:475 pp. Fischer, Verlag, Jena. **1971b.** Zur Kenntnis der Epicriidae Berlese, 1885 (Acarina, Anactinochaeta). *Zool. Anz.* 186(1/2):105–114. **1976.** Eine neue Milbengattung der Eviphididae Berlese, 1913 (Acarina, Parasitiformes). *Abh. Ber. Naturkundemus. Goerlitz* 50(3): 1–6. **1978a.** Zur Kenntnis der Milbengattung *Parasitus* Latreille, 1795 und *Trachygamasus* Berlese, 1906 (Acarina, Parasitiformes). *Zool. Jahrb. (Syst.)* 105:347–359. [PARASITIDAE] **1978b.** Die Gattung *Pseudoparasitus* Oudemans, 1902. *Mitt. Zool. Mus. Berl.* 54(2):203–212. [LAELAPIDAE] **1978c.** Zur Kenntnis der Milbengattungen *Hypoaspis, Androlaelaps* und *Reticulolaelaps* (Acarina, Parasitiformes, Dermanyssidae). *Zool. Jahrb. (Syst.)* 105:1–32. [LAELAPIDAE] **1979a.** Die Gattung *Hypoaspis* Canestrini, 1884 (Acarina, Parasitiformes). *Zool. Jahrb. (Syst.)* 106:65–104. [LAELAPIDAE] **1979b.** Zur Kenntnis der Raubmilbengattungen *Cheiroseius* Berlese, 1916, *Asca* v. Heyden, 1826 und *Halolaelaps* Berlese et Trouessart, 1889 (Acarina, Parasitiformes). *Mitt. Zool. Mus. Berl.* 55(2):251–267. [ASCIDAE, HALOLAELAPIDAE] **1980.** Die Raubmilbengattung *Lasioseius* Berlese, 1916. *Zool. Jahrb. (Syst.)* 107:344–367. **1983.** Verbreitung und Bedeutung von Raubmilben der Cohors Gamasina als Antagonisten von Nematoden. *Pedobiologia* 25:419–432. **1985.** Die Raubmilbengattung *Proctolaelaps* Berlese, 1923. *Zool. Jahrb. Syst.* 112:185–206. **Krantz, G. W. 1960.** A re-evaluation of the Parholaspinae Evans 1956 (Mesostigmata : Macrochelidae). *Acarologia (Paris)* 2(4):393–433. [PARHOLASPIDIDAE] **1978.** *A Manual of Acarology,* 2nd ed. Oregon State University, Corvallis, Oreg. 509 pp. **Krantz, G. W. and J. G. Wernz. 1979.** Sperm transfer in *Glyptholaspis americana* (Gamasida : Macrochelidae). Pp. 441–446 in J. G. Rodriguez (ed.). *Recent Advances in Acarology,* Vol. II. Academic Press, New York **Krauss, W. 1970.** Die europäischen Arten der Gattung *Macrocheles* Latrielle 1829 und *Geholaspis* Berlese 1918. Gangsystematik der Parasitiformes. *Acarologie (Nuernb.)—Schriftenr. Vergl. Milbenkd.* 14:2–43. [MACROCHELIDAE] **Lee, D. C. 1970.** The Rhodacaridae (Acari : Mesostigmata); classification, external morphology and distribution of genera. *Rec. S. Aust. Mus. (Adelaide)* 16(3):1–219. **1974.** Rhodacaridae (Acari : Mesostigmata) from near Adelaide, Australia. III. Behaviour and development. *Acarologia (Paris)* 16(1):21–44. **Lindquist, E. E. 1975.** *Digamasellus* Berlese, 1905, and *Dendrolaelaps* Halbert, 1915, with descriptions of new taxa of Digamasellidae

(Acarina : Mesostigmata). *Can. Entomol.* 107(1):1–43. **Lindquist, E. E. and G. O. Evans. 1965.** Taxonomic concepts in the Ascidae, with a modified setal nomenclature for the idiosoma of the Gamasina (Acarina : Mesostigmata). *Mem. Entomol. Soc. Can.* 47:64 pp. **Macfadyen, A. 1953.** Notes on methods for the extraction of small soil arthropods. *J. Anim. Ecol.* 22:65–77. **McBrayer, J. F., J. M. Ferris, L. J. Metz, C. S. Gist, B. W. Cornaby, Y. Kitazawa, T. Kitazawa, J. G. Wernz, G. W. Krantz and H. Jensen. 1977.** Decomposer invertebrate populations in U.S. forest biomes. *Pedobiologia* 17:89–96. **Merchant, V. A. and D. A. Crossley. 1970.** An inexpensive, high-efficiency Tullgren extractor for soil microarthropods. *J. Ga. Entomol. Soc.* 5(2):83–87. **Metz, L. J. and M. H. Farrier. 1969.** Acarina associated with decomposing forest litter in the North Carolina Piedmont. Pp. 43–52 in *Proc. 2nd Int. Congr. Acarology, Sutton Bonnington.* Akadémiai Kiadó, Budapest, Hungary. **Michael, A. D. 1892.** On the variations in the internal anatomy of the Gamasinae, especially in that of the genital organs and their mode of coition. *Trans. Linn. Soc. Lond.* 5(2):281–324. **Micherdzinski, W. 1969.** *Die Familie Parasitidae Oudemans 1901 (Acarina, Mesostigmata).* Zak. Zool. Syst. Polsk. Akad. Nauk, Krakow. 690 pp. **Oliver, J. H. 1971.** Parthenogenesis in mites and ticks (Arachnida : Acari). *Am. Zool.* 11(2):283–299. **Petrova-Nikitina, A. D. 1969.** Analysis of the family Parholaspidae Evans, 1956 (Parasitiformes, Gamasoidea). Pp. 187–190 in *Proc. 2nd Int. Cong. Acarology, Sutton Bonnington.* Akadémiae Kiadó, Budapest, Hungary. **Price, D. W. 1973.** Abundance and vertical distribution of microarthropods in the surface layers of a California pine forest soil. *Hilgardia* 42(4):121–148. **1975.** Vertical distribution of small arthropods in a California pine forest soil. *Ann. Entomol. Soc. Am.* 68(1):174–180. **Till, W. M. 1963.** Ethiopian mites of the genus *Androlaelaps* Berlese s. lat. (Acari : Mesostigmata). *Bull. Br. Mus. (Nat. Hist.) Zool.* 10(1):1–104. [LAELAPIDAE] **Westerboer, I. 1963.** Die Familie Podocinidae Berlese 1916. In H.-J. Stammer (ed.). *Beiträge zur Systematik und Ökologie mitteleuropäischer Acarina 2.* Geest und Portig, Leipzig, East Germany. *Mesostigmata* 1(4):179–450. [PODOCINIDAE, ASCIDAE] **Westerboer, I. and F. Bernhard. 1963.** Die Familie Phytoseiidae Berlese 1916. In H.-J. Stammer (ed.). *Beiträge zur Systematik und Ökologie mitteleuropäischer Acarina 2,* Geest und Portig, Leipzig, East Germany. *Mesostigmata* 1(5):451–777. [PHYTOSEIIDAE, AMEROSEIIDAE, ASCIDAE] **Witalinski, W. 1972.** Mites of the genus *Holoparasitus* Oudemans, 1936 (Acarina, Parasitidae). *Acta Zool. Cracov.* 17(9):217–237. **Woodring, J. P. and C. A. Galbraith. 1976.** The anatomy of the adult uropodid (*Fuscouropoda agitans* (Arachnida : Acari)), with comparative observations on other Acari. *J. Morphol.* 150(1):19–58. [UROPODIDAE]

Key to Abbreviations Used in Figures

ad anterodorsal seta
al anterolateral seta
an anal shield
arb arthrodial brush
ap apotele
av anteroventral seta
bc basis capituli
cam camerostome
cn corniculus
cs capitular seta
cx coxa
df fixed cheliceral digit
dg deutosternal groove
dm movable cheliceral digit
ds dorsal cheliceral seta
en endopodal shield
ep epistome
epg epigynial shield
ex exopodal shield
fm femur
fp fovea pedalis
fs femoral scale
gen genital aperture
gn genu
h_1, h_2, h_3 hypostomatic setae 1, 2, 3
hyp hypostome
im internal malae
j, *J* dorsal setae
mpd metapodal shield
mst metasternal shield
pa paradactylus
pd pilus dentilis
pfm palpfemur

Key to Abbreviations Used in Figures (*Continued*)

pl	posterolateral seta	*sp*	spermatodactyl
pp	adgenital pore plate	*st*	sternal shield
pst	presternal shield	st_1, st_2, st_3, st_4	sternal seta 1, 2, 3, metasternal seta
pt	peritreme		
ptr	palptrochanter	*stg*	stigma
pts	peritrematic shield	*sty*	salivary stylet
pv	posteroventral seta	*tb*	tibia
r, R	marginal setae	*trt*	tritosternum
s, S	lateral setae	*v*	ventral shield
sc	scabellum	*z, Z*	median setae

21

Acarina: Prostigmata (Actinedida)

JOHN KETHLEY

Morphological and biological diversity abound in the acarine suborder Prostigmata. It is divided into six cohorts containing 31 superfamilies and 136 families; considering only named taxa there are 1100 genera, and 14,000 or more species. Terrestrial prostigmatid mites of the soil–litter community are a significant subset of these totals, representing 60 families, 681 genera, and nearly 6400 species. The prostigmatid groups of greatest direct interest to soil biologists are rather small: free-living fungivores/algavores include nine families, 27 genera, and 86 species; free-living predators comprise 26 families, 251 genera, and nearly 1200 species. However, these free-living taxa must be distinguished from the following array of other prostigmatid mites which may be found in the sampling process: seven families, 69 genera, and 1000 or more species of obligate plant feeders; four families, 69 genera, and 1000 or more species of specialized fungus feeders having obligate relationships with other arthropods; and 11 families, 262 genera, and 3000 or more species of nymphal and adult predators having larvae specialized as parasites of invertebrates or vertebrates.

External Morphology

The body is divided into two regions. The anterior region or *gnathosoma* bears the mouthparts and palpus. The posterior region or *idiosoma* externally bears the legs, reproductive structures, and anus. Morphological plasticity abounds in prostigmatid mites and involves major modifications of both the gnathosoma and idiosoma from ancestral conditions.

Gnathosoma

The gnathosoma consists of paired palpi flanking the chelicerae which are above the dorsum of a subcapitulum. Primitively, each palpus was a leglike tactile structure comprised of five homogeneous articulating segments (trochanter, femur, genu, tibia, and tarsus). However, segments have become modified or fused in various groups and the function may have changed. The femur and genu are fused in Caeculidae (Fig. 21.119), and in the families of Eupodoidea and Tydeoidea (Figs. 21.42, 21.50, 21.60, 21.65, 21.74). A raptorial palp has been evolved in at least two different ways. One or more tibial setae have been modified into a clawlike seta with a tendency to displace the tarsus from a terminal to a subterminal (ventral) position (Figs. 21.119, 21.125, 21.131, 21.177), followed by a reduction in the size of the tarsus in the Anystina and derivative groups (Figs. 21.142, 21.161, 21.230). A second raptorial morphology has been achieved in Cunaxidae with the development of terminal tarsal spines and often mesally directed spines (Figs. 21.106, 21.107). Palpal reduction is seen in Tenuipalpidae (Fig. 21.247) and the Heterostigmata (Tarsocheylidae, Tarsonemidae, etc.; Figs. 21.254, 21.259).

Primitively, each chelicera comprised a basal segment, a fixed digit bearing two setae, and a movable digit (Fig. 21.18), and each chelicera moved independently. The reduction of the fixed digit and modification of the movable digit into a piercing structure has occurred at least three times (Figs. 21.51, 21.95, 21.136). The cheliceral bases are mesally adnate in Cryptognathidae (Fig. 21.200), partially fused in Caligonellidae (Fig. 21.172), to almost fused in Raphignathidae (Fig. 21.182), with the movable digit reduced to a styletiform piercing structure. In Tetranychoidea, the cheliceral bases are fused to each other to form a *stylophore*; the bases are capable of united movement over the subcapitulum, each styletiform movable digit is capable of independent motion, and the stylophore itself may be withdrawn into the body (see Figs. 21.250, 21.251). The cheliceral bases are completely fused to each other *and* to the hypostome to form a *stylophore capsule* in the Cheyletoidea and Heterostigmata (Fig. 21.230). In Erythraeidae, the chelicerae are retractile, the palps remain external to the body. Retraction of the entire gnathosoma into the body has independently evolved in Cryptognathidae (Fig. 21.200), Calyptostomatidae (Fig. 21.327), and Smarididae (Fig. 21.330). The ventral surface of the subcapitulum bears distal adoral setae and proximal subcapitular setae. The anterolateral margin of the subcapitulum bears a pair of strongly sclerotized and bladelike setae, the *rutella* (Fig. 21.39) in a few families of Endeostigmata; the rutellum is absent in other prostigmatid mites.

Primitively, stigmatal openings and associated canal-shaped surface features (peritremes) are absent. Where present, stigmata are located at the base of the gnathosoma between the chelicerae in most Eupodina, Anystina, and some Eleutherengona. Evolved peritremes can be located at the base of the gnathosoma (Fig. 21.124), on the prodorsum (Fig. 21.141), on the mesal surface of the cheliceral bases (Fig. 21.150), or on the dorsal surface of the partially fused (Figs. 21.172, 21.176), or completely fused cheliceral bases (Fig. 21.230); the stigmata and peritremes attributed to *Stigmalycus* (Alicorhagiidae) by Theron et al. (1970[M])† are in fact the podocephalic canal (see below). Stigmata are absent in juveniles and most males of Heterostigmata; however, enlarged stigmata on the anterolateral margins of the prodorsum are unique to females of Heterostigmata (Fig. 21.254).

Idiosoma

Subdivisions and General Morphology. Subdivisions of the idiosoma are designated in relationship to the leg-bearing and non-leg-bearing regions. Ancestrally, the *podosoma* is the anterior portion, including legs I–IV; the *opisthosoma* is the region posterior to legs IV (Fig. 21.5). The podosomal and opisthosomal separation is obliterated in most groups because dorsal elements of the opisthosoma have encroached above ventral elements of the podosoma that support legs III and IV. These ventral elements become "captured" by the opisthosoma, forming a new tagma, the *hysterosoma*. The remaining elements of the original podosoma comprise the *propodosoma*, and the division between these two secondary body regions is termed the *sejugal groove*. The term *proterosoma* has been used to denote a combined propodosoma and gnathosoma. In some groups a further body subdivision can be seen posterior to legs IV, which delimits an anterior region of the hysterosoma termed the *metapodosoma* (Fig. 21.134).

The presence and distinctness of these secondary subdivisions varies not only between groups, but also with ontogenetic stage. Secondary patterns of sclerotization can serve to sharply delimit body regions or obliterate them through fusion(s). The extent of body sclerotization is highly variable. In most groups the body integument is weakly sclerotized or possesses a series of discrete sclerites associated with dorsal setae. A single holonotal sclerite is present in adults of some Penthalodidae (Figs. 21.41, 21.47), Labidostommatidae (Fig. 21.111), Cryptognathidae (Fig. 21.198), Homocaligidae (Fig. 21.195), and some Stigmaeidae (Fig. 21.217). The cuticle of caeculid mites is extremely thick and leathery (Fig. 21.117). Sclerotization of the prodorsum may be quadrate (Figs. 21.121, 21.160), hypertrophied to extend over the gnathosoma (Fig. 21.270), or reduced to a linear form (Figs. 21.329, 21.330, 21.337).

Primitively, the dorsum of the propodosoma or *prodorsum* bears six pairs of structures of setal origin (Fig. 21.5). The propodosoma is considered to represent dorsally two to four insensibly fused tergites, but there is no

† Reference citations are keyed to the reference lists: [B], "Biology–Ecology," and [M], "Morphology–Taxonomy."

indication of the original segmental distribution of these six pairs of setae. Consequently, an assessment of the homologs of these setae throughout prostigmatid mites is based strictly on topographic equivalency. Numerous signatures have been used for the prodorsal setae, and it seems timely to propose uniform notations and equivalents for these structures (Appendix 1). The anterior two pairs of prodorsal setae are *verticals*; one pair (*vi*) is mesad or internal, and one pair (*ve*) is laterad or external, and either may be modified as trichobothria. The internal verticals, *vi*, are modified as *trichobothria* (Figs. 21.103, 21.120) in the families Adamystidae, Anystidae, Bdellidae, Cunaxidae, Erythraeidae, some Johnstonianidae, Smarididae, and Sphaerolichidae. The external verticals, *ve*, are modified as trichobothria (Fig. 21.13) in Bimichaeliidae, Lordalycidae, and Nanorchestidae. The middle two pairs of prodorsal setae are *scapulars*, of which one pair (*sci*) is internal, and one pair (*sce*) external. In many families of prostigmatid mites *sci* are modified as trichobothria (Fig. 21.103); the external pair, *sce*, is always setiform. The presence of one or two posterior pairs of setae common in the Oribatei, the *posterior exobothridials* (*exp* or *ei*) and the *interlamellars* (*in*), is restricted to the families Alicorhagiidae, Bimichaeliidae, Grandjeanicidae, Micropsammidae, Nanorchestidae, Oehserchestidae, Proteonematalycidae, Sphaerolichidae, and Terpnacaridae (Fig. 21.5). In addition to prodorsal setae, an anterior protuberance, the *naso* (Figs. 21.9, 21.13, 21.123), may be present. Primitively, the ventral surface of the naso bears a pair of eyes that become fused into a single median unpaired eye in some Endeostigmata and most Anystina. The naso is present but lacks the median unpaired eye in most Eupodoidea and Parasitengona. The naso is absent, but the median unpaired eye is present in Cytinae (Bdellidae), some Tydeidae (*Triophtydeus*), and Calyptostomatidae (Parasitengona). The naso and median eyes are absent in all Eleutherengona. One or two pairs of eyes may be present in the region of the vertical or scapular setae. One pair of cup-shaped structures (*urnulae*) is found in nymphs and adults of Balaustiinae (Erythraeidae) (Fig. 21.333). The products of one to three pairs of coxal glands, located dorsolaterally above the insertion of legs I, are carried to the gnathosoma via a pair of *podocephalic canals* (Fig. 21.120). Maximally, there are three pairs of supracoxal setae present, often as minute, fingerlike processes (Fig. 21.120); one pair (*ep*) is located dorsally at the base of the palpus, one pair (*eI*) dorsolaterally at the level of insertion of legs I, and one pair (*eII*) dorsolaterally at the level of insertion of legs II. Supracoxals II are present only in the deutonymph and subsequent active stages of Anystidae, terrestrial Parasitengona, and Bdellidae.

Anamorphosis. True metameric segmentation is absent from Prostigmata. Primitively, the hysterosoma reflects the remnants of ancestral segmentation through the presence of transverse rows of setae and associated proprioceptors (*cupules*) which form a series of segmental remnants.

The full adult complement of setae is acquired by a series of developmental additions to the posterior of the body. This process parallels closely the paraproctal addition of segments (anamorphosis) seen in Crustacea. The relative stability of these developmental patterns in the Oribatei led Grandjean (1939a[M]) to develop and apply a coherent system of notation for the segmental remnants, setae, and cupules. Lindquist (1977[M]) has shown the homologies for these opisthosomal structures for heterostigmatid mites. The same developmental pattern occurs in the primitive elements of prostigmatid mites, especially the Endeostigmata. The extension of this notation scheme to other prostigmatid groups has been delayed due to lack of adequate comparative developmental studies, as well as the penchant of many workers to develop and adhere to ad hoc schemes for taxa studied in isolation. A uniform scheme of chaetotactic notation is applied below, and a table of setal equivalents is supplied in Appendix 2.

The larva is the first active life stage in most prostigmatids. Primitively, the larval hysterosoma bears five transverse regions designated (anterior to posterior) *C, D, E, F,* and *H* (the letter *g* is reserved for the genital region), and the paraproctal segment (the valves encompassing the anus) are initially designated *PS*, for pseudanal (Fig. 21.120). Three segmental remnants are then added during development, one in each of the next three nymphal instars. Each paraproctal element is represented by a pair of anal valves displacing the previous segmental remnant toward the anterior. First, the protonymph adds the adanal segmental remnant (*AD*) posteriorly as valves between the anus and segmental remnant *PS*. Next, the deutonymph adds the anal segmental remnant (*AN*) posteriorly as valves between the anus and segmental remnant *AD*. The final addition occurs in the tritonymph with the peranal segmental remnant (*PA*) posteriorly inserted between the anus and segmental remnant *AN*. Each transverse region bears corresponding pairs of setae: C—c_1, c_2, c_3, c_4; D—d_1, d_2; E—e_1, e_2; F—f_1, f_2, f_3; H—h_1, h_2, h_3; PS—ps_1, ps_2, ps_3; AD—ad_1, ad_2, ad_3; AN—an_1, an_2, an_3; PA—pa_1, pa_2, pa_3, pa_4. Laterally on segmental remnants *D* to *AD* are the respective cupules: *D—ia*; *E—im*; *F—ip*; *H—ih*; *PS—ips*; *AD—iad*. In the added regions the appearance of the cupules is ontogenetic, but is delayed one instar relative to the respective segmental remnant. Thus *ia, im, ip,* and *ih* are larval; *ips* appears in the protonymph and *iad* in the deutonymph. The addition of cupule pair *ian* (typical of numerous oribatid tritonymphs) does not occur in prostigmatid mites, and the addition of *ips* and *iad* is restricted to some Endeostigmata. The maximum number of cupules in all other prostigmatid mites is four pairs (*ia, im, ip, ih*).

The primitive complement of hysteronotal segmental remnants (*C—PA*), setae (28 pairs), and cupules (six pairs) has been maintained in Terpnacaridae (Fig. 21.5) and some Micropsammidae. Other early derivative taxa that possess *C—PA* but lack one or more pairs of cupules are Bimichaeliidae, Grandjeanicidae, some Micropsammidae, some Oehserchestidae, and Alicorhagiidae (*Epistomalycus*). The primitive condition has been modified in other taxa by either a reduction or an increase in the number of setae. Reductions have involved both the regressive loss of postlarval structures and/or the suppression of structures that are typically larval in origin. Postlarval regressions first involve the terminal nymphal segmental remnant *PA*. The addition of numerous body setae (plethotrichous neotrichy) in Nanorchestidae prohibits an assessment of setal or segmental identities in this family.

The regression (i.e., nonappearance) of *PA* and associated setae has occurred in Alicorhagiidae (except *Epistomalycus*), Lordalycidae, Proteonematalycidae, Adamystidae, and *Stigmacheylus* [incorrectly placed in Pseudocheylidae by Baker and Atyeo (1964[M])]. The subsequent regression of segmental remnant *AN* (and *PA*) and associated setae has occurred in the remaining prostigmatid mites except Bdellidae. In some cases the regression may involve the setae, but not the segmental remnant. In Bdellinae, a single pair of setae is added to the anal valves of the deutonymph (Fig. 21.101c). A similar condition is seen in the regression of *AD* and associated setae in the superfamily Eupodoidea. The anal valves of larvae of all Eupodoidea bear either three or two pairs of *ps* setae. The protonymph of Rhagidiidae and Strandtmanniidae has three pairs of *ps* setae flanking the anal valves and one pair of *ad* setae on the valves. The protonymph of eupodids, penthalodids, and penthaleids has three or two pairs of *ps* setae flanking the nude anal valves, implying that the *AD* segmental remnant has been added, but not the *ad* setae. The protonymphal addition of *only AD* occurs in the following families; Anystidae, Caeculidae, Cunaxidae, all Eupodoidea, Labidostommatidae, Paratydeidae, Teneriffiidae, and some Tydeidae. Some taxa have a typical addition of the segmental remnants, but delay the addition of the corresponding setae. For example, species of *Stigmalycus* (Alicorhagiidae) possess a single pair of *ps* setae in the larva, while postlarval stages possess ps_{1-3}; *ad* setae are absent in the protonymph, but the deutonymph and adult possess ad_{1-3}; deutonymphs possess a single pair of *an* setae, while the adults possess an_{1-3}. The prostigmatid groups that do not exhibit additions to the body chaetome beyond the larval *ps* series are: Raphignathoidea, Cheyletoidea, Tetranychoidea, Eriophyoidea, Heterostigmata, and Parasitengona.

Setal suppressions have also occurred within the *c—ps* series. Setae, c_3, c_4, and f_3 have been lost in all prostigmatid groups except some Endeostigmata. Setae *d2* and *e2* are lost in most families of Anystina, Eupodoidea, Tydeoidea, Bdelloidea, Caligonellidae, Raphignathidae, and all Heterostigmata. The presence of three or more pairs of *c* and two or more pairs of *d* and *e* in larvae of highly derivative groups such as Stigmaeidae, Parasitengona, and Tetranychoidea is considered secondary.

A small number of groups exhibit developmental retardations or additions within the *c—ps* series. Pseudanal setae are absent in larvae of *Cryptognathus* (Cryptognathidae), *Xenocaligonellidus* (Xenocaligonellididae), and *Coptocheles* and *Neognathus* (Caligonellidae), but present in the respective protonymphs. The topographic equivalent of f_2 is added in the female-forming deutonymph of some Stigmaeidae (Figs. 21.206, 21.210). The equivalent of f_1 is added in the deutonymph of Raphignathidae (Figs. 21.180, 21.183). The equivalents of e_1 and f_1 are added in the deutonymphs of many Cheyletidae (Fig. 21.224).

The developmental addition to the hysterosoma of large numbers of setae (plethotrichous neotrichy) obscures recognition of the *c—pa* series. This addition has occurred independently in almost every cohort: most Bimichaeliidae and Nanorchestidae (Endeostigmata); some Anystidae, Caeculidae, and Pseudocheylidae (Anystina); Penthaleidae and Strandtmanniidae (Eupodina); Cheyletidae and Xenocaligonellididae (Eleutherengona); and all terrestrial Parasitengona.

Genital Region. In most prostigmatid mites the genital region is located ventrally, posterior to legs IV and anterior to the anal valves. Aggenital setae may flank the genital valves and genital setae. Primitively, a vestibule covered by paired genital valves contains the genitalia, genital papillae, and eugenital setae; females may have an eversible ovipositor (Fig. 21.154). The ultrastructure of the genital papillae parallels that of osmoregulatory organs in other arthropods (Alberti 1979[M]). The appearance of genital papillae is ontogenetic; three pairs of papillae are present in adults of Adamystidae, Bimichaeliidae, Grandjeanicidae, Oehserchestidae, Terpnacaridae (Endeostigmata), Anystidae (minute in females), Caeculidae, Pomerantziidae, Teneriffiidae (Anystina), Bdellidae (Eupodina), Trombidioidea, and Calyptostomatoidea (Parasitengona). Two or three pairs are present in adults of Alicorhagiidae, Lordalycidae, Nanorchestidae, Nematalycidae (Endeostigmata), and Neotrombidiidae (Parasitengona). Two pairs of genital papillae are present in adults of Micropsammidae, Sphaerolichidae, Eupodoidea, Tydeoidea (some Tydeidae lack papillae), Cunaxidae, Labidostommatidae (Eupodina), and Paratydeidae (Anystina). Genital papillae are absent in adults of all Eleutherengona, Heterostigmata, Pseudocheylidae (Anystina), and most Erythraeoidea (Parasitengona). Genital tracheae, often arising in the anterior portion of the body and terminating in the genital vestibule, are present in Sphaerolichidae (Grandjean 1939b[M]), many Bdellidae (Grandjean 1938[M]), and some species of *Speleorchestes* (Nanorchestidae) (Grandjean 1939b[M]), and Micropsammidae.

Sexual dimorphism is expressed in three trends. In most endeostigmatid mites, dimorphism is evidenced only by the eugenital setae; females often have fewer eugenital setae than males, a condition shared by the Caeculidae, Labidostommatidae, Eupodoidea, and most Tydeoidea (Figs. 21.98, 21.99). The accessory gland of Eupodoidea males is an enlarged saclike structure that is moderately well sclerotized (Fig. 21.56). A second trend is seen in the family Sphaerolichidae, where the male has hypertrophied eugenital setae (Fig. 21.26) and some sclerotization within the genital vestibule. This condition is elaborated in the Anystina and Bdellidae, where the male genitalia are heavily sclerotized and extremely complex (Figs. 21.102, 21.133, 21.138, 21.147, 21.159). A third trend is the suppression of the eugenital setae and genital papillae that is paralleled by coalescence of the genital and anal regions and the development of an eversible aedeagus in males in several lineages.

The shift from reproduction by spermatophores to direct insemination probably evolved in at least three lineages of prostigmatid mites. Cunaxid adults have genital papillae, but lack eugenital setae (except females of *Parabonzia*), and the males have an aedeagus (Fig. 21.109). Males possess an aedeagus in at least two genera of Tydeidae (*Proctotydeus* and *Homeopronematus*). Adults lack genital papillae and eugenital setae, and the males have an aedeagus in Raphignathoidea, Cheyletoidea, Heterostigmata, and Tetranychoidea (Figs. 21.178, 21.179, 21.184, 21.209, 21.229, 21.234) [except possibly in Tarsocheylidae Lindquist (1986[M])]. Within the cohort Eleutherengona, the only males to retain setae of genital origin are in the Tarsocheylidae (see Fig. 21.257 and Appendix 2, col. 28) and in most Tetranychoidea. Males of other Eleutherengona have the genital and anal regions coalesced, and only setae of paraproctal origin (*ps* series) and one, two, or no pairs of aggenital setae present (see Figs. 21.173, 21.178, 21.179, 21.183, 21.184, 21.196, 21.208, 21.209, 21.229, 21.261, 21.264; see also Appendix 2, cols. 19–27, 29–31, 35, 36). Theoretically, in species where seminal fluid is transferred via an aedeagus, the female should possess a bursa copulatrix and spermatheca. However, the presence of the bursa copulatrix and spermatheca has been demonstrated for only a few species of Tenuipalpidae (Castagnoli 1974[M], for example).

Legs. Primitively, larvae have six articulating leg segments beyond the coxa: trochanter–femur–genu–tibia–tarsus–pretarsus; and adults have at least one femur subdivided into articulating basi- and telofemora. The coxae are immovably fused to the body venter. The coxal remnants are represented by coxal fields delineated by internal apodemes from which intrinsic coxal musculature originates. The femora may be subdivided in most Endeostigmata, Eupodina, and terrestrial Parasitengona, and some Labidostommatidae exhibit a third subdivision of femur I. The trochanters, genua, and tibiae are rarely subdivided. The tarsi are often entire, but may have a single acrotarsal subdivision in Labidostommatidae, or numerous pseudosegments in some Anystidae, Teneriffiidae, and Eupodidae (especially *Linopodes*). Primitively, the pretarsi consist of paired claws and a setulate empodial pad (Fig. 21.4). The empodium may be modified as a claw (Fig. 21.7), a rod from which arise hairlike elements terminating in bifurcate tips (tenent hairs) (Fig. 21.215), a membranous pad (Figs. 21.162, 21.258, 21.279), or they may be absent. Paired claws have been lost independently in a number of groups. The adult number of pairs of legs is reduced to three in some Tenuipalpidae, and to only two pairs in all

Eriophyoidea. Other leg reductions have evolved independently in numerous groups of parasites. Legs IV may exhibit saltatorial modifications. Coxal fields IV have a large surface area and possess extensive internal musculature and apodemes in Granjeanicidae, Oehserchestidae, Sphaerolichidae, and some Terpnacaridae (*Terpnacarus*) (Grandjean 1939b[M]). Similar modifications, which also include a trochanteral–femoral apodeme of IV, occur in Nanorchestidae (Grandjean 1939a[M]), some Bimichaeliidae (*Pachygnathus ornithorhynchus*) (Grandjean 1939a[M]), and some Alicorhagiidae (*Epistomalycus*). The ability to jump correlates with hypertrophy of basifemur IV in some Eupodidae and Rhagidiidae.

The legs bear both tactile structures (setae) and chemoreceptors (solenidia, and setae modified as eupathidia). Cuticular structures of setal origin are optically active when viewed through polarized light, whereas solenidia are not optically active. Structures of setal origin are of four basic types: tactile setae, eupathidia, famuli, and trichobothria. Eupathidia are modified setae having a hollow core, are often blunt tipped, and have an enlarged alveolus. A famulus (*eta*, ε) may be umbellate (Fig. 21.17), stellate (Fig. 21.46), or peglike. Primitively, the famulus is restricted to tarsi I and II. The famulus of tarsus I is present in all prostigmatid mites except Eleutherengona, Heterostigmata, Nematalycidae (Endeostigmata), Paratydeidae, Pseudocheylidae and Teneriffiidae (Anystina), and Tydeoidea (Eupodina). The famulus of tarsus II is present in Micropsammidae, Nematalycidae, Sphaerolichidae, Terpnacaridae, variable in Bimichaeliidae and Nanorchestidae (Endeostigmata), Adamystidae, Anystidae, Caeculidae (Anystina), Bdellidae, Penthaleidae, Penthalodidae, Rhagidiidae, and Strandtmanniidae (Eupodina), Erythraeoidea, and Trombidioidea (Parasitengona). A famulus-like structure, termed *k,* is found dorsally on tibia I in Adamystidae, Anystidae, Caeculidae and Teneriffiidae (Anystina), Bdellidae, Cunaxidae and Labidostommatidae (Eupodina), most terrestrial Parasitengona, and all Heterostigmata. A famulus-like structure is present dorsally on genua I and II in Bdellidae, some Cunaxidae, Penthalodidae (Eupodina), most terrestrial Parasitengona, and many Raphignathoidea (Eleutherengona). Leg trichobothria are tactile setae with an enlarged, often recessed alveoli. These are on the dorsal leg surface in Bdellidae, Cunaxidae (Eupodina), Caeculidae (Anystina), Tetranychidae (Eleutherengona), and some Parasitengona; or they are ventral on basifemora I and II of Anystidae and Sphaerolichidae, and on basi-femora I–III of Adamystidae. Solenidia may be found on the genua, tibiae, and tarsi, but rarely on the femora. Solenidia are usually erect, surface eversions of the cuticle that lack alveoli. In most Eupodoidea, the solenidia of tarsi I and II are recumbent and slightly recessed on the segment surface (Figs. 21.67–21.69) [= rhagidial organ(s) of authors]. The solenidion of tibia I is completely recessed within the segment in Ereynetidae (= ereynetal organ of authors). The solenidion of tibia II is partially or completely recessed in Rhagidiidae (Fig. 21.62) and Strandtmanniidae (Fig. 21.69). Notations for solenidia are both segmental and developmental: femorals are *theta* (θ) genuals are *sigma* (σ), tibials are *phi* (ϕ), and tarsals are *omega* (ω). A notation of N_1, N_2, or N_3 implies the order of ontogenetic appearance. Leg chaetotaxy and solenidiotaxy are orthotrichous in most groups (i.e., topographic homologies can be established for individual structures, across developmental stages of a given species, and across species, genera, and families). Although not emphasized in the key to the families, leg chaetotaxy and solenidiotaxy are extremely useful in delineation of genera and species within the various families. Comparative studies also facilitate the recognition of the developmental stages, since setal additions and/or reductions during ontogeny are usually predictable.

Developmental Stages

The basic life-history pattern comprises an egg and six instars: prelarva–hexapod larva–three octopod nymphs–adult male and female. Immatures possess a prodorsal ecdysial line and lack functional genitalia, whereas adults lack an ecdysial line but possess functional genitalia.

Ancestrally, the prelarva was a normally formed, active, feeding instar. Among extant taxa, no prelarva is known to feed, and mobility is evidenced only in *Speleorchestes* (Nanorchestidae) and perhaps *Saxidromus* (Adamystidae). Prelarvae of *Speleorchestes* possess a well-developed body and leg chaetotaxy as well as pretarsi. A body chaetotaxy and pretarsi are also present in prelarvae of Adamystidae, but the leg setae are reduced to small spinose structures. Prelarvae possess legs, but the segments are ill defined, and setae and pretarsi are absent in Alicorhagiidae, Anystidae, Bdellidae, Caeculidae, Eupodidae, Lordalycidae, Micropsammidae, Nematalycidae, Rhagidiidae, Terpnacaridae, and many Parasitengona. Prelarvae completely lack legs in Bimichaeliidae (*Alycus*), Cunaxidae, Nanorchestidae (*Nanorchestes* and *Neonanorchestes*), Proteonematalycidae, Tetranychidae, Tydeidae, and some Parasitengona (*Balaustium*). Prelarvae should exist, but have not been described for any species of Raphignathoidea.

Primitively, prelarvae and larvae possess a pair of discoid or fingerlike structures (*urstigmata* = *Claparéde's organ*) laterad of coxal fields I and II. The presence of urstigmata in prelarvae and larvae, both of which lack genital papillae and genital valves, corresponds to the presence of genital papillae in nymphs and adults (the Oudemans–Grandjean rule). When present, the number of pairs of genital papillae in nymphs is developmental; protonymphs have a single pair and deutonymphs have two pairs present; three pairs of genital papillae are present in the tritonymph only if the adult possesses three pairs. In general, enough diagnostic characters are present on larvae to permit family-level identification.

In most groups the juveniles are homeomorphic with respect to the adults, differing in their leg and/or palpal chaetotaxy and the chaetome of the presumptive genital region. These ontogenetic differences among juveniles with respect to their adults may also reflect juvenile sexual dimorphism in a few groups. Sexual dimorphism in juveniles may be evidenced by differences within the genital vestibule in Bimichaeliidae, or differences in the solenidiotaxy, hysteronotal chaetotaxy, and chaetotaxies of the genital and aggenital regions in most Raphignathoidea, or by differences in the morphology of the larval pseudanal setae in some Tarsonemidae (Lindquist 1986a[M]).

A major departure from homeomorphic juveniles and adults occurs in the Parasitengona, where larvae are heteromorphic with respect to the nymphs and adults, and the protonymph and tritonymph are inactive, only partially expressed stages that develop within the cuticle of the preceding stages. A similar developmental pattern occurs in species of *Neonanorchestes* (Nanorchestidae), where the protonymph and tritonymph are also inactive, morphologically reduced stages. In other taxa, such as Labidostommatidae, some species of Rhagidiidae, and *Neognathus* (Caligonellidae), only the larva is a regressive, inactive, nonfeeding stage. The number of juvenile stages may be regressed in some groups: adults of Alicorhagiidae, Proteonematalycidae, Stigmocheylidae, at least one species of Bimichaeliidae (*Bimichaelia diadema*), and all known Raphignathoidea, Cheyletidae, and Tetranychoidea (except Tuckerellidae) are preceded by only two nymphal instars. Greater regression occurs in the Tarsonemida, where there are no active nymphal instars and females are either oviparous or larviparous. Kuznetzov (1980[M]) reported several other departures from the basic seven-stage life-history pattern which indicate great developmental plasticity within the family Tydeidae. Thus *Tydeus californicus* is larviparous, with three active nymphal stages preceding the adult; *Venilia liberta* has an inactive, incompletely developed tritonymph that is dormant from July until the following spring; and species of *Tydaeolus* and *Microtydeus* lack the tritonymphal stage.

Biology

Feeding

The prostigmatid mites to be found in the soil community have diverse feeding habits. Terminology for characterization of feeding habits applied principally to oribatid mites can be extended to certain groups of prostigmatid mites. Schuster (1956[B]) distinguished between consumers of microflora (i.e., algae, bacteria, and fungi) as microphytophages and consumers of vascular plant material as macrophytophages. Luxton (1972[B]) defined subgroups within the micro- and macrophytophage categories and defined categories for consumers of material of animal origin (i.e., zoophagy as feeding on a living animal, necrophagy as feeding on carrion, and coprophagy as feeding of fecal material). Unfortunately, Luxton's concept of macrophytophages and its subgroups has been applied primarily to consumers of "dead or dying plant tissues" (Krantz and Lindquist 1979[B], p. 122; Harding and Stuttard 1974[B]; Luxton and Petersen 1982[B]). Rather than redefine macrophytophagy *sensu* Luxton (1972[B]) to include consumers of tissue fluids of living vascular plants, Krantz and Lindquist considered phytophagy to include both higher plant tissue fluid feeders and herbivores (for consumers of entire living plants or plant parts).

The lack of sufficient detailed food preference studies of prostigmatid mites makes definitive characterization of feeding habits of specific groups highly speculative. However, categorization of feeding habits based on observations or rearings of related taxa, gut content analyses, or mouthpart morphology could provide a useful beginning. Prostigmatid feeding habits encompass all three subgroups of microphytophagy *sensu* Luxton (1972[B]) (mycophagy, bacteriophagy, and phycophagy), predators of arthropods and nematodes (zoophagy), plant tissue fluid feeders, and obligate invertebrate or vertebrate parasites. To date, no prostigmatids have been shown to be macrophytophages *sensu* Luxton (1972[B]), necrophages, or coprophages. With some exceptions noted below, the feeding patterns of prostigmatid mites appear to follow broad taxonomic groups (Table 21.1).

The biological plasticity within prostigmatid mites confounds unimodal characterization of feeding types. Within a given group different species can be consumers at two or more trophic levels. Different species within the families Tarsonemidae and Tydeidae belong to each of the subgroups within microphytophagy as well as predation, plant tissue feeding, and even parasitism. Most Stigmaeidae are considered to be predaceous, but species of *Eustigmaeus* feed on mosses (Gerson 1972[M]). The deutonymphs and adults of most terrestrial Parasitengona are considered to be predaceous, but species of *Balaustium* feed on pollen and plant tissue in addition to Hemiptera and Thysanoptera (Grandjean 1947[M], Newell 1963[B])! Species of Endeostigmata have been traditionally considered to be fungivorous. However, *Alycus roseus* (Bimichaeliidae) is strictly a predator of nematodes (Walter 1987a[B]), and one species of *Pachygnathus* in the *P. ornithorhynchus* group (Bimichaeliidae) is thought to be a predator of copepods and tardigrades (Schuster 1979[B]). Two species of *Alicorhagia* (Alicorhagiidae) are omnivores, feeding on algae and fungi but requiring nematodes in their diet to complete development (Walter 1987a,b[B]). Characterization of feeding habits within a single species may be quite difficult when the species has been shown to be either predaceous or fungivorous, as in caeculids [see Coineau (1974b[M] versus Crossley and Merchant (1971[B,M])] or in tydeids (Brickhill 1958[B]). Specific feeding patterns are unknown for even a single representative of several families. Consequently, a questionable assessment is given based on gnathosomal morphology (Pomerantziidae) or on taxonomic affinities (Barbutiidae, Xenocaligonellididae). The relatively close taxonomic relationship of the

Table 21.1. Feeding Pattern–Macrohabitat Relations Among Prostigmatid Mites Likely to Be Found or Encountered in the Soil Community[a]

	Feeding Pattern				Microhabitat			
	Micro-phyto-phages	Predator	Plant Feeder	Parasite Vert., Invert.	Aerial	Upper Litter	Humus	Soil below 10 cm
Anystoidea		+			+	+		
Barbutiidae		?				+		
Bdelloidea		+			+	+	+	+
Caeculidae	+	+				+		
Caligonellidae		+				+		
Camerobiidae		+			+	+		
Cheyletidae		+			+	+		
Cryptognathidae		+				+		
Endeostigmata	+	+			+	+	+	+
Ereynetidae		+		+		+		
Eriophyoidea			+1		+	+		
Eupalopsellidae		+2			+	+		
Eupodidae	+	+				+	+	+
Homocaligidae		?	?			+		
Labidostommatidae		+				+		
Parasitengona		+3		+4	+	+	+	+
Paratydeidae	+	+					+	+
Penthaleidae			+			+		
Penthalodidae	+		+			+	+	+
Pomerantziidae		?						+
Pyemotoidea				+5	+	+		
Pygmephoroidea	+	+			+	+		
Raphignathidae		+			+	+		
Rhagidiidae		+			+	+	+	+
Stigmaeidae		+	+		+	+		
Strandtmanniidae		+				+	+	
Tarsonemidae	+	+	+	+6	+	+	+	
Tetranychoidea			+1		+	+		
Tydeidae	+	+	+	?	+	+	+	+
Xenocaligonellididae		?			+			

[a] Entries given for a superfamily or higher category pertain to all families of that group; see the text for exceptions and discussion. 1, Obligate plant feeders; 2, predators of scale insect eggs; 3, predators of arthropods only as nymphs and adults; 4, parasites of vertebrates or invertebrates only as larvae; 5, obligate parasites of invertebrates; 6, parasites/parasitoids of insects; +, known feeding pattern and/or macrohabitat; ?, probable feeding pattern.

Homocaligidae to the stigmaeid genus *Eustigmaeus* as well as a strong affinity for *Sphagnum* mosses raises the possibility of plane tissue fluid feeding for this family. More detailed studies of specific taxa are required to answer these questions.

Reproductive and Developmental Biology

Data on mating patterns, modes of reproduction, fecundity, development periods, and longevity for specific prostigmatids are scarce (excluding arboreal plant feeders); however, some generalizations can be given. Primitively, mating is by spermatophores. Males produce stalked spermatophores individually but in the immediate absence of females. A more derived behavior involves the production of spermatophores in the immediate vicinity of females or the formation of a male–female pair. An elaborate mating sequence is known for some species of Adamystidae, where the male grasps the female, produces a stalked spermatophore, elevates and rotates the female 180°, and then forces the female down onto the spermatophore (Coineau 1976[M]). Spermatophore producing prostigmatids have traditionally been considered to be diplo-diploid, bisexual species. However, Helle et al. (1984[B]) demonstrated haplo-diploidy in a species of Ereynetidae, and thelytokous parthenogenesis has been demonstrated for some species of Terpnacaridae (*Terpnacarus*) (Theron 1976[M]) and Rhagidiidae (Ehrnsberger 1977[B]). Because males are unknown, parthenogenesis is suspected in Alicorhagiidae (*Alicorhagia* and *Stigmalycus*), Grandjeanicidae, Lordalycidae, Oehserchestidae, Nematalycidae, Pomerantziidae, Proteonematalycidae, Stigmocheylidae, and some Bimichaeliidae (*Bimichaelia diadema*). Males are either unknown or very rare (facultative parthenogenesis) in some Bdellidae and Cunaxidae (Bdelloidea). The production of haploid males from unfertilized eggs (arrhenotoky) is thought to be the general pattern for eleutherengones, where mating involves direct insemination (Oliver 1971[B], Helle et al. 1984[B]). However, some species of Tetranychidae and Tenuipalpidae are known to be thelytokous (Oliver 1983[B]), and thelytoky is strongly suspected in *Cheyletus eruditus* (Oliver 1971[B]).

Fecundity may be highly variable within a single family; for example, in Trombidiidae, females of *Georgia* produce as few as 10 eggs, females of *Eutrombidium* and *Trombidium* produce up to 3500 eggs (Robaux 1974[M]), and females of *Dinothrombium* produce between 80,000 and 100,000 eggs (Newell 1979[B]). Specific fecundity data for other prostigmatids are lacking except for tetranychoids and arboreal stigmaeids. It is clear that smaller prostigmatid females (i.e., less than 400 μm in body length) mature only one or two eggs at a time (Luxton 1981[B]). However, this physical constraint of the body volume required to mature an egg should not be taken to imply a low fecundity, since data on adult longevity and oviposition frequency are wanting.

The sparse life-history data of terrestrial prostigmatids can easily be summarized. Chada (1956[B]) reported that *Penthaleus major* (Penthaleidae) required 61 days for development, with adults living 37 days; data (in days) for individual stages were: egg 25, larva 12.5, protonymph 8.6, deutonymph 7.8, tritonymph 7.0; the mites complete two generations per year in north-central Texas feeding on grasses; the first generation develops in the fall from summer diapausing eggs, the second generation develops in the spring, and density peaks occurs in early February and early April. Brickhill (1958[B]) reared *Tydeus bakeri* (Tydeidae) on honeydew and found that 28–34 days were required for development from egg to adult; females began oviposition 3–4 days after eclosion and laid an average of 4 eggs per clutch. Brickhill also found that *Lorryia ferulus* (Tydeidae) required 26–30 days for development of egg to adult when fed honeydew. McCoy et al. (1969[B]) reared *Parapronematus acaciae* (Tydeidae) on fungi and found that 15–18 days were required for development of egg to adult; females produced an average of 6 eggs. Wahab et al. (1974[B]) reared *Tydeus californicus* on potato leaves and found that 19 days were required for development of larva to adult at 30°C and 50% relative humidity; females produced 19–29 larvae, averaging 2 larvae per day. Snetsinger (1956[B]) reported that *Spinibdella depressa* (Bdellidae) required 22–29 days for development of egg to adult at 70°F and 90% relative humidity when fed tetranychid mites; data (in days) for individual stages were: egg 8, larva 1–2, protonymph 2–3, deutonymph 5–8, tritonymph 6–8; mites failed to complete development at less than 75% relative humidity and temperatures above or below the range 60–70°F. Beer and Dailey (1956[B]) reported that *Cheyletomorpha lepidopterorum* (Cheyletidae) required an average of 21 days for development of egg to adult; data (in average number of days) for each stage were: egg 3, larva 4, protonymph 5, deutonymph 9; each female produced an average of 16 eggs over a 9-day period; females survived an additional 13 days after oviposition. Beer and Dailey also found that *Cheyletus eruditus* (Cheyletidae) required an average of 21 days for development of egg to adult; data (in average number of days) for each stage were: egg 3, larva 4, protonymph 5, deutonymph 9; each female produced an average of 22 eggs laid in two equal clutches over a 7-day period; females survived a maximum of 83 days.

The seasonal patterns of peak population density given in a number of studies (Franco et al. 1979[B]), Luxton, 1981[B], Loots and Ryke 1966[B], Athias 1975[B], Aoki et al. 1977[B], Ishikawa et al. 1977[B]) suggest that many of the smaller prostigmatid species (most Endeostigmata, some Eupodidae, Rhagidiidae, and Tydeidae (Eupodina) and Heterostigmata have only one or two generations per year.

Ecology

General Aspects of Distribution

Prostigmatid mites are broadly distributed geographically and occur throughout terrestrial ecosystems. Within a given habitat some are further restricted in microhabitat distribution, while others exhibit considerable movement between microhabitats. From the viewpoint of soil ecologists two basic types of movement are important, that between soil and plants and that between soil horizons.

This within-habitat movement may be of a seasonal, diurnal, or temporally random nature and may be obligatory, facultative, or accidental in terms of the life cycle of individual mite species. Those prostigmatid mites that accidentally occur in, and are sampled from the soil community are found there for a variety of reasons. Obligate plant tissue-fluid feeders, such as members of the Eriophyoidea and Tetranychoidea, can be abundant in samples, mostly as a result of dislodgment from vegetation. Predaceous prostigmatids occurring primarily on emergent vegetation, such as most Bdellidae and Anystidae, also occur in soil samples. This may also reflect an accidental dislocation from the vegetation, or it may be a result of a migration from the aerial habitat due to environmental conditions, or even a genuine affinity for the upper litter–humus region. Unlike other major groups of mites, a significant fraction of prostigmatids are found at depths well in excess of 10 cm in most non-forest ecosystems. Consequently, aerial bdellids must be distinguished from bdellids that are restricted to depths below 10 cm. Obligate invertebrate parasitoids such as Pyemotoidea, which can also be found in soils, may utilize hosts that exhibit considerable within-habitat movement. The soil–litter community is the contact zone for parasitic larvae of the terrestrial Parasitengona seeking vertebrate or invertebrate hosts that move across or through the habitat. Even those species restricted to specific aerial microhabitats may be encountered if the microhabitat itself becomes incorporated in the soil. Associations with lichens are common (Gerson and Seaward 1977[B], André 1979[B]), but lichen fragments or entire thalli may be dislodged from trees and fall to the surface of the soil. Other prostigmatid species may be associated with liverworts, mosses, leaves, twigs, bark and wood fragments, or other materials that continuously enter the soil system from above.

Such irregularities in distribution cause problems in interpreting structure of soil mite communities and the functional role of component species. Soil ecologists tend to distinguish between aerial grazing food webs and detritus food webs (Petersen and Luxton 1982[B]). Characterization of feeding habits and microhabitat occurrence (Table 21.1) will facilitate such a separation. However, additional studies are needed to detail microhabitat preference of individual taxa.

Knowledge of within-habitat movements may also show some commonly cited generalizations to be untenable. For example, a seemingly positive relationship between substrate organic content and a ratio of the number of oribatids to the number of prostigmatids (Loots and Ryke 1966[B]), has been suggested in a number of studies (di Castri 1973[B], Athias 1975[B], Dindal 1979[B]). However, the "oribatid/prostig ratio" may be extremely misleading for two reasons, one involving feeding pattern and the other vertical migration. Although the Oribatei constitute a relatively homogeneous group with respect to basic feeding pattern, prostigmatid mites are quite heterogeneous. The ecological significance of a ratio in which one element consists of species of the same trophic level (saprophagous oribatids), and the other consists of species in multiple trophic levels (fungivorous, predaceous, parasitic, or plant-feeding prostigmatids) appears questionable.

Seasonal differences in spatial distribution can dramatically alter oribatid/prostigmatid ratios at the same site. Although there is relatively little vertical migration of soil mites in the top 10 cm of an organic soil in forests (as summarized in Petersen 1982a[B]) vertical migration is quite pronounced in deserts (Edney et al. 1976[B], Franco et al. 1979[B]), grasslands (Leetham and Milchunas 1985[B]), temperate prairies (Lussenhop 1972[B]), tropical savannas (Athias 1975, 1976[B]), and mediterranean-type ecosystem soils (Price 1975[B]). A reanalysis of the data of Price (1975[B]) for the top 7.6 cm sampled yields on oribatid/prostigmatid ratio of 3.88 for samples taken in late March, but a very different value, 0.59, for samples taken in late July, due to the downward migration of 56% of the individuals of Oribatei.

General Characteristics of Populations

Meaningful comparisons of population parameters among different terrestrial ecosystems require comparable data; unfortunately, this is rarely possible. Within-site heterogeneity and differences among sampling and extracting procedures have long been recognized as significant factors impinging on our ability to develop comparable data sets. This may be especially true for prostigmatid mites, where Petersen (1982a[B]) noted that density data appearing unduly low for this group may in part be due to their extremely small size and/or differential recovery rates. Similarly, population density estimates through time may be misleading due to poor recovery of molting and/or resting stages that are resistant to extraction. This is considered a serious difficulty with data for the Oribatei because nearly 30% of the annual cycle may be spent in molting (Luxton 1982[B]); the equivalent problem has not been addressed for prostigmatid mites. The following discussion of density, biomass, and seasonality and Tables 21.2–21.4 does not include suspected underestimates.

Density and Diversity. Estimates of mean annual density of prostigmatid mites per square meter and percent of total mites represented by prostigmatids developed from the literature are organized by major ecosystem in Table 21.2. Moss-turf bank habitats in the antarctic have a low overall acarine density and diversity, but prostigmatids (principally species of Nanorchestidae, Eupodidae, and Ereynetidae) constitute over 90% of the mites (Table 21.2). Prostigmatids in desert habitats range from fewer than 1000 to 6000 per square meter and constitute from 40% to nearly 85% of the acarine density (Table 21.2). Dominant species represent the Bdelloidea, Caligonellidae, Nanorchestidae, Pygmephoridae, Tarsonemidae, and Tydeidae in North America; Nanorchestidae, Paratydeidae, and Pseudocheylidae in Australia.

Prostigmatids represent the dominant acarine group in herbaceous habitats in arctic–alpine, temperate (mesic), and tropical regions, where intensive studies indicate densities ranging from 39,000 to 63,000 per square meter, 31,000 to 95,000 per square meter and 10,000 to 18,000 per square meter, respectively (Table 21.2). Prostigmatids may represent as much as 85% of the total number of mites. Species of Eupodidae, Pygmephoridae (Pyemotidae of authors), Scutacaridae, Tarsonemidae, and Tydeidae are abundant in these herbaceous habitats. Species of Bdellidae and Nanorchestidae are also abundant in temperate (mesic) grasslands, while species of Stigmaeidae are abundant in tropical grasslands. The prostigmatid fauna of hydric temperate herbaceous habitats is depauperate, and densities range from nearly 1000 per square meter to less than 4000 per square meter.

Prostigmatids are abundant in conferous and deciduous forests, but reflect a range of only 20–30% of the total number of mites due to the increased abundance of Oribatei. Mean annual densities vary from 20,000 per square meter to nearly 120,000 per square meter in coniferous forests, but range only from 20,000 to 87,000 per square meter in deciduous forests (Table 21.2). Species of the following families are abundant in both coniferous (Price 1975[B]) and deciduous (Aoki et al. 1977[B], Ishikawa et al. 1977[B], Luxton 1981[B]) forests: Nanorchestidae, Eupodidae, Rhagidiidae, Tydeidae, Pygmephoridae (Pyemotidae of authors), Scutacaridae, and Tarsonemidae; species of Bimichaeliidae, Bdelloidea, Cryptognathidae, Caligonellidae, Raphignathidae, Stigmaeidae, and Paratydeidae are also abundant in western U.S. coniferous forest soils (Price 1975[B]); species of Alicorhagiidae and Bdellidae may be abundant in deciduous forests (Aoki et al. 1977[B]). In tropical forest soils, Chiba et. al (1975[B]) found that seven species distributed among the families Eupodidae (two species), Nanorchestidae, Rhagidiidae, Tarsonemidae, and Tydeidae represented 75% of all the prostigs encountered.

Biomass. Estimates of mean annual biomass are organized by major ecosystem in Table 21.3. Due to their small average body size, their contribution to the total acarine biomass of deciduous and coniferous forest

Table 21.2. Density/Diversity Relations of Prostigmatid Soil Mites

Habitat	Annal $\bar{x}$ no./m^2	Percent of Total Mites	Number of Prostig. Species	Sample Depth (cm)	Source
Moss-turf bank, Antarctica	7,515	93.1	6	6	Goddard (1979[B])
Moss-turf bank, Antarctica	7,700	94.5	6	24	Block (1980[B])
Warm Deserts/Arid Scrub					
Joshua tree desert, California, U.S.	245	10.7	6	—	Wallwork (1972a[B])
Sclerophyllous grassland, Australia	580	84.0	1–5	5	Wood (1971[B])
Desert steppe, Australia	850	70.2	1–5	5	Wood (1971[B])
Mojave desert, U.S.	1,832	44.5	25	30	Franco et al. (1979[B])
Low-layered woodland, Australia	2,430	65.6	1–5	5	Wood (1971[B])
Shrub steppe, Australia	2,890	66.9	1–5	5	Wood (1971[B])
Semiarid mallee scrub, Australia	6,130	73.4	1–5	5	Wood (1971[B])
Arctic/Alpine Herbaceous					
Meadow, Alaska, U.S.	1,100	16.0	16?	10	Douce and Crossley (1977[B])
Alpine sedge mat, Austria	1,631	4.9	—	1	Schatz (1981[B])
Alpine meadow, Austria	2,264	9.7	—	1	Schatz (1981[B])
Dryas–Graminium heath, Alaska, U.S.	38,700	83.0	16	10	Douce and Crossley (1977[B])
Dryas heath, Alaska, U.S.	63,200	83.0	16	10	Douce and Crossley (1977[B])
Temperate Herbaceous—Hydric					
Saltmarsh, U.K.	920	4.9	7	3	Luxton (1967a,b[B])
Juncus, U.K.	1,100	5.3	3	15	Macfayden (1952[B])
Deschampsia tussocks, U.K.	1,400	1.8	3	15	Macfayden (1952[B])
Nardus grassland, U.K.	1,460	1.8	—	3	Block (1965[B])
Eriophorum vaginatum moor, U.K.	3,100	9.1		6	Block (1966a[B])
Molinia fen, U.K.	3,500	2.6	3	15	Macfayden (1952[B])

Table 21.2. *(Continued)*

Habitat	Annal $\bar{x}$ no./m^2	Percent of Total Mites	Number of Prostig. Species	Sample Depth (cm)	Source
	Temperate Herbaceous—Mesic				
Mineral soils, U.K.	910	8.4	23	7.6	Davis (1963[B])
Limestone grassland, U.K.	3,905	9.2	1	3	Block (1965[B])
Pasture, Sweden	5,700	28.1	57	5	Weis-Fogh (1948[B])
Pasture, New Zealand	14,359	18.4	14	18	Adams (1971[B])
Prairie, Wisconsin, U.S.	31,920	46.3	21	9	Lussenhop (1972[B])
Pasture, New Zealand	41,110	35.1	19	15.2	McMillan (1969[B])
Grassland, Sweden	50,500	45.1	25	20–26	Persson and Lohm (1977[B])
Fell field, northern Finland	52,570	85.2	—	6	Huhta and Koskenniemi (1975[B])
Festuca–Agrostis grassland, U.K.	61,000	40.1	37	12	Wood (1967a[B])
Nardus grassland, U.K.	75,000	42.6	31	12	Wood (1967a[B])
Short-grass prairie, Colorado, U.S.	94,700	66.8	35	60	Leetham and Milchunas (1985[B])
	Tropical Grasslands				
Burned savanna, Ivory Coast	7,840	43.7	—	85	Athias (1975[B])
Unburned savanna, Ivory Coast	10,900	38.6	—	85	Athias (1975[B])
Savanna, Chile	12,500	29.7	—	60	Covarrubias et al. (1964[B])
Savanna woodland, Australia	19,640	69.1	1–5	5	Wood (1971[B])
Pasture soil, South Africa	21,700	68.7	44	5	Loots and Ryke (1966[B])
	Temperate Coniferous Forest				
Pinus forest, Tennessee, U.S.	20,310	24.2	—	6.4	Crossley and Bohnsack (1960[B])
Picea woodland, U.K.	35,000	35.4	—	7.7–15.4	Evans et al. (1961[B])
Larix woodland, U.K.	45,000	29.6	—	7.7–15.4	Evans et al. (1961[B])

Picea forest, Quebec, Canada	49,920	8.0	9	8	Behan et al. (1978[B])
Scots pine, 150 lb H_2SO_4/yr, 6 years, Sweden	52,780	22.1	—	10–12	Bååth et al. (1980[B])
Picea forest, northern Finland	61,200	35.1	—	6	Huhta and Koskenniemi (1975[B])
Evergreen montane forest, South Africa	65,800				van den Berg and Ryke (1967–1968[B])
Pinus sylvestris woodland, U.K.	68,000	27.3	—	7.7–15.4	Evans et al. (1961[B])
Pseudotsuga woodland, U.K.	69,000	21.1	—	7.7–15.4	Evans et al. (1961[B])
Pinus ponderosa woodland, California, U.S.	74,980	51.3	60	25.4	Price (1973[B])
Picea forest, southern Finland	91,800	28.9	—	6	Huhta and Koskenniemi (1975[B])
Pinus nigra woodland, U.K.	100,000	30.8	—	7.7–15.4	Evans et al. (1961[B])
Scots pine, 50 lb H_2SO_4/yr 6 years, Sweden	106,670	34.5	—	10–12	Bååth et al. (1980[B])
Scots pine, 23-yr regrowth, Sweden	117,340	37.2	—	10–12	Bååth et al. (1980[B])
Temperate Deciduous Forest					
Fagus woodland, Denmark	20,480	30.4	40	6	Luxton (1981[B], 1982[B])
Quercus woodland, Japan	22,300	30.1	42	15	Ishikawa et al. (1977[B])
Mixed temperate deciduous forest, Japan	22,550	21.7	50	15	Aoki et al. (1977[B])
Temperate deciduous forest, North Carolina, U.S	31,400	17.9	—	5	Seastedt and Crossley (1978[B])
Quercus–Fagus woodland, U.K.	78,000	18.1	—	7.7–15.4	Evans et al. (1961[B])
Quercus woodland, U.K.	87,000	21.5	—	7.7–15.4	Evans et al. (1961[B])

Table 21.2. (*Continued*)

Habitat	Annal $\bar{x}$ no./m^2	Percent of Total Mites	Number of Prostig. Species	Sample Depth (cm)	Source
Tropical Forest					
Lowland rain forest, Malaysia	1,520	8.4	45	15	Chiba et al. (1975[B])
Semideciduous forest, Nigeria	3,020	22.0	14	7.5–10	Lasebikan (1974[B])
5–8 years of bush regrowth, Nigeria	14,830	32.2	—	15	Critchley et al. (1979[B])
Sclerophyll mallee, Australia	18,380	52.7	1–5	5	Wood (1971[B])
Acacia scrub, South Africa	22,160	76.8	54	5	den Heyer and Ryke (1966[B])
Sclerophyll forest, Chile	38,574	38.1	—	60	Covarrubias et al. (1964[B])
Cultivated Habitats					
Banana field, northern India	4,120	47.5	—	22.5	Singh and Pillai (1975[B])
Citrus orchard, northern India	10,260	74.0	—	22.5	Singh and Pillai (1975[B])
Fallow land, northern India	10,370	74.4	—	22.5	Singh and Pillai (1975[B])
Fodder crop field, northern India	10,730	75.9	—	22.5	Singh and Pillai (1975[B])
Clear-cut, cultivated field, Nigeria	17,701	72.1	—	15	Critchley et al. (1979[B])

Table 21.3. Biomass and Percent of Total Acarine Biomass for Prostigmatid Soil Mites[a]

Habitat	mg/m²	Percent of Total Mite Biomass	Source
Tundra	10	11.1	Petersen (1982b[B])
Mojave desert, U.S.	4.4	48.8	Franco et al. (1979[B])
Temperate grassland	40	33.3	Petersen (1982b[B])
Grassland, Colorado, U.S	75.2	43.1	Crossley et al. (1975[B])
	136.0	32.6	Walter et al. (1987[B])[b]
Tropical grassland	50	62.5	Petersen (1982b[B])
Grassland, South Africa	63.7	30.5	Oliver and Ryke (1965[B])
Temperate coniferous forest	30	6.0	Petersen (1982b[B])
Picea forest, Southern Finland	28.8	4.2	Huhta and Koskenniemi (1975[B])
Picea forest, Northern Finland	20.8	3.6	Huhta and Koskenniemi (1975[B])
Temperature deciduous forest (mor)	10	1.1	Petersen (1982b[B])
Birch forest, Denmark	3.67	0.9	Luxton (1982[B])
Temperate deciduous forest (mull)	10	3.3	Petersen (1982b[B])

[a] Estimates based on mites recovered by high-gradient extraction, except Walter et al. (1987[B]).

[b] Values from Crossley et al. (1975[B]), adjusted to reflect extraction efficiencies obtained with heptane flotation.

systems is slight [0.9–4.2%, Luxton (1982[B]) and Huhta and Koskenniemi (1975 [B]), respectively], as well as that of total acarine respiration (2.5–18.3%, Luxton (1982[B]) and Huhta and Koskenniemi (1975[B]), respectively), even though the mean annual number of prostigmatids may reach values in excess of 100,000 per square meter. In contrast, prostigmatids represent the highest percent of total acarine biomass in deserts and grasslands (Table 21.3). Although these values are small relative to the total community biomass, the specific role of prostigmatid species in the community may be of great significance.

Seasonal Variation

Observed Patterns. Estimates of density for individual species through time are rare in the literature. Average density and peak density data for specific groups of prostigmatids are given in Table 21.4 for three ecosystems: a desert, a tropical grassland, and a temperate deciduous forest.

Luxton (1981[B]) found an overall mean monthly density of 20,480 prostigmatid mites per square meter in a Danish beechwood soil community; the peak density (40,140 per square meter) occurred in the winter (December); population density crashes occurred in May (7640 per square

Table 21.4. Density, Seasonality, and Biomass Data for Some Prostigmatid Mites in Three Ecosystems

Higher Category/ Wt Family	Genus	Annual $\bar{x}$ Nos./ m^2	Peak Nos./ m^2	Peak Months*	$\bar{x}$ No. Eggs in ♀♀	No. Gen./ yr	$\bar{x}$ Dry Wt (μg)
Mojave Desert, U.S. (Edney et al. 1976, Franco et al. 1979)							
Bdelloidea		144	238	Mar.*, Aug., Oct./Nov.	—	—	2.6
Caeculidae		11	32	Sept.*, Oct.	—	—	60.6
Erythraeidae		32	96	Jan., Mar.*	—	1	—
Nanorchestidae		338	601	Mar., May/June, Aug., Oct.*/Dec.	—	—	1.6
Raphignathoidea		142	224	June, Aug.*/Dec.	—	—	1.8
Trombidioidea		19	51	Feb.*/Mar., Nov.	—	1	117
		22	55	Feb.*/Mar., Nov.	—	—	22.2
Temperate Deciduous Forest, Denmark (Luxton 1981, 1982)							
Alicorhagiidae	*Alicorhagia*	360	800	May/June*, Dec.	1.1	2	0.36
Ereynetidae	*Ereynetes*	72	190	May, Aug.*, Nov.	—	2–3	0.19
Eupodidae	*Cocceupodes* sp.1	738	1,400	Apr., June*, Oct., Dec./Jan.	1.3	3	0.32
	Cocceupodes sp.2	283	850	Apr.*, July, Oct., Dec./Jan.	1.1	3	0.14
	Eupodes sp.1	2,542	3,500	Feb./Mar.*, June, Dec.*	1.3	1	1.6
	Eupodes sp.2	1,708	3,400	Apr.*, June, Dec., Feb.	1.3	2	0.36
	Linopodes	182	290	Aug.*, Oct., Dec.	2.8	2	0.63

Nanorchestidae	*Nanorchestes*	2,735	5,000	Feb./Mar., July, Oct., Dec.*	1.4	3	0.14
Pygmephoridae	*Neopygmephorus*	443	3,600	Oct.*	—	1	0.34
	Pseudopygmephorus	315	1,600	June, Dec.*	—	1	0.26
Rhagidiidae	*Coccorhagidia*	170	325	July*, Oct.*, Dec.	1.2	3	—
	Rhagidia	390	950	Aug., Dec.*	1.6	2–3	—
Scutacaridae	*Diversipes*	408	1,900	Feb.*	—	1	0.06
	Scutacarus sp.1	518	600	Oct.*	—	1	0.11
	Scutacarus sp.2	211	900	Apr., June*	—	2	0.08
Tarsonemidae	*Tarsonemus*	4,917	11,500	Feb.*, Aug., Dec.	1	3	0.05
Tydeidae	*Coccotydeus*	762	850	Dec.*	1	1	0.19
	Lorryia	80	220	Aug.*, Oct.*	1.2	2	0.63
	Microtydeus	3,160	4,100	Aug., Oct., Dec., Feb.*	1	4?	0.04
	Tydeus	95	280	May, Aug.*, Nov.	—	3	0.19
Tropical Grassland, South Africa (Oliver and Ryke 1966, Loots, and Ryke 1966)							
Caeculidae		69	494	Feb.*/Mar.	—	1	—
Cunaxidae	*Cunaxa*	208	733	Oct./Nov.*, Feb.	—	2?	14.0
Lordalychidae		7	51	Oct.*, May*	—	2	14.0
Nanorchestidae	*Nanorchestes*	2,408	6,307	Oct./Nov.*/Jan./Feb., Apr./July	—	3	2.0
	Speleorchestes	1,147	3,767	Nov.*, Dec., Feb.	—	2	2.0
Paratydeidae		235	478	Oct.*/Nov., May/Aug.	—	2	12.0
Rhagidiidae	*Rhagidia*	38	153	Nov./Dec., May, June*, July	—	2	2.0
Tarsonemidae	*Tarsonemus*	4,982	12,681	Nov./Feb., Apr./May*, June	—	2	1.5
Tydeidae	*Coccotydeus*	5,990	15,476	Nov./Feb., Apr./May*/Aug.	—	2	1.5
	Pronematus	1,146	2,829	Nov./Feb.*	—	2	—

Asterisks denote month(s) in which peak densities occurred.

meter) and September (7800 per square meter). Luxton suggested the winter peak was due to rainfall (November) stimulating egg hatching, with the emergence of juveniles paralleling a renewed substrate. A general pattern of winter population density peaks is consistent with the data of Franco et al. (1979[B]) and Edney et al. (1975[B]) for Mojave Desert populations, and Oliver and Ryke (1966[B] and Loots and Ryke (1966[B]) for South African grasslands, as indicated in Table 21.4. Edney et al. (1975[B]) and Franco et al. (1979[B]) noted population density increases and decreases paralleling periods of significant rainfall (i.e., January, July, October, and December).

In general, population density peaks of predaceous groups tend to coincide with those of their microphytophagous prey, but very few studies have documented the component species. For example, densities of the predaceous stigmaeid *Cheylostigmaeus scutatus* parallel those of its collembolan prey *Onychiurus debilis* in an English salt marsh (Luxton 1967a[B]); high numbers of winter-active predaceous groups (Anystidae, Bdellidae, and Rhagidiidae) were directly correlated with collembolan abundance in southern Canada (Aitchison 1979[B]).

Other predaceous groups, such as the Trombidiidae (Parasitengona), have highly specialized patterns of temporal occurrence. The deutonymphs and adults of the subfamilies Trombidiinae and Microtrombidiinae are hemiedaphic and euedaphic, respectively, and move into the upper litter or soil only briefly (Robaux 1974[M]). The active postlarval stages of *Dinothrombium* emerge in very large numbers for only a few hours following a rain, when they feed on sexual forms of termites, then return to solitary burrows 7–12 cm below the surface (Tevis and Newell 1962[B], *D. pandorae* in California; Newell 1979[B], *D. tinctorium* in Zimbabwe).

The time of peak recruitment even for closely related species within the same habitat may vary. Luxton (1981[B]) found peak egg production coinciding with peak adult population density levels (Table 21.4) for most of the prostigmatid species, except *Eupodes sp. 1,* where peak egg production occurred in June (population density peak in December), and *Eupodes sp.2,* where peak egg production occurred in October (population density peak in April), suggesting an extended egg maturation period for these species.

When taken in their entirety, prostigmatid population densities generally evidence a major seasonal peak in the winter period, as noted above. However, peak abundance for all prostigmatids does not occur at this time. Luxton (1981[B]) (see also Table 21.4) reported that individual species peaks in the late spring: *Eupodes sp.1* in March, *Cocceupodes sp.2* and *Eupodes sp.2* in April; or summer: *Alicorhagia, Cocceupodes sp.1* and *Scutacarus sp.2* in June, *Coccorhagidia* in July; or early fall: *Ereynetes, Linopodes, Lorryia,* and *Tydeus* in August, *Scutacarus sp.1* in October. Similar nonwinter peaks are indicated in Table 21.4 for a desert and a grassland system. Future detailed studies may reveal abundance relationships for the nonwinter months to represent a series of individual population density peaks occurring at different periods, which may reflect different roles of individual taxa within the soil community.

Possible Causes. Contemporary climactic patterns strongly influence decomposition and the movements of nutrients through a given ecosystem (Bourlière and Hadley 1970; Noy-Meir 1973[B], 1974[B]), and as such bear directly on prostigmatid densities and seasonal fluctuations. Little is known about direct involvement of abiotic factors in the life cycles of prostigmatids, but various species can be active under quite extreme conditions. For example, Aitchison (1979[B]) found nine species of prostigmatids to be active at ground level under subnivean (>20 cm snow thickness, average temperature of −4 to −5°C) winter conditions in southern Canada, and supercooling points of −31.0 and −23.1°C for two antarctic species *Stereotydeus villosus* (Penthalodidae) and *Nanorchestes antarcticus* (Nanorchestidae) were reported by Block (1982[B]). At the other extreme, Hutterer and Rack (1977[M]) found *Chaussieria vennustissima* (Anystidae), an extremely rapid-walking predator occurring on the surface of barren soil in Austria, to be most active between 39 and 41°C, and tolerating temperatures as high as 43°C.

Prostigmatids in both cold and hot desert ecosystems are basically water limited due to the short-term, highly seasonal aspect of available water. When available, water promotes a rapid pulse of primary production that represents a resource flush for microarthropods. Consequently, acarine populations must be capable of rapid growth ("*r*" selected, Wallwork 1980[B]). In part, this may account for the significant decrease in oribatid dominance in these habitats relative to forests, since most Oribatei have relatively low reproductive potentials. Grassland populations respond to a longer pulse of nutrient flow, but available water (as rainfall) is ultimately limiting. Even forest ecosystems, where the overall pattern of production is relatively constant, exhibit both temporal and spatial heterogeneity of production on a microhabitat level due to variability of litter fall and rates of decomposition for individual plant species (summarized in Dickinson and Pugh 1974[B]). Thus the prostigmatids in both coniferous and deciduous forests consist of a mix of taxa, some capable of a rapid response to seasonally restricted nutrient input, and others adapted to more stable, longer-term seasonal patterns. Characterization of individual mite taxa will require life-history data for individual species.

Paleontological Considerations

In part, the current general and within-habitat distribution of prostigmatid mites appears to reflect environmental constraints imposed during the early evolution of the group. The great age and early proliferation of the free-living elements of prostigmatid mites is attested to by the occurrence of a surprisingly diverse assemblage of Devonian fossils (Dubinin

1962[M]) from the Rhyne Chert formation in Scotland. These prostigmatids are considered to represent five extinct genera in four extant families; three of these are endeostigmatids [Alicorhagiidae, Bimichaeliidae (as Pachygnathidae), and Nanorchestidae] which are considered the earliest derivative prostigmatids, the fourth represents a more derived family, the Tydeidae. This deposit is considered by Kevan et al. (1975[B]) and Rolfe (1980[B]) to be of Siegenian/Emsian (Devonian) age. Consequently, a consideration of prevailing conditions during these periods may shed light on contemporary patterns of distribution.

The early terrestrial flora was simple, and during the Cambrian and Ordovician periods was at most a microflora of algae, fungi, and bacteria (Schopf 1970). The first vascular plants are generally thought to have arisen no earlier than the Downtonian (late Silurian) (Kevan et al. 1975[B], Edwards 1980[B]). These psilophytes and later trimerophytes "are usually considered to have lived along rivers, around lakes, or in coastal areas" (Edwards 1980[B], p. 75). The climactic conditions of the late Silurian and Gedinnian/Siegenian (lower Devonian) are thought to have been "typical of present-day arid or semi-arid regions" (Scott 1980[B], p. 90). Rolfe (1980[B], Scott (1980[B]), Chaloner (1970[B]), and Kevan et al. (1975[B]) have each noted the great importance of adaptations for reducing water loss in the evolution of early terrestrial arthropods.

Thus the endeostigmatids and their early derivatives probably evolved in relatively harsh environments, associated with the margin of bodies of water. An invasion of the fully terrestrial habitat may have followed coastal debris, possibly in the form of algal wrack, to solve the problem of water loss initially. Within microhabitats where organic matter was abundant but subject to periodic desiccation, vertical migration by mites to levels of adequate available water would have been a successful adaptation to avoid water loss. Mites that were wind dispersed inland into habitats with little or no organic matter might have survived by following moisture gradients in the soil. When viewed in light of the above, the studies of Nef (1971[B]), in which he documented that oribatids are geonegative above a humidity threshold of 20–25% of water and become geopositive only below that threshold, take on a new significance for understanding the biological basis of vertical migration in soil mites.

Contemporary distribution of early derivative prostigmatids may be relictual, reflecting this ancient geological history. Thus it is interesting to note that the endeostigmatid families Micropsammidae and Nematalycidae are known only from deep sand–sandy soils which are *all* of shoreline beach or alluvial origin, even though the habitat may now be found far inland. All of the endeostigmatid families and early derivative prostigmatids such as Adamystidae and Paratydeidae have their greatest diversity in arid or semiarid habitats. A number of these groups also occur at great depth (up to 100 cm) in sandy desert–grassland soils (Table 21.1), but additional studies are required to document the potential pervasiveness of these taxa.

Ecological Roles

Within the mites, microphytophagous prostigmatids appear to be the ecological equivalent of the Oribatei in soil ecosystems where there is a major decrease in dominance of the Oribatei (grasslands relative to forests) or poor representation (deserts and high arctic–antarctic habitats). Although the specific role of individual prostigmatids is unknown for most ecosystems, Santos et al. (1981[B]), and Santos and Whitford (1981[B]) have demonstrated a measurable impact by tydeids on bacteriophagous nematodes and a significant influence of pygmephorids (pyemotids of authors), and tarsonemids as fungal grazers stimulating fungal activity.

Impact of Perturbations

There is an extensive body of literature pertaining to environmental perturbations. Environmental manipulations have involved agricultural–silvicultural practices, numerous types of artificial substrates, applications of artificial "acid" rain, liming, various salts, and insecticides. Unfortunately, the majority of these studies lumped all prostigmatid mites into a single category (if considered at all). Within the insecticide impact studies, for example, one cannot distinguish any differential impact for taxa of different trophic levels; other studies attempted to partition fungivores and predators, but did so incorrectly and drew confusing conclusions. Curiously, the insecticide impact studies did not take advantage of the enormous number of studies in applied entomology where it has long been known that predaceous arthropods are very sensitive to insecticides and plant feeders are initially suppressed, but evidence a strong population rebound in the absence of decimated predators. Thus the results of Hoy and Shea (1981[B]), for example, are satisfying but not surprising: applications of pesticides to the soil generally depressed or eliminated the predaceous prostigmatids (Bdellidae, Cunaxidae, Eupodidae, and Raphignathidae) and resulted in a rebound effect for fungivorous–herbivorous prostigmatids (especially Tydeidae). They also noted some differential effects: Cryptognathidae were eliminated by carbaryl but unaffected by lindane, whereas Eupodidae were eliminated by lindane and only slightly affected by chlorpyrifos or carbaryl.

Techniques

Both dynamic and mechanical techniques have been used to collect soil mites. The most frequently used dynamic methods are Tullgren–Berlese funnels (or their numerous modifications), and high-gradient extractors. Most mechanical techniques involve flotation followed by concentration of the mites at a lipophilic–water interface. Edwards and Fletcher (1971[B]) compared the efficiency of a number of extraction methods and found dynamic methods, especially high-gradient techniques, to be more efficient than flotation procedures. Recently, flotation was "shown to extract twice as many species and 4 times as many individuals" from prairie soils as high-gradient procedures (Walter et al. 1987[B], p. 232). Flotation procedures also recover inactive stages. For strictly qualitative studies of the B horizon or deeper, flotation techniques have the advantage in processing a much greater volume of soil. This advantage in qualitative studies is balanced in using dynamic extractors only when the leaf litter–humus layer has been concentrated by the use of sifters prior to placing the sample in funnels. Ethyl alcohol (70%) is an ideal fluid preservative. Formalin should not be used.

Heavily sclerotized specimens or old, alcohol-preserved specimens usually require clearing prior to slide preparation. Lactophenol (lactic acid, 10 ml; phenol crystals, 3.6 g; distilled water, 5 ml) is a commonly used clearing agent. Specimens cleared in lactophenol must be carefully rinsed in a graded series of water–alcohol solutions prior mounting lest the specimens collapse or explode due to osmotic pressure differentials. Vitzthum's fluid (chloral hydrate, 11.5 g; phenol, 90 ml; distilled water, 1 ml) avoids some of the problems of lactophenol. Nesbitt's fluid (chloral hydrate, 80 g; distilled water, 50 ml; concentrated hydrochloric acid, 5 ml) is a more powerful clearing agent that allows direct transfer into aqueous mounting media. Most prostigmatid groups can be adequately studied as whole specimen mounts using standard gum arabic media. Identification of most adult terrestrial Parasitengona is greatly facilitated by dissection of the gnathosoma from the body. Strandtmann's (1949[M]) modification of Berlese–Hoyer's mounting medium uses a higher ratio of chloral hydrate to water plus a small portion of acetic acid. Ratios for the components of Strandtmann's fluid are: distilled water, 40 ml; gum arabic, 40 g; chloral hydrate, 350 g; glycerine, 25 ml; and glacial acetic acid, 15 ml, and *must* be fully dissolved in the sequence given *without* the application of heat. Slides should be ringed with a suitable moisture-impervious compound such as an underwater electrical paint. Additional clearing agents and mounting media are discussed in Singer (1967[M]).

Identification

The following key is artificial. It is applicable to all nominal families of prostigmatid mites found in the soil community on a world basis. Representatives of the following groups are not included in the key: parasites [Heterocheylidae (Heterocheyloidea), Ioliniidae (Tydeoidea), Podapolipidae (Tarsonemoidea), Pterygosomatidae (Pterygosomatoidea), and eight families of Cheyletoidea], some obligate plant feeders (families of Eriophyoidea), aquatic Parasitengona (water mites), and Halacaroidea (Eupodina, marine mites). Although not documented in the literature, at least one species in every family included in the key is known from North America.

Until experience is gained, and determinations verified, family-level identification requires the use of a compound microscope. To facilitate identification, the families (and subfamilies in part) of terrestrial Parasitengona are keyed separately as larvae and adults. The number of nominal species and genera for each family is based on a literature review through 1982 on a world basis. The higher category grouping of families into superfamilies, subcohorts, and cohorts follows the classification of Kethley (1982[M]), with the exception of placing Nematalycidae within the Endeostigmata. An extensive listing of the primary systematic literature through 1982 is provided for the serious student to facilitate identification at the generic level. The only extensive generic-level treatment for prostigmatid soil mites is given in Gilyarov (1978[M]), with species-level keys to the fauna of the USSR.

Illustrations have been modified from those published by H. M. André, W. T. Atyeo, E. W. Baker, Y. Coineau, E. A. Cross, D. A. Crossley, Jr., J. Den Heyer, Z. Feider, U. Gerson, R. H. Gonzalez, F. Grandjean, H. H. Keifer, E. E. Lindquist, I. M. Newell, R. A. Norton, A. C. Oudemans, D. W. Price, P. Robaux, H. O. R. Schubart, M. Shiba, R. W. Strandtmann, F. M. Summers, H. Witte, and T. G. Wood.

ACKNOWLEDGMENTS

A large number of people have assisted in different ways in this work. Gratitude is expressed to the numerous students of the Ohio State University Summer Institute of Acarology courses in soil acarology during the past 14 years who struggled with and helped correct earlier drafts of the key. I am grateful to numerous people who assisted in making critical literature and specimens available. The assistance of W. T. Atyeo, D. A. Crossley, Jr., E. E. Lindquist, W. W. Moss, V. Behan-Pelletier, and W. C. Welbourn for review and criticism of the manuscript is gratefully acknowledged. Special thanks and indebtedness are due R. A. Norton for critical assistance with an extensive reorganization of the manuscript. Last, but by far from least, illustrations were rendered by Ms. C. Richardson.

ARTIFICIAL KEY TO THE FAMILIES OF TERRESTRIAL ENDEOSTIGMATID AND PROSTIGMATID MITES

1a Only adult females (and males in **Tarsocheylidae**) with anterolateral prodorsal stigmatal openings (Fig. 21.254); only setae *e* and *f*, cupules *im* and *ip* present on a common sclerite, (Fig. 21.253), or *ip* absent (Fig. 21.259), other setae and cupules on other sclerites; palps usually reduced to two or three segments (Figs. 21.254, 21.272) (except **Tarsocheylidae**); Subcohort **HETEROSTIGMATA** **48**

1b Juveniles and adults with stigmatal openings at base of gnathosoma (Fig. 21.182), on gnathosoma (Fig. 21.176), or absent (Fig. 21.201); setae *e*, *f* and cupules *im*, *ip* not on a common sclerite or absent . . . **2**

2a (1) Prodorsal trichobothria present (Figs. 21.1, 21.13, 21.47, 21.103). . . **3**

2b Prodorsal trichobothria absent (Fig. 21.170) **28**

3a (2) Palpal tibia with at least one clawlike seta, usually arising from the distal margin of the tibia (Figs. 21.119, 21.125, 21.151, 21.169, 21.177, 21.202, 21.230, 21.235, 21.252) **4**

3b Palpal tibia without a clawlike seta (Figs. 21.24, 21.33, 21.42, 21.65, 21.106, 21.107, 21.114, 21.247) **9**

4a (3) Heavily sclerotized; spines present on anterior surface of legs I; Superfamily **CAECULOIDEA** (Figs. 21.117–21.119); 61 species in nine genera . . . . **Caeculidae**

4b Weakly sclerotized; spines absent on anterior surface of legs I . . . **5**

5a (4) Peritremes emergent or elaborated on anterior margin of prodorsum (Figs. 21.124, 21.141); body of nymphs and adults setose, but not densely covered with setae (Figs. 21.121, 21.127; *d, e, f, h* series each with rarely more than three pairs of setae); two to four pairs of cupules present; Superfamily **ANYSTOIDEA** (in part) **6**

5b Peritremes usually absent; if present, located between cheliceral bases; body of nymphs and adults usually densely covered with setae (Figs. 21.329, 21.335, 21.350; *d, e, f, h* series usually unrecognizable), cupules usually absent; Cohort **PARASITENGONA** **56**

6a (5) Pretarsi I–IV elongate, annulate; paired claws slender, minute; empodium membranous (Figs. 21.160–21.164); seven species in three genera . . . . **Pseudocheylidae**

6b Pretarsi not elongate (tarsi may be subdivided) or annulate; paired claws robust, not minute; empodium clawlike or absent. **7**

7a (6) Pretarsal claws II–IV of nymphs and adults linear, with dense setulate processes; pretarsal empodia III and IV absent (Figs. 21.369, 21.370); three species in one genus. . . . . **Stigmocheylidae**

7b Pretarsal claws II–IV of nymphs and adults with serrate, bipectinate, or nude margins; pretarsal empodia III and IV present, clawlike, or padlike **8**

8a (7) Peritreme sinuous, elaborated on anterolateral margins of prodorsum, distal margin not elevated above the body (Fig. 21.141); palp tarsus very short, not extending beyond the tip of the tibial clawlike seta (Fig. 21.142); pretarsal empodia I and II absent (Figs. 21.144, 21.145); see Figs. 21.141–21.147; eight species in two genera . . . . **Teneriffiidae**

8b Peritreme linear or slightly curved, distal margin elevated above the body (Fig. 21.124); palp tarsus extending beyond tip of tibial clawlike seta (Figs. 21.125, 21.126, 21.129, 21.131); pretarsal empodium present, clawlike, or padlike (Fig. 21.130); see Figs. 21.120–21.133; 40 species in 14 genera . . . . **Anystidae**

9a (3) Coxal fields I–IV expanded, rectangular, mesally contiguous, body heavily sclerotized; Superfamily **LABIDOSTOMMATOIDEA** (= **Nicoletiellidae**) (Figs. 21.110–21.116); 48 species in nine genera . . . . **Labidostommatidae**

9b Coxal fields I–IV not as above, body generally weakly sclerotized **10**

10a (9) Trichobothria present on at least tibia IV (Fig. 21.100), *or* tarsus I (Figs. 21.22, 21.29) **11**

10b Leg trichobothria absent **14**

11a (10) Trichobothrium present on tibia IV, absent from tarsus I; Superfamily **BDELLOIDEA**. **12**

11b Trichobothrium present on tarsus I, absent from tibia IV **13**

12a (11) Adults with three pairs of genital papillae, and two or more eugenital setae; palps long, antenniform, usually elbowed and with one or two elongate distal setae; leg trichobothria present on tarsi III and IV, tibiae I and IV (Figs. 21.100–21.102); 72 species in 14 genera . . . . **Bdellidae**

12b Adults with two pairs of genital papillae, and usually lacking eugenital setae; palps usually not elbowed, usually terminating in a tarsal spine; leg trichobothria present only on tibia IV (Figs. 21.103–21.109); 55 species in 14 genera . . . . **Cunaxidae**

13a (11) Adults with two pairs of genital papillae; two pairs of prodorsal trichobothria present (Figs. 21.22–21.28); 4 species in one genus. . . . . . **Sphaerolichidae**

13b Adults with three pairs of genital papillae; only one pair of prodorsal trichobothria present (Figs. 21.29, 21.30); two species in one genus **Grandjeanicidae**

14a (10) Two pairs of prodorsal trichobothria present **15**

14b Only one pair of prodorsal trichobothria present **18**

15a (14) Anterior pair of prodorsal trichobothria set in a communal depression; fixed cheliceral digit truncate distally (Figs. 21.1–21.4); seven species in one genus. . . . . **Lordalycidae**

15b Anterior pair of prodorsal trichobothria not set in a communal depression; fixed cheliceral digit not truncate. **16**

16a (15) Anterior pair of prodorsal trichobothria inserted on naso; peritreme present at the base of the chelicerae (Figs. 21.155–21.159); four species in two genera . . . . **Adamystidae**

16b Anterior pair of prodorsal trichobothria not inserted on naso; peritremes absent **17**

17a (16) Pretarsi I–IV tridactylate, each provided with two claws and an empodium; labrum unsclerotized (= **Pachygnathidae**) (= **Alycidae**) (Figs. 21.8–21.14); 34 species in seven genera . . . . **Bimichaeliidae**

17b Pretarsi I–IV monodactylate, paired claws absent, empodial claw present; unpaired, basally curved, sclerotized bladelike structure present externally under and between the cheliceral bases (sclerotized labrum) (Figs. 21.31–21.35); 30 species in five genera . . . . **Nanorchestidae**

18a (14) Pretarsi II–IV lack claws, only an empodium present; paired, linear, sclerotized bladelike structures often present internally in the region of the prodorsum (sclerotized esophagus) (Figs. 21.36–21.39); 5 species in three genera . . . . **Alicorhagiidae**

18b Pretarsi II–IV usually with paired claws and an empodium; sclerotization of esophagus (when present), not extending internally to the level of the prodorsum **19**

19a (18) Prodorsum with only three pairs of structures (two pairs of setae, one pair of trichobothria); peritreme present at base of chelicerae (Figs. 21.134–21.140); seven species in six genera . . . . . . **Paratydeidae**

19b Prodorsum with four or more pairs of structures present; peritreme absent **20**

20a (19) Rutellum present (Fig. 21.39); palps with five segments; opisthosomal chaetotaxy with four pairs in the *c* series, three pairs in each of the *f* and *h* series, three or four pairs in each of the *ad*, *an*, and *pa* series (Fig. 21.5) **21**

20b Rutellum absent (Figs. 21.51, 21.57, 21.74); palps with four segments (Figs. 21.42, 21.50, 21.65, 21.74); opisthosomal chaetotaxy with only two pairs in each of the *c*, *f*, and *h* series; at most one pair in the *ad* series (Figs. 21.40, 21.41, 21.54, 21.55, 21.72, 21.73); *AN* and *PA* absent; *or* opisthosomal chaetotaxy neotrichous, setal series *c*, *d*, *e*, *f*, and *h* obscured (Figs. 21.63, 21.64). **22**

21a **(20)** Pretarsus I tridactylate (Fig. 21.6); six pairs of prodorsal structures present (five pairs of setae and one pair of trichobothria) (Figs. 21.5–21.7), nine species in three genera . . . . **Terpnacaridae**

21b Pretarsus I monodactylate (Fig. 21.19); five pairs of prodorsal structures present (four pairs of setae and one pair of trichobothria) (Figs. 21.15–21.21), four species in one genus . . . . . . **Oehserchestidae**

22a **(20)** Tarsus I usually with at least one recumbent solenidion (Figs. 21.61, 21.67, 21.78); famulus present on tarsus I, often stellate-shaped (Fig. 21.78); naso usually present (Fig. 21.54); Superfamily **EUPODOIDEA** **23**

22b Solenidia of tarsus I erect, not recumbent (Fig. 21.90); famulus absent; naso absent; Superfamily **TYDEOIDEA** (in part) **27**

23a **(22)** Adults well sclerotized; dorsum of idiosoma ornamented in a pentagonal pattern with a distinctive "V"- or "Y"-shaped furrow *or* with a sclerotized anterior projection extending over the gnathosoma and usually bearing setae *vi*; cupules distinctly lyriform (Figs. 21.40–21.47); 35 species in four genera . . . . **Penthalodidae**

23b Body weakly sclerotized, unornamented, without V- or Y-shaped furrow; sclerotized prodorsal extension over the gnathosoma absent; cupules circular or apparently absent. **24**

24a **(23)** Anal opening dorsal or dorsoterminal; intercoxal regions I–IV with more than 15 pairs of setae present; palpal femorogenu (second proximal segment) with two to six setae (Figs. 21.48–21.53); 23 species in four genera . . . . **Penthaleidae**

24b Anal opening ventral or terminal; intercoxal regions I–IV with less than 10 pairs of setae present; palpal femorogenu with at most two setae (Fig. 21.60). **25**

25a **(24)** Tibia II solenidion inserted on the surface; subcapitulum with two pairs of setae [two pairs basal and none in the adoral region (Fig. 21.77), *or* one pair in the adoral region and one pair basal (Fig. 21.76)]; each chelicera with one seta; tarsus I with a maximum of two solenidia; coxal field III maximally with three or four setae (Figs. 21.72–21.80); 86 species in seven genera. . . . . **Eupodidae**

25b Tibia II solenidion completely recessed within a pit (Fig. 21.62); subcapitulum with four pairs of setae (Fig. 21.57) (two pairs in adoral region *and* two pairs basal); each chelicera with one (Fig. 21.58), or two setae (Fig. 21.59); tarsus I *usually* with three or more solenidia; coxal field III with four or more (never three) setae **26**

26a **(25)** Palpal tibia (third segment) with three setae; tibia I with two or more solenidia; prodorsal region with three pairs of setae and a pair of trichobothria; hysterosomal setal series *c*, *d*, *e*, *f*, and *h* clearly recognizable as such, with 2, 1, 1, 2, and 2 pairs of setae, respectively (Figs. 21.54–21.62); 127 species in 21 genera . . . . **Rhagidiidae**

26b Palpal tibia with one seta; tibia I with one solenidion; prodorsal region with four pairs of setae and a pair of trichobothria; hysteronotal chaetome neotrichous; setal series *c, d, e, f, h,* obscured, respective series with 6–10, 5, 4–5, and 3 pairs of setae (Figs. 21.63–21.71); two species in one genus, ***Pilorhagidia*** Strandtmann and Goff, 1977 (= ***Strandtmannia*** Zacharda, 1978) . . . . **Strandtmanniidae**

27a **(22)** Solenidion of tibia I when present, not recessed into a pit. Dorsal opisthosomal trichobothria absent (Figs. 21.87–21.91), over 200 species in 42 genera . . . . **Tydeidae**

27b Solenidion of tibia I recessed into an inverted saclike structure with a narrow duct opening near the distal margin of the segment; a pair of dorsal opisthosomal trichobothria may be present; netlike ornamentation usually present on the prodorsum and legs (Figs. 21.92–21.99); 62 species in 11 genera . . . . **Ereynetidae**

28a **(2)** Cheliceral bases completely separate mesally; genital papillae usually present in nymphs and adults; larvae with urstigma **29**

28b Cheliceral bases mesally adnate, partially or completely fused, or apparently absent; genital papillae usually absent in nymphs and adults, larvae without urstigma **33**

29a **(28)** Peritremes, often small and three- to five-chambered, present between cheliceral bases **30**

29b Peritremes absent **31**

30a **(29)** Rutellum absent; hysteronotal series *c–d–e–f–h* with 2–1–1–2–2 respective pairs of setae; coxal fields III and IV of nymphs and adults longitudinally approximate; females with an elongate evertable ovipositor bearing six or more eugenital setae; trochanters I–IV of adults each with at least one seta present; Superfamily **POMERANTZIOIDEA** (Figs. 21.148–21.154); five species in one genus. . . . . . . **Pomerantziidae**

30b Rutellum present, tridentate; hysteronotal series *c–d–e–f–h* with 4–2–2–2–3 respective pairs of setae; coxal fields III and IV of nymphs and adults longitudinally separate; female without an elongate evertable ovipositor, eugenital setae absent; trochanters I–IV of adults nude (Figs. 21.376–21.380); monotypic. . . . . **Proteonematalycidae**

31a **(29)** Prodorsal and opisthosomal sclerites absent; body linear to elongate; free-living in soil, sand, or interstitial subterranean waters. **32**

31b One or more dorsal sclerites present; body suboval or slightly elongate; aquatic . . . .

Palps usually with only three or four segments; four dorsal and four ventral sclerites present; genital papillae minute or absent; marine or estuarine; Superfamily **HALACAROIDEA;** 345 species in 30 genera . . . . **Halacaridae**

Palps usually with five free segments (only four segments in ***Hydrachna***); idiosomal sclerotization various, but not as above; genital papillae usually present; Principally fresh water (**Pontarachnidae** is intertidal); Cohort **PARASITENGONA** (in part) . . . . Water mites, or **HYDRACHNELLAE**

32a **(31)** Six pairs of prodorsal setae present; rutellum massive, heavily sclerotized, tridentate; adults with at least one seta present on trochanters III, IV, and genu IV, tibiae III and IV each with three setae, and femur IV with two setae (Figs. 21.371–21.375). . . **Micropsammidae**

32b At most seven prodorsal setae present, three pairs and an unpaired seta; rutellum minute, weakly sclerotized; adults without setae on trochanters III, IV, or genu IV, tibiae III and IV each with no more than two setae, and femur IV with no more than one seta (Figs. 21.81–21.86); eight species in five genera . . . . **Nematalycidae**

33a (28) Males and females with only two pairs of legs (Figs. 21.234–21.245); obligate plant feeders, over 2000 species in 50 genera and three families . . . . Superfamily **ERIOPHYOIDEA**

33b Males and females with three or four pairs of legs. **34**

34a (33) Stylophore retractile into idiosoma (Figs. 21.250, 21.251); associated peritremes present on anterior margin of prodorsum; movable cheliceral digit recurved; female genital opening transverse (Fig. 21.236); pretarsal claws with tenent hairs; Superfamily **TETRANYCHOIDEA** **35**

34b Stylophore not retractile into idiosoma; associated peritremes on dorsum of stylophore (Fig. 21.230), at base of stylophore (Fig. 21.182), or absent (Fig. 21.201); movable cheliceral digit linear; female genital opening longitudinal; pretarsal claws without tenent hairs (**Barbutiidae** is an exception) **39**

35a (34) Palps linear, tibial clawlike seta absent (Figs. 21.246, 21.247); 540 species in 21 genera . . . . **Tenuipalpidae**

35b Palpal tibial clawlike seta present. **36**

36a (35) Tarsus I with two short, bulbous solenidia (Fig. 21.233) **37**

36b Solenidia of tarsus I linear **38**

37a (36) Anal valve setae (ps_{1-3}), simple, nude; eyes absent (Figs. 21.231–21.236); five species in one genus . . . . **Linotetranidae**

37b Some anal valve setae, bifurcate or dendritic; eyes present (Figs. 21.237–21.239); 1 species . . . . **Allochaetophoridae**

38a (36) Setal row *c* with five or more pairs of setae; setae of posterior margin (row *h*), flagelliform or bipectinate (Figs. 21.240–21.242); 17 species in one genus . . . . **Tuckerellidae**

38b Setal row *c* with no more than four pairs of setae; setae of posterior margin not flagelliform or bipectinate (Figs. 21.248–21.252); over 900 species in 70 genera . . . . **Tetranychidae**

39a (34) Stylophore completely fused to subcapitulum; base of moveable digit contained within the stylophore capsule; male genital opening subdorsal or dorsal; genital and anal openings coalesced; Superfamily **CHEYLETOIDEA** (Figs. 21.220–21.230); over 200 species in 68 genera **Cheyletidae**

39b Cheliceral bases variously fused to each other, never completely fused to subcapitulum; base of movable digit located at the tip of the cheliceral base, not within stylophore; male genital opening ventral or terminal, rarely dorsal (**Cryptognathidae** is an exception); genital and anal openings of females contiguous, rarely coalesced; Superfamily **RAPHIGNATHOIDEA** **40**

40a (39) Entire gnathosoma elongate, retractile, enclosed within a dorsal sheath arising from the prodorsum (Figs. 21.198–21.200), 27 species in one genus . . . . **Cryptognathidae**

40b No prodorsal extension covering gnathosoma **41**

41a (40) Legs greatly elongate, genu less than one-fourth the length of the tibia; trochanters visible from the dorsal aspect (Figs. 21.185–21.189); 98 species in five genera. . . . . **Camerobiidae**

41b Legs not greatly elongate, genu greater than one-half the length of the tibia; trochanters not visible from the dorsal aspect **42**

42a (41) Idiosoma heavily sclerotized, hemispherical; prodorsum with transversely arched groove; paired flask- or funnel-shaped structures internal to the groove (Figs. 21.194–21.197); six species in two genera **Homocaligidae**

42b Prodorsum lacking a transversely arched groove and associated internal flask- or funnel-shaped structures; variously sclerotized, but not hemispheric in shape. **43**

43a (42) Peritremes and stigmatal openings associated with cheliceral base (Figs. 21.167, 21.172, 21.192) **44**

43b Peritremes and stigmata absent (Fig. 21.201). **47**

44a (43) Cheliceral bases adnate, three- to five-chambered peritreme associated with paraxial surface of cheliceral base; pretarsal claws with tenent hairs; coxal field II widely separated from coxal field III (Figs. 21.165–21.169); two species in one genus . . . . **Barbutiidae**

44b Cheliceral bases fused mesally; peritremes on the dorsal surface of the cheliceral bases (Figs. 21.172, 21.176), or slightly proximal to the cheliceral bases (Fig. 21.182); pretarsal claws nude, without tenent hairs; coxal field II approximate to coxal field III **45**

45a (44) Peritremes located posterior to fused cheliceral bases, on extreme anterior margin of prodorsum (Figs. 21.180–21.184); 24 species in two genera . . . . **Raphignathidae**

45b Peritremes associated with dorsum of fused cheliceral bases **46**

46a (45) At least six pairs of body setae as long as the entire idiosoma; body circular in outline; palpal tarsus greatly reduced (Figs. 21.190–21.193); 3 species in 2 genera . . . . **Xenocaligonellididae**

46b No body setae as long as the entire idiosoma; body linear, not circular in outline; palpal tarsus conspicuous (Figs. 21.170–21.179); 16 species in four genera. . . . . **Caligonellidae**

47a (43) Palpal tarsus elongate, $1\frac{1}{2}$–3 times the length of the tibia, *AND* palpal genu with only one seta; tenent hairs (Fig. 21.203), arising from the empodium (Figs. 21.201–21.205); 10 species in four genera . . . . . **Eupalopsellidae**

47b Palpal tarsus not elongate, less than $1\frac{1}{4}$ the length of the tibia (the tibia may be elongate), *AND* palpal genu usually with two setae; tenent hairs (Fig. 21.215), arising from an empodial axis (Figs. 21.206–21.219); over 200 species in 21 genera . . . . **Stigmaeidae**

48a (1) Cheliceral bases fused mesally into a subtrapezoidal stylophore, but separate from the subcapitular extremity; palps three-segmented (Fig. 21.255); genital opening a longitudinal slit; males with anterolateral stigmatal openings, Superfamily **TARSOCHEYLOIDEA** (Figs. 21.253–21.258); six species in two genera . . . . . . **Tarsocheylidae**

48b Cheliceral bases fused mesally to each other and ventrally to the subcapitulum (forming a stylophore capsule); palps reduced to no more than two nonarticulating segments, extending to or slightly beyond the subcapitular extremity; genital opening reduced; males without anterolateral stigmatal openings **49**

49a (48) Legs IV of females three-segmented, with femur and genu fused, and without claws or empodium; legs IV of males usually four-segmented, with a single sessile claw; Superfamily **TARSONEMOIDEA**; plant feeders, fungus feeders, insect associates (Figs. 21.259–21.262); over 350 species in 30 genera. . . . . **Tarsonemidae**

49b Femur and genu of legs IV of female separate, often with claws and a membranous empodium (Fig. 21.266); legs IV of males five-segmented (Fig. 21.264) **50**

50a (49) Trochanter IV subtriangular (similar to trochanter III), or, if compressed, wider than long; gnathosoma of male not reduced to a tubiform structure; insect or plant associates; Superfamily **PYEMOTOIDEA** **51**

50b Trochanter IV subquadrangular, generally longer than wide; gnathosoma of male tubiform, apparently nonfunctional for feeding; Superfamily **PYGMEPHOROIDEA** **54**

51a (50) Pretarsus I with two claws, 18 species in five genera . . . . **Dolichocybidae**

51b Pretarsus I with one claw **52**

52a (51) Coxal fields I and II with four to six pairs of setae; gnathosoma of male and female normally developed, 25 species in two genera . . . . . . **Pyemotidae**

52b Coxal fields I and II with three pairs of setae, or less; gnathosoma either very large, or reduced and fused into prodorsum. **53**

53a (52) Prodorsal trichobothria present; gnathosoma very large, wider than anterior margin of prodorsum; coxal fields I and II with three pairs of setae; two species in 1 genus . . . . **Caraboacaridae**

53b Prodorsal trichobothria absent; gnathosoma reduced, partly or completely fused into the prodorsum; coxal fields I and II with fewer than three pairs of setae; 10 species in three genera. . . **Acarophenacidae**

54a (50) Dorsal opisthosomal sclerites large, covering prodorsum completely, anterior margin striate, forming a roof over the gnathosoma; distance between insertions of legs II–III equal to the distance between insertions of legs III–IV (Figs. 21.270–21.280); over 400 species in 17 genera. . . . . **Scutacaridae**

54b Dorsal opisthosomal sclerites usually smaller and not covering entire prodorsum (if prodorsum is covered in dorsal aspect, then sclerite without striations on free margin); distance between the insertions of legs II and III at least twice distance between legs III and IV . . . **55**

55a (54) Prodorsum with two to three pairs of setae; femur I with three to five setae (if only three setae, then dorsal seta modified, not setiform) (Figs. 21.263–21.266); over 300 species in 15 genera . . . **Pygmephoridae**

55b Prodorsum with only one pair of setae; femur I with three setiform setae, dorsal seta unmodified (Figs. 21.267–21.269); 25 species in seven genera. . . . . **Microdispidae**

56a (5) Larvae (three pairs of legs present) **57**

56b Nymphs and adults (four pairs of legs present) **78**

57a (56) Coxal fields I and II contiguous or nearly so (Figs. 21.282, 21.285); urstigma and anal opening usually present; one or two pairs of prodorsal trichobothria present **58**

57b Coxal fields I and II usually distinctly separate (Figs. 21.322, 21.325); urstigma and anal opening usually absent; two pairs of prodorsal trichobothria present; Superfamily **ERYTHRAEOIDEA**. **73**

58a (57) Coxal fields I–III multisetose (15 or more setae in each field); palp genu with three setae (occasionally two or four); parasites of Diptera (Tipulidae and Limoniidae) world; Superfamily **CALYPTOSTOMATOIDEA** (Figs. 21.281–21.283). . . . . **Calyptostomatidae**

58b Coxal fields I and II, each with one or two setae (Figs. 21.291, 21.294); palp genu with less than three setae (Figs. 21.289, 21.304) **59**

59a (58) Palp genu with two setae; aquatic **PARASITENGONA,** over 4000 species in 320 genera, 47 families, and eight superfamilies. . . . . Water mites, or **HYDRACHNELLAE**

59b Palp genu with a single seta or none **60**

60a (59) Dorsum of opisthosoma with at least one sclerite bearing two or more setae (Figs. 21.284, 21.290). **61**

60b Dorsal sclerites of opisthosoma, if present, bearing at most a single seta (Figs. 21.295, 21.299) **65**

61a (60) Coxal field II with two setae (Fig. 21.287); **Trombidiidae** (in part) **62**

61b Coxal field II with one setae (Figs. 21.291, 21.294) **64**

62a (61) Femur I with six setae, genu II without microseta *k*. . . . . Subfamily **Neothrombiinae**

62b Femur I with five setae, genu II with microsetae *k*. **63**

63a (62) Palp femur and genu nude (Fig. 21.286); genu II and III each with two solenidia (Figs. 21.284–21.286). . . . . . Subfamily **Allothrombiinae**

63b Palp femur with one dorsal seta (Fig. 21.289), palp genu nude; genu II and III each with only one solenidion (Figs. 21.287–21.289) . . . . . Subfamily **Trombidiinae**

64a (61) Genu II and III each with two branched setae; femur I with six branched seta; claws of pretarsus III may be asymmetric (Figs. 21.290–21.292). . . . . Subfamily **Microtrombidiinae**
Includes **Georgiinae, Feideriinae, Eutrombidiinae,** *sensu* Vercammen-Grandjean, 1973 [M]

64b Genu II and III each with three or four branched setae; femur I with five branched setae; claws of pretarsus III symmetric; parasites of Homoptera; holarctic (Figs. 21.293, 21.294) . . . . . . **Podothrombiidae**

65a (60) Anal sclerite present with one (Fig. 21.298) or two (Fig. 21.305) pairs of setae **66**

65b Anal sclerite absent (Fig. 21.315). **69**

66a (65) Pretarsal claws I–III absent, empodium clawlike . . . . **Neotrombidiidae** (in part)

66b Pretarsal claws and empodium I–III present **67**

67a (66) Anal sclerite with only one pair of setae; host unknown; holarctic; **Johnstonianidae** (in part); Subfamily **Polydiscinae** (Figs. 21.295–21.298) . . . . ***Polydiscus***

67b Anal sclerite with two pairs of setae (Figs. 21.300, 21.305). **68**

68a (67) Femora I–III subdivided (Figs. 21.299–21.301) . . . . . **Chyzeriidae**

68b Femora I–III entire; parasites of Diptera; nearctic; **Johnstonianidae,** in part; Subfamily **Lasseniinae** (Figs. 21.305, 21.306) . . . . ***Lassenia***

69a (65) Genu II with three branched setae; prodorsal sclerite with a single anteromedian seta (unpaired *vi*); parasites of vertebrates; worldwide (Figs. 21.314–21.317) . . . . **Trombiculidae**

69b Genu II with four branched setae; prodorsal sclerite with a pair of anteromedian setae (paired *vi*), which may be modified into a second pair of trichobothria **70**

70a (69) Femora II and III subdivided; setae on anteromedian structure of prodorsal sclerite setiform or trichobothria (Figs. 21.302–21.304) **Johnstonianidae**

70b Femora II and III entire; setae on anteromedian structure of prodorsal sclerite setiform. **71**

71a **(70)** Femur I with seven branched setae; empodium present, pretarsal claws absent (Fig. 21.320); (Figs. 21.318–21.320) . . . . **Neotrombidiidae** (in part)

71b Femur I with six branched setae **72**

72a **(71)** Pretarsi with only paired claws *or* only clawlike empodium; parasites of arthropods (Figs. 21.309–21.313) . . . . **Trombellidae**

72b Pretarsi with paired claws *and* clawlike empodium; parasites of vertebrates, worldwide . . . (Figs. 21.307, 21.308) . . . **Leeuwenhoekiidae**

73a **(57)** Trichobothria present on tarsus I and tibia I, occasionally on genu I; tibia I with at least three (3) solenidia; paired pretarsal claws usually symmetrical (Figs. 21.324–21.326) . . . . **Smarididae**

73b Trichobothria absent from genu, tibia, and tarsus I; tibia I with no more than two solenidia; paired pretarsal claws usually asymmetrical (Figs. 21.321–21.323); **Erythraeidae** **74**

74a **(73)** Prodorsum with two pairs of eyes . . . . . . Subfamily **Erythraeinae**

74b Prodorsum with only one pair of eyes **75**

75a **(74)** Coxal fields II and III each with two setae; genuae II and III each with 11 or 12 setae . . . . Subfamily **Callidosomatinae** (in part)

75b Coxal fields II and III each with only one seta; genuae II and III each with less than 11 setae (usually 9) **76**

76a **(75)** Trochanters I, II, and palp genu each with only one seta. . . . . . . Subfamily **Leptinae**

76b Trochanters I, II, and palp genu each with two or more setae **77**

77a **(76)** Palp femur with two setae, palp genu with three setae; trochanter I with two setae, genuae II and III each with only one seta . . . . . . Subfamily **Callidosomatinae** (in part)

77b Palp femur with only one seta, palp genu with two setae; trochanter I with three setae, genuae II and III nude. . . Subfamily **Balaustiinae**

78a **(56)** Chelicerae capable of being completely withdrawn into the body; movable cheliceral digit usually linear in form **79**

78b All gnathosomal elements not capable of being retracted into the idiosoma; movable cheliceral digit curved, hooklike **86**

79a **(78)** Only one pair of prodorsal trichobothria present (Fig. 21.327); prodorsal sclerite reduced, bearing only 1 pair of trichobothria and devoid of other setae; two pairs of genital papillae present; cuticle crenulate; cosmopolitan (Figs. 21.327, 21.328), five species in one genus . . . . **Calyptostomatidae**

79b Two pairs of prodorsal trichobothria present (Figs. 21.329, 21.330, 21.337); prodorsal sclerite linear or platelike, bearing more than one pair of trichobothria and at least one seta; genital papillae usually absent (if present, the palps are external, and not capable of being withdrawn into the body); Superfamily **ERYTHRAEOIDEA** . . . **80**

80a **(79)** Entire gnathosoma retractile. Dorsal setae generally flattened, broad, serrate, and/or leaflike (Figs. 21.329, 21.330), 40 species in nine genera; **Smarididae** **81**

80b Palps external, only the chelicerae are retractile; body setae variously shaped, usually simple or pectinate; over 300 species in 34 genera; **Erythraeidae** 82

81a **(80)** One eye on each side of the prodorsum (Fig. 21.329). . . . . Subfamily **Hirstiosomatinae**

81b Two eyes on each side of the prodorsum (Fig. 21.330) . . . . Subfamily **Smaridinae**

82a **(80)** Two eyes on each of the prodorsum (Fig. 21.334) . . . . Subfamily **Erythraeinae**

82b One eye on each side of the prodorsum 83

83a **(82)** Prodorsal region with two cup-shaped structures (urnulae) located near the eyes (Figs. 21.332, 21.333) . . . . . Subfamily **Balaustiinae**

83b Prodorsal region without urnulae 84

84a **(83)** Eyes entirely in front of the prodorsal sclerite (Fig. 21.325) . . . . . . Subfamily **Myrmicotrombiinae**

84b Eyes between the trichobothrial bases or more posterior. 85

85a **(84)** Eyes anterior of the midlength of the prodorsal sclerite (Fig. 21.331). Subfamily **Leptinae**

85b Eyes posterior to the midlength of the prodorsal sclerite (Figs. 21.336, 21.337) . . . . Subfamily **Callidosomatinae**

86a **(78)** Prodorsum with only one pair of trichobothria; prodorsal sclerite usually well developed 87

86b Prodorsum with two pairs of trichobothria; prodorsal sclerite well developed or reduced 98

87a **(86)** Dorsal body setae trifurcate (Figs. 21.338–21.340), 24 species in five genera . . . . **Neotrombidiidae**

87b Dorsal body setae variously shaped, but not trifurcate 88

88a **(87)** Pregenital tubercle present; body setae simple; prodorsal sclerite well sclerotized (Fig. 21.342); **Johnstonianidae** (in part) . . . **Lasseniinae**

88b Pregenital tubercle absent; other characters variable 89

89a **(88)** Prodorsal trichobothria expanded; palp tibia usually with five or fewer setae (Figs. 21.347, 21.348); **Johnstonianidae** (in part) . . . . **Charadracarinae**

89b Prodorsal trichobothria flagelliform; palp tibia usually with more than 10 setae 90

90a **(89)** Dorsum with 2–10 pairs of elongate densely setose processes (Fig. 21.341); 15 species in six genera . . . . **Chyzeriidae**

90b Elongate, setose dorsal processes absent 91

91a **(90)** Palp tibia without spinose setae in addition to the clawlike seta (Figs. 21.362, 21.363); 470 species in 74 genera; **Trombidiidae** (in part) 92

91b Palp tibia with at least one spinose seta adjacent to the large clawlike seta (Figs. 21.349, 21.350, 21.365, 21.366) 93

92a **(91)** Anterior margin of prodorsal sclerite with up to four setae (Figs. 21.360–21.364) . . . . Subfamily **Trombidiinae**

92b Anterior margin of prodorsal sclerite with more than four setae (Fig. 21.359) . . . . Subfamily **Allothrombiinae**

93a **(91)** Palp tibia with only two distinct spines in tandem near the large clawlike seta; body usually constricted between legs II and III, giving body a figure-eight shape **94**

93b Palp tibia often with numerous spinose setae; if spinose setae appear to be in tandem, then more than three are present; body not constricted between legs II and III **95**

94a (93) Anterior margin of prodorsal sclerite with two, one, or no setae; setae on palp genu more numerous or equal to those on the palp femur (Figs. 21.349–21.352); over 1700 species in 74 genera . . . . **Trombiculidae**

94b Two setae present on the anterior margin of the prodorsal sclerite; setae on palp genu less numerous than on palp femur (Fig. 21.353); over 400 species in 26 genera . . . . **Leeuwenhoekiidae**

95a (93) Palp tibia with 5–20 spinose setae, not arranged in distinct rows; prodorsal sclerite weekly sclerotized anterior to the trichobothria; two setae present on the anterior margin of the prodorsal sclerite (Fig. 21.358); 20 species in two genera. . . . . **Podothrombiidae**

95b Palp tibia with spinose setae arranged in one or two distinct rows; antiaxial surface of palp tibia with zero to three large spinose setae; anterior margin of prodorsal sclerite variously developed, may be devoid of setae **96**

96a (95) Prodorsal sclerite narrow, poorly developed or absent; posterior of body with two or more sclerites present; body setae may be simple (Fig. 21.354); 46 species in 20 genera . . . . **Trombellidae** (in part)

96b Prodorsal sclerite well developed; posterior of body lacking sclerites; body setae variously shaped; **Trombidiidae** (in part). **97**

97a (96) Anterior margin of prodorsal sclerite with only two setae; body setae biramous (Figs. 21.360, 21.361). . . . . . Subfamily **Neothrombiinae**

97b Anterior margin of prodorsal sclerite with more than two setae; body setae various, but not biramous (Figs. 21.365–21.368) . . . . Subfamily **Microtrombidiinae**

Includes **Georgiinae, Feideriinae, Eutrombidiinae,** *sensu* Vercammen-Grandjean, 1973[M].

98a (86) Dorsum with a sclerotized band outlining the body; palp tibia with two large clawlike setae adjacent to each other; anterior margin of the prodorsum extending over the gnathosoma (Figs. 21.355–21.357); **Trombellidae** (in part) . . . . **Moyanellinae**

98b Dorsum without sclerotized band; two large clawlike setae of palp tibia not adjacent to each other; anterior extension of prodorsum over the gnathosoma absent (Figs. 21.343–21.346); 40 species in 11 genera **Johnstonianidae**

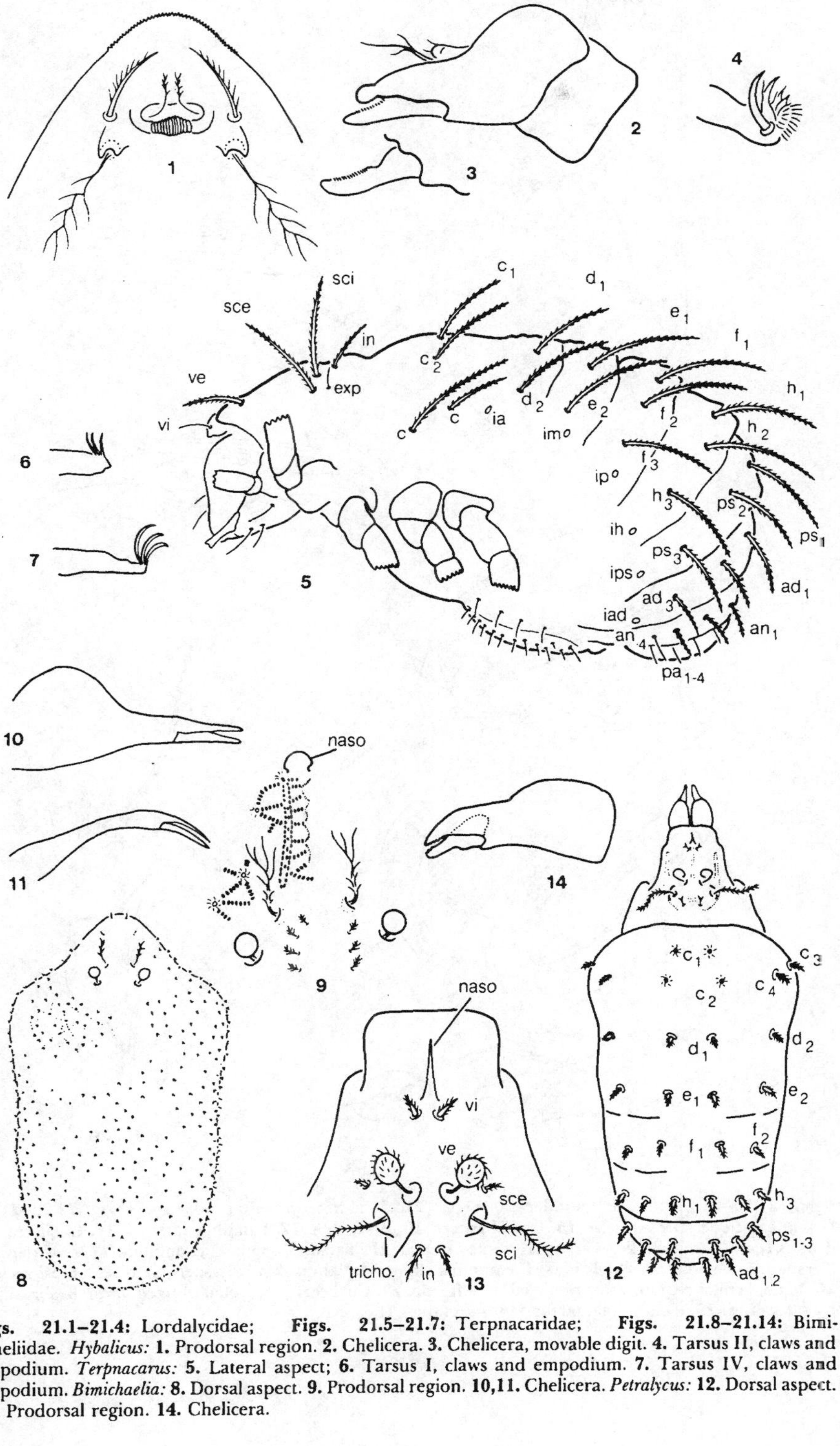

Figs. 21.1–21.4: Lordalycidae; **Figs. 21.5–21.7:** Terpnacaridae; **Figs. 21.8–21.14:** Bimichaeliidae. *Hybalicus:* **1.** Prodorsal region. **2.** Chelicera. **3.** Chelicera, movable digit. **4.** Tarsus II, claws and empodium. *Terpnacarus:* **5.** Lateral aspect; **6.** Tarsus I, claws and empodium. **7.** Tarsus IV, claws and empodium. *Bimichaelia:* **8.** Dorsal aspect. **9.** Prodorsal region. **10,11.** Chelicera. *Petralycus:* **12.** Dorsal aspect. **13.** Prodorsal region. **14.** Chelicera.

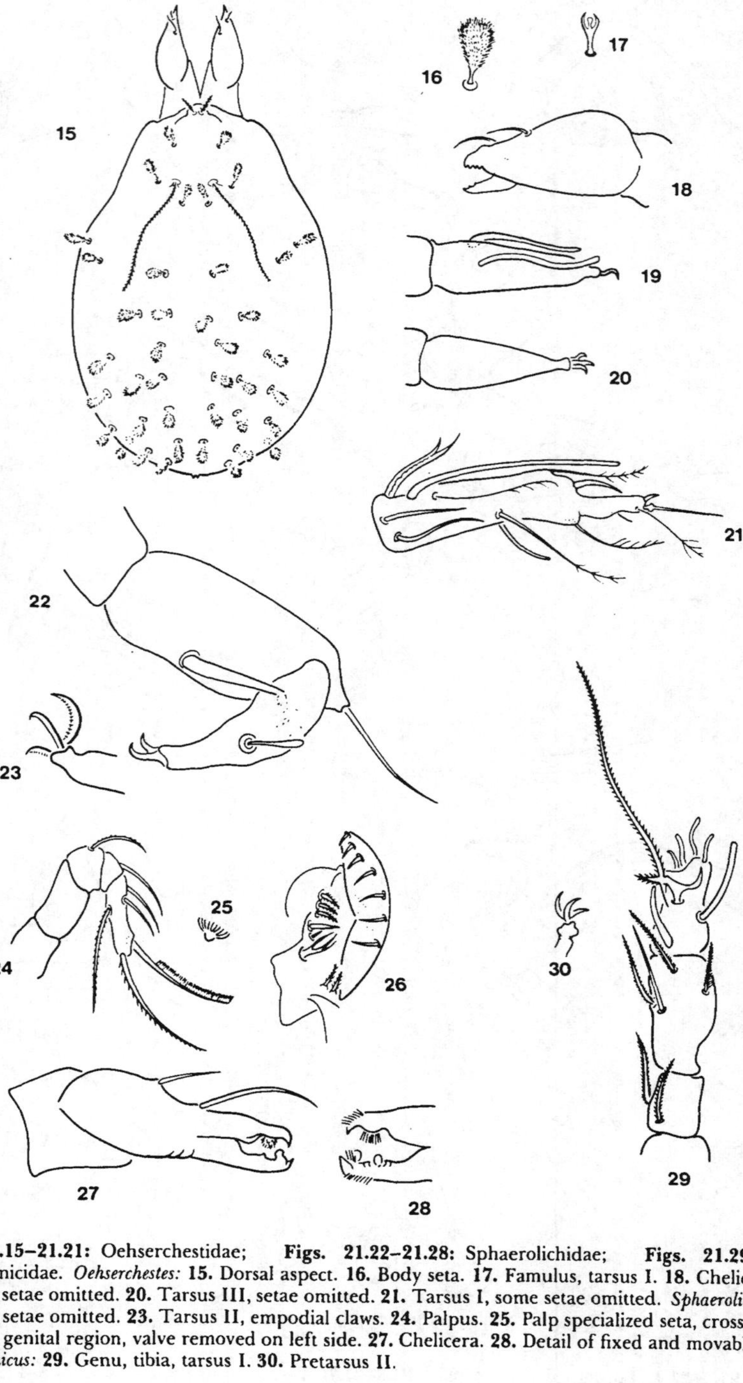

Figs. 21.15–21.21: Oehserchestidae; **Figs. 21.22–21.28:** Sphaerolichidae; **Figs. 21.29–21.30:** Grandjeanicidae. *Oehserchestes:* **15.** Dorsal aspect. **16.** Body seta. **17.** Famulus, tarsus I. **18.** Chelicera. **19.** Tarsus I, setae omitted. **20.** Tarsus III, setae omitted. **21.** Tarsus I, some setae omitted. *Sphaerolichus:* **22.** Tarsus I, setae omitted. **23.** Tarsus II, empodial claws. **24.** Palpus. **25.** Palp specialized seta, cross section. **26.** Male, genital region, valve removed on left side. **27.** Chelicera. **28.** Detail of fixed and movable digits. *Grandjeanicus:* **29.** Genu, tibia, tarsus I. **30.** Pretarsus II.

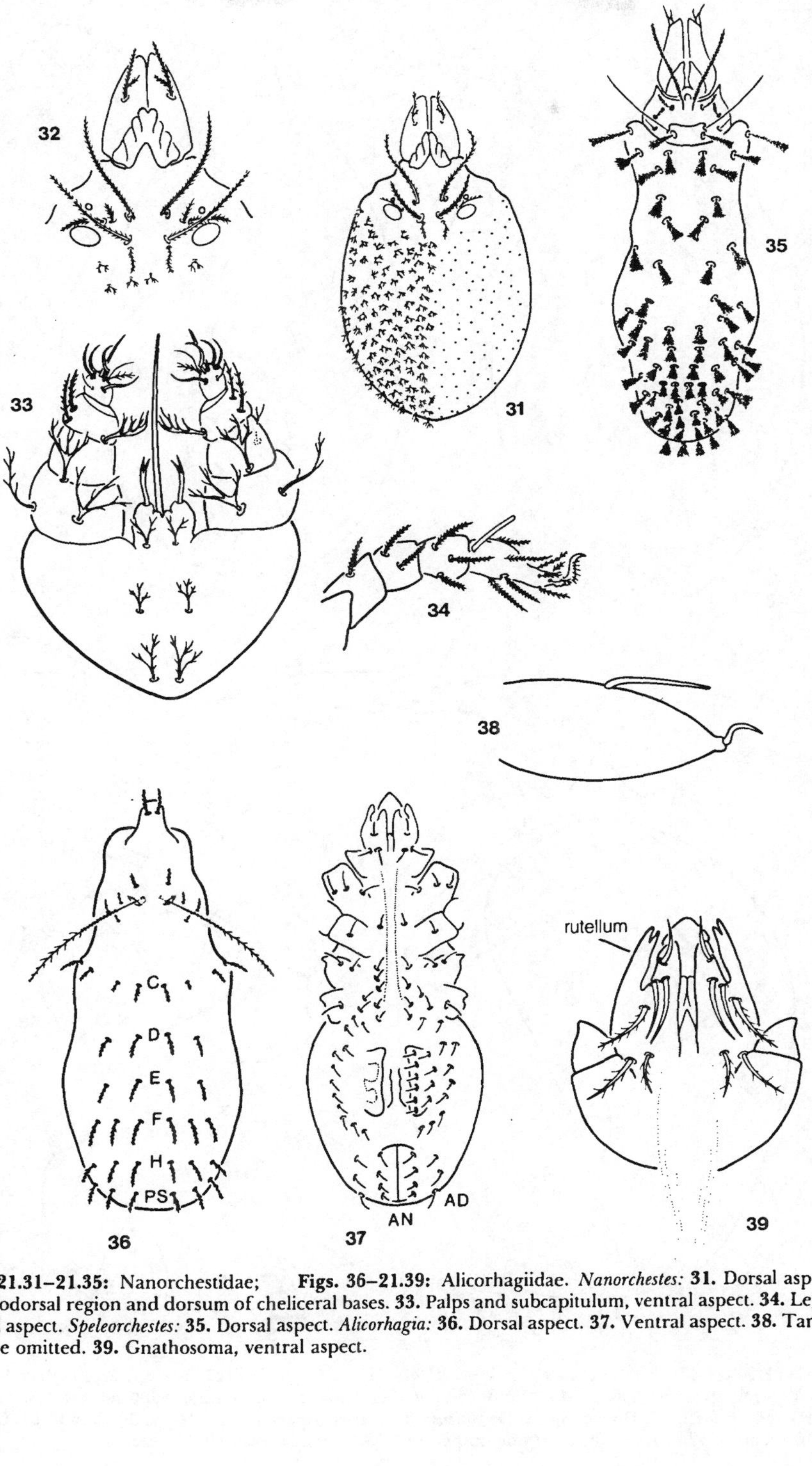

Figs. 21.31–21.35: Nanorchestidae; **Figs. 36–21.39:** Alicorhagiidae. *Nanorchestes:* **31.** Dorsal aspect. **32.** Prodorsal region and dorsum of cheliceral bases. **33.** Palps and subcapitulum, ventral aspect. **34.** Leg I, lateral aspect. *Speleorchestes:* **35.** Dorsal aspect. *Alicorhagia:* **36.** Dorsal aspect. **37.** Ventral aspect. **38.** Tarsus I, setae omitted. **39.** Gnathosoma, ventral aspect.

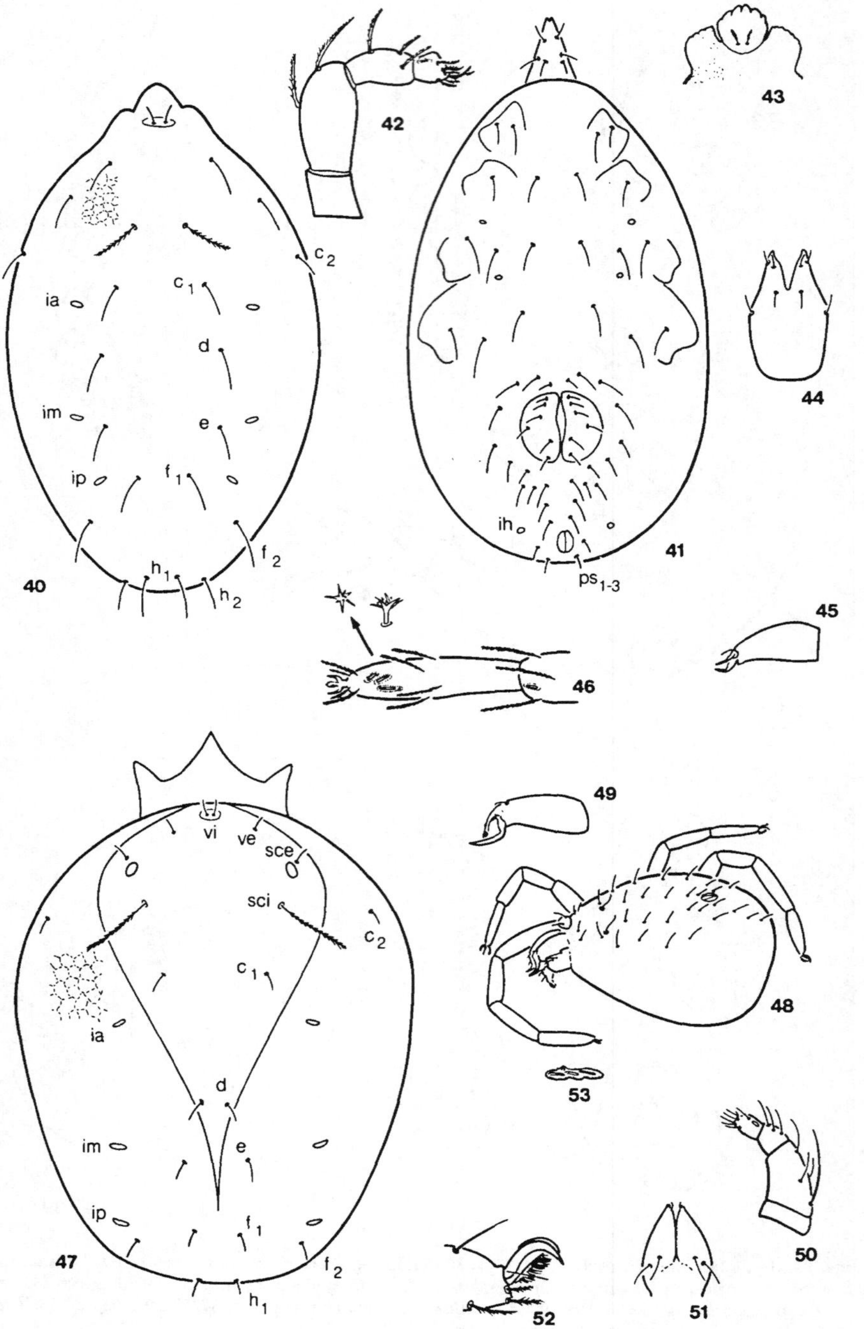

Figs. 21.40–21.47: Penthalodidae; **Figs. 21.48–21.53:** Penthaleidae. *Stereotydeus:* **40.** Dorsal aspect. **41.** Ventral aspect. **42.** Palpus. **43.** Sclerotized naso. **44.** Gnathosoma, ventral. **45.** Chelicera. **46.** Tarsus I, dorsal. *Penthalodes:* **47.** Dorsal aspect. *Penthaleus:* **48.** Lateral aspect. **49.** Chelicera. **50.** Palpus. **51.** Gnathosoma, ventral. **52.** Tarsus IV, claws and empodium. **53.** Tarsus I, recumbent solenidia.

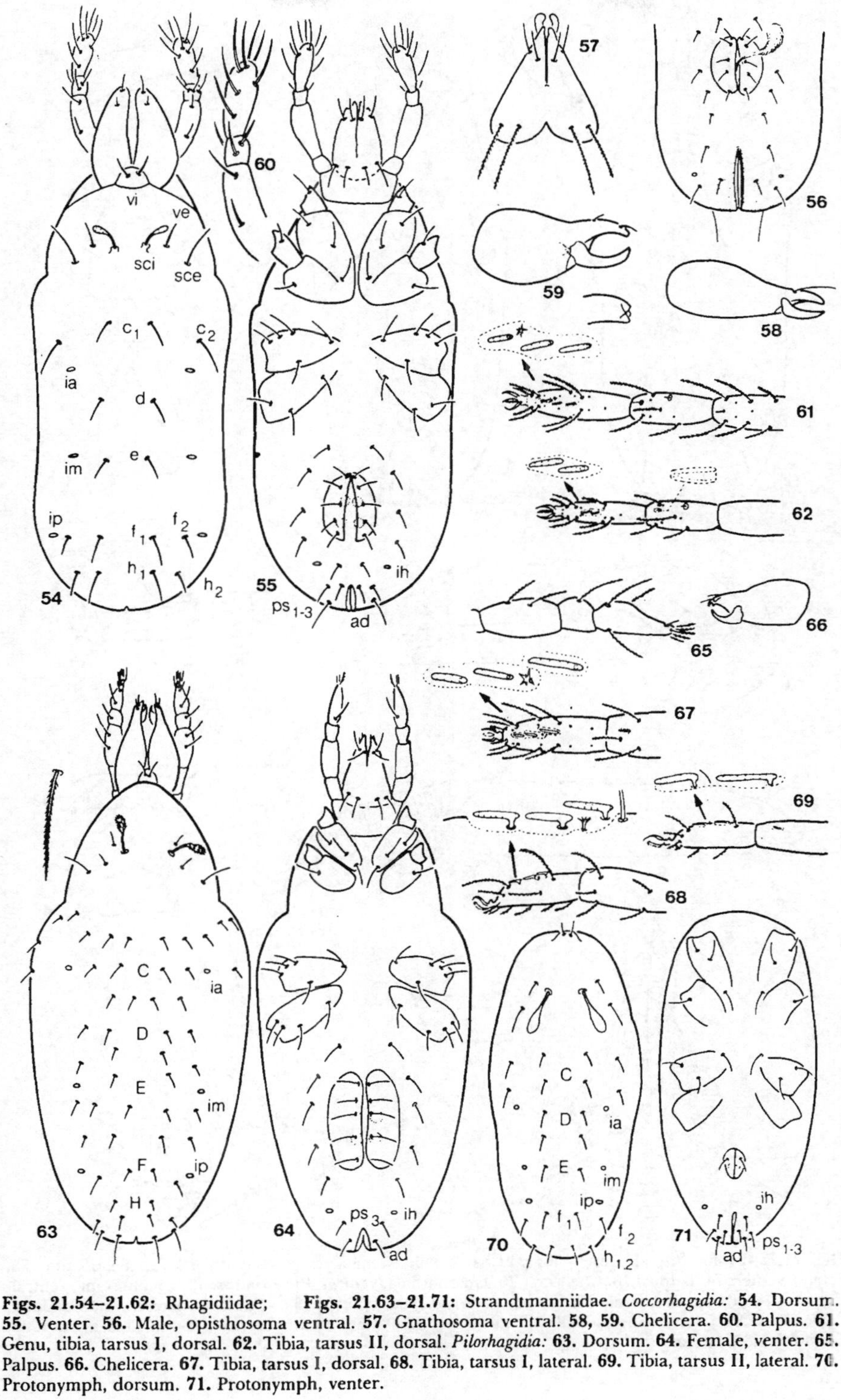

Figs. 21.54–21.62: Rhagidiidae; **Figs. 21.63–21.71:** Strandtmanniidae. *Coccorhagidia:* **54.** Dorsum. **55.** Venter. **56.** Male, opisthosoma ventral. **57.** Gnathosoma ventral. **58, 59.** Chelicera. **60.** Palpus. **61.** Genu, tibia, tarsus I, dorsal. **62.** Tibia, tarsus II, dorsal. *Pilorhagidia:* **63.** Dorsum. **64.** Female, venter. **65.** Palpus. **66.** Chelicera. **67.** Tibia, tarsus I, dorsal. **68.** Tibia, tarsus I, lateral. **69.** Tibia, tarsus II, lateral. **70.** Protonymph, dorsum. **71.** Protonymph, venter.

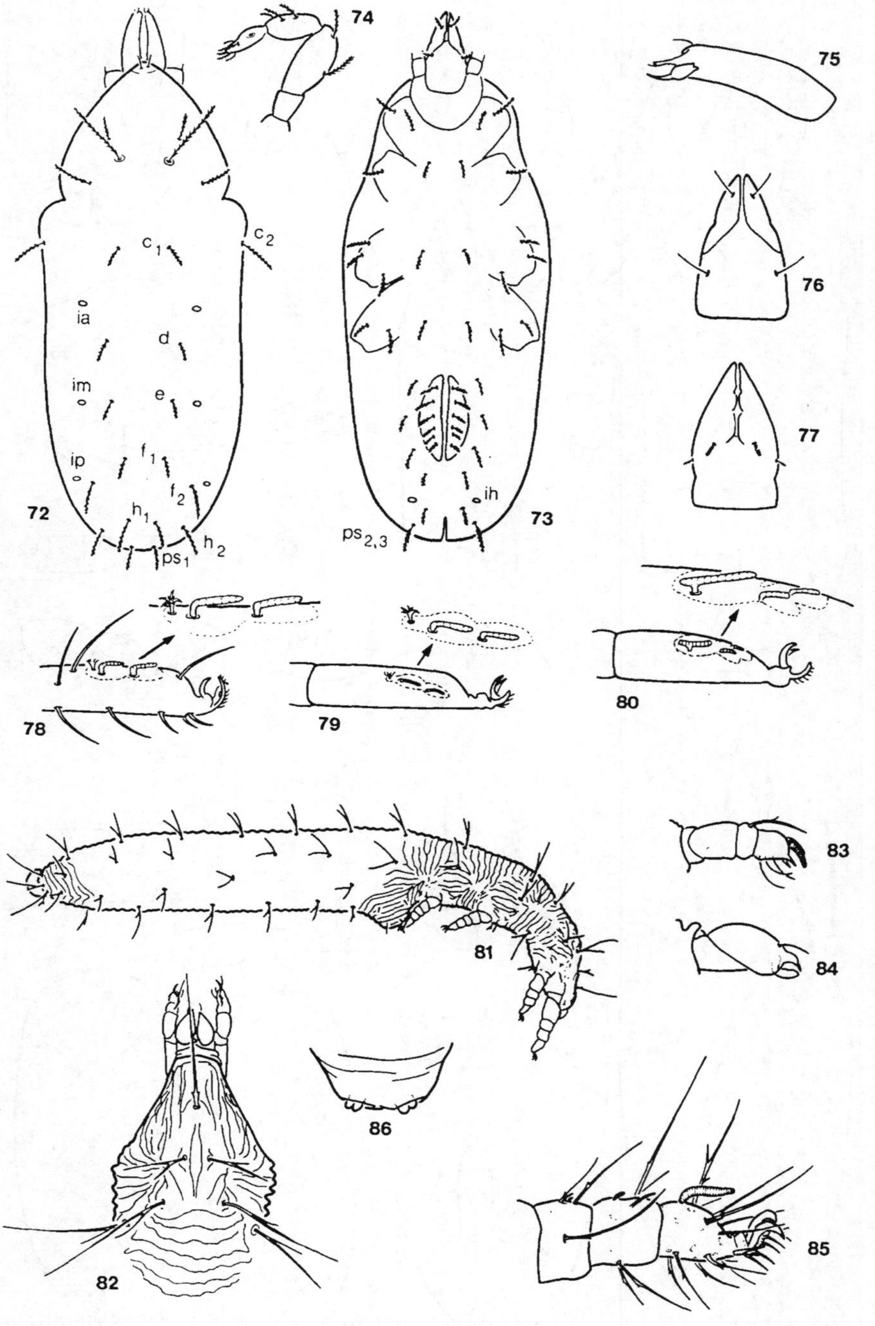

Figs. 21.72–21.80: Eupodidae; **Figs. 21.81–21.86:** Nematalycidae. *Eupodes:* **72.** Female, dorsum. **73.** Female venter. **74.** Palpus. **75.** Chelicera. **76.** Gnathosoma, ventral. *Protereunetes:* **77.** Gnathosoma, ventral. *Eupodes:* **78.** Tarsus I, lateral. **79.** Tarsus I, dorsal setae omitted. **80.** Tarsus II, lateral, setae omitted. *Psammolycus:* **81.** Lateral aspect. **82.** Prodorsal region. **83.** Palpus. **84.** Chelicera. **85.** Genu, tibia, tarsus I. **86.** Genital region, lateral.

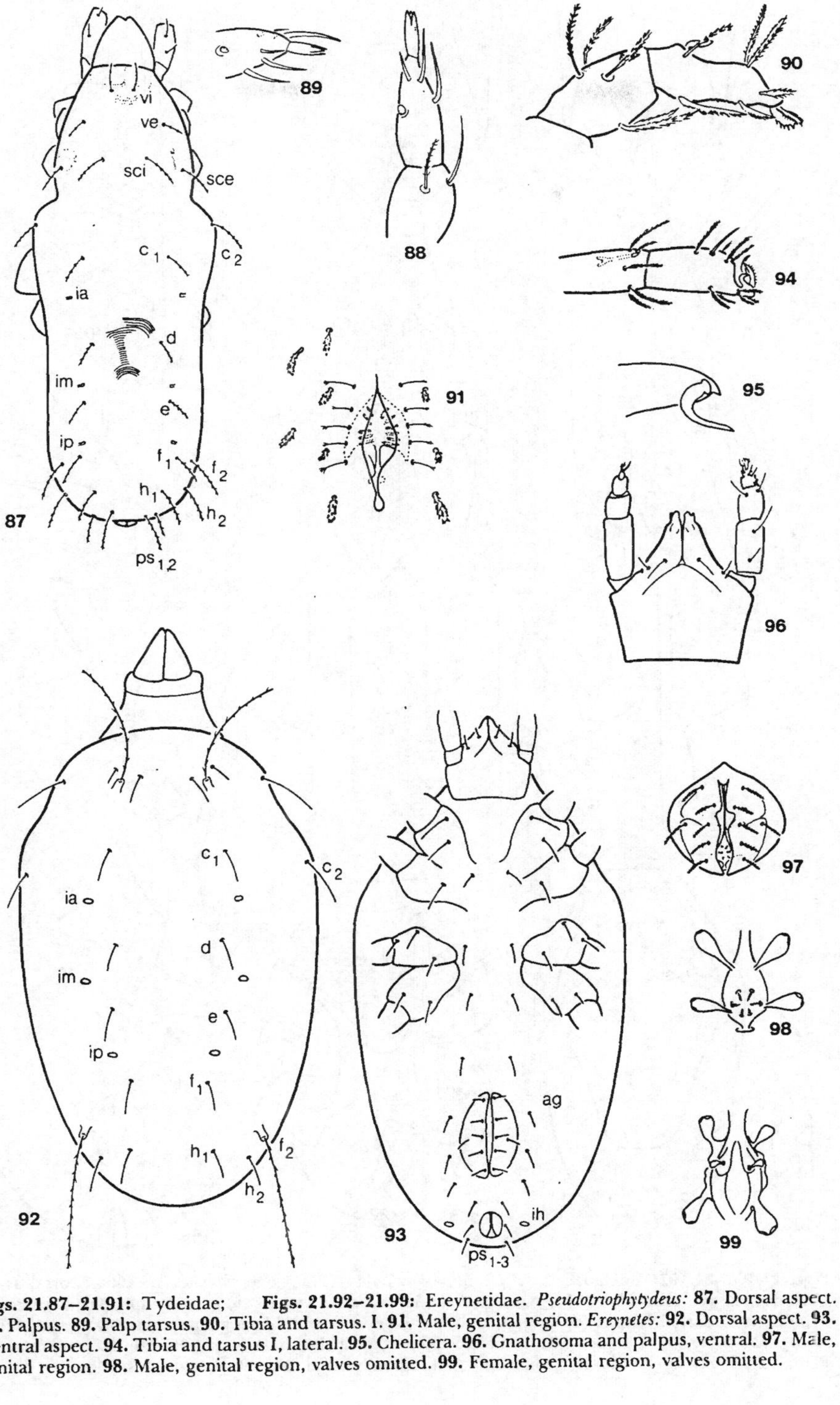

Figs. 21.87–21.91: Tydeidae; **Figs. 21.92–21.99:** Ereynetidae. *Pseudotriophytydeus:* **87.** Dorsal aspect. **88.** Palpus. **89.** Palp tarsus. **90.** Tibia and tarsus. I. **91.** Male, genital region. *Ereynetes:* **92.** Dorsal aspect. **93.** Ventral aspect. **94.** Tibia and tarsus I, lateral. **95.** Chelicera. **96.** Gnathosoma and palpus, ventral. **97.** Male, genital region. **98.** Male, genital region, valves omitted. **99.** Female, genital region, valves omitted.

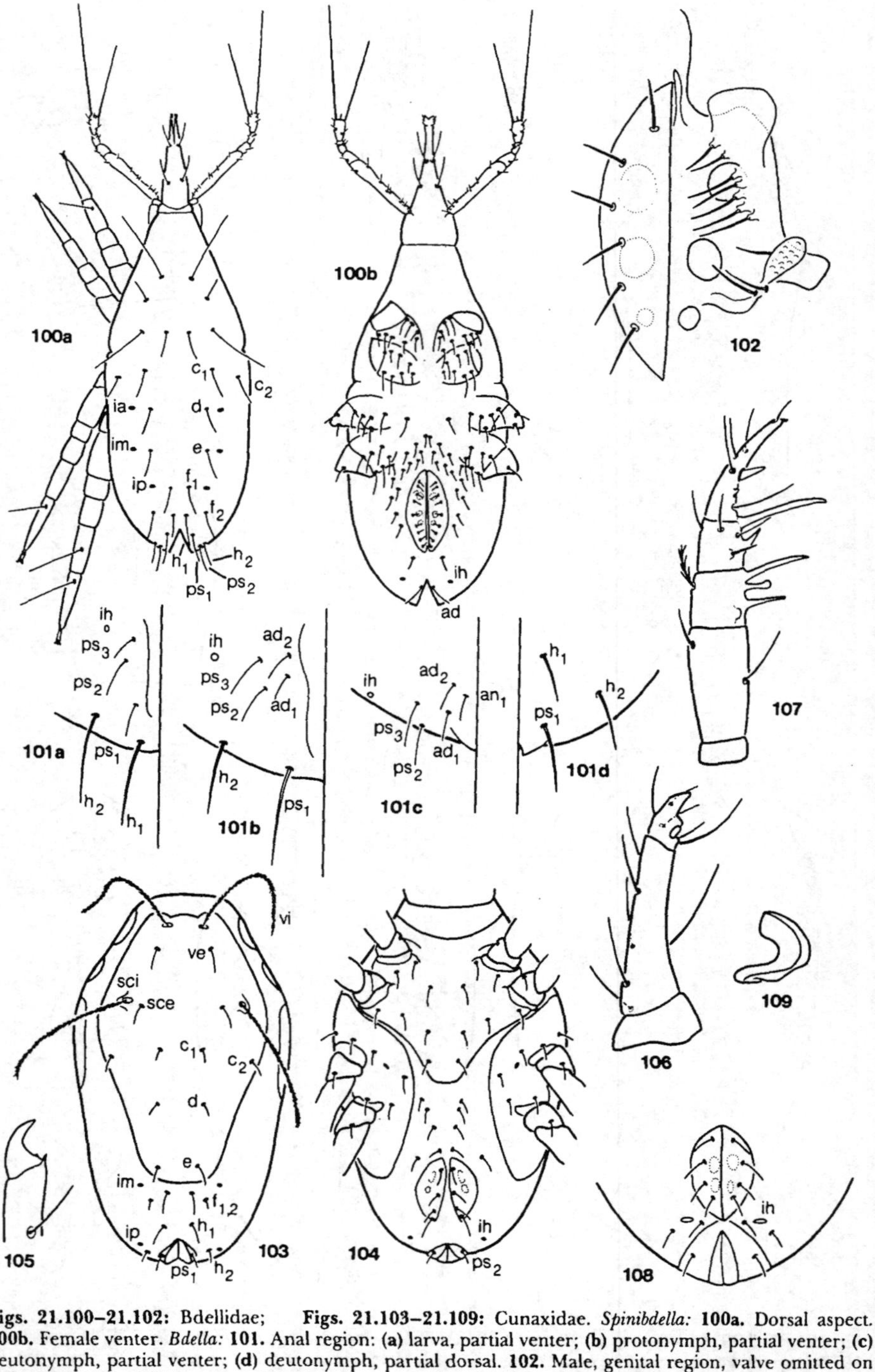

Figs. 21.100–21.102: Bdellidae; **Figs. 21.103–21.109:** Cunaxidae. *Spinibdella:* **100a.** Dorsal aspect. **100b.** Female venter. *Bdella:* **101.** Anal region: **(a)** larva, partial venter; **(b)** protonymph, partial venter; **(c)** deutonymph, partial venter; **(d)** deutonymph, partial dorsal. **102.** Male, genital region, valve omitted on right. *Coleoscirus:* **103.** Dorsum. **104.** Female, venter. **105.** Chelicera. *Pulaeus:* **106.** Palpus. *Armascirus:* **107.** Palpus. *Dactyloscirus:* **108.** Male, genital and anal region, ventral. **109.** Male, aedeagus, lateral.

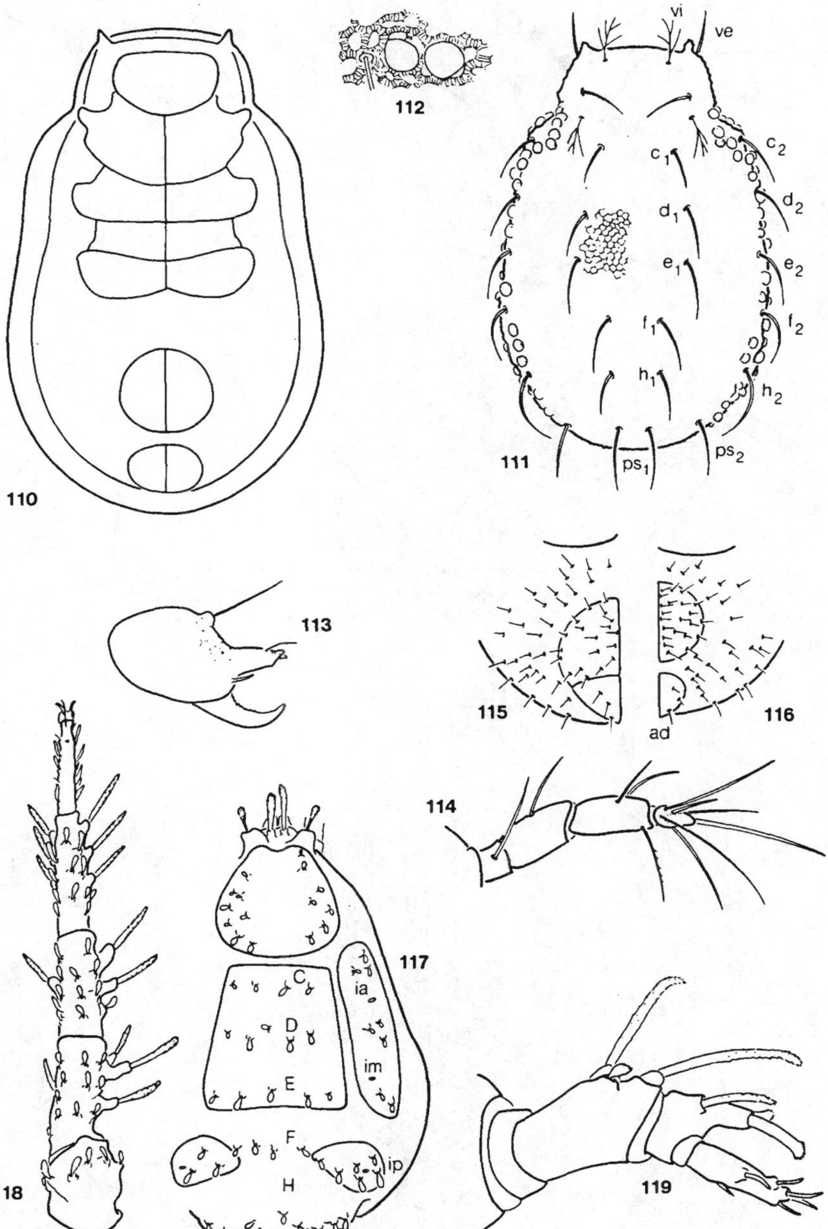

Figs. 21.110–21.116: Labidostommatidae; **Figs. 21.117–21.119:** Caeculidae. *Labidostomma:* **110.** Male, venter. **111.** Dorsum. **112.** Dorsal cuticle, detail. **113.** Chelicera. **114.** Palpus. **115.** Female, genital and anal regions. **116.** Male, genital and anal regions. *Procaeculus:* **117.** Dorsum. **118.** Leg I. **119.** Palpus.

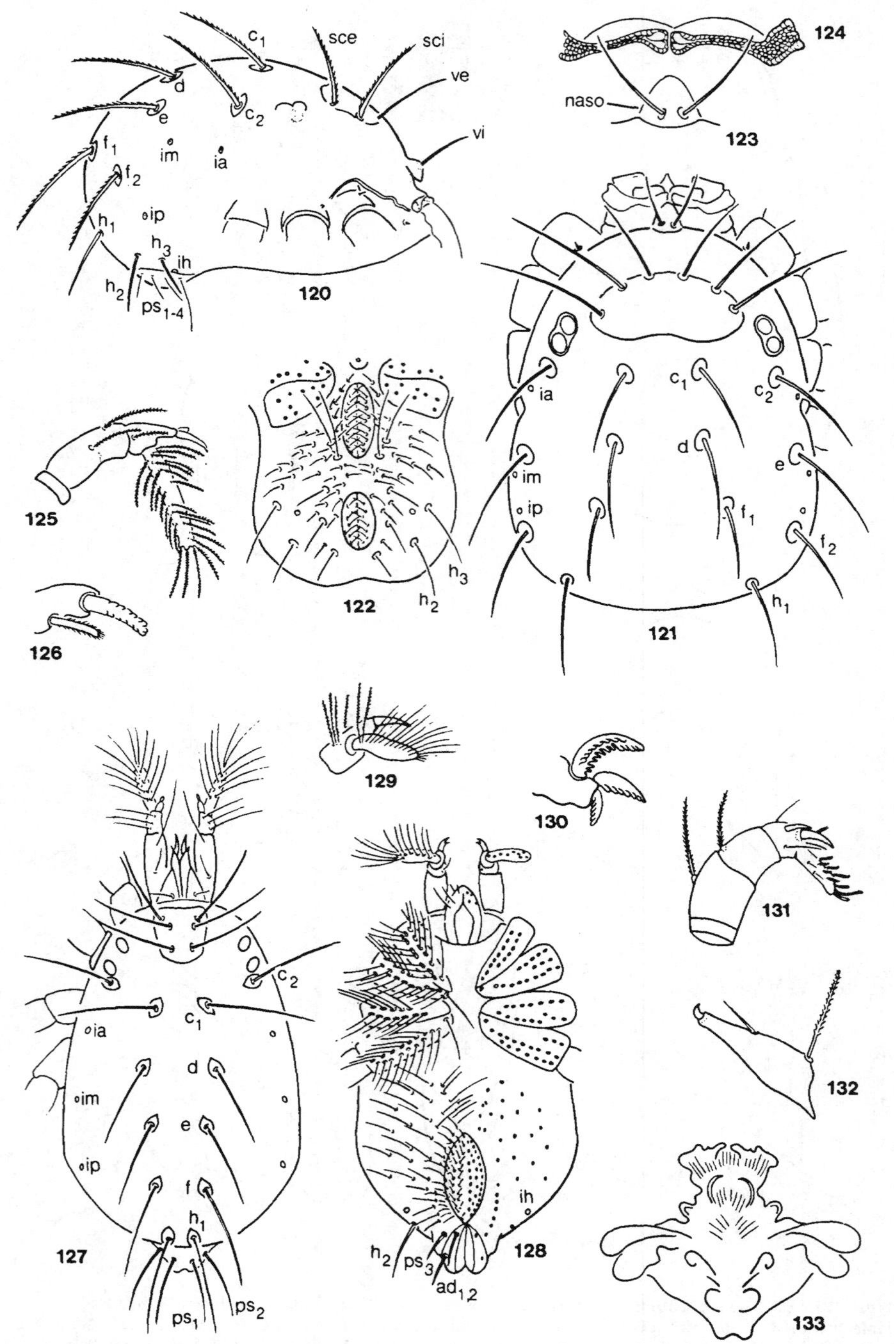

Figs. 21.120–21.133: Anystidae. *Anystis:* **120.** Larva, lateral aspect. **121.** Female, dorsum. **122.** Female, opisthosoma, ventral. **123.** Naso. **124.** Peritreme. *Chaussieria:* **125.** Palpus. *Erythracarus:* **126.** Palpal tibia. **127.** Dorsum. **128.** Venter. *Bechsteinia:* **129.** Palpal tibia and tarsus. **130.** Tarsus IV, claws and empodium. *Tarsotomus:* **131.** Palpus. **132.** Chelicera. **133.** Male, genital region, valves omitted.

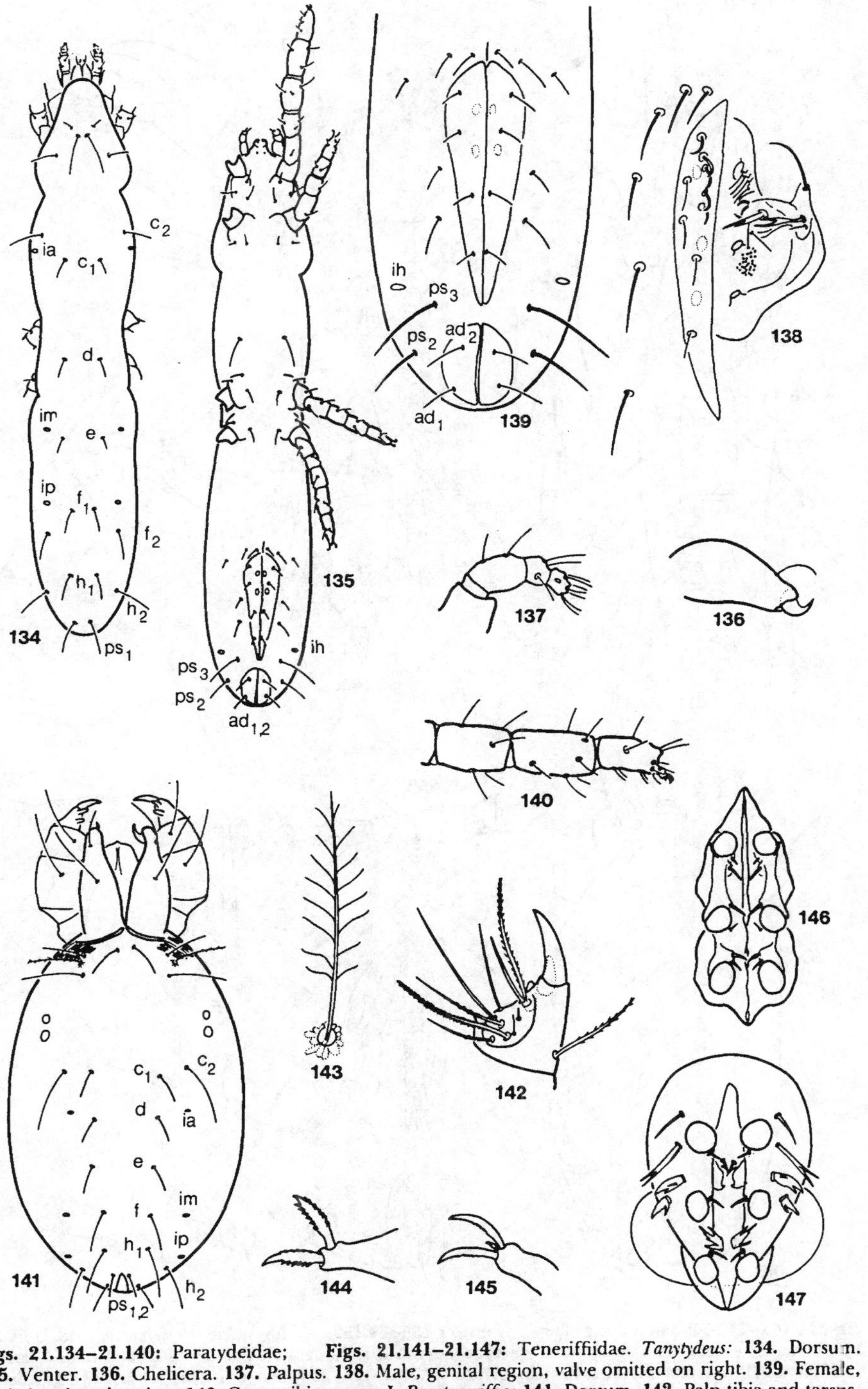

Figs. 21.134–21.140: Paratydeidae; **Figs. 21.141–21.147:** Teneriffiidae. *Tanytydeus:* **134.** Dorsum. **135.** Venter. **136.** Chelicera. **137.** Palpus. **138.** Male, genital region, valve omitted on right. **139.** Female, genital and anal region. **140.** Genu, tibia, tarsus I. *Parateneriffia:* **141.** Dorsum. **142.** Palp tibia and tarsus. **143.** Prodorsal trichobothrium. **144.** Tarsus I claws. **145.** Tarsus IV claws and empodium. **146.** Female, genital region, valves omitted. **147.** Male, genital region, valves omitted.

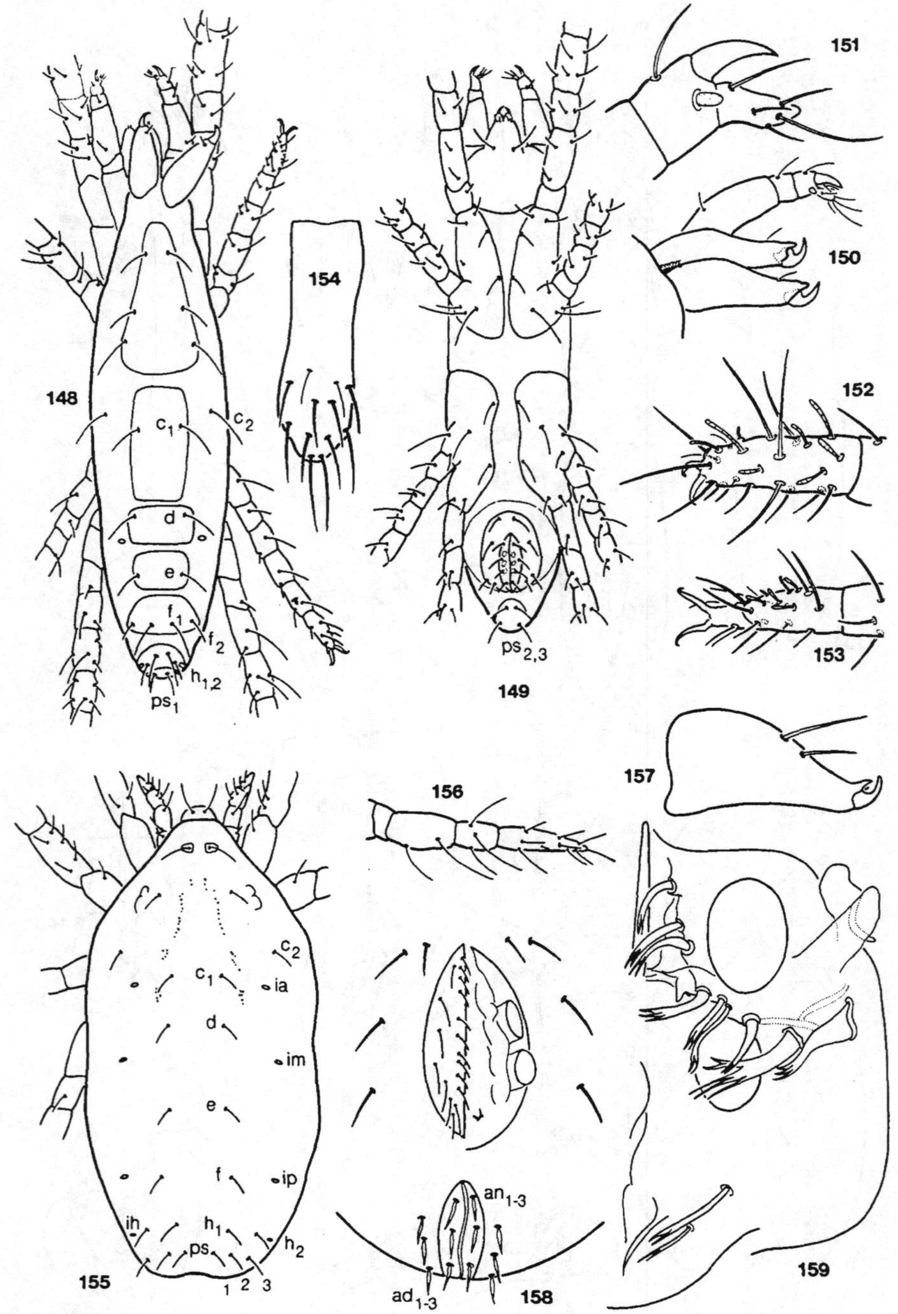

Figs. 21.148–21.154: Pomerantziidae; **Figs. 21.155–21.159:** Adamystidae. *Pomerantzia:* **148.** Dorsum. **149.** Female, venter. **150.** Palpus and chelicerae, lateral. **151.** Palpal tibia and tarsus. **152.** Tarsus I, lateral. **153.** Tarsus II, dorsolateral. **154.** Female, ovipositor. *Adamystis:* **155.** Dorsum. **156.** Palpus. **157.** Chelicera. **158.** Female, genital and anal regions, genital valve removed on right. **159.** Male, internal genitalia, left side.

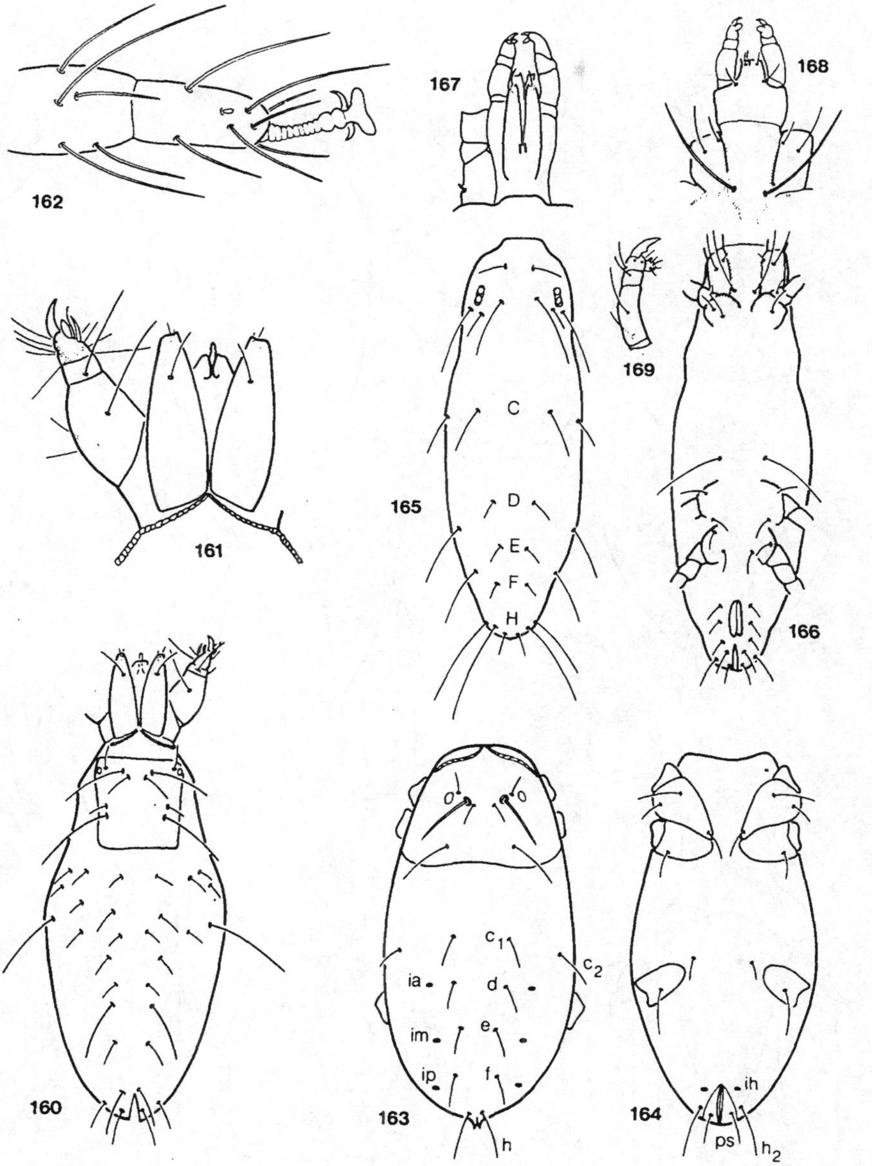

Figs. 21.160–21.164: Pseudocheylidae; **Figs. 21.165–21.169:** Barbutiidae. *Anoplocheylus:* **160.** Dorsum. **161.** Gnathosoma and prodorsal peritremes, dorsal aspect. **162.** Tibia and tarsus II, dorsal. **163.** Larva, dorsum. **164.** Larva, venter. *Barbutia:* **165.** Dorsum. **166.** Venter. **167.** Gnathosoma dorsal. **168.** Gnathosoma ventral. **169.** Palpus.

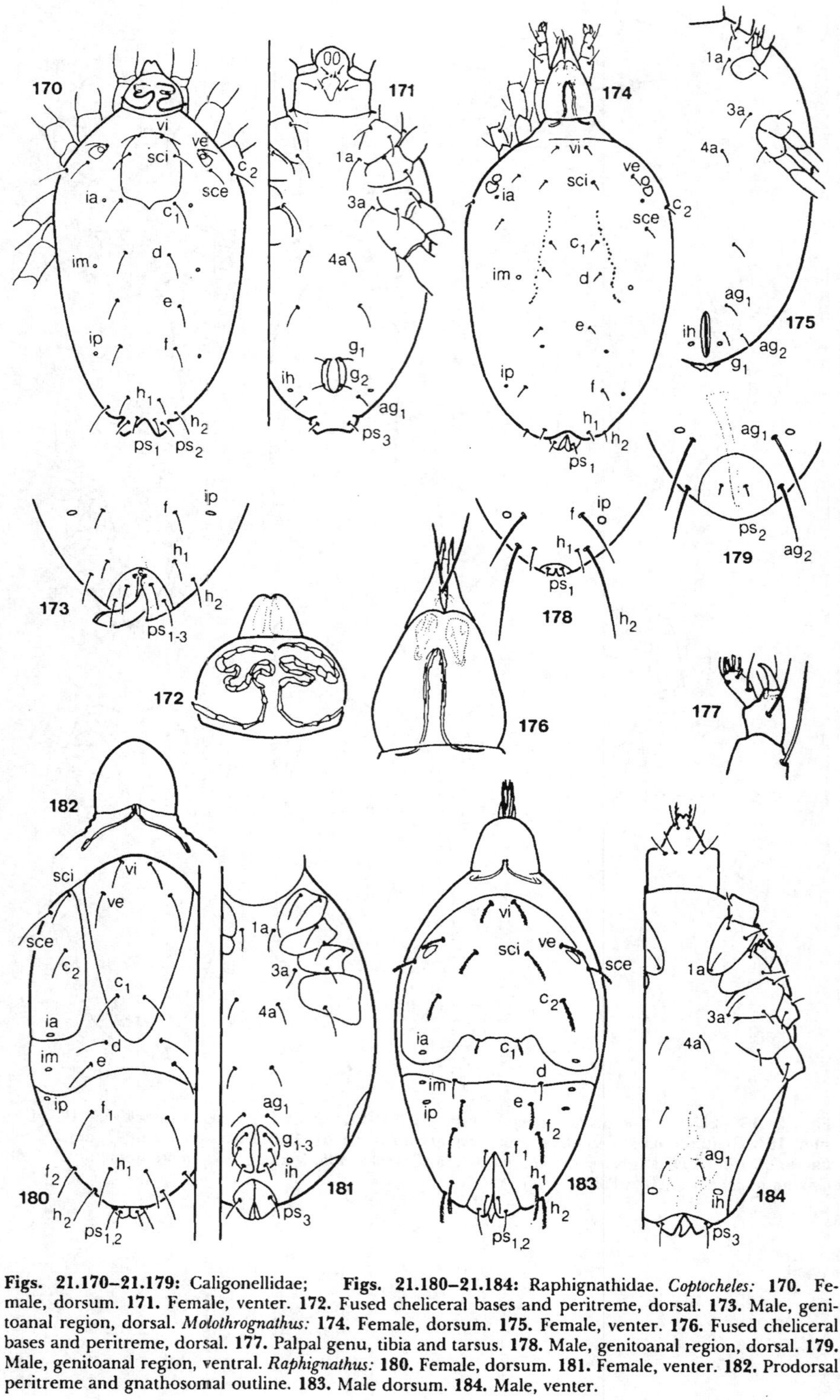

Figs. 21.170–21.179: Caligonellidae; **Figs. 21.180–21.184:** Raphignathidae. *Coptocheles:* **170.** Female, dorsum. **171.** Female, venter. **172.** Fused cheliceral bases and peritreme, dorsal. **173.** Male, genitoanal region, dorsal. *Molothrognathus:* **174.** Female, dorsum. **175.** Female, venter. **176.** Fused cheliceral bases and peritreme, dorsal. **177.** Palpal genu, tibia and tarsus. **178.** Male, genitoanal region, dorsal. **179.** Male, genitoanal region, ventral. *Raphignathus:* **180.** Female, dorsum. **181.** Female, venter. **182.** Prodorsal peritreme and gnathosomal outline. **183.** Male dorsum. **184.** Male, venter.

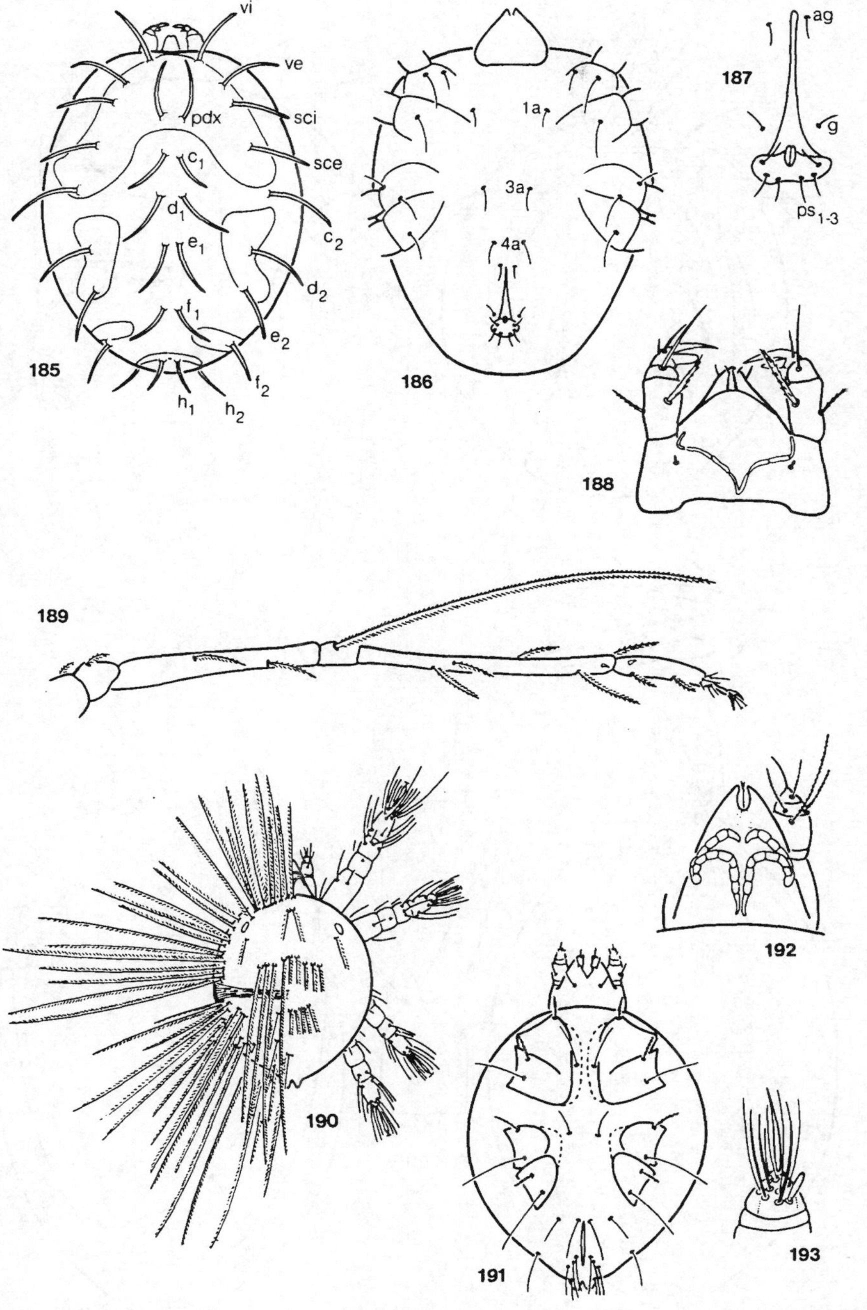

Figs. 21.185–21.189: Camerobiidae; **Figs. 21.190–21.193:** Xenocaligonellididae. *Neophyllobius:* **185.** Female, dorsum. **186.** Female, venter. **187.** Female, genitoanal region. **188.** Gnathosoma dorsal. **189.** Leg II lateral. *Dasythraeus:* **190.** Female, dorsal aspect. **191.** Female, venter. **192.** Fused cheliceral bases, peritreme and palpus, in part, dorsal. **193.** Palpal tibiotarsus.

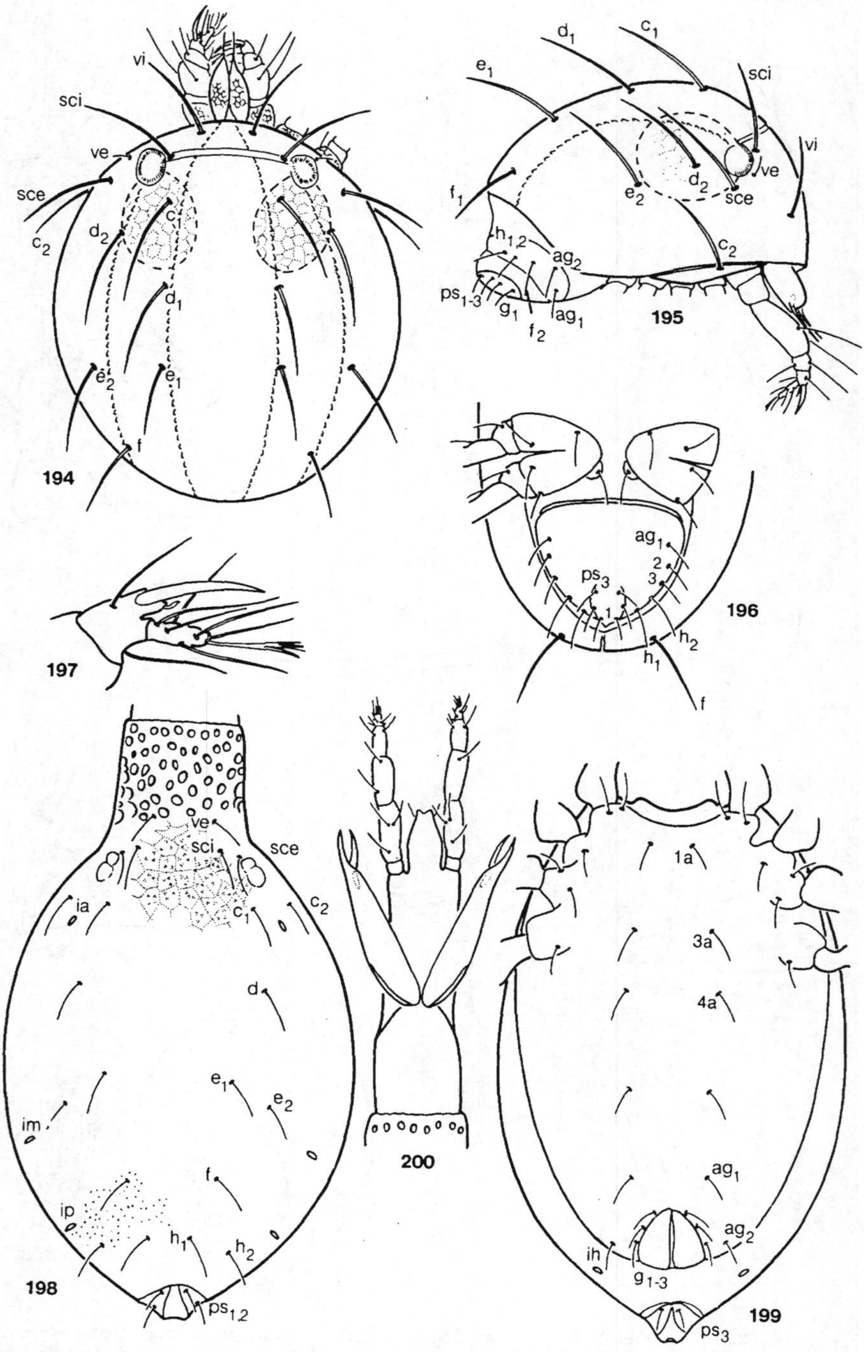

Figs. 21.194–21.197: Homocaligidae; **Figs. 21.198–21.200:** Cryptognathidae. *Annerosella:* **194.** Female, dorsum. **195.** Female, lateral aspects, legs omitted. **196.** Male, opisthosoma ventral. **197.** Palpal tibia and tarsus, lateral aspect. *Cryptognathus:* **198.** Female, dorsum. **199.** Female, venter. **200.** Gnathosoma dorsal.

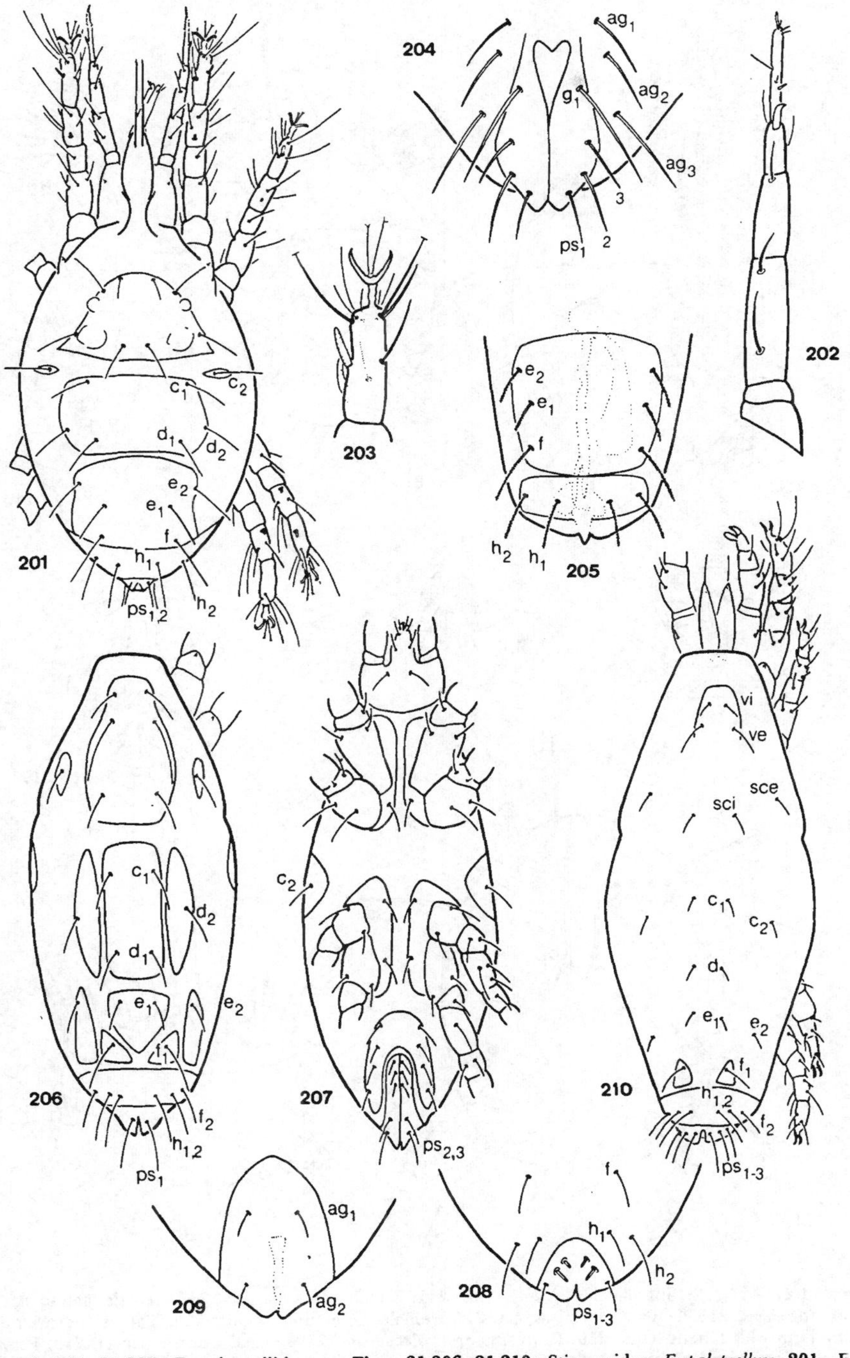

Figs. 21.201–21.205: Eupalopsellidae; **Figs. 21.206–21.210:** Stigmaeidae. *Eupalopsellus:* **201.** Female, dorsal aspect. *Eupalopsis:* **202.** Palpus. **203.** Male, tarsus I. **204.** Female, genitoanal region. **205.** Male, posterior of opisthosoma, dorsal. *Stigmaeus:* **206.** Female, dorsum. **207.** Female, venter. **208.** Male, genitoanal region, dorsal. **209.** Male, genitoanal region, ventral. **210.** Female, dorsum.

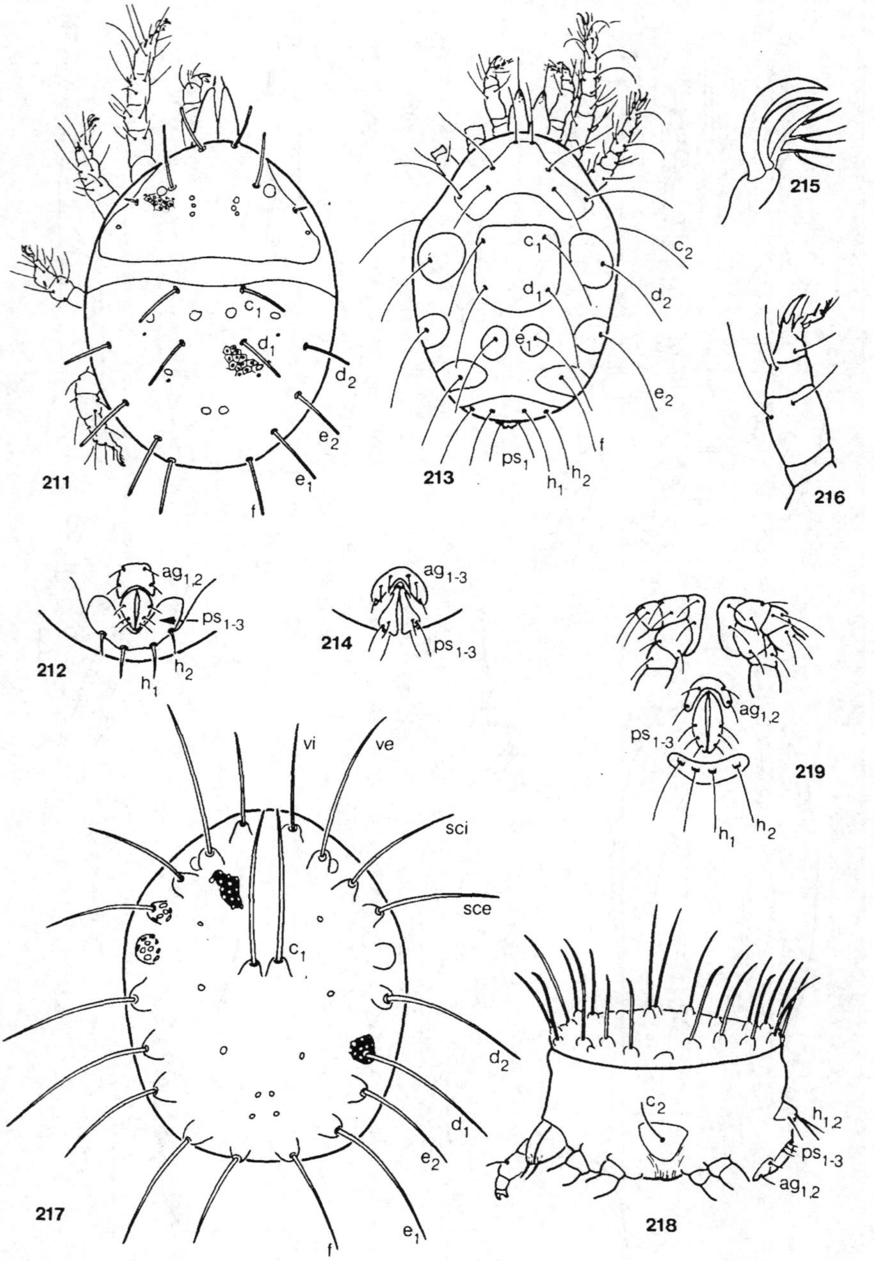

Figs. 21.211–21.219: Stigmaeidae. *Eustigmaeus:* **211.** Female, dorsal aspect. **212.** Female, genitoanal region. *Stigmaeus:* **213.** Female, dorsal aspect. **214.** Female, genitoanal region. **215.** Tarsus I, claws nude, empodium with tenent hairs. **216.** Palpus, lateral. *Mendanaia:* **217.** Female, dorsal aspect. **218.** Female, lateral aspect, legs omitted. **219.** Female, posterior of opisthosoma, ventral.

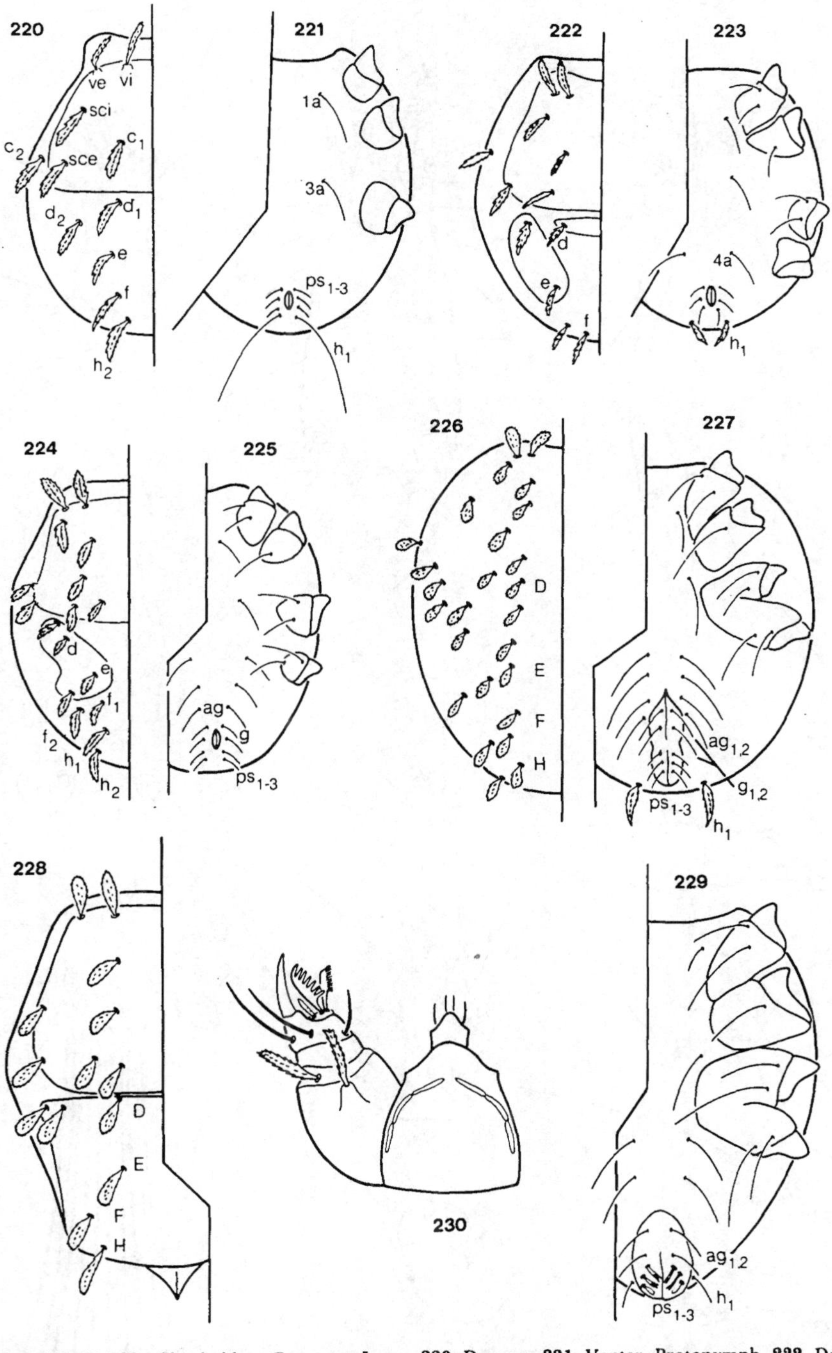

Figs. 21.220–21.230: Cheyletidae. *Eutogenes:* Larva, **220.** Dorsum. **221.** Venter. Protonymph, **222.** Dorsum. **223.** Venter. Deutonymph, **224.** Dorsum. **225.** Venter. Female, **226.** Dorsum. **227.** Venter. Male, **228.** Dorsum. **229.** Venter. **230.** Female, gnathosoma, dorsal.

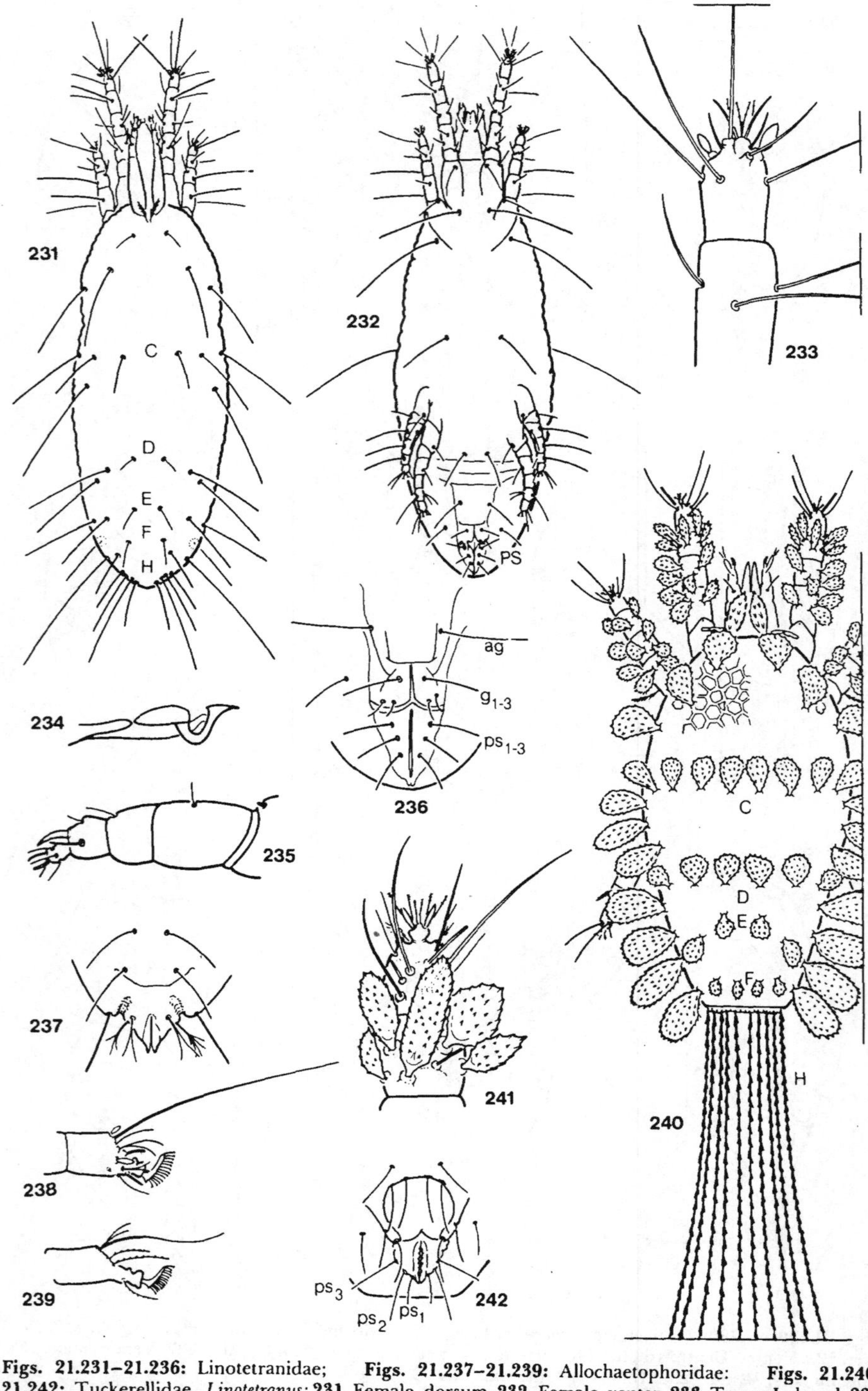

Figs. 21.231–21.236: Linotetranidae; **Figs. 21.237–21.239:** Allochaetophoridae: **Figs. 21.240–21.242:** Tuckerellidae. *Linotetranus:* **231.** Female, dorsum. **232.** Female, venter. **233.** Tarsus I, dorsal. **234.** Male aedeagus, lateral. **235.** Palpus, lateral. **236.** Female, genitoanal region. *Allochaetophora:* **237.** Nymph, genitoanal region. **238.** Nymph, tarsus II, lateral. **239.** Nymph, tarsus IV, lateral. *Tuckerella:* **240.** Female, dorsal aspect. **241.** Female, tibia and tarsus I, dorsal. **242.** Female, genitoanal region.

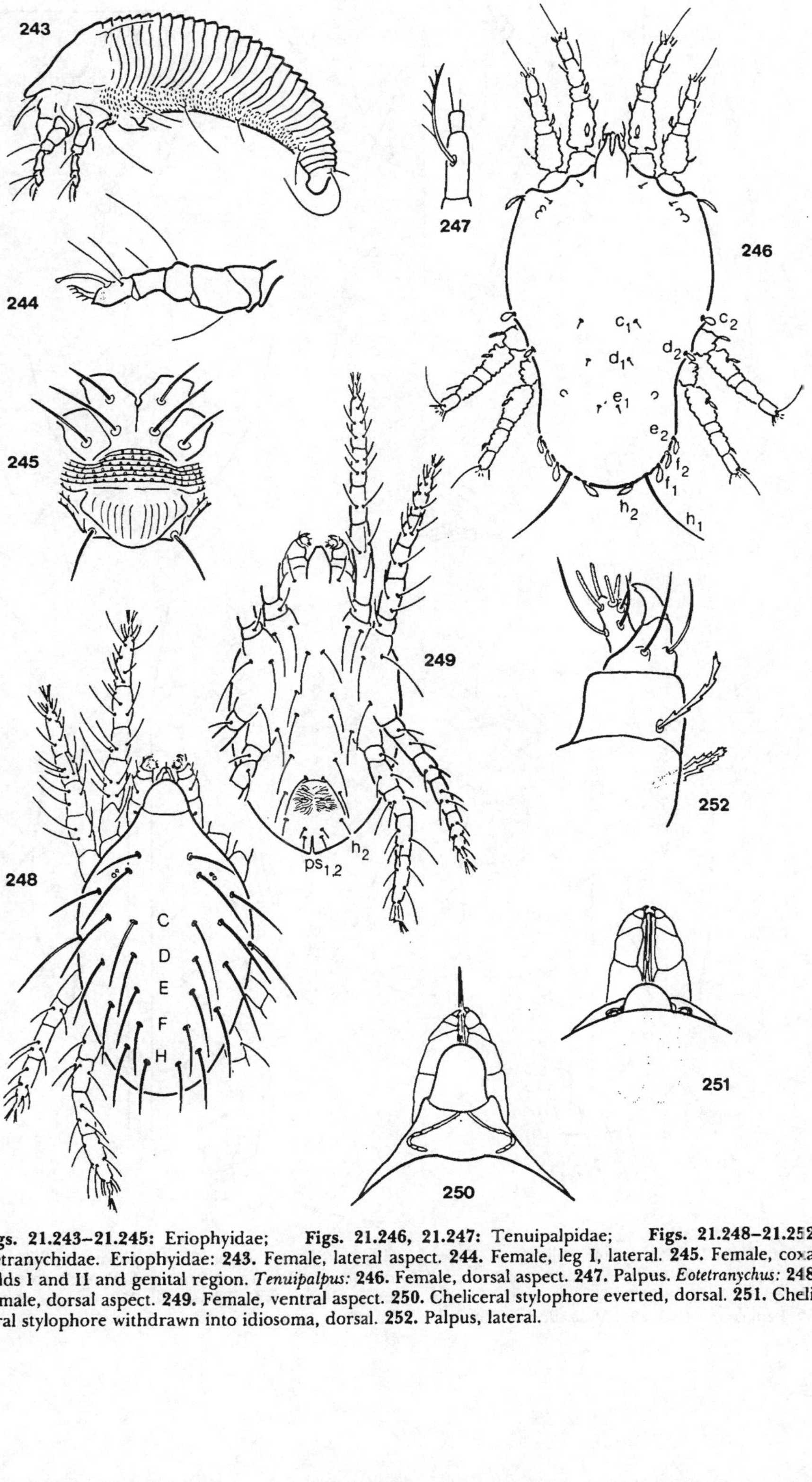

Figs. 21.243–21.245: Eriophyidae; **Figs. 21.246, 21.247:** Tenuipalpidae; **Figs. 21.248–21.252:** Tetranychidae. Eriophyidae: **243.** Female, lateral aspect. **244.** Female, leg I, lateral. **245.** Female, coxal fields I and II and genital region. *Tenuipalpus:* **246.** Female, dorsal aspect. **247.** Palpus. *Eotetranychus:* **248.** Female, dorsal aspect. **249.** Female, ventral aspect. **250.** Cheliceral stylophore everted, dorsal. **251.** Cheliceral stylophore withdrawn into idiosoma, dorsal. **252.** Palpus, lateral.

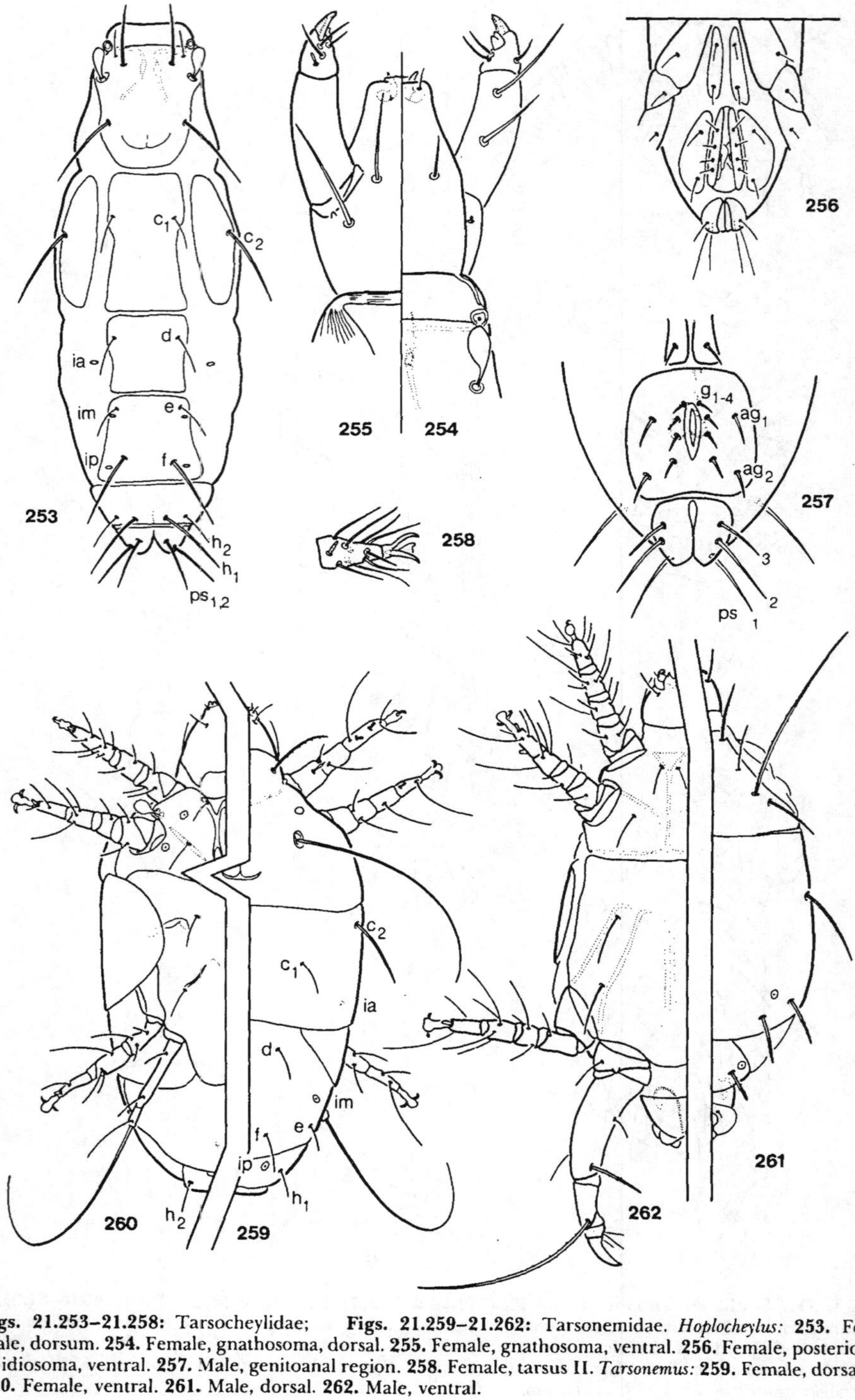

Figs. 21.253–21.258: Tarsocheylidae; **Figs. 21.259–21.262:** Tarsonemidae. *Hoplocheylus:* **253.** Female, dorsum. **254.** Female, gnathosoma, dorsal. **255.** Female, gnathosoma, ventral. **256.** Female, posterior of idiosoma, ventral. **257.** Male, genitoanal region. **258.** Female, tarsus II. *Tarsonemus:* **259.** Female, dorsal. **260.** Female, ventral. **261.** Male, dorsal. **262.** Male, ventral.

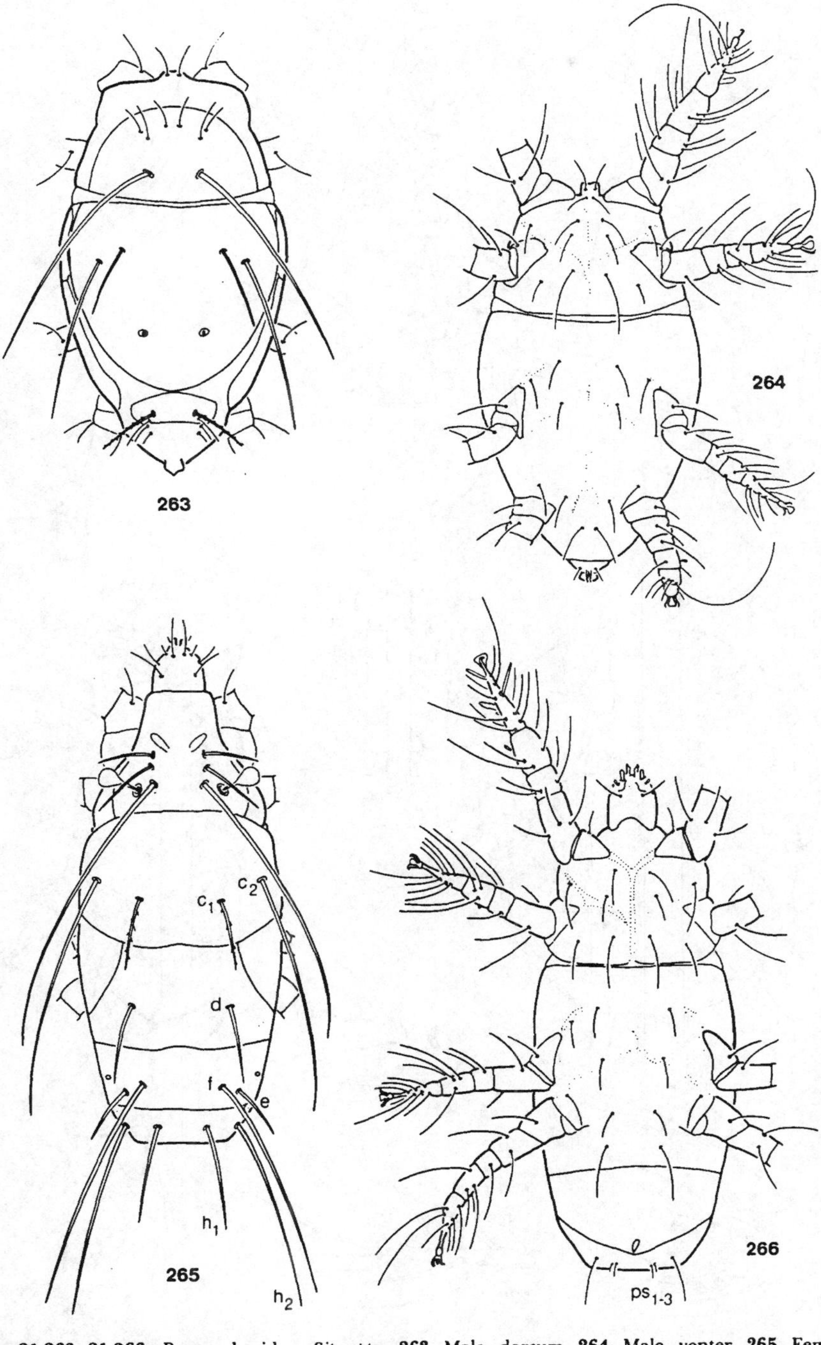

Figs. 21.263–21.266: Pygmephoridae. *Siteroptes:* **263.** Male, dorsum. **264.** Male, venter. **265.** Female, dorsum. **266.** Female, venter.

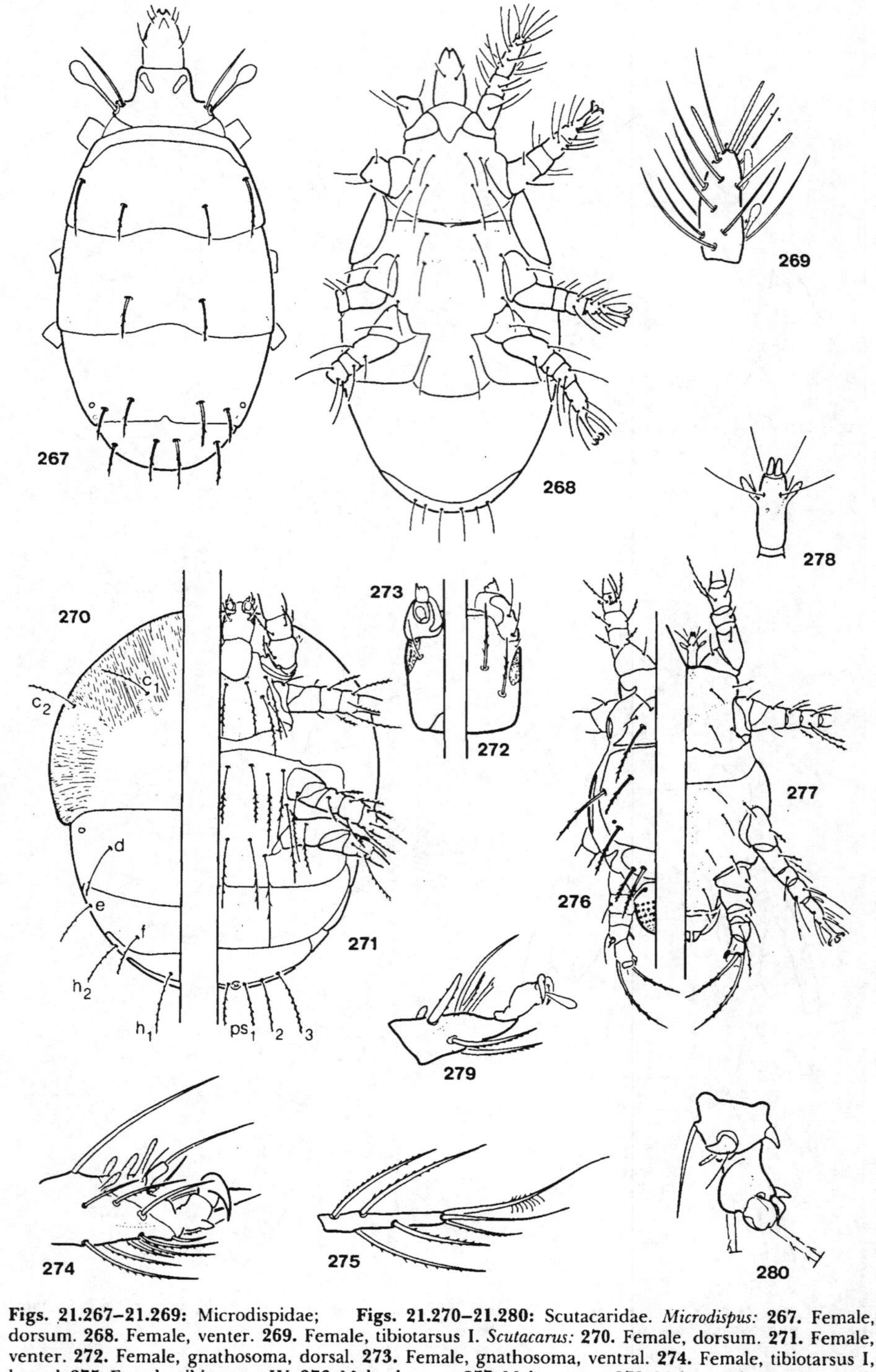

Figs. 21.267–21.269: Microdispidae; **Figs. 21.270–21.280:** Scutacaridae. *Microdispus:* **267.** Female, dorsum. **268.** Female, venter. **269.** Female, tibiotarsus I. *Scutacarus:* **270.** Female, dorsum. **271.** Female, venter. **272.** Female, gnathosoma, dorsal. **273.** Female, gnathosoma, ventral. **274.** Female, tibiotarsus I, lateral. **275.** Female, tibiotarsus IV. **276.** Male, dorsum. **277.** Male, venter. **278.** Male, gnathosoma, ventral. **279.** Male. tarsus I, lateral. **280.** Male, tarsus IV.

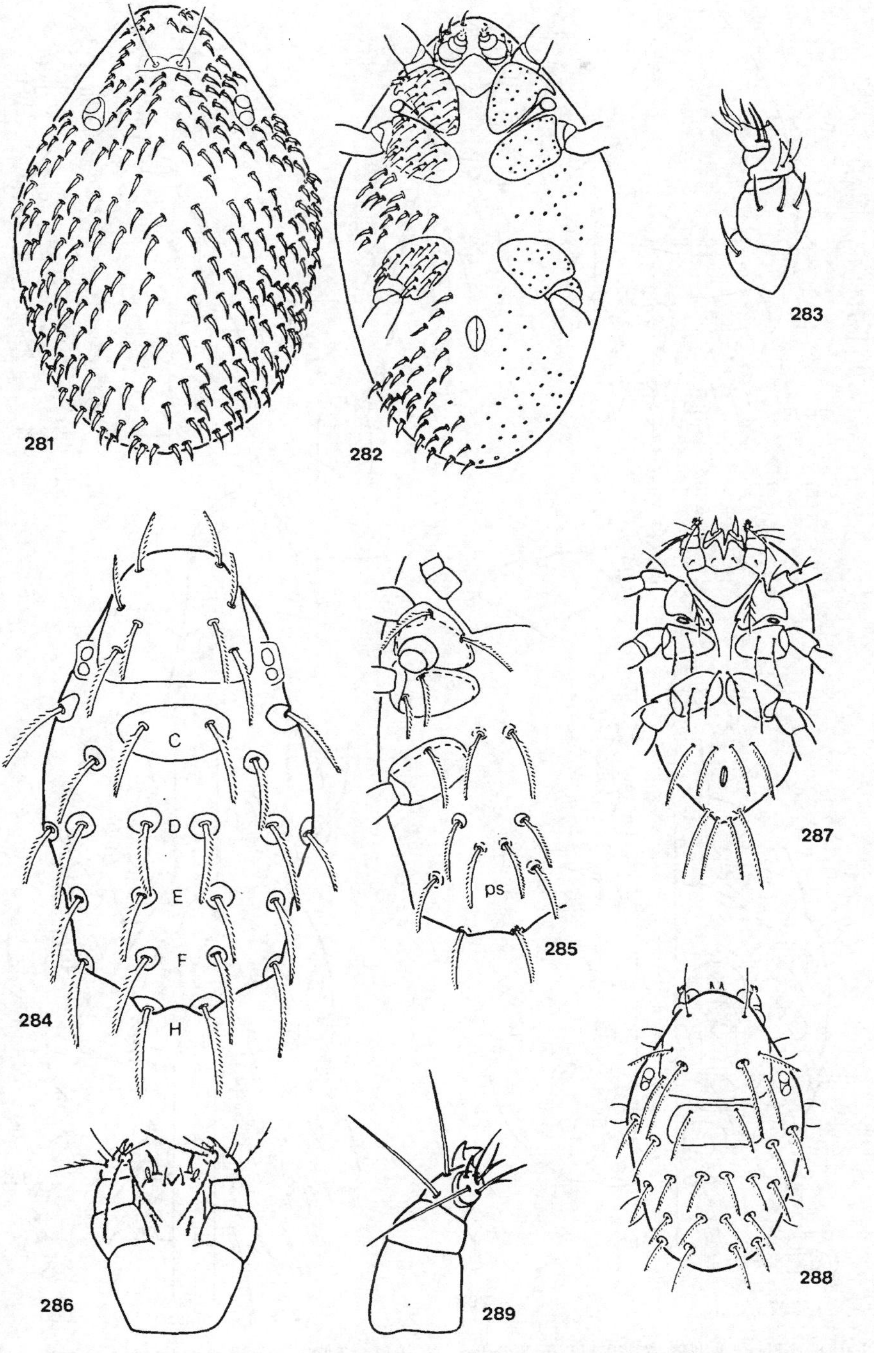

Larvae: **Figs. 21.281–21.283:** Calyptostomatidae; **Figs. 21.284–21.286:** Trombidiidae; **Figs. 21.287–21.289:** Trombidiinae. *Calyptostoma:* **281.** Dorsum. **282.** Venter. **283.** Palpus. Allothrombiinae—*Allothrombium:* **284.** Dorsum. **285.** Venter. **286.** Gnathosoma, ventral. Trombidiinae—*Trombidium:* **287.** Venter. **288.** Dorsum. **289.** Palpus.

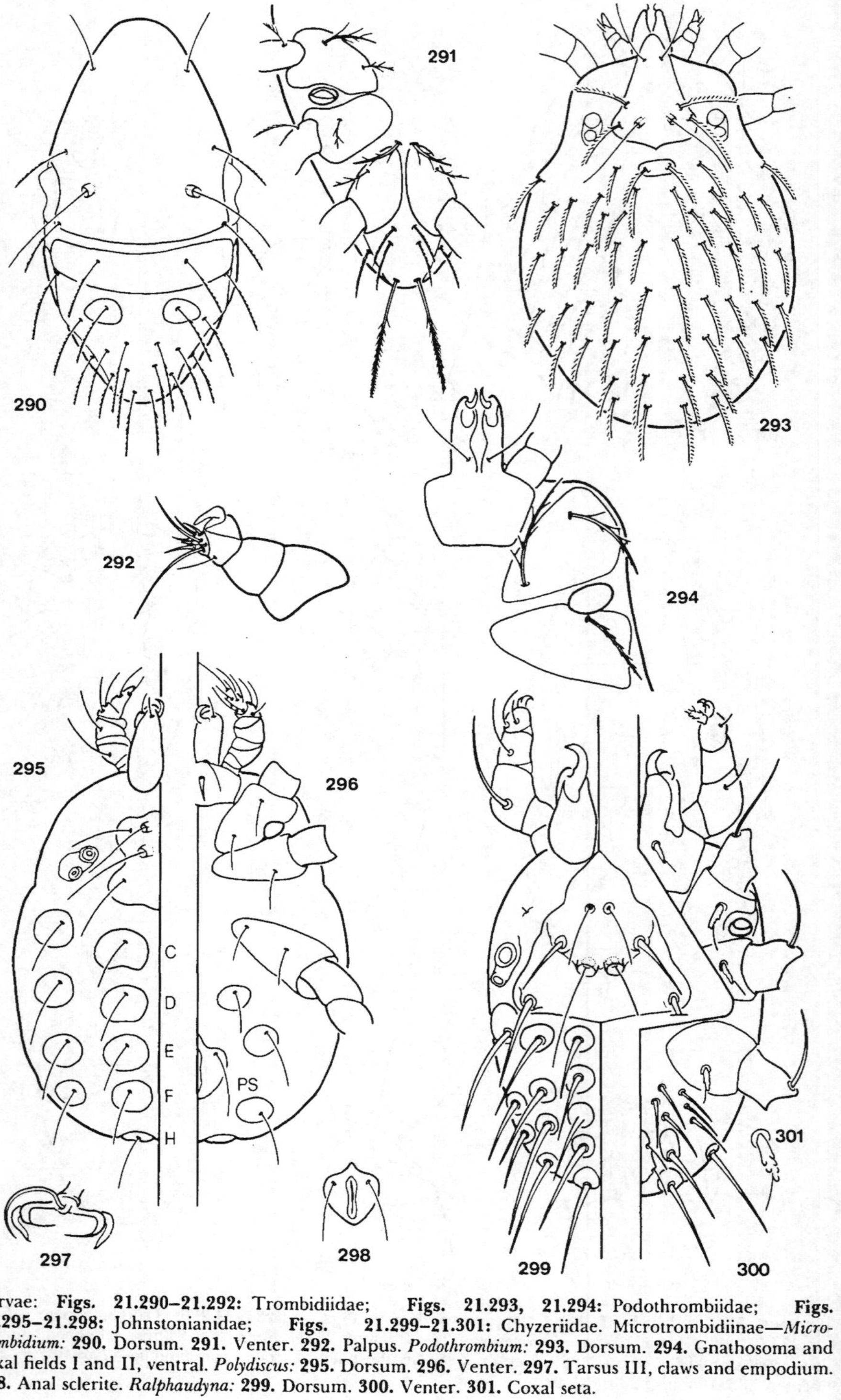

Larvae: **Figs. 21.290–21.292:** Trombidiidae; **Figs. 21.293, 21.294:** Podothrombiidae; **Figs. 21.295–21.298:** Johnstonianidae; **Figs. 21.299–21.301:** Chyzeriidae. Microtrombidiinae—*Microtrombidium:* **290.** Dorsum. **291.** Venter. **292.** Palpus. *Podothrombium:* **293.** Dorsum. **294.** Gnathosoma and coxal fields I and II, ventral. *Polydiscus:* **295.** Dorsum. **296.** Venter. **297.** Tarsus III, claws and empodium. **298.** Anal sclerite. *Ralphaudyna:* **299.** Dorsum. **300.** Venter. **301.** Coxal seta.

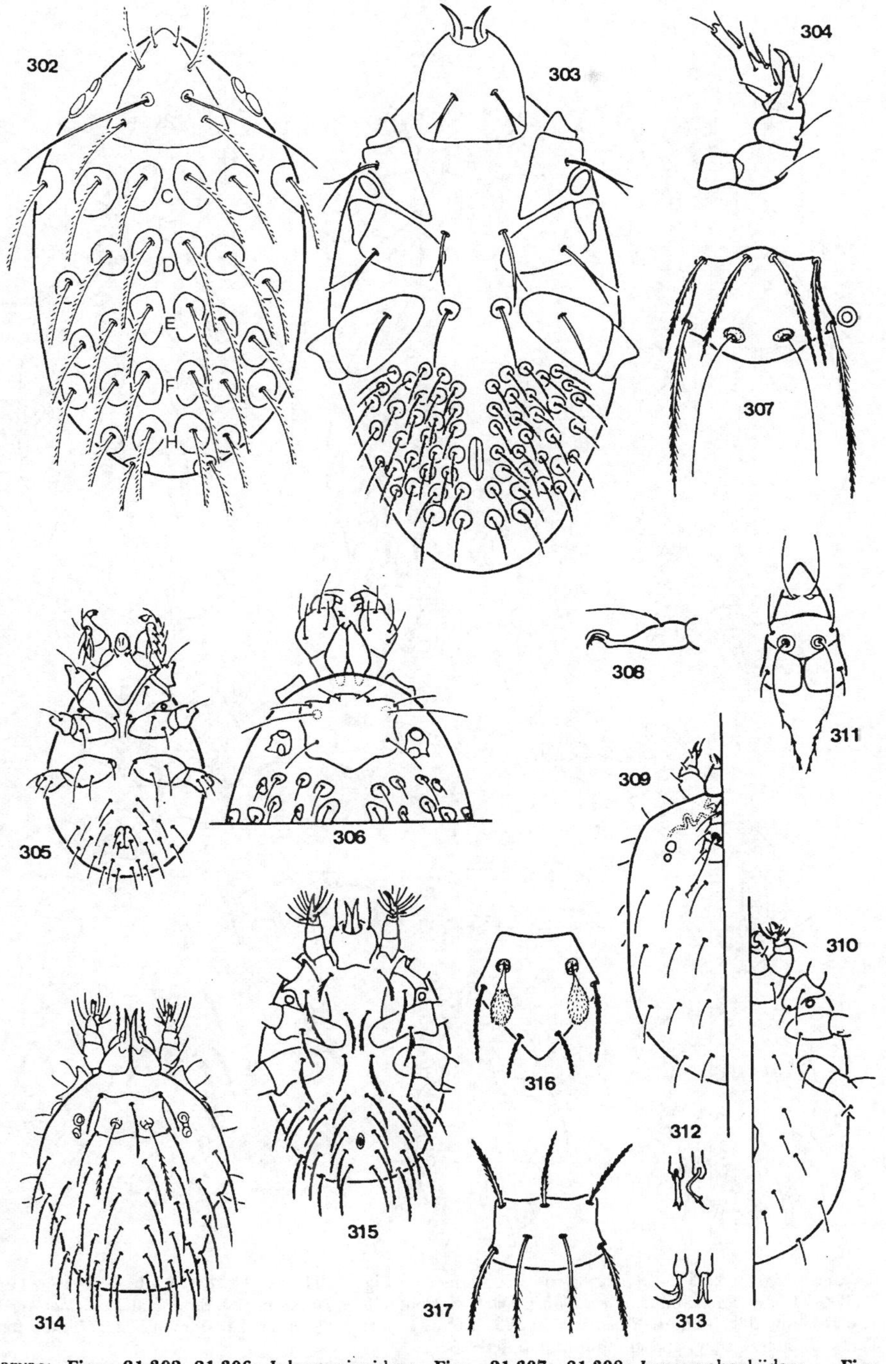

Larvae: **Figs. 21.302–21.306:** Johnstonianidae; **Figs. 21.307, 21.308:** Leeuwenhoekiidae; **Figs. 21.309–21.313:** Trombellidae; **Figs. 21.314–21.317:** Trombiculidae. *Diplothrombium:* **302.** Dorsum. **303.** Venter. **304.** Palpus. *Lassenia:* **305.** Venter. **306.** Dorsum. *Shunsennia:* **307.** Prodorsal sclerite. **308.** Tarsus III, setae omitted. *Durenia:* **309.** Dorsum. **310.** Venter. **311.** Prodorsal sclerite. **312.** Tarsus I empodial claw, dorsal, lateral. **313.** Tarsus III claws, dorsal, lateral. *Leptotrombidium:* **314.** Dorsum. **315.** Venter. *Walchia:* **316.** Prodorsal sclerite. *Eutrombicula:* **317.** Prodorsal sclerite.

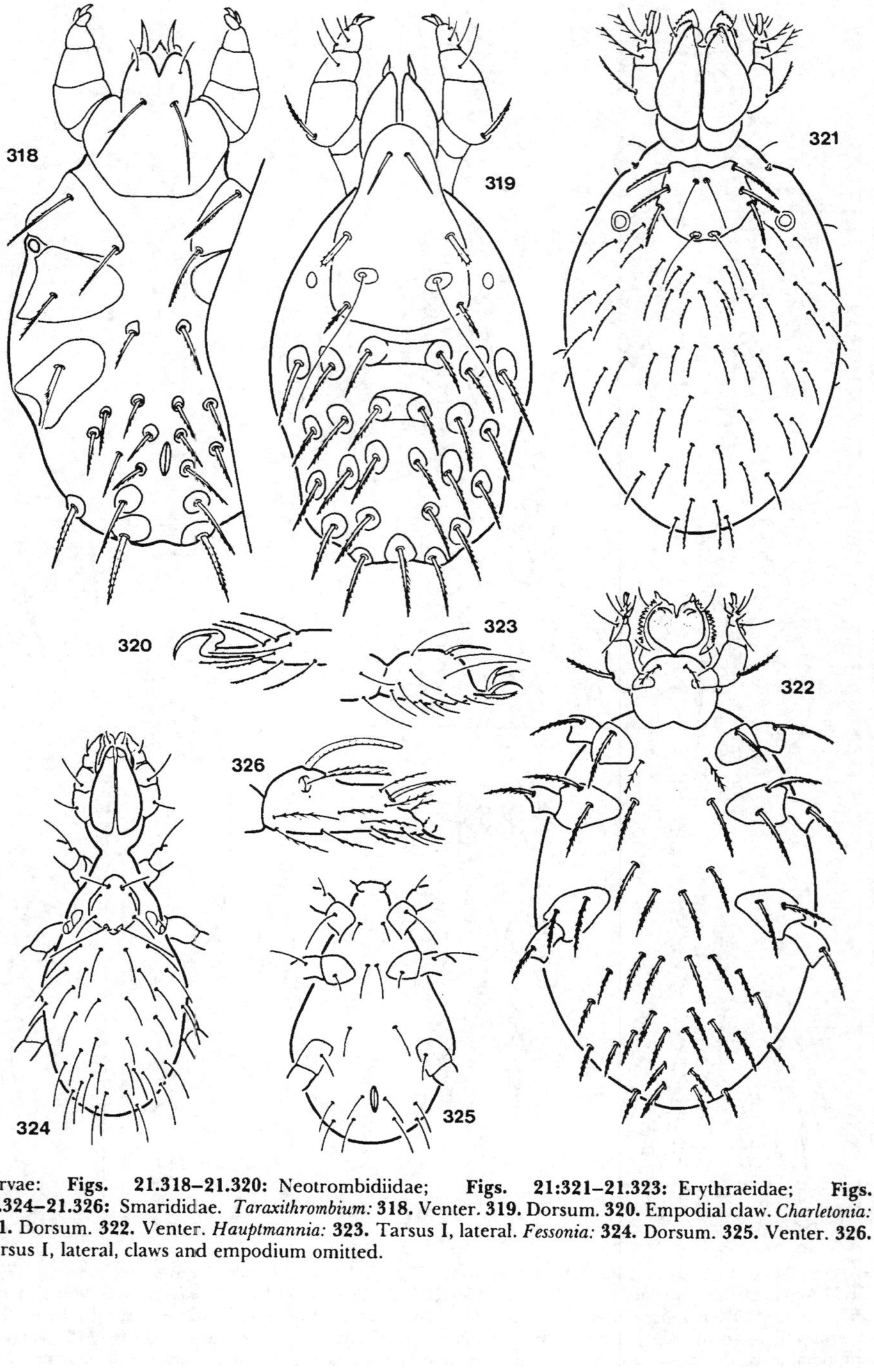

Larvae: **Figs. 21.318–21.320:** Neotrombidiidae; **Figs. 21:321–21.323:** Erythraeidae; **Figs. 21.324–21.326:** Smarididae. *Taraxithrombium:* **318.** Venter. **319.** Dorsum. **320.** Empodial claw. *Charletonia:* **321.** Dorsum. **322.** Venter. *Hauptmannia:* **323.** Tarsus I, lateral. *Fessonia:* **324.** Dorsum. **325.** Venter. **326.** Tarsus I, lateral, claws and empodium omitted.

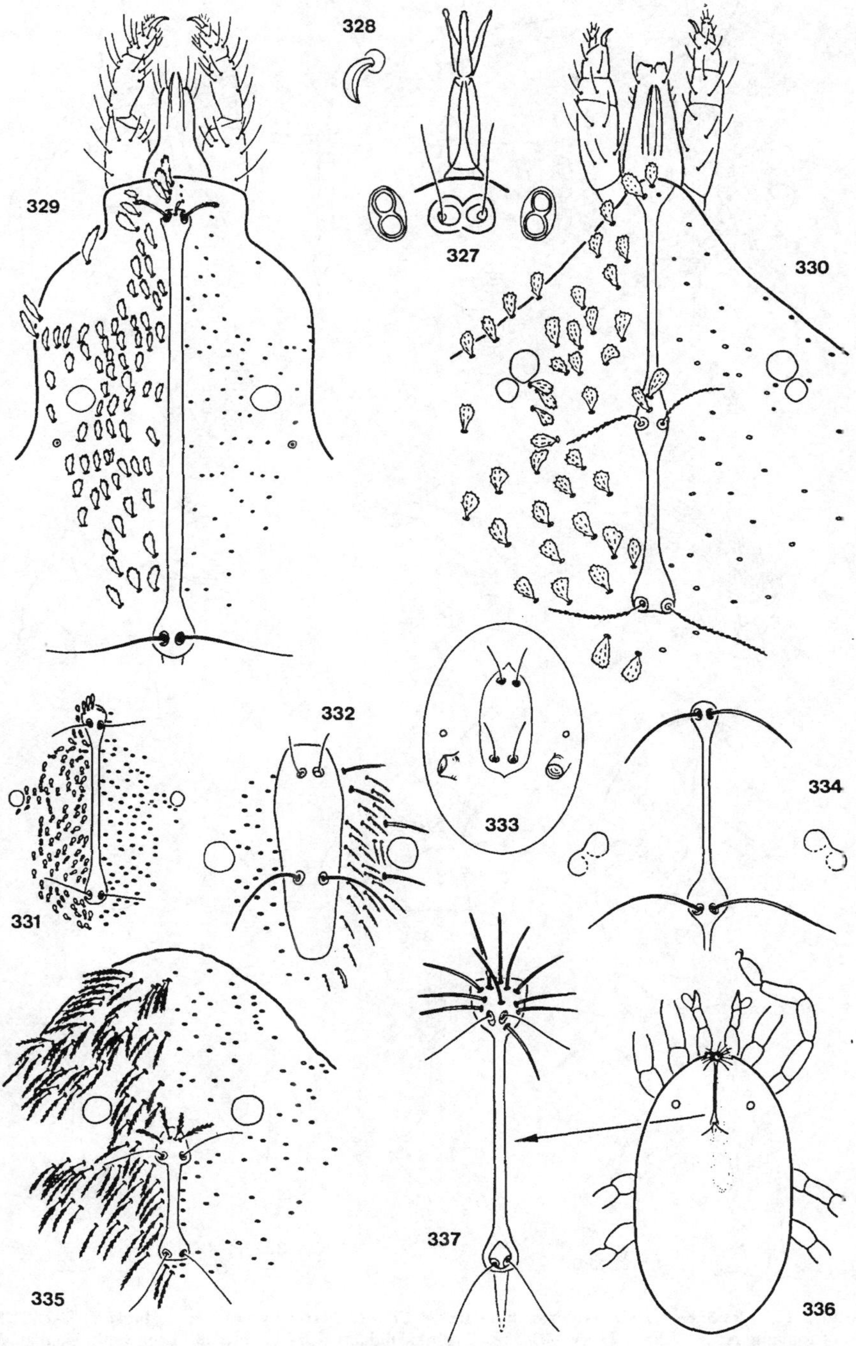

Adults: **Figs. 21.327, 21.328:** Calyptostomatidae; **Figs. 21.329, 21.330;** Smarididae; **Figs. 21.331–21.337:** Erythraeidae. *Calyptostoma:* **327.** Prodorsal sclerite and gnathosoma everted, diagrammatic. **328.** Body seta. Hirstiosomatinae—*Trichosmaris:* **329.** Prodorsal region and palps. Smaridinae—*Fessonia:* **330.** Prodorsal region and palps. Leptinae—*Leptus:* **331.** Prodorsal region. Balaustiinae—*Wartookia:* **332.** Prodorsal region. *Balaustium:* **333.** Prodorsal region. Erythraeinae—*Augustonella:* **334.** Prodorsal region. Myrmicotrombidiinae—*Myrmicotrombidium:* **335.** Prodorsal region. Callidosomatinae—*Charletonia:* **336.** Dorsum, schematic; **337.** Prodorsal sclerite.

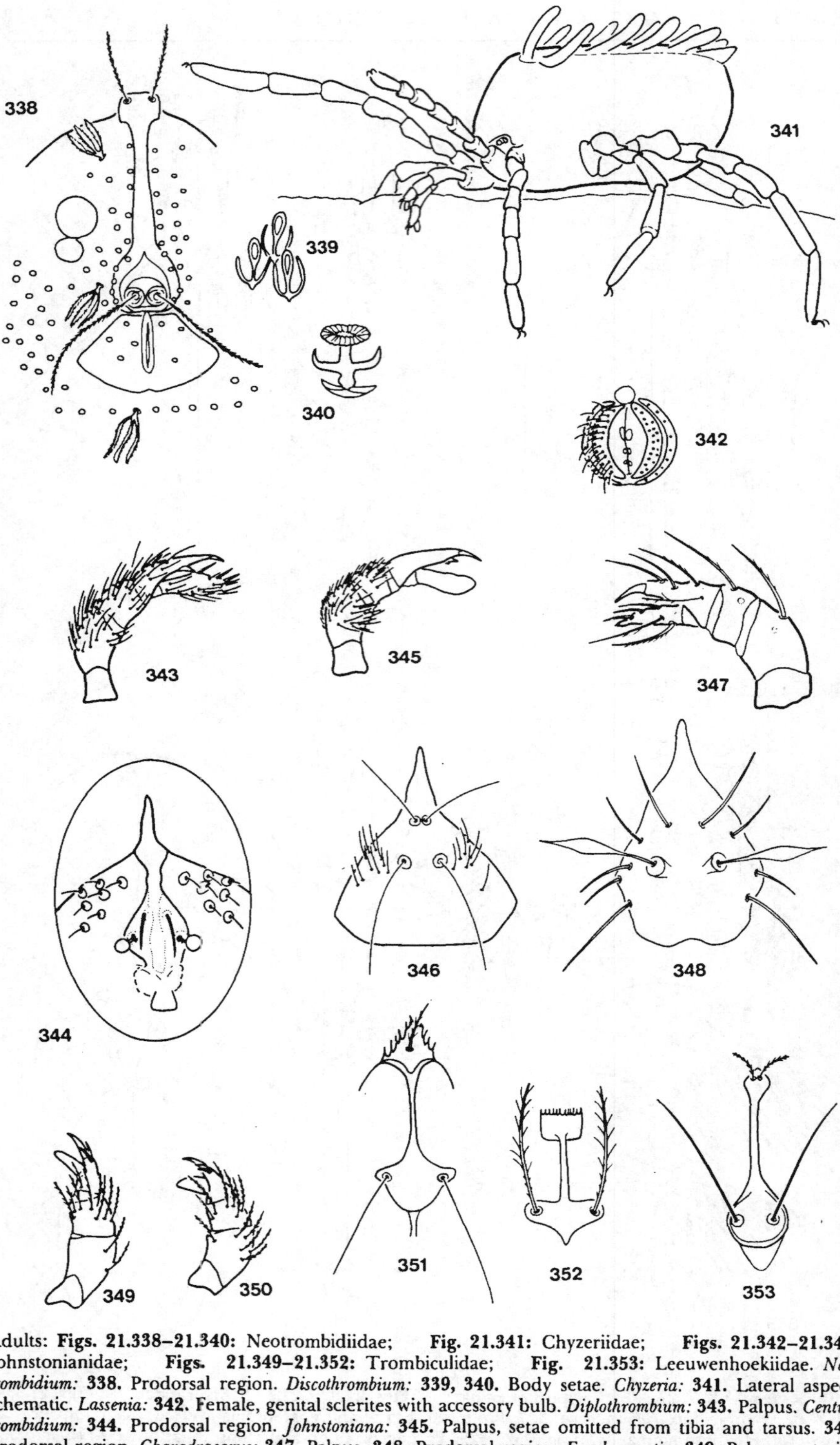

Adults: **Figs. 21.338–21.340:** Neotrombidiidae; **Fig. 21.341:** Chyzeriidae; **Figs. 21.342–21.348.** Johnstonianidae; **Figs. 21.349–21.352:** Trombiculidae; **Fig. 21.353:** Leeuwenhoekiidae. *Neotrombidium:* **338.** Prodorsal region. *Discothrombium:* **339, 340.** Body setae. *Chyzeria:* **341.** Lateral aspect, schematic. *Lassenia:* **342.** Female, genital sclerites with accessory bulb. *Diplothrombium:* **343.** Palpus. *Centrotrombidium:* **344.** Prodorsal region. *Johnstoniana:* **345.** Palpus, setae omitted from tibia and tarsus. **346.** Prodorsal region. *Charadracarus:* **347.** Palpus. **348.** Prodorsal region. *Euschongastia:* **349.** Palpus, anterior aspect. **350.** Palpus, posterior aspect. *Neotrombicula:* **351.** Prodorsal sclerite. *Walchia:* **352.** Prodorsal sclerite. *Hannemania:* **353.** Prodorsal sclerite.

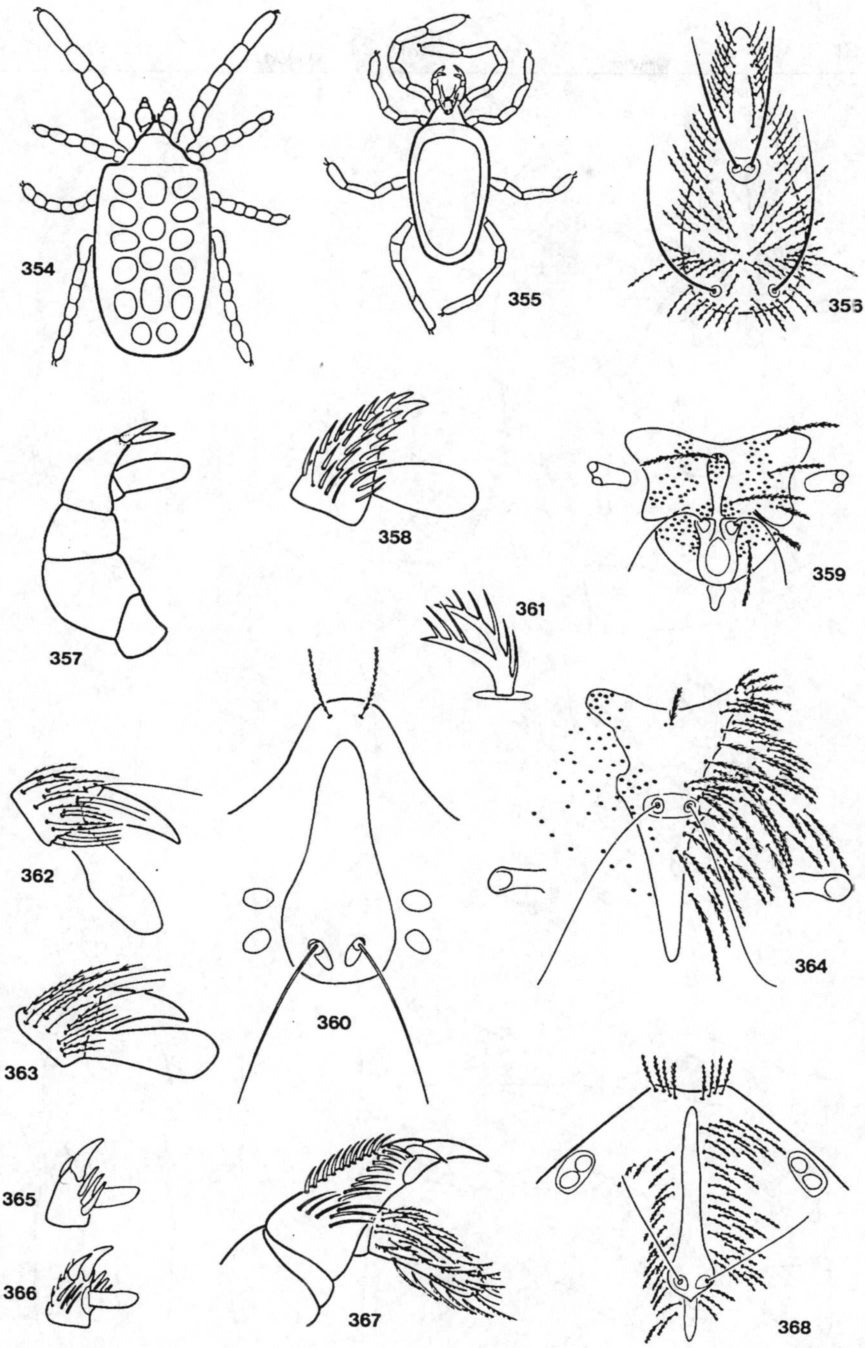

Adults: **Figs. 21.354–21.357:** Trombellidae; **Fig. 21.358:** Podothrombiidae; **Figs. 21.359–21.368:** Trombidiidae. Trombellinae—*Trombella:* **354.** Dorsal aspect, schematic. Moyanellinae—*Moyanella:* **355.** Dorsal aspect, schematic. **356.** Prodorsal region. **357.** Palpus, setae omitted. *Podothrombium:* **358.** Palp tibia and tarsus, setae omitted from tarsus. Allothrombiinae—*Allothrombium:* **359.** Prodorsal region. Neothrombiinae—*Neothrombium:* **360.** Prodorsal region, schematic; **361.** Biramous body setae. *Parathrombidium:* **362.** Palp tibia and tarsus, anterior aspect, setae omitted from tarsus. **363.** Palp tibia and tarsus, posterior aspect, setae omitted from tarsus. **364.** Prodorsal region. Microtrombidiinae—*Eutrombidium:* **365.** Palp tibia and tarsus, posterior aspect, setae omitted from tarsus. **366.** Palp tibia and tarsus, anterior aspect, setae omitted from tarsus. *Atractothrombium:* **367.** Palp tibia and tarsus. *Pedothrombidium:* **368.** Prodorsal region.

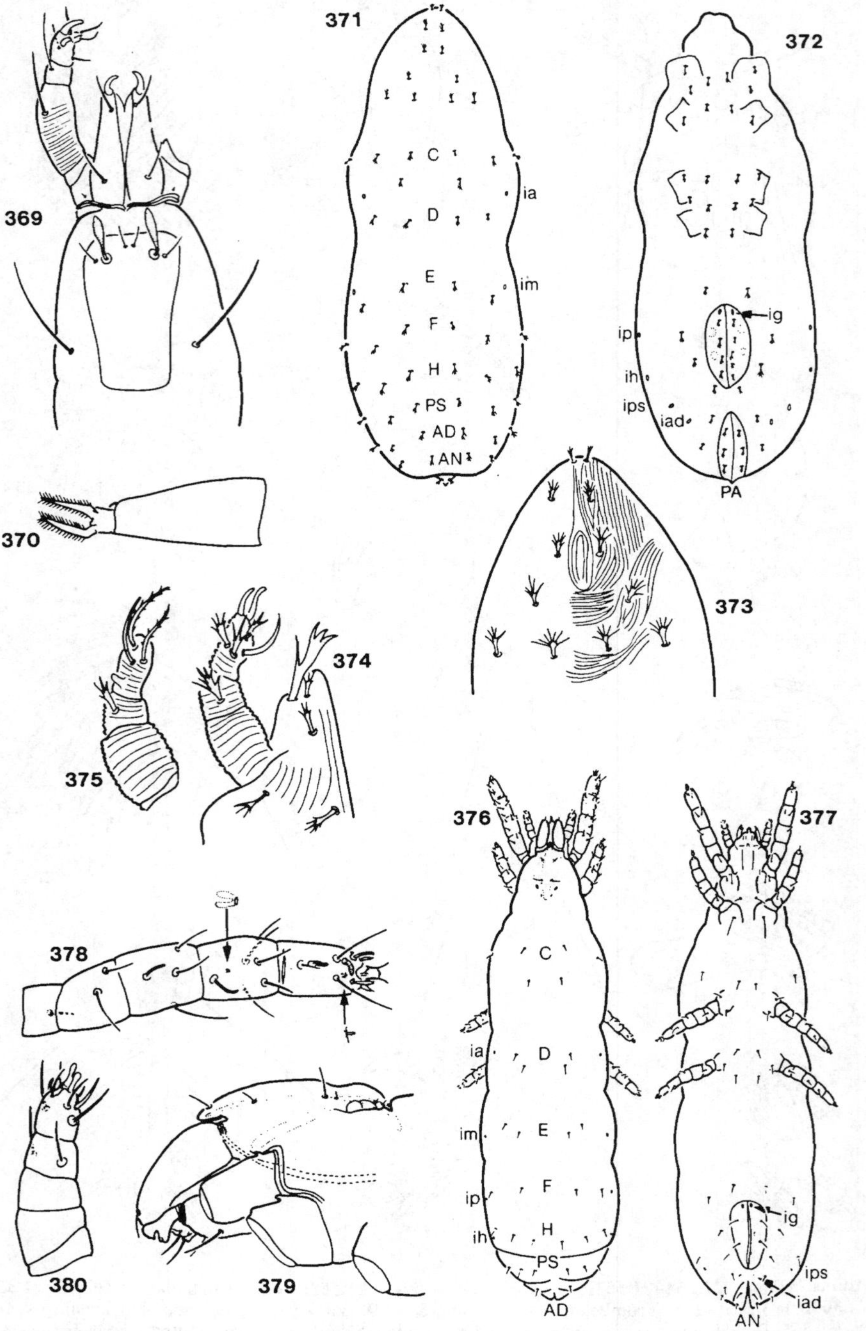

Figs. 21.369, 21.370: Stigmocheylidae; **Figs. 21.371–21.375:** Micropsammidae; **Figs. 21.376–21.380:** Proteonematalycidae. *Stigmocheylus:* **369.** Gnathosoma and prodorsal region. **370.** Tarsus IV, dorsal, setae omitted. Micropsammidae: **371:** Dorsal aspect. **372.** Ventral aspect. **373.** Prodorsal region. **374.** Subcapitulum, left side, ventral aspect. **375.** Palpus, dorsal. *Proteonematalycus:* **376.** Dorsal aspect. **377.** Ventral aspect. **378.** Leg I, dorsal aspect. **379.** Palpus, dorsal aspect. **380.** Gnathosoma and prodorsal region, lateral aspect, palps and legs omitted.

References

BIOLOGY–ECOLOGY

Abbott, D. T. and D. A. Crossley, Jr. 1982. Woody litter decomposition following clear-cutting. *Ecology* 63(1):35–42. **Adams, E. C. G. 1971.** Ecological studies of microarthropods in a New Zealand pasture with special reference to the collembola. *Pedobiologia* 11:321–337. **Aitchison, C. W. 1979.** Winter-active subnivean invertebrates in Southern Canada. III. Acari. *Pedobiologia* 19:153–160. **Anderson, J. M. 1975.** Succession, diversity and trophic relationships of some soil animals in decomposing leaf litter. *J. Anim. Ecol.* 44(2):475–495. **André, H. 1979.** Notes on the ecology of corticolous epiphyte dwellers. 1. The mite fauna of fruticose lichens. Pp. 551–557 in J. G. Rodriguez (ed.). *Recent Advances in Acarology,* Vol. 1. Academic Press, New York. **Aoki, J., K. Ishikawa and M. Shiba. 1977.** Soil mites of the Shrine Forest of Meiji-jingu in the central part of Tokyo. Pp. 81–107 in Y. Kitazawa (ed.). *Biomass of Animals of Terrestrial Ecosystems in Japan.* Rikkyo University, Tokyo. **Athias, F. 1975.** Données complémentaires sur l'abondance et la distribution verticale des Microarthropodes de la savane de Lamto (Côte d'Ivoire). *Bull. Mus. Natl. Hist. Nat. (Paris) Ser. 3* 308. *Ecol. Gen.* 24:1–28. **1976.** Recherche sur les microarthropodes du sol de la savane de Lamto (Côte d'Ivoire). *Ann. Univ. Abidjan Ser. E Ecol.* 9:191–271. **Bååth, E., B. Berg, U. Lohm, B. Lundgren, H. Lundkvist, T. Rosswall, B. Söderström and A. Wiren. 1980.** Effects of experimental acidification and liming on soil organisms and decomposition in a Scots pine forest. *Pedobiologia* 20:85–100. **Beer, R. E. and D. T. Dailey. 1956.** Biological and systematic studies of two species of cheyletid mites, with a description of a new species (Acarina, Cheyletidae). *Univ. Kans. Sci. Bull.* 37(1), no. 5:393–437. **Behan, V. M. and S. B. Hill. 1980.** Distribution and diversity of North American arctic soil Acari. Pp. 717–740 in D. Dindal (ed.). *Soil Biology as Related to Land Use Practices.* Proceedings of the 7th International Soil Zoology Colloquium, ISSS. **Behan, V. M., S. B. Hill and D. K. McE. Kevan. 1978.** Effects of nitrogen fertilizers, as urea, on Acarina and other arthropods in Quebec black spruce humus. *Pedobiologia* 18:249–163. **Belfield, W. 1956.** The arthropoda of the soil in a West African pasture. *J. Anim. Ecol.* 25:275–287. **Bengtson, S.-A., A. Fjellberg and T. Solhoy. 1974.** Abundance of tundra arthropods in Spitsbergen. *Entomol. Scand.* 5:137–142. **Block, W. 1965.** Distribution of soil mites (Acarina) on the Moor House National Nature Reserve, Westmorland, with notes on their numerical abundance. *Pedobiologia* 5:244–251. **1966a.** Seasonal fluctuations and distributions of mite populations in moorland soils, with a note on biomass. *J. Anim. Ecol.* 35:487–503. **1966b.** The distribution of soil acarina on eroding blanket bog. *Pedobiologia* 6:27–34. **1970.** Microarthropods in some Uganda soils. Pp. 195–202 in J. Phillipson (ed.). *Methods of Study in Soil Ecology.* UNESCO, Paris. **1980.** Aspects of the ecology of Antarctic soil fauna. Pp. 741–758 in D. Dindal (ed.). *Soil Biology as Related to Land Use Practices.* Proceedings of the 7th International Soil Zoology Colloquium, ISSS. **1982.** Supercooling points of insects and mites on the Antarctic Peninsula. *Ecol. Entomol.* 7:1–8. **Block, W. and W. B. Banage. 1968.** Population density and biomass of earthworms in some Uganda soils. *Rev. Ecol. Biol. Sol* 5:515–521. **Block, W. and S. R. Young. 1978.** Metabolic adaptations of Antarctic terrestrial micro-arthropods. *Comp. Biochem. Physiol. A Comp. Physiol.* 61:363–368. **Bourlière, F. and M. Hadley. 1980.** Combination of qualitative and quantitative approaches. Pp. 1–6 in D. E. Reichle (ed.). *Analysis of Temperate Forest Ecosystems. Ecological Studies* Vol. 1. Springer–Verlag, New York. **Brickhill, C. D. 1958.** Biological studies of two species of tydeid mites from California. *Hilgardia* 27(20):601–620. **Broadbent, A. B. and A. D. Tomlin. 1982.** Comparison of two methods for assessing the effects of Carbofuran on soil animal decomposers in cornfields. *Environ. Entomol.* 11(5):1036–1042. **Burges, A. and F. Raw (eds.). 1967.** *Soil Biology.* Academic Press, London. 532 pp. **Butcher, J. W., R. Snider and R. J. Snider. 1971.** Bioecology of edaphic Collembola and Acarina. *Annu. Rev. Entomol.* 16:249–288. **Chada, H. L. 1956.** Biology of the winter grain mite and its control in small grains. *J. Econ. Entomol.* 49(4):515–520. **Chaloner, W. G. 1970.** The rise of the first land plants. *Biol. Rev. Camb. Philos. Soc.* 45:353–377. **Chiba, S., T. Abe, J. Aoki, G. Imadate, K. Ishikawa, M. Kondoh, M. Shiba and H. Watanabe. 1975.** Studies on the productivity of soil animals in Pasoh Forest Reserve, West Malaysia. I. Seasonal change in the density of soil mesofauna: Acari, Collembola and others. *Sci. Rep. Hirosaki Univ.* 22(2):87–124. **Chiba, S., T. Abe, M. Kondoh and M. Shiba. 1976.** Studies on the productivity of soil animals in Pasoh Forest Reserve, West Malaysia. II. Seasonal change in the density of Nematoda and Enchytraeidae. *Sci. Rep. Hirosaki Univ.* 23:74–78. **Collins, N. J., J. H. Baker and P. J. Tilbrook. 1975.** Signy Island, Maritime Antarctic. In T. Rosswall and O. W. Heal (eds.). *Structure and Function of Tundra Ecosystems. Ecol. Bull.* (Stockholm) 20:345–374. **Coulson, J. C. and J. B. Whittaker. 1978.** The ecology of moorland animals. In O. W. Heal and D. F. Perkins (eds.). *Production Ecology of British Moors and Montane Grasslands. Ecol. Stud.* 27:52–93. **Covarrubias, R. R. F. and F. di Castri. 1964.** Observaciónes ecologico-cuantitativas sobre la fauna edafica de zonas semiaridae del norte de Chile. *Monogr. Ecol. Biogeogr. Chile* 2:1–109. **Critchley, B. R., A. G. Cook, U. Critchley, T. J. Perfect, A. Russell-Smith and R. Yeadon. 1979.** Effects of bush clearing and soil cultivation on the invertebrate fauna of a forest soil in the humid tropics. *Pedobiologia* 19:425–438. **Crossley, D. A., Jr. 1970.** Roles of microflora and fauna in soil systems. *Pesticides in the Soil* 30–35. **1977.** The role of arthropods in forest ecosystems. Pp. 49–56 in W. J. Mattson (ed.). *Proceedings in the Life Sciences.* Springer-Verlag, New York. **Crossley, D. A., Jr. and K. K. Bohnsack. 1960.** Long-term ecological study in the Oak Ridge Area. III. The oribatid fauna in pine litter. *Ecology* 41(4):628–638. **Crossley, D. A., Jr. and V. A. Merchant. 1971.** Feeding by caeculid mites on fungus demonstrated with radioactive traceers. *Ann. Entomol. Soc. Am.* 64(4):760. **Crossley, D. A., Jr., C. W. Proctor, Jr. and C. Gist. 1975.** Summer biomass of soil microarthropods of the Pawnee National Grassland, Colorado. *Am. Midl. Nat.* 93(2):491–495. **Davis, B. N. K. 1963.** A study of micro-arthropod communities in mineral soils near Corby, Northants. *J. Anim. Ecol.* 32:49–71. **Day, F. P., Jr. 1982.** Litter decomposition rates in the seasonally flooded Great Dismal Swamp. *Ecology* 63(3):670–678. **Den Heyer, J. and P. A. J. Ryke. 1966.** A mesofaunal investigation of the soil in a thorn-tree (*Acacia karroo*) biotope. *Rev. Biol.* 5:309–364. **Dhillon, B. S. and N. H. E. Gibson. 1962.** A study of the acarina and Collembola of agricultural soils. 1. Numbers and distribution in undisturbed grassland. *Pedobiologia* 1:189–209. **di Castri, F. 1963.** État de nos connaissances sur les biocenoses édaphiques du Chili. Pp. 375–385 in Doeksen and van der Drift (eds.). *Soil Organisms.* Amsterdam. **1973.** Soil animals in latitudinal and topographical gradients of Mediterranean ecosystems. Pp. 171–190 in F. di Castri and H. A. Mooney (eds.). *Mediterranean Type Ecosystems: Origin and Structure.* Springer-Verlag, Berlin. **Dickinson, C. H. and G. J. F. Pugh (eds.). 1974.** *Biology of Plant Litter Decomposition,* Vol. 2. Academic Press, New York. 775 pp. **Dindal, D. L. 1979.** Soil arthropod microcommunities of the Pine Barrens. Pp. 527–539 in R. T. T. Forman (ed.). *Pine Barrens: Ecosystem and Landscape.* Academic Press, New York. **Douce, G. K. 1976.** Biomass of soil mites (Acari) in arctic coastal tundra. *Oikos* 27:324–330. **Douce, G. K. and D. A. Crossley, Jr. 1977.** Acarina abundance and community structure in an arctic coastal tundra. *Pedobiologia* 17:32–42. **Edney, E. B., J. F. McBrayer, P. J. Franco and A. W. Phillips. 1976.** *Abundance and Distribution of Soil Microarthropods in Rock Valley, Nevada.* US/IBP Desert Biome Res. Memo 75–29. Utah State University Logan, Utah, pp. 39–46. **Edwards, C. A. 1965.** Effects of pesticide residues on soil invertebrates and plants. In *Ecology and the Industrial Society.* 5th Symposium of the British Ecological Society, pp. 239–261. **1980.** Early land floras. Pp. 55–86 in A. L. Panchen, (ed.). *The Terrestrial Environment and the Origin of Land Vertebrates.* Academic Press, New York. **Edwards, C. A. and K. E. Fletcher. 1971.** A comparison of extraction methods for terrestrial arthropods. Pp. 150–185. in J. Phillipson, (ed.). *Methods of Study in Quantiative Soil Ecology.* IBP Handb. 18. Blackwell Scientific Publications, Oxford. **Edwards, C. A., E. B. Dennis, and E. W. Empson. 1967.** Pesticides and the soil fauna: effects of aldrin and DDT in an arable field. *Ann. Appl. Biol.* 60:11–22. **Edwards, C. A., D. E. Reichle, and D. A. Crossley, Jr. 1969.** Experimental manipulation of soil invertebrate populations for trophic studies. *Ecology* 50(3):495–498. **1970.** The role of soil invertebrates in turnover of organic matter and nutrients. Pp. 147–172. in D. Reichle (ed.). *Ecological Studies. Analysis and Synthesis, Vol. 1.* Springer, Berlin. **Ehrnsberger, R. 1974.** Prälarval- und Larvalentwicklung bei Rhagidiiden (Acarina: Prostigmata). *Osnabrucker Naturwiss. Mitt.* 3:85–117. **1977.** Fortpflanzungsverhalten der Rhagidiidae (Acarina Trombidiformes). *Acarologia (Paris)* 19(1):67–73. **Elbadry, E. A., A. A. Yousef, and S. H. Metwally. 1976.** Life history studies on the soil mite *Scutacarus agypticus* with description of the immature stages (Acarina : Scutacaridae). *Pedobiologia* 16:167–170. **Evans, G. O., J. G. Sheals, and D. Mac Farlane. 1961.** *The Terrestrial Acari of the British Isles.* British Museum of Natural History, London. 219

pp. **Franco, P. J., E. B. Edney, and J. F. McBrayer. 1979.** The distribution and abundance of soil arthropods in the northern Mojave desert. *J. Arid Environ.* 2:137–149. **Garay, I., J. C. Da Fonseca, and P. Blandin. 1980.** The effects of trampling on the fauna of a forest floor. I. Microarthropods. Pp. 200–212. in D. Dindal, (ed.). *Soil Biology as Related to Land Use Practices.* Proceedings of the 7th International Soil Zoology Colloquium, ISSS. **Gerson, U. and M. R. D. Seaward. 1977.** Lichen–invertebrate associations. Pp. 69–119 in M. R. D. Seaward, (ed.). *Lichen Ecology.* Academic Press, New York. **Gist, C. S. and D. A. Crossley, Jr. 1975.** The litter arthropod community in a southern Appalachian hardwood forest: numbers, biomass and mineral element content. *Am. Midl. Nat.* 93:107–122. **Goddard, D. G. 1979.** The Signy Island terrestrial reference sites. XI. Population studies on the Acari. *Br. Antarct. Surv. Bull.* 48:71–92. **Goffinet, G. 1975.** Écologie édaphique des milieux naturals du Haut-Shaba (Zaire). I. Caractéristiques, écotopiques et synécologie comparée des zoocenoses intercaliques. *Rev. Ecol. Biol. Sol* 12:691–722. **1976.** Écologie édaphique des écosystèmes naturels du Haut-Shaba (Zaire). II. Phénologie et fluctuations démographiques au niveau des groupes zoologiques dominants et de quelques populations d'Arthropodes. *Bull. Ecol.* 7:335–352. **Grandjean, F. 1947.** Au sujet des Érythroides. *Bull. Mus. Nctl. Hist. Nat. (Paris)* 2(19):327–334. **Greenslade, P. J. M. and P. Greenslade. 1968.** Soil and litter fauna densities in the Solomon Islands. *Pedobiologia* 7:362–370. **Haarløv, N. 1960.** Microarthropods from Danish soils: ecology, phenology. *Oikos Suppl.* 3:1–176. **Håvar, S. and T. Amundsen. 1981a.** Effects of liming and artificial acid rain on the mite (Acari) fauna in coniferous forest. *Oikos* 37(1):7–20. **Hågvar, S. and B. R. Kjøndal. 1981a.** Effects of artificial acid rain on the microarthropod fauna in decomposing birch leaves. *Pedobiologia* 22:409–422. **1981b.** Succession, diversity and feeding habits of microarthropods in decomposing birch leaves. *Pedobiologia* 22:385–408. **Hanlon, R. D. G. and J. M. Anderson. 1980.** Influence of macroarthropod feeding activities on microflora in decomposing oak leaves. *Soil Biol. Biochem.* 12:255–261. **Harding, D. J. L. and R. A. Stuttard. 1974.** Microarthropods. Pp. 489–532. in C. H. Dickinson, and G. J. F. Pugh, (eds.). *Biology of Plant Litter Decomposition,* Vol. 2. Academic Press, New York. **Helle, W., H. R. Bolland, S. H. M. Jeurissen, and G. A. van Seventer. 1984.** Chromosome data on the Actinedida, Tarsonemida and Oribatida. Pp. 449–454 in D. A. Griffiths, and C. E. Bowman, (eds.). *Acarology VI,* Vol. 1. Ellis Horwood, Chichester, West Sussex, England. **Heugens, A. and E. van Daele. 1981.** The influence of salt concentration on the mite population in pine litter. *Pedobiologia* 22:39–51. **Hoy, J. B. 1980a.** Ecological impact of Lindane on a pine plantation soil microarthropod community. *Environ. Entomol.* 9(2):164–174. **1980b.** Effects of Lindane, Carbaryl, and Chlorpyrifos on non-target soil arthropod communities. Pp. 71–81 in D. Dindal, (ed.). *Soil Biology as Related to Land Use Practices.* Proceedings of the 7th International Soil Zoology Colloquium, ISSS **Hoy, J. B. and P. J. Shea. 1981.** Effects of lindane, chlorpyrifos, and carbaryl on a California pine forest soil arthropod community. *Environ. Entomol.* 10(5):732–740. **Huhta, V. 1976.** Effects of clear-cutting on numbers, biomass and community respiration of soil invertebrates. *Ann. Zool. Fenn.* 13:63–80. **Huhta, V. and A. Koskenniemi. 1975.** Numbers, biomass and community respiration in soil invertebrates in spruce forests at two latitudes in Finland. *Ann. Zool. Fenn.* 12:164–182. **Huhta, V., M. Nurminen, and A. Valpas. 1969.** Further notes on the effect of silvicultural practices upon the fauna of coniferous forest soil. *Ann. Zool. Fenn.* 6:327–334. **Huhta, V., E. Ikonen and P. Vilkamaa. 1979.** Succession of invertebrate populations in artificial soil made of sewage sludge and crushed bark. *Ann. Zool. Fenn.* 16:223–270. **Ishikawa, K., M. Shiba, and J. Aoki. 1977.** Soil mites of a secondary forest of Musashi-murayama in the western suburb of Tokyo. Pp. 109–308 in Y. Kitazawa, (ed.). *Biomass of Animals of Terrestrial Ecosystems in Japan.* Rikkyo University, Tokyo. **Kevan P. G., W. G. Chaloner, and D. B. O. Savile. 1975.** Interrelationships of early terrestrial arthropods and plants. *Paleontologie (Prague)* 18(2):391–417. **Koponen, S. and M.-L. Ojala. 1975.** Quantitative study of invertebrate groups in the soil and ground layer of the IBP sites at Kevo, northern Finland. *Ref. Kevo Subarct. Res. Stn.* 12:45–52. **Košir, M. 1975.** Ernährung and Entwicklung von *Pygmephorus mesembrinae* und *P. quadratus* (Pygmephoridae, Tarsonemini, Acari) und Bemerkungen über drei weitere Arten. *Pedobiologia* 15:313–329. **Krantz, G. W. and E. E. Lindquist. 1979.** Evolution of phytophagous mites (Acari). *Annu. Rev. Entomol.* 24:121–158. **Lasebikan, B. A. 1974.** Preliminary communication on microarthropods from a tropical rain forest in Nigeria. *Pedobiologia* 14:402–411. **1981.** Comparative studies of the arthropod fauna of the soil and a decaying log of an oil palm tree in a tropical forest. *Pedobiologia* 21:110–116. **Leetham, J. W. and D. G.**

Milchunas. 1985. The composition and distribution of soil microarthropods in the shortgrass steppe in relation to soil water, root biomass, and grazing by cattle. *Pedobiologia* 28:311–325. **Lindquist, E. E., B. D. Ainscough, F. V. Clulow, R. C. Funk, V. G. Marshall, H. H. J. Nesbitt, B. M. O'Connor, I. M. Smith and P. R. Wilkinson. 1979.** 12. Acari in H. V. Danks, (ed.). *Canada and Its Insect Fauna. Mem. Entomol. Soc. Can.* 108:252–291. **Lohm, U. and T. Persson, (eds.). 1977.** *Soil Organisms as Components of Ecosystems.* Proceedngs of the 6th VI International Soil Zoology Colloquium, ISSS. Ecol. Bull. (Stockholm) 25:1–614. **Loots, G. C. and P. A. J. Ryke. 1966.** A comparative, quantitative study of the micro-arthropods in different types of pasture soil. *Zool. Afr.* 2(2):167–192. **Lussenhop, J. 1972.** Distribution and phenology of a prairie soil arthropod population. Pp. 60–72 in J. H. Zimmerman, (ed.). *Proceedings of the 2nd Midwest Prairie Conference* 1970, University of Wisconsin, Madison, Wis. **Luxton, M. 1967a.** The ecology of saltmarsch Acarina. *J. Anim. Ecol.* 36:257–277. **1967b.** The zonation of saltmarsh Acarina. *Pedobiologia* 7:55–66. **1972.** Studies on the oribatid mites of a Danish beech wood soil. *Pedobiologia* 12:434–463. **1981.** Studies on the prostigmatic mites of a Danish beech wood soil. *Pedobiologia* 22:277–303. **1982.** The biology of mites from beech woodland soil. *Pedobiologia* 23:1–8. **Luxton, M. and H. Petersen. 1982.** Survey of the main animal taxa of the detritus food web. *Oikos* 39(3):293–294. **Macfadyen, A. 1952.** The small arthropods of a *Molinia* fen at Cothill. *J. Anim. Ecol.* 21:87–117. **Maclean, S. F., Jr. 1975.** Ecology of tundra invertebrates at Prudhoe Bay, Alaska. Pp. 114–123 in J. Brown, (ed.). *Ecological Investigations of the Tundra Biome in the Prudhoe Bay Region, Alaska.* Biol. Papers, Spec. Rep. 2 University of Alaska. **Maclean, S. F., Jr., G. K. Douce, E. A. Morgan, and M. A. Skeel. 1977.** Community organization in the soil invertebrates of Alaskan arctic tundra. in U. Lohm, and T. Persson, (eds.). *Soil Organisms as Components of Ecosystems.* Proceedings of the 6th International Soil Zoology Colloquium, ISSS. Ecol. Bull (Stockholm) 25:90–101. **Maldague, M. E. 1961.** Relations entre le couvert végétal et la microfaune. *Publ. Inst. Natl. Etude Agron. Congo Ser. Sci.* 90:122 pp. **1970.** Rôle des animeaux édaphiques dans la fertilité des sols forestiers. *Publ. Inst. Natl. Etude Agron. Congo Ser. Sci.* 112:245 pp. **Marshall, V. G. 1974.** Seasonal and vertical distribution of soil fauna in a thinned and urea-fertilized Douglas fir forest. *Can. J. Soil Sci.* 54:491–500. **McBrayer, J. F., D. E. Reichle, and M. Witkamp. 1974.** *Energy Flow and Nutrient Cycling in a Cryptozoan Food-Web.* EDFB-IBP-73-8. Oak Ridge National Laboratory, Oak Ridge, Tenn. **McBrayer, J. F., J. M. Ferris, L. J. Metz, C. S. Gist, B. W. Cornaby, Y. Kitazawa, T. Kitazawa, J. G. Wernz, G. W. Krantz, and H. Jensen. 1977.** Decomposer invertebrate populations in U.S. forest biomes. *Pedobiologia* 17:89–96. **McCoy, C. W., A. G. Selhime, and R. F. Kanavel. 1969. The feeding behavior and biology of** *Paraponematus acaciae* (Acarina: Tydeidae). *Fla. Entomol.* 52(1):13–19. **McMillan, J. H. 1969.** The ecology of the acarine and collembolan fauna of two New Zealand pastures. *Pedobiologia* 9:372–404. **Metz, L. J. 1971.** Vertical movement of Acarina under moisture gradients. *Pedobiologia* 11:262–268. **Moore, J. C., D. E. Walter, and H. W. Hunt. 1988.** Arthropod regulation of micro- and mesobiota in below-ground detrital food webs. *Annu. Rev. Entomol.* 33:419–439. **Murphy, P. W. (ed.). 1962.** *Progress in Soil Zoology.* Proceedings of the First International Soil Zoology Colloquium, ISSS. Butterworth, London. 398 pp. **Nef, L. 1971.** Influence de l'humidité sur le géotactisme des Oribates (Acarina) dans l'extracteur de Berlese-Tullgren. *Pedobiologia* 11:433–445. **Newell, I. M. 1963.** Feeding habits in the genus *Balaustium* (Acarina, Erythraeidae), with special reference to attacks on man. *J. Parasitol.* 49(3):498–502. **1979.** *Acarus tinctorius* Linnaeus 1767 (Trombidiidae). Pp. 425–428 in J. G. Rodriguez, (ed.). *Recent Advances in Acarology,* Vol. 2. Academic Press, New York. **Noy-Meir, I. 1973.** Desert ecosystems: environment and producers. *Annu. Rev. Ecol. Syst.* 4:25–51. **1974.** Desert ecosystems: higher trophic levels. *Annu. Rev. Ecol. Syst.* 5:195–214. **Oliver, J. H., Jr. 1971.** Parthenogenesis in mites and ticks (Arachnida: Acari). *Am. Zool.* 11:283–299. **1977.** Cytogenetics of mites and ticks. *Annu. Rev. Entomol.* 22:407–429. **1983.** Chromosomes, genetic variance and reproductive strategies among mites and ticks. *Bull. Entomol. Soc. Am.* 29(2):8–17. **Oliver, P. G. and P. A. J. Ryke. 1966.** Seasonal fluctuations of the mesofauna in soil under kikuyu grass. *Mem. Inst. Invest. Cient. Mocambique Ser. A. (Cienc. Biol.)* 7:235–280. **1969.** The influence of citricultural practices on the composition of soil Acari and Collembola populations. *Pedobiologia* 9:277–281. **Persson, T. and U. Lohm. 1977.** Energetical significance of the annelids and arthropods in a Swedish grassland ecosystem. *Ecol. Bull. (Stockholm)* 23:1–211. **Petersen, H. 1982a.** Basis for evaluation of the density and biomass data. *Oikos* 39(3):295–395. **1982b.** Structure and size of soil animal populations. *Oikos* 39(3):306–329. **1982c.** The total soil fauna biomass and its composition. *Oikos* 39(3):330–339. **Petersen, H. and M. Luxton. 1982.** Introduction. *Oikos* 39(3):291–292. **Price, D. W. 1973.** Abundance and vertical distribution of microarthropods in the surface layers of a California pine forest soil. *Hilgardia* 42(4):121–148. **1975.** Vertical distribution of small arthropods in a California pine forest soil. *Ann. Entomol. Soc. Am.* 68(1):174–180. **Price, D. W. and G. S. Benham, Jr. 1976.** Vertical distribution of pomerantziid mites (Acarina: Pomerantziidae). *Proc. Entomol. Soc. Wash.* 78(3):309–313. **Rolfe, W. D. I. 1980.** Early invertebrate terrestrial faunas. Pp. 117–157 in A. L. Panchen (ed.). *The Terrestrial Environment and the Origin of Land Vertebrates.* Academic Press, New York. **Ryke, P. A. J. and G. C. Loots. 1967.** The composition of the micro-arthropod fauna in South African soils. Pp. 538–545 in O. Graff, and J. E. Satchell, (eds.). *Progress in Soil Biology.* Vieweg, Braunschweig, West Germany. **Santos, P. F. and W. G. Whitford. 1981.** The effects of microarthropods on litter decomposition in a Chihuahuan desert ecosystem. *Ecology* 62(3):654–663. **Santos, P. F., E. DePree, and W. G. Whitford. 1978.** Spatial distribution of litter and microarthropods in a Chihuahuan desert ecosystem. *J. Arid Environ.* 1:41–48. **Santos, P. F., J. Phillips, and W. G. Whitford. 1981.** The role of mites and nematodes in early stages of buried litter decomposition in a desert. *Ecology* 62(3):664–669. **Schatz, H. 1981.** Abundanz, Biomasse und Respirationstrate der Arthropoden-Mesofauna im Hochgebirge (Obergurgl, Tiroler Zentralalpen). *Pedobiologia* 22:52–70. **Schopf, J. W. 1970.** Precambrian micro-organisms and evolutionary events prior to the origin of vascular plants. *Biol. Rev.* 45:319–353. **Schuster, R. 1956.** Der Anteil der Oribatiden an den Zersetzungsvorgängen im Boden. *Z. Morphol. Oekol. Tiere* 45:1–33. **1979.** Soil mites in the marine environment. Pp. 593–602 in J. G. Rodriguez, (ed.). *Recent Advances in Acarology,* Vol. 1. Academic Press, New York. **Scott, A. C. 1980.** The ecology of some upper Palaeozoic floras. Pp. 87–115 in A. L. Panchen, (ed.). *The Terrestrial Environmental and the Origin of Land Vertebrates.* Academic Press, New York. **Seastedt, T. R. and D. A. Crossley, Jr. 1978.** Further investigations of microarthropod populations using the Merchant–Crossley high-gradient extractor. *J. Ga. Entomol. Soc.* 13(4):339–344. **1981.** Microarthropod response following cable logging and clear-cutting in the southern Applachians. *Ecology* 62(1):126–135. **Seastedt, T. R., A. Kothari, and D. A. Crossley, Jr. 1980.** A simplified gelatine embedding technique for sectioning litter and soil samples. *Pedobiologia* 20:55–59. **Singh, J. and K. S. Pillai. 1975.** A study of soil micro-arthropod communities in some fields. *Rev. Ecol. Biol. Sol.* 12(3):579–590. **Snetsinger, R. 1956.** Biology of *Bdella depressa,* a predaceous mite. *J. Econ. Entomol.* 49(6):745–746. **Stegeman, L. C. 1964.** The effects of the carbamate insecticide carbaryl upon forest soil mites and Collembola. *J. Econ. Entomol.* 57(6):803–808. **Tevis, L., Jr. and I. M. Newell. 1962.** Studies on the biology and seasonal cycle of the giant red velvet mite, *Dinothrombium pandorae* (Acari, Trombidiidae). *Ecology* 43(3):497–505. **Turner, F. B. and R. M. Chew. 1981.** Production by desert animals. Pp. 199–260. in P. W. Goodall, and R. A. Perry, (eds.). *Arid Land Ecosystems,* Vol. 2. IBP 17. Cambridge University Press, Cambridge. **van den Berg, J. and P. A. J. Ryke. 1967–1968.** A systematic–ecological investigation of the acarofauna of the forest floor in Magoebaskloof (South Africa) with special reference to the Mesostigmata. *Rev. Biol.* 6(1–2):157–234. **Wahab, A. E. A., A. A. Yousef, and H. M. Hemeda. 1974.** Biological studies on the tydeid mite, *Tydeus californicus* (Banks) *Bull. Soc. Entomol. Egypte* 58:349–353. **Wallwork, J. A. 1970.** *Ecology of Soil Animals.* McGraw-Hill, London. 283 pp. **1972a.** Distribution patterns and population dynamics of the micro-arthropods of a desert soil in southern California. *J. Anim. Ecol.* 41:291–310. **1972b.** Mites and other microarthropods from the Joshua Tree National Monument, California. *J. Zool. (Lond.)* 168:91–105. **1980.** Desert soil microarthropods in an "r"-selected system. Pp. 759–769 in D. Dindal, (ed.). *Soil Biology as Related to Land Use Practices.* Proceedings of the 7th International Soil Zoology Colloquium, ISSS. **Walter, W. E. 1987a.** Belowground arthropods of semiarid grasslands. Pp. 271–290 in J. L. Capinera, (ed.). *Integrated Pest Management on Rangeland: A Shortgrass Prairie Perspective.* Westview Press, Boulder, Colo. **1987b.** Trophic behavior of "mycophagous" microarthropods. *Ecology* 68(1):226–229. **Walter, W. E., J. Kethley, and J. C. Moore. 1987.** A heptane flotation method for recovering microarthropods from semiarid soils, with comparison to the Merchant–Crossley high-gradient extraction method and estimates of microarthropod biomass. *Pedobiologia* 30:221–232. **Weis-Fogh, T. 1948.** Ecological investigations on mites and collemboles in the soil. *Nat. Jutl.* 1:139–277. **Whitford, W. G. and P. F. Santos. 1980.** Arthropod and detritus decomposition in desert ecosystems. Pp. 770–778 in D. Dindal, (ed.).

Soil Biology as Related to Land Use Practices. Proceedings of the 7th International Soil Zoology Colloquium, ISSS. **Wicklow, D. T. and W. F. Whittingham. 1974.** Soil microfungal changes among the profiles of disturbed conifer–hardwood forests. *Ecology* 55:3–16. **Wiegert, R. G. 1974.** Litterbag studies of microarthropod populations in three South Carolina old fields. *Ecology* 55:94–102. **Witkamp, M. and D. A. Crossley, Jr. 1966.** The role of arthropods and microflora in breakdown of white oak litter. *Pedobiologia* 6:293–303. **Wood, T. G. 1967a.** Acari and Collembola of moorland soils from Yorkshire, England. I. Description of the sites and their populations. *Oikos* 18:102–117. **1967b.** Acari and Collembola of moorland soils from Yorkshire, England. II. Vertical distribution in four grassland soils. *Oikos* 18:137–140. **1967c.** Acari and Collembola of moorland soils from Yorkshire, England. III. The micro-arthropod communities. *Oikos* 18:277–292. **1971.** The distribution and abundance of *Folsomides deserticola* (Collembola : Isotomidae) and other microarthropods in arid and semi-arid soils in southern Australia, with a note on nematode populations. *Pedobiologia* 11:446–468. **Zyromska-Rudzka, H. 1979.** Disappearance of Scutacaridae (Acarina) due to high concentrations of mineral nutrients in a meadow. *Pedobiologia* 19:9–17.

Morphology–Taxonomy

Alberti, G. 1979. Fine structure and probable function of genital papillae and Claparède organs of Actinotrichida. Pp. 501–508 in J. G. Rodriguez, (ed.). *Recent Advances in Acarology,* Vol. 2. Academic Press, New York. **Alberti, G. and R. Ehrnsberger. 1977.** Rasterelektronenmikroskopische Untersuchungen zum Spinnvermögen der Bdelliden und Cunaxiden (Acari : Prostigmata). *Acarologia (Paris)* 19(1):55–61. **André, H. M. 1979.** A generic revision of the family Tydeidae (Acari : Actinedida). I. Introduction, paradigms and general classification. *Bull. Ann. Soc. R. Belg. Entomol.* 108(3,4):189–208. **1980.** A generic revision of the family Tydeidae (Acari : Actinedida). IV. Generic descriptions, keys and conclusions. *Bull. Ann. Soc. R. Belge Entomol.* 116:104–168. **Athias-Henriot, C. 1961.** Nouveaux acariens phytophages d'Algérie (Actinotrichida, Tetranychoidea : Tetranychidae, Linotetranidae). *Ann. Ec. Natl. Agric. Alger* 3(3):1–10. **Atyeo, W. T. 1960.** A revision of the mite family Bdellidae in North and Central America. *Univ. Kans. Sci. Bull.* 40(8):345–499. **1961.** The taxonomic position of the genus *Neophyllobius* Berlese, 1886 (Acarina, Caligonellidae) with the description of a new genus and species. *Acarologia (Paris)* 3(2):153–158. **1963.** New and redescribed species of Raphignathidae (Acarina) and a discussion of the chaetotaxy of the Raphignathoidea. *J. Kans. Entomol. Soc.* 36(3):172–186. **Atyeo, W. T. and D. A. Crossley, Jr. 1961.** The Labidostommidae of New Zealand (Acarina). *Rec. Dom. Mus. (Wellington)* 4(4):29–48. **Baker, E. W. 1965.** A review of the genera of the family Tydeidae (Acarina). Pp. 95–1033 in J. A. Naegle, (ed.). *Advances in Acarology,* Vol. 2. Cornell University Press, Ithaca, N.Y. **Baker, E. W. and W. T. Atyeo 1964.** A review of the mites of the family Pseudocheylidae Oudemans, 1909 (Acarina, Prostigmata). *Bull. Univ. Nebr. State Mus.* 4(12):257–272. **Baker, E. W. and A. E. Pritchard. 1953.** The family categories of tetranychoid mites, with a review of the new families Linotetranidae and Tuckerellidae. *Ann. Entomol. Soc. Am.* 46(2):243–258. **Beer, R. E. and A. Nucifora. 1965.** Revisione dei generi della famiglia Tarsonemidae (Acarina). *Boll. Zool. Agrar. Bachic.* 2(7):19–43. **Bolland, H. R. 1986.** Review of the systematics of the family Camerobiidae (Acari, Raphignathoidea). I. The genera *Camerobia, Decaphyllobius, Tillandsobius,* and *Tycherobius. Tijdschr. Entomol.* 129:191–215. **Brennan, J. M. and M. L. Goff. 1977.** Keys to the genera of chiggers of the western hemisphere (Acarina : Trombiculidae). *J. Parasitol.* 63(3):554–566. **Castagnoli, M. 1974.** La spermateca in alcune specie de Brevipalpini. *Redia* 55:83–88. **Coineau, Y. 1964.** Une nouvelle espèce française de Labidostommida (Acariens, Prostigmates) *Eunicola travei* sp.n. *Vie Milieu* 15(1):153–175. **1974a.** Un type nouveau d'Acariens Prostigmates libres: les Saxidromoidea nouvelle super-famille. *C.R. Acad. Sci. (Paris) Ser. D. Sci. Nat.* 278:1059–1062. **1974b.** Éléments pour une monographie morphologique, écologique et biologique des Caeculidae (Acariens). *Mem. Mus. Natl. Hist. Nat. (Paris) Ser. A Zool.* 81:1–299. **1976.** Les pariades sexuelles des Saxidrominae Coineau 1974 (Acariens Prostigmates, Adamystidae). *Acarologia (Paris)* 18(2):234–240. **1977.** La première prélarve éllatostatique connue chez les Acariens. *Acarologia (Paris)* 19(1):46–54. **Coineau, Y. and P. Theron. 1983.** Les Micropsammidae, n. fam., d'acariens Endeostigmata des sables fins. *Acarologia (Paris)* 24(3):275–280. **Coineau, Y., A. Fize, and M. C. Delamare-Deboutteville. 1967.** Découverte en France des Acariens Nematalycidae

Strenzke, à l'occasion des travaux d'aménagement du Languedoc-Roussilon. *C.R. Acad. Sci. (Paris) Ser. D. Sci. Nat.* 265:685–688. **Cross, E. A. 1965.** The generic relationships of the family Pyemotidae (Acarina: Trombidiformes). *Univ. Kans. Sci. Bull.* 45:29–275. **Cross, E. A. and J. C. Moser. 1975.** A new dimorphic species of *Pyemotes* and a key to previously described forms (Acarina: Tarsonemoidea). *Ann. Entomol. Soc. Am.* 68(4):723–732. **Crossley, D. A., Jr. 1960.** Comparative external morphology and taxonomy of nymphs of the Trombiculidae (Acarina). *Univ. Kans. Sci. Bull.* 40(6):135–321. **Crossley, D. A., Jr. and V. Merchant. 1971.** Feeding by caeculid mites on fungus demonstrated with radioactive tracers. *Ann. Entomol. Soc. Am.* 64(4):760–762. **Den Heyer, J. 1977.** Bonzinae, a new subfamily of the Cunaxidae (Prostigmata: Acari). *Acarologia (Paris)* 29(4):601–618. **1978a.** A new cunaxid subfamily and the neotype designation of *Cunaxoides croceus* (Koch, 1838) (Prostigmata: Acari). *Acarologia (Paris)* 20(3):338–350. **1978b.** Coleoscirinae, a new cunaxid subfamily and two new South African species of *Coleoscirus* Berlese, 1916 (Prostigmata: Acari). *Acarologia (Paris)* 20(4):522–541. **1980.** A classification system for the family Cunaxidae (Actinedida: Acarida). *Publ. Univ. North Ser. A* 23:1–12. **1981.** Systematics of the family Cunaxidae Thor, 1902 (Actinedida: Acarida). *Publ. Univ. North Ser. A* 24:1–19. **Dubinin, V. B. 1962.** Acaromorpha. Pp. 447–473 in Yu. A. Orlov, (chief ed.), B. B. Rodendorf, (vol. ed.). *Fundamentals of Paleontology*, Vol. 9. Akademiya Nauk, Moscow. **Ehara, S. 1965.** Two new species of Teneriffiidae from Japan, with notes on the genera *Heteroteneriffia* and *Neoteneriffiola* (Acarina: Prostigmata). *Seto Mar. Biol. Lab.* 13(3):221–229. **1975.** Description of a new species of *Tuckerella* from Japan (Acarina: Tuckerellidae). *Int. J. Acarol.* 1(2):1–5. **Fain, A. 1979.** Idiosomal and leg chaetotaxy in Cheyletidae. *Acarologia (Paris)* 5(4):305–310. **Feider Z. 1955.** Fauna Republicii Populare Romine. Arachnida, Acarina, Trombidoidea. *Acad. Repub. Pop. Rom.* 5(1):1–186. **Feider, Z. and N. Vasiliu. 1969.** Revision critique de la famille Nicoletiellidae. *Proc. 2nd Intern. Cong. Acar.* 202–207. **Gerson, U. 1966.** A redescription of *Eupalopsis maseriensis* (Canestrini and Fanzago) (Acarina: Eupalopsellidae). *Isr. J. Zool.* 15:148–154. **1968.** Some raphignathoid mites from Israel. *J. Nat. Hist.* 2:429–437. **1972.** Mites of the genus *Ledermuelleria* (Prostigmata: Stigmaeidae) associated with mosses in Canada. *Acarologia (Paris)* 13(2):319–343. **Gilyarov, M. S. (ed.). 1978.** *A Key to the Soil-Inhabiting Mites. Trombidiformes.* Akademiya Nauk, Leningrad. 271 pp. (in Russian). **Goff, M. L., R. B. Loomis, W. C. Welbourn, and W. J. Wrenn. 1982.** A glossary of chigger terminology (Acari: Trombiculidae). *J. Med. Entomol.* 19(3):221–238. **Gonzalez, R. H. 1978a.** A new species of xenocaligonellid mite from the Galapagos Islands (Acari). *Proc. Entomol. Soc. Wash.* 80(2):191–196. **1978b.** *Annerosella knorri*, a new species of mite on water lettuce from Thailand (Acari: Homocaligidae). *Int. J. Acarol.* 4(3):221–225. **Gonzalez-Rodriguez, R. H. 1965.** A taxonomic study of the genera *Mediolata, Zetzellia*, and *Agistemus* (Acarina: Stigmaeidae). *Univ. Calif. Publ. Entomol.* 41:1–64. **Grandjean, F. 1937.** Le genre *Pachygnathus* Dugès (*Alycus* Koch) (Acariens) (Cinquième et dernière partie). *Bull. Mus. Natl. Hist. Nat. (Paris) Ser. 2.* 9:262–269. **1938.** Observations sur les Bdelles (Acariens). *Ann. Soc. Entomol. Fr.* 107:1–24. **1939a.** Les segments postlarvaires de l'hysterosoma chez les oribates (Acariens). *Bull. Soc. Zool. Fr.* 64:273–284. **1939b.** Quelques genres d'acariens appartenant au groupe des Endeostigmata. *Ann. Sci. Nat. Zool. Ser.* 11(2):1–22. **1939c.** Observations sur les Acariens (5e série). *Bull. Mus. Natl. Hist. nat. (Paris)* 2(11):394–401. **1942a.** Observations sur les Labidostommidae (2e série). *Bull. Mus. Natl. Hist. nat. (Paris)* 2(14):185–192. **1942b.** Observations sur les Acariens (7e série). *Bull. Mus. Natl. Hist. nat. (Paris)* 2(14):264–267. **1942c.** Quelques genres d'acariens appartenant au groupe des Endeostigmata (2e série). Première partie. *Ann. Sci. Nat. Zool. Ser.* 11(4):85–135. **1943a.** Le développement postlarvaire d' "*Anystis*" (Acarien). *Mem. Mus. Natl. Hist. Nat. (Paris) (N.S.)* 18(2):33–77. **1943b.** Quelques genres d'acariens appartenant au groupe des Endeostigmata (2e série). Deuxième partie. *Ann. Sci. Nat. Zool. Ser.* 11(5):1–59. **1944.** Observations sur les Acariens de la famille des Stigmaeidae. *Arch. Sci. Phys. Nat.* 5(26):103–131. **1947.** Étude sur les Smarisidae et quelques autres Érythroides (Acariens). *Arch. Zool. Exp. Gen.* 85(1):1–26. **Hunter, P. E. and E. A. Cross. 1968.** Key to North american Ereynetinae and description of a new species (Acarina: Ereynetidae). *J. Ga. Entomol. Soc.* 3(4):154–158. **Hutterer, R. and G. Rack. 1977.** Beitrag zur Morphologie und Biologie von *Chausseria venustissima* (Berlese, 1882) (Acarina, Trombidiformes, Anystidae). *Acarologia (Paris)* 18(3):421–441. **Kethley, J. B. 1977.** The status of *Hybalicus* Berlese, 1913 and *Oehserchestes* Jacot, 1939 (Acari: Acariformes: Endeostigmata). *Fieldiana Zool.* 72(4):59–64. **1982.** Acariformes-

Prostigmata. Pp. 117–145. in S. Parker (ed.). *Synopsis and Classification of Living Organisms,* Vol. 2. McGraw-Hill, New York. **1988a.** Proteonematalycidae, a new mite family (Acari: Acariformes), from fore-dune sand of the southern shores of Lake Michigan. *Int. J. Acarol.* 14: in press. **1988b.** The status of *stigmocheylus* Berlese (Acari: Prostigmata). *J. Entomol. Sci.* 23: in press. **Krantz, G. W. 1978.** *A Manual of Acarology,* 2nd ed. Oregon State University, Corvallis Oreg. 508 pp. **Kuznetzov, N. N. 1980.** Adaptive peculiarities of ontogenesis in the Tydeidae (Acariformes). *Zool. J. Moscow.* 59(7):1018–1023. Translation 1061523, Translation Bureau, Ottawa, Canada. **Lindquist, E. E. 1976.** Transfer of the Tarsocheylidae to the Heterostigmata, and reassignment of the Tarsonemina and Heterostigmata to lower hierarchic status in the Prostigmata (Acari). *Can. Entomol.* 108:23–48. **1977.** Homology of dorsal opisthosomal plates, setae, and cupules of heterostigmatic mites with those of other eleutherengone prostigmata (Acari). *Acarologia (Paris)* 19(1):97–104. **1978.** On the synonymy of *Tarsonemus waitei* Banks, *T. setifer* Ewing, and *T. bakeri* Ewing, with redescription of species (Acari: Tarsonemidae). *Can. Entomol.* 110:1023–1048. **1986a.** A means of determining sex in larval Tarsonemidae (Acari: Heterostigmata) based on external structure. *Exp. & Appl. Acarol.* 2:323–328. **1986b.** The world genera of Tarsonemidae (Acari: Heterostigmata): a morphological, phylogenetic and systematic revision, with a reclassification of family-group taxa in the Heterostigmata. *Mem. Entomol. Soc. Can.* 139:1–517. **Lindquist, E. E. and P. H. Vercammen-Grandjean. 1971.** Revision of the chigger-like larvae of the genera *Neotrombidium* Leonardi and *Monunguis* Wharton, with a redefinition of the subfamily Neotrombidiinae Feider in the Trombidiidae (Acarina: Prostigmata). *Can. Entomol.* 103:1557–1590. **Lindsay, K. L. 1972.** *Nanorchestes antarcticus* Strandtmann; a taxonomic description of the instars. *Acarologia (Paris)* 13(4):593–606. **Luxton, M. 1973.** Mites of the genus *Cryptognathus* from Australia, New Zealand and Niue Island. *Acarologia (Paris)* 15(1):53–75. **Mahunka, S. 1970.** Considerations of the systematics of the Tarsonemina and the description of new European taxa (Acari: Trombidiformes). *Acta Zool. Acad. Sci. Hung.* 16:137–174. **Marshall, V. G. 1970.** Tydeid mites (Acarina: Prostigmata) from Canada. I. New and redescribed species of *Lorryia. Ann. Entomol. Soc. Que.* 15(1):17–52. **McDaniel, B. and E. G. Bolen. 1981.** A new genus and two new species of Nanorchestidae from Padre Island, Texas (Acari: Prostigmata). *Acarologia (Paris)* 22(3):253–256. **McDaniel, B., D. K. Morihara, and J. K. Lewis. 1975.** A new species of *Tuckerella* from South Dakota and a key with illustrations of all known described species. *Acarologia (Paris)* 17(1):274–283. **1976.** The family Teneriffiidae Thor, with a new species from Mexico. *Ann. Entomol. Soc. Am.* 69(3):527–537. **Moser, J. C. and E. Z. Cross. 1975.** Phoretomorph: a new phoretic phase unique to the Pyemotidae (Acarina: Tarsonemoidea). *Ann. Entomol. Soc. Am.* 68:820–822. **Newell, I. M. 1957.** Studies on the Johnstonianidae (Acari, Parasitengona). *Pac. Sci.* 11:396–466. **Norton, R. A. 1977.** An example of phoretomorphs in the mite family Scutacaridae. *J. Ga. Entomol. Soc.* 12:185,186. **Norton, R. A. and G. S. Ide. 1974.** *Scutacarus baculitarsus agaricus* n. subsp. (Acarina: Scutacaridae) from commercial mushroom houses, with notes on phoretic behavior. *J. Kans. Entomol. Soc.* 47(4):527–534. **Oudemans, A. C. 1936.** Neues über Anystidae (Acari). *Arch. Naturgesch. N.F.* 5(3):364–446. **Pittard, D. A. 1971.** A comparative study of the life stages of the mite, *Stereotydeus mollis* W. & S. (Acarina). *Pac. Insects Monogr.* 25:1–14. **Price, D. W. 1975.** Notes on the genus *Pomerantzia* Baker with a description of a new species from Indiana (Acarina: Pomerantziidae). *Proc. Entomol. Soc. Wash.* 77(4):487–490. **Price, D. W. and G. S. Benham, Jr. 1976.** Vertical distribution of pomerantziid mites (Acarina: Pomerantziidae). *Proc. Entomol. Soc. Wash.* 78(3):309–313. **Pritchard, A. E. and E. W. Baker. 1955.** A revision of the spider mite family Tetranychidae. *Mem. Pac. Coast Entomol. Soc.* 2:1–472. **Rack, G. 1975.** Phoretisch auf Kleinsaugern gefundene Arten der Gattung *Pygmephorus* (Acarina, Pygmephoridae). *Mitt. Hamb. Zool. Mus. Inst.* 72:157–176. **Rack, G. and G. C. Eickwort. 1980.** Biology and description of a new Pygmephorid mite (Acarina: Tarsonemida) associated with the soil-nesting bee *Agapostemon nasutus* (Hymenoptera: Halictidae). *Acarologia (Paris)* 21(2):267,268. **Robaux, P. 1967.** Contribution à l'étude des acariens Trombidiidae d'Europe. *Mem. Mus. Natl. Hist. Nat. (Paris) Ser. A Zool.* 46:1–124. **1969.** Thrombidiidae d'Amérique du Sud. II. Chyzerini, Thrombidiinae; solénidiotaxie du palpe chez quelques Thrombidiidae; relations entre les *Podothrombium* Berlese 1910, *Variathrombium* n.g. et les Thrombidiinae s.l. *Acarologia (Paris)* 11(1):69–93. **1970.** Étude des larves de Thrombidiidae. III. La larve de *Johnstoniana errans* (Johnston) 1852, description de l'adulte et de la nymphe. *Acarologia (Paris)* 12(2):339–356. **1974.** Recherches sur le développement et la biologie des acariens "Trombidiidae."

Mem. Mus. Natl. Hist. Nat. (Paris) Ser. A. Zool. 85:1–186. **1975a.** Observations sur quelques Actinedida (= Prostigmates) du sol d'Amérique du Nord. I. Une nouvelle espèce de Caligonellidae (Acari-Raphignathoidea): *Coptocheles grandjeani* n. sp. *Acarologia (Paris)* 17(2):236–242. **1975b.** Observations sur quelques Actinedida (= Prostigmates) du sol d'Amérique du Nord. II. Redescription de *Neophyllobius farrieri* De Leon 1958 (Acari-Raphignathoidea-Camerobiidae). *Acarologia (Paris)* 17(2):248–256. **1975c.** Observations sur quelques Actinedida (= Prostigmates) du sol d'Amérique du Nord. III. Description de deus nouvelles espèces de *Cryptognathus.* (Acari-Raphignathoidea-Cryptognathidae). *Acarologia (Paris)* 17(2):257–269. **1975d.** Observations sur quelques Actinedida (= Prostigmates) du sol d'Amérique du Nord. V. Barbutiidae, une nouvelle famille d'acariens (Acari: Raphignathoidea) et description d'une espèce nouvelle appartenant au genre *Barbutia. Acarologia (Paris)* 17(3):480–488. **1976.** Observations sur quelques Actinedida (= Prostigmates) du sol d'Amérique du nord. VII. Sur deux espèces nouvelles de Raphignathidae (Acari). *Rev. Ecol. Biol. Sol* 13(3):505–516. **1977a.** Observations sur quelques Actinedida (= Prostigmates) du sol d'Amérique du Nord. VI. Sur deux espèces nouvelles de Labidostommidae (Acari). *Acarologia (Paris)* 18(3):442–461. **1977b.** Observations sur quelques Actinedida (= Prostigmates) du sol d'Amérique du Nord. VII. Sur deux formes larvaires nouvelles de Trombidioidea (Acari). *Acarologia (Paris)* 18(4):651–667. **Schubart, H. O. R. 1973.** The occurrence of Nematalycidae (Acari: Prostigmata) in central Amazonia with a description of a new genus and species. *Acta Amazonica* 3(3):53–57. **Schuster, R. and I. J. Schuster. 1977.** Ernährungs- und fortpflanzungsbiologische Studien an der Milbenfamilie Nanorchestidae (Acari: Trombidiformes). *Zool. Anz.* 199:89–94. **Shiba, M. 1969.** Taxonomic investigations on free living mites in the sub alpine forests of Shiga Heights, IBP Area. II. Prostigmata. *Bull. Natl. Sci. Mus. (Tokyo)* 12(1):65–115. **Singer, G. 1967.** A comparison between different mounting techniques commonly employed in acarology. *Acarologia (Paris)* 9(3):475–484. **Smiley, R. L. and J. O. Whitaker, Jr. 1981.** Studies on the idiosomal and leg chaetotaxy of the Cheyletidae (Acari) with descriptions of a new genus and four new species. *Int. J. Acarol.* 7(1–4):109–127. **Southcott, R. V. 1961.** Studies on the systematics of the Erythraeoidea (Acarina), with a critical revision of the genera and subfamilies. *Aust. J. Zool.* 9(3):367–610. **1972.** Revision of the larvae of the tribe Callidosomatini (Acarina: Erythraeidae) with observations on post-larval instars. *Aust. J. Zool. Suppl. Ser.* 13:1–84. **1963.** The Smarididae (Acarina) of North and Central America and some other countries. *Trans. R. Soc. S. Aust.* 86:159–245. **Strandtmann, R. W. 1949.** The blood-sucking mites of the genus *Haemolaelaps* (Acarina: Laelaptidae) in the United States. *J. Parasitol.* 35(3):325–352. **1967.** *Terrestrial Prostigmata. Entomology of Antarctica.* Antarct. Res. Ser. 10, Publ. 1574. American Geophysical Union, Washington, D.C. pp. 51–80. **1971.** The eupodoid mites of Alaska (Acarina: Prostigmata). *Pac. Insects* 13(1):75–118. **Strandtmann, R. W. and M. L. Goff. 1978.** The Eupodoidea of Hawaii (Acarina: Prostigmata). *Pac. Insects* 19(3,4):121–143. **Strandtmann, R. W. and P. J. Tilbrook. 1968.** Some Prostigmata (Acari) from Signy Island, South Orkney Islands, and Deception Island, South Shetland Islands. *Br. Antarct. Surv. Bull.* 17:51–57. **Summers, F. M. 1960.** *Eupalopsis* and Eupalopsellid mites (Acarina: Stigmaeidae, Eupalopsellidae). *Fla. Entomol.* 43(3):119–138. **1966.** Genera of the mite family Stigmaeidae Oudemans (Acarina). *Acarologia (Paris)* 8(2):230–250. **Summers, F. M. and W. M. Chaudhri. 1965.** New species of the genus *Cryptognathus* Krammer (Acarina: Cryptognathidae). *Hilgardia* 36(7):313–326. **Summers, F. M. and D. W. Price. 1970.** Review of the mite family Cheyletidae. *Univ. Calif. Publ. Entomol.* 61:1–153. **Summers, F. M. and E. I. Schlinger. 1955.** Mites of the family Caligonellidae. *Hilgardia* 23(12):539–561. **Theron, P. D. 1974.** Hybalicidae, a new family of endeostigmatic mites (Acari: Trombidiformes). *Acarologia (Paris)* 16(3):397–412. **1975.** Three new species of the genus *Alycosmesis* (Acari: Terpnacaridae) from South Africa. *J. Entomol. Soc. South. Afr.* 38(2):289–296. **1976.** New species of the genus *Terpnacarus* Grandjean (Acari: Terpnacaridae) with notes on the biology of one species. *J. Entomol. Soc. South. Afr.* 39(1):131–141. **1977.** New species of the genus *Petralycus* Grandjean (Acari: Endeostigmata) from South Africa. *Acarologia (Paris)* 19(1):38–45. **Theron, P. D. and P. A. J. Ryke. 1969.** The family Nanorchestidae Grandjean (Acari: Prostigmata) with descriptions of new species from South African soils. *J. Entomol. Soc. South. Afr.* 32(1):31–60. **1975a.** Three new species of the family Sphaerolichidae (Acari: Endeostigmata) from South Africa. *Acarologia (Paris)* 17(2):220–235. **1975b.** Five new species of the family Lordalychidae (Acari: Endeostigmata) from South Africa. *Acarologia (Paris)* 17(4):631–651. **Theron, P. D., M. K. P. Meyer, and P. A. J. Ryke. 1969.** Two new genera of the family Paratydeidae (Acari : Prostigmata) from South African soils. *Acarologia (Paris)* 11(4):697–710. **1970.** The family Alicorhagiidae Grandjean (Acari : Trombidiformes) with descriptions of a new genus and species from South African soils. *Acarologia (Paris)* 12(4):668–676. **Tseng, Y. H. 1977.** A contribution to the knowledge of Formosan cheyletid mites (Acarina : Prostigmata). *Proc. Natl. Sci. Counc. Repub. China* 10(2):1–52. **1982.** Mites of the family Stigmaeidae of Taiwan with key to genera of the world (Acarina : Prostigmata). *Phytopathol. Entomol. NTU* 9:1–52. **van der Hammen, L. 1969.** Notes on the morphology of *Alycus roseus* C. L. Koch. *Zool. Meded. (Leiden)* 43(15):177–202. **Vercammen-Grandjean, P. H. 1968.** *Chigger mites of the Far East (Acarina : Trombiculidae and Leeuwenhoekiidae).* Special Study. U.S. Army Medical Research and Development Command, Washington, D.C., 135 pp. **1973.** Sur les status de la famille des Trombidiidae Leach, 1815 (Acarina, Prostigmata). *Acarologia (Paris)* 15:102–114. **Vistorin-Theis, G. 1975.** Entwicklungszyklus der Calyptostomiden (Acari, Trombidiformes). *Acarologia (Paris)* 17(4):683–692. **1976.** Morphologisch-taxonomische Studien an der Milbenfamilie Calyptostomidae (Acari, Trombidiformes). *Sb. Osterr. Akad. Wiss. Math.-Nat. Kl. Abt. I* 185:55–89. **Volgin, V. I. 1969.** Acarina of the family Cheyletidae, world fauna. *Akad. Nauk SSSR Zool. Inst. Opred. Faune SSSR* 101:1–432. **Wainstein, B. A. 1965.** On the system of Hydrachnellae and their position in the suborder Trombidiformes. *Tr. Inst. Biol. Vnutr. Vod, Akad. Nauk SSSR* 11:66–83. **Wallace, M. M. H. and J. A. Mahon. 1972.** The taxonomy and biology of Australian Bdellidae (Acari). I. Subfamilies Bdellinae, Spinibdellinae and Cytinae. *Acarologia (Paris)* 14(4):544–580. **1976.** The taxonomy and biology of Australian Bdellidae (Acari). II. Subfamily Odontoscirinae. *Acarologia (Paris)* 18(1):65–123. **Welbourn, W. C. and O. P. Young. 1987.** New genus and species of Erythraeinae (Acari : Erythraeidae) from Mississippi with a key to the genera of North American Erythraeidae. *Ann. Entomol. Soc. Am.* 80:230–242. **Witte, H. 1978.** Die postembryonale Entwicklung und die funktionelle Anatomie des Gnathosoma in der Milbenfamilie Erythraeidae (Acarina, Prostigmata). *Zoomorphologie* 91(2):157–189. **Wood, T. G. 1964.** A new genus of Stigmaeidae (Acarina, Prostigmata) from New Zealand. *N. Z. J. Sci.* 7:579–584. **1967.** New Zealand mites of the family Stigmaeidae (Acari, Prostigmata). *Trans. R. Soc. N.Z. Zool.* 9(9):93–139. **1969.** The Homocaligidae, a new family of mites (Acari : Raphignathoidea), including a description of a new species from Malaya and the British Solomon Islands. *Acarologia (Paris)* 11(4):711–729. **1973.** Revision of Stigmaeidae (Acari : Prostigmata) in the Berlese Collection. *Acarologia (Paris)* 15(1):76–95. **Zacharda, M. 1979a.** The ecology of Rhagidiidae. Pp. 537–542 in J. G. Rodriguez, (ed.). *Recent Advances in Acarology,* Vol. 1. Academic Press, New York. **1979b.** Strandtmanniidae—a new family of Eupodoidea (Acarina : Prostigmata). *Vestn. Cesk. Spol. Zool.* 43(1):76–81. **1980.** Soil mites of the family Rhagidiidae (Actinedida : Eupodoidea), morphology, systematics, ecology. *Acta Univ. Carol. Biol.* 1978:489–785. **Zaher, M. A. and E. A. Gomaa. 1979.** Genus *Neophyllobius* in Egypt with description of three new species (Prostigmata-Neophyllobiidae). *Int. J. Acarol.* 5(2):123–130.

Appendix 1. Prodorsal Setal Equivalents for Prostigmatid Mites

Family	Setal Signature						Source
	vi	*ve*	*sci*	*sce*	*exp*	*in*	
Adamystidae	*bn**	*pa*	*bp**	*po*	–	–	Coineau (1974a[M])
Alicorhagiidae	*bf*	*ba*	*bb**	*be*	*br*	*bm*	Grandjean (1939b[M])
Anystidae	*sa**	*na*	*sp**	*nb*	–	–	Grandjean (1943a[M])
Bdellidae	ant.	lat.	post.	med.			
	sens.*	prop.	sens.*	prop.	–	–	Atyeo (1963[M])
Bimichaeliidae-1[a]	*bf*	*ba**	*bb**	*be*	*br*	*bm*	Grandjean (1943b[M])
Bimichaeliidae-2	*bf*	*ba**	*bb**	*be*	–	*bm*	Grandjean (1943b[M])
Caeculidae	*po*	*pa*	*bo**	*pm*	–	–	Coineau (1974b[M])
Cheyletidae	*vi*	*ve*	*sci*	*sce*	–	–	Present work
Cunaxidae	*PS1**	*dl l*	*PS2**	*dc1*	–	–	Den Heyer (1977[M])
Ereynetidae	*cc*	*ca*	*cb**	*cd*	–	–	Grandjean 1939c[M])
	v.p.	*v.a.*	+*	*sc*	–	–	Strandtmann and Tilbrook (1968[M])
Grandjeanicidae	*ro*	*le*	*bo**	*xa*	–	*in*	Theron (1974[M])
Ioliniidae	*vi*	*ve*	*sci**	*ce*	–	–	Present work
Labidostommatidae	*bo.a.*	*ga*	*bo.p.**	*ge*	–	–	Grandjean (1942[M])
Lordalycidae	–	*le**	*bo**	+	–	–	Theron and Ryke (1975b[M])
Nanorchestidae	*na*	*nf**	*nb**	*ne*	*nr*	*nm*	Grandjean (1942b[M])
Nematalycidae	*pf*	*pa*	*pb*	*pe*	–	–	Coineau et al. (1967[M])
Oehserchestidae	*ro*	*le*	*bo**	*xa*	–	*in*	Theron (1974[M])
Parasitengona-1[b]	*vi**	*ve*	*sci**	*sce*	–	–	Present work
Parasitengona-2	*vi*	*ve*	*sci**	*sce*	–	–	Present work
Paratydeidae	–	*ve*	*sci**	*sce*	–	–	Present work
Penthalodidae	*i.v.*	*e.v.*	*T**	*s.c.*	–	–	Pittard (1971[M])
Pomerantziidae	–	*ve*	*sci*	*sce*	–	–	Present work
Pseudocheylidae	*vi*	*ve*	*sci**	*sce*	–	–	Present work
Pygmephoridae	–	*pml/pr*	+*/*pml*	*pi*	–	–	Rack and Eickwort (1980[M])
Rhagidiidae	*i.v.*	*e.v.*	*tr.**	*s.c.*	–	–	Strandtmann, 1971[M])

Appendix 1. (*Continued*)

Family	Setal Signature						Source
	vi	*ve*	*sci*	*sce*	*exp*	*in*	
Scutacaridae	–	*pdl*	*s**/*pd3*	*pd2*	–	–	Norton and Ide (1974[M])
Sphaerolichidae	*bf**	*ba*	*bb**	*be*	*br*	*bm*	Grandjean (1939b[M])
	*le**	*ro*	*bo**	*xa*	*xp*	*in*	Theron and Ryke (1975a[M])
Stigmaeidae	*ae*	*be*	*ce*	*de*	–	–	Grandjean (1944[M])
Stigmocheylidae	*vi*	*ve*	*sci**	*sce*	–	–	Present work
Tarsocheylidae	*vi*	*ve*	*sci**	*sce*	–	–	Present work
Tarsonemidae	*v*1	*v*2	+*/+	*sc*2	–	–	Lindquist (1978[M])
Teneriffiidae	*v*	*a*	*s**	*p*	–	–	Ehara (1965[M])
Terpnacaridae	*bf*	*ba*	*bb**	*be*	*br*	*bm*	Grandjean (1939b[M])
Tetranychidae	*dp*1	*dp*2	*dp*3	*dp*4	–	–	Pritchard and Baker (1955[M])
Tydeidae	*P*1	*P*2	S*	*P*3	–	–	Baker (1965[M])

[a] Bimichaeliidae-1, = Bimichaeliidae excluding *Petralycus;* Bimichaeliidae-2 = *Petralycus.*

[b] Parasitengona-1, = Erythraeidae, Smarididae; Parasitengona-2 = all other Parasitengona.

* = Trichobothrium; + = present; – = absent; signature/signature = female/male.

Appendix 2. Body Chaetome Equivalents in Endeostigmata, Anystina, and Eupodina

1	2	3	4	5	6	7	8	9	10	11	12	13	14	15	16	17
vi	*bn**	*bn**	*sa**	*v*	*po*	—	*i.v.*	*iv*	*i.v.*	P_1	p_1	*cc*	*v.p.*	ant. sens.*	PS_1*	*bo. a**
ve	*pa*	*pa*	*na*	*a*	*pa*	+	*e.v.*	*ev*	*e.v.*	P_2	p_2	*ca*	*v.a.*	lat. prop.	dl_1	*ga*
sci	*bp**	*bp**	*sp**	*s**	*bo**	+	*tr.**	*tr**	*T**	*S**	*s**	*cb**	+*	post. sens.*	PS_2*	*bo. p**
sce	*po*	*po*	*nb*	*p*	*pm*	+	*s.c.*	*sc*	*s.c.*	P_3	p_3	*cd*	*s.c.*	med. prop.	dc_1	*ge*
c_1	a_1	a_1	*nc*	d_1	a_1	c_1	*i.h.*	*ih*	*i.h.*	D_1	d_1	*ce*	*h.i.*	int. hum.	dc_2	*gm*
c_2	a_2	a_2	*nd*	*l*	a_2	c_2	*e.h.*	*eh*	*e.h.*	L_1	l_1	*cf*	*h.e.*	ext. hum.	dl_2	*gr*
d_1	b_1	b_1	*ne*	d_2	b_1	*d*	d_1	d_1	$d._1$	D_2	d_2	*cg*	d_1	lat. prop.	dc_3	*da*
d_2	b_2	—	—	—	b_2	—	—	—	—	L_2	l_2	—	—	—	—	*la*
e_1	c_1	c_1	*nf*	d_3	c_1	*e*	d_2	d_2	$d._2$	D_3	d_3	*ch*	d_2	int. lumb.	dc_4	*db*
e_2	c_2	—	—	—	e_2	—	—	—	—	L_3	—	—	—	—	—	*lb*
f_1	d_1	d_1	f_1	d_4	d_1	f_1	*i.l.*	*il*	*i.l.*	D_4	d_4	*ci*	*l.i.*	int. sac.	dc_5	*dc*
f_2	—	—	f_2	—	d_2	f_2	*e.l.*	*el*	*e.l.*	L_4	l_4	*ck**	+*	ext. sac.	dl_5	*lc*
h_1	e_1	e_1	h_1	*i*	e_1	h_1	*i.s.*	*is*	*i.s.*	D_5	d_5	*cl*	*s.i.*	int. clun.	dc_6	*dd*
h_2	e_2	e_2	h_2	*o*	*h*	h_2	*e.s.*	*es*	*e.s.*	L_5	l_5	*cm*	*s.e.*	anal	dl_6	*ld*
h_3	—	—	h_3	—	e_2	—	—	—	—	—	—	—	—	—	—	—
ps_1	ps_1	ps_1	ps_1	*p′*	ps_1	ps_1	a_1	a_3	$a._3$	*A*	h_1	*cn*	a_2	post. a. (*p*)	dc_7 (*a*)	*de*
ps_2	ps_2	ps_2	ps_2	*a′*	ps_2	ps_2	a_2	a_2	$a._2$	—	h_2	+	a_1	anal (*p*)	dl_7 (*pa*)	*le*
ps_3	ps_3	ps_3	ps_3	±	ps_3	ps_3	a_3	a_1	$a._1$	—	—	—	—	anal (*p*)	*pa* (−)	+
ps_4	—	—	ps_4	—	—	—	—	—	—	—	—	—	—	—	—	—
ad_1	ad_1	ad_1	+	+	ad_1	—	a_4	a_4	†	†	*ps*	†	†	anal (*q*)	a_1 (−)	a_1
ad_2	ad_2	ad_2	+	+	ad_2	—	—	—	†	†	—	†	†	anal (*q*)	a^2 (−)	a_2
ad_3	ad_3	ad_3	+	+	—	—	—	—	†	†	—	†	†	anal (*q*)	—	a_3
ad_4	—	—	+	—	—	—	—	—	—	—	—	—	—	—	—	—
an_1	a_1	an_1	—	—	—	—	—	—	—	—	—	—	—	anal (*r*)	—	—
an_2	a_2	an_2	—	—	—	—	—	—	—	—	—	—	—	—	—	—
an_3	a_3	an_3	—	—	—	—	—	—	—	—	—	—	—	—	—	—
ag_1	+	1	+	+	ag_1	+	pgs_1	$pags_1$	pgs_1	PG_1	ag_1	+	pg_1	pg_1	pg_1	+
ag_2	+	2	+	+	ag_2	+	pgs_2	$pags_2$	pgs_2	PG_2	ag_2	+	pg_2	pg_2	pg_2	+
ag_3	+	3–5	—	+	ag_3	+	pgs_3	$pags_3$	pgs_3	PG_3	ag_3	+	pg_3	pg_3	pg_3	+
g_1	+	1	+	+	g_1	+	g_1	pgs_1	$g.s._1$	G_1	g_1	+	g_1	g_1	g_1	+

Appendix 2. *(Continued)*

1	2	3	4	5	6	7	8	9	10	11	12	13	14	15	16	17
g_2	+	2	+	+	g_2	+	g_2	pgs_2	$g.s._2$	G_2	g_2	+	g_2	g_2	g_2	+
g_3	+	3	+	+	g_3	+	g_3	pgs_3	$g.s._3$	G_3	g_3	+	g_3	g_3	g_3	+
g_4	+	4–7	+	+	g_4	+	g_4	pgs_4	$g.s._4$	G_4	g_4	+	g_4	g_4	g_4	+
1a	+	+	+	+	+	+	*ics*I	+	I-3	V_1	+	+	*ci*	+	+	+
3c	+	+	+	+	+	+	*ics*III	+	III-4	V_2	+	+	*c*iii	+	+	+
4a	+	+	+	+	+	+	*ics*IV	+	IV-3	V_3	+	+	*c*iv	+	+	+
4c	+	+	+	+	+	?	–	–	–	–	–	–	–	+	+	+
ia	+	+	+	+	+	+	+	ly_1	+	+	+	+	+	+	+	+
im	+	+	+	+	+	+	+	ly_2	+	+	+	+	+	+	+	+
ip	+	+	+	+	+	+	+	ly_3	+	+	+	+	+	+	+	+
ih	+	+	+	+	+	+	+	ly_4	+	+	+	+	+	+	+	+

Column 1. Modified from Grandjean (1939[M]). Cols. 2, 3. ADAMYSTIDAE (Coineau 1974a[M]); Col. 2, *Saxidromus;* Col. 3, *Adamystis.* Col. 4. ANYSTIDAE (Grandjean 1943a[M]). Col. 5. TENERIFFIIDAE (Ehara 1965[M]). Col. 6. CAECULIDAE (Coineau 1974b[M]). Col. 7. POMERANTZIIDAE (Lindquist 1977[M]). Cols. 8, 9. RHAGIDIIDAE: Col. 8 (Strandtmann 1971[M]). Col. 9 (Zacharda 1980[M]). Col. 10. PENTHALODIDAE (Pittard 1971[M]). Columns 11, 12. TYDEIDAE: Col. 11 (Baker 1965[M]), Col. 12 (Andrè 1980[M]). Cols. 13, 14. EREYNETIDAE: Col. 13 (Grandjean 1939c[M]), Hunter and Cross 1968[M]), Col. 14 (Strandtmann and Tilbrook 1968[M]). Col. 15. BDELLIDAE (Atyeo 1963[M]), Grandjean 1938[M]). Col. 16. CUNAXIDAE (Den Heyer 1977[M], 1980[M]). Col. 17. LABIDOSTOMMATIDAE Grandjean (1942a[M]), Coineau 1964[M]).

* = Trichobothrium; + = present; – = absent; (signature) = system used subsequently; † = only segment *AD* added, no setae present.

Body Chaetome Equivalents in Raphignathoidea

1	18	19	20	21	22	23	24	25	26	27
vi	*ae*	*ae*	–	l_1	+	*ae*	*ae*	*ae*	*ae*	*ae*
ve	*ce*	*ce*	*ae*	l_2	+	*be*	*be*	*be*	*me*	*ce*
sci	*be*	*be*	*pv*	l_3	+	*ce*	*ce*	–	*pm*	*be*
sce	+	*de*	*pr*	l_4	+	*de*	*de*	*de*	*ce*	*de*
pdx_1	–	–	–	$*mc_1$	–	–	–	–	–	–
c_1	*a*	*a*	*po*	mc_2	*cp*	*a*	*a*	*ce*	*a*	*a*
c_2	*he*	*he*	*sc*	l_5	*h*	*he*	*he*	*he*	*he*	*he*
d_1	*b*	*b*	*a*	mc_3	dc_1	*b*	*b*	*a*	*b*	*b*
d_2	–	–	–	l_6	s_1	*la*	*la*	*la*	*la*	*la*
e_1	*c*	*c*	*b*	mc_4	dc_2	*c*	*c*	*c*	*c*	*c*
e_2	–	–	*la*	l_7	s_2	*lm*	*lm*	*lm*	*lm*	*lm*
f_1	*d*	~*d*	*c*	mc_5	dc_3	*d*	*d*	*li*	*li*	*li*
f_2	–	*lr*	–	l_8	l_2	†*lr*/–	*lr*/–	–	–	–
h_1	*e*	*e*	*d*	mc_6	dc_5	e_1	e_1	*e*	*e* (*le*)	pg_5/*e*
h_2	*f*	*le*	*le*	l_9	dc_4	*e*2/*lr*	e_2	*le*	*le* (*e*)	pg_4/*le*
ps_1	ps_1	ps_1	*+	+	+	+/e_2	ps_1	g_1	+ (g_4)	ag_4/ag_3
ps_2	ps_2	ps_2	*+	+	+	*w*/+	ps_2	g_2	+ (g_3)	ag_3/ag_2
ps_3	ps_3	ps_3	*+	+	+	–/*w*	ps_3	g_3	+ (g_2)	ag_2/ag_1
ag_1	*v*	*u*/±	(*ve*)	+	+	*t*	*u*	+	+ (pg_1)	pg_1
ag_2	±	–	(*vf*)	–	–	*u*	*v*	+	+ (pg_2)	pg_2
ag_3	–	–	–	–	–	*v*	*w*	–	+ (pg_3)	pg_3
g_1	g_1/–	g_1/–	*g*/–	+/–	+/?	+/–	g_1/–	–	+ (g_1)/–	ag_1/–
g_2	g_2/–	g_2/–	*g*/–	–	+/?	–	g_2/–	–	–	–
g_3	–	g_3/–	*g*/–	–	–	–	g_3/–	–	–	–
1*a*	+	1*a*	(*va*)	+	+	+	1*a*	+	+	+
3*a*	+	3*a*	(*vb*)	+	+	+	3*a*	+	+	+
4*a*	*t*	4*a*	(*vc*)	+	+	+	4*a*	+	+	+
4*c*	*u*	*t*	(*vd*)	–	–	–	*t*	–	–	–
ia	+	+	+	–	–	–	–	–	–	–
im	+	+	+	–	–	–	–	–	–	–
ip	+	+	+	–	–	–	–	–	–	–
ih	+	+	+	–	–	–	–	–	–	–

Column 1. Modified from Grandjean (1939a[M]). Col. 18. CALIGONELLIDAE (Robaux 1975a[M]). Col. 19. RAPHIGNATHIDAE (Atyeo 1963[M]). Col. 20. CRYPTOGNATHIDAE (Summers and Chaudhri 1965[M], Robaux 1975c[M]). Col. 21. CAMEROBIIDAE (Gerson 1968[M]). Col. 22. XENOCALIGONELLIDAE (Gonzalez 1978a[M]). Col. 23. BARBUTIIDAE (Robaux 1975d[M]). Col. 24, 25. STIGMAEIDAE: 24, *Storchia* (Grandjean 1944[M]); 25, *Mullederia* (Wood 1964). Col. 26. EUPALOPSELLIDAE (Summers 1960[M], (Gerson 1966[M]. Col. 27. HOMOCALIGIDAE (Gonzalez 1978b[M]). Signature/signature = female/male; (signature) = system used subsequently; + = present; – = absent; ± = present or absent; * = seta added in protonymph; ~ = seta added in second nymph; † = seta added in adult female.

Body Chaetome Equivalents in Heterostigmata, Tetranychidae, and Cheyletidae

1	28	29	30	31	32	33	34	35	36	37
vi	+	–	v_1	–	–	dp_1	*LA*	*vi*	*vi*	*vi*
ve	+	*pml/pr*	v_2	pd_1	dp_1	dp_2	*LA*	*ve*	*ve*	*ve*
sci	+*	+*/*pml*	+*/+	s^*/pd_3	dp_2	dp_3	*LA*	*sci*	*sci*	*sci*
sce	+	*pi*	sc_2	pd_2	dp_3	dp_4	*LA*	*sce*	*sce*	*sce*
c_1	c_1	c_1	c_1	c_1	dc_1	dc_1	*LA*	d_1/d_2	d_2	d_3
c_2	c_2	c_2	c_2	c_2	dl_1	dl_1	*LA*	*h*	*h*	*h*
cx_1	–	–	–	–	*h*	*h*	N_2	$-/d_1$	d_1	d_2
cx_2	–	–	–	–	–	–	–	–	–	d_1
d_1	*d*	d_1	*d*	*d*	dc_2	dc_2	*LA*	$-/d_3$	d_3	d_4
d_2	–	–	–	–	dl_2	dl_2	*LA*	l_1	l_1	l_1
dx	–	–	–	–	–	sl_1	–	–	–	–
e_1	*e*	e_1	*e*	e_2	dc_3	dc_3	N_2	$-/l_3$	d_4	d_5
e_2	–	–	–	–	dl_3	sl_2	*LA*	l_2	l_2	l_2
ex	–	–	–	–	–	sl_2	–	–	–	–
f_1	*f*	e_2	*f*	e_1	s_1	s_1	N_2	l_4/l_5	d_5	l_4
f_2	–	–	–	–	s_2	s_2	*LA*	l_3/l_4	l_3	l_3
fx	–	–	–	–	–	sl_3	–	–	–	–
h_1	h_1	f_1	h_1	f_1	*c*	*c*	*LA*	d_5/g_7	$+/a_2$	a_1
h_2	h_2	f_2	h_2	f_2	*ppa*	*ppa*	*LA*	l_5/g_6	l_4/a_1	d_3
h_3	–	–	–	–	–	–	–	–	–	–
ps_1	ps_1	h_2	ps_1	h_1	a_2	a_3	*LA*	a_3/g_5	a_2/g_5	a_2
ps_2	ps_2	h_3	ps_2	h_2	a_1	a_2	*LA*	a_2/g_4	a_1/g_4	g_5
ps_3	ps_3	h_1	–	h_3	–	a_1	*LA*	a_1/g_3	g_4/g_3	g_4
ag_1	+	–	–	–	ag_1	ag_1	N_2	g_2	g_2	g_2
ag_2	+	–	–	–	*apa*	*apa*	N_2	g_3/d_5	l_5/l_4	g_3
ag_3	–	–	–	–	–	–	–	–	–	–
g_1	+	–	–	–	*g*	*g*	N_2	$g_4/-$	$g_3/-$	–
g_2	+	–	–	–	–	–	–	$g_5/-$	–	–
g_3	+	–	–	–	–	–	–	–	–	–
g_4	+	–	–	–	–	–	–	–	–	–
1*a*	+	2*a*	+	2*a*	+	+	*LA*	ic_1	ic_1	ic_1
3*a*	+	3*b*	+	3*a*	+	+	*LA*	ic_3	ic_3	ic_3
4*a*	+	4*a*	+	4*a*	+	+	N_2	ic_4	ic_4	ic_4
4*c*	+	4*b*	–	4*b*	+	+	N_1	g_1	g_1	g_1
ia	+	+	+	+	+	+	–	–	–	–
im	+	+	+	+	+	+	+	+	?	?
ip	+	+	+	+	+	+	+	+	+	+
ih	+	+	+	+	+	+	+	+	?	+

Column 1. Modified from Grandjean (1939a[M]). Col. 28. TARSOCHEYLIDAE (Lindquist 1977[M]), Col. 29. PYGMEPHORIDAE (Rack and Eickwort 1980[M]). Col. 30. TARSONEMIDAE (Lindquist 1978[M]. Col. 31. SCUTACARIDAE (Norton and Ide 1974[M]). Col. 32, 33. TETRANYCHIDAE (Pritchard and Baker 1955[M]): 32, Tetranychinae; 33, Bryobiinae. Cols. 34–37, CHEYLETIDAE: 34, current work; 35 (Fain 1979[M]) *Cheyletus*; 36, 37 Smiley and Whitaker (1981[M]; 36, *Camincheyletus*; 37, *Acaropsellina*.

* = Trichobothrium. 28–31, 35. Signature/signature = female/male, 36. Signature/signature = nymph 2/male. + = Present; – = absent; ? = presence not known; LA = larval seta; N 1 = protonymphal seta; N 2, = seta appears in second nymph.

22

Acarina: Astigmata (Acaridida)

JAMES R. PHILIPS

The acarine suborder Astigmata is almost entirely composed of mycophagous (Walter 1987), saprophagous, phytophagous, and parasitic mites which occur on their hosts, on stored food products, or in houses, nests, soil, and litter. Astigmatic mite populations vary widely in soil and litter, but generally comprise up to 2.4% of the microarthropods in forests, grasslands, old fields, and tundra sites (Pillai and Winston 1963, Oswald and Minty 1970, Weigert 1974, Douce and Crossley 1977, Fain and Wauthy 1979). Virtually none occur in California pine forests (Price 1973), but they represent one-third of the mite fauna in California desert soil (Wallwork 1972). High numbers have also been found in uncultivated, mineral, and pasture soil in Great Britain (Salt et al. 1948, Sheals 1957, Davis 1963), and these mites may be locally common in other sites.

Morphology

The Astigmata are weakly sclerotized, slow-moving, pale mites that lack respiratory stigmata and tracheae, and respire through the cuticle. Segmentation is not apparent, and the body is divided into two main regions, the gnathosoma and the idiosoma (Fig. 22.1), which are separated by a flexible membrane. The gnathosoma contains the chelicerae and two-segmented palps. The roof of the oral opening occurs below the chelicerae, and the ventral part of the gnathosoma is formed by the fusion of the palpal coxae. In most soil Astigmata, the chelicerae are chelate–dentate (Fig. 22.3), with fixed and movable digits.

The idosoma is usually divided into the propodosoma and the hystero-

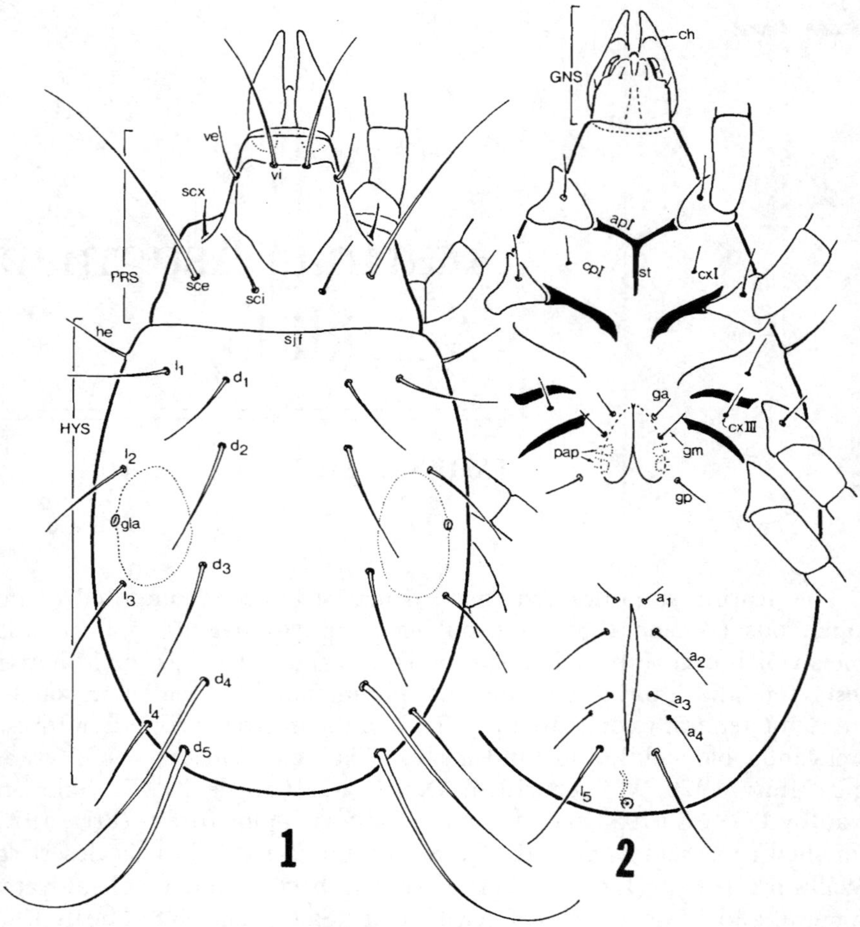

Fig. 22.1. Adult acarid female, dorsal (*PRS*, propodosoma; *HYS*, hysterosoma, *sjf*, sejugal furrow). **Fig. 22.2.** Adult acarid female, ventral (*GNS*, gnathosoma; *apI*, apodeme I; *ch*, chelicera, *cpI*, coxal plate I; *pap*, genital papillae; *st*, sternum).

soma by the sejugal furrow. A sclerotized plate called a shield covers part of the dorsal propodosoma in some species, and a separate hysterosomal shield also may occur. Dorsal sensory setae occur in varying forms and numbers on the idiosoma. Typically, there are four pairs of dorsal propodosomal setae: anterior internal and external vertical setae (*vi* and *ve*), and posterior internal and external scapular setae (*sci* and *sce*). A small, simple, or dendritic supracoxal seta (*scx*) may occur laterally on the propodosoma at the base of leg I. On the hysterosoma occur up to five pairs of dorsomedial setae (d_1–d_5) medially, five pairs of dorsolateral setae (l_1–l_5) and two pairs humeral setae, one lateral (*h*) and one ventral (*sh*). Ventrally, the idiosoma (Fig. 22.2) also bears sensory setae. Coxal setae (*cx* I, *cx* III) occur on the epimera, or coxisternal regions, of legs I and III. Three pairs of genital setae (*ga*, *gm*, and *gp*) surround the genital opening, and up to six pairs of anal setae surround the anus.

This setal nomenclature is based primarily on the system suggested by Atyeo and Gaud (1966). The setal terminology of Zakhvatkin (1941) and Hughes (1964) is also widely used in the literature. The main differences between the systems are: d_5 and l_4 are termed internal sacral setae (*sai*) and external sacral setae (*sae*); l_2 and l_3 are mediolateral setae *la* and *lp*; l_1 and *h* are anterolateral humeral setae *hi* and *he*; *sh* is *hv*; a_1 and a_2 are preanal setae (pra_1, pra_2) and a_3–a_6 and l_5 are postanal setae (pa_1–pa_5).

Between the mediolateral setae l_2 and l_3, laterodorsal hysterosomal gland openings are normally visible. These glands (*gla*) may be colorless or brownish, and appear as large sacs.

The primary genital opening occurs ventrally between the legs. In the female this aperture, the oviporus, is generally an inverted U, V, or Y, flanked by two pairs of knoblike or fingerlike structures termed genital papillae. Oviposition occurs through this aperture, but sperm are received through a small, circular opening, the bursa copulatrix, at the posterior end of the body. The male aedeagus (Figs. 22.4, 22.5) is supported by struts

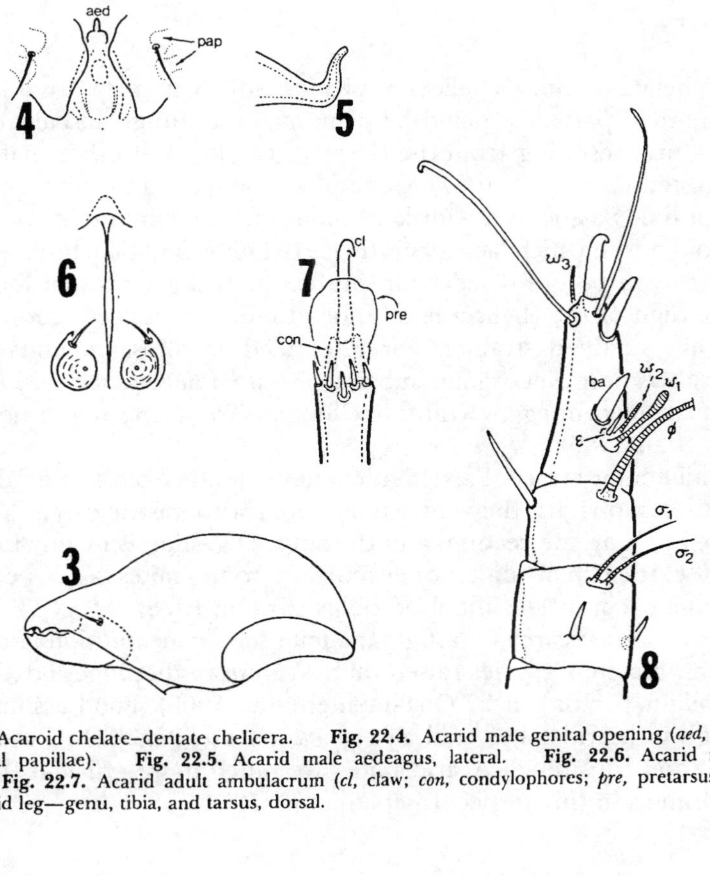

Fig. 22.3. Acaroid chelate–dentate chelicera. **Fig. 22.4.** Acarid male genital opening (*aed*, aedeagus; *pap*, genital papillae). **Fig. 22.5.** Acarid male aedeagus, lateral. **Fig. 22.6.** Acarid male anal region. **Fig. 22.7.** Acarid adult ambulacrum (*cl*, claw; *con*, condylophores; *pre*, pretarsus). **Fig. 22.8.** Acarid leg—genu, tibia, and tarsus, dorsal.

and the genital opening is flanked by two pairs of genital papillae, while the anal opening is often flanked by a pair of copulatory suckers (Fig. 22.6).

Two pairs of legs arise from the ventral propodosoma, and two pairs from the ventral hysterosoma. The legs are composed of five segments: trochanter, femur, genu, tibia, and tarsus. The basal coxae are completely fused with the ventral idiosoma, but their anterior edges may be visible as internal sclerotized apodemes, and apodemes I often fuse to form a median sternum. The posterior coxal edges may also be sclerotized, forming internal epimerites. A claw may be present at the apex of the tarsus, attached to a fleshy pretarsus and often supported by sclerotized condylophores in the pretarsus (Fig. 22.7). A pair of small copulatory suckers may be found on tarsus IV of some males.

Three kinds of sensory structures occur on the legs: setae, solenidia, and the famulus. Normal tactile setae contain actinochitin and are birefringent under polarized light. Chemosensory solenidia (ω, ϕ, σ) usually appear to be transversely striate, lack actinochitin, and do not exhibit birefringence. Solenidia are restricted to the tarsus (ω_1–ω_3), tibia (ϕ), and genu (σ_1, σ_2). The famulus (ε) is a small, generally conical birefringement chemosensor which contains actinochitin and occurs on tarsus I (Fig. 22.8).

Biology

The chelate–dentate chelicerae of most soil Astigmata are adapted to masticate food particles, including plant material, fungi, and algae. Liquefied products resulting from the decay of carrion and other material are also ingested (Wallwork 1967). Generally, high-protein food is preferred, although lipids associated with keratinous animal substances are ingested by acaroid mites, which lack a keratinase (Akimov and Schehur 1972), and *Carpoglyphus,* a pest of dried fruit, prefers high-sugar-content foods. The movable digit of the chelicerae of anoetid mites is reduced, and the palps are highly modified strainers for filter feeding. Most anoetids feed on microorganisms in wet organic substrates, but *Histiostoma murchiei* H. and J. feed on earthworm eggs (Krantz 1978), and *Tyrophagus* ingest nematodes (Walter et al. 1986).

The alimentary tract of astigmatic mites extends from a muscular pharynx to the esophagus, the ventriculus (often with gastric caeca), the intestine, the hindgut, the rectum, and the anus. Digestion is mainly intracellular, but extracellular digestion occurs in some mites as a peritrophic membrane encompasses the food bolus (Krantz 1978).

Information is scarce on astigmatic mite food consumption and defecation rates, but four species raised on a *Neurospora* diet ingested 10–50 μg fresh weight per day at 27°C (Pimentel et al. 1960). Food assimilation is very high in acarid mites, and *Tyrophagus putrescentiae* (Schr.) utilizes 40–60% of food consumed. Acarid mites are much more efficient than most oribatid mites in this respect (Stepien et al. 1971).

Astigmatic mites facilitate the humification process since they fragment material, providing a greater surface area for subsequent attack by other organisms. Some digestion of litter material occurs, but resistant material is concentrated in their fecal pellets and then utilized by other organisms (Harding and Stuttard 1974). Bacterial and fungal spores are transported on the exterior of astigmatic mites, and spores are also distributed in fecal pellets (O'Connor 1984). *Tyrophagus putrescentiae* has been observed to ingest soil as well as organic matter, altering the microstructure of the soil substrate (Robaux et al. 1977). Terrestrial arthropods typically have a tracheal system, but soil astigmatic mites lack trachea and are adapted to low-oxygen environments (Wallwork 1967). Since some astigmatic mite populations increase under conditions of nearly anaerobic decomposition, these species are useful as biological indicators in assessing changes in soil and litter conditions as a result of human perturbations (Dindal et al. 1977).

Sexual reproduction in astigmatic mites occurs through the direct transfer of sperm to the female bursa compulatrix by the male aedeagus. Usually, no packet of sperm is formed, but spermatophores have been observed in the acarid genera *Acarus* and *Lardoglyphus* (Griffiths and Boczek 1977). Parthenogenesis occurs in some anoetids, and both thelytokous and arrhenotokous strains of *Histiostoma feroniarum* (Dufour) have been cultured (Hughes and Jackson 1958).

Soil astigmatic mites can complete their life cycle in several days to several weeks. Females can lay up to 800 eggs in a month in the laboratory, and adults have an average longevity of 23–46 days under various culture conditions (Barker 1964, Rodriguez and Stepien 1973, Woodring 1969). The life cycle consists of a maximum of six stages: egg, larva, protonymph, heteromorphic deutonymph, tritonymph, and adult. In a few species, the larval, protonymph, or deutonymph stages form within the eggshell in the female.

The larval stage has three pairs of legs and lacks genital papillae. The protonymph has four pairs of legs, a rudimentary genital opening, and one pair of genital papillae. The tritonymph may develop directly from the protonymph, and has two pairs of genital papillae. Many species have a heteromorphic deutonymphal stage which may occur between the protonymph and tritonymph. This stage, also known as the hypopus (Fig. 22.9), is highly resistant to adverse environmental conditions, lacks mouthparts, and does not feed except in some endofollicular and subcutaneous parasitic species. The hypopus is also a transportation stage, and may be either active or inert. Active entomophilic forms possess a posteroventral sucker plate and are phoretic on insects. The typical sucker plate contains two central pairs of suckers and two lateral and posterior pairs of conoids: conical, soft, suckerlike modified setae. Species phoretic on mammals usually possess posteroventral ridged claspers (Fig. 22.10) which hold onto a hair. Inert hypopi have no phoretic adaptations, remain within the protonymphal skin, and are transported by wind. Hypopi possess two pairs of genital papillae near a rudimentary genital opening, just

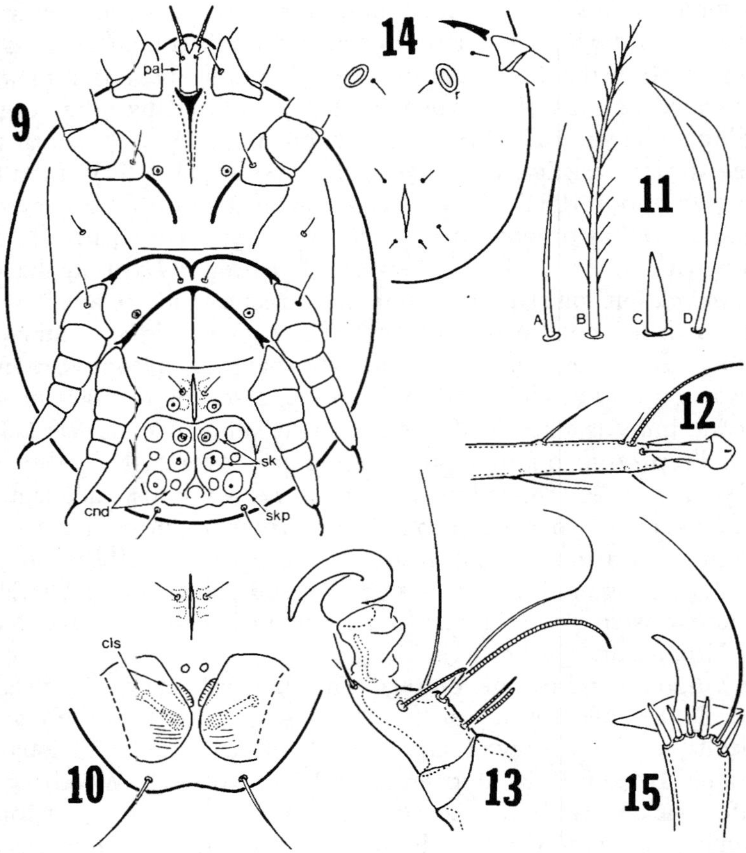

Fig. 22.9. Acarid hypopus, ventral (*cnd,* conoids; *pal,* palposoma; *sk,* suckers; *skp,* sucker plate). **Fig. 22.10.** Glycyphagid hypopus, posteroventral region (*cls,* claspers). **Fig. 22.11.** Setae: (A) simple; (B) pectinate; (C) spinose; (D) foliate. **Fig. 22.12.** Glycyphagid adult ambulacrum. **Fig. 22.13.** Chaetodactylid hypopus ambulacrum. **Fig. 22.14.** Adult anoetid hysterosoma, ventral (*r,* ring). **Fig. 22.15.** *Myianoetus* ambulacrum.

above the sucker plate or claspers in active forms. The gnathosoma is absent in all hypopi, but active hypopi usually have an anteroventral palposoma—a smooth, rectangular, or bilobed structure terminating anteriorly in a pair of long solendidia (α). A pair of ocelli occur on the propodosoma or hysterosoma in a few species.

The hypopus stage represents an adaptation unique to astigmatic mites, and it occurs in most soil astigmatic mites (O'Connor 1982). Pyroglyphid mites and the acarid genera *Tyrophagus, Suidasia,* and *Thyreophagus* apparently lack a hypopus, but many other acarid, anoetid, and saproglyphid mites are insect associates with entomophilic hypopi (Philips 1984). Glycyphagid mites are generally mammal associates, breeding in mammal nests with hypopi phoretic on mammals. However, both active and inert hypopi occur in the genera *Glycyphagus* and *Lepidoglyphus,* as well as in the acarid genus *Acarus.* Formation of the hypopus is influenced by such factors as population density, food availability, temperature, and humidity. The thick cuticle of the hypopus enables it to survive unfavorable conditions while it is transported to a more suitable environment by wind, arthropods, or mammals. This dispersal mechanism has greatly contributed to the success of the Astigmata, but hypopi of nidicolous and stored products species are occasionally found in soil as well as the more common litter species.

In addition to heteromorphic deutonymphs, some heteromorphic males are known, and male polymorphism in acarids comprises four types: homeotype homeomorphic, homeotype heteromorphic, bimotype bimorphic, and bimotype pleomorphic (Woodring 1969). Homeotype males have bodies and setal lengths resembling females; bimotype males are larger with longer setae. Homeomorphic and bimorphic males have legs resembling females. Heteromorphic and pleomorphic males use their modified legs to kill homeomorphic males, but the dominant type of male varies depending on the species and environment.

Astigmatic mite populations in some soils have significantly increased as a result of human activities, such as wheat cultivation (Kines and Sinha 1973), pipeline construction (Tomlin 1977) and applications of insecticides (Marshall 1979) and nitrogen fertilizer (Behan et al. 1978). Astigmatic mites serve as prey for many predatory mites and other invertebrates, and the increase in astigmatic mite populations in orchard soils after insecticide spraying was associated with a decline in predator populations (Karg 1978). Application of nitrogen fertilizer to black spruce humus increased the populations of microorganisms which are potential food sources of the mycophagous and saprophagous astigmatic mites (Behan et al. 1978).

The most important soil genera are *Tyrophagus, Schwiebea, Rhizoglyphus, Glycyphagus, Lepidoglyphus,* and *Histiostoma. Schwiebea* is widespread in soil and litter, but *Tyrophagus, Glycyphagus,* and *Lepidoglyphus* are also common in many additional habitats, including nests (Philips and Dindal 1979), carrion, dung, stored products, and houses. The genera *Acarus, Tyrophagus, Suidasia, Carpoglyphus,* and *Glycyphagus* can all cause contact dermatitis in people who handle infested materials (Baker et al. 1967). Acarid, glycyphagid, and pyroglyphid mites can also infest human lungs and the intestinal and urinary tracts when accidentally ingested or inhaled. Pyroglyphid mites are generally found in nests, stored products, and house dust and are the primary cause of house-dust allergies. They eat mainly skin scales and fungi (Wharton 1976, Lustgraaf 1978) and are occasionally found in soil and litter.

Very few soil astigmatic mites are phytophagous, but some species of *Tyrophagus* and *Rhizoglyphus* are significant plant pests (Garman 1921, Johnston and Bruce 1965). *Rhizoglyphus* attacks the bulbs and roots of many plants throughout the world (Manson 1971).

Anoetid mites, especially *Histiostoma,* are found in moist litter, decaying fungi, carrion, nests, and similar habitats containing wet, decomposing organic matter. Some species are aquatic, living in pitcher plants or water-filled tree holes (Hughes and Jackson 1958, Fashing 1973, Fashing and O'Connor 1984).

The families Chortoglyphidae, Fusacaridae, Guanolichidae, and Rosensteiniidae are not widespread soil species. *Chortoglyphus* is common in barns, stored products, houses, and nests, while *Fusacarus* inhabits mammal nests. Guanolichid mites are associated with bats and bat guano (Fashing 1980), as are most rosensteiniid mites. *Rosensteinia,* however, is associated with cockroaches (O'Connor and Reisen 1978).

The families Chaetodactylidae and Hemisarcoptidae are also insect symbionts. Chaetodactylid hypopi are phoretic on bees and the adults are found in their nests. Hemisarcoptid hypopi are phoretic on ladybeetles and the adults are predators of scale insects.

Collection and Study Techniques

Astigmatic mites in soil and litter are easily collected along with other arthropods by extraction in a modified Tullgren funnel (Murphy 1962). Funnel extraction involves placing a sample on a screen inside a funnel, below which is a collecting jar containing a preservative solution, such as alcohol and glycerin. A light bulb is often placed at the top of the funnel, and as the sample dries, the mites move down and fall into the collecting jar. High-gradient funnel extractors are more complex, and set up steep temperature and humidity gradients in the sample, resulting in a greater extraction efficiency (Krantz 1978). Samples should remain in funnels for a week or more, depending on the size and moisture content of the sample. During extraction, the desiccating environment may cause formation of hypopi, resulting in collection of many more hypopi in the jar than were present in the sample when it was collected. Slow-moving mites sensitive to desiccation are not efficiently extracted in funnels, and relatively few astigmatic mite larvae will be collected with this method. Insects in the sample should be examined carefully for phoretic hypopi. Large numbers of astigmatic mites may remain on the sides of the funnels, so after extraction, funnels should be washed with alcohol and the solution added to the jar.

If live specimens are needed, the mites can be collected in a jar containing moist paper at the base of the funnel, or the sample can be hand sorted directly under the microscope. Astigmatic mites, although small, are mainly white and show up well against dark substrates.

When funnel extraction is not possible, astigmatic mites as well as other invertebrates can be extracted from preserved samples through flotation. This method involves mixing solutions such as alcohol and kerosene with the sample. Debris precipitates in the alcohol, while mites float in the kerosene and that layer can be poured off.

Astigmatic mites can be cultured in small glass or plastic cells with plaster–charcoal bottoms. High humidity must be maintained, but many species of fungi are suitable foods, as well as wheat germ and other grains (Pimentel et al. 1960, Woodring 1969, Rodriguez and Stepien 1973, Hughes 1976). Anoetids thrive on wet decomposing plant and animal matter, and pyroglyphids grow well on skin scales and wheat germ.

Most astigmatic mites can be mounted without clearing, but guanine crystals may obscure the features of some specimens, such as anoetids. Immersion of the mite in Nesbitt's fluid for several days dissolves these crystals, but Nesbitt's fluid is a powerful clearing agent and specimens will shrivel if left in it too long. Permanent slide mounts of the mites can be made using Hoyer's medium. Recipes for making Nesbitt's fluid, Hoyer's medium, and other clearing and mounting media for mites are given by Krantz (1978). For species identification of astigmatic mites, slide mounts and phase contrast or interference contrast microscopy are necessary.

Identification

The classification of astigmatic mites involves primarily morphological characters found on the idiosoma and legs. Chaetotaxy is very important and the number, location, and lengths and shapes of the body and leg setae are very useful distinguishing features. Setae may be simple (Fig. 22.11*A*), pectinate (Fig. 22.11*B*), spinose (Fig. 22.11*C*), foliate (Fig. 22.11*D*), or other shapes. Macrosetae longer than half the length of the idiosoma occur on the tarsi of some hypopi. The position, lengths, and numbers of solenidia are similarly significant. The structure of the cuticle varies widely, and the idiosoma may be smooth, striate, wrinkled, scaly, spiny, tuberculate, or pitted. Sclerotization varies, and the idiosoma may be unsclerotized and pale, completely sclerotized and brown, or with discrete sclerotized shields–propodosomal or hysterosomal or both. The ventral apodemes may be fused or free. Claws and their supporting structures may be present, reduced, or absent on the tarsi. In the male, anal and tarsal suckers occur in some groups, while in the hypopus, posteroventral structures such as the sucker plate or ridged claspers are important.

The hypopus presents great taxonomic difficulties because it is a heteromorphic stage whose morphology greatly differs from that of the corresponding adult. Many hypopi that have been collected have not been reared to adults, and many genera are known only from the hypopus stage. Some genera that are distinctive as adults, such as *Sancassania, Rhizoglyphus,* and *Schweibea,* have hypopi which are very difficult to differentiate. Active hypopi of the genera *Glycyphagus* and *Lepidoglyphus* both were originally described as species of the genus *Dermacarus,* a large genus of hypopi, and the former two genera cannot yet be adequately characterized in a key to hypopi. With further rearing of hypopi, these taxonomic problems should be clarified.

ACKNOWLEDGMENT

I thank Dr. Roy A. Norton for his thorough review of this chapter.

KEYS TO FREE-LIVING TERRESTRIAL ASTIGMATA OF NORTH AMERICA

Key to Stages of Astigmatic Mites

1a Three pairs of legs . . . . **Larvae**
1b Four pairs of legs **2**
2a One pair of genital papillae; genitalia undeveloped. . . . . . . **Protonymph**
2b Two pairs of genital papillae **3**
3a Mouthparts absent; posteroventral suckers or claspers usually present . . . . **Hypopus**
3b Mouthparts present. **4**
4a Genitalia undeveloped. . . . . **Tritonymph**
4b Genitalia developed. **5**
5a Aedeagus present; suckers on tarsus IV and near anus present or absent . . . . **Male**
5b Oviporus and bursa copulatrix present; anal and tarsal suckers absent. . . . **Female**

Key to Families of Adults

1a Two pairs of sclerotized rings (Fig. 22.14) on the ventral hysterosoma; chelicerae dentate, movable digit vestigial . . . . **Anoetidae**
1b Ventral sclerotized rings absent; chelicerae chelate–dentate with movable digit **2**
2a Tarsi I and II without pretarsus and claw, terminating in short spines and a longer seta; genu II with one solenidion; associated with bats. . . . . **Guanolichidae, *Neoguanolichus***
2b Tarsi I and II with fleshy pretarsus or claw or both **3**
3a Tibia I with one solenidion (ϕ), without setae; claws absent; associated with scale insects . . . . **Hemisarcoptidae, *Hemisarcoptes***
3b Tibia I with one solenidion (ϕ) and one or two setae; claws usually present **4**
4a Tibia I with one seta and one solenidion; tarsus I with two apical solenidia; vertical setae (*vi, ve*) absent; idiosoma with simple striations. . . . . **Pyroglyphidae**
4b Tibia I with two setae and one solenidion; setae *vi* present; idiosoma usually lacking simple striae **5**
5a Tarsi with a pair of sclerotized condylophores supporting a well-developed empodial claw in a fleshy pretarsus (Fig. 22.7) **6**
5b Tarsi usually with indistinct condylophores supporting a claw in a fleshy pretarsal stalk (Fig. 22.12) **7**
6a Sejugal furrow absent; ventral gnathosoma usually with a pair of large lobe-like processes; males without anal and tarsal suckers; idiosoma usually rough

with pectinate setae; associated with cockroaches and bats . . . . **Rosensteiniidae**

6b Sejugal furrow present; ventral gnathosoma without large lobelike processes; males usually with a pair of anal suckers and a pair of suckers on tarsus IV; idiosoma usually smooth with simple setae . . . . **Acaridae**

7a Claws well developed 8

7b Claws reduced (Fig. 22.12) 9

8a Claws larger than pretarsal stalk; apodemes free; idiosomal setae serrate or pectinate; cuticle rough; associated with bees . . . . **Chaetodactylidae**

8b Claws smaller than pretarsal stalk; apodemes I and II united with the sternum; idiosomal setae simple; cuticle smooth . . . . **Carpoglyphidae, *Carpoglyphus***

9a Idiosomal setae usually pectinate; cuticle usually rough, with minute spines, scales, or tubercles; sejugal furrow usually absent; males usually without anal and tarsal suckers . . . . **Glycyphagidae**

9b Idiosomal setae simple, cuticle usually smooth 10

10a Sejugal furrow present; idiosoma unsclerotized; usually two solenidia on genu I . . . . **Saproglyphidae**

10b Sejugal furrow absent 11

11a Idiosoma sclerotized, brown; genu I with two solenidia; males without anal and tarsal suckers; in mammal nests . . . . **Fusacaridae, *Fusacarus***

11b Idiosoma unsclerotized, pale; genu I with one solenidion; males with one pair of anal suckers and one pair of suckers on tarsus IV . . . . **Chortoglyphidae, *Chortoglyphus***

Key to Families of Hypopi

1a Tarsus I with one solenidion; tibia I with two solenidia; legs III and IV generally pointing anteriorly . . . . **Anoetidae**

1b Tarsus I with two or three solenidia; tibia I with one solenidion; legs III and IV generally pointing posteriorly 2

2a Tarsi I–III with fleshy pretarsi and enlarged hooklike claws (Fig. 22.13); tarsus IV with one long terminal macroseta; phoretic on bees, ***Osmia*** and ***Xylocopa*** . . . . **Chaetodactylidae**

2b Tarsi I–III generally without enlarged hooklike claws 3

3a Tarsi I–III with terminal claw supported by stalklike pretarsus or claws absent; eyes generally present on propodosoma. 4

3b Tarsi I–IV with sessile claws or apodemes III free and at least six setae on tarsus IV or posteroventral sucker plate absent; eyes generally absent . . . 6

4a Tarsus IV without claw 5

4b Tarsus IV with claw; palposomal solenidion absent; mediolateral eyes present on propodosoma . . . . **Carpoglyphidae, *Carpoglyphus***

5a Palposomal solenidion absent; eyes absent; phoretic on ladybeetles, ***Chilocorus*** **Hemisarcoptidae, *Hemisarcoptes***

5b Palposomal solenidion generally well-developed; eyes generally present apically on the propodosoma . . . . **Saproglyphidae**

6a Posteroventral sucker plate present; scapular setae smooth. . . . . **Acaridae**

6b Suckers absent or scapular setae pectinate; posteroventral ridged claspers developed to varying degrees, occasionally absent . . . . **Glycyphagidae**

Key to Genera of Acaridae—Adults

1a Female claws bifid; males heteromorphic with bifid claws or single claws; male tarsus III with two large apical spines . . . . ***Lardoglyphus***

1b Tarsi without bifid claws; male tarsus III without two apical spines **2**

2a Cuticle with tubercles, scales, or wrinkles **3**

2b Cuticle smooth. **4**

3a Cuticle covered with tubercles; idiosomal setae pectinate; in decaying logs ***Fagacarus***

3b Cuticle with scales or scalelike wrinkles . . . . ***Suidasia***

4a External vertical setae (*ve*) well developed, situated anteriorly near the internal vertical setae (*vi*). **5**

4b External vertical setae very short or absent, when present situated mediolaterally on the propodosomal shield; hysterosomal chaetotaxy often reduced, with several pairs of setae absent **8**

5a External scapular setae (*sce*) as long or longer than internal scapular setae (*sci*) **6**

5b Internal scapular setae longer than external scapular setae; solenidion σ_1 on genu I less than three times as long as σ_2 . . . . **7**

6a Setae *ve* more than half the length of *vi,* solenidion σ_1 on genu I less than three times as long as σ_2 . . . . ***Aleuroglyphus***

6b Setae *ve* less than half the length of *vi;* solenidion σ_1 on genu I more than three times longer than σ_2. . . . . ***Acarus***

7a Setae *vi* more than four times as long as *ve;* setae *ve* smooth and located posteriorly to *vi*. . . . . ***Mycetoglyphus***

7b Setae *vi* much less than four times as long as *ve;* setae *ve* pectinate and located lateral or slightly posterolateral to *vi* . . . . ***Tyrophagus***

8a Seta *ba* on tarsi I and II in front of ω_1 large, thick, spinose (Fig. 22.8) . . . **14**

8b Seta *ba* on tarsi I and II in front of ω_1 slender, simple **9**

9a Setae *sci,* l_1, d_1–d_3 absent **10**

9b All idiosomal setae present. **11**

10a Setae *vi, sci,* l_1, d_1, d_2, and d_3 absent . . . . ***Michaelopus***

10b Setae *ve, sci,* l_1, d_1, d_2, d_3, l_2, and l_4 absent; posterior end of male hysterosoma with a shield terminating in a projecting semicircular lobe. . . . . ***Thyreophagus***

11a Supracoxal setae (*scx*) thin and long, with a few very short barbs **12**

11b Supracoxal setae thick with many long barbs; ventral setae of tibiae spinose ***Cosmoglyphus***

12a Ventral setae of tibiae spinose; one to three foliate setae present on tarsi I and II. . . . . ***Sancassania***

12b Ventral setae of tibiae simple; tarsi without foliate setae **13**

13a Setae d_3 and d_4 two or more times longer than setae d_1 and d_2 . . . . ***Acotyledon***

13b Setae d_3 and d_4 about as long as setae d_1 and d_2; associated with bees ***Horstia***

14a Genu of leg III with one solenidion and no setae; hysterosomal shield present in males and present or absent in females; setae *sci,* l_1, d_1, d_2, and d_5 absent; in water-filled tree holes . . . . ***Naiadacarus***

14b Genu of leg III with one solenidion and one seta **15**

15a Setae *ve* and *sci* usually present; setae d_1 and d_2 present; females with six pairs of anal setae; males without hysterosomal shield. . . . . ***Rhizoglyphus***

15b Setae *ve* absent; setae *sci*, d_1, and d_2 usually absent; females with at most two pairs of anal setae; males with hysterosomal shield. **16**

16a Setae *sce, h*, d_4, and l_4 more than 25% of the length of the idiosoma; males with a posterior hysterosomal shield terminating in a four-lobed, fanlike structure; setae *sci*, l_1, d_1, and d_2 absent. . . . . ***Histiogaster***

16b Setae *sce* and d_4 less than 25% of the length of the idiosoma; males' hysterosomal shield not projecting posteriorly, without lobes; setae *sci*, l_1, d_1, and d_2 usually absent. . . . . ***Schwiebea***

Key to Genera of Acaridae—Hypopi

1a Eyes present, located far apart on the anterolateral margins of the propodosoma . . . . ***Michaelopus***

1b Eyes absent or located close together at apex of propodosoma **2**

2a Leg IV without claw **3**

2b Leg IV with claw **4**

3a Palposoma present; associated primarily with dermestid beetles, *Dermestes* ***Lardoglyphus***

3b Palposoma absent; associated with bees and wasps . . . . ***Tortonia***

4a Eyes generally present; apodemes III and IV usually free . . . ***Histiogaster***

4b Eyes absent; apodemes III and IV usually united medially **5**

5a Genu I with two apical solenidia . . . . ***Schulzea***

5b Genu I with one solenidion **6**

6a Tarsus IV with one or two macrosetae greater than or equal to half the length of the idiosoma; basal setae of genua I and II pectinate. **7**

6b Tarsus IV without macrosetae; basal setae of genua I and II smooth . . . **9**

7a Tarsus IV with one macroseta; tarsus III without macrosetae; setae *sci* and *sce* usually long, reaching posterior edge of propodosomal shield. . . . ***Acarus***

7b Tarsus IV with two macrosetae, tarsus III with one macroseta **8**

8a Idiosoma dorsally punctate, without striations; setae *sci* and *sce* short, not reaching posterior edge of propodosomal shield . . . . ***Lackerbaueria***

8b Idiosoma dorsally punctate, with striations; associated with bees . . . . . . ***Horstia***

9a Setae d_1–d_5 well developed, thick; two solenidia on tarsus I **10**

9b Setae d_1–d_5 short, simple; usually three solenidia on tarsus I. **11**

10a Setae *sci, sce*, d_1–d_4 leaflike, ridged; setae l_2–l_4, d_5 thin, setiform; associated with the tenebrionid beetle ***Bolitotherus*** . . . . ***Boletoglyphus***

10b Setae *sci, sce* d_1–d_5, l_2–l_4 spinose; tarsi III and IV with heavy clawlike setae ***Konoglyphus***

11a Palposoma shorter than its basal width; associated with bees . . . . ***Kuzinia***

11b Palposoma much longer than its basal width **12**

12a Sucker plate far from posterior end of body; margins of body flattened, very wide; legs III and IV usually not visible from above; genua IV longer than femora IV. . . . . ***Cosmoglyphus***

12b Sucker plate near posterior end of body; body margins not flattened **13**

13a Famulus absent; associated with the fly ***Mallota*** and water-filled tree holes ***Naiadacarus***
13b Famulus small, spinose, located next to ω_1 . . . . **14**
14a Sucker plate reduced, with one pair of suckers and no conoids, or longer than wide, with two pairs of suckers and four small conoids in a transverse line ***Acotyledon***
14b Sucker plate wider than long, with 2 large pairs of suckers and four conoids not on a straight line **15**
15a Setae *vi* usually very small, about equal to *sci*. . . . . ***Schwiebea***
15b Setae *vi* usually much longer than *sci*. **16**
16a Setae *vi* 35–44 μm long, as long as or slightly shorter than palposomal solenidion α; epimera I and III with conoids (Fig. 9), tarsus I with eight setae ***Rhizoglyphus***
16b Setae *vi* generally less than 35 μm, shorter than palposomal solenidion α; epimera I and III with conoids or simple setae; tarsus I often with nine or more setae . . . . ***Sancassania***

Key to Genera of Anoetidae—Adults

1a Female genital opening transverse; male tarsi without pretarsal lobes, aedeagus not protruding and genital suckers lacking . . . . ***Histiostoma***
1b Female genital opening longitudinal; male tarsi with fleshy pretarsal lobes (Fig. 22.15) or genital suckers present 2
2a Posterior sclerotized rings far above anal opening; males with genital suckers ***Bonomoia***
2b Posterior sclerotized rings very near the anus, within a distance equal to the ring diameter; males without genital suckers 3
3a Dorsal setae short; female apodemes I not united; male aedeagus largely protruding and vertical setae short. . . . . ***Myianoetus***
3b Dorsal setae long, situated on tubercles; female apodemes I united; male aedeagus not protruding and vertical setae long, as long as the gnathosoma . . . ***Glyphanoetus***

Key to Genera of Anoetidae—Hypopi

1a Claws forked . . . . ***Myianoetus***
1b Claws undivided 2
2a Some hysterosomal setae short, thick, spinose . . . . ***Hexanoetus***
2b Hysterosomal setae filiform 3
3a Eyes present on anterolateral hysterosoma . . . . ***Bonomoia***
3b Eyes absent 4
4a All legs with claws 5
4b Leg IV without claw 6
5a Simple setae or setal insertions present on epimere I, conoids on epimere III; some hysterosomal setae long, anteriorly directed . . . . . . ***Glyphanoetus***
5b Simple setae, insertions, or conoids on both epimeres . . . . . . ***Histiostoma***
6a Genu and femur of legs III and IV partly fused, leg segments elongate, dorsum foveolate, phoretic on the fly ***Haematobia*** . . . . ***Ameranoetus***
6b Genu and femur of legs III and IV distinctly separate 7

7a Leg III without claw, phoretic on halictid bees . . . . *Anoetus*
7b Leg III with claw . . . . *Histiostoma*

Key to Genera of Chaetodactylidae—Adults

1a Propodosomal shield absent; apodemes III and IV fused; associated with *Osmia* . . . . *Chaetodactylus*
1b Propodosomal shield present; apodemes III and IV fused; associated with *Xylocopa* . . . . *Sennertia*

Key to Genera of Chaetodactylidae—Hypopi

1a Propodosomal shield absent; associated with *Osmia* . . . . . . *Chaetodactylus*
1b Propodosomal shield present; associated with *Xylocopa* . . . . . . *Sennertia*

Key to Genera of Glycyphagidae—Adults

1a Cuticle densely striate with large rough scales on dorsal idiosoma; body unsclerotized, pale; genu I with two apical solenidia; setae *vi* and *ve* far apart *Aeroglyphus*
1b Cuticle not striate, generally with spines, scales, tubercles, or pits 2
2a Body unsclerotized, pale, usually with minute spines or tubercles 3
2b Body sclerotized, brown, granular, usually with scales, tubercles, or pits 10
3a Two apical solenidia on genu I; claws and condylophores poorly developed 4
3b One spical solenidion on genu I 9
4a Large leaflike subtarsal scale present; setae *ve* posterolateral to *vi* . . . . . . *Lepidoglyphus*
4b Leaflike subtarsal scale absent 5
5a Cuticle with minute spines or smooth; setae d_5 long, barbed 6
5b Cuticle with small tubercles 8
6a Setae d_2 simple, short; propodosomal shield absent . . . . *Echimyopus*
6b Dorsal idiosomal setae all long, barbed 7
7a Narrow, elongate propodosomal shield present; cuticle with minute spines; setae *ve* posterolateral to *vi* . . . . *Glycyphagus*
7b Propodosomal shield absent; male cuticle smooth; setae *ve* anterolateral to *vi;* associated with marsupials . . . . *Marsupialichus*
8a Propodosomal shield absent; setae d_5 short, spinose; associated with muskrats . . . . *Zibethacarus*
8b Propodosomal shield present; setae d_5 barbed . . . . *Dermacarus*
9a Claws and condylophores not visible; pretarsus small; sejugal furrow absent; anal opening not bordering genital area . . . . *Blomia*
9b Claws and condylophores well developed, acaridlike; pretarsus large; sejugal furrow present; anal opening bordering genital area . . . . *Euglycyphagus*
10a Dorsal hysterosomal setae pectinate; one ventral seta on tibiae I and II. . . *Ctenoglyphus*
10b Dorsal hysterosomal setae smooth; two ventral setae on tibiae I and II 11
11a Cuticle pitted; hysterosoma rounded posteriorly, with fine setae 12

11b Cuticle with scales, tubercles, or small spines; hysterosoma lobed or rounded posteriorly, with thick spinose setae and occasionally fine setae **14**

12a Genu III with one solenidion and one seta; tibia IV with one solenidion and one seta **13**

12b Genu III with one seta and no solenidia; tibia IV with one solenidion and no setae . . . . ***Xenoryctes***

13a Genua and tibiae of legs with ridges and pectinate setae; legs I–IV similar, articulated ventrally . . . . ***Gohieria***

13b Legs not ridged, with spinose setae; legs I and II articulated dorsally, much thicker than legs III and IV; in tree holes . . . . ***Fusohericia***

14a Females with setae d_1, d_3, d_4, fine; hysterosoma with small spines and pits, posteriorly rounded; apodemes fused on each side to form two lateral sclerotized arcs next to the genital opening. Males with dorsal setae spinose; hysterosoma with scales, two pairs of anal setae. . . . . ***Orycteroxenus***

14b Dorsal setae spinose; posterior hysterosoma with lobes bearing small spines, most lobes with long spinose setae. Females with six pairs of anal setae; apodemes I and II free, III and IV fused. Males with four pairs of anal setae ***Labidophorus***

Key to Genera of Glycyphagidae—Hypopi

1a Tarsus III without claw **2**

1b Tarsus III with claw. **3**

2a Posteroventral hysterosoma with two pairs of ridged claspers and long narrow clasping folds attached basally and extending beyond the posterior end of the body; associated with kangaroo rats, ***Dipodomys*** . . . . ***Dipodomyopus***

2b Posteroventral hysterosoma with very small, sclerotized claspers; clasping folds not free, not extending beyond posterior end of body; associated with mice, ***Heteromys,*** and ***Perognathus*** . . . . ***Neolabidophorus***

3a Tarsus III with single claw **4**

3b Tarsus III with forked claw . . . . ***Ctenoglyphus***

4a Hysterosoma posteroventrally with two pairs of ridged claspers. **6**

4b Hysterosoma without posteroventral ridged claspers **5**

5a Claws with pretarsus and condylophores; sucker plate present, with a pair of very large posterior suckers and a pair of very small anterior suckers; sejugal furrow present; tarsi without macrosetae; propodosomal and most hysterosomal setae pectinate; associated with trogid beetles, ***Trox*** . . . ***Euglycyphagus***

5b Claws sessile; idiosoma without sucker plate; sejugal furrow absent; one macroseta on tarsus IV; in hair follicles, skin or nests of mammals. . . . . . . ***Echimyopus***

6a Genital papillae located laterally, adjacent to trochanter IV **7**

6b Genital papillae medially located near genital opening **8**

7a Genital papillae short, rounded; associated with ***Heteromys*** . . . . ***Mediolabidophorus***

7b Genital papillae long, generally directed posteriorly. **8**

8a Flattened barbed seta on tibia IV . . . . ***Metalabidophorus***

8b Cylindrical barbed seta on tibia IV; associated with beavers, ***Aplodontia***. . . ***Microlabidopus***

9a Setae *ve* absent . . . . ***Dermacarus***

9b Setae *ve* present 10
10a Claws IV much shorter than claws III 11
10b Claws III and IV equal or subequal 12
11a Claws I and III equal or subequal; associated with marsupials . . . . ***Marsupialichus***
11b Claws I much longer than claws III. . . . . ***Labidophorus***
12a Sides of hysterosoma with one or two pairs of recurved hooklike processes ***Orycteroxenus***
12b Sides of hysterosoma rounded 13
13a Tarsus IV with one macroseta, 120 μm long; associated with moles, ***Scalopus*** ***Scalopacarus***
13b Tarsus IV without macrosetae 14
14a Tarsi I and II with six setae; claspers very small and poorly developed; dorsal idiosoma covered with small pits connected by lines forming a network. . . ***Neoxenoryctes***
14b Tarsi I and II with eight setae; claspers well developed; dorsal idiosoma covered with pits, reticulate, or smooth . . . . ***Xenoryctes***

Key to Genera of Pyroglyphidae—Adults

1a Propodosomal shield anteriorly broad and truncate; external scapular setae (*sce*) much longer than internal scapulars (*sci*); two pairs of long posterior hysterosomal setae 3
1b Propodosomal shield anteriorly narrowing and elongate, scapular and hysterosomal setae short 2
2a Genu I with one solenidion; male with anal suckers . . . . ***Euroglyphus***
2b Genu I with two solenidia; male without anal suckers . . . . . . ***Pyroglyphus***
3a Females with dorsal hysterosomal shield; first pair of genital papillae of male anterior to genital opening; males with simple setae on tarsus III . . . . . . ***Sturnophagoides***
3b Females without dorsal hysterosomal shield; both pairs of genital papillae of male lateral to genital opening or male with two medial thick spinose setae on tarsus III 4
4a Tarsus I and generally tarsus II with chitinous spinose apical processes, males with simple setae medial on tarsus III; male genital papillae lateral to genital opening . . . . ***Dermatophagoides***
4b Tarsi I and II without chitinous spinose apical processes; male with two medial thick spinose setae on tarsus III; male genital papillae anterior to genital opening . . . . ***Hirstia***

Key to Genera of Saproglyphidae—Adults

1a Legs without claws 2
1b Legs with claws 3
2a Cuticle smooth . . . . ***Nanacarus***
2b Cuticle with scales . . . . ***Sapracarus***
3a Setae d_1–d_3 thick, spinose . . . . ***Kennethiella***
3b Setae d_1–d_3 simple 4
4a Three pairs of genital setae; three pairs of anal setae 5

4b Three pairs of genital setae; four pairs of anal setae . . . . . *Monobiacarus*
5a Setae *ve* present. . . . . *Vespacarus*
5b Setae *ve* absent. . . . . *Calvolia*

Key to Genera of Saproglyphidae—Hypopi

1a Tarsus III with three setae . . . . *Congovidiella*
1b Tarsus III with more than three setae. 2
2a Tarsus III with four setae. . . . . *Sapracarus*
2b Tarsus III with more than four setae 3
3a Tarsus III with five setae . . . . *Euvidia*
3b Tarsus III with more than five setae 4
4a Tarsus III with six setae. 5
4b Tarsus III with more than six setae 9
5a Eyes present. 6
5b Eyes absent . . . . *Crabrovidia*
6a Tarsus IV with three setae . . . . *Calvolia*
6b Tarsus IV with more than three setae. 7
7a Tarsus IV with four setae . . . . *Vespacarus*
7b Tarsus IV with five setae 8
8a Tarsus III with a ventral spinose process; palposomal setae present . . . . *Kennethiella*
8b Tarsus III without a spinose process; palposoma and palposomal solenidion absent. . . . . *Monobiacarus*
9a Tarsus III with seven setae. 10
9b Tarsus III with eight setae . . . . *Vidia*
10a Palposomal setae present. . . . . *Zethovidia*
10b Palposomal setae absent . . . . *Zethacarus*

References

Akimov, E. A. and L. E. Schehur. 1972. Features of the nutrition of the grain mites *Glycyphagus domesticus* (DeG.), *Tyrophagus putrescentiae* (Schr.) and the root mite *Rhizoglyphus echinopus* (Fum. and Rob.) on some proteinoids (in Russian, English summary). *Vestn. Zool.* 6(6):45–48. **Atyeo, W. T. and J. Gaud. 1966.** The chaetotaxy of sarcoptiform feather mites. *J. Kans. Entomol. Soc.* 39:337–346. **Baker, E. W. 1962.** Some Acaridae from bees and wasps (Acarina). *Proc. Entomol. Soc. Wash.* 64(1):1–10. **Baker, E. W. and D. A. Crossley, Jr. 1964.** A new species of mite, *Fusohericia lawrencei,* from an artificial tree-hole (Glycyphagidae). *Ann. Natal Mus.* 16:1–6. **Baker, E. W. and F. Cunliffe. 1960.** Notes on saproglyphid mites associated with solitary wasps. *Proc. Entomol. Soc. Wash.* 62:209–231. **Baker, E. W., T. M. Evans, D. I. Gould, W. B. Hull and H. L. Keegan. 1967.** *A Manual of Parasitic Mites of Medical or Economic Importance.* Natl. Pest Control Assoc. Tech. Publ. 170 pp. **Barker, P. S. 1964.** Bionomics of *Caloglyphus anomalus* Nesbitt (Acarina : Acaridae). *Manit. Entomol.* 8:41–47. **Behan, V. M., S. B. Hill and D. K. McE. Kevan. 1978.** Effects of nitrogen fertilizers, as urea, on Acarina and other arthropods in Quebec black spruce humus. *Pedobiologia* 18:249–263. **Cooreman, J. 1959.** Note sur le genre *Aeroglyphus* Zachvatkine, 1941 (Acaridiae, Glycyphagidae). *Bull. Inst. R. Sci. Nat. Belg.* 35(33):1–19. **Cross, E. A. 1968.** Descriptions of three new species of mites phoretic upon native bees. *Southwest. Nat.* 13(3):325–334. **Davis, B. N. K. 1963.** A study of micro-arthropod communities in mineral soils near Corby, Northants. *J. Anim. Ecol.* 32:49–71. **Delfinado, M. D. and E. W. Baker. 1976.** Notes on hypopi (Acarina) associated with bees and wasps. *J. N.Y. Entomol. Soc.* 84(2):76–90. **Dindal, D. L., J. P. Moreau and L. Theoret. 1977.** *Effect of Spray Irrigation of Municipal Waste-Water on Soil Invertebrate Populations and the Potential Influence on Physical Factors of Soil.* Tech. Rep. Proj. B-048-NY. Office of Water Research and Technology. U.S. Dept. of Interior, Washington, D.C. 16 pp. **Douce, G. K. and D. A. Crossley, Jr. 1977.** Acarine abundance and community structure in an arctic coastal tundra. *Pedobiologia* 17:32–42. **Fain, A. 1967.** Solenidiotaxy of leg I in the hypopi of the Acaridiae (Acari: Sarcoptiformes). *Rev. Zool. Bot. Afr.* 76(3–4):244–248. **1969.** Les deutonymphes hypopiales vivant en association phoretique sur les mammifères (Acarina: Sarcoptiformes). *Bull. Inst. R. Sci. Nat. Belg.* 45(33):1–262. **1972.** Notes sur les hypopes des Saproglyphidae. (Acarina: Sarcoptiformes). II. Redéfinition des genres. *Acarologia (Paris)* 14(2):225–249. **1977.** Caractères comparés des hypopes de *Rhizoglyphus echinopus* Fum. et Robin et de *Sancassania chelone* Oudemans (Astigmates: Acaridae). *Acarologia (Paris)* 19(1):105–111. **1977.** Notes sur le genre *Schwiebea* Oudemans, 1916 (Acarina, Astigmata, Acaridae). *Bull. Ann. Soc. R. Belg. Entomol.* 113:251–276. **Fain, A. and D. E. Johnston. 1974.** Three new species of hypopi phoretic on springtails (Collembola) in England (Acari: Acaridiae). *J. Nat. Hist.* 8:411–420. **Fain, A. and R. A. Norton. 1979.** *Fagacarus verrucosus* n.g., n.sp. (Acari, Astigmata) from decaying beechwood in the U.S.A. *Int. J. Acarol.* 5(1):5–8. **Fain, A. and J. R. Philips. 1977a.** Astigmatic mites from nests of birds of prey in the U.S.A. I. Description of four new species of Glycyphagidae. *Int. J. Acarol.* 3(2):105–114. **1977b.** Astigmatic mites from nests of birds of prey in the U.S.A. II. Two new species of the genera *Fusacarus* Michael and *Blomia* Oudemans (Glycyphagidae). *Acta Zool. Pathol. Antverp.* 69:155–162. **1978a.** Astigmatic mites from nests of birds of prey in the U.S.A. III. *Sapracarus tuberculatus* g.n., sp.n (Acari, Astigmata, Saproglyphidae). *Acta Zool. Pathol. Antverp.* 70:227–231. **1978b.** Astigmatic mites from nests of birds of prey in the U.S.A. IV. Description of the life-cycle of *Acotyledon paradoxa* Oudemans, 1903. *Zool. Meded. (Leiden)* 53(3):29–39. **1978c.** Notes on the genus *Suidasia* Oudemans, 1905 with description of a new species from Australia. *Int. J. Acarol.* 4(2):115–123. **1979.** Astigmatic mites from nests of birds of prey in the U.S.A. V. Four new species of Anoetidae. *Int. J. Acarol.* 5(2):147–153. **Fain, A. and G. Wauthy. 1979.** Les Acaridae (Acari, Astigmata) d'horizons hemiédaphiques dans trente forêts décidues de Belgiques. *Bull. Ann. Soc. R. Belg. Entomol.* 115:169–182. **Fain, A. and J. O. Whitaker. 1973.** Phoretic hypopi of North American mammals (Acarina: Sarcoptiformes, Glycyphagidae). *Acarologia (Paris)* 15(1):144–170. **Fashing, N. A. 1973.** The post-embryonic stages of a new species of *Mauduytia* (Acarina: Anoetidae). *J. Kans. Entomol. Soc.* 46:454–468. **1974.** A new subfamily of Acaridae, the Naiadacarinae, from water-filled treeholes (Acarina: Acaridae). *Acarologia (Paris)* 16(1):166–181. **1980.** *Neoguanolichus panamensis,* a new species of Guanolichidae from bat caves in Panama (Acarina: Sarcoptiformes). *Acarologia (Paris)* 21(1):108–116. **Fashing, N. J. and B. M. O'Connor. 1984.** *Sarraceniopus*—a new genus for histiostomatid mites inhabiting the pitchers of the *Sarraceniaceae* (Astigmata: Histiostomatidae). *Int. J. Acarol.* 10(4):217–227. **Garman, P. 1921.** *A study of the bulb mite. Conn. Agric. Exp. Stn. Bull.* 225:115–132. **Grandjean, F. 1939.** La chaetotaxie des pattes chez les Acaridiae. *Bull. Soc. Zool. Fr.* 64:50–60. **Griffiths, D. A. 1970.** A further systematic study of the genus *Acarus* L. 1758 (Acaridae, Acarina), with a key to species. *Bull. Br. Mus. (Nat. Hist.) Zool.* 19(2):85–118. **Griffiths, D. A. and J. Boczek. 1977.** Spermatophores of some acaroid mites (Astigmata: Acarina). *Int. J. Insect. Morphol. Embryol.* 6(516):231–238. **Harding, D. J. and R. A. Stuttard. 1974.** Microarthropods. Pp. 489–532 in C. H. Dickinson and G. J. F. Pugh (eds.). *Biology of Plant Litter Decomposition,* Vol. 2. Academic Press, New York. **Hughes, A. M. 1964.** The chaetotaxy of some Acaridei (Astigmata). Proceedings of the First International Congress on Acarology. *Acarologia (Paris)* 6(suppl):65–74. **1976.** *The mites of stored food and houses. Ministry Ag. Fish Food Tech. Bull.* 9:1–400. **Hughes, R. D. and C. G. Jackson. 1958.** A review of the Anoetidae (Acari). *Va. J. Sci. N.S.* 9(1):5–198. **Ide, G. S. and S. Mahunka. 1978.** *Ameranoetus haematobii* gen. et sp. nov. from the USA (Acari: Anoetidae). *Folia Entomol. Hung.* 31(1):47–49. **Johnston, D. E. and W. A. Bruce. 1965.** *Tyrophagus neiswanderi,* a new acarid mite of agricultural importance. *Ohio Agric. Res. Dev. Cent.* 977:3–17. **Karg, W. 1978.** Milben als Indikatoren zur Optimierung von Pflanzenschutzmassnahmen in Apfelintensivanlagen. *Pedobiologia* 18:415–425. **Kines, K. A. and R. N. Sinha. 1973.** A laboratory study of fauna and flora in an agricultural soil in Manitoba. *Manit. Entomol.* 7:59–66. **Krantz, G. W. 1978.** *A Manual of*

Acarology. Oregon State University, Corvallis, Oreg. 509 pp. **Linquist, E. L. 1977.** Acari. Pp. 252–290 in H. V. Danks (ed.). *Canada and Its Insect Fauna. Mem. Ent. Soc. Can.* 108. 573 pp. **Lustgraaf, B. V. D. 1978.** Ecological relationships between xerophilic fungi and house-dust mites (Acarida : Pyroglyphidae). *Oecologia (Berl.)* 33:351–359. **Mahunka, S. 1974.** Beiträge zur Kenntnis der an Hymenopteren lebenden Milben (Acari). II. *Folia Entomol. Hung.* 27(1):99–108. **Manson, D. C. M. 1971.** A contribution to the study of the genus *Rhizoglyphus Claparède 1869* (Acarina : Acaridae). *Acarologia (Paris)* 13(4):621–650. **Marshall, V. G. 1979.** Effects of the insecticide diflubenzuron on soil mites of a dry forest zone in British Columbia. Pp. 129–134 in J. G. Rodriguez (ed.). *Recent Advances in Acarology,* Vol. 1. Academic Press, New York. 631 pp. **Mostafa, A. R. 1970.** Saproglyphid hypopi (Acarina: Saproglyphidae) associated with wasps of the genus *Zethus* Fabricius. *Acarologia (Paris)* 12:168–192. **Murphy, P. W. 1962.** Extraction methods for soil animals. I. Dynamic methods with particular reference to funnel processes. Pp. 75–114 in P. W. Murphy (ed.). *Progress in Soil Zoology.* Butterworths, London. **O'Connor, B. M. 1979.** A new species of *Ctenoglyphus* (Acari: Glycyphagidae) from *Meriones unguiculatus* (Rodentia : Cricetidae) with notes on the subfamily Ctenoglyphinae. *J. Med. Entomol.* 15(3):265–271. **1982.** Evolutionary ecology of astigmatid mites. *Annu. Rev. Entomol.* 27:385–410. **1984.** Acarine–fungal relationships: the evolution of symbiotic associations. Pp. 354–381 in Q. Wheeler and M. Blackwell (eds.). *Fungus–Insect Relationships.* Columbia University Press, New York. 514 pp. **O'Connor, B. M. and W. K. Reisen. 1978.** *Chiroptoglyphus,* a new genus of mites associated with bats with comments on the family Rosensteiniidae. *Int. J. Acarol.* 4(3):179–194. **Ostaff, D. P. and R. Freitag. 1973.** Phenology of carabid beetles of Thunder Bay and their associated mite population. *Proc. Entomol. Soc. Ont.* 104:34–37. **Oswald, E. T. and L. W. Minty. 1970.** Soil acarine fauna of southeastern Manitoba. I. Forest soils. *Manit. Entomol.* 4:76–87. **1971.** Soil acarine fauna of southeastern Manitoba. II. Riparian communities. *Manit. Entomol.* 5:71–77. **Philips, J. R. 1984.** Acarine symbionts of trogid beetles. Pp. 552–556 in D. A. Griffiths and C. E. Bowman (eds.). *Acarology VI,* Vol. 1. Ellis Horwood, Chichester, West Sussex, England. 645 pp. **Philips, J. R. and D. L. Dindal. 1979.** The acarine community of nests of birds of prey. Pp. 559–562 in J. G. Rodriguez (ed.). *Recent Advances in Acarology,* Vol. 1. Academic Press, New York. 631 pp. **Philips, J. R. and R. A. Norton. 1978.** *Lardoglyphus falconidus* n.sp. (Acarina : Acaridae) from the nest of an American kestrel (*Falco sparverius* L.). *Acarologia (Paris)* 20(1):128–137. **Pillai, P. R. P. and P. W. Winston. 1963.** A preliminary study of the Acaridiae (Acarina, Sarcoptiformea) of Colorado. *Entomol. News* 74:49–55. **Pimentel, D., M. W. Rumsey and F. A. Streams. 1960.** Rearing tyroglyphid mites on *Neurospora. Ann. Entomol. Soc. Am.* 53(4):549. **Price, D. W. 1973.** Abundance and vertical distribution of microarthropods in the surface layers of a California pine forest soil. *Hilgardia* 42(4):121–144. **Robaux, P., C. Y. Jeanson and D. Barbier. 1977.** Microstructures construites par un acarien *Tyrophagus putrescentiae* dans une argile. Étude experimentale et microscopique. *Ecol. Bull.* 25:489–493. **Rodriguez, J. G. and Z. A. Stepien. 1973.** Biology and population dynamics of *Caloglyphus berlesei* (Michael) (Acarina : Acaridae) in xenic diet. *J. Kans. Entomol. Soc.* 46(2):176–183. **Salt, G., F. S. J. Hollick, F. Raw and M. V. Brian. 1948.** The arthropod population of pasture soil. *J. Anim. Ecol.* 17:139–150. **Scheucher, R. 1959.** Systematik and Ökologie der deutschen Anoetinen. *Beitr. Syst. Okol. Mitteleur. Acarina* 1(2):233–384. **Sheals, J. G. 1957.** The collembola and acarina of uncultivated soil. *J. Anim. Ecol.* 26:125–134. **Stepien, Z. A., W. Goscynski and J. Boczek. 1971.** The energy budget of *Tyrophagus putrescentiae.* Pp. 373–388 in M. Daniel and B. Rosicky (eds.). *Proceedings of the 3rd International Congress on Acarology.* Dr. W. Junk, The Hague, The Netherlands. 837 pp. **Tomlin, A. D. 1977.** Pipeline construction-impact on soil micro- and mesofauna (Arthropoda and Annelida) in Ontario. *Proc. Entomol. Soc. Ont.* 108:13–17. **Turk, E. and F. Turk. 1959.** Systematik und Ökologie der Tyroglyphiden Mitteleuropas. *Beitr. Syst. Okol. Mitteleur. Acarina* 1(1):3–321. **Wallwork, J. A. 1967.** Acari. Pp. 363–396 in A. Burges and F. Raw (eds.). *Soil Biology.* Academic Press, New York. 532 pp. **1972.** Distribution patterns and population dynamics of the microarthropods of a desert soil in southern California. *J. Anim. Ecol.* 41:291–310. **Walter, D. E. 1987.** Trophic behavior of "mycophagous" microarthropods. *Ecology* 68(1):226–229. **Walter, D. E., R. A. Hudgens and D. W. Freckman. 1986.** Consumption of nematodes by fungivorous mites, *Tyrophagus* spp. (Acarina, Astigmata, Acaridae). *Oecologia (Berl.)* 70(3):357–361. **Wharton, G. W. 1976.** House dust mites. *J. Med. Entomol.* 12(6):577–621. **Weigert, R. G. 1974.** Litterbag studies of microarthropod populations in three

South Carolina old fields. *Ecology* 55:94–102. **Woodring, J. P. 1966a.** North American Tyroglyphidae (Acari). I. New species of *Calvolia* and *Nanacarus,* with keys to the species. *Proc. La. Acad. Sci.* 29:76–84. **1966b.** North American Tyroglyphidae (Acari). III. The genus *Histiogaster,* with descriptions of four new species. *Proc. La. Acad. Sci.* 29:113–136. **1969.** Observations on the biology of six species of acarid mites. *Ann. Entomol. Soc. Am.* 62(1):102–108. **1973.** Four new anoetid mites associated with halictid bees. *J. Kans. Entomol. Soc.* 46(3):310–327. **Zakhvatkin, A. A. 1941.** *Fauna of the U.S.S.R. Arachnoidea 6(1), Tyroglyphoidea (Acari).* Zool. Inst. Acad. Sci. USSR, N.S. 28. Translated by American Institute of Biological Science. 573 pp.

Bibliography

Additional references of value to researchers and students are listed below.

Bååth, E., B. Berg, U. Lohm, B. Lundgren, H. Lundkvist, T. Rosswall, B. Soderström and A. Wirren. 1980. Effects of experimental acidification and liming on soil organisms and decomposition in a Scots pine forest. *Pedobiologia* 20:85–100. **Baker, G. T. 1983.** Observations on the biology of *Rhizoglyphus robini* Claparède (Acaridae). *J. Appl. Entomol.* 95(2):162–165. **Bolger, T. 1985.** The rate of disappearance of *Salix* litter on cutaway bog and the microarthropods associated with it. *Pedobiologia* 28:145–153. **Bolger, T., and J. P. Curry. 1980.** Effects of cattle slurry on soil arthropods in grassland. *Pedobiologia* 20:246–253. **Broadbent, A. B., and A. D. Tomlin. 1980.** Effects of carbofuran on the soil microcommunity in a cornfield. Pp. 82–93 *in* D. L. Dindal (ed.). *Soil Biology as Related to Land Use Practices.* Proc. VII Soil Zool. Coll. USEPA Office of Pesticide and Toxic Substances. Washington, D. C. 880 pp. **Critchley, B. R., A. G. Cook, U. Critchley, T. J. Perfect, A. Russell-Smith, and R. Yeadon. 1979.** Effects of bush clearing and soil cultivation on the invertebrate fauna of a forest soil in the humid tropics. *Pedobiologia* 19:425–438. **Fain, A. 1982.** Revision des genres *Thyreophagus* Rondani, 1874, et *Michaelopus* Fain & Johnston, 1974 (Acari, Acaridae) avec description de neuf espèces nouvelles. *Bull. Inst. roy. Sci. nat. Belg.* 54(7):1–47. **Fain, A. 1982.** Cinq espèces du genre *Schwiebea* Oudemans, 1916 (Acari, Astigmata) dont trois nouvelles découvertes dans des sources du sous-sol de la ville de Vienne (Autriche) au cours des travaux du Metro. *Acarologia* 23(4):359–371. **Fain, A. 1985.** Notes on two genera of mites (*Viedebanttia* and *Nanacarus*) (Acari: Astigmata) described by Oudemans. *Zool. Meded.* 59(23):275–282. **Fain, A. 1985.** Observations sur les genres *Suidasia* Oudemans, 1905, *Tyrophagus* Oudemans, 1924 et *Madaglyphus* Fain, 1971 (Acari, Acaridae). *Rev. Zool. afr.* 99:159–164. **Fain, A., and Y. Mumcuoglu. 1979.** *Glycyphagus* (*Myacarus*) *helveticus* sp. N. (Acari, Astigmata) vivant dans le nid d'*Arvicola terrestris* en Suisse. *Rev. Suisse Zool.* 86(2):455–461. **Fain, A., and J. R. Philips. 1981.** Astigmatic mites from nests of birds of prey in the USA. VI. The adult forms of *Echimyopus orphanus* Fain & Philips, 1977 and of *Dermacarus pilitarsus* Fain & Philips, 1977. *Intl. J. Acarol.* 7(1–4):235–237. **Fain, A., J. R. Philips, and N. Wilson. 1979.** Two new species of the genus *Dermacarus* Haller, 1880 (Acari, Glycyphagidae). *Intl. J. Acarol.* 5(1):73–76. **Fain, A., and E. J. Spicka. 1986.** The life cycle of *Lepidoglyphus hylandi* (Fain, 1969) comb. nov. (Acari, Glycyphagidae). *Acarologia* 27(4):317–324. **Fain, A., E. J. Spicka, G. S. Jones and J. O. Whitaker, Jr. 1985.** The life-cycle of two astigmatic mites, *Glycyphagus* (*Zapodacarus*) *newyorkensis* (Fain, 1979) comb. nov. and *Glycyphagus* (*Zapodacarus*) *zapus* sp. n. (Glycyphagidae). Acarologia 26(2):155–169. **Ferguson, C. A. 1985.** A new genus and species of mite (Acari: Acaridae) from Mexico. *Zool. J. Linn. Soc.* 85:275–290. **Franco, P. J., E. B. Edney and J. F. McBrayer. 1979.** The distribution and abundance of soil arthropods in the Northern Mojave desert. *J. Arid Envir.* 2:137–149. **Garay, I., J. C. DaFonseca and P. Blandin. 1980.** The effects of trampling on the fauna of a forest floor. I. Microarthropods. Pp. 200–212 *in* D. L. Dindal (ed.). *Soil Biology as Related to Land Use Practices.* Proc. VII Soil Zool. Coll. USEPA Office of Pesticide and Toxic Substances. Washington, D. C. 880 pp. **Gerson, U., S. Yathom, S. Capua and D. Thorens. 1985.** *Rhizoglyphus robini* Claparède (Acari: Astigmata: Acaridae) as a soil mite. *Acarologia* 26(4):371–380. **Gerson, U., and R. Schneider. 1982.** The hypopus of *Hemisarcoptes coccophagus* Meyer (Acari: Astigmata: Hemisarcoptidae). *Acarologia* 23(2):171–176. **Griffiths, D. A. 1979.** The morpho-species and its relationship to the biological species in the genus *Tyrophagus* (Acaridae, Acarina). Pp. 199–212 *in* J. G. Rodriguez (ed.). *Recent Advances in Acarology.* Vol. 1. Academic Press. New York. 631 pp. **Gould, S. J. 1982.** Solitary origins of some societies. *Nat. Hist.* 91(9):6–12. **Heungens, A., and E. van Daele. 1984.** The influence of some acids, bases and salts on the mite and Collembola population of a pine litter substrate. *Pedobiologia* 27:299–311. **Hurlock, E. T., D. M. Armitage and B. E. Llewellin. 1980.** Seasonal changes in mite (Acari) and fungal populations in aerated and unaerated wheat stored for three years. *Bull. ent. Res.* 70:537–548. **Hutson, B. R., and L. G. Veitch. 1983.** Mean annual population density of Collembola and Acari in the soil and litter of three indigenous South Australian forests. *Aust. J. Ecol.* 8:113–126. **Karg, W. 1987.** On the knowledge of the genus *Schwiebea* Oudemans, 1916. (Acarina, Sarcoptiformes). *Deut. Entomol. Zeit.* 34(1–3):141–148. **Lagerlof, J., and O. Andren. 1985.** Succession and activity of microarthropods and enchytraeids during barley straw decomposition. *Pedobiologia* 28:343–357. **Lombert, P. A. H. M., F. S. Lukoschus and B. M. O'Connor. 1982.** The life-cycle of *Cosmoglyphus inaequalis* Fain and Caceres, 1973, with comments on the systematic position of the genus. *Bull. Inst. roy. Sci. nat. Belg.* 54(10):1–17. **Loring, S. J., R. J. Snider and L. S. Robertson. 1981.** The effects of three tillage practices on Collembola populations. *Pedobiologia* 22:172–184. **Lukoschus, F. S., T. Woeltjes, E. A. Juckwer and A. Fain. 1979.** Life cycle of *Orycteroxenus galemys* sp. nov. (Astigmata: Glycyphagidae). *Intl. J. Acarol.* 5(1):29–38. **Luxton, M. 1981.** Studies on the astigmatic mites of a Danish beech wood soil. *Pedobiologia* 22:29–38. **Nesbitt, H. H. J. 1985.** A new mite from bromeliad leaf-axils from Costa Rica. *Intl. J. Acarol.* 11(3):209–214. **Newton, J., and M. P. Thomas. 1979.** The effects of trampling on the soil Acari and Collembola of a heathland. *Intl. J. Envir. Stud.* 13:219–223. **O'Connor, B. M. 1979.** Evolutionary origins of astigmatid mites inhabiting stored products. Pp. 273–278 *in* J. G. Rodriguez (ed.). *Recent Advances in Acarology.* Vol. 1. Academic Press. New York. 631 pp. **O'Connor, B. M. 1981.** *A systematic revision of the family-group taxa in the non-psoroptidid Astigmata.* Ph. D. Dissertation. Cornell University, Ithaca, New York. University Microfilms. 594 pp. **O'Connor, B. M. 1982.** Acari: Astigmata. Pp. 146–169 *in* S. B. Parker (ed.). *Synopsis and Classification of Living Organisms.* Vol. 2. McGraw-Hill. New York. **O'Connor, B. M. 1984.** Nomenclatorial status of some family-group names in the non-psoroptidid Astigmata (Acari: Acariformes). *Intl. J. Acarol.* 10(4):203–207. **Peterson, H., and M. Luxton. 1982.** A comparative analysis of soil fauna populations and their role in decomposition processes. *Oikos* 39(3):287–388. **Philips, J. R. 1981.** Mites (Acarina) from nest of Norwegian birds of prey. *Fauna Norv. Ser. B.* 28:44–47. **Philips, J. R., M. Root and P. DeSimone. 1983.** Arthropods from a saw-whet owl (*Aegolius acadicus*) nest in Connecticut. *Entomol. News* 94(2):60–64. **Purvis, G., and J. P. Curry. 1980.** Successional changes in the arthropod fauna of a new ley pasture established on previously cultivated arable land. *J. Appl. Ecol.* 17:309–321. **Purvis, G., and G. O. Evans. 1982.** Astigmatic mites of the genera *Schwiebea* Oudms. and *Nanacarus* Oudms. from S. E. Ireland. *J. nat. Hist.* 16:815–821. **Quintero, M. T., and A. Acevedo. 1984.** Studies on deep litter mites in farms in Mexico. Pp. 629–634 *in* D. A. Griffiths and C. E. Bowman (eds.). *Acarology VI.* Vol 1. Ellis Horwood Ltd. Chichester. 645 pp. **Sinha, R. N. 1979.** Role of Acarina in the stored grain ecosystem. Pp. 263–272 *in* J. G. Rodriguez (ed.). *Recent Advances in Acarology.* Vol. 1. Academic Press. New York. 631 pp. **Sinha, R. N. 1984.** Acarine community in the stored rapeseed ecosystem. Pp. 1017–1025 *in* D. A. Griffiths and C. E. Bowman (eds.). *Acarology VI.* Vol. 2. Ellis Horwood Ltd. Chichester. 1296 pp. **Spicka, E. J., and B. M. O'Connor. 1980.** Description and life-cycle of *Glycyphagus* (*Myacarus*) *microti* sp. n. (Acarina: Sarcoptiformes: Glycyphagidae) from *Microtus pinetorum* from New York, U.S.A. *Acarologia* 21(3–4):451–476. **Stepien, Z. A., and J. G. Rodriguez. 1982.** Caloric values of mites for bioenergetical studies. *Intl. J. Acarol.* 8(3):189–192. **Timms, S., D. N. Ferro and R. M. Emberson. 1982.** General biology and nomenclature of *Sancassania berlesei* (Michael). *Acarologia* 22(4):385–390. **Timms, S., D. N. Ferro and R. M. Emberson. 1982.** Andropolymorphism and its heritability in *Sancassania berlesei* (Michael) (Acari: Acaridae). *Acarologia* 22(4):391–398. **Whelan, J. 1985.** Seasonal fluctuations and vertical distribution of the acarine fauna of three grassland sites. *Pedobiologia* 28:191–201. **Whelan, J. 1986.** An analysis of the acarine communities in the herbage and soil of three adjoining grassland sites in Co. Kildare, Ireland. *Pedobiologia* 29:149–164.

23

Acarina: Oribatida

ROY A. NORTON

Oribatid mites (suborder Oribatida or Cryptostigmata), often called "moss mites" or "beetle mites," are one of the numerically dominant arthropod groups in the organic horizons of most soils, where their densities commonly reach several hundred thousand individuals per square meter. Of the major mite suborders, oribatid mites have remained tied most closely to this habitat, which is their ancestral one. The incredible variety of morphological, physiological and behavioral adaptations found among the suborders Astigmata, Prostigmata, and Mesostigmata, especially those lineages having developed close symbiotic relationships with other arthropods and vertebrates, is for the most part unknown in this group of mites. Their evolutionary history is apparently a long one. Most of the known specimens preserved in Oligocene Baltic amber can be reliably assigned to extant genera, and examples belonging to what are usually considered rather highly derived genera are known from the Jurassic.

One might expect such an ancient group, with habitat divergences restricted to minor forays into aquatic and hyperedaphic situations, to show little taxonomic diversity, but the contrary is true. About 1000 genera have been described and included in more than 150 families, with the fauna of much of the tropics and southern hemispheres hardly investigated. In temperate regions even cursory examination of organic soil horizons in a given forest stand or grassland soil will uncover examples of 30–50 species; intensive investigations may yield twice this number.

Basic External Morphology

This section is a very simplified introduction to the general external structures of adult oribatid mites, intended primarily as a springboard for the use of the family key. Our knowledge of the details of external morphology is almost exclusively a result of the investigations of the late F. Grandjean, whose works have been listed by Travé and Vachon (1975).

Size and General Facies

Adult oribatid mites range in size from approximately 150 μm to over 1.5 mm; a large majority are in the range 300–700 μm. Color is usually determined by the degree of sclerotization and melanization of the integument, ranging from white or very light yellow through various shades of brown to black. Rarely, red or orange pigments are evident, as in some members of the Brachychthoniidae. There is little or no sexual dimorphism; the following discussion and family keys can be applied to any adult.

Ancestral arachnid tagmata, the prosoma (containing all cephalic structures and supporting the appendages), and opisthosoma (containing the genital and anal openings, all primary reproductive structures and most of the digestive system), are not apparent in the body divisions of oribatid mites. Secondary body divisions are generally thought to have been derived by the ventral encroachment of parts of the prosoma (legs III and IV and supportive structures) on the opisthosoma. Accompanying this distortion the original prosoma–opisthosoma separation was obliterated, so that a *sejugal groove* divides the derived *proterosoma* (bearing the first two pairs of legs) from the *hysterosoma* (bearing the last two pairs of legs). If these two units freely articulate (i.e., if the sejugal groove is unsclerotized), the body form is known as *dichoid* (see Figs. 23.25, 23.26, *sjg*). If the sejugal region is sclerotized (proterosoma and hysterosoma are fused), the body form is *holoid* (see Fig. 23.43).

Several distinct body regions are recognizable in this derived body form. The dorsal and dorsolateral region of the sclerotized proterosoma is known as the *prodorsum* (Fig. 23.1, *pd*) (or *aspis,* Fig. 23.2, *asp,* if the lateral regions of the proterosoma are unsclerotized). The large dorsal sclerite of the hysterosoma is called the *notogaster* (Fig. 23.1, *ng*), which may or may not be completely defined by a *circumgastric scissure* (Fig. 23.1 *cgs*). In ventral aspect, the mouthparts (or *gnathosoma*) are located in the anterior region of the proterosoma, the *epimeral region,* including the area approximately delimited by the insertion of the eight legs, spans the sejugal groove, and the *anogenital region* lies posteriad of the last pair of legs.

A highly specialized *ptychoid* body form is found in certain families collectively known as "box mites." In these groups the aspis and epimeral region are freely articulated, both with each other and with the remainder of the body (Fig. 23.2). The epimera and legs can be drawn into the body and the aspis can close over and cover them; as a result, no leg or soft integument is exposed (Fig. 23.3).

Integument and Its Derivatives

In adults the integument proper is well sclerotized except in some of the most primitive oribatid mites. The outer surface may be smooth or have various types of general relief, such as granulation, reticulation, shallow pitting, or pock marks. Large or small expanses of integument may be porose, penetrated by many minute canals. In many families there is secreted a nonchitinous external layer, the *cerotegument,* which covers all or part of the body and is often loosened or damaged during the preparation of specimens. Cerotegument is usually partly composed of tubercles of a granular, columnar, or filamentous nature, or may have reticulate designs.

Various types of hairlike structures are derived from the integument. Normal setae of the body and legs serving a tactile function are usually simple, solid, and often externally barbed, but may be flagellate, branched, leaflike, fan-shaped, or have other unusual structure. Certain other setae, or setiform organs, on the legs are chemosensory, with internal protoplasmic extensions from the epidermis. Details of leg setation in the Oribatei are summarized elsewhere (Norton 1977).

Prodorsum (Figs. 23.1, 23.4)

In most oribatid mites the *rostrum* (*rm*), or anterior extremity of the prodorsum, extends anteriorly or anteroventrally to cover the gnathosoma. In a few soft-bodied primitive families the poorly sclerotized rostrum is limited in extent, exposing most of the chelicerae in dorsal aspect (see Fig. 23.16).

The surface of the prodorsum may be smooth or may have a variety of large tubercles, ridges, and other structures, in addition to the overall integumental relief mentioned above. In many groups there are paired longitudinal ridges located either medially or laterally. If the structures are relatively broad, ribbonlike, with at least one free edge in cross section, they are called *lamellae* (*lam*). Usually, lamellae will also have a free distal end, or *cuspis* (*cu*), not directly fused to the prodorsum. If the ridges are simply narrow longitudinal thickenings, without cusps or free edges, they are called *costulae* (Fig. 23.46, *cos*). Lamellae may be entirely separate from each other, fused for most or part of their length, or connected by a transverse ridge or thickening, the *translamella* (*tr*).

In less primitive groups there are five pairs of prodorsal setae. The *rostral setae* (*ro*) are the most anterior, inserted on or near the rostrum. The *lamellar setae* (*le*) are located medially; when lamellae are present the setae are usually at the end of the cuspis. The *interlamellar setae* (*in*) are inserted posteromedially, often just anterior to the sejugal groove. At each postero-lateral corner of the prodorsum is a modified sensory seta of rheotactile function called a *sensillus* (*ss*), which takes a variety of shapes but almost always is distinct from other prodorsal setae. The sensillus is inserted in a cuplike *bothridium* (*bo*), which may have a funnel-shaped rim. Laterally, below the sensillus in dorsal aspect, is the paired *exobothridial seta* (*ex*), which is often inconspicuous. In more primitive groups there are two pairs of exobothridial setae (Fig. 23.27, *exa, exp*), one behind or above the other.

Notogaster (Figs. 23.1, 23.5, 23.6)

The notogaster takes diverse forms. Most often it is more or less hemispherical, but it can be elongate, dorsoventrally flattened, rectangular, laterally compressed, or conical, and may be partially or completely fused to the prodorsum, obliterating the sejugal groove dorsally. In certain groups parts of the exuviae from immature instars are retained and carried on the notogaster. Compacted organic and mineral debris completely or partially covers the notogaster of some species, often in characteristic formations.

Certain groups have winglike anterolateral projections called *pteromorphs* (*ptm*), which extend beyond the general body outline and may curve ventrally over the bases of the legs. Pteromorphs may be movable (Figs. 23.1, 23.34) (able to be appressed to the sides of the body) or immovable (Fig. 23.36). When viewed laterally, movable pteromorphs display a hinge of weakly sclerotized integument near their base, and usually project noticeably from the general outline of the notogaster. The notogaster of some groups may have small, nonmovable angular projections extending from the anterolateral corners. These *humeral projections* (Fig. 23.40, *hp*) do not extend laterally, as pteromorphs do.

Less conspicuous on the notogaster of most higher oribatids is the octotaxic system of either porose areas or sacculi. Mediad of the pteromorphs is often a pair of well-defined regions (Fig. 23.5, *Aa*), usually circular, of concentrated integumental pores or canals called *areae porosae,* which appear lighter than the surrounding area in transmitted light. Usually, there are four such pairs, the other three (A_1, A_2, A_3) being arranged laterally in the posterior half of the notogaster. Care must be taken not to confuse these with internal points of muscle insertion, which also appear light colored in transmitted light and are often present in an irregular band near the lateral and posterior margins.

Sacculi (Fig. 23.6, *Sa*, S_1–S_3), which are homologous to areae porosae and are present in the same locations, are bag-shaped invaginations with porose walls which open to the surface through an inconspicuous pore. Sacculi usually appear darker than the surrounding integument, since transmitted light passes through more cuticle. These should not be confused with less conspicuous, simple, slit sense organs, three of which (*ia, im, ip*) are usually distributed as in Figs. 23.5 and 23.6. Both areae porosae and sacculi seem to be associated with ductless glands (Woodring and Cook 1962).

Another paired gland, the *opisthosomal gland,* opens mediolaterally on the notogaster of most oribatid mites. Both the outline of the gland and its external opening (*gla*) may be large and conspicuous, or very difficult to discern, depending on the group.

In certain more primitive families there are transverse scissures (*ts*) which in dorsal aspect seem to divide the notogaster into several sections (Figs. 23.21–23.24). These scissures have no counterparts on the ventral surface.

More primitive oribatid mites have 16 pairs of notogastral setae, but this number can be greatly reduced, or certain setae may be very inconspicuous or represented only by their alveolus.

Ventral Structures (Figs. 23.7, 23.8)

The epimeral region is the area approximately delimited by the insertions of the eight legs, and results from the fusion of four pairs of platelike coxisternal elements or epimeres. *Apodemes* (*ap*), sclerotized, lamellate invaginations that serve as attachment sites for certain muscles, may clearly and completely define each pair of epimera in transmitted light (Fig. 23.43), or certain of the apodemes may be incomplete medially (not reaching the midline, Fig. 23.8) or apparently absent. The number of epimeral setae varies, but most commonly epimera I–IV have 3–1–3–4 pairs of setae, respectively.

In primitive oribatid mite groups the anogenital region is comprised of four pairs of seta-bearing plates, which occupy essentially all of the ventral surface posteriad of epimere IV (Fig. 23.7). The *genital plates* (*gp*) enclose the genital aperture through which the elongate *ovipositor* (Fig. 23.9) (adult females) or short *genital sclerite* (Fig. 23.10) (adult males) can be extruded during reproductive activities. Laterad of each genital plate is an *aggenital plate* (*agp*). The anal opening is enclosed by a pair of *anal plates* (*anp*) and *adanal plates* (*adp*), respectively. This overall condition is known as the *macropyline* condition (i.e., the aggenital and adanal plates are independent). Many macropyline groups also have a small unpaired surface sclerite, the *preanal plate* (*pp*), between the genital and anal apertures. In less primitive groups a *brachypyline* condition is found (Fig. 23.8) where the aggenital and adanal plates have hypertrophied and fused, forming a so-called *ventral plate* (*vp*), in which the anal and genital plates are inserted. Adanal and aggenital setae are still distinguishable on the ventral plate. These differences are one basis for the traditional classification of oribatid mites into two major groups, the "lower oribatids," and the "higher oribatids" or Brachypylina.

Mouthparts

The mouthparts of groups with a well-developed rostrum are folded at rest into an anteroventral vestibule known as the *camerostome*. These mouthparts include the *subcapitulum*, or infracapitulum (Fig. 23.11), paired *palps* (Fig. 23.12), and paired *chelicerae.* The greater part of the ventral surface of the complex subcapitulum is formed, like the epimera, from the coxisternal elements of the palpal segment; the pharynx is enclosed in this structure and the mouth opens near its distal end. The palps have from two to five segments, with the latter number being the rule. Of most value for our purposes are the chelicerae, the great majority of which are *chelate–dentate* (Fig. 23.13) in form. A few families have modified chelicerae in which the distal portion is greatly narrowed and the chelae are small, with teeth indistinct or absent; these are called *peloptoid* or pelopsiform chelicerae (Fig. 23.14). The family Gustaviidae has unique chelicerae, the distal portion of which is almost acicular, with serrated margins (Fig. 23.15); there are no chelae.

Legs

Although there are many morphological characters associated with the legs of oribatid mites which are useful from a classificatory viewpoint, it will suffice in this treatment to mention that all oribatid mites have legs with five segments; *trochanter, femur, genu, tibia,* and *tarsus* (proximal to distal). In many groups trochanters I and II are very small and easily overlooked. The tarsi are monodactyl or tridactyl (rarely bidactyl).

Feeding Biology

Oribatid mites deviate from the typical arachnid mode of feeding (predation, with preoral digestion, followed by the ingestion of fluids) in that they are slow-moving particulate feeders. In most instances chelicerae and structures of the subcapitulum are used together to cut or tear particles into sizes suitable for swallowing. Soil microorganisms and decaying higher-plant material constitute the diet for the most part, but there are varying degrees of food specificity. Luxton (1972), in a modification of work by Schuster (1956), categorized oribatid mites in the following way, based on feeding habits.

Macrophytophages are those that feed strictly on higher-plant material, usually after some decomposition has taken place. Some of these, such as the "box mites" of the families Phthiracaridae and Euphthiracaridae, and certain Carabodidae, burrow into and feed on decaying wood of logs, branches, twigs, or leaf petioles. These woody tissue feeders are known as xylophages; those feeding on softer, nonvascular tissue of decaying leaves are categorized as phyllophages.

Microphytophages, such as one finds in the families Oppiidae, Eremaeidae, Damaeidae, and others, feed on certain elements of the soil microflora. Among them, mycophages (by far the most common) utilize fungi or yeasts, phycophages feed on algae, and some are believed to be bacteriophages, scraping bacterial films or clumps from decaying materials.

Panphytophages, including perhaps a majority of oribatid mites, are those that utilize both microbial and higher plant materials, either concurrently or at different stages in the life cycle. These general feeders include such common families as Nothridae, Camisiidae, Liacaridae, Oribatulidae, Galumnidae, and others.

Many oribatids have been shown to readily eat nematodes in culture (Muraoka and Ishibashi 1976), but the extent to which this and other incidental feeding habits, such as carrion feeding and coprophagy, occur in nature is unknown.

The presence of feeding specificity within groups of macrophytophages has not been convincingly demonstrated, although certain mites are known to prefer deciduous or coniferous leaf litter, or leaf petioles over twig tissue. Feeding specificity among microphytophagous groups is better documented. Laboratory preference tests (e.g., Hartenstein 1962) have shown that a given mycophage will eat certain fungi voraciously and ignore others, even if starved. In some instances these preferences have been confirmed by the examination of gut contents of field-collected specimens, and Luxton (1972) indicated that the vertical distribution of certain oribatid mites in the soil horizons corresponds to that of their preferred fungal food. However, Mitchell and Parkinson (1976) found that a fungus preferred in laboratory studies is not always the one eaten in natural surroundings, even if it is present in abundance. Also, it has been shown by a number of workers that mycophagous oribatid mites may feed readily and develop to adulthood on a given fungus food in the laboratory, yet be incapable of reproducing without a different nutrition source.

It has been proposed (Anderson 1978) that in deciduous forest habitats, specialized fungal feeding is most prevalent soon after leaf fall, when different litter components are highly discrete. After a certain amount of degradation and mixing, generalized feeding becomes more common.

When considering macrophytophages, the question is often posed as to whether they are deriving energy from the higher plant material itself or the microflora ingested with it. Early work on this problem indicated that the ratio of the two major structural components of the plant material changed during gut passage. Lignin, a recalcitrant, nondigestible material, increased in proportion to cellulose, indicating some cellulose removal. Enzymological investigations have shown the presence of a cellulase in all macrophytophages and panphytophages studied (Luxton 1972), but not in the microphytophages, as one might expect. The same work indicates that microphytophages and panphytophages possess an enzyme that breaks down trehalose, a dissacharide produced in significant amounts by fungi and yeasts, but very little by higher plants. The absence of this enzyme in macrophytophages suggests that they do not utilize the ingested fungal material.

The source of the cellulase in the digestive systems of macrophytophagous oribatid mites has not been determined, but considering how far removed this mode of feeding is from the fluid feeding of most arachnids, it is probable that it is produced by a symbiotic gut microflora. Stefaniak and Seniczak (1976) and Seniczak and Stefaniak (1978) demonstrated the presence of a well-developed gut flora in certain oribatid mites. It is composed primarily of bacteria, correlating well with the relatively high pH of the alimentary canal.

Reproductive Biology

Mating

Copulation is unknown among oribatid mites; rather, the male deposits a stalked spermatophore on the substrate by means of the complex genital sclerite, or penis (Woodring 1970, Cancela da Fonseca 1967). The female envelops the globular spermatophore head, containing rodlike, immobile sperm, with the genital vestibule. Although there is usually no male–female contact, laboratory studies have shown that in some species the onset and rate of spermatophore production by males depends on the local presence of females; in others, these functions are independent of female presence. One or two species are known in which the male performs a simple dance, perhaps leading the female over the spermatophore. Thelytokous parthenogenesis has been demonstrated in mites of some families, such as Nothridae, Camisiidae, Eremaeidae, Oppiidae, and Tectocepheidae, and is suspected in other groups because of the absence or rarity of males in collections.

Developmental Stages

The typical seven life stages of oribatid mites include the egg, prelarva, larva, protonymph, deutonymph, tritonymph, and adult. There are few variations to this scheme; the most important of these occurs in certain aquatic or semiaquatic groups which are viviparous, with the egg and prelarval instar being passed within the body of the female.

The eggs of most oribatid mites are white, but the egg chorion in certain groups becomes sclerotized, and therefore brownish in color, prior to oviposition. The number of eggs developed within a female at one time varies from one, in the small Brachychthoniidae, up to 16 or more. Eggs are pressed into crevices with the elongate ovipositor, or left on the surface of the substrate. Unlike certain other soil mites, oribatids do not usually camouflage eggs by piling on bits of debris. Eggs may be deposited one at a time in scattered locations, as in many Damaeidae, or deposited all at once in a batch, as in some Ceratozetidae. After oviposition, when the development of the egg is complete, the egg chorion partially splits, exposing the legless, immobile prelarval instar or "deutovum stage," which never leaves the egg. The larva is the first mobile feeding instar.

Oribatid mites undergo anamorphic development, in which body segments are added terminally at certain points in the life cycle. With few exceptions, this involves addition to the larval complement of two segments around the anus, one first present in the protonymph, the other in the deutonymph.

Although there are exceptions, the primitive, or lower oribatids, are essentially paurometabolous. The development from larva to adult involves anamorphosis and the predictable addition of setae to the body and appendages, but only slight, gradual changes in overall appearance. The Brachypylina, or higher oribatids, have a contrasting development, somewhat analogous to the holometabolous development of higher insects. In addition to anamorphosis and ontogenetic changes in setation, there is usually a striking change in overall appearance between the tritonymph and adult instars, with the adult usually being much more externally ornate and heavily sclerotized than the tritonymph. As in holometabolous insects, this metamorphosis not only affects appearance, but may also correspond to changes in feeding preferences and habitat selection; these changes are usually not incurred by more primitive oribatid mites.

In either group, the instars are usually readily distinguishable. The larva is the only six-legged instar; all others have eight legs. The nymphs are most easily distinguished by counting the number of paired genital papillae, seen just inside the genital opening in ventral aspect; protonymphs have one, deutonymphs two, and tritonymphs three pairs. Adults also have three pairs, but the female ovipositor or male genital sclerite can be seen inside the genital opening; adults are also usually more heavily sclerotized, except when recently eclosed.

Temporal Aspects of Life Cycle

Relatively little is known about the phenology and duration of oribatid mite life cycles in nature, although there is an abundance of laboratory data, most of which was obtained under optimal conditions, including unnaturally high constant temperatures. It is likely that a majority of medium-sized temperate-zone oribatid mites complete their life cycle in approximately one year. Larger mites may take two years. In boreal or arctic environments even longer periods may be needed; Mitchell (1977a) postulated that 3 years are necessary for the complete life cycle of a medium-sized member of the Ceratozetidae in Alberta.

The timing of egg maturation and oviposition appears to vary between groups. Mitchell (1977b), for example, suggested that certain oribatid mites which inhabit the surface litter tend to oviposit during the whole year, whereas those inhabiting the fermentation or humus horizons tend to be seasonal, ovipositing in spring or summer. There is increasing evidence that iteroparity, or repeated reproduction by the same female, is common among oribatid mites. This, combined with the fact that the time of oviposition may determine which instar overwinters (Weigmann 1975), leads to considerable overlapping of generations and moderately stable population densities.

Although feeding and development are curtailed when soils become frozen, various adaptations allow oribatid mite survival. Sømme and Conradi-Larsen (1977a,b) have shown that oribatids can undergo behavioral and physiological changes to avoid death due to freezing of the hemolymph. Food particles in the gut can serve as nucleating agents for the formation of ice crystals, but the oribatid mites studied stopped feeding at the onset of cold temperatures. Also, these mites have the ability to accumulate glycerol in the hemolymph during autumn and early winter, which serves as an antifreeze, allowing them to supercool. In spring, feeding resumes and glycerol levels decrease. These mites can also withstand considerable periods without oxygen at 0°C—a valuable adaptation since it is likely that anaerobic conditions develop in frozen soil.

Local Distributions in Soil

Certain oribatid mites are known to be obligatory or facultative inhabitants of nonsoil substrates, such as tree and rock surfaces, lichens, and mosses, sometimes with a considerable degree of microhabitat specificity (Aoki 1967, Hammer 1972, Travé 1963). Their greatest diversity and numbers, however, occur in the organic horizons of the soil. In general, one finds greater diversity in the most highly structured organic horizons. Thus soils of a mor humus type are more favorable than those of a mull humus type. Relatively few oribatid mites are known to prefer the euedaphic zone (i.e., the mineral layers of the soil). Members of the Epilohmanniidae, Eulohmanniidae, and certain Oppiidae are common, if not obligatory inhabitants of this region, and most of these are narrow, elongate, or very small mites adapted to the small pore spaces. In the hemiedaphic, or organic layers, greatest numbers and diversity of oribatid mites are found in the fermentation layer (Anderson 1978), where the overall biological activity is highest. Some studies have demonstrated seasonal migration of certain mites downward from litter and fermentation layers during the winter. According to Lebrun (1964), these usually represent medium to large species; smaller mites do not show such migration to the same extent.

Like many other soil arthropods, oribatid mites exhibit a clumped or contagious horizontal distribution within the soil, probably caused by a similar distribution of favored resources. The consequences of such a distribution on statistical analyses of populations has been discussed by Gerard and Berthet (1966).

Anthropocentric Significance

The fact that few, if any, oribatid mites are of direct economic significance is undoubtedly one of the major reasons behind the poor state of our taxonomic and biological knowledge. The instances where oribatid mites have been implicated in monetary-loss situations are generally quite isolated, and usually involve suspected feeding on fruits or leaves of cultivated plants, or limited occurrence in stored products. Whether or not the mite initiates damage in these instances, or whether they are feeding on exudates that result from damage by other sources or microflora developing on such damaged material, is usually not determined. Certain members of the Haplozetidae are known root feeders but of no economic significance.

Of indirect but well-known significance is the role of certain oribatid mites as intermediate hosts of cestodes. Mammalian tapeworms of the family Anoplocephalidae utilize members of the Brachypylina (particularly Galumnidae and Oribatulidae) for the development of cysticercoids (Sengbusch 1977). Of greatest economic significance is *Moniezia expansa* (Rudolphi), the sheep tapeworm. Oribatid mites which periodically climb into low vegetation during the most humid parts of the day may be ingested along with the grasses as sheep graze, transmiting mature cysticercoids to the sheep.

Ecological Significance

Oribatid mites have relatively low metabolic rates and contribute little to total soil metabolism directly, despite being present in high numbers in most noncultivated soils. Even though a significant portion of annual leaf fall in temperate forests may be ingested by these animals, very little of the energy content is extracted. But basing their ecological significance on metabolic contributions to the breakdown of organic matter would be misleading. Current research suggests that the principal function of this group in the soil system is one of regulating organic matter decomposition, most of which is accomplished by the soil microflora. More complete reviews can be found in Wallwork (1983) and Norton (1986) and references cited therein.

Rates of organic matter decomposition are affected in several ways. Feeding by microphytophagous mites may stimulate new growth and help prevent nutrient tie-up resulting from fungistasis and bacteriostasis. Macrophytophagous mites are responsible for much comminution of organic matter into smaller particles, with greater water content, more surface area for microbial attack, and a higher pH, which is more favorable for bacterial growth. Both types of feeders (and, of course, panphytophages) disseminate spores or other propagules, both externally and internally, and inoculate new substrates.

The composition of the microbial component of the soil community is also affected directly by selective oribatid feeding and indirectly by nonrandom dissemination of spores. Also, feeding on antibiotic-producing fungi may have significant indirect effects on the diversity of microbes by altering antibiotic output, which could influence the range of materials that can be decomposed by the system.

Much of the above may also be said for other saprophagous small arthropods, especially the Collembola, and the question is often raised as to the difference in ecological roles of these two groups. Collembola appear to be *r*-strategists, with relatively high metabolic and feeding rates and reproductive potential; these values tend to be much lower among the oribatid mites, suggesting that oribatids are *K*-strategists, relative to the Collembola. Crossley (1977a) has hypothesized that in soils Collembola may be the most dominant mycophages in periods of rapid growth of fungal mycelium. Their *r*-strategy allows them to respond quickly to new food resources. Oribatid mites (at least the mycophagous groups) may utilize senescent growths of fungi, perhaps parts where nutrients, especially calcium, are concentrated. Todd et al. (1975) have demonstrated that calcium is concentrated in the integument of oribatid mites, and this group may have direct as well as indirect effects on the cycling of this important nutrient (Cromack et al. 1977, Crossley 1977b).

Techniques

Most commonly used for the collection of oribatid mites are dynamic extractors such as the Berlese–Tullgren funnel and its many modifications, but certain flotation methods also work well if the soil has little organic matter (Murphy 1962a,b, Martin 1977). The nature of the dynamic extractor and the physical regime under which it is operated greatly affects extraction efficiency (Aucamp et al. 1964, Huhta 1972), but this efficiency varies from group to group. Adult oribatid mites are satisfactorily extracted in most commonly used apparati, but the extraction of soft-bodied immatures, usually slower moving and less resistant to desiccation than adults, is usually much less efficient. Immatures of groups such as the Carabodidae, Phthiracaridae, and Euphthiracaridae, which burrow into wood and other substrates, are very rarely collected by extractors. Such materials must be picked apart and examined by hand. Similarly, oribatid mites naturally subject to irregular periods of desiccation, such as those living in or on lichens, are less likely to respond effectively; certain of these mites are known to become immobile during periods of dryness.

Except by experienced workers, even family identifications require the use of a compound microscope. Observation of the external structures used in keys necessitates clearing the mites, usually with concentrated or dilute lactic acid, depending on the degree of sclerotization. Because of their often nearly spherical bodies and inflexible exoskeletons, oribatid mite specimens mounted whole on slides in standard gum arabic media usually become broken and useless in a relatively short time, as the medium shrinks in drying. Small pieces of paper, glass, plastic, or other material used as props under the cover glass may help, but usually a small degree of dissection is best if permanent or semipermanent slides are desired. One useful technique when working with brachypyline oribatids is to use fine needles to tear away the notogaster of cleared specimens at the cirgumgastric scissure, then mount it alongside the rest of the body under the same cover glass. Another observational method is the use of temporary mounts, in which a cavity slide is half covered with a cover glass, lactic acid is used as a medium, and the whole, cleared mite is introduced using a small loop or brush, so that it barely touches the slide and the cover glass. In this fashion a single specimen can be turned and studied from all aspects; it is then returned to a preservative, for which most standard alcohol solutions are acceptable. Details of these and other appropriate observational techniques can be found in works by Balogh (1972), Grandjean (1949), Krantz (1978), and Travé (1965).

Identification

The following key is a very simplified one written to allow beginners, with some practice to make family-level identifications of common adult oribatid mites from North America, north of Mexico. It is somewhat less applicable for subtropical regions, such as southern Florida, and for arboreal and other nonsoil habitats. Common aquatic families are not included. More serious students are referred to Balogh (1972), who dealt with the known world fauna at the generic level, or to Balogh (1963), the scope of which is restricted to the holarctic region. No monographs are currently available that treat all groups of North American oribatid mites at the species level, but references to works of lesser scope can be found in the catalogue by Marshall et al. (1987).

In this key no attempt is made to include reference to a suprafamilial classification; the simplification needed to produce the key, as well as the present controversy over higher categories, preclude this. However, families identified from couplets 1–22 are considered "lower oribatid mites" those identified at couplets 23 and beyond are included in the Brachypylina, or "higher oribatid mites". Wallwork (1969) has provided a key to major oribatid mite groups based on immatures.

ACKNOWLEDGMENTS

Some of the illustrations have been redrawn and modified from those published by J.-I. Aoki, F. Bernini, F. Grandjean, A. P. Jacot, J.-C. Lions, M. Sellnick, L. G. Sitnikova, and K. Suzuki. I am indebted to my colleague Dr. James R. Philips for his critical review of this manuscript. Information in this chapter was based on research and literature published through 1980, with some updated references through 1987.

KEY TO FAMILIES

1a Rostrum poorly developed; chelicerae mostly exposed in dorsal aspect (Fig. 23.16); usually white and soft-bodied **2**

1b Rostrum well developed (Fig. 23.4); chelicerae mostly hidden in dorsal aspect; color usually yellow, brown or black; moderate to heavy sclerotization **5**

2a Legs with six free segments (femur divided); usually, some notogastral setae very long, darkly pigmented. **3**

2b Legs with five free segments; no long, pigmented notogastral setae. . . . . **GEHYPOCHTHONIIDAE**

3a Tarsi bidactyl . . . . **PALAEACARIDAE**

3b Tarsi tridactyl **4**

4a Tarsi homotridactyl (empodial claw approximately equal in size to lateral claws); sensillus short, flattened . . . . **APHELACARIDAE**

4b Tarsi heterotridactyl (empodial claw distinctly smaller than lateral claws); sensillus elongate, not flattened . . . . **CTENACARIDAE**

5a Body ptychoid (Fig. 23.2); circular or laterally compressed in cross section **6**

5b Body not ptychoid; circular or dorsoventrally compressed in cross section **9**

6a Genital and anal plates relatively small, separately inserted on broad ventral plate (Fig. 23.17). . . . . **MESOPLOPHORIDAE**

6b Genital and anal plates relatively large, adjacent, covering most of the ventral surface; no distinctly separate ventral plate **7**

7a Genital and anal plates broad, adjacent but not fused together (Fig. 23.20) **PHTHIRACARIDAE**

7b Aggenital and adanal plates fused together, elongated (Figs. 23.18, 23.19) **8**

8a Genital plates fused with aggenital plates, and anal plates fused with adanal plates (Fig. 23.18). . . . . **EUPHTHIRACARIDAE**

8b Anal and adanal plates separated by a distinct scissure (Fig. 23.19); genital and aggenital plates separated or not . . . . **ORIBOTRITIIDAE**

9a Notogaster dorsally divided by one or more distinct transverse scissures (Figs. 23.21–23.24); opisthosomal gland absent. **10**

9b Notogaster without transverse scissures; opisthosomal gland present or absent. **13**

10a Notogaster with a single complete transverse scissure **11**

10b Notogaster with two or three complete transverse scissures **12**

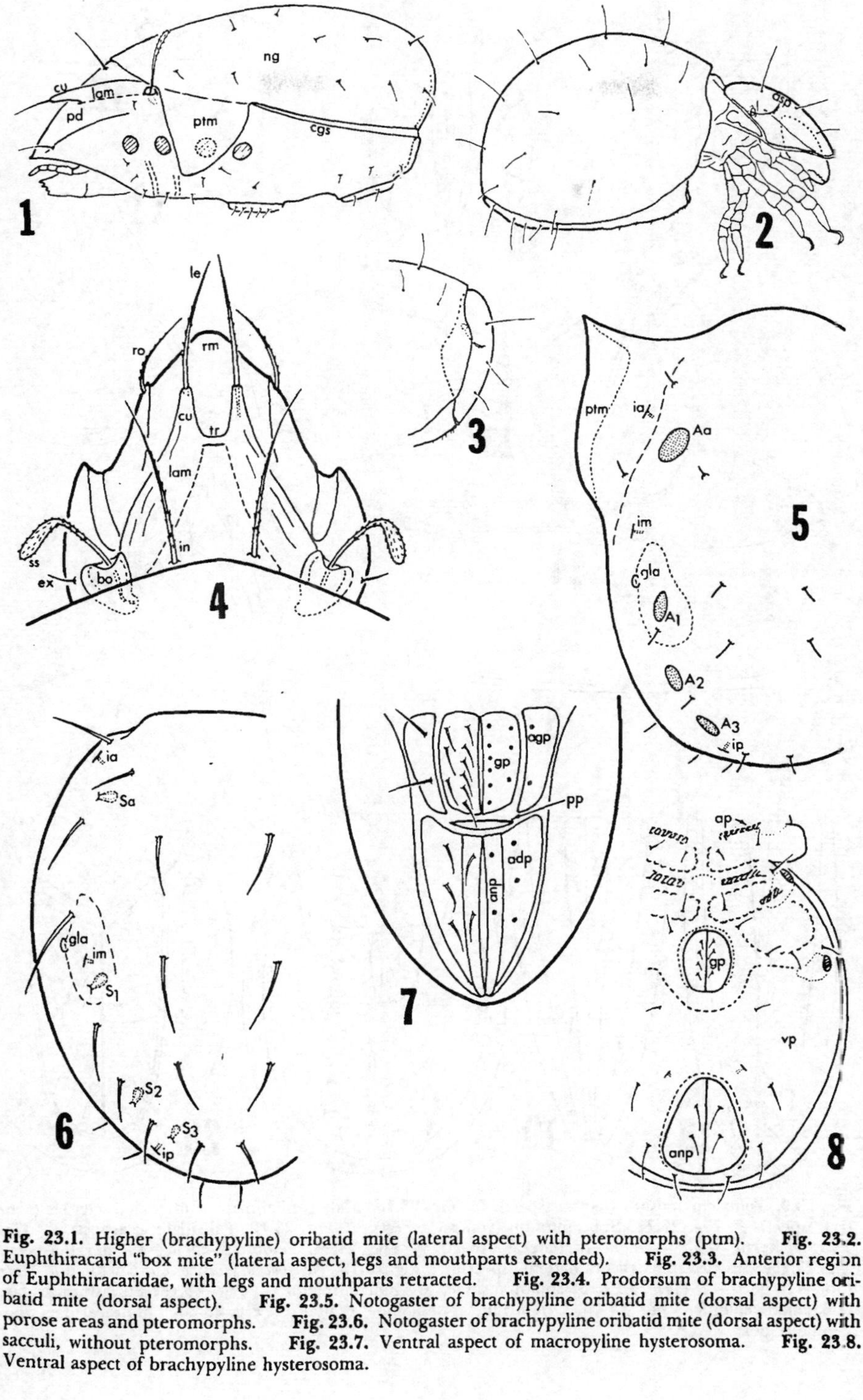

Fig. 23.1. Higher (brachypyline) oribatid mite (lateral aspect) with pteromorphs (ptm). **Fig. 23.2.** Euphthiracarid "box mite" (lateral aspect, legs and mouthparts extended). **Fig. 23.3.** Anterior region of Euphthiracaridae, with legs and mouthparts retracted. **Fig. 23.4.** Prodorsum of brachypyline oribatid mite (dorsal aspect). **Fig. 23.5.** Notogaster of brachypyline oribatid mite (dorsal aspect) with porose areas and pteromorphs. **Fig. 23.6.** Notogaster of brachypyline oribatid mite (dorsal aspect) with sacculi, without pteromorphs. **Fig. 23.7.** Ventral aspect of macropyline hysterosoma. **Fig. 23.8.** Ventral aspect of brachypyline hysterosoma.

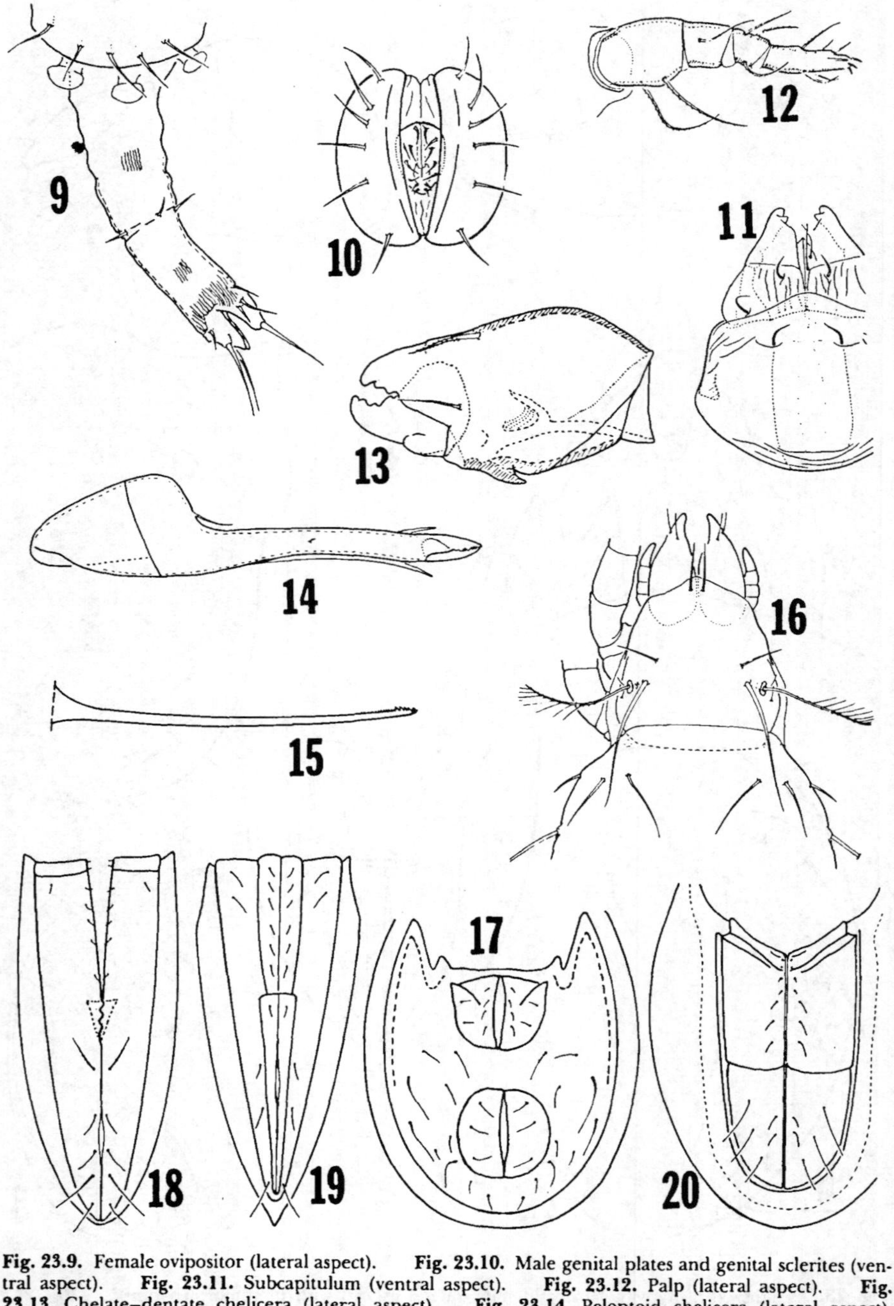

Fig. 23.9. Female ovipositor (lateral aspect). **Fig. 23.10.** Male genital plates and genital sclerites (ventral aspect). **Fig. 23.11.** Subcapitulum (ventral aspect). **Fig. 23.12.** Palp (lateral aspect). **Fig. 23.13.** Chelate–dentate chelicera (lateral aspect). **Fig. 23.14.** Peloptoid chelicera (lateral aspect). **Fig. 23.15.** *Gustavia* sp. chelicera (lateral aspect). **Fig. 23.16.** *Nehypochthonius porosus* Norton and Metz (dorsal aspect). **Fig. 23.17.** Mesoplophoridae, anogenital region. **Fig. 23.18.** Euphthiracaridae, anogenital region. **Fig. 23.19.** Oribotritiidae, anogenital region. **Fig. 23.20.** Phthiracaridae, anogenital region.

11a Transverse scissure distinctly anteriorly curved; anterior half of notogaster usually with appearance of a second, poorly defined, medially incomplete scissure (Fig. 23.22). . . . . **ENIOCHTHONIIDAE**

11b Transverse suture straight or slightly posteriorly curved; no apparent second scissure (Fig. 23.21) . . . . **HYPOCHTHONIIDAE**

12a Notogaster with two complete transverse scissure (Fig. 23.23) (very small species, usually less than 250 μm) . . . . **BRACHYCHTHONIIDAE**

12b Notogaster with three complete transverse scissures (Fig. 23.24) (mostly larger than 250 μm) . . . . **COSMOCHTHONIIDAE**

13a Body form dichoid (Figs. 23.25, 23.26); two pairs of exobothridial setae **14**

13b Body form holoid; one pair of exobothridial setae **17**

14a Prodorsum relatively broad, as broad as notogaster (Fig. 23.27); preanal plate present . . . . **LOHMANNIIDAE**

14b Prodorsum narrower than maximum width of notogaster (Fig. 23.28); preanal plate absent **15**

15a Genital plates transversely divided; separate paired aggenital plates distinguishable, bearing aggenital setae . . . . **PERLOHMANNIIDAE**

15b Genital plates entire; aggenital setae not on paired aggenital plates . . . **16**

16a A parabolic scissure present, passing between genital and anal plates (Fig. 23.26); ventral region without distinct longitudinal or transverse apodemes between either legs III or IV . . . . **EULOHMANNIIDAE**

16b Without such a scissure; ventral region with distinct longitudinal and transverse apodemes between both legs III and IV . . . **EPILOHMANNIIDAE**

17a Ventral region of the macropyline type; aggenital and adanal plates discrete, movable **18**

17b No discrete, movable aggenital or adanal plates **21**

18a Prodorsal bothridium undeveloped; sensillus small, simple seta (Fig. 23.29) **MALACONOTHRIDAE**

18b Bothridium well developed; sensillus of different shape than other prodorsal setae **19**

19a Aggenital setae present; notogaster usually flat or concave dorsally, with more or less parallel sides in dorsal aspect; posterior notogastral setae often inserted on large tubercles (Fig. 23.30); body usually totally or partially covered with adherent mineral or organic debris . . . . **CAMISIIDAE**

19b Aggenital setae absent; notogaster usually more or less convex dorsally, with convex sides (Fig. 23.31); posterior setae not on large tubercles; body without noticeable adherent debris **20**

20a Preanal plate present; four or five pairs of setae on epimera II; setae on genital plate not all concentrated medially; prodorsum covered with distinct areoles or reticulation . . . . **NOTHRIDAE**

20b Preanal plate absent; no more than one pair of setae on epimera II; genital setae all concentrated along medial border of plate; prodorsum without distinct areoles or reticulation . . . . **TRHYPOCHTHONIIDAE**

21a Ventral region fused to notogaster, only a pair of crescent-shaped scissures (SU) directed behind genital plates (Fig. 23.32) . . . . **NANHERMANNIIDAE**

21b Ventral region of the normal brachypyline type; ventral plate completely separated from the notogaster by a circumgastric scissure **22**

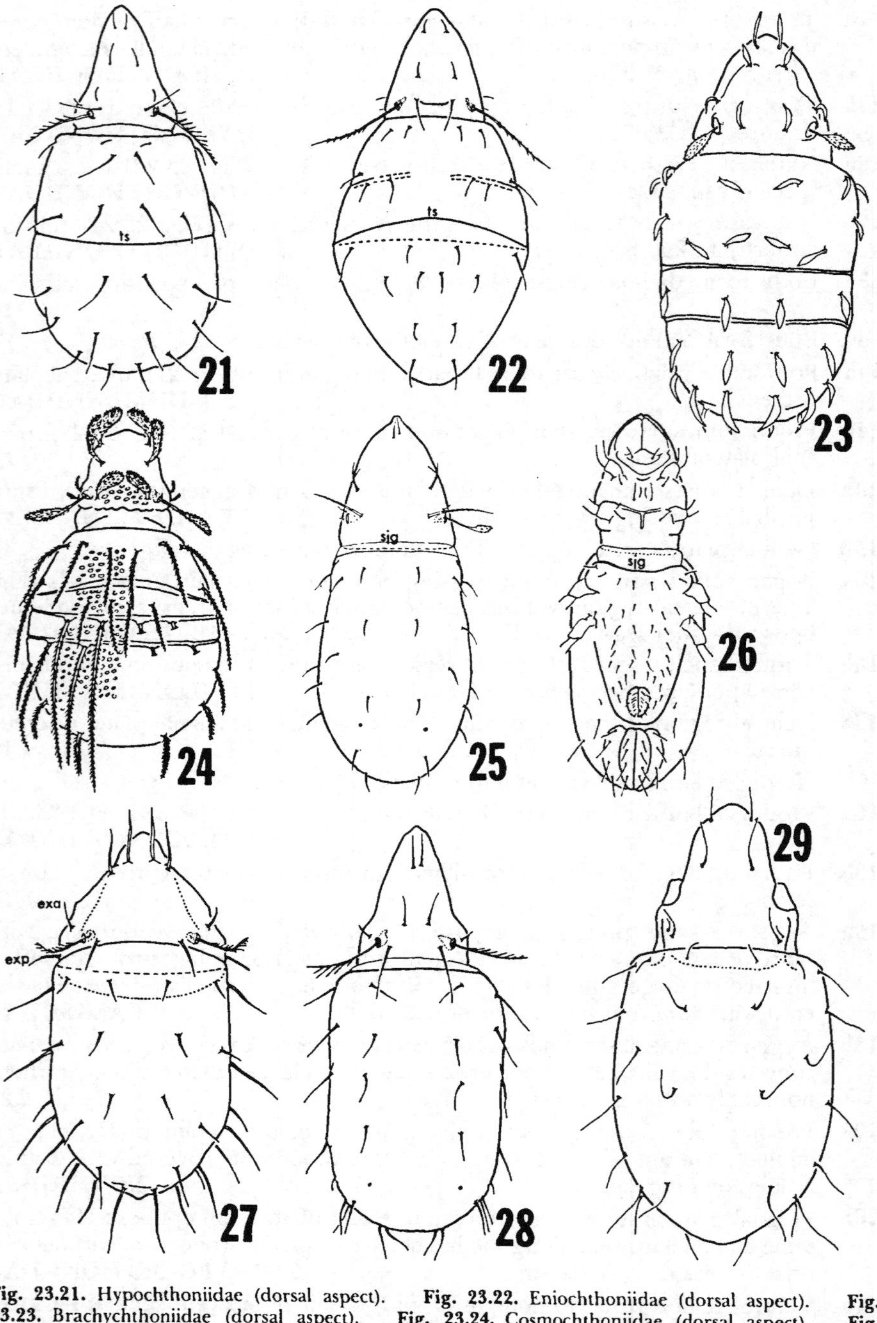

Fig. 23.21. Hypochthoniidae (dorsal aspect). **Fig. 23.22.** Eniochthoniidae (dorsal aspect). **Fig. 23.23.** Brachychthoniidae (dorsal aspect). **Fig. 23.24.** Cosmochthoniidae (dorsal aspect). **Fig. 23.25.** Epilohmanniidae (dorsal aspect). **Fig. 23.26.** Eulohmanniidae (ventral aspect). **Fig. 23.27.** Lohmanniidae (dorsal aspect). **Fig. 23.28.** Perlohmanniidae (dorsal aspect). **Fig. 23.29.** Malaconothridae (dorsal aspect).

22a A pair of distinct, funnel-shaped tubes present laterally on notogaster, through which the opisthosomal gland opens (Fig. 23.33). . . . . **HERMANNIELLIDAE**

22b Lateral tubes absent **23**

23a Notogaster with octotaxic system of porose areas or sacculi (Figs. 23.5, 23.6); pteromorphs usually present **24**

23b Notogaster without octotaxic system; pteromorphs usually absent **32**

24a Chelicerae peloptoid (Fig. 23.14); with well-developed cerotegument covering most of body . . . . **PHENOPELOPIDAE**

24b Chelicerae normal, chelate–dentate (Fig. 23.13); cerotegument mostly inconspicuous or absent **25**

25a Pteromorphs movable and auriculate, extending anteriorly and posteriorly well past articulation with notogaster (Fig. 23.34) **26**

25b Pteromorphs, if present and movable; not auriculate **27**

26a Notogaster usually with porose areas; six pairs of genital setae . . . . **GALUMNIDAE**

26b Notogaster usually with sacculi; five pairs of genital setae . . . . **PARAKALUMMIDAE**

27a Prodorsal lamellae broad (spanning most of prodorsum) and long (extending anteriorly almost to rostrum or beyond), and often medially fused for a portion of their length; pteromorphs large, immovable (Figs. 23.35, 23.36) **28**

27b Lamellae narrow, rarely touching medially (but may be joined by a translamella) and rarely approaching rostrum anteriorly **29**

28a Lamellar cusp rounded or with a single lateral point; lamellar seta inserted at mediodistal extremity of lamella or on its ventral surface; pteromorphs often with a long, knifelike anterior projection (Fig. 23.35);. . . . . **ACHIPTERIIDAE**

28b Lamella usually deeply incised terminally, producing medial and lateral points; lamellar setae inserted at base of incision (Fig. 23.36); pteromorphs without knifelike projection; . . . . **ORIBATELLIDAE**

29a Five or fewer pairs of genital setae; lamellae without free cusp; pteromorphs of various types, occasionally absent **30**

29b Six pairs of genital setae; lamellae usually with well-developed cusp; pteromorphs well developed, curving ventrally; movable or immovable **31**

30a Pteromorphs movable; triangular in lateral aspect. . . . **HAPLOZETIDAE**

30b Pteromorphs immovable (usually projecting laterally rather than curving ventrally) or absent . . . . **ORIBATULIDAE (*sensu lato*)**

31a Pteromorphs movable; anterior margin of notogaster overhangs dorsosejugal suture, connecting pteromorphs like a bridge (Fig. 23.37); translamella present . . . . **MYCOBATIDAE**

31b Pteromorphs movable or immovable, but without bridgelike connection; translamella present or absent . . . . **CERATOZETIDAE**

32a Prodorsum with lamellae **33**

32b Prodorsum without lamellae; narrow costulae may be present **42**

33a Integument thick, dark, distinctly sculptured with coarse ridges, tubercles, or areoles; lamellae usually broad, close to lateral margin of prodorsum. . . . **34**

33b Integument smooth, at most with fine, indistinct punctuation or tuberculation; lamellae of various form **35**

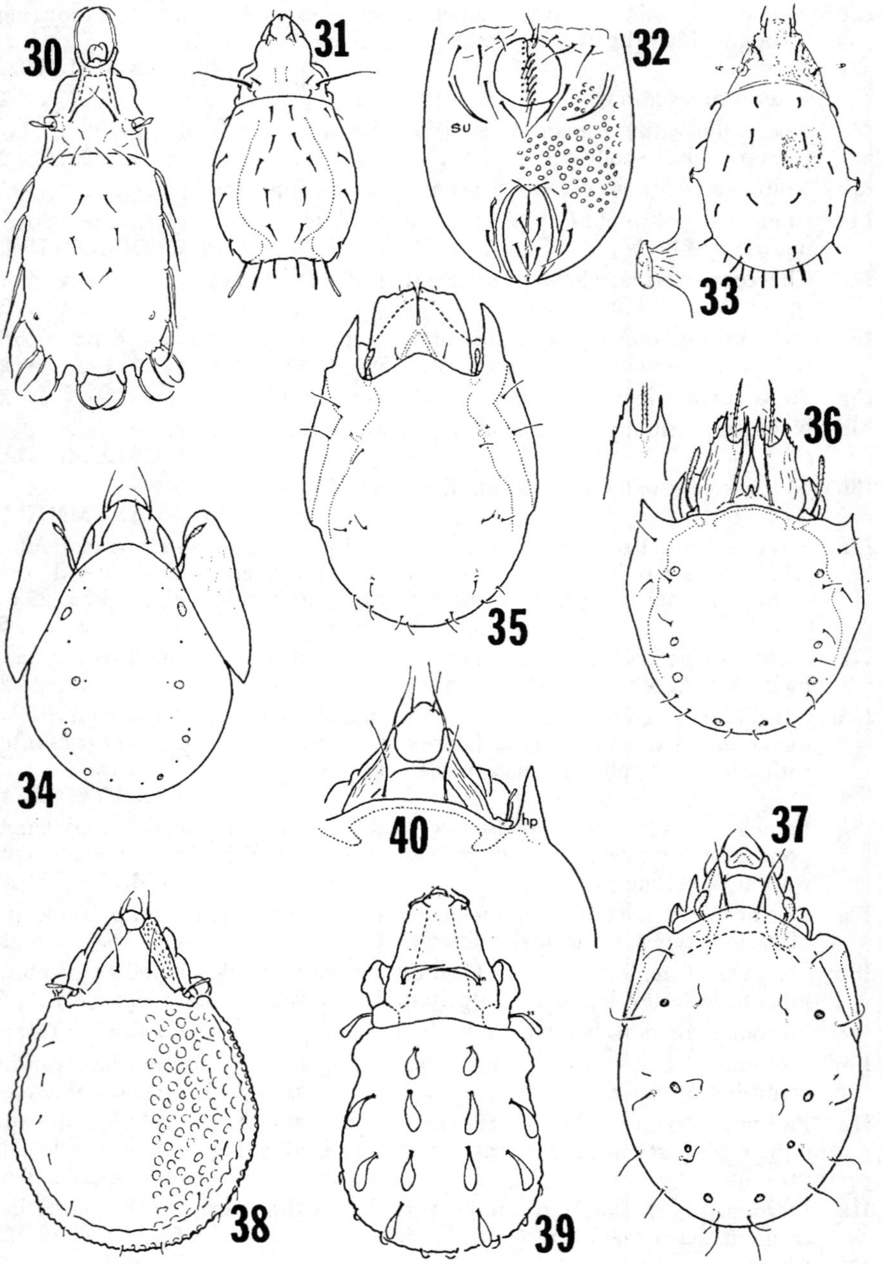

Fig. 23.30. Camisiidae (dorsal aspect). **Fig. 23.31.** Nothridae (dorsal aspect). **Fig. 23.32.** Nanhermanniidae, anogenital region. **Fig. 23.33.** Hermanniellidae (dorsal aspect) and enlargement of opisthosomal gland opening. **Fig. 23.34.** Galumnidae (dorsal aspect). **Fig. 23.35.** Achipteriidae (dorsal aspect). **Fig. 23.36.** Oribatellidae (dorsal aspect), with enlargement of lamellar cuspis. **Fig. 23.37.** Mycobatidae (dorsal aspect). **Fig. 23.38.** Cepheidae (dorsal aspect). **Fig. 23.39.** Carabodidae (dorsal aspect). **Fig. 23.40.** Tenuialidae, anterior region (dorsal aspect).

34a Usually with 6 pairs of genital setae; distance between genital and anal plates much less than length of the latter; notogastral setae mostly distributed in a pair of curved marginal rows (Fig. 23.38) . . . . **CEPHEIDAE**

34b Usually with four pairs of genital setae; genital and anal plates separated by at least the length of the latter; notogastral setae not distributed in a pair of marginal rows (Fig. 23.39) . . . . **CARABODIDAE**

35a Notogaster with large pointed humeral projection (Fig. 23.40) . . . . **TENUIALIDAE**

35b Humeral projection absent or inconspicuous **36**

36a Sejugal suture incomplete, notogaster anteriorly fused to prodorsum (Fig. 23.41); lamellae broad, marginally situated; relatively small species (usually less than 350 μm) . . . . **TECTOCEPHEIDAE**

36b Sejugal suture complete; lamellae usually situated more medially on prodorsum; usually larger species **37**

37a Chelicerae reduced, with relatively small diameter for most of their length **38**

37b Chelicerae normal, robust, chelate–dentate **39**

38a Chelicerae needlelike, without a movable digit; distally with small serrations (Fig. 23.15) . . . . **GUSTAVIIDAE**

38b Chelicerae peloptoid (Fig. 23.14). . . . . **METRIOPPIDAE**

39a Lamellae narrow, separated throughout their length; usually with very long cusp (Fig. 23.42); notogastral setae reduced to one to three pairs . . . . . . **CERATOPPIIDAE**

39b Lamellae broad, usually medially touching or fused, or with a short translamella; if completely separate, then with cusp very short or absent **40**

40a Distance between genital and anal plates less than length of the latter; mostly smaller species (usually no more than 350 μm) . . . . . . **ASTEGISTIDAE**

40b Distance between genital and anal plates greater than the length of the latter; mostly larger species **41**

41a Integument smooth, often shiny. . . . . **LIACARIDAE**

41b Integument with fine punctuation or tuberculation . . . . . **XENILLIDAE**

42a Epimera II–IV usually rectangular, entirely defined by transverse, medially fused apodemes (Fig. 23.43); notogaster with 16 pairs of conspicuous setae; large (about 1000 μm), dark species with round, convex notogaster . . . . **HERMANNIIDAE**

42b All paired apodemes usually not meeting medially; if so, epimera not all evenly rectangular in appearance (apodemes not transverse, or else indistinct or absent); notogaster with fewer than 16 pairs of setae **43**

43a Genital plates transversely divided (Fig. 23.44); notogaster carries exuviae of immature instars; notogastral setation reduced; at most three to five pairs of conspicuous setae, restricted to posterior third; large (about 1000 μm), dark species with sculptured integument. . . . . **LIODIDAE**

43b Genital plates entire; notogaster various; usually smaller species **44**

44a Notogaster flat-topped or concave in lateral aspect, elliptical in dorsal aspect; usually with heavy, ornamented cerotegument; notogastral setation reduced; setae absent from anterior half; tibia I with a pronounced dorsal tubercle which overhangs tarsus and bears a long sensory seta. . . . . **GYMNODAMAEIDAE**

44b Notogaster convex, with or without cerotegument; setae present on anterior half; tibia I with tubercle less pronounced or absent **45**

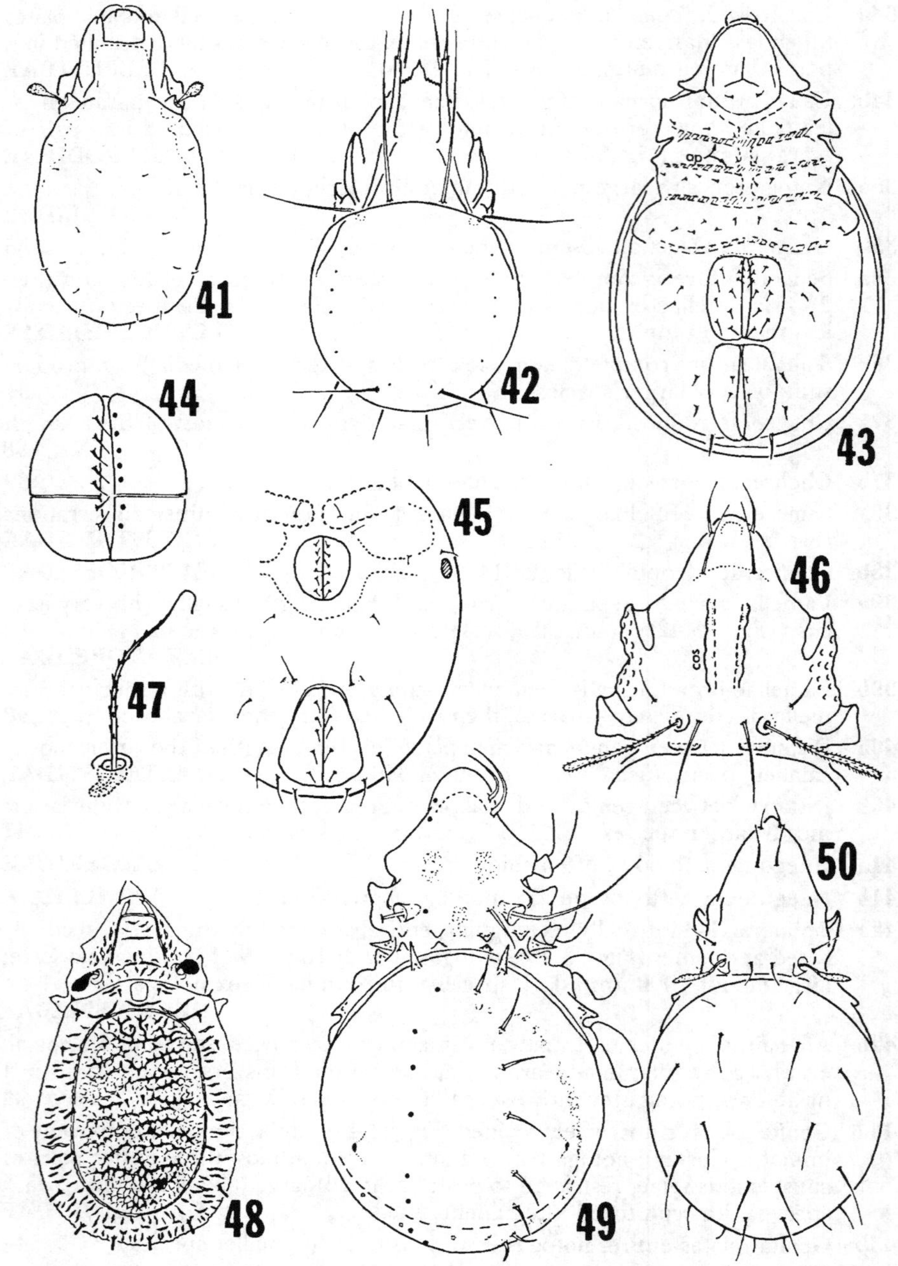

Fig. 23.41. Tectocepheidae (dorsal aspect). **Fig. 23.42.** Ceratoppiidae (dorsal aspect). **Fig. 23.43.** Hermanniidae (ventral aspect). **Fig. 23.44.** Liodidae, genital plates. **Fig. 23.45.** Eremaeidae, anogenital region (ventral aspect). **Fig. 23.46.** Eremaeidae, prodorsum (dorsal aspect). **Fig. 23.47.** Eremulidae, sensillus. **Fig. 23.48.** Cymbaeremaeidae (dorsal aspect). **Fig. 23.49.** Damaeidae (dorsal aspect). **Fig. 23.50.** Oppiidae (dorsal aspect).

45a Ventral plate with five or more pairs of setae (Fig. 23.45) (not including setae of epimere IV) 46

45b Ventral plate with four pairs of setae (Fig. 23.8) (aggenital and three adanals) 48

46a With three to five pairs of setae on anal plate; sensillus usually clavate or somewhat spatulate; prodorsum usually with costulae (Fig. 23.46) . . . . . **EREMAEIDAE**

46b With two pairs of anal setae; sensillus finely tapered or flagellate distally; prodorsum with or without costulae 47

47a Sensillus bent distally in a characteristic shepherd's crook (Fig. 23.47); costulae usually distinct and parallel . . . . **EREMULIDAE**

47b Sensillus often curved, but not as above; parallel costulae absent . . . . . **EREMOBELBIDAE**

48a Integument of notogaster highly sculptured with pitting or vermiform ridges; margin flat, central region convex (Fig. 23.48); a small, round or square light-colored area (lenticulus) usually present near anterior border of notogaster . . . . **CYMBAEREMAEIDAE**

48b Notogastral integument mostly smooth, shiny; usually evenly convex; lenticulus absent. 49

49a Chelicerae peloptoid (Fig. 23.14). . . . . **SUCTOBELBIDAE**

49b Chelicerae normal, chelate dentate (Fig. 23.13) 50

50a Costulae present 51

50b Costulae absent. 53

51a Genital and anal plates large; distance between them half the length of the latter or less . . . . **THYRISOMIDAE**

51b Genital and anal plates smaller; separated by a distance usually equal to or greater than the length of the latter 52

52a Anterior margin of notogaster truncate; costulae long, reaching well into anterior half of prodorsum . . . . **AUTOGNETIDAE**

52b Anterior margin of notogaster rounded; costulae usually shorter, restricted to midregion of prodorsum . . . . **OPPIIDAE** (in part)

53a All conspicuous notogastral setae (eight pairs) distributed in a pair of medial longitudinal rows (Fig. 23.49); epimere IV not well defined posteriorly. . . **DAMAEIDAE (*sensu latu*)**

53b Notogastral setae not in a pair of rows (Fig. 23.50); epimere IV defined posteriorly by a long, curving distinct border . . . . . **OPPIIDAE** (in part)

References

Anderson, J. M. 1978. Inter- and intra- habitat relationships between woodland Cryptostigmata species diversity and the diversity of soil and litter microhabitats. *Oecologia* (*Berl.*) 32:341–348. **Aoki, J.-I. 1967.** Microhabitats of oribatid mites on a forest floor. *Bull. Nat. Sci. Mus.* (*Tokyo*) 10:133–138. **Aucamp, J. L., G. C. Loots and P. A. J. Ryke. 1964.** Notes on the efficiency of funnel batteries for the extraction of soil micro-arthropods. *Tydskr. Natuurwet.* 4:105–126. **Balogh, J. 1963.** Identification keys of holarctic oribatid mites (Acari) families and genera. *Acta Zool. Hung.* 9:1–60. **1972.** *The Oribatid Genera of the World.* Akademiai Kiado, Budapest, Hungary. 188 pp. **Cancela da Fonseca, J. P. 1967.**

Le spermatophore de *Damaeus quadrihastatus* Markel et Mayer (acarien oribate). Pp. 227–232 in *Proceedings of the 2nd International Congress on Acarology,* Sutton Bonington. **Cromack, K., Jr., P. Sollins, R. L. Todd, D. A. Crossley, Jr., W. M. Fender, R. Fogel and A. W. Todd. 1977.** Soil microorganism–arthropod interactions: fungi as major calcium and sodium sources. Pp. 78–84 in W. J. Mattson (ed.). *The Role of Arthropods in Forest Ecosystems.* Springer-Verlag, New York. **Crossley, D. A., Jr. 1977a.** The roles of terrestrial saprophagous arthropods in forest soils: current status of concepts. Pp. 49–56 in W. J. Mattson (ed.). *The Role of Arthropods in Forest Ecosystems.* Springer-Verlag, New York. **1977b.** Oribatid mites and nutrient cycling. Pp. 71–85 in D. L. Dindal (ed.). *Biology of Oribatid Mites.* SUNY College of Environmental Science and Forestry, Syracuse, N.Y. **Gerard, G. and P. Berthet. 1966.** A statistical study of microdistribution of Oribatei (Acari). II. The transformation of the data. *Oikos* 17:142–149. **Grandjean, F. 1949.** Observation et conservation des très petits Arthropodes. *Bull. Mus. Natl. Hist. Nat. (Paris) Ser. 2* 21:363–370. **Hammer, M. 1972.** Microhabitats of oribatid mites on Danish woodland floor. *Pedobiologia* 12:412–423. **Hartenstein, R. 1962.** Soil Oribatei. I. Feeding specificity among forest soil Oribatei. *Ann. Entomol. Soc. Am.* 55:202–206. **Huhta, V. 1972.** Efficiency of different dry funnel techniques in extracting Arthropods from raw humus forest soil. *Ann. Zool. Fenn.* 9:42–48. **Krantz, G. W. 1978.** *A Manual of Acarology,* 2nd ed. Oregon State University, Corvalis, Oreg. 509 pp. **Lebrun, P. 1964.** Note sur les migrations des oribatides (Acari) de petite taille. *Bull. Ann. Soc. R. Entomol. Belg.* 100:69–77. **Luxton, M. 1972.** Studies on the oribatid mites of a Danish beech wood soil. *Pedobiologia* 12:434–463. **Marshall, V. G., R. M. Reeves and R. A. Norton.** A catalogue of the Oribata of Continental US and Canada. *Mem. Can. Entomol. Soc.* No. 139. 418 pp. **Martin, J. E. H. 1977.** *The Insects and Arachnids of Canada.* Part 1. *Collecting, Preparing and Preserving Insects, Mites, and Spiders.* Publ. 1643. Research Branch, Agriculture Canada. 182 pp. **Mitchell, M. J. 1977a.** Life history strategies of oribatid mites. Pp. 65–69 in D. L. Dindal (ed.). *Biology of Oribatid Mites.* SUNY College of Environmental Science and Forestry, Syracuse, N.Y. **1977b.** Population dynamics of oribatid mites (Acari, Cryptostigmata) in an aspen woodland soil. *Pedobiologia* 17:305–319. **Mitchell, M. J. and D. Parkinson. 1976.** Fungal feeding of oribatid mites (Acari: Cryptostigmata) in an aspen woodland soil. *Ecology* 57:302–312. **Muraoka, M. and N. Ishibashi. 1976.** Nematode-feeding mites and their feeding behaviour. *Appl. Entomol. Zool.* 11:1–7. **Murphy, P. W. 1962a.** Extraction methods for soil animals. I. Dynamic methods with particular reference to funnel processes. Pp. 75–114 in P. W. Murphy (ed.). *Progress in Soil Zoology.* Butterworth, London. **1962b.** Extraction methods for soil animals. II. Mechanical methods. Pp. 115–155 in P. W. Murphy (ed.). *Progress in Soil Zoology.* Butterworth, London. **Norton, R. A. 1977.** A review of F. Grandjean's system of leg chaetotaxy in the Oribatei and its application to the Damaeidae. Pp. 33–62 in D. L. Dindal (ed.). *Biology of Oribatid Mites.* SUNY College of Environmental Science and Forestry, Syracuse, N.Y. **1986.** Aspects of the biology and systematics of soil arachnids, particularly saprophagous and mycophagous mites. *Quaest. Entomol.* 21:523–541. **Schuster, R. 1956.** Der Anteil der Oribatiden an den Zersetzungsvorgangen im Boden. *Z. Morphol. Oekol. Tiere* 45:1–33. **Sengbusch, H. G. 1977.** Review of oribatid mite-anoplocephalan tapeworm relationships (Acari; Oribatei: Cestoda; Anoplocephalidae). Pp. 87–102 in D. L. Dindal (ed.). *Biology of Oribatid Mites.* SUNY College of Environmental Science and Forestry, Syracuse, N.Y. **Seniczak, S. and O. Stefaniak. 1978.** The microflora of the alimentary canal of *Oppia nitens* (Acarina, Oribatei). *Pedobiologia* 18:110–119. **Sømme, L. and E.-M. Conradi-Larsen. 1977a.** Cold-hardiness of collembolans and oribatid mites from windswept mountain ridges. *Oikos* 29:118–126. **1977b.** Anaerobiosis in overwintering collembolans and oribatid mites from windswept mountain ridges. *Oikos* 29:127–132. **Stefaniak, O. and S. Seniczak. 1976.** The microflora of the alimentary canal of *Achipteria coleoptrata* (Acarina, Oribatei). *Pedobiologia* 16:185–194. **Todd, R. L., D. A. Crossley, Jr. and J. C. Stormer, Jr. 1975.** Chemical composition of microarthropods by electron microprobe analysis—a preliminary report. In C. J. Arceneaux (ed.). *32nd Annual Proceedings of the Electron Microscopy Society of America,* St. Louis, Mo. **Travé, J. 1963.** Écologie et biologie des oribates (acariens) saxicoles et arboricoles. *Vie Milieu Suppl.* 14:1–267. **1965.** Quelques techniques de récolte, de tirage, d'observation et de conservation des oribates (acariens) et autres microarthropodes. *Rev. Ecol. Biol. Sol* 2:23–47. **Travé, J. and M. Vachon. 1975.** Francois Grandjean, 1882–1975 (notice biographique et bibliographique). *Acarologia (Paris)* 17:1–19. **Wallwork, J. A. 1969.** Some basic principles underlying the classification and identification of cryptostigmatid mites. Pp. 155–166 in J. G. Sheals (ed.). *The Soil Ecosystem.* Systematics Association Publ. 8, London. **1983.** Oribatids in forest ecosystems. *Annu. Rev. Entomol.* 28:109–130. **Weigmann, G. 1975.** Labor- und Freilanduntersuchungen zur Generationsdauer von Oribatiden (Acari : Oribatei). *Pedobiologia* 15:133–148. **Woodring, J. P. 1970.** Comparative morphology, homologies, and functions of the male system in oribatid mites (Arachnida : Acari). *J. Morphol.* 132:425–451. **Woodring, J. P. and E. F. Cook. 1962.** The internal anatomy, reproductive physiology, and molting process of *Ceratozetes cisalpinus* (Acarina Oribatei). *Ann. Entomol. Soc. Am.* 55:164–181.

24

Terrestrial Isopoda

WILLIAM B. MUCHMORE

The terrestrial isopods, also known as woodlice, sow bugs, pill bugs, and slaters, make up the suborder Oniscoidea of the crustacean order Isopoda. They are generally elliptical, dorsoventrally flattened animals with a prominent pair of antennae, seven pairs of legs, and (usually) obvious uropods. They range in length from 2 to 30 mm, are smooth or rough surfaced, and their color varies from dark gray in the larger, more vagile forms to white in the small, hypogean ones.

External Morphology

The body is divided into a *head* or *cephalon, thorax* or *pereion,* and *abdomen* or *pleon* (Fig. 24.1). The first section of the body is conveniently called the *head,* although it actually includes the first thoracic segment as well as the true head segments. Dorsally, it is covered by an undivided, somewhat oval sclerite, which may have anterolateral processes and may bear laterally a pair of *eyes.* The eyes are simple, composed of a single ocellus, or compound, composed of several or many closely associated ocelli. Anteriorly, the head bears a pair of very small, inconspicuous *first antennae* and a pair of larger, prominent *second antennae.* The latter are important sense organs and are usually referred to simply as the *antennae.* They are composed of five rather stout, basal segments, and two or more, slender, distal segments which constitute the *flagellum.* Ventrally, surrounding the mouth is a buccal mass consisting of the mouthparts: *mandibles, first maxillae, second maxillae,* and *maxillipeds.*

The pereion consists of seven broadly articulated segments, each with a tergite which is more or less expanded at the sides as epimera. The head is more or less set back into the first pereion segment, and the epimera of the last segment extend back over the pleon to various extents. Ventrally, each pereion segment bears a pair of well-developed *pereiopods* or *legs* (Fig. 24.2), typically composed of a coxa fused to the body and six free segments (*basis, ischium, merus, carpus, propodus* and *dactylus*). The legs are much alike, although the more posterior ones are usually larger and the anterior ones may be slightly modified in males. In the females there are medial, leaflike extensions of the bases of some of the pereiopods, called *oöstegites,* which form a brood pouch for the developing eggs and embryos.

Tapering back from the pereion are the six *pleon* segments, including the terminal *telson* or *pleotelson.* As in the pereion, there may be epimera extending posterolaterally, especially on segments 3, 4, and 5. Each of the pleon segments bears a pair of appendages, the *pleopods,* which consist of a short, basal *protopodite,* an inner branch or *endopodite,* and an outer branch or *exopodite.* Endopodites and exopodites of the first five pairs are usually flat and leaflike, and the exopodites often contain *tracheal tubules.* In males the first two pleopods are usually modified as more or less elongate copulatory organs. The appendages of the telson (*uropods*) usually have elongate inner and outer branches (*rami*), which serve as sensory organs; in forms that can roll into a ball, however, the rami may be reduced and flattened to conform to a smooth body outline.

Food

Most isopods are omnivores or scavengers, feeding mainly on vegetable and also on animal matter. Their mandibles are heavily sclerotized and capable of chewing fairly dense material. Most of their food seems to be already dead and decaying—in litter, under logs and rocks, in crevices, under bark, in dumps, wherever plant and animal refuse and detritus accumulates and moisture is present. On occasion they have been implicated in eating living plants; they may find optimal conditions in greenhouses or outdoors in the southern states and in their numbers cause noticeable damage to various plants.

Reproduction and Life History

Reproduction is quite straightforward in the terrestrial isopods. Females are fertilized by males using their modified first and second pleopods as a copulatory organ. The eggs (up to 100 or more) are retained in a brood pouch formed by oöstegites medial to the legs. Development is direct, giving rise to a juvenile with the general characteristics of the species, but imperfect in some details of segmentation, appendage development, and

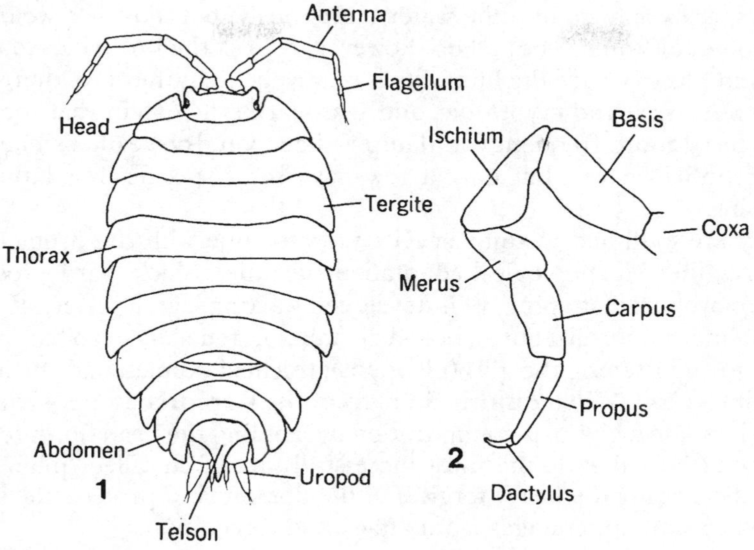

Figs. 24.1, 24.2. *Oniscus asellus,* dorsal view of entire animal, leg.

so on. These are soon liberated from the brood pouch and after four or five molts are like the adults except for size, proportions, pigmentation, and sexual characters.

Molting is peculiar in the isopods in that it is accomplished in two separate stages: first the posterior half of the skin is shed, then 2 or 3 days later the anterior half is shed. Thus animals in the process of molting may appear bicolored, either the back or front halves being lighter, depending on the stage of the process. Also, after shedding the posterior exuviae, the back half of the animal expands and becomes larger than the front half, until the anterior molt is accomplished.

Although most isopods reproduce normally sexually, a few may be parthenogenetic. Best known of these is *Trichoniscus pusillus,* which is widespread in northern Europe and has been imported into America. In most populations, only females are found, although males may occur occasionally, especially in the spring.

Ecology and Habits

The terrestrial isopods (Oniscoidea) are the only large group of Crustacea to become adapted for life on the land. A few amphipods can live in moist forest litter or on beaches, and a few crabs are successful land dwellers in tropical places, usually near the sea. But large numbers of oniscoid isopods have become quite independent of the water and may be found in many inland locations, even in some which are classified as deserts. Some littoral species may go into the water temporarily, but most live well away from bodies of water. They all do, however, restrict their activities to times when and places where the humidity is relatively high; they are, therefore, mainly nocturnal and cryptozoic, and seasonal to the extent that they are encountered more frequently in damp rather than dry weather. They are often found in caves, but only a few have been specifically adapted as troglobites.

They are modified in some special ways to cope with the problems of terrestrial life. Morphological adaptations include a thick, hard exoskeleton, dorsoventral flattening, well-developed walking legs, a variety of sense organs, internal fertilization, a brood chamber in females, a capillary water system for moistening the gills (pleopods), tracheal tubules, and the ability to roll into a ball. When disturbed, many of the terrestrial isopods can curl themselves into a globular configuration by bending the head down toward the telson (those that do this most successfully are often called "pill bugs"). This exposes only the heavy tergites of the dorsum and protects the softer underparts and appendages from attack and desiccation.

They are usually found associated with damp, decaying vegetation or other organic matter, and may be present in very large numbers in accumulations of such materials as occur in compost, manure, dump heaps, wrack, windrows, and tree holes. Many of the American oniscoids are anthropophilic and have apparently been introduced from Europe or elsewhere through human activity.

Only a few studies have been made to ascertain the importance of terrestrial isopods in their communities. The scanty evidence available suggests that under natural conditions they do little more than triturate a certain portion of the litter in which they live, and, of course, they serve as a food source for a variety of invertebrates and vertebrates.

Collection and Study

Terrestrial isopods are usually easily collected individually by hand from their hiding places under vegetation, logs, bark, rocks, boards, rubbish, and animal carcasses. They may also be separated from litter and debris by the use of a Berlese or Tullgren funnel.

They should be collected and preserved in 70–80% alcohol; it is permissible to add a little glycerol to prevent complete desiccation in case the alcohol should evaporate.

All the characters used in the key may be seen on whole specimens observed with a hand lens or a dissecting microscope. Frequently, however, the identification of species depends on detailed study of certain parts under a compound microscope. These parts, such as mouthparts, legs, and pleopods, may be mounted temporarily in glycerine or Hoyer's medium for such observation.

Identification

The following key is designed to identify a mature specimen to genus. Because juveniles look so much like adults, there is often a possibility of confusing juveniles of large species with adults of small species. Ideally, males with obviously modified first and second pleopods or females with brood pouches should be used. These will then usually be in the size ranges indicated for the genera (body length is measured from the front of the head to the tip of the telson on a flattened individual).

In life most terrestrial isopods are gray or brown, often mottled with white or yellow; some have reduced pigmentation and appear as tan, pink, or white. In many species there is much variation in color and pattern; also, the lighter colors, such as white, yellow, and pink, tend to be lost or obscured after preservation in alcohol. Therefore, colors have been indicated in the key only in a general way.

ACKNOWLEDGMENT

I thank Dr. George O. Schultz for reviewing this chapter and providing valuable suggestions.

KEY TO GENERA

1a Flagellum of second antenna with 10 or more articles (Fig. 24.3); rami of uropod very long and slender (Fig. 24.4); Family **Ligiidae** **2**

1b Flagellum of second antenna with five or fewer articles; rami of uropod rather short and stout (Fig. 24.1) **3**

2a (1) Inner angle of uropod produced, so that rami attach at different levels (Fig. 24.4); terrestrial . . . . ***Ligidium*** Brandt

Representatives of this genus are widespread in the eastern half of North America and on the west coast from California to Alaska. There are 8–10 named species, of which ***L. elrodii*** (Packard) is most common in the east and ***L. gracile*** (Dana) in the west. They live in moist litter, especially in swampy situations and along streams and springs. Body length up to 10 mm.

2b Inner angle of uropod not produced, and rami attach at same level (Fig. 24.5); littoral . . . . ***Ligia*** Fabricius

All members of this genus are littoral, and representatives are probably found along every coast; the common sea slater. The taxonomy is not stable, but the most common in the east is probably ***L. exotica*** Roux and in the west ***L. pallasii*** Brandt. They live on rocks and piles just above the water. Body length to 30 mm.

3a (1) Flagellum markedly more slender than basal segments of second antenna, composed of three to six indistinct articles (Fig. 24.6) **4**

3b Flagellum nearly as stout as basal segments of second antenna, composed of two to four distinct articles (Fig. 24.7) **14**

4a (3) Eye of one to three ocelli (Fig. 24.8) or absent; terrestrial; Family **Trichoniscidae** **5**

4b Eye of five or more aggregated ocelli; littoral; Family **Scyphacidae 12**

5a (4) Eye of three ocelli (Fig. 24.8) **6**

5b Eye of a single ocellus **7**

5c Eye absent; colorless, cavernicolous forms **11**

6a (5) Dorsal surface smooth and shiny. . . . . ***Trichoniscus*** Brandt

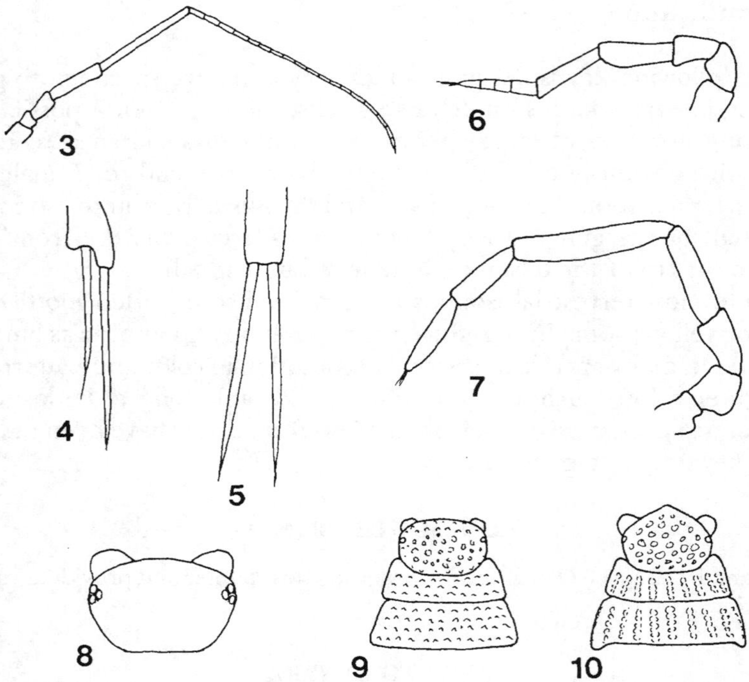

Fig. 24.3. *Ligia* sp., antenna. **Fig. 24.4.** *Ligidium* sp., uropod. **Fig. 24.5.** *Ligia* sp., uropod. **Fig. 24.6.** *Trichoniscus pusillus*, antenna. **Fig. 24.7.** *Porcellio* sp., antenna. **Fig. 24.8.** *Trichoniscus pusillus*, head, dorsal view. **Fig. 24.9.** *Miktoniscus* sp., head and pereionites 1 and 2. **Fig. 24.10.** *Haplophthalmus danicus*, head and pereionites 1 and 2.

Three European species have been recorded from North America. ***T. pusillus*** Brandt (= ***T. demivirgo*** Blake) is widespread in the east, north to Newfoundland, and has been reported from Washington; ***T. pygmaeus*** Sars is known from Newfoundland, New York, and Illinois; and ***T. provisorius*** Racovitza from Newfoundland only. They all live in moist woodland litter and in human refuse. Body length to 5 mm.

6b Dorsal surface covered with tiny tubercles. . . . ***Cordioniscus*** Graeve

In North America one species, ***C. stebbingi*** (Patience), has been found, only in greenhouses in Massachusetts and Washington. In Europe it lives in moist litter. Body length to 3.5 mm.

7a **(5)** Dorsal surface tuberculate. **8**

7b Dorsal surface smooth and shiny . . . . ***Hyloniscus*** Verhoeff

One European species, ***H. riparius*** (C. L. Koch), is found in several eastern states and north to Newfoundland. It lives in moist litter and refuse. Body length to 7 mm.

8a **(7)** Dorsal surface with tubercles distributed randomly or in transverse rows (Fig. 24.9); usually colored. **9**

8b Dorsal surface with tubercles in distinct longitudinal rows (Fig. 24.10); white . . . . ***Haplophthalmus*** Schöbl

One European species, ***H. danicus*** Budde-Lund, is locally common in eastern United States and north to Newfoundland; it has also been found in Utah. It inhabits moist litter and decaying wood. Body length to 4 mm.

9a **(8)** Rosy pink in color; endopodite of first pleopod of male long and spoon-shaped (Fig. 24.11) . . . . ***Androniscus*** Verhoeff

One European species, ***A. dentiger*** Verhoeff, has been found in greenhouses in Ontario and Newfoundland and outdoors in New York. It lives in damp litter and soil. Body length to 6 mm.

9b Brownish gray to white; endopodite of first pleopod of male not spoon-shaped **10**

10a (9) Exopodite of first pleopod of male with two slender processes posteriorly (Fig. 24.12) . . . . ***Trichoniscoides*** Sars

One European species, ***T. sarsi*** Patience, has been found in Newfoundland, under rocks near human habitation. Body length to 3 mm.

10b Exopodite of first pleopod of male without posterior processes (Fig. 24.13) . . . . ***Miktoniscus*** Kesselyak

Several species are known from the eastern United States. ***M. halophilus*** Blake lives in lowlands along the coast, while others occur in the mountains and the Mississippi valley. A few are found in caves and may be modified by loss of pigment and reduction or loss of eyes. They are usually found in moist litter. Body length to 5.5 mm.

11a (5) Endopodite of first pleopod of male a long, narrow blade, at least two-thirds as long as exopod (Fig. 24.14) . . . . . ***Amerigoniscus*** Vandel

Nine species are known from restricted localities in Virginia, Georgia, Tennessee, Oklahoma, Texas, New Mexico, and Oregon. All were found in caves except one from Oregon, which was under rocks and moss in dense forest. Body length to 5 mm.

11b Endopodite of first pleopod of male reduced, less than one-fourth as long as exopod (Fig. 24.15) . . . . ***Brackenridgia*** Ulrich

This genus (= ***Protrichoniscus*** Archangeli) is represented by three species in the United States, while others occur in Mexico. ***B. cavernarum*** Ulrich and ***B. reddelli*** (Vandel) live in caves in Texas, while ***B. heroldi*** (Archangeli) is found in damp humus in California. Body length to 5.5 mm.

12a (4) Eye small, of 5–10 ocelli; head with prominent anterolateral processes (Fig. 24.16). **13**

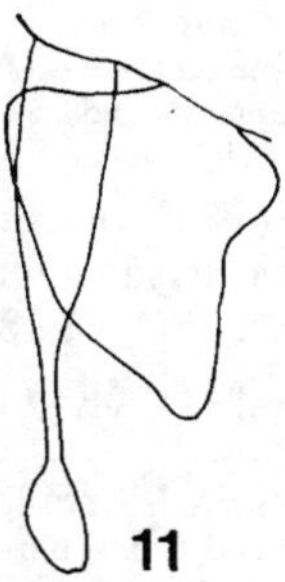
11

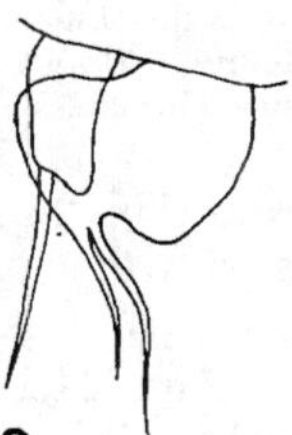
12

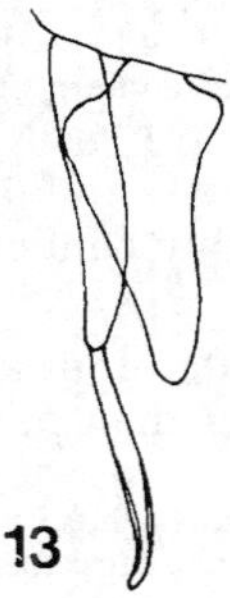
13

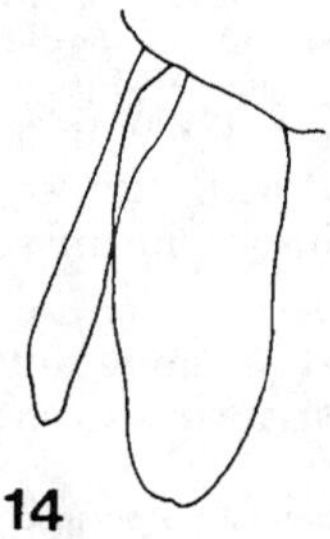
14

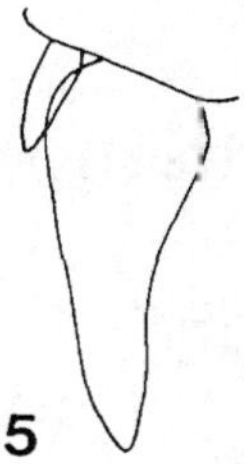
15

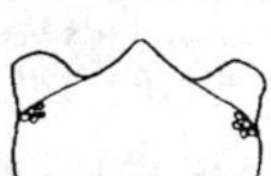
16

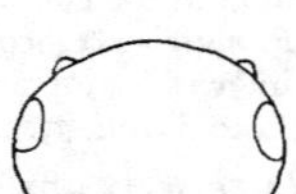
17

18

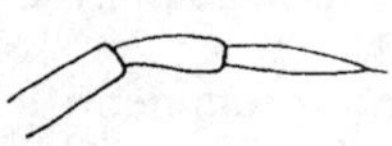
19

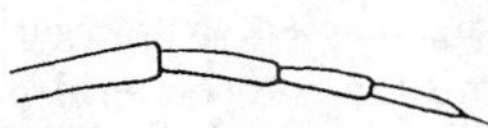
20

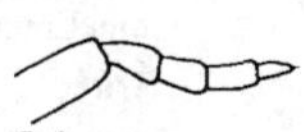
21

Fig. 24.11. *Androniscus dentiger*, ♂, pleopod 1. **Fig. 24.12.** *Trichoniscoides sarsi*, ♂, pleopod 1. **Fig. 24.13.** *Miktoniscus* sp., ♂, pleopod 1. **Fig. 24.14.** *Amerigoniscus paynei*, ♂, pleopod 1. **Fig. 24.15.** *Brackenridgia cavernarum*, ♂, pleopod 1. **Fig. 24.16.** *Armadilloniscus ellipticus*, head. **Fig. 24.17.** *Scyphacella arenicola*, head. **Fig. 24.18.** *Trichorhina triocis*, head. **Fig. 24.19.** *Porcellio* sp., flagellum of antenna. **Fig. 24.20.** *Oniscus asellus*, flagellum of antenna. **Fig. 24.21.** *Tylos* sp., flagellum of antenna.

12b Eye large, of many ocelli; head without prominent anterolateral processes (Fig. 24.17) . . . . ***Scyphacella*** Smith

One species, ***S. arenicola*** Smith, ranges along the east coast from Florida to Massachusetts. It is littoral, on sandy beaches. Body length to 4.5 mm.

13a (12) Body broad (elliptical) and flattened . . . . . ***Armadilloniscus*** Uljanin

Three species are known. ***A. ellipticus*** (Harger) occurs on the east coast from Florida to Massachusetts; ***A. holmesi*** Archangeli and ***A. lindahli*** (Richardson) are found on the west coast from California to Washington. All are littoral in habits, found mainly in moist sand. Body length to 4.5 mm.

13b Body narrow (oblong) and convex . . . . ***Detonella*** Lohmander

One species, ***D. papillicornis*** (Richardson), is found on the west coast from Washington to Alaska. It lives in moist sand on ocean beaches. Body length to 4 mm.

14a (3) Eye of one to three ocelli (Fig. 24.18) or absent; body white; Family **Squamiferidae** **15**

14b Eye compound, of many ocelli; body usually colored **16**

15a (14) Eye of one to three ocelli. . . . . ***Trichorhina*** Budde-Lund

Three species of this tropical genus have been recorded from the United States. ***T. triocis*** Mulaik occurs in southern Texas, while ***T. tomentosa*** Budde-Lund and ***T. heterophthalma*** Lemos de Castro have been found in Florida and Georgia, respectively. They live in leaf litter. Body length to 3.5 mm.

15b Eyes lacking; myrmecophilous . . . . ***Platyarthrus*** Brandt

Two species of this European genus have been found in the United States: ***P. hoffmannseggi*** Brandt in Connecticut and New York and ***P. schöbli*** Budde-Lund in Texas. They are always found in association with ants. Body length to 5.5 mm.

16a (14) Flagellum of second antenna of two articles (Fig. 24.19). **19**

16b Flagellum of three articles (Fig. 24.20); Family **Oniscidae** (in part) **17**

16c Flagellum of four articles (Fig. 24.21); Family **Tylidae**. ***Tylos*** Latreille

Three species of this tropical genus are known in the United States. ***T. latreillei*** Audouin and ***T. niveus*** Budde-Lund have been found in Florida and ***T. punctatus*** Holmes and Gay is common in southern California. They are all inhabitants of sandy ocean beaches. Body length to 20 mm.

17a (16) Head with distinct anterolateral processes (Fig. 24.22); body elliptical; outline of pleon smoothly continuous with that of pereion (see Fig. 24.1) **18**

17b Head without anterolateral processes (Fig. 24.23); body elongate; pleon abruptly narrower than pereion (Fig. 24.24) . . . . ***Philoscia*** Latreille

About a dozen species of this genus have been reported from the United States, but the systematics are still uncertain. Most species are littoral, and one or another probably occurs on all coasts, in seaweed and debris on beaches and in tidal marshes. ***P. vittata*** Say is apparently most common in the east and ***P. richardsonae*** Holmes and Gay in the west. One European form, ***P. muscorum*** (Scopoli), is anthropophilic and has been introduced into inland localities in Washington and New York, where it is found in moist litter and debris. Body length to 5 mm, greater for ***P. muscorum***.

18a (17) Body broad, rather flattened; anterolateral processes of head prominent; telson elongate (Fig. 24.25); terrestrial . . . . ***Oniscus*** Linnaeus

One species of this European genus, ***O. asellus*** Linnaeus, has been introduced into North America and has become widely distributed in northern United States and southern Canada. It is usually found in litter and refuse around human habitations, and sometimes occurs in large numbers. Body length to 18 mm.

18b Body strongly convex; anterolateral processes not prominent; telson short, rounded (Fig. 24.26); littoral, west coast only. . . . . ***Alloniscus*** Dana

Three species of this genus have been described from the west coast. Most common is ***A. perconvexus*** Dana, found under debris on sandy beaches from California to British Columbia. Body length to 16 mm.

19a (16) Uropods extend beyond telson, which is produced (Fig. 24.27); Family **Oniscidae** (in part). **20**

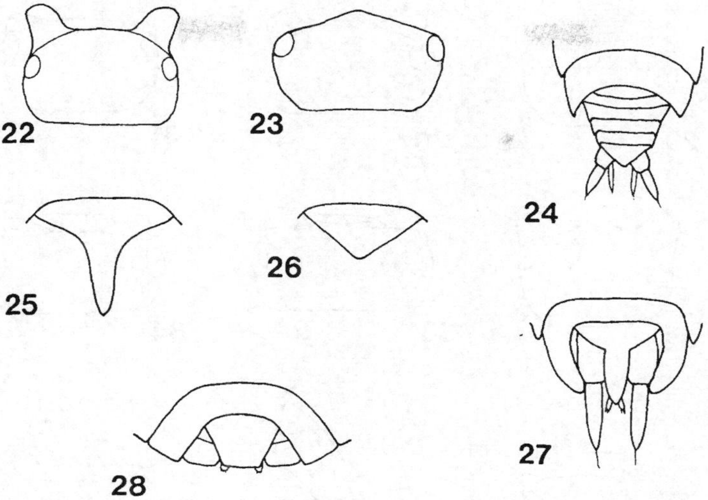

Fig. 24.22. *Oniscus asellus*, head. **Fig. 24.23.** *Philoscia muscorum*, head. **Fig. 24.24.** *Philoscia muscorum*, posterior end of pereion and pleon. **Fig. 24.25.** *Oniscus asellus*, telson. **Fig. 24.26.** *Alloniscus perconvexus*, telson. **Fig. 24.27.** *Cylisticus convexus*, posterior pleonites, telson and uropods. **Fig. 24.28.** *Armadillidium vulgare*, posterior pleonites, telson and uropods.

19b Uropods not extending beyond telson, which is broadly truncate (Fig. 24.28) **24**

20a (19) Front of head with prominent bulbous expansion (Fig. 24.29); distal article of flagellum of second antenna three times as long as basal article; small . . . . ***Rhyscotus*** Budde-Lund

One species of this tropical genus has been reported from southern Texas and Florida, namely, ***R. texensis*** (Richardson). It is found in moist litter. Body length to 6 mm.

20b Head without a prominent anterior expansion; articles of flagellum subequal in length; medium to large in size **21**

21a (20) Outline of pereion and pleon continuous; front margin of head produced into a median lobe (when viewed from above) (Fig. 24.30). . . **22**

21b Pleon distinctly narrower than pereion; front margin of head nearly straight (when viewed from above) (Fig. 24.31) . . . . ***Metoponorthus*** Budde-Lund

This genus (= ***Porcellionides*** Miers) is mainly represented in North America by one European species, ***M. pruinosus*** (Brandt). This is widespread through the United States and southern Canada, especially around human habitations. Three other species have been described from restricted localities in southern states. Body length to 12 mm.

22a (21) Median lobe of head sharp pointed (when viewed from above) (Fig. 24.30); body quite convex, can be rolled into a ball . . . . ***Cylisticus*** Schnitzler

One European species, ***C. convexus*** (DeGeer), is widespread in the northern half of the United States and southern Canada. It is found in a wide variety of litter and debris, mainly around human habitations. Body length to 15 mm.

22b Median lobe of head rounded (when viewed from above) (Fig. 24.32); body moderately flattened, cannot be rolled into a ball **23**

23a (22) Exopodites of all 5 pleopods with tracheae (white patches) . . . . . . ***Trachelipus*** Budde-Lund

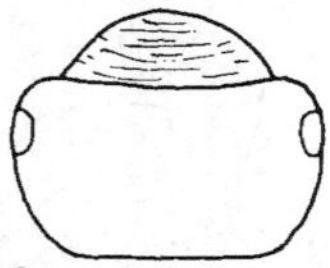
29

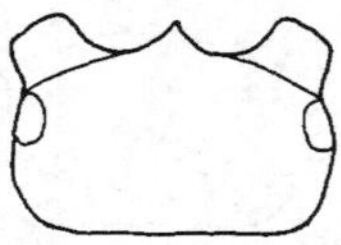
30

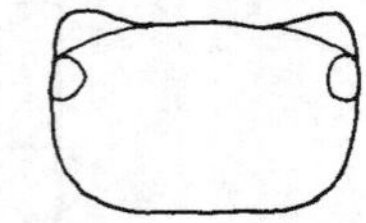
31

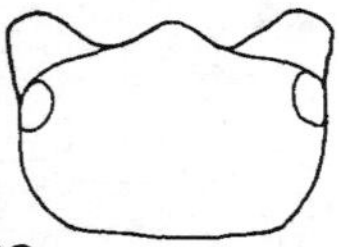
32

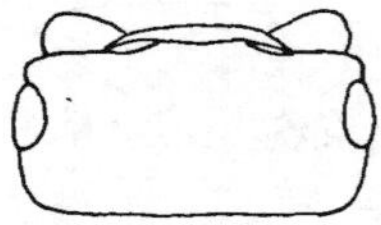
33

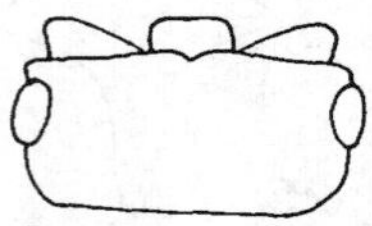
34

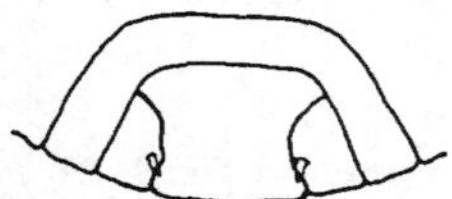
35

36

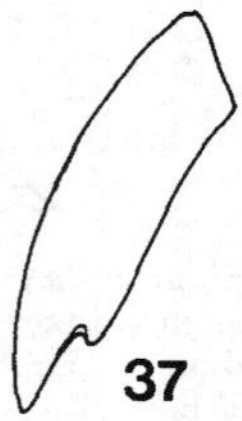
37

Fig. 24.29. *Rhyscotus texensis,* head. **Fig. 24.30.** *Cylisticus convexus,* head. **Fig. 24.31.** *Metoponorthus pruinosus,* head. **Fig. 24.32.** *Trachelipus rathkei,* head. **Fig. 24.33.** *Armadillidium vulgare,* head. **Fig. 24.34.** *Armadillidium nasatum,* head. **Fig. 24.35.** *Venezillo* sp., posterior pleonites, telson and uropods. **Fig. 24.36.** *Venezillo* sp., ventral view of lateral margin of pereion segment 1. **Fig. 24.37.** *Cubaris* sp., ventral view of lateral margin of pereion segment 1.

This European genus (= ***Tracheoniscus*** Verhoeff) is represented in North America by one species ***T. rathkei*** (Brandt). This is very common in the northern half of the United States and southern Canada. It is found in a wide variety of litter and debris and under the loose bark of trees. Body length to 15 mm.

23b Exopodites of only first two pleopods with tracheae. . . . . ***Porcellio*** Latreille

Three European species are widespread and common in the United States and Canada: ***P. scaber*** Latreille, ***P. laevis*** Latreille, and ***P. spinicornis*** Say. They are found in a variety of habitats, some fairly dry, and especially around human habitations. Another European species, ***P. dilatatus*** Brandt, has been reported from greenhouses in British Columbia, Washington, Oregon, Idaho, and Newfoundland. Two or three other species have been described from restricted localities in Texas. Body length to 18 mm.

24a **(19)** External ramus of uropod broad and flat, hinged to basal segment and contributing to posterior outline of body (Fig. 24.28); internal ramus small, attached ventrally; Family **Armadillidiidae**. . . . . ***Armadillidium*** Brandt

Two European species of this genus are found in North America, ***A. vulgare*** Latreille) and ***A. nasatum*** Budde-Lund. The former is the more widespread but both are locally common. They are easily distinguished by the shape of the front of the head (Figs. 24.33 and 24.34). Both are found in a wide variety of habitats, some very dry, and may be numerous in greenhouses. These are the common "pill bugs." Body length to 15 mm.

24b Both rami of uropod very small, attached dorsally and ventrally to basal segment, which itself is large and expanded and contributes to posterior outline of body (Fig. 24.35); Family **Armadillidae** **25**

25a (24) Posterior part of lateral margin of pereion segment 1 with a deep notch (when viewed from below) (Fig. 24.36) . . . . ***Venezillo*** Verhoeff

This tropical genus is represented in the United States by eight or nine poorly known species, reported from Florida, Georgia, Texas, Arizona, and California. They are found in damp litter. Body length to 10 mm.

25b Posterior part of lateral margin of pereion segment 1 entire, although a small notch may occur inside and below the margin (when viewed from below) (Fig. 24.37) . . . . ***Cubaris*** Brandt

One species, ***C. murina*** Brandt, of this tropical genus has been reported from Florida, where it is found in damp litter. Body length to 12 mm.

References

References to the early work on terrestrial isopods of North America may be found in the works of Van Name (1936, 1940, 1942). Below are listed most of the significant papers on these isopods published since 1941.

Alexenberg, M. L. 1962. A key to the sowbugs of Long Island, New York. *Turtox News* 40:22–23. **Bell, R. T. 1971.** *Handbook of the Malacostraca of Vermont and Neighboring Regions* (*Crayfish, Sowbugs and Their Relatives*). R. T. Bell, Burlington, Vt. pp. 1–65. **Brooks, J. L. 1942.** Notes on the ecology and the occurrence in America of the myrmecophilous sowbug, *Platyarthrus hoffmannseggi* Brandt. *Ecology* 23:427–437. **Causey, D. 1952.** The terrestrial Isopoda of Arkansas. *Proc. Arkansas Acad. Sci.* 5:25–30. **1953.** Additional records of terrestrial isopods in Arkansas. *Proc. Arkansas Acad. Sci.* 6:49–60. **Cole, G. 1959.** A summary of our knowledge of Kentucky crustaceans. *Trans. Ky. Acad. Sci.* 20:65–79. **Dexter, R. W., W. F. Hahnert and J. A. Beatty. 1988.** Distribution of the terrestrial Isopoda on islands of western Lake Erie. Pp. 102–110 in J. E. Downhower (ed.). *The Biogeography of the Island Region of Western Lake Erie.* Ohio State Univ. Press, Columbus. **Eberly, W. R. 1954.** The terrestrial isopods (Oniscoidea) of Indiana. *Proc. Indiana Acad. Sci.* 63:272–277. **Edney, E. B. 1953.** The woodlice of Great Britain and Ireland. *Proc. Linn. Soc. Lond.* 164:49–98. **1954.** Woodlice and the land habitat. *Biol. Rev. Camb. Philos. Soc.* 29:185–219. **1968.** Transition from water to land in isopod crustaceans. *Am. Zool.* 8:309–326. **Garthwaite, R. and C. Sassaman. 1985.** *Porcellionides floria,* new species, from North America; provinciality in the cosmopolitan isopod *Porcellionides pruinosus* (Brandt). *J. Crust. Biol.* 5:539–555. **Hamner, W. M., M. Smyth and E. D. Mulford, Jr. 1968.** Orientation in the sand-beach isopod *Tylos punctatus. Anim. Behav.* 16:405–409. **1969.** The behavior and life history of a sand-beach isopod, *Tylos punctatus. Ecology* 50:442–453. **Hatch, M. H. 1947.** The Chelifera and Isopoda of Washington and adjacent regions. *Univ. Wash. Publ. Biol.* 10:155–274. **1949.** Studies on the fauna of Pacific northwest greenhouses (Isopoda, Coleoptera, Dermaptera, Orthoptera, Gastropoda). *J. N.Y. Entomol. Soc.* 57:141–165. **Hatchett, S. P. 1947.** Biology of the Isopoda of Michigan. *Ecol. Monogr.* 17:47–79. **Hayes, W. B. 1970.** The accuracy of pitfall trapping for the sand-beach isopod *Tylos punctatus. Ecology* 51:514–516. **Heeley, W. 1941.** Observations on the life-histories of some terrestrial isopods. *Proc. Zool. Soc. Lond. B* 111:79–149. **Holsinger, J. R. 1967.** New data on the range of the troglobitic trichoniscid isopod, *Caucasonethes henroti. J. Tenn. Acad. Sci.* 42:15. **Judd, W. W. 1965.** Terrestrial sowbugs (Crustacea: Isopoda) in the vicinity of London, Ontario. *Can. Field-Nat.* 79:197–202. **Lemos de Castro, A. 1965.** On the systematics of the genus *Littorophiloscia* Hatch (Isopoda, Oniscidae). *Arq. Mus. Nac. Rio J.* 53:85–98. **Lindroth, C. H. 1957.** *The Faunal Connections between Europe and North America.* Wiley, New York. **Menzies, R. J. 1950.** Notes on California isopods of the genus *Armadilloniscus,* with the description of *Armadilloniscus coronacapitalis* n. sp. *Proc. Calif. Acad. Sci.* 26:467–481. **Muchmore, W. B. 1957.** Some exotic terrestrial isopods (Isopoda, Oniscoidea) from New York State. *J. Wash. Acad. Sci.* 47:78–83. **1964.** New terrestrial isopods of the genus *Miktoniscus* from eastern United States (Crustacea: Isopoda: Oniscoidea). *Ohio J. Sci.* 64:51–57. **1970a.** A new troglobitic trichoniscid isopod of the genus *Caucasonethes. J. Tenn. Acad. Sci.* 45:27–

28. **1970b.** New records of European woodlice in New York. *Entomol. News* 81:198–199. **Mulaik, S. B. 1960.** Contribución al conocimiento de los isópodos terrestres de México (Isopoda, Oniscoidea). *Rev. Soc. Mex. Hist. Nat.* 21:79–292. **Mulaik, S. and D. Mulaik. 1942.** New species and records of American terrestrial isopods. *Bull. Univ. Utah* 32(6), *Biol. Ser.* 6(7):1–11. **1943.** New Texas terrestrial isopods with notes on other species. *Bull. Univ. Utah* 34(3), *Biol. Ser.* 8(1):1–15. **Palmen, E. 1951.** A survey of the Oniscoidea (Isopoda terr.) of Newfoundland. *Ann. Zool. Soc. Zool.-Bot. Fenn. "Vanamo"* 14(1):1–27. **Paris, O. H. 1963.** The ecology of *Armadillidium vulgare* in California grassland: food, enemies, and weather. *Ecol. Monogr.* 33:1–22. **Sargent, J. E. 1973a.** Terrestrial isopods of Minnesota. *J. Minn. Acad. Sci.* 38:88–92. **1973b.** Albino terrestrial isopods. *Am. Midl. Nat.* 89:241. **Schultz, G. A. 1961a.** Distribution and establishment of a land isopod in North America. *Syst. Zool.* 10:193–196. **1961b.** *Cubaris murina* Brandt, an isopod crustacean new to the United States. *Crustaceana (Leiden)* 3:169–170. **1962.** *Miktoniscus grayi*, a new species of terrestrial isopod from North Carolina. *J. Elisha Mitchell Sci. Soc.* 78:47–57. **1963a.** *Venezillo evergladensis*, a new species of terrestrial isopod crustacean from Florida. *Trans. Am. Microsc. Soc.* 82:209–213. **1963b.** *Philoscia robusta*, a new species of terrestrial isopod crustacean from south-eastern United States. *J. Elisha Mitchell Sci. Soc.* 79:26–29. **1963c.** *Trichorhina donaldsoni*, new species, a terrestrial isopod crustacean from Florida. *Am. Midl. Nat.* 69:435–440. **1964a.** *Philoscia bonita*, a new species of isopod crustacean from Florida. *Am. Midl. Nat.* 71:156–160. **1964b.** Two additional data on terrestrial isopod crustaceans: *Ligidium blueridgensis*, sp. nov., from Georgia and a North Carolina cave location for *Miktoniscus linearis* (Patience, 1908). *J. Elisha Mitchell Sci. Soc.* 80:90–94. **1964c.** *Taxonomy and Ecology of Some American Terrestrial Isopods (Oniscoidea).* Ph.D. dissertation, Duke University, Durham, N.C. 362 pp. **1965a.** The distribution and general biology of *Hyloniscus riparius* (Koch) (Isopoda, Oniscoidea) in North America. *Crustaceana (Leiden)* 8:131–140. **1965b.** The reduction of *Philoscia vittata* Say, 1818, to a synonym of *Philoscia muscorum* (Scopoli, 1763). *Crustaceana (Leiden)* 8:107–108. **1965c.** Terrestrial isopods from caves and mines in Texas and northern Mexico with a description of *Venezillo tanneri* (Mulaik and Mulaik) allotype. *Tex. J. Sci.* 17:101–107. **1966.** *Philoscia miamiensis* n. sp., an isopod crustacean from Florida with ecological notes on the new species. *Trans. Am. Microsc. Soc.* 85:457–462. **1970a.** Descriptions of new subspecies of *Ligidium elrodii* (Packard) comb. nov. with notes on other isopod crustaceans from caves in North America (Oniscoidea). *Am. Midl. Nat.* 84:36–45. **1970b.** A review of the species of the genus *Tylos* Latreille from the New World (Isopoda, Oniscoidea). *Crustaceana (Leiden)* 19:297–305. **1971.** *Platyarthrus schoebli* Budde-Lund new to Texas and the New World with notes on *P. hoffmannseggi* Brandt (Crustacea, Oniscoidea, Squamiferidae). *Tex. J. Sci.* 23:297–299. **1972a.** The Armadillidae of Florida (Isopoda, Oniscoidea). *Q. J. Fla. Acad. Sci.* 35:1–4. **1972b.** A review of species of the family Scyphacidae in the New World (Crustacea, Isopoda, Oniscoidea). *Proc. Biol. Soc. Wash.* 84:472–488. **1974a.** The status of the terrestrial isopod crustaceans *Philoscia muscorum, P. vittata, P. robusta* and *P. miamiensis* in the New World (Oniscoidea, Philosciidae). *Crustaceana (Leiden)* 27:147–153. **1974b.** Terrestrial isopod crustaceans (Oniscoidea) mainly from the West Indies and adjacent regions. I. *Tylos* and *Ligia*. *Stud. Fauna Curacao Other Caribb. Isl.* 45:162–173. **1975.** Terrestrial isopod crustaceans (Oniscoidea) from coastal sites in Georgia. *Bull. Ga. Acad. Sci.* 34:185–194. **1976.** *Miktoniscus halophilus* Blake, *M. medcofi* (Van Name) and *M. morganensis* n. comb., reconsidered with notes on New World species of the genus (Crustacea, Isopoda, Trichoniscidae). *Am. Midl. Nat.* 95:28–41. **1977.** Terrestrial isopod crustaceans (Oniscoidea) from St. Catherines Island, Georgia. *Ga. J. Sci.* 35:151–158. **1982.** *Amerigoniscus malheurensis*, new species, from a cave in western Oregon (Crustacea: Isopoda: Trichoniscidae). *Proc. Biol. Soc. Wash.* 95:89–92. **1982.** Terrestrial isopods (Crustacea: Isopoda: Oniscoidea) from North Carolina. *Brimleyana* 8:1–26. **1983.** Disposition of three species of Oniscoidea from western Atlantic seashores (Crustacea: Isopoda: Halophilosciidae and Philosciidae). *Proc. Biol. Soc. Wash.* 96:440–451. **Schultz, G. A. and C. Johnson. 1984.** Terrestrial isopod crustaceans from Florida (Oniscoidea). Tylidae, Ligiidae, Halophilosciidae, Philosciidae, and Rhyscotidae. *J. Crust. Biol.* 4:154–171. **Vandel, A. 1941.** La faune Nord-Atlantique. *Rev. Franc. Entomol.* 16:1–11. **1950.** Isopodes terrestres recueillis par C. Bolivar et R. Jeannel (1928) et le Dr. Henrot (1946). *Arch. Zool. Exp. Gen.* 87:183–210. **1953a.** A new terrestrial isopod from Oregon, *Caucasonethes rothi* n. sp. *Pac. Sci.* 7:175–178. **1953b.** Remarques systématiques, morphologiques, et biogéographiques sur

un groupe de Trichoniscidae Nord-Atlantique (Crustacea: Isopodes terrestres). *Bull. Mus. Natl. Hist. Nat.* (*Paris*) Ser. 2 25:368–375. **1960.** Isopodes terrestres (Première partie). *Faune Fr.* 64:1–416. **1962.** Isopodes terrestres (Deuxième partie). *Faune Fr.* 66:417–931. **1965.** Les Trichoniscidae cavernicoles (Isopoda terrestria; Crustacea) de l'Amérique du Nord. *Ann. Speleol.* 20:347–389. **1977.** Les espèces appartenant au genre *Amerigoniscus* Vandel, 1950 (Crustacés, Isopodes, Oniscoides). *Bull. Soc. Hist. Nat. Toulouse,* 113:303–310. **Van Name, W. G. 1936.** The American land and fresh-water isopod Crustacea. *Bull. Am. Mus. Nat. Hist.* 71:1–535. **1940.** A supplement to the American land and fresh-water isopod Crustacea. *Bull. Am. Mus. Nat. Hist.* 77:109–142. **1942.** A second supplement to the American land and fresh-water isopod Crustacea. *Bull. Am. Mus. Nat. Hist.* 80:299–329. **Warburg, M. 1968.** Behavioral adaptations of terrestrial isopods. *Am. Zool.* 8:545–559. **White, J. J. 1968.** Bioenergetics of the woodlouse *Tracheoniscus rathkei* Brandt in relation to litter decomposition in a deciduous forest. *Ecology* 49:694–704.

25

Chilopoda

PETER MUNDEL

The organisms of the class Chilopoda are all generally termed centipedes. The five orders are morphologically and biologically quite distinct from one another, the orders themselves being rather homogeneous, with each considered to be of great antiquity. Centipedes are readily distinguished from all superficially similar organisms by the unique presence of the forcipules, the modified first trunk segment upon which the true head rests. The appendages of this segment are modified into a robust pincerlike structure, the ducts of the poison glands opening near their tips. The bite is potentially dangerous in only a very few large tropical scolopendromorph species. No dangerous forms occur anywhere in North America, and most species are incapable of inflicting a bite upon humans. The coxosternal plates of this segment form a structure termed the prosternum. Its configuration is taxonomically useful. The actual head capsule contains the structures typical of the other Mandibulata, a single pair of antennae, one pair of mandibles, a first and a second maxillae. The first pair of walking legs is borne on what is morphologically the second trunk segment. Each trunk segment bears a single pair of legs. There is always an odd number of pedal segments. Walking legs in all centipedes are basically composed of seven segments: coxa, trochanter, prefemur, femur, tibia, tarsus (which is often divided), and a pretarsus frequently bearing two accessory claws. Protection and structural integrity for each segment are provided by lightly sclerotized plates, tergites, sternites, and assorted pleurites, which are connected to one another by more delicate cuticle. The body is terminated by a coalescence of three segments, the pregenital, the genital, and the anal, which form a small postpedal region.

Two subclasses and five orders are generally recognized. The subclasses,

Anamorpha and Epimorpha, are characterized principally by their modes of development, but share other characters as well. The Epimorpha hatch with the same number of pedal segments as the adult: 21 or 23 in the order Scolopendromorpha, and 31 or more in the order Geophilomorpha. In both orders are one tergite, one sternite, and various pleurites associated with each pedal segment, although the Geophilomorpha possess an intercalary tergite, giving the appearance of having two tergites per segment. The young of the three orders of the subclass Anamorpha hatch with fewer segments and leg pairs than the adults, which always have 15 pedal segments. All three of these orders have 15 sternites, one per pedal segment. The Scutigeromorpha have only seven large tergites, Craterostigmomorpha have 21, and the Lithobiomorpha have 15.

The fauna of North America is rich and varied. The vast majority, of the species, and most of the higher taxa as well, were described by the single pioneering investigator Ralph V. Chamberlin. His first publications on centipedes appeared in 1902 and his last in 1967. Chamberlin's work is monumental in its scope and quantity. In too many cases, however, formal descriptions and diagnoses are, although sufficient to validate a new name, totally insufficient to permit the recognition of the taxon without recourse to the holotype. Regretably it is now apparent that synonymies are rampant among the North American fauna, and to further complicate matters, there remain a great many undescribed taxa at various levels.

In distribution the North American species are not as restricted in their occurrences as the literature might lead one to believe. This false view of restricted distribution is due to the large number of forms known only from their type series. The higher category systematics of the North American families and genera and indeed that of the rest of the world's fauna as well, is a shambles: it is hoped that this will soon be remedied.

Those investigators interested in ecology and behavior should not be discouraged by these details. Taxa are, in most cases, readily recognizable once the characters are learned and understood. As organisms, centipedes are fascinating creatures, and despite their abundance and frequent occurrence they are not at all well studied and thus present many opportunities for various investigations.

Preparation and Study

Centipedes are best preserved for study by first fixing in an aqueous solution consisting of 70% isopropyl alcohol and 1–3% glacial acetic acid for 12 hours and then transferred for permanent storage in 70% isopropyl alcohol. The fixing solution should be decanted off with care to avoid loss of commensal mites. The use of ethyl alcohol should be avoided, for it not only makes specimens hard and brittle but, additionally, causes them to die in rather contorted positions, making material very difficult to work with.

Live material can easily be narcotized for study by the use of either CO_2 gas or by brief submergence in soda water and revived without apparent ill effects.

Larger animals may be examined whole; smaller specimens frequently need to be slide mounted. Nearly all geophilomorphs require dissection of mouthparts for determination of even the family, and due to their size, slide mounting is preferred. Hoyer's mountant and its variants are excellent, although long-term storage of these preparations can be difficult. Slides must be properly dried, a process that can be difficult in certain climates. Slides may generally be examined in a few days. Slides, once dry, must be effectively sealed to prevent "fingering." Glyptol 1201 is excellent for this purpose. Kaiser's mountant and other glycerin jelly media are adequate for temporary mounts.

Lithobiomorpha

Adults range in size from less than 5 mm to over 50 mm, and always have 15 pedal segments. In distribution they are well represented in the temperate regions of both the northern and southern hemispheres. The southern hemisphere stocks are primarily of the families Henicopidae and Anopsobiidae. The northern stocks may be broadly interpreted as Lithobiidae, with a very few, but conspicuous, henicopids. Lithobiomorphs are rare and poorly represented in tropical regions. The family Lithobiidae has been subdivided by Chamberlin (1912–1925) into several families, and a lesser number have been added by European investigators. Chamberlin has described nearly 100 genera, largely North American. Many are difficult to recognize. At the species level there appear to exist many synonyms. Chamberlin made considerable use of leg spur placement (plectrotaxy) in species diagnoses. Although spur placement is of great utility, Chamberlin misapplied it. During successive molts adults can continue to add a few more spurs to a spur series, especially anteriorly. Spurs on the legs that undergo sexual modification, and the rearmost legs in general, tend to be rather variable. Hence many of the diagnoses in the literature are of no utility. Examinations of large series of individuals of single populations will occasionally show the full range of "species variation" in the entire genus. Many genera are monotypic; others, however, are not. Some large Chamberlinian genera are clearly artificial, united by superficial resemblances, such as the modification of a particular leg segment.

Identification of single individuals can be difficult, except for males of taxa that exhibit notable sexual modification of either one or both pairs of rear legs.

Females are readily distinguished from males by the presence in the postpedal region of prominent gonopods which are used for handling eggs and for picking up spermatophores. Eggs are laid, camouflaged with dirt and debris, and hidden individually. Several taxa are readily reared in culture.

Broadly interpreted, it is the family Lithobiidae *s.l.* which contains the vast bulk of the North American fauna in the order Lithobiomorpha. As mentioned earlier, the taxon has been divided into a larger number of families. Many of the genera of the southern United States, nearly all of those of Mexico, and a number of taxa from southern Europe as well, have females that exhibit smooth edges on the terminal claw of each gonopod. In contrast, the taxa from more northern regions and from cooler parts of tropical localities, such as mountain areas, have females with a prominent lateral tooth on each lateral claw edge, making the claw appear tripartite. Chamberlin considered the smooth claw diagnostic of the family Gosibiidae and the toothed claw diagnostic of the family Lithobiidae *sensu* Chamberlin. Other characters have been used to separate out a number of the more morphologically interesting taxa into separate families. The family name Watobiidae Chamberlin 1915 refers to a group of rather diminutive species from southern North America whose relationship to one another is uncertain. They are characterized principally by extremely reduced plectrotaxy. The family Ethopolidae Chamberlin 1925 is defined by the presence of sievelike coxal pores; the genus *Bothropolys* is very common in parts of northern North America. The interesting genus *Pseudolithobius* Stuxberg of the U.S. west coast has been placed in the family Pseudolithobiidae Matic 1973, along with a probably quite unrelated Turkish species, on the basis of their possession of coxal pores on the last five pairs of coxa instead of the unbiquitous arrangement of having them on only the last four pairs of coxa. The genus and family level system of the Lithobiidae *s.l.* is very poorly understood. Considerable revisionary work will be required before a stable and workable phylogeny can be formulated. There are nearly 100 described genera; a very large proportion of them are based on North American species, of which a large number are monotypic. A number of genera with multiple species have their species defined by characters known to vary greatly with successive molts, and hence many synonyms exist in widely distributed, frequently encountered North American "genera."

The Anopsobiidae Verhoeff 1907 are diminutive forms, eyeless or with a single pair of ocelli, the adults with anal gland retained, coxal productions prominent or at least always present, and spurs absent.

The genera referred to this family are nearly all restricted to the extreme southern hemisphere. There are isolated reports of species from the USSR (Zalesskaja 1979) and from the Indochina area (Attems 1953).

Members of the family Henicopidae Pocock 1902 comprise the dominant fauna of the temperate regions of the southern hemisphere. Several taxa are, however, nearly worldwide in distribution and frequently encountered. The cosmopolitan species *Lamyctes fulvicornis* Meinert is common in

North America and elsewhere. *Zygethobius* Chamberlin is apparently native to North America, living in mostly mountainous regions in the central Appalachian region in the east and in scattered locales in the west and Pacific northwest.

Scutigeromorpha

These are commonly called "house centipedes." The name is derived from the species *Scutigera coleoptrata* L., which is frequently encountered in dwellings in much of Europe and North America, and outdoors in mild climates. The majority of scutigeromorph taxa, however, occupy a variety of natural habitats, mostly in the world's warmer regions.

A notable feature of the order is the presence of an unusual respiratory system. There is a single dorsal "spiracle" at the posterior of each large tergite. The tracheae within form a surprisingly lunglike structure which directly aerates the hemolymph rather than having tracheae course through the body as do all other centipedes. The large, prominent, highly functional compound eye is similarly unique to this order. The tarsi of the legs are subdivided into many subsegments. Animals walk with an abundance of these tarsal segments on the substrate, yet they are remarkably agile and fast and capable of skillfully capturing a variety of small arthropods, upon which they feed.

Males and females are recognized by the presence of gonopods in both. In the female the gonopods are large, prominent, and fused in a single forceps-like structure. In males there are four small, somewhat fingerlike projections in the genital area.

There are perhaps more than a dozen genera worldwide with a little over 100 species. Some taxa are not at all well known and the relationships among the various taxa are not clear. Classic interpretations recognize two families, separable by use of the key herein.

Craterostigmomorpha

This unusual, monogeneric, relict order is restricted in distribution to Tasmania and New Zealand. Members have 21 tergites but only 15 pedal segments in the adult. There is little known of the life history and biology except that they are collected only seasonally and appear to have a preference for living in rotten wood. The females tend the eggs and young, which hatch with fewer than their adult complement of pedal segments. In males the postpedal region is conspicuously modified into a structure of yet undetermined function. There are perhaps three species.

Geophilomorpha

Adults vary in size from less than 1 cm to over 25 cm. All are eyeless and possess 14 antennal segments; the number of pedal segments varies from a minimum of 31 to over 180. The number in a given taxon is reasonably consistent. In distribution geophilomorphs are found nearly worldwide. All are burrowers or crevice exploiters of one sort or another; their small cross-sectional area makes them eminently suited for such an existence. Geophilomorphs are biologically and morphologically the most diverse centipedes, excellent subjects for comparative studies of various sorts. The systematics are not well understood; the North American fauna suffers from having been described largely by Chamberlin. There are many undescribed taxa and the distributions of known species is not well understood. Most species live preferentially in a particular microhabitat; there are species that characteristically exploit logs; others, various soil and humus layers. Deep-soil exploiters exist in many places and are likely to be highly susceptible to the adverse influences of soil compaction.

Females can be separated from males by the presence of small gonopods in the males. In a number of taxa males also possess larger or more inflated ultimate legs. There can, in some taxa, be sexual dimorphism in the presence or absence of certain pore fields and in the number of pedal segments.

For determinations, dissection of the head and mouthparts, and slide mounting, are required. The head capsule is removed from the forcipular segment by carefully incising the cuticle both at the rear of the capsule dorsally and as far rearward of the second maxillae as possible ventrally. For this purpose, minute pins are pushed into the ends of thin, wooden matchsticks, and then the very tips are drawn across a fine honing stone in such a manner as to produce a bladelike edge near the tip. Structures should be cut rather than torn or teased, for some of the structures associated with the maxillae are quite delicate. The first and second maxillae should be removed as a single unit. The mandibles thus exposed can be carefully removed by severing their connections. The maxillae and mandibles should be mounted under a single small coverslip at the left end of the slip. The head and body should be mounted under a single larger coverslip in the center of the slide, their ventral surfaces up. Cut the body into two or more pieces if it does not fit under the coverslip conveniently. Soak the body in distilled water to remove excess alcohol, which is not compatible with the mounting medium. Place the data label at the right of the slide with sufficient room below it for a determination label. The coverslip on the head and body may require a small weight to press it down slightly while the mountant cures. For larger specimens it may be necessary to use a small device to hold the coverslip parallel to the slip during curing.

Families can readily be identified by examination of the mandibles and head capsule. Three major mandibular structures of diagnostic utility can be encountered (see Figs. 25.1–25.4). The simplest structures are simple

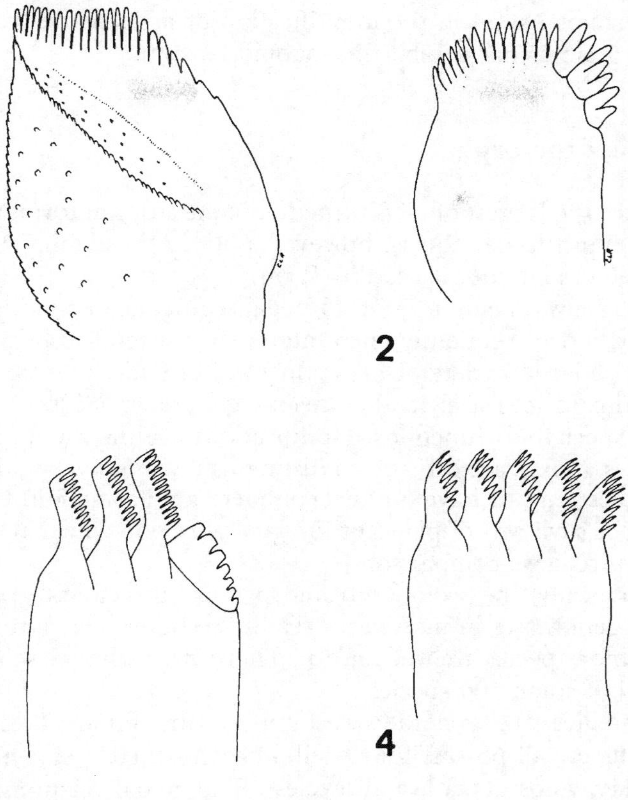

Figs. 25.1–25.4. Mandibular structures. **25.1.** Simple teeth. **25.2.** Dentate laminae with simple teeth. **25.3.** Dentate laminae with pectinate laminae. **25.4.** Pectinate laminae.

teeth, fine, often hyaline, evenly arrayed along the biting edge (Fig. 25.1). Two other structures of systematic utility are laminar structures. These are pectinate laminae, which are comblike arrays usually several projecting from the edge of the mandible (Figs. 25.3–25.4). The other, dentate laminae, are not likely to be confused with simple teeth. Dentate laminae are robust, well-chitinized structures with toothed margins except in the Mecistocephalidae, where the single dentate is virtually indistinguishable from the pectate laminae (Figs. 25.2, 25.3). These characters are used in separating the readily recognized generally accepted families. Characters of the labrum and clypeus are not as reliable or as easy to interpret correctly. Chamberlin established several families, notably Soniphilidae 1912, Sogonidae 1912, and Tampiyidae, which are without doubt close to one another and might be broadly interpreted as all belonging to the family Geophilidae *s.l.* All these "families" are distinguished by subtle characterizations of the labrum and clypeus which are of questionable value, although perhaps useful in the identification of at least the frequently encountered genera that exhibit the sogonid labrum.

Scolopendromorpha

These are the largest of all centipedes. Some tropical forms may be over 30 cm in length; other forms, however, notably those inhabiting soil, are much smaller, some not exceeding 2 cm.

There are always either 21 or 23 pedal segments, depending on genus. The larger, more frequently encountered forms tend to be active predators, taking a large variety of prey: they will in many cases scavenge.

The ultimate legs are, in the various genera, modified to perform a variety of specialized functions. Modifications are the same in both males and females. (Sexes are difficult to distinguish without very close examination. Females tend to have a more complete genital region. In some taxa there may be divisions of plates of the genital region which are easily seen; others require close comparison.)

Taxonomically, the scolopendromorphs are the easiest group to work with. The genera are in most cases readily recognizable, and many of the more common species are well known. There are perhaps a score of genera and a total of some 500 species.

Two families are generally recognized: the family Scolopendridae, whose members all possess four ocelli in each ocular area, and the family Cryptopidae, whose taxa are all eyeless. The actual relationship of these two families to one another and the interrelationship of the genera are not at all clear.

Identification

The following key to the centipede families of the world is largely artificial; the characters used are not in all cases the actual characters that define the phylogenetic lineage. The characters used are those which are readily observed in a particular taxa and separate it from remaining taxa. Thus characters must not be used in isolation of characters that occur in earlier couplets. Once formally identified, members of local faunas can in most cases be readily identified in the field by use of informal characters. These characters do not lend themselves well to use in keys, which by their nature must encompass potentially large numbers of taxa, some undescribed, from extensive and varied areas.

The higher-level systematics of centipedes is not currently well understood; a large number of taxa await both description and discovery. Several families previously synonymized in the literature are contained in this key, since use of the characters that define these groups does at least permit a number of taxa (genera) to be grouped together for the purpose of identification.

The bibliography is not complete but contains the citations of most major works of utility, useful reviews of taxa, references that contain keys to various groups in some cases not apparent from the title alone, and useful papers that deal with the general biology of centipedes. The references are annotated to increase their utility to investigators unfamiliar with both the animals and the literature. Publications containing only descriptions of new species are not listed, and a great many North American species are known only from the original descriptions. Some studies that appear in multiple parts are for brevity, indicated by citation of but a single title.

ACKNOWLEDGMENTS

I would like to thank Clara L. Richardson for the illustrations and the University of Wisconsin Zoological Museum for help at various stages of the preparation of this manuscript.

ARTIFICIAL KEY TO ORDERS AND FAMILIES OF THE CENTIPEDES OF THE WORLD

1a Adults with 15 pairs of legs, young with fewer; Subclass **Anamorpha** . . . 2

1b Adults with the same number of pairs of legs as their young, either 21 or 23, or 31 or more; Subclass **Epimorpha** 7

2a Spiracles single and middorsal, seven dorsal tergites, compound eyes; Order **Scutigeromorpha** (one to three families) 3

2b Spiracles paired on segments where present, the openings lateral. 4

3a Antennal segments much wider than long, male gonopods peglike . . . . . **Scutigeridae**

3b Antennal segments as long as wide or longer than wide, male gonopods flattened or leaflike . . . . **Pselliodidae**

4a With 21 tergites and 15 sternites . . . . Order **Craterostigmomorpha**
Monogeneric, Tasmania and New Zealand; possibly three species.

4b With 15 tergites and sternites; Order **Lithobiomorpha** (three to nine variously defined families, many genera) 5

5a Adults with coxal pores present on only the last two pairs of coxae, eyes absent or a single ocelli in each ocular area, largely restricted to the southern hemisphere; four genera . . . . **Anopsobiidae**

5b In adults coxal pores present on more than the last two pairs of coxae . . . 6

6a With no more than a single large ocellus in each ocular area, a few forms eyeless, spurs on legs absent, setae often well represented. . . **Henicopidae**

6b Ocelli generally more than four in each ocular area, a few eyeless forms, true spurs present on legs, as are setae . . . . **Lithobiidae sensu lato**

7a Twenty-one or 23 pairs of legs; Order **Scolopendromorpha** (2 or 3 families) 8

7b Thirty-one or more pairs of legs; Order **Geophilomorpha** (about a dozen family-level groups) 9

8a With four ocelli on either side of head . . . . **Scolopendridae**

8b Eyeless . . . . **Cryptopidae**

9a Forcipules with some segments fused, consisting of fewer than the four typical/usual segments **10**

9b Forcipules consisting of the usual four segments **11**

10a Forcipule consists of three segments, several genera South American and African . . . . **Aphilodontidae**

10b Forcipule consists of two segments . . . . **Eriphantidae**
Monotypic, Baja California.

11a Mandible complex, with complex lamellar structures, these either bearing toothlike prominences (dentate laminae) or being comblike (pectinate laminae), simple teeth present or absent (Figs. 25.2–25.4) **12**

11b Mandibles simple; singular, simple teeth arrayed along the biting edge (Fig. 25.1) **13**

12a With one or more clearly discerned dentate laminae (other teeth variable; Figs. 25.2, 25.3) **14**

12b With apparently all pectinate laminae only (Fig. 25.4) **16**

13a Labral teeth always projecting rearward, anterior portion of body not especially tapered, head of "normal size," midpiece of labrum always smaller than lateral pieces (Figs. 25.5, 25.6), only a few species brightly colored, worldwide, many genera, several "families"; **Geophilidae sensu lato**. **17**

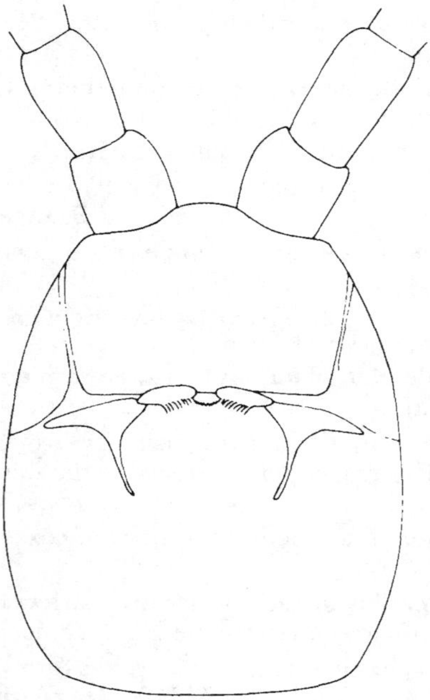

5

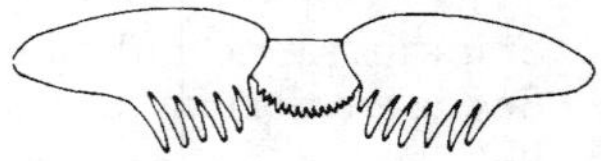

6

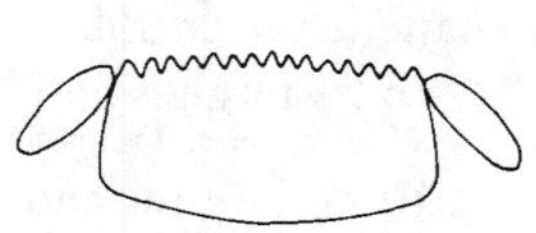

7

8

Figs. 25.5–25.8. Labrum. **25.5.** Location on head showing rearward projection of labral teeth. **25.6.** Rearward projection of labral teeth, clypeus–labrum suture apparent. **25.7.** Forward projection of labral teeth. **25.8.** Clypeus with rearward extension replacing medial labrum.

13b With labral teeth usually appearing to project forward (Fig. 25.7), anterior half of body notably tapered, head "small," midpiece of labrum always larger than side pieces, body color often red, orange, or pink . . . . **Dignathodontidae**

14a With both pectinate *and* dentate laminae prominant (Fig. 25.3) . . . . **Himantaridae**

14b With dentate laminae *and* simple mandibular teeth (Fig. 25.2) **15**

15a Antennae foliform, body color generally light, many genera, widespread . **Schendylidae**

15b Distal segments of antennae notably inflated, body color often dark, various genera, tropical and subtropical . . . . **Ballophilidae**

16a Head capsule about as long as wide, covers forcipules, tropical, various genera . . . . **Oryidae**

16b Head capsule much longer than wide, forcipules very large, several genera, mostly tropical . . . . **Mecistocephalidae**

17a With the pretarsus either possessing a bladelike structure or pretarsus much modified, very hooklike or toothed **18**

17b Legs normal, not as above; **Geophilidae sensu lato** **19**

18a New World, Mexican and Central American tropics, three genera . . . . . **Neogeophilidae**

18b Old World tropics, several genera . . . . **Gonibregmatidae**

19a With labrum fused to the clypeus partially or completely, the divisions of the three labrum pieces not readily evident. Often with a production of the medial portion of the clypeus extended rearward, tending to replace medial portion of the labrum (Fig. 25.8) . . . . **Sogonidae**

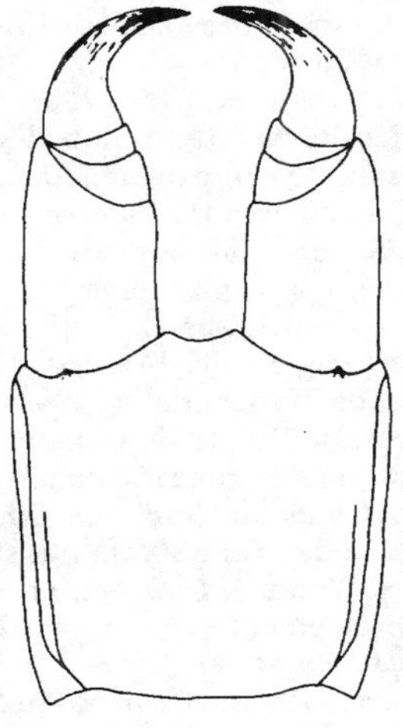

9

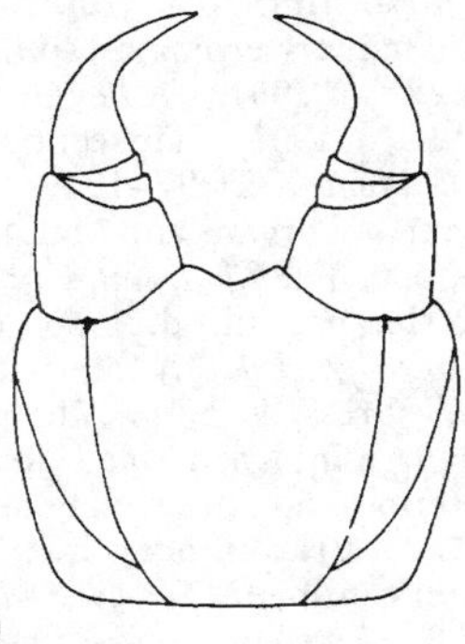

10

Figs. 25.9–25.10. Head capsule, sutures, and forcipules. **25.9.** Coxosternal and prosternal lateral sutures subparallel, forcipules very large relative to head capsule. **25.10.** Coxosternal sutures run obliquely forward converging toward anterior corner of prosterum, forcipules relatively small, as large or slightly larger than head.

19b Suture between clypeus and labrum usually apparent, the three pieces comprising the labrum usually distinct (Fig. 25.6) but midpiece occasionally rudimentary **20**

20a Prosternal lateral and coxosternal sutures subparallel to one another (Fig. 25.9), head capsule much longer than wide, forcipules usually much larger than head **21**

20b Coxosternal sutures run obliquely forward converging toward anterior corner of prosternum (Fig. 25.10), head capsule about as long as wide, or only a little longer than wide, forcipules small, not much larger than head, worldwide, many genera . . . . **Geophilidae sens. str.**

21a Second maxilla with a prominent sclerotized ridge on the coxosternum, worldwide, many genera . . . . **Chilenophilidae**

21b Without prominent ridge on coxosternum of second maxilla, worldwide, several genera . . . . **Pachymeriidae**

References

Andersson, G. 1976. Post-embryonic development of *Lithobuis forficatus* (L.). *Entomol. Scand.* 7:161–168. **Applegarth, A. G. 1952.** The anatomy of the cephalic region of a centipede, *Pseudolithobius megaloporus* Stuxberg (Chilopoda). *Microentomologica* 17:127–171. **Archey, G. 1937.** Revision of the Chilopoda of New Zealand. *Rec. Aukl. Inst. Mus.* 2(2):71–100. **Attems, C. von. 1914.** Die indo-australishen Myriopoden. *Arch. Naturgesch. Abt. A* 80(1914):1–398. **1926.** Progoneata und Chilopoda. In *Kükenthal und Krumbach's Handbuch der Zoologie,* Vol. 4. Walter de Gruyter, West Berlin. 402 pp. (An excellent review of all that was known of centipede biology up to that time.) **1928.** The Myriapoda of South Africa. *Ann. S. Afr. Mus.* 26:1–431. **1929.** Geophilomorpha. *Tierreich* 52(1):1–388. (Indispensable for the identification of members of this order. Extensive keys and descriptions, although many North American taxa are missing.) **1930.** Scolopendromorpha. *Tierreich* 54(2):1–308. (Thorough discussions and descriptions of the majority of the world's taxa.) **1947.** Neue Geophilomorpha des Wiener Museums. *Ann. Naturhist. Mus. Wien* 55:50–149. (An update of his 1929 work.) **1953.** Myriopoden von Indochina. *Mem. Mus. Natl. Hist. Nat. (Paris) Ser. A Zool.* 5(3):133–230. **Auerbach, S. 1949.** A preliminary ecological study on certain deciduous forest centipedes. *Am. Midl. Nat.* 42(1):220–227. **1951.** A key to the centipedes of the Chicago area. *Bull. Chic. Acad. Sci.* 9(6):109–114. **1951.** The centipedes of the Chicago area with special reference to their ecology. *Ecol. Monogr.* 21:97–124. **Bennett, D. S. and S. M. Manton. 1962(1963).** Arthropod segmental organs and Malphigian tubules with particular reference to their function in the Chilopoda. *Ann. Mag. Nat. Hist.* 13(5):545–556. **Blower, G. 1951.** A comparative study of the chilopod and diplopod cuticle. *Q. J. Microsc. Sci.* 92:141–161. **Blower, J. G. (ed.). 1974.** *Myriapoda. II.* International Congress on Myriapodology. *Symp. Zool. Soc. Lond.* 32. Academic Press, London. 712 pp. **Brölemann, H. 1930.** Chilopodes. *Fauna Fr.* 25(1):1–405. (An indispensable reference on the western European fauna.) **1932.** Tableaux de determination des chilopodes signales en Afrique du Nord. *Bull. Soc. Hist. Nat. Alger* 23(2):31–64. **Brunhumber, B. S. 1969.** The mode of the sperm transfer in the scolopendromorph centipede, *Cormocephalus anceps anceps* Porat. *J. Linn. Soc. Lond. Zool.* 48:409–420. **1970.** Egg laying, maternal care and development of young in the scolopendramorph centipede, *Cormocephalus anceps anceps* Porat. *J. Linn. Soc. Lond. Zool.* 49:225–234. **Bucherl, W. 1939.** Os Quilopodes do Brasil. *Mem. Inst. Butantan (São Paulo)* 13:49–362. **1941(1942).** Catálogos dos Quilopodes de zona neotrópica. *Mem. Inst. Butantan (São Paulo)* 15:251–372. **1959.** Kritische Untersuchungen der *Newportia*-Arten. *Beitr. Neotrop. Fauna* 1:229–242. **1974.** Die Scolopendramorpha der neotropischen Region. Pp. 99–133. In J. G. Blower (ed.). *Myriapoda. Symp. Zool. Soc. Lond.* 32. (A thorough review, most useful.) **Chamberlin, R. V. 1912a.** The Chilopoda of California. 3. *Pomona Coll. J. Entomol. Claremont* 4(1):651–672. **1912b.** The Geophiloidea of the southeastern states. *Bull. Mus. Comp. Zool.* 54(13):405–436. **1912c.** The Henicopidae of America north of Mexico. *Bull. Mus. Comp. Zool.* 57(1):1–36. (Chamberlin's eight-part monograph on the North American lithobiomorph fauna, although much outdated, contains the original descriptions of many of North America's more common species.) **1913.** The lithobiid genera *Nampabius, Garibius, Tidabius,* and *Sigibius. Bull. Mus. Comp. Zool.* 57(2):37–104. **1915a.** The genus *Watobius. Bull. Mus. Comp. Zool.* 57(3):104–112. **1915b.** New chilopods from Mexico and the West Indies. *Bull. Mus. Comp. Zool.* 59(8):493–541. **1916.** The lithobiid genera *Oabius, Kiberius, Paobius, Arebius, Nothembius,* and *Tigobius. Bull. Mus. Comp. Zool.* 57(4):113–201. **1918.** The Chilopods and Diplopods of the West Indies. *Bull. Mus. Comp. Zool.* 62(5):149–262. **1920.** The Myriapoda of the Australian region. *Bull. Mus. Comp. Zool.* 64(1):1–269. **1922a.** The centipedes of Central America. *Proc. U.S. Natl. Mus.* 60(2402):1–17. **1922b.** Further studies of North America Lithobiidae. *Bull. Mus. Comp. Zool.* 57(6):257–283. **1922c.** The Gosibiidae of America north of Mexico. *Bull. Mus. Comp. Zool.* 57(5):203–255. **1925a.** The Ethopolidae of America north of Mexico. *Bull. Mus. Comp. Zool.* 57(7):385–437. **1925b.** The genera *Lithobius, Neolithobius, Gonibius* and *Zinapolys* in America north of Mexico. *Bull. Mus. Comp. Zool.* 57(8):441–501. **1938.** On eighteen new Lithobiomorphus chilopods. *Ann. Mag. Nat. Hist.* 11(2):625–635. **1940.** Diagnoses of ten new chilopods with a new genus of Sogonidae and a key to the species of *Lophobius. Pan-Pac. Entomol.* 16(2):49–56. **1941a.** New genera and species of American lithobiid centipedes. *Bull. Univ. Utah. Biol. Ser.* 6(6):1–23. **1941b.** New genera and species of North American geophiloid centipedes. *Ann. Entomol. Soc.* 34:773–790. **1942.** A new American centipede of the genus *Scutigera* (Chilopoda: Scutigeridae). *Entomol. News* 53:10–11. **1943a.** On Mexican centipedes. *Bull. Univ. Utah Biol. Ser.* 7(3):1–55. **1943b.** Some records and descriptions of American chilopods *Proc. Biol. Soc. Wash.* 56:97–108. **1944.** Chilopods in the collection of Field Museum of Natural History. *Zool. Serv. Field Mus. Nat. Hist. Publ.* 588, 28(4):175–216. **1956.** The Chilopoda of Chile. *Acta. Univ. Lund Ark. N.S.* 51(5):1–61. **1966.** A new genus in the chilopod family Dignathodontidae with proposal of two subfamilies (Chilopoda: Geophilomorpha). *Proc. Biol. Soc. Wash.* 79:215–220. **Cloudsley-Thompson, J. 1958.** *Spiders, Scorpions, Centipedes and Mites: The Ecology of Woodlice, "Myriapods" and Arachnids.* Pergamon Press, London. 228 pp. **1960.** A new sound-producing mechanism in centipedes. *Entomol. Mon. Mag.* 96:110–113. **Crabill, R. 1953a.** Concerning a new genus *Dinocryptops,* and the nomenclatorial status of *Otocryptops* and *Scolopocryptops* (Chilopoda: Scolopendromorpha Cryptopidae). *Entomol. News* 64:96. **1953b.** The Schendylidae of northeastern North America. (Chilopoda: Geophilomorpha: Schendylidae). *J. N.Y. Entomol. Soc.* 61:93–98. **1958.** Report of *Geophilus proximus* in North America with a key to its eastern North American congeners (Chilopoda: Geophilomorpha: Geophilidae). *Entomol. News* 69:15–17. **1959.** A synonymical list of American Himantaridae, with a generic key and description of a new genus (Chilopoda: Geophilomorpha: Himantaridae). *Entomol. News* 70:117–126, 153–159. **1960.** A new American genus of cryptopid centipede, with an annotated key to the scolopendromorph genera from America north of Mexico. *Proc. U.S. Natl. Mus.* 111:1–15. **1961.** A catalogue of the Schendylinae of North America including Mexico, with a generic key and proposal of a new *Simoporus* (Chilopoda: Geophilomorpha: Schendylinae). *Entomol. News* 72:29–36. **1962.** Plectrotaxy as a systematic criterion in lithobiomorphic centipedes (Chilopoda: Lithobiomorpha). *Proc. U.S. Natl. Mus.* 113:399–412. **1979.** A new cryptopid genus, with a key to the genera known to occur in North America including Mexico (Chilopoda: Scolopendromorpha: Cryptopidae). *Proc. Entomol. Soc. Wash.* 79(3):346–349. **Crawford, C. S., W. A. Riddle and S. Pugach. 1975.** Overwintering physiology of the centipede *Scolopendra polymorpha. Physiol. Zool.* 48(3):290–294. **Eason, T. E. 1964.** *Centipedes of the British Isles.* Clowes London. 294 pp. (An excellent review of that fauna on an elementary level, very useful as an introduction, in English, to the taxonomic study of centipedes.) **Enghoff, H. 1975.** Notes on *Lamyctes coeculus* (Brölemann), a cosmopolitic, parthenogentic centipede (Chilopoda: Henicopidae). *Entomol. Scand.* 6:45–46. **Fahlander, K. 1938.** Beiträge zur Anatomie and Systematischen Einteilung der Chilopoden. *Zool. Bidr. Uppsala* A:1–148. **Jeekel, C. A. W. 1963–1964.** Beitrag zur Kenntnis der Systematik und Ökologie der Hundertfüsser (Chilopoda) Nordwestdeutschlandes. *Abhl. Verh. Naturwiss. Ver. Hamb.* (*N.F.*) 8:111–153. **Johansen, O. A. and F. H. Butt. 1941.** *Embryology of Insects and Myriapods.* McGraw-Hill, New York. 462 pp. **Johnson, B. M. 1952.** *The Centipedes and Millipedes of Michigan.* Univ. Mich. Diss. Publ. 3773. 488 pp. **Joly, R. and M. Descamps. 1968.** Étude comparative du complex endocrine céphalique chez les Myriapodes Chilopodes. *Gen. Comp. Endocrinol.* 10:364–375. **Jones, D. T. 1944.** List of publications of Dr. Ralph V. Chamberlin to June 1st, 1944. *Vintonia* 4:1–22. **Klingel, H. 1959.** Die Paarung des *Lithobius forficatus*. *Verh. Dtsch. Zool. Ges.* 34:326–332. **1960.** Vergleichende Verhaltensbiologie der Chilopoden *Scutigera coleoptrata* L. ("Spinnenassel") und *Scolopendra cingulata* Latr. (Skolopender). *Z. Tierpsychol.* 17:11–30. **Knoll, H. J. 1974.** Untersuchungen zur Entwicklungsgeschichte von *Scutigera coleoptrata* L. (Chilopoda). *Verh. Dtsch. Zool. Ges.* 67:126–130. **Kraus, O. (ed.). 1978.** Myriapoda. 3rd International Kongress für Myriapodologie. *Abh. Verh. Naturwiss. Ver. Hamb.* (*N.F.*) 21–22:1–377. **Lawrence, R. F. 1960.** Myriapodes Chilopodes. *Fauna Madagascar* 12:1–121. **LeMoili, F. 1977.** Ethogram of cleaning behavior in the chilopod *Scutigera coleoptrata* L. in maintenance activity and in conflict situations. *Atti. Soc. Ital. Sci. Nat. Mus. Civ. Stor. Nat. Milano* 119:77–84. **Lewis, J. G. E. 1961.** The life history and ecology of the littoral centipede *Strigamia* (= *Scolioplanes*) *maritima* (Leach). *Proc. Zool. Soc. Lond.* 137:221–248. **1962.** The ecology and taxonomy of the centipedes found on the shore in the Plymouth area. *J. Mar. Biol. Assoc. U.K.* 42:655–664. **1963.** On the spiracle structure and resistance to desiccation of four species of geophilomorph centipedes. *Entomol. Exp. Appl.* 6:89–94. **1965.** The food and reproductive cycles of the centipede, *Lithobius variegatus* and *L. forficatus* in a Yorkshire woodland. *Proc. Zool. Soc. Lond.* 144:269–283. **1966.** The taxonomy and biology of the centipede, *Scolopendra amazonica* in the Sudan. *J. Zool.* (*Lond.*) 149:188–203. **1967.** The scolopendromorph centipedes of the Sudan with remarks on taxonomic characters in the Scolopendridae. *Proc. Linn. Soc. Lond.* 178:185–207. **1968.** Individual variation in a population of the centipede *Scolopendra amazonica* from Nigeria and its implications for methods of taxonomic discrimination in the Scolopendridae. *J. Linn. Soc. Lond. Zool.* 47:315–326. **1969.** The variation of the centipede *Scolopendra amazonica* in Africa. *Zool. J. Linn. Soc.* 48:49–57. **1972a.** The population density and biomass of the centipede *Scolopendra amazonica* (Bucherl) (Scolopendramorpha, Scolopendridae) in sahel savanna in Nigeria. *Entomol. Mon. Mag.* 108:16–18. **1972b.** The life histories and distribution of the centipedes *Rhysida nuda togoensis* and *Ethmostigmus trigonopodus* (Scolopendramorpha, Scolopendridae). *J. Zool.* (*Lond.*) 167:399–414. **1973.** The taxonomy, distribution and ecology of centipedes of the genus *Asanda* (Scolopendramorpha, Scolopendridae) In Nigeria. *Zool. J. Linn. Soc.* 52(2):97–112. **(In Press).** *The Biology of Centipedes.* Cambridge University Press, Cambridge **Manton, S. M. 1965.** The evolution of arthropodan locomotory mechanisms. Part 8. *J. Linn. Soc. Lond. Zool.* 45:251–484. (Functional requirements and body design in Chilopoda, together with a comparative account of their skeletomuscular systems and an appendix on a comparison between burrowing forces of annelids and chilopods and its bearing on the evolution of the arthropodan hemocoel.) **1974.** Segmentation in Symphyla, Chilopoda and Pauropoda in relation to phylogeny in Myriapoda. *Symp. Zool. Soc. Lond.* 32:163–198. **1977.** *The Arthropoda; Habits, Functional Morphology, and Evolution.* Clarendon Press, Oxford. 527 pp. **Matic, Z. 1973.** Pseudolithobiidae fam nov: una nuova famiglia dell'ordine Lithobiomorpha. *Fragm. Entomol.* 9(3):135–142. **Mundel, P. 1979.** The centipedes (Chilopoda) of the Mazon Creek. Pp. 361–378 in M. Nitecki (ed.). *The Mazon Creek Fossils.* Academic Press, New York. 581 pp. **1981.** *A Review of the Lithobiomorph Centipedes of Mexico.* University Microfilms, Ann Arbor, Mich. 287 pp. **Palmen, E. and M. Rantala. 1954.** On the life history of ecology of *Pachymerium ferrugineum* (C. L. Koch) (Chilopoda : Geophilomorpha). *Ann. Zool. Soc. Zool.-Bot. Fenn. "Vanamo"* 16(3):1–44. **Rillig, G. 1960.** *Lithobius forficatus. Grosses Zool. Prakt.* 13b:1–136. (A very useful review of gross anatomy, copiously illustrated.) **Scheffel, H. 1963.** Zur Häutungsphysiologie der Chilopoden. *Zool. Jahrb.* (*Zool. Physiol.*) 70:284–290. **1969.** Utersuchungen über die hormonale Regulation von Häutung und Anamorphose von *Lithobius forficatus* (L.) (Myriapoda, Chilopoda). *Zool. Jahrb.* (*Allg. Zool.*) 74:436–505. **Schubart, O. 1967.** Diplopoda, Symphyla, Pauropoda, Chilopoda. *Tierwelt Mitteleur.* 2(3):1–51. **Summers, G. F. 1979.** An illustrated key to the chilopods of the northcentral states of the United States. *J. Kans. Entomol. Soc.* 52(4):690–700. **Summers, G. F. and C. W. Uetz. 1979.** Microhabitats of woodland centipedes in a stream side forest. *Am. Midl. Nat.* 102(2):346–352. **Summers, G. F., J. A. Beatty and N. Magnuson. 1980.** A checklist of Illinois centipedes (Chilopoda). *Great Lakes Nat.* 13(4):241–257. **Tobias, D. 1969.** Grundsätzliche Studien zur Art-Systematik der Lithobiidae. *Abh. Senckenb. Naturforsch. Ges.* 532:1–51. **Wurmli, M. 1974.** Systematic criteria in the Scutigeromorpha. *Blower* 197:89–98. **1975a.** Zur Systematik der Scutigeriden Europas und Kleinasiens (Chilopoda : Scutigeromorpha) Vorarbeiten zu einer Monographie der Scutigermorpha 1. *Ann. Naturhist. Mus. Wien* 77:399–408. **1975b.** Review der Hundertfüsser Gattung *Thereuonema* (Chilopoda : Scutigeromorpha). *Entomol. Ger.* 2(2):189–196. **1975c.** Systematische Kriterien in der Gruppe von *Scolopendra morsitans* Lin., 1758 (Chilopoda : Scolopendramorph; Scolopendridae). *Dtsch. Entomol. Z.* (N.F.) 22:201–206. **1976.** Contributo alla conoscenza del *Plutonium zwierleini* Cavanno, 1881 (Chilopoda : Scolopendramopha : Cryptopidae). *Anim. Catania* 2:209–213. **1977.** Zur Systematik der Gattung *Scutigera* (Chilopoda : Scutigeridae). *Abh. verh. naturwiss. Ver. Hamb.* (N.F.) 20:123–131. **1978.** Synopsis der neotropischen Pselliodidae (Chilopoda : Scutigeromorpha). *Stud. Neotrop. Fauna Environ.* 13(2):135–142. **Verhoeff, K. W. 1902–1925.** Chilopoda. Pp. 1–175 in H. G. Bronn (ed.). *Klassen und Ordnungen des Tierreichs,* Bd. 5. Leipzig, Leipzig, East Germany. (One of the most extensive monographs on centipedes available.) **1905.** Entwicklungsstufen der Steinlaufer, Lithobiiden. *Zool. Jahrb.* (*Anat.*) *Suppl.* 8:195–298. **1925.** Beiträge zur Kenntnis der Steinlaufer, Lithobiiden. *Arch. Naturgesch.* (*Berlin*) *Abt. A* 91, 9:124–158. **1934.** Beiträge zur Systematik und Geographie der Chilopoden. *Zool. Jahrb.* (*Syst.*) 66(1–2):1–112. **1937.** Zur Kenntnis der Lithobiiden. *Arch. Naturgesch.* (*Leipzig*) (N.F.) 6:171–257. **1938.** Chilopoden-studien, zur Kenntnis der Epimorpha. *Zool. Jahrb. Jena* (*Syst.*) 71:339–388. **Zalesskaja, N. T. 1979.** *The Lithobiomorpha of the USSR.* Fauna USSR. Nauka Publishers, Moscow. 211 pp. (Excellent figures of many of the commonly encountered palearctic forms.) **Zaunick, R. 1962.** Karl Wilhelm Verhoeff. *Lebensdarstell. Dtsch. Naturforsch.* 9:1–50. (His biography and a list of his publications.)

26

Diplopoda

RICHARD L. HOFFMAN

Millipedes† are common and conspicuous in the fauna of upper soil and litter horizons, particularly in tropical and temperate broadleaf forest regions of the world. Neglected for decades by all but a few systematists, and represented in many texts and references as a minor group of insignificant arthropods, the class Diplopoda actually may assert a claim to considerable importance and interest. The present content of 115 families and about 1700 genera is dispersed among no less than 15 orders and three subclasses, and the nearly 10,000 described species are probably less than a fifth of the actual number of living forms.

Interest in the systematics and biology of millipeds has greatly increased during the past three decades, and studies made during this time have confirmed intimations by earlier workers that the class has a remarkable potential as a source of meaningful discoveries, not only reflecting biological details peculiar to the group itself, but also shedding light on such broader subjects as biogeography and evolution. The synecological roles of millipeds in the soil and litter milieu are becoming appreciated as of basic importance, and knowledge of these animals as members of the soil communities becomes increasingly a necessity for investigators working in such areas.

The most fundamental problem, that of making accurate identifications of specimens, is particularly acute and reflects the retarded state of knowledge of the class, which is roughly at the point reached by entomology in

† Unabridged dictionaries in this country present the common name for this taxon as 1) millipede, 2) milliped. The spelling "millipede," with the terminal "e," is common in British literature. However, for decades, the term "milliped," without the terminal "e," has been used by diplopodologists in the technical literature from North America.

about 1850. Prior to the middle of this century, practically all of the taxonomic work on Diplopoda was the product of a small group of specialists (never more than five or six at one time), mostly amateurs working in isolation and publishing in several languages. The insulating effects of nationalism were exacerbated particularly among European specialists by personal idiosyncracies in taxonomic and nomenclatorial procedures. The greatest emphasis was placed on the description of new taxa, and attempts at revisionary synthesis were rare and superficial. Not only were classifications often highly contradictory and inconsistent, but even group names differed considerably: up to 1960 there were three different sets of ordinal names in current usage.

At the present time, only the relatively small fauna of northern Europe is well known, and even there numerous nomenclatorial adjustments remain to be effected. For the rest of the world, taxonomy remains in the "alpha" stage, and biological studies are in their infancy. During the past several decades great progress has been made in upgrading knowledge of the Nearctic fauna, but several large and important families still remain to be monographed, and the American species of Callipodida cannot even be referred to a particular family until the entire order has been revised. The soil biologist consulting this chapter will therefore discover very irregular coverage: a few families are in relatively good taxonomic condition, others are in the stage of being revised piecemeal, and for all too many there is simply no literature helpful in making identifications. This lack of balance will probably persist for the rest of this century. Eventual redress can be hastened by the cooperation of those who have much to gain from it: soil biologists willing to accumulate study material suitable for taxonomic purposes!

Definition

The class Diplopoda may be defined as a group of terrestrial, tracheate, mandibulate, oviparous, anamorphic Arthropoda (or if the phylum Arthropoda is considered to be polyphyletic, Labrata, Myriapoda, Uniramia, Atelocerata, or whatever restricted phylum name may appeal) characterized by simple, eight-segmented antennae; whole-limb mandibles; coalescence of embryonic maxillary appendages into a gnathochilarium; and undifferentiated trunk segments (diplosomites) composed of the fusion of two embryonic somites and generally bearing two pairs of legs; teloblastic segmentation, with the proliferation zone located just anterior to the last segment. Body segments are very variable regarding degree of fusion of tergal, pleural, and sternal sclerites. Trachea are diffusely branched, not anastomosing, opening through stigmata at lateral ends of sterna. Most species are adapted to some extent for rolling into a coil, spiral, or sphere. Except in a few orders, metazona of most body segments contain internal glands (ozadenes) secreting volatile allomones capable of repelling or discouraging predators and suppressing fungal and bacterial growth.

External Anatomy

Millipeds are homonomomously metameric, and are composed of a head, a variable number of similar trunk segments, and a terminal pygidial segment containing the anus. It is probable that originally all trunk segments bore two pairs of legs, although in all living species the first several following the head appear to have only a single pair.

Head

The head is a compact capsule composed of six embryonic somites and some elements derived from a seventh (first trunk somite). There is no external demarcation into definite parts, but a somewhat arbitrary recognition of topographic regions is usually followed. The uppermost area behind the antennae is the *epicranium,* normally with a *longitudinal median suture;* between and just below the antennae is the *frons,* below this is the *clypeus,* and the ventral margin of the head is the *labrum.* The lateralmost areas just below the antennal sockets are referred to as *genae.* In the order Spirobolida a postantennal lateral lobe (*parietal sclerite*) is set off by a suture line. The antennae are simple, composed of seven articles and a much reduced apical disk, sunk into the end of the seventh; this disk usually carries four sensory cones and is considered to represent an eighth antennomere. Sometimes one or more of the distal articles carries a distinct sensory pit on its outer surface. Most millipeds have light-sensitive *ocelli,* ranging from two or three tiny round dots to large and prominent ovoid or subtriangular clusters (*ocellaria*) of well-pigmented convex ocelli; immature specimens have fewer ocelli than adults and many cave species (as well as all species of Polydesmida, Platydesmida, and Siphonophorida) never have ocelli at any stage of life.

In most kinds of millipeds the *mandibles* are large and massive, attached to the cranium by an elongate hinge line, and composed of three segments. The basalmost (*cardine*) is largest and often subject to sexual dimorphism, being more prominent in males. The apical joint has several prominent blunt teeth and numerous rows of pectinate lamellae. The two pairs of maxillae characteristic of tracheate arthropods are represented in Diplopoda by a large, flat plate (*gnathochilarium*) forming the posteroventral side of the head, its exact derivation uncertain but presumably composed of elements from both pairs of maxillae. Of very constant shape for each order or suborder, the gnathochilarium is useful in defining higher taxa and representative types are illustrated (Figs. 26.14–26.18). Typically, there is a small-to-large triangular or pentagonal median sclerite (*mentum*), distal to which are two small elongate rectangular plates (*lingual lamellae*), and a larger stipe on each side; basally, a transverse *prementum* is often present (sometimes medially suturate). Just behind and often closely in contact with the prementum is another transverse sclerite, the *gula* (*hypostoma*), which is actually the sternum of the first trunk segment and not part of the gnathochilarium. The latter articulates with postoccipital apodemes of the cranium by means of a small cardine sclerite at each end of the prementum. The orders Polyzoniida and Siphonophorida have the head acutely prolonged, often into a fine beak, and in these groups the mandibles and gnathochilarium occur in reduced and atypical form, with individual sclerites fused, obscure, or wanting.

Segments

The first trunk segment (*collum*) is a transverse postcephalic plate of very variable form, often small, ellipsoidal, and flat, to large, broad, and hoodlike, covering much of the head in dorsal or lateral aspect. The sternal element (*gula*) is reduced to a small transverse sclerite without appendages or tracheal openings.

Trunk segments are composed of two embryonic somites, and all but the first four bear two pairs of legs. The appendages are located in the posterior half of the segment (the *metazonum*); the pair in the anterior half (*prozonum*) have been displaced posteriorly, permitting telescoping of the prozonum into the metazonum of the preceding segment. In some orders these two subsegments are superficially divided by a transverse constriction (*stricture*).

Segmental structure varies greatly among the different orders. In adult polydesmoids all segmental sclerites are coalesced into a solid ring without trace of sutures. In others (julids, spirostreptoids) each segment is composed of two dorsal elements (pleuroterga of the pro- and metazona) and two small but distinct imbricated sterna; no pleural sclerites are visible. In spiroboloids there are three transverse belts of sclerites divided by middorsal and lateral sutures (visible only in recently molted specimens), giving a total of 12 tergal plates, a distinct pleural sclerite on each side, and two sterna, a total of 15 parts. In both this and the preceding example, all elements are fused in adults even if suture lines are evident, so that the sterna are immovably fixed to the pleuroterga. In many orders (e.g., Platydesmida, Chordeumatida) the segment is functionally a pleurotergum in which the pleural sclerites are not distinguishable and the sterna are attached only by connective tissue, thus flexibly joined and freely movable; usually they are retracted into the body cavity. Stigmata are located at the lateral ends of the sterna, just at the lateral coxal condyle. The sterna may be very wide (Polydesmida) or narrow enough that the coxae are in contact medially. In the majority of diplopods each metazonum from the fifth or sixth segment to the penultimate carries an *ozopore* on each side, which is the opening of an internal gland (*ozadene*) secreting an allomone of variable composition and odor. A few groups appear never to have possessed ozadenes, while others (notably small polydesmoids) seem to have lost them secondarily.

In several groups (most Platydesmida and Polydesmida, some Chordeumatida) the metazona are produced laterally into *paranota* of variable size and width (in earlier literature referred to as "keels"), which when set high on the segment impart a distinctive "flat-backed" appearance to the animal. In those species adapted for rolling into a sphere, the paranota are laterally narrowed and directed ventrad, their apices accommodated usually by the greatly enlarged paranota of one of the anterior segments.

The last body segment is a legless ring (sometimes an incomplete ring lacking a sternum, with the ventral ends overlapping) containing the anal valves (*paraprocts*) and often produced dorsomedially into a projection (*epiproct*) of variable size and shape. In chordeumatids and callipodids the median posterior edge of this segment carries two or three pairs of small spinnerets, which produce a silk used by the young stages to produce a molting cocoon. Occasionally, the *hypoproct*, a transverse sclerite subtending the ventral ends of the paraprocts, may be indistinguishably fused with the posterior edge of the preceding sternum.

Legs

In most millipeds the legs are similar throughout the body and in both sexes, except that the anterior legs of males are often thicker than in females and sometimes provided with large internal glands opening through ventral knobs or processes. Frequently, one or more of the podomeres in males will be provided with an eversible membranous pad useful in holding to the female during mating. In some primitive forms the coxae have eversible sacs, thought to be used in water absorption. Typically, the leg is composed of six segments (*coxa, prefemur, femur, postfemur, tibia, tarsus*) and a tarsal claw, the latter simple and acute except in polyxenids, where multilobed, and in polyzoniids, where provided with a large ventral paronychium (Fig. 26.11). In some groups there is an accessory ring intercalated between coxa and prefemur.

External Genitalia

The male *gonopore* is located either on the coxa of the second pair of legs, or on a rudimentary intercalary structure between the second and third pairs, probably the remnant of the anterior legpair of the third trunk segment. This structure, informally referred to as the *penis,* may consist of paired sclerotized terminations of the vasa deferentia enclosed in a small basal framework, or may be reduced to a single eversible tube. In adult males the anterior legpair of the seventh segment is always modified into sperm transfer apparatus (*gonopods*), and in most groups the posterior pair of this segment is also modified (sometimes secondarily reduced or missing).

Very young specimens cannot easily be sexed, but in later nymphal instars the legs of the seventh segment differentiate into gonopods, either gradually through several stadia or fairly abruptly during the two last molts. Sometimes the gonopods are carried withdrawn into the body, or may be partly exposed, and in polydesmoids the single pair of gonopods is held almost completely outside the ventral surface of the segment. If two pairs are present, one usually functions to position or spread the female genitalia, the other to transfer a spermatophore or seminal fluid. In many groups the gonopods of at least one pair have a distinct prostatic groove on the distalmost of the two segments (normally, there is a coxal segment plus a telopodite which represents the remaining podomeres).

The female appendages (*cyphopods*) are specialized structures (considered by some to represent the vestigial coxae of otherwise missing legs) located behind the second pair of legs; they are somewhat bivalve in structure and contain the opening of the oviducts as well as seminal receptacles for sperm storage. In a few groups cyphopods have appreciable taxonomic importance.

Digestive System

The enteron of diplopods is a simple straight tube divided into three regions: *proenteron* (derived from embryonic stomadeum), *mesenteron* (from endoderm), and *metenteron* (proctodeal); the first is represented by the pharynx, the second by esophagus, stomach, and intestine, and the third by rectum. The mesenteron contains the usual peritrophic membrane. There is no distinct proventriculus. The rectum is often adapted for water recovery. Malpighian tubules enter the metenteron just beyond its junction with the mesenteron.

Most diplopods are detritivores, feeding opportunistically and generally on leaf litter and decomposing plant material, including the excrement of herbivore mammals. A few species will feed also on animal carrion and some are known to be preferential carnivores (chiefly in cave ecosystems).

Depending on the degree of adaptation to moist or dry habitats, the feces of millipedes vary greatly in composition: generally watery in forest-dwelling species of polydesmoids, for instance, but relatively dry and pelletized produced by spiroboloids.

The role of millipeds in contribution to the formation of mull humus remains inadequately documented, although generally considered to be appreciable. Unquestionably, these animals are important in assisting the mechanical and chemical degredation of litter into soil.

Reproductive Characteristics

Reproduction in Diplopoda is by the external development of internally fertilized eggs. The sexes are separate; paired gonads in each sex are located ventrad to the enteron and extend through most of the body cavity, often with transverse connectives, and open through gonopores located on or posterior to the coxae of the second pair of legs. In many groups the vasa deferentia terminates in sclerotized *"penes"* capable of some extrusion, and in all the oviducts open through cyphopods containing seminal receptacles. Mating is accomplished by prolonged clasping of the partners (up to 48 hours); there is usually little or no courtship, but in several species males are provided with glands secreting a substance that attracts females and/or distracts them while the male effects clasping. Spermatic material is extruded onto the gonopods either prior to clasping or after the genitalia are engaged, and then transferred to the seminal receptacles. In pentazonoid forms, in which gonopods are not developed, the male accomplishes spermatophore transfer with his mouthparts while holding the female with one or both of his telopods (modified last pair of legs).

Eggs are spherical, shelled, and heavily yolked; early development is similar to that of insects (meroblastic cleavage, flexion of the embryo, anamorphic posteclosion development). The number of eggs varies greatly (up to as many as 2000 tiny eggs among some polydesmoids, or as few as three or four only) and they are accorded different treatment in various taxa: laid in clusters and brooded by one of the parents (Platydesmida), scattered singly in the humus environment (Spirobolida), or enclosed in an igloo-shaped mud chamber built by the mother (Polydesmida). The young usually hatch with three pairs of legs and pass through anamorphic growth stages, with segments and legs added at each moult. Segmentation is teloblastic, the proliferation zone located in the penultimate segment. Development is gradual, hemimetabolous, without major changes in structure or appearance.

In some groups maturity is reached within a year and death may soon follow mating. In others, there may be a spring and fall (or fall and spring) mating period, and in more primitive taxa the life span may exceed several years, with continued growth and molting following the onset of sexual maturity, but without further increase in segment number. There are usually at least seven immature stadia, the first of which may be passed within the egg capsule. Functional external genitalia appear at the final molt in most groups, although maturation of the gonads may already have taken place. Mating usually occurs shortly after the final molt into sexual maturity.

Parthenogenesis is certainly not a common phenomenon among Diplopoda, but has been recorded for several species, and probably occurs in some others in which the known sex ratios are highly disparate (e.g., 1

♂ : 100 ♀♀). All of the established and presumed parthenogenetic species are quite small (less than 10 mm in length).

Habitat Preferences and Seasonal Occurrence

The great majority of Nearctic millipeds inhabit fairly moist biotopes, typically the upper horizons of litter in broadleaf mesic forests. Abundance and diversity seems correlated with the presence of calcareous substrates. In treeless areas the droppings of large mammals often provide both food and shelter. Many species are found in caves, both as troglophiles and obligate troglobites, and can be collected in baited pitfall traps or on organic material brought in for the purpose.

Immature stages are more prone to remain hidden in the deepest layers of litter and humus, and in mineral soil itself, and the presence of a given species in an area may be unsuspected until the time of maturity, when hundreds or thousands of specimens appear on the surface.

The chemical composition of the litter or soil has some effect on the character of the milliped fauna. In the eastern United States, pine forests have relatively few species, although polyxenids may occur in pine litter in great numbers. The acidic soil under ericaceous plants similarly harbors few species, but in the Appalachian region several genera are either confined to such habitat or are most abundant there. Most millipeds are nocturnal, but many can be seen out during the daylight hours in moist shady forest; a few even cross the highways in full sunshine, and large numbers of several species are prone to mass daylight movements.

Members of arctic–boreal taxa mature in the autumn months and may be found as adults during the winter only. More typically, maturity is reached in late spring or summer, at which time the older instars and new adults occur on the surface or in upper layers of litter.

Collection and Preservation

Millipeds may be picked up by hand and dropped into 70% isopropyl alcohol. If straight specimens are desired, they may be inserted into short lengths of plastic tubing, or rolled up in paper and tied with thread. Large specimens can be injected with preservative to prevent internal decomposition and subsequent fragmentation. Collection fluids that contain acetic acid or ethyl acetate should be avoided, as they decalcify the exoskeleton and greatly distend the body with internal bubbles of CO_2. Nor should formalin solutions be used, as they result in brittleness and loss of appendages.

In general, the greatest number of species and individuals occur in moist broadleaf forests developed in limy substrata. A careful regional survey should include other biotopes as well, as some species inhabit, almost unaccountably, places that seem wholly unsuitable. During hand collecting, the dry uppermost layer of leaves can be brushed aside and the moister horizons below carefully picked apart. The undersides of partly embedded stones and logs can be examined, as well as loose bark. Mechanical (funnel) techniques often are very successful in obtaining the smaller species in quantity, but rarely disclose anything that careful hand picking would overlook. Metal cans sunk in the soil, with some ethylene glycol as preservative, often yield good catches. Examination of the stomach contents of toads may be rewarding.

Most millipeds are nocturnal or crepuscular, and night collecting (especially when humidity is very high) often produces spectacular yields of the larger species which may not be found by hand picking at the same site during the day. Many species are seasonally restricted, so that year-round collecting is required to sample a local fauna adequately.

Since accurate identification usually depends on examination of male genitalia, collecting should be extended enough to obtain at least one adult male; the presence of gonopods can be ascertained easily even in the smallest species with a hand lens (and the unaided eye, with practice). Samples sent to a specialist for identification should include males and should always be labeled accurately with respect to locality, date, and collector (a vial of specimens with only a code number or symbol is an abomination and has little taxonomic value; some systematists will not identify such material).

A reference or study collection can be built up along various lines, the simplest techniques being by far the best. Material can be housed in homeopathic vials with neoprene stoppers, or in cotton-plugged shell vials stored in wide-mouthed jars. Arrangement can be systematic or follow a numerical collection-number sequence (the latter requires a card index for location of material). Jar lids of metal are strongly preferred to plastic lids, which break easily. In either case, the paper lid liner should be removed; the metal lid should be coated inside with spray enamel and a thin layer of paraffin added. This technique greatly retards the loss of alcohol by evaporation. Shell vials plugged with cotton should be inverted (cotton end down) so that if alcohol loss does occur, the vials will remain wet until virtually all has evaporated from the jar. Labels may be written with soft pencil, India ink, or typewriter, always on cotton fiber or rag content paper (cheaper grades of pulp paper will dissolve after a short immersion in fluid). Small, neat, detailed labels are preferable. Code numbers alone should *never* be the sole data in a vial because the corresponding data book frequently manages to become lost.

A binocular dissecting microscope with magnification of 90 or 120× is adequate to take a specimen of even the smallest species through the key to families that follows. Examination of the gonopods or gnathochilarium may require removal from the specimen and mounting in glycerine or a permanent medium, and use of a compound microscope. Once removed, however, such structures can be stored in microvials together with the body of the specimen.

It is usually necessary to have mature specimens, at least for initial identification of a species. Generally, maturity is indicated by (a) the presence of legs on all body segments, or (b) well-developed pigmentation and color pattern, or (c) completely sclerotized and calcified exoskeleton, or (d) the presence of sclerotized and functional external genitalia (even if largely concealed within the seventh segment). All of these points should be considered collectively, since the adults of some small species may be depigmented, and in some groups adults may have several legless posterior segments.

In general, the species of any given family look very much alike externally, and once a little experience has been gained in taking specimens through the key, it should be possible to recognize families by sight without having to examine the gnathochilarium or gonopods, or count segments. In cases of doubt at any particular couplet, try taking the specimen both ways: often some subsequent statement will rule out a given possibility. It is recommended that close attention be paid to statements of distribution. One would not expect to find species of Paeromopodidae in Florida, or Eurymerodesmidae in Oregon. Marking those families with local representation should greatly "shorten" the key for most users.

Classification of North American Diplopod Families

The key to families given in this chapter is based primarily on easily observable external characters common to both sexes, and as such does not reflect relationships as would a key based more on structure of the male reproductive system. The following list is provided to arrange the families in a systematic sequence which includes additional suprafamilial taxa names, as well as names (printed in italics) now considered to be synonymous or equivalent with those employed here, and references to recent literature which will be useful to anyone wishing to confirm an identification or pursue it to lower levels of classification.

Prior to 1980 there existed no reference work that endeavored to list all known generic and subgeneric names in the Diplopoda with taxonomic position and synonymy indicated. Such information is available for the North American fauna in the checklist compiled by Chamberlin and Hoffman (1958); this source is now largely obsolete and useful chiefly for its nearly complete bibliography up to 1955. I have prepared a provisional classification of the world fauna (Hoffman 1980) which contains no keys or descriptions but which should be useful in indicating the size and distribution of all taxa down to subgenus, as well as some idea of relationships. An updated catalog of the millipeds of North and Middle America is now in preparation, but an adequate manual for the identification of even the U.S. species cannot be expected until several large and difficult taxa have been revised. For the present, the user can only rely on the technical systematic papers cited in the bibliography of this chapter. For specimens taken in urban, agricultural, or otherwise disturbed areas, and therefore possibly introduced from Europe, reference to the regional manuals by Blower (1958) and Schubart (1934) will prove worthwhile.

Class Diplopoda [Attems 1926, Verhoeff 1926–1932, Hoffman 1980]

- Subclass Penicillata (*Pselaphognatha*)
  - Order Polyxenida (*Ancyrotricha, Schizocephala*)
    - Superfamily Polyxenoidea
      - Family Polyxenidae
- Subclass Pentazonia (*Opisthandria*)
  - Superorder Oniscomorpha
    - Order Glomerida (*Plesiocerata*)
      - Family Glomeridae
- Subclass Helminthomorpha (*Proterandria*)
  - Superorder Ommatophora
    - Order Polyzoniida (*Colobognatha* in part)
      - Suborder Polyzoniidea (Cook and Loomis 1928)
        - Family Polyzoniidae (Loomis 1971, *Polyzonium* only)
        - Family Siphonotidae
        - Family Hirudisomatidae
  - Superorder Anocheta
    - Order Spirobolida
      - Suborder Spirobolidea
        - Family Spirobolidae (Keeton 1960a, 1966)
        - Family Floridobolidae (Keeton 1959)
        - Family Atopetholidae (Hoffman and Orcutt 1960)
  - Superorder Diplocheta
    - Order Spirostreptida
      - Suborder Cambalidea (Loomis 1938)
        - Family Cambalidae (*Cambalopsidae, Nannolenidae, Leioderidae*)
      - Suborder Epinannolenidea
        - Family Choctellidae (Hoffman 1963)
      - Suborder Spirostreptidea
        - Family Spirostreptidae (Causey 1975)
    - Order Julida (*Symphyognatha, Zygocheta*)
      - Family Blaniulidae (*Nemasomatidae, Isobatidae, Zosteractiidae*)
      - Family Paeromopodidae (*Californiulidae*)
      - Family Parajulidae
      - Family Julidae

Superorder Typhlogena
Order Siphonophorida
Family Siphonorhinidae (Cook and Loomis 1928; *Illacme* only)
Family Siphonophoridae (Cook and Loomis 1928, *Siphonacme* only)
Order Platydesmida
Family Andrognathidae (Gardner 1975)
Superorder Coelocheta
Order Callipodida (Loomis 1937). [No American families are distinguished in this order, owing to the chaotic condition of the taxonomy of the group; Loomis's paper treats all Nearctic genera except three described after 1937: *Delophon* (revised by Shelley 1979a), *Tetracion* (Hoffman 1956), and *Texophon* (Chamberlin 1946). The genus that he included under the name *Spirostrephon* is now *Abacion*.]
Order Chordeumatida (*Chordeumida*) (Shear 1972)
Suborder Striariidea
Family Striariidae (Cook 1899)
Suborder Chordeumatidea
Superfamily Heterochordeumatoidea
Family Adritylidae (Shear 1971)
Family Conotylidae (Shear 1971, 1976)
Superfamily Cleidogonoidea
Family Cleidogonidae (Shear 1972)
Family Trichopetalidae
Superfamily Brannerioidea
Family Branneriidae
Family Tingupidae
Family Rhiscosomididae (Shear 1973)
Family Apterouridae (Shear 1977)
Superfamily position uncertain
Family Urochordeumatidae
Family Caseyidae
Superorder Mercheta
Order Polydesmida (*Proterospermophora, Polydesmoidea, Merocheta*)
Suborder Chelodesmidea
Superfamily Xystodesmoidea
Family Xystodesmidae (*Eurydesmidae, Fontariidae*)
Family Eurymerodesmidae
Superfamily Sphaeriodesmoidea
Family Sphaeriodesmidae (*Cyclodesmidae, Desmonidae*)
Superfamily Platyrhacoidea
Family Platyrhacidae (*Euryuridae*) (Hoffman 1979, *Euryurus*)
Suborder Paradoxosomatidea
Family Paradoxosomatidae (*Strongylosomidae*)

Suborder Polydesmidea
Superfamily Polydesmoidea
Family Polydesmidae (Hoffman 1962, *Scytonotus*)
Family Haplodesmidae (Blower 1980, *Prosopodesmus*)
Superfamily Trichopolydesmoidea
Family Macrosternodesmidae (*Nearctodesmidae*)
Superfamily Stylodesmoidea
Family Pyrgodesmidae (*Stylodesmidae, Chytodesmidae, Stiodesmidae*)

KEY TO NORTH AMERICAN ORDERS AND FAMILIES

1a Body wall membranous, without carbonate impregnation, body composed of head and 11 trunk segments, each of the latter except the last with a large cluster of setae on each side and two transverse rows of flattened trichomes; terminal segment with two dense clusters of whitish macrosetae; adults with 13 pairs of legs, tarsal claws large and laminate, males without legs modified for either clasping or sperm transfer; minute animals 2–3 mm in length strongly resembling the larvae of dermestid beetles (Fig. 26.1) . . . . Order **Polyxenida,** Family **Polyxenidae**

One genus, ***Polyxenus,*** with about six named species recorded for North America, their taxonomic status unsettled. Eastern United States, Texas, California, Washington, Oregon.

1b Body wall (except in immature and molting stages) usually sclerotized and reinforced with calcium salts; body composed of more than 11 segments; setae, if present, always single and simple, never in dense brushes or clusters; adults with minimally 17 pairs of legs, the tarsal claws simple and acute; at least one pair of legs modified in adult males to function during mating; usually larger animals, most Nearctic species longer than 4 mm. **2**

2a Body capable of enrolling into a compact sphere, composed of 13 apparent segments of which the second is broadened laterally to accommodate the tips of the following terga during volvation (Fig. 26.2); sterna divided and separated by the medially placed coxae; last pair of legs of males enlarged and modified into forcipulate structures for clasping the female during mating Order **Glomerida,** Family **Glomeridae**

Onomeris, two species in southeastern United States, and ***Glomeroides,*** one, vicinity of San Francisco Bay.

2b Body not capable of rolling into a sphere (except in the polydesmoid family **Sphaeriodesmidae**), composed of 19 or more segments in adults; sterna always intact and never laterad to coxae; last legs of males normal, but those of seventh (often eighth as well) segment modified as sperm transfer organs (gonopods) **3**

3a Body of adults with 18–20 segments, the segmental sclerites fused into a single solid ring usually without trace of sutures; ocelli always missing; gonopods formed from only anterior legpair of seventh segment; Order **Polydesmida** **4**

3b Body of adults with more than 20 segments, which are less completely coalesced: sterna and pleura either joined by membrane or at least distinguishable by existing suture lines; eyes normally present except in the orders **Platydesmida** and **Siphonophorida** and some cave-adapted species of **Chordeumatida** and **Spirostreptida;** gonopods usually composed of both pairs of

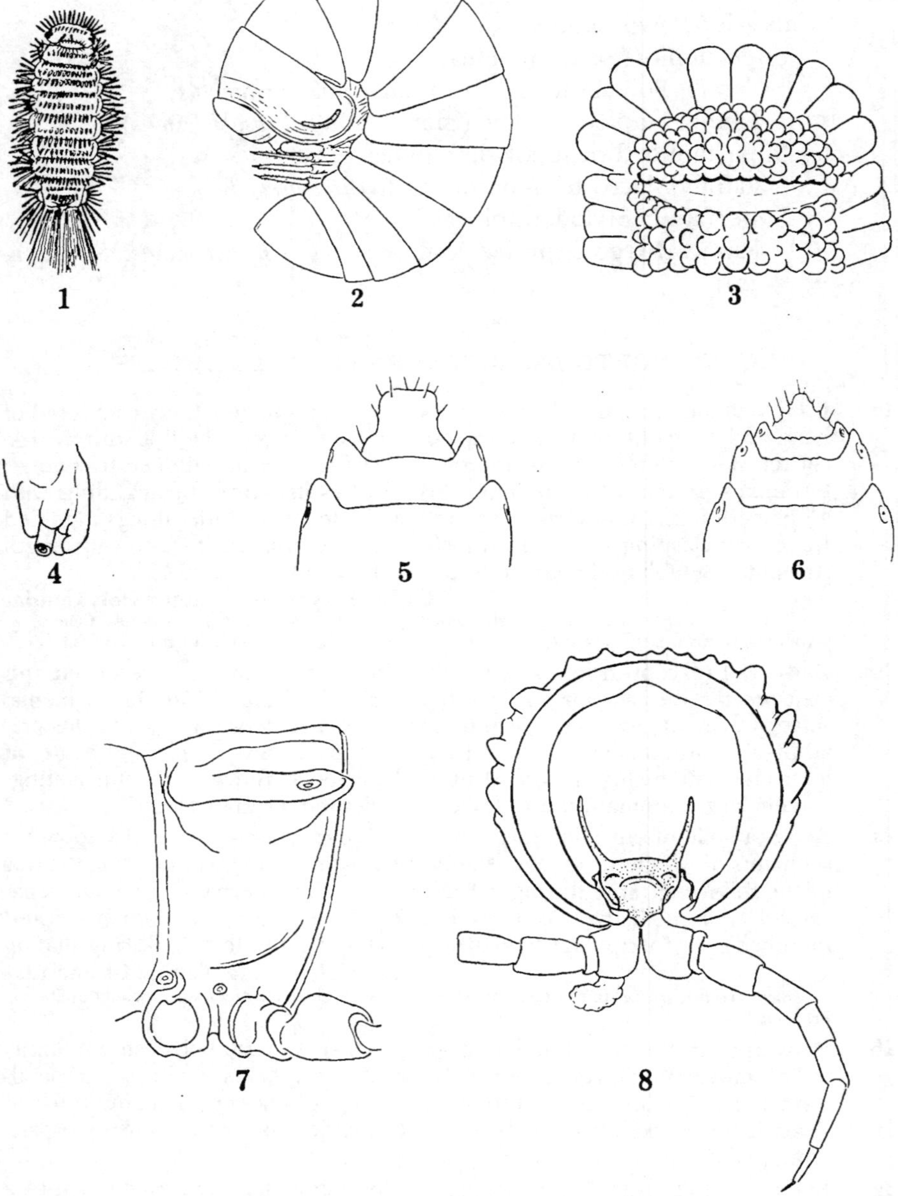

Fig. 26.1 *Polyxenus* (Polyxenidae), habitus of entire animal, dorsal aspect. **Figs. 26.2.** *Onomeris underwoodi* (Cook) (Glomeridae), entire animal partly enrolled, lateral aspect showing enlarged second segment modified for volvation. **Fig. 26.3.** *Myrmecodesmus* (Pyrgodesmidae), first two segments, dorsal aspect. **Fig. 26.4.** *Myrmecodesmus* (Pyrgodesmidae), right paranotum of midbody poriferous segment, dorsolateral aspect showing porostele. **Fig. 26.5.** *Euryurus maculatus* (Koch) (Platyrhacidae), posteriormost segments, dorsal aspect showing broadened and truncate epiproct. **Fig. 26.6.** *Pseudopolydesmus branneri* (Bollman) (Polydesmidae), posteriormost segments showing acute epiproct typical of most polydesmoids. **Fig. 26.7.** *Ergodesmus remingtoni* (Hoffman) (Macrosternodesmidae), midbody segment in lateral aspect showing oblique ridge dorsad to posterior leg base. **Fig. 26.8.** *Abacion magnum* (Loomis) (Callipodida: family uncertain). Midbody segment in anterior aspect, showing everted coxal sack and sternum (stippled) attached by pleurotergum only by membrane, also intercalary podomere between coxa and prefemur.

legs of the seventh segment, and anterior pair of eighth segment sometimes also modified; if only one pair, the sterna are not coalesced with pleuroterga and the latter are ornamented with longitudinal ridges 12

4a Body modified for volvation, paranota directed nearly ventral, those of second and third segments distinctly enlarged; sterna narrow, legs almost in contact; last segment enlarged and hoodlike; ozopores absent; segments 5–18 with deep round pit at base of anterior edge of paranota . . . . Family **Sphaeriodesmidae**

Desmonus, six species from Georgia and Kentucky to Texas. *Desmoniella*, one species, Oklahoma.

4b Body not modified for volvation, at most forming a flat coil or spiral, paranota directed more horizontally, those of second and third segments not enlarged; sterna much wider than length of a coxa; ozopores present; segments without pit at anterior base of paranota 5

5a Collum considerably enlarged and flabellate, concealing most or all of head in dorsal aspect (Fig. 26.3); at least part of the ozopore series borne on cylindrical whitish porosteles (Fig. 26.4) 6

5b Collum much smaller, covering only the posteriomost part of the head; ozopores opening directly on paranotal surface, never stalked 7

6a Metaterga with four or five transverse rows of subequal tubercules; coxae of gonopods small, not globose, carried mostly inside body cavity . . . . Family **Haplodesmidae**

One pantropical genus and species, *Prosopodesmus jacobsoni*, introduced into urbanized areas in south Florida and Louisiana.

6b Metaterga with at most three transverse rows of tubercles or flattened elevated areas, usually with two paramedian longitudinal rows of enlarged tubercles; gonopod coxae large, globose, carried almost entirely outside the body . . . . Family **Pyrgodesmidae**

Two native genera, *Psochodesmus* with one species widespread in Florida and *Myrmecodesmus* with two species along the Gulf Coast from west Florida to Texas; one genus, *Calyptodesmus*, with one species in Florida, doubtless introduced from South America.

7a Paranota of second segment lower than level of end of collum and paranota of third segment; metaterga of segments 5–17 smooth and polished, with distinct transverse sulcus . . . . Family **Paradoxosomatidae**

Two introduced genera, *Oxidus gracilis* being widespread and acclimatized over much of eastern United States, and *Asiomorpha coarctata*, so far confined to Florida and the Gulf coast states.

7b Paranota of second segment set either at same level as end of collum and paranota of third segment, or not as low as end of collum; metaterga of segments 5–17 either without trace of sulcus, or, if such is vaguely indicated, surface is textured into several transverse series of low polygonal areas. . . 8

8a Metaterga smooth and polished, without trace of sulcus; collum wider than head, and extended ventrolaterad behind mandibles 9

8b Metaterga with three to five transverse rows of polygonal areas (or, rarely, low tubercules), often with vague transverse sulcus between first and second series; collum flat, not wider than head and not turned ventrad laterally . . . 11

9a Epiproct apically broadened and truncate (Fig. 26.5) . . . . Family **Platyrhacidae**

Two genera in southeastern United States, north to Ohio and Minnesota: *Euryurus* with six species and *Auturus* with about four; the species of both genera partial to rotting wood (Hoffman 1978).

9b Epiproct apically acute (Fig. 26.6) **10**

10a Prefemora of legs with acute ventrodistal spine, except in a few genera confined to the Pacific northwest . . . . Family **Xystodesmidae**

Numerous genera, associated with specific geographic regions: central and southern California, ***Xystocheir, Amplocheir, Paimokia, Anombrocheir, Wamokia, Motyxia;*** northern California to Alaska and Montana, ***Harpaphe, Hybaphe, Isaphe, Tubaphe, Chonaphe, Montaphe, Metaxycheir, Orophe;*** New Mexico, ***Stenodesmus;*** east of the Great Plains, ***Boraria, Gyalostethus, Erdelyia, Cherokia, Caralinda, Pleuroloma, Pachydesmus, Dicellarius, Nannaria, Oenomoea, Apheloria, Rudiloria, Dixioria, Sigmoria, Cleptoria, Brachoria, Dynoria, Lyrranea, Deltotaria, Stelgipus, Cheiropus, Croatania, Semionellus.*** Many of these genera have been revised (Buckett et al. 1968a, b; Hoffman 1956–1967; Shelley 1977, 1978, 1980).

10b Prefemora of legs without distal spine . . . . . Family **Eurymerodesmidae**

Two genera in the region Iowa to Texas, east to Georgia: ***Eurymerodesmus*** with about 22 species, ***Paresmus*** with five.

11a Sides of metazona with oblique ridge above base of legs (Fig. 26.7); limbus usually unmodified; if these characters are difficult to observe then species will be less than 10 mm long and nearly colorless . . . .
Family **Macrosternodesmidae**

In Newfoundland, ***Ophiodesmus albonanus,*** introduced from Europe; eastern United States, ***Chaetaspis;*** Texas and New Mexico, ***Speodesmus;*** Arizona, ***Oodedesmus;*** southern California, ***Phreatodesmus, Harpagonopus;*** Nevada, ***Tidesmus;*** northwestern states, ***Nearctodesmus, Kepolydesmus, Ergodesmus,*** members of these last genera often bright red.

11b Sides of metazona without oblique ridges above base of legs; limbus usually laciniate; most species larger than 10 mm . . . . . . Family **Polydesmidae**

On the west coast, ***Bidentogon;*** in Texas and New Mexico, ***Speodesmus*** (cavernicolous); in New Mexico and Utah, ***Utadesmus;*** in eastern North America, ***Pseudopolydesmus, Polydesmus*** (introduced species only), and ***Scytonotus,*** the latter with species also in the northwestern states, western Canada, and Alaska.

12a Sterna, and in some cases also the pleura, not firmly attached to adjoining sclerites but more or less flexibly articulated by connective tissue (Fig. 26.8) **13**

12b Sterna, pleura, and terga coalesced into rigid cylindrical rings, the sclerites usually at least partly distinguishable by the persistence of the articular sutures **30**

13a Metaterga without median longitudinal suture **14**

13b Metaterga with well-defined median longitudinal suture **18**

14a Head small, conical, with two or three pairs of black ocelli; antennae nearly straight, the individual articles nearly all the same size and shape; body whitish to rose or light brown, smooth and polished, resembling a small slug or terrestrial flatworm . . . . Order **Polyzoniida. 15**

14b Head variable in form, but usually as broad as collum, without trace of ocelli; antennae relatively long, some of the distal articles enlarged and with round sensory pits . . . . Order **Siphonophorida 17**

15a Metaterga with two or three transverse rows of setae; tarsal claws with large prominent paronychium (Fig. 26.9); animals quick and active in life, usually bright pink . . . . Family **Siphonotidae**

One species, ***Siphonotus purpureus*** (Pocock), introduced from the West Indies into Florida and Louisiana.

15b Metaterga glabrous; tarsal claws simple, without paronychium; animals typically slow and deliberate, not bright pink **16**

16a Caudal edges of metaterga slightly but distinctly upturned (examine in lateral profile); pleuroterga of last segment fused ventrally, forming a complete ring enclosing paraprocts (Fig. 26.10); male gonopore opening through a small projection on ventral side of coxae of second pair of legs . . . .
Family **Hirudisomatidae**

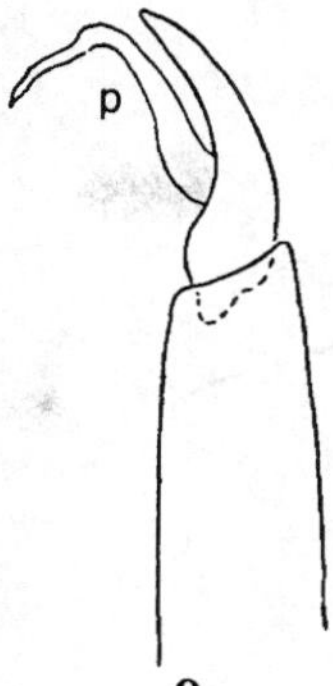

9

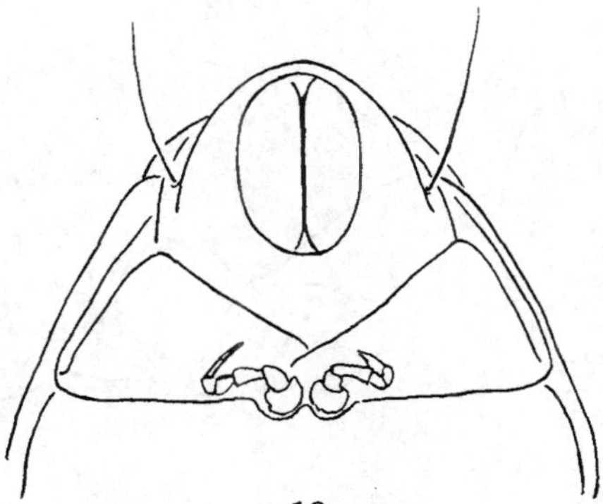

10

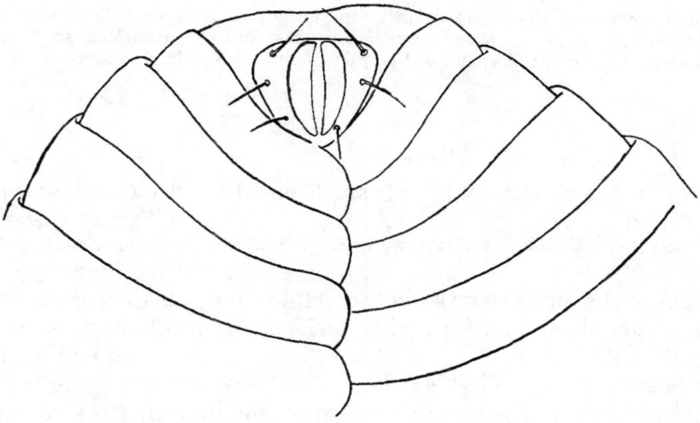

11

Fig. 26.9. *Rhinotus purpureus* (Pocock) (Siphonotidae), apex of tarsus of midbody leg showing tarsal claw and enlongated paronychium (*P*). **Fig. 26.10.** *Rhinotus purpureus* (Pocock) (Siphonotidae), posterior end of body, ventral aspect. **Fig. 26.11** *Polyzonium* sp. (Polyzoniidae), posterior end of body, ventral aspect.

Euzonium, Hypozonium, and ***Octoglena,*** each with species in California, ***Hypozonium*** north to Washington; an unnamed genus in the southern Appalachians. The group urgently requires revision.

16b Caudal edges of metaterga not raised; ventral ends of pleuroterga of last segment overlapping (Fig. 26.11); male gonopore on membranous "penes" originating in intersegmental membrane behind coxae of second pair of legs Family **Polyzoniidae**

Eastern United States: ***Polyzonium*** with about 11 nominal species; west coast states: ***Bdellozonium, Buzonium, Calzonium,*** and ***Pizonium,*** mostly monotypic.

17a Antennae nearly straight, extending directly in front of head, which is usually prolonged into an acute rostrum (Fig. 26.12); 5th antennomere usually much

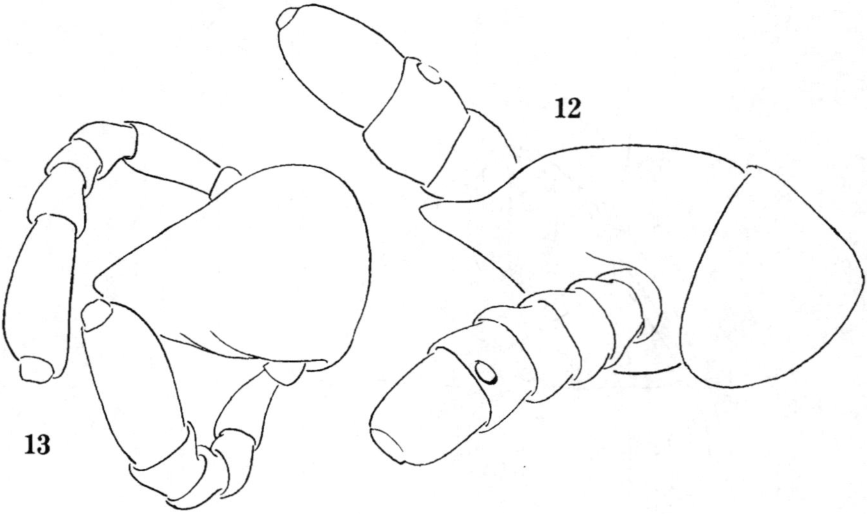

Fig. 26.12. *Siphonophora* sp. (Siphonophoridae), anterior end of body showing massive, nearly straight antennae with sensory organ on fifth antennomere. **Fig. 26.13.** *Siphonorhinus* sp. (Siphonorhinidae), head and antennae to show slender, elbowed posture of latter, which lack sensory organs.

enlarged, it and the sixth with small but distinct circular sensory pit near outer end . . . . Family **Siphonophoridae**

Two genera: ***Siphonophora*** with two species in southern Texas and ***Siphonacme*** with one species in Arizona.

17b Antennae distinctly elbowed between third and fourth articles, the sixth largest, neither it nor the fifth with sensory pits, head short, acutely pyriform (Fig. 26.13) . . . . Family **Siphonorhinidae**

One genus and species, ***Illacme plenipes*** Cook and Loomis, in southern California.

18a Head without ocelli, more or less acuminate, mouthparts somewhat reduced (gnathochilarium; Fig. 26.14); ozopores present but apparently nonfunctional; adults with more than 30 segments, metazona usually with distinct, often extremely broad paranota; coxae of most legs with eversible sacs; Order **Platydesmida** . . . . Family **Andrognathidae**

Andrognathus with a single species in southeastern United States, ***Ischnocybe, Mitocybe,*** and ***Gosodesmus*** with single species in California, ***Brachycybe*** with two eastern and four Californian species.

18b Ocelli present except in a few cavernicolous species; head and mouthparts of normal form, gnathochilarium as in Figs. 26.15 and 26.16; ozopores present or absent; coxal sacs, if present, confined to anterior end of body **19**

19a Body with more than 40 segments in adults, the metaterga with conspicuous sharp longitudinal ridges; ozopores present, the allomone with the odor of phenol; moderate to large (60 mm) active omnivores . . . . Order **Callipodida**

North American genera of this group are not at present allocated to families pending revision of the order. East of the Great Plains: ***Abacion*** with four species, ***Delophon*** with four (Shelley 1979a), ***Tetracion*** with two (in Alabama and Tennessee caves); ***Texophon,*** one species, in extreme southern Texas; ***Tynomma, Diactis, Heptium,*** and ***Colactis*** in California, the last also occurs in Mexico and Arizona.

19b Body with 26–30 segments in adults, metaterga with ridges only in one family; ozopores absent; only ninth and tenth pairs of legs with coxal sacs (in males); Order **Chordeumatida** **20**

20a Body segments with 10–12 prominent elevated thin longitudinal ridges, the surface between them densely granular; collum enlarged, hoodlike, largely concealing the head . . . . Family **Striariidae**

Striaria is widespread in eastern United States and on the west coast, *Vaferia* and *Speostriaria* occur in California (Loomis 1936).

20b Body segments smooth or variously textured but never with parallel ridges; collum small, concealing at most only basal part of head **21**

21a Mentum of gnathochilarium transversely divided, with a small triangular promentum set off (Fig. 26.15). **22**

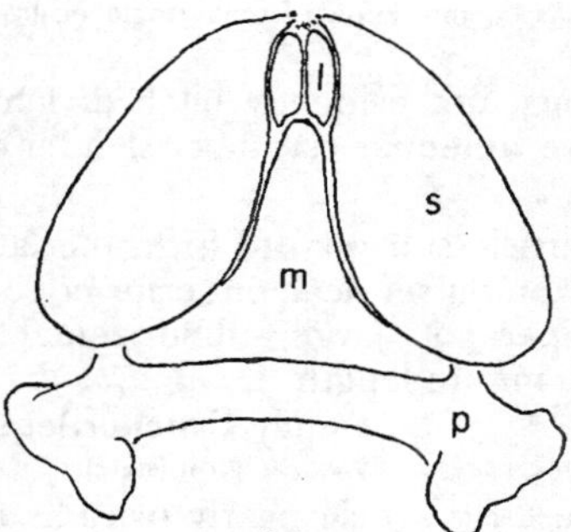

14

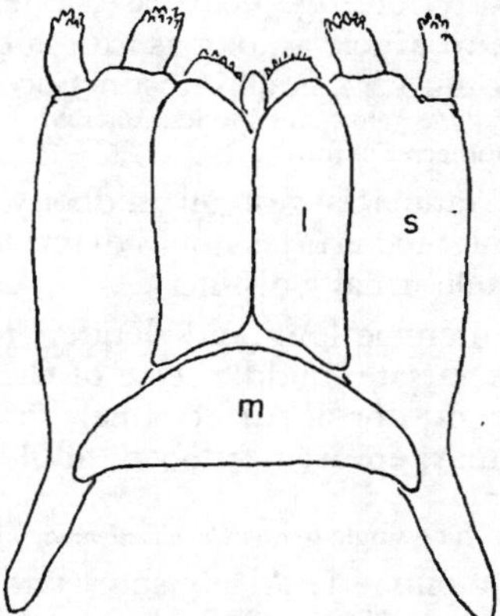

15

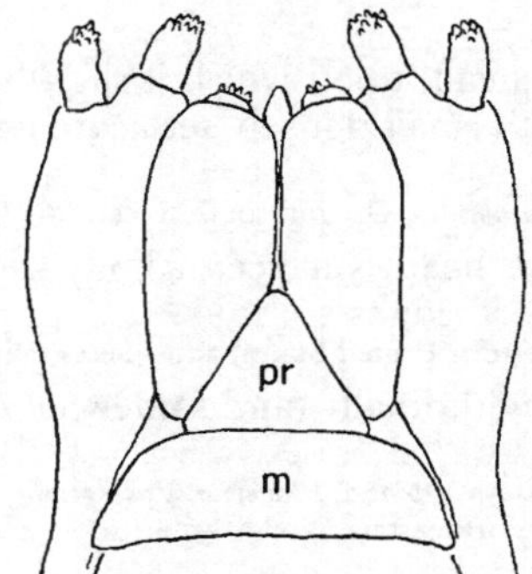

16

Fig. 26.14. *Brachycybe lecontii* (Wood) (Andrognathidae), gnathochilarium. *L*, lingual lamella; *S*, stipe, *M*, mentum, *P*, prebasilar plate. **Fig. 26.15.** *Conotyla melinda* Hoffman (Conotylidae), gnathochilarium. Symbols as in Fig. 26.14. **Fig. 26.16.** *Cleidogona major* Cook and Collins (Cleidogonidae), gnathochilarium. *Pr*, promentum; *M*, mentum.

21b Mentum of gnathochilarium entire, no promentum (Fig. 26.16) **29**

22a Adults with only 26 body segments . . . . Family **Branneriidae**

One species, ***Branneria carinata*** (Bollman), is widespread over southeastern United States.

22b Adults with 28 or 30 body segments **23**

23a Metazona of at least some body segments produced laterally into broad prominent flattened paranota; dorsal surface variously textured, not smooth **24**

23b Metazona not produced into true paranota, the segments either cylindrical or in a few cases angularly produced dorsolaterally by enlargement of the basal tubercules of dorsolateral setae; tergal surface smooth and polished (elevated into low, smooth tubercles in some species of ***Pseudotremia***, Family **Cleidogonidae**) **27**

24a Paranota largest on anterior segments, gradually reducing in size posteriorly to segment 7, each with acute anterior angle and smaller triangular lobe on posterior edge near base; antennae strongly clavate distad to fourth article, third article as long as fifth and fourth much shorter than either; less than a dozen very small indistinct ocelli . . . . Family **Apterouridae**

One genus and species, ***Apterourus horizontalis*** Loomis, known from a single locality in southern California.

24b Paranota largest on midbody segments, not evidently bilobed; lengths of antennal articles not as described above, antennae clavate or slender distally; ocelli usually distinct **25**

25a Antennae long and slender, fourth article subequal to fifth, the latter not incrassate; middle setae of each metazonum set near posterior edge of segment; dorsal surface finely granular; paranota large, subhorizontal, rectangular; epiproct trilobed; adults 15–20 mm in length . . . . Family **Urochordeumatidae**

The single genus ***Urochordeuma***, with two (one?) species in Washington (state).

25b Antennae relatively short and clavate, fifth article nearly twice as thick as fourth, latter much shorter than third or fifth; all three metazonal setae on each side of dorsum forming a transverse series near anterior edge; paranota laterally depressed; epiproct not trilobed; adults less than 10 mm in length **26**

26a Collum broader than head, upon which it slightly overlaps; surface of body segments textured with small densely set acute conic tubercles . . . . Family **Rhiscosomididae**

The single genus ***Rhiscosomides***, Oregon south to central California.

26b Collum narrower than head; surface of body segments set with numerous small sharp-edged short ridges . . . . Family **Tingupidae**

Tingupa, central and western United States, and ***Buotus***, Virginia and North Carolina.

27a Sides of metazona longitudinally (and somewhat obliquely) striate (best seen on dried specimen) . . . . Family **Caseyidae**

The nominal genera ***Caseya, Opiona, Placerna, Vasingtona***, and ***Zantona***, California to Washington; ***Underwoodia***, northeastern North America.

27b Sides of metazona smooth **28**

28a Adults greater than 10 mm in length; antennae distally slender; dorsal setae less than one-fourth of body diameter; epigaean species with more than 10 ocelli . . . . Family **Cleidogonidae**

Pseudotremia, about 40 species in the central Appalachians and Interior Lowlands, mostly in caves; ***Cleidogona***, about 30 species in eastern United States north to Indiana and Massachusetts; ***Tiganogona***, eight species, central United States, chiefly the Ozark region.

28b Adults less than 10 mm in length; antennae distally clavate; dorsal setae equal

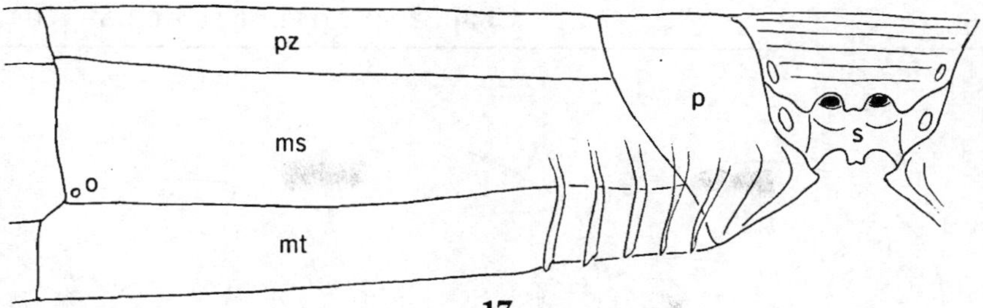

17

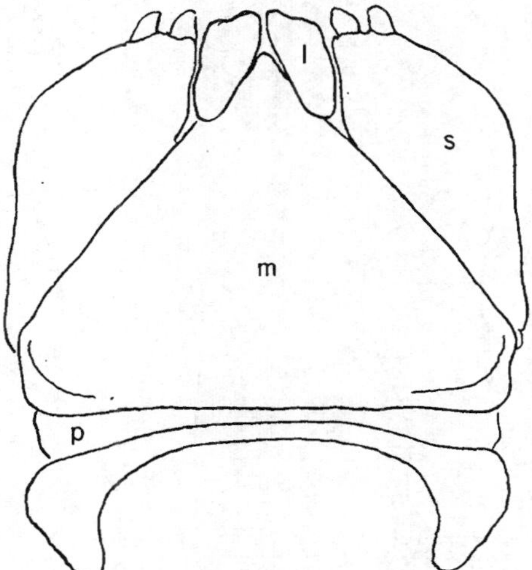

18

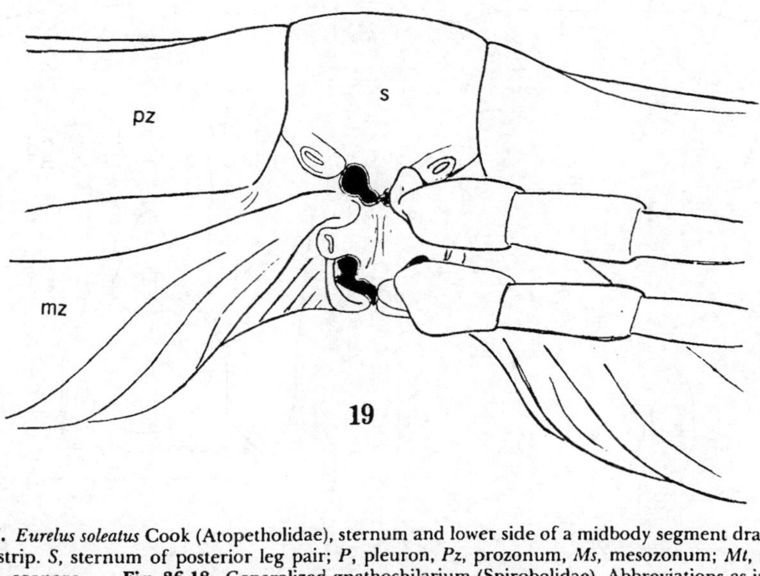

19

Fig. 26.17. *Eurelus soleatus* Cook (Atopetholidae), sternum and lower side of a midbody segment drawn as flattened strip. *S*, sternum of posterior leg pair; *P*, pleuron, *Pz*, prozonum, *Ms*, mesozonum; *Mt*, metazonum; *O*, ozopore. **Fig. 26.18.** Generalized gnathochilarium (Spirobolidae). Abbreviations as in Fig. 26.14. **Fig. 26.19.** Midventral region of body segment (Spirostreptidae). *S*, sternum of anterior legpair; *Pz*, prozonum; *Mt*, metazonum. Legs removed from right side; coxal cavities shown in black.

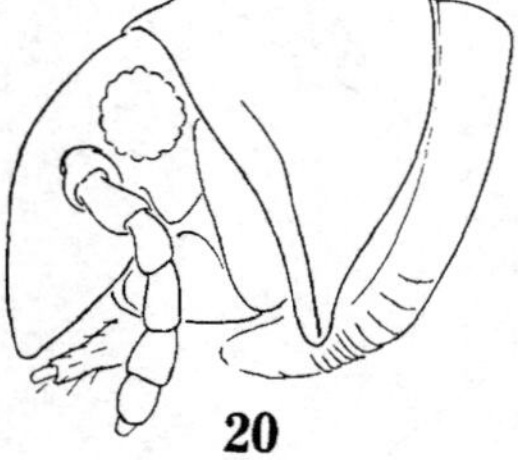

20

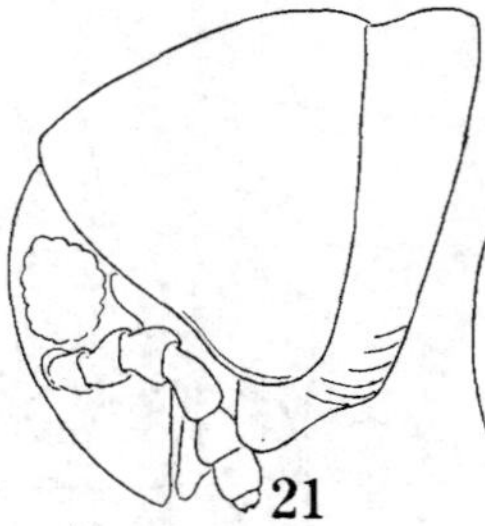

21

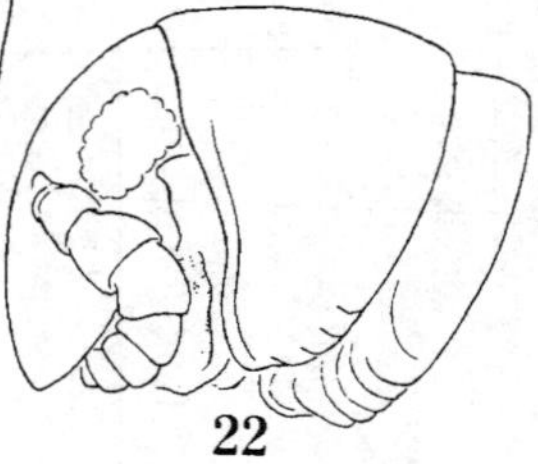

22

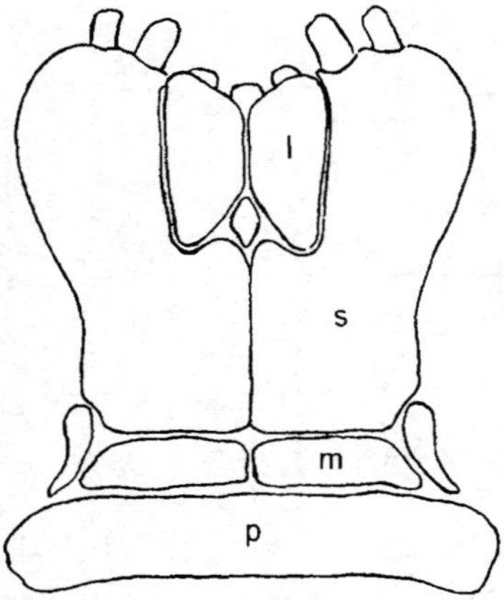

23

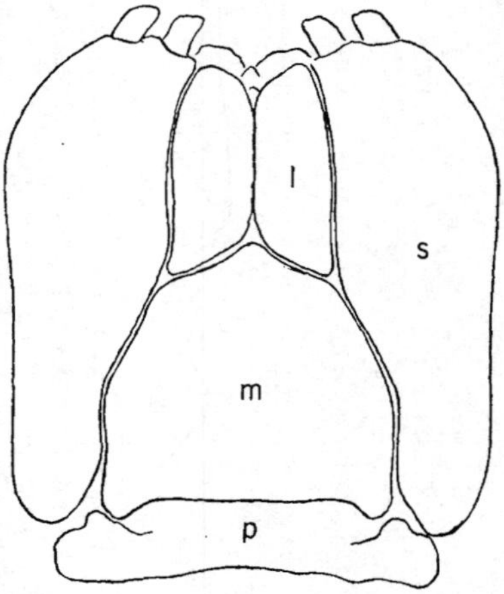

24

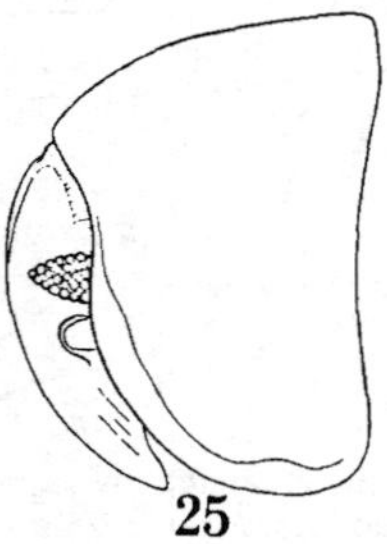

25

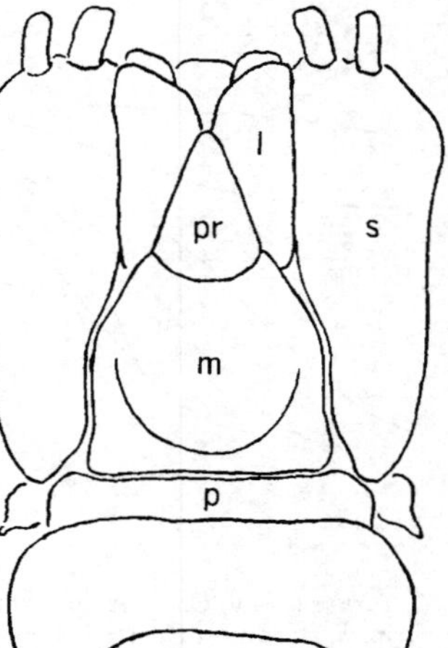

26

or exceeding body diameter; epigaean species with less than 10 ocelli and body colorless. . . . . Family **Trichopetalidae**

Trichopetalum and ***Trigenotyla,*** mostly epigaean, collectively with about a dozen species, central and northeastern North America; ***Scoterpes,*** exclusively troglobitic, with numerous species in central United States (Missouri, Alabama, Tennessee, Kentucky).

29a Length of adults 25–30 mm; telopodite of tenth legs of males much reduced, the coxae enlarged . . . . Family **Adritylidae**

The single genus ***Adrityla,*** with three species in the Rocky Mountain region from Utah to British Columbia.

29b Length of adults less than 25 mm; telopodite of tenth legs of males normal, coxae only slightly enlarged . . . . Family **Conotylidae**

Conotyla, 13 species in northeastern United States; ***Achemenides,*** one species, north-central states; ***Austrotyla,*** six species, Rocky Mountains and northern Mississippi valley; ***Idagona,*** Idaho, one species; ***Corypus, Brunsonia*** and ***Macromastus,*** one species each, Montana; ***Bollmanella*** and ***Taiyutyla,*** numerous species, northwest (Shear 1974).

30a Ocellaria basically rounded to round–oval, no transverse suture evident between them; pleural sclerites distinct (Fig. 26.17); a pair of legs present on segment 3; gnathochilarium as shown in Fig. 26.18; Order **Spirobolida** . . **31**

30b Ocellaria more often reniform or subtriangular, usually with a fine suture evident between their inner angles; pleural sclerites completely coalesced with lower ends of terga, no suture line remaining (Fig. 26.19); segment 3 without legs; gnathochilarium of a different form (Figs. 26.24, 26.26) . . . **33**

31a Collum acutely narrowed laterad, coming to a blunt point below level of second segment (Fig. 26.20); paraprocts strongly convex, their mesal edges meeting at a reentrant angle and lined with a fringe of fine short hairs; ventral setae of tarsi numerous, in two rows . . . . . Family **Atopetholidae**

Atopetholus, Watichelus, and ***Onychelus*** in southern California; ***Arinolus*** in Arizona; ***Piedolus*** in Utah; ***Anelus, Centrelus, Comanchelus, Eurelus, Toltecolus,*** and ***Uvaldia*** in Texas and New Mexico, all requiring revision to establish a number of valid species.

31b Collum broadly rounded laterally, not extending ventrad as far as ventrolateral angle of second segment; paraprocts slightly convex, their mesal edges compressed, not forming reentrant angle, no marginal setae; ventral tarsal setae sparse, usually in a single row 32

32a Ventrolateral region of second segment produced below and in front of lower end of collum; antennomeres subcylindrical, mostly longer than broad; lateral side of mandibular stipe not depressed (Fig. 26.21); paraprocts with distinctly compressed mesal margins; apex of mentum without convexity Family **Spirobolidae**

Narceus, two widespread species in eastern United States and a third in Florida; ***Chicobolus,*** monotypic, South Carolina to Key West; ***Hiltonius,*** four species in southern California; ***Tylobolus,*** seven species, Oregon and California, Utah, Kansas(?) (Keeton 1960a).

Fig. 26.20. *Atopetholus californicus* Chamberlin (Atopetholidae), head and first two body segments, lateral aspect, showing acutely pointed lateral end of collum typical of this family. **Fig. 26.21.** *Narceus americanus* (Palisot de Beauvois) (Spirobolidae), head and first two body segments, lateral view, showing anteroventral lobe of second segment subtending rounded end of collum. **Fig. 26.22.** *Floridobolus penneri* Causey (Floridobolidae), head and first two body segments, lateral view, showing short flattened antennomeres and depression in side of mandibular stipe. **Fig. 26.23.** Generalized gnathochilarium (Julidae). *S*, stipe; *L*, lingual lamella; *M*, mental sclerite; *P*, prebasilar plate. **Fig. 26.24.** Generalized gnathochilarium (Spirostreptidae), symbols as in Fig. 26.23. **Fig. 26.25.** *Choctella cummingsi* Chamberlin (Choctellidae), head and collum, lateral aspect to show concealment of antenna by anterior lateral curvature to collum edge. **Fig. 26.26.** *Cambala hubrichti* Hoffman (Cambalidae), gnathochilarium. Symbols as in Fig. 26.23; *Pr*, promentum; *M*, mentum.

32b Ventrolateral part of second segment lower than end of collum but not produced anteroventrad beneath it; antennomeres flattened, as wide as long; mandibular stipe depressed to accommodate antennae (Fig. 26.22); paraprocts without compressed mesal margins; apex of mentum with prominent median convexity . . . . Family **Floridobolidae**

The single species ***Floridobolus penneri*** in south central Florida (Keeton 1960b).

33a Stipes of gnathochilarium broadly in contact medially, with a small promental sclerite isolated in front of the median commissure and a larger transverse mental plate (or two plates) at the base (Fig. 26.23); Order **Julida** **34**

33b Stipes of gnathochilarium always widely separated by the large central mentum (Figs. 26.24, 26.26); Order **Spirostreptida**. **37**

34a Metazona with fine but distinct longitudinal striation entirely around segments. **35**

34b Metazonal striation, if present at all, confined to lower sides of segments, below ozopores **36**

35a Male genitalia large and partially exposed, sixth and seventh segments greater in diameter than fifth and eighth; ventral part of head long, face distinctly longer than broad in frontal view; most species more than 50 mm long . . . . Family **Paeromopodidae**

Two genera of uncertain content, ***Paeromopus*** with six nominal species in California, and ***Californiulus*** with about half a dozen species in Montana, Oregon, Idaho, and California; also a disjunct monotypic genus ***Aprosphylosoma,*** Oregon, possibly referable here.

35b Male genitalia concealed inside body, sixth and seventh segments not distinctly broader than fifth and eighth; face short, about as broad as long in frontal view; our species less than 50 mm in length . . . . . Family **Julidae**

All our species are introduced from Europe and most are successfully acclimatized, especially in northeastern North America. Several species of ***Allajulus*** (previously recorded as ***Cylindroiulus*** or ***Diploiulus***), one of ***Ophyiulus,*** and two of ***Brachyiulus,*** nearly all widespread in urban areas.

36a Body small and slender, about 20 times as long as broad; first pair of legs of males either similar to the others or reduced to minute hooklike vestiges; second segment of females not notably enlarged . . . . Family **Blaniulidae**

This family in its traditional sense has recently been divided into several families on the basis of small differences in male genitalia, visible only in microscope preparations. Perhaps subfamily status is more appropriate for these taxa. **Blaniulidae sen. str.,** would include ***Virgoiulus*** with two, and ***Okeanobates*** with one, species in eastern United States; **Nemasomatidae** with ***Orinisobates*** (three species), ***Chelojulus*** (one species), and ***Telsonemasoma*** (one species) chiefly in Montana, Idaho, and Washington; **Zosteractinidae,** with ***Zosteractis*** and ***Ameractis*** (one and two species, respectively) confined to caves in central and eastern United States.

36b Body stouter, at most 15 times as long as broad; first pair of legs of males greatly enlarged and very conspicuous; second segment of females thicker than third and fourth . . . . Family **Parajulidae**

A large and poorly understood group, much in need of revision. The nominal genera now recognized are, east of the Rockies, ***Aliulus, Aniulus, Arvechambus, Ethojulus, Gyniulus, Hakiulus, Illiulus, Okliulus, Oriulus, Pseudojulus, Ptyoiulus, Teniulus,*** and ***Uroblaniulus;*** in the Rockies (Arizona and New Mexico north to Canada), ***Apacheiulus, Gosiulus, Oriulus;*** on the west coast, north to Alaska, ***Codiulus, Bollmaniulus, Karteroiulus, Mulaikiulus, Saiulus, Simiulus,*** and ***Spathiulus.*** Many of these names are doubtlessly redundant.

37a Collum enlarged and broadly rounded laterally, covering most of antennae (Fig. 26.25); hypoproct fused with sternum of preceeding segment; color metallic bronzy in life; body relatively stout, about 10 times as long as broad Family **Choctellidae**

The single genus ***Choctella*** with two species in Alabama and Tennessee (Hoffman 1965).

37b Collum less enlarged, the antennae entirely exposed; hypoproct usually discrete; living specimens not metallic; body more slender, 15–20 times as long as broad **38**

38a Adults of U.S. species greter than 100 mm long; lateral ends of collum with several parallel engraved striations; lingual lamellae in contact, not separated by mentum or promentum (Fig. 26.24) . . . . Family **Spirostreptidae**

One genus, ***Orthoporus,*** with about three species in southwestern United States, Texas to Arizona.

38b Adults of U.S. species less than 60 mm long; ends of collum at most with one submarginal striation; lingual lamellae separated by anterior prolongation of mentum or promentum (Fig. 26.26) . . . . Family **Cambalidae**

Poorly known taxon badly in need of revision, most of the nominal genera listed are probably invalid. The genus ***Cambala*** has about five species in eastern United States; the following, mostly monotypic genera, have been proposed for species from California (Shelley 1979b): ***Nannolene, Paiteya, Titsona, Buwatia, Platydere, Pharodere, Endere, Leiodere, Odachurus, Alakene, Doilene,*** and ***Tigolene.***

References

Attems, C. 1926. Myriopoda. In Krumbach (ed.). *Handbuch der Zoologie.* 4(1). Walter de Gruyter, Berlin. 402 pp. **Blower, J. 1958.** *British Millipedes (Diplopoda). Linn. Soc. Lond. Synop. Br. Fauna.* (11):1–74 **Buckett, J. S. and M. R. Gardner. 1968a.** Revision of the millipede genus *Harpaphe* Cook from western North America (Polydesmida : Xystodesmidae). *Occas. Pap. Bur. Agric. (Calif.)* (11):1–51. **1968b.** Revision of the millipede genus *Wamokia* Chamberlin from the Sierra Nevada of central California (Diplopoda : Polydesmida : Xystodesmidae). *Proc. Biol. Soc. Wash.* 81:511–538. **Causey, N. B. 1975.** Desert millipedes (Spirostreptidae) of the southwestern United States and adjacent Mexico. *Occas. Pap. Mus. Texas. Tech. Univ.* (35):1–12. **Chamberlin, R. V. and R. L. Hoffman. 1958.** Checklist of the millipedes of North America. *Bull. U.S. Natl. Mus.* (212):1–236. **Cook, O. F. 1899.** The diplopod family Striariidae. *Proc. U.S. Natl. Mus.* 21:667–676. **Cook, O. F. and H. F. Loomis. 1928.** Millipedes of the order Colobognatha, with descriptions of six new genera and type species, from Arizona and California. *Proc. U.S. Natl. Mus.* 72(18):1–26. **Gardner, M. R. 1975.** Revision of the millipede family Andrognathidae in the Nearctic region. *Mem. Pac. Coast Entomol. Soc.* (5):1–61. **Hoffman, R. L. 1956–1967.** Revisions of xystodesmid millipede genera: *Proc. U.S. Natl. Mus.: Dixioria* 106:1–19, **1956;** *Pachydesmus,* 108:181–218, **1958;** *Cherokia,* 112:227–264, **1960;** *Deltotaria,* 113:15–35, **1961;** *Boraria* and *Gyalostethus,* 117:305–347, **1963;** *Cleptoria,* 124:1–27, **1967. 1962.** The milliped genus *Scytonotus* in eastern North America, with the description of two new species. *Am. Midl. Nat.* 67:241–249. **1965.** A second species in the diplopod genus *Choctella* (Spirostreptida : Choctellidae). *Proc. Biol. Soc. Wash.* 78:55–58. **1978.** North American millipeds of the genus *Euryurus* (Polydesmida : Platyrhacidae). *Trans. Am. Entomol. Soc.* 104:37–68. **1980.** *Classification of the Diplopoda.* Museum d'Histoire Naturelle, Geneva, Switzerland. 236 pp. **Hoffman, R. L. and B. S. Orcutt. 1960.** A synopsis of the Atopetholidae, a family of spiroboloid millipeds. *Proc. U.S. Natl. Mus.* 111:95–165. **Keeton, W. T. 1959.** A revision of the milliped genus *Brachoria* (Polydesmida : Xystodesmidae). *Proc. U.S. Natl. Mus.* 109:1–58. **1960a.** A taxonomic study of the milliped family Spirobolidae (Spirobolida : Diplopoda). *Mem. Am. Entomol. Soc.* (54):1–146. **1960b.** A new family for the diplopod genus *Floridobolus* (Spirobolida: Spirobolidea). *Bull. Brooklyn Entomol. Soc.* 54:1–7. **1966.** The species of the milliped genus *Tylobolus* (Diplopoda : Spirobolida). A reexamination. *Trans. Am. Entomol. Soc. (Phila.)* 92:17–28. **Loomis, H. F. 1936.** New millipede of the American family Striariidae. *J. Wash. Acad. Sci.* 26:403–409. **1937.** Crested millipedes of the family Lysiopetalidae [= order Callipodida in modern usage] in North America, with descriptions of new genera and species. *Proc. U.S. Natl. Mus.* 84:97–135. **1938.** The cambaloid millipedes of the United States, including a family new to the fauna and new genera and species. *Proc. U.S. Natl. Mus.* 86:27–66. **1971.** The diplopod genus *Polyzonium* in North America. *Fla. Entomol.* 52:245–251. **Schubart, O. 1934.** Tausendfüssler oder Myriapoda. I. Diplopoda, in F. Dahl, *Die Tierwelt Deutschlands und der angrenzenden Meersteile . . . ,* Part 28. Gustav Fisher, Jena, East Germany. 318 pp. **Shear, W. A. 1971.** The millipede family Conotylidae in North America, with a description of the new family Adritylidae (Diplopoda: Chordeumida). *Bull. Mus. Comp. Zool.* 141:55–97. **1972.** Studies in the millipede order Chordeumida (Diplopoda): a revision of the family Cleidogonidae and a reclassification of the order Chordeumida in the New World. *Bull. Mus. Comp. Zool.* 144:151–352. **1973.** The millipede family Rhiscosomididae (Diplopoda: Chordeumida). *Psyche (Camb.)* 80:189–203. **1974.** The millipede genus *Bollmanella* (Diplopoda, Chordeumida, Conotylidae). *Psyche (Camb.)* 81:134–146. **1976.** The millipede family Conotylidae (Diplopoda, Chordeumida). Revision of the genus *Taiyutyla,* with notes on recently proposed taxa. *Am. Mus. Novit.* 2600:1–22. **Shelley, R. M. 1977.** The milliped genus *Croatania* (Polydesmida: Xystodesmidae). *Proc. Biol. Soc. Wash.* 90:302–325. **1978.** Millipeds of the eastern Piedmont region of North Carolina, U.S.A. (Diplopoda). *J. Nat. Hist.* 12:37–79. **1979a.** A revision of the milliped genus *Delophon,* with the proposal of two new tribes in the subfamily Abacioninae (Callipodida: Caspiopetalidae). *Proc. Biol. Soc. Wash.* 92:533–550. **1979b.** A synopsis of the milliped genus *Cambala,* with a description of *C. minor* Bollman (Spirostreptida: Cambalidae). **Proc. Biol. Soc. Wash.** 92:551–571. **1980.** A revision of the milliped genus *Pleuroloma* (Diplopoda: Xystodesmidae). *Can. J. Zool.* 58:129–168. **Verhoeff, K. W. 1926–1932.** Diplopoda. Pp. 1–2084. In H. G. Bronn (ed.). *Klassen und Ordnungen des Tierreichs, wissenschlaftlich dargestellt in Wort und Bild.,* Vol. 5, abt. 2, buch 2. Leipzig, East Germany.

27

Pauropoda

ULF SCHELLER

Five families and more than 500 species of Pauropoda have been described, from various countries. Most probably the number of species is considerably higher. Only limited attention has been drawn to the Pauropoda so far because of their small body size, their susceptibility to environmental changes, and the patchiness in their local distribution.

Three families (italicized below) with about 70 species were recorded from North America.†

Order **Pauropoda**

- Suborder Hexamerocerata
  - Family Millotauropodidae
- Suborder Tetramerocerata
  - Infraorder Ectomorpha
    - Family *Pauropodidae*
    - Family Afrauropodidae
    - Family *Brachypauropodidae*
  - Infraorder Endomorpha
    - Family *Eurypauropodidae*

The Pauropoda are progoneate, dignathous, blind myriapods, generally 0.5–1.5 mm long with 8–11 leg-bearing trunk segments in the adult stage. The body is cylindrical and whitish in Hexamerocerata and Ectomorpha, oval and dorsoventrally flattened and brownish in Endomorpha. The North American species have 9 or 10 pairs of legs as adults. The Pauropo-

† While this book was in production the author described 26 new species, one subfamily, two new genera and one new subgenus from North America (Scheller 1984, 1985, 1986, 1988). He has also found 10 new species and one new subgenus from the Virgin Islands.

didae are more or less long-legged, most often speedy runners for short distances. The Brachypauropodidae and Eurypauropodidae have proportionately shorter legs and generally reduced agility. Their external morphology is briefly treated below under taxonomic anatomy.

The classic work on post-embryonic development of Pauropoda was produced by Tiegs (1947), while Verhoeff (1934) published early taxonomic works.

Digestive Characteristics

Gastrointestinal Mechanisms

The digestive tract is uniform throughout the group and hence offers no characters of taxonomic value. Most species have poorly developed mouthparts. The mouth cavity, pharynx, and esophagus form a short tube where the salivary glands open. Posteriorly follows the midgut, a straight tube running through almost the entire length of the body. It leads into a short rectum opening just anterior of the anal plate.

The Tetramerocerata are fungus eaters or feed on the semiliquid end products of decayed plant material. The Hexamerocerata, which are not known from North America, have grinding mandibles and may consume solid food.

Relationship to Humification

In most environments the occurrence of the Pauropoda is very patchy and the populations sparse. Their influence on the breaking down of vegetable litter therefore plays a role in modifying soils only very locally. Their incapability to burrow and their small size and narrow esophagus also indicate that their action on the organic content of soils is insignificant.

Excretory Characteristics

In the few species studied closely there is a pair of Malpighian tubes on either side of the midgut opening at the junction of the latter and the rectum. Observations indicate that the midgut epithelium continually discharges concretions into the lumen of the midgut, which are eventually extruded through the anus. The concretions have been shown to be a product of the animal's own metabolism. The feces seem to be droplets in liquid form. The fecal production rate is unknown.

Reproduction

The species are bisexual. The gonopores are on the ventral side between the second pair of legs, paired submedian in the male but unpaired median in the female. The sexes are easily distinguished: males have two conical penes, females a single almost concealed genital aperture. Sexual dimorphism is rare. Some species have very low sex ratios and there may sometimes be a lowered sexual reproduction. There are species in which males are unknown which may be parthenogenetic.

Male Morphology

Four hollow testes lie above the alimentary canal in the posterior part of the body. Their vasa deferentia pass down, two on either side, below the intestine, where they continue forward as four vesiculae seminales. The latter join and form two exit channels the anterior part of which constitutes the ejaculatory ducts, which pass to either side of the nerve cord to the penes.

Female Morphology

The reproductive organs are simple. There is in the hind part of the body a single unpaired ovary below the midgut. It is connected with the genital aperture by a single oviduct and a short vagina. In the latter a receptaculum seminis opens, forming a sac below the nerve cord, a sac that is usually spherical.

Life Cycle

Only the life cycle of the Pauropodidae is well known and is described below. The total length of the larval period of free-living animals is unknown but may last for 3–4 months.

Oviposition. The eggs are laid singly or in clumps in the soil or in the decaying material in which the animals live. They are spherical, whitish, about 0.05–0.15 mm in diameter. After about 2 weeks the outer membrane breaks and the embryo enters the pupoid phase.

Pupoid Phase. The pupoid phase is quiescent and remains for a few days. The pupal sheath is setose and may have outgrowths showing the position of the antennae and more or less the first three pairs of legs.

The following stages are all whitish (in Eurypauropodidae brownish) and actively moving. The descriptions below are restricted to some external characters of special importance for recognizing the stages.

First Instar Larva. The first instar larva has 7 abdominal segments, 3 pairs of legs, and 2 pairs of trichobothria. After about 3 weeks it molts and discloses the following stage.

Second Instar Larva. The second instar larva has 8 abdominal segments, the new segment being the seventh abdominal, which has developed between the sixth segment and the anal segment of the previous instar. There are 5 pairs of legs and 3 pairs of trichobothria.

Third Instar Larva. The third instar larva has 10 abdominal segments, 6 pairs of legs, and 4 pairs of trichobothria.

Fourth Instar Larva. The fourth instar larva has 11 abdominal segments, 8 pairs of legs, and 4 pairs of trichobothria.

Adult Animal. The adult has 12 abdominal segments, 9 (seldom 10) pairs of legs, and 4 pairs of trichobothria.

Ecological Characteristics

The Pauropoda inhabit a variety of soil types. They are sometimes found in litter and decaying logs and several species occur also under bark and moss carpets, where darkness, humidity, and temperature are suitable. Because they cannot burrow they are bound to crevices and tunnels already at hand. They are associated with Collembola, various types of mites, worms, and insect larva, and with Protura, Diplura, Symphyla, and Pseudoscorpionidea. They most often occur in dark sites rich in humus, in moist but not in wet or dry places, and rather in deciduous leaf litter than in coniferous litter, but Pauropoda may be found in a variety of environments.

They can sometimes be abundant. Locally up to several hundred specimens may occur in a square decimeter in damp deciduous forests. Preferably, they gather at a depth of a few centimeters but may occur deeper, sometimes at a depth of at least 0.75 m.

Collection and Microscopical Technique

Most pauropods are very small and fragile, not easy to deal with, and sometimes difficult to obtain. Sifting and hand sorting of debris is not a good method, but collecting individual specimens in situ may give good results. Larger specimens can be picked up from the underside of moist stones and branches with a fine brush. Individuals in the main part of the species move quickly and intermittently with a mouselike running motion and are therefore easy to distinguish from all other small soil-living arthropods except a few collembolans. However, reduced agility occurs, for example, in juvenile stages and in adults in Eurypauropodidae and Brachypauropodidae. These forms are difficult to discover and they are best extracted by funnel or flotation techniques.

Most funnels will provide results, but the Tullgren funnels, particularly as recently modified, are best for collecting agile forms. The slow-moving and strongly sclerotized Eurypauropodidae are most conveniently obtained by means of Winkler funnels. However, Pauropoda are most susceptible to the slightest provocation—too much moisture, too little moisture, too much light—so a gentle application of heat and light to extrude the animals from the sample is necessary. Because pauropods are unable to burrow, it may be advantageous to place the samples on gauze in an inverted positon without disturbance. Precautions should always be taken when using funnels to ensure that the laboratory atmosphere is not too dry, since the animals may be driven inward and not downward.

In general, juvenile stages are seldom obtained in funnel collection. Particularly for quantitative estimations and for information on life histories, a more adequate sampling technique is needed. Flotation is a method that can be more effective than funnels, particularly in soils with low contents of organic matter. Even a simple water flotation may give good results: in particular, it is easy to float off various stages of Pauropodidae species, whereas the strongly sclerotized Eurypauropodidae may not be obtained. The use of a more complicated apparatus or a technique requiring repeated procedures is of limited value, owing to the fragility of most species.

For systematic purposes the morphological details of both antennae and pygidium must be available for study. Because a proper orientation of the former and a good extension of the latter greatly facilitate the work, the animals should be fixed and preserved in fluids that keep them soft, transparent, and stretched. Cold strong ethyl alcohol often gives irregular contraction. If pure ethyl alcohol is to be used, it must not exceed 70%. The drawbacks of alcohol are eliminated if it is mixed with small amounts of acetic acid and formaldehyde.

Small species and juvenile stages may be studied in the collection fluid, after some lactic acid has been added. Large specimens, particularly strongly sclerotized ones, have to be cleared in pure lactic acid or a mixture of about 80% lactic acid, 15% glycerine, and 5% formaldehyde. Staining is unnecessary. Preferably the animals should be placed in one of the foregoing solutions on an object glass and a cork wedge (15 mm long, 1.5 mm at thickest end) is placed under one side of the cover glass. The wedge can be moved so that cover glass presses or rolls the animal so that it can be examined from all sides without damage. The use of phase contrast facilitates study of the setae, the claws, the anal plate, the trichobothria, and details in the antennal structure.

Taxonomic Structures

The general structure of parts useful in taxonomic work is described below. Internal anatomy is omitted. The characters considered to be of primary importance are depicted in Figs. 27.1–27.3 and should be studied before attempting to use the keys. The Hexamerocerata and the Afrauropodidae (Tetramerocerata) are not treated.

Head

The tergal side of the head is triangular with many setae, the arrangement, shape, and length of which are of value in separating species. Laterally located is a pair of temporal organs (Fig. 27.1). They are more or less ovoid. A mediotergal spatulate plate (Fig. 27.7*a*) may occur.

Antennae

The antennae (Fig. 27.2) consist of a slightly telescopic four-segmented stalk and two branches protruding from its last segment. One branch is tergal, *t*, and has a single flagellum, F_1; the other is sternal, *s*, and bears two flagella, F_2 and F_3, one or two globuli, *g*, and one or two setae, *q*. The last segment of the stalk has the four setae p, p', p'', and *r*, sometimes also one or two more, p''' and *u*. The shape, relative length, and diameter of these structures present good characters on both the generic and specific levels.

Trunk

There are 12 segments. The first has vestiges of a pair of legs and legs are lacking on the last segment (*Allopauropus* s.g. *Decapauropus*) or the last two segments (all other Tetramerocerata). The tergites are thin or very slightly sclerotized in Pauropodidae, partly sclerotized in Brachypauropodidae, and very strongly sclerotized in Eurypauropodidae. Their shape, degree of sclerotization, and division into sclerites are important on family and generic levels. They are provided with setae, spines, or other protuberances, the number, shape, and arrangement of which (Figs. 27.1, 27.4, 27.5) have proved taxonomically useful. There are five pairs of long trichobothria, T_1–T_5, more or less pubescent and usually increasing in length posteriorly (Fig. 27.1). The third pair, particularly, is important for species recognition. It often has an apical end swelling or may be thickened at the middle; or it is branched.

Pygidium

The pygidium (Fig. 27.3) is divided horizontally. The tergum has four or five pairs of setae: the a_1, a_2, and a_3 are often of subequal shape but of

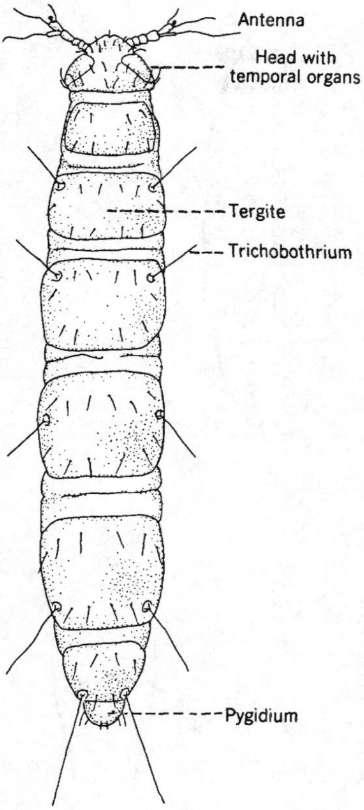

1

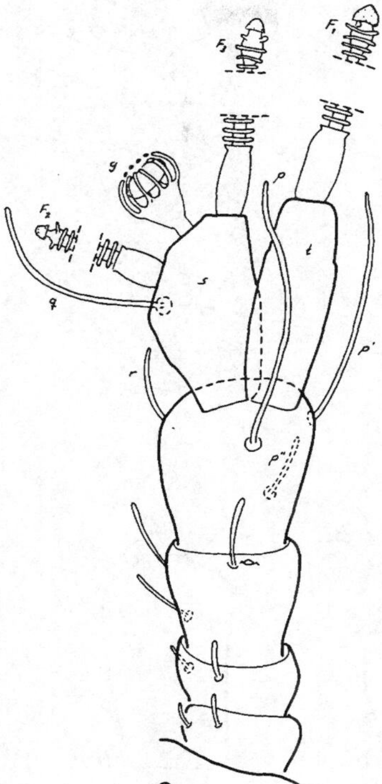

2

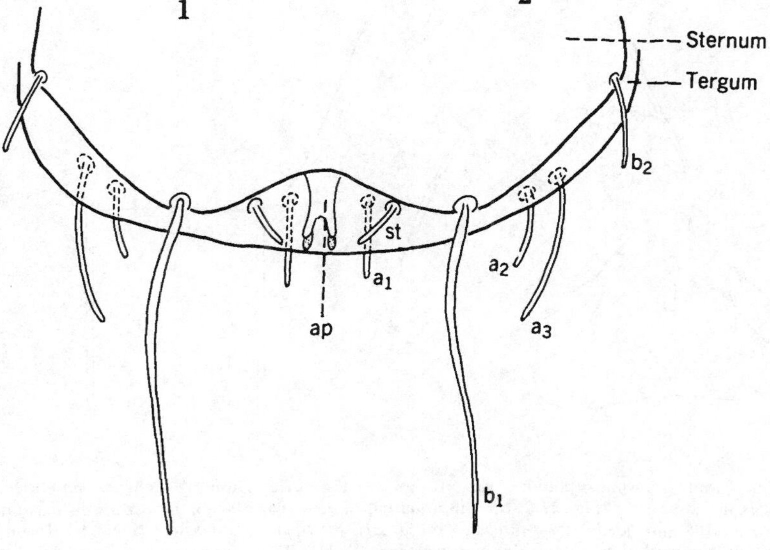

3

Fig. 27.1. Morphology of a pauropod. Family Pauropodidae, adult, tergal view. Legs not shown. **Fig. 27.2.** Antenna. Family Pauropodidae, tergal view. *t*, tergal branch; *s*, sternal branch; F_1, F_2, F_3, flagella *g*, globulus; *q*, seta on sternal branch; *p*, *p*′, *p*″, *r*, setae on fourth segment. **Fig. 27.3.** Pygidium. *Allopauropus vulgaris* (Hansen), ventral view. a_1, a_2, a_3, *st*, setae on tergum; *ap*, anal plate; b_1, b_2, setae on sternum.

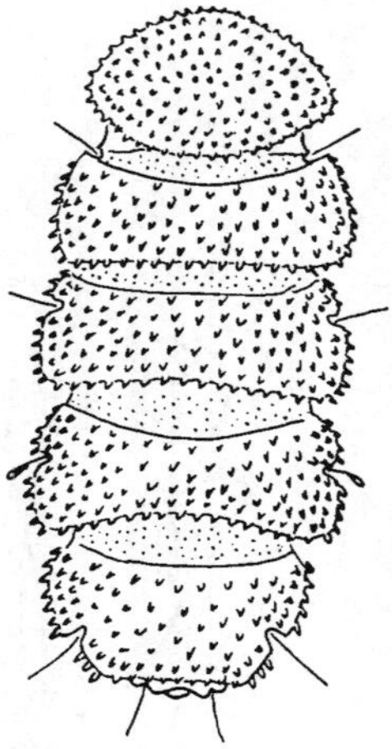

4

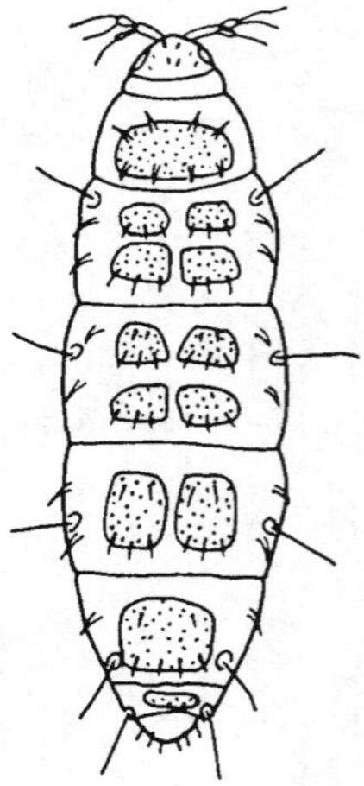

5

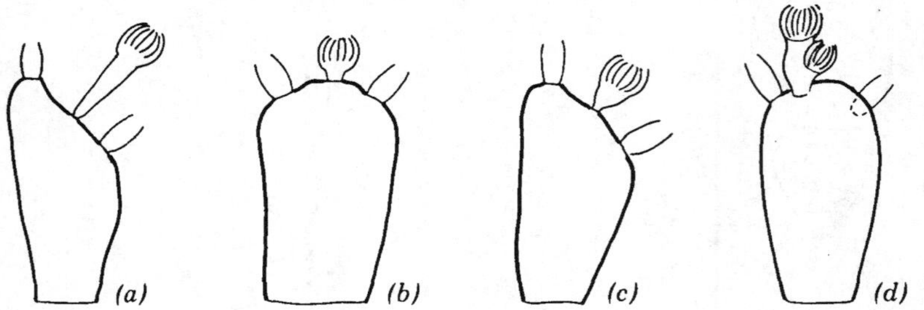

6

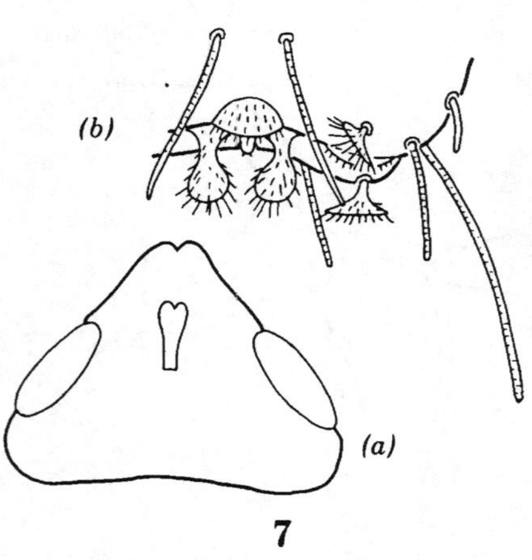

7

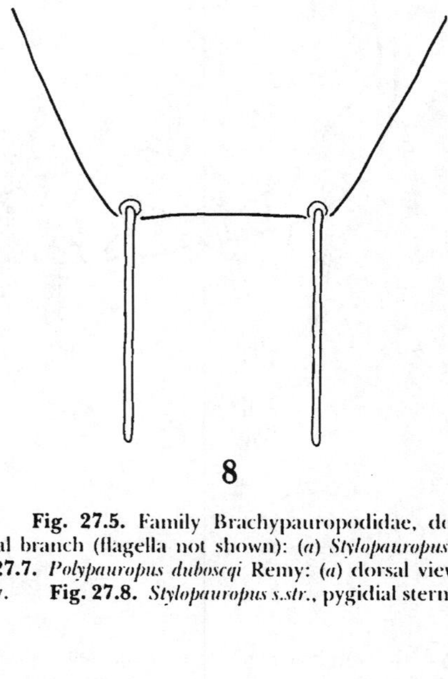

8

Fig. 27.4. Family Eurypauropodidae, dorsal view. **Fig. 27.5.** Family Brachypauropodidae, dorsal view. Legs not shown. **Fig. 27.6.** Sternal antennal branch (flagella not shown): (*a*) *Stylopauropus*; (*b*) *Pauropus;* (*c*) *Allopauropus;* (*d*) *Polypauropus*. **Fig. 27.7.** *Polypauropus duboscqi* Remy: (*a*) dorsal view of head with mediotergal plate; (*b*) pygidium, dorsal view. **Fig. 27.8.** *Stylopauropus s.str.*, pygidial sternum, ventral view (generalized).

different length; the styli, *st,* are often different and inserted on the posteroventral side, usually very near the anal plate. A fifth aberrant pair, t_1, is found in *Polypauropus* (Fig. 27.7*b*).

The sternum lies below the tergum and has a pair of posterior setae, b_1, and most often one or two pairs more of the b_2, b_3, and t_2. A most peculiar organ, the anal plate, is inserted on the posteromedian part. It is a small sclerotized plate often provided with branches and/or appendages. Each species, even the immature stages, has a plate that is characterized by its shape (form, size, cuticular structure) and, if any, its appendages (length, thickness, direction, surface structure, insertion points). The anal plate constitutes a unique system of characters with the help of which the species are most often immediately recognized.

The high distinctiveness shown by the anal plate in distinguishing species is supplemented by other good pygidial taxonomic characters easily studied together at the same time. Thus the shape and proportions of the pygidial setae are similar in all stages. However, molting specimens have to be dealt with cautiously and there are reasons to suspect a more complicated pygidial chaetotaxy during the postembryonic development in Eurypauropodidae.

Legs

Taxonomically valuable characters are the number of segments (five or six); the shape of the tarsus; the shape and length of the setae on tarsus, tibia, coxa, and trochanter; and the morphology of the apical organ of the tarsus.

KEY TO SPECIES

A key is given to the families, genera, and species known to occur in North America. The species described by Hilton are dubious and were omitted. Some doubtful records by earlier authors as well as some species from Canada and the United States recently reported by Scheller (1984, 1985, 1986, 1988), were also excluded.

1a Antennae with four-segmented stalk, scarcely telescopic; two antennal branches on fourth segment; 6 or 9–10 tergites; 9–10 leg-bearing segments; Suborder **Tetramerocerata** **2**

1b Antennae with six-segmented stalk, strongly telescopic; two antennal branches, one each on fifth and sixth segments; 12 tergites; 11 leg-bearing segments . . . . Suborder **Hexamerocerata**
Tropical, not yet known in North America.

2a First and last tergites large, covering at least head and anterior part of pygidium respectively; body flattened (Fig. 27.4); Family **Eurypauropodidae** **67**

2b First and last tergites small, leaving head and pygidium free **3**

3a Body fusiform; generally long-legged; tergites undivided, generally not sclerotized (Fig. 27.1); Family **Pauropodidae** **4**

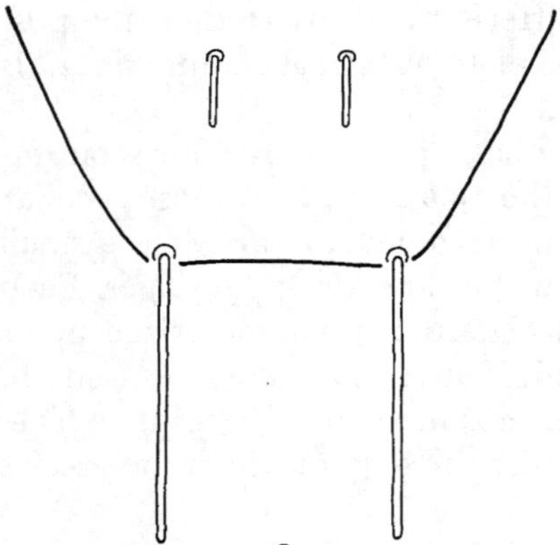

9

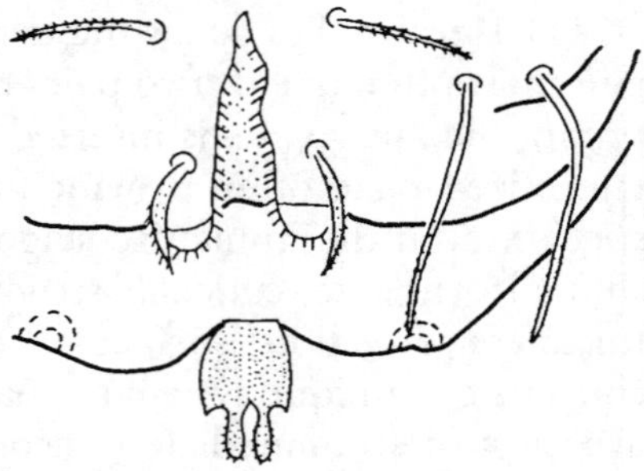

10

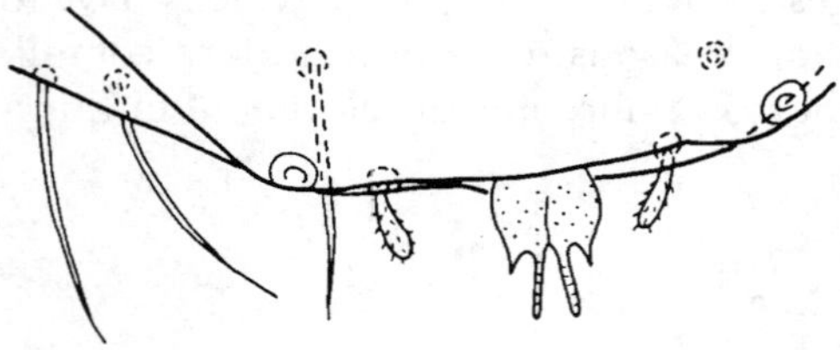

11

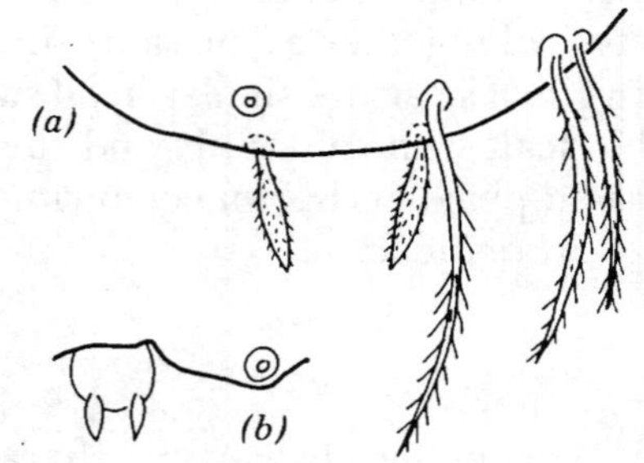

12

Fig. 27.9. *Stylopauropus* subg. *Donzelotauropus* Remy, pygidial sternum, ventral view (generalized). **Fig. 27.10.** *Stylopauropus (S.) sulcatus* Remy, pygidium, dorsal view; tergum with posteromedian furrow (above), sternum with anal plate (below). (After Remy.) **Fig. 27.11.** *Stylopauropus (S.) fratuelis* Remy, pygidium, ventral view. (After Remy.) **Fig. 27.12.** *Stylopauropus (S.) allmenae* Remy: (*a*) anal plate, ventral view; (*b*) pygidial tergum, dorsal view. (After Remy.)

3b Body oval; short-legged; tergites often divided (Fig. 27.5), generally distinctly sclerotized; Family **Brachypauropodidae** **62**

4a Sternal antennal branch with one globulus and one long seta q (Fig. 27.2); mediotergal plate of head lacking **5**

4b Sternal antennal branch with two globuli joined to a single stalk (Fig. 27.6*d*) and setae q and q'; mediotergal plate of head present (Fig. 27.7*a*) . . . . . ***Polypauropus*** Remy, 1932

One species, ***P. duboscqi*** Remy (Fig. 27.7)

5a Stalk of globulus distinctly longer than diameter of globulus (Fig. 27.6*a*); ***Stylopauropus*** Cook, 1896 . . . . **6**

5b Stalk of globulus shorter than diameter of globulus (Fig. 27.6*b,c*) **20**

6a Pygidial sternum with setae b_1 (Fig. 27.8); tergite I with 4 + 4 or 4 + 6 setae; Subgenus ***Stylopauropus* s.str.** **7**

6b Pygidial sternum with setae b_1 and b_3 (Fig. 27.9); tergite I with 4 + 4 setae; Subgenus ***Donzelotauropus*** Remy, 1957. . . . . **14**

7a Pygidial tergum cleft by a posteromedian pubescent furrow (Fig. 27.10); pygidial setae a_1 much shorter than a_2 . . . . ***Stylopauropus (Stylopauropus) sulcatus*** Remy, 1958

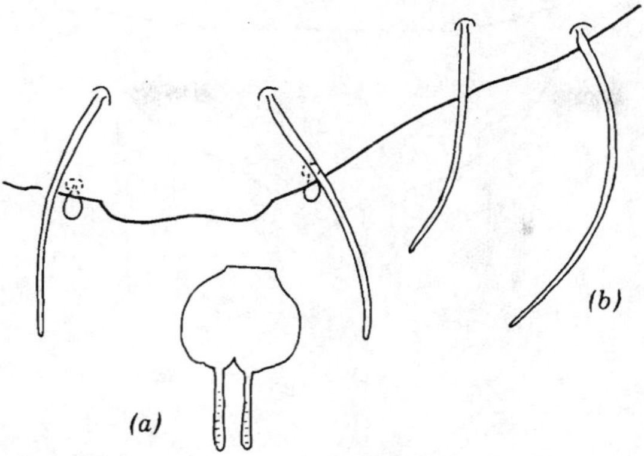

13

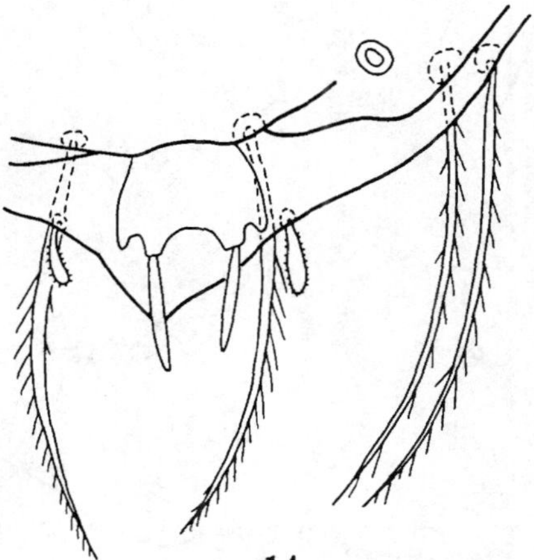

14

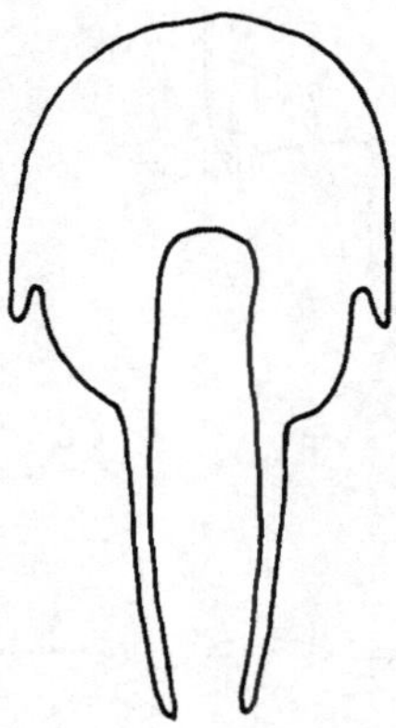

15

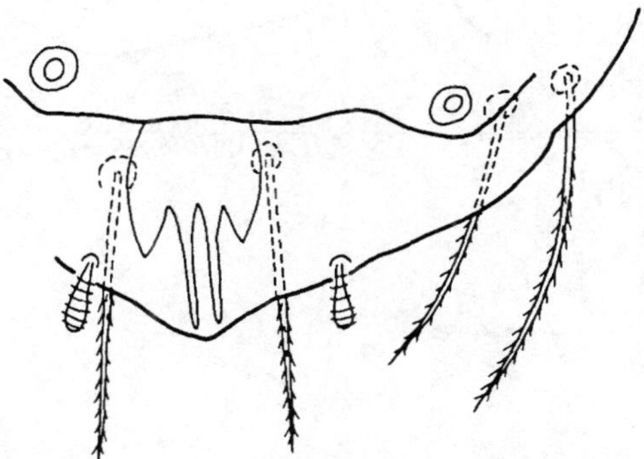

16

Fig. 27.13. *Stylopauropus (S.) dybasi* Remy: (*a*) anal plate; (*b*) pygidial tergum, dorsal view. (After Remy.) **Fig. 27.14.** *Stylopauropus (S.) senticosus* Remy, pygidium, ventral view. (After Remy.) **Fig. 27.15.** *Stylopauropus (S.) brito* Remy, anal plate. (After Chalupský.) **Fig. 27.16.** *Stylopauropus (S.) californianus* Remy, pygidium, ventral view. (After Remy.)

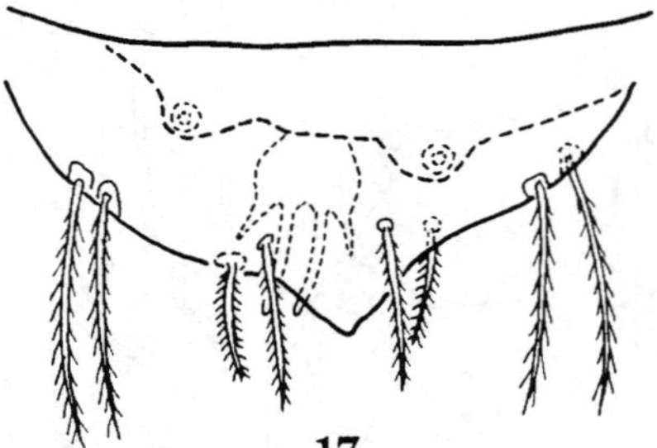

17

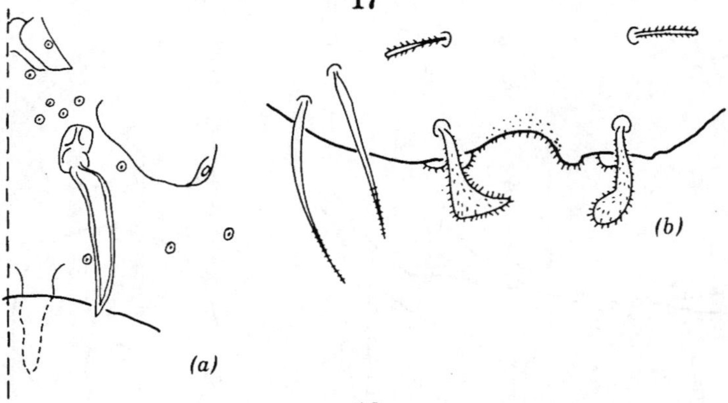

18

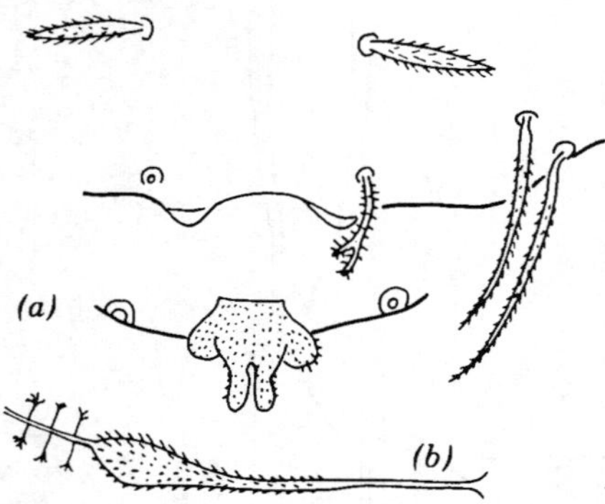

19

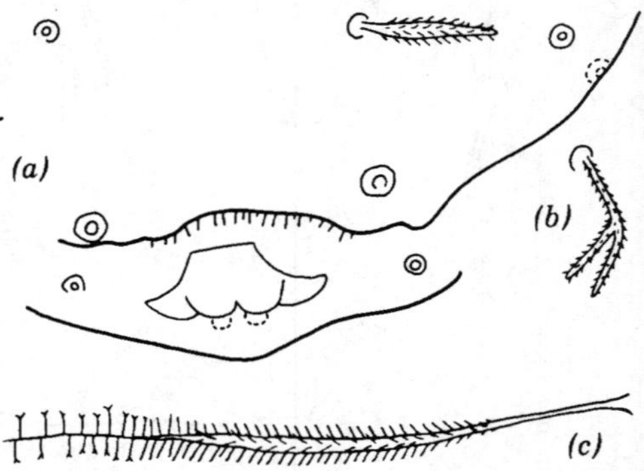

20

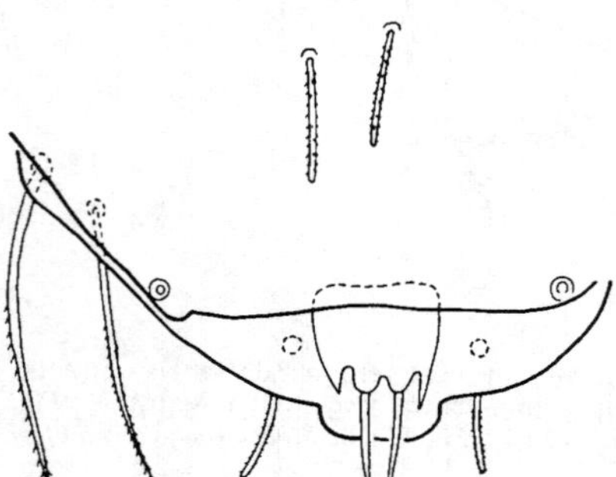

21

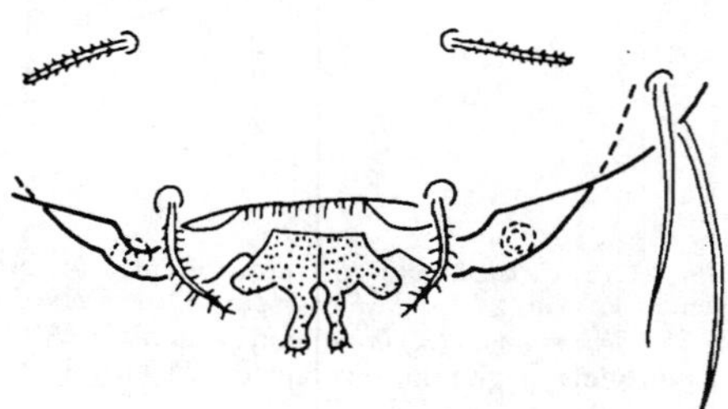

22

7b Pygidial tergum not cleft; pygidial setae a_1 about as long as or longer than a_2 . . . **8**

8a Posteromedian part of pygidial tergum straight, slightly rounded or with a low rounded bulge **9**

8b Posteromedian part of pygidial tergum with a triangular projection . . . **11**

9a Pygidial setae a_1 about as long as a_2; anal plate with distinct posterolateral corners (Fig. 27.11) . . . . ***S.(S.) fratuelis*** Remy, 1958

9b Pygidial setae a_1 distinctly longer than a_2; anal plate with rounded posterolateral corners **10**

10a Posterior margin of anal plate between appendages convex; pygidial setae a_3 shorter than a_2; *st* lanceolate (Fig. 27.12) . . . . ***S.(S.) allmenae*** Remy, 1958

10b Posterior margin of anal plate between appendages indented; pygidial setae a_3 longer than a_2; *st* short clavate (Fig. 27.13) . . ***S.(S.) dybasi*** Remy, 1956*b*

11a Anal plate with broad rounded indention between posterior appendages (Fig. 27.14) . . . . ***S.(S.) senticosus*** Remy, 1956*b*

11b Anal plate with narrow incision between posterior appendages **12**

12a Posterolateral corners of anal plate strongly reduced or lacking (Fig. 27.15) ***S.(S.) brito*** Remy, 1949

12b Posterolateral corners of anal plate protruding. **13**

13a Posterolateral corners of anal plate with triangular appendages; tergal antennal branch four to five times longer than wide (Fig. 27.16) . . . . ***S.(S.) californianus*** Remy, 1958

13b Posterolateral corners of anal plate with thin appendages (reduced in subsp. ***brevicornis***); tergal antennal branch seven times longer than wide (Fig. 27.17) . . . . ***S.(S.) pedunculatus*** (Lubbock, 1867)

14a Head with one pair of dorsal macrochaetae **15**

14b All dorsal setae on head of normal shape **17**

15a *st* club-shaped; head with a dorsal posteromedian furrow (Fig. 27.18) . . . ***Stylopauropus (Donzelotauropus) securiger*** Remy, 1958

15b *st* furcate; head without posteromedian furrow **16**

16a Proximal part of T_3 distinctly club-shaped (Fig. 27.19) . . . . ***S.(D.) gladiator*** Remy, 1958

16b Proximal part of T_3 slowly narrowing outward (Fig. 27.20) . . . . ***S.(D.) machaerophorus*** Remy, 1958

17a *st* thin cylindrical **18**

17b *st* clavate **19**

18a Anal plate with thin pointed appendages (Fig. 27.21) . . . . ***S.(D.) quadrisulcus*** Remy, 1958

18b Anal plate with broad lateral lobes and two thick irregular posterior appendages (Fig. 27.22) . . . . ***S.(D.) strigilarius*** Remy, 1958

Fig. 27.17. *Stylopauropus (S.) pedunculatus* (Lubbock), pygidium, ventral view. **Fig. 27.18.** *Stylopauropus (D.) securiger* Remy: (*a*) head, right side, dorsal view; (*b*) pygidial tergum, dorsal view. (After Remy.) **Fig. 27.19.** *Stylopauropus (D.) gladiator* Remy: (*a*) pygidium, dorsal view; (*b*) T_3, proximal part. (After Remy.) **Fig. 27.20.** *Stylopauropus (D.) machaerophorus* Remy: (*a*) pygidium, dorsal view; (*b*) *st.*; (*c*) T_3, proximal and middle part. (After Remy.) **Fig. 27.21.** *Stylopauropus (D.) quadrisulcus* Remy, pygidium, ventral view. (After Remy.) **Fig. 27.22.** *Stylopauropus (D.) strigilarius* Remy, pygidium, dorsal view. (After Remy.)

19a Anal plate appendages subcylindrical, blunt; posteromedian incision U-shaped (Fig. 27.23) . . . . ***S.(D.) hystriculus*** Remy, 1958

19b Anal plate appendages tapering, pointed; posteromedian incision V-shaped (Fig. 27.24) . . . . ***S.(D.) schusteri*** Remy, 1959

20a Tergites thin **21**

20b Tergites sclerotized . . . . ***Scleropauropus*** Silvestri, 1902 **61**

21a Anterior and posterior margins of sternal antennal branch of equal length (Fig. 27.6*b*). ***Pauropus*** Lubbock, 1867 **22**

21b Anterior margin of sternal antennal branch shorter than posterior margin (Fig. 27.6*c*) . . . . ***Allopauropus*** Silvestri, 1902 **26**

22a Submedian branches of anal plate triangular (Fig. 27.25) . . . . ***P. lanceolatus*** Remy, 1956

22b Submedian branches of anal plate subcylindrical, cut **23**

23a Pygidial setae a_3 longer than a_2; posterodistal appendages of anal plate cylindrical (Fig. 27.26) . . . . ***P. valens*** Remy, 1956*b*

23b Pygidial setae a_3 shorter than a_2; posterodistal appendages of anal plate lanceolate or clavate **24**

24a *st* with thickened base; distal part of pygidial setae b_3 slowly tapering (Fig. 27.27) . . . . ***P. occidentalis*** Remy, 1956*b*

24b *st* subcylindrical; distal part of pygidial setae b_3 thickened **25**

25a Anal plate with broad and deep incision between posteromedian branches; antennal branch *t* 2.5 times longer than wide (Fig. 27.28). . . . . ***P. amoenifer*** Remy, 1958

25b Anal plate with narrow and ground incision between posteromedian branches; antennal branch *t* at least five times longer than wide (Fig. 27.29) ***P. collignoni*** Remy, 1956*a*

26a Three pairs of setae on pygidial sternum: b_1, b_2, and b_3 (Fig. 27.30) . . . . Subgenus ***Allopauropus* s.str. 27**

26b Two pairs of setae on pygidial sternum: b_1 and b_2 (Fig. 27.31) . . . . Subgenus ***Decapauropus*** Remy, 1957 **31**

27a *st* thin, tapering, pointed; pygidial setae a_3 shorter than a_1 and a_2 (Fig. 27.32) ***Allopauropus (Allopauropus) crucifer*** Remy, 1958

27b *st* clavate or bladder-shaped; pygidial setae a_3 longer than a_1 and a_2 . . . **28**

28a *st* clavate; posteromedian margin of anal plate straight or convex with two short submedian appendages. **29**

28b *st* bladder-shaped; anal plate with ventral posteromedian indentation and four distal appendages: two long submedian and two short posterolateral **30**

29a Pygidial setae a_1 about one-fourth of length of a_2; globulus lengthened (Fig. 27.33). . . . . ***A.(A.) cantralli*** Remy, 1958

29b Pygidial setae a_1 a little shorter than a_2; globulus short (Fig. 27.34) . . . . . ***A.(A.) danicus*** (Hansen, 1902)

30a Pygidial setae a_1 only slightly longer than a_2; antennal globulus almost unstalked (Fig. 27.35). . . . . ***A.(A.) carolinensis*** (Starling, 1943)

30b Pygidial setae a_1 much longer than a_2; antennal globulus with distinct stalk (Fig. 27.36) . . . . ***A.(A.) brevisetus*** Silvestri, 1902

31a Trichobothria simple **32**

31b Trichobothria ramified (Fig. 27.37) . . . . ***Allopauropus (Decapauropus) bouini*** Remy, 1955

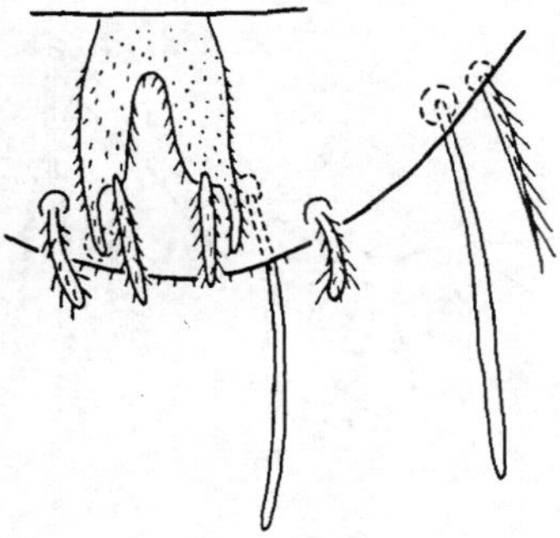

23

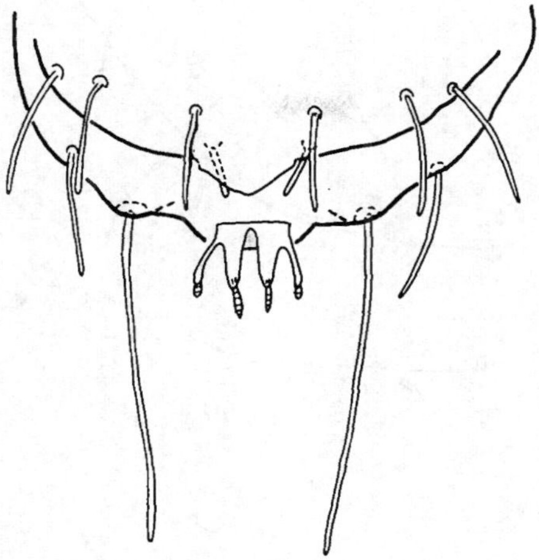

25

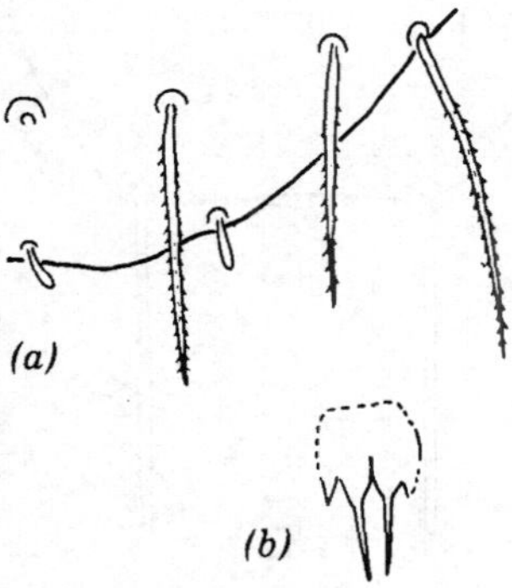

24

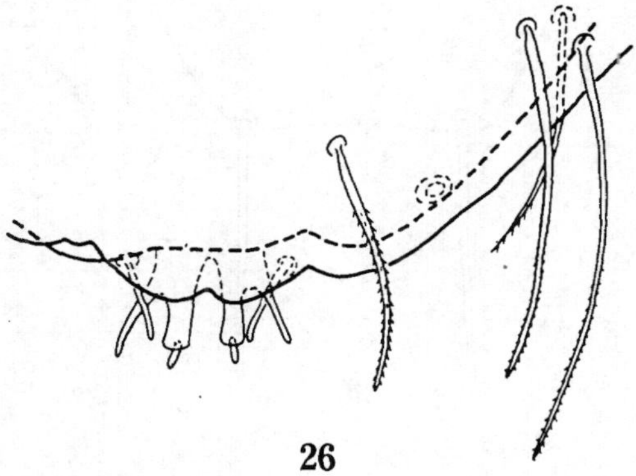

26

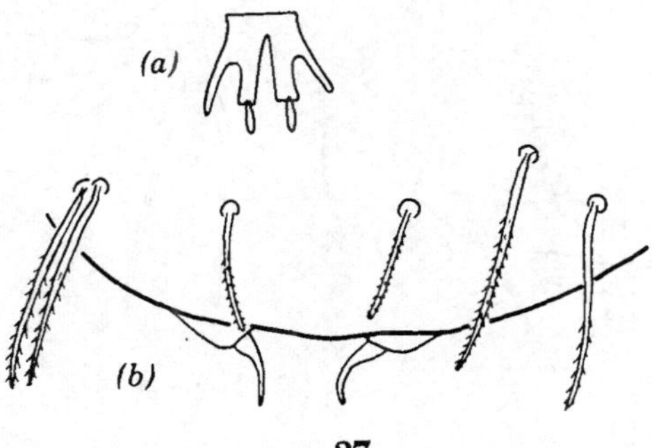

27

Fig. 27.23. *Stylopauropus (D.) hystriculus* Remy, pygidial sternum, ventral view. (After Remy.) **Fig. 27.24.** *Stylopauropus (D.) schusteri* Remy: (*a*) pygidial tergum, dorsal view; (*b*) anal plate. (After Remy). **Fig. 27.25.** *Pauropus lanceolatus* Remy, pygidial tergum, ventral view. **Fig. 27.26.** *Pauropus valens* Remy, pygidial tergum, dorsal view. (After Remy.) **Fig. 27.27.** *Pauropus occidentalis* Remy: (*a*) anal plate; (*b*) pygidial tergum, dorsal view. (After Remy.)

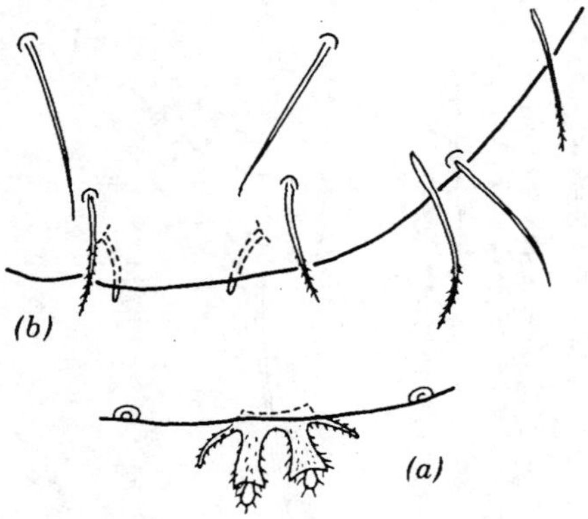

28

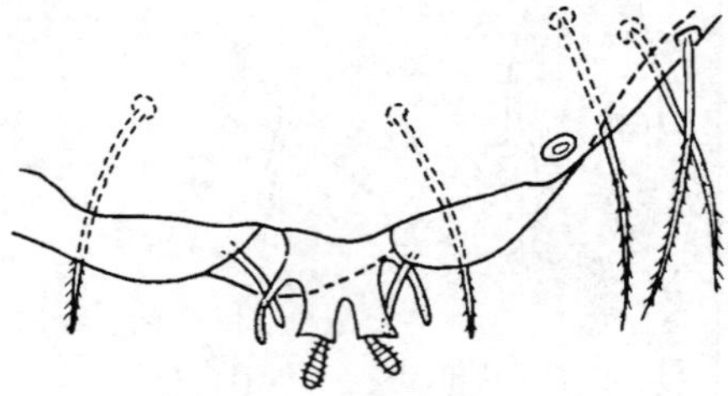

29

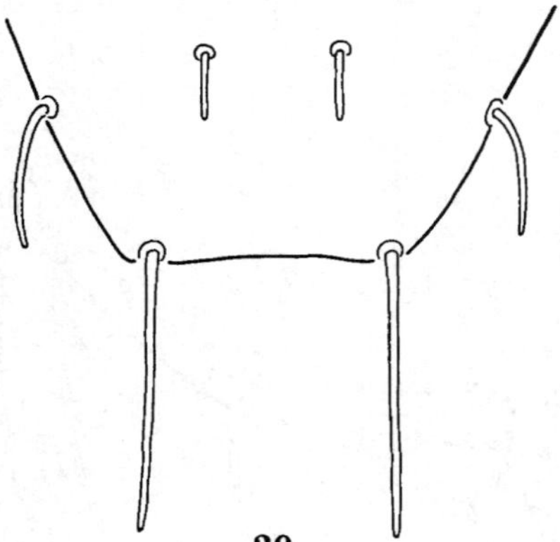

30

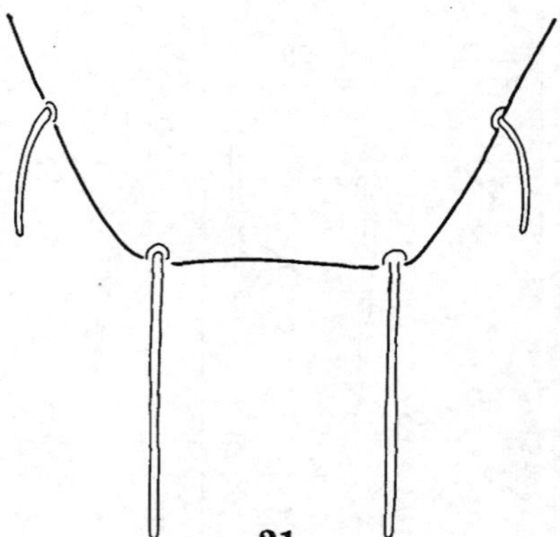

31

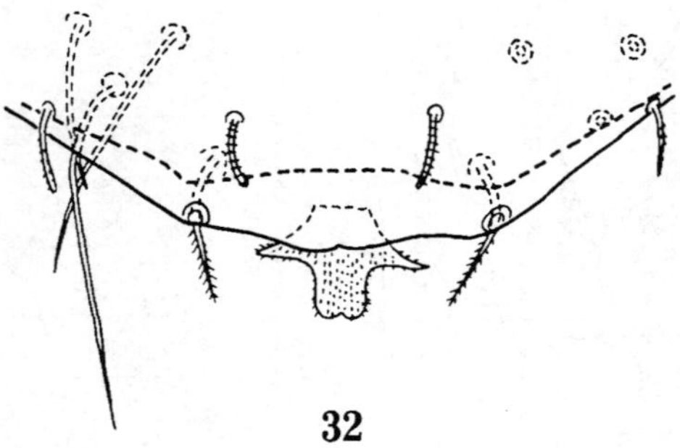

32

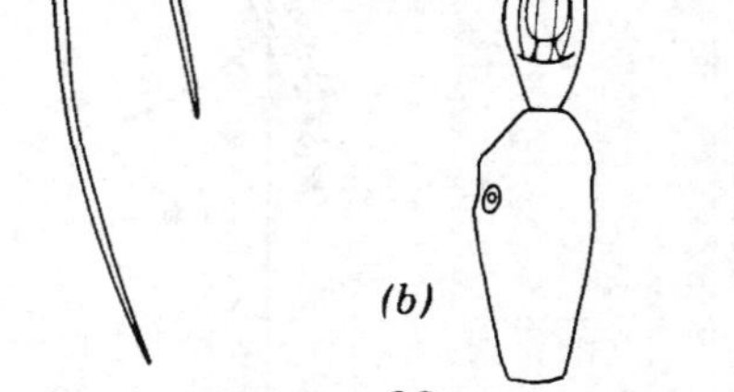

33

Fig. 27.28. *Pauropus amoenifer* Remy: (*a*) anal plate; (*b*) pygidial tergum, dorsal view. (After Remy.) **Fig. 27.29.** *Pauropus collignoni* Remy, pygidium, ventral view. (After Remy.) **Fig. 27.30.** *Allopauropus s.str.*, pygidial sternum, ventral view (generalized). **Fig. 27.31.** *Allopauropus* subg. *Decapauropus* Remy, pygidial sternum, ventral view (generalized). **Fig. 27.32.** *Allopauropus (A.) crucifer* Remy, pygidial tergum, ventral view. (After Remy.) **Fig. 27.33.** *Allopauropus (A.) cantralli* Remy: (*a*) pygidium, ventral view; (*b*) sternal antennal branch. (After Remy.)

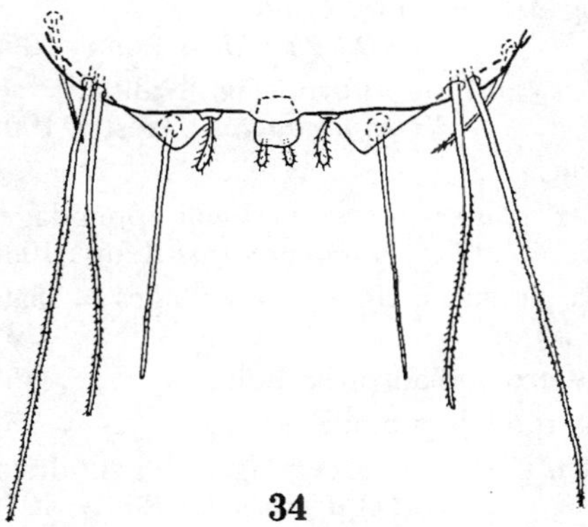

34

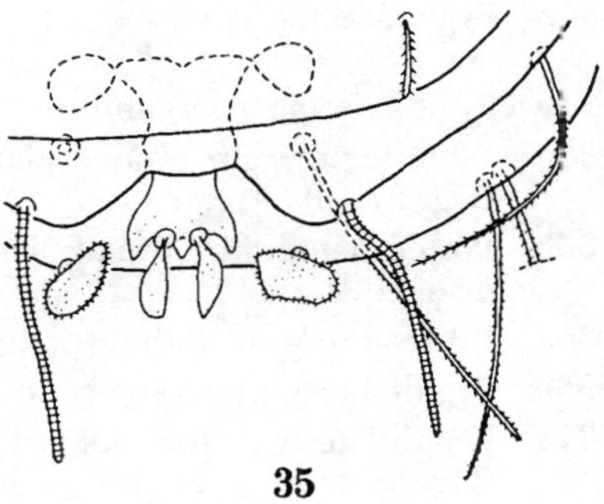

35

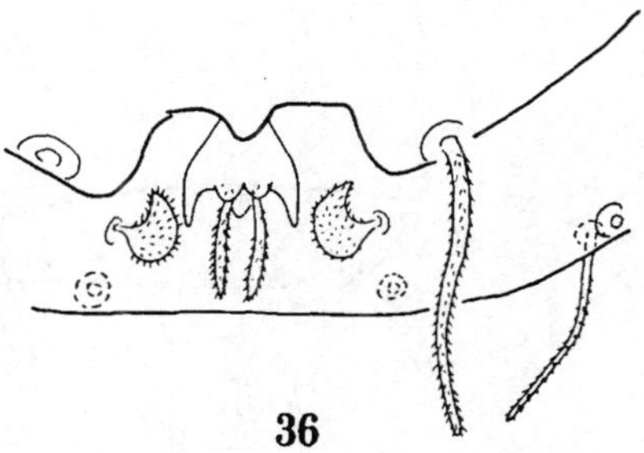

36

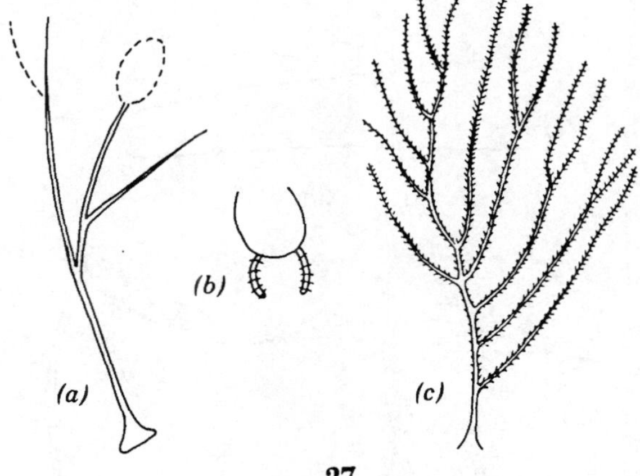

37

Fig. 27.34. *Allopauropus (A.) danicus* (Hansen), pygidium, tergal view. **Fig. 27.35.** *Allopauropus (A.) carolinensis* Starling, pygidium, ventral view. **Fig. 27.36.** *Allopauropus (A.) brevisetus* Silvestri, pygidium, ventral view. **Fig. 27.37.** *Allopauropus (D.) bouini* Remy: (*a*) T_3; (*b*) anal plate; (*c*) T_1. (After Remy.)

32a T_3 with end-swelling (Figs. 27.39*b*, 27.41*b*, 27.42*c*, 27.43*b*, 27.44*a*) **33**
32b T_3 without end-swelling (Fig. 27.1) **39**
33a Distal seta on tarsus of last pair of legs simple **34**
33b Distal seta on tarsus of last pair of legs furcate (Fig. 27.38) . . . . ***A.(D.) gaetulicus*** Remy, 1952
34a Pygidial setae a_1 very short, globular, a_2 and a_3 setiform (Fig. 27.39) . . . . ***A.(D.) productus*** Silvestri, 1902
34b Pygidial setae a_1, a_2 and a_3 all setiform **35**
35a Posterior margin of anal plate convex, plate with two vestigial appendages (Fig. 27.40) . . . . ***A.(D.) excornis*** Remy, 1958
35b Posterior margin of anal plate with a median incision, appendages of plate distinct **36**
36a Pygidial tergum with a triangular posteromedian projection **37**
36b Pygidial tergum with a rounded posteromedian bulge **38**
37a Anal plate with four long, cylindrical, posterior appendages; *st* cylindrical (Fig. 27.41) . . . . ***A.(D.) aegyptiacus*** Remy, 1950

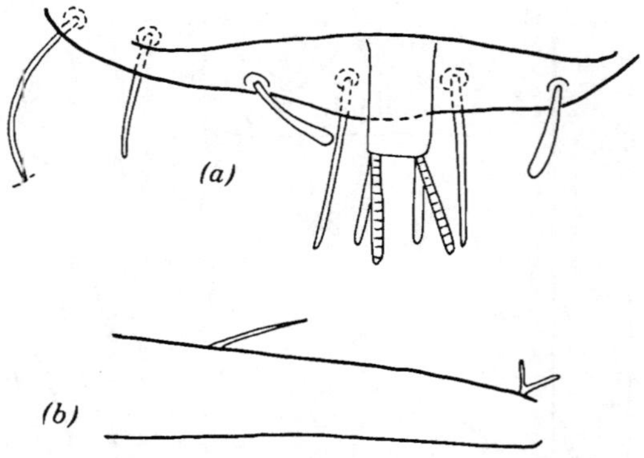

38

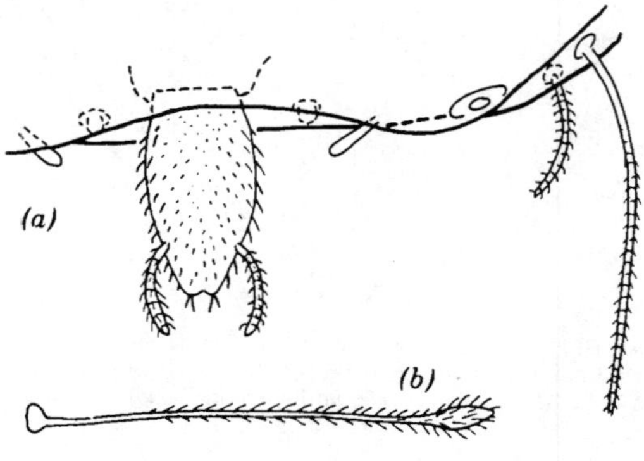

39

Fig. 27.38. *Allopauropus (D.) gaetulicus* Remy: (*a*) pygidial tergum, ventral view (after Remy); (*b*) tarsus of last pair of legs (rough drawing after Remy's text.) **Fig. 27.39.** *Allopauropus (D.) productus* Silvestri; (*a*) pygidium, ventral view, sternal setae not shown; (*b*) T_3.

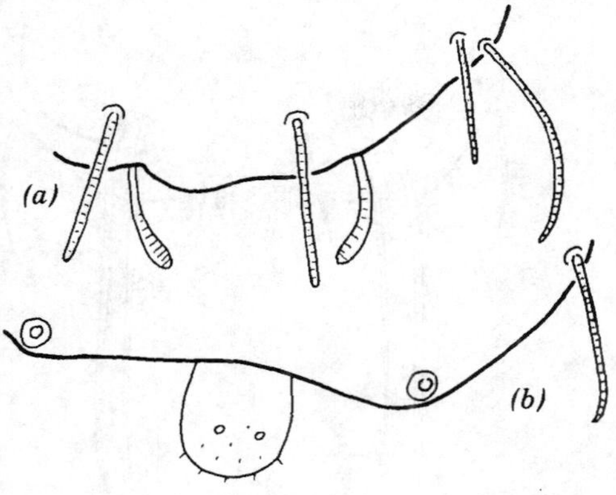

40

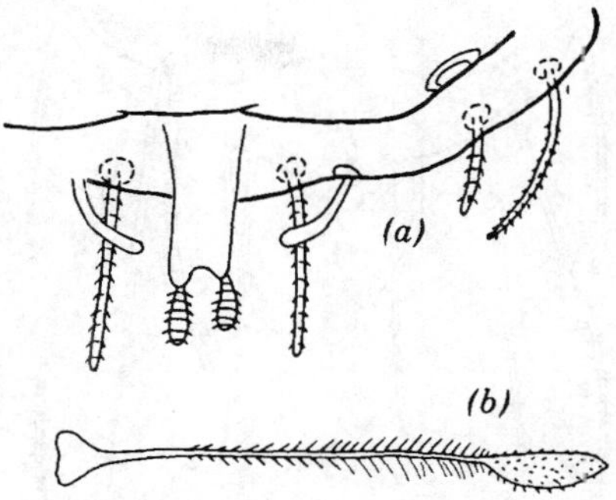

43

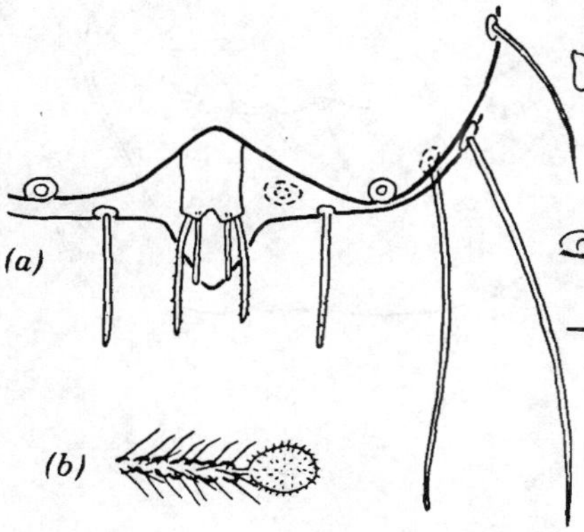

41

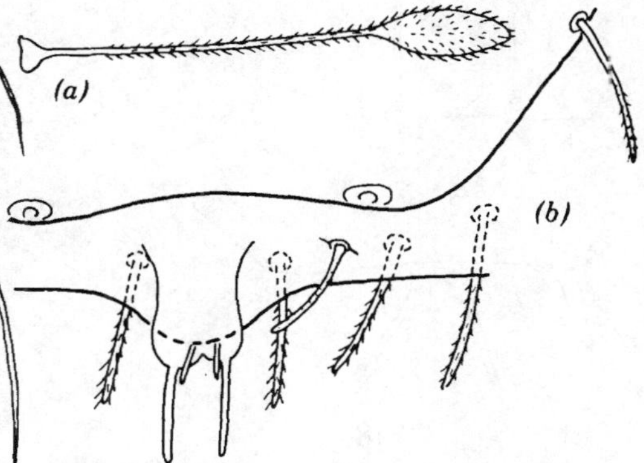

44

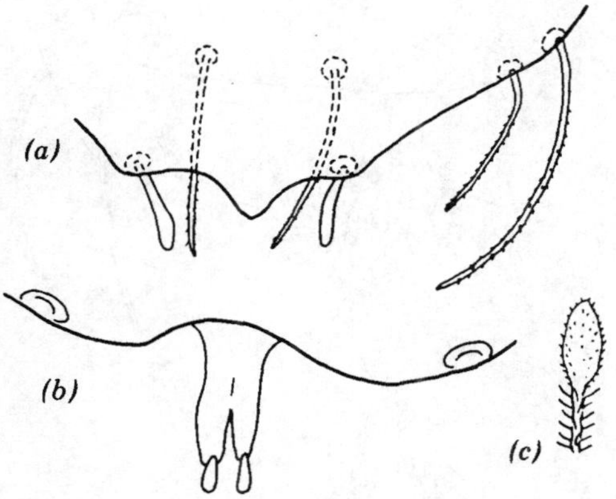

42

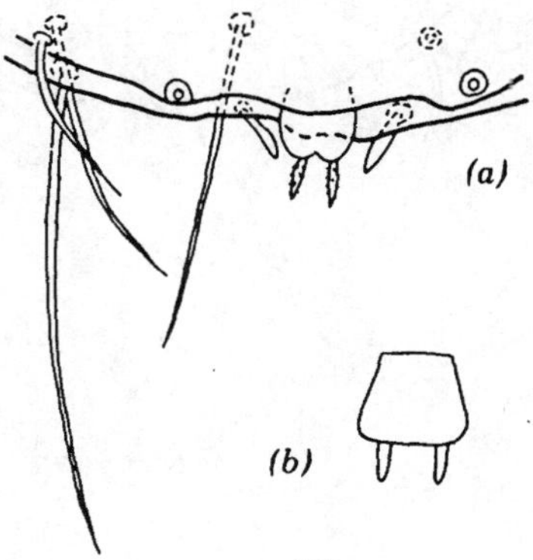

45

Fig. 27.40. *Allopauropus (D.) excornis* Remy: (*a*) pygidial tergum, dorsal view; (*b*) pygidial sternum, ventral view. (After Remy.) **Fig. 27.41.** *Allopauropus (D.) aegyptiacus* Remy: (*a*) pygidium, ventral view; (*b*) T_3, distal part. **Fig. 27.42.** *Allopauropus (D.) barroisi* Remy: (*a*) pygidial tergum, ventral view; (*b*) pygidial sternum, distal part; (*c*) T_3, distal part. (After Remy.) **Fig. 27.43.** *Allopauropus (D.) milloti* Remy: (*a*) pygidium, ventral view; (*b*) T_3. **Fig. 27.44.** *Allopauropus (D.) proximus* Remy: (*a*) pygidium, ventral view; (*b*) T_3. (After Remy.) **Fig. 27.45.** (*a*) *Allopauropus (D.) causeyae* Starling, anal plate (after Starling.) (*b*) *Allopauropus (D.) dybasi* Remy, pygidium, ventral view (after Remy).

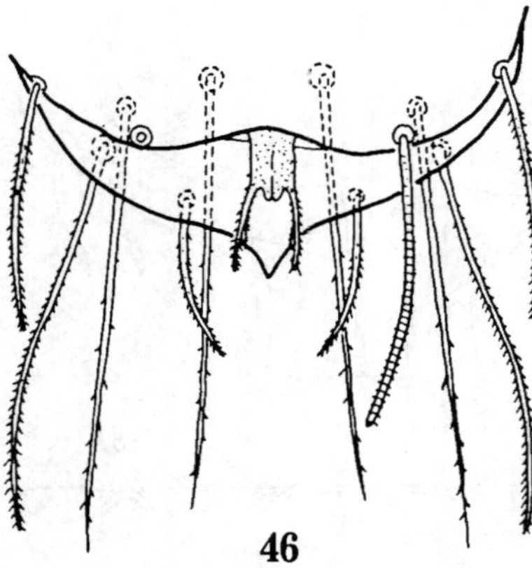

46

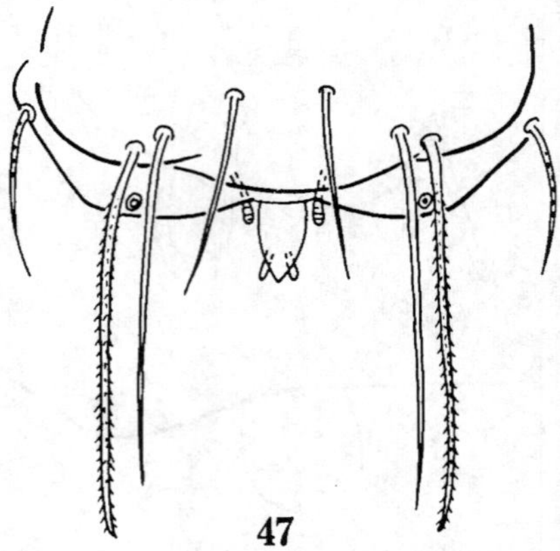

47

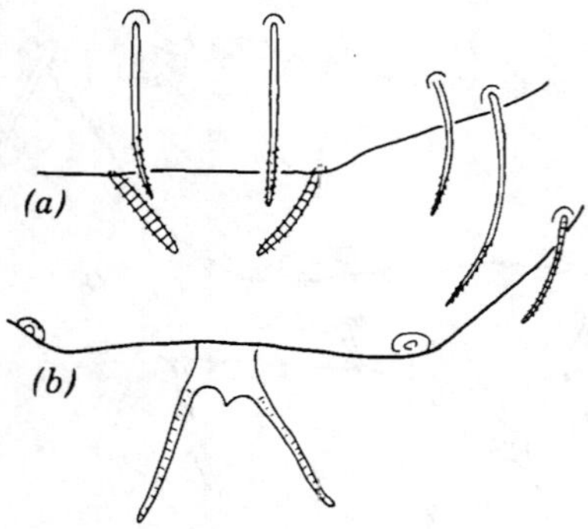

48

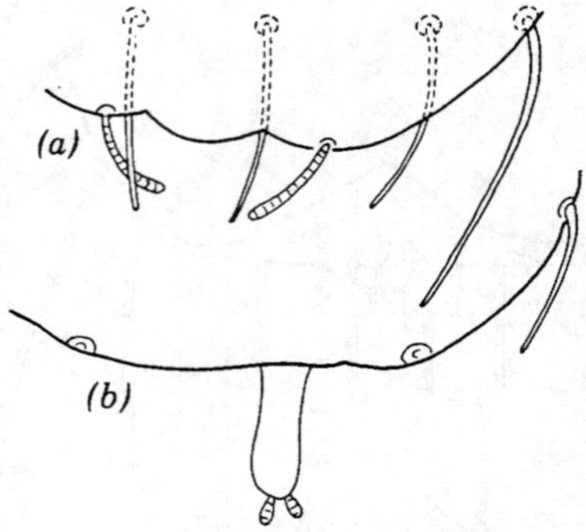

49

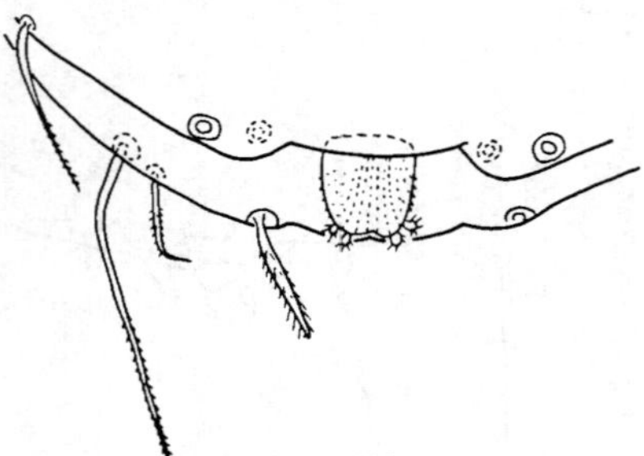

50

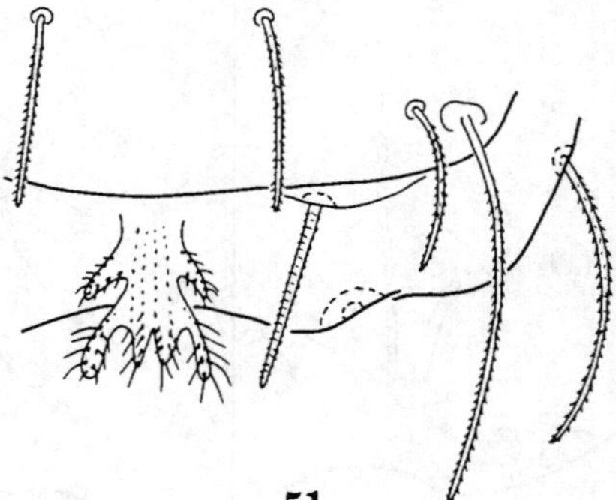

51

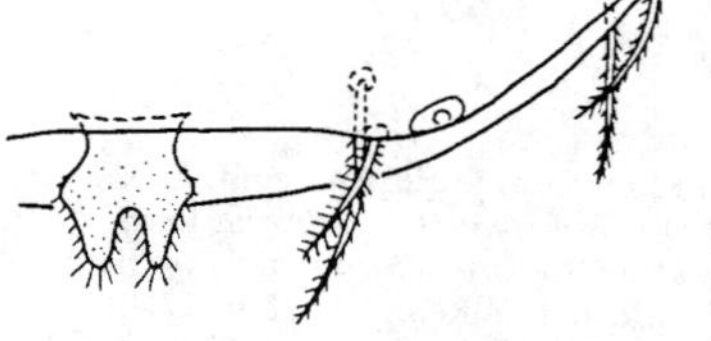

52

37b Anal plate with two short, clavate, posterior appendages; *st* clavate (Fig. 27.42) . . . . ***A.(D.) barroisi*** Remy, 1956
38a Pygidial setae a_2 much shorter than a_3; *st* clavate (Fig. 27.43) . . . . ***A.(D.) milloti*** Remy, 1948
38b Pygidial setae a_2 and a_3 subequal in length; *st* cylindrical (Fig. 27.44) . . . . ***A.(D.) proximus*** Remy, 1948
39a Posterior margin of anal plate straight (Fig. 27.45*b*). . . . . ***A.(D.) causeyae*** Starling, 1943
39b Posterior margin of anal plate with a posteromedian indentation or projection. **40**
40a Pygidial setae a_2 longer than a_3. . . . . **41**
40b Pygidial setae a_2 as long as or shorter than a_3 . . . . **42**
41a *st* clavate; anal plate with two fusiform posterior appendages (Fig. 27.45*a*) ***A.(D.) dybasi*** Remy, 1957
41b *st* subcylindrical, thin, pointed; anal plate with two cylindrical posterolateral appendages (Fig. 27.46) . . . . ***A.(D.) ribauti*** Remy, 1937
42a Posterior margin of anal plate convex or with a median projection . . . **43**
42b Posterior margin of anal plate with a median indentation **45**
43a *st* clavate; T_5 with long stiff pubescence hairs (Fig. 27.47) . . . . ***A.(D.) aristatus*** Remy, 1936
43b *st* subcylindrical; T_5 with short pubescence **44**
44a Appendages of anal plate twice longer than plate, cylindrical (Fig. 27.48). . ***A.(D.) mucronatus*** Remy, 1958
44b Appendages of anal plate about one-third of length of plate, clavate (Fig. 27.49) . . . . ***A.(D.) usingeri*** Remy, 1938
45a Anal plate strongly pubescent **46**
45b Anal plate glabrous or slightly granular **49**
46a *st* lanceolate (Fig. 27.50). . . . . ***A.(D.) intonsus*** Remy, 1956
46b *st* thin, tapering, pointed **47**
47a Anal plate with two lateral appendages on each side (Fig. 27.51) . . . . . . ***A.(D.) cuenoti*** (Remy, 1931)
47b Anal plate posteriorly bipartite or with two posterolateral appendages . . . **48**
48a Anal plate bipartite; posteromedian indentation of plate U-shaped, distinct (Fig. 27.52) . . . . ***A.(D.) distinctus*** Bagnall, 1936
48b Anal plate with two clavate posterolateral appendages; posteromedian indentation very small (Fig. 27.53) . . . . ***A.(D.) leconteorum*** Remy, 1958
49a *st* clavate **50**
49b *st* subcylindrical or very slightly thickened distally **51**

Fig. 27.46. *Allopauropus (D.) ribauti* Remy, pygidium, ventral view. (After Remy.) **Fig. 27.47.** *Allopauropus (D.) aristatus* Remy, pygidium, dorsal view. **Fig. 27.48.** *Allopauropus (D.) mucronatus* Remy: (*a*) pygidial tergum, dorsal view; (*b*) pygidial sternum, ventral view. (After Remy.) **Fig. 27.49.** *Allopauropus (D.) usingeri* Remy; (*a*) pygidial tergum, ventral view; (*b*) pygidial sternum, ventral view. (After Remy). **Fig. 27.50.** *Allopauropus (D.) intonsus* Remy, pygidium, ventral view. (After Remy). **Fig. 27.51.** *Allopauropus (D.) cuenoti* (Remy), pygidium, dorsal view. **Fig. 27.52.** *Allopauropus (D.) distinctus* Bagnall, pygidium, ventral view.

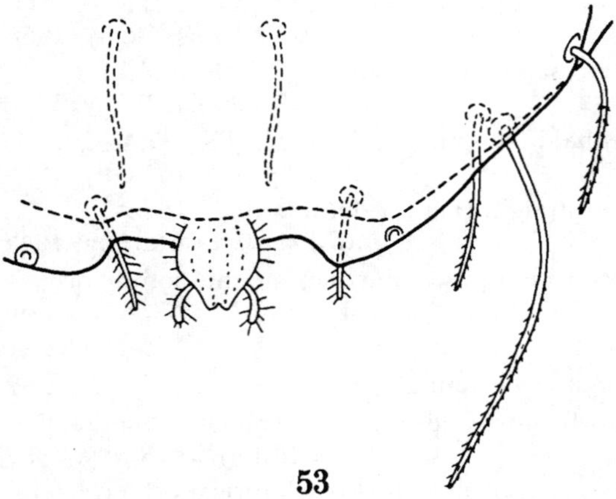

53

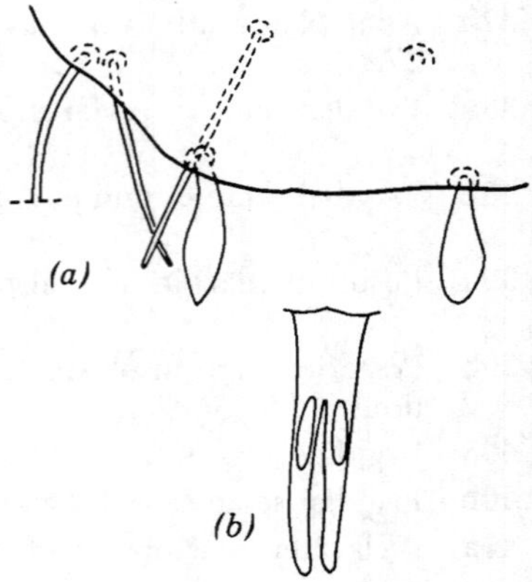

54

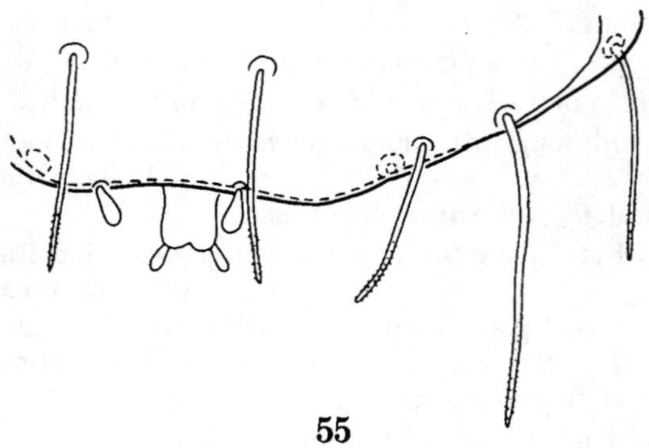

55

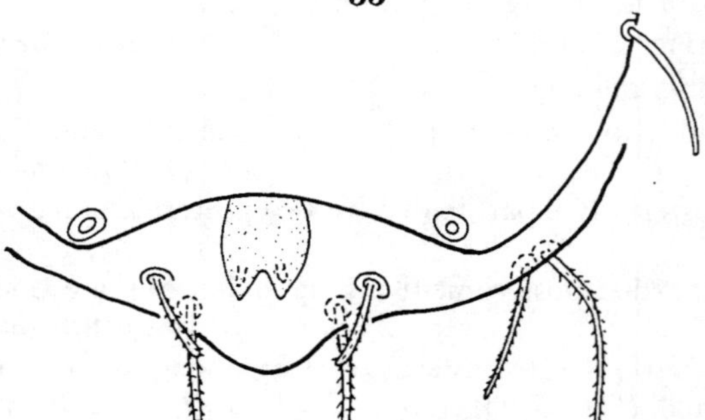

56

Fig. 27.53. *Allopauropus (D.) leconteorum* Remy, pygidium, ventral view. (After Remy.) **Fig. 27.54.** *Allopauropus (D.) pulcher* Remy: (*a*) pygidial tergum, ventral view; (*b*) anal plate, ventral view. (After Remy.) **Fig. 27.55.** *Allopauropus (D.) linsleyi* Remy, pygidium, dorsal view. (After Remy.) **Fig. 27.56.** *Allopauropus (D.) helveticus* (Hansen), pygidium, ventral view.

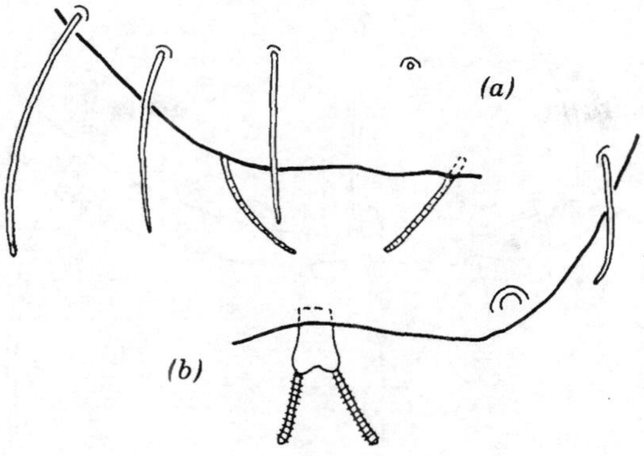

57

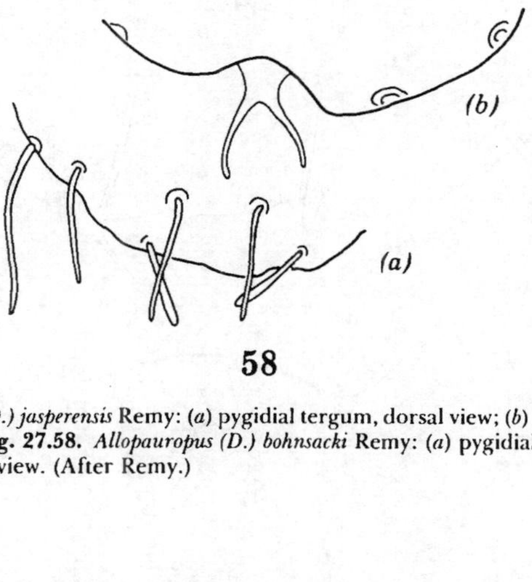

58

Fig. 27.57. *Allopauropus (D.) jasperensis* Remy: (*a*) pygidial tergum, dorsal view; (*b*) pygidial sternum, ventral view. (After Remy.) **Fig. 27.58.** *Allopauropus (D.) bohnsacki* Remy: (*a*) pygidial tergum, dorsal view; (*b*) pygidial sternum, ventral view. (After Remy.)

50a Anal plate narrow with two middle and two posterior clavate appendages (Fig. 27.54) . . . . ***A.(D.) pulcher*** Remy, 1956

50b Anal plate short with two posterolateral appendages (Fig. 27.55) . . . . . . ***A.(D.) linsleyi*** Remy, 1958

51a F_2 almost as long as F_3 . . . . **52**

51b F_2 about half as long as F_3 . . . . **54**

52a Anal plate convex laterally, posterior part cleft into two triangular lobes each with a short tergal appendage (Fig. 27.56) . ***A.(D.) helveticus*** (Hansen, 1902)

52b Anal plate concave laterally, posterior appendages longer than plate **53**

53a Anal plate with two low rounded posterior lobes with two diverging appendages; pygidial setae a_2 and a_3 subequal (Fig. 27.57) . . . . ***A.(D.) jasperensis*** Remy, 1957

53b Anal plate posteriorly forming a broad V; pygidial setae a_2 much shorter than a_3 (Fig. 27.58) . . . . ***A.(D.) bohnsacki*** Remy, 1957

54a Pygidial tergum with a distinct posteromedian lobe between *st* (Fig. 27.59) ***A.(D.) lobiger*** Remy, 1956

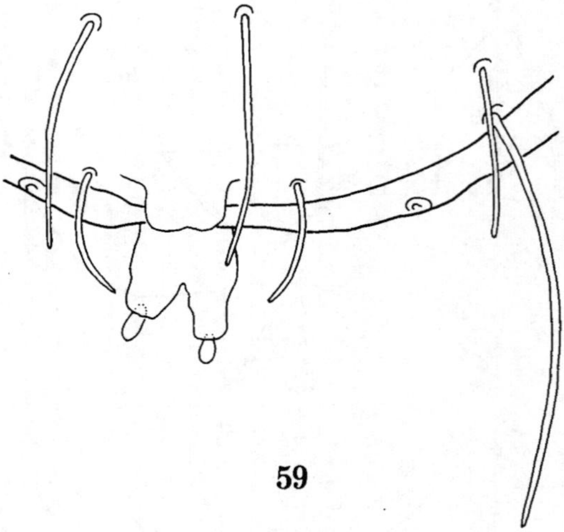

59

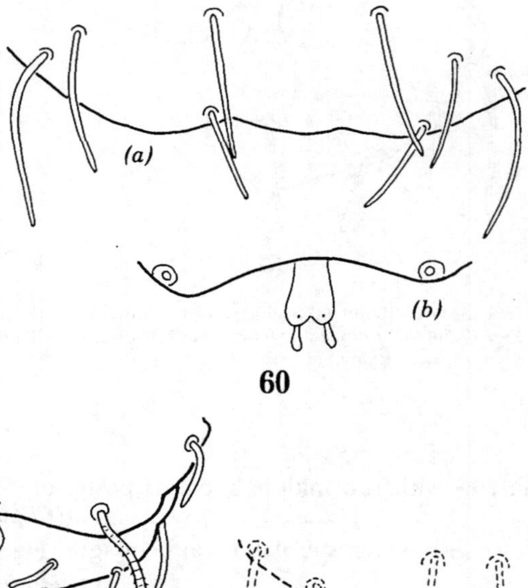

60

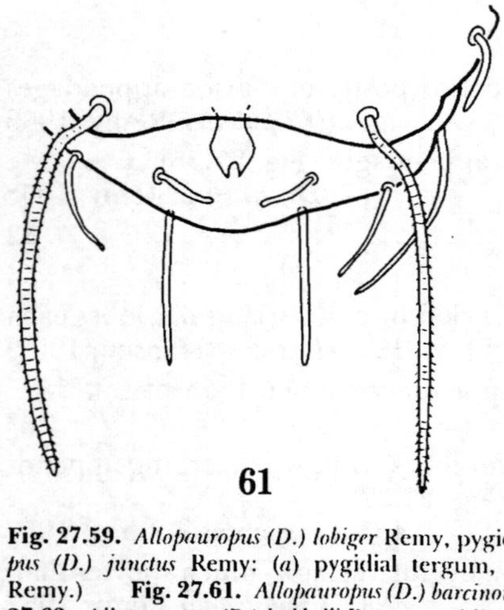

61

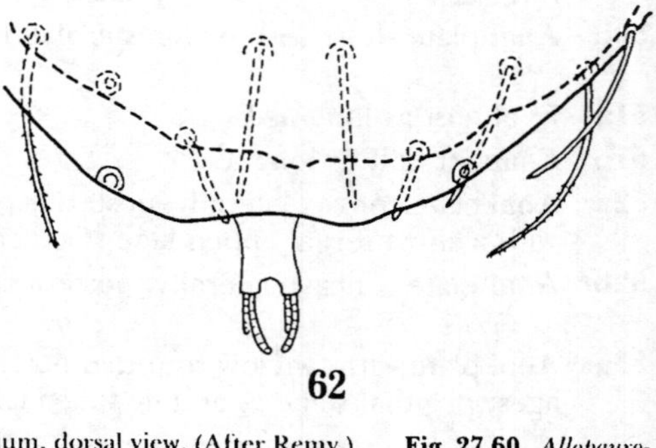

62

Fig. 27.59. *Allopauropus (D.) lobiger* Remy, pygidium, dorsal view. (After Remy.) **Fig. 27.60.** *Allopauropus (D.) junctus* Remy: (*a*) pygidial tergum, dorsal view; (*b*) pygidial sternum, ventral view. (After Remy.) **Fig. 27.61.** *Allopauropus (D.) barcinonensis* Remy, pygidium, ventral view. (After Remy.) **Fig. 27.62.** *Allopauropus (D.) hubbelli* Remy, pygidium, ventral view. (After Remy.)

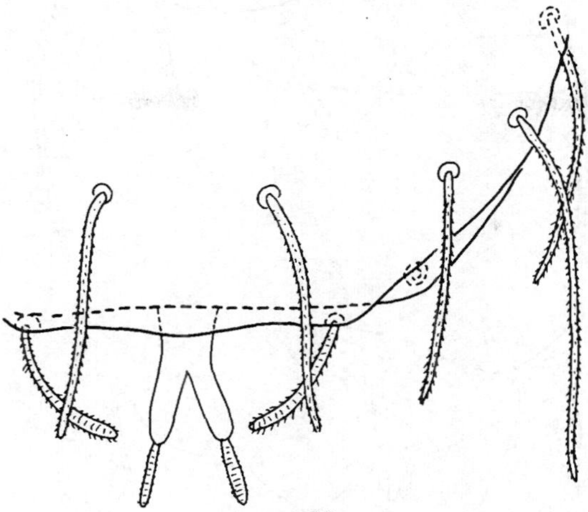

63

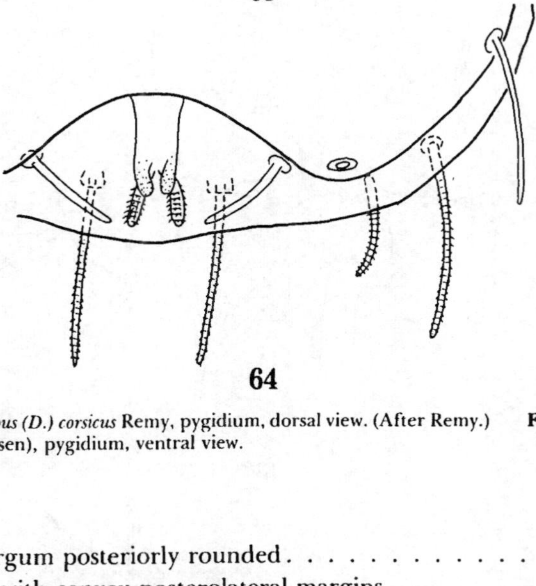

64

Fig. 27.63. *Allopauropus (D.) corsicus* Remy, pygidium, dorsal view. (After Remy.) **Fig. 27.64.** *Allopauropus (D.) vulgaris* (Hansen), pygidium, ventral view.

54b Pygidial tergum posteriorly rounded **55**

55a Anal plate with convex posterolateral margins **56**

55b Anal plate with concave lateral margins **57**

56a Anal plate with two short posterolateral rounded lobes and two clavate posterior appendages (Fig. 27.60) . . . . ***A.(D.) junctus*** Remy, 1956

56b Anal plate with two short posterior pointed projections; no appendages (Fig. 27.61) . . . . ***A.(D.) barcinonensis*** Remy, 1933

57a Distal end of T_3 subcylindrical (Fig. 27.62) . . . ***A.(D.) hubbelli*** Remy, 1956

57b Distal end of T_3 tapering **58**

58a Anal plate much longer than broad with deep posteromedian indentation **59**

58b Anal plate about as long as broad with ground posteromedian indentation **60**

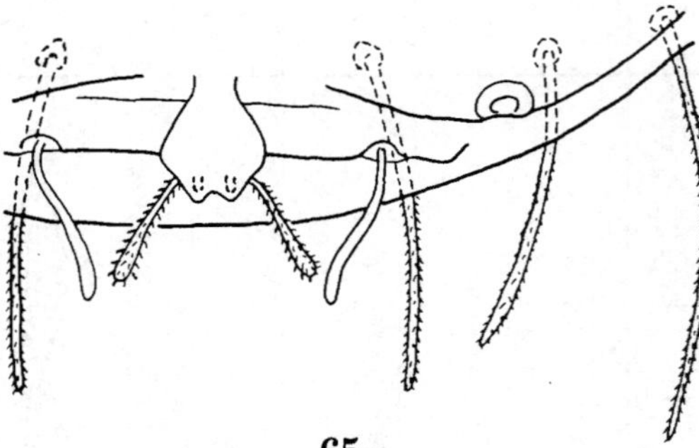

65

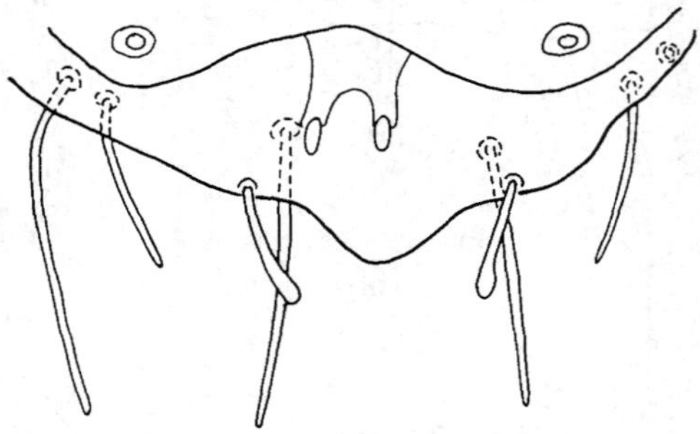

66

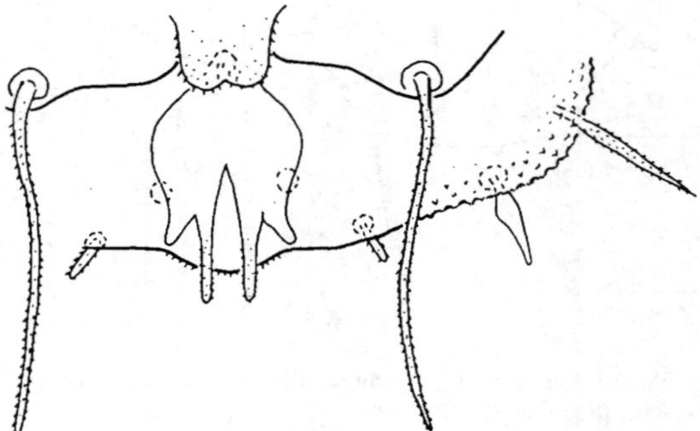

67

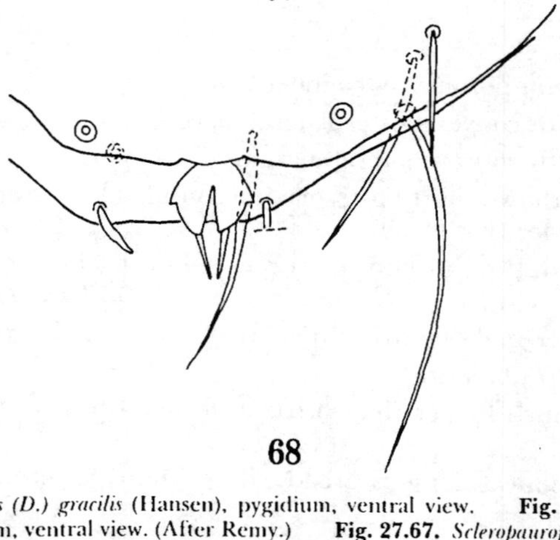

68

Fig. 27.65. *Allopauropus (D.) gracilis* (Hansen), pygidium, ventral view. **Fig. 27.66.** *Allopauropus (D.) cognatus* Remy, pygidium, ventral view. (After Remy.) **Fig. 27.67.** *Scleropauropus (Scleropauropus) lyrifer* Remy, pygidium, ventral view. **Fig. 27.68.** *Scleropauropus (Scleropauropoides) dukensis* (Starling), subad., pygidium, ventral view. (After Remy.)

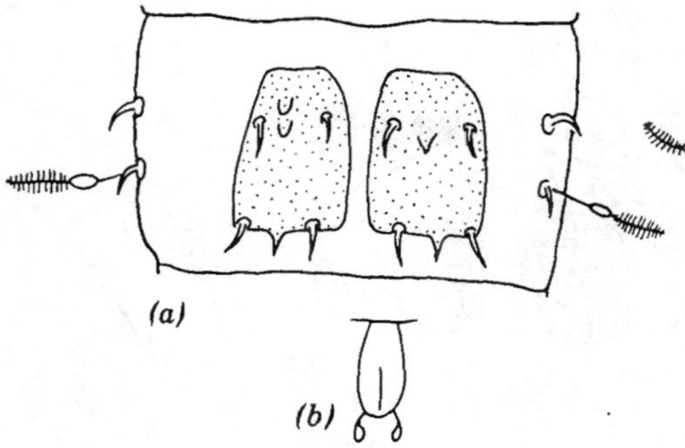

69

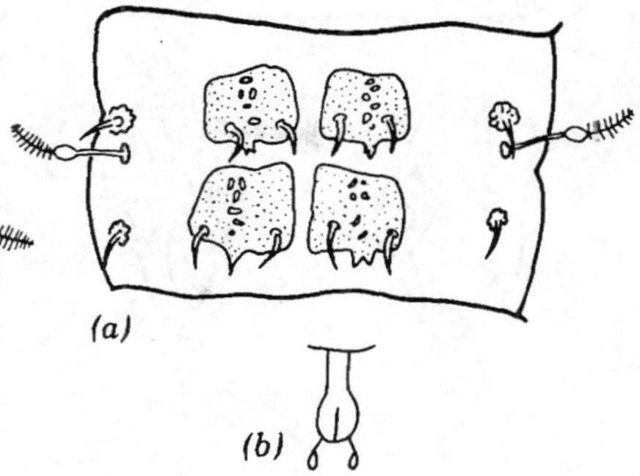

70

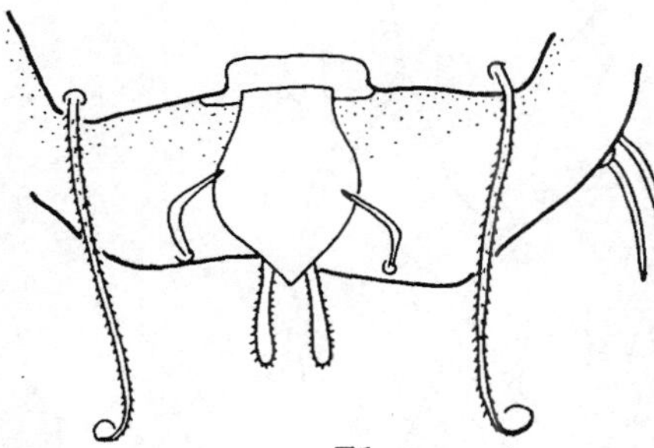

71

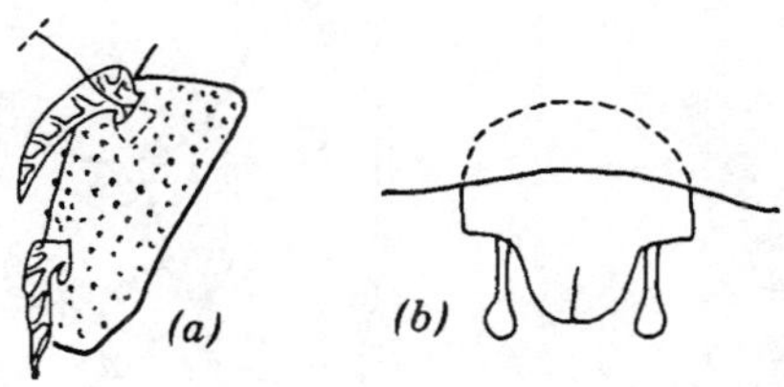

72

Fig. 27.69. *Aletopauropus lentus* MacSwain and Lanham, subad.: (*a*) tergite IV, dorsal view; (*b*) anal plate. (After MacSwain and Lanham.) **Fig. 27.70.** *Zygopauropus hesperius* MacSwain and Lanham: (*a*) tergite IV, dorsal view; (*b*) anal plate. (After MacSwain and Lanham.) **Fig. 27.71.** *Brachypauropus pearsei* Starling, pygidium, ventral view. (After Starling.) **Fig. 27.72.** *Deltopauropus macswaini* Remy: (*a*) tergite II, left side, lateral sclerite; (*b*) anal plate. (After Remy).

59a Posterior branches of anal plate deeply cleft, diverging; $s = q$ (Fig. 27.63)
A.(D.) corsicus Remy, 1940

59b Posterior branches of anal plate short, subparallel; $s > q$ (Fig. 27.64) . . . .
A.(D.) vulgaris (Hansen, 1902)

60a Anal plate with narrow base and four distal appendages: two long posterolateral, two short submedian on sternal side (Fig. 27.65). . . . .
A.(D.) gracilis (Hansen, 1902)

60b Anal plate broadest at base with two short clavate appendages posteriorly (Fig. 27.66) . . . . ***A.(D.) cognatus*** Remy, 1956*b*

61a Pygidial sternum with setae b_1 and b_3 (see Fig. 27.9). . . . .
Subgenus ***Scleropauropus* s.str.**

One species, ***S.(S.) lyrifer*** Remy, 1936 (Fig. 27.67)

61b Pygidial sternum with setae b_1 and b_2 (cf. Fig. 27.31) . . . .
Subgenus ***Scleropauropoides*** Remy, 1957

One species, ***S.(S.) dukensis*** (Starling, 1943) (Fig. 27.68)

62a Tergite IV with two sclerites (Fig. 27.69) . . . .
Aletopauropus MacSwain and Lanham, 1948

One species, ***A. lentus*** MacSwain and Lanham, 1948

62b Tergite IV with four sclerites (Fig. 27.70) **63**

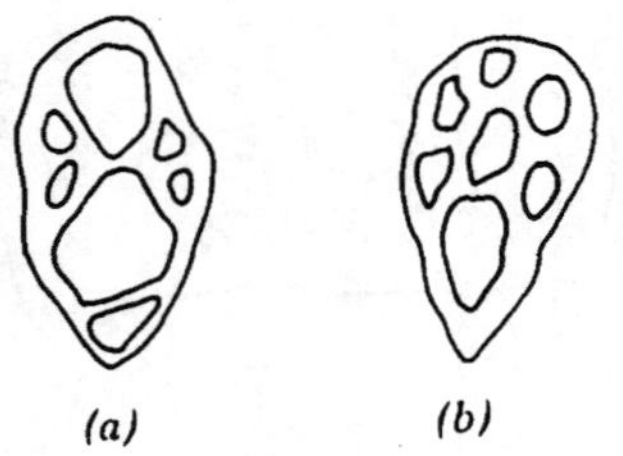

73

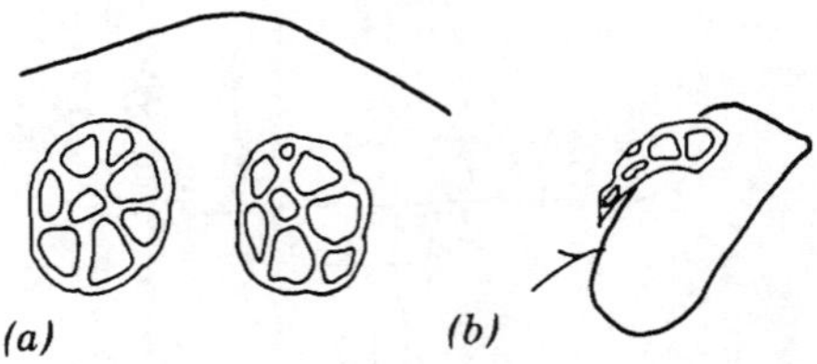

74

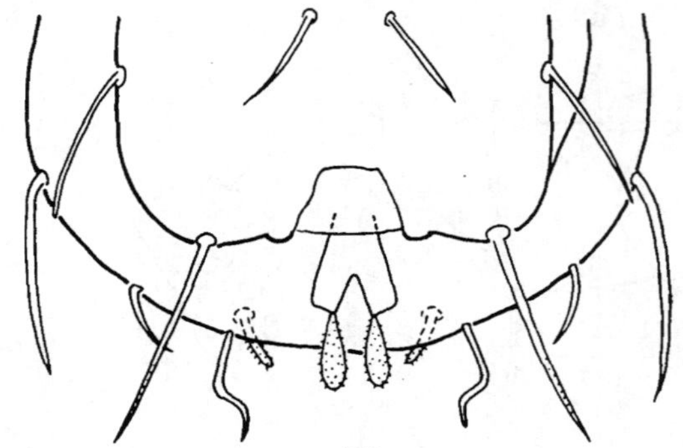

75

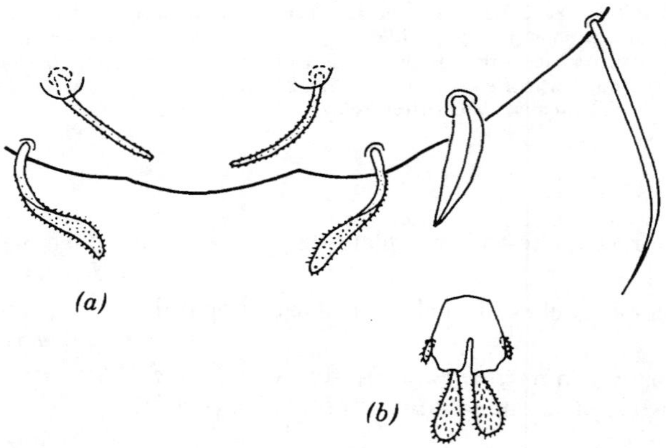

76

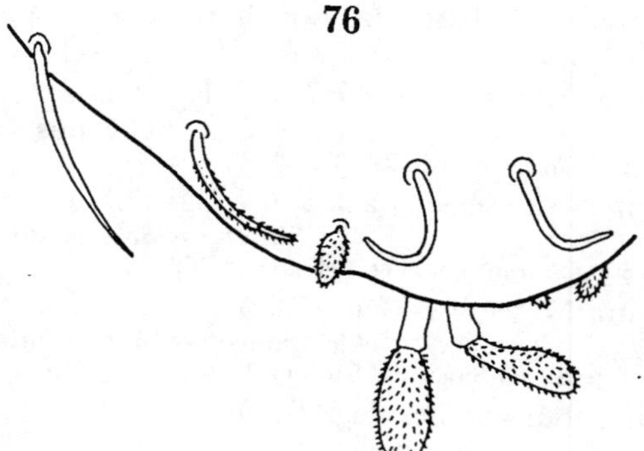

77

63a Tergite IV with four rudimentary lateral sclerites. . . . . *Zygopauropus* MacSwain and Lanham, 1948
One species, *Z. hesperius* MacSwain and Lanham, 1948 (Fig. 27.70)

63b Tergite IV with two rudimentary lateral sclerites. **64**

64a Dorsal sclerites with hooklike setae . . . . . . ***Brachypauropus*** Latzel, 1884
One species, ***B. pearsei*** Starling, 1943 (Fig. 27.71)

64b Dorsal sclerites with leaflike protuberances . . . . ***Deltopauropus*** MacSwain and Lanham, 1948, **65**

65a Lateral sclerite on tergite II with two leaflike protuberances (Fig. 27.72) ***D. macswaini*** Remy, 1956

65b Lateral sclerite on tergite II with one leaflike protuberance and one furcate seta. **66**

66a Mediodorsal leaflike protuberances of anterior row on tergite V oval (Fig. 27.73). . . . . ***D. luteus*** MacSwain and Lanham, 1948

66b Mediodorsal leaflike protuberances of anterior row on tergite V rounded (Fig. 27.74) . . . . ***D. magnus*** MacSwain and Lanham, 1948

67a All legs five-segmented; apical organ of tarsus consists of a main claw and an anterior secondary claw . . . . ***Gravieripus*** Remy, 1937
One species, **G. *armatus*** MacSwain and Lanham, 1948 (Fig. 27.75)

67b First and ninth pair of legs five-segmented, intervening pairs six-segmented; apical organ of tarsus consists of a main claw and two secondary claws, one anterior and one posterior; ***Eurypauropus*** Ryder, 1879 . . . . **68**

68a Tergites I-V with two pairs of areas clear of large spines; pygidial setae a_1 curved inward, converging; a_2 daggerlike; *st* narrowly foliate (Fig. 27.76) ***E. spinosus*** Ryder, 1879

68b Tergites I-V with one pair of areas clear of large spines; pygidial setae a_1 distally curved outwards; a_2 thin, tapering; *st* short, spatulate (Fig. 27.77) ***E. unciger*** Remy, 1956b
Another species, ***Eurypauropus*** sp. Remy (1954), is omitted because it is described from a juvenile specimen only.

References

Hansen, H. J. 1902. On the genera and species of the order Pauropoda. *Vidensk. Medd. Dan. Naturhist. Foren.* 1901:323–424. **MacSwain, J. W. and U. N. Lanham. 1948.** New genera and species of Pauropoda from California. *Pan-Pac. Entomol.* 24(2):69–84. **Remy, P. A. 1954.** Quelques Pauropodes du Michigan. *Rev. Fr. Entomol.* 21:125–131. **1956a.** Pauropus Collignoni n.sp. des environs de Carmel. *Bull. Soc. Hist. Nat. Moselle* 37:119–121. **1956b.** Pauropodes des Etats-Unis d'Amérique. *Mem. Soc. Natl. Sci. Nat. Math. Cherbourg* 47:1–48. **1957.** Pauropodes des Etats-Unis. *Rev. Fr. Entomol.* 24:82–87. **1958.** Pauropodes des Etats-Unis d'Amérique et de la Jamaique. *Mem. Soc. Natl. Sci. Nat. Math. Cherbourg* 48:1–77. **1959.** Stations Nord-Américaines de Pauro-

Fig. 27.73. *Deltopauropus luteus* MacSwain and Lanham, subad., tergite V, anterior submedian protuberances. (After MacSwain and Lanham.) **Fig. 27.74.** *Deltopauropus magnus* MacSwain and Lanham, subad.: (*a*) tergite V, anterior submedian protuberances; (*b*) tergite II, left side, lateral sclerite. (After MacSwain and Lanham.) **Fig. 27.75.** *Gravieripus armatus* MacSwain and Lanham, pygidium, ventral view. (After MacSwain and Lanham.) **Fig. 27.76.** *Eurypauropus spinosus* Ryder: (*a*) pygidial tergum, dorsal view; (*b*) anal plate, ventral view. (After Remy.) **Fig. 27.77.** *Eurypauropus unciger* Remy, pygidial tergum, dorsal view. (After Remy.)

podes, avec description de deux nouvelles espèces. *Bull. Soc. Sci. Nancy* 1959:183–194. **Scheller, U. 1984.** Pauropoda from Canada. *Can. J. Zool.* 62:2074–2091. **1985.** Taxonomic and distributional notes on pauropods from the United States. *Entomol. Scand.* 16:237–257. **1986.** Beringian Pauropoda. *Entomol. Scand.* 17:363–391. **1988.** *The Pauropoda of the Savannah River Plant, Aiken, South Carolina.* SRO-NERP 17:1–99. U.S. Dept. Energy, Aiken. **Starling, J. H. 1943.** Pauropoda from the Duke Forest. *Proc. Entomol. Soc. Wash.* 45:183–200. **Tiegs, O. W. 1947.** The development and affinities of the Pauropoda, based on a study of *Pauropus silvaticus*. I and II. *Q. J. Microsc. Sci.* 88:165–267, 275–336. **Verhoeff, K. W. 1934.** Pauropoda. Pp. 121–200 in H. G. Bronn (ed.). *Klassen und Ordnungen des Tierreichs,* Vol. 5, Abt. 2, Buch 3, Lief. 2. Leipzig, East Germany.

28

Symphyla

CLIVE A. EDWARDS

The Symphyla are a Class (Pocock 1893) and Order (Ryder 1880) of the phylum Arthropoda. Members of the order are extremely common inhabitants of soil in all parts of the world. They are important in breaking down organic matter, and some species are quite important pests, since they feed on the roots of many crops.

The first species of Symphyla was described in Scopoli in 1763 as *Scolopendra nivea* and believed to be a centipede. He did not realize that it possessed such distinctive characters as to justify it being placed in a separate class. The second species, *Scolopendrella notacantha,* was named by Gervais in 1839, from a site near Paris, France, and the third, *Scolopendrella immaculata,* by Newport (1845) from individuals collected near London, England. The latter species were classified as species of millipedes in the Geophilidae. Menge (1851) suggested that *Scolopendrella* was neither a centipede nor a millipede, but rather resembled the hexapod, *Campodea,* in many morphological characters and habits, and he thought that it should be considered as a transitional genus between the Lepismidae and Scolopendridae. In 1880, Ryder created a new order, the Symphyla, deriving this name from the affinity of these animals with both myriapods and insects, and this new group was later raised to the level of a class (Pocock 1893).

The first record of Symphyla in North America was by Packard (1873) who described a species he called *Scolopendrella americana.* He also believed the Symphyla to be an important phylogenetic connecting link between the Thysanura and myriapods, and considered that they had characters more closely related to the insects than to the myriapods. Grassi (1886b) published a thorough review, in which he discussed the morphology, physiology, and habits of the Symphyla in relation to other arthropods.

In the following years, a voluminous literature on the morphology and systematic position of the Symphyla was published, and there was very considerable controversy as to whether the Symphyla were phylogenetically ancestral to the insects. Pocock (1893) considered that because the differential characters between insects and centipedes are linked so well in Symphyla, they must be regarded as the living form that comes closest to being a hypothetical ancestor of the two main divisions of tracheate Arthropoda. Carpenter (1903) also postulated a Symphyla ancestry for insects, centipedes, and millipedes, and supported the ranking of Symphyla, Diplopoda, and Chilopoda as classes. This argument was developed by Tillyard (1931), who suggested an insectan ancestral group with many of the characters of the Symphyla, and Imms (1936), who published a paper on the ancestry of insects which he derived from a Symphylan ancestor.

Considering the taxonomic and ecological importance of the Symphyla, the scientific literature on the group is not voluminous, with no more than about 400 papers published to date. There have been several really excellent reviews of the group, particularly those of Grassi (1886a,b), which discussed not only taxonomy, anatomy, and physiological systems, but also habitats, and Hansen (1904), who reviewed the whole of the taxonomy to that date. Verhoeff (1933–34) later updated knowledge on the Symphyla.

The scientific literature on the ecology and biology of Symphyla is very sparse (Anglade 1967; Archey 1926, 1913; Edwards 1955, 1958, 1959a, 1961; Friedel 1928; Juberthie-Jupeau, 1963; Michelbacher 1938, 1942a, 1948, 1949; Williams 1907a,b; and Waterhouse 1967, 1968), which may be one reason why the ecological importance of the Symphyla is seldom appreciated by soil biologists and ecologists.

External Morphology

Symphyla are white, ranging from 0.2 to 1.5 cm in length, with 12 pairs of legs when adult (Figs. 28.1, 28.2).

The *head* (Fig. 28.3) is heart-shaped with biting mouthparts. There are no eyes, but there is a postantennal organ at the base of each antenna. There is a single pair of stigmas just below the postantennal organ. The antennae are long and multijointed, with sense organs at the tip, and up to 60 segments.

The adult *body* consists of 12 apparent segments and a posterior telson. On the dorsal surface, there are between 15 and 22 dorsal scuta, of different shapes and patterns in the different genera (Figs. 28.1, 28.2). They are progoneate, with the genital aperture located on the ventral side of the third body somite (Fig. 28.4). The sexes can be distinguished readily only in mounted cleaved specimens.

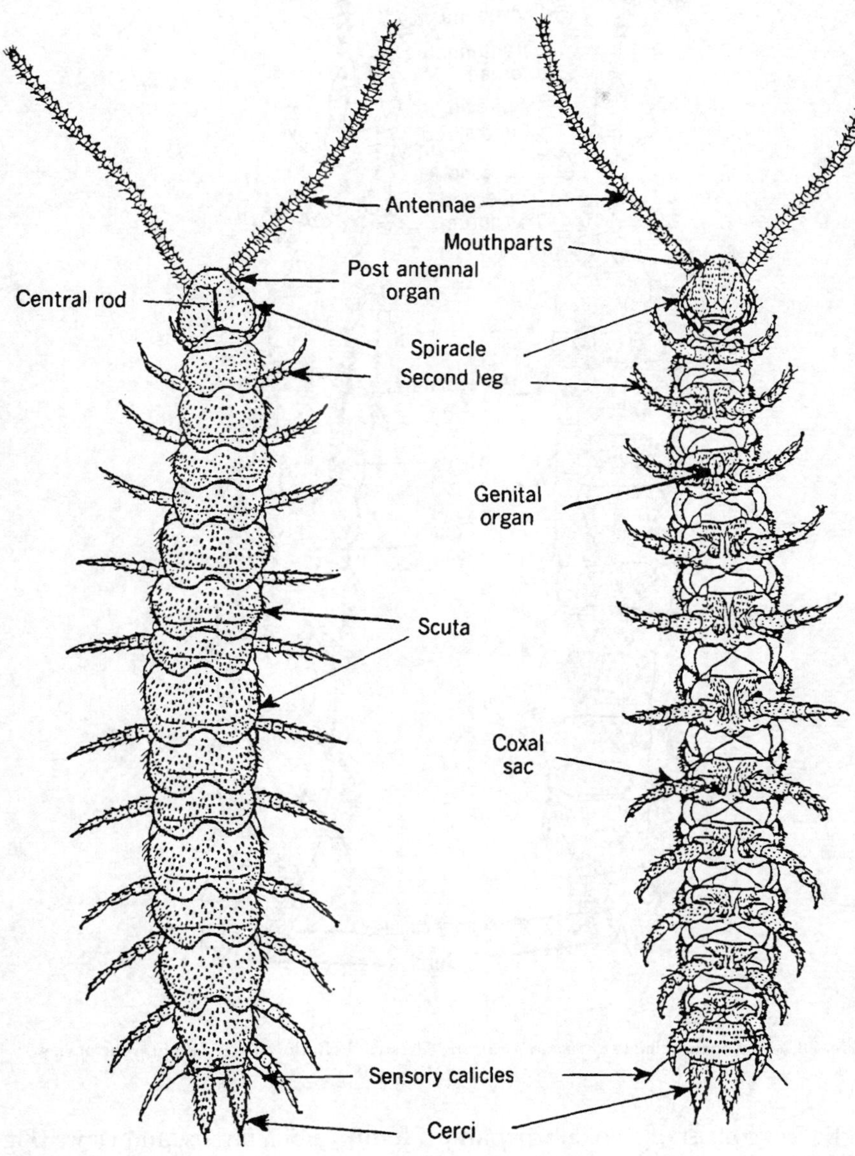

Fig. 28.1. Scutigerellidae (*Scutigerella immaculata* Neupo). Left, dorsal view. Right, ventral view.

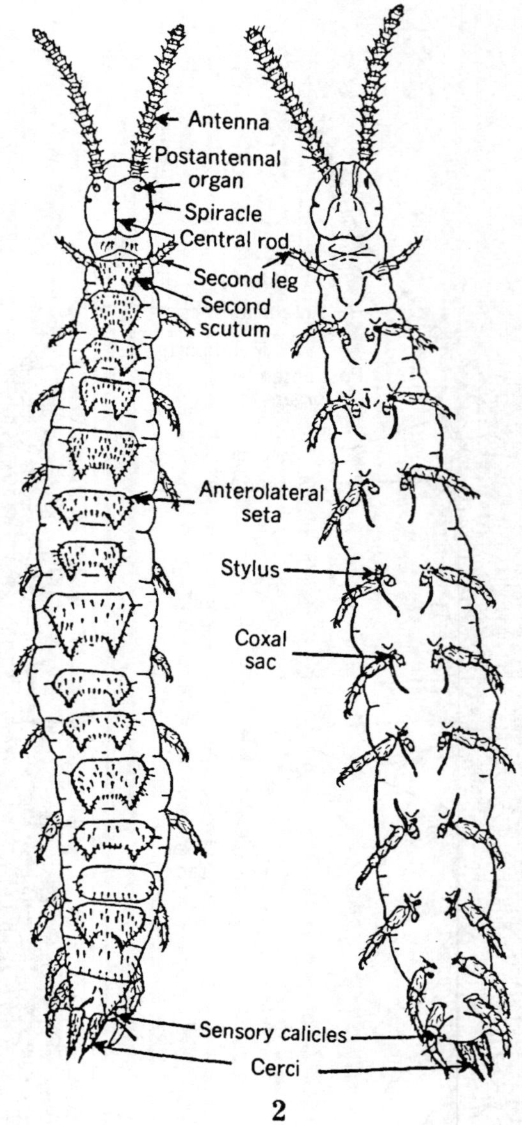

Fig. 28.2. Scolopendrellidae (*Symphylella vulgaris* Silvestri). Left, dorsal view. Right, ventral view.

The *legs* consist of a coxa, trochanter, femur, tibia, tarsus, and claws (Fig. 28.5). There are 12 pairs of legs of which a four-segmented first pair of legs is present only in Scutigerellidae, and is reduced or even vestigial in other genera. At the base of the third-to-last pair of legs is a sensory stylus whose shape varies in the different genera. Associated with the posterior of each coxal segment is a coxal sac. The coxal sacs are absent on the first pair of legs, and those associated with the second, eleventh, and twelfth pairs of legs are reduced, or vestigial in species of Scutigerellidae, and missing altogether from the last three pairs of legs in species of Scolopendrellidae.

Long *sensory hairs* grow from cup-shaped receptacles on either side of the last segment (Fig. 28.6). Pointed paired *cerci* occur at the rear of the animal. Each cercus has a *spinneret* at its tip which has a silk gland opening at the base of a long seta.

Food Habits

Most species of Symphyla are omnivorous and can feed on soft tissues of either plants or animals (Feytaud 1926, Larrieu 1934). Species belonging to the Scutigerellidae show some preference for living plant material, and can

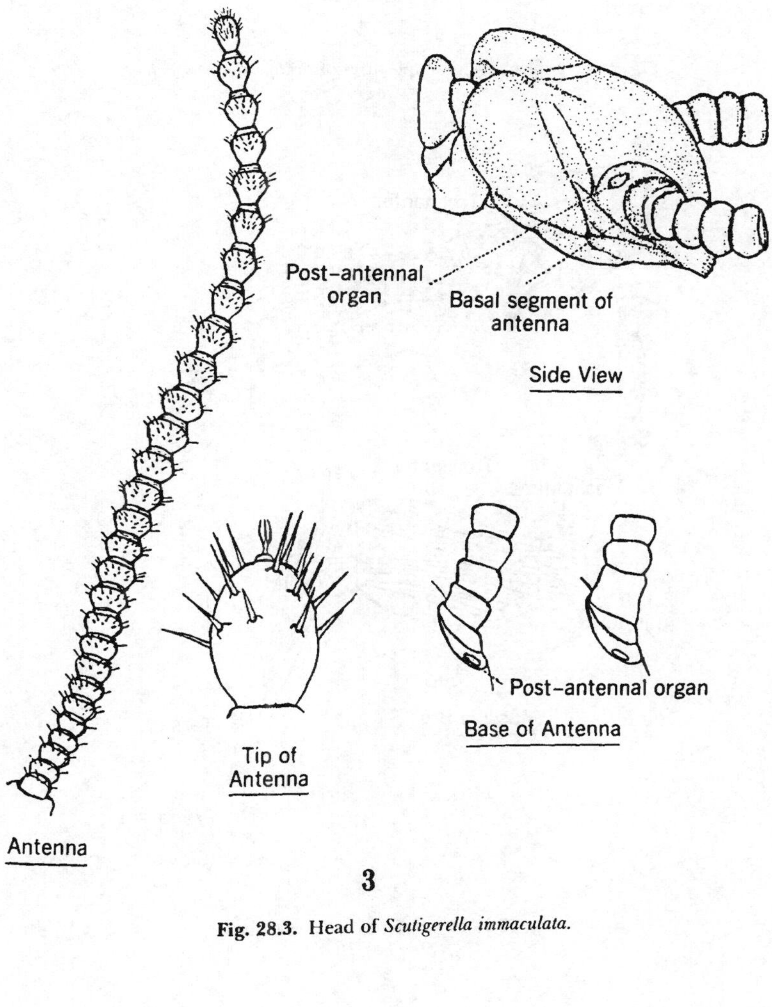

Fig. 28.3. Head of *Scutigerella immaculata.*

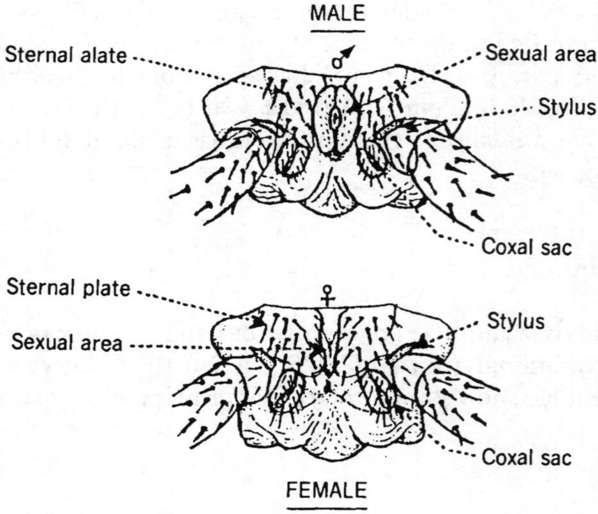

4

Fig. 28.4. External genital apertures of *Scutigerella immaculata*.

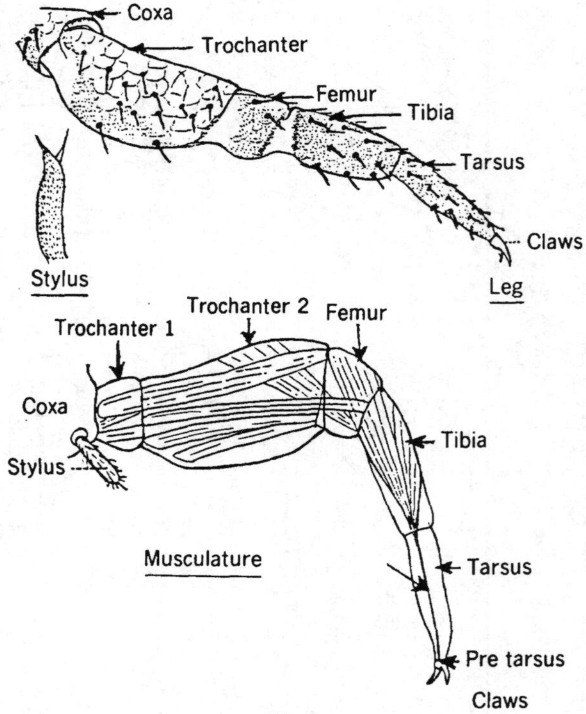

5

Fig. 28.5. Leg of *Scutigerella immaculata*.

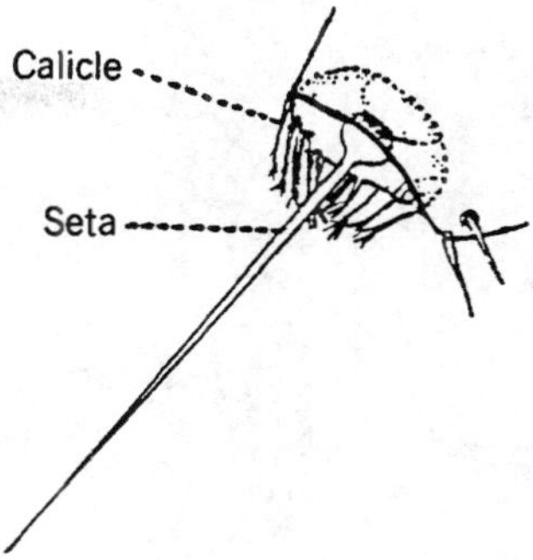

Fig. 28.6. Posterior sense calicle of *Scutigerella immaculata.*

be serious pests of an extremely wide range of plants due to their voracious feeding habits, often consuming more than 20 times their own weight of vegetable matter in 24 hours (Edwards 1959a). When fresh plant material is not available, they can live on microorganisms, decaying vegetable matter, or invertebrate corpses. Only rarely will species belonging to the Scolopendrellidae feed on fresh vegetable material, consuming mainly decaying organic matter. They, too, are extremely voracious and will attack decaying vegetable matter at an earlier stage of decomposition than will many other soil-inhabiting invertebrates. Their habits of moving rapidly up and down the soil profile distributes their feces widely through soil.

Reproduction, Development, and Life Cycle

The *eggs* are white and heavily ridged when laid, gradually darkening to a light tan color. They are laid in clusters of 9–25 and these are guarded by the female; if left, fungi invade and eggs do not hatch. They normally hatch in 8–28 days, dependent on temperature. The newly hatched *first instar young* (Fig. 28.7) has 10 dorsal scuta, 6 pairs of legs, 4 pairs of styli, and only 6 antennal segments. The posterior end of the body is enlarged and differs in structure from that of other instars. After about 2–3 days, the first instar young molt into *second instar* (Fig. 28.8), which have 10 dorsal scuta, 7 pairs of legs, 5 pairs of styli, and 13 segments to the antennae, and the spinnerets appear. After 4–7 days, these molt into *fourth instar,* which have 13 dorsal scuta, 9 pairs of legs, 7 pairs of styli, and 17 antennal segments. After 10–12 days, they molt into *fifth instar,* which have 14 dorsal scuta, 10 pairs of legs, 8 pairs of styli, and 19 antennal segments. After 11–15 days, they molt into *sixth instar,* which have 15 dorsal scuta, 11 pairs of legs, 9 pairs of styli, and 20–24 antennal segments. After 13–20 days, these molt into *seventh instar,* with the full complement of 15 dorsal scuta, 12 legs, 10 pairs of styli, and 24–27 antennal segments.

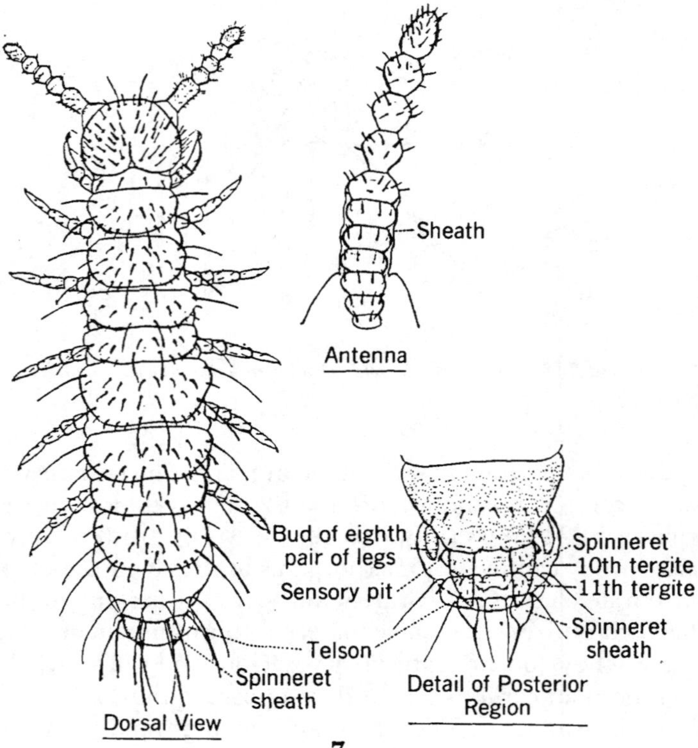

Fig. 28.7. Newly hatched *Scutigerella immaculata.*

These animals are still not sexually mature and they continue to molt at 1- to 3-month intervals. There are minor changes in the penultimate scuta up to the twelfth molt (Fig. 28.9), and about this time they become sexually mature. Molting continues throughout their life cycles and at every molt they add antennal segments (except when they are broken), up to a maximum of 59. In protected conditions such as cultures, they can live a very long time, undergoing more than 50 molts. Several workers have kept individuals for long periods, ranging from 2 years 4 months (Filinger 1931) to 4 years (Michelbacher 1939a,b) to 7 years (Verhoeff 1933–1934).

The mating behavior of *Scutigerella immaculata* is quite complex (Juberthie-Jupeau 1963). The male produces a thin stalk from the anterior sexual aperture, attaches a seminal droplet or spermatophore to this, and leaves it attached to some surface. When a female finds this she consumes it and deposits it in a special cheek pouch. She then removes an ovum from her vulva, places it in her mouth, where it is fertilized, and than attaches it to a support, such as a soil particle or plant appendage (Figs. 28.10, 28.11). There is no evidence as to whether parthenogenesis can occur.

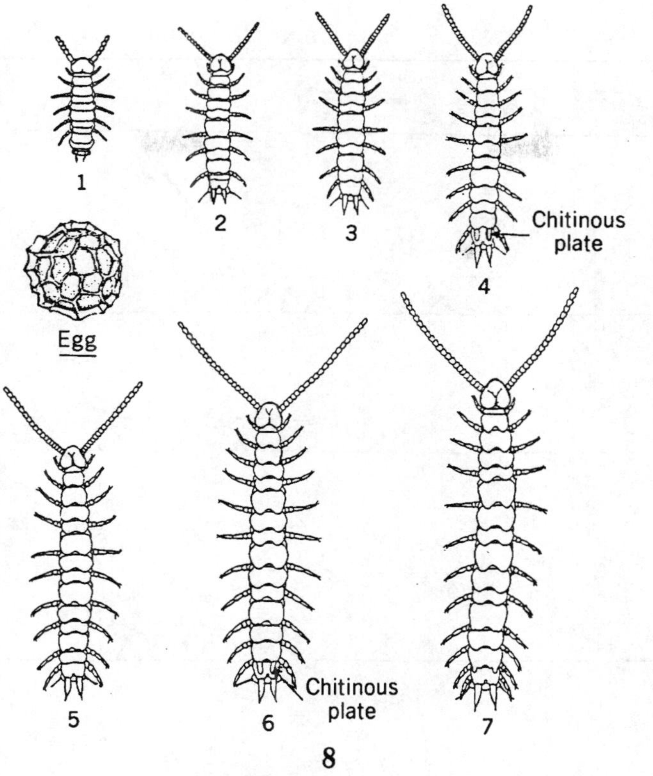

Fig. 28.8. Developmental stages of *Scutigerella immaculata*, seven successive instars.

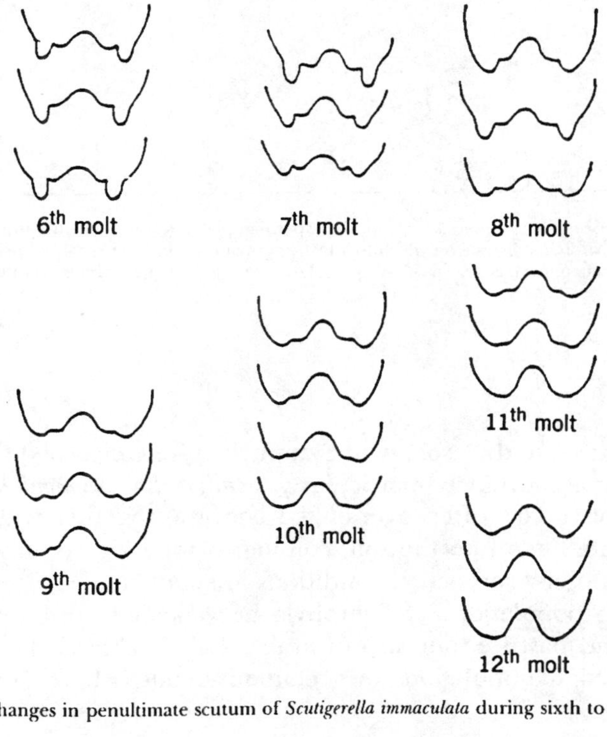

Fig. 28.9. Changes in penultimate scutum of *Scutigerella immaculata* during sixth to twelfth molts.

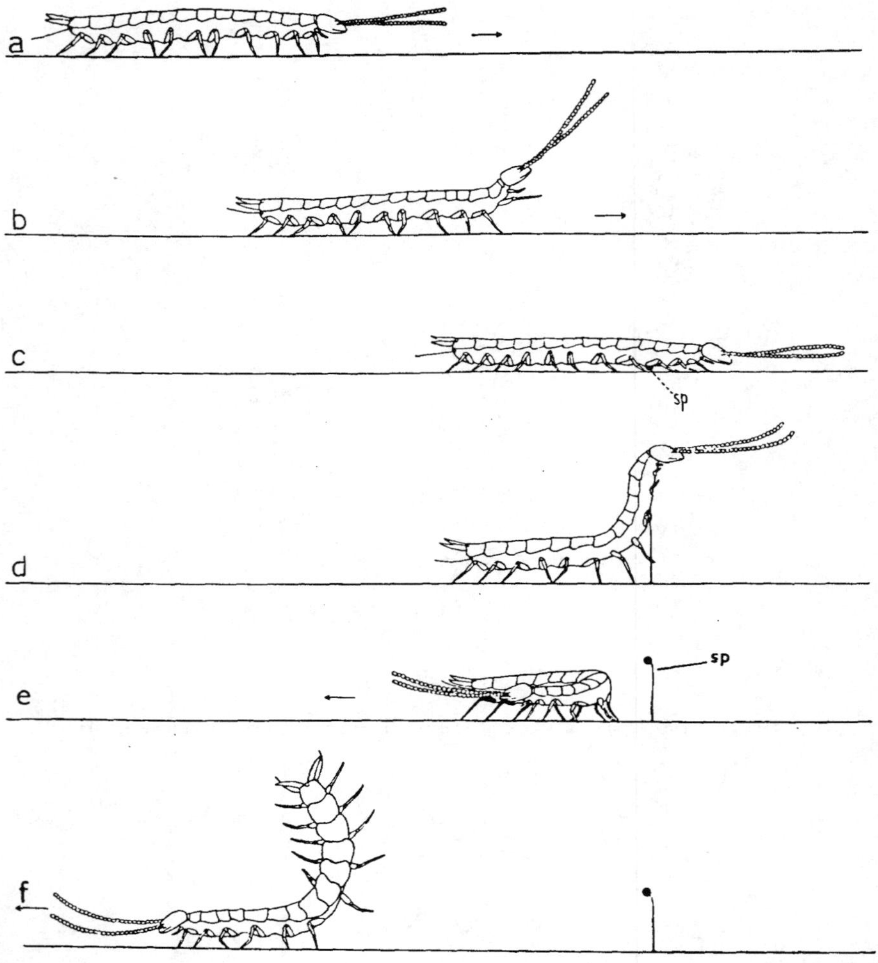

Fig. 28.10. Male *Scutigerella immaculata* depositing spermatophore (after Juberthie-Jupeau 1963). (*a–b*) male searches for proper substrate on which to attach spermatophore; (*c*) spermatophore (*sp*) is attached; (*d*) stalk with spermatophore is drawn out of genital opening; (*e–f*) male retreats from spermatophore.

Ecology

Most studies into the ecology of Symphyla have concerned those species of economic importance, particularly *Scutigerella immaculata* (Wymore 1924). Symphyla are often overlooked because, together with *Campodea,* they penetrate deep into the soil, coming to the surface only when they need food and environmental conditions are suitable.

Very high populations of Symphyla have been recorded not only in woodland and pasture, but also in arable soils (Table 28.1). They reach particularly high populations in greenhouse soils, where environmental conditions are ideal and relatively stable. Edwards (1958) sampled 415 locations in the United Kingdom and United States and found Symphyla present in 32% of grassland soils, 20% of forest soils, 44% of cultivated soils, and 53% of greenhouse soils. They were most common in loams, had lowest populations in sands and clays, with clay loams and sandy clay loams intermediate.

In most temperate climates, Symphyla have distinct seasonal changes in numbers and vertical distribution (Edwards 1959a). In North America,

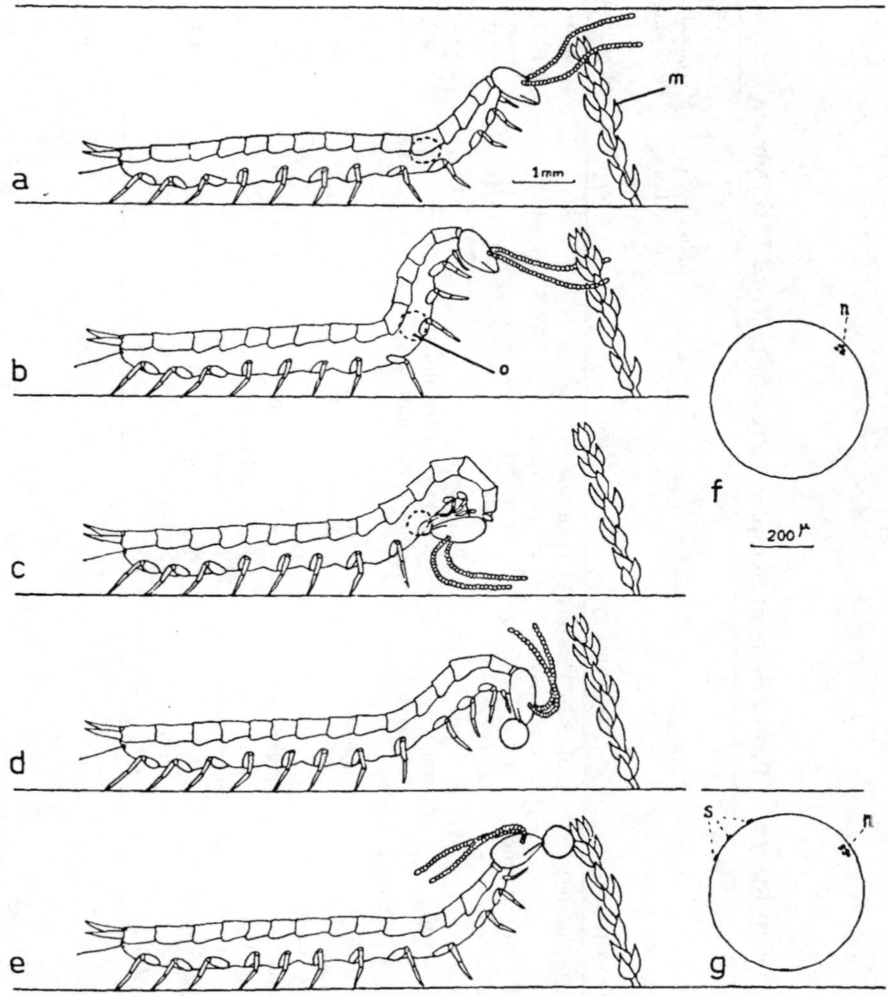

Fig. 28.11. Female *Scutigerella immaculata* fertilizing and depositing egg (after Juberthie-Jupeau 1963). The female locates the spermatophore, consumes it, and sperm collects in a cheek pouch; (a) she then searches for an appropriate moss plant (*m*); (b) the unfertilized ovum (*o* and *f*) moves to the genital opening; (c) the ovum, with nucleus (*n*), is removed from the opening by the mouth parts; (d) sperm (*s*) discharged from the cheek pouch is spread over the surface of the ovum (*g*) resulting in fertilization; (e) fertilized ovum is attached to moss plant and left unattended to hatch.

Table 28.1. Proportions of Symphyla Present in Total Invertebrate Arthropod Populations

Reference (Place)	Location	Collection Method	Number of Invertebrates (millions/acre)	Myriapoda as Percentage of Total Invertebrates	Symphyla	
					Percentage of Total Invertebrates	Percentage of Myriapoda
Morris 1922 (England)	Unmanured pasture	Hand sorted	3.62	24.2	1.8	7.3
	Manured pasture	Hand sorted	10.94	16.4	2.0	11.3
Thompson 1924 (England)	Pasture	Wet sieved and flotation	19.5	Small	Small	86.3
Morris 1927 (England)	Unmanured pasture	Wet sieved	1.21	7.9	0.3	3.1
	Manured pasture	Wet sieved	5.10	36.2	0.7	2.0
Sawa 1930 (Japan)	Arable	Hand sorted and flotation	—	—	0.8	79.7
	Grassland	Hand sorted and flotation	—	—	0.2	43.9
Van Zwaluwenberg 1930 (Hawaii)	Sugarcane	Hand sorted	70.0	25.2	5.6	—
	Fallow	Hand sorted	—	26.9	9.3	—
Baweja 1939 (England)	Arable	Flotation	67.6–111.8	18.0	0.06	0.3
Williams 1941 (U.S.)	Forest litter	Flotation (Ladell)	15.5–40.0	3.13–3.52	0.21–0.26	5.9–8.3
Starling 1944 (U.S.)	Forest litter	Tullgren funnel	—	—	—	16.6
Pearse 1946 (U.S.)	Forest litter	Tullgren funnel	—	—	—	37.1
Salt et al. 1948 (England)	Grassland	Sieved and flotation	1,068.8	2.1	1.5	69.5
	Grassland	Sieved and flotation	—	—	—	67.0
Salt 1952 (Africa)	Coffee plantation	Sieved and flotation	—	—	—	63.0

numbers are high close to the soil surface in spring, low in summer, with a second increase in autumn. Under greenhouse conditions, the spring migration from the subsoil to the surface occurs earlier than outdoors.

The factors controlling these migrations include soil moisture content, temperature, the availability of suitable food close to the soil surface, feeding cycles, molting cycles, egg-laying patterns, and diurnal rhythms. Symphyla require 100% relative humidity for extended survival. They have an optimum temperature range of 15–21°C and can survive temperatures only between 2 and 28°C. Because of their ability to penetrate deep into soils, they are able to survive adverse conditions for long periods and can recolonize the upper layers of soil when conditions are more favorable.

Collection and Population Sampling

Because Symphyla can penetrate so deeply into soil, they are difficult to collect or to assess populations. They are best sampled by taking deep soil cores (Edwards 1955). The animals can be extracted from the soil in several different ways. Hand sorting is unsatisfactory because many of these small arthropods may be overlooked, and they move very rapidly when disturbed, although their white color makes them relatively easy to see. Tullgren or Berlese funnels, which are usually efficient at extracting arthropods, are not very efficient for Symphyla, which do not emerge from soil readily in response to heat. Washing soil samples and then using a wet sieving technique, followed by flotation on a dense solution of magnesium sulfate is reasonably efficient, but some of these fragile arthropods may be damaged during the washing process. The best method seems to be to gently crumble soil samples in containers of water. Symphyla will float to the surface readily and can be picked off with a fine brush and transferred to culture containers containing plaster of Paris or some other culture medium.

Identification

Hansen (1903), recognized only two genera, *Scutigerella* and *Scolopendrella.* In 1913, Bagnall raised these genera to the rank of subfamilies, and placed three genera in each subfamily. Ribaut (1913) described a new genus, *Geophilella,* which he thought was sufficiently distinct to merit the erection of a third subfamily, the Geophilellinae, but in a later paper (1931), after a valuable study of the organization of the Symphyla, stated that he considered the most natural division to be into two families, the Scutigerellidae and the Scolopendrellidae, leaving *Geophilella* as a genus included in the latter family. However, Attems (1926), listed a separate family, the Geophilellidae, for this genus. Verhoeff (1933–1934) retained the two families of Ribaut, but divided the Scolopendrellidae into three subfamilies, Ribautiellinae, Geophilellinae, and Scolopendrellinae, according to the number of distinct dorsal scuta present. However, this scheme is not tenable since there is much overlapping of characters between these subfamilies. Furthermore, the differences in numbers of dorsal scuta results only from subdivision of a basic scutal pattern. In one genus, *Symphylellopsis,* subdivision of the third scutum occurs in one species and not another (Fisher et al. 1974). Such an unstable character does not seem to be a valid basis for erection of new subfamilies. Hence, only two families will be recognized, and no subfamilies are included in the keys given here.

No keys to genera were available after Bagnall's work in 1913, until I produced keys, by which time a further nine genera had been added to the six described by him, and one subgenus, *Scolopendrelloides,* warranted the status of genus. These were incorporated into a key to world genera (Edwards 1958). Of the 15 genera described in the literature, one, *Pseudoscutigerella* Hilton, was considered a synonym of *Geophilella* by H. E. Ewing (Michelbacher 1941), who examined the type specimen. Another genus, *Scolopendrellina* Hilton, was described insufficiently to be a recognizable genus, and although the type specimen was stated to be in the museum of Pomona College, California, it no longer appears to be in existence. Hence these two genera have not been considered in the keys. Another genus, *Neosymphyla,* was described subsequently by Edwards and Belfield (1967). The commonest genera appear to be *Scutigerella, Hanseniella, Symphylella* and *Symphylellopsis,* and for the first three of these genera, numerous species have been described.

No up-to-date keys to species of the Symphyla are available, with the exception of those for the genus *Scutigerella* (Michelbacher 1942b), and this is in need of revision because a number of new species have been described since it was published. A very large number of species have been described for the genera *Symphylella* and *Hanseniella,* many of which may be invalid, and specific determinations present considerable difficulty. Keys to British species of Symphyla have been made (Edwards 1959c).

There have not been many workers on Symphyla in the United States and only six genera and 26 species have been recorded to date. There is little doubt that a diverse Symphyla fauna exists in the United States, so it is suggested that initially the key to world genera, developed by Edwards (1959b), should be consulted, so that if genera not previously recorded are met they can be identified. However, for general ecological studies a key to these genera recorded from the United States to date is given and a list of species recorded by Hilton (1931, 1933) and others is provided.

It must be emphasized that the Symphyla vary considerably in characters, such as chaetotaxy and relative size of parts of the body. Great caution should be exercised in description of new species, especially where good series of adult specimens are not available. Mounting techniques are important in the accurate identification of specimens. Satisfactory temporary mounts can be made in lactic acid, and more permanent preparations can be obtained by the use of polyvinyl alcohol medium, type A3. The latter method has the merit of extending the legs into a position suitable for study. The best type of permanent mount requires more preparation and is made by first fixing the specimen in a flat position, pinning it beneath a coverslip, and then flooding with acetic acid. It is then cleared in 10% KOH, stained with 1% acetic acid fuchsin for 24–48 hours, and mounted in Canada balsam.

Some difficulties may arise in connection with several characters used in the keys. In particular, in most genera the first scutum, and the first pair of legs are smaller, and the first pair of legs possesses fewer segments than subsequent ones. In some species, the first pairs of legs may be virtually absent and only represented by small protuberances behind the head. Similarly, the first scutum may be so reduced that it is only represented by a row of setae. Furthermore, the sutures that separate some scuta may be relatively indistinct, so that it may be difficult to determine the actual number of scuta present.

KEY TO NORTH AMERICAN FAMILIES OF SYMPHYLA

The letters given in parentheses throughout the keys refer to Fig. 28.12.

1a Scuta large and with rounded posterior margins (*a,b*); styli at base of legs very large (*n*); paired posterior sense calicles, with many setae irregularly distributed around margin of pit (*l*); central rod of head capsule unbroken (*r*); first pair of legs shorter, but more than half as long as following pair (*h*); length usually more than 4 mm . . . . **Scutigerellidae** Bagnall

1b Scuta either with pointed posterior projections, or reduced in size (*c–g*); styli at base of legs small or reduced to slight protuberances; paired sense calicles with smooth margin to pit (*k*); central rod of head capsule with break in middle (*s*); first pair of legs usually less than half as long as the following pair (*i,j*), or absent; length usually less than 4 mm . . . . . . **Scolopendrellidae** Bagnall

KEY TO NORTH AMERICAN GENERA OF SCUTIGERELLIDAE

1a With deep cavity beneath middle of caudal margin of last scutum between cerci, overlapped by posterior margin of last scutum: last scutum more or less emarginate (*p*); anterolateral seta directed mainly backward (*a*) . . . . ***Scutigerella*** Ryder

1b Without cavity beneath caudal margin of last scutum q: last scutum not emarginate; anterolateral setae directed mainly forward or laterally (*b*) . . . . . . ***Hanseniella*** Bagnall

KEY TO NORTH AMERICAN GENERA OF SCOLOPENDRELLIDAE

1a Seventeen dorsal scuta or less; scuta usually large and distinct (*c–e*); cerci large, with numerous setae; first pair of legs with claws or vestigial (*i* or *j*) . . . **2**

1b Twenty-one dorsal scuta or more; scuta may be small and indistinct with few setae (*f*) or even reduced to small oval plaques; cerci small with sparse setae **3**

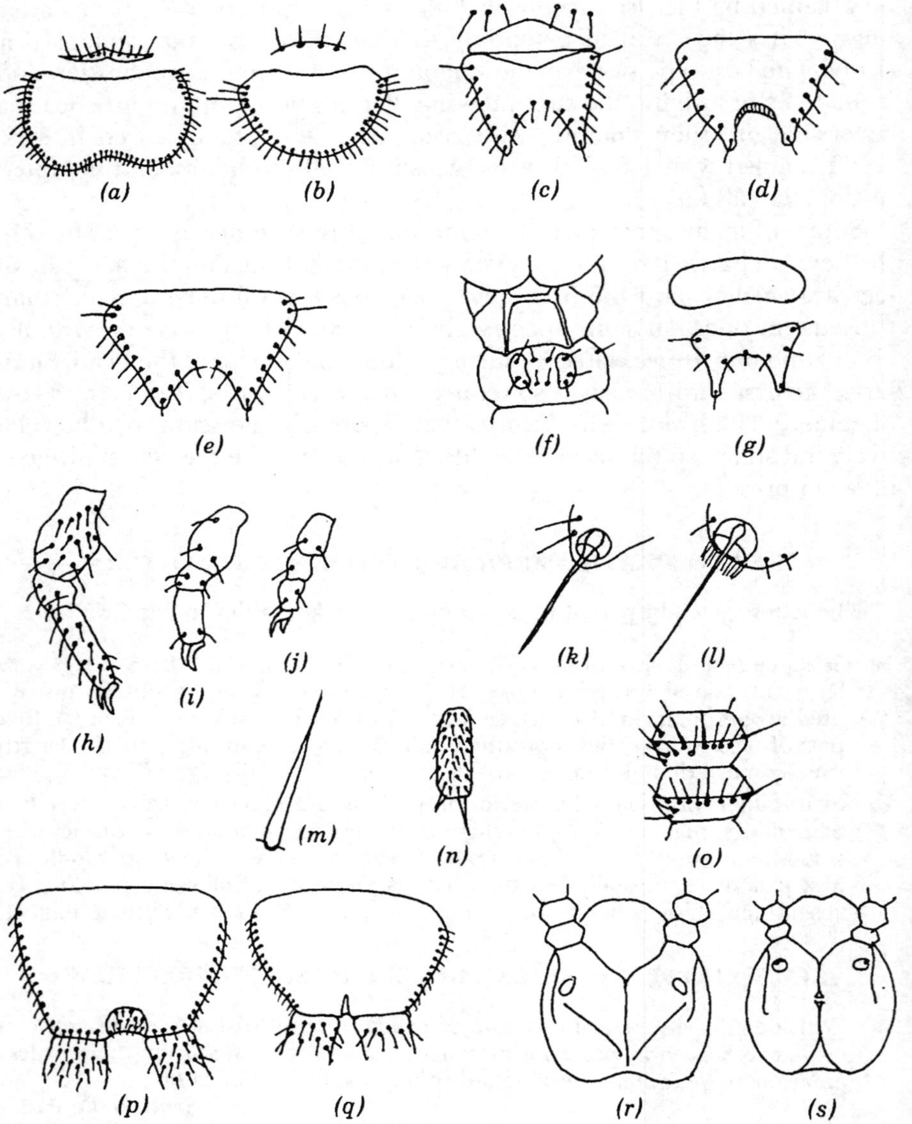

Fig. 28.12. Characters of Symphyla used in key to genera: (*a*) second dorsal scutum of *Scutigerella;* (*b*) second dorsal scutum of *Hanseniella;* (*c*) second dorsal scutum of *Symphylella;* (*d*) second dorsal scutum of *Scolopendrella;* (*e*) second dorsal scutum of *Scolopendrellopsis;* (*f*) second dorsal scutum of *Geophilella;* (*g*) second dorsal scutum of *Symphylellopsis;* (*h*) dorsal aspect of right first leg of *Scutigerella;* (*i*) dorsal aspect of right first leg of *Scolopendrella;* (*j*) dorsal aspect of right first leg of *Symphylellopsis;* (*k*) dorsal view of left sense calicle of *Symphylella;* (*l*) dorsal view of left sense calicle of *Scutigerella;* (*m*) stylus of *Hanseniella;* (*n*) scutal seta of *Hanseniella;* (*o*) antennal segments of *Hanseniella;* (*p*) last dorsal scutum of *Scutigerella;* (*o*) last dorsal scutum of *Hanseniella;* (*r*) dorsal view of head of *Scutigerella;* (*s*) dorsal view of head of *Symphylella.*

2a First pair of legs vestigial; represented only by small protuberances sometimes with terminal hair; coxal sacs at base of legs 3–9 only . . . . ***Symphylella*** Silvestri

2b First pair of legs large; more than two-thirds as long as second pair (*i*); hind margin of each scutum with a band of longitudinal striae between processes (*d*); cerci without transverse terminal ridges. . . . . . ***Scolopendrella*** Gervais

2c First pair of legs less than half as long as second pair; no band of striae on scuta (*e*); cerci with transverse terminal ridges. . . . . . ***Scolopendrellopsis*** Bagnall

3a Twenty-two dorsal scuta; first very large; other scuta represented only by paired oval peaques bearing few setae (*f*); cerci with longitudinal ridges on terminal area . . . . ***Geophilella*** Ribaut

3b Twenty-one or 22 dorsal scuta; first moderately large; some other scuta with narrow triangular projections on posterior margin (*g*); cerci with transverse ridges on terminal areas. . . . . ***Symphylellopsis*** Ribaut

North American Species of Symphyla

Scutigerellidae

Scutigerella immaculata Newport
Scutigerella silvestri Michelbacher
Scutigerella palmonii Michelbacher
Scutigerella nodicercus Michelbacher
Scutigerella verhoeffi Michelbacher
Scutigerella causeyae Michelbacher
Scutigerella linsleyi Michelbacher
Scutigerella inculta Michelbacher
Hanseniella vandykei Michelbacher
Hanseniella californica Hilton

Scolopendrellidae

Symphylella isabellae (Grassi)
Symphylella vulgaris Hansen
Symphylella brevipes Michelbacher
Symphylella subterranea Michelbacher
Symphylella essigi Michelbacher
Symphylella capitata Michelbacher
Symphylella longiseta Michelbacher
Symphylella oviceps Michelbacher
Symphylella sierrae Michelbacher
Symphylella capicola Michelbacher
Symphylella rossi Michelbacher
Symphylellopsis alba Michelbacher
Symphylellopsis geum Michelbacher
Geophilella americana Hilton
Scolopendrellopsis sensiferis Hilton
Scolopendrellopsis santa Hilton

North American Genera and Species Inadequately Described

Symphylella panama Hilton
Symphylella santa Hilton
Symphylella cubae Hilton
Scolopendrella notella Hilton

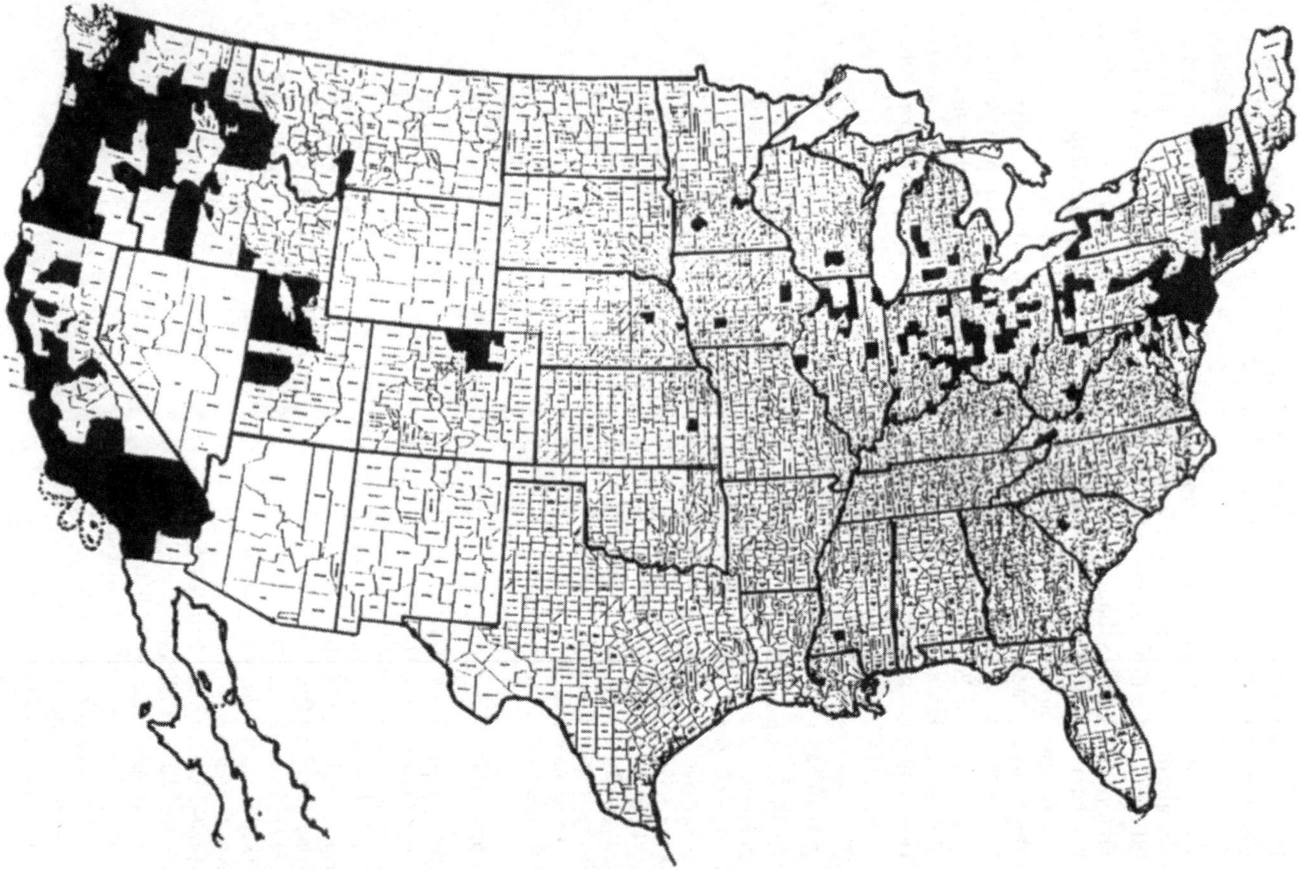

Fig. 28.13. Distribution of *Scutigerella immaculata* in the United States (after Waterhouse 1970). Solid shaded areas indicate counties where specimens have been reported.

Scolopendrella numexta Hilton
Scolopendrellina californica Hilton
Pseudoscutigerella americana Hilton (probably a synonym for *Geophilella*)

Only *Scutigerella immaculata* has been studied sufficiently for a distribution to be given (Waterhouse 1970) (Fig. 28.13). This probably represents more of distribution of crops attacked by this species than being a true distribution of the species. Most other records are for single locations.

References

Anglade, P. 1967. Étude de populations de Symphyles en sol cultivé et d'influence de traitements du sol. Pp. 372–381. in O. Graff. and J. E. Satchell (eds.). *Progress in Soil Biology.* North Holland Publishing Co., Amsterdam. **Archey, G. 1926.** Symphyla. Pp. 11–19 in Thilo. Krumbach (ed.). *Handbuch der Zoologie,* Vol. 4, No. 1. Walter de Gruyter, Berlin. **1913.** On the classification of the order Symphyla. *J. Linn. Soc. Lond. Zool.* 32:195–99. **Attems, C. 1926.** Myriopoda. In Thilo. Krumbach (ed.). *Handbuch der Zoologie.* 4(1). Walter de Gruyter, Berlin. 402 pp. **Baweja, K. D. 1939.** Studies of the soil fauna with particular reference to recolonization of sterilized soil. *J. Anim. Ecol.* 8:120–161. **Carpenter, G. H. 1903.** On the relationships between the classes of the Arthropoda. *Proc. R. Ir. Acad. Sect. B. Biol. Geol. Chem. Sci.* (1902–1904) 24:320–350. **Edwards, C. A. 1955.** Soil sampling for symphylids and a note on populations. Pp. 152–156 in D. K. McE. Kevan (ed.). *Soil Zoology.* Butterworths, Oxford. **1958.** The ecology of Symphyla. I. Populations. *Entomol. Exp. Appl.* 1:308–319. **1959a.** The ecology of Symphyla. II. Seasonal soil migrations. *Entomol. Exp. Appl.* 2:257–267. **1959b.** Keys to the genera of the Symphyla. *J. Linn. Soc. Lond.* 44:164–169. **1959c.** A revision of the British Symphyla. *Proc. Zool. Soc. Lond.* 133 pt. 3:403–439. **1961.** The ecology of Symphyla. III. Factors controlling soil distributions. *Entomol. Exp. Appl.* 4:239–256. **Edwards, C. A. and W. Belfield. 1967.** A new genus and species of Symphyla, *Neosymphyla ghanensis,* with comments on Symphyla segmentation. *Rev. Biol. Ecol. Sol* IV(3):517–521. **Feytaud, J. 1926.** Contribution à l'étude des symphyles dans leurs rapports avec l'agriculture. *Rev. Zool. Agric. Appl.* 25:1–11. **Filinger, G. A. 1931.** The garden centipede, *Scutigerella immaculata* Newport. *Ohio Agric. Exp. Stn. Bull.* 486:1–33. **Fisher, J. R., R. H. Langille and A. J. Keaster. 1974.** A note on the occurrence of *Scolopendrellopsis santa* in Missouri, U.S.A. *J. Kans. Entomol. Soc.* 47(2):259–261. **Friedel, H. 1928.** Ökologische und physiologische Untersuchungen an *Scutigerella immaculata* (Newp.). *Z. Wiss. Biol. Abt. A Morphol. Okol. Tiere* 10(4): 738–797. **Gervais, P. 1839.** Rapport sur un mémoire relatif aux insectes myriapodes. *C.R. Acad. Sci. (Paris)* 9:530–532. **Grassi, B. 1886a.** I progenitori degli insetti e dei Myriapodi, morfologia delle *Scolopendrelle. Mem. R. Accad. Sci. Torino Ser.* 2 37:593–624. **1886b.** Über Verwandtschaftsbeziehungen der Myriapoden. *Dtsch. Naturforsch. Aerzte Berl. Tagebl.* 59. Abstracted in: *J. R. Microsc. Soc. Ser. 2* (1):384–385. **Hansen, H. J. 1903.** The genera and species of the order Symphyla. *Q. J. Microsc. Sci. (Lond.)* 47:1–101. **Hilton, W. A. 1931.** Symphyla from North America. *Ann. Entomol. Soc. Am.* 24(3):537–552. **1933.** Western Symphyla. *Ann. Entomol. Soc. Am.* 26:557–559. **Imms, A. D. 1936.** The ancestry of insects. *Trans. Soc. Br. Entomol.* 3:1–32. **Juberthie-Jupeau, 1963.** Recherches sur la reproduction et la mue chez les Symphyles. *Arch. Zool. Exp. Gen.* 102(1):1–172. **Larrieu, G. 1934.** Contribution à l'étude de *Scutigerella immaculata* Newport. *Rev. Zool. Agric. Appl.* 33(9,10,11):129–174. **Menge, A. 1851.** Myriapoden der Umgegend von Danzig. *Neueste Schr. Naturforsch. Ges. Danzig* 4(Heft 4):1–22. **Michelbacher, A. E. 1938.** The biology of the garden centipede, *Scutigerella immaculata. Hilgardia* II(3):55–148. **1939a.** Notes on Symphyla with descriptions of three new species from California. *Pan-Pac. Entomol.* XV(1):21–28. **1939b.** Further notes on Symphyla with descriptions of three new species from California. Ann. Entomol. Soc. Am. XXXII:747–757. 1941. Two genera of Symphyla new to the United States with descriptions of three new species. *Ann. Entomol. Soc. Am.* XXXIV:139–150. **1942a.** Contributions towards a knowledge of the insect fauna of Lower California. 5. Symphyla. *Proc. Calif. Acad. Sci. Ser. 4*

XXIV(5):153–160. **1942b.** A synopsis of the genus *Scutigerella. Ann. Entomol. Soc. Am.* XXXV(3):267–288. **1948.** Movement and seasonal populations of the garden centipede in greenhouse soil. *J. Econ. Entomol.* 41(5):707–16. **1949.** The ecology of Symphyla. *Pan-Pac. Entomol.* 25:1–11. **Morris, H. M. 1922.** The insect and other invertebrate fauna of arable land at Rothamsted. *Ann. Appl. Biol.* 9:282–305. **1927.** The insect of other invertebrate fauna of arable land at Rothamsted, II. *Ann. Appl. Biol.* 14:442–464. **Newport, G. 1845.** Monograph of the class Myriopoda, order Chilopoda. *Trans. Linn. Soc. Lond.* 19:265–302, 349–439. **Packard, A. S. 1873.** Occurrence of rate and new myriapods in Massachusetts. *Boston Soc. Nat. Hist. Proc.* 16:111. **Pearse, A. S. 1946.** Observations on the microfauna of Duke Forest. *Ecol. Monogr.* 16:127–150. **Pocock, R. I. 1893.** On the classification of the tracheate Arthropoda. *Zool. Anz.* 16:271–275. **Ribaut, H. 1913.** Un genre nouveau de la classe des symphyles (myriopodes). *Bul. Soc. Hist. Nat. Toulouse* 46:77–84. **1931.** Observations sur l'organisation des symphyles. *Bull. Soc. Hist. Nat. Toulouse* 62:443–465. **Ryder, J. A. 1880.** *Scolopendrella* as the type of a new order of articulates (Symphyla). *Am. Nat.* 14:375–376. **Salt, G. 1952.** The arthropodan population in the soil of some E. African pastures. *Bull. Environ. Res.* 43:203–226. **Salt, G., F. S. Hollick, F. Rau and M. V. Brian. 1948.** The arthropodan population of pasture soil. *J. Anim. Ecol.* 17:139–150. **Sawa, R. 1930.** A preliminary survey of the arthropodan fauna of the University Farm of Komaba. *J. Coll. Agric. Tokyo Imp. Univ.* 10(5):329–345. **Scopoli, G. A. 1763.** *Entomologïa Carniolica Exhibens Insecta Carnioliae Indigena et Distributa in Ordines, Genera, Species, Varietates.* J. T. Trattner, Vindobonae, Italy. 420 pp. **Starling, J. H. 1944.** Ecological studies of the Pauropoda of the Duke Forest. *Ecol. Monogr.* 16:291–310. **Thompson, M. 1924.** The soil population. An investigation of the biology of the soil in certain districts of Aberystwyth. *Ann. Appl. Biol.* 11:349–394. **Tillyard, R. J. 1931.** The evolution of the class Insecta. *Pap. Proc. R. Soc. Tasmania* 1930:1–89. **Van Zwaluwenburg, R. H. 1930.** New myriapod records. *Proc. Hawaii. Entomol. Soc.* 7(3):394. **Verhoeff, K. W. 1933–1934.** Symphyla and Pauropoda. Pp. 1–120. in H. G. Bronns (ed.). *Klassen und Ordnungen des Tierreichs.* Band. 5, *Arthropoda,* Abt. 2 (Myriapoda), Buch 3, Lief. 1 (Symphyla). Leipzig, Leipzig, East Germany. **Waterhouse, J. S. 1967.** Population studies of the garden Symphylan. *Can. Entomol.* 99(7):696–702. **1968.** Studies on the garden Symphylan *Scutigerella immaculata. Can Entomol.* 100(2):172–178. **1970.** Distribution of the garden Symphylan in the United States. *J. Econ. Entomol.* 63(2):390–394. **Williams, E. C. 1941.** An ecological study of the floor fauna of the Panama rain forest. *Bull. Chic. Acad. Sci.* 6:63–124. **Williams, S. R. 1907a.** Habits and structure of *Scutigerella immaculata* Newport. *Boston Soc. Nat. Hist. Proc.* 33:461–485. **1907b.** *Notes on Scutigerella immaculata, Its Eggs and Larva.* 7th International Zoological Congress Proceedings, (Cambridge, Mass., pp. 656–659. **Wymore, F. H. 1924.** Biology and control of the garden centipede. *J. Entomol. Zool.* 16(3):73–88.

29

Insecta: Protura

T. P. COPELAND

GENTARO IMADATÉ

The Protura are minute, slender, wingless insects characterized by the absence of antennae and eyes of any kind, by entognathous mouthparts and a 12-segmented abdomen, the first three of which bear paired ventral appendages. The forms of the genitalia are unique among the arthropods, and the gonotreme position behind the penultimate segment is shared among the insects only with the Collembola.

Silvestri (1907)† described the first proturan species, followed soon thereafter by Berlese's monograph (1909) treating all aspects of their anatomy and biology. Condé (1945, 1947) first recognized the great systematic value of the foretarsal sensillae and gave them names. Tuxen (1955, 1956) refined the scheme by naming the setae and adapting the entire system to the eosentomids. These advances, along with the recognition of the importance of the female genitalia, made possible for the first time definitive descriptions of species.

Tuxen's monograph (1931) treating proturan morphology with notes on their systematics and ecology was preliminary to his comprehensive book *The Protura.* With its publication in 1964 the modern era of proturan study came of age. In this encyclopedia Tuxen redescribed, using proven taxonomic characters, all the known species and thoroughly summarized knowledge of their morphology, development, phylogeny, distribution,

† Throughout this chapter, a number of authors and dates shown are given only to assist the reader in locating references; generally, they are older citations that are not cited in the reference section because of space limitations. However, these older citations may be found by perusing the bibliographies in the selected references at the end of this chapter.

and ecology. Nosek (1973) followed with his monograph on the European fauna. Most of our knowledge of Asiatic and Pacific Island Protura has been derived from the many publications and book (1974) by Imadaté and from the papers of Yin (1963, 1965, 1974, 1977, 1979). The latter has not only described many Chinese species, but the unique species found have prompted a reassessment of proturan phylogeny, the significance of which probably has not yet been fully realized.

Reproductive Characteristics and Postembryonic Development

The habits of proturans, details of their life history, and their embryonic development largely remain a mystery, but the remarkable work on proturan eggs by Bernard (1976, 1979) may advance this frontier. Bizzari and Fratello (1971), Dallai (1972), Fratello (1972, 1975), and Fratello et al. (1976) have shown the diploid chromosome number in Protura to vary from 8 to 20. Absence of males and the uncoupling of the oocyte chromosomes in *Proturentomon minimum* indicate a parthenogenetic development for this species. In other protentomids (e.g., *Hesperentomon*), males are very rare, and this, too, indicates a probable parthenogenetic development.

Protura exhibit anamorphosis, and during their postembryonic development the abdomen increases in number of segments from nine to 12. The process was investigated in detail by Tuxen (1949), and his diagnosis of the larval stages—prelarva, larva I, larva II, maturus junior, and preimago—has been supported by all subsequent investigations. However, the first stage was not observed hatching from the egg. Bernard showed that the prelarva is indeed the first postembryonic stage.

Prelarva (Fig. 29.9). This stage is morphologically unique compared to all succeeding instars. The mouthparts (Fig. 29.10), foretarsi, abdominal appendages (Figs. 29.11–29.16), and so on, are not well developed. Foretarsal claws and empodia are absent in eosentomids (Fig. 29.18) but present in acerentomids. The chaetotaxy is similar to all known forms, and the abdomen consists of nine segments.

Larva I, the Second Instar. General appearance of the body, mouthparts, abdominal appendages, claws, and so on, are as in the imago, but the abdomen has only nine segments. The shapes and positions of the foretarsal sensillae are only slightly different from those of the adult. During development from larva I to adult, the chaetotaxy evolves progressively. Body seta number usually increases with each molt by adding new setae to those existing in the previous stage. This occurs in every developmental stage (see Figs. 29.33–29.38, 29.41–29.48). For example, in larva I all the principal pairs of posterior seta, *p* 1, 2, 3, 4, and 5 are present on abdominal tergites II–VI in all known forms (Figs. 29.41–29.45), and those five pairs are also present in all instars between larva I and imago (Figs. 29.42–29.44, 29.46–29.48).

Larva II. A new segment emerges between the eighth and terminal segments, giving a 10-segmented abdomen.

Maturus Junior. Two additional segments appear and the abdomen is now 12-segmented, as in the adult. Genitalia in both sexes are still absent.

Preimago. This instar is distinguishable from the maturus junior only in the males of acerentomids in that an undeveloped squama genitalis appears (Figs. 29.6–29.7).

Imago. Genitalia in both sexes are fully developed. Protura do not molt after attaining sexual maturity.

Ecology

Protura occur in all regions of the world where sufficient moisture supports plant growth except in the true arctic and antarctic. They are among the most abundant soil arthropods. Proturans are to be found in forest litter and humus; in tree cavities; under bark on dead trees; under stones; in peat bogs; in pasture, cultivated field, and grassland soils; on moss and lichens; in underground small mammal nests; and in caves.

These insects are especially abundant in woodland humus. In the moist deciduous forest soils of eastern Tennessee, 150–200 individuals, representing two or three species and genera, are commonly extracted from only the surface layers on a 1-ft^2 plot. In a California pine forest, Price (1973), using core samples extending to the 10-inch mineral subsoil level, estimated a population density of 4204 per square meter.

From the moist deciduous forests of eastern Tennessee and Virginia, the population density and diversity decrease northward toward colder climates and toward the drier soils to the east, south, and west. Dry sandy soils or those with little organic material, and soils from under evergreen trees and shrubs have fewer Protura than do those from beneath moist deciduous vegetation.

Protura are found to a surprising depth in the ground, especially when one considers that they have no obvious adaptations for burrowing. Price (1973) gives these data in percent of total numbers present by levels: litter–humus, 18%; and mineral subsoil 0–2 in. 23%; 2–4 in., 23.5%; 4–6 in., 16.6%; and 6–10 in., 18.9%. In this pine forest habitat, a significant portion of the total number was found below the 6-in. subsoil level. Most studies show a much higher percentage in the more superficial layers and fewer in the deeper ones, in every case depending on the composition of the soil.

Studies in Europe and Japan have shown that some species are ecologically tolerant and consequently have a wide distributional range, whereas others are intolerant and have restricted ones. These studies also indicate that some species with wide ranges are generally abundant, while others with equally broad ranges are less frequently encountered. The limited information available on North American forms makes generalizations hazardous, but from what is known, it appears that our fauna follows similar patterns. A very few species occur in both Europe and America or in Asia and America.

The food of almost all proturan species is unknown, but Sturm (1959) observed them feeding on mycorrhizal fungi; it may be advantageous to look for Protura at the roots of mycorrhiza-bearing trees (Tuxen 1977). Although these insects are abundant in soil, they probably make no significant contribution to humification because of their relatively small total biomass.

Distribution

The first proturans reported from North America were collected in New York and described by Silvestri in 1909. Twelve years passed before the second American species was recorded (Ewing 1921), but since then they have been found to be widely distributed over North America.

Protura have been reported from the tundra of Alaska and Canada (Weber 1950, Tuxen 1955, Nosek 1977), alpine Krummholz in Wyoming (Copeland, unpublished), and in the boreal forest of Canada with its extension along the Appalachian Mountains (Sailer, unpublished; Durey and Copeland 1968). Numerous state records exist for their occurrence in the deciduous forest biome from Massachusetts to Florida and westward to the biome limits (Ewing 1940; Bernard 1975; Bonet, 1950; Copeland 1964, 1978; Walker and Rust 1975). The grassland, Pacific coniferous forest, and more arid regions are represented by records from Arizona, California, Colorado, Indiana, Iowa, Kansas, Oklahoma, Oregon, South Dakota, Texas, and Utah (Ewing 1940; Hilton 1929, 1930, 1943; Chandler 1957; Price, 1960, 1973, 1975; von Wicklen 1962; Imadaté 1980). Although some 46 species are presently known from North America, the group has been inadequately surveyed.

Techniques

Protura are usually collected by use of Berlese funnels utilizing either heat and light or naphthalene to drive out the microfauna. They may also be found by examining plant roots, moss, humus, duff, and beneath bark on dead trees. For more critical ecological work, a flotation extraction may be more desirable. For taxonomic and distributional studies the authors prefer the funnels. The material is collected and preserved in 70–80% ethanol.

Because of their small size, the specimen must be cleared and mounted on microscope slides for examination under an oil-immersion lens. Copeland mounts directly from 70% ethanol into Salmon's (1951) type MA-2 polyvinyl alcohol lacto-phenol solution with no prior clearing necessary. Turtox CMCP-10 formula is used in the same manner and gives similar results. The slides are cured for several days in an oven at 55°C. Within 6 months the cover glasses should be ringed with two or three coats of epoxy, fingernail polish, or similar compound to make the preparations more permanent. Great care must be taken to ensure that at least one of the foretarsi is retained.

Morphology and Definition of Terms

Protura have three body regions: head, thorax, abdomen (Figs. 29.1, 29.2, 29.9); three pairs of thoracic legs; and a 12-segmented abdomen in the adult. The mouthparts are entognathous. Eyes and antennae are absent, the forelegs serving as antennae. Spiracles are present in all North American Eosentomoidea and absent in our Acerentomoidea. The first three abdominal segments bear ventrally a pair of one or two segmented appendages, meaning with or without terminal vesicle. Cerci are absent. Protura are among the smallest of insects, with body length ranging from approximately 0.4 to 2 mm fully extended.

Labrum and LR (Figs. 29.68, 29.69). The labrum may be short or long, and this is usually expressed as a ratio *LR,* head length (base of cranium to rostral setae 1) divided by labrum length.

Rostral Setae (Figs. 29.68, 29.69) A row of setae found across the anterior margin of the head capsule. The central pair is the rostral setae 1 (*RS* 1).

Mandibles (Figs. 29.68, 29.69). In the Acerentomoidea the mandibles are more or less stylet-shaped, at least in Protentomidae. In Eosentomidae they may be very long and extremely narrow, very long and more broad, intermediate, or broad and short. In Eosentomidae (except *Stylétoentomon*), they possess longitudinal striae and three or more tiny, apical teeth. Most species possess only three teeth and the number is constant for the species.

Maxillary and Labial Palpi (Figs. 29.3–29.5, 29.68). The former are longer and more conspicuous, whereas the latter are much smaller and more difficult to see. The labial palpi with their sensory structures are taxonomically useful in the Acerentomoidea.

Maxillary Gland Canal (Filamento di Sostegno). These open at the base of the lobi externi of the maxillae. The canal shows various modifications in Acerentomoidea and is regarded as an important feature among both genera and species (Figs. 29.28, 29.30, 29.32, 29.53–29.56). The canal usually has a heart-shaped widening (calyx) with or without peculiar appendices. In Eosentomidae (Fig. 29.70), the glands are small, spherical

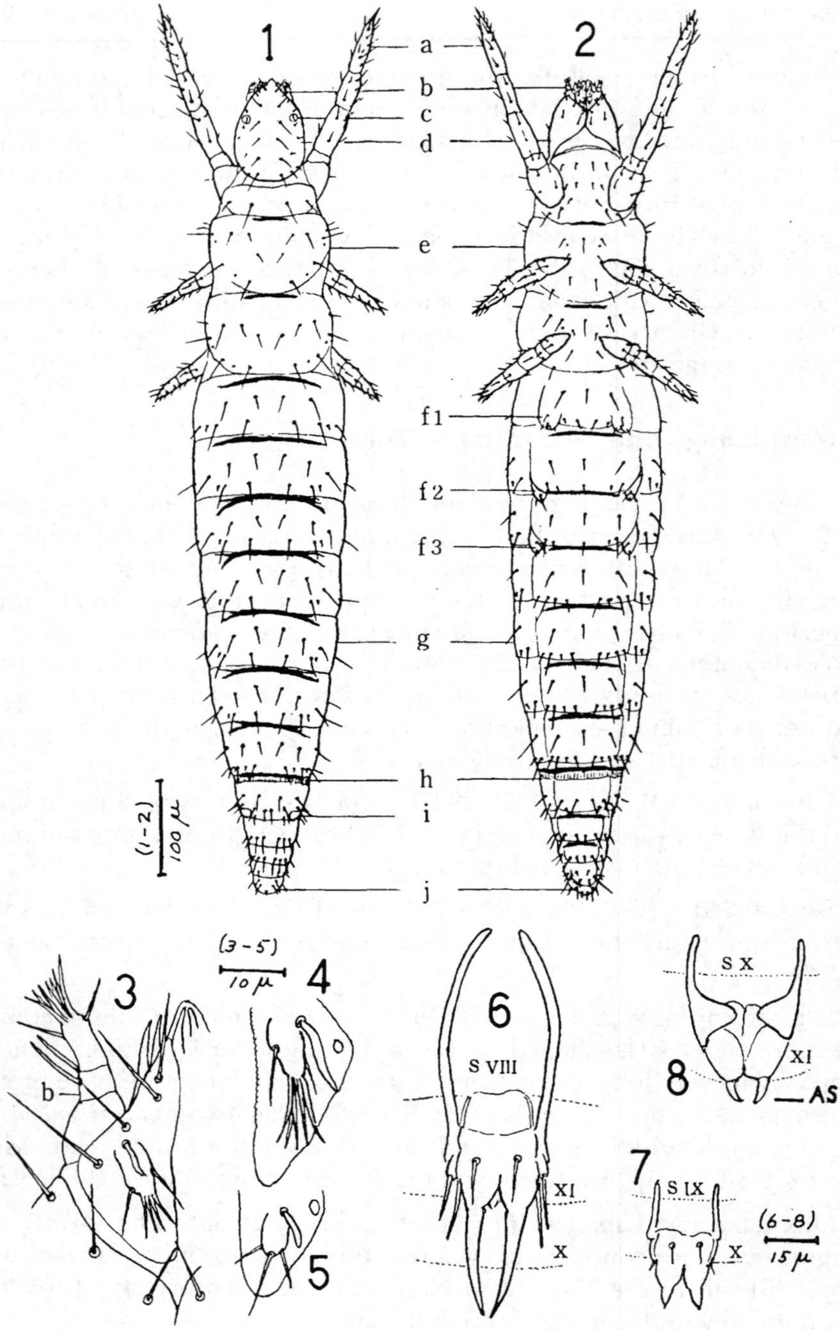

Figs. 29.1–29.8. 1,2. Dorsal and ventral view, *Amerentulus*. **3.** Mouthparts, ventral view, *Amerentulus*. **4,5.** Labial palpus, *Nipponentomon* (4) and *Berberentulus* (5). **6.** Male genital organ. **7.** The same, preimago. **8.** Female genital organ. **6–8.** *Nipponentomon*. *AS*, acrostylus; *a*, foretarsus; *b*, maxillary palpus; *c*, pseudoculus; *d*, head; *e*, mesothorax; *f*1 to *f*3, abdominal appendages I–III; *g*, abdomen V; *h*, striate band of abdomen VIII; *i*, comb (opening of abdominal gland); *j*, abdomen XII; *S* VIII–X, abdominal sternites VIII, IX, and X.

bodies located either in the pro- or mesothorax. The secretory canal is long, very fine, and undifferentiated. Location of the glands is constant within the species, but they are sometimes displaced during the mounting process.

Pseudoculi and PR (Fig. 29.1*c*). Pseudoculi are round or oval paired sensory structures on the dorsolateral surfaces of the head. The ratio *PR* is derived by dividing head length to rostral setae 1 by pseudoculus length. In Eosentomidae and Acerentomidae, the pseudoculi are fairly uniform, but in Protentomidae they are useful characters on the generic level. In *Proturentomon* and *Protentomon* these structures have distinct prolongations ("levers") proximally (Figs. 29.27, 29.29), and in *Hesperentomon* they possess a median opening (Fig. 29.31).

The only characters of general taxonomic importance on the thorax are the presence or absence of spiracles and the chaetotaxy. Spiracles (Fig. 29.70) are possessed by all North American Eosentomoidea but are absent in our Acerentomoidea. The setae consist of a median pair and an anterior and posterior row (Figs. 29.19, 29.70). See the section on chaetotaxy for the system used in their designation.

Foretarsus. This structure possesses characters of paramount taxonomic importance and a thorough mastery of its morphology is absolutely essential for species determination. The foretarsus can be visualized as a four-sided structure with an upper (anterior) surface containing the alpha (α) setae, the ventral with the beta (β) setae, the exterior with the gamma (γ) setae, and the inner face with the delta (δ) setae. The setae surround the sensillae† and serve to locate them. The size, shape, and position of the sensillae in relation to the setae are constant within the species but vary interspecifically.

In Eosentomidae (Figs. 29.74, 29.75), one finds eight α, nine β, eight δ, and five γ setae. On the dorsal surface, the claviform *t* 1 is easily identified. It is located near the center of the tarsus on the level of seta α 3′ or between it and α 3. The position of the *t* 1 determines the ratio BS, length of *basis* divided by *summit*. The *t* 1, *t* 2, and *t* 3 form a straight line along the tarsus with *t* 2 inserted near the level of α 4 and *t* 3 on the level of α 6. The pretarsal sensilla *s* is just beyond the *t* 3 and is usually capitate (spatulate), but in some species it is setiform.

The *t* 2, α 4, *d*, and *b*′2 form a quadrangle, the shape and size of which is fairly constant within the species. Sensilla *y* is roughly on the level of *b*′2 but on the opposite side of α 4. Not located on the dorsal face but always conspicuous in dorsal view is the *a*′. It is setiform–lanceolate and usually proximal and lateral to *t* 1.

† The authors know that the use of *sensilla* and *sensillae* for *sensillum* and *sensilla* is linguistically incorrect, but the former are universally used by authors treating the Protura, and for uniformity, we have followed their example.

On the exterior side near the middle of the tarsus, there is a transverse row of setae and sensillae, which is also a good reference point. It is composed of sensilla *b*, seta β 4, *x*, and *c* in that order. The row is on the level of or a little distal to *t* 1. Distal to and in line with β 4 are sensillae *e* and *g*. Both are spatulate in *Eosentomon* and either setiform or spiniform (Figs. 29.72, 29.73) in *Isoentomon*. Laterodistal from *e* and separated from it by γ 4 is the *f* 1; the *f* 2 is between the *f* 1 and *t* 3. Sensilla *b*′1, when present, is usually between setae δ 3′ and δ 4′. The *c*′ is on or near the level of *t* 3 and separated from it by seta α 6. Sensilla *a* is always near the midpoint between γ 1 and 2. The sensilla *b*′1 is absent in many species, with *c*′ rarely so.

In Acerentomoidea (Figs. 29.57–29.62), there are three dorsal foretarsal sensillae, *t* 1, *t* 2, *t* 3; seven exterior ones, *a–g*; and three interior sensillae, *a*′, *b*′, *c*′; seven dorsal setae, α 1–7; seven ventral setae, β 1–7; five exterior setae, γ 1–5; and six interior setae; δ 1–6.

Throughout the postembryonic development between larva I and imago, these sensillae and setae are fairly constant, but *b*′ and *c*′ may be exceptionally labile. One or both are often absent in the earlier stages, such as larva I and larva II, and appear in the later stages (larva II, maturus junior).

In Protentomidae the reduction of sensillae and setae is not rare (e.g., *t* 1 and *c* are lacking in some members of *Proturentomon* and *Protentomon*), and the extreme reduction is found in the Asian genus *Condeellum,* which has only seven foretarsal sensillae. On the other hand, *c*′ is doubled in *Hesperentomon,* and in Acerentomidae, *b*′ is often lacking.

Chaetotaxy. The system for naming the setae was devised by Tuxen (1949) and slightly modified in 1964. Figures 29.19, 29.20, 29.70, and 29.71 illustrate the method in Acerentomoidea and Eosentomoidea. The thoracic and abdominal setae generally occur in two rows, are numbered from the median line outward, and include the abdominal pleurals with the tergals. Setae in the anterior rows are prefixed with the letter *a* (*a* 1, *a* 2, etc.) and those in the posterior rows with *p* plus the numeral. In the posterior rows, one also finds the accessory setae, and these are identified with a prime symbol (e.g., *p* 1′, *p* 2′). For a more thorough explanation, see Tuxen (1964).

The chaetotaxy is a useful character in identifying the genera and species, but it must be used with caution. The chaetotaxy evolves successively during postembryonic development, and it is not rare that a particular pattern can be found only in the adult. Another problem is individual variation, but this is usually confined to the setae acquired in the last postembryonic stage. In other words, the setae follow the well-known biological principle that the earlier a structure makes its appearance in organismal development, the more stable it is. In most instances setal abnormalities are readily apparent because they involve the loss or gain of a single seta in a row, making it asymmetrical.

Chaetotaxic pattern at the earliest stage is similar within each suborder for all known forms. At the following stage newly added setae are usually common within a family or genus. The later the stage, the greater the chaetotaxic differentiation between species and/or genera (Figs. 29.33–29.38, 29.41–29.48).

Abdominal Appendages. In Eosentomidae (Figs. 29.14–29.16) all three pairs are similar, each appendage bearing five setae and a terminal vesicle. In Acerentomoidea, the appendages (Fig. 29.2) show differentiation and are very useful characters, especially among the genera.

In North American *Hesperentomon* all three pairs have terminal vesicles and four setae each (Figs. 29.25, 29.26). *Protentomon* and *Proturentomon* are characterized by the reduction of the third pair, which has no terminal vesicle (Figs. 29.23, 29.24). Only two setae are found on the third appendage.

In all members of Acerentomidae only the first pair of appendages possess terminal vesicles. The number and size of the setae on the second and third pairs are the important characters of acerentomid genera. *Amerentulus* seems to be the most common acerentomid in the eastern United States. This genus is characterized by three setae on the second and third abdominal appendages, of which the median apical is a microchaeta and often represented as little more than a pit. This structure is easily overlooked, especially if viewed from an oblique angle.

Striate Band. A striate band (Acerentomidae, Figs. 29.1, 29.2) or serrated transverse line (Protentomidae) is found on the anterior border of tergum VIII. The striate band is well developed in many genera (Figs. 29.63, 29.66), but is reduced in a few genera, such as *Baculentulus* and *Berberentulus* (Fig. 29.64). In the latter genera, striae are rarely found in the band, although distinct longitudinal striae are visible in the band of the *Baculentulus*. The band is absent in Eosentomidae.

Abdominal Glands. These open by large pores on the posterior lateral marginal or tergum VIII. The gland pore lid may have various pectinate ornamentations, called "comb on abdomen VIII." The shape of the comb is highly specialized in Acerentomidae and is of taxonomic importance to species and/or genera (Figs. 29.63–29.65). Lids are either absent or greatly reduced in Eosentomidae.

Other pectinate structures are often found on the abdominal segments in Acerentomidae. The extreme development among North American genera is found on abdomen IX–X of *Nipponentomon* (Figs. 29.66, 29.67).

External Genitalia. The male apparatus (Fig. 29.6) is similar in the species of each suborder and is not used in systematics. The female squama genitalis is of great taxonomic value at the species level, especially in eosentomids. For this reason, we give figures only to introduce the basic terminologies used in its description and to portray some of the shapes encountered (Figs. 29.8, 29.76–29.79). For full descriptions in both sexes, see Tuxen (1964).

ACKNOWLEDGMENTS

We would like to acknowledge the help of Dr. E. C. Bernard, Miss A. Gray, Dr. J. W. Hart, Dr. J. Nosek, Dr. J. A. Powell, Dr. J. Rusek, and Dr. P. Wygodzinsky, and of the reviewer, Dr. S. L. Tuxen.

Identification

The following key contains all the genera known from North America. Apart from phylogenetic meaning, it employs characteristics that will most readily distinguish the suborders, families, and genera represented in North America, not necessarily characteristics that would be useful on a worldwide basis. However, present knowledge of the proturan fauna of North America is not very satisfactory. A considerable number of genera new to North America will no doubt be recorded in the future, and then the key must be amended.

In ecological surveys young larvae such as larva I and II are very often collected, and frequently more than 70% of the total individuals captured are of young larvae. In the following key, characteristics that can be used through all the stages between larva I and imago are preferentially employed. Characteristics with an asterisk, however, are useful only for adults. The first stage, prelarva, cannot be determined to genus with certainty.

KEY TO FAMILIES OF PRELARVA

1a Foretarsus without claw and empodium (Fig. 29.18); abdominal appendages I–III with terminal vesicle (Figs. 29.14–29.16); Suborder **Eosentomoidea** . . Family **Eosentomidae**

1b Foretarsus with claw and empodium (Fig. 29.17); Suborder **Acerentomoidea 2**

2a At least first two pairs of abdominal appendages with terminal vesicle . . . . Family **Protentomidae**

2b Only abdominal appendage I with terminal vesicle (Figs. 29.11–29.13) . . . Family **Acerentomidae**

KEY TO SUBORDERS OF LARVA I TO ADULT

1a Spiracles present (Fig. 29.70); all three pairs of abdominal appendages similar (Figs. 29.14–29.16), each with terminal vesicle and five setae; abdomen without pectinated structures; lids on abdominal gland pores absent or very small and without comb; abdominal segment VIII without serrate band or striate border; female squama genitalis without acrostylus but usually with a processus sternalis* (Figs. 29.76–29.79). . . . . **Eosentomoidea**

1b Spiracles absent; abdominal appendages (Figs. 29.2, 29.11–29.16, 29.23–29.26) each with four or fewer setae; abdomen with or without pectinated

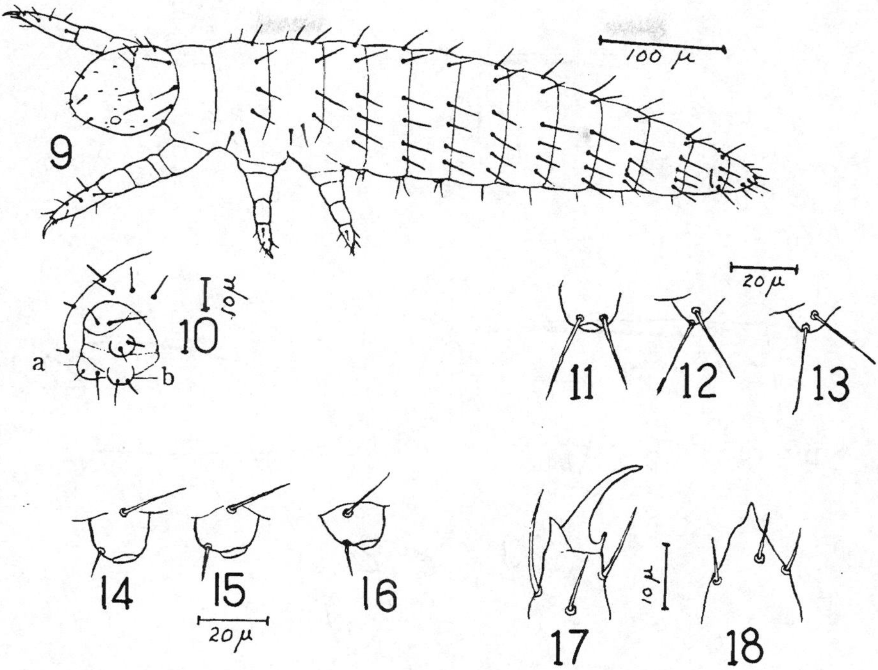

Figs. 29.9–29.18. 9–18. Prelarva. **10.** Mouthparts, ventral view (*a*, maxillary palpus; *b*, labial palpus). **11–13, 14–16.** Abdominal appendages I, II, and III. **17,18.** Apex of foretarsus. **9–13,17.** *Nipponentomon*. **14–16,18.** *Eosentomon*.

structures (Figs. 29.66, 29.67); lids of abdominal gland pores developed and often with comb (Figs. 29.63–29.66); abdominal segment VIII with striate band and often a serrate border (Figs. 29.63–29.66); female squama genitalis with acrostylus and without processus sternalis* (Fig. 29.8) . **Acerentomoidea**

Eosentomoidea: Key to North American Genera

1a Mandibles long and extremely slender, stylet-shaped (Fig. 29.68); labrum long and slender; foretarsal sensillae *e* and *g* spatulate, *b'* 1 present or absent (Figs. 29.74, 29.75) . . . . *Styletoentomon* Copeland

1b Mandibles not stylet-shaped, broader and always possessing three or more small, apical teeth (Fig. 29.69); labrum and mandibles short or long; foretarsal sensillae *e* and *g* spatulate, spiniform, or setiform 2

2a Foretarsal sensillae *e* (if present) and *g* short spiniform or long setiform (Figs. 29.72, 29.73), *b'* 1 present or absent; processus sternalis of female squama genitalis "dissolved" (Maya type of Tuxen 1964, Fig. 29.76)* . . . . *Isoentomon* Tuxen

2b Foretarsal sensillae *e* and *g* spatulate, *b'* 1 present or absent; processus sternalis of various forms, including Maya type (Figs. 29.76–29.79)* . . . . *Eosentomon* Berlese

Acerentomoidea: Key to the North American Families and Genera

1a At least first two pairs of abdominal appendages with terminal vesicle Family **Protentomidae 2**

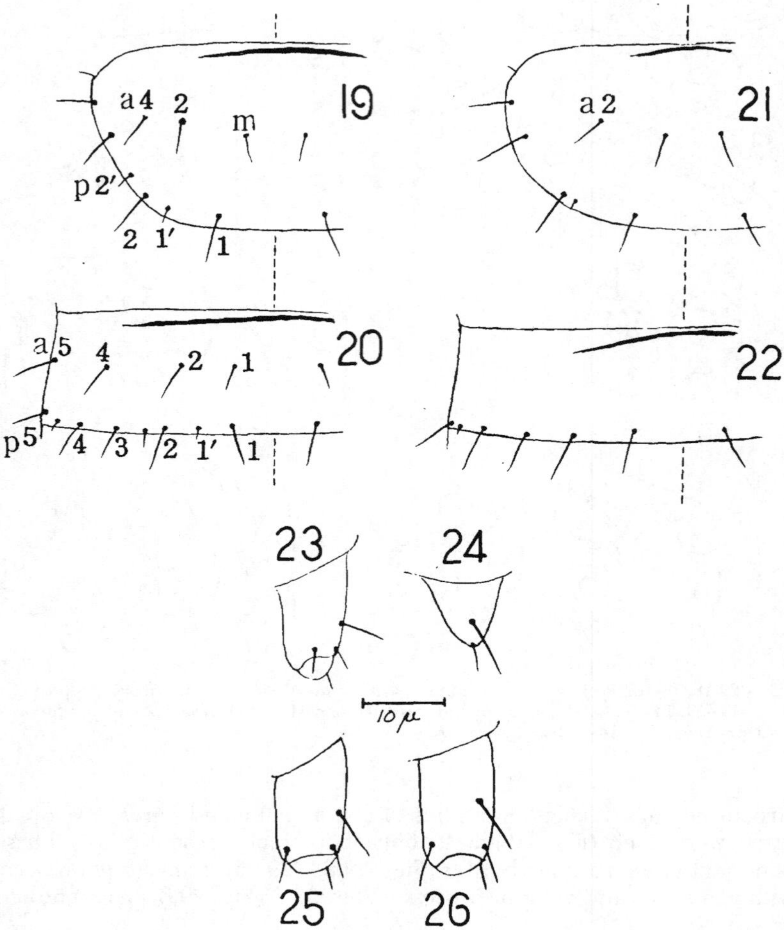

Figs. 29.19–29.26. 19,21. Tergal chaetotaxy, metathorax. **20,22.** Tergal chaetotaxy (including pleural), abdomen V. **23,25.** Abdominal appendage II. **24,26.** The same, III. **19,20,25,26.** *Hesperentomon.* **21–24.** *Proturentomon.*

1b Only abdominal appendage I with terminal vesicle . . . . Family **Acerentomidae 4**

2a (1) Canal of maxillary gland with long sausage-like dilatation (Fig. 29.32); pseudoculus with distinct median opening (Fig. 29.31); abdominal appendage III usually with terminal vesicle (Fig. 29.26); metathorax with dorsal anterior setae *a* 2 and 4 (Fig. 29.19)*; abdominal tergites II–VI with four pairs of anterior setae (Fig. 29.20)* . . . . Genus ***Hesperentomon*** Price

2b Canal of maxillary gland with heart-shaped or circular dilatation (Figs. 29.28, 29.30); pseudoculus without median opening; abdominal appendage III without terminal vesicle (Fig. 29.24); abdominal tergites II–VI without anterior setae (Fig. 29.22) or with only one pair of anterior setae *a* 1* **3**

3a (2) Pseudoculus with a large triangular prolongation proximally (Fig. 29.29); metathorax with dorsal anterior setae *a* 2 and without *a* 4 (Fig. 29.21)*. . . . . Genus ***Proturentomon*** Silvestri

3b Pseudoculus with a long parallel-sided prolongation proximally (Fig. 29.27); metathorax with dorsal *a* 2 and 4 or *a* 2 only* . . . . Genus ***Protentomon*** Ewing

Recently the Asian genus ***Neocondeellum*** Tuxen et Yin was newly recorded from Tennessee (Bernard, E. C. 1985). This runs to 2b in our key, but the genus is distinguishable from 3a ***Proturentomon*** or 3b ***Protentomon*** by the following characters: canal of maxillary gland with bladder like excrescence at the place of calyx, pseudoculus with short prolongation proximally, metathorax with dorsal anterior setae, a 2 and 4, and abdominal tergites II–VI with two pairs of anterior setae (a 1 and 2).

4a (1) On abdominal tergites II–VI, the third principal pair of posterior setae, *p* 3, situated a little anterior to the other posterior setae (Figs. 29.45–29.48) 5

The seta pair, *p* 3, may be misinterpreted as one of the anterior pairs. At the stages larva I and II, it is correctly interpreted (Figs. 29.45, 29.46), but in later stages, the pair is more anterior and is often mistaken as an anterior pair (Figs. 29.47, 29.48).

4b On abdominal tergites II–VI, *p* 3 located in the same row as the other posterior setae (Figs. 29.41–29.44); abdominal appendages II–III with two setae, the apical seta more than one-half the length of the subapical (Fig. 29.52); labial palpus with a tuft of setae apically (Fig. 29.4) . . . 12

5a (4) Abdominal appendages II–III with three setae 6

5b Abdominal appendages II–III with two setae 8

6a (5) Labial palpus with a few setae apically; striate band on abdomen VIII reduced (Figs. 29.5, 29.64) . . . . Genus ***Fjellbergella*** Nosek

The generic features of ***Fjellbergella*** need to be revised thoroughly, although we follow those mentioned in Nosek (1978).

6b Labial palpus ornamented with a tuft of setae apically (Fig. 29.3); striate band on abdomen VIII well developed (Fig. 29.63) 7

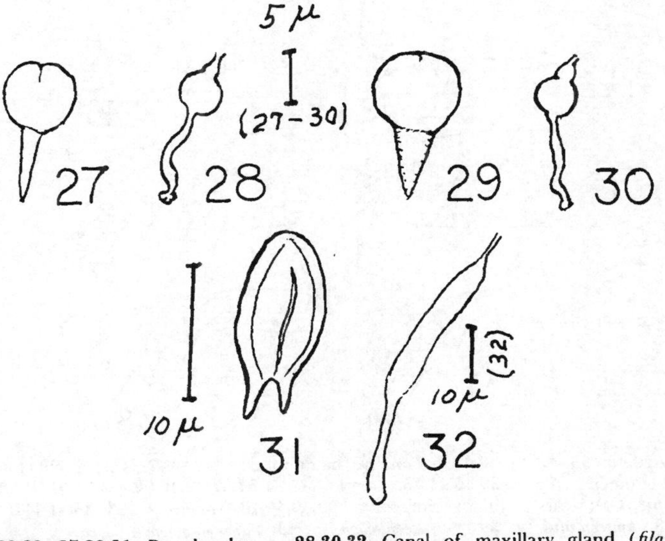

Figs. 29.27–29.32. 27,29,31. Pseudoculus. **28,30,32.** Canal of maxillary gland (*filamento di sostegno*). **27,28.** *Protentomon.* **29,30.** *Proturentomon.* **31,32.** *Hesperentomon.*

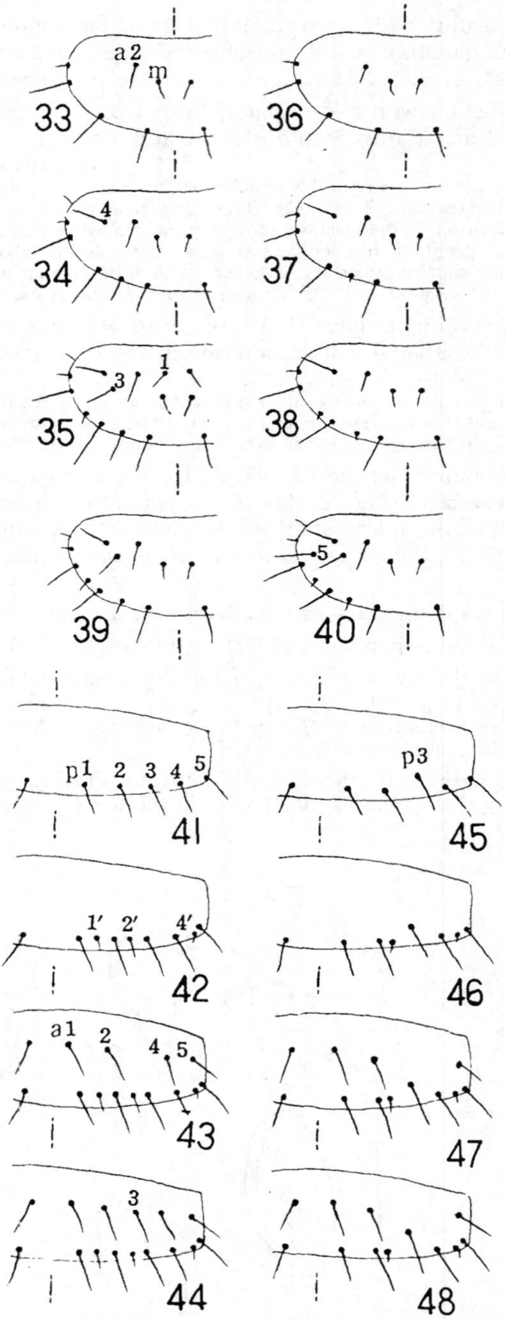

Figs. 29.33–29.48. 33–40. Tergal chaetotaxy, metathorax. **41–47.** Tergal chaetotaxy (including pleural), abdomen V. **33,36,41,45.** Larva I. **34,37,42,46.** Larva II. **43,47.** Maturus junior. **35,38.** Maturus junior—imago. **39,40,44,48.** Imago. **33–35,41–44.** *Nipponentomon.* **36–38,45–48.** *Amerentulus.* **39.** *Verrucoentomon.* **40.** *Yamatentomon.*

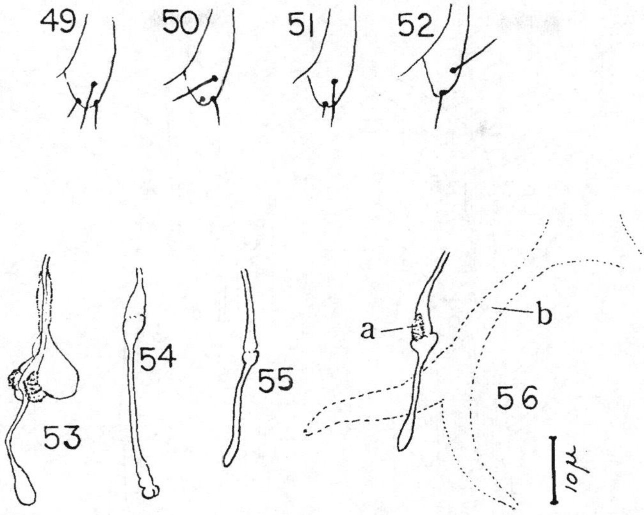

Figs. 29.49–29.56. 49–52. Abdominal appendage II. **53–56.** Canal of maxillary gland (*filamento di sostegno*) (*a*, racemose appendix; *b*, proximal arm of fulcrum). **49.** *Acerentulus.* **50,54.** *Amerentulus.* **51,55.** *Baculentulus.* **52,56.** *Nipponentomon.* **53.** *Vesciculentomon.* [After Rusek (1974).]

7a **(6)** Median apical seta on abdominal appendages II–III more than one-third of the lateral apical in length (Fig. 29.49); canal of maxillary gland with simple terminal dilatation only (Fig. 29.55) . . . . Genus ***Acerentulus*** Berlese

7b Median apical seta on abdominal appendages II–III represented only by a pit or a microchaeta (Fig. 29.50); canal of maxillary gland with a subterminal row of globules as well as terminal dilatation (Fig. 29.54) Genus ***Amerentulus*** Tuxen

In ***Amerentulus,*** the median apical seta on abdominal appendages II–III is so minute that it is easily overlooked.

8a Apical seta on abdominal appendages II–III more than one-half of the subapical in length **9**

8b Apical seta on abdominal appendages II–III less than one-half of the subapical in length (Fig. 29.51); labial palpus with a few setae apically (Fig. 29.5) **10**

9a **(8)** Labial palpus ornamented with a tuft of setae apically; striate band on abdomen VIII reduced . . . . Genus ***Tuxenentulus*** Imadaté

9b Labial palpus with a few setae apically; striate band on abdomen VIII well developed . . . . Genus ***Proacerella*** Bernard

10a **(8)** Striate band on abdomen VIII not reduced. . . . . Genus ***Gracilentulus*** Tuxen

10b Striate band on abdomen VIII reduced (Fig. 29.64) **11**

11a **(10)** Foretarsal sensilla *t* 1 claviform (Figs. 29.61, 29.62). . . . . Genus ***Berberentulus*** Tuxen

11b Foretarsal sensilla *t* 1 baculiform (Figs. 29.59, 29.60) . . . . Genus ***Baculentulus*** Tuxen

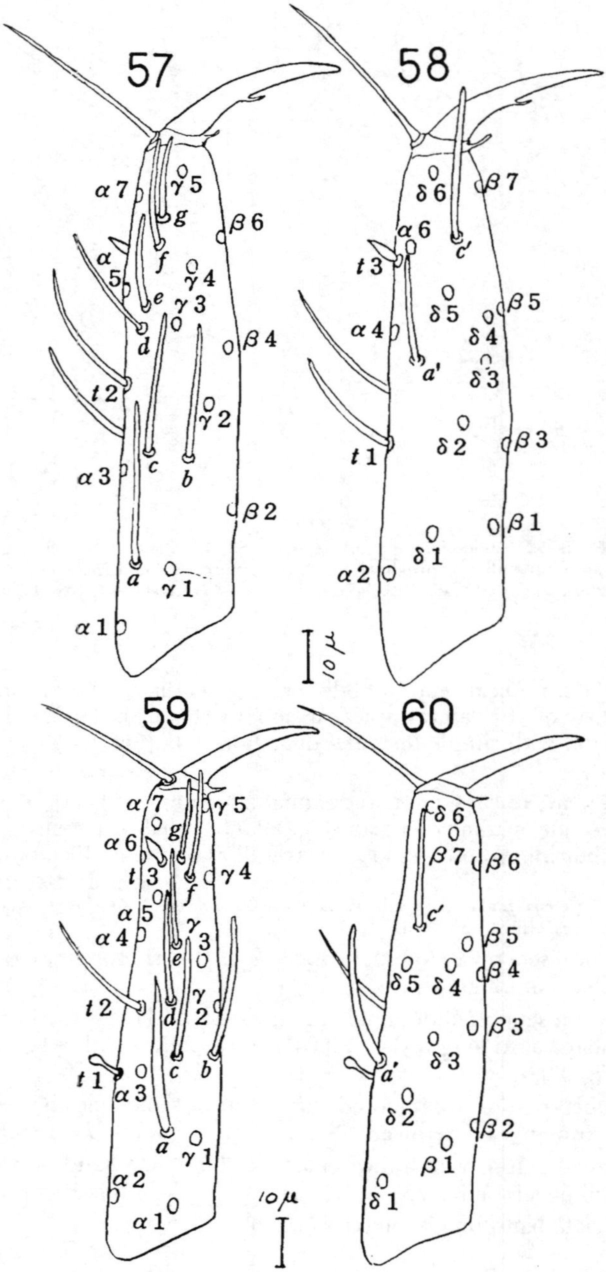

Figs. 29.57–29.67. 57–62. Foretarsal sensillae (**57, 59,** and **61,** exterior view; **58, 60,** and **62,** interior view). **63,64.** Abdominal tergite VIII. **65.** Comb on abdomen VIII. **66.** Abdomen VIII–XI, dorsal view. **67.** Abdomen IX and X, ventral view. [*m*, comb (opening of abdominal gland); *s*, striate band]. **57,58,66,67.** *Nipponentomon.* **59,60,65.** *Yamatentomon* **61–63.** *Amerentulus.* **64.** *Baculentulus.*

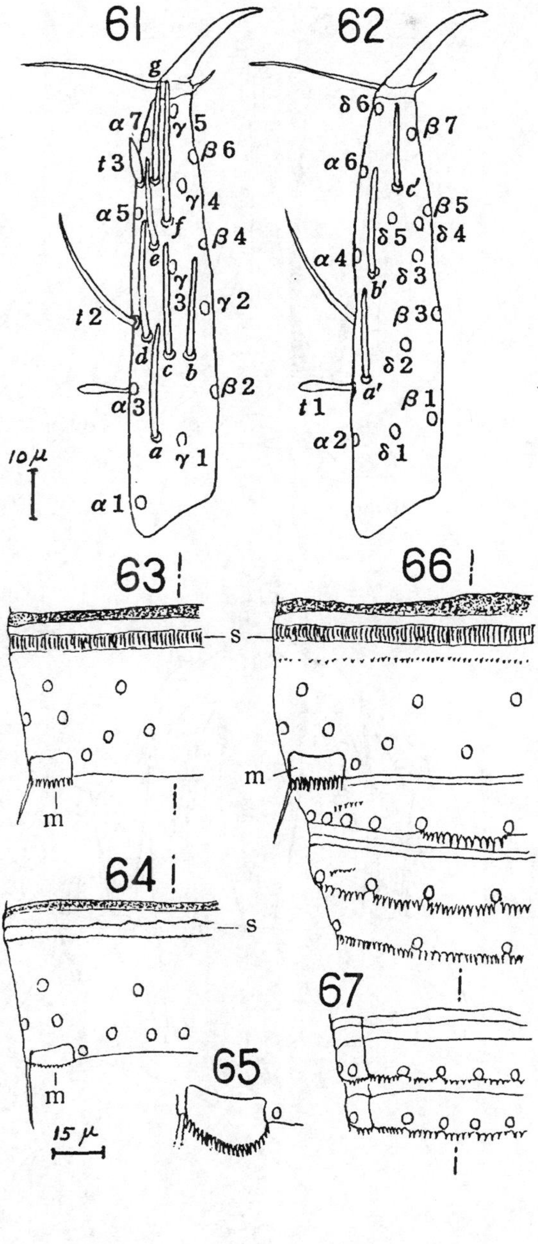
61
62
g
α7
γ5
β6
t3
γ4
α5
f
β4
e
γ
3
γ2
t2
d
c
b
β2
α3
a
γ1
10μ
α1
δ6
β7
α6
c′
β5
δ5
δ4
α4
δ3
b′
β3
δ2
a′
t1
β1
α2
δ1
63
66
s
m
64
67
65
15μ

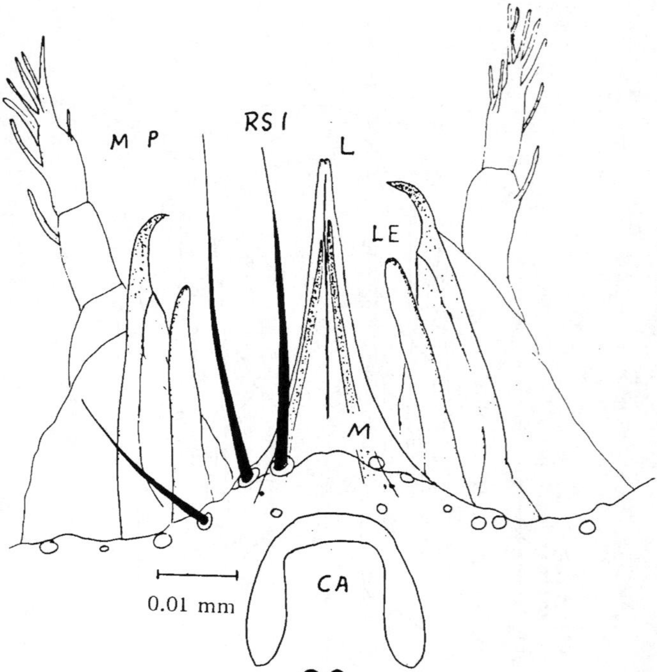

68

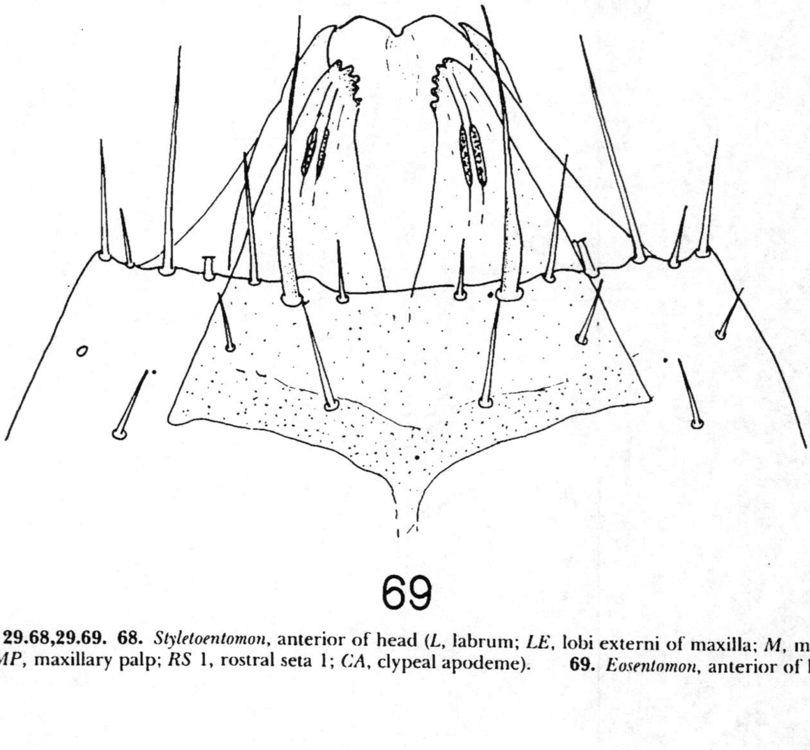

69

Figs. 29.68,29.69. 68. *Styletoentomon*, anterior of head (*L*, labrum; *LE*, lobi externi of maxilla; *M*, mandible; *MP*, maxillary palp; *RS* 1, rostral seta 1; *CA*, clypeal apodeme). **69.** *Eosentomon*, anterior of head.

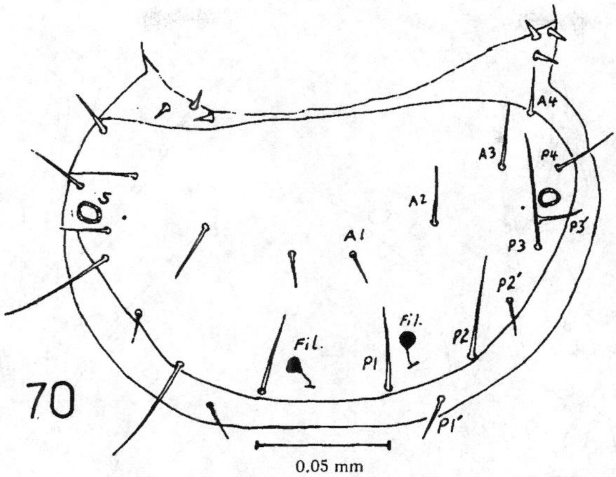

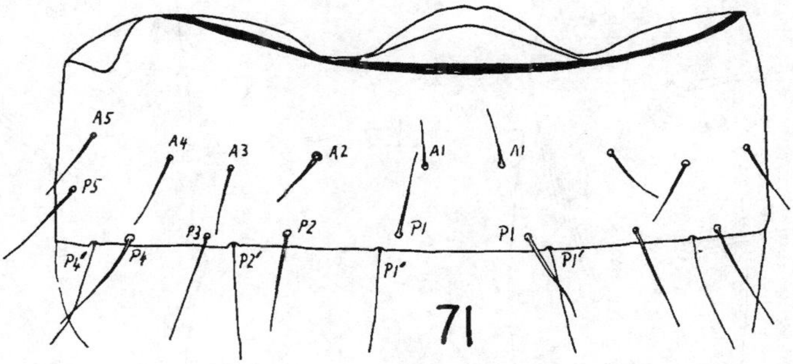

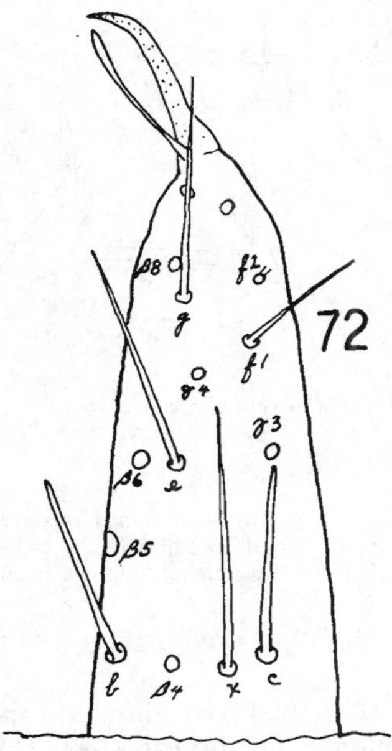

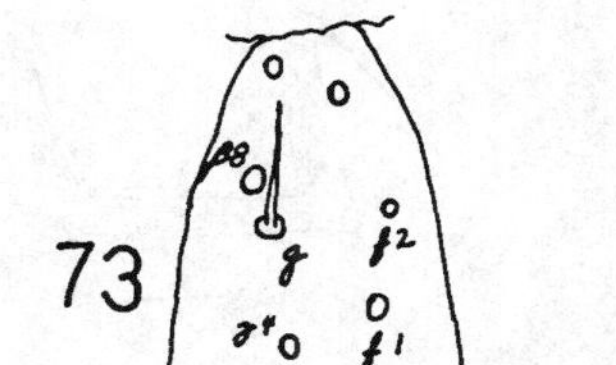

Figs. 29.70–29.73. 70. *Eosentomon*, mesothoracic tergum (*A*1–*A*4 and *P*1–*P*4, anterior and posterior setae; *Fil*, glands of *filamento di sostegno; S*, spiracle). **71.** *Eosentomon*, abdominal tergum IV. **72,73.** *Isoentomon;* foretarsus, distal–ventral end.

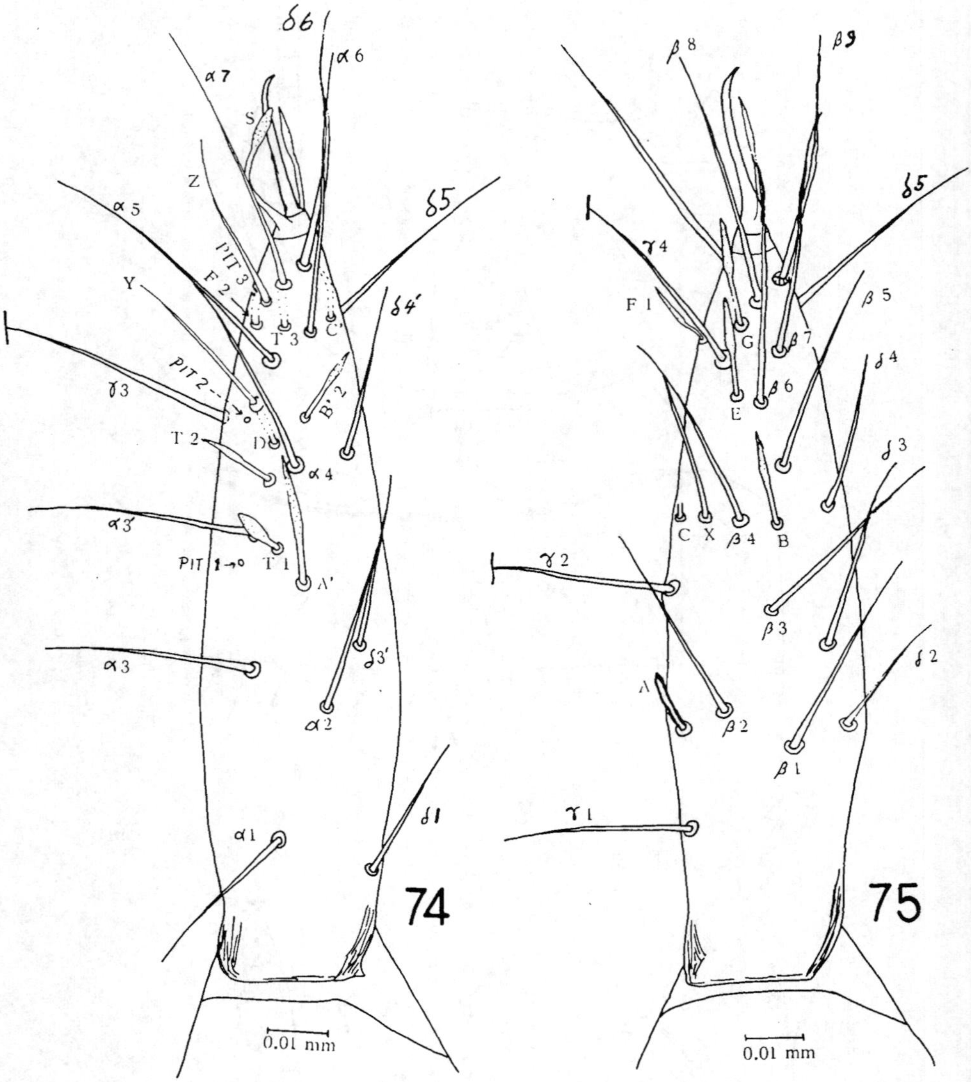

Figs. 29.74, 29.75. *Eosentomon;* foretarsus, dorsal and ventral views.

The eight acerentomid genera mentioned above have three common characteristics: Foretarsal sensilla *t* 1 claviform, baculiform, or spatulate and not filiform or spindle-shaped; metathorax with dorsal anterior setae *a* 2 and 4, without *a* 1, 3, and 5*; canal of maxillary gland without racemose appendix on the calyx.

12a (4) Foretarsal sensilla *t* 1 baculiform (Fig. 29.59) or claviform, not spindle-shaped or filiform **13**

12b Foretarsal sensilla *t* 1 spindle-shaped (Fig. 29.58) or filiform; canal of maxillary gland with racemose appendices on the calyx (Fig. 29.56) **15**

13a (12) Canal of maxillary gland with a large, smooth, vesicular appendix and racemose appendices on the calyx (Fig. 29.53); foretarsal sensilla *d* very close to *e* (Fig. 29.57); comb on abdomen VIII normal, not protruded posteriorly; metathorax with dorsal anterior setae *a* 1, 2, 3, and 4 (Fig. 29.35) or *a* 2, 3, and 4 (Fig. 29.39)*. . . Genus ***Vesiculentomon*** Rusek

13b Canal of maxillary gland with racemose appendices, but without a large, smooth, vesicular appendix on the calyx **14**

14a **(13)** Foretarsal sensilla *d* very close to *e; a'* in the same row as *t* 2; comb on

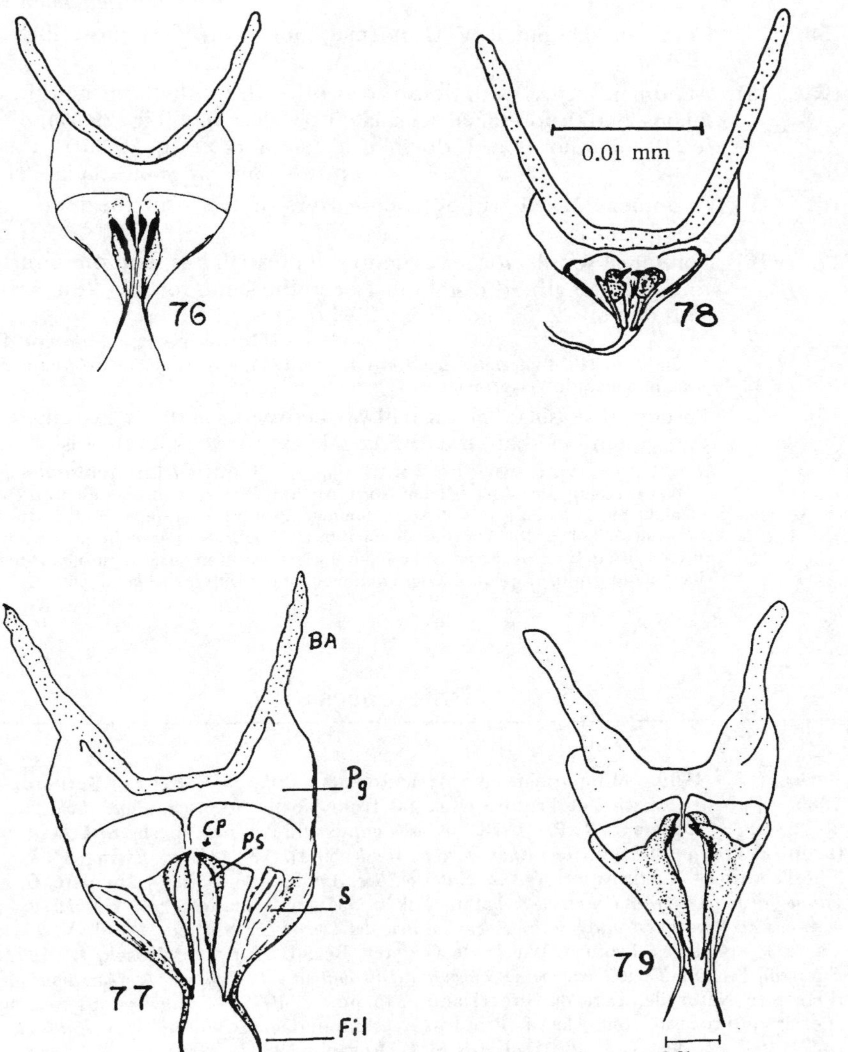

Figs. 29.76–29.79. **76.** *Isoentomon,* female genitalis. **77.** *Styletoentomon,* female genitalis. (*BA,* basal apodeme; *CP,* caput processus; *Fil,* filum processus; *Pg,* perigynium; *PS,* processus sternalis; *S,* stylus). **78,79.** *Eosentomon,* female genitalis.

abdomen VIII normal, not protruded posteriorly; metathorax with dorsal *a* 2, 3, and 4* . . . . Genus ***Nosekiella*** Rusek

14b Foretarsal sensilla *d* about halfway between *c* and *e* (Fig. 29.59); *a'* only a little distal to *t* 1 (Fig. 29.60); comb on abdomen VIII protruded posteriorly (Fig. 29.65); metathorax with dorsal *a* 2, 3, 4, and 5 (Fig. 29.40)* . . . . Genus ***Yamatentomon*** Imadaté

15a (12) Comb on abdomen VIII protruded posteriorly; foretarsal sensilla *d* about halfway between *c* and *e; a'* a little distal to *t* 1; metathorax with dorsal *a* 2, 3, 4, and 5 or *a* 1, 2, 3, 4, and 5 (Fig. 29.40)* . . . . Genus ***Filientomon*** Rusek

15b Comb on abdomen VIII normal, not protruded posteriorly (Fig. 29.66) **16**

16a (15) Abdomen IX–XI with dense rows of teeth on the hind margin (Figs. 29.66–29.67); foretarsal sensilla *d* very close to *e* (Fig. 29.57); *a'* distal to *t* 2; metathorax with dorsal *a* 1, 2, 3, and 4 (Fig. 29.35)*. . . . . . Genus ***Nipponentomon*** Imadaté et Yosii

16b Abdomen IX–XI without dense rows of teeth on the hind margin **17**

17a (16) Foretarsal sensilla *d* very close to *e; a* normal, less than one-fourth the foretarsal length; *a'* distal to *t* 1 or in the same row as *t* 2; metathorax with dorsal *a* 2, 3, and 4 (Fig. 29.39)*. . . . . Genus ***Verrucoentomon*** Rusek

In Rusek (1974), ***Acerella canadensis*** Tuxen, 1955, was erroneously placed in ***Nosekiella*** instead of in ***Verrucoentomon.***

17b Foretarsal sensilla *d* about halfway between *c* and *e; a* extremely long, more than one-half the foretarsal length; *a'* slightly distal to *t* 2; metathorax with dorsal *a* 2 and 4* . . . Genus ***Alaskaentomon*** Nosek

Very recently a new genus from North America, ***Paracerella,*** was described (Imadaté, G. 1980. A new genus of Acerentomidae (Protura) from Japan and North America. Kontyû, Tokyo, 48). This new genus runs to 17b ***Alaskaentomon*** in our key, but the former differs from the latter in the following features: foretarsal sensilla *a* normal, less than one-fourth the foretarsal length and metathorax with dorsal *a* 2, 3, and 4.

References

Berlese, A. 1909. Monografia dei Myrientomata. *Redia* 6:1–182. **Bernard, E. C. 1985.** Two new species of Protura (Insecta) from North America. *Proc. Biol. Soc. Wash.* 98:72–80. **Copeland, T. P. 1978.** A new genus and two new species of Eosentomoidea (Protura: Eosentomidae). *Proc. Entomol. Soc. Wash.* 80(4):473–484. **Ewing, H. E. 1940.** The Protura of North America. *Ann. Entomol. Soc. Am.* 33:495–551. **Imadaté, G. 1974.** *Fauna Japonica Protura (Insecta).* Keigaku, Tokyo. 351 pp. **Janetschek, H. 1970.** *Handbuch der Zoologie eine Naturgeschichte der Staemme des Tierreiches,* 2nd ed., Band IV, 2. Haelfte, Tiel 2/3; Spegielles, Protura. Walter de Gruyter, Berlin. 72 pp. **Nosek, J. 1973.** *The European Protura: Their Taxonomy, Ecology and Distribution with Keys for Determination.* Muséum d'Histoire Naturelle, Geneva, Switzerland. 345 pp. **1977.** A new genus and six new species of Protura from Alaska (Protura: Acerentomidae, Eosentomidae). *Entomol. Scand.* 8:271–284. **Nosek, J. and D. Keith McE. Kevan. 1984.** A species of Acerentomidae (Protura) new to Canada. *Rev. Entomol. Que.* 29:55–61. **1978.** Key and diagnoses of proturan genera of the world. *Annot. Zool. Bot.* Bratislava 122:1–54. **Szeptycki, A. 1985.** Polish Protura. III. *Eosentomon bohemicum* Rusek and related species, 1966. *Pol. Pismo Entomol.* 55:531–574. **Tuxen, S. L. 1931.** Monographie der Proturen. I. Morphologie nebst Bemerkungen über Systematik and Ökologie. *Z. Morphol. Oekol. Tiere* 22:671–

720. **1949.** Uber den Lebenszyklus und die postembryonale Entwicklung zweier danisher Proturengattungen. *K. Dan. Vidensk. Selsk. Biol. Skr.* 6(3):1–50. **1963.** Phylogenetical trends in the Protura. *Z. Zool. Syst. Evolution forsch.* 1:277–310. **1964.** *The Protura. A Revision of the Species of the World with Keys for Determination.* Hermann, Paris. 360 pp. **1984.** *Brasilidia* and a new genus and species of Protura (Insecta) from Argentina. Together with an improvised key to Acerentomidae. *Rev. Ecol. Biol. Sol* 21:283–295. **Yin, W.-Y. 1965.** Studies on Chinese Protura. II. A new family of the suborder Eosentomoidea. *Acta Entomol. Sin.* 14(2):186–195. **1984.** A new phylogeny of Protura with approach to its origin and systematic position. *Sci. Sin. Ser.* B (*Chem. Biol. Agric. Med. & Earth Sci.*) 27:149–150.

30

Insecta: Microcoryphia and Thysanura

LYNN M. FERGUSON

The thysanurans are small wingless insects of two principal types of structure represented by the Orders Thysanura (Bristletails, Silverfish, Firebrats) and Microcoryphia (Jumping Bristletails). They are similar to the pterygote insects in that they are ectognathus, that is, the mandibles and maxillae are fully exposed and the labium is a free appendage. Their antennae and external reproductive structures are also similar to those of the winged insects. Bristletails are found living in forest litter, under bark, in rock crevices, and under stones. Silverfish and firebrats are found in human dwellings and other buildings. Still others live in termite and ant nests and in caves.

The Order Thysanura previously included the silverfish, firebrats, jumping bristletails, and lesser known entities (some within the order and others that are now considered as not belonging, such as the diplurans). Remington (1954) proposed that the order be divided into two orders or suborders, corresponding to the former two superfamilies Machiloidea, the jumping bristletails, and the Lepismatoidea, including the silverfish and firebrats.

The silverfish and jumping bristletails have much in common morphologically and biologically, such as similarities in their mating behavior; therefore, in this chapter the two orders will be discussed together in regard to their morphology, biology, and collection. The keys illustrate the hierarchical ranking of the various categories of taxa found in the United States. Although six or seven families, containing eight subfamilies (Remington 1954), are presently recognized within these two orders, there are basically two types that make up most of the species commonly encountered: the Lepismatidae, which includes the silverfish (*Lepisma, Cteno-*

lepisma) and firebrat (*Thermobia*), and the Machilidae, which includes the jumping bristletails (*Petrobius, Machilis,* and others).

Morphological Characteristics

Primitively wingless, terrestrial insects (Figs. 30.1, 30.2). Body length, 4–20 mm; with or without pigmented scales; joints of antennal flagellum lacking intrinsic muscles; mouthparts ectognathus, simple; maxillae and labium with long palpi; thoracic segments with large paranotal lobes; legs all similar, coxae large, tarsi with two to five segments; most abdominal segments with a pair of styli on the coxopodites (Fig. 30.5*H,E*), some of which may also have posterior paired exsertile vesicles (often secondarily lost) (Fig. 30.5*E,I*); tip of abdomen with eleventh segment bearing two cerci and a median caudal filament; penis well developed (Fig. 30.5*H*); ovipositor rather large, composed of two pairs of gonapophyses.

Order Microcoryphia

Body form subcylindrical (Fig. 30.1), thorax strongly arched; head hypognathous; eyes large, contiguous, with hundreds of ommatidia (Fig. 30.6*A–D*); vertex greatly reduced; frons large, with median ocellus; lateral ocelli large; mandibles primitive, consisting of a basal part with a single point of articulation to the cranium and a distal part with separate molar and incisor processes (Fig. 30.5*G*); palpus of maxillae with seven segments; labium with palpus of three unequal segments, the distal not greatly expanded apically; thoracic terga with very large paranotal lobes against sides; all leg segments essentially cylindrical; meso- and metathoracic coxae typically with styli (may be absent on mesothoracic coxa or both); femur and tibia large; tarsus with three tarsomeres; abdominal segments II–VII covered ventrally by three distinct sclerites: an anteromedial sternite and two lateral coxopodites (Fig. 30.5*D,E*); coxopodites of abdominal segments II–IX each with a posterolateral stylus (Fig. 30.5*D,E,H*); posterior edge of segments I–VII typically with one or two pairs of exsertile vesicles mesad of the styli (commonly absent on segment I, absent on V–VII in one species, all absent in one genus) (Fig. 30.5*D,E*).

Order Thysanura

Body form more or less flattened dorsoventrally (Fig. 30.2), thorax never arched; head usually prognathous; eyes absent or rather small, laterally situated, far apart, with about 12 large facets which may be independent, simple eyes; ocelli usually absent; vertex and frons broad; mandible very different from Microcoryphia, having both anterior and posterior articulation points and with the molar and incisor areas confluent and not very different; palpus of maxillae with five segments (distal segment may consist of two pseudosegments) (Fig. 30.5*A,B*); labium with palpus of three segments, usually very broad apically (Fig. 30.3*G,I*); thoracic terga with strong paranotal lobes but much less than in Microcoryphia; coxae very large, strongly flattened dorsoventrally (Fig. 30.3*E*), never with styli; femur and tibia smaller than in Microcoryphia; tarsus with two to five tarsomeres (Fig. 30.4*B,E*); abdominal segments II–VII covered ventrally by a single undivided sclerite (possibly fused sternite and coxopodites) (Fig. 30.5*C*); posterolateral styli present on some segments; posterior edge of some abdominal segments with one pair of exsertile vesicles, but often entirely absent.

In addition to the morphological characters listed above for the two orders, *Machilis* and *Lepisma* have 9 and 10 pairs of spiracles, respectively. However, the general condition of the respiratory organs in the thysanurans is primitive, with little or no specialization. In the Machilidae the tracheae do not anastomose, and the system of each spiracle is separate from that of its neighbor. There is more elaboration in *Lepisma,* with a common longitudinal trunk running down each side, and a transverse trunk in each segment uniting the spiracles of opposite sides (Lawrence 1953). In both orders the spiracles are simple and without a closing apparatus.

It has been suggested that the exsertile vesicles (eversible sacs) found on the ventral surface of various segments of the abdomen represent primitive organs of respiration. In *Campodea* (Order Diplura) the exsertile vesicles function in water absorption (Drummond 1953). These thin-walled sacs are covered with a delicate membrane and can be distended with blood, the pressure of which serves to evaginate them. The vesicles are collapsible and can be withdrawn beneath the body surface by the aid of special retractor muscles attached to their walls. In the thysanurans the exsertile vesicles are associated with small stylelike appendages and are more numerous in *Machilis,* two pairs being found on abdominal sternites II–V and one pair on sternites VI and VII; *Lepisma* has none.

Sensory structures include compound eyes and ocelli. The compound eyes of *Machilis* are very large with approximately 1000 minute facets and in their outward appearance closely resemble the eyes of most pterygote insects. It is theoretically possible that both mosaic images and movement can be perceived by *Machilis.* The eyes of *Lepisma,* on the other hand, are much reduced and the ocelli are absent. There are only 12 strongly convex facets, larger than those of *Machilis* and resembling the eyes of the Isopoda, rather than those of flying insects. The eyes of *Lepisma* are obviously simpler than those of *Machilis* or the pterygote insects, having in the course of time undergone reduction and simplification of their elements. The primitive thysanuran *Tricholepidion gertschi* (Family Lepidotrichidae, Order Thysanura) still retains ocelli (Wygodzinsky 1961).

Cryptozoic animals are characterized by the poor development of their visual organs, the Machilidae being the exception. Tactile organs, on the other hand, are numerous and highly specialized. Of particular note are the unique and peculiar type of sensory hair represented by the trichobothrium, a specialized organ of tactile sense which is almost entirely confined to the cryptozoic orders (Lawrence 1953). Trichobothria are long, fine, delicate hairs set in large distinct pits in the cuticle. They may combine both auditory and tactile functions. In the Machilidae, Wygodzinsky (1941) has found trichobothria on the legs and maxillary palps of certain males, as well as on the thoracic segments. In addition to these, a tactile sense spine has been found between the ommatidia of the machilid *Petrobius*. And a paired or single "frontal organ" of unknown function is found in *Lepisma* and *Machilis*. The organ consists of an aggregation of sense cells near the eyes and are connected with the middle part of the brain by a narrow strand of nervous tissue. Only in *Lepisma* do the frontal organs still remain in contact with the cuticle.

For additional material on morphology, consult Matsuda (1965, 1970, 1976), Manton (1972), Snodgrass (1935, 1952), Boudreaux (1979), and Gupta (1979).

Feeding and Predation

Machilis and *Lepisma* are both browsers on a wide range of softer plant materials such as algae and lichens. Their mandibles, although different in form, both have masticating structures. *Lepisma saccharina,* a pest in domestic dwellings and warehouses, consumes paper, the starch coverings of book bindings, and various woven fabrics.

Predators of *Machilis* include salticid spiders and scorpions. A harvestman (Opiliones) was observed to eat the molted skin of a species of *Machilis;* however, they did not attack the living animals.

Escape movements by thysanurans include rapid running by *Lepisma* and the jumping ability of the Machilidae. The leaping movements of the Machilidae are somewhat similar to those of the Amphipoda. There is a sudden bending and release of the posterior part of the abdomen which, striking against the ground, enables the body to be catapulted upward and forward. The legs and cercal appendages play a minor role in leaping, since *Machilis* can still leap effectively when these appendges have been amputated. According to Denis (1949), *Petrobius* is able to leap a distance of 10 cm and 4–5 cm into the air.

Reproduction

Thysanurans initiate courtship by vibrating one another with their antennae and palps, particularly by the male to determine the readiness of the female for mating. This being accomplished, the male of the silverfish *Lepisma* quickly swings his tail back and forth against a wall or other elevation. He spins a few threads with his penis and under the threads he deposits a spermatophore on the ground. The female then runs under the threads, touching them with her raised bristlelike abdominal appendages. She then stops and searches for the sperm packet with her genital opening, finally taking it up. After the release of the spermatozoa, the empty spermatophore comes to rest between the genital valves and is eaten by the female.

In *Machilis* the male also spins a thread with his penis, which he attaches to the ground at one end; the other end he holds high in the air by raising his abdomen. Three or four sperm droplets are placed along the thread. The male wraps around the female, turning her so that her raised abdomen comes in contact with the thread. She then picks up the sperm droplets with the end of her ovipositor.

The courtship behavior in both thysanurans ensures a rapid transfer of the sperm to the female's reproductive tract. This external method of fertilization also indicates the primitiveness of these terrestrial arthropods and in part restricts them to a humid environment for the sake of reproduction.

After fertilization of her eggs, the female of *Machilis* searches for a crevice or depression on a stone with her flexible and sensitive ovipositor. When a suitable place is found, the depression is cleared further with the ovipositor before the eggs are laid, either singly or in small groups of 14–32 eggs. The eggs are then made inconspicuous with a covering of minute particles of soil or debris (Wygodzinsky 1941). *Ctenolepisma* lays its eggs in lots of 2–20, pushing them about 20 mm into crevices with its long ovipositor. The eggs of *Ctenolepisma* measure 1.15 mm × 0.83 mm (Lindsay 1940). It is probable that more than one batch of eggs is laid by a female during a summer.

In the Thysanura the sex ratios of *Ctenolepisma longicaudata* are about equal. In the Microcoryphia males are absent in at least two species of *Machilis* and numerous in others, and *Trigoniophthalmus alternatus* in North America is apparently parthenogenetic.

Life Cycle

The egg in the Machilidae remains at the embryological stage far longer than in other Apterygota, Wygodzinsky (1941) having recorded a dormant period of 9 months for *Machilis* and 11–12 months for *Trigoniophthalmus*. In *Machilis* the number of larval stages is six or more. During the first two or three stages there are no scales on the body or legs. The first stage is fairly active but no food is taken. The second stage larvae eat their own exuviae.

The Lepismatidae have the largest number of stages in the Apterygota, *Ctenolepisma* having 14 according to Lindsay (1940). In the first stage the larvae are very sluggish and no food is taken. In the second stage the larvae actively seek food. There is little difference among the various instars except that of size.

There is probably a long period of growth between the end of the larval phase and the attainment of full maturity, proceeding by small and almost imperceptible changes (Lawrence 1953). The number of molts undertaken before reaching maturity in *Machilis* is at least five and in *Ctenolepisma* about 13. Molting continues in the adults, with as many as 60 molts having been recorded. It has been suggested that the number of molts is of more than the usual importance for the cryptozoic fauna, since they are particularly liable to external infection from various molds and other fungi. Such attacks are extremely lethal, and those organisms which molt often, both before and after maturity, would be at an advantage in warding off the infection before it had taken a hold or penetrated the cuticle.

The duration of ecdysis or molting seems to be a matter of minutes, taking about 10–15 minutes for *Lepisma,* but only 4–5 minutes for *Machilis.* The cuticle or exoskeleton splits longitudinally down the middle of the back, including the head, thorax, and a number of abdominal segments. The exuviae are often found on the underside of large stones, although Wygodzinsky (1941) found that the molted skin was often eaten by various Machilidae and *Lepisma.* Lindsay (1940) reported that the silverfish *Ctenolepisma longicaudata* sometimes eats the molt. No special precautions seem to be taken to ensure safety and privacy at this critical time in the life cycle.

Thysanurans are particularly long-lived. The known duration of life for *Lepismachilis* is 3 years (Wygodzinsky 1941), and for *Ctenolepisma* 7–8 years (Lindsay 1940).

Collection

Specimens may be taken by means of a Berlese funnel or by actively searching under stones, logs, and other plant debris. They may be collected into small vials; however, a small net is useful for catching the jumping bristletails. Specimens should be preserved in 75–80% alcohol.

Identification

The keys that follow are based largely on the works of the late Pedro Wygodzinsky (1961, 1972, 1980), a recognized world authority on Thysanura.

KEYS TO THE GENERA AND TO SOME SPECIES OF THYSANURA (SENSU STRICTO AND MICROCORYPHIA) OF THE CONTIGUOUS UNITED STATES

1a Body cylindrical or laterally compressed; one mandibular condyle; compound eyes large, contiguous, ocelli present; coxae bear styli (Fig. 30.1); spiracles present on abdominal segments II–VIII; cerci shorter than caudal filament (Fig. 30.1) . . . . Subclass **Archaeognatha,** Order **Microcoryphia 18**

1b Body flattened; two mandibular condyles; compound eyes small or absent, ocelli often absent; coxae without styli; spiracles present on abdominal segments I–VIII; cerci about as long as caudal filament (Fig. 30.2) . . . . Subclass **Dicondylia,** Order **Thysanura 2**

2a (1) Compound eyes and ocelli absent . . . . Family **Nicoletiidae 4**

2b Eyes present. **3**

3a (2) Ocelli absent; scales present. . . . . Family **Lepismatidae 6**

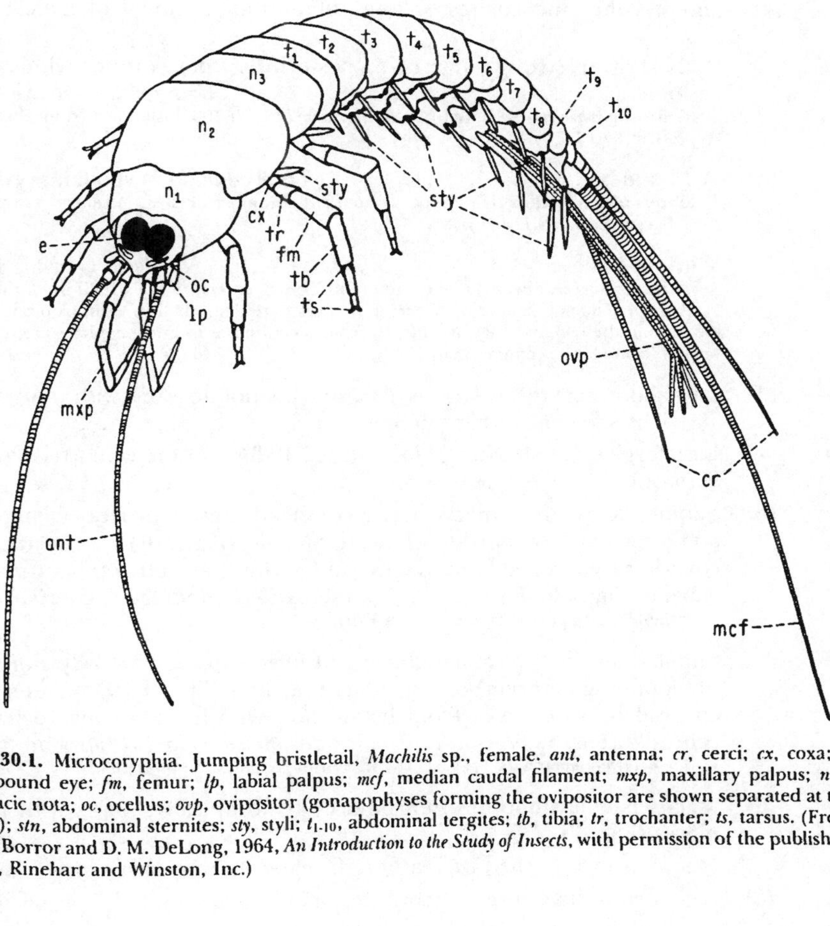

Fig. 30.1. Microcoryphia. Jumping bristletail, *Machilis* sp., female. *ant*, antenna; *cr*, cerci; *cx*, coxa; *e*, compound eye; *fm*, femur; *lp*, labial palpus; *mcf*, median caudal filament; *mxp*, maxillary palpus; $n_{1\text{-}3}$, thoracic nota; *oc*, ocellus; *ovp*, ovipositor (gonapophyses forming the ovipositor are shown separated at the apex); *stn*, abdominal sternites; *sty*, styli; $t_{1\text{-}10}$, abdominal tergites; *tb*, tibia; *tr*, trochanter; *ts*, tarsus. (From D. J. Borror and D. M. DeLong, 1964, *An Introduction to the Study of Insects*, with permission of the publisher, Holt, Rinehart and Winston, Inc.)

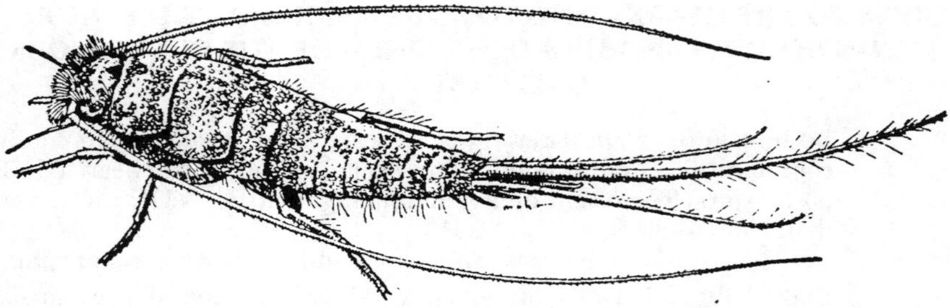

Fig. 30.2. Thysanura. *Thermobia domestica*, firebrat. (From Illinois Natural History Survey; [see H. H. Ross, *A Textbook of Entomology*, 3rd ed., 1965, John Wiley & Sons, Inc., Fig. 166, p. 219.]

3b Ocelli present; scales absent . . . . Family **Lepidotrichidae**
Only one living genus and species known, ***Tricholepidion gertschi*** Wygodzinsky 1961, found in leaf litter and under bark in a small area of northern California. Other members of this family are represented by fossils in Baltic amber.

4a (2) Body elongate, approximately parallel-sided; thorax slightly wider than abdomen; abdominal sterna II–VII subdivided into coxites and sternite, and coxites of ninth abdominal segment of male fused **5**

4b Body shorter, strongly narrowed posteriorly; thorax much wider than abdomen . . . . ***Grassiella*** Silvestri 1898
Grassiella wheeleri (Escherich) [= ***Battigrassiella*** Paclt 1963] is known from Texas (Wygodzinsky 1958).

5a (4) With scales . . . . ***Texoreddellia*** Wygodzinsky 1973
Texoreddellia texensis (Ulrich) occurs in a large number of caves in central Texas and the adjacent Edwards Plateau.

5b Without scales . . . . ***Nicoletia*** Gervais 1842
Nicoletia ruckeri Silvestri 1905 is found in Texas, ***N. tergata*** Mills 1940 in Florida. Parthenogenetic populations of ***N. meinerti*** Silvestri 1905 are known from tropical localities in the Old and New World and from greenhouses in Europe. Males are only known from Hawaii (Zimmerman 1948).

6a (3) Last abdominal tergite triangular, sharply pointed and with more than 1 + 1 bristle-combs (Fig. 30.3*A*). **7**

6b Last tergite not sharply pointed (Fig. 30.3*F*, *H*) or, when triangular, with only 1 + 1 bristle-combs **8**

7a (6) Submedian fields of macrochaetae on head capsule broadly triangular, attaining anterior border of head capsule (Fig. 30.3*C*); prosternum covering fore coxae from below, and without a central tuft of macrochaetae (Fig. 30.3*B*) . . . . ***Stylifera gigantea*** Escherich 1905
Free-living populations in southern Florida.

7b Submedian fields of macrochaetae of head capsule narrowly elongate, not attaining anterior border of head capsule (Fig. 30.3*D*); prosternum covered by fore coxae from below, and with a tuft of macrochaetae centrally (Fig. 30.3*E*). . . . . ***Acrotelsa collaris*** (Fabricius 1793)
[= ***Lepisma mucronata*** Packard 1873]; Florida, common in households.

8a (6) Macrochaetae smooth, often bifid or trifid apically; males with parameres (Fig. 30.5*H*) **9**

8b Macrochaetae barbed or feathered; males lacking parameres . . . **10**

9a (8) Last abdominal tergite truncate or faintly rounded apically (Fig.

30.3*F*); apical segment of labial palp wider than long (Fig. 30.3*G*); dorsum of undamaged living specimens uniformly slate gray. . . . . ***Lepisma saccharina*** Linnaeus 1758

Silverfish. Cosmopolitan in habitations; free-living populations found only in extreme southern Florida.

9b Last abdominal tergite triangular (Fig. 30.3*H*); apical segment of labial palp longer than wide (Fig. 30.3*I*); dorsum of undamaged living specimens dark brown . . . . ***Allacrotelsa spinulata*** (Packard 1873)

[= ***Stachisma mexicana*** Wygodzinsky 1949]; very common in southern and western states: Tennessee, South Carolina, Florida, Arkansas, Oklahoma, Texas, Arizona, California, Oregon, and Washington.

10a **(8)** Insects golden yellow; body length not more than 6 mm when mature; generally myrmecophilous; caudal appendages very short, not longer

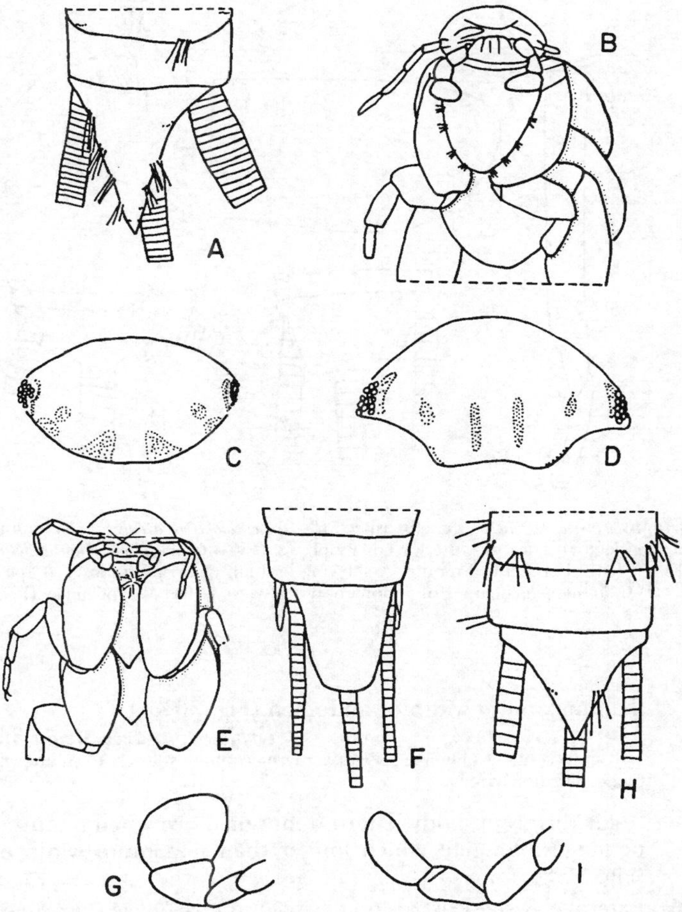

Fig. 30.3. (*A–C*) *Stylifera gigantea*: (*A*) apex of abdomen, seen from above and slightly lateral; (*B*) anterior portion of body, ventral view; (*C*) head capsule with fields of macrochaetae, schematic. (*D,E*) *Acrotelsa collaris*: (*D*) head capsule with fields of macrochaetae, schematic; (*E*) anterior portion of body, ventral view. (*F,G*) *Lepisma saccharina*: (*F*) apex of abdomen, dorsal view; (*G*) labial palp. (*H,I*) *Allacrotelsa spinulata*: (*H*) apex of abdomen, dorsal view; (*I*) labial palp. [After Wygodzinsky (1972).]

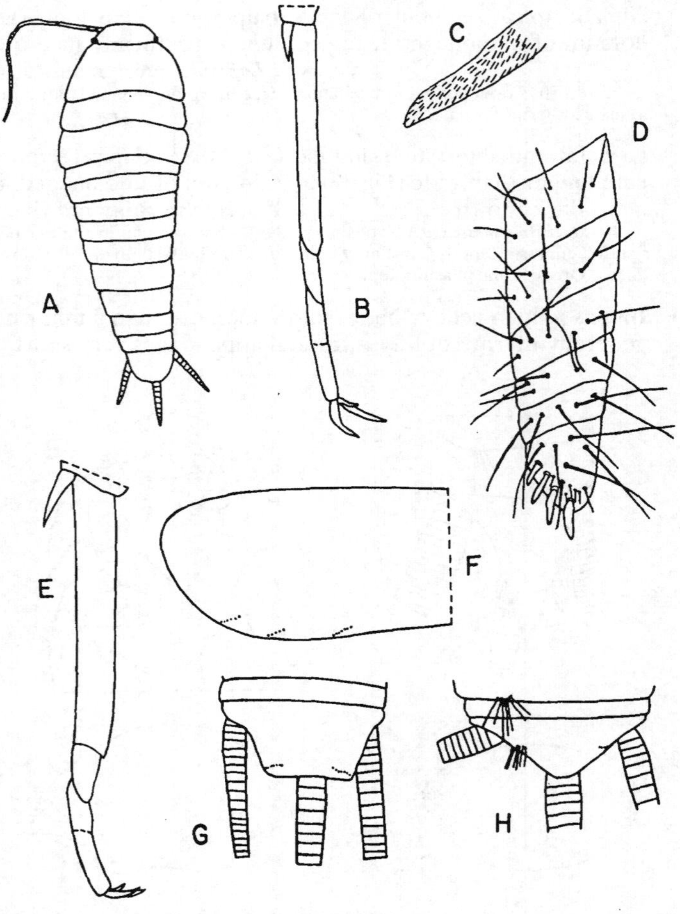

Fig. 30.4. (*A*) *Mirolepisma deserticola*, general aspect. (*B–D*) *Leucolepisma arenaria*; (*B*) hind tarsus; (*C*) median claw of hind leg; (*D*) apex of posterior gonapophysis. (*E–G*) *Ctenolepisma longicaudata*: (*E*) tarsus of hind leg; (*F*) left half of abdominal tergite V, showing location of bristle-combs; (*G*) apex of abdomen, dorsal view. (*H*) *C. lineata pilifera*, apex of abdomen, dorsal view. [After Wygodzinsky (1972).]

than maximum width of abdomen (Fig. 30.4*A*) . . . . **Mirolepisma deserticola** Silvestri 1938

Genus contains only one species, a myrmecophilous desert form known only from California and Arizona.

10b Color different; body length 6 mm or more when mature; caudal appendages elongate, much longer than maximum width of abdomen (Fig. 30.2) **11**

11a **(10)** Lateral claws at least as long as second tarsomere (Fig. 30.4*B*); median claw with conspicuous cuticular projections (Fig. 30.4*C*); ovipositor apically with fossorial spines (Fig. 30.4*D*) . . . . **Leucolepisma arenaria** Wall 1954

Genus contains a single species, a desert form found only in the southwest: California, Nevada, and Arizona.

11b Lateral claws shorter than second tarsomere (Fig. 30.4*E*); median claw smooth; ovipositor lacking fossorial spines **12**

12a (11) Several abdominal tergites with 3 + 3 bristle-combs (Fig. 30.4*F*); maxillary palps with five segments (Fig. 30.5*B*) **13**

12b Abdominal tergites at most with 2 + 2 bristle-combs; maxillary palps with five or six (Fig. 30.5*A*) apparent segments **17**

13a (12) Abdominal sternites each with one median bristle-comb (Fig. 30.5*C*); 3 + 3 bristle-combs on abdominal tergites II–V . . . .
Ctenolepisma targionii (Grassi and Rovelli 1889)

Only known from South Carolina in the western hemisphere; original range is the Mediterranean region.

13b Abdominal sternites without median bristle-combs; 3 + 3 bristle-combs on abdominal tergites II–V, II–VI, or II–VII **14**

14a (13) Only abdominal tergites II–V with 3 + 3 bristle-combs . . . .
Ctenolepisma diversisquamis Silvestri 1908

[= ***C. reducta*** Folsom 1923]; free-living in southern Florida.

14b Abdominal tergites II–VI or II–VII with 3 + 3 bristle-combs . . . **15**

15a (14) Only abdominal tergites II–VI with 3 + 3 bristle-combs; tenth abdominal tergite truncate or faintly emarginate posteriorly (Fig. 30.4*G*) **16**

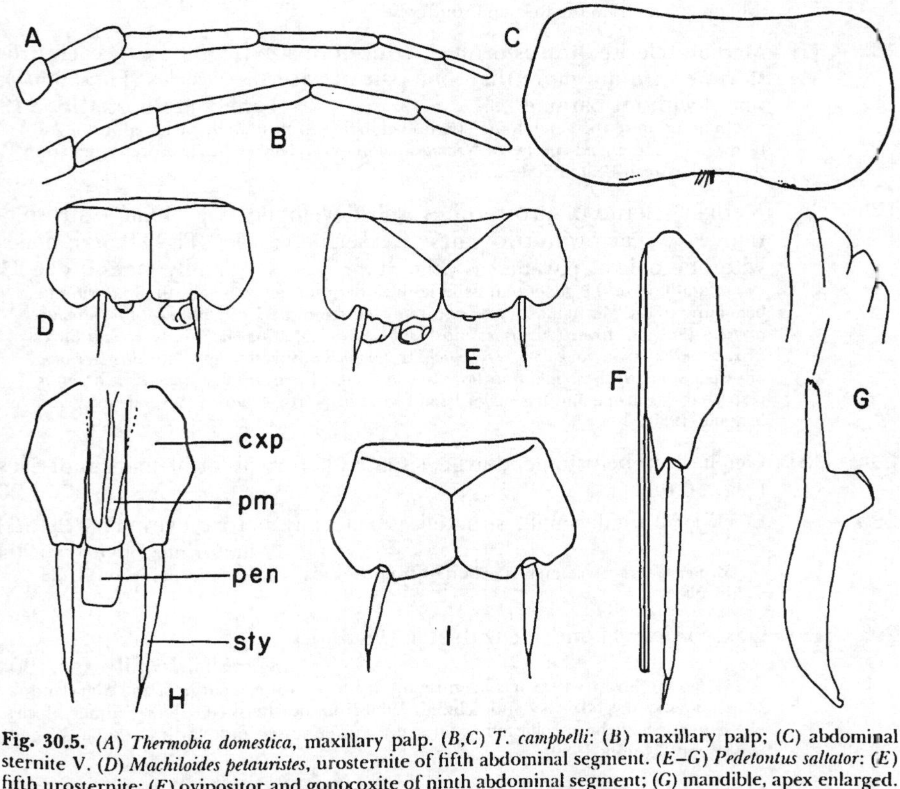

Fig. 30.5. (*A*) *Thermobia domestica*, maxillary palp. (*B,C*) *T. campbelli*: (*B*) maxillary palp; (*C*) abdominal sternite V. (*D*) *Machiloides petauristes*, urosternite of fifth abdominal segment. (*E–G*) *Pedetontus saltator*: (*E*) fifth urosternite; (*F*) ovipositor and gonocoxite of ninth abdominal segment; (*G*) mandible, apex enlarged. (*H,I*) *Petrobius maritimus*: (*H*) Genital region: *cxp*, coxopodite; *pen*, penis; *pm*, paramere; *sty*, stylus; (*I*) urosternite VIII of male. [(*A–C*) After Wygodzinsky (1972); (*D–I*) after Wygodzinsky and Schmidt (1980).]

15b Abdominal tergites II–VII with 3 + 3 bristle-combs; tenth abdominal tergite subtriangular (Fig. 30.4*H*). **Ctenolepisma lineata pilifera** (Lucas 1840)

Four-lined silverfish. [= ***Lepisma quadriseriata*** Packard 1873, ***L. rubro-violacea*** Schött 1897, ***L. reticulata*** Schött 1897]. Introduced from the Mediterranean region; very common in the southwest: Oregon, California, Nevada, and Arizona. Also common in the east: known from Maine to Tennessee and Georgia, and west to Missouri.

16a **(15)** Maximum length about 10 mm; antennae and caudal filaments not longer than body . . . . ***Ctenolepisma ciliata*** (Dufour 1831)

California; introduced from the Mediterranean area.

16b Length 13–15 mm; antennae and caudal appendages longer than body. . . . . ***Ctenolepisma longicaudata*** Escherich 1905

Long-tailed silverfish. [= ***C. urbana*** Slabaugh 1940]. Found in dwellings: North Carolina to Florida, west to Louisiana; Illinois, Missouri, and southern California.

17a **(12)** Maxillary palp with five segments (Fig. 30.5*B*); only one pair of stylets present; scales of undamaged living specimen uniformly gray dorsally ***Thermobia campbelli*** (Barnhart 1951)

Only known from two domestic populations in Pennsylvania and Ohio.

17b Maxillary palp with six apparent segments (Fig. 30.5*A*); two or three pairs of stylets present; scales of undamaged living specimen forming complex pattern. . . . . ***Thermobia domestica*** (Packard 1873)

Firebrat. Found throughout North America in dwellings where it is warm. Free-living populations in the west and southwest.

18a **(1)** Median sclerite of urosternites reduced in size (Fig. 30.5*D*); each urosternite with not more than one pair of exsertile vesicles (Fig. 30.5*D*); males without parameres . . . . Family **Meinertellidae 19**

In addition to the genera and species belonging to this family included in the key below, an undescribed species of ***Neomachilellus*** Wygodzinsky has been reported from the southeastern United States.

18b Median sclerite of urosternites well developed (Fig. 30.5*E*); urosternites II–V each with two pairs of exsertile vesicles (Fig. 30.5*E*); males with one pair of parameres (Fig. 30.5*H*). . . . Family **Machilidae 21**

In addition to the genera and species included in the key below, the following taxa belonging to the Machilidae have been reported from the United States: ***Petridiobus arcticus*** (Folsom) from Arkansas; and from California, ***Mesomachilis nearcticus*** Silvestri, ***Mixomachilis remingtoni*** Paclt, ***Neomachilis halophilus*** Silvestri, and ***Machilis strenuva*** Silvestri. Some of these identifications are in doubt; for example, ***Machilis*** is a genus restricted to Europe and has never been found in North America (Wygodzinsky and Schmidt 1980).

19a **(18)** Ocelli dumbbell-shaped, large, situated before anterior margin of eyes (Fig. 30.6*A*) **20**

19b Ocelli rounded, small, situated sublaterally before eyes (Fig. 30.6*B*) ***Machilinus*** Silvestri 1904

M. nevadensis Sweetman is reported from Nevada; ***M. aurantiacus*** (Schött) from California.

20a **(19)** Coxa of legs II and III with styli (Fig. 30.1). . . . . ***Machiloides*** Silvestri 1904

M. banksi Silvestri 1911 has been reported form northern Virginia and Maryland; ***M. petauristes*** Wygodzinsky and Schmidt 1980 from northern New Jersey; other species of the genus have been collected as far south as Tennessee and North Carolina (Wygodzinsky and Schmidt 1980).

20b Coxa of legs III only with styli. . . . . ***Hypomachilodes*** Silvestri 1911

H. texanus Silvestri from Texas.

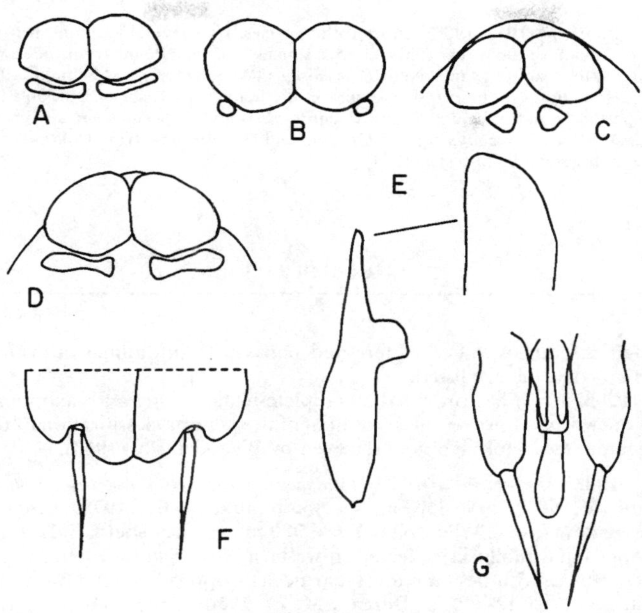

Fig. 30.6. (*A*) *Machiloides* sp., compound eyes and ocelli, front view. (*B*) *Machilinus* sp., compound eyes and ocelli, front view. (*C*) *Trigoniophthalmus alternatus*, compound eyes and ocelli, front view. (*D–G*) *Petrobius brevistylis*: (*D*) compound eyes and ocelli, front view; (*E*) mandible, apex enlarged; (*F*) posterior portion of coxites of urosternite VIII of male; (*G*) genital area of male. [(*A,B*) After Wygodzinsky (1955); (*C–G*) after Wygodzinsky and Schmidt (1980).]

21a **(18)** Ocelli subtriangular, located submedially (Fig. 30.6*C*); ovipositor completely covered by coxites of ninth segment. . . . . **_Trigoniophthalmus_** Verhoeff 1910

T. alternatus (Silvestri 1904) is found in the northeastern United States and throughout Europe except Scandinavia. No males have been found among the American specimens, and therefore, the American populations are probably parthenogenetic.

21b Ocelli shoesole-shaped, transverse, situated before anterior border of eyes (Fig. 30.6*D*); ovipositor of mature females extending to or beyond level of apex of styli of ninth segment (Fig. 30.5*F*) **22**

22a **(21)** Mandibles with four apical teeth (Fig. 30.5*G*); parameres of males segmented . . . . **_Pedetontus_** Silvestri 1911

P. saltator Wygodzinsky and Schmidt 1980 is known from the following New England states: Connecticut, Massachusetts, New Jersey, New York, and Pennsylvania.

22b Mandibles without distinct apical teeth (Fig. 30.6*E*); parameres of males entire (Fig. 30.5*H*) . . . . **_Petrobius_** Leach 1809 **23**

23a **(22)** Coxites of urosternite VIII of male each with a distinct lobe posteromedially (Fig. 30.6*F*); distal portion of penis elongate pear-shaped, narrowest at base (Fig. 30.6*G*) . . . **_Petrobius brevistylis_** Carpenter 1913

[= ***P. canadensis*** Paclt 1969]. Inhabits cliffs along rocky coasts, usually 5–20 ft above high-tide mark.

23b Coxites of urosternite VIII of male truncate posteromedially (Fig. 30.5*I*); apical portion of penis short, subrectangular, subequal in width throughout (Fig. 30.5*H*). . . . . . . **_Petrobius maritimus_** Leach 1809

Bousfield (1958, 1962) reported this species from several localities in Nova Scotia, and Schaefer (1965) reported a species similar to ***P. maritimus*** from the coast of Maine. All of these sightings may be of ***P. brevistylis*** (Wygodzinsky and Schmidt 1980).

In addition to the two species above, the following species of ***Petrobius*** have been reported from the United States: ***P. californicus*** and ***P. persquamosus*** from California, ***P. superior*** and ***P. submutans*** from Oregon, and ***P. calcaratus*** from Colorado. These species were described by Silvestri (1911).

References

Besides cited references, other papers and books with additional information on Thysanura and Microcoryphia are listed.

Juri Paclt (1963, 1967) has provided a complete bibliography and taxonomic keys to the Thysanura of the world. However, his treatment of the generic classification of the Thysanura (Family Nicoletiidae) was not favorably received by Wygodzinsky (1963).

Arnett, R. H. 1985. *American Insects: A Handbook of the Insects of America North of Mexico.* Van Nostrand Reinhold, New York. 850 pp. **Boudreaux, H. B. 1979.** *Arthropod Phylogeny with Special Reference to Insects.* Wiley, New York. 320 pp. **Bousfield, E. L. 1958.** Littoral marine arthropods and mollusks collected in western Nova Scotia, 1956. *Proc. N.S. Inst. Sci.* 24:303–325. **1962.** Studies on littoral marine arthropods from the Bay of Fundy region. *Nat. Mus. Can. Bull.* 183:42–62. **Delany, M. J. 1954.** Thysanura and Diplura. *Handbooks for the Identification of British Insects,* Vol. 1, No. 2. Royal Entomological Society, London. pp. 1–7. **Denis, J. 1949.** Ordre des Thysanoures. Pp. 209–275 in P.-P. Grassé (ed.). *Traité de Zoologie,* Vol. 9. Masson, Paris. **Drummond, F. H. 1953.** The eversible vesicles of *Campodea* (Thysanura). *Proc. R. Entomol. Soc. Lond.* 28:145–148. **Grzimek, B. 1975.** Insects. *Grzimek's Animal Life Encyclopedia,* Vol. 2. Van Nostrand Reinhold, New York. 643 pp. **Gupta, A. P. 1979.** *Arthropod Phylogeny.* Van Nostrand Reinhold, New York. 762 pp. **Lawrence, R. F. 1953.** *The Biology of the Cryptic Fauna of Forests.* A.A. Balkema, Cape Town, South Africa. 408 pp. **Lindsay, E. 1940.** The biology of the silverfish, *Ctenolepisma longicaudata,* with particular reference to its feeding habits. *Proc. Royal Soc. Victoria,* s.n. 52:83. **Manton, S. M. 1972.** The evolution of arthropodan locomotory mechanisms. 10. Locomotory habits, morphology and evolution of the hexapod classes. *Zool. J. Linn. Soc.* 51:203–400. **Matsuda, R. 1965.** Morphology and evolution of the insect head. *Mem. Am. Entomol. Inst. (Gainesville)* 4. 334 pp. **1970.** Morphology and evolution of the insect thorax. *Mem. Entomol. Soc. Can.* 76. 431. pp. **1976.** *Morphology and Evolution of the Insect Abdomen.* Pergamon Press, New York. 532 pp. **Packard, C. E. 1947.** Fun with the fire brat. *Turtox News* 25(9):1–2. **Paclt, J. 1963.** Thysanura fam. Nicoletiidae. *Genera Insectorum de P. Wytsman,* fasc. 216e. 58 pp. **1967.** Thysanura fam. Lepidotrichidae, Maindroniidae, Lepismatidae. *Genera Insectorum de P. Wytsman,* fasc. 218e. 86 pp. **Remington, C. L. 1954.** The suprageneric classification of the order Thysanura (Insecta). *Ann. Entomol. Soc. Am.* 47(2):277–286. **Schaefer, P. E. 1965.** *Bull. North. N. Engl. Acad. Sci.* (abstract). **Schaller, F. 1968.** *Soil Animals.* University of Michigan Press, Ann Arbor, Mich. 144 pp. **Silvestri, F. 1911.** Contribúto alla conoscènza dei Machilidae dell'America settentrionàle. *Boll. Lab. Zool. Portici* 5:324–350. **Slabaugh, R. E. 1940.** A new Thysanuran, and a key to the domestic species of Lepismatidae (Thysanura) found in the United States. *Entomol. News* 51:95–98. **Snodgrass, R. E. 1935.** *Principles of Insect Morphology.* McGraw-Hill, New York. 667 pp. **1952.** *A Textbook of Arthropod Anatomy.* Comstock Publishing Associates, Ithaca, N.Y. 363 pp. **Tomlin, A. D. 1979.** 22. Microcoryphia. 23. Thysanura. Pp. 304–305. In H. V. Danks (ed.). *Canada and Its Insect Fauna. Mem. Entomol. Soc. Can.* 108. **Wallwork, J. A. 1970.** *Ecology of Soil Animals.* McGraw-Hill, London, 283 pp. **Wygodzinsky, P. 1941.** Beiträge zur Kenntnis der Dipluren und Thysanuren der Schweiz. *Denkschr. Schweiz. Naturforsch. Ges.* 74(2):113–227. **1955.** Thysanura. In *South African Animal Life.* 2:83–190. Almquist and Wiksells Boktryckari AB, Uppsala, Sweden. **1958.** Sobre algunos "Nicoletiidae" americancos (Thysanura, Insecta) [*On Some American Nicoletiidae*]. *Acta Zool. Lilloana* 16:97–120. **1961.** On the surviving representative of the Lepictotrichidae (Thysanura). *Ann. Entomol. Soc. Am.* 54:621–627. **1963.** On Paclt's Nicoletiidae (Thysanura) in Genera Insectorum. *Ann. Mag. Nat. Hist. Ser. 13* 6:265–269. **1972.** A review of the silverfish (Lepismatidae, Thysanura) of the United States and Caribbean area. *Am. Mus. Novit.* (2481):1–26. **1973.** Description of a new genus of cave thysanuran from Texas (Nicoletiidae, Thysanura, Insecta). *Am. Mus. Novit.* (2518):1–8. **Wygodzinsky, P. and K. Schmidt. 1980.** Survey of the Microcoryphia (Insecta) of the northeastern United States and adjacent provinces of Canada. *Am. Mus. Novit.* (2701):1–17. **Zimmerman, E. C. 1948.** *Insects of Hawaii,* Vol. 2: Apterygota to Thysanoptera inclusive. Order Thysanura. Honolulu. Pp. 29–38.

31

Insecta: Diplura

LYNN M. FERGUSON

The diplurans (entotrophi) are an order of entognathus apterygote hexapods which are considered as a separate class by some zoologists and as an order of the class Insecta by others (Mackerras 1970). Of the entognathus hexapods, which also includes the Protura and Collembola, the Diplura are most like true insects, similar to the Thysanura with which they used to be included. Diplurans generally live in damp humus or soil and in caves. Although at least seven families are presently recognized, there are basically two types that make up most of the order. These are the japygids, with their forceps-like cerci, and the campodeids, with their long filamentous cerci.

Morphological Characteristics

Small to large, elongate, parallel-sided hexapods (Figs. 31.1, 31.2). Chewing mouthparts enclosed within the head capsule (entognathus); mandible often with a movable prostheca (Fig. 31.3). Antennae long, many-segmented, moniliform. Compound eyes and ocelli absent. Legs five-segmented with a single, undivided tarsus. Abdomen with 10 distinct segments and rudimentary structures of a possible eleventh; lateral styli and eversible vesicles on most of the first seven urosternites (Fig. 31.4); gonopore between segments VIII and IX. Cerci present, long and filamentous (Fig. 31.1), short and annulated (Fig. 31.5), or short and forceps-like (Fig. 31.2); sometimes containing the duct of a silk gland (Fig. 31.5). Median caudal filament absent. Tracheae relatively well developed, with two to

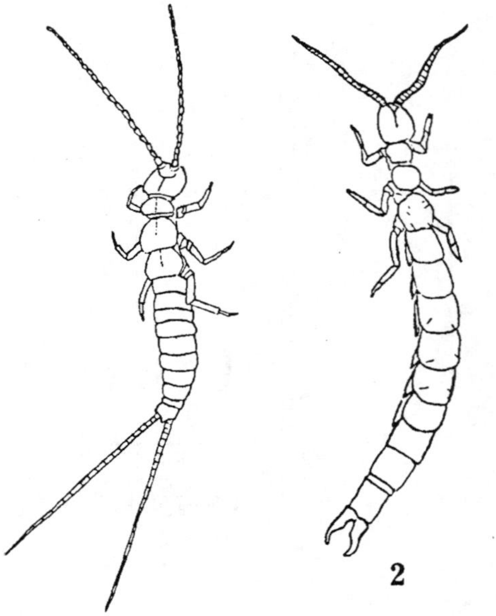

Figs. 31.1, 31.2. 1. Campodeid dipluran, setae omitted. [After Grzimek (1975).] **2.** Japygid dipluran. [After Grzimke (1975).]

four pairs of spiracles on thoracic pleura and none or seven pairs on abdominal pleura. Body covered with setae or, rarely, with scales.

Feeding and Predation

Japygid diplurans have been known to eat Collembola, Isopoda, Symphyla, campodeid diplurans, other japygids, Diptera larvae, beetles, vegetable debris, and fungal mycelia. They are therefore omnivorous, but they seem to prefer animal prey (Pagés 1951). Campodeid diplurans eat mites and other small arthropods, enchytraeid worms, fungal spores and mycelia, and detritus. Cavernicolous campodeids also ingest a large quantity of sediment, possibly feeding on bacteria and other organic matter present. Diplurans (campodeids) in turn are fed on by harvestmen, ants (Wilson 1955), carabid cave beetles, and japygids (Kosaroff 1935).

Reproduction

Campodeids reproduce by means of spermatophores placed at random by the males, in the presence or absence of females (Bareth 1964). Therefore, there is no well-developed intromittent male sex organ (Fig. 31.6). In

Campodea remyi each spermatophore consists of a globule 50–70 μm in diameter mounted on a peduncle 50–100 μm high, with a diameter of 7–12 μm (Fig. 31.7). One to four bundles of sperm are enclosed in a granular fluid within the globule. The sperm remain viable within the spermatophore for 2 days. A male may produce around 200 spermatophores in a week, some of which will be eaten by himself and other insects.

Insemination is dependent on a female with mature ova finding a recently placed spermatophore and picking it up with the valves that sur-

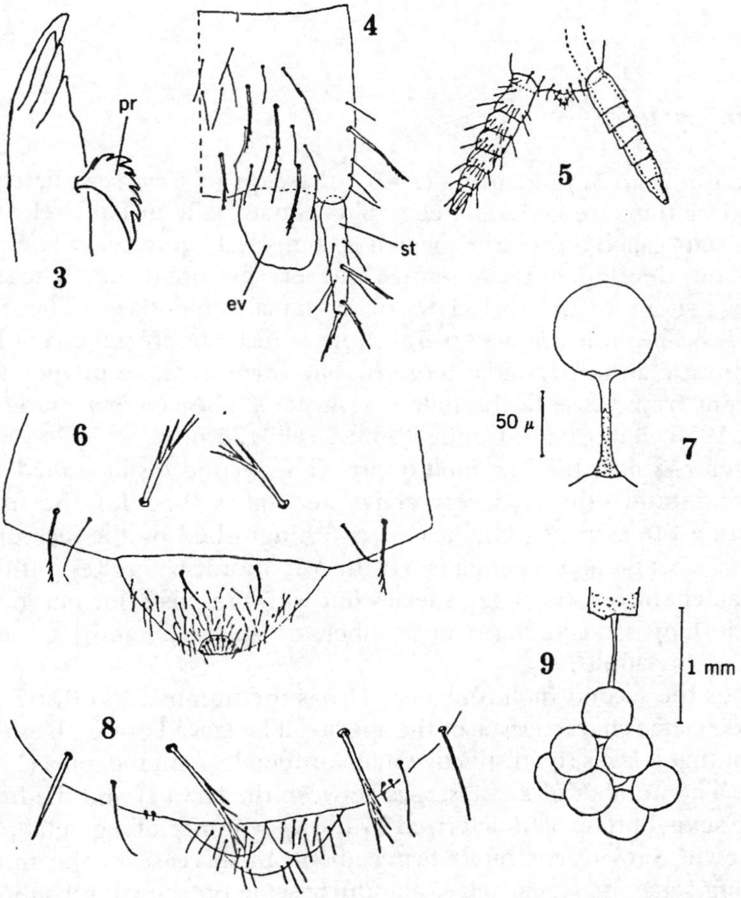

Figs. 31.3–31.9. **3.** Tip of left mandible of *Anajapyx menkei,* with lacinia mobilis or movable prostheca (*pr*) [After Smith (1960).] **4.** Left half of urosternite VI of a campodeid, showing stylus (*st*) and extended eversible vesicle (*ev*). **5.** Cerci of *Projapyx,* showing inner canal. [Modified from Smith (1960).] **6.** Male genital papilla and part of urosternite VIII of *Litocampa* sp. **7.** Spermatophore of *Campodea remyi.* [Modified from Bareth (1964).] **8.** Female genital papilla, consisting of two ventral valves and one dorsal valve, and part of urosternite VIII of *Litocampa* sp. **9.** Egg cluster and stalk of *Campodea remyi* attached to a stone. [After Wygodzinsky (1941).]

round her genital opening (Fig. 31.8). The female lays her fertilized eggs in a cavity in the soil. She attaches the egg cluster (Fig. 31.9) to a stalk produced by a glandular secretion, suspending the eggs so that they do not make contact with the soil. The egg cluster of *Campodea remyi* contains four to nine eggs which are 0.4 mm in diameter (Wygodzinsky 1941).

Japygids also suspend their egg cluster on stalks from stones or pebbles which line a cavity in the soil. The egg cluster of *Japyx* contains 25–35 eggs which are 0.75 mm in diameter (Silvestri 1928b). In the japygids, the female remains with the eggs, curling her body around them in a protective manner like that of certain female centipedes. She also remains with the newly hatched larvae until they develop their forceps-like cerci and can fend for themselves (Silvestri 1948).

Life Cycle

According to Wygodzinsky (1941), the eggs of *Campodea* hatch 12–13 days after they are laid. The egg splits equatorially in half, releasing the first instar (called a prelarva), which is completely quiescent. The prelarva is not fully developed: the antennae and cerci are not distinctly segmented, the legs show no trace of claws or pretarsal appendices. There are no spiracles. The cuticle bears several short, conical, finely pubescent hairs on the thoracic and abdominal tergites. The form of these phaneres is very different from those in the following instars. The venter is nude (Condé 1946, 1956; Bareth and Condé 1965; Orelli 1956).

After 1–2 days the first molt occurs. The second instar (called a larva I by Condé and others) is very active and takes food for the first time, including the exuviae. The larva I is distinguished by the lack of trichobothria on antennal segments III to VI, thoracic tergites with macrochaetae characteristic of the species but with the posterior marginal setae and clothing setae reduced in number, no spiracles, and legs similar to those of the adults.

After the second molt, the larva II has the number and distribution of macrochaetae characteristic of the species. The larva II resembles the adult except that it lacks the distinctive male or female genital papilla (Figs. 31.6, 31.8). The number of larval stages between the larva II and the first instar of the sexual phase, characterized by the appearance of a genital papilla, is unknown. Subsequent molts bring about an increase in the number of clothing setae, marginal setae, glandular setae on the lateral subcoxal appendages of urosternite I, and the number of setae on the genital papilla (Condé 1956).

The early development of *Japyx* is similar to that of the campodeids. According to Silvestri (1948), 15 days elapse between the laying of the eggs and their hatching. The first instar (larva I of Pagés 1951) is motionless except for some slow contractions and distentions. The first molt occurs after 5–6 days, and the second instar (larva II) appears. Larva II also performs some slow locomotory movements but is still immobile for the most part. Neither larva I nor larva II take any food. They form a more-or-less rounded cluster of individuals, hanging from the stalk in their brood chamber, embraced by their mother.

Larva I and II of the japygids are homologous to the prelarva of the campodeids (Condé 1956). The campodeid prelarva is less developed than

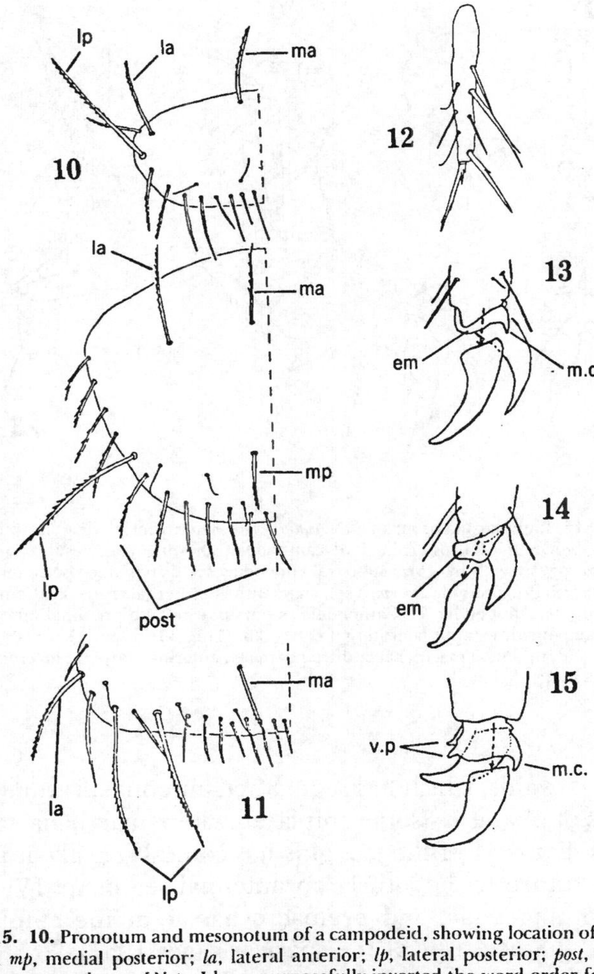

Figs. 31.10–31.15. 10. Pronotum and mesonotum of a campodeid, showing location of macrochaetae (*ma*, Medial anterior; *mp*, medial posterior; *la*, lateral anterior; *lp*, lateral posterior; *post*, posterior marginal setae). Clothing setae not shown. [*Note:* I have purposefully inverted the word order for the names of the designated macrochaetae, (e.g., lateral posterior instead of posteriolateral), so that the words will agree with the widely accepted French abbreviations for these macrochaetae.] **11.** Abdominal tergite of a campodeid, showing location of macrochaetae (abbreviations same as in Fig. 31.10). **12.** Stylus of a campodeid. **13.** Lateral view of pretarsus of *Teljapyx* sp. (*em*, empodium; *m.c.*, median claw). [After González (1964).] **14.** Lateral view of pretarsus of *Dinjapyx* sp. (*em*, empodium). [After González (1964).] **15.** Lateral view of pretarsus of *Heterojapyx* sp. (*m.c.*, median claw; *v.p.*, ventral projections). [After González (1964).]

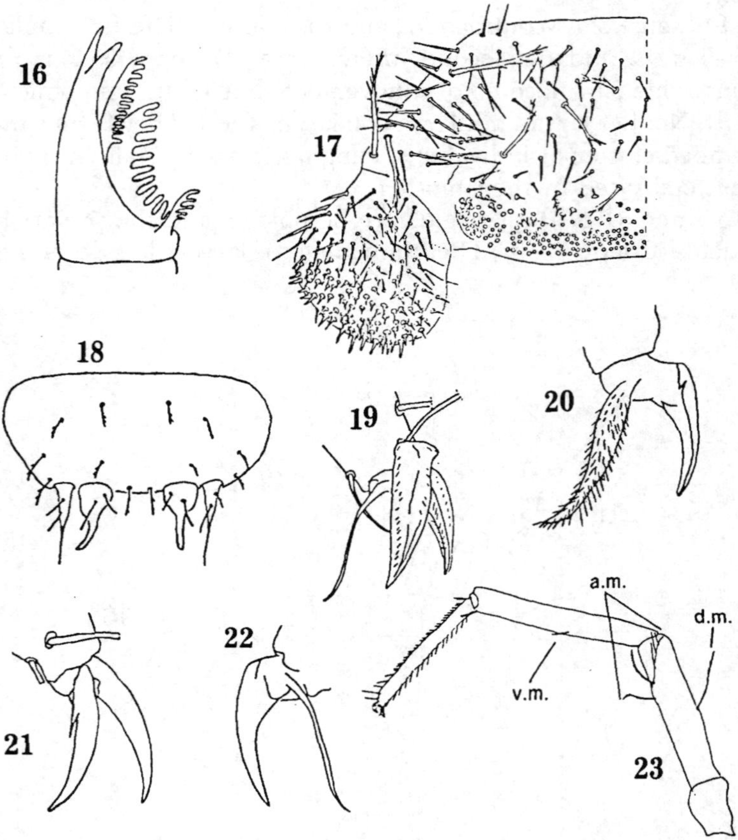

Figs. 31.16–31.23. **16.** Right lacinia of maxilla of *Anajapyx hermosus*; ventral view showing pectens. [After Smith (1960).] **17.** Left half of urosternite I of campodeid, showing lateral subcoxal appendage. **18.** Urosternite I of *Anajapyx* sp. with pyriform subcoxal appendages and styli; microsetae omitted on sternite. [After Smith (1960).] **19.** Pretarsus of *Litocampa* sp., claws with laterotergal crests. **20.** Campodeid pretarsus with flattened, pilose lateral appendix. **21.** Campodeid pretarsus: claw with proximal spiniform process. **22.** Pretarsal claw of campodeid, without laterotergal crests. **23.** Campodeid leg III (coxa omitted), showing major macrochaetae (*d.m.*, dorsal macrochaeta of femur; *a.m.*, anterior marginal macrochaetae of femur; *v.m.*, ventral macrochaeta of tibia).

larva I of the japygids, which has segmented antennae, rudimentary claws, and two types of phaneres (some emplaced where macrochaetae are found in the adults). Larva II of the japygids has conical sensilla at the emplacements of the future trichobothria on antennal segments IV to VI, more numerous clothing setae, and premacrochaetae at the emplacements of macrochaetae characteristic of the species (Pagés 1951).

The second molt occurs 10 days after the appearance of the larva II. The third instar (larva III) resembles the adults in having forceps almost totally developed and the same hairs and setae. The young presumably leave the brood chamber, or nest, at this time and seek food on their own.

Diplurans continue to molt throughout their life, increasing in size and/

or the number of clothing setae on the body. Accordingly, they are able to regenerate lost parts such as legs, antennae, and cerci. Indeed, the cerci of the campodeids are capable of autotomy, becoming detached at a line of weakness near their bases (Marten 1939). *Campodea* apparently eat their own exuviae (Wygodzinsky 1941).

Collection

Specimens may be collected by placing soil and humus in a Berlese funnel. By looking under stones and logs in damp soil and in caves, specimens may also be collected by hand with a small camel's hair brush dipped in alcohol, or with an aspirator. In caves, I have found baiting with Roquefort or blue cheese particularly helpful. Molasses traps used in Mammoth Cave, Kentucky, for collecting cave crickets also caught many *Litocampa cookei*. Specimens should be preserved in 70–75% alcohol.

Measurements and Characteristics

The arrangement and number of large setae (macrochaetae), particularly on the thoracic and abdominal tergites (Figs. 31.10, 31.11), are used in identification. Ratios of the lengths of thoracic macrochaetae and setae are used to help distinguish certain species. The occurrence of other phaneres, such as the trichobothria of the antennae, are also important. Other characters used are the number of antennal segments, the form of the pretarsal claws and appendices, the morphology of the first abdominal urosternite, the form of the cerci, and the particular form of the mandibles.

Identification

The following key to the subfamilies of the Diplura is basically that of Pagés (1959), with incorporated modifications by Smith (1960) and González (1964). In the key to the genera and subgenera of the Subfamily Campodeinae in the United States, three undescribed genera are included for the sake of completeness.

KEY TO THE FAMILIES AND SUBFAMILIES OF DIPLURA

1a Cerci one segmented, forcipiform (Fig. 31.2); mandible without lacinia mobilis (prostheca); with or without trichobothria on antennae; if present, on antennal segments IV–VI or IV–XV; two or four pairs of thoracic spiracles; abdominal spiracles present; styli without several distinctly developed setae; anal valvules not well developed, or lacking; Suborder **DICELLURATA,** Superfamily **JAPYGOIDEA** **2**

1b Cerci multisegmented, filiform (Fig. 31.1); mandible with lacinia mobilis (prostheca) (Fig. 31.3); trichobothria present, beginning on antennal segments III, IV, or V; two or three pairs of thoracic spiracles; abdominal spiracles present or absent; styli with several very distinctly developed setae (Fig. 31.12); anal valvules well developed; Suborder **RHABDURA** **7**

2a (1) Labial palps present; four pairs of thoracic spiracles; Family **JAPYGIDAE** Lubbock 1942 . . . . **4**

2b Labial palps absent; two pairs of thoracic spiracles **3**

3a (2) Antennae without trichobothria; lateral subcoxal appendages present on both sexes; pretarsus with empodium bent backwards dorsally, forming a median unpaired claw (unguiculus) (Fig. 31.13); forceps with glandular orifice on proximal tergal surface; small forms . . . . Family **PARAJAPYGIDAE** Pagés 1959

Cosmopolitan; species of two or three genera in the United States.

3b Trichobothria present on at least segments IV and XV; pretarsal empodium rounded off, without projections (Fig. 31.14); females without lateral subcoxal appendages; forceps without glandular orifice; gigantic forms. . . . Family **DINJAPYGIDAE** González 1964

Known only from Peru and Bolivia.

4a (2) Trichobothria on antennal segments IV–VI only; pretarsal empodium strongly bent backward dorsally, forming a median unpaired claw (unguiculus), and rounded ventrally (Fig. 31.13) **5**

4b Trichobothria on at least segments IV–XV; pretarsal empodium bent backwards dorsally, forming a median unpaired claw (unguiculus), and with two sharp, conical, ventral projections (Fig. 31.15) **Heterojapyginae,** Womersley 1939

No North American species; known from Australia, New Zealand, Madagascar, Pamir, and Tibet.

5a (4) Macrosetae smooth, no barbules . . . **Japyginae,** Womersley 1939

Cosmpolitan; five or six genera known from the United States.

5b Macrosetae densely barbed, plumose **6**

6a (5) Apical lamina of lacinia pectinate (Fig. 31.16); styli with two setae; forceps asymmetrical; body more than 5 mm in length . . . . . . **Evalljapyginae,** Silvestri 1948

North and Central America; ***Evalljapyx*** Silvestri 1911 is the dominant genus of japygids in the western United States; several species.

6b Apical lamina of lacinia without pectens; one seta on inner face of styli; forceps symmetrical; body less than 5 mm in length . . . . . **Provalljapyginae,** L. Smith 1962

Provalljapyx Silvestri 1949 reported from North America and Brazil.

7a (1) Trichobothria beginning on antennal segment III; maxilla without palpus, lacinia without pectens; urosternite I without styli, with one pair of subcoxal appendages (Fig. 31.17); styli present on urosternites II–VII; abdominal spiracles absent; cerci long or short, without inner canal, closed at apex; Superfamily **CAMPODEOIDEA 9**

7b Trichobothria beginning on antennal segments IV or V; palpus on maxilla, pectens on lacinia (Fig. 31.16); urosternite I with one pair of lateral posterior styli and one pair of sublateral posterior subcoxal appendages; styli present on urosternites I–VII; seven pairs of abdominal spiracles present; cerci short, with inner canal, opening apically (Fig. 31.5); Superfamily **PROJAPYGOIDEA** **8**

8a (7) Epicranial suture complete, anterior branches (frontal sutures) beaded; trichobothria on antennal segments IV–XXII, terminal segment with two sensory pits, one on anterior and one on posterior side; lacinia of maxilla simple; labial palpi well developed and subcylindrical; three pairs of thoracic spiracles; all sternal thoracic apodemes well developed, Y-shaped; subcoxal appendages of urosternite I cylindrical, pilose, with terminal glandular setae (Fig. 31.17); eversible vesicles absent on urosternites; Malpighian tubules absent; genital papilla of male of ***Projapyx*** type; genital papilla of female very reduced. . . . . Family **PROJAPYGIDAE** Cook 1899

Two genera: ***Projapyx*** Cook 1899, seven species known from West Africa (Guinea and Cameroon) and Brazil; ***Symphylurinus*** Silvestri 1909, 19 species known from Mexico, South America, Africa, Australia, China, and India.

8b Epicranial suture incomplete, reduced to its posterior branch (coronal suture); trichobothria on antennal segments V–X, and on even-numbered segments thereafter in some species; apex of terminal segment of antenna covered by large placoid sensilla; lacinia of maxilla bifid (Fig. 31.16); labial palpi very reduced; two pairs of thoracic spiracles; apodeme of prothoracic sternum apparently reduced to its posterior branch; subcoxal appendages of urosternite I pyriform (Fig. 31.18), very slightly pilose, without glandular setae; urosternites II–VII with pair of eversible vesicles (Fig. 31.4); 16 Malpighian tubules present; genital papilla of male similar to ***Campodea*** type (Fig. 31.6); genital papilla of female not reduced (Fig. 31.8). . . . Family **ANAJAPYGIDAE** (Paclt 1957)

Single genus, ***Anajapyx*** Silvestri 1903. ***A. hermosus*** L. Smith 1960 described from California. Four other species known from Mexico and Italy.

9a **(7)** Epicranial suture incomplete, reduced to proximal portion of posterior branch (coronal suture); trichobothria on antennal segments III–VII; terminal segment of antennae without cupuliform structure; labial palpi absent; two pairs (intermediate absent) of thoracic spiracles; urosternite I with one pair of very small subcoxal appendages; cerci short . . . Family **PROCAMPODEIDAE** Silvestri 1948

Procampodea macswaini Condé and Pagés 1956, described from California.

9b Epicranial suture complete, beaded on entire extent; trichobothria on antennal segments III–VI; terminal segment of antenae with cupuliform structure at its apex; labium with palpi in form of flattened protuberances with setigerous plates, and with one pair of internal palpiform processes (labial palpi of some authors); three pairs of thoracic spiracles; urosternite I with one pair of well developed subcoxal appendages; cerci long; Family **CAMPODEIDAE** Lubbock 1873. . . . . **10**

10a **(9)** Body with scales; lateral pretarsal claws simple, anterior claw similar to posterior; pretarsus with or without small median claw (unguiculus); lateral pretarsal appendices always present; posterior marginal setae of urosternite VIII may show sexual dimorphism. **11**

10b Body without scales; lateral pretarsal claws simple or with laterotergal crests, anterior claw similar to posterior or smaller; pretarsus with (***Eutrichocampa hispanica*** Silvestri 1932 only) or without small median claw (unguiculus); lateral pretarsal appendices absent, rudimentary, or well developed, subcylindrical or flattened, smooth or pubescent; posterior marginal setae of urosternite VIII similar on both sexes . . . . **Campodeinae,** Condé 1956

11a **(10)** Scales on entire trunk; pretarsus with (***Lepidocampa*** Oudemans 1890) or without (***Syncampa*** Silvestri 1931) small median claw (unguiculus); lateral pretarsal appendices flattened, with long pubescence; posterior marginal setae of urosternite VIII show sexual dimorphism (larger setae with numerous barbs on males) . . . . . . **Lepidocampinae,** Condé 1956

Found throughout the tropics and subtropics, and also at very high altitudes in the Andes and Himalayan mountains; all species are endogeans. Only ***Lepidocampa weberi*** Oudemans 1890 has been reported from the United States (California and Hawaii).

11b Scales on metathorax and abdomen (***Tritocampa*** Silvestri 1933) or on abdomen only (***Hemicampa*** Silvestri 1912); pretarsus without small median claw (unguiculus); lateral pretarsal appendices subcylindrical, smooth; posterior marginal setae of urosternite VIII not showing any sexual dimorphism . . . **Hemicampinae,** Condé 1956

Hemicampa boudreauxi Condé and Geeraert 1962 was described from Texas and Louisiana; other species of ***Hemicampa*** known from Mexico. One other genus, ***Tritocampa*** Silvestri 1933, of Guyana. All species are endogeans.

KEY TO GENERA AND SUBGENERA OF CAMPODEINAE IN THE UNITED STATES

1a Lateral pretarsal appendices well developed, subcylindrical (setiform) (Fig. 31.19) or flattened (Fig. 31.20) **5**

1b Lateral pretarsal appendices absent or reduced to spiniform processes attached proximally to the claws (Fig. 31.21) **2**

2a (1) Pretarsal claws simple, slightly unequal in length (Fig. 31.22) . . . **3**

2b Pretarsal claws with laterotergal crests (Fig. 31.19); posterior claw much larger than anterior . . . . ***Haplocampa*** Silvestri 1912

Endogeans and cavernicoles of midwestern and western United States and Canada (Ferguson 1975).

3a (2) Femur III with dorsal macrochaeta (Fig. 31.23); pronotum with 3 + 3 (*ma, la, lp*) macrochaetae; mesonotum with 4 + 4 (*ma, la,* 2 *lp*) macrochaetae (Fig. 31.10). . . . . . . ***Eumesocampa*** Silvestri 1933

Endogeans of Colorado (***E. danielsi*** Silvestri 1933 and ***E. lutzi*** Silvestri 1933) and of New York, Pennsylvania, Ohio, and Maryland [***E. frigillis*** (Hilton) 1936, reported by Condé and Geeraert (1962)], and cavernicoles (at least two undescribed species) of West Virginia, Illinois, and Missouri (Ferguson 1978).

3b Femur III without dorsal macrochaetae; pronotum with 2 + 2 (*ma, lp*) or 3 + 3 (*ma, la, lp*) macrochaetae; mesonotum with 0 or less than 4 + 4 macrochaetae; ***Metriocampa*** Silvestri 1911. . . . . **4**

Redefined by Condé (1956).

4a (3) Pronotum with 2 + 2 (*ma, lp*) macrochaetae . . . . Subgenus ***Metriocampa* s. str.** Silvestri 1911

Redefined by Silvestri (1933). Six species and one subspecies are recorded from the United States; all are endogeans.

4b Pronotum with 3 + 3 (*ma, la, lp*) macrochaetae . . . . Subgenus ***Tricampa*** Silvestri 1933

Considered a separate genus by Paclt (1957) and as a subgenus of ***Metriocampa*** by Bareth and Condé (1958) and by Condé and Geeraert (1962). Four species; endogeans and a few cavernicoles (troglophiles).

5a (1) Mesonotum with 4 + 4 (*ma, mp, la, lp*) macrochaetae; lateral pretarsal appendices of two forms: anterior appendix smooth, subcylindrical (setiform) (Fig. 31.19); posterior appendix pilose, flattened (Fig. 31.20) . . . . **Campodeid Genus A**

Highly modified cavernicole of Carlsbad Caverns and other caves of the Guadalupe Mountains of southeastern New Mexico and southwestern Texas.

5b Not as above **6**

6a (5) Lateral pretarsal appendices subcylindrical (setiform), covered with fine hairs or barbules. **7**

6b Not as above **8**

7a (6) Pretarsal claws with well-developed laterotergal crests; femur III with dorsal macrochaeta . . . . **Campodeid Genus B**

Cavernicole of Nevada.

7b Pretarsal claws simple, without laterotergal crests; femur III without dorsal macrochaetae . . . .
Campodea Subgenus ***Hypercampa*** Silvestri 1933

Only one species known: *C.* (*H.*) ***essigi*** Silvestri 1933 of California; an endogean.

8a (6) Pronotum with four pairs of macrochaetae (*ma*, 2 *la*, *lp*); pretarsal claws with well-developed laterotergal crests; lateral pretarsal appendices smooth, subcylindrical (setiform) . . . .
Campodeid Genus C

Cavernicole of the Smoky Mountains, Tennessee.

8b Pronotum with less than four pairs of macrochaetae; pretarsal claws with or without laterotergal crests; lateral pretarsal appendices smooth and subcylindrical (setiform), or flattened and possibly pubescent (Figs. 31.19, 31.20) **9**

9a (8) Pretarsal claws simple, lacking laterotergal crests; claws subequal in size **11**

9b Pretarsal claws with well-developed or slightly developed laterotergal crests; anterior claw smaller than posterior claw; ***Litocampa*** Silvestri 1933 . . . . **10**

10a (9) Abdominal tergites I–VII without macrochaetae; tergite VIII with 2 + 2 *lp* macrochaetae; abdominal segment IX with 4 + 4 macrochaetae (total) . . . . Subgenus ***Cocytocampa*** Paclt 1957

L.* (*C.*) *perkinsi (Silvestri 1934) of Oahu (Hawaiian Islands), an endogean. One other species properly belongs to this subgenus, ***L.* (*C.*) *solomonis*** Bareth and Condé 1972, of the Solomon Islands.

10b Macrochaetae on at least some of abdominal tergites I–VII (Fig. 31.11); tergite VIII with 3 + 3 or more *lp* macrochaetae; abdominal segment IX with 5 + 5 or more macrochaetae (total) . . . .
Subgenus ***Litocampa* s. str.** Silvestri 1933

Cavernicoles (19 or more species) in primarily the eastern and southern United States; elsewhere in the world, endogeans as well as cavernicoles occur.

11a (9) Femur III without dorsal macrochaetae. **12**

11b Femur III with dorsal macrochaeta . . . .
Podocampa (Silvestri 1932)

Former subgenus of ***Campodea***, elevated to generic rank by Condé (1956). Lateral pretarsal appendices may be either smooth, subcylindrical, or pubescent, flattened; therefore, according to Condé and Geeraert (1962), ***Podocampa*** includes several species formerly assigned to the genus ***Parallocampa*** Silvestri 1933. Five species reported from Texas and Louisiana; all are endogeans.

12a (11) Lateral pretarsal appendices flattened, pubescent . . . .
Meiocampa (Silvestri 1933)

Former subgenus of ***Parallocampa*** Silvestri 1933, raised to generic rank by Paclt (1957). Five species in California, Arizona, and Oregon; all are endogeans.

12b Lateral pretarsal appendices smooth, subcylindrical (setiform), except distal region which may be slightly enlarged and flattened; ***Campodea*** Westwood 1842 . . . . **13**

13a (12) Mesonotum and metanotum with respectively 1 + 1 (*la*) and 0 macrochaetae . . . . Subgenus ***Monocampa*** Silvestri 1932

C.* (*M.*) *morgani Silvestri 1911, an endogean described from North Carolina and Florida.

13b Mesonotum and metanotum with more than 1 + 1 and 0 macrochaetae **14**

14a **(13)** Mesonotum with 2 + 2 (*ma*, *la*) macrochaetae; metanotum with 1 + 1 (*ma*) macrochaetae. . . . . Subgenus ***Dicampa*** Silvestri 1932

One species reported from San Antonio, Texas, very similar to ***C.*** **(*D.*)** ***escalerai*** Silvestri 1932 of Spain; an endogean.

14b Mesonotum with 3 + 3 (*ma*, *la*, *lp*) macrochaetae; metanotum with 2 + 2 (*ma*, *lp*) or 1 + 1 (*ma*, rarely *lp*) macrochaetae. . . . . Subgenus ***Campodea*** **s. str.** Westwood 1842

Redefined by Silvestri (1932). Approximately 20 species and four subspecies have been reported from the United States; all are endogeans.

References

Important comprehensive works are Condé (1956) for campodeids, Pagés (1951) for japygids, and Pagés (1959) for general classification. Paclt (1957) provides an excellent bibliography. The other papers listed refer to major contributions to our knowledge of diplurans in the United States and to cited references.

Bareth, C. 1964. Structure et dépôt des spermatophores chez *Campodea remyi. C.R. Hebd. Seances Acad. Sci.* (*Paris*) 259(12):1572–1575. **Bareth, C. and B. Condé. 1958.** Campodéidés endogés de l'ouest des États-Unis (Washington, Oregon, Californie, Arizona), *Bull. Soc. Linn. Lyon* 27:226–248, 265–276, 297–304. **1965.** La prélarve de *Campodea* (*C.*) *remyi. Rev. Ecol. Biol. Sol* 2(3):397–401. **1972.** Diploures Campodéidés des Iles Salomon. *Rev. Ecol. Biol. Sol* 9(2):235–256. **Chandler, L. 1956.** The orders Protura and Diplura in Indiana. *Proc. Indiana Acad. Sci.* 66:112–114. **Condé, B. 1946.** À propos du développement postembryonnaire des Campodéidaé. *Bull. Soc. Entomol. Fr.* 51:69–71. **1949.** Campodéidés cavernicoles de la région Appalaches. *Notes Biospeol.* 4:125–137. **1956.** Matériaux pour une monographie des Diploures Campodéidés. *Mem. Mus. Natl. Hist. Nat.* (*Paris*) *Ser. A Zool.* 12(1955):1–202. **1973.** Campodéidés endoges de l'Est des Etats-Unis. *Bull. Soc. Linn. Lyon* 42(spécial):17–29. **Condé, B. and P. Geeraert. 1962.** Campodéidés endogés du centre des États-Unis. *Arch. Zool. Exp. Gen.* 101:73–160. **Condé, B. and J. Thomas. 1957.** Contribution à la faune des Campodéidés de Californie (Insectes Diploures). *Bull. Soc. Linn. Lyon* 26:81–96, 118–127, 142–155. **Ferguson, L. M. 1975.** Subterranean campodeids (Insecta : Diplura : Campodeidae) from the northwestern United States. *Bull. Assoc. Southeast. Biol.* (*Abstr.*) 22(2):53. **1978.** A preliminary report on the *Eumesocampa* (Insecta : Diplura : Campodeidae) of North America. *Va. J. Sci.*, 29(2):60 (abstract). **1981.** *Cave Diplura of the United States*, Proc. 8th Int. Cong. Speleol. (Bowling Green, KY).1:11–12. **Gardner, R. E. 1914.** Some notes on the distribution of Cinura in the vicinity of Claremont, with description of a new species. *J. Entomol. Zool.* (*Claremont*) 6:86–92. **González, R. H. 1964.** Japygoidea de Sud America. 6. Revisión de la Familia Dinjapygidae (Womersley, 1939) (Insecta : Diplura). *Acta Zool. Lilloana* 20:113–128. **Grzimek, B. 1975.** Insects. In *Grzimek's Animal Life Encyclopedia*, Vol. 2. Van Nostrand Reinhold, New York. 643 pp. **Hilton, W. A. 1932.** The *Campodea* of California. *J. Entomol. Zool.* (*Claremont*) 24:47–51. **1936.** *Campodea* from the United States. *J. Entomol. Zool.* (*Claremont*) 28:5–10. **Kosaroff, G. 1935.** Beobachtungen über die Ernährung der Japygiden. *Mitt. Naturwiss. Inst. Sofia* 8:181–185. **Lawrence, R. F. 1953.** *The Biology of the Cryptic Fauna of Forests.* A.A. Balkema. Cape Town, South Africa. 408 pp. **Mackerras, I. M. 1970.** Evolution and classification of the Insects. Pp. 152–167 in CSIRO, *The Insects of Australia.* Melbourne University Press, Canberra, Australia. **Marten, W. 1939.** Zur Kenntnis von *Campodea. Z. Morph. Oekol. Tiere* 36:41–88. **Orelli, M. 1956.** Untersuchungen zur postembryonalen Entwicklung von *Campodea. Verh. Naturforsch. Ges. Basel* 67:501–74. **Packard, A. S. 1888.** The cave fauna of North America, with remarks on the anatomy of the brain and origin of the blind species. *Mem. Natl. Acad. Sci.* 4:1–156. **Paclt, J. 1957.** Diplura. *Genera Insectorum de P. Wytsman*, fasc. 212E. 123 pp. **Pagés, J. 1951.** Contribution à la connaissance des diploures. *Bull. Sci. Bourgogne* 13, Suppl. 9. 97 pp., 12 plates. **1959.** Remarques sur la classification des diploures. *Trav. Lab. Zool. Fac. Sci. Dijon* 26. 25 pp. **Reddell, J. R. 1983.** A checklist and bibliography of the Japygoidea (Insecta: Diplura) of North America, Central America, and the West Indies. *Texas Mem. Mus., Pearce-*

Sellards Ser. (37):1–41. **Silvestri, F. 1912.** Nuòvi gèneri e nuòve spècie di Campodeidae (Thysanura) dell' Amèrica Settentrionàle. *Boll. Lab. Zool. Gen. Agrar. Portici* 6:5–25. **1928a.** Description of a new species of *Japyx* (Thysanura) from Potter Creek cave, Shasta County, California. *Univ. Calif. Publ. Entomol.* 4:335–340. **1928b.** On postembryonal development of Japygidae (Thysanura). *Trans. 4th Int. Congr. Entomol.* (*Ithaca*) 2:905–908. **1932.** Campodeidae (Thysanura) de España. Parte primera. *Eos* 8:115–164. **1933.** Quàrto contribúto alla conoscènza dei Campodeidae (Thysanura) del Nord Amèrica. *Boll. Lab. Zool. Gen. Agrar. Portici* 27:156–204. **1934a.** Campagne spéologique de C. Bolivar et R. Jeannel dans l'Amérique du Nord (1928). II. Campodeidae. Biospeologica LX. *Arch. Zool. Exp. Gen.* 76:379–383. **1934b.** Description of a new species of Thysanura (Campodeidae) from the Hawaiian Islands. *Proc. Hawaii. Entomol. Soc.* 8:519–522. **1947.** On some Japygidae in the Museum of Comparative Zoology (Dicellura). *Psyche* (*Camb.*) 54:209–229. **1948.** Japyginae (Japygidae: Insecta Diplura) della fauna Italiana finora note. *Boll. Lab. Entomol. Agrar. Portici* 8:236–296. **Smith, L. E. 1960.** The family Projapygidae and Anajapygidae (Diplura) in North America. *Ann. Entomol. Soc. Am.* 53:575–583. **Wilson, E.O. 1955.** Ecology and behavior of the ant *Belonopelta deletrix* Mann (Hymenoptera, Formicidae). *Psyche* (*Camb.*) 62:82–87. **Wygodzinsky, P. 1941.** Beiträge zur Kenntnis der Dipluren und Thysanuren der Schweiz. *Denkschr. Schweiz. Naturforsch. Ges.* 74(2):113–227. 5 pls.

32

Insecta: Collembola

KENNETH A. CHRISTIANSEN

The Collembola are small wingless hexapodous arthropods with distinct head, one pair of antennae, six abdominal segments, lacking true compound eyes, and with a ventral tube and often with a ventral jumping organ or furcula. Most consider them to be a separate class, but they were most commonly placed in the class Insecta subclass Apterygota.

A number of features readily distinguish the Collembola from other soil arthropods. The ventral tube or collophore is a cylindrical midventral appendage on the first abdominal segment. It is always present and is equipped with a pair of more or less eversible vesicles at the tip. The furcula, which may be reduced or absent, is a jumping organ on the third and fourth abdominal segment with a single basal segment and paired distal branches. The eyes are simple ocelli, eight plus eight or fewer, or a pair of lateral dark eyepatches. The legs have only a single tibiotarsus beyond the "knees" and a single claw or unguis at the tip, typically with a smaller opposing ventral unguiculus.

Adult Collembola range from 0.5 to 8 mm in length but most soil forms are 1–3 mm long as adults. There are many groups of similar-sized arthropods which are often confused with Collembola. Table 32.1 lists the most common confused forms and shows how they may be distinguished from Collembola.

Taxonomic Anatomy

Good general discussions of collembolan anatomy are available in Paclt (1956), *Biologie der primär flügellosen Insekten,* and Christiansen and Bel-

Table 32.1. Distinguishing Other Groups from Collembola

Group	Distinguished from Collembola by:
Acarina	No antenna, generally four pairs of legs, no collophore
First-instar Diplopoda	No collophore, no legs on segment behind head, large mandibles usually visible at side of head
First-instar Pauropoda	Branched antenna
Wingless Homoptera	More than six abdominal segments, long outstanding antenna, no collophore
Wingless Psocoptera	More than six abdominal segments, head vertical, long antenna, no collophore

linger (1979), *The Collembola of North America North of the Rio Grande.* The discussion below is limited to organs used in this key (see Fig. 32.1).

The cuticle of Collembola is covered with granules, knobs, or reticulations. These are often of taxonomic value. Differentiated areas occur. Simple smoother areas are called fovea, while those in the family Neelidae, equipped with specialized setae, are called sensory fields. Simple porelike circular areas occur in several groups. In the family Onychiuridae they are usually surrounded by cuticular rings and are called pseudocelli. These serve as autohemoragic devices. Cuticular outgrowths are sometimes in the form of large knobs but more commonly in the form of spines, setae, and/or scales. Spines occur mainly on the furcula, head, or posterior abdomen, and may be complex. The end of the abdomen in many poduromorphs and some Isotomidae bears a series of specialized spines called anal horns. Setae are most commonly smooth and conical but come in a wide array of shapes and ornamentations. Large setae or macrochaetae are often complex and are of great importance in identification. Scales occur only in the family Entomobryidae and are fusiform to oval or circular, and striate.

Pigment is located in the epidermal cells. Pigments of almost any color may be present, but a blue that serves as an indicator (turning red in alkali) is the most common. Elaborate patterns occur, particularly in the Entomobryidae and Sminthuridae.

The antennae are four segmented, but the first segment may be divided into two subsegments and the fourth segment is often annulate or subsegmented. The fourth antennal segment is usually equipped with a number of blunt, generally thicker "sensory" setae, much used in classification. The apex of the third antennal segment has a pair of laterodorsal thick blunt setae. This apical organ may be associated with complex organs or relatively simple ones.

The postantennal organ is found between the eyepatch and antennal base. In its simplest form the structure is an oval or rounded vesicle. In its most elaborate form it becomes a series of many tubercles variously arranged.

The mouthparts are somewhat similar to those of other insects. The labium bears an apical pair–complex of setaceous groups of papillae. The maxillae and mandibles are concealed in the head. The former are quite elaborate and varied in structure, generally with a mandiblelike ramus and a number of fringed lamellae. The mandibles typically have a basal molar area with many teeth and a distal disjunct row of two to three larger teeth. In many hypogastrurids the molar plate is lost and in some forms the mandibles and maxillae are reduced to a series of piercing cutting stylets.

The three thoracic segments have distinct seta-bearing tergites in the Poduromorpha, but in the Entomobryomorpha the first segment is reduced and lacks setae. In the Symphypleona the segmental boundaries are difficult to locate and the thorax and abdomen are not separate.

The legs may have one or two subcoxal regions indicated by weak incomplete external sclerites. Characteristic setal patterns and modified setae are found on all segments, particularly the trochanter and tibiotarsus. One or more outstanding blunt, clavate, or acuminate setae projecting over the unguis are called tenent hairs. The unguis or upper claw is universally present and consists of two to five lamellae ending in free teeth. Additional teeth may occur along the lamellae. A sheath or tunica may surround the unguis. The unguiculus or empodial appendage is a ventral lanceolate process with two to four lamellae which may bear teeth. It is reduced or absent in some forms.

The abdomen is typically divided into six segments, although the last is morphologically a telson. The abdominal segments of the Symphylpleona are fused into at most two masses (greater and lesser abdomen). In the Arthropleona the fifth and sixth, and less commonly the fourth through sixth segments are sometimes fused. The first abdominal segment bears the ventral tube, used for water intake, respiration, and adhesion. The third abdominal segment in furculate forms bears a small bifurcate organ called the tenaculum which serves to hold the furcula appressed to the body. The furcula arises from the fourth abdominal segment and consists of a basal manubrium, distal elongate dentes, and a pair of generally smaller toothed mucrones. The furcula is lacking or vestigial in the most Onychiuridae and is often reduced or absent in the Hypogastruridae and Isotomidae. It is always well developed in the Entomobryidae and the Symphypleona.

Biology

The Collembola live in a wide variety of habitats, but most species, except for littoral and neustonic forms, are at least occasionally found in the soil. Most species live here, where they, along with the mites, make up the majority of the arthropod fauna. They have a wide variety of diets, but most soil forms feed on decayed vegetation and/or microflora. Development is direct with adults differing from juveniles in proportion, size, pigment (usually juveniles are paler), and the absence of a genital aperture. In some genera a diapause occurs which may be associated with regressive modification of mouthparts and digestive system and even striking external modification of curticle and the development of spines (ecomorphosis). Collembola molt throughout life with instars ranging from four to more than 50.

Sperm transfer is indirect in soil forms. Males deposit spermatophores which are picked up by females.

Distribution

Collembola are found in all soil habitats, but they are most abundant and common in the litter layer. In desert regions they are limited to protected moist areas. They are extremely abundant in cold climates, and their proportion of the soil fauna increases as the annual temperature diminishes. Most genera are extremely widespread; however, there are a number of primarily arctic forms and a small number of thermophilic genera, largely of tropical origin.

Preservation and Identification

Collembola can be readily collected manually with the aid of an aspirator or a wetted brush. Tullgren or Berlese funnels or flotation are also good collecting methods. Specimens are best stored in 70% isopropyl alcohol. For examination many methods are available; however, a good one uses Marc Andre 1 liquid for clearing. The formula follows:

Distilled water	30 cm^3
Chloral hydrate	40 g
Glacial acetic acid	30 cm^3

Specimens can be placed in this from any standard preservative, but they should have been in the preservative for at least 2 weeks. Fresh specimens are likely to explode, although this is often good for displaying chaetotaxy. Specimens are kept in this medium until cleared. Very dark specimens may have to be bleached first in KOH or hydrogen peroxide. The specimens may be safely kept in Marc Andre 1 as long as 36 hours. After clearing, complete specimens are mounted in Marc Andre 2 solution:

Distilled water	50 cm^3
Chloral hydrate	200 g
Glycerine	40 cm^3
Gum arabic	20 g

Specimens are mounted on double coverslips loosely attached to a microscope slide and cured in an oven at 45–60°C for 24–36 hours. It is generally best to mount one specimen with a side and one with a dorsal view. In pigmented forms it is important to have one specimen still showing the pigment distribution.

Identification is best done using a compound microscope. The meanings of most descriptions of position will be clear. The terms "outer," "inner," "anterior," and "posterior" with reference to legs presume that these are extended straight laterally. "Dorsal" and "ventral" applied to the furcula refer to this position as the organ extends posteriorly. More detailed information about classification can be found in *The Collembola of North America* (Christiansen and Bellinger 1979).

ACKNOWLEDGMENT

Dr. Richard J. Snider graciously reviewed this manuscript.

KEY TO GENERA

1 or **120** Body linear; at least four abdominal segments clearly separated (Fig. 32.1); Suborder **Arthropleona** **2**

2 or **51** First thoracic segment dorsally distinct and with dorsal setae (Section **Poduromorpha**) (Fig. 32.2). **3**

3 or **10** Pseudocelli present, at least on antennal bases, eyes absent (Fig. 32.3); Family **Onychiuridae** **4**

4 or **5** Furcula with dens and mucro present (Fig. 32.4) . . . . ***Lophognathella*** Börner

This is the only genus of the family with blue or black pigment. There is a single species, ***L. choreutus***, known from North America; found only in Oregon and northern California. Maximum length about 2 mm.

5 or **4** Furcula reduced to a pair of knobs, or absent **6**

6 or **7** Only three antennal segments evident (third, and fourth segments fused) (Fig. 32.5) . . . . ***Sensiphorura*** Rusek

Color white. Body slender and elliptical. Pseudocelli not clearly visible with light microscope. Known from a single species, ***S. marshalli***, so far known only from Vancouver Island. Maximum length about 0.5 mm.

7 or **6** Last two antennal segments distinct **8**

8 or **9** Apical organ of third antennal segment with two clubs behind four to six integumentary papillae (Fig. 32.6) . . . . . . ***Onychiurus*** Gervais

White, pale yellow, or reddish. Body broadly elliptical to fusiform. Unguiculus present and generally well developed. Anal horns are present or absent. There are 47 species divided into three subgenera (***Archaphorura***, ***Protaphorura***, and ***Onychiurus* s.s.**). Many new species remain to be described. The members of the genus occur all over the new world and they are dominant forms in the euedaphon. Maximum sizes range from 0.7 to 4.0 mm with most species having maxima between 1 and 2 mm.

9 or **8** Apical organ of third antennal segment with one to three exposed sense clubs (Fig. 32.7) . . . . ***Tullbergia*** Lubbock

Without trace of pigment. Body elongate fusiform or elliptical. Unguiculus minute. These slender minute forms are dominant in the deeper layers of the euedaphon. There are 105 described species occurring all over the new world and probably many undescribed forms. Maximum lengths vary from 0.5 to 1.5 mm but most maxima are around 1 mm.

10 or **3** Pseudocelli absent; Family **Hypogastruridae** **11**

11 or **34** Mandibles with clear molar plate (Fig. 32.8); Subfamily **Hypogastrurinae**. **12**

12 or **15** Postantennal organ absent **13**

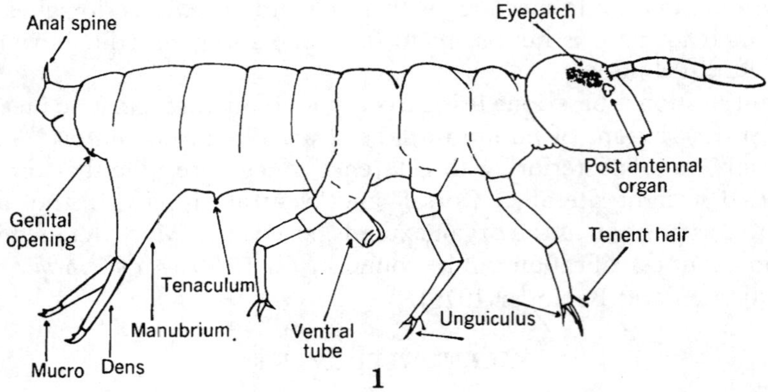

1

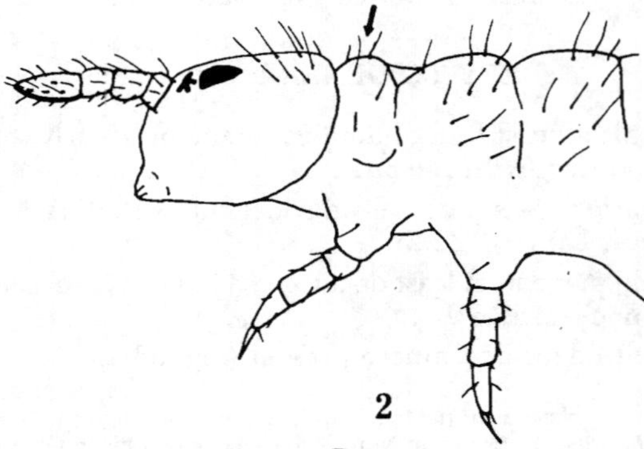

2

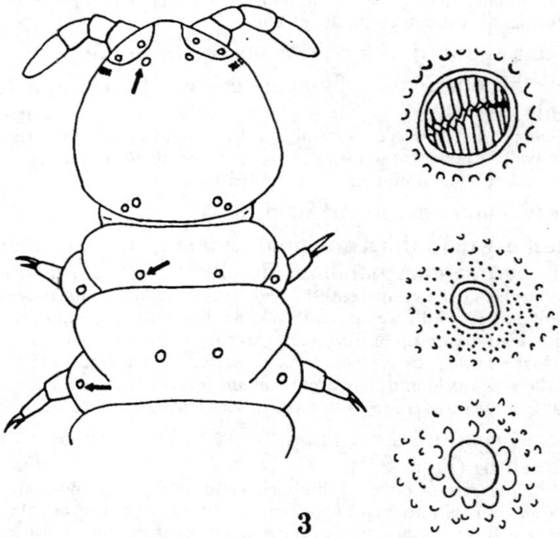

3

Fig. 32.1. General Arthropleona body shape. **Fig. 32.2.** Head and first two thoracic segments of schematic hypogastrurid. **Fig. 32.3.** *Onychiurus* composite dorsal view showing anterior end and different types of pseudocelli.

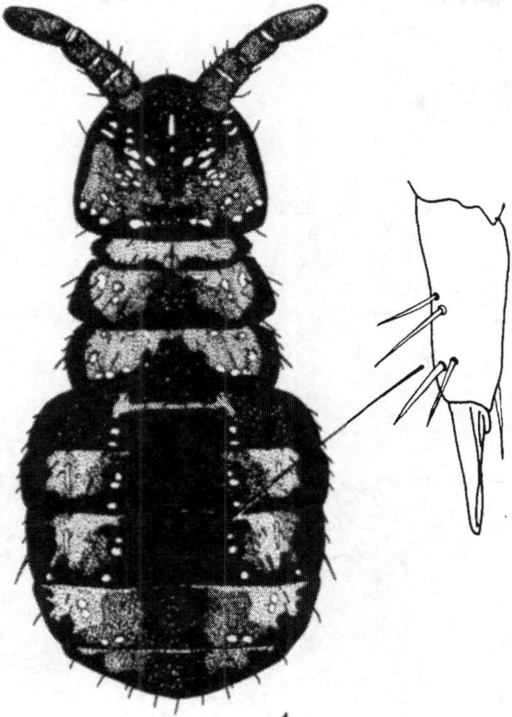

4

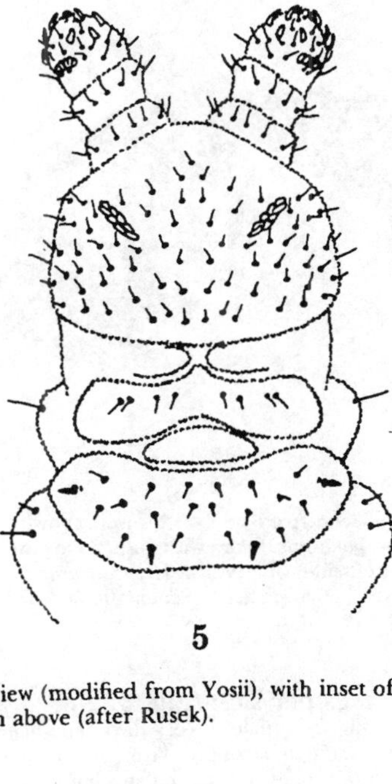

5

Fig. 32.4. *Lophognathella* dorsal view (modified from Yosii), with inset of furcula. **Fig. 32.5.** Anterior portion of *Sensiphorura* seen from above (after Rusek).

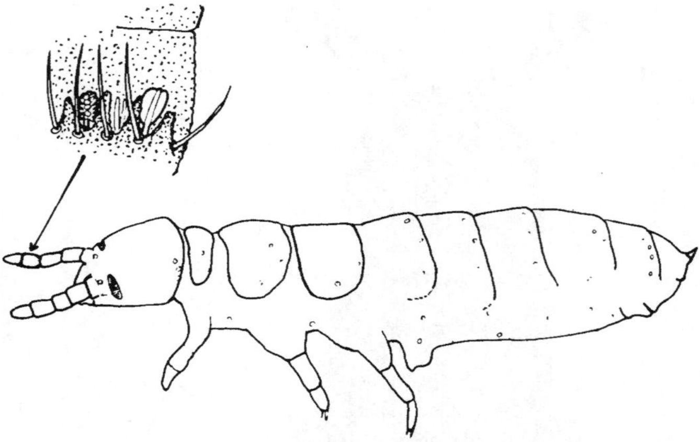

6

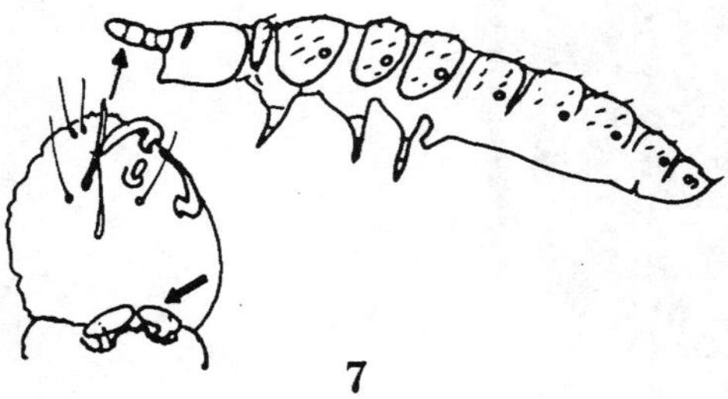

7

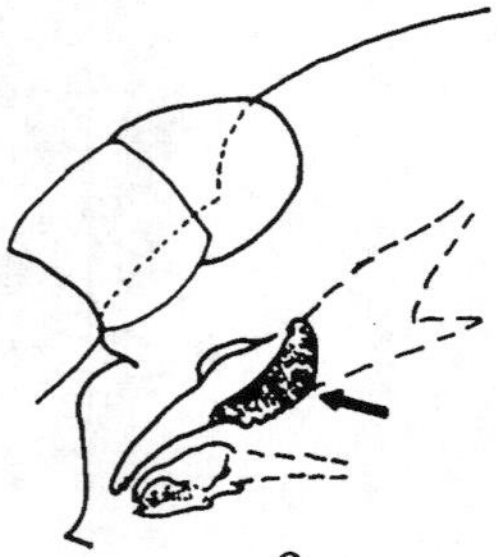

8

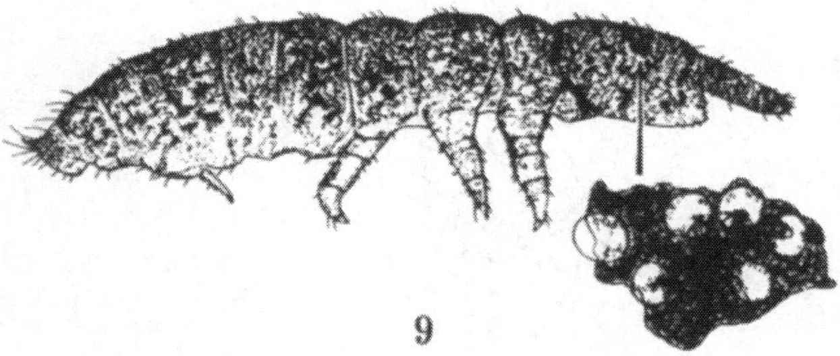

9

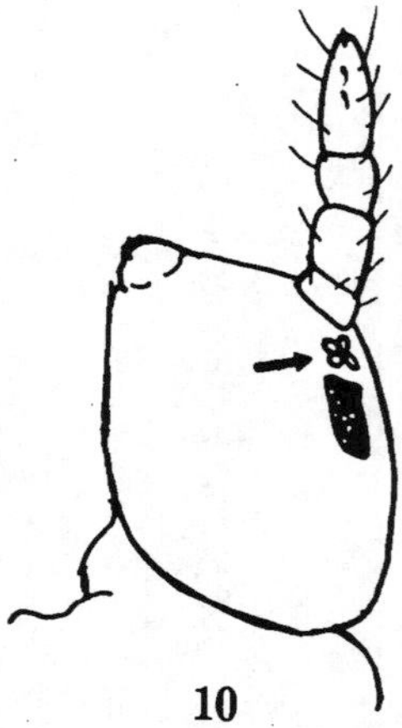

10

Fig. 32.6. Generalized *Onychiurus* seen from side, with insert showing the third antennal segment sense organ. **Fig. 32.7.** Habitus of *Tullbergia falca,* with insert showing third antennal segment sense organ. **Fig. 32.8.** Mouthparts (left side) of a typical *Hypogastrura.* **Fig. 32.9.** *Xenylla grisea* (after Linnianemi), with insert showing eyes. **Fig. 32.10.** Schematic podurid head showing postantennal organ.

13 or **14** Four to five eyes per side (Fig. 32.9) . . . . ***Xenylla*** Tullberg

Pigment varying from pale to very dark blue-black. Unguiculus absent and foot usually with two weakly to strongly clavate tenent hairs. In a few species the dens is strongly reduced or absent. Uncleared specimens may be confused with the subgenus ***Schoetella*** of the genus ***Hypogastrura.*** There are 15 Nearctic species found everywhere except for the high arctic. Maximum sizes between 1 and 2 mm.

14 or 13 Eyes and pigment absent . . . . ***Acherontiella*** Absolon

Body broadly fusiform without pigment eyes or postantennal organ. Unguiculus and furcula absent. Two clavate tenent hairs per foot. This Latin American genus is represented in the United States by a single species, ***A. sabina***. The only known collection is from a kangaroo rat's nest from New Mexico. Maximum length 1.75 mm.

15 or 12 Postantennal organ present (Fig. 32.10) **16**

16 or 17 Anal spines 3, dorsal dental setae 9 or more (Fig. 32.11) . . . . . . ***Triacanthella*** Schäffer

Postantennal organ of four simple lobes. Foot with well developed unguiculus and a single clearly clavate tenent hair. This genus has a single known Nearctic species, ***T. copelandi***. So far recorded only from Tennessee. Maximum length 1.5 mm.

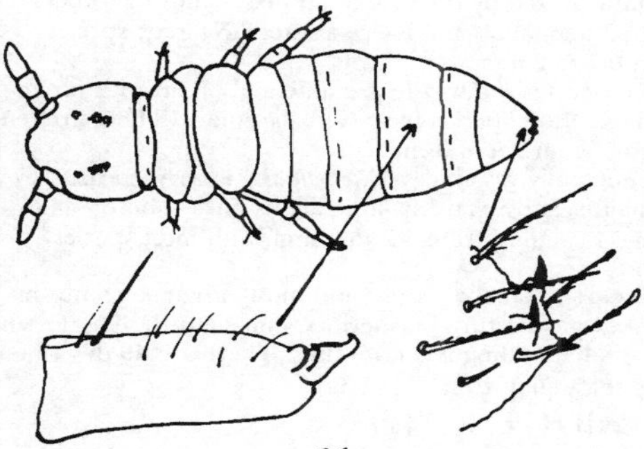

11

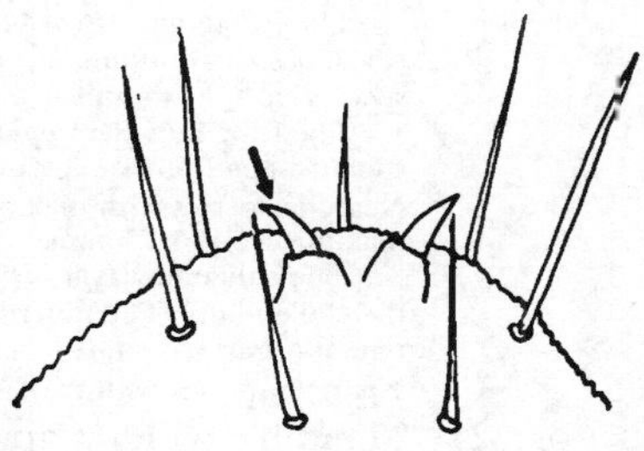

12

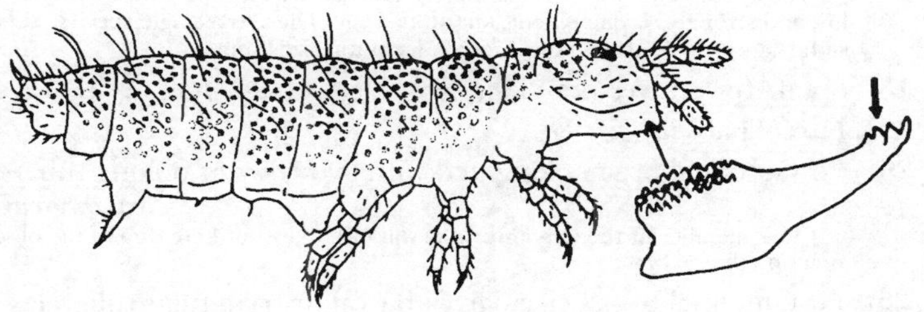

13

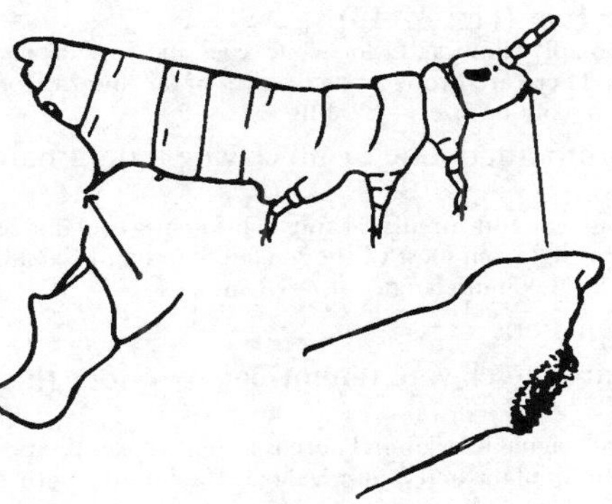

14

Fig. 32.11. Habitus *Triacanthella copelandi*, with inserts showing dentes and spines. **Fig. 32.12.** Typical hypogastrurid and spines seen from above. **Fig. 32.13.** Typical *Hypogastrura (Ceratophysella)* (after Stach), with insert showing mandibles with apical teeth. **Fig. 32.14.** Habitus *Stenogastrura hiemalis*, with inserts showing mandible lacking apical teeth and mucro.

17 or **16** Anal spines two or absent (Fig. 32.12) **18**

18 or **23** Eyes 8 + 8 . . . . **19**

19 or **20** Furcula absent . . . . ***Knowltonella*** Wray

Color deep purple-blue. Unguiculus absent. Four clavate tenent hairs on each foot. A single species ***K. idahoensis*** is known, collected only in Idaho. This species could easily be confused with ***Hypogastrura (Schoetella)*** if the furcula is overlooked. Maximum length 1.5 mm.

20 or **19** Furcula present **21**

21 or **22** Mandible with apical teeth (Fig. 32.13) . . . . ***Hypogastrura*** Bourlet

Color varied but most generally bluish or brownish. The postantennal organ varies in size and shape but is usually four-lobed. Furcula well developed and the parts always distinct. Many forms display ecomorphosis and may be mistaken for other genera. The genus is abundant all over North America. The four disparate subgenera.

The subgenus ***Schoetella*** is characterized by the lack of an unguiculus (all others have this), and a simple, small postantennal organ. There are four Nearctic species with maximum size ranging from 0.7 to 2 mm.

The subgenus ***Mitchellania*** is characterized by the postantennal organ with posterior lobes surrounding a central boss. Most species have cephalic spines. There are ten species with maximum sizes ranging from 1.5 to 3 mm.

The subgenus ***Ceratophysella*** generally resembles ***Mitchellania***, is characterized by a spoon-shaped mucro, a typical habitus, and relatively large anal spines. There are 26 Nearctic species with maximum sizes ranging from 1.2 to 3 mm, with most species maxima around 1.5 mm.

The subgenus ***Hypogastrura* s.s.** is by far the largest and most varied subgenus in the Collembola. The mucro is varied in structure but never spoon-shaped. The postantennal organ has short lobes never surrounding a central boss. There are 49 described Nearctic species with maximum size ranging from 1 to 3 mm.

22 or **21** Mandible without apical teeth (Fig. 32.14) . . . .
Stenogastrura Christiansen and Bellinger

Color deep blue with rounded paler spots and appendages pale. Unguis, unguiculus, and strongly clavate tenent hair all present. There is a single species, ***S. hiemalis***, so far known only from Nova Scotia. Maximum size 3 mm.

23 or **18** Six or fewer eyes per side. **24**

24 or **29** Furcula absent **25**

25 or **26** Postantennal organ elongated with two rows of oblong tubercles . . .
Sensiphorura Rusek

This member of the **Onychiuridae** will often key out here due to the obscure pseudocelli; see 6.

26 or **25** Postantennal organ circular with oval or triangular tubercles . . . **27**

27 or **28** Inner margin of unguis strongly toothed; with three clavate tenent hairs per foot (Fig. 32.15) . . . . ***Tafallia*** Bonet

Body broadly elliptical. Color white, eyes and furcula absent. Anal horns large and prominent. There are two Nearctic species so far known from Utah, New Mexico, and California. Maximum size 1–1.2 mm.

28 or **27** Unguis untoothed, one or no clavate tenent hair per foot (Fig. 32.16)
Willemia Börner

Eyes pigment and furcula lacking. This genus contains seven localized Nearctic species, recorded from most of the United States and Canada except for the southern tier of states. Maximum lengths 0.7–1.1 mm.

29 or **24** Furcula present **30**

30 or **31** More than one clavate tenent hair per foot (Fig. 32.17) . . . .
Mesachorutes Absolon

Eyes and pigment lacking. There is a single Nearctic species, ***M. thomomys***, so far known from Montana and South Dakota. Maximum length 1.2 mm.

31 or **30** One or no clavate tenent hair per foot **32**

32 or **33** Mandible with strong molar plate and several apical teeth (Fig. 32.18; see also Fig. 32.8) . . . . ***Schaefferia*** Absolon

Color white, yellow, or with scattered pigment. Often with dark eye patches. Eyes from 2 + 2 to 6 + 6. The subgenus ***Schaefferia* s.s.** is known from a single Tennessee

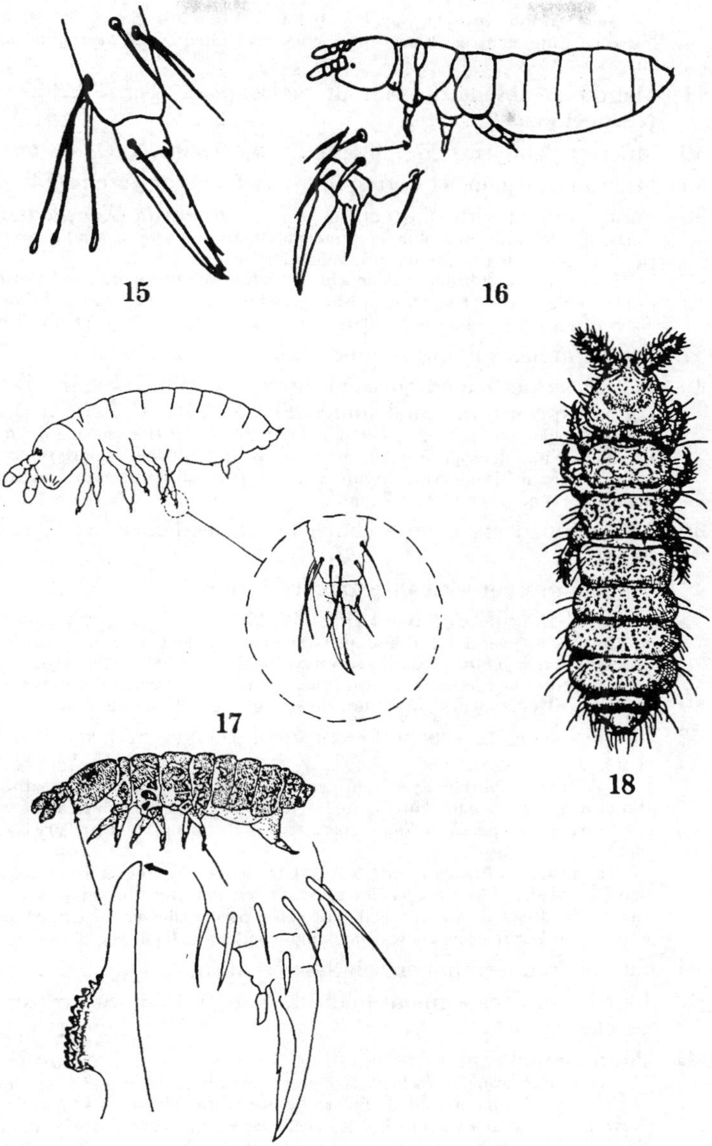

Fig. 32.15. Hind foot complex of *Tafallia insularis*. Fig. 32.16. Habitus of *Willemia dubia*, with insert showing foot complex. Fig. 32.17. Habitus of *Mesogastrura thomomys* (after Chamberlain) with insert showing hind foot complex. Fig. 32.18. Habitus of *Schaefferia* sp. (after Thibaud). Fig. 32.19. Habitus of *Microgastrura minutissma*, with inserts showing mandible and hind foot complex.

species ***S. cheoha,*** and the subgenus ***Typhlogastrura*** has three known Nearctic species. ***S. variablilis*** is a surface species recorded from Alaska and Wisconsin. There are two cave species found in Alabama and Arkansas. Maximum lengths 1.5–2.5 mm.

33 or 32 Mandible with reduced molar plate and without apical teeth (Fig. 32.19). . . . . ***Microgastrura*** Stach

Color reddish brown to gray. Eyes 6 + 6. There is a single species, ***M. minutissma,*** found in scattered areas all over the contiguous United States. Maximum size about 0.5 mm.

34 or **11** Mandibles absent or without molar plate (Fig. 32.20); Subfamily **Neanurinae** **35**

35 or **36** Mucro trilamellate (Fig. 32.21) . . . . ***Odontella (Odontella)*** Schäffer

36 or **35** Mucro not trilamellate, frequently reduced or absent (Fig. 32.22) **37**

37 or **38** Foot complex with unguiculus . . . . ***Odontella (Xenyllodes)*** Axelson

Body color white and blue. Postantennal organ in a deep pit and generally triradiate. Eyes generally 5 + 5 (one species has 2 + 2).

There are two subgenera ***Odontella* s.s.** with 8 Nearctic species and ***Xenyllodes*** with two Nearctic species. The genus is widespread in the United States and Canada but has not been recorded from the southern tier of states. Maximum lengths 1–2 mm.

38 or **37** Foot complex without unguiculus. **39**

39 or **40** Maxilla sickle-shaped, three or more anal spines or spine like setae, or (in one species) two anal spines (Fig. 32.23). . . ***Friesea*** Dalla Torre

Color blue. Eyes from 2 + 2 to 8 + 8. Furcula varies from having the mucro manubrium and dens all clear but small, to absent. Most species have three or five anal horns. There are 18 described Nearctic species from all over North America. Maximum size range from 0.7 to 1.7 mm.

40 or **39** Maxilla quadrate, narrow, lamellate, or needlelike (see Figs. 32.20 and 32.24). **41**

41 or **44** Furcula present with all segments distinct **42**

42 or **43** Head of maxilla quadrate (Fig. 32.25) . . . . ***Brachystomella*** Agren

Color blue. Eyes 8 + 8. The fourth antennal segment is fused to the third. This primarily tropic genus is readily recognized by the peculiar maxilla. There are six recorded Nearctic species. Except for one record from Alaska, all records are from the contiguous United States. Maximum sizes range from 0.8 to 1.2 mm.

43 or **42** Maxilla elongate, crochetlike or serrate (Fig. 32.26; see also Fig. 32.24 *A,B*) . . . . ***Pseudachorutes*** Tullberg

Color blue to purple. Eyes eight per side. Mandible two- to eight-toothed. Maxilla two-bladed and elongate. Furcula well developed. There are two Nearctic subgenera. The subgenus ***Pseudochorudina*** is characterized by having the maxillary blades with toothed inner margin.

The subgenus ***Pseudachorutes* s.s.** is characterized by crochetlike or apically toothed maxillary blades. There are 17 described species; however, the genus is poorly known and many species are undescribed. The genus occurs all over the United States and Canada but is rare everywhere. Maximum lengths run from 1 to 2 mm.

44 or **41** Furcula reduced (mucro always absent) or lacking **45**

45 or **48** Last abdominal segment bilobed, body usually with conspicuous tubercles **46**

46 or **47** Postantennal organ present (Fig. 32.27) . . . . . . ***Morulina*** Börner

Color blue, lavender, or very dark green. Mandibles large and 5- to 10-toothed. Maxillae vary from one-bladed and crochetlike to three-bladed and heavily serrate. Furcula reduced to a knob or knobs. There are seven Nearctic species found all over the United States and Canada. Maximum size range from 2 to 5 mm. These are typically inhabitants of the litter layer.

47 or **46** Postantennal organ absent (Fig. 32.28) . . . . ***Neanura*** MacGillivray

Color generally white or blue. Eyes from absent to 4 + 4. Body generally somewhat dorsoventrally compressed. Furcula absent or reduced to a tubercle. There are four Nearctic subgenera. The subgenus ***Crossodonthina,*** characterized by complex fringed mouthparts, is represented by three species found in Alaska, British Columbia, Washington, Idaho, Oregon, and northern California. The subgenus ***Lobella*** has a complex mandible and simple crochetlike maxilla. It is represented by a single Californian species, ***N. (Lobella) ambigua.*** The great majority of Nearctic ***Neanura*** fall into the subgenera ***Neanura*** and ***Lathriopygia,*** characterized by crochetlike maxillae and tridentate mandibles. There are 14 Nearctic species of these two subgenera. Maximum lengths range from 1.5 to 3.5 mm.

48 or **45** Last abdominal segment rounded. **49**

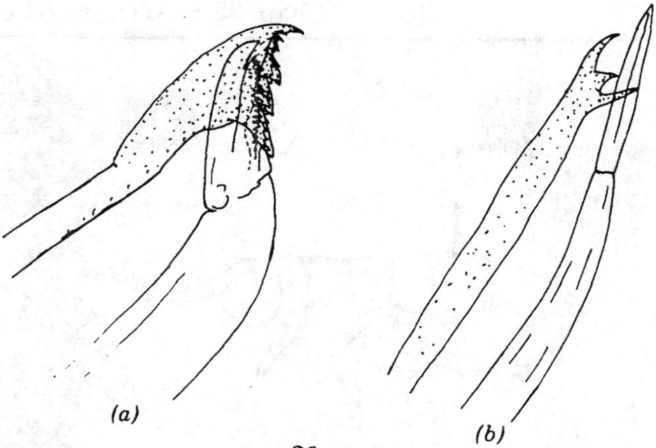

20

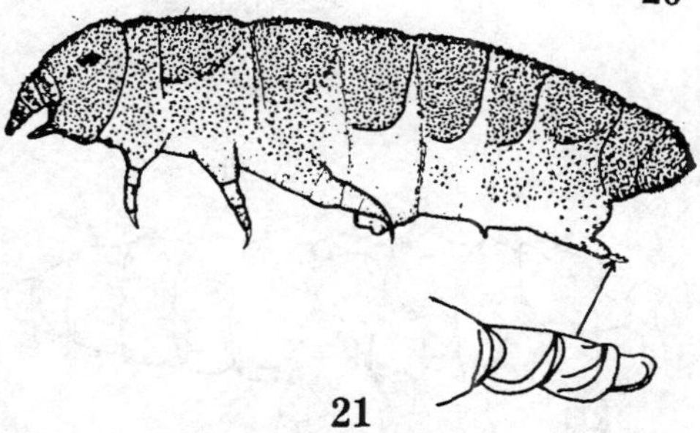

21

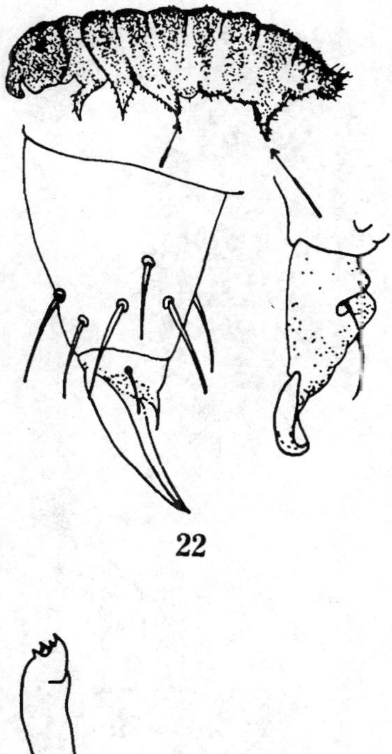

22

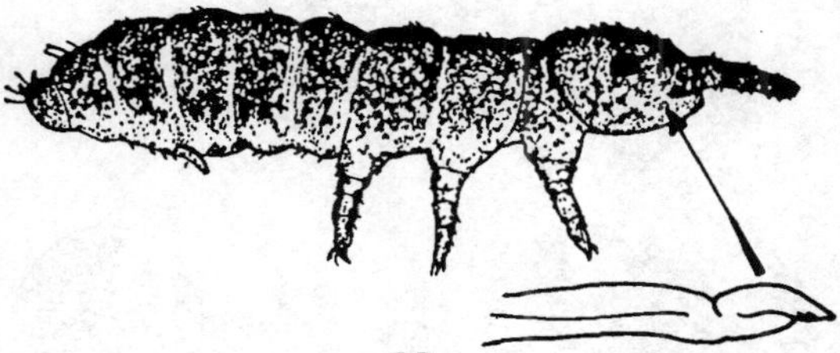

23

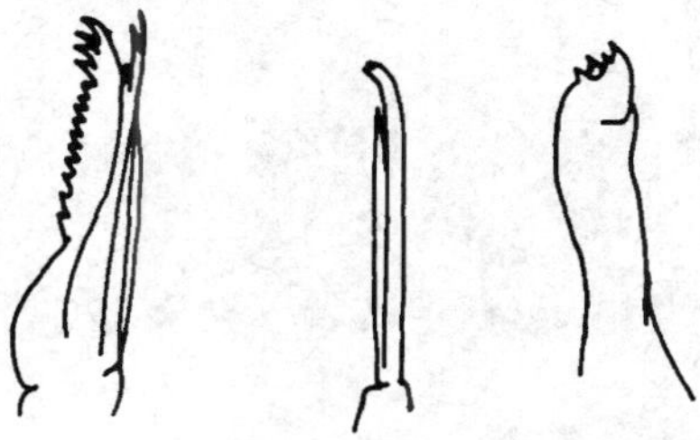

24

Fig. 32.20. Mouthparts of (*a*) *Anurida granaria* and (*b*) *Neanura muscorum*. Mandibles stippled. **Fig. 32.21.** Habitus of *Odontella (Odontella) shasta*, with insert showing mucro seen from above. **Fig. 32.22.** Habitus of *Odontella (Xenyllodes) armatus*, with inserts showing mucro and foot complex. **Fig. 32.23.** Habitus of *Friesea claviseta* (after Linnianemi), with insert showing maxilla. **Fig. 32.24.** Three neanurid maxillary types: (*a,b*) *Pseudachorutes;* (*c*) *Brachystomella*.

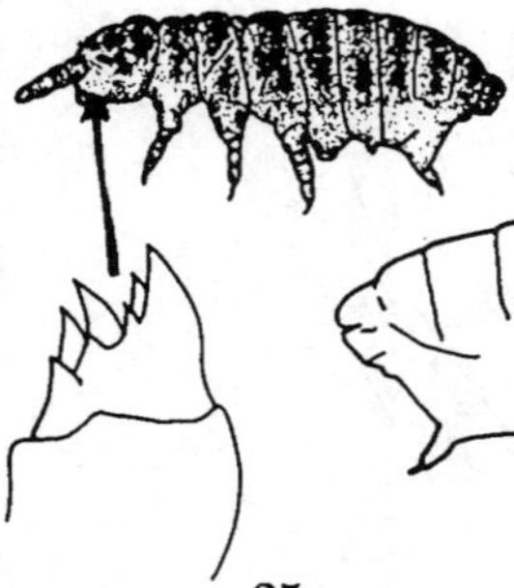

25

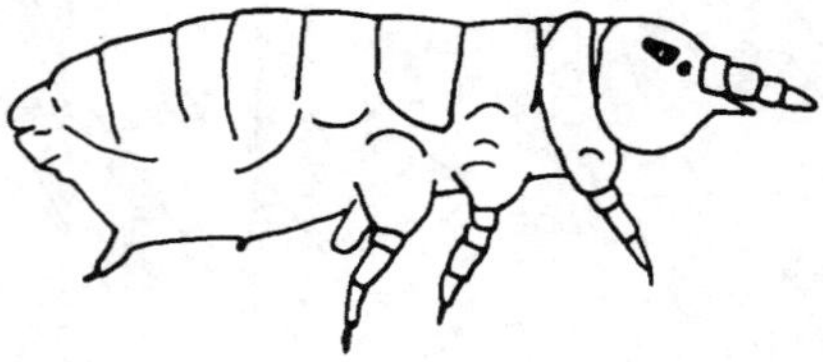

26

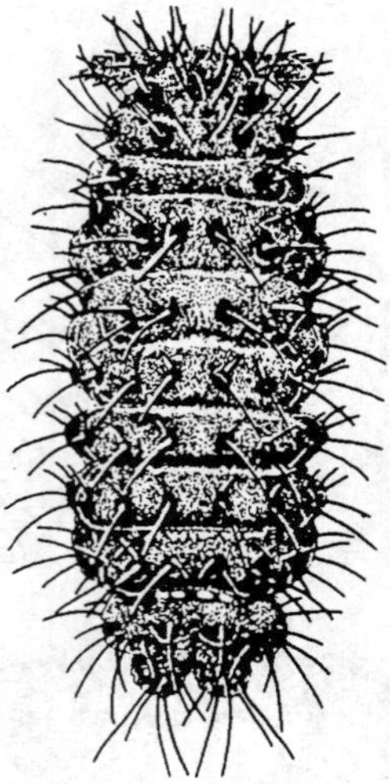

28

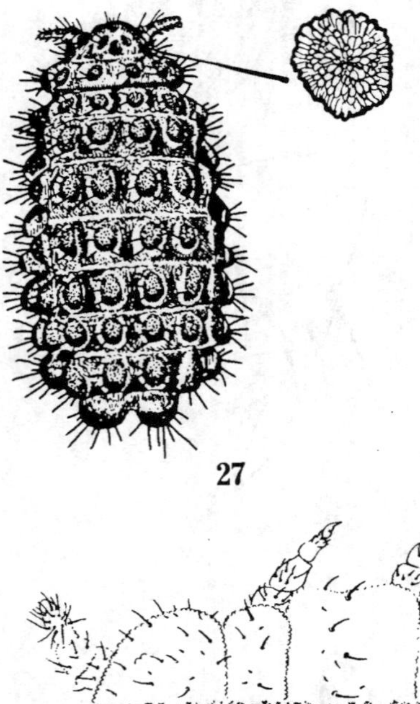

27

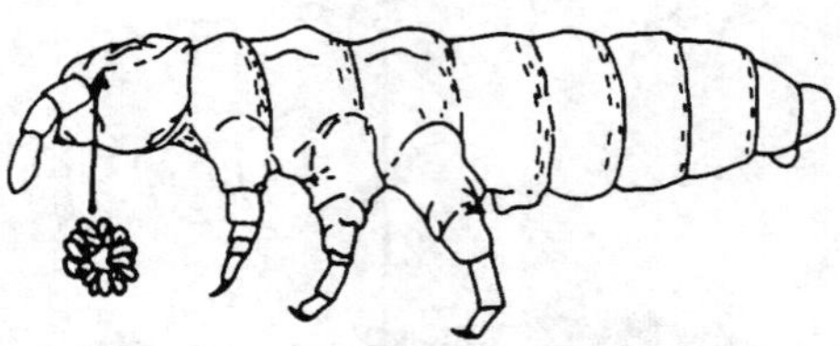

29

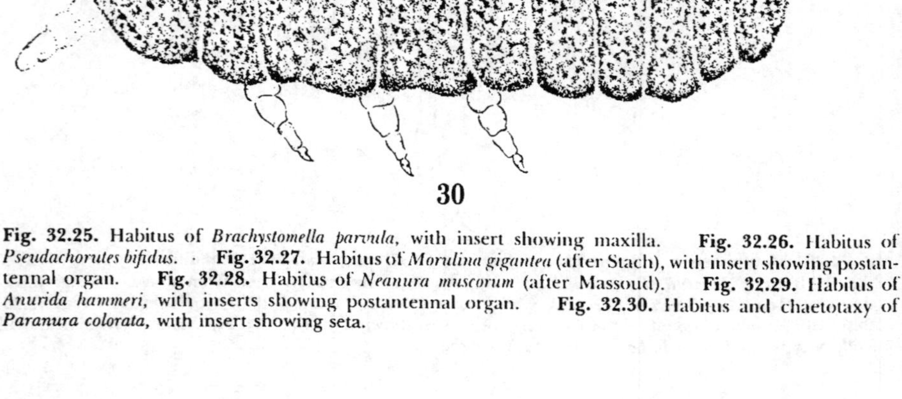

30

Fig. 32.25. Habitus of *Brachystomella parvula*, with insert showing maxilla. **Fig. 32.26.** Habitus of *Pseudachorutes bifidus*. **Fig. 32.27.** Habitus of *Morulina gigantea* (after Stach), with insert showing postantennal organ. **Fig. 32.28.** Habitus of *Neanura muscorum* (after Massoud). **Fig. 32.29.** Habitus of *Anurida hammeri*, with inserts showing postantennal organ. **Fig. 32.30.** Habitus and chaetotaxy of *Paranura colorata*, with insert showing seta.

49 or **50** Postantennal organ present (Fig. 32.29) . . . . ***Anurida*** Laboulbene

Color white or purplish. Mandible with 3–15 teeth. Maxilla varied. Eyes 0 to 8 + 8. Furcula usually absent. The soil fauna has three Nearctic subgenera. Only the subgenus ***Protachorutes*** has tenaculum and a vestige of the furcula. The single Nearctic species, ***P. bicolor,*** is known only from California. The subgenus ***Micranurida*** characterized by simple maxillae has 6 Nearctic species, not known south of Maryland and Illinois. The subgenus ***Anurida*** **s.s.** characterized by complex maxillae has 18 described Nearctic primarily northern species but are rarely found all over the United States and Canada. The maximum size of the subgenera ***Protachorutes*** and ***Micranurida*** range from 0.5 to 0.7 mm while those of ***Anurida*** **s.s.** are generally between 1 and 4 mm.

50 or **49** Postantennal organ absent (Fig. 32.30) . . . . . . ***Paranura*** Axelson

Color white or (one species) blue. Eyes 0 to 3 + 3. Mandibles tridentate and maxilla needlelike. Furcula reduced. Body generally similar to ***Neanura*** except for the absence of lobes or tubercles. There are five Nearctic species widely distributed in Canada and the United States. Maximum length 1–3 mm.

51 or **2** First thoracic segment without clear dorsal region or tergite or setae (see the section on **Entomobryomorpha**) (Fig. 32.31) **52**

52 or **53** Mucro hairy; fourth antennal segment much shorter than third (Family **Entomobryidae**) Subfamily **Tomocerinae** (Fig. 32.32) . . . . . . **Tomocerus** Nicolet

Color with scales silvery white to shiny black. With scales removed white or yellow, rarely brown. Eyes 6 + 6 to none. The genus is divided into six subgenera largely on a basis of chaetotaxy features. There are 16 species found all over North America. Maximum length 2–7 mm.

53 or **52** Mucro smooth or with at most two setae, third antennal segment at most only slightly longer than fourth **54**

54 or **57** Dentes with dentate spines, mucro subequal in length to dens. Family **Entomobryidae,** Subfamily **Oncopodurinae** **55**

55 or **56** Eyes and pigment absent (Fig. 32.33) . . . . ***Oncopodura*** Carl and Lebedinsky

Body silvery or white, occasionally with scattered flecks of pigment. Postantennal organ usually present. There are six Nearctic species found primarily in caves. They are so far known from the northern United States (Wisconsin and New Hampshire) south. Maximum lengths 1–2 mm.

56 or **55** Eyes present, adults pigmented (Fig. 32.34) . . . ***Harlomillsia*** Bonet

Body color of adults blue-gray and juveniles white. Apex of antenna with unique large apical spatulate seta. Postantennal organ present. There is a single species, ***H. oculata,*** so far recorded in the southeastern United States and Oregon. Maximum length 0.6 mm. This is an inhabitant of the upper soil layers.

57 or **54** Dentes without spines or with simple short smooth or finely ciliate spines **58**

58 or **83** Postantennal organ absent **59**

59 or **60** Body without scales, small body setae smooth and acuminate (Fig. 32.35) . . . . ***Isotomiella*** Bagnall

Fourth antennal segment with six large oval sensillae. Tenent hairs acuminate: Unguiculus well developed. Mucro tridentate. The species ***Folsomia (Folsomina) onychiurina*** will key out here, but it can readily be distinguished by its falcate mucro. The single Nearctic species has been recorded from Idaho and Massachussetts to Utah and North Carolina. Maximum length 1.1 mm. It is primarily an inhabitant of the upper soil layers, common in caves.

60 or **59** Body with scales and/or with some small body setae ciliate **61**

61 or **64** Dentes smooth (not crenulate) **62**

62 or **63** Mucro at least one-third as long as dens. Dens with large fringed scales, Subfamily **Cyphoderinae** (Fig. 32.36). . . ***Cyphoderus*** Nicolet

Similar to ***Pseudosinella*** in appearance. Color silvery white. Eyes and postantennal organ absent. Unguiculus with large wing tooth. This tropical genus has a single Nearctic species, ***C. similis,*** so far reported from ant nests in Massachussetts, New Jersey, Iowa, and Louisiana. Maximum size 1.5 mm.

63 or **62** Mucro less than one-fourth as long as dens. Subfamily **Paronellinae**

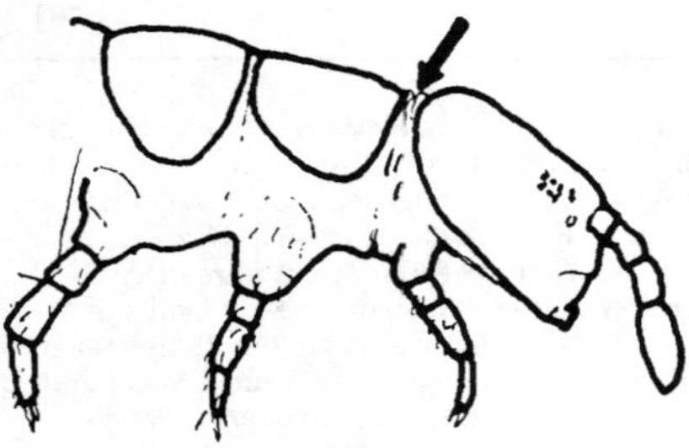

31

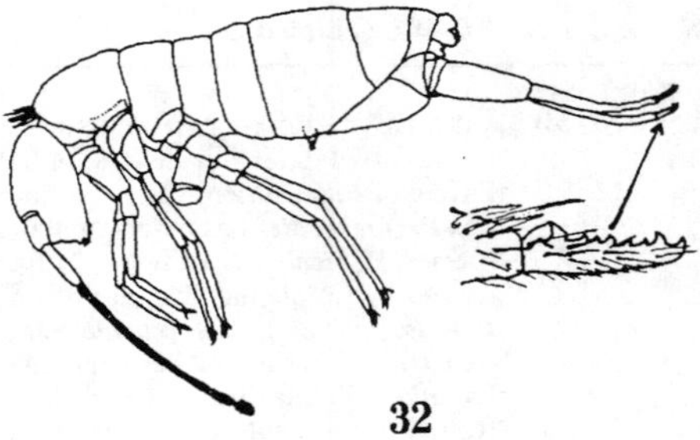

32

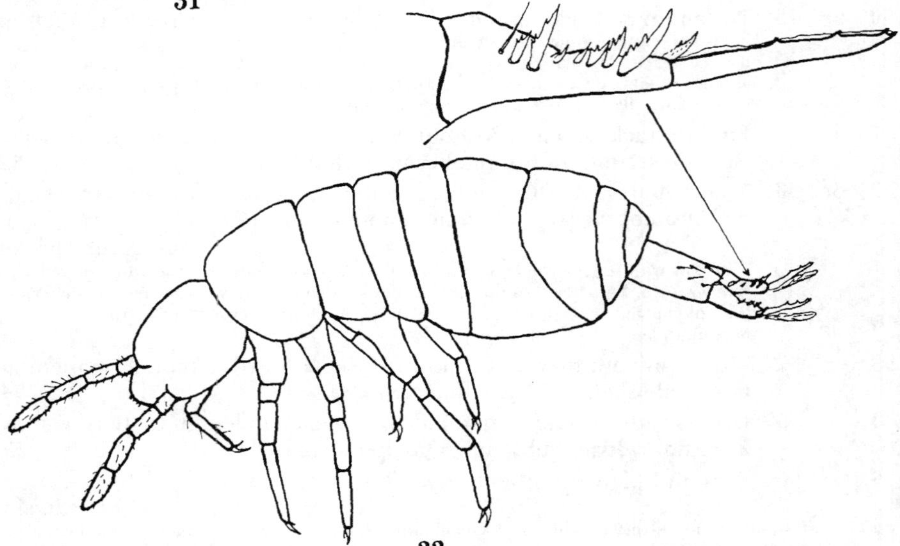

33

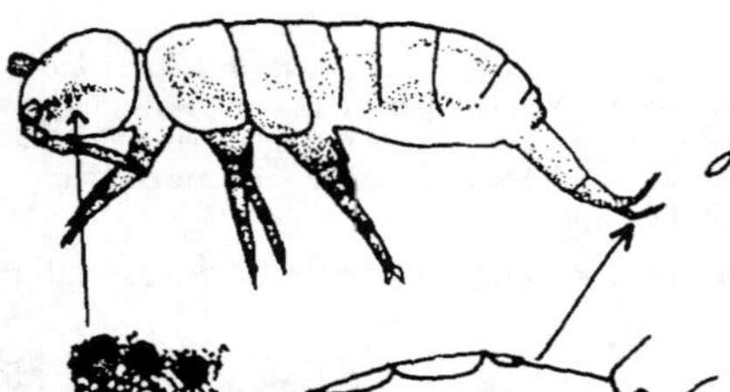

34

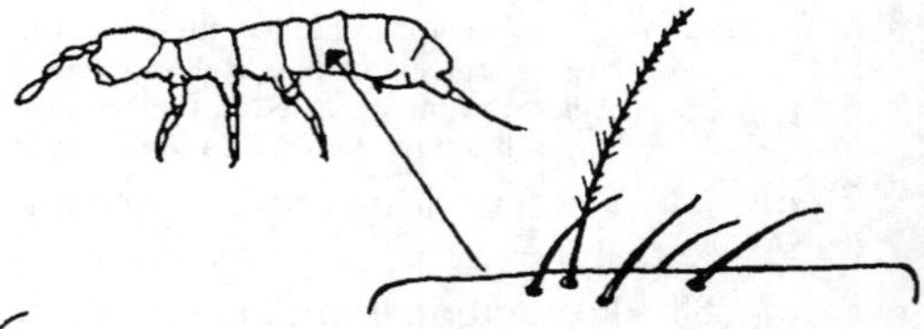

35

36

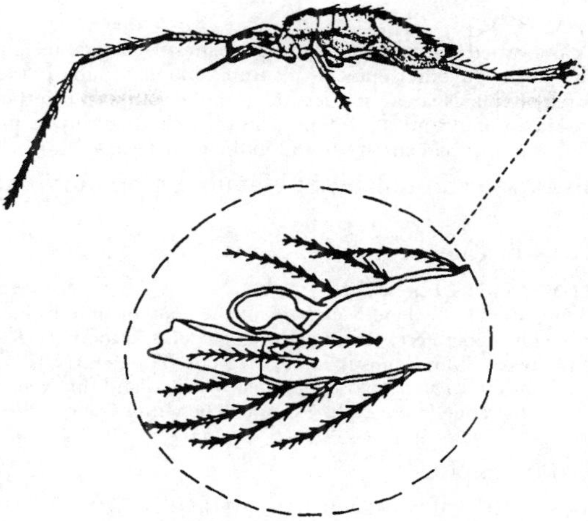

37

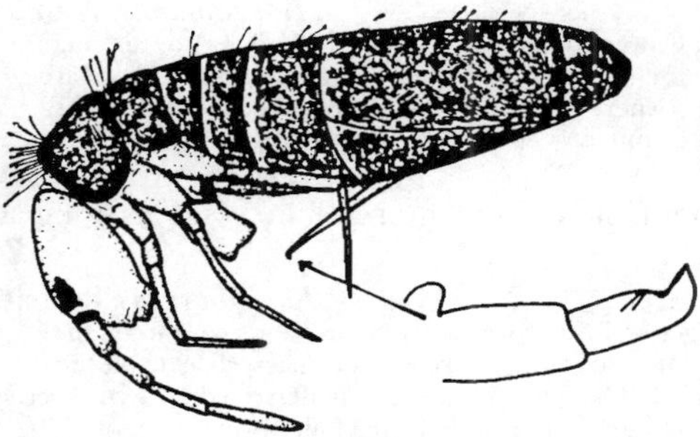

38

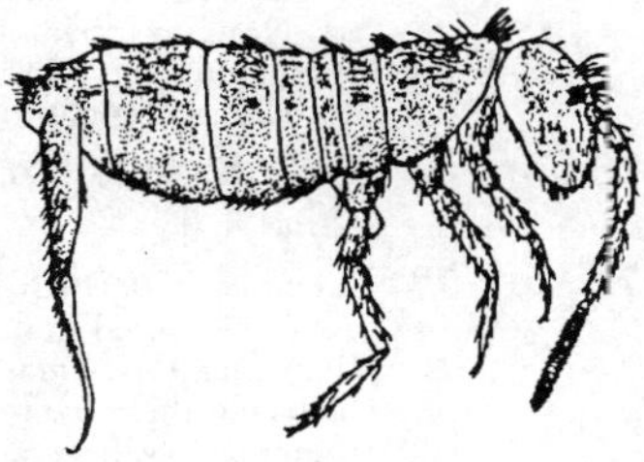

39

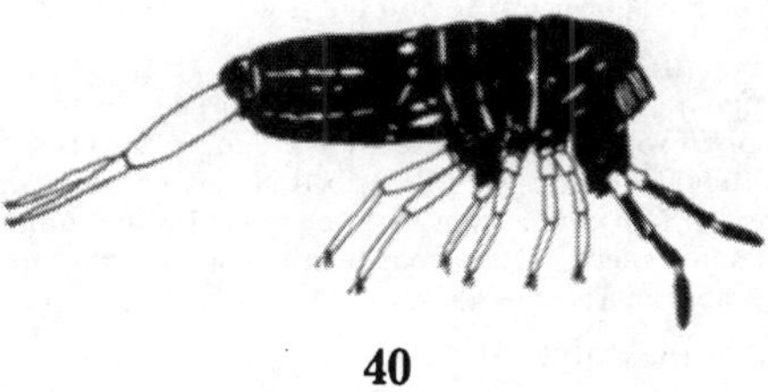

40

Fig. 32.37. Habitus of *Salina banksi* (after Pedigo), with insert of mucro, after Mills. **Fig. 32.38.** Habitus of *Seira bipunctata*, with insert showing mucro and apex of dens. **Fig. 32.39.** Habitus of *Heteromurus nitidus* (after Salmon). **Fig. 32.40.** Habitus of *Lepidocyrtus paradoxus* (after Snider).

Fig. 32.31. Schematic representation of entomobryid thorax. **Fig. 32.32.** Habitus of *Tomocerus flavescens* (after Folsom), with inserts showing mucro. **Fig. 32.33.** Habitus of *Oncopodura hoffi*, with insert showing mucro and dens. **Fig. 32.34.** Habitus of *Harlomillsia oculata* with inserts of dens and eyes and postantennal organ (after Mills). **Fig. 32.35.** Habitus of *Isotomiella minor* (after Gisin), with insert showing large setae. **Fig. 32.36.** Dens and mucro of *Cyphoderus similis* (after Folsom).

(Fig. 32.37) . . . . ***Salina*** MacGillivry

Colors varied (usually with a striking pattern). Ungual teeth are small. Scales are absent. This is the only genus of this tropical family found in the United States. There are three or four Nearctic species found in the southern two-thirds of the United States. Maximum lengths 2–3 mm. This genus is primarily an inhabitant of the vegetation but young specimens are found in the litter layer.

64 or **61** Dens dorsally crenulate; Subfamily **Entomobryinae** (see Fig. 32.45) . **65**

65 or **76** Body with scales **66**

66 or **67** Mucro falcate (Fig. 32.38) . . . . ***Seira*** Lubbock

Pigment generally blue over parts of the body or in a distinct pattern Eyes 8 + 8. Unguis with moderate to small teeth. Unguiculus untoothed. There are six or seven Nearctic species, found mostly in arid regions. The genus has been recorded from many southwestern states plus Louisiana, Florida, and the Nebraska Sand Hills. Maximum lengths range from 2.2 to 3.5 mm. The genus is largely found under rocks and debris.

67 or **66** Mucro bidentate **68**

68 or **73** Dentes with scales on ventral surface **69**

69 or **70** Fourth abdominal segment at midline less than twice as long as third (Fig. 32.39). ***Heteromurus*** Wankel

Color white to reddish with scattered pigment. Eyes 2 + 2 or 1 + 1 or absent. Unguiculus with small to moderate outer tooth. Tenent hairs acuminate. There are two Nearctic species, found everywhere except for the arctic and subarctic regions. Maximum length 2.5 mm. The commonest species, ***H. nitidus,*** appears to be introduced and occurs primarily in caves.

70 or **69** Fourth abdominal segment at midline more than twice as long as third **71**

71 or **72** Eyes 8 + 8 (Fig. 32.40). . . . . ***Lepidocyrtus*** Bourlet

Color varied with pigment generally uniform or in broad bands or (rarely) in simple patterns. Intact animals often metallic. There are few macrochaetae on the body and these are relatively small. There are 15 adequately described Nearctic species found everywhere except the high arctic. The thorax sometimes projects forward as shown for ***L. paradoxus.*** In most species it is much less prominent. Maximum length between 1 and 3.5 mm (mostly between 1 and 1.5 mm). The genus is primarily an inhabitant of the subcortical region but is also found in the litter layer.

72 or **71** Eyes 6 + 6 or fewer (Fig. 32.41) . . . . ***Pseudosinella*** Schäffer

Color generally silvery white, but some species are blue. The unguis varies but is often equipped with large basal teeth. The unguiculus is sometimes equipped with large basal teeth. There are 22 described Nearctic species occurring everywhere except the high arctic. Maximum lengths range from 1.5 to 4.5 mm. Most species are cave forms, but some species are primarily inhabitants of the upper layers of the soil and under rock and debris habitats.

73 or **68** Dens without scales **74**

74 or **75** Scales broad and clearly striate (Fig. 32.42) . ***Willowsia*** Shoebotham

Body uniform purple or blue or complexly patterned. Eyes 8 + 8. There are many large macrochaetae in addition to the scales. There are two species widely distributed over North America. Maximum lengths 1.5–2.0 mm. The genus is largely associated with manufactured structures. The species illustrated is the house collembolan in North America.

75 or **74** Scales narrow without clear markings (Fig. 32.43) . . . . ***Janetschekbrya*** Yosii

Habitus, eyes, and foot as in ***Willowsia.*** There is a single Nearctic species, ***J. arida,*** so far found in Arizona, New Mexico, and Texas. Maximum size 2.1 mm. This genus appears to inhabit vegetation and under rocks and debris in arid regions.

76 or **65** Body without scales. **77**

77 or **78** Eyes 4 + 4 or fewer (Fig. 32.44). . . . . ***Sinella*** Brook

Body color generally white, occasionally with scattering of pigment. Unguis with inner teeth sometimes large. Mucro bidentate or (one species) falcate. There are 16 Nearctic species from all over the continguous United States. Maximum size from 1.5 to 4.5 mm. Most of the species are cave forms; however, soil and greenhouse species also occur.

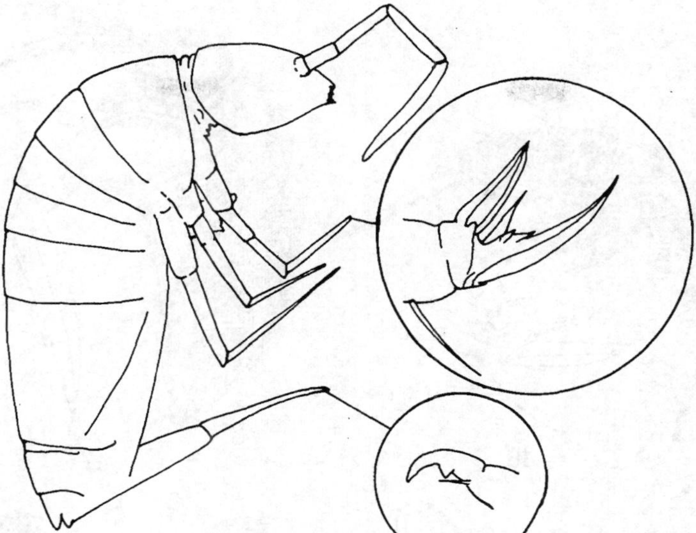

41

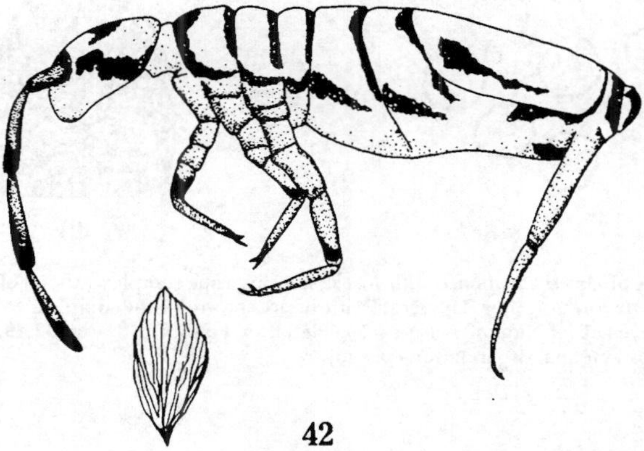

42

43

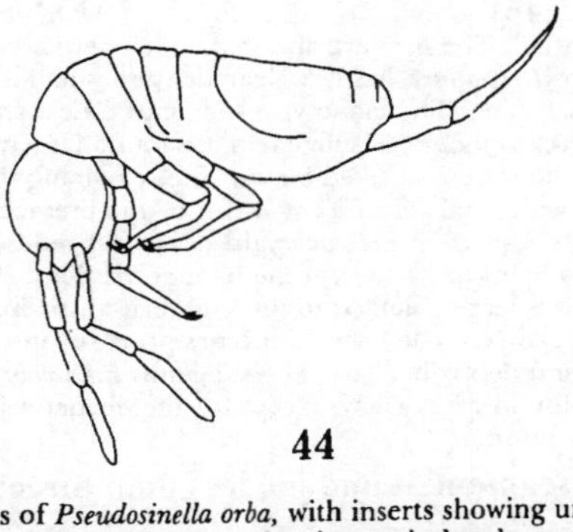

44

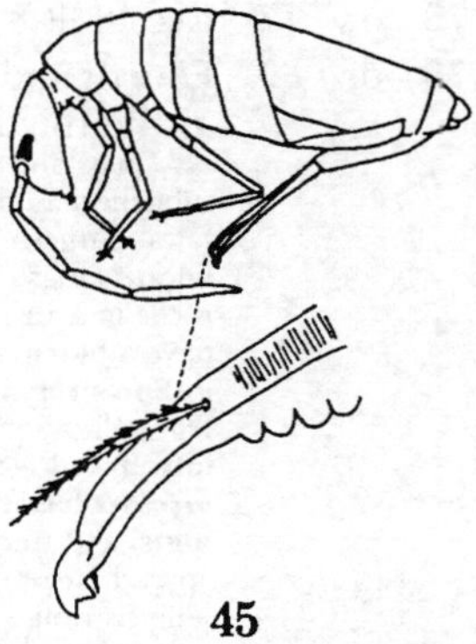

45

Fig. 32.41. Habitus of *Pseudosinella orba,* with inserts showing unguis and mucro. **Fig. 32.42.** Habitus of *Willowsia nigromaculata,* with insert showing typical scale. **Fig. 32.43.** Scales of *Janetschekbrya arida.* **Fig. 32.44.** Habitus of *Sinella basidens.* **Fig. 32.45.** Habitus of *Entomobrya (Entomobryoides) guthriei,* with insert showing mucro and dens.

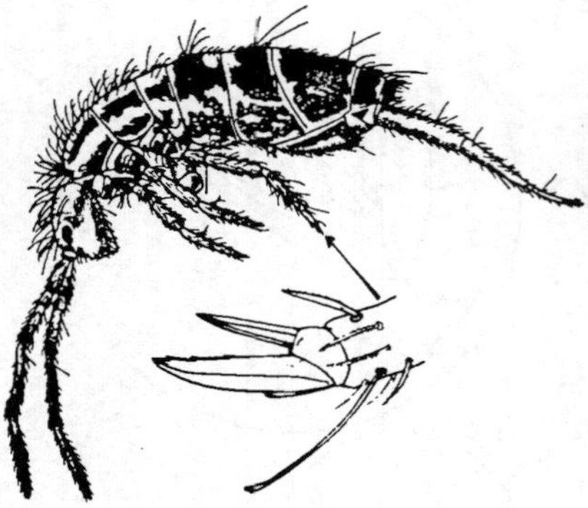

46

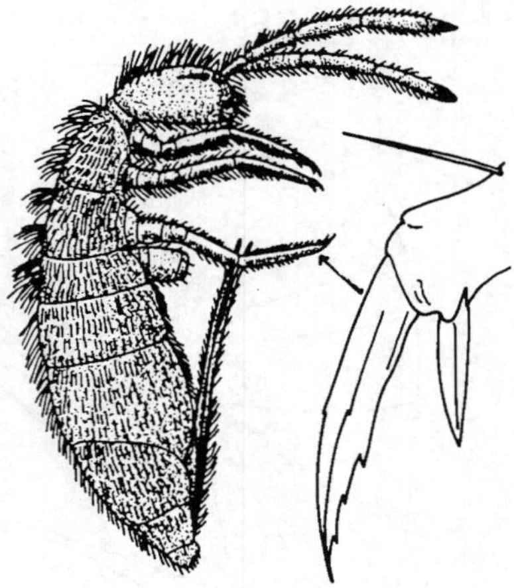

47

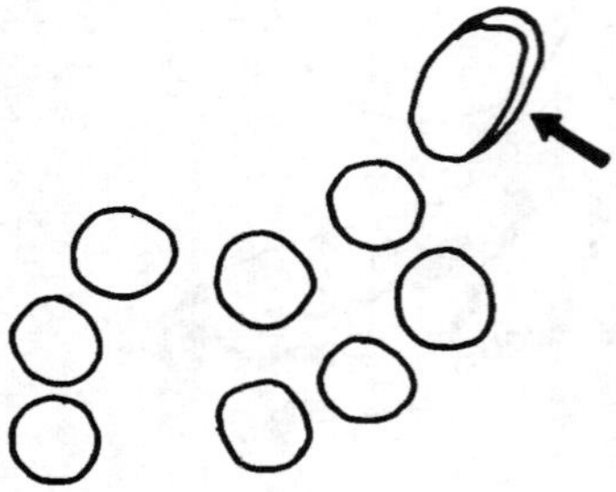

48

49

Fig. 32.46. Habitus of *Orchesella ainsliei*, with insert showing foot complex (after Folsom). **Fig. 32.47.** Habitus of *Corynothrix borealis* (after Tullberg), with insert showing foot complex. **Fig. 32.48.** Postantennal organ and eyes of left side of *Isotoma trispinata* (after Folsom). **Fig. 32.49.** Head of *Proisotoma immersa* showing postantennal organ (after Folsom).

78 or **77** Eyes 8 + 8 . . . . **79**

79 or **80** Fourth abdominal segment at midline more than three times as long as the third (Fig. 32.45) . . . . ***Entomobrya*** Rondani

Color and pattern varied. The eyes are always 8 + 8. There are five soil-living subgenera. The subgenus ***Drepanura*** has two Nearctic species and is characterized by a falcate mucro with a basal spine. The monotypic subgenus ***Calx*** is characterized by a falcate mucro without basal spine. Both subgenera are found largely under debris or rocks in arid regions of the western United States. The remaining three subgenera all have a bidentate mucro with basal spine. They differ by the presence of an apical bulb in ***Entomobrya* s.s.** and its absence in ***Entomobryoides***. The subgenus ***Homidia*** differ from all other subgenera by having spines at the base of the dens. ***Homidia*** has two introduced Nearctic species largely limited to the southern Mississippi valley. ***Entomobryoides*** has five widespread Nearctic forms. It occurs primarily in subcortical, ant nests, and under rocks and debris habitats. The subgenus ***Entomobrya*** has 26 widespread Nearctic species found everywhere except for the deeper soil layers. Maximum lengths range from 2 to 3.5 mm.

80 or **79** Fourth abdominal segment at midline less than three times as long as the third **81**

81 or **82** Unguis with two unpaired and paired inner teeth, first and second antennal segments subsegmented (Fig. 32.46) ***Orchesella*** Templeton

Color and patterns extremely varied. The eyes are always 8 + 8. The mucro is always bidentate with a basal spine. The genus has 14 Nearctic species mostly limited to the United States and southern Canada from the Rocky Mountains east. A few introduced species occur sporadically on the west coast. Maximum lengths range from 2 to 5 mm. These are largely inhabitants of the litter layer. They can often be spotted by their slow movement.

82 or **81** Unguis with three inner teeth, antennae not subsegmented (Fig. 32.47) . . . . ***Corynothrix*** Bourlet

Body color green to greenish yellow, evenly distributed. Eyes 8 + 8. Mucro bidentate with basal spine. There is a single Nearctic species, ***C. borealis,*** found only in the high arctic of Alaska and Canada. Maximum length 3 mm. It inhabits the litter layer and low vegetation.

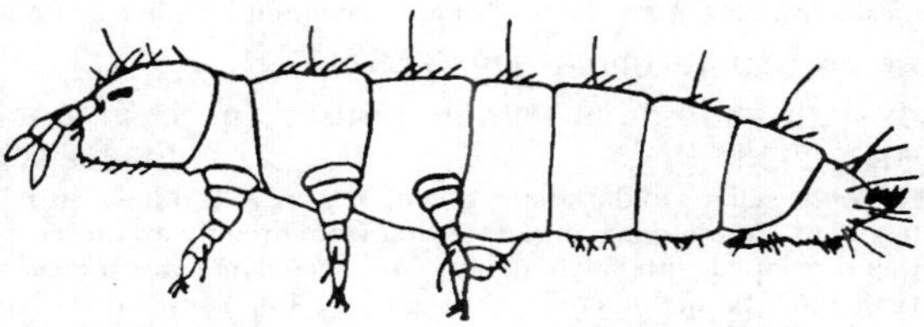

50

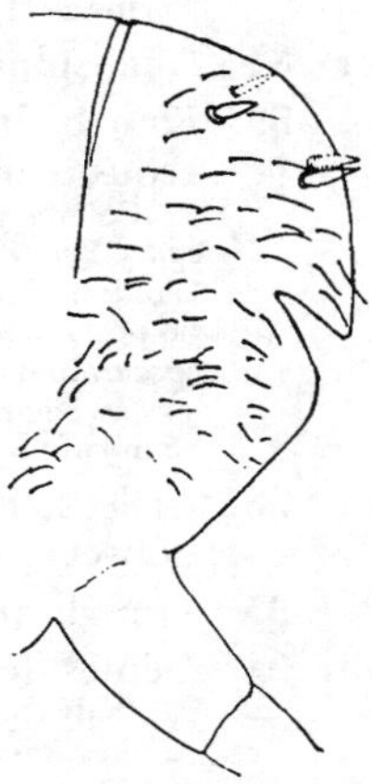

52

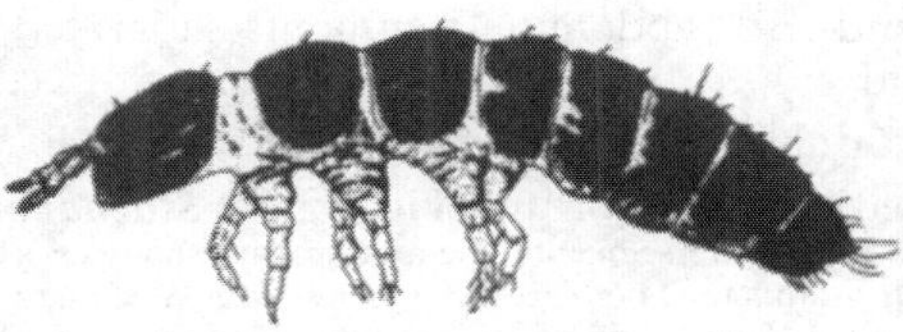

51

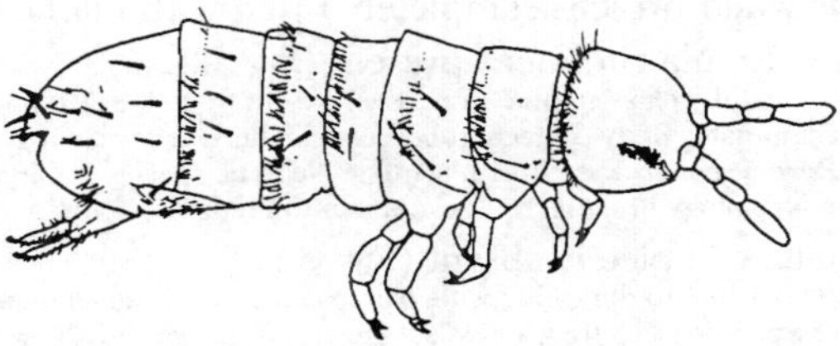

53

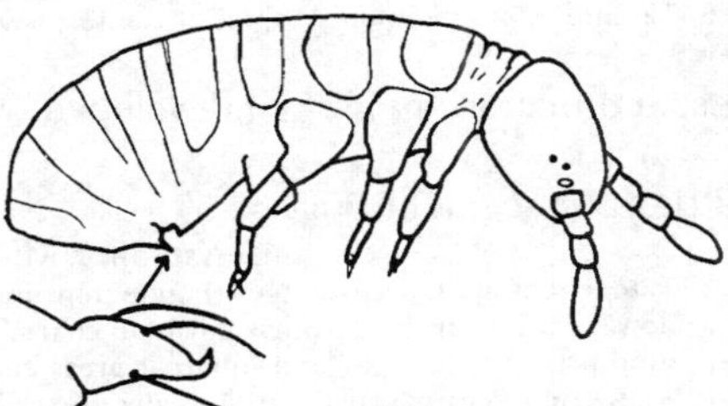

54

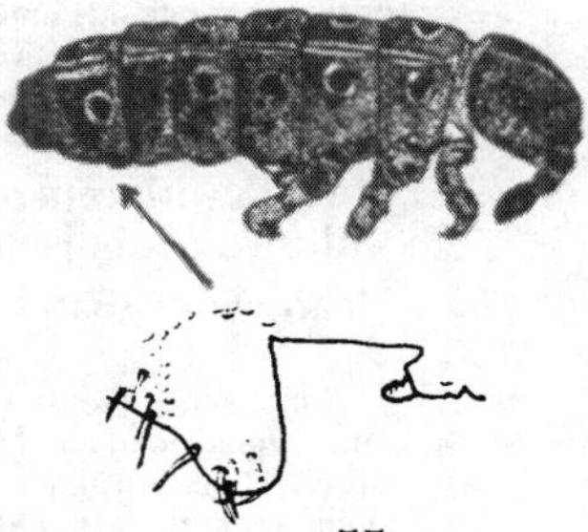

55

Fig. 32.50. Habitus of *Tetracanthella arctica* (after Folsom). **Fig. 32.51.** Habitus of *Uzelia setifera* (after Stach). **Fig. 32.52.** End of abdomen of *Proisotoma (Appendisotoma) aera* showing subterminal anal spines. **Fig. 32.53.** Habitus of *Folsomia alpha* with main setae drawn but minor setae omitted. **Fig. 32.54.** Habitus of *Stachanorema contra,* with insert showing furcula. **Fig. 32.55.** Habitus of *Coloberella octogenaria,* with insert showing furcula.

83 or **58** Postantennal organ present; Family **Isotomidae** (Figs. 32.48 and 32.49) **84**

84 or **87** Anal spines 2–4, terminal at end of abdomen **85**

85 or **86** Four anal spines (Fig. 32.50). . . . . ***Tetracanthella*** Schött

Color mottled medium to dark blue or blue gray. Eyes 8 + 8. Postantennal organ elongate elliptical. Unguis untoothed; unguiculus one-fourth to one-half as long as unguis. Cuticle heavily reticulate and showing clear sclerites. There are 6 Nearctic species ranging from Point Barrow, Alaska, to central California and Kentucky. Maximum lengths 1–1.8 mm. They are primarily inhabitants of moss and other soil vegetation. They are also found in soil litter.

86 or **85** Two anal spines (Fig. 32.51) . . . . ***Uzelia*** Absolon

Color dark blue. Unguiculus absent. Furcula reduced to a triangular flap or absent, otherwise as ***Tetracanthella.*** There are two Nearctic species, known from Alaska, Northwest Territories, Nova Scotia, and Pennsylvania. Maximum length 1.7 mm.

87 or **84** Anal spines absent or subterminal (Fig. 32.52) **88**

88 or **89** Fourth through sixth abdominal segments fused (no clear nonsetaceous bands) (Fig. 32.53) . . . . ***Folsomia*** Willem

Color white, reddish or yellow, to dark blue, green, or gray with numerous pale spots. Eyes 8 + 8 to absent. Postantennal organ varying from broadly oval to elongate elliptical. Furcula well developed with bidentate mucro. ***F. Folsomina onychiurina*** has no postantennal organ and a falcate mucro [see couplet 59 (60)]. There are 27 Nearctic species found all over North America. Maximum lengths 0.5–1.8 mm with most species being between 1.0 and 1.5 mm. These are primarily soil forms being most abundant in the upper layers.

89 or **88** At least fourth and fifth abdominal segments separated by a clear nonsetaceous band **90**

90 or **95** Furcula reduced or absent; mucro, if present, hooklike **91**

91 or **92** Dentes demarcated from manbrium (Fig. 32.54) ***Stachanorema*** Wray

Body color blue with pale intersegmental membranes and spots. Eyes 2 + 2 or 8 + 8. Integument strongly granulate. There are two species from Washington and California. Maximum length about 1 mm. Primarily inhabitants of the subcortical and litter habitats in forest areas.

92 or **91** Dens when present completely fused with manubrium **93**

93 or **94** Clear furcula rudiment present (Fig. 32.55). . . . ***Coloberella*** Latzel

Color dull yellow to blue-gray. Eyes absent or 8 + 8. All body setae smooth, short, and acuminate. Body coarsely tuberculate. There are two subgenera: ***Coluberella* s.s.** and ***Paranurophorus,*** each with a single Nearctic species. Little is known about the range or biology of either species. Maximum length 1.5 mm.

94 or **93** Furcula completely absent (Fig. 32.56) . . . . . ***Anurophorus*** Nicolet

Color white to dark blue with pale patches and intersegmental membranes. Eyes up to eight per side. There are two subgenera. ***Anurophorus* s.s.** with reticulate integument and six or more eyes per side has seven Nearctic species. ***Pseudanurophorus*** with one or two eyes per side and granulate integument has two Nearctic species. The genus is found all over the Nearctic region from Tennessee, Pennsylvania, and central California north. Maximum lengths 0.6–1.5 mm. Most commonly found in coniferous or mixed coniferous–decidous forests.

95 or **90** Furcula with mucro, dens, and manbrium all well developed. Mucro with two or more teeth **96**

96 or **97** Head wider than second thoracic segment (Fig. 32.57) . . . . ***Metisotoma*** Maynard

Color blue. Eyes 8 + 8. No clavate tenent hairs. Dens longer than manbrium. Mucro quadridentate. Macrochaetae vary from smooth to finely ciliate to coarsely ciliate. There is a single species, found primarily in arctic and subarctic areas and alpine regions. Scattered records occur from Pennsylvania and Montana north. Maximum length 2.2 mm. This is a carnivorous form.

97 or **96** Head subequal to, to narrower than, second thoracic segment . . **98**

98 or **99** Dentes with dorsal spines (Fig. 32.58) . . . . . ***Semicerura*** Maynard

Color dark green to purplish black. Eyes 8 + 8. Mucro tridentate. Macrochaetae of abdomen and leg bases unilaterally ciliate. There are two Nearctic species so far re-

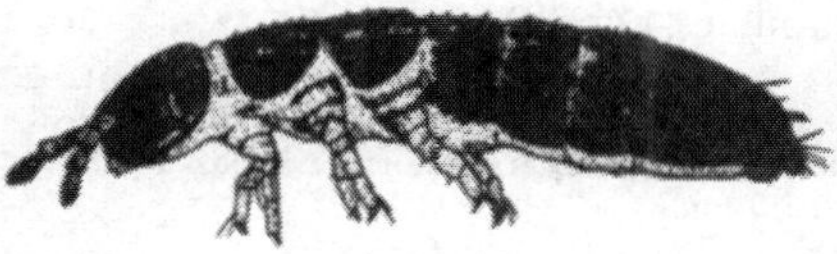

56

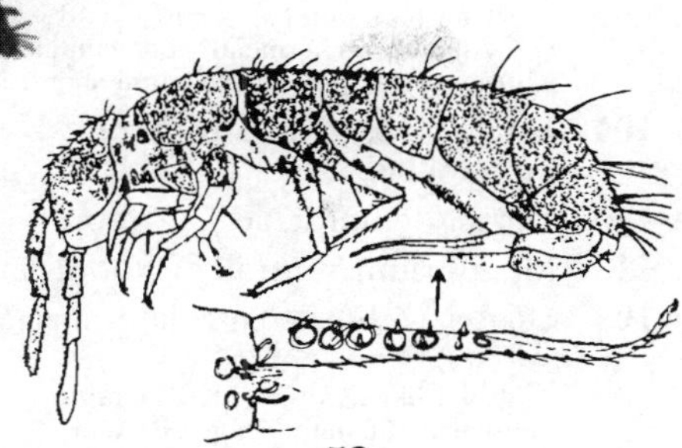

58

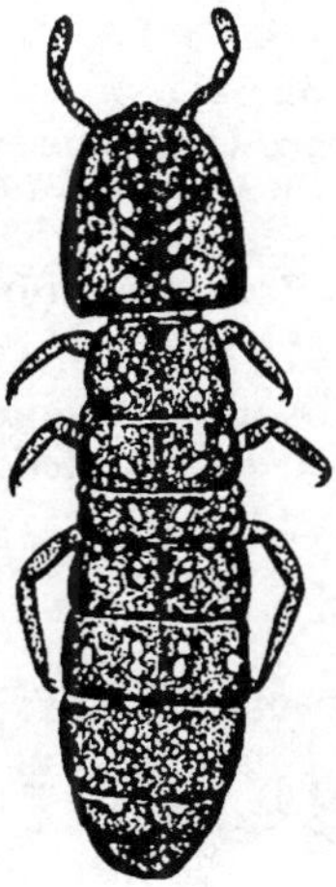

57

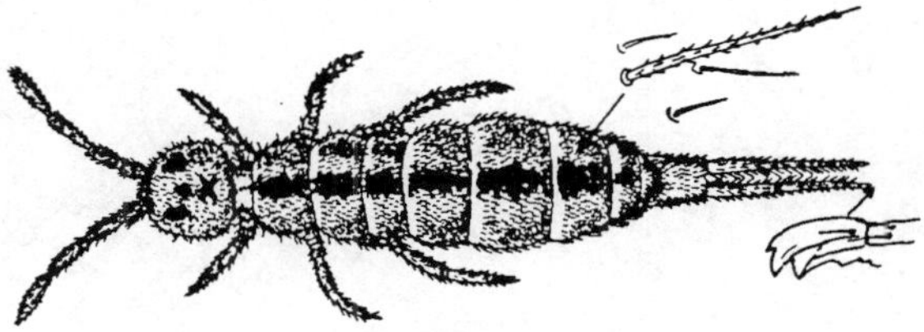

59

Fig. 32.56. Habitus of *Anrophorus septentrionalis* (after Stach). **Fig. 32.57.** Habitus of *Metisotoma grandiceps* (after Cassagnau). **Fig. 32.58.** Habitus of *Semicerura miltispinata* (after James), with insert showing dental spines. **Fig. 32.59.** Habitus of *Isotomurus tricolor* (after Folsom), with inserts showing large multilaterally ciliate and mucro.

corded from West Virginia, New York, and Quebec. The forms are probably much more widespread but rare. Maximum length 1.5 mm.

99 or **98** Dentes without spines **100**

100 or **101** Macrochaetae of abdomen mostly multilaterally finely ciliate (Fig. 32.59) . . . . ***Isotomurus*** Börner

Color yellow, green, blue, or purplish, often longitudinal stripes or transverse bands. Eyes 8 + 8. Tenaculum with 8–24 setae. Only the nominate subgenus is found in the soil. There are six Nearctic species recorded from all over the Nearctic except for the high arctic and Florida, Georgia, Alabama, Mississippi, and South Carolina. Maximum lengths 2.5–3 mm. This is generally a riparian form.

101 or **100** Macrochaetae absent or smooth or unilaterally coarsely ciliate . **102**

102 or **103** Center of manubrium of adult specimens with 12 or more setae (Fig. 32.60) . . . . ***Isotoma*** Bourlet

Color variable but rarely patterned. Eyes 0 to (in most cases) 8 + 8. Mucro tridentate or quadidentate (very rarely quinquedentate). The genus is large and varied. There are seven subgenera but ony four are common in the soil. The subgenus ***Isotoma*** (three species) is characterized by spinelike setae at the apex of the manubrium. The subgenera ***Vertagopus*** and ***Pseudisotoma*** (eight species) are characterized by clavate tenent hairs. Most species (42) belong to the subgenus ***Desoria.*** The genus is found all over the Nearctic region in all types of soil habitats. Maximum lengths range from 1.0 to 4.5 m.

103 or **102** Center of manubrium of adult specimen with 10 or fewer ventral setae **104**

104 or **105** Eight anal spines present on papillae (Fig. 32.61) . . . . ***Weberacantha*** Christiansen

Color blue with pale spots. Eyes 8 + 8. Mucro tridenate. Integument finely reticulate. All body setae smooth and acuminate. Known only from northern Alaska and the Northwest Territories. Maximum length 1.5 mm.

105 or **104** Anal spines generally absent. If present, fewer than eight and never on papillae. **106**

106 or **110** Dorsal dental setae 7 or more **107**

107 or **113** Manubrium ventrally without setae **108**

108 or **109** Dentes without ventrolateral setae (Fig. 32.62). . . . . ***Bonetrura*** Christiansen & Bellinger

Color blue. Eyes 8 + 8. Pronotum granulate. Known only from West Virginia and Louisiana. Maximum length 1.2 mm.

109 or **108** Dentes with ventrolateral setae.. . . . . ***Proisotoma*** Börner (in part)

110 or **106** Dorsal dental setae 3 or fewer. **111**

111 or **112** Abdomen with spinelike setae on sixth abdominal segment, dens with two setae on ventral, half (Fig. 32.63) . . . . ***Isotomodes*** Linnaniemi

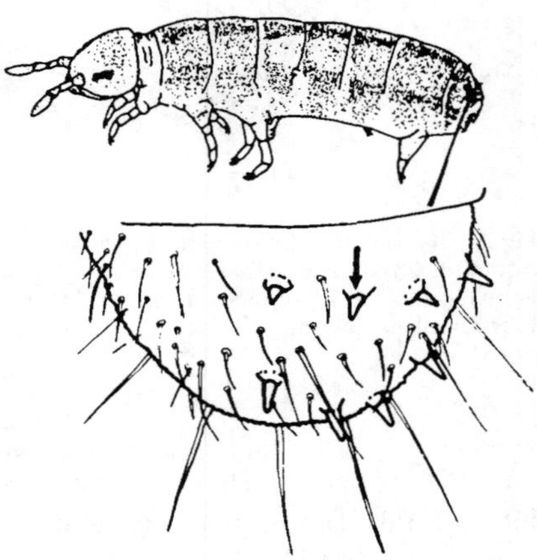

61

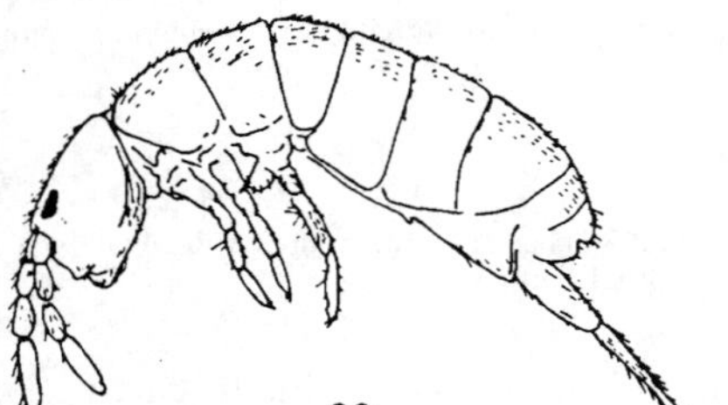

60

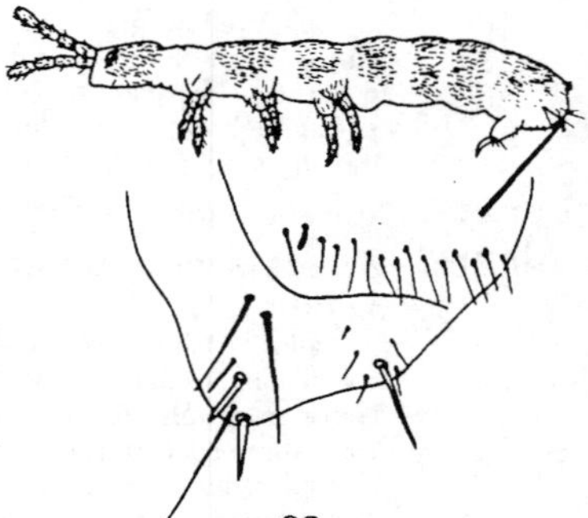

63

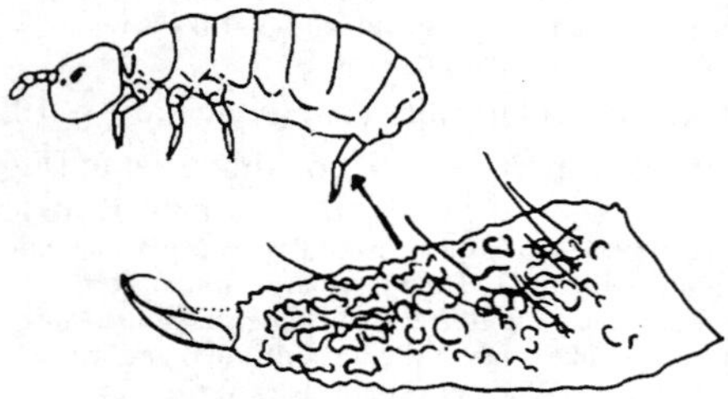

62

Fig. 32.60. Habitus of *Isotoma* (*Desoria*) sp. **Fig. 32.61.** Habitus of *Weberacantha octa*, with insert showing anal spines. **Fig. 32.62.** Habitus of *Bonetrura boneti* (after Yosii), with insert showing dens and mucro. **Fig. 32.63.** Habitus of *Istomodes falsus*, with insert showing apex of abdomen (after Gama).

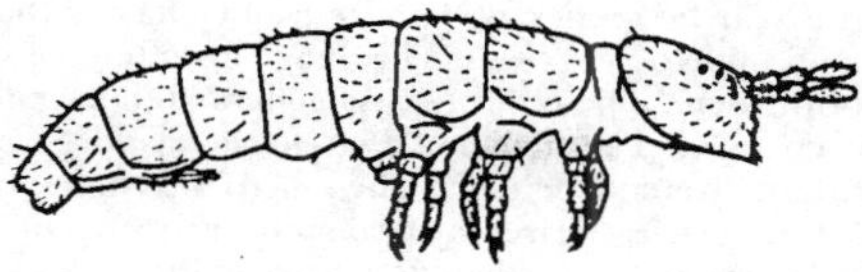

64

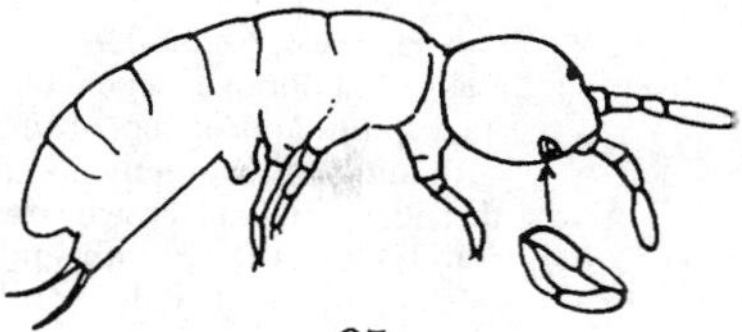

65

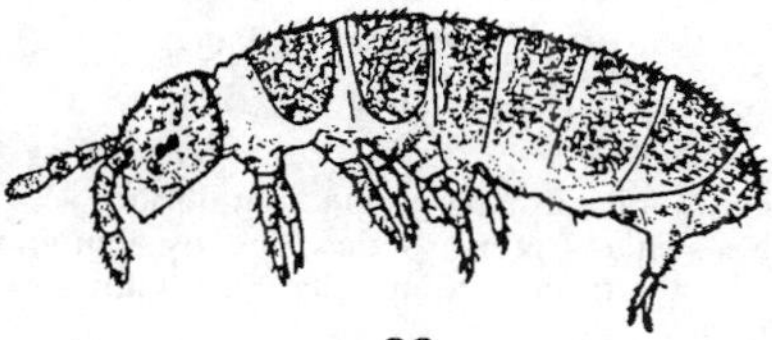

66

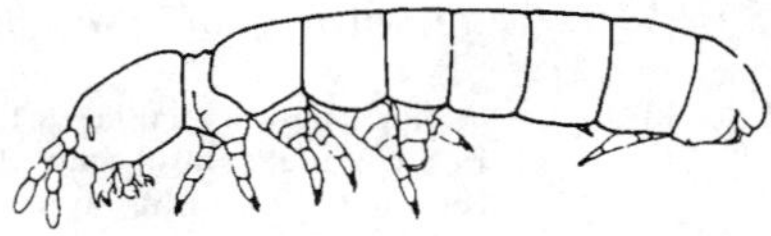

67

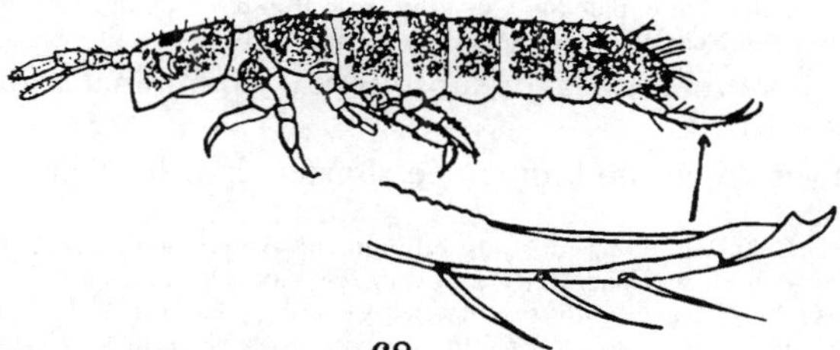

68

Fig. 32.64. Habitus of *Folsomides americanus* (after Stach). **Fig. 32.65.** Habitus of *Microisotoma achromata*, with insert showing postantennal organ. **Fig. 32.66.** Habitus of a typical *Proisotoma* (after Stach). **Fig. 32.67.** Habitus of *Dagmaea tenuis* (after Folsom). **Fig. 32.68.** Habitus of *Cryptopygus similis* with insert showing dens and mucro (after James).

Pigment and eyes absent. Mucro bidentate. There are four Nearctic species so far recorded from California, Iowa, Michigan, Washington, and New Jersey. Additional study will probably show them to be widespread. Maximum lengths vary from 0.8 to 1.2 mm. They are largely inhabitants of the middle soil layers.

112 or **111** Abdomen without posterior spinelike setae, dens with one or no setae on ventral half (Fig. 32.64) . . . . ***Folsomides*** Stach

Color white with scattered pigment. Eyes 8 + 8 to 0. Mucro bidentate. The body setae are all smooth and acuminate. There are five Nearctic species known from all over North America except Alaska and extreme arctic Canada. Maximum size from 0.6 to 1 mm.

113 or **107** Manubium ventrally with setae **114**

114 or **115** Postantennal organ with margin divided into quadrants (Fig. 32.65) ***Micrisotoma*** Bellinger

Pigment and eyes absent. Mucro bidentate. Fifth and sixth abdominal segments fused. Some of the large abdominal setae are unilaterally serrate or ciliate. There is a single species so far reported from Connecticut, Massachusetts, North Carolina, and Quebec.

115 or **114** Postantennal organ with margin not divided into quadrants. **116**

116 or **117** Mucro tridentate or mucro bidentate with fifth and sixth abdominal segments separated by a clear, nonsetaceous band (Fig. 32.66) . . . ***Proisotoma*** (in part)

Color white to gray or green to blue. Eyes generally 8 + 8, rarely reduced or absent. There are 0–10 ventral setae on the center of the manubrium. There are three subgenera. ***Appendisotoma,*** with eight Nearctic species, is characterized by having the last two abdominal segments fused and three dental teeth. This subgenus often has anal spines and/or sacs at the end of the dens. The other two subgenera both have the fifth and sixth abdominal segments separated. The subgenus ***Proisotoma*** has 11 Nearctic species and is characterized by a dorsally crenulate dens. The mucro has three or (rarely) 2 teeth. The subgenus ***Ballistura*** is characterized by a smooth, furrowed, or dorsally tuberculate dens and generally boat-shaped bidentate mucro and has 14 Nearctic species. The genus is found all over the Nearctic region. Maximum lengths range from 0.8 to 1.8 mm.

117 or **116** Mucro bidentate *and* last two abdominal segments fused **118**

118 or **119** Dens with six or fewer setae on ventral half (Fig. 32.67). . . . . . . ***Dagamea*** Yosii

Color white, sometimes with scattered pigment. Eyes absent. Manubrium with 3 + 3 (rarely 2 + 2) ventral setae. There is a single Nearctic species. The distribution is uncertain due to confusion in the extant literature. Maximum length 1.0 mm.

119 or **118** Dens with eight or more ventral setae (Fig. 32.68). . . . . ***Cryptopygus*** Willem

Color white, pale blue, or gray. Eyes 0 to 8 + 8. There are nine species found all over the Nearctic region. Maximum length 0.6 to 2.5 m.

120 or **1** Body broadly oval or subspherical, abdominal segments 1–4 more or less fused **121**

121 or **122** Eyes absent and antennae shorter than head. Family **Neelidae** (Fig. 32.69). . . . . ***Neelus*** Folsom

Color white or with scattered pigment. Antennae three- or four-segmented with large blunt sensillae. Body with very few setae all small smooth and acuminate. Body usually with a series of sensory fields bearing small variously formed sensillae. There are three subgenera: ***Megalothorax,*** with three Nearctic species characterized by three antennal segments; ***Neelus,*** with a single Nearctic species characterized by having the fourth antennal segment subequal to the third; and ***Neelides,*** with three Nearctic species characterized by having the third antennal segment wider than the fourth. The genus has been found all over the Nearctic region. Maximum lengths range from 0.3 to 0.7 mm. These are typically inhabitants of the deeper soil layers.

122 or **121** Eyes present and/or antennae longer than head. Family **Sminthuridae** **123**

123 or **124** Fourth antennal segment less than one-half as long as third. Subfamily **Dicyrtominae** (Fig. 32.70) . . . . ***Dicyrtoma*** Bourlet

Color varied, often with complex patterns. Eyes 8 + 8. There are four Nearctic subgenera: ***Bothriovulsus, Dicyrtoma, Dicyrtomina,*** and ***Ptenothrix,*** but only the last subgenus is regularly found in Nearctic soil habitats. There are 19 Nearctic species of the genus found all over North America from British Columbia south. Primarily an inhabitant of low vegetation but frequently under stones, debris, or bark.

124 or **123** Fourth antennal segment as long as the third **125**

125 or **130** Two or more clavate tenent hairs per foot **126**

126 or **127** Greater abdomen with heavy outstanding serrate setae . . . . ***Vesicephalus*** Richards

Background color yellow with clear pigment pattern. Eyes 8 + 8. Head with large interocular vesicles. Mucro with one edge serrate. There are two Nearctic species, one from Washington, California, and Oregon, the second recorded from Minnesota, Iowa, Missouri, and North Carolina. Maximum lengths about 2 mm.

127 or **126** Greater abdominal setae never large and serrate **128**

128 or **129** Tenent hairs thick, two to three, parallel to large axis of tibiotarsus, very strongly clavate (Fig. 32.72) . . . . ***Bourletiella*** Banks

Color varied, often with a striking pattern. Eyes 8 + 8. Greater abdomen and head without spinelike or serrate setae. Female subanal appendage well developed. There are five subgenera, two of which are neustonic. The remaining subgenera are largely inhabitants of vegetation; however, they are occasionally found in the litter layer. The subgenus ***Bourletiella*** has nine Nearctic species characterized by the extreme sexual dimorphism. The subgenus ***Deuterosminthurus*** with seven Nearctic species is characterized by the dorsally excavate abdomen of the females. The subgenus ***Prorastriopes*** has

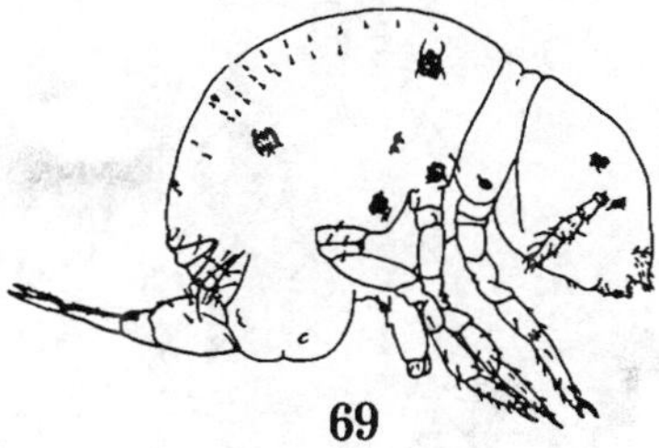

69

70

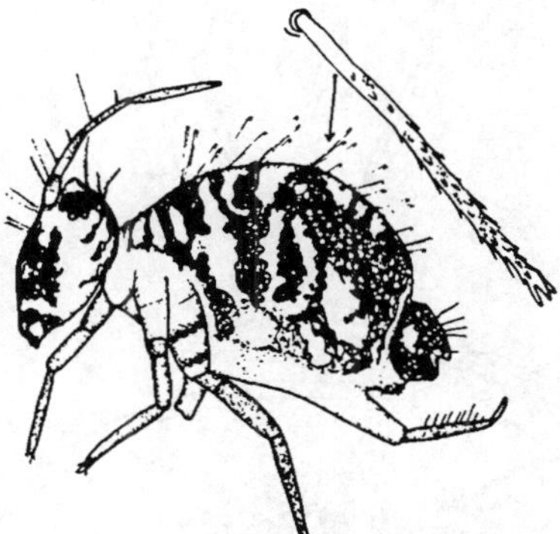

71

Fig. 32.69. Habitus of *Neelus (Megalothorax) minimus* (after Bonet). **Fig. 32.70.** Habitus of *Dicyrtoma (Ptenothrix) marmorata,* showing spinelike setae. **Fig. 32.71.** Habitus of *Vesicephalus occidentalis,* with insert showing body setae.

two Nearctic species, characterized by truncate tibiotarsal setae. Known from all over the Nearctic except for arctic and subarctic regions. Maximum lengths of males range from 0.6 to 0.9 mm, of females from 0.8 to 1.2 mm.

129 or **128** Tenent hairs slender, not parallel to long axis of tibiotarsus, weakly to moderately clavate, usually four or more per foot (Fig. 32.73). . . . ***Sminthurinus*** Börner

Color varied, often patterned. Eyes 8 + 8. Female subanal appendage well developed. There are three subgenera, but most forms found in the soil belong to ***Polykatianna,*** characterized by a forward-pointing basal mucronal tooth, and ***Sminthurinus,*** characterized by a forked female supranal seta. There are 13 Nearctic species found all over the Nearctic except for arctic and subarctic regions. Maximum in lengths 1–1.7 mm.

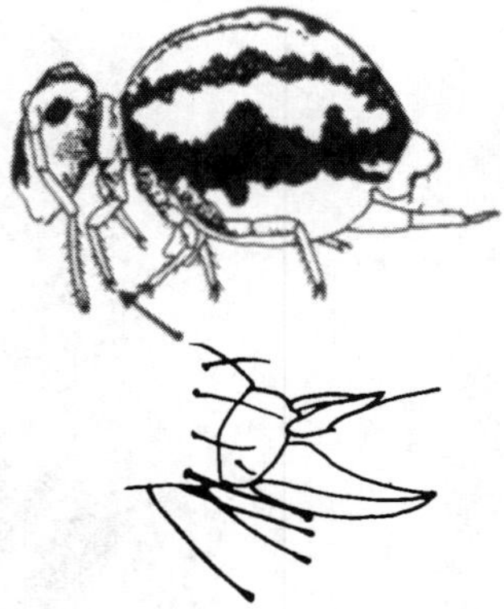

73

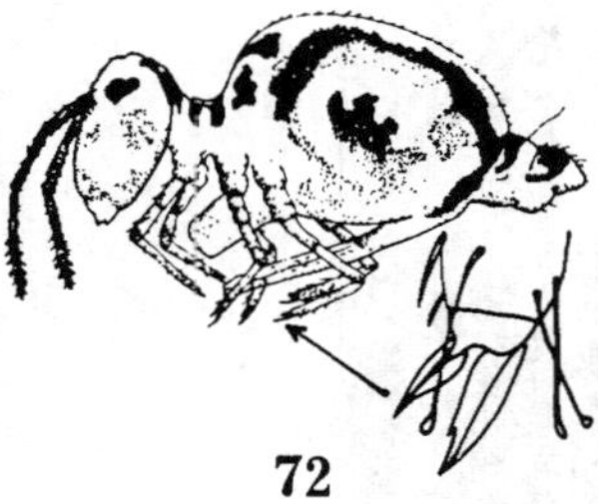

72

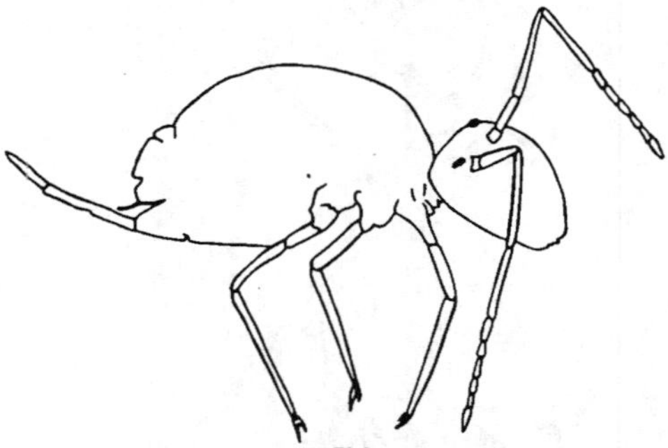

74

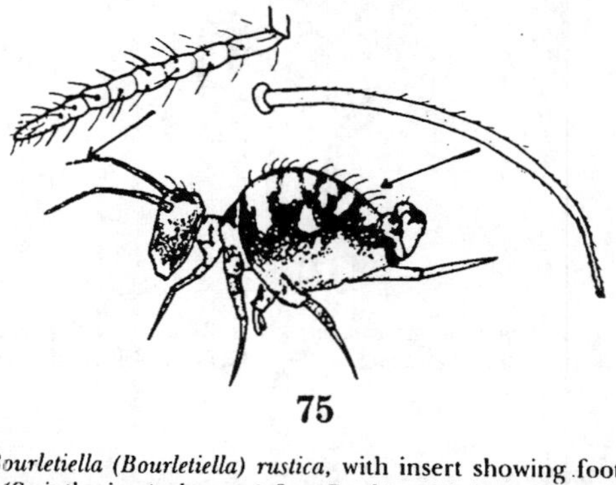

75

Fig. 32.72. Habitus of *Bourletiella (Bourletiella) rustica*, with insert showing foot complex. **Fig. 32.73.** Habitus of *Sminthurinus (Sminthurinus) elegans* (after Stach), with insert showing foot complex. **Fig. 32.74.** Habitus of *Arrhopalites clarus.* **Fig. 32.75.** Habitus of *Sphyrotheca*, with inserts showing antennae and abdominal setae.

130 or **125** Clavate tenent hairs absent (rarely one and weakly clavate) . . . **131**

131 or **132** Eyes 2 + 2 or fewer (Fig. 32.74) . . . . ***Arrhopalites*** Börner

Color generally white with scattering of reddish pigment. Fourth antennal segment usually subsegmented. Dentes with spines. Female with well-developed subanal appendage. There are 12 Nearctic species known from all regions except Florida and the high arctic. Maximum lengths from 0.7 to 1.8 mm. The genus is primarily a cave group but is also common in the litter layer and superficial soil of mixed or coniferous forests.

132 or **131** Eyes 4 + 4 or more **133**

133 or **136** Greater abdomen with many heavy usually clavate or truncate setae (often serrate or ciliate) **134**

134 or **135** Fourth antennal segment subsegmented (Fig. 32.75). . . . . ***Sphyrotheca*** Börner

Background color yellow to pink with patterned blue pigment. Eyes 8 + 8. Most species have conspicuous cephalic spinelike setae. There are three Nearctic species from scattered localities all over the United States and southern Ontario. Maximum length about 1 mm. Primarily found under bark, debris, and stones, and occasionally in litter.

135 or **134** Fourth antennal segment unsubsegmented (Fig. 32.76) . . . . ***Neosminthurus*** Mills

Background color yellow or white with dark blue or black pigment. Eyes 8 + 8. Large body setae truncate or clavate and serrate or heavily ciliate or grooved. Female subanal appendage well developed. There are three Nearctic species known from all over the United States except for the Pacific coast. Not recorded from Canada. Maximum length 1.2 mm.

136 or **133** Greater abdomen without heavy spinelike setae **137**

137 or **138** Antennal subsegments 10 or more (Fig. 32.77) . . . . ***Sminthurus*** Latreille

Pigment varying, generally with striking patterns in several colors. Eyes 8 + 8. There are two subgenera, ***Allacma,*** with one species characterized by palmate female subanal appendage, and ***Sminthurus,*** with 18 Nearctic species and acuminate subanal appendages. The genus is found all over the Nearctic region. Maximum lengths range from 2 to 3 mm. The genus is primarily an inhabitant of vegetation but is occasionally found under stones, debris, or loose bark.

138 or **137** Antennal subsegments 7 or fewer **139**

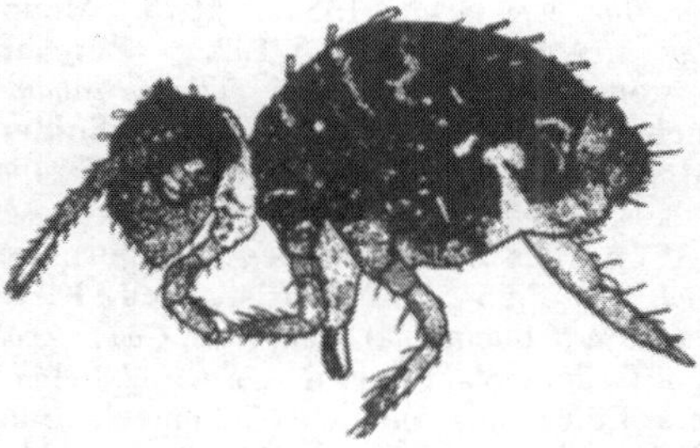

76

77

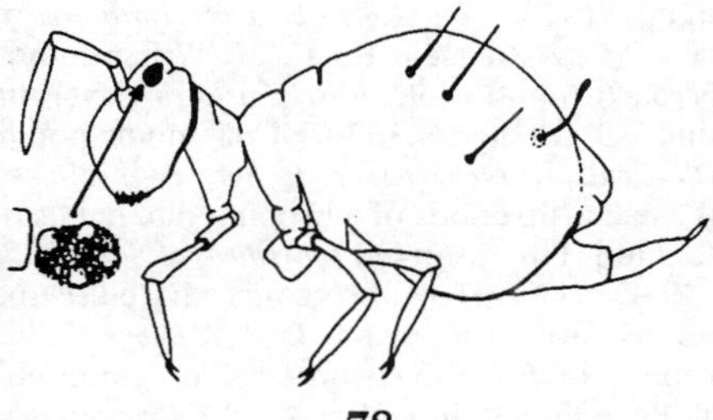

78

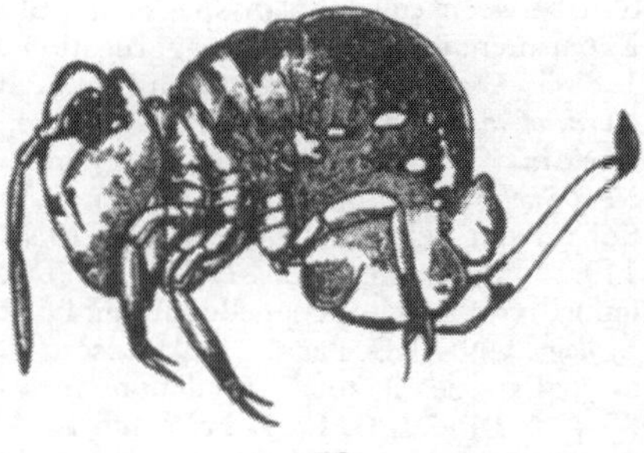

79

Fig. 32.76. Habitus of *Neosminthurus clavatus* (after Mills). **Fig. 32.77.** Habitus of *Sminthurus medialis* (after Betsch). **Fig. 32.78.** Habitus of *Collophora quadrioculata* (after Richards), with insert showing eyepatch. **Fig. 32.79.** Habitus of *Sminthurides aquaticus* (after Folsom).

139 or **140** Eyes 4 + 4 (Fig. 32.78) . . . . ***Collophora*** Richards

Color pale gray with scattered fine pigment granules. Fourth antennal segment with three to five weakly demarcated subsegments. There is a single Nearctic species so far known from Louisiana and Arizona. Maximum length 0.5 mm.

140 or **139** Eyes 6 + 6 or 8 + 8 (Fig. 32.79) . . . . ***Sminthurides*** Börner

Color varied but pigments are generally pink, blue-purple, or black. Complex patterns are sometimes present. Eyes usually 8 + 8 with 2 + 2 much smaller than others and possibly absent. Sexual dimorphism extreme, with females lacking the subanal appendage and the much smaller males having complex grasping antennae. Body setae are small and acuminate. There are four subgenera with 18 species, largely neustonic. Only the two species of the subgenus ***Sphaeridia*** are commonly found in soil habitats. Maximum lengths range from 0.6 to 1.5 mm. In soils, found primarily in the litter layer.

Bibliography

Adam, R. H. 1941. Stratification, diurnal and seasonal migration of the animals in a deciduous forest. *Ecol. Monogr.* 11:189–227. **Aitchison C. W. 1974.** A sampling technique for active subnivian invertebrates in southern Manitoba. *Manit. Entomol.* 8:32–36. **Arnett, P. M. 1969.** A study of collembolan populations associated with four seral stages leading to the beech–maple climax. *Proc. Indiana Acad. Sci.* 78:231–240. **Auerbach, S. I. 1958.** The soil ecosystem and radioactive waste disposal to the ground. *Ecology* 39:522–529. **Bellinger, P. F. 1954.** Studies of soil fauna with special reference to the Collembola. *Bull. Conn. Agric. Exp. Stn.* 583:1–67. **Bird, R. D. 1930.** Biotic communities of the aspen parkland of central Canada. *Ecology* 11:356–442. **Blake, I. H. 1927.** A comparison of the animal communities of coniferous and deciduous forests. *Ill. Biol. Monogr.* 10:371–520. **1931.** Further studies on deciduous forest animal communities. *Ecology* 12:508–527. **Bonet, F. 1943.** Sobre la clasification de los Oncopoduridae (Collembola) con descripción de especies nuevas. *An. Esc. Nac. Cienc. Biol. Mex.* 3:127–153. **1948.** Monografía de la familia Neelidae (Collembola). *Rev. Soc. Mex. Hist. Nat.* 8:131–192. **Butcher, J. W., E. Kirknel and M. Zabik. 1969.** Conversion of DDT to DDE by *Folsomia candida* (Willem). *Rev. Ecol. Biol. Sol* 6:291–298. **Butcher, J. W., R. Snider and R. J. Snider. 1971.** Bioecology of edaphic Collembola and Acarina. *Annu. Rev. Entomol.* 16:249–288. **Cassagnau, P. 1961.** Ecologie du sol dans les Pyrénées centrales. Les biocenoses des collemboles. *Actual. Sci. Ind.* 1283:1–225. **1971.** Les différents types d'ecomorphose chez les collemboles Isotomidae. *Rev. Ecol. Biol. Sol.* 8:55–57. **Christiansen, K. A. 1958.** The Nearctic members of the genus *Entomobrya* (Collembola). *Bull. Mus. Comp. Zool. Harv. Univ.* 118:437–545. **1964.** Bionomics of Collembola. *Annu. Rev. Entomol.* 9:147–178. **1965.** A revision of the Nearctic members of the genus *Tomocerus* (Collembola Entomobryidae). *Rev. Ecol. Biol. Sol* 1:639–677. **1966.** The genus *Arrhopalites* (Collembola: Sminthuridae) in the United States and Canada. *Int. J. Speleol.* 2:43–73. **1967.** Competition between collembolan species in culture jars. *Rev. Ecol. Biol. Sol* 4:439–462. **1970.** Experimental studies on the aggregation and dispersion of Collembola. *Pedobiologia* 10:180–198. **Christiansen, K. A. and P. F. Bellinger. 1979.** *The Collembola of North America North of the Rio Grande.* Grinnell College, Iowa. **Christiansen, K. A., M. Willson and J. Tecklin. 1961.** Preliminary study of the microarthropod ecology of Hunters Cave. *Bull. Natl. Speleol. Soc.* 23:63–70. **Clemen, R. B. and L. P. Pedigo. 1970.** Collembola populations from selected arable fields. *Proc. North Cent. Branch. Entomol. Soc. Am.* 25:115–119. **Coleman, D. C. 1970.** Food webs of small arthropods of a broomsedge field studied with radio-isotope-labelled fungi. Pp. 203–207 in J. Phillipson (ed.). *Methods of Study in Soil Ecology.* UNESCO, Paris. **Davidson, V. S. 1932.** The effect of seasonal variability upon animal species in total populations in a deciduous forest succession. *Ecol. Monogr.* 2:305–333. **Dindal, D. L., F. Folts and R. A. Norton. 1975.** Effects on DDT on community structure of soil microarthropods in an old field. Pp. 505–512 in J. Vanek (ed.). *Progress in soil zoology.* Academia, Prague. **Dindal, D. L., D. Schwert and R. A. Norton. 1975.** Effects of sewage effluent disposal on community structure of soil invertebrates. Pp. 419–426 in J. Vanek (ed.). *Progress in Soil Zoology* Academia, Prague, Czechoslovakia. **Folsom,T. W.**

1937. Nearctic Collembola, or springtails, of the family Isotomidae. *Bull. U.S. Natl. Mus.* 168:1–144. **Gisin, H. 1960.** *Colembolenfauna Europas, 1–312.* Museum d'Histoire Naturelle, Geneva, Switzerland. **Gama, M. M. da. 1963.** Monografia do gènero Isotomedes (Insects, Collembola), *Mem. Estud. Mus. Zool. Univ. Coimbra* 284:1–44. **Hammer, M. 1953.** Investigations on the microfauna of northern Canada. II. Collembola. *Acta Arct.* 6:1–180. **Jensen, P. and K. W. Corbin. 1966.** Some factors affecting aggregation of *Isotoma viridis,* Bourlet and *Arion fasciatus,* Nilson. *Ecology* 47:332–334. **Jensen, P., G. L. Jacobson and D. E. Willard. 1973.** Effects of mowing and raking on Collembola. *Ecology* 54:564–572. **Knight, C. B. 1961.** The Tomocerine (Collembola) in old field strands of North Carolina. *Ecology* 42:140–149. **1977.** Seasonal and microstratal dietary research on *Tomocerus* (Collembola) in two forested ecosystems of eastern North Carolina. *Rev. Ecol. Biol. Sol* 13:6950610. **Knight, C. B. and R. A. Angel. 1967.** A preliminary study of the dietary requirements of *Tomocerus* (Collembola). *Am. Midl. Nat.* 77:510–517. **Knight, C. B. and J. P. Chesson, Jr. 1966.** The effect of DDT on the forest floor Collembola of a loblolly pine stand. *Rev. Ecol. Biol. Sol* 3:129–139. **Knight, C. B. and V. Read. 1969.** Microstratification of Tomocerus (Collembola) in a pine–open field continuum. *Rev. Ecol. Biol. Sol* 6:221–234. **Lussenhop, J. 1973.** The soil arthropod community of a Chicago expressway margin. *Ecology* 54:1124–1137. **1976.** Soil arthropod response to prairie burning. *Ecology* 57:88–98. **Massoud, Z. 1967.** Monographie des Neanuridae, collemboles poduromophes à pièces buccales modifies. *Biol. Am. Aust.* 3:1–339. **Mills, H. B. 1934.** *A Monograph of the Collembola of Iowa.* Collegiate Press, Ames, Iowa. 143 pp. **Ostidiek, J. L. 1961.** Luctuation in populations of Collembola within leaf litter in the Patuxent Research Refuge, Maryland. *Biol. Stud. Cathol. Univ. Am.* 62:1–44. **Paclt, J. 1956.** Biologie der primär flügellosen Inseckten. G. Fischer, Jena, DDR. 258 pp. **Richards, W. R. 1968.** Generic classification, evolution, and biogeography of the Sminthuridae of the world (Collembola). *Mem. Entomol. Soc. Can.* 53:1–54. **Stach, T. 1947.** *The Apterygotan Fauna of Poland in Relation to the World-Fauna of This group of Insects. Family: Isotomidae.* Polska Akademija Umiejetnosci, Krakow, Poland. 488 pp. **1949a.** Ibid., *Families: Anuridae and Pseudachorutidae.* Polska Akademija Umiejetnosci, Krakow, Poland. 122 pp. **1949b.** Ibid., *Families: Neogastruridae and Brachystomellidae.* Polska Akademija Umiejetnosci, Krakow, Poland. 341 pp. **1954.** Ibid., *Family: Onychiuridae.* Polska Akademia Nauk, Krakow, Poland. 219 pp. **1956.** Ibid., *Family: Sminthuridae.* Panstwowe Wydawnictwo Naukowe, Krakow, Poland. 289 pp. **1957.** Ibid., *Families: Needlidae and Dicyrtomidae* Panstwowe Wydawnictwo Naukowe, Krakow, Poland. 113 pp. **1960.** Ibid., *Tribe: Orchesellini.* Panstwowe Wydawnictwo Naukowe, Krakow, Poland. 151 pp. **1963.** Ibid., *The apterygotan fauna of Poland in relation to the world-fauna of this group of insects. Tribe Entomobryini,* Panstwowe Wydawnictwo Naukowe, Krakow, Poland. 126 pp. **Waldorf, E. S. 1971.** The reproduction of biology of *Sinella curviseta* (Collembola: Entomobryidae) in laboratory culture. *Rev. Ecol. Biol. Sol* 8:451–463.

33

Insecta: Isoptera

W. L. NUTTING

Termites have aptly been called "social cockroaches" because of their ancestry; they might as well be considered microruminants for their role in nature. They are both primary consumers (herbivores) and decomposers and, as such, play an important role in the biological turnover of cellulosic materials and in the physical and chemical alteration of soils.

Wood (1978) reviewed the feeding habits of termites and has classified the variety of organic matter on which they feed as follows: living vegetation of all sorts, both aerial and subterranean; dead vegetation, either standing, fallen, or buried and in various stages of decomposition; dung; humus; fungi, either in rotting vegetation or cultivated within the nest (Old World); and special or incidental items such as algae, lichens, organic-rich nest materials (carton), and dried tissues of vertebrate corpses. Their predilection for cellulose qualifies them as pests of human society. In the United States and parts of Canada, termites are notorious for the destruction they wreak on homes, buildings, poles, and posts, but in warmer climates the list extends to timber, pastures, and crops. It is well known that they depend largely on symbiotic protozoans and bacteria for the digestion of this diet. LaFage and Nutting (1978) have written an extensive review of the complex nutrient dynamics of termites.

The Isoptera form a small (morphologically) homogeneous order of some 2200 species and all are social. They are primarily a tropical group, but 39 species occur in the lower 48 states and the southernmost parts of two Canadian provinces (Ontario and British Columbia). Of these, 22 live in, or at least occasionally enter, the soil. They belong to three families: one so-called "damp-wood" termite in the Kalotermitidae, another damp-wood and seven subterranean species in the Rhinotermitidae, and 13 subterranean species in the Termitidae. Large populations occur in parts of the southeast and locally elsewhere. They are active the year around where there is little or no freezing but during progressively shorter seasons with increasing elevation and latitude. It has been suggested that their distribution is ultimately limited by the length of the summer foraging period. It eventually becomes too short for colonies to recover population losses incurred during enforced hibernation and starvation several feet underground and out of reach of adequate food. They may also remain active through the year where nests are located in or under heated buildings.

Termites live in large long-lived colonies, fully comparable in their organization and complexity with ant societies. Most members of a colony—larvae, nymphs, and workers—are soft, delicate, unpigmented insects, generally blind and easily desiccated. Subterranean colonies are believed to have one or more nests or groups of chambers at varying depths in the ground, or sometimes in logs, which they occupy according to season. This closed environment is opened only to release winged reproductives during the annual flight or swarming season. Their foraging gallery system is dynamic, with shallow tunnels continually being abandoned and rerouted to locate items of wood and other litter which intermittently falls to the ground above them.

Termites share this resource with a variety of other soil invertebrates, fungi, and microorganisms. None of our species builds mounds; therefore, we rarely see them or any evidence of their activity even though they may be abundant. Nevertheless, we are beginning to realize that termites play a considerable role in some ecosystems through modification of habitat, their contribution to energy flow, and the cycling of nutrients (Wood and Sands, 1978). The following books together cover nearly every aspect of termite biology and ecology: Krishna and Weesner (1969, 1970), Lee and Wood (1971b), Wilson (1971), Brian (1978, 1983), and Hermann (1979). Annotated bibliographies of the termite literature in Snyder (1956, 1961, 1968), Ernst and Araujo (1986), and in *Termite Abstracts* (established in 1981) are definitely worth knowing about.

External Morphology

These very generalized insects have retained many characteristics of their cockroach ancestors. The flattened blattid body plan includes an opisthognathous head, a broad, protective pronotum and leathery forewings, long running legs, large compound eyes, slender antennae in front, and prominent sensory cerci behind. These features have been subtly altered in the termite body plan for a life-style of tunneling through wood and soil: a cylindrical form with prognathous head, small pronotum and delicate, narrow dehiscent wings, stout walking legs, eyes reduced or absent, and relatively short antennae and cerci (Fig. 33.3).

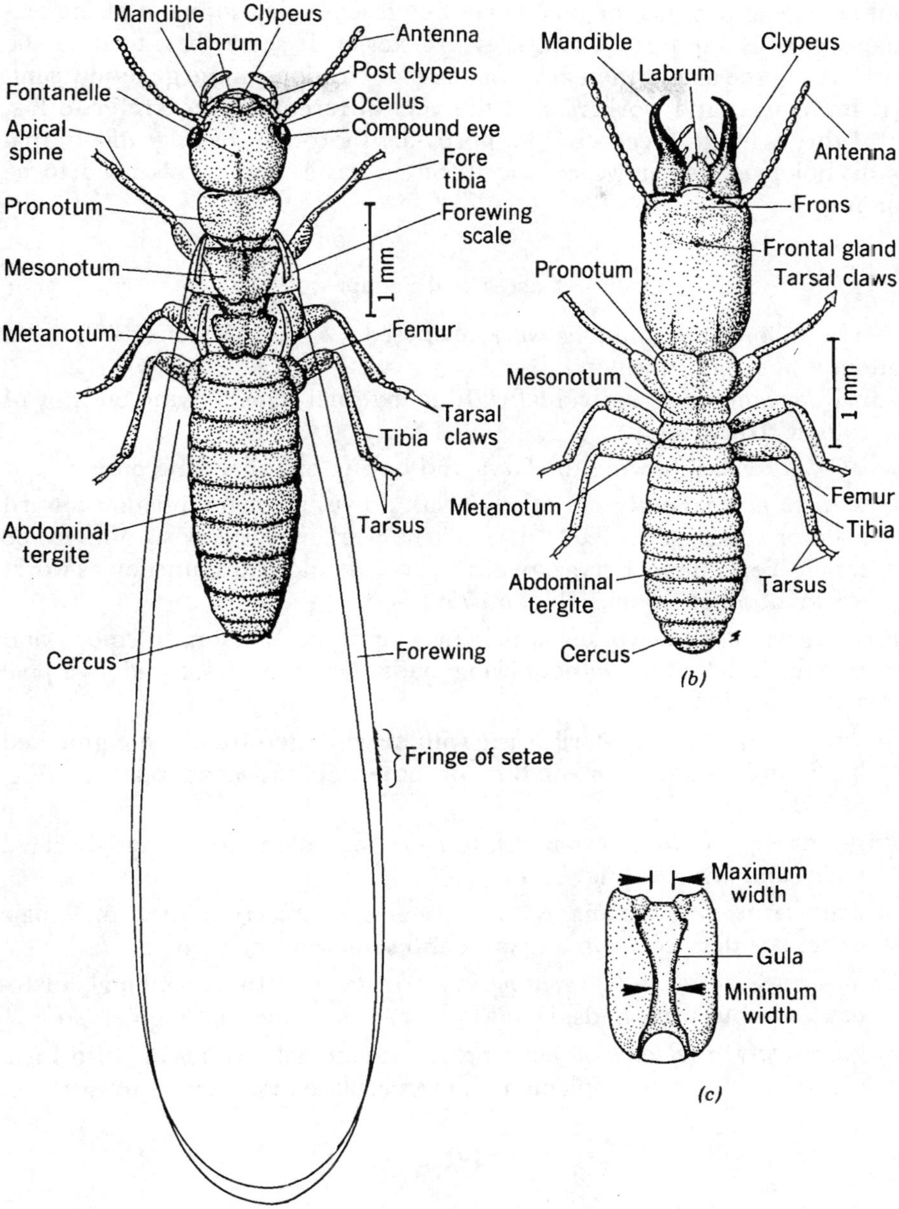

Fig. 33.1. External morphology of a typical termite (*Reticulitermes tibialis* Banks): (*a*) alate caste; (*b*) soldier caste; (*c*) maximum/minimum gular widths.

Consonant with the evolution of eusociality, a division of labor has appeared, with its attendant, morphologically distinct worker and soldier castes in addition to the usual winged reproductives (Fig. 33.15). However, the immatures, workers, and reproductives are sometimes so conservative structurally as to be difficult to identify to species. While workers and soldiers generally include members of both sexes, development of the sexual organs is suppressed and they are sterile. It seems best to treat the features in the alates, then any notable modifications in the generally similar immatures and workers, and the very different soldiers (refer to Fig. 33.1 throughout). Weesner (1969) provides thorough coverage of external morphology. First, however, the different castes and forms need to be defined.

Castes and Forms

The following definitions were proposed by E. M. Miller (1969) in an attempt at standardization.

Larva an immature individual with no external signs of wing buds or of soldier structures.

Nymph a form following the larva and having external wing buds.

Worker a permanently sterile caste with no visible differentiation toward alate or soldier. It is likely that a "true worker" caste exists only in the family Termitidae. Larvae, nymphs, and pseudergates function as workers in all other families (Fig. 33.7*b*).

Pseudergate a false worker which may undergo "stationary" molts and remain indefinitely without wing buds, between larval and nymphal stages (Fig. 33.3*c*).

Soldier a permanently sterile caste with heavily sclerotized and pigmented head and enlarged mandibles or other defensive adaptations (Fig. 33.1*b*).

First-form reproductive (*primary or imago*) the colony-founding, dealated males and females (Fig. 33.1*a*).

Second-form reproductive (*brachypterous neotenic, nymphoid*) a functional male or female derived from a nymph and somewhat pigmented.

Third-form reproductive (*apterous neotenic, ergatoid*) a functional male or female without wing buds, larval in form, and somewhat pigmented.

Supplementary or replacement reproductive functional second- or third-form reproductives may supplement or even replace first-form reproductives.

Head

The body of all castes is subcylindrical in cross section and flexible for easy maneuvering in tunnels. The head is nearly circular in dorsal outline and directed forward so that the mouthparts are strategically placed for carving and drilling wood or soil. The compound eyes begin to appear only in the alate line. All other forms are generally blind. Simple eyes or ocelli occur on most alates, one just above and anterior to each compound eye. A small spot occurs on top of the head of alates (and soldiers) in the middle and higher termites. It marks the opening of the frontal gland, but its function in this caste is not known. It lies approximately between the compound eyes and is no more than a raised pinpoint in the rhinotermitids (Fig. 33.1*a*) but is variously shaped and pigmented in the termitids (Fig. 33.16*a*).

The mouthparts are well developed and of the biting–chewing type in all forms but the soldiers. The mandibles of most soldiers are variously enlarged for defensive use and often armed with teeth on their inner margins (Fig. 33.3*b*). The nasute soldiers (Fig. 33.17*d*) of two rather uncommon species in the southwest have no obvious mandibles. They have the front of the head drawn out into a tapering snout or "nasus," with the frontal gland opening at its tip. They are thus equipped to eject a viscous, toxic secretion from their heads and so to entangle and disable their enemies. The midventral area of the soldier head, particularly those with elongate heads, is covered by a long, narrow plate, the gula (Fig. 33.1*c*), which is used in separating some of the *Reticulitermes* species.

Thorax

Although this part of the body is very complex, there are few structures that are used in identification. Of the three subdivisions, pro-, meso-, and metathorax, only the expanded dorsal plate, or pronotum of the first, need be mentioned. Its shape and dimensions are sometimes important in identifying alates or soldiers.

The other two thoracic segments bear the wings, which are fairly delicate and nearly equal in size and shape. They are deciduous along basal or humeral sutures (HS) (Fig. 33.2*a*), and lost generally soon after the dispersal flights. The small basal portion remaining is called the wing scale (WS). The wing membrane may bear numerous minute hairs (microtrichia), or faint patterns of either microscopic punctations or reticulations. It may be transparent or slightly opaque and, especially in some termitids, lightly pigmented.

The wing veins, from the anterior or costal margin (CM), are subcosta (SC), radius (R), radial sector (RS), media (M), and cubitus (Cu) (Fig. 33.2*a*). As a result of fusion or reduction, a complete set of veins is not present in all groups. Some veins may be conspicuously pigmented, whereas others are scarcely discernible. Expect some variation in venational details from one individual to another within species. In the nymph developing wings appear as posterior projections (wing buds, wingpads) on the lateral margins of meso- and metathorax.

The legs of termites are very conventional walking legs, and all species here have four-segmented tarsi. The tarsal claws are well developed, particularly in the kalotermitids. The number of spurs at the distal end of the tibia is useful for separating a few groups (Figs. 33.16*b*, 33.17*b*).

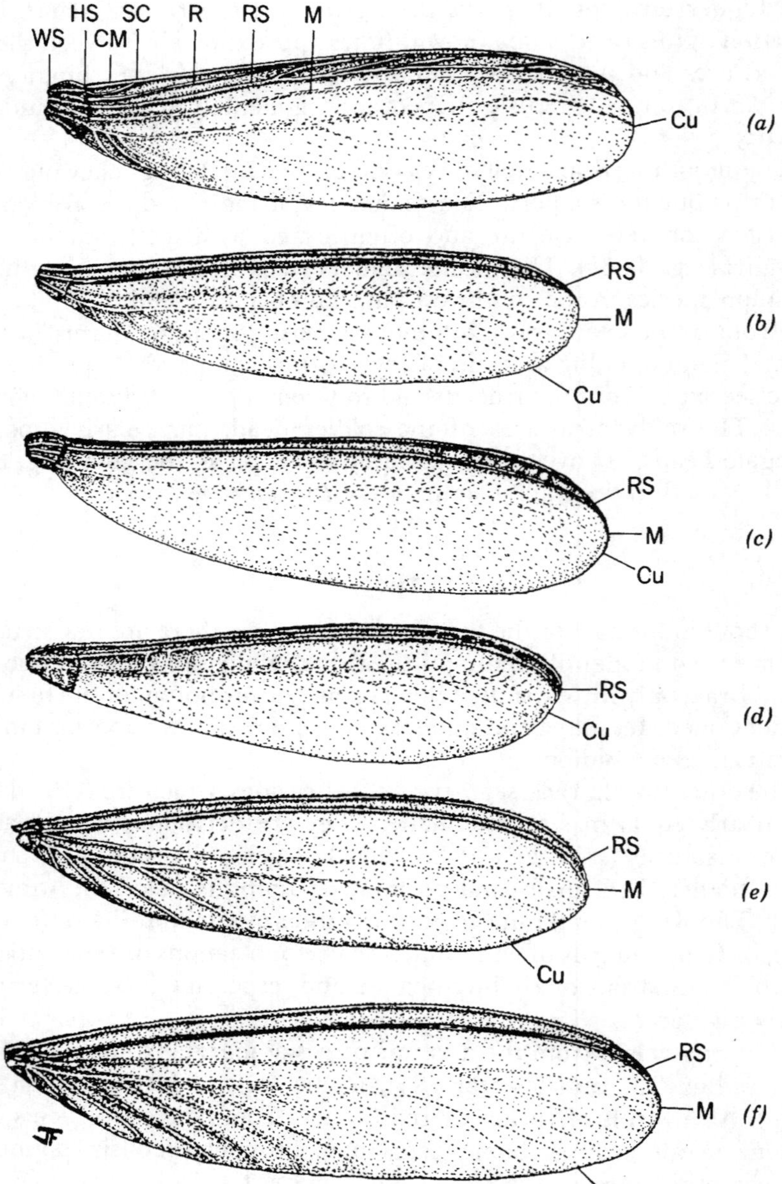

Fig. 33.2. Details of wing venation: (*a*) *Paraneotermes*; (*b*) *Heterotermes*; (*c*) *Reticulitermes*; (*d*) *Prorhinotermes*; (*e*) *Amitermes*; (*f*) *Gnathamitermes*. *WS*, wing scale; *CM*, costal margin; *HS*, humeral sutures. Wing veins as follows: *SC*, subcosta; *R*, radius; *RS*, radial sector; *M*, media; *Cu*, cubitus.

Abdomen

The abdomen is 10-segmented with conspicuous broad tergites and sternites. The first sternite is absent in both sexes, the tenth is absent in the male, and the eighth through tenth are absent or reduced in the female. Male imagoes are thus identifiable by the shorter seventh through ninth sternites and, in the lower families, by the conspicuous paired styli projecting posteriorly from the center of the free margin of the last sternite. Females are easily recognized by the much longer terminal (seventh) sternite. Larger, immature forms can be sexed by the shape of the posterior sternites, but this is not the case with the Termitidae. Although both sexes of all castes bear short cerci on either side near the tip of the abdomen, they are longer and more conspicuous in the lower termites.

The body surface is nearly naked, although there are short, delicate setae sparsely distributed over most sclerotized areas. Immatures are unpigmented, a dirty white or cream color, except for heavily sclerotized structures such as mandibles and maxillae. Alates are always pigmented, varying from pale browns to black. The lighter species usually fly at night, the darker ones during the day or twilight. Neotenic reproductives are generally amber colored. Like the immatures, workers of most species are unpigmented, although the head capsules are sometimes pale yellow or amber. An exception to this are the workers of the two species of *Tenuirostritermes*. Their bodies are medium brown and heads a darker brown. They forage without cover at night and on cloudy days. Experiments by Nel and Hewitt (1969) on similarly pigmented individuals of a South African hodotermitid showed that the pigmentation provides significant protection against solar radiation. Soldier heads are always heavily sclerotized, with head capsules ranging from pale yellow to dark chestnut, and mandibles usually darker browns to black.

Digestive Characteristics

The termite alimentary system is deceptively simple when its capabilities are considered (Noirot and Noirot-Timothée 1969). The digestive tube is divided into fore-, mid-, and hindgut, much like that of a cockroach. A small distended crop at the end of the foregut can hold small amounts of food before further grinding by the hardened cuticular teeth of the proventriculus or "gizzard." The midgut is fairly short and lined with a glandular epithelium. A variable number (4–15) of Malpighian tubules, excretory organs, empty into the tract at the junction of mid- and hindgut.

The hindgut contains five distinguishable segments, beginning with a short first segment and enteric valve which prevents hindgut contents from backing up into the midgut. The third segment, or paunch, is enlarged in all termites and holds the well-known symbiotic microbiota, which digest the diet of plant material (Honigberg 1970). Colon and rectum follow the paunch. It has been estimated that the microorganisms in the hindgut constitute about one-third of the termite's wet weight. The lower and middle termites harbor both flagellate protozoans and bacteria, with a preponderance of the former. In contrast, the Termitidae generally have no protozoans in the hindgut, but contain symbiotic bacteria instead.

Unique Gastrointestinal Mechanisms

Termites are nearly as well known for their basic ability to digest cellulose as for their consequent capability of destroying wood and wood products. Cellulose is probably the most abundant of the continuously cycled organic materials in the biosphere. Soil bacteria and fungi are the primary agents of its degradation. Among the relatively few animals that metabolize cellulose there are two groups that employ basically different strategies. Some protozoans, molluscs, and arthropods actually produce their own cellulolytic enzymes; the termites and ruminants depend upon a rich gut fauna and/or flora which metabolize the cellulose. However, there is increasing evidence that the termites belong to both groups (LaFage and Nutting 1978).

It appears that the lower termites depend primarily on their symbiotic flagellates for the digestion of cellulose. However, they also possess enzymes of their own for the digestion of soluble carbohydrates. Yamin and Trager (1979) have obtained the protozoan *Trichomitopsis termopsidis* (from *Zootermopsis* sp.) in axenic culture and found that it ferments labeled cellulose to CO_2, acetate, and H_2. They have also demonstrated that a hindgut fauna of this flagellate alone will prolong the survival of its termite host. Among the termitids there is not as yet sufficient evidence to support a case for cellulose digestion by either the intestinal bacteria or by the termites themselves (Waller and LaFage 1987).

Next to cellulose, the lignins are probably the most abundant group of continuously cycled organic compounds. The white-rot fungi are the only microorganisms with the capacity for complete decomposition of lignin. As a group, the lignins are less efficiently digested by the termites and their symbiotes than cellulose or the hemicelluloses. There is good evidence that several species of both lower and higher termites do attack and degrade lignin. However, there is no information to suggest how the termites and their symbiotes accomplish this. In studies on six species of termites and several different woods, assimilation efficiencies ranged from 54 to 93% on a dry weight basis (Wood 1978). Wood and Sands (1978) used an average value of 68% for overall assimilation efficiency in calculations of energy flow through termite populations in different ecosystems. This puts the termites in a class with land snails of the genus *Helix* and ruminants as surprisingly efficient assimilators of cellulose.

While the nitrogen content of plant material is highly variable, it is extremely low in wood. The ability of termites to survive and reproduce on diets containing so little nitrogen has been the subject of much speculation. Four sources of nitrogen, are open to them: new nitrogen from (a) their diet and (b) atmospheric nitrogen, and recycled nitrogen from (c) their symbiotes and (d) waste materials from the symbiotes and the termites themselves. The diet certainly provides a wide range, albeit in minimal amounts, of largely conventional nitrogenous compounds which are digested and utilized by both members of the partnership. There is considerable evidence to show that termite gut bacteria fix atmospheric nitrogen, which can be incorporated into termite tissues (Waller and LaFage 1987). Breznak and his group have continued investigations on the role of bacteria in hindgut fermentations, with emphasis on the nitrogen economy of termites (Breznak 1984).

Relationships to Humification

Termites enter the decomposition process at an early stage by their very choice of food. Relatively few North American species feed on living woody or herbaceous plants. There are many humivorous termites in the tropics, but we do not know whether any of our species have this habit. The majority favor sound dead or decaying wood and other plant materials, either standing, fallen or buried. In feeding they subdivide the detritus into particles of many sizes, some of which are never collected or eaten. Thus more surface area is progressively exposed to attack by free-living soil microorganisms.

In its passage through their guts, many of the polysaccharides and some of the lignin in the plant material are digested by the symbiotes in the hindgut. Termite fecal material contains some cellulose, variable amounts of lignin, demethoxylated cleavage products of lignin, and alkali- and water-soluble materials. The products of wood digestion in termite feces resemble the humic extracts from soil organic matter. Lee and Wood (1971b) reported that this material inhibited microbial activity.

While mound-building termites, restricted to the tropics, concentrate organic matter and associated breakdown products in and around their nests or mounds, subterranean species tend to leave them in their extensive workings and gallery systems. In the course of their attack on surface wood, rhinotermitids typically plaster their galleries with fecal material and pack abandoned voids with soil. Many termitids, especially species of *Amitermes* and *Gnathamitermes* in the United States, build delicate soil shelters over wood, plant debris, and dung as they work (Nutting et al. 1987). Subterranean termites nest in the soil at varying depths and, during favorable seasons, they are continually driving new tunnels through upper levels of the soil in search of new food items on the surface above them. As a result of these activities the subdivided detritus and fecal material become buried and mixed with the soil. Nutrients returned to the soil from the dead bodies of termites may not be trivial, even though termites are cannibalistic (LaFage and Nutting 1978). Values of termite tissues and feces returned to the soil are given in the next section.

Lee and Wood (1971b) thoroughly reviewed the literature on the effects of termite activity on the physical and chemical characteristics of soils, but there had been no data published on any North American species at that time. Studies on *Heterotermes aureus* and *Gnathamitermes perplexus* in a semiarid grassland of southern Arizona have since shown that both species add considerable amounts of clay, organic carbon, PO_4, Na, Ca, Mg, and HCO_3 to the surface soils. *Heterotermes* increased the amount of N (Kjeldahl), NH_4, and K, and *Gnathamitermes* the amount of NO_3 (Nutting et al. 1987). Schaefer and Whitford (1981) reported on nutrient cycling by large populations of *Gnathamitermes tubiformans* in the Chihuahuan Desert of New Mexico. Estimates of the amounts of litter and soil moved by termites are given in the section "Unique Ecological Characteristics."

Excretory Characteristics

It had been assumed that nitrogen excretion by termites followed the pattern common to insects in general; that is, cells of the Malpighian tubules transfer nitrogenous wastes by diffusion or active transport from blood to the lumen of the tubules. Water, certain salts, and other useful compounds are selectively reabsorbed through the lower end of the tubules and the hindgut, particularly by the specialized cells of the rectal pads or papillae. Being sparingly water soluble, uric acid and urates are precipitated and passed out with the feces. The major end product of protein metabolism in the symbiotic protozoans is probably NH_3. The composition of fecal material is given above under "Relationships to Humification."

Urinary Counterparts

Potrikus and Breznak (1980) have shown that both kalotermitids and rhinotermitids do form uric acid and can store large amounts of it in the fat body. However, they do not void significant amounts of this purine in their feces. They also harbor uricolytic bacteria in the hindgut, and Potrikus and Breznak (1981) have suggested that the uric acid serves as a nitrogen bank which can be drawn upon for internal recycling by both symbiotes and host (Breznak 1984).

Fecal Production Rate

High assimilation efficiency reduces fecal output to no more than 40–50% of the ingested food. Dry-wood termites form hard, dry pellets which they pack into abandoned chambers or cast outside the nest through dump holes. Subterranean species produce liquid or pasty feces with which they plaster the walls of their galleries and chambers, or occasionally deposit in unused cavities. Fecal material thus tends to be concentrated in areas where it may not be immediately available to plant roots. This situation is obviously not so extreme as it is in mound builders. Wood and Sands (1978) estimated returns via termite feces to the soil in three types of western African savanna to range from 0.8 g (15.5 kJ) to 12.4 g (233 kJ)/m^2. Returns via termite tissue (corpses) turned out to be 27.6–277 kJ/m^2. These authors review what little has been published on the decomposition of the corpses and fecal material of termites. An estimate of fecal production rate by a nonsubterranean, damp-wood termite (*Zootermopsis* sp.) might be made from data in Hungate's (1941) paper on nitrogen economy.

Unique Reproductive Characteristics

The success of the social insects is largely due to the development of a caste system in which typically sterile caste (workers and sometimes soldiers) work to assist a reproductive caste in rearing their sibs.

Normal

A termite colony is basically a family composed of the founding pair, or king and queen, and one or more broods of their offspring. Among the lower termites the larvae and nymphs serve the reproductives; among the middle and higher termites an emerging or true worker caste provides the labor force. Unlike the situation in the social ants, bees, and wasps, the male termite (king) does not die after courtship and mating, but remains with the female (queen) to assist her and mate with her periodically.

The termites, particularly the lower families, have the remarkable ability to produce neotenic replacement or supplementary reproductives (Howard and Haverty 1980, Myles and Nutting 1988). A few cases of multiple primary reproductives are known among the Termitidae. With such reproductive flexibility a colony can take advantage of increased abundance of resources (litter fall, for example) to grow, or even to divide, and mature more rapidly. Neotenic reproductives can also be produced to head fragments of colonies and so to survive after passive and gradual, or active and sudden, division of a diffusely organized colony. Results of preliminary studies by Howard et al. (1981) indicate that the fusion and division of colonies of *Reticulitermes flavipes* and *R. virginicus* provide great potential for colony growth and expansion into unoccupied areas.

Typical Life Cycle

A subterranean termite colony in a temperate climate probably reaches maturity in 5 or 10 years; dry-wood termites average perhaps a few years less. This means that a colony is sufficiently prosperous to produce all castes, including winged reproductives. Colonies release their alates seasonally in one or more dispersal flights, apparently in response to a combination of physical and meteorological factors. The lower termites fly under a wide range of weather conditions and stage many flights of relatively small numbers over a prolonged season. The higher termites generally fly under more specific conditions, producing fewer flights of larger numbers, although the season may still cover several months. Most of our species probably fly no more than a few hundred yards before settling.

Alates shed their wings soon after the flight. Males excitedly seek out the females, which typically assume a "calling" attitude with abdomen raised and, presumably, dispensing a sex pheromone. After meeting, the pair proceed in tandem to find a suitable nesting site either in wood or soil. The male usually assists the female in excavating a cell, and copulation is delayed until they are established within their nuptial chamber.

The life of a colony begins with the appearance of the first brood of eggs and larvae within days, sometimes weeks, after pairing. The incipient colony grows slowly at first, so that colonies of our fastest growing subterranean termitids probably contain no more than a few hundred individuals at the end of the first year. Information on colony development and maturation is scarce. Details from laboratory studies of some two dozen species, representing all families, have been summarized by Nutting (1969a).

Parthenogenesis

A number of cases has been reported but there is no evidence that parthenogenesis plays a significant role in the reproductive life of any termite species (Nutting 1969a). Howard et al. (1981) consider *Reticulitermes virginicus* females to be capable of parthenogenetic colony foundation.

Unique Ecological Characteristics

Subterranean termites are among the most abundant but cryptic of animals. Except for the tropical harvesters and mound builders, they are almost never seen. However, Wood and Sands's (1978) review convincingly shows that termites play an important role in tropical and warm temperate ecosystems, particularly with regard to habitat modification and their contributions to energy flow and nutrient cycling. A few examples here should be sufficient.

One of the few ecological studies on North American termites was made on a shrub-invaded, semiarid grassland in southern Arizona. The total population of five subterranean species was estimated at 1025 per square meter with a biomass of 0.414 g/m^2 (dry weight) (Haverty and Nutting 1975a). One prominent species (*Heterotermes aureus*) had a population of 431 per square meter or 0.148 g/m^2 (Haverty et al. 1975). In a year-long study, the maximum surface foraging density of the commonest species (*Gnathamitermes perplexus*) was estimated at 70.52 per square meter or 0.045 g/m^2 (LaFage et al. 1976). [Similar records cited by Wood and Sands (1978) range up to 6825 per square meter or 18.9 g/m^2 for all (four) species in a west African agroecosystem.] Continuing studies on the Arizona species in the same area are reviewed by Jones and Nutting (1987). Howard et al. (1982) have contributed important demographic information on two destructive *Reticulitermes* spp.

Annual production of alates by five species (four subterranean) in the same semiarid grassland amounted to 67,174 per hectare or 174 g/ha (Nutting and Haverty 1976). While this compares favorably with Lepage's (1972) estimate of alate production by *Trinervitermes trinervius* in Sahelian savana of 15,000 per hectare, it is rather less than Bouillon's (1969) estimate of 560,000 per hectare for *Cubitermes exiguus* in Zaire! Since very few alates succeed in founding new colonies, it is evident that these relatively stable insect societies are producing a seasonably dependable, energy-rich food which is eagerly eaten by a great variety of predators, vertebrate and invertebrate.

Even in semiarid, warm temperature areas, termite populations consume impressive amounts of woody litter and forage. For example, we have calculated that *Heterotermes* removed superficial dead wood at the rate of 78.9 kg/ha in a particular year, or 3.7% of the standing crop biomass (2127 kg/ha) and 17.5% of the annual production of dead wood (450 kg/ha per year) (Haverty and Nutting 1975b). By direct measurement, this termite consumed paper at a rate of at least 19.5, and possibly as much as 33.9 kg/ha per year. Unpublished, but fairly good estimates were made of woody litter consumption by all nine species (two dry-wood termites) in the same semiarid grassland in southern Arizona. Total consumption came to 414 kg/ha per year, 92% of the annual dead wood production of 450 kg/ha. Bodine (1973) estimated that *Gnathamitermes tubiformans* removed 360 kg grass/ha per year from a short-grass Texas rangeland or 12% of net primary productivity. The effect of termite grazing is occasionally so great that poorly managed rangelands are denuded of grass. Whitford and his group continue their studies on the role of termites in the detritus cycle of the Chihuahuan Desert ecosystem (MacKay et al. 1987).

The activity of our non-mound-building, subterranean termites may result in rates of soil accretion sufficient to influence soil formation—over hundreds or thousands of years in certain areas. Both *Heterotermes* and *Gnathamitermes* pack their workings with soil, as do many other termites. Referring again to the year-long study in the southern Arizona grassland, it was estimated that *Heterotermes* moved soil to the surface at a rate of 72 kg/

ha per year, *Gnathamitermes* at 559 kg/ha per year, and the two species together in the areas of overlap at 93 kg/ha per year. In sum, the two species deposited soil on the surface at a combined rate of 724 kg/ha per year (Nutting et al. 1987). Lee and Wood (1971a) reported that erosion of abandoned mounds at two different sites in Australia added soil to the ground surface at rates of 1150 and 4700 kg/ha per year.

Collection and Technique

The best way to determine the presence of subterranean termites is to walk through an area, turning over stones and every bit (above a few centimeters long) of woody litter, dung, cardboard, and paper. Since *Gnathamitermes* and some of the *Amitermes* build conspicuous tubes and broader shelters of soil, these should be watched for on dead plant material, sticks, logs, tree trunks, and fence posts. If any termites are present, a half-hour's search should uncover foraging groups if the weather is warm and the soil is damp, otherwise at least some vacant galleries or shelters. The listing of one or more termite and pest control operators in a local telephone directory is an indication that termites are a problem in the area. Lee and Wood (1971b) provide a very useful review of methods for sampling termite populations. While many techniques will yield termites, three have proven somewhat more effective than others: soil coring, baiting, and litter processing.

Soil core sampling is a common technique and was used by Ueckert et al. (1976) on shortgrass rangeland in west Texas. (These workers also used an automotive-type vacuum cleaner to collect termites from their surface shelters.) Variations on coring in the form of pits have been used by Sands (1965) and trenches by Kemp (1955). However, there are serious limitations to the technique, not the least of which are physical: hardpan and shallow or stony soils. Termites are also likely to be distributed below, or to escape, the reach of ordinary coring tools. Termites can be separated from the soil by hand, by flotation (Salt et al. 1948), or by modified Berlese–Tullgren funnel (Haarlov 1947).

Wooden stakes or blocks have been widely used for testing the relative intensity of termite activity and hence the effectiveness of chemicals against termite attack (Kemp 1955). A modification of the bait sampling method, using toilet paper rolls as bait, provided much information on the foraging populations of *Heterotermes* and *Gnathamitermes* and their behavior (LaFage et al. 1973, 1976; Haverty et al. 1975). The rolls are an easily handled, standard unit of practically pure cellulose and are readily accepted by termites. (They can also be set out to attract and collect termites for laboratory use.)

Finally, litter can be examined for termites and other invertebrates by sifting or the use of Berlese funnels. Litter bags of nylon, fiberglass, or plastic mesh have also been widely used for studying the effects of various organisms, including termites, on the rate of decomposition (Wood 1978). Items of woody litter have been sampled in large quadrats (Haverty and Nutting 1975a) or belt transects (Johnson and Whitford 1975). Each piece of woody litter was examined for evidence of termite activity, and all associated foraging groups were collected and counted. Subterranean termites can often be separated from wood (or soil) in metal or plastic trays by "papering" (Gay et al. 1955, McMahan and Watson 1977), that is, by covering the wood or soil with damp paper toweling. The termites leave the wood in favor of the paper, from which they are then easily brushed. The larger pieces of wood may have to be sawed into short lengths, split, and the termites, particularly kalotermitids (e.g., *Paraneotermes*), extracted by hand with forceps and brush.

It must be realized that termite populations may be highly aggregated, so that sampling patterns obviously must be worked out to suit their density and dispersion in the area under study. Such distributions are probably a result of the dispersion and abundance of suitable host materials.

Whatever sampling method is used, the worker caste is certain to be the most numerous. Soldiers are infrequent in the Rhinotermitidae and Termitidae, generally making up less than 2% of the individuals in foraging groups and even in very large associations containing reproductives. The Kalotermitidae maintain higher proportions; groups of *Paraneotermes* may contain up to about 15% soldiers. *Tenuirostritermes* is exceptional in foraging on the soil surface at night or on cloudy days. Foraging parties average about 20% soldiers, but they may run as high as 70% or more. Alates are produced and swarm seasonally, usually during mild, humid periods (Nutting 1969a,b, 1979). Numbers produced depend on the size and economy of the parent colony. Since they assemble in superficial chambers before swarming, they may occasionally be taken in soil samples.

For identification, termites should be collected and stored in 95% alcohol. They are so small and soft-bodied that they become shriveled and distorted when mounted on pins or points. Either soldiers or alates (or primary reproductives) are required for specific identification; both are even better.

Measurement and Identification

The caste system affords a double opportunity for the identification of most species of termites. Immatures and workers are generally impossible to distinguish below family level at most. Soldiers, however, bear many useful characters, especially on the head, associated with their highly specialized defensive behavior. The reproductive caste have fewer distinguishing features, owing to their relatively conservative morphological evolution.

A few species can be identified with a good eye after a bit of experience. Several more can be resolved with a 10× hand lens, but confident identification of some of the *Amitermes* species calls for a good stereomicroscope with magnifications of 25–50× and an ocular micrometer. Specimens are best studied submerged in alcohol. They can be held down by a dab of Vaseline or modeling clay, or with small pins in a shallow, wax-bottomed dish. Wings may be examined under alcohol against a light background, or on microscope slides, either dry or in glycerin.

Specific Characteristic Identifying Structures

Most of the structures used in the keys below are easily found for evaluation, counting or measurement. Examples of wing venation appear in Fig. 33.2. Important structures on the soldier and alate castes are illustrated in Fig. 33.1*a* and *b* except those few given special emphasis as follows: fore- and hindwing scales overlapping, Fig. 33.4*a*; not overlapping, Fig. 33.7*a*; two or three fore tibial spines, Figs. 33.17*b* and 33.16*b*; and maximum/minimum gular widths, Fig. 33.1*c*.

The only features that may be difficult to resolve are the fontanelle, when inconspicuous, and the number of fore tibial spines. Careful manipulation of the specimen in strong oblique light is necessary to reveal the fontanelle as a raised point, as in *Reticulitermes*, or as a depressed point, as in *Anoplotermes*. Again, careful manipulation of the fore tarsi is required to distinguish the dark brown spines at the apex of the fore tibia, one on the inner and two on the outer face in *Gnathamitermes* (Fig. 33.16*b*) and one on each face in *Tenuirostritermes* (Fig. 33.17*b*). Color can be a very useful character, but it is a very difficult one to put into words. Although coloration within species in an area varies little, specimens stored in alcohol tend to fade, for example, from black to dark brown after a few years.

Specific Measurements Needed

The identification of either caste requires very few careful measurements. Total length (TL) is often helpful, especially with very small or very large species. The measurements in the keys have been taken from the literature and from alcoholic specimens, with head extended, from tip of labrum or mandibles to wing tips, or tip of the abdomen for soldiers. TL does not usually vary more than ±1 mm in average-sized specimens. More exacting measurements need to be made for pronotal width and for com-

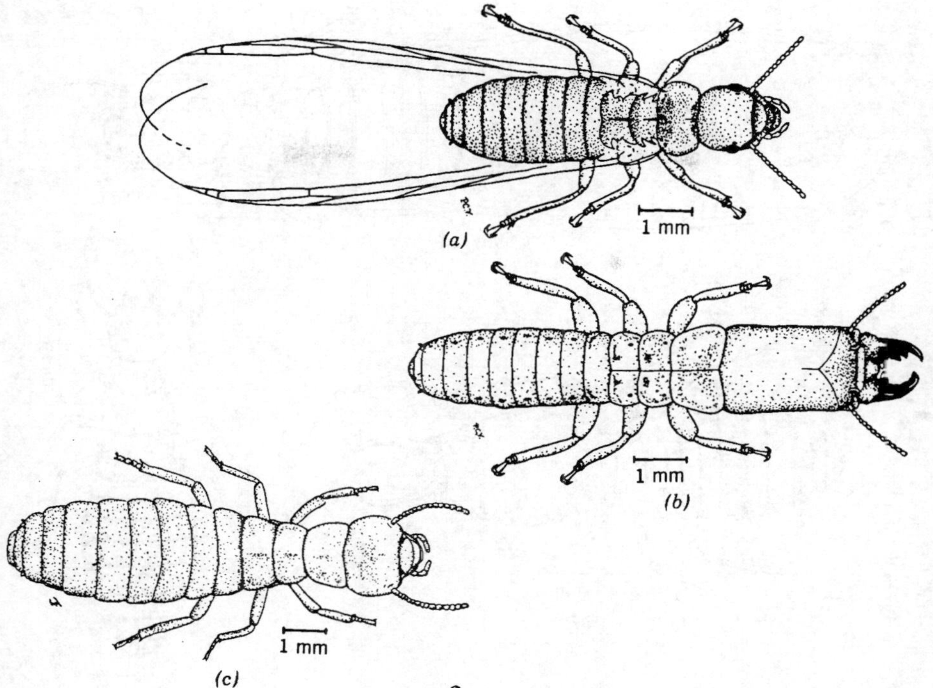

3

Fig. 33.3. *Paraneotermes simplicicornis* (Banks): (*a*) alate; (*b*) soldier; (*c*) pseudergate.

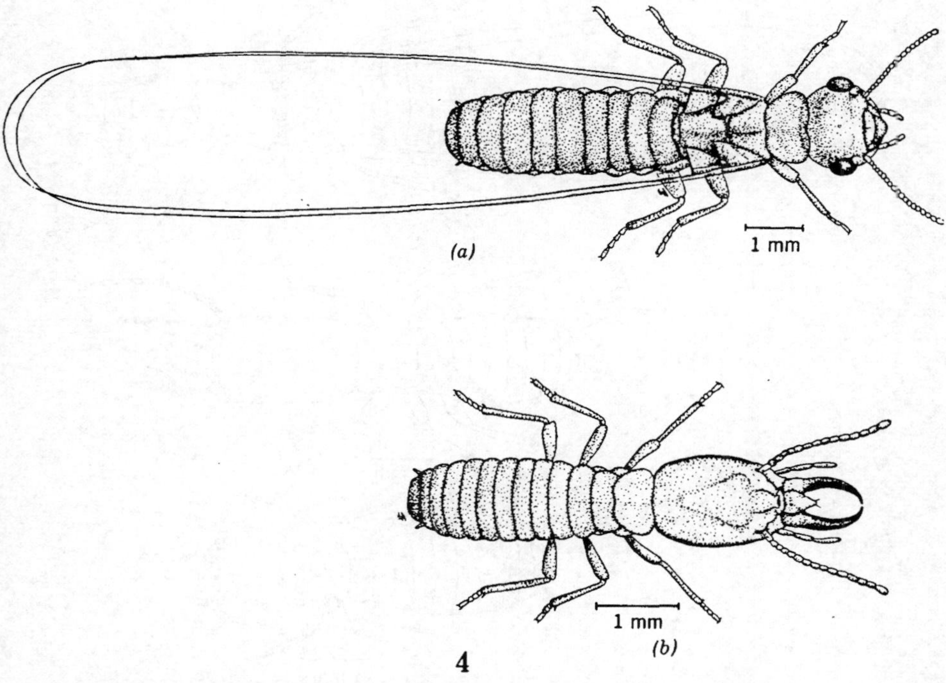

4

Fig. 33.4. *Coptotermes formosanus* Shiraki: (*a*) alate with fore- and hindwing scales overlapping; (*b*) soldier.

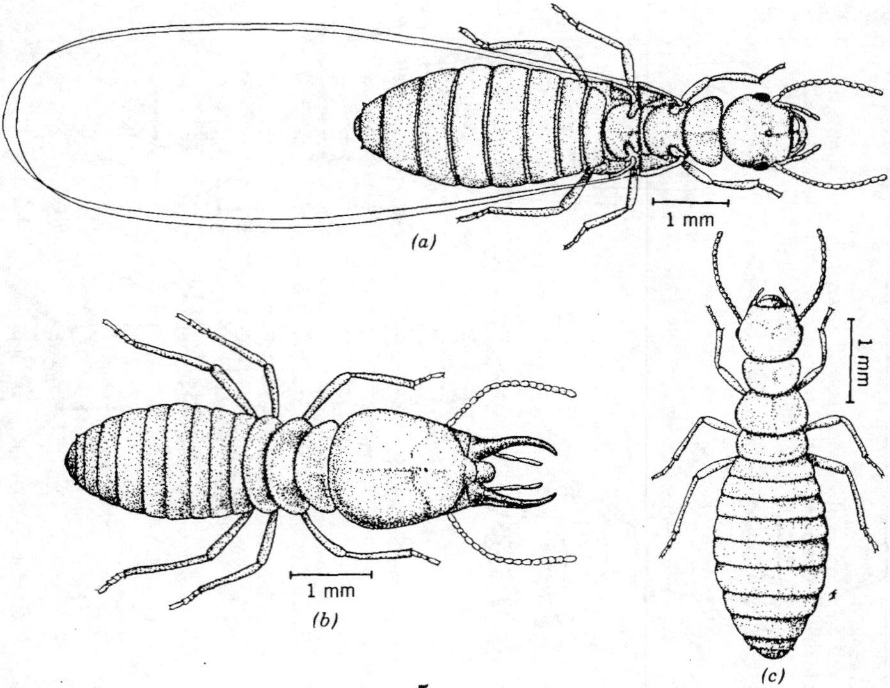

5

Fig. 33.5. *Prorhinotermes simplex* (Hagen): (*a*) alate; (*b*) soldier; (*c*) "worker".

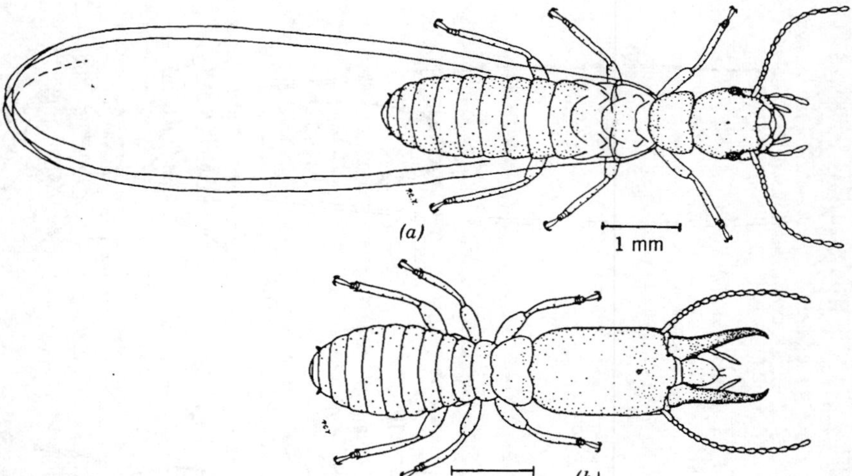

6

Fig. 33.6. *Heterotermes aureus* (Snyder): (*a*) alate; (*b*) soldier; (*c*) "worker".

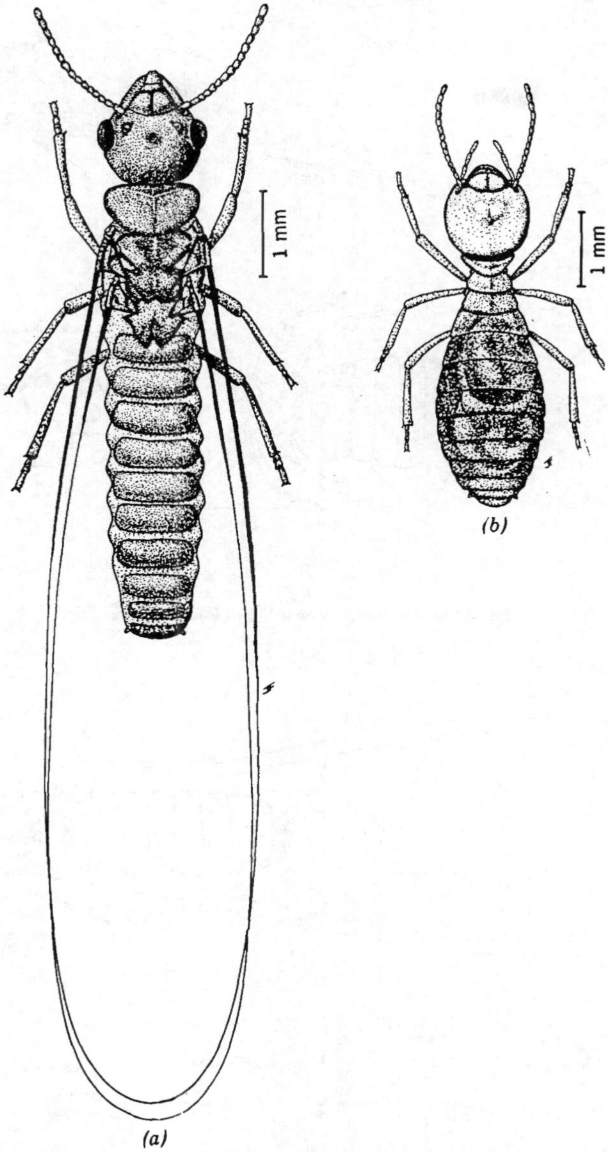

Fig. 33.7. *Anoplotermes fumosus* (Hagen): (*a*) alate with fore- and hindwing scales not overlapping; (*b*) worker.

parisons of head length and width for soldiers of some *Reticulitermes* species and head width for both alates and soldiers of *Amitermes* species. Comparisons of maximum and minimum gular width have also been used for separating soldiers of *Reticulitermes*. The number of antennal segments is easily counted and occasionally very useful.

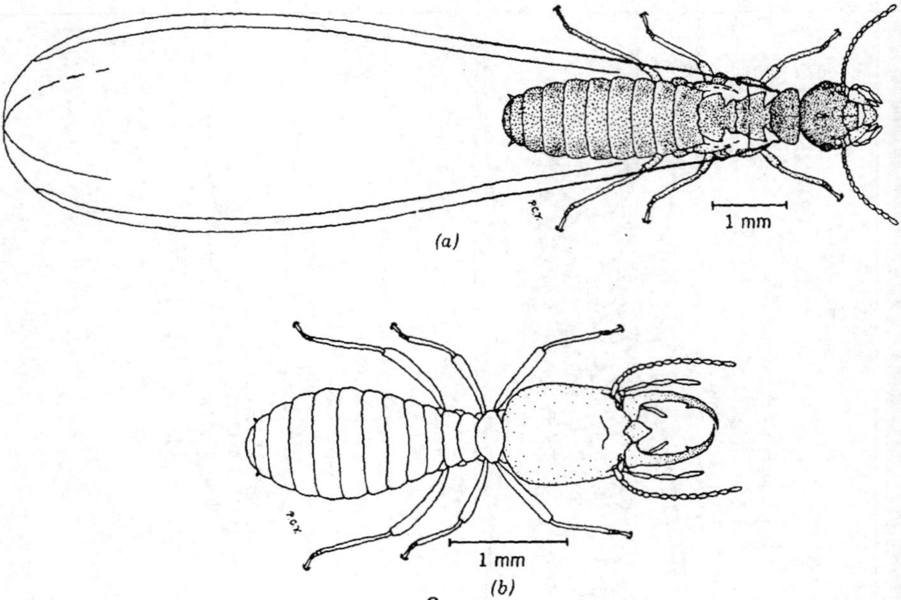

8

Fig. 33.8. *Amitermes emersoni* Light: (*a*) alate; (*b*) soldier.

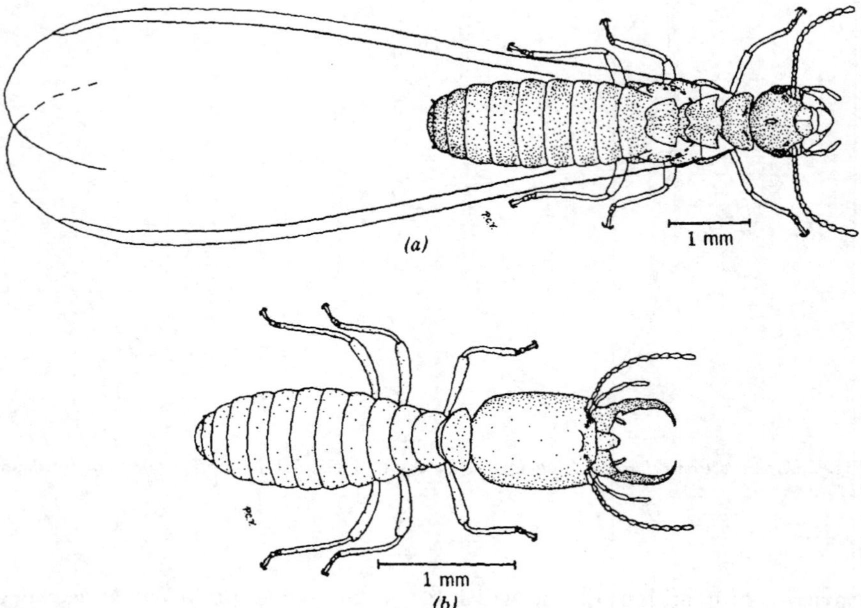

9

Fig. 33.9. *Amitermes coachellae* Light: (*a*) alate; (*b*) soldier.

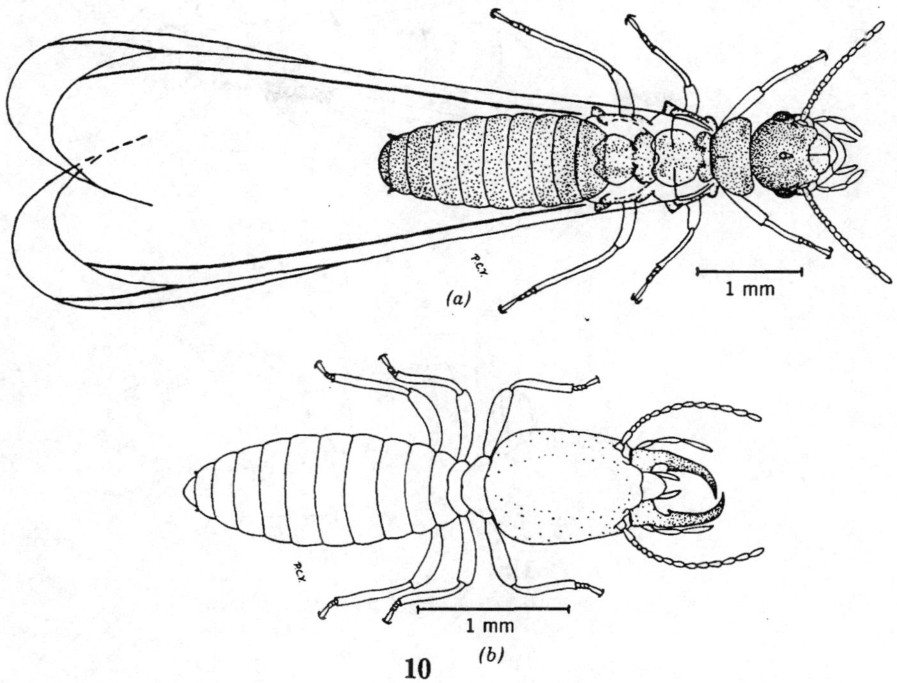

10

Fig. 33.10. *Amitermes minimus* Light: (*a*) alate; (*b*) soldier.

Introduction to Keys

In assembling the following keys, I have drawn on the keys and notes of Snyder (1954) and Weesner (1965). With considerable support from the illustrations, they are designed to lead the soil biologist at least to family and genus and, in most cases, to the species of any subterranean termite in America north of Mexico. This includes all of our termitids, which typically nest in the soil and work dead wood and vegetation on or in the soil. It also

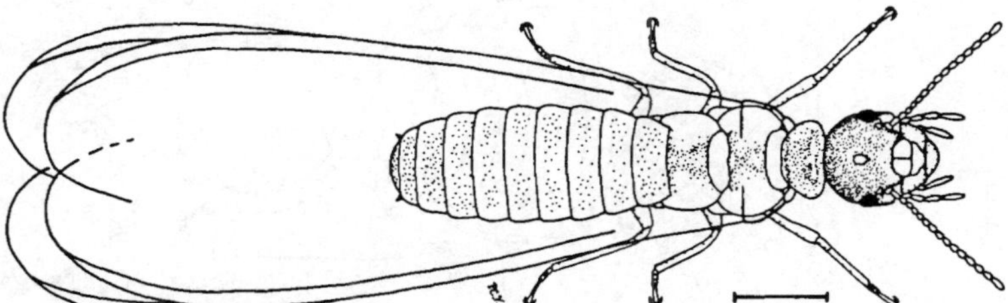

11

Fig. 33.11. *Amitermes pallidus* Light: alate (soldier and worker unknown).

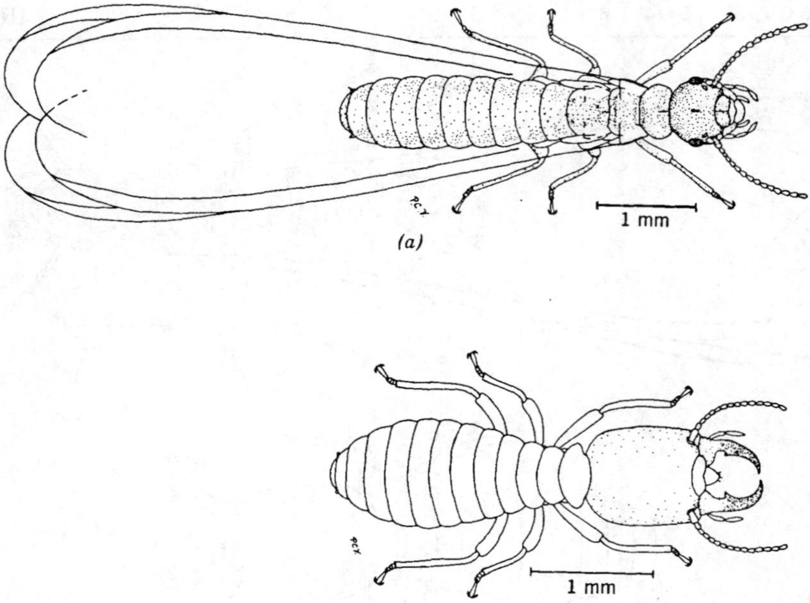

12

Fig. 33.12. *Amitermes parvulus* Light: (*a*) alate; (*b*) soldier.

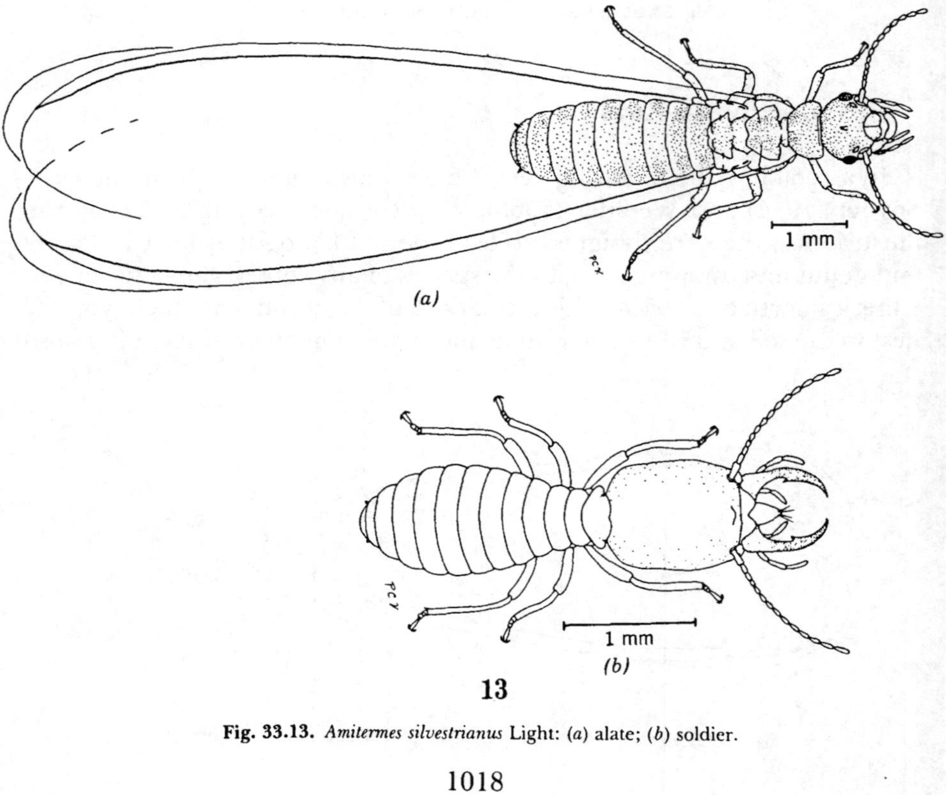

13

Fig. 33.13. *Amitermes silvestrianus* Light: (*a*) alate; (*b*) soldier.

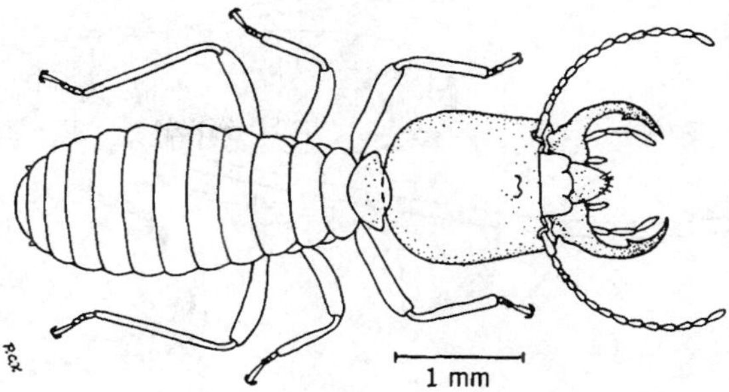

14

Fig. 33.14. *Amitermes snyderi* Light: soldier (alate unknown).

includes most of the rhinotermitids, which maintain their reproductives and brood either in the soil or in logs or stumps at or above the soil surface. *Prorhinotermes* is exceptional in nesting in "damp wood": standing dead trees and fairly solid stumps and logs in mangrove swamps and pine woods. It occasionally tunnels into adjacent soil, perhaps for moisture. *Paraneotermes*, the "subterranean" kalotermitid, presumably also nests in damp

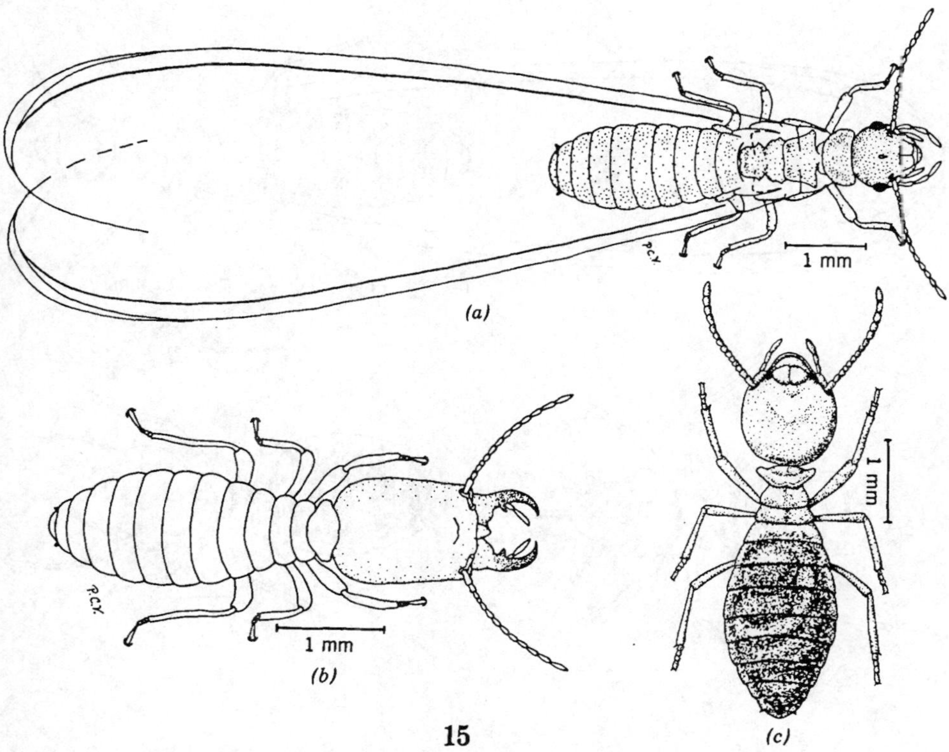

15

Fig. 33.15. *Amitermes wheeleri* (Desneux): (*a*) alate; (*b*) soldier; (*c*) worker.

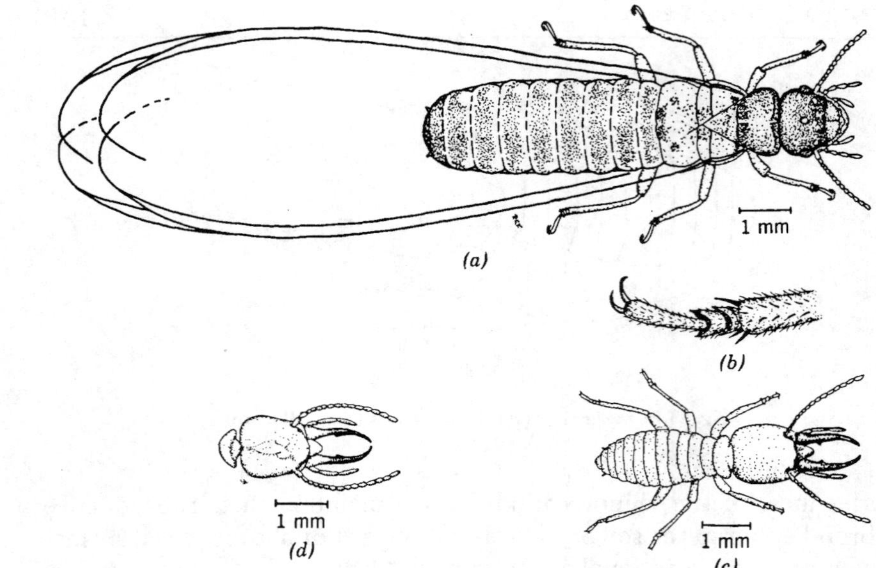

16

Fig. 33.16. *Gnathamitermes perplexus* (Banks): (*a*) alate; (*b*) fore tibia with three spines; (*c*) soldier. (*d*) *G. tubiformans* (Buckley), soldier.

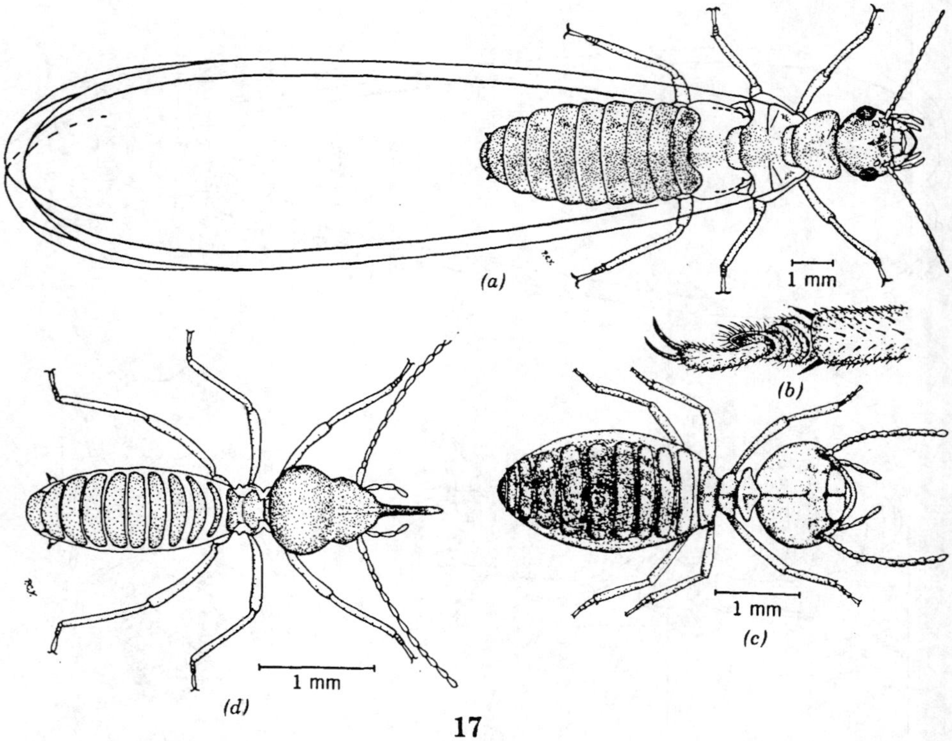

17

Fig. 33.17. *Tenuirostritermes tenuirostris* (Desneux): (*a*) alate; (*b*) fore tibia with two spines; (*c*) worker; (*d*) nasute soldier.

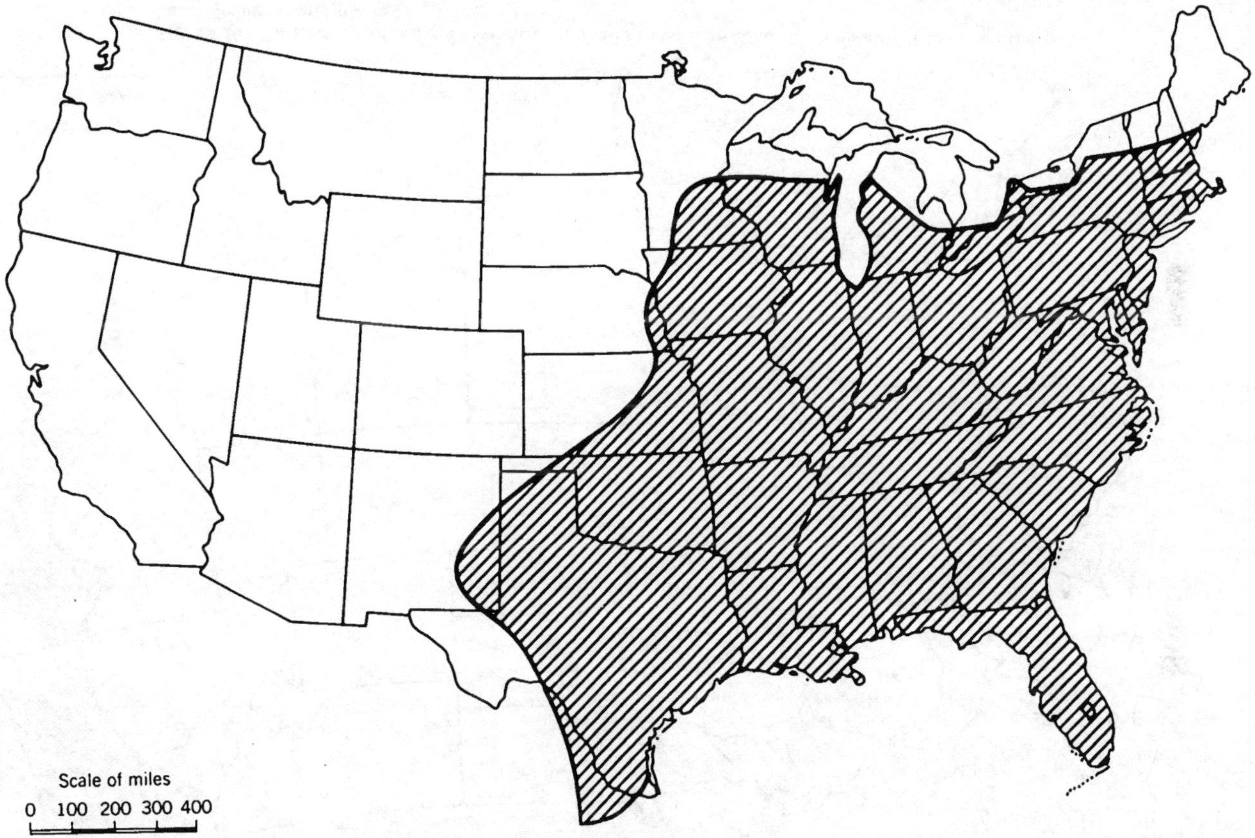

Fig. 33.18. Distribution of *Reticulitermes flavipes* Kollar.

Fig. 33.19. Distribution of *Reticulitermes hesperus* Banks (west coast) and *R. arenincola* Goellner (southern Lake Michigan sand dunes).

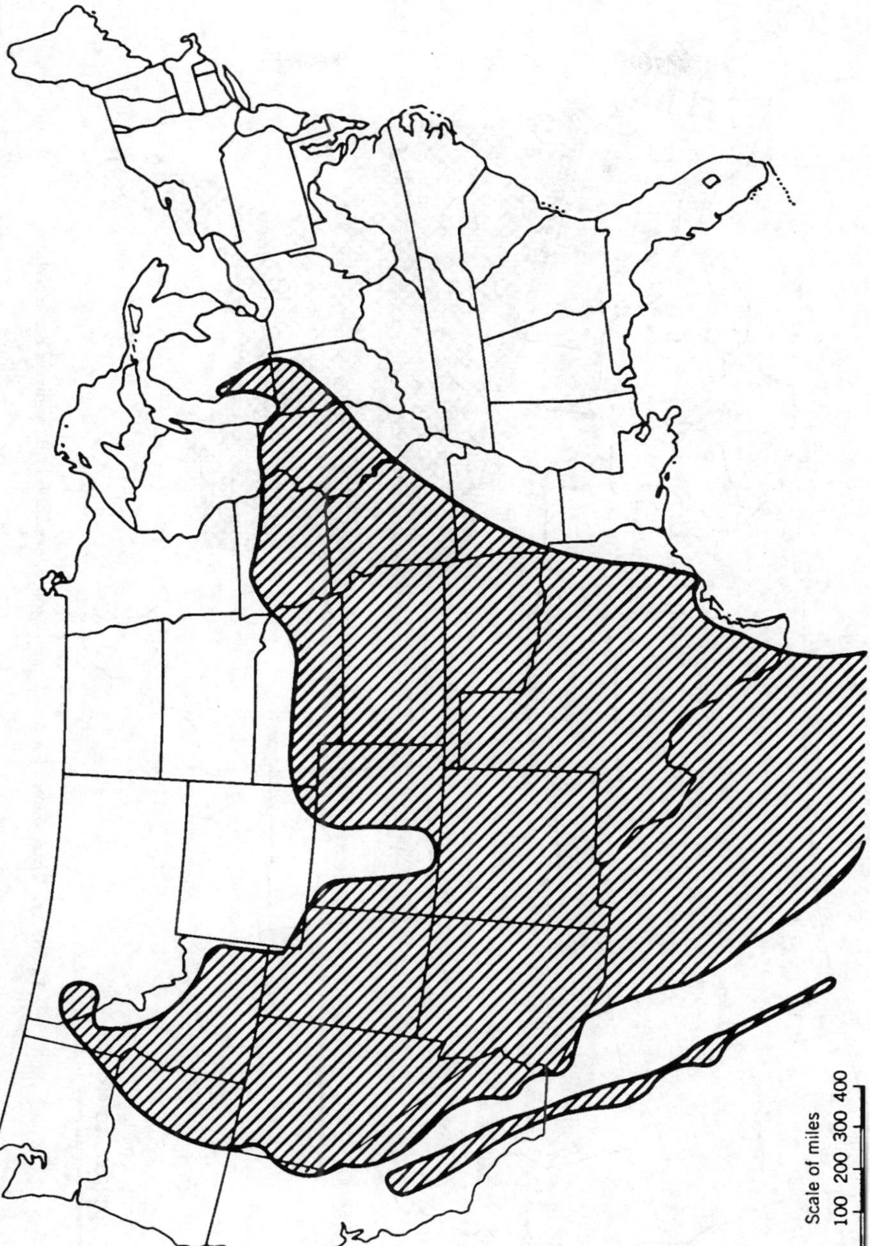

Fig. 33.20. Distribution of *Reticulitermes tibialis* Banks.

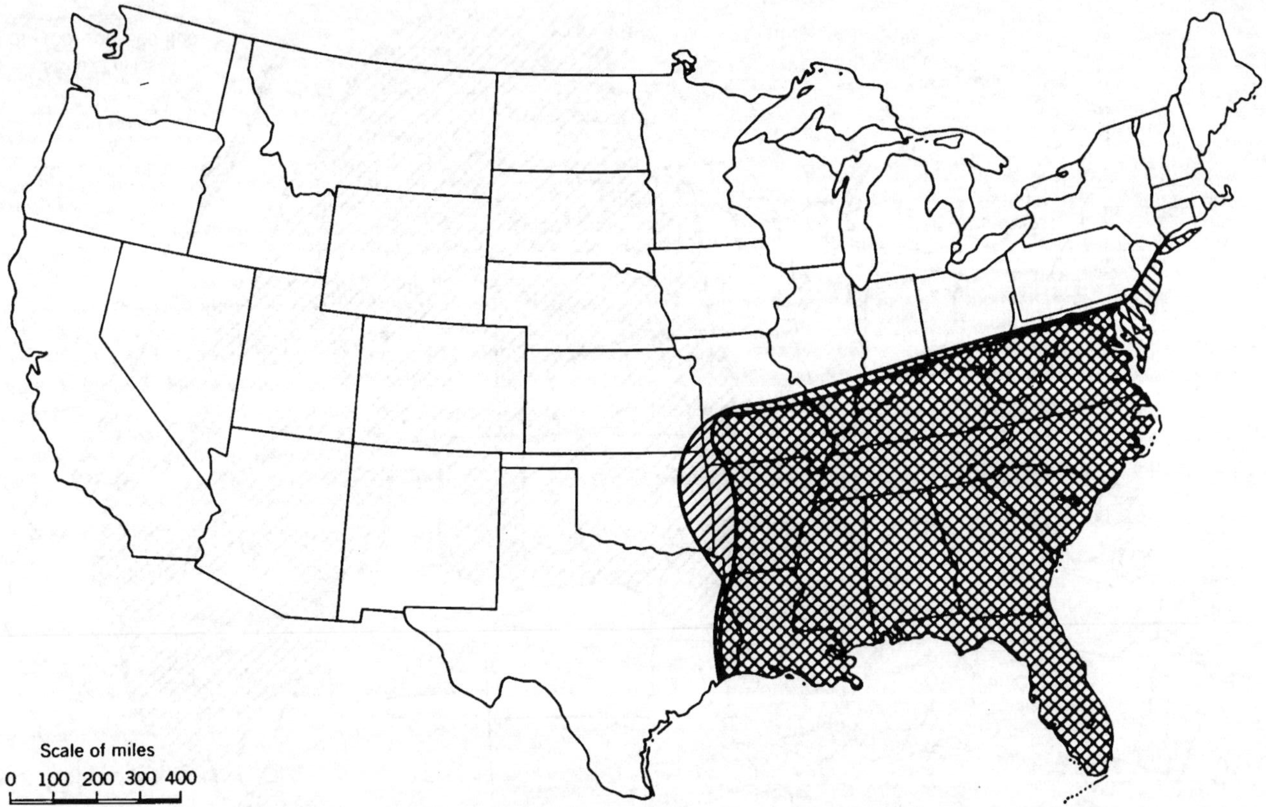

Fig. 33.21. Distribution of *Reticulitermes hageni* Banks ▨ and *R. virginicus* Banks ▧.

wood (stumps and roots) and tunnels through the soil to adjacent food sources.

It should be emphasized that all castes are not known for three species. *Anoplotermes* is the only North American genus, but one of several genera of soldierless termites that occur in various parts of the world. The alate is the only caste known for *Amitermes pallidus*; a colony has never been found. Conversely, soldiers and workers are the only castes known for *Amitermes snyderi*; foraging groups have been found, but never with reproductives or alates. Soldiers of *A. snyderi* are considered too large to be matched with *A. pallidus* alates (Light 1932, Nutting 1969b).

The species of *Reticulitermes* are especially difficult and greatly in need of revision. The keys by Miller (1949, Florida only), Snyder (1954, United States), and Weesner (1965, alates only) are the best that we have. [Weesner (1965, p. 37) would not even attempt a key to soldiers, as she felt that it would only be misleading.] In certain areas identification is fairly straightforward, but problems of variation arise from the realities of their catholic tastes, great adaptability, and wide, overlapping ranges (see maps, Figs. 33.18–33.21). There is some evidence for hybridization between *R. hesperus* and *tibialis*, *flavipes* and *virginicus*, and possibly, even *flavipes* and *lucifugus* in Europe. Alates are generally easier to distinguish than soldiers, and the odds of coming up with a "correct" determination are obviously increased with both soldiers and alates from the same colony. Couplets to the species of *Reticulitermes* in the following key are based on Snyder (1954), a longtime authority on the genus. Distributions and a few additional characteristics used by Miller (1949) and Weesner (1965) have also been incorporated. It must be admitted, however, that there is no key that will guarantee certain identifications of the species of *Reticulitermes* anywhere in North America.

ACKNOWLEDGMENTS

The writing of this contribution and some of the unpublished research reported herein were supported by State Research Project 144 in the Agricultural Experiment Station, University of Arizona, Tucson. For reading the manuscript, checking the keys, and contributing a number of helpful suggestions, I am much indebted to Drs. M. I. Haverty, J. P. LaFage, and F. G. Werner. I would like to acknowledge here the careful and conscientious work of Pamela Cope Vesterby and Joel Floyd, who made the drawings for the keys.

KEY TO TERMITE SPECIES

Key to Alates

1a Several dark, heavily sclerotized veins and diagonal branches between anterior (costal) margin and median vein; no frontal gland pore (fontanelle) on top of head: median and cubital veins approaching each other or joined distally; postclypeus flat and not divided at midline; head and body uniform dark brown; total length, with wings, (TL) 11–12 mm (Fig. 33.3); southwestern, frequent (**Kalotermitidae**) . . . ***Paraneotermes simplicicornis*** (Banks)

1b Only two prominent veins along leading edge of wing (costal margin and radial sector), with at most a few nearly perpendicular veins between near apex; fontanelle present on top of head (inconspicuous in ***Coptotermes***, ***Anoplotermes***, and ***Tenuirostritermes***); postclypeus convex and divided at midline **2**

2a **(1b)** Forewing scale about same size as hindwing scale, at least not overlapping; wings slightly opaque, veins posterior to radial sector (RS) lightly pigmented, conspicuous; pronotum of nonreproductives much narrower than head, raised in front, saddle-shaped; southwestern (**Termitidae**) **11**

2b Forewing scale distinctly larger than, and overlapping, hindwing scale; wings clear, veins posterior to RS unpigmented, difficult to distinguish; pronotum of nonreproductives normal, shieldlike; widespread (**Rhinotermitidae**) **3**

3a **(2a)** Fontanelle conspicuous; median vein absent, wing margin not fringed with fine setae; light yellow-brown; TL about 11 mm (Fig. 33.5); restricted to coastal southeastern Florida, frequent. . . . . ***Prorhinotermes simplex*** (Hagen)

3b Fontanelle inconspicuous, usually a raised or depressed point; median vein present **4**

4a **(3b)** Wings densely covered with very fine hairs (setae); antennae with more than 18 segments; light yellow-brown; TL about 14–15 mm (Fig. 33.4); introduced: restricted to Houston, Galveston, Beaumont, Texas; Lake Charles and delta area, Louisiana; Biloxi, Mississippi; Mobile, Alabama; Memphis, Tennessee; Charleston, South Carolina; coastal Broward and northern coastal Dade counties, Florida (R. H. Scheffrahn, personal communication), frequent. . . . . ***Coptotermes formosanus*** Shiraki

4b Wings not densely covered with fine setae; antennae with less than 18 segments; yellow-brown to black **5**

5a **(4b)** Pale yellowish brown; wings tinted brown with faint brownish veins; margins fringed with fine setae; TL about 9.0–10.5 mm (Fig. 33.6); southeastern California, south-central and southwestern Arizona (Sonoran Desert), common . . . . ***Heterotermes aureus*** (Snyder)

5b Yellowish brown to black; wings almost colorless to translucent brown or gray; not fringed with fine setae; most of six species widespread (***Reticulitermes***) **6**

6a **(5b)** Light yellowish brown; wings less than 6 mm long and 1.75 wide; TL about 8 mm; southeastern (Fig. 33.21), frequent . . . . ***Reticulitermes hageni*** Banks

6b Dark brown to black **7**

7a **(6b)** Tibiae plainly darkened or black, contrasting with pale tarsi . **10**

7b Tibiae pale as tarsi **8**

8a **(7b)** Ocelli about equal to, generally more than, one diameter away from compound eye; wings tinted gray to brown; forewing usually darker; TL about 10 mm, wing more than 6 mm long; most common termite, eastern half of United States (Fig. 33.18) . . . . . . ***Reticulitermes flavipes*** Kollar

8b Ocelli plainly less than one diameter from compound eye . . **9**

9a **(8b)** Smaller, TL about 8 mm; wings less than 6 mm long; brownish black to black; southeast, frequent (Fig. 33.21) . . . . ***Reticulitermes virginicus*** Banks

9b Larger, TL about 9 mm; wings more than 6.5 mm long, less than 2 mm wide; dark brown; vicinity Boston, Massachusetts, and dune areas southern shores Lake Michigan, frequent (Fig. 33.19) . . . ***Reticulitermes arenincola*** Goellner

10a **(7a)** Smaller, TL about 9 mm; brownish black; wing membrane brownish gray; tibiae slightly darkened; far west, common (Fig. 33.19) . ***Reticulitermes hesperus*** Banks

10b Larger, TL about 9.0–10.5 mm; brownish black; wing membrane almost colorless or tinged with brown; tibiae definitely blackish (Fig. 33.1); most common termite, midwest and western United States (Fig. 33.20) . . . . ***Reticulitermes tibialis*** Banks

11a **(2a)** Fontanelle conspicuous, lighter than head capsule; fore tibiae with three apical spines **12**

11b Fontanelle inconspicuous; fore tibiae with two apical spines . . **20**

12a **(11a)** Large species, TL about 13–16 mm; southwestern (***Gnathamitermes***) **13**

12b Small species, TL less than 13 mm; southwestern (***Amitermes***) . **14**

13a **(12a)** Shining mahogany brown; fontanelle roughly circular (Fig. 33.16*a,c*); southern California, Nevada, and Arizona, very common . . . . ***Gnathamitermes perplexus*** (Banks)

13b Dark brownish black; fontanelle oval (Fig. 33.16*d*); southern New Mexico and Texas, very common . . . . ***Gnathamitermes tubiformans*** (Buckley)

14a **(12b)** Light yellowish brown to dark brown **15**

14b Black or blackish **18**

15a **(14a)** Light yellowish brown or light brown; smaller, TL less than 10 mm **16**

15b Medium to dark brown; larger, more than 10 mm **17**

16a **(15a)** Light golden brown, head conspicuously darker, postclypeus yellow; head width more than 0.95 mm; TL about 9.5 mm; soldier and worker unknown (Fig. 33.11); south-central Arizona (Sonoran Desert), rare . . . . ***Amitermes pallidus*** Light

16b Light brown; head width less than 0.90 mm; TL about 7.5 mm (Fig. 33.12); known only from vicinity of Tucson and Grand Canyon (Colorado River at Cardenas Creek), Arizona, and San Antonio, Texas, very rare . . . . ***Amitermes parvulus*** Light

17a **(15b)** Larger, TL about 11.5–13.0 mm; head width more than 1.05 mm; postclypeus yellowish brown (Fig. 33.13); southern California and Arizona (Colorado and Sonoran Deserts), rare . . . . ***Amitermes silvestrianus*** Light

17b Smaller, TL about 10–11.5 mm; head width less than 1.0 mm; postclypeus pale yellow (Fig. 33.15); widespread southern California and Nevada to Brownsville, Texas, frequent . . . . ***Amitermes wheeleri*** (Desneux)

18a **(14b)** Very small, TL about 7.5–8.5 mm; head width 0.84–0.90 mm; dull grayish black (Fig. 33.10); widespread, southern California and Nevada to Brownsville, Texas, frequent ***Amitermes minimus*** Light

18b Larger, TL about 10 mm **19**

19a **(18b)** Smoky black; TL about 9.0–10.5 mm (Fig. 33.9); southern California, Nevada, and Arizona (Colorado, Mojave, and Sonoran Deserts), rare . . . . ***Amitermes coachellae*** Light

19b Dark, shining black; TL about 11–12 mm (Fig. 33.8); southern California and Arizona (Colorado and Sonoran Deserts), rare . . ***Amitermes emersoni*** Light

20a **(11b)** Large, TL about 13–14 mm; dark brown with pale brown wings; antennae 14-segmented, third segment very short; ocelli prominent, compound eyes hemispherical, protruding; mandibles long, narrow, protruding to give front of head a pointed look (Fig. 33.7); southern Texas, rare . . . . ***Anoplotermes fumosus*** (Hagen)

20b Very large, TL over 20 mm; dark brown, wings tinted brown; antennae 15-segmented, third segment longer than second; ocelli, eyes, mandibles normal; southwestern (***Tenuirostritermes***) . . . **21**

21a **(20b)** Ocelli less than one-half long diameter from compound eyes; TL about 20–23 mm (Fig. 33.17); southeastern Arizona and southwestern New Mexico, rare ***Tenuirostritermes tenuirostris*** (Desneux)

21b Similar; except ocelli fully one-half long diameter from compound eyes; central and southern Texas, rare . . . . ***Tenuirostritermes cinereus*** (Buckley)

Key to Soldiers

1a Head bulbous, drawn out in front of constriction into long snout or nasus; mandibles minute, hidden in dorsal view; soldiers (and workers) small, brown, with antlike behavior; southwestern; (**Termitidae**, in part) (***Tenuirostritermes***) **2**

1b Head without a long snout; mandibles conspicuous in dorsal view **3**

2a **(1a)** Smaller, TL about 2.8 mm (of worker about 3.5 mm); head blackish brown; antennae 12-segmented; central and southern Texas, rare . . . . ***Tenuirostritermes cinereus*** (Buckley)

2b Larger, TL about 3.0–3.5 mm (of worker about 4.0–4.5 mm); head reddish brown; antennae of largest soldiers 13-segmented (Fig. 33.17); southeastern Arizona, southwestern New Mexico, and western Texas, rare . . . ***Tenuirostritermes tenuirostris*** (Desneux)

3a **(1b)** Pronotum of soldiers (and workers) normal, shieldlike **4**

3b Pronotum of soldiers (and workers) small, saddle-shaped; (**Termitidae**) **5**

4a **(3a)** Mandibles with several teeth of varying size on left, fewer on right, short, heavy, black; head longer than wide, of variable proportions, heavily sclerotized, reddish brown; antennae 14-segmented, third segment not enlarged; TL about 8–11 mm (Fig. 33.3); southwestern, frequent (**Kalotermitidae**) . . . . ***Paraneotermes simplicicornis*** (Banks)

4b Mandibles with no teeth (**Rhinotermitidae**) **14**

5a **(3b)** Soldier caste absent, but worker fairly distinctive; total length (TL) 4–5 mm; postclypeus swollen and divided at midline; fore tibiae swollen; abdomen dull, dirty gray, "baglike"; head, thorax and legs off-white (Fig. 33.7); southern Texas, rare . . . . ***Anoplotermes fumosus*** (Hagen)

5b Soldier caste present, mandibles each with one prominent tooth **6**

6a **(5b)** Mandibles long, straight, slender, incurved at tips, marginal teeth relatively small and near middle; southwestern (***Gnathamitermes***) **7**

6b Mandibles short, strongly curved or bowed, teeth in various positions; southwestern (***Amitermes***) **8**

7a **(6a)** Acute teeth below middle of mandible, may be prominent, inconspicuous, unequal; TL about 5–6 mm; head pale yellow; abdomen mottled gray (Fig. 33.16*a,c*); southern California, Nevada, and Arizona, very common . . . . . . ***Gnathamitermes perplexus*** (Banks)

7b Obtuse teeth cut into margin near middle of mandible; otherwise similar to 7 (Fig. 33.16); southern New Mexico and Texas, very common. . . . . ***Gnathamitermes tubiformans*** (Buckley)

8a **(6b)** Teeth near base of mandibles. **9**

8b Teeth near or beyond middle of mandibles; barblike **12**

9a **(8a)** Teeth large, conical, directed inward; head width less than 1.0 mm; TL about 4–5 mm (Fig. 33.15); widespread, southern California and Nevada to Brownsville, Texas, frequent . . . . ***Amitermes wheeleri*** (Desneux)

9b Teeth acute, cut out of margin and set off from basal hump by deep notch. **10**

10a **(9b)** Large, head width more than 1.0 mm; TL about 3.5 mm (Fig. 33.9); southern California, Nevada, and Arizona (Colorado, Mojave, and Sonoran Deserts), rare . . . ***Amitermes coachellae*** Light

10b Small, head width less than 1.0 mm **11**

11a **(10b)** Head obviously longer than broad, head width 0.78–0.90 mm; mandibles long, slender; TL about 3.5 mm (Fig. 33.10); widespread, southern California and Nevada to Brownsville, Texas, frequent . . . . ***Amitermes minimus*** Light

11b Head nearly "square", head width 0.78–0.87 mm; mandibles short, broad; TL about 3.5 mm (Fig. 33.12); known only from vicinity Tucson, Arizona, and San Antonio, Texas, very rare . . . ***Amitermes parvulus*** Light

12a **(8b)** Teeth near middle of mandible; head width about 1.25 mm; TL about 3.5 mm (Fig. 33.13); southern California and Arizona (Colorado and Sonoran Deserts), rare . . ***Amitermes silvestrianus*** Light

12b Teeth two-thirds of distance from base to tip of mandible. . . **13**

13a **(12b)** Mandibles long, slender; head width less than 1.0 mm; TL about 4.0 mm (Fig. 33.8); southern California and Arizona (Colorado and Sonoran Deserts), rare. . . . . . . ***Amitermes emersoni*** Light

13b Mandibles short, heavy; head width 1.25–1.45 mm; TL about 4–6 mm; alate unknown (Fig. 33.14); southern California and Arizona (Colorado, Mojave, and Sonoran Deserts), rare . . . . ***Amitermes snyderi*** Light

14a **(4b)** Fontanelle opening beneath a short hood just above clypeus, head narrowed anteriorly; TL about 5.0–5.5 mm (Fig. 33.4); introduced: restricted to Houston, Galveston, Beaumont, Texas; Lake Charles and delta area, Louisiana, frequent . . . . ***Coptotermes formosanus*** Shiraki

14b Fontanelle one-fourth to one-third of head length behind clypeus **15**

15a **(14b)** Fontanelle conspicuous; head distinctly narrowed anteriorly; mandibles long, slender; labrum broadly rounded apically; TL about 5.5–6.5 mm (Fig. 33.5); restricted to coastal southeastern Florida, frequent . . . . ***Prorhinotermes simplex*** (Hagen)

15b Fontanelle inconspicuous; head parallel-sided, longer than wide; labrum acutely rounded **16**

16a **(15b)** Mandibles fairly long (usually longer than head width), slender, slightly incurved at tips; TL about 5.0–6.0 mm; (of worker 4.5–5.0 mm) (Fig. 33.6); southeastern California, south-central and southwestern Arizona (Sonoran Desert), common . . . . ***Heterotermes aureus*** (Snyder)

16b Mandibles short (usually shorter than head width), stout, rather strongly incurved at tips (left somewhat heavier than right); most of six species widespread (***Reticulitermes***) **17**

17a **(16b)** Gula less than twice as wide in front (maximum width) as in middle (minimum width); TL about 5.5–7.0 mm (Fig. 33.1); most common termite, midwest and western United States (Fig. 33.20) . . ***Reticulitermes tibialis*** Banks

17b Gula at least twice as wide in front as in middle **18**

18a **(17b)** Head at least twice as long as wide; TL about 5.5–6.5 mm; far west, common (Fig. 33.19) . . . . ***Reticulitermes hesperus*** Banks

18b Head less than twice as long as wide **19**

19a **(18b)** Head about 1.5 times as long as wide in front; pronotal width 0.81–1.10 mm; TL about 5.5–6.7 mm; most common termite, eastern half of United States (Fig. 33.18) ***Reticulitermes flavipes*** Kollar

19b Head about same proportions, but smaller; pronotal width 0.67–0.81 mm; TL less than 6.0 mm **20**

20a **(19b)** Distribution limited to vicinity Boston, Massachusetts, and dune areas southern shores Lake Michigan (Fig. 33.19), frequent; maximum gular width more than twice its minimum width; TL 4.5–4.9 mm . . . . ***Reticulitermes arenincola*** Goellner

21a **(20b)** Gula widens rather abruptly from middle, anteriorly, maximum width 2.5 times its minimum width; usually larger and with mandibles slightly more "hooked" than ***R. hageni***; TL about 4.5–5.5 mm; southeastern (Fig. 33.21), frequent ***Reticulitermes virginicus*** Banks

21b Gula widens gradually from middle, anteriorly, maximum width twice its minimum width; usually smaller and with mandibles less "hooked" than ***R. virginicus***; TL about 4.5–5.0 mm; southeast (Fig. 33.21), frequent . . . . ***Reticulitermes hageni*** Banks

References

Anon. 1980. *Termite Abstracts.* Taylor & Francis, London. **Bodine, M. C. 1973.** *The Impact of the Termite Gnathamitermes tubiformans on Rangeland Herbage Dynamics.* M.S. thesis, Texas Tech University, Lubbock, Tex. 39 pp. **Bouillon, A. 1969.** *Rev. Ecol. Biol. Sol* 6:469. **Breznak, J. A. 1984.** Biochemical aspects of symbiosis between termites and their intestinal microbiota. Pp. 173–203 in J. M. Anderson, A. D. M. Rayner and D. W. H. Walton (eds.). *Invertebrate–Microbial Interactions.* Cambridge University Press, Cambridge. **Brian, M. V. (ed.). 1978.** *Production Ecology of Ants and Termites.* Cambridge

University Press, Cambridge. 409 pp. **1983.** *Social Insects.* Chapman & Hall, London. 377 pp. **Ernst, E. and R. L. Araujo. 1986.** *A Bibliography of Termite Literature, 1966–1978.* Wiley, New York. 903 pp. **Gay, F. J., T. Greaves, F. G. Holdaway and A. H. Wetherly. 1955.** *Standard Laboratory Colonies of Termites for Evaluating the Resistance of Timber, Timber Preservatives, and Other Materials to Termite Attack.* Bull. 277. CSIRO, East Melbourne, Australia. 60 pp. **Haarlov, N. 1947.** *J. Anim. Ecol.* 16:115. **Haverty, M. I. and W. L. Nutting. 1975a.** *Environ. Entomol.* 4:480. **1975b.** *Insectes Soc.* 22:93. **Haverty, M. I., W. L. Nutting and J. P. LaFage. 1975.** *Environ. Entomol.* 4:105. **Hermann, H. R. (ed.). 1979.** *Social Insects,* Vol. 1. Academic Press, New York. 437 pp. **Honigberg, B. M. 1970.** Protozoa associated with termites and their role in digestion. Pp. 1–36 in K. Krishna and F. Weesner (eds.). *Biology of Termites,* Vol. 2. Academic Press, New York. **Howard, R. W. and M. I. Haverty. 1980.** *Environ. Entomol.* 9:458. **Howard, R. W., E. J. Mallette, M. I. Haverty and R. V. Smythe. 1981.** *Psyche (Camb.)* 88:75. **Howard, R. W., S. C. Jones, J. K. Mauldin and R. S. Beal. 1982.** *Environ. Entomol.* 11:1290. **Hungate, R. E. 1941.** *Ann. Entomol. Soc. Am.* 34:467. **Johnson, K. A. and W. G. Whitford. 1975.** *Environ. Entomol.* 4:66. **Jones, S. C. and W. L. Nutting. 1989.** Foraging ecology of subterranean termites in the Sonoran Desert. In J. O. Schmidt (ed.). *Interactions among Plants and Animals in the Western Deserts.* University of New Mexico Press, Albuquerque, N. Mex. (in press). **Kemp, P. B. 1955.** *Bull. Entomol. Res.* 46:113. **Krishna, K. and F. Weesner (eds.). 1969, 1970.** *Biology of Termites,* 2 Vols. Academic Press, New York. 598 and 643 pp. **LaFage, J. P. and W. L. Nutting. 1978.** Nutrient dynamics of termites. Pp. 165–232. in M. V. Brian (ed.). *Production Ecology of Ants and Termites.* Cambridge University Press, Cambridge. **LaFage, J. P., W. L. Nutting and M. I. Haverty. 1973.** *Environ. Entomol.* 2:954. **LaFage, J. P., M. I. Haverty and W. L. Nutting. 1976.** *Sociobiology* 2:155. **Lee, K. E. and T. G. Wood. 1971a.** *Pedobiologia* 11:376. **1971b.** *Termites and Soils.* Academic Press, New York. 251 pp. **Lepage, M. 1972.** *Terre Vie* 26:388. **Light, S. F. 1932.** Contribution toward a revision of the American species of Amitermes Silvestri. Pp. 355–414 in *Univ. Calif. Publ. Entomol.,* Vol. 5, No. 17. University of California, Berkeley, Calif. **MacKay, W. P., J. C. Zak and W. G. Whitford. 1989.** The natural history and role of subterranean termites in the northern Chihuahuan Desert. In J. O. Schmidt (ed.), *Interactions among Plants and Animals in the Western Deserts.* University of New Mexico Press, Albuquerque, N. Mex. (in press). **McMahan, E. A. and J. A. L. Watson. 1977.** *J. Aust. Entomol. Soc.* 16:455. **Miller, E. M. 1949.** *A Handbook of Florida Termites.* Tech. Series. University of Miami Press, Coral Gables, Fla. 30 pp. **1969.** Caste differentiation in the lower termites. Pp. 233–282 in K. Krishna and F. Weesner (eds.). *Biology of Termites,* Vol. 1. Academic Press, New York. **Myles, T. G. and W. L. Nutting. 1988.** *Q. Rev. Biol.* 63:1. **Nel, J. J. C. and P. H. Hewitt. 1969.** *Nature (Lond.)* 223:862. **Noirot, Ch. and C. Noirot-Timothée. 1969.** The digestive system. Pp. 49–88 in K. Krishna and F. Weesner (eds.). *Biology of Termites,* Vol. 1. Academic Press, New York. **Nutting, W. L. 1969a.** Flight and colony foundation. Pp. 233–282 in K. Krishna and F. Weesner (eds.). *Biology of Termites,* Vol. 1. Academic Press, New York. **1969b.** *Pan-Pac. Entomol.* 45:320. **1979.** *Sociobiology* 4:141. **Nutting, W. L. and M. I. Haverty. 1976.** *Sociobiology* 2:145. **Nutting, W. L., M. I. Haverty and J. P. LaFage. 1987.** *J. Arid Environ.* 12:233. **Potrikus, C. J. and J. A. Breznak 1980.** *Insect Biochem.* 10:19. **1981.** *Proc. Natl. Acad. Sci. USA* 78:4601. **Salt, G., F. S. Hollick, F. Raw and M. V. Brian. 1948.** *J. Anim. Ecol.* 17:180. **Sands, W. A. 1965.** *J. Anim. Ecol.* 34:557. **Schaefer, D. A. and W. G. Whitford. 1981.** *Oecologia (Berl.)* 48:277. **Snyder, T. E. 1954.** *Order Isoptera. The Termites of the United States and Canada.* Tech. Bull. National Pest Control Association, New York. 64 pp. **1956.** *Annotated, Subject-Heading Bibliography of Termites 1350 B.C. to 1954 A.D.* Smithsonian Miscellaneous Collections 130, Publ. 4258. Smithsonian Institution, Washington, D.C. 305 pp. **1961.** *Supplement to the Annotated, Subject-Heading Bibliography of Termites 1955–1960.* Smithsonian Miscellaneous Collections 143, Publ. 4463. Smithsonian Institution, Washington, D.C. 137 pp. **1968.** *Second Supplement to the Annotated, Subject-Heading Bibliography of Termites 1961–1965.* Smithsonian Miscellaneous Collections 152, Publ. 4705. Smithsonian Institution, Washington, D.C. **Ueckert, D. N., M. C. Bodine and B. M. Spears. 1976.** *Ecology* 57:1273. **Waller, D. A. and J. P. LaFage. 1987.** Nutritional ecology of termites. Pp. 487–532 in F. Slansky, Jr., and J. G. Rodriguez (eds.). *The Nutritional Ecology of Insects, Mites, and Spiders.* Wiley, New York. **Weesner, F. M. 1965.** *The Termites of the United States, a*

Handbook. National Pest Control Association, Elizabeth, N.J. 70 pp. **1969.** External anatomy. Pp. 19–47 in K. Krishna and F. Weesner (eds.). *Biology of Termites,* Vol. 1. Academic Press, New York. **Wilson, E. O. 1971.** *The Insect Societies.* Belknap Press of Harvard University Press, Cambridge, Mass. 548 pp. **Wood, T. G. 1978.** Food and feeding habits of termites. Pp. 55–80 in M. V. Brian (ed.). *Production Ecology of Ants and Termites.* Cambridge University Press, Cambridge. **Wood, T. G. and W. A. Sands. 1978.** The role of termites in ecosystems. Pp. 245–292 in M. V. Brian (ed.). *Production Ecology of Ants and Termites.* Cambridge University Press, Cambridge. **Yamin, M. A. and W. Trager. 1979.** *J. Gen. Microbiol.* 113:417.

34

Insecta: Psocoptera

ALFONSO NERI GARCÍA ALDRETE

To the present, 60 species of Psocoptera have been recorded from ground litter in the continental United States (see Table 34.1); they are distributed in 28 genera of the three recognized suborders within the group. This account is based on material available to me or known from the literature to be present in the United States. Some of the species here dealt with are new to science and need to be named; it is likely that the number of litter psocids will increase as more collecting is done, specially in the western part of the country. Consequently, the identification keys here provided will have to be modified to incorporate the new additions.

External Morphology

Psocids are free-living, exopterygote Neoptera, ranging in size from about 1 to 12 mm. The head is large and mobile, with long, filiform antennae; the number of flagellomeres ranges from nine to about 50. Winged adults have usually three ocelli, and a bulbous postclypeus is distinctive of the order (Figs. 34.1, 34.2, 34.5, 34.7, 34.12, 34.13, 34.19, 34.24). The mandibles are strong and asymmetrical, and the laciniae are modified to form maxillary "picks" (Figs. 34.3, 34.14, 34.18, 34.24). The maxillary palps are four-segmented (Figs. 34.3, 34.17), the labial palps are reduced, and the hypopharynx is modified, having two ovoid lingual sclerites, connected by filaments to a cup-shaped sitophore sclerite (Fig. 34.4). The wings are membranous, held either flat or rooflike over the back. Venation is reduced (Fig. 34.31), and the wing surface may be hairy, scaly, or bare, the hairs being mostly confined to the veins and margin. The tarsi are two-

Table 34.1. Species of Psocoptera Known from Litter in the Continental United States

Species[a]	Probable Status	Distribution
Thylacella cubana (Banks) (Tg)	Casual	Southeastern U.S.
Nepticulomima sp. (Tg)	Casual	Florida
Echmepteryx hageni (Packard) (Tg)	Casual	Eastern U.S.
Echmepteryx intermedius Mockford (Tg)	Casual	Florida
Echmepteryx madagascariensis (Kolbe) (Tg)	Casual	Florida
Echmepteryx sp. (Tg)	Casual	Florida
Pteroxanium kelloggi Ribaga (Tg)	Primary	Northwest coast
Lepolepis occidentalis Mockford (Tg)	Primary	Eastern U.S.
Lepolepis sp. (Tg)	Primary	Florida
Cerobasis guestfalica Kolbe (Tg)	Casual	Eastern U.S.
Lepinotus reticulatus Enderlein (Tg)	Primary	Southern U.S.
Myrmicodipnella aptera Enderlein (Tg)	Primary	California
Rhyopsocus micropterus Mockford (Tg)	Primary	California
R. phillipsae Sommerman (Tg)	Casual	Georgia, Florida, Texas
R. squamosus Mockford and Gurney (Tg)	Primary	Texas
Psyllipsocus oculatus Gurney (Tg)	Casual	Texas
Belaphotroctes alleni Mockford (T)	Casual	Texas, New Mexico
B. badonneli Mockford (T)	Casual	Florida, Texas
B. ghesquierei Badonnel (T)	Casual	Florida
B. hermosus Mockford (T)	Casual	Texas
Liposcelis bostrychophilus Badonnel (T)	Primary (part)	Georgia, Florida, Texas
L. corrodens Heymons (T)	Casual	Across northern U.S.
L. deltachi Sommerman (T)	Casual	Texas
L. entomophilus Enderlein (T)	Primary (Part)	Georgia, Florida, Illinois
L. formicarius Hagen (T)	Primary	California, Idaho, Colorado
L. liparus Broadhead (T)	Primary (part)	Arizona, Texas, New Mexico, Nebraska
L. nasus Sommerman (T)	Casual	Texas
L. prenolepidis Enderlein (T)	Primary	California
L. simulans Broadhead (T)	Casual	Indiana, Idaho, New Mexico, Colorado, Washington, California
L. terricolis Badonnel (T)	Casual	California, Texas, New York, Illinois
L. triocellatus Mockford (T)	Primary	California
L. villosus Mockford (T)	Primary	California
Nanopsocus oceanicus Pearman (T)	Primary (part)	Northeastern U.S.
Tapinella maculata Mockford and Gurney (T)	Casual	Texas
Sphaeropsocus sp. (T)	Primary	California
Epipsocus crosbyanus (Chapman) (P)	Primary	Eastern U.S.
E. lepicidinarius (Chapman) (P)	Casual	Eastern U.S.
Caecilius flavidus (Stephens) (P)	Secondary	Eastern U.S.
C. maritimus Mockford (P)	Primary (part)	California
C. nadleri Mockford (P)	Primary	Eastern U.S.

Table 34.1. (*Continued*)

Species[a]	Probable Status	Distribution
C. posticus Banks (P)	Primary	Eastern U.S.
C. sommermanae Mockford (P)	Secondary	Eastern U.S.
C. totonacus Mockford (P)	Primary	New Mexico, Colorado
Graphopsocus cruciatus (L.) (P)	Secondary	East and west coasts
Teliapsocus conterminus (Walsh) (P)	Secondary	Eastern U.S., Wyoming, Washington
Polypsocus corruptus (Hagen) (P)	Secondary	Eastern and Northwestern U.S.
Lachesilla pallida (Chapman) (P)	Primary	Eastern U.S.
L. pedicularia (L) (P)	Casual	Eastern U.S.
L. D-5 (P)	Casual	Florida
Ectopsocus briggsi McLachlan (P)	Secondary	West coast
E. meridionalis Ribaga (P)	Casual	Eastern U.S.
E. vachoni Badonnel (P)	Primary	Southern U.S.
Ectopsocopsis cryptomeriae Enderlein (P)	Secondary	Southern U.S.
Peripsocus alachuae Mockford	Casual	Florida
P. quadrifasciatus (Harris) (P)	Casual	Eastern U.S.
Archipsocus nomas Gurney (P)	Casual	Florida
Archipsocus panama Gurney (P)	Casual	Florida
Hemipsocus pretiosus Banks (P)	Casual	Florida
Trichadenotecnum alexanderae Somm. (P)	Casual	Eastern U.S.
Indiopsocus sp. (P)	Casual	Florida

[a] Tg, Trogiomorpha; T, Troctomorpha; P, Psocomorpha.

or three-segmented in adults and always two-segmented in nymphs; besides this character, nymphs can be recognized by the reduction of meso- and metathorax, by not having a recognizable external genitalia nor coxal organs and by having, in winged species, rigid, thickened, not articulated wing buds of variable length according to the instar. Frequently, the posterior coxae have a rasp presumed to be stridulatory. Cerci are absent and, generally, the abdominal terga 8–10 are fused to form a sclerotized semiring, the clunium. The abdomen ends in an epiproct and two paraprocts, regarded as remnants of the eleventh segment. In winged forms each paraproct has a sense tubercle with a variable number of trichobothria arising from it. Sexual dimorphism is common; it always affects the external genitalia, frequently affects eye size, males having larger eyes, and sometimes affects general body size and coloration, and occasionally involve wing dimorphism, one of the sexes being macropterous and the other being apterous or micropterous. Psocids are primitive hemipteroids, their nearest relatives being the Phthiraptera. Workers that have most contributed to the knowledge of the Psocoptera in this country are G. Enderlein, H. A. Hagen, N. Banks, P. J. Chapman, A. B. Gurney, K. M. Sommerman,

E. L. Mockford, and G. E. Eertmoed. Some important references for the study of this group are the following: Smithers (1–3), Badonnel (4), Chapman (5), Roesler (6), Günther (7), and New (8).

Reproduction

This subject was reviewed by New (9) and by Mockford (10). Psocids reproduce either sexually or asexually; parthenogenesis is common, some of its cases involving thelytokous populations of bisexual species. Obligatory and facultative thelytoky are known in psocids; these phenomena were reviewed by Mockford (11).

Courtship behavior has been studied in only a few species of psocids, so the spectrum of this behavior for the order is largely unknown. In some species it involves running about and wing fluttering by the male and antennal vibrating on the potential mate by one or both sexes; in some cases the male runs over the female, pushes under her from in front, and then copulate, facing in the same direction for a variable period of time. After this, in some species the female turns 180°, so that the copulating pair faces opposite directions (4,9,10,12,13). Copulation may last from a few seconds to about half an hour (10); longer periods may involve the transfer of a spermatophore (13). Copulatory positions end to end, male above female, and female above male are known in the order, the position seeming to be constant in any species (10). Klier (13) studied the courtship and copulation in several European species.

The life cycle of psocids consists, typically, of an egg stage, six immature stages, and the adult stage. The eggs are deposited singly or in groups; they are commonly cemented to the substrate and in some species are covered by threads spun by the female. In some cases, they are encrusted with debris. The duration of the different stages in the life cycle varies inversely with the temperature at which they have developed (4,14).

Ecology

Mockford (10) classified the principal habitats of psocids into six series: (a) living foliage, (b) dead foliage, (c) ground litter, (d) bark of trees, (e) rock surface, and (f) human habitation; he noted that the nests of birds and mammals fit into the second or third series; it seems that no psocids occur in the soil proper, being there only as strays from the litter above. It is possible sometimes to find adults and nymphs of litter species at several meters depth. New (15) divided the psocids found in leaf litter into three ecological groups: (a) primary litter dwellers, which spend their whole life history in the litter and do not frequent other habitats; (b) secondary litter dwellers, which have generations in two different habitats, one being litter;

and (c) casual litter dwellers, which do not normally breed in the litter and are infrequently found there.

The third category includes individuals of species that normally occur in other habitats and fall occasionally to the ground; there they may either return to their original habitat or die. The secondary litter dwellers are species that inhabit foliage and present a seasonal habitat shift. The last generation of the year is completed in the litter by individuals of various life-history stages that fall to the ground with deciduous leaves to overwinter or to lay overwintering eggs. The following spring the adults invade the new foliage of trees and shrubs after having completed the first generation in the litter. This category has been best studied in temperate regions. Primary litter dwellers often present morphological characteristics associated with the habitat: alary polymorphism occurs in higher frequency than in the order as a whole. The usual form involves males macropterous and females brachypterous, although in some cases the males are micropterous and the females are polymorphic. Associated with the brachypterous condition is the poor development of thoracic structures (notal lobes and sutures, mesoprecoxal bridges, and trochantins), ocelli, ctenidia of the tarsal segments, trichobothria of the sensory fields of paraprocts, as well as retention of nymphal duplex paraproctal spines and setae in the adults. Also, the coxal organ of the hind legs is generally absent. Thus the facies of primary litter dwellers is often highly neotenic. The probable status of the species occurring in litter in the United States is indicated in Table 34.1. Mockford (16) reviewed polymorphism in psocids and recorded that of 1400 species known in 1965, 88 showed alary polymorphism. The same author (10) presented data to show that selection must exist against development of macroptery in primary litter psocids; he also analyzed the distribution of primary litter species by higher taxonomic categories and the number of primary litter species in each category in relation to the total number of species known; he found that of about 1800 species of Psocoptera, 129 occur in litter, 92 of which are primary litter dwellers. The distribution of litter dwellers in the suborders of Psocoptera is as follows: in the Trogiomorpha, of a total of 161 species, 27 occur in litter, 16 being probably primary; 53 of the 217 species of Troctomorpha occur in litter, and 45 of them are believed to be primary dwellers. In the Psocomorpha, the largest suborder, of 1500 species, 53 have been recorded in litter, 30 of which are thought to be primary dwellers. Since the Trogiomorpha and the Troctomorpha may be more ancient than the Psocomorpha, and since they have relatively more primary litter dwellers, Mockford (10) suggested that for the whole order, ground litter is one of the primitive habitats. He also suggested that for some groups litter is a specialized habitat arrived at via dwelling in dried hanging leaves.

There are some associations of psocids with vertebrates; it is pertinent to cite them here since most of the species involved also occur in litter. Gurney (17) recorded psocids infesting chinchillas, puppies, and human hair; Pearman (18) recorded eight species of psocids collected on *Rattus* and *Mastomys* from Tanzania and St. Helena. Badonnel (19) mentioned 12 species of psocids from the fur or nests of nine species of mammals in Angola, one fissiped, one insectivore, six rodents, and one lemurid. In the United States, Mockford (20) cited four species of psocids from sleeping nests of *Neotoma fuscipes* in California. Processing of *Neotoma floridana* and squirrel nests in Tullgren funnels in Florida has yielded five and six species of psocids, respectively. Psocoptera from nests of birds have been recorded by Hicks (21), Rapp (22), Wlodarczyk (23), and Wlodarczyk and Martini (24). Mockford (25) recorded four species of psocids from the plumage of migratory birds. I have unpublished records of six species of psocids from hanging nests of icterids in southern Mexico. These associations are probably not obligate. The material used to construct the nests (dead leaves and branches principally) constitutes a suitable habitat for psocids, and they have taken advantage of such a situation, somewhat similar to the ground litter habitat. The phoretic implications of these relationships are obvious, as pointed out by Mockford (25). Concerning associations with other insects, some species are frequently present in wasps, ants, and termite nests (26; personal observation).

Psocids feed on algae, lichens, particles of organic debris, and fungal mycelium and spores. Some species can be cultured on dry yeast. Examination of the digestive tract contents of *Epipsocus crosbyanus* showed predominance of fungal hyphae and spores in addition to organic debris. Digestive tract contents of other litter species have not been investigated, nor has the role that psocids may play as litter decomposers.

Collection and Mounting Techniques

Psocids can be extracted from litter by placing it on a sifter and shaking it several times over a cloth; the individuals are then taken from the cloth either with an aspirator or a fine brush; if they are going to be preserved, they may be placed directly in 80% ethyl alcohol. Also, litter samples can be processed in a number of ways to extract the psocids (27–29). Edwards and Fletcher (29) compared several extraction methods, both mechanic and dynamic, for terrestrial arthropods and found that the best general method is a Tullgren funnel of the Macfadyen type, with a gradient of temperature and moisture through the sample. They compared several extraction methods on three soil types and three habitats and found that all methods tested were suitable to extract psocids on sandy and clay or loam soils, and recommended for peat soil the following methods: (a) simple funnel with heat, (b) Kempson infrared extractor, and (c) Macfadyen air-conditioned funnel. Although these methods refer to extraction from soil samples, the results can probably be extended to extraction from litter. New (15) found that a Kempson Bowl extractor was more efficient than Tullgren funnels to extract psocids from litter.

Identification of psocids requires that they be mounted temporarily or permanently for microscopic examination. Temporary mounts can be made placing the specimen on a bed of cotton strands covered by glycerine on a concave slide; after the examination is done, the specimen is returned to alcohol. More frequently, permanent mounts are necessary, and that requires dissection of the specimen, under the dissecting microscope usually at 70–80×, and utilizing fine needles and "watchmaker's" forceps. The choice of the mounting medium and stain depends on the worker; I find mounting in chloral gum medium quite convenient, since it does not require clearing of the parts in KOH or staining. The specimen is dissected in lactophenol and the parts are mounted directly in the medium; observation is greatly facilitated due to the low refractive index of the medium. It has the disadvantage that since it contains H_2O, evaporation of this substance through time may damage the preparation; it thus becomes of primary importance to seal the slides after drying them on a warming tray at 34°C for at least 24 hours. The following directions may be useful in making permanent slides: specimens preserved dry must be relaxed prior to dissection either by placing in warm water with a small amount of detergent during 24 hours to rehydrate and restore flexibility or by placing in a humid chamber. If the specimen is too hardened, it may be placed in acetic chloral (chloral hydrate, acetic acid, and water in proportions 4 : 3 : 3) during 12 hours or as needed, to be softened. The antenna, wings, and legs of one side of the body are separated from it and are transferred to a drop of the mounting medium on a slide, where they are arranged and covered with a coverslip. To remove the mouthparts it is convenient to separate the head from the body, orient it postclypeus down, and holding it in place with one needle, separate the labium, maxillae, and sitophore sclerite, cutting carefully with the other needle. The labrum-epipharynx and the mandibles are also separated from the head capsule, and the different pieces are transferred to a drop of medium on the same slide, arranged, and covered. The female genitalia are prepared as follows. The terminal third of the abdomen is sectioned and the subgenital plate is freed by inserting a needle between it and the ninth sternum and cutting at the sides. Then the clunium is cut longitudinally at either side next to the gonapophyses and ninth sternum. Thus the gonapophyses and ninth sternum form one piece and the clunium, epiproct, and paraprocts are separated as another. These parts are mounted in a third drop of medium on the same slide. Remains of the digestive tract, membranes, muscles, and the fecal pellet are removed. The male genitalia are similarly treated; the hypandrium is separated by cutting along each side and the phallosome is liberated from it and adjacent membranes. The clunium, epiproct, and paraprocts are flattened as in the case of the female. Thus for each sex the appendages, mouthparts, and genitalia are mounted separately on the same slide. Care should be taken to avoid excess medium and to flatten the pieces as much as possible without distorting them. Small, dorsoventrally flattened psocids (Liposcelidae, Pachytroctidae) can be mounted whole after clearing in lactophenol. Kühnelt (28) gives general preservation and mounting techniques for soil arthropods and includes the formula to prepare the chloral gum medium. New (8) described the procedure of dissection and permanent mounting of psocids, utilizing a choice of several stains and using Euparal as mounting medium. This technique also gives excellent results, except for the membranous parts, which are often deformed, due to dehydration.

Measurement and Identification

The parts to be measured must be held as horizontal as possible to minimize error; measurements of leg parts must be taken from condyle to condyle; the most important measures and their standard abbreviations are the following:

Head: IO is the minimum distance between the compound eyes in dorsal view of head; D is the greatest anteroposterior compound eye diameter and d is the greatest lateral compound eye diameter, all in dorsal view of head; PO is the ratio d/D; f_1, f_2, f_3, and so on, are lengths of the antennal flagellomeres.

Thorax: FW and HW are the lengths of fore- and hindwings, respectively; F, T, t_1, t_2, and so on, are the lengths, respectively of femur, tibia, and tarsomeres of posterior leg; t_1ct is the number of ctenidia on the first posterior tarsomere.

For the Liposcelidae the following measurements are also important: S I is the length of the longest lateral seta on the pronotum; S II is the length of the longest anterolateral marginal seta of mesonotum; Sa is the length of the longest dorsolateral seta of ninth abdominal tergum, and Se is the length of the longest seta on epiproct. Some important characters in psocid identification are the following:

Head: General color pattern, presence or absence of ocelli, shape of P_4 (fourth segment of maxillary palps); shape of labrum and presence or absence on it of sclerotized bands (compare Figs. 34.1, 34.5, 34.7, 34.12, and 34.13 with Fig. 34.24), and structure of the labium-hypopharynx (Fig. 34.4).

Thorax: General color pattern, position of the wings over the back, either flat or rooflike; wing shape and venation, wing color pattern, and presence or absence of scales or hairs on veins, membrane, and wing margins; leg structure, shape of femur, presence or absence of rows of setae on surface of femora and tibiae; shape of pretarsal claws, shape of pulvillus, and presence or absence of preapical denticles on claw.

Abdomen: By far, the external genitalia of both sexes, as well as the structure of epiproct and paraprocts provide the most useful and important characters in identification. The external genitalia show considerable variation among the three suborders of psocids, the maximum complexity and variability being presented within the Psocomorpha. For a general account of external genitalia in the Psocoptera, see Badonnel (30). The female has a subgenital plate that partially covers the gonapophyses. The latter are usually composed of three pairs of valves, but reduction from this number is common; the shape and configuration of all these structures, as well as the structure of the ninth sternum, have diagnostic value. In the male there is a simple or complex hypandrium overlying the phallosome. The latter is formed by the external and internal parameres and endophallus. The internal parameres generally articulate with the externals, and may join each other distally to form an aedeagus. There is great variation from this basic scheme, and here again, the peculiarities of all these pieces are valuable in species discrimination. In both sexes, the shape of the paraprocts and epiproct, presence or absence of prongs, chetotaxy, and number of trichobothria in the paraproctal sensory fields also aid in identification.

ACKNOWLEDGMENTS

I extend my gratitude to Dr. André Badonnel, Dr. Gary E. Eertmoed, and Dr. Edward L. Mockford, who all provided a critical review of the manuscript.

KEY TO SUBORDERS OF PSOCOPTERA

The three suborders of the Psocoptera are represented in the litter fauna of the continental United States. They can be separated with the following key.

1a Antennae with more than 20 segments, adults with three-segmented tarsi; pterostigma absent or not thickened; labial palps two-segmented, hypopharyngeal filaments separated throughout their length; gonapophyses reduced, only external valves well developed; paraprocts with a strong marginal spine (Fig. 34.21) . . . . **TROGIOMORPHA**

1b Antennae 13–17, if other than 13-segmented; distal flagellomeres secondarily ringed; hypopharyngeal filaments fused into a single filament before joining the sitophore; tarsi two- to three-segmented; gonapophyses variously developed; paraprocts without marginal spine **2**

2a Antennae 11- to 17-segmented, typically 15-segmented; some flagellar segments secondarily ringed; labial palps one- to two-segmented; tarsi three-segmented (seldom two-segmented); pterostigma not thickened. SGP (subgenital plate) with a T-shaped sclerite, sometimes reduced or absent **TROCTOMORPHA**

2b Antennae usually 13-segmented; labial pals one-segmented and reduced; tarsi two- to three-segmented; pterostigma thickened . . . . . **PSOCOMORPHA**

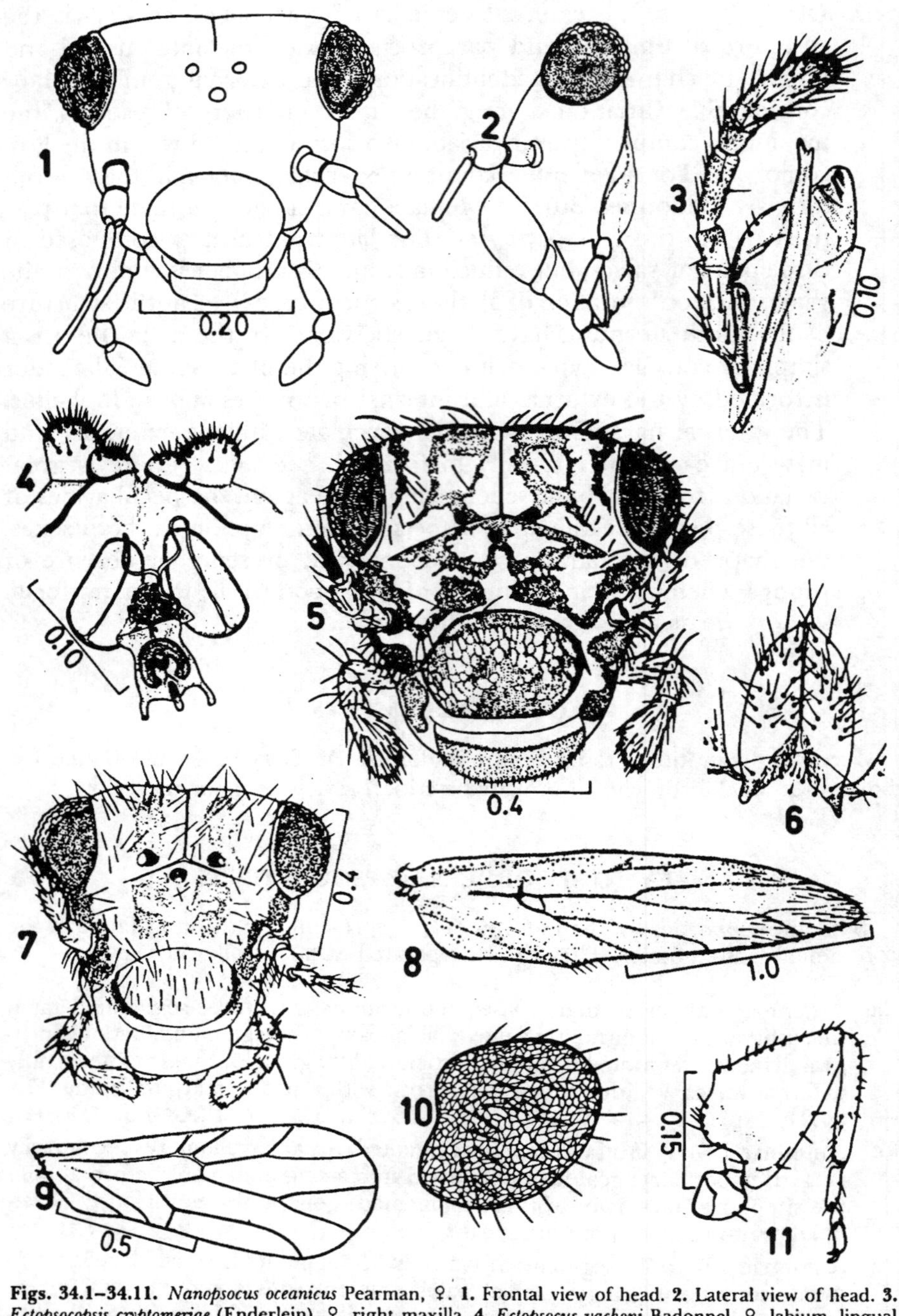

Figs. 34.1–34.11. *Nanopsocus oceanicus* Pearman, ♀. **1.** Frontal view of head. **2.** Lateral view of head. **3.** *Ectopsocopsis cryptomeriae* (Enderlein), ♀, right maxilla. **4.** *Ectopsocus vachoni* Badonnel, ♀, labium, lingual and sitophore sclerites. **5.** *Pteroxanium kellogi* (Ribaga), ♀, frontal view of head. **6.** *Lepolepis* sp., ♀, gonapophyses. **7.** *Lepolepis occidentalis* Mockford, ♀, frontal view of head. **8.** *L. occidentalis* Mockford, ♀, forewing. **9.** *Nanopsocus oceanicus* Pearman, ♀, forewing. 10. *Lepinotus reticulatus* Enderlein, ♀, forewing. **11.** *Liposcelis* sp., ♀, right posterior leg. Scales in mm. Figure 34.2 to same scale as Fig. 34.1. Figures 34.6 and 34.10 to same scale as Fig. 34.11.

Key to Species of Trogiomorpha

Sixteen species of Trogiomorpha are found in litter in the continental United States; seven are primary litter dwellers and the other species have been only casually found there (Table 34.1). Some of the species recorded are widely distributed (2); ***Myrmicodipnella aptera*** was described from a single specimen from an ant's nest in San Francisco, California (31), it may well be a species of ***Cerobasis***. The litter Trogiomorpha of the United States may be separated with the following key.

1a Body and wings clothed with scales or scalelike hairs; forewings usually elongated, pointed; tarsal claws with at least one preapical tooth 2

1b Body and wings not clothed with scales; forewings rounded; tarsal claws mostly without preapical teeth 10

2a Forewings short, broad, not reaching tip of abdomen; venation evanescent; if long winged with or without a small closed cell between M and Cu_1; hindwings with closed cell, if present, on distal end 3

2b Forewings reaching beyond tip of abdomen, venation distinct. Forewings with large closed cell above $M + Cu_1$; hindwings with closed cell, if present, on proximal end 5

3a Wing polymorphic; short winged forms without hindwings; scalelike forewings not reaching beyond fourth abdominal segment, tibiae without dark transverse bands 4

3b Long winged forms unknown; forewings broad, reaching beyond fourth abdominal segment, hindwings vestigial; tibiae with two wide dark bands; head marks (Fig. 34.5) (7,8) . . . . ***Pteroxanium kellogi*** (Ribaga)

4a Abdominal scales golden brown; head marks (Fig. 34.7); short- and long-winged forms present; forewing (Fig. 34.8); gonapophyses slender, narrowing distally (32). . . . . ***Lepolepis occidentalis*** Mockford

4b Abdominal scales dark brown, head marks not as above; only short-winged forms known; gonapophyses stout (Fig. 34.6) . . . . ***Lepolepis*** sp.

5a Gonapophyses slender, distally pointed 6

5b Gonapophyses stout; apices blunt (Fig. 34.15); body and wings with scalelike hairs; head with dark lateral bands from compound eyes to lower genae near postclypeus (33) . . . . ***Thylacella cubana*** (Banks)

6a Lacinial tips stout, trifid; hindwings without closed basal cell 7

6b Lacinial tips slender, bifid (Fig. 34.14); hindwings with small, closed basal cell; head fuscous, with wide U-shaped area on frons next to postclypeus ***Nepticulomima*** sp.

7a Tibiae banded 8

7b Tibiae unbanded 9

8a Head marks (Fig. 34.12); third valvulae long and slender (34) ***Echmepteryx hageni*** (Packard)

8b Head marks similar to ***E. hageni,*** but lacking a narrow band on each side paralleling margin of compound eye; third valvulae shorter than in ***E. hageni*** (34) . . . . ***Echmepteryx intermedia*** Mockford

9a Head marks (Fig. 34.13); forewings without distinct pigmented areas ***Echmepteryx*** sp.

9b Head unmarked, forewings with a large, dark longitudinal band; body pale (35) . . . . ***Echmepteryx madagascariensis*** (Kolbe)

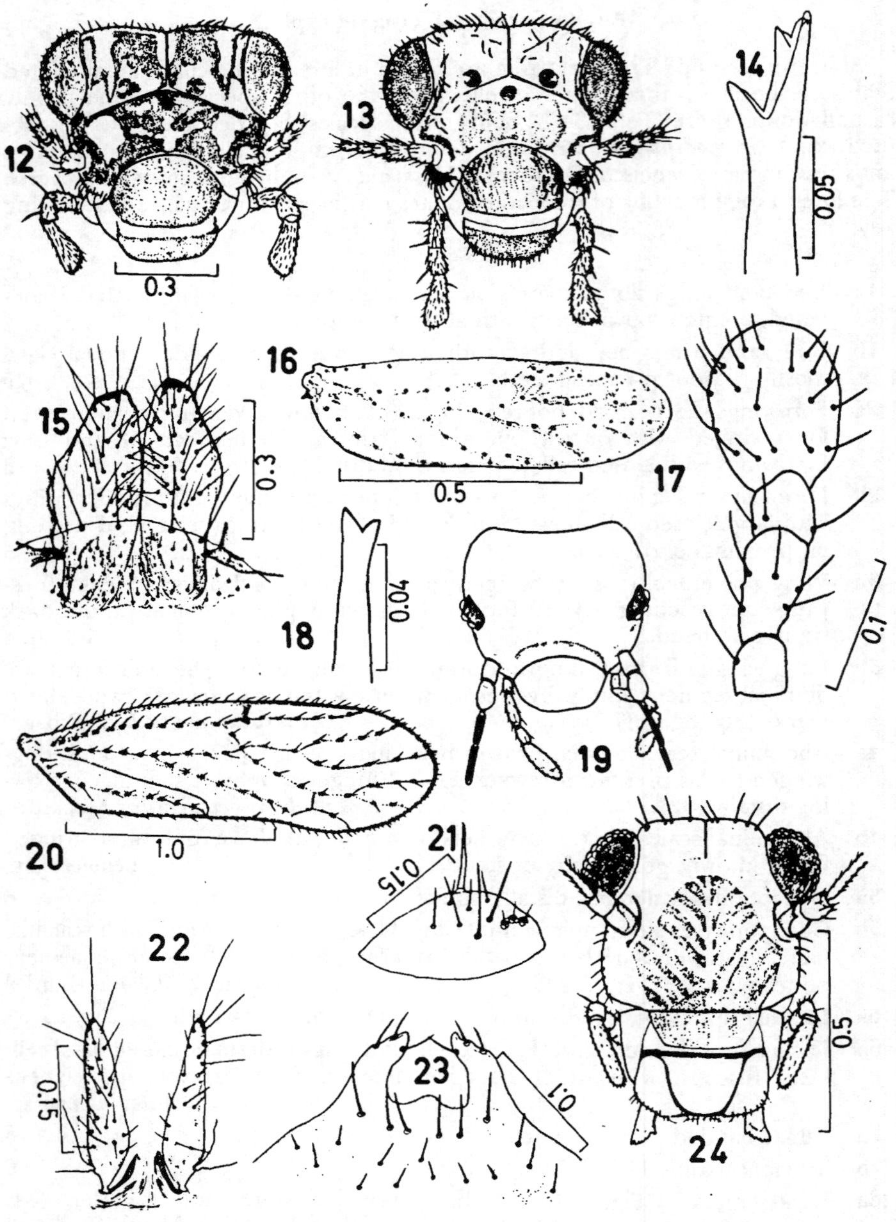

Figs. 34.12–24. 12. *Echmepteryx hageni* (Packard), ♀, frontal view of head. **13.** *Echmepteryx* sp., ♀, frontal view of head. **14.** *Nepticulomima* sp., ♀, lacinial tip. **15.** *Thylacella cubana* (Banks), ♀, gonapophyses. **16.** *Psyllipsocus oculatus* Gurney, ♀, forewing. **17.** *Belaphotroctes ghesquierei*, Badonnel, ♀, left maxillary palp. **18.** *Rhyopsocus squamosus* Mockford & Gurney, ♀, lacinial tip. **19.** *Liposcelis* sp., ♀, frontal view of head. **20.** *Hemipsocus pretiosus* Banks, ♀, forewing. **21.** *Lepolepis occidentalis* Mockford, ♀, paraproct. **22.** *Cerobasis guestfalica* (Kolbe), ♀, gonapophyses. **23.** *Ectopsocus briggsi* McLachlan, ♀, distal end of subgenital plate. **24.** *Epipsocus crosbyanus* (Chapman), ♀, frontal view of head. Scales in mm. Figure 34.13 to same scale as Fig. 34.12. Figure 34.19 to same scale as Figure 34.21.

10a Lacinial tips bifid (Fig. 34.18); frequently brachypterous or with short, flap-like forewings **11**
10b Lacinial tips trifid; apterous or winged **14**
11a P_4 distally expanded; hind tibiae with three strong apical spines **12**
11b P_4 elongate; hind tibiae with two strong apical spines; forewing (Fig. 34.10) setose, reticulated (7,8) . . . . ***Lepinotus reticulatus*** Enderlein
12a Male clunium extended posteriorly to form two broad, rounded, lateral lobes, and a median, spatulate projection; apex of hypandrium slightly convex; a narrow, sclerotized band extending anteriorly, from anterior border of hypandrium; macropterous and brachypterous (36) . . . . ***Rhyopsocus phillipsae*** Sommerman
12b Male clunium not as above; hypandrium not as above, brachypterous . . **13**
13a Compound eyes large, with many facets; apex of hypandrium slightly bi-lobed; accessory bodies of spermatheca without lateral indentations (37) . . ***Rhyopsocus squamosus*** Mockford and Gurney
13b Compound eyes small, with few facets; apex of hypandrium rounded; accessory bodies of spermatheca bean-shaped (20) . . . . ***Rhyopsocus micropterous*** Mockford
14a Second segment of maxillary palps with a sensillum on inner side; apterous **15**
14b Second segment of maxillary palps without a sensillum or strong seta; long-winged or brachypterous; forewing (Fig. 34.16) (38) . . . . ***Psyllipsocus oculatus*** Gurney
15a Head with an anchor-shaped mark between epicranial and epistomal sulci; hind tibiae with two strong, apical spines; gonapophyses (Fig. 34.22) (7) . . ***Cerobasis guestfalica*** (Kolbe)
15b Head pattern not as above; hind tibiae with three spines in addition to apical ones (31) . . . . ***Myrmicodipnella aptera*** Enderlein

Key to Species of Troctomorpha

Nineteen species of Troctomorpha, in five genera, are presently known from litter in the United States (Table 34.1). The species of ***Sphaeropsocus*** is as yet undescribed. These species may be separated with the following key, in which the species of ***Liposcelis*** are not included:

1a Body dorsoventrally flattened; hind femora dilated **2**
1b Body not flattened; hind femora not dilated **3**
2a P_4 at least slightly wider than other segments (Fig. 34.17); a slender transverse sclerite on mesonotum, anterior to the middle; hind femora without protuberances. . . . . Genus ***Belaphotroctes*, 5**
2b P_4 not as above; without transverse sclerites on meso-metanotum; hind femora each with a small outer protuberance (Fig. 34.11). . . . . . Genus ***Liposcelis***
3a Compound eyes with few ocelloids; forewings elytriform, venation incomplete; hind wings absent . . . . ***Sphaeropsocus*** sp.
3b Compound eyes well developed; wings not elytriform **4**
4a Body distinctly pigmented; pale red-brown bands from compound eyes to midpoint of clypeus, meeting there; a dark, red-brown band on each side of thorax, above coxal insertions; abdomen with six series of red-brown spots on segments 1–8. Females fully winged and apterous; males apterous (37) . . . ***Tapinella maculata*** Mockford and Gurney

4b Body pale brown, without pigmented spots; fully winged and apterous forms known (Figs. 34.1, 34.2, 34.9) (39,40) . . . . ***Nanopsocus oceanicus*** Pearman

5a Females with P_4 about 2× as wide in middle as next segment and not bearing rodlike or short sensilla on its lower surface (4) . . . . ***Belaphotroctes ghesquierei*** Badonnel

5b Females with P_4 either as above or 1.5× as wide in middle as next segment and bearing one or two groups of rodlike sensilla or scattered, short sensilla on lower surface **6**

6a Females with P_4 fully 2× as wide in middle as next segment and bearing on its lower surface two groups of closely set rodlike sensilla near tip (42) . . . . . ***B. badonneli*** Mockford

6b Females with P_4 about 1.5× as wide in middle as next segment and bearing on its lower surface one group of closely set rodlike sensilla, or scattered, short sensilla **7**

7a Females with one group of closely set rodlike sensilla on lower surface of P_4, (43) . . . . ***B. hermosus*** Mockford

7b Females with scattered rodlike sensilla on lower surface of P_4 (44) . . . . . . ***B. alleni*** Mockford

Key to Species of Psocomorpha

The Psocomorpha is the most diverse group. It is represented by 25 species in 13 genera, seven of the species being primary litter dwellers (Table 34.1). The following key can be used to separate the species of litter Psocomorpha in the continental United States.

1a Labrum with a longitudinal line-like sclerite on each half of anterior surface; females apterous, nymphlike; lacinial tips broad, with eight denticles, one set apart (Fig. 34.24) (5) . . . . ***Epipsocus crosbyanus*** (Chapman)

Males fully winged, extremely rare; only one specimen known so far. The closely related ***E. lepicidinarius*** (Chapman) found normally on rocks, occurs casually in litter.

1b Labrum not as above; wings variable; gonapophyses variable; lacinial tips not as above **2**

2a Forewings with areola postica **8**

2b Forewings without areola postica **3**

3a Pretarsal claws without preapical teeth; forewings with row of hairs on most veins; pterostigma rectangular. **5**

3b Pretarsal claws with preapical teeth, forewings glabrous, pterostigma wide distally. **4**

4a Apical process of female subgenital plate (SGP) approximately triangular, distally rounded; male clunial process with lateral flanges, almost triangular, bearing no denticles (45). . . . . ***Peripsocus alachuae*** Mockford

4b Apical process of female SGP subquadrate, distally concave; male clunial process without lateral flanges, straight posteriorly, bearing denticles. Forewing (Fig. 34.30) (5) . . . . ***P. quadrifasciatus*** (Harris)

5a Female SGP bilobed, gonapophyses complete; ninth tergite of male with transverse comb of spurs **6**

5b Female SGP (Fig. 34.41); gonapophyses reduced to external valves; spermathecal opening with sclerifications; ninth tergite of male modified to form a complex (46) sclerotized frame . . ***Ectopsocopsis cryptomeriae*** (Enderlein)

6a Lobes of female SGP long, slender; male phallosome with stout, complex, sclerotized "bridge" between lateral fingerlike apophyses **7**

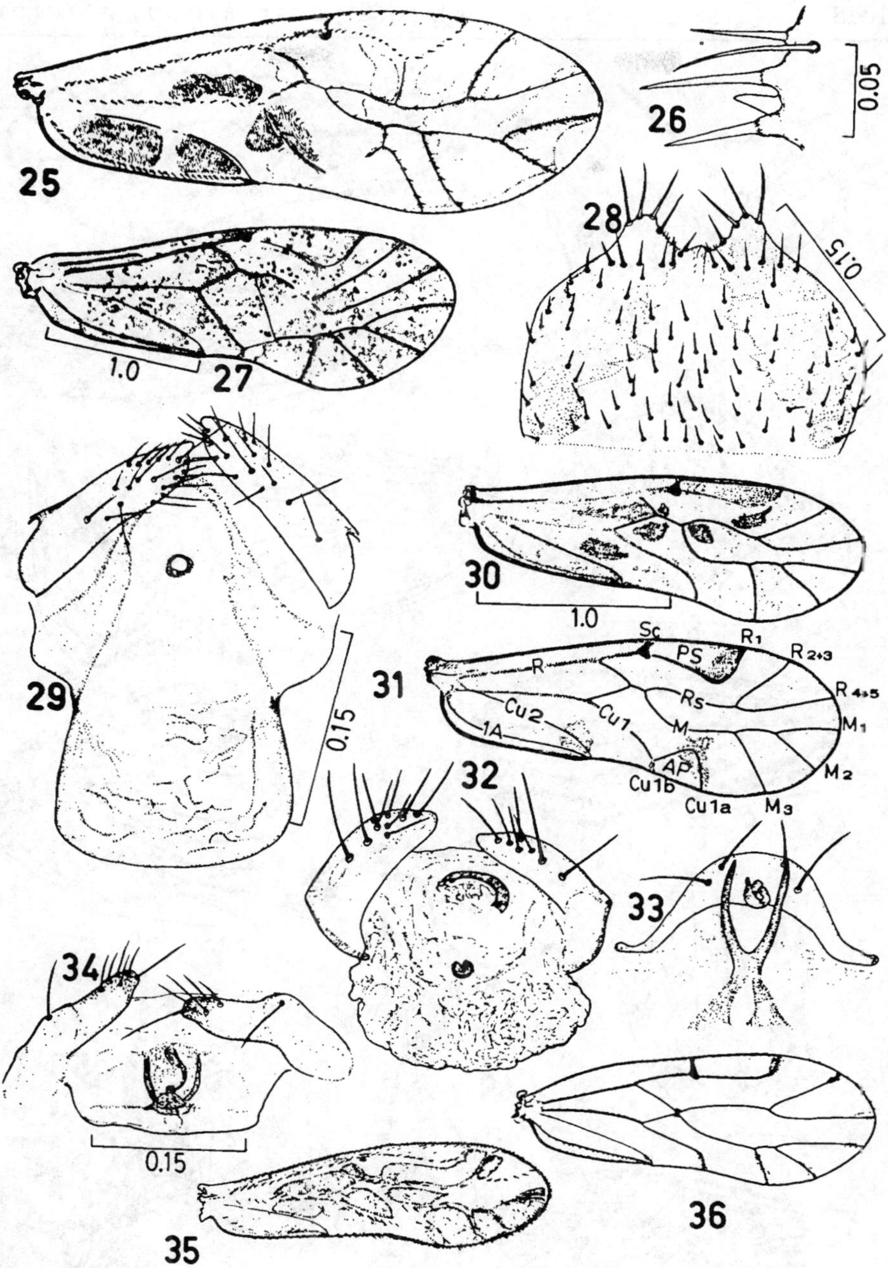

Figs. 34.25–36. 25. *Graphopsocus cruciatus* (L), ♀, forewing. **26.** *Teliapsocus conterminus* (Walsh), ♀, spines of border of paraproct. **27.** *Trichadenotecnum alexanderae* Sommerman, ♀, forewing. **28.** *Ectopsocus vachoni* Badonnel, ♀, subgenital plate. **29.** *Lachesilla* D-5, ♀, gonapophyses and ninth sternum. **30.** *Peripsocus quadrifasciatus* (Harris), ♀, forewing. **31.** *Lachesilla pallida* (Chapman), ♂, forewing venation; main veins labeled (*PS*, pterostigma, *AP*, areola postica). **32.** *L. pallida* (Chapman), ♀, gonapophyses and ninth sternum. **33.** *L. pallida* (Chapman), ♂, epiproct and apophysis of clunium. **34.** *L. pedicularia* (L.), ♀, gonapophyses and ninth sternum. **35.** *Caecilius posticus* Banks, ♂, forewing. **36.** *Ectopsocus briggsi* McLachlan, ♂, forewing. Scales in mm. Figures 34.25, 34.31, and 34.35 to same scale as Figure 34.30. Figure 34.32 to same scale as Figure 34.34. Figure 34.33 to same scale as Figure 34.29. Figure 34.36 to same scale as Figure 34.27.

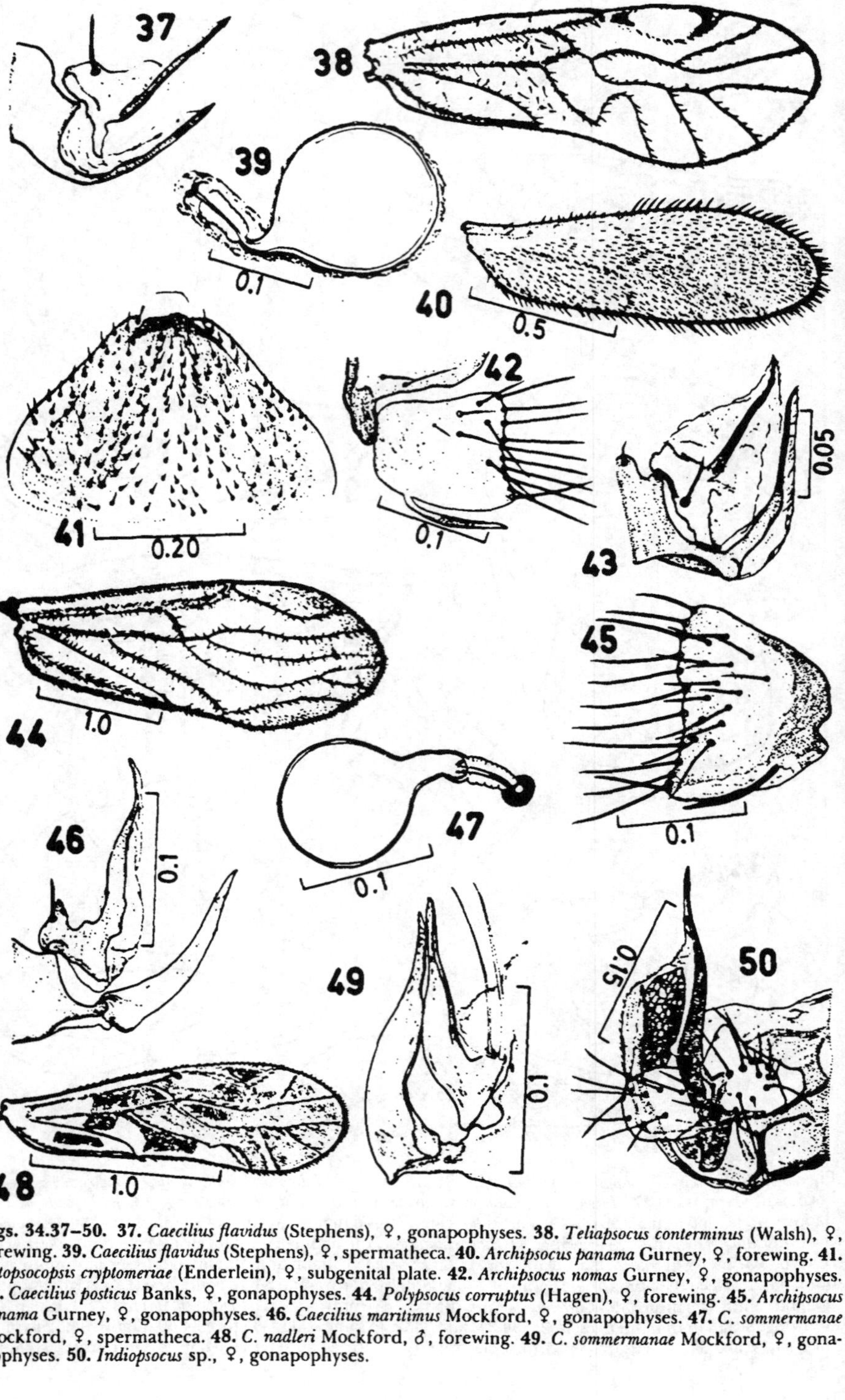

Figs. 34.37–50. 37. *Caecilius flavidus* (Stephens), ♀, gonapophyses. **38.** *Teliapsocus conterminus* (Walsh), ♀, forewing. **39.** *Caecilius flavidus* (Stephens), ♀, spermatheca. **40.** *Archipsocus panama* Gurney, ♀, forewing. **41.** *Ectopsocopsis cryptomeriae* (Enderlein), ♀, subgenital plate. **42.** *Archipsocus nomas* Gurney, ♀, gonapophyses. **43.** *Caecilius posticus* Banks, ♀, gonapophyses. **44.** *Polypsocus corruptus* (Hagen), ♀, forewing. **45.** *Archipsocus panama* Gurney, ♀, gonapophyses. **46.** *Caecilius maritimus* Mockford, ♀, gonapophyses. **47.** *C. sommermanae* Mockford, ♀, spermatheca. **48.** *C. nadleri* Mockford, ♂, forewing. **49.** *C. sommermanae* Mockford, ♀, gonapophyses. **50.** *Indiopsocus* sp., ♀, gonapophyses.

6b Lobes of female SGP short, stout (Fig. 34.28); male phallosome without bridge as above. Females wing-dimorphic, males nearly apterous (35,47) ***Ectopsocus vachoni*** Badonnel

7a Lobes of female SGP incurved (Fig. 34.23), duplex spine of paraproct margin with slightly subequal prongs; forewings faintly spotted (Fig. 34.36); bisexual (48) . . . . ***E. briggsi*** McLachlan

7b Lobes of female SGP posteriorly directed; duplex spine of paraproct margin with subequal prongs; forewings distinctly spotted; parthenogenetic (48) ***E. meridionalis*** Ribaga

8a Vein Cu_{1a} not joined to M (areola postica free). **12**

8b Vein Cu_{1a} fused to M or joined to it by a crossvein (areola postica fused or joined to M by a crossvein) **9**

9a Cu_{1a} fused to M; gonapophyses complete; external valves with a distinct prominence, posteriorly directed **11**

9b Cu_{1a} joined to M by a crossvein, gonapophyses complete or reduced to dorsal and ventral valves; if the former, no prominence on external valves. . . **10**

10a M in forewing three-branched; pterostigma widest distally; a crossvein between pterostigma and Rs (Fig. 34.25) (7). . . . ***Graphopsocus cruciatus*** (L)

10b M in forewing two-branched; pterostigma widest basally; a crossvein between pterostigma and Rs (Fig. 34.20); gonapophyses broad, complete (3). . . . . ***Hemipsocus pretiosus*** Banks (L)

11a First and second sections of Cu_{1a} in forewing, forming an almost straight line, distal section of Cu_{1a} somewhat recurved (Fig. 34.27) (49). . . . . ***Trichadenotecnum alexanderae*** Sommerman

11b First and second sections of Cu_{1a} in forewing with an angle between them; distal section of Cu_{1a} not recurved; gonapophyses (Fig. 34.50) (33) ***Indiopsocus*** sp.

12a Pretarsal claws without preapical teeth; two pairs of gonapophyses present **15**

12b Pretarsal claws with preapical teeth; gonapophyses reduced to external valves. **13**

13a Female SGP rounded; male clunium with one or two apophyses, directed posteriorly; epiproct with slender, acuminate projection **14**

13b Female SGP projected posteriorly; forewings hyaline; gonapophyses and ninth sternum (Fig. 34.29) (50) . . . . ***Lachesilla*** sp. D-5

14a Gonapophyses stout (Fig. 34.32); male clunium with a single forked apophysis (Fig. 34.33); forewing (Fig. 34.31) (5,51) . . . . . ***L. pallida*** (Chapman)

14b Gonapophyses slender (Fig. 34.34); male clunium with two slender apophyses (5,50,51). . . . . ***L. pedicularia*** (L)

15a Dorsal valves of gonapophyses slender, styliform; external valves broad, forewing ciliation dense, throughout the membrane in macropterous forms; venation reduced, evanescent at wing margin **16**

15b Gonapophyses not as above; forewing ciliation mostly along veins and margin; if the membrane is ciliated; cilia restricted to basal end or along posterior border of wing; venation distinct. **17**

16a f_1 longer than pedicel. Parameres not converging anteriorly. Gonapophyses (Fig. 34.45) (52) . . . . ***Archipsocus panama*** Gurney

16b f_1 shorter than pedicel. Parameres converging anteriorly. Gonapophyses (Fig. 34.42) (52) . . . . ***A. nomas*** Gurney

17a Forewings with cilia on membrane as well as along veins and margin . . **18**

17b Forewings with cilia along veins and margin; membrane bare, at most with microtrichia **19**

18a Cilia of forewing membrane in a band along posterior border, from distal end of pterostigma to base of areola postica; *M* two-branched, areola postica low, forming an elongated, horizontal cell; forewing (Fig. 34.44) (53) ***Polypsocus corruptus*** (Hagen)

18b Cilia of forewing membrane on basal end of wing; *M* three-branched; areola postica high. Forewing (Fig. 34.38). Duplex spine of paraprocts (Fig. 34.26) (53) . . . . ***Teliapsocus conterminus*** (Walsh)

19a Forewings mostly hyaline; body pale yellow **20**

19b Forewings fumose or with distinct marks; body red-brown; males macropterous, females micropterous and macropterous **22**

20a Dorsal valves of gonapophyses each with a basal seta; "neck" of spermatheca slender; pterostigma widest distally. **21**

20b Dorsal valves of gonapophyses each with two basal setae (Fig. 34.49) "neck" of spermatheca stout (Fig. 34.47); pterostigma slightly widened distally (34) ***Caecilius sommermanae*** Mockford

21a Gonapophyses (Fig. 34.37); spermatheca (Fig. 34.39); male mesothoracic tibiae swollen proximally and tapering distally (7,54) . ***C. flavidus*** (Stephens)

What appears in the literature as ***Caecilius aurantiacus*** (Hagen) is in reality ***C. flavidus*** (Stephens) (54).

21b Gonapophyses (Fig. 34.36); glandular area of spermatheca longer than in ***C. flavidus;*** male mesothoracic tibiae not as above (55) ***C. maritimus*** Mockford

22a Forewings of males and macropterous females fumose, with pale areas in the membrane (56) . . . . ***C. totonacus*** Mockford

22b Forewings of males and macropterous females with a distinct, colored pattern **23**

23a Male forewing with a spot through distal half of pterostigma (Fig. 34.48). Head pale brown; female SGP with lateral apophyses little protruded (56) ***C. nadleri*** Mockford

23b Male forewing with colorless pterostigma; wing pattern (Fig. 34.35). Head deep chocolate brown; gonapophyses (Fig. 34.43); female SGP with lateral apophyses more protruded than in ***C. nadleri*** (5,56) . . . ***C. posticus*** Banks

References

1. **Smithers, C. N. 1965.** *Aust. Zool.* 13:137–209.
2. **Smithers, C. N. 1967.** *Aust. Zool.* 14:1–145.
3. **Smithers, C. N. 1972.** *Aust. Mus. Syd. Mem.* 14:1–351.
4. **Badonnel, A. 1951.** Psocoptères. pp. 1301–1340 in P. P. Grassé (ed.). *Traité de Zoologie,* Vol. 10, fasc. 2. Masson, Paris.
5. **Chapman, P. J. 1930.** *J. N.Y. Entomol. Soc.* 38:219–290, 319–403.
6. **Roesler, R. 1944.** *Stettin. Entomol. Zeit.,* 105:117–166.
7. **Günther, K. K. 1974.** Stäublause, Psocoptera. Pp. 1–314 in *Die Tierwelt Deutschlands,* 61 Teil. Gustav Fischer, Jena, East Germany.

8. **New, T. R. 1974.** Pp. 1–102 in *Psocoptera, Handbooks for the Identification of British Insects,* Vol. 1, Part 7. Royal Entomological Society of London.
9. **New, T. R. 1971.** *Entomologist (Lond.)* pp. 59–79.
10. **Mockford, E. L.** *Bionomics of Litter and Soil Inhabiting Psocoptera.* Unpublished manuscript.
11. **Mockford, E. L. 1971.** *Am. Zool.* 11:327–339.
12. **Mockford, E. L. 1977.** *Southwest. Nat.* 22:21–29.
13. **Klier, E. 1956.** *Zool. Jahrb.* 75:207–286.
14. **Garcia Aldrete, A. N. 1973.** *Ciencia (Mex. City)* 28:73–77.
15. **New, T. R. 1969.** *Trans. Soc. Br. Entomol.* 18:169–180.
16. **Mockford, E. L. 1965.** *Proc. North Cent. Branch Entomol. Soc. Am.* 20:82–86.
17. **Gurney, A. B. 1950.** *Pest Control Technol. Entomol. Sect.* 131–163.
18. **Pearman J. V. 1960.** *Entomologist (Lond.)* 93:246–250.
19. **Badonnel, A. 1969.** *Publ. Cult. Cia. Diamantes Angola* 79:1–152.
20. **Mockford, E. L. 1971.** *Pan-Pac. Entomol.* 47:127–140.
21. **Hicks, E. A. 1959.** *Checklist and Bibliography of the Ocurrence of Insects in Bird's Nests.* Iowa State College Press, Ames, Iowa. 681 pp.
22. **Rapp, W. F. 1961.** *Entomol. News* 72:195.
23. **Wlodarczyk, J. 1963.** *Fragm. Faun. (Warsaw)* 10:361–366.
24. **Wlodarczyk, J. and J. Martini. 1969.** *Ecol. Pol. Ser. B.* 15:323–336.
25. **Mockford, E. L. 1967.** *Proc. Entomol. Soc. Wash.* 69:307–309.
26. **Badonnel, A. 1973.** *Publ. Cult. Cia. Diamantes Angola* 87:59–104.
27. **Southwood, T. R. E. 1966.** *Ecological Methods with Particular Reference to the Study of Insect Populations.* Methuen, London. 391 pp.
28. **Kühnelt, W. 1976.** *Soil Biology,* 2nd ed. Faber & Faber, London. 483 pp.
29. **Edwards, C. A. and K. E. Fletcher. 1971.** *A comparison of extraction methods for terrestrial arthropods.* In J. Phillipson (ed.). *Methods of Study in Quantitative Soil Ecology: Population, Production and Energy Flow.* IBP Handbook 18. Blackwell Scientific, Oxford. 297 pp.
30. **Badonnel, A. 1970.** Psocoptera. In S. L. Tuxen (ed.). *Taxonomist's Glossary of Genitalia in Insects,* 2nd rev. enlarged ed. Munksgaard, Copenhagen.
31. **Enderlein, G. 1909.** *Boll. Lab. Zool. Gen. Agrar. Portici* 3:329–339.
32. **Mockford, E. L. 1955.** *Am. Midl. Nat.* 53:436–441.
33. **Mockford, E. L. 1974.** *Entomol. Am.* 48:103–215.
34. **Mockford, E. L. 1974.** *Fla. Entomol.* 57:255–257.
35. **Kolbe, H. J. 1885.** *Berl. Entomol. Z.* 29:183–192.
36. **Sommerman, K. M. 1956.** *J. Wash. Acad. Sci.* 46:145–149.
37. **Mockford, E. L. and A. B. Gurney. 1956.** *J. Wash. Acad. Sci.* 46, 353–368.
38. **Gurney, A. B. 1943.** Ann. Entomol. Soc. Am. 36:195–220.
39. **Pearman, J. V. 1928.** *Entomol. Mon. Mag.* 64:133–137.
40. **Badonnel, A. 1977.** *Bull. Mus. Natl. Hist. Nat. (Paris) Ser.3* 478, *Zool.* 335:917–949.
41. **Badonnel, A. 1949.** *Bull. Inst. R. Sci. Nat. Belg.* 25:64 pp.
42. **Mockford, E. L. 1972.** *Fla. Entomol.* 55:153–163.
43. **Mockford, E. L. 1963.** *Ann. Entomol. Soc. Am.* 56:25–37.
44. **Mockford, E. L. 1978.** *Proc. Entomol. Soc. Wash.* 80:556–574.
45. **Mockford, E. L. 1971.** *Jour. N.Y. Entomol. Soc.* 79:89–115.
46. **Enderlein, G. 1907.** *Stettin. Entomol. Zeit.,* 68:90–106.
47. **Badonnel, A. 1945.** *Rev. Fr. Entomol.* 12:44.
48. **Mockford, E. L. 1959.** *Proc. Entomol. Soc. Wash.* 61:260–266.
49. **Sommerman, K. M. 1948.** *Proc. Entomol. Soc. Wash.* 50:165–173.
50. **Garcia Aldrete, A. N. 1974.** *Folia Entomol. Mex.* 27:88 pp.

51. **Sommerman, K. M. 1946.** *Ann. Entomol. Soc. Am.* 39:627–661.
52. **Gurney, A. B. 1939.** *J. Wash. Acd. Sci.* 29:501–515.
53. **Mockford, E. L. 1978.** *Trans. Am. Entomol. Soc.* 104:139–190.
54. **Lienhard, C. 1977.** *Res. Rech. Sci. Parc Nat. Suisse* 14:417–551.
55. **Mockford, E. L. 1965.** *Trans. Am. Entomol. Soc.* 91:121–166.
56. **Mockford, E. L. 1966.** *Trans. Am. Entomol. Soc.* 92:133–172.

35

Insecta: Coleoptera Carabidae Adults and Larvae

ROSS T. BELL

The family Carabidae, or ground beetles, contains over 2000 North American species. The limits of the family have not been generally agreed upon. As treated here, it includes the Rhysodidae, Cicindelidae, Trachypachidae, and Omophronidae, all of which have been recognized as separate families by some authorities. Carabidae in the broad sense includes all of the nonaquatic members of the suborder Adephaga. Adult Adephaga differ from other beetles in having the sternum of the first abdominal segment interrupted by the hind coxae, so that it appears in the intact specimen to be divided into three parts, one between the coxae, and the others at either side. In all Adephaga there is a notopleural suture separating the pronotum from the propleura. (This is in contrast to the dominant suborder Polyphaga, in which the notopleural suture is absent.) Most Adephaga have distinct propleura, separated from the prosternum by a sternopleural suture. In the Rhysodini and some Cychrini the sternopleural suture is reduced or absent, so that the propleura are fused to the prosternum, not to the pronotum as in Polyphaga.

The adults of Carabidae may be distinguished from those of the aquatic Adephaga (Dytiscidae, Haliplidae, Gyrinidae) by the absence of swimming hairs and other adaptations for aquatic existence. The beetles most often mistaken for Carabidae are the terrestrial Tenebrionidae or darkling beetles (suborder Polyphaga). In addition to the subordinal characters, Tenebrionidae may be recognized by the fact that the outer antennal segments are somewhat thickened, the anterior margin of the head is thickened, and the eyes are more or less emarginate anteriorly.

The larvae of suborder Adephaga differ from those of Polyphaga in having two segments below the knee joint (excluding the claw). In

Polyphaga, there is only one segment, excluding the claw. Larvae of Carabidae differ from those of the aquatic families in having open, functional spiracles on most of the abdominal segments. Most larvae of Carabidae have well-developed urogomphi ("tails") on the ninth abdominal segment. The tenth segment forms a uropod, a structure resembling a proleg, which serves to support the tip of the abdomen. Larvae of Cicindelinae, Rhysodini, and *Dromius*, and the mature larvae of Brachinini and *Lebia*, lack urogomphi. Of these groups, the Cicindelinae inhabit vertical burrows in the soil, the Rhysodini burrow in rotten wood, and most of the remaining groups are parasitoid forms found associated with the pupae of other beetles.

The typical terrestrial carabid larvae, with well-developed urogomphi, are superficially much like those of Staphylinidae (suborder Polyphaga). The two families are easily distinguished by the segmentation of the leg.

Ecology

Much of the available ecological information on Carabidae is summarized by Thiele (1977). Lindroth (1961–1969) included available ecological information with his taxonomic accounts of North American species, while Larochelle (1975) added much information about northeastern species.

Feeding

Adults of most Carabidae are primarily carnivorous, feeding on almost any small animal of suitable size. Many species feed extensively on dead or disabled arthropods, and thus act primarily as scavengers rather than predators. A large number of species include some plant material in the diet, primarily sprouts, berries, tender shoots, and flower parts. Some species have also been taken in large numbers on vertebrate carrion. The ecological impact of unspecialized, opportunistic Carabidae is thus very hard to assess.

Two large tribes, Harpalini and Amarini, are primarily phytophagous, although many of the species take some animal food. Many of these species feed on fruits or seeds, especially those of Umbelliferae and Cruciferae. Some species hoard seeds, and one European species is known to provide seeds for its larvae.

The tribes Cychrini and Licinini are specialized for predation on Gastropoda. The Cychrini have long, narrow mandibles, allowing them to feed on the snail through the aperture of the shell, while the Licinini have very short mandibles, adapted for crushing the shell.

Promecognathus and *Metrius* are specialized for predation on Diplopoda. *Helluomorphoides* is adapted for wrestling prey away from ants. *Notiophilus* is especially adept at catching Collembola, although it preys as well on many other small arthropods. Some species of *Dyschirius* feed mainly on the burrowing Staphylinid *Bledius*. Some species of *Bembidion* feed extensively on the eggs of insects. A number of other tribes and genera have specialized mouthparts, which indicate an unusual diet, as yet unstudied or poorly understood. These include *Leistus*, *Loricera*, *Panagaeus*, *Micrixys*, *Chlaenius* (subgenus *Anomoglossus*), *Chlaenius* (subgenus *Pseudanomoglossus*), and *Pterostichus* (subgenus *Cylindrocharis*).

Carabidae make extensive use of extraoral digestion. A majority of Carabidae bolt relatively large fragments of the prey and store them in a crop. These pieces are more finely comminuted in the proventriculus. Identifiable fragments of arthropod prey are found in the crop contents and can often be identified rather exactly. A second group of species grind the food much more finely in the jaws and rely much more heavily on extraoral digestion. In Carabini and Cicindelinae, this tendency has developed to the point that only liquids are imbibed, and the grinding apparatus of the proventriculus has become vestigial. In these groups there are no recognizable fragments of the prey in the contents of the alimentary canal.

Much less is known about the feeding habits of the larvae than about those of the adults. There is a general tendency for the larva to feed on the same type of food as the corresponding adult. Larvae of Amarini and Harpalini, for instance, are largely phytophagous, while those of Cychrini, like the adult, feed on snails. The feeding methods of the two stages are entirely different. Larvae have little or no ability to grind the food, so that they rely much more than does the adult on extraoral digestion. A large majority of larvae of Carabidae have narrow, sickle-shaped mandibles, adapted for piercing the prey. Probably most species of this type ambush and kill small arthropods. Larvae of *Nebria* are of this type, and at least some of the species are in the larval state in late fall and winter, when they feed extensively on arthropods that have been killed or inactivated by low temperatures. The largely phytophagous larvae of Harpalini and Amarini have the mandibles much shorter and more triangular than in other tribes. The larva of Cicindelini constructs a vertical burrow in the soil, from which it projects the anterior part of the body to seize passing arthropods. *Lebia* and *Brachinus* have parasitoid larvae, which feed on the pupae of other insects. *Lebia* larvae develop on the pupae of Chrysomelid beetles, while those of *Brachinus* develop on the pupae of water beetles of several families. In both genera, the first instar larva is a typical carabid larva which searches for the host, while the later instars have reduced appendages and sclerotization.

Life History

In the majority of ground beetles of the temperate zone, there is one generation per year. In a few subarctic and arctic species, two or more years are required to reach sexual maturity. In some species, a part of the adult population survives the subsequent winter and breeds a second time.

The eggs in most species are laid singly in the soil, in leaf litter, or in rotten wood. In *Chlaenius* (subgenus *Agostenus*) and in a few other groups, each egg is coated with mud and deposited on a plant. The eggs are white and elongate–oval, with few distinguishing features. The total egg production of a single female varies between 30 and 600. Lower fecundity is known from a few cave species which lay very large eggs.

Most species have three larval instars, although a few parasitoid species have more. Pupation is in the soil or in rotten wood. Morphological differences between pupae of different species are few and have been little studied.

Almost all Carabidae of the temperate zone have a short breeding season, synchronized by photoperiod, or by a combination of temperature and photoperiod. A majority of species breed in the spring. A few species copulate in the fall but delay egg laying until spring. An important minority of species breed and lay eggs in the fall. In those species that lay eggs in the spring, the larva completes development by the fall, and pupation follows immediately. The new adults in some of these species remain in the pupal cell until spring, while in others, they emerge and forage for a time before going into hibernation. In the fall breeders, the larva has a period of quiescence or a diapause in the winter and completes development in the spring.

Effects on Soils

Probably most Carabidae make some use of burrowing, both as adults and as larvae. In most species this is limited to slight modification of previously existing cavities or to burrowing in loose materials. In either case, the effect on the soil is minimal. A minority of genera are specialized burrowers. Both the mandibles and the legs are used in digging. Where such Carabidae are abundant, they probably have a significant effect on soil structure. The most specialized burrowers belong to the tribe Scaritini. Omophronini, Broscini, and some genera of Harpalini are also specialized burrowers.

Migration and Dispersal

Many Carabidae make a seasonal migration from a primary habitat to a winter refuge. This is especially true of species of shorelines and marshes, which are not suitable spots for overwintering. It is also true to some extent of species of open fields and cultivated land. Seasonal migration must be taken into account in ecological investigations, since in favorable spots winter visitors may actually outnumber permanent residents. For this reason, primarily riparian and arboreal genera have been included in the generic keys; only strictly cavernicolous genera have been omitted. Where the distances are not great, the migration may be accomplished on foot, either by the adult or the larva. In many species, however, the adult uses flight. The spring migration is made by day in many species and is thus much more conspicuous than the fall migration, which is generally nocturnal.

Adults of many species make frequent dispersal flights. These flights are nocturnal, and are reflected in the large numbers of Carabidae, which are attracted to lights. They are characteristic of species of temporary or widely dispersed habitats, such as the shores of temporary ponds and pools, disturbed soil, and cultivated ground.

Carabidae may be divided among three groups, depending on the development of the hind wings. In macropterous species, fully developed hind wings occur in all individuals. In brachypterous species, all individuals have vestigial wings, while in dimorphic species, some individuals have fully developed hind wings and others have vestigial ones. Macropterous species may be divided into two subgroups. A minority of species use the wings during the pursuit of prey or as a means of escaping predators. This group includes most species of *Cicindela* and *Lebia* and a few species of *Bembidion*. A much larger group of species use the wings only for dispersal movements or for seasonal migration. Brachypterous species are mostly forest species, and of course, have much more limited dispersal abilities than do the macropterous species.

Many species of Carabidae are dimorphic with respect to the hindwings. In some species the two morphs are nearly equally abundant; in others one morph may outnumber the other by several hundredfold. In fact, some species presently thought to be consistently macropterous or consistently brachypterous may actually prove to be dimorphic, with the rarer morph simply not having been found. In many species the relative abundance of the two morphs varies geographically. The macropterous morph is often predominant at the edge of the range. In some species the brachypterous morph predominates in areas thought to mark Pleistocene refugia, while the macropterous morph dominates regions covered by the last glaciation.

Collection

Diurnal Carabidae may be hand collected. An aspirator is convenient for catching the smaller species. Cicindelinae and some Bembidiini fly readily and are best stalked with an insect net. Many nocturnal Carabidae shelter for the day under superficial cover, such as stones, logs, and fallen leaves and can be collected by overturning these objects. The numbers of Carabidae under superficial cover varies tremendously in response to changes in temperature and humidity, and a species that appears abundant in the spring may appear to be absent in the hotter part of the summer. In reality, it may still be present in large numbers but may be sheltering by day in deep, inaccessible cavities in the soil. Littoral species may be collected in great numbers by splashing parts of the shoreline. Marsh and bog species hide in clumps of vegetation just above the water level. These species can be collected by tramping vegetation into the water, after which they will be found swimming at the surface. Smaller species from the forest floor may be collected by gathering large amounts of leaf litter and immersing it in water.

Pitfall trapping will accumulate large numbers of ground beetles, including some species that are rarely caught by other methods. Empty pitfall traps may be used to collect live Carabidae. Dead specimens can be obtained in pitfall traps which have a small amount of formaldehyde added to water. This is a particularly good method to use when traps cannot be checked every day. The formaldehyde preserves the beetles while deterring small mammals from eating them. Partly arboreal species can be collected by tying pieces of stiff cloth to tree trunks. The beetles shelter under the cloth as they would under loose bark.

Those Carabidae with functional hindwings can sometimes be caught in large numbers at light traps, using either incandescent or ultraviolet (blacklight) light. Ideal trapping occurs when the temperature is above 20°C, the air is nearly still, and the sky is overcast.

It is very difficult to quantify the abundance of particular species of Carabidae. Most species are both wary and mobile enough to evade many standard sampling techniques. The percentage of the beetles escaping may vary widely, depending on the alertness of the beetles, which is in turn a function of the temperature. The results of trapping methods depend, of course, on the mobility as well as the abundance of the beetles. Nevertheless, pitfall trapping provides a valuable perspective on the effectiveness of other methods of sampling.

Preservation

The best quality specimens of adult Carabidae are made by killing the beetles in a jar or vial containing ethyl acetate vapor. A few drops of ethyl acetate are added to the jar at intervals, as a certain amount of the vapor escapes each time the jar is opened. The jar is partly filled with hardwood sawdust, which absorbs any fluids given off by the beetles, and also prevents them from injuring one another. Crumpled paper towels can be used as a substitute. Specimens fixed in ethyl acetate are clean and relaxed. The internal genitalia, which are sometimes important in critical taxonomic work, are less likely to be distorted than by other methods. Specimens killed in ethyl acetate are then pinned or point mounted.

Seventy percent ethyl alcohol can also be used for killing and preserving Carabidae. This is quicker and more convenient when large numbers of specimens are taken during ecological studies. It has the further advantage that the internal organs are preserved, so that the stomach contents and the state of development of the gonads can be checked later. Major disadvantages are that the details of the surface sculpture are difficult to see in wet specimens, and that the internal sac of the penis may be distorted by the alcohol. Alcoholic specimens can later be pinned or point mounted. Often, however, grease or oil seeps from the specimen and partially disguises the surface sculpture and setal pattern, as well as providing an adhesive to which dust particles adhere. Grease can also alter the color pattern to some extent. Grease can be removed by soaking the specimen in xylene or carbon tetrachloride. Larvae can be fixed and preserved in 70% alcohol.

Species Identification

There are more than 2000 species of Carabidae in North America. The most recent checklist is by Erwin et al. (1977). By far the most useful publication for species identification is Lindroth's "The Ground Beetles (Carabidae, Excluding Cicindelinae) of Canada and Alaska" (published in six parts, 1961–1969). The title of this great work is somewhat misleading: Lindroth treated all North American species of many genera or subgenera; in others, he omitted only certain southern species-groups. For a few genera not revised in full by Lindroth, there are revisions by other modern authors; for a number of others, modern keys and descriptions have yet to be written. Because of the great number of species, it has proven to be impractical to present species keys here. In the generic keys below, for each genus the best available key or keys are cited. When Lindroth is cited, the page numbers are cited. In a few cases, when genera have recently been divided, the proper key couplet in Lindroth is also included. For the groups not treated by Lindroth, I have generally cited the most recent comprehensive key. Exceptions have been made in the case of Thomas L. Casey. This author described thousands of Carabidae in the period 1884–1918. His keys are not illustrated and are in a form that makes them almost unusable. Moreover, Casey did not admit the possibility of variation within a species. Of the nearly 900 Caseyan species investigated by Lindroth, fewer than 10% proved to be valid species. In groups not investigated by Lindroth or other modern authors, there are still many of Casey's names "on the books." In such cases I have not cited Casey's key, but have listed the best key previous to Casey.

In summary, the practicality of species identification of adult Carabidae in the United States diminishes in proportion to one's distance from the Canadian border. In those states near to Canada, virtually 100% of the specimens can be identified to species with confidence. The farther south one goes, the greater the difficulties encountered in groups not treated by Lindroth. Even in the most southern states, a majority of groups can be keyed to species, but the identifier will find it necessary to assemble a large number of scattered keys and isolated descriptions.

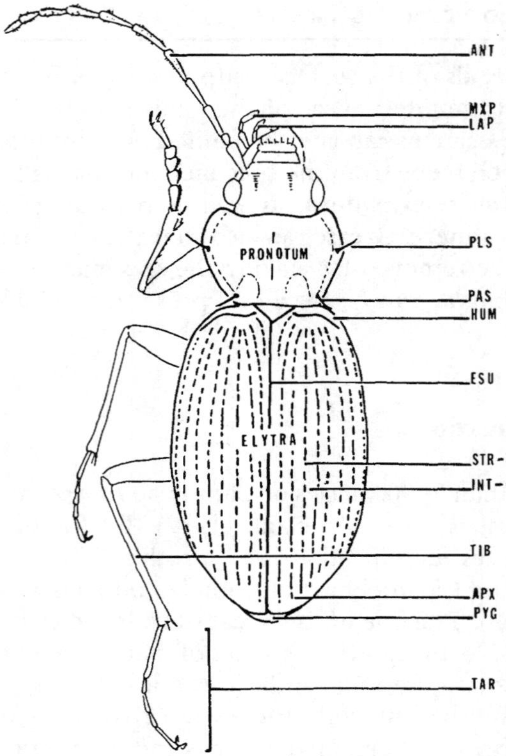

1

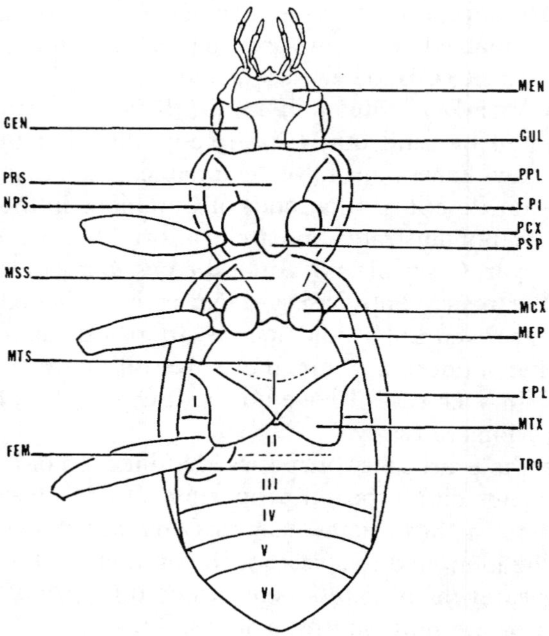

2

Fig. 35.1. Habitus, dorsal aspect. **Fig. 35.2.** Habitus, ventral aspect.

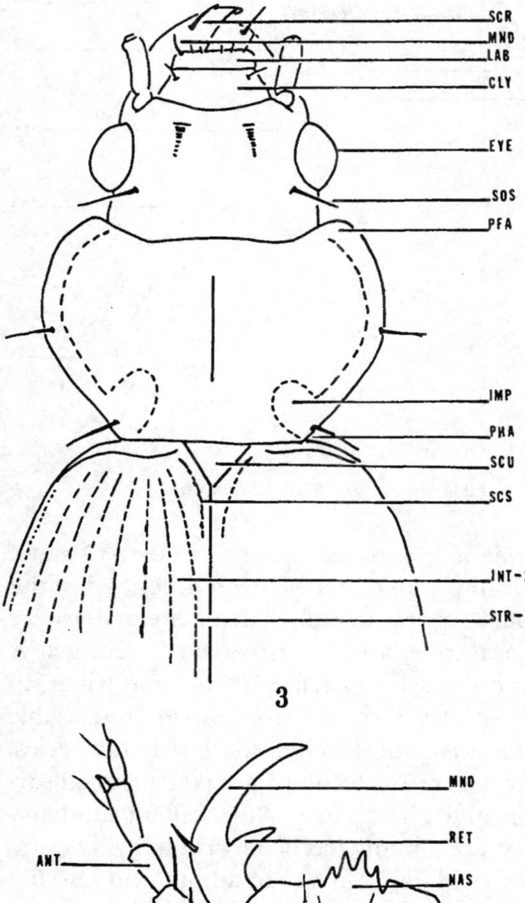

3

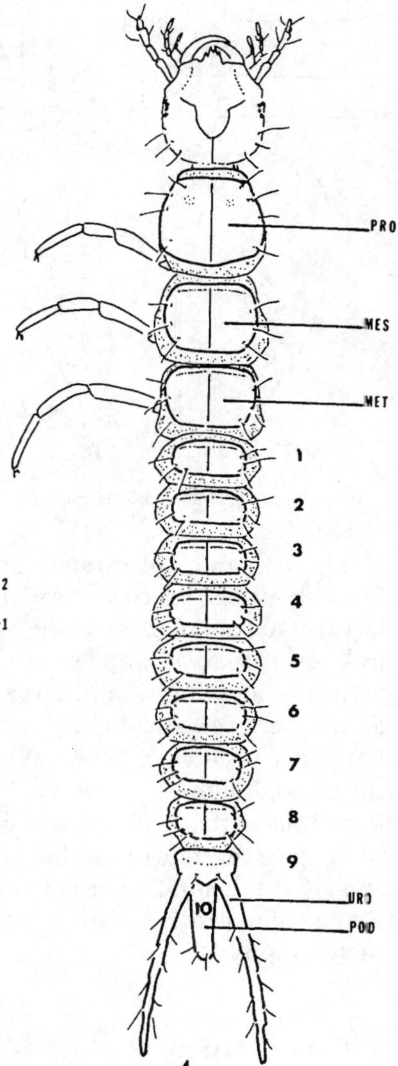

4

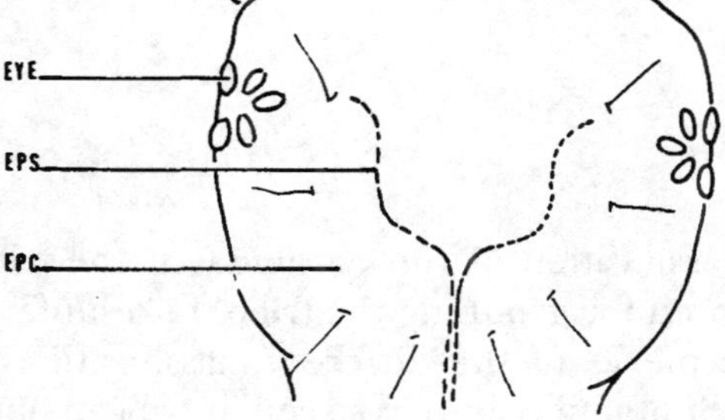

5

Fig. 35.3. Head, prothorax, anterior elytra, dorsal aspect. **Fig. 35.4.** Larva, dorsal aspect. **Fig. 35.5.** Head, larva, dorsal aspect.

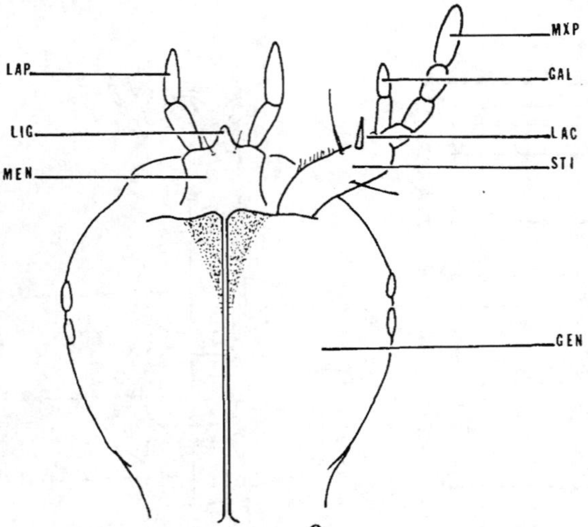

6

Fig. 35.6. Head, larva, ventral aspect.

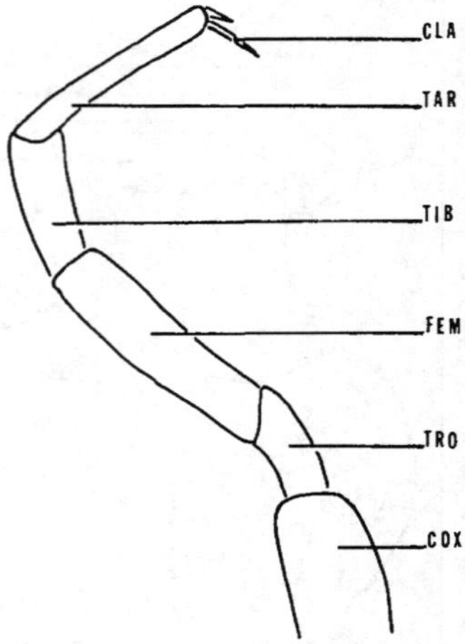

7

Fig. 35.7. Leg, larva.

The taxonomy of carabid larvae is in a much more rudimentary state. Only about 220 species of American Carabidae have been described in the larval state, about 10% of the total. Approximately half of the genera are unknown as larvae, and even several tribes are also unknown. Existing keys allow the identification of larvae only as far as tribe, or in some tribes, to genus. In a few small genera, most of the species are known, and usable keys have been cited below. When it is critical to identify larvae to species, the ecological investigator will have to resort to rearing larvae or to culturing adults so that they will lay eggs in the laboratory. An excellent summary of the present state of carabid larval taxonomy has been written by Thompson (1979). It summarizes methods of linking larvae to adults, and also has a more detailed key than is possible in this work, as well as an extensive bibliography.

Classification

The most stable unit in the classification of the Carabidae is the tribe. Diverse schemes have been proposed for grouping the tribes into subfamilies. A discussion of the relative merits of the different classifications is beyond the scope of this chapter. Subfamilies have consequently been omitted from the keys. Key characters (indicated in Figs. 35.1 to 35.7) have been selected for ease in identification, not for their phylogenetic significance; consequently, tribes close together in the key are not necessarily closely related. The keys are based on the fauna of the continental United States and Canada, and some extralimital species will not key correctly in them.

List of Abbreviations Used in Figures and Key Illustrations

ANT	antenna
APX	apex (elytra)
CLA	claw
CLY	clypeus
COX	coxa
EPC	epicranium
EPI	epipleuron (pronotum)
EPL	epipleuron (elytron)
EPS	epicranial suture
ESU	elytral suture
EYE	eye (lateral ocellus)
FEM	femur
GAL	galea
GEN	gena
GUL	gula
HUM	humerus
IMP	basal impression
INT	interval
LAB	labrum
LAC	lacinia
LAP	labial palpus
LIG	ligula
MCX	mesocoxa
MEN	mentum
MEP	metepimeron
MES	mesothorax
MET	metathorax
MND	mandible
MSS	mesosternum
MTE	metepisternum
MTS	metasternum
MTX	metacoxa
MXP	maxillary palpus
NAS	nasale
NPS	notopleural suture
PAS	angular seta (pronotum)
PCX	procoxa
PFA	front angle (pronotum)
PHA	hind angle (pronotum)
PLS	lateral seta (pronotum)
POD	uropod
PPL	propleuron
PRO	prothorax
PRS	prosternum
PSP	prosternal process
PUL	pulvillus
PYG	pygidium
RET	retinaculum
SCR	scrobal seta
SCS	scutellar striole
SCU	scutellum
SOS	supraorbital seta
STI	stipes
STR	stria
TAR	tarsus
TIB	tibia
TRO	trochanter
URO	urogomphi

KEY TO ADULTS

1a Mentum shorter than mandibles; mentum separated from head capsule laterally by cleft in which maxilla is visible; maxillary and labial palpi completely exposed (Fig. 35.8*a*); dorsal surface of head entire **2**

1b Mentum longer than mandibles, completely concealing latter in ventral view; mentum fused to head capsule laterally, completely concealing maxilla (Fig. 35.8*c*); dorsal surface of head divided into

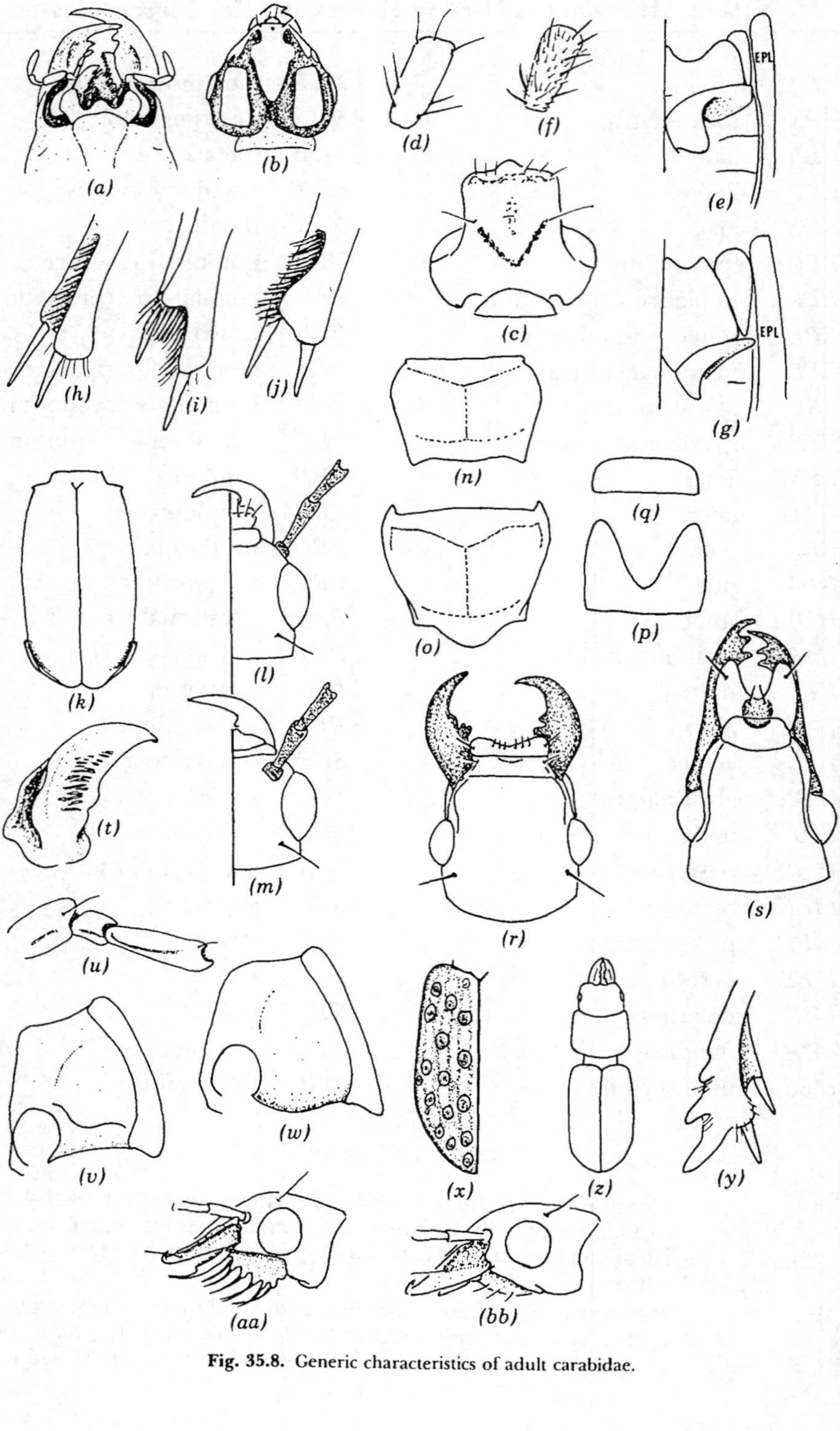

Fig. 35.8. Generic characteristics of adult carabidae.

three lobes (Fig. 35.8*b*); dark reddish-brown beetles 5.4–8.1 mm long; in rotten wood, including dead roots . . . . Tribe **RHYSODINI**

Two genera, eight species. Key: Bell (1970).

2a (1a) Outer antennal segments without short pubescence (Fig. 35.8*d*) (although a few tactile setae are present); coxa of hind leg bordering elytral epipleuron (Fig. 35.8*e*); oval beetles 3.7–6.1 mm long; Rocky Mountains and Pacific Coast. . . . . Tribe **TRACHYPACHINI,** ***Trachypachus*** Motschulsky

Four species. Key: Lindroth (1961, pp 1–4).

2b Outer antennal segments covered with short pubescence in addition to tactile setae (Fig. 35.8*f*); hind coxa either separated from elytral epipleuron or touching latter at one point **3**

3a (2b) Hind coxa in contact with medial margin of elytral epipleuron at one point only (Fig. 35.8*g*); hind coxae widely separated from one another; metasternum without median sulcus; minute beetles 1.7 mm long, with elytra truncate, last segment of palpus narrow, pointed, in streamside gravels, northern Rockies and westward Tribe **GEHRINGIINI,** ***Gehringia olympica*** Darlington

3b Hind coxa separated from elytral epipleuron (Fig. 35.9*p,q*); coxae contiguous or narrowly separated from one another **4**

4a (3b) Both spurs of anterior tibia at or near apex of tibia (second spur more posterior to than proximal to first spur (Fig. 35.8*h,j*) . . . **5**

4b Second spur clearly proximal to first spur (Fig. 35.8*i*) (by at least two-thirds of length of spur) or else second spur absent (some spp. of ***Lebia***); or else antennal cleaning organ in the form of an emargination of the inner margin of the tibia, between the spurs . . . **20**

5a (4a) Antennal cleaning organ in the form of an emargination of inner margin of cleaning organ, proximal to both spurs (Fig. 35.8*i,j*) **6**

5b Antennal cleaning organ in the form of a vertical groove on inner surface of tibia, with its distal end between the spurs (Fig. 35.8*h*), or else cleaning organ obsolete **7**

6a (5a) Elytron with prominent, outward-projecting ridge in apical fifth (Fig. 35.8*k*); species of Mexican border region . . . . Tribe **OZAENINI**

Four genera, four species. Key: Ball (1960).

6b Elytron without such a ridge; stout black beetles 9–13 mm long, Pacific Coast forests . . . . Tribe **METRIINI,** ***Metrius*** Eschscholt

Two species. Key: Lindroth (1961 p. 7).

7a (5b) Insertion of antenna dorsad to mandible (Fig. 35.8*l*); Tribe **CICINDELINI** **8**

7b Insertion of antenna posterior to base of mandible (Fig. 35.8*m*) **11**

8a (7a) Eyes large, prominent, separated by less than three times width of one of them; hind coxae contiguous; most species with functional wings; many species metallic; most species with pale marks on elytra **9**

8b Eyes small, separated by five or more times width of one of them; hind coxae slightly separated; hind wings vestigial; color black or dark brown **10**

9a **(8a)** Pronotum without obvious anterior angles (Fig. 35.8*n*); elytra with punctures fine or obsolete, not in rows; legs dark in almost all species. . . . . *Cicindela* Linnaeus
One hundred species. Keys: Wallace (1961), Leng (1902), Willis (1968).

9b Pronotum with prominent anterior angles (Fig. 35.8*o*); elytra with slightly confused rows of coarse punctures; legs pale; southern two-thirds of United States . . . . *Megacephala* Latreille
Three species. Key: Leng (1902).

10a **(8b)** Pronotal margin prominent, separating flattened disk from pronotal epipleuron; length 13–22 mm; Pacific coast forests . . . *Omus*
Twenty-six species. Key: Leng (1902).

10b Pronotal margin narrow; pronotal disk convex, curved almost evenly into pronotal epipleuron; western plains and deserts; length 30 mm or more. . . . . *Amblycheila* Say
Four species. Key: Vaurie (1955).

11a **(7b)** Labrum bilobed (Fig. 35.8*p*) **12**

11b Labrum truncate (Fig. 35.8*q*) or at most slightly emarginate **17**

12a **(11a)** Lateral lobes of labrum short, at most equaling median length of labrum; head massive (Fig. 35.8*r*); Tribe **CARABINI** **13**

12b Lateral lobes of labrum elongate, three or more times median length of labrum (Fig. 35.8*s*); head long, slender; Tribe **CYCHRINI** **14**

13a **(12a)** Third segment of antenna more than twice as long as second segment, with strong longitudinal keel on one side (Fig. 35.8*u*); mandibles transversely wrinkled near base (Fig. 35.8*t*). . . . . *Calosoma* Weber
Forty-five species. Key: Gidaspow (1959).

13b Third segment of antenna less than twice as long as second segment, without keel, or with very short one near base; mandibles smooth or punctate near base . . . . *Carabus* Linnaeus
Twelve species. Key: Lindroth (1961, p. 32).

14a **(12b)** Pleural suture of pronotum evident (Fig. 35.8*v*); tarsi pubescent above **15**

14b Pleural suture of pronotum absent (Fig. 35.8*w*); tarsi glabrous above; labrum with six setae **16**

15a **(14a)** Labrum with four setae; elytron with 10 or fewer striae or striae nearly absent; length less than 10 mm. . . . . . *Maronetus* Casey
Four species, southern Appalachians. Key: Schwart (1895).

15b Labrum with 6 setae; elytron with 11 or more striae; length 15 mm or more . . . . *Scaphinotus* Dejean
Fifty species. (Keys: Lindroth (1961, p. 15), Ball (1960) [subgenera], Valentine (1935) [SG. Steniridea], Gidaspow (1968, 1973), Van Dyke (1936, 1938).

16a **(14b)** Median depression of labrum not extended onto clypeus (Fig. 35.8*r*); elytra with shining tubercles but entirely without striae; length 15–27 mm; northern Rockies and Pacific coast forests. . . *Cychrus* Fabricius
Three species. Key: Lindroth (1961, p. 95).

16b Median depression of labrum extended onto clypeus (Fig. 35.8*s*); elytra with distinct striae; posterior parts of intervals in some species broken into tubercles; forests east of Great Plains. 10–19 mm *Sphaeroderus* Dejean
Five species. Key: Lindroth (1961, p. 26).

17a **(11b)** Elytra without striae, but with rows of large ocellate pits (Fig. 35.8*x*); two setae over each eye; riparian beetle 8.5–11 mm long, Rockies and west . . . .
Tribe **OPISTHIINI,** ***Opisthius richardsoni*** Kirby

17b Elytra striate, without ocellate pits; one seta above each eye; Tribe **NEBRIINI** **18**

18a **(17b)** Lateral surface of maxilla and gula with spinose hairs, each arising from tubercle (Fig. 35.8*aa*); lower part of lateral face of mandible forming broad lamella 7.5–10 mm; Pacific coast forests . . . . . .
Leistus Frölich

Two species. Key: Lindroth (1961, p. 56).

18b Lateral surface of maxilla and gula without tubercles or spinose hairs (Fig. 35.8*bb*); lower part of lateral face of mandible not forming prominent lamella **19**

19a **(18b)** Each elytron with 10 complete striae, without scutellar striole; 8–15 mm; subarctic . . . . ***Pelophila*** Dejean

Two species. Key: Lindroth (1961, p. 57).

19b Each elytron with scutellar striole and nine complete striae; 7.5–17 mm; mostly riparian or alpine . . . . ***Nebria*** Latreille

Forty-five species. Key: Lindroth (1961, p. 62).

20a **(4b)** Apex of anterior tibia with distal tooth which is at least equal in length to distal spur (Fig. 35.8*y*); mesothorax narrowed and elongate (body pedunculate) (Fig. 35.8*z*); lateral lobe of frons forming shelf dorsad to base of antenna; Tribe **SCARITINI** **21**

20b Apex of anterior tibia without distal tooth (though lateral tooth is present at apex in ***Euryderus***); body rarely pedunculate **28**

21a Basal segment of antenna longer than combined length of segments 2–4; one supraorbital seta; large beetles 16 mm or more in length, with enlarged, toothed mandibles **22**

21b Basal segment of antenna shorter than combined lengths of segments 2–4 combined; two supraorbital setae; length 2.5–8.5 mm **23**

22a Hind angles of pronotum rectangular (Fig. 35.9*a*); body broad, somewhat flattened; length 28–38 mm; grasslands, dry open woods . . . . ***Pasimachus*** Bonelli

Fourteen species. Key: Bänninger (1950).

22b Hind angles of pronotum denticulate (Fig. 35.9*b*); lateral portion of base oblique; body narrow, cylindrical; 16–30 mm; burrows under large stones in grasslands, cultivated ground . . . .
Scarites Fabricius

One or more species.

23a **(21b)** Dorsal surface of head with eight longitudinal carinae (Fig. 35.9*c*); 3.0–5.2 mm; riparian . . . . ***Schizogenius*** Putzeys

Twenty-four species. Key: Whitehead (1972).

23b Dorsal surface of head not longitudinally carinate **24**

24a Elytral epipleura interrupted by twisted fold opposite abdominal sternum VI (Fig. 35.9*d*) **25**

24b Elytral epipleura not interrupted, not appearing twisted (Fig. 35.9*e*) **26**

25a **(24a)** Clypeus squarely truncate (Fig. 35.9*f*); mandibles long, slender, 4–6 mm; eastern and southeastern United States, mostly floodplain forests . . . . ***Ardistomis*** Putzeys

Five species. Key: Leconte (1857).

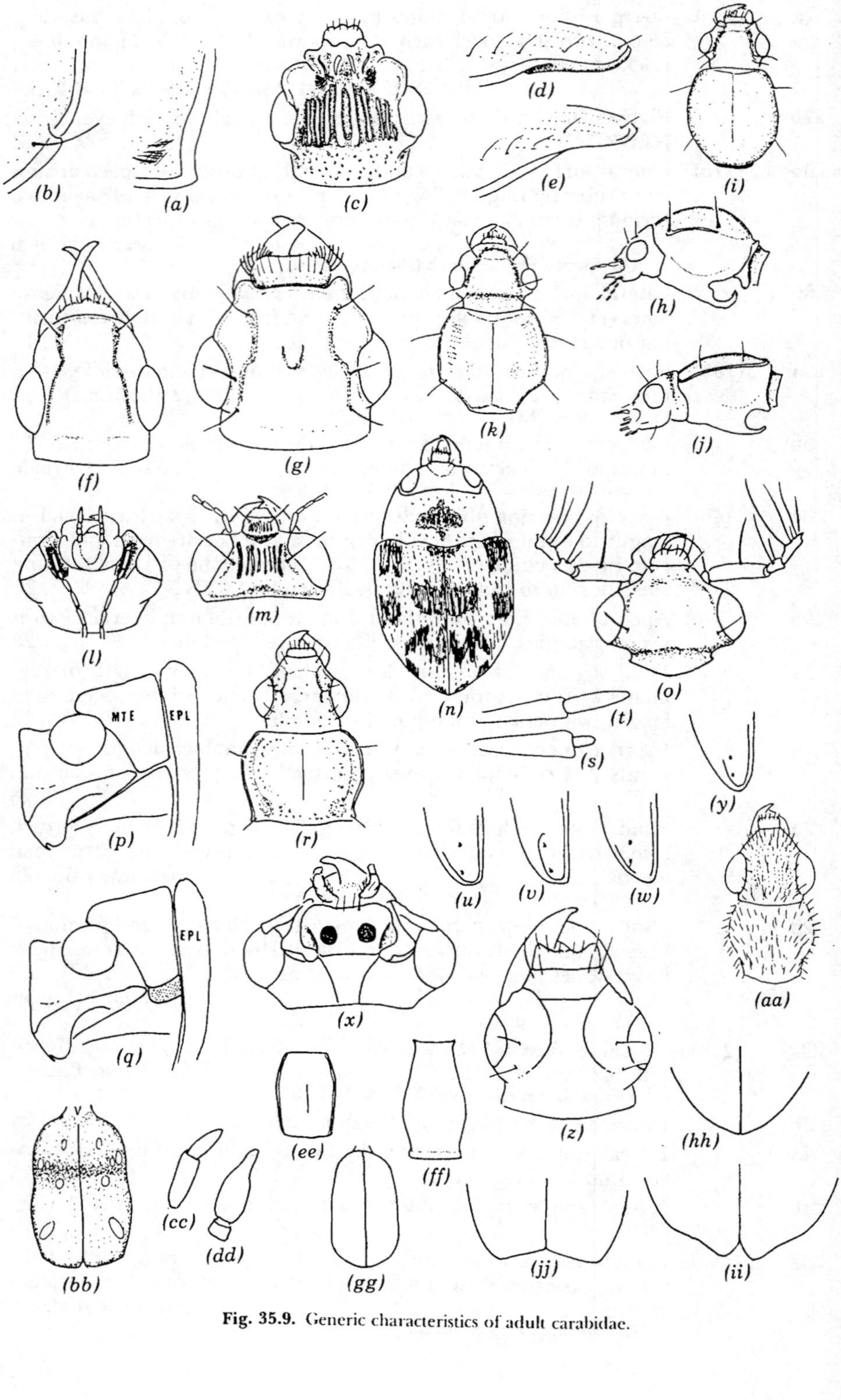

Fig. 35.9. Generic characteristics of adult carabidae.

25b Clypeus with a lateral projecting tooth or lobe (Fig. 35.9*g*); mandibles short, curved; riparian, eastern United States, 7 mm . . . . . ***Aspidoglossa subangulata*** Chaudoir

26a **(24b)** Apical segment of each palpus with attenuate point; length 2.5–3.0 mm; rufous, somewhat depressed; southern Florida . . . . ***Oxydrepanus rufus*** Putzeys

26b Apical segment of antenna not attenuate. **27**

27a **(26b)** Pronotum more or less globose; marginal rim of pronotum ending at or anterior to angular seta (Fig. 35.9*h,i*); 2.5–6.4 mm; most species riparian, but some in grasslands . . . . . . ***Dyschirius*** Panzer

Sixty species. Key: Lindroth (1961, p. 33).

27b Pronotum not globose; lateral margins straight; marginal rim of pronotum extended posterior to angular seta (Fig. 35.9*j,k*); length 4.7–8.5 mm; largely riparian or in cultivated ground . . . . ***Clivina*** Latreille

Twenty-five species. Keys: Lindroth (1961, p. 159), Leconte (1879a).

28a **(20b)** Deep groove present on ventral side of head between eye and maxilla, receiving first antennal segment when latter is folded posteriorly (Fig. 35.9*l*); margin of head expanded, forming shelf which conceals base of antenna and condyle of mandible in dorsal view; Gulf coast and southwestern states . . . . Tribe **PSEUDOMORPHINI,** ***Pseudomorpha*** Kirby

Eighteen species. Key: Notman (1925).

28b No groove for antenna on ventral surface of head; scape of antenna and dorsal condyle of mandible not concealed in dorsal view **29**

29a **(28b)** Frons and clypeus with 6–12 longitudinal carinae (Fig. 35.9*m*); second interval of elytron at least as wide as intervals 3 and 4 combined; elytron appearing to have two strioles; eyes very large; length 3.4–6 mm . . . . Tribe **NOTIOPHILINI,** ***Notiophilus*** Duméril

Fourteen species. (Key: Lindroth (1961, p. 92).

29b Frons and clypeus without longitudinal carinae (but irregularly longitudinally wrinkled in some species); second interval narrower than combined widths of intervals 3 and 4; elytron with one striole or without a striole **30**

30a **(29b)** Scutellum entirely concealed beneath (Fig. 35.9*n*) posterior edge of pronotum; elytron with 14–15 striae; form oval or circular; riparian; 4.8–7.8 mm. . . . . Tribe **OMOPHRONINI,** ***Omophron*** Latreille

Eleven species. Key: Benschoter and Cook (1956).

30b Scutellum not concealed; elytron with 12 or fewer striae. . . . **31**

31a **(30b)** Outer margin of maxilla, ventral surface of head and antennal segments 1–6 with many stiff spinose hairs; head strongly, suddenly constricted behind eye (Fig. 35.9*o*); elytron with 12 striae, without striole; 6.5–8.5 mm . . . . Tribe **LORICERINI,** ***Loricera*** Latreille

Three species. Key: Lindroth (1961, p. 122).

31b Outer margin of maxilla, ventral side of head with only the normal tactile setae; head constricted or not; elytron with 11 or fewer stria; striole present in most species **32**

32a **(31b)** Anterior tibia with second spur retracted 0.16 or less of distance to base of tibia; metepimeron not visible (Fig. 35.9*p*); not overlapping abdomen; Tribe **ELAPHRINI** **33**

32b Anterior tibia with second spur retracted 0.25–0.33 of distance to base of tibia; metepimeron visible as scalelike structure overlapping abdominal sternum I at side (Fig. 35.9*q*) **35**

33a **(32b)** Width across eyes as wide or wider than greatest width of pronotum; elytron without striae but with three rows of large ocellate foveae; length 5.5–11.2 mm; riparian . . . . . ***Elaphrus*** Fabricius

Fourteen species. Key: Lindroth (1961, p. 109).

33b Width of head across eyes less than width of pronotum; elytral striae distinct; elytron with foveate, punctures in intervals 3, 5, and 7 but these foveae not ocellate **34**

34a **(33b)** Pronotum with sides reflexed; frons with two longitudinal grooves on each side; inner and outer grooves connected by transverse groove at level of anterior supraorbital seta (Fig. 35.9*r*); marshes and bogs, rare, 10–18 mm . . . . ***Blethisa*** Bonelli

Five species. Key: Lindroth (1961, p. 105).

34b Pronotum with sides not reflexed; frontal grooves not connected by transverse groove; 7–9 mm; arctic . . . . . ***Diacheila*** Motschulsky

Two species. Key: Lindroth (1961, p. 102).

35a **(32b)** Last segment of each palpus much narrower and shorter (0.25 mm or less, usually 0.10 mm) than penultimate, latter clavate . . . **36**

35b Last segment of each palpus equal to or longer than penultimate segment; latter not clavate **49**

36a **(35a)** Elytra truncate; one supraorbital seta; minute soil beetles . . . . . Tribe **MICRATOPINI,** ***Micratopus*** Casey

Several species. No key available.

36b Elytra not truncate; two supraorbital setae; Tribe **BEMBIDIINI** **37**

37a **(36b)** Eyeless beetles 1–2 mm long, southeastern United States and Pacific coast forest . . . . ***Anillinus*** Casey

Five species. Key: Jeannel (1937).

37b Eyes present **38**

38a **(37b)** Scutellar striole present **39**

38b Scutellar striole absent **41**

39a **(38a)** Apical segment of maxillary palpus much narrower than anteapical segment, and about 0.1 to 0.25 as long as latter (Fig. 35.9*s*). . . . **40**

39b Apical segment of maxillary palpus conical, its base only slightly narrower than anteapical segment, apical about 0.33 as long as ante-apical (Fig. 35.9*t*); riparian, Pacific Coast forest . . . . . . ***Phrypeus ricksecken*** Hayward

40a **(39a)** Pronotum and elytron pubescent; elytra striation not evident; eyes very large; arctic, and one introduced species on cultivated land in east . . . . ***Asaphidion*** Gozis

Three species. Key: Lindroth (1963, p. 204).

40b Pronotum and elytron glabrous (in a few species, each elytral interval with one row of very long setae); at least bases of inner elytral striae evident; mostly riparian or alpine, but some soil species in forests and cultivated ground) . . . . ***Bembidion*** Latreille

Four hundred species. Keys: Lindroth (1963), Hayward (1897).

41a **(38b)** Sutural stria recurved at apex; eyes large **42**

41b Sutural stria not recurved at apex; eyes small; California . . . . . ***Lymneops*** Casey

Two species. Key: Casey (1918).

42a **(41b)** Recurrent stria nearly parallel to margin (Fig. 35.9*u*); flattened beetles usually found under bark. **43**

42b Recurrent stria divergent from margin (Fig. 35.9*v*) **44**

43a **(42a)** Pronotum with several marginal setae; pronotal margin reflexed; recurrent stria interrupted near elytral apex (Fig. 35.9*w*); apex of elytron pale; 1.5–1.8 mm . . . . ***Mioptachys flavicauda*** Say

43b Pronotum with 1 marginal seta; pronotal margin not reflexed; recurrent stria not interrupted (see **42a**); dorsal surface entirely black, dull; 2.5–3.2 mm . . . . ***Tachyta*** Kirby

Four species. Key: Erwin (1975).

44a **(42b)** Mentum with large pit organs (Fig. 35.9*x*); form varying; if body is strongly convex, then striae are strongly punctate **45**

44b Mentum without large pit organs; body very convex; elytral striae faintly punctate or impunctate; bare soil, riparian, some in ant nests; 1.9–3.2 mm . . . . ***Elaphropus*** Motschulsky

Thirty species. Key: Lindroth (1966, p. 410, couplet 2).

45a **(44a)** Recurrent stria short, not hooked around posterior dorsal puncture (Fig. 35.9*y*) **46**

45b Recurrent stria longer, hooked around posterior dorsal puncture (see **41b**) **48**

46a **(45a)** Marginal stria of elytron entire; dorsal surface pale brown, elytra with common dark spot; 2.5–3.2 mm; riparian, eastern United States . . . . ***Pericompsus ephippiatus*** Say

46b Marginal stria interrupted at middle or absent. **47**

47a **(46b)** Marginal stria of elytron interrupted at middle; length 2.8–3.2 mm; strongly convex, with several strongly punctured striae; in sawdust piles, introduced . . . . . . ***Porotachys bisulcatus*** Nicolai

47b Marginal stria entirely absent; body somewhat flattened; elytron with one impunctate stria; 1.3–1.7 mm . . . . ***Polyderis*** Motschulsky

Two species. Key: Lindroth (1966, p. 413, couplet 25).

48a **(45b)** Elytra strongly iridescent; 1.8–3.1 mm; mostly riparian. . . . . . ***Paratachys*** Casey

Nineteen species. Keys: Lindroth (1966, p. 413, couplet 19), Hayward (1900).

48b Elytra not iridescent; halophilous spp.; 2.0–2.7 mm . . . . ***Tachys*** Stephens

Twelve species. Keys: Lindroth (1966, p. 413, couplet 26); Hayward (1900).

49a **(35b)** Frons with deep grooves concentric on and distant from eyes; latter small (Fig. 35.9*z*); apical segment of maxillary palpus conical, longer than penultimate segment; 3.5–6.5 mm; Tribe **TRECHINI** . . . **50**

49b Frontal grooves, if well developed, not concentric on and distant from eyes; maxillary palpus not conical and elongate **51**

50a **(49a)** Elytra hairy; dorsal surface rufous with bluish cross-band near elytral apex; 4.4–5.5 mm; northeastern North America, in cultivated ground, especially in mouse burrows. ***Lasiotrechus discus*** Fabricius

50b Elytra glabrous; elytra without bluish cross-band; mostly in forests ***Trechus*** Latreille

Forty species. Keys: Lindroth (1961, p. 192), Barr (1962).

51a (**49b**) Head across eyes conspicuously broader than pronotum (Fig. 35.9*aa*) **52**

51b Head across eyes equal to or narrower than pronotum . . . **57**

52a (**51a**) Claws pectinate, south Texas . . . . Tribe **AGRIINI,** ***Agra oblongopunctatum*** Chevrolat

52b Claws not pectinate. **53**

53a (**52a**) Pronotum cordate (Fig. 35.9*aa*); elytra with long erect setae; Tribe **LACHNOPHORINI** **54**

53b Pronotum subcylindrical; elytra glabrous. **56**

54a (**53a**) Elytra with strong constriction at basal third (Fig. 35.9*bb*); small riparian species of southern states and Mexican border . . . . . . ***Ega*** Castelnau

Two species. Key: Schaupp (1879a).

54b Elytra without constriction at basal third **55**

55a (**54b**) Apical segment of labial palpus fusiform (Fig. 35.9*cc*); dorsal surface black; southern states, near ponds . . . . ***Euphorticus pubescens*** Dejean

55b Apical segment of labial palpus oval, with acuminate apex (Fig. 35.9*dd*); head and pronotum metallic green; elytra pale with dark marks; riparian; west coast and Mexican border . . . . ***Lachnophorus*** Dejean

Two species. Key: Schaupp (1879b).

56a (**53a**) Pronotum with base and apex of equal width (Fig. 35.9*ee*); apex of elytron entire; elongate beetle with black head; pronotum and inner intervals brown; outer elytral intervals yellow, eastern North America . . . . Tribe **CTENODACTYLINI,** ***Leptotrachelus*** Latreille

Three species. No key available.

56b Pronotum with base broader than apex (Fig. 35.9*ff*); apex of elytron obliquely truncate (Fig. 35.9*gg*). . . . . Tribe **ODACANTHINI,** ***Colliuris*** DeGeer

Four species. Lindroth (1969, p. 1008).

57a (**51b**) Tip of elytron either evenly rounded (Fig. 35.9*hh*) or obliquely sinuate (Fig. 35.9*ii*). **58**

57b Tip of elytron transversely or slightly obliquely truncate (Fig. 35.9*jj*) **128**

58a (**57a**) Head, including mandibles, 1.5 times longer than prothorax; elytra short, inflated, with only faint traces of striation (Fig. 35.10*a*); black beetles 8–16 mm; long, from Pacific coast forests . . . . Tribe **PROMECOGNATHINI,** ***Promecognathus*** Chaudoir

Two species. Key: Lindroth (1961, p. 127).

58b Without the foregoing combination of characters (if head is elongate, striae are distinct; if striae are obsolete, head is less elongate) **59**

59a (**58a**) Pronotum nearly spherical; except for its base, which is fitted to pedunculate mesothorax; Tribe **BROSCINI** **60**

59b Pronotum more or less flattened; mesothorax not markedly pedunculate. **62**

60a (**59a**) Angular seta present; elytral striae distinct; head constricted behind eyes (Fig. 35.10*b*); Pacific Coast forests **61**

60b Angular seta absent (Fig. 35.10*c*); elytral striae almost obliterated;

head not constricted behind eyes; dorsal surface feebly bronzed; 6–10 mm; xerophilous, Canada, northern New England, Rocky Mountains south to Colorado. . . . . . *Miscodera arctica* Paykull

61a **(60a)** Frons irregularly wrinkled (Fig. 35.10*b*); dorsal surface coppery or bronzed; 12–18 mm. . . . . *Zacotus matthewsi* Leconte

61b Frons not wrinkled (Fig. 35.10*d*); dorsal surface dull black; 9.0–9.4 mm . . . . *Broscodera insignis* Mannerheim

62a **(59b)** Mandible with scrobal seta **63**

62b Scrobal seta absent **69**

63a **(62a)** Head with distinct constriction posterior to eyes (Fig. 35.10*e*) **64**

63b Head without constriction posterior to eyes; length 5–9 mm; beetles associated with seashores or alkali lakes . . . . Tribe **POGONINI**

Four genera, five species. Key: Ball (1960).

64a **(63a)** Outer antennal segments oval, constricted basally, length less than 1.5 of greatest width (Fig. 35.10*f*); abdominal sterna III–V each with deep groove parallel to posterior margin; Tribe **NOMIINI** **65**

64b Outer antennal segments elongate, length more than twice greatest width, segments not suddenly constricted at base (Fig. 35.10*g*); abdominal sterna without grooves paralleling posterior margins of sterna III–V; Tribe **PATROBINI** **66**

65a **(64a)** Hind angle of pronotum with two teeth, separated by notch; one pair of clypeal setae; eye large, convex (Fig. 35.10*e*); in forests, under bark; 6.0–7.7 mm . . . . *Nomius pygmaeus* Dejean

65b Hind angle with tooth single; eye small, protuberant (Fig. 35.10*h*); in forests; two pairs of clypeal setae; 5.5–6.2 mm . . . . *Psydrus piceus* Leconte

66a **(64b)** Eye large, separated from base of head by 1.5 or less of length of eye (Fig. 35.10*i*); length 8 mm or more **67**

66b Eye small, separated from base of head by more than twice length of eye; 5.2–6.0 mm, barren streamside gravels, Pacific northwest *Platidiolous vandykei* Kurnakow

67a **(66a)** Head with deep, linear constriction behind eyes **68**

67b Head with shallow constriction behind eye; 8.5–17 mm; barren stream gravels, mountains of west and northeast . . . . *Diplous* Motschulsky

Four species. Key: Lindroth (1961, p. 187).

68a **(67a)** Elytron with three to four setiferous punctures in interval 3; pronotum with one lateral seta (Fig. 35.10*i*); 8–15 mm; hygrophilous, but often away from standing water . . . . *Patrobus* Dejean

Six species. Key: Lindroth (1961, p. 178).

68b Elytron with 10–12 setiferous punctures in interval 3, also several in anterior parts of intervals 1,5; pronotum with three to four marginal setae; 10–13 mm; in beaver houses . . . . *Platypatrobus lacustris* Darlington

69a **(62b)** Two orbital setae over each eye **70**

69b One orbital setae over each eye **95**

70a **(69a)** Last segment of maxillary palpus arising obliquely from penultimate segment (Fig. 35.10*j,k*); red and black beetles with head

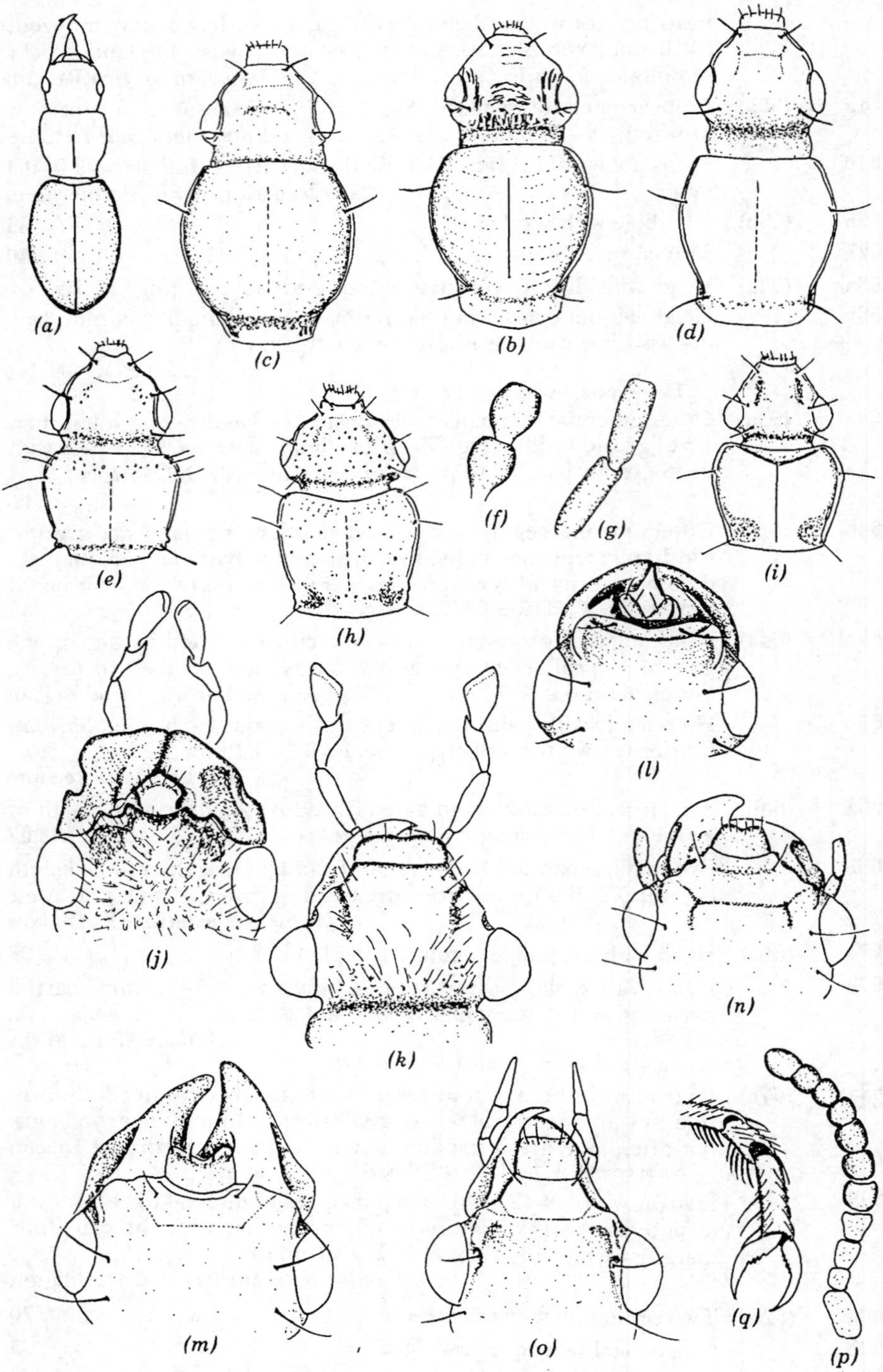

Fig. 35.10. Generic characteristics of adult carabidae.

constricted near base and dorsal surface hairy; Tribe **PANAGAEINI** **71**

70b Last segment of maxillary palpus arising from apex of penultimate segment **72**

71a **(70a)** Clypeus truncate; eyes protruding; head deeply constricted behind eyes (Fig. 35.10*k*) . . . . ***Panagaeus*** Latreille

Three species. Key: Leconte (1879b).

71b Clypeus with median notch, bounded on either side by tooth; head not markedly constricted behind eyes; eyes less protruding (Fig. 35.10*j*); Texas, New Mexico . . . . ***Micrixys distinctus*** Haldeman

72a **(70b)** Clypeus deeply emarginate at base, more or less exposing basal membrane of labrum; labrum bilobed. (Fig. 35.10*l,m*); TRIBE LICININI **73**

72b Clypeus truncate or very slightly emarginate; labrum truncate to shallowly emarginate (Fig. 35.10*n*) **75**

73a **(72a)** Labrum emarginate; neither left nor right mandible with dorsal notch (Fig. 35.10*m*); 9.5 mm. or more in length **74**

73b Labrum cleft nearly to base; either right or left mandible with deep dorsal notch (Fig. 35.10*l*); 3.8–6.5 mm . . . . . ***Badister*** Clairville

Thirteen species. Key: Lindroth (1969, p. 956).

74a **(73a)** Metepisternum longer than wide (see **32a**); interval 7 without carina. . . . . ***Diplocheila*** Brulle

Eight species. Keys: Ball (1959), Lindroth (1968, p. 940).

74b Metepisternum as wide as long; interval 7 carinate, at least near base . . . . ***Dicaelus*** Bonelli

Fourteen species. Key: Ball (1959).

75a **(72b)** Last segment of maxillary palpus acutely pointed (Fig. 35.10*o*); length 2 mm; elytron glabrous except for band of pubescence along lateral margin; in rotting vegetation. . . . .
Tribe **PERIGONINI,** ***Perigona nigriceps*** Castelnau

75b Last segment of maxillary palpus truncate or oval (Fig. 35.10*n*); length 4 mm or more **76**

76a **(75b)** Mandible short, strongly curved; distance between dorsal condyle and apex of mandible less than distance between dorsal condyles of right and left mandibles (Fig. 35.10*n*); Tribe **AMARINI**. . . . **77**

76b Mandible long, less curved; distance between dorsal condyle and apex of mandible much more than distance between the two dorsal condyles **78**

77a **(76a)** Penultimate segment of labial palpus with two large tactile setae; frontal furrows represented by shallow transverse impressions; 5.0–5.8 mm; forests, eastern Canada, northeastern United States
Disamara arenaria Leconte

77b Penultimate segment of labial palpus with three or more setae; frontal furrows better developed, in most species clearly longer than broad . . . . ***Amara*** Bonelli

Two hundred species. Keys: Lindroth (1968, p. 656), Hayward (1908).

78a **(76b)** Antennal segments 5–10 submonilliform, short, slightly compressed (Fig. 35.10*p*); tibiae distinctly flattened; length 11–17 mm; in rotten logs. . . . . Tribe **MORIONINI,** ***Morion*** Latreille

Two species. Key: Allen (1968).

78b Antennal segments longer, subcylindrical; tibiae not distinctly flattened; Tribe **PTEROSTRICHINI** **79**

79a (78a) Tarsi pubescent dorsally (Fig. 35.10*q*); length 13–18 mm; dorsal surface black, elytra with bluish sheen; introduced from Europe and Africa; restricted to port areas of west coast and eastern Canada . . . . ***Pristonychus*** Dejean

Two species. Key: Lindroth (1966, p. 549).

79b Tarsi not pubescent dorsally **80**

80a (79b) Tarsal claws denticulate or serrate (Fig. 35.10*q*) **81**

80b Tarsal claws smooth **82**

81a (80a) Pronotum oval, hind angles strongly rounded (Fig. 35.11*a*); 8.7–13.2 mm . . . . ***Synuchus*** Gyllenhall

Two species. Key: Lindroth (1966, p. 551).

81b Pronotum with hind angle either rectangular (Fig. 35.11*b*) or obtuse (Fig. 35.11*c*); 7–15 mm . . . . ***Calathus*** Bonelli

Eight species. Keys: Lindroth (1966, p. 542), Ball and Negre (1972).

82a (80b) Elytral epipleura interrupted by a twisted fold near apex (see **24a**) **83**

82b Epipleura not interrupted near apex (see **24b**) **90**

83a (82a) Apical segment of labial palpus broadly triangular (Fig. 35.11*d*); its width subequal to length of segment; stout oval spp.; 13–20 mm long; base of pronotum only slightly narrower than greatest width of pronotum; forests; black with faint purple reflections . . . . . ***Myas*** Dejean

Two species. Key: Lindroth (1966, p. 444).

83b Apical segment of labial palpus narrow, much longer than broad **84**

84a (83b) Median tooth of mentum entire (Fig. 35.11*e*); elytra iridescent **85**

84b Median tooth of mentum emarginate or cleft (Fig. 35.11*f*). . . **87**

85a (84a) Scutellar striole present; pronotum broadly margined; dorsum black with iridescence on elytra; southeastern United States . . . ***Pterostichus* (*Piesmus*) *monedulus*** Germar

85b Scutellar striole absent; pronotum narrowly margined **86**

86a (85b) Sides of pronotum strongly constricted near base; anterior tarsus of male symmetrical; southern Texas . . . . . ***Oxycrepis ulkei*** Horn

86b Sides of pronotum not or but slightly constricted near base (Fig. 35.11*g*); anterior tarsus of male very asymmetrically dilated (Fig. 35.11*h*); hygrophilous; mostly in marshes . . . ***Loxandrus*** Leconte

Sixty-three species. Key: Allen (1972).

87a (84b) Penultimate segment of labial palpus bisetose **88**

87b Penultimate segment of labial palpus with three or more setae; eastern United States, Great Plains, Arizona. . . . . ***Evarthrus*** Leconte

Forty-three species. Key: Freitag (1969).

88a (87a) Pronotum strongly cordate, base very strongly narrowed; sides strongly curved; hind angles very obtuse (Fig. 35.11*i*). . . . . . . ***Evarthrus***

A few species. Key: Freitag (1969).

88b Pronotum not strongly cordate (if somewhat cordate, then hind angles are obtuse) **89**

89a (88b) Pronotum with base very broad, scarcely narrower than greatest width (Fig. 35.11*j*); black, elytra at least slightly iridescent; antennal

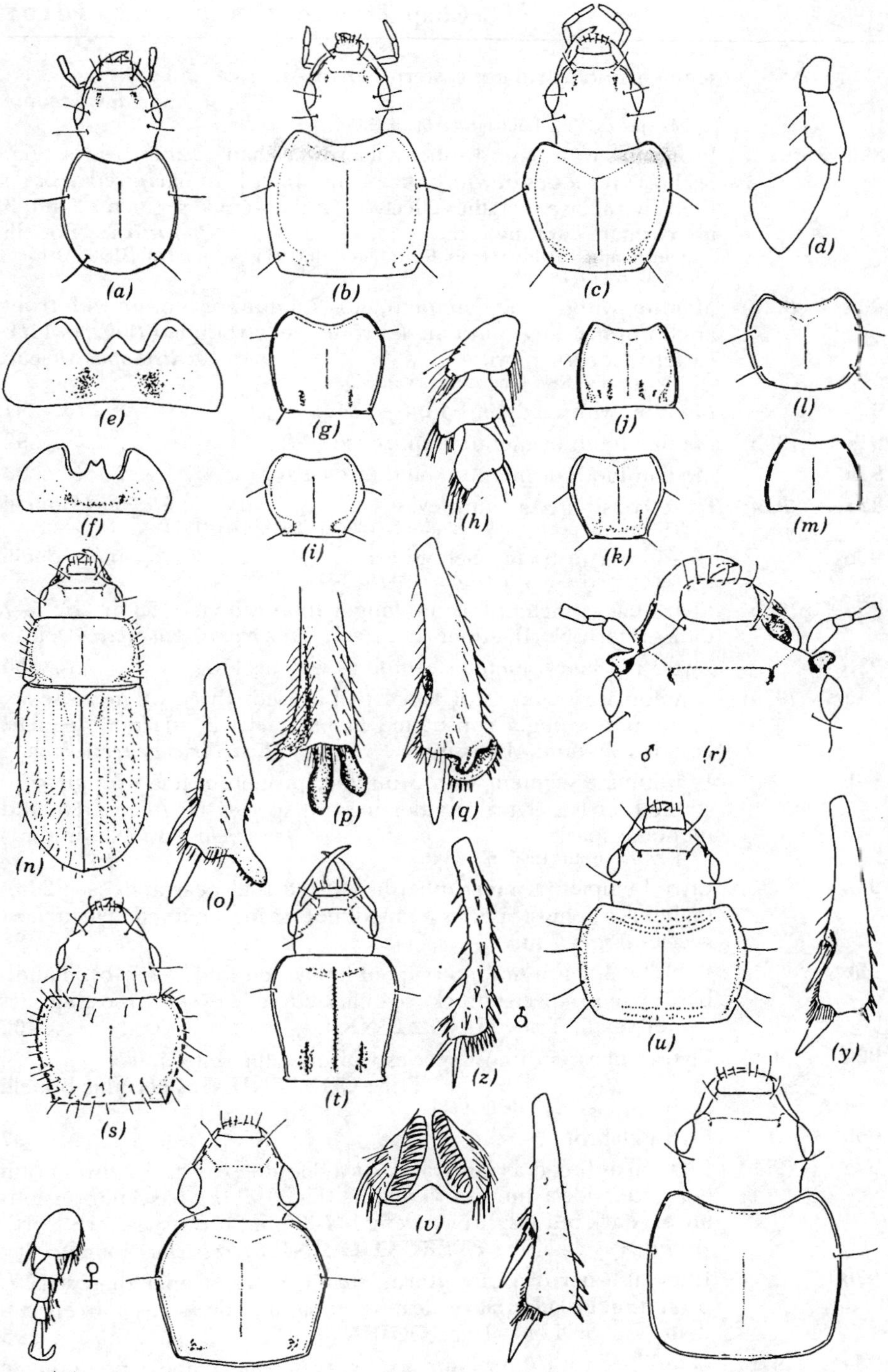

Fig. 35.11. Generic characteristics of adult carabidae.

segments not carinate; eastern North America . . . . **Abacidus** Leconte

Six species. Key: Lindroth (1966, p. 535).

89b Pronotum with base distinctly narrower than greatest width (Fig. 35.11*k*) (in species in which the pronotum is least narrowed at base; then elytra have metallic reflections and antennal segments 2 and 3 are carinate laterally) . . . . **Pterostichus** Bonelli

One hundred eighty species. Keys: Lindroth (1966, p. 448), Ball (1966), Hacker (1968), Barr (1971).

90a **(82b)** Mentum without median tooth; 4.0–7.5 mm; pronotum with front angles protruding, hind angles completely rounded (Fig. 35.11*l*); eastern North America . . . . **Olisthopus** Dejean

Two species. Key: Lindroth (1966, p. 533).

90b Mentum with median tooth (see **84a**). **91**

91a **(90b)** Median tooth of mentum entire (see **84a**) **92**

91b Median tooth of mentum bifid (see **84b**) **93**

92a **(91a)** Head constricted behind eyes . . . . **Platynus** Bonelli

Twenty-six species. Keys: Lindroth (1966, p. 557), Barr (1971).

92b Head not constricted behind eyes. . . . . **Agonum** Bonelli

Eighty species. Key: Lindroth (1966, p. 557).

93a **(91b)** Elytra hairy; metepisternum longer than wide (see **3a** or **2b**); 5–7 mm, eastern North America . . . . . . **Atranus pubescens** Dejean

93b Elytra not hairy; metepisternum as wide as long **94**

94a **(93a)** Penultimate segment of labial palpus with three or more setae; pronotum strongly constricted at base; elytral striation regular; length 12.8 mm; Mississippi. . . . . . . **Evarthrus gravesi** Freitag

94b Penultimate segment with two setae; pronotum less strongly constricted at base; elytral striation in some species irregular; arctic and montane species . . . . **Pterostichus** Bonelli (in part)

Key: Lindroth (1966, p. 448).

95a **(69b)** Elytral epipleura interrupted by twisted fold near apex (see **24a**); mandibles in most species pointed, not strongly curved; eye in most species large; 7 mm **96**

95b Elytral epipleura not interrupted by twisted fold (see **24b**); mandibles in most species thick, strongly curved; eyes in most species rather small; Tribe **HARPALINI** **102**

96a **(95a)** Elytra pubescent; most species with metallic reflections . . . . . . Tribe **CHLAENIINI, Chlaenius** Bonelli

Fifty species. Key: Bell (1960).

96b Elytra glabrous **97**

97a **(95b)** Interval as broad as interval 8; stria 8 as deep as stria 7; pronotum elongate, widest anterior to middle (Fig. 35.11*k*); basal impressions linear; dark reddish brown; length 7–9 mm; forests, eastern North America . . . Tribe **PTEROSTICHINI, Pterostichus honestus** Say

97b Interval 9 narrow, careniform; stria 8 much deeper than stria 7; pronotum broadest at or near base; basal impressions scarcely evident (Fig. 35.11*m*); Tribe **OODINI** **98**

98a **(97b)** Mandibles with dorsal surface excavated near apex; penultimate segment of labial palpus with two setae **99**

98b Mandible with dorsal surface not excavated near tip; penultimate segment of labial palpus without setae **100**

99a (98a) Elytral striae rather deep, coarsely punctate; stria 7 as distinct as others; southeast United States. . . . . . ***Evolenes exarata*** Dejean

99b Elytral striae very shallow; finely punctate; stria 7 fainter than stria 6 or 8; Louisiana . . . . ***Dercylinus impressus*** Leconte

100a (98b) Clypeal setae present; 7.5–14.4 mm; in marshes, semiaquatic; eastern United States . . . . ***Oodes*** Bonelli

Four species. Key: Lindroth (1969, p. 997).

100b Clypeal setae absent. **101**

101a (100b) Labrum with six setae; elytron with stria 7 as well developed as striae 1–6; eastern United States . . . . ***Anatrichis*** Leconte

Three species. Key: Horn (1870).

101b Labrum with three setae; elytron with stria 7 faint or absent; southern and eastern United States. . . . . ***Stenocrepis*** Chaudoir

Six species. Key: Horn (1870).

102a (95b) Prothorax elongate anterior to coxae; anterior coxal cavity about 0.33 of length of prothorax; elongate, brown beetles; southern United States. . . . . ***Stenomorphus*** Leconte

Key: Van Dyke (1943).

102b Pronotum not elongate anterior to coxae; anterior coxal cavity about 0.50 of length of prothorax; body rarely elongate . . . **103**

103a (102b) Anterior tibia with laterally directed tooth at apex (Fig. 35.11*o*); body very stout (Fig. 35.11*n*); sides of pronotum, uneven, unnumbered intervals of elytron with long setae; 10.16 mm; grasslands, central North America . . . . ***Euryderus grossus*** Say

103b Anterior tibia without laterally directed tooth. **104**

104a (103a) Anterior tibia with broadly rounded plate at apex (Fig. 35.11*q*); spurs of hind tibia spatulate (Fig. 35.11*p*); color pale; length 13–17 mm; eastern North America, burrows in sandy soil . . . . ***Geopinus incrassatus*** Dejean

104b Anterior tibia without expanded plate at apex; spurs of hind tibia not spatulate **105**

105a (104b) Apical spur of anterior tibia trifid. **106**

105b Apical spur of anterior tibia not trifid. **107**

106a (105a) Each elytral interval with two rows of setigerous punctures; 5–14 mm; western North America . . . . ***Dicheirus*** Mannerheim

Five species. Key: Noonan (1968).

106b Elytra with few setiferous punctures near apex of striae 3, 5, 7 (or with one or two of the rows missing); 9–14 mm; east of Rockies. . . . . ***Anisodactylus*** (Subgenus ***Gynandrotarsus***) Laferté

Seven species. Key: Lindroth (1968, p. 835, couplet 4).

107a (105b) Tooth present at side of frons above base of antenna; male with angulate process on right mandible (Fig. 35.11*r*); 7.5–11.5 mm; open fields, United States, east of Rockies. . . . . ***Cratacanthus dubius*** Beauvois

107b Frons without tooth above base of antenna; right mandible of male not modified **108**

108a (106b) Head with transverse row of setiferous punctures between eyes (Fig. 35.11*s*); pronotum and elytra with long, erect setae, 7–12 mm; grasslands, interior of North America . . . ***Piosoma*** Leconte

Two species. Key: Lindroth (1968, p. 745).

108b Head without transverse row of setae **109**

109a (108b) Angular seta of pronotum present; eyes with short pubescence; outer elytral intervals pubescent; 3.2–5.2 mm; Canada and Rocky Mountains and northern New England; dry, thin forest, grasslands. . . . . ***Trichocellus*** Ganglbauer

Two species. Key: Lindroth (1968, p. 875).

109b Angular seta absent; eye not hairy **110**

110a (109b) Mandibles very long, pointed; labrum very short (Fig. 35.11*t*); 4.2–4.8 mm; eastern North America . . . . ***Amerinus linearis*** Leconte

110b Mandible short, curved; labrum not especially short **111**

111a (110b) Penultimate segment of labial palpus with more than three setae **112**

111b Penultimate segment of labial palpus with two or rarely three setae **123**

112a (111a) Elytron with several setiferous punctures in or near striae 2, 5, and 7 . . . . **113**

112b Elytron with up to one setiferous puncture in stria 2 only . . **116**

113a (112a) Several lateral setae on pronotum (Fig. 35.11*u*) length 11–13 mm; color pale yellowish brown; sand bars and dunes, upper midwest. ***Hartonymus hoodi*** Casey

113b One lateral seta on pronotum; length 4–10 mm; color darker, often metallic or iridescent **114**

114a (113b) Pronotum with sides strongly narrowed, base almost as narrow as apex (Fig. 35.11*w*); male with segments 1–4 of anterior and middle tarsus dilated, with two rows of scales ventrally (Fig. 35.11*v*); female with segment 1 dilated (Fig. 35.11*x*), densely hairy ventrally; length 6.0–7.5 mm; eastern North America, under bark . . . . . ***Gyandropus hylacis*** Dejean

114b Pronotum with sides less rounded, base clearly wider than apex; males of some species with anterior and middle tarsi dilated, with two rows of scales, other species with tarsus of males not dilated; female with tarsi not dilated **115**

115a (114a) Anterior tibia with six to eight stout spines on outer margin (Fig. 35.11*y*); male with anterior tarsus not dilated; hind angles of pronotum obtusely rounded (Fig. 35.11*bb*); middle tibia of male strongly curved, with small tubercles on medial surface (Fig. 35.11*z*) . . . . ***Discoderus*** Leconte

Nineteen species. Key: Horn (1883).

115b Anterior tibia with five or fewer spines on outer margin (Fig. 35.11*aa*); male with anterior and middle tarsi dilated, with one row of scales ventrally; hind angles obtuse to rectangular; middle tibia of male straight, without tubercles . . . . . . ***Selenophorus*** Dejean

Forty-three species. Keys: Lindroth (1968, p.822), Horn (1880).

116a (112b) Hind tarsi more elongate; first segment of hind tarsus longer than second and third segments combined (Fig. 35.12*b*); male with front and middle tarsi dilated (Fig. 35.12*a*), densely hairy ventrally (Fig. 35.12*c*) **117**

116b Hind tarsi shorter; first segment of hind tarsus shorter than combined lengths of second and third segments; middle and front tarsi of male various **118**

117a **(116a)** Mentum with well-developed median tooth (see **84a**) . . . . ***Anisotarsus*** Chaudoir

Ten species. Keys: Lindroth (1968, p. 866), Van Emden (1953).

117b Mentum with median tooth absent or minute . . . . ***Anisodactylus*** Dejean

Thirty species. Key: Lindroth (1968, p. 835).

118a **(116b)** Front and middle tarsi of male not dilated, without ventral pubescence or scales; 11.0–13.5 mm; Alaska, northwest Canada . . . . ***Harpalobrachys leirodes*** Motschulsky

118b Front and middle tarsi of male dilated, and either densely pubescent (see **119a**) or else with two rows of scales ventrally (see **114a**) **119**

119a **(118b)** Front and middle tarsi of male densely pubescent below (Fig. 35.12*c*); frontal fovea prolonged toward eye . . . . ***Anisodactylus*** (Subgenus ***Anadaptus*** Casey)

Seven species. Key: Lindroth (1968, p. 837, couplet 27).

119b Front and middle tarsi of male with two rows of scales below (see **114a**) **120**

120a **(119b)** Frontal fovea small but deep, not at all prolonged toward eye **121**

120b Frontal fovea prolonged toward eye **122**

121a **(120a)** Antennal segments 1 or 1 and 2 pale; segments 3–11 black; black, with black legs; length 7.3–9.8 mm; pronotum with sides parallel except near apex, where they are strongly incurved. Canada and northern United States, south, in mountains . . . . ***Harpalellus basilaris*** Kirby

121b Outer segments of antenna paler than those at middle, or else antenna entirely pale; size, shape, and color extremely varied . . ***Harpalus*** Latreille

Sixty species. Key: Lindroth (1968, p. 751).

122a **(120b)** Elytra strongly iridescent; length 9–12 mm, eastern North America, deciduous forests . . . . ***Trichotichnus*** Morawitz

Three species. Key: Lindroth (1968, p. 818).

122b Elytra not iridescent; length 6.5–8 mm; eastern United States, open fields . . . . ***Episcopellus autumnalis*** (Fig. 35.12*d*) Say

123a **(111b)** Frontal grooves extended obliquely to the orbital puncture; 6 mm; southern Texas . . . . ***Pogonodaptus mexicanus*** Bates

123b Frontal grooves not extended to orbit, or else extended to orbit anterior to orbital seta **124**

124a **(123a)** Length 15–20 mm; head of male greatly enlarged (Fig. 35.12*e*) (compare with female); southern Arizona and New Mexico . . . . ***Polpochila*** Solier

Two species. Key: Negre (1963).

124b Length 12 mm or less; head of male not larger than that of female. **125**

125a **(124b)** Mentum with median tooth (see **84a**) **126**

125b Mentum without median tooth **127**

126a **(125a)** Segments 1–4 of front and middle tarsi of male with dense pubescence beneath; southern Arizona, northern Mexico . . . . ***Pelmatellus lucidus*** Casey

126b Segments 1–4 of front and middle tarsi of male with two rows of

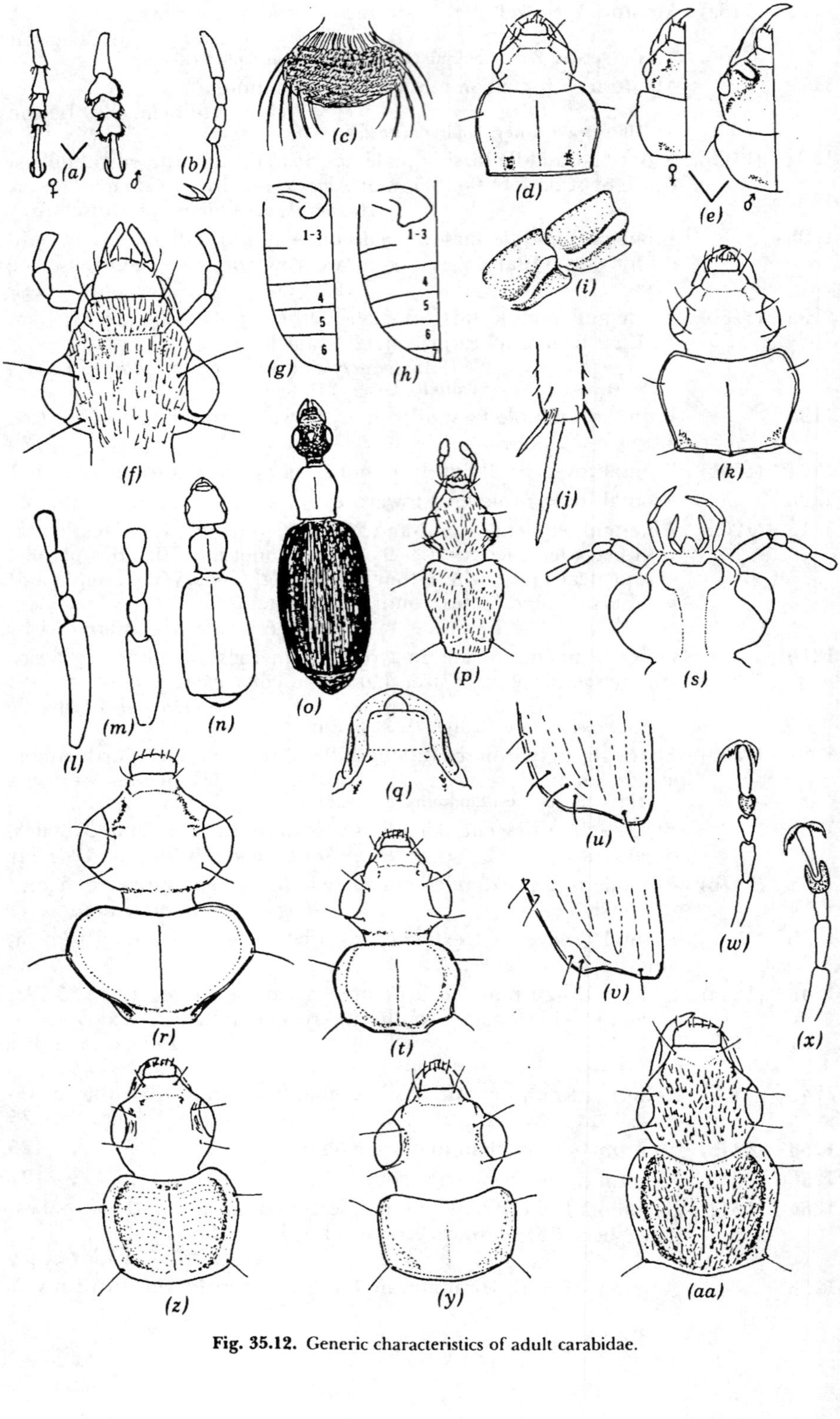

Fig. 35.12. Generic characteristics of adult carabidae.

scales beneath (see **114a**), or without any ventral vestiture; 2.8–7.0 mm . . . . ***Bradycellus*** Erichson
Twenty-five species. Key: Lindroth (1968, p. 880).

127a **(125b)** Posterior series of setiferous punctures in stria 8 continuous; 2.4–4.1 mm . . . . ***Acupalpus*** Latreille
Fifteen species. Key: Lindroth (1968, p. 925).

127b Posterior series of setiferous punctures in stria 8, divided into two groups of four each; 3.2–9.1 mm. . . . . . . ***Stenolophus*** Stephens
Twenty-two species. Key: Lindroth (1968, p. 905).

128a **(57b)** One orbital seta present; dorsal surface at least partly hairy . . . **129**

128b Two orbital setae; dorsal surface glabrous or hairy (Fig. 35.12*f*) **130**

129a **(128a)** Abdomen with six visible sterna (Fig. 35.12*g*); body entirely pale; length 10 mm or less . . . .
Tribe **DRYPTINI,** ***Pseudaptinus*** Castelnau
Seven species. Key: Leconte (1879d).

129b Abdomen with seven or eight visible sterna (Fig. 35.12*h*); elytra dark, in most species blue; head and pronotum pale in almost all species . . . . Tribe **BRACHININI,** ***Brachinus*** Weber
Forty-nine species. Key: Erwin (1970)

130a **(128b)** Outer antennal segments short, quadrangular, strongly flattened (Fig. 35.12*i*); anterior tibia strongly flattened; length 12 to 17 mm; elytra hairy. . . . . Tribe **HELLUONINI,** ***Helluomorphoides*** Ball
Seven species. Key: Ball (1956).

130b Outer antennal segments at most slightly flattened; anterior tibia round in cross section **131**

131a **(130b)** Second (inner) spur of hind tibia very long, almost equal in length to first segment of tarsus (Fig. 35.12*j*) **132**

131b Second spur of hind tibia much shorter than first segment of tarsus **133**

132a **(131a)** Hind tibial spurs serrate (Fig. 35.12*j*); head not constricted posteriorly to form a neck (Fig. 35.12*k*); elytra pale, in most species with pattern of dark spots; 4.5–6.0 mm; sandy areas. . . . .
Tribe **MASOREINI,** ***Tetragonoderus*** Dejean
Four species. Key: Lindroth (1968, p. 1010).

132b Hind tibial spurs not serrate; head constricted to form a neck; elytron dark with pale apex and pale subasal spot; 4.2–5.5 mm, eastern United States . . . .
Tribe **LEBIINI,** ***Nemotarsus elegans*** Leconte

133a **(131b)** First antennal segment longer than segments 2–4 combined (Fig. 35.12*l*); dorsal surface hairy; head with narrow neck, distinctly less than half width of head across eyes **134**

133b First antennal segment shorter than segments 2, 3, and 4 combined (Fig. 35.12*m*); dorsal surface hairy or glabrous; head either not narrowed behind eyes, or only moderately narrowed **135**

134a **(133a)** Entire dorsal surface pale; eyes small (Fig. 35.12*n*); length 5.2–6.0 mm . . . . Tribe **ZUPHIINI,** ***Zuphium*** Latreille
Five species. Key: Leconte (1879c).

134b Head and elytra black in most species; pronotum orange; eyes prominent (Fig. 35.12*o*); length 15–24 mm . . . .
Tribe **DRYPTINI,** ***Galerita*** Fabricius
Five species. Key: Lindroth (1969, p. 1091).

135a **(133b)** All segments of antenna densely pubescent; penultimate segment of labial palpus with three or more setae; color pale; 6–9 mm; dorsal surface hairy; southern and western United States (Fig. 35.12*p*) . . . . Tribe **DRYPTINI,** ***Pseudaptinus*** Castelnau

Seven species. Key: Leconte (1879d, p. 62).

135b Antennal segments 1–3 not pubescent (but with a few tactile hairs); penultimate segment of labial palpus with two setae **136**

136a **(135b)** Each elytral stria with two rows of punctures; elytral striae wider than intervals, latter costate; southern Texas . . . . Tribe **EUCHEILINI,** ***Inna texana*** Schaeffer

136b Stria narrower than intervals with one row of punctures or impunctate **137**

137a **(136b)** Mandible without scrobe; pronotum angulate at side **138**

137b Mandible with scrobe (Fig. 35.12*q*); pronotum with sides not angulate; Tribe **LEBIINI** **139**

138a **(137a)** Tarsal claws smooth; gular sutures indistinct (Fig. 35.12*s*) . . . . Tribe **PENTAGONICINI,** ***Pentagonica*** (Fig. 35.12*r*) Schmidt-Goebel

Four species. (Key: Lindroth (1969, p. 1013)

138b Tarsal claws pectinate; gular sutures distinct; Florida, on palmetto . . . . Tribe **LEBIINI** ***Onota floridana*** Horn

139a **(137b)** Elytron with 3 setigerous punctures forming triangle at outer apical angle (Fig. 35.12*v*); head strongly and suddenly constricted behind eyes (Fig. 35.12*t*); most species brightly colored or patterned; mostly planticolous. . . . . ***Lebia*** Latreille

Forty species. Keys: Madge (1967), Lindroth (1969, p. 1017).

139b Setigerous punctures not forming triangle at outer apical angle (Fig. 35.12*u*); head more gradually constricted or not constricted **140**

140a **(139b)** Labrum as long as wide or nearly so **141**

140b Labrum transverse at least twice as wide as long **142**

141a **(140a)** Dorsal surface of tarsus hairy; body metallic green or blue above; southern and central United States . . . . ***Coptodera*** (Fig. 35.12*y*) Dejean

Two species. Key: Horn (1882).

141b Dorsal surface of tarsus glabrous; pronotum partly or entirely orange; southeastern states . . . . . . ***Phloeoxena signata*** Dejean

142a **(140b)** Fourth segment of tarsus cylindrical or slightly emarginate (Fig. 35.12*w*) **143**

142b Fourth segment of tarsus bilobed (Fig. 35.12*x*) **155**

143a **(142a)** Total length 6 mm or more. **144**

143b Total length 5 mm or less. **151**

144a **(143a)** Mentum without a tooth; tarsal claws dentate (Fig. 35.12*w*). . **145**

144b Mentum with median tooth (see **84a**); tarsal claws smooth or dentate **146**

145a **(144a)** Antennal segments 1 and 2 pubescent; elytra iridescent; south Texas . . . . ***Eucaerus varicornis*** Leconte

145b Antennal segments 1 and 2 without pubescence (but with a few tactile setae); elytra not iridescent; dark brown, arboreal . . . . . ***Dromius piceus*** (Fig. 35.12*z*) Dejean

146a **(144b)** Tarsal claws smooth; dorsal surface hairy; elytra metallic green or blue; pronotum and head vary from rufous to black; 6–8 mm; Great Plains and west . . . . . ***Tecnophilus croceicollis*** Menetries

146b Tarsal claws serrate (see **142a**) **147**

147a **(146b)** Dorsal surface of tarsus densely hairy. **148**

147b Dorsal surface of tarsus glabrous or with a few scattered hairs . . . **149**

148a **(147a)** Dorsal surface of head, pronotum and elytra densely pubescent (Fig. 35.12*aa*). ***Cymindis*** Latreille

Sixteen species. Key: Lindroth (1969, p. 1072).

148b Dorsal surface glabrous . . . . ***Apenes*** Leconte

Five species. Key: Horn (1882).

149a **(147b)** Penultimate segment of labial palpus with three or more setae . . ***Pinacodera*** Schaum

Twelve species. Key: Horn (1882).

149b Penultimate segment of labial palpus with two setae **150**

150a **(149b)** Abdominal segments 4–6 each with long lateral setae (Fig. 35.13*b*); black not metallic; 6.5–9.5 mm; introduced, coastal. . . . . ***Plochionus*** **(s. str.)** ***pallens*** (Fig. 35.13*a*) Fabricius

150b Abdominal segments 4–6 without lateral setae; dorsal surface with faint to bright metallic reflections; southern and western North America . . . . ***Calleida*** (Sg. ***Philophuga*** Motschulsky)

Three species. Key: Larson (1969).

151a **(143b)** Elytral striae distinctly impressed; tarsal claws quite smooth; black, dorsal surface in most species with metallic reflections; riverside gravels; 3.2–4.5 mm . . . . ***Apristus*** Chaudoir

Fourteen species. Keys: Lindroth (1969, p. 1044), Horn (1882)

151b Elytral striae very shallow or faint; tarsal claws denticulate to serrate (see **142a**); dorsal surface rarely metallic **152**

152a **(151b)** Apical segment of labial palpus dilated, nearly as wide as long (Fig. 35.13*c*); elytron black with pale shoulder mark or entirely pale . . ***Axinopalpus*** Leconte

Nine species. No complete key available.

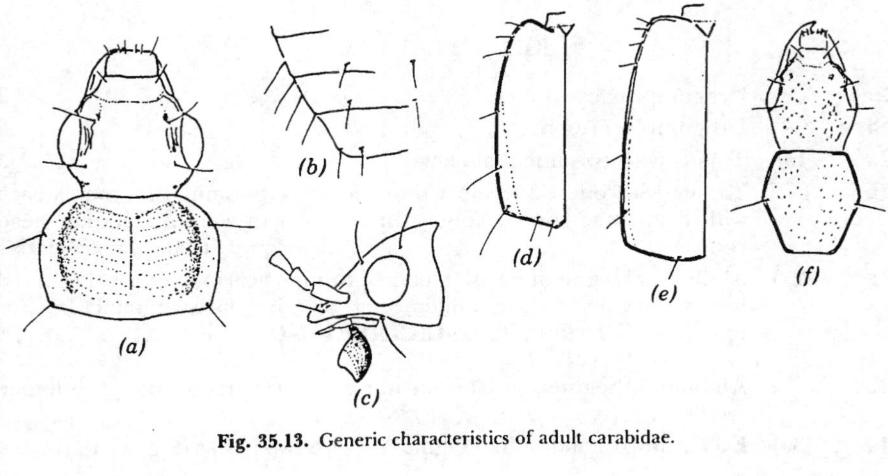

Fig. 35.13. Generic characteristics of adult carabidae.

152b Apical segment of labial palpus at least twice as long as wide . . . **153**

153a **(152b)** Pronotum truncate at base; pale with head black; east coast on beach grasses; 3.3–3.6 mm . . . . ***Dromius atriceps*** Leconte

153b Pronotum rounded at base; dorsal surface black **154**

154a **(153b)** Mentum with bifid tooth (see **84b**); antennal pubescence begins on segment 4; basal margin of elytron extends medially almost to suture; tip of elytron with medial half oblique (Fig. 35.13*d*); 2.7–3.5 mm.; on bare soils . . . . ***Metabletus americanus*** Dejean

154b Mentum without tooth; antennal pubescence begins on segment 3; basal margin of elytron not extending medial to base of stria 3; tip of elytron entirely transverse (Fig. 35.13*e*); 2.4–4.1 mm . . . . . ***Microlestes*** Schmidt-Goebel

Seven species. Key: Lindroth (1969; p. 1048).

155a **(142b)** Mentum without tooth **156**

155b Mentum with median tooth (see **84a**) **157**

156a **(155a)** Head not constricted behind eyes; head, pronotum coarsely punctate; pronotum as long as wide; south Texas . . . . ***Cylindronotum aeneum*** Putzeys

156b Head slightly constricted behind eyes; head and pronotum impunctate; pronotum wider than long; southern states . . . . ***Euproctinus*** Leng and Mutchler

Two species. Key: Wickham (1897).

157a **(155b)** Elytra entirely metallic green; pronotum narrow, only slightly wider than head (Fig. 35.13*f*), either red or metallic green . . . . ***Calleida*** **(s. str.)** Dejean

Twelve species. Key: Horn (1882).

157b Elytra brown or patterned with dark brown and orange . . . **158**

158a **(157b)** Middle and hind tarsi with middorsal groove on segments 1–3; anterior angles of pronotum distinct; Arizona. . . ***Lecalida*** Casey

Two species. Key: Casey (1920).

158b Middle and hind tarsi without middorsal groove; anterior angles of pronotum broadly rounded; eastern North America . . . . ***Plochionus*** (Subgenus ***Menidius*** Chaudoir)

Three species. Key: Horn (1882).

KEY TO LARVAE

1a Urogomphi absent **2**

1b Urogomphi present **9**

2a **(1a)** Tarsus with two unequal claws **3**

2b Tarsus with one claw; most thoracic and abdominal segments each with transverse row of spines; in burrows in wood, including dead roots . . . . Tribe **RHYSODINI**

3a **(2a)** Abdominal segment 5 with a pair of humps bearing anteriorly curved hooks (Fig. 35.14*a*); abdomen long, slender; larva inhabits burrow open to soil surface; Tribe **CICINDELINI**. **4**

Key: Hamilton (1925).

3b Abdominal segment 5 without humps or recurved hooks, abdomen swollen **7**

4a **(3a)** Each hump of abdominal segment 5 with two hooks (Fig. 35.14*a*) . . . **5**

4b		Each hump with three hooks; Pacific coast forest . . . . ***Omus***
5a	(4a)	Six lateral ocelli on each side of head **6**
5b		Four lateral ocelli on each side of head; Great Plains and western deserts. . . . . ***Amblycheila***
6a	(5a)	Both hooks of each abdominal hump point anteriorly; southern half of United States . . . . ***Megacephala***
6b		Inner hook short, straight, outer one very long, curved (Fig. 34.14*a*) ***Cicindela***
7a	(3b)	Ninth segment of abdomen forming concave terminal disk . . . . . Tribe **OZAENINI**
7b		Ninth abdominal segment not forming concave disk. **8**
8a	(7b)	Anterior claw largest in anterior leg; posterior claw largest in middle and hind legs; abdomen greatly swollen . . . . Tribe **PSEUDOMORPHINI**
8b		Anterior claw largest in all legs; abdomen not swollen. . . . . Tribe **LEBIINI** ***(Dromius)***
9a	(1b)	Tarsus with one claw **10**

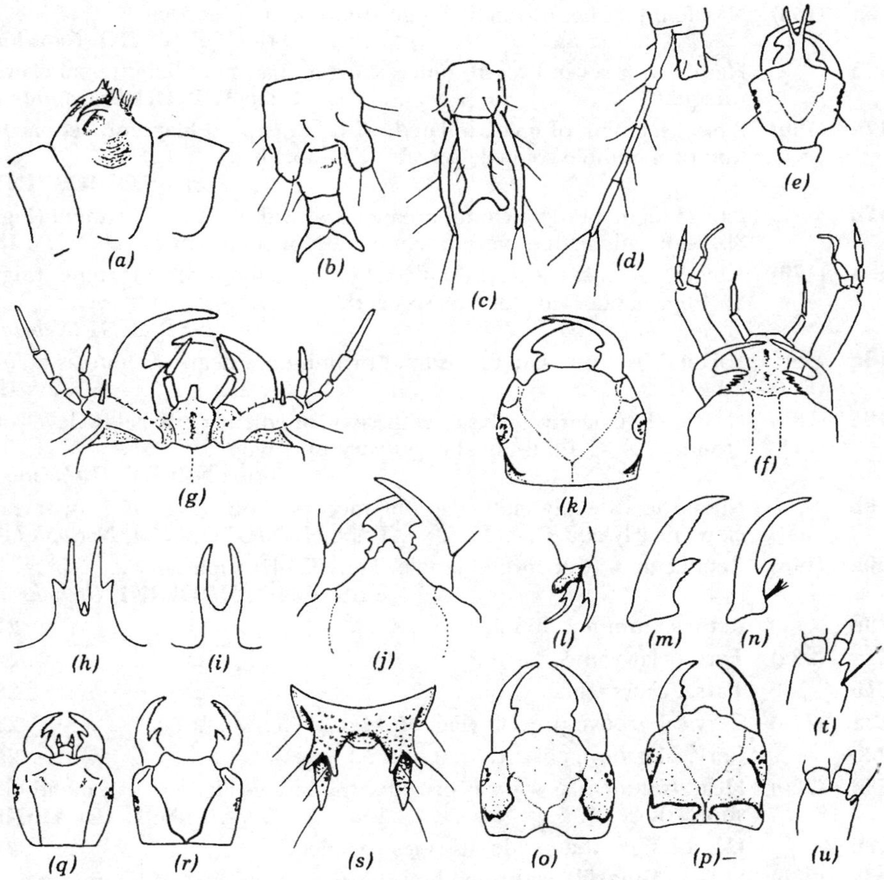

Fig. 35.14. Generic characteristics of larval carabidae.

9b Tarsus with two claws **14**

10a **(9a)** Urogomphus knoblike, half or less length of segment 10 (Fig. 35.14*b*) **11**

10b Urogomphus elongate, varying from slightly shorter than segment 10 to several times its length **12**

11a **(10a)** Ligula absent; antenna with four segments . . . . Tribe **BRACHININI,** ***Brachinus***

11b Ligula present; antenna with two segments . . . . Tribe **LEBIINI** ***(Lebia)***

12a **(10b)** First segment of antenna the longest . . . . Tribe **BROSCINI**

12b First segment of antenna not the longest **13**

13a **(12b)** Labial palpus with four segments. . . . . Tribe **TRECHINI**

13b Labial palpus with two segments . . . . Tribe **BEMBIDIINI** and **POGONINI**

14a **(9b)** Head constricted posteriorly to form narrow neck **15**

14b Head not or scarcely constricted posteriorly **20**

15a **(14a)** Urogomphi segmented (Fig. 35.14*d*) **16**

15b Urogomphi not segmented (Fig. 35.14*c*) **17**

16a **(15a)** Nasale bifurcate, distinctly longer than mandibles. (Fig. 35.14*e*) . . . Tribe **DRYPTINI,** ***Galerita***

16b Nasale with several teeth, much shorter than mandible; tarsal claws unequal . . . . Tribe **ZUPHIINI** ***(Zuphium)***

17a **(15b)** Apical segment of galea inflated, with long apical filament; retinaculum of mandible coarsely serrate (Fig. 35.14*f*) . . . . Tribe **LORICERINI**

17b Apical segment of galea not inflated and without apical filament (Fig. 35.14*g*); retinaculum very finely serrate or not serrate **18**

18a **(17b)** Lacinia of maxilla represented by medially directed spine (Fig. 35.14*g*); nasale with four or six teeth . . . . Tribe **NEBRIINI** ***(Nebria)***

18b Lacinia absent; nasale forms bifid prominence nearly as long as mandible **19**

19a **(18b)** Nasale cleft nearly to base, with small median tooth below level of others (Fig. 35.14*h*); tarsal claws very unequal . . . . Tribe **NEBRIINI** ***(Leistus)***

19b Nasale less deeply cleft, without median tooth (Fig. 35.14*i*); tarsal claws nearly equal . . . . Tribe **NOTIOPHILINI,** ***Notiophilus***

20a **(14b)** Retinaculum of mandile double (Fig. 35.14*j*); riparian . . . . Tribe **OMOPHRONINI,** ***Omophron***

20b Retinaculum not double **21**

21a **(20a)** Tarsal claws unequal **22**

21b Tarsal claws equal **28**

22a **(21a)** Cervical groove present (Fig. 35.14*k*) **23**

22b Cervical groove absent **26**

23a **(22a)** Head rounded, as wide as or wider than long (Fig. 35.14*k*); mandible short, thick . . . . Tribe **HARPALINI**

23b Head longer than wide, its sides parallel **24**

24a **(23b)** Stipes of maxilla with membranous area at middle of lateral aspect; two lateral ocelli on each side . . . . . Tribe **MORIONINI,** ***Morion***

24b Stipes with lateral aspect entirely schlerotized; Tribe **SCARITINI** **25**

25a **(24b)** Six lateral ocelli on each side . . . . *Dyschirius*

25b One lateral ocellus on each side . . . . *Clivina*

26a **(22b)** Antenna with five segments; urogomphi short, widely separated, with only five setae in all instars. . . . . Tribe **PTEROSTICHINI,** *Evarthrus*

27b Antenna with four segments; cerci long, slender, with more than five setae except in first instar **27**

27a **(26b)** Urogomphi long, slender, articulated with abdominal segment 9; riparian; western North America . . . . Tribe **OPISTHIINI,** *Opisthius richardsoni*

27b Urogomphi rigidly joined to Segment 9 . . . . Tribe **PERIGONINI,** *Perigona*

28a **(21b)** Urogomphi segmented (see **15a**) **29**

28b Urogomphi not segmented (see **15b**) **31**

29a **(28a)** Tarsal claw with basal tooth and pulvillus (Fig. 35.14*l*); lacinia of maxilla absent; cervical groove absent . . . . Tribe **LEBIINI**

29b Tarsal claw without basal tooth or pulvillus; lacinia vestigial to well developed; cervical groove present (but very reduced in first instar *Colliuris*) **30**

30a **(29b)** Blade and retinaculum of mandible denticulate (Fig. 35.14*m*) . . . . Tribe **ODACANTHINI,** *Colliurus*

30b Blade denticulate but retinaculum smooth . . . . Tribe **CTENODACTYLINI,** *Leptotrachelus*

31a **(28b)** Cervical groove present **32**

31b Cervical groove absent **37**

32a **(31a)** Antennae three times or more length of head . . . . Tribe **PANAGAEINI,** *Panagaeus*

32b Antennae approximately as long as head capsule **33**

33a **(32b)** Mandible with penicillus (Fig. 35.14*n*) (tuft of setae on medial margin near base); the second segment of either antenna or maxillary palpus is not the longest segment of that appendage **34**

33b Mandible without a penicillus (but in some species with one or two setae in its place); the second segment longest in both antenna and maxillary palpus **35**

34a **(33a)** Head capsule rounded, wider than long; mandible short, stout, with groove along lateral face . . . . Tribe **AMARINI,** *Amara*

34b Head capsule as long as wide, its sides nearly parallel; mandible long, slender, without lateral groove . . . . . . Tribe **PTEROSTICHINI**

35a **(33b)** Femora and tibiae without stout spines; cervical groove limited to sides of head . . . . Tribe **PATROBINI**

35b Femora and tibiae with stout spines; cervical groove not limited to side of head; Tribe **SCARITINI** **36**

36a **(35b)** Cervical groove bent anteriorly at medial end (Fig. 35.14*o*); urogomphi strongly convergent at tips . . . . *Pasimachus*

36b Cervical groove continued toward midline of head (Fig. 35.14*p*); urogomphi not convergent . . . . *Scarites*

37a **(31b)** Eyes absent; Pacific Coast, river sands . . . . Tribe **GEHRINGIINI,** *Gehringia olympica*

37b Eyes present 38

38a (37b) Three lateral ocelli on each side . . . . Tribe **PSYDRINI**

38b Six lateral ocelli on each side 39

39a (38b) Medial margin of blade of mandible serrate or crenulate. 40

39b Medial margin of mandible smooth. 43

40a (39b) Retinaculum of mandible serrate or crenulate; antennae much longer than head; on most species, head paler than body; Tribe **LICINI** 41

40b Retinaculum smooth; antennae not elongate; arctic species. . . . 42

41a (40a) Frontal plate forming more than half of rear margin of head in dorsal view (Fig. 35.14*q*) . . . . *Badister*

41b Frontal plate barely reaching hind margin of head (Fig. 35.14*r*). . . *Dicaelus*

42a (40b) Urogomphi fixed to segment 9. . . . Tribe **ELAPHRINI,** *Diacheila*

42b Urogomphi articulated to segment 9. . . . . Tribe **NEBRIINI,** *Pelophila*

43a Terga of body segments forming broad shelves on either side of body (giving the appearance of a sowbug); urogomphi very short and simple . . . . Tribe **CYCHRINI**

43b Terga not forming broad lateral shelves 44

44a (43b) Retinaculum serrate; antennae much longer than head . . . . Tribe **LICININI,** *Diplocheila*

44b Retinaculum smooth 45

45a (44b) Tarsal claw with ventral tooth and pulvillus . . . . Tribe **LEBIINI,** *Metabletus*

45b Tarsal claws simple, without tooth or pulvillus 46

46a (45b) Tarsal claws long, straight . . . . Tribe **MASOREINI,** *Tetragonoderus*

46b Tarsal claws of normal length, curved 47

47a (46b) Urogomphi much shorter than segment 10, broad, with four setae; lacinia absent . . . . Tribe **TRACHYPACHINI,** *Trachypachus*

47b Urogomphi varying from nearly equal to much longer than segment 10; lacinia of maxilla present 48

48a (47b) Urogomphi strongly tapered with one or two nodes or accessory points (Fig. 35.14*s*) . . . . Tribe **CARABINI**
Key: Burgess and Collins (1917).

48b Urogomphi not strongly tapered, not pointed 49

49a (48b) Urogomphi thick, curved, with many prominent setiferous nodes; head not conspicuously paler than body; Tribe **ELAPHRINI** 50
Key: Lindroth (1954).

49b Urogomphi either short and slender, without conspicuous nodes, or else articulated and very long . . . . Tribes **CHLAENIINI and OODINI**

50a (49a) Lacinia of maxilla distad to major seta (Fig. 35.14*t*); urogomphi each with about 50 major setae . . . . *Blethisa*

50b Lacinia with major seta of stipes arising from its tip (lacinia may be very small) (Fig. 35.14*u*); urogomphi each with about 15 major setae *Elaphrus*

References

Allen, R. T. 1968. A synopsis of the tribe Morionini in the western hemisphere with descriptions of two new species (Coleoptera: Carabidae). *Caribb. J. Sci.* 8(3,4):141–163. **1972.** A revision of the genus *Loxandrus* Leconte (Coleoptera: Carabidae) in North America. *Entomol. Am.* 46:1–184. **Ball, G. E. 1956.** A revision of the North American species of the genus *Helluomorphoides* Ball (Coleoptera, Carabidae, Helluonini). *Proc. Entomol. Soc. Wash.* 58(2):67–91. **1959.** A taxonomic study of North American Licini with notes on the Old World species of the genus *Diplocheila* (Coleoptera). *Mem. Am. Entomol. Soc.* 16:1–258. **1960–1971.** Carabidae. Pp. 55–182 in R. H. Arnett (ed.). *The Beetles of the United States.* University of Michigan Press, Ann Arbor, Mich. **1966.** A revision of the North American species of the subgenus *Cryobius* Chaudoir (*Pterostichus*, Carabidae, Coleoptera). *Opusc. Entomol. Suppl.* 28:1–166. **Ball, G. E. and J. Negre. 1972.** The taxonomy of the Nearctic species of the genus *Calathus* Bonelli (Coleoptera: Carabidae: Agonini). *Trans. Am. Entomol. Soc.* 98:412–533. **Bänninger, M. 1950.** The subtribe Pasimachina (Coleoptera, Carabidae, Scaritini). *Rev. Entomol. (Rio J.)* 21(3):481–511. **Barr, T. C., Jr. 1962.** The genus Trechus (Coleoptera: Carabidae: Trechini) in the southern Appalachians. *Coleopt. Bull.* 16:65–92. **1971.** The North American *Pterostichus* of the subgenus *Cylindrocharis* Casey (Coleoptera, Carabidae). *Am. Mus. Novit.* (2445):1–14. **Bell, R. T. 1960.** A revision of the genus *Chlaenius* Bonelli (Coleoptera, Carabidae) in North America. *Misc. Publ. Entomol. Soc. Am.* 1(3):98–166. **1970.** The Rhysodini of North America, Central America, and the West Indies (Coleoptera: Carabidae or Rhysodidae). *Misc. Publ. Entomol. Soc. Am.* 6(6):289–324. **Benschoter, C. A. and E. F. Cook. 1956.** A revision of the genus *Omophron* (Carabidae, Coleoptera) of North America and north of Mexico. *Ann. Entomol. Soc. Am.* 49:411–429. **Burgess, A. F. and C. W. Collins. 1917.** The genus Calosoma. *Bull. U.S. Dept. Agric.* 417:1–124. **Casey, T. L. 1918.** A review of the North American Bembidiinae. *Mem. Coleoptera* 8:1–223. **1920.** Random studies among the American Caraboidea. *Mem. Coleoptera* 9:133–299. **Erwin, T. L. 1970.** A reclassification of bombardier beetles and a taxonomic revision of the North and Middle American species (Carabidae: Brachinida). *Quaest. Entomol.* 6:4–215. **1975.** Studies of the subtribe Tachyina (Coleoptera: Carabidae: Bembidiini). III. Systematics, Phylogeny, and Zoogeography of the genus *Tachyta* Kirby. *Smithson. Contrib. Zool.* 208:1–68. **Erwin, T. L., D. R. Whitehead and G. E. Ball. 1977.** Family 4. Carabidae. Pp. 1–68 in Blackwelder and R. H. Arnett. *Checklist of the Beetles of Canada, United States, Mexico, Central America and the West Indies* (Yellow Version 4) **Freitag, R. 1969.** A revision of the species of the genus *Evarthrus* Leconte (Coleoptera: Carabidae). *Quaest. Entomol.* 5:89–212. **Gidaspow, T. 1959.** North American caterpillar hunters of the genera *Calosoma* and *Callisthenes* (Coleoptera: Carabidae). *Bull. Am. Mus. Nat. Hist.* 116(3):229–343. **1968.** A revision of the ground beetles belonging to *Scaphinotus*, subgenus *Brennus*. *Bull. Am. Mus. Nat. Hist.* 140(3):135–192. **1973.** Revision of ground beetles of American genus *Cychrus* and four subgenera of Genus *Scaphinotus* (Coleoptera, Carabidae). *Bull. Am. Mus. Nat. Hist.* 152(2):51–102. **Hacker, H. A. 1968.** The species of the subgenus *Leptoferonia* Casey (Coleoptera: Carabidae: *Pterostichus*). *Proc. U.S. Natl. Mus.* 124:1–61. **Hamilton, C. C. 1925.** Studies on the morphology, taxonomy and ecology of the larvae of holarctic tiger beetles (family Cicindelidae). *Proc. U.S. Natl. Mus.* 65(2530):1–87. **Hayward, R. 1897.** On the species of *Bembidion* of American north of Mexico. *Trans. Am. Entomol. Soc. (Phila.)* 24:32–158. **1900.** A study of *Tachys* of boreal America. *Trans. Am. Entomol. Soc. (Phila.)* 26:191–238. **1908.** Studies in *Amara*. *Trans. Am. Entomol. Soc. (Phila.)* 34:13–65. **Horn, G. H. 1870.** On the species of *Oodes* and allied genera of the United States. *Trans. Am. Entomol. Soc. (Phila.)* 3:105–109. **1880.** A review of the species of *Anisodactylus* and critical notes on the species of *Selenophorus* inhabiting the United States. *Proc. Am. Philos. Soc.* 19:162–183. **1882.** Synopsis of the tribe Lebiini. *Trans. Am. Entomol. Soc. (Phila.)* 10:126–183. **1883.** Synoptic table—*Discoderus*. *Bull. Brooklyn Entomol. Soc.* 5:52–53. **Jeannel, R. 1937.** Les Bembidiides endoges. *Rev. Fr. Entomol.* 4:1–23. **Larochelle, A. 1975.** *Les Carabidae du Quebec et du Labrador.* Bull. I. Départment de Biologie du College Bourget, Rigaud. 255 pp. **Larson, D. J. 1969.** A revision of the genera *Philophuga* Motschulsky and *Tecnophilus* Chaudoir with notes on the North American Callidina (Coleoptera: Carabidae). *Quaest. Entomol.* 5:15–84. **Leconte, J. L. 1857.** Synopsis of the species of *Clivina* and allied genera inhabiting the United States. *Proc. Acad. Nat. Sci. Phila.* 8:75–83. **1879a.** Synoptic table—*Clivina*. *Bull. Brooklyn Ento-*

mol. Soc. 2:32–33. **1879b.** Synoptic table—*Panagaeus. Bull. Brooklyn Entomol. Soc.* 2:59–62. **1879c.** Synoptic table—*Zuphium. Bull. Brooklyn Entomol. Soc.* 2:61,62. **1879d.** Synoptic table—*Diaphorus. Bull. Brooklyn Entomol Soc.* 2:62–85. **Leng, C. W. 1902.** Revision of the Cicindelidae of boreal America. *Trans. Am. Entomol. Soc. (Phila.)* 28:93–186. **Lindroth, C. H. 1954.** A revision of *Diachila* Motsch. and *Blethisa* Bon. *K. Fysiogr. Sallsk. Lund Forh. (N.S.)* 65:1–28. **1961–1969.** The ground beetles (Carabidae, excluding Cicindelinae) of Canada and Alaska. 1192 pp. *Opusc. Entomol. Suppl.* 20, 24, 29, 33, 34, 35. Part I. pp. I-XLVIII, 1969. Part II. pp. 1–200, 1961. Part III. pp. 201–408. 1963. Part IV. pp. 409–648, 1966. Part V. pp. 649–944, 1968. Part VI. pp. 945–1192, 1969 **Madge, R. B. 1967.** A revision of the genus *Lebia* Latreille in America north of Mexico (Coleoptera, Carabidae). *Quaest. Entomol.* 3:139–242. **Negre, J. 1963.** Revision du genre *Polpochila* Solier (Col., Carabidae). *Rev. Fr. Entomol.* 30:25–241. **Noonan, G. R. 1968.** A revision of the genus *Dicheirus* Mannerheim (Col., Carabidae). *Opusc. Entomol.* 33:281–304. **Notman, H. 1925.** A review of the beetle family Pseudomorphidae and a suggestion for a rearrangement of the Adephaga, with descriptions of a new genus and new species. *Proc. U.S. Natl. Mus.* 67(14):1–34. **Schaupp, F. G. 1879a.** Synoptic table—*Ega. Bull. Brooklyn Entomol. Soc.* 2:85. **1879b.** Synoptic table—*Lachnophorus. Bull. Brooklyn Entomol. Soc.* 2:86. **Schwartz, E. A. 1895.** Notes on *Nomaretus*, with descriptions of two new species. *Proc. Entomol. Soc. Wash.* 3:269–273. **Thiele, H. U. 1977.** *Carabid Beetles in Their Environments. A Study on Habitat Selection by Adaptations in Physiology and Behavior.* Zoophysiology and Ecology 10. Springer-Verlag, Berlin. 369 pp. **Thompson, R. G. 1979.** Larvae of North American Carabidae with a key to the tribes. Pp. 209–292. in T. L. Erwin, G. E. Ball and D. R. Whitehead (eds.). *Carabid Beetles: Their Evolution, Natural History and Classification.* Dr. W. Junk, The Hague, The Netherlands. **Valentine, J. M. 1935.** Speciation in *Steniridia*, a group of Cychrine beetles (Family Carabidae, Genus *Scaphinotus*). *J. Elisha Mitchell Sci. Soc.* 51(2):341–375. **Van Dyke, E. C. 1936.** A review of the subgenus *Nomaretus* Lec. of the genus *Scaphinotus* Dejean (Coleoptera–Carabidae). *Bull. Brooklyn Entomol. Soc.* 31:37–44. **1938.** A review of the genus *Scaphinotus*, subgenus *Scaphinotus* Dejean (Coleoptera, Carabidae). *Bull. Entomol. Soc. Am. (N.S.)* 18:93–133. **1943.** New species and subspecies of North American Carabidae. *Pan-Pac. Entomol.* 19:17–30. **Van Emden, F. 1953.** The Harpalini genus *Anisotarsus* Dejean (Col. Carab.). *Ann. Mag. Nat. Hist. Ser. 12* 6:513–547. **Vaurie, P. 1955.** Review of *Amblycheila. Am. Mus. Novit.* (1724):1–26. **Wallace, J. B. 1961.** *The Cicindelidae of Canada.* University of Toronto Press, Toronto, Ontario, Canada. 74 pp. **Whitehead, D. R. 1972.** Classification, phylogeny and zoogeography of *Schizogenius* Putzeys (Coleoptera, Carabidae, Scaritini). *Quaest. Entomol.* 8:131–348. **Wickham, H. F. 1897.** Coleoptera of the lower Rio Grande. *Bull. Lab. Nat. Hist. State Univ. Iowa* 4:96–115. **Willis, H. L. 1968.** Artificial key to the species of *Cicindela* of North America north of Mexico (Coleoptera : Cicindelidae). *J. Kans. Entomol. Soc.* 41:303–317.

36

Insecta: Coleoptera Ptiliidae†

HENRY S. DYBAS

The family Ptiliidae (featherwing beetles) contains the smallest known beetles, the majority of the species being 1 mm or less in length. The common name is derived from the structure of the wings (Fig. 36.72), which are believed to function primarily for passive flotation, as in the ciliate seeds of dandelions (Horridge 1956; but see Pringle 1957, pp. 77–80). The featherwings are normally folded out of sight under the wing covers or elytra, which in some genera are shortened and truncate. However, in preserved specimens the featherwings often project beyond the apex of the elytra and help (together with the minute body size) in the recognition of this family of beetles.

The family Ptiliidae is widely distributed throughout the world, especially in moist temperate and tropical regions. About 23 genera and 115 species have been listed from the United States and Canada (Arnett 1968); many more remain to be described and named, judging by existing collections. The beetles are abundant in many situations, but are easily overlooked because of their minute size and cryptic habits.

Habitats

Featherwing beetles live in habitats such as leaf litter on the ground, decaying logs, compost heaps, tree holes, decaying fungi, animal dung, under bark of dead trees, decaying algae on the seashore, and other moist, decaying organic matter suitable for growth of the molds and other fungi

† Send all technical inquiries on this family to Linda Dybas.

on whose spores and hyphae the beetles feed. Ptiliidae are frequently the most abundant beetles in such situations.

Life Cycle

Under favorable conditions, Ptiliidae appear to reproduce continuously, as evidenced by finding larvae together with both teneral and fully hardened adults at different times of the year. Only a single egg is accommodated and matured in the abdomen at a time, and this egg is relatively huge, nearly one-half the length of the female in species of average size in the family (Dybas 1978b). The developmental period seems to be short; the mean time from egg to adult in three British species of *Ptinella* is 32–45 days at 20°C, and the number of larval instars is three (Taylor 1975). The larvae (Fig. 36.71) of most species are slender and active and are white in color [Böving and Craighead (1931); but see Dybas (1976) for further details]. They occur together with the adults in the same situations and appear to have the same feeding habits. The pupal stage has been adequately described only for *Acrotrichis fascicularis* (Hinton 1941).

Collection

The most effective general method of collecting Ptiliidae is by using the Berlese or insect funnel (Borror et al. 1976, pp. 726–727). Decaying moist organic detritus is placed on a screen near the top of the funnel and subjected to gentle heat and drying (most commonly an electric light bulb is the heat source). The featherwing beetles and other soil arthropods gradually work their way down through the debris and accumulate in a jar of preservative at the apex of the funnel. The use of this simple apparatus permits the systematic sampling of the microhabitats in an area and the accumulation of adequate series for study. Other collecting methods are (a) hand picking in the field (e.g., collecting *Ptinella* species under bark); (b) sifting debris with a hand screen over a white cloth; and (c) waving a fine-mesh "aerial plankton" net in the late afternoon when featherwing beetles and other insects are often found flying or floating in the air. A low-power-head magnifier (about 2×) helps in detecting ptiliid beetles in the field and leaves the hands free.

Species of the genus *Actidium,* and a related undescribed genus, live on bare sand and gravel bars along streams (a seemingly unusual habitat for the family) and can best be collected by flotation. The moist surface sand and gravel is troweled into a bucket of water, stirred a few times, and the Ptiliidae, along with other beetles and arthropods, allowed to float to the surface. The ptiliids break through the surface film and rapidly take off from the surface. *Actidium* species may be only 0.5 mm long and are difficult to detect on the water surface, even in bright sunshine and with the help of a head magnifier.

The smallest featherwing beetles, and thus the most minute of all beetles, belong to *Nanosella* and allied genera. Adults and larvae live in the spore tubes on the under surface of fresh polypore fungi (Polyporaceae), where they feed on growing spores. When a host polypore is disturbed, the beetles leave the spore tubes, cross the under surface for a short distance, and disappear into other spore tubes. They can be collected by touching them with a wetted forceps point and transferring them to a vial of preservative. Here also a magnifier and good outdoor illumination are helpful.

Polymorphism

Many species of featherwing beetles are characterized by a striking polymorphism in which each sex is represented by two different forms: (a) a *normal morph* with well-developed eyes, wings, and body pigmentation; and (b) a *vestigial morph* in which eyes, wings, and body pigmentation are reduced or lacking (Dybas 1978b). The vestigial morph is always the more abundant form, comprising 90% or more of the individuals in a population. The two morphs occur together at the same time and place, the normal morph functioning as a dispersal form. The two forms are so different that they may appear to belong to different species or genera. Polymorphism of this kind occurs, among North American Ptiliidae, in *Ptinella, Pteryx, Ptinellodes,* and in some undescribed genera.

Parthenogenesis

Ptiliidae have an unusually high incidence among the Coleoptera of thelytokous parthenogenesis (the production of females only from unfertilized eggs). Species lacking a male sex are known in several genera in our fauna: *Acrotrichis, Ptinella,* and *Bambara.* In the bog ptiliid *Ptiliopycna moerens* (Matth.), males are known in only a few bog populations; throughout most of the range in the northeastern United States and adjacent Canada, only females are present. [For further details on parthenogenesis in the Ptiliidae, see Dybas (1966, 1978a,b).]

Family Diagnosis

Featherwing beetles are characterized by their small size, distinctive featherwings (Fig. 36.72), form of antenna (Fig. 36.49–36.52), distinctive pump apparatus (Fig. 36.65–36.70) of the spermatheca, reduced tarsal segmentation, and the structure of the head capsule, which lacks gular sutures and which has mouthparts of characteristic form (Fig. 36.73).

The smallest Ptiliidae, and therefore the most minute of all beetles, are undescribed tropical American fungus-spore inhabitants (related to *Nanosella*) that are only about 0.30 mm long. At the other extreme is the "giant" of the family, *Acrotrichis godmani* (Matth.), 2.0 mm or more in length, which occurs in Central America. Most Ptiliidae, however, are between 0.5 and 1.2 mm long.

The featherwings, which give the family its common name, are highly distinctive (Fig. 36.72). The wing membrane is very narrow and is borne on a single sclerotized stalk; the membrane is furnished with very long marginal hairs which form much the greater part of the expanse of the featherwing. At rest the wings are folded under the elytra or wing covers by means of several transverse folds, and the marginal hairs are closely appressed along the membrane margin (Dybas 1966), showing through the elytra in pale species as dark, longitudinal vittae. *Nossidium* and *Motschulskium* possess less specialized featherwings; they have two sclerotized basal struts, a much broader wing membrane, and much shorter marginal hairs. In repose this form of wing has longitudinal as well as transverse folds (Forbes 1926). The featherwings are lost in some species (e.g., "*Ptinella*" *mekura* Kubota) and are reduced or lacking in the vestigial morph of polymorphic species in the genera *Ptinella, Ptinellodes,* and *Pteryx.*

The antennae (Fig. 36.49–36.52) are usually slender and 11-segmented (10-segmented in some fungus spore-tube inhabitants). The first two segments are always considerably enlarged, and the terminal two or three are gradually or more abruptly enlarged to form a loose club.

The spermatheca (Fig. 36.65–36.70) is highly diagnostic in its unique pump apparatus, which is similar only to some members of the small related family Limulodidae (consisting of obligate ant guests found in the New World and the Australian region). The spermatheca is usually well sclerotized and shows extraordinary variability in the family (e.g., Figs. 36.65–36.70). In any one species, however, it is remarkably constant in form and therefore is extremely useful in identification.

The tarsal segments are reduced in number; there are usually only two or, indistinctly, three segments visible.

The head capsule and mouthparts exhibit little variation in the family. The gular sutures are invariably lacking, the posterior arms of the tentorium being attached on or near the ventral rim of the capsule. The lacinia and galea and maxillary palpi are characteristic in form throughout the family (Fig. 36.73).

In the members of the subfamily Acrotrichinae (in a restricted sense) the apex of the abdomen has a characteristic structure. North American genera of this group are *Nephanes, Smicrus, Ptiliopycna,* and the species-rich *Acrotrichis.* The two terminal terga or dorsal segments (morphological segments IX and X) are fused together in a characteristic way, the point of fusion being marked by a lateral tooth on each side (Figs. 36.53, 36.55). In the seashore genus *Actinopteryx,* of uncertain affinities, the terminal two terga are fused in the male and distinct in the female. In all other Ptiliidae, morphological segment X is demarcated from IX by a distinct suture. *Note:* The posterior margin of tergum VII has a characteristic fringe (Figs. 36.53, 36.55), termed the *dermal* or *palisade fringe,* in all Ptiliidae except for certain flightless forms; this structure serves as a useful landmark for establishing the identity of the terminal abdominal segments.

Preparation for Study

Because of their minute size, featherwing beetles are best prepared as whole mounts on microscope slides and studied with a compound microscope at magnifications of 100× to 400×. [See Borror et al. (1976, pp. 738–739) for preparation techniques and materials.] Many ptiliids are thinly sclerotized and lightly pigmented, and can be rapidly and conveniently mounted, ventral side up, directly from the preservative (usually 70% ethanol) into Hoyer's medium on a microscope slide. After 1 or 2 hours on a slide warming table at about 50°C, the specimen is usually cleared enough for examination of the external exoskeletal structures as well as sclerotized internal structures such as the aedeagus of the male and the spermatheca of the female. Darkly pigmented species (e.g., *Acrotrichis* spp.) require bleaching with hydrogen peroxide or other bleaching agents before mounting. In many cases, however, critical study of the exoskeletal features requires specimens from which soft tissues have been removed by maceration with substances such as potassium hydroxide or lactophenol (Borror et al. 1976, pp. 738–739).

Current State of Knowledge

The family Ptiliidae is one of the least known groups of Coleoptera. Few North American species can be identified correctly from the literature, and judging by existing collections, many species still remain to be named and described. The following key to genera reflects the current state of knowledge of the group. A number of the genera (e.g., *Acrotrichis, Ptilium, Microdium*) are treated in a broad sense (*sensu lato* or s.l.). These genera no doubt will have to be restricted in the future and additional genera established when the species are revised and their relationships understood better. A number of other genera, presumably undescribed, also occur in our fauna but are not included in the key. However, the 25 genera in the present key include the largest and most widely distributed genera and probably include the great majority of the species to be found in this region, judging from the collections at hand.

Unlike previous keys to the genera of North American Ptiliidae (Matthews 1884; Arnett 1960), the following key incorporates characters that require microscope slide preparations, greatly expanding the number of available characters for defining and recognizing the genera in our region.

ACKNOWLEDGMENTS

I thank Dr. Rupert L. Wenzel, Dr. Robert D. Waltz, and Dr. Fred G. Andrews for critical review of this chapter. My grateful appreciation is extended to Dr. Linda Dybas, without whom this manuscript would never have been completed.

Genera of North American Ptiliidae in Key

Acrotrichis Motsch.
Actidium Matth.
Actinopteryx Matth.
Bambara Vuillet
Cylindrosella Barber
Hydnosella Dybas
Micridium Motsch.
Millidium Motsch.
Motschulskium Matth.
Nanosella Motsch.
Nephanes Thoms.
Nossidium Erichs.
Oligella Motsch.
Porophila Dybas
Ptenidium Erichs.
Pteryx Matth.
Ptiliola Hald.
(= *Nanoptilium* Flach)
Ptiliolum Flach
Ptiliopycna Casey
Ptilium Erichs.
Ptinella Motsch.
Ptinellodes Matth.
Smicrus Matth.
Suterina Dybas
(= *Suterella* [Dybas])
Throscoptilium Barber

KEY TO GENERA OF PTILIIDAE FOUND IN NORTH AMERICA NORTH OF MEXICO†

1a Hind coxae contiguous, or nearly so (Figs. 36.21, 36.23–36.28); elytra complete, not shortened or truncate (Figs. 36.1–36.12) **2**
The abdomen may be extended well beyond the elytral apices in specimens in liquid or on microscope slides.

1b Hind coxae separated by at least one-eighth metasternal width (Figs. 36.29–36.41); elytra shortened or truncate. **10**

2a **(1)** Metasternal lines present (Figs. 36.25–36.28) (very short in ***Hydnosella***, Fig. 36.26); apex of abdomen (pygidium) with a single median tooth of variable form (e.g., acute, emarginate, or bifid, Figs. 36.57–36.58) **(Nanosellini)** **5**

2b Metasternal lines lacking (Figs. 36.21–36.24). Pygidium without a single median tooth (in ***Nossidium*** ♀ the pygidium is multidentate, Fig. 36.62) **3**

3a **(2)** Mesocoxae externally partly separated by a median mesosternal process that is keeled and sharply acuminate posteriorly (Fig. 36.24); eyes emarginate behind, dorsally with a backward directed flange which

† ***Dilineum*** Casey and ***Ptiliodina*** Casey are not included in the key. They may possibly be synonyms of ***Micridium* s.l.** and ***Pteryx***, respectively.

receives the anterior pronotal angles when head is retracted (Fig. 36.64); featherwing, with a basal scape consisting of one sclerotized strut; membrane narrow, its marginal hairs about 4 times as long as width of membrane (Fig. 36.72); body medium in size, about 0.5–0.6 mm from anterior margin of pronotum to apex of elytra (excluding head and apex of abdomen which may be variably exerted); body form as in Fig. 36.3 . . . . ***Bambara*** Vuillet (= ***Eurygyne*** Dybas)

Pantropical genus with eight species (six parthenogenetic) in the United States. Distribution: Florida and Gulf coast west to Texas (Dybas 1966, as ***Eurygyne***). Habitat: decaying litter on ground. One species (***B. fusca*** Dybas) extends north to Illinois and east to Maryland in sawdust piles. ***B. testacea*** (Britt.) (= ***lutea*** Dybas), yellow, parthenogenetic, cosmopolitan, is abundant in southeastern United States (spermatheca: Fig. 36.65).

3b Mesocoxae separated externally by a median mesosternal process that is subtruncate behind (Figs. 36.21–36.23); eyes not emarginate, without dorsal flange; featherwing with a basal scape of two sclerotized struts; membrane broad, its marginal hairs relatively short, no longer than width of membrane; body relatively large, 0.75–1.0 mm . . . **4**

4a (3) Underside with two sutures extending laterally from each mesocoxa; pronotum with sides evenly curved, and with a sharply defined dorsal groove along its explanate margin; body reddish or light brown in color, strongly convex, upper surface shining, bearing long, semierect setae; uncleared body color light brown or reddish (Figs. 36.1, 36.21, 36.22, 36.62, 36.67) . . . . ***Nossidium*** Er.

Three U.S. species. Habitat: forest floor debris. Frequent in eastern United States. Spermatheca of ***Nossidium*** sp.: Fig. 36.67.

4b Underside with one suture extending laterally from each mesocoxa; pronotum with sides sinuate posteriorly, without dorsal marginal groove; color dark; body flattened, with fine, silky, semirecumbent pubescence; uncleared body darkly colored (Fig. 36.23) . . . . ***Motschulskium*** Matth.

One species (***M. sinuatocolle*** Matth.) found under seaweed on seashore of west coast of North America.

5a (2) Head normal, about two-thirds as wide as pronotum (Fig. 36.4); habitat: underside and spore tubes of fresh shelf fungi (**Polyporaceae** and **Hydnaceae**) **6**

5b Head very narrow, about one-third as wide as pronotum (Fig. 36.6); size minute, about 0.45 mm; color yellow . . . . ***Suterina*** Dybas, (= ***Suterella*** Dybas)

One species ***S. microcephala*** (Dybas). Habitat: forest floor litter, eastern United States.

6a (5) Mesosternal elevation somewhat arrowhead shaped, broadest at middle, narrowed anteriorly and posteriorly (Figs. 36.25–36.27) **7**

6b Mesosternal elevation with sides parallel posteriorly and broadly overlapping medial parts of mesocoxae, strongly constricted anteriorly before it joins collar (Fig. 36.28). . . . . ***Throscoptilium,*** Barber

One species (***T. duryi*** Barber), eastern United States. Habitat: fresh polypore fungi (**Polyporaceae**), especially ***Phellinus gilvus*** (Schweinitz) Patouillard (formerly ***Polyporus gilvus***)

7a (6) Form extremely long and narrow, about five times as long as broad (Fig. 36.5) . . . . ***Cylindrosella,*** Barber

One species (***C. dampfi*** Barber) described from Mexico. An unidentified species of this genus occurs at higher elevations in southern Arizona. Another occurs in Kentucky and adjacent southern Indiana and Illinois in spore tubes on undersurface of the perennial polypore fungus ***Fomes robiniae*** (Murr.) Sacc. & D. Sacc., on living black locust trees (***Robinia pseudoacacia*** L.).

7b Form elongate–oval, not more than three times as long as broad (Fig. 36.4). **8**

8a Basal angle of prothorax with a cylindrical pit opening on ventral side (Figs. 36.26, 36.46) . . . . ***Hydnosella*** Dybas

One species (***H. globitheca*** Dybas), Indiana. Habitat: undersurface of fungus, ***Stecherinum*** sp. **(Hydnaceae).**

8b Prothorax without cylindrical pits **9**

9a **(8)** Lateral mesocoxal suture directed laterad for a distance before curving anteriorly (Fig. 36.25); pygidial spine a short lobe with a tooth on each side (Fig. 36.58); antennae 11-segmented (Fig. 36.52). . . . . ***Porophila*** Dybas

One species (***P. malkini*** Dybas). Oregon, California. Habitat: spore tubes on undersurface of fresh polypore fungi [***Fomes pinicola*** (Fr.) Cke.]

9b Mesocoxal suture curving anteriorly almost immediately (Fig. 36.27); pygidial spine acute or bifid (Fig. 36.57); antennae apparently 10-segmented; last segment with a distinct constricted apical part (Fig. 36.51) (see also Figs. 36.4 and 36.68) . . . . ***Nanosella*** Motsch.

Two nominal species in the eastern United Stages. Habitat: spore tubes on undersurface of fresh polypore fungi **(Polyporaceae).** Also several undescribed species, United States and neotropics. Spermatheca of ***Nanosella*** sp.: Fig. 36.68.

10a **(1)** Eyes absent, or reduced in size and pale in color; metathoracic wings reduced or lacking; body pigmentation pale (*Note:* If uncertain as to condition of eyes, wings, and body pigmentation, proceed to couplet 11) **25**

10b Eyes present, prominent, black (except in teneral specimens); metathoracic feather wings fully developed, the appressed marginal hairs showing through the elytra in pale species as dark longitudinal vittae on each side **11**

11a **(10)** Pygidium of characteristic form as in Figs. 36.53–36.55, with a lateral tooth on each side, and with median tooth present (Fig. 36.53) or absent (Fig. 36.55); aedeagus symmetrical in abdomen, furnished with two ventral hooks (Fig. 36.54); elytra shortened, truncate (Figs. 36.17–36.20); mesocoxal suture straight, never curved (Figs. 36.31–36.33, 36.36) **(Acrotrichinae)** **12**

11b Pygidium not as above, with or without median tooth, or pygidium serrate as in Figs. 36.59 and 36.62; never with only one tooth on each side. Aedeagus not as above; elytra truncate, or complete and rounded at apex; lateral mesocoxal suture straight or curved **16**

12a **(11)** Hind angles of pronotum distinctly produced posteriorly (Fig. 36.19); pygidium almost always with a median tooth in addition to the lateral tooth on each side (Fig. 36.53) (see also Figs. 36.31 and 36.54) . . . . ***Acrotrichis*** Motsch.

Acrotrichis **s.l.** (incl. subg. ***Ctenopteryx*** Flach) with about 40 nominal species in the Nearctic region, is the largest and most ubiquitous genus in North America. The species are mostly black and occur in decaying floor litter, dung, compost, decaying fungi, nests, under carrion, and so on. The genus is urgently in need of revision; few species, even the commonest, can be correctly identified from the literature. ***Acratrichis,*** commonly used in the United States, is a misspelling (Biström and Silfverberg 1979).

12b Posterior pronotal angles right-angled or obtusely rounded (Fig. 36.20), not produced posteriorly; median pygidial tooth present or absent (Figs. 36.53, 36.55) **13**

13a **(12)** Mesosternal keel widened anteriorly and joined to the mesosternal collar (Fig. 36.33); form convex; posterior angles of pronotum obtusely rounded (Fig. 36.20); pygidium with median tooth as well as a lateral tooth on each side . . . . ***Ptiliopycna*** Casey

One species, ***P. moerens*** (Matth.), in northeastern United States and adjacent Canada. Habitat: sphagnum and other mosses in bogs and swamp forests. Most populations parthenogenetic; males known from only a few bogs (Dybas 1978a). Spermatheca: Fig. 36.69.

13b Mesosternal keel not joined to mesosternal collar (Figs. 36.32, 36.36); form not highly convex; posterior angles of pronotum obtusely or sharply angled; pygidium without median tooth (Fig. 36.55). . . . 14

14a Fine transverse line delimiting mesosternal collar sharply anteriorly indented at middle, sinuate laterally (Fig. 36.36); pronotum with sides straight or slightly sinuate at posterior third (Fig. 36.18); body flattened, surface opaque, granulate; antennal segments 5–7 about $3\frac{1}{2}$–4 times as long as broad, not abruptly constricted apically (Fig. 36.49); male with trifoliate patch of short stout spines on underside of abdomen near apex; size 0.8–1.0 mm . . . . ***Smicrus*** Matth.

Two nominal species (? one widespread species) Habitat: floor litter, especially in damp situations.

14b Fine line of mesosternal collar straight or slightly sinuate at middle, not sharply indented; its sides straight or slightly sinuate (Fig. 36.32); body not opaque or strongly granulate **15**

15a (14) Antennal segments 5–8 abruptly constricted at apex (Fig. 36.50); spermatheca spiral in form as in Fig. 36.66 (see also Figs. 36.17, 36.32, and 36.55) . . . . ***Nephanes*** Thoms. **s. str.**

Six or seven nominal species in the United States, including ***N. titan*** Newn., described from Europe, which is common in stable sweepings and wet compost. ***Nephanes rugulodorsus*** Hatch should be excluded; it belongs instead in ***Ptiliolum*** or near.

15b Antennal segments 5–8 not abruptly constricted at apex; spermatheca not sprial in form . . . . ***Nephanes*** **s.l.**

Several species (? all undescribed) found in Illinois, Indiana, Florida, Texas, New Mexico, and other states are probably not congeneric with ***Nephanes titan*** Newn., the type species of ***Nephanes.*** Habitat: forest floor debris, basal tree holes.

16a (11) Prosternum with a well developed posteriorly directed intercoxal process (Fig. 36.45); upper surface polished; base of pronotum usually with four foveae (Fig. 36.2); mesosternum modified for reception of prosternal process (Fig. 36.35); most species large size, dark . . . . . ***Ptenidium*** Er.

About 13 nominal species in our region, generally distributed. Habitat: forest floor litter, compost, grass piles, tree holes, nests.

16b Prosternum with intercoxal process absent or only feebly developed; mesosternum otherwise **17**

17a (16) Pronotum with characteristic form as in Fig. 36.7, its constricted posterior portion narrower than anterior margin; middle coxae separated by an almost parallel prolongation of the mesosternum (Fig. 36.34) . . . ***Actidium*** Matth.

The six described species in North America are from California east to Texas, but the genus also occurs east to Florida. Common on bare gravel and sand bars along water courses (in Indiana three species occur together), and abundant in piles of decaying algae on the seashore (Florida). Also occurs in intertidal habitats (fissures in rocks) on the west coast.

17b Pronotum of different form **18**

18a (17) Elytral suture with about 10 small tubercles, humerus with a small tooth or angulation (Fig. 36.47); lateral mesocoxal suture straight, directed anteriorly at about a 45° angle (Fig. 36.37); body form as in Fig. 36.12 ***Ptiliola*** Hald., (= ***Nanoptilium*** Flach)

Abundant, widely distributed. Three nominal species in North America. Habitat: under decaying mushrooms, dung, and other "ripe" decaying organic matter.

18b Not as above **19**

19a (18) Collar and median elevation of mesosternum of characteristic form as in Fig. 36.29; mesocoxal sutures straight, angled anteriorly; body black; pronotum with hind angles produced posteriorly (Fig. 36.16); male aedeagus symmetrical in abdomen when viewed from beneath . . . . ***Actinopteryx*** Matth.

One species. ***A. fucicola*** Allib., European coasts and east coast United States. Habitat: algae on seashore Florida to? Virginia, also mud and debris in mangrove swamps. The terminal two dorsal abdominal segments fused in male, distinct in female.

19b Mesosternum not as above **20**

20a (19) Pygidium serrulate, middle tooth large (Fig. 36.59); body form as in Fig. 36.11; larger species $\geq$ 0.7 mm (see also Fig. 36.48) ***Ptiliolum*** Flach

Uncertain number of species, some probably listed under ***Ptilium*** in North American catalogs. Habitat: forest floor and similar. Widely distributed, especially in northern and western states and Canada.

20b Pygidium dentate or not, never serrulate **21**

21 (20) Pygidium with two apical teeth (Fig. 36.61); mesosternum with a sharply defined median triangular elevation (Figs. 36.39, 36.40); pronotum with (Fig. 36.8) or without sulci. **22**

21b Pygidium with or without a single apical tooth; mesosternum not elevated or feebly elevated and demarcated at middle (Figs. 36.37, 36.38, 36.41) **23**

22a (21) Triangular mesosternal elevation with a median fovea which has an internal extension laterally on each side (Fig. 36.40); pronotum with three well-developed longitudinal sulci. . . . . . . ***Millidium*** Motsch.

One introduced species, ***M. minutissimum*** (Ljungh), has been found in Illinois, Wisconsin, and Washington. Habitat: horse manure, stable sweepings, decaying vegetation, cut grass piles. Listed as a subgenus of ***Ptilium*** in most European works.

22b Triangular mesosternal elevation without a fovea (Fig. 36.39); Pronotum with or without longitudinal sulci (see also Figs. 36.8 and 36.61) ***Ptilium*** Er. **s.l.**

Several species (? undescribed) have been collected in Florida, Texas, Arizona, and elsewhere. The 13 nominal species listed in our U.S. catalogs probably belong elsewhere (e.g., ***Pteryx, Ptiliolum***).

23a (21) Longitudinal metasternal carinae present, partly or completely developed (Figs. 36.38, 36.41) **24**

23b Metasternal carinae absent; lateral mesocoxal sutures curved, directed laterally, thence curved anteriorly (Fig. 36.30); metasternal collar lacking; mesosternum tumid or with a weakly developed keel between mesocoxae; hind coxae separated by half metasternal width or more (Fig. 36.30); aedeagus a curved tube lying on its side in the abdomen, and appearing asymmetrical from beneath (Fig. 36.56); species polymorphic, the vestigial morph by far the most common morph, with eyes and metathoracic wings reduced or lacking, and body pale in color **25**

24a (33) Posterior coxae broadly lamellate, separated by about one-third of the metasternal width (Fig. 36.38); mesosternal humeri sharply angulate (Fig. 36.38); mesosternal elevation not sharply defined, vaguely trifoliate (Fig. 36.38); convex, yellow, pronotum longitudinally sulcate or not; metasternal margin between hind coxae comblike in males of some species (see also Figs. 36.9 and 36.60) . . . .
Micridium Motsch. **s.l.** (? = ***Dilinium*** Casey)

One or two nominal species, as well as several undescribed species abundant and widely distributed in eastern North America. Habitat: forest floor litter, one species (no pronotal sulci) common in tree holes, eastern North America.

24b Posterior coxae narrowly lamellate, separated by about one-sixth of the metasternal width (Fig. 36.41). Mesosternal humeri rounded (Fig.

36.41); mesosternal elevation not well defined, indistinctly triangular (Fig. 36.41), body linear, flattened. Pronotum with a shallow median longitudinal depression (Fig. 36.10) (see also Fig. 36.5) ***Oligella*** Motsch.

Not previously listed from the United States but several species (? undescribed, ? introduced) in our fauna appear to belong to this genus or near. Specimens have been collected mainly in forest floor litter and in stable sweepings, but also in organic thatch of ant nests in prairie. An undescribed, strongly polymorphic species of this or a related undescribed genus has been found in Arizona in leaf sheaths of dead or dying sotol plants (***Dasylirion*** sp.).

25a (23) Pronotum with posterior angles somewhat produced posteriorly (Fig. 36.15); female with two heavily sclerotized clawlike structures inside apex of abdomen (Fig. 36.63); strongly polymorphic, eyes and wings not detectable in vestigial morph (Fig. 36.70) . . . ***Ptinellodes*** Matth.

Two polymorphic species, southeastern United States. Habitat: ***P. lecontei*** G. & H (spermatheca: Fig. 36.63) occurs under bark, in logs, and in sawdust piles. ***P. suteri*** Dybas (spermatheca: Fig. 36.70) occurs mainly in forest floor litter (Dybas 1978).

25b Pronotum with posterior angles obtusely rounded or sharply angled, never produced posteriorly (Figs. 36.42–36.44) **26**

26a (25) Pronotum narrower anteriorly than at base, widest behind middle; posterior angles obtusely angulate, never sinuate and acute; sides narrowly explanate at middle, more widely so posteriorly (the explanate margin in slide preparations is delimited internally by an endoskeletal thickening) (Fig. 36.44); the eyes are pale and reduced in size in the vestigial morph (see also Fig. 36.14) . . . . ***Pteryx*** Matth.

Four nominal polymorphic species and at least five more in existing collections, widespread in United States. Habitat: forest floor debris, tree holes, logs, sawdust piles, sphagnum bogs. Genus needs revision before species can be correctly identified.

26b Pronotum as wide or wider anteriorly than at base, widest anterior to middle; posterior angles sinuate and sharply angled in some species, broadly rounded in others; margins not explanate (Figs. 36.42, 36.43; see also Figs. 36.13, 36.30, and 36.56) . . . . . . ***Ptinella*** Motsch. s l.

Four nominal species in the United States, several more in collections, generally distributed in the United States. Habitat: mainly under bark, also in tree holes, logs, and sawdust piles. Strongly polymorphic in both sexes except in ***"Ptinella" mekura*** Kubota, in which only vestigial females are known. This species was described from Japan and is known also from the Mediterranean region; it is widely distributed in our region and occurs in litter on the soil.

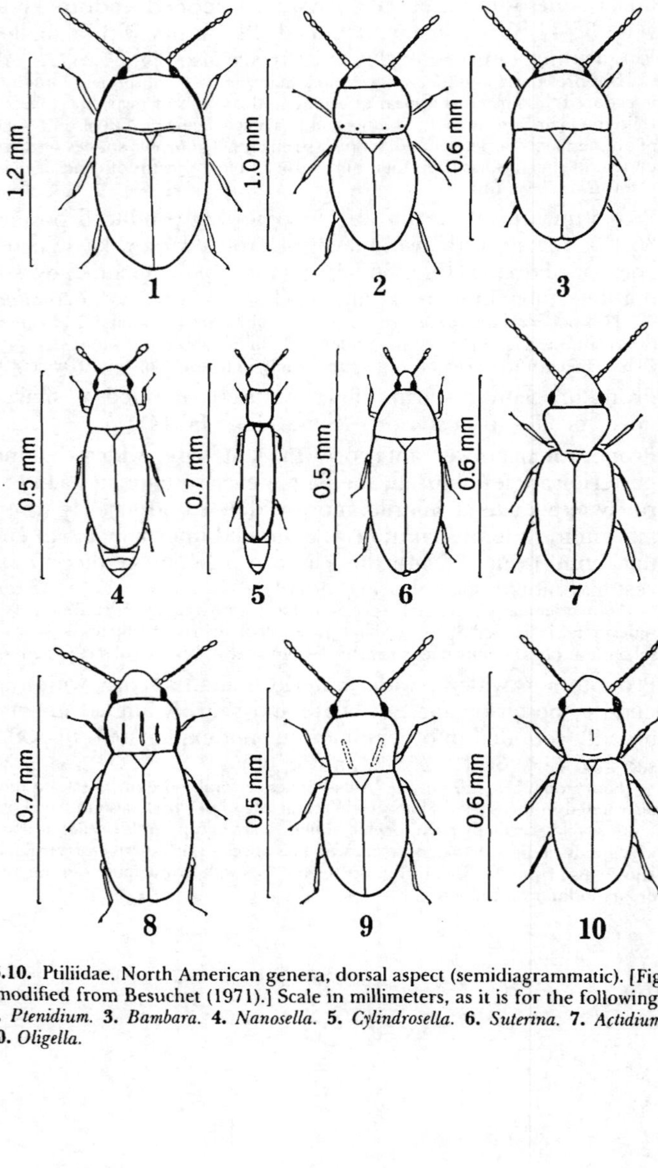

Fig. 36.1–36.10. Ptiliidae. North American genera, dorsal aspect (semidiagrammatic). [Figures 1, 2, 7–10 redrawn or modified from Besuchet (1971).] Scale in millimeters, as it is for the following illustrations. **1.** *Nossidium.* **2.** *Ptenidium.* **3.** *Bambara.* **4.** *Nanosella.* **5.** *Cylindrosella.* **6.** *Suterina.* **7.** *Actidium.* **8.** *Ptilium.* **9.** *Micridium.* **10.** *Oligella.*

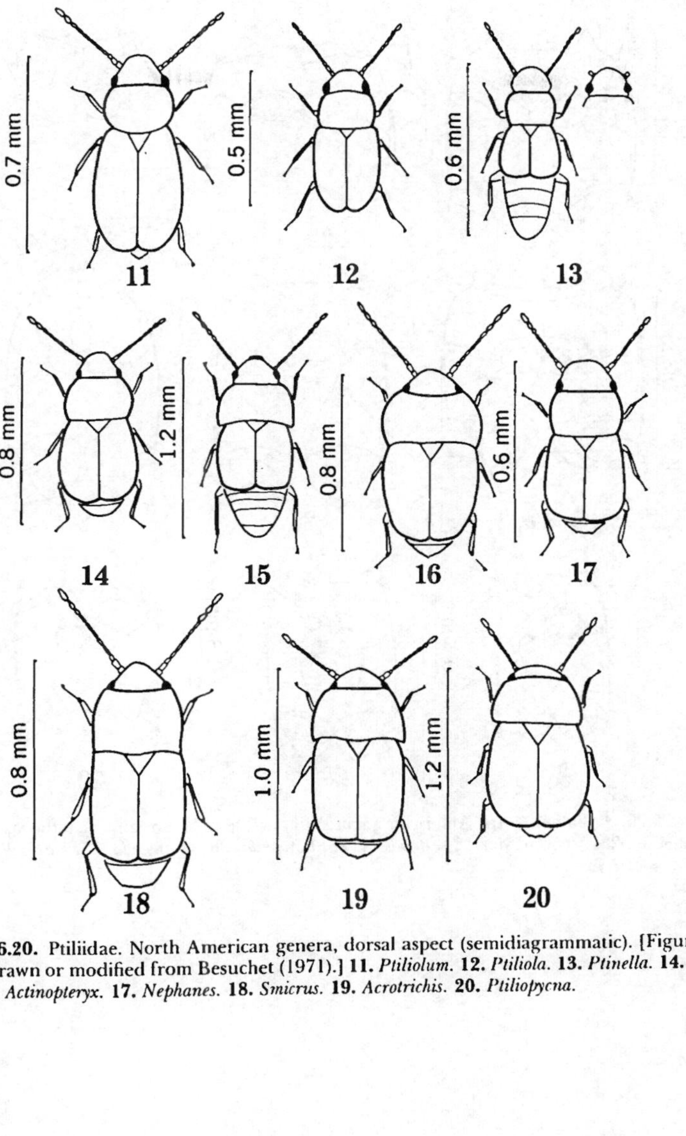

Figs. 36.11–36.20. Ptiliidae. North American genera, dorsal aspect (semidiagrammatic). [Figures 11, 12, 14, 16–19 redrawn or modified from Besuchet (1971).] **11.** *Ptiliolum.* **12.** *Ptiliola.* **13.** *Ptinella.* **14.** *Pteryx.* **15.** *Ptinellodes.* **16.** *Actinopteryx.* **17.** *Nephanes.* **18.** *Smicrus.* **19.** *Acrotrichis.* **20.** *Ptiliopycna.*

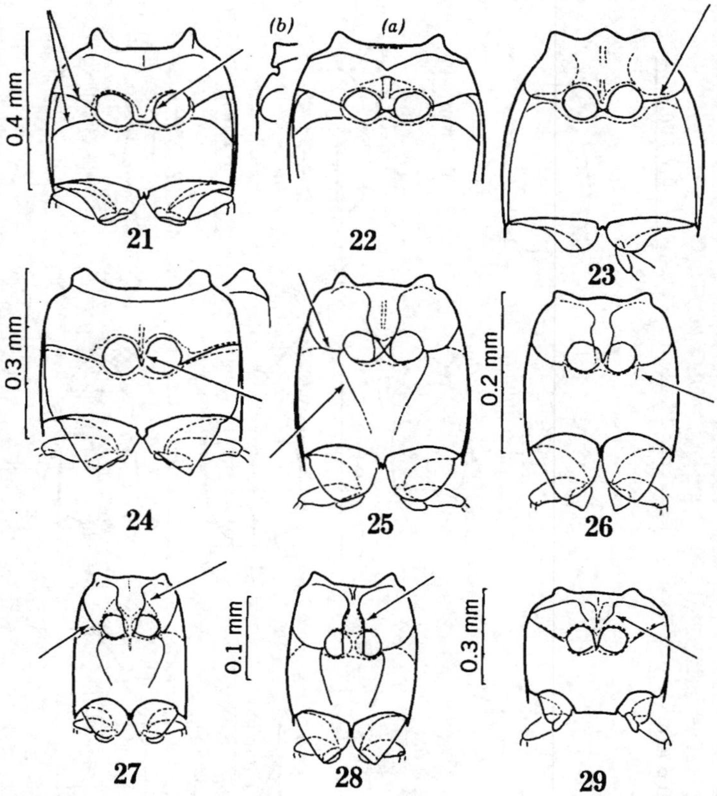

Fig. 36.21–36.29. Ptiliidae. North American genera; underside of meso- and metathorax. **21.** *Nossidium.* **22.** *Nossidium. s.l.* **23.** *Motschulskium.* **24.** *Bambara.* **25.** *Porophila.* **26.** *Hydnosella.* **27.** *Nanosella.* **28.** *Throscoptilium.* **29.** *Actinopteryx.*

Figs. 36.30–36.38. Ptiliidae. North American genera; underside of meso- and metathorax. **30.** *Ptinella.* **31.** *Acrotrichis.* **32.** *Nephanes.* **33.** *Ptiliopycna.* **34.** *Actidium.* **35.** *Ptenidium.* **36.** *Smicrus.* **37.** *Ptiliola.* **38.** *Micridium.*

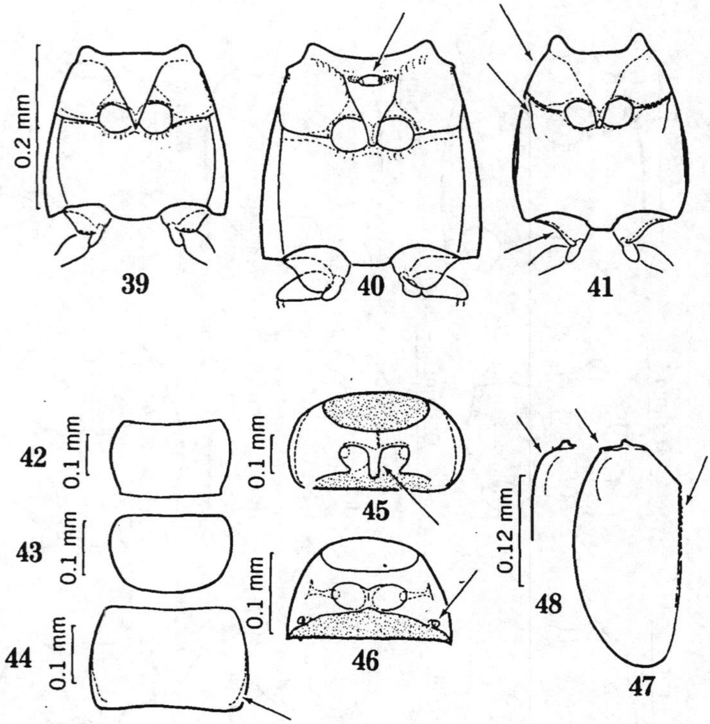

Figs. 36.39–36.48. Ptiliidae. **39–41:** Underside of meso- and metathorax (**39** *Ptilium.* s.l.; **40** *Millidium;* **41** *Oligella*). **42,43.** pronotum of *Ptinella* species. **44.** pronotum of *Pteryx* sp. **45.** *Ptenidium* sp., intercoxal prosternal process. **46.** *Hydnosella globitheca* Dybas, showing cylindrical pit on underside of basal angle of prothorax. **47.** *Ptiliola* sp. (= *Nanoptilium*), dentate elytral humerus and sutural tubercles. **48.** *Ptiliolum* sp., rounded elytral humerus.

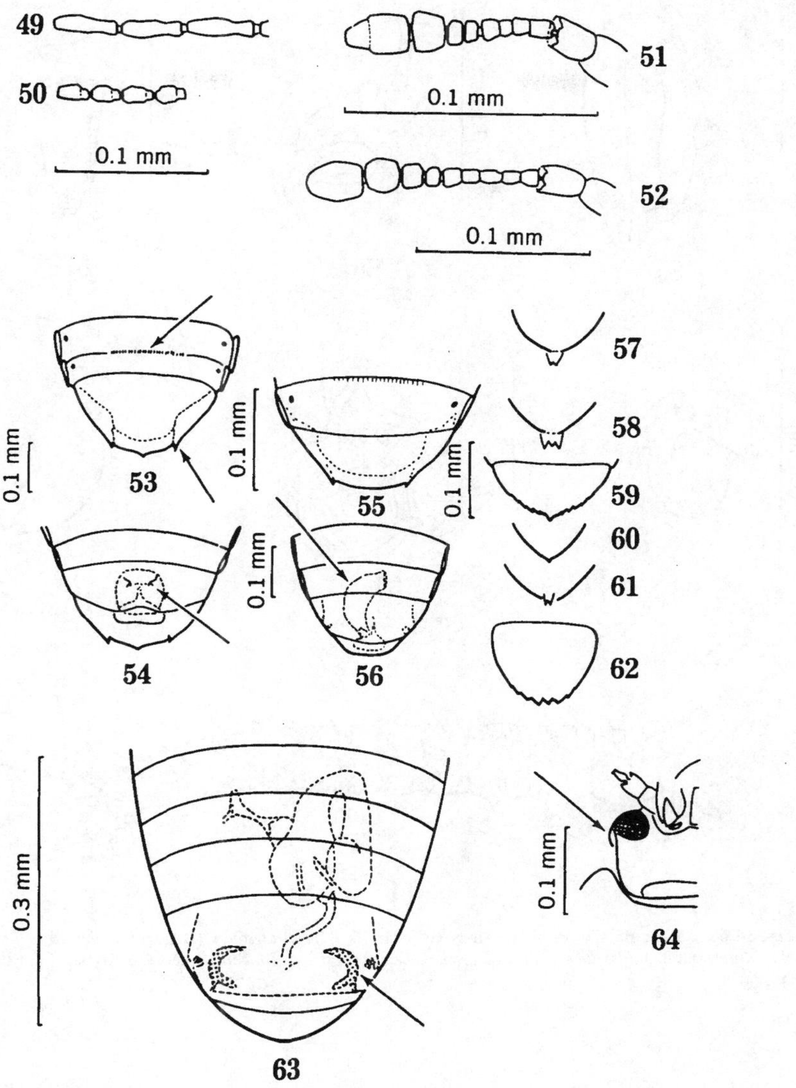

Figs. 36.49–36.64. Ptiliidae. **49.** *Smicrus* sp., antennal segments 5-7. **50.** *Nephanes* s. str., antennal segments 5-8. **51.** *Nanosella* sp., 10-segmented antenna. **52.** *Porophila malkini* Dybas, 11-segmented antenna. **53.** *Acrotrichis* sp., dorsal apex of abdomen showing terminal pygidium (fused terga IX and X) with two lateral and one median tooth (note palisade fringe on posterior margin of tergum VII, which is present in this position in all Ptiliidae except certain flightless forms. **54.** *Acrotrichis* sp., ventral aspect of male abdomen showing (dashed lines) symmetrical aedeagus with two ventral hooks and associated internal sterum (anteriorly stalked). **55.** *Nephanes* sp., dorsal aspect of apex of abdomen, apical median tooth lacking. **56.** *Ptinellodes lecontei* ventral aspect of apex of abdomen showing (dashed lines) male aedeagus lying asymetrically on its side and associated internal sternite IX (stalked anteriorly). **57–62.** Apex of abdomen: **57.** *Nanosella* sp. **58.** *Porophila malkini* Dybas. **59.** *Ptiliolum* sp. **60.** *Ptiliola* (= *Nanoptilium*) sp. **61.** *Ptilium* s.l. sp. **62.** *Nossidium* sp., female. **63.** *Ptinellodes*. **64.** *Bambara* sp., underside of head, showing dorsal flange of eye.

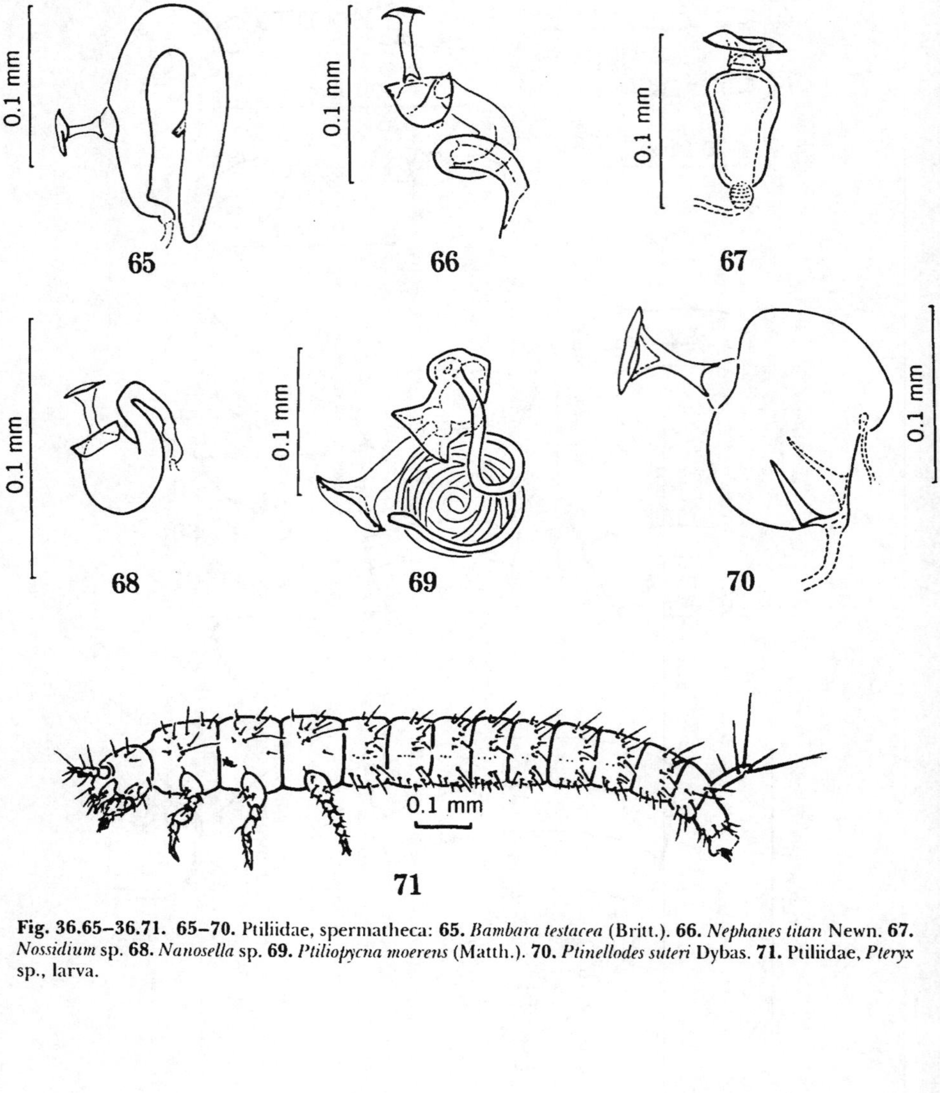

Fig. 36.65–36.71. 65–70. Ptiliidae, spermatheca: **65.** *Bambara testacea* (Britt.). **66.** *Nephanes titan* Newn. **67.** *Nossidium* sp. **68.** *Nanosella* sp. **69.** *Ptiliopycna moerens* (Matth.). **70.** *Ptinellodes suteri* Dybas. **71.** Ptiliidae, *Pteryx* sp., larva.

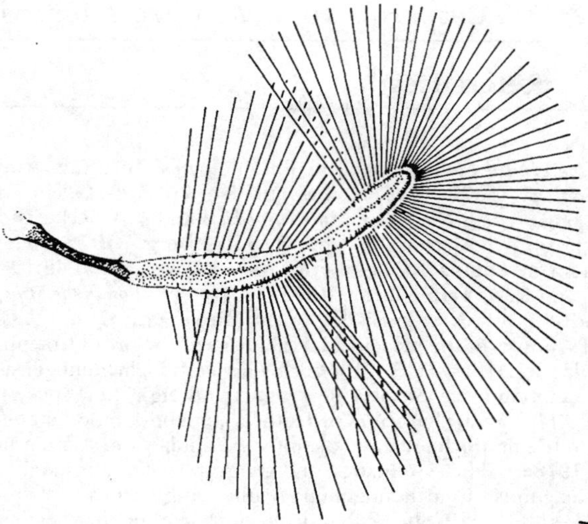

72

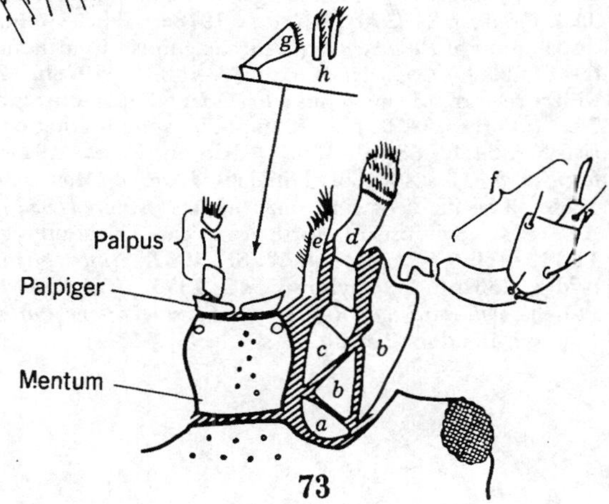

73

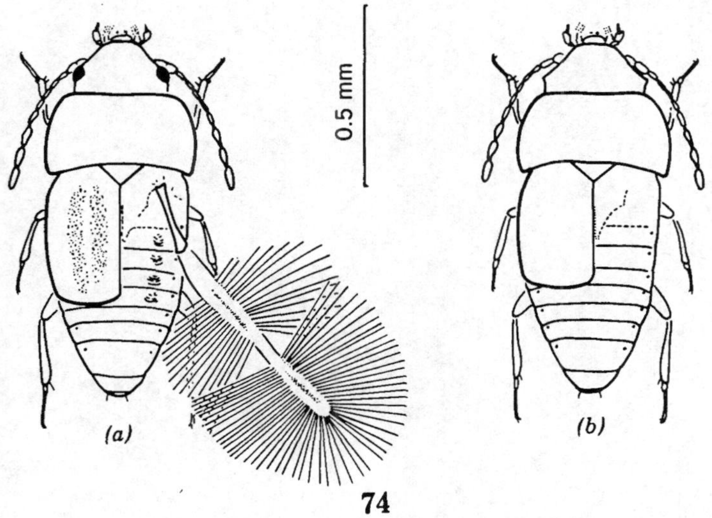

74

Figs. 36.72, 36.74. 72. Ptiliidae, *Bambara testacea* (Britt.), featherwing. **73.** Ptiliidae, *Acrotrichis* mouthparts (semi diagrammatic) on underside of head. (*a*) carod; (*b*) stipes; (*c*) subgalea; (*d*) galea; (*e*) lacinea; (*f*) palpus; (*g*) paraglossa; (*h*) hypopharynx at paraglossa. **74.** *Ptinellodes lecontei.* (*a*) normal morph; (*b*) vestigial morph.

References

Arnett, R. 1968. *The Beetles of the United States.* American Entomological Institute, Ann Arbor, Mich. 1112 pp. **Besuchet, C. 1971.** 21. Familie: Ptiliidae. Pp. 311–342 in H. Freude, K. W. Harde and G.A. Lohse (eds.). *Die Käfer Mitteleuropas,* Band 3. Goecke and Evers, Krefeld, West Germany. 365 pp. **Biström, O. and H. Silfverberg. 1979.** The type species of the European genera of Ptiliidae (Coleoptera). *Ann. Entomol. Fenn.* 45:12–15. **Borror, D. J., D. M. Delong and C. A. Triplehorn. 1976.** *An Introduction to the Study of Insects,* 4th ed. Holt, Rinehart and Winston, New York. 852 pp. **Böving, A. C. and F. C. Craighead. 1931.** An illustrated synopsis of the principal larval forms of the order Coleoptera. *Entomol. Am.* 11:1–351. 125 pls. **Dybas, H. S. 1966.** Evidence for parthenogensis in the featherwing beetles, with a taxonomic review of a new genus and eight new species. *Fieldiana Zool.* 51:11–52. **1976.** The larval characters of featherwing and limulodid beetles and their family relationships in the Staphylinoidea (Coleoptera: Ptiliidae and Limulodidae). *Fieldiana Zool.* 70:29–78. **1978a.** The systematics, and geographical and ecological distribution of *Ptiliopycna,* a Nearctic genus of parthenogenetic featherwing beetles (Coleoptera: Ptiliidae) *Am. Midl. Nat.* 99:83–100. **1978b.** Polymorphism in featherwing beetles, with a revision of the genus *Ptinellodes* (Coleoptera: Ptiliidae). *Ann. Entomol. Soc. Am.* 71:29–78. **Forbes, W. T. M. 1926.** The wing folding patterns of the Coleoptera. *J. N.Y. Entomol. Soc.* 34:42–68, 91–139. **Hinton, H. E. 1941.** The immature stages of *Acrotrichis fascicularis* (Herbst) (Col. Ptiliidae). *Entomol. Mon. Mag.* 77:245–250. **Horridge, G. A. 1956.** The flight of very small insects. *Nature (Lond.)* 178:1334–1335. **Matthews, Rev. A. 1884.** Synopsis of North American Trichopterygidae. *Trans. Am. Entomol. Soc. (Phila.)* 11:113–156. **Pringle, J. W. S. 1957.** *Insect Flight.* Cambridge University Press, Cambridge. 133 pp. **Taylor, V. A. 1975.** *The Biology of Feather-Winged Beetles of the Genus* Ptinella *with Particular Reference to Coexistence and Parthenogenesis.* Ph.D. thesis, University of London, London. 296 pp.

37

Insecta: Coleoptera Silphidae

and the Associated Families

Agyrtidae and Leiodidae

STEWART B. PECK

The beetle superfamily Staphylinoidea contains some 10 families, of which the Ptiliidae, Staphylinidae, and Pselaphidae are covered in other chapters of this book. These families all share an intimate relationship with soil-associated habitats along with the families Silphidae, Agyrtidae, and Leiodidae: the subjects of this chapter. General keys to beetle families and a coverage of general beetle morphology can be found in introductory entomology texts such as Borror et al. (1976) or in general works on beetles such as Arnett (1963), Crowson (1967), Dillon and Dillon (1961), Jacques (1951), and Hatch (1953, 1957). Recent research efforts have reexamined evolutionary relationships between these groups and the classification followed here is that recently proposed by Lawrence (1982) and Lawrence and Newton (1982).

These three beetle families collectively spend most of their adult and larval lives in association with soil and its extensions; they usually function as scavengers and decomposers. Many occur in association with carrion, dung, decaying plant material, and fungi. They are important in ecosystems in promoting recycling of nutrients. A few specialized scavengers have become guests in nests of social insects, in burrows and nests of ground-dwelling vertebrates, as ectoparasites of some mammals, and as blind and wingless inhabitants of deep soil and caves.

Members of these families are often collected through general sampling of soil, litter, and decomposing organic matter. A program using pitfall traps baited with various materials (Newton and Peck 1975, Southwood

1966) can yield data on food and habitat preferences, seasonality, reproductive activity, and so on. To collect most genera in these families, one must concentrate on their special habitats and habits.

The Family Silphidae

These are the large carrion beetles. They are 10–35 mm long, often brightly colored, and are commonly found with the bodies of dead animals. They are most closely allied to the family Staphylinidae (Lawrence and Newton 1982) from which they may be separated by their clavate or capitate 10- or 11-segmented antennae with dense pubescence on at least segments 9–11, by their large and conical front coxae, and enlarged pronotum. The elytra often bear red or orange markings, and the tarsi are always five-segmented. The members of the family vary much in shape, from almost circular to elongate–oblong, and from flattened to strongly convex. The elytra vary from short and truncate (exposing last three dorsal abdominal segments), to apically rounded or acute.

In many species the adults and larvae feed on carrion and in at least some cases on fly larvae. Some seem to be more associated with decaying fungi or vegetable matter, and one genus is strictly phytophagus. In Eurasia some genera are predatory specialists on snails and caterpillars. In all cases, though, the silphids are intimately tied in the completion of their life cycle to ground and soil habitats.

The attractive red or orange and black *Nicrophorus* (Fig. 37.1) are called burying beetles because they can rapidly excavate the soil from under a dead mouse or other small animal until the body sinks into the ground. They exhibit a very interesting parental care, and the larvae are fed by a parent through regurgitation (work of Pukowski, abstracted by Balduf 1935). Stridulation may be used for inter- and intraspecific communication (Bredohl 1984, Niemitz 1972, Schumacher 1973). Verification of these biologies and a comparative-evolutionary study of *Nicrophorus* biology has not been made for North American species (see Milne and Milne 1944, 1976).

It is amazing that so little is known in North America about the general biology, ecology, and natural history of these large, attractive, and easy-to-collect beetles. Some studies have examined the entire community of carrion insects and the position of silphids in the process of faunal succession (Johnson 1975, Payne, 1965, Payne and King 1970, Reed 1958, Walker 1957). Many studies have looked at limited aspects of the biology of all or a few silphids in a limited geographic area (Shubeck 1969, Shubeck et al. 1977). Older literature indicates that various species have preferences for the carrion of certain classes of vertebrates, but there now seems to be little basis to believe that the beetles choose between the carrion of homiotherm or poikilotherm vertebrates (Shubeck 1976). Olfactory sensitivity is of primary importance in locating carrion. Early studies were concerned with locating the sense organs (Dethier 1947, and references) but modern electrophysiological studies have precisely identified the sense receptors and some of the specific decay compounds that activate them (Waldow 1973). The effect of wind in locating carrion is important (Shubeck 1968). Silphids have reduced interspecific competition by "dividing" their carrion resource in several ways. There are differences in the season in which a species is most active (Anderson 1982a), in their diel periodicities (Shubeck 1971), in the age of the carrion that they can use, and in habitat preferences (Anderson 1982a) (e.g., such as upland forests, swamp forests, open grassy or steppe habitats, and boreal versus warm temperature forests). But so little is yet known of their present-day ecologies that although they can be locally common as fossils in Pleistocene deposits, they are not yet useful indicators of paleoecological or paleoenvironmental conditions (Miller and Peck 1979). The size of the carrion resource may also be important in resource partitioning. The Silphinae seem to be abundant as adults and larvae on large carcasses as well as on small ones, but the Nicrophorinae seem to use only smaller carcasses for reproductive purposes.

The beetles may generally act to maintain ecosystem health and productivity by aiding the recycling of nutrient matter. While doing so they reduce the frequency of occurrence and suitability of carrion as a site of fly oviposition and maggot development. Their feeding activities may also destroy potential foci of infection of pathenogenic bacteria or parasitic worms. Rabies virus (but not anthrax bacteria) is inactivated in the gut of *Nicrophorus* (Sidor 1970), which also phoretically carry *Poecilochirus* and several other mites (Springett 1968, Wilson 1983) and several species of nonpathogenic nematodes as commensals (Theodorides 1950). Detailed studies of *Nicrophorus* ecology have been done only on species in Michigan (Wilson and Fudge 1984, Wilson and Knollenberg 1984, Wilson et al. 1984).

Systematics

Much of the existing literature is difficult to use in making determinations of North American silphids. Because of the widespread and growing interest in them and the lack of comprehensive keys, the following is provided, which should allow determination of all species in North America north of Mexico, based on a modern understanding of their systematics. It is modified from a manuscript key of world genera and North American species of Silphidae by Dr. R. B. Madge, Commonwealth Institute of Entomology, London. The generic placement of North American species of Silphinae is somewhat confused in the older literature. The silphine genera used in the following key are not the traditional ones for North America, but do bring the fauna into evolutionary perspective with the world fauna, and should help establish stability in nomenclature. Anderson and Peck (1985) present keys and distribution maps for Canadian and Alaskan species. Distribution and bionomics of the U.S. species are summarized by Peck and Kaulbars (1988). Anderson (1982b) and Anderson and Peck (1985) give data on all known larvae. The fauna of Latin America is reviewed by Peck and Anderson (1985a) and phylogenetic relationships are proposed. A complete systematic catalog is that of Peck and Miller (1989).

Morphological Characters

Males usually have broadly expanded protarsal segments, while those of the female resemble the segments on the other legs. The outer and bent-under margins of the elytra are the epipleura (sing., epipleuron). The elytra may have one to three upraised ridges or costae (sing., costa). A raised area or callus may occur near the posterior end of the outermost costa. The elytral suture is where the two elytra meet at the midline. The elytra of *Nicrophorus* have broad and transverse red or yellow-orange bands called fasciae. Surface microsculpture and hairs may be difficult to see in worn, greasy, or dirty specimens (which should be cleansed in a warm detergent–ammonia solution).

ACKNOWLEDGMENTS

I thank Robert Anderson, J. M. Campbell, Ronald B. Madge, Scott E. Miller, Alfred F. Newton, Paul Shubeck, and Quentin D. Wheeler for sharing their research results and helping me to achieve an up-to-date understanding of the beetles covered in this chapter, and for checking part or all of the manuscript. Jarmila Kukalova-Peck lovingly prepared the illustrations. Field and laboratory research on soil and litter scavenging beetles has been supported by operating grants from the Natural Sciences and Engineering Research Council of Canada (NSERCC).

KEY TO SUBFAMILIES, GENERA, AND SPECIES OF U.S. AND CANADIAN SILPHIDAE†

1a Frontoclypeal suture absent (Figs. 37.3, 37.4); elytra usually tricostate and tips usually prolonged at suture; antennae distinctly 11-segmented (second segment about as long as third), gradually widening into club; gula strongly constricted medially (fifth abdominal tergum without a pair of stridulatory files); Subfamily **Silphinae** **2**

1b Frontoclypeal suture present (Fig. 37.5); elytra always smooth and truncate; antennae apparently 10-segmented (the second segment being very short and more or less hidden in the tip of the first, Fig. 37.5); last four segments abruptly widened into a distinctly swollen club; gula absent except for narrow triangular piece anteriorly (fifth abdominal tergum with a pair of midlongitudinal stridulatory files); Subfamily **Nicrophorinae**, Genus ***Nicrophorus*** Fabricius, 1775 (Fig. 37.1) **16**

† Key prepared by Ronald B. Madge and Stewart B. Peck.

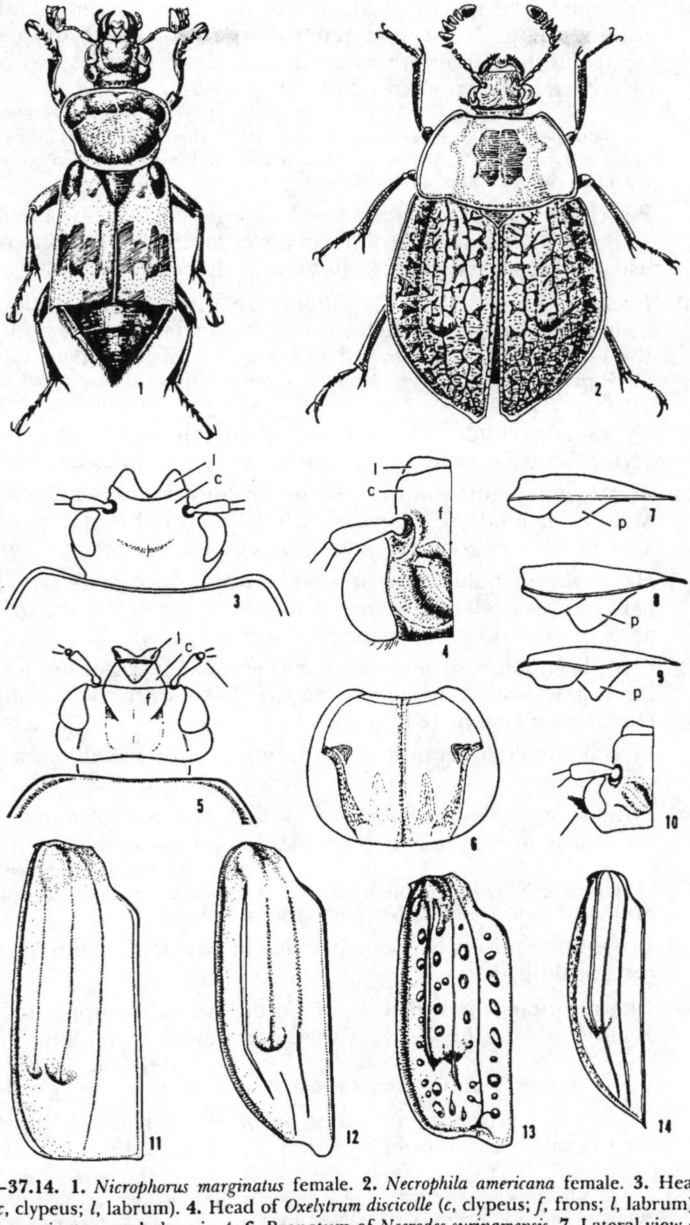

Figs. 37.1–37.14. 1. *Nicrophorus marginatus* female. **2.** *Necrophila americana* female. **3.** Head of *Aclypea bituberosa* (*c*, clypeus; *l*, labrum). **4.** Head of *Oxelytrum discicolle* (*c*, clypeus; *f*, frons; *l*, labrum). **5.** Head of *Nicrophorus marginatus*, symbols as in 4. **6.** Pronotum of *Necrodes surinamensis*. **7.** Lateral view of pronotum *Necrodes surinamensis* (*p*, pronotal postcoxal lobe). **8.** Lateral view pronotum *Thanatophilus trituberculatus*. **9.** Lateral view pronotum *Thanatophilus sagax*. **10.** Head *Oiceoptoma inaequale*. **11.** Elytron *Thanatophilus truncatus*. **12.** Elytron *Thanatophilus coloradensis*. **13.** Elytron *Thanatophilus lapponicus*. **14.** Elytron *Oxelytrum discicolle* female.

2a (1a) Postcoxal lobe of pronotum (seen in side view) short and broadly rounded (Fig. 37.7); moderate to large males (with expanded protarsi) with metafemora swollen and toothed toward apex; pronotum orbicular, widest toward middle (Fig. 37.6) . . . .
Genus ***Necrodes*** Leach, 1815

Our only species is ***N. surinamensis*** (Fabricius); widely distributed in the United States and southern Canada in farmland, prairie, and forests, often in great numbers. Ratcliffe (1972) has summarized the natural history.

2b Postcoxal lobe of pronotum well developed, usually angular (Figs. 37.8, 37.9); metafemora of males never toothed or swollen; pronotum usually widest toward base and not as markedly orbicular **3**

3a (2b) Eyes very large (Fig. 37.4); underside of elytra with many longer hairs in region of posterior callus (raised tubercle) (careful positioning of the light is required to see them) . . . Genus ***Oxelytrum*** Gistel, 1848

Several species occur in the Neotropics, but only the most widespread, ***O. discicolle*** (Brullé) (Fig. 37.14), has crossed the Rio Grande into extreme southern Texas.

3b Eyes smaller, underside of elytra without longer hairs in region of posterior callus (at most with a fine tomentum of microsetae) **4**

4a (3b) Head with a short row of prominently longer and more erect hairs on side just behind eyes (Figs. 37.2, 37.10) (styli of female genitalia with terminal setae extending along outer face, coxites curving down) **5**

4b Head without a short row of prominently longer, more erect hairs just behind eyes (styli of female genitalia with terminal setae restricted to apex or curving around apex; coxites variable) **10**

5a (4a) Elytral shoulders with small tooth (Fig. 37.19); posterior face of metafemora without carinae (female genitalia with styli small); Genus ***Oiceoptoma*** Leach, 1815 . . . . **6**

5b Elytral shoulders rounded; posterior face of metafemora with two carinae **8**

6a (5a) Pronotum bicolored, with a black disk and broad orange margins; metatibiae in female abruptly enlarged on apical third . . . .
O. noveboracense (Forster)

Widespread in more northerly parts of eastern and north-central United States and southern Canada; most active in spring and early summer.

6b Pronotum entirely black; metatibiae in female not abruptly enlarged on apical third **7**

7a (6b) Elytral epipleuron wide (i.e., on posterior half the upper oblique part is at least twice the width of the lower vertical part) (Fig. 37.17). . .
O. inaequale (Fabricius)

Widespread in eastern North America; active in spring and early summer.

7b Elytral epipleuron narrow (i.e., on posterior half the upper oblique part is subequal to lower vertical part) (Fig. 37.18). . . . .
O. rugulosum (Portevin)

This has a poorly understood distribution in the southeastern United States from southern Indiana to Florida and Texas; previously confused with foregoing species.

8a (5b) Pronotum black, entirely covered with a simple dense punctuation; (female genitalic styli short); Genus ***Heterosilpha*** Portevin, 1926 . . . **9**

8b Pronotum bicolored and with elongated punctures; (styli of female genitalia large, usually extending beyond apex of coxite) . . . . . .
Genus ***Necrophila*** Kirby and Spence, 1828

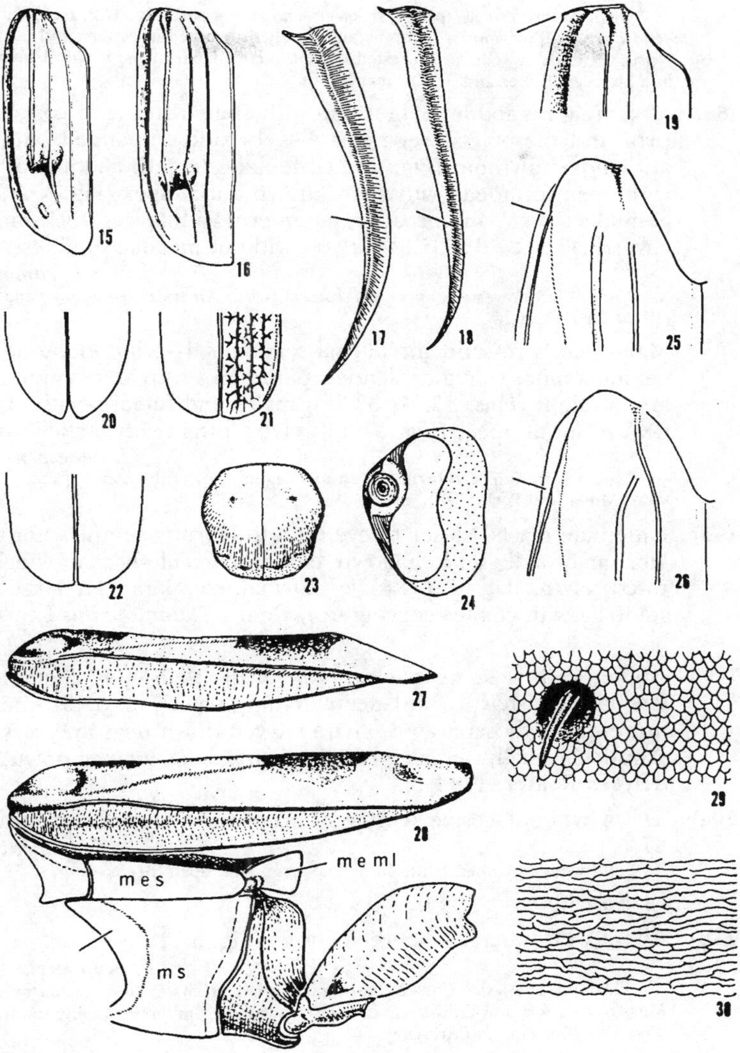

Figs. 37.15–37.30. **15.** Elytron of *Thanatophilus trituberculatus.* **16.** Elytron of *Thanatophilus sagax.* **17.** Epipleuron of *Oiceoptoma inaequale.* **18.** Epipleuron of *Oiceoptoma rugulosum.* **19.** Elytral humeral angle *Oiceoptoma inaequale.* **20.** Elytral tip *Heterosilpha ramosa* female. **21.** Elytral tip *Heterosilpha ramosa* male. **22.** Elytral tip *Heterosilpha aenescens* male and female. **23.** Pronotum *Nicrophorus carolinus.* **24.** Interior face tenth antennal segment *Nicrophorus investigator.* **25.** Elytral humeral angle *Nicrophorus pustulatus.* **26.** Elytral humeral angle *Nicrophorus orbicollis.* **27.** Epipleuron *Nicrophorus investigator.* **28.** Elytral epipleuron and part of thorax *Nicrophorus hybridus* (*meml*, metepimeral lobe; *mes*, metepisternum; *ms*, metasternum with anterior bald patch). **29.** Isodiametric or granular microsculpture of elytral disk *Nicrophorus vespilloides.* **30.** Transverse meshwork or brick-wall microsculpture elytral disk *Nicrophorus orbicollis.*

North America's sole species, ***N. americana*** (Linnaeus) (Fig. 37.2) has distinctive yellow pronotal margins; commonly and widely distributed in eastern and central North America in open and forested habitats, active from spring to fall. It may mimic bumble bees (Fisher and Tuckerman 1986).

9a (8a) Males (eighth abdominal tergum with apex truncate or concave) with pro- and mesotarsal segments 1–4 broadly expanded, and elytral apex not drawn out (Fig. 37.21); females (eighth abdominal tergum with apex rounded) with elytra drawn out at apex (Fig. 37.20); male genitalia thicker and broader, parameres with apices thick and downcurved (Figs. 37.33, 37.34); elytra without metallic shininess. . . . . ***H. ramosa*** (Say)

Widespread in western and north-central North America. See Brewer and Bacon (1975) for biology.

9b Males with pro- and mesotarsal segments 1–4 not expanded; male genitalia more thin and slender, parameres with apices more narrow and straight (Figs. 37.31, 37.32); males and females with elytra not drawn out at apex; (Fig. 37.22); elytra often with metallic shininess ***H. aenescens*** (Casey)

Ranges from Baja California north to Oregon, west of the Sierra–Cascade mountains (Miller and Peck 1979).

10a (4b) Antennae placed closer to eye than to labrum; temples not swollen; labrum broadly and shallowly emarginate; mesocoxae widely separated; elytra flat at apex; eyes normal, comparatively large (female genitalia with coxites curving up); Genus ***Thanatophilus*** Leach, 1815 **11**

10b Antennae as close to, or closer to, labrum than to eye; temples swollen; labrum sharply and deeply emarginate (Fig. 37.3); mesocoxae more narrowly separated; elytra curved down near apex; eyes comparatively small; (female genitalia with coxites curving down); Genus ***Aclypea*** Reitter, 1884 . . . . **15**

11a (10a) Elytra without costae (costae effaced), and abruptly truncate (Fig. 37.11) . . . . ***T. truncatus*** (Say)

Southwestern United States in desert, scrub, and open forests.

11b Elytra with costae **12**

12a (11b) Elytra with tubercles between costae (Fig. 37.13). . . . . ***T. lapponicus*** (Herbst)

Throughout Alaska, Canada, northeastern United States, and the southwest to New Mexico, Arizona, and California; in forest, grassland, and arctic–alpine tundra habitats. Also occurring in Eurasia.

12b Elytra without tubercles between costae. **13**

13a (12b) Inner two elytral costae absent (effaced) except at apex (Fig. 37.12); eighth abdominal segment orange . . . . ***T. coloradensis*** (Wickham)

Only at treeline and above in arctic–alpine tundra in the Rocky Mountains from New Mexico to Alaska (Peck and Anderson 1982).

13b Elytra with three fully developed costae; eighth abdominal segment black **14**

14a (13b) Outer elytral costa extending beyond apical callus (Fig. 37.16); postcoxal lobe of pronotum pointed (Fig. 37.9); female with sutural apices of elytra not prolonged but similar to those of male (Fig. 37.16). . . ***T. sagax*** (Mannherheim)

Pacific northwest, north-central states, central and northern Canada, and Alaska.

14b Outer elytral costae ending at apical callus (Fig. 37.15); postcoxal lobe of pronotum broadly rounded (Fig. 37.8); female with sutural apices of elytra prolonged (Fig. 37.15) . . . . *T. trituberculatus* (Kirby)

Central and northern Canada and northern Alaska.

15a **(10b)** Pronotum with narrower anterior margin near middle; margin at anterior angle flatter; often with impunctate (smooth) spot on anterior third behind the eye (in addition to other spots near midline and along base) . . . . *A. opaca* (Linnaeus)

A Eurasian species in the far north of Canada and Alaska (Anderson and Peck 1984).

15b Pronotum with wider anterior margin near middle; margin at anterior angle more upcurved; never with impunctate spot on anterior third behind eye (although other spots may be present) . . . . *A. bituberosa* (LeConte)

Phytophagous on native **Chenopodiaceae** and various garden and crop plants like spinach and sugar beets. In western prairie and mountain states; Nebraska to California, north to Northwest Territories. Active in early spring. See Cooley (1917) and Anderson and Peck (1984) for biology.

16a **(1b)** Pronotum (Fig. 37.23) without an anterior transverse impression; with extremely narrow lateral margins; and with distinct punctures along base . . . . *N. carolinus* (Linnaeus)

Widely distributed but most commonly found in the southeastern coastal plain northward to central plains states and adjacent Canada; usually in open (mesic to arid) habitats.

16b Pronotum (Fig. 37.1) with an anterior transverse impression, distinct at least at sides of disk; lateral margins not extremely narrow; base without distinct punctures **17**

17a **(16b)** Frons and pronotal disk red; abdominal sterna with setae along posterior margins much shorter on basal than on apical segments; tarsal empodium normally quadrisetose . . . . *N. americanus* Olivier

This, our largest species, was formerly widely distributed in the eastern United States and into extreme southern Ontario (in heavily forested habitats?). The species has been only rarely taken in the past few decades and may be endangered or extinct over most of its former range (Anderson 1982c).

17b Frons and pronotal disk black; abdominal sterna with marginal setae (especially at middle) long on both basal and apical segments; tarsal empodium bisetose **18**

18a **(17b)** Pronotum covered with dense golden pubescence except for two small spots on disk; seventh abdominal tergum covered with dense semierect pubescence . . . . *N. tomentosus* Weber

Widely distributed over eastern and central United States and southern Canada in both forested and open habitats. In behavior and appearance it seems to be a bumblebee mimic (Fisher and Tuckerman 1986).

18b Pronotum glabrous or with only sparse hairs anteriorly and on margins; seventh abdominal tergum with short, sparse, depressed hairs **19**

19a **(18b)** Posterior lobe of metepimeron (Fig. 37.28) with many long, golden hairs **20**

19b Posterior lobe of metepimeron with dark hairs or only a few short golden hairs, or glabrous **22**

20a **(19a)** Anterior face of procoxae with very long hairs on basal half; middle black band of elytra usually reaching the dorsal ridge of the epipleuron; inner face of metatibiae with a dense brush of hair. **21**

20b Anterior face of procoxae without very long hairs on basal half; middle black band of the elytra never reaching the dorsal ridge of the epipleuron; inner face of metatibiae with sparse hair . . . . ***N. marginatus*** Fabricius

Widely distributed across the United States and southern Canada; in open grassland, old-field areas, desert scrub, and sagebrush steppe habitats.

21a (20a) Next-to-last antennal segment with outer edge usually deeply emarginate (Fig. 37.36); basal segment of antennal club black; anterior black elytral band reaching epipleural ridge but not significantly crossing onto epipleuron; posterior elytral fascia reaching dorsal ridge of epipleuron . . . . ***N. obscurus*** Kirby

Distributed in prairies of central Canada and the United States.

21b Next-to-last antennal segment with outer edge usually only shallowly emarginate (Fig. 37.35); basal segment of antennal club red or black; *if* red, then anterior black elytral band crossing over epipleural ridge onto epipleuron; posterior elytral fascia may or may not reach dorsal epipleural ridge; *if* basal antennal segment black, then elytral markings very reduced or absent . . . . ***N. gutula*** Motschulsky

This species is widespread across the western United States and adjacent southern Canada; in grassland, shrub, and open montane forest habitats; usually all black in the Pacific northwest except for reddish elytral shoulders; populations (to which the name ***N. hecate*** Bland has been applied) inland from the Pacific coast and in the Rocky Mountain states have normal to widely expanded fasciae. Geographic variation is discussed by Anderson and Peck (1986).

22a (19b) Metatibiae curved; basal black band of elytra not reaching the epipleuron . . . . ***N. sayi*** Laporte

The species is widely distributed in the more temperate parts of eastern North America and northwest to Alberta; in forested habitats; usually active in very early spring.

22b Metatibiae straight; basal black band of elytra usually reaching the epipleuron **23**

23a (22b) Elytral disk with strong isodiametric microsculpture (appearing as equidimensional or granular units at magnification of about 100–150×) (Fig. 37.29) **24**

23b Elytral disk with microsculpture arranged in fine transverse lines or meshes (appearing as an irregular "brick-wall" pattern of units at magnification of about 100–150×) (Fig. 37.30) **27**

24a (23a) Metasternal pubescence brown; metepisternum (Fig. 37.28) with pubescence long and well developed; anterior margin of pronotum straight **25**

24b Metasternal pubescence golden; metepisternum with pubescence short and poorly developed; anterior margin of pronotum concave **26**

25a (24a) Elytral disk entirely black (very rarely with small red spots) . . . . . ***N. nigrita*** Mannerheim

The species is limited to a wide strip, in and west of the Sierra–Cascade mountain range, paralleling the Pacific, from southern British Columbia to southern California; usually in openly forested habitats.

25b Elytral disk distinctly bifasciate . . . . ***N. mexicanus*** Matthews

Distributed in the southern Rocky Mountain states southward; in sparsely forested habitats.

26a (24b) Base of epipleuron orange with a postbasal black spot; both upper and lower faces of antennal segments 9 and 10 with a patch of dense white hairs arranged in a figure-eight pattern across base (Fig. 37.24) ***N. vespilloides*** Herbst

Distribution across northern United States and throughout Canada and Alaska; in forested habitats that are colder, and wetter (usually marshy, boggy, or swampy); also occurring in Eurasia; commonly confused with the following species.

26b Base of epipleuron all black, antennal segments 9 and 10 without a patch of dense white hairs (figure-eight pattern may be present as faint vestige only) . . . . *N. defodiens* Mannerheim

Distribution in more northerly localities in the United States and across Canada, and in better drained or drier forests; from Alaska to central California in the west, and penetrating the southeast in the upper elevation forests of the Appalachian Mountains. Geographic variation is discussed by Anderson and Peck (1986).

27a **(23b)** Elytral epipleuron black **28**

27b Elytral epipleuron orange-red **29**

28a **(27a)** Dorsal ridge of epipleuron short, reaching anteriorly only to level of tip of scutellum (Fig. 37.26); elytra bifasciate . . . . *N. orbicollis* Say

One of the most common species in forested habitats in eastern and central North America.

28b Dorsal ridge of epipleuron long, reaching anteriorly to level of basal third of scutellum (Fig. 37.25); elytra with only small posterior fascia *N. pustulatus* Herschel

Widely distributed in forested habitats in eastern North America and northwest to Alberta.

29a **(27b)** Metasternum with elongate bald patch just behind mesocoxae (Fig. 37.28); with continuous multiple and irregular rows of stout and erect hairs running from anterior face of elytral shoulder to base of epipleural ridge; elytral epipleuron narrow, the dorsal ridge usually straight or curving gently upward at posterior end (Fig. 37.28); moderate to large males with free apical process of metatrochanter strongly bent back. . . . . *N. hybridus* Hatch and Angel

Distributed in northern plains and Rocky Mountain states into adjacent Canada; in open and sparsely forested habitats.

29b Metasternum without an elongate bald patch just behind mesocoxae; multiple irregular rows of hairs on elytral shoulder, but ending well before the base of epipleural ridge; elytral epipleuron wide, the dorsal ridge usually curving down at posterior end (Fig. 37.27); moderate to large males with free apical process of metatrochanter not strongly curved back . . . . *N. investigator* Zetterstedt

Widely distributed across Canada and southward along the Rockies to New Mexico; in forested habitats; also occurring in Eurasia. Geographic variation is discussed by Anderson and Peck (1986).

Family Agyrtidae

Members of this family have no common name. They have traditionally been included within the Silphidae, but recent study interprets them to have many adult and larval characters that separate them from the staphylinid-like silphids, and more closely allying them with the Leiodidae (Lawrence and Newton 1982). Very little is known of their general biology. They seem to be general scavengers and may frequently be baited with carrion. Their greatest diversity is in the Pacific Northwest, and only one species occurs in eastern North America. They are usually associated with litter and soil habitats, although most species of *Pteroloma* and *Apteroloma* seem connected with streamside or other cool, moist habitats in montane areas and may be locally common on snowfields at night, feeding on dead insects.

The family is most easily separated from the Silphidae by the nontruncate elytra, which bear 9 (or 10) punctate striae (Figs. 37.42–37.45). The antennae are filiform or only gradually clavate, and there are five or six

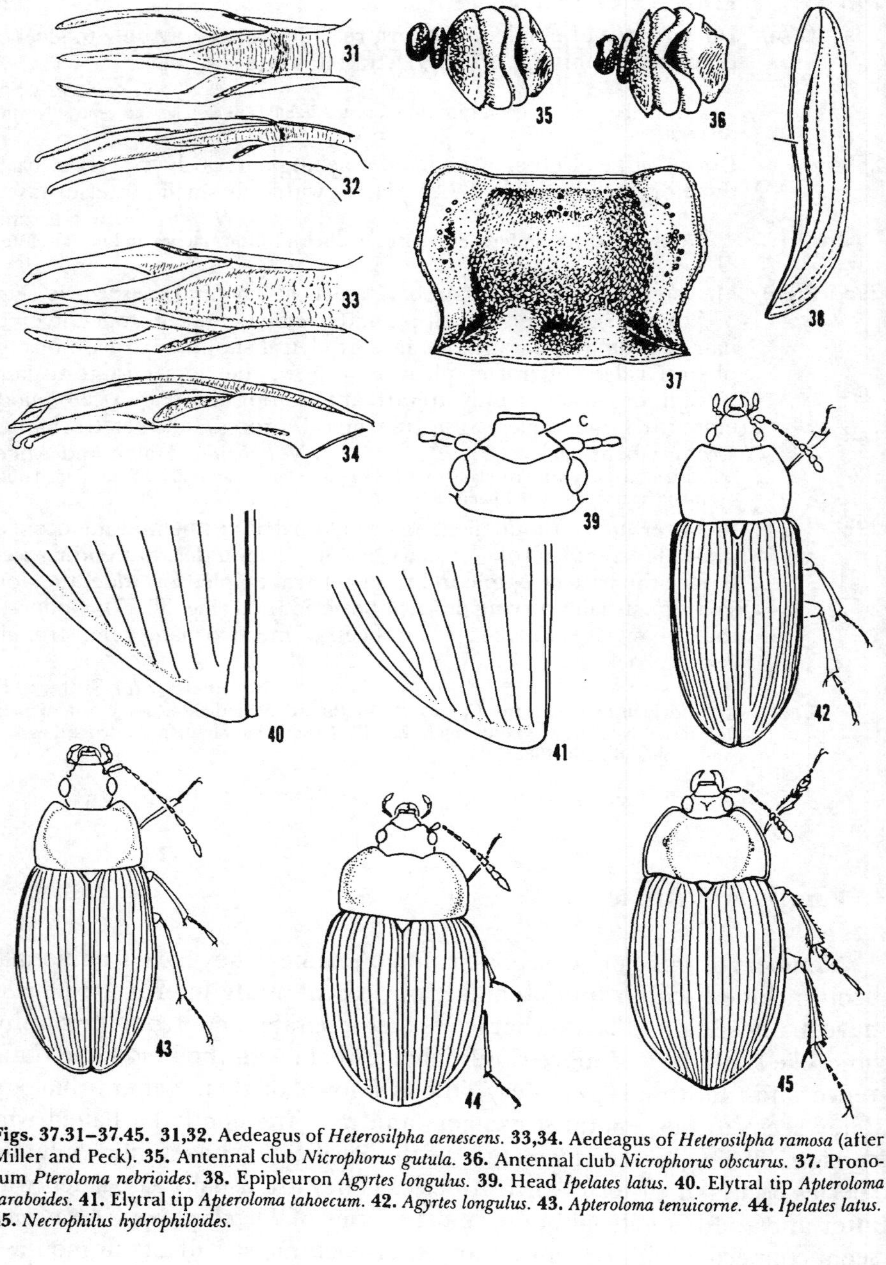

Figs. 37.31–37.45. 31,32. Aedeagus of *Heterosilpha aenescens*. **33,34.** Aedeagus of *Heterosilpha ramosa* (after Miller and Peck). **35.** Antennal club *Nicrophorus gutula*. **36.** Antennal club *Nicrophorus obscurus*. **37.** Pronotum *Pteroloma nebrioides*. **38.** Epipleuron *Agyrtes longulus*. **39.** Head *Ipelates latus*. **40.** Elytral tip *Apteroloma caraboides*. **41.** Elytral tip *Apteroloma tahoecum*. **42.** *Agyrtes longulus*. **43.** *Apteroloma tenuicorne*. **44.** *Ipelates latus*. **45.** *Necrophilus hydrophiloides*.

visible abdominal sternites (always six in the Silphidae). Additional description, characters, or detailed distributions are given by Lawrence (1982) and Lawrence and Newton (1982), and Anderson and Peck (1985).

Key to Genera and Species of U.S. and Canadian Agyrtidae†

1a Dorsal ridge of elytral epipleuron (in side view) depressed behind shoulder (Fig. 37.38); last segment of maxillary palp swollen; Genus ***Agyrtes*** Frölich, 1799 . . . . **2**

1b Dorsal ridge of elytral epipleuron evenly curved from shoulder to apex; last segment of maxillary palp not inflated **3**

2b **(1a)** Antennal club of four segments; posterior angles of pronotum sharp; elytral epipleuron glabrous; aedeagus in side view with pronounced bend . . . . ***A. longulus*** (LeConte)

Fig. 37.42. From southern California to Alaska, inland to Idaho; seemingly most active in winter months; presumably a forest litter scavenger (Hatch 1957, Peck 1974b).

2b Antennal club of five segments; posterior angles of pronotum rounded; elytral epipleuron hairy; aedeagus in side view relatively straight . . . . ***A. similis*** Fall

Limited to mountain forests in southern California; active in winter months; presumed scavenger (Peck 1974b).

3a **(1b)** Last five or six segments of antennae with uniform vestiture of microsetae (plus ring of long tactile setae); segments 1–5 or 6 with only a ring of tactile setae; tenth segment transverse; Genus ***Necrophilus*** Latreille, 1829. . . . . **4**

3b Antennal segments evenly pubescent (in addition to tactile setae), the pubescence gradually becoming shorter and denser apically; tenth segment usually elongate **5**

4a **(3a)** Winged; metasternum long; first abdominal spiracle large and eliptical ***N. hydrophiloides*** Guérin-Méneville

Fig. 37.45. In moist, forested sites from coastal southern California to southern Alaska; often carrion scavenger (Hatch 1957).

4b Wingless; metasternum short; first abdominal spiracle almost circular ***N. pettitii*** Horn

A deep litter and soil scavenger; from New York, Ontario, and Missouri south to Louisiana and northern Florida; most frequently found at high elevations in the southern Appalachians (Peck 1981a).

5a **(3b)** Posterior coxae distinctly separated; wingless; elytra completely without shoulders . . . . Genus ***Lyrosoma*** Mannerheim, 1953

The only North American species is ***L. opacum*** Mannerheim, known from Alaskan coastal and island sites, where it feeds on decaying seaweed in and above the tidal zone (Hatch 1957).

5b Posterior coxae contiguous; winged or wingless; elytra with well-marked shoulders **6**

6a **(5b)** Clypeus transversely "pentagonal" in shape (Figs. 37.39, 37.44); mandibles not toothed; preapical antennal segments with a sharp ridge around apex of each segment. . . . . Genus ***Ipelates*** Reitter, 1884

The only North American species is ***I. latus*** (Mannerheim) (Fig. 37.44); from northern California to Alaska and Alberta; a forest litter inhabitant, often under bark, and apparently feeding on fleshy fungi (Hatch 1957, Schwaller 1983).

† Prepared by Ronald B. Madge and Stewart B. Peck.

6b Clypeus rectangular or trapezoidal; mandibles toothed; preapical antennal segments without a sharp ridge around apex of each segment **7**

7a **(6b)** Pronotum with a strong depression in each posterior angle and at middle of base (Fig. 37.37); tibiae with a longitudinal dorsal ridge; frons usually with two pale "ocelli" or "eyespots"; protarsi of males scarcely widened; aedeagus with a single large paramere . . . . Genus ***Pteroloma*** Gyllenhal, 1827

The only North American species is ***P. nebrioides*** Brown of the Rocky Mountains of British Columbia, Alberta, and Montana (Hatch 1957).

7b Pronotum without basal impressions; tibiae without a longitudinal dorsal ridge; protarsi of males strongly widened; aedeagus without parameres; Genus ***Apteroloma*** Hatch, 1927 . . . . **8**

8a **(7b)** Elytral epipleuron without punctures **9**

8b Elytral epipleuron densely punctate along lower margin. **10**

9a **(8a)** Internal angle of elytral apex evenly rounded (Fig. 37.41); pronotum strongly transverse, base more than one and a half times the length along midline. . . . . ***A. tahoecum*** (Fall)

A scavenger of forest stream margins in the Sierra Nevada mountains of California and adjacent Oregon (Hatch 1957).

9b Internal angle of elytral apex sharp (often with broad tooth) (Fig. 37.40); pronotum only slightly transverse; base less than one and a third times the length along the midline . . . . . ***A. caraboides*** (Fall)

A rare scavenger from northern California to British Columbia and Idaho (Hatch 1957).

10a **(8b)** Posterior angles of pronotum obtusely angulate. . . . . ***A. arizonicum*** (Van Dyke)

A rare scavenger of mountain forest streamsides, from Arizona, New Mexico, and possibly mountains of west Texas; seemingly active in fall and winter.

10b Posterior angles of pronotum rounded (Fig. 37.43) . . . . ***A. tenuicorne*** (LeConte)

Often abundant in montane areas at streamsides or damp and wet habitats and at edges of (and at night upon) melting snowfields; from British Columbia and southern Alberta to California and Colorado (Hatch 1957).

The Family Leiodidae

The family Leiodidae is composed of small beetles, from 1 to 8 mm in length; usually of an ovoid shape; and usually with 11-segmented antennae with an interrupted five-segmented club, with segment 8 smaller than 7 and 9. Other antennal conditions occur but are less common. What are grouped here have been variously placed elsewhere under the names Catopidae, Cholevidae, Leptodiridae, Anisotomidae, Colonidae, Leptinidae, and Platypsyllidae. A consistent family feature (although difficult to observe) is the shared derived character of a usually nearly enclosed gutter ringing the antennal articulation on the dorsal face of segments 7, 9–10, or 8–10 or (in 10-segmented antennae) 7–9. These segments often have internal sensory vesicles opening into the gutter (Peck 1977a). Further characterization is given in Lawrence (1982) and Lawrence and Newton (1982). This section follows Newton's (1984) newly proposed higher classification within what now seems to be a monophyletic family.

The members of the family are for the most part scavengers; inhabiting decaying litter, leaves, fungi (see Newton 1984), moss and grass roots, and the subcortical region of dead trees. Some species live on carrion, may be attracted to dung, and occur in the nests of ants, in caves, and the burrows of ground-dwelling reptiles, birds, and mammals. The habits of the individual subfamilies are fairly diverse and are discussed separately below.

The Subfamily Leptininae

These are very dorsoventrally flattened, louselike beetles, 2–5 mm long, with eyes reduced or absent. They are called mammal nest beetles, and have been traditionally treated as a separate family. However, they have internal antennal vesicles on antennal segments 9 and 10, an advanced character for the leiodid assemblage, and represent a line of Cholevinae (Catopinae) that has become specialized for a life of ectoparasitism on small mammals. *Leptinus* is occasionally found in soil and leaf litter, but more frequently on various voles, moles, and shrews, and in their nests, and rarely, in old nests that have been occupied by bumblebees and *Vespula* wasps (Peck 1982). They may questionably be a guest in *Formica* ant nests.

The Subfamily Cholevinae

These have been variously placed under the names Catopidae and Leptodiridae, and Catopinae of the Leiodidae. Zwick (1979) has shown that the family group name Cholevinae has priority over Catopinae. These may be called the small carrion beetles. They are elongate–oval; pale brownish, to brown, to black; and 2–6 mm in length. They often have a pronotum and elytra with a finely granulate surface or with faint cross striations or strigae. Their eighth antennal segment is always shorter and smaller in diameter than the seventh and ninth segments. The tarsal formula is almost always 5–5–5. These fragile, rapidly running beetles are mostly found on vertebrate carrion, but may act as more general litter decomposers by feeding on the bodies of terrestrial arthropods and decaying plant material. Some are found most frequently on decomposing fungi, in caves on moist organic matter such as bat and cricket guano, and in the nests of ants, where they are probably scavengers. They may be collected in their special microhabitats, or by setting large or small dung and carrion baited pitfall traps (Newton and Peck 1975, Zwick 1979), and by sifting rotting leaves, forest litter, moss, and grass roots. Baited traps can be used to determine seasonality and habitat preferences of the species (Peck and Anderson 1985b).

The Subfamily Coloninae

Little is known of the biology and general natural history of this subfamily. All species are in the single genus *Colon*. The beetles have a characteristic facies that is easy to distinguish when once seen. The antennal club is gradually formed from the last four to six segments. Contrasted with other members of the family, segment 8 is not smaller than 7 and 9. [See Hatch (1933, 1957) for more details.] The beetles are taken in forest litter, humus, and moss, and often near tree bases or stumps, where they and their larvae may be fungal feeders. The use of flight-intercept traps, or the sweeping of low vegetation in forest glades and clearings on still, warm, and humid evenings and afternoons, can occasionally take adults.

The Subfamily Catopocerinae

This group is endemic to North America. There is no common name. It is characterized by widely separated hind coxae, an elongate prosternum, and lack of internal vesicles on the antennal segments, which bear periarticular gutters. Two genera are known, of which both are eyeless and inhabitants of deep forest litter or caves. They may be general scavengers but association with subterranean fungi is known in some (Fogel and Peck 1975).

The Subfamily Leiodinae

These may be collectively called the round fungus beetles. They are generally convex, shiny, oval beetles, 1.5–7 mm in length and from pale yellow-brown to reddish brown to black in color. The antennal club and tarsal formula vary with the genus. Many species can contract the head and prothorax under the body, conceal their appendages, and roll into a ball. The beetles occur in decomposing vegetation and in and on the soil, in fungi, under bark, and in decaying wood and leaves, where they feed on fungi and slime molds. Different groups show different feeding preferences. The Leiodini and Hydnobiini seem to feed on subterranean fungi, while the Pseudoleiodini feed on soft fungi and are more generally saprophagous, and the Agathidiini seem to be specialists on the plasmodia and fruiting bodies of slime molds or Myxomycetes (Fogel and Peck 1975, Newton 1984, Russell 1979, Wheeler 1979a,b). The beetles may be caught by sifting and Berlese extraction from their microhabitats, or in flight in forests, or by sweeping as they rest on vegetation toward dusk, or in Malaise and "window" or "flight intercept" traps (Peck and Davies 1980, Chandler and Peck 1989). Higher classification in the group was previously unsatisfactory, and the keys here use recently proposed advances (Newton 1984).

KEY TO SUBFAMILIES, TRIBES, GENERA, AND SOME SPECIES OF U.S. AND CANADIAN LEIODIDAE

1a Head with occipital carina or elevated crest (Fig. 37.50) (this may be hard to see if head is tightly retracted against pronotum; the crest is weak in ***Platycholeus*** which has a tarsal formula of 5–5–5 in males and 4–5–5 in females) **2**

1b Head without occipital carina or crest **14**

2a **(1a)** Occipital crest overlapping pronotum when head is in repose; cervical sclerites absent; procoxal cavities internally open behind; body dorso-ventrally flattened; eyes reduced or absent; Subfamily **Leptininae 3**

2b Occipital carina resting against front of pronotum when head is in repose; cervical sclerites present; procoxal cavities internally closed behind; body and eyes usually normal; Subfamily **Cholevinae 5**

3a **(2a)** Prosternum short and acute at apex, not extending between procoxae; length about 2 mm . . . . . . Genus ***Leptinus*** (Müller, 1817)

Distributed over much of eastern North America and the Pacific northwest; with three described species (Fig. 37.46), most frequently found in nests and fur of mice, shrews, and moles and occasionally in litter (Peck 1982). Their biology is not well known, although it has been studied for ***L. testaceus*** of Europe (Buckle 1976, Ising 1969).

3b Prosternum produced posteriorly, extending beyond procoxae and appearing to separate them, *or* forming a broad flat plate, ending in a median lobe fringed with long setae **4**

4a **(3b)** Prosternum produced posteriorly; separating the procoxae . . . . . Genus ***Leptinillus*** Horn, 1882

L. aplodontiae Ferris has a prosternum without apical setae; is 3.5 mm long; and is confined to the fur and nests of the mountain beaver ***Aplodontia rufa*** from California to Washington. ***L. validus*** (Horn) has a prosternum fringed with long, closely set setae extending to the mesosternum; is 4.5 mm long; and is confined to the lodges and fur of the beaver (***Castor canadensis***) throughout the host's range (see Parks and Barnes 1955, Wood 1965).

4b Prosternum forming a broad, flat plate, ending in a median lobe fringed with long setae . . . . . . Genus ***Platypsyllus*** Ritsema, 1869

The highly modified and flea-appearing ***P. castoris*** Ritsema is a true ectoparasite of beavers (***Castor***) in North America and Eurasia (see Wood 1965, Janzen 1963).

5a **(2b)** Posterior coxae somewhat separated, tarsi all five-segmented except for female protarsi with four segments . . . . Tribe **BATHYSCIINI,** Genus ***Platycholeus*** Horn, 1880

Two species, ***P. leptinoides*** Crotch and ***P. opacellus*** Fall are the only described North American representatives of this large and predominantly cavernicolous tribe of Eurasia. They occur in the Pacific northwest, in litter and under bark of logs and stumps, often with ants and termites (Hatch 1957).

5b Posterior coxae contiguous, all tarsi five-segmented **6**

6a **(5b)** Elytra with setal bases arranged as transverse or oblique strigae. . . **7**

6b Elytra with granular surface of irregularly arranged setal bases; strigae absent; Tribe **Cholevini** **11**

7a **(6a)** Hind margin of hind tibia with two inner long spines and an outer row or comb of short and equal spines only (Fig. 37.54); Tribe **Ptomaphagini** **8**

7b Hind margin of hind tibia with two long inner spines (and perhaps a comb of equal spines), but more importantly, also with about four longer outer spines (Fig. 37.53); Tribe **Nemadini** **10**

8a **(7a)** Form oval; with long and erect hairs as well as short recumbent hairs; mesosternal carina low, effaced, or absent; compact and modified for life as guests in ant nests . . . . . . Genus ***Echinocoleus*** Horn, 1885

Three species are known as inhabitants of nests of ***Pogonomyrmex*** and ***Novomessor*** harvester ants in the western United States (Peck 1976).

8b Form elongate oval (Fig. 37.47); with short recumbent hairs only; mesosternal carina present and usually well developed; may live with ants but not as highly modified myrmecophile. **9**

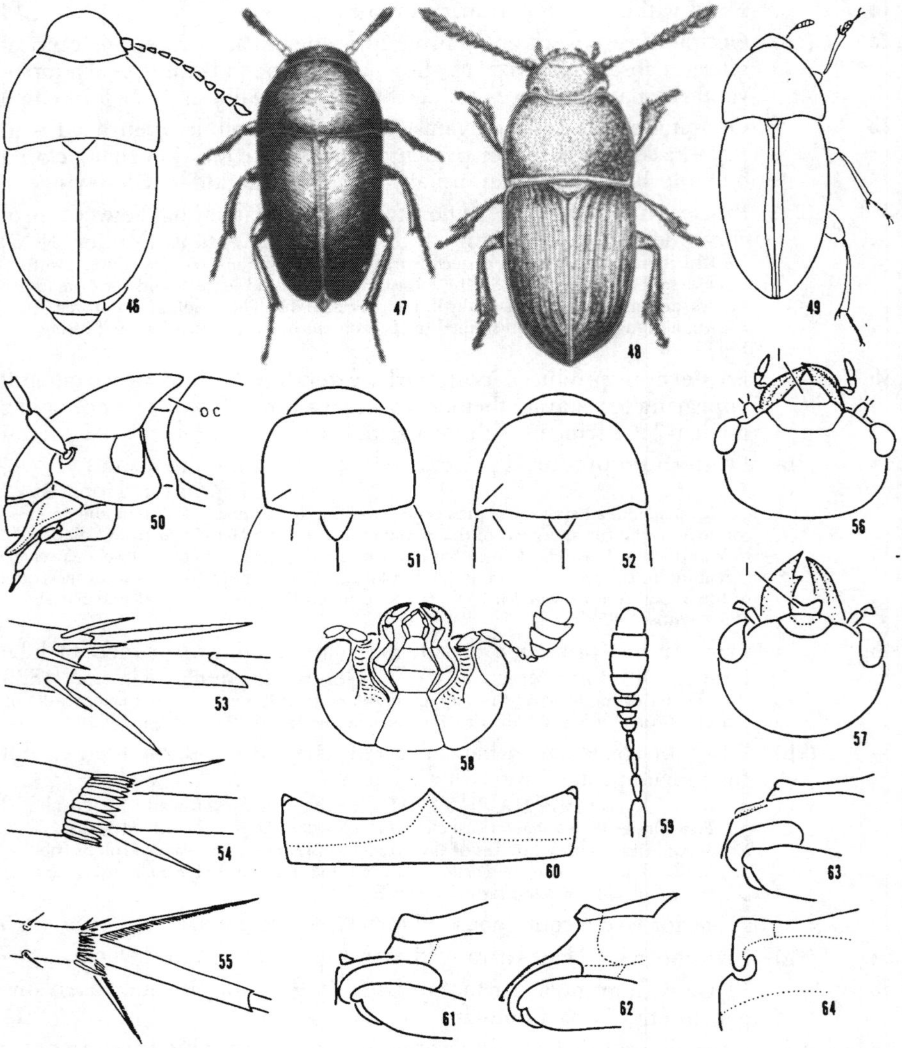

Figs. 37.46–37.64. 46. *Leptinus americanus*. **47.** *Ptomaphagus consobrinus*. **48.** *Catopocerus appalachianus*. **49.** *Colon* sp. **50.** *Ptomaphagus shapardi* (*oc*, occipital carina). **51.** Pronotum *Catops* sp. **52.** Pronotum *Sciodrepoides* sp. **53.** Metatibial spurs *Dissochaetus* sp. **54.** Metatibial spurs *Ptomaphagus* sp. **55.** Metatibial spurs *Prionochaeta opaca*. **56.** Head *Agathidium* sp. **57.** Head *Hydnobius* sp. **58.** Head with antennal grooves *Apheloplastus egenus*. **59.** Antenna *Colon* sp. **60.** First visible abdominal segment *Aglyptinus laevis*. **61.** Mesosternum *Cyrtusa luggeri*. **62.** Mesosternum *Leiodes* sp. **63.** Mesosternum *Colenis impunctata*. **64.** Mesosternum *Cainosternum imbricatum* (from Wheeler, in litt.).

9a **(8a)** Size medium (2.6 mm) to smaller; flight wings absent; eyes reduced to poorly defined collection of about 20 pigmented facets; aedeagal tip more elaborately sculptured, broader, and more blunt . . . .
Genus ***Adelopsis*** Portevin, 1907

Sixteen species, all forest litter and soil inhabiting scavengers, are known from the southeastern United States (Peck 1978b).

9b Size medium (2.6 mm) to larger; flight wings usually present and eyes usually normal (wings absent, and eyes smaller (Fig. 37.50) or reduced to unpigmented spot in most cavernicolous species); aedeagal tip simple, elongate and pointed . . . . . Genus ***Ptomaphagus*** Illiger, 1798

Twenty-five species are known from the United States and Canada (Peck 1973, 1978a), occupying a wide range of litter and soil habitats as scavengers, and extensions of these habitats such as ant nests and animal burrows. Eighteen additional species have become specialized for cave habitats in the southern United States (Peck 1973, 1984, 1986).

10a (7b) Basal segments of male (with expanded protarsi) midtarsi weakly dilated and spongy pubescent beneath; mesosternal carina more elevated and extending nearly to anterior margin of mesosternum. . . Genus ***Nemadus*** Thomson, 1867

Ten species are described from North America, where they are scavengers in forest litter and sometimes in caves and animal burrows; some species seem restricted to ant nests (Hatch 1933, Jeannel 1936, Fall 1937).

10b Basal segments of male (with expanded protarsi) midtarsi not expanded; mesosternal carina feebly elevated, sometimes confined to region between mesocoxae . . . . Genus ***Dissochaetus*** Reitter, 1884

Three species occur as scavengers in forest litter, in ant nests and on carrion in eastern, central, and southwestern North America (Hatch 1933, Jeannel 1936).

11a (6b) Segments 1–10 of antennae bipectinate; with anterior projections, giving sawtoothed appearance. . . Genus ***Catoptrichus*** Murray, 1856

One species, ***C. frankenhauseri*** (Mannerheim), occurs in forest litter in the Pacific northwest; habits unknown, probably scavenger (Hatch 1933).

11b Antennae normal, not bipectinate **12**

12a (11b) Tibial spurs long, serate, longest metatibial spur as long as first metatarsal segment (Fig. 37.55); first mesotarsal segment not dilated in males. . . . . Genus ***Prionochaeta*** Horn, 1880

One species, ***P. opaca*** (Say) occurs in forest litter and some caves in eastern North America; common on carrion (Peck 1977b).

12b Tibial spurs not long and serate; much shorter than first tarsal segment; first mesotarsal segment dilated in males. **13**

13a (12b) Pronotum with base sinuate on either side just within the more or less distinctly rectangular posterior angles (Fig. 37.52). . . . . Genus ***Sciodrepoides*** Hatch, 1933

Two subspecies of two Eurasian species, ***S. fumatus terminans*** (LeConte) and ***S. watsoni hornianus*** (Blanchard), are commonly found as scavengers in many forest and soil-related habitats across the forested parts of the northern half of the continent (Hatch 1933, Jeannel 1936).

13b Pronotum with base arcuate, the hind angles obtuse or more or less evidently rounded (Fig. 37.51) . . . . . Genus ***Catops*** Paykull, 1798

Ten described species occur in forest litter and related habitats across North America; carrion scavenger (Hatch 1933, Jeannel 1936, Peck, in preparation).

14a (1b) Antennae 11-segmented; club gradual, of four to six uninterrupted segments, segment 8 as large as segments 9–10 (Fig. 37.59) (and with periarticular gutter and internal vesicles); cervical sclerites absent; intersegmental membranes between abdominal sterna with brick-wall pattern of minute sclerites; female with four visible abdominal sterna . . . . . . Subfamily **Coloninae,** Genus ***Colon*** Herbst, 1797

Some 35 species are known from litter habitats in North America. Their shape (Fig. 37.49) and antennal club is distinctive; males often with tooth on hind femora; biology and larvae unknown (Hatch 1933, 1957).

14b Antenna 10- or 11-segmented; club usually interrupted when 11-segmented (segment 8 usually smaller than segments 9 and 10 and without periarticular gutter or internal vesicles); cervical sclerites present; abdominal intersegmental membrane without minute sclerites; females with five or six visible abdominal sternites. **15**

15a **(14b)** Hind coxae separated by about a third their width; prosternum in front of coxae longer than coxal width; always with no eyes; Subfamily **Catopocerinae** **16**

15b Hind coxae not separated; prosternum in front of coxae much shorter than coxal width; Subfamily **Leiodinae** **17**

16a **(15a)** Appendages short, body ovoid and flattened, pronotum and elytra of similar widths (Fig. 37.48) . . . .
Tribe **CATOPOCERINI**, Genus ***Catopocerus*** Motschulsky, 1869

Fourteen species are named in this widespread North American genus of forest soil and deep litter (Hatch 1957, Peck 1974); scavengers or feeders on subterranean fungi (Fogel and Peck 1975).

16b Appendages conspicuously long, body elongate and rounder, pronotum much narrower than elytra . . . .
Tribe **GLACICAVICOLINI,** Genus ***Glacicavicola*** Westcott, 1968

The single species, ***G. bathyscioides*** Westcott, a highly evolved cave inhabitant, occurs as a scavenger in lava tube and limestone caves in Idaho and Wyoming (Peck 1974, 1981; Westcott 1968).

17a **(15b)** Labrum deeply emarginate apically (Fig. 37.57) (obscure in ***Cyrtusa*** and some ***Hydnobius***) **18**

17b Labrum shallowly or not at all emarginate apically (Fig. 37.56) . . . **28**

18a **(17a)** Tarsal formula 5–5–5; Tribe **HYDNOBIINI**. **19**

18b Tarsal formula 5–5–4; Tribe **LEIODINI**. **20**

19a **(18a)** Antennal club three-segmented, segment 7 as small as 8 . . . . . .
Genus ***Triarthron*** Maerkel, 1840

Two species known; from Pennsylvania to Oregon (Hatch 1957).

19b Antennal club five-segmented, segment 8 smaller than 7 . . . . . .
Genus ***Hydnobius*** Schmidt, 1847

Sixteen species, mostly western, but several with a general northern distribution; subterranean fungi feeders (Hatch 1957).

20a **(18b)** Head without antennal grooves beneath; body not contractile . . . **21**

20b Head with antennal grooves beneath (Fig. 37.58); body sometimes contractile. **25**

21a **(20a)** Mesosternum oblique (Fig. 37.62). **22**

21b Mesosternum vertical between the coxae (Fig. 37.61) **23**

22a **(21a)** Mesosternum with median longitudinal carina. . . . .
Genus ***Leiodes*** Latreille, 1792

Some 29 generally distributed species known in North America; usually in forested habitats (Hatch 1957).

22b Mesosternum without median carina. . . . .
Genus ***Ecarinosphaerula*** Hatch, 1929

The single species, ***E. ecarina*** Hatch, is known from British Columbia and Nevada (Hatch 1957).

23a **(21b)** Mesosternum with a median longitudinal carina **24**

23b Mesosternum without median carina Genus ***Cyrtusa*** Erichson, 1842

The species ***C. luggeri*** Hatch is spread from Quebec and Michigan to British Columbia (Hatch 1957).

24a **(23a)** Antenna 10-segmented, club four-segmented . . . .
Genus ***Anogdus*** LeConte, 1866

Four species occur in the eastern United States and British Columbia (Hatch 1957).

24b Antenna 11-segmented, club five-segmented (segment 8 hidden by 7 and 9 when club is contracted) . . . Genus ***Neocyrtusa*** Brown, 1937

Eight species are known; widely distributed (Hatch 1957).

25a **(20b)** Antenna 11-segmented, club five-segmented. . . . . Genus ***Lionothus*** Brown, 1937

The genus is widespread in North America, but only the species ***L. ulkei*** Brown is recorded from the District of Columbia.

25b Antenna 10-segmented, club three- or four-segmented. **26**

26a **(25b)** Middle and hind tibiae distinctly wider than anterior tibiae; middle and hind tarsi wide, compressed; intervals between elytral striae nearly smooth. **27**

26b All tibiae relatively more slender and elongate; tarsi slender, cylindrical; elytral intervals with dense punctures, approaching size of strial punctures . . . . Genus ***Caenocyrta*** Brown, 1937

The species ***C. picipennis*** (LeConte) is distributed from the District of Columbia to Quebec, to British Columbia (Hatch 1957).

27a **(26a)** Antennal club four-segmented (Fig. 37.58) . . . . Genus ***Zeadolopus*** Broun, 1903

The species ***Z. egenus*** (LeConte) ranges from Quebec and Michigan to Georgia. Until recently, the genus was called ***Apheloplastus*** (Newton 1983).

27b Antennal club three-segmented . . . . Genus ***Isoplastus*** Horn, 1880

The species ***I. fossor*** ranges from Quebec and Michigan to the District of Columbia.

28a **(17b)** All tarsi three-segmented; abdominal sternum 3 (first visible) with transverse carina (Fig. 37.60) . . . . Tribe **SCOTOCRYPTINI,** Genus ***Aglyptinus*** Cockerell, 1906

The only species in the United States and Canada is ***A. laevis*** (LeConte), in the east from Canada to Louisiana; usually associated with fungi.

28b All tarsi of at least four segments; abdominal sternum 3 (first visible) without transverse carina **29**

29a **(28b)** Tarsal formula 5–4–4, segmentation not sexually dimorphic, elytra usually transversely striolate; male with enlarged (tenent) setae on protarsi only; tibiae without longitudinal carinae; Tribe **PSEUDOLEIODINI** **30**

29b Tarsi sexually dimorphic in segmentation, 5–5–4 in male, 5–4–4 or 4–4–4 in female; elytra not transversely striolate; male tenent setae usually on pro- and mesotarsi; tibiae with longitudinal carinae; Tribe **AGATHIDIINI**. **31**

30a **(29a)** Elytra without longitudinal rows of punctures; color pale (testaceous); mesosternal carina simple (Fig. 37.63) Genus ***Colenis*** Erichson, 1845

C. impunctata (LeConte) is the only North American species; in forests from eastern United States and Canada to British Columbia (Hatch 1957).

30b Elytra with longitudinal rows of punctures; color dark; mesosternal carina with transverse depression (Fig. 37.64) . . . . Genus ***Cainosternum*** Notman, 1931

Only one known species, ***C. imbricatum*** Notman from New York (Wheeler 1984).

31a **(29b)** Eighth antennal segment distinctly narrower than seventh; antennal club five-segmented; head narrowed behind the eyes; body form convex to hemispherical, contractile. . . Genus ***Anisotoma*** Panzer, 1797

Fifteen species are known; generally distributed in forests over much of Canada and the United States; feeders on plasmodia and fruiting bodies of slime molds (Russell 1979, Wheeler 1979a).

31b Eighth antennal segment not (or slightly) narrower than seventh, antennal club abruptly three-segmented; head often broad behind eyes ("postocular tempora" well developed); body form variable, sometimes oblong–eliptical and subdepressed, sometimes hemispherical and highly contractile. **32**

32a (31b) Elytra with nine complete, punctate striae; head narrowed behind eyes; body form oblong–eliptical, subdepressed; clypeus not emarginate. . . . . Genus ***Stetholiodes*** Fall, 1910

The only species, ***S. laticollis*** Fall, occurs in Indiana (Wheeler 1981).

32b Elytra with fewer than nine complete, punctate striae (or without striae); head often wide behind eyes; body form variable, often hemispherical and very contractile; clypeus contractile . . . . Genus ***Agathidium*** Panzer, 1797

Forty-six species are known; widely distributed across the United States and Canada; usually in forested habitats and on slime molds (Hatch 1957, Russell 1979).

References

Anderson, R. S. 1982a. Resource partitioning in the carrion beetle (Coleoptera : Silphidae) fauna of southern Ontario: ecological and evolutionary considerations. *Can. J. Zool.* 60:1314–1325. **1982b.** Burying beetle larvae: Nearctic *Nicrophorus* and Oriental *Ptomascopus morio* (Silphidae). *Syst. Entomol.* 7:249–264. **1982c.** On the decreasing abundance of *Nicrophorus americanus* Olivier (Coleoptera : Silphidae) in eastern North America. *Coleopt. Bull.* 36:362–365. **Anderson, R. S. and S. B. Peck. 1984.** Bionomics of Nearctic species of *Aclypea* Reitter: phytophagous "carrion" beetles (Coleoptera : Silphidae). *Pan-Pac. Entomol.* 60:248–255. **1985.** *The Insects and Arachnids of Canada.* 13. *The Carrion Beetles of Canada and Alaska (Coleoptera : Silphidae and Agyrtidae).* Publ. 1778. Agriculture Canada, Ottawa, Ontario, Canada. 121 pp. **1986.** Geographic patterns of colour variation in North American *Nicrophorus* burying beetles (Coleoptera : Silphidae). *J. Nat. Hist.* 20:283–297. **Arnett, R. H., Jr. 1963.** *The Beetles of the United States (A Manual for Identification).* Catholic University of America Press, Washington, D.C. 1112 pp. **Balduf, W. V. 1935.** *The Bionomics of Entomophagous Coleoptera.* John S. Swift Co., St. Louis, Mo. 220 pp. **Borror, D. J., D. M. DeLong and C. A. Triplehorn. 1976.** *An Introduction to the Study of Insects,* 4th ed. Holt, Rinehart, and Winston, New York. 852 pp. **Bredohl, R. 1984.** Zur Bioakustic mitteleuropäischer Totengräber (Coleoptera : Silphidae : *Necrophorus*). *Entomol. Gen.* 10:11–25. **Brewer, J. W. and T. R. Bacon. 1975.** Biology of the carrion beetle *Silpha ramosa* Say. *Ann. Entomol. Soc. Am.* 68:786–790. **Buckle, A. P. 1976.** Studies on the biology and distribution of *Leptinus testaceus* Müller within a community of mixed small mammal species. *Ecol. Entomol.* 1:1–6. **Chandler, D. and S. B. Peck. 1989.** Species richness and abundance of Leiodidae (Coleoptera) in an uncut and 40 year old forest in New Hampshire. *Can. J. Zool.* (in press). **Cooley, R. A. 1917.** The spinach carrion beetle. *J. Econ. Entomol.* 10:94–102. **Crowson, R. A. 1967.** *The Natural Classification of the Families of Coleoptera,* reprint ed. E. W. Classey, Faringdon, Oxfordshire, England. 214 pp. **Dethier, V. G. 1947.** The role of the antennae in the orientation of carrion beetles to odors. *J. N.Y. Entomol. Soc.* 65:285–293. **Dillon, E. S. and L. S. Dillon. 1961.** *A Manual of Common Beetles of Eastern North America.* Row, Peterson, Evanston, Ill. 884 pp. **Fall, H. C. 1937.** The North American species of *Nemadus,* with descriptions of new species (Coleoptera, Silphidae). *J. N.Y. Entomol. Soc.* 45:335–340. **Fisher, R. M. and R. D. Tuckerman. 1986.** Mimicry of bumble bees and cuckoo bumble bees by carrion beetles (Coleoptera : Silphidae). *J. Kans. Entomol. Soc.* 59:20–25. **Fogel, R. and S. B. Peck. 1975.** Ecological studies of hypogeous fungi. I. Coleoptera associated with sporocarps. *Mycologia* 67:741–747. **Hatch, M. H. 1933.** Studies on the Leptodiridae (Catopidae) with descriptions of new species. *J. N.Y. Entomol. Soc.* 41:187–239. **1953, 1957.** *The beetles of the Pacific Northwest.* I. *Introduction and Adephaga.* 340 pp. II. *Staphyliniformia.* 384 pp. Univ. Washington Publ. Biol. 16. University of Washington Press, Seattle, Wash. **Ising, E. 1969.** Zur Biologie des *Leptinus testaceus* Müller, 1817 (Insecta, Coleoptera). *Zool. Beitr.* 15:393–456. **Janzen, D. H. 1963.** Observations on populations of adult beaver-beetles, *Platypsyllus castoris* (Platypsyllidae : Coleoptera). *Pan-Pac. Entomol.* 39:215–228. **Jaques, H. E. 1951.** *How to Know the Beetles.* Wm. C. Brown, Dubuque, Iowa. 372 pp. **Jeannel, R. 1936.** Monographie des Catopidae. *Mem. Mus. Natl. Hist. Nat. (Paris) (N.S.)*, Vol. 1. 433 pp. **Johnson, M. D. 1975.** Seasonal and microseral variations in the insect populations on carrion. *Amer. Midl. Nat.* 93(1):79–90. **Lawrence, J. F. 1982.** Coleoptera. Pp. 482–553 in S. P. Parker (ed.). *Synopsis and Classification of Living Organisms*, Vol. 2. McGraw-Hill, New York. **Lawrence, J. F. and A. F. Newton, Jr. 1982.** Evolution and classification of beetles. *Annu. Rev. Ecol. Syst.* 13:261–290. **Miller, S. E. and S. B. Peck. 1979.** Fossil carrion beetles of Pleistocene California asphalt deposits, with a synopsis of Holocene California Silphidae (Insecta : Coleoptera : Silphidae). *Trans. San Diego Soc. Nat. Hist.* 19:85–106. **Milne, L. J. and M. J. Milne. 1944.** Notes on the behavior of burying beetles (*Nicrophorus* spp.). *J. N.Y. Entomol. Soc.* 52:311–327. **Milne, J. J. and M. Milne. 1976.** The social behavior of burying beetles. *Sci. Am.* 235(2):84–89. **Newton, A. F., Jr. 1983.** New generic synonymies, new combinations, and distributional comments on Leiodini (Coleoptera : Leiodidae). *Coleopt. Bull.* 37:173–176. **1984.** Mycophagy in Staphylinoidea. Pp. 302–351 in Q. Wheeler and M. Blackwell (eds.). *Fungus–Insect Relationships; Perspectives in Ecology and Evolution.* Columbia University Press, New York. 514 pp. **Newton, A. and S. B. Peck. 1975.** Baited pitfall traps for beetles. *Coleopt. Bull.* 29:45–46. **Niemitz, O. 1972.** Bioakustische, verhaltensphysiologische und morphologische Untersuchungen an *Necrophorus vespillo* (Fab.). *Forma Functio* 5:209–230. **Parks, J. J. and J. W. Barnes. 1955.** Notes on the family Leptinidae including a new record of *Leptinillus validus* (Horn) in North America (Coleoptera). *Ann. Entomol. Soc. Am.* 48:417–421. **Payne, J. A. 1965.** A summer carrion study of the baby pig *Sus scrofa* L. *Ecology* 45(5):592–602. **Payne, J. A. and E. W. King. 1970.** Coleoptera associated with pig carrion. *Entomol. Monthly Mag.* 105(1969):224–232. **Peck, S. B. 1973.** A systematic revision and the evolutionary biology of the *Ptomaphagus* (*Adelops*) beetles of North America (Coleoptera : Leiodidae : Catopinae), with emphasis on cave-inhabiting species. *Bull. Mus. Comp. Zool. Harv. Univ.* 145:29–162. **1974a.** The eyeless *Catopocerus* beetles (Leiodidae) of eastern North America. *Psyche (Camb.)* 81:377–397. **1974b.** A review of the *Agyrtes* (Silphidae) of North America. *Psyche (Camb.)* 81:501–506. **1974c.** Biology of the Idaho lava tube beetle, *Glacicavicola. Bull. Natl. Speleol. Soc.* 36:1–3. **1976.** The myrmecophilous beetle genus *Echinocoleus* in the southwestern United States (Leiodidae, Catopinae). *Psyche (Camb.)* 83:51–62. **1977a.** An unusual sense receptor in internal antennal vesicles of *Ptomaphagus* (Coleoptera : Leiodidae). *Can. Entomol.* 109:81–86. **1977b.** A review of the distribution and biology of the small carrion beetle *Prionochaeta opaca* of North America (Coleoptera : Leiodidae : Catopinae). *Psyche (Camb.)* 83:299–307. **1978a.** New montane *Ptomaphagus* beetles from New Mexico and zoogeography of southwestern caves (Coleoptera : Leiodidae : Catopinae). *Southwest. Nat.* 23:227–238. **1978b.** Systematics and evolution of forest litter *Adelopsis* in the southern Appalachians (Coleoptera : Leiodidae : Catopinae). *Psyche (Camb.)* 85:355–382. **1981a.** Distribution and biology of flightless carrion beetle *Necrophilus pettitii* in eastern North America (Coleoptera : Silphidae). *Entomol. News* 92:181–185. **1981b.** The Idaho cave beetle *Glacicavicola* also occurs in Wyoming. *Coleopt. Bull.* 35:141–142. **1982.** A review of the ectoparasitic *Leptinus* beetles of North America (Coleoptera : Leptinidae). *Can. J. Zool.* 60:1517–1527. **1984.** The distribution and evolution of cavernicolous *Ptomaphagus* beetles in the southeastern United States (Coleoptera; Leiodidae; Cholevinae) with new species and records. *Can. J. Zool.* 62:730–740. **1986.** Evolution of adult morphology and life-history characters in cavernicolous *Ptomaphagus* beetles. *Evolution* 40:1021–1030. **Peck, S. B. and R. S. Anderson. 1982.** The distribution and biology of the alpine–tundra carrion beetle *Thanatophilus coloradensis* (Wickham) in North America (Coleoptera : Silphidae). *Coleopt. Bull.* 36:112–115. **1985a.** Taxonomy, phylogeny, and biogeography of the carrion beetles of Latin America (Coleoptera : Silphidae). *Quaest. Entomol.* 21:247–317. **1985b.** Seasonal activity and habitat associations of adult small carrion beetles in southern Ontario (Coleoptera, Leiodidae, Cholevinae). *Coleopt. Bull.* 39:347–353. **Peck, S. B. and A. E. Davies. 1980.** Collecting small beetles with large-area "window" traps. *Coleopt. Bull.* 34:237–239. **Peck, S. B. and M. M. Kaulbars. 1987.** A synopsis of the distribution and bionomics of the carrion beetles (Coleoptera : Silphidae) of the conterminous United States. *Proc. Entomol. Soc. Ont.* 118:47–81. **Peck, S. B. and S. E. Miller. 1989.** *A Catalog of the Coleoptera of North America, Family*: Silphidae. U.S. Dept. Agric. Handb. 529–28 (in press). **Ratcliffe, B. C. 1972.** The natural history of *Necrodes surinamensis* (Fabr.) (Coleoptera; Silphidae). *Trans. Am. Entomol. Soc. (Phila.)* 98:359–410. **Reed,**

H. B., Jr. 1958. A study of dog carcass communities in Tennessee, with special reference to the insects. *Am. Midl. Nat.* 59:213–245. **Russell, L. K. 1979.** Beetles associated with slime molds (Mycetozoa) in Oregon and California (Coleoptera: Leiodidae, Sphindidae, Lathridiidae). *Pan-Pac. Entomol.* 55:1–9. **Schwaller, W. 1983.** Die Arten der Gattung *Ipelates (sensu lato)* (Coleoptera : Silphidae : Agyrtinae). *Rev. Suisse Zool.* 90:101–110. **Schumacher, R. 1973.** Beitrag zur Kenntnis der Stridulationsapparate einheimischer *Necrophorus*-Arten (*Necrophorus humator* Ol., *Necrophorus investigator* Zetterst., *Necrophorus vespilloides* Herbst) (Insecta, Coleoptera). *Z. Morphol. Tiere* 75:65–75. **Shubeck, P. P. 1968.** Orientation of carrion beetles to carrion: random or non-random? *J. N.Y. Entomol. Soc.* 76:253–265. **1969.** Ecological studies of carrion beetles in Hutcheson Memorial Forest. *J. N.Y. Entomol. Soc.* 77:138–151. **1971.** Diel periodicities of certain carrion beetles (Coleoptera: Silphidae). *Coleopt. Bull.* 25:41–46. **1976.** Carrion beetle responses to poikilotherm and homoiotherm carrion (Coleoptera: Silphidae). *Entomol. News* 87:265–269. **Shubeck, P. P., N. M. Downie, R. L. Wenzel and S. B. Peck. 1977.** Species composition of carrion beetles in a mixed-oak forest. *W. L. Hutcheson Mem. For. Bull.* 4:12–17. **Sidor, C. 1970.** Resultati ispitivanja uloge insekata grobara (*Necrophorus vespillo* L.) u prenosenju besnila. *Mikrobiologija* (*Belgr.*) 7:131–138. **Southwood, T. R. E. 1966.** *Ecological Methods with Particular Reference to the Study of Insect Populations.* Methuen London. 391 pp. **Springett, B. P. 1968.** Aspects of the relationship between burying insects *Necrophorus* spp. and the mite, *Poecilochirus necrophori* Vitz. *J. Anim. Ecol.* 37:417–424. **Theodorides, J. 1950.** Notes diverse sur les *Necrophorus* (Coleoptera, Silphidae). *Bull. Inst. R. Sci. Nat. Belg.* 26: no. 52. 20 pp. **Waldow, U. 1973.** Elektrophysiologie eins neuen Aasgeruchrezeptors und seine Bedeutung für das Verhaltern des Totengräbers (*Necrophorus*). *J. Comp. Physiol.* 83:415–424. **Walker, T. J., Jr. 1957.** Ecological studies of the arthropods associated with certain decaying materials in four habitats. *Ecology* 38:262–276. **Westcott, R. L. 1968.** A new subfamily of blind beetle from Idaho ice caves with notes on its bionomics and evolution (Coleoptera: Leiodidae). *Contrib. Sci.* (*Los Ang.*) 141, 14 pp. **Wheeler, Q. D. 1979a.** Slime mold beetles of the genus *Anisotoma* (Leiodidae): classification and evolution. *Syst. Entomol.* 4:251–309. **1979b.** Revision and cladistics of the middle American genus *Creagrophorus* Matthews (Coleoptera : Leiodidae). *Quaest. Entomol.* 15:447–479. **1981.** Diagnosis and phylogenetic relationships of the monotypic genus *Stetholiodes* (Coleoptera: Leiodidae). *Ohio J. Sci.* 81:165–168. **1986.** Rediscovery and cladistic placement of the genus *Cainosternum* (Coleoptera: Leiodidae). *Ann. Entomol. Soc. Am.* 79:377–383. **Wilson, D. S. 1983.** The effect of population structure on the evolution of mutualism: a field test involving burying beetles and their phoretic mites. *Am. Nat.* 121:851–870. **Wilson, D. S. and J. Fudge. 1984.** Burying beetles: intraspecific interactions and reproductive success in the field. *Ecol. Entomol.* 9:195–203. **Wilson, D. S. and W. G. Knollenberg. 1984.** Food discrimination and ovarian development in burying beetles (Coleoptera: Silphidae: *Nicrophorus*). *Ann. Entomol. Soc. Am.* 77:165–170. **Wilson, D. S., W. G. Knollenberg and J. Fudge. 1984.** Species packing and temperature dependent competition among burying beetles (Silphidae, *Nicrophorus*). *Ecol. Entomol.* 9:205–216. **Wood, D. M. 1965.** Studies on the beetles *Leptinillus validus* (Horn) and *Platypsyllus castoris* Ritsema (Coleoptera: Leptinidae) from beaver. *Proc. Entomol. Soc. Ont.* 95:33–63. **Zwick, P. 1979.** Contributions to the knowledge of Australian Cholevidae (Catopidae auct.: Coleoptera). *Aust. J. Zool. Suppl. Ser.* 70. 56 pp.

38

Insecta: Coleoptera Staphylinidae Adults and Larvae

ALFRED F. NEWTON, JR.

Part A: Adults

The family Staphylinidae is one of the largest families of beetles, with about 32,000 known species and over 200 new ones being described each year. It is a worldwide and ancient group, with a fossil history extending back to the early Jurassic period. In North America, the family is represented by nearly 400 genera and over 3000 named species. About half of these species are found in soil and litter, forming one of the most common and ecologically important insect components of the soil fauna.

Staphylinids are usually easily distinguished from other beetles by their short truncate elytra, which leave more than half of the rather flexible abdomen exposed. The closely related pselaphids, which also have short elytra, differ in having relatively broad, nonflexible abdomens and in having numerous foveae on the body surfaces. Earwigs (insect order Dermaptera) look superficially similar but have a pair of pinchers or cerci at the abdominal apex, and usually have more than 11 antennal segments.

Adult Morphology

Length 1–35 mm; form ovoid to very elongate (Figs. 38A.1, 38A.2); color yellowish to very dark; well sclerotized. Head prognathous to hypognathous, with or without distinct neck. Compound eyes present or absent, pair of ocelli sometimes present. Antenna of 11 (rarely 10 or 9) segments, slender or clavate, weakly or not at all clubbed (or club one-segmented). Labrum free, variable in shape. Mandibles usually projecting and acute, sometimes apically blunt or not visible. Maxilla with two distinct inner lobes (galea and lacinia) and four- (rarely five-) segmented palp. Labium variable, with a pair of usually three-segmented palps. Mentum present. Gular area well sclerotized, gular sutures separate, fused, or absent. Pair of cervical sclerites present.

Thorax: Pronotum variable in shape, usually laterally margined. Scutellum usually visible, triangular. Elytra truncate, usually very short and exposing five to six abdominal segments above, rarely longer; sometimes striate. Hind wings usually present, compactly folded beneath elytra. Coxae variable in size and shape, the coxae of each segment usually contiguous but sometimes moderately separated. Tarsi usually all of five segments, sometimes of four, three, or two segments or heteromerous.

Abdomen elongate, with six or seven segments visible ventrally and several visible dorsally, the exposed segments well sclerotized and usually margined at sides. Abdominal segments usually connected by relatively long membranes (one-fifth to one-half as long as each segment) which often are distinctly patterned and which allow the abdomen to flex and telescope greatly.

Sexual dimorphism usually subtle or absent externally and restricted to internal genitalia. Aedeagus variable in shape, basal piece reduced or absent, median lobe usually bulbous at base and with small foramen on one side for entry of seminal vesicle. Ovipositor usually not visible externally, short and nonpiercing.

Detailed and well-illustrated treatments of staphylinid external morphology are given by Blackwelder (1936), Coiffait (1972), Tikhomirova (1973), and Naomi (1987). Internal morphology is poorly known and not used in classification or identification, except for sclerotized parts of the genitalia and the presence and structure of abdominal defensive glands (Araujo 1978).

Biology and Ecology

A majority of staphylinids whose habits are known are agile predators, feeding on a variety of prey ranging from nematodes, mites, and Collembola for small species to larvae and adults of beetles and other insects for larger species. Feeding in many species, at least among the subfamilies Staphylininae and Paederinae, involves extra-oral digestion, so that the gut contains little or no solid material. Notable exceptions to predatory habits among soil dwellers are species of the subfamilies Osoriinae and Oxytelinae, whose guts contain quantities of decomposing organic matter on which they feed. At least some species of *Micropeplus* (Micropeplinae), *Holotrochus* and *Thoracophorus* (Osoriinae), *Coproporus, Sepedophilus,* and *Tachinus*

(Tachyporinae), and several genera of Aleocharinae, are known or suspected to feed on molds or other soil-inhabiting fungi (Newton 1984).

The life-cycles of very few staphylinids are known, but these show a common pattern. Mating takes place in the usual habitat of the adults and is brief. Incubation and larval development are rapid, being completed typically in a few weeks. Mature larvae may continue through a brief pupal stage (one to several weeks) and emerge as adults, or the larvae may remain quiescent and overwinter without change, pupating the following season. Adults are relatively long-lived, and overwintering occurs primarily in the adult stage. At least one, sometimes several, generations occur per year. Adults and immature stages may occur simultaneously in the same habitat. (Larvae are discussed in detail in Part B of this chapter.)

Most soil staphylinids are found in uncompacted, moist forest litter which is rich in decomposing organic material. Dry litter of exposed and desert areas has few species, but local moist sites in such areas may be densely populated. Deep mineral soil is known to be the preferred habitat of many Leptotyphlinae. Most staphylinids have fully developed flight wings and fly readily, but wing reduction and loss is common among litter-inhabiting species, and all members of one soil-dwelling subfamily (Leptotyphlinae) are flightless. Many staphylinids occur in soil-related microhabitats such as dung and carrion, higher fungi, vertebrate and invertebrate nests, and wet sites along streams and other bodies of water, and hence are occasionally found in soil.

General discussions of the biology and ecology of staphylinids are given by Mank (1923), Voris (1934), Balduf (1935), Coiffait (1972), and Tikhomirova (1973). References to notes on individual North American species may be found in Moore and Legner (1974a, 1975); more recent relevant studies concern species of *Osorius* (Smith et al. 1978) and *Paederus* (Frank and Kanamitsu 1987). There have been no detailed ecological studies on the soil and litter fauna of North America, but many such studies in Europe give useful data on staphylinid genera which also occur in North America: for example, Coiffait (1960), Dennison and Hodkinson (1984), Evans (1967), Focarile (1985), Friebe (1983). Kasule (1968), and Topp (1979).

Collection and Study Techniques

Staphylinidae may be directly hand-collected from soil habitats, or may be caught by use of pitfall traps set in the soil. More effective, especially for smaller species, is the collection of soil or litter samples from which the fauna is subsequently extracted by means of Berlese or Tullgren funnels, or Winkler/Moczarski extractors. Efficiency is improved by sifting to eliminate larger litter and rocks. Other standard methods of extracting soil fauna are also effective. Minute deep-soil species such as Leptotyphlinae are probably best collected by the flotation method described by Coiffait (1959, pp. 239–240), which has been little used in North America. Flying species may be very effectively sampled with flight intercept or "window" traps.

Adults may be killed and preserved in 80% ethanol or similar liquid preservative, or killed dry and pinned, pointed or carded by the use of standard entomological techniques. Whole dried or alcoholic specimens are usually identifiable to genus at least without further preparation, but dissection of male genitalia is sometimes required for reliable species determination; techniques are given in those taxonomic works that use genitalia. Leptotyphlinae and smaller Aleocharinae are best studied as slide preparations after clearing in 1 *N* potassium hydroxide or other clearing agent.

Most of the techniques mentioned above are discussed in detail by Martin (1977); see also Smetana (1971), Wheeler and McHugh (1987), and Besuchet et al. (1987) for litter sifting and extracting methods, and Peck and Davies (1980) for "window" trap design.

Classification

The classification of Staphylinidae at all levels, including family limits, is rather unstable and in need of revision, as illustrated by the contrasting family/subfamily systems of Coiffait (1972), Lawrence and Newton (1982), and Naomi (1985). The subfamily classification adopted here is a conservative one in widespread use by staphylinid taxonomists worldwide, and differs somewhat (mainly in the ranking of some tribes and subfamilies) from the more artificial system recently used for North America by Moore and Legner (1974b, 1975, 1979) and Arnett (1985). The primarily fungus-inhabiting families Scaphidiidae (several genera in North America) and Dasyceridae (*Dasycerus* only), the carrion-associated family Silphidae (see Chapter 37), and the large litter-inhabiting families Pselaphidae (see Chapter 39) and Scydmaenidae, are all closely related to and probably derived from Staphylinidae, and may eventually be combined with this family.

Identification

Keys to all genera of North American Staphylinidae, exclusive of the subfamily Aleocharinae, are given by Arnett (1968) and Moore and Legner (1974b, 1979). Seevers (1978) gives a key to North American genera of Aleocharinae, and Herman (1970a) presents a key to world genera of Oxytelinae.

North American staphylinid species are in general difficult to identify, since up-to-date revisionary studies with keys have not been published for most genera, and there are many undescribed species and unrecognized synonyms. Hatch (1957) includes keys to most species of the Pacific Northwest region (British Columbia, Washington, Oregon and Idaho). Frank (1986) gives a checklist of Florida species with references that often include keys. The most recent complete catalog of North American species is that of Moore and Legner (1974a, 1975) which gives references to most species-level keys published up to then. More recently published keys, and other significant publications, are cited where appropriate in the generic key below.

Three groups of staphylinids (subfamilies Aleocharinae and Leptotyphlinae, and tribe Xantholinini of Staphylininae) are difficult to identify even to genus without well-illustrated keys and, often, special specimen preparation. These groups are not keyed to genus here, but existing generic keys are cited where appropriate.

ACKNOWLEDGMENT

I thank M. K. Thayer and J. S. Ashe for reviewing the entire manuscript.

KEY TO SOIL GENERA (ADULTS)

The key that follows is limited to genera in which at least one species is frequently found in soil or organic litter of forested or nonforested areas. Genera with species primarily restricted to occurrence on dung, carrion, fruiting bodies of higher fungi, wet habitats such as bogs and edges of streams and lakes, and nests of vertebrates or invertebrates are excluded, although individuals of these species are occasionally found in soil. Habits of North American species, if unknown, are assumed to be similar to those of exotic congeners. The many genera for which no published information on habits exists were included or excluded on the basis of personal field observations and label data of museum specimens, supplemented by the *in litteris* comments of other specialists, who are here thanked for their cooperation: J. M. Campbell; P. M. Hammond; L. H. Herman, Jr.; E. R. Hoebeke; I. Moore; A. Smetana; M. K. Thayer; and L. E. Watrous. M. K. Thayer prepared the portion of the key dealing with **Omaliinae.**

All figures were made with the aid of a camera lucida but are not drawn to scale. The figures are semidiagrammatic, with setae and irrelevant structures usually omitted.

1a Head with pair of ocelli on vertex (Fig. 38A.4); **Omaliinae** (in part) . . . **2**

1b Head without ocelli (but vertex may have pair of tentorial pits as in Fig. 38A.2) **16**

2a Apical segment of maxillary palp much smaller than preapical segment, which is greatly swollen (Fig. 38A.21) **3**

2b Apical segment of maxillary palp about as wide as preapical segment, which is not swollen **4**

3a Gular sutures separate throughout their length (Fig. 38A.7) ***Subhaida*** Hatch

Seven species, western, usually high altitudes. Key: Campbell (1978).

3b Gular sutures confluent at middle (similar to Fig. 38A.8) . . . .
Eudectoides Campbell

One species, *E. crassicornis* (LeConte), southeastern, very rare. See Campbell (1978).

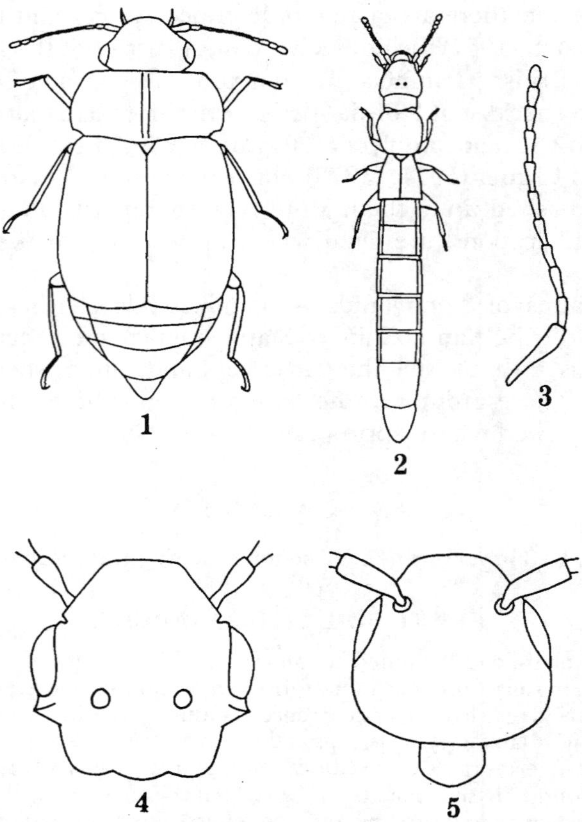

Fig. 38A.1–38A.5. **1.** *Megarthrus pictus* Motschulsky. **2.** *Xenotyphlus taylorensis* Coiffait. **3.** *Homaeotarsus bicolor* (Gravenhorst), right antenna, dorsal. **4.** *Deinopteroloma pictum* (Fauvel), head, dorsal. **5.** *Aleodorus* sp., head, dorsal.

4a Posterior tarsus with segment 5 (last segment) as long as or longer than segments 1–4 together **13**

4b Posterior tarsus with segment 5 shorter than segments 1–4 together . . . **5**

5a All antennal segments densely pubescent; apical segment of maxillary palp five times as long as preapical segment. . . . . ***Lesteva*** Latreille

Two species, eastern.

5b First two to four antennal segments not densely pubescent like more apical ones; apical segment of maxillary palp less than four times as long as preapical segment **6**

6a Head with neck impression absent or indistinct dorsally (Fig. 38A.4) . . **10**

6b Head with distinct impressed line or transverse row of punctures behind ocelli (Fig. 38A.6) **7**

7a Labrum distinctly bilobed (Fig. 38A.14); body narrow and parallel-sided . . . ***Xylodromus*** Heer

Three species, mostly northern.

7b Labrum with anterior margin straight, slightly convex, or at most slightly emarginate; various body shapes **8**

8a Apical segment of maxillary palp at least twice as long as preceding segment, usually longer; body ovoid in dorsal view; pronotum more transverse, at least 1.5 times as wide as long, usually wider; some species with all or nearly all of abdominal tergites covered by elytra . . . . ***Anthobium*** Samouelle (in part)
Thirteen species, mostly western.

8b Apical segment of maxillary palp no more than 1.75 times as long as preceding segment; body relatively elongate; pronotum only slightly transverse, no more than 1.5 times as wide as long; at least three abdominal tergites not covered by elytra **9**

9a Head and pronotum shining, without microsculpture, not pubescent; head with subocular ridge. . . . . ***Arpedium*** Erichson
Two species, eastern. Key: Campbell (1984).

9b Head and pronotum opaque, with dense granulate or reticulate microsculpture, pubescent; head without subocular ridge . . . . . ***Eucnecosum*** Reitter
Three species, mostly northern. Key: Campbell (1984).

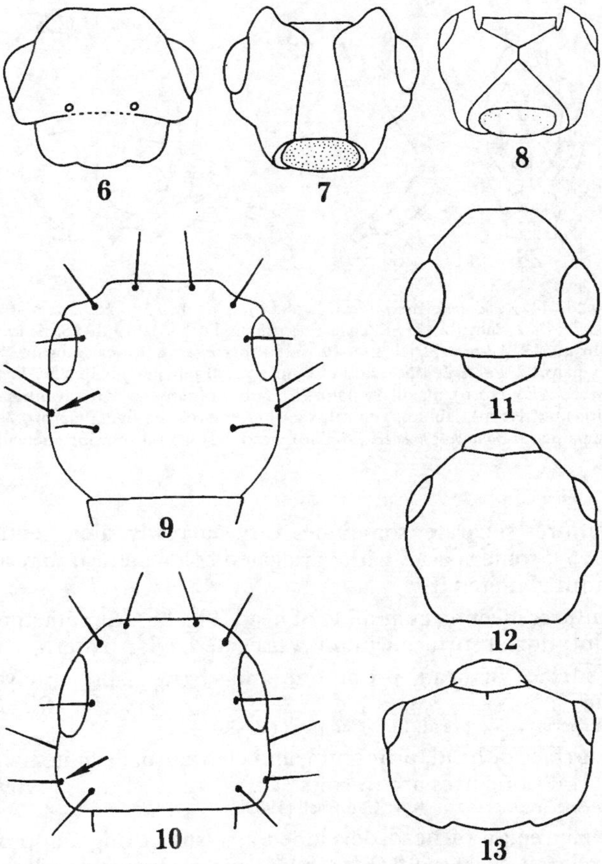

Fig. 38A.6–38A.13. 6. *Xylodromus concinnus* (Marsham), head, dorsal. **7.** *Subhaida ingrata* (Hatch), head, ventral. **8.** *Olophrum obtectum* Erichson, head, ventral. **9.** *Staphylinus badipes* LeConte, head, dorsal, showing postocular seta (arrow). **10.** *Platydracus viridanus* (Horn), head, dorsal, showing postocular seta (arrow). **11.** *Palaminus* sp., head, dorsal. **12.** *Derops divalis* (Sanderson), head, dorsal. **13.** *Coproporus ventriculus* (Say), head, dorsal.

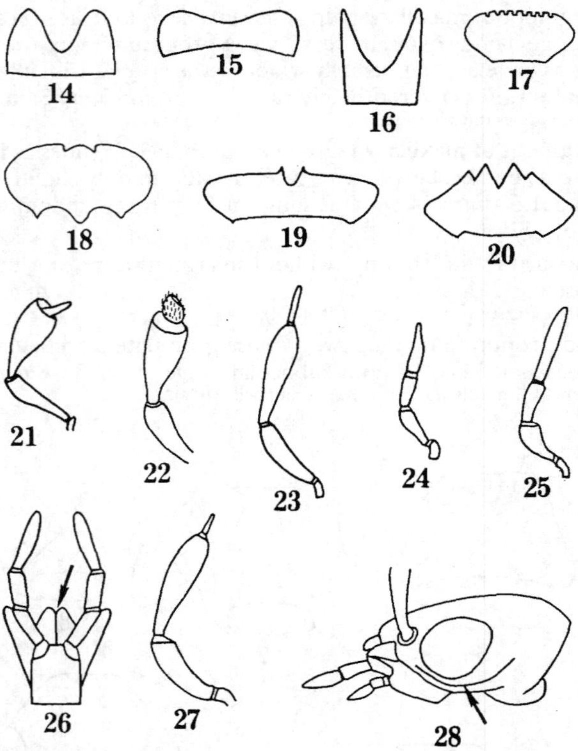

Fig. 38A.14–38A.28. **14.** *Xylodromus concinnus* (Marsham), labrum. **15.** *Syntomium confragosum* Mäklin, labrum. **16.** *Megalopinus* sp., labrum. **17.** *Edaphus americanus* Puthz, labrum. **18.** *Achenomorphus corticinus* (Gravenhorst), labrum. **19.** *Astenus* sp., labrum. **20.** *Echiaster brevicornis* (Casey), labrum. **21.** *Subhaida ingrata* (Hatch), maxillary palp. **22.** *Paederus obliteratus* LeConte, apex of maxillary palp. **23.** *Aleodorus* sp., maxillary palp. **24.** *Erichsonius patella* (Horn), maxillary palp. **25.** *Hesperus baltimorensis* (Gravenhorst), maxillary palp. **26.** *Platydracus viridanus* (Horn), labium, ventral, showing emarginate ligula (arrow). **27.** *Diochus schaumi* Kraatz, maxillary palp. **28.** *Bolitobius kremeri* (Malkin), head, left lateral showing subocular carina (arrow).

10a Gular sutures separate, sometimes very narrowly, along entire length (see Fig. 38A.7); some species with variegated color pattern may appear to have gular sutures united **12**

10b Gular sutures meeting at midline of head (Fig. 38A.8), sometimes united only at a point; dorsal surface without variegated color pattern **11**

11a Dorsal surface of head, pronotum, and elytra glabrous; elytra confusedly punctate. . . . . ***Olophrum*** Erichson

Seven species, widespread. Key: Campbell (1983b).

11b Dorsal surface of head, pronotum, and elytra usually quite hairy; if glabrous, then elytral punctures are in rows . . . . ***Acidota*** Stephens

Three species, northern. Key: Campbell (1982b).

12a Postocular region of head developed as a sharp ridge, appearing in dorsal view as a large sharp tooth (Fig. 38A.4); mesosternum usually carinate at least basally; elytra with light and dark brown color pattern (obscured in teneral specimens) . . . . ***Deinopteroloma*** Jansson

Two species, western. Key: Smetana (1985).

12b Postocular region with at most a very small "tooth" in dorsal view;

mesosternum not carinate; elytra not bicolored . . . . **Anthobium** Samouelle (in part)

See couplet 8.

13a Lateral margins of pronotum serrate and more or less biangulate (Fig. 38A.37) . . . . **Acrulia** Thomson

One species, *A. tumidula* (Mäklin), western.

13b Lateral margins of pronotum evenly rounded (may be sinuate, but not angulate), usually smooth but may be faintly serrulate 14

14a Elytral punctures in (sometimes slightly irregular) impressed rows; mesosternum not carinate medially . . . . **Elonium** Leach

Four species, widespread.

14b Elytra confusedly punctate, without impressions between punctures; mesosternum carinate or not 15

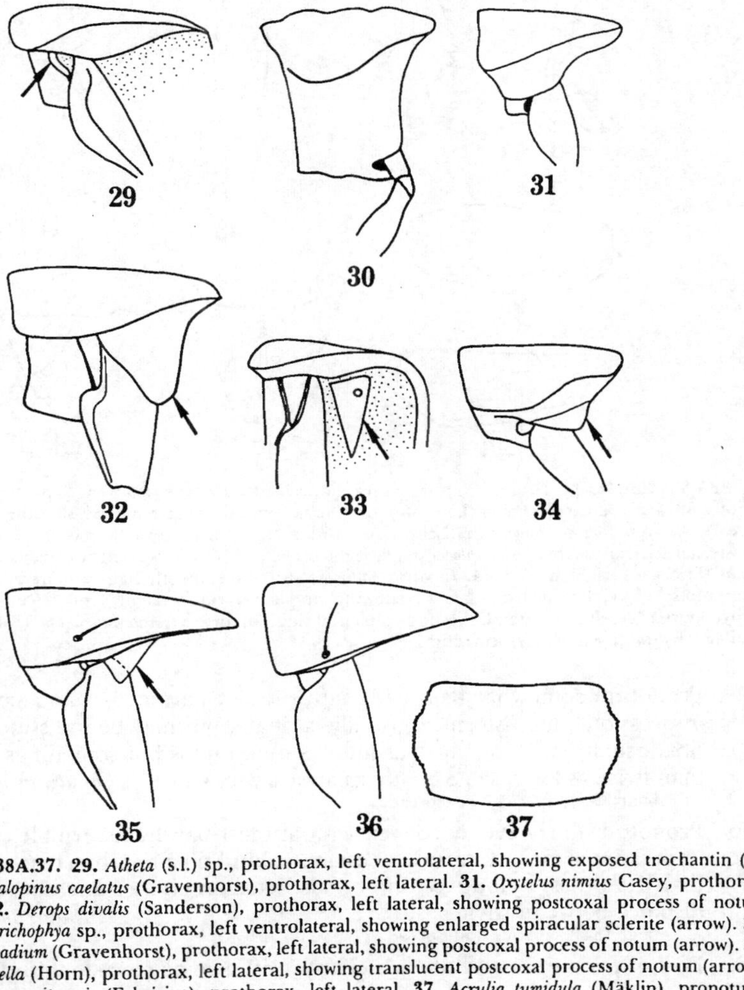

Fig. 38A.29–38A.37. 29. *Atheta* (s.l.) sp., prothorax, left ventrolateral, showing exposed trochantin (arrow). **30.** *Megalopinus caelatus* (Gravenhorst), prothorax, left lateral. **31.** *Oxytelus nimius* Casey, prothorax, left lateral. **32.** *Derops divalis* (Sanderson), prothorax, left lateral, showing postcoxal process of notum (arrow). **33.** *Trichophya* sp., prothorax, left ventrolateral, showing enlarged spiracular sclerite (arrow). **34.** *Homaeotarsus badium* (Gravenhorst), prothorax, left lateral, showing postcoxal process of notum (arrow). **35.** *Erichsonius patella* (Horn), prothorax, left lateral, showing translucent postcoxal process of notum (arrow). **36.** *Philonthus cyanipennis* (Fabricius), prothorax, left lateral. **37.** *Acrulia tumidula* (Mäklin), pronotum, dorsal.

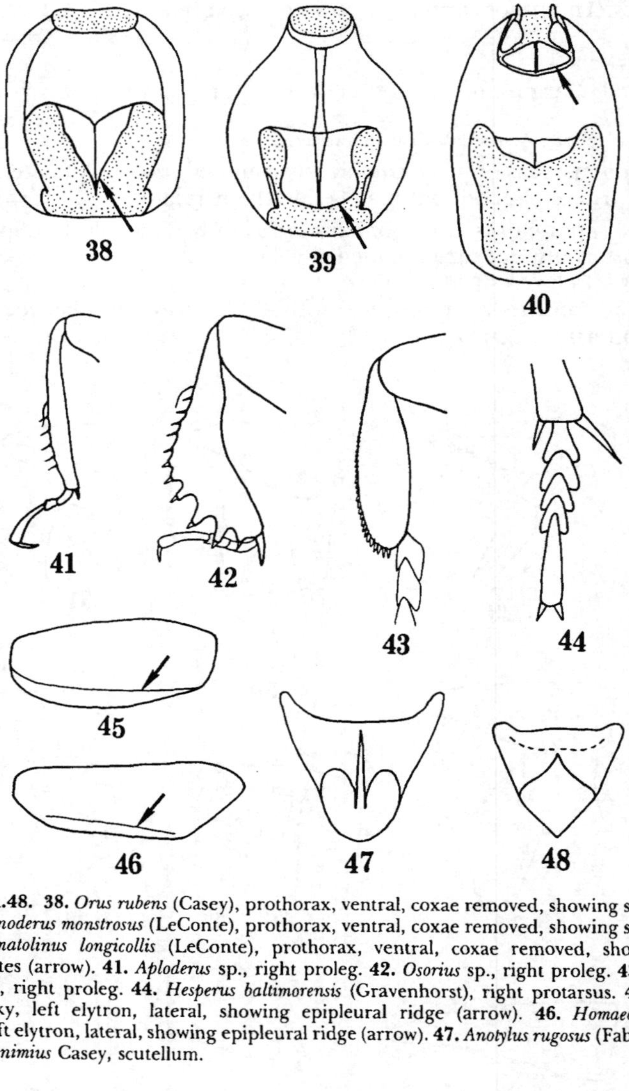

Fig. 38A.38–38A.48. **38.** *Orus rubens* (Casey), prothorax, ventral, coxae removed, showing sternal process (arrow). **39.** *Stamnoderus monstrosus* (LeConte), prothorax, ventral, coxae removed, showing sternal process (arrow). **40.** *Nematolinus longicollis* (LeConte), prothorax, ventral, coxae removed, showing pair of antesternal sclerites (arrow). **41.** *Aploderus* sp., right proleg. **42.** *Osorius* sp., right proleg. **43.** *Sepedophilus basalis* (Erichson), right proleg. **44.** *Hesperus baltimorensis* (Gravenhorst), right protarsus. **45.** *Megarthrus pictus* Motschulsky, left elytron, lateral, showing epipleural ridge (arrow). **46.** *Homaeotarsus badium* (Gravenhorst), left elytron, lateral, showing epipleural ridge (arrow). **47.** *Anotylus rugosus* (Fabricius), scutellum. **48.** *Oxytelus nimius* Casey, scutellum.

15a Pronotum somewhat flattened, with distinct paramedial and posterolateral impressions; mesosternum usually carinate, but may be indistinctly so; hind tibia usually twice or more as long as hind tarsus but sometimes slightly less than twice as long; setae on elytra always very short . . ***Omalium*** Gravenhorst

About 15–20 species, widespread.

15b Pronotum fairly evenly convex with at most barely discernible very shallow impressions; mesosternum not carinate; hind tibia less than twice (usually less than 1.5 times) as long as hind tarsus; setae on elytra sometimes longer than distance between their bases . . . . ***Hapalaraea*** Thomson

About 15–20 species, widespread.

16a Disk of pronotum and of each elytron with two or more longitudinal carinae **17**

16b Disk of pronotum and of elytron without carinae, but longitudinal grooves sometimes present 20

17a Most abdominal segments consisting of complete ring without joint between dorsal and ventral halves; **Osoriinae** (in part) . . ***Thoracophorus*** Motschulsky

About three species, eastern and southern.

17b Each abdominal segment with joint and one or two supplementary sclerites between dorsal and ventral halves 18

18a Antenna of nine segments, apical segment enlarged to form club; body compact, similar in outline to Fig. 38A.1; **Micropeplinae** . . ***Micropeplus*** Latreille

Twelve species, widespread. Key: Campbell (1973b).

18b Antenna of 11 segments, without well-developed club; body elongate, slender; **Pseudopsinae** 19

19a Apical segment of maxillary palp as wide as, and much longer, than, preapical segment . . . . ***Nanobius*** Herman

One species, ***N. serricollis*** (LeConte), western. See Herman (1977).

19b Apical segment of maxillary palp shorter and much narrower than preapical segment . . . . ***Pseudopsis*** Newman

Nine species, widespread. Key: Herman (1975).

20a Antennae inserted on surface of head behind a line drawn between anterior margins of eyes (Fig. 38A.5); last segment of maxillary palp much shorter and narrower than preapical segment (Fig. 38A.23); elytron without epipleural ridge 21

20b Antennae inserted at front or side margins of head anterior to a line drawn between anterior margins of eyes (Fig. 38A.4); last segment of maxillary palp variable in size; elytron with (Figs. 38A.45, 38A.46) or without epipleural ridge 22

21a Procoxa not enclosed, trochantin visible (Fig. 38A.29); hind coxae contiguous. . . . . **Aleocharinae** (most)

Nearly 200 genera and over 1300 species, widespread and common, many in litter and soil. Badly in need of revision at all levels! Key to genera: Seevers (1978). Keys to genera or species of certain soil-linked groups: cave-associated genera and species (Klimaszewski and Peck 1986); **Falagriini** (Hoebeke 1985); **Gyrophaenina** (Ashe 1984, Seevers 1951); ***Aleochara*** (Klimaszewski 1984); ***Myllaena*** (Klimaszewski 1982); and ***Seeversiella*** (Ashe 1986).

21b Procoxa enclosed at base, trochantin not visible (similar to Fig. 38A.30); hind coxae not contiguous; **Steninae** . . . . ***Stenus*** Latreille

Nearly 200 species, widespread, mostly in wet areas.

22a Abdomen ventrally with seven complete, articulated segments exclusive of small apical genital segment which may be visible at apex (Fig. 38A.52); eyes present; tarsi three-segmented; **Oxytelinae** (in part) 23

22b Abdomen ventrally with six complete, articulated segments exclusive of genital segment, which may be visible at apex (Fig. 38A.53); if genital segment enlarged to form apparent seventh segment (Fig. 38A.2) then eyes absent; tarsi of three, four, or five segments 28

23a Pronotum subquadrate, disc strongly convex, strongly shining, nearly impunctate . . . . ***Apocellus*** Erichson

Nine species, mostly southern.

23b Pronotum transverse, disk slightly convex to flattened, shining to dull, sparsely to densely punctate 24

24a Middle coxae subcontiguous (similar to Fig. 38A.54) 25

24b Middle coxae separated by metasternal process (Fig. 38A.49) 26

25a Protibia with longitudinal row of spines (Fig. 38A.41); procoxal fissure more or less open, exposing trochantin (see Figs. 38A.29,38A.32) . . . . ***Aploderus*** Stephens

Seven species, western.

25b Protibia without longitudinal row of spines; procoxal fissure closed, concealing trochantin (similar to Fig. 38A.31) . . . . ***Carpelimus*** Leach

About 80 species, widespread, mainly in wet habitats.

26a Trochantin of front legs exposed (similar to Fig. 38A.29). . . . . ***Platystethus*** Mannerheim

Three species, widespread, mainly on dung.

26b Trochantin of front legs concealed (Fig. 38A.31). **27**

27a Scutellum with diamond-shaped impression (Fig. 38A.48) . . . . ***Oxytelus*** Gravenhorst

Ten species, widespread, mainly on dung. Key to five Florida species: Frank and Thomas (1981).

27b Scutellum with trilobed impression (Fig. 38A.47) . . . . ***Anotylus*** Thomson

Seventeen species, widespread, mainly on dung.

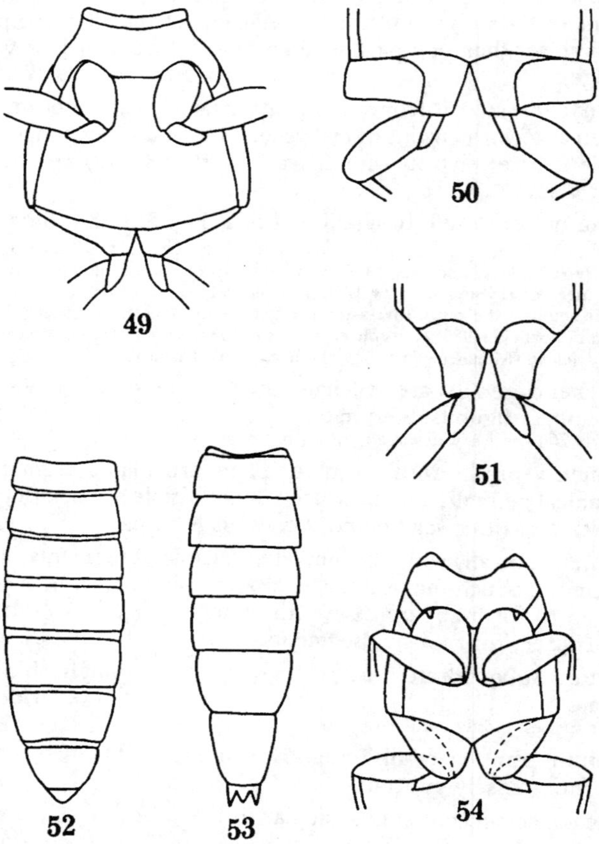

Fig. 38A.49–38A.54. 49. *Oxytelus nimius* Casey, meso- and meta-thorax, ventral. **50.** *Coproporus ventriculus* (Say), hind coxae, ventral. **51.** *Erichsonius patella* (Horn), hind coxae, ventral. **52.** *Oxytelus nimius* Casey, abdomen, ventral. **53.** *Erichsonius patella* (Horn), abdomen, ventral. **54.** *Habrocerus capillaricornis* (Gravenhorst), meso- and metathorax, ventral.

28a Front coxa enclosed basally, trochantin not or scarcely visible (Figs. 38A.30, 38A.31) **29**

28b Front coxa not enclosed basally, trochantin clearly visible (Figs. 38A.29, 38A.32, 38A.33, 38A.34) **37**

29a Labrum produced anteriorly into pair of spiniform processes (Fig. 38A.16); eyes very large, occupying entire side of head; **Megalopsidiinae** . . . . . . *Megalopinus* Eichelbaum

Two species, southeastern.

29b Labrum not as above, transverse; eyes smaller, occupying only part of side of head, or absent **30**

30a Most abdominal segments consisting of complete ring without joint between dorsal and ventral halves **31**

30b Each abdominal segment with joint and one to two supplementary sclerites between dorsal and ventral halves **33**

31a Mandibles short, not falciform; labrum small, emarginate, without teeth; tarsi three-segmented; **Osoriinae** (in part) . . . . *Espeson* Schaufuss

Two new species, Florida and North Carolina. Genus not previously recorded from North America.

31b Mandibles long, falciform; labrum very wide, entire, toothed; tarsi five-segmented; **Euaesthetinae** (in part) **32**

32a Elytra about as long as pronotum; antenna inserted much closer to side than to midline of head . . . . *Stictocranius* LeConte

One species, ***S. puncticeps*** LeConte, eastern.

32b Elytra much shorter than pronotum; antenna inserted as close to midline as to side of head . . . . *Fenderia* Hatch

One species, ***F. capizzii*** Hatch, western.

33a Eyes absent; form threadlike (Fig. 38A.2); tarsi three- or two-segmented; size minute, about 1 mm long . . . . **Leptotyphlinae**

Nine genera and 13 species, California, Alaska, and Florida, usually in deep litter and soil. Keys to genera and species: Coiffait (1962), California; Frank and Thomas (1984), Florida; and Smetana (1986), Alaska.

33b Eyes present; form less slender; tarsi five- or four-segmented; size larger **34**

34a Mandibles short, not falciform; labrum emarginate or convex, without teeth; tarsi five-segmented; **Oxytelinae** (in part) **35**

34b Mandibles long, falciform; labrum broad, finely toothed; tarsi four-segmented; **Euaesthetinae** (in part) **36**

35a Labrum emarginate (Fig. 38A.15); side margins of pronotum crenulate . . . *Syntomium* Curtis

Four species, northern. Key: Hatch (1957).

35b Labrum convex; side margins of pronotum smooth . . . *Mitosynum* Campbell

One species, ***M. vockerothi*** Campbell (1982a), New Brunswick.

36a Head above with pair of deep impressions (dorsal tentorial pits) between eyes . . . . *Edaphus* Motschulsky

Four species, eastern. Key: Puthz (1974).

36b Head above not impressed . . . . *Euaesthetus* Gravenhorst

About 18 species, widespread, mainly in wet habitats.

37a Most abdominal segments consisting of complete ring without joint between dorsal and ventral halves **38**

37b Each abdominal segment with joint and usually one to two supplementary sclerites between dorsal and ventral halves **41**

38a Head with distinct neck (Fig. 38A.11); **Paederinae** (in part) . . . . *Palaminus* Erichson

Fourteen species, mainly southern, mostly on vegetation.

38b Head without neck; **Osoriinae** (in part) **39**

39a Head nearly or quite as wide as pronotum; tibiae, especially front pair, strongly expanded externally and bearing large spines (Fig. 38A.42) . . . . ***Osorius*** Latreille

Nine species, mostly southern.

39b Head three-fourths or less as wide as pronotum; tibiae slender, without large external spines **40**

40a Prosternum, in lateral outline, flat in front of procoxae. . . . . ***Clavilispinus*** Bernhauer

Five species, widespread.

40b Prosternum, in lateral outline, tumid in front of procoxae . . . . (=***Neotrochus*** Blackwelder) ***Holotrochus*** Erichson

Four species, mostly southern.

41a Hind coxa strongly excavate throughout its length for retraction of hind femur (Fig. 38A.54) **42**

41b Hind coxa not excavate laterally, its posterior face vertical, oblique or horizontal (Figs. 38A.49, 38A.50, 38A.51) **43**

42a Antenna with three-segmented club, club segments twice or more as wide as preceding segments; elytra long, not more than three abdominal segments exposed dorsally; subfamily uncertain . . . . ***Empelus*** LeConte

One species, ***E. brunnipennis*** LeConte, western, rare. See Hatch (1957).

42b Antenna extremely slender, not at all clubbed; elytra short, about five to six abdominal segments exposed dorsally; **Habrocerinae** . . . . ***Habrocerus*** Erichson

Two species, widespread.

43a Trochantin of front legs (see Fig. 38A.29) with blunt, more or less straight external edge, not projecting (Figs. 38A.32, 38A.33); hind coxa variable, usually transverse, its apex not very prominent, its anterior edge straight to slightly curved (Figs. 38A.49, 38A.50) **44**

43b Trochantin of front legs with acute, convex, projecting external edge (Fig. 38A.34); hind coxa triangular, its apex strongly projecting posteriorly, its posterior face vertical, its anterior edge strongly convex (Fig. 38A.51) . . **63**

44a Head distinctly and abruptly constricted behind eyes to form broad neck, at least laterally (Fig. 38A.12) **45**

44b Head not or gradually narrowed behind eyes (Fig. 38A.13) **52**

45a Prothorax with large postcoxal process, spiracular sclerite relatively small and inconspicuous or not visible (Fig. 38A.32) **46**

45b Prothorax without distinct postcoxal process, but with large spiracular sclerite behind coxa (Fig. 38A.33) **50**

46a Apical segment of maxillary palp longer than preapical segment **47**

46b Apical segment of maxillary palp shorter than preapical segment **48**

47a Head as in Fig. 38A.12, eyes not prominent; all antennal segments elongate; length over 4 mm, form slender; **Tachyporinae** (in part) . . . . (= ***Rimulincola*** Sanderson) ***Derops*** Sharp

One species, ***D. divalis*** (Sanderson), Ozarks of Arkansas and Illinois, rare. See Smetana (1983).

47b Head not as in Fig. 38A.12, eyes prominent; preapical two or more antennal segments transverse; length under 3 mm, form broadly oval; **Omaliinae** (in part) . . . . ***Pycnoglypta*** Thomson

Two species, northern. Key: Campbell (1983a).

48a Abdomen very broad, about half again as wide as pronotum, with indistinct pair of lateral sclerites dorsally; head with pair of impressions between eyes;

Omaliinae (in part) . . . . ***Omalorphanus*** Campbell and Chandler
One species, ***O. aenigma*** Campbell and Chandler (1987), Oregon.

48b Abdomen about as wide as pronotum, with two distinct pairs of lateral sclerites dorsally; head without impressions between eyes; **Phloeocharinae 49**

49a Elytron with epipleural ridge (see Fig. 38A.45); size larger, over 3.5 mm . . . ***Dytoscotes*** Smetana and Campbell
One species, ***D. pacificus*** Smetana and Campbell (1980), Oregon.

49b Elytron without epipleural ridge; size smaller, 2.5 mm or less. . . . . ***Phloeocharis*** Mannerheim
One species, ***P. californica*** Smetana and Campbell (1980), California.

50a Elytron without epipleural ridge; tergum of first visible abdominal segment not covered by elytra; **Trichophyinae** . . . . ***Trichophya*** Mannerheim
Three species, widespread.

50b Elytron with epipleural ridge (Fig. 38A.45); tergum of first visible abdominal segment covered by elytra (Fig. 38A.1); **Proteininae** **51**

51a Pronotum with median longitudinal impressed line, hind angles emarginate (Fig. 38A.1) . . . . ***Megarthrus*** Curtis
Six species, mainly northern.

51b Pronotum without impressed line, hind angles entire . . ***Proteinus*** Latreille
Eight species, widespread. See Frank (1979).

52a Antenna 10-segmented; tarsi four-segmented; **Aleocharinae** (in part) ***Cypha*** Samouelle
Four species, widespread.

52b Antenna 11-segmented; tarsi five-segmented. **53**

53a Pronotum widest in anterior half, sides crenulate, disk impressed; **Oxytelinae** (in part) . . . . ***Coprophilus*** Latreille
Two species, northern, rare. Key: Campbell (1979b).

53b Pronotum widest in basal half, sides smooth, disk evenly convex; **Tachyporinae** (in part) **54**

54a Head without subocular carina. **55**

54b Head with subocular carina (Fig. 38A.28) **59**

55a Front tibia with external ctenidium (Fig. 38A.43); dorsum of head and pronotum densely pubescent . . . . ***Sepedophilus*** Gistel
Thirty species, widespread, many in fungi. Key: Campbell (1976).

55b Front tibia without ctenidium; dorsum of head and pronotum sparsely or not pubescent **56**

56a Mesosternum longitudinally carinate, at least posteriorly; elytral epipleuron acutely reflexed, not or scarcely visible in lateral view **57**

56b Mesosternum not carinate; elytral epipleuron not acutely reflexed, visible in lateral view (similar to Fig. 38A.45) **58**

57a Elytron bicolored, dark brown with apical portion and broad sutural and humeral longitudinal band yellow . . . . ***Cilea*** Jacquelin du Val
One species, ***C. silphoides*** (Linnaeus), widespread. See Campbell (1975).

57b Elytron unicolorous or bicolored with apical portion lighter in color . . . . ***Coproporus*** Kraatz
Nine species, widespread but mostly southern, often under bark or on fermenting materials. Key: Campbell (1975).

58a Apical segment of maxillary palp as wide as, and as long as or longer than, preapical segment . . . . ***Tachinus*** Gravenhorst
About 40 species, widespread but mainly northern, often in rotting fungi or other decaying materials. Keys: Campbell (1973a, 1988).

58b Apical segment of maxillary palp much narrower and shorter than preapical

segment . . . . ***Tachyporus*** Gravenhorst
About 30 species, widespread. Key: Campbell (1979a).

59a Apical segment of maxillary palp at most half as wide as, and shorter than, preapical segment . . . . ***Mycetoporus*** Mannerheim
About 22 species, widespread.

59b Apical segment of maxillary palp more than half as wide as, and about as long as, preapical segment. **60**

60a Apical segment of labial palp wider than preapical segment, with oblique excavation . . . . ***Bolitobius*** Leach
Four species, widespread.

60b Apical segment of labial palp not wider than preapical segment, without excavation **61**

61a Apical segment of labial palp much narrower than preapical segments; elytron usually with four or more longitudinal rows of setae on disc. . . . . . ***Bryoporus*** Kraatz
Several species, widespread.

61b Apical segment of labial palp subequal in width to preapical segments; elytron with a single longitudinal row of setae on disc **62**

62a Body narrowly fusiform, sides strongly tapering anteriorly and posteriorly from level of elytra; ocular puncture and seta (near dorsoposterior edge of eye) reduced or absent . . . . ***Lordithon*** Thomson
Twenty-four species, widespread, usually on fungi. Key: Campbell (1982c).

62b Body narrowly elongate, sides subparallel; ocular puncture and seta present ***Carphacis*** Des Gozis
Three species, widespread, often on fungi. Key: Campbell (1980).

63a Pronotum with large, opaque postcoxal process (Fig. 38A.34); intersegmental membranes of abdomen with "brick wall" pattern; **Paederinae** **64**

63b Pronotum without postcoxal process (Fig. 38A.36), or with small translucent process (Fig. 38A.35); intersegmental membranes of abdomen usually with rounded or triangular pattern, or pattern indistinct; **Staphylininae** . . . **82**

64a Antenna anteriorly flexile and strongly geniculate at first joint, basal segment very elongate (Fig. 38A.3) **65**

64b Antenna posteriorly flexile, not strongly geniculate, basal segment not very elongate **66**

65a Elytron with epipleural ridge near side margin (Fig. 38A.46) . . . . ***Homaeotarsus*** Hochhuth
About 36 species, mainly southern, mainly in wet habitats.

65b Elytron without epipleural ridge . . . . ***Biocrypta*** Casey
Two species, southern.

66a Apical segment of maxillary palp compressed, truncate and pubescent (Fig. 38A.22) . . . . ***Paederus*** Fabricius
Fifteen species, widespread.

66b Apical segment of maxillary palp not as above **67**

67a Apex of hind tibia with distinct ctenidium on both sides; neck never less than one-fourth as wide as maximum head width . . . . (including ***Lobrathium*** Mulsant and Rey) ***Lathrobium*** Gravenhorst
About 150 species, widespread, mainly in wet habitats. Partial keys: Watrous (1980, 1981).

67b Apex of hind tibia with ctenidium only on inner side; neck variable. . . **68**

68a Gular sutures separate throughout their length (see Fig. 38A.7) **69**

68b Gular sutures united, at least posteriorly **75**

69a Neck one-fifth to one-eighth as wide as maximum head width . . . . ***Scopaeus*** Erichson

About 40 species, widespread, mainly in wet habitats.

69b Neck one-fourth or more as wide as maximum head width **70**

70a Pronotum distinctly longer than wide (Fig. 38A.38) . . . . ***Orus*** Casey

Fifteen species, mostly western. Keys: Herman (1964, 1965).

70b Pronotum not or scarcely longer than wide **71**

71a Antenna with basal two segments enlarged, at least twice as wide as remaining segments, which are slender, not progressively wider toward apex . . . . . ***Thinocharis*** Kraatz

Five species, mainly southern.

71b Antenna normal, basal two segments less enlarged, remaining segments progressively wider toward apex **72**

72a Head and pronotum densely and very finely punctate, without larger punctures except at margins . . . . ***Lithocharis*** Dejean

Twelve species, widespread.

72b Head and pronotum with few or many larger punctures on the disk, sometimes densely punctate **73**

73a Labrum with small median tooth or rounded prominence (Fig. 38A.18) . . . ***Achenomorphus*** Motschulsky

One species, ***A. corticinus*** (Gravenhorst), eastern.

73b Labrum without median projection **74**

74a Gular sutures distinctly divergent posteriorly from before middle (see Fig. 38A.7) . . . . ***Sunius*** Curtis

About 28 species, mainly western.

74b Gular sutures most approximate along middle or at base of head . . . . . . ***Medon*** Stephens

About 50 species, mainly western.

75a Prosternum only moderately dilated over procoxae (Fig. 38A.38); neck often less than one-fourth as wide as maximum head width **76**

75b Prosternum strongly expanded over procoxae, either contiguous with sides of pronotum or narrowly separated from them (Fig. 38A.39); neck usually one-fourth or more as wide as maximum head width **79**

76a Neck at narrowest point not more than one-fifth as wide as maximum head width **77**

76b Neck at narrowest point at least one-fourth as wide as maximum head width **78**

77a Head with deep coarse punctures, intervals without ground sculpture and shining . . . . ***Rugilus*** Samouelle

Twelve species, widespread.

77b Head without deep coarse punctures, with dense ground sculpture, dull (= ***Stilicolina*** Casey) ***Eustilicus*** Sharp

Five species, mainly southern. Key: Herman (1970b).

78a Prothorax not, or only slightly, longer than wide; labrum with median denticle . . . . ***Deroderus*** Sharp

One species, ***D. tabacinum*** (Casey), Arizona. See Frania (1986).

78b Prothorax at least 1.1 times as long as wide; labrum without median denticle ***Stilocharis*** Sharp

At least one species, unnamed, Arizona. See Frania (1986).

79a Prosternum narrowly separated from pronotum behind coxae (Fig. 38A.39) **80**

79b Prosternum contiguous with pronotum behind coxae **81**

80a Prothorax not distinctly prolonged anteriorly . . . . ***Stilicopsis*** Sachse
Two species, southeastern.

80b Prothorax narrowly prolonged at middle anteriorly (Fig. 38A.39). . . . . . ***Stamnoderus*** Sharp
Three species, southeastern.

81a Labrum bidentate (Fig. 38A.19) . . . . ***Astenus*** Dejean
About 24 species, widespread, mostly in wet habitats.

81b Labrum quadridentate (Fig. 38A.20) . . . . ***Echiaster*** Erichson
Two species, southeastern.

82a Large median antesternal sclerite, or pair of contiguous sclerites, present anterior to prosternum (Fig. 38A.40). **83**

82b Median antesternal sclerite absent, or vague sclerotized area present . . **85**

83a Neck not more than about one-third as wide as maximum head width; elytra overlapping at suture, sutural edge arcuate . . . . **Xantholinini**
Twenty-four genera and 96 species, widespread, mostly in litter and soil. Key to genera and species: Smetana (1982).

83b Neck about half as wide as maximum head width; elytra contiguous at suture, sutural edge straight; **Othiini** **84**

84a Disc of pronotum with one pair of punctures . . . . ***Atrecus*** Du Val
Five species, widespread but mainly northern, mainly under bark. Key: Smetana (1982).

84b Disc of pronotum with at least four pairs of punctures. . . ***Parothius*** Casey
Two species, western, mainly under bark. Key: Smetana (1982).

85a Apical segment of maxillary palp minute (Fig. 38A.27); neck about a third as wide as head; **Diochini** . . . . ***Diochus*** Erichson
One species, ***D. schaumi*** Kraatz, eastern. See Smetana (1982).

85b Apical segment of maxillary palp larger, usually half or more as long as preapical segment; neck about half or more as wide as head. **86**

86a Head with subocular ridge (see Fig. 38A.28); disc of pronotum usually with only one to two pairs of punctures; **Quediini** **87**

86b Head without subocular ridge; disk of pronotum usually multipunctate; **Staphylinini** **88**

87a Apical segment of maxillary palp subulate, much smaller than preapical segment . . . . ***Heterothops*** Stephens
Seventeen species, widespread. Key: Smetana (1978).

87b Apical segment of maxillary palp subequal in length and width to preapical segment. . . . . ***Quedius*** Stephens
Ninety species, widespread. Keys: Smetana (1971, 1981).

88a Apical segment of labial palp greatly enlarged, more than twice length and width of preapical segment . . . . ***Tympanophorus*** Nordmann
One species, ***T. puncticollis*** Erichson, northern.

88b Apical segment of labial palp not enlarged. **89**

89a Ligula distinctly notched (Fig. 38A.26); pronotum densely punctate, punctures usually separated by less than their diameters; large species, length 12 mm or more **90**

89b Ligula entire; pronotum usually less densely punctate, punctures separated by more than their diameters; smaller species, length usually less than 12 mm. **91**

90a Base of postocular macroseta of head nearer to eye than to base of head (Fig. 38A.9). . . . . ***Staphylinus*** Linnaeus
Fourteen species, mainly northern, mainly in wet habitats. Introduced species: Newton (1987).

90b Base of postocular macroseta of head nearer to base of head than to eye (Fig. 38A.10). . . . . ***Platydracus*** Thomson

About 27 species, mainly eastern and southern, often on dung.

91a Largest lateral setigerous puncture of pronotum separated from lateral pronotal margin at most by little more than the width of the puncture (Fig. 38A.36). **92**

91b Largest lateral setigerous puncture of pronotum separated from lateral pronotal margin by about three times the width of the puncture or more (Fig. 38A.35). **94**

92a Apical segment of labial palp not or very little narrower than preapical segment, shape variable . . . . ***Philonthus*** Curtis

About 150 species, widespread, often on dung, carrion, and other decomposing matter.

92b Apical segment of labial palp much narrower than preapical segment, cylindrical **93**

93a Basal segments of protarsus slender, without spatulate setae below . . . . . ***Gabrius*** Curtis

About 26 species, mainly northern and western.

93b Basal segments of protarsus broad (see Fig. 38A.44), bearing spatulate setae below . . . . ***Gabronthus*** Tottenham

Two species, widespread. Key: Frank (1983).

94a Apical segment of maxillary palp subulate or fusiform (Fig. 38A.24) **95**

94b Apical segment of maxillary palp very elongate, subcylindrical (Fig. 38A.25). **96**

95a Translucent postcoxal process of pronotum present (Fig. 38A.35); mesosternum not carinate . . . . ***Erichsonius*** Fauvel

Eighteen species, widespread, mainly eastern. Keys: Frank (1975, 1981a).

95b Postcoxal process of pronotum not present; mesosternum transversely carinate . . . . ***Neobisnius*** Ganglbauer

Fourteen species, widespread. Key: Frank (1981b).

96a Basal segments of protarsus slender, without spatulate setae below . . . . . ***Belonuchus*** Nordmann

Eleven species, mostly southern, mostly on fermenting vegetable matter.

96b Basal segments of protarsus broad (Fig. 38A.44), bearing spatulate setae below. . . . . ***Hesperus*** Fauvel

Four species, eastern and southwestern. Key: Moore (1958).

References (Adults)

Araujo, J. 1978. Anatomie comparée des systèmes glandulaires de défense chimique des Staphylinidae. *Arch. Biol.* 89:217–250. **Arnett, R. H., Jr. 1968.** *The Beetles of the United States (A Manual for Identification).* The American Entomological Institute, Ann Arbor, Mich. 1112 pp. **1985.** *American Insects: A Handbook of the Insects of America North of Mexico.* Flora & Fauna Publications, Gainesville, Fla. 850 pp. **Ashe, J. S. 1984.** Generic revision of the subtribe Gyrophaenina (Coleoptera : Staphylinidae : Aleocharinae) with a review of the described subgenera and major features of evolution. *Quaest. Entomol.* 20:129–349. **1986.** *Seeversiella bispinosa,* a new genus and species of athetine Aleocharinae (Coleoptera : Staphylinidae) from North America. *J. N.Y. Entomol. Soc.* 94:500–511. **Balduf, W. F. 1935.** *The Bionomics of Entomophagous Coleoptera.* John S. Swift Co., St. Louis, Mo. 220 pp. **Besuchet, C., D. H. Burckhardt and I. Löbl. 1987.** The "Winkler/Moczarski" eclector as an efficient extractor for fungus and litter Coleoptera. *Coleopt. Bull.* 41:392–394. **Blackwelder, R. E. 1936.** Morphology of the coleopterous family Staphylinidae.

Smithson. Misc. Collect. 94(13):102 pp. **Campbell, J. M. 1973a.** A revision of the genus *Tachinus* (Coleoptera : Staphylinidae) of North and Central America. *Mem. Entomol. Soc. Can.* 90. 137 pp. **1973b.** New species and records of New World Micropeplinae (Coleoptera : Staphylinidae). *Can. Entomol.* 105:569–576. **1975.** A revision of the genera *Coproporus* and *Cilea* (Coleoptera : Staphylinidae) of America North of Mexico. *Can. Entomol.* 107:175–216. **1976.** A revision of the genus *Sepedophilus* Gistel (Coleoptera : Staphylinidae) of America North of Mexico. *Mem. Entomol. Soc. Can.* 99. 89 pp. **1978.** A revision of the North American Omaliinae (Coleoptera : Staphylinidae). 1. The genera *Haida* Keen, *Pseudohaida* Hatch, and *Eudectoides* new genus. 2. The tribe Coryphiini. *Mem. Entomol. Soc. Can.* 106. 87 pp. **1979a.** A revision of the genus *Tachyporus* Gravenhorst (Coleoptera : Staphylinidae) of North and Central America. *Mem. Entomol. Soc. Can.* 109. 95 pp. **1979b.** *Coprophilus castoris,* a new species of Staphylinidae (Coleoptera) from beaver lodges in eastern Canada. *Coleopt. Bull.* 33(2):223–228. **1980.** A revision of the genus *Carphacis* Des Gozis (Coleoptera : Staphylinidae) of North America. *Can. Entomol.* 112:935–953. **1982a.** *Mitosynum vockerothi,* a new genus and new species of Coleoptera (Staphylinidae : Oxytelinae) from eastern Canada. *Can. Entomol.* 114:687–691. **1982b.** A revision of the North American Omaliinae (Coleoptera : Staphylinidae). 3. The genus *Acidota* Stephens. *Can. Entomol.* 114:1003–1029. **1982c.** A revision of the genus *Lordithon* Thomson of North and Central America (Coleoptera : Staphylinidae). *Mem. Entomol. Soc. Can.* 119. 116 pp. **1983a.** A new species of *Pycnoglypta* Thomson (Coleoptera : Staphylinidae) from eastern Canada. *Can. Entomol.* 115:361–370. **1983b.** A revision of the North American Omaliinae (Coleoptera : Staphylinidae): the genus *Olophrum* Erichson. *Can. Entomol.* 115:577–622. **1984.** A revision of the North American Omaliinae (Coleoptera : Staphylinidae): the genera *Arpedium* Erichson and *Eucnecosum* Reitter. *Can. Entomol.* 116:487–527. **1988.** New species and records of North American *Tachinus* Gravenhorst (Coleoptera : Staphylinidae). *Can. Entomol.* 120:231–295. **Campbell, J. M. and D. S. Chandler. 1987.** *Omalorphanus aenigma,* an unusual new genus and species of Omaliinae (Coleoptera : Staphylinidae) from Oregon. *Can. Entomol.* 119:315–327. **Coiffait, H. 1959.** Monographie des Leptotyphlites (Col. Staphylinidae). *Rev. Fr. Entomol.* 26:237–437. **1960.** Les Coléoptères du Sol. *Vie Milieu Suppl.* 7, 204 pp. **1962.** Les Leptotyphlitae (Col. Staphylinidae) de Californie. *Rev. Fr. Entomol.* 29(2):154–165. **1972.** Coléoptères Staphylinidae de la Région Paléarctique Occidentale. I. Généralités; Sous-familles: Xantholininae et Leptotyphlinae. *Nouv. Rev. Entomol. Suppl.* 2(2):651 pp. **Dennison, D. F. and I. D. Hodkinson. 1984.** Structure of the predatory beetle community in a woodland soil ecosystem. V. Summary and conclusions. *Pedobiologia* 26:157–170. **Evans, M. E. G. 1967.** Notes on feeding in some predaceous beetles of the woodland floor. *Entomologist (Lond.)* 1967:300–303. **Focarile, A. 1985.** Le cenosi fitosaprobie di Coleotteri in lettièra di Ontano Verde (*Alnus viridis* Chaix) nelle Alpi occidentàli. *Boll. Mus. Reg. Sci. Nat. Torino* 3:79–126. **Frania, H. E. 1986.** Status of *Eustilicus* Sharp, *Trochoderus* Sharp, *Deroderus* Sharp, and *Stilocharis* Sharp (Coleoptera : Staphylinidae : Paederinae) with implications for classification of the Medonina and Stilicina. *Can. J. Zool.* 64:467–480. **Frank, J. H. 1975.** A revision of the New World species of the genus *Erichsonius* Fauvel (Coleoptera : Staphylinidae). *Coleopt. Bull.* 29(3):177–203. **1979.** A new species of *Proteinus* Latreille (Coleoptera : Staphylinidae) from Florida. *Fl. Entomol.* 62:329–340. **1981a.** A new *Erichsonius* species from Arizona with discussion on phylogeny within the genus (Coleoptera : Staphylinidae). *Coleopt. Bull.* 35:97–106. **1981b.** A revision of the New World species of the genus *Neobisnius* Ganglbauer (Coleoptera : Staphylinidae : Staphylininae). *Occas. Pap. Fl. State Collect. Arthropods* 1. 60 pp. **1983.** New records of Philonthini from the circum-Caribbean region (Coleoptera : Staphylinidae). *Fl. Entomol.* 66:473–481. **1986.** A preliminary checklist of the Staphylinidae (Coleoptera) of Florida. *Fl. Entomol.* 69:363–382. **Frank, J. H. and K. Kanamitsu. 1987.** *Paederus, sensu lato* (Coleoptera : Staphylinidae): natural history and medical importance. *J. Med. Entomol.* 24:155–191. **Frank, J. H. and M. C. Thomas. 1981.** *Oxytelus incisus* Motschulsky and *O. pennsylvanicus* Erichson (Coleoptera, Staphylinidae, Oxytelinae) in Florida. *Fl. Entomol.* 64:399–405. **1984.** *Cubanotyphlus largo,* a new species of Leptotyphlinae (Coleoptera : Staphylinidae) from Florida. *Can. Entomol.* 116:1411–1417. **Friebe, B. 1983.** Zur Biologie eines Buchenwaldbodens. 3. Die Käferfauna. *Carolinea* 41:45–80. **Hatch, M. H. 1957.** The beetles of the Pacific Northwest. II. Staphyliniformia. *Univ. Wash. Publ. Biol.* 16. 384 pp. **Herman, L. H., Jr. 1964.** A revision of *Orus* Casey. I. Subgenus *Leucorus* Casey and a new subgenus (Coleoptera : Staphylinidae). *Coleopt. Bull.* 18:112–121. **1965.** Revision of *Orus.* II. Subgenera *Orus, Pycnorus* and *Nivorus* (Coleoptera : Staphylinidae). *Coleopt. Bull.* 19:73–90. **1970a.** Phylogeny and reclassification of the genera of the rove-beetle subfamily Oxytelinae of the world (Coleoptera, Staphylinidae). *Bull. Am. Mus. Nat. Hist.* 142(5):343–454. **1970b.** The ecology, phylogeny and taxonomy of *Stilicolina* (Coleoptera, Staphylinidae, Paederinae). *Am. Mus. Novit.* 2412. 26 pp. **1975.** Revision and phylogeny of the monogeneric subfamily Pseudopsinae for the world (Staphylinidae, Coleoptera). *Bull. Am. Mus. Nat. Hist.* 155(3):241–318. **1977.** Revision and phylogeny of *Zalobius, Asemobius,* and *Nanobius,* new genus (Coleoptera, Staphylinidae, Piestinae). *Bull. Am. Mus. Nat. Hist.* 159:45–86. **Hoebeke, E. R. 1985.** A revision of the rove beetle tribe Falagriini of American north of Mexico (Coleoptera : Staphylinidae : Aleocharinae). *J. N.Y. Entomol. Soc.* 93:913–1018. **Kasule, F. K. 1968.** Field studies on the life-histories of some British Staphylinidae (Coleoptera). *Trans. Soc. Br. Entomol.* 18(4):49–80. **Klimaszewski, J. 1982.** Studies of Myllaenini (Coleoptera : Staphylinidae, Aleocharinae). 1. Systematics, phylogeny, and zoogeography of Nearctic *Myllaena* Erichson. *Can. Entomol.* 114:181–242. **1984.** A revision of the genus *Aleochara* Gravenhorst of America north of Mexico (Coleoptera : Staphylinidae, Aleocharinae). *Mem. Entomol. Soc. Can.* 129. 211 pp. **Klimaszewski, J. and S. B. Peck. 1986.** A review of the cavernicolous Staphylinidae (Coleoptera) of eastern North America. I. Aleocharinae. *Quaest. Entomol.* 22:51–133. **Lawrence, J. F. and A. F. Newton, Jr. 1982.** Evolution and classification of beetles. *Annu. Rev. Ecol. Syst.* 13:261–290. **Mank, H. G. 1923.** The biology of the Staphylinidae. *Ann. Entomol. Soc. Am.* 16:220–237. **Martin, J. E. H. 1977.** *The Insects and Arachnids of Canada.* 1. *Collecting, Preparing and Preserving Insects, Mites and Spiders.* Publ. 1643. Ottawa, Ontario, Canada. Agriculture Canada, 182 pp. **Moore, I. 1958.** The North American species of *Hesperus* Fauvel, with descriptions of two new species (Coleoptera; Staphylinidae). *Trans. San Diego Soc. Nat. Hist.* 12(18):311–318. **Moore, I. and E. F. Legner. 1974a.** Bibliography (1758 to 1972) to the Staphylinidae of America north of Mexico (Coleoptera). *Hilgardia* 42(16):511–547. **1974b.** Keys to the genera of the Staphylinidae of America north of Mexico exclusive of the Aleocharinae (Coleoptera : Staphylinidae). *Hilgardia* 42(16):548–563. **1975.** *A Catalogue of the Staphylinidae of America North of Mexico (Coleoptera).* Spec. Publ. 3015. University of California Division of Agricultural Sciences, Berkeley, Calif. 514 pp. **1979.** *An Illustrated Guide to the Genera of the Staphylinidae of America North of Mexico Exclusive of the Aleocharinae (Coleoptera).* Priced Publ. 4093. University of California Division of Agricultural Sciences, Berkeley, Calif. 332 pp. **Naomi, S.-I. 1985.** The phylogeny and higher classification of the Staphylinidae and their allied groups (Coleoptera, Staphylinoidea). *Esakia* 23. 27 pp. **1987.** Comparative morphology of the Staphylinidae and the allied groups (Coleoptera, Staphylinoidea). I. Introduction, head sutures, eyes and ocelli. *Kontyu (Tokyo)* 55:450–458. **Newton, A. F., Jr. 1984.** Mycophagy in Staphylinoidea (Coleoptera). Pp. 302–353 in Q. Wheeler and M. Blackwell (eds.). *Fungus/Insect Relationships. Perspectives in Ecology and Evolution.* Columbia University Press, New York. **1987.** Four *Staphylinus (sensu lato)* species new to North America, with notes on other introduced species (Coleoptera : Staphylinidae). *Coleopt. Bull.* 41:381–384. **Peck, S. B. and A. E. Davies. 1980.** Collecting small beetles with large-area "window" traps. *Coleopt. Bull.* 34:237–239. **Puthz, V. 1974.** A new revision of the Nearctic *Edaphus*-species and remarks on other North American Euaesthetinae (Coleoptera, Staphylinidae). *Rev. Suisse Zool.* 81(4):911–932. **Seevers, C. H. 1951.** A revision of the North American and European staphylinid beetles of the subtribe Gyrophaenae (Aleocharinae, Bolitocharini). *Fieldiana Zool.* 32:655–762. **1978.** A generic and tribal revision of the North American Aleocharinae (Coleoptera : Staphylinidae). *Fieldiana Zool.* 71:1–289. **Smetana, A. 1971.** Revision of the tribe Quediini of America north of Mexico (Coleoptera : Staphylinidae). *Mem. Entomol. Soc. Can.* 79. 303 pp. **1978.** Revision of the tribe Quediini of America north of Mexico (Coleoptera : Staphylinidae). Suppl. 4. *Can. Entomol.* 110:815–840. **1981.** Revision of the tribe Quediini of America north of Mexico (Coleoptera : Staphylinidae). Suppl. 5. *Can. Entomol.* 113:631–644. **1982.** Revision of the subfamily Xantholininae of America north of Mexico (Coleoptera : Staphylinidae). *Mem. Entomol. Soc. Can.* 120. 389 pp. **1983.** The status of the staphylinid genera *Derops* Sharp and *Rimulincola* Sanderson (Coleoptera). *Entomol. Scand.* 14:269–279. **1985.** Systematic position and review of *Deinopteroloma* Jansson, 1946, with descriptions of four new species (Coleoptera, Silphidae and Staphylinidae (Omaliinae)). *Syst. Entomol.* 10:471–499. **1986.** *Chionotyphlus alaskensis* n.g., n.sp., a Tertiary relict from unglaciated interior Alaska (Coleoptera, Staphylinidae). *Nouv. Rev. Entomol. (N.S.)* 3:171–187. **Smetana, A. and J. M. Campbell. 1980.** A new genus and two new phloeocharinae species from the Pacific coast of North America (Coleoptera : Staphylinidae). *Can. Entomol.*

112:1061–1069. **Smith, R. L., G. C. Lanzaro, J. E. Wheeler and A. Snyder. 1978.** Ecology and behavior of *Osorius planifrons*. *Ann. Entomol. Soc. Am.* 71:752–755. **Tikhomirova, A. L. 1973.** *Morphological and Ecological Features and Phylogeny of the Staphylinidae (with a Catalogue of the Fauna of the USSR)*. Academy of Sciences, USSR, A. N. Severtsov Institute of Evolutionary Morphology and Ecology of Animals, Moscow. 190 pp. (in Russian). **Topp, W. 1979.** Vergleichende Dormanzuntersuchungen an Staphyliniden (Coleoptera). *Zool. Jahrb. (Syst.)* 106:1–49. **Voris, R. 1934.** Biologic investigations on the Staphylinidae (Coleoptera). *Trans. Acad. Sci. St. Louis* 28:231–261. **Watrous, L. 1980.** *Lathrobium (Tetartopeus)*: natural history, phylogeny and revision of the nearctic species (Coleoptera, Staphylinidae). *Syst. Entomol.* 5:303–338. **1981.** Studies of *Lathrobium (Lobrathium)*: Revision of the *grande* species group (Coleoptera : Staphylinidae). *Ann. Entomol. Soc. Am.* 74:144–150. **Wheeler, Q. D. and J. V. McHugh. 1987.** A portable and convertible "Moczarski/Tullgren" extractor for fungus and litter Coleoptera. *Coleopt. Bull.* 41:9–12.

Part B: Larvae

Larvae of Staphylinidae, although less often encountered than adults, are also common components of the soil fauna. Larvae of most or all of the 1500 or more species of soil-associated adults probably also occur in soil, but most of these species have not yet been found and identified as larvae.

Most staphylinid larvae may be distinguished from other beetle larvae by possession of the combination of a pair of articulated appendages (urogomphi) at the apex of the ninth abdominal tergum (Fig. 38B.1); mandibles without contiguous molar lobes at their bases; and a maxilla with a single undivided inner lobe which is fixed (Figs. 38B.4, 38B.37, 38B.42) or articulated (Figs. 38B.2, 38B.43–38B.45). Larvae of some closely related families, such as Scaphidiidae, Pselaphidae, and Scydmaenidae, are difficult to separate from staphylinid larvae (see Kasule 1966; Newton 1989). Larvae of the staphylinid subfamilies Paederinae and Staphylininae superficially resemble larvae of several other beetle groups, but may be distinguished as follows: from Carabidae by having urogomphi articulated rather than fixed at base and by having only a single segment without claws (tarsungulus) beyond the tibia of the legs; from Hydrophilidae and Histeroidea by having an articulated appendage at the apex of the stipes of the maxilla (Figs. 38B.43–38B.45) rather than on the first apparent palpal segment; and from Cantharidae and other beetle larvae by the presence of articulated urogomphi.

Larval Morphology

Length about 0.5–25 mm; form elongate (Fig. 38B.1) to ovoid; color white to yellowish, dorsum sometimes with brown transverse bands or entirely brown, or variegated; head, thoracic and abdominal sclerites and appendages lightly to well sclerotized, with extensive exposed membranous areas between.

Head prognathous to (Micropeplinae) hypognathous, with or without distinct neck (Figs. 38B.2–38B.5). Compound eyes absent, one to six ocelli (usually black) present on each side of head, or ocelli absent. Dorsum of head with Y-shaped ecdysial lines (epicranial sutures) which end in or near the antennal foramina (Figs. 38B.3, 38B.5); stem of "Y" absent in *Micropeplus*. Antenna well developed, of three or four segments; penultimate segment with large sensory appendage usually on anterior (inner) side but sometimes dorsal, ventral, or posterior in position (Figs. 38B.3, 38B.5–38B.11). Epistomal suture absent. Labrum either (a) articulated, but less distinct than that of most beetle larvae, lacking internal apodemes (tormae) at the hind margins and usually accompanied by one to two pairs of lateral sclerites as in Fig. 38B.3; or (b) fused to head capsule to form nasale, which is usually toothed or spinose anteriorly as in Fig. 38B.5. Mandibles without contiguous molar lobes at base, their apices usually projecting and acute but sometimes multidentate or otherwise modified (Figs. 38B.2–38B.5, 38B.24–38B.36). Maxilla with a single distinct inner lobe (mala), which may be fixed (Figs. 38B.4, 38B.37, 38B.42) or small, fingerlike, and articulated at base (Figs. 38B.2, 38B.43–38B.45). Maxillary palp apparently three-segmented (Figs. 38B.37–38B.40, 38B.42–38B.44) or four-segmented (Figs. 38B.41, 38B.45), not counting minute, semicircular basal segment (palpifer). Labium with a median ligula of variable shape and a pair of two- or three-segmented palps (Figs. 38B.2, 38B.4, 38B.46–38B.48). Mentum present (Fig. 38B.4) or absent (Fig. 38B.2). Gular area variable, may be membranous with submental sclerite to completely sclerotized except for thin ecdysial line(s) (Figs. 38B.2, 38B.4). Posterior arms of tentorium attached directly to ventral surface of head rather than to transverse bridge.

Thorax: Tergum of each segment consisting of a single sclerite, with midlongitudinal ecdysial line. Terga two to three usually with transverse carina behind anterior margin. Pleura and sterna variable, membranous or with one or more sclerites present on each segment. Pair of large annular spiracles present laterally between segments 1 and 2. Legs well developed, usually five-segmented (coxa, trochanter, femur, tibia, and clawlike tarsungulus; trochanter and femur fused in *Micropeplus*); longest macroseta of leg on trochanter; tarsungulus usually bisetose.

Abdomen elongate, 10-segmented. Segments 1–8 usually similar in appearance; tergum and sternum of each segment usually of a single sclerite or divided midlongitudinally by membranous area; tergum and sternum with or without transverse carina behind anterior margin; pair of annular spiracles present in or just outside of lateral edges of tergum; one to two pairs of small pleural sclerites may be present. Segment 9 consisting of tergum and sternum with or without transverse carina; tergum with pair of appendages (urogomphi) at posterolateral edges; urogomphi usually two- or one-segmented (Figs. 38B.1, 38B.56, 38B.57), rarely unarticulated. Segment 10 cylindrical, transverse to elongate; apex with anus and eversible membranous lobes which usually bear fine teeth (four large teeth in many Aleocharinae) (Figs. 38B.1, 38B.56, 38B.57).

Sexual dimorphism absent. Internal morphology poorly known and not used in classification or identification.

Larval instars three, rarely two. Instars usually similar in structure and appearance, with minor differences in chaetotaxy and in shape and relative size of appendages. Instar 1 of some species with one or more pairs of fine teeth on dorsum of head, thorax or abdomen which probably function as egg-bursters.

The only reasonably detailed treatment of staphylinid larval morphology is that of Verhoeff (1919).

Biology and Ecology

The habits of staphylinid larvae are much more poorly known than the habits of the adults. In general, larvae occur in the same habitat and feed on the same substances as adults of the same species, hence comments on the trophic behavior of adults in the preceding section also apply to larvae.

Larval developmental time is usually short relative to adult longevity, and extensive areas of exposed membrane result in greater susceptibility to desiccation in larvae as compared with adults. Larval locomotion is probably less efficient (no wings and relatively shorter legs), with the result that larvae in general are less frequently encountered than adults, are less vagile, and do not participate in dispersal.

Larvae that occur in soil-related habitats such as dung and carrion, fungi, animal nests, and so on, usually leave these habitats to pupate in nearby soil. In *Aleochara* and allied genera of dung- and carrion-frequenting Aleocharinae, first-instar larvae enter soil to seek Diptera puparia, in which they complete their development as endoparasitoids.

Further information on larval biology is contained in the references cited in Part A of this chapter.

Collection and Study Techniques

Staphylinid larvae may be collected by the same methods used for adults, as discussed in Part A. They also may often be obtained by confining adults, with suitable food, in small rearing containers until eggs are laid. The eggs or newly emerged larvae of carnivorous species are best removed to individual containers for further development.

Larvae should be killed by immersion in near-boiling water or in a liquid preservative, and kept permanently in such a preservative. Peterson's KAAD mixture, Kahle's and similar fixatives, and a mixture of 95% ethanol (75%), water (20%), and glacial acetic acid (5%) all give good results, and plain 80% ethanol is satisfactory.

Whole larvae in fluid are often identifiable to genus without further preparation, but in most cases study is facilitated by removal of the head for ease of orientation. Critical study of head structures requires clearing the head in 1 *N* potassium hydroxide or other clearing agent and examining it as a temporary or permanent slide mount with a compound microscope. The number and arrangement of ocelli should be noted before clearing. Lightly sclerotized and unpigmented larvae, after clearing, should be stained in acid fuchsin, dehydrated in acetone and xylene, and examined in anhydrous media as temporary (mineral oil) or permanent (resinous) slide mounts.

The preservation and study techniques mentioned above are discussed in detail by Martin (1977) and Stehr (1987), while Steel (1970) describes a very successful staphylinid rearing method.

Identification

Larvae of members of fewer than a third of the North American genera and of only a few percent of the species have been described. Thus the only completely reliable method of species determination of field-collected larvae is by rearing the larvae through to the adult stage followed by identification of the adults, using the guidelines presented in Part A. The cast skin of the final larval instar may then be used to identify unreared larvae of the same species. Determined larvae may also be obtained by rearing from field-collected, identified adults (see above). Under certain conditions it may be possible to associate larvae with adults collected at the same time and place, but such association must be done very carefully to prevent misidentification, and the determination should be considered tentative until confirmed by other means.

Larvae not reared or reliably associated with adults may be tentatively identified to the subfamily and often generic level by use of keys constructed primarily on the basis of reared European species. Such keys include, in approximately increasing order of completeness, Böving and Craighead (1931), Paulian (1941), Pototskaya (1967), Kasule (1966, 1968, 1970), and Topp (1978). The cited works of Kasule are especially well illustrated and documented. Keys to known larval genera of the world have been published for two subfamilies, Proteininae (Steel 1966) and Omaliinae (Steel 1970). Newton (1982) presented a revised subfamily key to separate larvae of Osoriinae, Oxytelinae, and Piestinae, and reviewed character variation among oxyteline genera.

No comprehensive keys specifically intended for the North American fauna have yet appeared, but Frank (1989) includes a subfamily key and extensive discussion and description of North American genera and species. Mank (1923), Böving and Craighead (1931), and Voris (1939) have described a number of North American species. Moore and Legner (1974) give a general catalog of the few described immature stages of our fauna. The following works, not mentioned in that catalog, also describe larvae of North American species of the cited genera or higher groups: Aleocharinae: *Atheta* (Ashe and Watrous 1984), *Gyrophaenina* (Ashe 1986), and *Meronera* (Ashe 1985); Micropeplinae (Newton, in press); Osoriinae: *Thoracophorus* (Boháč 1978); Oxytelinae: *Carpelimus* (Moore and Legner 1973) and *Platystethus* (Legner and Moore 1977); Paederinae: *Lathrobium* (Watrous 1981) and genera of *Stilicina* and *Medonina* (Frania 1986); Staphylininae: *Belonuchus* (Silvestri 1945), *Philonthus* (Boller 1983), and *Staphylinus* and allied genera (Frank 1979, Lesage 1977, Orth et al. 1975); and Tachyporinae: *Sepedophilus* (Newton 1984).

Larvae of four small staphylinid subfamilies found in North America have not yet been described: Leptotyphlinae, Megalopsidiinae, Pseudopsinae, and Trichophyinae. Larvae of some members of the first three of these subfamilies are available and are included in the key that follows (formal descriptions of these larvae are in preparation by A. Smetana or A. Newton, for publication elsewhere).

KEY TO SOIL GENERA (LARVAE)

The key that follows is restricted to genera whose adults are considered to be litter or soil inhabitants according to the criteria put forth in Part A of this chapter. Staphylinid larvae normally found in other habitats may occasionally be found in soil, especially when about to pupate; most such larvae should key out at least to the correct subfamily.

The key is based largely on those of Kasule (1966, 1968, 1970) and Steel (1970) but is extensively modified by use of North American material. Because larvae of so few North American species and even genera are known, the key must be considered provisional and determinations made from it are tentative at best.

All figures were made with the aid of a camera lucida from cleared slide preparations. Except for those of **Omaliinae,** which are redrawn from Steel (1970), all figures are original and based on North and Central American species. Previously undescribed larvae were identified either by rearing or by carefully repeated association with adults in restricted habitats. All larvae illustrated are believed to be final (third) instar larvae.

The key to adults should be consulted for authors of generic names and information on the size and distribution of genera or higher taxa.

1a Body broad, flattened, terga with lateral projections; head without median epicranial suture; **Micropeplinae** . . . . *Micropeplus*

1b Body usually very elongate (Fig. 38B.1); terga without lateral projections; head with median epicranial suture (Figs. 38B.3, 38B.5) **2**

2a Segment 8 of abdomen with a median dorsal glandular structure (Fig. 38B.56); one ocellus present on each side of head . . . . **Aleocharinae** (in part)

2b Segment 8 of abdomen without glandular structure; number of ocelli variable, up to six on each side of head **3**

3a Labrum articulated (Fig. 38B.3); head without distinct neck (Fig. 38B.3), or rarely with very broad neck (similar to Fig. 38B.4); abdominal sterna usually entire **4**

3b Labrum completely fused to head capsule to form nasale (Fig. 38B.5); head usually with distinct neck (Figs. 38B.2, 38B.4, 38B.5); abdominal sterna often divided along midline **36**

4a Mandible with two or more subapical teeth, apically more or less widened and scooplike, inner edge not serrulate (Figs. 38B.28–38B.31); ligula transverse, truncate (Fig. 38B.49); four or fewer ocelli present on each side of head **5**

4b Mandible usually with a single tooth and one or two preapical teeth, rather narrow at apex, inner edge often serrulate (Figs. 38B.24–38B.27, 38B.32–38B.36); ligula variable, usually narrow or conical (Figs. 38B.46–38B.48); *or* if mandibles and ligula resemble first half of couplet, then five or six ocelli present on each side of head **13**

5a Left and right mandibles approximately symmetrical, with same number (usually two or one) of large subapical teeth (Fig. 38B.28); penultimate segment of maxillary palp as long as or shorter than preceding segment (Fig. 38B.38), *or* urogomphus falciform; urogomphus one-segmented; **Oxytelinae** **6**

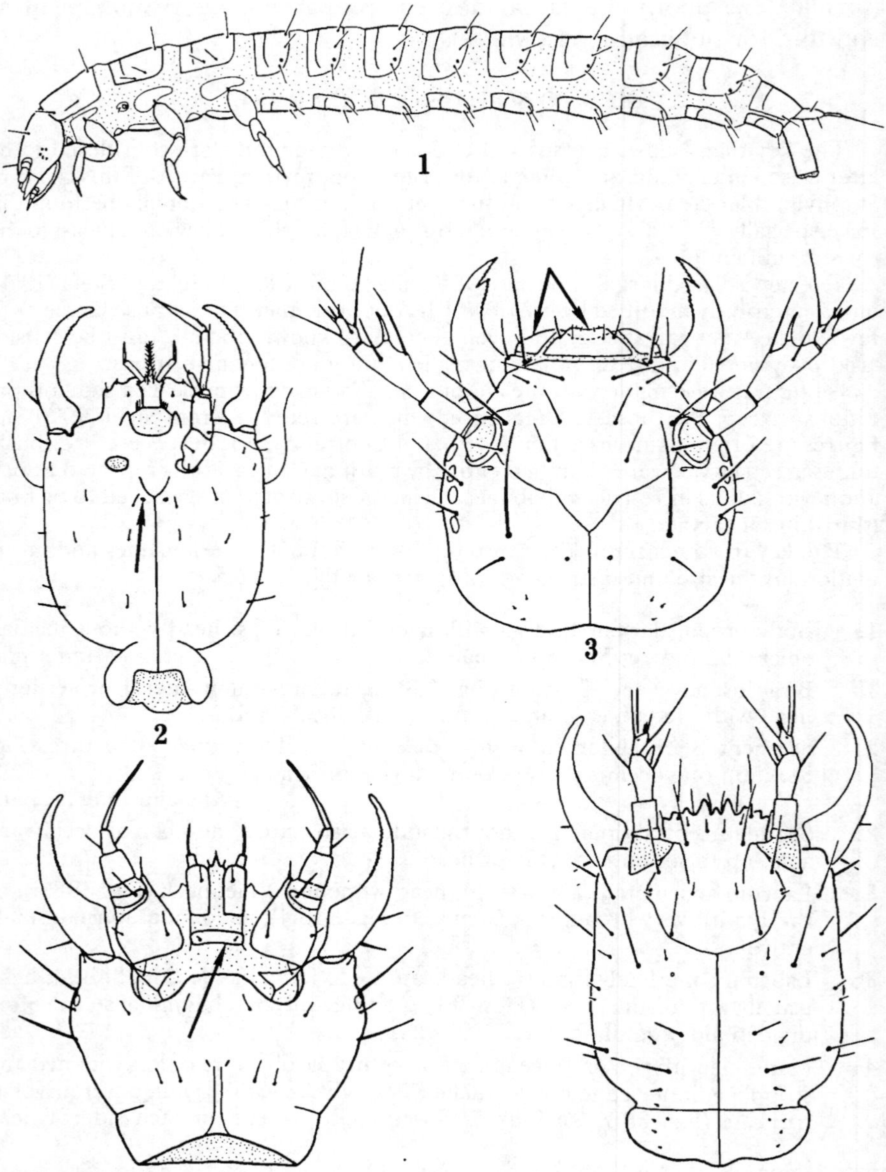

Figs. 38B.1–38B.5. 1. *Phloeocharis californica* Smetana and Campbell. **2.** *Sunius confluentus* (Say), head, ventral, showing tentorial pit (arrow). **3.** *Tachyporus californicus* Horn, head, dorsal, showing labrum (arrows). **4.** *Euaesthetus* sp. (Massachusetts), head, ventral, showing mentum (arrow). **5.** *Diochus* sp. (Mexico), head, dorsal.

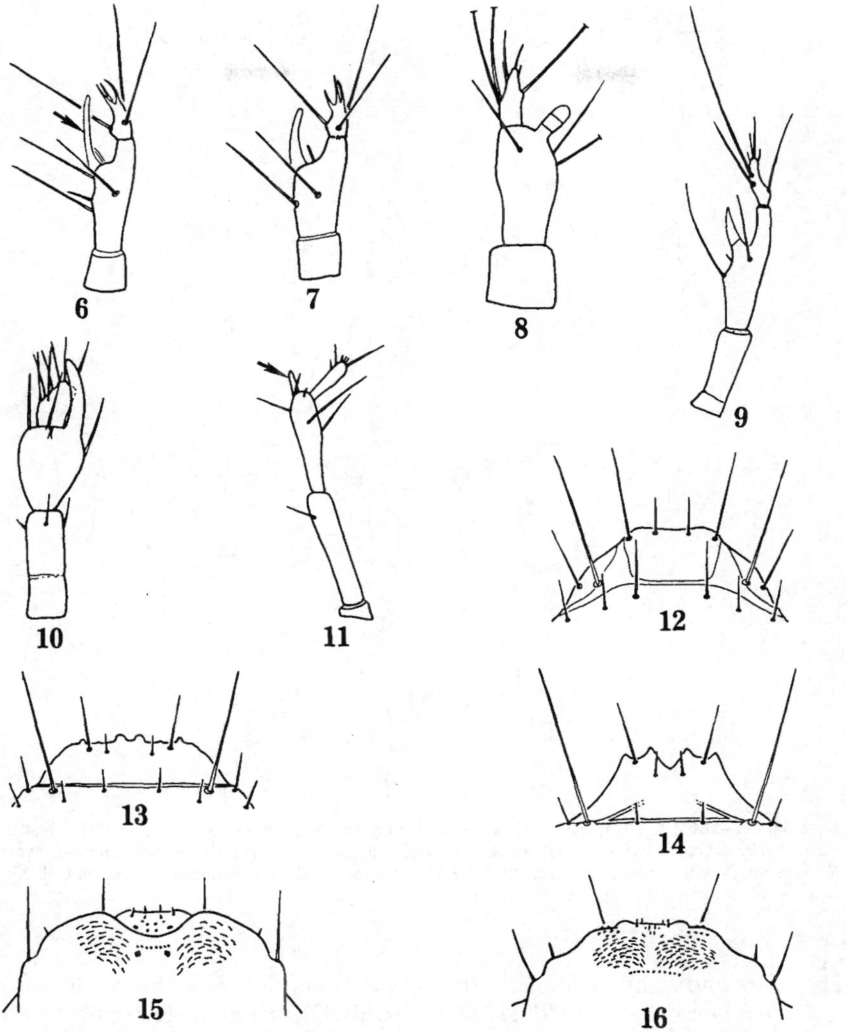

Figs. 38B.6–38B.16. 6. *Acrulia inflata* (Gyllenhal), right antenna, dorsal, showing sensory appendage (arrow). **7.** *Hapalaraea pygmaea* (Paykull), right antenna, dorsal. **8.** *Mycetoporus* sp. (Massachusetts), left antenna, dorsal. **9.** *Sepedophilus castaneus* (Horn), right antenna, dorsal. **10.** *Palaminus* sp. (Tennessee), left antenna, dorsal. **11.** *Homaeotarsus* sp. (California), left antenna, dorsal, showing sensory appendage (arrow). **12.** *Anthobium unicolor* (Marsham), labrum, dorsal. **13.** *Acrolocha minuta* (Olivier), labrum, dorsal. **14.** *Omalium rivulare* (Paykull), labrum, dorsal. **15.** *Tachyporus californicus* Horn, labrum, ventral. **16.** *Tachinus* sp. (New York), labrum, ventral. [Figures 6, 7, 12, 13, and 14 redrawn from Steel (1970).]

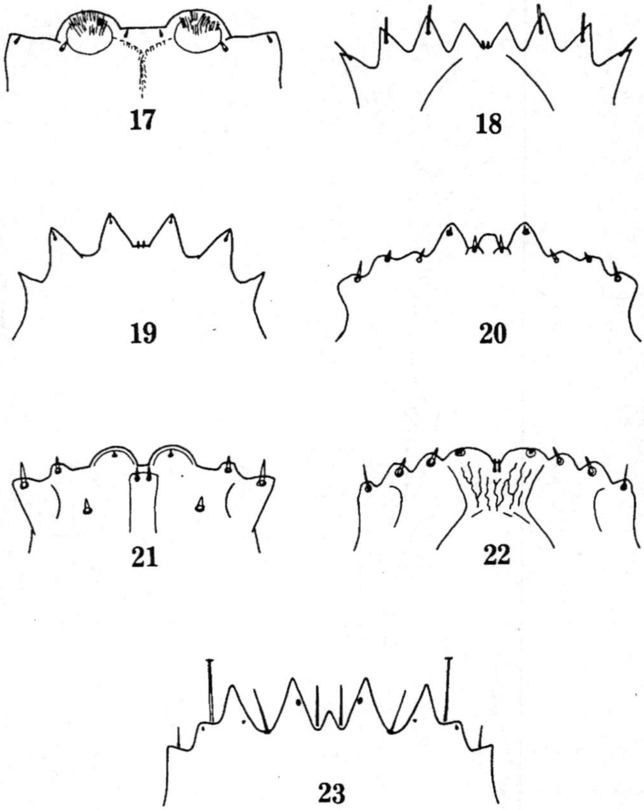

Figs. 38B.17–38B.23. **17.** *Homaeotarsus* sp. (California), nasale, ventral. **18.** *Astenus* sp. (Michigan), nasale, ventral. **19.** *Echiaster* sp. (Panama), nasale, ventral. **20.** *Lathrobium grande* LeConte, nasale, ventral. **21.** *Scopaeus* sp. (Mexico), nasale, ventral. **22.** *Achenomorphus corticinus* (Gravenhorst), nasale, ventral. **23.** *Philonthus* sp. (New Jersey), nasale, ventral.

5b Left and right mandibles with dissimilar dentition, or with scooplike or mola-like lobes near apex (Figs. 38B.29–38B.31); penultimate segment of maxillary palp longer than preceding segment (Fig. 38B.39); urogomphus one- or two-segmented, straight; **Osoriinae** **10**

6a Head with three or four ocelli on each side **7**

6b Head with one or no ocelli on each side. **8**

7a Body about half as wide as long, capable of rolling into a ball; urogomphus much shorter than last abdominal segment . . . . ***Syntomium***

7b Body much more elongate, not capable of rolling into a ball; urogomphus longer than last abdominal segment. . . . . ***Carpelimus***

8a Ocelli absent; urogomphus falcate; penultimate segment of maxillary palp longer than preceding segment. . . . . ***Coprophilus***

8b One ocellus present on each side of head; urogomphus straight; penultimate segment of maxillary palp shorter than preceding segment (Fig. 38B.38) **9**

9a Prementum less than twice as long as ligula; urogomphus two to three times as long as its width at base . . . . ***Platystethus***

9b Prementum about twice as long as ligula (Fig. 38B.49); urogomphus about four to five times as long as its width at base . . . . ***Oxytelus, Anotylus***

10a Three or four ocelli present on each side of head; left mandible with multidentate or bladelike mesoventral edge (Figs. 38B.29, 38B.30); thin, brown, transverse bands present on each abdominal segment **11**

10b One or no ocellus present on each side of head; left mandible with thick projection, bearing rough surface, on mesoventral edge (Fig. 38B.31); abdomen without brown transverse bands. **12**

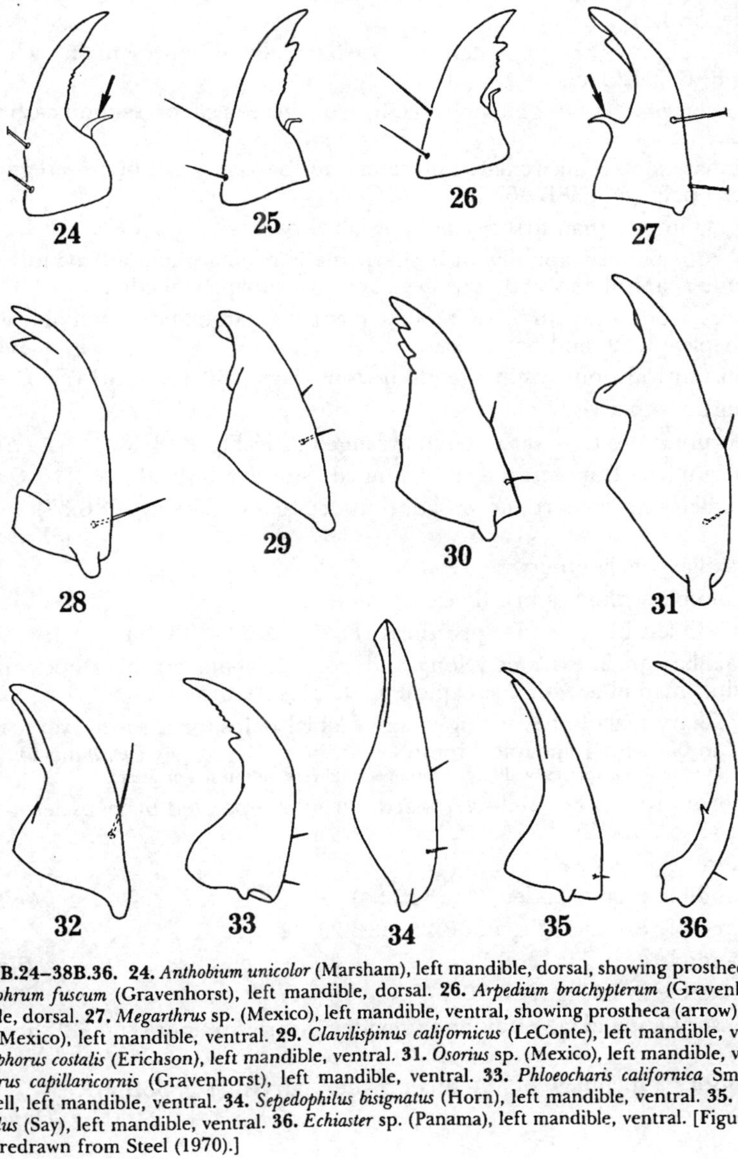

Figs. 38B.24–38B.36. 24. *Anthobium unicolor* (Marsham), left mandible, dorsal, showing prostheca (arrow). **25.** *Olophrum fuscum* (Gravenhorst), left mandible, dorsal. **26.** *Arpedium brachypterum* (Gravenhorst), left mandible, dorsal. **27.** *Megarthrus* sp. (Mexico), left mandible, ventral, showing prostheca (arrow). **28.** *Oxytelus* sp. (Mexico), left mandible, ventral. **29.** *Clavilispinus californicus* (LeConte), left mandible, ventral. **30.** *Thoracophorus costalis* (Erichson), left mandible, ventral. **31.** *Osorius* sp. (Mexico), left mandible, ventral. **32.** *Habrocerus capillaricornis* (Gravenhorst), left mandible, ventral. **33.** *Phloeocharis californica* Smetana and Campbell, left mandible, ventral. **34.** *Sepedophilus bisignatus* (Horn), left mandible, ventral. **35.** *Coproporus ventriculus* (Say), left mandible, ventral. **36.** *Echiaster* sp. (Panama), left mandible, ventral. [Figures 24, 25, and 26 redrawn from Steel (1970).]

11a Four ocelli present on each side of head; left mandible with bladelike mesoventral edge (Fig. 38B.29) . . . . ***Clavilispinus***

11b Three ocelli present on each side of head; left mandible with a few teeth on mesoventral (as well as mesodorsal) edge (Fig. 38B.30) . . . ***Thoracophorus***

12a One ocellus present on each side of head. . . . . ***Holotrochus***

12b Ocelli absent. . . . . ***Osorius***

13a Head with a single ocellus present on each side (rarely absent); urogomphus minute, shorter than apical abdominal segment which is usually small (Fig. 38B.56) . . . . **Aleocharinae** (in part)

13b Head with six, five, or three ocelli present on each side (very rarely absent); urogomphus usually longer than apical abdominal segment which is larger (Fig. 38B.1). **14**

14a Urogomphus one-segmented; five ocelli (rarely, zero) present on each side of head; **Omaliinae** **15**

14b Urogomphus two-segmented; six, five, or three ocelli present on each side of head **26**

15a Ligula short, semicircular, not extending beyond apex of first segment of labial palp (Fig. 38B.46) **16**

15b Ligula longer than first segment of labial palp **17**

16a Labrum bilobed apically with sharp median emargination; mandible with slender, apically forked prostheca and two subapical teeth . . . . . ***Acidota***

16b Labrum apically with broad arcuate median emargination; mandible without subapical teeth and prostheca . . . . ***Subhaida***

17a Labrum with four distinct teeth in front (Figs. 38B.13, 38B.14) **18**

17b Labrum otherwise **19**

18a Labrum more transverse, teeth arranged as in Fig. 38B.13. . . . . ***Elonium***

18b Labrum less transverse, teeth arranged as in Fig. 38B.14 . . . . . ***Omalium***

19a Maxillary mala short and broad, rounded externally (Fig. 38B.37) . . . . . ***Xylodromus***

19b Maxillary mala otherwise **20**

20a Mandible without a prostheca **21**

20b Mandible with a slender prostheca (Figs. 38B.24–38B.26) **22**

21a Maxillary mala extremely long and slender, about reaching apex of labial palps; mandible with 2 subapical teeth . . . . ***Lesteva***

21b Maxillary mala not reaching apex of labial palps, or if so not very slender; mandible with 1 subapical tooth . . . . ***Eusphalerum*** Kraatz

Twenty-seven species, widepread (adults on flowers, not in soil or litter).

22a Labrum with three well-sclerotized portions separated by less sclerotized areas (Fig. 38B.12) **23**

22b Labrum otherwise **25**

23a Mandible more slender (Fig. 38B.24) . . . . ***Anthobium***

23b Mandible broader (Figs. 38B.25, 38B.26). **24**

24a Mandible as in Fig. 38B.25 . . . . ***Olophrum***

24b Mandible as in Fig. 38B.26 . . . . ***Arpedium***

25a Sensory appendage of antenna slender, nearly straight (Fig. 38B.6) . . . . ***Acrulia***

25b Sensory appendage of antenna broader, curved (Fig. 38B.7) . . . . ***Hapalaraea***

26a Maxillary mala very long, slender (Fig. 38B.42); mandible with prostheca (Fig. 38B.27); **Proteininae** **27**

26b Maxillary mala less elongate; mandible without prostheca **28**

27a Six ocelli present on each side of head . . . . *Megarthrus*

27b Three ocelli present on each side of head. . . . . *Proteinus*

28a Head with subbasal carina, delimiting very broad neck dorsally and laterally (similar to Fig. 38B.4); mandible with large subapical projection, shaped like the blade of a guillotine; **Megalopsidiinae** . . . . *Megalopinus*

28b Head without subbasal carina, without neck (Fig. 38B.3); mandible variable, not as above **29**

29a Five ocelli present on each side of head; antenna three-segmented **30**

29b Six ocelli present on each side of head, or antenna four-segmented (Fig. 38B.9); **Tachyporinae** **31**

30a Ocelli arranged in a single, irregularly curved line; mandible as in Fig. 38B.32; **Habrocerinae**. . . . . *Habrocerus*

30b Ocelli arranged in two groups of three anterior, two posterior (Fig. 38B.1); mandible as in Fig. 38B.33; **Phloeocharinae** . . . . *Phloeocharis*

31a Ocelli arranged in circle; antenna short, three-segmented (Fig. 38B.8); abdominal spiracles placed in edges of terga . . . . *Mycetoporus, Bolitobius, Lordithon*

31b Ocelli arranged differently; antenna longer, more or less distinctly four-segmented (Fig. 38B.9); abdominal spiracles in membrane just outside terga **32**

32a Mandible chisel-shaped at apex (Fig. 38B.34); five ocelli present on each side of head . . . . *Sepedophilus* (in part)

32b Mandible not chisel-shaped (Figs, 38B.3, 38B.35); usually six ocelli present on each side of head **33**

33a Mandible with one or more teeth very near apex, inner edge not serrulate (Fig. 38B.35); ligula broad, blunt, one-segmented (Fig. 38B.47). . . . . . . *Coproporus, Cilea*

33b Mandible with single large preapical tooth, part of inner edge serrulate (Fig. 38B.3); ligula usually long, conical, two-segmented (Fig. 38B.48) **34**

34a Apical segment of antenna very elongate, innermost macroseta inserted more distally than other two macrosetae and about twice as long as other macrosetae (Fig. 38B.9) . . . . *Sepedophilus* (in part)

34b Apical segment of antenna less elongate, insertions of three macrosetae about equidistant from base of segment, innermost seta less than twice as long as other macrosetae (Fig. 38B.3) **35**

35a Labrum as in Fig. 38B.15; sensory appendage of antenna slender, acute (Fig. 38B.3) . . . . *Tachyporus*

35b Labrum as in Fig. 38B.16; sensory appendage of antenna broader and shorter. . . . . *Tachinus*

36a Mentum present (Fig. 38B.4); mala not articulated (Fig. 38B.4); urogomphus one-segmented or not articulated at base. **37**

36b Mentum absent (Fig. 38B.2); mala articulated (Figs. 38B.2, 38B.43–38B.45); urogomphus usually two-segmented, rarely one-segmented **39**

37a Ocelli absent; urogomphus not or indistinctly articulated . . . . **Leptotyphlinae**

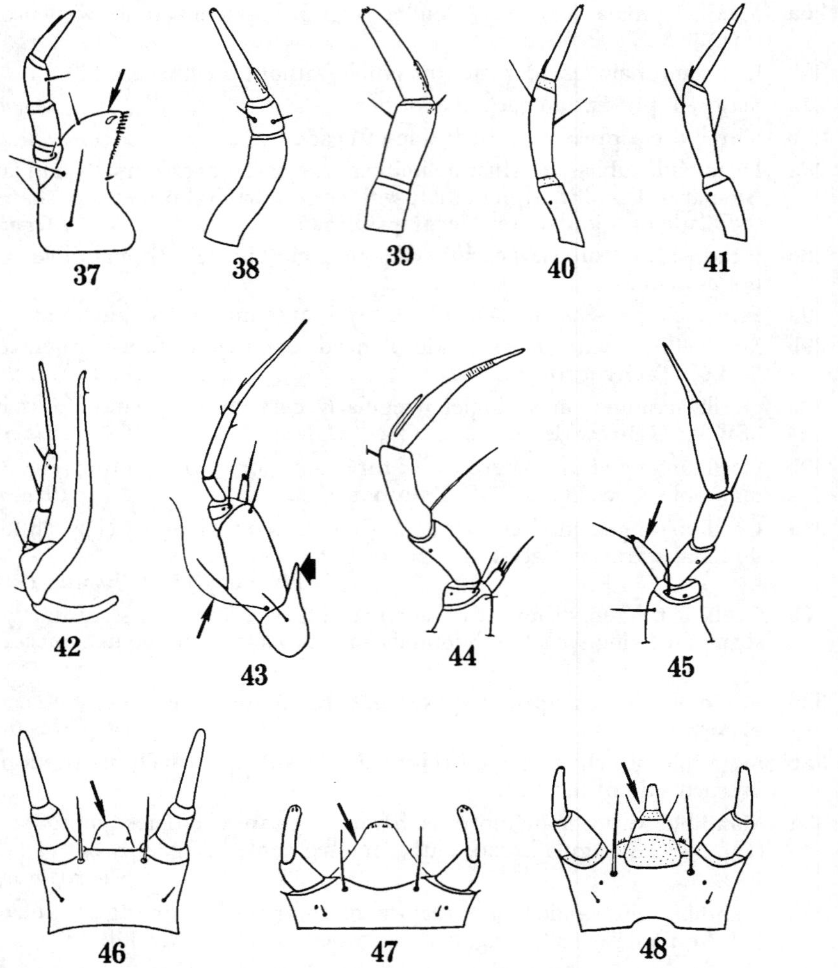

Figs. 38B.37–38B.48. 37. *Xylodromus concinnus* (Marsham), right maxilla, ventral, showing mala (arrow). **38.** *Oxytelus* sp. (Mexico), left maxillary palp, ventral. **39.** *Thoracophorus costalis* (Erichson), left maxillary palp, ventral. **40.** *Platydracus zonatus* (Gravenhorst), right maxillary palp, ventral. **41.** *Philonthus* sp. (Mexico), apex of right maxilla, ventral. **42.** *Megarthrus* sp. (Mexico), right maxilla, ventral. **43.** *Astenus* sp. (Michigan), right maxilla, ventral, showing trichobothrium (slender arrow) and spine on cardo (broad arrow). **44.** *Paederus* sp. (Mexico), apex of right maxilla, ventral. **45.** *Scopaeus* sp. (Mexico), apex of left maxilla, ventral, showing articulated mala (arrow). **46.** *Acidota crenata* (Fabricius), labium, ventral, showing ligula (arrow). **47.** *Coproporus ventriculus* (Say), labium, ventral, showing ligula (arrow). **48.** *Tachyporus californicus* Horn, labium, ventral, showing ligula (arrow). [Figures 37 and 46 redrawn from Steel (1970).]

37b Six ocelli present on each side of head; urogomphus articulated at base . . . 38

38a Antenna longer than mandible; ligula broad and bilobed; head capsule without neck; **Steninae** . . . . ***Stenus***

38b Antenna shorter than mandible; ligula small, acute (Fig. 38B.4); head capsule with weak neck constriction (Fig. 38B.4); **Euaesthetinae** . . . . ***Euaesthetus***

39a Antenna three-segmented (similar to Fig. 38B.6); abdominal terga and sterna each consisting of a single median sclerite; **Pseudopsinae** 40

39b Antenna four-segmented (Figs. 38B.10, 38B.11); abdominal terga and sterna divided along midline by membranous area **41**

40a Urogomphus two-segmented; six distinct ocelli on each side of head . . . . *Nanobius*

40b Urogomphus one-segmented; ocelli on each side of head closely clustered, not countable . . . . *Pseudopsis*

41a Usually five to six ocelli (rarely one) present on each side of head; trichobothria present on stipes (Figs. 38B.2, 38B.43), on head capsule near ocelli (Fig. 38B.2) and on pronotum; prosternal area as in Fig. 38B.50, cervicosternum transverse; **Paederinae**. **42**

41b Usually four, two, or one ocelli present on each side of head; trichobothria absent; prosternal area as in Fig. 38B.51, cervicosternum triangular; **Staphylininae** **50**

42a Penultimate segment of antenna about twice as wide as preceding segment, antenna indistinctly four-segmented (Fig. 38B.10); urogomphus shorter than last abdominal segment, similar to Fig. 38B.57. . . . . *Palaminus*

42b Penultimate segment of antenna about equal in width to preceding segment, antenna four-segmented (Fig. 38B.11); urogomphus longer than last abdominal segment **43**

43a Nasale without teeth, with pair of rounded projections bearing dense setal brush below (Fig. 38B.17); antenna with sensory appendage on outside of penultimate segment (Fig. 38B.11) . . . . *Homaeotarsus, Biocrypta*

43b Nasale with four or more teeth or rounded projections (Figs. 38B.18–38B.22); antenna with sensory appendage on dorsal or inner side of penultimate segment (similar to Figs. 38B.5, 38B.10) **44**

44a Cardo of maxilla with large tooth which is about as long as remainder of cardo (Fig. 38B.43); nasale with four pairs of acute teeth (Fig. 38B.18) *Astenus*

44b Cardo without tooth or with much smaller tooth (Fig. 38B.2); nasale with fewer pairs of acute teeth, or with some of teeth rounded (Figs. 38B.19–38B.22). **45**

45a Apical segment of maxillary and labial palps with transversely ribbed flexible region about one-third the distance from the apex (Fig. 38B.44); mala about half or less as long as first segment of maxillary palp (Fig. 38B.44) . . . . . *Paederus*

45b Apical segment of maxillary and labial palps with (Fig. 38B.45) or without (Fig. 38B.2) a single point of flexation; mala more than half as long as first segment of maxillary palp (Fig. 38B.45) **46**

46a Nasale with three pairs of acute teeth (Fig. 38B.19); mandible with single tooth on inner edge about one-third from base (Fig. 38B.36) . . . *Echiaster*

46b Nasale with rounded teeth or with a median tooth in addition to pairs of acute teeth (Figs. 38B.20–38B.22); mandible with numerous fine teeth on inner edge, or teeth absent. **47**

47a Nasale with a median tooth in addition to three to four pairs of lateral teeth, which may be rounded to acute (Fig. 38B.20). . . . . *Lathrobium*

47b Nasale without median tooth, with three to four pairs of rounded teeth, those of the innermost pair being much wider than long (Figs. 38B.21, 38B.22) **48**

48a Second antennal segment quadrate; nasale with two pairs of teeth lateral to median broadly rounded pair (Fig. 38B.21); last segment of maxillary palp transversely divided more than one-third before apex (Fig. 38B.45) . . . . *Scopaeus*

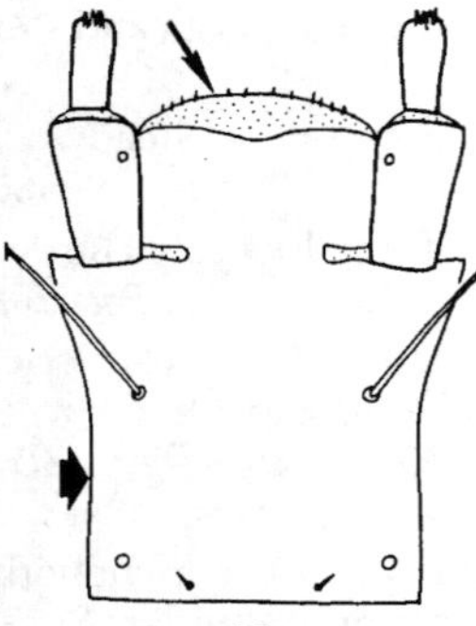

49

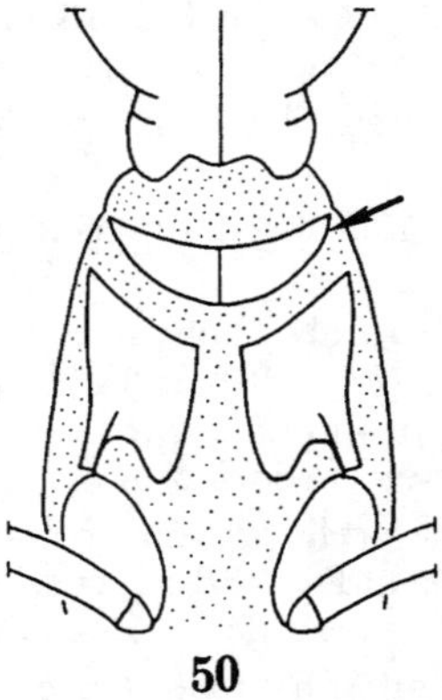

50

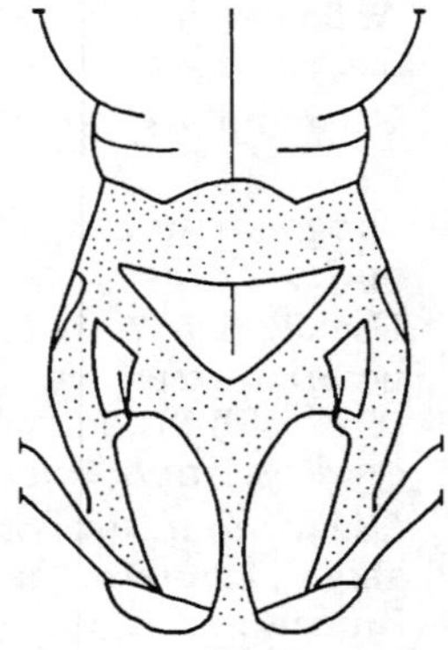

51

52

53

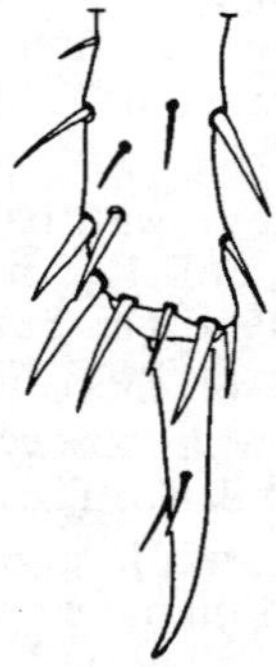

54

55

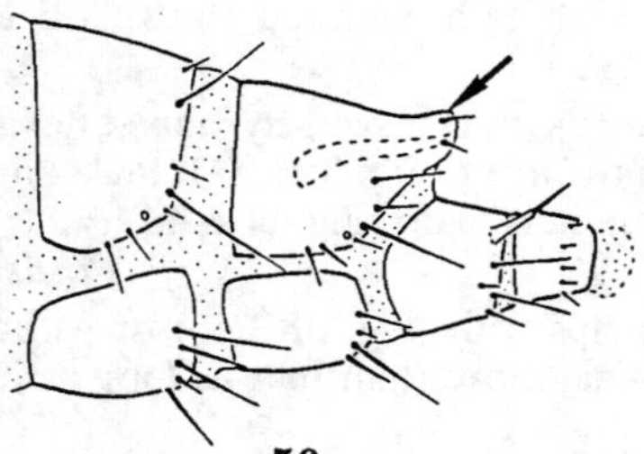

56

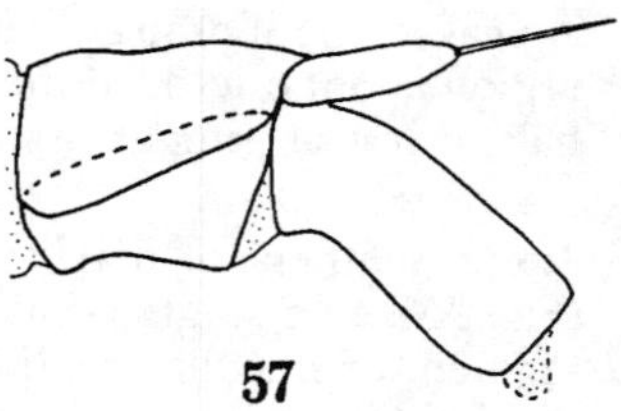

57

Figs. 38B.49–38B.57. 49. *Oxytelus* sp. (Mexico), labium, ventral, showing ligula (slender arrow) and prementum (broad arrow). **50.** *Homaeotarsus* sp. (California), prothorax, ventral, showing cervicosternum (arrow). **51.** *Staphylinus badipes* LeConte, prothorax, ventral. **52.** *Diochus* sp. (Mexico), apex of left proleg, anterior. **53.** *Staphylinus badipes* LeConte, apex of left proleg, anterior. **54.** *Platydracus zonatus* (Gravenhorst), apex of left proleg, anterior. **55.** *Platydracus zonatus* (Gravenhorst), abdominal tergal setae at high magnification. **56.** *Gyrophaena* sp. (Mexico), abdominal segments 7–10, left lateral, showing glandular structure on tergum 8 (arrow). **57.** *Pseudopsis* sp. (California), abdominal segments 9 and 10, left lateral (most setae omitted).

48b Second antennal segment at least twice as long as wide (as in Fig. 38B.11); nasale with three (rarely two) pairs of teeth lateral to median broadly rounded pair (Fig. 38B.22); last segment of maxillary palp with only minute apical subdivision. **49**

49a Ventral ecdysial lines of head ending at about level of tentorial pits . . . . ***Deroderus, Eustilicus, Rugilus, Stilocharis***

49b Ventral ecdysial lines of head extending anterior to level of tentorial pits (Fig. 38B.2) . . . . ***Achenomorphus, Lithocharis, Medon, Sunius***

50a One or two ocelli on each side of head; protibia with single row of simple setae on anterior face (Fig. 38B.52); tarsungulus with only two setae, in longitudinal row or very unequal in size (Fig. 38B.52); vestiture of simple setae only; maxillary palp four-segmented, labial palp three-segmented . . . **51**

50b Four ocelli in close group on each side of head; protibia without comb or, if comb present, it is composed of split setae (Fig. 38B.53) or not arranged in single row; tarsungulus with three setae or, if only two present, these are equal in size and about equidistant from base (Fig. 38B.54); vestiture, especially of dorsum of abdomen, usually with numerous blunt or frayed setae (Fig. 38B.55); maxillary palp of three or four segments, labial palp of two or three segments **52**

51a Two ocelli, large anterior and small posterior, on each side of head; tarsungular setae unequal in size, not in longitudinal row (Fig. 38B.52); **Diochini** . . . . ***Diochus***

51b One ocellus, which may not be pigmented, on each side of head; tarsungular setae equal in size, in longitudinal row on ventral surface. . . . . **Othiini, Xantholinini**

52a Urogomphus shorter than apical abdominal segment (similar to Figs. 38B.1, 38B.57); **Quediini** **53**

52b Urogomphus as long as or longer than apical abdominal segment; **Staphylinini** **54**

53a Protibial comb present, composed of split setae (as in Fig. 38B.53) . . . . . ***Quedius***

53b Protibial comb absent . . . . ***Heterothops***

54a Protibial comb present and composed of split setae (Fig. 38B.53), or split setae covering most of anterior face of protibia . . . . ***Staphylinus***

54b Protibial comb absent, or irregular and composed of a few simple setae . . **55**

55a Protibia without comb, all tibiae broad (Fig. 38B.54); tarsungulus with two setae only (Fig. 38B.54); maxillary palp three-segmented (Fig. 38B.40), labial palp two-segmented. . . . . ***Platydracus***

55b Protibia with or without comb, tibiae more slender; tarsungulus with three setae; maxillary palp four-segmented (Fig. 38B.41), labial palp three-segmented **56**

56a Nasale with seven teeth . . . . ***Gabrius***

56b Nasale with nine teeth (Fig. 38B.23). . . ***Philonthus, Belonuchus, Neobisnius***

References

Ashe, J. S. **1985.** Fecundity, development and natural history of *Meronera venustula* (Erichson) (Coleoptera : Staphylinidae : Aleocharinae). *Psyche (Camb.)* 92:181–204. **1986.**

Structural features and phylogenetic relationships among larvae of genera of gyrophaenine staphylinids (Coleoptera : Staphylinidae : Aleocharinae). *Fieldiana Zool. (N.S.)* No. 30. 60 pp. **Ashe, J. S. and L. E. Watrous. 1984.** Larval chaetotaxy of Aleocharinae (Staphylinidae) based on a description of *Atheta coriaria* Kraatz. *Coleopt. Bull.* 38:165–179. **Boháč, J. 1978.** Description of the larva and pupa of *Thoracophorus brevicristatus* (Coleoptera : Staphylinidae). *Acta Entomol. Bohemoslov.* 75:394–399. **Boller, F. 1983.** Zur Larvalmorphologie der Gattung *Philonthus* Curtis (Coleoptera, Staphylinidae). *Spixiana (Muench.)* 6:113–131. **Böving, A. G. and F. C. Craighead. 1931.** An illustrated synopsis of the principal larval forms of the order Coleoptera. *Entomol. Am. (N.S.)* 11:1–351. **Frania, H. E. 1986.** Larvae of *Eustilicus* Sharp, *Rugilus* Leach, *Deroderus* Sharp, *Stilocharis* Sharp, *Medon* Stephens (Coleoptera : Staphylinidae : Paederinae : Paederini), and their phylogenetic significance. *Can. J. Zool.* 64:2543–2557. **Frank, J. H. 1979.** Larval morphology and the classification of *Staphylinus (sensu lato)* (Col., Staphylinidae). *Entomol. Mon. Mag.* 114:235–238. **1989.** Staphylinidae (Staphylinoidea). In F. W. Stehr (ed.). *Immature Insects,* Vol. 2. Kendall/Hunt, Dubuque, Iowa (in press). **Kasule, F. K. 1966.** The subfamilies of the larvae of Staphylinidae (Coleoptera) with keys to the larvae of the British genera of Steninae and Proteininae. *Trans. R. Entomol. Soc. Lond.* 118(8):261–283. **1968.** The larval characters of some subfamilies of British Staphylinidae (Coleoptera) with keys to the known genera. *Trans. R. Entomol. Soc. Lond.* 120(4):115–138. **1970.** The larvae of Paederinae and Staphylininae (Coleoptera : Staphylinidae) with keys to the known British genera. *Trans. R. Entomol. Soc. Lond.* 122(2):49–80. **Legner, E. F. and I. Moore. 1977.** The larva of *Platystethus spiculus* Erichson (Coleoptera : Staphylinidae) and its occurrence in bovine feces in irrigated pastures. *Psyche (Camb.)* 84(2):158–164. **Lesage, L. 1977.** Stades immatures de staphylins. I. La larve et la nymphe de *Staphylinus cinnamopterus* Gravenhorst (Coleoptera : Staphylinidae). *Nat. Can. (Que.)* 104:235–238. **Mank, H. G. 1923.** The biology of the Staphylinidae. *Ann. Entomol. Soc. Am.* 16:220–237. **Martin, J. E. H. 1977.** *The Insects and Arachnids of Canada.* 1. *Collecting, Preparing and Preserving Insects, Mites and Spiders.* Publ. 1643. Agriculture Canada, Ottawa, Ontario, Canada. 182 pp. **Moore, I. and E. F. Legner. 1973.** The larva and pupa of *Carpelimus debilis* Casey (Coleoptera : Staphylinidae). *Psyche (Camb.)* 80(4):289–294. **1974.** A catalogue of the taxonomy, biology and ecology of the developmental stages of the Staphylinidae (Coleoptera) of America north of Mexico. *J. Kans. Entomol. Soc.* 47(4):469–478. **Newton, A. F., Jr. 1982.** A new genus and species of Oxytelinae from Australia, with a description of its larva, systematic position, and phylogenetic relationships (Coleoptera : Staphylinidae). *Am. Mus. Novit.* 2744:1–24. **1984.** Mycophagy in Staphylinoidea (Coleoptera). Pp. 302–353 in Q. Wheeler and M. Blackwell (eds.). *Fungus/Insect Relationships. Perspectives in Ecology and Evolution.* Columbia University Press, New York. **1989.** Micropeplidae (Staphylinoidea). In F. W. Stehr (ed.). *Immature Insects,* Vol. 2. Kendall/Hunt, Dubuque, Iowa (in press). **Orth, R. S., I. Moore, T. W. Fisher and E. F. Legner. 1975.** Biological notes on *Ocypus olens,* a predator of brown garden snail, with descriptions of the larva and pupa (Coleoptera : Staphylinidae). *Psyche (Camb.)* 82(3–4):292–298. **Paulian, R. 1941.** Les premiers états des Staphylinoidea [Coleoptera]. Étude de morphologie comparée. *Mem. Mus. Natl. Hist. Nat. (Paris) (N.S.)* 15:361 pp. **Pototskaya, V. A. 1967.** *Classification Key of the Larvae of* Staphylinidae *in the European Part of the USSR.* Academy of Sciences, USSR, A. N. Severtsov Institute of Evolutionary Morphology and Ecology of Animals, Moscow. 120 pp. (in Russian). **Silvestri, F. 1945.** Descrizióne e Biología del Coleottero Stafilinide *Belonuchus formosus* Grav. introdotto in Italia per la lòtta cóntro Ditteri Tripaneidi. *Boll. Lab. Entomol. Agrar. Portici* 5:312–326. **Steel, W. O. 1966.** A revision of the staphylinid subfamily Proteininae (Coleoptera). I. *Trans. R. Entomol. Soc. Lond.* 118(9):285–311. **1970.** The larvae of the genera of the Omaliinae (Coleoptera : Staphylinidae) with particular reference to the British fauna. *Trans. R. Entomol. Soc. Lond.* 122(1):1–47. **Stehr, F. W. 1987.** *Immature Insects.* Kendall/Hunt, Dubuque, Iowa. 754 pp. **Topp, W. 1978.** Bestimmungstabelle für die Larven der Staphylinidae. In B. Klausnitzer (ed.). *Ordnung Coleoptera (Larven).* Bestimmungsbücher zur Bodenfauna Europas, Lfg. 10. W. Junk, The Hague, The Netherlands. 378 pp. **Verhoeff, K. W. 1919.** Studien über die Organisation und Biologie der Staphylinoidea. IV. Zur Kenntnis der Staphyliniden-Larven. V. Zur Kenntnis der Oxyteliden-Larven. *Arch. Naturgesch. Abt. A* 85(6):1–111. **Voris, R. 1939.** The immature stages of the genera *Ontholestes, Creophilus* and *Staphylinus.* Staphylinidae (Coleoptera). *Ann. Entomol. Soc. Am.* 32(2):288–303. **Watrous, L. E. 1981.** Studies of *Lathrobium (Lobrathium):* Revision of the *grande* species group (Coleoptera : Staphylinidae). *Ann. Entomol. Soc. Am.* 74:144–150.

39

Insecta: Coleoptera Pselaphidae†

DONALD S. CHANDLER

Pselaphidae generally may be found in any situation where litter, low vegetation, or loose bark maintain a zone of high humidity. Species are most diverse and also most commonly collected in forest leaf or wood litter, but a few species may be abundant in certain grassland and desert habitats. The distribution of the family in North America reflects this dependence on moist leaf litter. Of the over 700 species, nearly half occur in or near the region of the Appalachian Mountains, with a second major assemblage in California and the Pacific northwest. There are relatively few species known from the Rocky Mountain and plains states, although this may be largely due to lack of adequate sampling.

The family is closely related to the Staphylinidae and shares many of their distinguishing characters, with one genus being notably staphyliniform (*Mayetia*). Pselaphidae can usually be recognized by a combination of the following characters: antennae with apical antennomere(s) enlarged, body generally stouter and more rigid, two to three tarsomeres, and presence of vertexal foveae.

Despite the large number of species, pselaphids are usually rarely encountered in collections. This is undoubtedly due in part to the small size of the species (0.7–4.5 mm) and to their often obscure habitats. A few of the larger species may be seen beneath bark or stones, but the majority of species are detected only by processing litter with specalized equipment, usually a Berlese funnel. Despite collecting difficulties, the family has a sound taxonomic basis and about half of the 98 genera have been revised within the past 30 years.

† Scientific Contribution 1268 of the New Hampshire Agricultural Experiment Station.

Biology

All species have well-developed mandibles and are predaceous on invertebrates such as annelids, Collembola, insect larvae, mites, and Symphyla (Engelmann 1956, Barr 1974). Engelmann (1956) and Reichle (1969) demonstrated differences in prey choice and feeding behavior between adults of several free-living species, and de Marzo (1986) documented the feeding behavior of the larva of a free-living species. The Clavigerinae and some species scattered through several tribes are obligate myrmecophiles and have specialized feeding habits. Species of Clavigerinae feed on living and dead ant larvae, and also are fed by trophallaxis (Park 1932, Akre and Hill 1973). A fair number of other species are synoeketes in ant nests, particularly in the northern latitudes.

The biological information known for most species usually consists of notes on the type of litter or the ant host with which the adults were collected. Only 12 larvae have been described for the world, three of these from North America (Böving and Craighead 1931, Wagner 1975). Eggs have not been observed and the pupae of two European species have been described (Besuchet 1956). The only note on larval habitat in North America is by Wagner (1975), who observed that larvae of *Euplectus confluens* LeConte are infrequently collected with the adults in tree-hole material.

The typical adult habitat is moist leaf or wood litter of the F layer. However, a few species or genera are most frequent in the L or H layers, with one genus (*Mayetia*) being primarily in mineral soil or in grass sods. Type of wood or leaf litter may influence the presence of beetle species (Chandler 1987), either through inherent chemical properties or differences in moisture retention.

Reproduction

Both sexes are known for all but a few species, and it appears that the typical reproductive pattern is heterosexual. In the only paper that deals with the pselaphid life-cycle (Park et al. 1953), observations are based entirely on the appearance of recently emerged adults, which are light in color. In this study of an Illinois prairie population, the authors found that there were two peaks in the emergence of teneral adults, one in the spring and the other in the fall. Individuals of five species at this site were found to have mature sperm and were observed copulating during the time of the spring peak (Mickey and Park 1956). The diploid number of chromosomes was found by Mickey and Park to be 28 for *Batrisodes globosus* (LeConte) and 30 for *Pselaphus fustifer* Casey.

Collection and Technique

The small size and obscure habitats of pselaphids render it necessary to use specialized collecting techniques. The preferred procedure is to select litter samples and process them with a Berlese funnel, since Pselaphidae are usually present in any litter in low numbers. Many collectors prefer to sift the litter through ¼-in. (about 1 cm^2) hardward cloth into a bag to decrease the volume of the sample, increase arthropod concentration, and sample from a larger area.

The principles and basic design of the Berlese funnel are presented in Borror et al. (1981). The usual design differences for beetle collectors are directed toward increasing the screen surface area to accommodate larger samples. Specimens should be collected in 70% alcohol or Barber's fluid [formula in Borror et al. (1981)].

Other species may be found beneath bark of dead trees or beneath partially embedded stones. Many species whose habitats are poorly known are attracted to white and ultraviolet lights.

Identification

The Pselaphidae are readily distinguished from all other families except the Staphylinidae by the five visible abdominal segments posterior to the abbreviated elytra. Pselaphidae are distinguished from the Staphylinidae by a combination of the following characters: enlarged apical antennomere(s), three or fewer tarsomeres, presence of vertexal foveae, and lack of ocelli.

Males are necessary for identification of most species, and even necessary to recognize a few genera. Males can be readily recognized with some practice by the presence of an additional sternite or a modified terminal sternite (flattened, concave). Genera can often be characterized by the location of the male cuticular modifications: projections of the tibiae, femora and trochanters; excavations and projections of the abdomen and head; and swellings, foveae, and projections of the antennomeres. Females either lack these modifications or they are feebly developed.

Species over 1 mm can usually be identified when glued to an insect point. Care must be taken to keep one side and the center free of glue, since key foveae and other characters can be easily obscured. Smaller species should be mounted on slides for identification. To prepare a slide mount, the specimens should first be cleared (5% KOH followed by 10% acetic acid), disarticulated in glycerine or alcohol, and examined in glycerine on a slide. After identification the beetle should be placed in a permanent mounting medium, such as Canada balsam, with small props of a broken cover glass inserted in the medium beneath the cover glass. Specimens may also be stored in a genitalia vial in glycerine.

External Anatomy

The body of a pselaphid is heavily sclerotized and elongate to ovoidal. The color varies from yellow to black, and may be strikingly bicolored. The pubescence is usually aciculate, but may be capitate, scalelike, elongate, and split (trichomes) or rarely absent from most of the body. The head is quadrate to triangular, abruptly constricted at the base to form a neck. Three to 11 antennomeres are present, usually with the apical one to three enlarged. Vertexal foveae are usually present. The maxillary palpi have three or four segments, with the first very small and the last enlarged.

The pronotum is narrower than the abdomen and constricted basally. The disk is flat to convex, often with antebasal sulci or foveae. The elytra are short, with the metawings usually present. The abdomen is often slightly constricted at the base, with five to seven segments visible. The legs have three tarsomeres, rarely reduced to two.

Acknowledgments

I thank Mr. Robert O. Schuster, Dr. John F. Burger, and Dr. R. Marcel Reeves for their valuable reviews and suggestions.

KEY TO THE NORTH AMERICAN GENERA OF PSELAPHIDAE

Both males and females may be identified except for the few genera for which males are required. Specimens mounted on points should be identifiable at a maximum of 72×.

1a Three antennomeres, first difficult to see, third much longer than other two together (Fig. 39.2); **Clavigerinae** **2**

1b Nine to 11 antennomeres, last antennomere at most as long as remaining segments combined **3**

2a (1) Eyes present; **Fustigerini** (2 species) . . . . ***Fustiger***

2b Eyes absent; **Adraniini** (6 species; Wickham 1901) . . . . . . ***Adranes***

3a (1) Mesotrochanters long, junction with femora less than one-half trochanter length (Fig. 39.3); **Pselaphinae** **4**

3b Mesotrochanters short, junction with femora more than one-half trochanter length (Fig. 39.1) **18**

4a (3) Basal two tarsomeres short, only distal tarsomere relatively long (Fig. 39.4), southwestern United States; **Arhytodini** (4 species; Chandler 1976). . . . . ***Caccoplectus***

4b Basal tarsomere short, distal two tarsomeres relatively long (Fig. 39.5). **5**

5a (4) Third segment of maxillary palpi with tuft of setae projecting from outer angle (Figs. 39.6–39.8) **6**

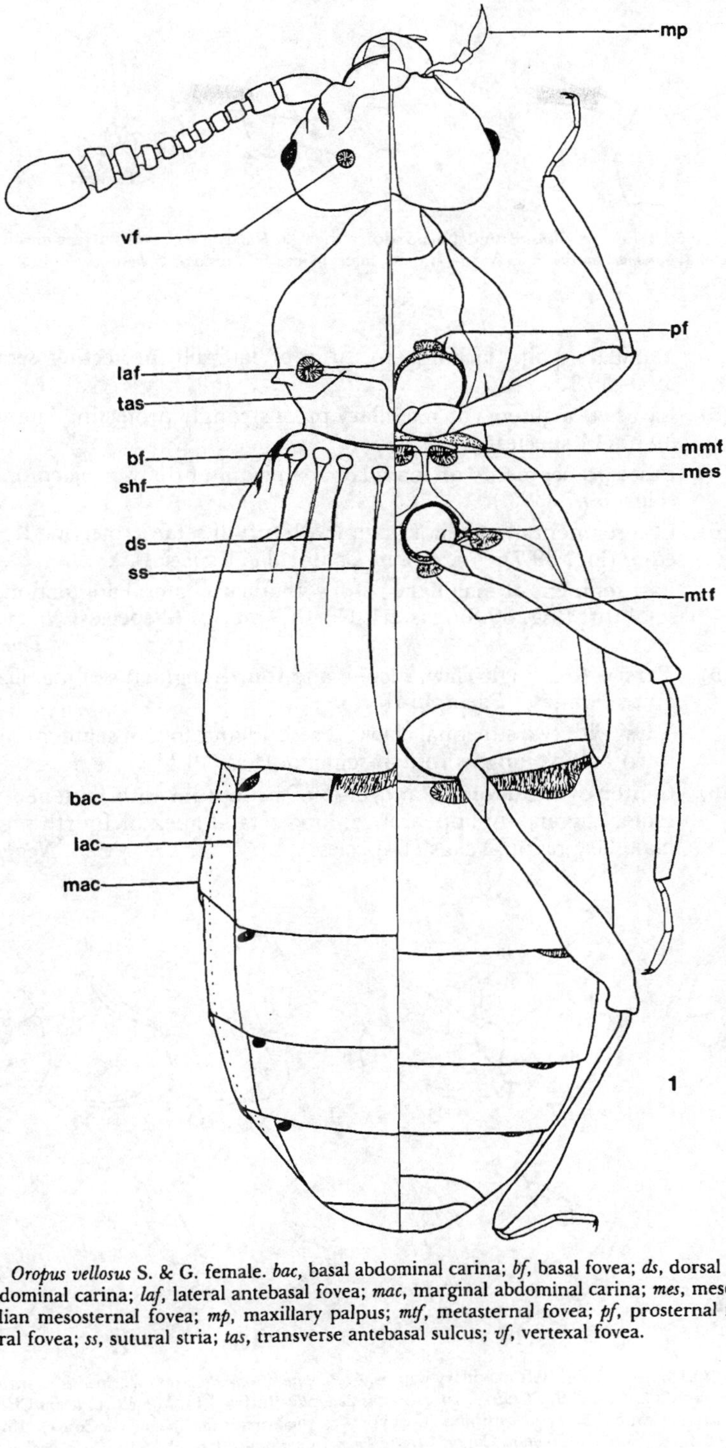

Fig. 39.1. *Oropus vellosus* S. & G. female. *bac*, basal abdominal carina; *bf*, basal fovea; *ds*, dorsal striae; *lac*, lateral abdominal carina; *laf*, lateral antebasal fovea; *mac*, marginal abdominal carina; *mes*, mesosternum; *mmf*, median mesosternal fovea; *mp*, maxillary palpus; *mtf*, metasternal fovea; *pf*, prosternal fovea; *shf*, subhumeral fovea; *ss*, sutural stria; *tas*, transverse antebasal sulcus; *vf*, vertexal fovea.

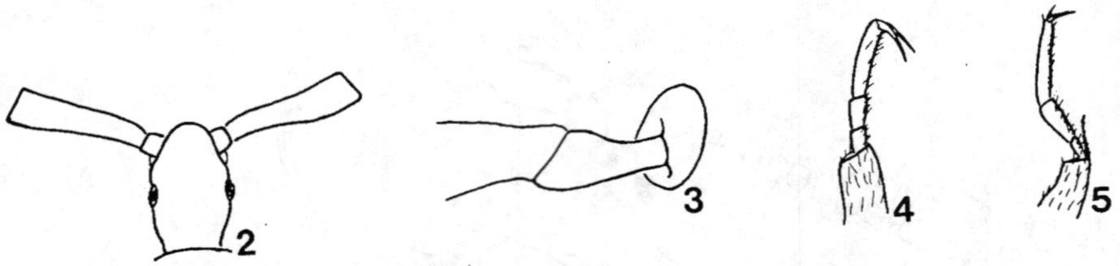

Figs. 39.2–39.5. **2.** *Fustiger fuchsi* Brendel head, dorsal view. **3.** *Hamotus nodicollis* Raffray mesotrochanter, ventral view. **4.** *Sonoma hespera* P. & W. metatarsus, lateral view. **5.** *Batrisodes globosus* (LeConte) metatarsus, lateral view.

5b Maxillary palpi lacking any tufts of laterally projecting setae (Figs. 39.9–39.13) 8

6a (5) Last two segments of maxillary palpi strongly projecting laterally (Fig. 39.6) (14 species) . . . . ***Pilopius***

6b Last two segments of maxillary palpi only briefly projecting laterally (Figs. 39.7–39.8). 7

7a (6) Last segment of maxillary palpi with tuft of setae projecting from outer edge (Fig. 39.7); Arizona to California (1 species) . . . . . . ***Ctenisis***

7b Last segment of maxillary palpi with abrupt lateral angulation, lacking setal tuft (Fig. 39.8); eastern North America (2 species). . . . . ***Tmesiphorus***

8a (5) Tarsi with a single claw; second and fourth segments of maxillary palpi quite elongate; **Pselaphini**. 9

8b Tarsi with two subequal claws; if second and fourth segments elongate, third half as long as fourth segment (Fig. 39.11). 10

9a (8) Center of head vertex and base of pronotum with flattened scalelike setae, "sugary" in appearance; long seta at apex of fourth segment of maxillary palpi; Texas (1 species) . . . . ***Neopselaphus***

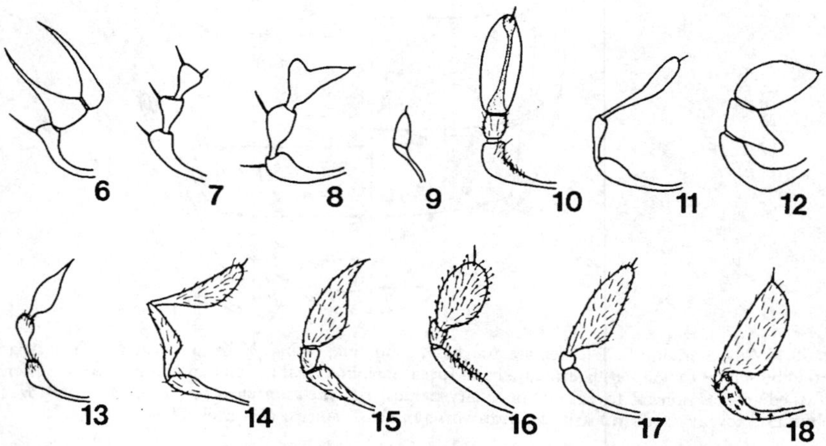

Figs. 39.6–39.18. Dorsal view left maxillary palpus. **6.** *Pilopius ocularis* (Casey). **7.** *Ctenisis raffrayi* Casey. **8.** *Tmesiphorus costalis* LeConte. **9.** *Biotus* sp. **10.** *Hamotus nodicollis* Raffray. **11.** *Mipseltyrus levini* Chandler. **12.** *Cedius ziegleri* LeConte. **13.** *Tyrus corticinus* (Casey). **14.** *Cylindrarctus longipalpis* (LeConte). **15.** *Batrisodes globosus* (LeConte). **16.** *Valda frontalis* Casey. **17.** *Machaerodes carinatus* (Brendel). **18.** *Pselaptrichus rothi* Park.

9b Head and pronotal base glabrous, shining, lacking setae, seta at apex of fourth maxillary palpi short, not easily visible (5 species) . . ***Pselaphus***

10a **(8)** At least head with flattened scalelike setae in foveae, sulci, and at head base, setae "sugary" in appearance; maxillary palpi small, three-segmented (Fig. 39.9) **11**

10b Body with only aciculate setae; maxillary palpi variable **12**

11a **(10)** Metatibiae flattened throughout length; Arizona to California (1 species) . . . . ***Biotus***

11b Metatibiae swollen near apex, cylindrical throughout length; southern United States (2 species) . . . . ***Atinus***

12a **(10)** Last antennomere barely larger than others, not forming a distinct club (Fig. 39.19); eastern North America; **Ceophyllini** (1 species) . . . . . ***Ceophyllus***

12b With distinct antennal club of one to three antennomeres; **Tyrini** . . **13**

13a **(12)** Last segment of maxillary palpi enlarged, with mesal longitudinal sulcus (Fig. 39.10) **14**

13b Last segment of maxillary palpi not much wider than penultimate segment, surface without sulcus (Figs. 39.11–39.13) **15**

14a **(13)** Antennal club consisting of last antennomere; eastern North America (1 species) . . . . ***Circocerus***

14b Antennal club consisting of last three antennomeres; southern United States (4 species; Chandler 1974) . . . . ***Hamotus***

15a **(13)** Pronotum with median tubercle at base, lacking antebasal foveae; Arizona (1 species) . . . . ***Anitra***

15b Pronotal base lacking tubercle, with antebasal foveae present . . . **16**

16a **(15)** Last three segments of maxillary palpi elongate (Fig. 39.11), together much longer than head (4 species) . . . . ***Mipseltyrus***

16b Segments of maxillary palpi not so elongate (Figs. 39.12–39.13), together much shorter than head **17**

17a **(16)** Maxillary palpi with third segment strongly projecting medially (Fig. 39.12); eastern North America (4 species; Park 1949) . . . . . ***Cedius***

17b Maxillary palpi with third segment elongate (Fig. 39.13) (5 species) . . ***Tyrus***

18a **(3)** Two or three tarsomeres, distal tarsomere longer than basal tarsomere(s) (Fig. 39.4); **Faroninae** **19**

18b Three tarsomeres, distal two much longer than basal tarsomere (Fig. 39.5) **22**

19a **(18)** Maxillary palpi resting in large glabrous cavity below antennal insertions; nine antennomeres; Arizona; **Pyxidicerini** (1 species) . . . . . ***Bythinoplectus***

19b Lacking glabrous cavity below antennal insertions; 10–11 antennomeres **20**

20a **(19)** Two tarsomeres; eyes absent; body elongate and slender, less than 1 mm; **Mayetiini** (18 species; Schuster et al. 1959, 1960) . . . . ***Mayetia***

20b Three tarsomeres; eyes present; body larger; **Faronini** **21**

21a **(20)** Antennal tubercles separated by sulcus extending anteriorly from frontal fovea (Fig. 39.20); Pacific Coast States (4 species; Schuster and Marsh 1958b) . . . . ***Megarafonus***

21b Area between antennal tubercles concave, but clearly not a sulcus extending from frontal fovea (Fig. 39.21) (25 species; Marsh and Schuster 1962) . . . . ***Sonoma***

22a (18) Abdomen rounded laterally, margins incomplete, marginal carinae present only on first tergite, lateral carinae may be present on first four tergites (Fig. 39.22); **Batrisinae** **23**

22b Abdomen angulate laterally, margins distinct, marginal and lateral carinae present on first four tergites (Fig. 39.1) **27**

23a (22) Head with cuticular tubercle in place of eyes **24**

23b Head with eyes present **25**

24a (23) With spine in place of eyes; eastern North America (31 species; Barr 1974) . . . . ***Arianops***

24b With knob in place of eyes; Texas (1 species) . . . . . . ***Texamaurops***

25a (23) Pronotum smooth, lacking tubercles; eastern North America and Arizona (5 species) . . . . ***Arthmius***

25b Pronotum with lateral tubercles adjacent to median antebasal fovea **26**

26a (25) Metatibiae with long apical spur of coalesced setae (Fig. 39.5) (74 species; Park 1947, 1956; Grigarick and Schuster 1962b) . . . ***Batrisodes***

26b Metatibiae with apical setae short and free, not forming a spur; eastern North America (14 species; Park 1965). . . . . ***Batriasymmodes***

27a (22) With long exposed neck obscured by ruff of long setae (Fig. 39.23); **Speleobamini** **28**

27b Neck short, without long obscuring setae (Fig. 39.1) **29**

28a (27) With eyes; foveae present on vertex, pronotum, and elytra; southeastern United States (2 species; Park 1956) . . . . ***Prespelea***

28b Lacking eyes; vertex, pronotum, and elytra lacking foveae; Alabama (1 species) . . . . ***Speleobama***

29a (27) Maxillary palpi with third segment elongate, more than half as long as fourth segment, inner margin swollen medially (Fig. 39.14); **Tychini**. **30**

29b Maxillary palpi with third segment less than half the length of fourth segment (in ***Basolum*** same length, but small, not projecting on inner margin) (Figs. 39.15–39.18) **34**

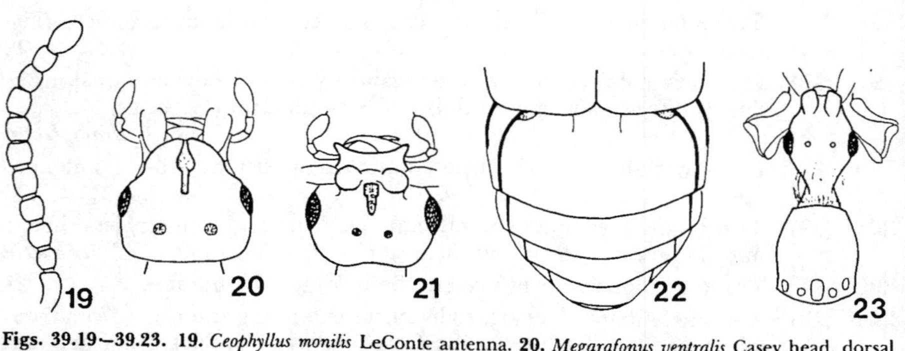

Figs. 39.19–39.23. 19. *Ceophyllus monilis* LeConte antenna. **20.** *Megarafonus ventralis* Casey head, dorsal view. **21.** *Sonoma hespera* P. & W. head, dorsal view. **22.** *Batrisodes globosus* (LeConte) abdomen, dorsal view. **23.** *Prespelea copelandi* Park head, dorsal view.

30a (29) Vertexal foveae within one eye facet diameter of eye margin (5 species) *Cylindrarctus*

30b Vertexal foveae four to six eye facet diameters from eye margin . **31**

31a (30) Penultimate segment of maxillary palpi sharply angulate on inner margin; eastern North America (7 species) . . . . *Custotychus*

31b Penultimate segment of maxillary palpi smoothly rounded on inner margin; all but one species from western North America **32**

32a (31) Male genitalia asymmetrical (13 species; Schuster and Marsh 1958a) . *Hesperotychus*

32b Male genitalia symmetrical **33**

33a (32) Male metasternum simple (13 species; Grigarick and Schuster 1962a) *Hylotychus*

33b Male metasternum with median tubercle (6 species; Park and Wagner 1962) . . . . *Lucifotychus*

34a (29) Last segment of maxillary palpi swollen, with erect capitate setae over surface (Fig. 39.16); California; **Valdini** (1 species) . . . . *Valda*

34b Maxillary palpi of various shapes, but lacking any obviously modified setae, all of them cylindrical **35**

35a (34) Maxillary palpi with fourth segment enlarged, three or more times wider than third (Figs. 39.17–39.18); first antennomere elongate, at least as long as next three together **Bythinini** **36**

35b Maxillary palpi with fourth segment not so enlarged (Fig. 39.1); first antennomere rarely as long as next three **40**

36a (35) Eyes absent. **37**

36b Eyes present **38**

37a (36) Pronotum with antebasal sulcus lightly impressed or absent; sutural stria of elytra distinct at least basally; southeastern United States (3 species, Park 1951). . . . . *Speleochus*

37b Pronotum with antebasal sulcus distinct; sutural stria of elytra lacking; southeastern United States (3 species; Park 1951) . . . . *Subterrochus*

38a (36) Maxillary palpi with second segment smooth (Fig. 39.17); eastern North America (1 species) . . . . *Machaerodes*

38b Second segment of maxillary palpi studded with tubercles (Fig. 39.18) **39**

39a (38) Pronotum with antebasal sulcus evenly arcuate between lateral foveae; male genitalia asymmetrical; eastern North America (5 species; Park 1953) . . . . *Tychobythinus*

39b Pronotum with antebasal sulcus angulate at middle; male genitalia symmetrical; western North America (37 species; Schuster and Marsh 1956) . . . . *Pselaptrichus*

40a (35) Metacoxae flat, usually widely separated, not conically produced over sternite at junction with trochanters (Fig. 39.24); **Brachyglutini** . . **41**

40b Metacoxae conically produced at junction with trochanters, projecting out over sternite, usually contiguous to narrowly separated (Fig. 39.25); **Euplectinae** **51**

41a (40) Body glabrous and polished; metacoxae narrowly separated; eastern United States (2 species; Park et al. 1976) . . . . *Eupsenius*

41b Body with setae present over surface; metacoxae widely separated **42**

42a (41) Elytra with distinct discal stria extending apically from basal foveae **43**

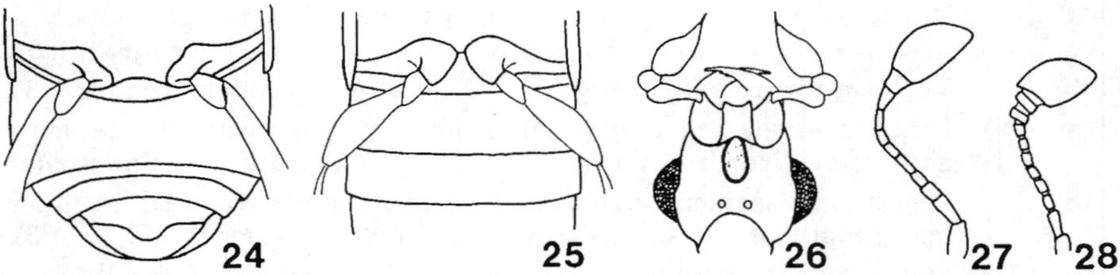

Figs. 39.24–39.28. 24. *Reichenbachia snowi* Fletcher abdomen, ventral view. **25.** *Euplectus acomanus* Casey abdomen, ventral view. **26.** *Decarthron abnorme* (LeConte) head, ventral view. **27.** *Pselaptus belfragei* LeConte antenna. **28.** *Scalenarthrus horni* LeConte antenna.

42b Elytra lacking discal striae 48

43a (42) Pronotum with deep transverse antebasal sulcus between antebasal foveae (13 species; Fall 1927) . . . . ***Rybaxis***

43b Pronotum lacking antebasal sulcus between foveae. 44

44a (43) Head lacking vertexal foveae; eastern North America. . . . . ***Nisaxis***

44b Head with vertexal foveae 45

45a (44) Pronotum with antebasal foveae subequal, large and pubescent (22 species) . . . . ***Brachygluta***

45b Pronotum with nude median antebasal fovea 46

46a (45) Pronotum with lateral antebasal foveae indistinct, nude; head venter with median oval fossa set off by carinoid edges (Fig. 39.26); 10 antennomeres (27 species; Park 1958). . . . . ***Decarthron***

46b Pronotum with lateral antebasal foveae large and pubescent; head venter with median carinoid ridge; eleven antennomeres 47

47a (46) Pronotum medially convex; first tergite with basal carinae (76 species; Grigarick and Schuster 1967) . . . . ***Reichenbachia***

47b Pronotum with median longitudinal sulcus; first tergite lacking basal carinae; Florida (1 species) . . . . ***Briaraxis***

48a (42) Pronotum constricted just before base, this area roughened laterally 49

48b Pronotum straight from lateral angles to base, lacking constriction, basal area with texture similar to pronotal disc. 50

49a (48) Elytra with three to four basal foveae; pronotum with median antebasal fovea distinct (1 species) . . . . ***Eutrichites***

49b Elytra with two basal foveae hidden by pronotal base; pronotum with median antebasal fovea small to almost absent; eastern North America (3 species) . . . . ***Anchylarthron***

50a (48) Last two antennomeres straight (Fig. 39.27); mesosternum strongly carinate between coxae; Texas to California (1 species) . . . ***Pselaptus***

50b Last antennomere meeting tenth at oblique angle (Fig. 39.28); mesosternum flat between coxae; Arizona to California (1 species). . . . . ***Scalenarthrus***

51a (40) Head with prominent frontal tubercle, antennae inserted near apex (Fig. 39.29). 52

51b Head with front flattened, lacking prominent frontal tubercle (Fig. 39.1) 53

52a (51) Pronotal disc with deep median longitudinal sulcus; Florida (1 species) ***Rhinoscepsis***

52b Pronotal disc flattened, with at most a feeble longitudinal sulcus; California (1 species) . . . . ***Morius***

53a (51) Antennae geniculate, first antennomere about half as long as remaining antennomeres combined; eastern United States (Fig. 39.30) (8 species) . . . . ***Rhexius***

53b Antennae straight, first antennomere rarely longer than next two together (Fig. 39.1) **54**

54a (53) Metacoxae widely separated by one-half metasternal length; northeastern United States (1 species) . . . . ***Trichonyx***

54b Metacoxae contiguous to separated by one-fourth metasternal length **55**

55a (54) Prosternum with median longitudinal carina **56**

55b Prosternum flat or with smooth raised median area **61**

56a (55) Pronotum with lateral antebasal foveae free, lacking antebasal sulcus; northeastern United States (1 species) . . . . ***Bibloporus***

56b Pronotum with antebasal sulcus connecting foveae **57**

57a (56) Elytra lacking subhumeral fovea; Pacific Northwest (1 species) . . . . ***Abdiunguis***

57b Elytra with subhumeral fovea **58**

58a (57) Mesocoxal cavities separated by greater than one-fourth cavity width; eastern North America (2 species) . . . . ***Thesium***

58b Mesocoxal cavities contiguous **59**

59a (58) Pronotum with sharply defined median sulcus from base to antebasal sulcus; Pacific northwest (1 species) . . . . ***Hatchia***

59b Pronotum with thin median carina from base to antebasal sulcus . **60**

60a (59) Head with nude vertexal foveae; western North America (2 species; Park and Schuster 1955) . . . . ***Trisignis***

60b Head with pubescent vertexal foveae; eastern North America (5 species; Park 1956) . . . . ***Eutyphlus***

61a (55) Elytra lacking subhumeral fovea **62**

61b Elytra with subhumeral fovea **71**

62a (61) Pronotum with sub-basal tooth on lateral margins; Pacific coast states (3 species) . . . . ***Tetrascapha***

62b Pronotum with sub-basal margins smooth, lacking tooth; eastern North America and Arizona **63**

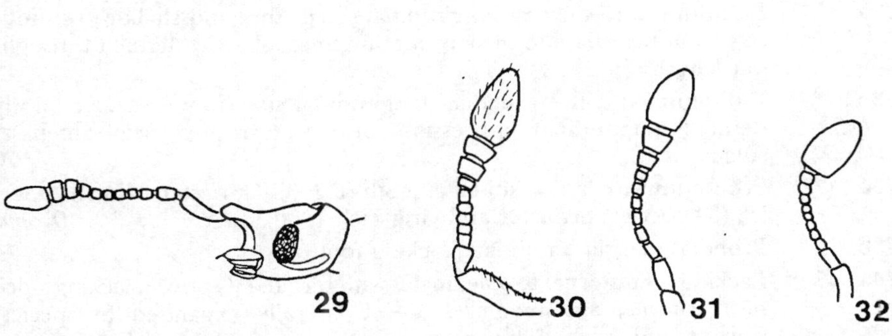

Figs. 39.29–39.32. **29.** *Rhinoscepsis bistriatus* LeConte head, lateral view. **30.** *Rhexius* sp. antenna. **31.** *Trimioplectus obsoletus* Brendel antenna. **32.** *Melba maja* Brendel antenna.

63a **(62)** Elytra with three antebasal foveae **64**
63b Elytra with two antebasal foveae **65**
64a **(63)** Head with eyes easily visible in dorsal view, distinct impression between vertexal foveae; lacking metasternal foveae; Pennsylvania (2 species; Bowman 1934) . . . . ***Trigonoplectus***
64b Head with eyes barely visible in dorsal view, convex between vertexal foveae; metasternal foveae present; Arizona (1 species; Chandler 1985) . . . . ***Trimioarcus***
65a **(63)** Last antennomere at most as long as previous four combined (Fig. 39.31); eastern North America (2 species; Park 1949). . ***Trimioplectus***
65b Last antennomere at least as long as previous six combined (Fig. 39.32) **66**
66a **(65)** Body conspicuously granulate; tergites I–III with basal carinae; Arizona (2 species; Chandler 1985) . . . . ***Zonaira***
66b Body smooth, shining; basal carinae only on tergite I **67**
67a **(66)** Last segment of maxillary palpi about as large as third; eastern North America (2 species). . . . . ***Basolum***
67b Last segment of maxillary palpi obviously larger than third **68**
68a **(67)** Tenth antennomere asymmetrical, triangular in lateral view; body 0.6–0.7 mm long; eastern North America (1 species). . . . . . ***Dalmosella***
68b Tenth antennomere symmetrical, trapezoidal to rectangular; body 0.8 mm or longer **69**
69a **(68)** First tergite half again as long as second; males with tubercle on vertex behind antennal insertions; eastern North America (3 species) . . . . ***Trimiomelba***
69b First two tergites subequal in length; males with vertex smooth, lacking tubercle **70**
70a **(69)** Metasternal foveae large, easily visible; male mesofemora inflated; eastern North America and Arizona (12 species) . . . . ***Melba***
70b Metasternal foveae small or lacking, difficult to see; male mesofemora as wide as profemora; Arizona (2 species; Chandler (1985) ***Allotrimium***
71a **(61)** Pronotal disk with faint to distinct median longitudinal sulcus or a distinct discal fovea (in ***Oropodes***, ***Euplecturga***, and ***Foveoscapha*** this is faint but distinct in almost all species, these genera only in Pacific coast states) **72**
71b Pronotal disk convex to flattened. **83**
72a **(71)** Pronotum with median longitudinal sulcus thin and shallow, continuous from near base to near pronotum apex, sharply defined throughout length (Fig. 39.1). **73**
72b Pronotum with deep median longitudinal sulcus or fovea, or faintly defined longitudinal impression, often interrupted near antebasal sulcus **76**
73a **(72)** Pronotum with lateral subbasal tooth (Fig. 39.1); western North America (34 species; Schuster and Grigarick 1960) . . . . ***Oropus***
73b Pronotal margins crenulate, lacking tooth **74**
74a **(73)** Lacking metasternal foveae; males with tergite IV convex, lacking sulci, head simple, antennomeres V–VI laterally expanded (4 species; Carlton and Allen 1986). . . . . ***Euboarhexius***
74b Two metasternal foveae; male antennomeres normal, not expanded **75**

75a (74) Tergite I with basal carinae; males with vertex modified behind antennal insertions, tergite IV convex, lacking sulci; eastern North America (5 species; Carlton 1983) . . . . *Conoplectus*

75b Tergite I lacking basal carinae; males with vertex simple, tergite IV with basolateral sulci; California (9 species; Schuster and Grigarick 1962) . . . . *Rhexidius*

76a (72) Anterior prosternal foveae present (Fig. 39.33) **77**

76b Anterior prosternal foveae lacking, ... **78**

77a (76) Head with nude vertexal foveae; labral apex concave to convex (13 species; Wagner 1975) . . . . *Euplectus*

77b Head with pubescent vertexal foveae; labral apex with U-shaped notch; eastern North America (1 species; Wagner 1975) . . . . *Leptoplectus*

78a (76) Head with pubescent vertexal foveae; eastern North America . . . **79**

78b Head with nude vertexal feveae; Pacific coast states **80**

79a (78) Vertexal foveae separated by less than one foveal width; tergite I lacking basal carinae (4 species). . . . . *Thesiastes*

79b Vertexal foveae separated by one and a half or greater foveal widths; tergite I with basal carinae (13 species; Wagner 1975) . . *Pycnoplectus*

80a (78) Metasternal foveae present. **81**

80b Metasternal foveae absent **82**

81a (80) Elytra with four antebasal foveae (1 species) . . . . . . *Foveoscapha*

81b Elytra with two antebasal foveae; Oregon (1 species; Grigarick and Schuster 1980) . . . . *Bontomtes*

82a (80) Antennomere X asymmetrical, triangular in lateral view; promesocoxal foveae present (Fig. 39.34) (3 species; Grigarick and Schuster 1976) . *Euplecturga*

82b Antennomere X symmetrical, quadrate; promesocoxal foveae absent (6 species; Grigarick and Schuster 1976) . . . . *Oropodes*

83a (71) Head with broad shallow sulci above and below eyes; Florida (1 species; Wagner 1975) . . . . *Acolonia*

83b Head lacking sulci above and below eyes. **84**

84a (83) Metacoxae separated by one-fourth metasternal length; Florida (1 species). . . . . *Lemelba*

84b Metacoxae contiguous to narrowly separated. **85**

85a (84) Prosternal foveae absent; Arizona (2 species; Chandler 1985) *Simplona*

85b Prosternal foveae present, pubescent. **86**

86a (85) Metasternal foveae absent **87**

86b Metasternal foveae present. **88**

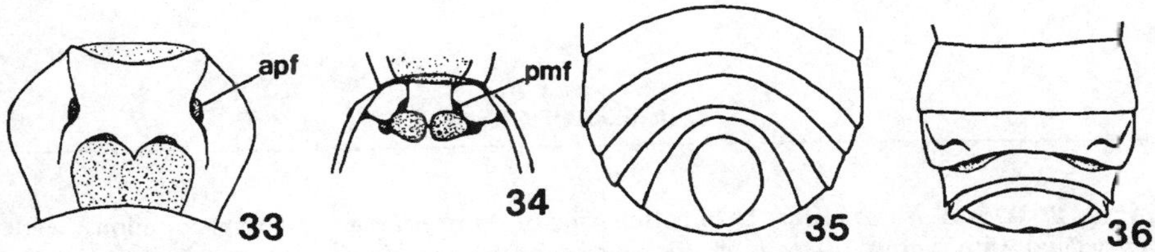

Figs. 39.33–39.36. 33. *Euplectus acomanus* Casey pronotum, ventral view; apf, anterior prosternal fovea. **34.** *Euplecturga* sp. sternal region; pmf, promesocoxal fovea. **35.** *Actium falcatum* G. & S. abdomen, ventral view. **36.** *Cupila multifossa* G. & S. abdomen, ventral view.

87a **(86)** Elytra with two basal foveae, body surface smooth; widespread (8 species; Grigarick and Schuster 1971) . . . . ***Actiastes***

87b Elytra with three basal foveae; body surface granulate; eastern North America (1 species; Grigarick and Schuster 1980). . . . . ***Liniolis***

88a **(86)** Pronotum with median antebasal fovea bisected by thin longitudinal carina; eastern North America (5 species) . . . . ***Bibloplectus***

88b Pronotum with median fovea lacking a bisecting carina **89**

89a **(88)** Tergite I with strong median basal impression. **90**

89b Tergite I lacking marked impression at base, convex **91**

90a **(89)** Median basal impression densely setate; two metasternal foveae; California to Oregon (2 species). . . . . ***Pilactium***

90b Median basal impression sparsely setate; single metasternal fovea; Arizona (1 species) . . . . ***Allobrox***

91a **(89)** Head across eyes close to or as wide as pronotum, head trapezoidal **92**

91b Head across eyes distinctly narrower than pronotum, head almost triangular. **94**

92a **(91)** Vertexal foveae small, nude; male mesofemora inflated; Arizona (1 species; Chandler 1985) . . . . ***Tomoplectus***

92b Vertexal foveae large, pubescent; male mesofemora about as wide as profemora; eastern North America **93**

93a **(92)** Tergites lacking any basal carinae (7 species; Park 1962) . . ***Ramecia***

93b Tergite I with basal carinae (4 species). . . . . ***Pseudactium***

94a **(91)** Promesocoxal foveae present (Fig. 39.34) (2 species; Chandler 1986) . ***Actizona***

94b Promesocoxal foveae lacking **95**

95a **(94)** Lacking median mesosternal foveae; Arizona to Texas (3 species; Chandler 1985) . . . . ***Saxet***

95b Median mesosternal foveae present **96**

96a **(95)** Males with seventh sternite large, circular to oval (Fig. 39.35); vertexal and lateral pronotal foveae often pubescent (35 species; Grigarick and Schuster 1971). . . . . ***Actium***

96b Males with seventh sternite small, transverse (Fig. 39.36); pubescence of foveae variable **97**

97a **(96)** Pronotum with median sulcus complete from base to antebasal sulcus; vertexal and lateral pronotal foveae pubescent; Pacific coast states (3 species; Grigarick and Schuster 1968) . . . . ***Cupila***

97b Pronotum lacking median sulcus from base to antebasal sulcus; vertexal and lateral pronotal foveae nude; southeastern United States (2 species; Schuster and Grigarick 1968) . . . . ***Pygmactium***

References

Akre, R. D. and W. B. Hill. 1973. Behavior of *Adranes taylori,* a myrmecophilous beetle associated with *Lasius sitkaensis* in the Pacific northwest. *J. Kans. Entomol. Soc.* 46:526–536. **Barr, T. C., Jr. 1974.** The eyeless beetles of the genus *Arianops* Brendel (Coleoptera : Pselaphidae). *Bull. Am. Mus. Nat. Hist.* 154(1):1–51. **Besuchet, C. 1956.** Larves et nymphes de Pselaphides (Coleopteres). *Rev. Suisse Zool.* 63:697–705. **Borror, D. J., D. M. DeLong and C. A. Triplehorn. 1981.** *An Introduction to the Study of Insects,* 5th ed. W.B.

Saunders, Philadelphia. 827 pp. **Böving, A. G. and F. C. Craighead. 1931.** An illustrated synopsis of the principal larval forms of the order Coleoptera. *Entomol. Am.* (1930) 11:1–351. **Bowman, J. R. 1934.** *The Pselaphidae of North America.* Published privately, Pittsburgh, Pa. 149 pp., keys to almost all genera. **Carlton, C. E. 1983.** Revision of the genus *Conoplectus* Brendel (Coleoptera : Pselaphidae). *Coleopt. Bull.* 37:55–80. **Carlton, C. E. and R. T. Allen. 1986.** Revision of the genus *Euboarhexius* Grigarick and Schuster (Coleoptera : Pselaphidae). *Coleopt. Bull.* 40:285–296. **Chandler, D. S. 1974.** The *Hamotus* of Arizona (Coleoptera : Pselaphidae). *Pan-Pac. Entomol.* (1973) 49:378–382. **1976.** A revision of the genus *Caccoplectus* (Coleoptera : Pselaphidae). *Coleopt. Bull.* (1975) 29:301–316. **1985.** The Euplectini of Arizona (Coleoptera : Pselaphidae). *Entomography* 3:107–126. **1986.** New Pselaphidae from New Hampshire (Coleoptera). *Psyche (Camb.)* 93:121–125. **1987.** Species richness and abundance of Pselaphidae (Coleoptera) in old-growth and 40-year-old forests in New Hampshire. *Can. J. Zool.* 65:608–615. **de Marzo, L. 1986.** Osservazióni etologiche sulle larve di *Batrisodes oculatus* Aube (Coleoptera : Pselaphidae). *Frustula Entomol.* 7–8:501–506. **Engelmann, M. D. 1956.** Observations on the feeding behavior of several pselaphid beetles. *Entomol. News* 67:19–24. **Fall, H. C. 1927.** The North American species of *Rybaxis. Psyche (Camb.)* 34:218–226. **Fletcher, F. C. 1932.** Undescribed North American species of Pselaphidae (Coleoptera), including a synopsis of the genus *Rhexidius* Casey. *Can. Entomol.* 64:29–35. **Grigarick, A. A. and R. O. Schuster. 1962a.** Notes on Tychini from western North America (Coleoptera : Pselaphidae). *Pan-Pac. Entomol.* 38:169–177. **1962b.** Species of the genus *Batrisodes* from the Pacific slope of western North America (Coleoptera : Pselaphidae). *Pan-Pac. Entomol.* 38:199–213. **1967.** *Reichenbachia* found in the United States west of the continental divide (Coleoptera : Pselaphidae). *Univ. Calif. Publ. Entomol.* 47:1–45. **1968.** A revision of the genus *Cupila* Casey (Coleoptera : Pselaphidae). *Pan-Pac. Entomol.* 44:38–44. **1971.** A revision of *Actium* Casey and *Actiastes* Casey (Coleoptera : Pselaphidae). *Univ. Calif. Publ. Entomol.* 67:1–56. **1976.** A revision of the genus *Oropodes* Casey (Coleoptera : Pselaphidae). *Pan-Pac. Entomol.* 52:97–109. **1980.** Discrimination of genera of Euplectini of North and Central America (Coleoptera : Pselaphidae). *Univ. Calif. Publ. Entomol.* 87:1–56. **Marsh, G. A. and R. O. Schuster. 1962.** A revision of the genus *Sonoma* Casey (Coleoptera : Pselaphidae). *Coleopt. Bull.* 16:33–56. **Mickey, G. H. and O. Park. 1956.** Cytological observations on pselaphid beetles. *Ohio J. Sci.* 56:155–164. **Park, O. 1932.** The myrmecocoles of *Lasius umbratus mixtus aphidicola* Walsh. *Ann. Entomol. Soc. Am.* 25:77–88. **1947.** Observations on *Batrisodes* (Coleoptera : Pselaphidae), with particular reference to the American species east of the Rocky Mountains. *Bull. Chic. Acad. Sci.* 8:45–132. **1949.** New species of Nearctic pselaphid beetles and a revision of the genus *Cedius. Bull. Chic. Acad. Sci.* 8:315–343. **1951.** *Cavernicolous Pselaphid Beetles of Alabama and Tennessee, with Observations on the Taxonomy of the Family.* Geol. Surv. Alabama, Museum Paper 31, pp. 1–107. **1953.** New or little known pselaphid beetles of the United States, with observations on taxonomy and evolution of the family Pselaphidae. *Bull. Chic. Acad. Sci.* 9:249–283. **1956.** New or little known species of pselaphid beetles from southeastern United States. *J. Tenn. Acad. Sci.* 31:54–100. **1958.** New or little known species of pselaphid beetles chiefly from southeastern United States. *J. Tenn. Acad. Sci.* 33:39–74. **1960.** Cavernicolous pselaphid beetles of the United States. *Amer. Midl. Nat.* 64:66–104 (key to cavernicolous species). **1962.** A review of the genus *Ramecia* (Coleoptera : Pselaphidae). *Ohio J. Sci.* 62:150–156. **1965.** Revision of the genus *Batriasymmodes* (Coleoptera : Pselaphidae). *Trans. Am. Microsc. Soc.* 84:184–201. **Park, O. and R. O. Schuster. 1955.** A new subtribe of pselaphid beetles from California. *Nat. Hist. Misc. (Chic.)* (148):1–6. **Park, O. and J. A. Wagner. 1962.** Family Pselaphidae. Pp. 4–31 in M. H. Hatch (ed.) *The Beetles of the Pacific Northwest.* Part III. *Pselaphidae and Diversicornia I.* (1961). Univ. Wash. Publ. Biol. 16. University of Washington, Seattle. 503 pp. **Park, O., S. Auerbach and M. Wilson. 1953.** Pselaphid beetles of an Illinois prairie: the population. *Ecol. Monogr.* 23:1–16. **Park, O., J. A. Wagner and M. W. Sanderson. 1976.** Review of the pselaphid beetles of the West Indies (Coleoptera : Pselaphidae). *Fieldiana. Zool.* 68:1–90. **Reichle, D. E. 1969.** Distribution and abundance of bog-inhabiting pselaphid beetles. *Trans. Ill. State Acad. Sci.* 62:233–264. **Schuster, R. O. and A. A. Grigarick. 1960.** A revision of the genus *Oropus* Casey (Coleoptera : Pselaphidae). *Pac. Insects* 2:269–299. **1962.** A revision of the genus *Rhexidius* Casey (Coleoptera : Pselaphidae). *Pan-Pac. Entomol.* 38:1–14. **1968.** A new genus of pselaphid beetle from southeast United States (Coleoptera : Pselaphidae). *Pan-Pac. Entomol.* 44:112–118. **Schuster, R. O., and G. A. Marsh. 1956.** A revision of the genus *Pselaptrichus* Brendel (Coleoptera : Pselaphidae).

Univ. Calif. Publ. Entomol. 11:117–158. **1958a.** A new genus of Tychini from California (Coleoptera : Pselaphidae). *Pan-Pac. Entomol.* 34:125–137. **1958b.** A study of the North American genus *Megarafonus* Casey (Coleoptera : Pselaphidae). *Pan-Pac. Entomol.* 34:187–194. **Schuster, R. O., G. A. Marsh and O. Park. 1959.** Present status of the tribe Mayetini in the United States—Part I (excluding California). *Pan-Pac. Entomol.* 35:117–127. **1960.** Present status of the tribe Mayetini in the United States—Part II California. *Pan-Pac. Entomol.* 36:15–24. **Wagner, J. A. 1975.** Review of the genera *Euplectus, Pycnoplectus, Leptoplectus,* and *Acolonia* (Coleoptera : Pselaphidae) including Nearctic species north of Mexico. *Entomol. Am.* 49:125–207. **Wickham, H. F. 1901.** Two new blind beetles, of the genus *Adranes,* from the Pacific coast. *Can. Entomol.* 33:25–28.

40

Insecta: Coleoptera Scarabaeidae Larvae

HARUO TASHIRO

The family Scarabaeidae, referred to as scarab or scarabaeid beetles, contain about 1280 North American species and vary greatly in size, color, and habits. The most distinguishing common feature in adults is the antennae, terminating in an oval or elongate club composed of three (rarely more) leaves or lamellae that can be folded closely together or spread apart to show the individual lamellae (Borror and DeLong 1970). This feature has given rise to the name lamellicorn beetles, and depending on their food habits, are divided into lamellicorn scavengers and lamellicorn leaf chafers (Comstock 1977). Another common feature is their fossorial front tibia adapted for digging in soil (Blatchley 1910).

The most distinguishing characteristic of scarabaeid larvae is their stout, wrinkled C-shaped configuration when at rest, with the tip of the abdomen touching the legs or nearly so. With exceptions in one subfamily there are three larval stages. Most descriptions refer to the third stage larvae but usually apply to the second stage as well (Ritcher 1966).

The superfamily Scarabaeoidea contains two other families that might be confused with Scarabaeidae since members of all three families occupy similar habitats and have similar appearances. Lucanidae have C-shaped larvae, but the adult antennae terminate in pectinate or comblike segments incapable of being folded into a compact club. Larvae of Passalidae are not C-shaped. Also, their metathoracic larval legs are rudimentary, whereas in most Scarabaeidae they are well developed.

Morphology

External Larval Morphology

Scarabaeid adults are mostly heavy-bodied, oval to elongate, usually convex beetles with 8- to 11-segmented lamellate antennae, the most useful character for distinguishing this family.

"White grub" is the common name that Peterson (1951) applies to all scarabaeid larvae. Blatchley (1910) reserves this name only to those larvae that live in the ground and damage roots, while Ritcher (1966) applies this term in particular to grubs of economic importance to agricultural crops. Larvae of the subfamily Melolonthinae containing many of the most injurious species of Scarabaeidae are known universally as "white grubs."

Figure 40.1, a third-stage larva of a June beetle, *Phyllophaga* spp., typifies a generalized family member. Some of the larvae that feed on decaying wood and other organic matter have noticeably distended caudal segments. The head is generally brown or black and the general body color is white or yellowish except for the caudal end, which may be dark due to the accumulated feces in the rectal sac and visible through the semi transparent cuticle (Peterson 1951).

Mouthparts are well developed for chewing with strongly sclerotized mandibles which are often asymetrical. Maxillae are well developed with three- to four-segmented maxillary palpi. Antennae are three- to four-segmented. The epipharynx, the inner surface of the labrum and clypeus,

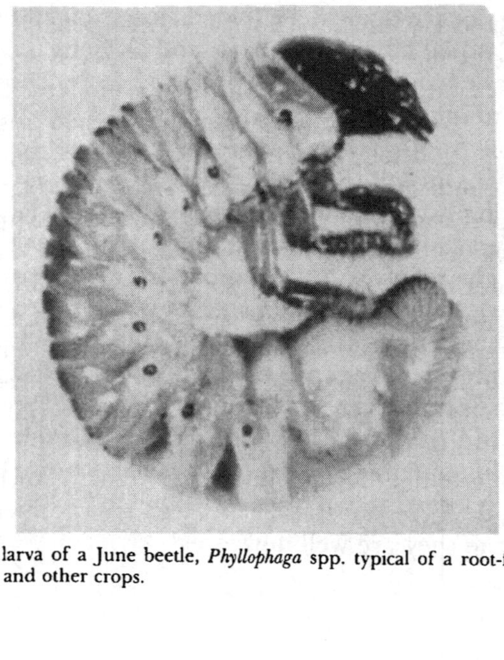

Fig. 40.1. Third-stage larva of a June beetle, *Phyllophaga* spp. typical of a root-feeding species highly destructive to turfgrass and other crops.

has numerous setae, spines, pits, and sclerotized processes useful in identification (Peterson 1951).

The three-segmented thorax generally bears four segmented legs, which usually terminate in pointed structures resembling claws. The abdomen consists of 9 or 10 segments, with each segment generally with three annulets in each of the first six or seven segments. Spiracles are present on the prothorax and the first eight abdominal segments. Ventrally, the caudal segment may bear groups of spines, scales, or stiff hairs, referred to collectively as the raster. Anal slit is either transverse, Y- or V-shaped. Both are useful in identification. Excellent treatises were published by Böving (1942) and Butt (1944) on the external larval morphology of the European chafer, *Rhizotrogus majalis* (Razoumowsky), recently changed from *Amphimallon* (Sutherland 1978).

Internal Larval Morphology

Internal morphology of scarabaeid larvae can also be represented by the European chafer (Fig. 40.2.) (Splittstoesser et al. 1973). The alimentary canal is distinctly divided into the stomodaeum (foregut), ventricles (midgut), and the proctodaeum (hindgut), with the latter consisting of the anterior intestine, rectal sac, and the rectum. The darkened accumulated feces and soil in the rectal sac generally has the color of the soil of the larval habitat in the case of subterranean forms. Cecal diverticula occur at the anterior and posterior areas of the ventriculus. There are 10 present at the junction of the stomodaeum and ventriculus and four just anterior to the junction of the ventriculus and proctodaeum. There are four Malpighian

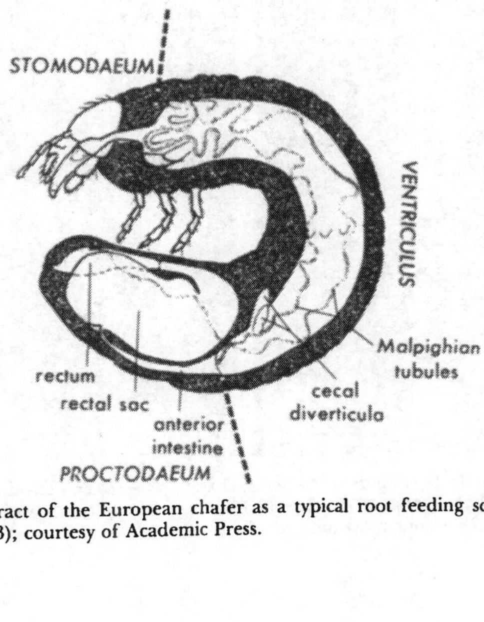

Fig. 40.2. Alimentary tract of the European chafer as a typical root feeding scarabaeid larvae. [After Splittstoesser et al. (1973); courtesy of Academic Press.

tubules, which extend forward the entire length of the midgut, then loop backward, and finally terminate on the ventral surface of the rectal sac (Menees 1958).

Larval sex can be distinguished in a number of scarabaeids by the appearance of a kidney-shaped terminal ampullae in the ventral median region of the ninth abdominal segment (Fig. 40.3*A*) in males (Tashiro et al. 1969). Ampullae are also present in females, in the posterior ventral portion of the eighth abdominal segment, but adipose tissue obscures their view through the cuticle (Fig. 40.3*B*). Vas deferens or lateral oviducts connect anteriorly to the ampullae and end in testes or ovaries in the seventh abdominal segments (Menees 1957).

Ecology

There are 14 recognized subfamilies of scarabaeids; food habits of 12 are given in Table 40.1 as modified from Ritcher (1958). Adult and larval food habits of subfamilies are so diverse that only generalizations hold true.

The subfamilies fall into two main groups, the lamellicorn scavengers, consisting mainly of the first four subfamilies, and the lamellicorn leaf chafers, consisting mainly of the last four subfamilies listed in Table 40.1. I have taken the liberty of placing the middle four subfamilies into a minor group since they consist of relatively few species whose biology is little

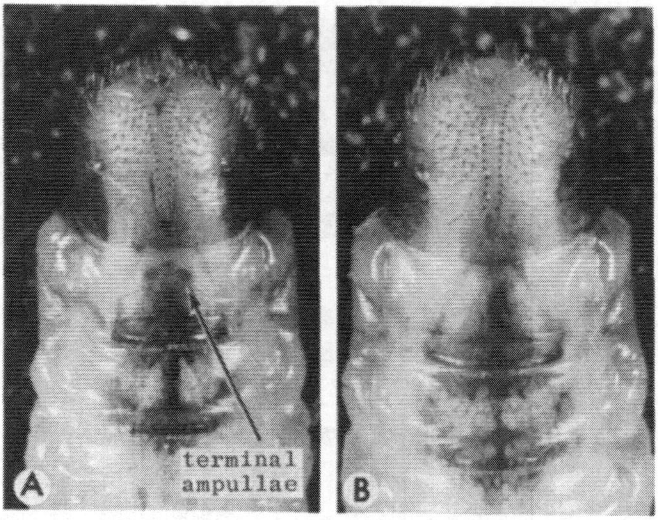

Fig. 40.3. Ventral abdominal segments of European chafer larvae to show terminal ampullae visible in the male (*A*) and invisible in the female (*B*). [After Tashiro et al. (1969); courtesy of Cornell University, New York State Agricultural Experiment Station.]

Table 40.1. Food Habits of Scarabaeidae[a]

Subfamily	Saprophagous: Carrion	Saprophagous: Dung	Saprophagous: Humus	Saprophagous: Decaying Vegetable Matter	Saprophagous: Duff	Saprophagous: Litter	Saprophagous: Wood	Phytophagous: Fungi	Phytophagous: Seed Plants
Coprinae (Scarabaeinae)	A L	A L		A				A	
Aphodiinae	A L	A L	L	A L				A	L
Geotrupinae	A	A L	L	L	L	L		A	
Troginae	A L								
Acanthocerinae							L		
Hybosorinae		L (?)							
Pleocominae									L
Glaphyrinae				L					L
Melolonthinae			L						A L
Rutelinae			L				L		A L
Dynastinae		L	L	L	L	L	L		A L
Cetoniinae		L	L				L		A

Source: Modified from Ritcher (1958).

[a] A, adults; L, larvae.

known. Condensations of mainly Ritcher's (1958) commentaries on primarily food habits follow.

Habits of Lamellicorn Scavengers

Dung is the major food of both adults and larvae of Coprinae (known as dung beetles and tumble bugs) and Aphodiinae (known as Aphodian dung beetles) but adults also feed on carrion, decaying vegetable matter, and fungi. Tumble bugs chew off a piece of dung, work it into a ball, roll it to a burial site, and upon burying, deposit an egg into the ball. Upon hatching the larva is assured of a food supply. The sacred scarab of ancient Egypt, *Scarabaèus sàcer* L., belongs to this group (Borror and DeLong 1970). Aphodiine larvae are primarily dung feeders, but a few feed on grass roots. The black turfgrass ataenius, *Ataenius spretulus* (Halderman), is currently a serious pest of golf course fairways (Niemczyk and Wegner 1979).

Geotrupinae (earth-boring dung beetle) adults are found beneath cow dung, horse manure, or carrion, but some occur in logs or decaying fungi. Larval food is all provisioned by their adults. Larvae of at least one species feeds on leaves, buds, and shoots, also provisioned by adults.

Troginae (or skin beetle) adults are found on old dry animal carcasses feeding on the hide, hairs, feathers, and dried tissues on the bone. Some species are found in owl pellets and beneath bark or roots (Borror and DeLong 1970). Larvae of most species feed in soil beneath carcasses of dead animals, while others are found in bird's nests and rodent burrows. Some species feed on grasshopper eggs.

Habits of Minor Subfamilies

Adults of Acanthocerinae, Hybosorinae, Pleocominae (rain beetles), and Glaphyrinae are not known to feed. Rain beetles are so named because flights usually follow rains when winged males emerged from the soil to seek burrows where wingless females reside (Ritcher 1966). Glaphyrine adults are very hairy and resemble bumblebees in flight (Ritcher 1966).

Larvae of Acanthocerinae feed on wood; pleocomine larvae feed on tree roots and some on grasses. Larvae of a glaphyrine species damage roots of cranberry.

Habits of Lamellicorn Leaf Chafers

To this group belong the subfamilies Melolonthinae (June beetles and chafers), Rutelinae (shining leaf chafers), Dynastinae (rhinoceros, hercules, and elephant beetles), and Cetoniinae (the flower beetles). Larvae of many species are agriculturally important because they feed on roots of many crops. Also, there are many saprophagous species.

Adults of nearly all Melolonthinae are vigorous feeders on plant tissue, devouring leaves, flowers, and fruits of trees and shrubs. Many are nocturnal. Larvae of this subfamily are the most damaging to roots of grasses, legumes, field crops, trees, and shrubs. Species of May or June beetles, *Phyllophaga* spp., cause widespread damage to pastures, corn, potatoes, strawberries, and tree seedlings. The European chafter and the Asiatic garden beetle, *Maladera castanea* (Arrow), are both introduced species whose larvae are major pests of turfgrasses. Neither adult is destructive.

Of the Rutelinae, the introduced Japanese beetle, *Popillia japonica* Newman, feeds on nearly 300 species of plants, while its larva feeds on turfgrass roots. Because of the combined adult and larval feeding habits and widespread distribution east of the Mississippi River, it is considered as one of the most serious of insect pests to agriculture. The Oriental beetle, *Anomala orientalis* Waterhouse, another introduced species, is a serious turfgrass pest. The spotted pelidnota, *Pelidnota punctata,* is a common ruteline whose larvae are found in decaying stumps or rotten wood and whose adults feed on leaves of cultivated and wild grapes.

Adults of Dynastinae usually attack stems and roots in search of liquid nourishment. Larvae of this subfamily have the most diverse food habits, feeding on dung and all decaying vegetable matter as well as roots of living plants. Larvae of the northern masked chafer, *Cyclocephala borealis* Arrow, and the southern masked chafer, *C. immaculata* (Oliver), are both destructive to pasture grasses and lawns. The latter is also found under cattle droppings and in moist soil near straw stacks and manure (Ritcher 1966).

Adults of Cetoniinae are liquid feeders on aboveground parts of plants, preferring nectar or sap or juices of ripening fruits and vegetables. Some feed on pollen. The larvae almost never feed on roots of living plants but prefer decaying vegetable matter in the soil, dung, or dead wood. When larvae of the green June beetle, *Cotinis nitida* (L.), damage young tobacco, clover pastures, and lawns in southeastern United States, it is a result of physical disturbances of the soil and roots, not larval feeding. These larvae have a peculiar habit of crawling on their backs, making no use of their small legs (Blatchley 1910).

Life Cycles

Numbers of Molts

With the exception of some members of Pleocominae, whose larvae molt 7–11 times with a molt each year, the scarabaeids have three larval instars.

Length of Life Cycle

Life cycles of the lamillicorn scavengers and the minor subfamilies are little known compared to the lamellicorn leaf chafers. In the latter group it may be 1, 2, or 3 years, and in more northern latitudes some Melolonthinae may require 4 or 5 years. Due to the economic importance of the lamellicorn leaf chafers, some of their life cycles have been worked out in great detail.

One-Year Life Cycle. Many of the most destructive species normally complete their life cycle in 1 year, but a small percentage may require 2 years (Tashiro et al. 1969). In the more northern latitudes, the Japanese beetle may require 2 years to complete their development from eggs deposited in late August (Hawley 1944).

The detailed life cycle of the European chafer, as shown in Fig. 40.4, and all stages and stadia present, as shown in Fig. 40.5, serve to illustrate a typical 1-year life cycle.

Normally, third-stage larvae overwinter. After disappearance of all frost, larvae migrate to near the soil surface to resume feeding on grass roots. Populations exceeding 10–20 per 0.1 m^2 can cause major damage, as shown in Fig. 40.6. In the northeast, larval feeding is completed by the latter half of May; they then construct earthen cells at a depth of 5–10 cm in which they become prepupae and pupae. Adults emerge during June into early July, make nightly mating flights following warm, sunny days, and mate as beetles rest in trees. Since adults essentially do not feed, they must depend on fat stored as larvae and normally live no more than 2 weeks.

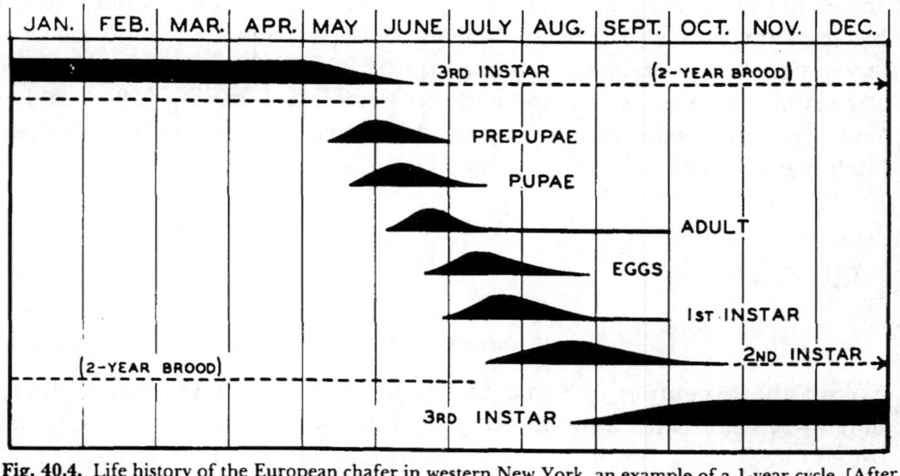

Fig. 40.4. Life history of the European chafer in western New York, an example of a 1-year cycle. [After Tashiro et al. (1969); courtesy of Cornell University, New York State Agricultural Experiment Station.]

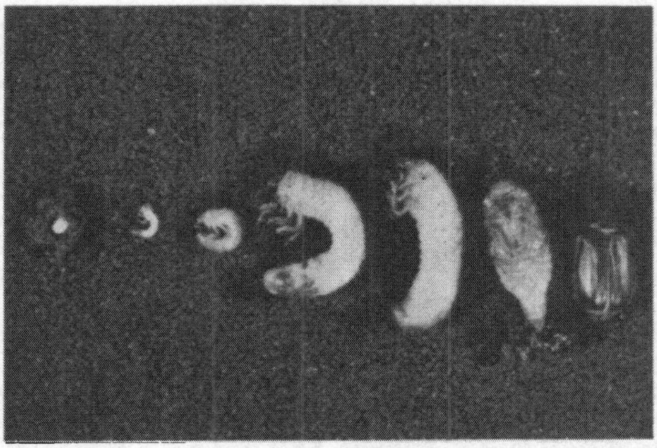

Fig. 40.5. Stages and stadia in the life cycle of the European chafer; egg, first, second, third-stage larvae, prepupa, pupa, and adult.

Fig. 40.6. Major damage to home lawn (*A*) and to winter wheat (*B*) caused by European chafer third-stage larvae. Equally severe damage can also develop in the fall.

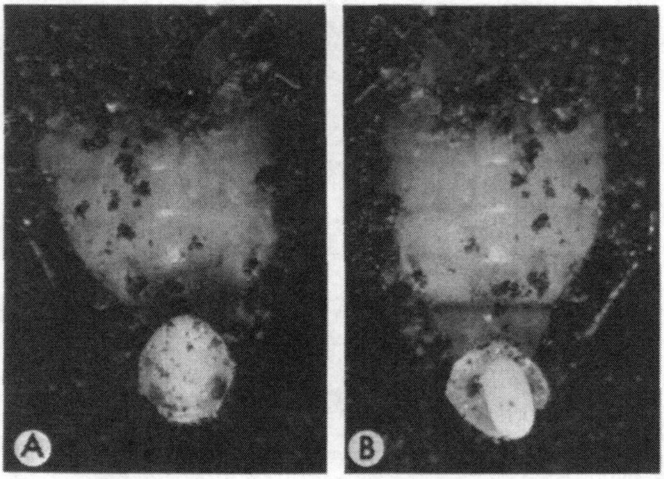

Fig. 40.7. Evagination of vaginal tract for preparing earthen cell by European chafer female (*A*). Egg is deposited simultaneously with invagination of vagina (*B*). [After Tashiro (1987)].

They return to the soil during their nightly flights, preferably to non-cultivated areas of pastures, weeds, and turfgrass, where females deposit eggs singly in earthen cells. Evagination of the vaginal tract into a balloon-like structure (Fig. 40.7*A*) compresses the moist soil and forms a cell (Tashiro 1987). An egg is deposited simultaneously with invagination of her vagina (Fig. 40.7*B*). Young eggs are ellipsoidal, but as they age become spherical. They hatch in 1 or 2 weeks.

From July into September larvae pass through the three larval stages, and by mid-September, with ample soil moisture, the third-stage larvae feed very actively within 2–3 cm of the soil surface. Summer soil moisture governs vertical distribution of larvae. As the surface soil dries, larvae migrate downward to seek moist soil at 12–15 cm or deeper. With sufficient water to moisten the soil continuously to the depths of the larvae, they will migrate to near the soil surface again within 24 hours. Feeding near the soil surface continues late into fall. As frost begins to form in surface soil, larvae will migrate downward for the winter.

Other related species whose larvae feed primarily on turfgrasses and underground stems have similar life cycles with minor deviations. Overwintering larvae of the Japanese beetle feed 3–4 weeks later during spring than those of the European chafer in the northeast; adults appear nearly a month later in late June and July. Individuals live from 4 to 6 weeks feeding and ovipositing during their entire adult life (Fleming 1972). During mid-September into October, like the European chafer, all the larvae are feeding very near the soil surface. In a mixed population, under identical environmental conditions, a much larger percentage of the Japanese beetle are second-stage larvae than the European chafer during fall. By the time of adult emergence the following summer, this difference is compensated for by the longer larval feeding period of the Japanese beetle in the spring. Also, the Japanese beetle larvae migrate downward at the first onset of cold weather a full month before the European chafer. Other well-known species destructive to turf with similar life cycles in the northeast are the Oriental beetle, the Asiatic garden beetle, and the northern masked chafer.

Three-Year Life Cycle. May or June beetles, genus *Phyllophaga,* of which there are some 152 species in North America north of Mexico (Luginbill and Painter 1953), have life cycles that may vary from 2 to 4 years in some species, depending largely on latitude. In the northeast and Canada there are many species that have a 3-year life cycle, as illustrated in Fig. 40.8.

Overwintering adults appear in early spring and are first noticed when they fly into screens or windows as they are attracted to light on warm evenings. Adults are most abundant during May and June, when they feed on the leaves of many common shade and ornamental trees at night. Eggs are deposited in soil at a depth of about 10 cm. They hatch in about a

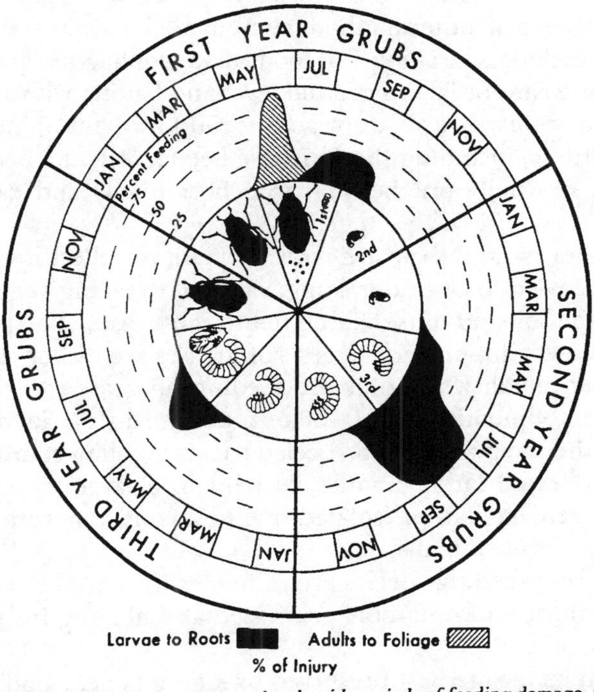

Fig. 40.8. Diagram of 3-year life cycle of a June beetle with periods of feeding damage. [After Hammond (1940). Reproduced from Agriculture Canada's Publication 668, *Farmers' Bulletin 86*, with the permission of the Minister, Supply and Services, Canada.]

month and by September are second-stage larvae, causing damage to grass roots before descending to lower levels to overwinter.

Second-stage larvae migrate to near the surface early during the second spring and after feeding become third-stage larvae. Underground feeding damage is greatest during July into September as third-stage larvae feed on roots of turfgrass or cultivated crops. They overwinter the second year in this stage.

During spring of the third year third-stage larvae migrate to near the soil surface to complete their feeding, pupate during July and August, become adults in September, and without emerging, overwinter as adults to complete their 3-year life cycle.

Collection

Adults of many species are attracted to light and hence are easily collected in light traps. This is primarily true of the more agriculturally important species whose larva are injurious to underground stems and grass roots. Woodruff (1973) indicates that blacklight (actually ultraviolet light) is much more attractive to beetles than is visible light. The advent of blacklight traps has resulted in more species and specimens collected and records obtained than all other methods combined. The less effective methods he refers to include searching for food of saprophagous groups or trapping by sinking cans or jars to ground level and baiting with such materials as proprionic or butyric acid, amyl acetate, and fermenting malt, molasses, and yeast. Others, including the Japanese beetle, Oriental beetle, and rose chafer, feed diurnally on their favorite host plants and can readily be picked off.

On the other hand, larvae are more difficult to collect in quantity since most are concealed in or under dung, carrion, decaying vegetable matter, wood, or in the soil and must be dug out in most cases. The phytophagous species of greatest importance as pests of grasses are rather easily collected during spring or fall when they are feeding on grass roots near the soil surface. When populations of 10–20 or more third-stage larvae per 0.1 m^2 are present, the sod may easily be peeled back, revealing countless numbers of larvae. Collected larvae should be held in a large volume of soil or medium of their habitat, or isolated one to a container with their habitat medium to prevent cannabalism.

Larvae of root feeding species are rather easily reared to adults by placing them in moist soil containing grass seed and holding for pupation and eclosion.

Scarabaeid larvae are best preserved by killing in near-boiling water and placing in 70% ethyl alcohol as practiced by Ritcher (1966) or 70% isopropyl alcohol as practiced by Woodruff (1973).

Identification

Taxonomic literature on Scarabaeidae is much more abundant for adults then for larvae. Therefore, larval identification may require rearing individuals to adults. Blatchley (1910) has the most complete description of adults of the family as a whole. Borror and DeLong (1970) have an adult key to subfamilies based on more recently accepted divisions. Woodruff's (1973) publication on the scarab beetles of Florida should not be overlooked in the study of six saprophagous subfamilies. His illustrations, "Stereoscan" (scanning electron microscopy) photos, and distribution maps, not only for Florida but the entire United States, adds much to his descriptions. Lugenbill and Painter's (1953) keys to May (or June) beetle adults are the most complete for this important genus.

Identification of Larvae to Subfamilies

For the most complete taxonomic work on the larvae of Scarabaeidae, one must turn to Ritcher (1966), whose work is the most comprehensive and recent. Ritcher's (1966) key to the subfamilies of Scarabaeidae based on characters of the third (ultimate) stage larvae is given with modifications. Figures 40.9–40.62 have been consolidated from Ritcher's (1966) plates. Use of his keys are recommended for larval identifications beyond the subfamily level.

Identification of Most Common Turfgrass Species

In the northeast, the most commonly found larvae in turfgrass are introduced species: the Japanese beetle, the European chafer, the Oriental beetle, and the Asiatic garden beetle. These are generally followed in order of abundance by the northern masked chafer, a native scarabaeid with a 1-year life cycle, and various June beetle, *Phyllophaga* spp., larvae. A key to field identification of these species and a *Phyllophaga* based on the rastral patterns and shape of anal slit is presented and supported by illustrations (Fig. 40.63). These characters are readily seen with the aid of a 10× hand lens.

KEY TO THE SUBFAMILIES OF SCARABAEID LARVAE†

1a Maxilla with galea and lacinia distinctly separated (Fig. 40.9) **2**

1b Maxilla with galea and lacinia entirely fused (Fig. 40.10), or fused proximally but distally free, or (rarely) not fused but fitting tightly together (Fig. 40.11) **10**

† Modified from Ritcher (1966).

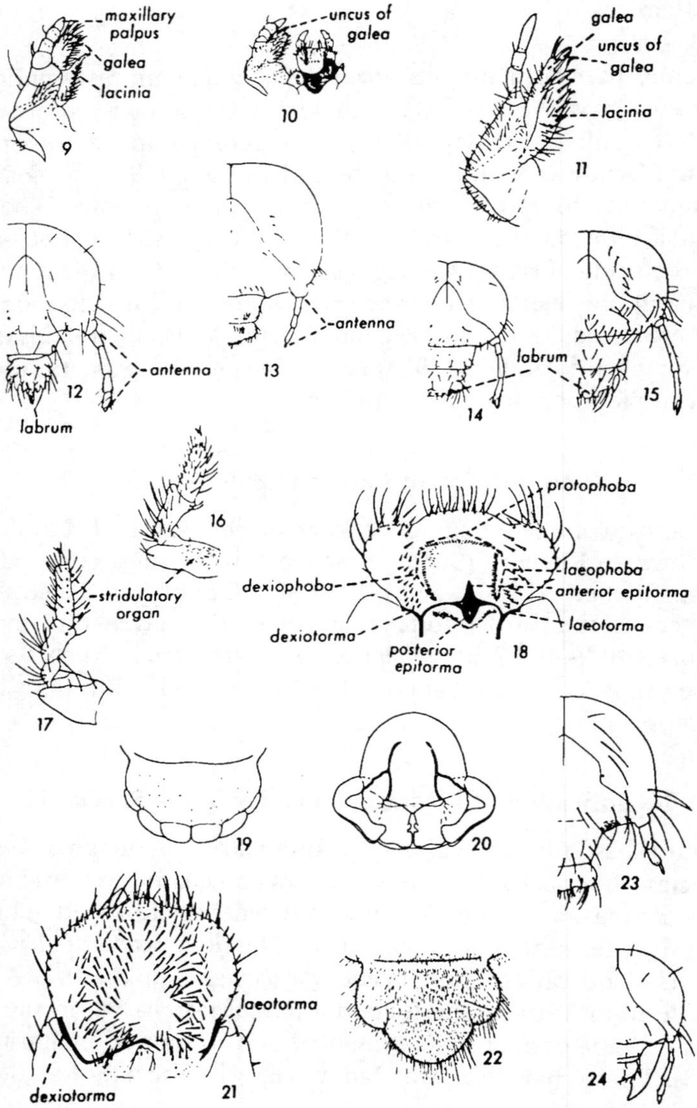

Figs. 40.9–40.24. Supportive illustrations for key to mainly subfamilies Hybosorinae and Acanthocerinae of Scarabaeidae larvae. [After Ritcher (1966); courtesy of Oregon State University Press.]

2a Distal segment of antenna little, if any, reduced in diameter compared to the other antennal segments (Fig. 40.12) **3**

2b Distal segment of antenna much reduced in size, often appearing as an appendage of the penultimate segment (Fig. 40.13) **5**

3a Anterior margin of labrum trilobed or broadly rounded; without serrations or truncate lobes . . . . a few **Aphodiinae**

3b Anterior margin of labrum serrate or with truncate lobes (Figs. 40.12, 40.14) **4**

4a Anterior margin of labrum with three truncate lobes (Figs. 40.14, 40.15); prothoracic and mesothoracic legs with stridulatory organs (Figs. 40.16, 40.17) . . . . **Hybosorinae**

4b Anterior margin of labrum strongly serrate (Fig. 40.12); stridulatory organs may be present on meso- and metathoracic legs . . . . . . **Acanthocerinae**

5a Epipharynx with tormae united mesally, anterior phoba present (Fig. 40.18); anal opening usually surrounded by fleshy lobes (Figs. 40.19, 40.20) . . . 6

5b Tormae of epipharynx not united mesally, anterior phoba absent (Fig. 40.21); last abdominal segment without fleshy lobes (Fig. 40.22) 9

6a Antenna three-segmented (Fig. 40.23). 7

6b Antenna four- or apparently five-segmented (Figs. 40.13, 40.24) 8

7a Legs two- or three-segmented (Figs. 40.25, 40.26); abdominal terga often divided into two distinct annulets (Fig. 40.27); body often hump-backed; meso- and metathoracic legs often with stridulatory organs (Figs. 40.25, 40.26) and latter often reduce in size (Fig. 40.25) . . . . **Geotrupinae**

7b Legs four-segmented; abdominal terga divided into three distinct annulets; body not hump-backed; legs never with stridulatory organs. . . . **Troginae**

8a Legs two-segmented (Fig. 40.28); claws very small or absent (Figs. 40.29, 40.30); if present, each with one or two terminal setae (Fig. 40.29); antenna four-segmented (Fig. 40.13); body strongly hump-backed (Figs. 40.31, 40.32);. . . . . **Scarabaeinae**

8b Legs four-segmented (Fig. 40.33); claws well-developed (Fig. 40.33); antenna four- or apparently five-segmented (Figs. 40.34, 40.24); body not hump-backed (Fig. 40.33) . . . . most **Aphodiinae**

9a Antenna three-segmented (Fig. 40.35); anal slit angulate (Fig. 40.36); body without a uniform covering of soft hairlike setae; meso- and metathoracic legs with stridulatory organs (Fig. 40.37); claws each bearing two setae (Fig. 40.37) . . . . **Pleocominae**

9b Antenna four-segmented (Fig. 40.38); anal slit transverse; body with a uniform dense covering of soft, hairlike setae (Fig. 40.22); meso- and metathoracic legs without stridulatory organs; claws each bearing four setae (Fig. 40.39) . . . . **Glaphyrinae**

10a Mandible with ventral stridulatory area absent, indistinct, or consisting of a patch of minute granules; anal opening usually angulate or Y-shaped; galea and lacinia fused proximally but separated distally or tightly fitted together; last antennal segment always with a single, large, oblong, dorsal, sensory spot (Fig. 40.40) . . . . **Melolonthinae**

10b Mandible with a ventral, oval, stridulatory area consisting of a few to many transverse, granular ridges (Fig. 40.41); anal opening a straight or slightly curved, transverse cleft (Fig. 40.42); galea and lacinia of maxilla fused forming a mala (Fig. 40.43); last antennal segment with one or more dorsal sensory spots (Figs. 40.44, 40.45) 11

11a Lacinia of maxilla with one or two well-developed terminal unci (Fig. 40.46) or uncus almost absent (Fig. 40.47); maxillary stridulatory teeth always with anteriorly projecting points, never truncate (Figs. 40.48, 40.49) 12

11b Lacinia of maxilla with three well-developed terminal unci (Fig. 40.50); maxillary stridulatory teeth truncate (Fig. 40.51) or with anteriorly projecting points (Fig. 40.52) 13

12a Labrum symmetrical, either entire (Fig. 40.53), bilobed (Fig. 40.54), or trilobed (Fig. 40.55); epizygum absent (Fig. 40.56); epipharynx always with a single nesium (sclerotized plate absent), plegmata absent (Fig. 40.56); hapto-

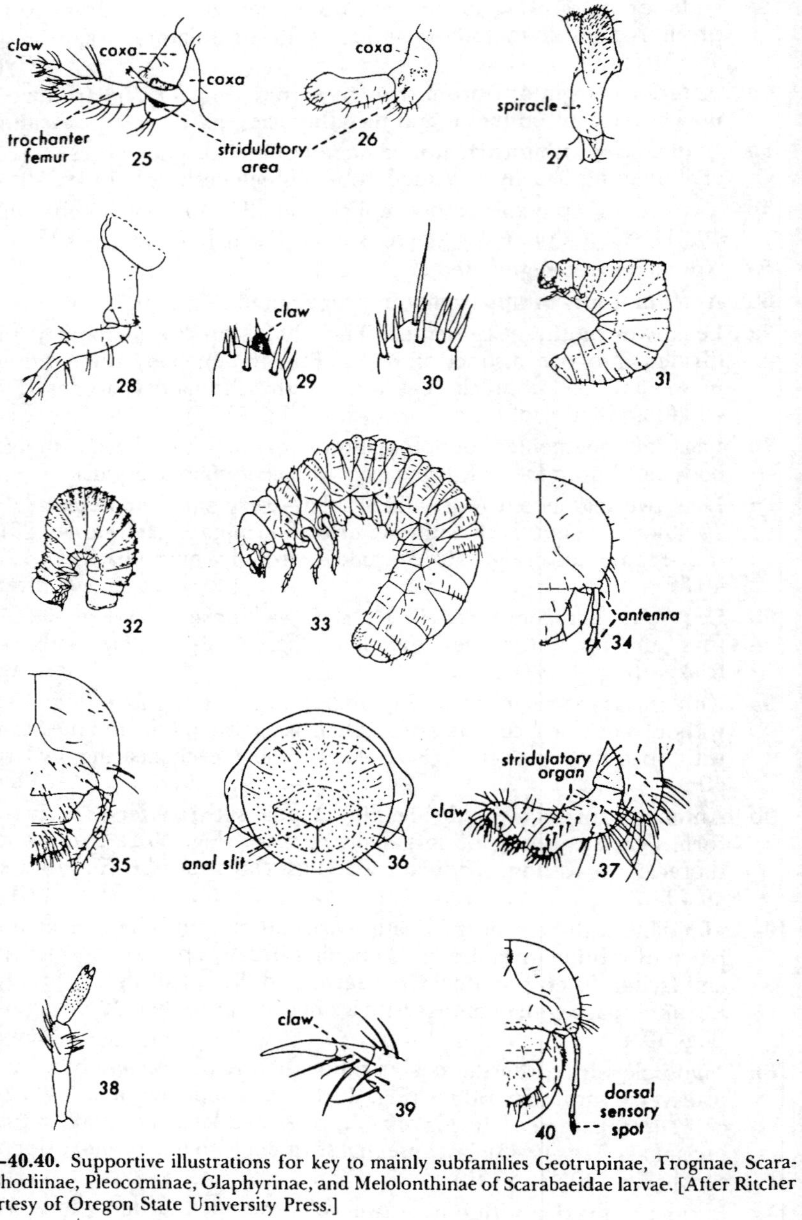

Figs. 40.25–40.40. Supportive illustrations for key to mainly subfamilies Geotrupinae, Troginae, Scarabaeinae, Aphodiinae, Pleocominae, Glaphyrinae, and Melolonthinae of Scarabaeidae larvae. [After Ritcher (1966); courtesy of Oregon State University Press.]

merum often with a conspicuous, transverse, curved row of stout setae (Figs. 40.55, 40.56) which may or may not be interrupted by a median process (Fig. 40.53); if the lacinia has two terminal unci, the two unequal in size (Fig. 40.57) . . . . **Cetoniinae**

12b Labrum usually asymmetrical; epizygum usually present (Fig. 40.58); epipharynx usually with two nesia and usually with plegmata (Fig. 40.58); haptomerum never with a conspicuous, transverse, curved row of stout setae; if the lacinia has two terminal unci, the two are usually equal in size (Figs. 40.59, 40.60) . . . . (Tribes **ANOMALINI** and **RUTELINI**) **Rutelinae**

13a Plegmatia present (Fig. 40.61); haptomerum of epipharynx with a dense transverse row of six or nine heli (Fig. 40.61); maxillary stridulatory teeth never truncate . . . . (Tribe **ADORETINI**) **Rutelinae**

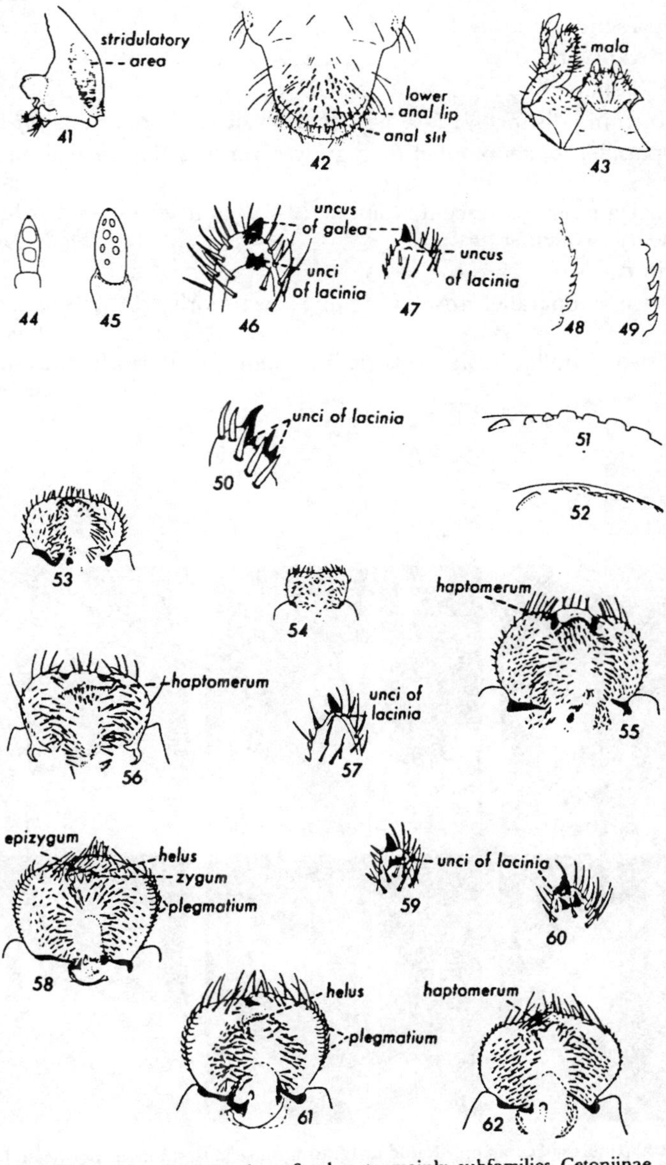

Figs. 40.41–40.62. Supportive illustrations for key to mainly subfamilies Cetoniinae, Rutelinae, and Dynastinae of Scarabaeidae larvae. [After Ritcher (1966); courtesy of Oregon State University Press.]

13b Plegmatia absent (Fig. 40.62); haptomerum of epipharynx with a raised, bilobed or toothlike, or ridgelike process (Fig. 40.62); maxillary stridulatory teeth usually truncate (Fig. 40.51). . . . . **Dynastinae**

FIELD IDENTIFICATION OF MOST COMMON SUBTERRANEAN SCARABAEID LARVAE IN THE NORTHEAST U.S.

1a Palida absent, teges scattered over lower anal lip and tenth segment *Northern masked chafer*

1b Palidia present **2**

2a Anal slit transverse **3**

2b Anal slit Y-shaped **4**

3a Palidia forming distinct V; 5–8 pali on each side. . . . . *Japanese beetle*

3b Palidia parallel to, subparallel or slightly diverging, 11–16 pali on each side *Oriental beetle*

4a Palidium transverse, crescent shaped, 20–27 stout closely set pali; maxillae with greatly swollen stipes. . . . . *Asiatic garden beetle*

4b Palidia in two rows **5**

5a Palidia two subparallel rows widening posteriorly; 14–22 pali each side *European chafer*

5b Palidia two parallel rows constricting slightly anteriorly and posteriorly *June beetle larvae*

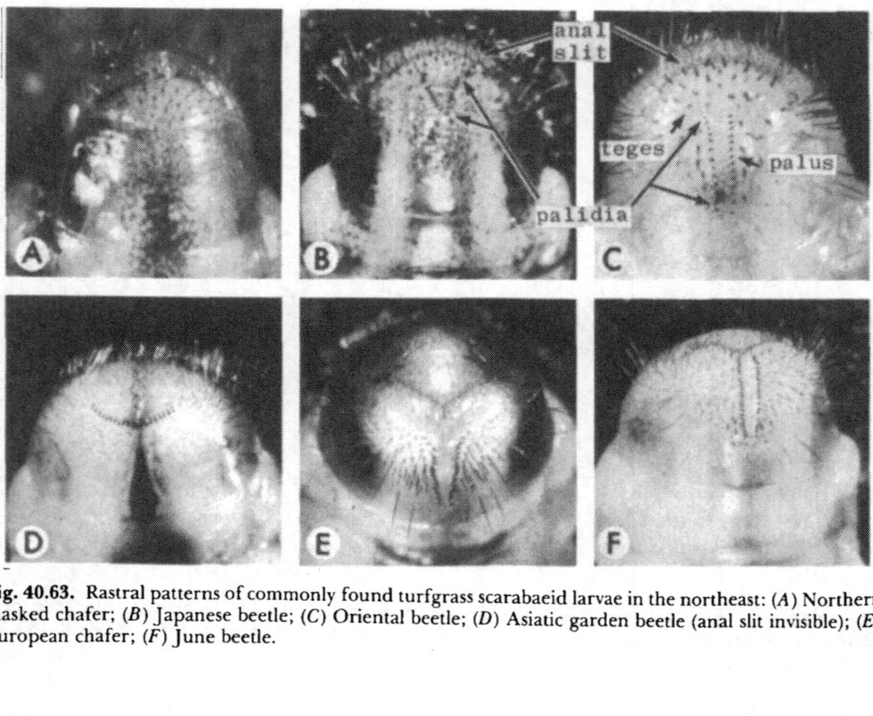

Fig. 40.63. Rastral patterns of commonly found turfgrass scarabaeid larvae in the northeast: (*A*) Northern masked chafer; (*B*) Japanese beetle; (*C*) Oriental beetle; (*D*) Asiatic garden beetle (anal slit invisible); (*E*) European chafer; (*F*) June beetle.

References

Blatchley, W. S. 1910. *An Illustrated Descriptive Catalog of the Coleoptera or Beetles (Exclusive of the Rhyncophora) Known to Occur in Indiana.* Nature Publishing Co., Indianapolis, Ind. 1386 pp. **Borror, D. J. and D. M. DeLong. 1970.** *An Introduction to the Study of Insects,* 3rd ed. Holt, Rinehart and Winston, New York. 812 pp. **Böving, A. G. 1942.** Description of the third-stage larva of *Amphimalon majalis* (Razoumowsky). *Proc. Entomol. Soc. Wash.* 44(6):111–121. **Butt, F. H. 1944.** External morphology of *Amphimallon majalis* (Razoumowsky) (Coleoptera, the European chafer). *Cornell Univ. Agric. Exp. Stn. Mem.* 266. 18 pp., 13 plates. **Comstock, J. H. 1977.** *An Introduction to Entomology,* 9th ed. Cornell University Press, Ithaca, N.Y. 1064 pp. **Fleming, W. E. 1972.** Biology of the Japanese beetle. *U.S. Dept. Agric. Tech. Bull.* 1449. 129 pp. **Hammond, G. H. 1940.** White grubs and their control in eastern Canada. *Dom. Canada Dept. Agric. Publ.* 668. 18 pp. **Hawley, I. M. 1944.** Notes on the biology of the Japanese beetle. *U.S. Bur. Entomol. Plant Quar.* E-615. 19 pp. **Lugenbill, P., Sr. and H. R. Painter. 1953.** May beetles of the United States and Canada. *U.S. Dept. Agric. Tech. Bull.* 1060. 102 pp., 78 plates. **Menees, J. H. 1957.** Sex identification in some larvae of Scarabaeoidea. *Bull. Brooklyn Entomol. Soc.* 52(4):97–100. **1958.** The anatomy and histology of the larval alimentary canal of the European chafer, *Amphimallon majalis* Razoumowsky (Scarabaeidae). *J. N.Y. Entomol. Soc.* 66:75–86. **Niemczyk, H. D. and G. S. Wegner. 1979.** *Life History and Control of the Black Turfgrass Ataenius.* Ohio Report on Research and Development, pp. 85–88. **Peterson, A. 1951.** *Larvae of Insects: An Introduction to Nearctic Species. II. Coleoptera, Diptera, Neuroptera, Siphonoptera, Mecoptera, Trichoptera.* Edwards Brothers, Ann Arbor, Mich. 416 pp. **Ritcher, P. O. 1958.** Biology of Scarabaeidae. In. E. A. Steinhaus (ed.) *Annual Review of Entomology,* Vol. 3. Annual Reviews, Palo Alto, Calif. 520 pp. **1966.** *White Grubs and Their Allies—A Study of North American Scarabaeoid Larvae.* Oregon State University Press, Corvallis, Oreg. 219 pp. **Splittstoesser, C. M., H. Tashiro, S. L. Lin, K. H. Steinkraus and B. J. Fiori. 1973.** Histopathology of the European chafer *Amphilmallon majalis,* infected with *Bacillus popilliae. J. Invertebr. Pathol.* 22:161–167. **Sutherland, D. W. S. 1978.** Common names of insects and related organisms (1978 revision). *Entomol. Soc. Am. Spec. Publ.* 78-1. 132 pp. **Tashiro, H. 1987.** *Turfgrass Insects of the United States and Canada.* Cornell University Press, Ithaca, N.Y. 391 pp. **Tashiro, H., G. G. Gyrisco, F. L. Grambrell, B. J. Fiori and H. Breitfeld. 1969.** Biology of the European chafer *Amphimallon majalis* (Coleoptera: Scarabaeidae) in northeastern United States. *New York State Agric. Exp. Stn. Bull.* 828. 71 pp. **Woodruff, R. E. 1973.** *Arthropods of Florida and Neighboring Land Areas.* Vol. 8. *The Scarab Beetles of Florida.* Florida Dept. Agric. Consumer Serv. Contrib. 260. 220 pp.

Insecta: Diptera Adults

J. F. McALPINE

The order Diptera (true flies) includes such diverse forms as crane flies, midges, horseflies, flower flies, blowflies, and many others. In number of species it is probably exceeded only by Coleoptera (Alexander 1969). Latest estimates (Hardy 1977) indicate that it contains some 200,000 different species throughout the world, nearly half of which are still undescribed (i.e., unnamed). Sabrosky (1966) estimated that Diptera make up about one-eighth of the world's insect species and one-fifth of those in the Nearctic region. The book *A Catalog of the Diptera of America North of Mexico* (Stone et al. 1965) recognizes 105 families, 1971 genera, and 16,130 species in the United States and Canada. In the *Manual of Nearctic Diptera,* 108 families are recognized and it is estimated that there are more than 25,000 Nearctic species representing about 2150 genera (McAlpine et al. 1981). Up to one-half of these species and genera are likely to be encountered in soil ecology studies.

The order consists of two suborders, the Nematocera, in which the antennal flagellum usually contains at least four freely articulated segments (Figs. 41.3, 41.28–41.40),† and the Brachycera, in which all the segments of the flagellum beyond the first one are more or less consolidated into a specialized style or arista (Figs. 41.1, 41.2, 41.4–41.6, 41.8, 41.9, 41.41–41.96). The suborders are divided into infraorders, each of which contains one or more superfamilies. The names and hierarchical ranking of all these taxonomic levels are shown in Table 41.1.

† The figures used throughout this chapter are taken from the *Manual of Nearctic Diptera,* Vol. 1 (McAlpine et al. 1981), and are reproduced here with the permission of the Minister of Supply and Services, Canada.

Table 41.1. Classificiation of Nearctic Diptera Showing Those Likely to be Encountered in Soil Biology Studies as Larvae (L) and Pupae (P)

Order DIPTERA

Suborder NEMATOCERA

Infraorder Tipulomorpha

Superfamily Tanyderoidea Tanyderidae (P)	Superfamily Tipuloidea Tipulidae (L, P)

Infraorder BLEPHARICEROMORPHA

Superfamily Blephariceroidea Blephariceridae	Superfamily Deuterophlebioidea Deuterophlebiidae	Superfamily Nymphomyioidea Nymphomyiidae

Infraorder AXYMYIOMORPHA

Superfamily Axymyioidea
Axymyiidae

Infraorder BIBIONOMORPHA

Superfamily Pachyneuroidea Pachyneuridae	Superfamily Bibionoidea Bibionidae (L, P)	Superfamily Sciaroidea Mycetophilidae (L, P) Sciaridae (L, P) Cecidomyiidae (L, P)

Infraorder PSYCHODOMORPHA

Superfamily Psychodoidea Psychodidae (L, P)	Superfamily Trichoceroidea Trichoceridae (L, P) Superfamily Scatopsoidea Scatopsidae (L, P) Synneuridae	Superfamily Anisopodoidea Anisopodidae (L, P)

Infraorder PTYCHOPTEROMORPHA

Superfamily Ptychopteroidea
Ptychopteridae (P)

Infraorder CULICOMORPHA

Superfamily Culicoidea Dixidae Chaoboridae Culicidae	Superfamily Chironomoidea Thaumaleidae Simuliidae Ceratopogonidae (L, P) Chironomidae (L, P)

Table 41.1. (*Continued*)

Suborder BRACHYCERA

Infraorder TABANOMORPHA

Superfamily Tabanoidea
- Pelechorhynchidae (L, P)
- Tabanidae (L, P)
- Athericidae (P)
- Rhagionidae (L, P)

Superfamily Stratiomyoidea
- Xylophagidae (L, P)
- Xylomyidae
- Stratiomyidae (L, P)

Infraorder ASILOMORPHA

Superfamily Asiloidea
- Therevidae (L, P)
- Scenopinidae (L, P)
- Vermileonidae (L, P)
- Mydidae (L, P)
- Apioceridae (L, P)
- Asilidae (L, P)

Superfamily Bombylioidea
- Acroceridae (P)
- Nemestrinidae (P)
- Bombyliidae (P)
- Hilarimorphidae

Superfamily Empidoidea
- Empididae (L, P)
- Dolichopodidae (L, P)

Infraorder MUSCOMORPHA

Section ASCHIZA

Superfamily Lonchopteroidea
- Lonchopteridae (L, P)

Superfamily Platypezoidea
- Platypezidae (L, P)
- Phoridae (L, P)

Superfamily Syrphoidea
- Syrphidae (L, P)
- Pipunculidae (P)

Section SCHIZOPHORA

Subsection ACALYPTRATAE

Superfamily Conopoidea
- Conopidae (P)

Superfamily Nerioidea
- Cypselosomatidae
- Micropezidae (L, P)
- Neriidae

Superfamily Diopsoidea
- Tanypezidae (L, P)
- Strongylophthalmyiidae (L, P)
- Psilidae (L, P)
- Diopsidae (L, P)

Superfamily Tephritoidea
- Lonchaeidae (L, P)
- Otitidae (L, P)
- Platystomatidae (L, P)
- Pyrgotidae (P)
- Tephritidae (P)
- Richardiidae (P)
- Pallopteridae
- Piophilidae (L, P)

Superfamily Opomyzoidea
- Clusiidae
- Acartophthalmidae
- Odiniidae
- Agromyzidae (P)
- Opomyzidae (P)
- Anthomyzidae (L, P)
- Aulacigastridae
- Periscelididae
- Asteiidae
- Milichiidae (L, P)
- Carnidae (L, P)
- Braulidae

Superfamily Sciomyzoidea
- Coelopidae (P)
- Dryomyzidae (L, P)
- Sciomyzidae (P)
- Ropalomeridae
- Sepsidae (L, P)

Table 41.1. *(Continued)*

	Suborder BRACHYCERA	
Superfamily Lauxanioidea	Superfamily Sphaeroceroidae	Superfamily Ephydroidea
Lauxaniidae (L, P)	Helomyzidae (L, P)	Curtonotidae (L, P)
Chamaemyiidae	Trixoscelididae (L, P)	Drosophilidae (L, P)
	Chyromyidae	Diastatidae (L, P)
	Rhinotoridae	Camillidae
	Sphaeroceridae (L, P)	Ephydridae (L, P)
		Chloropidae (L, P)
		Cryptochaetidae
		Tethinidae (L, P)
		Canacidae
	Subsection CALYPTRATAE	
Superfamily Muscoidea	Superfamily Oestroidea	Superfamily Hippoboscoidea
Scatophagidae (L, P)	Calliphoridae (P)	Hippoboscidae
Anthomyiidae (L, P)	Oestridae (P)	Nycteribiidae
Muscidae (L, P)	Sarcophagidae (P)	Streblidae
	Rhinophoridae (P)	
	Tachinidae (P)	

Total: 75/108 families

Many Diptera occupy, at some stage of their development, habitats that are intermediate between terrestrial and aquatic, on the one hand, and between highly organic soils and living or decaying organic matter on the other. For example, larvae of Tanyderidae live in sandy stream bottoms, but they move to the soil along the margin of the stream to pupate and emerge as adults. The larvae of many Tephritidae infest the stems, flowers, and fruits of plants, but they often pupate in the soil, whence the adults emerge. Consequently, a problem arises as to where to draw the line between those families which are to be included in this work and which ones are to be excluded. To some degree the decision has to be arbitrary.

At least 75 of the 108 families that occur in the Nearctic region are to some extent soil-related (Table 41.1). By this I mean that some stages of some member(s) of these 75 families are likely to be found in soil as it is defined in the Introduction (Chapter 1). Strictly aquatic groups (James 1959, Merritt and Schlinger 1984) (e.g., Blephariceroidea, Culicoidea, and most Chironomoidea) are excluded. Groups that are restricted to dead and living trees (Teskey 1976) (e.g., Axymyiidae, Pachyneuridae, Synneuridae, Xylomyidae, Clusiidae, Odiniidae, Aulacigastridae) are also excluded. Similarly, certain predatory and parasitic groups, such as Chamaemyiidae, Braulidae, Hippoboscidae, Nycteribiidae, and Streblidae, are very unlikely to be encountered in soil even as pupae, and they are excluded. On the other hand, parasitic groups, such as Tachinidae, Rhinophoridae, Sarcophagidae, and Oestridae, frequently pupate in the soil or soil litter, and

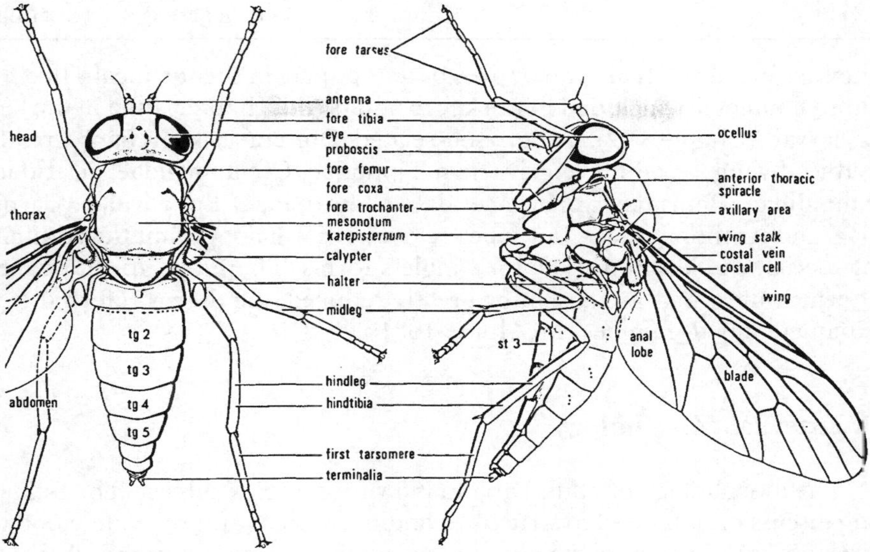

1 Symphoromyia montana ♀

2 Symphoromyia montana ♀

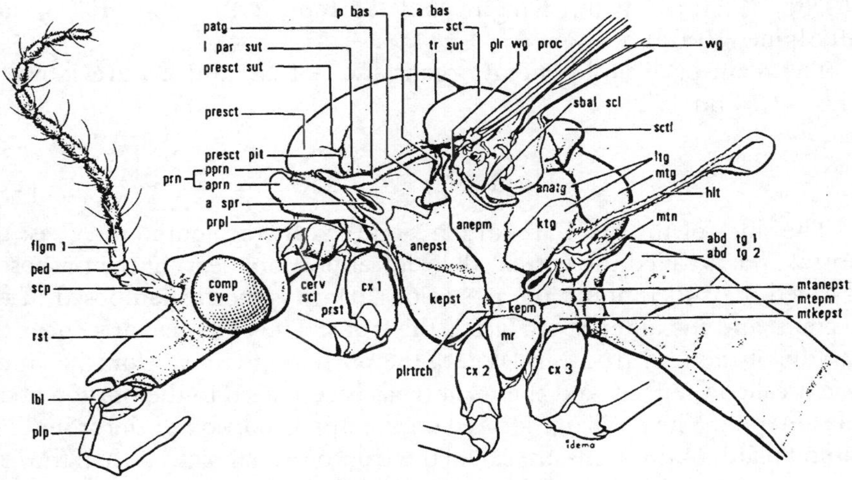

3 Tipula trivittata ♂

Figs. 41.1.–41.3. Main anatomical divisions, parts, and sclerites of generalized flies. **1,2.** *Symphoromyia montana* Aldrich, **Rhagionidae,** in dorsal and left lateral views, respectively. **3.** *Tipula trivittata* Say, **Tipulidae,** in left lateral view.

Abbreviations:

a bas, anterior basalare
abd tg, adominal tergite
anatg, anatergite
anepm, anepimeron
anepst, anepisternum
aprn, antepronotum
a spr, anterior spiracle
cerv scl, cervical sclerite
comp eye, compound eye
cx, coxa
flgm, flagellomere
hlt, halter
kepm, katepimeron
kepst, katepisternum
ktg, katatergite
lbl, labellum
l par sut, lateral parapsidal suture
ltg, laterotergite
mr, meron
mtanepst, metanepisternum
mtepm, metepimeron
mtg, mediotergite
mtkepst, metakatepisternum
mtn, metanotum
patg, paratergite
p bas, posterior basalare
ped, pedicel
plp, palpus
plrtrch, pleurotrochantin
plr wg proc, pleural wing process
pprn, postpronotum
presct, prescutum
presct pit, prescutal pit
presct sut, prescutal suture
prn, pronotum
prpl, propleuron
prst, prosternum
rst, rostrum
sbal scl, subalar sclerite
scp, scape
sct, scutum
sctl, scutellum
st, sternite
tr sut, transverse suture
wg, wing.

they are included. Inclusion of groups that pupate in the soil means that 24 more families are included in the key to adults (this chapter) than in the key to larvae (Chapter 42). A few soil-related Diptera lack or have greatly reduced wings as adults (e.g., certain Tipulidae, Cecidomyiidae, Sciaridae, Empididae, Phoridae, Sphaeroceridae, Chloropidae, Ephydridae, Carnidae, and Anthomyzidae), but space restrictions will not permit the addition of special couplets in the key for wingless forms. Therefore, for those flies in which the wings are absent or greatly reduced, the user is referred to a complete key to families (McAlpine 1981b).

External Morphology

The morphology of adult Diptera is a very complex subject. The following discussion is limited to structures and terms that must be understood to use the key. For more details the reader is referred to comprehensive general treatments given by such workers as Crampton (1942), Hennig (1973), Knight (1970), Knight and Laffoon (1970a,b,c, 1971), and McAlpine (1981a).

The main parts and general organization of an adult fly are shown in Figs. 41.1 and 41.2.

Head (Figs. 41.1–41.7)

The sides of the head are largely occupied by the compound eyes; the dorsal area is called the vertex; the ventral portion, bearing the proboscis, is called the subcranium; the posterior surface is mostly composed of the occiput, and the anterior surface is the frons. The antennae arise near the middle of the true frons; in Diptera the portion of frons below the antennae is called the face; and the term frons is restricted to the portion above the antenna. The ocelli are located on the upper portion of the frons. The lateral plates of the frons are called the fronto-orbital plates; when they are transversely or obliquely divided (Fig. 41.160), as in many acalyptrate families, the upper part is called the orbital plate and the lower part is called the frontal plate. In holoptic forms (Fig. 41.6), mostly males, the frons is largely obliterated by the enlarged eyes. The face is bounded laterally by the parafacials and anteriorly by the clypeus. The gena lies below the compound eye and extends from the vibrissal angle, anteriorly, to the postgena posteriorly. The mouthparts or proboscis consist of a labrum-epipharynx above and a labium and hypopharynx below. Those groups with typical biting mouthparts, such as Ceratopogonidae (Fig. 41.13), also have a pair of styliform mandibles and a similar pair of maxillae. Those groups with lapping mouthparts, such as Calliphoridae (Fig. 41.14), lack bladelike mandibles and maxillae. Usually, each maxilla bears a maxillary palpus. In Nematocera it contains three to five segments (Figs. 41.3, 41.13),

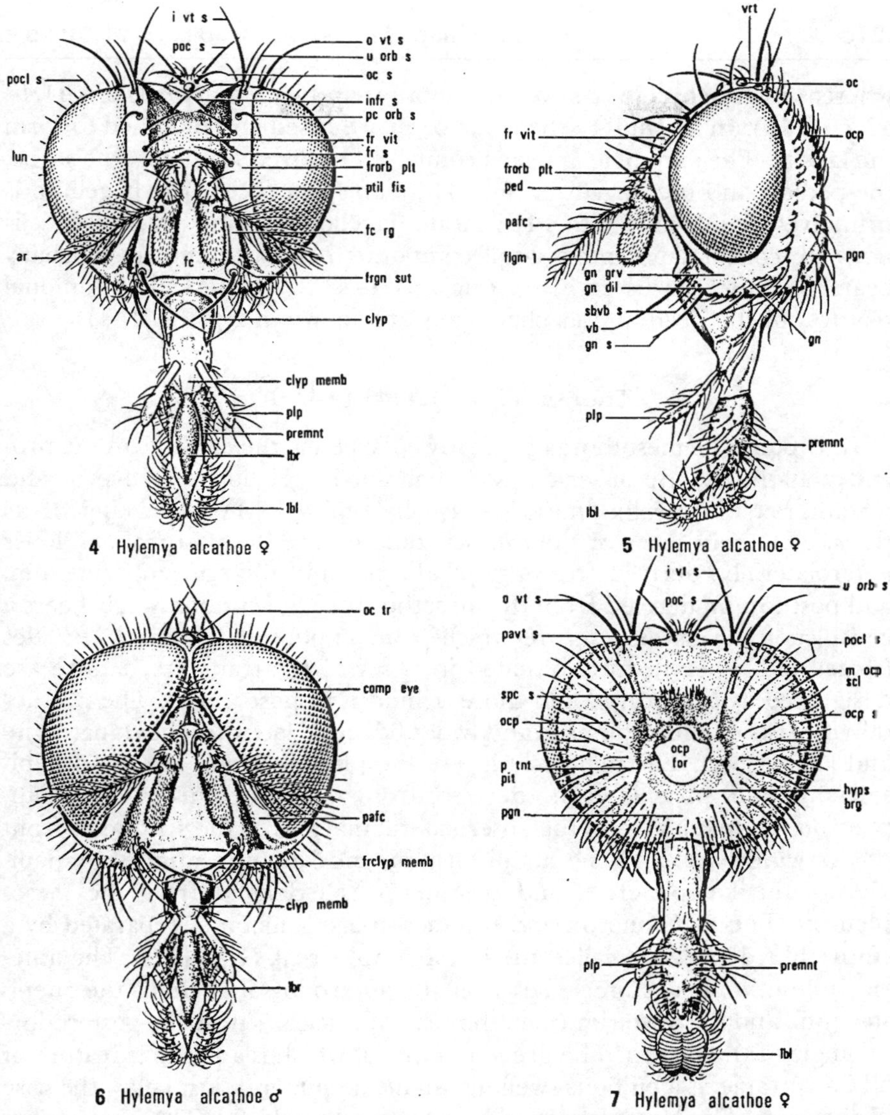

Figs. 41.4.–41.7. Head of *Hylemya alcathoe* (Walker), **Anthomyiidae. 4.** Anterior. **5.** Left lateral view of female. **6.** Anterior view of male. **7.** Posterior view of female. Note dichoptic condition of female (4) and holoptic condition of male (6).

Abbreviations:

ar, arista
clyp, clypeus
clyp memb, clypeolabral membrane
comp eye, compound eye
fc, face
fc rg, facial ridge
flg, flagellum
flgm, flagellomere
frclyp memb, frontoclypeal membrane
frgn sut, frontogenal sucture
frorb plt, fronto-orbital plate
fr s, frontal seta
fr vit, frontal vitta
gn, gena
gn dil, genal dilation
gn grv, genal groove
gn s, genal seta
hyps brg, hypostomal bridge
infr s, interfrontal seta
i vt s, inner vertical seta
lbl, labellum
lbr, labrum
lun, lunule
m ocp scl, median occipital sclerite
oc, ocellus
ocp, occiput
ocp for, occipital foramen
ocp s, occipital seta
oc s, ocellar seta
oc tr, ocellar triangle
o vt s, outer vertical seta
pafc, parafacial
pavt s, paravertical seta
pc orb s, proclinate orbital seta
ped, pedicel
pgn, postgena
plp, palpus
pocl s, postocular seta
poc s, postocellar seta
premnt, prementum
ptil fis, ptilinal fissure
p tnt pit, posterior tentorial pit
sbvb s, subvibrissal setula
spc s, supracervical setae
u orb s, upper orbital seta
vb, vibrissa
vrt, vertex

whereas in Brachycera it never has more than two segments (Figs. 41.4–41.7, 41.14). In all Diptera the labial palpi are fused and modified to form the labella. The antennae are each composed of three segments: the scape, the pedicel and the flagellum (Fig. 41.3). In Nematocera the flagellum is primitively subdivided into 14 separate flagellomeres. In Brachycera it is normally consolidated into a single compound basal segment which usually bears a slender stylus or arista (Figs. 41.1–41.2, 41.4–41.9). Additional features of the head and its chaetotaxy are shown in Figs. 41.4–41.9.

Thorax (Figs. 41.3, 41.8, 41.9)

In Diptera the mesothorax is greatly enlarged at the expense of the pro- and metathorax. Because it is so dominant in relation to the pro- and metathorax, the prefix "meso" is usually omitted when speaking of its parts. The main parts of the thoracic dorsum are the postpronotal lobes anterolaterally, derived from the prothorax, and the scutum, scutellum, and postnotum, derived from the mesothorax. The scutum usually bears a conspicuous transverse suture which is an important landmark. Bristles (setae) and other features situated in front of the transverse suture are designated as presutural, and those behind it as postsutural. The postnotum consists of a mediotergite and two subdivided laterotergites (anatergite and katatergite). The main sclerites of the pleural region are the proepisternum and proepimeron, derived from the propleuron; the notopleuron, a sunken area at the anterolateral margins of the scutum in front of the wing base, and the anepisternum, the katepisternum, the anepimeron, the katepimeron, and the meron, all derived from the mesopleuron. The katepimeron and the meron are sometimes separated by a suture-like depression called the coxopleural streak (Fig. 41.9). The anterior (mesothoracic) spiracle lies near the anterodorsal angle of the anepisternum, and the posterior (metathoracic) spiracle lies near the posterodorsal angle of the meron. The greater ampulla, which is a peculiar feature of all Calyptratae, is a bulbous swelling on the anepimeron just below the base of the wing. The metapleuron is usually greatly reduced. The prosternum usually occurs as a distinct, separate sclerite between the anterior coxae; sometimes it is fused laterally with the proepisternum forming a precoxal bridge. Both the mesosternum and metasternum are usually invaginated in Diptera and not visible externally. All these and additional features, including the thoracic chaetotaxy, are shown in Figs. 41.3, 41.8, and 41.9.

Wing and Halter

In Diptera, only the first or mesothoracic pair of wings is developed as functional flight organs; the metathoracic pair is reduced to a pair of small, clublike balancing organs, the halteres (Figs. 41.1–41.3, 41.149). The axillary area of each wing (Fig. 41.11) contains three sclerotized axillary plates.

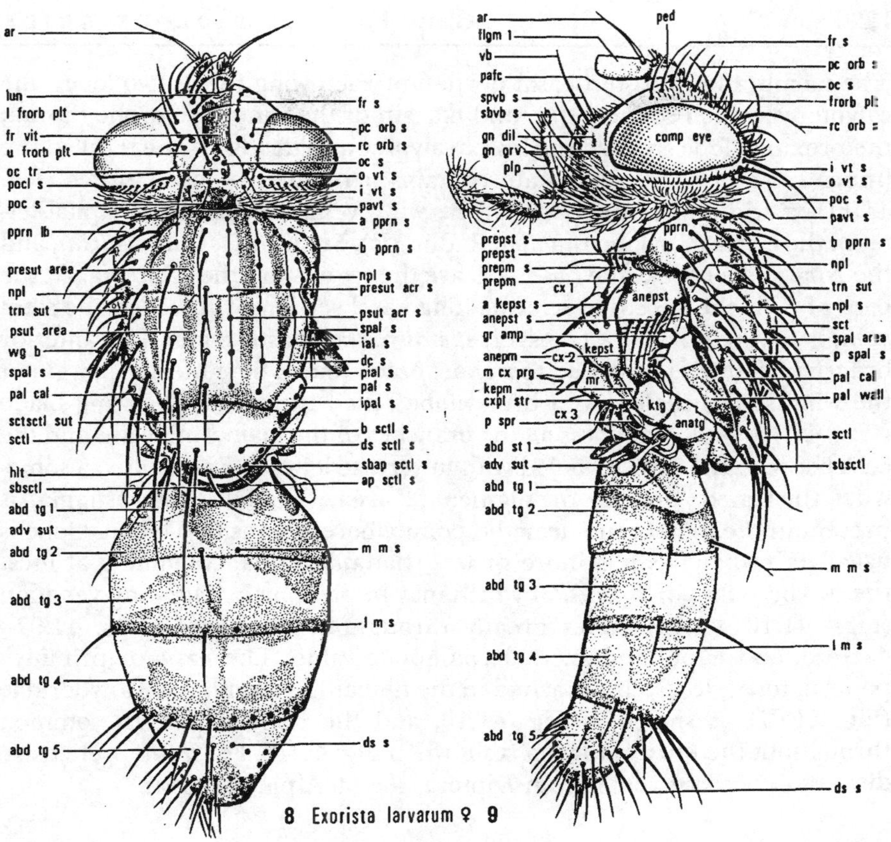

8 Exorista larvarum ♀ 9

Figs. 41.8.–41.9. Morphology, chaetotaxy, and terminology of body of *Exorista larvarum* (Linnaeus), **Tachinidae,** in **8.** dorsal and **9.** lateral views.

Abbreviations:

abd st, abdominal sternite
abd tg, abdominal tergite
acr s, acrostichal seta
adv sut, adventitious suture
a kepst s, anterior katepisternal seta
anatg, anatergite
anepm, anepimeron
anepst, anepisternum
anepst s, anepisternal seta
a pprn s, anterior postpronotal seta
ap sctl s, apical scutellar seta
ar, arista
b pprn s, basal postpronotal seta
b sctl s, basal scutellar seta
comp eye, compound eye
cx, coxa
cxpl str, coxopleural streak
dc s, dorsocentral seta
ds s, discal seta
ds sctl s, discal scutellar seta
flgm, flagellomere
frorb plt, fronto-orbital plate
fr plt, frontal plate
fr s, frontal seta
fr vit, frontal vitta
gn dil, genal dilation
gn grv, genal groove
gr amp, greater ampulla
hlt, halter
ial s, intra-alar seta
ipal s, intrapostalar seta
i vt s, inner vertical seta
kepm, katepimeron
kepst, katepisternum
ktg, katatergite
l m s, lateral marginal seta
lun, lunule
m cx prg, mid coxal prong
m m s, median marginal seta
mr, meron
npl, notopleuron
npl s, notopleural seta
oc s, ocellar seta
oc tr, ocellar triangle
orb plt, orbital plate
o vt s, outer vertical seta
pafc, parafacial
pal cal, postalar callus
pal s, postalar seta
pal wall, postalar wall
pavt s, paravertical seta
pc orb s, proclinate orbital seta
ped, pedicel
pial s, postintra-alar seta
pocl s, postocular setae
poc s, postocellar seta
pprn, postpronotum
pprn lb, postpronotal lobe
prepm, proepimeron
prepm s, proepimeral seta
prepst, proepisternum
prepst s, proepisternal seta
presut acr s, presutural acrostichal seta
presut area, presutural area of scutum
p spal s, posterior supra-alar seta
p spr, posterior spiracle
psut acr s, postsutural acrostichal seta
psut area, postsutural area
rc orb s, reclinate orbital seta
sbap sctl s, subapical scutellar seta
sbsctl, subscutellum
sbvb s, subvibrissal seta
sct, scutum
sctl, scutellum
sctsctl sut, scutoscutellar suture
spal area, supra-alar area
spal s, supra-alar seta
spvb s, supravibrissal seta
tr sut, transverse suture
vb, vibrissa
wg b, wing base.

The membranous posterobasal portion of each wing forms two lobes, the calypteres, which connect the hind margin of the wing base to the thorax; the proximal lobe is called the lower calypter and the distal one is called the upper calypter. The wing stalk contains a number of basal plates (Fig. 41.11) in addition to the bases of the wing veins (labeled with capital letters): the Costa (*C*), Subcosta (*Sc*), Radius (*R*), Media (*M*), Cubitus (*Cu*), and the Anal veins (*A*). At the base of *C* are the tegula and the basicosta. At the base of *Sc* is the subcostal sclerite. The basal section of *R*, called the stem vein, is marked off by a transverse, suturelike construction. The anterior branch of *M*(*MA*) is reduced to a short brace vein. The posterior margin of the wing stalk usually has a broad lobe called the alula. The wing blade (Figs. 41.10, 41.12) contains all the branches of the main wing veins and the cells between the veins (labeled with lowercase letters). Frequently, a somewhat thickened, opaque or pigmented area, called the pterostigma, is present in the membrane near the point where *R* joins *C*. The anal lobe is usually a rather flexible, more or less triangular area, containing at most the two branches of *A*, both of which may be atrophied. The wing venation (Figs. 41.10, 41.12) varies greatly throughout the order (Figs. 41.97–41.148), and is of fundamental taxonomic value. The basic or primitive pattern, most closely approached in the nematocerous family Tanyderidae (Fig. 41.97), is shown in Fig. 41.10, and the derived pattern common throughout the Muscomorpha is shown in Fig. 41.12. For a comprehensive discussion of wing venation in Diptera, see McAlpine (1981a).

Legs

Each leg consists of a coxa, trochanter, femur, tibia, and tarsus (Figs. 41.1, 41.2). These segments differ in the foreleg, midleg, and hindleg. The different leg surfaces are ascertained by imagining that all the legs are fully extended laterally at right angles to the main axis of the body. Thus each segment of each leg has an anterior, posterior, dorsal, and ventral surface. Intermediate surfaces are named on the same basis (e.g., anterodorsal and posteroventral). The tarsus of almost all Diptera is subdivided into five tarsomeres on all legs. On the distal end of the fifth tarsomere is the acropod (Figs. 41.15–41.17). Because it is so closely associated with the fifth tarsomere it is usually wrongly regarded as an integral part of that segment. Distally it usually bears a pair of claws, and in addition it may have a pair of pulvilla and an unpaired median empodium. The presence or absence of pulvilla or an empodium, or both, together with the form of the empodium, serve as important key characters in Diptera.

Abdomen

The abdomen of a fly (Figs. 41.18–41.22) is primitively composed of 11 segments. Each abdominal segment consists of a dorsal tergite and a ven-

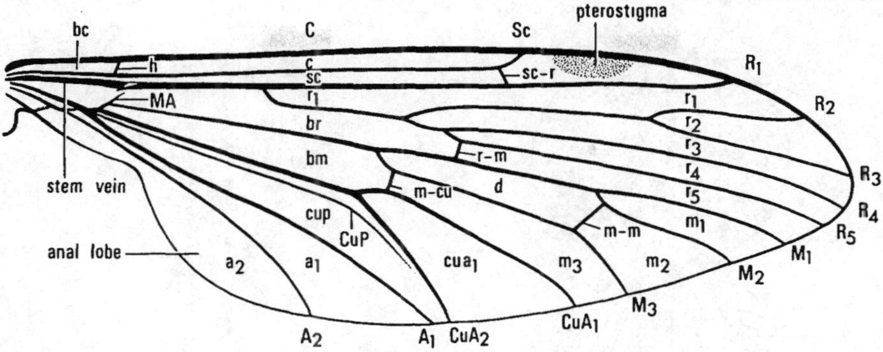

10 Ground-plan of Diptera wing

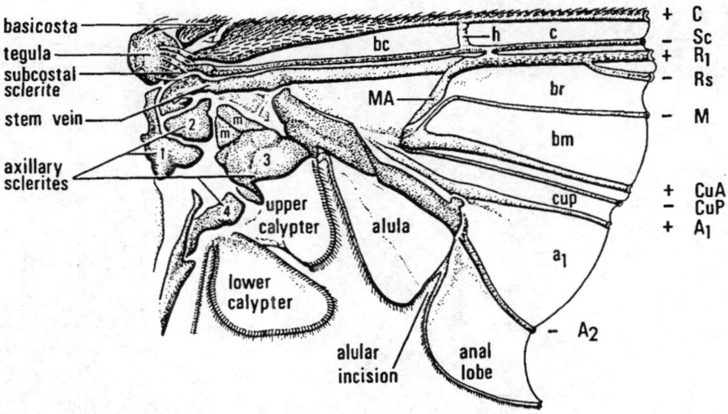

11 Tabanus americanus ♀

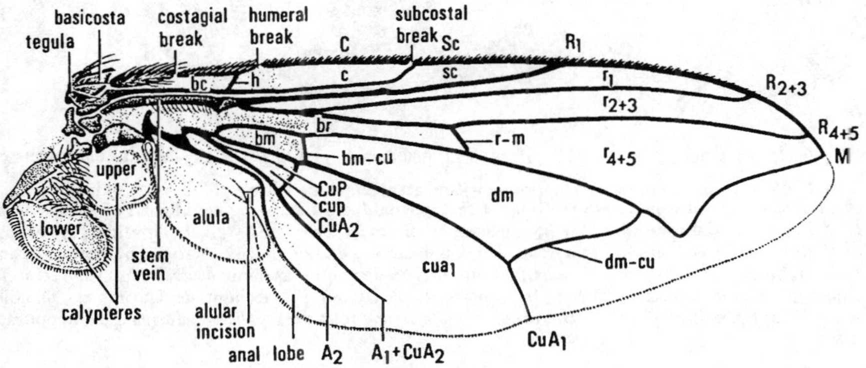

12 Paralucilia wheeleri

Figs. 41.10.–41.12. Dorsal views of right wing to show structure and venation. **10.** Wing stalk and blade of hypothetical primitive Diptera. **11.** Axillary area and wing stalk of *Tabanus americanus* Forster, **Tabanidae.** **12.** Complete wing of *Paralucilia wheeleri* (Hough), **Calliphoridae. Veins:** A_1, A_2, branches of anal veins; *C*, costa; *Cu*, cubitus; CuA_1, CuA_2, branches of anterior branch of cubitus; *CuP*, posterior branch of cubitus; *M*, media; M_1, M_2, M_3, posterior (sectoral) branches of media; *MA*, anterior branch of media; *R*, radius; R_1, anterior branch of radius; R_2, R_3, R_4, R_5, posterior (sectoral) branches of radius. **Cells:** a_1, a_2, anal; *bc*, basal costal; *bm*, basal medial; *br*, basal radial; *c*, costal; cua_1, anterior cubital (cubital fork); *cup*, posterior cubital; *d*, discal (lm_2); *dm*, discal medial; m_1, m_2, m_3, medial; r_1, r_2, r_3, r_4, r_5, radial. Crossveins: *bm-cu*, basal medial–cubital; *dm-cu*, discal medial–cubital; *h*, humeral; *m-m*, medial; *r-m*, radial medial.

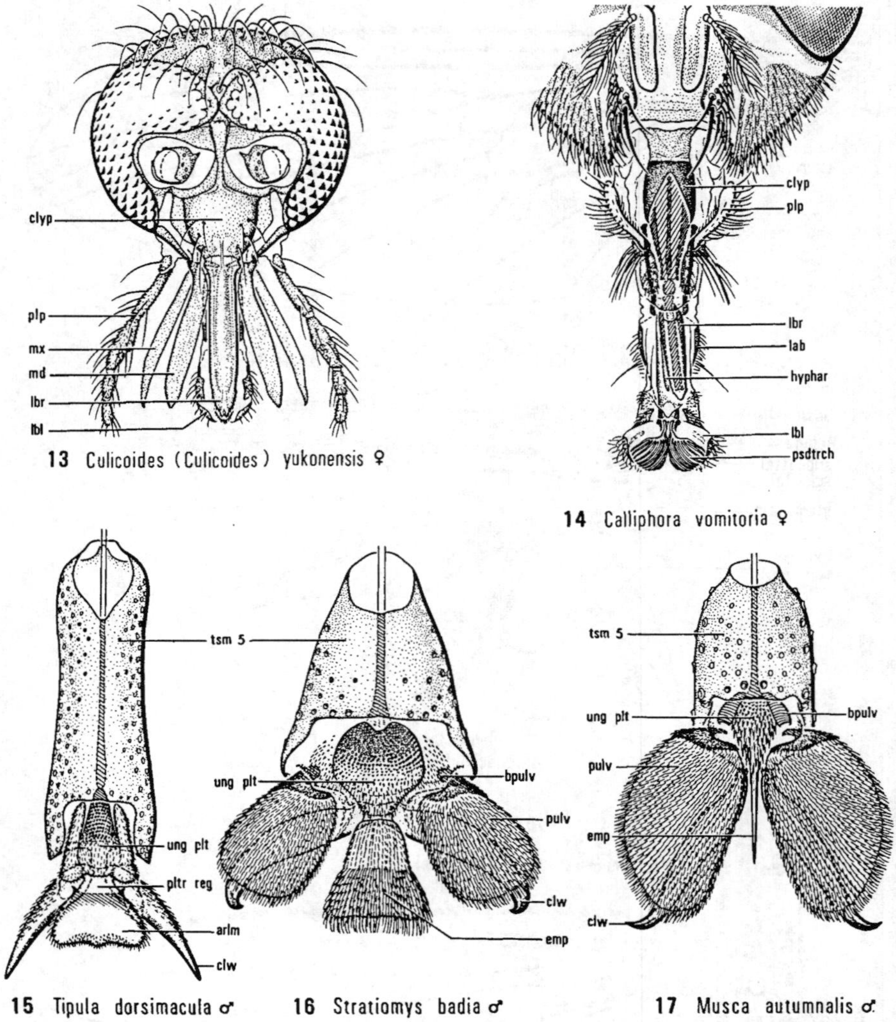

Figs. 41.13.–41.17. Mouthparts (anterior view) and acropods (ventral views) of flies. **13.** Biting mouthparts of *Culicoides (Culicoides) yukonensis* Hoffman, **Ceratopogonidae. 14.** Lapping mouthparts of *Calliphora vomitoria* (Linnaeus), **Calliphoridae. 15.** Simple acropod of *Tipula dorsimacula* Walker, **Tipulidae. 16.** Acropod of *Stratiomys badia* Walker, **Stratiomyidae,** with a pulvilliform empodium. **17.** Acropod of *Musca autumnalis* DeGeer, **Muscidae,** with a bristlelike empodium. *Abbreviations: arlm,* arolium; *bpulv,* basipulvillus; *clw,* claw; *clyp,* clypeus; *emp,* empodium; *hyphar,* hypopharynx; *lab,* labium; *lbl,* labellum; *lbr,* labrum; *md,* mandible; *mx,* maxilla; *plp,* palpus; *pltr reg,* plantar region; *psdtrch,* pseudotrachea; *pulv,* pulvillus; *ung plt,* unguitractor plate.

tral sternite, which are connected laterally by a pleural membrane. Tergite 1 is almost always shortened and closely associated with tergite 2. The abdomen is usually longer and more slender and has more of the terminal segments exposed in the Nematocera (Figs. 41.18, 41.21) and lower Brachycera than in the Muscomorpha (Figs. 41.19, 41.20, 41.22). Shorten-

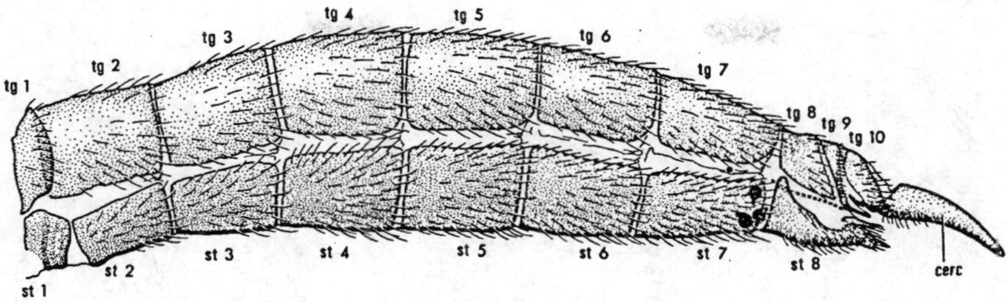

18 Trichocera columbiana ♀

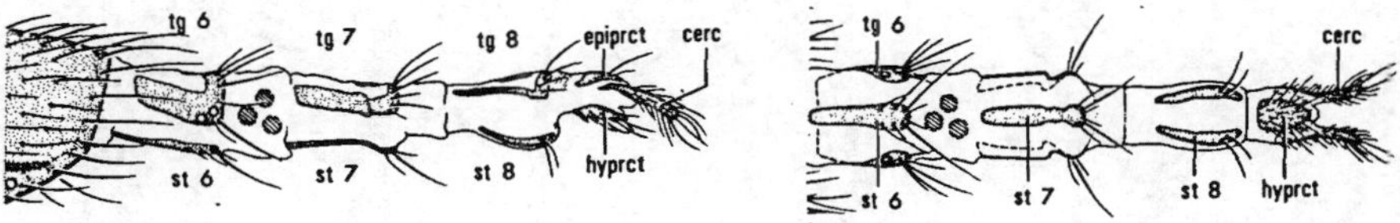

19 Anthomyia pluvialis ♀

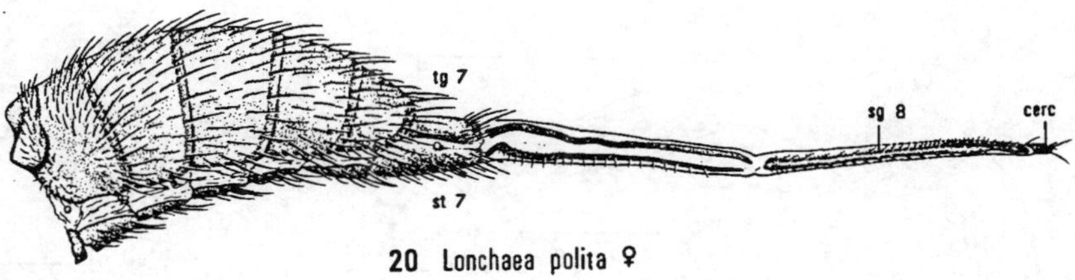

20 Lonchaea polita ♀

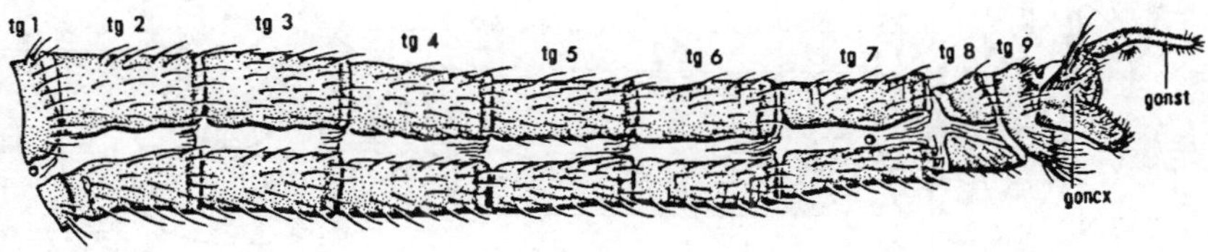

21 Trichocera garretti ♂

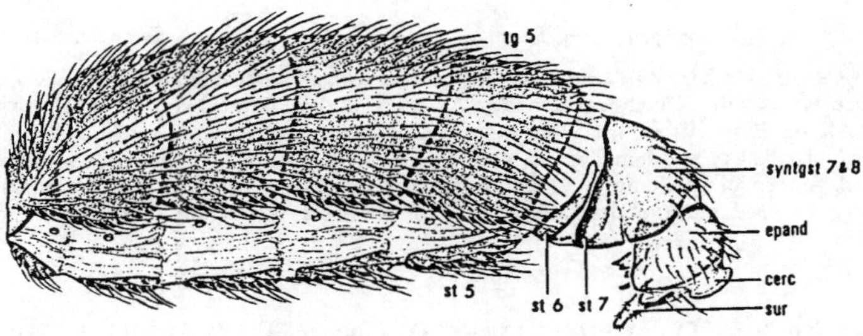

22 Neopiophila setaluna ♂

Figs. 41.18.–41.22. Female **18–20.** and male **21,22.** abdomens, respectively. of: **18.** *Trichocera columbiana* Alexander, **Trichoceridae; 19.** *Anthomyia pluvialis* (Linnaeus), **Anthomyiidae; 20.** *Lonchaea polita* Say, **Lonchaeidae; 21.** *Trichocera garretti* Alexander, **Trichoceridae;** and **22.** *Neopiophila setaluna* McAlpine, **Piophilidae.** *cerc,* cercus; *epand,* epandrium; *epiprct,* epiproct; *goncx,* gonocoxite; *gonst,* gonostylus; *sg,* segment; *st,* sternite; *syntgst,* syntergosternite; *tg,* tergite.

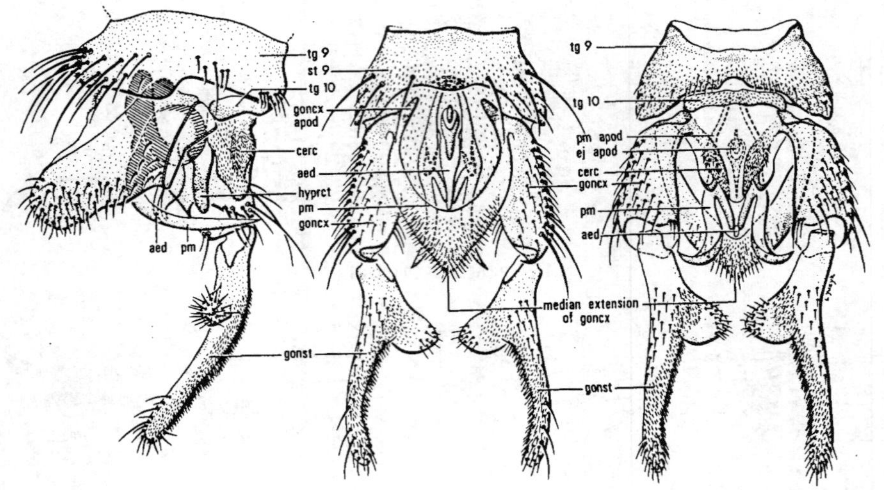

23 Trichocera garretti ♂ **24** Trichocera garretti ♂ **25** Trichocera garretti ♂

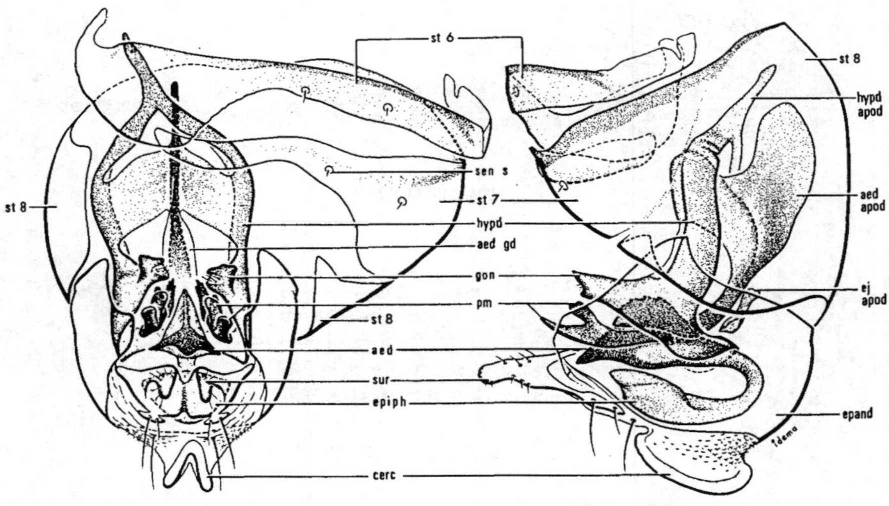

26 Neopiophila setaluna ♂ **27** Neopiophila setaluna ♂

Figs. 41.23.–41.27. Male terminalia: **23–25.** ventral, left lateral and dorsal views respectively, of *Trichocera garretti* Alexander, **Trichoceridae. 26,27.** ventral and left lateral views respectively of *Neopiophila setaluna* McAlpine, **Piophilidae.** *aed,* aedeagus; *aed apod,* aedeagal apodeme; *aed gd,* aedeagal guide; *cerc,* cercus; *ej apod,* ejaculatory apodeme; *epand,* epandrium; *epiph,* epiphallus; *gon,* gonopod; *goncx,* gonocoxite; *goncx apod,* gonocoxal apodeme; *gonst,* gonostylus; *hypd,* hypandrium; *hypd apod,* hypandrial apodeme; *pm,* paramere; *pm apod,* parameral apodeme; *sen s,* sensory seta; *st,* sternite; *tg,* tergite.

ing takes place by coalescence of the basal segments, reduction and coalescence of the terminal segments, telescopic retraction of the terminal segments into the more proximal segments, and downward and forward flexing of the terminal segments. The segments that are modified for copulation and oviposition are referred to as the terminalia. In Nematocera and lower Brachycera usually only the segments beyond segment 7 are modified for reproduction. In Muscomorpha, especially the males, all segments beyond segment 4 are frequently specially adapted for reproductive purposes. In both male and female terminalia the segments before the genital opening (between 8 and 9 in females, and between 9 and 10 in males) are referred to as pregenital segments; those behind the genital opening are called postgenital segments. The terminal portion of both male and female terminalia, consisting of rudiments of segment 11, a pair of cerci, and the anus is called the proctiger. The number of abdominal spiracles varies from eight in primitive forms to several pairs in specialized forms. The main parts of the female terminalia are shown in Figs. 41.18–41.20. In some groups, especially the Tephritoidea the terminalia are adapted for piercing (Fig. 41.20); in such groups the base of the ovipositor is formed by the seventh segment, the main shaft by the eighth segment, and the apex by the proctiger, which is comprised mainly of the fused cerci.

The main parts of the male terminalia (Figs. 41.21–41.27) are the epandrium (tergite 9) and associated lobes called surstyli (derived from tergite 10), the hypandrium and associated lobes called the gonopods (two-segmented in most Nematocera and lower Brachycera, one-segmented in practically all Muscomorpha), and the parameres (situated between the dorsomedial base of the gonopod and the posterolateral base of the aedeagus), the aedeagus, sternite 10, the epiproct, the hypoproct, the cerci, and the anus. In some Nematocera and lower Brachycera the male terminalia are permanently rotated through 180° and therefore inverted. In all Muscomorpha they are rotated through 360° and therefore circumverted (Fig. 41.22); at the same time they are usually flexed forward and stored under the pregenital segments for protection. For a comprehensive treatment of both male and female terminalia of Diptera, see McAlpine (1981a) in the *Manual of Nearctic Diptera,* or Van Emden and Hennig (1970).

ACKNOWLEDGMENTS

The typescript was reviewed by Dr. H. J. Teskey and Dr. J. R. Vockeroth; their time and valuable suggestions were very much appreciated.

KEY TO FAMILIES

1a Antennal flagellum with four or more freely articulated flagellomeres, with apical segments not consolidated into a stylus or arista (Figs. 41.3, 41.28–41.40); palpus usually with three to five segments; Suborder **Nematocera** 2

1b Antennal flagellum consolidated into a single compound segment bearing a stylus or arista (Figs. 41.1, 41.2, 41.41–41.96); palpus with not more than two segments; Suborder **Brachycera** 14

2a Halter with a basal appendage, the prehalter (Figs. 41.40, 41.149) **Ptychopteridae**

2b Halter without a prehalter (Fig. 41.3) 3

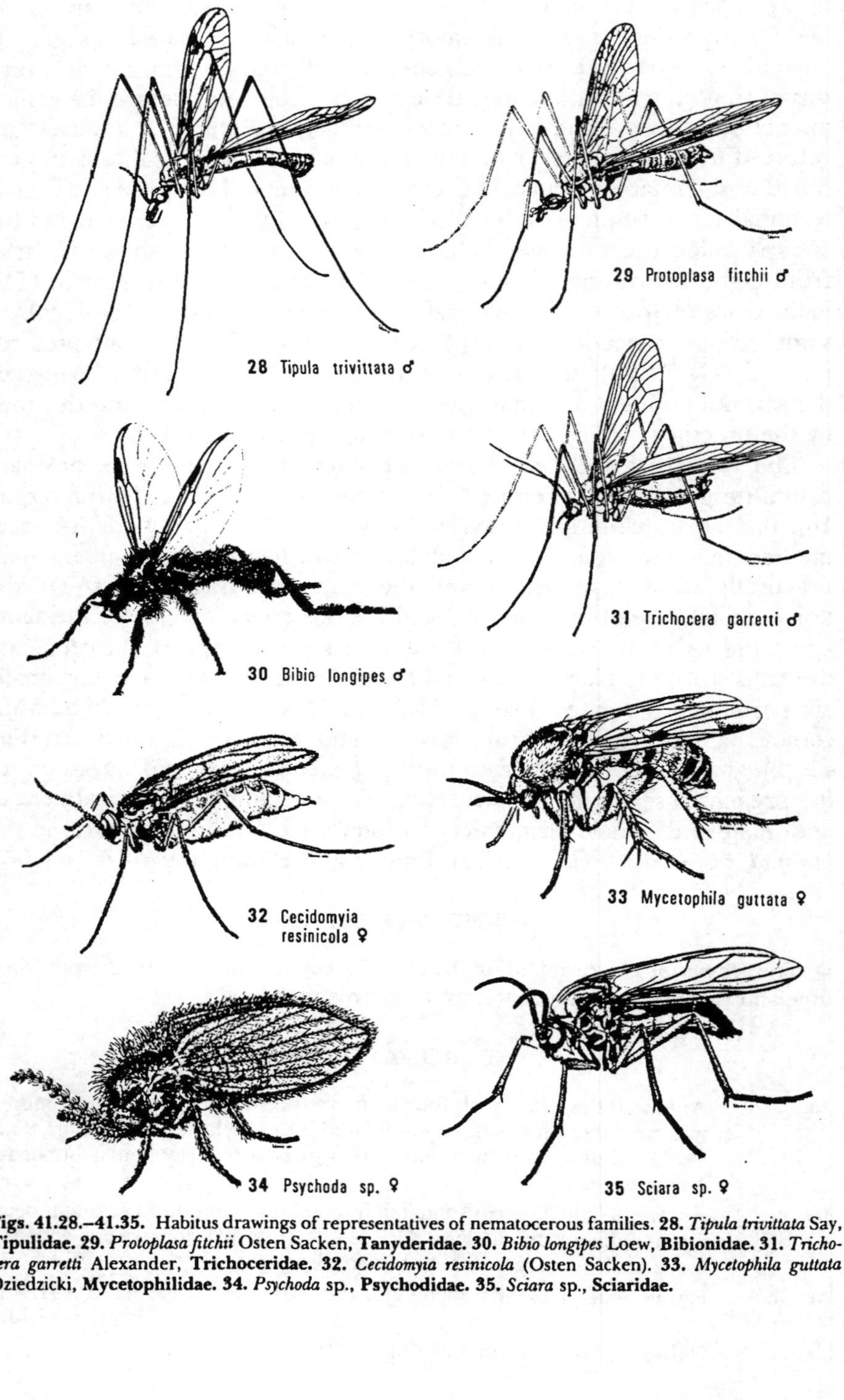

Figs. 41.28.–41.35. Habitus drawings of representatives of nematocerous families. **28.** *Tipula trivittata* Say, **Tipulidae. 29.** *Protoplasa fitchii* Osten Sacken, **Tanyderidae. 30.** *Bibio longipes* Loew, **Bibionidae. 31.** *Trichocera garretti* Alexander, **Trichoceridae. 32.** *Cecidomyia resinicola* (Osten Sacken). **33.** *Mycetophila guttata* Dziedzicki, **Mycetophilidae. 34.** *Psychoda* sp., **Psychodidae. 35.** *Sciara* sp., **Sciaridae.**

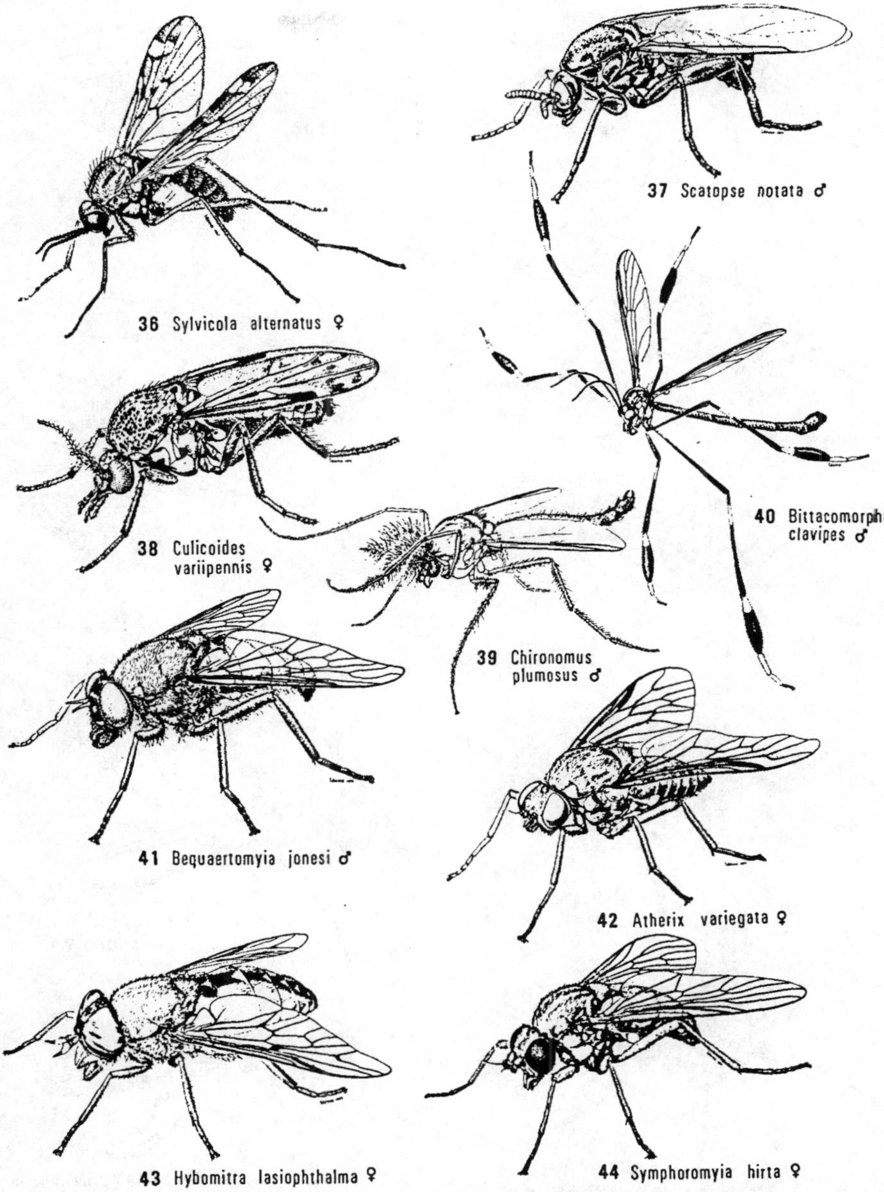

Figs. 41.36.–41.44. Habitus drawings of representatives of families of Nematocera and orthorrhaphous Brachycera. **36.** *Sylvicola alternatus* (Say), **Anisopodidae. 37.** *Scatopse notata* (Linnaeus), **Scatopsidae. 38.** *Culicoides (Monoculicoides) variipennis* (Coquillet), **Ceratopogonidae. 39.** *Chironomus plumosus* (Linnaeus), **Chironomidae. 40.** *Bittacomorpha clavipes* (Fabricius), **Ptychopteridae. 41.** *Bequaertomyia jonesi* (Cresson), **Pelecorhynchidae. 42.** *Atherix variegata* Walker, **Athericidae. 43.** *Hybomitra lasiophthalma* (Macquart), **Tabanidae. 44.** *Symphoromyia hirta* Johnson, **Rhagionidae.**

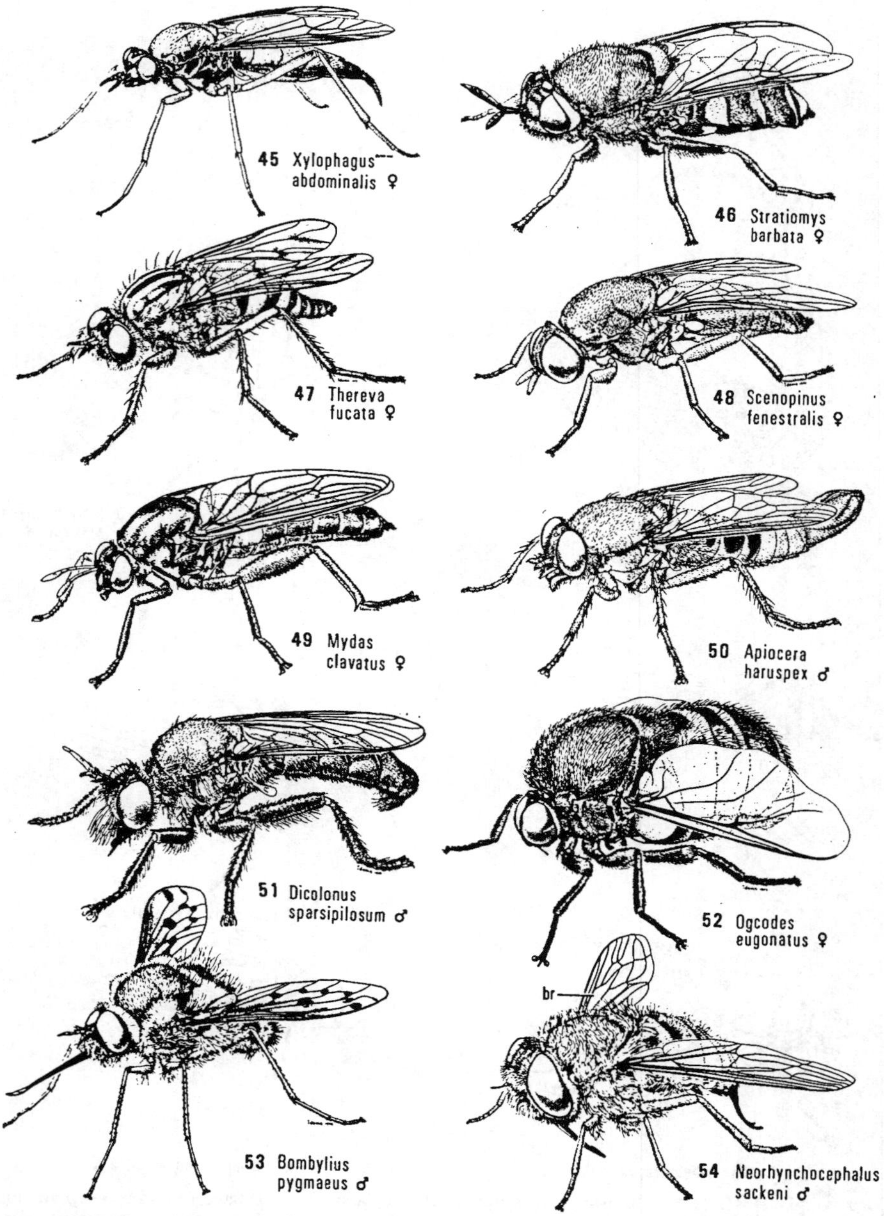

Figs. 41.45.–41.54. Habitus drawings of representatives of families of orthorrhaphous Brachycera. **45.** *Xylophagus abdominalis* Loew, **Xylophagidae. 46.** *Stratiomys barbata* Loew, **Stratiomyidae. 47.** *Thereva fucata* Loew, **Therevidae. 48.** *Scenopinus fenestralis* (Linnaeus), **Scenopinidae. 49.** *Mydas clavatus* (Drury), **Mydidae. 50.** *Apiocera haruspex* Osten Sacken, **Apioceridae. 51.** *Dicolonus sparsipilosum* Back, **Asilidae. 52.** *Ogcodes eugonatus* Loew, **Acroceridae. 53.** *Bombylius pygmaeus* Fabricius, **Bombyliidae. 54.** *Neorhyncocephalus sackenii* (Williston), **Nemestrinidae.**

3a Wing with two strong anal veins, A_1 and A_2, reaching wing margin (Figs. 41.99, 41.100). **4**

3b Wing with, at most, A_1 reaching wing margin, A_2 at least, fading out before wing margin **5**

4a Ocelli absent (Fig. 41.3); A_2 usually more than half as long as A_1 and relatively straight (Fig. 41.99); scutum with a complete V-shaped suture (Fig. 41.3, 41.28) . . . . **Tipulidae**

4b Ocelli present (Fig. 41.31); A_2 much less than half as long as A_1, and strongly curved at apex (Fig. 41.100); scutum with V-shaped suture incomplete in middle (Fig. 41.31) . . . . **Trichoceridae**

5a *R* with five free branches; cell *d* present; anal lobe well developed (Fig. 41.97); resemble crane flies (Fig. 41.29) . . . . **Tanyderidae**

5b *R* rarely with five branches, but if so (some **Psychodidae** Fig. 41.98), then cell *d* absent and anal lobe undeveloped; form variable **6**

6a *C* continuing around wing, although weaker along hind margin (Figs. 41.98, 41.103) **7**

6b *C* ending just beyond insertion of last branch of *R*, usually near apex of wings (Figs. 41.101, 41.102, 41.104–41.108) **8**

7a Wing with abundant, long, coarse hairs; usually pointed at apex, and with 10 veins reaching wing margin (Fig. 41.98); small mothlike flies that usually hold wings rooflike over body (Fig. 41.34) . . . . **Psychodidae**

7b Wing with shorter, sparser, finer hairs; rounded at apex, and with three to seven veins reaching wing margin (Fig. 41.103); small, delicate gnatlike flies that usually hold wings flat over body (Fig. 41.32). . . . **Cecidomyiidae**

8a Ocelli absent. **9**

8b Ocelli present **10**

9a Two branches of *M* present, M_2 usually distinct, sometimes weakened at base; never more than two branches of *R* reaching wing margin, R_{2+3} if present, crossvein-like and forming a closed cell r_1 (Fig. 41.102); mouthparts of female usually with bladelike mandibles (Fig. 41.13); postnotum usually without a longitudinal groove; tiny biting midges (Fig. 41.38) . . . . **Ceratopogonidae**

9b One branch of *M* present, M_2 never apparent; usually three branches of *R* reaching wing margin, R_{2+3} if present, reaching wing margin (Fig. 41.104); mouthparts of both sexes lacking functional mandibles; postnotum usually with a longitudinal groove; small to moderately large, nonbiting midges (Fig. 41. 39). . . . . **Chironomidae**

10a Wing with cell *d* present and *M* with three branches (Figs. 41.36, 41.107). . . . . **Anisopodidae**

10b Wing with cell *d* absent and *M* never with more than two branches (Figs. 41.101, 41.105, 41.106, 41.108) **11**

11a Wing with two closed basal cells, that is, with cells *br* and *bm* separate and closed distally (Fig. 41.101); pulvilli and empodium strongly and equally developed, acropod with three relatively similar pads (Fig. 41.30) . . . . **Bibionidae**

11b Wing with one closed basal cell, that is, either with cells *br* and *bm* confluent (Fig. 41.108) or with cell *bm* open to wing margin (Figs. 41.105, 41.106); pulvilli at most weakly developed, acropod never with three similar pads **12**

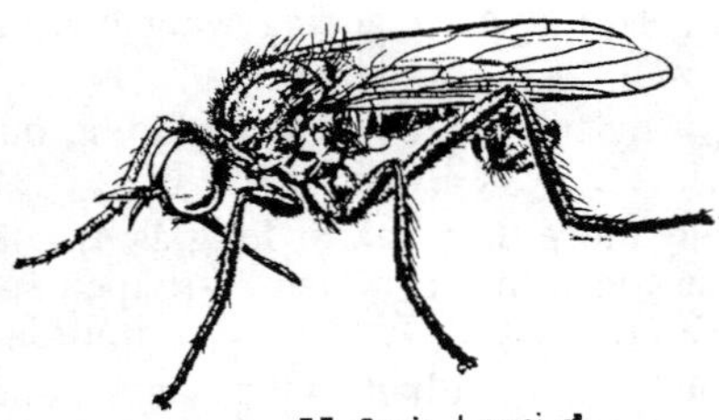

55 Empis browni ♂

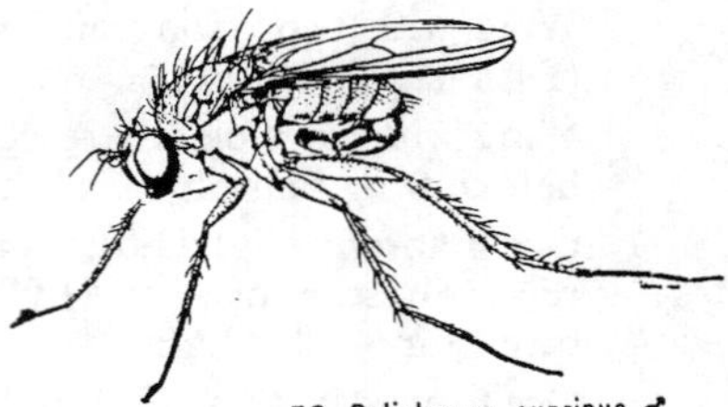

56 Dolichopus cuprinus ♂

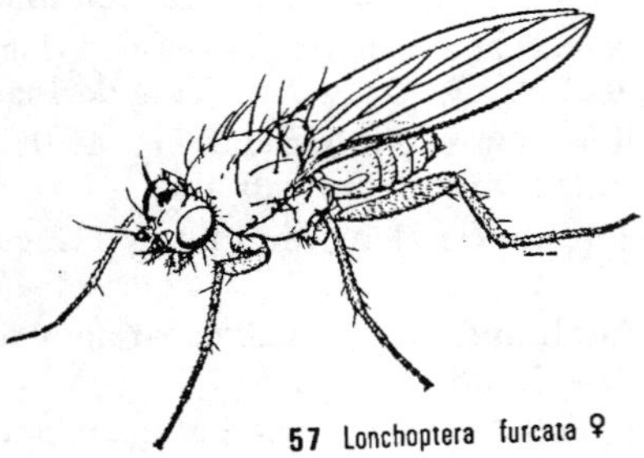

57 Lonchoptera furcata ♀

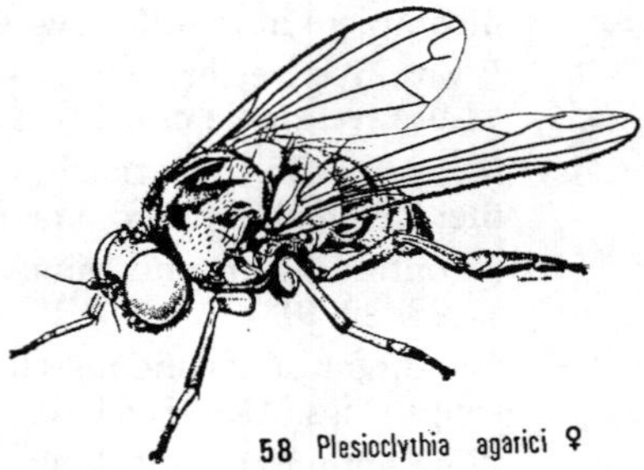

58 Plesioclythia agarici ♀

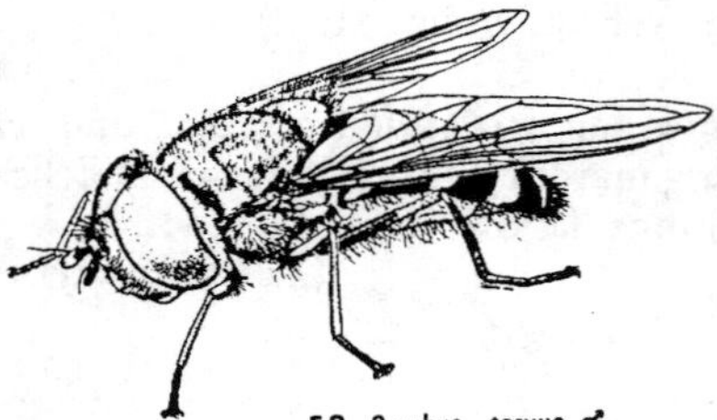

59 Syrphus torvus ♂

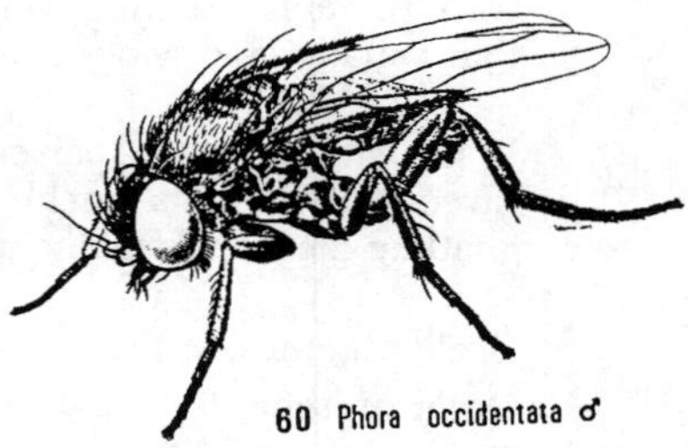

60 Phora occidentata ♂

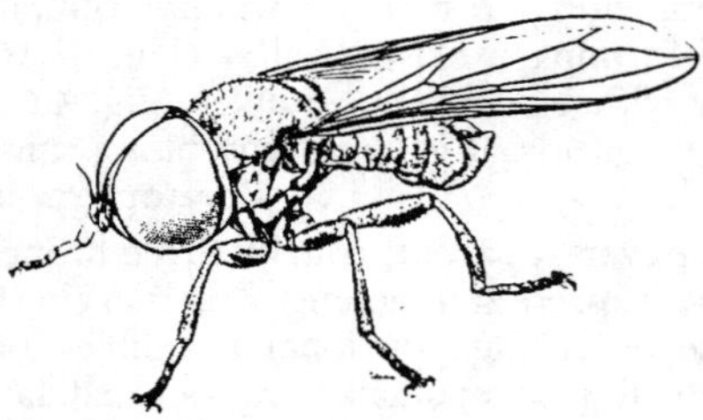

61 Pipunculus ater ♂

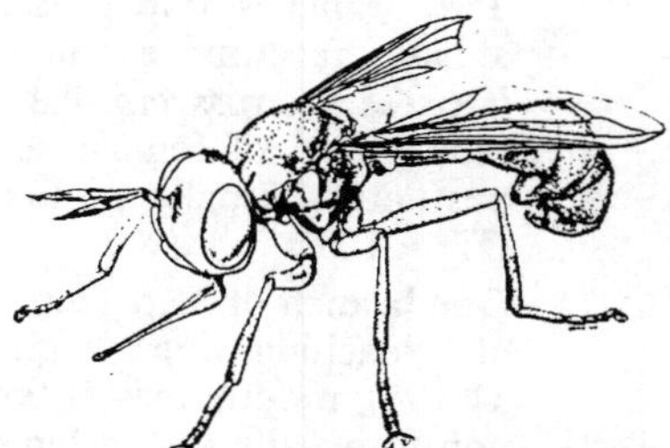

62 Physoconops brachyrhynchus ♂

Figs. 41.55.–41.62. Habitus drawings of representatives of families of orthorrhaphous and cyclorrhaphous Brachycera. **55.** *Empis browni* Curran, **Empididae. 56.** *Dolichopus cuprinus* Wiedemann, **Dolichopodidae. 57.** *Lonchoptera furcata* (Fallén), **Lonchopteridae. 58.** *Plesioclythia agarici* (Willard), **Platypezidae. 59.** *Syrphus torvus* Osten Sacken, **Syrphidae. 60.** *Phora occidentata* Malloch, **Phoridae. 61.** *Pipunculus ater* Meigen, **Pipunculidae. 62.** *Physoconops brachyrhynchus* (Macquart), **Conopidae.**

12a Tibiae without apical spurs; wing with *C* ending far before wing apex, and usually with only *C* and *R* darkly pigmented (Figs. 41.37, 41.108) **Scatopsidae**

12b Tibiae with apical spurs; wing with *C* ending at or near wing apex, and usually with all veins darkly pigmented (Figs. 41.105, 41.106) . . . **13**

13a Eyes meeting in a narrow eye bridge above antennae (Fig. 41.35) (except in ***Pnyxia*** in which eyes are reduced in both sexes and wings are lacking in female); wing with stem and fork of *M* subequal in length, and with fork distinctly bell-shaped (Fig. 41.106); all mesothoracic pleural sclerites bare (Fig. 41.35) . . . . **Sciaridae**

13b Eyes never meeting above antennae (Fig. 41.33); wing usually with fork of *M* much longer than stem, and lanceolate rather than bell-shaped (Fig. 41.105); usually, some mesothoracic pleural sclerites at least weakly haired (Fig. 41.33) . . . . **Mycetophilidae**

14a Ptilinal fissure and lunule absent (Figs. 41.41–41.61); CuA_2 usually long, often reaching wing margin or joining A_1 near its apex (Figs. 41.109–41.131); **Orthorrhaphous Brachycera** and **Muscomorpha, Aschiza** **15**

14b Ptilinal fissure and lunule present (Figs. 41.4, 41.62–41.96); CuA_2 usually short, usually joining A_1 near wing base (Figs. 41.132–41.148); **Muscomorpha, Schizophora** **36**

15a Empodia pulvilliform, each acropod with three relatively similar pads below tarsal claws (Fig. 41.16); CuA_2 free (Fig. 41.111) or meeting A_1 in an acute angle near wing margin (Figs. 41.109, 41.110, 41.112) . . **16**

15b Empodia usually setiform (Fig. 41.17) or absent; if acropod with three pads then empodium much narrower than pulvilli, if empodia somewhat pulvilliform (some water-striding **Empididae** and **Dolichopodidae**), then CuA_2 joining A_1 at an obtuse angle far from wing margin (Figs. 41.123, 41.125) **24**

16a Head unusually small, rarely more than half as wide as thorax; eyes holoptic or nearly so in both sexes; lower lobe of calypter extremely large, wider than head (Fig. 41.52) . . . . **Acroceridae**

16b Head more than half as broad as thorax; eyes never holoptic in female; lower lobe of calypter smaller, not as wide as head **17**

17a Venation peculiar: branches of *R* and *M* more or less converging to apex of wing; branches of *M* curving forward and ending before or scarcely beyond apex of wing, and with a composite "diagonal vein" running from distal end of cell *br* to posterior margin of wing (Fig. 41.54) . . . . **Nemestrinidae**

17b Venation more normal than above; branches of *R* and *M* diverging to apical wing margin; branches of *M* joining margin far behind apex of wing, and without a composite "diagonal vein" (Figs. 41.109–41.112) **18**

18a *C* ending near apex of wing (Figs. 41.111, 41.112). **19**

18b *C* continuing around wing, though usually weaker along posterior margin (Figs. 41.109, 41.110) **20**

19a *C* usually ending well before, rarely nearly at, wing apex; branches of *R* more or less crowded anteriorly and all ending in margin well before wing apex; cell *d* (or *dm*) short, usually little longer than wide (Fig. 41.112); tibial spurs usually absent (Fig. 41.46) . . . . **Stratiomyidae**

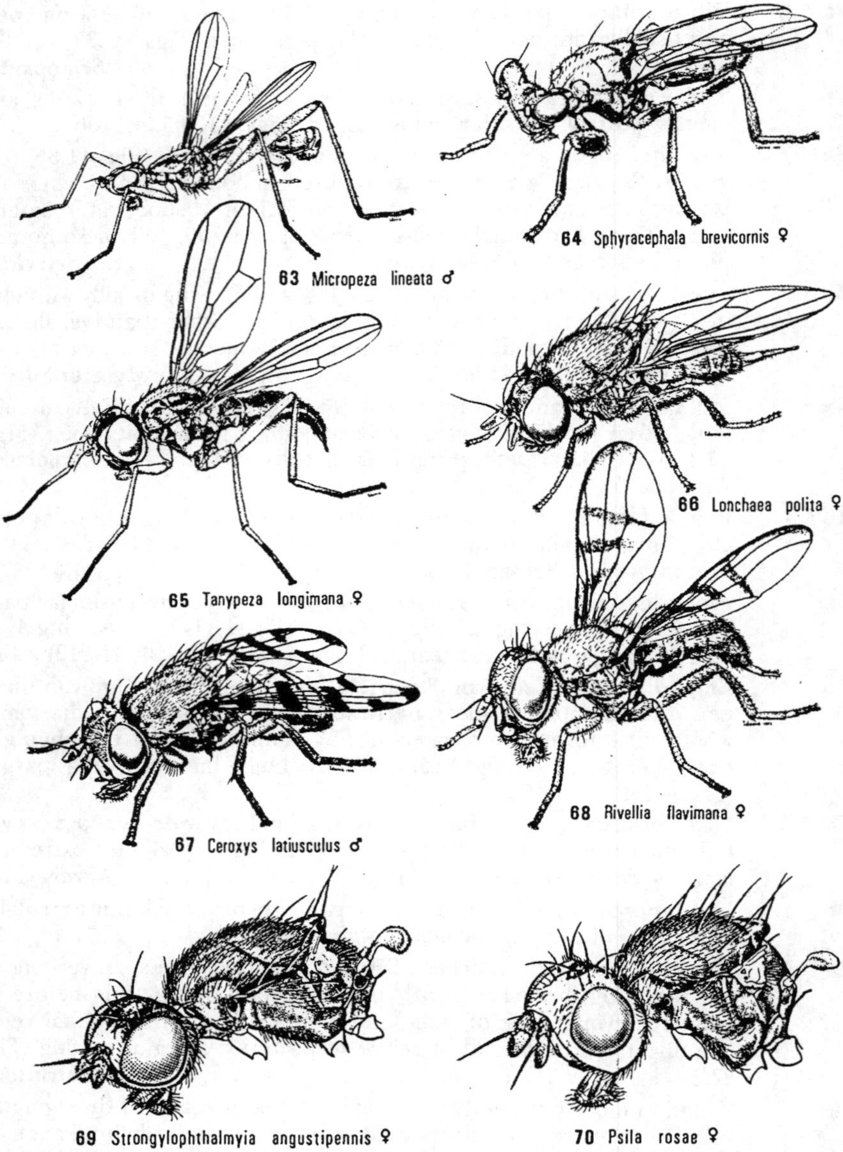

Figs. 41.63.–41.70. Habitus drawings of represenatives of families of acalyptrate Muscomorpha. **63.** *Micropeza lineata* Van Duzee, **Micropezidae. 64.** *Sphyracephala brevicornis* (Say), **Diopsidae. 65.** *Tanypeza longimana* Fallén, **Tanypezidae. 66.** *Lonchaea polita* Say, **Lonchaeidae. 67.** *Ceroxys latiusculus* (Loew), **Otitidae. 68.** *Rivellia flavimana* Loew, **Platystomatidae 69.** *Strongylophthalmyia angustipennis* Melander, **Strongylophthalmyiidae. 70.** *Psila rosae* (Fabricius), **Psilidae.**

19b C extending beyond wing apex; branches of R not crowded anteriorly, with R_5 ending at or beyond wing apex; cell d at least two times as long as broad (Fig. 41.111); tibial spurs present on mid and hind tibiae (Fig. 41.45) . . . . **Xylophagidae**

20a Subscutellum strongly developed; posterior thoracic spiracle with a scalelike elevation immediately behind it (Fig. 41.151) (reduced in ***Glutops***) **21**

20b Subscutellum absent or very weakly developed; posterior thoracic spiracle never with a scalelike elevation immediately behind it. . . **23**

21a Flagellum with a slender, nonannulated arista (Fig. 41.42); wing with cell r_1 closed by vein R_{2+3} meeting C at end of R_1 (Fig. 41.110). . . . **Athericidae**

21b Flagellum with a coarse, annulated stylus (Figs. 41.41, 41.43); wing with cell r_1 open, and with vein R_{2+3} meeting C far beyond end of R_1 (Fig. 41.109) **22**

22a Both upper and lower calypteres large, subequal in size (Fig. 41.151); first abdominal tergite deeply notched in middle of posterior margin, and with a median suture (Fig. 41.152) . . . . **Tabanidae**

22b Upper calypter moderately large; lower one scarcely developed; first abdominal tergite without a median notch or suture (Fig. 41.41) . . . **Pelecorhynchidae**

23a Wing with anal lobe and alula undeveloped; alular incision absent (Fig. 41.117); scutellum small, with flat bare disk. . . . . . **Vermileonidae**

23b Wing with anal lobe and alula developed; alular incision present (Figs. 41.1, 41.2); scutellum large, with convex hairy disk (Figs. 41.1, 41.2) **Rhagionidae**

24a CuA_2 reaching wing margin near A_1 (Figs. 41.118, 41.120) or joining A_1 near wing margin (Figs. 41.113–41.115, 41.130, 41.131); if joining A_1, then CuA_2 at least 1.5 times longer than apical section of A_1, except in a few **Bombyliidae** (***Apolysis*** and ***Prorates*** spp.) **25**

24b CuA_2 absent (Figs. 41.124, 41.129), vestigial (Fig. 41.123) or joining A_1 far from wing margin (Figs. 41.116, 41.121–41.123, 41.128); if joining A_1, then CuA_2 not or scarcely longer than apical section of A_1, except in a few **Empididae** (***Brachystoma*** spp.) and **Platypezidae** (***Polypivora*** spp.) **31**

25a Branches of M peculiarly curved forward, more or less parallel to posterior wing margin; at least M_1 ending freely in wing margin before wing apex (Fig. 41.113); large asilidlike flies **26**

25b Branches of M not curved as above, but if forwardly bent, then not ending freely in wing margin (Fig. 41.131); size and form variable **27**

26a Flagellum unusually long, with a slender stalklike base and strongly clubbed apex; stalk at least twice as long as combined length of scape and pedicel (Fig. 41.49); a single ocellus present (Fig. 41.49) . . . . . **Mydidae**

26b Flagellum shorter than head, without a stalklike base; entire flagellum not much longer than combined length of scape and pedicel (Fig. 41.50); three ocelli present (Fig. 41.50) . . . . **Apioceridae**

27a Spurious vein evident as a strong veinlike fold between Rs and M; M_1 curved forward more or less in line with crossvein $dm–cu$ and joining

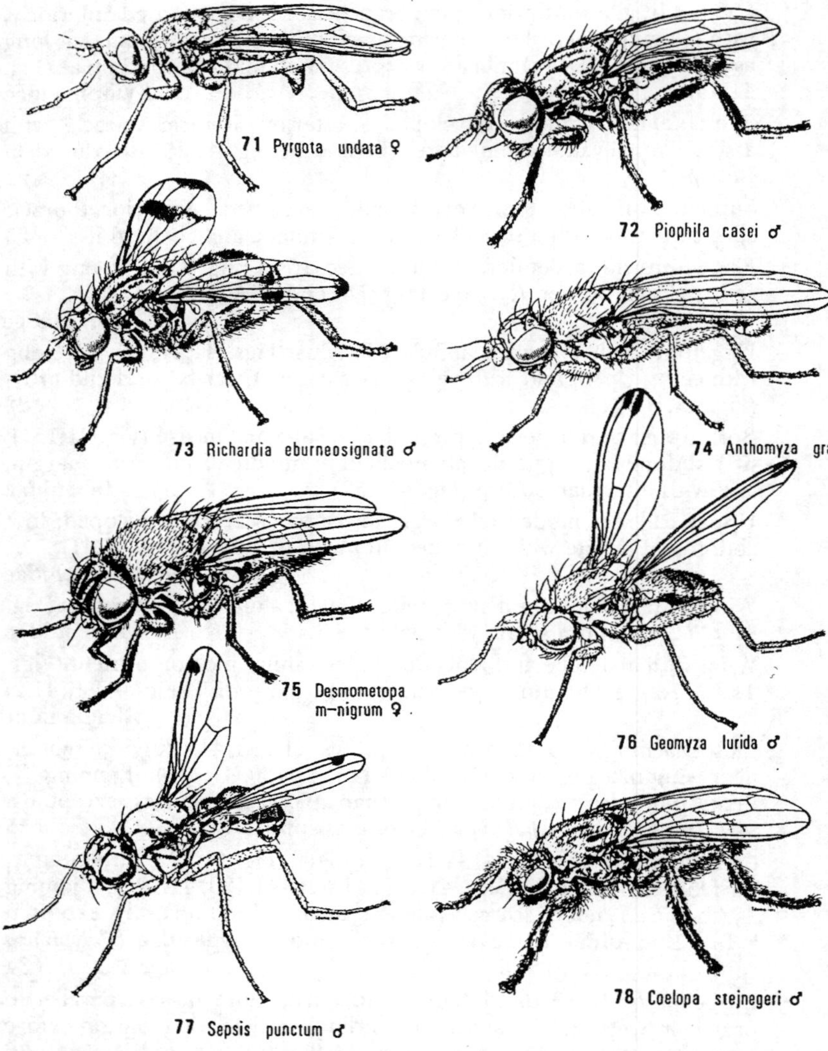

Figs. 41.71.–41.78. Habitus drawings of representatives of families of acalyptrate Muscomorpha. **71.** *Pyrgota undata* Wiedemann, **Pyrgotidae. 72.** *Piophila casei* (Linnaeus), **Piophilidae. 73.** *Richardia eburneosignata* Hennig, **Richardiidae. 74.** *Anthomyza gracilis* Fallén, **Anthomyzidae. 75.** *Desmometopa m-nigrum* (Zetterstedt), **Milichiidae. 76.** *Geomyza lurida* (Loew), **Opomyzidae. 77.** *Sepsis punctum* Linnaeus, **Sepsidae. 78.** *Coelopa stejnegeri* Aldrich, **Coelopidae.**

an unbranched R_{4+5} in a crossvein-like manner (Figs. 41.59, 41.131) **Syrphidae**

27b Spurious vein undeveloped; M_1 usually not curved forward in a crossvein-like manner, but if so (some **Bombyliidae**), then joining a furcate R_{4+5} (Fig. 41.118) **28**

28a Head strongly hemispherical, with compound eyes unusually large and almost meeting on midline both above and below antenna (Fig. 41.61);

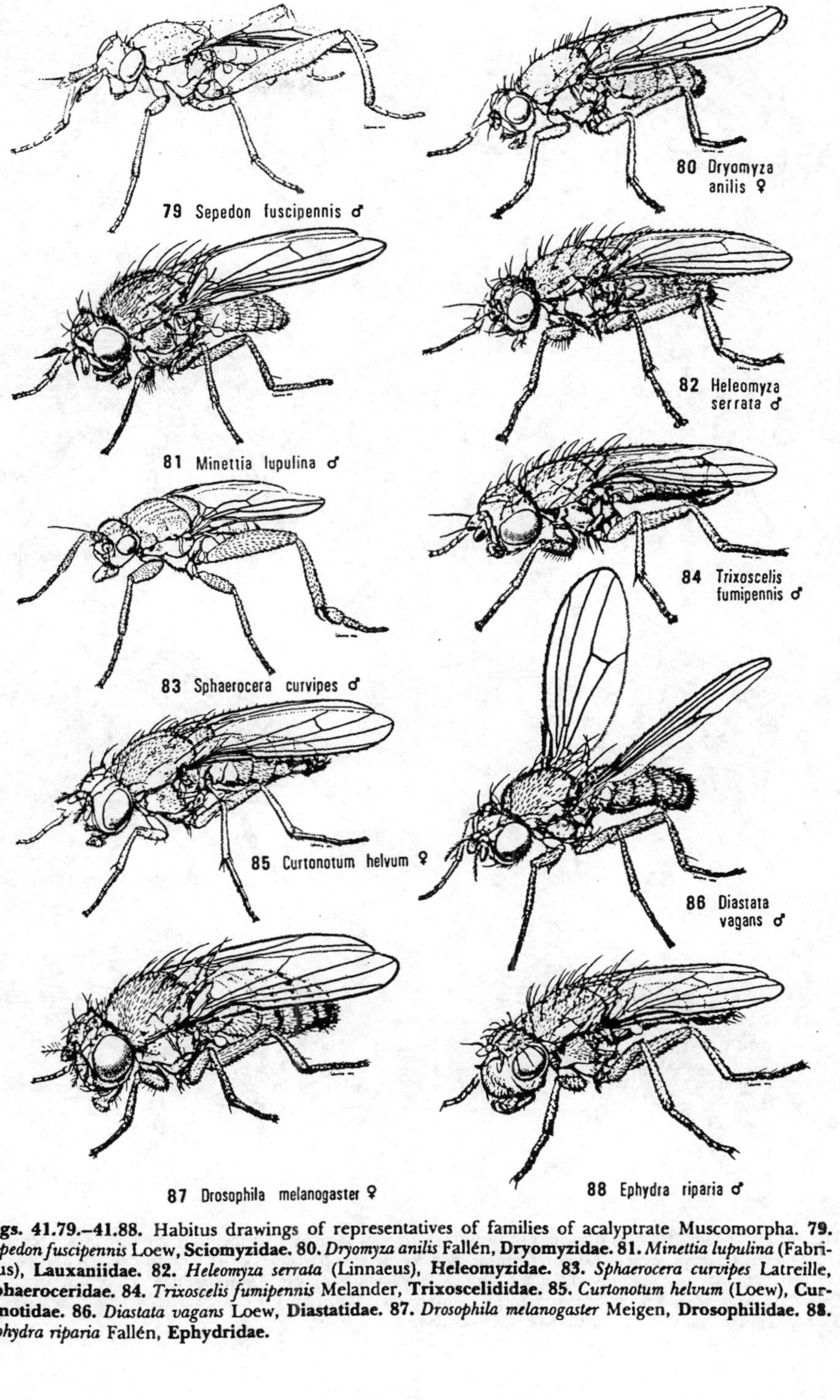

Figs. 41.79.–41.88. Habitus drawings of representatives of families of acalyptrate Muscomorpha. **79.** *Sepedon fuscipennis* Loew, **Sciomyzidae. 80.** *Dryomyza anilis* Fallén, **Dryomyzidae. 81.** *Minettia lupulina* (Fabricius), **Lauxaniidae. 82.** *Heleomyza serrata* (Linnaeus), **Heleomyzidae. 83.** *Sphaerocera curvipes* Latreille, **Sphaeroceridae. 84.** *Trixoscelis fumipennis* Melander, **Trixoscelididae. 85.** *Curtonotum helvum* (Loew), **Curtonotidae. 86.** *Diastata vagans* Loew, **Diastatidae. 87.** *Drosophila melanogaster* Meigen, **Drosophilidae. 88.** *Ephydra riparia* Fallén, **Ephydridae.**

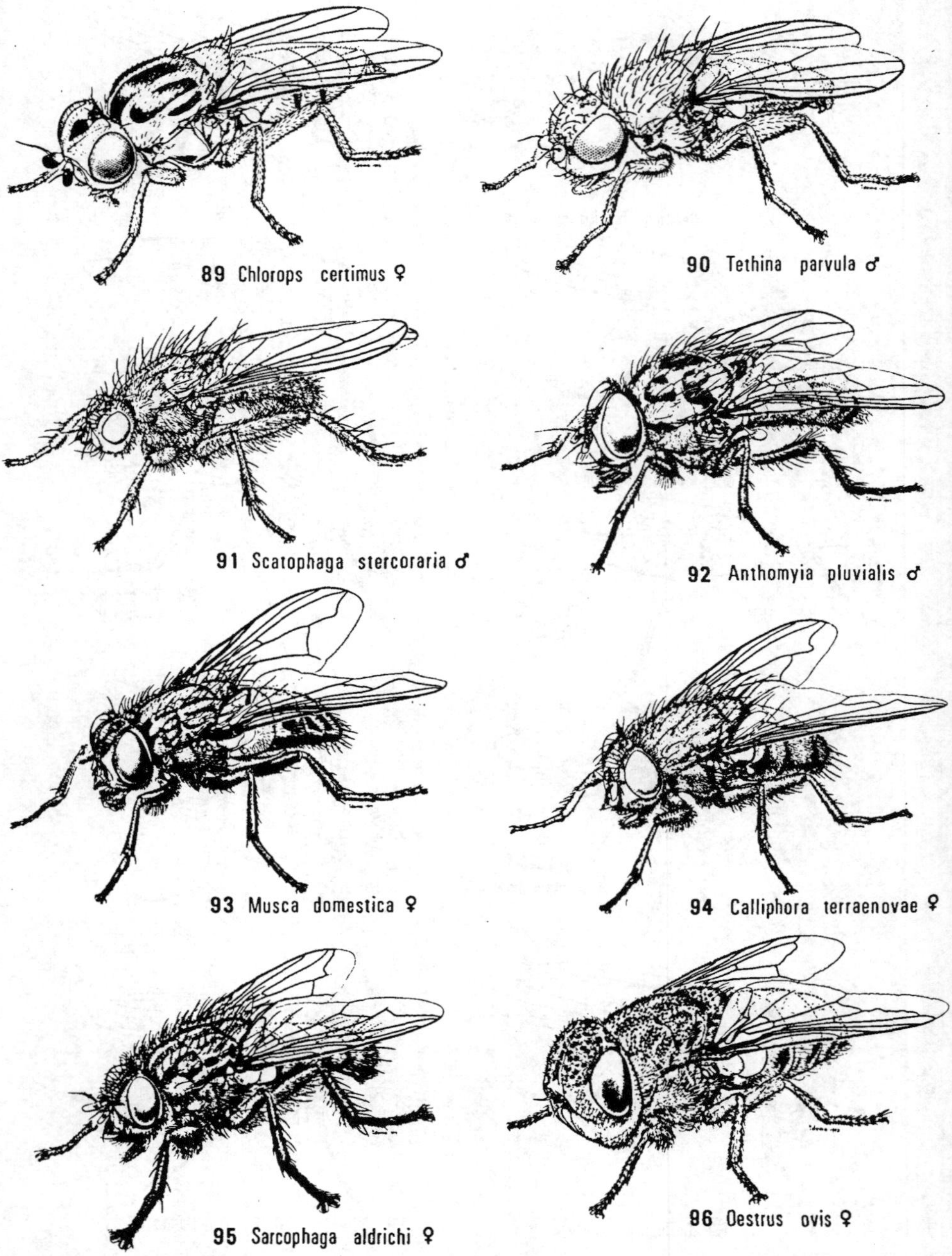

Figs. 41.89.–41.96. Habitus drawings of representatives of families of acalyptrate and calyptrate Muscomorpha. **89.** *Chlorops certimus* Adams, **Chloropidae. 90.** *Tethina parvula* (Loew), **Tethinidae. 91.** *Scatophaga stercoraria* (Linnaeus), **Scatophagidae. 92.** *Anthomyia pluvialis* Malloch, **Anthomyiidae. 93.** *Musca domestica* Linnaeus, **Muscidae. 94.** *Calliphora terraenovae* Macquart, **Calliphoridae. 95.** *Sarcophaga aldrichi* Parker, **Sarcophagidae. 96.** *Oestrus ovis* Linnaeus, **Oestridae.**

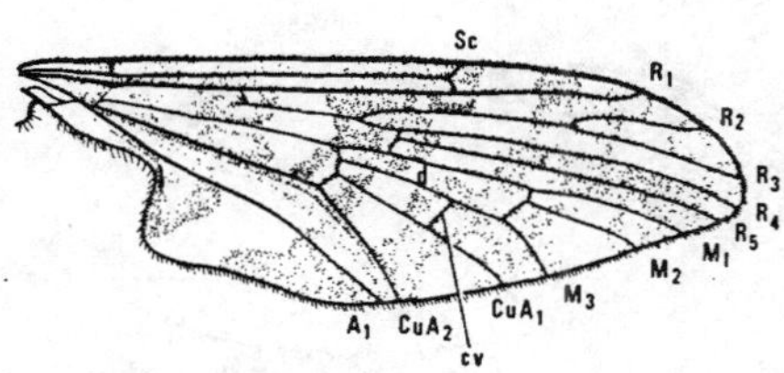

97 Protoplasa fitchii ♂

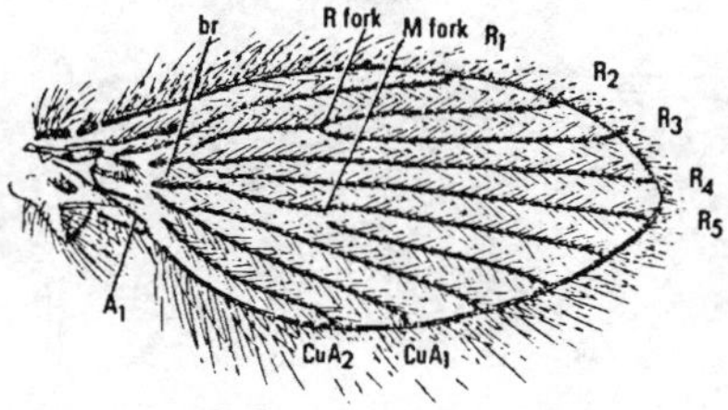

98 Pericoma marginalis ♂

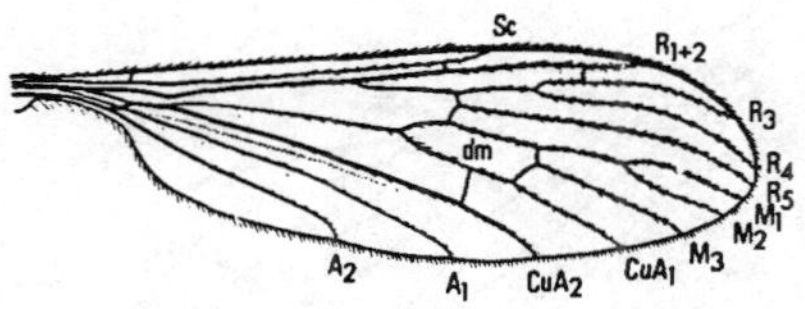

99 Prolimnophila areolata ♂

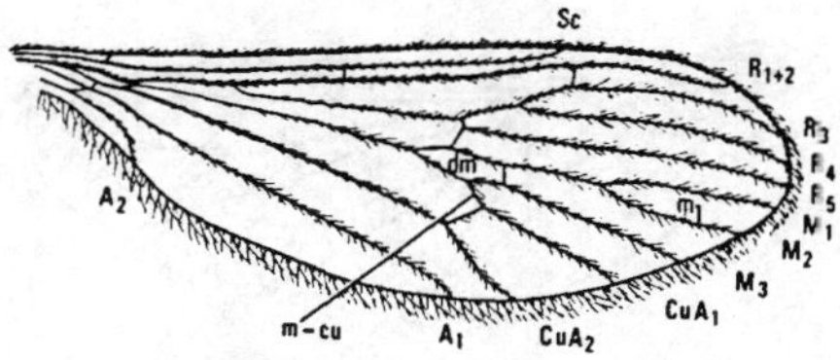

100 Paracladura trichoptera ♂

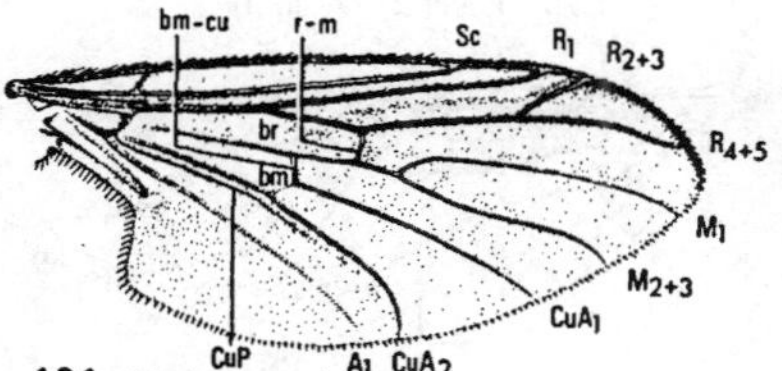

101 Plecia americana ♂

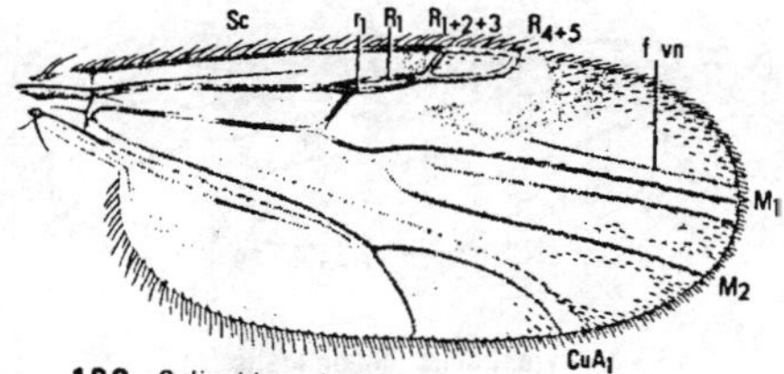

102 Culicoides insignis ♀

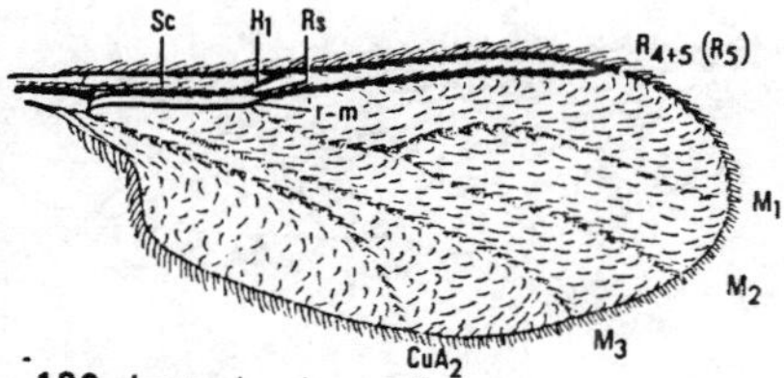

103 Lestremia cinerea ♂

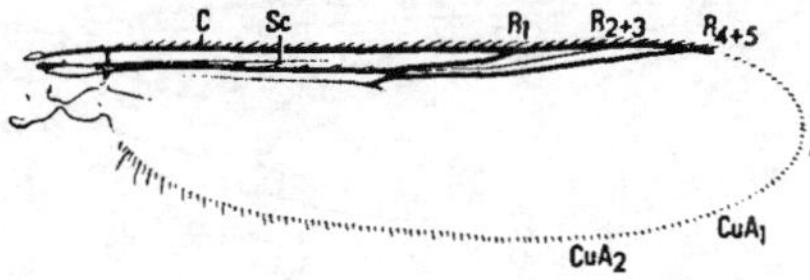

104 Smittia sp. ♂

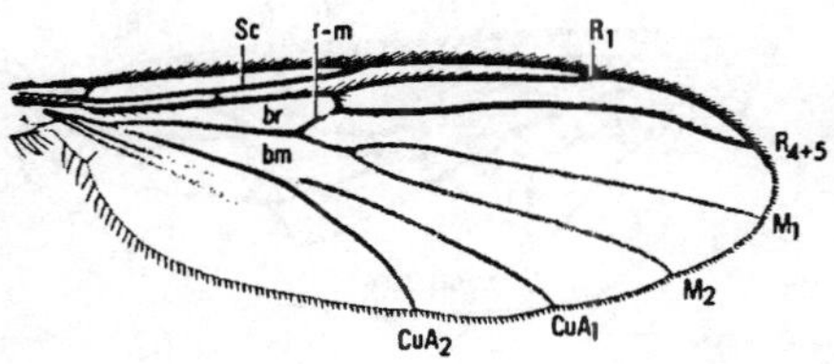

105 Boletina sp. ♂

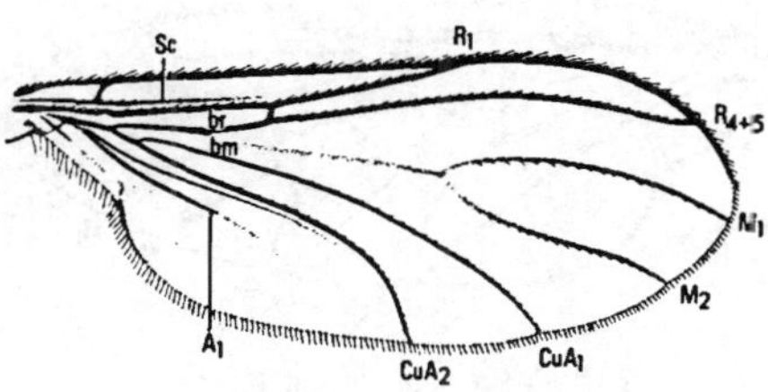

106 Sciara humeralis ♂

Figs. 41.97–41.106. Wings of Nematocera. **97.** *Protoplasa fitchii* Osten Sacken, **Tanyderidae. 98.** *Pericoma marginalis* (Banks), **Psychodidae. 99.** *Prolimnophila areolata* (Osten Sacken), **Tipulidae. 100.** *Paracladura trichoptera* (Osten Sacken), **Trichoceridae. 101.** *Plecia americana* Hardy, **Bibionidae. 102.** *Culicoides insignis* Lutz, **Ceratopogonidae. 103.** *Lestremia cinerea* Macquart, **Cecidomyiidae. 104.** *Smittia* sp., **Chironomidae. 105.** *Boletina* sp., **Mycetophilidae. 106.** *Sciara* sp., **Sciaridae.**

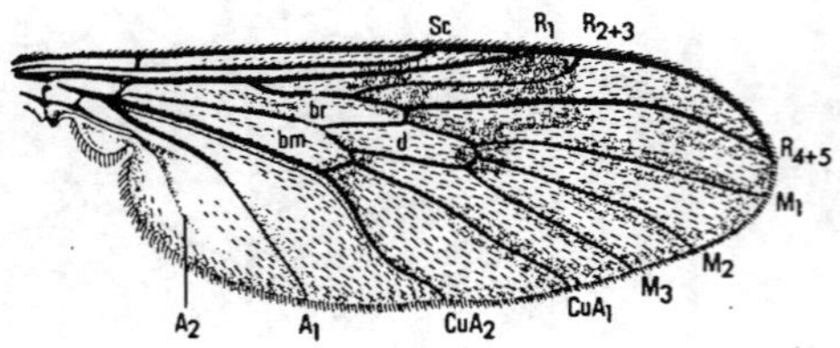

107 Sylvicola fenestralis ♂

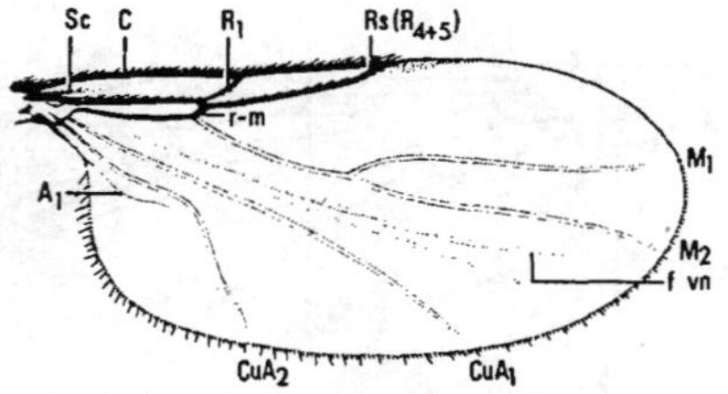

108 Coboldia fuscipes ♂

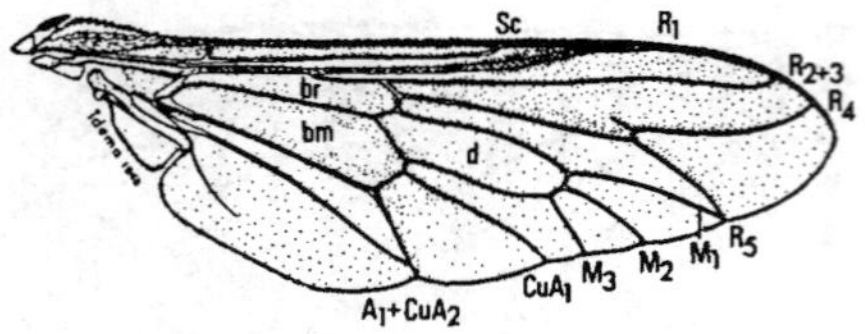

109 Esenbeckia delta ♂

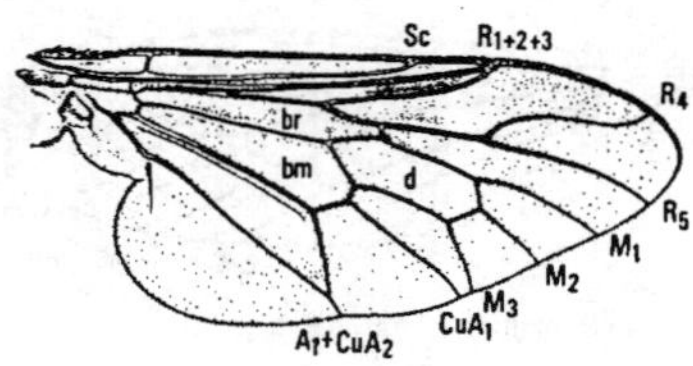

110 Atherix variegata ♀

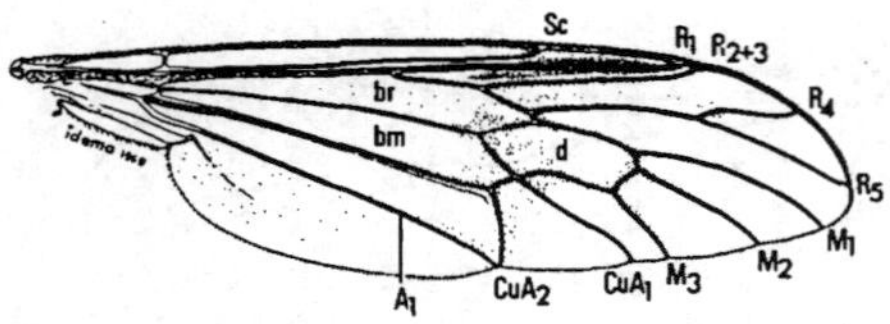

111 Xylophagus abdominalis ♀

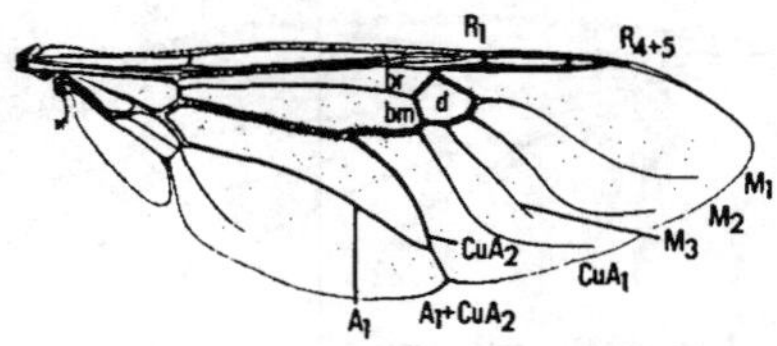

112 Anoplodonta nigrirostris ♀

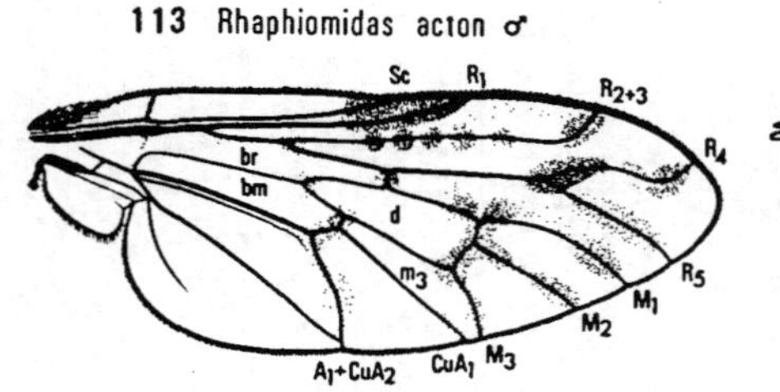

113 Rhaphiomidas acton ♂

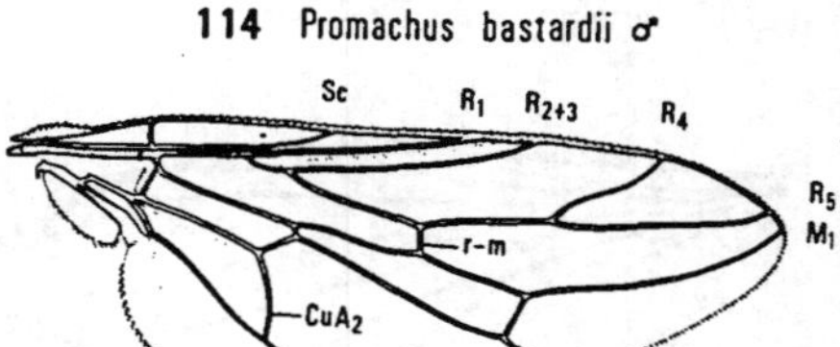

114 Promachus bastardii ♂

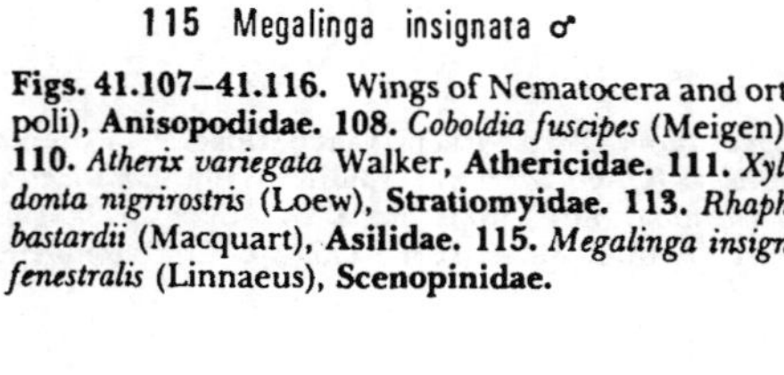

115 Megalinga insignata ♂

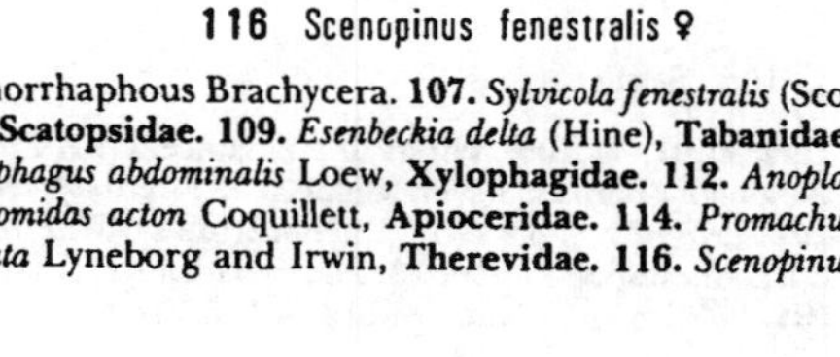

116 Scenopinus fenestralis ♀

Figs. 41.107–41.116. Wings of Nematocera and orthorrhaphous Brachycera. **107.** *Sylvicola fenestralis* (Scopoli), **Anisopodidae. 108.** *Coboldia fuscipes* (Meigen), **Scatopsidae. 109.** *Esenbeckia delta* (Hine), **Tabanidae. 110.** *Atherix variegata* Walker, **Athericidae. 111.** *Xylophagus abdominalis* Loew, **Xylophagidae. 112.** *Anoplodonta nigrirostris* (Loew), **Stratiomyidae. 113.** *Rhaphiomidas acton* Coquillett, **Apioceridae. 114.** *Promachus bastardii* (Macquart), **Asilidae. 115.** *Megalinga insignata* Lyneborg and Irwin, **Therevidae. 116.** *Scenopinus fenestralis* (Linnaeus), **Scenopinidae.**

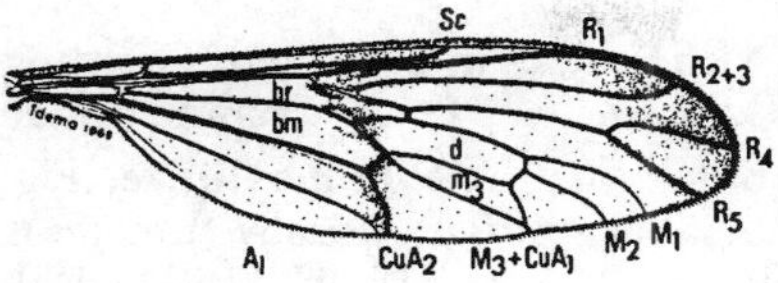

117 Vermileo tibialis ♂

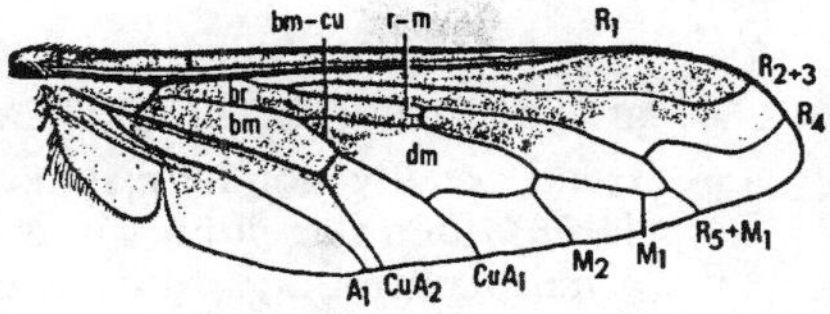

118 Bombylius major ♂

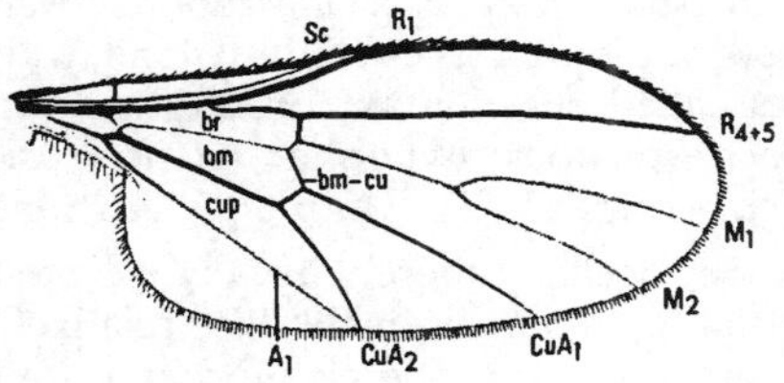

119 Empidideicus humeralis ♀

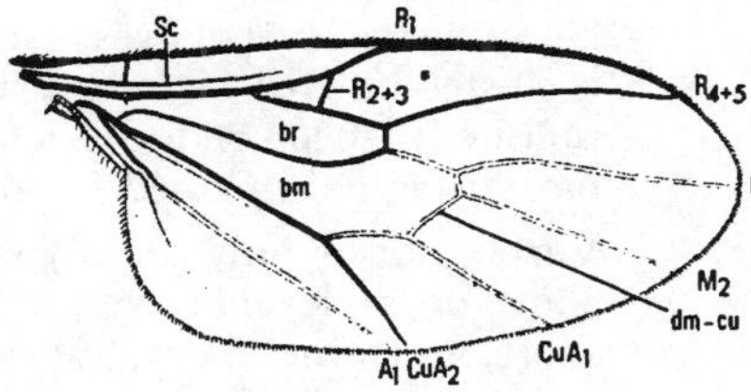

120 Glabellula crassicornis ♀

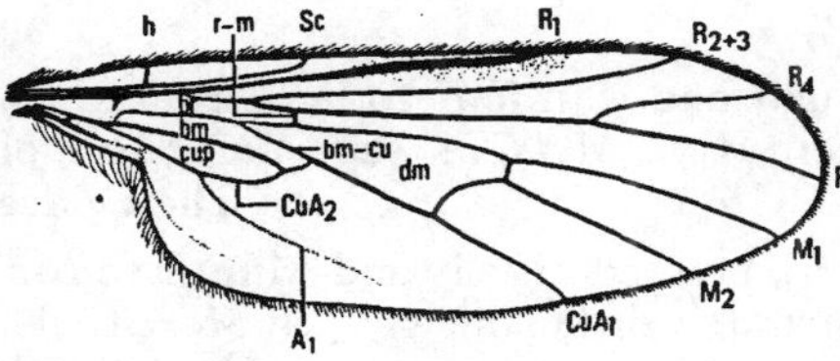

121 Hilara femorata ♂

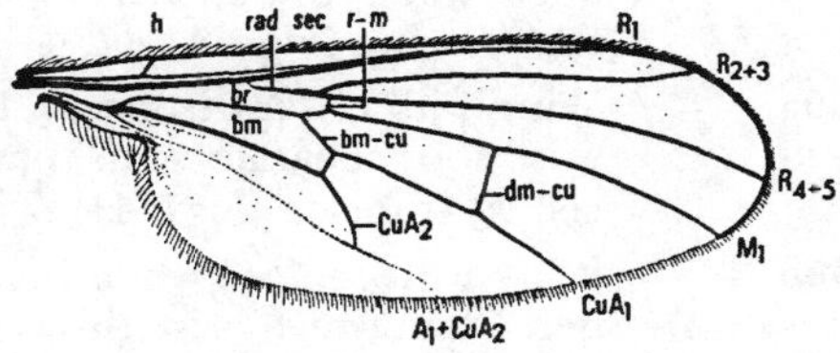

122 Hybos reversus ♀

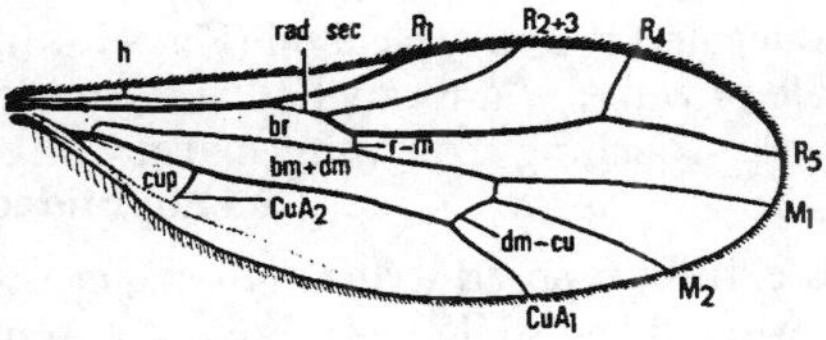

123 Neoplasta scapularis ♂

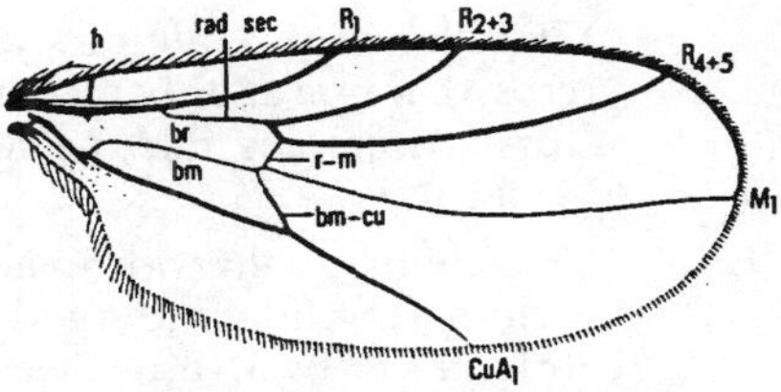

124 Microempis testacea ♀

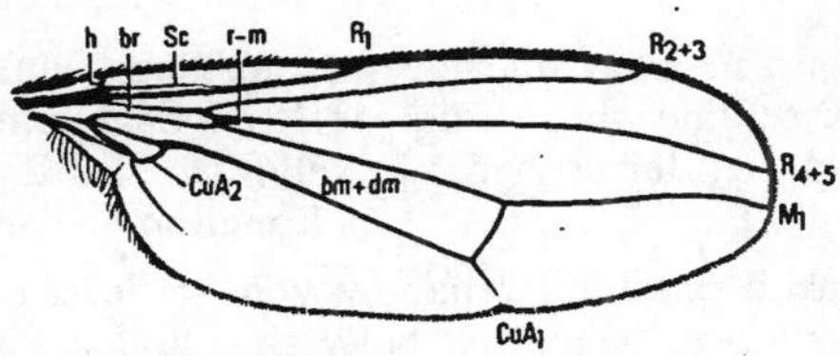

125 Hydrophorus intentus ♂

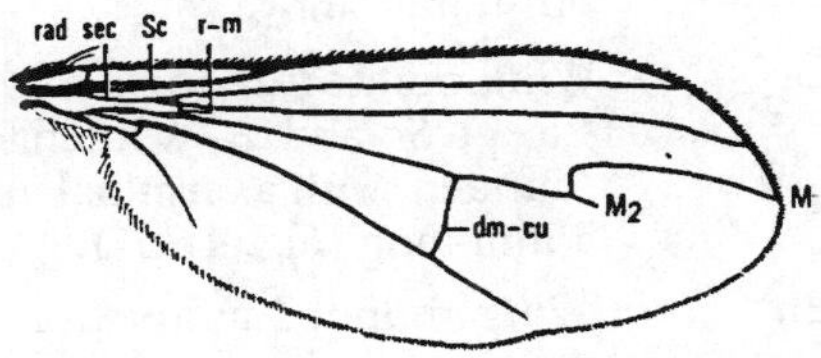

126 Dolichopus cuprinus ♂

Figs. 41.117–41.126. Wings of orthorrhaphous Brachycera. **117.** *Vermileo tibialis* (Walker), **Vermileonidae. 118.** *Bombylius major* Linnaeus, **Bombyliidae. 119.** *Empidideicus humeralis* Melander, **Bombyliidae. 120.** *Glabellula crassicornis* (Greene), **Bombyliidae. 121.** *Hilara femorata* Loew, **Empididae. 122.** *Hybos reversus* Walker, **Empididae. 123.** *Neoplasta scapularis* (Loew), **Empididae. 124.** *Micrempis testacea* Melander, **Empididae. 125.** *Hydrophorus intentus* Aldrich, **Dolichopodidae. 126.** *Dolichopus cuprinus* Wiedemann, **Dolichopodidae.**

flagellum with a dorsal arista (Fig. 41.61); R_{4+5} unbranched; *C* ending at wing apex (Fig. 41.130) . . . . **Pipunculidae**

28b Head usually not hemispherical, but if so (some **Bombyliidae,** Fig. 41.53), then flagellum without a dorsal arista. R_{4+5} usually furcate; *C* usually continuing around wing (Figs. 41.114, 41.115, 41.118–41.120) **29**

29a Vertex usually distinctly excavated between eyes; ocellar tubercle well below dorsal level of eyes; eyes never holoptic; face relatively long, with a cluster or row of long bristles called the mystax; proboscis stout, polished; labella reduced and inconspicuous; hypopharynx protrusible, strongly developed for piercing (Fig. 41.51). . . . . **Asilidae**

29b Vertex not or only slightly concave; ocellar tubercle usually elevated above dorsal level of eyes; eyes usually holoptic in male; face relatively short, sometimes hairy, but without a mystax; proboscis short and stout to long and slender, usually dull pruinose; labella usually well developed and conspicuous; hypopharynx not protrusible, not developed for piercing (Figs. 41.47, 41.53) **30**

30a Wing with cell *bm* truncate distally and with four corners from which arise four separate veins, namely M_{1+2}, M_3, CuA_1, and CuA_2; base of cell m_3 truncate (Figs. 41.47, 41.115) . . . . **Therevidae**

30b Wing with cell *bm*, when present, pointed distally and with three corners from which arise three separate veins, namely M_{1+2}, $M_3 + CuA_1$, and CuA_2 (Figs. 41.53, 41.118–41.120); cell m_3, when present, with base pointed (as in Fig. 41.117) . . . . **Bombyliidae**

31a Flagellum with a minute stylus concealed in a subapical pit (Fig. 41.48); crossvein *r–m* at or beyond middle of wing; *M* unbranched and curved forward, joining R_{4+5} or closely approaching it in wing margin (Fig. 41.116) . . . . **Scenopinidae**

31b Flagellum usually with an elongate, fully exposed stylus or arista; crossvein *r–m* well before middle of wing (Figs. 41.121–41.128) or absent (Fig. 41.129); *M* branched (Figs. 41.121, 41.123, 41.126–41.129) or unbranched (Figs. 41.122, 41.124, 41.125), sometimes curved forward but not as above **32**

32a Wing pointed at apex, and with peculiar linear venation; main veins except *Sc* and R_3 with black setulae above (Fig. 41.127); flagellum rounded, with a terminal arista; slender brownish or yellowish flies, 2–5 mm long (Fig. 41.57). . . . . **Lonchopteridae**

32b Wing rounded at apex, and with radiating venation; veins at least in posterior half of wing not setulose; flagellum, body size, and color variable **33**

33a Wing with branches of *R* strongly thickened and crowded into anterior base, and with four other weak and peculiarly aligned veins in remainder of wing blade; *C* ending near middle of anterior margin (Fig. 41.129); small humpbacked flies, 1–4 mm long (Fig. 41.60). . . . . . **Phoridae**

33b Wing with branches of *R* not strongly thickened and crowded anterobasally, and with other veins normal; *C* extending at least to wing apex **34**

34a Wing with both A_1 and *Sc* reaching wing margin, and with cell *cup* acute at posterior apex (Fig. 41.128); hind tarsus, at least in male, with one or more basal tarsomeres expanded and flattened; arista three-segmented, terminally situated (Fig. 41.58) . . . . **Platypezidae**

34b Wing rarely with A_1 reaching wing margin (some **Empididae**), but if so either *Sc* incomplete or cell *cup* obtuse or rounded at posterior apex (as in Fig. 41.121); hind tarsus not modified as above; arista or stylus two-segmented, terminally or dorsally situated **35**

35a *Rs* originating at or near level of crossvein *h*, distal to crossvein *h* by, at most, length of crossvein *h*; crossvein *r–m* in basal fourth of wing; cells *bm* and *dm* confluent, that is, crossvein *bm–cu* absent (Figs. 41.125, 41.126); *Sc* usually abruptly curved posteriorly and fused with R_1 (Fig. 41.126), except in **Hydrophorinae** (Fig. 41.125); body (Fig. 41.56) usually with metallic color . . . . **Dolichopodidae**

35b *Rs* originating well distal to level of crossvein *h*, usually distal to it by more than length of crossvein *h*; crossvein *r–m* distal to basal fourth of wing; cell *bm* usually separated from cell *dm* (when cell *dm* present) by crossvein *bm–cu*; *Sc* usually joining *C* or ending freely, never abruptly joining R_1 (Figs. 41.121–41.124); body (Fig. 41.55) usually without metallic color . . . . **Empididae**

36a Greater ampulla present as a bulbous swelling below wing base (Fig. 41.9); vibrissa usually present (Figs. 41.5, 41.9); antennal pedicel always with a complete dorsal seam (Fig. 41.150); **Calyptratae** . . . **37**

36b Greater ampulla usually absent, but if indistinctly present (some **Tephritidae** and **Psilidae**), then vibrissa absent; antennal pedicel usually without a complete dorsal seam; **Acalyptratae** **50**

37a Large heavy-bodied hairy but bristleless flies, 9–25 mm long, resembling honeybees, carpenter bees, or bumblebees (Fig. 41.96); head inflated, with reduced or atrophied mouthparts, no vibrissae, and small sunken antennae; meron (Fig. 41.9) usually with a cluster of long hairs; larvae exclusively obligate parasites of mammals . . . . . . **Oestridae**

37b Usually smaller trimmer ordinarily bristled flies, 1–15 mm long; head almost always with well-developed mouthparts, vibrissae, and antennae; meron bare or with a row of bristles, sometimes with scattered fine hairs in addition to bristles; larvae rarely obligate parasites of mammals **38**

38a Meron with a row of bristles, and sometimes with additional hairs (Fig. 41.9) **39**

38b Meron without bristles or hairs **46**

39a Subscutellum strongly developed, prominently and evenly convex in profile (Fig. 41.9) . . . . **Tachinidae**

39b Subscutellum absent or weakly developed (Fig. 41.153), if weakly developed **(Rhinophoridae),** upper half more or less membranous and concave in profile **40**

40a Abdomen and usually thorax with a distinct metallic blue or green luster; palpus usually orange-yellow (Fig. 41.94) . . . . most **Calliphoridae**

40b Abdomen usually mostly dull, variegated gray, brown, or black; sometimes shining black, but never metallic blue or green. Palpus blackish to yellowish **41**

41a Thorax with silky wavy yellowish hairs in addition to normal black bristles and hairs . . . . **Calliphoridae (*Pollenia* spp.)**

41b Thorax without silky yellowish hairs **42**

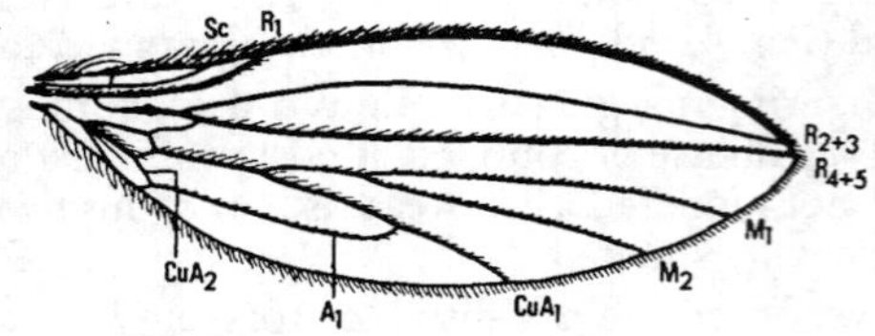

127 Lonchoptera furcata ♀

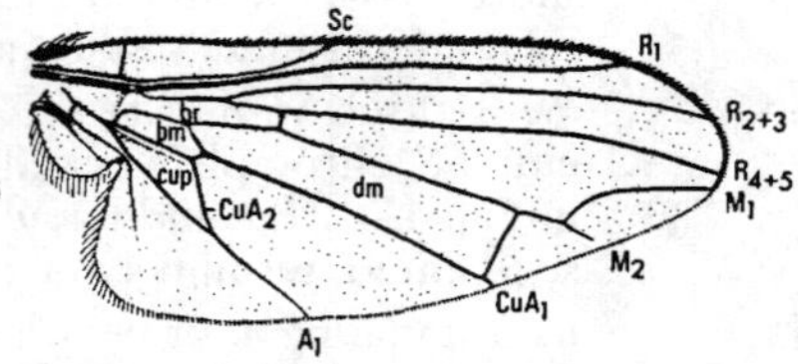

128 Platypeza consobrina ♂

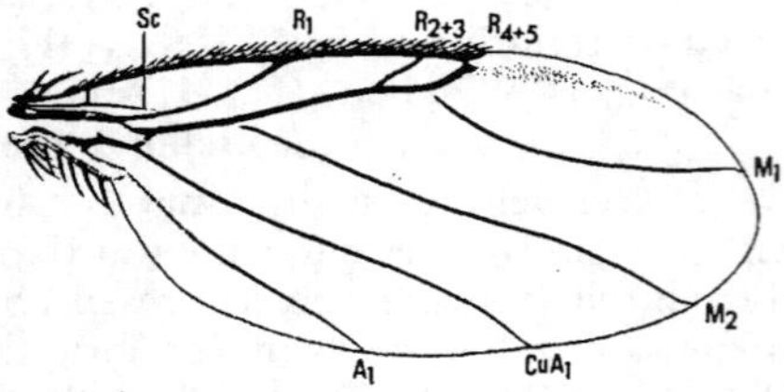

129 Lecanocerus compressiceps ♂

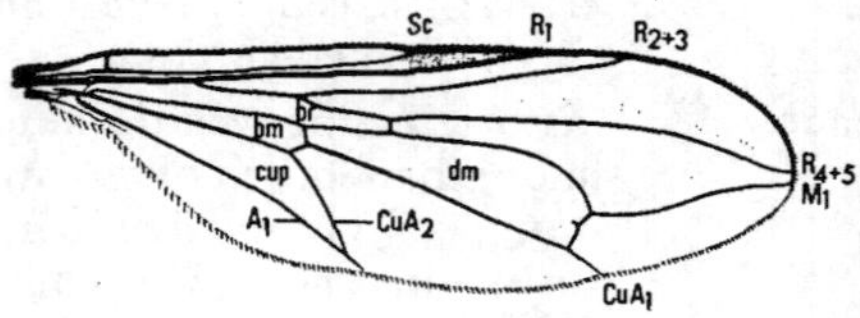

130 Pipunculus fuscus ♂

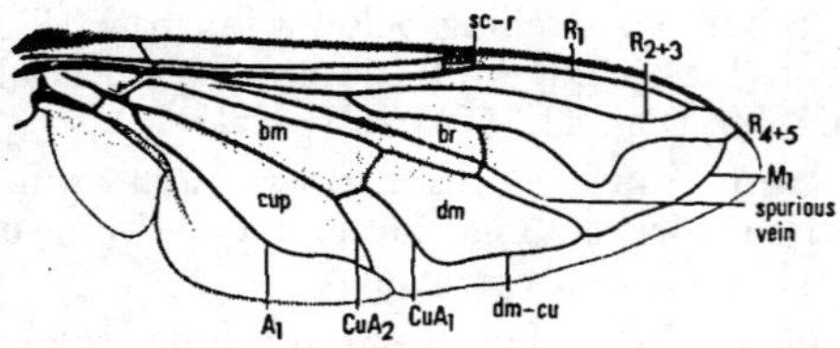

131 Eristalis tenax ♂

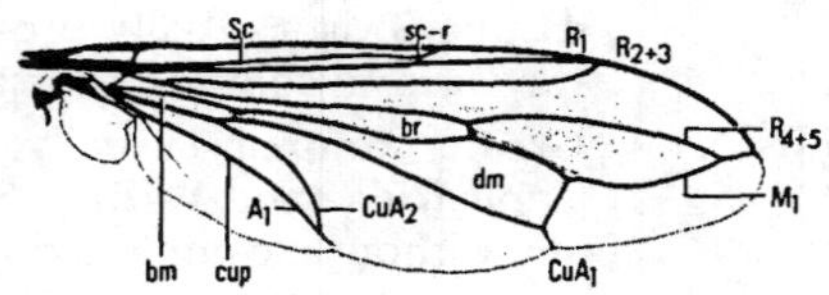

132 Physocephala texana ♀

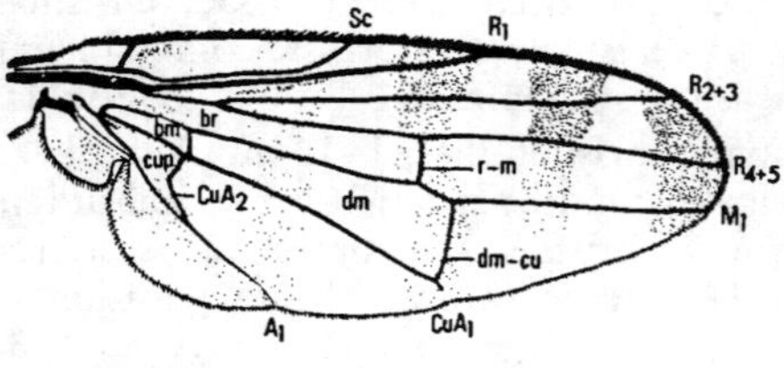

133 Pseudotephritus cribellum ♀

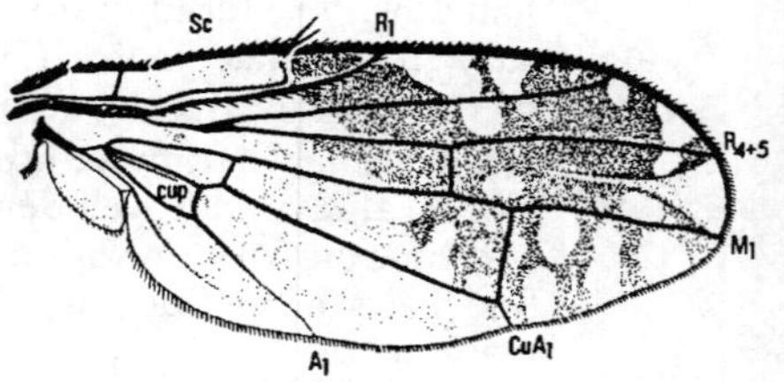

134 Tephritis subpura ♀

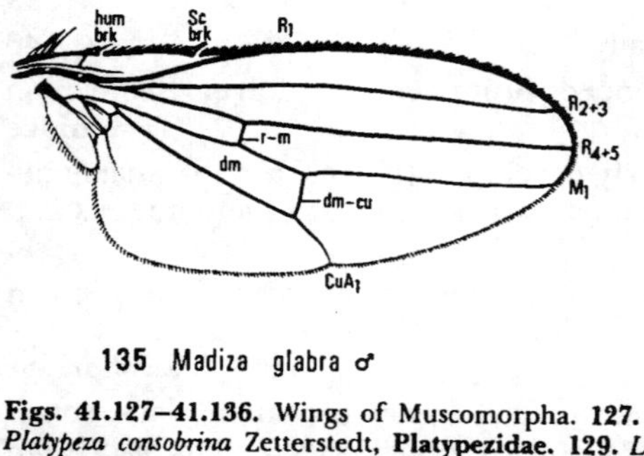

135 Madiza glabra ♂

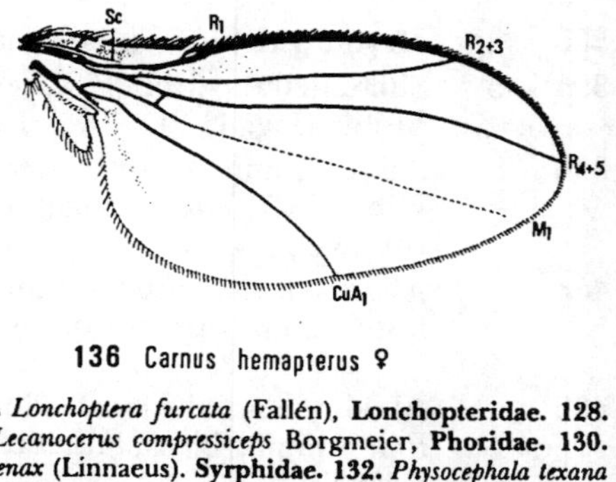

136 Carnus hemapterus ♀

Figs. 41.127–41.136. Wings of Muscomorpha. **127.** *Lonchoptera furcata* (Fallén), **Lonchopteridae. 128.** *Platypeza consobrina* Zetterstedt, **Platypezidae. 129.** *Lecanocerus compressiceps* Borgmeier, **Phoridae. 130.** *Pipunculus fuscus* Loew, **Pipunculidae. 131.** *Eristalis tenax* (Linnaeus). **Syrphidae. 132.** *Physocephala texana* (Williston), **Conopidae. 133.** *Pseudotephritis cribellum* (Loew). **Otitidae. 134.** *Tephritis subpura* (Johnson), **Tephritidae. 135.** *Madiza glabra* Fallén, **Milichiidae. 136.** *Carnus hemapterus* Nitzsch, **Carnidae.**

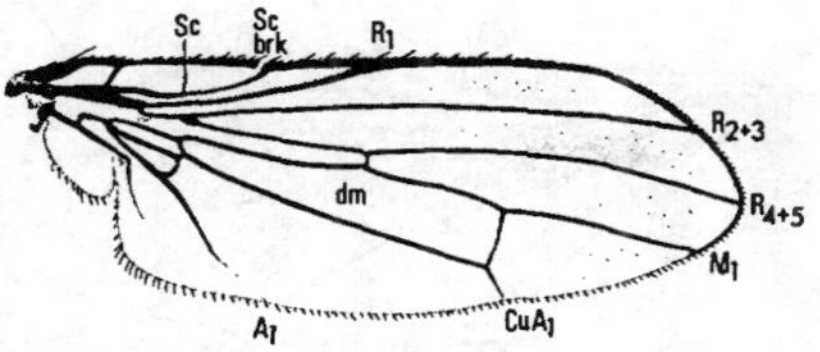

137 Tephrochlamys rufiventris ♀

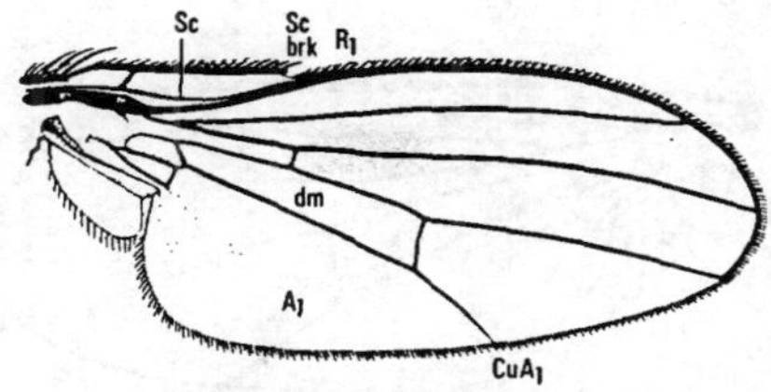

138 Tethina horripilans ♂

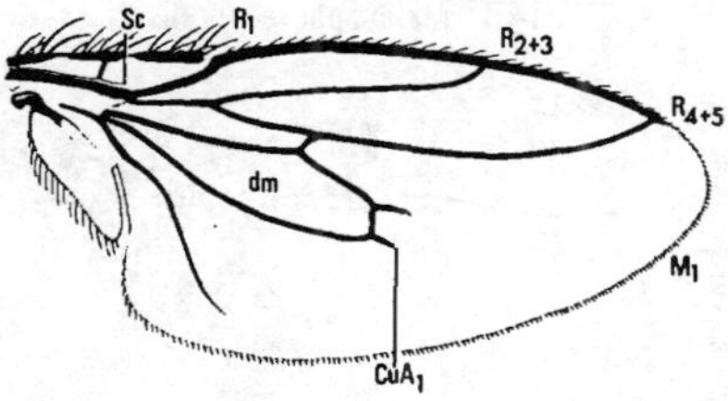

139 Leptocera fontinalis ♀

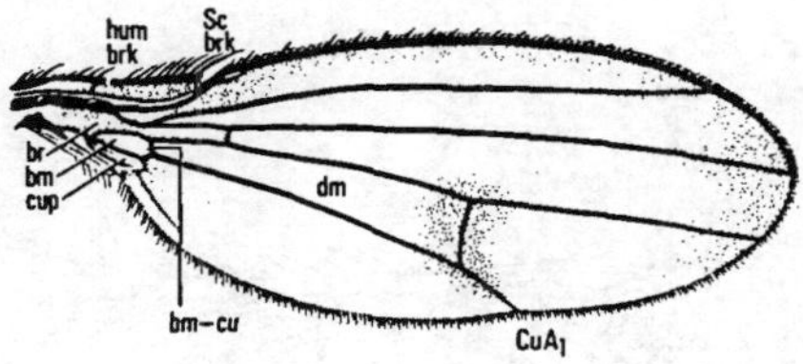

140 Diastata vagans ♂

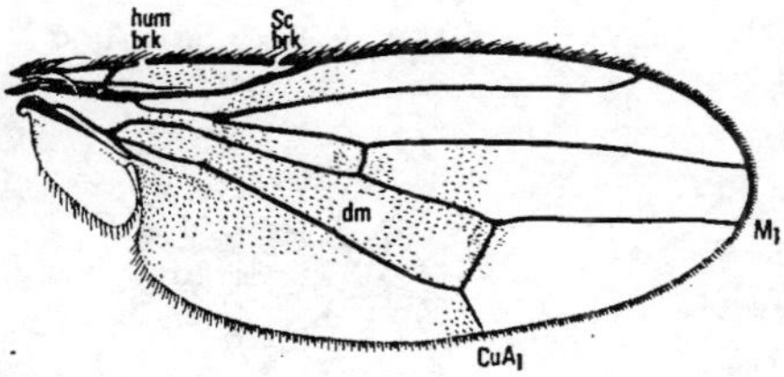

141 Parydra aquila ♀

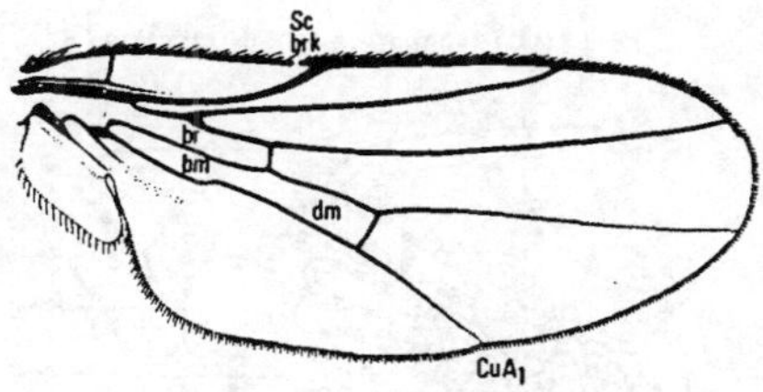

142 Chlorops certimus ♀

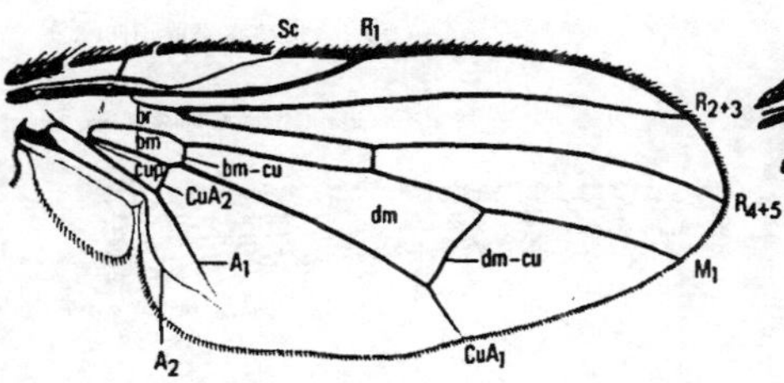

143 Fannia canicularis ♀

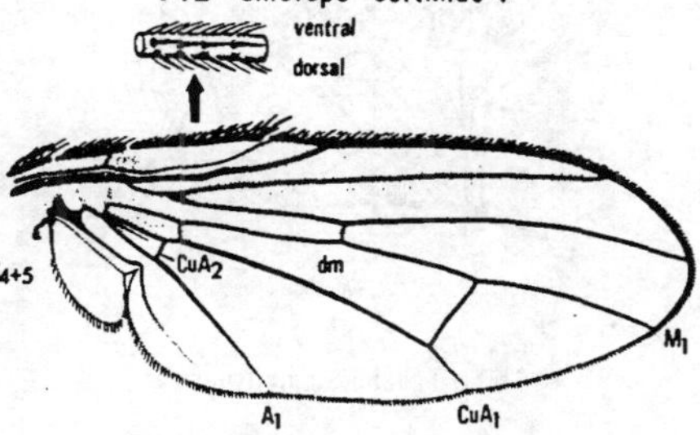

144 Botanophila spinidens ♀

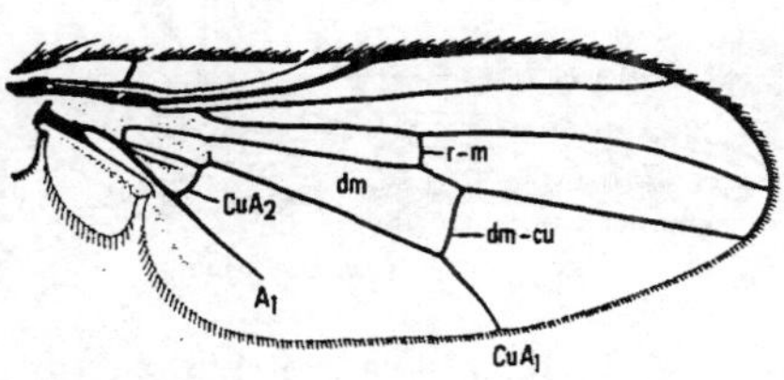

145 Acridomyia sp. ♂

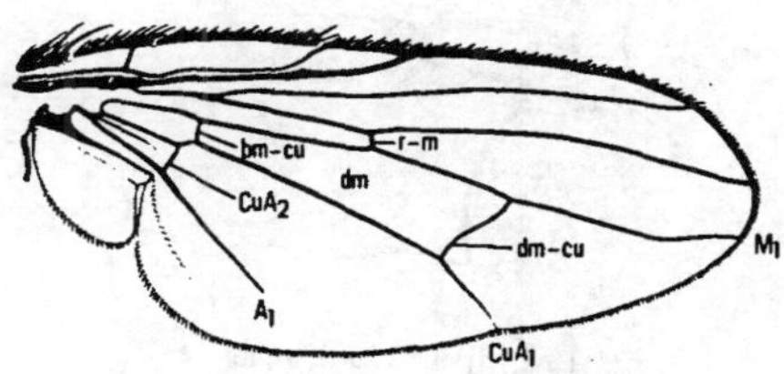

146 Muscina assimilis ♀

Figs. 41.137–41.146. Wings of Muscomorpha. **137.** *Tephrochlamys rufiventris* (Meigen), **Heleomyzidae. 138.** *Tethina horripilans* (Melander), **Tethinidae. 139.** *Leptocera fontinalis* (Fallén), **Sphaeroceridae. 140.** *Diastata vagans* Loew, **Diastatidae. 141.** *Parydra aquila* (Fallén), **Ephydridae. 142.** *Chlorops certimus* Adams, **Chloropidae. 143.** *Fannia canicularis* (Linnaeus), **Muscidae. 144.** *Botanophila spinidens* (Malloch), **Anthomyiidae. 145.** *Acridomyia* sp., **Anthomyiidae. 146.** *Muscina assimilis* (Fallén), **Muscidae.**

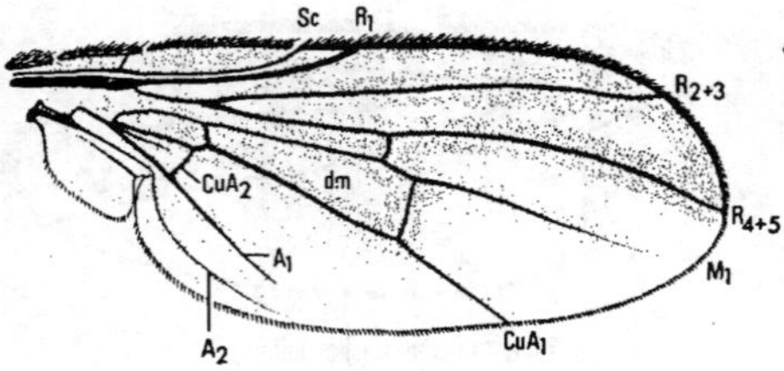

147 Bezzimyia sp. ♀

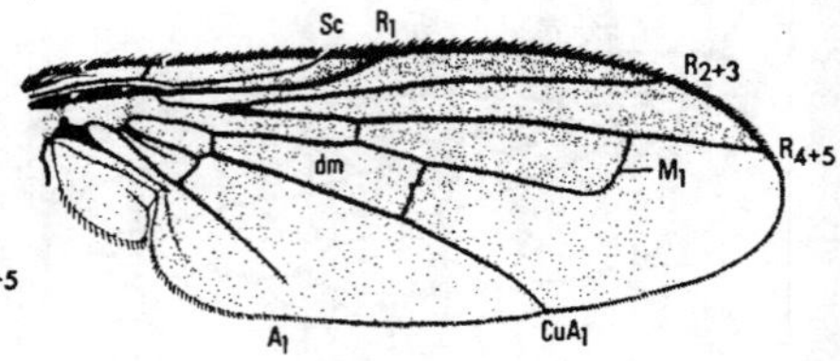

148 Melanophora roralis ♀

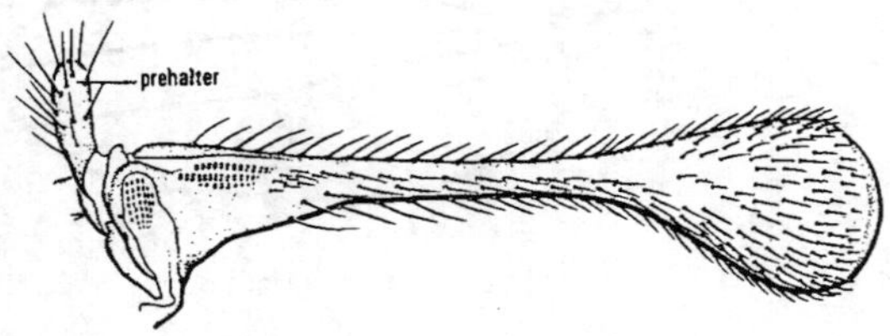

149 Ptychoptera quadrifasciata ♂

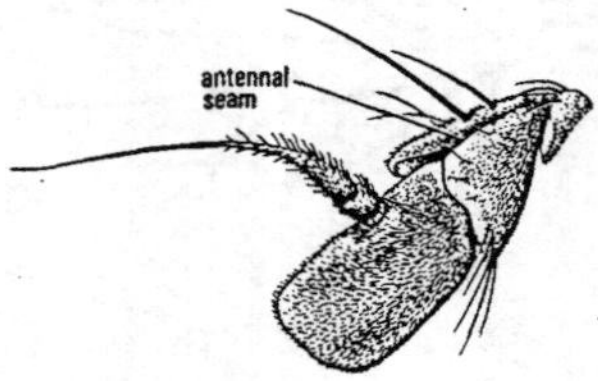

150 Cordilura ustulata ♂

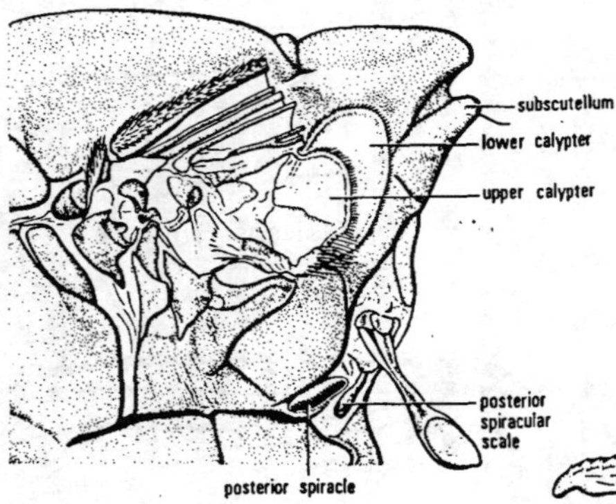

151 Tabanus sulcifrons ♀

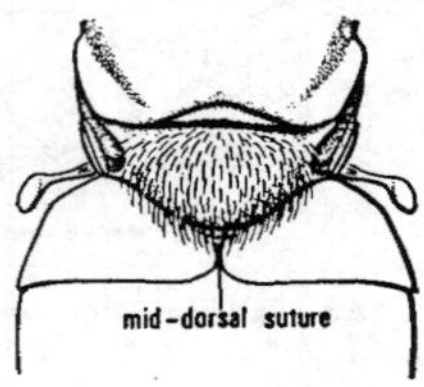

152 Tabanus sulcifrons ♀

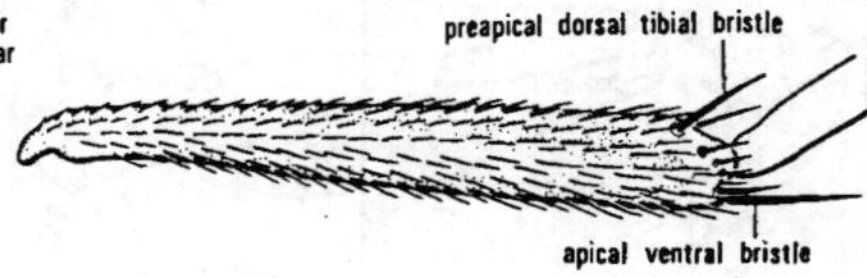

154 Clusiodes sp. ♀

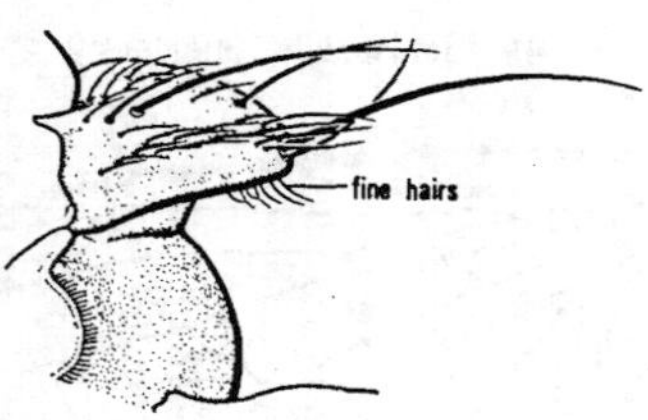

153 Hylemya alcathoe ♂

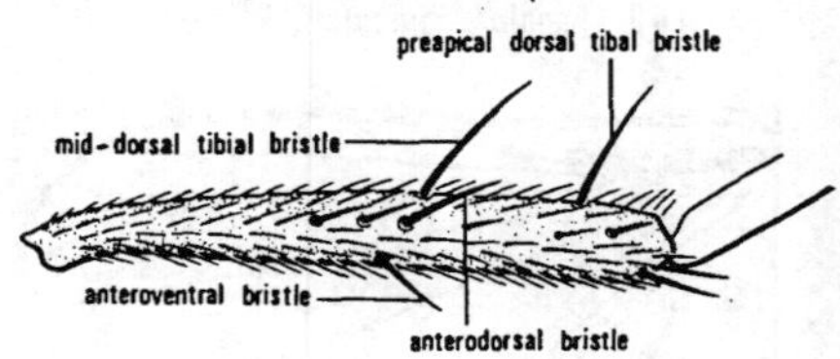

155 Fannia canicularis ♂

Figs. 41.147–41.155. Various parts of flies. **147.** Wing of *Bezzimyia* sp., **Tachinidae. 148.** Wing of *Melanophora roralis* (Linnaeus), **Rhinophoridae. 149.** Halter of *Ptychoptera quadrifasciata* Say, **Ptychopteridae. 150.** Antenna of *Cordilura ustulata* Zetterstedt, **Scatophagidae. 151.** Details of calypteres and posterior thoracic spiracle of *Tabanus sulcifrons* Macquart, **Tabanidae. 152.** Scutellum and abdominal tergite 1 of *Tabanus sulcifrons*, **Tabanidae. 153.** Scutellum of *Hylemya alcathoe* Walker, **Anthomyiidae,** left lateral view. **154.** Preapical dorsal and apical ventral bristles of left middle tibia of *Clusiodes* sp., **Clusiidae. 155.** Bristles of left hind tibia of *Fannia canicularis* (Linnaeus), **Fanniirae, Muscidae.**

42a Scutum usually with three conspicuous black stripes on a gray background (Fig. 41.95); notopleuron usually with three or four bristles; hind coxa usually with hairs on posterior surface; arista usually plumose (Fig. 41.95) . . . . **Sarcophagidae (Sarcophaginae)**

42b Scutum with or without black stripes; notopleuron always with only two bristles (as in Figs. 41.8, 41.9); hind coxa always without hairs on posterior surface; arista bare to plumose **43**

43a Coxopleural streak absent . . . . **Calliphoridae (Angioneurini)**

43b Coxopleural streak present (Fig. 41.9) **44**

44a Second abdominal sternite overlapping margins of second abdominal tergite (Fig. 41.159); lower calypter with inner base expanded, impinging on scutellum (Fig. 41.157) . . . **Sarcophagidae (Miltogramminae)**

44b Second abdominal sternite not overlapping margins of second abdominal sternite (Fig. 41.158); lower calypter with inner base not expanded, widely removed from the scutellum (Fig. 41.156) **45**

45a *M* complete and joining R_{4+5} (Fig. 41.148) . . . . . . **Rhinophoridae**

45b Apical portion of *M* atrophied (Fig. 41.147) . . . . **Tachinidae (*Bezzimyia*)**

46a (37) A_2 curved forward beyond apex of A_1, with course of A_1 (if extended) intersecting A_2 before wing margin; *Sc* always nearly straight on apical two-thirds (Fig. 41.143); hind tibia always with a strong bristle near middle of dorsal surface, in addition to preapical dorsal bristle (Fig. 41.155). . . . . **Muscidae (Fanniinae)**

46b A_2 not curved as above, with course of A_1 not intersecting A_2 before wing margin; *Sc* usually distinctly curved forward on apical half or less (Figs. 41.144–41.146); hind tibia without a strong bristle near middle of dorsal surface (though a similar bristle called the calcar sometimes present on posterodorsal surface) **47**

47a Palpus absent; crossveins *r–m* and *dm–cu* separated by a distance not or scarcely greater than length of crossvein *dm–cu*; crossvein *bm–cu* partially atrophied (Fig. 41.145) . . . . **Anthomyiidae (*Acridomyia* spp.)**

47b Palpus present; crossveins *r–m* and *dm–cu* usually more widely separated; crossvein *bm–cu* usually complete (Figs. 41.144, 41.146). . . **48**

48a A_1 never attaining wing margin even as a fold (Fig. 41.146); lower calypter broad, never linear (Fig. 41.93) . . . . most **Muscidae**

Coenosopsia prima Malloch **(Anthomyiidae)** from Texas and Florida keys out here; it differs from **Muscidae** in having an outstanding seta at lower base of first tarsomere of hind leg and two spiracles on sixth abdominal segment of female.

48b A_1 usually traceable to wing margin at least as a creaselike fold (Fig. 41.144), but if not (some **Scatophagidae**), then lower calypter linear **49**

49a Scutellum usually with fine pale hairs on apicoventral surface (Fig. 41.153), and frons usually narrower in male than in female (Figs. 41.4, 41.6); if scutellum bare on ventral surface and frons wide in male **(Fucelliinae,** some **Myopinini),** then frons with strong interfrontal bristles (Fig. 41.4); occiput never with fine pale hairs; with coarse black bristles and hairs only; katepisternum usually with two to four bristles. Lower calypter linear or broad (Fig. 41.92) . . . . . . **Anthomyiidae**

49b Scutellum bare on ventral surface; frons wide in both sexes and without interfrontal bristles; occiput almost always with rather numerous fine pale hairs, sometimes interspersed with additional black bristles

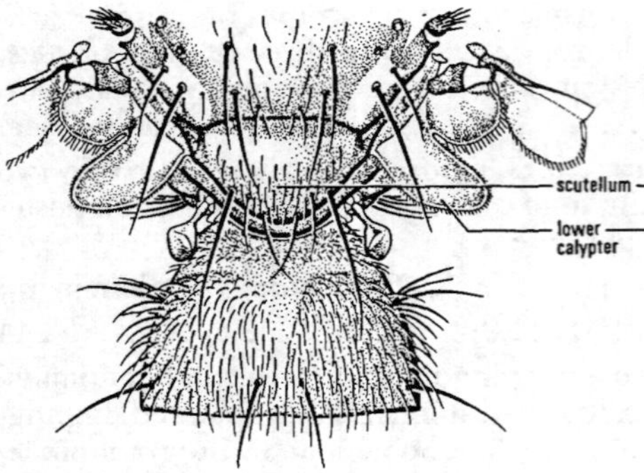

156 Melanophora roralis ♂

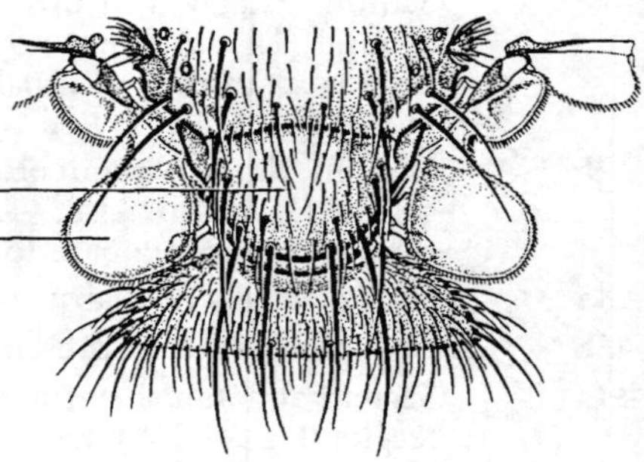

157 Anicia campestris ♀

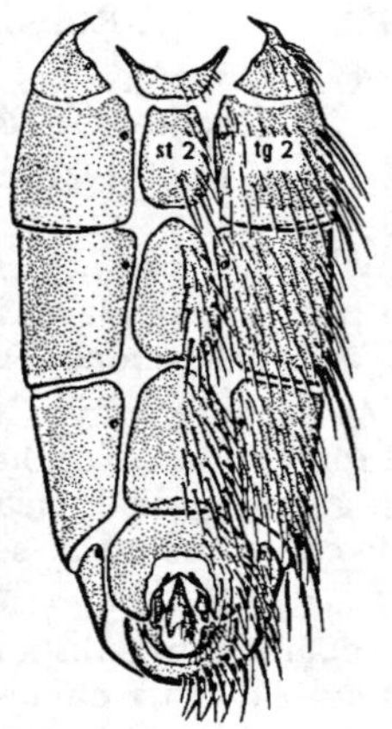

158 Melanophora roralis ♂

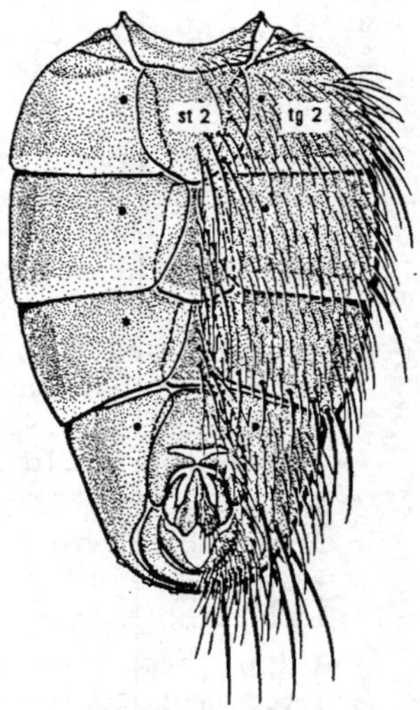

159 Anicia campestris ♂

Figs. 41.156–41.159. Calypteres and abdomens of Rhinophoridae and Sarcophagidae. Relative positions of lower calypter in *Melanophora roralis* (Linnaeus), **Rhinophoridae, 156;** and in *Anicia campestris* (Fallén), **Sarcophagidae, 157.** Relative positions of abdominal sternite 2 in *Melanophora roralis*, **158;** and *Anicia campestris*, **159.**

and hairs; katepisternum usually with one bristle; lower calypter always linear (Fig. 41.91) . . . . **Scatophagidae**

50a (36) Proboscis usually very long and slender, frequently two or more times longer than head; pedicel usually longer than first flagellomere (Fig. 41.62); crossvein *sc–r* present or cell *cup* distinctly longer than cell *bm*, or with both characters present simultaneously; *M* joining R_{4+5} or closely approaching it (Fig. 41.132) . . . . **Conopidae**

50b Proboscis usually shorter and stouter, not longer than head except in some Milichiidae, where it is elbowed (Fig. 41.75); pedicel usually shorter than first flagellomere, but if longer (most **Sciomyzidae,** Fig. 41.79, and some **Piophilidae,** Fig. 41.72, e.g., males of ***Prochyliza***), then crossvein *sc–r* absent and cell *cup* shorter than cell *bm*; *M* approaching R_{4+5} or not (Figs. 41.133–41.142). **51**

51a Ocelli absent in Nearctic species; medium-to-large flies, with heavily marked wings (Fig. 41.71). **Pyrgotidae**

51b Ocelli present; size variable, with wings marked or unmarked . . . **52**

52a Eye at end of a stout stalk; antennae widely displaced laterally; scutellum with only two subapical bristles, each arising from an elongate process (Fig. 41.64) . . . . **Diopsidae**

52b Eye not on a stalk; antennae not displaced; scutellum usually with four bristles not arising from elongate processes **53**

53a Hind tarsus with first tarsomere distinctly swollen and usually shorter than second tarsomere (Fig. 41.83); *Sc* always incomplete; CuA_1 usually incomplete (Fig. 41.139) . . . . **Sphaeroceridae**

53b Hind tarsus with first tarsomere not swollen and longer than second tarsomere; *Sc* complete or incomplete; CuA_1 usually complete . . . **54**

54a *Sc* abruptly bent forward at nearly 90°, weakened beyond the bend and ending at subcostal break; wing almost always with markings (Fig. 41.134); pedicel with a dorsal cleft; vibrissae absent (Fig. 41.160); greater ampulla (see Fig. 41.9) usually weakly distinguishable; anepimeron (see Fig. 41.9) always with bristles or hairs or with both . . . . **Tephritidae**

54b *Sc* less abruptly bent; disagreeing with above in one or more of remaining characters **55**

55a Slender-bodied flies, with long slender legs (Figs. 41.63, 41.65); R_{4+5} and *M* meeting or strongly converging at wing margin; *Sc* always complete (Figs. 41.63, 41.65) **56**

55b Usually stouter flies, with shorter more robust legs; R_{4+5} and *M* subparallel or divergent at wing margin; *Sc* complete or incomplete . . . **57**

56a Eyes very large, distinctly higher than long, and closer together in male than in female; ocellar bristles present; R_1 setose above; katepisternum evenly pilose, but without outstanding bristles or hairs (Fig. 41.65) . **Tanypezidae**

56b Eyes smaller than above, not much higher than long, and equidistant in both sexes; ocellar bristles absent; R_1 not setose; katepisternum never evenly pilose, always with one to many outstanding bristles or hairs (Fig. 41.63) . . . . **Micropezidae**

57a *Sc* complete or nearly so, ending in *C* or just short of it and free from R_1 distally (Figs. 41.133, 41.137, 41.138) **58**

57b *Sc* incomplete, not reaching *C*, often fusing with R_1 distally (Figs. 41.136, 41.139–41.142) **75**

58a *C* without a subcostal break (Fig. 41.133) **59**

58b *C* with a subcostal break (Figs. 41.137, 41.138). **65**

59a Posterior thoracic spiracle with one or more fine bristles on lower margin; form usually antlike; head subspherical; palpus vestigial (except in ***Orygma***); abdomen usually elongate and basally constricted (Fig. 41.77) . . . . **Sepsidae**

59b Posterior thoracic spiracle without bristles or outstanding hairs; form usually not antlike; palpus usually well developed **60**

60a Metepisternum setose below posterior thoracic spiracle; thorax distinctly flattened; squatty and strongly bristled flies (Fig. 41.78) . . . . **Coelopidae**

60b Metepisternum bare; thorax not very flattened; form and bristling variable **61**

61a Some or all tibiae with a preapical dorsal bristle (as in Fig. 41.154) . . **62**

61b All tibiae without a preapical dorsal bristle. **64**

62a Postocellar bristles distinctly convergent (as in Fig. 41.165); wing membrane always somewhat milky (Fig. 41.81) . . . . **Lauxaniidae**

62b Postocellar bristles divergent (as in Fig. 41.8), almost parallel (as in Fig. 41.160), or absent; wing membrane variable **63**

63a Clypeus large and prominent, in profile distinctly bulging beyond lower margin of face (Fig. 41.161); pedicel always short (Figs. 41.161, 41.80). . . . . **Dryomyzidae**

63b Clypeus small and withdrawn, in profile more or less concealed under lower margin of face (Fig. 41.79); pedicel usually elongate (Fig. 41.79) **Sciomyzidae**

64a Postocellar bristles distinctly developed, divergent (Fig. 41.67); katepisternal bristle present (except in ***Myrmecothea,*** a strikingly antlike genus with two scutellar setae) . . . . **Otitidae** (in part)

64b Postocellar bristles absent or very weakly developed; katepisternal bristles absent, though hair or pile commonly present; form never especially antlike, always with four scutellar setae (Fig. 41.68). . . . . . . **Platystomatidae**

65a Vibrissa present (Figs. 41.72, 41.75, 41.82, 41.84, 41.85), but sometimes weak in **Tethinidae** (Fig. 41.90) **66**

65b Vibrissa absent (Figs. 41.66, 41.67, 41.73), but subvibrissal bristles sometimes vibrissa-like (e.g., **Lonchaeidae**) (Fig. 41.66) **73**

66a *C* with a distinct humeral break in addition to subcostal break (Figs. 41.135, 41.136). **67**

66b *C* without humeral break, with subcostal break only (Figs. 41.137, 41.138). **69**

67a Arista broadly plumose; *C* extensively and strongly spinose in Nearctic species; all tibiae with a strong preapical dorsal bristle; medium-sized yellowish flies, with strongly arched scutum (Fig. 41.85) . . . . **Curtonotidae**

67b Arista bare or shortly haired; *C* not spinose; tibiae without a preapical dorsal bristle; small to very small, mostly shining black flies, with normally arched scutum **68**

68a Gena broad, and with a row of bristles in middle; proboscis with stout bulbous prementum and short inconspicuous labella (Fig. 162); wing with M_1 weak and unpigmented (Fig. 41.136) . . . **Carnidae** (in part)

68b Gena usually narrow, but if broad, bristles confined to lower margin; proboscis with slender prementum and long conspicuous labella folded back along prementum (Fig. 41.75); wing with M_1 normal (Fig. 41.135) **Milichiidae** (in part)

69a Postocellar bristles divergent (Fig. 41.163) **70**

69b Postocellar bristles convergent (Figs. 41.82, 41.84, 41.164–41.167) or absent (Fig. 41.90) **71**

70a Frons always with three to five strong inclinate frontal bristles that are relatively similar to the two reclinate orbital bristles in size and spacing (Fig. 41.163). . . . . **Agromyzidae** (in part)

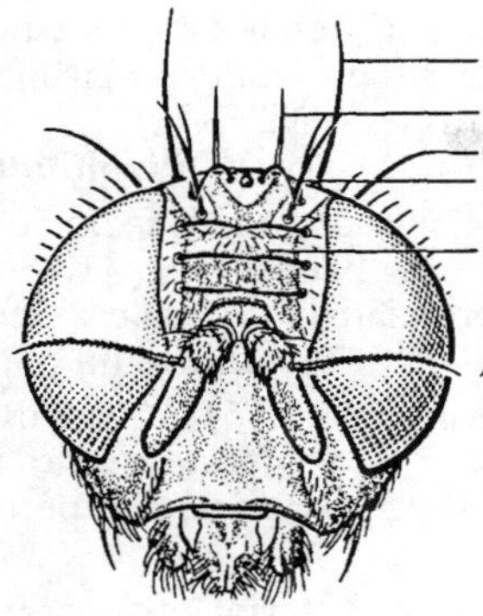

160 Rhagoletis pomonella ♀

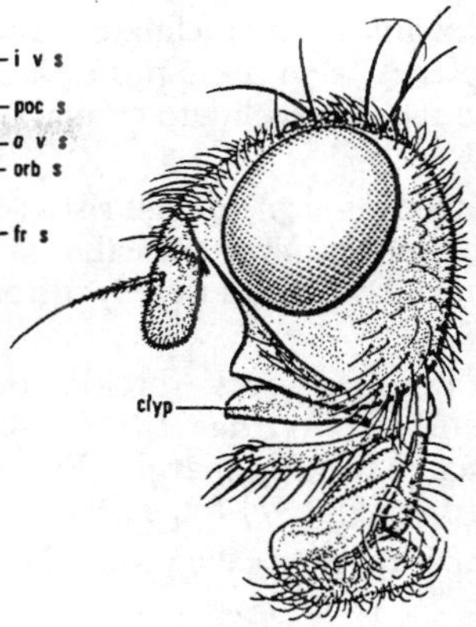

161 Dryomyza anilis ♀

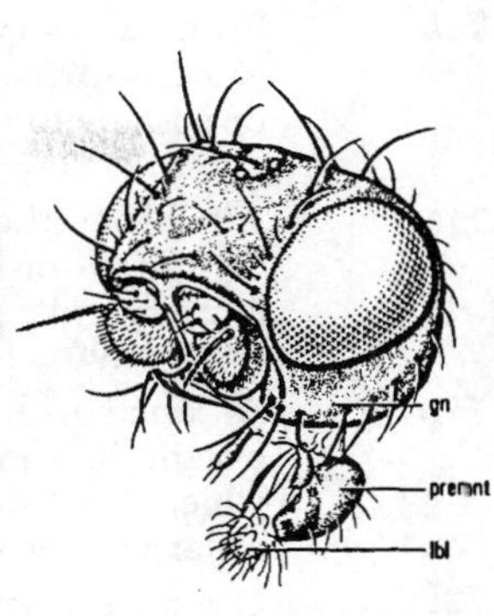

162 Carnus hemapterus ♀

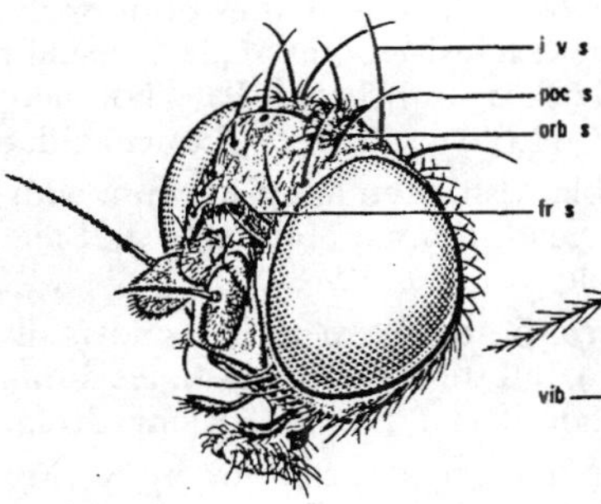

163 Agromyza pseudoreptans ♀

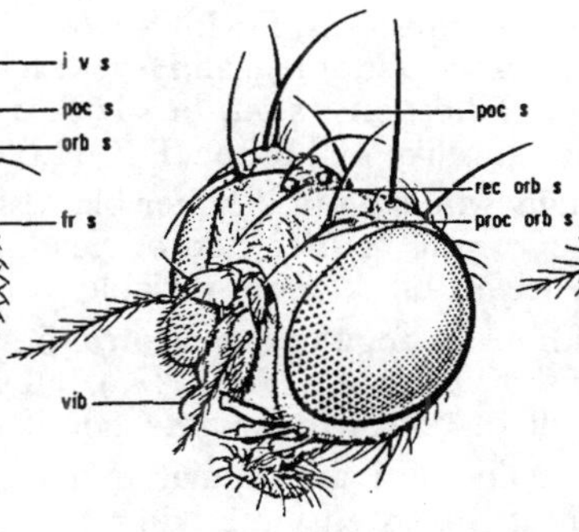

164 Diastata vagans ♂

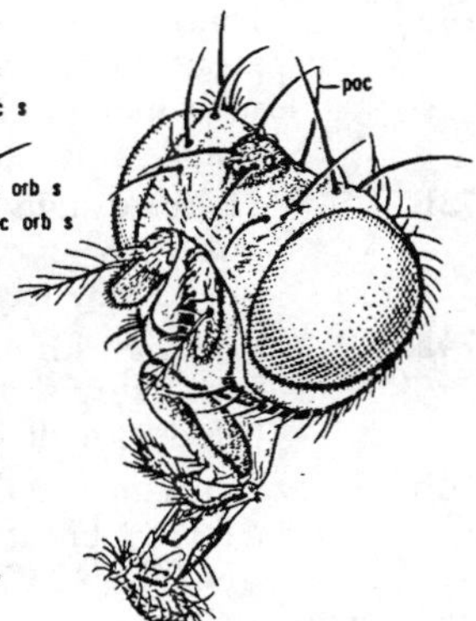

165 Drosophila colorata ♀

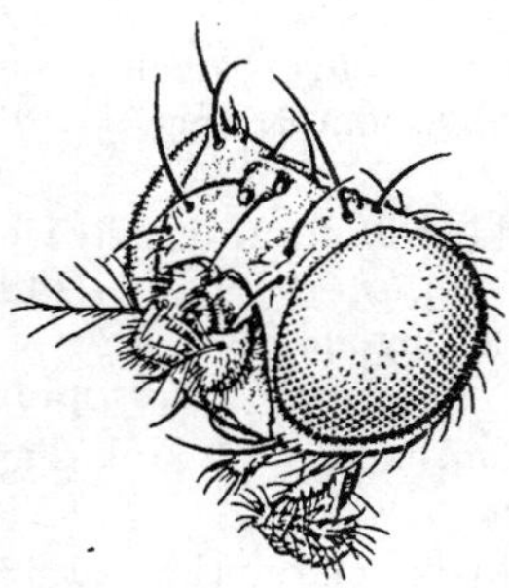

166 Microdrosophila quadrata ♀

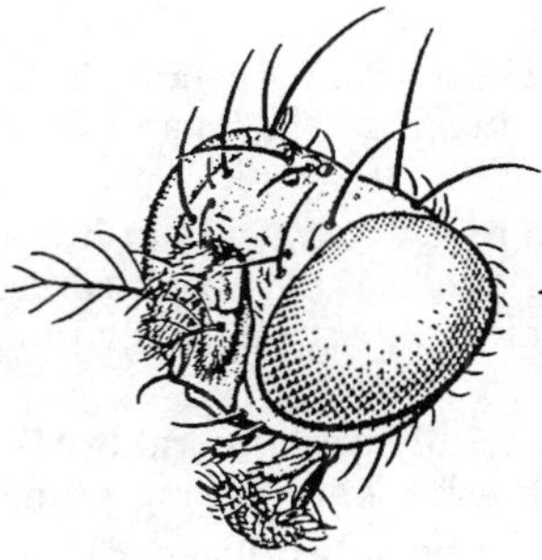

167 Chymomyza amoena ♀

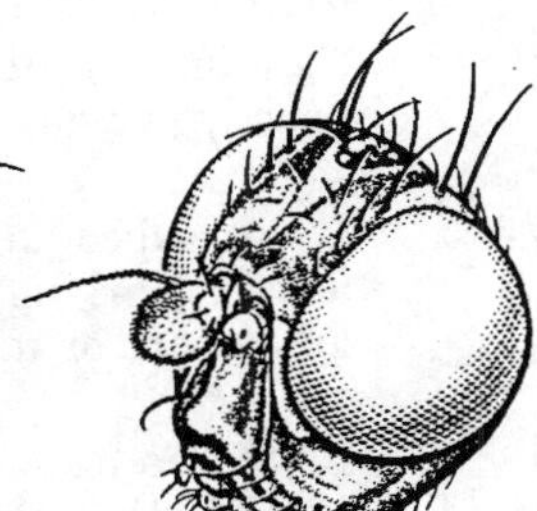

168 Cinderella lampra ♀

Figs. 41.160–41.168. Heads of acalyptrate Muscomorpha. **160.** *Rhagoletis pomonella* (Walsh), **Tephritidae. 161.** *Dryomyza anilis* Fallén, **Dryomyzidae. 162.** *Carnus hemapterus* Nitzsch, **Carnidae. 163.** *Agromyza pseudoreptans* Nowak, **Agromyzidae. 164.** *Diastata vagans* Loew, **Diastatidae. 165.** *Drosophila colorata* Walker, **Drosophilidae. 166.** *Microdrosophila quadrata* (Sturtevant), **Drosophilidae. 167.** *Chymomyza amoena* (Loew), **Drosophilidae. 168.** *Cinderella lampra* Steyskal, **Heleomyzidae.** *Abbreviations: clyp,* clypeus; *fr s,* frontal seta; *i vt s,* inner vertical seta; *orb s,* orbital seta; *o vt s,* outer vertical seta; *proc orb s,* proclinate orbital seta; *poc s,* postocellar seta; *rec orb s,* reclinate orbital seta; *vib,* vibrissa.

70b Frons seldom with any strong inclinate frontal bristles, but if one or two present (***Actenoptera***), then these positioned distant from and inclined oppositely to the two lateroclinate orbital bristles (Fig. 41.72) . . . . . **Piophilidae**

71a All tibiae without a preapical dorsal bristle; *C* not spinose; A_1 fading out on apical third or more, never traceable to wing margin (Fig. 41.138); vibrissa and postocellar bristles weakly differentiated to virtually absent (Fig. 41.90). . . . . **Tethinidae** (in part)

71b All tibiae with a preapical dorsal bristle (as in Fig. 41.154); *C* spinose, except in some **Heleomyzidae** (***Borboropsis*** and ***Oldenbergiella***) in which A_1 is traceable to wing margin (Fig. 41.137); vibrissa and postocellar bristles strong (Figs. 41.82, 41.84) **72**

72a Ocellar bristles arising on ocellar triangle above anterior ocellus (Fig. 41.82) . . . . **Heleomyzidae**

72b Ocellar bristles arising just outside ocellar triangle beside or slightly below anterior ocellus (Fig. 41.84) . . . . **Trixoscelididae**

73a Halter black. Anepisternum with a row of strong bristles posteriorly; frons, face, thorax, abdomen, and legs (exclusive of tarsi) black, usually with metallic reflections; stout bristly flies, with broad flat abdomen; female with lancelike ovipositor (Fig. 41.66). . . . . . . **Lonchaeidae**

73b Halter usually whitish, but if halter blackish, then anepisternum without a row of strong bristles, and/or head, thorax, abdomen, and legs not as extensively black; form variable **74**

74a Cell *cup* without an angular extension at posterior apex; frons normally with a single orbital bristle (Fig. 41.73), but this bristle sometimes duplicated in male of ***Omomyia hirsuta*** Coquillett . . . . . . **Richardiidae**

74b Cell *cup* usually with an angular extension at posterior apex (Fig. 41.133), but if this extension lacking, then frons with two orbital bristles (Fig. 41.67). . . . . **Otitidae**

75a (57) CuA_2 and A_1 absent or vestigial, never forming a closed cell *cup*; crossvein *bm–cu* absent, making cells *bm* and *dm* confluent (Figs. 41.136, 41.141, 41.142). **76**

75b CuA_2 and A_1 present, always forming a closed cell *cup*; crossvein *bm–cu* usually present, making cells *bm* and *dm* usually separate (Fig. 41.140) **78**

76a *C* with subcostal break only, without humeral break; CuA_1 usually with a kink marking position of atrophied crossvein *bm–cu* (Fig. 41.142); arista bare or micropubescent, never plumose (Fig. 41.89) . . . . . . **Chloropidae**

76b *C* with both subcostal and humeral breaks; CuA_1 without a kink (Figs. 41.136, 41.141); arista bare to long plumose **77**

77a Face strongly convex; subcranial cavity usually very large (Fig. 41.88); postocellar bristles, if present, divergent; M_1 strong and normally pigmented (Fig. 41.141) . . . . **Ephydridae**

77b Face concave; subcranial cavity normal (Fig. 41.162); postocellar bristles, if present, subparallel; M_1 weak and unpigmented (Fig. 41.136) **Carnidae (*Carnus, Meoneura*)**

78a Vibrissa absent (Figs. 41.69, 41.70); katepisternum without an outstanding bristle, though frequently setulose (Figs. 41.69, 41.70); mostly slender rather fragile flies, with elongate wings **79**

78b Vibrissa present (Figs. 41.164–41.168); katepisternum always with one or more outstanding bristles; form variable **80**

79a Notopleuron usually with two bristles; anepisternum with an outstanding bristle in addition to fine setulae (Fig. 41.69); ocelli situated far forward, the anterior one about midway between vertex and antennae (Fig. 41.69) . . . . **Strongylophthalmyiidae**

79b Notopleuron with one bristle, the posterior one; anepisternum without a bristle, with fine setulae only (Fig. 41.70); ocelli situated near vertex, the anterior one on dorsal third of frons (Fig. 41.70) . . . . **Psilidae**

80a *C* with both a humeral and a subcostal break (Figs. 41.135, 41.136, 41.140) **81**

80b *C* without a humeral break, with a subcostal break only (Fig. 41.138) **85**

81a Long axis of antenna strongly elbowed, with flagellum distinctly decumbent; arista frequently long plumose (Figs. 41.164–41.167) . . **82**

81b Long axis of antenna nearly straight, with flagellum more or less porrect; arista bare or practically bare (Figs. 41.162, 41.163, 41.168). . . **83**

82a *C* with small, evenly spaced, erect spinules beyond subcostal break (Fig. 41.140); proclinate orbital bristle arising dorsolaterally to lowermost reclinate orbital bristle (Fig. 41.164) . . . . **Diastatidae**

82b *C* without erect spinules, with ordinary reclinate setulae only; proclinate orbital bristle usually arising ventromedially to reclinate orbital bristle (Fig. 41.165); if arising above reclinate orbital bristle (***Chymomyza***, Fig. 41.167), then dorsomedially to it; if arising laterally to it (***Microdrosophila***, Fig. 41.166), then ventrolaterally to it . . . . **Drosophilidae**

83a Face tuberculate (Fig. 41.168); midtibia with a preapical dorsal bristle (as in Fig. 41.154) . . . . **Heleomyzidae (*Cinderella*)**

83b Face not tuberculate; all tibiae lacking a preapical dorsal bristle . . **84**

84a Gena broad, with a row of bristles in middle; proboscis with stout bulbous prementum and short inconspicuous labella (Fig. 41.162) . . **Carnidae** (in part)

84b Gena usually narrow, but if broad, then bristles confined to lower margin; proboscis with slender prementum and long conspicuous labella folded back along prementum (Fig. 41.75) . . . **Milichiidae** (in part)

85a Anepimeron setulose; frons with a single orbital bristle (Fig. 41.76) **Opomyzidae**

85b Anepimeron usually bare, but if setulose, then frons with more than one orbital bristle **86**

86a Postocellar bristles divergent (Fig. 41.163), always present . . . . . . **Agromyzidae** (in part)

86b Postocellar bristles convergent or absent **87**

87a Anepisternum bare; fore femur usually with a strongly developed ctenidial spine (Fig. 41.74). . . . . **Anthomyzidae**

87b Anepisternum setulose (Fig. 41.90); fore femur without a distinct ctenidial spine . . . . **Tethinidae** (in part)

References

Alexander, C. P. 1969. Baron Osten Sacken and his influence on American dipterology. *Annu. Rev. Entomol.* 14:1–18. **Crampton, G. C. 1942.** · Guide to insects of Connecticut. VI. The Diptera or true flies of Connecticut. First Fascicle. The external morphology of the Diptera. *Bull. Conn. St. Geol. Nat. Hist. Surv.* 64:10–165. **Hardy, D. E. 1977.** *The Present and Future Status of our Knowledge of the Diptera Faunas of the World.* Communicated at 14th Congress of Entomology, Washington, D.C. **Hennig, W. 1973.** Ordnung Diptera (Zweiflugler). *Handb. Zool. 4(2) (Lfg.20):1–337.* **James, M. T. 1959.** Diptera. Pp. 1057–1079 in *Fresh Water Biology.* W. T. Edmondson (ed.). Wiley, New York. **Knight, K. L. 1970.** A mosquito taxonomic glossary. I. Adult head (external). *Mosq. Syst. News Lett.* 2:23–33. **Knight, K. L. and J. L. Laffoon. 1970a.** A mosquito taxonomic glossary. II. Adult head (internal). *Mosq. Syst. News Lett.* 2:69–81. **1970b.** A mosquito taxonomic glossary. III. Adult thorax. *Mosq. Syst. News. Lett.* 2:132–148. **1970c.** A mosquito taxonomic glossary. IV. Adult thoracic appendages. *Mosq. Syst. News Lett.* 2:165–177. **1971.** A mosquito taxonomic glossary. V. Abdomen (except female genitalia). *Mosq. Syst. News Lett* 3:8–25 **McAlpine, J. F. 1981a.** Morphology and terminology—adults. Pp. 9–63 in J. F. McAlpine et al. (eds.). *Manual of Nearctic Diptera,* Vol. 1. Agric. Can. Monogr. 27. Canadian Government Publishing Centre, Hull, Quebec 678 pp. **1981b.** Key to families—adults. Pp. 89–124 in J. F. McAlpine et al. (eds.). *Manual of Nearctic Diptera,* Vol. 1. Agric. Can. Monogr. 27. Canadian Government Publishing Centre, Hull, Quebec. 678 pp. **McAlpine, J. F., B. V. Peterson, G. E. Shewell, H. J. Teskey, J. R. Vockeroth and D. M. Wood. 1981.** *Manual of Nearctic Diptera,* Vol. 1. Agric. Can. Monogr. 27. Canadian Government Publishing Centre, Hull, Quebec. 678 pp. **Merritt, R. W. and E. I. Schlinger. 1984.** Aquatic Diptera. 2. Adults of aquatic Diptera. Pp. 467–490. *in* R. W. Merritt and K. W. Cummins (eds.). *An Introduction to the Aquatic Insects of North America.* Kendall/Hunt Publishing Co., Dubuque, Iowa. **Sabrosky, C. W. 1966.** *A Catalogue of the Diptera of Americas South of the United States.* 1. *Introduction.* Pp. I–XIII. Department of Zoology, Secretary of Agriculture, Sao Paulo, Brazil. **Stone, A., C. W. Sabrosky, W. W. Wirth, R. L. Foote and J. R. Coulson. 1965.** *A Catalog of the Diptera of America North of Mexico.* U.S. Dept. Agric. Handb. 276. 1696 pp. **Teskey, H. J. 1976.** Diptera larvae associated with trees in North America. *Mem. Entomol. Soc. Can.* 100:1–53. **Van Emden, F. and W. Hennig. 1970.** Diptera. Pp. 215–359. in S. L. Tuxen (ed.). *Taxonomists' Glossary of Genitalia of Insects.* Munksgaard, Copenhagen, Denmark.

42

Insecta: Diptera Larvae

H. J. TESKEY

Diptera larvae are in general moisture loving, and few can survive desiccation for prolonged periods. Many representatives of the family are aquatic or semiaquatic. Parasitic and carnivorous larvae are bathed in ample moisture in the bodies of their hosts. The same is true of burrowing phytophagous larvae. Even those larvae feeding externally on plants, in addition to the moisture obtained in feeding, are in the higher-humidity gradient of the transpiring plant. Following the same trend, most Diptera larvae associated with soil are found in wet-to-damp situations and usually in soils with abundant organic matter. The organic matter provides not only food for the many saprophagous larvae to be found here, but also a higher moisture retention level. Relatively few Diptera larvae inhabit drier soils, and almost all of these are predatory and have integumentary adaptations for the retardation of moisture loss.

The composition of the soil fauna is difficult to define. The distinction from an aquatic fauna and the fauna of decaying trees are particularly difficult to set. The decisions can only be arbitrary and are here considered to include those larvae inhabiting soil of either or both mineral and organic composition that is normally exposed above the water level and extensively decayed trees in close contact with the soil that have lost all recognizable bark. If difficulty is experienced in identifying larvae from this boundary situation, recourse can be made to keys identifying aquatic larvae (James 1959, Teskey 1978) and larvae from decaying trees (Teskey 1976). The only phytophagous larvae that are included in the following key are those feeding externally on the roots of plants and on fungi. Larvae infesting fungi are known primarily from their association with the exposed sporophore, but it is assumed that the larvae of fungicolous Diptera may also feed on the mycelium of fungi and would be, in most cases, true soil dwellers. Larvae associated with nests and burrows of animals are also included, as are larvae inhabiting fecal matter. Many larvae feeding on emergent vegetation fall to the ground and burrow into the soil to pupate. Their life in soil before pupation is so brief that it is unlikely that they would be recovered in soil samples. Therefore, it is impractical to include them in the key. Larvae representing families of Diptera that are parasitic on other soil-inhabiting invertebrates (e.g., Bombyliidae, Nemestrinidae, Acroceridae, and Tachinidae) are also not included. Such larvae are hidden from view within their host and would not be directly encountered in the normal course of sampling soil. Adults of these parasites which might be obtained in emergence traps or from individual rearings of the host are identified in Chapter 41.

Diptera larvae, in common with the immature stages of most other insects, are inadequately known. No more than 5 to 10% of the species have been described, many in insufficient detail for diagnosis. The deficiency is particularly great among the Muscomorpha (cyclorrhaphous Brachycera) and is probably even more deficient with respect to the soil fauna because of increased difficulties associated with observation in this habitat. Hence some caution must be exercised in the use of the following key.

Biological information on the larvae is also deficient. What is known is often of a very generalized nature. Brief statements summarizing our present knowledge, with respect to a soil habitat, are included in the key. For those wishing further information on the larval systematics and biology of individual families, some selected references are included in the key following the family name. Further information may be obtained in the indispensable publications of Hennig (1948, 1950, 1952). Another important source of systematic and biological information of both adults and larvae of all families of Diptera is McAlpine (1981, 1987). Other references identifying the larvae of terrestrial or soil inhabiting Diptera are by Brauns (1954) and Ghilarov (1964).

It is always advantageous to rear larvae to the adult stage to confirm identification. But so that identifications of larvae may be improved, associated larval material should be retained and preserved for future study and description by interested workers. This can be done either by preserving identical larvae to the one that is reared or by finding and preserving the last exuvium of the larva being reared as soon as possible after the larva has pupated. Preferably both procedures should be done, as the larval skin serves to confirm the correct association of the preserved larva, while the latter provides better material for description. In the Muscomorpha the last larval exuvium forms the characteristic puparium in which pupation and adult development take place. All larval material should be preserved in 70–80% alcohol, with appropriate labels identifying the habitat and the adult association. The adult should be pinned. To prevent shriveling and darkening, the larvae should first be killed in hot water or an appropriate fixative such as KAAD (a mixture of kerosene, acetic acid, 95% ethyl or isopropyl alcohol, and dioxane in approximate proportions of 1 : 3 : 30 : 1) before transfer to the lower concentration of alcohol.

External Morphology

Larvae of Diptera are extremely variable in form. The only diagnostic character common to all larvae is the lack of jointed thoracic legs. Head structures range from a completely exposed heavily sclerotized head capsule usually having well-developed mouthparts, including opposed mandibles (Figs. 42.6, 42.7, 42.16–42.31) (most Nematocera), through various degrees of reduction of the head capsule, yielding forms with several slender rods that are partially retracted within the thorax (Figs. 42.8–42.10, 42.13) (e.g., some Tipulidae) and an accompanying rotation of the mandibles to the vertical plane (Figs. 42.36, 42.39, 42.40, 42.42, 42.49, 42.51, 42.53, 42.57) (orthorrhaphous Brachycera). In the cyclorraphous Brachycera (Muscomorpha), there is a complete loss of the external sclerotized parts of the head capsule and compensatory growth of the internal tentorial and pharyngeal sclerite, forming a characteristically shaped, completely internal cephalopharyngeal skeleton (Figs. 42.62, 42.63, 42.78).

The body of most larvae is slender and composed of 11 or 12 segments, but may be variously swollen in part or overall, or dorsoventrally compressed, or the segmentation may be completely obscured by a profusion of wrinkles or lines on the segments (Fig. 42.59). The segments may be secondarily subdivided (Figs. 42.24, 42.25, 42.35) or combined. The body integument is usually soft and pliable. In some species it is densely covered with short fine hairs or pubescence, or the hairs may be scattered and sparse and these may be placed singly or in tufts or fringes. Some segments may bear sclerotized plates. Small fleshy processes or tubercles may be present on one or more body segments (Figs. 42.60, 42.67), but more often these are restricted to the terminal segment surrounding the posterior spiracles (Figs. 42.70, 42.73, 42.79, 42.84). Oval or round, tuberclelike projections in the form of prolegs may be present on the prothorax (Figs. 42.28, 42.30) or on one or more abdominal segments (Figs. 42.27, 42.41). These bear apical spinules or small curved spines (crochets) that assist in locomotion. Raised transverse ridges or creeping welts bearing spinules serve the same purpose in other larvae (Figs. 42.45, 42.55, 42.68, 42.70), particularly those of the Muscomorpha.

Of considerable use in larval systematics is the arrangement of the spiracles. Spiracles may be absent (apneustic) or present, usually laterally, on various of the segments except the mesothorax, according to the systems and nomenclature shown in Fig. 42.1. Spiracles on the prothorax are called the anterior spiracles, while the eighth pair of abdominal spiracles, usually

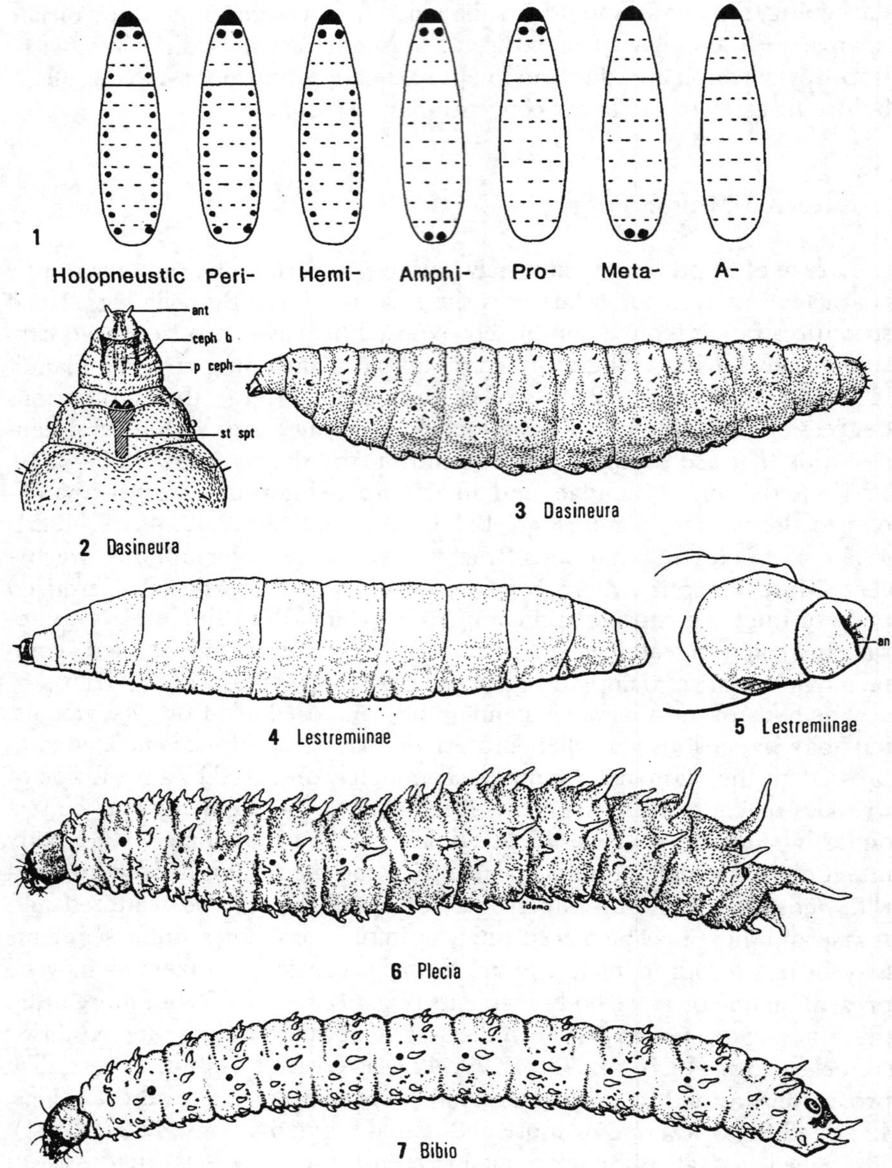

Figs. 42.1–42.7. **1.** Schematic representation of spiracular systems of Diptera larvae. **2–5. Cecidomyiidae:** **2,** ventral view of head and thoracic segments of *Dasineura* sp.; **3,** lateral view of larva of same; **4,** lateral view of larva of Lestremiinae; **5,** oblique posterior view of same. **6,7. Bibionidae:** **6,** lateral view of larva of *Plecia nearctica;* **7,** same of *Bibio* sp. *an,* anus; *ant,* antenna; *cephb,* cephalic bar; *pceph,* postcephalic segment; *st spt,* sternal spatula.

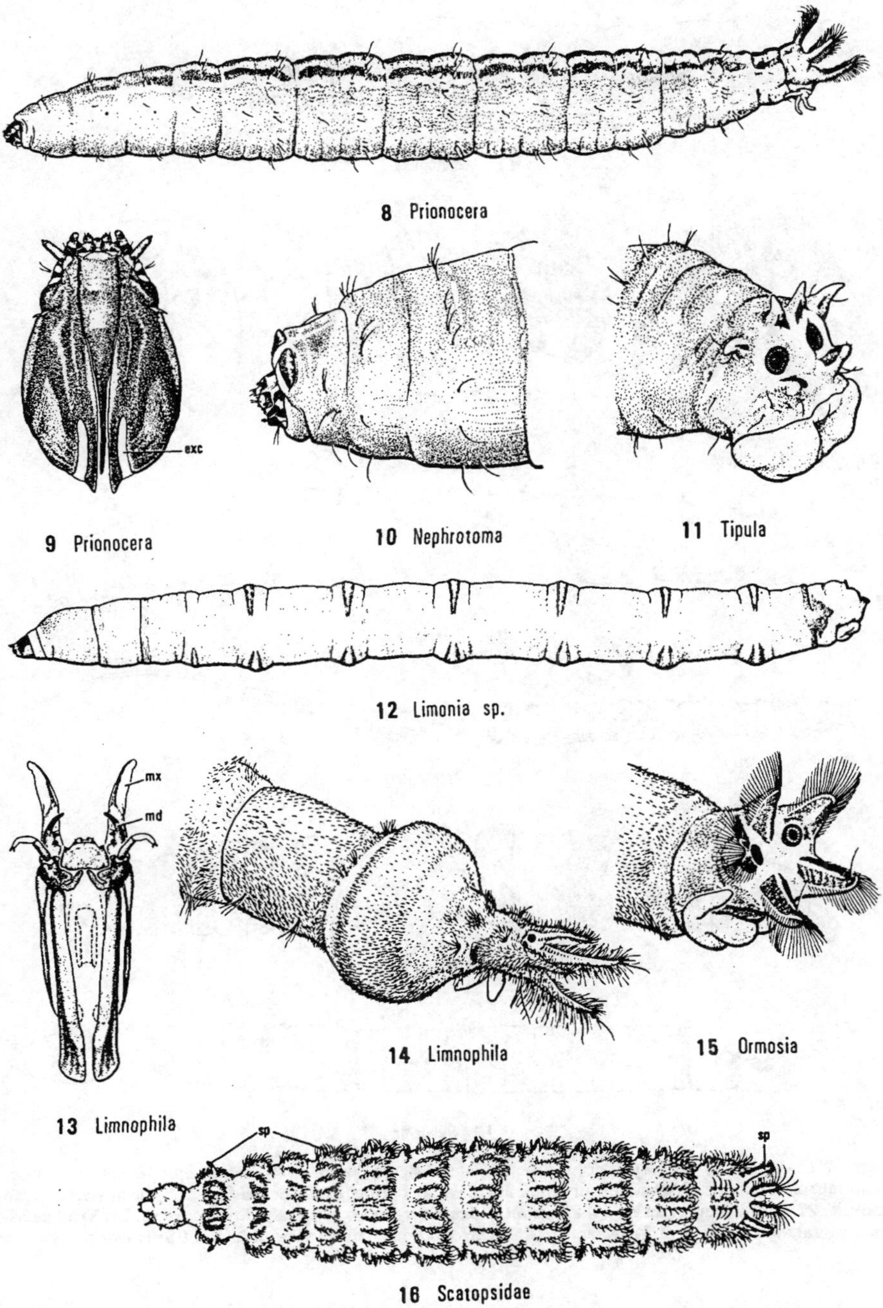

Figs. 42.8–42.16. 8–15. Tipulidae: 8, lateral view of larva of *Prionocera* sp.; **9,** dorsal view of head capsule of same; **10,** retracted head capsule and thoracic segments of *Nephrotoma* sp.; **11,** oblique posterior view of terminal segments of *Tipula trivittata* Say; **12,** lateral view of larva of *Limonia* sp.; **13,** dorsal view of head capsule of *Limnophila* sp.; **14,** oblique posterior view of terminal segment of same; **15,** same of *Ormosia* sp. **16. Scatopsidae,** dorsal view of larva. *exc,* excavation; *md,* mandible; *msc,* maxilla; *sp,* spiracle.

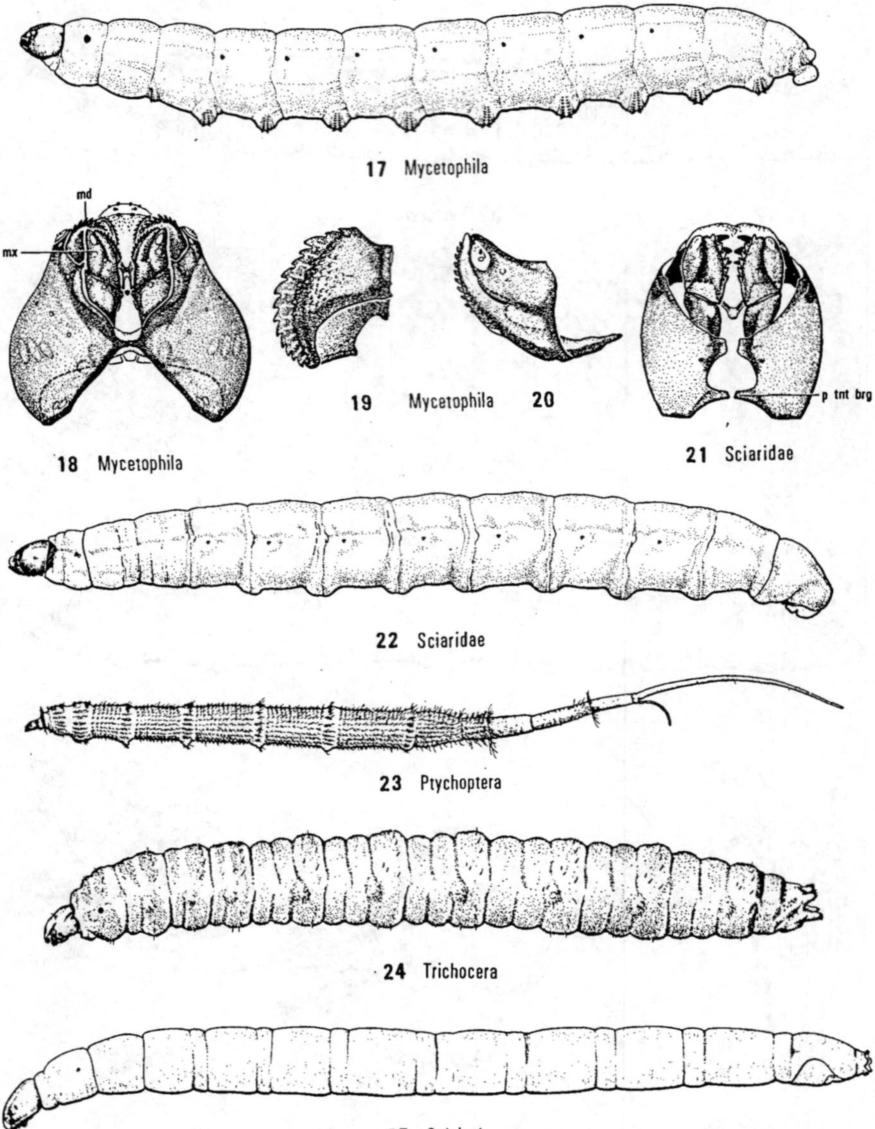

Figs. 42.17–42.25. 17–20. Mycetophilidae: 17, lateral view of larva of *Mycetophila* sp.; **18,** ventral view of head capsule of same; **19,** mandible of same; **20,** maxilla of same. **21,22. Sciaridae: 21,** ventral view of head capsule; **22,** lateral view of larva. **23. Ptychopteridae,** lateral view of larva of *Ptychoptera* sp. **24. Trichoceridae,** lateral view of larva of *Trichocera* sp. **25. Anisopodidae,** lateral view of larva of *Sylvicola* sp.

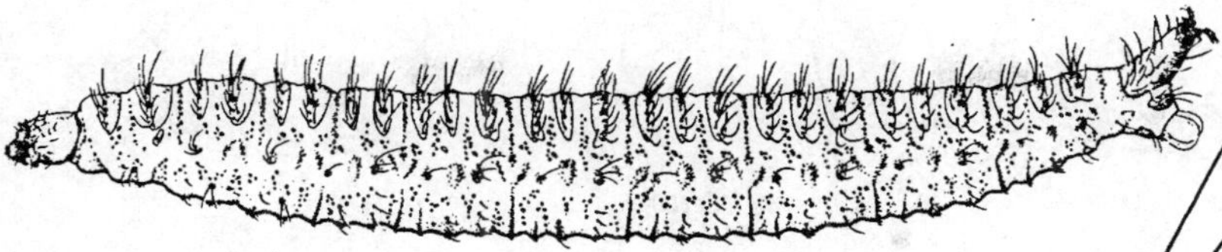

26 Pericoma

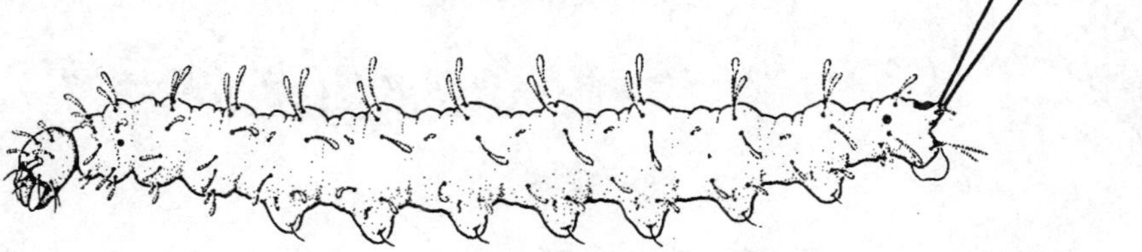

27 Phlebotomus

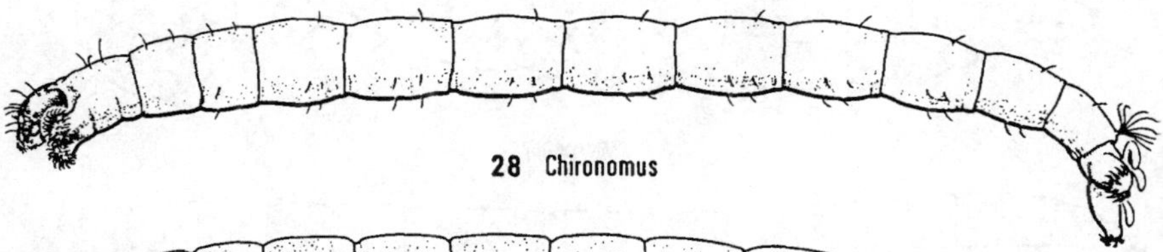

28 Chironomus

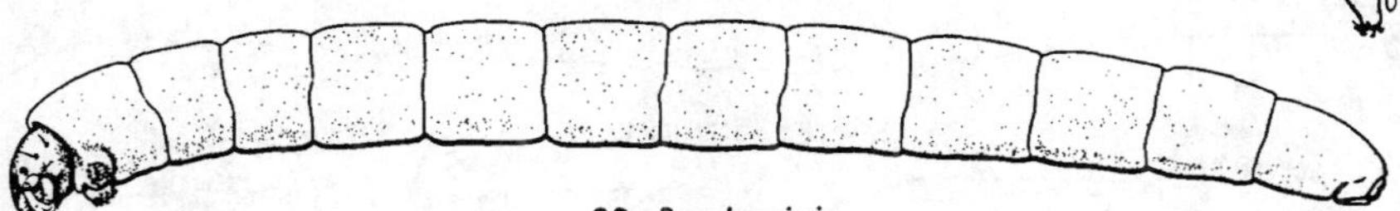

29 Pseudosmittia

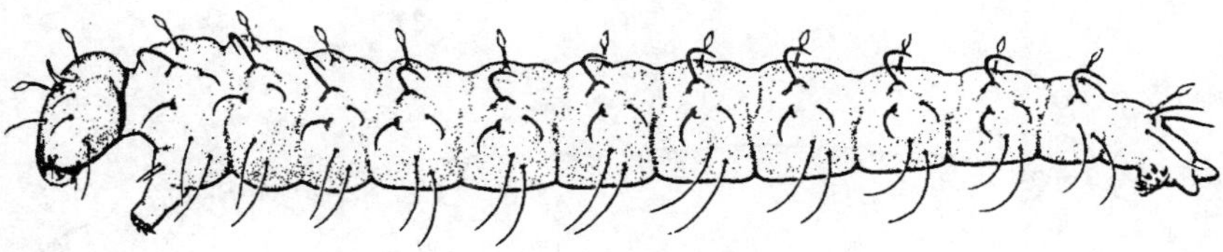

30 Forcipomyia

31 Culicoides

Figs. 42.26–42.31. 26,27. Psychodidae: 26, lateral view of *Pericoma* sp.; **27,** same of *Phlebotomus* sp. **28,29. Chironomidae: 28,** lateral view of *Chironomus* sp.; **29,** same of *Pseudosmittia* sp., **30,31. Ceratopogonidae: 30,** lateral view of *Forcipomyia* sp.; **31,** dorsal view of *Culicoides* sp.

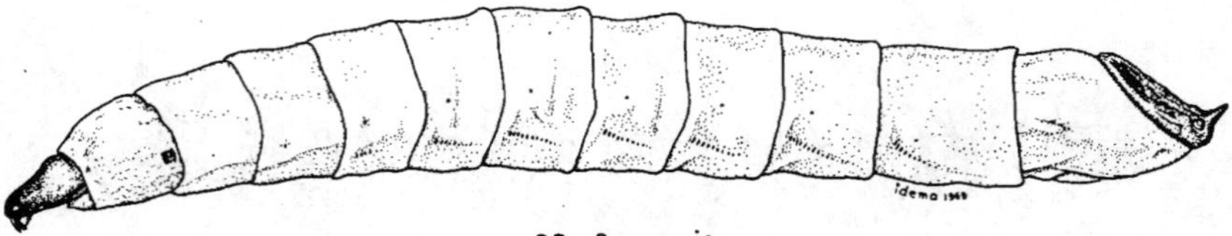

32 Coenomyia

33 Vermileo

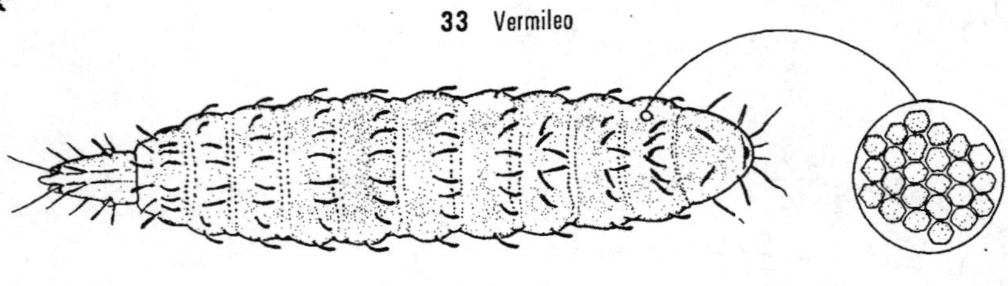

34 Microchrysa

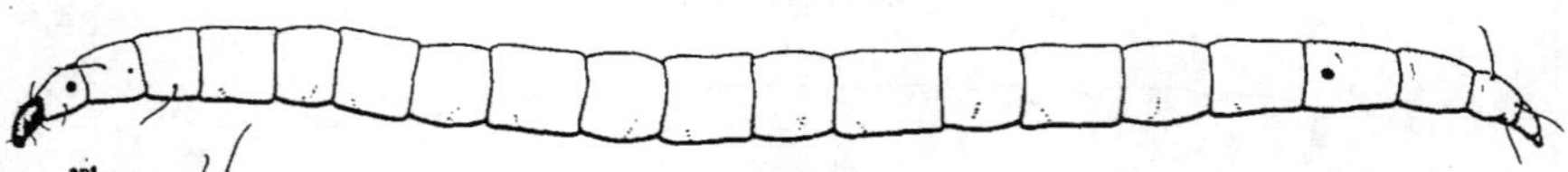

35 Scenopinus

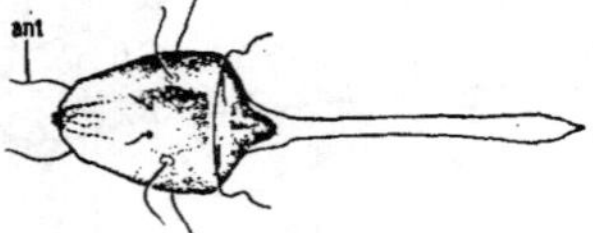

36

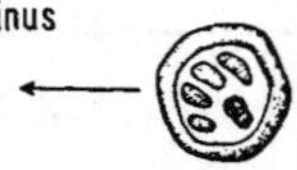

37

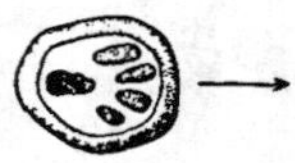

38

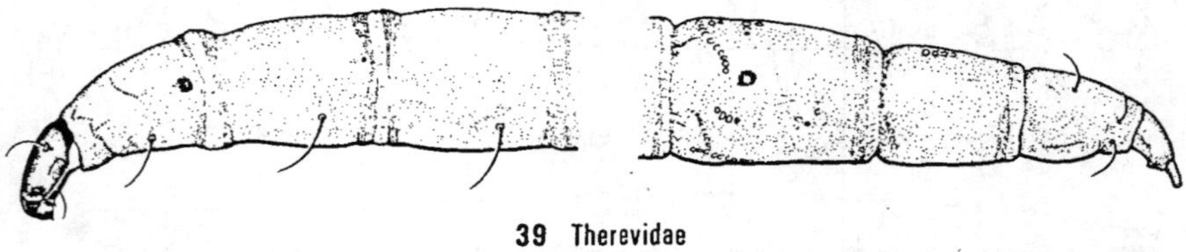

39 Therevidae

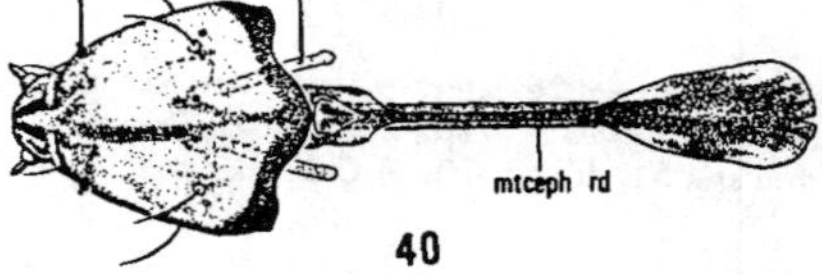

40

Figs. 42.32–42.40. 32. Xylophagidae, lateral view of *Coenomyia ferruginea* (Scopoli). **33. Vermileonidae,** lateral view of *Vermileo comstocki* Wheeler **34. Stratiomyidae,** dorsal view of *Microchrysa polita* (Linnaeus) with enlargement of calcareous integumental plaques. **35–38. Scenopinidae: 35,** lateral view of *Scenopinus* sp.; **36,** dorsal view of head capsule of same; **37,38** anterior and posterior spiracle of same. **39–40. Therevidae; 39,** lateral view of anterior and posterior segments; **40,** dorsal view of head capsule. *ant,* antenna; *mtceph rd,* metacephalic rod; *tnt,* tentorial arm.

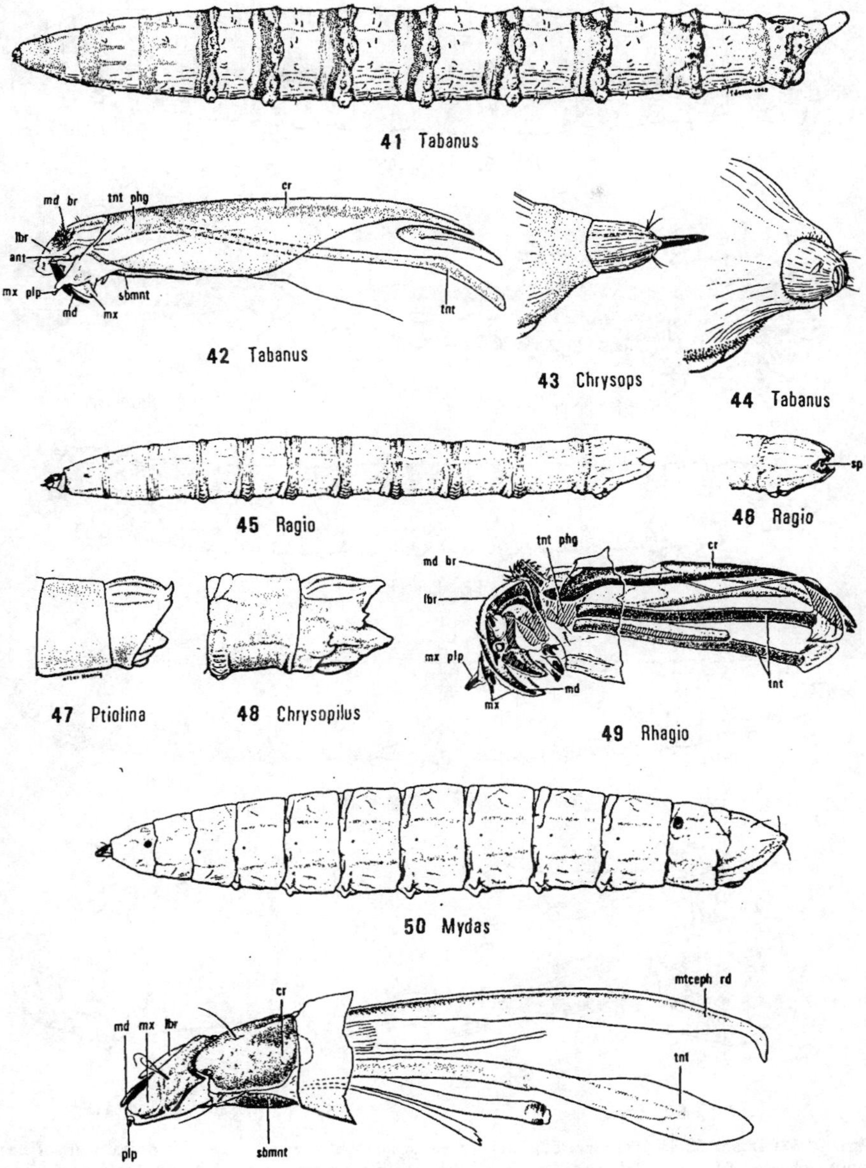

Figs. 42.41–42.51. 41–44. Tabanidae: 41, lateral view of *Tabanus reinwardtii* Wiedemann; **42,** head capsule of same; **43,** respiratory siphon with retractile spine of *Chrysops cincticornis* Walker; **44,** respiratory siphon of *Tabanus marginalis* Fabricius. **45–49. Rhagionidae: 45,** lateral view of *Rhagio* sp; **46,** oblique posterior view of terminal segment of same; **47,48** lateral views of terminal segments of *Ptiolina* sp. and *Chrysopilus flavibarbus* Adams; **49,** head capsule of *Rhagio* sp. **50,51. Mydidae: 50,** lateral view of *Mydas clavatus* (Drury); **51,** head capsule of same. *ant,* antenna; *cr,* cranium; *lbr,* labrum; *md,* mandible; *md br,* mandibular brush; *mtceph rd,* metacephalic rod; *mx,* maxilla; *mx plp,* maxillary palp; *sbmt,* submentum; *tnt,* tentorial arm; *tnt phg,* tentorial phragma.

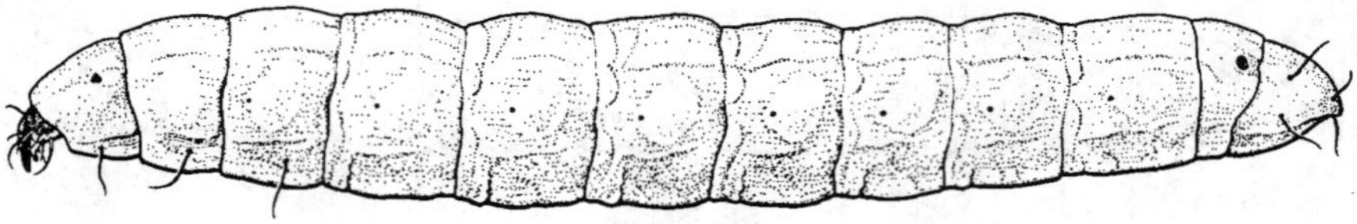

52 Promachus

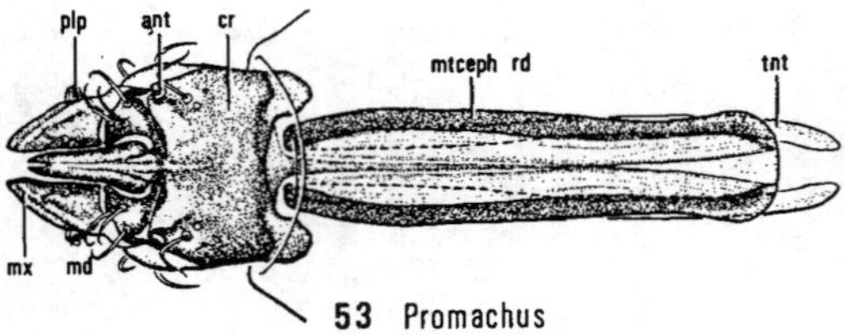

53 Promachus

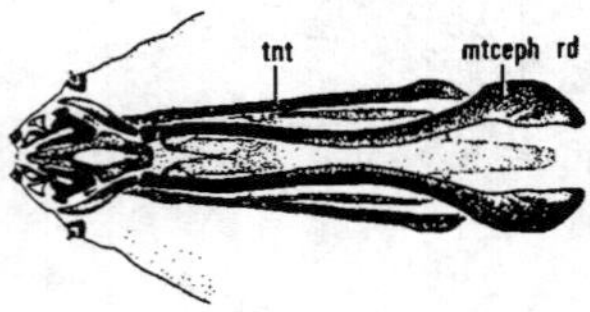

54 Rhaphium

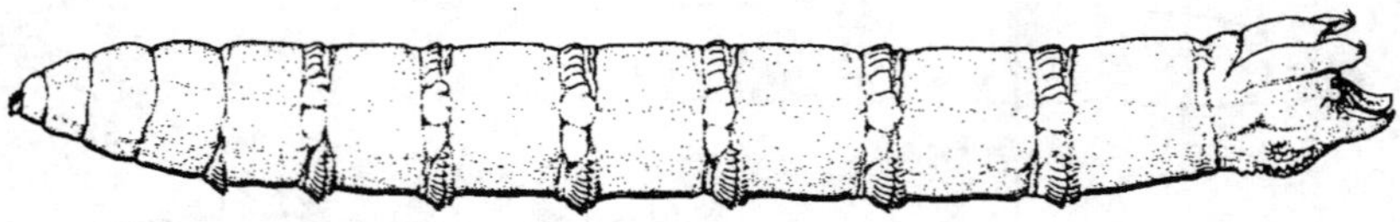

55 Rhaphium

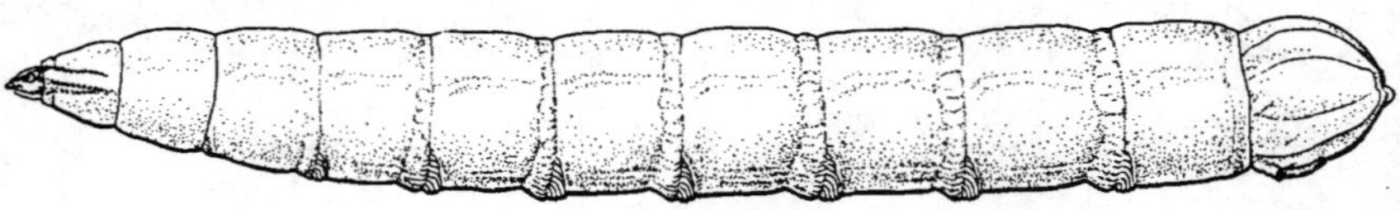

56 Rhamphomyia

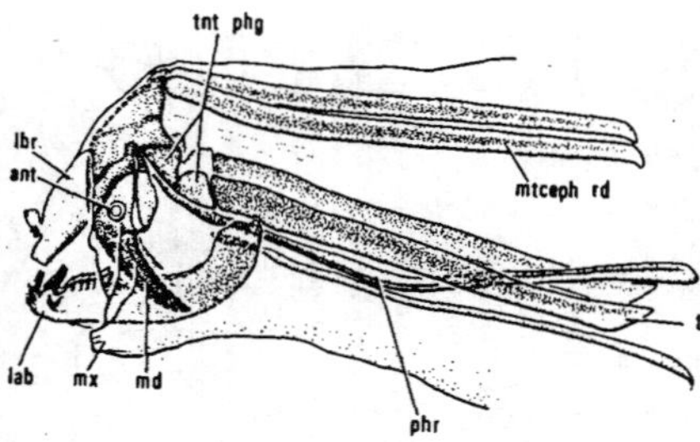

57 Empidinae

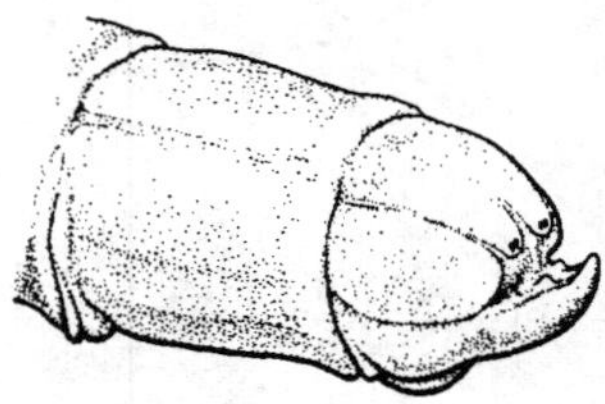

58 Phyllodromia

Figs. 42.52–42.58. 52,53. Asilidae: 52, lateral view of larva of *Promachus* sp.; 53, dorsal view of head capsule of same. **54, 55. Dolichopodidae: 54,** dorsal view of head capsule of *Rhaphium slossonae* (Johnson); **55,** lateral view of larva of same. **56–58. Empididae: 56,** lateral view of *Rhamphomyia sp.*; **57,** lateral view of head capsule of Empidinae; **58,** oblique posterior view of *Phyllodromia* sp. Abbreviations are as for Figs. 42.41 to 42.51.

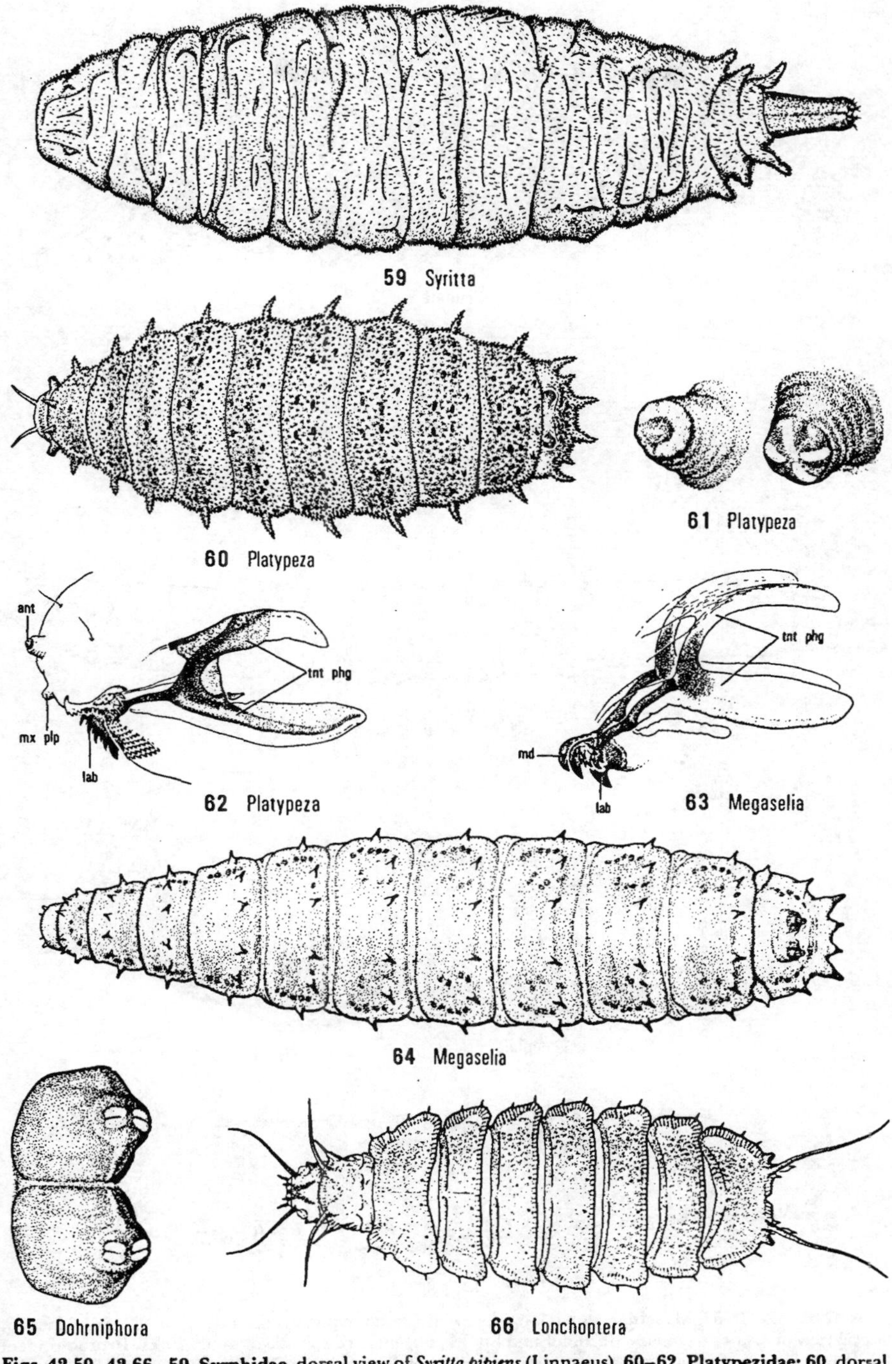

Figs. 42.59–42.66. 59. Syrphidae, dorsal view of *Syritta pipiens* (Linnaeus). **60–62. Platypezidae: 60,** dorsal view of *Platypeza* sp.; **61,** anterior and posterior spiracle of same; **62,** cephalopharyngeal skeleton of same. **63–65. Phoridae: 63,** cephalopharyngeal skeleton of *Megaselia* sp; **64,** dorsal view of larva of same; **65,** posterior spiracles of *Dohrniphora* sp. **66. Lonchopteridae,** dorsal view of larva of *Lonchoptera* sp. *ant.* antenna; *lab,* labium; *mx plp,* maxillary palp; *tnt phg,* tentorial phragma.

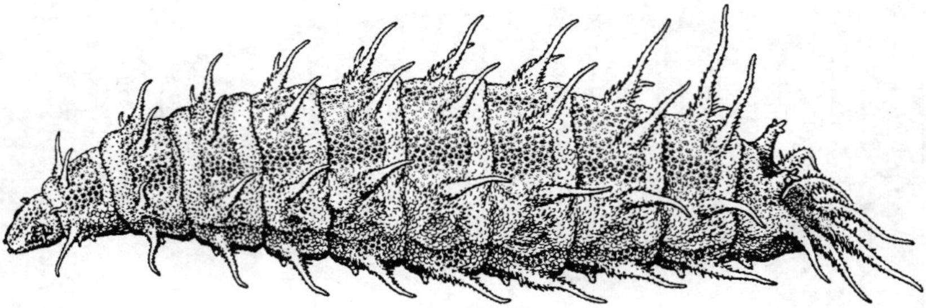

67 Fannia

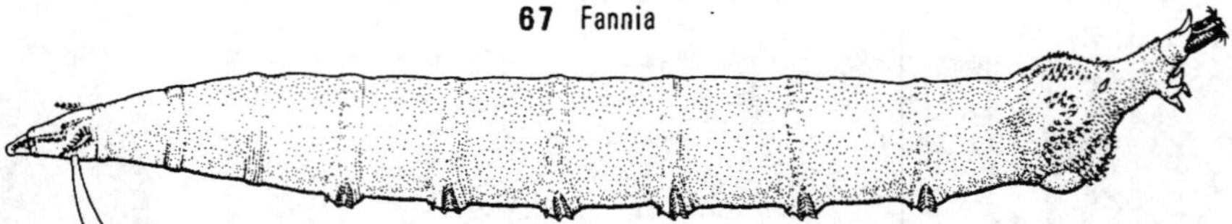

68 Sepsis

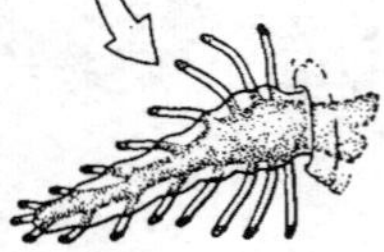

69

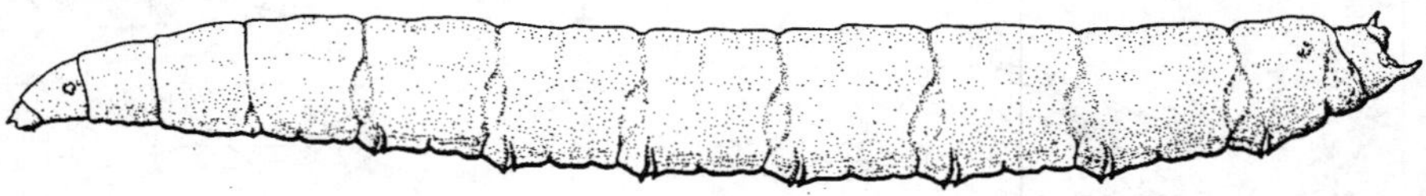

70 Drosophila

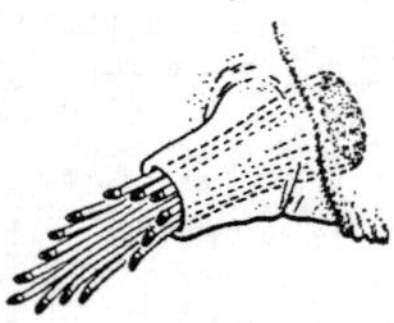

71 Drosophila

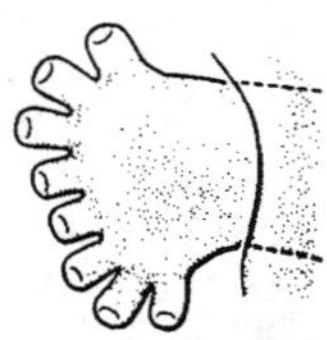

72 Lasiopiophila

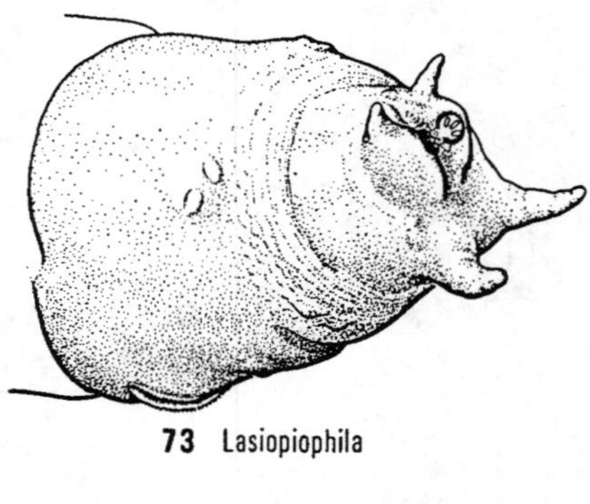

73 Lasiopiophila

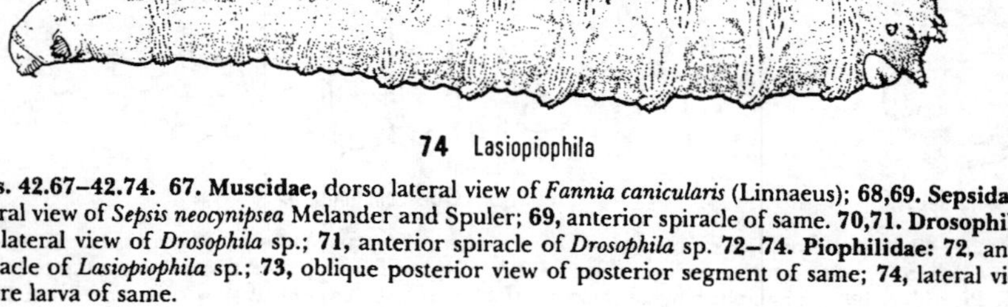

74 Lasiopiophila

Figs. 42.67–42.74. 67. Muscidae, dorso lateral view of *Fannia canicularis* (Linnaeus); **68,69. Sepsidae: 68,** lateral view of *Sepsis neocynipsea* Melander and Spuler; **69,** anterior spiracle of same. **70,71. Drosophilidae: 70,** lateral view of *Drosophila* sp.; **71,** anterior spiracle of *Drosophila* sp. **72–74. Piophilidae: 72,** anterior spiracle of *Lasiopiophila* sp.; **73,** oblique posterior view of posterior segment of same; **74,** lateral view of entire larva of same.

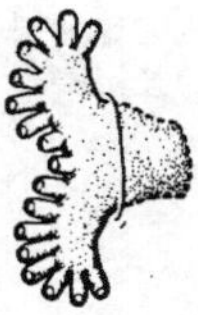

75 Scatophaga

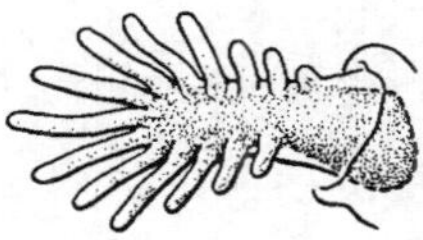

76 Desmometopa

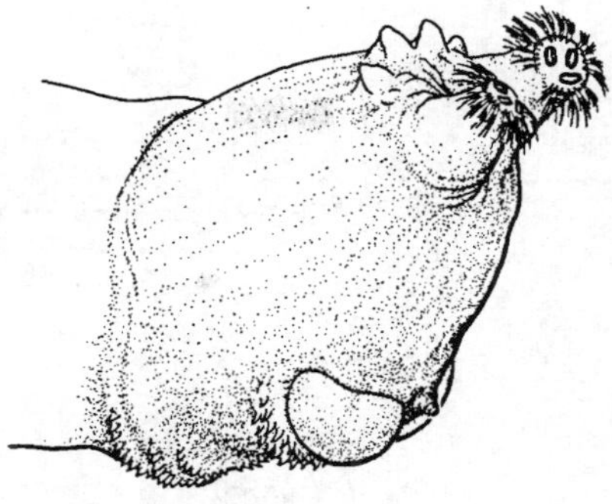

77 Desmometopa

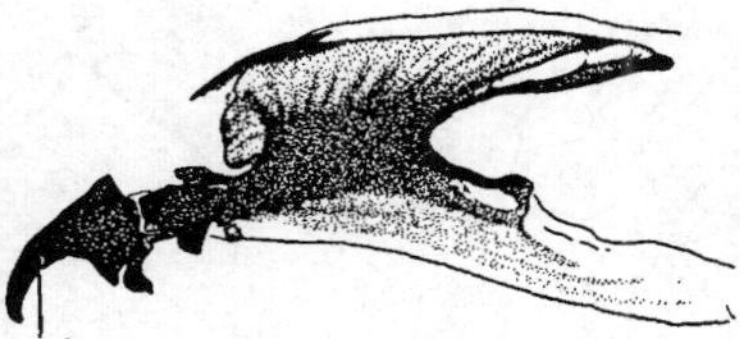

78 Sarcophaga

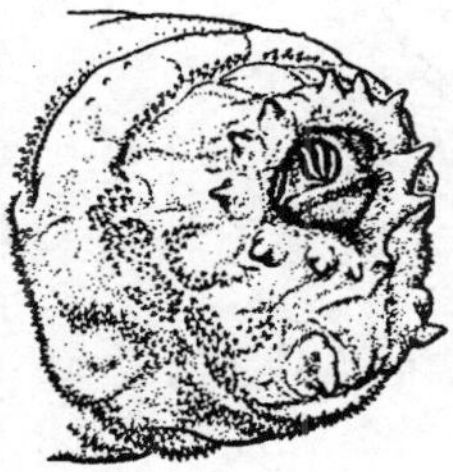

79 Sarcophaga

80 Sarcophaga

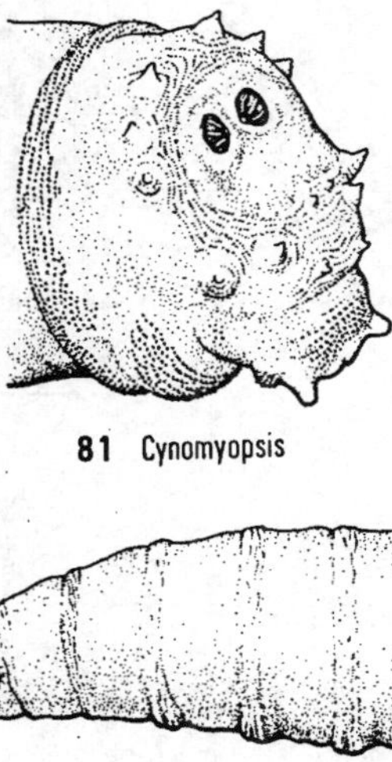

81 Cynomyopsis

82 Cynomyopsis

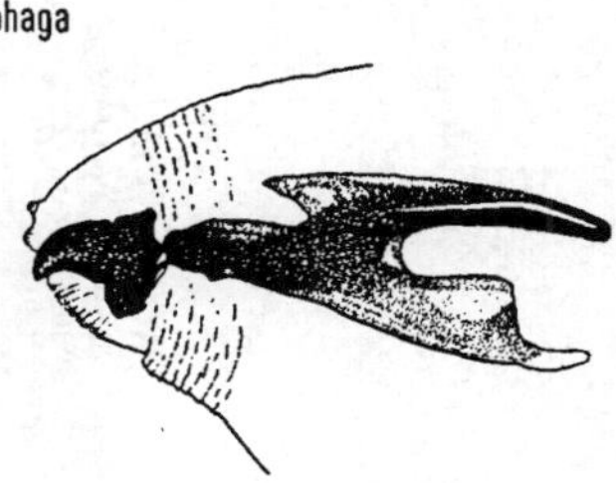

83 Hylemya

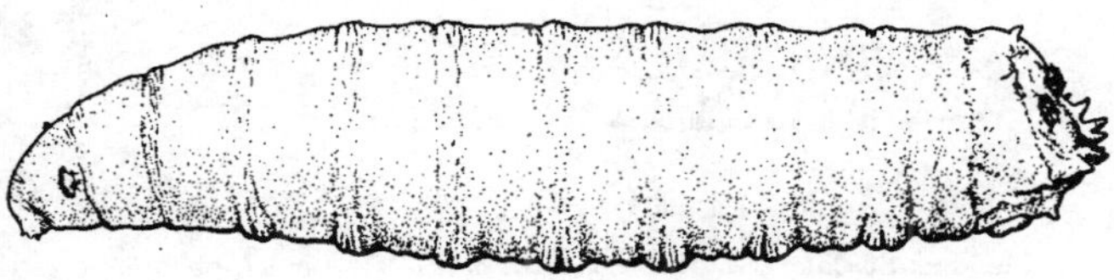

84 Hylemya

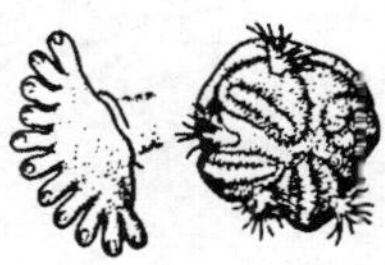

85 Hylemya

Figs. 42.75–42.85. 75. Scatophagidae, anterior spiracle of *Scatophaga* sp. **76,77. Milichiidae: 76,** anterior spiracle of *Desmometopa m-nigrum* (Zetterstedt); **77,** oblique posterior view of terminal segment of same. **78–82. Sarcophagidae: 78,** cephalopharyngeal skeleton of *Sarcophaga guerula* Walker; **79,** oblique posterior view of terminal segment of same; **80,** posterior spiracles of same; **81,** oblique posterior view of terminal segment of *Cynomyopsis cadaverina* (Robineau-Desvoidy); **82,** posterior spiracle of same. **83–85. Anthomyiidae: 83,** cephalopharyngeal skeleton of *Hylemya brassicae* Bouché; **84,** lateral view of larva of same; **85,** anterior and posterior spiracle of same.

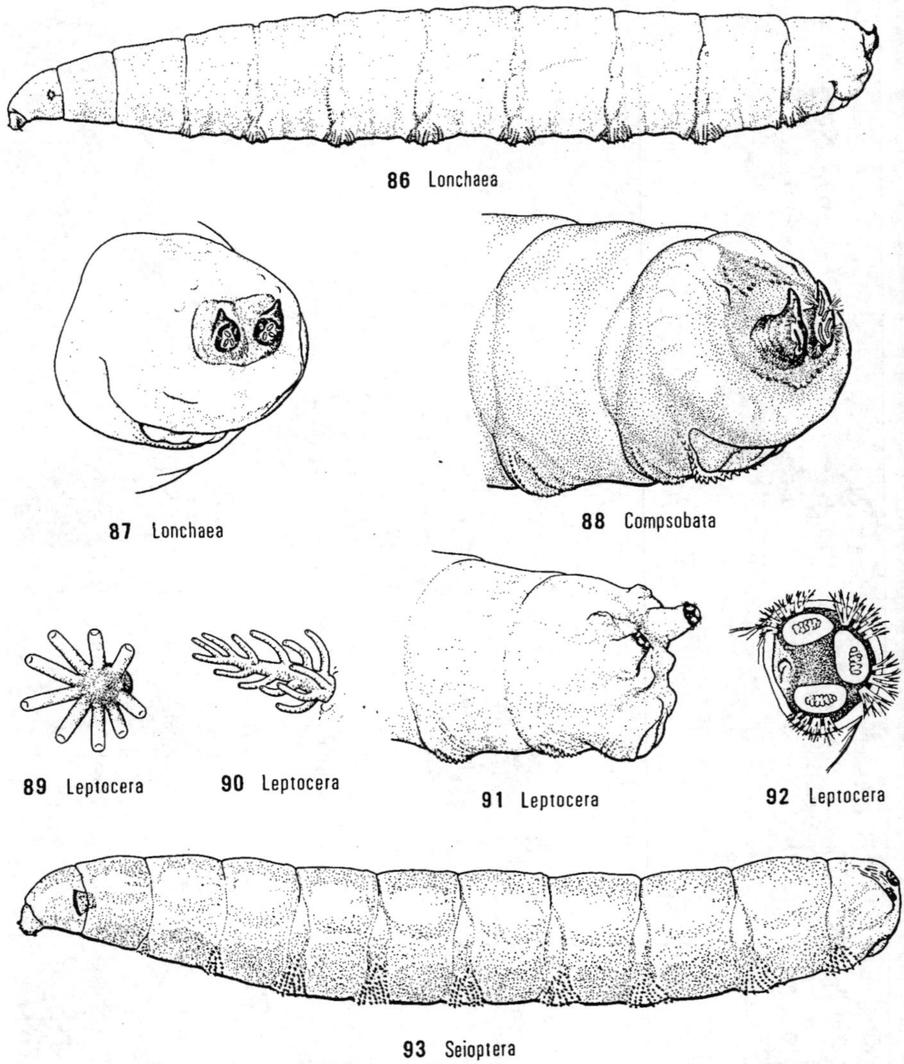

Figs. 42.86–42.93. 86,87. Lonchaeidae: 86, lateral view of *Lonchaea corticis* Taylor; **87,** oblique posterior view of terminal segment of same. **88. Micropezidae,** oblique posterior view of *Compsobata univitta* (Walker). **89–92. Sphaeroceridae: 89, 90,** anterior spiracles of two species of *Leptocera;* **91,** oblique posterior view of terminal segments of *Leptocera* sp; **92,** posterior spiracle of same. **93. Otitidae,** lateral view of *Seioptera vibrans* (Linnaeus)

located on the terminal segment, are called the posterior spiracles. Intermediate spiracles take the number or name of the segment on which they occur. Spiracles vary considerably in form, particularly in number and arrangement of spiracular openings and in the presence of accessory structures. The spiracle normally comprises a spiracular plate, which is usually bordered by a sclerotized rim and bears the spiracular opening (Fig. 42.82). The posterior spiracles of Muscomorpha larvae normally occur on the

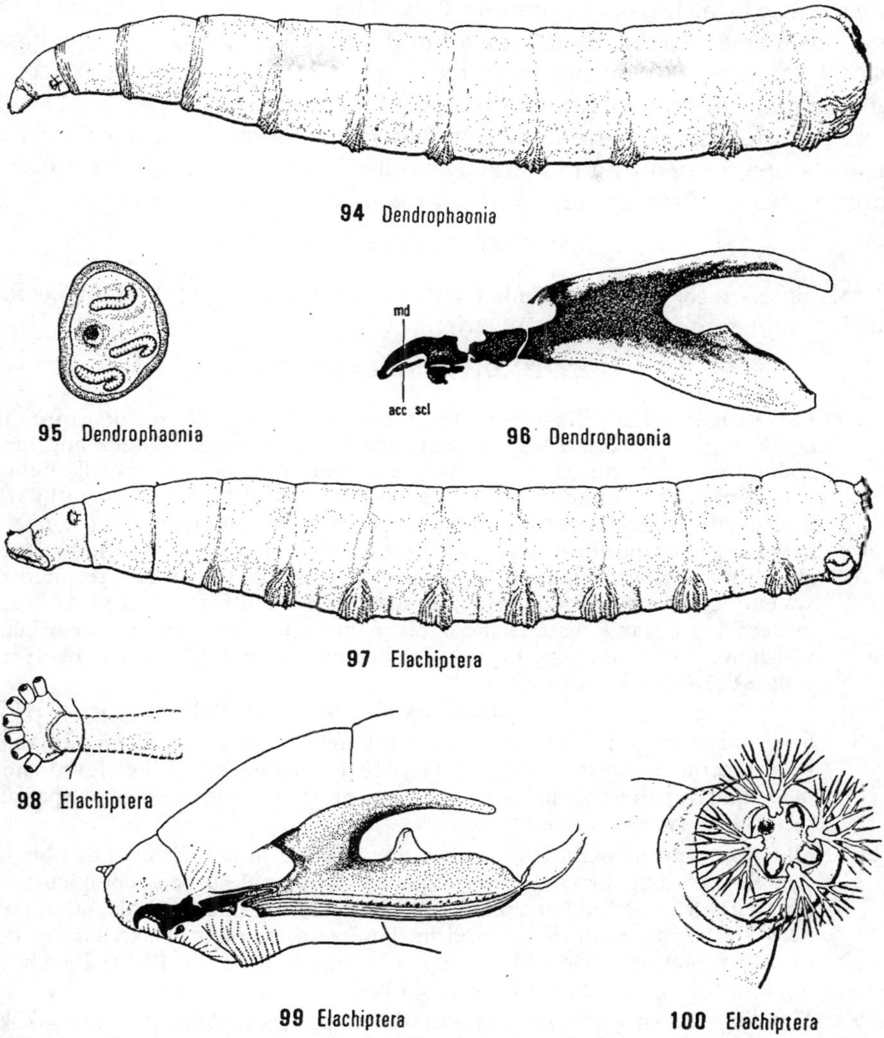

Figs. 42.94–42.100. 94–96. Muscidae: 94, lateral view of *Dendrophaonia* sp.; **95,** posterior spiracle of same; **96,** cephalopharyngeal skeleton of same. **97–100. Chloropidae: 97,** lateral view of *Elachiptera* sp.; **98,** anterior spiracle of same; **99,** cephalopharyngeal skeleton of same; **100,** posterior spiracle of same. *acc scl,* accessory oral sclerite; *md,* mandible.

caudal, often somewhat flattened face of the terminal segment, which is referred to as the spiracular disk. In some other larvae a portion of the terminal segment bearing the spiracles is lengthened into a short or long, sometimes sclerotized, respiratory siphon (Figs. 42.23, 42.26, 42.41, 42.59, 42.68).

The anterior spiracles of Muscomorpha differ quite widely from the posterior spiracles. They are commonly projecting, with the spiracular openings situated individually on short to relatively long papillae or filaments that branch from a common stalk. The branching pattern may take various forms, usually resembling a fan (Figs. 42.72, 42.85) or a tree (Figs. 42.69, 42.76, 42.90), or may be tufted (Fig. 42.71). The openings are commonly more numerous than on the posterior spiracle.

The anus occurs ventrally or caudally on the terminal segment and is usually surrounded by a moderately swollen anal pad of slightly different appearance and surface texture than the adjacent integument.

ACKNOWLEDGMENTS

My sincere recognition is extended to Drs. C. D. Dondale and J. F. McAlpine for their thorough review of this manuscript.

KEY TO DIPTERA FAMILIES

1a Larva small, often yellow to reddish; a small postcephalic segment present between prothorax and head capsule in addition to three thoracic and nine abdominal segments (Figs. 42.2–42.4); head capsule very small, lightly sclerotized, and rounded or only slightly longer than broad; mouthparts indistinctly differentiated; antenna two- or three-segmented; paired cephalic bars projecting posteriorly from lateral margins (Fig. 42.2); prothorax usually with a variously shaped sternal spatula ventrally (Fig. 42.2); respiratory system peripneustic or apneustic; most larvae phytophagous but many living in decaying organic matter; the latter include the Subfamily **Lestremiinae,** which are recognizable by having anus terminal (Fig. 42.5) rather than ventral; **NEMATOCERA** (in part) . . . .
Cecidomyiidae (Mamaev and Krivosheina 1965)

1b Size and color various; postcephalic segment usually not distinguishable; head capsule or internal cephalopharyngeal skeleton with well-differentiated mouthparts; paired cephalic bars lacking; prothorax without sternal spatula; respiratory system variable **2**

2a Mandibles normally moving against one another in a horizontal or oblique plane, rarely hooklike or sickle-shaped; head capsule usually complete and permanently exserted (eucephalic) (Figs. 42.6, 42.7, 42.16–42.41), but if partially retracted within thorax and incomplete as a result of excisions in the capsule posteriorly (**Tipulidae,** couplet 3, Figs. 42.8–42.10, 42.13), then lacking tentorial arms; most **NEMATOCERA** **3**

2b Mandibles moving parallel to one another in a vertical plane, usually hooklike or sickle-shaped (Figs. 42.42, 42.49, 42.57, 42.78); head capsule variously reduced posteriorly and partially or almost entirely retracted within thorax (hemicephalic) or replaced by an internal cephalopharyngeal skeleton (Figs. 42.42, 42.49, 42.51, 42.53, 42.54, 42.57, 42.62, 42.63, 42.78, 42.96, 42.99), but if appearing complete and permanently exserted, then with a slender metacephalic rod extending into prothorax (Fig. 42.36, 42.40); **BRACHYCERA** **15**

3a Head capsule longitudinally excised dorsolaterally to varying degrees, in extreme cases reduced to several slender rods (Figs. 42.9, 42.13); capsule retracted more than half its length within thorax (Figs. 42.8, 42.10, 42.12); respiratory system usually metapneustic, only rarely apneustic; posterior spiracular disk usually with two to six marginal lobes (Figs. 42.8, 42.11, 42.14, 42.15); predominantly aquatic or semiaquatic in soil and organic matter, with some species in relatively dry habitats . . . . . **Tipulidae** (Alexander 1920)

3b Head capsule complete and retracted within thorax; respiratory system usually not as above **4**

4a Respiratory system holopneustic or peripneustic, with eight pairs of abdominal spiracles; posterior spiracles usually conspicuously larger than other seven pairs of abdominal spiracles **5**

4b Respiratory system not as above; if hemipneustic, with only seven pairs of abdominal spiracles, then all abdominal spiracles of equal size (Figs. 42.17, 42.22) **6**

5a Respiratory system holopneustic; all segments usually with tuberculous or spinous processes; posterior spiracles sessile on lateral surface of terminal or penultimate segment, (Figs. 42.6, 42.7); scavengers in soils rich in humus, sometimes feeding on plant roots . . . . . . **Bibionidae** (Krivosheina 1962)

5b Respiratory system peripneustic; only terminal abdominal segment having tuberculous processes, involving a pair of dorsally sclerotized lobes or a broad sclerotized shelf behind the anus and below the posteriorly directed spiracles; preceding abdominal segments dorsally with abundant short setae arranged in a regular pattern (Fig. 42.16); in many kinds of decaying plant and animal materials, including excrement . . . . **Scatopsidae**

6a Respiratory system hemipneustic, rarely propneustic or apneustic; mandible or maxilla or both flattened or lamellate, with a serrate inner margin (Figs. 42.18–42.21) **7**

6b Respiratory system amphipneustic, metapneustic, or apneustic; mandible and maxilla not flattened, without a serrate inner margin **8**

7a Posterior tentorial bridge complete or nearly complete (Fig. 42.18) (usually visible beneath integument within occipital cavity in preserved specimens without special treatment); abdominal creeping welts without sclerotized spicules (Fig. 42.22); feeding on decaying plant material, animal excrement, and fungi . . . . **Sciaridae**

7b Posterior tentorial bridge absent (Fig. 42.21) or, if bridge partially formed, then abdominal creeping welts with sclerotized spicules (Fig. 42.17); feeding on fungi . . . . **Mycetophilidae** (Madwar 1937)

8a Posterior spiracles mounted at end of a slender respiratory siphon that is nearly or fully as long as the body when completely extended (Fig. 42.23); in wet to saturated organic soils . . . . **Ptychopteridae**

8b Posterior spiracles mounted on no more than a short siphon, much shorter than body, or spiracles absent **9**

9a Prothorax with one proleg or a pair of prolegs ventrally; respiratory system apneustic (Figs. 42.28–42.30). **10**

9b Prothorax lacking prolegs; respiratory system apneustic, amphipneustic, or metapneustic **11**

10a All body segments with prominent tubercles or setae or both (Fig. 42.30); in damp situations associated with decaying organic matter . . . . **Ceratopogonidae (Forcipomyiinae)**

10b Body segments except sometimes the terminal segment lacking prominent tubercles and setae (Figs. 42.28, 42.29); primarily aquatic but many Orthocladiinae in terrestrial habitats. . . . . **Chironomidae** (Strenzke 1950)

11a Respiratory system apneustic; larva slender, smooth, with uniform beadlike segments; long setae present only on terminal abdominal segment (Fig. 42.31); in damp soils, often on coastal and inland beaches . . . . **Ceratopogonidae (Leptoconopinae, Ceratopogoninae)**

11b Respiratory system amphipneustic or metapneustic; larva usually somewhat wrinkled, with secondarily divided segments; distinctive setation or sclerotized plaques may be present on some segments **12**

12a Secondary segmentation of prothorax and abdominal segments evident, with these segments having a distinct narrow annulus anteriorly (Fig. 42.25); posterior spiracles sessile on surface of terminal abdominal segment and placed either laterally or apically, with apically positioned spiracles surrounded by five small lobes (Fig. 42.25); saprophagous in moist decaying organic material . . . . **Anisopodidae** (Brindle 1962, Keilin and Tate 1940)

12b Secondary segmentation either not apparent or, more usually, with thoracic and first abdominal segments subdivided into two sections and remaining abdominal segments subdivided into three elementary sections; posterior spiracles either mounted on a respiratory siphon or as above, but if sessile and apical in position, then surrounded by two or four lobes **13**

13a Posterior spiracles and a pair of fanlike setal brushes borne dorsally at posterior margin of a sclerotized plate on terminal abdominal segment or at apex of a short sclerotized respiratory siphon projecting posterodorsally from the terminal segment; a sclerotized plaque or plaques evident dorsally on one or more of the secondary segmental divisions (Fig. 42.26); in decaying organic material . . . . **Psychodidae (Psychodinae)** (Quate 1955)

13b Posterior spiracles not borne on a respiratory siphon; sclerotized plaques absent **14**

14a Posterior spiracles situated laterally on terminal or penultimate abdominal segment; setae on integument prominent, often with some variously modified and others on dorsum of terminal segment very long (Fig. 42.27); breeding in soil, often in semidesert areas **15**

14b Posterior spiracles situated on apex of terminal abdominal segment and surrounded by four fleshy lobes (Fig. 42.24); setae variable, but none very long; in decaying vegetable matter . . . .
Trichoceridae (Brindle 1962, Keilin and Tate 1940)

15a Sclerotized portions of cranium always present and usually partially exposed externally (Figs. 42.32–42.35, 42.39, 42.41, 42.45, 42.50, 42.52, 42.55, 42.56); labrum, mandibles, or maxillae readily recognized; **orthorrhaphous BRACHYCERA** **16**

15b Externally exposed, sclerotized portions of cranium completely lacking; only a membranous cephalic segment anterior to prothorax remaining, this normally bearing two pairs of papillalike projections considered to be vestiges of antennae and maxillary palpi (Figs. 42.62, 42.83, 42.99); characteristically shaped cephalopharyngeal skeleton retracted completely within thorax or almost entirely absent in some; species usually parasitic (Figs. 42.62, 42.63, 42.78, 42.83, 42.96, 42.99); homologs of labrum and maxillae not clearly definable; **cyclorrhaphous BRACHYCERA (MUSCOMORPHA)** **27**

16a Head capsule of a rather smooth conical shape with oral cavity opening ventrally near apex and not visible in anterior view; posterodorsal surface of terminal abdominal segment with a large, oval sclerotized plate bearing spiracles anteriorly and a pair of pointed projections near posterior margin (Fig. 42.32); predaceous in soil and decaying wood . . **Xylophagidae (*Coenomyia*)**

16b Head capsule not as uniformly conical and oral cavity partially visible in anterior view; terminal abdominal segment lacking such a sclerotized plate **17**

17a Body dorsoventrally compressed; integument hardened by calcareous deposits accumulated in small roundish or hexagonal facets that impart a shagreened appearance; head capsule permanently partially exposed and capable of only slightly independent movement (Fig. 42.34); in soils rich in decaying plant and animal materials . . . **Stratiomyidae** (McFadden 1967)

17b Body of various forms; integument not hardened and not faceted with calcium deposits, sometimes tough and leathery; head capsule capable of much independent movement **18**

18a Body long and slender, eel-like, with 20 apparent segments; posterior spiracles situated laterally on fourth segment from end (Figs. 42.35, 42.39); head capsule appearing to be complete and permanently exserted, flexibly articulated posteriorly with a slender or apically expanded metacephalic rod lying within thorax (Figs. 42.36, 42.40) **19**

18b Body not eel-like, with no more than 12 apparent segments; posterior spiracles on terminal or penultimate abdominal segment; head capsule more or less reduced, especially posteroventrally, and partially retracted within thorax, with or without a single broad or nonspatulate metacephalic rod lying within thorax, or with two such rods present. **20**

19a Metacephalic rod expanded apically (Fig. 42.40); antenna minute and peglike; setae on each side of thoracic segments shorter than diameter of segments and situated ventrolaterally (Fig. 42.39); predaceous in soil and decaying wood . . . . **Therevidae**

19b Metacephalic rod slender throughout (Fig. 42.36); antenna long and filamentous; setae on each side of thoracic segments as long as diameter of segments, and filamentous; setae on each side of thoracic segments as long as diameter of segments, and mesothoracic seta situated higher on segment than prothoracic and metathoracic setae (Fig. 42.35); predaceous on termites and insects in nests and burrows of birds and animals . . . . **Scenopinidae**

20a Body with terminal segments distinctly enlarged; integument wrinkled and warty; abdominal segment 1 with a ventral proleg; abdominal segment 7 with a dorsal row of spinelike tubercles along posterior margin (Fig. 42.33); posterior spiracles situated dorsally on abdominal segment 8; constructs pitfall traps in sandy soil for capturing prey . . . **Vermileonidae** (Wheeler 1930)

20b Body usually cylindrical and relatively smooth-skinned; prolegs, if present, situated on several segments; abdominal segment 7 lacking dorsal row of tubercles; posterior spiracles, if present, situated caudally on terminal segment or dorsolaterally on penultimate or terminal segment **21**

21a Each mandible with a brush of backwardly curved spines situated above base and attached to mandible such that when it strikes downward the brush of spines is pulled forward from beneath a semitransparent membranous covering and the spines are erected (Figs. 42.42, 42.49); portion of cranium lying within thorax continuous with anterior exposed portion without apparent break between the two areas, although desclerotization may suggest bilateral division (Fig. 42.49); tentorial arms solidly attached anteriorly (Fig. 42.42), except in ***Rhagio*** (Fig. 42.49) **22**

21b Mandible without such a brush of spines; posterior portion of cranium (metacephalic rod or rods) lying within thorax separated from anterior exposed portion by a clear seam, allowing independent flexibility between the two portions (Figs. 42.51, 42.53, 42.54, 42.57); tentorial arms also flexibly attached anteriorly **23**

22a Posterior spiracular openings lying within fissures to either side of a pair of abutting vertically linear bars (Fig. 42.44) or a retractable laterally compressed spine (Fig. 42.43); spiracles situated at apex of a short respiratory siphon or a comparable slightly domed area delimited caudally on terminal segment and siphon; terminal segment lacking tubercles; several or all of anterior seven abdominal segments with an encircling row of projections sometimes bearing apical spicules and functioning as prolegs (Fig. 42.41); submental plate present (Fig. 42.42); in wet to drier, arable soils . . . . . . **Tabanidae** (Teskey 1969)

22b Posterior spiracular openings exposed (Fig. 42.46); spiracle circular or oval; tracheal trunks distinctly separated caudally; terminal segment either deeply cleft caudally to form two or four liplike lobes (Figs. 42.45, 42.46, 42.48), or bearing a pair of sclerotized hornlike processes dorsally and a pair of rounded lobes ventrally (Fig. 42.47); posterior spiracles on caudal face of dorsal lobes; first seven abdominal segments with ventral creeping welts; submentum absent; in soils rich in organic matter. . . . . **Rhagionidae** (Brindle 1959, Krivosheina 1960)

23a Head capsule with a single relatively narrow or broader metacephalic rod that is sometimes split almost to base; sclerotized submentum present ventrally on head capsule; maxilla large and heavily sclerotized, more prominent than slender mandible (Figs. 42.51, 42.53); nine abdominal segments present; respiratory system functionally amphipneustic, although remnants of spiracles forming a holopneustic system usually visible; posterior spiracles situated laterally on abdominal segment 8 (Figs. 42.50, 42.52) **24**

23b Head skeletonized in appearance, with two slender metacephalic rods and two tentorial arms the most prominent of features (Figs. 42.54, 42.57); submentum absent; maxilla sometimes appearing to be absent, and never heavily sclerotized nor more prominent than mandible; eight abdominal segments present; posterior spiracles, when present, located caudally on terminal segment; respiratory system amphipneustic, metapneustic, or apneustic **26**

24a Maxillae laterally compressed, tending to cup mandibles, and similar in length to mandibles; maxillary palpus situated apically (Fig. 42.51); habitat terrestrial usually in decaying wood . . . . **Mydidae**

24b Maxillae more or less dorsoventrally compressed, often toothed apically and concave ventrally to form gougelike digging structures, usually much longer than mandibles; maxillary palpus situated laterally (Figs. 42.53) **25**

25a Abdominal segment 8 about twice as long as wide; posterior spiracles lying laterally near anterior margin of this eighth segment; predacious on other insects in soil . . . . **Apioceridae**

25b Abdominal segment 8 no longer than half its diameter; posterior spiracles situated dorsolaterally in posterior half of eighth segment (Fig. 42.52); commonly predacious on other insects in dry, arable soils **Asilidae** (Melin 1923)

26a Metacephalic rods moderately expanded to spatulate apically (Fig. 42.54); terminal abdominal segment either with four (or rarely two ventral) primary lobes surrounding posterior spiracles (Fig. 42.55), or evenly rounded as in plant-mining forms; in wet to damp organic soils . . . . . **Dolichopodidae**

26b Metacephalic rods slender throughout (Fig. 42.57); terminal abdominal segment usually bearing a single median protuberance below posterior spiracles in nonaquatic larvae (Figs. 42.56, 42.68); in wet to damp soils or decaying vegetation . . . . **Empididae**

27a Posterior spiracles elevated on a single short to moderately long prominence, with spiracular plates fused along median margin (Fig. 42.59); body with dense pubescence and usually with systematically arranged spicules or tubercles on one or more segments; in wet-to-damp soils with abundant organic matter . . . . **Syrphidae** (Dixon 1960, Hartley 1961, Heiss 1938)

27b Posterior spiracles either sessile or elevated above surface of terminal abdominal segment; spiracular plates normally well separated **28**

28a Body distinctly flattened dorsoventrally, with thin striated lateral margins on tergal plates of all segments; long filamentous processes present on terminal abdominal segment and on first two thoracic segments (Fig. 42.66); in decaying vegetation . . . . **Lonchopteridae**

28b Body not dorsoventrally flattened, or if so, then lacking thin striated lateral margins on tergal plates; long filamentous processes when present not restricted to terminal abdominal segment and first two thoracic segments. . . **29**

29a Anterior spiracle simple, each with one to several sessile spiracular openings arrayed peripherally at apex of a short tubular or conical projection (Fig. 42.61); body often somewhat dorsoventrally flattened; all body segments usually bearing several systematically arranged spicules or tubercles, usually with those situated laterally most prominent (Figs. 42.60, 42.64, 42.67); tentoropharyngeal and hypopharyngeal sclerites of cephalopharyngeal skeleton finely constructed and fused together; hypopharyngeal sclerite usually continuous anteriorly with a single or multitoothed median labial sclerite, or with paired mandibles, or with both structures (Figs. 42.62, 42.63) **30**

29b Anterior spiracle either absent or bearing two or more short papillae or long filaments branching from apex of a spiracular stalk (Figs. 42.69, 42.71, 42.72); tentoropharyngeal and hypopharyngeal sclerites often more strongly constructed than above and distinctly separated; hypopharyngeal sclerite fused to hooklike labial sclerite only in first instar larva of some species . . **31**

30a Posterior spiracles each on short conical apically sclerotized spiracular supports, with four spiracular openings on each spiracle arranged radially around spiracular scar (Fig. 42.61); all known larvae feed on fungi. . . . . **Platypezidae** (Kessel, Buegler and Keyes 1973)

30b Posterior spiracles variously supported, each with spiracular openings arranged in two pairs placed one behind the other (Fig. 42.65); parasitic, carnivorous, or saprophagic in a variety of decaying animal or plant materials **Phoridae**

31a Spiculate or setiferous tubercles present on several body segments preceding terminal one (Fig. 42.67); in fungi and decaying plant and animal materials **Muscidae (*Fannia*)**

31b Tubercles entirely absent, or situated only on terminal abdominal segment **32**

32a One or more body segments densely clothed with minute setulae or spicules (Fig. 42.68), or terminal abdominal segment bearing a characteristic complement of one or more pairs of symmetrically placed papillae or tubercles (Figs. 42.68, 42.70, 42.74, 42.77, 42.79, 42.81, 42.84) **33**

32b Body segments lacking abundant setulae, spicules, papillae, or tubercles, and generally featureless with exception of spicules on creeping welts that may occasionally encircle anterior margins of a few segments. **42**

33a Terminal abdominal segment more or less tapered posteriorly because of close proximity or basal union of spiracular prominences (Figs. 42.68, 42.70), or the basal union sometimes attenuated into a respiratory tube (Figs. 42.62–42.64); shorter spiracular prominences usually each with a tubercle present dorsally or dorsolaterally at base; spicules or setulae may extensively cover integument of one or more segments 34

33b Terminal abdominal segment rather truncate; posterior spiracular prominences distinctly separate and lacking tubercles on the base of each prominence; spicules or setae usually present only on margins of segments. . . . 38

34a Anterior spiracle with a basal stalk terminating in many long filamentous processes (Fig. 42.71), and capable of being entirely retracted within the body; saprophagous in fermenting organic matter . . . . **Drosophilidae (Drosophilinae)**

34b Anterior spiracle absent or constructed differently than above, but if in form of an elongate stalk that can be retracted within the body, then apically bearing short serially arranged lateral buds (Fig. 42.64) 35

35a Terminal abdominal segment attenuated posteriorly into a more or less elongate respiratory tube lacking tubercles dorsally and usually capable of some invagination; spicules or setae entirely covering all abdominal segments; in mud, sand, grassy shores, and sedge meadows . . . . **Ephydridae**

35b Terminal abdominal segment tapered, either with a very short respiratory tube, or with closely placed or basally fused conical spiracular prominences; spiracular prominences each with a tubercle present dorsally at or near base; spicules or setulae not covering all aspects of every abdominal segment. . . 36

36a Spicules and pubescence extensively covering terminal abdominal segment only (Fig. 42.68); posterior spiracles usually with well-developed spiracular setae; anterior spiracle with buds projecting to either side of a more or less elongate central axis (Fig. 42.69); in excrement . . . . **Sepsidae**

36b Spicules present only at segmental margins of terminal abdominal segment, or extensively covering other segments besides the terminal one; posterior spiracles with spiracular setae inconspicuous or absent; anterior spiracle with buds projecting fanlike (except in some **Lauxaniidae**) (Fig. 42.72). . . . 37

37a Posterior spiracles situated on median sloping faces of spiracular prominences and appearing capable of retraction on one another (Fig. 42.23); segments immaculate except for tubercles on terminal segment and spicules on anterior ventral creeping welts of abdominal segments (Fig. 42.74); saprophagous in decaying plant and animal matter . . . . **Piophilidae**

37b Posterior spiracles situated squarely on apices of spiracular prominences; spicules on abdominal segments normally much more extensive than above; saprophagous on decaying leaves . . . . **Lauxaniidae** (Miller and Foote 1975, 1976)

38a Posterior spiracular openings arranged with longitudinal axes aligned such that two openings are nearly parallel to each other and the third is nearly at a right angle (Fig. 42.77); each spiracular opening often isolated on its own papilla-like projection; terminal segment often with transverse ridge of three or four small tubercles on dorsum near base of spiracular prominences (Fig. 42.77); saprophagous or coprophagous in carrion, excrement, or decaying plant material . . . . **Milichiidae** (in part)

38b Posterior spiracular openings normally rather symmetrically radiating from a spiracular scar; anal segment lacking a ridge of tubercles at base of spiracular prominences 39

39a Posterior spiracles sessile on surface of anal segment (Figs. 42.79, 42.81), and either lacking a sclerotized peritreme or with spiracular openings slitlike and positioned vertically or on a ventral median incline (Figs. 42.80, 42.82). . . 40

39b Posterior spiracles more or less distinctly elevated above plane of terminal segment (Fig. 42.84), and usually differing in other of above characters . . 41

40a Posterior spiracles in a deep spiracular cavity (Fig. 42.79); spiracular openings inclined more or less vertically; spiracular scar usually not visible; peritreme incompletely encircling spiracular plate (Fig. 42.80); feed almost exclusively on fresh or decaying animal matter . . . . **Sarcophagidae**

40b Posterior spiracles exposed at apex of terminal abdominal segment (Fig. 42.81); spiracular openings obliquely inclined; spiracular-scar present; peritreme completely encircling spiracular plate (Fig. 42.82); in fresh or decaying animal matter or excrement . . . . **Calliphoridae**

41a Anterior spiracle bicornuate, with buds arranged on two more or less distinct diverging arms (Fig. 42.75); cephalopharyngeal skeleton lacking parastomal bars (as in Figs. 42.78, 42.83); coprophagous, phytophagous **Scatophagidae**

41b Anterior spiracle fan-shaped or treelike (Figs. 42.85, 42.89, 42.90), or parastomal bars present in cephalopharyngeal skeleton, or both features present saprophagous, phytophagous, or coprophagous. . . . .
Heleomyzidae (in part)
Anthomyiidae
Sphaeroceridae (in part)

42a Posterior spiracles sessile on surface of terminal segment (Fig. 42.93); phytophagous or in decaying organic matter. . . . . . . **Otitidae** (in part)
Chloropidae (in part)

42b Posterior spiracles distinctly elevated on short supporting structures (Figs. 42.87, 42.88, 42.94, 42.97) 43

43a Posterior spiracular plate with one or more bordering spines or lobes, or with a sharp ridge along the dorsal margin (Figs. 42.86, 42.88) 44

43b Posterior spiracular plate without lobes, spines, or ridges along the margin 45

44a Posterior spiracular openings short and oval, aligned at almost right angles to one another (Fig. 42.87); saprophagous in dung and rotting vegetables. . .
Lonchaeidae, (in part)

44b Posterior spiracular openings radiating from spiracular scar at distinctly less than right angles, or irregularly or peripherally arranged (Fig. 42.88); saprophagous. . . . . **Micropezidae**

45a Anterior spiracle with buds projecting semicircularly (Fig. 42.31) or treelike (Fig. 42.90); coprophagous . . . . **Sphaeroceridae** (in part)

45b Anterior spiracle otherwise. 46

46a Cephalopharyngeal skeleton with accessory oral sclerites (Fig. 42.96), or posterior spiracles either with serpentine spiracular openings or openings distinctly angled (Fig. 42.95), or both features present; scatophagous, saprophagous, zoophagous . . . . **Muscidae** (in part)

46b None of these characteristics evident . . . . **Chloropidae** (in part)
Diopsidae
Heleomyzidae (in part)
Lonchaeidae (in part)
Otitidae (in part)
Milichiidae (in part)

References

Alexander, C. P. 1920. The crane-flies of New York. II. Biology and phylogeny. *Cornell Univ. Agric. Exp. Stn. Mem.* 38:695–1133. **Brauns, A. 1954.** *Untersuchungen zur angewandten Bodenbiologie.* Vol. 1. *Terricole Dipterenlarven.* Gottingen, Frankfurt, West Germany. 179 pp. **Brindle, A. 1959.** Notes on the larvae of the British Rhagionidae and Stratiomyidae with a key to the genera. *Entomol. Rec.* 71:126–133. **1962.** Taxonomic notes on the larvae of British Diptera. II. Triehoceridae and Anisopodidae. *Entomologist (Lond.)* 95:285–288. **Dixon, T. J. 1960.** Key to and descriptions of the third instar larvae of some species of Syrphidae (Diptera) occurring in Britain. *Trans. R. Entomol. Soc. Lond.* 112:345–379. **Ghilarov, M. S. 1964.** *Manual for Identification of the Soil Inhabiting Larvae of Insects.* Nauka, Moscow. 919 pp. (in Russian). **Hartley, J. C. 1961.** A taxonomic account of the larvae of some British Syrphidae. *Proc. Zool. Soc. Lond.* 136:505–573. **Heiss, E. M. 1938.** A classification of the larvae and puparia of the Syrphidae of Illinois exclusive of aquatic forms. *Ill. Biol. Monogr.* 16(4):1–142. **Henning, W. 1948, 1950, 1952.** *Die Larvenformen der Dipteren.* Akademie-Verlag, Berlin. Pt. 1, 185 pp.; Pt. 2, 458 pp.; Pt. 3, 628 pp. **James, M. T. 1959.** Diptera. Pp. 1057–1116, W. T. Edmondson (ed.). in *Ward and Whipple's Freshwater Biology.* Wiley, New York. **Keilin, P. and P. Tate. 1940.** Early stages of the families Trichoceridae and Anisopodidae. *Trans. R. Entomol. Soc. Lond.* 90:39–62. **Kessel, E. L., M. E. Buegler and P. M. Keyes. 1973.** A survey of the known larvae and puparia of Platypezidae, with a key to ten genera based on immature stages (Diptera). *Wasmann J. Biol.* 31:233–261. **Krivosheina, N. P. 1960.** Peculiarities of the earth-living larvae of Rhagionidae, Dolichopodidae and Empididae (Diptera). *Entomol. Obozr.* 39:144–155 (in Russian). **1962.** The European larvae of the Bibionidae (Diptera, Nematocera) with keys to several species. *Pedobiologia* 1:210–227 (in Russian). **Madwar, S. 1937.** Biology and immature stages of Mycetophilidae (Diptera, Nematocera). *Philos. Trans. R. Soc. Lond.* 227. 110 pp. **Mamaev, B. M. and N. P. Krivosheina. 1965.** *Larvae of gallmidges. Diptera, Cecidomyiidae.* Akademii Nauk SSSR, Moscow. **McAlpine, J. F. (ed.). 1981.** *Manual of Nearctic Diptera* 1:1–674. Monogr. 27. Research Branch, Agriculture Canada, Ottawa. **1987.** *Manual of Nearctic Diptera* 2:675–1332. Monogr. 28. Research Branch, Agriculture Canada, Ottawa. **McFadden, M. W. 1967.** Soldier fly larvae in America north of Mexico. *Proc. U.S. Natl. Mus.* 121(3569):1–72. **Melin, D. 1923.** Contributions to the knowledge of the biology, metamorphosis and distribution of the Swedish asilids in relation to the whole family of asilids. *Zool. Bidr. Uppsala* 8. 317 pp. **Miller, R. A. and B. A. Foote. 1975.** Biology and immature stages of eight species of Lauxaniidae (Diptera). *Proc. Entomol. Soc. Wash.* 77:308–328. **1976.** Biology and immature stages of eight species of Lauxaniidae (Diptera). II. Descriptions of immature stages and discussion of larval feeding habits and morphology. *Proc. Entomol. Soc. Wash.* 78:16–37. **Quate, L. W. 1955.** A revision of the Psychodidae (Diptera) in America north of Mexico. *Univ. Calif. Publ. Entomol.* 10(3):103–273. **Strenzke, K. 1950.** Systematik, Morphologie und Okologie der terrestrischen Chironomiden. *Arch. Hydrobiol.* 18:207–414. **Teskey, H. J. 1969.** Larvae and pupae of some eastern North American Tabanidae (Diptera). *Mem. Entomol. Soc. Can.* 63. 147 pp. 1976. Diptera larvae associated with trees in North America. *Mem. Entomol. Soc. Can.* 100:1–53. **1978.** Aquatic diptera, larvae of aquatic diptera. Chap. 18, Part 1 in R. W. Merritt and K. W. Cummins (eds.). *Introduction to the Aquatic Insects of North America.* Kendall/Hunt, Dubuque, Iowa. 441 pp. **Wheeler, W. M. 1930.** *Demons of the Dust.* W. W. Norton, New York.

43

Insecta: Hymenoptera Formicidae

GEORGE C. WHEELER

JEANETTE WHEELER

The Ants of North America (Bull. Mus. Comp. Zool. Harvard Coll. 104; 585 pp., 57 pl.) by the late William Steel Creighton (1902–1973) is still the bible for myrmecologists of the region. It was inevitable, however, that such an excellent work would immediately begin to generate its own obsolescence by making additions and revisions possible. Hence, although we have used it as a basis for our keys, it has been necessary to append the changes that have been made in the intervening 30 years—not so many, however, as would have been anticipated.

Creighton himself admitted (*ibid.,* p. 10) the inappropriateness of the title, because it does not include Mexico and Central America. "Nearctic Ants" would have been more suitable, because "our extensive Nearctic ant fauna is largely confined to regions north of Mexico and the few representatives of this fauna which occur in Mexico are strictly limited to the higher mountain ranges."

Identification and How to Use the Keys

Since the Nearctic ant fauna comprises some 550 species in 74 genera, it is impossible to key to species; indeed, we are hard pressed to key to genera in the space allotted. Only one feature of our keys requires explanation. The accompanying illustrations are labeled only by half-couplets (i.e., the numeral and letter beside a figure is the same as that of the half-couplet illustrated). If only a part of the figure is illustrative, that part is marked with an arrow.

If one wishes to key to species, they can still use most of Creighton's keys, provided that they heed our appendix after the key to each subfamily.

In our appendices we have given citations (titles omitted) for each reference. Thus we have avoided the necessity for a "Literature Cited," thereby saving considerable space.

Our keys are for worker ants only; frequently, the largest workers must be used because they are the only ones that show the characters. Queens may sometimes fit the keys, but males are so different that it is impossible to use worker keys.

ACKNOWLEDGMENT

We thank Roy R. Snelling for reading this manuscript and making many useful suggestions.

Key Morphological Characters

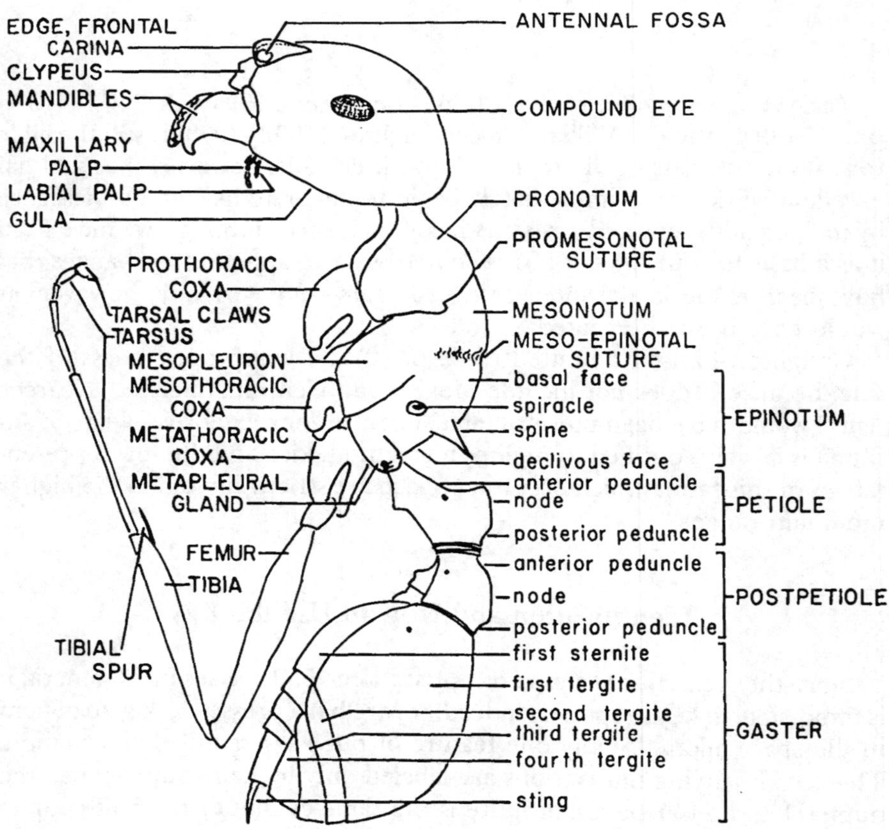

Fig. 43.1. Lateral aspect.

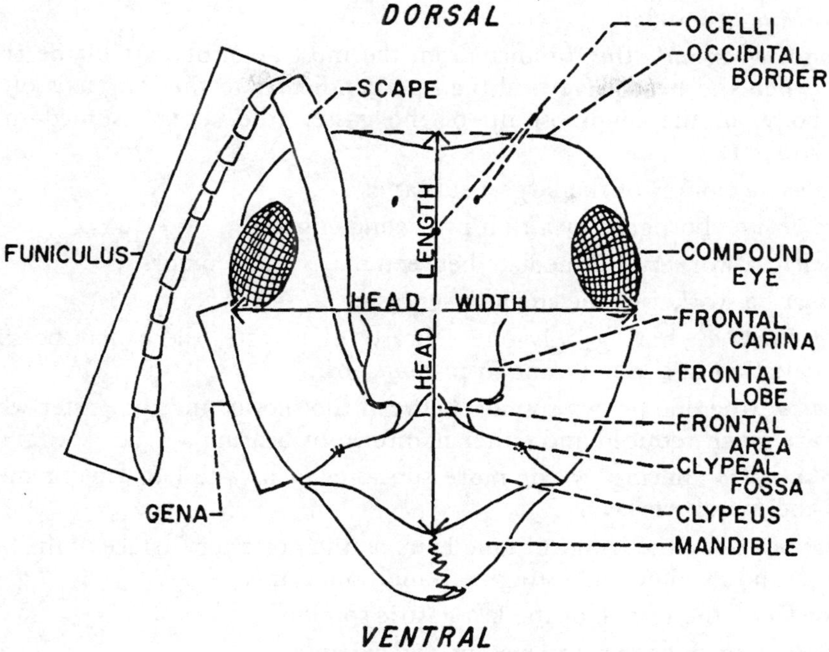

Fig. 43.2. Head characteristics.
Formicidae Figs. 43.1a to 43.7b
Ponerinae Figs. 43.1a to 43.14b
Dorylinae Figs. 43.1a to 43.2b
Myrmicinae Figs. 43.1a to 43.25a
Myrmicinae Figs. 43.23b to 43.36a
Dolichoderinae Figs. 43.1a to 43.5b
Formicinae Figs. 43.1a to 43.9b

Glossary

Carina a keel-like structure.

Caste among social insects, the various forms of each sex of mature individuals.

Concolorous uniformly colored.

Flexor surfaces surfaces that are brought closer together when a jointed structure is bent at a joint.

Frons (= front) the facial area above the clypeus and between the frontal carinae; dorsally, it passes without a definite boundary into the vertex.

Granulose roughened with small pelletlike elevations.

Hair a term used loosely in entomology; in general, a hair is isolated, somewhat stiff, and more or less erect. A downy covering of fine soft hairs, which generally lie against the integument, is called "pubescence."

Humerus shoulder.

Length of an ant the distance from the most anterior part of the frons (when the head is vertical, i.e., perpendicular to the long axis of the body) to the hindmost tip of the gaster (the latter assumed to be straight).

Major a worker of the largest subcaste.

Marginate bordered by a thin projecting edge.

Media a worker intermediate between majors and minors.

Minor a worker of the smallest subcaste.

Monomorphic having only one subcaste of a caste; there may be some variation in size but none in proportions.

Pedicel the one or two somites between the thorax and the gaster which are much reduced and either nodiform or bearing a scale.

Polymorphic having two or more subcastes of a caste (e.g., major media and minor worker).

Psammophore the fringe of long hairs on the posterior surface of the head (found in some harvesting ants and honey ants).

Pygidium the tergite of the last gastric somite.

Scrobe a groove that receives an appendage.

Subcastes the various forms of mature individuals of a caste; subcastes of the worker caste in ants are termed major (or solider), media, and minor (or minim).

Thorax a term that, when applied to ants, should be understood to mean the second major subdivision of the body; it includes the prothorax, mesothorax, metathorax, and epinotum.

Vertex the top of the head below the occiput and above the eyes and the frons.

KEY TO SUBFAMILIES OF FAMILY FORMICIDAE

1a Eyes absent or vestigial; pedicel usually of two segments in worker (one in female and male); clypeus short; frontal carinae short and vertical, not covering antennal insertions; antennae usually short; epinotum usually unarmed **Dorylinae**

1b Without this combination of characters **2**

2a Pedicel of one segment **3**

2b Pedicel of two segments. **7**

3a With a conspicuous constriction between first and second gastric somites. . . **4**

3b Without a constriction between first and second gastric somites **5**

4a Elongate, slender, and subcylindrical; scapes usually short and stout; antennal fossa more or less encircled by a lateral carina on the cheek (rarely obsolete); posterior surface of head usually with a distinct carina running ventrally from

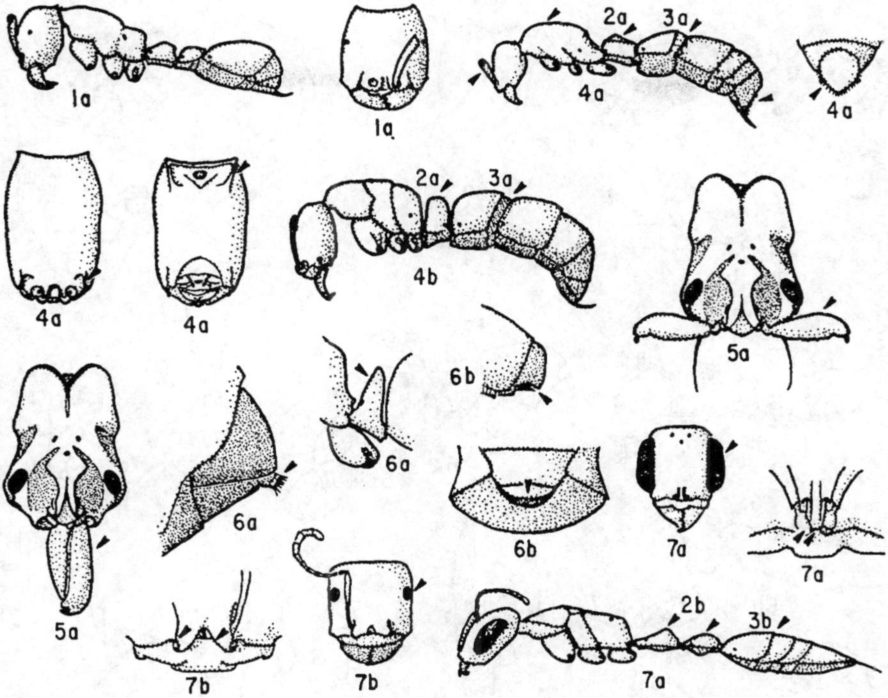

Formicidae Figs. 1a–7b. The numerals and letters on figures refer to numerals and letters of couplets in formicidae key.

each dorsolateral corner; dorsal surface of thorax with sutures indistinct or absent; pygidium margined laterally and posteriorly with a row of large or small (but always distinct) spines . . . . **Cerapachyinae**

4b Without this combination of characters . . . . **Ponerinae**

5a Mandibles articulated near middle of ventral border of head; when closed, parallel to each other; when fully opened, they form together a straight line parallel to ventral border of head; Tribe **Odontomachini** in the . . **Ponerinae**

5b Mandibles articulated to ventral corners of head 6

6a Opening at posterior end of gaster (acidipore) terminal, circular and usually surrounded by a fringe of hairs; sting vestigial; petiole usually scalelike **Formicinae**

6b Opening at posterior end of gaster (cloacal orifice) slitlike . . . . **Dolichoderinae**

7a Elongate, often very slender; eyes very large and elongate; clypeus with rounded upper margin, not prolonged upward between frontal carinae; frontal carinae usually close together, usually narrow and not expanded laterally to cover antennal insertions; antennae short . . . . **Pseudomyrmecinae**

7b Without this combination of characters; frontal carinae usually large, nearly always covering antennal insertions and nearly always well separated . . . . **Myrmicinae**

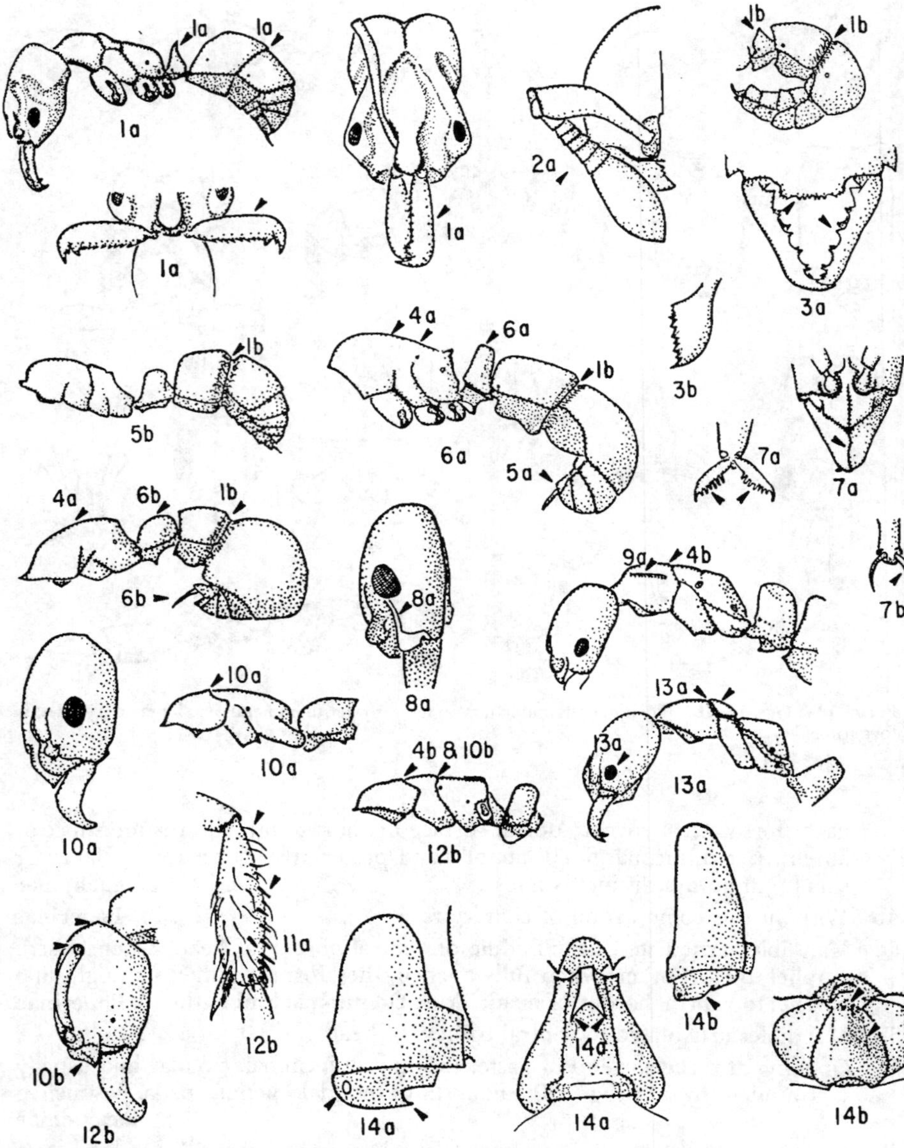

Ponerinae Figs. 1a–14b. The numerals and letters on figures refer to numerals and letters of couplets in the Ponerinae key.

KEY TO GENERA OF SUBFAMILY PONERINAE

1a Gaster without a distinct constriction between first and second somites; node and petiole conical with spine above; mandibles linear and inserted near midline of head; when closed, parallel to each other; forming a straight line when fully open (line parallel to ventral border of head). . . ***Odontomachus***

1b Gaster with distinct constriction between first and second somites; node of petiole blunt or rounded above; mandibles inserted at sides of head . . . **2**

2a Antenna short, stout and with nine segments . . . . *Discothyrea*
2b Antenna with more than nine segments 3
3a Ventral border of clypeus denticulate; mandible with a row of coarse bidentate teeth . . . . *Amblyopone*
3b Ventral border of clypeus variously shaped but never denticulate; mandibular teeth, when present, single 4
4a Thoracic dorsum without sutures, at most a shallow impression at point where suture should be, usually without impression. 5
4b Thoracic dorsum with at least promesonotal suture present, and usually mesoepinotal also 7
5a Apex of gaster directed ventrally or anteroventrally when major axis of gaster is in line with that of thorax; head and thorax punctatogranulose or punctatorugose; gaster smooth but with numerous piligerous punctures 6
5b Apex of gaster directed to rear when major axis of gaster is in line with that of thorax; head, thorax, and gaster covered with even, straight, longitudinal rugae . . . . *Gnamptogenys*
6a Petiole scalelike, with front and rear faces flattened; ventral border of clypeus not projecting in middle . . . . *Proceratium*
6b Petiole nodiform, low, and much rounded above; clypeus with a narrow median lobe projecting strongly beyond remainder of ventral border. . . . *Sysphincta*
7a Tarsal claws distinctly pectinate; mandible without distinct teeth . . . . . . *Leptogenys*
7b Tarsal claws simple; mandibular teeth distinct. 8
8a Cheek with a distinct carina extending from eye to clypeus . . . *Neoponera*
8b Cheek without a carina 9
9a Pronotum marginate on either side . . . . *Pachycondyla*
9b Pronotum not marginate on either side 10
10a Thoracic dorsum with only promesonotal suture present; clypeus flat, suture that separates it from front of head indistinct . . . . *Platythyrea*
10b Thoracic dorsum with promesonotal and mesoepinotal sutures; clypeus with projecting median lobe, suture separating clypeus from front of head distinct 11
11a Outer surface of middle tibia with stout stiff hairs in addition to normal flexible hairs 12
11b Middle tibia lacking distinct stiff hairs on outer surface 13
12a Larger ants (4.9–5 mm); mesopleuron distinctly striate; antennal scape reaching occipital border; dark brown to black . . . . *Mesoponera*
12b Smaller ants (3–3.5 mm); mesopleuron not distinctly striate; antennal scape failing to reach occipital by an amount greater than its greatest diameter; light to dark reddish brown . . . . *Cryptopone*
13a Mesonotum blisterlike and surrounded by a distinctly impressed suture; eye moderately large, facets distinct. . . . . *Brachyponera*
13b Without foregoing combination of characters 14
14a Subpetiolar process with anterior corner blunt or rounded and bearing a distinct fenestra, bidentate posteriorly; maxillary palp two-segmented . . . *Ponera*
14b Subpetiolar process a simple lobe with fenestra vestigial or absent, posterior teeth lacking; maxillary palp one-segmented . . . . *Hypoponera*

Appendix for Ponerinae

Creighton (1950) can still be used for keying species if the following changes are noted:

Subgenus *Brachyponera;* change to genus *Brachyponera.*
Euponera gilva change to *Cryptopone gilva.*
Euponera stigma change to *Mesoponera stigma.*
Odontomachus haematoda clarus change to *Odontomachus clarus.*
Subgenus *Parectatomma* change to genus *Gnamyptogenys.*
Ponera coarctata pennsylvanica change to *Ponera pennsylvanica.*
Ponera ergatandria change to *Hypoponera punctatissima.*
Ponera inexorata change to *Hypoponera inexorata.*
Ponera oblongiceps change to *Hypoponera gleadowi.*
Ponera opaciceps change to *Hypoponera opaciceps.*
Ponera trigona opacior change to *Hypoponera opacior.*
Genus *Stigmatomma* change to genus *Amblyopone.* ADD key to species from **Brown 1960,** *Bull. Mus. Comp. Zool. Harv. Unv.* 122:191–192.

ADD:

Discothyrea testaceus (see **Smith and Wing 1954,** *J. N.Y. Entomol. Soc.* 62:103–112).
Odontomachus brunneus (see **Brown 1976.** *Stud. Entomol.* 19:67–171).
Ponera exotica (see **Smith 1962,** *Acta Hymen.* 1:377–382).
Proceratium californicum (see **Snelling 1967,** *Nat. Hist. Mus. Los Ang. Cty. Contrib. Sci.* 124, 10 pp).

KEY TO GENERA OF SUBFAMILY DORYLINAE

1a Tarsal claws with teeth **2**
1b Tarsal claws without teeth . . . . ***Neivamyrmex***
2a Scape thick, its apical width greater then one-third its length. . . . . ***Nomamyrmex***
2b Scape slender, its apical width less than one-third its length . . . . . ***Labidus***

Appendix for Dorylinae

To key to the species of Dorylinae, see **Watkins 1976,** *The Identification and Distribution of New World Army Ants,* Baylor University, Waco, Tex., 102 pp., and ADD *Neivamyrmex nyensis* (see **Watkins 1977,** *Southwest. Nat.* 22:421).

Subfamily Cerapachyinae

Cerapachys augustae ranges from western Texas to southern Arizona; workers and females have 11-segmented antennae. *C. davisi* (western Texas) is known only

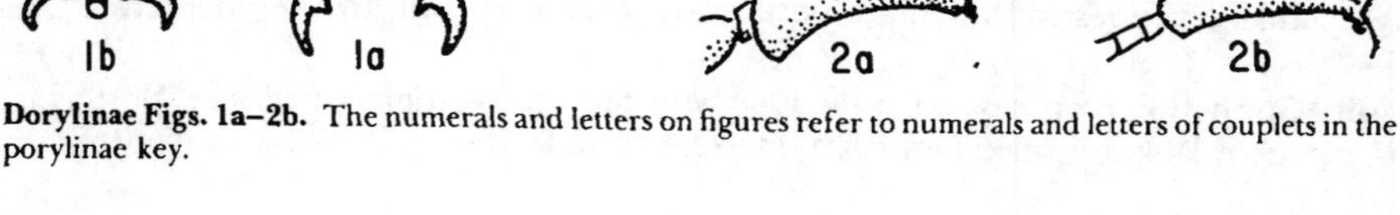

Dorylinae Figs. 1a–2b. The numerals and letters on figures refer to numerals and letters of couplets in the porylinae key.

from males. *Acanthostichus texanus* is known only from females collected in the southern tip of Texas; they have 12-segmented antennae.

Appendix for Subfamily Pseudomyrmecinae

Only one genus (*Pseudomyrmex*) occurs in the New World and it is essentially neotropical, but in the United States it is found in the southern Atlantic and Gulf coast states and westward to California. Use **Creighton 1955,** *J. N.Y. Entomol. Soc.* 63:17–20.

KEY TO GENERA OF SUBFAMILY MYRMICINAE

1a Antenna with four segments; pilosity mostly squamiform . . . . . ***Epitritus***
1b Antenna with more than four segments 2
2a Antenna with six segments 3
2b Antenna with more than six segments 4
3a Mandible long and slender, apex with two large teeth (one anterior to other) and smaller subapical tooth . . . . ***Strumitgenys***
3b Mandible much shorter and armed with numerous teeth . . . ***Smithistruma***
4a Antenna with seven segments, scape triangular with broad base; pubescence squamiform; erect hairs thickened (clavate or spatulate) . . . . ***Eurhopalothrix***
4b Antenna with more than seven segments and lacking combination of other characters 5
5a Frontal carina expanded to form overhanging rim at edge of head, behind which is deep scrobe for antenna; head of major appearing like shallow saucer . . . . ***Zacryptocerus***
5b Antennal carina and head shape not as above 6
6a Postpetiole attached to dorsal surface of first gastric somite; gaster convex ventrally and flattened dorsally, acutely pointed posteriorly . . . . ***Crematogaster***
6b Postpetiole attached to anterior end of first gastric somite; gaster about equally convex above and below 7
7a Antenna with 10 segments, the last two forming a distinct club . . . . ***Solenopsis***
7b Antenna with more than 10 segments; club, if present, only rarely of two segments 8
8a Antenna with 11 segments 9
8b Antenna with 12 segments 19
9a Dorsum of pro-, meso-, and epinotum with spines, teeth, bosses, or prominent ridges; antennal fossa bordered by delicate carina which runs diagonally inward from insertion of mandible past medial border of eye 10
9b Spines and teeth, when present, confined to epinotum; antennal fossa rarely bordered by diagonal carina (if present, worker length not exceeding 2 mm) 14
10a Frontal carina projecting ventrally over clypeus, largely or entirely concealing lateral portion when head is in full face view; thoracic spines, if present, short and dentiform 11
10b Frontal carina shorter, at most projecting over dorsal half of clypeus, full width of clypeus visible in full face view; thoracic spines long and prominent 12

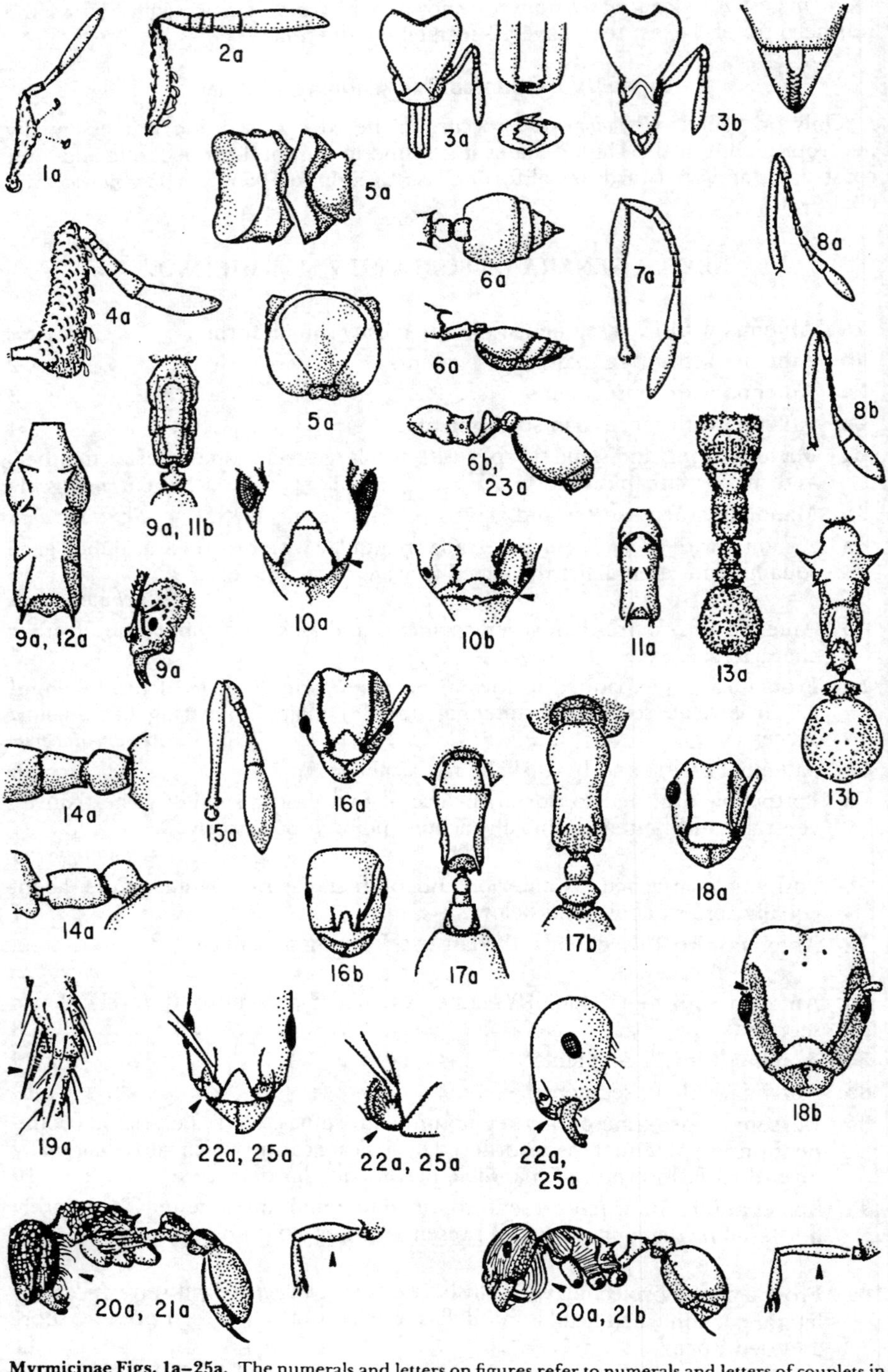

Myrmicinae Figs. 1a–25a. The numerals and letters on figures refer to numerals and letters of couplets in the Myrmicinae key.

11a Pro-, meso-, and epinotum each with a pair of short teeth or denticles; body hairs at least in part erect, gastric hairs, arising from small but distinct tubercles . . . . *Mycetosoritis*

11b Pro-, meso, and epinotum with bosses or carinae but without denticles; body hairs scalelike, appressed; gaster not tuberculate . . . . *Cyphomyrmex*

12a Promesonotum armed with two pairs of spines; large species, highly polymorphic, workers 2–12 mm long . . . . *Atta*

12b Promesonotum with at least three pairs of spines; smaller species, monomorphic or less polymorphic, workers 6 mm or less **13**

13a Entire insect (including scapes and legs) tuberculate; frontal carina extending almost to dorsal corner of head; occipital emargination shallow. . . . . *Trachymyrmex*

13b Tubercles largely confined to gaster, postpetiole and tops of occipital lobes; frontal carina indistinct dorsally and not extending to occipital corner; occiput deeply emarginate in largest worker . . . . *Acromyrmex*

14a Petiole subcylindrical, without node; epinotum unarmed. . . . *Xenomyrmex*

14b Node of petiole well developed; epinotum armed with spines or teeth . . . **15**

15a Antennal club very distinct, of two segments which are notably broader and longer than preceding seven small segments. . . . . *Oligomyrmex*

15b Antennal club, if present, of more than two segments and usually not separated abruptly from remainder of funiculus **16**

16a Frontal carina extending at least two-thirds of the distance to the occipital corner and bordered by a shallow scrobe for reception of scape, the latter flattened **17**

16b Frontal carina short, no antennal scrobe, scape not flattened . . . . *Leptothorax*

17a Postpetiole about as broad as long; thoracic humeri markedly angular; length 1.5–2 mm . . . . *.Wasmannia*

17b Postpetiole strongly transverse; thoracic humeri rounded; length 3.2 or more **18**

18a Antennal scrobe narrow, usually obliterated by heavy sculpture except immediately under overhanging edge of frontal carina; head, thorax, and petiolar node strongly reticulorugose . . . . *Xiphomyrmex*

18b Antennal scrobe broad, extending almost to medial border of eye; sculpture delicate . . . . *Harpogoxenus*

19a Middle and hind tibial spurs finely pectinate, with teeth distinct and regular but usually too small to see at magnifications of less than 100× **20**

19b Middle and hind tibial spurs simple or absent, very rarely with a few barbules but never pectinate. **24**

20a Thoracic dorsum with sutures obsolescent or absent; thorax not, or only slightly, impressed between meso- and epinotum; psammophore usually present **21**

20b At least mesoepinotal suture present and faint to distinct on thoracic dorsum; thoracic at least feebly impressed at mesoepinotal suture; psammophore absent **22**

21a Head and thorax extensively and coarsely reticulorugose; femora strongly incrassate; base of epinotal spines connected by a conspicuous keel; postpetiole (in profile) with large bulbous projection, deeper than long; psammophore feebly developed . . . . *Ephebomyrmex*

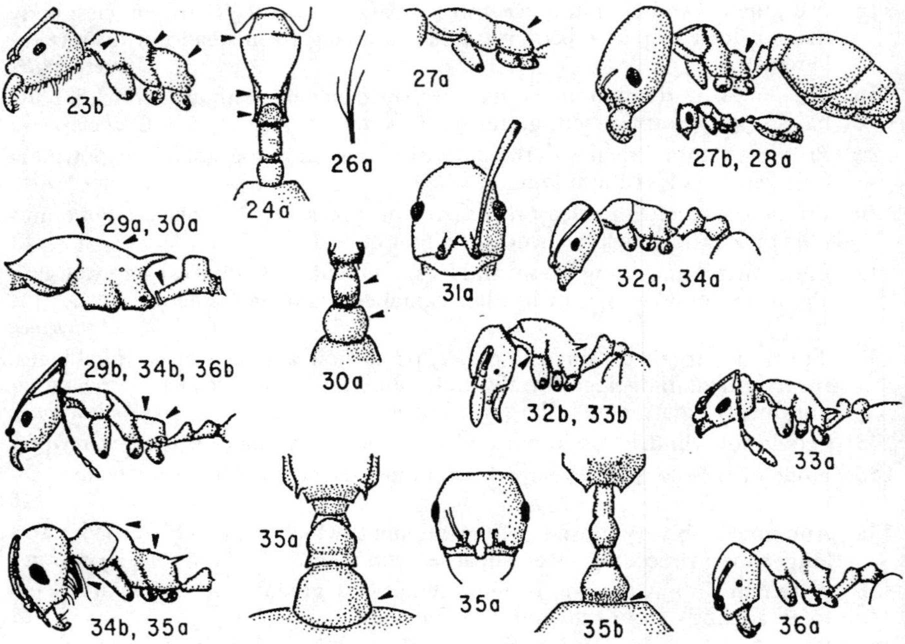

Myrmicinae Figs. 26a–36a.

21b Head and thorax not extensively and coarsely reticulorugose; femora not strongly incrassate; base of epinotal spines (when present) not connected by a conspicuous keel; postpetiole (in profile) smaller, ventral projection not bulbous, its depth little greater than its length; psammophore strongly developed. . . . . ***Pogonomyrmex***

22a Lateral protion of clypeus raised dorsally into a narrow ridge or carina which forms an abrupt semicircular boundary at ventral edge of antennal fossa ***Tetramorium***

22b Lateral portion of clypeus not raised; antennal fossa opening onto clypeus without elevated bounary. **23**

23a Epinotum armed with spines or teeth; promesonotal suture absent on thoracic dorsum; mesoepinotal suture moderately impressed . . . . . ***Myrmica***

23b Epinotum not armed with spines of teeth, usually evenly rounded but rarely with blunt protuberances; promesonotal suture visible on thoracic dorsum but faint; mesoepinotal suture strongly impressed . . . . ***Manica***

24a Thoracic humeri strongly angular; a pair of short denticles anterior to epinotal spines; petiole and postpetiole of nearly same diameter (in dorsal view) ***Myrmecina***

24b Without this combination of characters. **25**

25a Lateral portion of clypeus raised into narrow ridge or carina, forming a prominent semicircular boundary ventral to antennal fossa **26**

25b Clypeus not as above **27**

26a Many of the numerous erect body hairs branched or trifid . . . . ***Triglyphothrix***

26b Erect body hairs simple . . . . ***Tetramorium***

27a Epinotum unarmed, basal face at same level as dorsum of mesonotum . . . ***Monomorium***

27b Epinotum usually armed with spines or teeth, but if unarmed, basal face distinctly below level of dorsum of mesonotum. **28**

28a Worker caste dimorphic (rarely polymorphic) with head of major disproportionately large . . . . ***Pheidole***

28b Worker caste monomorphic, or if polymorphic head of major not disproportionately large **29**

29a Thoracic dorsum with mesoepinotal suture absent or very faintly indicated **30**

29b Thoracic dorsum with mesoepinotal suture well marked **34**

30a Totally lacking erect hairs; antennal club three-segmented; middle and hind tibiae without spurs; petiole with long cylindrical peduncle, node broadly oval; postpetiolar node 2× width of petiolar node . . . . . ***Cardiocondyla***

30b Without foregoing combination of characters **31**

31a Large species, length 10–12 mm; antennal scape projecting well beyond occipital border . . . . ***Novomessor***

31b Small species (not more than 4 mm); antenna usually not surpassing occipital border and never extending much beyond it. **32**

32a Thoracic dorsum flat or feebly convex; anterior peduncle of petiole short, thick, and not sharply set off from node; epinotal spines short to moderately long . . . . ***Leptothorax***

32b Thoracic dorsum distinctly convex; without definite sutures **33**

33a Antennal club three-segmented, shorter than remainder of funiculus; eyes with about 50 facets; smaller ants (2–2.6 mm long) . . . . . . ***Macromischa***

33b Antennal club three-segmented, robust, longer than remainder of funiculus; eyes with 8–12 pigmented facets; pronotum with anteroventral angle . . . ***Rogeria***

34a Thoracic dorsum in profile with epinotum as high as promesonotum, dorsum usually forming an unbroken plane, rarely with deep mesoepinotal impression; antennal club three-segmented . . . . ***Leptothorax***

34b Epinotum depressed well below level of pronotum, in profile mesonotum forming sloping declivity between them; antennal club indistinct, of four or five segments **35**

35a Postpetiole only slightly constricted behind, node low and not sharply set off from thick posterior peduncle; head quadrate, not notably narrower dorsal to eyes than vertrally; psammophore often present . . . . ***Veromessor***

35b Postpetiole strongly constricted posteriorly; psammophore never present **36**

36a Small ants (2.3–4 mm long); scape not reaching occipital border; promesonotum rather strongly convex, promesonotal suture indistinct or absent; eyes usually small to vestigial . . . . ***Stenamma***

36b Larger ants (3.25–7 mm long); scape surpassing occipital border; promesonotal suture usually distinct; eyes prominent. . . . . ***Aphaenogaster***

Appendix for Subfamily Myrmicinae

Creighton (1950) can still be used for keying species if the following revisions are noted:

Aphaenogaster. ADD *megommata* from **Smith 1963;** *J. N.Y. Entomol. Soc.* 71:244–246.
Atta. ADD *mexicana* from **Smith 1963,** *Proc. Entomol. Soc. Wash.* 65:299–302.
Cardiocondyla. ADD *ectopia* from **Snelling 1974,** *J. N.Y. Entomol. Soc.* 82:76–81.
Crematogaster (Acrocelia) is now *C. (Crematogaster)*; REVISION by **Buren 1968,** *J. Ga. Entomol. Soc.* 3:91–121.
Ephebomyrmex is now regarded as a genus. See under *Pogonomyrmex*.
Epipheidole inquilina is now *Pheidole inquilina*.
Epoecus pergandei is now *Monomorium pergandei*.
Erebomyrma longi is now *Oligomyrmex longi*.
Leptothorax. *Symmyrmica chamberlini* is now *L. chamberlini*.
Leptothorax canadensis is now called *L. muscorum*. ADD *carinatus* from **Cole 1957,** *J. Tenn. Acad. Sci.* 32:213–215; *hispidus*. from **Cole, 1957;** *J. Tenn. Acad. Sci.* 32:42–45; *stenotyle* from **Cole 1956,** *J. Tenn. Acad. Sci.* 31:28–31 and 214; *obliquicanthus* from **Cole 1953,** *J. Entomol. Soc. Wash.* 55:27–30; and *tuscaloosae* from **Wilson 1950,** *Psyche (Camb.)* 57:128–130.
Manica aldrichi is a synonym of *M. hunteri*.
Myrmica includes *Paramyrmica*. ADD *colax* from **Cole 1957,** *J. Tenn. Acad. Sci.* 32:37–42; *striolagaster* from **Cole 1953,** *J. Tenn. Acad. Sci.* 28:81–82 and **Gregg 1961,** *J. N.Y. Entomol. Soc.* 69:211; *mexicana* from **Gregg 1961,** *J. N.Y. Entomol. Soc.* 69:211; and *rugiventris* from **Gregg 1961,** *J. N.Y. Entomol. Soc.* 69:209–220.
Pheidole. REVISION: **Gregg 1958,** *J. N.Y. Entomol. Soc.* 66:7–48. *Sympheidole elecebra* becomes *Pheidole elecebra*.
Epipheidole inquilina becomes *Pheidole inquilina*.
Pogonomyrmex. REVISION: **Cole 1968,** *Pogonomyrmex Harvester Ants*. University of Tennessee Press, Knoxville, Tenn., 222 pp. ADD **Snelling 1981,** *Bull. South. Calif. Acad. Sci.* 80:97–112.
Solenopsis. ADD *invicta* from **Buren 1972,** *J. Ga. Entomol. Soc.* 7:126.
Stenamma. REVISION: **Smith 1957,** *Am. Midl. Nat.* 57:133–174; **Snelling 1973,** *Nat. Hist. Mus. Los Ang. Cty. Contrib. Sci.* 245, 38 pp.
Strumigenys. In Creighton = subgenus *Strumigenys*. ADD *eggersi*, *gundlachi* and *silvestrii* from **Brown 1962,** *Psyche (Camb.)* 69:238–267 (key, p. 257).
Trachymyrmex. ADD *neomexicanus* from **Cole 1952,** *J. Tenn. Acad. Sci.* 27:159–162; and *nogalensis* from **Byars 1951,** *Proc. Entomol. Soc. Wash.* 53:109–111.
Veromessor. ADD *lariversi* from **Smith 1951,** *Great Basin Nat.* 11:94–96; and *smithi* from **Cole 1963,** *Ann. Entomol. Soc. Am.* 56:678–682.

GENERA ADDED:
Eurhopalothrix. **Brown and Kempf 1960,** *Stud. Entomol. (Rio J.)* 3:207–208 and 246–247.
Quadristruma. **Brown 1949,** *Trans. Am. Entomol. Soc. (Phila.)* 75:47–51.
Rogeria. **Snelling 1973,** *Nat. Hist. Mus. Los Ang. Cty. Contrib. Sci.* 236, 7 pp
Smithistruma. **Brown 1953,** *Am. Midl. Nat.* 50:46–92; **1964,** *Trans. Am. Entomol. Soc. (Phila.)* 89:185.

KEY TO GENERA OF SUBFAMILY DOLICHODERINAE

1a Declivous face of epinotum very strongly concave; integument stiff and brittle; epinotum and often much of remainder of thorax, heavily sculptured . . . . ***Dolichoderus***

1b Declivous face of epinotum straight or nearly so; integument thin and flexible; sculpture everywhere fine. **2**

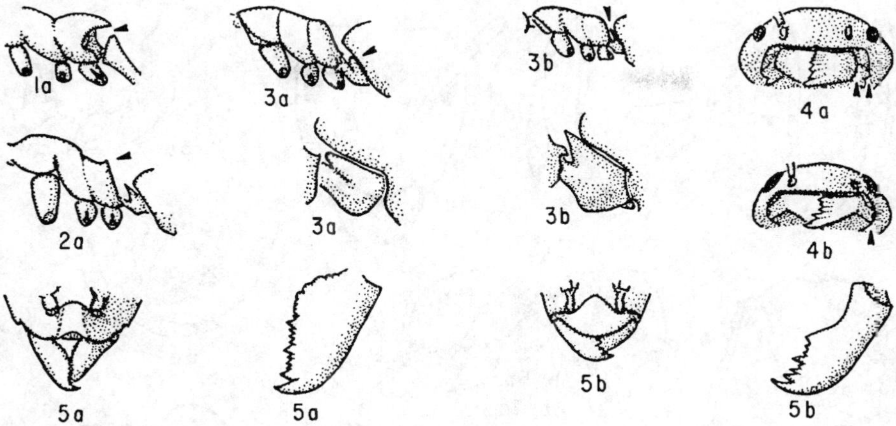

Dolichoderinae Figs. 1a–5b. The numerals and letters on figures refer to numerals and letters of couplets in the Dolichoderinas key.

2a Epinotum with prominent cone projecting upward . . . . *Conomyrma*

2b Without such a cone 3

3a Scale of petiole vestigial, strongly inclined forward and adnate to anterior peduncle; without hairs on thorax or head (except clypeus) . . . . *Tapinoma*

3b Scale of petiole distinct and erect or semierect, though often small; at least a few hairs on head and/or thorax 4

4a Worker caste moderately polymorphic (2.5–6 mm); ocelli present (at least in larger workers); gaster clothed with very dense pubescence which imparts a grayish or ashy tinge; a distinct gap between lateral margin of clypeus and mandible . . . . *Liometopum*

4b Worker caste monomorphic; without ocelli; other characters not as above 5

5a Masticatory border of mandible long and curved into basal border; masticatory border with four or five large teeth, with intercalary denticles present except between apical and preapical teeth; middle of ventral border of clypeus with a thick beveled triangle . . . . *Iridomyrmex*

5b Masticatory border of mandible short and sharply set off from basal border; masticatory border with four teeth, with denticles present only between teeth 3 and 4 and basally; ventral border of clypeus sharp and evenly curved . . . . *Forelius*

Appendix for Subfamily Dolichoderinae

Conomyrma was formerly called *Dorymyrmex*. *D. pyramicus* is now called *C. insana; D. p. bicolor* is now *C. bicolor; D. p. flavopectus* is now *C. flavopectus.*

According to **Snelling (in litt.)** there are only two species of *Iridomyrmex* (both introduced) in the United States, *humilis* and *iniquus; Iridomyrmex pruinosus* becomes *Forelius pruinosus.*

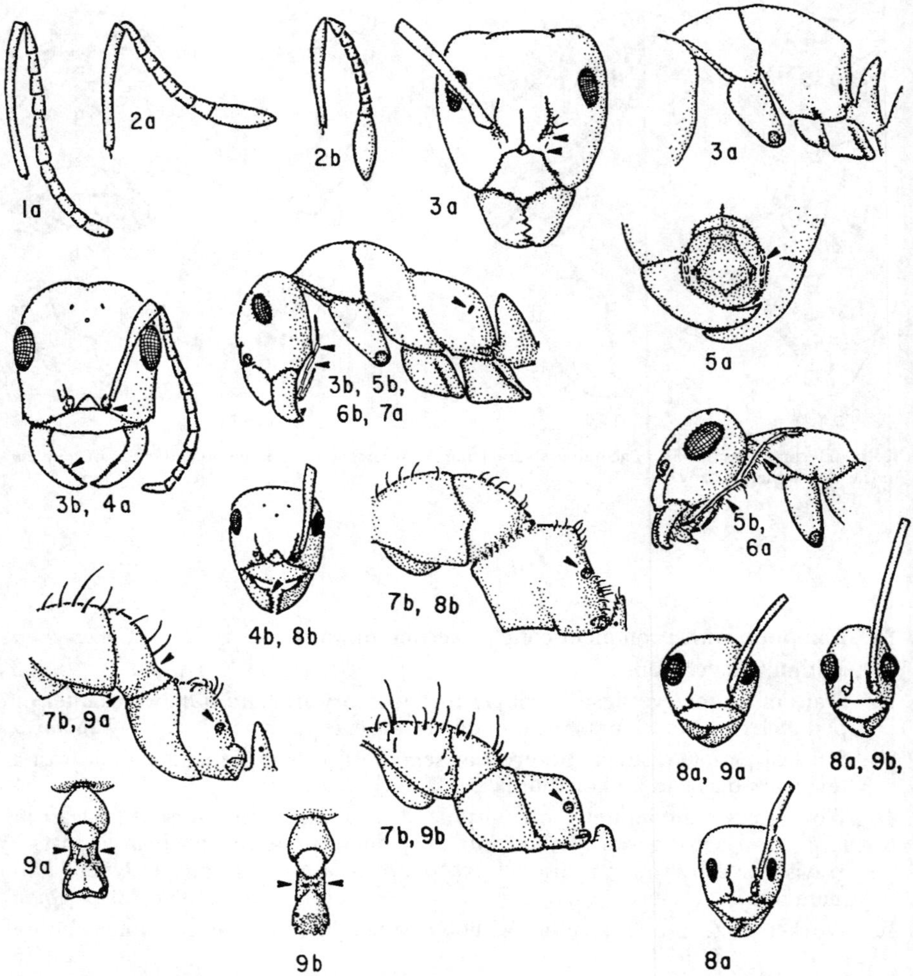

Formicinae Figs. 1a–9b. The numerals and letters on figures refer to numerals and letters of couplets in the Formicinae key.

KEY TO GENERA OF SUBFAMILY FORMICINAE

1a Antenna with 12 segments 3

1b Antenna with fewer than 12 segments 2

2a Antenna with 9 segments . . . . ***Brachymyrmex***

2b Antenna with 10 segments . . . . ***Acropyga***

3a Profile of thoracic dorsum evenly convex, epinotum not depressed below level of mesonotum; mesoepinotal suture not impressed or very slightly impressed; antenna inserted well above dorsal edge of clypeus . . . . ***Camponotus***

3b Profile of thoracic dorsum with epinotum distinctly depressed below level of mesonotum; meso-epinotal suture always distinct, often deep; antenna inserted at or near dorsal border of clypeus 4

4a Mandible sickle-shaped, its medial border finely serrate with very minute teeth . . . . *Polyergus*
4b Mandible triangular, its inner border distinctly dentate. 5
5a Maxillary palp very short and consisting of three segments. . . . . *Acanthomyops*
5b Maxillary palp longer and consisting of six segments 6
6a Maxillary palp longer than head, segments 3 and 4 very long, each as long or longer than two terminal segments combined; psammophore present . . . . *Myrmecocystus*
6b Maxillary palp not longer than head, segments 3 and 4 not unusually long; psammophore absent 7
7a Larger ants (2.5–9 mm long, usually (4.5–9 mm); epinotal spiracle a narrow slit; frontal carina prominent, its lateral margin slightly raised; ocelli conspicuous . . . . *Formica*
7b Smaller ants (2–4.5 mm long, usually 2–3.5 mm); epinotal spiracle round; frontal carina feebly developed, its lateral margin rounded or nearly flat; ocelli small or absent 8
8a Antennal scape surpassing occipital margin by at least one-third its length, usually much longer; erect body hairs coarse, long and usually brown or black 9
8b Antennal scape surpassing occipital margin by less than one-third its length, often much shorter; erect body hairs not coarse, but short and golden. . . . *Lasius*
9a Thorax strongly constricted and subcylindric at mesothorax, swollen in front of and behind constriction; most or all of eye above middle of side of head *Prenolepis*
9b Thorax only slightly constricted at mesothorax; most or all of eye below middle of side of head . . . . *Paratrechina*

Appendix for Subfamily Formicinae

Brachymyrmex. ADD *musculus* from **Wheeler and Wheeler 1978,** *Entomol. News* 89:189–190.

Camponotus (C.) pennsylvanicus ferrugineus is now *C. ferrugineus. C. p. modoc* is now *C. modoc.* ADD *quercicola* from **Smith 1953,** *J. N.Y. Entomol. Soc.* 61:211–214.

C. (Colobopsis). ADD *papago* from **Creighton 1952.** *Psyche (Camb.)* 59:148–162.

C. (Myrmaphaenus). C. (M.) bruesi is a synonym of *C. (M.) ulcerosus.* DELETE subgenus *Manniella.*

C. (Myrmentoma). C. (M.) rasilis is a synonym of *C. (M.) sayi.*

C. (Tanaemyrmex). ADD *dumetorum* from **Snelling 1970,** *Proc. Entomol. Soc. Wash.* 72:390–397. *C. (T.) fumidus festinatus* is now *C. (T.) festinatus. C. (T.) maccooki* is now *C. (T.) semitestaceus.*

Polyergus. Change *P. rufescens* to *P. breviceps* and synonymize into it all subspecies of *rufescens.*

REVISIONS:

Acanthomyops—**Wing 1968,** Cornell Univ. Agric. Exp. Stn. Mem. (405). 173 pp.

Paratrechina. **Trager 1984,** *Sociobiology* 9:49–162.

Lasius. **Wilson 1955,** *Bull. Mus. Comp. Zool. Harv. Univ.* 113:1–205. ADD *nevadensis* from **Cole 1956,** *J. Tenn. Acad. Sci.* 31:26–27, and *atopus* from *ibid.* **1958,** 33:75–77.

Myrmecocystus. **Snelling 1976,** *Nat. Hist. Mus. Los Ang. Cty. Sci. Bull.* 24, 163 pp.

CHANGES IN ***Formica:*** We now use the term "species-group" *neogagates* for *lasioides, limata,* and *neogagates,* which were in the subgenus *Proformica,* and *bradleyi, manni, oregonensis,* and *perpilosa,* which were in the subgenus *Raptiformica.* Creighton's keys may still be used.

Creighton's key to the *rufa* "species-group" may still be used if one makes a few changes. Combine *F. integra haemorrhoidalis, F. i. tahoensis* and *F. obscuripes ravida* into a single species *F. haemorrhoidalis.* ADD *calviceps* **Cole 1954,** *J. Tenn. Acad. Sci.* 29:164–165. All subspecies of *integroides* are now regarded as species.

Without females it is often impossible to distinguish the *microgyna* group from the *rufa* group. In fact, in future publications we will combine the two and make one key for an enlarged *rufa* group. *F. rasilis* is now a synonym of *densiventris.*

The *fusca* "species-group." Use **Francoeur's 1973** revision *(Mem. Soc. Entomol. Que.* (3), 316 pp. and/or **Wheeler and Wheeler's 1976** *("North Dakota Ants Updated,"* Desert Research Institute, Reno, Nev. pp. 11–16) revision of Francoeur. ADD *moki* from Creighton's *Neoformica* as *xerophila.*

The *pallidefulva* "species-group" is all that remains of Creighton's subgenus *Neoformica: archboldi, schaufussi,* and *pallidefulva.*

Creighton's subgenus *Raptiformica* minus the first five couplets of the key constitute our *sanguinea* "species-group." *F. sanguinea subnuda* is now *F. subnuda.*

ADD GENERA:

Acropyga: See **Snelling 1973,** *Nat. Hist. Mus. Los Angeles Cty. Contrib. Sci.* 236, 7 pp.
Plagiolepis: See **Smith 1957,** *J. N.Y. Entomol. Soc.* 65:195–198.

Index

This index is an extensive as well as an intensive reflection of the wealth of terms, ideas and knowledge presented by the contributors of this book. In this sense, it is an aid to locating specific taxon names, general taxonomic terms, ecological terms, methods of study, plus a variety of key characters and other general biological words that are applicable to soil ecology.

Page numbers presented in Roman typeface indicate entries found in text material, while bold face indicates terms illustrated in figures or referred to in tables, and italics show words that are found in identification keys. Synonyms and highly similar terms are presented in parentheses. Also, cross-references are given after many terms when appropriate.

Under each major taxon, which is treated as a book chapter, the first page of the chapter is given, followed by the inclusive pages of the identification keys for the group, and finally, the pages where the general illustrations of the taxon are found. Chapters were indexed further to the levels of specificity for taxon identification as provided by the individual authors. Pages listed under given taxon names may also include references to adjectival forms of the term, such as "acarid" listed under "Acaridae". Terms found in footnotes are presented with an "n" after the page number.

The index to subjects, in a number of cases, is presented using a conceptual approach. For example, the conceptual organization within the index relating to the ecological term "Microhabitats", includes all the factors-physical, chemical and biological (including food items)-of the microenvironment with which certain life forms are associated. Since these microhabitats support specific groups of organisms the microcommunity concept is also implied.

Due to the voluminous literature citations in this book, terms from reference sections are not included in this Index.